The
HARE
and the
TORTOISE

ALSO BY DAVID P. BARASH

The Whisperings Within

Sociobiology and Behavior

Aging: An Exploration

Stop Nuclear War! A Handbook
(with Judith Eve Lipton)

The Caveman and the Bomb
(with Judith Eve Lipton)

DAVID P. BARASH

The HARE *and the* TORTOISE

Culture, Biology, and Human Nature

VIKING

VIKING
Viking Penguin Inc., 40 West 23rd Street, New York, New York 10010, U.S.A.
Penguin Books Ltd, Harmondsworth, Middlesex, England
Penguin Books Australia Ltd, Ringwood, Victoria, Australia
Penguin Books Canada Limited, 2801 John Street, Markham, Ontario, Canada L3R 1B4
Penguin Books (N.Z.) Ltd, 182–190 Wairau Road, Auckland 10, New Zealand

First published in 1986 by Viking Penguin Inc.
Published simultaneously in Canada

Grateful acknowledgment is made for permission to reprint excerpts from the following copyrighted material:

"Sunday Morning" from *The Collected Poems of Wallace Stevens*. Copyright 1923 and renewed 1951 by Wallace Stevens. Reprinted by permission of Alfred A. Knopf, Inc.

"The Bobolink" by Edna St. Vincent Millay from *Collected Poems*, Harper & Row. Copyright 1928, 1955 by Edna St. Vincent Millay and Norma Millay Ellis.

"The Impact on Children and Adolescents of Nuclear Developments" by William Beardslee and John Mack in *Psychological Aspects of Nuclear Developments*, APA Task Force Report #20. Washington, D.C., American Psychiatric Association, copyright 1982. Reprinted with permission.

"Supernatural Songs, XII: Meru" from *The Poems* by W. B. Yeats, edited by Richard J. Finneran. Copyright 1934 by Macmillan Publishing Company, renewed 1962 by Bertha Georgie Yeats. Reprinted with permission of Macmillan Publishing Company and A. P. Watt Ltd.

A speech delivered by Charles Sade at Broome County Community College. Reprinted by permission of Charles M. S. Sade.

LIBRARY OF CONGRESS CATALOGING IN PUBLICATION DATA
Barash, David P.
The Hare and the tortoise.
Bibliography: p.
Includes index.
1. Social evolution. 2. Man—Animal nature.
3. Nature and nurture. 4. Sociobiology. I. Title.
GN360.B37 1986 304.5 85-40797
ISBN 0-670-81025-8

Printed in the United States of America by
The Book Press, Brattleboro, Vermont
Set in Baskerville

For Nanelle Rose

Contents

It is dangerous to show man too clearly how much he resembles the beast without at the same time showing him his greatness. It is also dangerous to allow him too clear a vision of his greatness without his baseness. It is even more dangerous to leave him in ignorance of both. But it is very profitable to show him both.

—PASCAL

Man is the only animal that laughs and weeps; for he is the only animal that is struck with the difference between what things are, and what they ought to be.

—HENRY HAZLITT

The HARE *and the* TORTOISE

CHAPTER ONE

THE SIAMESE SACK-RACE

It had been the world's first murder. The ape-man exultantly threw his club (actually the leg-bone of a zebra) into the air . . . and as it spun, it turned into an orbiting space station. In this stunning image from *2001: A Space Odyssey*, millions of moviegoers saw the human dilemma in microcosm: we carry the unmistakable signs of our animalness, and yet we do things that have carried us far from the realm of the merely organic. Ape-men all, we are the product of biological evolution—a slow and natural process—and yet we are also enmeshed in our own cultural evolution, which is fast and somehow "unnatural." As the ape-man's club traveled through the air, and ultimately, into outer space, four million years of biological and cultural evolution were collapsed into five seconds.

The journey of the human species dwarfs that of any astronaut, and it is a continuing trip. We are all time travelers, with one foot thrust into the cultural present and the other stuck in our biological past. If we are uncomfortable in this rather awkward posture, no one should be surprised. The human problem is relatively easy to describe but much more difficult to solve: as Pascal recognized so clearly, we are animals, yet we are also much more. The transition from ape to human was dwarfed by the transition from club to missile. Most anthropologists would agree that biologically a prehistoric ape probably wasn't all that far removed from Homo sapiens; it seems a much longer way, on the other hand, from a zebra's leg-bone to a space station. Yet, that remarkable and dizzying transformation occurred almost overnight as evolution reckons these things, and we, we alone, made it happen.

The little hyphen in ape-man is therefore the longest line imaginable, connecting two radically different worlds. Just as Janus the two-faced god was chosen to represent the first month of the Roman calendar because he had one face looking back to the year just past and the other looking ahead to the one forthcoming, our entire species is fundamentally two-faced, one face looking back into our own evolutionary past and the other looking ahead to a rapidly advancing future, all the while delicately balanced in a very transitory present. This book will attempt to make sense of that present, and to suggest a general theory for why it is so confusing, so dangerous, and yet such a hopeful moment in the human adventure.

Our uniqueness as human beings is obvious in nearly every aspect of life, from our stupendous constructions to our most casual conversations. But permeating it all, like the cartoonist's little devil perched on our shoulder and whispering in our ear, is our biological heritage. We have bodies and we have evolved, no less than the ill-fated zebra that our unpleasant ape-man ancestor found so useful. In short, we are biological beings as well as human beings. We bleed, we eat, we defecate, we reproduce, we die. Yet we also dream the most sublime odes, build the most remarkable machines, unlock the atom, imagine eternity and the deity. We are not only *part of* nature, but strangely *outside of* it as well, creatures who in many ways have transcended our organic selves, to think and do things that no other animal ever thinks or does. We can point with pride to our profound accomplishments, and with dismay to the difficulties they pose for us, and for the planet. We are not only Janus-faced, but Janus-souled, riven with a deep-seated dualism that is unique among the earth's creatures. It is both our glory and our curse.

In his *Essay on Man*, Alexander Pope lamented that

He hangs between, in doubt to act, or rest,
In doubt to deem himself a God, or beast;
In doubt his mind or body to prefer;
Born but to die, and reasoning but to err.

Pope wrote over a hundred years before Darwin, and now we know what the poet did not: We are both God *and* beast, and we do not so much hang between the two as we are stuck in both, simultaneously and irretrievably. Pope concluded that we were a creature of preeminent paradox:

Created half to rise and half to fall;
Great lord of all things, yet a prey to all;
Sole judge of truth, in endless error hurled:
The glory, jest, and riddle of the world!

Whatever the glory and whatever the jest, Homo sapiens is above all a riddle, the root of which must be proclaimed: the clash between our two fundamental characteristics, culture and biology. This essential dichotomy between the hare and the tortoise, between our galloping culture and our slow-moving biology, is the salient fact of human existence as well as the underlying basis for most of our problems. It is also the basis for this book.

To understand the conflict between culture and biology, we must first review their origins. Our essential bodily characteristics—and presumably, our emotional and mental ones as well—have developed by a gradual process of organic or biological evolution. Although the exact mechanism of this process is still being debated, the essential *fact* of evolution is no longer questioned. Thus, there is "the theory of evolution," and there are "theories of evolution." The various theories of evolution concern the possible mechanisms whereby the evolutionary process is fine-tuned (the role of geological catastrophes, the significance of neutral traits, etc.). But despite the claims of certain Christian fundamentalists, evolution itself is not really a "theory" at all, in the sense of unsupported speculation or a dressed-up hunch; rather, it is as close to fact as science has ever come, analogous to "cell theory," "atomic theory," "gravitational theory," or "the theory of relativity."

As to precisely what role evolution plays in influencing our behavior, there is room for debate. But if evolution is "only a theory," then it is only a theory that the earth is round.

The word "theory" derives from the Greek *theoria*, meaning "viewing or contemplating." A scientific theory is a coherent set of propositions that helps make sense out of facts that would otherwise seem chaotic. It is not a direct line to Truth, but neither is it a wild guess. When it comes to explaining the basic patterns of the living world, evolution has no peers among scientific theories; it does not even have any serious competitors.

Like our biology, our culture has also evolved. The process of *cultural evolution*, however, differs in fundamental ways from the biological evolutionary process that has shaped all life. Our capacity for culture was itself a product of biological evolution, and in this sense, human culture is a direct descendant of our biology. But like an errant child—or Frankenstein's monster—culture developed a momentum of its own and proceeded quite independently of the natural process that originally spawned it. This is because unlike biological evolution, cultural evolution has the capacity to take off on its own, to reproduce, to mutate, and to spread faster and more effectively than any "natural" system. While our biological nature remains shackled by genetics, lumbering along at a tortoise's pace—never faster than one generation at a step, and typically much slower even than that—our culture has been sprinting. In Aesop's fable, the tortoise eventually wins, because the hare is foolish, overconfident, and easily distracted, whereas the tortoise (although slow) is persistent. In the real world, culture and biology differ in speed, but they are equally foolish and equally persistent. And most important, they will both cross the finish line together, because despite their differences they are inextricably tied to each other. We might be tempted to sit back, amused, and watch the entertaining spectacle, sort of a comical, cosmic sack-race featuring two mismatched Siamese twins . . . except that we are part of the show.

The conflict between biology and culture is not a restatement of the ages-old nature/nurture controversy, although it has some similar elements. Biologists, psychologists, and others recognized some time ago that both nature (our genetic and biological heritage) and nurture (our experiences) combine inextricably to produce our behavior. We can no more claim primacy of the one over the other than we can claim that in the structure of a coin, heads are more important than tails. This is equally true for the hare and the tortoise. But although biology and culture are necessarily involved in producing our behavior, they do not necessarily have to get along. When they do, our private lives and our public societies are likely to go smoothly; in such cases, lacking any need for a remedy, most of us are unlikely to notice any need for an explanation.

Fortunately, there can be considerable harmony between our culture and our biology, largely because our biology is so flexible, like a "one size fits all" garment, able to conform to many different shapes and sizes. But not everything fits. Sometimes things just aren't right, and this is distressingly true for the human experience, which has more than its share of warts and rough edges. So whereas all human behavior derives from both biology and culture, both nature and nurture, it does not necessarily follow that biology and culture are always comfortably adjusted to each other. Furthermore, human attention is more likely to focus on conflict than to celebrate harmony, for the same reasons that the daily news reports do not tell us about the things that went well today. Just as we must look to the interaction between nature and nurture for the sources of our behavior, we can look to the conflict between nature and nurture for the sources of our difficulties. A useful rule in murder mysteries—so useful that it became a cliche—is *cherchez la femme* (look for the woman); when Homo sapiens is having trouble, a useful rule—not yet a cliche—would be to look for possible conflict between the hare and the tortoise.

To change the metaphor: two huge continents have drifted apart and now these great tectonic plates, culture and biology, grind together. The results, as we shall see, range from nearly trivial squeaks and wriggles, such as our troublesome sweet tooth or some of our sexual pecadilloes, to the most portentous quakes, such as nuclear war; while in between lie a host of middle sized tremors such as alienation, environmental abuse, and overpopulation. The conflict between culture and biology, the Siamese sack-race between the hare and the tortoise, is an event of paradoxical proportions, ranging from the seismic to the microscopic, from whole societies (indeed, the whole planet and its past, present, and future), to individual people and their likes and dislikes.

Before examining that conflict, in the next two chapters we shall examine the participants, reviewing briefly the anatomy of the tortoise and the hare.

CHAPTER TWO

ANATOMY OF THE TORTOISE

We live in an old chaos of the sun,
Or old dependency of day and night,
Or island solitude, unsponsored, free,
Of that wide water, inescapable.
—WALLACE STEVENS,
"Sunday Morning"

Biological evolution is the ultimate natural process. It is the great flow of events that created all living things and also the thread of continuity that links them together even now. For us, as for all living things, it is an old dependency. And it is inescapable. Julian Huxley once pointed out that the human species is evolution becoming aware of itself. However, most Homo sapiens didn't partake readily of this awareness, in part because the subtlety and slow pace of biological evolution make it difficult to identify, in part because of discomfort over a scientific truth that ran counter to established religious doctrine, and in part out of reluctance to admit a unity of human beings with "bestial" nature. "Descended from monkeys?" exclaimed the wife of the Bishop of Worcester in the mid–nineteenth century, "My dear, let us hope that it isn't true! But if it is, let us hope that it doesn't become widely known!"

Such reactions have a long history, and they often came gift-wrapped in elaborate intellectual paraphernalia as well. For example, after Copernicus had upset the comfortable, ego-pleasing, and universally accepted notion that the earth was the center of the universe, the Danish astronomer Tycho Brahe did his best to see to it that a more congenial view became widely known, even if it wasn't true. Brahe proposed that the five identified planets rotated about the sun, as Copernicus had demonstrated, but that the whole business, in turn, went around the earth! Brahe was not consciously seeking to deceive his contemporaries; he sought a view of the universe that was more in accord with what he wanted to be true.

Such a proposal was not, strictly speaking, a compromise, analogous to "splitting the difference" between contending parties. Rather, the Brahean solution was an ingenious but ultimately unsuccessful example of seeking to accommodate to new and awkward facts by accepting those that are undeniable—so long as they are marginal to the most deeply held beliefs—while clinging tenaciously to what is dearest to us, even if it may be incorrect.

Human beings periodically seek Brahean solutions. We try to adapt to new data, while nonetheless keeping our fundamental orientation intact: we grant the laws of physics, while leaving room for free will; we grant the finite limits to the earth's resources, while continuing to exploit them; we grant that nuclear weapons are unusable, while continuing to build more and more. Our view of human nature has also been rather Brahean: most people today accept evolution, but still maintain a conception of themselves as separate, unique, and distinct, God's special representatives on earth, if not in the universe. Like the inventive Tycho Brahe, we grudgingly accept certain unavoidable facts—such as our biological connectedness to other living things, notably reflected in paleontology, anatomy, embryology, and physiology—while strenuously resisting the heresy that in our behavior, too, we may be similarly connected to the other inhabitants of our planet.

Perhaps we should relax, since in fact our uniqueness is comfortably assured, even though on first glance it may seem otherwise. Our planet, admittedly, is insignificant compared to the rest of the universe. Not only do we circle the sun rather than vice versa, but even old Sol is rather small, out of the way, and second-rate by astronomical measures. And even here, on the third planet from the sun, not only are we animals, but as one species among millions, Homo sapiens is about as insignificant as the earth is among the heavens. However, when we examine the realm of life itself, the telescope is inverted: the universe, cold and perhaps utterly lifeless, is no longer vast and imposing, but instead, trivial compared to the marvelously living, breathing creatures that grace the earth. And in a sense, all these creatures pale by contrast with the human species, whose consciousness and culture make him and her something special indeed, something really new under this sun and perhaps all suns.

Lewis Mumford has written that "without man's cumulative capacity to give symbolic form to experience, to reflect upon it and refashion it and project it, the physical universe would be as empty of meaning as a handless clock; its ticking would tell nothing. The mindfulness of man makes the difference."

Our "mindfulness" is also a product of evolution, undoubtedly the greatest triumph of organization and complexity that a fundamentally unmindful, unsponsored process has yet created. "If this property of complexity could somehow be transformed into visible brightness," writes molecular biologist John Rader Platt, in *The Step to Man*,

> the biological world would become a walking field of light compared to the physical world . . . an earthworm would be a beacon . . . human beings would stand out like blazing suns of complexity, flashing bursts of meaning to each other through the dull night of the physical world between.

Perhaps Brahean solutions to our self-perception and especially our relation to the physical and biological world are less needed today than ever before, as we become increasingly aware of the universe as a handless clock, and, by contrast, of the mind-boggling complexity of life in general and of the human mind in particular. Such awareness is due at least partly to the welcome fact that our perception of "nature" is more sophisticated than ever before. The modern (and presumably, more accurate) view of nature is that it is neither the mysterious "élan vital" of an Henri Bergson, nor the rigid, clockwork automaticity of a Descartes, or a world of Newtonian billiard balls. Rather, in a curious merger of wisdom East and West—the world's great mystical traditions as well as modern physics and ecology—science today increasingly sees nature as a dynamic, ever-changing, permeable state of exchange and equilibrium; that is, nature is increasingly recognized to be neither a fixed, linear sequence of set levers and pulleys, nor a mystical bowl of amoeboid mush, but rather, a *process*. Joining the melee, recognizing ourselves as part of the process, we are neither diminished nor expanded, just described. So the need for a Brahean solution is thereby reduced, and we emerge freer than ever before to see ourselves not as others see us or as we might wish to see ourselves, but as, perhaps, we really are.

***

DARWIN did not "discover" evolution. It had been described and speculated upon by numerous authors before him, including his own grandfather. Rather, his great contribution was to identify a mechanism whereby evolution can plausibly be seen to have occurred: natural selection. Like so many great intellectual discoveries, natural selection is a logical necessity, given certain basic facts. Darwin's particular genius lay in recognizing the significance of commonplace, commonsense observations, just as it took a special mind for Isaac Newton, generations earlier, to divorce himself from the commonplace observation that things fall, and to view gravity as something worth examining. "Of course," commented Thomas Huxley upon reading *The Origin of Species*, "how stupid of me not to have thought of that!"

Natural selection—and hence, evolution—is simply a logical consequence of the way the biological world is constructed. In fact, it probably cannot be avoided. To maintain a constant population, individuals of any sexually reproducing species (with one male and one female as parents) must simply produce two surviving offspring, thereby replacing themselves. Fewer than this will eventually cause extinction; more will cause an increase in numbers, and the increase would itself continue to increase. Certain species such as the lynx and snowshoe hare in northern Canada or the lemming of Scandinavia experience cyclic increases in numbers followed by spectacular decreases. But by and large, the natural populations of most animals are balanced, showing only minor fluctuations in numbers. This indicates that in most cases, parents simply replace themselves, although most living things have the capacity to produce many more offspring than this. Thus, a female cod may produce a million eggs at a single spawning; if all these were to survive and then each of the million reproduce in turn, the oceans would soon be too small to hold all the codfish alone.

Even a slower breeder such as the robin, producing say four eggs at a sitting, can generate sixteen offspring in just four years, and if each of these offspring also reproduces comparably, our original pair will have been responsible for 104 direct descendants in this same four years. Or, the uncontrolled reproduction of houseflies would soon theoretically produce a ball of flies larger than the earth. But such numbers are hypothetical. These statements are mathematically correct; the fact that such dire predictions do not come true is testimony to the high mortality of living things in nature, and to the fact that reproductive potential is only rarely realized. All the potential offspring of most species do not survive. In fact, as we have seen, just two young of every species will eventually live to replace their parents, if the population is to remain constant, as most populations do. This means that 999,998 aspiring cod must perish annually along with 102 potential robins every four years, for two of each species that survive.

What separates the winners from the losers? Those that are better adapted to succeed will do so. In fact, this is what we mean by "better adapted." They will be *selected* by *nature* as the chosen representatives to propagate the next generation: in short, natural selection. As Darwin recognized, natural selection thus follows inevitably from the capacity of living things to reproduce in great numbers, combined with the fact that very few of the potential newcomers are actually successful.

This process could not produce change, however, unless the offspring from which nature selects differ genetically from each other. Thus, even a continual winnowing cannot achieve anything new if the components are all fundamentally the same. Only in the case of identical twins are two individuals of a sexual species genetically identical. With these rare exceptions, each individual is genetically distinct, each with its own particular makeup, its own particular contribution to make, and its own probability of failing. Its ultimate success depends on whether it is "fit"—that is, well adapted to survive and reproduce—and this fitness will be influenced by the unique gene combination each particular individual is carrying. Successful gene combinations are thereby selected for, eventually replacing their less successful relatives.

When Darwin first described natural selection, it was almost universally *mis*understood: it was thought to operate by violence and death. Evolutionary change was seen as the necessary outcome of Tennyson's "nature red in tooth and claw." "Social Darwinism" became a convenient credo of late-nineteenth-century laissez-faire capitalism. Under the excuse that "survival of the fittest" provided biological sanction for the most abhorrent social practices, domination of the weak by the strong was proclaimed as natural and right, a law of nature and hence, a dictate from God. Actually, the outcome of aggressive fights—waged either by tooth or stock option—is almost irrelevant to natural selection. As Darwin himself seemed to recognize, individuals are not selected for, *genes* are. To some extent then, the victor in a fight to the death might enjoy a selective advantage over an opponent, but only insofar as he or she would be more likely to leave successful offspring as a result. And even more important, advocates of social Darwinism fell victim to what David Hume had earlier called the "naturalistic fallacy": what is, ought to be. The social Darwinists misread not only science, but ethics as well.

Natural selection is not a prescription of how the world ought to be; it is a *description* of how it works and of the major way in which evolutionary change comes about. Natural selection is simply differential reproduction, primarily of individuals and their genes. Within any species, individuals leaving a larger number of offspring are being selected for—or rather, their genes are being selected, since they will enjoy a greater representation in succeeding generations. Those with fewer offspring, and fewer genetic copies projected into the future, are selected against. It's that simple. Evolution by natural selection favors any genetically influenced characteristic that increases the chances of its carriers leaving more offspring and other genetic relatives. These characteristics may include success in personal conflicts, but more likely, ability to prosper in their respective environments. Thus, the ability to find food, avoid enemies, attract a mate, obtain shelter, withstand climatic stress, and cooperate with others may all enjoy natural selection's favor, because possessors of these traits would be more likely to leave progeny that would carry similar tendencies and that would be more successful in turn, thereby increasing the frequency of these genes in the population. Natural selection thus proceeds inexorably from the simple fact that organisms are capable of great overproduction, while population size generally remains relatively constant; then, given genetic differences among individuals, selection will favor those traits that contribute most to reproduction in the particular environment in question.

The sources of these genetic differences, however, have only recently been identified. The fundamental building block of genetic variety is now known to be mutation, basically a stenographic error in the cell's copying machinery. The genetic composition of every living thing is coded by the specific arrangement of atoms in complex organic molecules, the nucleic acids (so named because they are most abundant in the nucleus). These nucleic acids—DNA in most animals—are unique to each species and must be copied precisely every time a cell divides, to insure that the daughter cells will be like the parent and ultimately to guarantee that watermelons produce watermelons while people produce people.

Usually this copying is remarkably exact, but occasionally an error makes the copied DNA very slightly different from the original, like a typed manuscript in which the wrong key was struck. In most cases, a mutation reduces the fitness of the living thing that carries it, just as an error generally reduces the quality of a typed work. Or imagine a delicate, carefully balanced machine: it is very unlikely that a random change will improve its operation. But occasionally, the error may actually improve the manuscript, by suggesting a better word than the original. Very rarely a mutation may actually improve performance and therefore be selected for. Perhaps one mutation in a thousand will be beneficial, while the mutations themselves occur about one in a *million*; nature is a very competent typist. At this point, however, the typewriter analogy breaks down, because mutations may involve not only the substitution of the wrong letter in a complex manuscript, but in fact the creation of an entirely new letter, thus potentially expanding the genetic repertoire, the "alphabet" of the species.

Mutations, like most other errors, are random in that specific instances cannot be predicted in advance. Like other errors, however, they are not entirely unpredictable either. Certain genes are more likely to mutate than others and although the exact time of any given mutation cannot be predicted, statistical estimates can be made. Thus we may say that gene "X" will mutate, on the average, once in one million copies, while gene "Y" will mutate once in ten million, just as insurance analysts will estimate different accident frequencies for different industrial operations. Furthermore, and once more like other accidents, mutation rates are influenced by external factors, in this case, certain chemicals

ENDANGERED SPECIES: BLACK RHINOCEROS

©ANDY WARHOL, 1983
silkscreen, 38″ × 38″
Ronald Feldman Fine Arts, Inc., NY, NY

Vulnerable or threatened by extinction, the African Black Rhinoceros has been reduced by 90% in ten years due to demand for its horn. Estimate: 10,000–30,000 remain.

(mustard gas for example), excessive heat, radiation (ultraviolet light, X-rays, or nuclear radiation), or even the presence of special genes that influence the mutation rate of other genes.

During their evolutionary history, most living things did not experience high-energy radiation; not surprisingly, therefore, they are vulnerable to its effects. This susceptibility of the genetic copying process to high-energy radiation is the basis for biologists' concern about excessive medical X-rays or radioactive nuclear fallout. In the absence of these artificially induced goads to higher mutation, enough genetic copying errors occur naturally, presumably in response to the normal imperfection of the material world or to natural factors in the environment, to provide evolution with a constant supply of genetic variation from which natural selection can select. It is sobering for human perfectionists to consider that all biological advance is ultimately based on error, nature's very essential fallibility.

Evolution by natural selection is a painfully slow process. Modern-day ferns, for example, are basically unchanged over hundreds of millions of years. The same goes for horseshoe crabs, turtles, and crocodiles. Even those species that have evolved "rapidly," such as horses, elephants, or human beings, took many hundreds of thousands and more often millions of years to progress from horselike, elephantlike, and humanlike ancestors to the forms we recognize today. Compared to the rate of cultural change, everything is a "living fossil."

Evolution would be even slower yet if it had to rely on mutation alone, since mutations are so very rare. Most living things, human beings included, long ago stumbled upon a wonderful device for producing a great variety of genetic combinations in every generation: sex. With sexual reproduction, genes from each parent are combined and recombined, organized and reorganized every generation, producing a unique genetic makeup for every offspring. This explains the seeming contradiction that whereas like always begets like—human beings never give birth to giraffes—children are never truly identical to their parents.

This, then, is the crucial biological consequence of sex: not reproduction per se (after all, many living things reproduce asexually), but rather, the creation of a vast store of genetic variation upon which natural selection can operate. Sexual reproduction shuffles and reshuffles the cards before every child is dealt his or her genetic "hand." In this way, mutations first appearing thousands of generations ago can be retested in different contexts until either a winning combination is reached or the hand is declared a failure. If successful, a combination will be favored by natural selection, selected *for*, and evolution will proceed by increasing its representation in the population; carriers of that combination will leave more offspring. Holders of losing hands, in turn, will leave fewer offspring. They will be selected *against*, and their combinations will eventually retire from the game, unable to compete with the winners. If everyone in the species is dealt a bad hand, or if the rules are changed too quickly, the species goes extinct.

Genetic variation increases the range of diversity shown for any characteristic. Such diversity can be seen, for example, in the wide range of human height from short Eskimo to tall Watusi, or even the varying statures of children from the same parents. We vary in all traits—from shoe size to intelligence—and this variability is in part an expression of the genetic variation that distinguishes each individual.

But why should such variation exist at all? With natural selection constantly pruning out the less successful variants, one might expect each species to eventually consist only of "super(wo)men," perfectly adapted to their environment. To some extent this is prevented by the very fact of error itself: mistakes do happen, ultimately beneficial or not. But beyond this, optimum characteristics for a species are always relative to its environment; so characteristics advantageous now may be of no value or even become liabilities if the environment changes . . . and eventually, it always does. Thus, keen eyesight might seem universally advantageous, but not to an animal inhabiting caves where it is perpetually dark. For such animals, eyes are not only useless but an actual hindrance since they are delicate and prone to injury and infection. Those fish and salamanders whose ancestors took up cave dwelling as a way of life have profited from the fact that they retained the genetic variety to produce eyeless forms when necessary.

Since selection tends to eliminate the unfit, and ultimately even the *less* fit, a greater diversity of types is present in each species before selection operates than after. Each species therefore experiences conflicting forces: mutation and sexual recombination acting to increase the range of variability, and natural selection generally tending to narrow this range. The pruning action of selection alone, without the disruptive effect of mutation and recombination, would tend to produce a somewhat uniform population, well adapted to a particular environment but vulnerable to change. Mutation and recombination, without the narrowing effects of selection, would produce an array of freaks, poorly equipped to be superior anywhere but perhaps with some representatives capable of surviving almost anywhere. Each species therefore arrives at its particular evolutionary strategy, combining varying amounts of short-term success with long-term insurance. There are numerous complex techniques for preserving hidden variability, keeping it from the probing eye of natural selection. For example, genes are normally found in pairs, of which one may be "dominant," overshadowing the influence of the other. In this way, a deleterious "recessive" mutation may be providentially retained in the population, only to prove advantageous—or alternatively, to be selected against—in the future. Among human beings, for instance, the tendency to produce blood that clots normally is controlled by a dominant gene; hemophilia is produced by its alternative, recessive form. Two seemingly normal people can therefore produce a hemophilic child if they are both carrying a recessive gene, masked by a dominant, and if the unlucky child gets a dose of the recessive gene from each parent.

If we reproduced asexually, those of us with normally clotting blood would be assured of producing offspring whose blood also clotted normally, since our offspring would be identical to ourselves. But by turning our backs on sexuality, we would also have foregone the opportunity of combining our genes with those of another person, an adventure whose outcome can be offspring that are new and guaranteed to be different, possibly less "fit" than ourselves, but possibly more so. Most of the higher animals, our own species included, have taken the plunge. We have opted for sexuality, with its consequent genetic variety and promise of evolutionary flexibility. By contrast, certain living things have been more conservative, mortgaging the future for a high degree of present-day fitness. The common yellow dandelion, for example, has foresaken sexual reproduction in return for immediate success. But when the environment changes, dandelions may well lack the genetic reserves needed to adapt to a new situation. In this regard, sexual abstention is in fact a profligacy of sorts, one for which the indulgent must eventually pay the price. Other living things, such as daphnia (water fleas) or aphids (plant lice), attempt an intermediate strategy. They rely largely on asexual reproduction, but they interject an occasional bout of sexuality each year, to reshuffle the cards just in case.

Ultimately, even a very sexy species may find itself unable to adapt to a changing environment and thus go extinct. In such situations the species is "overspecialized," analogous to the overspecialized worker whose technical training leads to unemployment. Overspecialization, in fact, is probably the most common cause of extinction. For example, saber-toothed "tigers" weren't really tigers but large, heavy-bodied scavengers whose enormous canine teeth probably served to pierce the thick hide of dead mastodons, mammoths, and titanotheres on which they fed. By the time their prey disappeared, these remarkable cats, saber teeth and all, apparently had gone too far down the road of specialization to adopt a new way of life. They were trapped, organisms without an environment, and are now extinct. Similarly, the sleek and graceful Everglades kite (a bird) feeds only on one species of snail, found in southern Florida. As Everglades National Park suffers increasing drought and other water problems compounded by human interference, the snail will probably disappear, and with it will go the Everglades kite.

By contrast, Homo sapiens is a relatively unspecialized animal, capable of exploiting a tremendous variety of life styles, "niches" to the professional ecologist. As we engineer rapid and far-reaching changes in the world's environments, we can look ahead to the likely disappearance of the world's specialized organisms—koala bear, condor, giant panda, and tiger—and continued success of the unspecialized: the raccoon, starling, rat, housefly, and cockroach. Keep in mind, however, that our own generalized biology combines with a highly specialized reliance on an extraordinary characteristic: culture. Some specializations, like the saber-tooth's canines, boxed their possessors into a cul-de-sac. They were tickets to doom. Others, like the peculiar reptiles that developed their forelimbs into wings and their scales into feathers, eventually giving rise to the wonderfully diverse order of birds, were the start of something big: a whole line of evolutionary success. Our cultural specialization has clearly taken flight; it remains to be seen whether we shall soar smoothly and safely like the descendants of *Archaeopteryx* or follow instead the precedent of Icarus, child of Daedalus, who flew too near the sun on his manmade wings, melted their wax, and crashed in ruin.

It is important to remember that evolution proceeds by the gradual accumulation, substitution, and exchange of favorable genes within individuals. Note that individuals do not evolve; species do. An individual is born with a particular genetic constitution and cannot change it, any more than the Everglades kite can shed its single-minded passion for snails. Only by reproduction are new genetic possibilities created, from which natural selection will choose some to bask in the future. The rate of evolutionary change will therefore depend on several different factors, notably the diversity present within a given population, the extent to which this diversity is a reflection of underlying genetic diversity, and the "selection pressure," the extent to which individuals and genes of one sort are reproductively favored over others. Even under optimum conditions for rapid change, evolution must await the passing of many, many generations as environmental pressures gradually knead and shape the gene pool of a species.

⁂

How evolution works is one thing; what it has done is another. No one really knows how life first appeared on earth. Because the event is shrouded in about six billion years of time past, it is entirely possible that we shall never know. Conflicting testimony at any legal proceeding will reveal that the more distant an event in time, the greater the disagreement about its details. As one of the most ancient events, the origin of life is no exception. However, we can skip the complex chemical formulations and describe one hypothetical sequence of events that has been accepted by a large number of experts. The early atmosphere probably consisted largely of hydrogen, nitrogen, water vapor, methane, and ammonia. By placing these substances in a constantly circulating system, applying occasional electric sparks (simulating ancient lightning) and ultraviolet radiation (simulating the sun), a great variety of complex organic molecules can be produced in the laboratory. Among these organic molecules are several amino acids, which are the basic components of proteins and even the precursors of nucleic acids themselves. This experiment may re-create the basic steps that actually took place so long ago; at least, it shows that inorganic compounds plus energy can produce organic compounds. As more and more complex molecules were formed, they would have interacted with each other, until a virtual "soup" of organic chemicals was produced. Given the rich consommé, it would only be necessary for one array to develop the ability to reproduce itself (as the DNA molecule does now in each of us) for life to have appeared.

The primitive life forms that were first able to reproduce were presumably surrounded by other organic molecules, generated by similar forces, but not quite at the "living" stage. This rich broth would have provided an abundant food source for any of our molecular ancestors that could take advantage of the energy accumulated in their structure. In a sense, therefore, the first living things were animallike in that they obtained their nourishment by "eating" other, dead things.

As the early animals began to slurp up their organic soup, natural selection presumably began to favor genetic variants that could produce their own food, using only the bare resources of carbon dioxide, water, and the energy of light: the first plants. As they diversified, plants not only provided an enhanced food supply for the increasingly hungry animals but by the process of photosynthesis (chemically joining carbon dioxide and water to produce glucose, a simple sugar) they added a new and important component to the atmosphere: oxygen. The effect of oxygen was twofold. By absorbing the sun's abundant ultraviolet radiation, oxygen was converted into ozone, which, present in the upper atmosphere today, screens out a high percentage of the ultraviolet radiation reaching the earth. Without this ozone layer, high levels of ultraviolet light would produce an increased frequency of skin cancer, blindness, and genetic mutations.

The second major impact of atmospheric oxygen was to provide a basic chemical environment in which life could be much more lively. Without oxygen, organisms would have to utilize a rather inefficient energy pathway, one that is employed today only by certain primitive organisms—such as yeast, which conveniently produce alcohol as a by-product—and by higher animals only for short periods at a time. An athlete's muscles can produce energy without oxygen during intense stress; accumulation of the end product of this process, lactic acid, is felt as a muscle cramp. When exercise is over, rapid breathing pays off the "oxygen debt" built up earlier. Limited to an oxygenless environment, animals were restricted to an uninspiring, sluggish existence. Freed from these metabolic strictures, however, they were ready for a riot of activity: individual and evolutionary, and given enough time, cultural as well as biological.

Early life was probably limited to the primitive seas. The first vertebrates evolved several hundred million years after the earliest invertebrates and were initially dwarfed by their highly evolved but boneless cousins. Trilobites (resembling present-day horseshoe crabs) and occasional giant eurypterids (enormous ocean-going scorpions) dominated the early oceans while small, inconspicuous vertebrate fish probably skulked in the shadows. These early fish even lacked movable jaws; in this respect they resembled modern lampreys and hagfish, unloved parasitic forms that attach themselves to larger game fish such as lake trout. But the primitive fish probably weren't parasitic, since there was little for them to parasitize. Rather, they almost certainly swam mournfully along the bottom, sucking up debris and rotten organic matter into their jawless mouths. Rather an unprepossessing beginning for what was to become the crowning glory of evolution.

These early vertebrates quickly "discovered" the value of armor plating, probably to protect themselves from the marauding invertebrates. (In evolutionary terms, individuals having a greater tendency to develop bony armor had a selective advantage over those that did not, producing more successful offspring until eventually—read here: millions of years later—a large proportion of the population was protected in this way.) Armor made increased size possible, since the defended fish could now hold their own in undersea competition. At last, possessors of jaws found themselves at an advantage, and fish resembling our modern forms began to appear. Efficient, ferocious jaws may then have rendered the cumbersome external armor unnecessary or even a liability. In any event, among the resulting jawed fish was a peculiar side-group, unimportant in themselves, but destined to play a mighty role in the earth's future and our own.

Take a modern fish such as a trout and perch it on dry land: in the water it swam readily, largely by undulations of its body.

But on land these beautifully coordinated movements are useless and the poor animal flops helplessly about. In water, the fish "breathes" easily by taking water into its mouth, past its gills, and out again through slits on each side of the head. Oxygen from the water is transferred into the fish's blood, while carbon dioxide goes the other way. But in air this apparatus is useless and the stranded creature quickly suffocates, a fish out of water indeed. Unlike their aquatic relatives, land animals also generally have muscular limbs enabling them to move about, as well as blind-ended pouches (lungs) where oxygen/carbon dioxide exchange takes place. About 300 million years ago, a peculiar and seemingly unimportant side-group of fish (the "crossopterygians") differed from their cousins in having fins joined to their bodies with thick, muscular attachments that provided strength and flexibility of movement: just the prerequisites for walking on land. In addition, they had simple, balloonlike pouches enabling them to gulp air when this was necessary for them to survive. It must have tasted to them like a good breath of fresh water.

It is easy to think of these animals as bold pioneers, courageously setting forth on the world's greatest adventure: the conquest of the land. Actually, they were nothing of the sort. Living in freshwater ponds and swamps, many of the early crossopterygian fish must have died as the climate changed and their aquatic environment dried up. The survivors were those equipped with primitive lungs, enabling them to stay alive by gulping air during the parched, anguished times between cooling rains. (Similar forms persist today, in the lungfish of South America, Africa, and Australia, true fish that nonetheless regularly remain out of water for weeks at a time.) When times got hard those fish equipped with both lungs and fleshy fins—again, like our modern lungfish—could wriggle over the land in search of another pond. Most likely, they simply dragged their bellies through the mud, up and down drying river beds. If they had any thoughts at all, our fishy ancestors were doubtless yearning to recapture the only way of life they had known: a wet one. And natural selection was favoring those who were most adept at doing it. With a strongly conservative intent, one of life's most radical adventures had begun.

Since plants had preceded them onto land, the early terrestrial animals found a rich food source waiting for them, as well as a generally unexploited environment. It must have been a relative Eden, since most of the direct competition among animals was back in the ponds and swamps, and most of the thorns and poisons—subsequently evolved by plants as defense against animals—had not yet been called into existence. A group of animals thus luxuriated on the shore, as increasingly successful landed immigrants. However, they were not entirely adapted to life on land, since they still had to get their feet wet once again in order to reproduce: their soft, jellylike eggs, adapted to water rather than land, would otherwise dry up and perish in the hot sun. These were the amphibians, represented by the frogs, toads, and salamanders of today. It then remained for certain of these amphibians to break the chain binding them to the water by evolving a hard-shelled egg within which they could carry a protective watery environment far from pond, stream, or swamp. Enter, the reptiles. (In a similar way, the earliest multicellular animals had derived some independence from their environment by enclosing themselves and carrying around a substitute for sea water, something we now call blood.)

With their newfound independence, the early reptiles swarmed over the primitive landscape, evolving into a great diversity of forms, including the mighty dinosaurs.

All the while, early in this great adaptive radiation of reptiles, a small sidebranch developed once again, once again carrying with it a number of peculiar, distinguishing characteristics, traits that would have seemed innocuous enough at the time, but would ultimately be revealed as evolutionary blessings. For example, some of these reptiles had their elbows rotated backward and their knees forward, so that instead of suspending themselves between four pillars as present-day lizards do, their bodies were elevated directly above their arms and legs. Their teeth also became increasingly specialized, some for grinding (molars), some for stabbing (canines), and others for cutting (incisors). This dentist's delight reflected their capacity for a more wide-ranging, omnivorous diet. By contrast, the teeth of modern reptiles are monotonously similar to each other: a cautious glimpse into a crocodile's mouth should convince the skeptic.

The internal anatomy was further rearranged in this side-branch, as it had been in the earlier transition from fish to amphibian, and again from amphibian to reptile. Reptilian scales gave way to insulating hair as these animals stumbled upon the trick of maintaining a constant internal body temperature in spite of the vagaries of the external environment. After all, air temperature changes much more than does water temperature. "Warm-bloodedness" therefore provided yet further independence from the outside, making these creatures more flexibly adapted to the land than their reptile ancestors had been. It also opened up regions of climatic extremes inaccessible to the dinosaurs and their relations. In addition, whereas the reptiles were limited to activity only at certain times of the day when the temperature was right, these new forms were free to design their own schedules.

These animals also made yet another major evolutionary discovery: developing young could be harbored within the mother and nourished by her blood supply. In this way, they were afforded greater protection and a constant temperature. And even after birth, nourishment was insured by feeding the young a fat- and protein-rich secretion produced by specialized sweat glands of the females: the mammary glands or breasts. Other important changes took place, such as building certain jaw bones in the middle ear, providing better hearing. These animals—the mammals—were initially rather small and unassuming, resembling modern shrews in appearance. They probably led fearful, inconspicuous lives in the shadow of the reptile giants, not unlike their vertebrate forebears in the primitive seas, perhaps occasionally snatching an unprotected egg. There was little opportunity for dramatic evolutionary expansion, since the dinosaurs and their reptile allies were already successful and flourishing, and there was little room left on the evolutionary stage while these monsters still held sway.

But then, about 100 million years ago and after an uncontested reign of more than 80 million years, the dinosaurs rapidly dwindled and disappeared. The causes of this massive die-off are unclear and still debated today. Perhaps those patiently gnawing mammals, with their superior reproductive methods and cunning ways, ate too many of the Goliath's eggs. More likely, the environment changed, leaving the dinosaurs too highly specialized to adapt. Maybe things became too dry, or too cold, and perhaps this was abetted by the impact of a large meteor, whose debris darkened the sky and cooled the ground for years on end.

In any event, the dinosaurs evidently lacked the flexibility necessary for long-term evolutionary survival. With their primary competitors eliminated, the field was thus opened for a riot of evolutionary experimentation among the early mammals. Flying, swimming, burrowing, grazing, browsing, drowsing, hopping, hanging, and predatory mammals filled the available niches; the surviving reptiles were restricted to relatively few representatives, concentrated in the tropics. Among the early mammals, one group took to the trees and retained many of the primitive features of their shrewlike precursors. These nondescript, apparently unspecialized little creatures deserve special watching, for among them—once again—are our ancestors.

Life in the trees exerted its own peculiar pressures. For one thing, it was necessary to move along branches rather than the flat ground. A grasping hand was more useful for this than a hoof or a claw, so these tree-dwelling forms—the primates—evolved long fingers and toes, with thumbs that could be opposed to them, providing a firm grip on cylindrical, three-dimensional objects. Thin, flat nails provided extra options without impeding the delicate work of the fingertips.

Further, most mammals had already become (and still are) "nose animals," guided about by a superbly developed sense of smell. But smells don't last long in the branches: treetops are windy places, with no continuous substrate on which a scent trail can be deposited or followed. So animals possessing a genetic tendency for a well-developed olfactory apparatus were simply wasting their time and energy, compared to others relying more on other senses, notably vision. The primates thus developed keen eyesight, particularly stereoscopic (three-dimensional) vision. With it, they could judge accurately the position of a branch or vine so as to grab it while leaping fifty feet or more above the forest floor. Stereoscopic vision, in turn, required that the eyes be rotated forward so they could focus independently on the same object, thus providing depth perception. This meant that the eyes had to lie in the same plane, like those of owls, rather than one on each side of the face, like dogs'. And this rotation required in turn that the doglike muzzle, still found in the terrestrial baboons for example, be greatly reduced in size, since it got in the way. Conveniently, our ancestors had pretty much given up being nose animals by this time anyway, so losing our muzzles was no great loss. Besides, our dextrous hands enabled us to explore the world manually, and to hold things up to be examined by our improving vision. The muzzle was thus expendable, along with the characteristic mammalian nose bristles. We became hand-eye animals.

Just when it looked like we would spend our future peacefully, if uneventfully, in the treetops, for some peculiar reason we decided to come down. This decision, in a sense, was no less momentous than the initial anonymous conquest of the land several hundred million years before. (It should be noted in passing that those expert tree climbers, the squirrels, descend a trunk head first, by adroitly rotating the bones of their wrists. Primates, by contrast, descend head up and rear end first. It is tempting to conclude that we have been backing into our future ever since.)

If our style was inauspicious, our motives—what biologists would call selective advantages—for resuming the terrestrial life were also unclear. Perhaps we had become too heavy to negotiate successfully the passage from tree to tree, since the weight-bearing capacity of branches decreases with distance from the trunk. We may have found ourselves forced to run from tree to tree along the ground. (But why, then, did we get so big in the first place?) Or perhaps our bold descent to the ground was but another would-be conservative venture that only appears daring in retrospect, like the first forays of the freshwater crossopterygan fish long before. Maybe the climate became increasingly dry, causing our forest home to thin out, gradually changing from lush jungle to sparse savannah as we find in much of east and south Africa today. Even now, the African savannah teems with wildlife, potential food for the omnivorous primate. Perhaps we were simply attracted to the better hunting down on the ground. On the other hand, and this notion is less pleasing to the human ego, perhaps we were forced down by other primates that were better adapted to an arboreal existence. Maybe some ancestor of the gorilla or chimpanzee, whose descendants today are losing out to the human presence in modern Africa, temporarily defeated us in the trees several million years ago, whereupon we moved reluctantly, in defeat and perhaps with no small resentment, to the flat, featureless, and seemingly marginal ground floor; refugees from the trees, pushed onto the plains.

Once established on the ground, we put our arboreal adaptations to good use. By flattening our feet and modifying our skeleton and leg musculature, we eventually stood straight. This upright stance may have caused certain large predatory animals to hesitate before attacking us. It may have given us a better view over the waving savannah grasses. But more important, it freed our arms and hands for other functions. Hands that had evolved for grasping branches could grasp them still, but to wield them as tools for digging up roots, or as weapons against would-be predators, prey, or each other.

The keen stereoscopic vision, so essential in the treetops, also permitted the precise hand-eye coordination that was needed for weapons to be used successfully. It is an important example of an early evolutionary teamwork, between the biology of what we *are* and the culture of what we *do*. When the team worked well together, the combination was unbeatable, as shown by the spectacular success of Homo sapiens in evolving and populating the planet.

Even with our biology and our newfound culture nicely complementing each other, we were still relative weaklings on the African savannah, still living precariously. So, those of us able to make efficient use of arms and hands were undoubtedly more successful in compensating for our physical weakness compared to the lion or the elephant. To a degree that may have never been true before in the history of life, we began to profit from what we *did* no less than from what we *were*. Digging, gathering and storing food, making rude shelters and clothing, killing other animals for food, as well as defending ourselves and our children from marauding beasts including other members of our own species: these things, as much as the structure of our lungs or our teeth, became a major arena for evolutionary success.

Not that our anatomy didn't continue to evolve. Bipedalism—walking upright on the hind legs—may have been, appropriately enough, a crucial first "step" toward becoming human, since it helped free our hands for tool use, tool making, and communication, rather than simple locomotion as in the other mammals. We quickly exploited an ability to adjust and improvise, traits that served a weak-bodied, large-brained animal so well in its new savannah home. Something crucial had begun: although biological evolution continued (just as it continues today), for the first time in the history of life, a new evolutionary factor had become significant, a factor whose importance was to challenge and eventually to overtake our biology. We started to become cultural animals.

The stage of human evolution described here is represented by a series of fossils known as Australopithecines, "southern apes," from south and east Africa. These early members of the human family were represented by *Australopithecus afarensis*, the earliest of which lived about four million years ago. They walked upright, were about four feet tall, and had brains that were about one pint in volume; approximately the size of modern chimpanzees'. In another two million years, there were at least three species, two heavy boned, slow moving, strictly vegetarian types (*Australopithecus robustus* and *A. boisei*), both of which apparently went extinct without leaving significant descendants, and a more slightly built, omnivorous form to which all of us trace our ancestry. Although we have no way of telling what these man-apes looked like (whether they were covered with hair like modern apes or virtually naked like us, for example) certain characteristics can be determined from their skeletons.

They stood somewhat less than five feet tall when adult, weighed less than a hundred pounds, and walked fully erect on their hind limbs. They had receding chins and foreheads, prominent eyebrow ridges, and teeth that look strikingly human. They also made simple chiseled tools and had brains that were nearly twice as large as those of the most primitive Australopithecines, roughly halfway between chimpanzees' and modern *Homo sapiens'*. They are the first representatives of the genus *Homo*, and have been given the species name *habilis*; *Homo habilis* in English would be something like "handyman."

Many anthropologists now agree that the early Australopithecines were meat eaters—if not exclusively so, at least for a large portion of their diet. And *Homo habilis* was literally "handy" as well, using his hands to wield tools and weapons. Skulls of medium and large animals, apparently killed by hard objects, have been unearthed from caves where *Homo habilis* lived and ate. The tree-dwelling crossopterygian fish had come once again to the land, and once again, they were making good.

We are justly proud of our enormous brain. Throughout the Australopithecine stage, however, that brain, although large by most animal standards, was still nothing to get excited about: no larger than a modern gorilla's. But beginning with those early, savannah dwelling, hunter/gatherer/scavengers, our brain enlarged at a fantastic pace, attaining human proportions (roughly twice the volume of a gorilla's) in a mere two million years. Although this may seem like a terribly long time, in evolutionary terms it is astonishingly short. For the brain to double in size within that timespan bespeaks the action of enormously powerful selective forces. The African man-apes were clearly living in an environment that placed a special premium on brainpower: contrary to what is commonly thought, our use of tools and weapons, and indeed, our remarkable development of nonbiological (cultural) extensions of our bodies, was not the *result* of a large brain, but rather, its *cause*. The gorilla-brained Australopithecines found themselves fortuitously preadapted to primitive tool use by virtue of their arboreal past. The cleverest of them were most successful in using tools and weapons to obtain food, to triumph over their enemies, and to support their families. Such individuals left more offspring. After all, a naked ape on the African savannah had to have something going for it in order to be successful; given our relatively weak backs, evolution strongly favored those of us with strong minds.

But the big-brained animal we call ourselves did not evolve solely in response to tool use. Other pressures were operating, all pushing in the same direction. Communication between individuals was of great advantage, whether to coordinate a hunt, plan a berry-picking, nut-gathering, or root-digging expedition, arrange for unified group defense, describe the locations of prey, enemies, water holes, or prospective shelter, consider alternative courses of action, or instruct offspring in the increasingly complex skills necessary for successful living. With the development of language, enormous horizons opened: we could discuss options, employ abstractions, evaluate the past, admire the present, plan the future. Individuals possessing these capacities enjoyed tremendous advantages, and so the evolution of the human brain received another forward thrust.

In addition to tool use, it therefore seems likely that natural selection smiled upon the early ability to interact effectively with others. This received additional impetus *from* brainpower, and in turn, provided additional pressure *for* the evolution of yet brainier specimens, especially those who could "handily" employ other individuals as "tools" for their own success.

Tool use, communication, and brain evolution thus interacted to produce a positive feedback ("vicious circle") loop. The ability to utilize tools and to coordinate our activities exerted initial selective pressures for increased brain capacity. This in turn made increased tool use and interpersonal collaboration possible, while the availability of increasingly sophisticated devices and opportunities further enhanced the desirability of a large brain. When *Homo habilis* evolved into *Homo erectus* ("Peking man"), about 1.5 million years ago, this process was well under way. By 350,000 years before Christ, *H. erectus* was using fire and ocher pigments, and by 60,000 years ago, *Homo sapiens* was on stage. Anthropologists are divided over whether the Neanderthal people were among our direct ancestors, or simply distant cousins. But it is generally acknowledged that *Homo sapiens neanderthalensis* were human beings, just a different subspecies or race from the modern form, *Homo sapiens sapiens.* The Neanderthalers buried their dead with flowers and almost certainly engaged in primitive religious rites. Archaeologists have found the skulls of great cave bears, enshrined by the Neanderthalers on rudimentary altars; something rather interesting must have been going on within the Neanderthalers' skulls as well. Cro-Magnon people were making beautiful paintings on cave walls, as well as carved figurines, 25,000 years ago, by which time, biologically, we were much the same as we are now.

Biological evolution is the most fundamental force for change acting on human beings; it is also the slowest, the most resistant, and perhaps as a result, the one most likely to be ignored. It is our species-wide purloined letter, responsible for those aspects of our bodies and our behavior that are so widespread and familiar that hardly anyone notices.

The long climb out of the organic soup had ended; or rather, our biology had produced what we are today, what we see when we look in the mirror. But to some extent it has been out of the soup and into the fire. Twenty-five thousand years ago, our metaphoric tortoise had pretty much reached where it is right now; human biology had arrived. But the human experience, by contrast, was just beginning. The last few steps (especially since we descended from the trees) may have been somewhat faster than they were before, but biological evolution has never been known for its speed . . . at least, not compared with its cultural cousin. Imagine a tortoise running (or rather, plodding) a marathon: virtually that entire route, from the origin of life to Cro-Magnon times—the long, winding journey through the jawless fish to the amphibians, the reptiles, the early mammals, and so forth—would be equivalent to all but the last foot or so. The last little bit of the tortoise's journey, from the Cro-Magnon days to the 1980s, would all be compressed into just a tiny fraction of the trip. But during that brief interval, while the tortoise takes yet another laborious step, an awful lot has been happening.

CHAPTER THREE

ANATOMY OF THE HARE

God took man as a creature of indeterminate nature, and, assigning him a place in the middle of the world, addressed him thus: "Neither a fixed body nor a form that is peculiar to thyself have we given thee, Adam; to the end that according to thy longing and according to thy judgment thou mayest have and possess what abode, what form, and what functions thou shalt desire. The nature of all other things is limited and constrained within the bounds of laws prescribed by us. Thou, constrained by no limits . . . shalt ordain for thyself the limits of thy nature . . . As the maker and molder of thyself in whatever shape thou shalt prefer, thou shalt have the power to degenerate into lower forms of life, which are brutish. Thou shalt also have the power, out of thy soul and judgment, to be reborn into the higher forms, which are divine.

—COUNT GIOVANNI PICO DELLA MIRANDOLA,
Italian Renaissance humanist, in
De Dignitate Hominis
(On the Dignity of Man)

Maker and molder of ourselves; constrained by no limits; choosing by our own judgment what form and function shall be ours: in short, God—according to Pico—bequeathed us cultural evolution. According to evolutionary biologists, the process worked somewhat differently: the oversized human brain, combined with hands capable of dextrous manipulation, helped us achieve ways of living and doing that went far beyond our bodies, and which very quickly overwhelmed the pace of biological change. Either way you look at it, early in human history we came under the influence of cultural evolution as well as biological evolution. And whereas sometimes the two operate harmoniously, sometimes they do not.

To some people, the word "culture" summons images of symphonies and painting, poetry and the Kennedy Center, opera and the weightier aspects of educational television. For our purposes, however, culture is much more general, widespread, and essential. Whereas human beings could survive without chamber music, they could not survive without culture; they could not survive without those various extensions of their bodies that they universally create, and without organizing themselves and orchestrating their lives in meaningful, symbolic, and profoundly extrabiologic ways. The culture of a particular human group might be defined in general terms as the sum total of all ways of living practiced by that group. The British anthropologist Edward Burnett Tylor first suggested this more inclusive concept in his book *Primitive Culture*, published in 1871. Tylor defined culture as "that complex whole which includes knowledge, belief, art, morals, law, custom and any other capabilities and habits acquired by man as a member of society."

Some major cultural inventions of the human species include tools and weapons, fire, agriculture, animal domestication, cities, smelting of iron, copper, and other metals, the wheel, gunpowder, the compass, the steam engine, the internal combustion engine, the airplane, penicillin, computers, and atomic energy, not to mention peanut butter, Hula Hoops, and underarm deodorants. Other major cultural developments, less likely to leave physical artifacts, but no less important to the human story, include language, writing, religion, political systems, laws, patterns of economic exchange, and science. Of course, many more factors could be added; to make the list complete, nearly every complex detail of human behavior would have to be included.

Cultures have progressed in different ways among different groups. Although certain phenomena may be universal—such as language, tools, and the recognition of some patterns of kinship relations among individuals—others are limited to specific local populations. The wheel, for example, was essentially unknown to New World inhabitants before the arrival of Europeans (although, interestingly, wheeled children's toys have been found among Aztec artifacts), while the practice of mate swapping, long characteristic of Eskimos, only recently appeared in certain segments of modern American culture, and less than ten years later, it seems to have disappeared again.

*
**

REGARDLESS of its details, however, all human culture shares something essential, a crucial characteristic that distinguishes its evolution from the slow, plodding biological evolution that preceded it: cultural evolution is independent of changes in genetic makeup. Whereas every individual is the *product* of biological evolution, he or she cannot evolve biologically. On the other hand, each of us can experience a variety of cultures during a single lifetime, and not only by visiting them. An aborigine brought from central Australia to western Europe can leap hundreds if not thousands of generations of cultural evolution within a few years. But the pace of cultural evolution is such that even the sedentary human being, just by living a few decades, is guaranteed to see a remarkable array of cultural practices come and go. New customs and inventions can spread from people to people, like lightning. Just like conductors of electricity, in fact, people can transmit currents of culture, as quickly as they are generated. But in another parallel to conductors of electricity, the people themselves will not really be changed, any more than a copper wire "evolves" during normal use.

This disparity between people, biologically, and people, culturally, occurs because culture is transmitted by a new mechanism, in which individuals may directly acquire characteristics that were initially obtained by another. It is a new process, quite divorced from normal biological events. With elaborate and sophisticated new technology, we can sometimes introduce the genes of one living thing into another; left to our own devices, this does not happen. Once formed as fertilized eggs, we are impenetrable to each other's genes. But once formed as human beings, we are readily penetrated by each other's ideas and cultures.

Whereas biological evolution is now acknowledged to proceed largely by natural selection (Darwinism), cultural evolution progresses by Lamarckism, "the inheritance of acquired characteristics." Herein lies the crucial difference between the tortoise and the hare. First proposed by Jean Baptiste Pierre Antoine de Monet, Chevalier de Lamarck, notably in his *Philosophie Zoologique* (1809), Lamarckism has since been generally discredited as a mechanism for biological evolution. But when it comes to culture, Lamarck was on the mark. His theory was based in large part on the effects of use and disuse: when a structure is used, it often grows, when ignored it atrophies. If these changes are then transmitted to one's offspring, a mechanism for long-term change is available, at least in theory.

Lamarckism suggests, for example, that the long necks of giraffes evolved because ancestral generations of giraffes kept stretching their necks, trying to reach food high in the African treetops. Similarly, a weight lifter could insure large biceps for his offspring simply by developing his own muscles. Unfortunately for Lamarck's reputation among biologists, the living world just doesn't work this way; there appears to be an almost complete separation between the somatic tissues of a living creature (its body) and its genes. Information flows from the DNA to the proteins, from genes to bodies, but not vice versa. Thus, biologists have cut the tails off many generations of adult mice, only to find each new generation is no more tailless than the ones before. And after thousands of years of circumcision, Jewish boys are still born with foreskins. Biological evolution depends on changes in *genes*, and genes are not changed by experiences acquired during the lifetime of the individual.

But culture is different. When medieval Japan was introduced to Western ideas and technology in the latter decades of the nineteenth century, it was able to absorb much of that culture with incredible speed. Enough to defeat Russia in the Russo-Japanese War and immediately establish itself as a major world power in the twentieth century. Rather than waiting for mutation and recombination to blunder onto the appropriate characteristics (even with the impossible assumption that such traits could be produced by biological evolution) and then waiting for their advantage to be recognized by natural selection, the Japanese people could immediately swallow whatever aspects of Western culture they considered desirable—just as the Russian people accepted the precepts of Marx and Lenin only a few decades later, and transformed their culture with equal speed. All this in less than one generation.

Similarly, when Alexander Graham Bell invented the telephone, it rapidly spread throughout the population because of its great usefulness. Imagine a biological evolutionary invention of equal value: relying upon genetic mechanisms of formation and transmission, it would spread much more slowly. Even the extraordinarily rapid evolution of human brain size, for example, proceeded at less than a snail's pace compared with the evolution of the telephone. If using a telephone depended on the existence of telephone genes, then Bell's children might possess that capacity (assuming it was a dominant trait), and perhaps 150 direct descendants would currently be using telephones. Instead, the technique was transferred by cultural evolution, and several billion people are telephone users today. Moreover, the external shape and internal design of the telephone has also been radically modified. During this time, essentially no biological evolution has occurred.

In addition, biological evolutionary change depends as we have seen on the gradual accumulation of many small steps. This is because a gross rearrangement of an organism's genetic system might well kill it, or at least cause so much disruption that it would be at a strong selective disadvantage. And the larger this disruption, the greater the impact. Biological evolution is thus forced to proceed by small steps because each step is essentially random. Natural selection sifts and chooses from the array of variations available to it, so that given enough time, order appears out of potential chaos.

The circumstances surrounding the emergence of cultural innovations are actually less well known than their biological counterparts. Like biological innovations, cultural innovations sometimes seem to appear randomly; it is almost impossible to predict the debut of human discoveries, innovations, or changes in customs. And who can describe the events that precede and give rise to an original idea? Like biological mutation, certain environments seem to stimulate more cultural innovation than others. The Renaissance was a "mutagenic" cultural environment, for example, whereas by contrast, the thousand years or so preceding it were not.

Whatever the causes of cultural innovations, they can spread not only independent of our genes, but in response to conscious decisions. When Europeans first tasted Oriental and Caribbean spices, for example, they were enchanted and wanted more. New trade routes were opened, new cities and a new merchant class became powerful, and human society—as well as human cooking technique—was changed in just a few decades. The Old World, similarly, incorporated tomatoes, potatoes, and corn from the New World, just as the New World quickly incorporated gunpowder and horses from the Old. Cultural diffusion can even outrun the actual migration of people themselves: a tiny band of Spanish adventurers conquered the much more numerous (and in many ways, more sophisticated) South American civilization largely because horses were unknown to the Incas and the Aztecs. A few hundred years later, the white conquerors of North America encountered

Plains Indians who were expert horsemen, but who had never met a European.

The spread of Valley Girl talk, skateboards, and Hamburger Helper may appear to be almost random and undirected, but in fact, there is often method to such apparent madness; new cultural patterns can be incorporated by the direct design and intent of human beings ("Mr. Watson, come here, I need you!") rather than under the control of a blind, opportunistic natural process. And whereas some such innovations can be trivial, others can entail major reorganizations of the entire social, technological, and personal fabric, for better or worse. As Leibnitz put it, *nature non facit saltus* (nature does not make leaps). But culture does. Sometimes it tries to look before it leaps, and this is another difference between cultural and biological evolution, since the latter is "blindly opportunistic," as the great geneticist Theodosius Dobzhansky pointed out. At other times, cultural evolution seems no more preplanned than its biological counterpart. Unlike nature, *cultur facit saltus.*

Human beings have developed Hinduism, the Renaissance, Christianity, the Industrial and Scientific Revolutions, Victorianism, New Deals, Five-Year Plans, Democracy, Communism, nuclear weapons, TV dinners, X-rated movies, video games, and birthday parties. Cultures may add, delete, or exchange major (or minor) components, transforming themselves within a fraction of a generation, as occurred in turn-of-the-century Japan and indeed, most cultures in contact with modern Western technology. Nearly everyone in the world has heard a transistor radio, although most do not know what a transistor is, and Burmese farmers often listen to rock and roll while plowing their fields by buffalo. In Ecuador, bananas are piled by hand into dugout canoes, as they probably have been for thousands of years, and then off they go, powered by outboard motors. Sometimes, as in these cases, one culture borrows traits from another. Often, the exchange is mutual, as with the fur parkas, snowshoes, and igloo design adopted by Americans and Europeans from the Eskimos, and the rifles and snowmobiles adopted, in turn, by modern Eskimos. In other cases, aspects of one culture are forcibly inserted into another, as with missionary Christianity or Islam.

The nature of cultures can be debated endlessly; they can be considered to be large, complex, and interdependent systems, like organisms, or the product of numerous discrete units analogous to genes in biological systems. Whereas cultures may undergo gradual transition, especially as a result of new environmental challenges (such as the Ice Age, the spread of industrialization, or resource shortages), they are also capable of gross mutation. The frequency of such convulsive change has been increasing in recent years, reaching a torrid pace during this century. And it shows every sign of gathering even more speed in the future.

Having distinguished biological from cultural evolution, we might further subdivide cultural evolution into two major components: social evolution and technological evolution. Social evolution includes forms of law and government, economics, the basic structuring of nation, family, work, and environment, music, art, literature, religion. Social evolution occurs in a time frame of decades, more often hundreds or even thousands of years. Although very fast and flexible compared with biological evolution, social evolution is nonetheless very slow compared with the other great pillar of cultural evolution, technological change. Of the two, technological evolution is by far the most rapid, bringing forth innovation at a rate that would be staggering for societal change, and unimaginable for biological change.

Thus, societies evolve with amazing speed by standards of natural selection, but slowly by standards of technology. Compare life in the United States today with 150 years ago, for example. The technology is dramatically different, including electrical appliances, automobiles, medicine, and nuclear weapons, to mention just a few. The societies, by contrast, are also different, but not wildly so: there may be fewer biparental households, but there are still households; the work week may be shorter, but there is still a work week; the government may be larger, but it is still a recognizable government, indeed, basically the same government—even the same Constitution—as existed in Andrew Jackson's time. By contrast to the revolutionary changes in technology and the moderate changes in society, the American of the Mexican-American War would be biologically indistinguishable from the American of today.

Technology proceeds by discovering and implementing methods by which human beings can act on the inanimate universe—on metal and stone, plastics and electronics, rocket ships and microwave ovens—or on the living, but nonhuman world: the planting and breeding of crops, the maintenance or exploitation of other animals, the care of our own innards. Technological evolution follows scientific discovery, and the basic insights of mathematics, physics, chemistry, and biology. It uncovers the laws of nature, and seeks to manipulate them; but, unlike social evolution, it does not write these laws.

Technology, for better or worse, has been progressing. It has allowed us to manipulate more of the world, and with less effort, than ever before. In its own way, biological evolution also involves progress of a sort, notably greater independence from one's environment (note the "progression" from naked genes, to bodies, to semiterrestrial amphibians, to warm-blooded mammals, etc.). Biological evolution also involves the accumulation of increasingly precise adaptations, such that highly evolved living things tend to "fit" their environment rather closely. However, there is very little evidence for progress in social evolution. Democracy, as practiced by the Greeks, has not been noticeably improved upon 3000 years later. Experiments with utopian societies come and go, with almost depressing regularity, and dictatorship—often with religious underpinnings as in the Ayatollah's Iran, or based upon secular theologies such as in the Soviet Union, are at least as old as the Age of the Pharaohs. Unlike technology or biology, societies do not so much evolve as revolve.

Are modern philosophers and theologians any farther advanced than Aristotle, Plato, Gautama Buddha, or Christ? Are today's Pulitzer or Nobel Prize winners better than Homer or Chaucer? Slavery is legally abolished worldwide, but forms of serfdom still exist and apartheid isn't very far behind (or ahead). Royalty and other forms of totalitarianism come and go, as does democracy. Religion oscillates from fundamentalism and absolutism to liberalism and "secular humanism" under various guises. Nontechnological, social culture may be characterized by circularity, or perhaps perturbations from a baseline, with repetitive reversions to that baseline, followed by deviations in the opposite direction; *Plus ça change, plus c'est la même chose* (the more things change, the more they stay the same). Perhaps the changes that characterize social evolution are due less to progress as such, than to the restless intelligence, interpersonal and intersocietal competition, and random mood swings writ large, which combine to make nontechnical social traits unstable or at least with a stability that seldom exceeds several hundred years. Thousand-Year reichs, of any sort, are rare indeed—and perhaps blessedly so.

It seems most likely that social evolution is characteristically unstable and unprogressive in large part because it involves the relationship of human beings to other human beings. The laws of nature can be revealed, but never repealed; the laws of human society, by contrast, are ours to write, erase, and rewrite. We can violate the latter, never the former. And we are much more likely to "get somewhere" with technological evolution since it is nearly always faster than social evolution.

We can better understand the differential pace of biological, societal, and technological evolution by considering those factors that *impede* change. Technological change is limited mainly by ideas, and also, by physical opportunity: access to the necessary materials, capital, support systems, etc. Societal change, by contrast, is limited by the presence of human beings themselves, which act like a buffer in a chemical solution. Thus, we may innovate, copy, and modify styles of clothing, for example, but human beings—by their very nature, it seems—insist on having clothing of some sort. Biological change, finally, is the most sluggish of all, since here the ballast is provided not by ideas, or even by human values, attitudes, and preferences, but rather by the human gene pool itself, which is the least modifiable of all human traits.

Much of the discordance in human existence may be due to the conflict between technological evolution and social evolution, both of these processes interacting within the cultural sphere; this particular conflict increasingly occupies sociologists, anthropologists, and specialists in the social management of technology. It well deserves such attention. For our purposes, however, it shall be useful to combine all fundamentally nonbiological factors (including both social and technological evolution) under a single rubric, and to designate them as "cultural evolution" as opposed to "biological evolution." In the concord and discord between these two, we shall search for the roots of ourselves, that which makes us human, happy, magnificent, and miserable.

These alternative levels of human evolution are similar to the anatomical distinction between our higher, cerebral functioning and lower, more primitive brain systems. Dr. Paul D. MacLean, former head of the Laboratory of Brain Evolution and Behavior of the National Institute of Mental Health, has emphasized a three-tiered conception of the human brain:

> Man finds himself in the predicament that Nature has endowed him essentially with three brains which, despite great differences in structure, must function together and communicate with one another. The oldest of these brains is basically reptilian. The second had been inherited from the lower mammals, and the third is a late mammalian development, which . . . has made man peculiarly man. Speaking allegorically of these three brains within a brain, we might imagine that when the psychiatrist bids the patient to lie on the couch, he is asking him to stretch out alongside a horse and a crocodile.

The distinction between cultural and biological evolution is not quite the same as that between the cerebrum on the one hand, and the paleomammalian and reptilian remnants on the other. Rather, it is between the entire human brain, in all its biological parts, and the most conspicuous *product* of that brain: culture.

Even this distinction is somewhat arbitrary, since as we have already seen, human biological and cultural evolution have been intimately connected, mutually supportive, and in a sense, inextricable. We are cultural animals. Culture is as "natural" to Homo sapiens as hooves to a horse or scales to a crocodile. A human being without culture would be as bizarre as a naked peacock, or a porcupine without quills.

In fact, culture is not even unique to our species. Chickadees in Britain, for example, quickly learned that they could sip the cream from unhomogenized milk by removing the foil caps; a tradition developed, with birds following milkmen and descending on the bottles before the homeowners did. Rats avoid poisoned bait, and red squirrels learn how to open acorns, once again without biological evolution taking place. Habitual methods of food preparation have been innovated by a single adult female Japanese macaque monkey, from whom they spread in a truly Lamarckian fashion to other troop members. She developed the now-famous tradition of dipping sweet potatoes in the ocean, to salt them; another time, the same individual discovered that she could separate wheat from sand by dropping handfuls into a stream, whereupon the wheat would float and could easily be scooped up. Other Japanese macaques picked up the behavior, without their genes changing in the slightest.

This renowned spread of new cultural traits among Japanese monkeys has even given rise to a new, culturally elaborated myth, the so-called "hundredth monkey phenomenon." According to this tale, once the hundredth monkey does a particular thing, the trait in question spreads rapidly throughout the population; this story is often told in the hope of encouraging the average citizen to become politically active, to "become the hundredth monkey." Although the hundredth monkey story is merely a well-intentioned fabrication, it is useful for our purposes, exemplifying another level of cultural evolution: the cultural evolution of an account of cultural evolution. (Mathematically inclined readers might call it "cultural evolution squared.")

When human beings invented tools and conceptual language, discovered the uses of fire and of domesticated animals, these were gradual—even slow and tedious—processes by modern standards. And in a sense, they were still at least minimally concordant with our biological evolution. But as the process has continued and the pace of cultural evolution has increased exponentially, the connectedness with biological evolution has grown more and more tenuous. A human being can walk to a bicycle, then pedal to an automobile, then get on a jet plane, or even a rocket ship without at any one point definitively severing his or her connectedness from the shambling bear or the sprinting cheetah. But the transition is not really a seamless web. When the quantitative difference becomes great enough, a qualitative difference has effectively been reached, even though there is no clear point of transition: a cheetah could fly as a passenger to the moon, and an astronaut stalk gazelles on the African savannah, but each would nonetheless be out of place. Yet, we still get occasional, nearly pathetic reminders of our animalness. There is something almost comic, for instance, about high-tech spacesuits designed with little tubes permitting primitive, smelly organic fluids to pass in and out, and even about the suits themselves, with their ungainly forked-stick shape, necessary to accommodate the two-legged space voyager who is still the same bipedal primate who once-upon-a-time walked about on the African savannah.

***

CULTURE and biology need not always be opposed. Indeed, it may be that most cultural practices are either neutral with regard to biological significance, or even strongly adaptive, such as the tenacious cultural defense of some form of marital bond, or indeed, most things that human beings do in their daily lives. This is not surprising, since cultures themselves succeed each other just like populations or species. As we have seen, a process analogous to natural selection determines their spread, although the mechanism is of course quite different. Natural selection does not involve any conscious choice on the part of those genes and gene combinations that experience maximum reproductive success. By contrast, culture often proceeds by the intentional selection of specific practices from among a large array of those available. In this sense, then, it is "teleological" or goal directed in a way that biological evolution never is. For example, Marxist ideologists have been able to design and carry out particular cultural practices in many nations within the past fifty years, just as capitalist ideologists have done for several hundred years before that. Although the results may not always have been exactly as intended, and much of human cultural evolution (like much of biological evolution) may be essentially random, the impact of conscious, end-directed choice in human cultural affairs certainly cannot be denied.

It may be argued that most human beings throughout most of human history have never had any real options for choice of cultural practices. Thus, the culture in which one is raised has a distinct and often pervasive influence, predisposing one's future decisions, and typically ruling others out altogether. Even if we were entirely free to decide upon our own cultural path, independent of our childhood experiences, the outlook of most people has always been narrowly restricted by the cultural models available to them. We simply do not have much chance to achieve drastic societal or cultural change. (Although, of course, we have even less chance to achieve biological change, radical or not.) Just as biological evolution must await the production of genetic diversity by mutation and sexual recombination, cultural evolution requires both exposure to new cultural practices, either through invention or observation, and the ability to adopt them.

For thousands of years, even while it was highly changeable

compared to our biology, human culture nonetheless changed slowly by modern standards. But then things began speeding up. The scientific revolution in particular began feeding upon itself, producing innovation at a greater and greater pace, while improved communication and transportation techniques enabled rapid dissemination of cultural change. Like when an asexual species suddenly becomes sexual, whole new avenues of (cultural) evolution were opened up and we have been rushing along them ever since, like a hare outrunning a tortoise.

Genetic combinations favored by natural selection are almost always adaptive; that is, they represent characteristics that tend to help their possessors do a better job of living and reproducing. If they didn't, they wouldn't be selected for. (There are some exceptions, as when a disadvantageous characteristic is naturally selected because it is somehow linked to another, advantageous one, whose benefit outweighs the other's detriment. But such events appear to be rare.) By contrast, success in cultural evolution is only indirectly determined by adaptive advantage. Thus, human culture may sometimes spread by physical superiority, as with Genghis Khan overrunning Asia or the white man overpowering the red in both North and South America. (The "physical superiority" here is of cultures, not necessarily of individuals; Genghis Khan's success was in fact due largely to the invention of stirrups, and the white man's, to horses and gunpowder.) But as mentioned earlier, success in physical combat is only one small aspect of total fitness, and conquering armies have often found themselves adaptively inferior to local populations in terms of their ability to deal with local conditions. The conquerors may thus absorb much of the "defeated" culture, coming to resemble it more than their own. In such cases, the losers may indeed have been beaten as individuals, but some aspects of their culture can still be victorious.

Biological hybridization is the union of two individuals from substantially different gene pools, usually members of different species. Sometimes when the hybridizers are similar but not too similar, the offspring show "hybrid vigor," and are hardier than either parent. When hybrids are made between different species, however, the results are usually adaptively inferior, since the combinations occur essentially at random, and each genetic system is adapted to itself, not to others. For example, given two well-constructed automobiles of different makes, it is unlikely that an unthinking exchange of parts between them would produce a better product. But if that exchange is performed by a competent mechanic, who consciously selects the best spark plugs while perhaps leaving the same carburetor, a real hot rod might result. Similarly with an exchange between cultures. Instead of the unplanned mixing characteristic of genetic hybridization, cultural hybridization often results in selection of the most appropriate combinations from both systems, leading to a more viable product than either was separately.

A successful culture may be superior in only a limited number of ways, as for example the recent successes of Western technological culture. The last 200 years have seen a drastic worldwide decline in the diversity of local cultures as "Westernization" has circled the globe. It is unlikely that this indicates true adaptive superiority. More likely, it reflects the pervasive influence of our superficially impressive and (perhaps superficially) effective science and technology. Only time will tell whether this system is adaptive over the long haul.

Culture may also spread by appeal to higher mental faculties, especially in combination with ascendant technology.* The rapid dissemination of Islam was facilitated by the former, and of missionary Christianity, the latter. Once again, the extent to which cultural practices are adaptive in the evolutionary sense can only be told by some historians of the distant future. Intellectual or

*In some cases, superior technology as such was not the major factor leading to the success of one culture over another. For example, missionary Christianity owed much of its successful spread to the fact that Caucasians introduced European diseases such as measles and smallpox, to which native populations lacked immunity.

ideological appeal has never been a direct factor in biological evolution, and such appeal may have no bearing whatever upon true adaptive value. On the other hand, an argument can be made that at least to a limited extent, cultural values are given emotional salience in proportion as they meet at least some of our biological needs. Success in evolutionary terms must be measured by continued existence over time; the absence of future humans to pass judgment upon a culture—or the entire species—will be ample judgment in itself.

One of the pitfalls of biological evolutionary change is that its reliance upon immediate adaptive advantage necessitates a rather shortsighted mechanism. Thus, short-term success may be achieved at the cost of long-term inflexibility and ultimate extinction. For example, an environment that is homogeneously light colored—such as occurs in White Sands National Monument, in New Mexico—selects for light-colored rodents, which are camouflaged from their predators. But when and if the background environment becomes darker, the light-colored mice—successful in the earlier environment—will quickly be selected against. Human cultures may also be susceptible to this capacity for overspecialization, finding themselves unable to adjust to changing conditions. They may achieve enduring success by anticipating, modifying, and compensating for the contingencies of the future. Or they may reveal that our unique teleological capacities are to us like the large canines were to the saber-tooth: useful for a time, but no ultimate security and maybe even a liability.

Unlike modern technology, some cultural practices have apparently existed in one form or another for thousands of years. They have demonstrated their adaptive utility and would presumably merit the imprimatur of biological evolution. For example, the Mosaic prohibition against eating pork may well have been prompted by recognition of the dangers of trichinosis. Whatever its theological underpinnings, it is thus good biology. Ironically, however, such prohibitions have in recent years been rendered largely unnecessary by further cultural practices such as hygienic hog-raising and careful cooking. Similarly, the east Asian and African practice of agricultural fertilization with human feces is poor biology because of the role of fecal matter in disease transmission, but good ecology in regions where soil is poor, scarce, or overused due to excessive human population.

In the absence of agricultural and culinary technology, and given sufficient time, one might expect the prohibition against pig-eating to become genetically incorporated within the human species, since those who refrained would have a lesser tendency of contracting disease and hence, a selective advantage over those that did not. Eventually—perhaps after 10,000 years or so—selection should therefore favor a genetic tendency to avoid pork, rendering cultural prohibitions unnecessary. The use of night soil fertilizers would ultimately depend on the balance struck in each local population between the abundance of disease organisms and the ecological demands on arable land. In both these cases, however, cultural evolution has overwhelmed whatever potential biology may have possessed.

Take another example: drinking milk. For the great majority of the world's human beings, it would seem as bizarre for an adult to drink the milk of a cow as it seems to us when the Masai drink the blood of their livestock. Most adult human beings lack the enzyme lactase, needed to digest the milk sugar, lactose. In fact, there is some correlation between the genetic ability to manufacture lactase and the complex cultural practices of dairying. Does this mean that there is a gene "for" dairying? Almost certainly not. But it does suggest that cultural practices—even complex and highly ritualistic or technologic practices—may be ultimately connected to our biology, often in unexpected ways.

It is extremely difficult to identify the degree to which human behaviors are genetically influenced, and to some extent, it is foolish to try. If a person stands five feet tall, it is meaningless to suggest that three feet and four inches of her stature is due to her genetic makeup, with the remaining one foot and eight inches due to her environment (nutrition, health, etc.). Rather, every inch of her height results from the interaction of her genes and her experiences. Certain human behaviors are nonetheless more rigidly constrained by genotype, while others are more flexible. Behavior that appears almost universal to all cultures, that has not varied during recorded history, and that is clearly of biological advantage, would seem likely to have genetic underpinnings. It is among such behavior that possible conflict between our biologically given inclinations and our culturally generated realities may be found.

The biological evolution of Homo sapiens has not stopped, even today. Nonetheless, it was virtually complete—that is, the human being we know today had already arrived—when human culture was really just beginning. Although the interval between the domestication of fire, for example, and the discovery of agriculture (perhaps 375,000 years) seems like an eternity to modern people, in fact it was a very brief time as biological evolution reckons these things. And whereas the interval between Gutenberg's invention of the printing press and Marconi's invention of the radio (about 500 years) is almost literally insignificant in biological time, it is immense in cultural time. Choose any 500 years of human history prior to about 1000 B.C.: although the cultural changes occurring during those years are bound to have been immense compared to the biological changes during the same time, such a timespan would be uneventful compared to the 500 years that our species has just witnessed.

The reason is the extraordinary pace of cultural evolution, freed from its biologic bonds. There is a general pattern for the rate of cultural change, differing in detail but generally accurate for most systems: cultural change has become approximately exponential. That is, the rate of change has itself been increasing, like a falling object accelerating by gravity, or—more to the point—like a rocket taking off. The general shape of human cultural evolution is therefore like this, ever-accelerating but becoming steeper as we approach modern times:

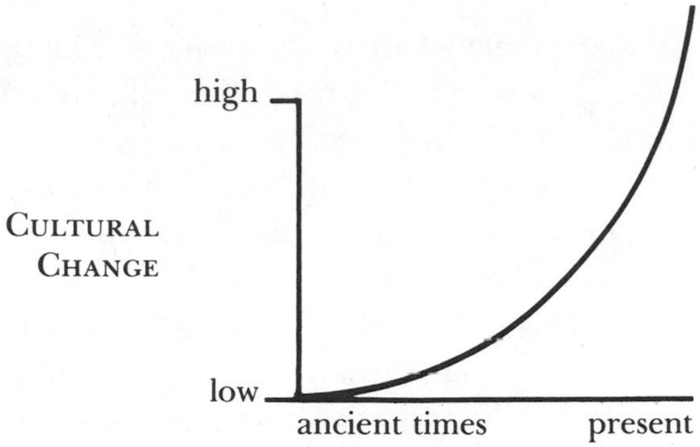
high
CULTURAL
CHANGE
low
ancient times
present

As just one example, let us consider power, defined by physicists as the ability to do work. For hundreds of thousands, or even millions of years, human power was limited by our own muscles. With the domestication of animals, perhaps 6000 years ago, we acquired ox-, camel-, and horsepower, as well as the basic simple machines such as the lever, inclined plane, wedge, wheel and axle, and screw. Then, we began harnessing the wind and the water, via sailboats, waterwheels, and windmills. Chemical power was unlocked after that, through the discovery of gunpowder. Electricity and the steam engine came on the scene only last century, followed shortly by the internal combustion engine and, just a few decades ago, the splitting of the atom with its release of colossal, unimaginable energy. More and more energy has been harnessed, in time intervals that were successively shorter and shorter.

A similar story could be told about most human cultural advances. Transportation: from walking and running, to riding a horse, to railroad trains and steamships, to automobiles and trucks, to airplanes, to supersonic jets, with the last advances all happening during the twentieth century alone. Communication: from the invention of language, to writing, to the printing press, to Morse code, to radio, to television, to personal telecommunications—again, with nearly as much change since the year 1900 as during the preceding 1900 years. Food technology: from hunting, scavenging, and foraging, to simple agriculture, to the appearance of cities made possible by agricultural surplus, to irrigation, to mechanized and chemically assisted agribusiness. The list goes on, including similar patterns in the technology of warfare and killing, and ironically its mirror-image technology in medicine and life-preservation; the complexity of human social organization; independence from outside weather; as well as the growth of human population itself. Most of the time, human beings have managed to keep their balance and their poise, at least partly because even changes that appear rapid when viewed from a distance, often seem slow compared to the pressing moment by moment events of our daily lives. But in fact, the pace of the human cultural experience has been extraordinary, measured by any other standards.

If the time that human beings have experienced in truly rapid cultural change appears as a mere surface gloss upon the deep underlying substructure of millions of years of biological evolution, this is not to devalue its importance. Rather, quite the opposite: despite the fact that we have spent only a few moments as cultural creatures after a long previous lifetime as biological beasts, it is also a fact that we are now entirely enmeshed in that "gloss." It is also the point. In an instant of brilliant illumination, the lightning bolt of cultural innovation has suddenly transfixed and drastically changed our landscape. It may have been brief, but it has also been potent; indeed, its brevity highlights its intensity. For while Homo sapiens is a slow-moving, painfully evolving, tortoiselike Darwinian creature, the speed of Lamarckian cultural evolution is almost like the speed of light. If the interval between the invention of writing and the computer (perhaps 7000 years) seems long, it is nearly undetectable when measured against the interval between the "invention" of jaws from the old vertebrate branchial arches and the "invention" of binocular vision by our arboreal ancestors (perhaps 700 million years); and yet, the former may be no less important than the latter.

In exploring the paradox of Homo sapiens, we shall therefore be exploring the strangeness of time itself. Time is a duration of something, or perhaps an interval between things. When nothing is happening, we say that time stands still; when a lot is going on, time seems to speed up. A second is approximately the space between heartbeats. A year is about what it takes for a dog to grow up. When we say that it takes nine months for a person to develop from fertilized egg to newborn child, we mean that it takes less than one rotation of the earth about the sun, about 270 rotations of the earth about its own axis (causing alternating periods of light and dark on its surface), and perhaps 23 million times longer than it takes to blow your nose.

Einstein has shown that time itself appears to slow down for objects—or, in theory, people—traveling very fast relative to others who are, comparatively speaking, standing still. But for all of our history, all the time, there is no reason to think that the nature of time itself has changed; what has changed is the amount of goings-on that has occupied a given stretch of time, comparing say 50,000 years ago, and today. This is not to say that day to day events didn't happen then, or that lives weren't full. Rather, the occurrences of one life did not carry through substantially into succeeding lives. Experiences didn't accumulate. Each generation had to learn things pretty much anew, and pretty much the same things each time: how to make a kidney, a fire, a digging stick.

"The whole succession of mankind," wrote Pascal, "in the course of so many centuries must be considered as one selfsame man who exists always and learns continuously." In existing "always" (perhaps 50,000 years) Homo sapiens has indeed been learning continuously, and as we have seen, the amount that we have learned has been increasing exponentially. And during that time, we have essentially remained "one selfsame man" and woman, at least in our basic biology.

Our experience on earth, while we have persisted as one, selfsame species, can been divided into intervals of human lifetimes. Taking the biblical threescore and ten (seventy) years as our yardstick, the human experience has lasted roughly 700 lifetimes. Of these, our first 500 were spent without agriculture, without domesticated animals, without metals, living in caves or huddled under trees. Cities appeared only in the last thirty lifetimes, and for the great majority of the world's people, only the last two or three. Only in the last ten such intervals has one lifetime been able to transmit its wisdom to another, nonoverlapping one, by writing. Electricity and steam power were effectively harnessed only two lifetimes ago, about the same time as the Industrial Revolution. And automobiles, computers, jet planes, nuclear weapons, air conditioning, plastics, antibiotics, washing machines, department stores, the Superbowl, world wars, pesticides, assembly lines, and widespread literacy—experiences of a lifetime, all—are uniquely the cultural furniture of just our most recent one, the 700th lifetime.

We evolved rapidly as a species, squeezing more into our 700 lifetimes as Homo sapiens than in thousands that had come before, and more than any other species in the course of its existence. But even then, people of the past could be forgiven for not noticing that much had changed during their lifetimes. Because probably not much had. Now, it is impossible to miss it.

The zebra's leg-bone really hasn't been up in the air very long. An ape-man threw it up, and when it comes down, it is still an ape-man who must catch it. The hare and the tortoise, as different as can be, are nonetheless the same creature.

The interaction between human culture and biology is not simple or easily described. Culture supports biology, in such obvious ways as the encouragement of physical health, production of food, or of shelter. In other, more subtle ways, culture seems to mimic biology. As anthropologist William Durham has emphasized, human cultural patterns may incline toward activities that are fitness enhancing, even though the participants may not consciously be aware of the connection. It is also possible, as physicist Charles Lumsden and sociobiologist Edward O. Wilson have argued mathematically, that our genes hold culture on a leash, so that even our complex cultural and social arrangements are somehow restrained, ultimately, by our DNA. (If so, however, it should be obvious that the genetic leash is very long.) In addition, culture may occasionally go its own stubborn way, following patterns of development and elaboration that are responsive only to itself, and to certain rules of learning and cultural inheritance, as geneticists M. Feldmann and L. Cavalli-Sforza have argued.

Human culture and human biology are both so vast and multifaceted that they are bound to intersect in many ways, in various different points and along oddly projecting edges and planes. Accordingly, it seems naive to expect that any single theory will account for the nature of that contact. Given the peculiar in-between-ness of our species, it even seems naive to expect that we can ever be complete enough to know anything for sure, about either ourselves or the world around us. "Cultural man proposes," writes anthropologist Weston La Barre in his book *The Ghost Dance*,

> but reality disposes, for man is only another kind of animal. In a universe where the very stars whirl and wheel in places we discern only light-years later, and even the great planets wander exquisitively responsive to bodies we incompletely descry in the night, a necessary wisdom is humility, a knowledge that we do not know but can only say that, blessed and burdened with our current projective fantasies, this is the way it seems, now.

One generalization that seems safe, however, is that given the radically different rates at which they develop and the dramatically different processes which they embody, culture and biology—although forced together by their conjunction in the same organism, Homo sapiens—must only rarely be in perfect harmony. Or at least, to echo La Barre, this is the way it seems, now.

Human societies, throughout history and continuing into the present, have eagerly seized on many biologically influenced traits and extended them, nonbiologically, into the cultural sphere. If males are more likely to fight than females, then cultures are likely to dictate that men are *expected* to fight more than women. If people are biologically inclined toward nepotism (treating relatives more favorably than nonrelatives), then culture is likely to reify this inclination, too, prescribing detailed patterns of appropriate and inappropriate behavior. In short, despite the remarkable flexibility that is its hallmark, culture can also lead to hardening of the categories.

Culturally demanded traits have often been exaggerated beyond anything produced by biological evolution, leading to an important, widespread phenomenon that we might call "cultural hyperextension." Rather than setting itself in opposition to innate human inclinations, culture often seeks to mimic and extend these inclinations, in the process outdoing nature itself and going too far. In some societies, accordingly, it is not only expected and required that males be aggressive, and more aggressive than women, but this difference is hyperextended beyond anything that natural selection would condone. Among the Plains Indians of North America, for example, a young man had to prove himself by going without food or water for days, seeking visions. An aspiring Masai warrior had to kill a lion. Among some of the New Guinea tribes, only a human head would do. Although sometimes women were also aggressive in these societies, it seems clear that any existing male-female differences were hyperextended by their cultures. In many cultures today, they still are.

Ornithologist S. Dillon Ripley has described "aggressive neglect" among certain birds, and a similar situation occurs among fish. In such cases, adults spend so much time defending their territories and squabbling with their neighbors that they do not adequately provision their young, keep the eggs warm or (in the fish example) free of fungus infections. As a result, their own offspring are less likely to survive. Aggressive neglect is almost certainly a pathological condition; it would be selected against, biologically. But pathologies can readily "evolve" in human cultural systems, especially when they represent seemingly adaptive hyperextensions of real situations and legitimate concerns. Today, one might argue that America has been subjected to a nuclear age example of aggressive neglect, the Caspar Weinberger disease. Defending one's offspring has long been adaptive; likewise for defending one's group. It has even been appropriate to sacrifice other possible benefits—even, sometimes, one's life—in the pursuit of security from competitors or predators. But it is only by a grossly maladaptive cultural hyperextension of such tendencies that the United States invests enormous quantities of its national treasure in a futile attempt to coerce its adversary, while in the process undermining its strength and the real security of its own children.

Efforts to "biologize" human behavior, whether successful or not, have typically been attempts to identify the biological underpinnings of what we do. By contrast, biologically based critiques of human behavior have sought to focus on presumed tendencies on the part of human culture to inhibit or to fall short of our biological inclinations. Culture then stands accused of being insufficiently biological, and the message, typically, is that if only we could structure our societies more in accord with our biology, then all would be well. But in searching for the likely causes of our malaise, discomfort, and danger, and in fact, just seeking to clarify the human situation, it seems likely that one culprit is cultural hyperextension, the hare's tendency to run for miles in a direction that the tortoise has just taken a single step. In short, sometimes the problem is not that culture is insufficiently biological, but rather that it is excessively so.

Given the likely disparity between culture and biology, it is surprising that earlier thinkers have so often been inclined to accept the way things *are* as the way nature intended them, the way they have always been, or should always be. In his *Politics*, for example, Aristotle wrote, "It is evident that the polis [city-state] belongs to the class of things that exist by nature, that man is by nature an animal intended to live in a polis." Not surprisingly, Aquinas said the same thing about the medieval, Holy Roman Empire, and to many in the present day, the nation-state appears to be self-evidently the highest and most biologically appropriate political organization for our species. A perspective on human biological as well as cultural evolution may help us break free from such limiting "chronocentrism," the notion that current times are the key to all time.

As opposed to "culture," the word "civilization" derives from the Latin "civis," which means citizen of a city; it typically refers to more elaborate, advanced, and technologic systems. The first civilizations therefore began even more recently than the first human cultures, and there is little doubt that the entire structure of civilization has developed during a time period so brief it is unlikely that we ourselves changed biologically during that interval. Perhaps our biological natures are so fluid, so flexible, so infinitely malleable and adaptable that we are able to coexist comfortably with any cultures that we create, and any civilizations as well. In short, perhaps we have essentially no genetically prescribed human nature; if so, then we literally are what our culture makes of us, and by definition, we couldn't be uncomfortable, except perhaps insofar as cultural systems impose their own unique and distressing pain or anxiety, or subject us to the din of inequity.

But on the other hand, if there is a human nature, a Darwinian tortoise no matter how diffuse underneath the cultural trappings of modern human beings, then we can be virtually guaranteed that it coexists uneasily, at best, with our Lamarckian hare. The hare will almost certainly undershoot our biology, or hyperextend it. As we have already seen, there would be little if any difficulty exchanging a Cro-Magnon and a modern infant but great incongruity in making the same switch with adults of both cultures. This incongruity (literally: an inability to fit), between our biology, which has evolved by the laborious process of natural selection, and our culture, which has appeared with explosive speed through cultural evolution, is the root of nearly every human difficulty. It will be the focus of the remainder of this book.

CHAPTER FOUR

SEX

From Procreation to Recreation

The sexual life of the camel is deeper than anyone thinks.
In a moment of amorous passion, he tried to make love to the Sphinx.
But the Sphinx's posterior arrangement lies deep in the sands of the Nile,
Which accounts for the hump on the camel
And the Sphinx's inscrutable smile.

—Anonymous

Anyone watching monkeys in a zoo might think that they were oversexed. In fact, Lord Solly Zuckerman, a highly respected British biologist, once concluded that the social behavior of a troop of baboons was almost entirely determined by sex: frequent mountings by the rapacious males matched only by nearly constant solicitation from the nymphomaniac females. Sex seemed to be the social glue that held monkey life together.

Within recent decades, however, biologists, anthropologists, and psychologists have conducted numerous studies of the behavior of free-living primates in their natural habitats. This work has revealed sex to be a relatively unimportant part of their normal behavior. In fact, all female mammals have a short period of time, ovulation, when they produce one or more eggs that are capable of being fertilized and producing offspring. Among most animals, such as deer or birds, ovulation is limited to certain seasons of the year (spring in birds, autumn in deer). Among others—mice and men, for example—it continues throughout the year in a regular, predictable cycle. In either case, conception can occur only during these restricted times, separated by relatively long periods of infertility.

Females of most species are sexually receptive to the males only during this brief ovulation period and in fact, they may react quite aggressively to any sexual overtures outside this time. In almost all cases, however, the males are also selective in their advances, coming into season ("rut" in deer and elk) in synchrony with the time of ovulation ("heat") by the females. In addition, among most species the males are able to determine when the females are maximally receptive by being sensitive to chemicals produced by the females at this time. The extraordinary interest shown by male dogs in the rear ends of bitches in heat attests to this sensitivity, as does their interest in the urine of other animals.

Because sexual interest is generally limited to periods of fertility and animals can precisely identify these times, sexual intercourse is actually quite rare among animals in nature. This makes sense, because copulation involves a great expenditure of energy, and energy—the basic currency of life—is not expended without good reason. Furthermore, copulating animals are generally rather preoccupied with each other and are therefore more susceptible to being attacked by a predator. In many cases, when the animals involved are highly aggressive or predatory, there is even danger that the prospective lover will be attacked by its mate. So sex may be not only wasteful, but downright dangerous.

Among human beings, however, sexuality is unique in that despite its dangers and disadvantages, copulation is *not* limited to times when reproduction is likely. (In fact, under many circumstances it is limited precisely to times when it is *unlikely*.) This seeming anomaly is explained when we realize that sexual behavior in Homo sapiens has been liberated from the purely reproductive function that it serves in nearly every other animal. Just as George Bernard Shaw once said that youth is so wonderful it's a shame that it's wasted on the young, human beings have found sex too wonderful, or useful, to be wasted on reproduction alone. It has been modified to serve a "higher" function: maintaining and strengthening the bond between adults.

In Robinson Jeffers' magnificent, brutal epic poem "Roan Stallion," we are given the image of a woman who seeks something more gratifying than her disappointing relationship with an insensitive husband. She literally falls in love with a glorious stallion, and that relationship, although sexual, is also deeply suffused with religious awe and mystical transcendence. At the end, the stallion kills the woman's husband, and the woman, in a final recognition of her fidelity to the human race, kills the stallion. For her, killing the roan stallion was almost like killing God: as a human being, her sexuality had been modified to serve a higher function, not just a physical act, and certainly not just procreation, but the essence of relationship. The stallion, by contrast, is capable of no such leap, either emotionally or cognitively. Human beings, alone among animals, are.

To understand this discontinuity, let us go from the sublime to the anatomic, and consider the peculiar anatomy of our species. In assuming an upright posture we diverged radically from the traditional body plan of other mammals. Most mammals are supported by four limbs, with the internal body organs basically suspended like salamis on a butcher's rack from a long backbone that is normally parallel to the ground. But by getting up on our hind legs we forced a radical redesign of our skeleton, especially our pelvis, not only to accommodate a new attachment for the hip muscles, but also to provide a basinlike scoop supporting our internal organs. This modification restricted the space available for our birth canal, thus making childbirth in human beings a rather trying affair when contrasted with other mammals. The newborn's head is exceptionally large relative to the rest of its body and in fact, would probably be larger yet if not for the restrictions imposed by a narrow exit tunnel. This limitation of human brain size at birth has in turn necessitated a prolonged period of brain development following birth. (It also necessitates a lot of protein in the first few years of childhood. Protein deficiencies at this time can deprive the brain of needed development and result in permanent retardation. The low protein diets of many underdeveloped nations as well as the starvation in sub-Saharan Africa are a particularly serious problem for this reason, even for those who survive.)

The natural delay of much brain growth and maturation until after birth has further necessitated a prolonged period of infant dependence among human beings, during which time substantial adult attention is needed, not only to protect the helpless infants but also to provide necessary training and guidance. This is where sex comes in. What mechanism would keep the parental pair together, thus providing maximum security and training for the young? If sex were limited to ovulation only—as it is with free-living baboons today, or with the roan stallion's more traditional mates—it is unlikely that the family as a relatively stable male-female union would have evolved. Instead, our social system might resemble that of the modern baboon, with numerous males and females but no permanent association between them, only the transient union of male-female "consort pairs" when the females are ovulating and receptive. Or perhaps we would be emulating the wild horses, with a dominant stallion maintaining a harem of mares who respond sexually, but probably do not love.

Human beings are unique among animals in that we participate extensively in nonreproductive sex. In order to enhance its effectiveness in cementing a bond between adults, the physical act has been reinforced by emotional factors generally identified as love. Whether love can operate extra-sexually is a debatable point, since even Platonic relationships such as "love" of God, country, or parent may involve sublimation of sexual impulses.

We are also unusual, if not unique, in the degree to which ovulation is hidden. Even casual zoo visitors can determine the estrus state of a female monkey, typically by the enormously enlarged and brightly colored perineal region. Women not only lack anything comparable, but in most cases, the exact time of ovulation is even concealed altogether; delicate thermometers or chemical tests are in fact required to detect what in other animals is usually immediately apparent. Sociobiologist Nancy Burley has suggested a novel explanation for this paradox, based on a confluence of evolution and human consciousness. Thus, she suggests that early in human evolution, the connection between ovulation and pregnancy may well have been recognized, along with the connection between childbirth and pain, and also mortality. This latter awareness, among women, could have inclined them to *avoid* pregnancy by refraining from sex when they were ovulating. Now, here comes the interesting twist: women who were conscious of their own ovulation would have been selected against, whereas those whose ovulation was concealed would have left more offspring—unintentionally, to be sure—thereby selecting for what they did not know. According to this hypothesis, then, it is not so much biology and culture that are in conflict here, but biology and consciousness; and in Burley's scheme at least, biology won.

It seems likely that fear of pregnancy inhibited female sexual behavior, once the connection between sexual intercourse and reproduction was understood. It is also a truism that with the invention of cheap and available contraceptives, sexual behavior has been more liberated from its biological consequences than ever before.

Evidence for the unique role of sex in human life comes from study of its physical components. Ejaculation by the male involves the expulsion of sperm in a rich, nutrient liquid suspension (semen), providing the microscopic sperm with food, energy, and a fluid medium through which they might swim to their ultimate rendezvous with a ripe egg within the female's body. Vaginal fluid is acidic, potentially lethal to sperm; male ejaculate therefore compensates by being appropriately alkaline. Discharge of the necessary fluid requires vigorous muscular contractions by the tubes that carry it, and the pleasurable sensation accompanying its release is closely connected with the release of tension as the tubes are physically emptied. It is thus interesting that the physiology of male orgasm does not differ significantly from humans to most lower mammals. The "need" for sexual release and satisfaction upon its accomplishment can be explained in evolutionary terms as due to the selective advantage of sending one's reproductive products on their way. (Individuals who copulate successfully are obviously more likely to leave offspring.)

The female version is a different story. There is no compelling evidence for a female orgasm in any animal other than Homo sapiens, although recent data suggests it as a possibility in certain monkey species and in the dwarf chimpanzees. It seems that in this regard we are again unique, or nearly so. By a great diversity of mechanisms, all initiated and then strengthened by the evolutionary force of natural selection, females of every species are variously motivated to copulate at the appropriate times. But in nearly all cases, there is no indication that they ever "get" anything out of it, other than simply to satisfy the motivating factors that led them to copulate in the first place. And ultimately, of course, getting pregnant as a result.

Among human beings, however, women are capable of experiencing orgasms of their own, although quite different from male ejaculation. Skinnerian psychologists might describe this phenomenon as "positive reinforcement," in that once experienced, it tends to increase the probability that the individual will repeat the behavior immediately preceding it. Orgasm, then, gives women a direct stake in copulation, making regular sexual activity more likely and with it, the increased coordination of male-female behavior that is the evolutionary payoff.

It is also worth noting that female orgasm is more difficult to achieve than its male counterpart. Thus, whereas men can often reach orgasm in a few minutes or even less, women typically take longer. Among certain animals it appears that dominant males often take longer to ejaculate than do subordinates. The mounting of a female grizzly bear by a dominant male is rather a leisurely affair, whereas subordinate males, obviously hurried because they are harried, spend most of their mating time looking over their shoulder, alert for the possible appearance of the dominants. Not surprisingly, they ejaculate very quickly, before they could be interrupted. Female grizzlies, to our knowledge, do not have orgasms, but it seems likely that even if they did, the sexual style of subordinate males would be unlikely to elicit very much response.

Assume now that sex among human beings serves social bonding as well as reproduction—recreation as well as procreation. Then it is not surprising that female orgasm, which may well be related to bonding, is more likely when the sex act itself is prolonged, and hence, when the male partner is more likely to be successful, effective, and desirable, the human equivalent of social dominance among animals. It remains to be seen, of course, whether "latency to ejaculation"—the time between penetration and ejaculation—in men is related to experience, self-esteem, and something analogous to social dominance. But it may be significant that premature ejaculation is particularly a complaint of young, inexperienced males, that ejaculation takes longer among older males, and that sexual responsiveness among women is much greater after lengthy sex—characteristic of dominant male animals—rather than rushed and harried sex, more typical of younger subordinates.

Many human physical characteristics attest to the importance of sex and indicate our potential for a high level of sexual fulfillment. Thus, all mammals feed their offspring milk, made from their mammary glands. But only human beings have breasts. Among all other mammals, the milk-making glands are trivial and inconspicuous except when the females are actively lactating. By contrast, human beings are unique in possessing well-developed, protuberant breasts even among nonreproductive individuals. There seems little doubt that this development is related to our exaggerated sexuality and to our elaboration of nonreproductive sex as well. By exaggerating a trait that is biologically associated with successful reproduction, natural selection may well signal reproductive competence; otherwise, there is no biological reason for female breasts to become enlarged at the onset of menstruation; they could just as well wait until pregnancy and lactation, as in all other mammals. But in human beings, sexuality has developed alluring aspects, quite independent from reproduction.

One needn't be a trained evolutionary biologist to speculate in this manner. Stephen Dedalus, the youthful hero of James Joyce's *A Portrait of the Artist as a Young Man*, suggests the following hypothesis while musing with his friends on the nature of female beauty (as perceived by men):

> [E]very physical quality admired by men in women is in direct connection with the manifold functions of women for the propagation of the species [modern biologists would here substitute "propagation of their own genes"]. It may be so. The world, it seems, is drearier than even you, Lynch, imagined. For my part I dislike that way out. It leads to eugenics rather than to esthetic. It leads you out of the maze into a new gaudy lectureroom where MacCann, with one hand on *The Origin of Species* and the other hand on the New Testament, tells you that you admired the great flanks of Venus because you felt that she would bear you burly offspring and admired her great breasts because you felt that she would give good milk to her children and yours.

Dreary or not, there may well be a good deal of truth in such notions.

It is noteworthy, however, that prelactational breast size among human beings actually does not correlate with milk production, so that the "great breasts" of Venus to which Stephen Dedalus refers may be potentially false and misleading advertisements. The nonlactating female breast is composed almost entirely of fat, whereas it is glandular tissue—which develops during pregnancy—that produces the milk. Large, fatty breasts may therefore be biologically deceitful, giving the illusion of mammalian plenty, but when push comes to shove, not necessarily any more substance.

Just as women have the largest breasts of any mammal, men have the longest penis of all primates, a characteristic that is as distinctive as our large brain, but to some, less elevating. Maybe the oversized penis of Homo sapiens is an adaptation to deposit sperm deeply into the female reproductive tract, necessitated by the tilting of the female pelvis that resulted from walking upright. Or perhaps there is some other reason; in any event, men seem to worry about their penis size about as much as women worry about their breast size. (Ironically, such concern on the part of men is also about as irrelevant, biologically, as the breast obsession of Western societies: once erect, there is very little difference in penis dimensions.) Regardless of our biology per se, it is clear that we have achieved substantial cultural hyperextension in our sexually related phobias and preoccupations.

As if all this hyperdevelopment of breasts and penises was not enough, our hairless body, covered with a great number of sensitive touch receptors, provides additional opportunity for enhanced sensual input. And the great dexterity of our fingers and lips (again, surpassing that of all other animals) also enables delicate and varied stimulation of the partner. We are clearly the sexiest animals on earth, by virtue of our biology alone.

Nonreproductive sex is an exceptional human attribute and evolution has obviously conspired to provide us with the opportunity and motivation for such activities, emancipated from reproduction. It is therefore ironic that the Catholic Church, for example, specifically considers such behavior to be immoral because it is "bestial" and "catering to our animal instincts." *Animals* are indeed limited by their instincts. They engage in sex only when it serves reproduction. People, by contrast, are in fact being uniquely human when they engage in sex for its own sake, for the higher relationship it helps achieve between a man and a woman. Sex limited to reproduction, on the other hand, is the norm for animals but not for people.

This is perhaps the most glaring example of conflict between the biological and cultural sides of sex. There are others, however. No biological phenomenon, in fact, has been more heavily overlain by cultural embroidery.

It is impossible to say which of the many marital arrangements practiced by different human cultures is the most biologically "natural." It may well be that local diversity in such arrangements is the biologically "correct" situation. Among monkeys and apes, different species practice different social systems and often, the same species may differ from place to place. In such cases, there seems to be adaptive significance to each particular arrangement. Living things have been selected to engage in different social systems depending on which ones are most successful in any given ecological situation. Yellow baboons forage through the dry east African savannah in large social groups with the males providing maximum protection for the females and young. By contrast, mandrill baboons of moist, forested west Africa travel in small nuclear family groups that appear to be monogamous. Among human beings, social and marital arrangements are determined by cultural and especially religious convention, without obvious regard to the regionally optimum ecological arrangement. Thus, the Moslem's maximum of four wives on the one hand, the Judeo-Christian policy of only one, on the other, may or may not be biologically sound, but both were probably arrived at for reasons other than their ecological and evolutionary utility.

The "double standard," allowing greater liberty for sexual experimentation and dalliance to men than to women, is well established in most human cultures. In general, men are more easily stimulated than women, and are the sexual aggressors. Rape is characteristically done by males to females, not vice versa, and this is not simply a result of the former's greater physical strength. Rather, it reflects a basic evolutionary consequence of maleness versus femaleness. Similarly, it is not just coincidental that prostitutes are usually females, visited by males—only rarely vice versa—

and that pornographic magazines are regularly purchased by millions of men but have much less market among women. The cultural differences, with all their complex embroidery, are almost certainly overlain on some clear-cut biological differences as well.

Females of every species produce eggs, each of which may be thousands of times larger than the male's sperm. Among such animals as fish, amphibians, reptiles, and birds this contrast in size is especially striking, since the egg is easily visible to the naked eye, whereas all sperm are microscopic. Among mammals the egg is considerably smaller than in lower forms, since the developing young are nourished from the mother's blood and the egg need not contain great food reserves. Nonetheless, the nutrient-laden egg dwarfs the tiny sperm. The production of eggs in all animals, including ourselves, clearly involves a much greater investment of metabolic energy than does the manufacture of sperm, especially since once fertilized, an egg must receive massive subsequent investments from the mother's body. Some birds may produce eggs totaling fully one third of their own body weight. An adult woman may produce about 400 eggs during her lifetime, whereas a man will discharge 100 to 300 *million* viable sperm at a single ejaculation. Because of their small size, the latter can be produced in fantastic numbers. In addition—and perhaps, most important responsibility for both pre- and postnatal care of young almost always falls upon the female. Among birds, prenatal care involves the production of large, well-supplied eggs, whereas among mammals, the female nourishes developing young from her own bloodstream and then later, by the milk she produces.

Seen from any angle, in most species the female has a greater stake in the outcome of any one copulation than does the male. She has invested more metabolic energy in egg making and she produces far fewer eggs. Hence, she stands a greater chance of losing out entirely if she is unsuccessful in securing proper fertilization for them. She literally has all her eggs in a limited number of baskets and—just as literally—she must bear the responsibility of a mistake. By contrast, the male's biology gives a degree of sexual license. It is no accident, then, that males are characteristically more easily aroused and more prone to be sexually forward than are females. Thus, even though male sexuality is typically cued to female receptivity, during the breeding season males are normally less discriminating than females.

Certain species of orchids get themselves pollinated by producing flowers that mimic female wasps. The sexually aroused male wasps try to copulate with them, get dusted with pollen, then proceed to the next acquiescent and deceptive flower, whereupon they unknowingly transfer the pollen between their successive floral "lovers." It is noteworthy that we don't find flowers masquerading as male wasps, simply because no self-respecting female would be so easily misled into copulating with a flower; only with another wasp. After all, she has large and precious eggs at stake; he has only some cheap and easily replaced sperm. If those creatures he finds so alluring were female wasps instead of flowers, he would be spreading his genes, and at little cost to himself. As it is, the wasp's maleness, with its eagerness and lack of discrimination, is readily exploited by the flowers.

In many other species, males will court almost anything, whereas the females are more demure and fussy about their choice of mates. Male crabeater seals, for example, congregate around females and their unweaned pups, perched on ice floes. The males try to mate; the females resist. The males bite the females; the females bite back. Both may get covered with blood, as the male seeks to induce the female to accept his sperm, while the female resists, presumably because it is not in her evolutionary interest to get pregnant too soon, since that would require weaning her pup before it is adequately grown. The course of true love among animals doesn't always run smoothly. Animal courtship often involves breaking down the resistance of the female, with stimulation of the male being relatively unimportant, because it is so easily achieved. Once again, it is easily achieved because as with the wanton wasps, the cost of male error is small and the potential benefit of success, large.

In at least one species of insect, the "Mormon cricket," the tables are turned: males are shy and females are sexually pushy. In this species the male transfers a large, gooey, nutrient-rich structure called a spermatophylax as part of mating; the spermatophylax represents a metabolic investment comparable to that of the female's eggs in many other species. Not surprisingly, therefore, in this case the male acts "femalelike" and vice versa. Female Mormon crickets must climb on top of the males before the latter will consent to mate with them, and the precious spermatophylax will only be transferred if the females in question are heavy enough; heavier females make more eggs, so in this unusual (but understandable) case, the male holds out for the mate with the biggest dowry.

Most of the time, however, it is the males who are eager to transfer their abundant sperm and the females who jealously guard access to their precious eggs.

The usual male-female difference in sexual style is also good biological evolutionary strategy. When males produce large numbers of sperm and can easily replace any that are discharged, natural selection would favor "playing fast and loose." By freely expending sperm on the slightest pretext, they would stand a greater chance of leaving offspring in the future.

In contrast, an error by a female may seriously handicap her reproductive future. Although it is to the male's advantage to maximize the dissemination of his sperm, good female strategy calls for optimization. Successful matings between males and females of different species produce offspring—hybrids—that are generally inferior to either parent in a variety of ways. As a result, the careless female who falls for the charms of a male from the wrong species has expended her precious genes (or, more to the point, her valuable reproductive investment) on a loser. Since offspring from such females are less likely to be successful, females with a genetic tendency to be less discriminating will leave fewer offspring. Eventually such types will decline in numbers, replaced by those of the more careful females who trusted their genes to male partners more likely to bring them success.

For example, domestic dogs are all members of the same species; hence they can exchange genes with one another. But artificial breeding programs have divided *Canis domesticus* into dozens of breeds, some of them dramatically different from each other. Take a female Pekinese or a tiny Chihuahua: in heat, she will readily accept any male dog, even a St. Bernard or Great Dane.* By accepting such an oversized mate, the miniaturized canine mother will have almost certainly signed her death warrant: her developing offspring will be too large to pass through her birth canal, and both bitch and litter will likely die unless someone performs a cesarean.

This is clearly a very artificial situation. Under natural conditions, a single species such as the domestic dog would not have differentiated into such very different shapes and sizes. Under current conditions, the small bitch who allows herself to be inseminated by a giant male is an evolutionary loser; such lack of discrimination would be strongly selected against. For her paramour, however, the situation would not be tragic at all. Dogs do not form long-lasting pair bonds, so if his mate dies in childbirth, the undiscriminating Lothario has not lost very much. This may seem overly harsh judgment, and certainly unfair in its punishment of the female while the equally "guilty" male gets off scot free. But evolution is not concerned with adjudicating morality, but rather with differential reproductive success.

We would expect therefore that under natural conditions, selection would tend to produce choosy females and relatively undiscriminating males. In many cases, a female whose mating preferences are overly lax may produce offspring that fail to develop normally and are aborted. Or the offspring may be perfectly healthy, but sterile, such as a mule which—true to its reputation—stubbornly refuses to pass on its parents' genes.

*In such cases, the physical mismatch in size between penis and vagina may make mating very difficult, however.

*
**

THE PRECEDING is not intended as a justification for the double standard in human society, or as a plea for its further elaboration. But it may help to explain an otherwise puzzling basic difference in the sexual behavior of men and women, even acknowledging that the human situation is much more complicated than it is among romantically inclined French poodles. Opportunities for disadvantageous hybridization may have existed among our Australopithecine ancestors, and even Cro-Magnons may have had the potential to interbreed with the Neanderthals. In fact, fossil skeletons at Mt. Carmel, in modern Israel, suggest that this may have occurred. But modern Homo sapiens is not seriously subject to any such temptations. Nature's scarlet letter, meted out automatically to errant animals, is handled by cultural means in the human species.

What does all this have to do with the conflict between biology and culture among Homo sapiens? As we have seen, natural selection, working through the advantage gained by a strong bond between parents, evolved mechanisms to achieve and maintain that bond, most notably the liberation of sex from its purely reproductive function. This may well have been achieved in part by endowing women with the capacity for orgasm, as well as the attendant emotional attachment of both partners. But by enhancing the female's sexual responsiveness, evolution has made the female receptive to males other than just her chosen mate. Thus, although it is characteristic of most animals (including humans) for the male to be more easily excitable, seduceable, and hence, likely to philander, the phenomenon of intentional female philandering is somewhat unusual in nature, but conspicuously well developed in human beings. Ironically, while selecting for female responsiveness as a means of maintaining the pair bond, evolution may also have set in motion forces that among many human cultures have traditionally acted to weaken that pair bond.

Different societies respond to a woman's sexual infidelity in almost every way imaginable, ranging from the death sentence to indifference to (very rarely) approval. Philandering by the male elicits equally varied responses, although among most societies it is generally regarded as less serious than comparable female behavior. This may well be due to the fact that male promiscuity has a firmer biological basis. The woman who gets pregnant has a substantial reminder of her behavior, and men whose mates get pregnant have lost some of their own potential fitness. The outrage of the cuckolded husband, often abetted and even encouraged by societal norms that may condone brutality or even murder as a response, represent a heavy-handed dose of cultural hyperextension.

Evolution, however, has set a potential trap for the dallying man as well. Among nonhuman animals there is little or no evidence of emotional concomitants to sexual behavior by the male, excepting the physical urgency associated with orgasm itself. But selection for the pair bond in human beings has also had subtle effects on the man as well as the woman, endowing him with a capacity for strong emotional attachments that may well equal hers. Clearly, such attachments are not universally felt, either by all human cultures or by all individuals within any given culture, or even by a single individual in all cases. But the "eternal triangle" is a fairly common occurrence, in which a man may love two women; or a woman, two men. Biological evolution may contribute to attachments that are often unacceptable to the standards of cultural evolution.

Prior to Western colonialism and Judeo-Christian social imperialism, the vast majority of human societies were in fact polygynous: the preferred marital system involved one successful man mated to several women. Given the biology of maleness and femaleness, it is possible for such arrangements to enhance the evolutionary fitness of all—except for the excluded bachelors. On the other hand, polyandry is predictably rare: there are very few societies in which one woman is typically mated to several men. Such an arrangement would likely reduce the fitness of the participating males, since it would lower the probability of any single one of them being a father; at the same time, it would do little for the fitness of the woman, since her maximum reproductive success is unlikely to be greatly enhanced. She can only have one child at a time, no matter how many husbands.

Human biology, moreover, is entirely consistent with the suggestion that Homo sapiens is a mildly polygynous species: men are somewhat larger and constitutionally more aggressive than women, and women become sexually mature earlier than men. This pattern is widely found among polygynous species, in which males are selected to delay engaging in vigorous male-male competition until they are old and large enough to have some chance of success. Bull elk, for example, do not compete for a harem of females until they are several years older than the females with whom they will eventually mate.

In this regard, it may appear that culture is the villain, often unreasonably frustrating the natural, healthy, biologically evolved proclivities of our species. But what is "natural" is not ipso facto good. As human beings, we have the right, perhaps the obligation, to choose monogamy over polygyny, or a uniform sexual standard over the double one. Many things that are biological, and hence, very "organic," are not necessarily very pleasant: typhoid, forest fires, gangrene, and hookworms, for example. If the disharmony between biology and culture is ultimately responsible for most human dilemmas, this does not imply that biology should therefore be permitted to win. If Homo sapiens is to win, however, then it might help to become more truly "sapient" about this fundamental conflict.

The biological connection between sex and love is almost certainly a result of the fact that the initial "goal" of evolution was to maintain and strengthen the pair bond, for the ultimate benefit of the children and hence, the fitness of the parents. Insofar as cultural restrictions serve to discourage adulterous or promiscuous behavior, thereby preventing erosion of the parental pair bond, such restrictions may ultimately be consistent with the thrust of biology as well. Furthermore, it may well be that excessive sexual freedom is ultimately doomed to impair the emotional integrity of its participants . . . not because it defies certain fundamental ethical or religious (that is, cultural) precepts, but rather because it ignores the basic connection between sex and love promoted by biological evolution.

Given that natural selection has succeeded in adding a potent emotional and pair-bonding function to the straightforward reproductive role of sex, it would be foolhardy to expect the two functions to be easily separated. Human beings achieve enormous personal, emotional benefits from sexual behavior, but these benefits can become hindrances to "pure, natural, free-love" when emotional ties are not only unsought but also unwanted. When culture prescribes sex without such ties, physical relationships can usually be made just that: physical, and devoid of personal bonding. But the ultimate cost may be great in terms of decreased ability to establish warm, loving relationships in the future. The classic case of sex without love is of course the prostitute, who by severing the biological connection and concentrating on the former, may suffer diminished capacity for the latter as well.

Biological evolution undoubtedly has had a profound impact upon human sexual behavior, selecting for its pair-bonding functions in addition to its purely reproductive effects. However, we must remember that biological evolution is a terribly slow process, easily outdistanced by cultural change. When natural selection began operating on the genetic systems of Homo sapiens and our precursors, fashioning our response to sexual encounters, human culture was very rudimentary, almost nonexistent. To primitive human beings, with virtually no complex society to surround and nourish us, the maintenance of a strong male-female pair was of crucial importance to the survival and success of children. Since that time, our culture has evolved so much that many of our biological characteristics may be anachronisms, as out of place as the bones of *Tyrannosaurus rex* in midtown Manhattan. But whereas those bones reside in the American Museum of Natural

History, some of our current biologic fossils still lurk within ourselves.

Evolution's payoff is in successful reproduction and, in turn, the successful reproduction of one's offspring and other relatives. A fatherless prehistoric child was at a real disadvantage, unlikely to be well provided with food or protected from enemies. Unless immediately adopted, an orphan was as good as dead. But in modern society, a successful paternal bond is only indirectly related, if at all, to the ultimate reproductive success of its offspring. Unwed parenthood is common, and the wealthy movie star or welfare mother will both probably produce reasonably normal, healthy children, eventually capable of having children themselves. Many cultures admittedly continue to discriminate against bastard children, and their parents, while supporting institutions such as religion that encourage relatively permanent male-female bonding. Thus, cultural defense of marriage may have developed in large part in support of the important biological function of marriage, the production of successful offspring. But the cultural innovations that are rendering the biologically evolved bonding of parents outmoded may well be undercutting the need for these cultural controls as well.

The institution of marriage also developed in support of cultural practices, notably the oppression of women, which are themselves facilitated by male-female differences. But other cultural developments such as schools, personal wealth, and government assistance, are drastically reducing the importance of the preexisting biological function. As a biologically necessary phenomenon, the male-female bond may be increasingly outmoded, even as we retain a deep-seated tendency nonetheless to respond strongly to real and potential partners. Parental bonding, in short, presumably began as a means to an evolutionary end, and it quickly incorporated sex as a useful tactic. Now, the end—successful reproduction—may well be achievable without the means, but we nonetheless pursue social and sexual bonds at least as much as reproduction itself.

Human beings still possess a coccyx, or tail-bone, as a remnant of our mammalian ancestry. In modern Homo sapiens, this structure has decreased since it is no longer selectively advantageous. Similarly, the appendix and tonsils are probably rudimentary compared to a time in our evolutionary past when they conferred some distinct advantage. As biological characteristics cease to be of value, they cease to be included in the genetic constitution of living things. Essentially, this is the Second Law of Thermodynamics at work: any complex structure, whether coccyx or the federal Constitution, requires an input of energy to keep it operating. Without natural selection or human consciousness providing regular input, nonrandom structures tend to decay: as the physicists put it, their entropy increases. But this process takes time, and living things, no less than governments, carry a lot of deadwood around with them. We might anticipate that as a result of present cultural trends emancipating a child's biological success from the success of its parents' social relationship, selection for biological factors maintaining that relationship would eventually decline. Biological evolution for the male-female bond may in fact be about to reverse itself. Clearly, its adaptive advantage (the production of successful children) is less important now than during our evolutionary past.

The precise role of biology in this entire business is difficult if not impossible to assess. Although impatient youth may bewail the conservatism of cultural change, our institutions are more likely to respond promptly to such drastically altered facts of life than is our genetic makeup. How they do, and indeed whether they do, is much more up to us than to our DNA. Ultimate success of male-female pair bonding in Homo sapiens may well depend upon the direction of cultural evolution and upon whether this bonding will be seen as worth maintaining for its own peculiar values. At one time, not only did the ends justify the means, they produced the means; now, the means may well have to stand, if they are to stand at all, as ends in themselves.

It is ironic that even as biological evolution may be inclined to eliminate the male-female bond because of its newly diminished utility, early cultural dissolution of it for the same reason may be bedeviled by the biological factors which are still tuned to maintaining the bond. It is certainly a very complicated story. Human culture has been exposing the adult pair bond to an increasing array of stresses. For example, we have already seen that if we did not differ significantly from other animals we might expect nothing from sex but procreation. But our evolution has provided other possibilities and accordingly, sexual behavior is now expected automatically to produce profound, "meaningful" relationships, experiences that are enriching almost to the point of mystical significance. We are in many ways a uniquely self-conscious animal, acting and simultaneously aware of ourselves as we do so. When it comes to experience as intense as sexuality, it is thus not surprising that we appraise performance, both our own and our partner's.

Such evaluative self-awareness may constructively improve any situation. But applied to sex, for example, it may also create temporary impotence, frigidity, and/or chronic dissatisfaction, the feeling that one is somehow missing something. Ironically, the proliferation of literate, scientific sex manuals has probably exacerbated the situation by heightening public awareness of what is "normal," and what, and how often, they should expect from their partner, themselves, and their experiences.

Another significant strain is imposed by the culturally evolved frequency of human encounters and interactions. Our genetic systems undoubtedly evolved through hundreds of thousands if not millions of years during which normal social interactions were limited to the other members of our primitive hunting and gathering bands. These probably contained no more than several dozen members each. Individuals within each band were certainly well known to each other. Even the occasional interaction with members of a different social unit probably constituted either a battle or a reunion of old acquaintances, since for better or worse, groups probably maintained a regular orientation relative to each other.

Thus, while the choice of sex partners may have been limited, so was temptation.

Today's human being may personally encounter literally hundreds, even thousands, of other people every day, many of whom may be quite attractive. This is something quite new to our experience as a species, although it is less obvious than computers, nuclear weapons, or nation-states. But whether identified or not, our genetic system almost certainly has not had time to evolve biological defenses against such hyperstimulation. Modern advertising barrages us with the images of our most alluring specimens in constant effort to generate the desired longing, envy, expectations, associations or simply to attract our attention. And given the effect of novelty on sexual behavior, attract our attention it does.

For nearly all species, sexual performance declines upon repeated experience with the same partner; however, among animals as diverse as rats and cattle, it can be returned to high levels upon exposure to a new one. It is said that when Calvin Coolidge and wife were separately touring a model farm, Mrs. Coolidge was impressed with the frequency with which the rooster mated: "Please point that out to Mr. Coolidge," she suggested. The President, in turn, asked whether the rooster was always mating the same hen. No, he was told, many different ones. "Please point *that* out to Mrs. Coolidge," he ostensibly replied. Students of animal behavior now speak of the "Coolidge effect," whereby sexual behavior, especially of males, is increased with a new partner. We obviously cannot say whether animals feel "bored" with a well-known mate, but they are clearly invigorated by a new one. To some extent at least, a similar phenomenon may occur among human beings. The classic "seven-year itch" reflects a time in marriage when other stimuli may be sought, and the marriage of old (usually wealthy or prominent) men to much younger women is often accompanied—if not precipitated—by the increased sexual vigor that usually ensues . . . if only for a short time. There is little known about the effect of novelty on sexual behavior in female animals, or on women either. It is likely to be similar to the male, although perhaps somewhat less intense. This is because, returning once again to the biology of maleness and femaleness, males (of most species) can increase their biological success by mating with a variety of partners and being relatively indiscriminate. A copulation with someone new may result in pregnancy, to the male's advantage, especially if he can avoid providing any subsequent resources or assistance. We are therefore biologically susceptible to, and evolutionarily unprepared for, such new stimuli, while at the same time our culture bombards us with them. It is not surprising that divorce rates are highest among the subculture most subjected to such disruptive sexual stimuli, the movie-star community.

Students of animal behavior have long recognized that many species possess a genetically determined susceptibility to particular stimuli that are often provided by another member of the species. Upon appearance of the correct signals, the other animal responds automatically with some basically instinctive behavior. The stimuli that elicit this behavior are referred to as "releasers" because they appear to release behavior that is fully developed within the animal, just waiting for the proper combination to let it out. In the model proposed by the famous ethologist Konrad Lorenz, a releasing stimulus permits a particular behavior to flow just like pulling the chain releases water accumulated above an old-fashioned toilet bowl.

For example, male European robins will attack a tuft of red feathers mounted on a stick and placed inside their territory, while ignoring a much more realistic stuffed robin that lacks the color red, which apparently releases robin aggressiveness. Crows will sometimes attack someone carrying a bit of black cloth, apparently because the unwitting victim has stimulated the "crow in distress" releasing mechanism normally reserved for predators such as hawks or owls. These reactions are automatic, not rational. This is why they can be produced by simple models with only the vaguest resemblance to a live animal, so long as the appropriate releaser is present. Similarly, among the North American woodpeckers known as yellow-shafted flickers, the males and females look almost identical except that the male sports a black streak on each side of his face, looking for all the world like a mustache. If an equivalent streak is painted onto a captured female, she will be attacked viciously by her mate. Possessing the releaser that says "male," she stimulates automatic aggressive behavior from her own consort.

Biologists have also noted that if the appropriate characteristics that make up a releaser are artificially exaggerated by an experimenter, animals will often prefer them to the naturally occurring signal, or will perform their particular behavior more intensely or for a longer time. These exceptionally successful manmade signals are called "supernormal" releasers. Many nesting birds, for example, will prefer to sit on the biggest egg possible, ignoring their own in favor of a larger experimental model. The oystercatcher, a common robin-sized shorebird, will apparently forget her own eggs and perch ridiculously on top of a huge artificial egg the size of a watermelon.

Another example: among birds that produce helpless young, such as robins and sparrows, parental housekeeping chores often include removing the shiny fecal sacks produced by the nestlings. If bird banders place the traditional shiny metal ring around a young bird's leg, the parent may literally attempt to throw the baby out with the bathwater, despite the protesting chirps from their own offspring.

As to our own species, if certain physical characteristics serve as releasers to human beings, their exaggeration—as for instance the 40-inch chest of the topless dancer or Playboy bunny—represents a culturally developed supernormal releaser. Our susceptibility to releasers is probably expressed in many other ways, going far beyond sexual stimuli. For example, Lorenz has pointed out that we tend to regard as "cute" and "cuddly" any picture of a child or animal with unusually large head and eyes, and an absence of protruding ears or nose: just look at any child's doll, or the cartoonist's representations of children. In such cases, the designer is playing upon our response to supernormal releasers by exaggerating those stimuli that normally release nurturant behavior in human beings.

Culturally produced supernormal releasers may provide enhanced pleasure to human beings who intelligently manipulate them. Ironically, however, they are also especially insidious precisely because they can be overridden. They lack the obvious automaticity—sometimes comic and absurd—of an incubating oystercatcher, a pugnacious robin, or a nest-cleaning sparrow. And therefore, we may well be susceptible to subliminal supernormal releasers, without realizing that they are acting upon us. If instead, like the engaging Mr. Toad of *The Wind in the Willows,* we became maniacally insane and uncontrollable when exposed to certain releasers (like that unhappy amphibian's "motor-cars"), then we could at least be on guard. But we are not directly at the mercy of our own mental mouse-trap response patterns. And as a result, paradoxically, we are almost defenseless against the subtle, hidden persuaders of our own making.

It might be argued that the evidence for genetically mediated susceptibility to releasers in human beings is thin indeed. Although female breasts have sexual significance in Western society, for example, a good case can be made that they are much less erotic in certain African societies where they are not provocatively hidden as a matter of course. It might therefore be best to modify the concept as it applies to humans, and to identify "cultural releasers"—stimuli that come to have specific behavioral significance largely as a result of cultural practices—with only possibly a faint hint of genetic underpinning. The particular stimuli would then vary from culture to culture.

Because of its peculiar quality of intentional directedness, distinguishing it from purely biological creations, human culture has a further potential: after generating culturally defined releasers, presumably by building on existing biological predispositions, societies can then proceed to exaggerate or hyperextend certain characteristics, thus eliciting a heightened response. The result? Another major source of potentially disruptive sexual stimuli, notably in the exaggeration of body features associated with sexuality. Thus, red lips have been emphasized with lipstick, eyes with eye shadow and mascara, and silicone implants can literally hyperextend the female chest. We are constantly subjected to these supernormal characteristics. Alternatively, we subject ourselves to these culturally generated characteristics *because* we are so susceptible to sexuality. In any event, not only do we produce cultural releasers, but we make them supernormal, and they both amuse and bedevil us.

⁂

In matters of sex, everything you can possibly imagine has occurred and much that you cannot imagine.

—Alfred Kinsey

We are among the few animals that show substantial variety in lovemaking; copulation among most animals is usually stereotyped and characteristic of each particular species, although some animals such as gorillas are quite inventive. The possibilities of copulation are clearly limited by the physical structure of our bodies and our genitals, but beyond that, human cultural ingenuity is almost boundless. Despite the great variety of lovemaking postures demonstrated by Homo sapiens, however, the ventral-ventral position appears to be the most common and generally popular. By contrast, the most common copulatory position among vertebrate animals involves the male climbing upon the female's back, the celebrated "doggy position." We are also unusual in employing the ventral-ventral position at all. In addition, we are unique, of course, in having such cultural "how to" books as the *Kamasutra* or *The Joy of Sex.* But if such sexual smorgasbords were part of our genetically mandated repertoire, we probably wouldn't need the books.

Among the many changes necessitated by our insistence on walking upright was elaboration of the gluteus maximus, the rear-end muscle. This may have hindered "normal" dorso-ventral copulation, possibly resulting in some ventral rotation of the vagina, thus facilitating face-to-face lovemaking. This basic postural change may also have been related to the very important connection between sex and emotional attachment, which was evolving at the same time. In short, face-to-face sex is personal sex. Shakespeare's "beast with two backs" is not really a beast at all, but rather, two human beings who are likely to know each other and who are in the process of getting to know each other better yet.

It is instructive to observe copulation (usually dorso-ventral) in the monkeys and apes: the female is strikingly uninvolved in the proceedings. If she happened to be eating when the male first mounted, she may continue to chew unconcernedly throughout the act. In fact, among the baboons and rhesus monkeys, females have been observed sexually soliciting an adult male who is eating something desirable. While the male is then occupied with his mounting, the simian Siren steals the food. With ventral-ventral copulation, such nonchalance is less likely and intercourse is more personal. The hands, for example, are available for fondling and manipulation of the partner, but for little else, while the eyes are facing each other, not gazing around for friends, enemies, food, or whatever. As ethologist Desmond Morris has emphasized, ventral copulation may thus be another genetically influenced mechanism for eliminating the emotional disinterest of animal matings and insuring the uniquely human sex/love relationship.

On the other hand, it is noteworthy that dorso-ventral copulation, even among human beings, is still the most likely to lead to conception, although it is definitely not the most popular. This may itself reflect the extent to which sex has been emancipated from its strictly reproductive function.

Not only are we unusual in making love belly-to-belly, but we also prefer privacy when we do it. In fact, this insistence is so ingrained that we may be surprised to learn that it also is unusual among animals. Copulating pairs of monkeys and apes do not appear to value their privacy or make a particular effort to hide their activities from the other group members, although among some species, consortships are formed during which male and female take a mini-honeymoon together, away from the others. The tendency for private and concealed lovemaking seems particularly strong among human beings, and in nearly all cultures.

The adaptive advantage of such behavior should be apparent. The deeply personal involvement so characteristic of human lovemaking—whether ventral-ventral or not—precludes a watchful eye for possible predators or competitors. Hence, early prehumans making love in an exposed savannah during broad daylight would be especially vulnerable, and less likely to rear successfully the eventual products of their intimacy. Secrecy would thus enhance safety, and success. Furthermore, mating is often physically demanding in our species, and both partners are generally less capable of vigorous physical exertion immediately following orgasm. This further jeopardizes the safety of public copulators, although of course, the same may hold true for other animals. But the intense orgasmic involvement of both sexes in humans exceeds that of all other animals and one might therefore expect the physical vulnerability following sexual intercourse to be greatest in human beings as well.

A final advantage to private sex derives from our general capacity for nonseasonal sex and the ease with which we can be stimulated. Thus, as Freud pointed out, human societies surround themselves with sex images and symbols reflecting both subliminal urges and conscious efforts at titillation. Highly sexed as we are, the sight (or sounds, or in fact, any reminder) of sexual activity is especially exciting to Homo sapiens and the public copulator runs the risk of stimulating immediate competition from among the onlookers.

It is not surprising that private sex is generally characteristic of modern human beings. Along with this clearly adaptive characteristic there have developed countless supportive cultural practices. One might expect this situation to represent another example of adaptive cultural characteristics, but then again, our culture and our biology do not perfectly complement each other. Within Western culture it is common for parents to overcompensate in defense of their sexual privacy, thereby imparting confusion to their children, and a feeling that sex is "dirty." Most Christian religions contain various sexual prohibitions, which may be seen as attempted defense of sexual privacy. But even among parents not strongly influenced by organized religion, a deep shyness often pervades and even prevents discussion of sex between parent and child. The result can be an adolescent and eventually an adult with a complex of guilt, confusion, and possible sexual inadequacy.

Another avenue to neurosis is provided by culturally mediated opportunities for massive overcompensation, resulting in exhibitionism, either through drugs or financial reward. In the latter case, a recent cartoon is especially significant: a naked man and woman are in bed being filmed for a movie scene, surrounded by lights, cameras, and countless people. The man, betraying an easily understood and clearly adaptive response, is complaining, "Somehow I just don't feel like it!"

Another basically adaptive characteristic of human sexual behavior that is ancient, widespread, and probably influenced by our genetic makeup (our "nature" as well as our "nurture"), is the so-called incest taboo. The horror of Oedipus may well be universal among human society. To understand the evolutionary advantage of prohibitions against matings of close relatives, we must first examine some basic facts of genetics.

Most genes occur in pairs, one set provided by the father and one by the mother. Both parents may contribute identical genes for a given trait or each may supply a different variant, in which case the offspring would be, in effect, a genetic mosaic for that particular characteristic. This does not necessarily result in an actual mosaic appearance for the characteristic itself; rather, one gene may "dominate" the other and be largely responsible for the final appearance of the trait, or the result may be somehow intermediate between either pure appearance, or even entirely different from each. Individuals carrying two differing genes for a particular trait tend to be healthier, more vigorous, and thus more fit than those possessing two identical genes. There are many possible reasons for this. One of the simplest is that disadvantageous genes arise from mutations and are generally recessive, not showing themselves in the trait because they are masked by their partner, which is then considered the dominant gene. In cases of genetic nonidentity, deleterious recessives can thus be masked by the corresponding dominant. But when both genes are identical, two recessives may occur together, and as a result the less advantageous recessive trait is expressed, and therefore the individual in question is less fit.

Considering the advantages of genetic variety, we would expect evolution to have devised numerous tricks for discouraging genetic identity among any pair of chromosomal traits within each individual, and one of the best ways to guard against such occurrences is to insist that mated pairs be related distantly, if at all, while prohibiting reproduction by close relatives. For example, imagine an individual containing identical genes for a particular trait; these genes may be represented by "GG." A relative would stand a good chance of being genetically the same (the chances would increase the closer the family relationship), thus producing "GG" offspring. A complete stranger, however, may have the genetic makeup "gg." Mating between them would thus produce the usually superior "Gg" offspring, one gene having been contributed by each parent.

Long before the genetic basis for this phenomenon was discovered, animal and plant breeders had known that continual mating of closely related individuals would eventually reduce the quality of the stock. "New blood," they knew, had to be brought in every once in awhile, otherwise the population would decline in health and vigor. Geneticists now term the deleterious results of extensive breeding among close relatives "inbreeding depression," something that is avoided among most animals by dispersal, the tendency of juveniles to leave home and seek their fortune elsewhere, away from their close genetic relatives. Among prehistoric human beings, however, dispersal was probably uncommon, since our species is so highly social that solitary, dispersing individuals would be unlikely to survive. So other mechanisms were necessary to prevent inbreeding. The common prohibition against mating of close relatives thus not only is a cultural practice with clearly adaptive biological significance, it may also be supported by a genetic tendency resulting from the direct pressures of natural selection upon our ancestors.

Israeli anthropologist Joseph Shepher has documented that children growing up in a kibbutz, who have been treated as though they were members of one large family, tend to avoid marrying each other, despite the fact that social pressures actually favor such unions. The young people themselves report that marrying a member of one's early play group would feel like "marrying a sister (brother)." Even though the individuals in this case are not biologically related, it seems likely that their underlying biological predispositions are inducing them to avoid "incest," using a social cue which is normally accurate but which, in the particular cultural environment of a kibbutz, is no longer reliable. Thus, by avoiding matings within one's play group, "biological" Homo sapiens has minimized the danger of incest and inbreeding, for many generations, whereas "cultural" Homo sapiens has outsmarted himself.

It may be objected that the incest taboo has not been absolutely universal among human beings. The well-known Egyptian pharaoh Tutankhamen and his wife, Queen Nefertiti, for example, were brother and sister. An extensive line of European royalty was plagued with hemophilia because of intermarriage among close relatives. Matings of first cousins have long been common in east European societies. Sociologist Pierre van den Berghe, among others, has pointed out that incestuous matings, especially between brother and sister, tend to occur in very special social situations: when enormous power, wealth, and/or spiritual values are to be concentrated within a very select circle. In such cases, cultural rules predominate over biological predisposition. (Also, at least the men often wind up having additional children via concubines, to whom they are not closely related; the royal women, by contrast, are more likely to be stuck in a situation that is, for them, biologically maladaptive.)

When biologists suggest that a trait is genetically influenced, they do not imply invariant, exact correspondence in all cases. We do not inherit blue eyes or tall stature as such, although both are under genetic influence. Rather, we inherit a potential, a range of possible expressions that will vary within certain limits, often rather wide limits at that, depending on the specific environment in which we live. Second-generation Japanese-Americans, with a gene pool that is indistinguishable from that of their parents, are nonetheless considerably taller than their parents because of better nutrition available to them in the United States. We may well have a biological predisposition for copulating with strangers, but certainly, we can meet strangers without copulating with them. And conversely, a biological predisposition for incest avoidance does not guarantee that incest will never occur.

To take another simple biological example, there is a genetically controlled characteristic in rabbits known as "Himalayan," in which the animals have patches of black on the tips of their forepaws, hind feet, and ears. The rest of the fur is pure white. The extremities of all animals are somewhat colder than their more central regions (witness our human problem of cold hands, feet, and ears) and rabbits possessing the Himalayan genes respond to cold temperatures by growing black fur. This can be experimentally tested by shaving off a patch of normal white fur from the side of a Himalayan rabbit. As expected, it grows back white again. Now, however, shave some fur and apply an ice pack during the regrowth period: the new fur comes in black. Clearly, black fur itself is not inherited in Himalayan rabbits, but rather the *ability* to produce black fur if the temperature falls below a certain threshold.

A more complex, but analogous example occurs in humans, with the "inheritance" of intelligence. There is clearly some genetic basis for differing IQ levels, but the possession of all the right genes for great brilliance will not make a genius out of a child raised in complete isolation, or one even moderately deprived of the necessary opportunities for mental exercise and growth. We inherit a range of possibilities, but the specific realization of these potentials depends on the particular environmental factors to which the individual is exposed. Similarly, a genetically mediated inclination for ventral, private copulation with nonrelatives and with a diversity of partners—perhaps more so for males than for females—could easily be swamped by cultural factors including religion, conscience, technology, and habit, with little regard to its ultimate biological utility.

When incest occurs in situations not involving royalty, it is most frequent between father and daughter, rare between siblings, and almost unheard of between mother and son. Moreover, by far the most frequent example of incest involves a stepfather with his stepdaughter, cases that are often closer to child abuse and that do not meet the biological definition of "incest" at all.

As we shall see in more detail in the following chapter, males and females are different, not only in their fundamental biology, but in certain behaviors as well. Not surprisingly, the behavioral differences are consistent with the more strictly biological. Males, in brief, are the inseminators; females, the inseminated. Males—as expected—are therefore the sexual aggressors and among many animal species, they are sometimes rapists as well. Through studies of ducks, geese, fish, and even insects, biologists have been compiling impressive catalogs of rape: the common factor seems to be that under certain conditions, notably when they are relatively unsuccessful otherwise, males attempt to maximize their fitness by forcing copulations with females.

A recent study by Randy and Nancy Thornhill has suggested that fitness considerations may be relevant to rape among human beings as well. They found that rape victims tend to be primarily women of childbearing age, whereas murder victims, by contrast, tend to be more evenly distributed across the population at large. If rape was simply a crime of violence against women, it would presumably be directed equally against all women. Moreover, the Thornhills also found, as predicted from evolutionary theory, that young men are overrepresented among rapists, and that these men tend to be relatively low in socioeconomic status, and thus, likely to be unattractive as mates if they followed socially sanctioned sexual strategies instead of rape.

Men are not only the most frequent transgressors in incest and rape, they are also far more likely than women to respond violently to adultery by their mate. In fact, the most frequent cause of violence among spouses, from wife-beating to murder, from Westchester to Mozambique, is sexual jealousy . . . especially male sexual jealousy. Among many societies, adultery is considered a crime only when it is performed by a wife, not by a husband. And moreover, it is often specifically designated a crime against the "offended husband," who may then, with the approval of society, wreak vengeance against his wife and/or her lover. There is little good evidence allowing us to compare female sexual jealousy with its male counterpart, but it is almost certain that among the great majority of human societies, women are much less enraged by their husbands' philandering than men are by the same behavior on the part of their wives. Much of this difference may be attributable to cultural traditions, which prescribe greater female tolerance and a sexual double standard. However, these cultural patterns, important as they are, also seem likely to emanate from biological patterns that are no less important: women get pregnant, not men. Accordingly, sexual dalliance by a wife is much more likely to compromise the fitness of her husband than vice versa.

With recent cultural advances in the technology of birth control as well as abortion, the strictly biological consequences of extramarital infidelity have been substantially diminished. As with the biological significance of the pair bond, the widespread cultural hyperextension of the double standard as well as male sexual jealousy therefore appears to be more inappropriate than ever. But not surprisingly, our "all too human" emotional responses—often intolerant, jealous, and sometimes even violent—still lag behind, caught in an evolutionary time-warp.

As intelligent, moralizing creatures, we can look at the seamier side of our own behavior, and establish cultural norms and codes that seek to outlaw and prevent actions that we judge to be intolerable. However, we do a disservice to the cause of humane understanding and ultimately, social betterment if we willfully ignore or deny the likely biological component of precisely those behaviors which we find so offensive. The likelihood that an act is in some sense "biological" does not make it good, just as being "cultural" does not make it bad. Few things are more "organic" than typhoid; when medical science seeks to understand the disease, this is not to support or encourage the typhoid bacillus. Similarly, when evolution offers some insights into rape and sexual violence, this is not to condone the behavior, but rather to help us understand something that outrages our cultural norms.

In the words of Saki (Hector Hugh Munro), "It takes all sorts to make sex," some sorts that we appreciate and enjoy, others that we find uncomfortable, embarrassing, and disagreeable, and yet others that are simply despicable. Underlying them, for better or worse, are all sorts of biology and all sorts of culture, variably dependent, sometimes opposed, and yet always intertwined.

CHAPTER FIVE

FEMINISM

Gorillas to Goldman to Gilligan

Marriage is primarily an economic arrangement, an insurance pact. It differs from the ordinary life insurance agreement only in that it is more binding, more exacting . . . If, however, woman's premium is a husband, she pays for it with her name, her privacy, her self-respect, her very life, "until death doth part." Moreover, the marriage insurance condemns her to life-long dependency, to parasitism, to complete uselessness, individual as well as social.

—Emma Goldman, "Marriage and Love" (1910)

Feminism is not unique to the 1980s or even the 1970s; its precursors can be identified at almost any stage of human history. What is unique, however, is its intensity, breadth of support, and depth of commitment in modern Western culture. It is clearly a cultural phenomenon, prominent in some societies, absent in others. It is almost certainly nongenetic in origin and is most commonly associated with a whole constellation of social and political values, in the Western world generally leftist and anti-establishment, since the Western "establishment" tends to be male dominated, politically conservative, and rather sexist. The correlation between political-social conservatism and sexism is so strong that it is often taken for granted. Here, as elsewhere, those things that we take for granted are most likely to yield insights into our innermost inclinations. In addition, it is not usually recognized that at the center of the feminist struggle—between woman and man, between woman and sexist society, and within many women themselves—is the persistent conflict between cultural and biological evolution.

Men and women are different, and not just in anatomy. Although societies may differ in what is recommended or even acceptable for either sex, and despite the fact that "men's work" in one society may be "women's work" in another, it remains true that men's behavior tends to differ consistently from women's behavior, in all human groups. In addition, a distinct pattern can be identified: women, for example, are universally more likely to be engaged in child care and men are universally more likely to go hunting and to kill each other. Whereas the evidence is accumulating that these differences are based at least to some degree on deep-seated biological distinctions between the sexes, we seem to have created cultural differences that go farther yet. Insofar as these differences are not commensurate with each other, we have a great source of irritation and injustice.

For some time, feminists resisted any efforts to identify differences between the sexes, and rightly so perhaps. Patronizing attitudes toward women as the "weaker sex," with implications of greater emotionality, lesser rationality, and generally less competence, have bolstered a wide array of institutions that have oppressed women for generations, and which to a large extent still do.* Locked in a vigorous struggle for equality, it seemed appropriate to resist as "sexist" any implication that women are not in every way "equal" to men. Early in any movement, it is convenient—often necessary—to simplify issues in the interest of clarity

*At a speech following the 1984 elections, Geraldine Ferraro pointed out to the great amusement of her (mostly female) audience that during their televised debate, Vice President Bush had shown dramatic mood swings, serious one moment then almost giddy the next. Ms. Ferraro wondered aloud whether this was hormonally based, and whether the nation was well advised to leave its direction to men, with their apparent biological instabilities.

and to prevent divisive disputation and destructive self-doubts. Fortunately, it seems that feminist thought has now matured, just as social and political thought generally tends to mature when its initial revolutionary goals begin to be achieved. It can therefore be hoped that feminists are now prepared to deal with, and to profit from, the insights of evolutionary biology.

Women have long been oppressed, in a variety of ways by most human cultures. Despite the conservative trend in the United States during the 1980s, it seems unlikely that women will much longer be denied equal opportunities, status, and respect, in both the workplace and the home, and equal pay for equal work. Like environmentalism before it, and perhaps the nuclear freeze movement in the near future, feminism has finally been transforming itself from a seemingly radical departure to an accepted way of thinking. In the process, it is at last becoming intellectually respectable to confront certain truths that were in fact always true, but were also inconvenient. Feminism as a legitimate claimant to social change will in the long run be strengthened by attention to biology and to the conflict between biological and cultural evolution, no less than by attention to economics and the conflict between private needs and male-dominated state policy.

Everyone has known for a very long time that men and women are structurally different, that women make eggs and men make sperm, that men have penises and women vaginas, that women have babies and lactate while men do not. These facts are not sexist; they are just facts. On the other hand, it may indeed be sexist to interpret some of these differences from a biased male perspective and suggest, as Freud did, that little girls perceive themselves as missing something that little boys have, suffering therefore from penis envy. One could as well suggest that when little boys grow up and get themselves covered with blood in fights or wars, as they often do in their sexist way, they are suffering from menstrual envy.

Aside from the obvious sexual differences, there is a clearcut distinction between men and women when it comes to physical size and strength. Although there are exceptions, men tend to be larger and stronger, and these differences are due to biological rather than cultural evolution. This may seem surprising. Given the fundamental distinction in their biological roles, one might expect the females to be generally larger. After all, men supply the sperm; women, the egg. The man's contribution is transient and relatively trivial in mass and energy; the woman's is long lasting and immensely costly to her. Since it is the woman who must literally carry the biological weight of reproduction, it would seem logical for her to be sturdier than the man. In a sense, she is. Thus, although muscular strength belongs especially to the male, females are biologically stronger as shown by their almost universally longer life span.

On the other hand, male superiority in muscular strength is also a product of biological evolution. The reason? Defense and competition. If we look at a social monkey now inhabiting the same African savannah that was probably the birthplace of modern Homo sapiens, we can see some of the selective pressures at work that almost certainly acted in shaping us. The common baboons of east Africa live in social units consisting of many adult males, females, juveniles, and young. The males are very powerful and often twice the size of the females. They also possess enormous canine teeth and an aggressive temperament, both of which become apparent when one male challenges another for social dominance and mating rights, or when a predator such as a leopard or cheetah appears on the scene. In the former case, the adult males threaten and if necessary fight; in the latter, they often do the same thing.

A similar correlation between size and aggressiveness among males holds true for the other terrestrial primates, and in fact, for most mammals. Even among the "lordly" lions, in which, incidentally, the males virtually never exert themselves to make a kill, they still function in defense of the pride. In this case they protect the kill especially from appropriation by hyenas, which have been known to chase an unattended lioness from her hard-earned meal. In other respects, the male lion is a virtual parasite of the female, since she does all the hunting. But when it comes to defense of the group, especially defending his own status from other males, the male lion is no slouch.

Male lions compete vigorously with each other for "ownership" of a pride. To the victor goes the opportunity of copulating with several females, and thus, reproducing by them. To the losers, a resentful career as bachelors, periodically trying to achieve social, sexual, and thus, evolutionary success.

Defending the group is a dangerous occupation and it might seem logical that evolution should have chosen the relatively expendable male for the task, while preserving the more important female. However, so far as we know, evolution does not act to benefit the species or even the group as such; rather, whenever some individuals leave more offspring than others, or when some genes leave more copies of themselves than do others, natural selection is taking place. Any species benefit is therefore likely to be purely incidental to the evolutionary jousting of individuals and their genes. Analogously, a nation's Gross National Product is the incidental result of individuals, corporations, and governmental entities seeking to maximize their own functioning; the GNP will generally go up when each subnational component does as well as possible, but not because enhancement of the GNP is the goal or the level at which competition is taking place. Similarly, species benefit is the result of selection acting to maximize the performance of each lower level component, not enhancement of the species per se.

Male lions defend their group against others, both hyenas and other lions, because there is something in it for them: by doing so, they are more likely to be evolutionarily successful. Presumably, the same fundamental pressures acted for thousands of generations on primitive Homo sapiens as well. Given the biology of sperm-making versus egg-making, it seems likely that such defense was primarily defense of the individual male's reproductive success; that is, it was probably self-serving, and in no way particularly laudable, gentlemanly, or gallant.

Unless females actively competed for males, there would be no equivalent selection pressure operating on them. If, on the other hand, there was a shortage of females, competition would ensue with natural selection favoring those females whose characteristics help them secure a mate and breed successfully. Among the common North American shorebirds known as spotted sandpipers, the usual pattern of sexual politics is reversed; females are notably malelike. Larger and more aggressive than their mates, female spotted sandpipers arrive on their breeding territories while the males are still loitering down south. These pushy, competitive females fight it out with each other, after which the males arrive and demurely settle down on one territory or another, typically several males within each female's "harem." Each of these males is subordinate to his territory-owning female, and each one incubates a clutch of eggs laid by the dominant female, who busies herself defending her boundaries and attending to the world of affairs while he cares for the offspring, unaided.

A similar pattern is found among those peculiar creatures, the sea horses, in which the females transfer their eggs to the males, who incubate them in special brood pouches. Significantly, female sea horses—like female spotted sandpipers—tend to be larger, brightly colored, and aggressive, while the males are small, drab, retiring, and sexually coy. These are all exceptions, although interesting exceptions to be sure, since they help prove the rule that the sex investing more (generally females) tends to be less aggressive and to be the subject of sexual competition on the part of the sex investing less (generally the males).

Among some animals, success in sexual competition may well involve physical strength. Among others, such as human beings, traits of little or no immediate survival value may nonetheless be of reproductive advantage if they are attractive to members of the opposite sex. There is much current debate among evolutionary biologists as to whether such apparently nonadaptive traits really are preferred, and if so, whether they are really nonadaptive. For example, it is at least possible that the large antlers of deer are a liability rather than an asset, since they require a substantial metabolic investment, and may make their bearers more conspicuous to predators. On the other hand, if an impressive rack of antlers indicates that its carrier is capable of surviving despite such a handicap, then females may preferentially mate with such an individual . . . thereby rendering the "handicap" an advantage! In any case, there are often strong selective pressures favoring large, impressive, and aggressive males who can defeat or dominate other males, while defending themselves and their dependents against attack.

Something similar to this phenomenon was first noted by Charles Darwin, who named it "sexual selection," mistakenly believing that because it often depended on the choice of attractive mating partners by the females—who were somehow possessed of an intuitive aesthetic sense—it was quite different from *natural* selection. We now recognize that selection is indeed occurring in these cases, but the selecting agent is the entire environment of the species, not just the proclivities of its females. Males do indeed strive to impress females, but perhaps even more, they attempt to outcompete other animals and particularly, other males of the same species. When the dominant silver-backed adult male gorilla repels the advances of another silver-back from another troop, he succeeds not only in male-male competition, but also in assuring himself the sexual attentions of his females.

Again, it might seem that evolution would still have produced larger and more powerful females by this same mechanism: females adept at defending the group from predators and capable of defeating possible competitors from other groups would naturally leave a greater number of successful offspring. This would cause selection for greater size and strength ("male characteristics") among females as well. And to some extent this has happened. But the fundamental biological distinction between male and female would foil the extensive development of female Amazonism. Thus, although a highly successful male could make a major contribution to the next generation's genetic makeup, the contribution of even a very successful female is limited by her biology. Reproduction in mammals requires a lengthy pregnancy, resulting in only a few offspring at a time; in primates it usually requires an even longer postnatal period of maternal care and attention. So, even though individual females, more than individual males, determine the reproductive output and ultimately the evolutionary success or failure of their species, even a hypothetical superfemale can make at best only a small genetic contribution to the next generation. By contrast, a successful male can fertilize many females, and therefore have a much greater evolutionary impact. The upshot of all this is a further tendency toward physical as well as behavioral differentiation between men and women among our ancestors.

This does not imply that males on balance have an advantage over females. In fact, the occasional male that is wildly successful is exactly balanced by many males who are not. Females, by contrast, are less likely to strike it rich via natural selection, but they are also less likely to strike out altogether. Being a male, then, is inherently risky, since among many species males are either successes or failures; being a female, by contrast, is more conservative, since nearly all females will breed, and the disparity between success and failure, biologically measured, is usually not very great.

But nothing in biology is gained for nothing. Most advantages are obtained at the expense of some disadvantages: A larger, stronger body requires more nourishment, and in times of food shortage, it is undoubtedly advantageous to have less bulk to support. This would provide a compensating advantage to smallness. Moreover, brightly colored males are necessarily more conspicuous, and thus, more susceptible to predation. And finally, whereas dominant males may be successful in evolutionary terms, they typically have a shorter life span than the evolutionary failures, which spend less time fighting and copulating, and more time eating and keeping out of harm's way. A mountain sheep ram, for example, or a bull elk, may well be exhausted and emaciated at the end of the rut, especially if he has been successful in maintaining his harem. Not surprisingly, females live longer than males among most animals, just as women live longer than men.

As a consequence of natural selection acting somewhat differently on males and females, the two sexes are therefore somewhat different, both physically and behaviorally. These biological differences also influence not only the proclivities of each sex, taken separately, but also the ways in which males and females, men and women, interact with each other. Males clearly dominate females among the monkeys and apes. Gorilla bands are led by a dominant silver-backed male; a clear progression of rank among dominant males can be identified among rhesus monkey bands and an oligarchy of several males often rules baboon troops. Frequently, the females have a distinct social hierarchy all their own, but any randomly chosen female is generally subordinate to any randomly chosen male. In addition, whereas male hierarchies are usually stable, changing only rarely when high-ranking members die or get very old, the female social ranks are constantly shifting, as a result of changes in the estrous cycle, presence or absence of dependent young, and possible consort association with a high-ranking male. Such biologically generated instability makes female dominance over males highly unlikely. In no cases, therefore, are primate groups consistently dominated by females. This is almost certainly a direct result of the biological distinctions between the sexes, based on hormonal and anatomical differences, which in turn are attributable to selection for male success in defense and competition, and for female success in cooperation and nurturance.

The characteristics of physique and temperament that make for aggressive success and were likely the object of intense selection among our ancestors seem to have generally led to social dominance of females by males. This correlation is not 100 percent, but common experience reveals how often domination goes along with physical size, strength, and aggressiveness. It is certainly true in most animals, and human beings do not appear to be an exception.

Male dominance in human beings might thus seem to be the biologically "correct" state of affairs. If we were purely biological beings, the issue would be closed, and in fact, would never have been raised! But we are unique in serving two, sometimes rather disparate masters, biology *and* culture, and this changes things altogether. Let us grant, if only for the sake of argument, that the general dominance of men over women derives from physical and behavioral differences due largely to selection for male aggressiveness, competitiveness, and defense of the social unit. There was probably great need for such characteristics as recently (in evolutionary terms) as 30,000 years ago. Since that time, however, cultural evolution has completely remodeled the contingencies of human survival and social life, with biology, as usual, lagging far behind. Men today only rarely function in direct defense of their genetic legacy; given the presence of nuclear weapons, in fact, those who do participate in such "defense" and who urge it upon the rest of us may ironically be endangering not only themselves but also those they claim to defend. They may in fact be the greatest threats to long-term evolutionary survival that our species has ever confronted.

The human tendency for a "nuclear family" with a strong male-female pair bond further reduces sexual selection for enhanced male characteristics. Thus, with roughly equal numbers of males and females in the population, and with (culturally imposed) monogamy, virtually every male gets to reproduce, instead of a few extraordinarily endowed specimens doing the lion's share of the breeding. There seems little doubt that given human biology and culture, selection for male-female differences will be decreasing. But there is equally little doubt that given the rate of biological evolution, such biological equalization must probably be measured in thousands of years. Where does that leave us?

We are stuck, saddled with an outdated biological system, rendered anachronistic by our rapidly evolving culture, and unacceptable by our expanding social consciousness. We possess physical characteristics and behaviors such as male dominance and aggressiveness that are offensive, inappropriate, and often downright dangerous. Although human culture, ironically, is itself responsible for this discrepancy, much of the difficulty in readjusting society stems from the cultural—not biological—supports to male dominance that have been erected by men. Many traditional behaviors, including such diverse things as retaining surname at marriage, identification of most political and economic activity as a male province, expected man-woman roles in home maintenance and wage earning, smoking cigars, or access to seats on a crowded bus, are purely social conventions. Admittedly, they may have been largely developed in response to the biological tendency for male protectiveness, competitiveness, and resulting patterns of dominance. But, such dominance is now irrelevant. Combined with widespread awareness of this fact, as well as the (culturally mediated) flexibility of human behavior, we can look forward to their elimination. But at the same time, cultural hyperextension of female suppression will not be dismantled quickly or easily. It may also require a hyperextended awareness of the need for such remedies.

There is a delicate titration between biological predispositions and social prescriptions. In such cases as male-female differences, it seems especially likely that society, seizing on a degree of evolutionary reality, has hyperextended it, making awkward, unjust, and offensive cultural mountains out of what may, in essence, be biological molehills. Men and women are indeed different, but in most cases they are less different than human social traditions have demanded them to be. Give cultural evolution a hand in such cases, and it takes the whole arm, ultimately, perhaps, to the detriment of all concerned. In other situations, the opposite may occur: evolutionarily given realities may be diminished rather than augmented by the action of cultural evolution. When this involves socialization to reduce otherwise potentially abrasive consequences (for example, of male aggressiveness), this cultural sandpaper may be quite beneficial, so long as it is applied gently and with sensitivity, as well as persistently. In other cases, cultural practices may run directly counter to evolutionary inclinations, and the result, while presumably in accord with conscious intent, may nonetheless cause great distress.

When it comes to male-female role differentiation, some biology remains clear. Among all mammals, females get pregnant, not males, and moreover they are equipped with milk-producing glands by which the infants are nourished after birth. Nearly all female mammals also possess genetically influenced behavior patterns that provide for care of the young. The new mother must generally remove the fetal membranes, usually eating them for the added nourishment and/or hormones they provide. Among many species, the newborn must be licked profusely in order for normal urination and defecation to occur. Adults often brood their young, warming them if they are cold, shading them if they are hot, removing fecal pellets to prevent fouling of the nest, and providing a steady supply of food. Such provisioning may or may not involve assistance by the father, but it always involves nursing by the mother. Female rats and mice will automatically retrieve their young if they have been scattered about, collecting them in a pile where they will be kept warm and safe. In most cases, this complex of maternal behaviors is activated by hormones produced by the mother herself, particularly the hormones oxytocin and prolactin, associated with lactation.

The situation in humans is undoubtedly more complex than among any animal. A good generalization in the study of animal behavior, in fact, is that among progressively smarter animals, the role of genetically influenced and hormonally activated behavior declines as the cerebral component increases. We have already seen, for example, that sexual behavior in human beings has been liberated from the imperious chemical domination by hormonal cycles that is found in all other mammals. Unquestionably, maternal behavior in Homo sapiens has also been greatly emancipated from the control often exerted by hormones and genetics. There is, in fact, no good evidence that mothers who are bottle feeding their infants have weaker maternal drives than do breast feeding mothers, in whom the maternal hormones are more prominent.

However, breast feeding has an undeniable side effect on the nursing mother's physiology: it tends to inhibit the next ovulatory cycle. This so-called lactational amenorrhea is entirely biological and is adaptive in reducing the likelihood that a mother will become pregnant again while still nursing an infant. In fact, among nontechnological societies, there is a correlation between the length of normal lactation, the length of postpartum sex taboos (socially mandated restrictions on sexual intercourse following the birth of a child), and the availability of protein in the maternal diet: lower protein levels tend to be associated with longer nursing and other culturally elaborated practices that reduce the probability of pregnancies following, maladaptively, too close one after the other. When corporations selling artificial infant "formula"—such as Nestlé in the recent past—successfully market their product in Third World nations, they disrupt this delicate synchrony between biology and culture, thereby removing an adaptive inhibition against excessive, unwanted pregnancies. In addition, whereas mothers' milk is generally free of pathogens, artificial infant formula made with local, contaminated water contributes significantly to chronic diarrhea and dysentery, major causes of death among newborns.

There is no unimpeachable evidence that childcare and nurturance must necessarily be women's work rather than men's. There is, in fact, overwhelming evidence that men are capable of substantial, and effective "mothering." The expression of any particular nurturant behavior is obviously very susceptible to cultural influences, especially social expectations and economic situation, as well as personal preferences, often born of one's own idiosyncratic experiences. However, it seems unlikely that a behavioral system so important to evolution as care of the young would not have some genetic component, even in that most liberated of species, the human animal. The system may be flexible and easily influenced by culture, but something is almost certainly there nonetheless. The well-known sensitivity of menstrual cycles to stress as well as the subtle and problematic but nonetheless likely effect of these cycles on behavior suggest that Homo sapiens is not immune to the effects of biology, in this case, our reproductive hormones.

Women's fear of identifying behavioral differences between themselves and men, combined with men's rather shortsighted and sometimes even churlish insistence that "men" somehow represent the entire human species, has resulted in some limited views of normal human nature. In her justly influential book *In a Different Voice: Psychological Theory and Women's Development,* Harvard professor Carol Gilligan has effectively outlined some of these misunderstandings. For example, girls have traditionally been thought to lag behind boys in the most widely acknowledged measures of moral development, devised by (male) psychologist Lawrence Kohlberg. This is because boys tend to evaluate situations by reference to abstract laws and ethical principles, whereas girls tend to focus on social connections and interpersonal relationships. There is, moreover, no arbitrary standard by which we can judge the former relative to the latter, although an appreciation of evolutionary biology can help us understand why the former is primarily a male pattern and the latter, female.

Gilligan emphasizes that female moral preferences tend to place "an ethic of responsibility as the center of women's moral concern, anchoring the self in a world of relationships and giving rise to activities of care," whereas by contrast, the highest stage of moral development, according to accepted psychological dogma, favors recognition of universal rights rather than personal responsibility. "The morality of rights," as Gilligan points out, "differs from the morality of responsibility in its consideration of the individual rather than the relationship as primary." It is probably no coincidence that a morality of rights rather than responsibility also happens to be the typical moral preference of males, and that the designers and interpreters of these systems also happen to be predominantly men.

Developmental psychologists have noted that when children

are playing, boys' games tend to last longer than girls', because boys quickly learn to adjudicate disputes by reference to abstract rules and principles. Gilligan points out that

> By participating in controlled and socially approved competitive situations, they learn to deal with competition in a relatively forthright manner—to play with their enemies and to compete with their friends—all in accordance with the rules of the game. In contrast, girls' play tends to occur in smaller, more intimate groups, often the best-friend dyad, and in private places. This play replicates the social pattern of primary human relationships in that its organization is more cooperative.

By contrast to boys, girls are more likely to stop the game when a disagreement arises, because they value the relationship among the participants more than abstract, blind justice, or the game itself. Although Gilligan refrains from suggesting the origin of the difference she so eloquently describes, its compatibility with biological evolution suggests that natural selection has been involved.

The biological evolutionary task of both males and females is to succeed in projecting copies of their genes into the future, to maximize their fitness. But as we have seen, success is likely to be achieved somewhat differently among the two sexes. Male success is typically achieved by effective competition; female success, by relationship, especially with their own offspring and other relatives. Thus, for boys and men, morality is at its most ideal and alluring when it is a morality of justice, of theoretical principles that place restraints upon aggressive, competitive, self-serving tendencies; for girls and women, on the other hand, morality is suffused with images of relationship, of caring, and of taking care of others. Male morality, as Gilligan describes it, is an ethic of *inhibiting* one's nasty self; female morality, in contrast, emphasizes *releasing* the caring self.

The classic developmental task for young boys, recognized by psychiatrists and psychoanalysts, is differentiation and individuation. It is a task that is appropriate to their biological task as well. Boys must separate themselves, physically and emotionally, from their primary caretaker (who is typically the mother), and become something different: a man, and a father. Becoming "your own man" means becoming different from your mommy. Young girls, however, are behaving more in concert with their ultimate biological needs if they model after their mother and achieve comparable relationships. In a world oriented toward achieving separateness and individuality, attachments appear as hindrances, as impediments to maturity. Erik Erikson has suggested similarly that for boys, identity must precede intimacy, whereas for girls, identity is found through relationships with others.

As Gilligan points out: "Since masculinity is defined through separation while femininity is defined through attachment, male gender identity is threatened by intimacy while female gender identity is threatened by separation. Thus males tend to have difficulty with relationships, while females tend to have problems with individuation." True to their biologically appropriate roles, men orient themselves toward career and success in the competitive world, entering readily into a series of hierarchically arranged systems, while women, true to theirs, orient themselves within networks of relationships. So when Freud identified the highest goals of mental health as "the ability to love and to work," he was not, in fact, describing a dual accomplishment to which both sexes equally aspire. Whereas men seek to be alone at the top, fearing in turn that others might get too close, women seek to be embedded in a network of human relationship and fear being isolated. As Gilligan puts it:

> The images of hierarchy and web . . . convey different ways of structuring relationships and are associated with different views of morality and self . . . As the top of the hierarchy becomes the edge of the web and as the center of a network of connection becomes the middle of a hierarchical

> progression, each image marks as dangerous the place which the other defines as safe.

Men fear failure; many women fear success. More than a century ago, suffragette Elizabeth Cady Stanton was so frustrated with women's proclivity for nurturance and sacrifice rather than assertion and worldly accomplishment that she urged a radical realignment in women's values. "Put it down in capital letters," she told a reporter, "SELF-DEVELOPMENT IS A HIGHER DUTY THAN SELF-SACRIFICE." Gilligan's work helps us understand that for women, self-development requires more effort than does self-sacrifice, and a look at our evolutionary history helps us understand why.

Following Stanton's injunction, women in growing numbers have finally begun to seek self-development, and to chafe at the slow pace at which society has permitted them to do their duty. Angered by the unresponsiveness of their own societies, many women found themselves in conflict with resistant and rigidified male-dominated and male-oriented social and governmental structures. Women began to seek and obtain freedom from sexist roles and from their inferior social status, roles and status that seem to have been originally suggested by our biology but were greatly magnified by our culture. At the same time, they were embroiled in another conflict, this one within themselves, between human biology and culture. Even with the heartlessness and social insensitivity that has characterized American federal politics in the 1980s, the fact is that women are more liberated from the tyranny of their biology than at any time in the long history of our species. Culture can now provide virtually all the minimal requirements for successful child-rearing—baby sitters, clinics, day care centers, schools—making the woman's contribution less and less necessary.

We have therefore gone beyond the simple biological concerns of the male as hunter, protector, selfish competitor, and (sometimes) provider, and the female as gardener, forager, sexual object, nursemaid, and educator. Women therefore find themselves caught in a double-bind between opportunities and desires for independence on the one hand, and the old biologically inspired yearnings for the reproductive role that evolution has already mapped out—and upon which society has built, and men capitalized—on the other. Small wonder women are ambivalent about their roles and their lives.

Men, by contrast, are generally having an easier time, since the old qualities of aggressiveness and daring, once so useful on the savannah, can be transferred more readily to the "outside" world of work, business, and professional competition. The crunch of culture upon biology therefore makes itself felt in every woman who debates the merits of career versus motherhood and in every man who finds himself called upon to relinquish some of the outmoded perquisites of dominance.

There can be no doubt that marriage, as it has been practiced by the great majority of human beings for most of our biological and cultural history, has been an institution that oppressed women. There can also be no doubt that at least in the West, some changes are under way. It would be ironic, though, if in the process of realizing their true potential as human beings at last, women find that rather than divesting themselves from the yoke of domestic slavery, they are simply permitted to assume new responsibilities—in the marketplace of work and competition—while still being burdened by the same old biological baggage as before. In her book *The Hearts of Men,* Barbara Ehrenreich, no less than Emma Goldman in the epigraph to this chapter, recognizes the shortcomings of marriage based on male dominance, female dependency, and economic bondage. Yet she also points out the tendency of men to take advantage of feminist striving for liberation by liberating themselves from husbandly and paternal responsibility and relationship, following perhaps their own untrammeled biology. This has left women stuck with *their* biology and simultaneously deprived of the culturally constructed "safety net," which, despite its shortcomings, has been one of traditional marriage's saving graces. Maybe its only saving grace.

As men have become increasingly aware of the possibilities of their own liberation, and of the debilitating stresses—emotional, cardiac, ulcer-inducing—of being success objects, women find themselves under attack from a new direction, caught between the Scylla of marital dependency and the Charybdis of second-class citizenship in the workplace (with childcare and family responsibilities often undiminished). Like Gilligan, Ehrenreich shies away from underlying causes. But the flight of American men from marital and paternal commitment that she so deplores and that growing numbers of men find so attractive is also a flight from a culturally imposed system to a seeming Shangri-La of self-realization, made all the more tantalizing by its congruence with male biology. It may be true, as Dr. Helen Caldicott likes to point out, that the woman most in need of liberation is the woman within every man, forced to play out a difficult, stressful, and often downright unpleasant macho role. But there is also a biological male within every man, and he, at least, may well resonate with the prospects of liberation from monogamy and the commitment to wife and children that it usually involves. After all, polygyny is biologically "natural" to our species, only inhibited by cultural proscriptions. (Whether is it "right" is another question.) And another biological consequence of maleness as opposed to femaleness is that whereas women are guaranteed that their children are biologically theirs, men are not. For the vast majority of mammals, male fitness is achieved by consorting with as many females as possible, while offering little or nothing in the way of paternal assistance.

* * *

GREAT blue herons are large, regal-looking birds that inhabit marshes, eat fish, and mate monogamously. Douglas Mock of the University of Oklahoma has found that within a minute after his mate leaves the nest area to begin foraging, the male great blue heron begins courting other females. These "extramarital" courtship efforts are not in any way disreputable, and certainly they are consistent with great blue heron biology: occasionally, a female finds a better partner and deserts her mate, and when and if that happens, the first male is better off if he has a replacement waiting in the wings (or, on the wing). Great blue herons—both male and female—apparently feel very little personal commitment to each other, but substantial commitment to their personal biological success. By contrast, extramarital endeavors among human beings may or may not be disreputable (depending on our cultural evaluation of such behavior), just as they may or may not be fitness enhancing.

Female stickleback fish deposit their eggs in a nest constructed by the male. The females then swim away, entirely free of domestic responsibilities, leaving their mates to guard the young. Once again, biological femaleness and maleness, operating in the context of each species' particular situation, dictate clear-cut roles for these animals; there is no such thing as a "liberated" stickleback fish, an oppressed one, or any that are notably ambivalent.

We, no less than the great blue heron or the stickleback fish, are bequeathed our maleness and femaleness by biology, and there is little that we can do about it. Manliness and womanliness, on the other hand, are judgments rendered by our cultures and by ourselves, and here, presumably, there is much that we can do. Whereas maleness among great blue herons or sticklebacks leads in a straightforward, uncomplicated way to male behavior, maleness among human beings need not necessarily lead to manliness, any more than femaleness always produces womanliness. Society, no less than biology, dictates manliness and womanliness, and such expectations may or may not be paralleled by our own inclinations. Sometimes the transition is less than smooth, because sometimes the perceptions, expectations, and restrictions of culture do not accord with those of biology. Out of this crucible of conflict arise some of our most frustrating impasses, and some of the most exciting opportunities to define ourselves as not only fully sexual but also fully liberated, committed to ourselves but also to each other, and as a result, fully human no less than other animals can be fully stickleback, or fully heron.

CHAPTER SIX

OF FAMILY AND FRIENDS

Altruistic Genes Playing Selfish Games

A hen is just an egg's way of making more eggs.
—Samuel Butler

In a sense, evolution is terribly selfish. Since selection favors characteristics that lead to differential reproduction, it favors behavior that helps each individual at the expense of all others. We can assume that in most cases at least, there is "room" in the environment for only a limited number of individuals of each species, or similarly, room for only a limited number of genes, competing for chromosomal space. An individual may thus promote his or her success not only by acquiring characteristics that are personally advantageous, but also by actively discouraging the success of competitors; insofar as life is a "zero-sum game," that means just about everyone else.

Seemingly opposed to this logic is the simple fact that many animals are not loners, fiercely competing with all others for a place in the evolutionary sun. Rather, they are social and often quite cooperative. Aristotle suggested that man is also a political animal, by which he did not mean that we are instinctively Democrats or Republicans, but rather, that we insist on associating with others, sometimes for cooperation, sometimes for competition, but nearly always by choice. Yet, there is other ancient wisdom on this score, notably *Homo homini lupus* (Man is a wolf to other men). We must also note, however, that wolves hunt in well-organized packs, within which each wolf has its place. Honeybees construct elaborate hives that may easily house thousands of workers, bison live in large herds, and baboons enjoy a complex social life based entirely on group membership. Animal sociableness, however, is actually not quite the anomaly that it may seem. Each individual within these species is better off as part of a group than it would be alone. There are no solitary baboons, or at least not for long. Although a single baboon is relatively defenseless, an organized troop can hold its own, making even a leopard change its mind and look for easier prey.

In addition to group defense, many animals gain added alertness to a predator's approach by enlisting the eyes, ears, and noses of other group members. Some animals, such as termites, actually produce a particular environment (within the nest) that is more conducive to survival than anything that could be made by a solitary termite. Social life also provides greater opportunity for learning and for the passing of traditions, as when members of a rat pack learn to avoid poisoned bait because the leader avoids it. In short, there are many advantages to cooperative behavior and thus, many reasons why animals live in apparent social harmony. In such cases, the social life is of selective advantage to each individual who lives in this manner, because it increases his or her personal chances of survival and ultimately, reproduction. Evolution will thus favor seemingly selfless behavior, but for selfish reasons.

This analysis becomes more revealing of the human situation when we consider behavior that helps another but does nothing to enhance the survival of the individuals doing the act. Perhaps the most obvious example would be parental behavior. Reproduction is so common in nature that we tend to take it for granted. What possible selfish advantage accrues to the dutiful progenitor?

In terms of personal survival, the answer, nearly always, is "none." Because sexual reproduction requires that two different individuals get together, intimately, it may be a time-consuming, difficult, and sometimes even dangerous activity. Thus, the female black widow spider is so named because of her occasional and unsavory habit of devouring her mate. Praying mantises sometimes do the same (once again, the female consuming the male) and in this case, there is even some evidence that the male copulates more vigorously after he has "lost his head" over a female—the cerebral ganglia tend to inhibit reflexes encoded in the lower nervous system, so a decorticate male, beheaded by his mate who may be more hungry for protein than for sex, will sometimes copulate successfully with his executioner.

Animals will often expend considerable effort, and run substantial risks, in order to reproduce: Pacific salmon batter themselves against rocks and swim upstream many miles, exhausting and often killing themselves even before they reach their spawning areas. Red-winged blackbirds stake out territories in the spring, and then fight fiercely to defend them—not for a place to eat, or to sleep, but to reproduce. In virtually every way, their lives would be simpler, safer, and probably longer if they simply looked after themselves and didn't bother trying to breed.

The antics of courting and copulating animals are notorious, and conspicuous. Roosters crow, elk bugle, elephants roar, and whales sing their eerie arias, sometimes for hours at a time. At such occasions, it seems that love-starved Romeos and Juliets are especially likely to be noticed by their food-starved predators. One of the most dramatic examples of the cost of reproducing comes from the common North American fireflies. These animals use their flashing patterns to announce their species and sex, and attract a mate; different species use different patterns, each one a distinctive Morse code. Thus, the male flashes his identity, the female responds with hers, and after a number of reciprocal reassurances, they fly away together into the moonlight. But in at least one species of firefly, the sexual code has been broken, and by a predator: in firefly courtship, three is definitely a crowd. This unwanted third party gives the flashing pattern used by the female of another species, whereupon a male of that species is attracted . . . and ends up being a meal rather than a mate. Entomologist James Lloyd of the University of Florida, who discovered this lethal intrigue, refers to "firefly femmes fatales."

Even once the various hurdles are surmounted and reproduction is well under way, the attentive parent often expends considerable energy raising its offspring: certain warblers have been observed to make 1000 trips per day, bringing insect food to their hungry nestlings. Carnivores such as wolves must do extra hunting to feed young pups. Nursing mothers of all species require extra food in order to support their dependent young. Furthermore, by consuming a portion of whatever food is available, the infants become potential competitors with their parents during times of food scarcity or drought and not only that, but when they grow up they are likely to compete even more seriously.

In addition to experiencing food and energy problems, the parent—just like the courter or copulator—may be more susceptible to predators than an "unattached" individual. Pregnant females, especially just before giving birth, are slower and less capable of outrunning their enemies. Many animals, including all primates, carry their newborn young, and the burden of heavy, dependent passengers could make a crucial difference if escape is called for. Given all these drawbacks to parenthood, how could it have been selected for? In a sense, the answer is obvious: if individuals declined to reproduce, then their species would last no longer than the life span of those individuals currently alive. But this is not in itself an explanation of why reproductive behavior occurs. Individuals do not perform with an eye on the ultimate good of the species (human beings, on occasion, may be exceptions here). If reproduction ceased because of immediate advantage obtained by the nonreproducers, the species would indeed soon go extinct. But the possibility of extinction cannot influence the evolutionary strategies of living things, since there is no way that organisms can reach ahead into the potential future, and then modify their current behavior accordingly.

But animals try hard to reproduce, despite the disadvantages it may entail. They do so because they are the descendants of others who did so, and moreover, who succeeded. Although bodies may be altruistic in a sense, looking out for other bodies—notably, those special bodies we call children—genes are "selfish" in that they look out for themselves. In another sense, however, they are "altruistic" too, in that they look out for other genes so long as those other genes are copies of themselves, just temporarily housed in another's body. Only those individuals with such altruistic propensities reproduced themselves in the past, and that's why every living thing that is here is here, and also why such things seek to reproduce.

Technically, individuals are not selected for, genes are; and by producing offspring as carriers of the parents' genetic material, living things enter the evolutionary arena. Since genes are being selected, not individuals, reproductive and parental behavior would be highly advantageous if it led to a maximum number of successful offspring. Behaviors that result in leaving more offspring would be selected for, and selection would cause a greater number of those genes (in this case, for effective parental behavior) to be represented in the future population. Animals will therefore expose themselves to all the rigors, dangers, and—at least in humans—irritations and inconveniences of having children in order to reproduce their genes. Not only will animals reproduce, but many will submit to considerable risk if necessary, in rearing and defending their offspring. The leaving of successful offspring is "worth" the chance of personal disaster.

This is not to suggest, incidentally, that living things seek to reproduce because of a conscious desire to bask in the warmth of natural selection's smile. Rather, selection has produced a host of short-term gratifications that lead animals to reproduce. Finding a mate or a nest site, copulating, caring for young, etc., all become, in effect, little gratifications in themselves. Animals go about satisfying one short-term need after the other, often being reinforced in the psychological sense, scratching the various itches that natural selection has established and that ultimately lead to selective advantage. It is not necessary for a reproducing animal to be aware of reproduction as its ultimate goal, any more than it is necessary for a tree to know that it will be more successful if it flowers in the spring instead of in the fall. Producing a maximum number of surviving genetic replicates is the ultimate goal, whether living things know it or not, and indeed, they can do a credible job of achieving their goal regardless of whether or not they are capable of acknowledging or even conceiving of its existence. Similarly, even a beginner can play a surprisingly good game of chess if he or she concentrates on the various possible subroutines—threatening two pieces simultaneously with a knight, containing the opposing queen, etc.—without necessarily orienting every move toward the ultimate goal of capturing the other side's king.

There is a limit to such a system. Although parents of many species will generally defend their young, a strategic retreat is usually preferred to certain death when the outcome is otherwise unavoidable for both parents and young. In such cases, individuals can ultimately leave more offspring by giving up on the present brood and escaping to raise another one. He or she who breeds and runs away may live to breed another day.

But, natural situations are rarely so cut and dried as this. Defense of a litter may or may not be successful, depending on such variable factors as the nature of the attacker, intensity of its assault, age and condition of the defenders, and degree of development of the young. A nesting pair of blackbirds will defend their young against marauding blue jays, but beat a hasty retreat before an approaching hawk. Each situation presents a different probability of success and hence, each must be assessed separately by the defending parent. This evaluation need not be conscious and rational, however. For example, marsupials differ from mammals in that their young are born in a very immature state, after which they develop in the mother's marsupium, or pouch.

One consequence of this arrangement is that the actual nourishment of marsupials is less efficient than in the case of most mammals, whose placenta permits a more effective exchange of nutrients. But there are some advantages for the marsupial as well: when chased by a predator, some species of wallabies and kangaroos have actually been known to abort a relatively large joey, thereby increasing the probability that the mother—now significantly lighter—will be able to run away. By contrast, placental mammals can hardly pause in mid-chase, undergo a convenient abortion, then keep running. Biologists speak of marsupial and placental mammals as each practicing a different reproductive "strategy," which simply means that they go about obtaining their evolutionary success in different, well-organized ways, whether they realize it or not.

The reproductive future of the adult also seems to be included in the evolutionary calculus. Thus, individuals who will not reproduce again or for whom each offspring represents a great investment of time and energy, will be more likely to risk all in defense of their young. Less persistence is expected from parents with a long reproductive life ahead of them or "easy breeders" with a negligible investment in any particular offspring.

Admittedly, it may seem cold hearted and insufferably clinical to interpret parental love in such "materialistic" evolutionary terms. But there can be no question of its validity among animals, and human beings, while very special, are very much animals as well.

Among the higher animals such as ourselves, hormones and genes are increasingly supplemented, modified, and even replaced by mental control of behavior. A female rat, deprived of the hormone prolactin, will not nurse and care for her babies. A female human being, similarly deprived, will not lactate, but can do a perfectly good job with a bottle; unlike the case with lower mammals, there is no evidence that maternal tendencies among Homo sapiens are evoked by maternal hormones. Our brain does the trick. Similarly with sexual receptivity, aggressive behavior, and so forth: among human beings, the highly developed brain makes decisions about behavior that among lower animals is controlled almost automatically by hormones and instincts. Not surprisingly, therefore, learning—and lots of it—is essential in our species. For the brain to be in control, alternative courses of action must be stored, this information must be readily retrievable, and there must be some ability to modify responses depending on what worked in similar situations in the past. If we relied on hormones and genes, we could simply trust our success to preconstructed chemical messengers, acting in concert with prewired neural circuits. But because our evolutionary strategy emphasizes a brainy flexibility, we are obliged to feed this brain, and not just with nutrients. And this provisioning of experience must begin early in life.

Prolonged dependence of the young is therefore the rule. The extraordinarily long period of infant helplessness among human beings requires continuous attention from the parents. During this time, children must not only be taught the basic requirements of successful living, as well as be defended from possible enemies, they must also be fed, cleaned, and cared for. The lower animals meet these requirements by relatively simple, automatic behaviors. Many birds respond instinctively to nestlings gaping for food. Students of animal behavior have found that many animals are automatically sensitive to certain—often exaggerated—signals in their environment, which release instinctive behavior. We can imagine the behavior (whether fleeing, fighting, or reproducing) as existing somehow within the animal, waiting only for the appropriate signal, whereupon it is released. The parent bird does not "love" its offspring; it simply follows the dictates of its genes to fill, mindlessly, wide-open mouths of the appropriate color and shape. In fact, certain birds, such as North America's brown-headed cowbird and red-headed duck, as well as the European cuckoo, regularly take advantage of this by laying their eggs in the nests of other species. These foreign young are treated like the parents' own, as long as they possess the appropriate releasers.

Similarly, the releasing of simple pent up physical pressure in the breast is probably more instrumental in stimulating nursing by female mammals than is conscious solicitude for the infants' welfare. Among most lower animals, in fact, parental behavior is sufficiently simple and short-lived to be safely relegated to specific automatic mechanisms, evolved by natural selection. For example, many animals learn the physical characteristics of their young when they give birth, during an extraordinary process of "one trial learning," known as "imprinting." As a result, females of many species learn irrevocably to recognize their young. The small tropical jewelfish, for one, does not instinctively know what her offspring look like; she carries her fertilized eggs in her mouth, incubating them there, and does not set eyes on them until they emerge as small, distinctively colored fry. Then, she becomes imprinted onto them, and will accept only them. If, however, an enterprising biologist substitutes young of another species before the mother jewelfish has ever set eyes on her own, first mouthful of offspring, she will subsequently prefer these foreigners to her own brood.

A mother goat will reject her own kid if she is denied the opportunity of smelling it within a few minutes after birth; a very brief exposure normally fixes the kid's identity in the mother's memory. If the newborn is experimentally removed immediately after birth and replaced by a different animal, the mother will become imprinted onto it, rejecting her natural offspring from then on. These behaviors might seem to indicate poor planning on evolution's part, since such blind instinctive obedience appears to leave much room for error. But actually, malfunctions of the type just described are limited almost entirely to cases of human intervention. When it occurs in nature, imprinting provides a simple, almost foolproof way of assuring parental recognition without overburdening the gene pool with unnecessary information.

Human beings pose a different situation. Since we require an exceptionally long training period, since we are dependent upon our parents for a remarkably long time, and since we behave in such complex ways as to have required intelligent and flexible responses from our ancestors, reliance upon simple releasers or imprinting would clearly be unsatisfactory. In our case evolution needed a particular mechanism, not previously used by animals, for maintaining the necessary close and attentive relationship between parent and offspring. The answer was love.

We encountered this word when discussing another novel human innovation, the intense male-female pair bond. The use of one term for these rather different phenomena is interesting and revealing. The love of two adults for one another is clearly different in kind from the love of parent for child. But our use of the same word for both may suggest an underlying similarity between the two relationships. Both carry strong emotions. So powerful is their hold on us, so pervasive their influence, that we generally find it impossible to step back and view them as *phenomena* that are almost certainly part of our biological makeup and therefore likely to be of some significance to evolution. Both are also heavily invested with cultural significance.

Both parental and sexual love also serve the same ultimate function, the maintenance of a close, long-lasting, well-coordinated relationship between individuals. Both forms of love serve an essential evolutionary purpose. To some extent, both are founded upon a rather banal sounding, but nonetheless significant occurrence: simple physical proximity and the familiarity among individuals that results.

Now admittedly, familiarity itself does not necessarily produce love (under certain circumstances, we are told, it can even "breed contempt"), but it can go a long way. There is increasing evidence from animal studies that simple exposure can produce increased tolerance, and eventually attraction. When a female song sparrow is ready to breed, she cautiously approaches the territorial boundaries of a likely mate. Up until now, he has been busy establishing his own special area, regularly defending it from intruding males who would dearly love to carve away a parcel for themselves, or appropriate the whole thing. The resident male is therefore very aggressive and shows no gallantry toward the female, who after all looks like just another trespasser and is attacked unmercifully. (There is virtually no difference in the appearance of male and female song sparrows.) But instead of either retreating or fighting as an intruding male would, the romantically inclined female simply flutters nearby, returning again and again, just "hanging around" near her attacker. At last, her patience is rewarded as the aggressiveness of the territory owner finally abates. To the observer, it seems that he has gradually gotten "used to" her presence. If they were human beings, we would say that they have fallen in love.

A simple yet fascinating experiment has demonstrated the effect of repeated exposure in producing attraction and even preference in mammals. Laboratory rats were divided into three groups: all were treated equally except that one group was given regular exposure to recordings of several Mozart symphonies; the second group heard the dissonant music of the twentieth-century composer Arnold Schoenberg. The third group was not exposed to either composer, so as to reveal whether inexperienced animals had any initial musical preference, and therefore, whether the preferences of the experimental groups could be attributed to their earlier music-listening experiences. After nearly two months, the rats were tested for their musical preferences. The third group, which had not been exposed to any music, did not show a strong preference, although on balance, they chose Mozart over Schoenberg. (Thereby revealing, one might conclude, the fundamental good judgment of untrammeled biology.) The first group, the one that had been initially exposed to the Mozart, showed an even stronger preference for his music when given a choice. The Schoenberg-exposed group, however, turned out to like Schoenberg. Each of the two experimental groups thus developed a definite preference for whatever it had been exposed to earlier, even when—as in the case of Schoenberg—this stimulus was held in disfavor by inexperienced animals. Familiarity, we must conclude, does not always breed contempt. For most living things, in fact, it produces preference.

It is a long way from musical rats and courting song sparrows to love in human beings, but the analogy seems to hold. Among male-female consort pairs in baboons and human beings, the partners spend long periods of time in each other's presence, often doing "nothing special," simply strengthening their affiliation by their mutual presence. In fact, sexual relations between human beings are often considered degrading if they are just "one night stands," lacking the extensive precopulatory period of increasing association and affection that somehow makes a relationship "meaningful." We can always rationalize this phenomenon by pointing out that getting to know each other enables deeper understanding, and hence, appreciation and ultimately love. But that's just the point.

Parents nearly always love their children. Why? Insofar as evolution is concerned, because children are a major route to biological success. As to the means of achieving this adaptive end, prolonged exposure is probably the key, coupled with the child's possession of the appropriate physical releasers: large head compared to the rest of the body, "cute" little nose and ears, the indescribable appeal of adult traits in miniature, unsteady gait—all of these are characteristics that evoke cuddliness and care-taking in Homo sapiens. We clearly do not succumb to immediate imprinting as does the nanny goat. A parent's first view of its newborn baby, somewhat discolored and possibly with misshapen head due to stresses during birth, often causes more disappointment than devotion. But ugly or cute, boy or girl, scrawny or fat, a child "grows on you" until quite soon it is the object of considerable love and devotion. The child need not even be the biological offspring of the prospective parents, as the great success of adoption has shown. Given the appropriate physical characteristics, evolution has arranged that simple familiarity will generate love, with all its attendant selective advantages.

It is always possible that the ultimate evolutionary significance of male-female and parent-child love relationships is coincidental or at most, tangential to their occurrence. We all accept the notion that human beings have a capacity and in fact, a need for love. "People who need people," we are told, "are the luckiest people in the world." We have a capacity for eating food as well, and indeed, a need for it, but no one considers him or herself especially blessed as a consequence. No one questions the evolutionary utility of food: why should love be different? Thus, loving relationships clearly satisfy a powerful and necessary urge within us, no less real than the adaptive urge to eat when hungry or to scratch when itchy. And just as we can develop pathologies of eating, from obesity to anorexia, we are also subject to pathologies of loving, although most of us do a remarkably good job of balancing our nutrient intake and our loving as well. In the wild state, animals such as goats must be prepared to move quickly right after birth; consequently, they are at some risk of misidentifying their own offspring, since these offspring are so precocious that they are walking and running when just a few hours old. The rigid imprinting between nanny and kid therefore makes good evolutionary sense.

For human beings under prehistoric situations, just as among nontechnologic people today, the chances of a newborn infant being attributed to the wrong mother were probably close to zero. However, in modern maternity hospitals, with sometimes literally dozens of newborns being "processed" daily, mix-ups can occur, and significantly, when they do, we are virtually as helpless as our own neonates.

Other animals show differences in their ability to recognize their offspring, differences that help illuminate our own otherwise puzzling limitations. Consider two species of birds, the rough-winged swallow and the bank swallow. Both dig themselves burrows where they nest in sand quarries and natural clay banks, but whereas pairs of rough-winged swallows nest in isolation from one another, bank swallows nest in dense colonies that may include hundreds of other birds. Ethologist Mike Beecher has been able to show that although bank swallows quickly learn to recognize their own young, and will refuse to feed a stranger who happens to land at their burrow entrance, rough-winged swallows apparently cannot tell one young bird from another. Because they breed in such dense colonies, adult bank swallows during their evolutionary history were probably often faced with the opportunity of caring for young that were not their own biological offspring. Those that did so were selected against relative to those that only invested in their own offspring. By contrast, adult rough-winged swallows normally do not encounter any offspring other than their own, so there is no reason for them to imprint onto the characteristics of their young, nor do they possess a genetic knowledge of them. So they innocently feed anyone they find at home, and sometimes they can be fooled. Natural selection, in a sense, can count on the likelihood that any youngsters in a rough-winged swallow's nest will be the ones that the mother has deposited there herself.

Human beings are probably more like rough-winged swallows than like bank swallows. That is, because of our preceding thousands of generations during which the probability of parents associating with the wrong offspring was very low, we lack innate recognition mechanisms, thereby rendering us susceptible to the maternity ward mix-up. This is an admittedly rare liability, one that is uniquely a result of medical technology and assembly-line maternity procedures. But along with it comes a rather substantial advantage: because we lack such recognition mechanisms, we have the capacity of adopting children not biologically our own, and of developing responses to them that are every bit as profound as those toward our genetic children. In short, we can have "cultural children" instead of, or in addition to, our biological children.

Once again, however, the picture is complicated. Our biologically mediated capacity to accept children that are not biologically our own, combined with the growing (culturally mediated) propensity for parents to divorce, places increasing numbers of children among nonbiological parents. Fortunately, stepfamilies can "work," in large part because we lack automatic rejection mechanisms, and because of the familiarity effect, as well as the fact that we are also conscious creatures who presumably enter into relationships with caution and often, a benevolent desire to make the best of situations. But stepfamilies are also susceptible to the ill effects of biological selfishness, since stepparents and stepchildren are usually intensely aware of their "step" relationship. It may be no coincidence that evil stepmothers and stepfathers are so often the villains of fairy tales: a disproportionate frequency of child abuse and child molestation involves the stepfather, whose involvement with the child comes as a result of his affiliation with the child's mother, not with the child itself. This pattern, although not excusable just because it is congruent with evolutionary expectation, is at least more understandable as a result.

Clearly, not all nonbiologic parents are child molesters or abusers, and stepfamilies can be wonderfully successful. But a glimpse of this increasingly widespread culturally mediated phenomenon through the lens of biological evolution provides a perspective that may be a healthy corrective to the frequent misconception that everything should always go smoothly . . . and then, the guilt and recriminations when it does not.

Both the male-female and parent-child relationships are commonplace situations that are almost universally taken for granted and rarely scrutinized. When considered in the light of evolutionary theory, in fact, they initially appear out-of-place and strange. What had seemed so natural is now puzzling, since the immediate costs are real, but the benefits, diffuse. Parental care seems such an obvious thing to provide, yet when the manifold personal disadvantages are considered, it is suddenly hard to explain. But upon further analysis, these behaviors once again assume a legitimate biological role, and with the added insight, we may well be better prepared to deal with culturally inspired perturbations so that the many faces of love can all make beautiful sense once more. It is said that before one studies Zen, the mountains are merely mountains and the flowers, flowers. To the earnest monk struggling to unravel their deeper meaning, the mountains and flowers are no longer what they had been. But when true enlightenment is finally achieved, the mountains are once again mountains and the flowers are flowers once again.

⁂

Around every person there is a circle or group of kindred of which such person is the center, the Ego, from whom the degree of the relationship is reckoned, and to whom the relationship itself returns . . . A formal arrangement of the more immediate blood kindred into lines of descent, with the adoption of some method to distinguish one relative from another, and to express the value of the relationship, would be one of the earliest acts of human intelligence.

—L. H. MORGAN (1871)

PARENTAL behavior would seem to be an obvious example of selfless behavior with a clear evolutionary function. We can also extend the biological explanation to cover such "altruistic" traits as the famous broken-wing display of the killdeer. If a fox approaches the nest of this common little shorebird, the adult will flutter away a short distance, seemingly in distress and feigning a broken wing. As the fox follows, the killdeer flaps away a bit further, seeming always to muster just barely enough strength to escape at the last instant, thus deceiving the fox into thinking that it will be easy prey. By this ruse, the predator is lured some distance from the nest at which point the crafty killdeer, its wing now miraculously mended, flies to safety. Such behavior is not without danger to the killdeer itself. But the benefits of increasing the chances that its offspring will survive have apparently been great enough for broken-wing feigning to have become firmly implanted in the species' biology.

Among many animals, the first one seeing potential danger will immediately give an alarm call, alerting others in the area. In some cases, the calling individual may not have any offspring at the time and thus could not be conferring any survival advantage upon them. In addition, the animal sounding the alarm may actually be at a disadvantage relative to the others, who gain by the alarm caller's apparent altruism. By sounding an alarm the sentry warns others of danger, at the same time making itself more conspicuous and hence, in greater danger than if it had simply kept quiet. It would have been more self-serving to flee silently or hide inconspicuously, leaving its fellows to find out for themselves that danger threatens. So why are alarm calls so common?

This knotty problem—and indeed, the whole paradox of animal altruism—was clarified if not entirely resolved by the British geneticist W. D. Hamilton, who pointed out that reproduction is only a special case of the more general phenomenon of natural selection acting on genes rather than individuals, groups, or species. Parents are especially concerned with their offspring because those offspring are a primary vehicle for the parents' own genes. In the same way, other genetic relatives are also carriers of one's genes; the closer the relationship, the greater the probability that a copy of a particular gene, present in one individual, will be present, by virtue of their common descent, in another. In fact, the preceding sentence can be turned around and made even more accurate: we define closeness of genetic relationship by the probability that copies of genes present in one individual will be present in another.

Hamilton's insight—commonly known as kin selection, or inclusive fitness theory—has emerged as one of the key concepts in sociobiology, the branch of evolutionary biology that is particularly concerned with the study of animal and human social behavior. The probability of shared genes turns out to be crucial evolutionary currency, and a major determinant of social behavior, in human beings as well as other animals.

Many species live in groups containing close relatives and the same explanation, in effect, can account for altruistic behavior within those groups as for parental behavior, provided that some sort of family relation exists between the altruist and the beneficiaries.

Picture a gene "X" that induces its carrier to utter alarm calls, as opposed to a gene "Y" that causes it to remain silent, selfishly. By calling an alarm, an individual carrying gene "X" would save the lives of some of its relatives, thus increasing the gene's frequency. At the same time, by exposing its carrier to higher predation, "X" would tend to decline in abundance. The ultimate balance would depend on three factors: the danger to the altruist, the benefit to the recipient, and the genetic closeness of the two individuals. The closer the relationship, the greater the probability that an alarm-calling gene will be warning itself, and therefore, the higher the risk that will be run until finally, with the very close parent-offspring relationship, risks may be very high indeed and yet still acceptable. On the other hand, natural selection would not generally promote altruistic self-sacrifice on behalf of distant relatives or total strangers.

Anthropologists have long recognized, and long been puzzled by, the fact that human beings the world over tend to organize their social life around systems of kinship. Some form of kin recognition and nepotism is found in every human society. We are obsessed with genes, even those among us who don't know DNA from dynamite. It is quite possible that the entire gamut of human kinship systems, from toleration of an undersirable house guest simply because he's a relative, to nepotism in business and government, is a by-product of this fundamental evolutionary phenomenon. Why do we bother identifying our relatives? Why do we treat them any differently from perfect strangers? On an immediate level, the answer is simple enough: we generally love our relatives, spend time with them, and trust them, at least, more than total strangers. Familiarity, perhaps. On an evolutionary level, such behavior is also in accord with natural selection.

Unlike most of the other examples discussed here, the relation between biological and cultural evolution in the case of human kin selection appears to be mutually supportive rather than conflicting. Insofar as the complex superstructure of kinship systems provides opportunities for appropriately directed biological altruism, cultural frameworks of this kind seem to be adaptive in the evolutionary sense. It is striking to consider that our common behavior toward relatives may have a direct basis in biological evolution. But some problems have arisen, owing largely to the rapid cultural advance of the past 30,000 years. For example, when our ancestors lived in small hunting-gathering bands that probably did not exceed 100 individuals, it is likely that kinship relations linked most of the membership. Selection would thus favor genes for altruistic defense of the group: analogous to gene "X" in our example of alarm calling. Such defense would be especially likely among the males, who are more directly engaged in competition, as well as better adapted for combat.

This could occur even if the defenders did not have offspring, or if immediate benefit to their family was not apparent. It is tempting to fall into the trap of arguing that such behavior is simply a result of "human nature" and that an evolutionary explanation is therefore neither relevant nor necessary. But as in the case of human love, discussed earlier, when we ask why nepotism is part of our behavioral repertoire, we are led to ask what has been the effect, if any, of culture upon this adaptive biological system.

The biological and behavioral system of nepotism was probably quite satisfactory before culture began its very rapid change. With the growth of tribes, communities, city-states, and eventually nations, we have been increasingly forced to interact with nonrelatives and on behalf of those we do not know or especially care about. We are expected to concern ourselves with the survival of perfect strangers. Along with this increased association with nonrelatives has come a concomitant decline in the association of relatives with each other. The common living arrangement of many nontechnologic people is an extended family in which grandparents, uncles, aunts, and cousins often live together, frequently under the same roof. Such arrangements are almost unheard of in modern America, and the consequences of such a culturally mandated deviation from human biology may well be severe, in terms of subtle psychological stress.

Child rearing, for example, is a different endeavor when it falls entirely on the shoulders of a small, nuclear family, as opposed to being a communal responsibility in which other, nonparental relatives are intimately involved. It is a basic principle of engineering that stress is reduced on any single point in proportion as it is spread across a large area. Similarly, the concentration of child rearing on just two adults (or worse yet, just one) seems almost a sure prescription for stress. And older adults, as well as younger nonreproductive ones, may be denied an otherwise satisfying outlet for their basic predispositions when they are denied regular interactions with grandchildren, nieces, or nephews.

Once upon a time, in short, an individual's evolutionary interests were adequately served if she simply insured the well-being of herself and her immediate associates. (As for himself and his immediate associates, this may have been less true, because of the phenomenon of male-male competition, discussed earlier.) But virtually overnight, things have changed. Our technology and economic system have made us dependent upon literally millions of people we will never know, tying our well-being to the fate of a large, rather arbitrary unit, the nation (and increasingly, the world) whereas our fundamental focus remains local: neighbors, family, and self. Cultural evolution has seen to it that "no man is an island," but we may lack the biological vision to appreciate this fact.

Our culturally determined social units have thus suddenly expanded to encompass more than just our relatives, while our biology has been especially primed for protective behavior only toward them and our close associates. Defense of these new larger units and cooperation within them have accordingly become a real problem. The defending individual must somehow be motivated to behave altruistically toward a unit that exceeds anything for which our biological nature has been prepared, and the task therefore falls upon culture to generate support for its own institutions. Human ethical systems are diverse, as expected in a thoroughly cultural phenomenon. They are also very similar, however, in that each seeks to bridge the gap between human biology and human culture: the former seeking selfish benefit, the latter demanding self-restraint at least, and often self-denial or even altruistic self-sacrifice.

On one level, then, we expect that human beings will be altruists; that is, although their actions may endanger or otherwise mortify their bodies, these acts (such as parenthood or nepotism) are in fact selfish on a genetic, and hence, an evolutionary scale. As society views us, however, human beings must appear to be profoundly and dangerously selfish, for the same reason: we tend to look out only for our own personal benefit or that of our relatives. Social psychologist Donald T. Campbell has suggested that there may be a deep quasi-evolutionary wisdom in the various systems of conventional religions and ethical morality that circumscribe the boundaries of acceptable human behavior, since without such restrictions the competitive urges of a normally selfish creature could be terribly destructive to society.

The English philosopher Thomas Hobbes suggested just this more than 300 years ago, when he pointed out that human beings tend naturally to a "warre of each against each," unless restrained by the power of the state, the *Leviathan.* Historically, such restrictions have been seen as necessary and indeed, laudable, although to be sure, without understanding their likely biological basis. In recent years, it has become fashionable to criticize such arrangements as artificial, unduly restrictive, and generally harmful. "Do your own thing," "If it feels good, do it," and other slogans of personal liberation were spawned by a sense of societal oppression, combined—especially during the 1960s—with a sense that established society, with its pollution, its vicious war in Vietnam, its untrammeled materialism, had no right to suppress the individual.

But ironically, those people most likely to support personal rebelliousness of this sort are also most likely to maintain that society itself has certain responsibilities, which go beyond the pursuit of private gain. It may be that an understanding of Homo sapiens' underlying selfishness—enshrined but not necessarily legitimized by the New Right in the 1980s—may once again swing the pendulum toward acceptance of society's rights and obligations vis-à-vis its otherwise unruly members. In the process, however, we had best be alert to the potential for abuse, something that nation-states—whether motivated by ideology of the far right or the far left—seem to find deeply compelling. Certainly, the continuing tension between human inclinations (in large part the product of biological evolution) and society's restrictions (the product of cultural evolution) can be destructive as well as restorative.

When it is widely perceived that individuals are unruly and troublesome, people seem more likely to support social functions such as legal restraints, taxation, police systems, and government structure in general, all aspects of organized society that appear to be fundamentally benevolent. However, there is a danger lurking here. By proclaiming that humanity is fundamentally flawed, riddled with a kind of original sin (whether generated by natural selection or God), we help lay the intellectual groundwork for a whole range of repressive and often despotic social processes. If we assume that human beings are "bad" at heart, then society must be protected against its own members. Following a period of worker discontent and near revolt in East Germany during 1953, the authorities announced that they were "disappointed" with the people, which prompted Bertolt Brecht to suggest that perhaps the government should therefore disband the people and elect another! Brecht, a dedicated Communist and supporter of the rights of society against those of the individual, nonetheless saw the danger of perceiving society to be more valuable than its members. The next step is a fateful one: toward the "national security" state, in which the security of that society, rather than the safety of its component individuals, becomes the only legitimate goal of collective action. Even democratic societies can, and have, walked down this potentially destructive road, especially in the modern era of looming nuclear holocaust. En route, other societies take on the appearance of evil incarnate, thereby justifying attitudes, policies, and the possession of weapons which in their own right would otherwise be seen as totally unacceptable.

Our tendencies to view each other with distrust may also be a direct outgrowth of our tendency to distrust ourselves, and then to project such fear onto others. Writing in *Psychology and Religion* in 1937, Carl Jung put it as follows:

> This terrifying power which nobody and nothing can check is mostly explained as fear of the neighbouring nation, which is supposed to be possessed by a malevolent fiend. Since nobody is capable of recognizing just where and how much he is himself possessed and unconscious, he simply projects his own conditions upon his neighbour, and thus it becomes a sacred duty to have the biggest guns and the most poisonous gas. The worst of it is that he is quite right. All one's neighbours are in the grip of some uncontrollable fear, just like oneself. In lunatic asylums it is a well-known fact that patients are far more dangerous when suffering from fear than when moved by rage or hatred.

Culture can have a useful, modulating effect on human fear and selfishness, leavening our nastier inclinations for the benefit of the group of which we are a part, and upon which we depend. As groups have grown, however, they have arrogated greater destructive power unto themselves, and at the same time have increasingly distanced themselves from the interests of the individuals who comprise them. Environmental degradation and nuclear weapons are cases in point: national leaders now find it expedient to follow policies that could lead to permanent pollution and/or resource shortage, as well as even threaten the destruction of life on earth in order to advance the "national interest." At such a juncture, the various techniques by which culture has overridden our biological proclivities for selfishness cease being benign and become malignant, possibly lethal.

Individuals do not usually give up their autonomy without a struggle. The tendency for self-preservation is strong, like that of reproduction and for about the same reason. In most animals, overcoming the instinct for self-preservation therefore requires potent evolutionary recompense. Respect for the smooth functioning of society is one thing. Permitting it to destroy oneself is another. How, then, does the modern nation-state recruit support for policies that are at best life-threatening and at worst, life-annihilating?

Most social monkeys are capable of organized group defense; in fact, this is one of the main reasons they are social in the first place. Although such defense may consist largely of bluff and show, it can be exceptionally ferocious, with vigorous coordinated attacks in which the enthusiasm of each defender is stimulated and magnified by the behavior of his comrades. The capacity for organized group aggressiveness in these animals may well have a genetic component, and some representation in our own species as well. It is known that human beings in groups, whether organized (an army) or unorganized (a mob), are capable of aggressive activities only rarely perpetrated by solitary individuals. We tend to lose our individuality, our conscience, and our restraints once we become part of a group, and the result can be violent in the extreme. And it seems that we positively yearn to make such an association, to lose ourselves while identifying with a larger collectivity, perhaps because in doing so, we "find ourselves," satisfying primitive adaptive wishes for family and tribal identities in the larger social unit. By exploiting this yearning through speeches, slogans, martial music, and all the trappings of national chauvinism, combined with an intensive program of group indoctrination beginning with the youngest children, societies have been remarkably successful in recruiting willing supporters, especially in times of real or imagined external threats.

Despite the fact that the nation is not, in fact, a biologically meaningful entity, nationalism has been able to achieve a remarkably strong hold on Homo sapiens, apparently because of the strong human tendency to establish social bonds, bonds of the sort that on a personal level are ultimately in our biological self-interest. Hence, nationalistic pseudo-relationships are generated via appeals to fictive kinship (*fraternité*) within the "fatherland" or "motherland." Ironically, in a world of nuclear weapons, the biologically appropriate, pro-life inclination of human beings to establish adaptive relationships with others has been manipulated by and directed toward an artificial, oversized, culturally defined, and increasingly unresponsive political unit—the nation—that threatens human life far more than nourishes it.

Similar conflicts between biology and culture are also apparent in more direct disputes regarding the personal defense of the society: specifically, soldiering. Animals don't behave as if they want to die. And neither do most people. They are unwilling to endanger their lives unless the benefits somehow outweigh the risks. When the risky behavior will result in a generally higher frequency of genes for such behavior, living on in other bodies that we call relatives, this behavior will typically be maintained because evolution favors it, via kin selection. But when the behavior confers only a very diffuse benefit for the genes concerned (the relatives), and when the risks are comparatively high, evolution will select against such tendencies. Defense of a nation, as opposed to a family, can be just such a case, in which we would expect individuals to balk at the more extreme demands of nationalism.

Both historically and biologically, soldiering may have been adaptive for the population as a whole. However, the advantage, if any, to a warrior is often low compared to the risk encountered, especially under conditions of modern warfare. Insofar as natural selection is concerned, individuals typically are motivated by individual benefit, not that of the group; evolution would judge it a bad bet, and select against such tendencies. Furthermore, under conditions of large social units, the individual who "selfishly" stays home and refuses to fight would seem to be in a position to leave more offspring than the "altruist" who marches off to war and may not return.

Primitive appeals to nationalism are used to generate popular support for policy agenda, and to recruit soldiers. But although the nation-state can silence dissent, force acquiescence, and in some cases even generate ardent enthusiasm, it is another thing for the nation-state's military apparatus to be assured a continuing supply of willing cannon fodder, since the preservation of one's own life is, after all, a pretty strong evolutionary imperative. One might expect this conflict between biology and culture to be ultimately resolved in favor of biology. But this is to ignore the importance of culture to the physical survival of our species, and to underestimate its resourcefulness in perpetuating itself. It must be recalled that cultures themselves will tend to develop adaptive characteristics through a process of competition with other cultures analogous to natural selection among strictly biological entities. For example, Napoleon's great battlefield successes were due at least partly to his invention of the *levée en masse*, the first time that the manpower of an entire nation was mobilized for military (or any other) purpose. The other great states of continental Europe learned their lessons well, and before the end of the nineteenth century, Germany in particular had a very efficient military draft. In such cases, however, the characteristic is assumed by the culture as a whole and is not part of the genetic makeup of individuals. Although biological selection operating at the level of groups is inefficient compared to the same process operating at the level of individuals or genes, cultural selection operating at the group level can be very potent indeed.

In his book *The Parable of the Tribes*, theologian Andrew Bard Schmookler examined the problem of power in human social evolution, reasoning as follows: "Imagine a group of tribes living within reach of one another. If all choose the way of peace, then all may live in peace. But what if all but one choose peace . . . ?" The result, as he sees it, is a system driven to ever greater accumulation of power—military, technologic, economic—because of competition among societies. In such cases, individuals are ground under by the desperate competitive machinations of their larger social units.

Forced conscription, as we have seen, is probably society's

simplest and crudest tool for insuring defensive altruism in a nonbiologically motivating situation, when salaries are inadequate to compensate for the risk and other costs of military service. Tax resistance, nonviolent civil disobedience, counterculture life styles, resistance to a military draft: all these and more are thus in the final analysis among the many possible manifestations of the conflict between biology and culture. More generally, resistance to "the system," when it becomes indifferent to life, or worse yet, life denying or life destroying, is often a *cri de coeur* abetted by any number of cultural techniques and prescriptions, but fundamentally a cry from the heart of life itself.

*
**

If there were only two men in the world, how would they get on? They would help one another, harm one another, flatter one another, slander one another, fight one another, make it up; they could neither live together nor do without one another.

—VOLTAIRE, *Philosophical Dictionary*

THE HUMAN capacity for group-oriented behavior is heightened by another characteristic of many social animals: rejection of foreigners. Most bees, wasps, and ants, for example, carry a particular odor, specific to their own hive, which permits other hive members to identify them. Foreigners, or residents that have been treated with foreign odors, are generally driven away or killed. Rats live in organized packs whose members are identified and tolerated. Woe to the strange rat who is introduced into the midst of a harmonious pack. The large ground-dwelling monkeys (macaques and baboons), whose behavior may be so suggestive of our own a few million years ago, live in tightly organized groups in which strangers are only rarely tolerated and aggression between groups often occurs. Once again, this is a pattern that makes sense in terms of biological evolution, since members of these relatively closed groups are often genetic relatives as well.

Among many tribal peoples, the word for human being is the name of the tribe: therefore, members of a different tribe are by definition not human beings. It is no coincidence that among many of the head-hunting tribes of the Amazon, killing a fellow tribesman is murder, whereas killing someone else is "hunting." By defining only their friends and relatives as truly human, members of the in-group are free to behave toward members of the out-group in ways that would not be socially acceptable toward fellow tribesmen, or biologically acceptable toward genetic relatives.

Killing of a fellow tribe member is generally prohibited, but killing someone from another tribe may be encouraged. After all, a member of a strange tribe is not a human being. This is not mere sophistry; it is a fundamental fact in the lives of many people, and it speaks eloquently of a world view in which we can identify evolution's oft-unpleasant hand. Kin selection is relevant to this murderous double standard, since a fallen stranger is less likely to be carrying genes in common with his murderer.

In proposing a biological tendency for xenophobia, we are not necessarily predicting 100 percent expression of this trait in all people or all groups: remember that genes define a range of possible expression, rather than rigidly determining a precise characteristic. This seeming equivocation is especially true for behavioral traits that can easily be modified by culture. It should be clear, however, that insofar as xenophobic tendencies exist, they generate troublesome susceptibility to propaganda picturing foreigners as criminal, immoral, not quite human, and certainly untrustworthy.

The human tendency to form groups with insiders and against outsiders is reflected in many aspects of our lives, acting within cultures as well as between them. It begins with children "ganging up" on one another and progresses through exclusive clubs, fraternities, sororities, union locals, service clubs, and political parties. In addition to such purely cultural affiliations that are achieved largely by choice and by personal effort, there are cultural units into which one is born and for which choice, although possible, is experienced only rarely. Religion, ethnic group, and nationality would be prime examples. Beyond this, conspicuous physical differences with a genetic basis provide a convenient rallying point for human discriminatory tendencies. Racial differences in skin color are a prominent case. And when such differences do not exist, we manufacture them, via clothing style, language, accent, secret passwords, astrological sign, or other totemic association.

We human beings have a notable inclination to exclude individuals who are conspicuously different from us in any way. Such behavior in its relatively primitive untrammeled form is probably biologically adaptive, for the following reason: among most animals, disease is a prominent cause of mortality, probably more important than is generally realized. Since many diseases can be transmitted by infected individuals, it would be advantageous if diseased animals were somehow prevented from associating closely with the healthy ones. Thus, in many animal societies, diseased or disfigured individuals are often mercilessly hounded and excluded from the group. As a general rule, those that are different get ostracized.

Unfortunately, "differentness" in human beings is much more often a function of opportunities, ideas, and inclinations (i.e., culture) than a biological condition. Loners, eccentrics, men with long hair and beards, people who go barefoot or wear beads, women without bras, "kooks" of all kinds become the subjects of society's antagonism. Only firm cultural insistence upon tolerance, motivated by recognition that ultimately society is best served by maintaining freedom and diversity, can save us from the stifling homogenization that would result from such evolutionarily generated xenophobia run wild.

At evolution's behest, we thus tend to defend and protect our offspring, favor our relatives, identify with groups, respond to mob psychology with a propensity for violence, and distrust outsiders and anyone who is different. The interaction of our biological and cultural heritage leaves us enmeshed in a complex mosaic of the bestial and the beautiful, of problems and possibilities.

The biologically evolved human capacity for benevolence and cooperation is not limited to patterns of genetic relatedness or immediate self-interest. There is yet another mechanism by which natural selection can get a handle on altruism. Known as "reciprocal altruism," or more simply, "reciprocity," it suggests how apparent altruism could evolve between individuals that are completely unrelated; indeed, even between members of different species.

The crucial requirement is that the beneficiary eventually reciprocates the favor, so that the altruist ends up ahead in the long run. Once again, as with kin selection, biological evolution can select for altruism so long as it is not really altruism at all, but rather, selfishness. For example, among the inhabitants of coral reefs are small, brightly colored "cleaner fish," notably wrasses of several different species. These little iridescent animals enter the gills and even the mouths of larger fish, including moray eels, removing various external parasites. The animals being cleaned profit from the service, and the cleaner fish get a meal. In a sense, the former animals are being altruistic, in not eating the cleaner when they could easily do so, just as the cleaner is being altruistic in not taking a bite out of its host. So each participant gives a little, and in the long run, both are better off.

Now, imagine a prehuman Australopithecine who has just killed an antelope. After stuffing himself and allowing his mate and his relatives to eat their fill, there may be little cost to sharing the remainder with other members of the group, especially if by doing so, the fellow doing the sharing makes it more likely that he will benefit from someone else's largesse next time around. If the cost of the altruistic act is not too great, and the opportunity for reciprocation is sufficiently high, it would not matter whether the participants are related or not (although, in fact, it would help).

Human beings are quite sensitive to reciprocation, typically responding with what has been called "moralistic aggression" when someone fails to return a favor. It may even be that without this capacity, reciprocal altruism would not evolve. This is because reciprocal systems are highly susceptible to being disrupted by cheaters, individuals who accept altruism from others but then refrain from paying them back later.* Such noncooperators enhance their fitness by finding it more blessed to receive than to give, thereby making chumps out of the altruists, who suffer a reduction in their fitness by giving without getting.

As a result, human reciprocal systems are best developed among individuals who know each other well and are likely to interact frequently in the future. We are more likely to trust a neighbor than a stranger, knowing that she is more likely to return a favor. In many cases, bonds of genetic relatedness are overlain with bonds of reciprocity, but among human beings at least, either one or the other can be sufficient to produce cooperation that may ultimately be beneficial to both. Conflicts arise under two circumstances: when the expected reciprocity is not forthcoming, and when suspicions of nonreciprocity interfere with cooperation that might otherwise be helpful or even necessary for the success of both parties. In the latter case in particular, such conflicts also involve biology and culture.

This problem has been recognized by mathematicians, psychologists, and political scientists for some time, and is called the Prisoner's Dilemma. The dilemma is as follows: imagine a simplified situation in which two individuals are interacting, each having the option of doing one of two things, either cooperate (altruistically) or cheat (selfishly). The payoff that each receives depends not only on what he or she does, but also on what the other one does, and yet each is forced to decide independently, not knowing whether the partner will be altruistic or selfish. We can represent the situation as follows, with the words inside the boxes indicating the payoffs received by individual 1:

*There is at least one species of fish, the saber-toothed blenny, which resembles the cleaner wrasse in both appearance and behavior, except that when the larger fish open their gill covers to be cleaned, the blenny takes a quick bite, and then darts away!

		INDIVIDUAL 2: cooperate	INDIVIDUAL 2: cheat
INDIVIDUAL 1	cooperate	reward	sucker
	cheat	temptation	punishment

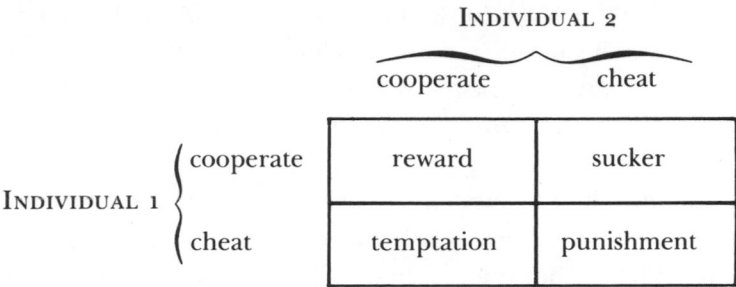

(Of course, a mirror image payoff would apply to individual 2.)

Cooperating altruistically brings a *reward* so long as the other individual also cooperates; cheating selfishly and refusing to cooperate brings a *punishment,* so long as the other individual is also selfish; cooperating altruistically when the other individual is selfish makes one a *sucker*; and finally, being selfish and refusing to cooperate yields the *temptation* to cheat if the other individual is cooperative (and also gets suckered in turn). Depending on the payoff values, there may be no dilemma at all. For example, if the payoff for mutual cooperation is high enough and the temptation to cheat is low enough, everyone should happily cooperate. Other patterns can also be analyzed. The most interesting one, however, is when temptation is high, the cost of being suckered is very great, and the reward of cooperation and the punishment for cheating are intermediate, with the former higher than the latter. Under such conditions, the participants are prisoners of a dilemma.

Here it is. Individual 1 would be best off if he cheated and individual 2 cooperated; he would then reap the temptation to cheat. However, individual 2 is in the same position. The two of them would be best off if they both cooperated, since the reward for cooperation is higher than the punishment for mutual cheating. But each is prevented from cooperating by fear that if he did so, the other might cheat, leaving the would-be altruist a sucker.

Look at it this way, again from the perspective of individual 1: he is trying to decide what to do, knowing that individual 2 will either cooperate or cheat. He reasons as follows. "If #2 cooperates, then the best strategy for me is to cheat, getting the *temptation* and leaving him a *sucker*. On the other hand, what if #2 cheats? Then once again the best strategy for me is to cheat, in which case admittedly I get the *punishment*, which may be bad, but less bad than the *sucker's* payoff which is otherwise in store for me." So, whatever individual 2 does, individual 1—following impeccable logic—is forced to cheat. Individual 2, following the same logic, does the same thing. The outcome of the Prisoner's Dilemma is that both individuals find themselves cheating, and receiving the *punishment* whereas if they had both cooperated and behaved altruistically, then both would have received the *reward*. Each one, fearing that the other will cheat and that he will be made a sucker, is forced to be a cheater himself. And so both wind up worse off than they would otherwise be.

"The reasonable man adapts himself to the world," wrote George Bernard Shaw in *Man and Superman*. "The unreasonable one persists in trying to adapt the world to himself. Therefore all progress depends on the unreasonable man." When two reasonable men meet, neither attempts to adapt the other to himself; that is, both cooperate and both receive a high payoff, the reward of mutual cooperation. But when an unreasonable man meets a reasonable one, the unreasonable one persists in trying to adapt the other to himself—that is, he cheats. In the short term, the result may be a kind of progress: the unreasonable man does better than the reasonable man because the unreasonable man receives the temptation for cheating while the reasonable one gets the sucker's payoff. But over time, the system retrogresses rather than progresses. Reasonable people become unreasonable, or else they disappear and are replaced by the unreasonable, who are more resilient because they do not get suckered. The result of such a system, therefore, is a population of all unreasonable people, none of whom cooperate. Admittedly, none of them get suckered, so in a sense their behavior is not entirely unreasonable, but they don't do any better than the relatively poor and punishing payoff of mutual cheating. Just as bad money drives out good, unreasonable or cheating behavior in a Prisoner's Dilemma tends to drive out reasonable or cooperating behavior. (The dilemma, once again, is that all participants would come out ahead if they could only figure out some way of being mutually reasonable, and cooperating.)

Human beings, even the illogical and nonmathematical among us, have probably been shaped by interactions of the Prisoner's Dilemma type. This is not to suggest that we necessarily run through the sometimes tortuous calculations or that our ancestors did. Rather, natural selection did the analysis and as a result, we tend to shy away from cooperating if there is too great a chance that we will end up being suckered. We therefore tend to be suspicious, especially when dealing with strangers who might cheat us. Moreover, we often tend to view the world as a Prisoner's Dilemma, even when it isn't.

Fortunately, however, there are many ways out of the Prisoner's Dilemma. As political scientist Robert Axelrod has described in his important book *The Evolution of Cooperation*, one of the most effective such routes involves "Tit for Tat," the exchange of small reciprocations each of which is relatively cost-free, so that a pattern of mutual trust and mutual benefit result. (It is also necessary, it seems, to retain the option of responding with a "cheat" if the partner/opponent does so.) Another way out is to recognize that mutually advantageous solutions can be available regardless of what the other side does, and that systems that under primitive, biological conditions were Prisoner's Dilemmas may yield different outcomes in modern times. For example, in cases such as overpopulation, resource depletion, criminal justice, or nuclear war, the worst outcome is not being suckered while the other side cheats, but rather, the mutual *punishment* for mutual cheating. Because of relatively recent cultural developments, in a variety of areas, the payoffs for cooperation have increased, along with the costs of cheating. Heedless of these charges, however, the primitive bio-psychology of Prisoner's Dilemma still holds sway.

As the world grows smaller and even strangers become reciprocators if not relatives, the tendency to see the world as a Prisoner's Dilemma becomes itself a dilemma, a dilemma that originates in the discordance between our biology and our culture.

CHAPTER SEVEN

AGGRESSION, KILLING, AND WAR

THE ARTS OF DEATH AND THE HEARTS OF MEN

I have examined Man's wonderful inventions and I tell you that in the arts of life Man invents nothing; but in the arts of death he outdoes Nature herself, and produces by chemistry and machinery all the slaughter of plague, pestilence and famine . . . In the arts of peace, Man is a bungler . . . His heart is in his weapons.

—GEORGE BERNARD SHAW, *Man and Superman*

It is difficult, perhaps impossible, to remain coolly cerebral when discussing aggression, killing, and war, particularly in the closing decades of the war-torn, holocaust-haunted twentieth century. The human mind, capable of marvelous creativity, keen insight, and soaring grandeur, is also capable of the most notorious excesses of ruthlessness, barbarity, and unmitigated horror. It can be the best of things and the worst of things, and never worse than when it hurts and kills. It is also frightening and confusing, not only in what it can do, but even in our uncertainty about what it *is*. Listen to British historian Hugh Trevor-Roper, describing the mind of Adolf Hitler (in his introduction to *Hitler's Secret Conversations, 1914–1944*):

> A terrible phenomenon, imposing indeed in its granite harshness and yet infinitely squalid in its miscellaneous

> cumber—like some huge barbarian monolith, the expression of giant strength and savage genius, surrounded by a festering heap of refuse—old tins and dead vermin, ashes and eggshells and ordure—the intellectual detritus of centuries.

In this chapter, we shall examine some of this detritus which makes up the human species, and which has accumulated not over centuries but rather, millennia. Are we really bunglers in the arts of peace? Do we really wear our hearts on our military arms? And if so, why?

People love to speculate about whether human beings possess an "aggressive instinct." Two rather distinct sides have emerged, one group convinced that we have a genetically mandated tendency to kill our fellows, and the other equally persuaded that human aggressiveness is simply a function of our environments and in no way related to the DNA of our animal past. Most likely, both are wrong. To understand this important controversy, which has generated considerable aggression in itself, it would be helpful to review some history.

Ethology, the science of animal behavior, is a relatively new discipline, owing much of its development to a union of old-time natural history with modern biology beginning largely in the 1930s. It is similar in some ways to the psychological study of comparative behavior, but with many differences in orientation, emphasis, and technique. These differences have produced rather divergent outlooks. For one thing, ethology is still largely a European specialty (although a vigorous following has developed in the United States), whereas comparative psychology is basically an indigenous American movement. Ethologists are biologists, primarily concerned with the study of behavior in an evolutionary context, and they tend to study animals for their own sakes.

By contrast, comparative psychologists are ultimately concerned with human behavior. When they study animals, it is with an eye toward our own species and with the expectation that by studying behavior occurring in both animals and Homo sapiens, they will gain insight into ourselves. Because of their emphasis on the animals for their own sakes, ethologists tend to conduct their studies in natural habitats, where the adaptive value of animal behavior will be more readily appreciated. On the other hand, on account of their emphasis upon experimental manipulation and control, most research by comparative psychologists is conducted in the laboratory.

Even the animals studied are different. Ethologists tend to investigate a wide variety of animals, in order to gain perspective on the diverse results of evolution. They concentrate on birds, fish, and insects. By contrast, comparative psychologists work largely with mammals, which bear greater resemblance to human beings. They have concentrated almost exclusively on studies of the white laboratory rat, an animal whose similarities to Homo sapiens can at least be debated. In fact, the comparative psychologists' preoccupation with white rats has been so great that a leading American behaviorist, Frank Beach, wrote an article in a prominent psychology journal urging his colleagues to widen their range of study animals. In his paper, a cartoon showed a reversal of the Pied Piper myth: A horde of lab-coated scientists were eagerly following a large white rat into the river outside of Hamelin Town.

Ethologists have specialized in studies of species-typical behaviors that are often unique to each species and that generally have a genetic basis. Comparative psychologists have concentrated on the more modifiable behaviors that are characteristic of ourselves. Chief among these, of course, is learning.

Given these underlying differences, it is not surprising that ethologists generally attribute a stronger genetic component to behavior than do psychologists, a difference in outlook that extends to the question of human aggressiveness as well. Thus, the advocates of a hereditary tendency for human aggressiveness have been led by ethologists such as Konrad Lorenz, Niko Tinbergen, Karl von Frisch, and Irenaus Eibl-Eibesfeldt. Within the past few decades, the American playwright Robert Ardrey and the British biologist Desmond Morris have popularized the ethological viewpoint in a series of provocative books. The opposing view, that aggressiveness is largely a result of specific environmental factors ("learning" in the broadest sense) and that peaceful societies are attainable with appropriate manipulation of the environment, has been urged by the American psychologist John Paul Scott and the anthropologist Ashley Montagu, among others. Psychiatrists tend to fall on both sides of the argument. Given their primary orientation toward human beings rather than animals, they have an understandable bias toward modifiable traits, and hence, those due to culture and early experience. On the other hand, the Freudian tradition of unconscious factors, as well as a sensitivity to genetics and biological underpinnings tends to make biological evolution also acceptable to psychiatry as a fount of human behavior.

We are impatient creatures wanting quick answers, especially to such a momentous question as the origin of human aggressiveness. But unfortunately, the best answer at this time is that we simply don't know which viewpoint is correct. It seems most likely that the truth lies somewhere in between. There is good evidence that many animals have a built-in "need" to discharge aggressiveness and that if an appropriate outlet is not available, they may injure their mates or even themselves. Animals frequently indulge in elaborate "appeasement" behaviors to defuse the aggressiveness of another, often a prospective mate. Aggressiveness, particularly between males, is frequently elicited automatically by the simple appearance of particular releasers, like the black mustache of a flicker, mentioned earlier. But many psychologists have also presented evidence for the role of experience in determining aggressiveness.

Scott was able to produce aggressive mice who were sure winners in any bout with an opponent, even a larger one, by previously exposing his candidate to a graded series of fights which were all "fixed" to insure its victory. The mouse that has fought and won in the past is much more likely to fight and win in the future. The implications are that animals learn to fight by fighting, just as they learn to win by winning. Further results also indicate that they learn *not* to fight simply by not fighting. This might suggest that among our own species, strict prohibitions against aggression would ultimately result in a nonaggressive temperament. In addition, Scott's research indicates that fighting is often a result of breakdown in the prevailing social system. One effect of social organization in many species thus appears to be the maintenance of order. Accordingly, by avoiding such breakdown in our own societies, we might help keep the peace.

Other suggestions for environmental control of aggression have also been made. The influential Yale psychiatrist J. Dollard and colleagues proposed long ago that frustration is a major cause of human aggression. Raising children in a frustration-free environment would thus appear to be a logical way to avoid unwanted aggression. (On the other hand, as Konrad Lorenz has pointed out, "nonfrustration" children frequently prove to be intolerable brats, whatever their level of aggression.) It has also been observed that animals often fight in direct response to painful stimuli. If two rats are kept together in a cage outfitted with an electrified grid, they will begin fighting with each other in response to a shock. This so-called "reflexive fighting" provides further undeniable evidence for the role of experience in influencing aggression. Indeed, all these independent hypotheses mesh nicely together, suggesting to some psychologists that a well-organized human society that eliminated frustration, pain, and the opportunities for fighting among children would have eliminated fighting and war among adults as well.

On the other hand, fighting in response to pain is also adaptive, and may well have been strongly selected by evolution. An animal experiencing an attack and the pain that comes with it might be well advised to respond by fighting. The electrified rat may "think" that the other rat is somehow responsible for the pain it has just felt, whereupon it responds—appropriately—by fighting. This argument does not rule out the potential significance of pain (or similarly, frustration). However, it does introduce an important new wrinkle; namely, the adaptive value of animal aggression, whether self-generated or in response to experience.

This was the major point of Lorenz's provocative book *On Aggression.* It is significant that the original German title was *Das sogenannte Böse*—literally, "the so-called evil"—implying that although violent human aggression may be a great evil, aggression itself is a basically adaptive characteristic, and one that evolution has fostered in animals, and possibly in human beings as well. In addition to facilitating simple defense, the capacity for aggression enables animals to assure adequate room between themselves and competitors. Individuals who are sufficiently aggressive are likely to be more fit than those who are wimps. Lorenz in particular has also emphasized that the very important male-female pair bond is supported in many species through the sublimation and redirection of each partner's aggressiveness toward the other, as well as the sharing of aggressiveness directed at outsiders. Given these advantages, the "so-called evil" appears increasingly desirable. It seems reasonable to suppose, therefore, that evolution may have somehow engineered it into the genetic makeup of human beings.

Anthropologists are fond of pointing triumphantly to a few basically nonaggressive human societies, such as the African pygmies, as "proof" that our species lacks a genetically mediated tendency for interpersonal violence. But again, the presence of apparent exceptions does not conclusively prove anything, since genes produce potentiality, not certainty. It is of some significance, moreover, that nonaggressive human societies exhibit great gusto in such other behaviors as eating, drinking, playing, laughing, and sex. This might actually support Lorenz's original idea (now nearly five decades old) that aggression builds up as a spontaneous drive, leading to the proposal that humanity's aggressiveness could be reduced by redirecting it, thereby encouraging more acceptable outlets for our pent-up energies. Furthermore, many seemingly nonaggressive people such as the Kung Bushmen, Australian aborigines, and certain Eskimo groups, have been forced into their peaceful stance after being defeated by other, more aggressive people. The salient fact remains that we are an overwhelmingly aggressive species, and this remains true whether our aggressiveness is generated out of whole cloth from within ourselves, or as a response to personal background, social situations, or frustration.

Lorenz and others have proposed more competitive athletics, on both the participant and spectator levels. Healthy competition—local, national, and international—could also be encouraged in such potentially beneficial fields as space and undersea exploration, socio-economic development, medical research, pure and applied science, and the arts. There is, in fact, some evidence that simply watching an aggressive act serves to decrease the antagonism present in an observer. In experiments, a group of subjects were insulted and thus made angry and aggressive: their blood pressure and pulse rate increased perceptibly. When they were subsequently shown films of boxing matches, automobile accidents, or other violent events, their pulse and blood pressure quickly returned to normal. It therefore appears that simple vicarious experiencing of aggressive release tends to diminish pent-up aggression. Most advocates of environmental control of human aggressiveness would disagree with these proposals, calling instead for the exact opposite: minimizing the opportunities for expressing aggression and violence, rather than encouraging its release even if in a "harmless" manner. They also have their studies to point to: research showing that exposure to explicit violence, especially on television, may well stimulate violence. Whereas the witness to real-life violence is often revolted and appalled, the witness to artificial violence, particularly if repeatedly experienced, seems to develop different norms for acceptable conduct as a result, as well as inaccurate perceptions regarding the actual consequences of violence.

It seems clear that too much aggressiveness is disadvantageous, and just as likely to be selected against among the ancestors of human beings as among any other living things. Since there are possible costs associated with aggressiveness (notably, the danger of being injured or killed) as well as potential benefits, it seems undeniable that being too aggressive is no less maladaptive than being insufficiently so. Beyond the danger of self-injury, hyper-aggressive individuals might harm their mates or relatives, waste time and energy that could be better spent on self-maintenance, and even lead to the phenomenon of aggressive neglect, in which offspring receive insufficient care because the parents are preoccupied with their own threatening, posturing, or fighting.

The "innate" level of human aggressiveness might ultimately be illuminated by the approach of sociobiology, which emphasizes the use of evolutionary biology, and especially the concept of maximum fitness—reproductive success of genes—to interpret and predict behavior. Rather than consider human beings to be either innately aggressive or innately nonaggressive, a sociobiological view suggests that we have been selected to behave aggressively under certain conditions and nonaggressively under others, depending on the consequences of such aggressiveness or nonaggressiveness for our evolutionary success. In short, aggressive behavior that is adaptive under one condition may be maladaptive under others; aggressive behavior that is adaptive for one individual (say, an adult male) may be maladaptive for another (say, a juvenile female). Like a chemist carefully measuring the right amount of an acid necessary to neutralize an explosive alkali, we can expect that in most cases, living things will titrate their high-risk behavior with some precision. Continuing the analogy, we can expect this to be done with much less precision when the substances involved have been newly created, so that their effects are not well known. Natural selection typically assures that living things "know" the reaction ranges of their behavior; but natural selection has not had much opportunity to act on modern weaponry.

In recent years, concern about human aggression run wild has been heightened by our development of nuclear weaponry and the capacity for self-annihilation. By the massive elaboration of our potential for killing, cultural evolution has rendered a possibly adaptive trait highly dangerous. Consider the progression of the "cides": from suicide, the killing of oneself, we can expand to homicide, the killing of someone else. Although certain biblical injunctions called for the next step, genocide—the killing of an entire people—it is only in more recent times that such a potential has been readily available and efficiently acted upon. Shortly afterward, we coined the term ecocide, for the killing of an entire ecosystem; and now, with nuclear weapons, we can look ahead to the latest and most challenging prospect of all, omnicide.

Animals do not view their own aggression as good or bad; it is simply part of their lives, like sleeping, eating, cheating, cooperating, or copulating. Human aggression, on the other hand, readily lends itself to ethical judgment, most often to condemnation. We must therefore concern ourselves both with the situations causing the actual appearance of aggression, and with its results, the ways in which human aggression is expressed. As is so often the case—and contrary to much popular opinion—the problem derives not so much from the instincts we *have*, but rather, from those we *lack*. And both the elicitation of human aggression and its actual expression are strongly influenced by the conflict between culture and biology.

Human beings may possess an instinct for spontaneous aggressive behavior, although it seems unlikely. Conceivably, all human violence results from particular environmental factors such as poor rearing conditions, frustration, social disorganization, personal neuroses or psychoses, etc. But even if human aggression is not generated *de novo* by our genetic constitution, the capacity for such behavior must ultimately derive from our genetic makeup, the results of biological evolution.

For an analogy, consider behavior for which genetic factors are unquestionably responsible: the capacity for complex learning. With sufficient training, a human being can learn to solve difficult problems in differential calculus, whereas even the most intelligent chimpanzee cannot. The distinction between their abilities is very great, almost entirely because of differences in their genetic makeup. Admittedly, an untutored human being cannot do calculus, but even an intensively coached chimp is utterly hopeless in this regard whereas any moderately intelligent Homo sapiens can learn the technique. This is not to say that we possess genes that specifically control the solution of calculus problems, while the chimpanzee does not. Rather, the human genetic constitution includes the capacity for elaborate symbolic and abstract thought whereas the chimp's does not. The fact that African bushmen do not regularly solve calculus problems does not mean that they are incapable of doing so. Their normal environment simply does not provide the appropriate circumstances. At minimum, then, human aggression must similarly be based on a *capacity* for aggressive behavior, which emerges under the "appropriate" circumstances.

Granted, then, that human aggression must derive at least indirectly from our biological makeup, one aspect of the aggression problem can be seen as springing from the interaction of our culture with this biology, since most human cultures provide situations that actually stimulate human aggression. Given our biological capacity (if not need) for aggression, many human cultures may even make its expression more likely than it would be if we lived with fewer cultural complexities. All this, in social systems that may avowedly seek to discourage aggression.

Violent incidents are rare among animal societies, in part because the order of social precedence is generally well known and adhered to, based on a biological capacity for establishing and maintaining such relationships. A hen in a flock, a cow in a herd, and a baboon in a troop all know their position relative to their colleagues, and fights are generally avoided—or if unavoidable, minimized—by the lower-ranking animal's deference to the higher. The system works so well that even such seemingly mild indications of dominance as threat or displacement of the subordinate by the dominant are rarely seen, because the former judiciously avoids confrontations with his or her superiors.

We apparently lack a well-developed biological capacity for social harmony attained in this manner; the despotism of such relationships among animals often seems viscerally unpleasant to us. We have thus substituted cultural rules for biological imperatives, requiring courts, jails, and police to accomplish what most animals have unconsciously achieved under evolution's guidance. Laws and police are uniquely human institutions; the baboon's policeman is his own biology.

This is not to claim that animal social systems always work perfectly or that tumult and sometimes even violence are unknown; rather, there is a predictable and consistent pattern to the ordering of animal social life. When disruption occurs, it is typically to the ultimate benefit of the disrupter. A simple experiment showed that among the mountain-dwelling Gelada baboons of Ethiopia, for example, individuals are remarkably calculating about whether or not to "make trouble," thereby showing considerable enlightened self-interest. Thus subordinate, mid-ranking, and high-ranking males were each exposed separately to different male-female pairs. High-ranking males typically intervened and tried to win the female for themselves; given their high social status, they were generally successful. Mid- and low-ranking males, on the other hand, typically did not intervene, so long as the behavior of the female in the male-female pair showed that she was closely affiliated with her male. In such a situation, a would-be sexual usurper would not only have to deal with the outraged "husband," but also with an uncooperative "wife," whose loyalty might be difficult to achieve. But when the female seemed bored, disaffected, or otherwise inattentive to her male, then even a mid- or low-ranking male would try to win her away.

Once established, social hierarchies among animals may be very long lasting. The "alpha" or number one animal often remains the uncontested dominant long after his physical powers have waned. He rules by reputation. Human beings similarly establish hierarchies based on reputation, but they are considerably less inviolate; indeed, they are constantly being tested. The result is almost continual emotional stress and occasional physical violence. Even if we did have a biological tendency to respect a social order once it is established, we would rarely have the opportunity to do so because of our advanced cultural systems: we encounter complete strangers all the time, people whose relation to ourselves cannot readily be determined.

Because of the efficiency with which animal social organization generally keeps the peace under natural conditions, students of animal behavior often have difficulty identifying the relative social ranking of particular individuals. A good technique for solving this problem is to introduce small amounts of some desirable commodity such as food, and watch them fight over it, or (more likely) see who defers to whom. With the savannah-dwelling baboons this is especially effective, since these animals are normally spread over a wide area, eating things like grass or roots that are evenly distributed over their range. Aggression may be induced among such animals by providing a situation of extreme competition for some peanuts or an apple.

Woodchucks are rather surly, independent, and almost solitary creatures, which become sociable only briefly, at mating. Most of the time, they keep away from each other, eating grasses and seeds. The males will sometimes fight, especially during the breeding season, but most woodchucks simply avoid one another. After all, their food is widely distributed, over a broad area, and usually there is nothing much to fight about. This is changed, however, when someone plants a small garden: suddenly, many animals are attracted to a limited area, where they rub shoulders in close proximity, and get mad at each other. In addition, they now have something to fight about: dozens of succulent pea plants, for example, in just ten feet of carefully cultivated row. Like the Greek gods and goddesses who began bickering among themselves when Eris, the goddess of discord, presented them with a golden apple inscribed "for the fairest," woodchucks are more likely to fight, wound, and kill each other when they have something clear-cut to fight over.

According to Greek mythology, the exploits of Eris gave rise ultimately to the Trojan War. In real life, comparable situations have persistently generated discord, in abundance. Our culture today places us in direct and continuing competition with each other for a limited amount of an easily identifiable resource: money, as well as social status. These are culturally mediated pressures that a small band of hunter-gatherer Australopithecines would never have experienced. But it is not necessary to look only at technological civilization to see examples of culturally inspired aggression. The struggle for simple physical possessions is a frequent cause of human violence, and the possession of objects of one sort or another is a universal human trait. Thus, the "crime" of stealing, with its frequent companion—violence—could not occur unless there was something to be stolen. Ownership of things external to our bodies is well developed in human beings, and is a direct outgrowth of our use of tools. But it is not unknown in some animals as well, and when ownership occurs, it too is often a source of conflict.

Many hunting animals fight to defend their kill from others, particularly from scavengers such as hyenas or jackals who try to steal it. Among birds, the Antarctic skua and tropical frigate bird are specialized air pirates that regularly get their meals by stealing the catch of such expert fishermen as gulls and cormorants. The gulls, in turn, commonly steal nest material and eggs from other unwary gulls. Male scorpion flies hunt for small prey items, which they present to females as part of courtship; sometimes, a male without such a prey item alights next to another that has one. Acting like a courting female, the empty-handed male attempts to steal the prize, which is normally passed to the female during mating. When he discovers that he has been tricked, the "wealthy" male tries to take back his possession, and a tug-of-war often ensues, sometimes leading to a brief fight as well.

Squabbles may break out over a prized food item among the monkeys and apes, but the disputed object is quickly consumed, and with it, the passions of the dispute. Human beings are unique among primates, however, in that we experience prolonged material ownership. This custom creates long-lasting grudges, and persistent interpersonal violence.

Most primates eat fruits, vegetables, or small invertebrates. In such cases, the food items are usually small and equally abundant over a wide area, thus making competitive struggles uneconomical. The higher monkeys and apes such as baboons, gorillas, and chimpanzees will eat medium-sized animals such as young gazelles and other monkeys when they can get them. The meat of larger animals is rich in protein and highly prized when available. However, a well-established social system generally determines priority at the kill, preventing undue aggression even then.

Among our partially carnivorous ancestors, it is likely that occasions for fighting over a kill were considerably more frequent than they are now among the monkeys and apes. It is therefore possible that we evolved a genetically influenced capacity for establishing and maintaining peacekeeping hierarchies, and that we still carry a remnant of this capacity. Thus, many of our cultural institutions, including systems of royalty, church hierarchies, armed forces ranks, and the various personnel levels on the corporate and governmental ladder may satisfy a tendency for such graded social organization. But the specifics in each case are clearly determined by cultural factors, not genetics. Our social institutions, even at their most despotic, lack the mousetrap guarantee of many of the rigidly instinctive behaviors shown by some of our animal cousins. If such systems were more firmly rooted in our biology, then the various obnoxious devices of tyrannical governments the world over would be unnecessary. (Or alternatively, they would seem less obnoxious!) Torture chambers, secret police, revolution, and counterrevolution all attest to the frequent discordance between our cultures and our biology. It is unclear precisely why human beings have experienced such a relaxation in genetic influence on their social organization. Presumably such flexibility helped us inhabit a wide range of environments, and deal with a wide range of circumstances even when we only occupied the African savannah. In any event, the pattern is consistent with the evolutionary decline in biological control and concomitant increase in cultural control of our behavior generally.

Regardless of whether we possessed, or still possess, a biological capacity for avoiding aggressive competition over food or other items, our culture has vastly increased the opportunities for similar aggression. It has been proposed that instead of Homo sapiens, we should have named ourselves *Homo faber*—Man, the maker—in recognition of our propensity to build and to use tools. No animal constructs as many different things as we do, and none values them so highly. Although a male stickleback fish will vigorously defend his territory, just as a female gull will defend her nest, no animal invests its possessions with the time and labor that human beings commonly do. To expend effort on something is to increase its value, and we put great labor into our things, valuing them very highly indeed, enough to fight and kill over them.

More than twenty years ago, in his popular book *The Territorial Imperative,* Robert Ardrey reviewed some of the ethologists' literature on territorial behavior in animals and used this as "proof" that we are territorial as well. He was both damned and applauded for these efforts. Basically, his arguments are too facile; it simply does not follow, for example, that because certain dragonflies are territorial, so are we. What occurs in animals need not necessarily occur among us.

True, human beings often demonstrate rigid conventions in their social use of space, but these patterns are culturally determined, varying greatly from one society to the next. The anthropologist Edward Hall has shown that significant differences exist, for example, between the conversational distances maintained by Arabs and Americans. Thus, the Syrian or Egyptian characteristically puts his face very close to the person with whom he or she is speaking. A friendly conversation requires that each participant be bathed in the odors and even the body heat of the other. By contrast, Americans and Western Europeans maintain greater interpersonal distance. These varying cultural styles can produce serious misunderstandings at international gatherings such as the United Nations, where persons from "short distance" countries (most of Latin America and many Mediterranean nations may be included here) sometimes appear pushy and overly forward to their "long distance" counterparts. On the other hand, Americans and others influenced by the social conventions of Western Europe may appear cold and indifferent to people accustomed to closer interactions.

Similar cultural differences occur with regard to living space, trespassing, and concepts of privacy. Far from being genetically tied to a territorial imperative, we appear to be susceptible to whatever mores prevail in our particular society.

It has also been claimed that our supposed territorial nature is partly responsible for our aggressiveness. This is doubly unlikely, since one consequence of territorial behavior—as in the case of social hierarchies—is a *reduction* in aggressive incidents. Among territorial species, each individual is necessarily aware of the established rights of his or her neighbors and generally respects them. The exceptions often occur early in the breeding season or at other transition times when territories are being set up. But such occasions are short-lived and almost always decided by bluff and threat, with the territorial owner at a clear advantage. Territoriality includes a tendency both to establish one's own piece of real estate and to respect the ownership of others. In fact, territorial ownership often confers near invincibility upon the proprietor. The boundaries may be remembered by experience, but respect for them is biologically rigid.

If we were truly territorial, fighting over space would be greatly reduced, especially once the boundaries were established. It is precisely because we lack much biological respect for territory that we fight so often over it. We want our own space, sometimes enough to fight for it, but at the same time, we are disinclined to respect very profoundly the territory of others. In most of the Western world we commonly stake out spatial claims, using picket fence, stone wall, barbed wire, printed sign, or locked door, all of which are themselves testimony not so much to territoriality but to its absence. And in doing so, we are following our culture, not our genes. Because these artificially inspired boundaries lack the support of biological evolution, they are rather insecure and subject to constant dispute. When cultural evolution presents situations for which biological evolution has not prepared us, look for trouble.

Despite the likelihood that human beings are not biologically territorial, let us reverse our field for a moment and assume for the sake of argument that we *are* territorial animals after all. A territorial biology would likely produce some sort of regular spacing between individuals and almost certainly, between families. In some cases, this could be accomplished by the widely spaced distribution of farms in the Midwest or the oppressive regularity of a typical suburban subdivision. Both patterns occur today, and the distances involved vary greatly. But, if there is an appropriate pattern of human spacing that best satisfies our genetic needs, it may be fifty feet apart, as in modern Levittown, or fifty miles, as in parts of Alaska. Not knowing, we cannot arrange our living to coincide with such possible needs. Moreover, the existing diversity of patterns suggests that even if we had such needs, they must be violated by the vast majority of living arrangements, since the variation is so great and the determining factors seem unlikely to be biological appropriateness.

The human genetic makeup—whatever it may be—is poured, like concrete, into molds of territorial use that are almost entirely determined by political, social, economic, and possibly random factors. But unlike concrete, genetic influences are not infinitely malleable. They cannot comfortably assume an infinite array of configurations, and if forced into inappropriate shapes, they may be stressed to the point of cracking. If we have any genetic propensity for territorial organization, then it is likely to be distorted and stressed by the wide diversity of contemporary human culture. Alternatively, if we lack such a propensity, then as we have seen, trouble also follows because we lack the stability that comes with biologically mandated adherence.

When it comes to questions of territory, the real crunch of culture upon biology could be expected to occur, once again, in our large urban areas. Just as a genetic tendency for social ranking, even if present, would be difficult to maintain in a big city, any purported territorial tendencies must be greatly strained in the urban crush. The motives for city dwelling are varied and complex. But economic factors clearly provide a major inducement to the urban dweller, with subsidiary incentives of excitement, novelty, sociocultural diversity, and simple inertia (having been born into the situation). It seems likely that even if territorial needs exist within us, powerful cultural factors with immediate appeal could readily seduce us to act contrary to them.

If human aggression derives in part from our culture acting upon a relative *absence* of biological characteristics, what sort of environmental determinants of aggression can be postulated? In *The Republic,* Plato sought to identify the proper form of human government, based on the underlying nature of the human soul. Although successful as philosophy, such efforts have not been notably successful in practice, perhaps because when it comes to human preferences for social organization, there is no single, underlying human soul. Our natures may be as insubstantial as the famous shadows dancing on the wall of Plato's cave.

If indeed we lack a genetic tendency for establishing any one particular kind of social system, our various forms of social organization must be strongly influenced by culture, perhaps entirely so. Human political systems are very diverse, ranging through tribalism, absolute monarchy, military dictatorship, republican democracy, fascism, socialism, and communism. Family structure, in turn, runs the full gamut from the private nuclear family to extended clans, and social patterning runs from densely urban and sedentary to pastoral and nomadic. Perhaps most important of all is the basic, diffuse social fabric of personal relationships, from the superstructure of political organization to the foundations of daily, family life. Here, the diversity of human relationships and organizations is truly staggering.

This varied social architecture has almost certainly developed without corresponding blueprints in our genes. For systems as diverse as communism and democracy, the guiding force has been a specific social, economic, and political philosophy, held initially by a few individuals. Such ideas have either prospered or withered, depending on the cogency with which they were expounded and the social, economic, and psychological climate among each group of people that was exposed to them. Social infrastructure, by contrast, reaching down ultimately to basic family relationships, is probably more resistant to alteration by a momentarily attractive ideology. In China, the human family has proved resistant to early Maoist efforts at drastic modification, just as in the United States human compassion has resisted the New Right's efforts to repeal New Deal entitlements and organize personal relations around selfish disregard of others. Although the fundamental social fabric of human behavior experienced by nearly all people is likely to be deeply biologic, the superficial patterns encountered by most of us seem to be largely a result of inertia and conformism; that is, we practice the social organization that our parents and relatives did, and that we see exhibited by our contemporaries, perhaps with a dollop of biological assistance.

To some extent the mosaic of social organizations exhibited by human beings is probably adaptive as well. We might therefore expect that the optimum social organization differs among tropical rain forest, prairie, seashore, and arctic dwellers, although the ideal arrangement for each particular case cannot be identified. But regardless of the forces producing specific patterns of social organization, such patterns once established would probably be continued by the joint actions of simple inertia and conformism (provided, of course, these patterns are not so grossly maladaptive as to produce extinction of the group, or so uncomfortable as to generate revolt).

Regardless of the methods of transmission, and irrespective of their adaptive values, it is likely that a large proportion of human social institutions lack much support from our genetic makeup. We can therefore anticipate a fair degree of instability in such systems, and once again, psychologists tell us that such instability is a prime breeding ground for aggression. In short, our social organizations (culturally evolved) seem likely to break down frequently since they lack corresponding support from our genes (biologically evolved). The result? More aggression.

It seems likely that we lack a genetic propensity for any particular form of social organization, and that in fact we have biologically evolved a capacity for utilizing many such cultural systems, just as we speak many different languages. Culture itself is a kind of specialization; we have specialized in being generalists. By not restricting us to a limited number of systems, biological evolution has provided one of the essential keys for successful exploitation of a great variety of habitats. Nonetheless, insofar as we are jacks of all trades but masters of none, we would still continue to lack the stability born of strong biological support for a particular way of life. Honeybee society, for example, is well organized and peaceful: with a place for everyone and everyone occupying his or her place. The workers work, the queen lays eggs, and the drones drone on. The fact remains that social breakdown is a prime ingredient in aggression, and any human culture—developed for whatever reason—that lacks a firm genetic substrate such as the honeybee enjoys can be relatively susceptible to such breakdowns.

All this should not be mistaken for an argument in favor of any particular type of social, political, or economic organization, in the forlorn hope that such an organization would be more congenial to our genes. In fact, it is exactly the opposite: since our genes apparently do not specify a social system for Homo sapiens, we are free to choose what we wish, or more likely, to suffer whatever is imposed on us. As the Grand Inquisitor pointed out in *The Brothers Karamazov,* freedom of choice can be an awful burden, and when it comes to social systems, the weight of that burden is especially real and can be especially troublesome, since there may be no system in which we are truly comfortable. Like the expatriate or the dedicated internationalist who can live with equal ease in a variety of countries but never feel that he or she is at home anywhere, Homo sapiens can live in a great variety of social systems without ever really being at home.

As we did when considering human territoriality, let us now

make the opposite assumption; namely, that we *do* possess a biologically evolved predisposition for some particular pattern of social organization. Here again, as with territorial behavior, the enormous heterogeneity of human cultural arrangements guarantees that many, if not most of them, are somehow inappropriate for us. It is significant that whereas the disruption of social organization may well be a major cause of aggression, much human violence seems to be directed toward disrupting existing cultural subsystems, rather than to be a result of their disorganization. Both of the above suppositions could act independently, and both lead to the same result. If we lacked genetic tendencies in support of our various cultural systems, we would clearly also lack any great reluctance to destroy them. On the other hand, if we possessed a genetic predisposition for some particular type of social organization and were denied fulfillment of it because of economic or ecologic conditions, or just simple historical accident, we might be inclined to rebel. In either case, aggression would ultimately be very likely: damned if we do and damned if we don't.

For a final example of this paradox, consider some consequences of the fact that we evolved in the tropics. The fossil record and our basic physiology both strongly suggest this. Our cold tolerance, for example, is extremely poor. Unclothed and without some basic technology, we could not survive outside the tropics. Yet clearly we do. In fact, we have spread over the globe, experiencing the widest geographic range of any species, and we have achieved this fantastic expansion by virtue of our master adaptation, culture. Human culture has thus allowed us to explore otherwise inaccessible places, settle in otherwise inhospitable lands, and fill a variety of ecological niches ranging from completely vegetarian to meat- and even insect-eating. Culture has been our necessary support and without it we would be lost; indeed, we would cease to be human. It frequently happens, however, that in relying so heavily on something that it becomes essential, we surrender some autonomy to it. We profit greatly from language, for example, and rely on it; at the same time, as linguist Benjamin Whorf has emphasized, our perception of the world is colored by the linguistic rules and vocabulary that we experience. We take advantage of mechanized agriculture to feed our immense population, and in the process, become dependent on pesticides, and millions die when the rains fail to come on schedule.

Assuming now that we possess some biological basis for our behavioral systems, we might also assume that these systems would be most appropriate for use in the tropics, where most of our early biological evolution took place. Our innate tendencies, under the watchful eye of natural selection, would then be attuned to systems that are adaptive for a tropical existence, whatever systems they may be. But as our culture began its rapid evolution, we developed the capacity to exist in regions far removed from our ancestral habitat. We rode our culture to mountain and desert, to subtropical and ultimately temperate and arctic regions. In so doing, we were probably forced by the exigencies of new and sometimes rather strenuous environments to employ cultural practices that were at least minimally adaptive. We had no choice but to deny, or at least ignore, any tropical biology that might still whisper within us. So the real culprit may be not culture, or biology, but the human wanderlust that forced us to adopt cultural styles that may be at variance with our biology.

Arnold Toynbee has suggested that the technological development of temperate zone people is due to the stimulating effect of seasonal change, combined with a certain degree of environmentally imposed necessity. We might expect to find Homo sapiens increasingly dependent on culture as we inhabit environments that are increasingly removed from our biological heritage. In this regard it would be interesting to see if temperate and arctic societies exhibit higher levels of neurosis, personal alienation, and aggression than do their counterparts in the tropics.

***

FIGHTING and winning makes an animal more likely to fight again. The same has been proposed for human beings. If so, it might also be that the vicarious experiencing of aggression—in other words, seeing or hearing it—would reduce inhibitions and foster aggressiveness in the spectator, just the opposite of the predictions generated from a strictly ethological perspective. Throughout history, human cultures have shown great ingenuity and expended considerable effort providing spectacles of aggression for their members. The ultimate motivation behind these performances may be the belief that vicarious experience of aggression would have a cathartic effect, like the Aristotelian concept of Greek tragedy, causing the observer somehow to discharge pent-up energies and "get the anger out." Perhaps it was expected that such experiences would render him or her less likely to threaten society with violence. Hence, the Roman fondness for "bread and circuses."

On the other hand, public display of aggression may also satisfy a profound need in our species, and not just in our rulers or in those poised to profit financially from such events. Primitive societies often hold ceremonial gatherings at which aggressive posturing takes place, with dancing, fighting, and even killing, often of sacrificial animals. In the worst case, a vicious circle ensues, with aggression feeding on itself as it demands public representation, which in turn generates more personal aggressiveness, which requires yet more public representation, and so on. Western society has produced the Roman gladiators, Christians eaten by lions or nailed to crosses, bullfights, public executions of all sorts, boxing matches, demolition derbies, and football games. With the advent of warfare, most recently abetted by radio and television, the details of organized human aggressiveness have been made apparent even to the noncombatant. Unlike the above examples, however, televised warfare is presumably not designed specifically for home entertainment. And in fact, televising of the Vietnam War seems to have increased nationwide repugnance for that conflict, whereas by contrast one of the dangers of artificial televised violence may be that it insulates the viewer from the real consequences of violent aggression. With that war finally ended, and with the result unsatisfying to boot, the American public was then free to indulge its fantasies of martial triumph, vicariously, through Rambo and other safe, celluloid escapades of wish fulfillment.

Psychologists have also proposed "frustration" as a major environmental cause of aggression. Webster's Dictionary defines frustration as "being prevented from achieving an objective or from gratifying certain impulses and desires, either conscious or unconscious." If we postulate a genetic tendency for certain behavior in human beings, such as territoriality or the establishment of a particular kind of social hierarchy, culture emerges as a major frustration in that it must often keep us from achieving our unconscious objectives. If Freud and Hobbes are correct, and civilization acts to frustrate and impede our otherwise unacceptable tendencies—indeed, if civilization is only possible because of this impedance—then culture may well be in the ironic position of causing aggression by the very act of preventing it. Again it seems that we just can't win.

One of the great benefits of culture, and indeed, one of the primary reasons for its existence, is that it empowers human beings to fulfill many of their desires. A certain degree of mastery over nature, or at least, temporary insulation from it, is made possible by technology which provides food, clothing, and shelter. Some leisure time is also generally assured, as well as numerous opportunities for self-expression through artistic media and (for the fortunate) through one's work. Theoretically at least, anyone with sufficient inclination and capacity to inquire into the workings of the natural world can become a scientist. Freed, in part, from those biological constraints that tended to make life nasty, brutish, and short, human beings have become the most playful animal on earth: Dutch historian Johan Huizinga once proposed the moniker *Homo ludens,* Man, the player. All this, and more, we owe to culture. It has been proposed that the relatively nonviolent African bushmen owe their low level of interpersonal aggression to a penchant for vigorous, energetic playfulness. And among animals, a playful mood is a distinctly nonaggressive one.

Our desires for social companions, of a particular sort, at particular times, doing particular things, can to some degree be satisfied by our social systems. Human culture may even be defined as a complex nongenetic system that satisfies human objectives, impulses, and desires. So despite the restrictions that it imposes, culture is also, ironically, an enormous nonfrustration device.

But beyond the satisfaction of such basic animal needs as food, warmth, sleep, elimination, reproduction, and sex, can anyone identify what a human being really wants? Culture doubtless provides much that we could not otherwise achieve. And yet it is no contradiction to point out that culture also generates many desires that must remain unfulfilled; that is, it also promotes frustration of a different sort. While satisfying many needs, it creates want. Someone living in today's highly evolved technological culture is likely to be more full of "wants" than his or her primitive ancestor, living at less remove from biology. Modern Homo sapiens, living atop an enormous superstructure of cultural opportunities and expectations, often finds him and herself assailed by vague stirrings of dissatisfaction. A potent source of such feelings is that old bugaboo: envy.

Modern culture has no monopoly on producing such feelings. It is likely that one of our early human ancestors, eyeing the results of a successful hunt, dearly wished that he and not his colleague was chewing on the giraffe's bloody haunches, or that she could have as easy a labor and delivery as her neighbor. And we have probably coveted our neighbor's wife (or husband) since the pair bond was first liberated from the genetic inflexibility found in most animals. But the scale at which modern culture now generates such emotions is probably unprecedented in the evolutionary history of our species.

Advertisement and the mass media also tease us with the images of some of our species's most attractive products (both human and material). Little wonder frustration is rampant. Once human culture progressed beyond the point of mere sustenance, we began surrounding ourselves with things. The keeping of possessions can easily lead not only to aggressive competition, but also to aggression via frustration. Conspicuous consumption is more than just a cliche for American status seekers. It is usually assumed that people keep up with the Joneses in order to raise their status; by contrast, our concern is not so much with the motivations of the conspicuous consumer as with his or her impact on that neighbor. One of the most subtle and powerful ways in which culture generates frustration is by exposing individuals to glimpses of another's possessions and accomplishments. By generating desires that cannot readily be gratified, and that may never be, culture is frustrating no less than gratifying.

Culturally induced frustration is especially apparent in most of the world's big cities, where rich, poor, and middle class almost literally rub shoulders—and sensibilities—every day. It is probably even more potent among many Third World nations, where the "revolution of rising expectations" is fueled by newfound awareness of the affluence of other people. Poverty is never easy to bear, but it was more tolerable before magazines, radio, television, and tourism made the impoverished aware of how badly off they really are: not just in absolute terms, but compared to the relative affluence of the Western world.

***

FINALLY, let us consider war. The relationship between war and aggression remains moot: war may be a *result* of aggression, simply its best organized and most deadly manifestation. Or conversely, war may be a *cause* of human aggression, the product of high-level governmental decisions which are then translated into personal resolve and emotion by the various stimulative and exploitative devices at society's command. Viewed either way, culture makes a major contribution to aggression by providing numerous opportunities for conflict between societies.

Alone among all animals, we provide the unique spectacle of individuals killing other members of the same species in order to "convert" them to some cultural practice. The cultural distinctions between people, whether adherence to different religions, economic systems, or ways of gaining a living (e.g., farming vs. herding) have always loomed large in humankind's reasons for going to war. We are all one species underneath, but we come packaged in many different cultures, thereby providing numerous options for distinguishing Us from Them. National identification, which typically aspires to be family and kinship identification writ large, facilitates both a defensive sensitivity and a dehumanization of the opponent.

Despite relatively high levels of aggression, animals only rarely kill or even seriously injure others of the same species. Recent field studies have shown that free-living animals are not nearly as idyllic as had been believed until the mid-1970s: wolves occasionally kill other wolves, lions sometimes kill other lions—typically, outsiders—and infanticide is quite frequent, especially when a male takes over an existing harem. In general, however, no living species practices the degree of intraspecies killing that we do, as measured either by its frequency or its ferocity. Ethologists have long emphasized that aggressive restraint among animals is typically achieved by a combination of behavior traits, nearly all of which have been evolved through natural selection and which accordingly rely on genetically coded mechanisms. When two rivals meet in an aggressive situation, bluff and threat generally ensue. In fact, the vast majority of such contests are immediately settled in this manner, with no recourse even to physical contact, let alone injury. By displaying themselves to best advantage, each contestant tries to impress the other with his prowess, so that the intimidated rival acknowledges inferiority or simply goes away.

Each species generally possesses a unique set of behaviors used for such occasions. The common features usually include exposure of teeth, claws, or whatever weapons the animal may possess, plus making itself appear as large as possible. Among fish, this may involve puffing up, spreading the fins, and orienting laterally to the opponent, while mammals commonly make their hair stand on end; all create the illusion of greater size. It is clearly better for disputes to be settled this way than by a prolonged, exhausting fight, with the threat of possible injury or death.

Should the two rivals appear evenly matched, a ritualized form of fighting may ensue, often resembling stylized posturing more than an actual struggle. Fish of many species will lock jaws, pushing and pulling each other in a combined effort to assess the others' strength while emphasizing their own. This is the basis for the famous and titanic struggles of bull elk or deer of many species, in which the males lock antlers and strain mightily against each other.

If the intent was to kill, such antlers could be used much more effectively against the unprotected sides of a rival. But this almost never happens. Instead, the competitors avoid such potentially lethal attacks, turning aside from their opponents' vulnerable flank and waiting for the opportunity to lock horns in ceremonial struggle once again. So ritualized are these behaviors that the pioneer European ethologists who have studied them in detail describe them as "tournaments" rather than true fights. Marine iguanas push against each other, locked forehead to forehead until the loser acknowledges defeat by dropping to his belly while the victor remains stiff-legged and threatening until his defeated rival scurries away.

Significantly, when males have lethal weapons and females do not, the latter typically lack the stylized aggressive tournaments so characteristic of the former. And those species that lack lethal weaponry altogether also tend to lack inhibitions against attacking one who is vulnerable.

Rattlesnakes fight frequently, and they are not immune to rattlesnake venom. (They are able to eat their own venom-injected prey only because the poison is broken down by the digestion; if bitten by another rattlesnake, they can die.) It is thus interesting that fighting rattlesnakes scrupulously avoid biting each other.

Instead, they engage in peculiar, highly stylized wrestling matches in which each one tries to push the rival over on its back. The two adversaries face each other with their heads and about one third of their bodies raised vertically above the ground, bobbing and weaving, while sometimes rubbing their ventral scales together. In making contact and straining against one another, they may raise themselves several feet above the ground, possibly giving rise to the caduceus symbol of the medical profession in the process. The victor pins the vanquished with the weight of his body for just a few seconds, and then the loser glides away, defeated but unbitten and very much alive.

Other animals with similarly lethal weapons can be counted on to refrain, typically, from using them against members of the same species. Thus, the Arabian oryx—equipped with long razor sharp horns—uses these weapons to push sideways and harmlessly against a rival. And giraffes, with strong and potentially dangerous hooves, harmlessly butt heads during aggressive confrontations. But just as the rattlesnake does not hesitate to use its venomous fangs against a wood rat, the oryx will use its horns and the giraffe, its hooves, against lions.

It is to an animal's selective advantage if it can efficiently dispatch either its would-be predator or its prey. But there is no real advantage to killing members of one's own species if the same result can be accomplished more easily and with less danger. After all, in many cases the loser in a "tournament" situation may well be a relative. If so, killing such a rival would be a Pyrrhic victory indeed, likely to be discouraged by kin selection. If success can be achieved without death or even severe injury to the opponent, then the victor could even gain some benefits from the other's company, without having to pay the price of fitness-lowering competition. Finally, interactions with prey or predator are comparatively one-sided as opposed to the double-edged sword of murderous intraspecies fighting. Selection for lethal attack on another species member could easily boomerang, creating the chances of either contestant being killed if both possess the appropriate genes, analogous to mutual defection in the Prisoner's Dilemma. Students of mathematical game theory have also demonstrated that under certain conditions, behaving as a "dove" can be more successful than behaving as a "hawk," even if hawks prevail in dove-hawk contests. (This is because hawks, as they become more numerous, are more likely to encounter other hawks and then to suffer damaging fights, whereas doves are more likely to live to fight another day.)

The avoidance of within-species killing, although not universal, is relatively well established for most animals, and is usually achieved by elaborate biologically evolved behaviors that are especially pronounced among those animals with a greater potential for lethality. A rabbit or a robin is clearly less likely to assassinate a colleague accidentally, or to be killed in turn, than is a leopard, eagle, or rattlesnake. Lethally armed animals, not surprisingly, have evolved effective and relatively harmless behavioral substitutes for murderous aggression; namely, displays and tournaments. Even when such measures are seemingly unable to prevent a "real" fight, a last fail-safe mechanism is usually available. For example, Konrad Lorenz has pointed out that the most "vicious" fight between two wolves in nature generally will not result in the loser being killed. Once a victor has clearly emerged, the loser does something that automatically inhibits the winner from pressing home its attack. Among wolves this may involve turning the neck to expose the soft underside, exactly the most vulnerable part of the animal's body; a similar response is seen when domestic dogs roll over on their backs, exposing their belly to a larger, dominant dog, or to a human being. Instead of killing his hapless opponent, the victor may then snap his jaws harmlessly in the air. Once another animal shows itself to be subordinate through this stereotyped behavior, it is unlikely to be killed. In short, the capacity for killing has generated controls which prevent its misuse. (Actually, we now know that wolves do occasionally kill each other, mainly when individuals from different packs come into conflict; the basic principle of restraint still holds, however, at least for fights within the wolf pack.)

In most cases in which animals have been found to kill other members of their species, the individuals had been in captivity. Often they are either prevented by the poverty of their artificial environments from exercising their full behavioral repertoire or they are confined in such small cages or tanks that simple retreat, the most common recourse of the defeated animal, is denied them. Rattlesnakes do not kill other rattlesnakes. But throughout recorded history, men have killed other men, and often, women and children as well. Perhaps the most persistent sound, echoing through our own evolutionary past as well as recent times, has been the beat of war drums.

The question of morals really isn't at issue here, especially since morality and ethics are basically society's effort to substitute cultural control of behavior for the biological control that we generally lack. The rattlesnake is responding to its genetic rather than its ethical system when it refrains from killing a rival. Despite the moral authority of the Mosaic prohibition against killing other human beings, we conveniently ignore that commandment without denying our biological heritage. Society finds it useful to inhibit killing when it might disrupt the smooth functioning of human social organization. But when society considers killing to be in its best interests, generally the killing of deviant individuals (executions) or of perfectly normal members of another society (warfare), the culturally invoked restrictions are easily lifted. Just as there has been virtually no effective biological evolution among human beings during the past two thousand years, there has been no effective moral evolution either. By A.D. 600, our species had already displayed the wisdom of not only Moses and Christ, but also Lao-tse, Confucius, and the Buddha. In the twentieth century, we had Hitler and Stalin . . . so it is not at all clear whether progress has been made. On the other hand, during the same interval we had enormous "progress" in weaponry, from the lance and broadsword to nuclear missiles.

One fundamental question remains to be answered: why do we lack effective, guaranteed, biologically evolved inhibitions against killing? The answer lies once more in the disparity between our biological and cultural evolution. In fact, it provides one of the clearest examples.

We lack genetically mediated killing inhibitions because natural selection didn't have much reason to endow us with any. After all, a naked, unarmed human being really isn't a very dangerous adversary to another, similar human being, unless he or she has had special (modern) combat training. It is extremely difficult for one person to kill another using only the devices with which biological evolution has equipped us. Our hands and feet are very inefficient weapons and our reduced teeth in their muzzleless, flat face present little threat of a killing bite such as found in the dog, cat, or weasel families, or even the modern reptiles and fish. Thus, throughout most of human evolution we were like the robin, rabbit, or female deer: lacking weapons capable of killing our fellows, we also lacked inhibitions against such killing, since we were virtually incapable of it anyhow.

Then came the rapid explosion of cultural evolution and the successive discovery of stone, club, knife, spear, blowgun, bow and arrow. These weapons probably first served to facilitate killing of prey or defense against predators. But defense and attack against other people were probably not far behind. Intraspecies fighting became the main function of human armaments, and with a great rush we invented swords, muskets, cannons, machine guns, poison gas, battleships, tanks, bombers, submarines, guided missiles, and finally, nuclear weapons. When the stag evolved his antlers or the rattlesnake its poison fangs, time was abundant—measured in millions of years—and the biological world stayed relatively unchanged for millennia. Each stage of development was accompanied by the corresponding evolution of genetically mediated behaviors carefully titrated to the biologically evolved equipment at hand. Otherwise, elk and rattlesnakes would probably have gone extinct long ago. The extraordinary development of human weaponry has occurred in a span of *thousands* of years, and our really crowning achievement in fiendishly destructive capacity is a product of the mid-twentieth century alone.

Albert Einstein once noted that whereas he had been taught that modern times began with the fall of the Roman Empire, in fact they began with the fall of the bomb that destroyed Hiroshima. In any event, such a time scale is much too short for natural selection to have operated effectively, or indeed, at all. We possess armament that is far more deadly than the most dangerous animal, but since it has developed by cultural rather than biological evolution, we lack the necessary behavioral controls that natural selection would have assured. Biological evolution would never have allowed such a souped-up hot rod to leave the assembly line without a good set of brakes. Biologically we are still rather innocuous apes, while our unrestrained culture makes us the greatest potential (and actual) killers the world has ever known.

It is conceivable that human beings are somewhat susceptible to appeasement or subordination behaviors by their fellow humans. And this may even have a weak genetic basis. We are generally less likely to kill a submissive, defeated individual who kneels at our feet. But the bowed head of the prospective victim has never in itself stayed the guillotine or the headsman's axe; human history abounds with the overt killing of supplicants. Admittedly, we do show a certain reluctance to kill women and children, somewhat preferring to slaughter our fellow *man.* This is probably of modest selective value, since as we have seen, males of most species compete with each other for access to females, as vehicles for perpetuating their (the males') genes. Killing the men and raping the women is thus a partial expression of our biology. In fact, many cultures take advantage of this apparent inhibition against killing women and children by using them as emissaries of peace or surrender, and also by distributing them as booty. But witness the slaughter of entire Cheyenne families by the American cavalry at Sand Creek, or the equivalent butchery at My Lai.

Maybe if we only possessed the right repertoire of subordination behaviors and automatic responses to them, then all would be well. But unfortunately even the unlikely development of biological controls to match our galloping culture would avail us little, because technology has provided us with increasingly efficient weapons that do their deadly work from increasingly greater distances. Subordination behaviors are only effective at the personal level: even if a villager is engaging in an exceptionally effective appeasement display, she cannot be noticed by the bombardier 20,000 feet up, or by the politician on another continent with his finger on the nuclear trigger. The whole trend of developments in human weaponry has been for effectiveness at ever increasing distances, from the club at three feet away to the cannon with a range of several miles, to the ICBM at 10,000 miles. This in itself makes biological inhibition a virtual impossibility.

Is there a solution? It is often easier to identify a problem than to solve it, and this seems particularly true of human aggression. Our culture, by the breathtaking speed of its advance, has placed us in a very dangerous position. But the capacity for culture is part of our genetic heritage and we could no sooner give it all up than forego use of our thumbs because they occasionally get hit by a hammer. Just as we must learn instead to use a hammer wisely, we had better develop proper control over our cultural productions. Perhaps we should also accept that certain things are too dangerous for us to use safely, or even to keep around at all. By their rambunctious primitiveness, human beings sometimes seem like incautious little children, who must be protected from their own immature inclinations; but they are the adults now, and there is no safe medicine chest in which we can safely store the household poison.

It should be emphasized that evolutionary considerations do not lead irrevocably to an accusatory posture in which Homo sapiens is castigated for being irrevocably tainted with a kind of biologically based original sin. First of all, nothing in human behavior is irrevocable. Although we irrevocably carry the imprint of evolution, this does not mean that we must irrevocably behave in any particular way. And second, our biological failings are more those of omission than of commission. We are threatened more by the genetic traits that we lack than by those we possess. Species self-loathing has been a surprisingly popular passion among human beings, and it has led more to an occasionally gratifying intellectual hair shirt than to useful insights. Among such protestations of disgust, some of the most widely disseminated have accompanied the recognition of our likely carnivorous ancestry, as reflected, for example, in these outraged observations by anthropologist Raymond Dart, in his fascinating and controversial *Adventures with the Missing Link:*

> The creatures that have been slain and the atrocities that have been committed . . . from the altars of antiquity to the abattoirs of every modern city proclaim the persistently bloodstained progress of man. He has either decimated and eradicated the world's animals or led them as domesticated pets to his slaughterhouses.

Mistaking predatory behavior for intraspecies aggression and war, Dart discerned the origin of human warfare in those torrents of animal blood:

> The loathsome cruelty of mankind to man . . . is explicable only in terms of man's carnivorous and cannibalistic origin. The blood-spattered, slaughter archives of human history from the earliest Egyptian and Sumerian records down to the most recent unspeakable atrocities of World Wars I and II, accord with universal cannibalism . . . in proclaiming this common blood-lust differentiator, this mark of Cain, that separates man dietetically from his anthropoid relatives.

Aldous Huxley expressed a similar, and not entirely unjustified feeling:

> The leech's kiss, the squid's embrace,
> The prurient ape's defiling touch:

And do you like the human race?
No, not much.

Let's be fair to our species; after all, it is the only one we shall ever be. We are neither gods nor devils, and with a bit of insight, we can even aspire to that state of secular grace urged upon us by Albert Camus, at which we shall be neither victims nor executioners. There may well be too much feisty competitiveness built into the human spirit for us ever to lie down together like lambs. At the same time, there may well be just enough evolutionary leavening in the form of enlightened self-interest, for us to reach a middle ground in the human tempest, something between the brutish barbarism of a Caliban and the lofty cerebralism of a Prospero. The word "enemy" derives from the Latin *in* (not) plus *amicus* (friendly), and it implies a state of hostility, often involving hatred for one's opponent. By contrast, the word "rival" originated with the Latin *rivus* (stream), and literally means "one who uses a stream in common with another." Rivals are therefore competitors; this much may be inevitable. They need not, however, be enemies.

Near Southbridge, Massachussetts, on the Connecticut border south of Worcester, there is a lovely lake with the extraordinary Mohican name of Chaubunagungamaug. In English it means "You Fish on Your Side, I Fish on My Side, Nobody Fishes in the Middle: No Trouble." The early residents of Lake Chaubunagungamaug, we may assume, were rivals but not enemies.

It is not too late for human beings to treat each other as rivals rather than enemies. It is too late, however, for biology to achieve this transformation, acting alone. The time is too short, and the need too urgent. We have traveled too far down the road of culture; having surrendered ourselves to its power, excitement, and uncertainty, we must ultimately find our salvation in that route as well. Once we disrupt a natural system—for example, by planting a field of corn—we are forced to keep relying on human input through cultivation, weeding, irrigation, and possibly even pesticides and herbicides, just to keep that system from falling apart. Similarly, we desperately require a stringent cultural system that will respect the exigencies of our evolutionary past while protecting us in our perilous, but irrevocable, departure from our biology in the present and future.

CHAPTER EIGHT

THE NEANDERTHAL MENTALITY

CAVEMAN CONSCIOUSNESS IN THE NUCLEAR AGE

When you look into the abyss, the abyss also looks into you.
—FRIEDRICH NIETZSCHE

Looking into the nuclear abyss, we can see nothing less than the end of the world, something that prophets have warned about for millennia, but which now—thanks to the possibility of worldwide environmental catastrophe known as nuclear winter—has at least become a potential reality. And looking into ourselves, we find a twentieth-century caveperson, wielding unheard of powers but still a primitive at heart.

"The splitting of the atom has changed everything but our way of thinking," wrote Einstein, adding, "and hence, we drift toward unparalleled catastrophe." Now, almost forty years later, that "drift" has become a powerful tide. We are confronted with something much harder than a mere physics problem: a *psychology* problem wrapped up in modern politics, driven at least in part by our evolutionary past. As Einstein himself recognized, psychology and its public manifestation, politics, are much more difficult—and important—than physics. It is crucial, therefore, that we begin to understand our way of thinking (and sometimes, not thinking) about nuclear weapons. We might also inquire into why our thought processes haven't changed, even though with the splitting of the atom, everything else has.

Nothing concentrates the mind, observed Samuel Johnson, like the prospect of being hanged in the morning. Similarly, nothing ought to concentrate the human mind like the growing prospect of nuclear war. But unfortunately, today's caveman is as ill-equipped to deal creatively with modern weapons as he or she is well equipped and often eager to swing a club—be it zebra bone or nuclear armed.

There is nothing unusual about possessing lethal weapons; the porcupine, for example, has prospered with its dangerous quills. But whereas a porcupine's mental and physical traits developed synchronously, ours are increasingly out of phase. When it comes to defending itself against predators, a porcupine relies on its sharp quills, and the combination is effective, because neither its behavior nor its quills have changed in millennia. The porcupine's quills are appropriate to its reliance on them because the two evolved in concert, the formidable weapons along with the appropriate instinctive and readily learned porcupine behavior patterns.

Not so, however, for modern human beings. It seems likely that for nearly all of our evolutionary history, we were rather porcupinelike ourselves; that is, our behavior and our capabilities were more or less in tune with one another. But in the last few thousand years, this cozy relationship has gone awry and nowhere is this disparity more spectacular or more dangerous than in the realm of nuclear weapons. Manipulating the brave new world of culture—in particular, the brave new nuclear weapons—is the same old biological Us. A Neanderthal has his finger on the button.

But, we reassure ourselves, no rational person would start a nuclear war, and can't we count on the rationality of Homo sapiens? On the contrary, "Only part of us is sane," wrote Rebecca West,

> only part of us loves pleasure and the longer day of happiness, wants to live to our nineties and die in peace, in a house

> that we built, that shall shelter those who come after us. The other half of us is nearly mad. It prefers the disagreeable to the agreeable, loves pain and its darker night despair, and wants to die in a catastrophe that will set life back to its beginnings and leave nothing of our house save its blackened foundations. Our bright natures fight in us with this yeasty darkness, and neither part is commonly quite victorious, for we are divided against ourselves . . .

As the gap between the hare and the tortoise has grown, the gap between human survival and oblivion has narrowed. More than twenty years ago, social psychologist Charles E. Osgood coined the phrase "Neanderthal mentality" in his influential discussions of the psychology of the arms race. Although clearly it is not anthropologically accurate (Neanderthalers may not even have been on the path of human descent), the term speaks eloquently of tendencies that are widespread and primitive. It is also appropriately pejorative. Unfortunately, the Neanderthal mentality is all too evident when erstwhile Homo sapiens confronts, or fails to confront, the problems posed by nuclear weapons.

What, then, is the Neanderthal mentality when applied to nuclear weapons? Basically, it is the tendency to employ prenuclear mental processes to a totally new, nuclearized world. To evolution, 1945 is less than yesterday. Despite a number of temporary setbacks, and a growing array of problems, the Neanderthal mentality served us well, for 99.999 percent of our evolutionary history, during which time it generated behavior that led on balance to social and biological success. But thanks to Einstein himself, and the Manhattan Project, the scene has suddenly changed. Thanks to biological evolution, however, the actors keep reading from the same outmoded script.

There is a problem, then, not only with the nuclear "hardware" but also with the human software. As Harvard negotiator Roger Fisher has pointed out, the mere possession of nuclear weapons does not in itself guarantee unparalleled catastrophe. France and Britain, for example, are both nuclear armed and have been mortal enemies for centuries. Yet strategic planners in London do not lie awake nights, anxiously anticipating a preemptive first strike from Paris. Closer to home, the United States Army, Navy, and Air Force are all immensely powerful, antagonistic, and armed to the teeth with nuclear weapons, yet they restrict their competition to the football field and Congressional appropriations committees.

Here are four aspects of the Neanderthal mentality that encapsulate the nuclear dilemma as resulting from the conflict between culture and biology.*

The first concern revolves around the question of aggressiveness and getting one's way in a pushy world. There is primitive appeal, for example, in the notion that we make ourselves safer by placing our opponents at risk. For generations, this may well have worked. If so, it was likely promoted by natural selection. But it is no longer appropriate to a world in which security, if it is to exist at all, must be mutual. As Roger Fisher points out, there is an American at one end of a rowboat and a Russian at the other, and the strategy of each seems to be that he will make his end of the boat more secure by making the other fellow's end more tippy. Following this "logic," we each deploy missiles theoretically capable of destroying the other side's missiles; this, in turn, makes Them less secure and is therefore supposed to make Us more secure. But in fact, the more nervous and jumpy we make them (and/or they make us), the more likely it becomes that one side or the other, fearing that it will not be able to retaliate if attacked, winds up attacking instead. And the more each side threatens the other, the more likely it becomes that either side will mistake a false alarm for the real thing. As a result, the frail craft we call the earth is more and more likely to capsize. And yet, to the Neanderthal, the ritual of reciprocal risk makes a primitive kind of sense.

For a very long time, having *more* made our primitive ances-

*For a much fuller development of the Neanderthal mentality applied to nuclear weapons, see *The Caveman and the Bomb: Human Nature, Evolution and Nuclear War*, by David P. Barash and Judith Eve Lipton, McGraw-Hill, 1985.

tors safer. As nations feel increasingly insecure in the nuclear age, then, the answer seems simple enough: more clubs, more warriors, more bows and arrows, more guns, more tanks, more bombers. But the circumstances have changed, and getting more does not make anyone safer; in fact, quite the opposite. But try telling that to today's nuclear Neanderthal. The more insecure he becomes, the more tightly he clutches the very sources of that insecurity, getting more and more weapons, and in turn getting more and more insecure when the other side follows suit, all the while looking more and more like a muscle-bound idiot in a Chinese finger puzzle.

Closely related to "more is better" is "fewer is worse": fear of having fewer nuclear weapons than one's opponent, no matter how meaningless the so-called nuclear balance. Accordingly, despite the fact that the United States has always been ahead in the nuclear arms race (and still is ahead today) we have continually been stampeded by an array of fictitious gaps; since the early 1950s, we have been subjected to The World According to Gap, and have responded every time by rushing to escalate the arms race, reducing everyone's safety in the process. We have reached massive and insane levels of "overkill." Overkill, it should be noted, is not a biologically meaningful concept. It has been defined by James Real as "pouring a bucket of gasoline on a baby who is already burning quite nicely."

The modern Neanderthal, it seems, is an intellectual descendant of Procrustes, that unpleasant Greek who adjusted the anatomy of hapless travelers to fit his special iron bed. When confronted with anything new, today's Neanderthal immediately reverts to Procrustean thinking, trying to fit the new problem onto his old conceptual framework. Hence, more nuclear weapons are better than fewer, the nuclear missile becomes just artillery or the catapult writ large, nuclear wars can usefully be threatened, fought, limited, and won, just like conventional wars always were.

Living things like to win, and there is no reason to think that primitive human beings were any different. Like football coaches who claim that a tie is about as satisfying as kissing your sister, our Stone Age cousins did not aim for a draw. Those who did were probably defeated by others who played to win. Evolution tends to be a zero-sum game: if one side wins, the other side often loses proportionately, with the total of all payoffs equal to zero. This is because there is only a limited amount of "niche space" available, and as we have already seen, most populations remain stable over time. So, my gain was your loss, and if you profited, it was ultimately at my expense unless, of course, we were relatives or potential reciprocators. Once more, the ground rules have changed, but the Neanderthal within us still goes by the old version, the one that prescribes victory.

Aggressiveness also raises the question of the adaptive potential of fighting. Animals don't fight all the time, and neither do human beings. We don't *need* to fight in the same way that we need to eat or to sleep. However, throughout most of our evolutionary history, we almost certainly did fight on some occasions, notably over food, living space, or mates. Primitive "warfare," for example, is common among nontechnologic peoples such as the Yanomamo of the upper Amazon or the Tsembaga Maring of highland New Guinea; in such cases, however, the mortality rate is relatively low and "war" is much closer to skirmishing. In addition, success in battle can bring success in life: animal protein, status, *lebensraum*, and often, women.

Today the danger is clear: there is simply no way that nuclear war can be anything but a losing proposition, and yet, we readily respond to frustration, competition, and threat by manning our battle stations, as though the world hasn't changed and warfare might still be worthwhile. No less than other living things, we are masterful cost/benefit strategists, doing something—such as fighting—when its benefits outweigh the costs. As we have seen, Homo sapiens' evaluation of such circumstances was fixed during our long evolutionary childhood, when war worked. Now the cost/benefit equation has changed drastically, and nuclear war can only bring disaster, but we nonetheless find war (any war) strangely attractive, and the prospect of victory, nearly irresistible.

The second aspect of prenuclear mental processes revolves around limitations in our ability to perceive danger. When the Neanderthal felt threatened, he generally was: a charging herd of mastodons, a forest fire, a bellicose fellow Neanderthal. And by the same token, when he felt safe, he generally was. (The masculine pronouns are employed here intentionally because of the high probability that *male* Neanderthals were the major instigators of and respondents to aggression and threat.) But once again, times have changed. As psychiatrist Jerome Frank has pointed out, nuclear weapons have no "psychological reality": they can not be seen, smelled, heard, or felt, and so the Neanderthal feels safe . . . even though he isn't. The threat is there, and is very real, but it lacks the tangible reality of a mugger with a knife, for example, something to which the Neanderthal is well attuned but which is actually far less threatening to us all. Concern about nuclear weapons vanished almost entirely for a time following 1963, when the limited test-ban treaty prohibited weapons testing in the atmosphere (and in space, and underwater). Such testing actually continued, however, and at a more rapid pace than ever, but it was moved underground: out of sight and hence, out of mind for most modern Neanderthals who, like the caricatured ostrich, feel safe so long as they can no longer see the danger.

During the Vietnam War, millions were motivated to end a conflict whose reality was brought home in daily newscasts. By contrast, today's nuclear Neanderthal is generally unmoved by a conflict that hasn't yet begun, and which will be over long before he or she has the opportunity to protest. One of the great challenges facing today's peace movement, then, in addition to overcoming the deep-seated human tendency to misperceive the uses and abuses of aggressiveness in a nuclear age, is a need to overcome the equally deep-seated tendency to restrict our perception of danger to those situations that connoted danger in the past, and which—thanks to the invention of nuclear weapons—are no longer valid hallmarks of risk today.

In the same vein, consider another interesting suggestion from Roger Fisher: because of the bloodless high technology and "nukespeak" surrounding nuclear weapons and their command and control, a president could order their use without really understanding, deep in his guts, what he is doing. Accordingly, perhaps we should replace the computer-coded little black briefcase with a small capsule, implanted next to the heart of a trusted aide. Then, to send forth the Emergency Action Message, the president would have to do something more human and hence *more real* than reciting "Execute SIOP Plan 1-GA4Z, PDQ." Instead, he would be required to cut open the aide's chest, dipping his own hands in human blood. Bright red human blood might be just what's needed, to shock even the most insulated Neanderthal back to reality.

Pain is a useful warning device, telling us that something is wrong. Whether a toothache or a stubbed toe, pain is an unpleasant and hence, effective way of getting our attention. Therefore, we are inclined—almost by definition—to avoid pain. But pain can be emotional as well as physical, and few things are as emotionally painful as facing the danger of nuclear annihilation. "Have a nice day!" proclaim the mindless, yellow smiley-face stickers that became emblematic of the 1970s; thinking about nuclear war will not help you to have a nice day. So the twentieth-century nuclear Neanderthal avoids pain by avoiding the issue. Ironically, once again, a "defense mechanism" that was useful in other contexts helps propel us toward unparalleled catastrophe in the modern age.

Other aspects of our prenuclear mentality further conspire to insulate us from even perceiving the nuclear threat. The very enormity of nuclear war, the literal amounts of power and energy involved, leave us uncomprehending. For the caveman, "hot" is 100 degrees in the shade, or perhaps the temperature of boiling water, or a fire. Maybe molten metal, at the extreme. But not 100 million degrees, the temperature inside a thermonuclear fireball. And what would hundreds of millions of deaths be like? Or a nuclear winter? We are unable to incorporate these realities into our caveman consciousness. It is not just that we don't want to entertain such thoughts, but rather, we are not able to do so: like visitors from another planet, whose antennae are not adjusted to the wavelengths of this new, nuclear world, we wander numb and unresponsive.

For the Neanderthal, as for Ecclesiastes, there was nothing much new under the sun. Even humanity's cultural advance, rapid by biological criteria, must have seemed excruciatingly slow and uneventful to those living through it, day by day. For most of us, even now, the past is a good guide to the future. If it hasn't happened yet, it is generally a good bet that it never will. Accordingly, we quickly reassure ourselves with the observation that nuclear war hasn't happened yet, in the four decades since 1945. Deterrence, we are told, works.

Few of us would take the fact that we are now alive as evidence that our death will never happen; and yet, many of us are reassured that nuclear war will not happen, because it hasn't happened yet. Freud suggested that the human capacity for "denial" is essential if we are to function normally from day to day. After all, our personal death is inevitable, so why obsess about it? Nuclear war, on the other hand, is not inevitable—but neither is it impossible or even unlikely, especially if our biology has its way and those most likely to oppose it are also most likely to ignore the whole issue, leaving the field clear to those military officers, politicians, and industrial contractors who profit personally from the arms race, and hence, are perfectly willing to gratify that aspect of their prenuclear mentalities.

Then there is habituation, the simplest and most primitive of all learning processes: learning *not* to respond. It is clearly maladaptive for an animal to respond to every stimulus that comes its way, and not surprisingly, then, even animals as simpleminded as flatworms stop responding to irrelevant stimuli after awhile. The progressive accumulation of nuclear weapons in the world has certainly not been gradual as measured in evolutionary time, but nonetheless the growth of superpower arsenals has taken an entire generation, a rather long time for quick-moving, conscious creatures. Almost insensibly, we have built more and more, never entirely registering what we were doing and thus, never adequately outraged.

The human nervous system is sensitive to short-term changes in stimulation, but adapts quickly to constancy or gradual change. Hence, we notice (adaptively) a new smell or a sudden sound, but we become literally insensitive to that same smell or sound if it is continued for a period of time without obvious consequence. Since nuclear weapons haven't been used directly against people since 1945, we have become almost as unaware of their presence as we are of the constant drone of a refrigerator motor, or our own smells in the bathroom.

It is said that if a frog is dropped into boiling water, it will jump out and save itself. But if placed in cold water that is then heated gradually, degree by degree, it will never notice the difference and boil to death. Our temperature has been rising, but most of us haven't been noticing.

There was widespread shock in the United States when the *Lusitania* was sunk by German submarines in 1917 and several hundred innocent civilians lost their lives. Then, the Italian use of poison gas in Ethiopia; the German bombing of the Spanish town of Guernica (immortalized by Picasso's painting); the killing of 35,000 Dutch civilians in Rotterdam; 100,000 or more in Dresdan, Hamburg, Tokyo, and finally, Hiroshima. If anything, our outrage has, over time, diminished, and we are no longer shocked by things that once seemed appalling. We have habituated to horrors.

After German Zeppelins had dropped some bombs over London in 1914, George Bernard Shaw wrote to the *London Times* proposing that the London County Council look into constructing air-raid shelters for the city's children, in the event that such attacks become more frequent. The newspaper reproved Shaw editorially, pointing out that it was inconceivable that so civilized a country as Germany—even when at war—would ever stoop to something so barbaric!

Another aspect of prenuclear thinking revolves around patterns of group identification, notably nationalism. The primitive human being relies heavily upon his fellows for success and survival, just like any other social animal, and in fact, more than most. For the many thousands of generations that preceded modern times, our ancestors sought and obtained safety and ultimately, reproductive success, in groups. And for most of that time, the larger the group the better. Families and tribal bands consisted of relatives and/or reciprocating colleagues, whereas by contrast, other families and other tribes were likely to be competitors and often antagonistic as well. Accordingly, there was an evolutionary payoff in what modern sociologists call "in-group amity, out-group enmity," mediated in part, perhaps, by kin selection.

Once again, however, a biologically adaptive tendency seems to have gone awry, leaving us at the mercy of "supernormal releasers," those exaggerated traits to which we are so peculiarly sensitive. Our allegiance is now readily seduced by culturally hyperextended groups known as nations, which offer much of the primitive gut-level satisfaction of family and tribe. Unlike rooting for the Detroit Lions, however, or belonging to the Benevolent and Protective Order of Elks, the modern patriot clings to a social unit that is *not* benign, and which in fact has elevated itself above the individual, threatening him and her with destruction for its own, national, "security."

When we review the annals of human nastiness, we tend to think of murder, rape, torture, arson, or assault, and yet, by far the greatest weight of human evil against fellow human beings has not been committed by the solitary sociopath, but rather, by Homo sapiens acting in a group. It is an excess of devotion, of subordination of the self to the group, not an excess of self-seeking, that leads us most dangerously astray. In human history, more harm has come from obedience than from disobedience (although we are typically warned against the latter, not the former). Similarly, more harm comes from excessively zealous group orientation than from individualism. The nation as a social unit has no fundamental, biological legitimacy, and yet it retains a deep human allegiance, typically because it mimics the primitive Neanderthal yearnings for deeply meaningful group associations.

As Freud pointed out in his *Thoughts for the Times on War and Death*,

> [the] state has forbidden to the individual the practise of wrong-doing, not because it desires to abolish it, but because it desires to monopolize it . . . A belligerent state permits itself every such misdeed, every such act of violence, as would disgrace the individual . . . Well may the citizen of the civilized world . . . stand helpless in a world that has grown strange to him.

Unfortunately, the world in its barbarity has not grown entirely strange to the human beings in whose name we prepare the most hideous atrocities. For so long as these acts are to be perpetrated in the name of "national security," which the nuclear Neanderthal erroneously equates with personal security, eons of evolution conspire to pave the way and legitimize the most illegitimate of plans and justify the most unjustifiable behavior.

Such collective self-deception also leads to "evil empires," and to a self-righteous certainty that whatever We do is good, and whatever They do is bad, even though such actions are often mirror images of each other. For example, when the USSR (a diverse group of Homo sapiens inhabiting eastern Europe and west-central Asia, largely having murderously displaced the indigenous inhabitants) invaded Afghanistan in support of a locally unpopular foreign government threatened by revolution, we denounced that as aggression. But when the United States (a diverse group of Homo sapiens inhabiting North America, largely having murderously displaced the indigenous inhabitants) invaded Vietnam in support of a locally unpopular government threatened by revolution, that was proclaimed a noble cause.

Neither side is able to see the world as the other side sees it. No surprise there; after all, during the 99.999 percent of our evolution that preceded modern times, such a leap of empathy was not needed. It was enough to take care of ourselves, secure in the knowledge that we were right, and they were wrong and in fact not really human anyhow. Precisely that tendency, to dehumanize one's opponents, greased the way for behaviors toward the Other that would never be permissible within the social group. For generations, human beings have defined themselves and their fellow group-members as "human," and considered that non-group-members are so alien as to be literally inhuman. Dehumanizing slang is typically directed toward such aliens, especially when the two sides are at war. Moreover, such attitudes facilitate the declaration of war in the first place. Kin selection may well be operating here, since it is a primitive biological tendency to behave benevolently toward relatives, competitively toward nonrelatives. This vestige of our evolution, adaptive in the distant past, could well facilitate a disregard of our shared humanity which in fact is not "inhuman," but rather, all too human.

Fortunately, dehumanization of the enemy, although perhaps encouraged by biological tendencies, is a delusion. It can readily be dispelled, simply by the recognition of our shared humanity. In *Homage to Catalonia*, his account of the Spanish Civil War, George Orwell described the homely situation that led to such a recognition on his part:

> At this moment a man, presumably carrying a message to an officer, jumped out of the trench and ran along the top of the parapet in full view. He was half-dressed and was holding up his trousers with both hands as he ran. I refrained from shooting at him. It is true that I am a poor shot and unlikely to hit a running man at a hundred yards . . . Still, I did not shoot partly because of that detail about the trousers. I had come here to shoot at "Fascists"; but a man who is holding up his trousers isn't a "Fascist," he is

visibly a fellow creature, similar to yourself, and you don't feel like shooting at him.

Similarly, Western politicians—arguing for additional nuclear weapons as "bargaining chips"—like to point out how dangerous it would be if they had to go into latest round of nuclear negotiations "naked." But perhaps that is precisely how it should be done. Maybe the world's people (in whose name, after all, such negotiations are being conducted) should require that high-level discussions of this sort be carried out only by naked human beings. Devoid of artifice, fully revealed as vulnerable, biological creatures, maybe they would laugh at each other and at their own pretensions; maybe they would find the wisdom and humility to act in the service of life instead of death.

Finally, let us consider another aspect of our prenuclear outlook that impedes antinuclear action. There was a certain primordial, adaptive wisdom by which the Neanderthal apportioned his or her energy. We avoid tackling problems that are too large and whereas we love projects, we abhor those that can't be completed successfully. Thus, we derive our day-to-day satisfactions from events that are on a personal, human scale; family, friends, work, recreation, etc. It simply does not pay to wrestle with a hurricane, a volcano, or an earthquake. At the very least, it is a waste of time to tackle oversized opponents. It may also be dangerous, and through the course of our evolution, natural selection has almost certainly acted against the Don Quixotes among us, who dreamed impossible dreams and broke their lances against unyielding windmills.

Unquestionably, nuclear war is the unyielding, oversized opponent par excellence, not only in the magnitude of its effects, but also in the size of the forces that must be tackled: bureaucratic, military, political, economic, etc. In short, the problem is big and each of us is small, so we readily answer the primitive urgings within, and leave the problem to someone else, to our nuclear priesthood and our parent/leaders, who, after all, are wiser and stronger than us, and whose benevolent attention to such things permits us to go about living our own little lives, reaping whatever primitive, personal-sized, and biologically appropriate rewards we can achieve.

> The hawk that motionless above the hill
> In the pure sky
> Stands like a blackened planet
> Has taught us nothing.

writes Edna St. Vincent Millay, in "The Bobolink."

> seeing him shut his wings and fall
> has taught us nothing at all.
> In the shadow of the hawk we feather our nests.

And so we go on, feathering our nests, oblivious to the shadow of the hawk—or perhaps seeing it but looking the other way because such sights are inconvenient, interfering as they do with the primitive, universal yearning to go on with our little day-to-day lives.

There may also be another factor, one that virtually everyone will deny, but which may nonetheless be worth examining. And this is the deep fascination that some people find with nuclear weapons themselves, precisely because they are so extraordinary, so powerful, and so impressive. They offer to Homo sapiens, a weak-bodied and anatomically unimpressive biological creature, the opportunity to identify with a level of force—and to some, of beauty—that is potently seductive. For example, consider this account, by Brigadier General Thomas Farrell, of the world's first atom bomb test, at Alamogordo, New Mexico:

> The effects could well be called unprecedented, magnificent, beautiful, stupendous and terrifying. No man-made phenomenon of such tremendous power had ever occurred

> before. The lighting effects beggared description. The whole country was lighted by a searing light with the intensity many times that of the midday sun. It was golden, purple, violet, gray and blue. It lighted every peak, crevasse and mountain range with a clarity and beauty that cannot be described but must be seen to be imagined. It was the beauty the great poets dream about but describe most poorly and inadequately. Thirty seconds after the explosion came, first the air blast pressing hard against people and things, to be followed almost immediately by the strong, sustained, awesome roar which warned of doomsday and made us feel that we puny things were blasphemous to dare tamper with the forces heretofore reserved to The Almighty. Words are inadequate tools for the job of acquainting those not present with the physical, mental and psychological effects. It had to be witnessed to be realized.

Thirty-six years later, psychologist Nicholas Humphrey delivered a nationally broadcast lecture on the BBC, titled "Four Minutes to Midnight," in which he gave a different perspective on the human potential of such power and beauty:

> "Do it beautifully!" says Hedda Gabler to Lövborg, as she hands him the gun. Oh yes, we'll do it beautifully. What more beautiful way to do it than in the way that poets dream about, but describe most poorly and inadequately? But the gun goes off by accident, and Lövborg dies miserably, shot not through the heart but through the balls.

The above only briefly evokes the perils of the nuclear Neanderthal as reflected in aggressiveness and success, perceptions of risk, group identification, and finally, impediments to activism. There are many other components as well, such as a tendency to use habitual, noncognitive behavior under stress (after all, it was quick reflexes rather than creative problem solving that saved our ancestors from the crouching saber-tooth); an inclination to imbue leaders with superhuman qualities of wisdom to match their superhuman qualities of power; a tendency to assume the worst in everyone else (paranoia), often employing a reflection of our own unpleasant traits and in the process, generating a self-fulfilling prophecy that produces precisely the effect we most fear; a perverse insistence on organizing the post-Hiroshima world around an intellectual system of primitive threat and punishment (i.e., deterrence), which, although wrapped in arcane intellectual pretensions, is fundamentally one of the simplest, most primitive ways in which one animal coerces another, and among complex animals, one of the least effective.

Once confronted with the threat posed by nuclear weapons—even after the denial, the habituation, the disordered priorities of national strength and national security, of risk, danger, error, and misperception, of enemies and of personal powerlessness are finally penetrated—even then, human beings typically reach for Brahean solutions: for example, seeking to make deterrence "more secure," rather than seeking to abolish nuclear weapons altogether. We prefer to tinker assiduously but superficially, toying with self-gratifying cosmetic adjustments that may help somewhat at the margins, while in fact maintaining our self-perceived interest in keeping things fundamentally as they are and at the same time gratifying our felt need to act "responsibly." Our failure to respond to radical problems with radical solutions is not surprising to anyone familiar with the general history of Homo sapiens, but in the case of nuclear weapons, it is almost exactly equivalent to rearranging the deck chairs on the *Titanic*.

*
* *

OUR PENCHANT for Brahean solutions is perhaps most dramatically revealed in the latest Star Wars schemes of the 1980s. Here we see a merger of humankind's pervasive faith in technology—the higher "tech" the better—with our penchant for responding to anxiety engendered by nuclear weapons by clutching yet more firmly the weapons themselves. In his book *Technics and Civiliza-*

tion, Lewis Mumford pointed out that "the belief that the social dilemmas created by the machine can be solved merely by inventing more machines is today [1934] a sign of half-baked thinking which verges close to quackery." Writing more than a decade before the first nuclear weapons were used, Mumford certainly did not have Star Wars in mind, but no better description of this very dangerous fantasy has ever been proposed.

Ever since President Reagan's now-famous Star Wars speech of March 1983, in which he proposed the development of a system to render offensive nuclear weapons "impotent and obsolete," growing attention has focused on the technological, economic, and strategic implications of such a plan. Analyses have proliferated as to whether or not Star Wars (or, as the Administration prefers to call it, the Strategic Defense Initiative) is technically feasible, economically affordable, or strategically desirable. Such debate and such concern are appropriate.

However, there has been surprisingly little attention to the psychological appeal of Star Wars. Whatever its merits or demerits technically, economically, or strategically, Star Wars is brilliant, psychologically. It offers a textbook portrait of prenuclear thought processes being applied to the fastest moving of today's hares. Star Wars is psychologically brilliant because it plays right into our primitive yearnings, precisely at a time when the old reliance on nuclear weapons for security is beginning to look increasingly threadbare to millions of people. At the same time, it is phenomenally obtuse because it is itself an example of those primitive yearnings, and at the highest government level.

Star Wars supporters claim that it would be morally superior to deterrence, since it will defend people, rather than threaten an opponent, or if necessary avenge oneself. In this respect, it is interesting that nuclear hawks, many of whom have made a career out of defending the use of planetary threats—that is, deterrence—are suddenly rushing to attack it, while nuclear doves, for many of whom deterrence is profoundly distasteful, find themselves defending it.

It is not difficult to see why some people support Star Wars:

high-level officials eager to remain in the good graces of their boss, military officers whose careers have suddenly become hitched to its research, testing, and eventual deployment, and high-tech military contractors who can smell a lucrative contract miles away. ("Dollars from heaven," gushed the *Wall Street Journal.*) There is something primitive, and understandable about that old motivator, greed. Among others, there is the appeal to some other deep-seated caveman inclinations, notably the feeling that security can be achieved by more weapons, certainly not by fewer. The attractiveness is an old one, the hope of having one's cake and eating it too: having one's missiles and hiding from them too. Or we might call it the Humpty-Dumpty syndrome: by the very nature of nuclear weapons, our national security has taken a great fall, and now the king is saying that he can *too* put it together again, if we only give him enough horses and enough men.

We must also not forget the seductiveness of high-tech itself, as Mumford warned. Despite ecological consciousness, small-is-beautiful, and so forth, it remains true that many of us have been mesmerized by a continuing, naive faith in technology, and a hope that science will save us, that machines will save us from machines. Pie-in-the-sky may be old hat, but armor-in-the-sky, a genuine astrodome from sea to shining sea is a potent lure, especially because it excuses us from dealing with the really difficult but ultimately unavoidable problem of working out negotiated, political settlements with our adversaries. After all, our technological skills are generally superior to our social skills, so when in doubt, we turn with relief to technology.

And finally, there is the appeal of unilateralism. After all, even ardent Cold Warriors are becoming jaded about somehow "beating" the Russians in the accumulation of offensive overkill. Star Wars offers them a new Holy Grail, an arena of strategic competition that plays to America's strength—technology—with the prospect that we'll just go ahead and solve this problem like real men . . . by ourselves.

Behind it all, it may be that the most cogent psychological prop for Star Wars is a cynical high-level recognition that it will guarantee an indefinite arms race, since as long as they are threatened with the prospect of a potential American Star Wars system, the Soviets will never agree to arms limitation, never mind reductions.

*
**

"MOM, can I have my birthday party early this year?" Nothing very unusual in this, coming from an impatient and precocious five-year-old. A bit more surprising, however, was the reason for the request: " 'cause I don't want to miss mine if there's a nuclear war."

A colicky baby, or a teenager with a drug problem demands our attention. A child with leukemia, anorexia, or even a bad case of acne quickly mobilizes parental concern and assistance. The alienated child, who tunes out of adult society and turns on to acid rock and drugs, is no less frightening or motivating for most parents. Some of this alienation has doubtless occurred as long as there were mammals and their awkward adolescent offspring, just beginning to surge with hormones and to seek their place in the social group. But as if this wasn't troublesome enough, the last half of the twentieth century has added something new: fear of nuclear war, that is, not so much fear that the future is uncertain, or likely to be difficult or even perhaps unpleasant, but rather, fear that there may not be a future at all.

Since the dawn of human consciousness, we have struggled with the reality of our own demise. It comes with the territory, in a sense, one of the bitter fruits of awareness. And yet, it is still difficult enough to talk with our own kids about what Kurt Vonnegut calls "plain old death." No wonder it is even harder to confront nuclear war, with its image of utter annihilation. Many parents cannot even talk with their children about sex, perhaps in part because of a desire for sexual privacy, discussed earlier. However difficult the facts of life, many more people shrink from discussing nuclear war, the facts of death. And yet, just as children simply cannot grow up without encountering sex, they also cannot grow up today without encountering nuclear war, in thought if not reality.

Most parents are frustrated, bewildered, shaken, and confused when faced with their children's anxieties about nuclear war. After all, we pride ourselves in "taking care" of our children, and nuclear war—more than anything else—confronts us with the harsh reality that our web of parentally provided security is really an illusion. We can invest in piano lessons and computer camps, and insist on daily dental flossing, so as to be good parents, and yet, if the bombs go off we have failed, utterly. Moreover, the fear and estrangement in some children is such that even if there isn't a nuclear war, we have failed anyhow.

The simple fact is that our children *are* threatened, as never before, and they know it. Never before have young members of the species Homo sapiens had serious reason to doubt the continuation of their species. Never before have we had a mushroom-clouded future. The "doomsday clock" of the *Bulletin of the Atomic Scientists* has been moved ahead to just three minutes before midnight, and responsible experts warn increasingly of the dwindling likelihood that we shall avoid nuclear war before the next century. To some extent, the current nuclear peace movement is responsible for the epidemic of nuclear anxiety among our youth; that is, our children are worrying because they are being told that they have something to worry about. The reality is, however, that blaming the peace movement for the widespread alienation and nuclear fear is like blaming the person who shouts "Fire," or who turns in the alarm, rather than the arsonist. The challenge for today's parents is to help their children respond effectively to that alarm, not to ignore, deny, or cover it up.

We may never know the precise psychological cost of living under the Bomb of Damocles, but it seems likely that an earlier generation has paid its share. Thus, many of today's parents are themselves veterans of "duck and cover" drills of the 1950s, and the strontium-90 still in their bones mixes all too well with a long-suppressed anxiousness in their hearts. Meanwhile, the current generation is not exempt. Just as the danger of nuclear war is probably higher now than at any time since the Cuban Missile Crisis, nuclear anxiety almost certainly has been keeping pace. A recent study by psychiatrists William Beardslee and John Mack has shown that one half of American children become aware of nuclear issues before age twelve. Among older children, one half thought this awareness affected their plans for marriage and the future. The Beardslee and Mack study, based on interviews with hundreds of school-age children, shows that children are "deeply disturbed" about the threat of nuclear war, profoundly pessimistic, and often, just plain scared.

What's wrong with being scared? According to psychoanalyst Sybille Escalona,

> profound uncertainty about whether or not mankind has a foreseeable future exerts a corrosive and malignant influence upon important development processes in normal and well-functioning children.

After all, "Young people come to terms with the adult world as long as it holds out a reasonable promise for fulfillment in some spheres of living." For many of our children, that "promise" looks more and more like a lie, and accordingly, children may be having more and more difficulty coming to terms with the adult world. A recent study of teenage California "stoners"—who had viewed the murdered body of one of their friends, yet been unmoved by the experience—revealed that these youngsters all felt hopeless about their own future, as though they were being denied any "promise for fulfillment" by the threat of nuclear annihilation.

The problem, however, is not simply how adults can best assuage the fear of our children, how we can minister to their worries, calm their anxieties, and help them to overcome their growing nuclear fear. The issue is not simply a need for juvenile reassurance, any more than a child caught in a burning house needs reassurance; reassurance be damned, the need is for rescuing. The problem is real, not imaginary; one not of attitude but of situation. The first step for parents wanting to help diminish their children's nuclear fear is therefore to recognize that the fear is legitimate—not only because it is real within them, and hence partaking of the "reality" of a psychosomatic complaint—but also because it is based on a real threat outside of them.

It is said that before shipping their children off to Babi Yar, thousands of Polish parents were very careful to see that they had brushed their teeth, combed their hair, and buttoned up their overcoats. Such fastidiousness may have made the parents feel good at the time, but cannot (in retrospect, admittedly) qualify as "good parenting." Similarly, good parenting in the Nuclear Age involves more than simply ministering to the fears of children, as though the fear itself is the problem. It must also seek to diminish the causes of such fear. That is, good parents in the 1980s will not only try to make their children *feel secure*, but also help them to *be safe*. Fortunately, doing one may well be the best way of doing the other. And in the process, curing some alienation in and between both generations.

Victorian sexual taboos did not make sex go away, just as parental reluctance to talk about it does not make it easier for children to become fully sexual adults. And just as it is especially difficult for parents to talk about sex unless they are both intellectually and emotionally comfortable with it themselves, parents will likely have difficulty talking with their children about nuclear war unless and until they have informed themselves about the issues.

In their report on children's attitudes toward nuclear war, Beardslee and Mack note that

> At each stage of development, the child mitigates disappointments by looking ahead and building a vision of the future in which he or she may possess what cannot now be had, or in which it is possible to become what he or she is

> incapable of being now. A healthy ego ideal builds out of possible goals or standards that are both realizable and worth struggling to achieve. But the building of such values, or of an ego ideal, depends on a present life that is perceived as stable and enduring and a future upon which the adolescent can, at least to some degree, rely.

The two psychiatrists go on to ask:

> But what happens to the ego ideal if society and its leaders are perceived cynically and the future itself is uncertain? Furthermore, how does it affect the ego ideal when the reason for that uncertainty is readily perceived to be the folly or "stupidity" of the adults around the adolescent who, because of perceived incompetence, greed, aggressiveness, lust for power, or ineffectualness can leave their children no future other than a planet contaminated by radiation and on the verge of incineration through the holocaust of nuclear war. In such a world, planning seems pointless, and ordinary values and ideals appear naive. In such a context, impulsivity, a value system of "get it now," the hyperstimulation of drugs, and the proliferation of apocalyptic cults that try to revive the idea of an afterlife while extinguishing individuality or discriminating perception, seem to be natural developments.

The best antidote for nuclear despair and alienation is not group therapy or chic attempts to "get the fear out," but rather, activism; this applies to children no less than adults. Moreover, just as children in the 1980s seem otherwise likely to grow up (if they do so at all) with a heavy dose of hopelessness and anger toward the adults who have created their world and now seem relatively uninvolved in seeking to improve it, children can profit enormously from images of their parents as active, courageous, concerned, and powerful people. Children are strongly influenced by the adult role models around them. Sybille Escalona notes that "growing up in a social environment that tolerates and ignores the risk of total destruction through voluntary human action tends to foster those patterns of personality functioning that can lead to a sense of powerlessness and cynical resignation."

In addition to the intrapsychic and personality costs, there may also be long-term social costs, borne ultimately by a society that will someday look to these children for leadership: "Growing up with the full knowledge that there may be no future and that the adult world seems to be unable to combat the threat can render the next generation less well equipped to avert actual catastrophe than it would be if the same threat existed in a different social climate."

On the other hand, growing up in an environment that fosters a loving, caring, and activist orientation toward all that is beautiful and special about this planet can encourage those patterns of personality functioning that lead to a sense of joy, power, and self-worth.

Let us also remember that we and our children are in this together, for better or worse. It is, in fact, the ultimate in togetherness: sharing the same fate on a small and endangered planet.

There have not as yet been any studies of children who grow up with an activist orientation toward nuclear war, but such children will probably not become stoners, nor will they feel helpless, hopeless, or despairing. They may, however, feel quite alienated from "the system" or a government that embraces nuclear weapons, and rightly so. But they will almost certainly feel connected to their family and to the human race. They may even develop a renewed faith in democracy, or at least, a renewed hope for its potential. Erik Erikson has suggested that infancy establishes a foundation of hope and "basic trust," followed by childhood and its rudimentary training in will, purpose, initiative, and skill, and then adolescence with its basic grounding in "some system of fidelity." And the greatest fidelity, perhaps, is not to an imagined god, or political system, a spouse, or even to oneself, but rather, fidelity to the planet.

It is ironic indeed that some parents are more upset that their children are upset about nuclear war, than about nuclear war itself. Many people, of course, do not accept personal responsibility for our nation's military and geopolitical policy. Insofar as that is so, the nation's failure to provide a safe world for them and their children is often less troublesome for them personally, since it is not seen as a personal failure on *their* part. But most parents, by contrast, do accept responsibility for the psychic wholeness of their children, not to mention their physical safety; accordingly, many adults have become active in the nuclear peace movement because of the despair and fear they have seen in their children (not to mention their own fear *for* those children.) These parents recognize implicitly the failure of the parents of Babi Yar.

What they may not realize, however, is that by becoming active themselves, they are almost certainly helping their children psychologically as well. To a child, parents are immensely powerful, and almost by definition, successful at what they attempt. So the active, committed parent, alienated from the nuclear nation-state but deeply connected to life itself, may be not only the best long-term hope for a child, but also the best short-term confidence builder. In a recent discussion among 17 students in a second grade class, 16 said they worried about nuclear war. Only one had hope for the future. He explained why: "I know there won't be a nuclear war because my daddy goes to meetings all the time to prevent it."

* * *

AS MODERN primitives, we are influenced by our evolutionary past. But this is not to say that we are prisoners of that past; indeed, we are in thrall only so long as we remain unaware of that influence. Once we recognize the outmoded source of our inclinations, the vestigial Neanderthal mentality like a swollen appendix now threatening to burst, we become empowered to excise it.

We are, after all, the most adaptable creatures on earth. We have given up slavery, the divine right of kings, human sacrifice, and dueling, all of which were at one time considered indelible reflections of "human nature." We can learn all sorts of things, like languages, music, or even respect for the rights of others, and we can inhibit inclinations that we recognize as inappropriate, like our muscle-headed Neanderthalism and reflex resort to Procrustean thinking. When it comes to the Neanderthal mentality and nuclear weapons, the conflict between biology and culture in human affairs is more pronounced and more dangerous than in any other realm. And yet, there is hope. Having recognized a problem and correctly identified a threat, Homo can be wonderfully sapient, correcting dangerous situations even if this includes correcting himself.

As Sigmund Freud saw it, culture has a responsibility to win back humanity from the ascendancy of our unleashed instincts:

> The fateful question for the human species is whether and to what extent their cultural development will succeed in mastering the disturbance of their communal life by the human instinct of aggression and self-destruction. It may be that in this respect precisely the present time deserves a special interest. Men have gained control over the forces of nature to such an extent that with their help they would have no difficulty in exterminating one another to the last man.
>
> —*Civilization and Its Discontents* (1930)

More than fifty years later, the problem is if anything more acute than Freud described, and somewhat different as well: our human "instincts" may well be the problem, but so is culture. Thus, it is precisely the runaway elaboration of nuclear culture, and its uncoupling from basic biological wisdom and inclinations, that so endangers the entire planet today. But Freud was also on target: We cannot wait for biology to save the day. Biological evolution is simply too slow. The responsibility cannot be foisted off on natural selection, any more than on government leaders. No, the responsibility is ours. Cultural evolution, acting in recent historical time, caused our nuclear mess. What is needed today, as Gunnar Myrdal put it, is "not the courage of illusory optimism but the courage of almost desperation."

According to Greek mythology, the gods punished Prometheus—who had impudently given fire to human beings—by chaining him to a great mountain, whereupon he was visited daily by a vulture, who chewed on his liver. Modern human beings, biological creatures acting not in deliberate evolutionary time but in a cultural frenzy, have unleashed a much more dangerous fire than Prometheus could ever have imagined. And this fire is made all the more lethal by the fact that, deep inside, we really aren't very "modern" ourselves. In *Prometheus Bound*, Aeschylus asks:

> Prometheus,
> Prometheus, hanging upon Caucasus,
> Look upon the visage
> Of yonder vulture:
> Is it not thy face,
> Prometheus?

Two thousand years later, Pogo said it more simply: "We has met the enemy and it is us."

CHAPTER NINE

POPULATION

Psychotic Rats, Open Faucets, and Closed Minds

Europe is over-populated, the world will soon be in the same condition, and if the self-reproduction of man is not "rationalized," as its labor is beginning to be, we shall have war. In no other matter is it so dangerous to rely upon instinct. Antique mythology realized this when it coupled the goddess of love with the god of war.

—Henri Bergson (1935)

Living things love to reproduce, not only figuratively but literally as well. That is, love is a means of reproducing, and human beings, no less than other animals and plants, tend to be very good at it. A sexually reproducing species need only leave two surviving offspring for each breeding pair for its population to remain constant from one generation to the next. This is more or less what usually happens, not because the individuals in question are especially concerned about the fate of their species, but rather, because competition among individuals and between species only rarely gives a pronounced advantage to anyone. If various factors in the environment did not act to reduce its numbers, each species would generate a population explosion of unimaginable proportions. If just two adults of any plant or animal reproduce unimpeded during their lifetime, and their offspring follow suit, then after a million years (a very short time in evolutionary terms) the entire earth and in fact, all of the visible universe would be packed with the quivering, living substance of our hypothetical organism. It should cause no surprise, therefore, to learn that living things do not achieve their full reproductive capacity. Their numbers are constantly depleted, both as adults and more commonly as immatures, embryos, or simple sperm and eggs. Some are depleted more than others; this, of course, is the stuff of natural selection.

In discussing natural selection, we concentrated on the survivors, and on those characteristics that confer genetic survival value. Now we must look at the losers. Many factors, separately or in combination, can cause living things to die or fail to reproduce. They can run out of food, or some necessary vitamins or minerals. They can fail to find a mate or a suitable place to live or raise young. They can succumb to the vagaries of climate, such as wind, rain, sun, or drought. They can fall to disease caused by bacteria, protozoa, or virus. They can be eaten while alive (by parasites such as the tapeworm or botfly) or first be killed and then eaten (by predators such as hawk or wolf or human being). Although these are hard fates for the individual victim, they also have the unintended benefit of preventing uncontrolled and potentially catastrophic population increases.

One famous case of misguided human altruism illustrates the importance of natural mortality in maintaining a healthy population. The Kaibab Plateau is an enormous wilderness in north-central Arizona. It supported a reasonably large deer population and a goodly number of predators, all in healthy balance with the environment. But the kind-hearted, deer-loving public, especially the hunters, wanted more deer and felt that this could be accomplished by killing their "enemies."

So, a systematic program of slaughter was initiated. From 1907 to 1923, 11 wolves, 600 cougars, and 3000 coyotes were removed from the Plateau. Before this, the deer population had been about 4000 healthy individuals, their numbers kept in check by the predators. With this check removed, the deer population increased spectacularly, to about 100,000. But long before that number was reached, something was clearly wrong. The deer were obviously running out of food, looking thin and sickly. Trees were being killed by overbrowsing and all vegetation had been stripped bare up to a height of about eight feet, the maximum that a starving deer can reach, standing on its hind legs. A great famine began in 1925 and more than half the deer starved to death in the next two years. By 1940, starvation was still killing more deer than the predators ever had, and the population was down to about 10,000.

More serious yet, the range itself had been gravely injured. Before predator removal it could probably have supported about 30,000 deer, but after the trappers and hunters had done their work, the extreme pressures of an unnaturally high population had greatly reduced the Plateau's ecological vigor. The Kaibab deer were victims of their own uncontrolled population. They not only suffer greatly from starvation, but seriously reduced the capacity of their environment to sustain life in the future.

The story of the Kaibab Plateau deer is in many ways an allegory of the planet earth and modern Homo sapiens. Our history has been one of progressively increasing numbers, first very gradually for thousands if not millions of years, then building up to a mighty roar in the last century. The worldwide human population at the discovery of agriculture, perhaps 10,000 years ago, was about 5 million. By the birth of Christ, there were about 200 million of us. We didn't reach our first billion until around 1850—approximately 52,000 years after Homo sapiens arrived on the scene. Our second billion appeared by around 1930, just eighty years later, and our third billion arrived around 1960, a mere thirty years after that. Billion number four was reached by 1975.

It should be clear that not only have our numbers been increasing at a phenomenal rate, but, even more frightening, the *rate of increase* has been increasing in many countries. Thus, even the phrase "population explosion," often used by people who are alarmed about our numbers, is inadequate. In an explosion, things fly very rapidly out from the source, but with decreasing speed farther from the epicenter, and as time goes on. There is no English word that describes the human population "explosion," an unprecedented event in which both the numbers and the rate have been increasing over time. It is more like an avalanche.

Like all animals, human beings are capable of a very high rate of reproduction, theoretically about twenty children per couple, but in practice more often six to ten in cases of unrestrained childbearing. This high birth rate was adaptive for our ancestors, among whom mortality was high. Because many children would normally die of disease, famine, exposure, or predation, primitive human beings required large families simply to keep even, and for individual parents, the cost of additional children was generally less than the benefit. As the Red Queen told Alice, it was necessary for us to run, just to remain where we were. To get anywhere, we had to go faster yet. To guarantee at least two surviving offspring, on the average, it was necessary to have six or seven, and among societies in which special value was ascribed to one sex (typically male) it was also necessary to have many children to ensure that at least one of the survivors would be a boy.

The explosive increase in human population has paralleled our explosive cultural evolution, in three major successive stages. The first probably occurred at the dawn of human evolution and corresponded to our initial invention of rudimentary culture (language, fire, etc.), which among its various effects also enabled greater survival for our offspring. The second, about 10,000 years ago, coincided with the invention of agriculture. This was a cultural advance of fundamental significance, transforming the bulk of humankind from a hunter-gatherer to an agrarian economy. From dependence on nature's whims, we achieved substantial control over our nourishment. Agriculture gave us predictable food supplies and made possible the concentration of population into cities, where the surpluses allowed individuals to specialize in other, nonagricultural pursuits. The population zoomed.

The third and greatest spurt began with the scientific revolution of the sixteenth and seventeenth centuries and has continued up to the present, having been fueled by the Industrial Revolution and more recently, the great hygienic and medical advances of the last 100 years. The human population is currently on a collision course with our ability to maintain social harmony in our overcrowded world, and with the basic physical and biological resources of the earth.

The Kaibab deer experienced their drastic population increase because their most important source of mortality, their predators, had been eliminated. Their reproductive capacity had always been high, geared to an equally high death rate. Like in an automobile that had always been driven with the brake on, the accelerator was also floored, just to keep the car moving. With the brake suddenly released and the gas pedal still pressed to the floor boards, the vehicle went out of control and crashed.

We have a high natural capacity for reproduction, also geared to a primitively high death rate. In human history, the most important natural braking agents have probably been starvation and disease rather than predation. When our ancestors began perfecting culture, becoming more efficient hunters, gatherers, and food preparers, they greatly reduced their losses from starvation. The invention of agriculture had a similar effect, while also permitting human survival in previously uninhabitable or difficult areas. In the last 100 years phenomenal strides have been made in our long battle against disease. Just as with the Kaibab deer, we are eliminating our cougars and wolves, only for us they are spelled malaria, cholera, and typhus. And just as with the Kaibab deer, our numbers are now skyrocketing; but with life just as with gravity, what goes up has a habit of coming down.

Population biologist Paul Ehrlich has emphasized that the Western world in particular has been practicing and exporting death control. On balance, human mortality has been greatly reduced, especially in the Third World, where it had always been appallingly high. There is, of course, nothing wrong with this; it is laudable. Starvation and disease are still two of our ugliest enemies and eliminating them must rank among the noblest human endeavors. But because the human population has increased beyond the earth's ability to provide consistently and safely for Homo sapiens, we witness the periodic, devastating famines such as those wracking Africa in the 1980s. Moreover, the long-range prospect for supporting an artificially inflated human population is bleak at best, and perhaps utterly impossible. To substitute one form of death for another, resulting from ultimate overpopulation and therefore likely to affect an even larger number of people, is irresponsible. The alternative, fortunately, is simple enough: continue our humanitarian efforts toward death control (in fact, increase them) but also simultaneously export *birth control.* As we ease up on the brake, our only hope for a smooth ride is to gradually release the accelerator as well.

We are faced with a social problem that would not occur if we weren't cultural as well as biological creatures. Thus, while our culture provides death control, our biology is still tuned by evolution to the production of a large number of offspring, in anticipation of the high mortality which is now largely averted. Under conditions of natural selection, analogous changes in mortality might eventually be reflected in an altered reproductive pattern. But this would take time on an evolutionary scale—thousands of years at least—whereas drastic cultural developments are occurring in decades, or less. As with the problem of human aggression, our population problem was born when cultural evolution left biological evolution far behind, and it must now be resolved by cultural forces acting alone. We simply can't wait for our genes to do it.

Fortunately, birth control might actually cost less than the present technology of death control while ultimately being no less important. A successful worldwide birth control program would have to overcome the opposition of certain religious groups as well as the rather paranoid (but sometimes justified) suspicions of certain ethnic groups—notably, many American blacks—who see birth control as a clever guise for racial genocide. Furthermore, it may have to buck a genetically motivated human desire for large families, originally instilled by natural selection long before our culture discovered death control. But when it comes to behavior, our biology is frequently subordinate to our culture, and fortunately it appears that strong culturally induced motivations for limiting family size can overcome any such biological tendency. Cultural evolution in this case offers not only the problem but also the solution. Birth control pills, condoms, IUDs, diaphragms, and sterilization surgery are all the products of human culture, which—if we can summon the requisite wisdom—permit us to thumb our collective noses at our genes.

The liberation of human sexuality from its purely reproductive function, discussed earlier, should facilitate use of contraception. In addition, as economic development proceeds, having too many children becomes an increasing social and financial burden for individual families. A hopeful and consistent phenomenon in recent years has been a tendency for birth rates to decline when socioeconomic conditions have improved. This is a purely cultural happening, the so-called "demographic transition," and it appears to be based on enlightened self-interest. A worldwide effort under the auspices of UNICEF has begun making significant progress in controlling juvenile diarrhea, the number one killer of young children. A double benefit seems to result: not only is suffering reduced immediately, but with their children no longer dying so readily, adults are more inclined to use birth control, thereby reducing suffering in the long term as well.

A similar shift often occurs among animals, when their situation changes such that parents are more successful producing a relatively small number of offspring and investing more heavily in each one, rather than producing a large number of offspring, and giving each one only a very small headstart. In peasant or hunter-gatherer societies, additional children are additional farmers, hunter-gatherers, and/or warriors. And when they die—as they often do—they are replaced. Among technologically advanced societies, by contrast, additional children represent a substantial drain on family resources, since each one requires heavy investments by way of medical care, clothing, and education, while there is relatively little immediate domestic productivity that they can contribute.

If our culture fails to respond to the threat of overpopulation, or if that response is inadequate, our biology just may do the job for us. The populations of many animals are limited by factors that ecologists call "density-dependent." This simply means that as the population increases, a higher number—or more effective yet, a higher percentage—of individuals die. It's a bit like the graduated income tax, with population density substituting for annual income and mortality substituting for the tax liability. Under density-dependent control, mortality increases as the population density increases and it decreases as the density declines. Some species, however, such as grasshoppers, appear to be limited by "density-independent" factors: death rate in such cases varies without regard to the actual population size. Thus, grasshopper numbers may be determined by environmental factors like drought or rainstorms that kill a relatively constant number of individuals, regardless of the total population. But it is the density-*dependent* species that are of greatest interest to us. These are the living things that seem to regulate their own population size. Analogous to a progressive tax structure, species of this sort experience a progressive death structure: "wealthier" (more abundant) populations suffer a proportionately higher mortality.

Density-dependent systems are examples of what engineers call "negative-feedback loops." The idea is really very simple. A positive feedback (the opposite) is what we colloquially call a "vicious circle," a situation in which some deviation from the norm results in a further deviation, with that deviation causing one larger yet, and so on. "The rich get richer" is an example of positive feedback ("and the poor get children"). A nuclear chain reaction is another example. Negative feedback, however, is more intellectually appealing, involving as it does a wonderfully balanced system, preset to absorb deviations and promptly return the system to an acceptable level. For a mundane and frequently overlooked example, consider the home thermostat. It may be set for a certain range of temperature, say sixty-five to sixty-seven degrees: if the house gets colder than sixty-five, the furnace turns on and warms it up to the acceptable range. Should the temperature exceed that range, the furnace shuts off, allowing the house to cool down. A really sophisticated negative feedback system might also incorporate central air conditioning, which would be activated if the temperature got too high.

Biological systems possess a vast array of negative feedback mechanisms, all acting to guarantee the precise physical and chemical conditions needed to maintain life. For example, our blood and body fluids must be kept within extremely narrow limits of acid and alkali if we are to avoid dying in convulsions. A more obvious example would be the maintenance of internal temperatures by a series of mechanisms similar to a household thermostat. Because we are warm-blooded animals, a change in body temperature of more than a few degrees in either direction could mean a coma and death. We unconsciously raise our temperature by shutting off the small blood vessels that lead to the skin, thus preventing heat loss, while rapidly contracting our muscles—shivering—to generate heat. If on the other hand we are getting overheated, more blood is pumped to the skin, where excess warmth is radiated into the air. At the same time, we also exude water onto our body surface—perspiration—which evaporates and cools us even more.

Analogous negative feedback processes operate in density-dependent control of animal populations. Flour beetles thrive in a mixture of dried cereals and grains, as the wholesale grocer is well aware. Left untouched in a feed bin, a small number will multiply rapidly. Eventually, the population will stop growing, in part because their own accumulating wastes are harmful to the animals; as the numbers increase the amount of waste increases likewise. There is always the outside possibility that something similar could occur among Homo sapiens. Thus, as toxic products such as mercury, lead, and PCBs accumulate in our environment, they could eventually interrupt our reproductive processes. And the industrial disaster in Bhopal, India, where thousands died, shows that even without war or widespread ecological collapse, modern civilization has some appalling "downside risks."

Aside from doing themselves in via their own toxic byproducts, flour beetles also kill each other directly. They can be cannibals, eating the young larval mealworms when they get the chance. Normally, in a sparse population, they don't get that chance very often; but as the population grows, individuals are more and more likely to bump into each other, with lethal consequences for the juveniles.

Something similar happens among grizzly bears. Male grizzlies, in particular, kill cubs. This is why the females chase them away before the birth of their young and why a sow bear with cubs is so dangerous to humans. She is highly protective of her young and evolution has told her that they are in danger. When the grizzly population is low, males only rarely get a chance to destroy many cubs, but should the population become inordinately high, we might expect this density-dependent mechanism to take its toll of the young bears, which in turn would restrict further population growth. It seems unlikely, by the way, that male grizzlies—any more than adult flour beetles—are consciously seeking to reduce the local population. Rather, they are probably just taking personal advantage of whatever opportunities come their way. To some extent they may also be following the dictates of selection, which would reward individuals who eliminate potential competitors, despite the fact that in some cases they may be killing their own genes. The overall population effect of such behavior is probably incidental, although no less real as a result.

Both these examples portray density-dependent population control in which adults attacked young. For a final example, in which adults attack other adults, let us journey to the cold waters at the bottom of the North Sea, and look briefly at the lives of the hard-shelled crabs that crawl about down there. Each animal is normally protected from its fellows by a strong external armor. But periodically it must molt, leaving the soft body dangerously exposed. If the sea-bottom population density is sufficiently low, a freshly molted crab will survive its brief susceptible condition. But if the animals are crowded, soft-bodied individuals will be found by their more numerous and invulnerable colleagues, and many will be eaten. High local populations thus tend to decrease.

It is unlikely that such mechanisms have ever operated to control human populations, and certainly no one predicts that cannibalism will exert any real effect on human numbers in the future. On the other hand, aggressive violence between people may be at least partially a function of population pressures. If higher population resulted in more violence, which in turn increased the death rate, we would have a straightforward density-dependent system that could (at least in theory) produce a relatively constant population size. There are several possible pathways linking population density and aggression.

Aggression is more likely to be expressed in a dense population, in which people are constantly bumping into the potential objects of their aggression. Any possible territorial instinct, however unlikely, would be severely stressed by increased crowding. Population density also seems likely to make social harmony more elusive than ever. It is probably no coincidence that our largest urban area, New York City, is generally acknowledged to be ungovernable. And as we have seen, the breakdown of social order has often been cited as a cause of aggression. It is significant that statistics for violent crime tend to be highest where population is the most dense. Furthermore, high populations tend to exaggerate demands for public services, while accentuating the discrepancies between rich and poor. Frustration rises, and with it once more, the tendency for aggression.

Despite these apparently tidy correlations, there is still no hard evidence for an aggressively mediated density-dependent control of human population and even if it was an imminent possibility, it wouldn't be desirable. In fact, none of the density-dependent control mechanisms inculcated in different species by biological evolution can hold a candle to the possibilities of cultural control, because such control would ideally involve *preventing* the problem, rather than dealing violently with it.

Although ecologists are divided as to the full significance of density-dependent controls in nature, they are generally agreed that to some extent at least, the size of a population influences the number that die, even when the lethal factor seems to be independent of density. This also applies to human populations. For example, if a drought or some other vagary of climate restricts the amount of food so that only 50 deer can survive on an isolated island, and 51 are present, one unfortunate must die. Similarly, if there are 200 deer then 150 will starve. This is density-dependent in a sense, in that the greater the population, the greater the number of individuals that die. The human parallel is appalling: if during conditions of periodic famine, a nation can support only 50 million people, but has been artificially inflated to a population of 200 million through massive death control without commensurate birth control, we can expect 150 million deaths when push ultimately comes to shove. And the larger the population, the more people affected.

It is also possible—and in some cases, more likely—that the costs will simply be spread throughout the population, with the survivors all tightening their belts somewhat. Everything we know about human behavior suggests that to some extent this will happen; on the other hand, everything we know about human behavior also suggests that sacrifices imposed by resource scarcity will not simply be distributed equally.

For many animals the environment dictates that only a limited number of individuals will be successful. In a territorial species requiring, for example, two acres per breeding pair, a ten-acre plot can accommodate only five pairs. The surplus, if any, is out of luck. Bobwhite quail generally have a hard time getting through the winter, but those that can obtain adequate shelter usually survive. Since there are usually more birds each fall than their environments can safely accommodate, a certain excess can be expected to die off each year. The environment sets the numbers that can survive and any excess usually perishes. The larger the population, the more animals are forced into suboptimal habitats and the higher the rate of mortality. When such displaced individuals eventually die, it is frequently "because" of predators: in the case of bobwhite quail, great horned owls or foxes. The predators are therefore often blamed for the mortality, whereas actually it is the prey species's environment, relative to its population size, that determines how many will die each year.

Sadly, a similar process occurs in human beings, although minus the predators. In the early 1970s, a typhoon in East Pakistan (now Bangladesh) claimed perhaps 500,000 lives. This terrible human loss may have seemed at the time to be a direct result of natural climatic factors, not in itself related to population size. But in fact, the Bengalis were victims of their own population density: the treacherous Gangetic Delta should never have been populated by so many people and indeed, would not have been if the whole area was not so horribly overcrowded. Like excess quail in winter, doomed to an inadequate habitat, millions of people were forced onto treacherous terrain by the very existence of their compatriots. Similarly with drought-stricken Africa, in which the population density, while less than that of Bangladesh, is nonetheless too high for the resources available.

Among most animals, competition tends to cull out the genetically less fit—since the "better" individuals can be expected to secure a safe place for breeding and survival—just as predators are more likely to eliminate the old and the sickly. But among humans, with our effective and ever increasing cultural buffering, there is virtually no chance for such selection to take place. Given the enormous potential present in every human being, and the equal enormity of personal pain and sorrow, only a gene-obsessed ghoul can take comfort from the notion that natural selection is somehow improving the species by such disasters. But even if such changes were theoretically possible, and even disregarding outraged ethics and morality, human living arrangements have virtually made such efforts biologically impossible: Imagine a child with fortuitously excellent gene combinations, born to an impoverished peasant family where she is one of nine children. There is virtually no real opportunity for this child to escape the perilous life on the Delta or on the arid Ethiopian plateau to which population pressures have driven her family. When the flood comes or the famine descends, she will be swept away, part of a human surplus, at the mercy of nature and an awful kind of density-dependence. How many heirs to Tagore were lost in that typhoon? How many Kenyattas have already starved?

Among many animals, numbers are also restricted by disease and parasites. Our peculiarly anthropocentric viewpoint often overlooks the fact that microbial pathogens are not a problem for our species alone. They are often the major factors controlling population size, especially among most free-living primates, for example. The effect here is once again somewhat density-dependent, since contagion is more easily spread in a dense population than in a sparse one, and overcrowded, malnourished individuals are more susceptible to disease.

If you were a parasite or disease-causing organism, death of your host would be a serious affair. You and your offspring would die along with your former benefactor unless there were opportunity to infect others. Disease outbreaks are therefore most likely when the susceptible population is large. The infection "runs its course" until the remaining individuals are those with generally higher resistance, and the population has thinned out sufficiently so that they are unlikely to contaminate others. There is a strong correlation between local population density in human beings and the chances of contracting disease. Who hasn't been exposed to the annoying cough and sneeze in a crowded bus or theater? And all parents know that their children are more likely to get sick when they attend nursery or public school, where they are exposed to other children. Many parasitic diseases such as liver flukes and roundworms are spread through human feces and as population density rises, so does the incidence of such debilitating diseases, unless strict public health measures are enforced. Because of the artificially high local population densities made possible by our culture, we are sitting very high on a precarious pinnacle of sanitation and food-producing technology, quite possibly on the verge of some drastic density-dependent controls. Temporary failures of our complex technology, due to floods, hurricanes, earthquakes, war, or simple excessive demands because of overcrowding itself, bring immediate local threats of such diseases as typhoid and cholera.

We must not overlook the effectiveness of modern medicine. But as human population increases, the need for medicine to counter this potentially density-dependent factor will clearly increase. We also should not overlook another culturally inspired aggravation, due to worldwide transportation. Because of the ubiquity of airplane, railroad, boat, and automobile, a disease originating in some obscure corner of the world, where the local population may be somewhat immune, can quickly spread throughout the globe.* Such relatively minor Old World diseases as measles and chicken pox, carried by explorers and missionaries, devastated many of the nonresistant native people of the Pacific and the New World. More recently, "Hong Kong flu" and other illnesses have spread far beyond the regions that would have experienced them in earlier days. And whereas AIDS, for example, may conceivably be eliminated by prompt attention and Herculean efforts, it is also possible that it will spread throughout the world, whereas in previous eras it would probably have remained geographically isolated for generations.

There is also growing evidence that many animals respond to increases in their population density with changes in their behav-

*Syphilis was unknown in the Old World until 1495, when Columbus's crew—and quite possibly, the Great Explorer himself—docked in Genoa, Italy, and began assiduously infecting Europe with the fruits of their various adventures in the New World.

ior and physiology that ultimately have the effect of reducing their numbers. Once again, there is no reason to believe that such density-dependence represents foresight on the part of individuals or their species; rather, individuals (and individual genes) may maximize their personal reproductive success under conditions of crowding by behavior that, incidentally, tends to reduce the population. The relevance of all this to human beings is unclear, but challenging. John C. Calhoun experimented with rats, observing their responses to different population sizes. He placed five pregnant rats in a quarter-acre outdoor pen and provided sufficient food and water for the expanding population. During the ensuing two-year period, the five pregnant females could theoretically have produced 50,000 offspring, but the population never came close to that. In fact, it never exceeded 200, and eventually stabilized at about 150 animals. In a laboratory setup, with rats isolated from each other in small, self-contained cages, 50,000 animals and more could easily have been maintained in the same area. The big question then: how did these rats in a semiwild state keep their numbers so low?

The answer was found in their social behavior. They organized themselves into a dozen bands of about twelve or thirteen members each, with a single dominant adult male leading each band. Fighting between bands was common, and often disrupted normal care of the young. This in turn caused a high mortality among the pups and prevented the population from increasing. In fact, even the final population of 150 was an uncomfortably high number, maintained only by artificial feeding and the inability of the animals to escape and thus spread out. In nature, the population of an equivalent area would have been considerably smaller yet, not necessarily as a result of disease, predation, or cannibalism, but rather, simply because of the social behavior of healthy animals.

The rats in this experiment had apparently engineered a rather precise form of density-dependent population control that kept their numbers below the level at which more serious disruptions would be expected. But exactly what sort of disruptions? What would their behavior be like if they were constantly bumping into each other? To test this, Calhoun began his most famous experiment, in which most of the rats were eventually exposed to a population density about twice that which had been experienced by the penned animals in the previous test.

The results were striking. Under these high densities, the generally cohesive organization of normal rat social life broke down almost completely and with this collapse came heightened aggression, physical disabilities, and high mortality of adults (particularly females) and young. An epidemic of tail biting erupted, and younger animals were often seriously injured. Most important, normal sexual behavior and care of the young were greatly disrupted. Rats normally make hollowed-out nests in which the young are born. The crowded animals, by contrast, generally made inadequate nurseries, usually because they failed to collect the proper nesting materials. They were also sloppy housekeepers, often dropping their bedding and arranging it poorly. Because of this, the young frequently became scattered at birth and very few survived.

The crowded females were also very poor mothers, nursing inadequately and taking indifferent care of their offspring. For example, mother rats will normally retrieve their young if they are scattered about, usually transferring them to a new nest if the old one is disturbed. The crowded animals, by contrast, showed very little retrieval. They often dropped their young while transferring them, and frequently left them where they fell. These unfortunate pups were then commonly eaten by marauding bands of male rats. Such destructive cannibalism is never found among a troop of free-living animals, whose population invariably remains at a more manageable level.

Sexual behavior among rats normally entails a predictable sequence with mating success dependent on careful following of the correct routine. Males identify females by their odor, and are particularly sensitive to whether they are in heat. When the time is right, the male chases the female, who runs into her burrow, only to peek outside and watch the male, who responds by doing a little jig. When this is finished, the female reemerges and copulation finally occurs. Males in the overcrowded situation often failed to observe the basic amenities of rat courtship, following females into their burrows and chasing them in groups, behavior reminiscent of gang rape in human beings. With their sensitive reproductive mechanism fouled up, abortion and death among females was unusually high. This also retarded further population growth.

It is particularly interesting for students of human mental illness that the "sick" behavior that developed among these overcrowded animals was not homogeneously distributed throughout the population. Different individuals developed particular pathologies; four different categories of abnormalities were observed. The few dominant and moderately aggressive males appeared relatively normal. Others became "hyperactive," harrassing females and eating their young. Some became pansexual and mounted anything: males, females, young, old, receptive or unreceptive. One group was passive, taking no interest in sex or fighting, while the last group withdrew entirely and traveled about only while the others slept.

Calhoun's rats may be sending us a message: when social order breaks down, it seems likely that our worst biologic potential will express itself. Unnaturally dense populations can produce social abnormalities that ultimately bring down the population size. Assuming now that it is valid to extrapolate from rats to people—and this is something that must be done cautiously, although it is done routinely in biomedical research—then such density-dependent biological controls may be painting a grim future for a world of unrestrained human population growth, even if we are successful in avoiding catastrophic population collapse. If we are sufficiently wise, however, or sufficiently lucky, we will never know if our biology has endowed us with the capacity for similar responses.

We do not know the psychological basis for the various behaviors of Calhoun's rats. However, other studies have suggested a general mechanism that may have wide relevance. Biologist John Christian discovered that conditions of social stress, such as those brought on by local overpopulation, result in excessive demands on the adrenal glands. Sitting quietly above the kidneys, these small structures regulate a variety of essential chemical processes within the body, including mobilization of sugar reserves and resistance to infection during stress. If called into constant use, the adrenals grow. The demands placed on the body by their excessive functioning can result in shock and eventually adrenal exhaustion and death. The best example of this comes from a study that is now several decades old, but unsurpassed in its completeness and in its possible message for an increasingly overcrowded world:

In 1916 a few deer were introduced onto St. James Island, several hundred uninhabited acres in Chesapeake Bay. Their numbers soon increased to about 300, clearly an unnaturally high population. Fortuitously, Christian was already studying these deer when over half the animals died in 1958. The dead animals seemed paradoxically healthy except for one thing: their adrenal glands were greatly enlarged. By 1959, with the population at more normal levels, adrenal size was also back to normal.

It would be interesting to compare the adrenal size of human urbanites with that of rural residents. On the other hand, given the capacity of human beings to adapt, it is entirely possible that the resident of Chicago is no more stressed by his or her environment than the inhabitant of Maine's north woods; moreover, we seem to have a remarkable aptitude for stressing ourselves, no matter how or where we live.

If increased population size increases the stress on individuals, ultimately causing overwork of the adrenal glands and a higher death rate, this wholly biological mechanism could operate as a density-dependent check on population growth. The catalyst here would be personal stress, though, not population size as such. A population that is numerically dense could conceivably avoid stimulating the adrenal mechanism so long as the individuals concerned were not stressed by the presence of their fellows. Similarly, an objectively low population density could activate the adrenal stress mechanism if the individuals involved were somehow very sensitive to each other. It's all in the perception of crowding, not crowding itself. Christian discovered that adrenal size among woodchucks, for example, was greatest not when the population was highest, but rather when aggressiveness among individuals was most intense. This occurred early in the spring, at the mating season, and then again later in the summer, when the dispersing young were seeking to carve out living space at the expense of the adults.

We might anticipate that the symptoms of stress would be greatest among individuals at the bottom of the social hierarchy. After all, they are being "dumped on" by everybody else while unable to retaliate. In fact this has been shown repeatedly: there is a consistent correlation in many animal species between low social rank and excessive adrenal size. Again, possible correlations among humans have not been investigated, although a study of stress-related illnesses among employees of the Bell Telephone Company found that individuals lower in the company hierarchy, such as linesmen and operators, had a higher frequency of such illnesses than did members of the Board of Directors. We might have to revise our conventional wisdom regarding the "harried executive" and the happy-go-lucky workman.

Many animal species show another interesting phenomenon, relating social status to stress and ultimately in some cases to the regulation of population size. Low-ranking individuals are frequently denied access to females, either because they have been unable to establish adequate territories and are therefore not among the eligible reproductives, or because the females simply show a preference for high-ranking mates. This is especially true among animals such as grouse or turkeys that are polygynous (one male servicing many females). Under these conditions, subordinate males are truly personae non grata and may never copulate; in fact, they may actually be unable to mate even given the opportunity. Studies of subordinate males have revealed that among many species, testosterone levels are considerably reduced and testes are greatly shrunken by comparison with the dominant, actively copulating males. Ethologists refer to this phenomenon as "psychological castration."

Among the monkeys and apes, arrangements vary from species to species, making generalizations hazardous, but a clear analogy of psychological castration is not usually found. In fact, overt sexual rivalry among males is rare, and even among the despotic baboons, subordinates get to copulate. During the peak of estrus, however, when a female is most fertile and intercourse is most likely to produce offspring, copulation rights are largely monopolized by the dominant adult males.

Humans differ from most of the well-studied primates in living monogamously and ostensibly pairing for life. Since there are approximately equal numbers of men and women, our family system virtually assures that each male can have the opportunity not only to copulate, but actually to reproduce. There is therefore little chance for psychological castration among humans. However, the increasing problem of "psychological impotence" may be very similar in that it is often brought on by the stresses of modern life in which people are subordinated to goals and systems, if not to individuals as well.

Two British investigators, A. S. Parkes and H. M. Bruce, once noticed that when a recently pregnant female mouse was exposed to a strange male, her pregnancy was often interrupted by a premature abortion. It was subsequently discovered that the actual presence of the male was not necessary for this effect: a pregnant female will abort if she is placed in a cage where a strange male has been. This suggested that odor was responsible, a hypothesis that was later confirmed when females had their olfactory lobes surgically removed. Pregnancy in these nonsmelling animals was no longer interrupted by strange males.

It is not known to what extent such a pregnancy block mechanism operates among free-living animals, if indeed it does at all. However, it could indicate another density-dependent population factor, since as a population rises, the chances of pregnant females encountering strange males—or at least their chemical calling cards in urine or feces—would rise as well. For such individuals, it would probably be adaptive to refrain from a breeding attempt when the social system is disrupted and behavioral pathologies such as we have just described are likely. Given that most mammals live in well-defined social networks, the presence of a strange adult male could indicate that such disruption is under way.

Operation of the "Bruce effect" clearly requires a good sense of smell. Human beings, however, are unusual among mammals in possessing a poorly developed sense of smell. It is therefore very unlikely that odor-induced abortions would be activated by increases in human population. In fact, our relative insensitivity to smell has probably given us the potential of enduring denser populations more easily than most mammals. The stress generated by population density must be relayed to the animal by its senses. And smell, so well developed in most mammals, probably serves as a major avenue informing each individual of the immediate degree of crowding. Since trees are airy, windy places, long-distance olfactory communication has not been perfected among primates, and anatomically speaking, we are rather ordinary primates (discounting our enormous thinking cap, of course). Without a good sense of smell, we can probably experience moderate congestion with comparatively little stress. With it, we might well find auditoriums, elevators, buses, and probably our cities as well, utterly intolerable.

The various biological density-dependent controls to overpopulation such as cannibalism, aggression, automatic abortion, and stress cannot be acceptable to human beings. But for most animals and their component genes, they are ways of making the best out of a bad situation. That's why they have evolved. We could possibly derive comfort from the thought that such density-dependent factors might not operate among our species. But if we truly lack these controls, and also fail to exercise alternative cultural controls, we may be in even greater peril, since the various density-*independent* controls are in the long run no more pleasant or acceptable. Once again, we may be more endangered by the instincts we lack—or that we possess but can easily override—than by those few that we possess.

Lemmings often do something drastic when their population gets high. However, their famous "suicidal" rush to the sea is misunderstood. Lemmings do not actually commit suicide; rather, when the population density becomes excessive, they emigrate in large numbers, often swimming rivers if necessary. If instead of a river, lemmings reach the ocean, they treat this lethal barrier as though it is just another surmountable one. Perhaps if we were more subject to density-dependent effects, or if our culture had not progressed so far ahead of our biology, we would have a stable population today. As it is, we unfortunately resemble the lemmings, not in rushing to the sea, but in treating new conditions as though they are just like other ones, which we have always surmounted in the past. Suicide need not be intentional to be real.

Some animals do seem to practice an effective form of birth control. In the northern Arctic, snowy owls and the equally predatory, gull-like jaegers apparently adjust their clutch size to the amount of prey available. These animals feed on small mammals such as lemmings and mice, which as we have seen experience periodic fluctuations in numbers. The birds produce more offspring, sometimes even "double clutching" in times of plenty (a lemming outbreak), and fewer when food is more scarce. The small English birds known as "tits," relatives of our chickadees, feed on insects and require an abundant summertime prey population to obtain enough food for their ravenous young. Accordingly, they also adjust their family size to the amount of food available. Their eggs must be laid in the spring, however, when very few insects are out, in anticipation of conditions later in the summer, when the young have hatched and must be fed. How is this prognostication achieved? Insect populations later in the year are determined by temperature and moisture conditions during the preceding spring, and apparently these clever birds use spring weather as a cue, telling them how many eggs to lay. The schooling, once again, is by natural selection, which promotes the successful and fails the dunces.

These are appealing solutions to potential overpopulation, because they involve preventing overproduction rather than eliminating an excess. However, they are clearly automatic and biological, part of the animals' hereditary constitution. There is no reason to believe that we are similarly endowed. If we are to be similarly efficient, then we must do it via our culture.

We might consider a species to be a bathtub, with the water level indicating population size. The faucet is on and water is flowing in (births). But the drain is also open and some is going out (deaths). When the two are balanced, the water level remains constant; engineers refer to this type of system as a "steady state." Human culture has given us the partial ability to close off the drain: since we have not also turned down the faucet to compensate, the water is rising. Dangerously. Many species possess automatic faucet adjusters, or safety drains (density-dependent mechanisms) that keep the water from getting too high. If, as appears to be the case, we lack these devices, then the tub will certainly overflow unless we use cultural means to turn down the faucet. Paul Ehrlich has suggested that we can administer an intelligence test by seeing how someone responds when confronted with an overflowing bathtub: does he or she rush about with bricks and mortar, building up the sides, or, instead, turn off the faucet?

Increases in food production, medical care, pollution control, and improved housing are clearly helpful but are not permanent solutions. They are based on increase and growth, whereas ultimately we must attain equilibrium. Nothing else will do. We are adjusted for short-term success under conditions of resource abundance and high mortality, not for long-term equilibrium under conditions of resource shortage and low mortality. Moreover, because of our long evolutionary history as expansive creatures with an expanding population, we respond poorly to equilibrium as a conscious goal. Those who warn of the limits to growth seem dour, gloomy, and glum, as opposed to the optimists who cheerfully espouse "growth, growth and more growth," as Senator Paul Laxalt announced to the Republican National Convention in 1984.

Nowhere in the natural world is unrestrained growth a prescription for ultimate success, although it is a good description of cancer. Carried far enough, growth, growth, and more growth can lead only to death, death, and more death.

Under strictly biological conditions, different species can be imagined pushing as hard as they can against each other, and similarly for individuals within each species. With everyone pushing hard, anyone who lets up loses out. Moreover, no one suffers ill-effects for behaving as though unrestrained growth is the goal, since it can never be achieved. But thanks to cultural evolution, we have eliminated much of the natural resistance that our own potentially expanding numbers would otherwise have encountered. Our cultural evolution has been notably lopsided in this respect, since it has given us not only the science and technology to prevent early death, but also the religious and other cultural prescriptions to "be fruitful and multiply" as God is reputed to have enjoined Adam. Moreover, given our evolutionary history (both biological and cultural) as a species, we are especially susceptible to such advice. As a result, there is very little at the moment for us to push against, so that if we don't ease up—and wise up—we must eventually fall on our faces.

CHAPTER TEN

ENVIRONMENT

From Monkey Manure to Ecological Ethic

From now on it will no longer be enough to ask if man can do something. We must also ask whether he ought to.

—David Brower

As a description of our snowballing, deeply technologic, and not very thoughtful style of galloping cultural reliance, the word "harebrained" is particularly apt. If we don't destroy ourselves by war, our harebrained schemes may bring down the curtain on the human adventure simply by making our planet uninhabitable during peace, with an unhealthy assist from overpopulation. We are in danger of choking on the millions of tons of air pollution that darken our skies; certain rivers are so grossly contaminated they have been declared fire hazards; toxic waste dumps alone contain enough poisons to end life on earth. The litany of human-induced environmental hazards goes on: salinization, erosion, deforestation, desertification, acidification, artificial climatic changes from nuclear winter and ozone depletion to an overheated greenhouse effect with melting of the polar ice caps. We cannot feed the people now on the earth, yet the population grows ever larger. Wilderness—once an enemy, now a beleaguered friend—shrinks daily, never to be reborn. Endangered species slip away, forever.

And even if by some miracle, we are able to maintain a quantity of human life in the future, the quality of that life seems ineluctably diminished as we diminish the quality of the natural world. There is a special value in forests and mountains, running brooks, and quiet ponds, a butterfly in the spring and an elk bugling in the fall: these are the stuff of life, and that means human life, too. Let us therefore briefly examine the origins of our present environmental crisis, because it is a deeply human crisis as well.

Living things do not live in a vacuum. Neither do they evolve in one. The circumstances of existence for all animals and plants relate them profoundly to each other, so that the evolutionary advantage or disadvantage of any characteristic must be evaluated in terms of the organism *and* its entire environment. Thus, the speed of an arctic wolf is relevant particularly to the speed of its prey, the caribou, while the high-crowned tooth patterns of a horse are attuned to the abrasive nature of grasses, its preferred food. The intricate harmony of the natural world is due to elimination of misfits, combined with elegant elaboration of the successful forms and their interconnections. By most measures, Homo sapiens has been among the most successful of living things. Our population, after all, is high. Our mortality rates are very low. We occupy an extraordinary range of environments. We manipulate vast quantities of energy and materials. We offer an image of glowing evolutionary success. But present-day success is very different from future persistence, as the dinosaurs would tell us if they could.

Those dinosaurs, so often derided as failures, were great evolutionary successes in their time. And that time lasted more than a hundred million years, much longer than our own ascendancy so far. Moreover, despite their ultimate extinction, it seems that the dinosaurs were somehow "natural" in a way that we aren't. Most organisms, dinosaurs included, fit into their environment because those that didn't were promptly eliminated, as they still are today. Those that are left do not only belong, they are *part of* their environments.

Of course, this is not to say that species are immortal; they clearly aren't. Over long periods of time, extinction is the rule, as species are unable to keep up with changing environments. Something analogous can be seen in a fairly short timespan as well: consider, for example, the progressive changes occurring when a hardwood forest in the Northeast is cleared by lumbering or fire. The initial plant growth consists of grasses and various annual "weeds," followed by larger herbs and then shrubby vegetation. Finally after many years, an oak forest may appear, but since the large trees screen out much of the sunlight, the ground is dark, which prevents successful growth of young oaks to replace the old-timers. But maple seedlings are comparatively "shade tolerant" and are able to grow in the dark forest floor. They may eventually replace the oaks and a mature maple forest finally results. This final stage, ecologists refer to it as the "climax" vegetation, is self-perpetuating and will continue until human beings or some other disaster strikes once more. Then it all begins again.

Each of the progressive "successional" stages through which our forest passed possess characteristic plants and animals that eventually make conditions unsuitable for their continuance and so are replaced by the next stage. They literally carry the seeds of their own destruction. Of course, this elimination occurs on a local level only and is not true extinction. The various species involved generally survive, perhaps in another successional stage somewhere else. If the human species is creating an increasingly unlivable environment for itself, perhaps the same thing is happening to us, but with the notable exception that if the whole planet is rendered uninhabitable, we must all go extinct. Space fantasies aside, there is nowhere else to go.

We are unlikely to look on ourselves with equanimity as a passing stage in the progression of life, even though we have only been here a very short time, by the evolutionary calendar. Even aside from our very understandable emotional involvement in our own fate, something rings false about human extinction—and the human-caused extinction of other living things—as "natural." For one thing, there appears to be something peculiarly *un*natural about what we are doing to ourselves and to our world. We don't quite fit into the neat, interlocking system that governs other living things and their destinies. When we disrupt the lives and the futures of other living things, it therefore seems very different and much less acceptable than when they disrupt themselves. But why? In what sense are we "less biological" than a dinosaur or an oak?

The answer seems obvious: we are every bit as biological as any living thing, in that we were produced by natural selection and are subject to certain basic laws of the organic world. At the same time, however, we are also creators and creatures of cultural evolution. Once again, our glory is also our problem. And it is the crux of our present environmental crisis.

In a sense, we destroy our environment because we are able to. No other species of animal or plant poses a threat to the integrity of the earth, because none has the means to do so; any that made their environment uninhabitable for themselves are therefore no longer around today. As with weaponry, our destructive capacity does not derive from our simple biological characteristics. As animals we make only a small dent in natural communities, and, except for our extraordinary numbers, from the neck down we are interesting but unremarkable animals. But with the addition of tools, division of labor, language, and higher rational and technological capabilities, we have been *exploiting* nature in a way never seen before. Without cultural evolution, this would not be possible. With it, we have accomplished most of the things we prize as making us particularly human. And with it, like unruly Samsons, we threaten to bring down the earthly temple, crashing upon our heads.

Whenever animals and plants went extinct in the course of geological time, they did so because they could not adapt to a changing environment, not because they evolved themselves into extinction. Species suicide is unknown in nature, quite simply because the characteristics of living things are the result of biological evolution. Any genetically influenced tendencies that lead to decreased reproductive success would necessarily be selected against in the course of evolution, to be replaced by more benign traits. In the course of biological evolution, then, extinction occurs only as a result of happenings beyond the species' control.

There once were deer ("Irish elk") that possessed enormous antlers, exceeding ten feet in length. According to an earlier view, these animals went extinct because their antlers grew too large, getting hopelessly entangled in vegetation or perhaps so weighting their heads that the animals couldn't see where they were going and bumped into one another or walked over cliffs. This is very unlikely. If certain individuals had begun developing antlers that were disadvantageously large, they would have left fewer offspring than their more appropriately endowed brothers, and the average antler size would have decreased (a sort of negative feedback, evolution-style). More likely, environmental changes did in the Irish elk.

By contrast, human environmental abuse—whether indirectly as a result of population growth, or directly as a result of destructive and polluting technology—is not subject to the typical biological controls. It is mediated by our culture, not by our genes. Furthermore, even if the tendency for environmental destruction was under genetic control, as it would be in animals subject only to biological evolution, the pace of our culturally induced environmental destruction is much too fast for differential reproduction to have any hope of restoring an equilibrium. Cultural evolution has provided us, almost overnight, with the tools to destroy the earth, while once more (as in the case of aggression) denying us the kinds of inhibitions needed to restrict their use. Biology, acting alone, would never have granted us this awesome capability without also including protective controls; at least, not for long. But with cultural evolution acting unimpeded, it's a whole new ball game.

In general, biological evolution proceeds by relatively small steps, rather than large, discontinuous leaps.* The consequences

*Some biologists, such as Stephen Jay Gould, have emphasized that evolutionary steps may occasionally be larger than had been thought. But this is a matter of the degree of fine tuning. We are still talking of steps, not leaps.

of a particular trait, such as antler size, can therefore be evaluated by natural selection as it develops incrementally. By contrast, the enormous advances of human cultural evolution have provided us with great chunks of new characteristics, upon which evolution has had no opportunity to ruminate, and which in many cases must be either accepted or rejected as momentous technologic quantum leaps: the steam-powered locomotive, electronics, nuclear power, long-lasting pesticides, strip mining, acid rain. If some Paleolithic forebear had evolved a propensity to do something that was ultimately harmful to itself or its genes, natural selection would have nipped it in the bud. But if the trait was beneficial to itself, even if harmful to others, then it would have been favored. Moreover, if like nuclear weapons, it had the capacity of destroying the world, then it could hardly be selected against, since natural selection can only act on occurrences, not possibilities.

By a tragic irony of fate, we have once again been equipped with *fewer* biological inhibitions upon environmentally destructive behavior than have most animals. This may be because we are primates. As a group, primates generally do not make permanent living quarters. Those animals that do, such as most birds, are generally imbued with instinctive prohibitions against fouling their own nest. The young either back up to the edge of the nest and "let fly" over the side, or they produce special fecal pellets that are presented to the parent, who dutifully carries them away. By contrast, most monkeys and apes sleep in a different place each night. Like hikers unconcerned about leaving a dirty campsite because they will not be returning, our near-relatives show little reluctance to defecate in their own beds.

Ground-dwelling animals such as woodchucks or wolves are very careful to keep their living chambers clean and therefore free of possible disease. But urine and feces pass immediately out of a monkey's arboreal world and become someone else's problem. What, we worry?

Terrestrial animals can be located by predators who use their noses to trap the individual whose toilet manners were untidy. Not so for the primates. Monkeys thus make terrible pets, in part because they are notoriously difficult to housebreak. As recent immigrants to the terrestrial world, people—like monkeys—have few qualms about fouling their nest. It should cause no surprise, then, that for all our intelligence, it is harder to toilet train a human being than a dog.

*
**

Every new source from which man has increased his power on earth has been used to diminish the prospects of his successors. All of his progress has been made at the expense of damage to his environment which he cannot repair and which he could not foresee.

—C. D. DARLINGTON

MOST animals are not renowned for their foresight. But on the other hand, their capabilities of doing harm are also limited; their power is limited to their bodies, and hence their ability to damage their environment is quite limited. Predatory animals, for example, have to work hard for a living. They survive only at the expense of other lives and their victims can be expected to exert themselves to their utmost, in the hope of staying alive. Catching prey is thus not without risk; at the very least, it takes time and energy. Predators therefore tend to be conservative hunters, killing only what they need. By contrast, the fruit-, leaf-, and insect-eating primates evolved in the tropics, where different edible species are available at different times. Food is often "there for the taking," involving little risk or difficulty, and with little to discourage gluttony. In addition, primates do not store food against times of scarcity, presumably because they rarely experience substantial scarcities, and also because their preferred foods do not store well. It is therefore additionally likely that human beings have evolved with very few biological inhibitions when it comes to exploitation of natural resources, or planning ahead for a rainy day.

Biologist Garrett Hardin wrote a very influential scientific paper nearly twenty years ago, titled "The Tragedy of the Commons." In it, he pointed to the situation in Britain, when sheep-owners grazed their flocks on public lands known as the Commons. Although everyone suffered if the Commons was overgrazed, no one felt personal responsibility for maintaining it. Each shepherd also preferred to graze his sheep on the Commons rather than on his own, private land, and moreover, each shepherd recognized that if he refrained, out of concern for the common good, then someone else would simply overgraze. As a result, the would-be altruist found himself a victim of a type of Prisoner's Dilemma, forced to cheat (that is, graze on the Commons rather than cooperatively refrain), because if he didn't, he would be victimized in the long run by the selfish, overgrazing, noncooperating exploiters. So everyone felt constrained to go for all he could get, at the expense of the welfare of the environment generally. The tragedy of the Commons was not only the pressure that it exerted toward personal selfishness and egocentricity, but also the destruction of potentially productive land.

It is interesting to compare the human situation in this regard with that of two closely related members of the weasel family, both of which are superbly efficient exploiters of their natural environment, but neither of which possess the human capacity for environmental destruction or for conceptualizing their potential for triumph or tragedy. The otter is a wonderfully capable hunter of fish and invertebrates. Except where human activities have greatly reduced their prey populations, otters rarely go hungry. They can generally catch more food than they need, but they don't do so. Instead, they have become notoriously playful creatures, swimming after worried fish just to satisfy their frolicsome nature. But what they kill, they eat.

The mink, on the other hand, is also a masterful hunter but without the otter's playfulness. Mink find killing very easy, in fact hard to resist. They appear to violate evolution's logic by often slaughtering more than they need. However, their marsh environment is highly productive and in no danger of being overexploited. The mink's behavior has been unchanged for millennia and is fully integrated into the natural community. Crow, coyote, fox, and hawk thus take advantage of the situation, often consuming the mink's uneaten victims, gaining life by the mink's excesses.

Human beings are much more extravagant than the mink in our resource use. Our waste materials are generally used by organisms we consider nasty or otherwise disagreeable, such as flies, cockroaches, rats, mice, and the algae that pollute our lakes and rivers, feeding on the nutrients added as a result of human hyperactivity. These outbreaks are a far cry from the beautifully balanced system supported in part by the mink's behavior, largely because human sewage, garbage, and other such delights have not been around long enough for such an intricate biological web to have evolved.

Our penchant for excess is undeniable. "Man's chief difference from the brutes," wrote William James in *The Will to Believe*,

> lies in the exuberant excess of his subjective propensities—his pre-eminence over them simply and solely in the number and in the fantastic and unnecessary character of his wants physical, moral, aesthetic, and intellectual. Had his whole life not been a quest for the superfluous, he would never have established himself as inexpugnably as he has done in the necessary.

As James saw it, however, this was cause for celebration, not regret: "And from the consciousness of this he should draw the lesson that his wants are to be trusted; that even when their gratification seems furthest off, the uneasiness they occasion is still the best guide of his life, and will lead him to issues entirely beyond his present power of reckoning. Prune down his extravagance, sober him, and you undo him." It remains to be seen, however, whether we shall undo ourselves first, through the results of this extravagance.

Perhaps the most pressing environmental danger now facing our planet (nuclear war aside), is the increasingly rapid destruction of the world's tropical forests. These ecosystems are the most intricately balanced, species-rich, and poorly understood on earth. They are also irreplaceable, yet they are being destroyed at an extraordinary rate, partly as a result of lumbering but even more by clearing in order to raise beef, which in turn is sold to American consumers in fast-food restaurants such as McDonald's or Burger King. Thanks to the extravagance of human need and greed, Third World countries are permitting their precious tropical forests to be destroyed, while in fact, even the newly cleared grazing land does not retain its vitality for more than a few years before being degraded to cementlike "laterite," which is useless both for making Quarter-Pounders and for regenerating a lush forest.

The primitive slash-and-burn agriculture that preceded such large-scale deforestation was nearly harmless by contrast. During hundreds and sometimes thousands of years of such primitive environmental perturbations, the small areas that were cut down had ample opportunity to regrow. The large-scale environment was scarcely affected. But now, thanks to our penchant for excess combined with the technologic capacity to act excessively, we threaten permanent harm.

A small—or even a moderate—quantity of something may be innocuous, or even healthy. Excessive amounts, however, readily go beyond diminishing returns and often become troublesome. In small doses, such elements as zinc, cadmium, or nickel are necessary for life; in excess, they are poisons. Food is also necessary for life; in excess, it produces obesity. We would all die without oxygen; we would also die if we were forced to breathe 100 percent oxygen. Exercise promotes health; in excess, it produces injury. A smokestack may be a sign of jobs and economic vitality; a forest of smokestacks may be a sign of air pollution and an environment rendered toxic. Agriculture is a marvelous way of providing food for people; unrestrained agriculture—complete with deforestation, water table depletion, and widespread use of pesticides—could sterilize the planet. There is, in short, much to be said for the Platonic "golden mean." But human cultural evolution seems to lead eventually not to moderation but rather, to excess.

Basically, our environmental crisis has been fueled by three fundamental factors: First, the capability of producing environmental destruction (technology in a broad sense); second, the lack of inhibitions and integrated control regarding the exercise of this capability; and third, a particular attitude toward nature. Animals, of course, probably have no "attitude" toward nature, and it would make little difference if they did. Their mental abilities are too poorly developed for such abstract considerations and in any case, their lack of culture prevents them from acting effectively upon any "world view." By contrast, human beings typically do have an attitude toward nature and furthermore, they act on it, often very effectively.

This attitude is generally antagonistic and exploitative, at least in the Western world. Our capacity to exploit may even have diminished our need—and hence, our ability—to live peacefully with other Homo sapiens. Since we have the ability to "defeat" nonhuman nature even as we may be defeated by other human beings, then we may have taken the path of least resistance during much of our experience as a species. After a conflict, if the loser leaves and succeeds elsewhere, his and her descendants may eventually have inherited the earth, but part of that inheritance may include relatively little capacity or inclination for working things out with our fellow human beings. It is therefore at least possible that our success in subduing nonhuman nature, and thereby colonizing new areas, has empowered us to be (at least for the short term) the masters of our planet, while at the same time depriving us of the much needed ability to live more harmoniously with each other.

We see the earth as something to be conquered rather than to be appreciated, as challenging and threatening rather than nourishing and protecting. Wilderness is to be tamed, not savored. Wildlife is a "resource," rather than a legitimate fellow inhabitant of our shared planet. And once you have seen one redwood tree, you have seen them all. Caught in a fundamentally schizophrenic division between culture and biology, we also tend to dichotomize the world, seeing it in arbitrarily dualistic terms: subject vs. object, good guy vs. bad guy, man vs. nature. We live in two worlds, the biological and the cultural, and we therefore feel somehow separated from ourselves. We respond by feeling separated from nature as well.

"The ordinary city-dweller," writes philosopher Susanne Langer,

> knows nothing of the earth's productivity; he does not know the sunrise and rarely notices when the sun sets; ask him what phase the moon is in, or when the tide in the harbor is high, or even how high the average tide runs, and likely as not he cannot answer you. Seed time and harvest are nothing to him. If he has never witnessed an earthquake, a great flood or a hurricane, he probably does not feel the power of nature as a reality surrounding his life at all. His realities are the motors that run elevators, subway trains, and cars, the steady feed of water and gas through the mains and of electricity over the wires, the crates of food-stuff that arrive by night and are spread for his inspection before his day begins, the concrete and brick, bright steel and dingy woodwork that take the place of earth and waterside and sheltering roof for him . . . Nature, as man has always known it, he knows no more.

Although this isolation from our environment may be largely unavoidable, a natural result of our capacity for cultural development, to some extent such attitudes are also encouraged by our peculiar cultural systems. Thus, the Western world view is strongly colored by Judeo-Christian religious concepts, which are themselves dualistic. Strict alternatives are at the core of Western religion: God vs. His creation, sin vs. redemption, heaven vs. hell.

We are only rarely sensitive to the unity between organism and environment. Nor are we readily inclined to act so as to preserve the integrity of the system as a whole. Under the traditional view, we were given "dominion" over all the earth, with explicit instructions to multiply and subdue it. We face life with a chip on our shoulder.

Not only do we feel isolated from and antagonistic toward nature, we also "use" our environment to provide various complex gratifications: leisure, recreation, the experience of speed, of overblown material abundance, of control. An animal asks only for the necessities, in return for which it provides necessities to others. The primitive Stone Age human being and the enlightened Eastern mystic are alike in taking from nature only the basic necessities while refraining from grossly destructive pursuits. The latter does so because of his greater understanding, the former because he is unable to do anything different. The rest of us lean on cultural levers of great power to pry from nature our various personal and collective needs and wants.

As Max Weber pointed out, to the Easterner, rationalism means rational adjustment *to* nature; to most Westerners, by contrast, it means rational mastery *over* nature, a mastery that would be impossible without the levers of culture . . . and that may well be ultimately impossible in any event. Seeking to "win" in our self-proclaimed battle against our environment, to wrest satisfactions from a biological world that is equally intense—but often less successful—in seeking to retain its own structural integrity, we may yet defeat our environment. But if so, we shall lose ourselves.

Comforts and luxuries become "necessities" because they are accessible through technology and because our colleagues have them. In addition, the simple facts of economic survival often require environmentally destructive behavior; the modern world "makes a living" by mining the environment, sometimes literally. Doing so, we are committing an economic sin, if not a theological one: dipping into our capital rather than living off the interest.

In our violent treatment of the earth we often reflect our own aggressiveness. Ethologists identify something called "redirected behavior," as when an angry man pounds a table or slams a door because he is inhibited from attacking the real object of his aggression. We may similarly redirect much of our frustrated interpersonal aggressiveness toward our environment. It may be no coincidence that some of the world's most aggressive people, notably Americans and western Europeans, are also among the most environmentally destructive. In the process we are doing violence to each other, and ourselves.

Echoing David Brower's injunction, which appeared as an epigraph for this chapter, Nobel Prize–winning physicist Murray Gell-Mann pointed out:

> [I]t used to be true that most things that were technologically possible were done but that certainly in the future this cannot and must not be so. As our ability to do all kinds of things, and the scale of them, increase—for the scale is planetary for so many things today—we must try to realize a smaller and smaller fraction of all the things that we can do. Therefore an essential element of engineering from now on must be the element of choice.

Fortunately, there is some good news. Human beings, intelligent primates that we are, can exercise choice. We can overcome our primitive limitations and shortsightedness. We can learn all sorts of difficult things, once we become convinced that they are important, or unavoidable. We can even learn to do things that go against our nature. A primate that can be toilet trained could possibly even be planet trained someday.

CHAPTER ELEVEN

TECHNOLOGY

Plasm on Brass?

Oh, what a world of profit and delight,
Of power, of honor, of omnipotence . . .
All things that move between the quiet poles
Shall be at my command . . .
A sound magician is a mighty god.
—Christopher Marlowe, *Doctor Faustus*

To a weak-bodied, biological creature, limited to the productions of sinew and bone, any magician—sound or unsound—is a mighty god. And with our biology increasingly augmented by our culture, our abilities have become increasingly magical. It is another question, however, whether along with the power, we have gained honor or delight, never mind omnipotence. (Profit, at least for a few, is another story.)

The word "technology" derives from the Greek *techne*, meaning art or skill. It has come to refer especially to industrial arts, applied science, and practical engineering, although at least one definition, in the Random House Dictionary, is "the sum of the ways in which a social group provide themselves with the material objects of their civilization." Except for its emphasis on physical objects, this definition is not far removed from a definition of culture itself.

Technologies are assumed to be complex. In the late twentieth century, technology has come to be virtually synonymous with the latest in scientific applications, as in "high-tech" electronics (e.g., computers), chemical engineering (e.g., plastics), metallurgy, medicine, and/or nuclear energy. But we can speak equally of the technology of Paleolithic hand axes, basket-weaving technology among the Navaho, or the technology of propaganda.

Technology arises from the artful and organized application of tools. Tools, then, are the building blocks of technology, and are generally considered to be relatively simple instruments, usually hand-held, for performing mechanical operations. For a time, it was thought that human beings were distinguished from other animals by our use of tools, and the term *Homo faber* (Man, the maker) was even proposed as our scientific name. But as we have learned more and more about the lives of animals, this presumed region of human uniqueness has been progressively encroached upon, since we now realize that many different animal species employ tools. Some even make them.

The woodpecker finch of the Galapagos Islands selects a cactus spine, breaks it off to the right length, then uses it to probe for insects. (It doesn't have a long, thin beak like a woodpecker, so it has to manufacture an artificial one.) Egyptian vultures drop rocks on ostrich eggs, breaking their shells. Sea otters dive for abalones and also for a flat rock. While floating luxuriously on their backs, the ocean-going gourmets then position the rock on their chests and open the tasty abalones by pounding them against the mollusc-opening tool. Chimpanzees go fishing for termites with a long thin stick or blade of grass—the simple tool is inserted into a termite mound, termites grab hold of it, and the instrument is then withdrawn, the termites eaten, and the process repeated.

In some cases, such as termite consumption by chimpanzees, the tools are luxuries. In others—for example, the termites' reliance on the elaborate cooling, protecting, and humidifying properties of their complicated mound—the tools are absolutely essential. In the former cases (tools as luxuries among animals), we can easily imagine the creature without its tools. A chimpanzee is still very much a chimpanzee if it never goes fishing for termites just as a sea otter can still be a sea otter even if it dines exclusively on crabs instead of abalones. In the latter cases (tools as essential to the lives of animals), the tools are an indispensable part of the animals' existence, inseparable from other aspects of the creatures' biology. It is impossible to be a termite without a termite mound or comparable structure, although we do not consider that termites possess a mound-building technology.

Human beings, not surprisingly, are different. We have embedded ourselves in a heavy matrix of tools and technology, and yet, although we are utterly dependent on this aspect of our culture—no less than the termite—we and our technology are not inseparable. Whereas a termite is inconceivable without its mound, a person can readily be imagined without a lathe, a spinning jenny, a Linotype machine, or a home computer. In short, although we have produced elaborate technology on which we are increasingly dependent, our biology and our technology have remained largely independent of each other.

Traditional wisdom among anthropologists and biologists is that the use of tools was central to the evolution of modern Homo sapiens. There is a minority viewpoint, however, that holds otherwise. In *The Myth of the Machine*, Lewis Mumford argued forcefully that tools and primitive technology actually had a more limited role in human biological evolution than is customarily supposed. Mumford suggested that the crucial developments that made us uniquely human were "soft" and hence, not easily fossilized: language, religious belief, compassion and empathy, social organization, conscience and consciousness, moral and ethical systems. In short, it is at least possible that art preceded utility, and meaning supersedes mechanism. "The burial of the body," wrote Mumford, "tells us more about man's nature than would the tool that dug the grave."

The traditional focus on the formative role of tools and early technology in human evolution may well reflect a modern obsession with tools and technology, a felt need to justify and rationalize the Technological Person of the twentieth century. There is an old, corny song with the refrain: "You made me what I am today; I hope you're satisfied." If tools and technology literally made us what we are today, then perhaps we ought to be satisfied—with what we are, no less than with our present-day technology, since the latter is only the most recent incarnation of that which made us. But if, on the other hand, our fundamental essence is not the child of Pleistocene technology, then perhaps the proliferation of today's more recent technology is not necessarily the natural and appropriate outcome of our early apprenticeship as a species, but rather, perhaps, a mere aberration, a gratuitous and rather dangerous hyperextension of a capability which, although important, does not warrant the worship that it commonly receives. Bedazzled by our ability to build, maybe we have fallen victim to a new neurosis: the edifice complex.

More seriously, if technology often seems to make a monkey of us these days, it may be because in our hearts (or more important, in our brains and our genes) we are still monkeys.

A word is considered to be an example of onomatopoeia if—like "buzz" or "plop"—it was formed by imitating some external event that served as a direct referent. In an analogous sense the first tools can be considered as onomatopoetic. That is, they were probably conceived initially as simple extensions of the human body: the club a stylized and more powerful hand and fist; the bowl and pouch more efficient cupped hands; the flint scraper a heavy-duty fingernail; the knife a stronger, more maneuverable tooth.* If so, then special plaudits are due the first, unknown inventor of a tool that was truly independent of our own biological

*Psychoanalysts might also suggest that the spear is an aggressive penis substitute, and the basket a vagina. After all, the former is typically used by men, the latter, by women. But it is worth remembering that when Saint Sigmund himself was queried as to the significance of his fondness for cigars, he replied: "Sometimes a cigar is just a cigar."

equipment: perhaps the blowgun, or the bow and arrow.

Freed from their imitative and hence, limited role, tools led eventually to machines: from the Greek *machana*, for pulley (pulleys are one of the famous "six simple machines" and are made, in turn, from an arrangement of such tools as the wheel and a piece of rope). Dazzled as many of us now are by the discoveries of the Industrial Revolution as well as the high-tech of the twentieth century, it is easy to overlook the many and important early technologies which have loomed large in our civilized history, if not our evolutionary formation: painting, the potter's wheel, the loom, musical instruments, the plow, writing, kitchen utensils, the horse collar, watermills and windmills, and plumbing.

Lewis Mumford made a useful distinction between three levels of technology: eotechnic, paleotechnic, and neotechnic. Eotechnic (from the Greek *eos* or dawn) represents the first, primitive stirrings, the dawn of technology itself. The materials of eotechnic advances were stone, hide, wood, cloth, and simple metals. Paleotechnic advances were particularly associated with the Industrial Revolution; they involved the harnessing of enormous amounts of power, often with the combustion of fossil fuels: steam engines, iron smelters, and the traditional smokestack industries are all basically paleotechnic. Finally, neotechnology relies heavily on electricity, miniaturization, and efficiency; it is less obviously power-oriented and polluting.

The transition from eotechnology to paleotechnology to neotechnology has involved a progression of more work being done with less human labor. At the same time, it is interesting to note that the actual skill required may be less: it is easier to operate a machine that makes chairs than to carve them by hand, easier to tell time with a digital clock than with a clock with hands, easier to navigate by automatic pilot than by compass, and easier yet by compass than by the stars. As we learn new skills and techniques, we also lose old ones. Some of them may be no loss at all: if there are fewer and fewer people alive who remember how to start a car by cranking it, this hardly matters since cars are no longer built with cranks. And even if children no longer learn to tie their shoes, perhaps this too is no real loss, since now there is Velcro. But if cake mixes replace the skills of baking, and voice-activated computers make writing (or even typing) obsolete, there may be some difficulties if it is ever really necessary to cook at home, or if the electric power is interrupted.

In one of the most important American literary and sociological documents, *The Education of Henry Adams,* the scion of one of the nineteenth century's most illustrious political families developed the view that human history has been caused not so much by people acting on other people, as by forces acting on people. The force that according to Henry Adams personified the Modern Age was represented by a gleaming, powerful dynamo, displayed at the 1900 Chicago Exposition. Just as the Virgin symbolized the force acting on people during the Middle Ages, the dynamo represented to Adams the force of the modern, technologic age. "May the Force be with you," we were advised in *Star Wars.* To Adams, who lived during the transition from paleotechnic to neotechnic, the force is inseparable from us, and is not entirely benevolent, since it is responsible for a breakdown in personal relationships.

Gleaming gears, belching smokestacks, or sparking wires are not prerequisites for depersonalization. The highly organized slave societies of ancient Egypt and Babylon showed us that hundreds of thousands of laborers can be forcibly conscripted and highly organized, deprived of their individual rights and depersonalized in the extreme, with no more than eotechnology. Mumford identified such depersonalization as the most powerful and pernicious machine of all, which he named the "mega-machine." The myth of the mega-machine was twofold: that it was irresistible, possessed of Pharaonic deity as well as the momentum of history, and that it was ultimately benevolent. The myth of today's technology is no different.

⁂

> *Only let the human race recover the right over nature which belongs to it by divine bequest, and let power be given it; the exercise thereof will be governed by sound reason and true religion.*
>
> —FRANCIS BACON, *Novum Organum*

FROM Francis Bacon onward—and to some degree in fits and starts even before him—the Western expectation was that technology would solve our problems. Descartes pointed out that by the appropriate attitudes and actions, we shall "render ourselves the lords and possessors of nature," and speaking for the nineteenth-century Age of Progress as well as for his own, optimistic America, the poet Longfellow urged his fellow citizens: "Let us act, that each tomorrow finds us further than today."

The goal of progress, that place where tomorrow is to find us smiling, was sometimes stated to be the elimination of work and drudgery, as witnessed by Aristotle's expressed wish:

> If every instrument, at command, or from foreknowledge of its master's will, could accomplish its special work . . . if the shuttle thus would weave and the lyre play of itself, then neither would the chief workman want assistants nor the master slaves.

For others, the concept of technology was identified more diffusely with progress, which in turn, may be one of the most important and unappreciated legacies of Christianity. In contrast to the Eastern religions, paganism, or Greco-Roman philosophy, Christianity introduced the notion that the world was going somewhere, like an arrow shot from a bow rather than a static paralysis or an endlessly turning and returning wheel. The concept of progress is seductive and easy enough to grasp, perhaps in part because individuals progress through successive stages in their own biological, social, and intellectual development. Since Christianity teaches that human beings are born into sin, from which they must redeem their souls, progress seems all the more natural as well as desirable. Indeed, it is our human responsibility, the reason we were put on earth. "The education of the human race," wrote St. Augustine in *The City of God,* "represented by the people of God, has advanced, like that of an individual, through certain epochs . . . so that it might gradually rise from earthly to heavenly things, and from visible to invisible."

In short, the intellectual underpinnings were already in place when Francis Bacon, a thousand years later, heralded the Industrial Revolution and the Renaissance by extolling the goal of human existence in more secular terms: "the enlarging of the bounds of human empire to the effecting of all things possible." Progress, and progress alone (bound now inextricably, it seemed, to technology and the machine) became a new Western religion. It is likely, in fact, that the fervor with which progress/technology was embraced during the Renaissance and the following Age of Rationality was due at least partly to the collapse of the medieval belief in divine order and perfection. If the City of God was largely slums, and perhaps without even a city-planner, perhaps then science and technology could take its place. "The human race is continually advancing in civilization and culture as its natural purpose," according to Immanuel Kant, preeminent German rationalist of the eighteenth century, "so it is continually making progress for the better in relation to the moral end of its existence, and . . . this progress, although it may sometimes be interrupted, will never be entirely broken off." This new faith—faith in rationalism, science, and technologic progress—had replaced theologic faith as the prime draft-horse of Western civilization. But the underlying fact was unchanged: the hare, offering not so much eternal bliss in heaven as secular joys in the form of technology and progress, would still be humankind's savior.

There is no reason to conclude that such attitudes ended 200 years ago. Herman Kahn, the savant who gave us thinkable nuclear wars, also extolled "the curative possibilities inherent in technological and economic progress," prophesying a world of plenty if we would only embrace technology with yet more confidence and less restraint than we have shown so far.

Even some theologians have converted. Notable among them is the French priest/paleontologist Pierre Teilhard de Chardin, whose work has generated something of a cult following, especially since his publication of *The Phenomenon of Man.* Teilhard shows that you don't have to be a physicist or even a supply-sider businessman to love technology. He develops a humanist and theological argument that ends up celebrating technology as a manifestation of the power of collective human endeavor, such that the future human being will be part of a merging of the biological and the cultural, into something new on earth, the "Noosphere," a "domain of interwoven consciousness" in which all people will merge as a "single, hyper-conscious arch molecule." Urging us not to fight technology, but to join it so as to help bring about this final stage of human evolution, Teilhard rejects the "nightmares of brutalization and mechanization which are conjured up to terrify us and prevent our advance." He extolls the Manhattan Project for having showed that "nothing in the universe can resist the converging energies of a sufficient number of minds sufficiently grouped and organized," and excoriates those who "had the temerity to assert that the physicists, having brought their researches to a successful conclusion, should have suppressed and destroyed the dangerous fruits of their invention. As though it were not every man's duty to pursue the creative forces of knowledge and action to their uttermost end!"

Physicists such as Herman Kahn or Edward Teller recommend a growing technologic horizon simply because it is there, and because we can do it (moreover, if we don't surely the Russians will . . . and where would we be then?), with little concern for its impact on the human soul. By contrast, Teilhard urges us to achieve a mystical union and in the process, to expand our selves.

For others, however, the growing reach of technology offers not so much a promise as a threat. Simple tools and utensils—even simple machines—are, as we have seen, basically extensions of the human body. As such, they are unlikely to take on a life of their own. But in the progression from eotechnic to paleotechnic, and perhaps even more from paleotechnic to neotechnic, the hare confronts the tortoise with devices that are increasingly foreign and autonomous. Because of this growing independence, our creations seem increasingly to be a Frankenstein's monster, or a Sorcerer's Apprentice: forces outside our bodies, acting by their own power, and perhaps, even, by their own will. The computer is the apotheosis of such a transformation, notably HAL in the movie *2001: A Space Odyssey*, which is not only autonomous, but malevolent as well.

For those lacking the secular faith of Kahn or the religious faith of Teilhard, the creations of Homo sapiens are often less than entrancing. Although it is a product of our own actions, technology—by its fundamental strangeness from our biology—can readily feel as though it is imposed upon us rather than emanating from us. In any event, and as a result of consciousness as well as our technologic inventiveness, human beings are damned (or blessed) to live forever at the brink of Awe: for the primitive, awe at nature; for the modern-day technicist, awe at his and her own productions. During the dark days of the late 1930s, Jewish parents in Germany were often dismayed to see their children goose-stepping and "heil-Hitlering" with their playmates. Even in the concentration camps, some inmates became as brutal as the guards. Psychiatrists call it "identification with the aggressor," and the critic cannot help wondering whether our planetary embrace of technology does not also include some aspects of this, especially as technology increasingly takes on the mantle of a separate, autonomous entity . . . and moreover, a potential aggressor.

Unlike the moon, which only exposes its bright side to the earth, technology has another aspect. Let us now turn, therefore, to those who view its dark side.

⁂

Even now in the enthusiasm for new discoveries, reported public interviews with scientists tend to run increasingly toward a future

> *replete with more inventions, stores of energies, babies in bottles, deadlier weapons. Relatively few have spoken of values, ethics, art, religion—all those intangible aspects of life which set the tone of a civilization and determine, in the end, whether it will be cruel or humane; whether, in other words, the modern world, so far as its interior spiritual life is concerned, will be stainless steel like its exterior, or display the rich fabric of genuine human experience.*
>
> —LOREN EISELEY

TECHNOLOGY has done wonders to lighten the load of suffering humanity: for example, we can cure and often prevent diseases such as diphtheria, cholera, polio, typhoid, and whooping cough, whereas others, such as smallpox, have been eliminated altogether. So let's not undervalue the lowly technologist. Without the stonemason's craft and the glazier's skill, there would be no Chartres Cathedral, and without the instrument maker, no Bach. Even in that extreme example of technology-run-wild, the nuclear arms race, technology has yielded some benefits as well: thanks to satellite surveillance, for instance, both the U.S. and the USSR can verify compliance with arms control treaties. The average twentieth-century human being, if fortunate enough to be at least moderately prosperous and to live in a technologically advanced nation, can be assured a reasonable chance of surviving infancy; growing up, he or she can contemplate the brilliant achievements of the world's past civilizations, and visit parts of the earth that were never available to our ancestors. We have more material abundance, with less work, than Homo sapiens has ever known before.

On the other hand, "goods" in the sense of material objects are not necessarily the same as "goods," the opposite of bads. (It is revealing that we use the same word for what should be two distinct concepts.) In widening a highway to adjust for excessive traffic, we often wind up stimulating more traffic so that instead of jammed two-lane roads we have jammed superhighways. In feeding an unprecedented number of people on our planet, the "green revolution" has spawned overpopulation that periodically leads to famines and ultimately, to environmental and human degradation. In producing the ultimate weapon, to keep the peace, we run a growing risk of losing everything, in the event of war. In short, there is a "flip side" to technology that isn't really very flip at all.

Although Philip II of Spain considered all inventors and innovators to be heretics, organized Western religion generally had less of a bone to pick with technology than with science. Thus, the Catholic Church resisted the heliocentric universe of Copernicus and Galileo, just as Protestantism opposed Charles Darwin; on the other hand, religion was more congenial to technology. In fact, it has been argued that the basic workplace virtues of thrift and punctuality were first developed in the very practical Benedictine monasteries of medieval Europe, from which they spread to the rest of the Western world. Those who dissented from technology's mandate—and its apparently overwhelming victory—were generally motivated by more secular concerns.

William Blake warned of the "dark Satanic mills" that accompanied the Industrial Revolution in Britain; paleotechnology in general, with its horrendous pollution, child labor, barbaric practices of near-slavery, and indifference to basic hygiene and fundamental human values, was not altogether beneficent in its impact on Homo sapiens, or on the rest of the natural world. As Lewis Mumford put it, while perfecting the mechanical art of multiplication, we neglected the moral and ethical art of division.

Even beyond the physical costs and dangers, and the social issue of justice and fairness, dissenters worried about the impact of technology on the human soul. Semiscientific visionaries, like H. G. Wells, foresaw a world in which the values of the machine depersonalized our species and ultimately destroyed it. *The Time Machine* painted a world in which romantic and humanistic values were embodied in the childlike and helpless Eloi, preyed upon by the cruel, rapacious Morlocks, machine-oriented troglodytes who ate human flesh. In their simplicity and passivity, the Eloi are reminiscent of de Tocqueville's warning, half a century earlier, that "a kind of virtuous materialism may ultimately be established in the world, which will not corrupt, but enervate the soul, and noiselessly unbend the springs of action." But in Wells's fantasy, the Morlocks no less than the Eloi were the victims of technology, condemned as they were to a brutal and cheerless underground existence as slaves to their machines.

Machines have grown in power and efficiency, becoming ubiquitous in the modern world. But although they have conferred many advantages, contrary to Aristotle's prediction they have not rendered slavery obsolete. In fact, the cotton gin actually increased the demand for slaves in the American South, and even with legal emancipation, the wage-slave is very much a reality today, consistent with Herbert Marcuse's observation that economic freedom should mean freedom *from* the economy. Moreover, the modern-day industrial worker, although not quite reduced to the nightmare of Wells's Morlocks, is increasingly becoming a latter-day pastoralist, a shepherd transformed into a machine-herd. "May not man himself become a sort of parasite upon the machine?" asks cyberneticist Norbert Wiener, "An affectionate machine-tickling aphid?"

The nineteenth-century liberal Russian intellectual and political troublemaker Alexander Herzen once predicted that technological advance in Russia would eventually produce "Genghis Khan with a telegraph." He was quite right: no better description has been written of Josef Stalin. But now, things are infinitely worse: Genghis Khan has nuclear weapons. Worse yet, Genghis Khan is transnational, found in Washington no less than in the Kremlin, in Belfast no less than in Beirut.

We have unleashed not only the atom, but the aggressive, inventive, and insatiable curiosity of Homo sapiens upon a world that previously knew only the feeble ploddings and proddings of strictly biological creatures.

Our reach, according to Robert Browning, should exceed our grasp, "or what's a heaven for?"—the poet enjoins us to strive to accomplish more than we are capable of achieving. Ironically, the situation has now reversed itself: our grasp has exceeded our reach. We have a technologic tiger by the tail, right here on earth. As for the Sorcerer's Apprentice, things are getting out of hand; or, like Genghis Khan with a telegraph, a Gulag, napalm, nuclear weapons, or a bulldozer, the artificially extended reach of our hands has exceeded our ability to achieve the necessary hand-mind co-ordination. We may well have accomplished more than we are capable of mastering. Like Halvard Solness, the tragic master builder in Ibsen's play of the same name, we have built higher than we can climb.

Sometimes, it all threatens to come tumbling down, and not just in war or intentional cruelty alone. Instead of "energy too cheap to meter," we have nuclear power plants that are economic basket cases, generating radioactive waste that will remain fiendishly toxic for many times longer than human civilization has yet existed. Instead of making the world a "global village," worldwide communications now make potentially lethal misunderstandings instantaneous; being able to talk with other nations has not given us anything more to say. The modern airplane, giving us convenient travel, has given us the bomber as well. The ability to transport food quickly and over long distances has resulted in the centralization of bakeries, for example. So whereas the average Frenchman can buy fresh-baked *petit pain* at a *boulangerie* within walking distance of his or her home, the average American has the privilege of buying prepackaged white bread, baked perhaps hundreds of miles away, and laced with preservatives to help the product survive its transportation from point of manufacture to wholesale warehouse, to supermarket, to breadbox. And when it comes to the daily grind of getting to work, although it is undeniable that our high-speed people-moving technology moves more people, more quickly than ever before, it is uncertain whether in fact the average commuter actually spends less time commuting than his or her counterpart several hundred years ago; having given ourselves the ability to travel long distances quickly, we also choose to live proportionately farther from our workplace.

At other times, it really does come tumbling down. The industrial disaster at Bhopal, India, in December 1984 is a classic, tragic example. At least 2500 people were killed, perhaps as many as 10,000. A few weeks earlier, liquid natural gas had exploded in Mexico City, with about 500 fatalities. Sixty thousand people had to be evacuated from the vicinity of the Three Mile Island nuclear power plant in Pennsylvania, in the aftermath of a leak and potential meltdown during 1979. Toxic wastes have ruined lives and threatened many more in Seveso, Italy, and Love Canal (near Niagara Falls, New York). In 1971, cargoes of Mexican wheat and United States barley, distributed in Basra, Iraq, had been treated with a mercury compound to prevent spoiling. The wheat was intended for use as seed, but the recipients, misunderstanding its purpose, consumed large quantities of the poisoned grain: as many as 6000 may have died. And there are asbestos, PCBs, Agent Orange, oil spills, and sewage leaks, to mention just a few. The disastrous explosion of the space shuttle Challenger was a particularly spectacular example of technology gone awry. It was troublesome for many Americans, not only because they personalized the catastrophe, but also because it represented the failure of a high-tech program that Americans had begun to take for granted.

In each case, the usual response is to blame operator error, mechanical design, faulty communications, insufficient local oversight, and so forth—certainly not the conflict between cultural and biological evolution, or between human capabilities and human need on the one hand, and technology and human greed on the other. According to political scientist Robert Engler, one historian estimated reassuringly that of those killed at Bhopal, one half "would not have been alive today if it were not for that plant and the modern health standards made possible by wide use of pesticides" of the sort manufactured there. The gods of technology giveth, and they taketh away.

Just as psychoanalysis has been described as treatment for the worried rich, so has concern for the environment and anxiety about technology. On this view, interestingly, the far right and the far left have sometimes converged, the former eager to justify maximum profits and to defuse public criticism and scrutiny of corporate irresponsibility, and the latter eager to enhance employment in today's smokestack industries. It is therefore noteworthy that the real victims of technology run wild, those most likely to be taken by this unleashed devil, are in fact the hindmost of society.

The severest industrial tragedies have not occurred in the United States, but in Third World countries, which often have environmental and worker protection laws that are notably lax compared to those of America, in an understandable effort to lure foreign investment and jobs. These industrial plants are typically located in slums, where land and labor are cheap. And where the effluent doesn't bother the affluent.

But eventually, the chickens must come home to roost, for the wealthy nations as well as the wealthy within each nation. "Of 600 registered generic pesticides in common use today," writes Jonathan Lash in *A Season of Spoils,* "79 to 84 percent have not been adequately tested for their potential to cause cancer, 60 to 70 percent have not been tested for their potential to cause birth defects, and 90 percent have not been tested for genetic mutations." The public has a legitimate, gut fear of catastrophe, and will hold its collective breath in fascination over towering infernos, nuclear meltdowns, and Titanic disasters, whether real or celluloid. By contrast, slow-acting substances, which accumulate over time and may exert their toll only over a span of years, simply don't register very high on humankind's Richter scale of ground-shaking events.

There is a real ambivalence here, not just an ambivalence of attitude, but one of substance as well. We are like the victims of a bad marriage, who nonetheless love each other: we can't live without technology, but we are having more and more trouble living with it as well. (Similarly, we can't live without our biological self, but find it difficult living with it as well.) Contemplate a sleek Lamborghini sports car, a finely tuned Jaguar XKE, or a smoothly humming Cadillac El Dorado, all triumphs of high technology—complete with electronic ignition, integrated circuits, computer-assisted carburetion, high-impact plastics, and space-age alloys. And then consider that they each rely on squashed dinosaur guts (that is, petroleum) to keep them going.

Technology is culture incarnate and yet it relies fundamentally on biology, no less than we—ourselves the incarnation of cultural evolution—rely on our own creaturely strengths and weaknesses. More than fifty years ago, W. F. Ogburn wrote *Living with Machines,* proposing the concept of "cultural lag." According to Ogburn, human subjective values (art, religion, ethics, emotion, even politics) have failed to keep pace with the innovations of machines. As Ogburn saw it, our human responsibility was to reduce that lag by speeding up our adjustment.* More recently, theologian Harvey Cox wrote that "the secular city signifies that point where man takes responsibility for directing the tumultuous energies of his time." Since technology has rendered most social and political structures obsolete, it is our job, as inhabitants of the "Secular City" rather than the City of God, to weave a "political harness to steer and control our technical centaurs."

But the problem is not just technology alone, or the tendency of human culture to lag behind a technology that is growing increasingly inhuman. Rather, it is the tendency of culture to outrun biology. Moreover, as we have seen, technology is a two-edged sword, offering benefits as well as extracting costs. We clearly need more than a dyspeptic rejection of modernity, we need instead an insistence that technology be measured by how much it contributes to life, rather than evaluating life by how much it contributes to the furthering of technology. Technology itself is not the enemy. The difficulty lies precisely with the creators and wielders of

*In "Sociology and the Atom," Ogburn gave an example of what he meant. He proposed that society's response to the challenge of the Nuclear Age—our way of diminishing the cultural lag that had been created—should be to dissolve our urban civilization and reconstitute the United States as a semirural society based on thousands of small towns dispersed around the countryside. It was a preview of the "crisis relocation" plans of the early 1980s.

technology; that is, with Homo sapiens, that confused and confusing creature of whom we are so justly proud and about whom we are so legitimately worried.

Earlier, we briefly considered the relationship of technological evolution to social evolution, pointing out that the latter tends to be much more sluggish, undirectional, sometimes circular, and perhaps as often regressing as advancing. As the "centaur" that has provided much of the brute force for human societies, technology has not been notably controlled or intelligent, but rather, something of a muscle-bound idiot. There is a more hopeful vision, however, since technology does progress, and—at least in theory—it can be harnessed and directed as we, the drivers, see fit. We cannot realistically accelerate biological evolution, nor can we count on altering nontechnological social evolution by contemplative philosophy, or fervent appeals to morality. But we can alter and orient technology; indeed, we do so every day. Population control can be achieved by appropriate uses of technology, and it seems likely that even the human desire for large numbers of children can be altered by technological advances that render child mortality reliably low. Polluting technology can be replaced by nonpolluting alternatives, and we can even engineer microorganisms with the genetic capability of degrading oil spills, plastics, and noxious chemicals, and particle accelerators to deplete plutonium and uranium-235. More efficient houses, industries, and automobiles, based on recycling and renewable resources, can be designed; to some extent, this is already being done, although on a small scale.

No matter how much our biology and even some of our social structures may resist, it is at least possible that we can use technology artfully and intelligently to orient our lives in healthy ways. If so, then today's manifest conflicts between technology, nontechnological culture, and biology may be transient and in the long run surmountable. Perhaps it is not too late for a reconciliation, with advances in science and technology serving as one possible fulcrum around which such harmony could be established.

Technology alone won't save us, any more than technology alone will destroy us. The missing piece in the puzzle is the human being: biologically evolved yet culturally and socially conditioned. And that cultural and social conditioning, by definition, is something over which we have power.

Many people have written about the "technological imperative," the notion that if something can be done, it must be. The implied image is of a large genie, arms akimbo, scowling and tapping his foot impatiently while waiting for us to do what eventually we must. Perhaps, however, we might occasionally show that genie who is the boss.

We have often assumed a confrontational attitude toward nature, frequently to a fault. We climb mountains because they are there, but we certainly haven't accepted the legitimacy of something—whether social injustice or smallpox—for the same reason. Why, then, should we accept the dominance of technology and machinery, just because they are there? The irony is all the more pronounced when we consider that unlike nature, technology exists only because we have created it. Rather than accept technology as an immutable given, and seek to rectify cultural lag by establishing human institutions that are more congenial to technology, perhaps we might consider establishing technologies that are more congenial to human beings, and to the rest of the planet.

"The Walker," in Robert Frost's poem, comes upon a nest of turtle eggs on a railroad track. He muses that

> The next machine that has the power to pass,
> Will get this plasm on its polished brass.

The polished brass of a railroad engine is a universe away from soft and squishy turtle-plasm. It is built of sterner stuff and is indifferent to the latter's fate. Only because of the intervention of another kind of squishy stuff—human-plasm—are the two coming into tragic contact; without that human-plasm, the polished brass wouldn't exist at all. So the world of polished brass and the world of plasm may not really be that far apart after all. We can look the other way when the train hurtles past, or shake our fists in impotent rage, but perhaps our responsibility lies deeper, and is more demanding: to climb aboard, and be a prudent engineer.

CHAPTER TWELVE

ALIENATION

Homo extraneus

I, a stranger and afraid
in a world I never made.
—A. E. Housman

Homo sapiens could also be called the alienated animal, estranged from much of his world, and from himself. We are probably unique in nature, the only animal that is somehow out of place in its own surroundings. Twentieth-century literature, poetry, theater, and cinema all reflect a rising tide of alienation. But whereas the strangeness of human beings in their own environment has probably been on the rise in recent years, it has actually been with us for a very long time, once again a result of our uncoordinated culture and biology. We are animals, living in a man-made environment. Small wonder we *feel* strange: we are strange.

There are many symptoms. Mental illness, anomie, frustration, boredom, antagonism, withdrawal, insensitivity: all these and more derive at least in part from a disharmony with the world we inhabit. (Including in this world, of course, our fellow human beings.) The human environment is a product of our culture, and within this complex edifice, a biological creature must live. Insofar as we possess any genetic basis for our behavioral tendencies and inclinations, this biological foundation must be largely appropriate to a pretechnological if not precultural environment. In effect, we entered a time machine just after we discovered tools, fire, agriculture, technology, and the rest of its offspring . . . and the dizzying pace of our trip has rendered us increasingly disoriented ever since.

"The heart of our difficulty," to historian Arnold Toynbee, "is the difference in pace between hare-swift movement of the scientific intellect, which can revolutionize our technology within the span of a single life-time and the tortoise-slow movement of the subconscious." In sketching the origins of alienation in the conflict between the hare and the tortoise, we shall consider first the discordance between human beings and the environments that our culture has enabled us to produce. Then, we shall look briefly at the alienation engendered by consciousness itself.

In a sense, virtually any human environment, even the most pastoral, must strictly be considered artificial in that it shows the unmistakable impress of ourselves and our culture. But perhaps no human environment compares with our cities when it comes to blatant disregard for certain aspects of our biology. Dependent upon the countryside for food, raw materials, and amelioration of their pollution, cities nevertheless have a life of their own. They are wholly manmade, where people are often crowded beyond belief and without relief, and in which one could easily spend a lifetime without walking on the earth or sitting under a tree. Needless to say, this is a peculiar setting for a living, breathing, sweating, biologically evolved creature.

It is also a very recent development: by 1800, only 50 cities in the world had populations of 100,000 or more. By 1985, there are more than 1500, of which more than 150 have one million or more residents. Mexico City alone is predicted to have 30 million inhabitants by the year 2000. To Plato, the size of a city was limited by the number of people who could hear the voice of a single orator; thanks to electronics and telecommunications, that number is now infinite. But it is another question whether our capacity to tolerate such numbers and such densities matches our ability to communicate and accumulate.

Although many people seem well adjusted to cities and would be loathe to leave, the truth is that our cities have enormous problems, most of them resulting from the disparity between our cultural creations and our biological needs. The problems of aggression and social disorganization were discussed earlier; while generally applicable to the human situation, they are also especially appropriate to humans in cities. There are, in addition, other alienating factors that are more specific to cities.

We all know the popular stereotype of the country bumpkin who is hypnotized and bedazzled by all the hustle, bustle, and bright lights of the Big Town. But we are all country hicks at heart. Cities present situations of excessive social stimuli—sights, smells, sounds—all ever-present, ever-changing, ever-demanding, urgently assaulting the senses. And there is little escape. One recourse of the city dweller is a retreat into what the religious philosopher Martin Buber called an "I-it" attitude toward our surroundings and our fellows. The alternative, "I-Thou," is a deeper, loving relationship in which the meaning of one partner is predicated upon the other, both achieving a transcendence of their otherwise limited selves. By contrast, the I-it relationship implies a thoroughly objectified attitude in which the individual sees him or herself as completely encapsulated within his or her skin, always distant from the other, from "it."

The city's characteristic bigness, noisiness, constant flux, and impersonality make I-Thou relationships unlikely. The average urbanite will encounter literally hundreds or thousands of people every day, and almost every one of them will be a stranger. This is a commonplace observation, but in a sense, a highly significant one. We meet strangers every day, not only passing them on the street, but crushing against them in public conveyances and doing business with them all the time. Visit a primitive human culture today in one of the backwaters of the world where homogenizing

Western culture has not yet penetrated: the people may be frightened, shy, antagonistic, or intensely interested, but they certainly won't ignore you.

You can be sure that the average tribesman in the mountains of New Guinea or the deserts of the Kalahari does not often meet with strangers. Regular dealings are almost entirely with relatives, friends, or at least acquaintances, and the chances are good that it was the same for our ancestors. The contrast with the average city dweller is extreme, and so are the results. Knowing a person is different from knowing a fact, a distinction that is oddly absent in English but well expressed in French, with the different verbs *connaître* (to know someone) and *savoir* (to know something). The former takes longer, is more difficult and more meaningful. When people know each other, they act differently than when they first meet. Formalities decline, protective mechanisms are relaxed. There is a mild but definite strain that comes with meeting a stranger. His or her attitudes and reactions are unknown. Although the circumstances of meeting generally indicate nonhostile intent, and often provide much additional information about what to expect, a certain biologically appropriate suspicion may well generate initial malaise. And this must become a constant part of the city dweller's life.

Even when introduced to someone at an intimate social gathering where personal interaction is expected, the overwhelming majority of people have great difficulty remembering their names, typically because they are stressed and preoccupied seeking to respond to the stranger(s).

One way to lessen this strain is to keep others outside the protective envelope of our existence. Actually, we cannot possibly know everyone in our path when we walk down a busy urban street. We cannot personally greet everyone on a crowded bus. We cannot reach out to the humanity that flourishes around us, breaking upon us like the ocean upon a rock. We cannot respond in the deeply human way for which biological evolution has undoubtedly prepared us. Rather, we must keep the world and each other at arm's length; although Buber saw it somewhat differently, I-itness is a defense mechanism, born of sheer necessity and helping us keep our emotional balance in a world made unstable and chaotic. So we surround ourselves with an impermeable barrier of indifference, sidestepping our fellows in silence, ever careful to avoid recognition of the other as a human being. In the New York City subways, no one looks anyone in the eye. Out on the street, we may even step over a fallen body, or observe an attempted murder with a cold indifference and reluctance to "get involved."

"Men, it has been well said, think in herds," we learn from a nineteenth-century treatise, quaintly titled *Extraordinary Popular Delusions and the Madness of Crowds.* "They go mad in herds. While they only recover their senses slowly and one by one."

In a sense, the urban dweller cannot be faulted for such behavior. It is forced upon him and her by the insane nature of the crowded madmade environment. Cities differ with respect to their encouragement of this attitude: New York City, for example, is one of the worst. Because it is almost impossible to find a comfortable place to sit, a drinking fountain, or bathroom facilities, its inhabitants are forced to use the city as a place between goals, a means to more private ends. They rush about, stone-faced and seemingly always in a hurry, callous to whatever or whomever they may encounter.

Many animals indulge in greeting and so-called "contact" signals, which also serve to reassure subordinates and appease the dominants. A flock of sparrows, like a crowd of people, is a noisy affair. Each bird periodically makes a short vocal call which informs the others of its presence and helps maintain optimum spacing among individuals. It also probably serves to reduce aggression between neighbors, informing them that all are bona fide members of the group. Among ground squirrels and prairie dogs, strangers sniff each other's faces in a stereotyped greeting ceremony. Dolphins chatter to each other almost constantly and chimpanzees extend their hands. Humans also shake hands when the meeting is formal or direct; when simply passing an acquaintance we generally utter a brief "hi" or equivalent, and/or smile and nod our head. These greeting ceremonies may appear inane or useless, but a brief experiment in withholding such signals will quickly reveal their very necessary function. We arouse quick suspicion and antagonism if we fail to greet our friends and acquaintances. Why? And what relevance does all this have to alienation and urban life?

The meeting of two animals produces immediate tension, especially if the encounter is unexpected. Most animals have therefore evolved means of reducing this tension, thereby preventing undue disruption of normal behaviors. It is also desirable to indicate association with others when such exists, so friends greet each other. Although the exact nature of these greetings varies among people depending on the specific culture, the general behavior is universal. Walk down a street of any small town: you will greet or be greeted by virtually everyone you meet. This is because with a small population, everyone knows everyone. Now walk along a busy city street: everyone studiously refrains from engaging anyone else. To do so would be physically impossible as well as physically dangerous. We can only speculate on the unresolved stresses generated by such unfulfilled encounters as we withdraw into ourselves. Protecting ourselves in one way, we stress ourselves in another.

One obvious result of the city's enforced anonymity is its crime rate. Unlike impulsive violence, in which the victim and perpetrator are often members of the same family or at least well known to each other, premeditated robbery or burglary is almost always directed at a stranger. A small-town resident doesn't rob the corner grocer; everyone knows nice old Mr. McPherson. But if McPherson is a nameless, familyless, disembodied, and anonymous spirit in a big city, he can be attacked with relative ease. Beyond this, the trend toward major corporate or conglomerate ownership makes the individual units all the more vulnerable. No one knows Mr. Safeway, but McPherson is a person. As our culture forces us together in such unnatural densities and proximities, the same defense mechanisms that defend us by maintaining emotional distance and indifference deny our natural inhibitions the opportunity to protect us.

It appears that as population density increases, the value attached to each individual decreases proportionately. We move from what in German is known as a *Gemeinschaft*, a society based on face-to-face personal ties, to a *Gesellschaft*, one characterized by impersonal, commercial relations.

The specter of galloping Gesellschaft provides a cogent argument for population control and for reducing urban concentrations, or at least encouraging the development of local neighborhoods where a healthy sense of community and of personal worth can still survive. It is ironic, for example, that one way for people to become suddenly very important is to go into the wilderness, away from people. On a backpacking trip where companions are rare, a stranger encountered on a lonely trail can be a delightful and valuable experience. There is often a lot to discuss: destinations and starting points, trail conditions, weather prospects, the location of good camping and watering sites, wildlife observed, etc. The same people, meeting at Fifth Avenue and Forty-second Street, would hardly respond to each other's presence. A good reason for getting away from the crush of people, if only temporarily, is to counter the city-bred alienation of people from each other.

Given all the disadvantages of city life, why do so many choose it? In most cases, there seems to be less a conscious selection of the city per se than a following of jobs, convenience, and economics, all of which may lead to the city. There is also, of course, the simple accident of those born to parents who were themselves previously lured by one of these circumstances. But other positive inducements exist as well. Earlier, we discussed the ethologists' concept of releasers and their potent artificial counterpart, supernormal releasers. These are manmade stimuli that elicit an exceptionally strong response by exaggerating certain characteristics of the normally occurring releaser. The city may well constitute a supernormal releaser for human beings, a cultural hyperextension of our fundamental sociality.

Like most other terrestrial primates, we are gregarious creatures, although our primitive social units were undoubtedly much smaller than the modern metropolis. We banded together for hunting, mating, child rearing, food gathering, defense, the transmission of culture, affection, and companionship. Evolution must have selected strongly against solitary tendencies, as it still does today among most primates, while favoring the social creature who was attracted to communal gatherings. The asocial person, like the solitary baboon, did not survive long and left few descendants. People would doubtless gather at a safe camping place, much as baboons do today at protected sleeping trees, or around the spoils of a successful hunt. For some animals, such as fish in a school or possibly primates in a troop, there may have been safety in numbers for purely statistical reasons as well. If a predator is likely to take, say, the first three individuals that he or she encounters, there would be a benefit to huddling closely together, not so much out of fondness, but in hope that as a result, someone else will be chosen rather than yourself. Not surprisingly, then, a school of fish becomes more tightly bunched when a barracuda swims near, and a flock of ducks flies even closer together when a peregrine falcon swoops overhead. British biologist W. D. Hamilton, who elucidated the mathematics of grouping for safety, called his scientific paper "Geometry for the Selfish Herd."

It may be an overstatement to describe a human city as a selfish herd, and certainly, most urban dwellers don't flock together out of conscious fear that a leopard might leap out of a dark alley, especially eager to make off with an easy, solitary target. But there is no doubt that we find safety, and reassurance in numbers. And if groupies were more fit than loners, the various signs of a group may well have been made attractive to us by biological evolution. However, within the last few thousand years, cultural evolution, impelled by the discovery of agriculture and following the path of least economic resistance, has spawned enormous gatherings of people, greatly surpassing anything we had "naturally" produced. The color, the noise, the variety, and the sheer excitement are fascinating and almost irresistible. Like moths to a flame, we have been drawn to this oversized stimulus, this supernormal releaser.

John C. Calhoun's experiments with overcrowded rats also suggest a far-reaching interpretation of human city-seeking. His rats, showing a wide gamut of hideous responses to excessive population density, were not actually forced to live so closely together. They *chose* it. The experiment was designed so that they had to feed from central bins from which each animal could only obtain a small amount of food at a time. Consequently, they spent a long time feeding and since there was room for many rats to feed together, they soon became accustomed to feeding alongside each other. Soon they became "conditioned" to each other's presence, actively seeking the larger group, even though the resulting overcrowding produced what are (at least from our viewpoint) very unpleasant results. Calhoun labeled the end product a "behavioral sink"—the unhealthy connotations were intentional—and he referred to his animals' intensive sociality as "pathological togetherness."

Yet there is hope. Eric Hoffer emphasized that many of our most valuable social and cultural advances came from cities. And as shown especially by Anne Whiston Spirn in her recent book *The Granite Garden,* many urban problems are in fact soluble. The Eternal City is doubtless a self-serving myth; the infernal city, however, need never become a reality, if we recognize cities as real environments, with the potential of being planned, humanized (sometimes even naturalized), and made tolerable and at times, even delightful. In our haste to produce supernormal releasers, it may be that we have too quickly given up on cities as environments. Spirn details the benefits that can result once we take account of the water dynamics, plant and animal life, soil compositions, wind patterns, and heat vectors that operate within so special an environment. Even ancient Rome met its citizens' needs for water, and modern-day Stuttgart has done a remarkable job adjusting human-made industry to nature-made airflow so as to minimize air pollution. After all, times have changed: once, we took our environments as we found them, or went elsewhere. Now, we have the capacity—indeed, the obligation—to make our own environments. When we begin doing this seriously, and not simply with an eye to economics, we may find ourselves feeling less like strangers in a strange, artificial land.

Aside from the alienation induced by our living places, notably our cities, there are the problems arising from the fact that we have surrounded ourselves with the increasingly strange products of our own creativity. When scientists wish to deal with devices they do not understand, they call them "black boxes." We know what goes into a black box and what comes out of it (inputs and outputs, to the engineer), but we don't know what goes on inside. Most psychologists, for example, treat the brain as a black box, with stimuli going in and behavior coming out. With the advent of increasingly elaborate technology, human beings are finding themselves increasingly surrounded by black boxes. We wake up in the morning, flick a switch (input), and somehow a light goes on (output). Pull the lever and the toilet flushes; turn a key and the car starts (usually). Not only in the big issues—having to do with the relationships among nations and the structure of the human experience writ large—but also in our everyday lives we have been surrendering our autonomy to things we comprehend only vaguely. Modern Homo sapiens lives increasingly removed from the primitive realities of rock and earth, water and wind, bird and leaf: things we can feel and understand primitively, if not intellectually.

In Greek mythology, the giant Antaeus drew his strength from the earth. He was invincible in combat so long as his feet were on the ground, but was eventually killed by Hercules, who strangled him while holding the hapless giant up in the air. We are also being lifted out of our element, and perhaps with similar results.

At one of Long Island's most exclusive country club beach resorts, ultra high heel shoes have long been de rigueur, worn daily by the socially active, jet-setting members. The most common injuries on the beach were not drownings or even abdominal cramps but rather, ruptured Achilles tendons in these high society women, suffered while walking barefoot on the beach. The requisite footwear for an advanced social life apparently caused shrinkage of the sturdy tendon leading to the heel. Hyperextending the heel, as a result of taking off the shoes and walking flat-footed in the soft, pockmarked sand, abruptly stretched the shrunken tendon and caused it to snap. Poetic justice, perhaps.

To the nonscientific mind, most things are either mystery or miracle. To the average twentieth-century inhabitant of any advanced Western nation, nothing has changed, except that we are expected to understand the mysteries and miracles on which we rely every day. Failing that understanding, emotionally uprooted by our dependence on things that are *from* us but not *of* us, our sense of connectedness is severely undercut.

Awakened by an alarm clock, he or she eats some packaged breakfast food and rides to work in a bus, train, or automobile. Once there our hero may sit behind a desk, conducting business over a telephone with people perhaps a thousand miles away, whom she doesn't even know. Or perhaps he performs some repetitious factory work, the end product of which he rarely sees. As tangible reward for such labors, there are pieces of paper that can be exchanged for things needed or wanted, the products of someone else's work. Small wonder that on Sunday afternoon he sets fire to some charcoal in his backyard and eagerly plunges his hands into a few pounds of raw hamburger meat, letting the blood squeeze out between his fingers; or she sways her body rhythmically and rapidly to high-amplitude sound, accompanied by large numbers of others, similarly disposed. At least this is real.

Not only has our culture placed us at a greater and greater distance from animal reality, also it subjects us daily to worries and stresses over which we have little if any control. Through television or newspapers we are informed of major political, military, social, and economic events which we feel individually powerless to influence, but which will clearly influence us. We are free to rage, worry, or approve, but the very size and complexity of our cultural apparatus makes it difficult for us to act effectively and see the fruition of such action.

During a recent eclipse of the sun, more people watched on television than in person, although it could have been viewed safely right outside. Shackles on the mind can be even more potent than on the body (or the Achilles tendon) and it is sobering to realize that culture is rendering us physically incapable of directly meeting the world, or mentally unwilling to do so.

Interestingly, outdoor activities such as hiking, canoeing, wildlife viewing, cross-country skiing, and mountain climbing are all excellent tonics for the complex culturally inspired disease of alienation. Such activities may provide their own kind of stress, but the stress is physical, direct, straightforward, and understandable. The simple realities of boot on ground, hand on rock, paddle in water—all are right there, without the interposition of bureaucracies, technologies, or ideologies. Small wonder interest in such activities has been growing exponentially in modern America.

And despite the intensive news coverage surrounding moon landings and space shuttles, children are still more likely to play cowboys and indians than astronauts. Not only has general interest in the space program fallen off rapidly—this was more or less expected—but the image of the astronaut has become remarkably dull and unexciting, while the cowboy rides on. The hero of our culture vs. the hero of our biology, and the latter is winning. The contrast is striking, the reason simple yet telling. Cowboys have a crucial something that astronauts lack: personal autonomy, direct physical control over events. The horse and gun, good guys and bad guys: these are simple realities, the stuff of cowboys and of cops and robbers and detective shows as well. And we long to make them our stuff too. Our fantasized spacemen such as Flash Gordon, Buzz Corbett, or Captain Video all participated in *personal* adventures in which strength, speed, courage, or intellect were successfully exercised in a straightforward manner.

By contrast, the reality of space travel is much nearer the reality of modern cultural-technological life. The erstwhile spaceman depends on an enormous support system, a "team" equipped with elaborate communicative, maintenance, and control devices. In fact, the astronaut is a virtual robot, pressing crucial buttons in response to orders from "Mission Control." If something goes wrong or a new course must be plotted, she must await orders from the computer and the engineers at home. The astronaut has virtually no real control over her environment, and aside from personal courage and technical competence, shows little that warrants emulation. As "doing" animals, we admire mastery over situations, while in her dependence on a myriad of imposing but somehow artificial cultural constructs, the astronaut reminds us too much of our own lives.

Greater mastery over space exploration is doubtless achieved by the concerted efforts of modern technology than could ever be accomplished by some rocket-mounted Buffalo Bill. But it is the culture's triumph, not the individual's. Our culture, in fact, has won many convincing victories. These triumphs, however, although not defeats for the individual, often leave us hollow and are sometimes Pyrrhic victories at best. For example, the institutions of mechanized agribusiness, combined with corporate distributorships, make food available to us in profusion and convenience. A great triumph of the "system." Yet what we gain in selection and facility we lose in satisfaction. It is rarely a very meaningful experience to buy some tomatoes at the supermarket. It's a different thing to feast upon a plump, red beauty you planted, tended, and watched grow. Not only that, but the home gardener has a known quantity; he has controlled its development. If for example, he doesn't want pesticides on his food, he can simply "make" it without them instead of depending on an unknown, unresponsive corporation, mass-producing a product perhaps thousands of miles away.

The technology of medicine has progressed similarly. There is a hollowness that accompanies many of our common practices, a feeling that the patient is being deprived not only of personal identity but of the right to meet life with the dignity of a sovereign being. Antidepressants and tranquilizers are prescribed with increasing frequency and unconcern, indicative of our preoccupation with symptoms rather than causes. We overmedicate nearly everything, with resulting natural selection for increasingly resistant disease organisms. Many women are still drugged into semiconsciousness during childbirth, admittedly sparing them some pain, but also cutting them off from the reality of one of the most intense experiences they can ever know. As with the increased interest in organic and home-grown foods, the growing popularity of "natural" childbirth reflects a growing rejection of the technologically mediated alienation doled out in modern obstetrical practice.

We are often rebuked for being a grossly materialistic culture, and yet in a sense, we are antimaterialist to an extreme. The philosopher Alan Watts once commented that a really materialist culture would have more respect for *material,* and would never tolerate the plastics, the veneer, the mass-produced throwaway garbage and shoddy workmanship that typically surround our lives. Similarly, although we are a nation of fat people—or at least, a nation of dieters—there is reason to doubt that we really value food. If we did, we wouldn't have consumed billions of McDonald's hamburgers. Instead, we seem to be captives of our own capabilities, impaled on our own hyperextended technology.

For example, one of the major innovations of the industrial revolution has been the factory. To understand its importance we must compare the theory behind factory operation with its more "primitive" predecessor, the artisan. When an individual's work is directed toward a whole, finished product, the activity can be deeply satisfying. Of course, the artisan must be well trained, and both apprenticeship and actual production are likely to be time consuming. Each worker in a factory, by contrast, specializes in a relatively simple operation, a very small part of the whole. Training takes less time and the product is completed more quickly since each step is handled by a different "specialist," usually with the aid of considerable mechanization as well. Job satisfaction is typically low. Potential satisfaction—the feeling of belonging and rapport not only with one's colleagues but also with one's work—has been sacrificed to efficiency and a sort of group mastery that can leave the individual isolated and unfulfilled.

On balance, and despite many attempts to break out of the situation, things appear to be getting more desperate as technology marinates in its own positive feedback. Holding a tiger by the tail, we fear to let go because we have traveled so far from the primitive state that we could not survive without our ferocious ally. Yet it carries us farther and farther into a land that is fundamentally strange, and therefore alienating as well as downright dangerous. And the more we struggle, the greater the alienation we produce, since our struggles necessarily involve the use of yet more artifice and artifact, which are basically the problem, not the solution.

Alvin Toffler's *Future Shock* describes another symptom of modern culture-induced alienation. A traumatic experience such as serious injury or the witnessing of some terrible event can result in "shock," a medical condition that may lead to death (see the discussion of stress-induced adrenal shock, earlier). Modern Homo sapiens—according to Toffler—is in a state of virtually continuous shock, produced by our inability to assimilate change at its new frenetic pace. We are disoriented, and it is getting worse as change comes faster and faster.

We are also an extraordinarily mobile society. It has become rare for children to make their adult lives near their birthplaces, and almost unheard of for families beyond the unit of husband-wife-and-children to live together. This likely deprives us of a great sense of continuity, to be found in living out our lives in the surroundings of our childhood memories. When people live in one place for a long time, they are said to have "put down roots." A plant gets nourishment and strength from its roots, and a sense of permanence and relatedness to the earth. By contrast, modern Western Homo sapiens lives in a state of seemingly permanent transience, rarely living in—or even remembering—our childhood homes, changing jobs and friends, schools and doctors, seeing only one very narrow side (the work-related one) of perhaps 99 percent of the people we meet. (In a term that never caught on, but deserved to, Toffler suggested that most working relationships were of such brief duration that instead of bureaucracy, we should identify "ad-hocracy.")

To some extent the seeming perpetual motion of modern Homo sapiens is a direct result of our cultural system. We go where the jobs are, where we are transferred, or in times past, where the draft could be avoided. In such cases, we seem to be helplessly dragged about by the riptide of events our culture has created. On the other hand, much of our mobility is intentional. Culture has given us the opportunity to better our lot. We can actively seek a better job, more pleasing climate, specialized schooling or medical care; mobility may well be a valuable tool for widening our horizons and making us happier. But regardless of whether our rootlessness is forced upon us or comes as a result of free selection from a newly opened range of possibilities, it is rootlessness nonetheless, and its effect on our psyche remains the same: we become strangers to our own land, our fellows, and to ourselves.

As a result of this estrangement, many people are also likely to feel estranged from society's customs, norms, and expectations. Earlier, we developed the notion that if Homo sapiens has a biologically predisposed preference for a particular kind of social structuring, such a preference is likely to be frustrated by the great multiplicity of human societies. We also suggested a more likely prospect, that our species lacks such a preference, and that as a result, we are also subject to disruption of our cultural systems and feeling alienated from them, since we lack the animal's innate patterns of social adherence, acceptance, and stability.

Let us assume that the human species, despite its rigid physical body, is basically amoebalike, able to ooze into almost any culturally defined shape. The role of culture, then, would be crucial in stabilizing our fundamental formlessness, providing structure and direction to what would otherwise be confusion and chaos. If, in the beginning, human nature is without form, and void, and darkness is upon the face of our being, then we owe a great debt to the Spirit of Culture, which moves over the face of the waters and says, "Let there be this way of living, and that." And "Thou shalt do this, and not that." In a sense, the genesis of human culture therefore unburdens us of the options of too many choices and too little human nature.

What happens, then, when alienation increases to the point of undermining confidence in one's culture? Unprecedented freedom can become confusing, even paralyzing. Independence can be transformed into fragmentation and disorientation. Individuals can find themselves wayward, lost, angry, and unfulfilled, uncertain what to choose for their lives, or even how to formulate such choices. When the past seems unacceptable as a guide and the future unpromising, the present assumes an overwhelming weight, which it cannot bear without collapsing.

We have a physical structure, our skeleton; in other respects, however, we cry out for structure, and yet our culture often provides us with structures that are profoundly inadequate and alienating, because they are too rigid, too uncaring, or too downright dangerous. Increasingly alienated from their cultures, people turn to other cultural practices that promise relief: primitive religion, violent terrorism, psychoanalysis, sociopathy, or involuted indifference.

In the years before Sigmund Freud, psychiatrists were commonly known as "alienists." And not surprisingly, the stuff of alienation is still the stuff of mental illness. As Freud pointed out in his *Civilization and Its Discontents,* civilization requires to some degree the frustration of basic human tendencies: aggressiveness, cruelty, violence, sexual debauchery, excessively selfish individualism of all sorts must be constrained if human beings are to live amicably with each other.

Just as civilizations can mimic and exaggerate biological traits through the process we have labeled cultural hyperextension, they are also prone to overreact to perceived human tendencies, tendencies that may be either lacking altogether or actually benign, but pathologic when frustrated. For example, in his book *Touching: The Human Significance of the Skin,* anthropologist Ashley Montagu described the various pernicious ways in which human societies have denied the natural, harmless, and healthy human need of children and adults for touching and being touched. The child's inclination for self-expression often leads to repressive school discipline; sexual yearnings lead to Victorian taboos, and so forth, all combining to produce a human being who is at best alienated and neurotic, and in whom any vestiges of mental health are testimony to an extraordinary constitutional toughness.

Radical psychiatrist R. D. Laing may have been exaggerating only a little when he wrote, in *The Politics of Experience:* "From the moment of birth, when the Stone Age baby confronts the twentieth century mother, the baby is subjected to these forces of violence, called love, as its mother and father, and their parents before them, have been. These forces are mainly concerned with destroying most of its potentialities, and on the whole this enterprise is successful."

To theologian-author Andrew Bard Schmookler, this frustration is specifically designed to produce rage and lust for power: "Power demands better servants than human beings are, as nature created them. Civilized societies need power, and thus they are compelled to re-create man. Socialization practices are the instruments by which social demands are translated into psychological structure." His explanation, though ingenious, seems overly contrived. More likely, the anger and frustration that follow from the practice of socialization are simply a result of the discordance between biological and cultural factors. Sometimes, our cultural practices demand too much of human beings; sometimes too little. Only very rarely can we expect them to hit it (that is, us) just right.

The strains generated by conflict between Christian theologic doctrine and biologic inclination were well portrayed by Saint Paul, when he wrote in anguish: "I delight in the law of God after the inward man: but I see another law in my members, warring against the law of my mind, and bringing me into the captivity to the law of sin which is in my members. O wretched man that I am!" (Romans 7: 22–24). Or as Hamlet might have said, O cursed spite, that ever culture tries to make biology right.

We began this chapter by looking at alienation as something that results from our culturally created environments, and have subsequently been moving toward examining alienation as something that also results from our unique consciousness, a level of self-awareness that apparently exists only in Homo sapiens. Apes and chimpanzees have a "self-image," or at least they will seek to wipe off grease paint applied to their faces once they see themselves in a mirror. But no animal can look at itself and be sad or disappointed. No animal obsesses about the gulf between the way things could be and the way they are. No animal is aware of its own, unavoidable death, and yet at the same time, aware that nothing can be done about it. Animals may be sad—sometimes downright miserable—but no animal feels sorry for itself, or feels worried about the fact that it is sad or disconnected.

Insofar as there are certain "inalienable" aspects to human nature, the more culture and human nature fail to correspond, the more strain will be produced, and the more alienated will be the human products. Eventually one or the other may crack: either the offending cultures will then change, or the offended individuals will develop neuroses or even psychoses.

Modern ecologists, Eastern mystics, and drug-stimulated

Westerners have recognized, apparently independently, that organism and environment are not separable, that the human skin, for example, does not isolate us from the rest of the world but rather, joins us to it. Flying in the face of this recognition, however, is the alienation of Homo sapiens from the manmade environment, from other human beings, and sometimes, from himself. A brilliant and sensitive man—also a former mental patient—Charles M. S. Sade (no relative to the notorious Marquis de) describes the experience of schizophrenia as resulting from a profound alienation between the biological self and the cultural whole. Sade writes that the schizophrenic

> has chosen himself, his limited individuality, his ego, to be his whole world. He usually seeks himself in a gradually enlarging interiority of mind. He is self-conscious, brooding, introspective, often passive-dependent externally, often hearing voices, which convey subtle and sensitive condemning or glorifying meanings. I don't think his ego is split or gone; in fact, it is probably magnified with delusions of grandeur which compensate for the real loss, the real split from his truer self that is being at one with things, people, and the world. To the extent possible he becomes an isolated fragment, alone in a mind, often tormented by guilt for his refusal to accept the larger reality. He nurses his secret hurt and in defiant pride, to affirm his isolated self, projects a solipsistic metaphysic, seeing the spirits of his inner world as ministering either to or against him and him alone, while what is left of the outer world means only in relation to him. Since he cannot actually completely withdraw from the outer world, though he seeks it, it impinges upon him as wholly other, mostly negative, blind to him or totally against him. So in some little place he has found to sit he asserts his kingship as the most important, most deeply persecuted person. His body means little to him and like a burden it weighs him down. He becomes finally a lonely, completely impotent god.

The loneliness of the schizophrenic is the loneliness of the human being isolated from his or her legitimate connections. The impotent god is transfixed by the disparity between his biological creatureliness and his culturally mediated alienation—often with an assist from some chemical imbalance in his very biological brain.

It is unclear whether the frequency or severity of mental illness is any greater in modern times. However, it seems likely that during the past few decades there have been significant increases in the use of mind-altering drugs, accompanied by a legitimate and growing concern on the part of parents, educators, and law enforcement people, who unfortunately are often terribly misinformed as to their effects and causes. Psychedelic drugs provide both escape from reality and a feeling of enhanced experience. ("Reality is only a crutch," reads a bumper sticker, "for people who can't handle drugs.") It does not really matter for our purposes whether these experiences are imagined or real, that is, whether one's sensitivity and perception are actually heightened or whether seeming revelations are just hallucinations, imaginary perceptions carrying no true insight into underlying realities. The important thing is that it seems that way to the user.

Use of marijuana, LSD, and more recently, cocaine, has become particularly prevalent among middle- and upper-class youth. In fact, it was the explosive increase of use among these groups that finally prompted today's massive social concern and even the rumblings of political action. These drugs have gained a wide audience because of the enhancement of perceptions and feelings they offer. They are not particularly new discoveries. They are not the latest wonder drugs, or recent contributions to "better living through chemistry." Marijuana and mescaline have been around for a very long time, although perhaps never achieving the popularity and the notoriety they enjoy today. They have been rediscovered overnight, by a populace weary of their alienated, secondhand approach to life and eager to be turned on and tuned into the vibrant sensations of a reality of which they sense they are being cheated. Drugs are a means to an end.

Actually, despite the possible health hazards involved, the use of mind-expanding drugs may be one of society's least dangerous efforts at overcoming alienation. For some, it appears that a true sense of living can only be achieved through acts of violence, most notably the killing of someone else. "Thrill murders" and serial killings are frequently perpetrated by deeply alienated people whose psychotic boredom fairly screams for behavior of high intensity, in order to make them know they are alive. While such behavior is clearly pathological, similar feelings, although less extreme, underlie many common features of everyday behavior. Snowmobiles, downhill skiing, driving cars, boats, and motorcycles too fast: all these are socially acceptable means of declaring a measure of personal mastery and autonomy. By seeking thrills with a degree of danger, technological man seeks to insert himself back into the world, to feel once again the pulse of real, biological life.

The need to feel such life seems especially acute for a species—the only one, as far as we can tell—that is aware of its eventual death. Awareness of death can add a certain spice to life, or it can drown existence in hopelessness and a deep, existential alienation. It can also lead to religions and cults, in a primitive search to achieve via cultural constructs the pseudo-community and secure rootedness of which knowledge of death has deprived our biological selves. If we were intelligent computers instead of human beings, then perhaps awareness of corrosion, short circuits, and the possibility of power outages would lead to an existential electric angst and along with it, a deep-seated alienation from our electronic selves and the disadvantages and perils that necessarily follow from this fact. Instead, we are intelligent animals, and therefore doomed to being aware of the limitations to our creaturely existence, limitations that are ultimately unavoidable even as we may perceive them to be tragic. We may elect to go gentle into that good night, or to rage against the dying of the light—but one way or the other, go we must.

In Plato's *Phaedo,* Socrates defined the fundamental wisdom of philosophy as "being mindful of death." He did not say, however, that we had to like it. On the one hand, mindfulness could lead to a fuller existence, if not to happiness. On the other, mindfulness in biological creatures cannot but lead to a painful awareness of our predicament, circumscribed as we are by our organic bodies and the certainty of our eventual death. The fundamental alienation of a creature doomed to die is thus a reassertion of our inescapable biology.

As Arthur Koestler emphasized, we are not "programmed" to conceive of our own deaths. When a computer confronts something for which it is not programmed,

> it is either reduced to silence, or it goes haywire. The latter seems to have happened, with distressing repetitiveness, in the most varied cultures. Faced with the intractable paradox of consciousness emerging from the pre-natal void and drowning in the post-mortem darkness, their minds went haywire and populated the air with the ghosts of the departed, gods, angels and devils, until the atmosphere became saturated with invisible presences which at best were capricious and unpredictable, but mostly malevolent and vengeful. They had to be worshipped, cajoled and placated by elaborately cruel rituals, including human sacrifice, Holy Wars and the burning of heretics.

Alienated from the facts of our lives, we perpetrated death on others.

Within eight years of writing these words, the brilliant Koestler, nearly eighty years old and terminally ill, took his own life.

CHAPTER THIRTEEN

THE BIOLOGICAL FUTURE

ADVICE FROM THE CHESHIRE CAT

The horseman serves the horse,
The neatherd serves the neat,
The merchant serves his purse,
The eater serves his meat;
'Tis the day of the chattel,
Web to weave, and corn to grind
Things are in the saddle,
And ride mankind.
—RALPH WALDO EMERSON

And what does biological evolution have in store for us dual creatures? Will we ride our culture or vice versa? It is inconceivable that we shall become in any way less biological in our nature and yet, equally unlikely that culture will somehow exert a lesser influence than it does today. In fact, despite occasional minority pushes for simplification in life style, we can expect a continuing increase in the technologic, abiologic aspects of human existence. Of course, we will also remain subject to natural selection. So long as some people have more offspring than others (for whatever reason), selection will occur, with the population's gene pool shifting in their direction.

But not only will this selection be—as usual—too slow to

counteract or even effectively harmonize with cultural evolution, it will also become increasingly less "natural." As birth control information becomes more widely disseminated, the major factor controlling reproductive success will be the rational intent of the prospective parents. Religious affinities, ethical and moral judgments, financial considerations, political and social atmosphere, and a host of personal factors will be the ultimate criteria, so that the evolutionary fitness of each individual and family will probably have very little genetic basis. Even the evolutionary fitness of genes will have little relationship to the characteristics of the genes themselves. This conscious, extragenetic influence on biological selection is unique in the history of evolution and there is no predicting where it will lead.

But it is nothing new for our species. "There is no stopping place in this life," observed the German theologian Meister Eckhart nearly 700 years ago, "no, nor was there ever one for any man, no matter how far along his way he'd gone. This above all, then, be ready at all times for the gifts of God, and always for new ones." We can substitute cultural evolution for God, if we wish, but the message remains. Thanks in part to nuclear weapons, environmental pollution, resource depletion, and overpopulation, the future isn't what it used to be. But it still is.

In the twentieth century, Alfred North Whitehead observed that it has always been the business of the future to be dangerous. Maybe so. Never before, however, has the future been doing its job quite so assiduously.

Insofar as biology goes, the human evolutionary future, in fact, is neither completely obscure nor demoralizing. We can assume that Homo sapiens will carry an increased load of deleterious genes (i.e., become somewhat "inferior") and will therefore come to rely increasingly upon supportive technology. This seemingly glum prophecy is not really so dire, or even so undesirable, as it might sound, however. It simply reflects our growing proficiency in the healing arts. To take one example, sufferers from diabetes had little hope just one hundred years ago; today, with artificially administered insulin, they can live virtually normal lives. Without medical treatment, diabetics are much less likely to have offspring than are healthy individuals; with it, there may be no difference in reproductive rates.

Since diabetes is genetically influenced, we can expect an increase in the frequency of associated genes, the natural selection pressure against them having been relaxed by an important cultural practice. Of course, such an increase in diabetes is less than catastrophic, since we have the capability to control the disease; among diabetics and their relatives, the merits of insulin injection can hardly be questioned. (Someday soon, we shall probably have the technology for pancreas transplants.) Ethically, we have no choice but to persist with cultural developments of this sort, even though they force the human species higher and higher up a slippery pyramid of dependence on cultural evolution. It is often easier to climb up than to get safely back down.

Among our biological characteristics, eyesight is a good example of the modification of "natural" selection by cultural evolution. Good vision must have been extremely important among our ancestors, as it still is for nontechnologic people today. It is obviously desirable to spot prey at a distance, ideally, before they can spot you, as well as to be aware of predators. Individuals with severe myopia would be at a selective disadvantage, and we might expect genetic combinations for good vision to be maintained at a high level in populations exposed to purely biological evolution. But, thanks to the optician's art, 20-20 vision is now bestowed upon anybody who can afford it, granting equal chances of survival and reproduction regardless of genetic constitution. When it comes to evolutionary change, culture has become the great equalizer. Even if culture was somehow unable to correct artificially for genetic inequalities in vision, other aspects of our culture have rendered visual acuity increasingly less important. Thus, possibly aside from increased danger of being hit by a car or walking into an open manhole, the visually incompetent in modern society would probably experience no severe reproductive difficulties.

The necessities of survival and reproduction are available for most Americans, and for many human beings the world over. "Superior," fortunate, or simply ruthless persons may often avail themselves of more comforts, luxuries, and satisfactions, but at least in America it is unlikely that they will have larger families. In fact, the reverse may be true. The "right" of all people to have children is not at question here: like all other rights, it is granted by society and is presumably beneficial to society as a whole. Nonetheless, our culture is tampering with the fundamentals of our biological evolution, and we should be prepared for the consequences, or rather, for the fact that there will be consequences, even if we cannot now anticipate them.

We were practicing *artificial* selection for thousands of years, long before the genetic basis of its success was understood. We have created distinct strains of dogs, for example, by choosing those individuals with characteristics we wanted, and causing them to breed with each other while prohibiting "undesirable" matings. By thus interceding between our domestic animals and natural selection, we have produced such divergent forms as the Chihuahua and the St. Bernard. From cow to chicken, our common farm animals also differ greatly from their wild relatives, upon whom only biological evolution had operated. Nearly all our food plants are likewise the result of artificial selection, by which we have "designed" strains with higher yields, greater disease resistance, and so forth. There is nothing new, then, in artificial selection; when it comes to biological evolution, Homo sapiens is in the habit of playing god.

But although our domestic plants and animals are "better" (for our purposes) than their wild relatives, they are also generally inferior to them in direct competition. For example, compare the barnyard pig with its probable ancestor, the wild boar. The farmer's pig has comparatively dull senses: keen vision, hearing, and smell are not of particular selective value in the barnyard environment, while they are clearly important for the free-living boar, a vibrant, alert, and energetic animal. The pig is slow moving and lethargic, a selective advantage to the farmer, who prefers his animals to gain weight rather than waste a lot of calories running around. The contrast between the domestic and the free-living is striking.

There may be a parallel between the evolutionary changes brought about by domestication and those potentially induced in ourselves by civilization. In both cases, biological entities are isolated from the rigorous pruning of natural selection. Instead, human beings have become the gardeners—and in our gardens, we cultivate ourselves.

In popular stereotype, the future human being is pictured with an enormous head perched upon a feeble body. This is a misconception. It is probably based on unconscious Lamarckian thinking. Under this view, traits acquired during a lifetime are passed on to one's children. Thus, it is reasoned that since we seem to use our brains more and more—reflecting the increased pace of cultural evolution—our successors will somehow be born with larger and larger heads. For this to happen, however, brains (like muscles) would have to get larger when they are used. Moreover, there would have to be a selective advantage to large brains, as there was when culture was first developing. In other words, for biological evolution to proceed in this way, people with greater intelligence would have to be producing more children than those with less intelligence.

As mentioned above, there is some reason to believe that if anything, the opposite may be occurring, with culture now operating to make possible equal genetic contributions from all, regardless of the biological qualities of each. There is even some potential for modern cultural evolution to reverse our earlier biological trend toward increased brain size. Thus, enlightened individuals with greater access to contraception and greater sensitivity to its importance are probably more likely to have smaller families.

Actually, we must carefully refrain from judgments of "superior" or "inferior" in this regard, since for biological evolution these terms simply mean the capacity to produce successful offspring and other carriers of one's genes. As far as evolution goes, lower intelligence may well have become superior. The distinction between biological and cultural values is underlined by the fact that Leonardo, Newton, Beethoven, George Washington, and Jesus Christ were all failures of natural selection, since none produced any offspring.

The situation is actually cloudier yet. Although it is fairly safe to say that socioeconomic level correlates inversely with family size—wealthier people generally have fewer children—there is virtually no hard evidence showing a consistent correlation between socioeconomic level and intelligence, and demographic trends are notoriously variable. The Kennedy mystique made large families fashionable; ecological consciousness caused a reversal; no one knows what will come next. Also, it is not necessarily true that brain size and intelligence are very closely related, at least within the current normal human range. Albert Einstein had an unusually small brain.

Although selective pressures on the human population have changed during the course of cultural evolution, so that we are exposed less and less to selection based on "traditional" biological qualities, natural selection is still occurring; only the framework has changed. One thousand years hence (or maybe even one hundred) biological evolution may be selecting for abilities to withstand the new pressures of human existence. Just as the use of DDT selected for resistant flies by killing off the vulnerable ones, increasing air and water pollution may select for people with inherent resistance to various contaminants. Just as the discovery of insulin has slowed down selection against diabetes, maybe increased smog levels will select for greater respiratory resistance, by killing off people with a tendency toward emphysema or lung cancer. (It is said that there are only good drivers in New York City; all the bad ones have been killed off.)

When an organism is placed in a contaminated environment, selection favors those individuals that can live and reproduce in spite of it. Perhaps the future human beings will be those who can live and reproduce despite their new internal environment of lead in their blood, mercury in their brains, strontium-90 in their bones, and PCBs in their fat. Beyond this, cultural evolution is providing a whole gamut of new, rather subtle environmental changes, which may also exert novel selective effects on Homo sapiens in the future. Our successors, for example, will have to cope with high noise levels, crowding, and the increased pace and attendant strain of modern technological life.

While this analysis may sound logically correct, it could equally well be all wrong. We are talking about many years and many generations. Given the present rate of cultural evolution, no one can predict the human environment just a single generation from now, never mind the hundreds required for substantial biological change. Modern physical contaminants and social environments may be entirely gone in a century or so, replaced by a whole new set of culturally generated conditions. Organisms in nature eventually reach a sort of equilibrium with their environment, so long as the environment itself remains more or less constant. But if the surroundings won't hold still, there is little chance for the living things to catch up. This may happen to us.

It is curiously satisfying to imagine a future in which biological evolution has somehow attuned us to the productions of our culture. But it is unlikely. Not only will culture always outstrip biology, but strangely enough, our own physical hardiness and short-term adaptability must ultimately get in the way of a lasting biological accommodation. Natural selection is generally concerned with an organism only *until* it has reproduced. This is why most living things do not live very long after breeding: having produced their offspring, and nursed and weaned them to independence, evolution is no longer very much interested in them. More accurately, deleterious mutations would quickly be selected against if they expressed themselves in a way that interfered with fitness; after one's genes have been sent on their way, the fate of the progenitors exerts relatively little effect, any more than a rocket designer is interested in the fate of the booster stage after the satellite has been launched.

Human beings, possibly whales and elephants, and perhaps some of the other primates are exceptions. They experience a fairly lengthy postreproductive life span because among animals for whom learning and judgment are important, the greater experience of older individuals is of selective advantage. With the invention of writing, the printing press, microfilm, and computer data processing, the value to society of elderly and knowledgeable people might seem to have progressively declined. If the pace of cultural change continues to accelerate, causing experience to be a detriment insofar as it is rooted in a different and hence, irrelevant past, we could also expect this trend to accelerate as well. (It seems noteworthy that this potential change in the usefulness of the elderly may be more true for technology than for nontechnical aspects of culture, such as religion, diplomacy, history, and so forth. And even with regard to the latest in path-breaking technology, graybeards may be of special value, in helping us recognize how new developments depart from previous experience, how on the other hand such developments may be simply reinventing the wheel, and/or how they may offer dangers as well as promises, based on similar events in the past.)

In any case, for natural selection to be effective in biologically adapting the human population to its culture, our new environments must somehow influence reproductive performance. This in itself is unlikely, since we are sufficiently sturdy that most culturally induced disease and illness do not seriously affect us until middle or old age. Thus, although acute mercury and lead poisoning may strike young children, most of the stresses produced by cultural evolution result in so-called degenerative diseases that have little influence on reproduction. Such increasingly important illnesses as heart disease, emphysema, and cancer seem to be caused by an accumulation of environmental "insults" like stress, anxiety, and a whole range of irritating and dangerous substances, chemicals, and situations. Since they generally don't affect us until after we've reproduced, however, there is little opportunity for those with greater resistance to leave more offspring. It is therefore unlikely that we could evolve much resistance even if the rate of cultural change was drastically slowed down.

But even if evolution doesn't care about the increased prominence of degenerative disease, we should. As part of the higher aspects of our mental evolution, we have developed a profound sensitivity to the survival and suffering of others. Perhaps this is an outgrowth of kin selection or reciprocity, or perhaps it is a purely cultural creation. In any case, we have reason to care, if only because we can easily imagine ourselves in the situation of others, and this capacity of anticipating the future is by itself another uniquely human attribute. By their nature, however, degenerative diseases build up slowly and often they are not clearly associated in the public mind with the environmental factors that are actually to blame. This makes alleviation of the problem yet more difficult.

Heart disease is probably our most serious degenerative illness; it has achieved almost epidemic proportions in the United States. This example is particularly interesting since it relates to the rest of the big-brained stereotype, the deteriorating body, while it also provides one of the most striking examples of the conflict between biological and cultural evolution. Our frighteningly high incidence of heart disease results especially from four combined factors: stress, eating habits, lack of exercise, and the use of certain drugs, notably tobacco and alcohol. All four have important biocultural aspects. We have already considered the likely contribution of cultural evolution to stress. Let us now consider the remaining three.

There is no doubt that being overweight is hard on one's heart. Our eating habits leave much to be desired. We seem to like things that aren't good for us. Why?

Here is a possible scenario:

Our ancestors probably ate lots of fruit. Modern primates still do. Fruit is nutritious and also high in sugar. It therefore became selectively advantageous for primates to like such foods, so enjoyment of sweets became part of our biological makeup; that is, individuals with such preferences left more successful descendants. As cultural evolution progressed, we began creating our own culinary environments, instead of simply depending on whatever nature had provided. We processed our own food, and, given our primate predisposition for sweets, we learned to make a whole array of "taste-tempting" delights. These tend to be very high in sugar and carbohydrates, and often little else. Human cultural evolution has thus taken a basically adaptive biological characteristic and perverted it, much to our immediate pleasure and ultimate detriment. Like our preference for big-headed cartoon children and big-busted silicone women, our sweet tooth is another culturally exploited supernormal releaser, a form of cultural hyperextension that has been paying special dividends to the confectionery and mortuary industries, and more recently, to physical fitness parlors.*

It seems "natural" for us to like sweet things, and indeed it is, just as it is natural for a raccoon to like crayfish or an anteater, ants. Our fondness for the taste of sugar is predicated upon its typical occurrence in nutritious foods present in the early human environment. Interestingly, the chemical lead acetate also tastes sweet to us, but is a deadly poison. Lead acetate is not normally found in our environment, and never was, certainly not in any abundance. If it had been, we would have evolved the ability to distinguish its taste from that of sugar, or we would never have considered both to be "sweet," or we would have gone extinct long ago.

In addition to sweets, we are also victimized by our fondness for "rich" foods, high in cholesterol. And this may be another ex-

*In *Sweetness and Power,* anthropologist Sidney Mintz has recently detailed how our biologically influenced fondness for sugar has become interwoven with a wide array of complex cultural practices, including slavery, imperialism, power politics, and industrialization.

ample of supernormal releasers and cultural hyperextension, due this time to the carnivorous habits of our Australopithecine ancestors. Wild game is usually quite lean, and animal fat, with its high energy value, was therefore undoubtedly prized by early hunters, as it still is among nontechnologic people. Our domestic animals currently produce enormous quantities of fat, and the beef industry gives special attention to well-marbled meat, which we find tasty and highly desirable. Unless restrained by new-found cholesterol consciousness, we gobble it up even as it lethally clogs our arteries.

Insufficient exercise is another prong on our trident of factors causing high rates of heart disease. To see its relevance to our theme, we must go back once again to the African savannah. The world in which human beings evolved was intensely physical. We had to run from our enemies, run to catch prey, walk long distances for water or shelter. Our physiology and anatomy, in particular our circulatory system, evolved with the expectation of vigorous stimulation for growth and proper maintenance. But although exercise may have become necessary for our health, we didn't have to like it.

One of the most notable achievements of cultural evolution has been the steady replacement of human labor by machines. Transportation has been particularly revolutionized, to the point that although an Australian aborigine may think nothing of covering twenty miles a day on foot, the average American would be horrified at the thought of walking a mile or two. We do almost all our traveling on our backsides, and our electrocardiograms have been showing it.

But if we evolved a fondness for sweet things "because" they were good for us, why didn't we also evolve a natural fondness for exercise, which is equally beneficial? Perhaps physical exercise—unlike ripe fruit—was unavoidable for our ancestors, so there was no need to generate a predisposition for it. Furthermore, it was certainly advantageous for the early hominids to take shortcuts and use cooperation or primitive tools to save energy wherever possible. We may therefore have evolved both a need for exercise and a tendency to be lazy.

Finally, what about tobacco and alcohol? Human beings are stimulus-hungry animals, and not only for sights and ideas but for things that we eat, drink, and breathe as well. Even today, perhaps the best assurance of a nutritionally adequate diet is one that is varied. We possess—and sometimes we appear to be possessed by—a powerful drive to stimulate our taste, smell, and other senses with sensations including those provided by alcohol, tobacco, and spices. It is not widely appreciated that prior to 1492, smoking tobacco was unknown to the rest of Homo sapiens, and that what we might call the Red Man's Revenge has been quite spectacular if not complete: it is almost certain that however many Native Americans were slaughtered directly by the invading Caucasians, and indirectly by their alcohol, even more Caucasians have been killed by tobacco in the years following.*

Freud once referred to the human species as a prosthetic god. With culture hyperextending that prosthesis, we are able to employ our artificial limbs to great effect. But unlike gods, we can also get "stressed out," paunchy, flabby, atherosclerotic, hypertensive, and we smoke and drink too much.

If biological evolution did occur by the inheritance of acquired characteristics, our proverbial weak-bodied human would undoubtedly be the *Homo* of the future. However, there is presently no evidence that weak-bodied people leave more successful offspring than do sturdier individuals; evolution of this sort by natural selection is therefore highly unlikely. So the best we can say is that whereas culture is undoubtedly having an enormous influence on our current biology, the shape of our future—and indeed, of *us*, in that future—is virtually unknowable.

Genetic diversity is essential to biological evolution and to the survival of ecological systems. When a new environment provides

*Poetic justice, perhaps, like the observation that in North America, the buffalo are coming back, while the railroads are going extinct!

new situations, natural selection chooses the best from among the available variety. The greater the variety, the greater the chance of good adaptation, and the lesser the danger of extinction. And yet, cultural evolution appears to be drastically reducing the worldwide biological and cultural diversity, thereby jeopardizing the future.

The last few hundred years have been a time of cultural homogenization, as diverse human societies—generally "primitive" in their technology—have disappeared, to be replaced by imitations of Western culture. Visit any of the world's large cities: the languages may differ but the life styles are essentially the same. The same clothing is worn in Nairobi, Bogotá, New York, Tokyo, and Brussels. When a human culture is gone, it is virtually lost forever, almost like a species that has gone extinct.

One of the dangers of a "monoculture" such as a cornfield is that it is unstable, and at the mercy of environmental change. Cultivated monocultures may do beautifully for a while, given constant human attention, but when a pest such as corn blight comes along, the whole crop can be wiped out. If all the corn plants are from the same strain, as they frequently are in modern "scientific" agriculture, there is no diversity to provide a reservoir of resistant individuals. In addition, healthy diverse systems have a self-damping effect on perturbations, analogous to negative feedback. Epidemics, for example, do not readily spread when susceptible individuals are surrounded by others in whom the disease cannot take hold. The parallel to human societies is not mere analogy.

In natural communities, each organism may produce descendants, and not only individuals but species as well thus tend to replace themselves sequentially. Imagine leaves in a deciduous forest during autumn: each falls separately, and is replaced in turn. To be sure, there are connections and crucial interdependencies, but also enough separation that failure for one is not disaster for all. With the advent of modern worldwide culture, however, we have not only homogenization but a tendency toward convergence as well, as though the fate of all the leaves in a forest was depending on the success of just one tree. The world, in short, has become very tightly wired: what happens in one place affects others. Court intrigues among the Incas 1500 years ago had virtually no effect on the simultaneous goings-on at Camelot. But now, when Beirut, Moscow, or Tokyo sneezes, someone on the other side of the world is bound to say *Gesundheit.*

Different human societies have different requirements, survive in different ways, and contain differing reservoirs of competence in dealing with the world's realities. With only one "world culture," we would be greatly limited in our ways of making a living. With diversity wiped smooth, we would all be equally susceptible to environmental perturbations such as epidemics or resource shortages. Furthermore, as local cultures go extinct in favor of the highly technological society of the West, all of the world's people come to rely more and more upon technology's elaborate superstructure. This may cause a temporary improvement in living standards, but at a dangerous cost, not only in the aesthetics of human variety, environmental abuse, alienation, and resource consumption, but also in social flexibility and adaptability.

Aside from all the personal, rootless insecurity that comes from a loss of cultural identity, such rapid change often transforms a society that had been coexisting harmoniously with its environment into one that is antagonistic, no longer fitting in with its land or each other. Furthermore, excessive worldwide reliance on a complex technology places the human species at the mercy of breakdowns in that technology.* Imagine the Eskimo, having abandoned his dogsled generations ago in favor of the snowmobile, who suddenly finds parts and fuel unavailable, for any of a variety of reasons. The dogsled required only what the arctic environment naturally provided: wood and sinews for the sled itself, fish or caribou meat to feed the dogs.

*A cartoon once depicted two Eskimos turning to each other, as the nuclear missiles arched back and forth overhead, saying, "Well, that's the end of civilization as *they* know it." Increasingly, it is civilization as the Eskimos know it, too.

Given their native cultures, representatives of the human species in the far north, for example, could conceivably survive a wide range of cataclysms, perhaps even nuclear holocaust itself. But once they become reliant upon a single culture, the fate of the entire human species becomes concentrated into a very few hands. Even now, humanity is composed of a diversity of people, living in different ways, in places ranging from deserts to tropical rain forests, to mountain meadows, ocean shores, and the Arctic. If we contain the nuclear Neanderthals among and within us, we stand a reasonable chance that some of us at least will experience a long evolutionary history. But as we become a homogenized monoculture, we put all our eggs in one basket, and a fragile one at that.

The danger in reducing human diversity goes beyond our destruction of indigenous cultures. It extends to the more fundamental genetic level of biological characteristics as well. Thus, each human racial group possesses certain distinctive genetic combinations, some of which are ultimately responsible for our recognizing the different races in the first place. Although there is no basis whatever for comparing the different races on any global scale of relative quality, it is certainly possible that each race is better adapted than any other to its particular environment, because at least in the past, natural selection has operated on the inhabitants of each given region, adjusting the surviving occupants to local conditions and vagaries. It is often difficult to identify the specific advantages of each characteristic, but the black African's skin is probably an adaptation preventing dangerous sunburn and aiding the radiation of excess heat, while the very pale Scandinavian complexion may be a reverse adaptation, enabling the cold-climate inhabitant to gain maximum advantage from the limited available sunshine. The short, squat Eskimos are well adapted to conserve body heat in their cold environment, and the tall Zulus to radiate it.

Isolated human populations often possess distinct frequencies of blood types and also, differences in disease resistance. The

American Indians and Eskimos were decimated by pneumonia and tuberculosis, to which they had little resistance. Similarly with the native Hawaiians and measles, a disease that was rarely fatal to the Western missionaries who brought it as a bonus along with the New Testament. When Charles Darwin made his famous voyage around the world, collecting much of the information he would later use to buttress his theory of evolution by natural selection, he visited Tierra del Fuego. These bleak and stormswept islands at the extreme southern tip of South America present a formidable environment for human beings. Yet they were inhabited by two tribes, the Onas and the Yaghans, who lived very simple lives of fishing, hunting, and gathering, entirely isolated from the "civilized" world of the nineteenth century. Today, both groups have been civilized to extinction.

To the hard-hearted objectivist, with no more empathy than knowledge of evolution, this may even seem like a good thing, resulting in "improvement" of the species. But if anything, the opposite is true. Although we have thus far considered local extinction as the elimination of "unfit" types (e.g., the demise of a people *lacking* certain disease resistance), these people may have also possessed various unique genetic attributes. Potentially "desirable" traits are lost forever along with their carriers. The Onas and Yaghans, for example, had an almost superhuman resistance to cold and physical privation, enabling them to live near-naked in virtually constant thirty-degree wind and sleet. But we shall never know how they managed it.

Although there is a particular poignancy about the loss of certain human biological traits, such as the Fuegians' cold tolerance, other vestiges of our biology are less appealing. Sickle cell anemia, for example, is a serious genetically transmitted disease of American and African blacks. It is caused by a double dose of a gene that is prevalent in equatorial West Africa, among other places, the original homeland of most of the American black population. The disease causes malformation of the blood cells ("sickling"), clogging small blood vessels and causing great pain and debility, while interfering with the transport of oxygen to the body tissues. Considering the severity of sickle cell anemia, one might expect natural selection to have eliminated it long ago from the African gene pool. But the gene has flourished, sometimes reaching a frequency as high as 40 percent. This is because a single dose of the sickling gene causes partial immunity to malaria, which is frequent where the sickling gene also prospers. Thus, despite its strong disadvantage in double dose, its advantage in single dose has kept the gene around. Once malaria is eliminated or adequate preventive drugs are available, or the people afflicted live somewhere else, there is no more "saving grace" to sickle cell anemia. There is no ethical argument for saving the sickle cell gene. But it should at least be clear that few things in nature are all bad or all good.

"Would you tell me, please, which way I ought to go from here?," Alice, lost in Wonderland, asked the Cheshire Cat.

> "That depends a good deal on where you want to get to," said the Cat.
>
> "I don't much care where—" said Alice.
>
> "Then it doesn't matter which way you go," said the Cat.
>
> "—so long as I get *somewhere,*" Alice added as an explanation.
>
> "Oh, you're sure to do that," said the Cat, "if you only walk long enough."

It is "natural" for a species with a well-developed culture to tamper with its biology, and unnatural to refrain. British scientist Dennis Gabor once suggested that our job was to invent the future. And one way or another, as the Cheshire Cat says, we are sure to do that. We may not know who is in the saddle, or exactly where we are going, but we're certainly on our way.

CHAPTER FOURTEEN

THE CULTURAL FUTURE

NED LUDD VERSUS THE THREE FACES OF FAUST

Civilization is hooped together, brought
Under a rule, under the semblance of peace
By manifold illusion: but man's life is thought,
And he, despite his terror, cannot cease
Ravening through century after century,
Ravening, raging, and uprooting that he may come
Into the desolation of reality.

—W. B. YEATS

In 1779, a simpleton named Ned Ludd lived in the British village of Leicestershire. As was—and still is—the way of boys, young Ned was often teased by the other village children. One day, he chased one of his tormentors into a house, but was unable to catch his quarry; the house contained two weaving machines, however, recently developed for the manufacture of stockings, and Ned Ludd vented his frustration and anger by destroying one of the machines. From then on, whenever a machine broke in Leicestershire, Ludd was blamed. Several decades passed, and by the end of 1811, the war with Napoleon was adding severe economic hardship to the already difficult conditions of newly industrialized Britain. Bands of rioters spread throughout Nottingham and its neighboring districts, destroying machinery in Yorkshire, Lancashire, Derbyshire, and Leicestershire, under the leadership of one "General Ludd," who may well have been mythical.

In any event, the "Luddites" angered the authorities, and struck fear into the newly emerging barons of the Industrial Revolution. The Luddites, too, were driven by anger and fear: anger at the Industrial Revolution itself, which was destroying British cottage industry, raising fear of widespread unemployment as a result of automation, as well as forcing those "lucky" enough to be employed to experience hideous working conditions. The rioting Luddites were harshly suppressed. Following the battle of Waterloo and the defeat of Napoleon, however, Britain suffered a depression, and in 1816, the rioting began again. This time, it spread across all of England, although once more it was eventually put down.

However much sympathy one might feel for the Luddites, we shall gain nothing by mindlessly smashing machinery, neither literal machinery nor the "machinery" of modern society. On the other hand, not all innovations and technologies are good, and not all of our cultural creations deserve respect, or even defense. Some warrant reproach, others (like nuclear weapons) deserve only to be dismantled and perhaps smashed to boot, whereas others (like vaccines, contraception, or literacy) warrant the widest possible dissemination. The quandary of biological Homo sapiens caught in a web of his and her own cultural creation demands a response that is worthy of our designation *sapiens* (the wise) not the inarticulate fury of young Master Ludd.

Unlike Rousseau, we have no reason to extol the noble savage, who probably wasn't all that noble anyhow. And unlike Thoreau, we cannot even retire to Walden Pond (which, incidentally, is now a popular tourist area). We must make the best of our divided existence as cultural animals. Just as we cannot deny our biology, we cannot deny our biological predisposition for culture and the fact that we shall never be able to turn aside from cultural evolution.

But with a clearer vision of what we want for ourselves and

our world, perhaps we can walk more carefully, stubbing our toe here and there, but at least staying on the high ground. Paul Valéry captured the shock, cynicism, and confusion of his generation following World War I when he wrote: "We hope vaguely, but dread precisely; our fears are infinitely clearer than our hopes." In reconciling the hare and the tortoise, one challenge is to fashion hopes that are no less precise, and hence no less attainable, than our fears. With a triumphant gasp of cultural awareness possibly we can even transcend ourselves, assess our position and its causes, and resolve to proceed with a clearer view of ourselves, our needs and our problems. We can evaluate our cultural inventiveness in the light of both our biology and our unique ethical perceptions, and then resolve not to confuse change with progress. After all, it is as absurd to claim that we cannot turn back the clock as it is to harrumph reflexively in opposition to everything that is new. We can, in fact, turn back a clock (even a digital one!) if we elect to do so. Whereas time itself is unstoppable, we can use the time allotted to us as we ourselves see fit. But we cannot turn back on our commitment to cultural solutions to our problems without turning our backs on being human.

Culture is not on trial here, only our uses of it. Given the flexibility of culture, it should actually be more adaptable than our biology alone. As we have seen, the problem is that often, culture is too flexible, serving as a lever that amplifies our biologically given inclinations, hyperextending our capabilities into situations that would be much less momentous if our biology alone were at work. This is why we now push so uninhibitedly on the cultural lever: for thousands of generations, most of today's cultural hyperextenders didn't exist and so our behavior was not harmful. In order for the interface between culture and biology to be benevolent, human behavior—motivated at least in part by biology and acting via culture—must not only *feel* right but also *be* right.

The cultural genie cannot be stuffed back into the bottle, but he can, perhaps, be made to obey. Is it hopelessly utopian and unreasonably idealistic to ask that we learn to measure our culture in terms of human beings rather than measure human beings in terms of machines or institutions? We cannot change "human nature," but we can seek to harmonize our culture with ourselves, always mindful however that what is biological is not necessarily what is good, and knowing also that we may never truly know ourselves.

Each situation must be considered on its own merits in attempting to prescribe remedies. In some cases, notably aggression and overpopulation, our species is necessarily committed to a cultural solution if there is to be any rational solution at all. Of course, all solutions are actually "cultural" in that they must be consciously determined and implemented. The question is whether we should try to move closer to our biology or yet farther away, and the answer must vary according to the tendency in question. Whatever the answers, a clear view of the hare and the tortoise might help us formulate the right questions.

In suggesting possible avenues of reconciliation between our biology and our culture, it is not surprising that being trained as a biologist inclines me to respect the former somewhat more and to criticize the latter. Thus, I would generally have us climb down from our dizzying heights of cultural and technological performance—modulate if not abandon our edifice complex—a recommendation that has been made by others, notably E. F. Schumacher, who has emphasized the value of "intermediate technology," "small is beautiful," and the merits of considering "economics as if people mattered." "Any third-rate engineer or researcher can increase complexity," writes Schumacher. "It takes a certain flair of real insight to make things simple again." In doing so, we might also make ourselves human again.

It is sometimes said among psychotherapists that the patient cannot be brought to a greater level of mental health than that of his or her therapists, just as a guru cannot be expected to lead disciples to more enlightenment than he or she has already attained. Similarly, we can use our diverse cultural innovations for good, but to no more good than we, as human beings, are capable of directing. We design and build societies and machines in the hope of achieving a degree of independence from, and control over, nature. But in the process, we empower ourselves with all the potent leverage of cultural evolution's febrile inventiveness, unleavened by any biologically based restraint. Disciples of Schumacher's approach, and those persuaded, perhaps, by the argument of this book, might argue that for human beings to be psychically sound and materially healthy, and for them to maintain themselves on their planet and with each other on a sustainable basis, they must integrate themselves more comfortably with their cultural evolution. Since we cannot speed up biological evolution, this requires careful scrutiny and often, simplification, of "progress."

On the other hand, many would recommend climbing higher yet. Psychologist B. F. Skinner, for example, while apparently recognizing the awkwardness of our present position, rather predictably takes this alternative view. In his book *Beyond Freedom and Dignity,* he suggested a massive increase in our reliance on cultural artifice, to result ultimately in a "technology of behavior" that will somehow harmonize humanity with its unique self-made environment. Depending on whom one listens to, we therefore might help ourselves by weakening the influence of culture, or by strengthening it. Either way, it seems incumbent on Homo sapiens to change things selectively.

The hare and the tortoise: culture and biology have been racing within us throughout our history. In recent years the hare has been accelerating mightily and the gap between the two has been widening. Reworking the analogy: we have one foot on the tortoise and the other on the hare. As they get farther apart, we are stretched more and more. It's getting awfully uncomfortable.

In *The Future as History,* Robert Heilbroner warned that

> When we estrange ourselves from history we do not enlarge, we diminish ourselves, even as individuals. We subtract from our lives one meaning which they do in fact

> possess, whether we recognize it or not. We cannot help living in history. We can only fail to be aware of it. If we are to meet, endure and transcend the trials and defeats of the future—for trials and defeats there are certain to be—it can only be from a point of view which, seeing the future as part of the sweep of history, enables us to establish our place in that immense procession in which is incorporated whatever hope humankind may have.

Replace "history" with "biology," and you have a reasonable summary of humanity's problem and its promise.

***

> *We are on Mr. Toad's Wild Ride, and though we may become "velocitized" and insensitive to our speed, or bored, or content to putter about in the apparently motionless comfort of the back seat, at some point we must dare again to look out the window and perceive that we are careening beyond the very human scales of space and time.*
>
> —Walter A. McDougall

According to legend, during the latter part of the sixteenth century there lived in Germany a man named Georg Faust, a magician and conjurer, who blasphemed, bragged about trafficking with the devil, and alternately amazed and irritated his contemporaries. His story was ultimately embellished in theater, opera, and literature, notably by the English playwright Christopher Marlowe, the German poet Johann Wolfgang von Goethe, and the novelist Thomas Mann. In one form or another, the Faust legend has had an extraordinary hold on the Western imagination, and no wonder. Marlowe's *Doctor Faustus* portrays the most widely known version, the tale of a brilliant polymath who turns ultimately to necromancy and thence to a pact with the devil, offering his soul in return for a life of power and voluptuous sensuality. He has his day (actually, twenty-four years) but in the end, pleading and despairing, Faust pays for his excess as he is borne away to eternal damnation by a company of devils.

Mann's *Doctor Faust* is considerably more modern in outlook, not surprising for a novel written in 1947. It tells of Adrian Leverkuhn, an arrogant, sickly musical genius who once again trades his soul, this time in return for creative accomplishment. In Mann's treatment, Leverkuhn's rise clearly parallels the rise of Nazism and the tortured megalomania of overwrought grandiosity and ultimate, inescapable doom.

Both Fausts are cautionary tales, Marlowe's focusing on the excesses of personal ambition and Mann's, on the social sanction of unwarranted power. In each, the central character is a tragic victim in the classic, Aristotelian sense: a hero laid low by a fundamental, inner flaw. And both stories seem to accord with our own flaw as well: Homo sapiens has made a comparable bargain, although the terms were never made explicit nor was the contract ever consciously drawn up. It was, however, sealed with our blood. Faust practiced diabolic hyperextension; ours has been merely cultural. But we have nonetheless been living in the Faust lane.

There is a third face of Faust, however: the one depicted in Goethe's magnificent theater/poem. It presents a different and rather more optimistic vision of Faust, and thus, of humanity. At the play's outset, God and Mephistopheles are debating the merits of Homo sapiens, the latter claiming

> The little god o' the world sticks to the same old way,
> And is as whimsical as on Creation's day.
> Life somewhat better might content him,
> But for the gleam of heavenly light which Thou has lent him:
> He calls it Reason—thence his power's increased,
> To be far beastlier than any beast.

As to the difference between humanity and other animals, Mephistopheles observes

> . . . he to me
> A long-legged grasshopper appears to be,
> That springing flies, and flying springs,
> And in the grass the same old ditty sings.

Would he still lay among the grass he grows in!
Each bit of dung he seeks, to stick his nose in.

A wager is made, with Mephistopheles betting that Faust will succumb to his diabolic temptations, and God betting that ultimately, "I shall lead him to a clearer morning." And of course, this eventually happens. After the furious passions of the Walpurgis Night, a love affair with Helen of Troy, myriad intellectual and scientific accomplishments, political and social success, and great martial triumphs, Faust turns finally to a recognition of human social obligations, whereupon he envisions a community of free people, under his tutelage and protection, living out their lives on free, reclaimed soil. Thereupon, he expresses his satisfaction and his desire that such a moment would last forever. This is what Mephistopheles was waiting for and the terms of his original contract: Faust had not been so much power-mad, as restless, a grasshopper compulsively sticking his nose into "each bit of dung." According to his deal with Mephistopheles, when Faust finally achieved true satisfaction on earth, then and only then would he be damned. Yet, when that finally occurs and Faust dies, God intervenes and has him transported to heaven as reward for his ultimate recognition of the highest goal of human existence, responsibility for others.

The limits of human potential are glimpsed, if at all, with much greater difficulty than are the outlines of our folly. In seeking to become angels, warned Pascal, we risk becoming less than men. His warning can be inverted as well: in seeking to become angels, we cannot become more than men or women. Perhaps Mephistopheles and a cynical reading of human biological evolution are correct, and we are doomed eternally to sing "the same old ditty," big-brained grasshoppers without even the good grace to recognize and act within our nature. On the other hand, perhaps we can reach for a deeper humanity, thumb our noses at Mephistopheles and at the grasshopper within us, saying "no" to the siren whisperings from our genes—or alternatively, to certian blandishments of our culture—whenever this seems necessary to achieve a truly sapient *Homo.*

Any effort to reveal human nature must unavoidably grapple with and acknowledge fundamental uncertainties. At minimum, we seem doomed to a kind of existential agnosticism, never knowing—and perhaps, never able to know—ourselves. What, then, should we do? It would be foolhardy if not irresponsible to let "nature" take its course, either our own biological nature, or the nature of human culture as currently constituted, because nature isn't necessarily good, and moreover, nature at this point has no course independent of the one we plot. We are also deeply ignorant both of our limitations and our capabilities. For a useful recommendation, we might turn from Faust to another German, Hans Vaihinger, who developed the philosophy of "*Als Ob*": As If. Vaihinger emphasized that we make the world meaningful indeed, tolerable—by living with certain essential fictions. We behave As If we are not personally going to die; we behave As If the sensory impressions that we receive are accurate reflections of an underlying reality; we behave As If a just and benevolent god will reward virtue. (Or for some of us, As If it carries its own reward, and for others As If such concerns can be ignored altogether.) We might also want to consider behaving As If we are free to say "no" or "yes" to our biology and our culture, as we choose, depending on whatever values we embrace. Wishful thinking cannot hold off personal death, or manipulate the insensate world, or create god from indifferent chaos. But when applied to human action on our human-dominated planet, behaving As If can become a self-fulfilling prophecy, something that empowers us in proportion as we believe it.

It remains to be seen, of course, whether Homo sapiens really does want to seize the future, and to do so in a uniquely human way, not just as the agents of natural selection acting around us and through us. "One of the most striking things about the struggle for freedom from intentional control is how often it has been lacking," wrote B. F. Skinner, in his *Beyond Freedom and Dignity.*

"Many people have submitted to the most odious religious, governmental, and economic controls for centuries." The theme recurs in some of our most renowned works of fiction as well. In "The Grand Inquisitor" chapter of Dostoevsky's *The Brothers Karamazov,* for example, Christ returns to earth during the Spanish Inquisition and is informed that "nothing has ever been more insupportable for a man and a human society than freedom." And in describing the effects of the Inquisition, we are told that "men rejoiced that they were again led like sheep, and that the terrible gift that had brought them such suffering was, at last, lifted from their hearts." At least, according to the Inquisitor.

On the other hand, that "terrible gift," human freedom, has motivated some of the most courageous and persistent actions of all history, from political revolution to scientific discovery.

Our challenge is to use our unique freedom as human beings to humanize culture, technology, science, and society; in doing so, our reward is that we shall have humanized ourselves. The struggle to maintain selfhood in the face of such dangers is awesome, and important, but it is not altogether new. After all, the effort to avoid being dehumanized in the twentieth century is not all that different from the earlier effort to avoid being "de-humanized" by a saber-tooth. We won that battle; we can also win this one.

There are, of course, some important differences between the threat of the saber-tooth and the threat of technology run wild: whereas de-humanization by the saber-tooth meant death of the body, dehumanization by our own creations means death of the soul. The danger in the former case was physical, immediate, and easily recognized, not different from threats that our primitive ancestors faced—successfully—for hundreds of millions of years. Appropriate responses were therefore easily evoked. The danger in the latter case is no less real, but nonetheless diffuse and arguable, often clothed in alluring Faustian promises of greater material abundance, personal accomplishment, or physical power. As with the problem of supernormal releasers, whose dangers to us are amplified, paradoxically, by their diffuseness, the threat of modern-day dehumanization is enhanced by its deceptive allure and subtle nuance.

Goethe's version is widely acknowledged to be the greatest Faust of all. But it is up to us whether Faust will be our text, and if so, which Faust. In this book, we have sought to trace the history and terms of humanity's bargain with evolution, both biological and cultural. We began with an image of murder, many eons ago. How appropriate, then, that we should end with the possibility of redemption.

Notes and References

ONE of the nice things about reading (and writing) a "trade" book as opposed to an academic tome is that you don't have to get bogged down by excessive references. So, when a poem or book's title was mentioned in the text, I have not repeated it here. On the other hand, one pleasure of reading any book is to use it as a jumping-off place for more information, just as it is another pleasure to recommend favorite sources to anyone who will listen. So, in this spirit, I offer the following notes and references: not in compliance with a scholarly ritual, but as one might introduce old friends to new ones, in the hope that they will enjoy each other's company.

CHAPTER TWO

More information regarding Tycho Brahe can be found in John Gade's *The Life and Times of Tycho Brahe* (1947, Princeton University Press: Princeton, New Jersey). There have been an enormous number of books written about evolution, many of them quite good, and some truly excellent. For scholarly textbook–type introductions—but nonetheless readable ones—my own favorites are *Evolution* (1977, W. H. Freeman: San Francisco) by the great Russian-born geneticist Theodosius Dobzhansky, *The Theory of Evolution* (1966, Penguin: Harmondsworth, England) by the British mathematical ecologist John Maynard Smith, and *Processes of Organic Evolution* (1977, Prentice-Hall: Englewood Cliffs, New Jersey) by the American botanist G. Ledyard Stebbins. Stebbins's *Darwin to DNA: Molecules to Humanity* (1982, W. H. Freeman: San Francisco) is more recent and has more material. These books emphasize the process of evolution, notably natural selection. For more focus on the history of the concept, try Loren Eiseley's *Darwin's Century* (1958, Doubleday: Garden City, New York). Peter Bowler's *Evolution: The History of an Idea* (1984, University of California Press: Berkeley) is—not surprisingly—less readable than Eiseley, but more scholarly. Two beautifully written classics that treat evolution in relation to human beings are Eiseley's *The Immense Journey* (1957, Random House: New York) and Garrett Hardin's *Nature and Man's Fate* (1959, Rinehart: New York). For a solid introduction to the actual history of evolution itself—notably, the process of evolutionary change among our distant ancestors—try Edwin H. Colbert's justly celebrated *Evolution of the Vertebrates* (1980, John Wiley: New York).

The literature on human evolution, and on primates, is also enormous. For starters, you might want to try Elwyn L. Simons's *Primate Evolution: An Introduction to Man's Place in Nature* (1972, Macmillan: New York) and also Alison Jolly's *The Evolution of Primate Behavior* (1972, Macmillan: New York). There has been much speculation but—not surprisingly—few facts discovered regarding the actual origin of life on earth; what is known is ably reviewed in *The Nature and Origin of the Biological World* (1982, Halsted: New York) by E. J. Ambrose. My favorite treatment of human history has always been H. G. Wells's *The Outline of History,* which, as the subtitle indicates, is no less than a "plain history of life and mankind" (1920, Macmillan: New York). It's also a treat to look through Hilaire Belloc's companion to Wells (1926, Sheed & Ward: London). Belloc, incidentally, summarized the West's reliance on culture—or rather, technology—when he wrote the following trenchant lines at the time India was seeking to overthrow British rule:

> Whatever happens, we have got
> The Maxim gun, and they have not.

CHAPTER THREE

For more on Lamarck, try *The Spirit of System: Lamarck and Evolutionary Biology* (1977, Harvard University Press: Cambridge, Mass.) by Richard W. Burkhardt. Paul MacLean's evocation of the three-part human brain occurs in his "A Triune Concept of the Brain and Behavior," part of a congress held at Queen's University in Kingston, Ontario, and published by the University of Toronto Press in 1973. Alvin Toffler's *Future Shock* (1970, Random House: New York) contains some themes similar to the present book, although he does not identify the culture/biology conflict as the principle cause, and he is much more enthusiastic about the human-made future than I am. For more on the history and consequences of our mobility, I recommend *Human Migrations: Patterns and Policies* (1978, Indiana University Press: Bloomington) edited by William McNeill and Ruth S. Adams.

For basic material on sociobiology, I immodestly recommend my own textbook *Sociobiology and Behavior* (1982, Elsevier: New York), and my trade book *The Whisperings Within* (1979, Harper & Row: New York). Edward O. Wilson's massive, now-classic compilation was titled *Sociobiology: The New Synthesis* (1975, Harvard University Press: Cambridge, Mass.) and was followed by his *On Human Nature* (1978, Harvard University Press: Cambridge, Mass.) As with most efforts to "biologize" human behavior, however, sociobiology has also been controversial; for two critiques—which to my mind unfairly lump sociobiology with racist eugenics and misguided social Darwinism—see Richard C. Lewontin, S. Rose, and L. J. Kamin's *Not in Our Genes* (1984, Pantheon: New York) and Stephen Jay Gould's *The Mismeasure of Man* (1981, W. W. Norton: New York). An intellectually challenging, mathematical argument for the connection between human evolution and culture is given by Charles Lumsden and Edward O. Wilson in their *Genes, Mind and Culture: The Coevolutionary Process* (1981, Harvard University Press: Cambridge, Mass.). A more readable treatment, for nonspecialists, is *Promethean Fire: Reflections on the Origin of the Mind* (1983, Harvard University Press: Cambridge, Mass.) by the same formidable pair. Robert Boyd and Peter J. Richerson have recently presented a series of models of *Culture and the Evolutionary Process* (1985, University of Chicago Press: Chicago), in which they analyze those situations in which natural selection would favor different capacities for cultural transmission. Finally, the Weston La Barre quote is from his magisterial *The Ghost Dance: Origin of Religion* (1970, Doubleday: Garden City, New York). For what it is worth, La Barre is my favorite anthropologist, a brilliant and iconoclastic scholar whose work deserves more attention than it has received.

CHAPTER FOUR

For a traditional ethological approach to the subject, see Margaret Bastock's *Courtship* (1967, Aldine: Chicago) and the relevant chapters of Irenaus Eibl-Eibesfeldt's well-illustrated text, *Ethology: The Biology of Behavior* (1975, Holt, Rinehart & Winston: New York). Sociobiologic views of sexual behavior can be found in my own *Sociobiology and Behavior* (1982, Elsevier: New York), and in Martin Daly and M. Wilson's *Sex, Evolution and Behavior* (1978, Duxbury: North Scituate, Mass.). A recent volume of scientific papers, edited by Cambridge University's Patrick Bateson, is also available: *Mate Choice* (1983, Cambridge University Press: New York). For the biology of maleness, see Robert L. Smith, ed., *Sperm Competition and the Evolution of Animal Mating Systems* (1984, Academic Press: Orlando, Fla.). Material on orchid/wasp mimicry, as well as other similar systems, is presented in Wolfgang Wickler's beautifully illustrated *Mimicry in Plants and Animals* (1968, McGraw-Hill: New York).

Regarding a sociobiologic view of human pair bonding, see my own *The Whisperings Within* (1979, Harper & Row: New York), and *The Evolution of Human Sexuality* (1979, Oxford University Press: New York) by anthropologist Donald Symons. Compare these treatments with reports by nonbiological writers, such as *A History of Courting* (1955, E. P. Dutton: New York) by Ernest Sackville Turner, or the more recent *Hands and Hearts: A History of Courtship* (1984, Basic: New York) by Ellen K. Rothman. German ethologist Irenaus Eibl-Eibesfeldt provides a traditional ethological perspective on human pair bonding in his *Love and Hate* (1972, Holt, Rinehart & Winston: New York). Probably the most influential—and popular—attempt to biologize human behavior, including a good exposition of our penchant for face-to-face lovemaking, was *The Naked Ape* (1965, McGraw-Hill: New York) by ethologist Desmond Morris. For inbreeding depression, see the massive (965-page!) *Genetics of Human Populations* (1971, W. H. Freeman: San Francisco) by Luigi L. Cavalli-Sforza and William F. Bodmer. For a sociobiologic perspective on incest: the best source is Joseph Shepher's *Incest: A Biosocial View* (1983, Academic Press: New York). And Randy and Nancy Thornhill present their research in "Human Rape: An Evolutionary Perspective," which appeared in the journal *Ethology and Sociobiology* (1983, 7: 137–173).

CHAPTER FIVE

Emma Goldman's "Marriage and Love" appears in her *Anarchism and Other Essays* (1911, Mother Earth Publishing Co.: New York). Friedrich Engels, collaborator with Karl Marx, wrote a treatment of the human family and the origin of female oppression that is remarkably "sociobiologic" in orientation: *The Origin of the Family, Private*

Property and the State (1972, International Publishers: New York). For the sociobiology of male-female differences, see my *Sociobiology and Behavior* (1982, Elsevier: New York) or Edward O. Wilson's *Sociobiology: The New Synthesis* (1975, Harvard University Press: Cambridge, Mass.). Anyone interested in feminism should consult the following if he/she hasn't already: *The Second Sex* (1952, Alfred A. Knopf: New York) by Simone de Beauvoir; *The Feminine Mystique* (1963, W. W. Norton: New York) by Betty Friedan; *The Female Eunuch* (1971, McGraw-Hill: New York) by Germaine Greer; and *Sexual Politics* (1970, Doubleday: Garden City, New York) by Kate Millett. It is noteworthy that Greer has recently paid homage to biology in her essay on the struggles between feminism and reproduction, *Sex and Destiny: The Politics of Human Fertility* (1984, Harper & Row: New York). Sociobiologist—and feminist—Sarah Blaffer Hrdy (not a misprint, no "a") provides a refreshing perspective on male-female evolution in her *The Woman That Never Evolved* (1981, Harvard University Press: Cambridge, Mass.).

Before Gilligan, the standard and rather male-centered view of human moral development was that of Lawrence Kohlberg, ably and influentially expressed in his *The Philosophy of Moral Development: Moral Stages and the Idea of Justice* (1981, Harper & Row: New York). In her *The Hearts of Men: American Dreams and the Flight from Commitment* (1983, Doubleday: Garden City, New York), Barbara Ehrenreich, like Gilligan, makes an argument that could have been designed by an evolutionary biologist, but wasn't.

CHAPTER SIX

Richard Alexander has written very effectively on the evolutionary aspects of human social organization, in his *Darwinism and Human Affairs* (1979, University of Washington Press: Seattle). For James Lloyd's work on fireflies, see his article "Aggressive Mimicry in *Photuris* Firefly *Femmes Fatales*," which appeared in the journal *Science* (1965, 149: 653–654). A fine review of the phenomenon of imprinting is available in W. Sluckin's *Early Learning in Man and Animals* (1970, Allen & Unwin: London). The experiment regarding musical preferences of rats is recounted in a chapter by social psychologist Robert Zajonc (rhymes with "science") titled "Attraction, Affiliation and Attachment," in *Man and Beast: Comparative Social Behavior,* edited by John F. Eisenberg and Wilton S. Dillon (1971, Smithsonian Institution Press: Washington, D.C.). The double standard among swallows is described in an article by Mike and Inger Beecher, "Sociobiology of Bank Swallows: Reproductive Strategy of the Male," in the journal *Animal Behaviour* (1979, 205: 1282–1285).

Excellent research on animal alarm calling has been reported by Paul W. Sherman, "Nepotism and the Evolution of Alarm Calls," in *Science* (1977, 197: 1246–1253). I also heartily recommend Donald T. Campbell's article "On the Conflicts between Biological and Social Evolution and between Psychology and Moral Tradition," appearing in the journal *American Psychologist* (1975, 30: 1103–1126). Sociologist Pierre van den Berghe treats human group identification from an evolutionary perspective in his prize-winning *The Ethnic Phenomenon* (1981, Elsevier: New York). The primitive biology underpinning modern-day nationalism is explored in *The Caveman and the Bomb* (1985, McGraw-Hill: New York) by David P. Barash and Judith Eve Lipton. And for an introduction to the very serious game of Prisoner's Dilemma—as well as a way out—see Robert Axelrod's *The Evolution of Cooperation* (1984, Basic: New York).

CHAPTER SEVEN

Frank Beach's warning to comparative psychologists, with the intriguing title "The Snark Was a Boojum," appeared in the journal *American Psychologist* (1950, 5: 115–124). The founder of modern ethology, Konrad Z. Lorenz, tells some marvelous animal stories—many of them related to aggression—in his justly famous *King Solomon's Ring* (1952, T. Y. Crowell: New York). Lorenz's *On Aggression* (1966, Harcourt, Brace and World: New York), although a bit dated in its "good of the species" arguments, nonetheless does a fine job of presenting the ethological view of animal and human nastiness. In part to counter such notions, the renowned American anthropologist Ashley Montagu edited *Man and Aggression* (1973, Oxford University Press: New York), which argued, aggressively, that we aren't instinctively aggressive at all. In his book *Aggression* (1958, University of Chicago Press: Chicago), animal psychologist John Paul Scott brought together data on the role of environmental factors—especially social disruption—in producing aggression among animals. Albert Bandura's *Aggression* (1973, Prentice-Hall: Englewood Cliffs, New Jersey) is an acclaimed social learning approach, and in *Frustration and Aggression* (1961, Yale University Press: New Haven, Conn.), psychiatrist John Dollard and others develop the thesis that frustration leads to aggressive behavior. For some biological views of aggression,

Robert Ardrey's books (*African Genesis,* 1961; *The Territorial Imperative,* 1966; and *The Social Contract,* 1970, all published by Atheneum: New York) have been criticized as glib and a bit naive; they are certainly entertaining, however, and worth reading nonetheless. For a somewhat more scholarly biological view of human war, representing the ethological perspective, see I. Eibl-Eibesfeldt's *The Biology of Peace and War* (1979, Viking: New York). And for a sociobiological flavor, try William Durham's article "Resource Competition and Human Aggression, Part I: A Review of Primitive War," in the journal *Quarterly Review of Biology* (1976, 51: 385–415).

A marvelous, accessible treatment of human spacing behavior was presented by anthropologist Edward T. Hall in *The Hidden Dimension* (1969, Doubleday: Garden City, New York). Microbiologist-humanist Rene J. Dubos has written about the range of human adaptability in *Man Adapting* (1965, Yale University Press: New Haven, Conn.) and *Of Human Diversity* (1974, Crown: New York). Johan Huizinga's *Homo ludens: A Study of the Play Element in Culture,* was published in 1950 by Beacon Press (Boston).

CHAPTER EIGHT

The imbalance between human biology and human culture is most pronounced—and most perilous—in the arena of nuclear weapons. For an elaboration of the hare and tortoise applied to nuclear war and the arms race, see *The Caveman and the Bomb: Human Nature, Evolution and Nuclear War* (1985, McGraw-Hill: New York) by David P. Barash and Judith Eve Lipton. A notable treatment of the arms race from a social psychological perspective was written by psychiatrist-psychologist Jerome Frank in 1965, then reprinted in 1982 as *Sanity and Survival in the Nuclear Age* (Random House: New York). Psychologist–public official Ralph White's *Fearful Warriors* (1984, Free Press: New York) ably describes the impact of fear and pride in international affairs, highlighting the psychological dimension in superpower relations. For another view of the psychological impact of nuclear weapons on human thought, see Joel Kovel's *Against the State of Nuclear Terror* (1984, South End Press: Boston). The Rebecca West quotation is from her *Black Lamb and Grey Falcon* (1940, Viking Penguin: New York), and the research by William Beardsley and John Mack is discussed in their chapter "The Impact on Children and Adolescents of Nuclear Developments," in *Psychological Aspects of Nuclear Developments,* Task Force Report #20, American Psychiatric Association, Washington, D.C. A powerful argument against the technological, strategic, and economic feasibility of Star Wars can be found in *The Fallacy of Star Wars* (1984, Vintage: New York) by physicists Richard Garwin, K. Gottfried, and H. Kendall. Perhaps the best treatment of "nuclear despair" is by Joanna Rogers Macy in *Despair and Personal Power in the Nuclear Age* (1983, New Society: Philadelphia).

CHAPTER NINE

Probably the most influential book on the threat and reality of overpopulation was Paul Ehrlich's nifty little *The Population Bomb* (1968, Ballantine: New York). For more meat, try Paul Ehrlich and Anne Ehrlich's *Population, Resources, Environment* (W. H. Freeman: San Francisco). Among the many books written about this issue, I particularly recommend Lester R. Brown's *In the Human Interest* (1974, W. W. Norton: New York) and *Ecocide and Population* (1971, St. Martin's: New York), edited by M. E. Adelstein and J. G. Pivall. The other side—an optimistic view that I consider misleading and dangerous—is represented by Julian L. Simon and Herman Kahn, eds., *The Resourceful Earth* (1984, Basil Blackwell: New York). The sad story of the Kaibab deer can be found in John P. Russo's *The Kaibab North Deer Herd,* a 1964 publication of the State of Arizona Game and Fish Department, Phoenix. For more on the demographic transition among human beings, see *Population Pressure and Cultural Adjustment* (1979, Human Sciences Press: New York) by Virginia Abernathy. And perhaps the best source for examples of density-dependence in animals is V. C. Wynne Edwards's *Animal Dispersion in Relation to Social Behavior* (1962, Hafner: New York). Professor Wynne Edwards attributed these effects to selection acting at the level of groups—not a popular viewpoint these days—but his work is a rich source of references and data, regardless of interpretation.

John Calhoun's work on Norway rats is recounted in his *The Ecology and Sociology of the Norway Rat* (1963, U.S. Public Health Service Publication #1008, Bethesda, Maryland). John Christian and David E. Davis review the relationship of adrenal size to stress and population density in their article "Endocrines, Behavior, and Population," which was published in the journal *Science* (1964, 146: 1550–1560). H. M. Bruce discusses "the Bruce effect" in "Smell as an Exteroceptive Factor," in the *Journal of Animal Science* (1966, supplement #25: 83–89). And for the classic account of lemming behavior and ecology, without the Walt Disney myths, try *Voles, Mice and Lemmings* (1942, Clarendon Press: Oxford) by the noted ecologist Charles S. Elton.

CHAPTER TEN

Two very important papers on the relationship of Homo sapiens to its environment are Lynn White's "The Historical Roots of Our Ecological Crisis" (reprinted in *Ecocide and Population,* referenced in Chapter 9) and Garrett Hardin's "The Tragedy of the Commons," in the journal *Science* (1961, 162: 1243–1248). The quote from Susanne Langer appears in her *Philosophy in a New Key* (1951, Harvard University Press: Cambridge, Mass.). When it comes to the possibility of establishing a healthy practical rapport between people and their economic needs, perhaps the most insightful thinker has been Ernst Friedrich Schumacher, especially his *Small Is Beautiful: Economics as if People Mattered* (1973, Harper & Row: New York) and *Good Work* (1979, Harper & Row: New York). The C. D. Darlington quote is from his *The Evolution of Man and Society* (1969, Simon & Schuster: New York). Readers concerned about the extinction of animal and plant species might want to consult Paul and Anne Ehrlich's *Extinction: The Causes and Consequences of the Disappearance of Species* (1981, Random House: New York). And for a chilling review of the prospects for nuclear winter, the most devastating environmental disaster of all, read *The Cold and the Dark* (1983, W. W. Norton: New York), edited by Paul Ehrlich, Carl Sagan, Donald Kennedy, and Walter Orr Roberts.

CHAPTER ELEVEN

Noted biologist John Tyler Bonner, an expert on the embryology and behavior of slime molds, has also written *The Evolution of Culture in Animals* (1980, Princeton University Press: Princeton, New Jersey). Pierre Teilhard de Chardin's enthusiasm for technology is expressed in his *The Phenomenon of Man* (1959) and *The Future of Man* (1964), both published by Harper (New York). Readers wanting more of Alexander Herzen might try consulting his sometimes provocative *From the Other Shore* (1956, George Braziller: New York). Writing in that wonderful leftist stalwart *The Nation,* on April 27, 1985, Robert Engler provided a very useful and disturbing review of human technology, "Technology Out of Control." In his important book *Normal*

Accidents (1984, Basic: New York), C. Perrow developed the thesis that overly complex technology leads inevitably to breakdowns and errors, some of them potentially catastrophic. Similar sentiments, but more gloomy and abstract, oriented more around a philosophical and theological perspective, can be found in the varied writings of France's Jacques Ellul, notably his *The Technological Society* (1964, Alfred A. Knopf: New York) and *The Betrayal of the West* (1978, Seabury: New York). Prolific and versatile historian William McNeill recently wrote what is probably the best account of the history of technology and armed force, *The Pursuit of Power* (1983, Basil Blackwell: New York). For a fine, level-headed, and well-balanced treatment of the role of technology in recent human history, see Victor Ferkiss, *Technological Man* (1969, George Braziller: New York), but if you want startling brilliance and erudition combined, the master is Lewis Mumford, who has specialized in architecture, written a critical study of Herman Melville, and produced several classics on technology and human society, including *Technics and Civilization* (1963) and *The Myth of the Machine* (1938), both published by Harcourt, Brace and World (New York).

CHAPTER TWELVE

To my mind, the most effective introduction to existentialism, the philosophy of alienation, is still William Barrett's *Irrational Man* (1958, Doubleday: Garden City, New York). Martin Buber's *I and Thou* (1970, Charles Scribner's Sons: New York) should not be missed, nor should Ian McHarg's *Design with Nature* (1971, Doubleday: Garden City, New York). Biologist W. D. Hamilton's article "Geometry for the Selfish Herd" appeared in the *Journal of Theoretical Biology* (1971, 31: 295–311). For a study of alienation as a major component of Karl Marx's analysis of modern society, see *Alienation, Praxis and Techne in the Thought of Karl Marx* (1976, University of Texas Press: Austin) by K. Axelos. Alienation is also very much on the mind of many specialists, from many different fields, as they survey the human condition. For philosophical perspectives, see I. Feuerlicht's *Alienation: From the Past to the Future* (1978, Greenwood Press: Westport, Conn.) and Morton Kaplan's *Alienation and Identification* (1976, Free Press: New York). Brian Baxter has reviewed the issue of alienation at the workplace in his *Alienation and Authenticity* (1982, Tavistock: New York). And for the impact of alienation on adolescents and the young in American Society, see Kenneth Kenniston's *The Uncommitted* (1965, Harcourt, Brace and World: New York).

CHAPTER THIRTEEN

The Emerson epigraph is from his "Ode, inscribed to W. H. Channing," which can be found in *The Selected Writings of Ralph Waldo Emerson* (1950, Modern Library: New York), edited by Brooks Atkinson. I have reviewed the biology (as well as other aspects) of aging in my *Aging: An Exploration* (1983, University of Washington Press: Seattle). Anthropologist Ruth Benedict, renowned for her *Patterns of Culture,* turned her attention to the problem of the human races in collaboration with Gene Weltfish in *Race: Science and Politics* (1959, Viking: New York). For sickle-cell anemia, try A. Cerami and E. Washington, *Sickle Cell Anemia* (1977, The Third Press: New York). The quote from economist-historian Robert L. Heilbroner is from his *The Future as History* (1960, Harper: New York).

CHAPTER FOURTEEN

For more on the Luddites, try either Frank O. Darvall's *Popular Disturbances and Public Order in Regency England* (1934, Oxford University Press: Oxford) or Frank Peel's *The Risings of the Luddites, Chartists and Plug-drawers* (1968, Cass: London). The latter has the advantage of an introduction by the brilliant and acerbic British social historian E. P. Thompson, who is also founder of the END (European Nuclear Disarmament) campaign. Finally, for a gentle yet stirring evocation of the human potential, try Rene J. Dubos's *Beast or Angel?* (1974, Charles Scribner's Sons: New York).

Index

Acknowledgments

Writers of nonfiction books often have a long list of people to thank, frequently after pointing out that books are rarely written by the author alone. Well, for better or worse, this one pretty much was. I even typed it myself. On the other hand, Drs. Barbara and Morris Lipton made helpful comments, and Dan Frank, my editor at Viking, and his assistant, André Bernard, prodded me beneficently to clarify arguments and prune unneeded verbiage. Nanelle Rose Barash (whose first four months overlapped the last four of this book) did much to distract me at critical junctures, thereby doubtless keeping the flow of ideas from becoming excessive.

Chart 40-1

GENETICS IN NURSING PRACTICE

Musculoskeletal Disorders

When assessing a patient with musculoskeletal complaints, nurses must not overlook the possibility of a genetic component to the patient's problems, such as the following:

- Achondroplasia
- Congenital talipes equinovarus (clubfoot)
- Developmental dysplasia of the hip (DDH) (congenital hip dysplasia)
- Ehlers-Danlos syndrome
- Marfan syndrome
- Stickler syndrome
- Osteogenesis imperfecta
- Osteoporosis
- Scoliosis

Nursing Assessments

Family History Assessment

- Assess for other similarly affected family members in the past three generations.
- Assess for the presence of other related genetic conditions (e.g., hematologic, cardiac, integumentary conditions).
- Determine the age at onset (e.g., fractures present at birth such as osteogenesis imperfecta, hip dislocation present at birth in DDH, or early-onset osteoporosis).

Patient Assessment

- Assess stature for general screening purposes (unusually short stature may be related to achondroplasia; unusually tall stature may be related to Marfan syndrome).
- Assess for disease-specific skeletal findings (e.g., pectus excavatum, scoliosis, long fingers [Marfan syndrome], osteoarthritis of the hip, and waddling gait [DDH]).
- Assess for disease-specific skin findings (e.g., velvety texture with unusual scarring and/or thin fragile skin [Ehlers-Danlos syndrome]).
- Assess for other common disease-specific findings (e.g., vision impairment [Stickler syndrome, Marfan syndrome], blue/gray sclerae, opalescent dentin, hearing impairment [osteogenesis imperfecta]).

Management Issues Specific to Genetics

- Inquire whether DNA gene mutation or other genetic testing has been performed on affected family members.
- If indicated, refer patient for further genetic counseling and evaluation so that family members can discuss inheritance, risk to other family members, and the availability of genetic testing and gene-based interventions.
- Offer appropriate genetic information and resources.
- Assess patient's understanding of genetic information.
- Provide support to families with newly diagnosed genetics-related musculoskeletal disorders.
- Participate in management and coordination of care of patients with genetic conditions and people predisposed to develop or pass on a genetic condition.

Genetics Resources

See Chapter 8, Chart 8–6 for genetics resources.

physical examination maneuvers that facilitate diagnosis of specific bone, muscle, and joint disorders. The extent of assessment depends on the patient's physical complaints, health history, and physical clues that warrant further exploration. The nursing assessment is primarily a functional evaluation, focusing on the patient's ability to perform activities of daily living.

Techniques of inspection and palpation are used to evaluate the patient's posture, gait, bone integrity, joint function, and muscle strength and size. In addition, assessing the skin and neurovascular status is an important part of a complete musculoskeletal assessment. The nurse should also understand and be able to perform correct assessment techniques on patients with musculoskeletal trauma. When specific symptoms or physical findings of musculoskeletal dysfunction are apparent, the nurse carefully documents the examination findings and shares the information with the primary provider, who may decide that a more extensive examination and a diagnostic evaluation are necessary.

Posture

The normal curvature of the spine is convex through the thoracic portion and concave through the cervical and lumbar portions. Common deformities of the spine include **kyphosis**, which is an increased forward curvature of the thoracic spine that causes a bowing or rounding of the back, leading to a hunchback or slouching posture. The second deformity of the spine is referred to as **lordosis**, or swayback, an exaggerated curvature of the lumbar spine. A third deformity is **scoliosis**, which is a lateral curving deviation of the spine (Fig. 40-4). Kyphosis can occur at any age and may be caused by degenerative diseases of the spine (e.g., arthritis or disk degeneration), fractures related to osteoporosis, and injury or trauma (Swartz, 2010). It may also be seen in patients with other neuromuscular disease. Lordosis can affect persons of any age. Common causes of lordosis include tight low back muscles, excessive visceral fat, and pregnancy as the woman adjusts her posture in response to changes in her center of gravity. Scoliosis may be congenital, idiopathic (without an identifiable cause), or the result of damage to the paraspinal muscles (e.g., muscular dystrophy).

During inspection of the spine, the entire back, buttocks, and legs are exposed. The examiner inspects the spinal curves and trunk symmetry from posterior and lateral views. Standing behind the patient, the examiner notes any differences in the height of the shoulders or iliac crests. Shoulder and hip symmetry, as well as the line of the vertebral column, are inspected with the patient erect and with the patient bending forward (flexion). Scoliosis is evidenced by an abnormal lateral curve in the spine; shoulders that are not level; an asymmetric waistline; and a prominent scapula, which is accentuated by bending forward. The examiner should then instruct the patient to bend backward (extension) with the examiner supporting the patient by placing hands on the posterior iliac spine (Bickley, 2009). Older adults experience a loss in height due to loss of vertebral cartilage and osteoporosis-related vertebral compression fractures. Therefore, an adult's height should be measured during each health screening.

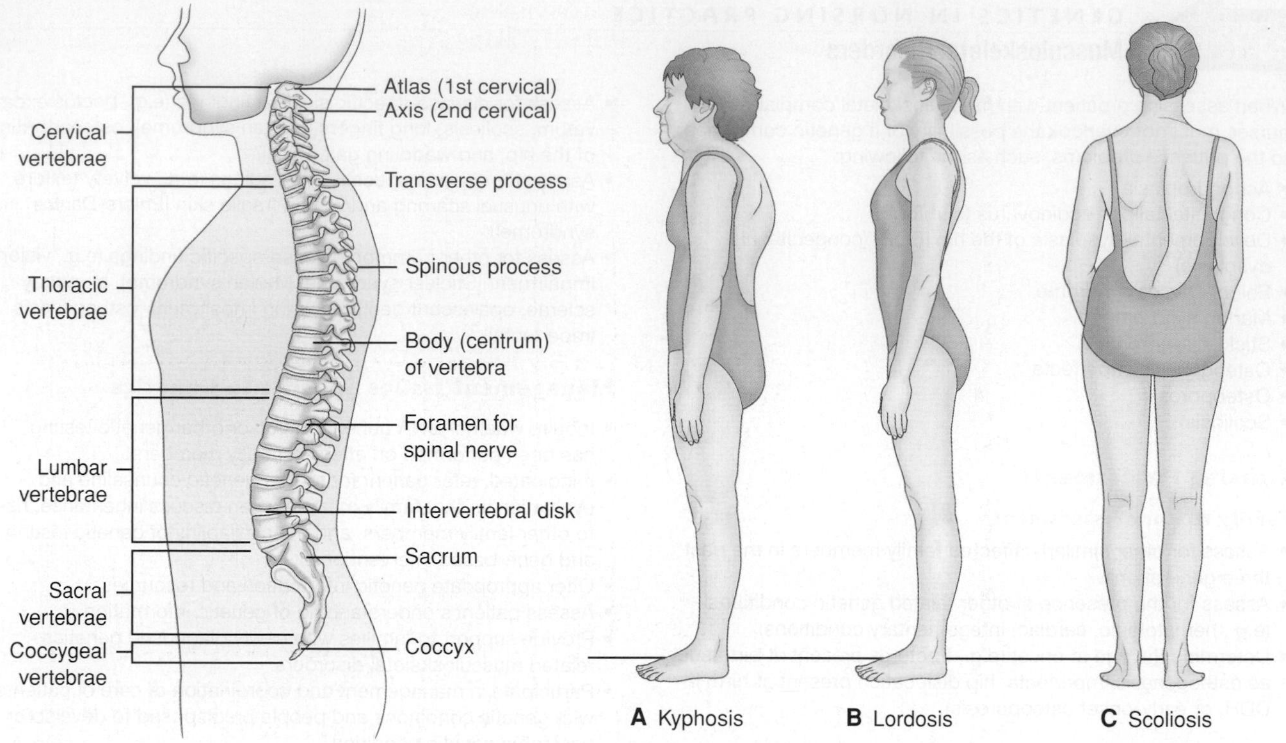
Atlas (1st cervical)
Axis (2nd cervical)
Cervical vertebrae
Transverse process
Spinous process
Thoracic vertebrae
Body (centrum) of vertebra
Foramen for spinal nerve
Lumbar vertebrae
Intervertebral disk
Sacrum
Sacral vertebrae
Coccygeal vertebrae
Coccyx
A Kyphosis
B Lordosis
C Scoliosis

FIGURE 40-4 • A normal spine and three abnormalities. **A.** Kyphosis: an increased convexity or roundness of the spine's thoracic curve. **B.** Lordosis: swayback; exaggeration of the lumbar spine curve. **C.** Scoliosis: a lateral curvature of the spine.

Gait

Gait is assessed by having the patient walk away from the examiner for a short distance. The examiner observes the patient's gait for smoothness and rhythm. Any unsteadiness or irregular movements (frequently noted in older adult patients) are considered abnormal. A limping motion is most frequently caused by painful weight bearing. In such instances, the patient can usually pinpoint the area of discomfort, thus guiding further examination. If one extremity is shorter than another, a limp may also be observed as the patient's pelvis drops downward on the affected side with each step. The knee should be flexed during normal gait; therefore, limited joint motion may interrupt the smooth pattern of gait. Evaluation of the knee involves the joints, bones, ligaments, tendons, and cartilage, and may include tests for the anterior and collateral ligaments, medial and lateral ligaments, and medial meniscus (Bickley, 2009). In addition, a variety of neurologic conditions are associated with abnormal gait, such as a spastic hemiparesis gait (stroke), steppage gait (lower motor neuron disease), and shuffling gait (Parkinson's disease).

Bone Integrity

The bony skeleton is assessed for deformities and alignment. Symmetric parts of the body, such as extremities, are compared. Abnormal bony growths due to bone tumors may be observed. Shortened extremities, amputations, and body parts that are not in anatomic alignment are noted. Fracture findings may include abnormal angulation of long bones, motion at points other than joints, and **crepitus** (a grating sound) at the point of abnormal motion. Movement of fracture fragments must be minimized to avoid additional injury. The nurse should include the following observations (Bickley, 2009):

- If the affected part is an extremity, how does its overall appearance compare to the unaffected extremity?
- Can the patient move the affected part? If an extremity is involved, does each toe or finger have normal sensation and motion (flexion and extension), and is the skin warm or cool?
- What is the color of the part distal to the affected area? Is it pale? Dusky? Mottled? Cyanotic?
- Does rapid capillary refill occur? (The nurse can gently squeeze a nail until it blanches, then release the pressure. The amount of time for the color under the nail to return to normal is noted. Color normally returns within 3 seconds. The return of color is evidence of capillary refill.)
- Is a pulse distal to the affected area palpable? If the affected area is an extremity, how does the pulse compare to the pulse of the unaffected extremity?
- Is edema present?
- Is any constrictive device or clothing causing nerve or vascular compression?
- Does elevating the affected part or modifying its position affect the symptoms?

Joint Function

The articular system is evaluated by noting range of motion, deformity, stability, tenderness, and nodular formation. Range of motion is evaluated both actively (the joint is moved by the muscles surrounding the joint) and passively (the joint is moved by the examiner). The examiner is familiar with the normal range of motion of major joints. Precise measurement of range of motion can be made by a goniometer (a protractor designed for evaluating joint motion). Limited range of motion may be the result of skeletal deformity, joint pathology, or **contracture** (shortening of surrounding joint structures) of the surrounding muscles, tendons, and joint capsule. In older adult patients, limitations of range of motion associated with osteoarthritis may reduce their ability to perform activities of daily living.

If joint motion is compromised or the joint is painful, the joint is examined for **effusion** (excessive fluid within the capsule), swelling, and increased temperature that may reflect active inflammation. An effusion is suspected if the joint is swollen and the normal bony landmarks are obscured. The most common site for joint effusion is the knee. If large amounts of fluid are present in the joint spaces beneath the patella, it may be identified by assessing for the balloon sign and for ballottement of the knee (Fig. 40-5). If inflammation or fluid is suspected in a joint, consultation with a specialist (e.g., orthopedic surgeon or rheumatologist) is indicated.

Joint deformity may be caused by contracture, dislocation (complete separation of joint surfaces), subluxation (partial separation of articular surfaces), or disruption of structures surrounding the joint. Weakness or disruption of joint-supporting structures may result in a weak joint that requires an external supporting appliance (e.g., brace).

Palpation of the joint while it is moved passively provides information about the integrity of the joint. Normally, the joint moves smoothly. A snap or crack may indicate that a ligament is slipping over a bony prominence. Slightly roughened surfaces, as in arthritic conditions, result in crepitus (grating, crackling sound or sensation) as the irregular joint surfaces move across one another.

The tissues surrounding joints are examined for nodule formation. Rheumatoid arthritis, gout, and osteoarthritis may produce characteristic nodules. The subcutaneous nodules of rheumatoid arthritis are soft and occur within and along tendons that provide extensor function to the joints. The nodules of gout are hard and lie within and immediately adjacent to the joint capsule itself. They may rupture, exuding white

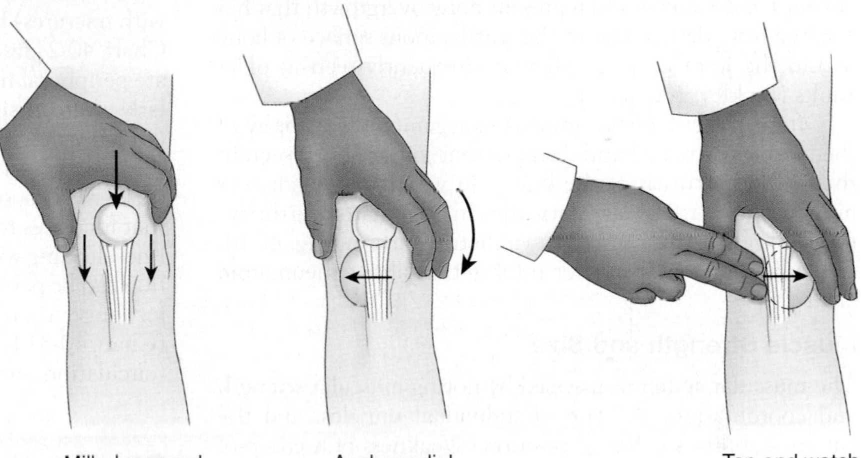
Milk downward
Apply medial pressure
Tap and watch for fluid wave
A

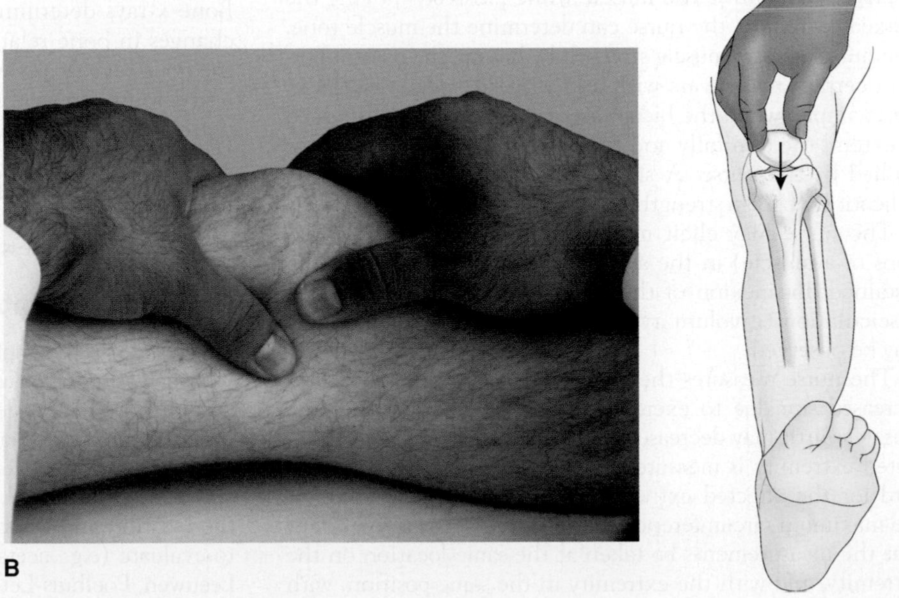
B

FIGURE 40-5 • Tests for detecting fluid in the knee. **A.** Technique for balloon sign. The medial and lateral aspects of the extended knee are milked firmly in a downward motion, which displaces any fluid downward. The examiner feels for any fluid entering the space directly inferior to the patella. When larger amounts of fluid are present, the subpatellar region feels as if it is "ballooning," and the balloon sign test is positive. **B.** Technique for ballottement sign. The medial and lateral aspects of the extended knee are milked firmly in a downward motion. The examiner pushes the patella toward the femur and observes for fluid return to the region superior to the patella. When larger amounts of fluid are present, the patella elevates, there is visible return of fluid to the region directly superior to the patella, and the ballottement test is positive. Photograph used with permission from Bickley, L. S. (2009). *Bates' guide to physical examination and history taking* (10th ed.). Philadelphia: Lippincott Williams & Wilkins.

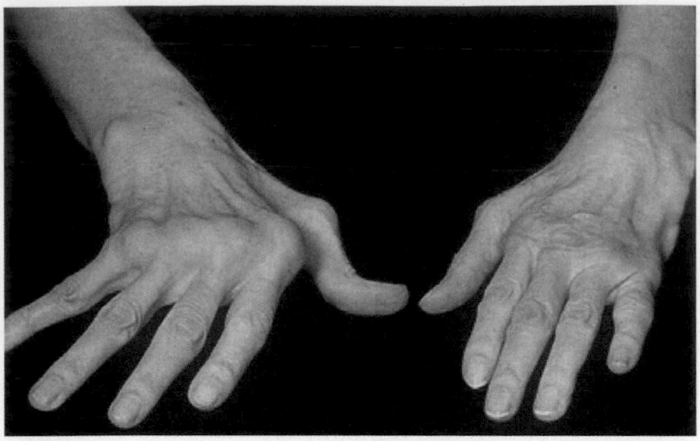

FIGURE 40-6 • Rheumatoid arthritis joint deformity with ulnar deviation of fingers and "swan neck" deformity of fingers (i.e., hyperextension of proximal interphalangeal joints with flexion of distal interphalangeal joints).

uric acid crystals onto the skin surface. Osteoarthritic nodules are hard and painless and represent bony overgrowth that has resulted from destruction of the cartilaginous surface of bone within the joint capsule. They are frequently seen in older adults (Bickley, 2009).

Often, the size of the joint is exaggerated by atrophy of the muscles proximal and distal to that joint. This is seen in rheumatoid arthritis of the knees, in which the quadriceps muscle may atrophy dramatically. In rheumatoid arthritis, joint involvement assumes a symmetric pattern (Fig. 40-6). (See Chapter 39 for further information about rheumatoid arthritis.)

Muscle Strength and Size

The muscular system is assessed by noting muscular strength and coordination, the size of individual muscles, and the patient's ability to change position. Weakness of a group of muscles may indicate a variety of conditions, such as polyneuropathy, electrolyte disturbances (particularly potassium and calcium), myasthenia gravis, poliomyelitis, and muscular dystrophy. By palpating the muscle while passively moving the relaxed extremity, the nurse can determine the muscle tone. The nurse assesses muscle strength by having the patient perform certain maneuvers with and without added resistance. For example, when the biceps are tested, the patient is asked to extend the arm fully and then to flex it against resistance applied by the nurse. A simple handshake may provide an indication of grasp strength.

The nurse may elicit muscle **clonus** (rhythmic contractions of a muscle) in the ankle or wrist by sudden, forceful, sustained dorsiflexion of the foot or extension of the wrist. **Fasciculation** (involuntary twitching of muscle fiber groups) may be observed.

The nurse measures the girth of an extremity to monitor increased size due to exercise, edema, or bleeding into the muscle. Girth may decrease due to muscle atrophy. The unaffected extremity is measured and used as the reference standard for the affected extremity. Measurements are taken at the maximum circumference of the extremity. It is important that the measurements be taken at the same location on the extremity, and with the extremity in the same position, with the muscle at rest. Distance from a specific anatomic landmark (e.g., 10 cm below the medial aspect of the knee for measurement of the calf muscle) should be indicated in the patient's record so that subsequent measurements can be made at the same point. For ease of serial assessment, the nurse may indicate the point of measurement by marking the skin. Variations in size greater than 1 cm are considered significant.

Skin

In addition to assessing the musculoskeletal system, the nurse inspects the skin for edema, temperature, and color. Palpation of the skin may reveal whether any areas are warmer, suggesting increased perfusion or inflammation, or cooler, suggesting decreased perfusion, and whether edema is present. Cuts, bruises, skin color, and evidence of decreased circulation or inflammation can influence nursing management of musculoskeletal conditions.

Neurovascular Status

The nurse must perform frequent neurovascular assessments of patients with musculoskeletal disorders (especially of those with fractures) because of the risk of tissue and nerve damage. Chart 40-2 describes methods the nurse may use to evaluate peripheral nerve function. The nurse needs to be particularly aware of signs and symptoms of compartment syndrome (which is described in detail later in this unit) when assessing the patient with a musculoskeletal injury. This neurovascular problem is caused by pressure within a muscle compartment that increases to such an extent that microcirculation diminishes, leading to nerve and muscle anoxia and necrosis. Function can be permanently lost if the anoxic situation continues for longer than 6 hours. Assessment of neurovascular status (Chart 40-3) is frequently referred to as assessment of CMS (circulation, motion, and sensation).

Diagnostic Evaluation

X-Ray Studies

Bone x-rays determine bone density, texture, erosion, and changes in bone relationships. X-ray study of the cortex of the bone reveals any widening, narrowing, or signs of irregularity. Joint x-rays reveal fluid, irregularity, spur formation, narrowing, and changes in the joint structure. Multiple x-rays, with multiple views (e.g., anterior, posterior, lateral), are needed for full assessment of the structure being examined. Serial x-rays may be indicated to determine the status of the healing process.

Computed Tomography

A computed tomography (CT) scan, which may be performed with or without the use of oral or intravenous (IV) contrast agents, shows a more detailed cross-sectional image of the body. It may be used to visualize and assess tumors; injury to the soft tissue, ligaments, or tendons; and severe trauma to the chest, abdomen, pelvis, head, or spinal cord. It is also used to identify the location and extent of fractures in areas that are difficult to evaluate (e.g., acetabulum) and not visible on x-ray (Van Leeuwen, Poelhuis-Leth, & Bladh, 2011).

Chart 40-2 ASSESSMENT

Assessing for Peripheral Nerve Function

Assessment of peripheral nerve function has two key elements: evaluation of sensation and evaluation of motion. The nurse may perform one or all of the following during a musculoskeletal assessment.

Nerve	Test of Sensation	Test of Movement
Peroneal	Prick the skin midway between the great and second toe. 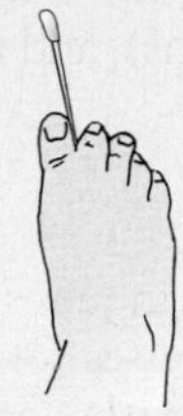	Ask the patient to dorsiflex the foot and extend the toes. 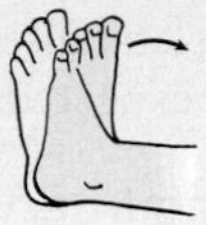
Tibial	Prick the medial and lateral surface of the sole. 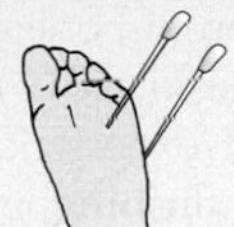	Ask the patient to plantar flex toes and foot. 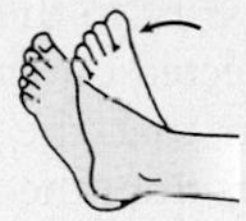
Radial	Prick the skin midway between the thumb and second finger. 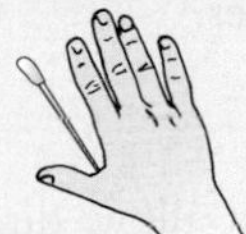	Ask the patient to stretch out the thumb, then the wrist, and then the fingers at the metacarpal joints. 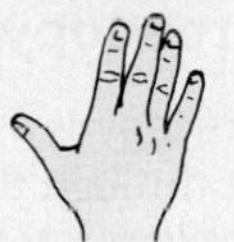
Ulnar	Prick the distal fat pad of the small finger. 	Ask the patient to abduct all fingers. 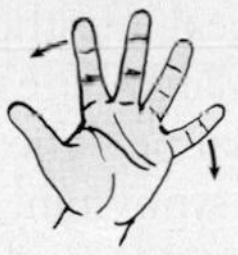
Median	Prick the top or distal surface of the index finger. 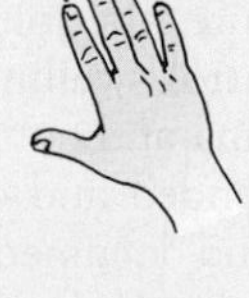	Ask the patient to touch the thumb to the little finger. In addition, observe whether the patient can flex the wrist. 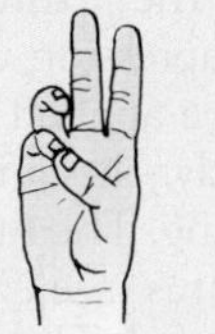

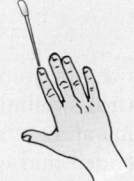

Chart 40-3

Indicators of Peripheral Neurovascular Dysfunction

Circulation

Color: Pale, cyanotic, or mottled
Temperature: Cool
Capillary refill: More than 3 seconds

Motion

Weakness
Paralysis

Sensation

Paresthesia
Unrelenting pain
Pain on passive stretch
Absence of feeling

Magnetic Resonance Imaging

Magnetic resonance imaging (MRI) is a noninvasive imaging technique that uses magnetic fields and radio waves to create high-resolution pictures of bones and soft tissues. It can be used to visualize and assess torn muscles, ligaments, and cartilage; herniated disks; and a variety of hip or pelvic conditions. The patient does not experience any pain during the procedure. The MRI scanner is noisy, and it may take 30 to 90 minutes to complete the test. Because an electromagnet is used, patients with any metal implants, clips, or pacemakers are not candidates for MRI (Van Leeuwen et al., 2011).

Quality and Safety Nursing Alert

Jewelry, hair clips, hearing aids, credit cards with magnetic strips, and other metal-containing objects must be removed before the MRI is performed; otherwise, they can become dangerous projectile objects or cause burns. Credit cards with magnetic strips may be erased, and nonremovable cochlear devices can become inoperable. In addition, transdermal patches (e.g., nicotine patch [NicoDerm], nitroglycerin transdermal [Transderm-Nitro], scopolamine transdermal [Transderm Scop], clonidine transdermal [Catapres-TTS]) that have a thin layer of aluminized backing must be removed before MRI because they can cause burns. The primary provider should be notified before the patches are removed.

To enhance visualization of anatomic structures, an IV contrast agent may be used. Patients who experience claustrophobia may be unable to tolerate the confinement of closed MRI equipment without sedation. Open MRI systems are available, but they use lower-intensity magnetic fields, which produce lower-quality images. Advantages of open MRI include increased patient comfort, reduced problems with claustrophobic reactions, and reduced noise.

Arthrography

Arthrography is used to identify the cause of any unexplained joint pain and progression of joint disease. A radiopaque contrast agent or air is injected into the joint cavity to visualize the joint structures such as the ligaments, cartilage, tendons, and joint capsule. The joint is put through its range of motion to distribute the contrast agent while a series of x-rays are obtained. If a tear is present, the contrast agent leaks out of the joint and is evident on the x-ray image.

Bone Densitometry

Bone densitometry is used to evaluate bone mineral density (BMD). This can be performed through the use of x-rays or ultrasound. The most common modalities used include dual-energy x-ray absorptiometry (DXA or DEXA), quantitative computed tomography (QCT), and quantitative ultrasound (QUS). DXA measures BMD and predicts fracture risk through accurate monitoring of bone density changes in patients with osteoporosis who are undergoing treatment. The density of bones in the spine, hip, and wrist may be calculated, as well as the total body. Peripheral dual-energy x-ray absorptiometry (pDXA) may be an alternative test that measures BMD of the forearm, finger, or heel, although its ability to project hip or spine fracture risk is less accurate than DXA.

Bone density may vary among different skeletal areas; therefore, BMD results may be normal at one site but low at another. Because these tests only measure density at specific sites, they may miss abnormal findings in other skeletal areas. Thus, although the BMD of the heel can be used to diagnose and monitor osteoporosis, predicting bone fracture risk related to osteoporosis is best achieved through DXA of the hip and spine. Hence, DXA is the most commonly prescribed diagnostic test for determining BMD (NOF, 2009; Van Leeuwen et al., 2011). (See Chapter 42 for a further discussion of osteoporosis risks.)

Nursing Interventions

Before the patient undergoes any of the imaging studies described previously (i.e., x-rays, CT scans, MRIs, arthrography, bone densitometry), the nurse prepares the patient. For these studies, the patient must lie still. During an MRI study, the patient may hear a knocking sound. In addition, the nurse assesses for conditions that may require special consideration during the study or that may be contraindications to the study (e.g., pregnancy; claustrophobia; inability to tolerate required positioning due to age, debility, or disability; metal implants). If contrast agents will be used for the CT scan, MRI, or arthrography, the patient is assessed for possible allergies (Van Leeuwen et al., 2011).

The patient having an arthrogram may feel some discomfort or tingling during the procedure. After the arthrogram, a compression elastic bandage may be applied if prescribed, and the joint is usually rested for 12 hours. Strenuous activity should be avoided until approved by the primary provider. The nurse provides additional comfort measures (e.g., mild analgesia, ice) as appropriate and explains to the patient that it is normal to experience clicking or crackling in the joint for 24 to 48 hours after the procedure until the contrast agent or air is absorbed.

Bone Scan

A bone scan is performed to detect metastatic and primary bone tumors, osteomyelitis, some fractures, and aseptic necrosis, and to monitor the progression of degenerative bone diseases. A bone scan may accurately identify bone disease before it can be detected on x-ray; as such, it may diagnose a stress fracture in a patient who continues to experience pain after x-ray findings are negative (Tabloski, 2010). A bone scan requires the injection of a radioisotope through an IV line; the scan is performed 2 to 3 hours afterward. At this point, distribution and concentration of the isotope in the bone are measured. The degree of nuclide uptake is related to the metabolism of the bone; areas of abnormal bone formation will appear brighter. An increased uptake of the isotope is seen in primary skeletal disease (osteosarcoma), metastatic bone disease, inflammatory skeletal disease (osteomyelitis), and fractures that do not heal as expected.

Nursing Interventions

Prior to the bone scan, the nurse inquires about possible allergies to the radioisotope and assesses for any condition that would contraindicate performing the procedure (e.g., pregnancy, breast-feeding). The nurse should educate the patient about why the bone scan may be indicated and explain that it can assist in the identification of bone disease before it can be detected on an x-ray (Tabloski, 2010). The nurse should explain that the patient may experience moments of discomfort from the isotope (e.g., flushing, warmth) but provide reassurance that the radionuclide poses no radioactive hazard (Van Leeuwen et al., 2011). In addition, the patient is encouraged to drink plenty of fluids to help distribute and eliminate the isotope. Before the scan, the nurse asks the patient to empty the bladder, because a full bladder interferes with accurate scanning of the pelvic bones.

Arthroscopy

Arthroscopy is a procedure that allows direct visualization of a joint through the use of a fiberoptic endoscope. Thus, it is a useful adjunct to diagnosing joint disorders. Biopsy and treatment of tears, defects, and disease processes may be performed through the arthroscope. The procedure takes place in the operating room under sterile conditions with either injection of a local anesthetic agent into the joint or general anesthesia. A large-bore needle is inserted, and the joint is distended with saline. The arthroscope is introduced, and joint structures, synovium, and articular surfaces are visualized. After the procedure, the puncture wound is closed with adhesive strips or sutures and covered with a sterile dressing. Complications are rare but may include infection, hemarthrosis, neurovascular compromise, thrombophlebitis, stiffness, effusion, adhesions, and delayed wound healing.

Nursing Interventions

After the arthroscopic procedure, the joint is wrapped with a compression dressing to control swelling. In addition, ice may be applied to control edema and enhance comfort. Frequently, the joint is kept extended and elevated to reduce swelling. The nurse monitors and documents the neurovascular status (see Chart 40-3). Analgesic agents are administered as needed. The patient is instructed to avoid strenuous activity of the joint, and exercises must be approved by the primary provider. The patient and family are instructed to monitor for signs and symptoms of complications (e.g., fever, excessive bleeding, swelling, numbness, cool skin) and the importance of notifying the primary provider should any of these occur (Van Leeuwen et al., 2011).

Arthrocentesis

Arthrocentesis (joint aspiration) is carried out to obtain synovial fluid for purposes of examination or to relieve pain due to effusion. Examination of synovial fluid is helpful in the diagnosis of septic arthritis and other inflammatory arthropathies and reveals the presence of hemarthrosis (bleeding into the joint cavity), which suggests trauma or a bleeding disorder. Normally, synovial fluid is clear, pale, straw colored, and scanty in volume. Using aseptic technique, the physician inserts a needle into the joint and aspirates fluid. Anti-inflammatory medications may be injected into the joint. A sterile dressing is applied after aspiration. There is a risk of infection after this procedure.

Nursing Interventions

The nurse should review the procedure with the patient and its indications. Hair may need to be removed from the site before the procedure. Pain may be a concern; telling the patient that analgesics may be administered to alleviate discomfort during the procedure may help decrease anxiety. Ice may be prescribed for the first 24 to 48 hours postprocedure; the patient should be educated about why it may be indicated (i.e., to diminish edema formation and pain). If antibiotics are prescribed postprocedure, the patient must be educated about their use and reminded to take medications as prescribed. The patient and family are educated about the possible signs and symptoms of complications, particularly infection and bleeding (e.g., fever, excessive bleeding, swelling, numbness, cool skin) and the importance of promptly notifying the health care provider if any of these occur (Van Leeuwen et al., 2011).

Electromyography

Electromyography (EMG) provides information about the electrical potential of the muscles and the nerves leading to them. The test is performed to evaluate muscle weakness, pain, and disability. The purpose of the procedure is to determine any abnormality of function and to differentiate muscle and nerve problems. An EMG can be used to identify the extent of damage if nerve function does not return within 4 months of an injury. Needle electrodes are inserted into selected muscles, and responses to electrical stimuli are recorded on an oscilloscope. Warm compresses may relieve residual discomfort after the study.

Nursing Interventions

Before the patient undergoes an EMG, the nurse inquires if the patient is taking any anticoagulant medications and assesses for any active skin infection. An EMG is usually contraindicated in persons receiving anticoagulant therapy (e.g., warfarin) because the needle electrodes may cause bleeding within the muscle. EMG also may be contraindicated in persons with extensive skin infections due to the risk of spreading infection from the skin to the muscle. The patient needs to be instructed to not use any lotions or creams on the day of the test (Van Leeuwen et al., 2011). If it is discovered that the patient is taking an anticoagulant or has a skin infection, this should be reported to the primary provider.

Biopsy

Biopsy may be performed to determine the structure and composition of bone marrow, bone, muscle, or synovium to help diagnose specific diseases. It involves excising a sample of tissue that can be analyzed microscopically to determine cell morphology and tissue abnormalities.

Nursing Interventions

The nurse educates the patient about the procedure and assures the patient that analgesic agents will be provided. The nurse monitors the biopsy site for edema, bleeding, pain, hematoma formation, and infection. Ice is applied as prescribed to control bleeding and edema. In addition, antibiotics and analgesic agents are administered as prescribed. The patient is instructed to report signs of redness, bleeding, or pain at the biopsy site as well as fever or chills to the primary provider (Van Leeuwen et al., 2011).

Laboratory Studies

Examination of the patient's blood and urine are used to identify the presence and amount of chemicals and other substances. The results may indicate a primary musculoskeletal problem (e.g., Paget's disease of the bone), a developing complication (e.g., infection), the baseline for instituting therapy (e.g., anticoagulant therapy), or the response to therapy, as well as possible causes of bone loss. Before surgery, coagulation studies are performed to detect bleeding tendencies (because bone is vascular tissue).

Serum calcium levels are altered in patients with osteomalacia, parathyroid dysfunction, Paget's disease, metastatic bone tumors, or prolonged immobilization. Serum phosphorus levels are inversely related to calcium levels and are diminished in osteomalacia associated with malabsorption syndrome. Acid phosphatase is elevated in Paget's disease and metastatic cancer. Alkaline phosphatase is elevated during early fracture healing and in diseases with increased osteoblastic activity (e.g., metastatic bone tumors). Bone metabolism may be evaluated through thyroid studies and determination of calcitonin, PTH, and vitamin D levels. Serum enzyme levels of creatine kinase and aspartate aminotransferase become elevated with muscle damage. Serum osteocalcin (bone GLA protein) indicates the rate of bone turnover. Urine calcium levels increase with bone destruction (e.g., parathyroid dysfunction, metastatic bone tumors, multiple myeloma) (National Guidelines Clearinghouse, 2011).

Specific urine and serum biochemical markers can be used to provide information about bone formation, as well as to document the effects of therapeutic interventions prescribed for patients diagnosed with musculoskeletal disorders. These include urinary *N*-telopeptide of type 1 collagen (N-Tx) and deoxypyridinoline (Dpd), both of which reflect increased osteoclast activity and increased bone resorption. Conversely, elevated serum levels of bone-specific alkaline phosphatase (ALP), osteocalcin, and intact *N*-terminal propeptide of type 1 collagen (P1NP) reflect increased activity of osteoblasts and enhanced bone remodeling activity (Singer & Eyre, 2008).

Critical Thinking Exercises

1 pq A 65-year-old woman presents to the primary provider's office with complaints of progressively worsening back pain over the course of 2 weeks that is unresponsive to self-administered ibuprofen (Motrin) and heat packs. She states that the pain is so bad at times that she cannot ambulate without her husband's assistance. She states that she does not believe she has had any recent back injury or trauma. She is in obvious discomfort and is unable to sit on the examination table. What are the initial priority questions and assessments that you want to perform in order to evaluate this patient? What is the evidence base indicating that this woman may be at increased risk for a particular musculoskeletal disorder? What recommendations might be made for appropriate testing in this patient?

2 pq A 29-year-old woman arrives at the emergency department (ED) after falling off of her mountain bike while at the park approximately 30 minutes ago. You are the triage nurse in the ED when the woman enters the waiting room. She is holding her right arm and shoulder, and you note that she is pale and shaking. There is a small cut on her arm, as well as a few abrasions, but no active bleeding. The right wrist and forearm appear swollen, and deformity is noted. What is your priority intervention for this patient? After the patient is brought back into the ED, what is your first physical assessment activity? What needs to be continually monitored on this patient? How would you assess the stability of her right arm? What diagnostic tests are most likely indicated? What is of most concern to you when treating a patient with musculoskeletal trauma? What comfort measures can be provided for this patient until her diagnosis is confirmed?

3 ebp You are a home health care nurse assigned to care for a 72-year-old man who has been discharged from a rehabilitation facility following a fractured hip from a fall. His current weight is 105 pounds and his height is 5 feet 2 inches. He has a medical history of chronic obstructive pulmonary disease, hypertension, 1 pack/day smoking history, and drinks one glass of wine a night. He is able to ambulate at this time with a walker. He lives independently but states that he has a daughter who lives close by to help him out if needed. He does not feel that he will need her assistance and is not quite sure why you have been assigned to his case because he "feels completely fine" since discharge. What is the evidence base that supports strategies for educating and planning care for this patient? In particular, what strategies can be implemented in order to educate this patient about fall prevention and maximize his musculoskeletal health? What is the strength of the evidence of the effectiveness of these strategies?

Brunner Suite Resources

Explore these additional resources to enhance learning for this chapter:

- **NCLEX-Style Questions and Other Resources on thePoint, http://thePoint.lww.com/Brunner13e**
- **Study Guide**
- **PrepU**
- **Clinical Handbook**
- **Handbook of Laboratory and Diagnostic Tests**

References

Books

Bickley, L. S. (2009). *Bates' guide to physical examination and history taking* (10th ed.). Philadelphia: Lippincott Williams & Wilkins.

Meiner, S. (2011). *Gerontological nursing* (4th ed.). St. Louis: Elsevier.

National Guidelines Clearinghouse. (2011). *Diagnosis and treatment of osteoporosis*. Bloomington, MN: Institute for Clinical Systems Improvement.

National Osteoporosis Foundation. (2010). *Clinician's guide to prevention and treatment of osteoporosis*. Washington, DC: Author.

Porth, C. M. (2011). *Essentials of pathophysiology* (3rd ed.). Philadelphia: Lippincott Williams & Wilkins.

Swartz, M. (2010). *Textbook of physical diagnosis: History and examination* (6th ed.). Philadelphia: Saunders.

Tabloski, P. (2010). *Gerontological nursing* (2nd ed.). Upper Saddle River, NJ: Pearson Education.

U.S. Department of Health and Human Services. (2012). *The surgeon general's report on bone health and osteoporosis: What it means to you*. Rockville, MD: Author.

Van Leeuwen, A., Poelhuis-Leth, D., & Bladh, M. (2011). *Davis's comprehensive handbook of laboratory and diagnostic tests with nursing implications* (4th ed.). Philadelphia: F. A. Davis.

Journals and Electronic Documents

Centers for Disease Control and Prevention. (2012). *Arthritis: Meeting the challenge of living well*. Available at: www.cdc.gov/chronicdisease/resources/publications/AAG/arthritis.htm

Cummings, S. R., Martin, J. S., McClung, M. R., et al. (2009). Denosumab for prevention of fractures in postmenopausal women with osteoporosis. *New England Journal of Medicine, 361*(8), 756–765.

National Osteoporosis Foundation. (2009). *Making a diagnosis*. Available at: www.nof.org/aboutosteoporosis/detectingosteoporosis/diagnosing

Singer, F., & Eyre, D. (2008). Using biochemical markers of bone turnover in clinical practice. *Cleveland Clinic Journal of Medicine, 75*(10), 739–750.

Resources

American College of Sports Medicine (ACSM), www.acsm.org

National Association of Orthopaedic Nurses (NAON), www.orthonurse.org

National Institute of Arthritis and Musculoskeletal and Skin Diseases, www.niams.nih.gov

National Osteoporosis Foundation, www.nof.org

Chapter 41

Musculoskeletal Care Modalities

Learning Objectives

On completion of this chapter, the learner will be able to:

1. Identify the preventive and health education needs of the patient with a cast, splint, or brace.
2. Describe the nursing management of the patient with a cast, splint, or brace.
3. Describe the various types of traction and the principles of effective traction.
4. Identify the preventive nursing care needs of the patient with an external fixator or in traction.
5. Describe the nursing management of the patient with an external fixator or in traction.
6. Compare the nursing needs of the patient undergoing total hip arthroplasty with those of the patient undergoing total knee arthroplasty.
7. Use the nursing process as a framework for care of the patient undergoing orthopedic surgery.

Glossary

abduction: movement away from the center or median line of the body
adduction: movement toward the center or median line of the body
avascular necrosis: death of tissue due to insufficient blood supply
brace: externally applied device to support the body or a body part, control movement, and prevent injury
cast: rigid external immobilizing device molded to contours of body part
cast syndrome: psychological (claustrophobic reaction) or physiologic (superior mesenteric artery syndrome) responses to confinement in body cast
continuous passive motion (CPM) device: a device that promotes range of motion, circulation, and healing
edema: soft tissue swelling due to fluid accumulation
external fixator: external metal frame attached to bone fragments to stabilize them
fracture: a break in the continuity of the bone
heterotopic ossification: misplaced formation of bone
neurovascular status: neurologic (motor and sensory components) and circulatory functioning of a body part
open reduction with internal fixation (ORIF): open surgical procedure to repair and stabilize a fracture
osteolysis: lysis of bone from inflammatory reaction against polyethylene particulate debris
osteomyelitis: infection of the bone
osteotomy: surgical cutting of bone
paresthesia: an abnormal sensation of tingling or numbness
petaling: smoothing the rough edges of a cast with smooth material (e.g., moleskin) so the underlying skin does not become abraded
sling: bandage used to support an arm
splint: device designed specifically to support and immobilize a body part in a desired position
traction: application of a pulling force to a part of the body
trapeze: overhead assistive device to promote patient mobility in bed

The management of musculoskeletal injuries and disorders frequently includes the use of casts, splints, braces, traction, surgery, or a combination of these. Because of the prevalence of musculoskeletal injuries and disorders, nurses practicing in many settings across the continuum of care, whether in the home, office, hospital, long-term care facility, or rehabilitation facility, must have an understanding of these modalities. Patient education is essential for optimal outcomes. The nurse prepares the patient for immobilization with casts or traction, and for surgery, when indicated. Nursing care is planned to maximize the effectiveness of these treatment modalities and to prevent potential complications associated with each of the interventions. The patient is instructed to manage care at home and how to safely resume activities.

The Patient in a Cast, Splint, or Brace

Casts

A **cast** is a rigid external immobilizing device that is molded to the contours of the body. A cast is used specifically to immobilize a reduced **fracture** (a break in the continuity of the bone), to correct or prevent a deformity (e.g., clubfoot, hip displacement), to apply uniform pressure to underlying soft tissue, or to support and stabilize weakened joints. Generally, casts permit mobilization of the patient while restricting movement of the affected body part.

Casting is the mainstay of treatment for most fractures (Boyd, Benjamin, & Asplund, 2009). The most common casting materials consist of fiberglass or plaster. The choice of material depends on several factors, which include the condition being treated, availability, and costs (Satryb, Wilson, & Patterson, 2011). Generally, the joints proximal and distal to the area immobilized are included in the cast. However, with some fractures, cast construction and molding may allow movement of a joint while immobilizing a fracture (e.g., three-point fixation in a patellar tendon weight-bearing cast).

Generally, casts can be divided into three main groups: arm casts, leg casts, and body or spica casts:

Short-arm cast: Extends from below the elbow to the palmar crease, secured around the base of the thumb. If the thumb is included, it is known as a thumb spica or gauntlet cast.

Long-arm cast: Extends from the axillary fold to the proximal palmar crease. The elbow usually is immobilized at a right angle.

Short-leg cast: Extends from below the knee to the base of the toes. The foot is flexed at a right angle in a neutral position.

Long-leg cast: Extends from the junction of the upper and middle third of the thigh to the base of the toes. The knee may be slightly flexed.

Walking cast: A short- or long-leg cast reinforced for strength

Body cast: Encircles the trunk

Shoulder spica cast: A body jacket that encloses the trunk, shoulder, and elbow

Hip spica cast: Encloses the trunk and a lower extremity. A double hip spica cast includes both legs.

Figure 41-1 illustrates long-arm and long-leg casts and areas in which pressure problems commonly occur with these casts.

Fiberglass Casts

Fiberglass casts are composed of polyurethane resins that have the versatility of plaster but are lighter in weight, stronger, water-resistant, and more durable than plaster. Fiberglass has the benefit of reaching full rigidity within 30 minutes of application (Satryb et al., 2011). Because they tend to be more difficult to contour and mold, fiberglass casts are more commonly used for simple fractures of the upper and lower extremities. They consist of an open-weave, nonabsorbent fabric that requires tepid water for activation. Heat is given off (an exothermic reaction) while the cast is applied. The heat given off during this reaction can be uncomfortable, and the nurse should prepare the patient for the sensation of increasing warmth so that the patient does not become alarmed. The heat given off from these casts does not place the patient's skin at risk for thermal injury like plaster casts (Walsh, 2009).

Some fiberglass casts use a waterproof lining (Gore-Tex), which permits the patient to shower, swim, or engage in hydrotherapy (the use of water for treatment). When the cast is wet, the patient is instructed to shake or drain water out of it; thorough drying is important to prevent skin breakdown, infection or irritation. The best results are achieved with casts that can easily drain, such as short-arm casts. Heels and elbows encased in wet casts may become macerated from the trapped water and therefore are associated with more skin breakdown.

Plaster Casts

Casts made of plaster are less costly and achieve a better mold than fiberglass casts; however, they are heavy, not water resistant, and can take up to 24 to 72 hours to dry postapplication, which can be inconvenient for the patient (Walsh, 2009). Rolls of plaster of Paris–impregnated bandages are dipped in cool water and applied smoothly to the body. An exothermic reaction occurs, similar to that seen with fiberglass casts. Serious burns may occur if hot water is used to dip the plaster, if multiple layers of padding and plaster are applied, or if the extremity is placed on a plastic coated mat or hospital pillow, because the heat generated by the chemical reaction cannot escape (Deignan, Iaquinto, Eskildsen, et al., 2011).

The time that it takes for a plaster cast to dry completely depends on its size, thickness, and location, as well as environmental drying conditions. A freshly applied cast should be exposed to circulating air to dry and supported on a firm

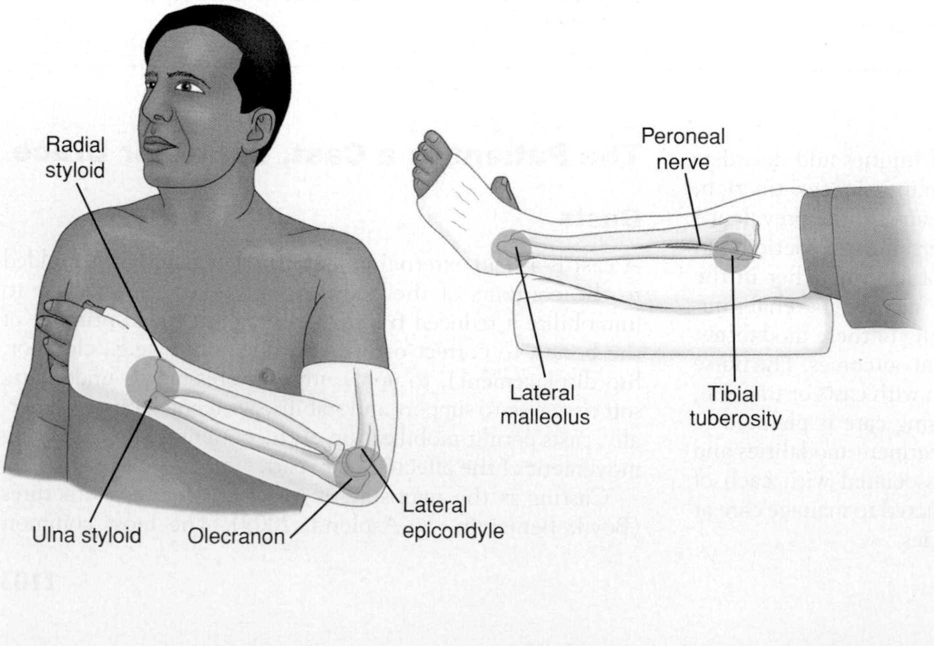
Radial styloid
Ulna styloid
Olecranon
Lateral epicondyle
Peroneal nerve
Lateral malleolus
Tibial tuberosity

FIGURE 41-1 • Pressure areas in common types of casts. (*Left*) Long-arm cast. (*Right*) Long-leg cast.

and smooth surface; it should not be placed on a hard surface or one with sharp edges. If elevation is requested to reduce swelling, a cloth-covered pillow is preferred to one covered in plastic, which could retain heat and prevent drying. When moving a patient with a wet plaster class, only the palms of the hands should be used so that indentations in the cast may be prevented; indentations can result in areas of pressure on the skin (Lynn, 2012). A wet plaster cast feels damp, appears dull and gray, sounds dull on percussion, and smells musty. The cast is dry when it feels hard and firm, has a white and shiny appearance, is resonant to percussion, and odorless.

On occasion, the plaster cast may have rough edges, which can crumble and cause skin irritation. **Petaling** the cast resolves this problem if the underlying stockinette does not cover the edges of the cast. To prevent skin breakdown, moleskin can be used over any rough area of the cast that may rub against the patient's skin.

Splints and Braces

Many injuries that were treated previously with casts may now be treated with other immobilization devices such as splints and braces. Splinting is more commonly used than casting in outpatient settings for acute and definitive management of orthopedic injuries (Boyd et al., 2009).

Splints are often used for simple and stable fractures, sprains, tendon injuries, and other soft tissue injuries. They offer many advantages over casts in that they are faster and easier to apply. They are also noncircumferential and thus do not compromise circulation when the natural swelling during the inflammatory phase of injury occurs. Pressure-related complications (e.g., skin breakdown, necrosis, compartment syndrome [see later discussions]) are more prevalent when soft tissue swelling occurs within a contained space (e.g., a circumferential cast). Splints are easily removed, facilitating inspection of the injury site. In addition, splints can be indicated to provide initial stability for fractures that are unstable while awaiting definitive care (Boyd et al., 2009).

Contoured **splints** of plaster or pliable thermoplastic materials may be used for conditions that do not require rigid immobilization, for those in which swelling may be anticipated, and for those that require special skin care. Splints made of thermoplastics are warmed and molded to custom-fit the affected body part (e.g., hand and thoracolumbosacral orthotics [TLSOs], clamshell-type back braces). The splint needs to immobilize and support the body part in a functional position, and it must be well padded to prevent pressure, skin abrasion, and skin breakdown. The splint is overwrapped with an elastic bandage applied in a spiral fashion and with pressure uniformly distributed so that circulation is not restricted.

Braces (i.e., orthoses) are used to provide support, control movement, and prevent additional injury. They are custom-fitted to various parts of the body; thus, they tend to be indicated for longer-term use than splints. The orthotist adjusts the brace for fit, positioning, and motion so that movement is enhanced, any deformities are corrected, and discomfort is minimized.

Many splints and braces are prefabricated and fastened with Velcro straps. They may be made of plastic and other materials such as cloth, leather, metal, and elastic. Knee immobilizers, ankle stirrups, and cock-up wrist splints are types of prefabricated splints and braces. Both splints and braces may be either custom-made or standard "off the shelf." Splints and braces are generally less compliant and permit more motion at the injury site than casts, which can be a serious disadvantage in that underlying injuries are not as well stabilized (Boyd et al., 2009).

Nursing Management of a Patient in a Cast, Splint, or Brace

Before the cast, splint, or brace is applied, the nurse completes an assessment of the patient's general health, presenting signs and symptoms, emotional status, understanding of the need for the device, and condition of the body part to be immobilized. Physical assessment of the part to be immobilized must include a thorough assessment of the skin and **neurovascular status** (i.e., neurologic and circulatory functioning), including the degree and location of swelling, bruising, and skin abrasions. In addition, the nurse gives the patient or family information about the underlying pathologic condition and the purpose and expectations of the prescribed treatment regimen. This knowledge promotes the patient's active participation in and adherence to the treatment program. The nurse prepares the patient for the application of the cast, splint, or brace by describing the anticipated sights, sounds, and sensations (e.g., heat from the hardening reaction of the fiberglass or plaster) that he or she may experience. Asking the patient and family what they know about the application and care of the cast can help determine opportunities for education (Walsh, 2009). The patient needs to know what to expect during application and the reason the body part must be immobilized (Chart 41-1).

The main nursing concern following the application of an immobilization device is assessment and prevention of neurovascular dysfunction or compromise of the affected extremity (Lynn, 2012). Assessments are performed at least every hour for the first 24 hours and every 1 to 4 hours thereafter to prevent neurovascular compromise related to **edema** (soft tissue swelling due to fluid accumulation) and/or the device. Neurovascular assessment includes the assessment of peripheral circulation, motion, and sensation of the affected extremity, assessing the fingers or toes of the affected extremity, and comparing them with those of the opposite extremity. When assessing peripheral circulation, the nurse must check peripheral pulses as well as capillary refill response (within 3 seconds), edema, and the color and temperature of the skin. While assessing motion, the nurse should note any weakness or paralysis of the injured body part. While assessing sensation, the nurse monitors for **paresthesia** (numbness or tingling) or absence of feeling in the affected extremity, which could indicate nerve damage (Satryb et al., 2011).

The "five Ps" indicative of symptoms of neurovascular compromise are *p*ain, *p*allor, *p*ulselessness, *p*aresthesia, and *p*aralysis (Johnston-Walker & Hardcastle, 2011). Early recognition of diminished circulation and nerve function is essential to prevent loss of function. The nurse adjusts the extremity so that it is above the level of the heart during the first 24 to 48 hours postapplication to enhance arterial

Chart 41-1 GUIDELINES

Guidelines for Applying a Cast

Equipment

• Drape for patient • Knitted material (e.g., stockinette) • Nonwoven roll padding • Casting material • Water and basin • Cast knife or cutter

Implementation

Procedure	Rationale
1. Support extremity or body part to be casted.	1. Minimizes movement; maintains reduction and alignment; increases comfort
2. Position and maintain part to be casted in position indicated by physician during casting procedure.	2. Facilitates casting; reduces incidence of complications (e.g., malunion, nonunion, contracture)
3. Drape patient.	3. Avoids undue exposure; protects other body parts from contact with casting materials
4. Wash and dry part to be casted.	4. Reduces incidence of skin breakdown
5. Place at least three layers of knitted material* (e.g., stockinette) over part to be casted. • Apply in smooth and nonconstrictive manner. • Allow additional material.	5. Protects skin from casting materials Protects skin from pressure Folds over edges of cast when finishing application; creates smooth, padded edge; protects skin from abrasion
6. Wrap soft, nonwoven roll padding* smoothly and evenly around part. • Use additional padding around bony prominences to protect superficial nerves (e.g., head of fibula, olecranon process).	6. Protects skin from pressure of cast Protects skin at bony prominences Protects superficial nerves
7. Apply plaster or fiberglass casting material evenly on body part. • Choose appropriate-width bandage. • Overlap preceding turn by half the width of the bandage. • Use continuous motion, maintaining constant contact with body part. • Use additional casting material (splints) at joints and at points of anticipated cast stress.	7. Creates smooth, solid, well-contoured cast Facilitates smooth application Creates smooth, solid, immobilizing cast Shapes cast properly for adequate support Strengthens cast
8. "Finish" cast. • Smooth edges. • Trim and reshape with cast knife or cutter.	8. Protects skin from abrasion Allows full range of motion of adjacent joints
9. Remove particles of casting materials from skin.	9. Prevents particles from loosening and sliding underneath cast
10. Support cast during hardening. • Handle hardening casts with palms of hands. • Support cast on firm, smooth surface. • Do not rest cast on hard surfaces or on sharp edges. • Avoid pressure on cast.	10. Casting materials begin to harden in minutes. Maximum hardness of nonplaster cast occurs in minutes. Maximum hardness of plaster cast occurs with drying (24–72 hours), depending on size, thickness, and location of cast and environment. Avoids denting of cast and development of pressure areas
11. Promote drying of cast. • Leave cast uncovered and exposed to air. • Turn patient every 2 hours, supporting major joints. • Fans may be used to increase airflow and speed drying.	11. Facilitates drying

*Nonabsorbent materials are used with nonplaster casts.

perfusion and control edema and notifies the primary provider at once if signs of compromised neurovascular status are present.

The nurse must carefully evaluate pain associated with the musculoskeletal condition, asking the patient to indicate the exact site and to describe the character and intensity of the pain using a pain rating scale (see Chapter 12). Most pain can be relieved by elevating the involved part, applying ice or cold packs, and administering analgesic agents as prescribed.

Quality and Safety Nursing Alert

A patient's unrelieved pain must be reported immediately to the primary provider to avoid necrosis, neuromuscular damage, and possible paralysis.

Pain associated with the underlying condition (e.g., fracture) is frequently controlled by immobilization. Pain due to edema that is associated with trauma, surgery, or bleeding into the tissues can frequently be controlled by elevation and, if prescribed, intermittent application of ice or cold packs. Ice bags (one third to one half full) or cold application devices are placed on each side of the cast, if prescribed, making sure not to indent or wet the cast.

Unrelieved or disproportionate pain may indicate complications. Pain associated with compartment syndrome (see Chapter 43 and later in this chapter) is relentless and is not controlled by modalities such as elevation, application of ice or cold, and usual dosages of analgesic agents. Severe burning pain over bony prominences, especially the heels, anterior ankles, and elbows, warns of an impending pressure ulcer. This may also occur from too-tight elastic wraps used to hold splints in place.

Quality and Safety Nursing Alert

The nurse must never ignore complaints of pain from the patient in a cast because of the possibility of problems, such as impaired tissue perfusion or pressure ulcer formation.

To promote healing, any skin lacerations and abrasions that may have occurred as a result of the trauma that caused the fracture must be treated before the cast, brace, or splint is applied. The nurse thoroughly cleanses the skin and treats it as prescribed. The patient may require a tetanus booster if the wound is dirty and if the last known booster was administered more than 5 years ago. Sterile dressings are used to cover the injured skin. If the skin wounds are extensive, an alternative method (e.g., external fixator) may be chosen to immobilize the body part. While the cast is on, the nurse observes the patient for systemic signs of infection, which include an unpleasant odor from the cast, splint, or brace, and purulent drainage staining the cast. Infection is more common from an open wound, but the moist, warm environment of a splint or cast can be an ideal conduit for infection (Boyd et al., 2009). If the infection progresses, a fever may develop. The nurse must notify the primary provider if any of these signs occur.

Finally, some degree of joint stiffness is an inevitable complication of immobilization. Every joint that is not immobilized should be exercised and moved through its range of motion to maintain function. The nurse encourages the patient to move all fingers or toes hourly when awake to stimulate circulation.

Monitoring and Managing Potential Complications

It is important to assess for potential complications resulting from casts, splints, and braces that can be serious and life threatening, such as compartment syndrome, pressure ulcer formation, and disuse syndrome.

Compartment Syndrome

Compartment syndrome—the most serious complication of casting and splinting—occurs when increased pressure within a confined space (e.g., cast, muscle compartment) compromises blood flow and low tissue perfusion occurs, most often in an extremity (Boyd et al., 2009). Ischemia and potentially irreversible neuromuscular damage can occur within a few hours if action is not taken (Fig. 41-2). Tight casts or constrictive splints are associated with this complication.

Clinical manifestations include dusky, pale appearance of the exposed extremity; cool skin temperature; delayed capillary refill; paresthesia; and unrelenting pain not relieved by position changes, ice, or analgesia. A hallmark sign is pain that occurs or intensifies with passive range of motion (Johnston-Walker & Hardcastle, 2011). The patient may complain that the cast, brace, or splint is too tight. The primary provider must be notified immediately.

If the complication is due to a cast or tight splint, the splint may be loosened or removed and the cast bivalved (cut in half longitudinally) to release constriction and allow for inspection of the skin. The nurse assists in maintaining limb alignment, and the extremity must then be elevated no higher than heart level to maintain arterial perfusion (Chart 41-2). If pressure is not relieved and circulation is not restored, a fasciotomy may be necessary to relieve the pressure within the muscle compartment. The nurse closely monitors the patient's response to conservative and surgical management of compartment syndrome. The nurse records frequent neurovascular responses and promptly reports changes to the primary provider. (See Chapter 43 for further discussion of compartment syndrome.)

Pressure Ulcers

Casts or inappropriately applied splints can put pressure on soft tissues, causing tissue anoxia and pressure ulcers. Lower extremity sites most susceptible are the heel, malleoli, dorsum of the foot, head of the fibula, and anterior surface of the patella. The main pressure sites on the upper extremity are located at the medial epicondyle of the humerus and the ulnar styloid (see Fig. 41-1).

If pressure necrosis occurs, the patient typically reports a very painful "hot spot" and tightness under the cast. The cast

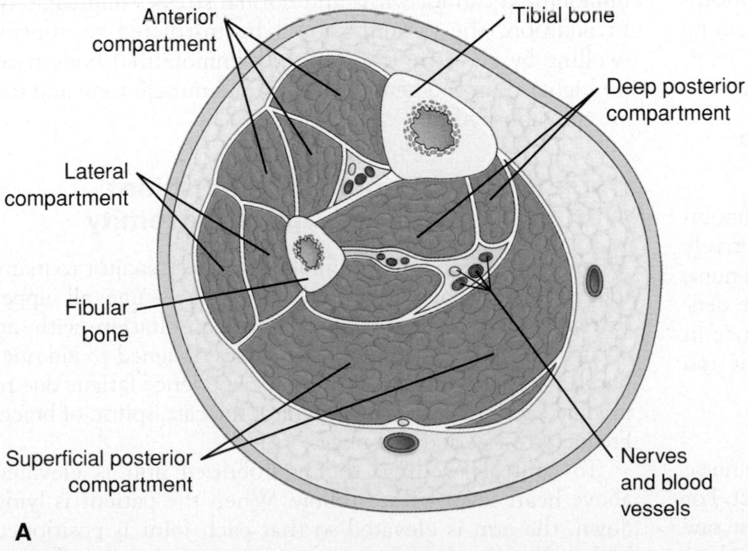
Anterior compartment
Tibial bone
Deep posterior compartment
Lateral compartment
Fibular bone
Superficial posterior compartment
Nerves and blood vessels
A
B

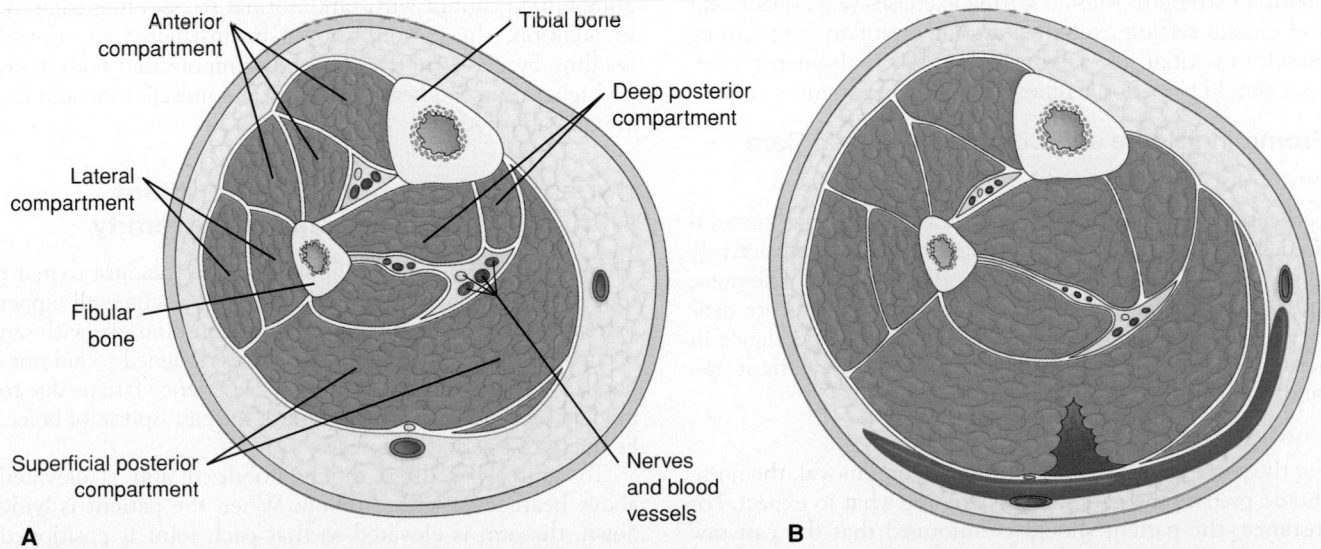

FIGURE 41-2 • **A.** Cross-section of normal lower leg with muscle compartments. **B.** Cross-section of lower leg with compartment syndrome. Swelling of muscles causes compression of nerves and blood vessels.

Chart 41-2 Procedure for Bivalving a Cast

The following procedure is used when a cast is bivalved:

1. With a cast cutter, a longitudinal cut is made to divide the cast in half.
2. The underpadding is cut with scissors.
3. The cast is spread apart with cast spreaders to relieve pressure and to inspect and treat the skin without interrupting the reduction and alignment of the bone.
4. After the pressure is relieved, the anterior and posterior parts of the cast are secured together with an elastic compression bandage to maintain immobilization.
5. To control swelling and promote circulation, the extremity is elevated (but no higher than heart level) to minimize the effect of gravity on perfusion of the tissues.

may feel warmer in the affected area, suggesting underlying tissue erythema. Drainage may stain the cast or splint and emit an unpleasant odor. Even if discomfort does not occur, there may still be extensive loss of tissue with skin breakdown and tissue necrosis. To assess for pressure ulcer development, the physician may bivalve or cut an opening (window) in the cast to allow for inspection, access, and possible treatment. A dressing may be applied over the exposed skin, and the cutout portion of the cast is replaced and held in place by an elastic compression dressing or tape. This prevents "window edema" from occurring, which is the swelling or bulging of the underlying soft tissue through the window opening.

Disuse Syndrome

Immobilization in a cast, splint, or brace can cause muscle atrophy and loss of strength, and can place patients at risk for disuse syndrome, which is the deterioration of body systems as a result of prescribed or unavoidable musculoskeletal inactivity (NANDA International, 2012). To prevent this, the nurse instructs the patient to tense or contract muscles (e.g., isometric muscle contraction) without moving the underlying bone. Isometric exercises, such as instructing the patient with a leg or arm cast to splint or brace to "push down" the knee or to "make a fist," respectively, helps reduce muscle atrophy and maintain strength. Muscle setting exercises (e.g., quadriceps and gluteal setting exercises) are important in maintaining muscles essential for walking (Chart 41-3). Isometric exercises should be performed hourly while the patient is awake.

Promoting Home and Community-Based Care

Educating the Patient About Self-Care

Self-care deficits occur when a portion of the body is immobilized. The nurse encourages the patient to participate actively in personal care and to use assistive devices safely. The nurse must assist the patient in identifying areas of self-care deficit and in developing strategies to achieve independence in activities of daily living (ADLs) (Chart 41-4). Patient and family education is also described in Chart 41-4.

Continuing Care

For the patient with a cast that is ready for removal, the nurse should prepare the patient by explaining what to expect. For instance, the patient should be informed that the cast saw uses an oscillating blade that vibrates but does not spin; thus, it cuts through the outer cast layer but does not penetrate deeply enough to injure the patient's skin (Satryb et al., 2011). The padding is then cut with scissors to ensure that the patient's skin will not be cut.

The formerly immobilized body part will be weak from disuse, stiff, and may appear atrophied. As the cast or splint is removed, the affected body part should be supported to prevent injury. The skin, which is usually dry and scaly from accumulated dead skin, is vulnerable to injury from scratching. The skin needs to be washed gently and lubricated with an emollient lotion. The patient should be instructed to avoid rubbing and scratching the skin, because doing so can cause damage to newly exposed skin.

The nurse and physical therapist educate the patient to resume activities gradually within the prescribed therapeutic regimen. Exercises prescribed to help the patient regain joint motion are explained and demonstrated. Because the muscles are weak from disuse, the body part that has been immobilized cannot withstand normal stresses immediately. In addition, the patient should be instructed to control swelling by elevating the formerly immobilized body part, no higher than the heart, until normal muscle tone and use are reestablished.

Chart 41-3 Muscle Setting Exercises

Isometric contractions of the muscle maintain muscle mass and strength and prevent atrophy.

Quadriceps Setting Exercise

- Position patient supine with leg extended.
- Instruct patient to push knee back onto the mattress by contracting the anterior thigh muscles.
- Encourage patient to hold the position for 5–10 seconds.
- Let patient relax.
- Have patient repeat the exercise 10 times each hour when awake.

Gluteal Setting Exercise

- Position patient supine with legs extended, if possible.
- Instruct patient to contract the muscles of the buttocks.
- Encourage patient to hold the contraction for 5–10 seconds.
- Let the patient relax.
- Have patient repeat the exercise 10 times each hour when awake.

Nursing Management of the Patient With an Immobilized Upper Extremity

The patient whose arm is immobilized must readjust to many routine tasks. The unaffected arm will assume all upper extremity activities. The nurse, in consultation with an occupational therapist, suggests devices designed to aid one-handed activities. The patient may experience fatigue due to modified activities and the weight of the cast, splint, or brace. Frequent rest periods are necessary.

To control swelling, the immobilized arm is elevated above heart level with a pillow. When the patient is lying down, the arm is elevated so that each joint is positioned higher than the preceding proximal joint (e.g., elbow higher than the shoulder, hand higher than the elbow).

Chart 41-4 HOME CARE CHECKLIST
The Patient With a Cast, Splint, or Brace

At the completion of home care education, the patient or caregiver will be able to:	PATIENT	CAREGIVER
• Describe techniques to promote cast drying (e.g., do not cover cast; expose cast to circulating air; handle damp plaster cast with palms of hands; do not rest the cast on hard surfaces or sharp edges that can dent soft cast).	✔	✔
• Describe approaches to controlling swelling and pain (e.g., elevate immobilized extremity to heart level, apply intermittent ice bag if prescribed, take analgesic agents as prescribed).	✔	✔
• Report pain uncontrolled by elevating the immobilized limb and by analgesic agents (may be an indicator of impaired tissue perfusion—compartment syndrome or pressure ulcer).	✔	✔
• Demonstrate ability to transfer (e.g., from a bed to a chair).	✔	
• Use mobility aids safely.	✔	
• Avoid excessive use of injured extremity; observe prescribed weight-bearing limits.	✔	
• Manage minor skin irritations (e.g., for skin irritation from edge of cast, splint, or brace, pad rough edges with tape or moleskin; to relieve itching, blow cool air from hair dryer; do not insert foreign objects inside the cast, splint, or brace).	✔	✔
• Demonstrate exercises to promote circulation and minimize disuse syndrome.	✔	
• State indicators of complications to report promptly to primary provider (e.g., uncontrolled swelling and pain; cool, pale fingers or toes; paresthesia; paralysis; purulent drainage staining cast; signs of systemic infection; cast, splint, or brace breaks).	✔	✔
• Describe care of extremity following cast, splint, or brace removal (e.g., skin care, gradual resumption of normal activities to protect limb from undue stresses, management of swelling).	✔	✔

A **sling** may be used when the patient ambulates. To prevent pressure on the cervical spinal nerves, the sling should distribute the supported weight over a large area of the shoulders and trunk, not just the back of the neck. The nurse encourages the patient to remove the arm from the sling and elevate it frequently.

Circulatory disturbances in the hand may become apparent with signs of cyanosis, swelling, and an inability to move the fingers. One serious effect of impaired circulation in the arm is Volkmann contracture, which is a specific type of compartment syndrome (Marshall & Browner, 2012). Contracture of the fingers and wrist occurs as the result of obstructed arterial blood flow to the forearm and hand. The patient is unable to extend the fingers, describes abnormal sensation (e.g., unrelenting pain, pain on passive stretch), and exhibits signs of diminished circulation to the hand. Irreversible damage develops within a few hours if action is not taken (see Chapter 43). This serious complication can be prevented with nursing surveillance and proper care.

Nursing Management of the Patient With an Immobilized Lower Extremity

The application of a leg cast, splint, or brace imposes a degree of immobility on the patient. Casts may include short-leg casts, extending to the knees, or long-leg casts, extending to the groin. Hinged knee braces and immobilizers typically extend from ankle to groin.

The patient's leg must be supported on pillows to the level of the heart to control swelling. Cold therapy or ice packs should be applied as prescribed over the fracture site for 1 to 2 days. The patient is taught to elevate the immobilized leg when seated. The patient should also assume a recumbent position several times a day with the immobilized leg elevated to promote venous return and control swelling (Lynn, 2012).

The nurse assesses circulation by observing the color, temperature, and capillary refill of the exposed toes. Nerve function is assessed by observing the patient's ability to move the toes and by asking about the sensations in the foot. Numbness, tingling, and burning may indicate peroneal nerve injury resulting from pressure at the head of the fibula.

Quality and Safety Nursing Alert

Injury to the peroneal nerve as a result of pressure is a cause of footdrop (the inability to maintain the foot in a normally flexed position). Consequently, the patient drags the foot when ambulating.

The nurse and physical therapist instruct the patient how to transfer and ambulate safely with assistive devices (e.g., crutches, walker) (see Chapter 10). The gait to be used depends on whether the patient is permitted to bear weight. If weight bearing is allowed, the cast, splint, or brace is reinforced to withstand the body weight. A cast boot or shoe, which is worn over the casted foot, provides a broad, nonskid walking surface.

Nursing Management of the Patient With a Body or Spica Cast

Casts that encase the trunk of the body (body cast) and portions of one or two extremities (spica cast) require special nursing strategies. Body casts are used to immobilize the spine. Hip spica casts are used for some femoral fractures and after some hip joint surgeries, and shoulder spica casts are used for some humeral neck fractures. These casts can remain in place between several weeks to several months, depending on the reason for its use (Reed, Carroll, Baccari, et al., 2011).

Nursing responsibilities include preparing and positioning the patient, assisting with skin care and hygiene, and monitoring for cast syndrome (see later discussion). Explaining the casting procedure helps reduce the patient's apprehension about being encased in a large cast. The nurse reassures the patient that several people will provide care during the application, support for the injured area will be adequate, and care providers will be as gentle as possible. Patients immobilized in large casts may develop **cast syndrome**, which may include psychological or physiologic manifestations (Satryb et al., 2011). The psychological component is similar to a claustrophobic reaction. The patient exhibits an acute anxiety reaction characterized by behavioral changes and autonomic responses (e.g., increased respiratory rate, diaphoresis, dilated pupils, increased heart rate, elevated blood pressure). The nurse needs to recognize the anxiety reaction and provide an environment in which the patient feels secure. The administration of pain and antianxiety medications prior to the procedure may help to reduce this reaction. Superior mesenteric artery syndrome is the physiologic manifestation associated with immobilization from a body cast. With decreased physical activity, gastrointestinal motility decreases, intestinal gases accumulate, intestinal pressure increases, and an ileus may occur. The patient exhibits abdominal distention and discomfort, nausea, and vomiting. As with other instances of adynamic ileus, the patient is treated conservatively with decompression (nasogastric intubation connected to suction) and intravenous (IV) fluid therapy until gastrointestinal motility is restored. Rarely, the abdominal distention can place added pressure on the superior mesenteric artery, reducing the blood supply to the bowel, which can result in gangrenous bowel. The descending aorta may also sustain pressure, as it may be compressed between the spine and pressure of abdominal distention, which results in ischemia. Placing a window in the abdominal portion of the cast or bivalving the cast may be sufficient to prevent or relieve pressure on the duodenum.

Quality and Safety Nursing Alert

The nurse monitors the patient in a large body cast for potential physiologic cast syndrome, noting bowel sounds every 4 to 8 hours, and reports abdominal discomfort and distention, nausea, and vomiting to the primary provider.

The patient with a body or spica cast is often cared for at home. The nurse educates the family about how to care for the patient, which includes providing hygienic and skin care, ensuring proper positioning, preventing complications, and recognizing symptoms that should be reported to the primary provider.

The Patient With an External Fixator

External fixators are used to manage complex open fractures with soft tissue damage. Complicated fractures of the humerus, forearm, femur, tibia, and pelvis are managed with external skeletal fixators. They are also used to correct defects, treat nonunion, and lengthen limbs. Their use has increased in recent years with advances in orthopedic trauma care (Tuttle, Smith, Wade, et al., 2009). The fixator provides skeletal stability for severe comminuted (crushed or splintered) fractures while permitting active treatment of extensive soft tissue damage (Fig. 41-3).

External fixation involves the surgical insertion of pins through the skin and soft tissues into and through the bone. An external metal apparatus is attached to these pins and is designed to hold the position of the fracture in proper alignment (Timms, Vincent, Santy-Tomlinson, et al. 2011). Benefits of external fixation, as opposed to other modes of treatment, include immediate fracture stabilization, minimization of blood loss (as compared to use of internal fixation), increased patient comfort, improved wound care, promotion of early mobilization and weight-bearing on the affected limb, and active exercise of adjacent uninvolved joints (Carroll & Koman, 2011). The disadvantage associated with the use of external fixators is an increased risk for pin site infections (Cavusoglu, Er, Inal, et al., 2009).

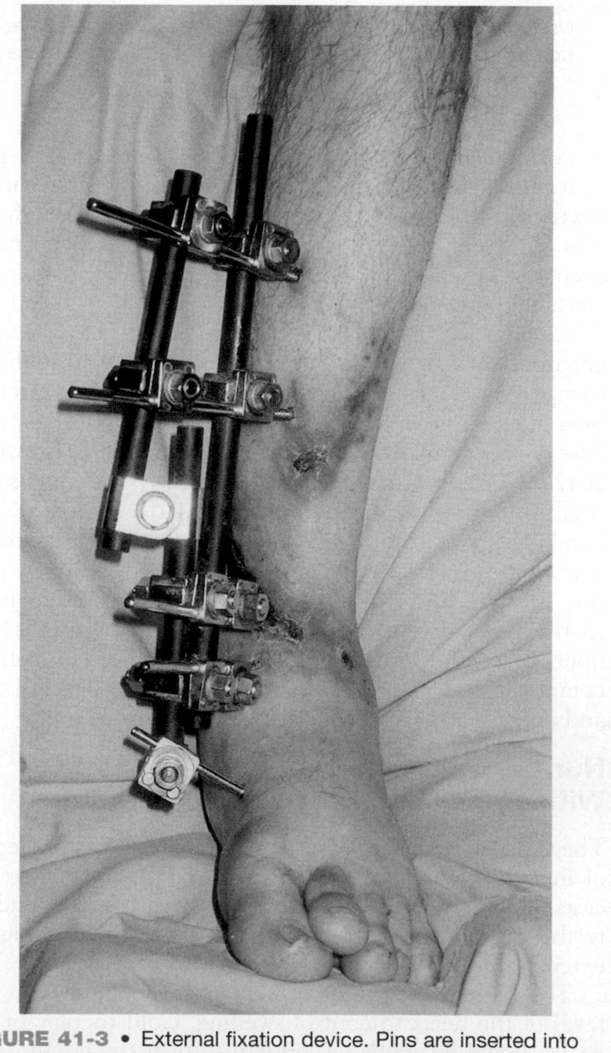

FIGURE 41-3 • External fixation device. Pins are inserted into bone. The fracture is reduced and aligned and then stabilized by attaching the pins to a rigid portable frame. The device facilitates treatment of soft tissue damaged in complex fractures.

Nursing Management

Patients must be prepared psychologically for application of the external fixator, as they may be at risk for an altered body image related to the overwhelming size and bulk of the apparatus. To promote acceptance of the device, patients should be given comprehensive information about the frame, reassurance that the discomfort associated with the device is minimal, and that early mobility is anticipated. Clothing and other materials may need to be altered or used to cover the device.

After the external fixator is applied, the extremity is elevated to the level of the heart to reduce swelling, if appropriate. Any sharp points on the fixator or pins are covered with caps to prevent device-induced injuries. The nurse must be alert for potential problems caused by pressure from the device on the skin, nerves, or blood vessels and for the development of compartment syndrome (see Chapter 43). The nurse monitors the neurovascular status of the extremity every 2 to 4 hours and promptly reports changes to the primary provider (Lynn, 2012). Because the pins are inserted externally, particular attention is focused on the pin sites for signs of inflammation and infection. The end goal is to avoid **osteomyelitis** (infection of the bone) (see Chapter 42 for a discussion of osteomyelitis). The nurse assesses each pin site at least every 8 to 12 hours for redness, swelling, pain around the pin sites, warmth, and purulent drainage, because these are the most common indicators of pin site infections (Timms et al., 2011). In the first 48 to 72 hours postinsertion, some serous drainage and mild redness at the pin sites is expected (Santy-Tomlinson, Vincent, Glossop, et al., 2011).

The nurse performs pin site care as prescribed, using aseptic technique. There is wide practice variation on pin care; however, evidence-based recommendations include cleansing each pin site separately to avoid cross-contamination with nonshedding material (e.g., gauze, cotton-tip swab) by using a chlorhexidine 2 mg/mL solution (Timms et al., 2011). Pin sites should be cleaned and dressed as prescribed unless there is copious drainage, the dressing becomes wet, or infection is suspected, in which case cleaning and dressing may be more frequent. If signs of infection are present or if the pins or clamps seem loose, the nurse notifies the primary provider.

Quality and Safety Nursing Alert

The nurse never adjusts the clamps on the external fixator frame. It is the physician's responsibility to do so.

If activity is restricted, the nurse encourages isometric exercises as tolerated to prevent complications of mobility (e.g., thrombus formation). When the swelling subsides, the nurse helps the patient become mobile within the prescribed weight-bearing limits (non–weight bearing to full weight bearing). Adherence to weight-bearing instructions minimizes the chance of loosening of the pins when stress is applied to the bone–pin interface. The external fixator may be removed once the soft tissue heals and there are no signs of infection. The fracture may require additional stabilization by a cast, molded orthosis, or internal fixation while healing.

Ilizarov fixation is a specialized type of external fixator consisting of numerous wires that penetrate the limb and are attached to a circular metal frame. This device is used to correct angulation and rotational defects, to treat nonunion (failure of bone fragments to heal), and to lengthen limbs. The device gently pulls apart the cortex of the bone and stimulates new growth through daily adjustment of the telescoping rods. The nurse must educate the patient about adjusting the telescoping rods and caring for the pin sites and apparatus, because this fixator can be in place for many months. When discharge is anticipated, the nurse educates the patient or caregiver about how to perform pin site care according to the prescribed protocol (clean technique can be used at home) (Cavusoglu et al., 2009) and to promptly report any signs of pin site infection. The nurse also instructs the patient or family to monitor neurovascular status and report any changes promptly. The patient or family members are instructed to check the integrity of the fixator frame daily and to report loose pins or clamps. A physical therapy referral is helpful in educating the patient how to transfer, use ambulatory aids safely, and adjust to weight-bearing limits and altered gait patterns (Chart 41-5).

The Patient in Traction

Traction uses a pulling force to promote and maintain alignment to an injured part of the body. The goals of traction include decreasing muscle spasms and pain, realignment of bone fractures, and correcting or preventing deformities. The type of traction, amount of weight, and whether traction can be removed for nursing care must be determined to obtain its therapeutic effects.

At times, traction needs to be applied in more than one direction to achieve the desired line of pull. When this is done, one of the lines of pull counteracts the other. These lines of pull are known as the vectors of force. The actual resultant pulling force is somewhere between the two lines of pull (Fig. 41-4). The effects of traction are evaluated with x-ray studies, and adjustments are made if necessary.

Traction is used primarily as a short-term intervention until other modalities, such as external or internal fixation, are possible. These modalities reduce the risk of disuse

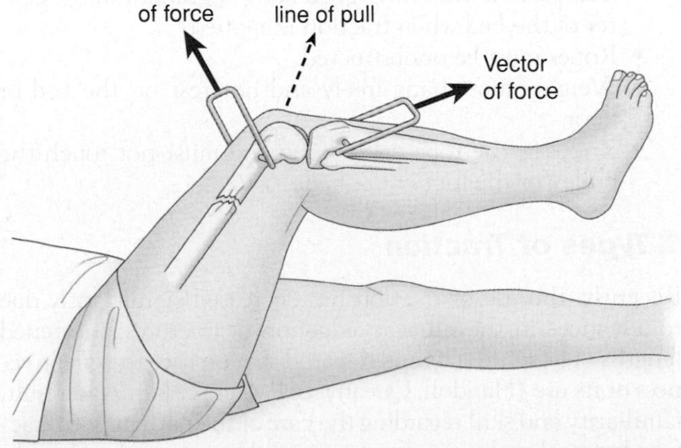
Vector
of force
Resultant
line of pull
Vector
of force

FIGURE 41-4 • Traction may be applied in different directions to achieve the desired therapeutic line of pull. Adjustments in applied forces may be prescribed over the course of treatment.

Chart 41-5 HOME CARE CHECKLIST

The Patient With an External Fixator

At the completion of home care education, the patient or caregiver will be able to:	PATIENT	CAREGIVER
• Demonstrate prescribed pin site care.	✔	✔
• State signs of pin site infection (e.g., redness, tenderness, increased or purulent pin site drainage) to be reported promptly.	✔	✔
• Describe approaches to controlling swelling and pain (e.g., elevate extremity to heart level, take analgesic agents as prescribed).	✔	✔
• Report pain uncontrolled by elevation and analgesic agents (may be an indicator of impaired tissue perfusion, compartment syndrome, or pin tract infection).	✔	
• Demonstrate ability to transfer.	✔	
• Use mobility aids safely.	✔	
• Avoid excessive use of injured extremity; observe prescribed weight-bearing limits.	✔	
• State indicators of complications to report promptly to primary provider (e.g., uncontrolled swelling and pain; cool, pale fingers or toes; paresthesia; paralysis; purulent drainage; signs of systemic infection; loose fixator pins or clamps).	✔	✔
• Describe care of extremity after fixator removal (e.g., gradual resumption of normal activities to protect limb from undue stresses).	✔	✔

syndrome and minimize hospital lengths of stay, often allowing the patient to be cared for in the home setting.

Principles of Effective Traction

Whenever traction is applied, countertraction must be used to achieve effective results. Countertraction is the force acting in the opposite direction. Usually, the patient's body weight and bed position adjustments supply the needed countertraction.

The following are additional principles to follow when caring for the patient in traction:

- Traction must be continuous to be effective in reducing and immobilizing fractures.
- Skeletal traction is *never* interrupted.
- Weights are not removed unless intermittent traction is prescribed.
- Any factor that might reduce the effective pull or alter its resultant line of pull must be eliminated.
- The patient must be in good body alignment in the center of the bed when traction is applied.
- Ropes must be unobstructed.
- Weights must hang freely and not rest on the bed or floor.
- Knots in the rope or the footplate must not touch the pulley or the foot of the bed.

Types of Traction

Recently, the use of traction has decreased significantly due to advances in the surgical reduction of fractures, shortened lengths of hospital stay, and research that queries the effectiveness of its use (Handoll, Queally, & Parker, 2011). As a result, familiarity and skill regarding the care of the patient with traction has declined among nursing staff. However, a basic working knowledge of the use of traction is necessary, because some orthopedic surgeons still prescribe traction for their patients.

There are several types of traction. *Straight* or *running traction* applies the pulling force in a straight line with the body part resting on the bed. The countertraction is provided by the client's body, and movement of the patient's body can alter the traction provided. Buck's extension traction (Fig. 41-5) is an example of straight traction. *Balanced suspension traction* (Fig. 41-6) supports the affected extremity off the bed and allows for some patient movement without disruption of the line of pull. With this traction, the countertraction is produced by devices such as slings or splints.

Traction may be applied to the skin (*skin traction*) or directly to the bony skeleton (*skeletal traction*). The mode of application is determined by the purpose of the traction. Traction can be applied with the hands (*manual traction*). This is temporary traction that may be used when applying a cast, giving skin care under a Buck's extension foam boot, or adjusting the traction apparatus.

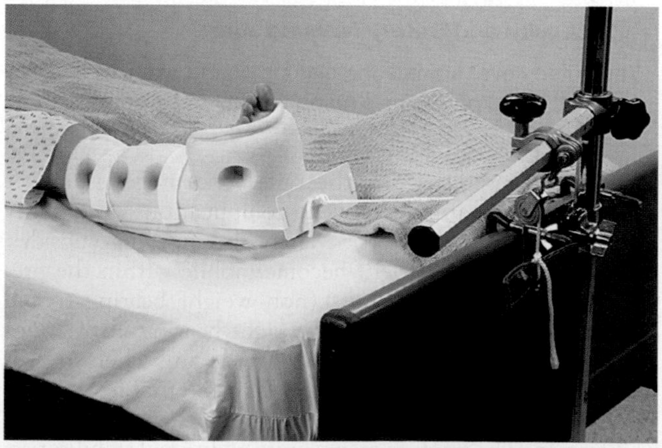

FIGURE 41-5 • Buck's extension traction. Lower extremity in unilateral Buck's extension traction is aligned in a foam boot and traction applied by the free-hanging weight. The Heelift® Traction Boot is shown here. Photo courtesy of DM Systems, Inc.

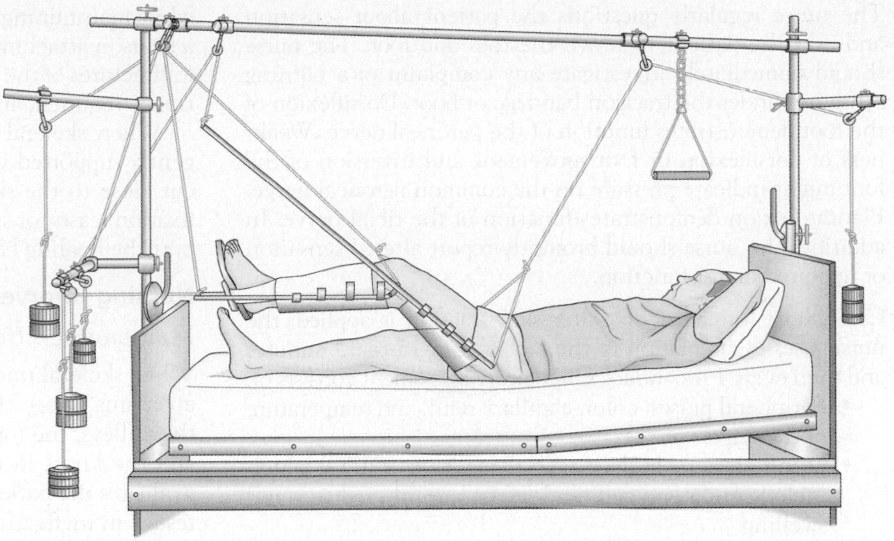

FIGURE 41-6 • Balanced suspension skeletal traction with Thomas leg splint. The patient can move vertically as long as the resultant line of pull is maintained. Note the use of overhead trapeze.

Skin Traction

Skin traction is used less frequently than in years past; indeed, a recent systematic review reported that there was no reduction in pain or complications associated with its use compared with other standard interventions (Abou-Setta, Beaupre, Dryden, et al., 2011). Nonetheless, skin traction may be prescribed for short-term use to stabilize a fractured leg, control muscle spasms, and immobilize an area before surgery. The pulling force is applied by weights that are attached to the client with Velcro, tape, straps, boots, or cuffs. The amount of weight applied must not exceed the tolerance of the skin. No more than 2 to 3.5 kg (4.5 to 8 lb) of traction can be used on an extremity. Pelvic traction is usually limited to 4.5 to 9 kg (10 to 20 lb), depending on the weight of the patient.

Types of skin traction used for adults include Buck's extension traction (applied to the lower leg) (described next), the chin halter strap (occasionally used to treat chronic neck pain), and the pelvic belt (sometimes used to treat lower back pain).

Buck's Extension Traction

Buck's extension traction (unilateral or bilateral) is skin traction to the lower leg. The pull is exerted in one plane when partial or temporary immobilization is desired (see Fig. 41-5). It is used to immobilize fractures of the proximal femur and hip before surgical fixation.

Before the traction is applied, the nurse inspects the skin for abrasions and circulatory disturbances. The skin and circulation must be in healthy condition to tolerate the traction. The extremity should be clean and dry before the foam boot or traction tape is applied.

To apply Buck's traction, the extremity is elevated and supported under the patient's heel and knee while the foam boot is placed under the leg, with the patient's heel in the heel of the boot. Next, the Velcro straps are secured around the leg. Traction tape that is overwrapped with an elastic bandage in a spiral fashion may be used instead of the boot. Excessive pressure is avoided over the malleolus and proximal fibula during application to prevent pressure ulcers and nerve damage. The rope is then affixed to the spreader or footplate over a pulley fastened to the end of the bed and attaches the prescribed weight—usually 5 to 8 pounds—to the rope. The weight should hang freely, not touching the bed or the floor (Lynn, 2012).

Nursing Interventions

Ensuring Effective Skin Traction

To ensure effective skin traction, it is important to avoid wrinkling and slipping of the traction bandage and to maintain countertraction. Proper positioning must be maintained to keep the leg in a neutral position. To prevent bony fragments from moving against one another, the patient should not turn from side to side; however, the patient may shift position slightly with assistance.

Monitoring and Managing Potential Complications

Skin Breakdown. During the initial assessment, the nurse identifies sensitive, fragile skin (common in older adults). The nurse also inspects the skin area that is in contact with tape, foam, or shearing forces, at least every 8 hours, for signs of irritation or inflammation. The nurse performs the following procedures to monitor and prevent skin breakdown:

- Removes the foam boots to inspect the skin, the ankle, and the Achilles tendon three times a day. A second person is needed to support the extremity during the inspection and skin care.
- Palpates the area of the traction tapes daily to detect underlying tenderness
- Provides frequent repositioning to alleviate pressure and discomfort, because the patient who must remain in a supine position is at increased risk for development of a pressure ulcer
- Uses higher-specification foam mattresses rather than standard hospital foam mattresses to prevent pressure ulcers (McInnes, Jammali-Blasi, Bell-Syer, et al., 2011)

Nerve Damage. Skin traction can place pressure on peripheral nerves. Care must be taken to avoid pressure on the peroneal nerve at the point at which it passes around the neck of the fibula just below the knee when traction is applied to the lower extremity. Pressure at this point can cause footdrop.

The nurse regularly questions the patient about sensation and asks the patient to move the toes and foot. The nurse should immediately investigate any complaint of a burning sensation under the traction bandage or boot. Dorsiflexion of the foot demonstrates function of the peroneal nerve. Weakness of dorsiflexion or foot movement and inversion of the foot might indicate pressure on the common peroneal nerve. Plantar flexion demonstrates function of the tibial nerve. In addition, the nurse should promptly report altered sensation or impaired motor function.

Circulatory Impairment. After skin traction is applied, the nurse assesses circulation of the foot within 15 to 30 minutes and then every 1 to 2 hours. Circulatory assessment consists of:

- Peripheral pulses, color, capillary refill, and temperature of the fingers or toes
- Manifestations of deep vein thrombosis (DVT), which include unilateral calf tenderness, warmth, redness, and swelling

The nurse also encourages the patient to perform active foot exercises every hour when awake.

Quality and Safety Nursing Alert

Neurovascular assessments are priority nursing interventions for patients in traction.

Skeletal Traction

Skeletal traction is often used when continuous traction is desired to immobilize, position, and align a fracture of the femur, tibia, and cervical spine. It is used when skin traction is not possible; greater weight (11 to 18 kg [25 to 40 lb]) is needed to achieve the therapeutic effect. Skeletal traction involves passing a metal pin or wire (e.g., Steinmann pin, Kirschner wire) through the bone (e.g., proximal tibia or distal femur) under local anesthesia, avoiding nerves, blood vessels, muscles, tendons, and joints. Traction is then applied using ropes and weights attached to the end of the pin. Alternatively, skeletal traction may involve the application of tongs to the head (e.g., Gardner-Wells or Vinke) that are fixed to the skull to immobilize cervical fractures (see Chapter 68).

The surgeon applies skeletal traction using surgical asepsis. The insertion site is prepared with a surgical scrub agent such as povidone-iodine solution. A local anesthetic agent is administered at the insertion site and periosteum. The surgeon makes a small skin incision and drills the sterile pin or wire through the bone. The patient feels pressure during this procedure and possibly some pain when the periosteum is penetrated.

After insertion, the pin or wire is attached to the traction bow or caliper. The ends of the pin or wire are covered with caps to prevent injury to the patient or caregivers. The weights are attached to the pin or wire bow by a rope and pulley system that exerts the appropriate amount and direction of pull for effective traction. The weights applied initially must overcome the shortening spasms of the affected muscles. As the muscles relax, the traction weight is reduced to prevent fracture dislocation and to promote healing.

Often, skeletal traction is balanced traction, which supports the affected extremity, allows for some patient movement, and facilitates patient independence and nursing care while maintaining effective traction. The Thomas splint with a Pearson attachment is frequently used with skeletal traction for fractures of the femur (see Fig. 41-6). Because upward traction is required, an overbed frame is used.

When skeletal traction is discontinued, the extremity is gently supported while the weights are removed. The pin is cut close to the skin and removed by the surgeon. Internal fixation, casts, or splints are then used to immobilize and support the healing bone.

Nursing Interventions

Maintaining Effective Skeletal Traction

When skeletal traction is used, the nurse checks the traction apparatus to see that the ropes are in the wheel grooves of the pulleys, the ropes are not frayed, the weights hang freely, and the knots in the rope are tied securely. The nurse also evaluates the patient's position, because slipping down in bed results in ineffective traction.

Quality and Safety Nursing Alert

The nurse must never remove weights from skeletal traction unless a life-threatening situation occurs. Removal of the weights defeats their purpose and may result in injury to the patient.

Maintaining Positioning

The nurse must maintain alignment of the patient's body in traction as prescribed to promote an effective line of pull. The nurse positions the patient's foot to avoid footdrop (plantar flexion), inward rotation (inversion), and outward rotation (eversion). The patient's foot may be supported in a neutral position by orthopedic devices (e.g., foot supports).

If the patient reports severe pain from muscle spasm, the weights may be too heavy or the patient may need realignment. Pain must be reported to the primary provider if body alignment fails to reduce discomfort. Opioid and nonopioid analgesics may be used to control pain. Muscle relaxants may be prescribed to relieve muscles spasms as needed.

Preventing Skin Breakdown

The patient's elbows frequently become sore, and nerve injury may occur if the patient repositions by pushing on the elbows. In addition, patients frequently push on the heel of the unaffected leg when they raise themselves. This digging of the heel into the mattress may injure the tissues. Therefore, the nurse should protect the elbows and heels and inspect them for pressure ulcers. Transparent film, hydrocolloid dressings, or skin sealants may be applied to bony prominences (such as elbows) to decrease friction (Institute for Clinical Systems Improvement [ICSI], 2012). To encourage movement without using the elbows or heel, a **trapeze** can be suspended overhead within easy reach of the patient (see Fig. 41-6). The trapeze helps the patient move about in bed and move on and off the bedpan.

Specific pressure points are assessed for irritation and inflammation at least every 8 hours. Patients at high risk for skin breakdown (e.g., older adults, malnourished patients) may need to be assessed more frequently (ICSI, 2012). Areas that are particularly vulnerable to pressure caused by a traction apparatus applied to the lower extremity include the ischial tuberosity, popliteal space, Achilles tendon, and heel. If the patient is not permitted to turn on one side or the other, the nurse must make a special effort to provide back care and to keep the bed dry and free of crumbs and wrinkles. The patient can assist by holding the overhead trapeze and raising the hips off the bed. If the patient cannot do this, the nurse can push down on the mattress with one hand to relieve pressure on the back and bony prominences and to provide for some shifting of weight. Higher-specification foam and pressure-relieving mattresses rather than standard hospital foam mattresses should be used to reduce the risk of pressure ulcer (McInnes et al., 2011).

For change of bed linens, the patient raises the torso while caregivers on both sides of the bed roll down and replace the upper mattress sheet. Then, as the patient raises the buttocks off the mattress, the sheets are slid under the buttocks. Finally, the lower section of the bed linens is replaced while the patient rests on the back. Sheets and blankets are placed over the patient in such a way that the traction is not disrupted.

Monitoring Neurovascular Status

The nurse evaluates the body part to be placed in traction and compares its neurovascular status (e.g., color, temperature, capillary refill, edema, pulses, ability to move, and sensations) to the unaffected extremity every hour for the first 24 hours after traction is applied and every 4 hours thereafter. The nurse instructs the patient to report any changes in sensation or movement immediately so that they can be promptly evaluated. DVT is a significant risk for the immobilized patient. The nurse encourages the patient to do active flexion–extension ankle exercises and isometric contraction of the calf muscles (calf-pumping exercises) 10 times an hour while awake to decrease venous stasis. In addition, anti-embolism stockings, compression devices, and anticoagulant therapy may be prescribed to help prevent thrombus formation.

> ***Quality and Safety Nursing Alert***
>
> **The nurse must promptly investigate every report of discomfort expressed by the patient in traction. Prompt recognition of a developing neurovascular problem is essential so that corrective measures can be instituted promptly.**

Providing Pin Site Care. The wound at the pin insertion site requires attention. The goal is to avoid infection and development of osteomyelitis (see Chapter 42). For the first 48 hours after insertion, the site is covered with a sterile absorbent nonstick dressing and a rolled gauze or Ace-type bandage. After this time, a loose cover dressing or no dressing is recommended (a bandage is necessary if the patient is exposed to airborne dust). Evidence-based recommendations for pin site care include the following (Holmes & Brown, 2005):

- Pins located in areas with considerable soft tissue should be considered at greatest risk for infection.
- At sites with mechanically stable bone-to-pin interfaces, pin site care should be performed on a daily or weekly basis (after the first 48 to 72 hours postinsertion, when drainage may be heavy). The frequency of pin care needs to be increased if mechanical looseness of pins or early signs of infection are present.
- Chlorhexidine 2 mg/mL solution is recommended as the most effective cleansing solution. If chlorhexidine is contraindicated (due to known sensitivity or skin reaction), saline should be used for cleansing (Timms et al., 2011).

The nurse must inspect the pin sites every 8 hours for reaction (i.e., normal changes that occur at the pin site after insertion) and infection. Signs of reaction may include redness, warmth, and serosanguineous drainage at the site. These signs subside after 72 hours. Signs of infection may mirror those of reaction but also include the presence of purulent drainage, pin loosening, tenting of skin at pin site, odor, and fever. Prophylactic broad-spectrum IV antibiotics may be administered for 24 to 48 hours postinsertion to prevent infection. Minor infections may be readily treated with antibiotics, and infections that result in systemic manifestations may additionally warrant pin removal until the infection resolves (Holmes & Brown, 2005).

> ***Quality and Safety Nursing Alert***
>
> **The nurse must inspect the pin site at least every 8 hours for signs of inflammation and evidence of infection.**

Due to a lack of evidence-based research findings, controversy remains about frequency of pin care, showering, and the use of massage to release skin adherence to pins (Holmes & Brown, 2005; Timms et al., 2011). Some evidence suggests that crusting at the pin site should not be removed, because the crust provides a natural barrier from bacteria (Timms et al., 2011). The patient and family should be taught to perform any prescribed pin site care prior to discharge from the hospital and should be provided with written follow-up instructions that include the signs and symptoms of infection.

Promoting Exercise. Patient exercises, within the therapeutic limits of the traction, assist in maintaining muscle strength and tone, and in promoting circulation. Active exercises include pulling up on the trapeze, flexing and extending the feet, and range-of-motion and weight-resistance exercises for noninvolved joints. Isometric exercises of the immobilized extremity (quadriceps and gluteal setting exercises) are important for maintaining strength in major ambulatory muscles (see Chart 41-3). Without exercise, the patient will lose muscle mass and strength, and rehabilitation will be greatly prolonged.

Nursing Management

Assessing Anxiety

The nurse must consider the psychological and physiologic impact of the musculoskeletal problem, traction device, and immobility. Traction restricts mobility and independence. The equipment often looks threatening, and its application can be frightening. Confusion, disorientation, and behavioral problems may develop in patients who are confined in a limited space for an extended time. Therefore, the nurse must assess and monitor the patient's anxiety level and psychological responses to traction.

Assisting With Self-Care

Initially, the patient may require assistance with self-care activities. The nurse helps the patient eat, bathe, dress, and toilet. Convenient arrangement of items such as the telephone, tissues, water, and assistive devices (e.g., reachers, overbed trapeze) may facilitate self-care. With resumption of self-care activities, the patient feels less dependent and less frustrated and experiences improved self-esteem. Because some assistance is required throughout the period of immobility, the nurse and the patient can creatively develop routines that maximize the patient's independence.

Monitoring and Managing Potential Complications

Immobility-related complications may include pressure ulcers (see Chapter 10), atelectasis, pneumonia, constipation, loss of appetite, urinary stasis, urinary tract infections, and venous thromboembolism (VTE) formation. Early identification of preexisting or developing conditions facilitates prompt interventions to resolve them.

Atelectasis and Pneumonia

The nurse auscultates the patient's lungs every 4 to 8 hours to assess respiratory status and educates the patient about performing deep-breathing and coughing exercises to aid in fully expanding the lungs and clearing pulmonary secretions. If the patient history and baseline assessment indicate that the patient is at risk for development of respiratory complications, specific therapies (e.g., the use of an incentive spirometer) may be indicated. If a respiratory complication develops, prompt institution of prescribed therapy is needed.

Constipation and Anorexia

Reduced gastrointestinal motility results in constipation and anorexia. A diet high in fiber and fluids may help stimulate gastric motility. If constipation develops, therapeutic measures may include stool softeners, laxatives, suppositories, and enemas. To improve the patient's appetite, the patient's food preferences are included, as appropriate, within the prescribed therapeutic diet.

Urinary Stasis and Infection

Incomplete emptying of the bladder related to positioning in bed can result in urinary stasis and infection. In addition, the patient may find the use of a bedpan uncomfortable and may limit fluids to minimize the frequency of urination. The nurse monitors the fluid intake and the character of the urine. Adequate hydration is important; therefore, the nurse instructs the patient to consume adequate amounts of fluid and to void every 3 to 4 hours. If the patient exhibits signs or symptoms of urinary tract infection, the nurse notifies the primary provider.

Venous Thromboembolism

Venous stasis that predisposes the patient to VTE occurs with immobility. The nurse educates the patient about how to perform ankle and foot exercises within the limits of the traction therapy every 1 to 2 hours when awake to prevent DVT. Involving family members in performing these exercises can increase adherence and promote the family's involvement with the patient's care (Palamone, Brunovsky, Groth, et al., 2011). The patient is encouraged to drink fluids to prevent dehydration and associated hemoconcentration, which contribute to stasis. The nurse monitors the patient for signs of DVT, including unilateral calf tenderness, warmth, redness, and swelling (increased calf circumference). The nurse promptly reports findings to the primary provider for evaluation and therapy.

During traction therapy, the nurse encourages the patient to exercise muscles and joints that are not in traction to prevent deterioration, deconditioning, and venous stasis. The physical therapist can design bed exercises that minimize loss of muscle strength. During the patient's exercise, the nurse ensures that traction forces are maintained and that the patient is properly positioned to prevent complications resulting from poor alignment.

The Patient Undergoing Orthopedic Surgery

Many patients with musculoskeletal dysfunction undergo surgery to correct the condition. Conditions that may be corrected by surgery include unstabilized fracture, deformity, joint disease, necrotic or infected tissue, and tumors. Frequent surgical procedures include **open reduction with internal fixation (ORIF)** and closed reduction with internal fixation (fracture fragments are not surgically exposed) for fractures; arthroplasty, meniscectomy, and joint replacement for joint conditions; amputation for severe extremity conditions (e.g., gangrene, massive trauma); bone graft for joint stabilization, defect filling, or stimulation of bone healing; and tendon transfer for improving motion. The goals include improving function by restoring motion and stability and relieving pain and disability. Chart 41-6 describes common orthopedic surgeries.

Chart 41-6 Common Orthopedic Surgical Procedures

Open reduction: correction and alignment of the fracture after surgical dissection and exposure of the fracture
Internal fixation: stabilization of the reduced fracture by the use of metal screws, plates, wires, nails, and pins
Arthroplasty: repair of joint problems through the operating arthroscope (an instrument that allows the surgeon to operate within a joint without a large incision) or through open joint surgery
Hemiarthroplasty: replacement of one of the articular surfaces (e.g., in a hip hemiarthroplasty, the femoral head and neck are replaced with a femoral prosthesis—the acetabulum is not replaced)
Joint arthroplasty or replacement: replacement of joint surfaces with metal or synthetic materials
Total joint arthroplasty or replacement: replacement of both articular surfaces within a joint with metal or synthetic materials
Meniscectomy: excision of damaged joint fibrocartilage
Amputation: removal of a body part
Bone graft: placement of bone tissue (autologous or homologous grafts) to promote healing, to stabilize, or to replace diseased bone
Tendon transfer: insertion of tendon to improve function
Fasciotomy: incision and diversion of the muscle fascia to relieve muscle constriction, as in compartment syndrome, or to reduce fascia contracture

Indications for a surgical procedure are based on the patient's age, underlying orthopedic condition, and general physical health and the impact of joint disability on daily activities. Timing of these procedures is important to ensure maximum function. In general, surgery should be performed before surrounding muscles become contracted and atrophied and serious structural abnormalities occur.

A blood loss of up to 1,500 mL during the procedure may be anticipated; therefore, several units of typed and cross-matched blood should be available (Spahn, 2010). Because most orthopedic surgeries are elective procedures, many patients donate their own blood during the weeks preceding their surgery. Autologous blood donations are cost-effective and eliminate many of the risks of transfusion therapy (see Chapter 32).

Blood is conserved during surgery to minimize loss. During orthopedic surgery on a limb (e.g., total knee arthroplasty [TKA]), a pneumatic tourniquet may be used intraoperatively to prevent intraoperative and postoperative bleeding and to maintain a clean surgical field. Although this technique may decrease the time needed for the operation and enables the surgeon to visualize the surgical field, the evidence is conflicting about how and when to use the tourniquet and whether bleeding is reduced (Tai, Lin, Jou, et al., 2011). Intraoperative blood salvage with reinfusion is used when a large volume of blood loss is anticipated. Postoperative blood salvage with intermittent autotransfusion also reduces the need for blood transfusion.

Joint Replacement

Patients with severe joint pain and disability may undergo joint replacement. Conditions contributing to joint degeneration include osteoarthritis, rheumatoid arthritis, trauma, and congenital deformity. Some fractures (e.g., femoral neck fracture) may cause disruption of the blood supply and subsequent **avascular necrosis** (death of tissue due to insufficient blood supply); management with joint replacement may be elected over ORIF. Joints frequently replaced include the hip, knee (Fig. 41-7), and finger joints. More complex joints (shoulder, elbow, wrist, ankle) are replaced less frequently.

Joint arthroplasty refers to the surgical removal of an unhealthy joint and replacement of joint surfaces with metal or synthetic materials. Total joint arthroplasty, also known as total joint replacement, involves the replacement of all components of an articulating joint. Most joint replacements consist of metal (e.g., cobalt-chromium, titanium) and high-density polyethylene components. There are two techniques that can be used to secure the components of the artificial joint. The first technique involves using implants cemented in the prepared bone with polymethylmethacrylate (PMMA), a bone-bonding agent that has properties similar to bone. However, loosening of prostheses due to cement–bone interface failure is a common cause of prostheses failures. The second technique uses press-fit, ingrowth prostheses (porous-coated, cementless artificial joint components). These allow the patient's bone to grow into and securely fix the prosthesis in the bone. Preliminary research has reported excellent fixation and improved long-term survival with porous-coated implants as compared to cemented

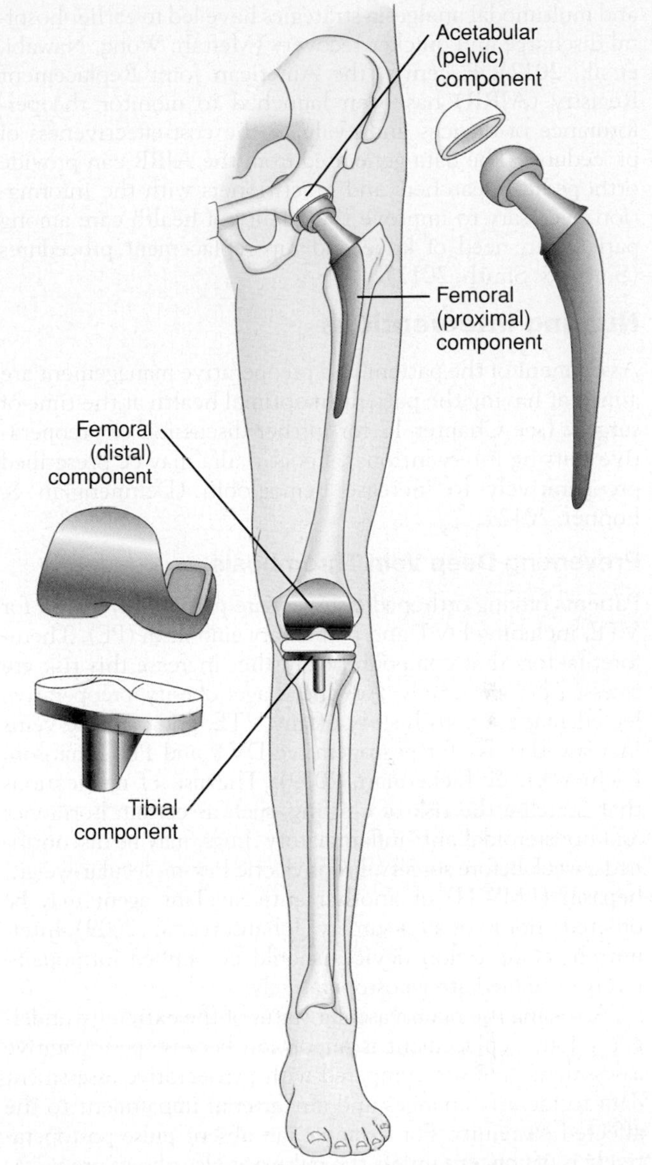
Acetabular
(pelvic)
component
Femoral
(proximal)
component
Femoral
(distal)
component
Tibial
component

FIGURE 41-7 • Examples of hip and knee replacement.

prostheses (Deirmengian & Lonner, 2012). Accurate fitting and the presence of healthy bone with adequate blood supply are important considerations in the use of cementless components (Whiteside, 2011). Much progress has been made in reducing prosthesis failure rates through improved techniques, improved materials, and the use of bone grafts; however, there is insufficient evidence to support whether the cemented or pressed-fit technique is superior (Gandhi, Tsvetkov, Davey, et al., 2009; Nakama, Peccin, Almeida, et al., 2012).

With joint replacement, patients may expect pain relief, return of joint motion, and improved functional status and quality of life. The scope of these improvements depends in part on patients' preoperative soft tissue condition and general muscle strength. Serious complications seldom occur, and recent innovations in total joint replacement surgery have made this a safer and more routinely performed surgery (Deirmengian & Lonner, 2012). In fact, minimally invasive surgical techniques, postoperative rehabilitation protocols, and multimodal analgesia strategies have led to earlier hospital discharge and quicker recovery (Meftah, Wong, Nawabi, et al., 2012). Recently, the American Joint Replacement Registry (AJRR) has been launched to monitor the performance of devices and evaluate the cost-effectiveness of procedures. The data generated from the AJRR can provide orthopedic researchers and practitioners with the information necessary to improve the quality of health care among patients in need of knee and hip replacement procedures (Smith & Smith, 2012).

Nursing Interventions

Assessment of the patient and preoperative management are aimed at having the patient in optimal health at the time of surgery (see Chapter 17 for further discussion of preoperative nursing interventions). Epoetin alfa may be prescribed preoperatively to increase hemoglobin (Deirmengian & Lonner, 2012).

Preventing Deep Vein Thrombosis

Patients having orthopedic surgery are particularly at risk for VTE, including DVT and pulmonary embolism (PE). Therefore, factors that compound or further increase this risk are assessed preoperatively. Advanced age, obesity, preoperative leg edema, previous history of any VTE, and varicose veins increase the risk for postoperative DVT and PE (Johanson, Lachiewicz, & Lieberman, 2009). The use of medications that increase the risk of clotting, such as certain hormones and nonsteroidal anti-inflammatory drugs, may be discontinued a week before surgery. Prophylactic low-molecular-weight heparin (LMWH) or another anticoagulant agent may be ordered prior to or after surgery (Johanson et al., 2009). Intermittent compression devices should be applied intraoperatively or immediately postoperatively.

Assessing the neurovascular status of the extremity undergoing joint replacement is important, because postoperative assessment data are compared with preoperative assessment data to identify changes and any arterial impairment to the affected extremity. For example, an absent pulse postoperatively is of concern unless the pulse was also absent preoperatively. Nerve palsy could occur as a result of surgery.

Preventing Infection

Preoperative assessment of the patient for recent or active infections, including urinary tract infection, is necessary because of the risk for postoperative infection. Any infection presenting 2 to 4 weeks before planned surgery may result in postponement of surgery. Preoperative skin preparation, such as showers with antiseptic soap, frequently begins 1 or 2 days before the surgery (Kamel, McGahan, Polisena, et al., 2012).

Research findings suggest that prophylactic broad-spectrum antibiotics administered minutes prior to incision are effective in preventing postoperative infection (Smith, Jacobs, Rodier, et al., 2011). If antibiotics are given too early (60 or more minutes before surgery), the patient's risk for postoperative infection may increase (Stefansdottir, Robertsson, W-Dahl, et al., 2009).

Culture of the joint during surgery may be important in identifying and treating subsequent infections. If osteomyelitis develops, it is difficult to treat (see Chapter 42). Persistent infection at the site of the prosthesis usually requires removal of the implant and joint revision. It is not always possible to achieve a functional joint when the reconstruction procedure has to be repeated.

Managing Pain

Assessment of the patient's pain preoperatively and any cultural and value preferences are important components related to the control of pain following joint surgery. Assessing the patient's level of understanding of the surgery and explaining what to expect in the postoperative period (e.g., incentive spirometry, pain control methods, activity limits) can improve outcomes. In particular, research findings suggest that patients who attend structured preoperative education classes report feeling better prepared for surgery and better able to control their pain after surgery (Kearney, Jennrich, Lyons, et al., 2011) (Chart 41-7).

Total Hip Arthroplasty

Total hip arthroplasty (THA; total hip replacement) is the replacement of a severely damaged hip with an artificial joint. Indications for this surgery include osteoarthritis, rheumatoid arthritis, femoral neck fractures (i.e., hip fracture), failure of previous reconstructive surgeries (failed prosthesis, **osteotomy**), and conditions resulting from developmental dysplasia or Legg-Calvé-Perthes disease (avascular necrosis of the hip in childhood). A variety of total hip prostheses are available. Most consist of a metal femoral component topped by a spherical ball made of metal, ceramic, or plastic that is fitted into a plastic or metal acetabular socket (see Fig. 41-7).

The surgeon selects the prosthesis that is best suited to the individual patient, considering various factors including skeletal structure and activity level. The patient has irreversibly damaged hip joints, and the potential benefits, including improved quality of life, outweigh the surgical risks. With the advent of improved prosthetic materials and operative techniques, the life of the prosthesis has been extended, and today younger patients with severely damaged and painful hip joints are undergoing total hip replacement.

Nursing Interventions

The nurse must be aware of and monitor for specific potential complications associated with THA. Complications that may occur include dislocation of the hip prosthesis, excessive wound drainage, VTE, infection, and heel pressure ulcer (Chart 41-8). The nurse also monitors for complications associated with immobility. Long-term complications include **heterotopic ossification** (formation of bone in the periprosthetic space), avascular necrosis, and loosening of the prosthesis.

Gerontologic Considerations

The older adult patient who has had THA merits special postoperative care considerations. Early THA surgery for hip fractures (within 24 to 36 hours) is recommended for most patients once a medical assessment has been made and the patient's condition has been stabilized appropriately. If there are no contraindications (e.g., history of a bleeding disorder), these patients should receive LMWH for DVT prophylaxis; mechanical devices should be used for patients in whom

Chart 41-7 NURSING RESEARCH PROFILE

Preoperative Education and Patient Outcomes After Joint Replacement Surgery

Kearney, M., Jennrich, M.K., Lyons, S., et al. (2011). Effects of preoperative education on patient outcomes after joint replacement surgery. *Orthopaedic Nursing*, *30*(6), 391–396.

Purpose

The limited data published on preoperative education to patients undergoing knee or hip replacement show mixed results and are insufficient to support or refute its use. Questions remain regarding patients' perceived preparation and their postoperative recovery in terms of pain, mobility, and complications. This descriptive, comparative study explored the experiences and outcomes of patients who attended a hospital-based preoperative total joint education class with those who did not attend. Specific research questions were as follows: (1) Do patients who attend the class feel better prepared for surgery, have more realistic expectations, and better adhere to measures to prevent complications? (2) Are there differences in discharge disposition or ability to control pain and participate in therapy? (3) Are there differences in pain severity, ambulation distance, length of stay, or complications?

Design

One hundred and fifty patients undergoing total hip or total knee replacement completed a self-administered survey about their preparation and expectations for surgery and their experiences after surgery. Eligible participants were identified as belonging in the experimental or control group after their surgery was finished based on whether they attended the educational class or not. Charts were reviewed for length of stay, complications, ambulation distance, and pain score. Postoperative complications were assessed by a follow-up phone call 30 days after discharge.

Findings

Of the 150 participants in the study, 77 attended the preoperative class and 73 did not attend. Participants who attended the structured preoperative class felt significantly better prepared for surgery ($p = .002$), and they also felt better able to control their pain after surgery ($p = .001$). No significant differences were found in postoperative pain scores, ambulation distances, or length of stay. Discharge disposition was not different between the two groups ($p = .425$). After the follow-up phone call, it was also determined that there was no difference in complication rates based on class attendance.

Nursing Implications

Participants in this study who attended a structured preoperative joint education class felt better prepared for surgery and better able to control their pain postoperatively than those who did not. Improvement in these two factors may increase patients' comfort and sense of control in the immediate postoperative period, improving their overall experience. Customizing preoperative education to meet specific patient needs may be an important factor that can optimize key clinical outcomes.

anticoagulants and antiplatelet agents are contraindicated. Providing an appropriate postoperative analgesic regimen for older adults can be challenging in the presence of impaired cognition, medical comorbidities, and possible drug interactions. Consulting with a pain management specialist to specifically tailor the analgesic type and dose may be helpful (see Chapter 12).

All older adult patients post-THA should be placed on a higher-specification, foam pressure-relieving mattress rather than a standard hospital mattress (Jenson, Cameron, & Lyn, 2010). A major goal following surgery in this patient population is early mobilization, in an effort to prevent the complications associated with prolonged immobility and to return the patient to functional activity (Egol & Straus, 2009). Early assisted mobilization and ambulation (begun within 48 hours of surgery) accelerates functional recovery and is associated with more direct discharges to home and less discharges to high-level care in previously community dwelling individuals (Jenson et al., 2010).

Preventing Dislocation of the Hip Prosthesis

For patients undergoing a posterior or posterolateral approach for THA, maintenance of the femoral head component in the acetabular cup is essential. The risk for dislocation is more common with this approach and may occur when the hip is in full flexion, adducted (legs together), and internally rotated. Therefore, correct positioning is maintained at all times. The patient should be in a supine position with his or her head slightly elevated and the affected leg in a neutral position. The use of an abduction splint, a wedge pillow (Fig. 41-8), or two or three pillows placed between the legs prevent adduction beyond the midline of the body. A cradle boot may be used to prevent leg rotation and to support the heel off the bed, preventing development of a pressure ulcer. When the nurse turns the patient in bed to the unaffected side, it is important to keep the operative hip in **abduction** (movement away from the center or median line of the body). The patient should not be turned to the operative side, which could cause dislocation, unless specified by the surgeon.

The patient's hip is never flexed more than 90 degrees. When using a fracture bedpan, the nurse instructs the patient to flex the unaffected hip and to use the trapeze to lift the pelvis onto the pan. The patient is also reminded not to flex the affected hip.

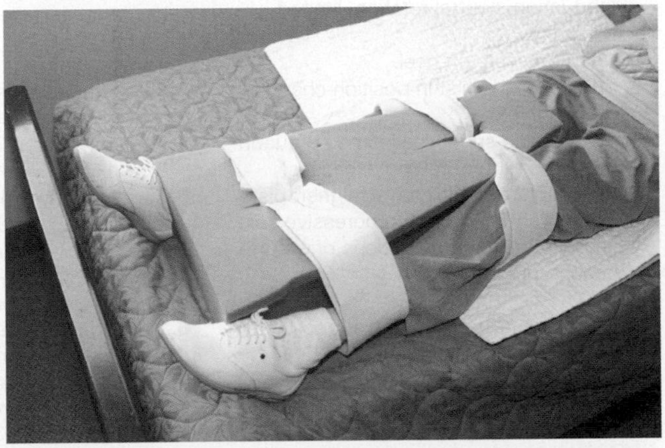

FIGURE 41-8 • An abduction pillow may be used after a total hip arthroplasty to prevent dislocation of the prosthesis.

(text continues on page 1123)

Chart 41-8

PLAN OF NURSING CARE
The Patient With a Total Hip Arthroplasty

NURSING DIAGNOSIS: Acute pain related to total hip arthroplasty
GOAL: Relief of pain

Nursing Interventions	Rationale	Expected Outcomes
1. Assess patient for pain using a standard pain intensity scale.	1. Pain is expected after a surgical procedure because of the surgical trauma and tissue response. Muscle spasms occur after total hip replacements. Immobility causes discomfort at pressure points.	• Describes discomfort • Expresses confidence in efforts to control pain • States pain is reduced; pain intensity scores are decreasing • Appears comfortable and relaxed • Uses physical, psychological, and pharmacologic measures to reduce pain and discomfort
2. Ask patient to describe discomfort.	2. Pain characteristics may help to determine the cause of discomfort. Pain may be due to complications (hematoma, infection, dislocation). Pain is an individual experience—it means different things to different people.	
3. Acknowledge existence of pain; inform patient of available analgesic agents or muscle relaxants.	3. The nurse can reduce the stress experienced by patient by communicating concern and availability of assistance to help the patient deal with the pain.	
4. Use pain-modifying techniques.	4.	
a. Administer analgesic agents as prescribed.	a. Patient will require parenteral opioids during the first 24–48 hours and then will progress to oral analgesic agents.	
b. Change position within prescribed limits.	b. The use of pillows to provide adequate support and relief of pressure on bony prominences assists in minimizing pain.	
c. Modify environment.	c. Interactions with others, distractions, and sensory overload or deprivation may affect pain experience.	
d. Notify surgeon about persistent pain.	d. Surgical intervention may be necessary if pain is due to hematoma or excessive edema.	
5. Evaluate and record discomfort and effectiveness of pain-modifying techniques.	5. Effectiveness of action is based on experience; data provide a baseline about pain experiences, pain management, and pain relief.	

NURSING DIAGNOSIS: Impaired physical mobility related to positioning, weight-bearing, and activity restrictions after total hip arthroplasty
GOAL: Achieves pain-free, functional, stable hip joint

Nursing Interventions	Rationale	Expected Outcomes
1. Maintain proper positioning of hip joint (abduction, neutral rotation, limited flexion).	1. Prevents dislocation of hip prosthesis.	• Maintains prescribed position • No heel pressure • Assists in position changes • Shows increased independence in transfers • Exercises hourly • Participates in progressive ambulation program • Actively participates in exercise regimen • Uses ambulatory aids correctly and safely
2. Keep pressure off heel.	2. Prevents pressure ulcer on heel.	
3. Instruct and assist in position changes and transfers.	3. Encourages patient's active participation while preventing dislocation.	
4. Instruct and supervise isometric quadriceps and gluteal setting exercises.	4. Strengthens muscles needed for walking.	
5. In consultation with physical therapist, instruct and supervise progressive safe ambulation within limitations of weight-bearing prescription.	5. Amount of weight bearing depends on patient's condition and prosthesis; ambulatory aids are used to assist the patient with non–weight-bearing and partial weight-bearing ambulation.	
6. Offer encouragement and support exercise regimen.	6. Reconditioning exercises can be uncomfortable and fatiguing; encouragement helps patient comply with exercise program.	
7. Instruct and supervise safe use of ambulatory aids.	7. Prevents injury from unsafe use and prevents falls.	

Chart 41-8

PLAN OF NURSING CARE

The Patient With a Total Hip Arthroplasty (continued)

COLLABORATIVE PROBLEMS: Hemorrhage; neurovascular compromise; dislocation of prosthesis; deep vein thrombosis; infection related to surgery
GOAL: Absence of complications

Nursing Interventions	Rationale	Expected Outcomes
Hemorrhage		
1. Monitor vital signs, observing for shock.	1. Changes in pulse, blood pressure, and respirations may indicate development of shock. Blood loss and stress of surgery may contribute to development of shock.	• Vital signs stabilize within normal limits • Amount of drainage decreases • No bright-red bloody drainage • Hematology values are within normal limits.
2. Note character and amount of drainage.	2. Within 48 hours, bloody drainage collected in portable suction device should decrease to 25–30 mL per 8 hours. Excessive drainage (>250 mL in first 8 hours after surgery) and bright-red drainage may indicate active bleeding.	
3. Notify surgeon if patient develops shock or excessive bleeding, and prepare for administration of fluids, blood component therapy, and medications.	3. Corrective measures need to be instituted.	
4. Monitor hemoglobin and hematocrit values.	4. Anemia due to blood loss may develop. Blood replacement or iron supplementation may be needed.	
Neurovascular Dysfunction		
1. Assess affected extremity for color and temperature.	1. The skin becomes pale and feels cool with decreased tissue perfusion. Venous congestion may produce cyanosis.	• Color normal • Extremity warm • Normal capillary refill • Moderate edema and swelling; tissue not palpably tense • Pain controllable • No pain with passive dorsiflexion • Normal sensations • No paresthesia • Normal motor abilities • No paresis or paralysis • Pulses strong and equal
2. Assess toes for capillary refill response.	2. After compression of the nail, rapid return of pink color indicates good capillary perfusion.	
3. Assess extremity for edema and swelling. Report patient complaints of leg tightness.	3. The trauma of surgery will cause edema. Excessive swelling and hematoma formation can compromise circulation and function.	
4. Elevate lower extremity. Keep elevated extremity lower than hip when in chair.	4. Minimizes dependent edema. Hip is never flexed more than 90 degrees to prevent dislocation.	
5. Assess for deep, throbbing, unrelenting pain.	5. Surgical pain can be controlled; pain due to neurovascular compromise is not relieved by treatment.	
6. Assess for pain on passive flexion of foot.	6. With nerve ischemia, there will be pain on passive stretch. Additionally, pain or tenderness may indicate deep vein thrombosis.	
7. Assess for change in sensations and numbness.	7. Diminished pain and sensory function may indicate nerve damage. Sensation in web between great and second toe—peroneal nerve; sensation on sole of foot—tibial nerve.	
8. Assess ability to move foot and toes.	8. Dorsiflexion of ankle and extension of toes indicate function of peroneal nerve. Plantar flexion of ankle and flexion of toes indicate function of tibial nerve.	
9. Assess pedal pulses in both feet.	9. Indicator of extremity circulation	
10. Notify surgeon if altered neurovascular status is noted.	10. Function of extremity needs to be preserved.	

(continues on page 1122)

Chart 41-8

PLAN OF NURSING CARE

The Patient With a Total Hip Arthroplasty (continued)

Nursing Interventions	Rationale	Expected Outcomes
Dislocation of Prosthesis		
1. Position patient as prescribed.	1. Hip component positioning (femoral component in acetabular component) needs to be maintained.	• Prosthesis not dislocated • Adheres to recommendations to prevent dislocation
2. Use abductor splint or pillows to maintain position and to support extremity.	2. Keeps hip in abduction and in a neutral rotation to prevent dislocation.	
3. Support leg and place pillows between legs when patient is turning and side-lying; turn to the unaffected side.	3–5. Prevent dislocation.	
4. Avoid acute flexion of hip (head of bed ≤60 degrees).		
5. Avoid crossing legs.		
6. Assess for dislocation of prosthesis (extremity shortens, internally or externally rotated, severe hip pain, patient unable to move extremity).	6. Findings may indicate dislocation of prosthesis.	
7. Notify surgeon of possible dislocation.	7. Joint dislocations compromise neurovascular status and future function of extremity.	
Deep Vein Thrombosis		
1. Use anti-embolism stocking or sequential compression device as prescribed.	1. Aids in venous blood return and prevents stasis.	• Wears anti-embolism stocking; uses compression device • No skin breakdown • Pulses equal and strong • Skin temperature normal • No calf pain or tenderness • Changes position with assistance and supervision • Participates in exercise regimen • Well hydrated • No chest pain; lungs clear to auscultation; no evidence of pulmonary emboli
2. Remove stocking for 20 minutes twice a day and provide skin care.	2. Skin care is necessary to avoid breakdown. Extended removal of stocking defeats purpose of stocking.	
3. Assess popliteal, dorsalis pedis, and posterior tibial pulses.	3. Pulses indicate arterial perfusion of extremity.	
4. Assess skin temperature of legs.	4. Local inflammation will increase local skin temperature.	
5. Assess for unilateral calf pain or tenderness every 8 hours.	5. Pain or tenderness may indicate deep vein thrombosis.	
6. Avoid pressure on popliteal blood vessels from equipment (e.g., abductor splint straps, sequential compression stockings) or pillows.	6. Compression of blood vessels diminishes blood flow.	
7. Change position and increase activity as prescribed.	7. Activity promotes circulation and diminishes venous stasis.	
8. Supervise ankle exercises hourly.	8. Muscle exercise promotes circulation.	
9. Monitor body temperature.	9. Body temperature increases with inflammation.	
10. Encourage fluids.	10. Dehydration increases blood viscosity.	
Infection		
1. Monitor vital signs.	1. Temperature, pulse, and respirations increase in response to infection. (Magnitude of response may be minimal in older adults.)	• Vital signs normal • Well-approximated incision without drainage or excessive inflammatory response • Minimal discomfort; no hematoma • Tolerates antibiotics
2. Use aseptic technique for dressing changes and emptying of portable drainage.	2. Avoids introducing organisms.	
3. Assess wound appearance and character of drainage.	3. Red, swollen, draining incision is indicative of infection.	
4. Assess complaints of pain.	4. Pain may be due to wound hematoma—a possible locus of infection—that needs to be surgically evacuated.	
5. Administer prophylactic antibiotics if prescribed, and observe for side effects.	5. Infected prosthesis is avoided.	

Chart 41-8

PLAN OF NURSING CARE

The Patient With a Total Hip Arthroplasty (continued)

NURSING DIAGNOSIS: Impaired home maintenance related to total hip arthroplasty
GOAL: Cares for self at home

Nursing Interventions	Rationale	Expected Outcomes
1. Assess home environment for discharge planning. **2.** Encourage patient to express concerns about care at home; explore together possible solutions to the problem. **3.** Assess availability of physical assistance for health care activities. **4.** Educate caregiver in home health care regimen. **5.** Instruct patient on posthospital care: **a.** Activity limitations (hip precautions, weight-bearing limits) **b.** Exercise instructions **c.** Safe use of ambulatory aids **d.** Wound care **e.** Measures to promote healing **f.** Medications, if any **g.** Potential problems **h.** Continuing health care supervision and management	**1.** Physical barriers (especially stairs, bathrooms) may limit patient's ability to ambulate and care for self at home. **2.** Patient may have special problems that need to be identified and resolved. **3.** Because of limitation of mobility and limited hip range of motion, patient may require some assistance in routine health care. **4.** Understanding of rehabilitative regimen is necessary for compliance. **5.** Lack of knowledge and poor preparation for care at home contribute to patient anxiety, insecurity, and nonadherence to therapeutic regimen.	• Home is accessible for patient at time of discharge • Appears relaxed and develops strategies to deal with identified problems • Personal assistance is available. • Demonstrates ability to provide necessary assistance within therapeutic prescription • Adheres to home care program • Keeps follow-up health care appointments

Limited flexion is maintained during transfers and when sitting. When the patient is initially assisted out of bed, an abduction splint or pillows are kept between the legs. The nurse encourages the patient to keep the affected hip in extension, instructing the patient to pivot on the unaffected leg with assistance by the nurse, who protects the affected hip from **adduction** (movement toward the center or median line of the body), flexion, internal or external rotation, and excessive weight bearing.

High-seat (orthopedic) chairs, semireclining wheelchairs, and raised toilet seats are used to minimize hip joint flexion. When sitting, the patient's hips should be higher than the knees. The patient's affected leg should not be elevated when sitting. The patient may flex the knee.

The nurse educates the patient about protective positioning, which includes maintaining abduction and avoiding internal and external rotation, hyperextension, and acute flexion, as described previously. At no time should the patient cross his or her legs or bend at the waist (e.g., to put on shoes and socks). Occupational therapists can provide the patient with devices to assist with dressing below the waist. Hip precautions should be enforced for 4 months or longer after surgery (Chart 41-9). A patient who has had an anterior surgical approach may not need these precautions.

Dislocation may occur with positioning that exceeds the limits of the prosthesis. The nurse must monitor for signs and symptoms of dislocation of the prosthesis, which include:

- Increased pain at the surgical site, swelling, and immobilization
- Acute groin pain in the affected hip or increased discomfort
- Shortening of the affected extremity
- Abnormal external or internal rotation of the affected extremity
- Restricted ability or inability to move the leg
- Reported "popping" sensation in the hip

If any of these clinical manifestations occur, the nurse (or the patient, if at home) immediately notifies the surgeon, because the hip must be reduced and stabilized promptly so that the leg does not sustain circulatory and nerve damage. After closed reduction, the hip may be stabilized with Buck's traction or a brace to prevent recurrent dislocation. As the muscles and joint capsule heal, the chance of dislocation diminishes. Stresses to the new hip joint should be avoided for the first 8 to 12 weeks, when the risk of dislocation is greatest.

Promoting Ambulation

Patients post-THA begin ambulation with the assistance of a walker or crutches within a day after surgery. The nurse and the physical therapist assist the patient in achieving the goal of independent ambulation. At first, the patient may be able to stand for only a brief period because of orthostatic hypotension. Specific weight-bearing limits on the prosthesis are based on the patient's condition, the procedure, and the fixation method. Usually, patients with cemented prostheses can proceed to weight bearing as tolerated. If the patient has a press-fit, cementless, ingrowth prosthesis, weight bearing immediately after surgery may be limited to

Chart 41-9 PATIENT EDUCATION

Avoiding Hip Dislocation After Arthroplasty With Posterior or Posterolateral Approach

Until the hip prosthesis stabilizes after hip replacement surgery, it is necessary to follow instructions for proper positioning so that the prosthesis remains in place. Dislocation of the hip is a serious complication of surgery that causes pain and loss of function and necessitates reduction under anesthesia to correct the dislocation. Desirable positions include abduction, neutral rotation, and flexion of less than 90 degrees. When you are seated, the knees should be lower than the hip.

Methods for avoiding displacement include the following:

- Keep the knees apart at all times.
- Put a pillow between the legs when sleeping.
- Never cross the legs when seated.
- Avoid bending forward when seated in a chair.
- Avoid bending forward to pick up an object on the floor.
- Use a high-seated chair and a raised toilet seat.
- Do not flex the hip to put on clothing such as pants, stockings, socks, or shoes. Positions to avoid after total hip replacement are shown in the illustrations.

Affected leg should not cross the center of the body

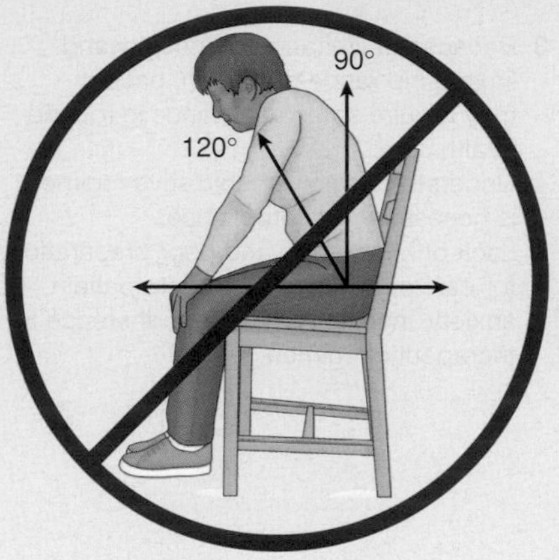
90°
120°

Hip should not bend more than 90 degrees

Affected leg should not turn inward

minimize micromotion of the prosthesis in the bone (Deirmengian & Lonner, 2012). As the patient is able to tolerate more activity, the nurse encourages transferring to a chair several times a day for short periods and walking for progressively greater distances.

Monitoring Wound Drainage

Fluid and blood accumulating at the surgical site, which could contribute to discomfort and provide a source for infection, may be drained with a closed suction portable suction device. Drainage of 200 to 500 mL in the first 24 hours is expected; by 48 hours postoperatively, the total drainage in 8 hours usually decreases to 30 mL or less, and the suction device is then removed; drains that remain in place for longer than 24 hours are at an increased risk for contamination, and infection may occur (McMorrow, 2011). The nurse promptly notifies the surgeon of excessive or foul-smelling drainage. If extensive blood loss is anticipated after total joint replacement surgery, an autotransfusion drainage system (in which the drained blood is filtered and reinfused into the patient during the immediate postoperative period) may be used to decrease the need for homologous blood transfusions.

Preventing Deep Vein Thrombosis

In the absence of prophylactic therapy, which includes mechanical (e.g., intermittent compression devices) and pharmacologic prophylaxis (e.g., anticoagulant medications), the risk of postoperative VTE is particularly high after orthopedic surgery. The incidence of DVT varies from 42% to 57% after hip arthroplasty and from 41% to 85% after knee arthroplasty (Geerts, Nergqvist, Pineo, et al., 2008). Without prophylaxis, DVT formation can develop within 7 to 14 days following surgery and lead to PE, which can be fatal. Early identification of the patient's VTE risk, ensuring that the patient receives the appropriate prophylaxis, instituting preventive measures, and monitoring the patient closely for clinical signs of the development of DVT and PE are key nursing responsibilities (Fitzgerald, 2010).

Signs of DVT include pain or tenderness over the area of the clot, swelling or tightness of the calf, and skin discoloration localized to one site; PE symptoms may include acute onset of dyspnea, tachycardia and pleuritic chest pain (Hiltz, 2011).

In order to prevent VTE, evidence-based guidelines recommend instituting mechanical and chemical (i.e., anticoagulant medication) prophylaxis (Johanson et al., 2009). Intermittent compression devices should be applied either intraoperatively or immediately postoperatively; these devices must remain on the legs at all times, even when the patient is out of bed. Patients should be instructed to dorsi- and plantar flex the ankles and the toes 10 to 20 times every half hour while awake. In addition, patients who are post-THA should be mobilized as soon as possible to assist with decreasing venous stasis; even patients with epidural catheters should stand and ambulate when they are physically able (Johanson et al., 2009).

Contemporary clinical practice guidelines support the use of anticoagulants as the primary effective DVT prophylaxis (Geerts et al., 2008; Johanson et al., 2009). The decisions about which medications to use are based on patient risk for

PE as well as risk for bleeding, and the dosage is based on patient weight. The use of aspirin, LMWH (e.g., enoxaparin [Lovenox], dalteparin [Fragmin]), and synthetic pentasaccharides (fondaparinux [Arixtra]) are recommended as prophylaxis for VTE after THA (Johanson et al., 2009) and should continue for at least 6 weeks following surgery (Watts, Howie, & Hughes, 2012).

Preventing Infection

Infection—a serious complication of THA—may necessitate removal of the prosthesis. Patients who are older, obese, poorly nourished, smoke cigarettes, or use corticosteroid medications (e.g., prednisone) and patients who have diabetes, rheumatoid arthritis, concurrent infections (e.g., urinary tract infection, dental abscess), or hematomas are at high risk for infection (Willis-Owen, Knoyves, & Martin, 2010).

Over time, one in five patients with THA will undergo revision of the prosthesis, most commonly because of aseptic loosening, infection, instability, or a mechanical complication. In particular, infection rates after THA vary from 0.4% to 1.4%, with most infections occurring during the first year (Biau, Leclerc, Marmor, et al., 2012). Because these joint infections are difficult to treat, strategies for preventing infections should be implemented at various steps of the process of care. These strategies may include ensuring that the patient's health is optimal preoperatively, preventing perioperative infections, ensuring that prophylactic antibiotics are administered as prescribed, and avoiding potential sources of infection (Biau et al., 2012; Willis-Owen et al., 2010).

If indwelling urinary catheters are used, they are removed as soon as possible to avoid infection (Wenger, 2010). If portable wound suction devices are used, they should also be removed as described previously. Current research findings suggest that drains are not necessary in elective joint replacement and should be removed as soon as possible to lower the risk of infection (McMorrow, 2011; Willis-Owen et al., 2010).

Acute infections may occur within 3 months after surgery and are associated with progressive superficial infections or hematomas. Delayed surgical infections may appear 4 to 24 months after surgery and may cause return of discomfort in the hip. Prophylactic antibiotics may be prescribed if the patient needs any future surgical or invasive procedures, such as tooth extraction or cystoscopic examination (American Academy of Orthopaedic Surgeons [AAOS] and the American Dental Association [ADA], 2012). Infections occurring more than 2 years after surgery are attributed to the spread of infection through the bloodstream from another site in the body. If an infection occurs, antibiotics are prescribed. Severe infections may require surgical débridement or removal of the prosthesis.

Promoting Home and Community-Based Care

Educating the Patient About Self-Care. Before the patient prepares to leave the acute care setting, the nurse provides thorough education to promote continuity of the therapeutic regimen and active participation in the rehabilitation process (Chart 41-10). The patient may be discharged to his or her home, a rehabilitation unit, a transitional care unit, or a long-term care facility. The nurse

Chart 41-10 Providing Home Care After Total Hip Arthroplasty

Considerations

- Pain management
- Wound care
- Mobility
- Self-care (activities of daily living)
- Potential complications

Nursing Interventions

Discuss with patient the following methods to reduce pain:

- Periodic rest
- Distraction and relaxation techniques
- Medication therapy (e.g., nonsteroidal anti-inflammatory drugs, opioid analgesic agents): actions of medications, administration, schedule, side effects

Instruct patient in the following:

- Keeping incision clean and dry
- Cleansing incision daily with soap and water and changing the dressing
- Recognizing signs of wound infection (e.g., pain, increased redness, swelling, purulent drainage, fever)

Explain that sutures or staples will be removed 10–14 days after surgery.

Educate patient about the following:

- Safe use of assistive devices
- Weight-bearing limits
- How to change positions frequently
- Limitations on hip flexion and adduction (e.g., avoid acute flexion and crossing legs)
- How to stand without flexing hip acutely
- Avoidance of low-seated chairs
- Sleeping with pillow between legs to prevent adduction
- Gradual increase in activities and participation in prescribed exercise regimen
- Use of important medications such as warfarin (Coumadin) and aspirin

Assess home environment for physical barriers.

Instruct patient to use elevated toilet seat and to use reachers to aid in dressing.

Encourage patient to accept assistance with activities of daily living during early convalescence until mobility and strength improve.

Arrange services and accommodations to address the patient's disability or illness, as appropriate.

Assess patient for development of potential problems, and instruct patient to report signs of potential complications:

- Dislocation of prosthesis (e.g., increased pain, shortening of leg, inability to move leg, popping sensation in hip, abnormal rotation)
- Deep vein thrombosis (e.g., calf pain, swelling, redness)
- Wound infection (e.g., pain, increased redness, swelling, purulent drainage, fever)
- Pulmonary emboli (e.g., shortness of breath, tachypnea, pleuritic chest pain)

Discuss with patient the need to continue regular health care (routine physical examinations) and screenings.

advises the patient of the importance of a daily exercise program in maintaining the functional motion of the hip joint and strengthening the abductor muscles of the hip, and reminds the patient that it will take time to strengthen and retrain the muscles.

The patient will need physical therapy to regain mobility. Assistive devices (crutches, walker, or cane) may be used for a time. After sufficient muscle tone has developed to permit a normal gait without discomfort, these devices are not necessary. In general, by 3 months, the patient can resume routine ADLs. Stair climbing may resume within 3 to 6 weeks following surgery (AAOS, 2011). Some discomfort with activity and at night is common for several weeks. Frequent walks, swimming, and the use of a high rocking chair are excellent for hip exercises. Restrictions must be kept in mind when resuming sexual activity. Sexual intercourse can be resumed based upon surgeon recommendation (typically 3 to 6 months postoperatively) and should be carried out with the patient in the dependent position (flat on the back) to avoid excessive adduction and flexion of the new hip. Attention to positioning and comfort may enhance the intimacy of the experience.

At no time during the first 4 months should the patient cross the legs or flex the hip more than 90 degrees. Assistive devices should be used for dressing, such as long-handled shoehorns or dressing sticks for on putting on shoes and socks. The patient should avoid low chairs and sitting for longer than 45 minutes at a time. These precautions minimize hip flexion and the risks of prosthetic dislocation, hip stiffness, and flexion contracture. Driving requires sufficient range of motion and muscle strength; most patients are given permission to drive 4 to 6 weeks postoperatively. Traveling long distances should be avoided unless frequent position changes are possible. Other activities to avoid include tub baths, jogging, lifting heavy loads, and excessive bending and twisting (e.g., lifting, shoveling snow, forceful turning). The primary provider may give the patient a card indicating that he or she has had a joint replacement; this card may be used to alert security personnel who use screening devices at airports or malls.

Continuing Care. A home care nurse may assess the patient's home for potential problems and monitor wound healing (see Chart 41-11 later in this chapter). The nurse, physical therapist, or occupational therapist assesses the home environment for physical barriers that may impede the patient's rehabilitation. In addition, the nurse or therapist may need to assist the patient in acquiring devices such as reachers and long-handled shoehorns or tongs to help with dressing or a toilet seat extender to elevate the toilet seat. After successful surgery and rehabilitation, the patient can expect a hip joint that is free or almost free of pain, has good motion, is stable, and permits normal or near-normal ambulation and function.

Total Knee Arthroplasty

TKA (total knee replacement), like THA, is considered for patients whose joint pain cannot be managed by nonsurgical treatment and who have severe pain and functional disability related to destruction of joint surfaces by osteoarthritis, rheumatoid arthritis, or posttraumatic (osteonecrotic) arthritis. When activity and mobility severely prevent patients from participating in ADLs, TKA is a successful, cost-effective, low-risk therapy that offers significant pain relief and restores quality of life and function (Parker, 2011). The femoral joint component can be metal or ceramic; the tibial component has a polyethylene surface that is either fixed to the metal tray attached to the tibia or slides freely (Bartlett, 2012). If the patient's ligaments have weakened, a fully constrained (hinged) or semiconstrained prosthesis may be used to provide joint stability. A nonconstrained (components are not linked) prosthesis depends on the patient's ligaments for joint stability.

Nursing Interventions

Postoperatively, the knee is dressed with a compression bandage. Ice or cold packs may be applied to control edema and bleeding. The nurse assesses the neurovascular status (movement, sensation, color, pulse) of the affected leg every 2 to 4 hours. It is important to encourage active flexion of the foot every hour when the patient is awake. Postoperative efforts are directed at preventing complications (VTE, peroneal nerve palsy, infection, bleeding, limited range of motion).

A wound suction drain may be present to remove fluid accumulating in the joint. If a drain is used, it is usually left in place for only 24 hours to reduce the risk of infection. Antibiotics are given prophylactically and continued for 24 hours postoperatively (Bartlett, 2012). If extensive bleeding is anticipated, an autotransfusion drainage system may be used during the immediate postoperative period. All blood must be transfused within 6 hours of collection (American Association of Blood Banks, 2009). The color, type, and amount of drainage are documented, and any excessive drainage or change in characteristics of the drainage is promptly reported to the primary provider.

A **continuous passive motion (CPM) device** may be prescribed to promote range of motion, circulation, and healing, and to prevent scar tissue from forming in the knee, which could decrease mobility and increase postoperative pain. The patient's leg is usually placed in this device immediately after surgery. The degree of flexion and extension of the joint and the cycle rate (the number of revolutions per minute) are prescribed by the surgeon, but it is often the responsibility of the nurse to maintain the device and monitor the patient's response to the therapy (Lynn, 2011) (Fig. 41-9).

Despite the widespread use of CPM, recent research findings suggest that CPM has no influence on immediate functional recovery, drainage, pain, or complications in post-TKA patients; in fact, postoperative knee swelling may persist longer (Alkire & Swank, 2010; Manier, Baviskar, Singhi, et al., 2012). CPM has also been used for preventing thrombosis in patients after TKA, although there is no evidence that this therapy is effective (He, Xiao, Lei, et al., 2012).

The physical therapist supervises exercises for strength and range of motion and educates the patient about how to use assistive devices based on weight-bearing restrictions. To avoid flexion contractures, patients are instructed to limit positions of flexion of the knee and to avoid a knee Gatch position and pillows behind the knee. If satisfactory flexion is not achieved, gentle manipulation of the knee joint under general anesthesia may be necessary about 2 weeks after surgery.

Patients who are postoperative for TKA should mobilize and ambulate by the first postoperative day (Johanson et al., 2009).

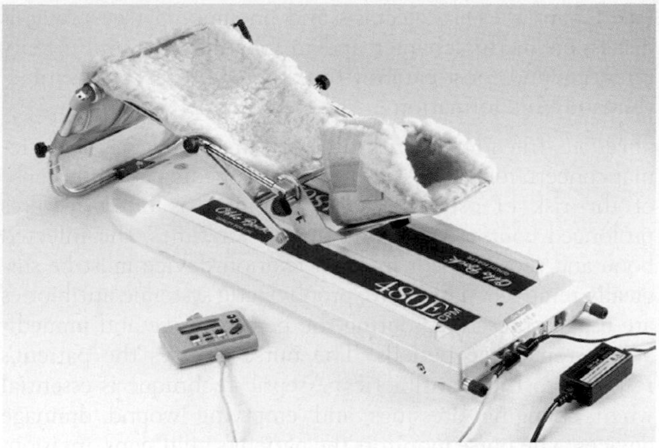

FIGURE 41-9 • Lower-limb continuous passive motion (CPM) device. The Otto Bock 480E Knee CPM is 11 kg (24 lb) and combines durable construction with portability and ease of operation. CPM is best applied immediately after surgery and continued, uninterrupted, for up to 6 weeks as prescribed by the physician. Photo courtesy of Otto Bock Healthcare, Minneapolis, MN.

The patient's weight-bearing status is determined by the orthopedic surgeon. The knee is usually protected with a knee immobilizer and is elevated when the patient sits in a chair. The typical requirements for discharge to home may include evidence of wound stability (e.g., no erythema, discharge, or redness), appropriate anticoagulation status by lab results (i.e., international normalized ratio [INR] between 1.5 and 2), progress toward physical therapy goals (e.g., appropriate use of walker), and satisfactory pain control with oral medications (Cook, Warren, Ganley, et al., 2008).

Acute rehabilitation usually takes about 1 to 2 weeks; length of time and discharge destination (e.g., home, acute rehabilitation unit) depends on the age and tolerance of the patient. If discharge is to home, the patient may continue to use the CPM device and may undergo physical therapy on an outpatient basis. Total recovery takes 6 weeks or longer, especially for those older than 75 years. Late complications that may occur include polyethylene-induced infection (**osteolysis**) and aseptic loosening of prosthetic components.

More than 95% of patients who have TKA will still have a functioning prosthesis 15 to 20 years after surgery (Hwang, Kong, Nam, et al., 2010). TKA is a viable option for improving the patient's health-related quality of life (Deirmengian & Lonner, 2012). Patients usually can achieve a pain-free, functional joint and participate more fully in life activities than before the surgery.

NURSING PROCESS

Postoperative Care of the Patient Undergoing Orthopedic Surgery

Assessment

After orthopedic surgery, the nurse reassesses the patient's needs related to pain, neurovascular status, health promotion, mobility, and self-esteem. Skeletal trauma and surgery performed on bones, muscles, or joints can produce significant pain, especially during the first 1 or 2 postoperative days. Tissue perfusion must be monitored closely, because edema and bleeding into the tissues can compromise circulation and result in compartment syndrome. Inactivity contributes to venous stasis and the development of VTE that may include DVT or PE. (See Chapter 19 for further discussion of postoperative nursing care, and Chapters 23 and 30, respectively, for discussion of DVT and PE.)

VTE is one of the most common and most dangerous of all complications occurring in the postoperative orthopedic patient. Advanced age, venous stasis, lower extremity orthopedic surgery, and immobilization are significant risk factors. The nurse assesses the patient daily for unilateral calf swelling, tenderness, warmth, and redness. The nurse promptly reports abnormal findings to the primary provider.

In addition, fat embolism syndrome (FES) (see Chapter 43) may occur with orthopedic surgery. The nurse must be alert to any signs and symptoms that may suggest the development of FES. These may include respiratory distress; onset of delirium or any acute change in level of consciousness; and development of unusual skin rashes, especially a papular rash on the upper torso.

Diagnosis

Nursing Diagnoses

Based on the assessment data, major nursing diagnoses after orthopedic surgery may include the following:

- Acute pain related to the surgical procedure, swelling, and immobilization
- Risk for peripheral neurovascular dysfunction related to swelling, constricting devices, or impaired circulation
- Impaired physical mobility related to pain, edema, or the presence of an immobilizing device (e.g., splint, cast, or brace)

(See Chapter 19 for additional applicable nursing diagnoses for patients after surgery.)

Collaborative Problems/Potential Complications

Potential complications may include any of those noted previously in Chapter 19, with particular attention to the following:

- Infection
- VTE, including DVT or PE

Planning and Goals

The major goals for the patient after orthopedic surgery mirror those for any patient having surgery (see Chapter 19). The specific major goals for the patient after orthopedic surgery may include relief of pain, adequate neurovascular function, improved mobility, and absence of complications.

Nursing Interventions

Relieving Pain

Pain can be intense following orthopedic surgery. Edema, hematomas, and muscle spasms contribute to the pain. The nurse assesses the patient's level of pain, evaluates the patient's response to therapeutic measures, and makes every effort to relieve the pain and discomfort. Patient and primary provider communication, thorough patient education, and accurate pain assessments are critical to optimal pain management (Gillaspie, 2010). Pain assessment must occur on an ongoing basis and take place at least as often as vital signs are assessed.

Multiple pharmacologic approaches to pain management exist, including the use of patient-controlled and epidural analgesia to relieve the pain. (See Chapter 12 for further discussion of the use of these devices and indications for analgesic medications.)

In addition to pharmacologic approaches to controlling pain, elevation of the operative extremity and application of ice and cold packs, if prescribed, help control edema and pain. Surgical drains inserted in the wound decrease fluid accumulation and hematoma formation. The nurse may find that nonpharmacologic interventions such as relaxation, distraction, healing touch, and guided imagery may help in reducing the patient's pain (Hardwick, Pulido, & Adelson, 2012).

The nurse should report increasing and uncontrollable pain to the orthopedic surgeon for evaluation. Pain should diminish rapidly after the initial postoperative period. After 2 to 3 days, most patients require only occasional oral analgesia for residual muscle soreness and spasm.

Maintaining Adequate Neurovascular Function

The nurse monitors the neurovascular status of the involved body part and notifies the primary provider promptly of any indications of diminished tissue perfusion. The patient is reminded to perform muscle setting, ankle, and calf-pumping exercises hourly while awake to enhance circulation.

Improving Physical Mobility

Patients are frequently reluctant to move after orthopedic surgery. Preoperative education about the planned postoperative treatment regimen promotes patient adherence to an optimal rehabilitation regimen (Kearney et al., 2011). Patients often increase their mobility once they have been reassured that movement within therapeutic limits is beneficial, the nurse will provide assistance, and discomfort can be controlled.

Metal pins, screws, rods, and plates used for internal fixation are designed to maintain the position of the bone until ossification occurs. They are not designed to support the body's weight, and they can bend, loosen, or break if stressed. The estimated strength of the bone, the stability of the fracture, reduction, and fixation, and the amount of bone healing are important considerations in determining weight-bearing limits. Although the incision may appear healed, the underlying bone requires more time to repair and regain normal strength. Some orthopedic procedures require weight-bearing restrictions. The orthopedic surgeon will prescribe the weight-bearing limits and the use of protective devices (orthoses), if necessary, after surgery.

The physical therapist tailors the rehabilitation program to each patient's needs. The goal is the patient's return to the highest level of function in the shortest time possible. Rehabilitation involves progressive increases in the patient's activities and exercises. Assistive devices (crutches, walker) may be used for postoperative mobility. Preoperative practice with assistive devices helps the patient use them appropriately postoperatively. The nurse makes sure that the patient uses these devices safely. (See Chapter 10 for discussions of crutch walking and the use of a walker.)

Monitoring and Managing Potential Complications

The patient undergoing orthopedic surgery is at risk for many postoperative complications that include hypovolemic shock (see Chapter 14), atelectasis and pneumonia (see previous discussion in this chapter and in Chapter 23), and urinary retention and constipation (see Chapter 19), as well as infection and VTE formation.

Infection. Infection is a risk after any surgery, but it is of particular concern for the postoperative orthopedic patient because of the risk of osteomyelitis. Osteomyelitis often requires prolonged courses of IV antibiotics. At times, the infected bone and prosthesis or internal fixation device must be surgically removed. Therefore, prophylactic systemic antibiotics are usually prescribed during the perioperative and immediate postoperative periods. The nurse assesses the patient's response to these antibiotics. Aseptic technique is essential when changing dressings and emptying wound drainage devices. The nurse monitors the patient's vital signs, incision, and drainage. The nurse monitors the patient for signs of urinary tract infection. Prompt assessment of and treatment for infection are essential.

Venous Thromboembolism and Deep Vein Thrombosis. Prevention of DVT requires the use of ankle and calf-pumping exercises, anti-embolism stockings, and intermittent compression devices applied to the patient's legs at all times. Adequate hydration and early mobilization are equally important. Aspirin and prophylactic fondaparinux, LMWH (e.g., enoxaparin, dalteparin), warfarin (Coumadin), or low-dose unfractionated heparin may be prescribed in the immediate postoperative period and up to 6 weeks postoperatively (Hiltz, 2011). The nurse monitors the patient for signs of DVT as described previously and promptly reports findings to the primary provider for management.

Concept Mastery Alert

There are two major methods used for DVT prophylaxis. Pharmacologic methods involve the use of drugs such as LMWH. Mechanical methods include the use of early ambulation, anti-embolism stockings, and intermittent compression devices.

Promoting Home and Community-Based Care

Educating the Patient About Self-Care. Because the hospital length of stay after orthopedic surgery is usually short, most convalescence and rehabilitation take place at home or in a nonacute care setting. The nurse educates the patient and family to recognize complications that must be reported promptly to the orthopedic surgeon. The patient must understand the prescribed medication regimen. The nurse should demonstrate proper wound care. The patient gradually resumes physical activities and adheres to weight-bearing limits. The patient must be able to perform transfers and use mobility aids safely. If the patient has a cast or other immobilizing device, family members are educated about how to assist the patient in a way that is safe for the patient and for the family member (e.g., using proper body mechanics when assisting the patient). Specific exercises need to be taught and practiced before discharge. The nurse discusses recovery and health promotion, emphasizing a healthy lifestyle and diet (Chart 41-11).

Continuing Care. If special equipment or home modifications are needed for safe care at home, they must be in place

Chart 41-11 HOME CARE CHECKLIST

The Patient Who Has Had Orthopedic Surgery

At the completion of home care education, the patient or caregiver will be able to:	PATIENT	CAREGIVER
• Describe wound care.	✔	✔
• State indicators of wound infections (e.g., redness, swelling, tenderness, purulent drainage, fever).	✔	✔
• Consume a healthy diet to promote wound and bone healing.	✔	
• Participate in prescribed exercise regimen to promote circulation and mobility.	✔	
• Use mobility aids safely.	✔	
• Observe prescribed weight-bearing and activity limits.	✔	
• Take prescribed therapeutic and prophylactic medications (e.g., antibiotics, anticoagulants, analgesic agents).	✔	
• State indicators of complications to report promptly to primary provider (e.g., uncontrolled swelling and pain; cool, pale fingers or toes; paresthesia; paralysis; purulent drainage; signs of systemic infection; signs of deep vein thrombosis or pulmonary embolism).	✔	✔
• Identify modifications of home environment to promote safe environment and independence during recovery and rehabilitation.	✔	✔

before the patient is discharged home. Discharge planning begins before surgery. The nurse, physical therapist, and social worker can assist the patient and family in identifying their needs and getting ready to care for the patient at home.

Frequently, home health nursing and home physical therapy are part of the discharge plan of care. These referrals provide resources and help the patient and family cope with the demands of care during recovery and rehabilitation. The nurse assesses the patient's progress and monitors for possible complications. Regular medical follow-up care after discharge needs to be arranged. The nurse reminds the patient and family about the importance of continuing health promotion and screening practices.

Evaluation

Expected patient outcomes may include:

1. Reports decreased level of pain
   a. Uses multiple approaches to reduce pain
   b. Uses oral analgesic medication as needed to control discomfort
   c. Elevates extremity to control edema and discomfort
   d. Moves with greater comfort
2. Exhibits adequate neurovascular function
   a. Exhibits normal color and temperature of skin
   b. Has warm skin
   c. Has normal capillary refill response
   d. Demonstrates intact sensory and motor function
   e. Demonstrates reduced swelling
3. Maximizes mobility within the therapeutic limits
   a. Requests assistance when moving
   b. Elevates edematous extremity after transfer
   c. Uses immobilizing devices as prescribed
   d. Complies with prescribed weight-bearing limitation
4. Exhibits absence of complications
   a. Does not experience shock
   b. Maintains normal vital signs and blood pressure
   c. Has clear lung sounds
   d. Demonstrates wound healing without signs of infection
   e. Does not experience urinary retention
   f. Voids clear urine
   g. Exhibits no signs of DVT or PE
   h. Does not experience constipation

Critical Thinking Exercises

1 pq A 62-year-old man has had a right THA for osteoarthritis. On his second postoperative day, he is resting comfortably in his bedside chair. When it is time to get him back to bed, you and the physical therapist assist in helping him transfer from the bed to the chair. He moves too quickly and rolls into bed onto the affected hip. He cries out in pain and claims that he felt a "popping" sound when he moved. What is your priority for assessment at this time? What is the possible cause of this pain? How could this be avoided in the future? What can the nurse say to the patient to assist him at this time?

2 pq A 55-year-old man has had a repair of a fracture of his left tibia and fibula. He has a long plaster cast that was applied last evening. He has had adequate pain control in this extremity with cold packs and oral analgesics. It is now the morning of his second postoperative day, and he reports pain unrelieved by the analgesic medication and a feeling of tightness in his calf. What additional assessments might you make at this time? What is your priority nursing diagnosis and plan of care?

3 ebp You are an experienced orthopedic nurse. You note that different orthopedic surgical groups have different indwelling urinary catheter practices for patients undergoing THA who come to the medical surgical floor where you work. One orthopedic group maintains an indwelling catheter for 3 to 4 days postoperatively, whereas another prefers intermittent catheterization for urinary retention. What are your concerns about leaving the indwelling catheter in place? What is the strength of the evidence that supports the best way to reduce this variation in practice to promote the best patient outcomes?

4 A 62-year-old woman was discharged from the hospital 4 days ago after a right TKA. She is moderately overweight, has extensive varicose veins, and has a history of smoking. She has been readmitted for pain and edema in her right calf and has a tentative diagnosis of DVT. She states that she was feeling fine and was ambulating around her house with her walker. She is questioning what is going on and does not understand why she is at risk for VTE. How would you respond to her? Describe the physical assessment that should be conducted for this patient. What are her risk factors for this disease process?

Brunner Suite Resources

Explore these additional resources to enhance learning for this chapter:

• NCLEX-Style Questions and Other Resources on thePoint, http://thePoint.lww.com/Brunner13e
• Study Guide
• PrepU
• Clinical Handbook
• Handbook of Laboratory and Diagnostic Tests

References

*Asterisk indicates nursing research.

Books

American Association of Blood Banks. (2009). *Guidance for the standards of perioperative autologous blood collection and administration* (4th ed.), Bethesda, MD: Author.

Lynn, P. (2012). *Taylor's clinical nursing skills: A nursing process approach* (3rd ed.). Philadelphia: Lippincott Williams & Wilkins.

Marshall, S. T., & Browner, B. D. (2012). Emergency care of musculoskeletal injuries. In C. M. Townsend, R. D. Beauchamp, B. M. Evers, et al. (Eds.). *Sabiston textbook of surgery* (19th ed.). Philadelphia: Saunders Elsevier.

NANDA International. (2012). *Nursing diagnoses: Definitions and classifications 2012–2014*. Philadelphia: Author.

Journals and Electronic Documents

Abou-Setta, A. M., Beaupre, L. A., Dryden, D. M., et al. (2011). *Pain management interventions for hip fracture* (Comparative Effectiveness Reviews, No. 30.). Rockville, MD: Agency for Healthcare Research and Quality. Available at: www.ncbi.nlm.nih.gov/books/NBK56675/

*Alkire, M., & Swank, M. (2010). Use of inpatient continuous passive motion versus no CPM in computer-assisted total knee arthroplasty. *Orthopaedic Nursing*, *29*(1), 79–85.

American Academy of Orthopaedic Surgeons. (2011). *Total hip replacement*. Available at: orthoinfo.aaos.org/topic.cfm?topic=A00377

American Academy of Orthopaedic Surgeons and the American Dental Association. (2012). *Prevention of orthopaedic implant infection in patients undergoing dental procedures guideline*. Available at: www.aaos.org/Research/guidelines/PUDP/PUDP_guideline.pdf

Bartlett, D. (2012). What you need to know about total knee arthroplasty. *OR Nurse*, *6*(1), 16–25.

Biau, D. J., Leclerc, P., & Marmor, S., et al. (2012). Monitoring the one year postoperative infection rate after primary total hip replacement. *International Orthopaedics*, *36*(6), 1155–1161.

Boyd, A. S., Benjamin, H. J., & Asplund, C. (2009). Principles of casting and splinting. *American Family Physician*, *79*(1), 16–22.

Carroll, E. A., & Koman, L. A. (2011). External fixation and temporary stabilization of femoral and tibial trauma. *Journal of Surgical Orthopedic Advances*, *20*(1), 74–81.

Cavusoglu, A. T., Er, M. S., Inal, S., et al. (2009). Pin site care during circular external fixation using two different protocols. *Journal of Surgical Orthopedic Trauma*, *23*(10), 724–730.

Cook, J. R., Warren, M., Ganley, K. J., et al. (2008). A comprehensive joint replacement program for total knee arthroplasty: A descriptive study. *BMC Musculoskeletal Disorders*, 9, 154–161. Available at: www.biomedcentral.com/content/pdf/1471-2474-9-154.pdf

Deignan, B. J., Iaquinto, J. M., Eskildsen, S. M., et al. (2011). Effect of pressure applied during casting on temperatures beneath casts. *Journal of Pediatric Orthopaedics*, *31*(7), 791–797.

Deirmengian, C. A., & Lonner, J. H. (2012). What's new in adult reconstructive knee surgery. *Journal of Bone and Joint Surgery*, *94*, 182–188.

Egol, K. A., & Straus, E. J. (2009). Perioperative considerations in geriatric patients with hip fracture. *Journal of Orthopaedic Trauma*, *23*(6), 386–394.

Fitzgerald, J. (2010). Venous thromboembolism: Have we made headway? *Orthopaedic Nursing*, *29*(4), 226–234.

Gandhi, R., Tsvetkov, D., Davey, J. R., et al. (2009). Survival and clinical function of cemented and uncemented prostheses in total knee replacement: A meta-analysis. *Journal of Bone and Joint Surgery, British Volume*, *91*(7), 889–895.

Geerts, W. H., Nergqvist, D., Pineo, G. F., et al. (2008). Prevention of venous thromboembolism: American College of Chest Physicians Evidence-based Clinical Practice Guidelines (8th ed.). *Chest*, *133*(6 Suppl.), 381S–453S.

*Gillaspie, M. (2010). Better pain management after total joint replacement surgery: A quality improvement approach. *Orthopaedic Nursing*, *29*(1), 20–24.

Handoll, H. H., Queally, J. M., & Parker, M. J. (2011). Pre-operative traction for hip fractures in adults. *Cochrane Database of Systematic Reviews*, (12), CD000168; doi: 10.1002/14651858.CD000168.pub3.

*Hardwick, M. E., Pulido, P. A., & Adelson, W. S. (2012). Nursing intervention using healing touch in bilateral total knee arthroplasty. *Orthopaedic Nursing*, *31*(1), 5–11.

He, M. L., Xiao, Z. M., Lei, M., et al. (2012). Continuous passive motion for preventing venous thromboembolism after total knee arthroplasty. *Cochrane Database of Systematic Reviews*, (1), CD008207; doi: 10.1002/14651858.pub2.

Hiltz, N. (2011). *Practice points: Thromboembolic disease. National Association of Orthopaedic Nurses*. Available at: www.orthonurse.org/p/do/sd/sid=417&type=0

Holmes, S. B., & Brown, S. J. (2005). Skeletal pin site care: National Association of Orthopaedic Nurses guidelines for orthopaedic nursing. *Orthopaedic Nursing*, *24*(2), 99–107.

Hwang, S. C., Kong, J. Y., Nam, D. M., et al. (2010). Revision total knee arthroplasty with a cemented posterior stabilized, condylar constrained or fully constrained prosthesis: A minimum 2-year-follow-up analysis. *Clinics in Orthopaedic Surgery*, *2*(2), 12–18.

Institute for Clinical Systems Improvement. (2012). *Health care protocol: Pressure ulcer prevention and treatment protocol* (3rd ed.). Available at: https://www.icsi.org/_asset/6t7kxy/PresUlcerTrmt-Interactive0112.pdf

Jenson, C. S., Cameron, I. D., & Lyn, M. (2010). Evidence-based guidelines for the management of hip fractures in older persons: An update. *Medical Journal of Australia*, *192*(1), 37–41.

Johanson, N. A., Lachiewicz, P. F., & Lieberman, J. R., et al. (2009). Prevention of symptomatic pulmonary embolism in patients undergoing total hip or knee arthroplasty. *Journal of the American Academy of Orthopaedic Surgeons*, *17*(3), 183–196.

Johnston-Walker, E., & Hardcastle, J. (2011). Neurovascular assessment in the critically ill patient. *Nursing in Critical Care*, *16*(4), 170–177.

Kamel, C., McGahan, L., Polisena, J., et al. (2012). Preoperative skin antiseptic preparations for preventing surgical site infections: A systematic review. *Infection Control and Hospital Epidemiology*, *33*(6), 608–617.

*Kearney, M., Jennrich, M. K., Lyons, S., et al. (2011). Effects of preoperative education on patient outcomes after joint replacement surgery. *Orthopaedic Nursing*, *30*(6), 391–396.

*Linari, L. R., Schofield, L. C., & Horrom, K. A. (2011). Implementing a bowel program: Is a bowel program an effective way of preventing constipation and ileus following elective hip and knee arthroplasty surgery? *Orthopaedic Nursing*, *30*(5), 317–321.

*Madsen, L., Herb, D., Magor, C., et al. (2010). Comparison of two bowel treatments to prevent constipation in post-surgical orthopaedic patients. *International Journal of Orthopaedic and Trauma Nursing*, *14*(2), 75–81.

Manier, R., Baviskar, J. V., Singhi, T., et al. (2012). To use or not to use continuous passive motion post–total knee arthroplasty. *Journal of Arthroplasty, 27*(12), 193–200.e1.

McInnes, E., Jammali-Blasi, A., Bell-Syer, S. E., et al. (2011). Support surfaces for pressure ulcer prevention. *Cochrane Database of Systematic Reviews*, (4), CD001735; doi: 10.1002/14651858.CD001735.pub4.

McMorrow, C. (2011). *Practice points: Nosocomial surgical site infection*. Available at: www.orthonurse.org/p/do/sd/sid=416&type=0

Meftah, M., Wong, A., Nawabi, D., et al. (2012). Pain management after total knee arthroplasty using a multimodal approach. *Orthopedics, 35*(5), e660–e664.

Nakama, G. Y., Peccin, M. S., Almeida, G. J., et al. (2012). Cemented, cementless or hybrid fixation options in total knee arthroplasty for osteoarthritis and other non-traumatic diseases. *Cochrane Database of Systematic Reviews*, (10), CD006193; doi: 10.1002/14651858.CD006193.pub2.

Palamone, J., Brunovsky, S., Groth, M., et al. (2011). "Tap and twist": Preventing deep vein thrombosis in neuroscience patients through foot and ankle range-of-motion exercises. *Journal of Neuroscience Nursing, 43*(6), 308–314.

Parker, R. J. (2011). Evidence-based practice: Caring for a patient undergoing total knee arthroplasty. *Orthopaedic Nursing, 30*(1), 4–8.

Reed, C., Carroll, L., Baccari, S., & Shermont, H. (2011). Spica cast care: A collaborative staff-led education initiative for improved patient care. *Orthopaedic Nursing, 30*(6), 353–358.

*Santy-Tomlinson, J., Vincent, M., Glossop, N., et al. (2011). Calm, irritated or infected? The experience of the inflammatory states and symptoms of pin site infection and irritation during external fixation: A grounded theory study. *Journal of Clinical Nursing, 20*(21–22), 3163–3173.

Satryb, S. A., Wilson, T. J., & Patterson, M. M. (2011). Casting: All wrapped up. *Orthopaedic Nursing, 30*(1), 37–41.

*Schneider, M. A. (2012). Prevention of catheter-associated urinary tract infections in patients with hip fractures through education of nurses to specific catheter protocols. *Orthopaedic Nursing, 31*(1), 12–18.

Smith, M., & Smith, T. (2012). The American Joint Replacement Registry. *Orthopaedic Nursing, 31*(5), 296–299.

Smith, M. A., Jacobs, L., Rodier, L., et al. (2011). Clinical quality indicators: Infection prophylaxis for total knee arthroplasty. *Orthopaedic Nursing, 30*(5), 301–304.

Spahn, D. R. (2010). Anemia and patient blood management in hip and knee surgery: A systematic review of the literature. *Anesthesiology, 113*(2), 482–495.

Stefansdottir, A., Robertsson, O., W-Dahl, A., et al. (2009). Inadequate timing of prophylactic antibiotics in orthopedic surgery. We can do better. *Acta Orthopaedica, 80*(6), 633–638.

Tai, T. W., Lin, C. J., Jou, I. M., et al. (2011). Tourniquet use in total knee arthroplasty: A meta-analysis. *Knee Surgery, Sports Traumatology, Arthroscopy, 19*(7), 1121–1130.

Timms, A., Vincent, M., Santy-Tomlinson, J., et al. (2011). *Royal College of Nursing guidance on pin site care: Report and recommendations from 2010 consensus project on pin site care*. Available at: www.rcn.org.uk/__data/assets/pdf_file/0009/413982/004137.pdf

Tuttle, M. S., Smith, W., Wade, R., et al. (2009). Safety and efficacy of damage control external fixation versus early definitive stabilization for femoral shaft fractures in the multiple-injured patient. *Journal of Trauma-Injury Infection & Critical Care, 67*(3), 602–605.

Walsh, C. (2009). Sign off on casting. *OR Nurse, 3*(5), 45–51.

Watts, A., Howie, C., & Hughes, H. (2012). Venous thrombosis and pulmonary embolism associated with hip and knee joint replacement over a 10-year period: A population-based study. *Journal of Bone & Joint Surgery, British Volume, 94*(Suppl. II), 85.

*Wenger, J. E. (2010). Cultivating quality: Reducing rates of catheter-associated urinary tract infection. *American Journal of Nursing, 110*(8), 40–45.

Willis-Owen, C. A., Knoyves, A., & Martin, D. K. (2010). Factors affecting the incidence of infection in knee and hip replacement: An analysis of 5,277 cases. *Journal of Bone & Joint Surgery, 92*(8), 1128–1133.

Whiteside, L. A. (2011). Does fixation matter: Cementless fixation for primary TKA. *Techniques in Knee Surgery, 10*(3), 129–135.

Resources

American Academy of Orthopaedic Surgeons/American Association of Orthopaedic Surgeons (AAOS), www.aaos.org

National Association of Orthopaedic Nurses (NAON), www.orthonurse.org

National Institute of Arthritis and Musculoskeletal and Skin Diseases, National Institutes of Health, www.niams.nih.gov

Chapter 42

Management of Patients With Musculoskeletal Disorders

Learning Objectives

On completion of this chapter, the learner will be able to:

1. Describe the nursing management, rehabilitation, and health education needs of the patient with low back pain.
2. Describe common musculoskeletal disorders of the hand, wrist, shoulder, and foot, and nursing care of the patient undergoing surgery to correct these disorders.
3. Explain the pathogenesis, prevention, and management of osteoporosis.
4. Use the nursing process as a framework for care of the patient with osteoporotic vertebral fracture.
5. Identify the causes and related management of patients with osteomalacia, Paget's disease, and septic arthritis.
6. Use the nursing process as a framework for care of the patient with osteomyelitis.
7. Describe the causes and related management of the patient with a primary or metastatic bone tumor.

Glossary

bursitis: inflammation of a fluid-filled sac in a joint
contracture: abnormal shortening of muscle or fibrosis of joint structures
involucrum: new bone growth around a sequestrum
osteopenia: low bone mineral density
osteoporosis: degenerative disease of the bone characterized by reduced mass, deterioration of matrix, and diminished architectural strength
radiculopathy: disease of a nerve root that may result in pain that radiates down the leg
sciatica: sciatic nerve pain; pain travels down back of thigh into foot
sequestrum: dead bone in abscess cavity
tendonitis: inflammation of muscle tendons

Musculoskeletal disorders, particularly impairment of the back and spine, are leading health problems and causes of disability. The functional and psychological limitations imposed on the patient may be severe. Nurses provide care for patients with these disorders in both inpatient and outpatient settings; because of this, nurses should be cognizant of these limitations and the effects they may have on these patients. The economic costs in terms of loss of productivity, medical expenses, and other costs that are not compensated are among the highest for any medical diagnosis (Bach & Holten, 2009).

Low Back Pain

The number of visits to health care providers for low back pain is second only to the number of visits for upper respiratory illnesses (Bach & Holten, 2009). Most low back pain is caused by one of many musculoskeletal problems, including acute lumbosacral strain, unstable lumbosacral ligaments and weak muscles, intervertebral disk problems, and unequal leg length. Depression (Karp, Weiner, Dew, et al., 2010), obesity, and stress are frequent comorbidities. Generally, back pain due to musculoskeletal disorders is aggravated by activity, whereas pain due to other conditions is not.

Older patients may experience back pain associated with osteoporotic vertebral fractures, osteoarthritis of the spine, and spinal stenosis. Higher numbers of areas of pain are associated with a higher level of disability (Buchman, Shah, Leurgens, et al., 2010). Other nonmusculoskeletal causes of back pain beyond the scope of this chapter include kidney disorders, pelvic problems, retroperitoneal tumors, and abdominal aortic aneurysms.

Pathophysiology

The spinal column can be considered an elastic rod constructed of rigid units (vertebrae) and flexible units (intervertebral disks) held together by complex facet joints, multiple ligaments, and paravertebral muscles. Its unique construction allows for flexibility while providing maximum protection for the spinal cord. The spinal curves absorb vertical shocks from running and jumping. The abdominal and thoracic muscles are important in lifting activities, working together to minimize stress on the spinal units. Disuse weakens these supporting muscular structures. Obesity, postural problems, structural problems, and overstretching of the spinal supports may result in back pain (Porth & Matfin, 2009).

The intervertebral disks change in character as a person ages. A young person's disks are mainly fibrocartilage with a gelatinous matrix. Over time, the fibrocartilage becomes dense and irregularly shaped. Disk degeneration is a common cause of back pain. The lower lumbar disks, L4-5 and L5-S1, are subject to the greatest mechanical stress and the greatest degenerative changes. Disk protrusion or facet joint changes can cause pressure on nerve roots as they leave the spinal canal, which results in pain that radiates along the nerve (Porth & Matfin, 2009). (See Chapter 70 for discussion of the management of intervertebral disk disease.)

Clinical Manifestations

The typical patient reports either acute back pain (lasting fewer than 3 months) or chronic back pain (3 months or longer without improvement) and fatigue. The patient may report pain radiating down the leg, which is known as **radiculopathy** or **sciatica**; presence of this symptom suggests nerve root involvement. The patient's gait, spinal mobility, reflexes, leg length, leg motor strength, and sensory perception may be affected. Physical examination may disclose paravertebral muscle spasm (greatly increased muscle tone of the back postural muscles) with a loss of the normal lumbar curve and possible spinal deformity.

Assessment and Diagnostic Findings

The initial evaluation of acute low back pain includes a focused history and physical examination, including general observation of the patient, gait evaluation, and neurologic testing (described in Chapter 40). The findings suggest either nonspecific back symptoms or potentially serious problems, such as sciatica, spine fracture, cancer, infection, or rapidly progressing neurologic deficit.

The diagnostic procedures described in Chart 42-1 may be indicated for the patient with potentially serious or prolonged low back pain. Red flags that trigger prescribing these studies include suspected spinal infection, severe neurologic weakness, urinary or fecal incontinence, and a new onset of back pain in a patient with cancer (Chou, Quaseen, Owens, et al., 2011). The nurse prepares the patient for these studies, provides the necessary support during the testing period, and monitors the patient for any adverse responses to the procedures.

Chart 42-1 Diagnostic Procedures for Low Back Pain

X-ray of the spine: may demonstrate a fracture, dislocation, infection, osteoarthritis, or scoliosis
Bone scan and blood studies: may disclose infections, tumors, and bone marrow abnormalities
Computed tomography (CT) scan: useful in identifying underlying problems, such as obscure soft tissue lesions adjacent to the vertebral column and problems of vertebral disks
Magnetic resonance imaging (MRI) scan: permits visualization of the nature and location of spinal pathology
Electromyogram (EMG) and nerve conduction studies: used to evaluate spinal nerve root disorders (radiculopathies)
Myelogram: permits visualization of segments of the spinal cord that may have herniated or may be compressed (infrequently performed; indicated when MRI scan is contraindicated)
Ultrasound: useful in detecting tears in ligaments, muscles, tendons, and soft tissues in the back

Medical Management

Most back pain is self-limited and resolves within 4 weeks with analgesics, rest, and avoidance of strain. Based on initial assessment findings indicating nonspecific back symptoms, the patient is reassured that the pain is not due to a serious condition and x-rays or other imaging modalities are not necessary (Chou et al., 2011). Management focuses on relief of discomfort, activity modification, and patient education.

Nonprescription analgesics such as acetaminophen (Tylenol) and nonsteroidal anti-inflammatory drugs (NSAIDs) and short-term prescription muscle relaxants (e.g., cyclobenzaprine [Flexeril]) are effective in relieving acute low back pain. Tricyclic antidepressants (e.g., amitriptyline [Elavil) and the newer dual-action serotonin-norepinephrine reuptake inhibitors (e.g., duloxetine [Cymbalta]) (Karp et al., 2010) or atypical seizure medications (e.g., gabapentin [Neurontin], which is prescribed for pain from radiculopathy) are used effectively in chronic low back pain. Opioid medications are indicated only short term for acute moderate to severe cases of low back pain, except in older adults, those with kidney disease, or those who must avoid chronic NSAID exposure because of its adverse gastric effects. Systemic corticosteroids are not considered effective in alleviating low back pain (Johnson, Neher, & St. Anna, 2011).

Effective nonpharmacologic interventions include thermal applications (hot or cold) and spinal manipulation (e.g., chiropractic therapy). Lumbar support belts are not recommended to treat acute low back pain but may be effective devices for preventing low back pain in occupational health settings (Donnelly, Callaghan, & Durkin, 2009). Cognitive-behavioral therapy (e.g., biofeedback), exercise regimens, spinal manipulation, physical therapy, acupuncture, massage, and yoga are all effective nonpharmacologic interventions for treating chronic low back pain (National Guideline Clearinghouse, 2010).

Most patients need to alter their activity patterns to avoid aggravating the pain. They should avoid twisting, bending, lifting, and reaching—all of which stress the back. The patient is taught to change position frequently. Sitting should be limited to 20 to 50 minutes based on level of comfort. Absolute bed rest is no longer recommended; typical activities of daily living (ADLs) should be resumed as soon as possible (Bach & Holten, 2009; Dahm, Brurberg, Jamtvedt, et al., 2010). A quick return to normal activities and a program of low-stress aerobic exercise are recommended. Conditioning exercises for both back and trunk muscles are begun after about 2 weeks to help prevent recurrence of pain.

Nursing Assessment

The nurse asks the patient to describe the discomfort (e.g., location, severity, duration, characteristics, radiation, weakness in the legs). Descriptions of how the pain occurred, such as with a specific action (e.g., opening a garage door) or with an activity in which weak muscles were overused (e.g., weekend gardening), and how the patient has dealt with the pain often suggest areas for intervention and patient education.

If back pain is a recurring problem, information about previous successful pain control methods helps in planning current management. Information about work and recreational activities helps identify areas for back health education. Because stress and anxiety can evoke muscle spasms and pain, the nurse assesses environmental variables, work situations, and family relationships. In addition, the nurse assesses the effect of chronic pain on the emotional well-being of the patient. Referral to a mental health professional for assessment and management of stressors contributing to the low back pain and related depression may be appropriate.

During the interview, the nurse observes the patient's posture, position changes, and gait. Often, the patient's movements are guarded, with the back kept as still as possible. The patient should be directed to a chair of standard seat height with arms for support. The patient may sit and stand in an unusual position, leaning away from the most painful side, and may need assistance when undressing for the physical examination.

On physical examination, the nurse assesses the spinal curve, any leg length discrepancy, and pelvic crest and shoulder symmetry. The nurse palpates the paraspinal muscles and notes spasm and tenderness. When the patient is in a prone position, the paraspinal muscles relax and any deformity caused by spasm can subside. The nurse asks the patient to bend forward and then laterally, noting any discomfort or limitations in movement. It is important to determine the effect of these limitations on ADLs. The nurse evaluates nerve involvement by assessing deep tendon reflexes, sensations (e.g., paresthesia), and muscle strength. Back and leg pain on straight-leg raising (with the patient supine, the patient's leg is lifted upward with the knee extended) suggests nerve root involvement.

Nursing Management

The major nursing goals for the patient include relief of pain, improved physical mobility, the use of back-conserving techniques of body mechanics, improved self-esteem, and weight reduction (as necessary) (Chart 42-2).

The nurse assesses the patient's response to analgesic agents. As the acute pain subsides, medication dosages are reduced. The nurse evaluates and notes the patient's response to various pain management modalities (as discussed in Chapter 12). The nurse cautions the patient with severe pain not to remain on bed rest because extended periods of inactivity are not effective and result in deconditioning. A medium to firm, nonsagging mattress (a bed board may be used) is recommended; there is no evidence to support the use of a firm mattress (National Guideline Clearinghouse, 2010). Lumbar flexion is increased by elevating the head and thorax 30 degrees by using pillows or a foam wedge and slightly flexing the knees supported on a pillow. Alternatively, the patient can assume a lateral position with knees and hips flexed (curled position) with a pillow between the knees and legs and a pillow supporting the head (Fig. 42-1). A prone position should be avoided because it accentuates lordosis. The nurse instructs the patient to get out of bed by rolling to one side and placing the legs down while pushing the torso up, keeping the back straight (National Institute of Neurological Disorders and Stroke (NINDS), 2011).

As the patient achieves comfort, an exercise program is gradually initiated with low-stress aerobic exercises, such as short walks or swimming. The physical therapist designs an exercise program for the patient to reduce lordosis, increase flexibility, and reduce strain on the back. It may include hyperextension exercises to strengthen the paravertebral muscles, flexion exercises to increase back movement and strength, and isometric flexion exercises to strengthen trunk muscles. Each 30-minute daily exercise period begins and ends with relaxation.

The nurse encourages the patient to adhere to the prescribed exercise program. Some patients may find it difficult to

Chart 42-2 HEALTH PROMOTION

Strategies for Preventing Acute Low Back Pain

Prevention

- Weight reduction as needed: Modify diet to achieve ideal body weight.
- Avoid high heels.
- Walk daily and gradually increase the distance and pace of walking.
- Avoid jumping and jarring activities.
- Stretch to enhance flexibility. Do strengthening exercises.

Body Mechanics

- Practice good posture.
- Avoid twisting, lifting above waist level, and reaching up for any length of time.
- Push objects rather than pull them.
- Keep load close to your body when lifting.
- Lift with the large leg muscles, not the back muscles.
- Squat while keeping the back straight when it is necessary to pick something up off the floor.
- Bend your knees and tighten abdominal muscles when lifting.
- Avoid overreaching or a forward flexion position.
- Use a wide base of support.

Work Modifications

- Adjust height of work area to avoid stress on back.
- Adjust height of chair using a footstool to position knees higher than hips.
- Use lumbar support in straight back chair with arm rests.
- Avoid prolonged standing and repetitive tasks.
- Avoid bending, twisting, and lifting heavy objects.
- When standing for any length of time, rest one foot on a small stool or box to relieve lumbar lordosis.
- Avoid work involving continuous vibrations.

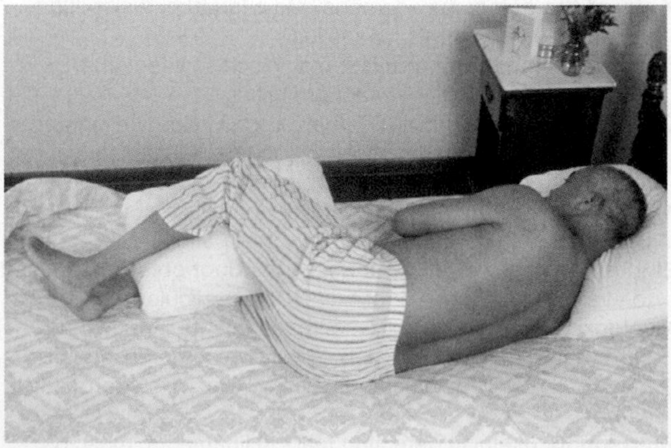

FIGURE 42-1 • Positioning to promote lumbar flexion. Photo by B. Proud.

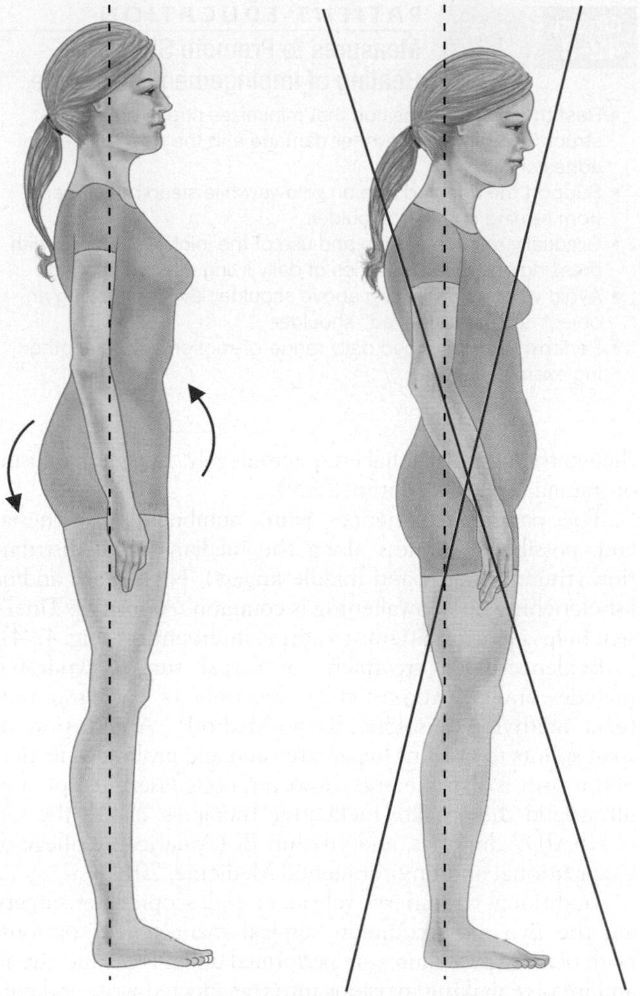

FIGURE 42-2 • Proper and improper standing postures. *(Left)* Abdominal muscles contracted, giving a feeling of upward pull, and gluteal muscles contracted, giving a downward pull. *(Right)* Slouch position, showing abdominal muscles relaxed and body out of proper alignment.

do so for a long period; in these instances, alternating activities may help facilitate adherence to the regimen (NINDS, 2011). Activities should not cause excessive lumbar strain or twisting, with avoidance of activities such as horseback riding and weight lifting.

Good body mechanics and posture are essential to avoid recurrence of back pain. The patient must be taught how to stand, sit, lie, and lift properly (Fig. 42-2). Providing the patient with a list of suggestions helps in making these long-term changes (see Chart 42-2). The patient who wears high heels is encouraged to change to low heels with good arch support. The patient who is required to stand for long periods should shift weight frequently and rest one foot on a low stool, which decreases lumbar lordosis. Standing on a foot cushion made of foam or rubber can be helpful. The proper posture can be verified by looking in a mirror to see whether the chest is up, the abdomen is tucked in, and the shoulders are down and relaxed. Locking the knees when standing is avoided, as is bending forward for long periods.

When sitting, the knees and hips should be flexed, with the knees level with the hips or higher to minimize lordosis.

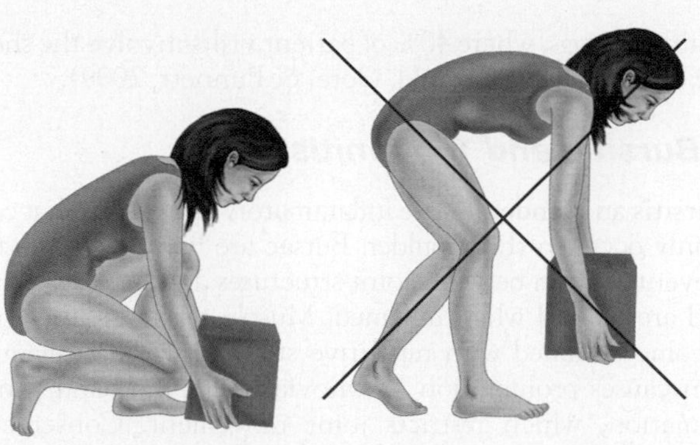

FIGURE 42-3 • Proper and improper lifting techniques. *(Left)* Correct position for lifting. This person is using the long and strong muscles of the arms and legs and holding the object so that the line of gravity falls within the base of support. *(Right)* Incorrect position for lifting. In this position, pull is exerted on the back muscles, and leaning causes the line of gravity to fall outside the base.

The feet should be flat on the floor or supported on a raised surface. Patients should avoid sitting on stools or chairs that do not provide firm back support.

The nurse instructs the patient in the safe and correct way to lift objects—using the strong quadriceps muscles of the thighs, with minimal use of weak back muscles (Fig. 42-3). With feet placed hip-width apart to provide a wide base of support, the patient should bend the knees, tighten the abdominal muscles, and lift the object close to the body with a smooth motion, avoiding twisting and jarring motions. To prevent recurrence of acute low back pain, the patient should avoid lifting more than one third of his or her weight without help.

Role-related responsibilities may have to be modified with the onset of low back pain (e.g., carrying children). As recovery progresses, the patient may resume them; however, if these activities contributed to the development of low back pain, resuming them may lead to the development of chronic low back pain, with associated disability. If the patient experiences secondary gains associated with low back disability (e.g., workers' compensation, easier lifestyle or workload, increased emotional support), a "low back neurosis" may develop. Antidepressants and counseling may be needed to assist the person in resuming a full, productive life. Specialized back clinics use multidisciplinary approaches to help the patient with chronic pain resume role-related responsibilities.

Obesity contributes to back strain by overtaxing the relatively weak back muscles in the absence of abdominal muscle support. Exercises are less effective and more difficult to perform when the patient is overweight. Weight reduction through diet modification is imperative to prevent recurrence of back pain. A sound nutritional plan that includes a change in eating habits and low-impact activities is vital. Noting achievement of weight reduction and providing positive reinforcement facilitate adherence. Back problems may resolve as optimal weight is achieved (NINDS, 2011).

Common Upper Extremity Problems

The structures in the upper extremities are frequently the sites of painful syndromes. This is especially true in occupational health settings, where 40% of patient visits involve the shoulder, wrist, and hand (Gold, Gore, & Punnett, 2009).

Bursitis and Tendonitis

Bursitis and **tendonitis** are inflammatory conditions that commonly occur in the shoulder. Bursae are fluid-filled sacs that prevent friction between joint structures during joint activity and are painful when inflamed. Muscle tendon sheaths also become inflamed with repetitive stretching. The inflammation causes proliferation of synovial membrane and pannus formation, which restricts joint movement. Conservative treatment includes rest of the extremity, intermittent ice and heat to the joint, and NSAIDs to control the inflammation and pain. Newer therapies that include extracorporeal shock waves, pulsed magnetic fields, laser phototherapy, and radiofrequency ablation may accelerate tendon healing, although further research is needed to determine their overall effectiveness. These modalities are expensive and are therefore not readily accessible to the uninsured. Arthroscopic synovectomy may be considered if shoulder pain and weakness persist. Corticosteroid injections remain more effective than most other interventions for short-term, rapid improvement (Gaujoux-Viala, 2009).

Loose Bodies

Loose bodies ("joint mice") may occur in a joint space as a result of articular cartilage wear and bone erosion. These fragments can interfere with joint movement ("locking the joint"). Loose bodies are removed by arthroscopic surgery if they cause pain or mobility issues.

Impingement Syndrome

Impingement syndrome is a general term that describes impaired movement of the rotator cuff of the shoulder. Impingement usually occurs from repetitive overhead movement of the arm or from acute trauma resulting in irritation and eventual inflammation of the rotator cuff tendons or the subacromial bursa as they grate against the coracoacromial arch. Early manifestations of this syndrome are characterized by edema from hemorrhage of these structures, pain, shoulder tenderness, limited movement, muscle spasm, and eventual disuse atrophy. The process may progress to a partial or complete rotator cuff tear (see Chapter 43).

Medications used to treat early impingement syndrome include oral NSAIDs or intra-articular injections of corticosteroids. Application of superficial cold or heat may subjectively improve patients' symptoms; however, a therapeutic exercise program (see Chapter 43) is required to improve outcomes, including reduction of pain and improved shoulder function (Chart 42-3).

Chart 42-3 PATIENT EDUCATION

Measures to Promote Shoulder Healing of Impingement Syndrome

- Rest the joint in a position that minimizes stress on the joint structures to prevent further damage and the development of adhesions.
- Support the affected arm on pillows while sleeping to keep from turning onto the shoulder.
- Gradually resume motion and use of the joint. Assistance with dressing and other activities of daily living may be needed.
- Avoid working and lifting above shoulder level or pushing an object against a "locked" shoulder.
- Perform the prescribed daily range-of-motion and strengthening exercises.

Carpal Tunnel Syndrome

Carpal tunnel syndrome is an entrapment neuropathy that occurs when the median nerve at the wrist is compressed by a thickened flexor tendon sheath, skeletal encroachment, edema, or a soft tissue mass. It frequently occurs in women between 30 and 60 years of age. Commonly caused by repetitive hand and wrist movements, it is also associated with rheumatoid arthritis, diabetes, acromegaly, hyperthyroidism, or trauma (Porth & Matfin, 2009).

The patient experiences pain, numbness, paresthesia, and, possibly, weakness along the median nerve distribution (thumb, index, and middle fingers). Night pain and/or fist clenching upon awakening is common. A positive Tinel's sign helps identify patients requiring intervention (Fig. 42-4).

Evidence-based treatment of carpal tunnel syndrome includes oral or intra-articular injections of corticosteroids (e.g., methylprednisolone [Depo-Medrol]). Application of wrist splints to prevent hyperextension and prolonged flexion of the wrist is also effective; however, yoga, laser therapy, and ultrasound therapy are ineffective therapies, as are the use of NSAIDs, diuretics, and vitamin B_6 (American College of Occupational and Environmental Medicine, 2011).

Traditional open nerve release or endoscopic laser surgery are the two most common surgical management options. Both of these procedures are performed under local anesthesia and involve making incisions into the affected wrist and cutting the carpal ligament so that the carpal tunnel is widened. Smaller incisions are made with the endoscopic laser procedure, resulting in less scar formation and a shorter recovery time than with the open method. Following either procedure,

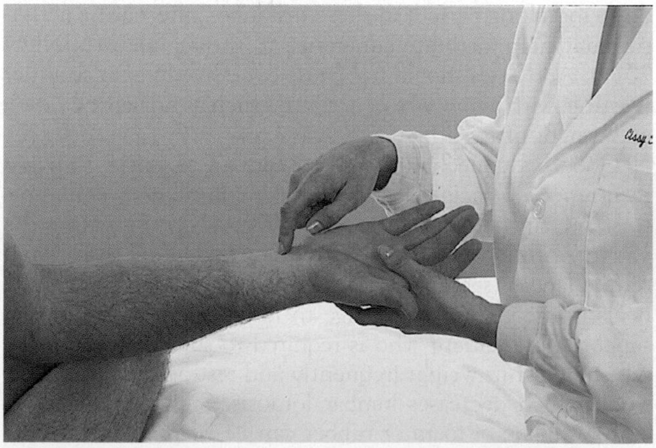

FIGURE 42-4 • Tinel's sign may be elicited in patients with carpal tunnel syndrome by percussing lightly over the median nerve, located on the inner aspect of the wrist. If the patient reports tingling, numbness, and pain, the test for Tinel's sign is considered positive. From Weber, J. W., & Kelley, J. (2010). *Health assessment in nursing* (3rd ed.). Philadelphia: Lippincott Williams & Wilkins. Photo by B. Proud.

the patient wears a hand splint and limits hand use during healing. The patient may need assistance with personal care. Full recovery of motor and sensory function after either type of nerve release surgery may take several weeks or months.

Ganglion

A ganglion—a collection of neurologic gelatinous material near the tendon sheaths and joints—appears as a round, firm, cystic swelling, usually on the dorsum of the wrist. It frequently occurs in women younger than 50 years (Porth & Matfin, 2009). The swelling is locally tender and may cause an aching pain. When a tendon sheath is involved, weakness of the finger occurs. Treatment may include aspiration, corticosteroid injection, or surgical excision. After treatment, a compression dressing and immobilization splint are used.

Dupuytren's Disease

Dupuytren's disease results in a slowly progressive **contracture** of the palmar fascia that causes flexion of the fourth, fifth, and, sometimes, middle finger, rendering these fingers more or less useless (Fig. 42-5). It is linked to an inherited autosomal dominant trait and occurs most frequently in men of Scandinavian or Celtic heritage who are older than 50 years (Black & Blazar, 2011). Dupuytren's disease is also associated with arthritis, diabetes, gout, cigarette smoking, and alcoholism. Starting as a nodule, it may or may not progress, producing a contracture of the fingers and palmar skin changes. The patient may experience dull and aching discomfort, morning numbness, and stiffness in the affected fingers. This condition starts in one hand, but eventually both are affected. Finger-stretching exercises or intra-nodular injections of corticosteroids may prevent contractures. With loss of movement, palmar and digital fasciectomies are performed to improve function.

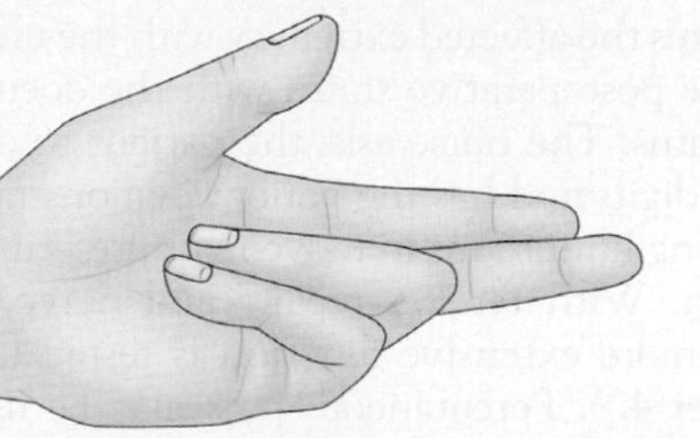

FIGURE 42-5 • Dupuytren's contracture, a flexion deformity caused by an inherited trait, is a slowly progressive contracture of the palmar fascia, which severely impairs the function of the fourth, fifth, and, sometimes, middle finger.

Nursing Management of the Patient Undergoing Surgery of the Hand or Wrist

Surgery of the hand or wrist, unless related to major trauma, is generally an ambulatory procedure. Before surgery, the nurse assesses the patient's level and type of discomfort, as well as limitations in function, caused by the impairing condition.

Hourly neurovascular assessment of the exposed fingers for the first 24 hours following surgery is essential for monitoring function of the nerves and perfusion. This is especially important if an intraoperative tourniquet technique was used, which is implicated in neurovascular deficits (Chart 42-4). The nurse

Chart 42-4 NURSING RESEARCH PROFILE

Neurovascular Assessment Practices of Orthopedic Nurses

Murphy, S., & O'Conner, C. (2010). So what! If a pneumatic tourniquet is used intraoperatively: A study of neurovascular assessment practices of orthopaedic nurses. *International Journal of Orthopaedic and Trauma Nursing*, *14*(1), 48–54.

Purpose

Pneumatic tourniquets are commonly used during surgery on extremities to provide better visualization of structures. However, many postoperative adverse effects can occur from their use, including superficial injuries (e.g., skin blisters, abrasions, lacerations) and deep tissue injuries (e.g., edema, pain, compartment syndrome). Safe and effective postoperative assessment parameters have not been codified in orthopedic nursing practice. This study aimed to codify practices of expert orthopedic nurses in assessing patients postoperatively who had pneumatic tourniquets applied on an extremity perioperatively.

Design

Using a descriptive exploratory design, questionnaires were distributed to 69 experienced orthopedic nurses in the Republic of Ireland with familiarity of the use of pneumatic cuffs. The Likert scale questions used were derived from previously conducted focus group interviews and pilot studies. Data collected included perceptions of who was responsible for discussing the use of perioperative tourniquets with patients and knowledge of where tourniquets are applied, duration of their use, and required postoperative care.

Findings

Thirty-nine percent of nurses believed that informing patients about the placement of the tourniquet and potential complications was the role of the surgeon, 20% believed that this was the role of the nurse, 9% believed that this was the role of the anesthesia provider, and 28% chose none of these options, which could mean that they felt the role should be shared. Most nurses (92%) agreed that neurovascular assessments were a critical part of their practice in the postoperative period. However, several common manifestations of tissue and vascular impairment were not noted by the nurses as something they considered during their assessments (e.g., pain, swelling). More than 63% admitted that they could not identify specific complications from the use of tourniquets.

Nursing Implications

Findings from this study suggest that even experienced orthopedic nurses are uncertain how intraoperative tourniquets may complicate patients' postoperative recovery. Hence, early warning signs and symptoms of complications can go unrecognized and underreported. Nurses should be educated about how the use of devices such as pneumatic tourniquets perioperatively may cause postoperative complications. They should be educated about how to assess for these complications in the postoperative period so that decisive, early interventions can be made.

compares the affected extremity with the unaffected extremity and the postoperative status with the documented preoperative status. The nurse asks the patient to describe sensations in the digits and has the patient demonstrate mobility, while enforcing limitations in movement prescribed by the patient's surgeon. With tendon repairs and nerve, vascular, or skin grafts, more extensive function is tested (see Chart 40-2 in Chapter 40). Percutaneous pins may be used to hold bones in position. These pins serve as potential sites of infection. Patient instructions concerning aseptic wound and pin care may be necessary.

Dressings provide support but should be nonconstrictive. Intermittent use of ice packs to the surgical area during the first 24 to 48 hours may be prescribed to control edema. Unless contraindicated, active extension and flexion of the fingers to promote circulation are encouraged, even though movement is limited by the bulky dressing.

Generally, pain and discomfort can be controlled by the use of oral analgesic agents. Patient education concerning the risk of falls and impaired cognition is important. Pain out of proportion to what is expected, particularly if it is accompanied with compromised neurovascular functioning, needs to be evaluated as an indication of compartment syndrome (see Chapter 43). Pain may be related to surgery, edema, hematoma formation, or restrictive bandages. To control swelling that may increase the patient's discomfort, the nurse instructs the patient to elevate the hand to heart level with pillows. If the patient is ambulatory, the arm is supported in a conventional sling with the hand elevated at heart level (see Fig. 43-10 in Chapter 43).

During the first few days after surgery, independent self-care is impaired. The patient may need to arrange for assistance with feeding, bathing, dressing, and toileting. Within a few days, the patient develops skills in one-handed ADLs and is usually able to function with minimal help and assistive devices. The nurse encourages the patient to use the involved hand, unless contraindicated, within the limits of discomfort. As rehabilitation progresses, the patient resumes use of the limb. Physical or occupational therapy–directed exercises may be prescribed. The nurse emphasizes compliance with the plan.

Promoting Home and Community-Based Care

Educating Patients About Self-Care

After the patient has undergone surgery, the nurse instructs the patient how to monitor neurovascular status and the signs of complications that need to be reported to the surgeon (e.g., paresthesia, paralysis, uncontrolled pain, coolness of fingers, extreme swelling, excessive bleeding, purulent drainage, foul odor, fever). The nurse discusses prescribed medications and their side effects with the patient. In addition, the nurse instructs the patient how to elevate the extremity and to apply ice (if prescribed) to control swelling. The use of assistive devices is demonstrated if such devices would be helpful in promoting accomplishment of ADLs. For bathing, the nurse instructs the patient to keep the dressing dry by covering it with a secured plastic bag. Generally, the wound is not redressed until the patient's follow-up visit with the surgeon (Chart 42-5).

Common Foot Problems

Disorders of the foot may be caused by poorly fitting shoes, which distort normal anatomy while inducing deformity and pain. Dermatologic problems commonly affect the feet in the form of fungal infections and plantar warts. Several systemic diseases affect the feet. Patients with diabetes are prone to develop corns and peripheral neuropathies with diminished sensation, leading to ulcers at pressure points of the foot. Patients with peripheral vascular disease and arteriosclerosis complain of burning and itching feet, resulting in scratching and skin breakdown. Foot deformities may occur with rheumatoid arthritis. Obesity can cause a host of foot anomalies, including arches that fall over time and plantar fasciitis.

Chart 42-5 HOME CARE CHECKLIST
Hand or Foot Surgery

At the completion of home care education, the patient or caregiver will be able to:	PATIENT	CAREGIVER
• Demonstrate how to assess neurovascular status.	✔	✔
• State abnormal findings (e.g., unrelenting pain; paralysis; paresthesia; cool, nonblanching digits) to report to primary provider promptly.	✔	✔
• Demonstrate control of edema by elevating extremity and applying ice intermittently if prescribed.	✔	✔
• Identify signs and symptoms of infection (e.g., elevated temperature, purulent drainage).	✔	✔
• Demonstrate exercises to promote circulation, unless contraindicated.	✔	
• Describe methods to prevent wound infection (e.g., keeping dressing clean and dry during activities of daily living).	✔	✔
• Describe the use of prescribed medications.	✔	✔
• Verbalize the need to keep appointment with surgeon for initial dressing change.	✔	✔
• Describe the prescribed weight-bearing limits, as applicable.	✔	✔
• Demonstrate the use of assistive devices, if appropriate.	✔	

The discomforts of foot strain are treated with rest, elevation, physiotherapy, supportive taping, and orthotic devices (American College of Foot and Ankle Surgeons [ACFAS], 2010). The patient must inspect the foot and skin under pads and orthotic devices for pressure and skin breakdown daily. If a "window" is cut into shoes to relieve pressure over a bony deformity, the skin must be monitored daily for breakdown from pressure exerted at the window area. Active foot exercises promote circulation and help strengthen the feet. Walking in properly fitting shoes is considered the ideal exercise.

Plantar Fasciitis

Plantar fasciitis, an inflammation of the foot-supporting fascia, presents as an acute onset of heel pain experienced with the first steps in the morning. The pain is localized to the anterior medial aspect of the heel and diminishes with gentle stretching of the foot and Achilles tendon. Management includes stretching exercises, wearing shoes with support and cushioning to relieve pain, orthotic devices (e.g., heel cups, arch supports, night splints), and corticosteroid injections (Lee, McKeon, & Hertel, 2009). Unresolved plantar fasciitis may progress to fascial tears at the heel and eventual development of heel spurs.

Corn

A corn is an area of hyperkeratosis (overgrowth of a horny layer of epidermis) produced by internal pressure (the underlying bone is prominent because of a congenital or acquired abnormality, commonly arthritis) or external pressure (ill-fitting shoes). The fifth toe is most frequently involved, but any toe may be involved.

Corns are treated by a podiatrist by soaking and scraping off the horny layer, by application of a protective shield or pad, or by surgical modification of the underlying offending osseous structure. Soft corns are located between the toes and are kept soft by moisture. Treatment consists of drying the affected spaces and separating the affected toes with lamb's wool or gauze. A wider shoe or wider toe box (i.e., part of shoe where toes are placed) may be helpful (ACFAS, 2009).

Callus

A callus is a thickened area of the skin that has been exposed to persistent pressure or friction. Faulty foot mechanics usually precede the formation of a callus. Treatment consists of eliminating the underlying causes and having a painful callus treated by a podiatrist. A keratolytic ointment may be applied and a thin plastic cup worn over the heel if the callus is on this area. Felt padding with an adhesive backing is also used to prevent and relieve pressure. Orthotic devices can be made to remove the pressure from bony protuberances, or the protuberance may be excised (ACFAS, 2009, 2010).

Ingrown Toenail

An ingrown toenail (onychocryptosis) is a condition in which the free edge of a nail plate penetrates the surrounding skin. A secondary infection or granulation tissue may develop. This painful condition is caused by improper self-treatment, external pressure (tight shoes or stockings), internal pressure (deformed toes, growth under the nail), trauma, or infection. Trimming the nails properly (clipping them straight across and filing the corners consistent with the contour of the toe) can prevent this problem. Active treatment consists of washing the foot twice a day and relieving the pain by decreasing the pressure of the nail plate on the surrounding soft tissue (ACFAS, 2010). Warm, wet soaks help drain an infection. A toenail may need to be excised by the podiatrist or primary provider if there are recurrent infections.

Hammer Toe

Hammer toe is a flexion deformity of the interphalangeal joint, which may involve several toes (Fig. 42-6A). Tight socks or shoes may push an overlying toe back into the line of the other toes. The toes usually are pulled upward, forcing the metatarsal joints (ball of the foot) downward. Corns develop on top of the toes, and tender calluses develop under the metatarsal area. Treatment consists of conservative measures: wearing open-toed sandals or shoes that conform to

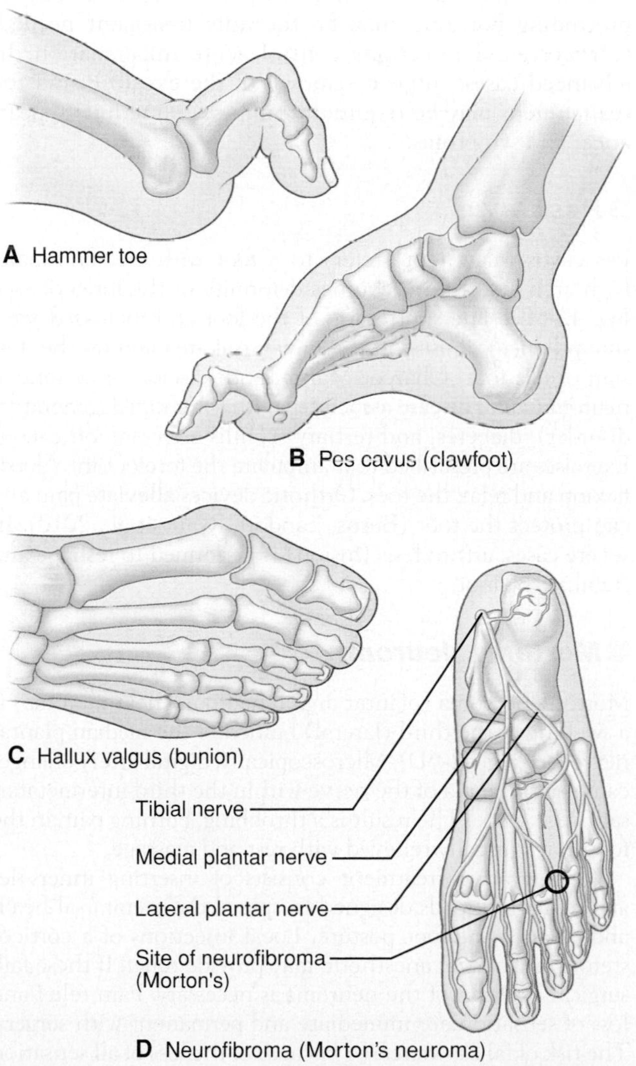
A Hammer toe
B Pes cavus (clawfoot)
C Hallux valgus (bunion)
Tibial nerve
Medial plantar nerve
Lateral plantar nerve
Site of neurofibroma (Morton's)
D Neurofibroma (Morton's neuroma)

FIGURE 42-6 • Common foot deformities.

the shape of the foot, carrying out manipulative exercises, and protecting the protruding joints with pads. Surgery (osteotomy) may be used to correct a resulting deformity. There is little evidence to support treatment of hammer toe when the patient does not report pain or other symptoms (ACFAS, 2009).

Hallux Valgus

Hallux valgus (bunion) is a deformity in which the great toe deviates laterally (see Fig. 42-6C). There is a marked prominence of the medial aspect of the first metatarsophalangeal joint. There is also osseous enlargement (exostosis) of the medial side of the first metatarsal head, over which a bursa may form (secondary to pressure and inflammation). Acute bursitis symptoms include a reddened area, edema, and tenderness.

Factors contributing to bunion formation include heredity, ill-fitting shoes, osteoarthritis, and the gradual lengthening and widening of the foot associated with aging. Treatment depends on the patient's age, the degree of deformity, and the severity of symptoms. In uncomplicated cases, wearing a shoe that conforms to the shape of the foot, or that is molded to the foot to prevent pressure on the protruding portions, may be the only treatment needed. Corticosteroid injections control acute inflammation. In advanced cases, surgical removal of the exostosis and toe realignment may be required to improve function, appearance, and symptoms.

Pes Cavus

Pes cavus (claw foot) refers to a foot with an abnormally high arch and a fixed equines deformity of the forefoot (see Fig. 42-6B). The shortening of the foot and increased pressure produce calluses on the metatarsal area and on the dorsum of the foot. Charcot-Marie-Tooth disease (a peripheral neuromuscular disease associated with a familial degenerative disorder), diabetes, and tertiary syphilis are common causes. Exercises are prescribed to manipulate the forefoot into dorsiflexion and relax the toes. Orthotic devices alleviate pain and can protect the foot (Burns, Landorf, Ryan, et al., 2010). In severe cases, arthrodesis (fusion) is performed to reshape and stabilize the foot.

Morton's Neuroma

Morton's neuroma (plantar digital neuroma, neurofibroma) is a swelling of the third (lateral) branch of the median plantar nerve (see Fig. 42-6D). Microscopically, digital artery changes cause an ischemia of the nerve within the third intermetatarsal (web) space. The result is a throbbing, burning pain in the foot that is usually relieved with rest and massage.

Conservative treatment consists of inserting innersoles and metatarsal pads designed to spread the metatarsal heads and balance the foot posture. Local injections of a corticosteroid and a local anesthetic may provide relief. If these fail, surgical excision of the neuroma is necessary. Pain relief and loss of sensation are immediate and permanent with surgery. The risk of falls is increased because of the loss of all sensation (ACFAS, 2010).

Pes Planus

Pes planus (flatfoot) is a common disorder in which the longitudinal arch of the foot is diminished. It may be caused by congenital abnormalities or associated with bone or ligament injury, excessive weight, muscle fatigue, poorly fitting shoes, or arthritis. Signs and symptoms include a burning sensation, fatigue, clumsy gait, edema, and pain. Exercises to strengthen the muscles and to improve posture and walking habits are helpful. A number of foot orthoses are available to give the foot additional support (Burns et al., 2010).

Nursing Management of the Patient Undergoing Foot Surgery

Surgery of the foot may be necessary because of various conditions, including neuromas and foot deformities (bunion, hammer toe, clawfoot) (Chart 42-6). Generally, foot surgery is performed on an outpatient basis. Before surgery, the nurse assesses the patient's gait and balance, as well as the neurovascular status of the foot. Additionally, the nurse considers the availability of assistance at home and the structural characteristics of the home in planning for care during the days after surgery.

Postoperative and home care follows the same principles as discussed earlier for hand surgery (Chart 42-5). After surgery, neurovascular assessment of the exposed toes (every 1 to 2 hours for the first 24 hours) is essential to monitor the function of the nerves and the perfusion of the tissues. If the patient is discharged within several hours after the surgery, the nurse educates the patient and family about how to assess for edema and neurovascular status (circulation, motion, sensation). Compromised neurovascular function can increase the patient's pain (see Chart 40-3 in Chapter 40).

Pain experienced by patients who undergo foot surgery is related to inflammation and edema. Formation of a hematoma may contribute to the discomfort. To control the anticipated edema, the foot should be elevated on several pillows when the patient is sitting or lying. Support of the entire limb under the knee is preferable. Ice packs applied intermittently to the surgical area during the first 24 to 48 hours may be prescribed to control edema and provide some pain relief. As activity increases, the patient may find that dependent positioning of the foot is uncomfortable. Simply elevating the foot often relieves the discomfort. Oral analgesic agents may be used to control the pain. The nurse instructs the patient and family about appropriate use of these medications.

After surgery, the patient will have a bulky dressing on the foot, protected by a light cast or a special protective boot. Limits for weight bearing on the foot will be prescribed by the surgeon. Some patients are allowed to walk on the heel and progress to weight bearing as tolerated; other patients are restricted to non–weight-bearing activities. Assistive devices (e.g., crutches, walker) may be needed. The choice of the devices depends on the patient's general condition and balance and on the weight-bearing prescription. Safe use of the assistive devices must be ensured through adequate patient education and practice before discharge (see Chapter 10). Strategies to move around the house safely while using assistive devices are also discussed with the patient. As healing progresses, the patient gradually resumes ambulation within

Chart 42-6 HOME CARE CHECKLIST
Osteoporosis

At the completion of home care education, the patient or caregiver will be able to:	PATIENT	CAREGIVER
Adolescents and Young Adults		
• List risk factors for osteoporosis.	✔	✔
• Identify calcium- and vitamin D–rich foods.	✔	
• Consume diet with adequate calcium (1,000–1,300 mg/day) and vitamin D.	✔	
• Engage in weight-bearing exercise daily.	✔	
• Modify lifestyle choices—avoid smoking, alcohol, caffeine, and carbonated beverages.	✔	
Menopausal and Postmenopausal Women (in addition to above)		
• Discuss calcium supplements.	✔	
• Engage in exercise that improves balance to reduce risk of falls.	✔	
• Demonstrate good body mechanics.	✔	
• Discuss pharmacologic agents to maintain and enhance bone mass.	✔	
• Review concurrent medical conditions and medications with primary provider to identify factors that contribute to bone mass loss.	✔	✔
• Assess home environment for hazards contributing to falls.	✔	✔
Men (in addition to above)		
• List risk factors associated with osteoporosis in men, including medications (e.g., corticosteroids, antiseizure medications, aluminum-containing antacids), chronic diseases (e.g., kidney, lung, gastrointestinal), and undiagnosed low testosterone levels.	✔	✔
• Participate in screening for osteoporosis.	✔	
• Talk with primary provider about the use of medications (e.g., alendronate) to enhance bone mass or to correct testosterone deficiency.	✔	

prescribed limits. The nurse emphasizes compliance with the therapeutic regimen.

The immobility of lower extremity surgery increases the risk of venous thromboembolism (VTE) development. (See Chapter 30 for VTE risk assessment and treatment.) Other postoperative complications may include limited range of motion, paresthesias, tendon injury, and recurrence of deformity (ACFAS, 2009). In addition, if percutaneous pins were used to hold bones in position, these pins may serve as potential sites of infection. Care must be taken to protect the surgical wound from dirt and moisture. When bathing, the patient can secure a plastic bag over the dressing to prevent it from getting wet. Patient instructions concerning aseptic wound care and pin care may be necessary. (See Chapter 41 for further discussion on pin care and infection prophylaxis.)

Metabolic Bone Disorders

Osteoporosis

Osteoporosis is the most prevalent bone disease in the world. More than 1.5 million osteoporotic fractures occur every year. Fractures requiring hospitalization have risen significantly over the past two decades (National Osteoporosis Foundation [NOF], 2013). More than 10 million Americans have osteoporosis, and an additional 33.6 million have **osteopenia** (i.e., low bone mineral density [BMD])—the precursor to osteoporosis. The consequence of osteoporosis is bone fracture. It is projected that one of every two Caucasian women and one of every five men will have an osteoporosis-related fracture at some point in their lives (NOF, 2013).

Prevention

Peak adult bone mass is achieved between the ages of 18 and 25 years in both women and men and is affected by genetic factors, nutrition, physical activity, medications, endocrine status, and general health (NOF, 2013). Risk factors for osteoporosis and their effects on bone remodeling and maintenance are noted in Figure 42-7.

Primary osteoporosis occurs in women after menopause (usually by age 51) and in men later in life, but it is not merely a consequence of aging. Failure to develop optimal peak bone mass and low vitamin D levels contribute to the development of osteopenia without associated bone loss. Early identification of at-risk teenagers and young adults, increased calcium and vitamin D intake, participation in regular weight-bearing exercise, and modification of lifestyle (e.g., reduced use of caffeine, cigarettes, carbonated soft drinks, and alcohol) are interventions that decrease the risk of fractures and associated disability later in life (NOF, 2013).

Secondary osteoporosis is the result of medications or diseases that affect bone metabolism. Men are more likely than women to have secondary causes of osteoporosis, including the use of corticosteroids (especially if they receive doses in

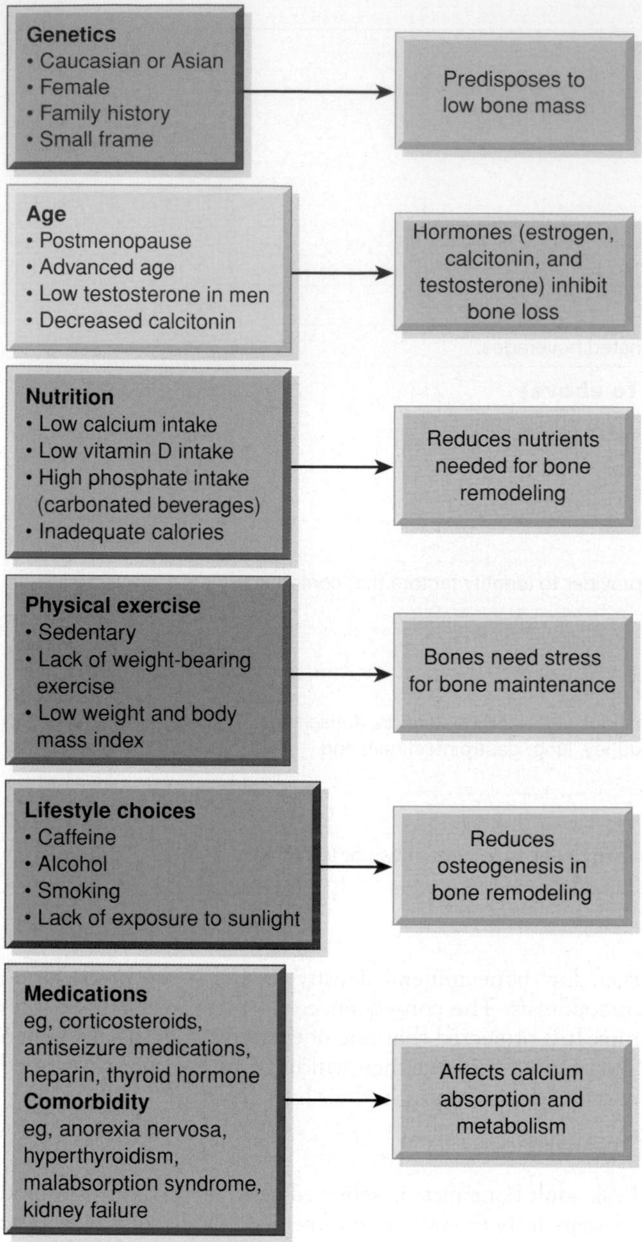
Genetics
• Caucasian or Asian
• Female
• Family history
• Small frame
Predisposes to low bone mass
Age
• Postmenopause
• Advanced age
• Low testosterone in men
• Decreased calcitonin
Hormones (estrogen, calcitonin, and testosterone) inhibit bone loss
Nutrition
• Low calcium intake
• Low vitamin D intake
• High phosphate intake (carbonated beverages)
• Inadequate calories
Reduces nutrients needed for bone remodeling
Physical exercise
• Sedentary
• Lack of weight-bearing exercise
• Low weight and body mass index
Bones need stress for bone maintenance
Lifestyle choices
• Caffeine
• Alcohol
• Smoking
• Lack of exposure to sunlight
Reduces osteogenesis in bone remodeling
Medications
eg, corticosteroids, antiseizure medications, heparin, thyroid hormone
Comorbidity
eg, anorexia nervosa, hyperthyroidism, malabsorption syndrome, kidney failure
Affects calcium absorption and metabolism

FIGURE 42-7 • Risk factors for osteoporosis and the effects of these factors on bone.

excess of 5 mg of prednisone daily for more than 3 months) and excessive alcohol intake. Specific disease states (e.g., celiac disease, hypogonadism) and medications (e.g., antiseizure agents, thyroid replacement agents, medroxyprogesterone [Depo-Provera], selective serotonin reuptake inhibitors, and proton pump inhibitors) that place patients at risk need to be identified and therapies instituted to reverse the development of osteoporosis (NOF, 2013). The degree of bone loss is related to the duration of medication therapy. When the drugs are discontinued or the metabolic problem is corrected, the progression is halted but restoration of lost bone mass may not occur.

Gerontologic Considerations

The prevalence of osteoporosis in women older than 80 years is 50%. The average 75-year-old woman has lost 25% of her cortical bone and 40% of her trabecular bone. Most residents of long-term care facilities have a low BMD and are at risk for bone fracture. One third of all hip fractures occur among men, and such fractures tend to be more lethal than those seen in women. It is estimated that the number of hip fractures and their associated costs will at least double by the year 2040 because of the projected aging of the U.S. population (NOF, 2013).

Routine vertebral fracture screenings are not recommended. However, 80% to 90% of these fractures can be seen incidentally on chest x-rays taken for other purposes. An older adult with a vertebral fracture has a 20-fold increased risk for a future osteoporotic fracture (Ensrud, 2011).

Older people absorb dietary calcium less efficiently and excrete it more readily through their kidneys. Postmenopausal women and older adults need to consume approximately 1,200 mg of daily calcium; quantities larger than this may place patients at heightened risk of renal calculi or cardiovascular disease (United States Preventive Services Task Force [USPSTF], 2013).

Pathophysiology

Osteoporosis is characterized by reduced bone mass, deterioration of bone matrix, and diminished bone architectural strength. Normal homeostatic bone turnover is altered; the rate of bone resorption that is maintained by osteoclasts is greater than the rate of bone formation that is maintained by osteoblasts, resulting in a reduced total bone mass. The bones become progressively porous, brittle, and fragile; they fracture easily under stresses that would not break normal bone. This occurs most commonly as compression fractures (Fig. 42-8) of the thoracic and lumbar spine, hip fractures, and Colles fractures of the wrist. These fractures may be the first clinical manifestation of osteoporosis (NOF, 2010).

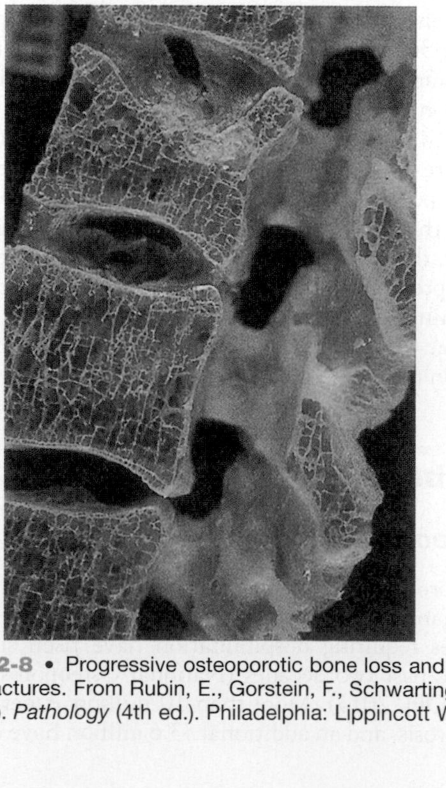

FIGURE 42-8 • Progressive osteoporotic bone loss and compression fractures. From Rubin, E., Gorstein, F., Schwarting, R., et al. (2004). *Pathology* (4th ed.). Philadelphia: Lippincott Williams & Wilkins.

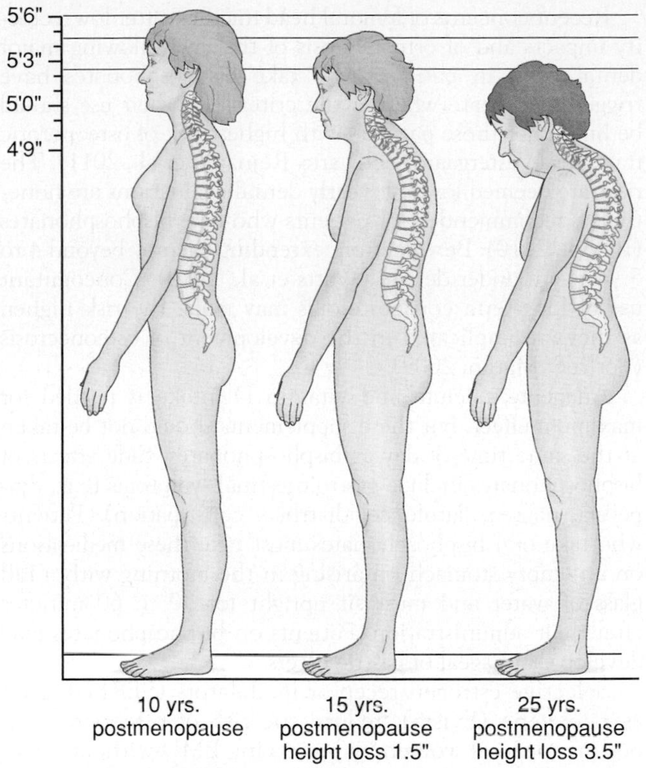
5'6"
5'3"
5'0"
4'9"
10 yrs. postmenopause
15 yrs. postmenopause height loss 1.5"
25 yrs. postmenopause height loss 3.5"

FIGURE 42-9 • Typical loss of height associated with osteoporosis and aging.

The gradual collapse of a vertebra may be asymptomatic. With the development of kyphosis (i.e., dowager's hump), there is an associated loss of height (Fig. 42-9). The postural changes result in relaxation of the abdominal muscles and a protruding abdomen. The deformity may also produce pulmonary insufficiency and increase risk for falls related to balance issues.

Age-related loss begins soon after the peak bone mass is achieved (i.e., in the fourth decade). Calcitonin, which inhibits bone resorption and promotes bone formation, is decreased. Estrogen, which inhibits bone breakdown, also decreases with aging. On the other hand, parathyroid hormone (PTH) increases with aging, thus increasing bone turnover and resorption. The consequence of these changes is net loss of bone mass over time.

The withdrawal of estrogens at menopause or with oophorectomy causes an accelerated bone resorption within the first 5 years after cessation of menses. Most women lose 10% of their bone mass. More than half of all women older than 50 years show evidence of osteopenia (Watts, Bliezikian, Camacho, et al., 2010).

Risk Factors

Small-framed, Asian, and Caucasian women are at greatest risk for osteoporosis (see Fig. 42-7). African American women generally are less susceptible to osteoporosis. Men have a greater peak bone mass and do not experience a sudden midlife estrogen reduction; as a result, osteoporosis occurs in men at a lower rate and at an older age (about one decade later) (Ducharme, 2010). It is believed that testosterone and estrogen are important in achieving and maintaining bone mass in men. Risk for osteoporosis rises with increasing age for both genders (Chart 42-6).

Nutritional factors contribute to the development of osteoporosis. A diet that includes adequate calories and nutrients needed to maintain bone, calcium, and vitamin D must be consumed. Patients who have had bariatric surgery are at increased risk for osteoporosis as the duodenum is bypassed, which is the primary site for absorption of calcium. Patients who have gastrointestinal diseases that cause malabsorption (e.g., celiac disease, alcoholism) may benefit from additional magnesium supplements (Watts et al., 2010).

Bone formation is enhanced by the stress of weight and muscle activity. When immobilized by casts, general inactivity, paralysis, or other disability, the bone is resorbed faster than it is formed, and osteoporosis results (Porth & Matfin, 2009). Resistance and impact exercises are most beneficial in developing and maintaining bone mass. Obesity has been recently recognized as a risk factor. Immobility contributes to the development of osteoporosis. Most overweight individuals are less active and may not consume key nutrients, which results in more risk for bone demineralization than the benefit of bone stress, leading to increased bone density, from extra weight (Bredella, Torianni, Ghomi, et al., 2011).

Assessment and Diagnostic Findings

Osteoporosis may be undetectable on routine x-rays until there has been 25% to 40% demineralization, resulting in radiolucency of the bones (Ensrud, 2011). When the vertebrae collapse, the thoracic vertebrae become wedge shaped and the lumbar vertebrae become biconcave. Osteoporosis is diagnosed by dual-energy x-ray absorptiometry (DEXA), which provides information about BMD at the spine and hip (see Chapter 40). The DEXA scan data are analyzed and reported as T-scores (the number of standard deviations above or below the average BMD value for a young, healthy Caucasian woman). T-scores that are gender specific for men are not yet available (Ducharme, 2010).

Baseline DEXA testing is recommended for all women older than 65 years, for postmenopausal women older than 50 years with osteoporosis risk factors, and for all people who have had a fracture thought to occur as a consequence of osteoporosis (NOF, 2013). BMD studies are also useful in assessing response to therapy and are recommended 3 months post any osteoporotic fracture. There is no evidence to support basic screening of men or for the optimal time interval to repeat studies following a normal baseline report.

Fracture risk can be estimated using the World Health Organization (WHO) Fracture Risk Assessment Tool (FRAX). Treatment is now reserved for those with a 10-year risk of more than 3% for hip fracture or 20% risk for other major fractures. The risks scores are based on BMD, personal and family history of fractures, body mass index, gender, age, and secondary factors such as medication use, smoking, and history of rheumatoid disease (NOF, 2013). Diabetes is not used in FRAX calculations, but recent evidence indicates a higher 10-year risk probability for osteoporotic hip fractures among patients with diabetes (Giangregorio, Lix, Johansson, et al., 2012).

Laboratory studies (e.g., serum calcium, serum phosphate, serum alkaline phosphatase [ALP], urine calcium excretion, urinary hydroxyproline excretion, hematocrit, erythrocyte sedimentation rate [ESR]), and x-ray studies are used to exclude other possible disorders (e.g., multiple myeloma, osteomalacia, hyperparathyroidism, malignancy) that contribute to bone loss. In men, low testosterone levels may be part of the cause (Ducharme, 2010).

Medical Management

A diet rich in calcium and vitamin D throughout life, with an increased calcium intake during adolescence and the middle years, protects against skeletal demineralization. Such a diet includes three glasses of skim vitamin D–enriched milk or other foods high in calcium (e.g., cheese and other dairy products, steamed broccoli, canned salmon with bones) daily. A cup of milk or calcium-fortified orange juice contains about 300 mg of calcium. The recommended adequate intake level of calcium for adults is 1,000 to 1,300 mg daily (USPSTF, 2013). The recommended vitamin D intake for adults 50 years and older is 800 to 1,000 IU daily (Ducharme, 2010).

Regular weight-bearing exercise promotes bone formation. Recommendations include 20 to 30 minutes of aerobic, bone-stressing exercise daily (e.g., not swimming). Weight training stimulates an increase in BMD. In addition, exercise improves balance, reducing the incidence of falls and fractures.

Pharmacologic Therapy

The first-line medications used to treat and prevent osteoporosis include calcium and vitamin D supplements and bisphosphonates for both genders (Ducharme, 2010). To ensure adequate calcium intake, a calcium supplement (e.g., Caltrate, Citracal) with vitamin D may be prescribed and taken with meals or with a beverage high in vitamin C to promote absorption. The recommended daily dose should be split and not taken as a single dose. Common side effects of calcium supplements are abdominal distention and constipation. Calcium from foods is better absorbed, but calcium supplements may be necessary for patients who are lactose intolerant.

Meta-analysis findings demonstrate that vitamin D taken alone is not effective in primary prevention of fractures (Avenell, Gillespie, Gillespie, et al., 2009). Patients who have vitamin D deficiency do benefit from taking ergocalciferol (vitamin D_2). Monitoring to keep levels greater than 30 mg/mL is recommended (Avenell et al., 2009).

Bisphosphonates that include daily or weekly oral preparations of alendronate (Fosamax), risedronate (Actonel), monthly oral preparations of ibandronate (Boniva), or yearly intravenous (IV) infusions of zoledronic acid (Reclast) all increase bone mass and decrease bone loss by inhibiting osteoclast function (Watts et al., 2010). However, primary prevention of fractures is limited mostly to vertebral bones; hip fracture prevention is less evident (Wells, Cranney, Peterson, et al., 2011). Longitudinal studies indicate an added 5 years of life occurs when fractures are prevented. Only alendronate and risedronate are approved by the U.S. Food and Drug Administration (FDA) for use in men (Ducharme, 2010).

Recent concerns of femoral head fractures after low-velocity impacts and of osteonecrosis of the jaw following major dental work in patients who take bisphosphonates have triggered a debate whether the criteria for their use should be limited to those patients with highest risk of osteoporotic fractures (Vestergaard, Schwartz, Rejnank, et al., 2011). The risks are deemed low, but yearly dental evaluations are nonetheless recommended for patients who take bisphosphonates (Brandi, 2010). Benefits from extending therapy beyond 4 to 5 years are under debate (Watts et al., 2010). Concomitant use of long-term corticosteroids may make the risk higher, as they are implicated in the development of osteonecrosis (Porth & Matfin, 2009).

Adequate calcium and vitamin D intake is needed for maximum effect, but these supplements should not be taken at the same time of day as bisphosphonates. Side effects of bisphosphonates include gastrointestinal symptoms (e.g., dyspepsia, nausea, flatulence, diarrhea, constipation). Patients who take oral bisphosphonates must take these medications on an empty stomach on arising in the morning with a full glass of water and must sit upright for 30 to 60 minutes after their administration. Patients on bisphosphonates may develop esophageal or gastric ulcers.

Selective estrogen receptor modulators (SERMs), such as raloxifene (Evista), reduce the risk of osteoporosis in postmenopausal women by preserving BMD without estrogenic effects on the uterus. They are indicated for both prevention and treatment of osteoporosis. SERMS are contraindicated in women with a history of VTE (Watts et al., 2010).

Teriparatide (Forteo) is an anabolic agent used for both genders that is administered subcutaneously once daily for 2 years. It is reserved for those with high fracture risk and long-term corticosteroid use. As a recombinant PTH, it stimulates osteoblasts to build bone matrix and facilitates overall calcium absorption. This drug is implicated in the development of osteosarcomas in laboratory animals (Watts et al., 2010). Therefore, the FDA cautions that it cannot be used in patients who have had radiation therapy or have other metabolic bone diseases.

The newest class of nuclear factor-kappa B ligand (RANKL) targeted therapies are indicated for persons who cannot tolerate or do not respond fully to the other medications and remain at high risk. These monoclonal antibody medications (e.g., denosumab [Prolia]) increase BMD and reduce the porosity of cortical bone by inhibiting the effects of tissue necrosis factor on the surface of osteoclasts, thereby reducing their activity (Lipton & Goessl, 2011). Denosumab has also demonstrated effects on reducing vertebral and wrist fractures.

Fracture Management

Fractures of the hip that occur as a consequence of osteoporosis are managed surgically by joint replacement or by closed or open reduction with internal fixation (e.g., hip pinning) as described in Chapters 41 and 43, respectively. Management of Colles fractures is also described in Chapter 43.

Osteoporotic compression fractures of the vertebrae are managed conservatively. Patients with these findings should be referred to an osteoporosis specialist. Most patients who experience these fractures are asymptomatic and do not require acute care management; for those who experience pain, acute care management is indicated as outlined in the following Nursing Process section. Percutaneous vertebroplasty or kyphoplasty (injection of polymethylmethacrylate [PMMA] bone cement into the fractured vertebra, followed by inflation of a pressurized balloon to restore the shape of the affected vertebra) can provide rapid relief of acute pain and improve quality of life (NOF, 2013). Patients who have not responded to first-line approaches to the treatment of vertebral compression fracture can be considered for these procedures. They are contraindicated in the presence of infection, multiple old fractures, and certain coagulopathies. The benefits of these expensive interventions are now questioned because of their association with increased rates of future fractures related to altered spinal mechanics (Ensrud, 2011).

NURSING PROCESS

The Patient With a Spontaneous Vertebral Fracture Related to Osteoporosis

Assessment

Recognition of risks and problems associated with osteoporosis form the basis for nursing assessment. The health history focuses on family history, previous fractures, dietary consumption of calcium, exercise patterns, onset of menopause, and the use of certain medications (e.g., corticosteroids), as well as alcohol, smoking, and caffeine intake. Any symptoms the patient is experiencing, such as back pain, constipation, or altered body image, are explored.

Physical examination may disclose localized pain, kyphosis of the thoracic spine, or shortened stature. Problems in mobility and breathing may exist as a result of changes in posture and weakened muscles.

Nursing Diagnoses

Based on the assessment data, major nursing diagnoses may include the following:

- Deficient knowledge about the osteoporotic process and treatment regimen
- Acute pain related to fracture and muscle spasm
- Risk for constipation related to immobility or development of ileus (intestinal obstruction)
- Risk for injury: additional fractures related to osteoporosis

Planning and Goals

The major goals for the patient may include knowledge about osteoporosis and the treatment regimen, relief of pain, improved bowel elimination, and absence of additional fractures.

Nursing Interventions

Promoting Understanding of Osteoporosis and the Treatment Regimen

Patient education focuses on factors influencing the development of osteoporosis, interventions to arrest or slow the process, and measures to relieve symptoms. The nurse emphasizes that people of any age need sufficient calcium, vitamin D, and weight-bearing exercise to slow the progression of osteoporosis. Patient education related to medication therapy as described previously is important. Patients must understand that having one fracture increases the probability of sustaining another (NOF, 2013).

Relieving Pain

Relief of back pain resulting from compression fracture may be accomplished by short periods of resting in bed in a supine or side-lying position. The mattress should be supportive. Knee flexion increases comfort by relaxing back muscles. Intermittent local heat and backrubs promote muscle relaxation. The nurse instructs the patient to move the trunk as a unit and to avoid twisting. When the patient is assisted out of bed, a trunk orthosis (e.g., lumbosacral corset) may be worn for temporary support and immobilization, although such a device is frequently uncomfortable and is poorly tolerated by many older adults (NOF, 2013). The patient gradually resumes activities as pain diminishes.

Improving Bowel Elimination

Constipation is a problem related to immobility and medications. Early institution of a high-fiber diet, increased fluids, and the use of prescribed stool softeners help prevent or minimize constipation. If the vertebral collapse involves the T10–L2 vertebrae, the patient may develop a paralytic ileus. The nurse therefore monitors the patient's intake, bowel sounds, and bowel activity.

Preventing Injury

Physical activity is essential to strengthen muscles, improve balance, prevent disuse atrophy, and retard progressive bone demineralization. Isometric exercises can strengthen trunk muscles. The nurse encourages walking, good body mechanics, and good posture. Daily weight-bearing activity, preferably outdoors in the sunshine to enhance the body's ability to produce vitamin D, is encouraged. Sudden bending, jarring, and strenuous lifting are avoided.

Gerontologic Considerations. Older adults fall frequently as a result of environmental hazards, diminished senses and cardiovascular responses, and responses to medications. The patient and family need to be included in planning for care and preventive management regimens. For example, the home environment should be assessed for elimination of potential hazards (e.g., scatter rugs, cluttered rooms and stairwells, toys on the floor, pets underfoot). A safe environment can then be created (e.g., well-lighted staircases with secure hand rails, grab bars in the bathroom, properly fitting footwear).

Evaluation

Expected patient outcomes may include:

1. Acquires knowledge about osteoporosis and the treatment regimen
   a. States relationship of calcium and vitamin D intake and exercise to bone mass
   b. Consumes adequate dietary calcium and vitamin D
   c. Takes prescribed medications, following instructions for administration
   d. Increases level of exercise
   e. Adheres to prescribed screening and monitoring procedures

2. Achieves pain relief
    a. Experiences pain relief at rest
    b. Experiences minimal discomfort during ADLs
    c. Demonstrates diminished tenderness at fracture site
3. Demonstrates usual pattern of bowel elimination
    a. Has active bowel sounds
    b. Reports regular pattern of bowel movements
4. Experiences no new fractures
    a. Maintains good posture
    b. Uses good body mechanics
    c. Engages in weight-bearing exercises (walks daily)
    d. Creates a safe home environment
    e. Accepts assistance and supervision as needed

Osteomalacia

Osteomalacia is a metabolic bone disease characterized by inadequate mineralization of bone. As a result, the skeleton softens and weakens, causing pain, tenderness to touch, bowing of the bones, and pathologic fractures. On physical examination, skeletal deformities (spinal kyphosis and bowed legs) give patients an unusual appearance and a waddling gait. These patients may be uncomfortable with their appearance and are at risk for falls and pathologic fractures, particularly of the distal radius and the proximal femur (Porth & Matfin, 2009).

Pathophysiology

The major defect in osteomalacia is a deficiency of activated vitamin D, which promotes calcium absorption from the gastrointestinal tract and facilitates mineralization of bone. The supply of calcium and phosphate in the extracellular fluid is low and does not move to calcification sites in bones.

Osteomalacia may result from failed calcium absorption or from excessive loss of calcium from the body (e.g., kidney failure). Gastrointestinal disorders (e.g., celiac disease, chronic biliary tract obstruction, chronic pancreatitis, small bowel resection) in which fats are inadequately absorbed are likely to produce osteomalacia through loss of vitamin D (along with other fat-soluble vitamins) and calcium, the latter being excreted in the feces with fatty acids. In addition, liver and kidney diseases can produce a lack of vitamin D because these are the organs that convert vitamin D to its active form.

Severe renal insufficiency results in acidosis. The body uses available calcium to combat the acidosis, and PTH stimulates the release of skeletal calcium in an attempt to reestablish a physiologic pH. During this continual drain of skeletal calcium, bony fibrosis occurs, and bony cysts form. Chronic glomerulonephritis, obstructive uropathies, and heavy metal poisoning result in a reduced serum phosphate level and demineralization of bone.

Hyperparathyroidism leads to skeletal decalcification and thus to osteomalacia by increasing phosphate excretion in the urine. Prolonged use of antiseizure medication (e.g., phenytoin [Dilantin], phenobarbital) poses a risk of osteomalacia, as does insufficient vitamin D (dietary, sunlight).

Osteomalacia that results from malnutrition (deficiency in vitamin D often associated with poor intake of calcium) is a result of poverty, poor dietary habits, and lack of knowledge about nutrition. It occurs most frequently in parts of the world where vitamin D is not added to food, where dietary deficiencies exist, and where sunlight is rare (Porth & Matfin, 2009).

Gerontologic Considerations

A nutritious diet is particularly important in older adults. Adequate intake of calcium and vitamin D is promoted. Because sunlight is necessary for synthesizing vitamin D, patients should be encouraged to spend some time in the sun. Prevention, identification, and management of osteomalacia in older adults are essential to reduce the incidence of fractures. When osteomalacia is combined with osteoporosis, the risk of fracture increases.

Assessment and Diagnostic Findings

On x-ray studies, generalized demineralization of bone is evident. Studies of the vertebrae may show a compression fracture with indistinct vertebral end-plates. Laboratory studies show low serum calcium and phosphorus levels and a moderately elevated ALP. Urine excretion of calcium and creatinine is low. Bone biopsy demonstrates an increased amount of osteoid, a demineralized, cartilaginous bone matrix that is sometimes referred to as pre-bone.

Medical Management

Physical, psychological, and pharmaceutical measures are used to reduce the patient's discomfort and pain. If the underlying cause of osteomalacia is corrected, the disorder may resolve. If kidney disease prevents activation of absorbed vitamin D, then supplementation requires the activated form (calcitriol). If osteomalacia is caused by malabsorption, increased doses of vitamin D, along with supplemental calcium, are usually prescribed. Exposure to sunlight may be recommended; ultraviolet radiation transforms a cholesterol substance (7-dehydrocholesterol) present in the skin into vitamin D.

If osteomalacia is dietary in origin, the interventions are akin to those discussed previously in the discussion on osteoporosis. Long-term monitoring of the patient is appropriate to ensure stabilization or reversal of osteomalacia. Some persistent orthopedic deformities may need to be treated with braces or surgery (e.g., osteotomy may be performed to correct long bone deformity).

Paget's Disease of the Bone

Paget's disease (osteitis deformans) is a disorder of localized rapid bone turnover, most commonly affecting the skull, femur, tibia, pelvic bones, and vertebrae. The disease occurs in about 2% to 3% of the population older than 50 years. The incidence is slightly greater in aging men than in women. A family history has been noted, with siblings often developing the disease (Falchetti, Marini, Mosi, et al., 2010). The cause of Paget's disease is not known (Brandi, 2010).

Pathophysiology

In Paget's disease, a primary proliferation of osteoclasts occurs, which induces bone resorption. This is followed by a compensatory increase in osteoblastic activity that replaces the bone. As bone turnover continues, a classic mosaic (disorganized) pattern of bone develops. Because the diseased bone is highly vascularized and structurally weak, pathologic fractures occur. Structural bowing of the legs causes malalignment of the hip, knee, and ankle joints, which contributes to the development of arthritis and back and joint pain (Brandi, 2010).

Clinical Manifestations

Paget's disease is insidious. Some patients do not experience symptoms but only have skeletal deformity. The condition is most frequently identified on x-ray studies performed during a workup for another problem. Sclerotic changes and cortical thickening of the long bones occur.

In most patients, skeletal deformity involves the skull. The skull may thicken, and the patient may report that a hat no longer fits. In some cases, the cranium, but not the face, is enlarged. This gives the face a small, triangular appearance. Most patients with skull involvement have impaired hearing from cranial nerve compression and dysfunction. Other cranial nerves may also be similarly affected.

The femurs and tibiae tend to bow, producing a waddling gait. The spine is bent forward and is rigid; the chin rests on the chest. The thorax becomes immobile during respiration. The trunk is flexed on the legs to maintain balance and the arms are bent outward and forward, appearing long in relation to the shortened trunk (Porth & Matfin, 2009).

Tenderness and warmth over bones may be noted due to increased bone vascularity. Patients with large, highly vascular lesions may develop high-output cardiac failure due to the increased vascular bed and metabolic demands (Porth & Matfin, 2009). The pain is mild to moderate, deep, and aching; it increases with weight bearing. Pain and discomfort may precede skeletal deformities of Paget's disease by years and are often wrongly attributed by the patient to old age or arthritis (Brandi, 2010).

Assessment and Diagnostic Findings

Elevated serum ALP concentration and urinary hydroxyproline excretion reflect increased osteoblastic activity. Higher values suggest more active disease. Patients with Paget's disease have normal blood calcium levels. X-rays confirm the diagnosis of Paget's disease by revealing local areas of demineralization and bone overgrowth in the characteristic mosaic patterns. Bone scans demonstrate the extent of the disease. Bone biopsy may aid in the differential diagnosis with other bone diseases (Porth & Matfin, 2009).

Medical Management

Pain usually responds to NSAIDs. Gait problems from bowing of the legs are managed with walking aids, shoe lifts, and physical therapy. Weight is controlled to reduce stress on weakened bones and misaligned joints. Asymptomatic patients may be managed with diets adequate in calcium and vitamin D and periodic monitoring.

Fractures, arthritis, and hearing loss are complications of Paget's disease. Fractures are managed according to location. Healing occurs if fracture reduction, immobilization, and stability are adequate. Severe degenerative arthritis may require total joint replacement; however, the afflicted "soft" bones do not make ideal surgical sites and are thus prone to complications. Loss of hearing is managed with hearing aids and communication techniques used with hearing-impaired people (e.g., speech reading, body language) (see Chapter 64).

Pharmacologic Therapy

Patients with moderate to severe disease may benefit from specific antiosteoclastic therapy. These medications reduce bone turnover, reverse the course of the disease, relieve pain, and improve mobility.

Bisphosphonates are the cornerstone of Paget therapy in that they stabilize the rapid bone turnover. Their use may not suppress all Paget symptoms (Brandi, 2010), but they reduce serum ALP and urinary hydroxyproline levels. (See earlier discussion on bisphosphonates).

Plicamycin (Mithracin), a cytotoxic antibiotic, may be used to control the disease. This medication is reserved for severely affected patients with neurologic compromise and for those whose disease is resistant to other therapy. This medication has dramatic effects on pain reduction and on serum calcium, ALP, and urinary hydroxyproline levels; however, there are significant side effects. It is administered by IV infusion; hepatic, kidney, and bone marrow function must be monitored during therapy. Clinical remissions may continue for months after the medication is discontinued.

Gerontologic Considerations

Because Paget's disease tends to affect older adults, patients and their families and caregivers should be educated about how to compensate for altered musculoskeletal functioning with an emphasis on the risk of falls. The home environment is assessed for safety to prevent falls and to reduce the risk of fracture. Strategies for coping with a chronic health problem and its effect on quality of life need to be developed. If age-related hearing loss is exacerbated by Paget's disease, alternative communication devices (e.g., text telephone, telecommunication device for the deaf) and home safety alarms may be indicated.

Musculoskeletal Infections

Osteomyelitis

Osteomyelitis is an infection of the bone that results in inflammation, necrosis, and formation of new bone. Osteomyelitis is classified as:

- Hematogenous osteomyelitis (i.e., due to bloodborne spread of infection)
- Contiguous-focus osteomyelitis, from contamination from bone surgery, open fracture, or traumatic injury (e.g., gunshot wound)

- Osteomyelitis with vascular insufficiency, seen most commonly among patients with diabetes and peripheral vascular disease, most commonly affecting the feet (Porth & Matfin, 2009)

Patients who are at high risk for osteomyelitis include older adults and those who are poorly nourished or obese. Other patients at risk include those with impaired immune systems, those with chronic illnesses (e.g., diabetes, rheumatoid arthritis), those receiving long-term corticosteroid therapy or immunosuppressive agents, and IV drug users.

Postoperative surgical wound infections typically occur within 30 days after surgery. They are classified as incisional (superficial, located above the deep fascia layer) or deep (involving tissue beneath the deep fascia). If an implant has been used, deep postoperative infections may occur within a year. Osteomyelitis may become chronic and may affect the patient's quality of life.

Pathophysiology

More than 50% of bone infections are caused by *Staphylococcus aureus* and increasingly of the variety that is methicillin resistant (i.e., methicillin-resistant *Staphylococcus aureus* [MRSA]) (Miller & Kaplan, 2009). Other pathogens include the gram-positive organisms streptococci and enterococci, followed by gram-negative bacteria, including pseudomonas.

The initial response to infection is inflammation, increased vascularity, and edema. After 2 or 3 days, thrombosis of the local blood vessels occurs, resulting in ischemia with bone necrosis. The infection extends into the medullary cavity and under the periosteum and may spread into adjacent soft tissues and joints. Unless the infective process is treated promptly, a bone abscess forms. The resulting abscess cavity contains dead bone tissue (the **sequestrum**), which does not easily liquefy and drain. Therefore, the cavity cannot collapse and heal, as it does in soft tissue abscesses. New bone growth (the **involucrum**) forms and surrounds the sequestrum. Although healing appears to take place, a chronically infected sequestrum remains and produces recurring abscesses throughout the patient's life. This is referred to as chronic osteomyelitis.

Clinical Manifestations

When the infection is bloodborne, the onset is usually sudden, occurring often with the clinical and laboratory manifestations of sepsis (e.g., chills, high fever, rapid pulse, general malaise). The systemic symptoms at first may overshadow the local signs. As the infection extends through the cortex of the bone, it involves the periosteum and the soft tissues. The infected area becomes painful, swollen, and extremely tender. The patient may describe a constant, pulsating pain that intensifies with movement as a result of the pressure of the collecting purulent material (i.e., pus). When osteomyelitis occurs from spread of adjacent infection or from direct contamination, there are no manifestations of sepsis. The area is swollen, warm, painful, and tender to touch. The patient with chronic osteomyelitis presents with a nonhealing ulcer that overlies the infected bone with a connecting sinus that will intermittently and spontaneously drain pus (Conterno & da Silva Filho, 2009).

Assessment and Diagnostic Findings

In acute osteomyelitis, early x-ray findings demonstrate soft tissue edema. In about 2 to 3 weeks, areas of periosteal elevation and bone necrosis are evident. Radioisotope bone scans, particularly the isotope-labeled white blood cell (WBC) scan, and magnetic resonance imaging (MRI) help with early definitive diagnosis. Blood studies reveal leukocytosis and an elevated ESR. Wound and blood culture studies are performed, although they are only positive in 50% of cases. Therefore, treatment with antibiotics may be prescribed without definitively isolating the offending organism (Conterno & da Silva Filho, 2009).

With chronic osteomyelitis, large, irregular cavities, raised periosteum, sequestra, or dense bone formations are seen on x-ray. Bone scans may be performed to identify areas of infection. The ESR and the WBC count are usually normal. Anemia, associated with chronic infection, may be evident. Cultures of blood specimens and drainage from the sinus tract are frequently unreliable for isolating the organisms involved. Antibiotic therapy is many times prescribed presumptively (Conterno & da Silva Filho, 2009).

Prevention

Prevention of osteomyelitis is the goal. Elective orthopedic surgery should be postponed if the patient has a current infection (e.g., urinary tract infection, sore throat). During surgery, careful attention is paid to the surgical environment. Prophylactic antibiotics, administered to achieve adequate tissue levels at the time of surgery and for 24 hours after surgery, are helpful (Gillespie, 2009). Urinary catheters and drains are removed as soon as possible to decrease the incidence of hematogenous spread of infection.

Aseptic postoperative wound care reduces the incidence of superficial infections and osteomyelitis. Prompt management of soft tissue infections reduces extension of infection to the bone or hematogenous spread. When patients who have had joint replacement surgery undergo extensive dental procedures or other invasive procedures (e.g., cystoscopy), prophylactic antibiotics are recommended before the procedure (Little, Jacobson, & Lockhart, 2010).

Medical Management

The initial goal of therapy is to control and halt the infective process. General supportive measures (e.g., hydration, diet high in vitamins and protein, correction of anemia) are instituted. The area affected with osteomyelitis is immobilized to decrease discomfort and to prevent pathologic fracture of the weakened bone

Pharmacologic Therapy

Bone infections are more difficult to eradicate than soft tissue infections because the infected bone is mostly avascular and less accessible to the body's natural immune response. Because there is decreased penetration by medications, antibiotic therapy is longer term than with other infections; typically it continues for 3 to 6 weeks. After the infection appears to be controlled, the antibiotic may be administered orally. However, there is little evidence to support optimal length of therapy (Conterno & da Silva Filho, 2009).

Surgical Management

If the infection is chronic and does not respond to antibiotic therapy, surgical débridement is indicated. The infected bone is surgically exposed, the purulent and necrotic material is removed, and the area is irrigated with sterile saline solution. A sequestrectomy (removal of enough involucrum to enable the surgeon to remove the sequestrum) is performed. In many cases, sufficient bone is removed to convert a deep cavity into a shallow saucer (saucerization). All dead, infected bone and cartilage must be removed before permanent healing can occur. A closed suction irrigation system may be used to remove debris. Wound irrigation using sterile physiologic saline solution may be performed for a week.

The wound is either closed tightly to obliterate the dead space or packed and closed later by granulation or possibly by grafting. The débrided cavity may be packed with cancellous bone graft to stimulate healing. With a large defect, the cavity may be filled with a vascularized bone transfer or muscle flap (in which a muscle is moved from an adjacent area with blood supply intact). These microsurgery techniques enhance the blood supply. The improved blood supply facilitates bone healing and eradication of the infection. These surgical procedures may be staged over time to ensure healing. Because surgical débridement weakens the bone, internal fixation or external supportive devices may be needed to stabilize or support the bone to prevent pathologic fracture (Gillespie, 2009).

NURSING PROCESS

The Patient With Osteomyelitis

Assessment

The patient reports an acute onset of signs and symptoms (e.g., localized pain, edema, erythema, fever) or recurrent drainage of an infected sinus with associated pain, edema, and low-grade fever. The nurse assesses the patient for risk factors (e.g., older age, diabetes, long-term corticosteroid therapy) and for a history of previous injury, infection, or orthopedic surgery. The gait may be altered as the patient avoids pressure and movement of the area. In acute hematogenous osteomyelitis, the patient exhibits generalized weakness due to the systemic reaction to the infection.

Physical examination reveals an inflamed, markedly edematous, warm area that is tender. Purulent drainage may be noted. The patient has an elevated temperature. With chronic osteomyelitis, the temperature elevation may be minimal, occurring in the afternoon or evening.

Nursing Diagnoses

Based on the assessment data, nursing diagnoses may include the following:

- Acute pain related to inflammation and edema
- Impaired physical mobility related to pain, use of immobilization devices, and weight-bearing limitations
- Risk for infection: bone abscess formation
- Deficient knowledge related to the treatment regimen

Planning and Goals

The patient's goals may include relief of pain, improved physical mobility within therapeutic limitations, control and eradication of infection, and knowledge of the treatment regimen.

Nursing Interventions

Relieving Pain

The affected part may be immobilized with a splint to decrease pain and muscle spasm. The nurse monitors the skin and neurovascular status of the affected extremity. The wounds are frequently very painful, and the extremity must be handled with great care and gentleness. Elevation reduces swelling and associated discomfort. Pain is controlled with prescribed analgesic agents and other pain-reducing techniques.

Improving Physical Mobility

Treatment regimens restrict weight-bearing activity. The bone is weakened by the infective process and must be protected by avoidance of stress on the bone. The patient must understand the rationale for the activity restrictions. The joints above and below the affected part should be gently moved through their range of motion. The nurse encourages full participation in ADLs within the prescribed physical limitations to promote general well-being.

Controlling the Infectious Process

The nurse monitors the patient's response to antibiotic therapy and observes the IV access site for evidence of phlebitis, infection, or infiltration. With long-term, intensive antibiotic therapy, the nurse monitors the patient for signs of superinfection (e.g., oral or vaginal candidiasis, loose or foul-smelling stools). The nurse carefully monitors the patient for the development of additional sites that are painful or sudden increases in body temperature.

If surgery is necessary, the nurse takes measures to ensure adequate circulation to the affected area (wound suction to prevent fluid accumulation, elevation of the area to promote venous drainage, avoidance of pressure on the grafted area), to maintain needed immobility, and to ensure the patient's adherence to weight-bearing restrictions. The nurse changes dressings using aseptic technique to promote healing and to prevent cross-contamination.

Promoting Home and Community-Based Care

Educating Patients About Self-Care. The patient and family are educated about the importance of strictly adhering to the therapeutic regimen of antibiotics. Patients and families often need to learn to maintain and manage the IV access and IV administration equipment in the home. Education includes the medication name, dosage, frequency, administration rate, safe storage and handling, adverse reactions, and necessary laboratory monitoring. In addition, the nurse provides instruction on aseptic dressing and warm compress techniques.

Continuing Care. The patient must be medically stable and physically able and motivated to adhere strictly to the therapeutic regimen of antibiotic therapy. The home care environment needs to be conducive to the promotion of health and to the requirements of the therapeutic regimen.

If warranted, the nurse completes a home assessment to determine the patient's and family's abilities regarding continuation of the therapeutic regimen. If the patient's support system is questionable or if the patient lives alone, a home care

Chart 42-7 HOME CARE CHECKLIST
Osteomyelitis

At the completion of home care education, the patient or caregiver will be able to:	PATIENT	CAREGIVER
• Describe osteomyelitis.	✔	✔
• Relieve pain with pharmacologic and nonpharmacologic interventions.	✔	
• State weight-bearing and activity restrictions.	✔	✔
• Demonstrate safe use of ambulatory aids and assistive devices.	✔	
• Describe the use of prescribed medications.	✔	✔
• Comply with antibiotic regimen.	✔	
• Promote healing through aseptic dressing changes.	✔	✔
• Demonstrate proper wound care.	✔	✔
• Report signs and symptoms of continuing infection or superinfection.	✔	✔

nurse may be needed to assist with IV administration of the antibiotics, monitoring for response to the treatment and evaluation of signs and symptoms of superinfections, and adverse drug reactions. The nurse stresses the importance of follow-up health care appointments (Chart 42-7).

Evaluation

Expected patient outcomes may include:

1. Experiences pain relief
   a. Reports decreased pain at rest
   b. Experiences no tenderness at site of previous infection
   c. Experiences minimal discomfort with movement
2. Increases in safe physical mobility
   a. Participates in self-care activities within restrictions
   b. Maintains full function of unimpaired extremities
   c. Demonstrates safe use of immobilizing and assistive devices
   d. Modifies environment to promote safety and to avoid falls
3. Shows absence of infection
   a. Takes antibiotic as prescribed
   b. Reports normal temperature
   c. Exhibits no edema
   d. Reports absence of drainage
   e. Laboratory results indicate normal WBC count and ESR.
   f. Wound cultures are negative.
4. Adheres to therapeutic plan
   a. Takes medications as prescribed
   b. Protects weakened bones
   c. Demonstrates proper wound care
   d. Reports signs and symptoms of complications promptly
   e. Consumes a healthy diet
   f. Keeps follow-up health care appointments

Septic (Infectious) Arthritis

Joints can become infected through spread of pathogens from other parts of the body (hematogenous spread) or directly through trauma or surgical instrumentation, causing septic arthritis. People at greatest risk include older adults, particularly those older than 80 years; people with comorbid conditions such as diabetes, rheumatoid arthritis, skin infection, or alcoholism; and people with a history of a joint replacement or other joint surgery or IV drug abuse (Goldenberg & Sexton, 2012). *S. aureus* is the most common cause of joint infections in all age groups, followed by other gram-positive bacteria, including streptococci (Mathews, Weston, Jones, et al., 2010). Gonococcal infection may cause septic arthritis through hematogenous spread; when seen, it tends to afflict young adults and is a rare occurrence in all but developing nations (Mathews et al., 2010).

Single knee or hip joints are most commonly infected in patients with septic arthritis, although up to 20% of cases involve more than one joint (i.e., polyarticular disease) (Mathews et al., 2010). Prompt recognition and treatment of an infected joint are important because accumulating purulent material may result in chondrolysis (destruction of hyaline cartilage), and continued hematogenous spread may lead to sepsis and death (see Chapter 14 for further discussion of sepsis). The overall mortality rate is about 11% and approaches 50% in patients with polyarticular disease, which may be attributed to the fact that many patients with septic arthritis are older with significant comorbidity and/or immunocompromise (Mathews et al., 2010).

Clinical Manifestations

The patient with acute septic arthritis presents with a warm, painful, swollen joint with decreased range of motion. Systemic chills, fever, and leukocytosis are sometimes present (Horowitz, Katzap, Horowitz, et al., 2011). Although any joint may be infected, 50% of cases involve a knee (Garcia-De La Torre & Nava-Zavala, 2009).

Assessment and Diagnostic Findings

An assessment for the source and cause of infection is performed. Diagnostic studies include aspiration, examination, and culture of the synovial fluid. Computed tomography (CT) and MRI may reveal damage to the joint lining. Radioisotope scanning may be useful in localizing the infectious process. There may not be any external wound or reported recent trauma.

Medical Management

Prompt treatment is essential and may save the prosthesis for patients who have had joint replacement surgery or avert severe sepsis. Broad-spectrum IV antibiotics are started promptly and then changed to organism-specific antibiotics after culture results are available (Gillespie, 2009). The IV antibiotics are continued until symptoms resolve. The synovial fluid is aspirated and analyzed periodically for sterility and decrease in WBCs.

The primary provider may aspirate the joint with a needle to remove excessive joint fluid, exudate, and debris. This promotes comfort and decreases joint destruction caused by the action of proteolytic enzymes in the purulent fluid. Occasionally, arthrotomy or arthroscopy is used to drain the joint and remove dead tissue (Garcia-De La Torre & Nava-Zavala, 2009).

The inflamed joint is supported and immobilized in a functional position by a splint that increases the patient's comfort. Analgesic agents are prescribed to relieve pain. The patient's nutrition and fluid status is monitored. Progressive range-of-motion exercises are prescribed as soon as the patient can begin movement without exacerbating symptoms of acute pain. If septic joints are treated promptly, recovery of normal function is expected. If the articular cartilage was damaged during the inflammatory reaction, joint fibrosis and diminished function may result. The patient is assessed periodically for recurrence over the next year.

Nursing Management

The nurse educates the patient and family about the septic arthritis physiologic process and explains the importance of supporting the affected joint, adhering to the prescribed antibiotic regimen, inspecting the skin under any splints that may be prescribed, and observing weight-bearing and activity restrictions. The patient is also educated that recurrence of infection in the near and far future is possible. The same interventions used for the patient with osteomyelitis are planned for the patient with septic arthritis (see previous discussion).

Bone Tumors

Neoplasms of the musculoskeletal system are of various types, including osteogenic, chondrogenic, fibrogenic, muscle (rhabdomyogenic), and marrow (reticulum) cell tumors as well as nerve, vascular, and fatty cell tumors. They may be primary tumors or metastatic tumors from cancers elsewhere in the body (e.g., breast, lung, prostate, kidney). Metastatic bone tumors are more common than primary bone tumors (Unni & Inwards, 2010).

Types

Benign Bone Tumors

Benign tumors of the bone and soft tissue are more common than malignant primary bone tumors. Benign bone tumors generally are slow growing, well circumscribed, and encapsulated; present few symptoms; and are not a cause of death. Benign masses include osteochondroma, enchondroma, bone cysts, osteoid osteoma, rhabdomyoma, and fibroma. Some benign tumors have the potential to become malignant.

Osteochondroma is the most common benign bone tumor. It usually occurs as a large projection of bone at the end of long bones (at the knee or shoulder), developing during growth. It then becomes a static bony mass. In fewer than 1% of patients, the cartilage cap of the osteochondroma may undergo malignant transformation after trauma, and a chondrosarcoma or osteosarcoma may develop (Unni & Inwards, 2010).

Bone cysts are expanding lesions within the bone. Aneurysmal (widening) bone cysts are seen in young adults, who present with a painful, palpable mass of the long bones, vertebrae, or flat bone. Unicameral (single-cavity) bone cysts occur more often in the first two decades of life and cause mild discomfort and possible pathologic fractures of the upper humerus and femur, which may heal spontaneously.

An osteoid osteoma is a painful tumor that occurs in children and young adults. The neoplastic tissue is surrounded by reactive bone formation that can be identified by x-ray. Enchondroma is a common tumor of the hyaline cartilage that develops in the hand, femur, tibia, or humerus. Usually, the only symptom is a mild ache. Pathologic fractures may occur in both types of tumors.

Giant cell tumors (osteoclastomas) are benign for long periods but may invade local tissue and cause destruction. They occur in young adults and are soft and hemorrhagic. Eventually, giant cell tumors may undergo malignant transformation and metastasize (Unni & Inwards, 2010).

Malignant Bone Tumors

Primary malignant musculoskeletal tumors are relatively rare and arise from connective and supportive tissue cells (sarcomas) or bone marrow elements (multiple myeloma; see Chapter 34). Malignant primary musculoskeletal tumors include osteosarcoma, chondrosarcoma, Ewing's sarcoma, and fibrosarcoma. Soft tissue sarcomas include liposarcoma, fibrosarcoma of soft tissue, and rhabdomyosarcoma. Bone tumor metastasis to the lungs is common.

Osteosarcoma is the most common and most often fatal primary malignant bone tumor. Prognosis depends on whether the tumor has metastasized to the lungs at the time the patient seeks health care. It appears most frequently in children, adolescents and young adults (in bones that grow rapidly), in older people with Paget's disease of the bone, and in persons with a prior history of radiation exposure. Clinical manifestations typically include localized bone pain that may be accompanied by a tender, palpable soft tissue mass. The primary lesion may involve any bone, but the most common sites are the distal femur, the proximal tibia, and the proximal humerus (Unni & Inwards, 2010).

Malignant tumors of the hyaline cartilage are called *chondrosarcomas*. These tumors are the second most common primary malignant bone tumor. They may grow and metastasize slowly or very fast, depending on the characteristics of the tumor cells involved (i.e., grade). Patients with low-grade chondrosarcomas tend to have a much better prognosis than those with high-grade chondrosarcomas (see Chapter 15 for a discussion of tumor grades). The usual tumor sites include the pelvis, femur, humerus, spine, scapula, and tibia. Metastasis to the lungs occurs in fewer than half of patients. These tumors may recur after excision (Unni & Inwards, 2010).

Metastatic Bone Disease

Metastatic bone disease (secondary bone tumor) is more common than primary bone tumors. Tumors arising from tissues elsewhere in the body may invade the bone and produce localized bone destruction (lytic lesions) or bone overgrowth (blastic lesions). The most common primary sites of tumors that metastasize to bone are the kidney, prostate, lung, breast, ovary, and thyroid. Metastatic tumors most frequently attack the skull, spine, pelvis, femur, and humerus and often involve more than one bone (polyostotic) (Porth & Matfin, 2009).

Pathophysiology

A tumor in the bone causes the normal bone tissue to react by osteolytic response (bone destruction) or osteoblastic response (bone formation). Adjacent normal bone responds to the tumor by altering its normal pattern of remodeling. The bone's surface changes and the contours enlarge in the tumor area.

Malignant bone tumors invade and destroy adjacent bone tissue. Benign bone tumors, in contrast, have a symmetric, controlled growth pattern and place pressure on adjacent bone tissue. Malignant bone tumors invade and weaken the structure of the bone until it can no longer withstand the stress of ordinary use; pathologic fracture commonly results.

Clinical Manifestations

Patients with metastatic bone tumor may have a wide range of associated clinical manifestations. They may be symptom free or have pain that ranges from mild and occasional to constant and severe, varying degrees of disability, and, at times, obvious bone growth. Weight loss, malaise, and fever may be present. The tumor may be diagnosed only after pathologic fractures occur.

With spinal metastasis, spinal cord compression may occur. It can progress rapidly or slowly. Neurologic deficits (e.g., progressive pain, weakness, gait abnormality, paresthesia, paraplegia, urinary retention, loss of bowel or bladder control) must be identified early and treated with decompression laminectomy to prevent permanent spinal cord injury.

Assessment and Diagnostic Findings

The differential diagnosis is based on the history, physical examination, and diagnostic studies, including CT, bone scans, myelography, arteriography, MRI, biopsy, and biochemical assays of the blood and urine. A surgical biopsy is performed for histologic identification. Extreme care is taken during the biopsy to prevent seeding and resultant recurrence after excision of the tumor. Chest x-rays are performed to determine the presence of lung metastasis. Surgical staging of musculoskeletal tumors is based on tumor grade and site (intracompartmental or extracompartmental), as well as on metastasis. Staging is used for planning treatment.

Serum ALP levels are frequently elevated with osteogenic sarcoma or bone metastasis. Hypercalcemia is also present with bone metastases from breast, lung, or kidney cancer. Symptoms of hypercalcemia include muscle weakness, fatigue, anorexia, nausea, vomiting, polyuria, cardiac dysrhythmias, seizures, and coma. Hypercalcemia must be identified and treated promptly.

During the diagnostic period, the nurse explains the diagnostic tests and provides psychological and emotional support to the patient and family. The nurse assesses coping behaviors and encourages the use of support systems. Because the terminology associated with benign and malignant growths sound similar, the nurse can clarify the meaning of these in terms of treatment and prognosis and may allay fears.

Medical Management

Primary Bone Tumors

The goal of primary bone tumor treatment is to destroy or remove the tumor rapidly (Unni & Inwards, 2010). This may be accomplished by surgical excision (ranging from local excision to amputation and disarticulation), radiation therapy if the tumor is radiosensitive, and chemotherapy (preoperative, intraoperative [neoadjuvant], postoperative, and adjunctive for possible micrometastases). Major gains are being made in the use of wide bloc excision with restorative grafting technique. Survival and quality of life are important considerations in procedures that attempt to save the involved extremity; however, surgical removal of the tumor may require amputation of the affected extremity, with the amputation extending well above the tumor to achieve local control of the primary lesion (see Chapter 43).

If possible, limb-sparing (salvage) procedures are used to remove the tumor and adjacent tissue. A customized prosthesis, total joint arthroplasty, or bone tissue from the patient (autograft) or from a cadaver donor (allograft) replaces the resected tissue. Soft tissue and blood vessels may need grafting because of the extent of the excision. Complications may include infection, loosening or dislocation of the prosthesis, allograft nonunion, fracture, devitalization of the skin and soft tissues, joint fibrosis, and recurrence of the tumor.

Because of the danger of metastasis with malignant bone tumors, chemotherapy is started before and continued after surgery in an effort to eradicate micrometastatic lesions. The goal of combined chemotherapy is greater therapeutic effect at a lower toxicity rate with reduced resistance to the medications. Soft tissue sarcomas are treated with radiation, limb-sparing excision, and adjuvant chemotherapy (see Chapter 15).

Secondary Bone Tumors

The treatment of advanced metastatic bone cancer is palliative. The therapeutic goal is to relieve the patient's pain and discomfort while promoting quality of life. If the bone is weakened, structural support and stabilization are needed to prevent pathologic fracture. Bones are strengthened by prophylactic internal fixation, arthroplasty, or PMMA (bone cement) reconstruction. Patients with metastatic disease are at higher risk than other patients for postoperative pulmonary congestion, hypoxemia, VTE, and hemorrhage.

Hematopoiesis is frequently disrupted by tumor invasion of the bone marrow or by treatment (chemotherapy, surgery, or radiation). Blood component therapy restores hematologic factors. Pain can result from multiple factors. Pain must be assessed accurately and managed with adequate interventions. External-beam radiation to involved metastatic sites may be used. Patients with multiple bony metastases may achieve pain control with systemically administered "bone-seeking" isotopes (e.g., strontium 89). (See Chapter 12 for more information about pain management.)

Additional therapies are used to treat the original cancer. Radiation, chemotherapy, and hormonal therapy may also be effective in promoting healing of osteolytic lesions (see Chapter 15). Bisphosphonates are effective in stabilizing bone and may prevent cancer spread, as are the RANK-L drugs (Lipton & Goessl, 2011) (see discussion of these medications in the Osteoporosis section).

Nursing Management

The nurse asks the patient about the onset and course of symptoms. During the interview, the nurse assesses the patient's understanding of the disease process, how the patient and the family have been coping, and how the patient has managed the pain. The nurse limits palpation of the mass to decrease any potential seeding process. The mass size and associated soft tissue swelling, pain, and tenderness are noted. Assessment of the neurovascular status and range of motion of the extremity provides baseline data for future comparisons. Evaluation of the patient's mobility and ability to perform ADLs is also documented.

The nursing care of a patient who has undergone excision of a bone tumor is similar in many respects to that of other patients who have had skeletal surgery. Explanation of diagnostic tests, treatments (e.g., wound care), and expected results (e.g., decreased range of motion, numbness, change of body contours) helps the patient deal with the procedures and changes and adhere to the therapeutic regimen. The nurse can most effectively reinforce and clarify information provided by the primary provider by being present during these discussions.

Accurate pain assessment and the use of pharmacologic and nonpharmacologic pain management techniques are used to relieve pain and increase the patient's comfort level. Oncology-associated bone pain is recognized as difficult to control. The nurse works with the patient in designing the most effective pain management regimen, thereby increasing the patient's self-efficacy.

Bone tumors weaken the bone to a point at which normal activities or even position changes can result in fracture. During nursing care, the affected extremities must be supported and handled gently. External supports (e.g., splints) may be used for additional protection. Surgery (e.g., open reduction with internal fixation, joint replacement) may be done in an attempt to prevent pathologic fracture. Prescribed weight-bearing restrictions must be followed. The nurse and physical therapist educate the patient about using assistive devices safely and strengthening unaffected extremities.

The nurse encourages the patient and family to verbalize their fears, concerns, and feelings. They need to be supported as they deal with the impact of the malignant bone tumor. Details of nursing interventions for oncology patients covered in Chapter 15 are appropriate to initiate with these patients. Referral to a psychiatric advanced practice nurse, psychologist, counselor, or spiritual advisor may be indicated for specific psychological help and emotional support.

Monitoring and Managing Potential Complications

Delayed Wound Healing

Wound healing may be delayed because of tissue trauma from surgery, previous radiation therapy, inadequate nutrition, or infection. The nurse minimizes pressure on the wound site to promote circulation to the tissues. An aseptic, nontraumatic wound dressing promotes healing. Monitoring and reporting of laboratory findings facilitate initiation of interventions to promote homeostasis and wound healing.

Repositioning the patient at frequent intervals reduces the incidence of skin breakdown and pressure ulcers. Special therapeutic beds or mattresses may be needed to prevent skin breakdown and to promote wound healing after extensive surgical reconstruction and skin grafting.

Inadequate Nutrition

Because loss of appetite, nausea, and vomiting are frequent side effects of chemotherapy and radiation therapy, adequate nutrition must be provided for healing and health promotion. Antiemetic agents and relaxation techniques reduce the adverse gastrointestinal effects of chemotherapy. Stomatitis is controlled with anesthetic or antifungal mouthwash (see Chapter 15). Adequate hydration is essential. Nutritional supplements or parenteral nutrition may be prescribed to achieve adequate nutrition.

Osteomyelitis and Wound Infections

Prophylactic antibiotics and strict aseptic dressing techniques are used to diminish the occurrence of osteomyelitis and wound infections. During healing, other infections (e.g., upper respiratory infections) need to be prevented so that healing efforts are not divided between the cancer and the new, acute process. If the patient is receiving chemotherapy, the nurse monitors the WBC count and instructs the patient to avoid contact with people who have colds or other infections.

Hypercalcemia

Hypercalcemia is a dangerous complication of bone cancer or any process involved with breakdown of bone. Symptoms include muscular weakness, incoordination, anorexia, nausea and vomiting, constipation, electrocardiographic changes (e.g., shortened QT interval and ST segment, bradycardia, heart blocks), and altered mental states (e.g., confusion, lethargy, psychotic behavior). Treatment includes hydration with IV administration of normal saline solution, diuresis, mobilization, and medications such as bisphosphonates. Because inactivity leads to additional loss of bone mass and increased calcium in the blood, the nurse assists the patient to increase activity and ambulation. (See Chapter 13 for further discussion of hypercalcemia and its management.)

Promoting Home and Community-Based Care

Educating Patients About Self-Care

Preparation for and coordination of continuing health care are begun early as a multidisciplinary effort. Patient education addresses medication, dressing changes, treatment regimens,

Chart 42-8 HOME CARE CHECKLIST
Bone Tumor

At the completion of home care education, the patient or caregiver will be able to:	PATIENT	CAREGIVER
• Describe tumor growth process.	✔	✔
• Control pain with pharmacologic and nonpharmacologic interventions.	✔	✔
• Support affected musculoskeletal area.	✔	
• Describe the use of prescribed medications.	✔	✔
• Comply with medication regimen.	✔	
• Consume diet to promote healing and health.	✔	
• State weight-bearing and activity restrictions.	✔	✔
• Demonstrate safe use of ambulatory aids and assistive devices.	✔	
• Identify complications of tumor and therapy.	✔	✔
• Report signs and symptoms of complications promptly.	✔	✔
• Engage in exercise that improves balance to reduce risk of falls.	✔	
• Use effective coping strategies.	✔	✔
• Assess home environment for hazards contributing to falls.	✔	✔
• Maintain role performance.	✔	✔

and the importance of physical and occupational therapy programs. The nurse educates the patient about weight-bearing limitations and special handling to prevent pathologic fractures. The patient and family must be educated about the signs and symptoms of possible complications as well as resources available for continuing care (Chart 42-8).

Continuing Care

Arrangements may be made with a home health care agency for home care supervision and follow-up. The home care nurse assesses the patient's and family's abilities to meet the patient's needs and determines whether the services of other agencies are needed. The nurse advises the patient to have telephone numbers readily available so that providers can be contacted in case concerns arise.

The nurse emphasizes the need for long-term health supervision to ensure cure or to detect tumor recurrence or metastasis and the need for recommended health screening. If the patient has metastatic disease, end-of-life issues may need to be explored. Referral for hospice care is made if appropriate.

Critical Thinking Exercises

1 ebp You are a staff nurse employed at a family practice clinic. A 30-year-old construction worker has been seeking treatment at the clinic for chronic low back pain after an injury a few months ago. He tells you that his mother recommended he ask for prescriptions to cover the "medical necessity" of obtaining the firmest bed mattress on the market, oral steroid medications, and acupuncture. What is the strength of the evidence for each of these potential therapies or interventions?

2 pq You are a volunteer registered nurse with a local ski patrol. A skier arrives at your Aid Station and requests help for his right great toe, which keeps sliding forward in his borrowed boots. On examination, you find a red, swollen toe tip with some darkening that resembles old blood under his nail. What is your priority assessment of this skier? What advice might you share with him for how he should conduct the rest of his day and for any future ski trips?

3 ebp At the geriatric medical clinic where you work as a nurse, a 74-year-old man presents with persistent low-grade back pain along his upper lumbar spine. He has a history of heavy cigarette smoking and alcohol consumption. He also takes levothyroxine supplement for hypothyroidism. He tells you that he has not had a DEXA scan, although his postmenopausal spouse has had a DEXA scan. What is the current evidence-based recommendation for osteoporosis prevention for this older male patient? Discuss the strength of the evidence that supports any risk factor reduction strategies that you consider implementing.

Brunner Suite Resources

Explore these additional resources to enhance learning for this chapter:
- NCLEX-Style Questions and Other Resources on thePoint, http://thePoint.lww.com/Brunner13e
- Study Guide
- PrepU
- Clinical Handbook
- Handbook of Laboratory and Diagnostic Tests

References

*Asterisk indicates nursing research.

Books

Gillespie, W. (2009). Bone and joint infections. In M. Loeb, F. Smaill, & M. Smieja (Ed.): *Evidence-based infectious diseases* (2nd ed.). Hoboken, NJ: Wiley-Blackwell.

National Osteoporosis Foundation. (2013). *Clinician's guide to prevention and treatment of osteoporosis*. Washington, DC: Author.

Porth, C. M., & Matfin, G. (2009). *Pathophysiology: Concepts of altered health states* (8th ed.). Philadelphia: Lippincott Williams & Wilkins.

Unni, K., & Inwards, C. (2010). *Dahlin's bone tumors* (6th ed.). Philadelphia: Lippincott Williams & Wilkins.

Journals and Electronic Documents

American College of Foot and Ankle Surgeons. (2009). Clinical practice guideline for the diagnosis and treatment of forefoot disorders. *Journal of Food and Ankle Surgery*, 48(2), 230–272.

American College of Foot and Ankle Surgeons. (2010). Diagnosis and treatment of heel pain. *Journal of Foot and Ankle Surgeons*, 49(3 Suppl.), S1–S19.

American College of Occupational and Environmental Medicine. (2011). *Carpal tunnel syndrome*. Available at: www.guideline.gov/content.aspx?id=34436&search=carpal+tunnel+syndrome

Avenell, A., Gillespie, W. J., Gillespie, L. D., et al. (2009). Vitamin D and vitamin D analogues for preventing fractures associated with involutional and post-menopausal osteoporosis. *Cochrane Database of Systematic Reviews*, (2), CD000227.

Bach, S., & Holten, K. (2009). Guideline update: What's the best approach to acute low back pain? *Journal of Family Practice*, 58(1), E1–E3.

Black E., & Blazar, P. (2011). Dupuytren disease: An evolving understanding of an age-old disease. *Journal of the American Academy of Orthopaedic Surgeons*, 19(12), 746–757.

Brandi, M. (2010). Current treatment approach for Paget's disease of the bone. *Discovery Medicine*, 10(52), 209–212.

Bredella, M., Torianni, M., Ghomi, R., et al. (2011). Determinants of bone mineral density in obese premenopausal women. *Bone*, 48(4), 748–754.

Buchman, A., Shah, R., Leurgans, S., et al. (2010). Musculoskeletal pain and incident disability in community dwelling older adults. *Arthritis Care and Research*, 62(9), 1287–1293.

Burns, J., Landorf, K. B., Ryan, M. M., et al. (2010). Interventions for the prevention and treatment of pes cavus. *Cochrane Database of Systematic Reviews*, (4), CD006154.pub2.

Chou, R., Quaseen, A., Owens, D., et al. (2011). Diagnostic imaging for low back pain: Advice for high-value health care from the American College of Physicians. *Annals of Internal Medicine*, 154(3), 181–189.

Conterno L., & da Silva Filho, C. R. (2009). Antibiotics for treating chronic osteomyelitis in adults. *Cochrane Database of Systematic Reviews*, (3), CD004439.pub2.

Dahm, K. T., Brurberg, K. G., Jamtvedt, G., et al. (2010). Advice to rest in bed versus advice to stay active for acute low-back pain and sciatica. *Cochrane Database of Systematic Reviews*, (6), CD007612.

Donnelly, C., Callaghan, C., & Durkin, J. (2009). The effect of an active lumbar support system on the seating comfort of officers in police fleet vehicles. *International Journal of Occupational Safety and Ergonomics*, 15(3), 295–307.

Ducharme, N. (2010). Male osteoporosis. *Clinics in Geriatric Medicine*, 26(2), 301–309.

Ensrud, K. (2011). Vertebral fractures. *New England Journal of Medicine*, 364(17), 1634–1642.

Falchetti, A., Marini, F., Mosi, L., et al. (2010). Genetic aspects of Paget's disease of the bone. *European Journal of Clinical Investigation*, 40(7), 655–667.

Garcia-De La Torre, I., & Nava-Zavala, A. (2009). Gonococcal and nongonococcal arthritis. *Rheumatic Disease Clinics of North America*, 35(1), 63–73.

Gaujoux-Viala, C. (2009). Efficacy and safety of steroid injections for shoulder and elbow tendonitis: A meta-analysis of randomized controlled trials. *Annals of the Rheumatic Diseases*, 68(12), 1843–1849.

Giangregorio, L., Lix, L., Johansson H., et al. (2012). FRAX underestimates fracture risk in patients with diabetes. *Journal of Bone and Mineral Research*, 27(2), 301–308.

Gold, J., Gore, R., & Punnett, L. (2009). Specific and non-specific upper extremity musculoskeletal disorder syndromes in automobile manufacturing workers. *American Journal of Industrial Medicine*, 52(2), 124–132.

Goldenberg, D. L., & Sexton, D. J. (2012). Septic arthritis in adults. *UpToDate*. Available at: www.uptodate.com/contents/septic-arthritis-in-adults

Horowitz, D. L., Katzap, E., Horowitz, S., et al. (2011). Approach to septic arthritis. *American Family Physician*, 84(6), 653–660.

Johnson, M., Neher, J., & St. Anna, L. (2011). How effective—and safe—are systemic steroids for acute low back pain? *Journal of Family Practice*, 60(5), 297–298.

Karp, J., Weiner, D., Dew, M., et al. (2010). Duloxetine and care management treatment of older adults with comorbid major depressive disorder and chronic low back pain: Result of an open-label pilot study. *International Journal of Geriatric Psychiatry*, 25(6), 633–642.

Lee, S. Y., McKeon, P., & Hertel, J. (2009). Does the use of orthoses improve self-reported pain and function measures in patients with plantar fasciitis? A meta-analysis. *Physical Therapy in Sport*, 10(1), 12–18.

Lipton, A., & Goessl, C. (2011). Clinical development of anti-Rank-L therapies for the treatment and prevention of bone metastasis. *Bone*, 48(1), 96–99.

Little, J., Jacobson, J., & Lockhart, P. (2010). The dental treatment of patients with joint replacements: A position paper from the American Academy of Oral Medicine. *Journal of the American Dental Association*, 141(6), 667–671.

Mathews, C. J., Weston, V. C., Jones, A., et al. (2010). Bacterial septic arthritis in adults. *Lancet*, 375(9717), 846–855.

Miller, L., & Kaplan, S. (2009). Staphylococcus aureus: A community pathogen. *Infectious Disease Clinics of North America*, 23(1), 35–52.

*Murphy, S., & O'Connor, C. (2009). So what! If a pneumatic tourniquet is used intraoperatively: A study of neurovascular assessment practices of orthopaedic nurses. *International Journal of Orthopaedic and Trauma Nursing*, 14(1), 48–54.

National Guideline Clearinghouse. (2010). *Guideline for the evidence-informed primary care management of low back pain*. Available at: www.guideline.gov/content.aspx?id=15668&search=low+back+pain#Section442

National Institute of Neurological Disorders and Stroke. (2011). *Low back pain fact sheet*. Available at: www.ninds.nih.gov/disorders/backpain/detail_backpain.htm

United States Preventive Services Task Force (USPSTF). (2013). Menopausal hormone therapy for the primary prevention of chronic conditions. *Annals of Internal Medicine*, 158(1), 47–54.

Vestergaard, P., Schwartz, T., Rejnank, L., et al. (2011). Risk of femoral shaft and subtrochanteric fracture among users of bisphosphonates and raloxifene. *Osteoporosis International*, 22(3), 993–1001.

Watts, N., Bliezikian, J., Camacho, P., et al. (2010). American Association of Clinical Endocrinologists medical guideline for clinical practice for the diagnosis and treatment of postmenopausal osteoporosis (Executive Summary). *Endocrine Practice*, 16(6), 1016–1019.

Wells, G. A., Cranney, A., Peterson, J., et al. (2011). Alendronate for the primary and secondary prevention of osteoporotic fractures in postmenopausal women. *Cochrane Database of Systematic Reviews*, (1), CD001155.pub 2.

Resources

National Institute of Arthritis and Musculoskeletal and Skin Diseases, www.niams.nih.gov

National Osteoporosis Foundation, www.nof.org

Paget Foundation, www.paget.org

Chapter 43

Management of Patients With Musculoskeletal Trauma

Learning Objectives

On completion of this chapter, the learner will be able to:

1. Differentiate between contusions, strains, sprains, dislocations, and subluxations.
2. Identify the signs and symptoms of an acute fracture.
3. Describe the treatment procedures of fracture reduction, fracture immobilization, and management of open and intra-articular fractures.
4. Describe the prevention and management of immediate and delayed complications of fractures.
5. Describe the rehabilitation needs of patients with fractures of the upper and lower extremities, pelvis, and hips.
6. Use the nursing process as a framework for care of the older adult patient with a fracture of the hip.
7. Identify sports- and occupation-related musculoskeletal disorders and their signs, symptoms, and treatments.
8. Describe the rehabilitation and health education needs of the patient who has had an amputation.
9. Use the nursing process as a framework for care of the patient with an amputation.

Glossary

allograft: tissue harvested from a donor for use in another person

amputation: removal of a body part, usually a limb or part of a limb

arthroscope: surgical scope injected into the joint to examine or repair

autograft: tissue harvested from one area of the body and used for transplantation to another area of the same body

avascular necrosis: death of tissue secondary to a decrease or lack of perfusion

contusion: blunt force injury to soft tissue

crepitus: a grating sound or sensation made by rubbing bony fragments together

débridement: surgical removal of contaminated and devitalized tissues and foreign material

delayed union: prolongation of expected healing time for a fracture

disarticulation: amputation through a joint

dislocation: complete separation of joint surfaces

fracture: a break in the continuity of a bone

fracture reduction: restoration of fracture fragments into anatomic alignment

malunion: healing of a fractured bone in a malaligned position

nonunion: failure of fractured bones to heal together

phantom limb pain: pain perceived in an amputated section

RICE: acronym for *r*est, *i*ce, *c*ompression, *e*levation

sprain: an injury to ligaments and muscles and other soft tissues at a joint

strain: a musculotendinous stress injury

subluxation: partial separation of joint surfaces

Unintentional traumatic injury is the fifth leading cause of death in the United States (National Center for Health Statistics [NCHS], 2013). Traumatic unintentional injuries are commonly called *accidents* by laymen. The term *accident* is considered inaccurate by trauma professionals, however, because "accidents *do not* happen." Indeed, implementation of evidence-based primary prevention policies (e.g., mandatory use of seat belts for drivers and passengers in motor vehicles) may prevent many unintentional injuries from occurring.

The leading causes of hospitalizations for patients who experience unintentional traumatic injuries include injuries to the musculoskeletal system, with fractures accounting for more than half of these. Various other musculoskeletal injuries, including contusions, sprains, and strains, collectively account for another 8% of causes for hospitalization (NCHS, 2009). Therefore, nurses who work in emergency departments, critical care units, and inpatient medical-surgical units frequently encounter patients who have experienced musculoskeletal trauma. However, upon discharge from the hospital, many of these patients require extensive periods of rehabilitation and follow-up. Thus, nurses who work in rehabilitation centers, long-term care facilities, ambulatory surgery centers, occupational health settings, and primary care clinics may all care for patients with musculoskeletal injuries.

Contusions, Strains, and Sprains

A **contusion** is a soft tissue injury produced by blunt force, such as a blow, kick, or fall, causing small blood vessels to rupture and bleed into soft tissues (ecchymosis, or bruising). A hematoma develops from bleeding at the site of impact, leaving a characteristic "black and blue" appearance. Approximately 48 hours later, breakdown of blood pigments changes the discoloration to appear yellow. Local symptoms (pain, swelling, and discoloration) are controlled with intermittent application of cold packs applied to the site and elevation of the extremity (Emergency Nurses Association [ENA], 2013). Most contusions resolve in 1 to 2 weeks.

Injury to a muscle or tendon from overuse, overstretching, or excessive stress may cause **strain** (National Association of Orthopaedic Nurses [NAON], 2007). Strains are graded along a continuum based on postinjury symptoms and loss of function and reflect the degree of injury. Three types of strain are recognized:

- A first-degree strain is mild stretching of the muscle or tendon. Signs and symptoms may include minor edema, tenderness, and mild muscle spasm, without noticeable loss of function.
- A second-degree strain involves partial tearing of the muscle or tendon. Signs and symptoms include loss of load-bearing strength with accompanying edema, tenderness, muscle spasm, and ecchymosis.
- A third-degree strain is severe muscle or tendon stretching with rupturing and tearing of the involved tissue. Signs and symptoms include significant pain, muscle spasm, ecchymosis, edema, and loss of function. An x-ray should be obtained to rule out bone injury, because an avulsion fracture (in which a bone fragment is pulled away from the bone by a tendon) may be associated with a third-degree strain (NAON, 2007). Magnetic resonance imaging (MRI) will reveal a third-degree strain, but x-rays do not reveal injuries to soft tissue or muscles, tendons, or ligaments.

A **sprain** is an injury to the ligaments and tendons that surround a joint. It is caused by a twisting motion or hyperextension (forcible) of a joint (NAON, 2007). The function of a ligament is to stabilize a joint while permitting mobility. A torn ligament causes a joint to become unstable. Blood vessels rupture and edema occurs; the joint is tender, and movement of the joint becomes painful. The degree of disability and pain increases during the first 2 to 3 hours after the injury because of the associated swelling and bleeding, especially if treatment is delayed. Sprains are graded in a manner similar to the grading system used for strains (NAON, 2007):

- A first-degree sprain is caused by stretching the ligamentous fibers, resulting in minimum damage. It is manifested by mild edema, local tenderness, and pain that is elicited when the joint is moved.
- A second-degree sprain involves partial tearing of the ligament. It results in increased edema, tenderness, pain with motion, joint instability, and partial loss of normal joint function.
- A third-degree sprain occurs when a ligament is completely torn or ruptured. A third-degree sprain may also cause an avulsion of the bone. Symptoms include severe pain, tenderness, increased edema, and abnormal joint motion.

Nursing Management

Treatment for contusions, strains, and sprains consists of resting and elevating the affected part, applying cold, and using a compression bandage. (The acronym **RICE**—*r*est, *i*ce, *c*ompression, *e*levation—is helpful for remembering treatment interventions.) Rest prevents additional injury and promotes healing. Intermittent application of moist or dry cold packs for during the first 72 hours after injury produces vasoconstriction, which decreases bleeding, edema, and discomfort (NAON, 2007). Care must be taken to avoid skin and tissue damage from excessive cold. An elastic compression bandage controls bleeding, reduces edema, and provides support for the injured tissues. Elevation at or above the level of the heart controls the swelling (NAON, 2007). If the sprain or strain is third degree, surgical repair or immobilization by a splint, brace, or cast may be necessary so that the joint will not lose its stability. The neurovascular status (circulation, motion, sensation) of the injured extremity is monitored at frequent intervals (e.g., every 15 minutes for the first 1 to 2 hours after injury) and then at lesser intervals (e.g., every 30 minutes) until stable. Decreases in sensation or motion and increases in pain level should be documented and reported to the patient's primary provider immediately so that compartment syndrome can be prevented (see later discussion).

After the acute inflammatory stage (e.g., 24 to 72 hours after injury), heat may be applied intermittently as needed to relieve muscle spasm and to promote vasodilation, absorption, and repair. Depending on the severity of injury, progressive passive and active exercises may begin in 2 to 5 days. Severe sprains and strains may require 3 to 6 weeks of immobilization before exercises are initiated. Strains and sprains take weeks or months to heal because ligaments and tendons have minimal blood supply. Splinting may be used to maintain stability at the injury site (NAON, 2007).

Joint Dislocations

A **dislocation** of a joint is a condition in which the articular surfaces of the distal and proximal bones that form the joint are no longer in anatomic alignment. A **subluxation** is a partial dislocation and does not cause as much deformity as a complete dislocation. In complete dislocation, the bones are literally "out of joint." Traumatic dislocations are orthopedic emergencies because the associated joint structures, blood supply, and nerves are displaced and may be entrapped with extensive pressure on them. If a dislocation or subluxation is not reduced immediately, **avascular necrosis** (AVN) may develop. AVN of bone is caused by ischemia, which leads to necrosis or death of the bone cells.

Signs and symptoms of a traumatic dislocation include acute pain, change in positioning of the joint, shortening of the extremity, deformity, and decreased mobility. X-rays of both the affected and symmetrical unaffected joint confirm the diagnosis and reveal any associated fracture (NAON, 2007).

Medical Management

The affected joint needs to be immobilized at the scene and during transport to the hospital. The dislocation is promptly reduced, and displaced parts are placed back in proper anatomic position to preserve joint function. Analgesia, muscle relaxants, and possibly anesthesia are used to facilitate closed reduction. The joint is immobilized by splints, casts, or traction and is maintained in a stable position. Neurovascular status is assessed at a minimum of every 15 minutes until stable. After reduction, if the joint is stable, gentle, progressive, active and passive movement is begun to preserve range of motion (ROM) and restore strength. The joint is supported between exercise sessions.

Nursing Management

Nursing attention is geared to frequent assessment and evaluation of the injury, including complete neurovascular assessment with proper documentation and communication with the primary provider. The patient and supportive family members are educated regarding proper exercises and activities as well as danger signs and symptoms to look for, such as increasing pain (even with analgesic agents), numbness or tingling, and increased edema in the extremity. These signs and symptoms may indicate compartment syndrome; if compartment syndrome is not identified and communicated to the primary provider, it may lead to disability or loss of the extremity (see later discussion).

Injuries to the Tendons, Ligaments, and Menisci

Rotator Cuff Tears

A rotator cuff tear is a tear in a tendon that connects one of the rotator muscles to the humeral head. The rotator cuff stabilizes the humeral head and is composed of four muscles and their tendons that include the supraspinatus, infraspinatus, teres minor, and subscapularis (NAON, 2007).

Rotator cuff tears may result from an acute injury or from chronic joint stresses. Patients complain of dull, aching pain that increases when they try to lift the affected arm above shoulder level, along with some limited ROM, and joint dysfunction, including muscle weakness (Bilal, Duffy, Shafi, et al., 2011). In many cases, patients with a rotator cuff tear experience night pain and cannot sleep on the involved side. The acromioclavicular joint is tender. X-rays are helpful in evaluating the joint. Arthrography and MRI or ultrasound are used to determine soft tissue pathology and the extent of the rotator cuff tear (Bilal et al., 2011).

Initial conservative management includes the use of nonsteroidal anti-inflammatory drugs (NSAIDs); rest with modification of activities; injection of a corticosteroid into the shoulder joint; and progressive stretching, ROM, and lengthening exercises. Some rotator cuff tears require arthroscopic **débridement** (removal of devitalized tissue) or arthroscopic or open acromioplasty with tendon repair. Postoperatively, the shoulder is immobilized for several days to 4 weeks. Physical therapy with shoulder exercises is begun as prescribed, and the patient is instructed in how to perform the exercises at home. The course of rehabilitation is lengthy (i.e., 6 to 12 months); functionality post rehabilitation depends on the patient's dedication to the rehabilitation regimen (NAON, 2007).

Epicondylitis

Epicondylitis is a chronic, painful condition that is caused by excessive, repetitive extension, flexion, pronation, and supination motions of the forearm. These motions result in inflammation (tendinitis) and minor tears in the tendons at the origin of the muscles on the lateral or medial epicondyles. Lateral epicondylitis (i.e., tennis elbow) is frequently identified in someone who repeatedly extends the wrist or frequently pronates and supinates the forearm. Pain develops over the lateral epicondyle and in the extensor muscles. If action is continued, pain continues to increase (Clinton & Murthi, 2008). Medial epicondylitis (i.e., golfer's or pitcher's elbow) is consistent with repetitive wrist flexion. Extreme tenderness occurs at the medial epicondyle. Pain greatly increases with wrist flexion against resistance.

Application of ice and administration of NSAIDs usually relieve the pain. In some instances, the arm is immobilized in a molded splint or cast. Because of its degenerative effects on tendons, local injection of a corticosteroid is traditionally reserved for patients with severe pain who do not respond to NSAIDs and immobilization; however, its efficacy is questionable, based upon findings from at least one recent randomized controlled trial (Coombs, Bisset, Brooks, et al. 2013). After pain subsides, rehabilitation exercises include gentle and gradual increased stretching of the tendons (Clinton & Murthi, 2008). A counterforce strap that limits extension of the elbow may be prescribed when activity is resumed.

Lateral and Medial Collateral Ligament Injury

Lateral and medial collateral ligaments of the knee (Fig. 43-1) provide stability lateral and medial to the knee.

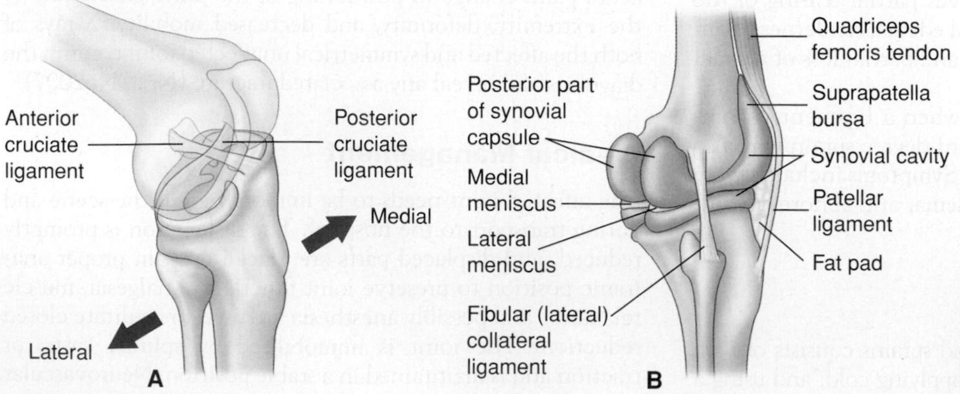
Anterior cruciate ligament
Posterior cruciate ligament
Medial
Lateral
A
Posterior part of synovial capsule
Medial meniscus
Lateral meniscus
Fibular (lateral) collateral ligament
Quadriceps femoris tendon
Suprapatella bursa
Synovial cavity
Patellar ligament
Fat pad
B

FIGURE 43-1 • Knee ligaments, tendons, and menisci. **A.** Anterolateral view. **B.** Posterolateral view.

Injury to these ligaments occurs when the foot is firmly planted and the knee is struck—either medially, causing stretching and tearing injury to the lateral collateral ligament, or laterally, causing stretching and tearing injury to the medial collateral ligament. The patient experiences an acute onset of pain, point tenderness, joint instability, and inability to walk without assistance.

Medical Management

Early management includes RICE. The joint is evaluated for fracture. Hemarthrosis (bleeding into the joint) may develop, contributing to the pain; should this occur, the joint fluid may be aspirated to relieve pressure.

Treatment depends on the severity of the injury. Conservative management includes limited weight bearing and the use of a protective brace. Early physical therapy with continuous passive ROM exercise and early weight bearing are encouraged, because these interventions are associated with superior functional outcomes (Logerstedt, Snyder-Mackler, Ritter, et al., 2010). The patient's return to full activities, including sports, depends on return of motion, functional stability of the joint, and muscle strength.

Surgical reconstruction may be indicated for some third-degree injuries (e.g., those that occur concomitant with bone avulsion); surgery may be performed immediately or may be delayed. The leg is immobilized for approximately 6 to 8 weeks. A progressive rehabilitation program helps restore the function and strength of the knee. Rehabilitation occurs over many months, and the patient may need to wear a derotational brace while engaging in sports to prevent reinjury.

Nursing Management

The nurse instructs the patient about proper use of ambulatory devices, the healing process, and activity limitation to promote healing. Education addresses pain management, the use of analgesic agents, the use of a brace, wound care, signs and symptoms of possible complications (e.g., altered neurovascular status, infection, skin breakdown), and self-care (Hegmann, 2011).

Cruciate Ligament Injury

The anterior cruciate ligament (ACL) and the posterior cruciate ligament (PCL) of the knee stabilize anterior and posterior motion of the tibia articulating with the femur (see Fig. 43-1). These ligaments cross each other in the center of the knee. Injury occurs when the foot is firmly planted and the leg sustains direct force, either forward or backward. If the force is forward, the ACL suffers the impact from the force, whereas backward force places force on the PCL. The injured person may report feeling and hearing a "pop" in the knee with this injury. If the patient exhibits significant swelling of the joint within 2 hours after the injury, the ACL or PCL may be torn. A torn cruciate ligament produces pain, joint instability, and pain with weight bearing. Immediate postinjury management includes RICE and stabilization of the joint until it is evaluated for a fracture. Severe joint effusion and hemarthrosis may require joint aspiration and wrapping with an elastic compression dressing (NAON, 2007).

Treatment depends on the severity of the injury and the effect of the injury on daily activities. Early treatment involves application of a brace and physical therapy. Surgical ACL or PCL reconstruction may be scheduled after near-normal joint ROM is achieved and includes tendon repair with grafting. This is typically performed as ambulatory arthroscopic surgery, a procedure in which the surgeon uses an **arthroscope** to visualize and repair the damage. The best surgical candidates include patients who are young and physically active. Older and less active patients, particularly with concomitant osteoarthritis, tend to benefit from nonsurgical therapy (NAON, 2007). After surgery, the patient is instructed to control pain with oral analgesic medications and cryotherapy (a cooling pad incorporated in a dressing). The patient and family are educated about monitoring the neurovascular status of the leg, wound care, and signs of complications that need to be reported promptly to the surgeon. Exercises (ankle pumps, quadriceps sets [see Chart 41-3 in Chapter 41], and hamstring sets) are encouraged during the early postoperative period. The patient must protect the graft by complying with exercise restrictions. The physical therapist supervises progressive ROM and weight bearing (as permitted). Continuous passive motion may be helpful in restoring full ROM (NAON, 2007).

Meniscal Injuries

Two crescent-shaped (semilunar) cartilages in the knee, called *menisci*, are located on the right and left side of the proximal tibia, between the tibia and the femur (see Fig. 43-1). These structures act as shock absorbers in the knee. Normally, little twisting movement is permitted in the knee joint. Twisting of the knee or repetitive squatting and impact may result in either tearing or detachment of the cartilage from its attachment to the head of the tibia. The peripheral third of the menisci have a small amount of blood flow, which allows that portion to heal if torn.

These injuries leave loose cartilage in the knee joint that may slip between the femur and the tibia, preventing full extension of the leg. If this happens during walking or running, the patient often describes the leg as "giving way." The patient may hear or feel a click in the knee when walking, especially when extending the leg that is bearing weight. When the cartilage is attached to the front and back of the knee but torn loose laterally (bucket-handle tear), it may slide between the bones to lie between the condyles and prevent full flexion or extension. As a result, the knee "locks."

When a meniscus is torn, the synovial membrane secretes additional synovial fluid due to the irritation, and the knee becomes very edematous. Initial conservative treatment includes immobilization of the knee, the use of crutches, anti-inflammatory agents, analgesic agents, and modification of activities to avoid those that cause the symptoms. MRI is the diagnostic tool used to detect a torn meniscus. Damaged cartilage is surgically removed (meniscectomy) arthroscopically. After surgery, a pressure dressing is applied. The most common complication is an effusion into the knee joint, which produces pain. The patient is instructed to continue quadriceps setting and ROM exercises (NAON, 2007).

Rupture of the Achilles Tendon

The Achilles tendon attaches the soleus and gastrocnemius muscles to the os calcis (i.e., the heel). Traumatic rupture of the Achilles tendon, generally within the tendon sheath, occurs during activities when there is a sudden contraction of the calf muscle with the foot fixed firmly to the floor or ground. The patient experiences sharp pain and cannot plantar flex the foot because the Achilles tendon is the plantar flexor for the ankle (Cline, Ma, Cydulka, et al., 2012). This most commonly occurs in adults older than 30 years who engage in sports (ENA, 2013).

MRI or ultrasound is indicated to determine the extent of the injury. A non–weight-bearing cast may be indicated, with the affected foot fixed in plantar flexion (see Chapter 41 for further discussion of casts). The cast is typically in place for 6 to 8 weeks. After immobilization with the cast, a heel lift is worn and progressive physical therapy to promote ankle ROM and strength is begun. Surgical repair is typically reserved for young, healthy athletes, because surgery is associated with fewer cases of re-rupture (NAON, 2007). After surgery, a cast is used to immobilize the joint, as described previously (NAON, 2007).

Fractures

A **fracture** is a complete or incomplete disruption in the continuity of bone structure and is defined according to its type and extent. Fractures occur when the bone is subjected to stress greater than it can absorb (Buckley & Panaro, 2012). Fractures may be caused by direct blows, crushing forces, sudden twisting motions, and extreme muscle contractions. When the bone is broken, adjacent structures are also affected, which may result in soft tissue edema, hemorrhage into the muscles and joints, joint dislocations, ruptured tendons, severed nerves, and damaged blood vessels. Body organs may be injured by the force that caused the fracture or by fracture fragments.

Types of Fractures

A *complete fracture* involves a break across the entire cross-section of the bone and is frequently displaced (removed from its normal position). An *incomplete fracture* (e.g., greenstick fracture) involves a break through only part of the cross-section of the bone; these more commonly occur in children (NAON, 2007). A *comminuted fracture* is one that produces several bone fragments. A *closed fracture* (simple fracture) is one that does not cause a break in the skin. An *open fracture* (compound, or complex, fracture) is one in which the skin or mucous membrane wound extends to the fractured bone. Open fractures are graded according to the following criteria (Schaller, 2012):

- Grade I is a clean wound less than 1 cm long.
- Grade II is a larger wound without extensive soft tissue damage or avulsions.
- Grade III is highly contaminated and has extensive soft tissue damage. It may be accompanied by traumatic amputation and is the most severe.

Fractures may also be described according to the anatomic placement of fragments. Specific types of fractures are reviewed in Figure 43-2.

An *intra-articular fracture* extends into the joint surface of a bone. Because each end of a long bone is cartilaginous, if the fracture is nondisplaced, x-rays will not always reveal the fracture because cartilage is nonradiopaque. MRI or arthroscopy will identify the fracture and confirm the diagnosis. The joint is stabilized and immobilized with a splint or cast, and no weight bearing is allowed until the fracture has healed. Intra-articular fractures often lead to posttraumatic arthritis (Buckley & Panaro, 2012).

Clinical Manifestations

The clinical signs and symptoms of a fracture include acute pain, loss of function, deformity, shortening of the extremity, crepitus, and localized edema and ecchymosis (NAON, 2007). Not all of these are present in every fracture.

Pain

The pain is continuous and increases in severity until the bone fragments are immobilized. The muscle spasms that accompany a fracture begin within 20 minutes after the injury and result in more intense pain than the patient reports at the time of injury. The muscle spasms can minimize further movement of the fracture fragments or can result in further bony fragmentation or malalignment (Porth & Matfin, 2009).

Loss of Function

After a fracture, the extremity cannot function properly because normal function of the muscles depends on the integrity of the bones to which they are attached. Pain contributes to the loss of function. In addition, abnormal movement (false motion) may be present.

Deformity

Displacement, angulation, or rotation of the fragments in a fracture of the arm or leg causes a deformity that is detectable when the limb is compared with the uninjured extremity.

Shortening

In fractures of long bones, there is actual shortening of the extremity because of the compression of the fractured bone. Sometimes, muscle spasms can cause the distal and proximal site of the fracture to overlap, causing the extremity to shorten (Porth & Matfin, 2009).

Crepitus

When the extremity is gently palpated, a crumbling sensation, called **crepitus**, can be felt or may be heard. It is caused by the rubbing of the bone fragments against each other.

Localized Edema and Ecchymosis

Localized edema and ecchymosis occur after a fracture as a result of trauma and bleeding into the tissues. These signs may not develop for several hours after the injury or may develop within an hour, depending on the severity of the fracture.

Emergency Management

Immediately after injury, if a fracture is suspected, the body part must be immobilized before the patient is moved. Adequate splinting is essential. Joints proximal and distal to the fracture also must be immobilized to prevent movement of fracture fragments. Immobilization of the long bones of

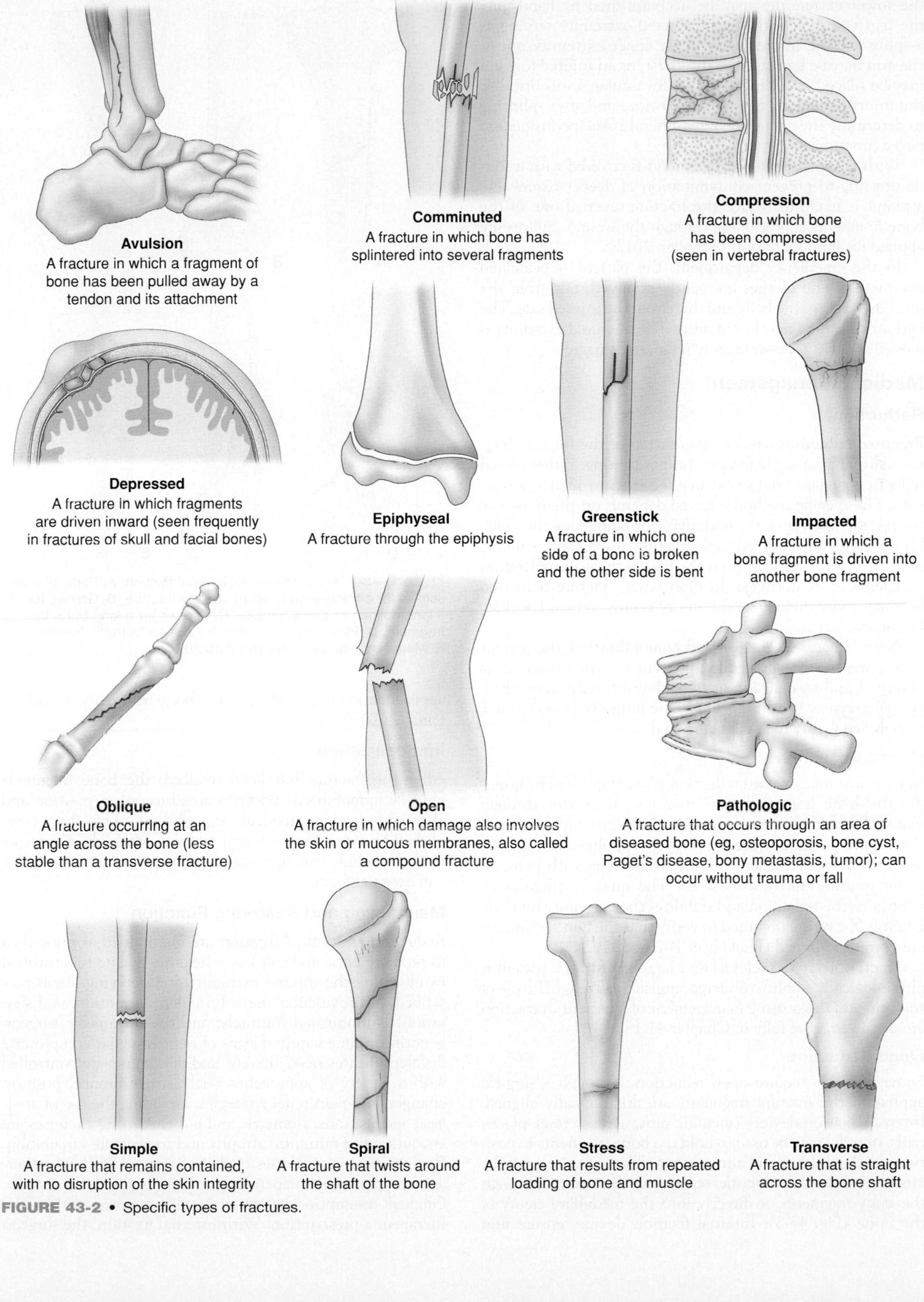
Avulsion
A fracture in which a fragment of bone has been pulled away by a tendon and its attachment
Comminuted
A fracture in which bone has splintered into several fragments
Compression
A fracture in which bone has been compressed (seen in vertebral fractures)
Depressed
A fracture in which fragments are driven inward (seen frequently in fractures of skull and facial bones)
Epiphyseal
A fracture through the epiphysis
Greenstick
A fracture in which one side of a bone is broken and the other side is bent
Impacted
A fracture in which a bone fragment is driven into another bone fragment
Oblique
A fracture occurring at an angle across the bone (less stable than a transverse fracture)
Open
A fracture in which damage also involves the skin or mucous membranes, also called a compound fracture
Pathologic
A fracture that occurs through an area of diseased bone (eg, osteoporosis, bone cyst, Paget's disease, bony metastasis, tumor); can occur without trauma or fall
Simple
A fracture that remains contained, with no disruption of the skin integrity
Spiral
A fracture that twists around the shaft of the bone
Stress
A fracture that results from repeated loading of bone and muscle
Transverse
A fracture that is straight across the bone shaft

FIGURE 43-2 • Specific types of fractures.

the lower extremities may be accomplished by bandaging the legs together, with the unaffected extremity serving as a splint for the injured one. In an upper extremity injury, the arm may be bandaged to the chest, or an injured forearm may be placed in a sling. The neurovascular status distal to the injury should be assessed both before and after splinting to determine the adequacy of peripheral tissue perfusion and nerve function.

With an open fracture, the wound is covered with a sterile dressing to prevent contamination of deeper tissues. No attempt is made to reduce the fracture, even if one of the bone fragments is protruding through the wound. Splints are applied for immobilization (Schaller, 2012).

In the emergency department, the patient is evaluated completely. The clothes are gently removed, first from the uninjured side of the body and then from the injured side. The patient's clothing may be cut away. The fractured extremity is moved as little as possible to avoid more damage.

Medical Management

Reduction

Fracture reduction refers to restoration of the fracture fragments to anatomic alignment and positioning. Either closed reduction or open reduction may be used to reduce a fracture. The specific method selected depends on the nature of the fracture; however, the underlying principles are the same. Usually, the physician reduces a fracture as soon as possible to prevent loss of elasticity from the tissues through infiltration by edema or hemorrhage. In most cases, fracture reduction becomes more difficult as the injury begins to heal (Buckley & Panaro, 2012).

Before fracture reduction and immobilization, the patient is prepared for the procedure; consent for the procedure is obtained, and an analgesic agent is administered as prescribed. Anesthesia may be administered. The injured extremity must be handled gently to avoid additional damage.

Closed Reduction

In most instances, closed reduction is accomplished by bringing the bone fragments into anatomic alignment through manipulation and manual traction. The extremity is held in the aligned position while the physician applies a cast, splint, or other device. Reduction under anesthesia with percutaneous pinning may also be used. The immobilizing device maintains the reduction and stabilizes the extremity for bone healing. X-rays are obtained to verify that the bone fragments are correctly aligned (Buckley & Panaro, 2012).

Traction (skin or skeletal) may be used until the patient is physiologically stable to undergo surgical fixation. (The use of traction and the nursing management of a patient in traction are discussed more fully in Chapter 41.)

Open Reduction

Some fractures require open reduction. Through a surgical approach, the fracture fragments are anatomically aligned. Internal fixation devices (metallic pins, wires, screws, plates, nails, or rods) may be used to hold the bone fragments in position until solid bone healing occurs. These devices may be attached to the sides of bone, or they may be inserted through the bony fragments or directly into the medullary cavity of the bone (Fig. 43-3). Internal fixation devices ensure firm approximation and fixation of the bony fragments (Buckle & Panaro, 2012).

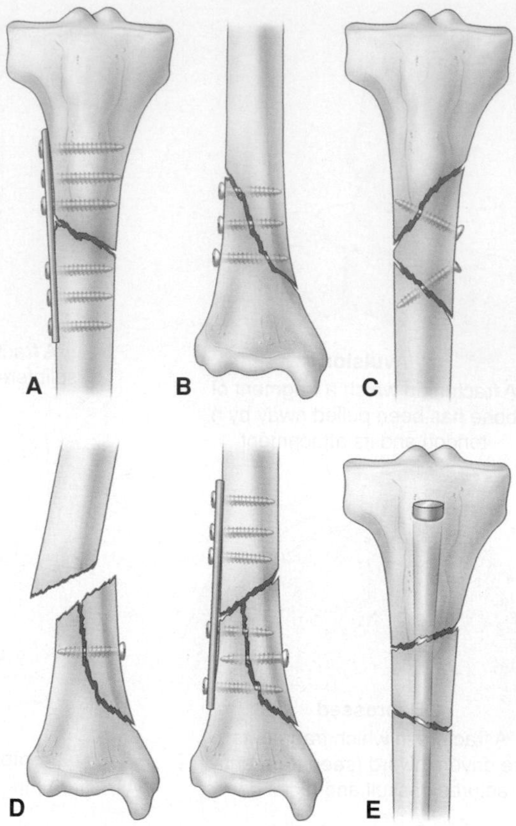
A
B
C
D
E

FIGURE 43-3 • Techniques of internal fixation. **A.** Plate and six screws for a transverse or short oblique fracture. **B.** Screws for a long oblique or spiral fracture. **C.** Screws for a long butterfly fragment. **D.** Plate and six screws for a short butterfly fragment. **E.** Medullary nail for a segmental fracture.

Immobilization

After the fracture has been reduced, the bone fragments must be immobilized and maintained in proper position and alignment until union occurs. Immobilization may be accomplished by external or internal fixation. Methods of external fixation include bandages, casts, splints, continuous traction, and external fixators.

Maintaining and Restoring Function

Reduction and immobilization are maintained as prescribed to promote bone and soft tissue healing. Edema is controlled by elevating the injured extremity and applying ice as prescribed. Neurovascular status (circulation, motion, and sensation) is monitored routinely, and the orthopedic surgeon is notified immediately if signs of neurovascular compromise develop. Restlessness, anxiety, and discomfort are controlled with a variety of approaches, such as reassurance, position changes, and pain relief strategies, including the use of analgesic medications. Isometric and muscle setting exercises are encouraged to minimize atrophy and to promote circulation. Participation in activities of daily living (ADLs) is encouraged to promote independent functioning and self-esteem. Gradual resumption of activities is promoted within the therapeutic prescription. With internal fixation, the surgeon determines the amount of movement and weight-bearing stress the extremity can sustain and prescribes the level of activity (NAON, 2007). (See Chapter 41 for more information about caring for patients who have a cast, are in traction, or are undergoing surgery.)

Nursing Management

Patients With Closed Fractures

The patient with a closed fracture has no opening in the skin at the fracture site. The fractured bones may be nondisplaced or slightly displaced, but the skin is intact. The nurse educates the patient regarding the proper methods to control edema and pain (Chart 43-1). It is important to educate about exercises to maintain the health of unaffected muscles and to increase the strength of muscles needed for transferring and for using assistive devices such as crutches, walkers, and special utensils. The patient is also educated to use assistive devices safely. Plans are made to help patients modify the home environment as needed and to ensure safety, such as removing floor rugs or anything that obstructs walking paths throughout the house. Patient education includes self-care, medication information, monitoring for potential complications, and the need for continuing health care supervision. Fracture healing and restoration of strength and mobility may take an average maximum of 6 to 8 weeks, depending on the quality of the patient's bone tissue (NAON, 2007).

Patients With Open Fractures

In an open fracture, there is a risk for osteomyelitis, tetanus, and gas gangrene. The objectives of management are to prevent infection of the wound, soft tissue, and bone and to promote healing of bone and soft tissue. Intravenous (IV) antibiotics are administered immediately upon the patient's arrival in the hospital along with tetanus toxoid if needed.

Wound irrigation and débridement are initiated in the operating room as soon as possible. The wound is cultured, and bone grafting may be performed to fill in areas of bone defects. The fracture is carefully reduced and stabilized by external fixation, and the wound is usually left open (see Chapter 41). If there is any damage to blood vessels, soft tissue, muscles, nerves, or tendons, appropriate treatment is implemented.

With open fractures, primary wound closure is usually delayed, particularly with higher-grade fractures. Heavily contaminated wounds are left unsutured and dressed with sterile gauze to allow for edema and wound drainage. Wound irrigation and débridement may be repeated, removing infected and devitalized tissue and increasing vascularity in the region (Schaller, 2012).

The extremity is elevated to minimize edema. Neurovascular status must be assessed frequently. Temperature is monitored at regular intervals, and the patient is monitored for signs of infection. Bone grafting may be necessary to bridge bone defects and to stimulate bone healing (Schaller, 2012).

Fracture Healing and Complications

Weeks to months are required for most fractures to heal. Many factors influence the time frame of the healing process (Chart 43-2). With a comminuted fracture, fragments must be properly aligned to attain the best healing possible. It is essential for the fractured bone to have blood supply to the area to facilitate the healing process. In general, fractures of flat bones (pelvis, sternum, and scapula) heal rapidly. A complex, comminuted fracture may heal slowly. Fractures at the ends of long bones, where the bone is more vascular and cancellous, heal more quickly than do fractures in areas where the bone is dense and less vascular (midshaft). Weight bearing, when prescribed, stimulates healing of

Chart 43-1 HOME CARE CHECKLIST
Closed Fracture

At the completion of home care education, the patient or caregiver will be able to:	PATIENT	CAREGIVER
• Describe approaches to control swelling and pain (e.g., elevate extremity to heart level; take analgesics as prescribed).	✔	✔
• Report pain uncontrolled by elevation and analgesics (may be an indicator of impaired tissue perfusion or compartment syndrome).	✔	✔
• Describe management of immobilizing device or care of incision.	✔	✔
• Consume diet to promote bone healing.	✔	
• Demonstrate ability to transfer.	✔	
• Use mobility aids and assistive devices safely.	✔	
• Avoid excessive use of injured extremity; observe prescribed weight-bearing limits.	✔	
• State indicators of complications to report promptly to primary provider (e.g., uncontrolled swelling and pain; cool, pale fingers or toes; paresthesia; paralysis; signs of local and systemic infection; signs of venous thromboembolism; problems with immobilization device).	✔	✔
• State possible delayed complications of fractures (i.e., delayed union; nonunion; avascular necrosis; complex regional pain syndrome, formally called *reflex sympathetic dystrophy syndrome;* heterotopic ossification).	✔	✔
• Describe gradual resumption of normal activities when medically cleared, and discuss how to protect fracture site from undue stresses.	✔	✔

Chart 43-2 Factors That Affect or Inhibit Fracture Healing

Factors That Enhance Fracture Healing

- Immobilization of fracture fragments
- Maximum bone fragment contact
- Sufficient blood supply
- Proper nutrition
- Exercise (weight bearing for long bones)
- Hormones (growth hormone, thyroid, calcitonin, vitamin D, anabolic steroids)
- Electric potential across fracture

Factors That Inhibit Fracture Healing

- Extensive local trauma
- Bone loss
- Weight bearing prior to approval
- Malalignment of the fracture fragments
- Inadequate immobilization
- Space or tissue between bone fragments
- Infection
- Local malignancy
- Metabolic bone disease (e.g., Paget's disease of the bone)
- Irradiated bone (radiation necrosis)
- Avascular necrosis
- Intra-articular fracture (synovial fluid contains fibrolysins, which lyse the initial clot and retard clot formation)
- Age >40 years
- Corticosteroids, nonsteroidal anti-inflammatory drugs
- Smoker
- Diabetes

Adapted from Buckley, R., & Panaro, C. D. (2012). General principles of fracture care. *Medscape.* Available at: emedicine.medscape.com/article/1270717-overview; Porth, C. M., & Matfin, G. (2009). *Pathophysiology: Concepts of altered health states* (8th ed.). Philadelphia: Lippincott Williams & Wilkins.

stabilized fractures of the long bones in the lower extremities (NAON, 2007).

If fracture healing is disrupted, bone union may be delayed or stopped completely. Factors that can impair fracture healing include inadequate fracture immobilization, inadequate blood supply to the fracture site or adjacent tissue, extensive space between bone fragments, interposition of soft tissue between bone ends, displacement of fracture fragments or ends, infection, and metabolic problems (NAON, 2007).

Complications of fractures may be either acute or chronic. Early complications include shock, fat embolism, compartment syndrome, and venous thromboembolism (VTE; deep vein thrombosis [DVT], pulmonary embolism [PE]). Delayed complications include delayed union, malunion, nonunion, AVN of bone, complex regional pain syndrome (CRPS, formerly called *reflex sympathetic dystrophy*), and heterotopic ossification.

Early Complications

Shock

Hypovolemic shock resulting from hemorrhage is more frequently noted in trauma patients with pelvic fractures and in patients with a displaced or open femoral fracture in which the femoral artery is torn by bone fragments. Treatment for shock consists of stabilizing the fracture to prevent further hemorrhage, restoring blood volume and circulation, relieving the patient's pain, providing proper immobilization, and protecting the patient from further injury and other complications (NAON, 2007). (See Chapter 14 for a discussion of shock.)

Fat Embolism Syndrome

After fracture of long bones or pelvic bones, or crush injuries, fat emboli frequently form (in as many as 88% of all fractures); however, their formation is believed to only infrequently cause morbid complications (in as few as 0.3% to 1.3% of these fractures). Fat embolism syndrome (FES) occurs when fat emboli cause morbid clinical manifestations (Tzioupis & Giannoudis, 2011). Most frequently, these occur in adults younger than 40 years and in men. FES is also more common in patients with multiple fractures (Stein, Yaekoub, Matta, et al., 2008). At the time of fracture, fat globules may diffuse from the marrow into the vascular compartment. The fat globules (i.e., emboli) may occlude the small blood vessels that supply the lungs, brain, kidneys, and other organs. The onset of symptoms is rapid, typically within 12 to 72 hours of injury (Tzioupis & Giannoudis, 2011).

Clinical Manifestations. The classic triad of clinical manifestations of FES include hypoxemia, neurologic compromise, and a petechial rash (NAON, 2007), although not all signs and symptoms manifest at the same time (Tzioupis & Giannoudis, 2011). The typical first manifestations are pulmonary and include hypoxia and tachypnea. The patient may exhibit a spectrum of pulmonary manifestations that may or may not include crackles, wheezes, precordial chest pain, cough, large amounts of thick white sputum, and tachycardia. The chest x-ray may show nothing of significance or a classic "snowstorm" infiltrate. Acute pulmonary edema, acute respiratory distress syndrome (ARDS), and heart failure may develop (Tzioupis & Giannoudis, 2011).

Neurologic manifestations typically include cerebral disturbances (due to hypoxia and the lodging of fat emboli in the brain) and manifest as mental status changes varying from headache and mild agitation to delirium and coma. These manifestations typically develop after the respiratory manifestations of FES develop, although they may develop concomitant with the onset of respiratory symptoms or occasionally precede them. Infrequent neurologic manifestations may also include focal impairments, including hemiplegia, aphasia, and acute changes in vision (Tzioupis & Giannoudis, 2011).

Quality and Safety Nursing Alert

Subtle personality changes, restlessness, irritability, or confusion in a patient who has sustained a fracture are indications for immediate arterial blood gas studies.

Petechiae formation tends to occur within 48 to 72 hours postinjury and almost always occurs after pulmonary and neurologic manifestations occur (Tzioupis & Giannoudis, 2011). The cause of the petechiae is not entirely clear, although it may be due to a transient thrombocytopenia. Classically, these petechiae are noted in the buccal membranes and conjunctival sacs, on the hard palate, and over the chest and anterior axillary folds. The petechiae are not present on any posterior body surfaces (Tzioupis & Giannoudis, 2011). The patient develops a fever greater than 39.5°C (103°F). Free fat may be found in the urine if emboli are filtered by the renal tubules. Acute tubular necrosis and kidney failure may develop (NAON, 2007).

Prevention and Management. Immediate immobilization of fractures, including early surgical fixation, minimal fracture manipulation, and adequate support for fractured bones during turning and positioning, and maintenance of fluid and electrolyte balance are measures that may reduce the incidence of fat emboli.

Prompt initiation of respiratory support, assessment, and monitoring is essential. The objectives of management are to support the respiratory system, to prevent respiratory failure, and to correct homeostatic disturbances. Acute pulmonary edema and ARDS are the most common causes of death from FES. Respiratory support is provided with high-flow oxygen. Controlled-volume ventilation with positive end-expiratory pressure may be used to prevent or treat pulmonary edema. Corticosteroids, such as methylprednisolone (Solu-Medrol) have been traditionally administered IV to treat the inflammatory lung reaction and to control cerebral edema, although their efficacy is questionable (Tzioupis & Giannoudis, 2011) (see Chapter 21 for care of the patient on a ventilator and Chapter 23 for the nursing management of respiratory failure). Vasopressor medications to support cardiovascular function are administered IV to prevent and treat hypotension, shock, and interstitial pulmonary edema. Accurate fluid intake and output records facilitate adequate fluid replacement therapy.

Compartment Syndrome

An anatomic compartment is an area of the body encased by bone or fascia (e.g., the fibrous membrane that covers and separates muscles) that contains muscles, nerves, and blood vessels. The human body has 46 anatomic compartments, and 36 of these are located in the extremities (Fig. 43-4). Compartment syndrome in an extremity is a limb-threatening condition that occurs when perfusion pressure falls below tissue pressure within a closed anatomic compartment (Bueche, 2010).

Acute compartment syndrome involves a sudden and severe decrease in blood flow to the tissues distal to an area of injury that results in ischemic necrosis if prompt, decisive intervention does not occur. The patient complains of deep, throbbing, unrelenting pain, which continues to increase despite the administration of opioids and seems out of proportion to the injury. A hallmark sign is pain that occurs or intensifies with passive ROM (e.g., pain intensifies with dorsiflexion of the wrist of the affected extremity) (Johnston-Walker & Hardcastle, 2011). This pain can be caused by (1) a reduction in the size of the muscle compartment because the enclosing muscle fascia is too tight or a cast or dressing is constrictive or (2) an increase in compartment contents because of edema or hemorrhage from the fracture site. The lower leg is most frequently involved, but the forearm is also at risk (see Fig. 41-2 in Chapter 41 for an illustration of compartment syndrome of the lower leg). The pressure within a muscle compartment may increase to such an extent that microcirculation diminishes, causing nerve and muscle anoxia and necrosis. Permanent function can be lost if the anoxic situation continues for longer than 4 hours (Bueche, 2010).

Assessment and Diagnostic Findings. Frequent assessment of neurovascular function after a fracture is essential and focuses on the "five Ps": *p*ain, *p*allor, *p*ulselessness *p*aresthesias, and *p*aralysis. Sensory deficits include deep, throbbing, escalating pain that increases with passive stretching. Paresthesia (burning or tingling sensation) and numbness are early signs of nerve involvement. Motion is evaluated by asking the patient to flex and extend the wrist or plantar flex and dorsiflex the foot. With continued nerve ischemia and edema, the patient experiences sensations of hypoesthesia (diminished sensation followed by complete numbness). Motor weakness may occur as a late sign of nerve ischemia. No movement (paralysis) indicates nerve damage (Bueche, 2010).

Peripheral circulation is evaluated by assessing color, temperature, capillary refill time, edema, and pulses. Cyanotic (i.e., blue-tinged) nail beds suggest venous congestion. Pallor or dusky and cold fingers or toes and prolonged capillary refill time suggest diminished arterial perfusion. Edema may obscure the function of arterial pulsation, and Doppler ultrasonography may be used to verify a pulse (Bickley, 2009). Pulselessness is a very late sign that may signify lack of distal tissue perfusion, but it is possible to have compartment syndrome with a weak pulse to the extremity (NAON, 2007).

Palpation of the muscle, if possible, reveals it to be swollen and hard. The orthopedic surgeon may measure tissue pressure by inserting a tissue pressure-monitoring device, such as a Wick catheter, into the muscle compartment (normal pressure is 8 mm Hg or less) (Fig. 43-5). Nerve and muscle tissues deteriorate as compartment pressure increases. Prolonged pressure of more than 30 mm Hg can result in compromised microcirculation (Bueche, 2010).

Medical Management. Prompt management of acute compartment syndrome is essential. The surgeon needs to be notified immediately if neurovascular compromise is suspected. Delay in treatment may result in permanent nerve and muscle damage or even necrosis and amputation.

> *Quality and Safety Nursing Alert*
>
> **Compartment syndrome is managed by maintaining the extremity at the heart level (*not above heart level*), and opening and bivalving the cast (see Chart 41-2 in Chapter 41) or opening the splint, if one or the other is present.**

If conservative measures do not restore tissue perfusion and relieve pain within 1 hour, a fasciotomy (surgical decompression with excision of the fascia) is indicated to relieve the constrictive muscle fascia. After fasciotomy, the wound is not sutured but is left open to allow the muscle tissues to expand; it is covered with moist, sterile saline dressings or with artificial skin. Alternatively, a vacuum dressing may be used to remove fluids and hasten wound closure (Bueche, 2010). The affected arm or leg is splinted in a functional position and elevated to heart level, and prescribed intermittent passive ROM exercises are usually performed. In 3 to 5 days, when the swelling has resolved and tissue perfusion has been restored, the wound is débrided and closed (possibly with skin grafts) (NAON, 2007). Complications that may occur after fasciotomy include AVN and infection.

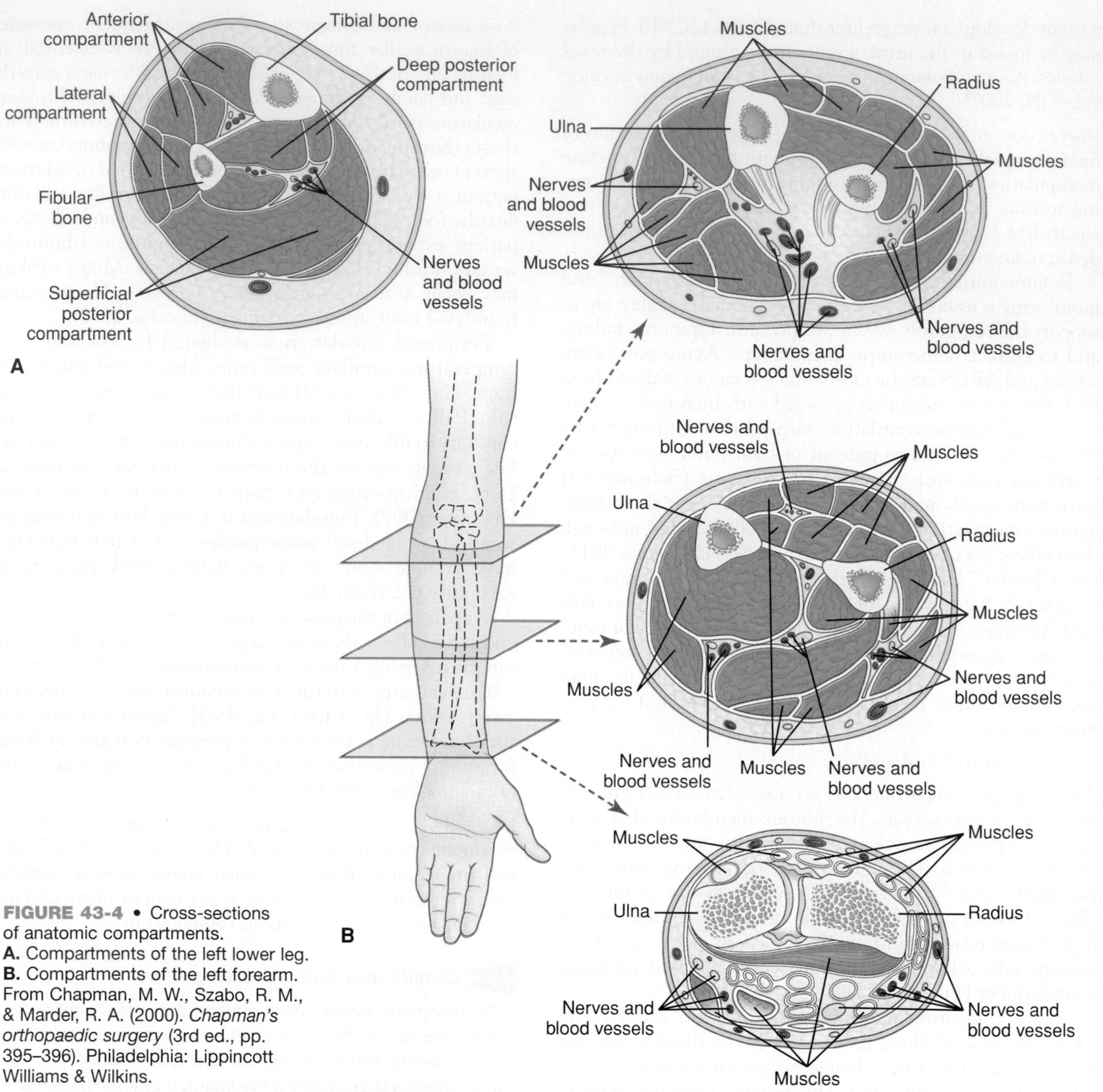
Anterior compartment
Tibial bone
Deep posterior compartment
Lateral compartment
Fibular bone
Nerves and blood vessels
Superficial posterior compartment
A
Muscles
Radius
Ulna
Muscles
Nerves and blood vessels
Muscles
Nerves and blood vessels
Nerves and blood vessels
Nerves and blood vessels
Muscles
Ulna
Radius
Muscles
Nerves and blood vessels
Muscles
Nerves and blood vessels
Muscles
Nerves and blood vessels
Muscles
Muscles
Ulna
Radius
Nerves and blood vessels
Nerves and blood vessels
Muscles
B

FIGURE 43-4 • Cross-sections of anatomic compartments. **A.** Compartments of the left lower leg. **B.** Compartments of the left forearm. From Chapman, M. W., Szabo, R. M., & Marder, R. A. (2000). *Chapman's orthopaedic surgery* (3rd ed., pp. 395–396). Philadelphia: Lippincott Williams & Wilkins.

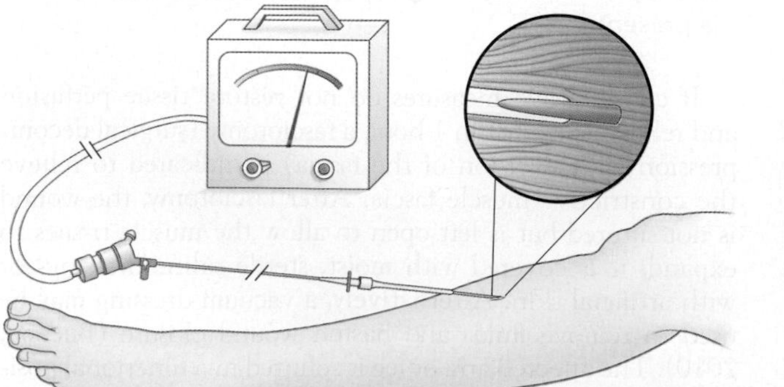

FIGURE 43-5 • The Wick catheter is inserted into a muscle compartment and continuously monitors compartment pressure. From Chapman, M. W., Szabo, R. M., & Marder, R. A. (2000). *Chapman's orthopaedic surgery* (3rd ed., p. 401). Philadelphia: Lippincott Williams & Wilkins.

Other Early Complications

VTE, including DVT and PE, are associated with reduced skeletal muscle contractions and bed rest. Patients with fractures of the lower extremities and pelvis are at high risk for VTE. PE may cause death several days to weeks after injury. (See Chapter 23 for a discussion of PE and Chapter 30 for a discussion of DVT and VTE.)

Disseminated intravascular coagulation (DIC) is a systemic disorder that results in widespread hemorrhage and microthrombosis with ischemia. Its causes are diverse and can include massive tissue trauma. Early manifestations of DIC include unexpected bleeding after surgery and bleeding from the mucous membranes, venipuncture sites, and gastrointestinal and urinary tracts. (See Chapter 33 for discussion of treatment for DIC.)

All open fractures are considered contaminated and are treated as soon as possible with copious irrigation, débridement, and IV antibiotics (Schaller, 2012). Surgical internal fixation of fractures carries a risk of infection. The nurse must monitor and instruct the patient regarding signs and symptoms of infection, including tenderness, pain, redness, swelling, local warmth, elevated temperature, and purulent drainage.

Delayed Complications

Delayed Union, Malunion, and Nonunion

Delayed union occurs when healing does not occur within the expected time frame for the location and type of fracture. Delayed union may be associated with distraction (pulling apart) of bone fragments, systemic or local infection, poor nutrition, or comorbidity (e.g., diabetes, autoimmune disease). The healing time is prolonged, but the fracture eventually heals (NAON, 2007).

Nonunion results from failure of the ends of a fractured bone to unite, whereas **malunion** is the healing of a fractured bone in a malaligned position (Fig. 43-6). In both of these instances, the patient complains of persistent discomfort and abnormal movement at the fracture site. Nonunion occurs most commonly in tibial fractures, whereas malunion occurs most commonly in fractures of the hand (or fingers). Factors contributing to nonunion and malunion include infection at the fracture site, interposition of tissue between the bone ends, inadequate immobilization or manipulation that disrupts callus formation, excessive space between bone fragments, limited bone contact, and impaired blood supply resulting in AVN. In nonunion, fibrocartilage or fibrous tissue exists between the bone fragments; no bone salts have been deposited. A false joint (pseudarthrosis) often develops at the site of the fracture (Porth & Matfin, 2009).

Medical Management. Malunion results in cosmetic deformities that do not resolve with surgical interventions. If functional impairments occur, however, then surgical correction may be an option (Lakshmanan, Damodaran, & Sher, 2012). Nonunion may be treated with internal fixation, bone grafting, electrical bone stimulation, or a combination of these therapies (Patel, McCarthy, & Herzenberg, 2012). Internal fixation stabilizes the bone fragments and ensures bone contact.

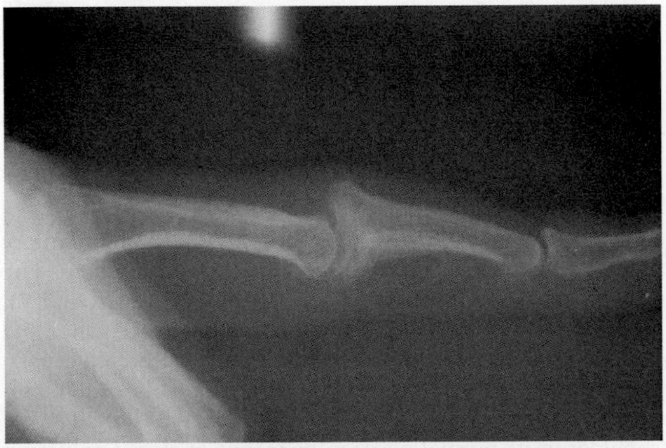

FIGURE 43-6 • Eight-month-old ring finger malunion in a 19-year-old female patient. From Strickland, J. W., & Graham, T. J. (2005). *Master techniques in orthopaedic surgery: The hand* (2nd ed.). Philadelphia: Lippincott Williams & Wilkins.

Bone grafts promote osteogenesis, osteoconduction, and osteoinduction. *Osteogenesis* (bone formation) occurs after transplantation of bone because the graft contains osteoblasts, which build bony matrix. Building of this structural bony matrix promotes *osteoconduction*, which is the growth of blood vessels and osteoblasts within the matrix. *Osteoinduction* is the stimulation of host stem cells to differentiate into osteoblasts by several growth factors, including bone morphogenetic proteins (BMPs), particularly BMP-2, BMP-4, and BMP-7 (Laurencin, Magge, & Khan, 2012).

Grafted bone undergoes a reconstructive process that results in a gradual replacement of the graft with new bone. During surgery, the bone fragments are débrided and aligned, infection (if present) is removed, and a bone graft is placed in the bony defect. The bone graft may be an **autograft** (tissue, frequently from the iliac crest, harvested from the patient for his or her own use) or an **allograft** (tissue harvested from a donor) (NAON, 2007). The bone graft fills the bone gap and provides a lattice structure for invasion by bone cells and actively promotes bone growth. The type of bone selected for grafting depends on function—cortical bone is used for structural strength, cancellous bone for osteogenesis, and corticocancellous bone for strength and rapid incorporation. Free vascularized bone autografts are grafted with their own blood supply, allowing for primary fracture healing.

After grafting, immobilization and non–weight-bearing exercises are required while the bone graft becomes incorporated and the fracture or defect heals. Depending on the type of bone grafted and the age of the patient, healing may take from 6 to 12 months or longer. Bone grafting complications include wound or graft infection, fracture of the graft, and nonunion. Specific problems associated with autografts include a limited quantity of bone available for harvest and harvest site pain that may persist for up to 2 years after harvest. Infrequent specific allograft complications include partial acceptance (lack of host and donor histocompatibility, which retards graft incorporation), graft rejection (rapid and complete resorption of the graft), and transmission of disease (rare) (Porth & Matfin, 2009).

Osteogenesis may be stimulated by electrical impulses; the effectiveness is similar to that of bone grafting. The use of electrical impulses is not effective with large bone gaps. The electrical stimulation modifies the tissue environment, making it electronegative, which enhances mineral deposition and bone formation that promotes bone growth. In some situations, pins that act as cathodes are inserted percutaneously, directly into the fracture site, and electrical impulses are directed to the fracture continuously. This method cannot be used when infection is present (Galkowski, Petrisor, Drew, et al., 2009).

Another method for stimulating osteogenesis is noninvasive inductive coupling. Pulsing electromagnetic fields are delivered to the fracture for approximately 10 hours each day by an electromagnetic coil over the nonunion site (Fig. 43-7). During the electrical stimulation treatment period, which takes 3 to 6 months or longer, rigid fracture fixation with adequate

FIGURE 43-7 • Bone healing stimulator applied to the arm. Reprinted with permission from EBI Medical Systems, Parsippany, NJ.

support is needed (Galkowski et al., 2009). Other osteogenic methods under investigation include using ultrasound and extracorporeal shock waves (Galkowski et al., 2009).

Nursing Management. The patient with a nonunion has experienced an extended time in fracture treatment and may become frustrated with prolonged therapy. The nurse provides emotional support and encouragement to the patient and encourages adherence to the treatment regimen. The orthopedic surgeon evaluates the progression of bone healing with periodic x-rays.

Nursing care for the patient with a bone graft includes pain management and monitoring the patient for possible complications. The nurse needs to reinforce educational information concerning the objectives of the bone graft, immobilization, non–weight-bearing exercises, wound care, monitoring for signs of infection, and the importance of follow-up care with the orthopedic surgeon (NAON, 2007).

Nursing care for the patient using bone stimulation devices focuses on patient education that addresses immobilization, weight-bearing restrictions, and correct daily use of the stimulator as prescribed.

Avascular Necrosis of Bone

AVN occurs when the bone loses its blood supply and dies. It may occur after a fracture with disruption of the blood supply to the distal area. It is also seen with dislocations, bone transplantation, prolonged high-dose corticosteroid therapy, chronic kidney disease, sickle cell anemia, and other diseases. The devitalized bone may collapse or reabsorb. The patient develops pain and experiences limited movement. X-rays reveal loss of mineralized matrix and structural collapse. Treatment generally consists of attempts to revitalize the bone with surgical decompression, bone grafts, prosthetic replacement, or osteotomy (Tofferi & Gilliland, 2012).

Complex Regional Pain Syndrome

CRPS is a painful sympathetic nervous system problem. It occurs infrequently, but when it does occur, it is most often in an upper extremity after trauma (e.g., Colles fracture; see later discussion) and is seen more frequently in women, typically between the ages of 30 and 55 years. Clinical manifestations of CRPS include severe burning pain, local edema, hyperesthesia, stiffness, discoloration, vasomotor skin changes (i.e., fluctuating warm, red, dry and cold, sweaty, cyanotic), and trophic changes that may include glossy, shiny skin and increased hair and nail growth. This syndrome is frequently chronic, with extension of symptoms to adjacent areas of the body. Disuse muscle atrophy and bone deossification (osteoporosis) may occur with persistent CRPS (Kale, 2012).

Nursing Management. Prevention may include elevation of the extremity after injury or surgery and selection of an immobilization device (e.g., external fixator) that allows for the greatest ROM and functional use of the rest of the extremity. Early effective pain relief is the focus of management. Pain may need to be controlled with analgesic agents. NSAIDs, corticosteroids, and muscle relaxants also may be used because of their analgesic properties. The nurse helps the patient cope with CRPS manifestations and explores multiple ways to control pain (see Chapter 12). A mental health referral may be appropriate.

Quality and Safety Nursing Alert

The nurse avoids using the affected extremity for blood pressure measurements and venipuncture in the patient with CRPS.

Heterotopic Ossification

Heterotopic ossification (myositis ossificans) is the abnormal formation of bone, near bones or in muscle, in response to soft tissue trauma or fracture after blunt trauma or total joint replacement. The muscle is painful, and normal muscular contraction and movement are limited. Early mobilization may prevent its occurrence. Usually, the bone lesion resorbs over time, but the abnormal bone eventually may need to be excised if symptoms persist (Banovac & Speed, 2011).

Concept Mastery Alert

For further review of fractures, visit thePoint for an Interactive Tutorial.

Fractures of Specific Sites

Clavicle

Fracture of the clavicle (collar bone) is a common injury that results from a fall or a direct blow to the shoulder. The clavicle helps maintain the shoulder in the upward, outward, and backward position from the thorax. Therefore, when the clavicle is fractured, the patient assumes a protective position, slumping the shoulders and immobilizing the arm to prevent shoulder movements. The treatment goal is to align the shoulder in its normal position by means of closed reduction and immobilization (NAON, 2007).

Most of these fractures occur in the middle third of the clavicle. A clavicular strap, also called a *figure-eight bandage*

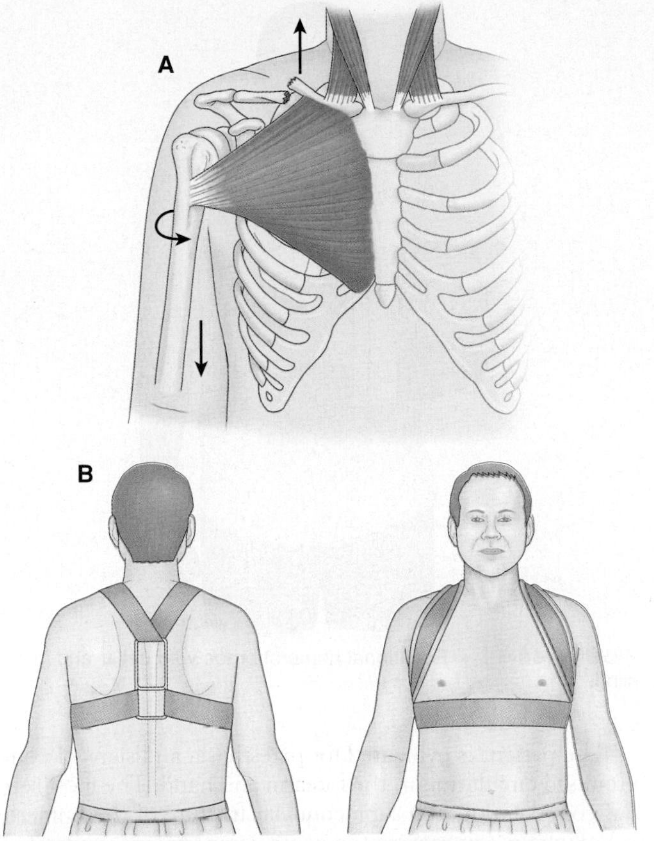
A
B

FIGURE 43-8 • Fracture of the clavicle. **A.** Anteroposterior view shows typical displacement in midclavicular fracture. **B.** Immobilization is accomplished with a clavicular strap.

(Fig. 43-8), may be used to pull the shoulders back, reducing and immobilizing the fracture. The nurse monitors the circulation and nerve function of the affected arm and compares it with the unaffected arm to determine variations, which may indicate disturbances in neurovascular status. A sling may be used to support the arm and relieve pain. The patient may be permitted to use the arm for light activities within the range of comfort (Cline et al., 2012).

Fracture of the distal third of the clavicle, without displacement and ligament disruption, is treated with a sling and restricted motion of the arm. When a fracture in the distal third is accompanied by a disruption of the coracoclavicular ligament that connects the coracoid process of the scapula and the inferior surface of the clavicle, the bony fragments are frequently displaced. This type of injury may be treated by open reduction with internal fixation (ORIF) (see Chapter 41 for discussion of ORIF).

The nurse cautions the patient not to elevate the arm above shoulder level until the fracture has healed (about 6 weeks) but encourages the patient to exercise the elbow, wrist, and fingers as soon as possible. When prescribed, shoulder exercises are performed to obtain full shoulder motion (Fig. 43-9). Vigorous activity is limited for approximately 3 months.

Complications of clavicular fractures include trauma to the nerves of the brachial plexus, injury to the subclavian vein or artery from a bony fragment, pneumothorax, nonunion,

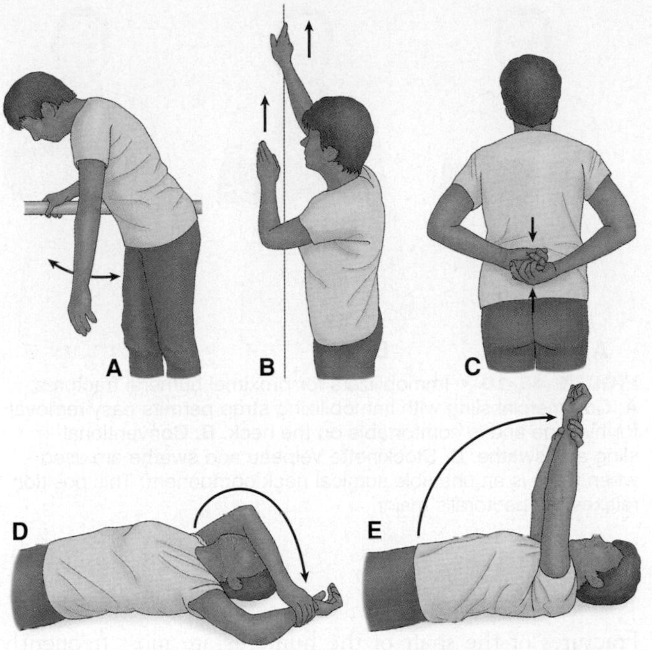
A
B
C
D
E

FIGURE 43-9 • Exercises that promote shoulder range of motion include pendulum exercise **(A)** and wall climbing **(B).** The unaffected arm is used to assist with internal rotation **(C),** external rotation **(D),** and elevation **(E).** In C, D, and E, the unaffected arm is used for power.

malunion, and posttraumatic arthritis (Mouzopoulos, Morakis, Stamatakos, et al., 2009).

Humeral Neck

Fractures of the proximal humerus may occur through the neck of the humerus. Impacted fractures of the surgical neck of the humerus are seen most frequently in older women after a fall on an outstretched arm. Active middle-aged patients who are injured in a fall may suffer severely displaced humeral neck fractures with associated rotator cuff damage (ENA, 2013).

The patient presents with the affected arm hanging limp at the side or supported by the uninjured hand. Neurovascular assessment of the extremity is essential to evaluate the full extent of injury and the possible involvement of the nerves and blood vessels of the arm.

Many impacted fractures of the surgical neck of the humerus are not displaced and do not require reduction. The arm is supported and immobilized by a sling and swathe that secure the supported arm to the trunk (Fig. 43-10). Limitation of motion and stiffness of the shoulder occur with disuse. Therefore, pendulum exercises begin as soon as tolerated by the patient. In pendulum or circumduction exercises, the physical therapist instructs the patient to lean forward and allow the affected arm to hang in abduction and rotate. These fractures require approximately 4 to 10 weeks to heal, and the patient should avoid vigorous arm activity for an additional 4 weeks. Residual stiffness, aching, and some limitation of ROM may persist for 6 months or longer (NAON, 2007).

When a humeral neck fracture is displaced, treatment consists of closed reduction, ORIF, or a total shoulder replacement. Exercises are begun after an adequate period of immobilization (NAON, 2007).

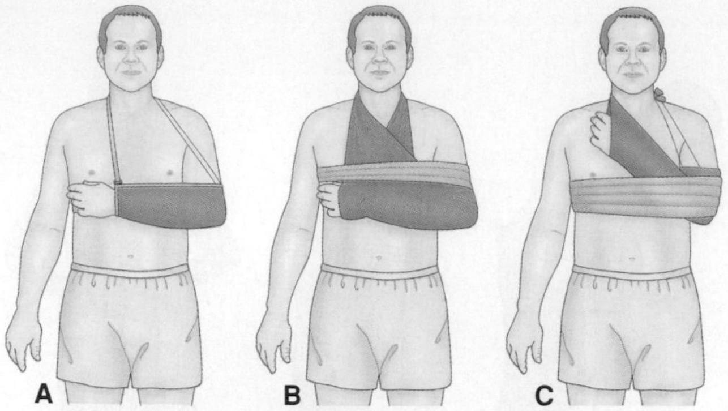
A
B
C

FIGURE 43-10 • Immobilizers for proximal humeral fractures. **A.** Commercial sling with immobilizing strap permits easy removal for hygiene and is comfortable on the neck. **B.** Conventional sling and swathe. **C.** Stockinette Velpeau and swathe are used when there is an unstable surgical neck component. This position relaxes the pectoralis major.

Humeral Shaft

Fractures of the shaft of the humerus are most frequently caused by (1) direct trauma that results in a transverse, oblique, or comminuted fracture or (2) an indirect twisting force that results in a spiral fracture. The nerves and brachial blood vessels may be injured with these fractures, so neurovascular assessment is essential to monitor the status of the nerve or blood vessels. Damage to either requires immediate attention.

Well-padded splints are used to initially immobilize the upper arm and to support the arm in 90 degrees of flexion at the elbow. A sling or collar and cuff support the forearm. The weight of the hanging arm and splints put traction on the fracture site. External fixators are used to treat open fractures of the humeral shaft (see Chapter 41). ORIF of a fracture of the humerus is necessary with nerve palsy, blood vessel damage, comminuted fracture, or displaced fracture (ENA, 2013).

Functional bracing is another form of treatment used for these fractures. A contoured thermoplastic sleeve is secured in place with interlocking fabric (Velcro) closures around the upper arm, immobilizing the reduced fracture. As swelling decreases, the sleeve is tightened, and uniform pressure and stability are applied to the fracture. The forearm is supported with a collar and cuff sling (Fig. 43-11). Functional bracing allows active use of muscles, shoulder and elbow motion, and good approximation of fracture fragments. Pendulum shoulder exercises are performed as prescribed to provide active movement of the shoulder, thereby preventing a "frozen shoulder." Isometric exercises may be prescribed to prevent muscle atrophy. The callus that develops is substantial, and the sleeve can be discontinued in about 8 weeks. Complications that are seen with humeral shaft fractures include delayed union and nonunion because of decreased blood supply in that area.

Elbow

Fractures of the distal humerus result from motor vehicle crashes, falls on the elbow (in the extended or flexed position), or a direct blow. These fractures may result in injury to the median, radial, or ulnar nerves (ENA, 2013).

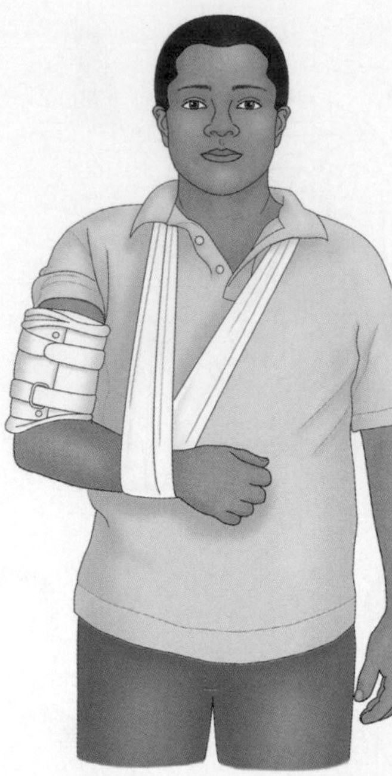

FIGURE 43-11 • Functional humeral brace with collar and cuff sling.

The patient is evaluated for paresthesia and signs of compromised circulation in the forearm and hand. The most serious complication of a supracondylar fracture of the humerus is Volkmann contracture (an acute compartment syndrome), which results from antecubital swelling or damage to the brachial artery (Chart 43-3). The nurse needs to monitor the patient regularly for compromised neurovascular status and signs and symptoms of acute compartment syndrome. Other potential complications are damage to the joint articular surfaces and hemarthrosis (i.e., blood in the joint), which may be treated by needle aspiration by the physician to relieve the pressure and pain.

The goal of therapy is prompt reduction and stabilization of the distal humeral fracture, followed by controlled active motion after swelling has subsided and healing has begun. If the fracture is not displaced, the arm is immobilized in a cast or posterior splint with the elbow at 45 to 90 degrees of flexion and placed in a sling. A thermoplastic splint is used to support the fracture (NAON, 2007).

Chart 43-3 Volkmann Contracture

- Observe the distal part of the extremity for swelling, skin color, nail bed capillary refill, and temperature. Compare affected and unaffected hands.
- Assess radial pulse.
- Assess for paresthesia (tingling and burning sensations) in the hand, which may indicate nerve injury or impending ischemia.
- Evaluate the patient's ability to extend and flex all fingers.
- Explore the intensity and character of the pain.
- Directly measure tissue pressure as prescribed.
- Report indications of diminished nerve function or diminished circulatory perfusion promptly before irreparable damage occurs; fasciotomy may become necessary.

Adapted from Kare, J. A. (2012). Volkmann contracture. *Medscape*. Available at: emedicine.medscape.com/article/1270462-overview

Usually, a displaced fracture is treated with ORIF. Excision of bone fragments may be necessary. Additional external support with a splint is then applied. Active finger exercises are encouraged. Gentle ROM exercise of the injured joint is begun about 1 week after internal fixation. Motion promotes healing of injured joints by producing movement of synovial fluid into the articular cartilage. Active exercise to prevent residual limitation of motion is performed as prescribed (NAON, 2007).

Radial Head

Radial head fractures are common and are usually produced by a fall on an outstretched hand with the elbow extended. If blood has collected in the elbow joint, it is aspirated to relieve pain and to allow early active elbow and forearm ROM exercises. Immobilization for nondisplaced fractures is accomplished with a splint that is applied for about 3 to 7 days (American College of Occupational and Environmental Medicine [ACOEM], 2007). The patient is instructed not to lift with the arm for approximately 4 weeks. If the fracture is displaced, surgery is typically indicated, with excision of the radial head when necessary. Postoperatively, the arm is immobilized in a posterior plaster splint and sling, and an appropriate exercise regimen is prescribed (NAON, 2007).

Radial and Ulnar Shafts

Fractures of the shaft of the bones of the forearm occur more frequently in children than in adults. The radius or the ulna may be fractured at any level. Frequently, displacement occurs when both bones are broken. The forearm's unique functions of pronation and supination must be preserved with proper anatomic alignment (NAON, 2007).

If the fragments are not displaced, the fracture is treated by closed reduction with a long-arm cast applied from the upper arm to the proximal palmar crease. Circulation, motion, and sensation of the hand are assessed before and after the cast is applied. The arm is elevated to control edema. Frequent finger flexion and extension are encouraged to reduce edema. Active motion of the involved shoulder is essential. The reduction and alignment are monitored closely by x-rays to ensure proper alignment. The fracture is immobilized for about 12 weeks; during the last 6 weeks, the arm may be in a functional forearm brace that allows exercise of the wrist and elbow. Lifting and twisting are avoided.

Displaced fractures are managed by ORIF, using a compression plate with screws, intramedullary nails, or rods. The arm is usually immobilized in a plaster splint or cast. Open and displaced fractures may be managed with external fixation devices. The arm is elevated to control swelling. Neurovascular status is assessed and documented. Elbow, wrist, and hand exercises are begun when prescribed by the physician.

Wrist

Fractures of the distal radius (Colles fracture) are common and are usually the result of a fall on an open, dorsiflexed hand. This fracture is frequently seen in older women with osteoporotic bones and weak soft tissues that do not dissipate the energy of the fall. The patient presents with a deformed wrist, pain, swelling, weakness, and limited finger ROM, as well as complaints of "tingling" in the affected hand (Altizer, 2008).

Treatment usually consists of closed reduction and immobilization with a short-arm cast. For fractures with extensive comminution, ORIF, arthroscopic percutaneous pinning, or external fixation is used to achieve and maintain reduction. The wrist and forearm are elevated for 48 hours after reduction to control swelling (Altizer, 2008).

Active motion of the fingers and shoulder should begin promptly to reduce swelling and prevent stiffness (Chart 43-4).

The fingers may swell due to diminished venous and lymphatic return. The nurse assesses the sensory function of the median nerve by pricking the distal aspect of the index finger. The motor function is assessed by the patient's ability to touch the thumb to the little finger. Diminished circulation and nerve function must be treated promptly (see previous discussion of compartment syndrome).

Chart 43-4 PATIENT EDUCATION

Encouraging Exercise after Treatment for Wrist Fracture

The nurse encourages active motion of the fingers and shoulder. The patient is instructed to perform the following exercises to reduce swelling and prevent stiffness:

- Hold the hand at the level of the heart.
- Move the fingers from full extension to flexion. Hold and release. (Repeat at least 10 times every hour when awake.)
- Use the hand in functional activities.
- Actively exercise the shoulder and elbow, including complete range-of-motion exercises of both joints.

Hand

Trauma to the hand is a frequent reason patients seek care in emergency departments, accounting for approximately 16 million injuries annually in the United States (Alke & Schraga, 2011). The most common type of metacarpal fracture in adults is referred to as boxer's fracture, which occurs when a closed fist bangs against a hard surface, fracturing the neck of the fifth finger. Falls and occupational injuries (e.g., machinery injuries, crushes) are the most common cause of phalangeal injury in adults (ENA, 2013). When any of the bones of the hand are fractured, the objectives of treatment are to regain maximum function of the hand and minimize cosmetic deformities. X-rays are the diagnostic studies of choice (Alke & Schraga, 2011).

For a nondisplaced fracture of the phalanx (finger bone), the finger is splinted for 3 to 4 weeks to relieve pain and to protect the finger from further trauma. Splinting sometimes consists of "buddy taping" a fractured finger to an adjoining nonfractured finger. Serial x-rays may be done to monitor healing. Displaced fractures and open fractures may require ORIF, using wires or pins. If the fracture is open, or if a fingernail is avulsed, antibiotics may be prescribed (Alke & Schraga, 2011).

The neurovascular status of the injured hand is evaluated and documented. Swelling is controlled by elevation of the hand. Functional use of the uninvolved portion of the hand is encouraged. Assistive devices might be recommended to aid the patient in performing ADLs until the hand has healed and functional status returns.

Pelvis

The sacrum, ilium, pubis, and ischium bones form the pelvis—a fused, stable, bony ring in adults (Fig. 43-12). Falls, motor vehicle crashes, and crush injuries can cause pelvic fractures. These are more commonly seen in younger and middle-aged adults, who engage in riskier behaviors (e.g., drive motor vehicles at high rates of speed) (NAON, 2007). There is a high mortality rate associated with unstable pelvic fractures (as high as 30%), primarily related to hemorrhage, although pulmonary complications, fat emboli, thromboembolic complications, and infection are also implicated (ENA, 2013). Management of severe, life-threatening pelvic fractures is coordinated with the trauma team (see Chapter 72).

Signs and symptoms of pelvic fracture include ecchymosis; tenderness over the symphysis pubis, anterior iliac spines, iliac crest, sacrum, or coccyx; local edema; numbness or tingling of the pubis, genitals, and proximal thighs; and inability to bear weight without discomfort. Computed tomography (CT) scanning of the pelvis helps determine the extent of injury by demonstrating sacroiliac joint disruption, soft tissue trauma, pelvic hematoma, and fractures. Neurovascular assessment of the lower extremities is completed to detect any injury to pelvic blood vessels and nerves (Russell, Jarrett, & Routt, 2012).

Hemorrhage and shock are two of the most serious consequences that may occur. Bleeding arises mainly from the laceration of veins and arteries by bone fragments and possibly from a torn iliac artery. The peripheral pulses, especially the dorsalis pedis pulses of both lower extremities, are palpated; absence of a pulse may indicate a tear in the iliac artery or one of its branches. Peritoneal lavage or abdominal CT may be performed to detect intra-abdominal hemorrhage. The patient is handled gently so that bony fragments are not displaced, which may exacerbate bleeding and shock (Russell et al., 2012).

The nurse assesses for injuries to the bladder, rectum, intestines, other abdominal organs, and pelvic vessels and nerves. To assess for urinary tract injury, the patient's urine is analyzed for blood. A voiding cystourethrogram and an IV urogram may be performed by a urologist. Laceration of the urethra is suspected in males with anterior fracture of the pelvis and blood at the urethral meatus. Females rarely experience a lacerated urethra (Russell et al., 2012). A urinary drainage catheter should not be inserted until the status of the urethra is known. Diffuse and intense abdominal pain, hyperactive or absent bowel sounds, and abdominal rigidity and resonance (free air) or dullness to percussion (blood) suggest injury to the intestines or abdominal bleeding.

Numerous classification systems have been used to describe pelvic fractures in relation to anatomy, stability, and mechanism of injury. Some fractures of the pelvis do not disrupt the pelvic ring; others disrupt the ring, which may be rotationally or vertically unstable. The severity of pelvic fractures varies. Long-term complications of pelvic fractures include malunion, nonunion, residual gait disturbances, back pain from ligament injury, and dyspareunia and erectile dysfunction (Russell et al., 2012).

Stable Pelvic Fractures

Stable fractures of the pelvis (Fig. 43-13) include fracture of a single pubic or ischial ramus, fracture of ipsilateral pubic and ischial rami, fracture of the pelvic wing of the ilium (Duverney fracture), and fracture of the sacrum or coccyx. If injury results in only a slight widening of the pubic symphysis or the anterior sacroiliac joint and the pelvic ligaments are intact, the disrupted pubic symphysis is likely to heal spontaneously with conservative management. Most fractures of the pelvis heal rapidly because the pelvic bones are mostly cancellous bone, which has a rich blood supply.

Stable pelvic fractures are treated with a few days of bed rest and symptom management until discomfort is controlled. Fluids, dietary fiber, ankle and leg exercises, anti-embolism stockings to aid venous return, logrolling, deep breathing, and skin care reduce the risk of complications and increase the patient's comfort. The patient with a fractured sacrum is at risk for paralytic ileus; therefore, bowel sounds should be monitored.

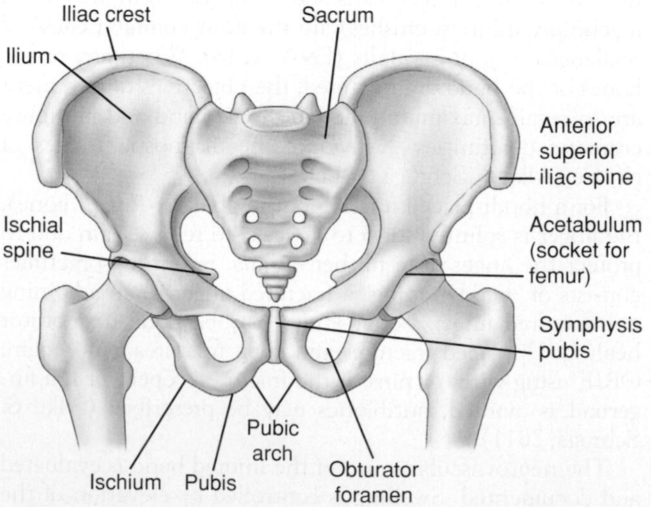
Iliac crest
Sacrum
Ilium
Anterior superior iliac spine
Ischial spine
Acetabulum (socket for femur)
Symphysis pubis
Pubic arch
Ischium
Pubis
Obturator foramen

FIGURE 43-12 • Pelvic bones.

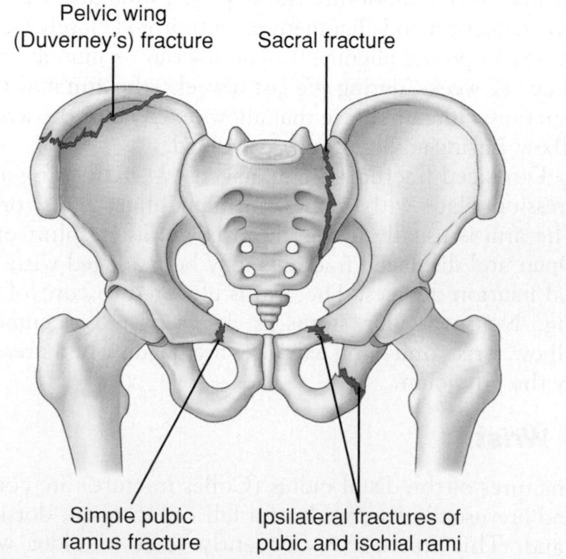
Pelvic wing (Duverney's) fracture
Sacral fracture
Simple pubic ramus fracture
Ipsilateral fractures of pubic and ischial rami

FIGURE 43-13 • Stable pelvic fractures.

The patient with a fracture of the coccyx experiences pain when sitting and when defecating. Sitz baths may be prescribed to relieve pain, and stool softeners may be given to ease defecation. As pain resolves, activity is gradually resumed with the use of assistive mobility devices. Early mobilization reduces problems related to immobility (NAON, 2007).

Unstable Pelvic Fractures

Unstable fractures of the pelvis (Fig. 43-14) may result in rotational instability (e.g., the "open book" type, in which a separation occurs at the symphysis pubis with sacroiliac ligament disruption), vertical instability, or a combination of both. Lateral or anteroposterior compression of the pelvis produces rotationally unstable pelvic fractures. Vertically unstable pelvic fractures occur when force is exerted on the pelvis vertically, as may occur when the patient falls onto extended legs or is struck from above by a falling object. Vertical shear pelvic fractures involve the anterior and posterior pelvic ring with vertical displacement, usually through the sacroiliac joint. There is generally complete disruption of the posterior sacroiliac, sacrospinous, and sacrotuberous ligaments.

Immediate treatment in the emergency department for a patient with an unstable pelvic fracture includes stabilizing the pelvic bones and compressing bleeding vessels with a pelvic girdle, which is an external binding and stabilizing device. If major vessels are lacerated, the bleeding may be stopped through embolization using interventional radiology techniques prior to surgery. More than 10% of deaths in patients with unstable pelvic fractures occur because of frank hemorrhage (ENA, 2013). Therefore, these patients are at risk for hemorrhagic shock (see Chapter 14 for nursing management of the patient in shock). When the patient is hemodynamically stable, treatment generally involves external fixation or ORIF. These measures promote hemostasis, hemodynamic stability, comfort, and early mobilization.

Acetabulum

Acetabular fractures are a type of intra-articular fracture. The typical mechanism of injury is that an external force drives the femoral shaft into the hip joint, fracturing the acetabulum. This may be caused by high-speed motor vehicle crashes (e.g., knees driven into dashboard, pedals forcibly driven upward into legs) or from falls from heights (Thacker, Tejwani, & Thakkar, 2012). Treatment depends on the pattern of fracture. Stable, nondisplaced fractures may be managed with traction and protective weight bearing so that the affected foot is placed on the floor only for balance. Displaced and unstable acetabular fractures are treated with open reduction, joint débridement, and internal fixation or arthroplasty. Internal fixation permits early non–weight-bearing ambulation and ROM exercise. Complications seen with acetabular fractures include malunion, nerve palsy, heterotopic ossification, and posttraumatic arthritis (Thacker et al., 2012).

Hip

The number of hip fractures in the United States is increasing each year with the aging of the American population. The lifetime risk of hip fracture at age 50 years is estimated at 15.8% (Kondo, Zierler, & Hagino, 2011). Hip fractures cause more than 300,000 hospitalizations annually among adults older than 65 years (NAON, 2007). Weak quadriceps muscles, slowed reflexes, decreased bone tensile strength, general frailty due to age, and conditions that produce decreased cerebral arterial perfusion (transient ischemic attacks, anemia, emboli, cardiovascular disease, effects of medications) contribute to the incidence of falls. Mortality rates are as high as one third of all patients by 1 year post hip fracture (NAON, 2007).

There are two major types of hip fracture. *Intracapsular fractures* are fractures of the neck of the femur. *Extracapsular fractures* are fractures of the trochanteric region (between the base of the neck and the lesser trochanter of the femur) and of the subtrochanteric region (Fig. 43-15). Fractures of the neck of the femur may damage the vascular system that supplies blood to the head and the neck of the femur, and the bone may become ischemic. For this reason, AVN is common in patients with femoral neck fractures (NAON, 2007).

Extracapsular intertrochanteric fractures have an excellent blood supply and heal more rapidly. However, extensive soft tissue damage may occur at the time of injury. It is not uncommon for the fracture to be comminuted and unstable. Older adults are particularly vulnerable to intertrochanteric fractures and do not have a prognosis as favorable as younger patients (NAON, 2007).

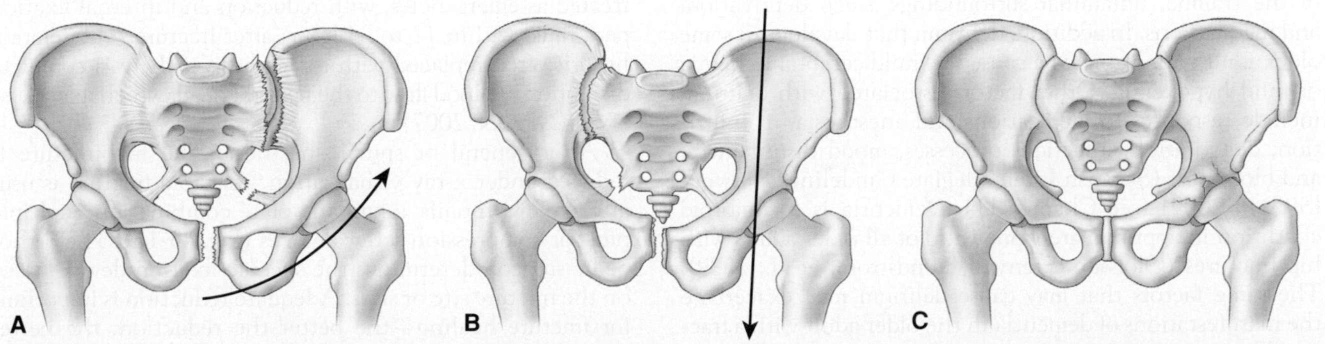
A
B
C

FIGURE 43-14 • Unstable pelvic fracture. **A.** Rotationally unstable fracture. The symphysis pubis is separated and the anterior sacroiliac, sacrotuberous, and sacrospinous ligaments are disrupted. **B.** Vertically unstable fracture. The hemipelvis is displaced anteriorly and posteriorly through the symphysis pubis, and the sacroiliac joint ligaments are disrupted. **C.** Undisplaced fracture of the acetabulum.

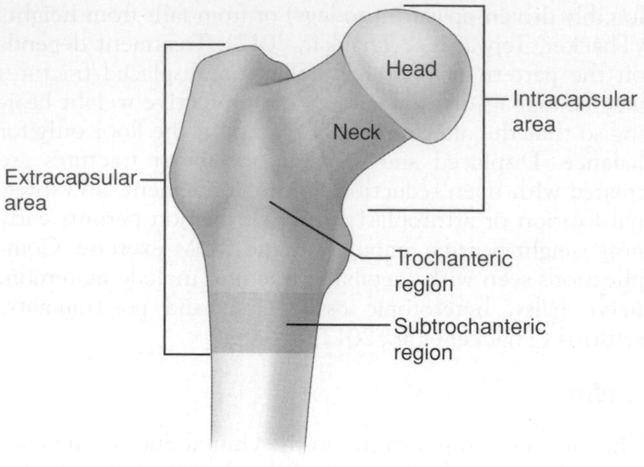
Head
Intracapsular area
Neck
Extracapsular area
Trochanteric region
Subtrochanteric region

FIGURE 43-15 • Regions of the proximal femur.

Clinical Manifestations

With fractures of the femoral neck, the leg is shortened, adducted, and externally rotated. The patient reports pain in the hip and groin or in the medial side of the knee. With most fractures of the femoral neck, the patient cannot move the leg without a significant increase in pain. The patient is most comfortable with the leg slightly flexed in external rotation. Impacted intracapsular femoral neck fractures cause moderate discomfort (even with movement), may allow the patient to bear weight, and may not demonstrate obvious shortening or rotational changes. With extracapsular femoral fractures of the trochanteric or subtrochanteric regions, the extremity is significantly shortened, externally rotated to a greater degree than intracapsular fractures, exhibits muscle spasm that resists positioning of the extremity in a neutral position, and has an associated area of ecchymosis. The diagnosis is confirmed by x-ray (NAON, 2007).

Gerontologic Considerations

Older adults (particularly women) who have low bone density from osteoporosis and who tend to fall frequently have a high incidence of hip fracture. Stress and immobility related to the trauma predispose the older adult to atelectasis, pneumonia, sepsis, VTE, pressure ulcers, and reduced ability to cope with other health problems. Many older adults hospitalized with hip fractures exhibit delirium as a result of stress of the trauma, unfamiliar surroundings, sleep deprivation, and medications. In addition, delirium that develops in some older adult patients may be caused by mild cerebral ischemia or mild hypoxemia. Other factors associated with delirium include responses to medications and anesthesia, malnutrition, dehydration, infectious processes, mood disturbances, and blood loss (Scottish Intercollegiate Guidelines Network [SIGN], 2009) (see Chapter 11). Dementia is a comorbid condition in approximately one third of all older adults with hip fractures (Olofsson, Stenvall, Lundstrom, et al., 2009). The same factors that may cause delirium may exacerbate the manifestations of dementia in the older adult with a fractured hip.

To prevent complications, the nurse must assess the older patient for chronic conditions that require close monitoring. Examination of the legs may reveal edema due to heart failure or absence of peripheral pulses from peripheral vascular disease. Similarly, chronic respiratory problems may be present and may contribute to the possible development of atelectasis or pneumonia. Coughing and deep-breathing exercises are encouraged. Frequently, older adults take cardiac, antihypertensive, or respiratory medications that need to be continued. The patient's responses to these medications should be monitored.

Dehydration and poor nutrition may be present. At times, older adults who live alone cannot call for help at the time of injury. A day or two may pass before assistance is provided, and as a result, dehydration occurs. Dehydration contributes to hemoconcentration and predisposes the patient to the development of VTE. Dehydration, inadequate nutritional intake, and immobility contribute to the development of pressure ulcers. Therefore, the patient needs to be encouraged to consume adequate fluids and a healthy diet (SIGN, 2009).

Muscle weakness may have initially contributed to the fall and fracture. Bed rest and immobility cause an additional loss of muscle strength unless the nurse encourages the patient to move all joints except the involved hip and knee. Patients are encouraged to use their arms and the overhead trapeze to reposition themselves. This strengthens the arms and shoulders, which facilitates walking with assistive devices.

Medical Management

Buck's extension traction, a type of temporary skin traction, was traditionally applied because it was believed to reduce muscle spasm, to immobilize the extremity, and to relieve pain. Its efficacy had never been established in clinical trials, however, so its routine prescription is not advocated (SIGN, 2009). The goal of surgical treatment for hip fractures is to obtain a satisfactory fixation so that the patient can be mobilized quickly and avoid secondary medical complications. Surgical treatment consists of (1) open or closed reduction of the fracture and internal fixation, (2) replacement of the femoral head with a prosthesis (hemiarthroplasty), or (3) closed reduction with percutaneous stabilization for an intracapsular fracture. Surgical intervention is carried out as soon as possible after injury. The preoperative objective is to ensure that the patient is in as favorable a condition as possible for the surgery. Displaced femoral neck fractures are treated as emergencies, with reduction and internal fixation performed within 12 to 24 hours after fracture. The femoral head is often replaced with a prosthesis if there is complete disruption of blood flow to the femoral head, which may cause AVN (NAON, 2007).

After general or spinal anesthesia, the hip fracture is reduced under x-ray visualization. A stable fracture is usually fixed with nails, a nail and plate combination, multiple pins, or compression screw devices (Fig. 43-16). The orthopedic surgeon determines the specific fixation device based on the fracture site or sites. Adequate reduction is important for fracture healing—the better the reduction, the better the healing.

Total hip arthroplasty (see Chapter 41) may be used in selected patients with acetabular defects (NAON, 2007).

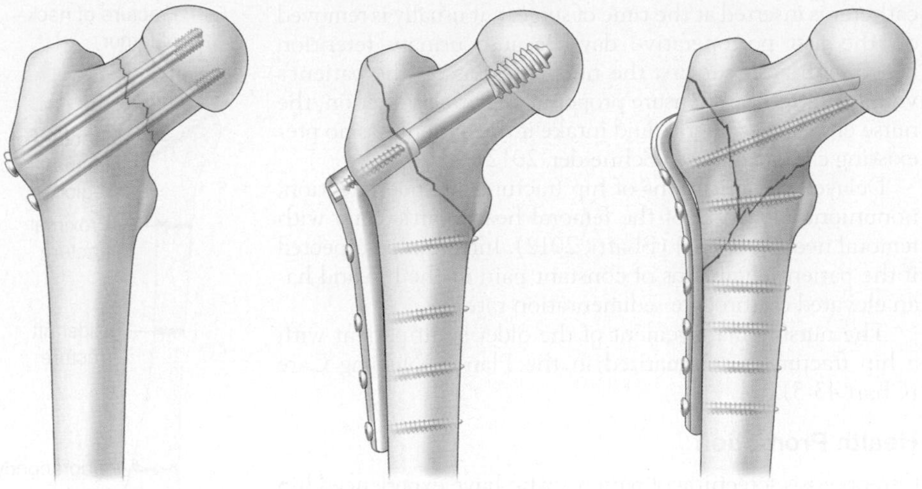
Cannulated screw fixation
Compression hip screw and side plate
Blade plate fixation

FIGURE 43-16 • Examples of internal fixation for hip fractures. Internal fixation is achieved through the use of screws and plates specifically designed for stability and fixation.

Nursing Management

The immediate postoperative care for a patient with a hip fracture is similar to that for other patients undergoing major surgery (see Chapters 19 and 41). Attention is given to pain management, prevention of secondary medical problems, and early mobilization of the patient so that independent functioning can be restored (see Chart 43-5).

During the first 24 to 48 hours, relief of pain and prevention of complications are important, and continuous neurovascular assessment is essential. The nurse encourages deep breathing and dorsiflexion and plantar flexion exercises every 1 to 2 hours. Thigh-high anti-embolism stockings or pneumatic compression devices are used, and anticoagulants are administered as prescribed to prevent the formation of VTE. The nurse administers prescribed analgesic medications and monitors the patient's hydration, nutritional status, and urine output (SIGN, 2009).

Repositioning the Patient

The most comfortable and safest way to turn the patient is to turn to the uninjured side. The standard method involves placing a pillow between the patient's legs to keep the affected leg in an abducted position. Proper alignment and supported abduction are maintained while turning (NAON, 2007).

Promoting Exercise

The patient is encouraged to exercise as much as possible by means of the overbed trapeze. This device helps strengthen the arms and shoulders in preparation for protected ambulation (e.g., toe touch, partial weight bearing). On the first postoperative day, the patient transfers to a chair with assistance and begins assisted ambulation. The amount of weight bearing that can be permitted depends on the stability of the fracture reduction. The physician prescribes the degree of weight bearing. In general, hip flexion and internal rotation restrictions apply only if the patient has had a hemiarthroplasty or total arthroplasty (NAON, 2007) (see Chapter 41). Physical therapists work with the patient on transfers, ambulation, and the safe use of assistive devices.

The patient can anticipate discharge to home or to an extended care facility with the use of assistive devices (see Chapter 10). Some modifications in the home may be needed, such as installation of elevated toilet seats and grab bars.

Monitoring and Managing Potential Complications

Neurovascular complications may occur from direct injury or edema in the area that causes compression of nerves and blood vessels. With hip fracture, bleeding into the tissues and edema are expected. Monitoring and documenting the neurovascular status of the affected leg are vital.

To prevent DVT, the nurse encourages intake of fluids and ankle and foot exercises. Anti-embolism stockings, pneumatic compression devices, and prophylactic anticoagulant therapy are indicated and should be prescribed (SIGN, 2009). Intermittent assessment of the patient's legs for signs of DVT, which may include unilateral calf tenderness, warmth, redness, and swelling, is indicated.

Pulmonary complications (e.g., atelectasis, pneumonia) are a threat to older patients undergoing hip surgery. Coughing and deep-breathing exercises, intermittent changes of position, and the use of an incentive spirometer may help prevent respiratory complications. Pain must be treated with analgesic agents, typically opioids; otherwise, the patient may not be able to cough, deep breathe, or engage in prescribed activities. The nurse assesses breath sounds to detect adventitious or diminished sounds.

Skin breakdown is often seen in older patients with hip fracture. Blisters caused by tape are related to the tension of soft tissue edema under a nonelastic tape. An elastic hip wrap dressing or elastic tape applied in a vertical fashion may reduce the incidence of tape blisters. In addition, patients with hip fractures tend to remain in one position and may develop pressure ulcers. Proper skin care, especially on the bony prominences, helps to relieve pressure. High-density foam mattress overlays may provide protection by distributing pressure evenly.

Loss of bladder control (incontinence or retention) may occur. In general, the routine use of an indwelling catheter is avoided because of the high risk of urinary tract infection. If a catheter is inserted at the time of surgery, it usually is removed on the first postoperative day. Because urinary retention is common after surgery, the nurse must assess the patient's voiding patterns. To ensure proper urinary tract function, the nurse encourages liberal fluid intake if the patient has no preexisting cardiac disease (Schneider, 2012).

Delayed complications of hip fractures include infection, nonunion, and AVN of the femoral head (particularly with femoral neck fractures) (Bhatti, 2012). Infection is suspected if the patient complains of constant pain in the hip and has an elevated erythrocyte sedimentation rate.

The nursing management of the older adult patient with a hip fracture is summarized in the Plan of Nursing Care (Chart 43-5).

Health Promotion

Osteoporosis screening of patients who have experienced hip fracture is important for prevention of future fractures. With dual-energy x-ray absorptiometry (DXA or DEXA) scan testing, the risk of additional fracture can be predicted. Specific patient education regarding dietary requirements, lifestyle changes, and weight-bearing exercise to promote bone health is needed. Specific therapeutic interventions need to be initiated to slow bone loss and to build bone mineral density (see Chapter 42). Prevention of falls is also important and may be achieved through exercises to improve muscle tone and balance and through the elimination of environmental hazards (Registered Nurses' Association of Ontario, 2011).

Femoral Shaft

Considerable force is required to break the shaft of a femur in adults. Most femoral fractures occur in young adults who have been involved in motor vehicle crashes or who have fallen from heights. Frequently, these patients have associated multiple traumatic injuries (Cline et al., 2012).

The patient presents with an edematous, deformed, painful thigh and cannot move the hip or the knee. The fracture may be transverse, oblique, spiral, or comminuted. Frequently, the patient develops shock, because the loss of 1,000 mL of blood into the tissues is common with these fractures (ENA, 2013). The diameter of the thigh should be closely monitored because expansion may indicate continued bleeding (Cline et al., 2012). Types of femoral fractures are illustrated in Figure 43-17A.

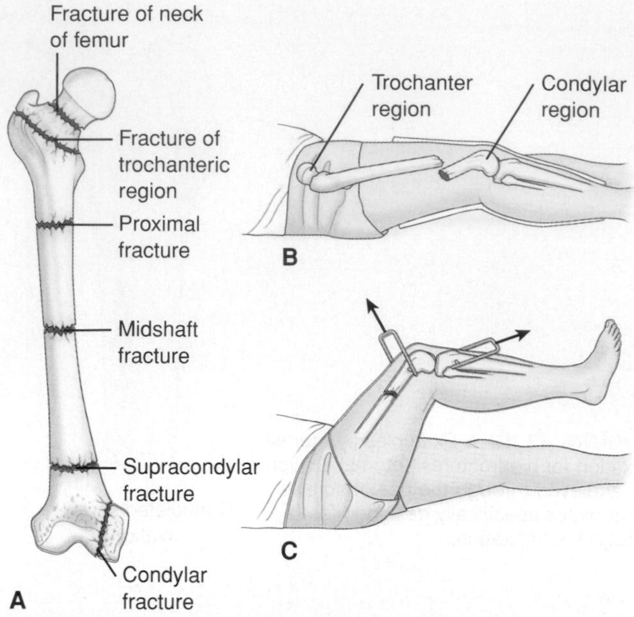
Fracture of neck of femur
Fracture of trochanteric region
Proximal fracture
Midshaft fracture
Supracondylar fracture
Condylar fracture
A
Trochanter region
Condylar region
B
C

FIGURE 43-17 • A. Types of femoral fractures. **B.** Example of deformity on admission to hospital. **C.** Adequate reduction is achieved when additional wire is inserted in the lower femoral fragment and vertical lift is secured.

Assessment and Diagnostic Findings

Assessment includes checking the neurovascular status of the extremity, especially circulatory perfusion of the lower leg and foot (popliteal, posterior tibial, and pedal pulses and toe capillary refill time), and comparing with the unaffected leg. A Doppler ultrasound may be indicated to assess blood flow. X-rays are used to confirm the diagnosis and determine the extent of injury (NAON, 2007). Dislocation of the hip and knee may accompany these fractures. Knee effusion suggests ligament damage and possible instability of the knee joint.

Medical Management

Continued neurovascular monitoring and documentation are important. The fracture is immobilized so that additional soft tissue damage does not occur. Generally, skeletal traction (Fig. 43-17B,C) or splinting is used to immobilize fracture fragments until the patient is physiologically stable and ready for ORIF procedures. IV opioid analgesic agents (e.g., morphine) are typically administered to treat pain; an anxiolytic agent (e.g., lorazepam [Ativan]) may also be prescribed (Keany & McKeever, 2011).

Internal fixation usually is carried out within 24 hours after injury (Scalea, 2008). Intramedullary locking nail devices are typically used (Keany & McKeever, 2011). Internal fixation permits early mobilization, which is associated with improved outcomes and recovery (Keany & McKeever, 2011). A thigh cuff orthosis may be used for external support. To preserve muscle strength, the patient is instructed to exercise the hip and the lower leg, foot, and toes on a regular basis. Active muscle movement enhances healing by increasing blood supply and electrical potentials at the fracture site. Prescribed weight-bearing limits are based on the type and location of the fracture and treatment approach. Physical therapy includes ROM and strengthening exercises, safe use of assistive devices, and gait training (NAON, 2007).

Infrequently, because of patient risk associated with anesthesia and surgery, these fractures may be managed with skeletal traction, prescribed with weights that maintain normal leg length (Keany & McKeever, 2011). This is a temporary intervention, however, until such time as the patient is stable and may tolerate surgical intervention (Keany & McKeever, 2011).

Open femoral fractures require immediate and extensive irrigation and débridement in the operating suite (see previous discussion of treatment for open fractures). Depending on needs for continued débridement, intramedullary nailing may be delayed (Keany & McKeever, 2011).

A common complication after fracture of the femoral shaft is restriction of knee motion. Active and passive knee

(text continues on page 1182)

Chart 43-5

PLAN OF NURSING CARE

Care of the Older Adult Patient With a Fractured Hip

NURSING DIAGNOSIS: Acute pain related to fracture, soft tissue damage, muscle spasm, and surgery
GOAL: Relief of pain

Nursing Interventions	Rationale	Expected Outcomes
1. Assess type and location of patient's pain whenever vital signs are obtained and as needed.	**1.** Pain is expected after fracture; soft tissue damage and muscle spasm contribute to discomfort; pain is subjective and is best evaluated on a pain scale of 0 to 10 and through description of characteristics and location, which are important for identifying cause of discomfort and for proposing interventions. Continuing pain may indicate development of neurovascular problems. Pain must be assessed periodically to gauge effectiveness of continuing analgesic therapy.	• Patient describes and rates pain on scale of 0 to 10 • Expresses confidence in efforts to control pain • Expresses comfort with position changes • Expresses comfort when leg is positioned and immobilized • Minimizes movement of extremity before reduction and fixation • Uses physical, psychological, and pharmacologic measures to reduce discomfort • Describes acceptable pain level that does not interfere with ability to participate in rehabilitation activities within 24–48 hours after surgery • Requests pain medications and uses pain relief measures early in pain cycle • States that positioning provides comfort • Appears comfortable and relaxed • Moves with increasing comfort as healing progresses
2. Acknowledge existence of pain; inform patient of available analgesic agents; record patient's baseline discomfort.	**2.** Reduces stress experienced by the patient by communicating concern and availability of help in dealing with pain. Documentation provides baseline data.	
3. Handle the affected extremity gently, supporting it with hands or pillow.	**3.** Movement of bone fragments is painful; muscle spasms occur with movement; adequate support diminishes soft tissue tension.	
4. Apply Buck's traction if prescribed; use trochanter roll.	**4.** Immobilizes fracture to decrease pain, muscle spasm, and external rotation of hip.	
5. Use pain-modifying strategies.	**5.** Pain perception can be diminished by distraction and refocusing of attention.	
a. Modify the environment.	**a.** Interaction with others, distraction, and environmental stimuli may modify pain experiences.	
b. Administer prescribed analgesic agents as needed.	**b.** Analgesics reduce the pain; muscle relaxants may be prescribed to decrease discomfort associated with muscle spasm.	
c. Encourage patient to use pain relief measures to relieve pain.	**c.** Mild pain is easier to control than severe pain.	
d. Evaluate patient's response to medications and other pain reduction techniques.	**d.** Assessment of effectiveness of measures provides basis for future management interventions; early identification of adverse reactions is necessary for corrective measures and care plan modifications.	
e. Consult with primary provider if relief of pain is not obtained.	**e.** Change in treatment plan may be necessary.	
6. Position for comfort and function.	**6.** Alignment of body facilitates comfort; positioning for function diminishes stress on musculoskeletal system.	
7. Assist with frequent changes in position.	**7.** Change of position relieves pressure and associated discomfort.	

NURSING DIAGNOSIS: Impaired physical mobility related to fractured hip
GOAL: Achieves pain-free, functional, stable hip

Nursing Interventions	Rationale	Expected Outcomes
1. Maintain neutral positioning of hip.	**1.** Prevents stress at the site of fixation	• Patient engages in therapeutic positioning
2. Use trochanter roll; roll to uninjured side.	**2.** Minimizes external rotation	• Uses pillow between legs when turning
3. Place pillow between legs when turning.	**3.** Supports leg; prevents adduction	• Assists in position changes; shows increased independence in transfers
4. Instruct and assist in position changes and transfers.	**4.** Encourages patient's active participation while preventing stress on hip fixation	• Exercises every 2 hours while awake

(continues on page 1178)

Chart 43-5

PLAN OF NURSING CARE

Care of the Older Adult Patient With a Fractured Hip (continued)

Nursing Interventions	Rationale	Expected Outcomes
2. Consider pre-injury blood pressure values and management of coexisting hypertension, if present.	2. Necessary for interpretation of current blood pressure determinations	• Exhibits stable postoperative hemoglobin and hematocrit values • Patient has clear breath sounds. • Breath sounds are present in all fields. • Exhibits no shortness of breath, chest pain, or elevated temperature
3. Note character and amount of drainage.	3. Excessive drainage and bright red drainage may indicate active bleeding.	
4. Notify surgeon if patient develops shock or excessive bleeding.	4. Corrective measures need to be instituted.	
5. Note hemoglobin and hematocrit values, and report decreases in values.	5. Anemia due to blood loss may develop; bleeding into tissues after hip fracture may be extensive; blood replacement may be needed.	
Pulmonary Complications		
1. Assess respiratory status: respiratory rate, depth, and duration; breath sounds; sputum. Monitor temperature.	1. Anesthesia and bed rest diminish respiratory effort and cause pooling of respiratory secretions. Adventitious breath sounds, pain on respiration, shortness of breath, blood tinged sputum, cough may indicate pulmonary dysfunction.	• Vital signs are stabilized within normal limits. • Patient has clear breath sounds. • Breath sounds are present in all fields. • Exhibits no shortness of breath, chest pain, or elevated temperature
2. Report adventitious and diminished breath sounds and elevated temperature.	2. Elevated temperature in the early postoperative period may be due to atelectasis or pneumonia.	• Arterial oxygen saturation (SaO_2) on room air is within normal limits. • Performs respiratory exercises; uses incentive spirometer as instructed
3. Supervise deep-breathing and coughing exercises. Encourage the use of incentive spirometer if prescribed.	3. Deep-breathing and coughing exercises promote optimal ventilation. Coexisting respiratory conditions diminish lung expansion.	• Changes position frequently • Consumes adequate fluids
4. Administer oxygen as prescribed.	4. Reduced ventilatory efforts may diminish SaO_2 when patient is breathing room air.	
5. Turn and reposition patient at least every 2 hours. Mobilize patient (assist patient out of bed) as soon as possible.	5. Promotes optimal ventilation; diminishes pooling of respiratory secretions	
6. Ensure adequate hydration.	6. Liquefies respiratory secretions; facilitates expectoration	
Peripheral Neurovascular Dysfunction		
1. Assess affected extremity for color and temperature.	1. The skin becomes pale and feels cool with decreased tissue perfusion. Venous congestion may cause cyanosis.	• Patient has normal color and the extremity is warm. • Demonstrates normal capillary refill response
2. Assess toes for capillary refill response.	2. After compression of the nail, rapid return of pink color indicates good capillary perfusion.	• Exhibits moderate swelling; tissue not palpably tense • States pain is tolerable
3. Assess affected extremity for edema and swelling.	3. The trauma of surgery will cause swelling; excessive swelling and hematoma formation can compromise circulation and function; edema may be due to coexisting cardiovascular disease.	• Reports no pain with passive dorsiflexion • Reports normal sensations and no paresthesia • Demonstrates normal motor abilities and no paresis or paralysis
4. Elevate affected extremity.	4. Minimizes dependent edema	• Has strong and equal pulses
5. Assess for deep, throbbing, unrelenting pain.	5. Surgical pain can be controlled; pain due to neurovascular compromise is refractory to treatment with analgesic medications.	
6. Assess for pain on passive flexion of foot.	6. With nerve ischemia, there will be pain on passive stretch.	
7. Assess for sensations and numbness.	7. Diminished pain and paresthesia may indicate nerve damage. Sensation in web between great and second toe—peroneal nerve; sensation on sole of foot—tibial nerve.	

Chart 43-5

PLAN OF NURSING CARE

Care of the Older Adult Patient With a Fractured Hip (continued)

Nursing Interventions	Rationale	Expected Outcomes
8. Assess ability to move foot and toes.	8. Dorsiflexion of ankle and extension of toes indicate function of peroneal nerve. Plantar flexion of ankle and flexion of toes indicate functioning of tibial nerve.	
9. Assess pedal pulses in both feet.	9. Indicates circulatory status of extremities	
10. Notify surgeon if diminished neurovascular status occurs.	10. Function of extremity needs to be preserved.	
Deep Vein Thrombosis		
1. Apply thigh-high anti-embolism stockings and/or sequential compression device as prescribed.	1. Compression aids venous blood return and prevents stasis.	• Wears thigh-high anti-embolism stockings • Uses sequential compression device • Experiences no more warmth than usual in skin areas • Exhibits no increase in calf circumference • Demonstrates no evidence of calf tenderness, warmth, redness, or swelling • Changes position with assistance and supervision • Participates in exercise regimen • Experiences no chest pain; has lungs clear to auscultation; presents no evidence of pulmonary emboli • Exhibits no signs of dehydration; has normal hematocrit • Maintains normal body temperature
2. Remove stockings for 20 minutes twice a day, and provide skin care.	2. Skin care is necessary to avoid skin breakdown. Extended removal of stocking or device defeats purpose.	
3. Assess popliteal, dorsalis pedis, and posterior tibial pulses.	3. Pulses indicate arterial perfusion of extremity. With coexisting arteriosclerotic vascular disease, pulses may be diminished or absent.	
4. Assess skin temperature of legs.	4. Local inflammation increases local skin temperature.	
5. Assess calf intermittently for tenderness, warmth, redness, and swelling.	5. Unilateral calf tenderness, warmth, redness, and swelling may indicate deep vein thrombosis.	
6. Measure calf circumference daily.	6. Increased calf circumference indicates edema or altered perfusion.	
7. Avoid pressure on popliteal blood vessels from appliances or pillows.	7. Compression of blood vessels diminishes blood flow.	
8. Change patient's position and increase activity as prescribed.	8. Activity promotes circulation and diminishes venous stasis.	
9. Supervise ankle exercises hourly while patient is awake.	9. Muscle exercise promotes circulation.	
10. Ensure adequate hydration.	10. Older adults may become dehydrated because of low fluid intake, resulting in hemoconcentration.	
11. Monitor body temperature.	11. Body temperature increases with inflammation (magnitude of response minimal in older adults).	

Nursing Interventions	Rationale	Expected Outcomes
Pressure Ulcers		
1. Monitor condition of skin at pressure points (e.g., heels, sacrum, shoulders); inspect heels at least twice a day.	1. Older adults are more prone to skin breakdown at points of pressure because of diminished subcutaneous tissue.	• Patient exhibits no signs of skin breakdown • Skin remains intact • Repositions self frequently • Uses protective devices
2. Reposition patient at least every 2 hours. Avoid skin shearing.	2. Avoids prolonged pressure and trauma to the skin.	
3. Administer skin care, especially to pressure points.	3. Immobility causes pressure at bony prominences; position changes relieve pressure.	
4. Use special care mattress and other protective devices (e.g., heel protectors); support heel off the mattress.	4. Devices minimize pressure on skin at bony prominences.	
5. Institute care according to protocol at first indication of potential skin breakdown.	5. Early interventions prevent tissue destruction and prolonged rehabilitation.	

(continues on page 1182)

Chart 43-5

PLAN OF NURSING CARE

Care of the Older Adult Patient With a Fractured Hip (continued)

NURSING DIAGNOSIS: Readiness for enhanced self-health management related to fractured hip and impaired mobility
GOAL: Exhibits self-health management behaviors

Nursing Interventions	Rationale	Expected Outcomes
1. Assess home environment for discharge planning.	**1.** Physical barriers (especially stairs, bathrooms) may limit patient's ability to ambulate and care for self at home.	• Home is accessible for patient at time of discharge. • Patient appears relaxed and develops strategies to deal with identified problems. • Has personal assistance available • Demonstrates ability to use necessary assistive devices within therapeutic prescription • Adheres to home care program; keeps follow-up health care appointments
2. Encourage patient to express concerns about care at home; explore with patient possible solutions to problems.	**2.** Patient may have special problems that need to be identified so that solutions might be identified.	
3. Assess availability of physical assistance for activities of daily living (ADLs) and health care activities.	**3.** Because of limitation of mobility, patient requires some assistance in ADLs and routine health care.	
4. Educate the caregiver about the home health care regimen.	**4.** Understanding of rehabilitative regimen is necessary for compliance.	
5. Educate patient and caregiver in posthospital care: **a.** Activity limitations **b.** Reinforcement of exercise instructions **c.** Safe use of ambulatory aids **d.** Wound care **e.** Measures to promote healing (nutrition, wound care) **f.** Medications **g.** Potential problems **h.** Continuing health care supervision	**5.** Lack of knowledge and poor preparation for care at home contribute to patient and caregiver anxiety, insecurity, and nonadherence to therapeutic regimen.	

exercises begin as soon as possible, depending on the stability of the fracture and knee ligaments. Other long-term complications may include malrotation, malunion, delayed union, and nonunion (Keany & McKeever, 2011).

Tibia and Fibula

Fractures of the tibia and fibula often occur in association with each other and tend to result from a direct blow, falls with the foot in a flexed position, or a violent twisting motion. Most of these fractures tend to be more distal than proximal; distal fractures may extend into the ankle joint (i.e., distal fractures of the tibia that extend into the joint are collectively referred to as pilon fractures) (Horn, Price, & Van Aman, 2011). The patient presents with pain, deformity, obvious hematoma, and considerable edema.

Assessment and Diagnostic Findings

The peroneal nerve is assessed; if damaged, the patient cannot dorsiflex the great toe and has diminished sensation in the first web space. The tibial artery is assessed for damage by evaluating pulses, skin temperature, and color and by testing the capillary refill response. The affected leg and ankle are compared with the unaffected leg and ankle. X-rays are indicated to determine the location, type, and extent of the fracture (NAON, 2007).

Medical Management

Most closed, nondisplaced fractures that do not involve the ankle joint (i.e., extra-articular fractures) are treated with closed reduction and immobilization in a non–weight bearing short-leg cast or brace. The leg is elevated to control edema. Partial weight bearing is usually prescribed after 7 to 10 days, depending on the type of fracture. Activity decreases edema and increases circulation. Fracture healing takes 4 to 6 weeks (NAON, 2007).

Displaced, open, or articular fractures may be treated with skeletal traction, internal fixation with intramedullary nails or plates and screws, or external fixation. External support may be used with internal fixation. Hip, foot, and knee exercises are encouraged within the limits of the immobilizing device. Partial weight bearing is begun when prescribed and is progressed as the fracture heals in 6 to 10 weeks (NAON, 2007).

Continued neurovascular evaluation is important. The development of acute compartment syndrome requires prompt recognition and communication to the orthopedic surgeon. Other complications include nonunion, delayed union, infection, and impaired wound edge healing due to limited soft tissue (Patel et al., 2012).

Rib

Uncomplicated fractures of the lower ribs occur frequently in adults of all ages, typically from either motor vehicle crashes or falls, and usually result in no impairment of function. They are typically diagnosed based on clinical presentation and confirmed with x-rays or ultrasound scans (Melendez & Doty, 2012). Because these fractures cause pain with respiratory effort, the patient tends to decrease respiratory excursions and refrains from coughing. As a result, tracheobronchial secretions are not mobilized, aeration of the lung is diminished, and a predisposition to atelectasis and pneumonia results. To help the patient cough and take deep breaths and use an incentive spirometer (see Chapter 21), the nurse may splint the chest with his or her hands, or may educate the patient on using a pillow to temporarily splint the affected site. NSAIDs may be prescribed to provide analgesic relief. Occasionally, an anesthesia care provider administers intercostal nerve blocks to relieve pain and to improve respiratory function (Melendez & Doty, 2012).

Chest strapping to immobilize the rib fracture is not used, because decreased chest expansion may result in atelectasis and pneumonia. The pain associated with rib fracture diminishes significantly in 3 or 4 days, and the fracture heals within 6 weeks. In addition to atelectasis and pneumonia, complications may include a flail chest, pneumothorax, and hemothorax (Melendez & Doty, 2012). The assessment and management of patients with these conditions are discussed in Chapter 23.

Thoracolumbar Spine

More than 150,000 people sustain fractures of the vertebral column annually in the United States; of these, most involve the thoracolumbar spine (Vinas, 2011). Fractures of the thoracolumbar spine may involve (1) the vertebral body, (2) the laminae and articulating processes, and (3) the spinous processes or transverse processes. The T12 to L2 area of the spine, called the *thoracolumbar junction*, is most vulnerable to fracture (Leahy & Rahm, 2012). Fractures generally result from indirect trauma caused by excessive loading, sudden muscle contraction, or excessive motion beyond physiologic limits. Osteoporosis contributes to vertebral body collapse (compression fracture) (Vinas, 2011).

Stable spinal fractures are caused by flexion, extension, lateral bending, or vertical loading. The anterior structural column (vertebral bodies and disks) or the posterior structural column (neural arch, articular processes, ligaments) is disrupted. Unstable fractures occur with fracture dislocations and involve disruption of both anterior and posterior structural columns.

The patient with a spinal fracture presents with acute tenderness, swelling, paravertebral muscle spasm, and change in the normal curves or in the gap between spinous processes. Pain is greater with moving, coughing, or weight bearing. Immobilization is essential until initial assessments have determined if there is any spinal cord injury and whether the fracture is stable or unstable (see Chapter 68). X-rays are initially indicated to confirm the fracture(s), and CT scans or MRI studies are then indicated to precisely determine the extent of injury (Vinas, 2011). If spinal cord injury with neurologic deficit does occur, it usually requires immediate surgery (laminectomy with spinal fusion) to decompress the spinal cord.

Stable spinal fractures are treated conservatively with limited bed rest. The head of the bed is elevated less than 30 degrees until the acute pain subsides (several days). Analgesic medications are prescribed for pain relief. The patient is monitored for a transient paralytic ileus caused by associated retroperitoneal hemorrhage. Sitting is avoided until the pain subsides. A spinal brace or plastic thoracolumbar orthosis may be applied for support during progressive ambulation and resumption of activities. X-rays are taken periodically to monitor the healing process. The patient is typically advised to restrict activities for 6 months (e.g., no lifting more than 20 pounds, no contact sports activities) (Leahy & Rahm, 2012).

The patient with an unstable fracture is treated with bed rest, possibly with the use of a special turning device or bed to maintain spinal alignment. Within 24 hours after fracture, open reduction, decompression, and fixation with spinal fusion and instrument stabilization are usually accomplished. Neurologic status is monitored closely during the preoperative and postoperative periods. Postoperatively, the patient may be cared for on the turning device or in a bed with a firm mattress. Progressive ambulation is begun a few days after surgery, with the patient using a body brace orthosis. Patient education emphasizes good posture, good body mechanics, and, after healing is sufficient, back-strengthening exercises. (See Chapter 68 for discussion of spinal cord injury.)

Sports-Related Injuries

Sport activities are very common, and, unfortunately, sports-related injuries are also common consequences. Table 43-1 displays common sports injuries, their mechanisms of injury, assessment findings, and acute care management.

Management

Patients who have experienced sports-related injuries are often highly motivated to return to their previous level of activity. Adherence to restriction of activities and gradual resumption of activities need to be reinforced. Injured athletes are at risk for reinjury and require follow-up and monitoring. With recurrence of symptoms, athletes need to diminish their level and intensity of activity to a comfortable level. The time required to recover from a sports-related injury can be as short as a few days or as long as 12 weeks, depending on the severity of the injury. The injured body part should have at least 90% of the strength of its uninjured symmetrical body part before sports activities are resumed. Increasing activities gradually to acclimate the muscles, tendons, and joints to the sport motions will assist in recovery and rehabilitation. It is generally advisable to resume sporting activities at half the previous level and gradually increase activity by an additional 10% to 15% on a weekly basis (American College of Sports Medicine [ACSM], 2011).

Prevention

Sports-related injuries can often be prevented by using proper equipment (e.g., running shoes for joggers, wrist guards for skaters) and by effectively training and conditioning the body. Specific training needs to be tailored to the person and the sport. Stretching prior to engaging in sports or exercise had long been recommended; however, studies done a decade ago suggest that stretching may not prevent injury. More recent studies note that stretching exercises are not detrimental, however, unless the stretching is prolonged (i.e., longer than a minute) (Kay & Blazevich, 2012).

TABLE 43-1 Common Sports Injuries

Anatomic Area	Mechanism of Injury	Assessment Findings	Sports Activity	Acute Management
Clavicle fracture	Fall on shoulder or outstretched arm Direct blow to the clavicle	Crepitus Holds arm closely to body Unable to raise affected arm above head Can feel movement of both ends of clavicle	Football Rugby Hockey Wrestling Gymnastics	Sling or shoulder immobilizer Ice NSAIDs
Dislocated shoulder	*Anterior:* Some combination of hyperextension, external rotation, and abduction Anterior blow to shoulder *Posterior:* Fall on flexed and adducted arm Direct axial load to humerus	Pain Lack of motion May feel empty shoulder socket Uneven posture in comparison to other shoulder Affected arm appears longer Abduction limited	Rugby Hockey Wrestling Skiing	Closed reduction Immobilizer Pendulum exercises
Dislocated elbow	Falling on a hand with a flexed elbow Elbow overextended	Intense pain Edema Limited motion Deformity Ecchymosis	Football Gymnastics Squash Wrestling Cycling Skiing	Immobilization Ice ROM exercises
Wrist sprain or fracture	Falling on an outstretched arm	Pain Edema Ecchymosis Deformity Limited motion	Skating Hockey Wrestling Skiing Soccer Handball Horseback riding	Ice Elevation Immobilization Gentle ROM for 4–6 weeks (for sprain only)
Knee sprain	Twisting injury that produces incomplete tear of ligaments and capsule around the joint	Pain Limited motion Edema Ecchymosis Tenderness over joint Joint appears stable	Basketball Football High jump	Ice Elevation Compression wrap Active ROM exercises Isometric exercises May immobilize
Knee strain	Sudden forced motion causing muscle to be stretched beyond normal capacity	Pain Limited motion Pain aggravated by activity	Soccer Swimming Skiing	Ice Elevation Rest Gradual return to activities
Meniscal tears of knee	Sharp, sudden pivot Direct blow to knee Forced internal rotation Wear from repetitive squatting or climbing Torsional weight-bearing force	Edema *Medial tear:* Pain occurs with hyperflexion, hyperextension, and turning in of knee with knee flexed. *Lateral tear:* Pain occurs with hyperflexion and hyperextension and internal rotation of foot with knee flexed. *Displaced fragment:* Inability to extend knee; "locked" Positive McMurray's sign*	Hockey Basketball Football	*Conservative:* RICE Exercising of quadriceps and hamstrings Resistive exercising NSAIDs Physical therapy *Surgical:* Arthroscopy
Ankle sprain	Foot is twisted, causing stretching or tearing of ligaments.	Pain Edema Limited motion Ecchymosis	Tennis Basketball Football Skating	Immobilization in cast or brace Ice Elevation Rest
Ankle strain	Sudden forced motion, stretching muscles beyond normal capacity	*Acute:* Severe pain *Chronic:* Achy pain	Running All ball sports	Immobilization in cast or brace Ice Elevation Rest
Ankle fracture	Inward turning on sole of foot and front of foot Supination with internal rotation Pronation with external rotation	Pain Edema Deformity Inability to bear weight	Contact sports Tennis Basketball	Ice Elevation Cast (4–6 weeks) Surgery if fracture is displaced or unstable
Metatarsal stress fracture	Occurs with repeated loading of bone; often in an unconditioned extremity	Forefoot pain that progressively worsens with activity Minimal or no forefoot swelling	Running Dance Skating	Rest Stop sports-related activity for 6 weeks Ice Weight bearing as indicated

NSAIDs, nonsteroidal anti-inflammatory drugs; ROM, range of motion; RICE, rest, ice, compression, elevation.
*McMurray's sign—manipulation of tibia while knee flexed produces audible "click."
Reprinted with permission from National Association of Orthopedic Nurses. (2007). *Core curriculum for orthopaedic nursing* (6th ed.). Boston: Pearson.

Occupation-Related Musculoskeletal Disorders

According to the U.S. Department of Labor, occupation-related musculoskeletal disorders (MSDs) are injuries or illnesses of the muscles, nerves, tendons, joints, cartilage, and bones that occur because of exposure to work-related risks. In 2011, occupation-related MSDs, also called *ergonomic injuries*, accounted for 33% of all injury and illness cases; six occupations accounted for 26% of these cases. Registered nurses and nursing assistants were included as two of these six occupations that reported high rates of occupation-related MSDs (U.S. Department of Labor, Bureau of Labor Statistics, 2012).

The most frequent types of MSDs that occurred among workers in all occupations in 2011 and that required days away from work were sprains, strains, and tears (38%) and soreness and pain (12%), with smaller percentages of other MSDs including fractures, amputations, and multiple injuries with fractures accounting for other reported cases (U.S. Department of Labor, 2012).

Prevention of Occupation-Related Musculoskeletal Disorders in Nursing Personnel

Nursing is consistently ranked among the top 10 occupations that are most involved in occupation-related injuries and lost days of work. Most of these injuries result in MSDs and are linked to patient handling and movement activities. Although the mechanism of injury for some of these MSDs may be acute, mounting evidence suggests that cumulative or "incremental" effects of repeated stress may be the root cause for many MSDs among nursing personnel (e.g., nursing assistants, licensed professional/vocational nurses, and registered nurses) (Watters, 2008). The American Nurses Association (ANA) launched a "Handle with Care" campaign in 2004, which was aimed at reducing occupation-related MSDs among nurses. Yet, results from a recent ANA survey of 4,600 nurses revealed that the majority still indicated that the fear of incurring an occupational-related MSD was one of their top three safety concerns. Survey results also revealed that 8 out of 10 nurses worked despite experiencing musculoskeletal pain and that 13% were injured at least three times on the job within the past year (ANA, 2012). As a response to these statistics, the ANA officially endorses the implementation of policies that may eliminate manual patient handling by nurses; in particular, the ANA advocates for the following (ANA, 2013):

- The implementation of education and training programs about safe patient handling for all nursing personnel
- Increased usage of assistive equipment and safe patient-handling devices
- Development of federal and state ergonomics policies for nursing personnel

Amputation

Amputation is the removal of a body part, often a limb. Amputation may be necessary because of the consequences of vascular diseases, which, in turn, is often a sequela of diabetes (see Chapter 30). Fulminating gas gangrene, trauma (crushing injuries, burns, frostbite, electrical burns, explosions, ballistic injuries), congenital deformities, chronic osteomyelitis, or malignant tumor may also necessitate an amputation. Of all these causes, vascular diseases account for 82% of all amputations (i.e., dysvascular amputations); 97% of these result in lower limb amputations. Males and African Americans are at heightened risk of having dysvascular amputations. Amputation of an upper limb occurs less frequently than that of a lower limb and is most often necessary because of either traumatic injury or a malignant tumor. However, upper limb trauma-related amputations account for 68% of all trauma-related amputations. Males are at higher risk for trauma-related amputations than females, and risk of having traumatic amputations increases with increasing age. It is estimated that 1.7 million Americans have had some type of limb loss and that 1 of every 200 Americans has had an amputation (National Limb Loss Information Center, 2008).

Amputation is used to relieve symptoms, to improve function, and to save or improve the patient's quality of life. If the health care team communicates a positive attitude, the patient adjusts to the amputation more readily and actively participates in the rehabilitative plan, learning how to modify activities and how to use assistive devices for ADLs and mobility (Chart 43-6).

Levels of Amputation

Amputation is performed at the most distal point that will heal successfully. The site of amputation is determined by two factors: circulation in the part and functional usefulness (i.e., meets the requirements for the use of a prosthesis).

The circulatory status of the limb is evaluated through physical examination and diagnostic studies. Muscle and skin perfusion is important for healing. Doppler flow studies with duplex ultrasound, segmental blood pressure determinations, and transcutaneous partial pressure of arterial oxygen (PaO_2) of the limb are valuable diagnostic aids. Angiography is performed if revascularization is considered an option.

The objective of surgery is to conserve as much limb length as needed to preserve function and possibly to achieve a good prosthetic fit. Preservation of knee and elbow joints is desirable. Figure 43-18 shows the levels at which a limb may be amputated. Most amputations involving limbs eventually can be fitted with a prosthesis.

The amputation of toes and portions of the foot can cause changes in gait and balance. A Syme amputation (modified ankle **disarticulation** amputation) is performed most frequently for extensive foot trauma and aims to produce a durable residual limb that can withstand full weight bearing. Below-knee amputation (BKA) is preferred to above-knee amputation (AKA) because of the importance of the knee joint and the energy requirements for walking. Knee disarticulations are most successful with young, active patients who can develop precise control of the prosthesis. When AKAs are performed, all possible length is preserved, muscles are stabilized and shaped, and hip contractures are prevented to maximize ambulatory potential (NAON, 2007). Most people who have a hip disarticulation amputation must rely on a wheelchair for mobility.

Chart 43-6 ETHICAL DILEMMA

Should an Older Adult Be Allowed to Refuse Treatment When That Treatment Is Likely to Extend Life?

Case Scenario

The patient is an 88-year-old man with an extensive cardiac history who has been a resident of a skilled nursing facility for the past 3 years. Comorbid conditions include type 2 diabetes (15 years) and severe peripheral vascular disease. For 6 months, the patient has suffered from a serious left leg wound that has not responded to treatment and has become much worse over the past 2 months. At this point, the leg is gangrenous to above the ankle, and the normal course of treatment would be to amputate the leg. However, the patient is refusing surgery and states that he wants "to die with all of my limbs intact." The patient has been diagnosed with mild dementia and has named his daughter as his durable power of attorney for health care. The nurses on the unit, the daughter, and the primary provider all agree that an amputation is necessary to save the patient's life. However, after hearing about the need for surgery and the likely outcome without the surgery, the patient adamantly continues to state, "I want to die with all of my limbs intact." The daughter and some of the nurses who care for the patient question his capacity to make such a decision competently.

Discussion

Several ethical issues are relevant to the resolution of the situation. The obligation to respect the patient's autonomy in the decision to refuse an amputation puts his life at risk. Although he is diagnosed with mild dementia, such a diagnosis does not necessarily mean that he lacks the capacity to make a competent decision.

Analysis

- Identify the ethical principles that are in conflict in this case (see Chart 3-3 in Chapter 3).
- What arguments would you offer *against* the surgery?
- What arguments would you offer *in favor of* the surgery?
- What resources are available to help you, your professional colleagues, the patient, and the daughter reach consensus on charting the best treatment plan for the patient?
- Assume that the patient is determined competent to make his own decisions and his surgery is not scheduled. As he begins to decline and loses consciousness, his daughter then wants to have him scheduled for an emergent amputation. How might this change or not change the ethical decision-making process and the patient's outcome?

Resources

See Chapter 3, Chart 3-6 for ethics resources.

Upper limb amputations are performed with the goal of preserving maximal functional length. The prosthesis is fitted early to ensure maximum function.

A "staged" amputation may be used when gangrene and infection exist. Initially, a guillotine amputation (e.g., nonclosed residual limb) is performed to remove the necrotic and infected tissue. The wound is débrided and allowed to drain. Sepsis is treated with systemic antibiotics. In a few days, after the infection has been controlled and the patient's condition has stabilized, a definitive amputation with skin closure is performed.

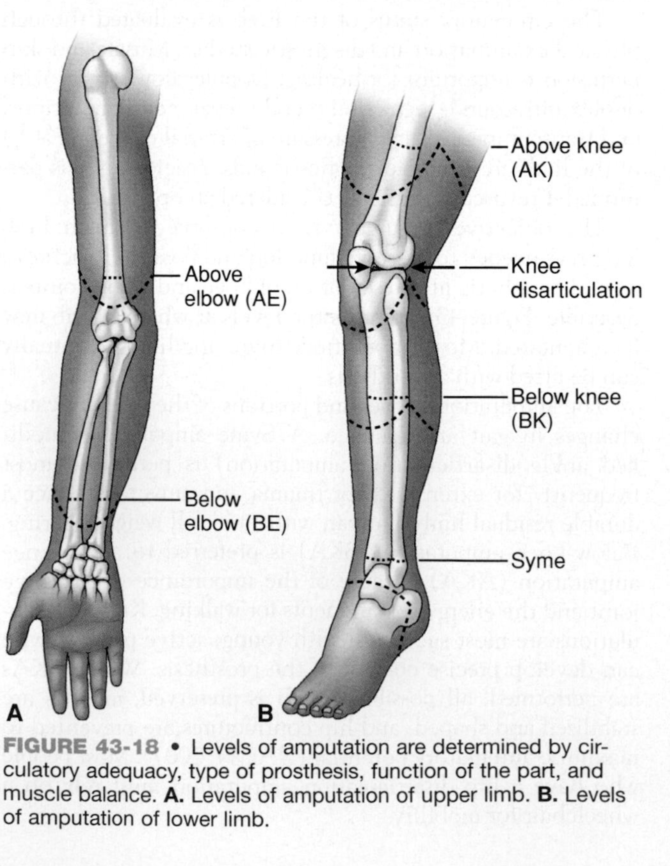

Above elbow (AE)
Below elbow (BE)
Above knee (AK)
Knee disarticulation
Below knee (BK)
Syme
A
B

FIGURE 43-18 • Levels of amputation are determined by circulatory adequacy, type of prosthesis, function of the part, and muscle balance. **A.** Levels of amputation of upper limb. **B.** Levels of amputation of lower limb.

Complications

Complications that may occur with amputation include hemorrhage, infection, skin breakdown, phantom limb pain, and joint contracture. Because major blood vessels have been severed, massive bleeding may occur. Infection is a risk with all surgical procedures. The risk of infection increases with contaminated wounds after traumatic amputation. Skin irritation caused by the prosthesis may result in skin breakdown. **Phantom limb pain** (pain perceived in the amputated section) is caused by the severing of peripheral nerves. Joint contracture is caused by positioning and a protective flexion withdrawal pattern associated with pain and muscle imbalance (NAON, 2007).

Medical Management

The objective of treatment is to achieve healing of the amputation wound, the result being a nontender residual limb with healthy skin for prosthetic use. Healing is enhanced by gentle handling of the residual limb, control of residual limb edema through rigid or soft compression dressings, and the use of aseptic technique in wound care to avoid infection (Department of Veteran Affairs, Department of Defense [VA/DoD], 2007).

A closed rigid cast dressing or an elastic residual limb shrinker that covers the residual limb may be used to provide uniform compression, to support soft tissues, to control pain, and to prevent joint contractures. Immediately after surgery, a sterilized residual limb sock is applied. Padding is placed over pressure-sensitive areas (VA/DoD, 2007).

For the patient with a lower limb amputation, the cast may be equipped to attach a temporary prosthetic extension (pylon) and an artificial foot. This rigid dressing technique is used as a means of creating a socket for immediate postoperative prosthetic fitting. The length of the prosthesis is tailored to the individual patient. Early minimal weight bearing using this device typically produces little discomfort. The cast is changed in about 10 to 14 days. A fever, severe pain, or a loose-fitting cast may necessitate earlier replacement (VA/DoD, 2007).

A removable rigid dressing may be placed over a soft dressing to control edema, to prevent joint flexion contracture, and to protect the residual limb from unintentional trauma during transfer activities. This rigid dressing is removed several days after surgery for wound inspection and is then replaced to control edema. The dressing facilitates residual limb shaping.

A soft dressing with or without compression may be used if there is significant wound drainage and frequent inspection of the residual limb is required. An immobilizing splint may be incorporated in the dressing. Residual limb wound hematomas are controlled with wound drainage devices to minimize infection.

Rehabilitation

The multidisciplinary rehabilitation team (patient, nurse, physician, social worker, physical therapist, occupational therapist, psychologist, prosthetist, vocational rehabilitation worker) helps the patient achieve the highest possible level of function and participation in life activities (Fig. 43-19). Prosthetic clinics and support groups for people with amputations facilitate this rehabilitation process (McFarland, Choppa, Betz, et al., 2010).

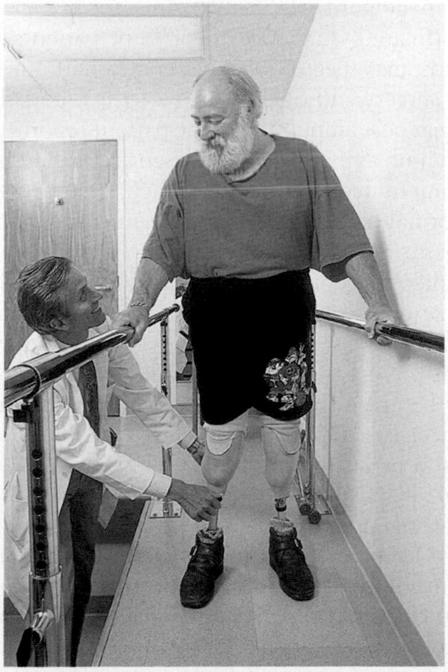

FIGURE 43-19 • Many patients with amputations receive prostheses soon after surgery and begin learning how to use them with the help and support of the rehabilitation team, which includes nurses, physicians, physical therapists, and others.

Patients who undergo amputation need support as they grieve the loss and change in body image. Their reactions can include anger, bitterness, and hostility. Psychological issues (e.g., denial, withdrawal) may be influenced by the type of support the patient receives from the rehabilitation team and by how quickly ADLs and the use of the prosthesis are learned. Knowing the full options and capabilities available with the various prosthetic devices can give the patient a sense of control over the resulting disability (Peerdeman, Boere, Witteveen, et al., 2011).

Patients who require amputation because of severe trauma can be young and healthy. These patients heal rapidly and are physically able to participate in a vigorous rehabilitation program. Because the amputation is the result of an injury, the patient needs psychological support in accepting the sudden change in body image and in dealing with the stresses of hospitalization, long-term rehabilitation, and modification of lifestyle.

Within the past decade, numerous U.S. soldiers have lost limbs because of ballistic injuries received while fighting in Iraq and Afghanistan. In order to best meet the complex needs of these young, previously healthy men and women, the U.S. Army instituted both a specialized treatment center for patients with amputations and a database registry to facilitate their long-term treatment and management. Treatment for these injured soldiers addresses not only their physical rehabilitation needs but also their emotional needs (Chart 43-7). Behavioral health services are considered key components to their therapy (McFarland et al., 2010).

NURSING PROCESS

The Patient Undergoing an Amputation

Assessment

Before surgery, the nurse must evaluate the neurovascular and functional status of the limb through history and physical assessment. If the patient has experienced a traumatic amputation, the nurse assesses the function and condition of the residual limb. The nurse also assesses the circulatory status and function of the unaffected limb. If infection or gangrene develops, the patient may have associated enlarged lymph nodes, fever, and purulent drainage. A culture and sensitivity test is obtained to determine the appropriate antibiotic therapy.

The nurse evaluates the patient's nutritional status and develops a plan for nutritional care in consultation with a dietitian or metabolic support team, if indicated. A diet with adequate protein and vitamins is essential to promote wound healing.

Any concurrent health problems (e.g., dehydration, anemia, cardiac insufficiency, chronic respiratory problems, diabetes) need to be identified and treated so that the patient is in the best possible condition to withstand the surgical procedure. The use of corticosteroids, anticoagulants, vasoconstrictors, or vasodilators may influence management and prolong or delay wound healing.

The nurse assesses the patient's psychological status. Evaluation of the patient's emotional reaction to amputation is important. Grief responses to permanent alterations in body image, function, and mobility are likely. Professional counseling can help the patient cope in the aftermath of amputation surgery.

Chart 43-7

NURSING RESEARCH PROFILE

Psychological Changes after Posttraumatic Amputation Among Veterans

Benetato, B. B. (2011). Posttraumatic growth among Operation Enduring Freedom and Operation Iraqi Freedom amputees. *Journal of Nursing Scholarship, 43*(4), 412–420.

Purpose

It is well established that psychosocial adjustment following traumatic major life events can cause depressive symptoms, including posttraumatic stress disorder (PTSD). However, recent research findings suggest that some people faced with traumatic major life events may experience positive changes post event, a dynamic experience called *posttraumatic growth* (PTG). Whereas PTSD has been studied in war veterans, PTG has not. Therefore, the purpose of this study was to examine PTG and its relationship to social support, rumination (i.e., as a measure of cognitive processing), and length of time from injury among Operation Enduring Freedom (OEP) and Operation Iraqi Freedom (OIF) veterans who experienced battle-related traumatic amputations.

Design

Following Institutional Review Board approval, a survey was mailed to all eligible OEF and OIF veterans who had a traumatic battle-related amputation and were listed in a Department of Veterans Affairs database. Survey instruments included the PTG Inventory, the Postdeployment Social Support Scale, and the Rumination Inventory. Of the 210 eligible veterans, 56 consented to participate in the study and completed the survey instruments.

Findings

The mean PTG scores in this sample of war veterans who had experienced traumatic battle-related amputations were generally higher than those reported by participants in other studies who faced major illnesses; however, there was much variance in these results. This finding supports the belief that PTG is an individualized process. This belief was further supported by the finding that there was no correlation between PTG scores and time since injury. In other words, the time it takes to adjust psychosocially following traumatic injury does not follow a predictable trajectory. There was a significant association between rumination and PTG: $r = .43$, $p = .001$, and between social support and PTG: $r = .24$, $p = .039$.

Nursing Implications

Both rumination (as a type of cognitive processing) and social support were associated with PTG in this sample of war veteran participants who had experienced traumatic amputation. Nurses are in unique positions in that they spend a lot of time with patients following traumatic amputation in acute care, rehabilitation, long-term care, and home health settings. Encouraging patients' ruminations about their traumatic experiences can improve their PTG. Furthermore, nurses should assess these veterans' extant social support networks and provide links to specific veterans' programs in order to foster their PTG experience.

Diagnosis

Nursing Diagnoses

Based on the assessment data, major nursing diagnoses may include the following:

- Acute pain related to amputation
- Impaired skin integrity related to surgical amputation
- Disturbed body image related to amputation
- Grieving and/or risk for complicated grieving related to loss of body part and resulting disability
- Self-care deficit: feeding, bathing, dressing, feeding, or toileting, related to amputation
- Impaired physical mobility related to amputation

Collaborative Problems/Potential Complications

Potential complications may include the following:

- Postoperative hemorrhage
- Infection
- Skin breakdown

Planning and Goals

The major goals of the patient may include relief of pain, including phantom limb pain, wound healing, acceptance of altered body image, resolution of the grieving process, independence in self-care, restoration of physical mobility, and absence of complications.

Nursing Interventions

Relieving Pain

Pain may be incisional or may be caused by inflammation, infection, pressure on a bony prominence, hematoma, or phantom limb pain. Muscle spasms may add to the patient's discomfort. Surgical pain can be effectively controlled with opioid analgesics that may be accompanied with evacuation of a hematoma or accumulated fluid. Changing the patient's position or placing a light sandbag on the residual limb to counteract the muscle spasm may improve the patient's level of comfort. Evaluation of the patient's pain and responses to interventions is an important component of pain management.

It is estimated that 60% to 80% of patients who have amputations may begin to experience phantom limb pain soon after surgery, although some patients report this pain as long as 1 year post amputation. The patient reports pain in the amputated limb as if it were still present. The pain is described as if the amputated limb feels crushed, cramped, or twisted in an abnormal position; this pain is sometimes accompanied by numbness or tingling. Phantom pain occurs intermittently and unpredictably; episodes of pain may last seconds or hours. When a patient describes phantom pains or sensations, the nurse acknowledges these feelings as real and encourages the patient to verbalize when in pain so that effective treatment may be given. Phantom pain typically diminishes over time for most patients, with episodes gradually becoming less frequent and of lesser duration (Wolff, Vanduynhoven, van Kleef, et al., 2011).

The pathogenesis of the phantom limb phenomenon is believed to involve changes in both peripheral and central neuronal mechanisms. Disruption in neuronal pathways are thought to cause neuroplastic changes that in turn result in changes in cortical representation, or the proprioceptive, tactile, and visual image of body parts as perceived by the cerebral cortex (Wolff et al., 2011). Although it had been thought that early intensive rehabilitation, desensitization therapy, and massage therapy were effective interventions in mitigating the intensity and duration of phantom limb pain, research-based evidence does not confirm that any of these interventions are effective. Opioid analgesics may be effective in relieving pain. In addition, beta-blockers may relieve dull, burning discomfort; antiseizure medications control stabbing and cramping pain; and tricyclic antidepressants may not only alleviate phantom pain but may also be prescribed to improve mood and coping ability. When medications are not effective in relieving pain, pulsed radiofrequency therapy (PRF) may be tried, although the evidence that supports its effectiveness is weak at present (Wolff et al., 2011). PRF delivers radiofrequency waves in short pulses directly to the residual limb by electrodes.

Promoting Wound Healing

The residual limb must be handled gently. Whenever the dressing is changed, aseptic technique is required to prevent wound infection and possible osteomyelitis.

Quality and Safety Nursing Alert

If the cast or elastic dressing inadvertently comes off, the nurse must immediately wrap the residual limb with an elastic compression bandage. If this is not done, excessive edema will develop in a short time, resulting in a delay in rehabilitation. The nurse notifies the surgeon if a cast dressing comes off so that another cast can be applied promptly.

Residual limb shaping is important for prosthesis fitting. The nurse instructs the patient and family to apply elastic wraps on the residual limb. Using elastic wraps on the residual limb is discouraged because they may apply inconsistent pressure on the residual limb, causing problems with shaping it to fit a prosthetic. After the incision is healed, the patient is instructed how to care for the residual limb (VA/DoD, 2007).

Enhancing Body Image

Amputation is a procedure that alters the patient's body image. The nurse who has established a trusting relationship with the patient is better able to communicate acceptance of the patient who has experienced an amputation. The nurse encourages the patient to look at, feel, and care for the residual limb. It is important to identify the patient's strengths and resources to facilitate rehabilitation. The nurse helps the patient regain the previous level of independent functioning. The patient who is accepted as a whole person is more readily able to resume responsibility for self-care; self-concept improves, and body image changes are accepted. Even with highly motivated patients, this process may take months.

Helping the Patient to Resolve Grieving

The loss of a limb (or part of one) may come as a shock even if the patient was prepared preoperatively. The patient's behavior (e.g., crying, withdrawal, apathy, anger) and expressed feelings (e.g., depression, fear, helplessness) reveal how the patient is coping with the loss and working through the grieving process.

The nurse creates an accepting and supportive atmosphere in which the patient and family are encouraged to express and share their feelings and work through the grief process. The support from family and friends promotes the patient's acceptance of the loss. The nurse helps the patient deal with immediate needs and become oriented to realistic rehabilitation goals and future independent functioning. Mental health and support group referrals may be appropriate (McFarland et al., 2010).

Promoting Independent Self-Care

Amputation affects the patient's ability to provide adequate self-care. The patient is encouraged to be an active participant in self-care. The patient needs time to accomplish these tasks and must not be rushed. Practicing an activity with consistent, supportive supervision in a relaxed environment enables the patient to learn self-care skills. The patient and the nurse need to maintain positive attitudes and minimize fatigue and frustration during the learning process.

Independence in dressing, toileting, and bathing depends on balance, transfer abilities, and physiologic tolerance of the activities. The nurse works with the physical therapist and occupational therapist to educate and supervise the patient in these self-care activities.

The patient with an upper limb amputation has self-care deficits in feeding, bathing, and dressing. Assistance is provided only as needed; the nurse encourages the patient to learn to do these tasks, using assistive feeding and dressing aids when needed. The nurse, therapists, and prosthetist work with the patient to achieve maximum independence.

Helping the Patient to Achieve Physical Mobility

Proper positioning prevents the development of hip or knee joint contracture in the patient with a lower limb amputation. Abduction, external rotation, and flexion of the lower limb are avoided. The residual limb may be placed in an extended position or elevated for a brief period after surgery (VA/DoD, 2007).

Quality and Safety Nursing Alert

The residual limb should not be placed on a pillow because a flexion contracture of the hip may result.

The nurse encourages the patient to turn from side to side and to assume a prone position, if possible, to stretch the flexor muscles and to prevent flexion contracture of the hip. The patient is encouraged not to sit for long periods of time to prevent flexion contracture. The legs should remain close together to prevent an abduction deformity. The nurse encourages the patient to use assistive devices to more readily perform self-care activities and to identify what home modifications, if any, should be made to perform these activities in the home environment.

Postoperative ROM exercises are started early because contracture deformities develop rapidly. ROM exercises include hip and knee exercises for patients with BKAs and hip exercises for patients with AKAs. It is important that the patient understands the importance of exercising the residual limb (VA/DoD, 2007).

The upper limbs, trunk, and abdominal muscles are exercised and strengthened. The extensor muscles in the arm and the depressor muscles in the shoulder play an important part in crutch walking. The patient uses an overbed trapeze to change position and strengthen the biceps. The patient may flex and extend the arms while holding weights. Doing push-ups while seated strengthens the triceps muscles. Exercises (such as hyperextension of the residual limb), conducted under the supervision of the physical therapist, also aid in strengthening muscles as well as increasing circulation, reducing edema, and preventing atrophy.

Because a patient who has had an upper limb amputated uses both shoulders to operate the prosthesis, the muscles of both shoulders are exercised. A patient with an above-elbow amputation or shoulder disarticulation is likely to develop a postural abnormality caused by loss of the weight of the amputated limb. Postural exercises are helpful.

Strength and endurance are assessed, and activities are increased gradually to prevent fatigue. As the patient progresses to independent use of the wheelchair, the use of ambulatory aids, or ambulation with a prosthesis, the nurse emphasizes safety considerations. Environmental barriers (e.g., steps, inclines, doors, throw rugs, wet surfaces) are identified, and methods of managing them are implemented. It is important to anticipate, identify, and manage problems associated with the use of the mobility aids. Proper instructions in using assistive devices will help prevent these problems.

Amputation of the leg changes the center of gravity; therefore, the patient may need to practice position changes (e.g., standing from sitting, standing on one foot). The patient is taught transfer techniques early and is reminded to maintain good posture when getting out of bed. A well-fitting shoe with a nonskid sole should be worn. During position changes, the patient should be guarded and stabilized with a transfer belt at the waist to prevent falling.

As soon as possible, the patient with a lower limb amputation is assisted to stand between parallel bars to allow extension of the temporary prosthesis to the floor with minimal weight bearing. How soon after surgery the patient is allowed to bear full body weight on the prosthesis depends on the patient's physical status and wound healing. As endurance increases and balance is achieved, ambulation is started with the use of parallel bars or crutches. The patient learns to use a normal gait, with the residual limb moving back and forth while walking with the crutches. To prevent a permanent flexion deformity from occurring, the residual limb should *not* be held up in a flexed position (VA/DoD, 2007).

The patient with an upper limb amputation is taught how to carry out ADLs with one arm. The patient is started on one-handed self-care activities as soon as possible. The use of a temporary prosthesis is encouraged. The patient who learns to use the prosthesis soon after the amputation is less dependent on one-handed self-care activities.

The patient with an upper limb amputation may wear a cotton T-shirt to prevent contact between the skin and shoulder harness and to promote absorption of perspiration. The prosthetist advises about cleaning the washable portions of the harness. Periodically, the prosthesis is inspected for potential problems.

The residual limb must be conditioned and shaped into a conical form to permit accurate fit, maximum comfort, and function of the prosthetic device. Elastic bandages, an elastic residual limb shrinker, or an air splint is used to condition and shape the residual limb. The nurse educates the patient or a member of the family about the correct method of bandaging.

Bandaging supports the soft tissue and minimizes the formation of edema while the residual limb is in a dependent position. The bandage is applied in such a manner that the remaining muscles required to operate the prosthesis are as firm as possible. An improperly applied elastic bandage contributes to circulatory problems and a poorly shaped residual limb.

Effective preprosthetic care is important to ensure proper fitting of the prosthesis. The major problems that can delay prosthetic fitting during this period are (1) flexion deformities, (2) nonshrinkage of the residual limb, and (3) abduction deformities of the hip.

The physician usually prescribes activities to condition or "toughen" the residual limb in preparation for a prosthesis. The patient begins by pushing the residual limb into a soft pillow, then into a firmer pillow, and finally against a hard surface. The patient is taught to massage the residual limb to mobilize the surgical incision site, decrease tenderness, and improve vascularity. Massage is usually started once healing has occurred and is first performed by the physical therapist. Skin inspection and preventive care are taught; mirrors may be used to visualize the skin on the residual limb (VA/DoD, 2007).

The prosthesis socket is custom-molded to the residual limb by the prosthetist. Prostheses are designed for specific activity levels and patient abilities. Types of prostheses include those that are hydraulic, pneumatic, biofeedback controlled, myoelectrically controlled, and synchronized. Adjustments of the prosthetic socket are made by the prosthetist to accommodate the residual limb changes that occur during the first 6 months to 1 year after surgery.

Some patients are not candidates for a prosthesis and are thus nonambulatory patients with amputations. If the use of a prosthesis is not possible, the patient is instructed in safe use of a wheelchair to achieve independence. A special wheelchair designed for patients who have had amputations is recommended. Because of the decreased weight in the front, a regular wheelchair may tip backward when the patient sits in it. In wheelchairs designed for patients who have had amputations, the rear axle is set back about 5 cm (2 inches) to compensate for the change in weight distribution.

Monitoring and Managing Potential Complications

After any surgery, efforts are made to reestablish homeostasis and to prevent complications related to surgery, anesthesia, and immobility. The nurse assesses body systems (e.g., respiratory, hematologic, gastrointestinal, genitourinary, skin) for problems associated with immobility (e.g., atelectasis, pneumonia, DVT, PE, anorexia, constipation, urinary stasis, pressure ulcers).

Massive hemorrhage due to a loosened suture is a potentially life-threatening problem. The nurse monitors the patient for any signs or symptoms of bleeding and monitors the patient's vital signs and suction drainage.

Quality and Safety Nursing Alert

Immediate postoperative bleeding may develop slowly or may take the form of massive hemorrhage resulting from a loosened suture. A large tourniquet should be in plain sight at the patient's bedside so that if severe bleeding occurs, it can be applied to the residual limb to control the hemorrhage. The nurse immediately notifies the surgeon in the event of excessive bleeding.

Infection is a common complication of amputation. Patients who have undergone traumatic amputation have contaminated wounds. The nurse administers antibiotics as prescribed. The nurse must monitor the incision, dressing, and drainage for indications of infection (e.g., change in color, odor, or consistency of drainage; increasing discomfort). The nurse also assesses for systemic indicators of infection (e.g., elevated temperature, leukocytosis with an increase of more than 10% bands on the differential) and promptly reports indications of infection to the surgeon.

Skin breakdown may result from immobilization or pressure from various sources. The prosthesis may cause pressure areas to develop. The nurse and the patient assess for breaks in the skin. Careful skin hygiene is essential to prevent skin irritation, infection, and breakdown. The healed residual limb is washed and dried (gently) at least twice daily. The skin is inspected for pressure areas, dermatitis, and blisters. If they are present, they must be treated before further skin breakdown occurs. Usually, a residual limb sock is worn to absorb perspiration and to prevent direct contact between the skin and the prosthetic socket. The sock is changed daily and must fit smoothly to prevent irritation caused by wrinkles. The socket of the prosthesis is washed with a mild detergent, rinsed, and dried thoroughly with a clean cloth. It must be thoroughly dry before the prosthesis is applied (VA/DoD, 2007).

Promoting Home and Community-Based Care

Educating the Patient About Self-Care. Before the patient is discharged to home or to a rehabilitation facility, the patient and family are encouraged to become active participants in care. They participate in care of the skin, residual limb, and prosthesis as appropriate. The patient receives ongoing education and practice sessions to learn to transfer and to use mobility aids and other assistive devices safely. The nurse explains the signs and symptoms of complications that must be reported to the primary provider (Chart 43-8).

Continuing Care in the Home and Community. After the patient has achieved physiologic homeostasis and has demonstrated achievement of major health care goals, rehabilitation continues either in a rehabilitation facility or at home. Continued support and evaluation by the home care nurse are essential.

The patient's home environment should be assessed prior to discharge. Modifications are made to ensure the patient's continuing care, safety, and mobility. An overnight or weekend experience at home may be tried to identify problems that were not identified on the assessment visit. Physical therapy and occupational therapy may continue in the home or on an outpatient basis. Transportation to continuing health care appointments must be arranged. The social service department of the hospital or the home health agency may be of great assistance in securing personal assistance and transportation services.

During follow-up health visits, the nurse evaluates the patient's physical and psychosocial adjustment. Periodic preventive health assessments are necessary. An older adult spouse may not be able to provide the assistance required if needed at home. Modifications in the plan of care are made on the basis of such findings. Often, the patient and family find involvement in a postamputation support group to be of

Chart 43-8 HOME CARE CHECKLIST
Amputation

At the completion of home care education, the patient or caregiver will be able to:	PATIENT	CAREGIVER
• Describe approaches to controlling pain (e.g., take analgesics as prescribed; use nonpharmacologic interventions).	✔	✔
• Report pain that is uncontrolled by analgesics and other pain management techniques.	✔	
• Describe care of residual limb and conditioning for prosthesis.	✔	✔
• Consume healthy diet to promote wound healing.	✔	
• Demonstrate ability to transfer.	✔	
• Use mobility and activity aids safely.	✔	
• Participate in rehabilitation program to regain functional independence.	✔	
• State indicators of complications to report promptly to primary provider (e.g., uncontrolled pain; signs of local or systemic infection; residual limb skin breakdown).	✔	✔
• Identify professionals and community agencies to help with transition to home.	✔	✔
• Identify support group to facilitate rehabilitation.	✔	✔
• Describe effects of amputation on self-image.	✔	
• Acknowledge grieving as part of coping process.	✔	✔
• Identify modifications of home environment to promote safe environment and independence during rehabilitation.	✔	✔
• Identify the importance of keeping follow-up appointments and participating in health screening and health promotion activities, including exercises.	✔	✔

value; here, they can share problems, solutions, and resources. Talking with those who have successfully dealt with a similar problem may help the patient develop a satisfactory solution.

Because patients and their family members and health care providers tend to focus on the most obvious needs and issues, the nurse reminds the patient and family about the importance of continuing health promotion and screening practices, such as regular physical examinations and diagnostic screening tests. Accessible facilities for screening, health care, and exercise are identified. Patients are instructed about their importance and are referred to appropriate health care providers.

Evaluation

Expected patient outcomes may include:

1. Experiences no pain, including phantom limb pain
   a. Appears relaxed
   b. Verbalizes comfort
   c. Uses measures to increase comfort and mitigate pain
   d. Participates in self-care and rehabilitative activities
   e. Reports diminished phantom sensations
2. Achieves wound healing
   a. Controls residual limb edema
   b. Exhibits healed, nontender, nonadherent scar
   c. Demonstrates residual limb care
3. Demonstrates improved body image and effective coping
   a. Acknowledges change in body image
   b. Participates in self-care activities
   c. Demonstrates increasing independence
   d. Projects self as a whole person
   e. Resumes role-related responsibilities
   f. Reestablishes social contacts
   g. Demonstrates confidence in abilities
4. Exhibits resolution of grieving
   a. Expresses grief
   b. Works through feelings with family and friends
   c. Focuses on future functioning
   d. Participates in support group
5. Achieves independent self-care
   a. Asks for assistance when needed
   b. Uses aids and assistive devices to facilitate self-care
   c. Verbalizes satisfaction with abilities to perform ADLs
6. Achieves maximum independent mobility
   a. Avoids positions contributing to contracture development
   b. Demonstrates full active ROM
   c. Maintains balance when sitting and transferring
   d. Increases strength and endurance
   e. Demonstrates safe transferring technique
   f. Achieves functional use of prosthesis
   g. Overcomes environmental barriers to mobility
   h. Uses community services and resources as needed
7. Exhibits absence of complications of hemorrhage, infection, or skin breakdown
   a. Does not experience excessive bleeding
   b. Maintains normal blood values
   c. Is free of local or systemic signs of infection
   d. Repositions self frequently
   e. Is free of pressure-related problems
   f. Reports any skin discomfort and irritations promptly

Critical Thinking Exercises

1 ebp You are a home health nurse and are assigned a new patient to your caseload—a National Guard veteran who stepped on a land mine and was released from duty after having a BKA of the right leg and an AKA of the left leg 2 months ago. Identify this patient's unique nursing, medical, physical therapy, occupational therapy, social, and emotional needs. Devise a nursing plan of care that addresses these needs. What is the strength of the evidence that supports your plan of care for this veteran?

2 You are a nurse who works in a college health clinic. A 19-year-old female student presents with complaints of right elbow pain that she has felt "on-again, off-again" but with worsening pain within the past 48 hours. She has no history of trauma to her elbow and has not been taking any medication to treat her pain. She receives scholarship money for being on the varsity tennis team. She tells you that she is worried about her pain because she is right-handed and it is the middle of the tennis season. How would you proceed with your assessments? What interventions do you think may be indicated? Might it be possible for her to continue to play tennis and not jeopardize her scholarship or her health?

3 pq You are a staff nurse on an orthopedic unit and are assigned to care for a 30-year-old man who had an ORIF for an unstable pelvic fracture. His surgical course was unremarkable, and it is now his second postoperative day. On your initial rounds, he appeared stable, had normal vital signs, and had no particular complaints. An hour later, his wife frantically calls you to his room, saying, "He is suddenly acting confused and is breathing heavy!" Before you reach the patient's room, what complication or complications may explain these sudden clinical manifestations? Describe your focused priority assessments and interventions.

Brunner Suite Resources

Explore these additional resources to enhance learning for this chapter:

- NCLEX-Style Questions and Other Resources on thePoint, http://thePoint.lww.com/Brunner13e
- Study Guide
- PrepU
- Clinical Handbook
- Handbook of Laboratory and Diagnostic Tests

References

*Asterisk indicates nursing research.

Books

Bickley, L. S. (2009). *Bates' guide to physical assessment* (10th ed.). Philadelphia: Lippincott Williams & Wilkins.

Cline, D. M., Ma, O. J., Cydulka, R. K., et al. (2012). *Tintinalli's emergency medicine manual*. New York: McGraw-Hill.

Department of Veterans Affairs, Department of Defense. (2007). *VA/DoD clinical practice guideline for rehabilitation of lower limb amputation*. Washington, DC: Author.

Emergency Nurses Association. (2013). *Sheehy's manual of emergency care*. St. Louis: Mosby.
National Association of Orthopedic Nurses. (2007). *Core curriculum for orthopaedic nursing* (6th ed.). Boston: Pearson.
Porth, C. M., & Matfin, G. (2009). *Pathophysiology: Concepts of altered health states* (8th ed.). Philadelphia: Lippincott Williams & Wilkins.

Journals and Electronic Documents

Alke, J., & Schraga, E. D. (2011). Hand fracture. *Medscape*. Available at: emedicine.medscape.com/article/825271-overview
Altizer, L. L. (2008). Colles' fracture. *Orthopaedic Nursing, 27*(2), 140–145.
American College of Occupational and Environmental Medicine. (2007). *Elbow disorders*. Elk Grove Village, IL: Author. Available at: www.guideline.gov/content.aspx?id=10883&search=elbow+disorders
American College of Sports Medicine. (2011). *ACMS information on … return to play—a coach's guide*. Available at: www.acsm.org/docs/brochures/return-to-play—a-coach%27s-guide.pdf
American Nurses Association. (2012). *News release: ANA leads initiative to develop national safe patient handling standards: Multidisciplinary group seeks to establish evidence-based guidelines to address deficiency*. Silver Spring, MD: Author.
American Nurses Association. (2013). *Safe patient handling and mobility*. Available at: www.nursingworld.org/MainMenuCategories/WorkplaceSafety/SafePatient?css=print
Banovac, K., & Speed, J. (2011). Heterotopic ossification. *Medscape*. Available at: emedicine.medscape.com/article/327648-overview
*Benetato, B. B. (2011). Posttraumatic growth among Operation Enduring Freedom and Operation Iraqi Freedom amputees. *Journal of Nursing Scholarship, 43*(4), 412–420.
Bhatti, N. S. (2012). Hip fracture. *Medscape*. Available at: emedicine.medscape.com/article/87043-overview
Bilal, R. H., Duffy, P. J., Shafi, B. B., et al. (2011). Rotator cuff pathology. *Medscape*. Available at: emedicine.medscape.com/article/1262849-overview
Buckley, R., & Panaro, C D. (2012). General principles of fracture care. *Medscape*. Available at: emedicine.medscape.com/article/1270717-overview
Bueche, K. L. (2010). Avoid the pressure of compartment syndrome. *OR Nurse Journal, 4*(1), 42–47.
Clinton, R. E., & Murthi, A. M. (2008). Lateral epicondylitis. *Current Orthopaedic Practice, 19*(6), 612–615.
Coombs, B. K., Bisset, L., Brooks, P., et al. (2013). Effect of corticosteroid injection, physiotherapy, or both on clinical outcomes in patients with unilateral epicondylalgia: A randomized controlled trial. *Journal of the American Medical Association, 309*(5), 461–469.
Galkowski, V., Petrisor, B., Drew, B., et al. (2009). Bone stimulation for fracture healing: What's all the fuss? *Indian Journal of Orthopaedics, 43*(2), 117–120.
Hegmann, K. T. (Ed.). (2011). Knee disorders. In *Occupational medicine practice guidelines. Evaluation and management of common health problems and functional recovery in workers* (3rd ed.). Elk Grove Village, IL. Available at: www.guideline.gov/content.aspx?id=36632&search=knee+disorders
Horn, P. L., Price, M. C., Van Aman, S. E. (2011). Orthopaedic trauma: Pilon fractures. *Orthopaedic Nursing, 30*(5), 293–298.
Johnston-Walker, E., & Hardcastle, J. (2011), Neurovascular assessment in the critically ill patient. *Nursing in Critical Care, 16*(4), 170–177.
Kale, S. (2012). Reflex sympathetic dystrophy surgery. *Medscape*. Available at: emedicine.medscape.com/article/1269453-overview
Kay, A. D., & Blazevich, A. J. (2012). Effect of acute static stretch on maximal muscle performance: A systematic review. *Medicine & Science in Sports & Exercise, 44*(1), 154–164.
Keany, J. E., & McKeever, J. E. (2011). Femur fracture. *Medscape*. Available at: emedicine.medscape.com/article/824856-overview
*Kondo, A., Zierler, B. K., & Hagino, H. (2011). The timing of hip fracture surgery and mortality within 1 year: A comparison between the United States and Japan. *Orthopaedic Nursing, 30*(1), 54–61.
Lakshmanan, P., Damodaran, P. R., & Sher, L. (2012). Malunion of hand fracture. *Medscape*. Available at: emedicine.medscape.com/article/1243899-overview
Laurencin, C. T., Magge, A., & Khan, Y. (2012). Bone graft substitute materials. *Medscape*. Available at: emedicine.medscape.com/article/1230616-overview
Leahy, M., & Rahm, M. (2012). Thoracic spine fractures and dislocations. *Medscape*. Available at: emedicine.medscape.com/article/1267029-overview
Logerstedt, D. S., Snyder-Mackler, L., Ritter, R. C., et al. (2010). Knee stability and movement coordination impairments: Knee ligament sprain. *Journal of Orthopaedic & Sports Physical Therapy, 40*(4), A1–A37.
McFarland, L. V., Choppa, A. J., Betz, K., et al. (2010). Resources for wounded warriors with major traumatic limb loss. *Journal of Rehabilitation Research & Development, 47*(4, App. III), 1–13.
Melendez, S. L., & Doty, C. I. (2012). Rib fracture treatment and management. *Medscape*. Available at: emedicine.medscape.com/article/825981-overview
Mouzopoulos, G., Morakis, E., Stamatakos, M., et al. (2009). Complications associated with clavicular fracture. *Orthopaedic Nursing, 28*(5), 217–224.
National Center for Health Statistics. (2009). *Injury in the United States: 2007 chartbook*. Available at: www.cdc.gov/nchs/ppt/injury/2007chartbook/injury_figures_PPT.htm
National Center for Health Statistics. (2013). *Accidents or unintentional injuries*. Available at: www.cdc.gov/nchs/fastats/acc-inj.htm
National Limb Loss Information Center. (2008). *Fact sheet: Amputation statistics by cause: Limb loss in the United States*. Knoxville, TN: Amputee Coalition of America. Available at: www.amputee-coalition.org
*Olofsson, B., Stenvall, M., Lundstrom, M., et al. (2009). Mental status and surgical methods in patients with femoral neck fracture. *Orthopaedic Nursing, 28*(6), 305–313.
Patel, M., McCarthy, J. J., & Herzenberg, J. (2012). Tibial nonunions. *Medscape*. Available at: emedicine.medscape.com/article/1252306-overview
Peerdeman, B., Boere, D., Witteveen, H., et al. (2011). Myoelectric forearm prostheses: State of the art from a user-centered perspective. *Journal of Rehabilitation Research & Development, 48*(6), 719–738.
Registered Nurses' Association of Ontario. (2011). *Prevention of falls and injuries in the older adult*. Toronto, ON: Author. Available at: www.guideline.gov/content.aspx?id=34756
Russell, G. V., Jarrett, C. A., & Routt, M. L. (2012). Pelvic fractures. *Medscape*. Available at: emedicine.medscape.com/article/1247913-overview
Scalea, T. M. (2008). Optimal timing of fracture fixation: Have we learned anything in the past 20 years? *Journal of Trauma, 65*(2), 253–260.
Schaller, T. M. (2012). Open fractures. *Medscape*. Available at: emedicine.medscape.com/article/1269242-overview
Schneider, M. A. (2012). Prevention of catheter associated urinary tract infections in patients with hip fractures through education of nurses to specific catheter protocols. *Orthopaedic Nursing, 31*(1), 12–18.
Scottish Intercollegiate Guidelines Network. (2009). *Management of hip fracture in older people. A national clinical guideline*. Edinburgh, Scotland: Author. Available at: www.guideline.gov/content.aspx?id=15206&search=hip+fracture
Stein, P. D., Yaekoub, A. Y., Matta, F., et al. (2008). Fat embolism syndrome. *American Journal of the Medical Sciences, 336*(6), 472–477.
Thacker, M. M., Tejwani, N., & Thakkar, C. (2012). Acetabulum fractures. *Medscape*. Available at: emedicine.medscape.com/article/1246057-overview
Tofferi, J. K., & Gilliland, W. (2012). Avascular necrosis. *Medscape*. Available at: emedicine.medscape.com/article/333364-overview
Tzioupis, C. C., & Giannoudis, P. V. (2011). Fat embolism syndrome: What have we learned over the years? *Trauma, 13*(4), 259–281.
U.S. Department of Labor, Bureau of Labor Statistics. (2012). *Nonfatal occupational injuries and illnesses requiring days away from work, 2011*. Available at: www.bls.gov/news.release/osh2.nr0.htm
Vinas, F. C. (2011). Lumbar spine fractures and dislocations. *Medscape*. Available at: emedicine.medscape.com/article/1264191-overview
Watters, C. (2008). Safe patient handling—are we there yet? *Orthopaedic Nursing, 27*(1), 38–41.
Wolff, A., Vanduynhoven, E., van Kleef, M., et al. (2011). Phantom pain. *Pain Practice, 11*(4), 403–413.

Resources

American College of Sports Medicine (ACSM), www.acsm.org
American Amputee Foundation (AAF), www.americanamputee.org/
Amputee Coalition of America, www.amputee-coalition.org/
Disabled American Veterans (DAV), www.dav.org
National Amputation Foundation (NAF), www.nationalamputation.org
National Association of Orthopaedic Nurses (NAON), www.orthonurse.org
National Institute for Occupational Safety and Health (NIOSH), www.cdc.gov/niosh/
National Institute of Arthritis and Musculoskeletal and Skin Diseases, www.niams.nih.gov
U.S. Department of Labor, Occupational Safety and Health Administration (OSHA), www.osha.gov
Wounded Warrior Project, www.woundedwarriorproject.org

Chapter 44

Assessment of Digestive and Gastrointestinal Function

Learning Objectives

On completion of this chapter, the learner will be able to:

1 Describe the structure and function of the organs of the gastrointestinal (GI) tract.
2 Describe the mechanical and chemical processes involved in digesting and absorbing nutrients and eliminating waste products.
3 Use assessment parameters appropriate for determining the status of GI function.
4 Discriminate between normal and abnormal GI function.
5 Identify the appropriate preparation, patient education, and follow-up care for patients who are undergoing diagnostic evaluation of the GI tract.

Glossary

absorption: phase of the digestive process that occurs when small molecules, vitamins, and minerals pass through the walls of the small and large intestine and into the bloodstream

achalasia: absence of peristalsis of the lower esophagus resulting in difficulty swallowing, regurgitation, and sometimes pain

amylase: an enzyme that aids in the digestion of starch

anus: last section of the gastrointestinal (GI) tract; outlet for waste products from the GI system

chyme: mixture of food with saliva, salivary enzymes, and gastric secretions that is produced as food passes through the mouth, esophagus, and stomach

digestion: phase of the digestive process that occurs when digestive enzymes and secretions mix with ingested food and when proteins, fats, and sugars are broken down into their component smaller molecules

dyspepsia: indigestion; upper abdominal discomfort associated with eating

elimination: phase of the digestive process that occurs after digestion and absorption, when waste products are evacuated from the body

esophagus: collapsible tube connecting the mouth to the stomach, through which food passes as it is ingested

fibroscopy (gastrointestinal): intubation of a part of the GI system with a flexible, lighted tube to assist in diagnosis and treatment of diseases of that area

hydrochloric acid: acid secreted by the glands in the stomach; mixes with chyme to break it down into absorbable molecules and to aid in the destruction of bacteria

ingestion: phase of the digestive process that occurs when food is taken into the GI tract via the mouth and esophagus

intrinsic factor: a gastric secretion that combines with vitamin B_{12} so that the vitamin can be absorbed

large intestine: the portion of the GI tract into which waste material from the small intestine passes as absorption continues and elimination begins; consists of several parts—ascending segment, transverse segment, descending segment, sigmoid colon, and rectum; also known as the colon

lipase: an enzyme that aids in the digestion of fats

pepsin: a gastric enzyme that is important in protein digestion

small intestine: longest portion of the GI tract, consisting of three parts—duodenum, jejunum, and ileum—through which food mixed with all secretions and enzymes passes as it continues to be digested and begins to be absorbed into the bloodstream

stomach: distensible pouch into which the food bolus passes to be digested by gastric enzymes

trypsin: enzyme that aids in the digestion of protein

Abnormalities of the gastrointestinal (GI) tract are numerous and represent every type of major pathology that can affect other organ systems, including bleeding, perforation, obstruction, inflammation, and cancer. Congenital, inflammatory, infectious, traumatic, and neoplastic lesions have been encountered in every portion and at every site along the length of the GI tract. As with all other organ systems, the GI tract is subject to circulatory disturbances, faulty nervous system control, and aging.

Apart from the many organic diseases to which the GI tract is susceptible, many extrinsic factors can interfere with its normal function and produce symptoms. Stress and anxiety, for example, often find their chief expression in indigestion, anorexia, or motor disturbances of the intestines, sometimes producing constipation or diarrhea. In addition to the state of mental health, physical factors such as fatigue and an inadequate or abruptly changed dietary intake can markedly affect the GI tract. When assessing and instructing the patient, the nurse should consider the variety of mental and physical factors that affect the function of the GI tract.

Anatomic and Physiologic Overview

Anatomy of the Gastrointestinal System

The GI tract is a pathway 7 to 7.9 m (23 to 26 ft) in length that extends from the mouth to the esophagus, stomach, small and large intestines, and rectum, to the terminal structure, the **anus** (Fig. 44-1). The **esophagus** is located in the mediastinum, anterior to the spine and posterior to the trachea and heart. This hollow muscular tube, which is approximately 25 cm (10 in) in length, passes through the diaphragm at an opening called the *diaphragmatic hiatus*.

The remaining portion of the GI tract is located within the peritoneal cavity. The **stomach** is situated in the left upper portion of the abdomen under the left lobe of the liver and the diaphragm, overlaying most of the pancreas (see Fig. 44-1). A hollow muscular organ with a capacity of approximately 1,500 mL, the stomach stores food during eating, secretes digestive fluids, and propels the partially digested food, or chyme, into the small intestine. The gastroesophageal junction is the inlet to the stomach. The stomach has four anatomic regions: the cardia (entrance), fundus, body, and pylorus (outlet). Circular smooth muscle in the wall of the pylorus forms the pyloric sphincter and controls the opening between the stomach and the small intestine.

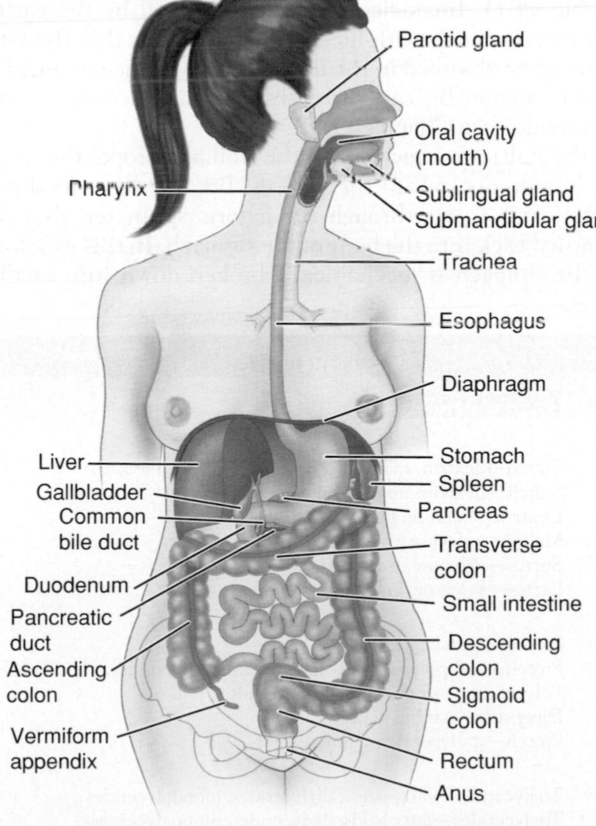
Parotid gland
Oral cavity
(mouth)
Pharynx
Sublingual gland
Submandibular gland
Trachea
Esophagus
Diaphragm
Liver
Stomach
Gallbladder
Spleen
Common
bile duct
Pancreas
Transverse
colon
Duodenum
Small intestine
Pancreatic
duct
Descending
colon
Ascending
colon
Sigmoid
colon
Vermiform
appendix
Rectum
Anus

FIGURE 44-1 • Organs of the digestive system and associated structures.

The **small intestine** is the longest segment of the GI tract, accounting for about two thirds of the total length. It folds back and forth on itself, providing approximately 70m (230 ft) of surface area for secretion and **absorption**, the process by which nutrients enter the bloodstream through the intestinal walls. It has three sections: The most proximal section is the duodenum, the middle section is the jejunum, and the distal section is the ileum. The ileum terminates at the ileocecal valve. This valve, or sphincter, controls the flow of digested material from the ileum into the cecal portion of the large intestine and prevents reflux of bacteria into the small intestine. Attached to the cecum is the vermiform appendix, an appendage that has little or no physiologic function. Emptying into the duodenum at the ampulla of Vater is the common bile duct, which allows for the passage of both bile and pancreatic secretions.

The **large intestine** consists of an ascending segment on the right side of the abdomen, a transverse segment that extends from right to left in the upper abdomen, and a descending segment on the left side of the abdomen. The sigmoid colon, the rectum, and the anus complete the terminal portion of the large intestine. A network of striated muscle that forms both the internal and the external anal sphincters regulates the anal outlet.

The GI tract receives blood from arteries that originate along the entire length of the thoracic and abdominal aorta and veins that return blood from the digestive organs and the spleen. This portal venous system is composed of five large veins: the superior mesenteric, inferior mesenteric, gastric, splenic, and cystic veins, which eventually form the vena portae that enters the liver. Once in the liver, the blood is distributed throughout and collected into the hepatic veins that then terminate in the inferior vena cava. Of particular importance are the gastric artery and the superior and inferior mesenteric arteries. Oxygen and nutrients are supplied to the stomach by the gastric artery and to the intestine by the mesenteric arteries (Fig. 44-2). Venous blood is returned

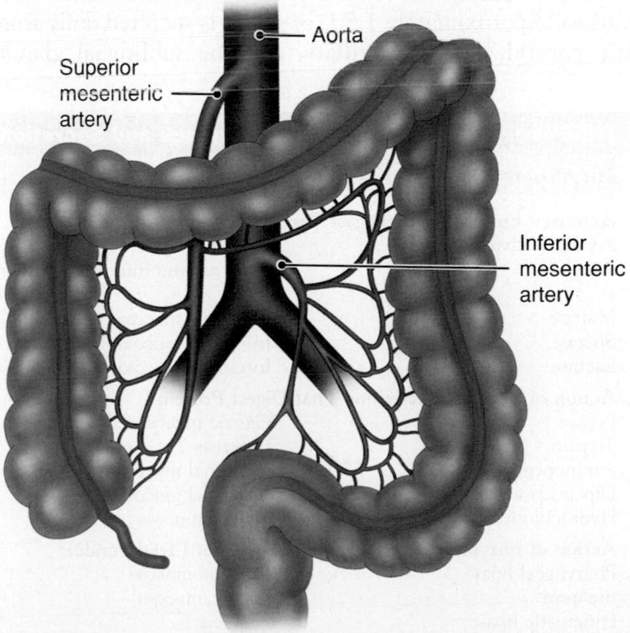
Aorta
Superior
mesenteric
artery
Inferior
mesenteric
artery

FIGURE 44-2 • Anatomy and blood supply of the large intestine.

from the small intestine, cecum, and the ascending and transverse portions of the colon by the superior mesenteric vein, which corresponds with the distribution of the branches of the superior mesenteric artery. Blood flow to the GI tract is about 20% of the total cardiac output and increases significantly after eating.

Both the sympathetic and parasympathetic portions of the autonomic nervous system innervate the GI tract. In general, sympathetic nerves exert an inhibitory effect on the GI tract, decreasing gastric secretion and motility and causing the sphincters and blood vessels to constrict. Parasympathetic nerve stimulation causes peristalsis and increases secretory activities. The sphincters relax under the influence of parasympathetic stimulation, except for the sphincter of the upper esophagus and the external anal sphincter, which are under voluntary control (Porth & Matfin, 2009).

Function of the Digestive System

All cells of the body require nutrients. These nutrients are derived from the intake of food that contains proteins, fats, carbohydrates, vitamins, minerals, and cellulose fibers and other vegetable matter, some of which has no nutritional value. Major functions of the GI tract include:

- Breakdown of food particles into the molecular form for **digestion**
- Absorption into the bloodstream of small nutrient molecules produced by digestion
- **Elimination** of undigested unabsorbed foodstuffs and other waste products

After food is ingested, it is propelled through the GI tract, coming into contact with a wide variety of secretions that aid in its digestion, absorption, or elimination from the GI tract.

Chewing and Swallowing

The process of digestion begins with the act of chewing, in which food is broken down into small particles that can be swallowed and mixed with digestive enzymes. Eating—or even the sight, smell, or taste of food—can cause reflex salivation. Approximately 1.5 L of saliva is secreted daily from the parotid, the submaxillary, and the sublingual glands. Ptyalin, or salivary amylase, is an enzyme that begins the digestion of starches. Water and mucus, also contained in saliva, help lubricate the food as it is chewed, thereby facilitating swallowing.

Swallowing begins as a voluntary act that is regulated by the swallowing center in the medulla oblongata of the central nervous system (CNS). As a bolus of food is swallowed, the epiglottis moves to cover the tracheal opening and prevent aspiration of food into the lungs. Swallowing, which propels the bolus of food into the upper esophagus, thus ends as a reflex action. The smooth muscle in the wall of the esophagus contracts in a rhythmic sequence from the upper esophagus toward the stomach to propel the bolus of food along the tract. During this process of esophageal peristalsis, the lower esophageal sphincter relaxes and permits the bolus of food to enter the stomach. Subsequently, the lower esophageal sphincter closes tightly to prevent reflux of stomach contents into the esophagus.

Gastric Function

The stomach, which stores and mixes food with secretions, secretes a highly acidic fluid in response to the presence or anticipated **ingestion** of food. This fluid, which can total 2.4 L/day, can have a pH as low as 1 and derives its acidity from **hydrochloric acid** (HCl) secreted by the glands of the stomach. The function of this gastric secretion is twofold: to break down food into more absorbable components and to aid in the destruction of most ingested bacteria. **Pepsin**, an important enzyme for protein digestion, is the end product of the conversion of pepsinogen from the chief cells (Table 44-1). **Intrinsic factor**, also secreted by the gastric mucosa, combines with dietary vitamin B_{12} so that the vitamin can be absorbed in the ileum. In the absence of intrinsic factor, vitamin B_{12} cannot be absorbed, and pernicious anemia results (see Chapter 33).

Peristaltic contractions in the stomach propel the stomach's contents toward the pylorus. Because large food particles cannot pass through the pyloric sphincter, they are churned back into the body of the stomach. In this way, food in the stomach is mechanically broken down into smaller

TABLE 44-1 The Major Digestive Enzymes and Secretions

Enzyme/Secretion	Enzyme Source	Digestive Action
Action of Enzymes That Digest Carbohydrates		
Ptyalin (salivary amylase)	Salivary glands	Starch→dextrin, maltose, glucose
Amylase	Pancreas and intestinal mucosa	Starch→dextrin, maltose, glucose
		Dextrin→maltose, glucose
Maltase	Intestinal mucosa	Maltose→glucose
Sucrase	Intestinal mucosa	Sucrose→glucose, fructose
Lactase	Intestinal mucosa	Lactose→glucose, galactose
Action of Enzymes/Secretions That Digest Protein		
Pepsin	Gastric mucosa	Protein→polypeptides
Trypsin	Pancreas	Proteins and polypeptides→polypeptides, dipeptides, amino acids
Aminopeptidase	Intestinal mucosa	Polypeptides→dipeptides, amino acids
Dipeptidase	Intestinal mucosa	Dipeptides→amino acids
Hydrochloric acid	Gastric mucosa	Protein→polypeptides, amino acids
Action of Enzymes/Secretions That Digest Fat (Triglyceride)		
Pharyngeal lipase	Pharynx mucosa	Triglycerides→fatty acids, diglycerides, monoglycerides
Steapsin	Gastric mucosa	Triglycerides→fatty acids, diglycerides, monoglycerides
Pancreatic lipase	Pancreas	Triglycerides→fatty acids, diglycerides, monoglycerides
Bile	Liver and gallbladder	Fat emulsification

→, converts to

TABLE 44-2 The Major Gastrointestinal Regulatory Substances

Substance	Stimulus for Production	Target Tissue	Effect on Secretions	Effect on Motility
Neuroregulators				
Acetylcholine	Sight, smell, chewing food, stomach distention	Gastric glands, other secretory glands, gastric and intestinal muscle	Increased gastric acid	Generally increased; decreased sphincter tone
Norepinephrine	Stress, other various stimuli	Secretory glands, gastric and intestinal muscle	Generally inhibitory	Generally decreased; increased sphincter tone
Hormonal Regulators				
Gastrin	Stomach distention with food	Gastric glands	Increased secretion of gastric juice, which is rich in HCl	Increased motility of stomach, decreased time required for gastric emptying Relaxation of ileocecal sphincter Excitation of colon Constriction of gastroesophageal sphincter
Cholecystokinin	Fat in duodenum	Gallbladder	Release of bile into duodenum	
		Pancreas	Increased production of enzyme-rich pancreatic secretions	
		Stomach	Inhibits gastric secretion somewhat	Slows gastric emptying
Secretin	pH of chyme in duodenum below 4–5	Stomach	Inhibits gastric secretion somewhat	Inhibits stomach contractions
		Pancreas	Increased production of bicarbonate-rich pancreatic juice	
Local Regulator				
Histamine	Unclear; substances in food	Gastric glands	Increased gastric acid production	

HCl, hydrochloric acid.

particles. Food remains in the stomach for a variable length of time, from 30 minutes to several hours, depending on the volume, osmotic pressure, and chemical composition of the gastric contents. Peristalsis in the stomach and contractions of the pyloric sphincter allow the partially digested food to enter the small intestine at a rate that permits efficient absorption of nutrients. This partially digested food mixed with gastric secretions is called **chyme**. Hormones, neuroregulators, and local regulators found in the gastric secretions control the rate of gastric secretions and influence gastric motility (Table 44-2).

Small Intestine Function

The digestive process continues in the duodenum. Duodenal secretions come from the accessory digestive organs—the pancreas, liver, and gallbladder—and the glands in the wall of the intestine itself. These secretions contain digestive enzymes: amylase, lipase, and bile. Pancreatic secretions have an alkaline pH due to their high concentration of bicarbonate. This alkalinity neutralizes the acid entering the duodenum from the stomach. Digestive enzymes secreted by the pancreas include **trypsin**, which aids in digesting protein; **amylase**, which aids in digesting starch; and **lipase**, which aids in digesting fats. These secretions drain into the pancreatic duct, which empties into the common bile duct at the ampulla of Vater. Bile, secreted by the liver and stored in the gallbladder, aids in emulsifying ingested fats, making them easier to digest and absorb. The sphincter of Oddi, found at the confluence of the common bile duct and duodenum, controls the flow of bile. Hormones, neuroregulators, and local regulators found in these intestinal secretions control the rate of intestinal secretions and also influence GI motility. Intestinal secretions total approximately 1 L/day of pancreatic juice, 0.5 L/day of bile, and 3 L/day of secretions from the glands of the small intestine. Tables 44-1 and 44-2 give further information about the actions of digestive enzymes and GI regulatory substances.

Two types of contractions occur regularly in the small intestine: segmentation contractions and intestinal peristalsis. *Segmentation contractions* produce mixing waves that move the intestinal contents back and forth in a churning motion. *Intestinal peristalsis* propels the contents of the small intestine toward the colon. Both movements are stimulated by the presence of chyme.

Food, ingested as fats, proteins, and carbohydrates, is broken down into absorbable particles (constituent nutrients) by the process of digestion. Carbohydrates are broken down into disaccharides (e.g., sucrose, maltose, and galactose) and monosaccharides (e.g., glucose, fructose). Glucose is the major carbohydrate that tissue cells use as fuel. Proteins are a source of energy after they are broken down into amino acids and peptides. Ingested fats become monoglycerides and fatty acid through emulsification, which makes them smaller and easier to absorb. Chyme stays in the small intestine for 3 to 6 hours, allowing for continued breakdown and absorption of nutrients.

Small, fingerlike projections called *villi* line the entire intestine and function to produce digestive enzymes as well as to absorb nutrients. Absorption is the major function of the small intestine. Vitamins and minerals are absorbed essentially unchanged. Absorption begins in the jejunum and is accomplished by active transport and diffusion across the intestinal wall into the circulation. Nutrients are absorbed at specific locations in the small intestine and duodenum, whereas fats, proteins, carbohydrates, sodium, and chloride are absorbed in the jejunum. Vitamin B_{12} and bile salts are absorbed in the ileum. Magnesium, phosphate, and potassium are absorbed throughout the small intestine.

Colonic Function

Within 4 hours after eating, residual waste material passes into the terminal ileum and slowly into the proximal portion of the right colon through the ileocecal valve. With each peristaltic wave of the small intestine, the valve opens briefly and permits some of the contents to pass into the colon.

Bacteria, a major component of the contents of the large intestine, assist in completing the breakdown of waste material, especially of undigested or unabsorbed proteins and bile salts. Two types of colonic secretions are added to the residual material: an electrolyte solution and mucus. The electrolyte solution is chiefly a bicarbonate solution that acts to neutralize the end products formed by the colonic bacterial action, whereas the mucus protects the colonic mucosa from the interluminal contents and provides adherence for the fecal mass.

Slow, weak peristalsis moves the colonic contents along the tract. This slow transport allows for efficient reabsorption of water and electrolytes, which is the major function of the colon. Intermittent strong peristaltic waves propel the contents for considerable distances. This generally occurs after another meal is eaten, when intestine-stimulating hormones are released. The waste materials from a meal eventually reach and distend the rectum, usually in about 12 hours. As much as one fourth of the waste materials from a meal may still be in the rectum 3 days after the meal was ingested.

Waste Products of Digestion

Feces consist of undigested foodstuffs, inorganic materials, water, and bacteria. Fecal matter is about 75% fluid and 25% solid material. The composition is relatively unaffected by alterations in diet because a large portion of the fecal mass is of nondietary origin, derived from the secretions of the GI tract. The brown color of the feces results from the breakdown of bile by the intestinal bacteria. Chemicals formed by intestinal bacteria are responsible in large part for the fecal odor. Gases formed contain methane, hydrogen sulfide, and ammonia, among others. The GI tract normally contains approximately 150 mL of these gases, which are either absorbed into the portal circulation and detoxified by the liver or expelled from the rectum as flatus.

Elimination of stool begins with distention of the rectum, which initiates reflex contractions of the rectal musculature and relaxes the normally closed internal anal sphincter. The internal sphincter is controlled by the autonomic nervous system; the external sphincter is under the conscious control of the cerebral cortex. During defecation, the external anal sphincter voluntarily relaxes to allow colonic contents to be expelled. Normally, the external anal sphincter is maintained in a state of tonic contraction. Thus, defecation is seen to be a spinal reflex (involving the parasympathetic nerve fibers) that can be inhibited voluntarily by keeping the external anal sphincter closed. Contracting the abdominal muscles (straining) facilitates emptying of the colon. The average frequency of defecation in humans is once daily, but this varies among people.

TABLE 44-3 Age-Related Changes in the Gastrointestinal System

Structural Changes	Implications
Oral Cavity and Pharynx	
• Injury/loss or decay of teeth • Atrophy of taste buds • ↓ Saliva production • Reduced ptyalin and amylase in saliva	Difficulty chewing and swallowing
Esophagus	
• ↓ Motility and emptying • Weakened gag reflex • ↓ Resting pressure of lower esophageal sphincter	Reflux and heartburn
Stomach	
• Degeneration and atrophy of gastric mucosal surfaces with ↓ production of HCl • ↓ Secretion of gastric acids and most digestive enzymes • ↓ Gastric motility and emptying	Food intolerances, malabsorption, or ↓ vitamin B_{12} absorption
Small Intestine	
• Atrophy of muscle and mucosal surfaces • Thinning of villi and epithelial cells	↓ Motility and transit time, which lead to complaints of indigestion and constipation
Large Intestine	
• ↓ Mucus secretion • ↓ Elasticity of rectal wall • ↓ Tone of internal anal sphincter • Slower and duller nerve impulses in rectal area	↓ Motility and transit time, which lead to complaints of indigestion and constipation ↓ Absorption of nutrients (dextrose, fats, calcium, and iron) Fecal incontinence

↓, decreased; HCl, hydrochloric acid.

Gerontologic Considerations

Although an increased prevalence of several common GI disorders occurs in the older adult population, aging per se appears to have minimal direct effect on most GI functions, in large part because of the functional reserve of the GI tract. Normal physiologic changes of the GI system that occur with aging are identified in Table 44-3. Careful assessment and monitoring of signs and symptoms related to these changes are imperative. Although irritable bowel symptoms decrease with aging, there seems to be an increase in many GI disorders of function and motility. The gastroenterologist frequently encounters older adult patients who report dysphagia, anorexia, dyspepsia, and disorders of colonic function (Chait, 2010; Eliopoulos, 2010; Poh, Navarro-Rodriguez, & Fass, 2010).

Assessment of the Gastrointestinal System

Health History

A focused GI assessment begins with a complete history. Information about abdominal pain, dyspepsia, gas, nausea

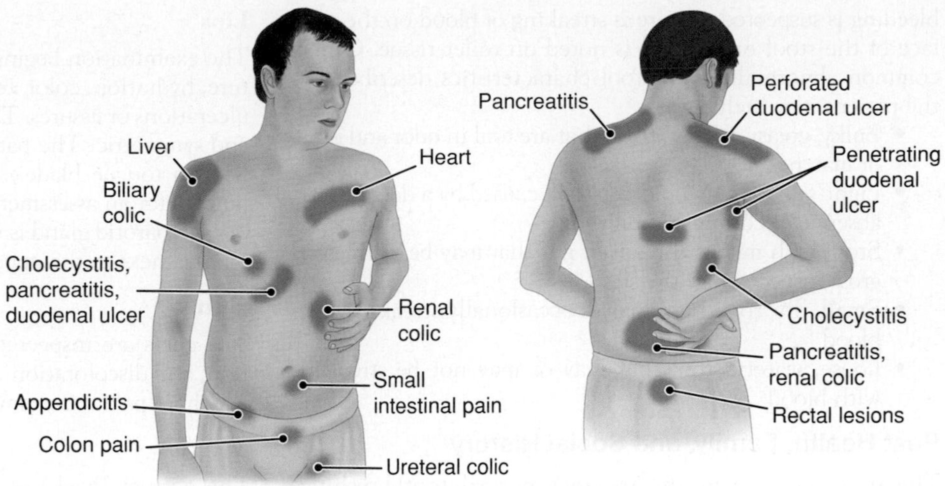
Liver
Biliary colic
Cholecystitis, pancreatitis, duodenal ulcer
Appendicitis
Colon pain
Heart
Renal colic
Small intestinal pain
Ureteral colic
Pancreatitis
Perforated duodenal ulcer
Penetrating duodenal ulcer
Cholecystitis
Pancreatitis, renal colic
Rectal lesions

FIGURE 44-3 • Common sites of referred abdominal pain.

and vomiting, diarrhea, constipation, fecal incontinence, jaundice, and previous GI disease is obtained (Bickley, 2009).

Common Symptoms

Pain

Pain can be a major symptom of GI disease. The character, duration, pattern, frequency, location, distribution of referred pain (Fig. 44-3), and time of the pain vary greatly depending on the underlying cause. Other factors such as meals, rest, activity, and defecation patterns may directly affect this pain (Bickley, 2009).

Dyspepsia

Dyspepsia, upper abdominal discomfort associated with eating (commonly called *indigestion*), is the most common symptom of patients with GI dysfunction. Indigestion is an imprecise term that refers to a host of upper abdominal or epigastric symptoms such as pain, discomfort, fullness, bloating, early satiety, belching, heartburn, or regurgitation; it occurs in approximately 25% of the adult population (Harmon & Peura, 2010). Typically, fatty foods cause the most discomfort because they remain in the stomach for digestion longer than proteins or carbohydrates. Salads, coarse vegetables, and highly seasoned foods may also cause considerable GI distress.

Intestinal Gas

The accumulation of gas in the GI tract may result in belching (expulsion of gas from the stomach through the mouth) or flatulence (expulsion of gas from the rectum). Usually, gases in the small intestine pass into the colon and are released as flatus. Patients often complain of bloating, distention, or feeling "full of gas" with excessive flatulence as a symptom of food intolerance or gallbladder disease.

Nausea and Vomiting

Nausea is a vague, uncomfortable sensation of sickness or "queasiness" that may or may not be followed by vomiting. It can be triggered by odors, activity, medications, or food intake. The emesis, or vomitus, may vary in color and content and may contain undigested food particles, blood (hematemesis), or bilious material mixed with gastric juices. The causes of nausea and vomiting are many; they may result from (1) visceral afferent stimulation (i.e., dysmotility, peritoneal irritation, infections, hepatobiliary or pancreatic disorders, mechanical obstruction); (2) CNS disorders (i.e., vestibular disorders, increased intracranial pressure, infections, psychogenic disorder); and/or (3) irritation of the chemoreceptor trigger zone from radiation therapy, systemic disorders, and antitumor chemotherapy medications (Wolfe, 2006).

Change in Bowel Habits and Stool Characteristics

Changes in bowel habits may signal colonic dysfunction or disease. Diarrhea, an abnormal increase in the frequency and liquidity of the stool or in daily stool weight or volume, commonly occurs when the contents move so rapidly through the intestine and colon that there is inadequate time for the GI secretions and oral contents to be absorbed. This physiologic function is typically associated with abdominal pain or cramping and nausea or vomiting. Constipation—a decrease in the frequency of stool, or stools that are hard, dry, and of smaller volume than typical—may be associated with anal discomfort and rectal bleeding. (See Chapter 48 for further discussion of diarrhea and constipation.)

The characteristics of the stool can vary greatly. Stool is normally light to dark brown; however, specific disease processes and ingestion of certain foods and medications may change the appearance of stool (Table 44-4). Blood in the stool can present in various ways and must be investigated. If blood is shed in sufficient quantities into the upper GI tract, it produces a tarry-black color (melena), whereas blood entering the lower portion of the GI tract or passing rapidly through it will appear bright or dark red. Lower rectal or anal

TABLE 44-4 Foods and Medications that Alter Stool Color

Altering Substance	Color
Meat protein	Dark brown
Spinach	Green
Carrots, beets, red gelatin	Red
Cocoa	Dark red or brown
Senna	Yellow
Bismuth, iron, licorice, charcoal	Black
Barium	Milky white

bleeding is suspected if there is streaking of blood on the surface of the stool or if blood is noted on toilet tissue. Other common abnormalities in stool characteristics described by the patient may include:

- Bulky, greasy, foamy stools that are foul in odor and may or may not float
- Light-gray or clay-colored stool, caused by a decrease or absence of conjugated bilirubin
- Stool with mucus threads or pus that may be visible on gross inspection of the stool
- Small, dry, rock-hard masses occasionally streaked with blood
- Loose, watery stool that may or may not be streaked with blood

Past Health, Family, and Social History

The nurse asks about the patient's normal toothbrushing and flossing routine; frequency of dental visits; awareness of any lesions or irritated areas in the mouth, tongue, or throat; recent history of sore throat or bloody sputum; discomfort caused by certain foods; daily food intake; the use of alcohol and tobacco, including smokeless chewing tobacco; and the need to wear dentures or a partial plate. For information about denture care, see Chart 44-1.

Past and current medication use and any previous diagnostic studies, treatments, or surgery are noted. Current nutritional status is assessed via history; laboratory tests (complete metabolic panel including liver function studies, triglyceride, iron studies, and complete blood count [CBC]) are obtained. History of the use of tobacco and alcohol includes details about type, amount, length of use, and the date of discontinuation, if any. The nurse and patient discuss changes in appetite or eating patterns and any unexplained weight gain or loss over the past year. The nurse also asks questions about psychosocial, spiritual, or cultural factors that may be affecting the patient.

Physical Assessment

The physical examination includes assessment of the mouth, abdomen, and rectum and requires a good source of light, full exposure of the abdomen, warm hands with short fingernails, and a comfortable and relaxed patient with an empty bladder.

Oral Cavity Inspection and Palpation

Dentures should be removed to allow good visualization of the entire oral cavity.

Chart 44-1 HEALTH PROMOTION

Denture Care

- Brush dentures twice a day.
- Remove dentures at night, and soak them in water or a denture product. (Never put dentures in hot water, because they may warp.)
- Rinse mouth with warm salt water in the morning, after meals, and at bedtime.
- Clean well under partial dentures, where food particles tend to get caught.
- Consume nonsticky foods that have been cut into small pieces; chew slowly.
- See dentist regularly to assess and adjust fit.

Lips

The examination begins with inspection of the lips for moisture, hydration, color, texture, symmetry, and the presence of ulcerations or fissures. The lips should be moist, pink, smooth, and symmetric. The patient is instructed to open the mouth wide; a tongue blade is then inserted to expose the buccal mucosa for an assessment of color and lesions. Stensen's duct of each parotid gland is visible as a small red dot in the buccal mucosa next to the upper molars.

Gums

The gums are inspected for inflammation, bleeding, retraction, and discoloration. The odor of the breath is also noted. The hard palate is examined for color and shape.

Tongue

The dorsum (back) of the tongue is inspected for texture, color, and lesions. A thin white coat and large, vallate papillae in a "V" formation on the distal portion of the dorsum of the tongue are normal findings. The patient is instructed to protrude the tongue and move it laterally. This provides the examiner with an opportunity to estimate the tongue's size as well as its symmetry and strength (to assess the integrity of the 12th cranial nerve [hypoglossal nerve]).

Further inspection of the ventral surface of the tongue and the floor of the mouth is accomplished by asking the patient to touch the roof of the mouth with the tip of the tongue. Any lesions of the mucosa or any abnormalities involving the frenulum or superficial veins on the undersurface of the tongue are assessed for location, size, color, and pain. This is a common area for oral cancer, which presents as a white or red plaque, an indurated ulcer, or a warty growth.

A tongue blade is used to depress the tongue for adequate visualization of the pharynx. It is pressed firmly beyond the midpoint of the tongue; proper placement avoids a gagging response. The patient is told to tip the head back, open the mouth wide, take a deep breath, and say "ah." Often, this flattens the posterior tongue and briefly allows a full view of the tonsils, uvula, and posterior pharynx. These structures are inspected for color, symmetry, and evidence of exudate, ulceration, or enlargement. Normally, the uvula and soft palate rise symmetrically with a deep inspiration upon saying "ah"; this indicates an intact vagus nerve (10th cranial nerve).

A complete assessment of the oral cavity is essential because many disorders, such as cancer, diabetes, and immunosuppressive conditions resulting from medication therapy or acquired immunodeficiency syndrome, may be manifested by changes in the oral cavity, including stomatitis.

Abdominal Inspection, Auscultation, Percussion, and Palpation

The patient lies supine with knees flexed slightly for inspection, auscultation, percussion, and palpation of the abdomen. For the purposes of examination and documentation, the abdomen can be divided into either four quadrants or nine regions (Fig. 44-4).

Consistent use of one of these mapping methods results in a thorough evaluation of the abdomen and appropriate documentation. The four-quadrant method involves the

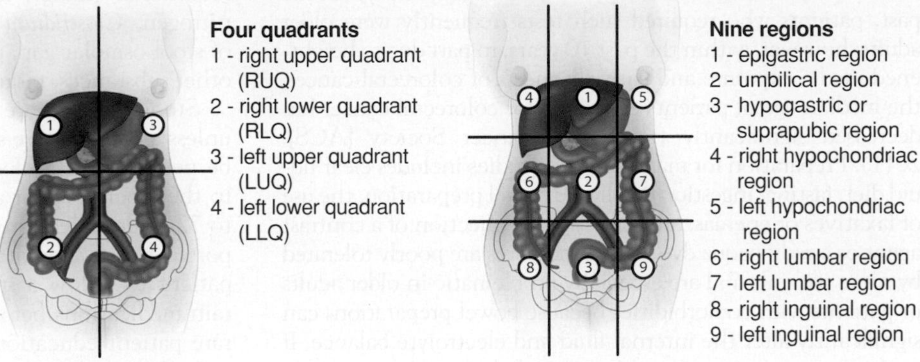
Four quadrants
1 - right upper quadrant (RUQ)
2 - right lower quadrant (RLQ)
3 - left upper quadrant (LUQ)
4 - left lower quadrant (LLQ)
Nine regions
1 - epigastric region
2 - umbilical region
3 - hypogastric or suprapubic region
4 - right hypochondriac region
5 - left hypochondriac region
6 - right lumbar region
7 - left lumbar region
8 - right inguinal region
9 - left inguinal region

FIGURE 44-4 • Division of the abdomen into four quadrants or nine regions.

use of an imaginary line drawn vertically from the sternum to the pubis through the umbilicus and a horizontal line drawn across the abdomen through the umbilicus. Inspection is performed first, noting skin changes, nodules, lesions, scarring, discolorations, inflammation, bruising, or striae. Lesions are of particular importance, because GI diseases often produce skin changes. The contour and symmetry of the abdomen are noted, and any localized bulging, distention, or peristaltic waves are identified. Expected contours of the anterior abdominal wall can be described as flat, rounded, or scaphoid.

Auscultation always precedes percussion and palpation, because they may alter sounds. Auscultation is used to determine the character, location, and frequency of bowel sounds and to identify vascular sounds. Bowel sounds are assessed using the diaphragm of the stethoscope for high-pitched and gurgling sounds (Gu, Lim, & Moser, 2010). The frequency and character of the sounds are usually heard as clicks and gurgles that occur irregularly and range from 5 to 35 per minute. The terms *normal* (sounds heard about every 5 to 20 seconds), *hypoactive* (one or two sounds in 2 minutes), *hyperactive* (5 to 6 sounds heard in less than 30 seconds), or *absent* (no sounds in 3 to 5 minutes) are frequently used in documentation, but these assessments are highly subjective (Li, Wang, & Ma, 2012). Using the bell of the stethoscope, any bruits in the aortic, renal, iliac, and femoral arteries are noted. Friction rubs are high pitched and can be heard over the liver and spleen during respiration. Borborygmi ("stomach growling") is heard as a loud prolonged gurgle.

Percussion is used to assess the size and density of the abdominal organs and to detect the presence of air-filled, fluid-filled, or solid masses. Percussion is used either independently or concurrently with palpation because it can validate palpation findings. All quadrants are percussed for overall tympani and dullness. Tympani is the sound that results from the presence of air in the stomach and small intestines; dullness is heard over organs and solid masses. The use of light palpation is appropriate for identifying areas of tenderness or muscular resistance, and deep palpation is used to identify masses. Testing for rebound tenderness is not performed by many examiners because it can cause severe pain; light percussion is used instead to produce a mild localized response when peritoneal irritation is present.

Rectal Inspection and Palpation

The final part of the examination is evaluation of the terminal portions of the GI tract, the rectum, perianal region, and anus. The anal canal is approximately 2.5 to 4 cm (1 to 1.6 in) in length and opens into the perineum. Concentric rings of muscle, the internal and external sphincters, normally keep the anal canal securely closed. Gloves, water-soluble lubrication, a penlight, and drapes are necessary tools for the evaluation. Although the rectal examination is generally uncomfortable and often embarrassing for the patient, it is a mandatory part of every thorough examination. For women, the rectal examination may be part of the gynecologic examination. Positions for the rectal examination include knee-chest, left lateral with hips and knees flexed, or standing with hips flexed and upper body supported by the examination table. Most patients are comfortable on the right side with knees brought up to the chest. External examination includes inspection for lumps, rashes, inflammation, excoriation, tears, scars, pilonidal dimpling, and tufts of hair at the pilonidal area. The discovery of tenderness, inflammation, or both should alert the examiner to the possibility of a pilonidal cyst, perianal abscess, or anorectal fistula or fissure. The patient's buttocks are carefully spread and visually inspected until the patient has relaxed the external sphincter control. The patient is asked to bear down, thus allowing the ready appearance of fistulas, fissures, rectal prolapse, polyps, and internal hemorrhoids. Internal examination is performed with a lubricated index finger inserted into the anal canal while the patient bears down. The tone of the sphincter is noted, as are any nodules or irregularities of the anal ring. Because this is an uncomfortable part of the examination for most patients, the patient is encouraged to focus on deep breathing and visualization of a pleasant setting during the brief examination.

Diagnostic Evaluation

GI diagnostic studies can confirm, rule out, stage, or diagnose various disease states. In particular, the diagnosis and treatment of cancer brings psychological distress to patients and their families and can cause anxiety, depressive symptoms, and adjustment disorders (Faul, Jim, Williams, et al., 2010; Greer, Solis, Temel, et al., 2011). After the diagnosis, time should be allotted for discussion with the patient, in addition to offering resource materials for information.

Many modalities are available for diagnostic assessment of the GI tract. The majority of these tests and procedures are performed on an outpatient basis in special settings designed for this purpose (e.g., endoscopy suite or GI laboratory). In the past, patients who required such tests frequently were older adults; however, within the past 10 years, in part due to heightened media exposure and early diagnosis of colorectal cancer, the median age of patients evaluated for colorectal cancer has decreased significantly (American Cancer Society [ACS], 2011). Preparation for many of these studies includes clear liquid diet, fasting, ingestion of a liquid bowel preparation, the use of laxatives or enemas, and ingestion or injection of a contrast agent or a radiopaque dye. These measures are poorly tolerated by some patients and are especially problematic in older adults or patients with comorbidities because bowel preparations can significantly alter the internal fluid and electrolyte balance. If further assessment or treatment is needed after any outpatient procedure, the patient may be admitted to the hospital.

Specific nursing interventions for each test are provided later in this chapter. General nursing interventions for the patient who is undergoing a GI diagnostic evaluation include:

- Establishing the nursing diagnosis
- Providing needed information about the test and the activities required of the patient
- Providing instructions about postprocedure care and activity restrictions
- Providing health information and procedural education to patients and significant others
- Helping the patient cope with discomfort and alleviating anxiety
- Informing the primary provider of known medical conditions or abnormal laboratory values that may affect the procedure
- Assessing for adequate hydration before, during, and immediately after the procedure, and providing education about maintenance of hydration

Serum Laboratory Studies

Initial diagnostic tests begin with serum laboratory studies, including but not limited to CBC, complete metabolic panel, prothrombin time/partial thromboplastin time, triglycerides, liver function tests, amylase, and lipase; possibly, more specific studies may be indicated, such as carcinoembryonic antigen (CEA), cancer antigen (CA) 19-9, and alpha-fetoprotein, which are sensitive and specific for colorectal and hepatocellular carcinomas, respectively. CEA is a protein that is normally not detected in the blood of a healthy person; therefore, when detected it indicates that cancer is present, although not what type of cancer is present. Primary providers can use CEA results to determine the stage and extent of the disease and the patient's prognosis for cancer, especially GI and, in particular, colorectal cancer (Bones, Byrne, O'Donoghue, et al., 2010). CA 19-9 is also a protein that exists on the surface of certain cells and is shed by tumor cells, making it useful as a tumor marker to follow the course of the cancer. CA 19-9 levels are elevated in most patients with advanced pancreatic cancer, but they may also be elevated in other conditions such as colorectal, lung, and gallbladder cancers; gallstones; pancreatitis; cystic fibrosis; and liver disease (Korkmaz, Unal, Selcuk, et al., 2010).

Stool Tests

Basic examination of the stool includes inspecting the specimen for consistency, color, and occult (not visible) blood. Additional studies, including fecal urobilinogen, fecal fat, nitrogen, *Clostridium difficile*, fecal leukocytes, calculation of stool osmolar gap, parasites, pathogens, food residues, and other substances, require laboratory evaluation.

Stool samples are usually collected on a random basis unless a quantitative study (e.g., fecal fat, urobilinogen) is to be performed. Random specimens should be sent promptly to the laboratory for analysis; however, the quantitative 24- to 72-hour collections must be kept refrigerated until transported to the laboratory. Some stool collections require the patient to follow a specific diet or refrain from taking certain medications before the collection. Thorough and accurate patient education regarding a specific stool study prior to collection greatly increases the accuracy of study results (Kumaravel, Hayden, Hall, et al., 2011).

Fecal occult blood testing (FOBT) is one of the most commonly performed stool tests. It can be useful in initial screening for several disorders, although it is used most frequently in early cancer detection programs. FOBT can be performed at the bedside, in the laboratory, or at home. Probably the most widely used in-office or at-home occult blood test is the Hemoccult II. It is inexpensive, noninvasive, and carries minimal risk to the patient. However, it should not be performed when there is hemorrhoidal bleeding. In addition, red meats, aspirin, nonsteroidal anti-inflammatory drugs, turnips, and horseradish should be avoided for 72 hours prior to the study, because they may cause a false-positive result; ingestion of vitamin C from supplements or foods can cause a false-negative result. Therefore, a careful assessment of the patient's diet and medication regimen is essential to avoid incorrect interpretation of results. A small amount of the specimen is applied to the guaiac-impregnated paper slide. If the test is performed at home, the patient mails the slide to the physician in an envelope provided for that purpose. Other occult blood tests that may yield more specific and more sensitive readings include Hemoccult SENSA and HemoQuant (Brady, 2012).

Fecal immunologic tests use monoclonal or polyclonal antibodies to detect the globin protein in human hemoglobin. There is no reaction with nonhuman hemoglobin or with foods that contain peroxidase activity; these tests are therefore more specific than guaiac tests (Khalid-de Bakker, Jonkers, Sanduleanu, et al., 2011; Kumaravel et al., 2011). Hematoporphyrin assays detect the broadest range of blood derivatives, but a strict dietary protocol is essential. Quantitative fecal immunochemical tests may be more accurate than guaiac testing and useful for patients who refuse invasive testing (Khalid-de Bakker et al., 2011).

Stool DNA testing is a relatively new means to detect certain DNA known to be related to colon cancer. More research is needed to determine how often the test needs to be performed. The stool DNA test does not require any dietary or medication restrictions and can detect neoplasia anywhere in the colon. The stool sample can be collected at home (Ned, Melillo, & Marrone, 2011).

Breath Tests

The hydrogen breath test was developed to evaluate carbohydrate absorption, in addition to aiding in the diagnosis of bacterial overgrowth in the intestine and short bowel syndrome. This test determines the amount of hydrogen expelled in the breath after it has been produced in the colon (on contact of galactose with fermenting bacteria) and absorbed into the blood.

Urea breath tests detect the presence of *Helicobacter pylori*, the bacteria that can live in the mucosal lining of the stomach and cause peptic ulcer disease. After the patient ingests a capsule of carbon-labeled urea, a breath sample is obtained 10 to 20 minutes later. Because *H. pylori* metabolizes urea rapidly, the labeled carbon is absorbed quickly; it can then be measured as carbon dioxide in the expired breath to determine whether *H. pylori* is present. Prior to urea breath testing, the patient is instructed to avoid antibiotics or bismuth subsalicylate (Pepto-Bismol) for 1 month before the test; sucralfate (Carafate) and omeprazole (Prilosec) for 1 week before the test; and cimetidine (Tagamet), famotidine (Pepcid), and ranitidine (Zantac) for 24 hours before the test. *H. pylori* also can be detected by assessing serum antibody levels.

Abdominal Ultrasonography

Ultrasonography is a noninvasive diagnostic technique in which high-frequency sound waves are passed into internal body structures, and the ultrasonic echoes are recorded on an oscilloscope as they strike tissues of different densities. It is particularly useful in the detection of an enlarged gallbladder or pancreas, the presence of gallstones, an enlarged ovary, an ectopic pregnancy, or appendicitis. Most recently, this technique has proven useful in diagnosing acute colonic diverticulitis (Reitz, Slam, & Chambers, 2011).

Advantages of abdominal ultrasonography include an absence of ionizing radiation, no noticeable side effects, relatively low cost, and almost immediate results. It cannot be used to examine structures that lie behind bony tissue, because bone prevents sound waves from traveling into deeper structures. Gas and fluid in the abdomen or air in the lungs also prevent transmission of ultrasound. An ultrasound produces no ill effects. However, some patients, typically pregnant women, have concerns regarding the energy emitted by the probe.

Endoscopic ultrasonography (EUS) is a specialized enteroscopic procedure that aids in the diagnosis of GI disorders by providing direct imaging of a target area. A small high-frequency ultrasonic transducer is mounted at the tip of the fiberoptic scope, which display images that are of higher-quality resolution and definition than regular ultrasound imaging. For patients undergoing complex endoscopic procedures, moderate sedation may be indicated (Scholten, 2012) (see Chapter 18 for further discussion of moderate sedation). EUS may be used to evaluate submucosal lesions, specifically their location and depth of penetration. In addition, EUS may aid in the evaluation of Barrett's esophagus, portal hypertension, chronic pancreatitis, suspected pancreatic neoplasm, biliary tract disease, and changes in the bowel wall due to ulcerative colitis. EUS is a safe and accurate test to use when selecting patients who are suspected of having biliary obstructive disease for therapeutic endoscopic retrograde cholangiopancreatography (ERCP) (Reitz et al., 2011). Intestinal gas, bone, and thick layers of adipose tissue that hamper conventional ultrasonography are not problems when EUS is used.

Nursing Interventions

The patient is instructed to fast for 8 to 12 hours before ultrasound testing to decrease the amount of gas in the bowel. If gallbladder studies are being performed, the patient should eat a fat-free meal the evening before the test. If barium studies are to be performed, they should be scheduled after ultrasonography; otherwise, the barium could interfere with the transmission of the sound waves. Patients who receive moderate sedation are observed for about 1 hour to assess for level of consciousness, orientation, and ability to ambulate. Patients treated on an outpatient basis are given instructions regarding diet, activity, and how to monitor for complications and may be discharged provided they have an escort (Scholten, 2010).

DNA Testing

Researchers have refined methods for genetics risk assessment, preclinical diagnosis, and prenatal diagnosis to identify people who are at risk for certain GI disorders (e.g., gastric cancer, lactose deficiency, inflammatory bowel disease, colon cancer) (Chart 44-2). In some cases, DNA testing allows clinicians to prevent (or minimize) disease by intervening before its onset and to improve therapy (Offit, 2011). People who are identified as being at risk for certain GI disorders may choose to have genetic counseling to learn about the disease and options for preventing and treating the disease, and to receive support in coping with the situation. (See Chapter 8 for further discussion of genetic counseling.)

Imaging Studies

Numerous minimally invasive and noninvasive imaging studies, including x-ray and contrast studies, computed tomography (CT), magnetic resonance imaging (MRI), positron emission tomography (PET), scintigraphy (radionuclide imaging), and virtual colonoscopy are available today.

Upper Gastrointestinal Tract Study

An upper GI fluoroscopy delineates the entire GI tract after the introduction of a contrast agent. A radiopaque liquid (e.g., barium sulfate) is commonly used; however, thin barium, diatrizoate sodium (Hypaque), and at times water are used due to their low associated risks. The GI series enables the examiner to detect or exclude anatomic or functional disorders of the upper GI organs or sphincters. It also aids in the diagnosis of ulcers, varices, tumors, regional enteritis, and malabsorption syndromes. The procedure may be extended to examine the duodenum and small bowel (small bowel follow-through). As the barium descends into the stomach, the position, patency, and caliber of the esophagus are visualized, enabling the examiner to detect or exclude any anatomic or functional derangement of that organ. Fluoroscopic examination next extends to the stomach as its lumen fills with barium, allowing observation of stomach motility, thickness of the gastric wall, the mucosal pattern, patency of the pyloric valve, and the anatomy of the duodenum. Multiple x-ray films are obtained during the procedure, and additional images may be taken at intervals for up to 24 hours to evaluate the rate of gastric emptying. Small bowel x-rays taken while the barium is passing through that area allow for observation

Chart 44-2

GENETICS IN NURSING PRACTICE

Digestive and Gastrointestinal Disorders

Several digestive and gastrointestinal disorders are associated with genetic abnormalities. Some examples are:

- Cleft lip and/or palate
- Familial adenomatous polyposis
- Hereditary nonpolyposis colorectal cancer
- Hirschsprung disease (aganglionic megacolon)
- Inflammatory bowel disease (e.g., Crohn's disease)
- Pyloric stenosis

Nursing Assessments

Family History Assessment

- Careful family history assessment for other family members with a similar condition (e.g., cleft lip/palate, pyloric stenosis)
- Assess for other family members in several generations with early-onset colorectal cancer
- Inquire about other family members with inflammatory bowel disease
- Assess family history for other cancers (e.g., endometrial, ovarian, kidney)

Patient Assessment

Assess for presence of other clinical conditions:

- With clefting—congenital heart defect, other birth defects suggestive of a genetic syndrome
- With familial adenomatous polyposis—congenital hypertrophy of retinal pigment epithelium

Management Issues Specific to Genetics

- Inquire whether any affected family member has had DNA mutation testing
- If indicated, refer for further genetic counseling and evaluation so that family members can discuss inheritance, risk to other family members, availability of genetic testing, and gene-based interventions
- Offer appropriate genetic information and resources
- Assess patients' understanding of genetic information
- Provide support to families with newly diagnosed genetic digestive disorders
- Participate in management and coordination of care for patients with genetic conditions and for those who are predisposed to develop or pass on a genetic condition

Genetics Resources

See Chapter 8, Chart 8-6 for genetics resources.

of the motility of the small bowel. Obstructions, ileitis, and diverticula can also be detected.

Variations of the upper GI study include double-contrast studies and enteroclysis. The double-contrast method of examining the upper GI tract involves administration of a thick barium suspension to outline the stomach and esophageal wall, after which tablets that release carbon dioxide in the presence of water are administered. This technique has the advantage of showing the esophagus and stomach in finer detail, permitting signs of early superficial neoplasms to be noted.

Enteroclysis is a very detailed, double-contrast study of the entire small intestine that involves the continuous infusion (through a duodenal tube) of 500 to 1,000 mL of a thin barium sulfate suspension; after this, methylcellulose is infused through the tube. The barium and methylcellulose fill the intestinal loops and are observed continuously by fluoroscopy and viewed at frequent intervals as they progress through the jejunum and the ileum. This process (even with normal motility) can take up to 6 hours and can be quite uncomfortable for the patient. The procedure aids in the diagnosis of partial small bowel obstructions or diverticula. The value of these x-ray screening studies has diminished as better technology has emerged (Graca, Freire, & Brito, 2010).

Nursing Interventions

Instruction regarding dietary changes prior to the study should include a clear liquid diet, with nothing by mouth (NPO) from midnight the night before the study (Tennyson & Semrad, 2011). Polyethylene glycol is considered the most effective bowel cleansing preparatory agent (Belsey, Crosta, Epstein, et al., 2012; Scholten, 2010). The nurse advises against smoking, chewing gum, and using mints because they can stimulate gastric motility. Typically, oral medications are withheld on the morning of the study and resumed that evening, but each patient's medication regimen should be evaluated on an individual basis. When a patient with insulin-dependent diabetes is NPO, his or her insulin requirements will need to be adjusted accordingly (see Chapter 51).

Follow-up care is provided after the upper GI procedure to ensure that the patient has eliminated most of the ingested barium. Fluids may be increased to facilitate evacuation of stool and barium.

Lower Gastrointestinal Tract Study

Visualization of the lower GI tract is obtained after rectal installation of barium. The barium enema can be used to detect the presence of polyps, tumors, or other lesions of the large intestine and demonstrate any anatomic abnormalities or malfunctioning of the bowel. After proper preparation and evacuation of the entire colon, each portion of the colon may be readily observed. The procedure usually takes about 15 to 30 minutes, during which time x-ray images are obtained.

Other means for visualizing the colon include double-contrast studies and a water-soluble contrast study. These tests are still occasionally used because they are relatively inexpensive and simple. A double-contrast or air-contrast barium enema involves the instillation of a thicker barium solution, followed by the instillation of air. The patient may feel some cramping or discomfort during this process. This test provides a contrast between the air-filled lumen and the barium-coated mucosa, allowing easier detection of smaller lesions.

If active inflammatory disease, fistulas, or perforation of the colon is suspected, a water-soluble iodinated contrast agent (e.g., diatrizoic acid [Gastrografin]) can be used. The procedure is the same as for a barium enema, but the patient must first be assessed for allergy to iodine or contrast agent. The contrast agent is eliminated readily after the procedure, so there is no need for postprocedure laxatives. Some diarrhea may occur in a few patients until the contrast agent has been totally eliminated.

Nursing Interventions

Preparation of the patient includes emptying and cleansing the lower bowel. This often necessitates a low-residue diet 1 to 2 days before the test, a clear liquid diet and a laxative the evening before, NPO after midnight, and cleansing enemas until returns are clear the following morning. The nurse makes sure that barium enemas are scheduled before any upper GI studies. If the patient has active inflammatory disease of the colon, enemas are contraindicated. Barium enemas also are contraindicated in patients with signs of perforation or obstruction; instead, a water-soluble contrast study may be performed. Active GI bleeding may prohibit the use of laxatives and enemas.

Postprocedural patient education includes information about increasing fluid intake, evaluating bowel movements for evacuation of barium, and noting increased number of bowel movements, because barium, due to its high osmolarity, may draw fluid into the bowel, thus increasing the intraluminal contents and resulting in greater output.

Computed Tomography

A CT scan provides cross-sectional images of abdominal organs and structures. Multiple x-ray images are taken from numerous angles, digitized in a computer, reconstructed, and then viewed on a computer monitor. As the sensitivity and specificity of CT scans have increased in recent years, so has its use. CT is a valuable tool for detecting and localizing many inflammatory conditions in the colon, such as appendicitis, diverticulitis, regional enteritis, and ulcerative colitis, as well as evaluating the abdomen for diseases of the liver, spleen, kidney, pancreas, and pelvic organs, and structural abnormalities of the abdominal wall. Because the adequacy of detail in the test depends on the presence of fat, this diagnostic tool is not useful for patient who are very thin or cachectic (Zamboni & Raptopoulos, 2010). The CT procedure is completely painless, but radiation doses are considerable. Continuous-motion (helical or spiral), three-dimensional CT that provides very detailed pictures of the GI organs and vasculature is also available (Zamboni & Raptopoulos, 2010).

Nursing Interventions

A CT scan may be performed with or without oral or intravenous (IV) contrast, but the enhancement of the study is greater with the use of a contrast agent. Any allergies to contrast agents, iodine, or shellfish, the patient's current serum creatinine level, and pregnancy status in females must be determined before administration of a contrast agent. Patients allergic to the contrast agent may be premedicated with a corticosteroid and antihistamine. Risk factors include an elevated serum creatinine, a history of drug allergy or contact hypersensitivity, and previous nonimmediate hypersensitivity reaction (Brockow & Ring, 2012; Mishra, Heavner, & Day, 2012).

In addition, kidney protective measures include the administration of IV sodium bicarbonate 1 hour before and 6 hours after IV contrast and oral acetylcysteine (Mucomyst) before or after the study. Both sodium bicarbonate and Mucomyst are free radical scavengers that sequester the contrast by-products that are destructive to kidney cells (Brown & Thompson, 2010; Mautone & Brown, 2010).

Magnetic Resonance Imaging

MRI is used in gastroenterology to supplement ultrasonography and CT. This noninvasive technique uses magnetic fields and radio waves to produce images of the area being studied. The use of oral contrast agents to enhance the image has increased the application of this technique for the diagnosis of GI diseases. It is useful in evaluating abdominal soft tissues as well as blood vessels, abscesses, fistulas, neoplasms, and other sources of bleeding.

The physiologic artifacts of heartbeat, respiration, and peristalsis may create a less-than-clear image; however, newer, fast-imaging MRI techniques help eliminate these physiologic motion artifacts. MRI is not totally safe for all people. Any ferromagnetic objects (metals that contain iron) can be attracted to the magnet and cause injury. Items that can be problematic or dangerous include jewelry, pacemakers, dental implants, paperclips, pens, keys, IV poles, clips on patient gowns, and oxygen tanks. MRI is contraindicated in patients with permanent pacemakers, artificial heart valves and defibrillators, implanted insulin pumps, or implanted transcutaneous electrical nerve stimulation devices, because the magnetic field could cause malfunction. MRI is also contraindicated for patients with internal metal devices (e.g., aneurysm clips), intraocular metallic fragments, or cochlear implants. Foil-backed skin patches (e.g., nicotine [NicoDerm], nitroglycerin [Transderm-Nitro], scopolamine [Transderm-Scop], clonidine [Catapres-TTS]) should be removed before an MRI because of the risk of burns; however, the patient's primary provider should be consulted before the patch is removed to determine whether an alternate form of the medication should be provided.

Nursing Interventions

Prestudy patient education includes NPO status 6 to 8 hours before the study and removal of all jewelry and other metals. The patient and family are informed that the study may take 60 to 90 minutes; during this time, the technician will instruct the patient to take deep breaths at specific intervals. The close-fitting scanners used in many MRI facilities may induce feelings of claustrophobia, and the machine will make a knocking sound during the procedure. Patients may choose to wear a headset and listen to music or wear a blindfold during the procedure. Open MRIs that are less close fitting eliminate the claustrophobia that many patients experience, although they produce lower-resolution images.

Positron Emission Tomography

PET scans produce images of the body by detecting the radiation emitted from radioactive substances. The radioactive substances are injected into the body IV and are usually tagged with a radioactive atom, such as carbon-11, fluorine-18, oxygen-15, or nitrogen-13. The atoms decay quickly, do not harm the body, have lower radiation levels than a typical x-ray or CT scan, and are eliminated in the urine or feces. The scanner essentially "captures" where the radioactive substances are in the body, transmits information to a scanner, and produces a scan with "hot spots" for evaluation by the radiologist or oncologist.

Scintigraphy

Scintigraphy (radionuclide testing) relies on the use of radioactive isotopes (i.e., technetium, iodine, and indium) to reveal displaced anatomic structures, changes in organ size, and the presence of neoplasms or other focal lesions such as cysts or abscesses. Scintigraphic scanning is also used to measure the uptake of tagged red blood cells and leukocytes. Tagging of red blood cells and leukocytes by injection of a radionuclide is performed to define areas of inflammation, abscess, blood loss, or neoplasm. A sample of blood is removed, mixed with a radioactive substance, and reinjected into the patient. Abnormal concentrations of blood cells are then detected at 24- and 48-hour intervals. Tagged red cell studies are useful in determining the source of internal bleeding when all other studies have returned a negative result.

Gastrointestinal Motility Studies

Radionuclide testing also is used to assess gastric emptying and colonic transit time. During gastric emptying studies, the liquid and solid components of a meal (typically scrambled eggs) are tagged with radionuclide markers. After ingestion of the meal, the patient is positioned under a scintiscanner, which measures the rate of passage of the radioactive substance from the stomach. This is useful in diagnosing disorders of gastric motility, diabetic gastroparesis, and dumping syndrome.

Colonic transit studies are used to evaluate colonic motility and obstructive defecation syndromes. The patient is administered a capsule containing 20 radionuclide markers and instructed to follow a regular diet and usual daily activities. Abdominal x-rays are taken every 24 hours until all markers are passed. This process usually takes 4 to 5 days; in the presence of severe constipation it may take as long as 10 days. Patients with chronic diarrhea may be evaluated at 8-hour intervals. The amount of time that it takes for the radioactive material to move through the colon indicates colonic motility.

Endoscopic Procedures

Endoscopic procedures used in GI tract assessment include fibroscopy/esophagogastroduodenoscopy (EGD), colonoscopy, anoscopy, proctoscopy, sigmoidoscopy, small bowel enteroscopy, and endoscopy through an ostomy.

Upper Gastrointestinal Fibroscopy/ Esophagogastroduodenoscopy

Fibroscopy of the upper GI tract allows direct visualization of the esophageal, gastric, and duodenal mucosa through a lighted endoscope (gastroscope) (Fig. 44-5). EGD is valuable when esophageal, gastric, or duodenal disorders or inflammatory, neoplastic, or infectious processes are suspected. This procedure also can be used to evaluate esophageal and gastric motility and to collect secretions and tissue specimens for further analysis.

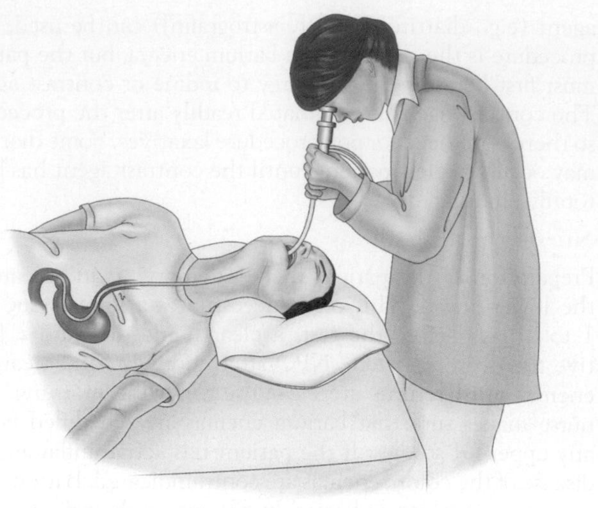

FIGURE 44-5 • Patient undergoing gastroscopy.

In EGD, the gastroenterologist views the GI tract through a viewing lens and can obtain images through the scope to document findings. Electronic video endoscopes also are available that attach directly to a video processor, converting the electronic signals into pictures on a television screen. This allows larger and continuous viewing capabilities, as well as the simultaneous recording of the procedure.

PillCam ESO, a pill-sized instrument equipped with two cameras, is available. Each camera takes seven photographs per second and transmits them wirelessly to a nearby storage device (Sieg, 2011; White & Kilgore, 2009). This technique is gaining popularity with patients and clinicians alike as a comfortable, convenient alternative to endoscopy. Two major drawbacks to this method of endoscopy are that it evaluates only the esophagus; in addition, the instrument may become lodged in a previously anastomosed section of bowel and require further endoscopic or surgical intervention for removal.

ERCP uses the endoscope in combination with x-rays to view the ductal structures of the biliary tract (Rabago, Ortega, Chico, et al., 2011). The side-viewing flexible scopes are used to visualize the common bile duct and the pancreatic and hepatic ducts through the ampulla of Vater in the duodenum. ERCP is helpful in evaluating jaundice, pancreatitis, pancreatic tumors, common bile duct stones, and biliary tract disease. ERCP is described further in Chapter 50.

Upper GI fibroscopy also can be a therapeutic procedure when combined with other procedures. Therapeutic endoscopy can be used to remove common bile duct stones, dilate strictures, and treat gastric bleeding and esophageal varices. Laser-compatible scopes can be used to provide laser therapy for upper GI neoplasms. Sclerosing solutions can be injected through the scope in an attempt to control upper GI bleeding.

After the patient is sedated, the endoscope is lubricated with a water-soluble lubricant and passed smoothly and slowly along the back of the mouth and down into the esophagus. The gastroenterologist views the gastric wall and the sphincters and then advances the endoscope into the duodenum for further examination. Biopsy forceps to obtain tissue specimens or cytology brushes to obtain cells for microscopic study can be passed through the scope. The procedure usually takes about 30 minutes.

The patient may experience nausea, gagging, or choking. The use of topical anesthetic agents and moderate sedation makes it important to monitor and maintain the patient's oral airway during and after the procedure. Finger or ear oximeters are used to monitor oxygen saturation, and supplemental oxygen may be administered if needed. Precautions must be taken to protect the scope, because the fiberoptic bundles can be broken if the scope is bent at an acute angle. The patient wears a mouth guard to keep from biting the scope.

Nursing Interventions

The patient should be NPO for 8 hours prior to the examination. Before the introduction of the endoscope, the patient is given a local anesthetic gargle or spray. Midazolam (Versed), a sedative that provides moderate sedation with loss of the gag reflex and relieves anxiety during the procedure, is administered. Atropine may be administered to reduce secretions, and glucagon may be administered to relax smooth muscle. The patient is positioned in the left lateral position to facilitate clearance of pulmonary secretions and provide smooth entry of the scope.

After gastroscopy, assessment includes level of consciousness, vital signs, oxygen saturation, pain level, and monitoring for signs of perforation (i.e., pain, bleeding, unusual difficulty swallowing, and rapidly elevated temperature). Temporary loss of the gag reflex is expected; after the patient's gag reflex has returned, lozenges, saline gargle, and oral analgesic agents may be offered to relieve minor throat discomfort. Patients who were sedated for the procedure must remain in bed until fully alert. After moderate sedation, the patient must be transported home with a family member or friend if the procedure was performed on an outpatient basis. Someone should stay with the patient until the morning after the procedure. Because of sedation, many patients will not remember postprocedure instructions. For this reason, discharge and follow-up instructions are provided to the person accompanying the patient home, as well as to the patient. In addition, many endoscopy suites have a program in which a nurse telephones the patient the morning after the procedure to find out if the patient has any concerns or questions related to the procedure.

Fiberoptic Colonoscopy

Historically, direct visualization of the bowel was the only means to evaluate the colon, but virtual colonoscopy (also known as CT colonography) has brought a more patient-friendly approach to this study. Virtual colonoscopy provides a computer-simulated endoluminal perspective of the air-filled distended colon using conventional CT scanning (Graca et al., 2010; Johnson, 2009).

Direct visual inspection of the large intestine (anus, rectum, sigmoid, transcending and ascending colon) is possible by means of a flexible fiberoptic colonoscope (Fig. 44-6). These scopes have the same capabilities as those used for EGD but are larger in diameter and longer. Still and video recordings can be used to document the procedure and findings.

This procedure is used commonly as a diagnostic aid and screening device. It is most frequently used for cancer screening and for surveillance in patients with previous colon cancer or polyps. (See Table 15-3 in Chapter 15 for details on the American Cancer Society's screening guidelines.) In addition, tissue biopsies can be obtained as needed, and polyps can be removed and evaluated. Other uses of colonoscopy include the evaluation of patients with diarrhea of unknown cause, occult bleeding, or anemia; further study of abnormalities detected on barium enema; and diagnosis, clarification, and determination of the extent of inflammatory or other bowel disease.

Therapeutically, the procedure can be used to remove all visible polyps with a special snare and cautery through the colonoscope. Many colon cancers begin with adenomatous polyps of the colon; therefore, one goal of colonoscopic polypectomy is early detection and prevention of colorectal cancer. This procedure also can be used to treat areas of bleeding or stricture. The use of bipolar and unipolar coagulators and heater probes, as well as injections of sclerosing agents or vasoconstrictors, are possible during this procedure. Laser-compatible scopes provide laser therapy for bleeding lesions or colonic neoplasms. Bowel decompression (removal of intestinal contents to prevent gas and fluid from distending the coils of the intestine) can also be completed during the procedure.

Colonoscopy is performed while the patient is lying on the left side with the legs drawn up toward the chest.

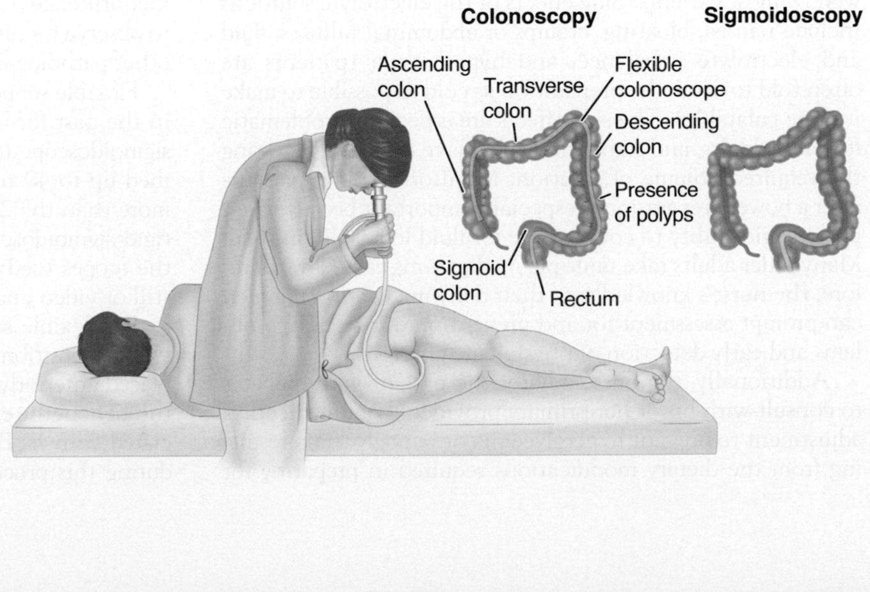
Colonoscopy
Sigmoidoscopy
Ascending colon
Transverse colon
Flexible colonoscope
Descending colon
Presence of polyps
Sigmoid colon
Rectum

FIGURE 44-6 • Colonoscopy and flexible fiberoptic sigmoidoscopy. For the colonoscopy, the flexible scope is passed through the rectum and sigmoid colon into the descending, transverse, and ascending colon. For the flexible fiberoptic sigmoidoscopy, the flexible scope is advanced past the proximal sigmoid and then into the descending colon.

The patient's position may be changed during the test to facilitate advancement of the scope. Biopsy forceps or a cytology brush may be passed through the scope to obtain specimens for histology and cytology examinations. Complications during and after the procedure can include cardiac dysrhythmias and respiratory depression resulting from the medications administered, vasovagal reactions, and circulatory overload or hypotension resulting from overhydration or underhydration during bowel preparation. The patient's cardiac and respiratory function and oxygen saturation are monitored continuously, with supplemental oxygen used as necessary. Typically, the procedure takes about 1 hour, and postprocedure discomfort results from instillation of air to expand the colon and insertion and movement of the scope during the procedure.

Nursing Interventions

The success of the procedure depends on how well the colon is prepared and on adequate sedation (Koornstra, 2009). Adequate colon cleansing provides optimal visualization and decreases the time needed for the procedure. Cleansing of the colon can be accomplished in various ways. The physician may prescribe a laxative for two nights before the examination and a Fleet's or saline enema until the return is clear the morning of the test. However, more commonly, polyethylene glycol electrolyte lavage solutions (GoLYTELY, CoLyte, and NuLYTELY) are used as intestinal lavages for effective cleansing of the bowel. The patient maintains a clear liquid diet starting at noon the day before the procedure. Then the patient ingests the lavage solution orally at intervals over 3 to 4 hours. If necessary, the nurse can give the solution through a feeding tube if the patient cannot swallow. Patients with a colostomy can receive this same bowel preparation. The use of lavage solutions is contraindicated in patients with intestinal obstruction or inflammatory bowel disease.

A sodium phosphate tablet (OsmoPrep, Visicol) can be used for colon cleansing prior to colonoscopy. Dosing consists of 32 tablets: 20 tablets (4 tablets every 15 minutes) with 8 ounces of any clear liquid (water, any clear carbonated beverage, or juice) on the evening prior to the examination, and 12 tablets (taken in the same manner) on the morning of the examination.

With the use of lavage solutions, bowel cleansing is fast (rectal effluent is clear in about 4 hours) and is tolerated fairly well by most patients. Side effects of the electrolyte solutions include nausea, bloating, cramps or abdominal fullness, fluid and electrolyte imbalance, and hypothermia (patients are often told to drink the preparation as cold as possible to make it more palatable). The side effects are especially problematic for older adults, and sometimes they have difficulty ingesting the required volume of solution. Monitoring older patients after a bowel preparation is especially important because their physiologic ability to compensate for fluid loss is diminished. Many older adults take multiple medications each day; therefore, the nurse's knowledge of their daily medication regimen can prompt assessment for and prevention of potential problems and early detection of physiologic changes.

Additionally, the nurse advises the patient with diabetes to consult with his or her primary provider about medication adjustment to prevent hyperglycemia or hypoglycemia resulting from the dietary modifications required in preparing for the test. The nurse also instructs all patients, especially older adults, to maintain adequate fluid, electrolyte, and caloric intake while undergoing bowel cleansing.

Special precautions must be taken for some patients. Implantable defibrillators and pacemakers are at high risk for malfunction if electrosurgical procedures (i.e., polypectomy) are performed in conjunction with colonoscopy. A cardiologist should be consulted before the test is performed, and the defibrillator should be turned off. These patients require careful cardiac monitoring during the procedure.

Colonoscopy cannot be performed if there is a suspected or documented colon perforation, acute severe diverticulitis, or fulminant colitis. Patients with prosthetic heart valves or a history of endocarditis require prophylactic antibiotics before the procedure.

Informed consent is obtained by the practitioner before the patient is sedated. Before the examination, an opioid analgesic agent or sedative (e.g., midazolam) is administered to provide moderate sedation and relieve anxiety during the procedure. Glucagon may be administered, if needed, to relax the colonic musculature and to reduce spasm during the test. Older or debilitated patients may require a reduced dosage of the analgesic agent or sedative to decrease the risks of oversedation and cardiopulmonary complications.

During the procedure, the patient is monitored for changes in oxygen saturation, vital signs, color and temperature of the skin, level of consciousness, abdominal distention, vagal response, and pain intensity. After the procedure, patients are maintained on bed rest until fully alert. Some patients have abdominal cramps caused by increased peristalsis stimulated by the air insufflated into the bowel during the procedure.

Immediately after the test, the patient is monitored for signs and symptoms of bowel perforation (e.g., rectal bleeding, abdominal pain or distention, fever, focal peritoneal signs). Because of the amnesic effects of midazolam, the patient may be unable to recall verbal information and should receive written instructions. If the procedure is performed on an outpatient basis, someone must transport the patient home. After a therapeutic procedure, the nurse instructs the patient to report any bleeding to the physician.

Anoscopy, Proctoscopy, and Sigmoidoscopy

Endoscopic examination of the anus, rectum, and sigmoid and descending colon is used to evaluate chronic diarrhea, fecal incontinence, ischemic colitis, and lower GI hemorrhage and to observe for ulceration, fissures, abscesses, tumors, polyps, or other pathologic processes.

Flexible scopes have largely replaced the rigid scopes used in the past for routine examinations. The flexible fiberoptic sigmoidoscope (see Fig. 44-6) permits the colon to be examined up to 40 to 50 cm (16 to 20 in) from the anus, much more than the 25 cm (10 in) that can be visualized with the rigid sigmoidoscope. It has many of the same capabilities as the scopes used for the upper GI study, including the use of still or video images to document findings.

For flexible scope procedures, the patient assumes a comfortable position on the left side with the right leg bent and placed anteriorly. It is important to keep the patient informed throughout the examination and to explain the sensations associated with it. Biopsies and polypectomies can be performed during this procedure. Biopsy is performed with small biting forceps introduced through the endoscope; one or more small pieces of tissue may be removed. If polyps are present, they may be removed with a wire snare, which is used to grasp the pedicle, or stalk. An electrocoagulating current is then used to sever the polyp and prevent bleeding. It is extremely important that all excised tissue be placed immediately in moist gauze or in an appropriate receptacle, labeled correctly, and delivered without delay to the pathology laboratory for examination.

Nursing Interventions

These examinations require only limited bowel preparation, including a warm tap water or Fleet's enema until returns are clear. Dietary restrictions usually are not necessary, and sedation usually is not required. During the procedure, the nurse monitors vital signs, skin color and temperature, pain tolerance, and vagal response. After the procedure, the nurse monitors the patient for rectal bleeding and signs of intestinal perforation (i.e., fever, rectal drainage, abdominal distention, and pain). On completion of the examination, the patient can resume his or her regular activities and diet.

Small Bowel Studies

There are several methods available for visualization of the small intestine, including capsule endoscopy and double-balloon endoscopy. Capsule endoscopy allows the noninvasive visualization of the mucosa throughout the entire small intestine. It is particularly useful in the evaluation of obscure GI bleeding. The technique consists of the patient swallowing a capsule embedded with a wireless miniature camera, a light source, and an image transmission system. The capsule is the size of a large vitamin pill (26 mm long, 11 mm wide, and 3.7 g in weight). It is propelled through the intestine by peristalsis. Images are transmitted from the end of the capsule to a recording device worn on a belt. Typically, the capsule passes from the rectum in 1 or 2 days. This diagnostic procedure is limited by its inability to allow for obtaining tissue samples for histology and for providing endoscopic therapy (Raphael, Patel, & Warren, 2010).

Double-balloon enteroscopy, also known as push-and-pull enteroscopy, has made it possible to visualize the mucosa of the entire small bowel as well as carry out diagnostic and therapeutic interventions (Fireman, 2010; Raphael et al., 2010). This endoscope is comprised of two balloons, one attached to the distal end of the scope and the other attached to the transparent overtube that slides over the endoscope. The endoscope is advanced using a push-and-pull technique that involves alternately inflating and deflating the balloons, causes telescoping of the small intestine onto the overtube. As a result of this telescoping, the endoscope can visualize much more of the small intestine than the length of the scope itself. The procedure takes between 1 and 3 hours and requires moderate sedation. Nursing interventions are similar to those for other endoscopic procedures.

Endoscopy Through an Ostomy

Endoscopy through an ostomy stoma is useful for visualizing a segment of the small or large intestine and may be indicated to evaluate the anastomosis for recurrent disease, or to visualize and treat bleeding in a segment of the bowel. Nursing interventions are similar to those for other endoscopic procedures.

Manometry and Electrophysiologic Studies

Manometry and electrophysiologic studies are methods for evaluating patients with GI motility disorders. The manometry test measures changes in intraluminal pressures and the coordination of muscle activity in the GI tract with the pressures transmitted to a computer analyzer.

Esophageal manometry is used to detect motility disorders of the esophagus and the upper and lower esophageal sphincter. Also known as esophageal motility studies, these studies are very helpful in the diagnosis of **achalasia**, diffuse esophageal spasm, scleroderma, and other esophageal motor disorders. The patient must refrain from eating or drinking for 8 to 12 hours before the test. Medications that could have a direct effect on motility (e.g., calcium channel blockers, anticholinergic agents, sedatives) are withheld for 24 to 48 hours. A pressure-sensitive catheter is inserted through the nose and is connected to a transducer and a video recorder. The patient then swallows small amounts of water while the resultant pressure changes are recorded. Evaluation of a patient for gastroesophageal reflux disease typically includes esophageal manometry.

Gastroduodenal, small intestine, and colonic manometry procedures are used to evaluate delayed gastric emptying and gastric and intestinal motility disorders such as irritable bowel syndrome or atonic colon. This is often an ambulatory outpatient procedure lasting 24 to 72 hours. Anorectal manometry measures the resting tone of the internal anal sphincter and the contractibility of the external anal sphincter. It is helpful in evaluating patients with chronic constipation or fecal incontinence and is useful in biofeedback for the treatment of fecal incontinence. It can be performed in conjunction with rectal sensory functioning tests. Dibasic sodium (Phospho-soda) or a saline cleansing enema is administered 1 hour before the test, and positioning for the test is either the prone or the lateral position.

Rectal sensory function studies are used to evaluate rectal sensory function and neuropathy. A catheter and balloon are passed into the rectum, with increasing balloon inflation until the patient feels distention. Then the tone and pressure of the rectum and anal sphincter are measured. The results are especially helpful in the evaluation of patients with chronic constipation, diarrhea, or incontinence.

Electrogastrography, an electrophysiologic study, also may be performed to assess gastric motility disturbances and can be useful in detecting motor or nerve dysfunction in the stomach. Electrodes are placed over the abdomen, and gastric electrical activity is recorded for up to 24 hours. Patients may exhibit rapid, slow, or irregular waveform activity.

Defecography measures anorectal function and is performed with very thick barium paste instilled into the rectum. Fluoroscopy is used to assess the function of the rectum and anal sphincter while the patient attempts to expel the barium. The test requires no preparation. The nurse educates the patient about what to expect during these procedures.

Gastric Analysis, Gastric Acid Stimulation Test, and pH Monitoring

Analysis of the gastric juice yields information about the secretory activity of the gastric mucosa and the presence or degree of gastric retention in patients thought to have pyloric or duodenal obstruction. It is also useful for diagnosing Zollinger-Ellison syndrome or atrophic gastritis.

The patient is NPO for 8 to 12 hours before the procedure. Any medications that affect gastric secretions are withheld for 24 to 48 hours before the test. Smoking is not allowed on the morning of the test because it increases gastric secretions. A small nasogastric tube with a catheter tip marked at various points is inserted through the nose. When the tube is at a point slightly less than 50 cm (21 in), it should be within the stomach, lying along the greater curvature. Once in place, the tube is secured to the patient's cheek and the patient is placed in a semireclining position. The entire stomach contents are aspirated by gentle suction into a syringe, and gastric samples are collected every 15 minutes for the next hour.

Important diagnostic information to be gained from gastric analysis includes the ability of the mucosa to secrete HCl. This ability is altered in various disease states, including:

- *Pernicious anemia:* Patients with this disease secrete no acid under basal conditions or after stimulation.
- *Severe chronic atrophic gastritis or gastric cancer:* Patients with these diseases secrete little or no acid.
- *Peptic ulcer:* Patients with this disease secrete some acid.
- *Duodenal ulcers:* Patients with this disease usually secrete an excess amount of acid.

The gastric acid stimulation test usually is performed in conjunction with gastric analysis. Histamine or pentagastrin is administered subcutaneously to stimulate gastric secretions. It is important to inform the patient that this injection may produce a flushed feeling. The nurse monitors the patient's blood pressure and pulse frequently to detect hypotension. Gastric specimens are collected after the injection every 15 minutes for 1 hour and are labeled to indicate the time of specimen collection after histamine injection. The volume and pH of the specimen are measured; in certain instances, cytologic study by the Papanicolaou technique may be used to determine the presence or absence of malignant cells.

Esophageal reflux of gastric acid may be diagnosed by ambulatory pH monitoring (Nusrat, Roy, & Bielefeldt, 2011). The patient is NPO for 6 hours before the test, and all medications affecting gastric secretions are withheld for 24 to 36 hours before the test. A probe that measures pH is inserted through the nose and into position about 12.7 cm (5 in) above the lower esophageal sphincter. It is connected to an external recording device and is worn for 24 hours while the patient continues usual daily activities. The end result is a computer analysis and graphic display of the results. This test allows for the direct correlation between chest pain and reflux episodes (Nusrat et al., 2011).

The Bravo pH monitoring system offers the advantage of pH monitoring of the esophagus without the transnasal catheter. The clinician, by means of endoscopy, attaches a capsule (approximately the size of a gel cap) to the patient's esophageal wall. Data related to pH are transmitted from the capsule to a pager-sized receiver that the patient wears. Data are collected for up to 48 hours and then downloaded and analyzed. The capsule spontaneously detaches from the esophagus in 7 to 10 days and then is passed through the patient's digestive system. The accuracy of this method of pH testing is greater than methods in which a catheter is used because the patient can eat normally and continue typical activities during the testing. The system reliably distinguishes reflux conditions from functional conditions if the study is performed with the patient off medications (Fletcher, Goutte, Slaughter, et al., 2011).

Laparoscopy (Peritoneoscopy)

With the tremendous advances in minimally invasive surgery, diagnostic laparoscopy is efficient, cost-effective, and useful in the diagnosis of GI disease. After a pneumoperitoneum (injecting carbon dioxide into the peritoneal cavity to separate the intestines from the pelvic organs) is created, a small incision is made lateral to the umbilicus, allowing for the insertion of the fiberoptic laparoscope. This permits direct visualization of the organs and structures within the abdomen, permitting visualization and identification of any growths, anomalies, and inflammatory processes. In addition, biopsy samples can be taken from the structures and organs as necessary. This procedure can be used to evaluate peritoneal disease, chronic abdominal pain, abdominal masses, and gallbladder and liver disease. However, laparoscopy has not become an important diagnostic modality in patients with acute abdominal pain, because less invasive tools (e.g., CT and MRI) are readily available. Laparoscopy usually requires general anesthesia and sometimes requires that the stomach and bowel be decompressed. Gas (usually carbon dioxide) is insufflated into the peritoneal cavity to create a working space for visualization. One of the benefits of this procedure is that after visualization of a problem, excision (e.g., removal of the gallbladder) can then be performed at the same time, if appropriate.

Critical Thinking Exercises

1 pq You are caring for a 60-year-old man with a history of diabetes, pancreatitis, alcohol abuse, and abdominal pain associated with meals. He states that he sometimes has nausea and vomiting after eating dinner. Identify questions that should be asked when taking the patient's history. What assessment parameters would be addressed? What are your priority assessments? What diagnostic tests would you expect to be ordered?

2 ebp You are working in a medical internist's office. A 56-year-old female patient comes into the office for a routine examination. She states that she did not get her recommended colonoscopy examination because no one in her family has colon cancer, and she does not believe that this type of screening is of any benefit to her. What education would you provide for this patient? What is the evidence base that supports the education? Discuss the strength of the evidence for the education. Identify the criteria used to evaluate the strength of the evidence.

Brunner Suite Resources

Explore these additional resources to enhance learning for this chapter:

- NCLEX-Style Questions and Other Resources on thePoint, http://thePoint.lww.com/Brunner13e
- Study Guide
- PrepU
- Clinical Handbook
- Handbook of Laboratory and Diagnostic Tests

References

Books

American Cancer Society. (2011). *Cancer facts & figures 2011*. Atlanta: Author.

Bickley, L. S. (2009). *Bates' guide to physical examination* (10th ed.). Philadelphia: Lippincott Williams & Wilkins.

Eliopoulos, C. (2010). *Gerontological nursing* (7th Ed). Philadelphia: Lippincott Williams & Wilkins.

Porth, C. M., & Matfin, G. (2009). *Pathophysiology. Concepts of altered health states* (8th ed.). Philadelphia: Lippincott Williams & Wilkins.

Wolfe, M. M. (Ed.). (2006). *Therapy of digestive disorders* (2nd ed.). Philadelphia: W. B. Saunders.

Journals and Electronic Documents

Belsey, J., Crosta, C., Epstein, O., et al. (2012). Meta-analysis: The relative efficacy of oral bowel preparations for colonoscopy 1985-2010. *Alimentary Pharmacological Therapy, 35*(2), 222–237.

Bones, J., Byrne, J., O'Donoghue, N., et al. (2010). Glycomic and glycoproteomic analysis of serum from patients with stomach cancer reveals potential markers arising from host defense response mechanisms. *Journal of Proteome Research, 10*(3), 1246 1265.

Brady, P. (2012). Fecal occult blood testing: Where do we stand? *Southern Medical Association, 105*(7), 362–363.

Brockow, K., & Ring, J. (2012). Anaphylaxis to radiographic contrast media. *Current Opinion in Allergy and Clinical Immunology, 11*(4), 326–331.

Brown, J., & Thompson, C. A. (2010). Contrast-induced acute kidney injury: The at-risk patient and protective measures. *Current Cardiology Report, 12*(5), 440–445.

Chait, M. (2010). Gastroesophageal reflux disease: Important considerations for the older patients. *World Journal of Gastrointestinal Endoscopy, 2*(12), 388–396.

Faul, L., Jim, H., Williams, C., et al. (2010). Relationship of stress management skill to psychological distress and quality of life in adults with cancer. *Psycho-Oncology, 19*(1), 102–109.

Fireman, Z. (2010). Capsule endoscopy: Future horizons. *World Journal of Gastrointestinal Endoscopy, 2*(9), 305–307.

Fletcher, K., Goutte, M., Slaughter, J., et al. (2011). Significance and degree of reflux in patients with primary extraesophageal symptoms. *Laryngoscope, 121*(12), 2561–2565.

Graca, B., Freire, P., & Brito, J. (2010). Gastroenterologic and radiologic approach to obscure gastrointestinal bleeding: How, why, and when? *Radiographics, 30*(1), 235–252.

Greer, J., Solis, J., Temel, J., et al. (2011). Anxiety disorders in long-term survivors of adult cancers. *Psychosomatics, 52*(5), 417–423.

Gu, Y., Lim, H., & Moser, M. (2010). How useful are bowel sounds in assessing the abdomen? *Digestive Surgery, 27*(5), 422 426.

Harmon, R., & Peura, D. (2010). Evaluation and management of dyspepsia *Therapeutic Advances in Gastroenterology, 3*(2), 87–98.

Johnson, C. (2009). CT colonography: Coming of age *American Journal of Roentgenology, 193*(5), 1239–1242.

Khalid-de Bakker, C., Jonkers, D., Sanduleanu, S., et al. (2011). Test performance of immunological fecal occult blood testing and sigmoidoscopy compared with primary colonoscopy screening for colorectal advanced adenomas. *Cancer Prevention Research, 4*(10), 1563–1571.

Koornstra, J. (2009). Bowel preparation before small bowel capsule endoscopy: What is the optimal approach? *European Journal of Gastroenterology & Hepatology, 21*(10), 1107–1109.

Korkmaz, M., Unal, H., Selcuk, H., et al. (2010). Extraordinarily elevated serum levels of CA 19-9 and rapid decrease after successful therapy: A case report and review of literature. *Turkish Journal of Gastroenterology, 21*(4), 461–463.

Kumaravel, V., Hayden, S., Hall, G., et al. (2011). New fecal occult tests may improve adherence and mortality rates. *Cleveland Clinic Journal of Medicine, 78*(8), 515–520.

Li, B., Wang, J., & Ma, Y. (2012). Bowel sounds and monitoring gastrointestinal motility in critically ill patients. *Clinical Nurse Specialist, 26*(1), 29–34.

Mautone, A., & Brown, J. (2010). Contrast-induced nephropathy in patients undergoing elective and urgent procedures. *Journal of Interventional Cardiology, 23*(1), 78–85.

Mishra, R., Heavner, J., & Day, M. (2012). Prevalence of adverse reactions to radiopaque contrast reported by patients presenting for interventional pain procedure. *Pain Practice, 13*(3), 182–190.

Ned, R., Melillo, S., & Marrone, M. (2011). Fecal DNA testing for colorectal cancer screening: The ColoSure test. *PLoS Currents, 22*(3), RRN1220.

Nusrat, S., Roy, P., & Bielefeldt, K. (2011). Wireless PH studies: Manometric or endoscopic guidance? *Diseases of the Esophagus, 25*(1), 26–32.

Offit, K. (2011). Personalized medicine: New genomics, old lessons. *Human Genetics, 130*(1), 3–14.

Poh, C., Navarro-Rodriguez, T., & Fass, R. (2010). Review: Treatment of GI reflux in the elderly. *American Journal of Medicine, 123*(6), 496–501.

Rabago, L., Ortega, A., Chico, I., et al. (2011). Intraoperative ERCP: What role does it have in the era of laparoscopic cholecystectomy? *World Journal of Gastrointestinal Endoscopy, 3*(12), 248–255.

Raphael, M., Patel, R., & Warren, B. (2010). Updates in small bowel imaging and endoscopy. *Journal of the American Osteopathic Association, 110*(12), 721–724.

Reitz, S., Slam, K., & Chambers, L. (2011). Biliary, pancreatic, and hepatic imaging for the general surgeon. *Surgical Clinics of North America, 91*(1), 59–92.

Scholten, S. (2010). Endoscopy: A guide for the registered nurse. *Critical Care Nursing Clinics of North America, 22*(1), 19–32.

Sieg, A. (2011). Capsule endoscopy compared with conventional colonoscopy for detection of colorectal neoplasms. *World Journal of Gastrointestinal Endoscopy, 3*(5), 81–85.

Tennyson, C., & Semrad, C. (2011). Advances in small bowel imaging. *Current Gastroenterology Reports, 13*(5), 408–417.

White, C., & Kilgore, M. (2009). PillCam ESO versus esophagogastroduodenoscopy in esophageal variceal screening: A decision analysis. *Journal of Clinical Gastroenterology, 43*(10), 975–981.

Zamboni, G. A., & Raptopoulos, V. (2010). CT enterography. *Gastrointestinal Endoscopic Clinics of North America, 20*(2), 347–366.

Resources

American Cancer Society, www.cancer.org

American Society for Gastrointestinal Endoscopy (ASGE), www.asge.org

Society of Gastroenterology Nurses and Associates (SGNA), www.sgna.org

Chapter 45

Digestive and Gastrointestinal Treatment Modalities

Learning Objectives

On completion of this chapter, the learner will be able to:

1. Describe the purposes and types of enteral and parenteral nutrition access devices.
2. Identify the purposes, indications for, and administration techniques of enteral and parenteral nutrition formulas.
3. Discuss nursing management of the patient who has a nasally placed feeding tube, stomal tube, or intravenous catheter.
4. Use the nursing process as a framework for care of the patient receiving enteral or parenteral nutrition support.
5. Describe the nursing measures used to prevent complications from enteral and parenteral nutrition support.

Glossary

antireflux valve: valve that prevents return or backward flow of fluid

aspiration: removal of substance by suction or the inhalation of fluids or foods into the trachea and bronchial tree

bolus: a feeding administered into the stomach in large amounts and at designated intervals

central venous access device (CVAD): a device designed and used for administration of sterile fluids, nutrition formulas, and medications into central veins

cyclic feeding: periodic infusion of feedings given over 8 to 18 hours

decompression (gastric/intestinal): removal of gastric or intestinal contents to prevent gas and fluid distention

dumping syndrome: physiologic response to rapid emptying of gastric contents into the small intestine, manifested by nausea, weakness, sweating, palpitations, syncope, and possibly diarrhea; occurs in patient who have had a rapid infusion of fluids and nutrition formulas directly into a tube accessing the small intestine

duodenum: the first part of the small intestine, which arises from the pylorus of the stomach and extends to the jejunum

enteral access device: device inserted into the gastrointestinal tract for either decompression and drainage or the infusion of nutrition formulas, fluid, and medications

enteral nutrition: nutritional formula feedings infused through a tube directly into the gastrointestinal tract

gastroparesis: partial paralysis of the stomach that results in decreased gastric motility and emptying

gastrostomy: surgical creation of an opening into the stomach for the purpose of administering fluids, nutrition formulas, and medications or for decompression and drainage of stomach contents

intravenous fat emulsion (IVFE): an oil-in-water emulsion of oils, egg phospholipids, and glycerin; also referred to as intravenous lipid emulsion

intubation: the insertion or placement of a tube into a body structure or passageway

jejunostomy: surgical creation of an opening into the jejunum for the purpose of administering fluids, nutrition formulas, and medications

jejunum: second portion of the small intestine, which extends from the duodenum to the ileum

lavage: flushing of the stomach with water or other fluids with a gastric tube to clear it

lumen: the channel within a tube or catheter

nasoduodenal tube: tube inserted through the nose into the proximal portion of the small intestine (i.e., duodenum)

nasoenteric tube: tube inserted through the nose into the stomach and beyond the pylorus into the small intestine

nasogastric (NG) tube: tube inserted through the nose into the stomach

nasojejunal tube: tube inserted through the nose into the second portion of the small intestine (i.e., jejunum)

orogastric tube: tube inserted through the mouth into the stomach

osmolality: ionic concentration of fluid

parenteral nutrition (PN): method of supplying nutrients to the body by an intravenous route

percutaneous endoscopic gastrostomy (PEG): a feeding tube inserted endoscopically into the stomach

peripherally inserted central catheter (PICC): a device inserted into a peripheral vein and designed and used for administration of sterile fluids, nutrition formulas, and medications into central veins

peristalsis: wavelike movement that occurs involuntarily in the alimentary canal

pH: the degree of acidity or alkalinity of a substance or solution

radiopaque: can be easily localized on x-ray

stoma: artificially created opening between a body cavity (e.g., stomach or intestine) and the body surface

stylet: a stiff wire placed in a catheter or other tube that allows the tube to maintain its shape during insertion

total nutrient admixture (TNA): an admixture of lipid emulsions, proteins, carbohydrates, electrolytes, vitamins, trace minerals, and water

This chapter presents several topics related to gastrointestinal (GI) intubation, including managing the care of patients with nasogastric (NG), nasoenteric, and gastrostomy and jejunostomy tubes; providing enteral tube feedings; gastric drainage; and education points related to home health care. In addition, parenteral nutrition (PN) is presented, including general indications for this nutritional modality and nursing care of patients receiving such support measures.

Gastrointestinal Intubation

GI **intubation** is the insertion of a flexible tube into the stomach, or beyond the pylorus into the **duodenum** (the first section of the small intestine) or the **jejunum** (the second section of the small intestine). The tube may be inserted through the mouth, the nose, or the abdominal wall. The tubes are of various diameters (French [Fr] size) and lengths, depending on their intended use. GI intubation may be performed in order to:

- Decompress the stomach and remove gas and fluid
- **Lavage** (flush with water or other fluids) the stomach and remove ingested toxins or other harmful materials
- Diagnose GI disorders
- Administer tube feedings, fluids, and medications
- Compress a bleeding site
- Aspirate GI contents for analysis

Tube Types

Orogastric and Nasogastric Tubes for Decompression, Drainage, Aspiration, and Lavage

A variety of tubes are used for decompression, drainage, **aspiration** (removal of substances by suction), and lavage of the stomach. An **orogastric tube** is a large-bore tube inserted through the mouth into the stomach and contains a wide outlet for removal of gastric contents; it is used primarily in the emergency department or an intensive care setting (see Chapter 72). A **nasogastric (NG) tube** is introduced through the nose into the stomach, often before or during surgery or at the bedside to remove fluid and gas from the upper GI tract by the process known as **decompression (gastric/intestinal)**. Other tubes, such as the Sengstaken-Blakemore tube, are used to treat bleeding esophageal varices (see Chapter 49). Commonly used oro- and NG tubes include the Levin tube and the gastric (Salem) sump tube.

Levin Tube

The Levin tube has a single **lumen** and is made of plastic or rubber. This tube is connected to low intermittent suction (30 to 40 mm Hg) to avoid erosion or tearing of the stomach lining, which can result from adherence of the tube's lumen to the mucosa of the stomach.

Gastric (Salem) Sump

The gastric (Salem) sump tube is a **radiopaque** (easily seen on x-ray), clear plastic, double-lumen NG tube. The inner, smaller lumen (known as the blue port) vents the larger suction-drainage tube to the atmosphere by means of an opening at the distal end of the tube. The sump tube can protect fragile gastric mucosa by maintaining a low (25 mm Hg) continuous force of suction at the drainage opening. The suction lumen may be irrigated to maintain patency.

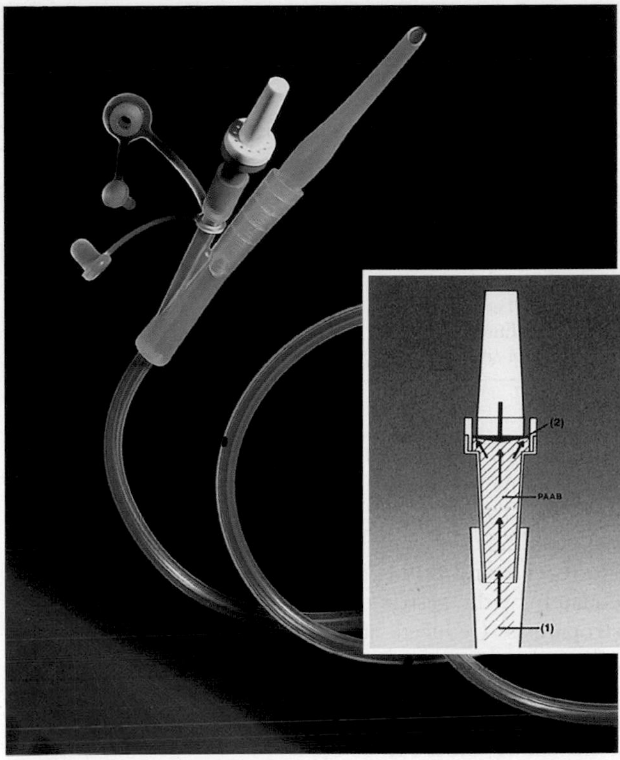
(2)
PAAB
(1)

FIGURE 45-1 • Gastric (Salem) sump tube equipped with a one-way valve that allows air to enter and can prevent reflux of gastric contents. The antireflux valve is designed with a pressure-activated air buffer (PAAB). The buffer is activated *(1)* and the valve closes *(2)* when pressure from gastric contents enters the tubing. Argyle Silicone Salem Sump Tube with preattached Argyle Salem Sump Anti-Reflux Valve courtesy of Sherwood Medical, St. Louis, MO.

The blue vent lumen should be kept above the patient's waist to prevent reflux of gastric contents through it; otherwise, it acts as a siphon. A one-way **antireflux valve** (prevents return or backward flow of fluid) seated in the blue pigtail can prevent the reflux of gastric contents out the vent lumen (Fig. 45-1). The valve is removed for irrigation of the suction lumen. To reestablish a buffer of air between the gastric contents and the valve, 20 mL of air is injected through the blue vent and the valve is reinserted.

Nasogastric and Nasoenteric Tubes for Administration of Tube Feedings, Fluids, and Medications

Some enteral tubes are manufactured to administer tube feedings, fluids, and medications. These tubes are made of various materials, including rubber, polyurethane, and silicone. They vary in length, diameter (Fr size), purpose, and placement in the GI tract (Table 45-1). They are smaller bore (generally 5 to 12 Fr) than the tubes made for gastric decompression and drainage, which lessens patient discomfort and nasal irritation. A nasally placed feeding tube is for short-term use and should stay in place for no more than 4 weeks before being replaced with a new tube (Bankhead, Boullata, Brantley, et al., 2009) (see later discussion on

TABLE 45-1 Nasogastric and Nasoenteric Tubes

Tube Type	Length (cm)	Size (French)	Lumen	Other Characteristics
Nasogastric Tubes				
Levin (plastic or rubber)	125	14–18	Single	Circular markings at intervals along the tube serve as guidelines for insertion.
Gastric sump or Salem (plastic)	120	12–18	Double	Smaller lumen acts as a vent
Moss	90	12–16	Triple	Contains both a gastric decompression lumen and a duodenal lumen for postoperative feedings
Sengstaken-Blakemore (rubber)			Triple	Two lumens are used to inflate the gastric and esophageal balloons, and one tube is reserved for suction or drainage.
Nasoenteric Feeding Tube				
Dobbhoff or EnteraFlo (polyurethane or silicone rubber)	160–175	8–12	Single	Tungsten-weighted tip, radiopaque, stylet

stomal tubes). Any feeding solution administered through a tube is poured through a syringe, delivered by gravity drip, or regulated by an electric pump.

NG feeding tubes are used for patients who have the ability to receive and process nutrition, fluids, and medications adequately by the gastric route. For those patients who have **gastroparesis** (reduced stomach motility), severe gastroesophageal reflux disease, impaired glottic closure, or undergone partial or total gastrectomy, or otherwise are at risk for **aspiration** (inhalation of substances into the airways), nasoenteric feeding tubes can be used.

Nasoenteric tubes placed in the duodenum are called **nasoduodenal tubes**, whereas those placed in the jejunum (the portion of the small intestine distal to the duodenum) are **nasojejunal tubes**. They can be inserted at the bedside, during surgery, fluoroscopically, or endoscopically. Nasally inserted feeding tubes are soft and pliable; therefore, they may kink when a **stylet** is not used, particularly if the patient is unable to swallow. However, caution is required when inserting feeding tubes with a stylet because there is a risk of tissue puncture or placement error.

During nasal insertion, the tip of the tube is initially directed toward the stomach, then can be further advanced through the pylorus into the small intestine if warranted. Fluoroscopic techniques may be used to visually direct feeding tubes into the stomach, duodenum, or jejunum. Several bedside techniques may be used to facilitate tube tip placements into the small intestine. These techniques include using air insufflation or manipulating the tube itself as it is being inserted (e.g., using a "corkscrew" technique). Prokinetic agents can be administered to facilitate movement of the feeding tube by **peristalsis** (involuntary wavelike movement) into the duodenum. In particular, administering the medication metoclopramide (Reglan) facilitates passage of non-weighted feeding tubes (Lord, Weiser-Maimone, Pulhamus, et al., 1993). An external magnet device may be used to help guide a postpyloric feeding tube that has a magnetized tip (Gabriel & Ackermann, 2004), and a bedside external electromagnetic feeding tube placement device can provide a visual guide during insertion and appears to preclude the need for postinsertion x-ray film (Koopman, Kudsk, Szokowski, et al., 2011; Powers, Luebbehusen, Spitzer, et al., 2011). Any of these techniques may be used alone to facilitate correct tube placement or in combination with the other techniques. Although still widely used in practice, tungsten-weighted tips do not facilitate migration of the tube's tip from the stomach into the intestine (Rees, Payne-James, King, et al., 1988).

Nursing Management

Preparing the Patient

The nurse explains the purpose of the tube to the patient prior to insertion to promote cooperation during the procedure. The general activities related to inserting the tube are then reviewed, including the fact that the procedure may cause gagging until the tube has passed beyond the throat.

Inserting the Tube

Before inserting the tube, the nurse determines the length that will be needed to reach the stomach or the small intestine. A mark is made on the tube to indicate the desired length. This length is traditionally determined by (1) measuring the distance from the tip of the nose to the earlobe and from the earlobe to the xiphoid process, and (2) adding up to 15 cm (6 in) for NG placement or at least 20 to 25 cm (8 to 10 in) or more for intestinal placement, although studies do not necessarily confirm that this is a reliable technique (Stepter, 2012) (Chart 45-1; Fig. 45-2).

During insertion, the patient usually sits upright with a towel or other protective barrier spread in a biblike fashion over the chest. The nostril may be swabbed or the oropharynx sprayed with an anesthetic agent to numb the nasal passage and suppress the gag reflex. To make the tube easier to insert, it should be lubricated with a water-soluble lubricant unless it has a dry coating (hydromer), which, when moistened, provides its own lubrication. Gloves should be worn during the procedure. The nostrils are inspected for any obstruction, and the more patent nostril is selected for use. The tip of the patient's nose is tilted upward, and the tube is aligned to enter the nostril. When the tube reaches the nasopharynx, the patient is instructed to lower the head slightly and, if able, to begin to swallow as the tube is advanced. The patient may also be encouraged to sip water through a straw to facilitate advancement of the tube if this action is not contraindicated. The oropharynx is inspected to ensure that the tube has not coiled in the pharynx or mouth.

Confirming Placement

To ensure patient safety, it is essential to confirm that the tube has been placed correctly. The tube tip may be in the esophagus, stomach, or small intestine, or inadvertently inserted in the lungs, most commonly in the right main bronchus. Inappropriate placement may occur in patients with decreased levels of consciousness, confused mental states, poor or

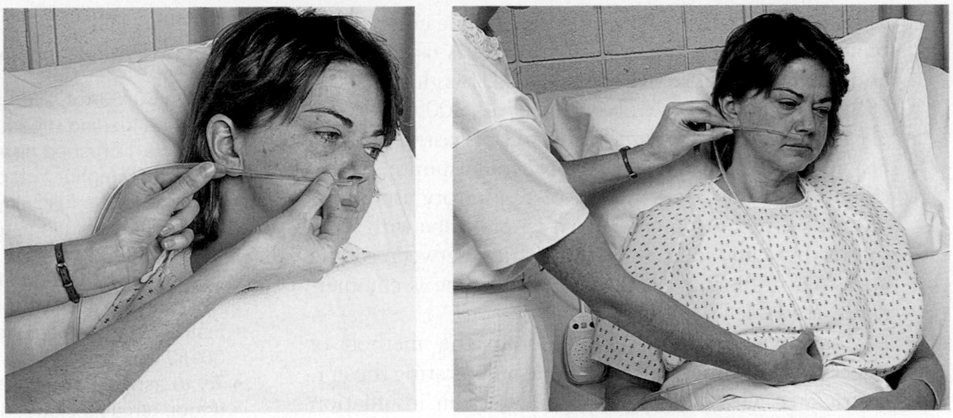

Measuring distance from nostril to tip of earlobe.

Measuring distance from earlobe to tip of xiphoid process.

Have the patient sit in a neutral position with head facing forward. Place the distal tip of the tubing at the tip of the patient's nose (N); extend tube to the tragus (tip) of the ear (E), and then extend the tube straight down to the tip of the xiphoid (X). The tube is placed up to 15 cm beyond that measured length.

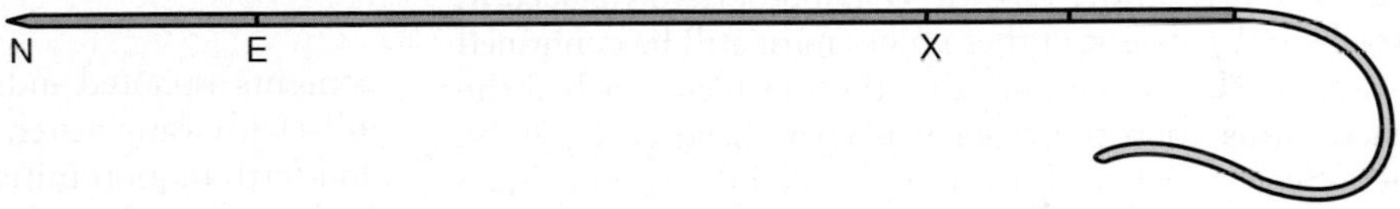
N
E
X

FIGURE 45-2 • Measuring length of nasogastric tube for placement into stomach.

absent cough and gag reflexes, or agitation during insertion. The presence of an endotracheal tube or tracheostomy also increases the risk of inadvertent placement of the tube in the lung. Initially, an x-ray should be used to confirm tube placement (Simons & Abdallah, 2012; Stepter, 2012). However, each time liquids or medications are administered, as well as once a shift for continuous feedings, the tube must be checked to ensure that it remains properly placed. The use of a combination of the following methods is recommended (Simons & Abdallah, 2012):

- *Measurement of tube length:* It is necessary to measure the length of the exposed portion of the tube and document the length. Every shift, the nurse measures this length and compares it with the original measurement. An increase in the length of exposed tube may indicate outward migration and possible malposition of the tube tip from the small intestine to the stomach or from the stomach into the esophagus. A decrease in length of exposed tube may indicate inward migration and the possibility that a tube placed in the stomach has migrated into the small intestine.
- *Visual assessment of aspirate color and volume:* Gastric aspirate is most frequently cloudy and green, tan, off-white, or brown and may be of large volume. Intestinal

Chart 45-1 NURSING RESEARCH PROFILE

Care of the Patient With Enteral Tube Feeding

Kenny, D., & Goodman, P. (2010). Care of the patient with enteral tube feeding: An evidence-based practice protocol. *Nursing Research*, *59*(1 Suppl.), S22–S31.

Purpose

Care of patients with enteral feeding tubes is often based on tradition and can vary widely between and within institutions. The purpose of this study was to develop, implement, and evaluate an evidence-based best practice protocol for managing patients receiving enteral tube feedings.

Design

Data collection tools included three domains: documentation of patient procedures, nursing knowledge of each of the specific procedures, and environment of care. Outcomes were measured at three levels: patient, nursing, and organization. Data were collected before and after implementation of the protocol (pretest–posttest design).

Findings

There was overall improvement in most measures consistent with best practices postimplementation. These included staff knowledge of how to unclog tubes, danger of using blue dye in feeding solution, medication administration practices, and patient position during feedings. In addition, overall appropriateness of documentation improved, as was the availability of suction equipment postimplementation.

Nursing Implications

The implementation of protocols such as this may presumably decrease rates of adverse events (e.g., aspiration) and improve outcomes. In addition, implementation of these types of evidence-based protocols translate nursing research and theory into practice, and improve collaboration between nurse scientists, nurse administrators, and nurses in direct caregiving roles.

aspirate is primarily clear and yellow to bile colored and typically smaller volume. Lung aspirate is usually clear and of small volume (Metheny, Reed, Berglund, et al., 1994; Metheny, Stewart, Nuetzel, et al., 2005).

- *pH measurement of aspirate:* The **pH** of gastric aspirate is acidic (1 to 5). The pH of intestinal aspirate is typically 6 or higher, and the pH of respiratory aspirate is more alkaline (7 or greater). An enteral tube with a pH sensor that can facilitate distinguishing between gastric and small intestinal placement of the tube is commercially available.
- *Air auscultation:* Studies have found the method of injecting air through the tube while auscultating the epigastric area with a stethoscope to detect air insufflation used alone is an unreliable indicator of gastric placement (Metheny, Dettenmeier, Hampton, et al., 1990).

Each of these methods of confirming tube placement has its advantages and disadvantages. Some new feeding tubes are now equipped with devices that more reliably determine tube location and prevent inadvertent tracheal malposition. One of these uses an end-tidal carbon dioxide detector that changes color when the tube tip is inappropriately placed in the trachea. Placement of these tubes must still be confirmed by x-rays, however, to ensure that the tips have reached the stomach rather than the esophagus (Bankhead et al., 2009; Munera-Seeley, Ochoa, Brown, et al., 2008). Another tube features a copper wire coiled around its stylet that generates an electromagnetic signal. This signal may be detected via a device that is placed over the patient's xiphoid process and visualized by computer imaging and is 99.5% as reliable as advanced radiographic techniques in confirming placement (Powers et al., 2011).

After the correct position of the tube tip has been confirmed, the feeding tube is secured to the nose (Fig. 45-3). A liquid barrier may be applied to the skin. The area can then be covered with a strip of hypoallergenic tape or a transparent adhesive dressing and then a second piece of tape that is partially split longitudinally can be placed over the initial piece to secure the tube.

Clearing Tube Obstruction

If it is difficult to instill or withdraw warm water into a clogged feeding tube, several declogging steps can be taken, including warm water penetration, milking the tube, infusing digestive enzymes, and employing mechanical declogging devices. When a tube is first noted to be clogged, a 30- to 60-mL syringe should be attached to the end of the tube and any contents aspirated and discarded. Then, the syringe should be filled with warm water, attached to the tube again, and a back-and-forth motion initiated to help loosen the clog (Marcuard & Stegall, 1990). If this is not effective in dislodging the clog, then the warm water should be left in place for 5 to 15 minutes to penetrate the clog (Arriola, Hatashima, & Klang, 2010). The tube can also be milked with the fingers from the insertion site outward to help dislodge the clog. If these techniques are not effective, then a solution of digestive enzymes activated with sodium bicarbonate can be tried (Chart 45-2).

A premanufactured declogging kit, such as the Clog Zapper, contains a syringe filled with enzymatic powder that is activated by pulling in water. An elongated hollow catheter is attached to this syringe so that the declogging solution gains close proximity to the clog. Inserting and twisting endoscopy or cytology brushes and commercial mechanical declogger devices into the feeding tube can only be used with larger-bore tubes and should be performed only by experienced providers. Of note, even though cola and cranberry juice are sometimes used to declog tubes, these fluids are not advocated because their acidic nature has been shown to worsen formula clogs by causing precipitation of proteins (Metheny, Eisenberg, & McSweeney, 1988; Nicalou & Davis, 1990; Wilson & Hayes-Johnson, 1987).

Chart 45-2 Procedure for Declogging a Feeding Tube

- With a 30–60-mL syringe, aspirate as much liquid as possible from the feeding tube and discard it.
- Dissolve 1 crushed non–enteric-coated sodium bicarbonate tablet (650 mg) or ¼ tsp baking soda in 10 mL water.
- Add one of the following forms of pancrelipase to the sodium bicarbonate solution and allow to dissolve:
  a. A 10,440-U crushed pancrelipase tablet (Viokace)
  b. An opened 12,000-U pancrelipase capsule (Creon)
  (*Note:* Mixture will turn a light brown color.)
- Instill declogging solution into feeding tube and clamp for 30–60 minutes.
- Try to aspirate and flush with warm water.
- If tube remains clogged, repeat steps once more.
- If tube remains clogged, notify primary provider to replace tube.

Adapted from Arriola, T. A. D., Hatashima, A., & Klang, M. G. (2010). Evaluation of extended-release pancreatic enzyme to dissolve a clog. *Nutrition in Clinical Practice*, *25*(5), 563–564; and Lord, L. M. (2003). Restoring and maintaining patency of enteral feeding tubes. *Nutrition in Clinical Practice*, *18*(5), 422–426.

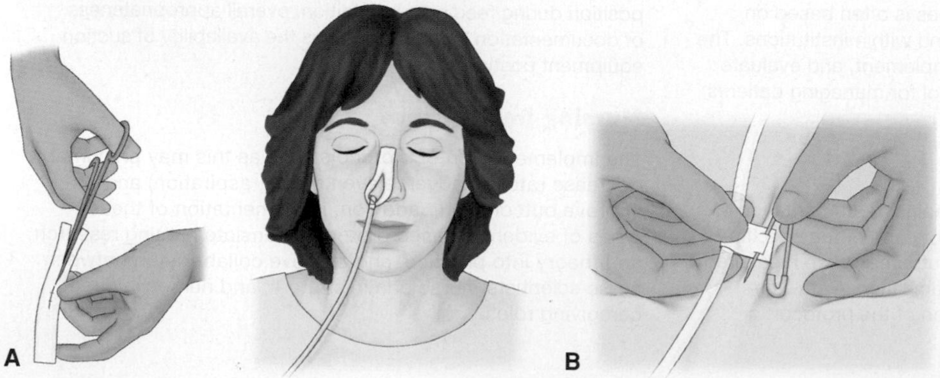
A
B

FIGURE 45-3 • Securing nasogastric (NG) tubes. **A.** The NG tube is secured to the nose with tape to prevent injury to the nasopharyngeal passages. **B.** The tubing is secured to the patient's gown with tape attached to a safety pin to prevent tension on the tube.

Quality and Safety Nursing Alert

Feeding tubes are more successfully declogged when intervention is initiated immediately after the obstruction is noted.

Monitoring the Patient and Maintaining Tube Function

NG tubes used for decompression and drainage are connected to a suction machine or a collection bag. Tubes used for **enteral nutrition** are attached to enteral delivery tubing or a syringe that contains the feeding formula. They can be connected continuously for pump feedings or intermittently so that the end of the tube is plugged between feedings. Malposition or dislodgement of the tube may be caused by tension on the tube with patient movement, coughing, suctioning, or airway intubation. If the nasally placed tube is removed inadvertently in a patient who has undergone esophageal or gastric surgery, it is usually replaced by the surgeon with care to avoid trauma to the suture line.

The nurse must keep an accurate record of all fluid intake, feedings, and irrigation volumes. To maintain patency, the tube is irrigated with water after every feeding and medication delivery and every 4 to 6 hours during continuous feedings, or if the tube is set to gravity drainage or suction. Normal saline or water can be used as an irrigant, depending on the patient's electrolyte levels. The nurse records the amount, color, and type of any drainage. When double- or triple-lumen tubes are used, each lumen is labeled according to its intended use for drainage, medication delivery, or feeding.

Providing Oral and Nasal Hygiene

Regular, conscientious oral and nasal hygiene is a vital part of patient care because the tube causes discomfort and pressure and may be in place for an extended length of time. The nose is inspected daily for skin irritation, and the nasal tape is changed every 2 to 3 days. If the nasal and pharyngeal mucosa are excessively dry, steam or cool vapor inhalations may be beneficial. Throat lozenges, an ice collar, chewing gum, or sucking on hard candies (if permitted) and limiting talking also assist in relieving patient discomfort.

Monitoring and Managing Potential Complications

Patients with NG or nasoenteric intubation are susceptible to a variety of problems, including fluid volume deficit, pulmonary complications, and tube-related irritations. These potential complications require careful ongoing assessment.

Symptoms of fluid volume deficit include dry skin and mucous membranes, decreased urinary output, lethargy, lightheadedness, hypotension, and increased heart rate. Assessment involves maintaining an accurate record of intake and output (I&O). This includes measuring fluid intake from tube feeding and flushes, oral liquids, and intravenous (IV) fluids. Output of urine, emesis, NG drainage, diarrhea, ostomies, fistulas, and drainage tubes should also be measured. The nurse assesses 24-hour fluid balance and reports negative fluid balance (output greater than intake, increased NG output, interruption of IV therapy, or any other disturbance in fluid I&O). (See Chapter 13 for further discussion of fluid volume deficit.)

Pulmonary complications from NG intubation can occur because coughing and clearing of the pharynx are impaired. Tubes may become malpositioned, retracting the distal end above the esophagogastric sphincter, which places the patient at increased risk of aspiration. Prokinetic agents such as erythromycin and metoclopramide may be prescribed to promote gastric emptying, presumably diminishing the likelihood of aspiration. Signs and symptoms of pulmonary complications include coughing during the administration of foods or medications, difficulty clearing the airway, tachypnea, and fever. Assessment includes regular auscultation of lung sounds and monitoring of vital signs. The nurse also carefully confirms the proper placement of the tube before instilling any fluids or medications.

Reducing the Risk for Aspiration

Aspiration pneumonia occurs when regurgitated stomach contents or enteral feedings from an improperly positioned feeding tube are instilled into the pharynx or the trachea or when oral secretions are aspirated. Feeding patients through tubes placed beyond the pylorus or using prokinetic agents can decrease the frequency of feeding regurgitation and aspiration. In addition, feedings and medications should always be administered with the patient in the semi-Fowler's position, and the patient's head should be elevated at least 30 to 45 degrees to reduce the risk of reflux and pulmonary aspiration. This position is maintained at least 1 hour after completion of an intermittent tube feeding and is maintained at all times for patients receiving continuous tube feedings. A reverse Trendelenburg position can be considered when it is not possible or advisable to elevate the head of the patient's bed (Bankhead et al., 2009).

Concept Mastery Alert

When administering continuous or cyclic tube feedings, preventing aspiration is a priority nursing intervention.

Removing the Tube

Before removing an enteral tube, the nurse may intermittently clamp it for a trial period of several hours to ensure that the patient does not experience nausea, vomiting, or distention. Before any tube is removed, it is flushed with 10 mL of water or normal saline to ensure that it is free of debris and away from the gastric lining. Gloves are worn when removing the tube. The tube is withdrawn gently and slowly for 15 to 20 cm (6 to 8 in) until the tip reaches the esophagus; the remainder is withdrawn rapidly from the nostril. If the tube does not come out easily, force should not be used, and the problem should be reported to the primary provider. As the tube is withdrawn, it is concealed in a towel to prevent secretions from soiling the patient or nurse. After the tube is removed, the nurse provides oral hygiene.

Administering Tube Feedings

Tube feedings are given to meet nutritional requirements when oral intake is inadequate or not possible and the GI tract is functional. The feedings are delivered to the stomach, duodenum, or proximal jejunum and help preserve GI integrity by preserving normal intestinal and hepatic metabolism.

TABLE 45-2 Conditions That May Require Enteral Therapy

Condition or Need	Examples
Preoperative bowel preparation	After administration of larger-volume cathartics
Gastrointestinal problems	Fistula, short-bowel syndrome, mild pancreatitis, Crohn's disease, ulcerative colitis, nonspecific maldigestion or malabsorption
Cancer therapy	Radiation, chemotherapy
Convalescent care	Surgery, injury, severe illness
Coma, semiconsciousness*	Stroke, head injury, neurologic disorder, neoplasm
Hypermetabolic conditions	Burns, trauma, multiple fractures, sepsis, acquired immunodeficiency syndrome, organ transplantation
Alcoholism, chronic depression, anorexia nervosa*	Chronic illness, psychiatric or neurologic disorder
Debilitation*	Disease or injury
Maxillofacial or cervical surgery	Disease or injury
Oropharyngeal or esophageal paralysis*	Disease or injury, neoplasm, inflammation, trauma, respiratory failure

*Because some patients with these conditions are at risk for regurgitating or vomiting and aspirating administered formula, each condition must be considered individually.

Tube feedings have several advantages over PN: They are lower in cost, safer, usually well tolerated by the patient, and easier to use both in extended care facilities and in the patient's home.

Nasoduodenal or nasojejunal feeding is indicated when the esophagus and stomach need to be bypassed or when the patient is at risk for aspiration. For tube feedings longer than 4 weeks, gastrostomy or jejunostomy tubes are preferred for administration of medications or food. Indications for enteral nutrition are summarized in Table 45-2.

Osmolality

The **osmolality** of normal body fluids is approximately 300 mOsm/kg. The body attempts to keep the osmolality of the contents of the stomach and intestines at this level. Osmolality is an important consideration for patients receiving tube feedings through the duodenum or jejunum because feeding formulas with a high osmolality may lead to undesirable effects. For example, when a concentrated solution of high osmolality entering the intestines is taken in quickly or in large amounts, water moves rapidly into the intestinal lumen from fluid surrounding the organs and the vascular compartment. The patient may have feelings of fullness, nausea, cramping, dizziness, diaphoresis, and osmotic diarrhea, collectively termed **dumping syndrome**. This can lead to dehydration, hypotension, and tachycardia. Patients fed by the small intestinal route vary in the degree to which they tolerate the effects of high osmolality; the nurse needs to be knowledgeable about the patient's formula and take steps to prevent this undesired effect. The small intestines may be able to adapt to a formula of high osmolality if it is initiated at a low hourly rate that is advanced slowly (Bankhead et al., 2009).

Formulas

The choice of formula to be delivered by tube feeding is influenced by the status of the GI tract and the nutritional needs of the patient. Formula characteristics that are considered include the chemical composition of the nutrient source (protein, carbohydrates, fat), caloric density, osmolality, fiber content, vitamins, minerals, electrolytes, and cost. A wide variety of containers, delivery systems, and enteral pumps are available for use with tube feedings.

Various tube feeding formulas are available commercially. Polymeric formulas are the most common; are composed of protein, carbohydrates, and fats in a high-molecular-weight form; and require that the patient has normal digestive function. Chemically defined or "predigested" formulas contain easier-to-absorb nutrients. Modular products contain only one major nutrient, such as protein, and are used to enhance commercially prepared products. Disease-specific formulas are available as adjuncts to treat various conditions. Fiber in some formulas helps bulk the stool to decrease the occurrence of both diarrhea and constipation. Some feedings are given as supplements, and others are administered to meet the patient's total nutritional needs. Dietitians and certified nutrition support clinicians collaborate with primary providers and nurses to determine the best formula for each patient. The volume of formula delivered varies depending on the caloric density of the formula and the energy needs of the patient. The overall goal is to achieve positive nitrogen balance and weight maintenance or gain without producing discomfort or diarrhea.

Administration Methods

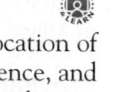

The tube feeding method chosen depends on the location of the tube in the GI tract, patient tolerance, convenience, and cost. Large-bore (larger than 12 Fr) nasally inserted tubes are uncomfortable; their usefulness for tube feedings is limited. Small-bore tubes manufactured for tube feedings are better tolerated; however, they require diligent monitoring and frequent flushing to remain patent.

Bolus and intermittent drip tube feeding methods are practical and inexpensive options for the tube-fed patient residing at home or in a long-term care facility, but they may be poorly tolerated in the acutely ill patient. **Bolus** feedings typically are divided into three to four feedings daily and can be administered into the stomach through a large syringe with a plunger or by gravity (with plunger removed) (Fig. 45-4). Bolus feedings can be delivered as quickly as the patient can tolerate them but are normally initiated slowly, increasing the rate as tolerated. With gravity feedings, raising or lowering the syringe above the abdominal wall regulates the rate of flow. The amount and flow rate is often determined by the patient's reaction. If the patient feels full, it may be desirable to slow the delivery time or give smaller volumes more frequently. The intermittent gravity drip feeding method requires administering feedings over 30 minutes or longer at designated intervals by a reservoir enteral bag and tubing, with the flow rate regulated by a roller clamp or automated pump.

Continuous feeding is the delivery of feedings incrementally by a slow drip over long periods. Slow drip feedings may reduce aspiration rates, distention, nausea, vomiting, and diarrhea in patients with poor gastric emptying or who are receiving hypertonic feeding solutions, as well as patients

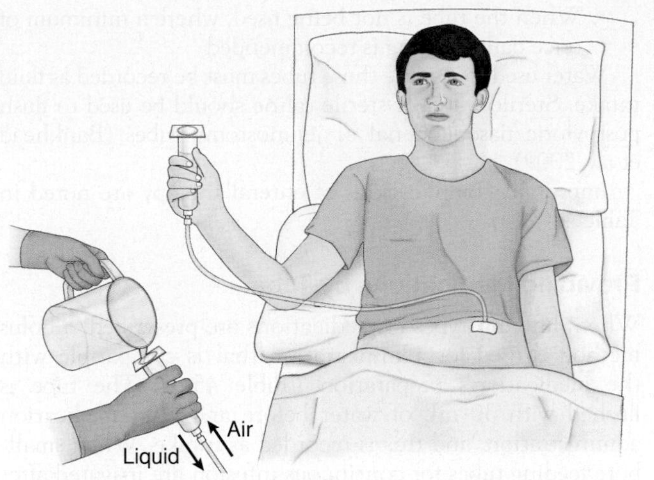
Air
Liquid

FIGURE 45-4 • Bolus gastrostomy feeding by gravity. Syringe is raised perpendicular to the abdomen so that feeding can enter by gravity.

with severe reflux or altered mental status. This method may also be used to administer tube feedings into the small intestine. Enteral feeding pumps control the delivery rate of the formula (Fig. 45-5). They allow for a constant flow rate and can infuse a viscous formula through a small-diameter feeding tube. However, the pumps are relatively heavy and must be attached to an IV pole, and they are expensive. Furthermore, they allow the patient less flexibility than intermittent feedings. Portable lightweight enteral pumps that are easier to handle are available for home use. In addition, most feeding pumps have built-in alarms that signal when the bag is empty, the battery is low, or the tube is occluded. The patient and caregiver need to be aware of these alarms and know how to "troubleshoot" the pump.

An alternative to the continuous infusion method is **cyclic feeding**, in which the infused feeding is given by an enteral feeding pump over 8 to 18 hours. Feedings may be infused at night to avoid interrupting the patient's lifestyle. Cyclic infusions may be appropriate for patients who are being weaned from tube feedings to an oral diet, for patients who cannot eat enough and need supplements, and for patients at home who need daytime hours free from the pump.

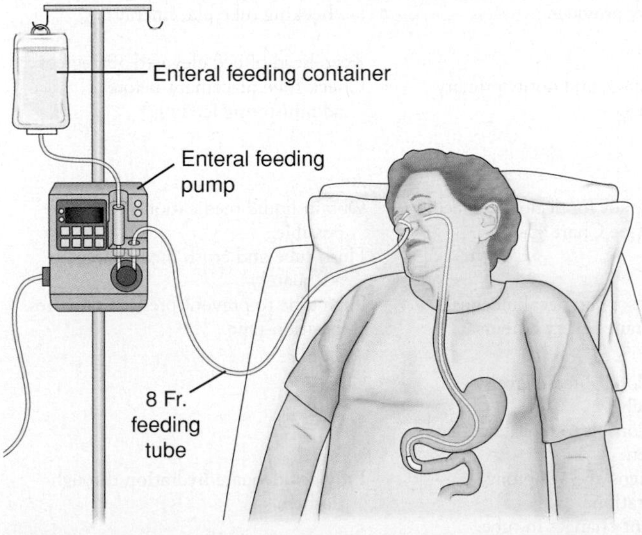
Enteral feeding container
Enteral feeding pump
8 Fr. feeding tube

FIGURE 45-5 • Nasoenteric tube feeding by continuous controlled pump. The head of the bed should be elevated to prevent aspiration.

Chart 45-3 ASSESSMENT

Assessing Patients Receiving Tube Feedings

Be alert for the following assessment findings:

- Tube placement, patient's position (head of bed elevated 30–45 degrees), and formula flow rate
- Patient's ability to tolerate the formula; observe for fullness, bloating, distention, nausea, vomiting, and stool pattern
- Clinical responses, as noted in laboratory findings (blood urea nitrogen, serum protein, prealbumin, electrolytes, kidney function, hemoglobin, hematocrit)
- Signs of dehydration (dry mucous membranes, thirst, decreased urine output)
- Amount of formula actually taken in by the patient
- Elevated blood glucose level, decreased urinary output, sudden weight gain, and periorbital or dependent edema
- Infection control practices: Replace any formula administered by an open system every 4 hours with fresh formula; change tube feeding container and tubing every 24 hours
- Check gastric residual volume before each feeding or, in the case of continuous feedings, every 4 hours; return the aspirate to the stomach
- Intake and output
- Weekly weights
- Recommendations made on dietitian consult

Adapted from Bankhead, R., Boullata, J., Brantley, S., et al. (2009). A.S.P.E.N. enteral nutrition practice recommendations. *Journal of Parenteral and Enteral Nutrition*, *33*(2), 122–167. Available at: pen.sagepub.com/content/early/2009/01/27/0148607108330314

Key assessment findings for patients receiving tube feedings are noted in Chart 45-3.

Maintaining Feeding Equipment and Nutritional Balance

The temperature and volume of the feeding, the flow rate, and the patient's total fluid intake are important factors to consider when tube feedings are administered. The schedule of tube feedings, including the correct quantity and frequency, is maintained. The nurse must carefully monitor the drip rate and avoid administering fluids too rapidly.

Gastric residual volumes (GRVs) are measured before each intermittent feeding and every 4 hours during continuous feedings. In non–critically ill patients, the interval for checking GRV may be increased to every 6 to 8 hours after the tube feeding goal rate is achieved (Bankhead et al., 2009). Any aspirated fluid is readministered to the patient to prevent a loss of fluids, electrolytes, nutrients, and medications that may be administered via the tube. High GRV should alert the nurse to examine the patient more closely but not necessarily stop the delivery of tube feedings. Recent literature provides data supporting the continuation of enteral tube feedings despite GRV as high as 500 mL if no signs or symptoms of gastrointestinal intolerance are present (e.g., nausea, emesis, abdominal distention). Techniques that may be employed to lower the gastric fluid volume and prevent aspiration include keeping the head of bed elevated between 30 and 45 degrees, maintaining a continuous drip tube feeding schedule, administering prokinetic agents as prescribed and converting to small bowel tube feedings (Lord, 2012).

Quality and Safety Nursing Alert

Although a residual volume of 200 mL or greater is generally considered a cause for concern in patients at high risk for aspiration, feedings do not necessarily need to be withheld in all patients.

Maintaining tube function is an ongoing responsibility of the nurse, patient, primary provider, and caregiver. To ensure patency and to decrease the chance of bacterial growth, sludge build-up, or occlusion of the tube, at least 30 mL of water is administered in each of the following instances (Bankhead et al., 2009)

- Before and after intermittent tube feeding and medication administration (with at least 5 mL of water in between each individual medication)
- After checking for gastric residuals and gastric pH
- Every 4 hours with continuous feedings
- When the tube feeding is discontinued or interrupted for any reason
- When the tube is not being used, where a minimum of once daily flushing is recommended

Water used to irrigate these tubes must be recorded as fluid intake. Sterile water or sterile saline should be used to flush postpyloric nasoduodenal or jejunostomy tubes (Bankhead et al., 2009).

Important complications of enteral therapy are noted in Table 45-3.

Providing Medications by Tube

When different types of medications are prescribed, a bolus method is used for administration that is compatible with the medication's preparation (Table 45-4). The tube is flushed with 30 mL of water before and after medication administration, and this is recorded as intake. When small-bore feeding tubes for continuous infusion are irrigated after administration of medications, a 30-mL or larger syringe is

TABLE 45-3 Complications of Enteral Therapy

Complications	Causes	Selected Nursing Interventions: Therapeutic	Selected Nursing Interventions: Preventive
Gastrointestinal			
Diarrhea	Hyperosmolar feedings Rapid infusion/bolus feedings Cold formula	Assess fluid balance and electrolyte levels; report findings. Implement changes in tube feeding formula or rate.	Appropriate rate of infusion and temperature of formula. Avoid multiple elixirs and prokinetic medications.
	Medications, especially antibiotic therapy	Review medications.	
Nausea/vomiting	Change in formula or rate Inadequate gastric emptying	Review medications.	Check residuals; if ≥200 mL, reinstill and recheck; report if residual is consistently high.
Gas/bloating/cramping	Air in tube Excess fiber	Notify primary provider if persistent.	Keep tubing free of air.
Constipation	Lack of fiber Inadequate fluid intake/dehydration Opioid use	Check fiber and water content; report findings.	Administer adequate amount of hydration as flushes. Consider cathartic.
Mechanical			
Aspiration pneumonia	Improper tube placement Vomiting with aspiration of tube feeding Flat in bed	Assess respiratory status and notify primary provider.	Implement reliable method for checking tube placement. Keep head of bed elevated 30 degrees.
Tube displacement	Excessive coughing/vomitus Tension on the tube or unsecured tube Tracheal suctioning Airway intubation	Stop feeding, and notify primary provider.	Check tube placement before administering feeding.
Tube obstruction	Inadequate flushing/formula rate Inadequate crushing of medications and flushing after administration	Follow policy for declogging feeding tubes (see Chart 45-2).	Obtain liquid medications when possible. Flush tube and crush medications adequately.
Nasopharyngeal irritation	Tube position/improper taping Use of large tubes	Assess nasopharyngeal mucous membranes every 8 hours.	Tape tube to prevent pressure on nares. Reposition tape.
Metabolic			
Hyperglycemia	Glucose intolerance High carbohydrate content of the feeding	Check blood glucose levels routinely. Dietary consult to reevaluate feeding regimen.	
Dehydration and azotemia (excessive urea in the blood)	Hyperosmolar feedings with insufficient fluid intake	Report signs and symptoms of dehydration. Implement changes in tube feeding formula, rate, or ratio to water.	Provide adequate hydration through flushes.

TABLE 45-4 Preparing Medication for Delivery by Feeding Tube

Medication Form	Preparation
Liquid	None
Simple compressed tablets	Crush and dissolve in water.
Buccal or sublingual tablets	Administer as prescribed.
Soft gelatin capsules filled with liquid	Make an opening in capsule, and squeeze out contents.
Enteric-coated tablets	Do not crush; change in form is required.
Timed-release tablets	Do not crush tablets, because doing so may release too much drug too quickly (overdose); check with pharmacist for alternative formulation.
Timed-release capsules or sustained-release capsules	Some can be opened and contents added to tube-feeding formula; *always* check with pharmacist before doing this.

used because the pressure generated by smaller syringes could rupture the tube.

Quality and Safety Nursing Alert

Administering medications through postpyloric enteric tubes may adversely affect their absorption; therefore, this should be avoided if possible. In addition, to avoid nutrient and drug interactions, medications should not be mixed with the feeding formulas.

Maintaining Delivery Systems

Tube feeding formula is delivered to patients by either an open or a closed system. The open system is packaged as a liquid or a powder to be mixed with water that is either poured into a feeding container or administered by a large syringe. The feeding container (which is hung on a pole) and the tubing used with the open system should be changed every 24 hours (Bankhead et al., 2009). The open system can be used for bolus feedings, intermittent feedings, or continuous drip feedings and can be delivered by push (with a syringe and plunger), gravity (syringe with plunger removed or gravity bag with roller clamp), or pump. To avoid bacterial contamination, the formula hang time in the bag at room temperature should never exceed what the formula manufacturer recommends, which is usually no more than 4 to 8 hours. Closed delivery systems use a prefilled, sterile container of about 1 L of formula that is spiked with enteral tubing and allows a hang time of 24 to 48 hours at room temperature. The closed delivery system must always use a pump to control formula rate in order to avoid dispensing a large formula volume in a short period of time.

Maintaining Normal Bowel Elimination Pattern

Patients receiving NG or **nasoenteric tube** feedings can experience diarrhea or constipation. Possible causes of diarrhea include:

- Malnutrition: A decrease in the intestinal absorptive area can cause diarrhea.
- Medication therapy:
  - Elixir-based medications—often contain sorbitol, which can act as a cathartic
  - Magnesium—acts as a cathartic
  - Antibiotics—thought to alter normal intestinal flora, allowing pathogenic bacteria to flourish
- *Clostridium difficile* (*C. difficile*) colitis: Can result after antibiotic use alters normal intestinal flora and promotes the abnormal growth of this potentially dangerous microbe. *C. difficile* colitis occurs most commonly in hospitalized patients treated with antibiotics and causes significant, potentially lethal diarrhea.
- Zinc deficiency: Zinc is lost with diarrhea, and zinc deficiency can then cause continued diarrhea.
- Concomitant lactose intolerance
- Concomitant hyperthyroidism
- Dumping syndrome: Formula is infused into the small intestine quickly or formula bypasses the stomach too readily into the small intestine and causes expansion of the intestinal wall. This leads to bloating, cramping, diarrhea, dizziness, diaphoresis, and weakness. Measures for managing the GI symptoms associated with dumping syndrome are presented in Chart 45-4.
- Contamination of the formula and feeding equipment with diarrhea-causing pathogens

Possible causes for constipation include:

- Inadequate water intake: Tube feedings typically do not meet total fluid needs and additional water needs to be administered.
- Administration of fiber-free tube feeding formulas
- Concomitant use of opioids

Maintaining Adequate Hydration

The nurse carefully monitors hydration because in many cases the patient cannot communicate the need for water. Water flushes are administered every 4 hours and after feedings to prevent hypertonic dehydration. The feeding may be initially given as a continuous drip in order to help the patient develop tolerance, especially for hyperosmolar solutions. Key nursing interventions include observing for signs of dehydration (e.g., dry mucous membranes, thirst, decreased urine output); administering water routinely; and monitoring I&O, residual volume, and fluid balance.

Chart 45-4 Preventing Dumping Syndrome

The following strategies may help prevent some of the uncomfortable signs and symptoms of dumping syndrome related to tube feeding:

- Slow the formula instillation rate to provide time for carbohydrates and electrolytes to be diluted.
- Administer feedings at room temperature, because temperature extremes stimulate peristalsis.
- Administer feeding by continuous drip (if tolerated) rather than by bolus, to prevent sudden distention of the intestine.
- Advise the patient to remain in semi-Fowler's position for 1 hour after the feeding; this position prolongs intestinal transit time by decreasing the effect of gravity.
- Instill the minimal amount of water needed to flush the tubing before and after a feeding, because fluid given with a feeding increases intestinal transit time.

Promoting Coping Ability

The psychosocial goal of nursing care is to support and encourage the patient to accept physical changes and to convey hope that daily progressive improvement is possible. If the patient is having difficulty adjusting to the treatment, the nurse intervenes by encouraging self-care within the parameters of the patient's activity level. In addition, the nurse reinforces an optimistic approach by identifying indicators of progress (daily weight trends, electrolyte balance, absence of nausea and diarrhea, improvement in plasma proteins).

Promoting Home and Community-Based Care

Educating Patients on Self-Care Management

Patients who require long-term tube feedings may have had recent surgery, dysphagia due to a neuromuscular disease or radiation or other types of trauma to the throat, an obstruction of the upper GI tract, or decreased level of consciousness. For a patient to be considered for tube feeding at home, the patient should:

- Be medically stable and successfully tolerating at least 60% to 70% of the feeding regimen
- Be capable of self-care or have a caregiver willing to assume the responsibility
- Have access to supplies and interest in learning how to administer tube feedings at home

Preparation of the patient for home administration of enteral feedings begins while the patient is still hospitalized. Ideally, the nurse educates the patient and caregiver while administering the feedings so that they can observe the mechanics and participate in the procedure, ask questions, and express any concerns. Before discharge, the nurse provides information about the equipment needed, formula purchase and storage, and administration of the feedings and water flushes (frequency, quantity, rate of instillation).

Family members who will be active in the patient's home care are encouraged to participate in education sessions. Available printed information about the equipment, the formula, and the procedure is reviewed. Arrangements are made to obtain the equipment and formula and have it ready for use before the patient's discharge.

Continuing Care

Referral to a home care agency is important so that a nurse can supervise and provide support during the first tube feedings at home. Additional visits will depend on the skill and comfort of the patient or caregiver in administering the feedings. During all visits, the nurse monitors the patient's physical status (weight, hydration status, vital signs, activity level) and the ability of the patient and family to administer the tube feedings correctly and assess the **enteral access device** and site. Enteral access devices require periodic replacement, and the nurse should be sure that the patient and caregiver have the necessary information to set up these tube replacement appointments. In addition, the nurse assesses for any complications. The patient or caregiver is encouraged to record times and amounts of feedings and water flushes, bowel patterns, and any symptoms that occur. The nurse can review the record with the patient and caregiver during home visits.

Gastrostomy and Jejunostomy

A **gastrostomy** is a procedure in which an opening is created into the stomach for the purpose of administering foods, fluids, and medications via a feeding tube or for gastric decompression in the setting of gastroparesis, gastroesophageal reflux disease, or intestinal obstruction. A gastrostomy is preferred over a nasally inserted tube to deliver enteral nutrition support longer than 4 weeks (Bankhead et al., 2009). Gastrostomy is also preferred over NG feedings in the patient who is comatose because the gastroesophageal sphincter remains intact, making regurgitation and aspiration less likely.

Balloon and non-balloon gastrostomy tubes (G tubes) may be placed surgically, endoscopically, or fluoroscopically. Each technique requires an abdominal incision, and either a permanent gastric **stoma** is created surgically that can be accessed with a feeding tube (Janeway gastrostomy) or a gastric stoma is established that remains open as long as it remains intubated (i.e., a tube remains in place). Insertion of a **percutaneous endoscopic gastrostomy (PEG)** requires the services of a provider skilled in endoscopy and utilizes moderate sedation. A lighted endoscope is inserted via the patient's mouth toward the stomach and then the stomach is inflated with air. The PEG tube is guided down the esophagus, into the stomach, and out through the abdominal incision. An internal fixation bolster is pulled snug against the stomach wall. An external retention bolster (crossbar, circular, or star shaped) is threaded down the tube and positioned snug to the skin. The tension between the external and internal fixation bolsters keeps the tube in place (Fig. 45-6A). G tubes can also be placed fluoroscopically by an interventional radiologist when an endoscope cannot be passed through a strictured or obstructed esophagus.

The initial G tube can be removed and replaced once the tract is well established, typically 6 weeks to 3 months after initial insertion. Routine replacement is indicated every 3 to 6 months for a balloon G tube and every 6 to 12 months for a non-balloon G tube. Replacement is also indicated for a tube that has clogged or fractured or has a ruptured balloon. The external G tube retention bolster should be fitted snugly to the stoma to prevent leakage of gastric secretions and is maintained in place through gentle traction between the internal and anchoring devices. Enough space should remain between the skin and the external anchor to allow a gauze pad to be placed between them (Fig. 45-7).

An alternative to G tubes that are bulky (e.g., they are usually coiled under an elastic binder or secured to the abdomen with tape or some type of attachment device) are low-profile gastrostomy devices (LPGDs) (see Fig. 45-6B). LPGDs may be inserted 6 weeks to 3 months after initial G tube placement or placed as the initial G tube. These devices are flush with the skin, eliminate the possibility of inward tube migration, have antireflux valves to prevent gastric leakage, and do not require tape or other securement devices. Patients requiring enteral nutrition support are able to conceal the feeding tube access site under their clothing. LPGDs require special connection tubing so they can be attached to the feeding container. Patients must be instructed to bring this connection tubing with them when traveling, going to the

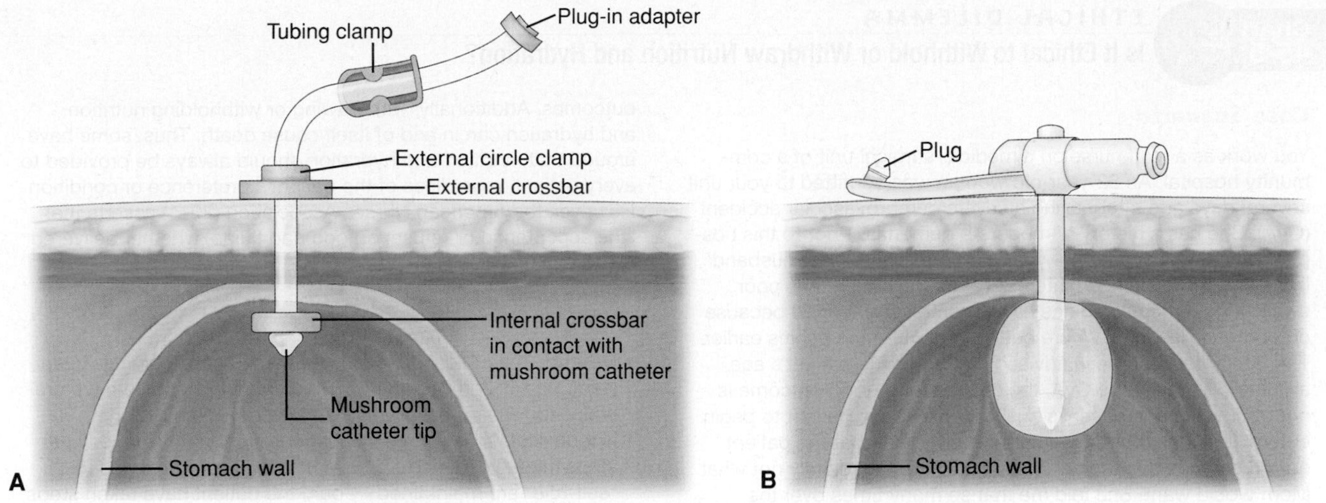
Plug-in adapter
Tubing clamp
External circle clamp
External crossbar
Internal crossbar in contact with mushroom catheter
Mushroom catheter tip
Stomach wall
A
Plug
Stomach wall
B

FIGURE 45-6 • **A.** A detail of the abdomen and the percutaneous endoscopic gastrostomy tube showing catheter fixation. **B.** A detail of the abdomen and the nonobturated low-profile gastrostomy device showing balloon fixation.

emergency department or hospital, or undergoing diagnostic procedures that require access into the GI tract.

A **jejunostomy** is a surgically placed opening into the jejunum for the purpose of administering food, fluids, and medications. A jejunostomy tube (J tube) is indicated when the gastric route is not accessible or to lower aspiration risk when the stomach is not functioning adequately to process and empty food and fluids.

The small intestine can also be accessed by placing a jejunal extension tube through an existing G tube and manipulating it through the pylorus into the small intestine endoscopically, fluoroscopically, or during a surgical procedure. A transgastric J tube can be placed that contains both a gastric and jejunal port so that both the stomach and small intestine can be accessed. There are also low-profile jejunostomy devices (LPJDs) that are placed via a gastric stoma; the distal end is positioned in the small intestine via passage through the pylorus. These devices have the same advantages as the LPGDs described previously.

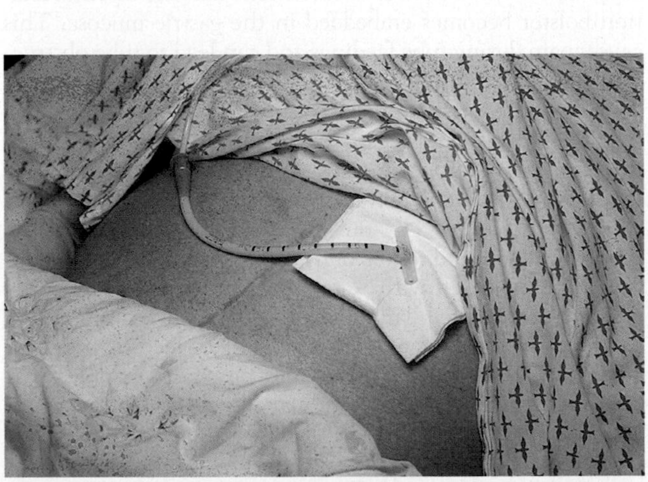

FIGURE 45-7 • Protection at the gastrostomy site. A percutaneous endoscopic gastrostomy tube may be protected by a dressing that allows access to the tube but covers the exit site.

NURSING PROCESS

The Patient With a Gastrostomy or Jejunostomy

Assessment

The focus of the preoperative assessment is to determine the patient's ability to understand and cooperate with the procedure. The nurse assesses the ability of both patient and family to adjust to a change in body image and to participate in self-care. There are multiple medical and ethical issues that the patient, the caregivers, and the physician should discuss together (see Chart 45-5).

The purpose of the procedure and expected postoperative course should be explained. The patient needs to know that the feeding tube will bypass the mouth and esophagus so that liquid feedings can be administered directly into the stomach or intestine. If the feeding tube is expected to be permanent, the patient should be made aware of this. If the procedure is being performed to relieve discomfort, prolonged vomiting, debilitation, or an inability to eat, the patient may find it more acceptable.

In the postoperative period, the patient's fluid and nutritional needs are assessed to ensure proper intake and GI function. The nurse inspects the tube for proper maintenance and the incision for any drainage, skin breakdown, or signs of infection. As the nurse evaluates patients' responses to the change in body image and their understanding of the feeding methods, interventions are identified to help them cope with the tube and learn self-care measures.

Diagnosis

Nursing Diagnoses

Based on the assessment data, major nursing diagnoses may include the following:

- Imbalanced nutrition: less than body requirements
- Risk for infection related to presence of wound and tube
- Risk for impaired skin integrity at tube insertion site
- Disturbed body image related to presence of tube

Chart 45-5 ETHICAL DILEMMA

Is It Ethical to Withhold or Withdraw Nutrition and Hydration?

Case Scenario

You work as a staff nurse on a medical-surgical unit of a community hospital. An 82-year-old woman was admitted to your unit several days ago after having a severe cerebrovascular accident (CVA). She has a history of moderate dementia. Prior to this hospitalization, she and her 84-year-old cognitively intact husband lived with their adult daughter. Because the patient has poor swallowing, enteral feedings have been recommended because outcomes following CVA are better when nutrition begins earlier. The family has also been advised that given the patient's age and the severity of the CVA, the chance for a good outcome is not optimistic. The patient's husband wants the patient to begin enteral feedings, but the daughter is adamant that the patient not commence this therapy, stating, "This is absolutely not what Mom would want! She told me that so many times over the years! Dad—we can't do this to her!"

Discussion

It is generally agreed that patients (or their designated decision makers) can refuse lifesaving treatment, particularly if the means of treatment are extraordinary (e.g., ventilators, dialysis machines, extracorporeal oxygenators). *Extraordinary means* include medications, treatments, and procedures that can be obtained only at excessive cost, pain, or inconvenience and offer no reasonable hope of benefit.

Ordinary means are those medications, treatments, and procedures that offer a reasonable hope of benefit and can be obtained without excessive expense, pain, or inconvenience. Nutrition and hydration therapy are perceived as ordinary means by many health care professionals, and good nutrition improves outcomes. Additionally, withdrawing or withholding nutrition and hydration can in and of itself cause death. Thus, some have argued that nutrition and hydration should always be provided to every patient, regardless of the patient's preference or condition. However, the American Nurses Association (2011) asserts that "the acceptance or refusal or food and fluids, whether delivered by normal or artificial means must be respected" (p. 1).

Analysis

- Describe the ethical principles that are in conflict in this case (see Chart 3–3 in Chapter 3). Which principle should have preeminence as you proceed to work with this family?
- Is it possible to preserve this patient's autonomy? If so, what steps might you take to ensure that the patient's rights to "self-rule" are maintained? Could the patient have taken steps to make her wishes known and legally binding in this type of situation before she became cognitively impaired?
- What resource might be available to you in your hospital to help you, the physician, and the patient and family determine what is in the patient's best interests?

Reference

American Nurses Association. (2011). *Forgoing nutrition and hydration*. Available at: www.nursingworld.org/MainMenuCategories/EthicsStandards/Ethics-Position-Statements/prtetnutr14451.pdf

Resources

See Chapter 3, Chart 3–6 for ethics resources.

Collaborative Problems/Potential Complications

Potential complications may include the following:

- Wound infection, cellulitis, and leakage
- GI bleeding
- Premature dislodgement of the tube

Planning and Goals

The major goals for the patient may include achieving nutritional requirements, preventing infection, maintaining skin integrity, adjusting to changes in body image, and preventing complications.

Nursing Interventions

Meeting Nutritional Needs

The first fluid nourishment is administered soon after tube insertion and can consist of a sterile water or normal saline flush of at least 30 mL. Formula feeding can begin as prescribed, typically within 2 to 24 hours post tube insertion. The infusion rate or bolus amount administered is gradually increased.

If the tube has been placed for gastric drainage, it can be connected to either low intermittent suction or to a gravity drainage bag. This drainage should be measured and recorded because it is a significant indicator of GI function. A decrease in the amount of drainage may indicate that the tube can be clamped for periods of time, allowing greater freedom of movement. High output can result in significant fluid and electrolyte losses.

Preventing Infection and Providing Skin Care

A thin gauze dressing should be applied between the tube insertion site and the G tube. It is normal to see scant serous drainage at the site for a couple of days post insertion. After this drainage ceases, the site may be left open to air. The nurse rotates the tube once daily to prevent skin breakdown and buried bumper syndrome. Buried bumper syndrome can occur when there is excessive traction on the G tube from the external retention bolster to the extent that the internal fixation bolster becomes embedded in the gastric mucosa. This causes pain during tube feedings and can lead to tube obstruction and peritonitis.

The skin surrounding a gastrostomy or jejunostomy requires special care because it may become irritated from the enzymatic action of gastric or intestinal juices that may leak around the tube. Left untreated, the skin becomes macerated and painful. The nurse washes the area around the tube with soap and water daily, removes any encrustation, rinses the area well with water, and pats it dry. Skin at the exit site is evaluated daily for signs of breakdown, irritation, excoriation, and the presence of drainage, bleeding or hypertrophic tissue growth or scattered, raised red papules that could indicate a yeast or candidal infection. Candida may appear in warm moist areas of the body; the area beneath the G tube external retention bolster is a common location for it to develop and spread. The nurse encourages the patient and family members to participate in this evaluation and in hygiene activities.

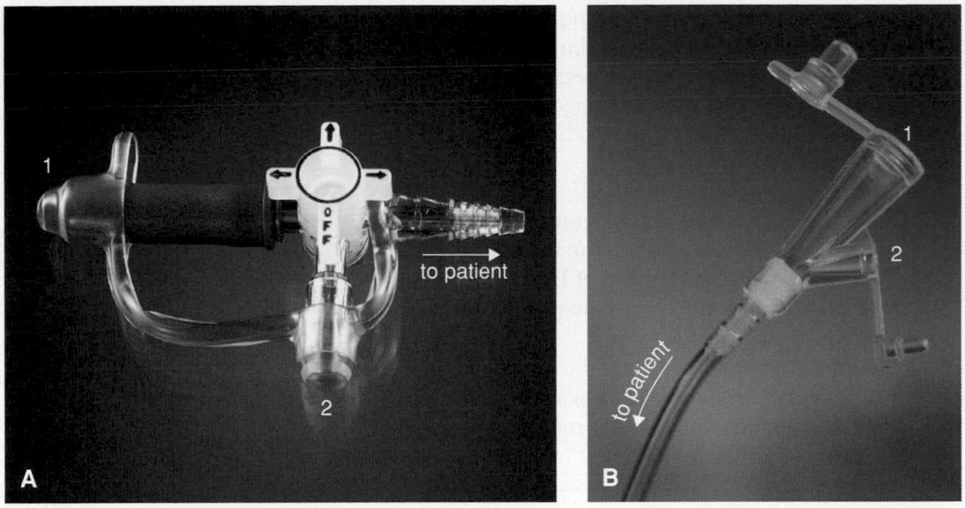
1
2
to patient
A
1
2
to patient
B

FIGURE 45-8 • The Lopez valve (**A**) and Distal Y connector (Bard) (**B**) allow feeding to continue through the main port *(1)* while providing a second port for flushing and medication delivery *(2)*. Part A reprinted with permission from ICU Medical. Part B reprinted with permission from Bard, Inc.

Enhancing Body Image

Eating is a major physiologic and social function, and the patient with a gastrostomy has experienced a major change in body image. The patient is also aware that gastrostomy as a therapeutic intervention is performed only in the presence of a major, chronic, or perhaps terminal illness. It is necessary to evaluate the existing family support system, because adjustment takes time and is facilitated by family acceptance. G tubes, transgastric J tubes, and G tubes with jejunal extensions can be transitioned to LPGDs and LPJDs as described previously and may be indicated to minimize the bulkiness and visibility of these tubes.

Monitoring and Managing Potential Complications

During the postoperative course, the most common complications are wound infection or cellulitis at the exit site, bleeding, excessive tightness of external retention bolster, and dislodgement. Because many patients who receive tube feedings are debilitated and have compromised nutritional status, any signs of infection are promptly reported to the primary provider so that appropriate therapy can be instituted. Bleeding from the insertion site in the stomach can also occur and should be reported promptly. The nurse closely monitors the patient's vital signs and observes all operative site drainage, vomitus, and stool for evidence of bleeding. If an external retention bolster, tape, securement device, or sutures are present, they are evaluated for adequate tension and securement. Excessive tension of the external retention bolster can cause excruciating pain and will lead to skin breakdown and ulceration. The nurse should notify the primary provider if excessive pain occurs at the incision site post insertion. Dislodgement of a recently inserted tube requires immediate attention because the tract can close within 4 to 6 hours if the tube is not replaced promptly.

Promoting Home and Community-Based Care

Educating the Patient About Self-Care. The patient with a G or J tube in the home setting must be capable of maintaining patency of the tube or have a caregiver who can do so. The nurse assesses the patient's level of knowledge and interest in learning about the tube, as well as an ability to understand how to flush, provide site care, and administer feedings or facilitate decompression and drainage. Education is similar to that described earlier. To facilitate self-care, the nurse encourages the patient to participate in flushing the tube, administering medications and tube feedings during hospitalization, and establishing as normal a routine as possible.

Adapters are available that can be secured to the end of the tube to create a "Y" site for ease of flushing, suction, or medication delivery (Fig. 45-8). The flushing equipment is cleaned with warm, soapy water and rinsed after each use. The tube can be marked at skin level to provide the patient with a baseline for later comparison. The patient or caregiver should be advised to monitor the tube's length and to notify the physician or home care nurse if the segment of the tube outside the body becomes shorter or longer.

Continuing Care. Referral to a home care agency is important to ensure initial supervision and support for the patient and caregiver. The home care nurse assesses the patient's status and progress and evaluates the care of the tube and healing status of the tube insertion site. Further instruction and supervision in the home setting may be required to help the patient and caregiver adapt to a physical environment and equipment that are different from the hospital setting. The nurse also reviews with the patient and caregiver what complications to report and assists the patient and family in establishing as normal a routine as possible.

Evaluation

Expected patient outcomes may include:

1. Achieves nutrition goals
   a. Attains weight goal
   b. Tolerates tube feeding prescription without nausea, emesis, cramping, abdominal pain, or feelings of early satiety
   c. Has acceptable bowel movements without constipation or large-volume liquid stools
   d. Has normal plasma protein, glucose, vitamin, and mineral levels
   e. Has normal electrolyte values
2. Is free of infection at enteral access site
   a. Is afebrile
   b. Has no induration, redness, pain, or purulent drainage
   c. Has no scattered papules indicative of a yeast infection

3. Has dry, intact skin surrounding enteral access site
   a. No evidence of excessive drainage or bleeding
   b. No skin breakdown or hypertrophic tissue growth
4. Adjusts to change in body image
   a. Is able to discuss expected changes
   b. Verbalizes concerns
5. Demonstrates skill in tube care
   a. Handles equipment competently
   b. Demonstrates how to maintain tube patency
   c. Keeps an accurate record of I&O
   d. Demonstrates how to gently wash tube site daily and keep clean and dry
6. Avoids other complications
   a. Exhibits adequate wound healing
   b. Tube remains intact and is routinely replaced for the duration of therapy

TABLE 45-5 Indications for Parenteral Nutrition

Condition or Need	Examples
Insufficient oral or enteral intake	Severe burns, malnutrition, short-bowel syndrome, acquired immunodeficiency syndrome, sepsis, cancer
Impaired ability to ingest or absorb food orally or enterally	Paralytic ileus, Crohn's disease, short gut, postradiation enteritis, high-output enterocutaneous fistula
Patient unwilling or unable to ingest adequate nutrients orally or enterally	Major psychiatric illness (e.g., severe anorexia nervosa)
Prolonged preoperative and postoperative nutritional needs	Extensive bowel surgery, acute pancreatitis

Adapted from A.S.P.E.N. (2009). Clinical guidelines for the use of parenteral and enteral nutrition in adult and pediatric patients, 2009. *Journal of Parenteral and Enteral Nutrition, 33*(3), 255–259.

Parenteral Nutrition

Parenteral nutrition (PN) is a method of providing nutrients to the body by an IV route. The nutrients are a complex admixture containing proteins, carbohydrates, fats, electrolytes, vitamins, trace minerals, and sterile water in a single container. The goals of PN are to improve nutritional status, establish a positive nitrogen balance, maintain muscle mass, promote weight maintenance or gain, and enhance the healing process (Alexander, Corrigan, Gorski, et al., 2010; Smith & Walsh, 2012).

Establishing Positive Nitrogen Balance

Most IV fluids do not provide sufficient calories or protein to meet the body's daily requirements. PN solutions can provide enough calories and nitrogen to meet the patient's daily nutritional needs. The patient with fever, trauma, burns, major surgery, or hypermetabolic disease requires additional daily calories (Mirtallo & Patel, 2012). When highly concentrated dextrose is administered, caloric requirements are satisfied and the body uses amino acids for protein synthesis rather than for energy. Additionally, electrolytes such as calcium, phosphorus, magnesium, and sodium chloride are added to the solution to maintain proper electrolyte balance and to transport glucose and amino acids across cell membranes.

The volume of fluid necessary to provide these calories peripherally can surpass fluid tolerance. To provide the required calories in a smaller volume, it is necessary to increase the concentration of nutrients and use a route of administration that rapidly dilutes incoming nutrients to the proper levels of body tolerance. Typically, a large, high-flow vein such as the superior vena cava (at the right atriocaval junction) is the preferred site.

Clinical Indications

The indications for PN include an inability to ingest adequate oral food or fluids within a 7 to 10 day timeframe (Krzywda & Meyers, 2010). Enteral nutrition should be considered before parenteral support because it assists in maintaining gut mucosal integrity and improved immune function and is typically associated with fewer complications. In both the home and hospital setting, PN is indicated in the situations listed in Table 45-5.

Formulas

A total of 1 to 3 L of solution is administered over a 24-hour period. The label of the solution is verified with the prescription. **Intravenous fat emulsions (IVFEs)** may be infused simultaneously with PN through a Y connector close to the infusion site and should not be filtered. Usually, 500 mL of a 10% IVFE or 250 mL of 20% IVFE is administered over 6 to 12 hours, one to three times a week. IVFEs can provide up to 30% of the total daily calorie intake.

Quality and Safety Nursing Alert

Before PN infusion is administered, the solution must be inspected for separation, oily appearance (also known as a "cracked solution"), or any precipitate (which appears as white crystals). If any of these are present, it is not used.

IVFEs can be mixed by the pharmacy staff with other components of PN to create a "three-in-one solution" commonly called a **total nutrient admixture (TNA)**. Whereas a filter is not used with IVFE, a special final filter (1.2-mcm filter) is used with TNA to prevent the administration of a precipitate (i.e., calcium, phosphorus, incompatibilities) that cannot be seen due to the opacity of the solution. Advantages of TNA over PN are cost savings in preparation and equipment, decreased risk of catheter or nutrient contamination, decreased nursing time, and increased patient convenience and satisfaction (Barber & Sacks, 2012). Ideally, the pharmacist, nutritionist, and primary provider should collaborate to determine the specific formula needed.

Initiating Therapy

PN solutions are initiated slowly and advanced gradually each day to the desired rate as the patient's fluid and dextrose tolerance permits. The patient's laboratory test results and response to PN therapy are monitored on an ongoing basis by the primary provider. Standing orders are initiated for weighing the patient; monitoring I&O and blood glucose; and baseline and periodic monitoring of complete blood count, platelet count, and chemistry panel, including serum carbon dioxide, magnesium, phosphorus, and triglycerides. A 24-hour urine nitrogen determination may be performed for analysis of nitrogen balance. In most hospitals, the PN solutions are prescribed on a daily standard PN order form. The formulation of the PN solutions is calculated carefully each day to meet the complete nutritional needs of the individual patient.

Administration Methods

Various vascular access devices are used to administer PN solutions in clinical practice. PN may be administered through either peripheral or central IV lines, depending on the patient's condition and the anticipated length of therapy. An infusion pump is always used for administration of PN.

Peripheral Method

To supplement oral intake, peripheral parenteral nutrition (PPN) may be prescribed. PPN is administered through a peripheral vein; this is possible because the solution is less hypertonic than a full-calorie PN solution. PPN formulas are not nutritionally complete because of their low dextrose content. Lipids are administered simultaneously to buffer the PPN and to protect the peripheral vein from irritation. The usual length of therapy using PPN is 5 to 7 days.

Quality and Safety Nursing Alert

Formulations with dextrose concentrations of more than 10% should not be administered through peripheral veins because they irritate the intima (innermost walls) of small veins, causing chemical phlebitis.

Central Method

Because central parenteral nutrition (CPN) solutions have five or six times the solute concentration of blood (and exert an osmotic pressure of about 2,000 mOsm/L), they are administered into the vascular system through a catheter inserted into a high-flow, large blood vessel (e.g., ideally at the superior vena cava/right atriocaval junction). Concentrated solutions are then very rapidly diluted to isotonic levels by the blood in this vessel.

Four types of **central venous access devices (CVADs)** are available: nontunneled (or percutaneous) central catheters, peripherally inserted central catheters (PICCs), tunneled catheters, and implanted ports.

Nontunneled Central Catheters

Nontunneled central catheters are used for short-term (less than 6 weeks) IV therapy in acute care settings. The subclavian vein is the most common vessel accessed because the subclavian area provides a stable insertion site to which the catheter can be anchored, is easily compressible (facilitating control of hemorrhage), allows the patient freedom of movement, and provides easy access to the dressing site. However, according to the Centers for Disease Control and Prevention (CDC; 2011), the subclavian access site should be avoided in patients with advanced kidney disease and those on hemodialysis to prevent subclavian vein stenosis. The second most common access sites include the basilic, brachial, or cephalic veins in the arm followed by the jugular vein. The femoral vein should be avoided for this purpose and should only be used as a last resort (CDC, 2011). For a patient with limited IV access, a triple-lumen catheter can be used because it offers three ports for various uses (Fig. 45-9). The use of a single-lumen catheter dedicated for the administration of PN is not typically feasible, because most patients require administration of medications and fluids in addition to PN and the line used to administer PN cannot be used for other purposes. According to the CDC (2011), no specific lumen type is recommended as best for PN administration.

When a patient requires IV access for PN, the insertion procedure is first explained so that the patient is aware of what to expect. The patient is placed supine in the Trendelenburg position to produce dilation of neck and shoulder vessels, which makes entry easier and decreases the risk of air embolus. The skin is cleansed with chlorhexidine to remove surface oils. To afford maximal accuracy in the placement of the catheter, the patient is instructed to turn his or her head away from the site of venipuncture and to remain motionless while the catheter is inserted and the wound is dressed.

The nurse assists during the insertion process by maintaining the sterile field and supporting the patient throughout the procedure. Maximal barrier precautions mandate that full-body sterile drapes are applied and sterile gloves, cap, gown, and masks are donned to reduce risk of catheter bloodstream infection (Institute for Healthcare Improvement [IHI], 2012) (see Chart 14-2 in Chapter 14). Lidocaine

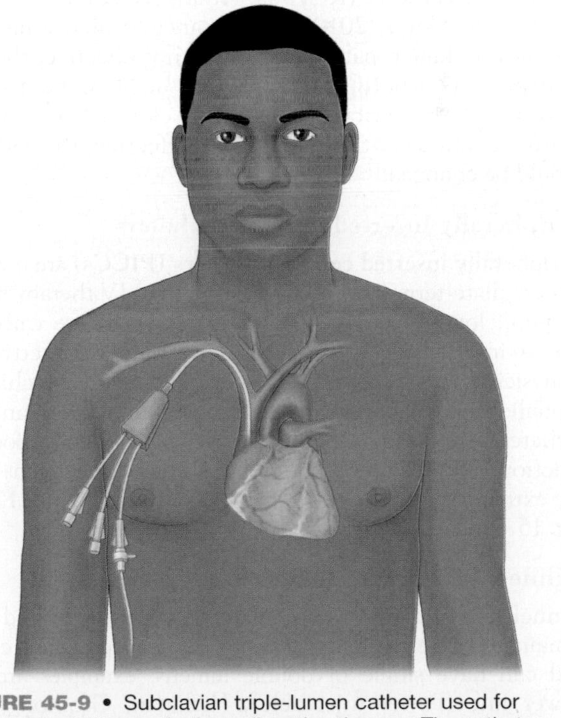

FIGURE 45-9 • Subclavian triple-lumen catheter used for parenteral nutrition and other adjunctive therapy. The catheter is threaded through the subclavian vein into the vena cava/right atriocaval junction. Each lumen is an avenue for solution administration. The lumens are secured with threaded needleless adapters or Luer-Lok–type caps when the device is not in use.

is injected to anesthetize the skin and underlying issues. A large-bore needle on a syringe is inserted and moved parallel to and beneath the clavicle until it enters the vein. A radiopaque wire is inserted through the needle into the vein. The catheter is then advanced over the wire, the needle is withdrawn, and the hub of the catheter is attached to the IV tubing. Until the syringe is detached from the needle and the catheter is inserted, the patient may be asked to perform the Valsalva maneuver. The patient is instructed to take a deep breath, hold it, and bear down with the mouth closed to produce a positive phase in central venous pressure, thereby lessening the possibility of air being drawn into the circulatory system (air embolism). The catheter is sutured to the skin. A chlorhexidine-impregnated gauze with a transparent covering (CDC, 2011) is applied using strict sterile technique (Timsit, Schwebel, Bouadma, et al., 2009).

The position of the tip of the catheter is checked with x-ray or fluoroscopy to confirm its location in the superior vena cava at the junction of the right atrium and to rule out a pneumothorax resulting from inadvertent puncture of the pleura. Once the catheter's position is confirmed, the prescribed CPN solution can be started. The initial rate of infusion is usually low, and the rate is gradually increased to the target rate.

An injection cap is attached to the end of each central catheter lumen, creating a closed system. IV infusion tubing is connected to the insertion cap of the central catheter with a threaded needleless adapter or Luer-Lok device. To ensure patency, all lumens are initially flushed with normal saline or diluted heparin (10 U/mL) after each intermittent infusion and after blood drawing; this flushing is necessary daily when the catheter is not in use. Force is never used to flush the catheter (Krzywda, Andris, & Edmiston, 2012; Krzywda & Meyer, 2010). If resistance is met, aspiration may restore lumen patency; if this is not effective, the primary provider is notified. Low-dose tissue plasminogen activator may be prescribed to dissolve a clot or fibrin sheath. If attempts to clear the lumen are ineffective, the catheter should be changed.

Peripherally Inserted Central Catheters

Peripherally inserted central catheters (PICCs) are used for intermediate-term (several days to months) IV therapy in the hospital, long-term care, or home setting. These catheters may be inserted at the bedside or in the outpatient setting by a physician or specially trained nurse. The basilic, brachial, or cephalic vein is accessed above the antecubital space, and the catheter is threaded to the superior vena cava/right atriocaval junction. Taking of blood pressure and blood specimens from the extremity with the PICC is avoided. (See Chapter 13 and Fig. 15-6 in Chapter 15.)

Tunneled Central Catheters

Tunneled central catheters are for long-term use and may remain in place for many years. These catheters are cuffed and can have single or double lumens; examples are the Power line (Power injectable), Hickman, Groshong, and Permacath. These catheters are inserted surgically. They are threaded under the skin (reducing the risk of ascending infection) to the subclavian vein and advanced into the superior vena cava.

Implanted Ports

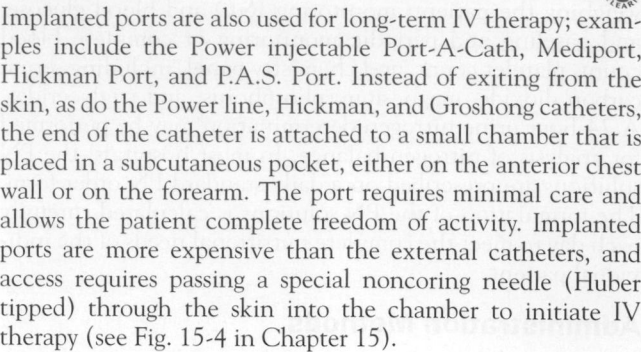

Implanted ports are also used for long-term IV therapy; examples include the Power injectable Port-A-Cath, Mediport, Hickman Port, and P.A.S. Port. Instead of exiting from the skin, as do the Power line, Hickman, and Groshong catheters, the end of the catheter is attached to a small chamber that is placed in a subcutaneous pocket, either on the anterior chest wall or on the forearm. The port requires minimal care and allows the patient complete freedom of activity. Implanted ports are more expensive than the external catheters, and access requires passing a special noncoring needle (Huber tipped) through the skin into the chamber to initiate IV therapy (see Fig. 15-4 in Chapter 15).

Discontinuing Parenteral Nutrition

The PN solution is discontinued gradually to allow the patient to adjust to decreased levels of glucose. If the PN solution is abruptly terminated, isotonic dextrose can be administered at the same rate the PN solution was infusing for 1 to 2 hours to prevent rebound hypoglycemia. Symptoms of rebound hypoglycemia include weakness, faintness, sweating, shakiness, feeling cold, confusion, and increased heart rate. Once all IV therapy is completed, the nontunneled central venous catheter or PICC is removed and an occlusive dressing is applied to the exit site. Tunneled catheters and implanted ports are removed only by the physician.

NURSING PROCESS

The Patient Receiving Parenteral Nutrition

Assessment

The nurse assists in identifying patients unable to tolerate oral or enteral feedings who may be candidates for PN. Indicators include significant weight loss (10% or more of usual weight), a decrease in oral food intake for more than 1 week, muscle wasting, decreased tissue healing, abnormal urea nitrogen excretion, and persistent vomiting and diarrhea (Mirtallo & Patel, 2012; Smith & Walsh, 2012). The nurse carefully monitors the patient's hydration status, electrolyte levels, and calorie intake.

Diagnosis

Nursing Diagnoses

Based on the assessment data, major nursing diagnoses may include the following:

- Imbalanced nutrition: less than body requirements related to inadequate oral intake of nutrients
- Risk for infection related to contamination of the central catheter site or infusion line
- Risk for imbalanced fluid volume related to altered infusion rate
- Risk for activity intolerance related to restrictions because of the presence of IV access device

Collaborative Problems/Potential Complications

The most common complications are pneumothorax, air embolism, a clotted or displaced catheter, sepsis, hyperglycemia, fluid overload, and rebound hypoglycemia. These problems and the associated collaborative interventions are described in Table 45-6.

TABLE 45-6 Complications of Parental Nutrition

Complications	Cause	Nursing Actions and Collaborative Interventions: Treatment	Nursing Actions and Collaborative Interventions: Prevention
Pneumothorax	Improper catheter placement and inadvertent puncture of the pleura	Place patient in Fowler's position. Offer reassurance. Monitor vital signs. Prepare for thoracentesis or chest tube insertion.	Assist patient to remain still in Trendelenburg position during catheter insertion.
Embolism	Disconnected tubing Cap missing from port Blocked segment of vascular system	Replace tubing immediately and notify physician. Replace cap and notify primary provider. Turn patient on left side and place in the head-low position. Notify primary provider.	Examine all tubing connection sites for their security.
Clotted catheter line	Inadequate/infrequent saline/heparin flushes Disruption of infusion	At direction of primary provider, flush with thrombolytic medication as prescribed.	Flush lines per established protocols. Monitor infusion rate hourly and inspect integrity of the line.
Catheter displacement and contamination	Excessive movement, possibly with a nonsecured catheter Separation of tubing and contamination	Stop the infusion, and notify the primary provider.	Examine all tubing connection sites. Avoid interrupting the main line or piggybacking other lines.
Sepsis	Separation of dressings Contaminated solution Infection at insertion site of catheter	Reinforce or change dressing quickly using aseptic technique. Discard. Notify pharmacist. Notify primary provider. Monitor vital signs.	Maintain sterile technique when changing tubing, dressing, or PN bag. Scrub the hub for 15 seconds prior to accessing line for any reason; air-dry prior to use.
Hyperglycemia	Glucose intolerance	Notify primary provider; addition of insulin to PN solution may be prescribed.	Monitor glucose levels (blood and urine). Monitor urine output. Observe for stupor, confusion, or lethargy.
Fluid overload	Fluid infusing rapidly	Decrease infusion rate. Monitor vital signs. Notify primary provider. Treat respiratory distress by sitting patient upright and administering oxygen as needed, if prescribed.	Use infusion pump. Verify correct infusion rate ordered.
Rebound hypoglycemia	Feedings stopped too abruptly	Monitor for symptoms (weakness, tremors, diaphoresis, headache, hunger, and apprehension); notify primary provider.	Gradually wean patient from PN.

PN, parenteral nutrition.
Adapted from Bankhead, R., Boullata, J., Brantley, S., et al. (2009). A.S.P.E.N. enteral nutrition practice recommendations. *Journal of Parenteral and Enteral Nutrition, 33*(2), 122–167. Available at: pen.sagepub.com/content/early/2009/01/27/0148607108330314

Planning and Goals

The major goals for the patient may include optimal level of nutrition, absence of infection, adequate fluid volume, optimal level of activity (within individual limitations), knowledge of and skill in self-care, and absence of complications.

Nursing Interventions

Maintaining Optimal Nutrition

A continuous, uniform infusion of PN solution over a 24-hour period is desired. However, in some cases (e.g., home care patients), cyclic PN may be appropriate. Cyclic PN is infused during a set period of time. The time periods for infusion are sufficient to meet the patient's nutritional and pharmacologic needs. Ideally, cyclic PN is infused over a 10- to 15-hour period that continues through the night. The cyclic PN is titrated up during the beginning of the infusion cycle and down at the conclusion of the infusion to prevent hyperglycemia and hypoglycemia, respectively.

The patient is initially weighed daily (this may be decreased to two or three times per week once stable) at the same time of the day under the same conditions for accurate comparison. Under the PN regimen, satisfactory weight maintenance or gain can usually be achieved. It is important to keep accurate I&O records and calculations of fluid balance. A calorie count is kept of any oral nutrients. Trace elements (copper, zinc, chromium, manganese, and selenium) are included in PN solutions and are individualized for each patient.

Preventing Infection

The high dextrose and fat content of PN solutions makes them an ideal culture medium for bacterial and fungal growth, and CVADs provide a port of entry. Gram-positive cocci, gram-negative bacilli, and *Candida* species are frequently isolated as causes of catheter-related bloodstream infections. Common organisms include *Staphylococcus aureus*, *Staphylococcus epidermidis*, *Pseudomonas aeruginosa*, *Acinetobacter* species, and *Klebsiella pneumoniae*.

Quality and Safety Nursing Alert

Meticulous aseptic technique is essential to prevent infection any time the IV line setup is manipulated.

The major sources of microorganisms for catheter-related bloodstream infections are the skin and the catheter hub. The catheter site is covered with an occlusive gauze dressing that is usually changed every 48 hours using sterile technique (Infusion Nurses Society [INS], 2011). Alternatively, the site may be covered with a transparent dressing that is changed weekly. Besides allowing frequent examination of the catheter site without changing the dressing, the transparent dressing also adheres well and is more comfortable for the patient. The CDC (2011) recommends changing CVAD dressings not more than every 7 days unless the dressing is damp, bloody, loose, or soiled. During dressing changes, the nurse and patient wear masks to reduce the possibility of airborne contamination. Sterile technique is used (e.g., the nurse wears sterile gloves). The area is checked for leakage; bloody or purulent drainage; a kinked catheter; and skin reactions such as inflammation, redness, swelling, or tenderness.

When an extension set is used with a central catheter, it is considered an extension of the catheter itself. It is not routinely changed with dressing or tubing changes. The dressing and tubing are labeled with the date, time of insertion, time of dressing change, and initials of the person who carried out the procedure; this information is also documented in the patient's medical record.

The catheter is another major source of colonization and infection. The use of chlorhexidine/silver sulfadiazine- or minocycline/rifampin-impregnated catheters is recommended for a patient whose catheter is expected to remain in place for longer than 5 days if there is concern over a possibility of the patient acquiring a central line–associated bloodstream infection (CLABSI) (CDC, 2011) (see Chapter 14 for further discussion of CLABSI prevention). Other mechanisms for prevention of infection include prophylactic antibiotic therapy, antithrombolytic agents, various exit-site dressings, and various disinfectants for cleansing catheter exit sites.

Maintaining Fluid Balance

To maintain an accurate rate of PN administration, an infusion pump is necessary. A designated rate is set in milliliters per hour, and the rate is checked every 3 to 4 hours. The infusion rate should not be increased or decreased to compensate for fluids that have infused too quickly or too slowly. If the solution runs out, 10% dextrose and water is infused at the same rate until the next PN solution is available from the pharmacy.

If the rate is too rapid, hyperosmolar diuresis can occur. Excess glucose is excreted by the renal tubules, pulling large volumes of water into the tubules via osmosis, resulting in higher-than-normal urine output and intravascular fluid volume deficit. If the flow rate is too slow, the patient does not receive the maximal benefit of calories and nitrogen. I&O is recorded every 8 hours so that fluid imbalance can be readily detected.

Encouraging Activity

Activities and ambulation are encouraged when the patient is physically able. With a central catheter, the patient is free to move the extremities, and normal activity should be encouraged to maintain good muscle tone. If applicable, the education and exercise program initiated by occupational and physical therapists is reinforced.

Promoting Home and Community-Based Care

Educating Patients About Self-Care. Successful home PN requires educating the patient and family in specialized skills using an intensive training program and follow-up supervision in the home. This is best accomplished through a team effort. Initiation of a home program facilitates the patient's discharge from the hospital.

Ideal candidates for home PN are patients who have a reasonable life expectancy after return home, have a limited number of illnesses other than the one that has resulted in the need for PN, and are highly motivated and fairly self-sufficient. Ethical dilemmas occur when the patient and family, as well as the caregiver, do not thoroughly understand what is involved in home PN. In addition, the ability to learn, availability of family interest and support, adequate finances, and physical plan of the home are factors that must be assessed when the decision about home PN is made (Chart 45-6).

Many home health care agencies have developed education brochures and videos for home PN treatment. Topics include catheter and dressing care, the use of an infusion pump, administration of fat emulsions, and instillation of heparin flushes. Education begins in the hospital and continues in the home or ambulatory infusion center (Chart 45-7).

Continuing Care. The home care nurse should be aware that the typical patient needs several instruction sessions for assessment of learning and reinforcement. More information about home patient education is presented in Chart 45-8. Special considerations for older adult patients who go home with nutrition support are presented in Chart 45-9.

Chart 45-6 ASSESSMENT

Assessing for Home Nutrition Support

Be alert to the following assessment findings:

- *Water:* Water is necessary for hand hygiene and cleaning of work areas.
- *Electricity:* A reliable power source is needed to provide proper lighting and charging of pumps.
- *Refrigeration:* Refrigeration must be adequate for accommodation of several bags of parenteral nutrition solution.
- *Telephone:* A telephone is necessary for contacting home health personnel, arranging for prompt delivery of supplies, and for emergency purposes.
- *Environment:*
  - Should be free of rodents and insects
  - Should have storage that is not accessible to pets and small children
  - Should be assessed for stairs, carpets, and inaccessible areas, which can limit mobility with infusion pumps if the patient has a disability

Adapted from Ireton-Jones, C., DeLegge, M., Epperson, L., et al. (2003). Management of the home parenteral nutrition patient. *Nutrition in Clinical Practice*, *18*(4), 310–317.

Chart 45-7

PATIENT EDUCATION

Educating Patients About Home Parenteral Nutrition

An effective home care education program prepares the patient to store solutions, set up the infusion, flush the line with heparin, change the dressings, and troubleshoot for problems. The most common complication is sepsis. Strict aseptic technique is taught for hand hygiene, handling equipment, changing the dressing, and preparing the solution.

Troubleshooting Mechanical Difficulties

Mechanical problems can arise with the infusion pump or catheter site. The patient should measure the external length of the catheter; this measurement is used as a comparison if dislodgement is suspected. The patient should receive a list of instructions on how to recognize catheter problems, including leakage, loose cap, blood clot, and dislodgement, and what to do for each problem.

Recognizing Metabolic Complications

The patient is given a list of signs and symptoms that indicate metabolic complications (neuropathies, mentation changes, diarrhea, nausea, skin changes, decreased urine output) and directions on how to contact the home health care nurse or physician if any of these complications occur. The patient is instructed to have routine serum chemistry and hematology tests as well.

Obtaining Psychosocial Support

The psychosocial aspects of home parenteral nutrition are as important as the physiologic and technical concerns. Patients must cope with the loss of eating and with changes in lifestyle brought on by sleep disturbances related to frequent urination during nighttime infusions. Major psychosocial reactions include depression, anger, withdrawal, anxiety, and impaired self-image. A successful home parenteral nutrition program depends on the patient's and family's motivation, emotional stability, and technical competence.

Chart 45-8

HOME CARE CHECKLIST

The Patient Receiving Parenteral Nutrition

At the completion of home care education, the patient or caregiver will be able to:	PATIENT	CAREGIVER
• Discuss goal and purpose of parenteral nutrition (PN) therapy.	✔	✔
• Discuss basic components of PN solution.	✔	✔
• List emergency phone numbers.	✔	✔
• Demonstrate how to handle PN solutions and medications correctly.	✔	✔
• Demonstrate how to operate infusion pump.	✔	✔
• Demonstrate how to prime tubing and filter.	✔	✔
• Demonstrate how to connect and disconnect PN infusion.	✔	✔
• Demonstrate how to perform catheter dressing changes.	✔	✔
• Demonstrate how to flush central line.	✔	✔
• Identify possible PN complications and interventions.	✔	✔

Chart 45-9

Home Parenteral and Enteral Nutrition

Age-related conditions that affect home nutrition support goals include the following:

- Arthritis—possible decreased hand dexterity and fine motor coordination
- Sensory impairment—inability to hear pump alarms; vision loss may affect ability to see pump menus or fill syringes
- Constipation—decreased overall bowel tone, which can cause intolerance of enteral feedings; water and fiber intake should be assessed
- Impaired thirst—may require close clinical management of fluid needs
- Obesity—decreased basal metabolic rate increases tendency toward weight gain; may require a kilocalorie reduction to compensate
- Diabetes—increased insulin resistance, which makes glucose control during PN infusion more challenging
- Depression/dementia—mood and memory disorders, which may present as low motivation to learn and adhere to the nutrition support regimen
- Multiple medications—conversion to an appropriate route or form required

Adapted from White, J., Brewer, D., Stockton, M., et al. (2003). Nutrition in chronic disease management in the elderly. *Nutrition in Clinical Practice*, *18*(1), 3–11.

Evaluation

Expected patient outcomes may include the following:

1. Attains or maintains nutritional balance
2. Is free of catheter-related infection
   a. Is afebrile
   b. Has no purulent drainage from the catheter insertion site
3. Is hydrated, as evidenced by good skin turgor
4. Achieves an optimal level of activity, within limitations
5. Demonstrates skill in managing PN regimen
6. Prevents complications
   a. Maintains proper catheter and equipment function
   b. Maintains metabolic balance within normal limits

Critical Thinking Exercises

1 You prepare to administer a patient's medications into his PEG tube but meet resistance when you attempt to instill water. You think that his tube may be clogged. Described the steps you will take to assess the patency of his tube and try to declog it as necessary.

2 ebp A patient who is malnourished and has Crohn's disease is recently admitted to the hospital with severe abdominal pain and exacerbation of his disease process. The patient is found to have a high output enterocutaneous fistula and will require bowel rest and PN. A PICC for TPN administration is inserted. Maximal barrier precautions are used during the PICC insertion, as well as chlorhexidine skin preparation and the use of a chlorhexidine disk at the insertion site. What evidence-based knowledge would you use to explain the rationale for PN to the patient? What is the strength of that evidence?

3 pq A male patient who is severely malnourished has recently been diagnosed with a complete small bowel obstruction related to a duodenal mass. He has lost 12% of his body weight over the past month. He is admitted to the hospital to begin preoperative PN and is scheduled for surgery in 7 to 10 days. What are your priority assessments of this patient as you admit him to your unit to begin his prescribed course of PN?

4 A 68-year-old woman who is obese had a G tube placed 2 days ago. She is now complaining of itching around the insertion site. You see a cluster of red pustules and think she may have a yeast infection. What are your next steps? Why is this patient at risk for a yeast infection? Could nursing interventions have prevented it from occurring?

Brunner Suite Resources

Explore these additional resources to enhance learning for this chapter:

- **NCLEX-Style Questions and Other Resources on thePoint, http://thePoint.lww.com/Brunner13e**
- Study Guide
- PrepU
- Clinical Handbook
- Handbook of Laboratory and Diagnostic Tests

References

*Asterisk indicates nursing research.
**Double asterisk indicates classic reference.

Books

Alexander, M., Corrigan, A., Gorski, L., et al. (2010). *Infusion nursing: An evidence-based approach* (3rd ed.). St. Louis, MO: Saunders Elsevier.

Barber, J., & Sacks, G. (2012). Chapter 15: Parenteral nutrition formulations. In C. M. Mueller (Ed.). *The A.S.P.E.N. nutrition support core curriculum: A case-based approach—the adult patient* (pp. 245–264). Silver Springs, MD: American Society for Parenteral and Enteral Nutrition.

Krzywda, E., Andris, D., & Edmiston, C. (2012). Chapter 16: Parenteral access devices. In C. M. Mueller (Ed.). *The A.S.P.E.N. nutrition support core curriculum: A case-based approach—the adult patient* (pp. 165–283). Silver Springs, MD: American Society for Parenteral and Enteral Nutrition.

Krzywda, E., & Meyer, D. (2010). Chapter 17: Parenteral nutrition. In M. Alexander, A. Corrigan, L. Gorski, et al. (Eds.). *Infusion nursing: An evidence-based approach* (3rd ed., pp. 316–350). St. Louis, MO: Saunders Elsevier.

Mirtallo, J., & Patel, M. (2012). Chapter 14: Overview of parenteral nutrition. In C. M. Mueller (Ed.). *The A.S.P.E.N. nutrition support core curriculum: A case-based approach—the adult patient* (pp. 234–244). Silver Springs, MD: American Society for Parenteral and Enteral Nutrition.

Journals and Electronic Documents

Gastrostomies, Nasogastric, and Nasoenteric Intubation and Feeding

Arriola, T. A. D., Hatashima, A., & Klang, M. G. (2010). Evaluation of extended-release pancreatic enzyme to dissolve a clog. *Nutrition in Clinical Practice, 25*(5), 563–564.

Bankhead, R., Boullata, J., Brantley, S., et al. (2009). A.S.P.E.N. enteral nutrition practice recommendations. *Journal of Parenteral and Enteral Nutrition, 33*(2), 122–167. Available at: pen.sagepub.com/content/early/2009/01/27/0148607108330314

**Gabriel, S. A., & Ackermann, R. J. (2004). Placement of nasoenteral feeding tubes using external magnetic guidance. *Journal of Parenteral and Enteral Nutrition, 28*(2), 119–122.

*Kenny, D., & Goodman, P. (2010). Care of the patient with enteral tube feeding: An evidence-based practice protocol. *Nursing Research, 59*(1 Suppl.), S22–S31.

Koopman, M. C., Kudsk, K. A., Szokowski, M. J., et al. (2011). A team-based protocol and electromagnetic technology eliminate feeding tube placement complications. *Annals of Surgery, 253*(2), 287–302.

Lord, L. M. (2012). Minimizing tracheobronchial aspiration in the tube-fed patient, part 2. *Nurse Practitioner, 37*(1), 8–10.

**Lord, L. M., Weiser-Maimone, A., Pulhamus, M., et al. (1993). Comparison of weighted vs unweighted enteral feeding tubes for efficacy of transpyloric intubation. *Journal of Parenteral and Enteral Nutrition, 17*(3), 271–273.

**Marcuard, S. P., & Stegall, K. S. (1990). Unclogging feeding tubes with pancreatic enzyme. *Journal of Parenteral and Enteral Nutrition, 14*, 198–200.

**Marcuard, S. P., Stegall, K. S., & Trogdon, S. (1989). Clearing obstructed feeding tubes. *Journal of Parenteral and Enteral Nutrition, 13*(1), 81–83.

* **Metheny, N., Dettenmeier, P., Hampton, K., et al. (1990). Determinant of inadvertent respiratory placement of small-bore feeding tubes. A report of 10 cases. *Heart and Lung, 19*(6), 631–638.

* **Metheny, N. Eisenberg, P., & McSweeney, M. (1988). Effect of feeding tube properties and three irrigants on clogging rates. *Nursing Research, 37*, 165–169.

* **Metheny, N., Reed, L., Berglund, B., et al. (1994). Visual characteristics of aspirates from feeding tubes as a method for predicting tube location. *Nursing Research, 43*(5), 282–287.

*Metheny, N., Stewart, J., Nuetzel, G., et al. (2005). Effect of feeding-tube properties on residual volume measurements in tube-fed patients. *Journal of Parenteral and Enteral Nutrition, 29*(3), 192–197.

Munera-Seeley, V., Ochoa, J. B., Brown, N., et al. (2008). Use of a colorimetric carbon dioxide sensor for nasoenteric feeding tube placement in critical care patients compared with clinical methods and radiography. *Nutrition in Clinical Practice, 23*(3), 318–321.

**Nicalou, D. P., & Davis, S. K. (1990). Carbonated beverages as irrigants for feeding tubes. *Annals of Pharmacotherapy, 24*, 840.

Powers, J., Luebbehusen, M., Spitzer, T., et al. (2011). Verification of an electromagnetic placement device compared with abdominal radiograph to predict accuracy of feeding tube placement. *Journal of Parenteral and Enteral Nutrition, 35*(4), 535–539.

**Rees, R. G., Payne-James, J. J., King, C., et al. (1988). Spontaneous transpyloric passage and performance of 'fine bore' polyurethane feeding tubes: A controlled clinical trial. *Journal of Parenteral and Enteral Nutrition, 12*(5), 469–472.

Simons, S. R., & Abdallah, L. M. (2012). Bedside assessment of enteral tube placement: Aligning practice with evidence. *American Journal of Nursing, 112*(2), 40–46.

Stepter, C. R. (2012). Maintaining placement of temporary enteral feeding tubes in adults. A critical appraisal of the evidence. *Medsurg Nursing, 21*(2), 61–68, 102.

**Wilson, M. F., & Hayes-Johnson, V. (1987). Cranberry juice or water? A comparison of feeding-tube irrigants. *Nutrition Support Service, 7*(7), 23–24.

Parenteral Nutrition

Centers for Disease Control and Prevention. (2011). *2011 guidelines for the prevention of intravascular catheter-related infections*. Available at: www.cdc.gov/hicpac/BSI/BSI-guidelines-2011.html

Infusion Nurses Society. (2011). Infusion nursing standards of practice. *Journal of Intravenous Nursing, 3*(Suppl. 1), 1S.

Institute for Healthcare Improvement. (2012). *How-to guide: Prevent central line-associated bloodstream infection*. Available at: www.ihi.org/knowledge/Pages/Tools/HowtoGuidePreventCentralLineAssociatedBloodstreamInfection.aspx

Smith, N., & Walsh, K. (2012). Parenteral nutrition, total. *CINAHL nursing guide*. Available at: web.ebscohost.com/nrc/detail?vid=6&hid=19&sid=f1023639-c814–4989–98b8-e06c2356580e%40sessionmgr15&bdata=JnNpdGU9bnJjLWxpdmU%3d#db=nrc&AN

Timsit, J. F., Schwebel, C., Bouadma, L., et al. (2009). Chlorhexidine-impregnated sponges and less frequent dressing changes for prevention of catheter-related infections in critically ill adults: A randomized controlled trial. *Journal of the American Medical Association, 301*, 1231–1241.

Resources

American Cancer Society, www.cancer.org
American Society for Nutrition (ASN), www.nutrition.org
American Society for Gastrointestinal Endoscopy (ASGE), www.asge.org
American Society for Parenteral and Enteral Nutrition (A.S.P.E.N.), www.nutritioncare.org
Infusion Nurses Society (INS), www.ins1.org
Oley Foundation, www.oley.org
Society of Gastroenterology Nurses and Associates (SGNA), www.sgna.org

Chapter 46

Management of Patients With Oral and Esophageal Disorders

Learning Objectives

On completion of this chapter, the learner will be able to:

1. Describe the nursing management of patients with conditions of the oral cavity.
2. Describe the relationship of dental hygiene and dental problems to nutrition and to disease.
3. Describe the nursing management of patients with abnormalities of the oral cavity, jaw, and salivary glands.
4. Describe the nursing management of patients with cancer of the oral cavity.
5. Use the nursing process as a framework for care of patients undergoing neck dissection.
6. Use the nursing process as a framework for care of patients with various conditions of the esophagus.
7. Describe the various conditions of the esophagus and their clinical manifestations and management.

Glossary

achalasia: absent or ineffective peristalsis (wavelike contraction) of the distal esophagus accompanied by failure of the esophageal sphincter to relax in response to swallowing

apicoectomy: an excision of the apex of the tooth root; performed when an infection develops despite having a root canal

Boerhaave syndrome: spontaneous esophageal rupture due to forceful vomiting

dysphagia: difficulty swallowing

dysplasia: abnormal change in cells

esophagogastroduodenoscopy (EGD): passage of a fiberoptic tube through the mouth and throat into the digestive tract for visualization of the esophagus, stomach, and small intestine; biopsies can be performed

Frey syndrome: a rare syndrome characterized by undesirable sweating and flushing occurring on the cheek, temporal region, and behind the ears after eating certain foods

gastroesophageal reflux (GERD): backflow of gastric or duodenal contents into the esophagus

halitosis: foul odor form the oral cavity; in laymen's terms, "bad breath"

hernia: protrusion of an organ or part of an organ through the wall of the cavity that normally contains it

lithotripsy: the use of shock waves to break up or disintegrate stones

odynophagia: pain on swallowing

parotitis: inflammation of the parotid gland

periapical abscess: abscessed tooth

pyrosis: heartburn

sialadenitis: inflammation of the salivary glands

stomatitis: inflammation of the oral mucosa

temporomandibular disorders: a group of conditions that cause pain or dysfunction of the temporomandibular joint and surrounding structures

vagotomy syndrome: dumping syndrome; gastrointestinal symptoms, such as diarrhea and abdominal cramping, resulting from rapid gastric emptying

xerostomia: dry mouth

Because digestion normally begins in the mouth, adequate nutrition is related to good dental health and the general condition of the mouth. Any discomfort or adverse condition in the oral cavity can affect a person's nutritional status. Changes in the oral cavity can influence the type and amount of food ingested as well as the degree to which food particles are properly mixed with salivary enzymes. Disease of the mouth or tongue can interfere with speech and thus affect communication and self-image. Esophageal problems related to swallowing can also adversely affect food and fluid intake, thereby jeopardizing general health and well-being. Given the close relationship between adequate nutritional intake and the structures of the upper gastrointestinal (GI) tract (lips, mouth, teeth, pharynx, esophagus), health education can help prevent disorders associated with these structures (Fig. 46-1).

DISORDERS OF THE ORAL CAVITY

Oral health is a very important component of a person's physical and psychological sense of well-being. Periodontal disease is the most common cause of tooth loss among adults (U.S. Department of Health and Human Services [HHS], 2010). Approximately 8.52% of adults between the ages of 20 and 64 years have periodontal disease, with 5.08% of adults having moderate to severe periodontal disease. Those individuals at risk for periodontal disease are older, black or Hispanic, current smokers, and have lower incomes and less education (HHS, 2010). Periodontal disease can be connected to variety of other systemic disease, such as cardiovascular disease, diabetes, and rheumatoid disease (American Academy

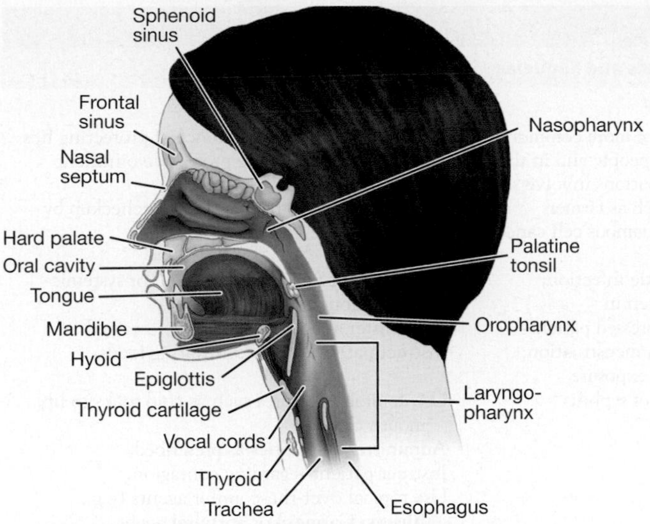
Sphenoid sinus
Frontal sinus
Nasal septum
Hard palate
Oral cavity
Tongue
Mandible
Hyoid
Epiglottis
Thyroid cartilage
Vocal cords
Thyroid
Trachea
Esophagus
Nasopharynx
Palatine tonsil
Oropharynx
Laryngo-pharynx

FIGURE 46-1 • Anatomy of the head and neck.

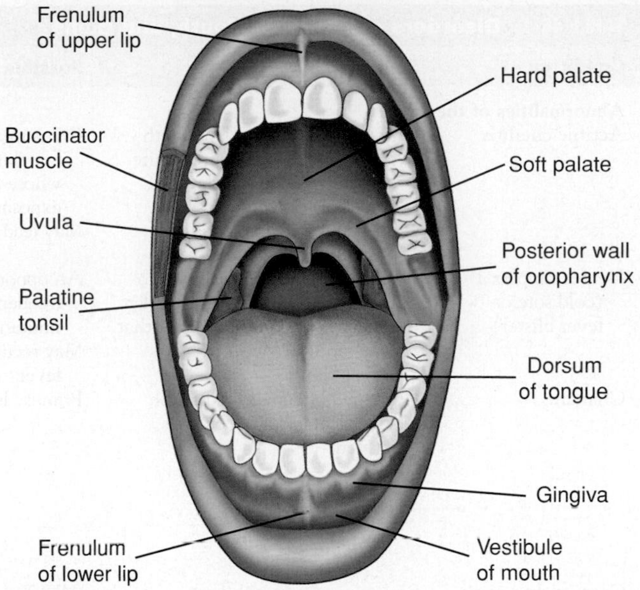
Frenulum of upper lip
Hard palate
Buccinator muscle
Soft palate
Uvula
Posterior wall of oropharynx
Palatine tonsil
Dorsum of tongue
Gingiva
Frenulum of lower lip
Vestibule of mouth

FIGURE 46-2 • Structures of the mouth, including the tongue and palate.

of Periodontology [AAP], 2008). Table 46-1 reviews common abnormalities of the oral cavity, their possible causes, and nursing considerations. Figure 46-2 illustrates structures of the oral cavity.

Dental Plaque and Caries

Tooth decay is an erosive process that begins with the action of bacteria on fermentable carbohydrates in the mouth, which produces acids that dissolve tooth enamel. Tooth enamel is the hardest substance in the human body, but dental erosion occurs for several reasons. Contributing factors include nutrition, soft drink consumption, and genetic predisposition. In addition, the extent of damage to the teeth may be related to the following:

- Presence of dental plaque, which is a gluey, gelatinlike substance that adheres to the teeth. The initial action that causes damage to a tooth occurs under dental plaque.
- Length of time acids are in contact with the teeth
- Strength of acids and the ability of the saliva to neutralize them
- Susceptibility of the teeth to decay

Dental decay begins with a small hole, usually in a fissure (a break in the tooth's enamel) or in an area that is hard to clean. Left unchecked, the decay extends into the dentin. Because dentin is not as hard as enamel, decay progresses more rapidly and in time reaches the pulp of the tooth.

Dentists can determine the extent of damage and the type of treatment needed using x-ray studies. Treatment for dental caries includes fillings, dental implants, or extraction, if necessary. In general, dental decay can occur in anyone. Contributing factors can be as simple as not brushing teeth on a regular basis to an improper diet. Older adults are subject to decay from drug-induced or age-related oral dryness (Chart 46-1).

Prevention

Measures used to prevent and control dental caries include practicing effective mouth care, reducing the intake of starches and sugars (refined carbohydrates), applying fluoride to the teeth or drinking fluoridated water, refraining from smoking, controlling diabetes, and using pit and fissure sealants. Regular dental visits are another method of preventive dental maintenance (HHS, 2012).

Mouth Care

Healthy teeth must be effectively cleaned on a daily basis. Brushing and flossing are particularly effective in mechanically breaking up the bacterial plaque that collects around teeth.

Normal mastication (chewing) and the normal flow of saliva also aid greatly in keeping the teeth clean. Because many ill patients do not consume adequate nutrients, they produce less saliva, which in turn reduces this natural tooth-cleaning process. The nurse may need to assume the responsibility for brushing the patient's teeth. Merely wiping the patient's mouth and teeth with a swab is ineffective. The

Chart 46-1 **Oral Conditions in the Older Adult**

Many medications taken by older adults cause dry mouth, which is uncomfortable, impairs communication, and increases the risk of oral infection. These medications include the following:

- Diuretic agents
- Antihypertensive medications
- Anti-inflammatory agents
- Antidepressant medications

Poor dentition can exacerbate problems of aging, such as:

- Decreased food intake
- Loss of appetite
- Social isolation
- Increased susceptibility to systemic infection (from periodontal disease)
- Trauma to the oral cavity secondary to thinner, less vascular oral mucous membranes

TABLE 46-1 Disorders of the Lips, Mouth, and Gums

Condition	Signs and Symptoms	Possible Causes and Sequelae	Nursing Considerations
Abnormalities of the Lips			
Actinic cheilitis	Irritation of lips associated with scaling, crusting, fissure; white overgrowth of horny layer of epidermis (hyperkeratosis) Considered a premalignant squamous cell skin cancer	Exposure to sun; more common in fair-skinned people and in those whose occupations involve sun exposure, such as farmers May lead to squamous cell cancer	Educate patient on importance of protecting lips from the sun by using protective ointment such as sun block. Instruct patient to have a periodic checkup by physician.
Herpes simplex 1 (cold sore or fever blister)	Symptoms may be delayed up to 20 days after exposure; singular or clustered painful vesicles that may rupture	An opportunistic infection; frequently seen in immunosuppressed patients May recur with menstruation, fever, or sun exposure	Use acyclovir (Zovirax) ointment or systemic medications as prescribed. Administer analgesic agents as prescribed. Instruct patient to avoid irritating foods.
Chancre	Reddened circumscribed lesion that ulcerates and becomes crusted	Primary lesion of syphilis	Use comfort measures such as cold soaks to lip, mouth care. Administer antibiotics as prescribed. Instruct patient regarding contagion. Use topical over-the-counter agents (e.g., Blistex, Carmex) or antiviral agents (e.g., acyclovir, penciclovir [Denavir]) as prescribed.
Contact dermatitis	Red area or rash; itching	Allergic reaction to lipstick, cosmetic ointments, or toothpaste	Instruct patient to avoid possible causes. Administer corticosteroids as prescribed.
Abnormalities of the Mouth			
Leukoplakia	White patches; may be hyperkeratotic; usually in buccal mucosa; usually painless	Fewer than 2% are malignant, but may progress to cancer Common among tobacco users	Instruct patient to see a physician if leukoplakia persists >2 weeks. Eliminate risk factors, such as cigarettes, smokeless tobacco.
Hairy leukoplakia	White patches with rough hairlike projections; typically found on lateral border of the tongue	Possibly viral; smoking and the use of tobacco Often seen in people who are HIV positive	Instruct patient to see the primary provider if condition persists >2 weeks.
Lichen planus	White papules at the intersection of a network of interlacing lesions; usually ulcerated and painful	Recurrences are common. May lead to a malignant process Unknown cause	Apply topical corticosteroids such as fluocinolone acetonide gel (Orabase). Avoid foods that irritate. Administer corticosteroids systemically or intralesionally as prescribed. Instruct the patient of need for follow-up if condition is chronic.
Candidiasis (moniliasis/ thrush)	Cheesy white plaque that looks like milk curds; when rubbed off, it leaves an erythematous and often bleeding base.	*Candida albicans* fungus; predisposing factors include diabetes, antibiotic therapy, and immunosuppression.	Antifungal medications such as nystatin (Mycostatin), amphotericin B, clotrimazole, or ketoconazole may be prescribed; these may be taken in pill form or as a suspension; when used as a suspension, instruct the patient to swish vigorously for at least 1 minute and then swallow.
Aphthous stomatitis (canker sore)	Shallow ulcer with a white or yellow center and red border; seen on the inner side of the lip and cheek or on the tongue; it begins with a burning or tingling sensation and slight swelling; painful; usually lasts 7–10 days and heals without a scar.	Associated with emotional or mental stress, fatigue, hormonal factors, minor trauma (e.g., biting), allergies, acidic foods and juices, and dietary deficiencies Associated with HIV infection May recur	Instruct the patient in comfort measures (e.g., saline rinses) and a soft or bland diet. Antibiotics or corticosteroids may be prescribed. Use over-the-counter benzocaine (Zilactin) as indicated.
Nicotine stomatitis (smoker's patch)	Two stages—begins as a red stomatitis; over time, the tongue and mouth become covered with a creamy, thick, white mucous membrane, which may slough, leaving a beefy red base.	Chronic irritation by tobacco	Cessation of tobacco use; if condition exists >2 weeks, a primary provider should be consulted and a biopsy may be needed.
Erythroplakia	Red patch on the oral mucous membrane	Nonspecific inflammation; more frequently seen in older adults	
Kaposi's sarcoma	Appears first on the oral mucosa as a red, purple, or blue lesion; may be singular or multiple; may be flat or raised	HIV infection	Instruct patient regarding side effects of planned treatment.

Condition	Signs and Symptoms	Possible Causes and Sequelae	Nursing Considerations
Stomatitis	Mild redness (erythema) and edema; severe forms include painful ulcerations, bleeding, and secondary infection.	Chemotherapy; radiation therapy; severe drug allergy; myelosuppression (bone marrow depression)	Prophylactic mouth care, including brushing, flossing, and rinsing, for any patient receiving chemotherapy or radiation therapy Educate patient about proper oral hygiene, including the use of a soft-bristled toothbrush and nonabrasive toothpaste; for painful ulcers, oral swabs with spongelike applicators can be used in place of a toothbrush; avoid alcohol-based mouth rinses and hot or spicy foods. Apply topical anti-inflammatory, antibiotic, and anesthetic agents as prescribed.
Abnormalities of the Gums			
Gingivitis	Painful, inflamed, swollen gums; usually, the gums bleed in response to light contact	Poor oral hygiene: food debris, bacterial plaque, and calculus (tartar) accumulate; the gums may also swell in response to normal processes such as puberty and pregnancy.	Educate patient about proper oral hygiene; toothbrushing, flossing, rinsing, dental appointments every 3–6 months.
Necrotizing gingivitis (trench mouth)	Gray-white pseudomembranous ulcerations affecting the edges of the gums, mucosa of the mouth, tonsils, and pharynx; halitosis; painful, bleeding gums; swallowing and talking are painful.	Poor oral hygiene; bacterial infection, inadequate rest, overwork, emotional stress, smoking, and poor nutrition may contribute to development.	Educate patient about proper oral hygiene; see Chart 46-2. Irrigate with 2%–3% hydrogen peroxide or normal saline solution. Avoid irritants such as smoking and spicy foods.
Herpetic gingivostomatitis	Burning sensation with the appearance of small vesicles 24–48 hours later; vesicles may rupture, forming sore, shallow ulcers covered with a gray membrane.	Herpes simplex virus; occurs most frequently in people who are immunosuppressed; may occur in other infectious processes such as streptococcal pneumonia, meningococcal meningitis, and malaria	Apply topical anesthetics as prescribed; may need opioids if pain is severe. Saline or 2%–3% hydrogen peroxide irrigations Antiviral agents such as acyclovir may be prescribed.
Periodontitis	Little discomfort at onset; may have bleeding, infection, gum recession, and loosening of teeth; later in the disease, tooth loss may occur.	May result from untreated gingivitis Poor or inadequate dental hygiene and inadequate diet contribute to development.	Instruct patient in proper oral hygiene. Instruct patient to consult a dentist or periodontist.

HIV, human immunodeficiency virus.

most effective method is mechanical cleansing (brushing). If brushing is impossible, it is better to wipe the teeth with a gauze pad and then have the patient swish an antiseptic mouthwash several times before expectorating into an emesis basin. A soft-bristled toothbrush is more effective than a sponge or foam stick. To prevent drying, the lips may be coated with a water-soluble gel.

Diet

Dental caries may be prevented by decreasing the amount of sugar and starch in the diet. Patients who snack should be encouraged to choose less cariogenic alternatives, such as fruits, vegetables, nuts, cheeses, or plain yogurt. In addition, frequent brushing, especially after meals, is necessary. Flossing should be performed daily.

Fluoridation

Fluoridation of public water supplies has been found to decrease dental caries. Some areas of the country have natural fluoridation; other communities have added fluoride to public water supplies. It is a cost-effective method; for every dollar spent on water fluoridation, there is $38 in savings each year from fewer cavities being treated (Centers for Disease Control and Prevention [CDC], 2010a). Yet only 27 states met the national health goal that 75% of their citizens have access to fluoridated public water systems. Studies suggest that by instituting a community water fluoridation program, tooth decay is reduced by as much as 18% to 40% (CDC, 2010a).

Fluoridation may also be achieved by having a dentist apply a concentrated gel or solution to the teeth; adding fluoride to home water supplies; using fluoridated toothpaste or mouth rinse; or using sodium fluoride tablets, drops, or lozenges.

Pit and Fissure Sealants

The occlusal surfaces of the teeth have pits and fissures—areas that are prone to caries. Some dentists apply a special coating to fill and seal these areas from potential exposure to cariogenic processes. These sealants can last 5 to 10 years, depending on how dry the tooth surface is prior to application. Dental sealants prevent 60% of tooth decay in treated teeth (CDC, 2010a).

Dental Health and Disease

Studies are ongoing that show the link between oral health and chronic disease such as diabetes, heart disease, low birth weight, premature births, and stroke. As early as 1996, it was posited that bacteria, specifically gram-negative bacteria, were the culprit that link periodontal disease to other diseases in the body, specifically heart diseases (Williams, 2008). More recently, it was confirmed that these bacteria cause an inflammatory response that initiates an increase in inflammatory markers such as C-reactive protein, white blood cells, and fibrinogen (Holmlund, Holm, & Lind, 2010). These markers are associated with an increased risk of cardiovascular disease. Data further suggest that if periodontal disease is treated, other chronic inflammatory conditions may be mitigated (CDC, 2010b).

The World Health Organization (WHO) Global Oral Health Programme (2012) espouses a global focus on oral health promotion and disease prevention. Methods that include developing water fluoridation programs, identifying those individuals at risk for poor oral hygiene, providing technical support for those nations that lack programs that target promotion of oral health, and building strong oral health programs will improve oral health as well as control chronic disease.

Dentoalveolar Abscess or Periapical Abscess

Periapical abscess, more commonly referred to as an abscessed tooth, involves a collection of pus in the apical dental periosteum (fibrous membrane supporting the tooth structure) and the tissue surrounding the apex of the tooth (where it is suspended in the jaw bone). The abscess may be acute or chronic. Acute periapical abscess is usually secondary to a suppurative pulpitis (a pus-producing inflammation of the dental pulp) that arises from an infection extending from dental caries. The infection of the dental pulp extends through the apical foramen of the tooth to form an abscess around the apex.

Chronic dentoalveolar abscess is a slowly progressive infectious process. In contrast to the acute form, a fully formed abscess may occur without the patient's knowledge. The infection eventually leads to a "blind dental abscess," which is actually a periapical granuloma. It may enlarge to as much as 1 cm in diameter. It is often discovered on x-ray films and is treated by extraction or root canal therapy, often with **apicoectomy** (excision of the apex of the tooth root).

Clinical Manifestations

The abscess produces a dull, gnawing, continuous pain, often with a surrounding cellulitis and edema of the adjacent facial structures, and mobility of the involved tooth. The gum opposite the apex of the tooth is usually swollen on the cheek side. Swelling and cellulitis of the facial structures may make it difficult for the patient to open the mouth. There may also be a systemic reaction, fever, and malaise.

Management

In the early stages of an infection, a dentist or oral surgeon may perform a needle aspiration or drill an opening into the pulp chamber to relieve pressure and pain and to provide drainage. Usually, the infection will have progressed to a periapical abscess. Drainage is provided by an incision through the gingiva down to the jawbone. Purulent material escapes under pressure. This procedure may be performed in the dentist's office, an outpatient surgery center, or a same-day surgery department. After the inflammatory reaction has subsided, the tooth may be extracted or root canal therapy performed. Antibiotics and opioids may be prescribed (Robertson & Smith, 2012).

The patient is assessed for bleeding after treatment and is instructed to use a warm saline or warm water mouth rinse to keep the area clean. The patient is also instructed to take antibiotic and analgesic agents as prescribed, to advance from a liquid diet to a soft diet as tolerated, and to keep follow-up appointments.

DISORDERS OF THE JAW

Abnormal conditions affecting the mandible (jaw) and the temporomandibular joint (which connects the mandible to the temporal bone at the side of the head in front of the ear) include congenital malformation, fracture, chronic dislocation, cancer, and syndromes characterized by pain and limited motion. Temporomandibular disorders and jaw surgery (a treatment common in many structural abnormalities or cancer of the jaw) are presented in this section.

Temporomandibular Disorders

Temporomandibular disorders are categorized as follows (National Institute of Dental and Craniofacial Research [NIDCR], 2011):

- Myofascial pain—a discomfort in the muscles controlling jaw function and in neck and shoulder muscles
- Internal derangement of the joint—a dislocated jaw, a displaced disk, or an injured condyle
- Degenerative joint disease—rheumatoid arthritis or osteoarthritis in the jaw joint

Diagnosis and treatment of temporomandibular disorders remain somewhat ambiguous, but the condition is thought to affect about 10 million people in the United States (NIDCR, 2011). Misalignment of the joints in the jaw and other problems associated with the ligaments and muscles of mastication are thought to result in tissue damage and muscle tenderness. Suggested causes include arthritis of the jaw, head injury, trauma or injury to the jaw or joint, stress, and malocclusion (although research does not support malocclusion as a cause).

Clinical Manifestations

Patients have jaw pain ranging from a dull ache to throbbing, debilitating pain that can radiate to the ears, teeth, neck muscles, and facial sinuses. They often have restricted jaw motion and locking of the jaw. There also may be a sudden change in the way the upper and lower teeth fit together. The patient may hear clicking, popping, and grating sounds when the mouth is opened, and chewing and swallowing may be difficult. Symptoms such as headaches, earaches, dizziness, and hearing problems may sometimes be related to temporomandibular disorders (Buescher, 2007; NIDCR, 2011).

Assessment and Diagnostic Findings

Diagnosis is based on the patient's report of pain, limitations in range of motion, dysphagia, difficulty chewing, difficulty with speech, or hearing difficulties. Magnetic resonance imaging and x-ray studies are generally only used for severe or chronic symptoms.

Medical Management

Signs and symptoms improve over time for the majority of patients with temporomandibular joint disorders, with or without treatment. Some practitioners think the role of stress in these disorders is overrated, but patient education in stress management may be helpful (to reduce grinding and clenching of teeth) (Sidebottom, 2009). Patients may also benefit from range-of-motion exercises. Pain management measures may include nonsteroidal anti-inflammatory drugs, with the possible addition of opioids, muscle relaxants, or mild antidepressants. Occasionally, intraoral orthotics (a plastic guard worn over the upper and lower teeth) may be worn to reposition the condyle head in the joint space to a more normal position, which in turn relieves the stress and pressure on the tissues of the joint (Ingawale & Goswami, 2009). This allows the tissues to heal. Conservative and reversible treatment is recommended (Scrivani, Keith, & Kaban, 2008). If irreversible surgical options are recommended, patients are encouraged to seek a second opinion.

Jaw Disorders Requiring Surgical Management

Correction of mandibular structural abnormalities may require surgery involving repositioning or reconstruction of the jaw. Simple fractures of the mandible without displacement, resulting from a blow to the chin, and planned surgical interventions, as in the correction of long or short jaw syndrome, may require treatment by these means. Jaw reconstruction may be necessary in the aftermath of trauma from a severe injury or cancer, both of which can cause tissue and bone loss.

Mandibular fractures are usually closed fractures. Rigid plate fixation (insertion of metal plates and screws into the bone to approximate and stabilize the bone) is the current treatment of choice in many cases of mandibular fracture and in some mandibular reconstructive surgery procedures (Ingawale & Goswami, 2009). Bone grafting may be performed to replace structural defects using bones from the patient's own ilium, ribs, or cranial sites. Rib tissue may also be harvested from cadaver donors (Scrivani et al., 2008).

Nursing Management

The patient who has had rigid fixation should be instructed not to chew food in the first 1 to 4 weeks after surgery. A liquid diet is recommended, and dietary counseling should be obtained to ensure optimal caloric and protein intake.

Promoting Home and Community-Based Care

The patient needs specific guidelines for mouth care and feeding. Any irritated areas in the mouth should be reported to the primary provider. The importance of keeping scheduled appointments to assess the stability of the fixation appliance is emphasized.

Consultation with a dietitian may be indicated so that the patient and family can learn about foods that are high in essential nutrients and ways in which these foods can be prepared so that they can be consumed through a straw or spoon while remaining palatable. Nutritional supplements may be recommended.

DISORDERS OF THE SALIVARY GLANDS

The salivary glands consist of the parotid glands, one on each side of the face below the ear; the submandibular and sublingual glands, both in the floor of the mouth; and the buccal gland, beneath the lips. About 1,200 mL of saliva is produced daily and swallowed. The glands' major functions are lubrication, protection against harmful bacteria, and digestion.

Parotitis

Parotitis (inflammation of the parotid gland) is the most common inflammatory condition of the salivary glands, although inflammation can occur in the other salivary glands as well. Mumps (epidemic parotitis), a communicable disease caused by viral infection and most commonly affecting children, is an inflammation of a salivary gland, usually the parotid.

People who are older, acutely ill, or debilitated with decreased salivary flow from general dehydration or medications are at high risk for parotitis. The infecting organisms travel from the mouth through the salivary duct. The organism is usually *Staphylococcus aureus* (except in mumps). The onset of this complication is sudden, with an exacerbation of both fever and the symptoms of the primary condition. The gland swells and becomes tense and tender. The patient feels pain in the ear, and swollen glands interfere with swallowing. The swelling increases rapidly, and the overlying skin soon becomes red and shiny.

Medical management includes maintaining adequate nutritional and fluid intake, good oral hygiene, and discontinuing medications (e.g., tranquilizers, diuretic agents) that can diminish salivation. Antibiotic therapy is necessary, and analgesics may be prescribed to control pain. If antibiotic therapy is not effective, the gland may need to be drained by a surgical procedure known as parotidectomy. This procedure may be necessary to treat chronic parotitis. The patient is advised to have any necessary dental work performed prior to surgery.

Sialadenitis

Sialadenitis (inflammation of the salivary glands) may be caused by dehydration, radiation therapy, stress, malnutrition, salivary gland calculi (stones), or improper oral hygiene. The inflammation is associated with infection by *S. aureus*, *Streptococcus viridans*, or *pneumococci*. In hospitalized or institutionalized patients, the infecting organism may be methicillin-resistant *Streptococcus aureus* (MRSA). Symptoms include pain, swelling, and purulent discharge. Antibiotics are used to treat infections. Massage, hydration, warm compresses, and corticosteroids frequently cure the problem. Chronic sialadenitis with uncontrolled pain is treated by surgical drainage of the gland or excision of the gland and its duct.

Salivary Calculus (Sialolithiasis)

Sialolithiasis, or salivary calculi (stones), usually occur in the submandibular gland. Salivary gland ultrasonography or sialography (x-ray studies filmed after the injection of a radiopaque substance into the duct) may be required to demonstrate obstruction of the duct by stenosis. Salivary calculi are formed mainly from calcium phosphate. If located within the gland, the calculi are irregular and vary in diameter from 3 to 30 mm. Calculi in the duct are small and oval.

Calculi within the salivary gland itself cause no symptoms unless infection arises; however, a calculus that obstructs the gland's duct causes sudden, local, and often colicky pain, which is abruptly relieved by a gush of saliva. This characteristic symptom is often disclosed in the patient's health history. On physical assessment, the gland is swollen and quite tender, the stone itself can be palpable, and its shadow may be seen on x-ray films.

The calculus can be extracted fairly easily from the duct in the mouth. Sometimes, enlargement of the ductal orifice permits the stone to pass spontaneously. Occasionally, **lithotripsy**, a procedure that uses shock waves to disintegrate the stone, may be used instead of surgical extraction for parotid stones and smaller submandibular stones. Lithotripsy requires no anesthesia, sedation, or analgesia. Side effects can include local hemorrhage and swelling. Surgery may be necessary to remove the gland if symptoms and calculi recur repeatedly.

Neoplasms

Although they are uncommon, neoplasms (tumors or growths) of almost any type may develop in the salivary gland; 50% of these are benign, and 70% to 80% arise in the parotid gland (Mendenhall, Werning, & Pfister, 2011). The incidence of salivary gland tumors is similar in men and women. Risk factors include prior exposure to radiation to the head and neck. Diagnosis is based on the health history and physical examination and the results of fine-needle aspiration biopsy.

Early-stage salivary gland tumors are usually curable with surgery alone. The prognosis is favorable if a major gland such as the parotid gland is involved (National Cancer Institute [NCI], 2012a). Dissection is carefully performed to preserve the seventh cranial nerve (facial nerve), although it may not be possible to do so if the tumor is extensive. Complications from surgery may involve facial nerve dysfunction and Frey syndrome. **Frey syndrome** is also known as auriculotemporal syndrome (De Bree, van der Waal, & Leemans, 2008). Symptoms include undesirable sweating and flushing occurring on the cheek, temporal region, and behind the ears after eating certain foods (De Bree et al., 2008). Successful treatments have occurred with botulinum toxin A injections.

If the tumor is malignant, radiation therapy may follow surgery. Radiation therapy alone may be a treatment choice for tumors thought to be localized or if there is risk of facial nerve damage from surgical intervention. Chemotherapy is usually used for palliative purposes. Recurrent tumors usually are more aggressive than initial tumors. Tumors of the salivary gland are associated with an increased incidence of second primary cancers (Gillison, 2007).

CANCER OF THE ORAL CAVITY AND PHARYNX

Cancers of the oral cavity and pharynx, which can occur in any part of the mouth or throat, are curable if discovered early. Risk factors for cancer of the oral cavity and pharynx include cigarette, cigar, and pipe smoking; the use of smokeless tobacco; excessive use of alcohol; and infection with human papillomavirus (HPV). Oral cancers are often associated with the combined use of alcohol and tobacco; these substances have a synergistic carcinogenic effect. Patient education directed toward avoiding high-risk behaviors is critical to prevent oral cancers.

The incidence of non-HPV–associated cancers of the oral cavity and pharynx is greatest in men older than 50 years. In general, it is almost twice as high in men as it is in women. Cancers of the oral cavity and pharynx occur more often in African Americans than in Caucasians (American Cancer Society [ACS], 2012).

HPV-associated cancers of the oropharynx behave differently than those that are not associated with HPV infection and tend to affect young males (Adelstein & Rodriguez, 2010). Survival rates for patients that have HPV-associated oropharyngeal cancers are usually quite good, but prognosis is worse if there is a history of tobacco abuse (Adelstein & Rodriguez, 2010). Approximately 35,000 new cases of oral cavity and oropharyngeal cancer occur annually in the United States. For the past 20 to 40 years, the number of new cases and death rate has been decreasing. Of patients with cancer of the oral cavity and oropharynx, 84% survive at least 1 year after diagnosis. Regardless of the stage of cancer at diagnosis, the 5-year relative survival rate is 59%, and the 10-year survival rate is 48% (ACS, 2012).

Pathophysiology

Malignancies of the oral cavity are usually squamous cell cancers, including those that are HPV associated (Adelstein & Rodriguez, 2010). Any area of the oropharynx can be a site of malignant growths, but the lips, the lateral aspects of the tongue, and the floor of the mouth are most commonly affected.

Clinical Manifestations

Many oral cancers produce few or no symptoms in the early stages. Later, the most frequent symptom is a painless sore or mass that does not heal. It may bleed easily and may present as a red or white patch that persists (ACS, 2012). A typical lesion in oral cancer is a painless indurated (hardened) ulcer with raised edges. As the cancer progresses, the patient may complain of tenderness; difficulty in chewing, swallowing, or speaking; coughing of blood-tinged sputum; or enlarged cervical lymph nodes.

Assessment and Diagnostic Findings

Diagnostic evaluation consists of an oral examination as well as an assessment of the cervical lymph nodes to detect possible metastases. Biopsies are performed on suspicious lesions (those that have not healed in 2 weeks). In people who use snuff or smoke cigars or pipes, high-risk areas include the buccal mucosa and gingiva. In those who smoke cigarettes and drink alcohol, high-risk areas include the floor of the mouth, the ventrolateral tongue, and the soft palate complex (soft palate, anterior and posterior tonsillar area, uvula, and the area behind the molar and tongue junction).

Human Papillomavirus Prevention

Currently, there are two prophylactic HPV vaccines available that are marketed to females between the ages of 9 and 26 years to prevent HPV-associated precancerous lesions of the uterine cervix. These vaccines protect against HPV-16, which is the common culprit not only in cervical cancer but also in oropharyngeal cancer (Adelstein & Rodriguez, 2010). It seems plausible that this vaccine could also prevent oropharyngeal cancer. Therefore, some advocate offering this vaccine universally not only to girls and young women but also to boys and young men (Adelstein & Rodriguez, 2010).

Medical Management

In patients diagnosed with oropharyngeal cancer, management varies with the nature of the lesion, the preference of the physician, and patient choice. Surgical resection and radiation therapy are standard treatment. Addition of chemotherapy may be useful for advanced disease (ACS, 2012).

In cancer of the lip, small lesions are usually excised liberally. Radiation therapy may be more appropriate for larger lesions involving more than one third of the lip because of superior cosmetic results. The choice depends on the extent of the lesion and what is necessary to cure the patient while preserving the best appearance. Tumors larger than 4 cm often recur.

In cancer of the tongue, treatment with radiation therapy and chemotherapy may preserve function and maintain quality of life. A combination of radioactive interstitial implants (surgical implantation of a radioactive source into the tissue adjacent to or at the tumor site) and external-beam radiation may be used. Surgical procedures include hemiglossectomy (surgical removal of half of the tongue) and total glossectomy (removal of the tongue). Glossectomy remains the principal treatment of advanced-stage or base of tongue cancers (Van Leirop, Basson, & Fagan, 2008). However, it remains controversial due to low cure rates and associated functional deficits that are associated with it (Van Leirop et al., 2008).

Often, cancer of the oral cavity has metastasized through the extensive lymphatic channel in the neck region, requiring a neck dissection and reconstructive surgery of the oral cavity. One common reconstructive technique involves the use of a radial forearm free flap (a thin layer of skin from the forearm along with the radial artery).

Nursing Management

The nurse assesses the patient's nutritional status preoperatively, and a dietary consultation may be necessary. The patient may require enteral (through the GI tract) or parenteral (intravenous [IV]) feedings before and after surgery to maintain adequate nutrition (see Chapter 45). Continual assessment and reevaluation are necessary. If a radial graft is to be performed, an Allen test on the donor arm must be performed to ensure that the ulnar artery is patent and can provide blood flow to the hand after removal of the radial artery. The Allen test is performed by asking the patient to make a fist and then manually compressing the ulnar artery. The patient is then asked to open the hand into a relaxed, slightly flexed position. The palm is pale. Pressure on the ulnar artery is released. If the ulnar artery is patent, the palm flushes within about 3 to 5 seconds.

Verbal communication may be impaired by radical surgery for oral cancer. It is therefore vital to assess the patient's ability to communicate in writing before surgery. Pen and paper are provided postoperatively to patients who can use them to communicate. A communication board with commonly used words or pictures is obtained preoperatively and given after surgery to patients who cannot write so that they may point to needed items. A speech therapist is also consulted postoperatively.

Postoperatively, the nurse assesses for a patent airway. The patient may be unable to manage oral secretions, making suctioning necessary. If grafting was part of the surgery, suctioning must be performed with care to prevent damage to the graft. The graft is assessed postoperatively for viability. Although color should be assessed (white may indicate arterial occlusion, and blue mottling may indicate venous congestion), it can be difficult to assess the graft by looking into the mouth. A Doppler ultrasound device may be used to locate the radial pulse at the graft site and to assess graft perfusion.

Nursing Management of the Patient With Conditions of the Oral Cavity

Promoting Mouth Care

The nurse instructs the patient in the importance and techniques of preventive mouth care. If a patient cannot tolerate brushing or flossing, an irrigating solution of 1 tsp of baking soda to 8 oz of warm water, half-strength hydrogen peroxide, or normal saline solution is recommended (Perry, 2008). The nurse reinforces the need to perform oral care and provides such care to patients who cannot provide it for themselves.

If a bacterial or fungal infection is present, the nurse administers the prescribed medications and instructs the patient how to administer the medications at home. The nurse monitors the patient's physical and psychological response to treatment.

Xerostomia (dryness of the mouth) is a frequent sequela of oral cancer, particularly when the salivary glands have been exposed to radiation or major surgery. It is also seen in patients who are receiving psychopharmacologic agents, patients with human immunodeficiency virus (HIV) infection, and patients who cannot close the mouth and as a result become mouth breathers. To minimize this problem, the patient is advised to avoid dry, bulky, and irritating foods and fluids, as well as alcohol and tobacco. The patient is also encouraged to increase intake of fluids (when not contraindicated) and to use a humidifier while sleeping. The use of synthetic saliva, a moisturizing antibacterial gel such as Biotene Oral Balance, or a saliva production stimulant such as pilocarpine (Salagen) may be helpful.

Stomatitis, a type of oral mucositis, which involves inflammation and breakdown of the oral mucosa, is often a side effect of chemotherapy or radiation therapy. Prophylactic mouth care is started when the patient begins receiving treatment; however, mucositis may become so severe that a break in treatment is necessary. If a patient receiving radiation therapy has poor dentition, extraction of the teeth before radiation treatment in the oral cavity is often initiated to prevent infection. Many radiation therapy centers recommend the use of fluoride treatments for patients receiving radiation to the head and neck. (See Chapter 15 for more information about stomatitis.)

Ensuring Adequate Food and Fluid Intake

The patient's weight, age, and level of activity are recorded to determine whether nutritional intake is adequate. A daily calorie count may be necessary to determine the exact quantity of food and fluid ingested. The frequency and pattern of eating are recorded to determine whether any psychosocial or physiologic factors are affecting ingestion. Based on the disorder and the patient's preferences, the nurse recommends changes in the consistency of foods and the frequency of eating. Consultation with a dietitian may be helpful. The goal is to help the patient attain and maintain desirable body weight and level of energy, as well as to promote the healing of tissue.

Supporting a Positive Self-Image

A patient who has a disfiguring oral condition or has undergone disfiguring surgery may experience an alteration in self-image. The patient is encouraged to verbalize the perceived change in body appearance and to realistically discuss changes or losses. The nurse offers support while the patient verbalizes fears and negative feelings (withdrawal, depressed mood, anger). The nurse listens attentively and determines the patient's needs and individualizes the plan of care. The patient's strengths and achievements are reinforced.

The nurse should determine the patient's concerns about relationships with others. Referral to support groups, a psychiatric liaison nurse, a social worker, or a spiritual advisor may be useful in helping the patient to cope with anxieties and fears. The patient's progress toward development of positive self-esteem is documented. The nurse should be alert to signs of grieving and should document emotional changes. By providing acceptance and support, the nurse encourages the patient to verbalize feelings.

Minimizing Pain and Discomfort

Oral lesions can be painful. Strategies to reduce pain and discomfort include avoiding foods that are spicy, hot, or hard (e.g., pretzels, nuts). A soft or liquid diet may be preferred. The patient is instructed about mouth care. The use of a soft toothbrush may prevent secondary trauma. It may be necessary to provide the patient with an analgesic agent such as viscous lidocaine (Xylocaine Viscous 2%) or opioids, as prescribed. Topical medications such as sucralfate (Carafate) and aluminum-magnesium liquid antacids may provide relief. The nurse can reduce the patient's fear of pain by providing information about pain control methods.

Preventing Infection

Leukopenia (a decrease in white blood cells) may result from radiation, chemotherapy, acquired immunodeficiency syndrome, and some medications used to treat HIV infection. Leukopenia reduces defense mechanisms, increasing the risk of infections. Malnutrition, which is also common among these patients, may further decrease resistance to infection. If the patient has diabetes, the risk of infection is further increased.

Laboratory results should be evaluated frequently and the patient's temperature checked every 4 to 8 hours for an elevation that may indicate infection. Visitors who might transmit microorganisms are prohibited if the patient's immunologic system is depressed. Sensitive skin tissues are protected from trauma to maintain skin integrity and prevent infection. Aseptic technique is necessary when changing dressings. Desquamation (shedding of the epidermis) is a reaction to radiation therapy that causes dryness and itching and can lead to a break in skin integrity and subsequent infection.

Signs of wound infection (redness, swelling, drainage, tenderness) are reported to the primary provider. Antibiotics may be prescribed prophylactically.

Promoting Home and Community-Based Care

Educating Patients About Self-Care

The patient who is recovering from treatment of an oral condition is instructed about mouth care, nutrition, prevention of infection, and signs and symptoms of complications (Chart 46-2). Methods of preparing nutritious foods that are seasoned according to the patient's preference and at the preferred temperature are explained to the patient and family. For some patients, it may be more convenient (but also more expensive) to use commercial baby foods than to prepare liquid and soft diets. The patient who cannot take foods orally may receive enteral or parenteral nutrition; the administration of these feedings is explained and demonstrated to the patient and the caregiver.

For patients with oral cancer, instructions are provided in the use and care of any dentures. The importance of keeping dressings clean and the need for conscientious oral hygiene are emphasized.

Continuing Care

The need for ongoing care in the home depends on the patient's condition. The patient, family members, and others health care team members responsible for home care (e.g., nurse, speech therapist, nutritionist, and psychologist) work together to prepare an individual plan of care.

Chart 46-2 HOME CARE CHECKLIST
The Patient With an Oral Condition

At the completion of home care education, the patient or caregiver will be able to:	PATIENT	CAREGIVER
• Demonstrate the use of suction equipment if indicated.	✔	✔
• State rationale for humidification.	✔	✔
• Identify foods necessary to meet caloric needs and dietary needs (i.e., change in consistency, seasoning limitations, supplements).	✔	✔
• Demonstrate effective oral hygiene.	✔	✔
• Demonstrate care of incision.	✔	✔
• State when next medical/dental follow-up appointment will be scheduled.	✔	✔

If suctioning of the mouth or tracheostomy tube is required, the necessary equipment is obtained and the patient and caregivers are taught how to use it. Considerations include the control of odors and humidification of the home to keep secretions moist. The patient and caregivers are educated to assess for obstruction, hemorrhage, and infection, as well as what actions to take if they occur. The home care nurse may provide physical care, monitor for changes in the patient's physical status (e.g., skin integrity, nutritional status, respiratory function), and assess the adequacy of pain control measures. The nurse also assesses the patient's and family's ability to manage incisions, drains, and feeding tubes and the use of recommended strategies for communication. The ability of the patient and family to accept physical, psychological, and role changes is assessed and addressed.

Follow-up visits to the primary provider are important to monitor the patient's condition and to determine the need for modifications in treatment and general care. Because patients and their family members, as well as health care providers, tend to focus on the most obvious needs and issues, the nurse reminds the patient and family about the importance of continuing health promotion and screening practices and refers them to appropriate practitioners. The nurse also reinforces instructions in an effort to promote the patient's self-care and comfort.

NECK DISSECTION

Malignancies of the head and neck include those of the oral cavity, oropharynx, hypopharynx, nasopharynx, nasal cavity, paranasal sinus, and larynx. (Laryngeal cancer is presented in Chapter 22.) These cancers account for fewer than 5% of all cancers (ACS, 2012). Depending on the location and stage, treatment may consist of radiation therapy, chemotherapy, surgery, or a combination of these modalities. Deaths from malignancies of the head and neck are primarily attributable to local-regional metastasis to the cervical lymph nodes in the neck. This often occurs by way of the lymphatics before the primary lesion has been treated. This local-regional metastasis is not amenable to surgical resection and responds poorly to chemotherapy and radiation therapy.

A radical neck dissection involves removal of all cervical lymph nodes (Fig. 46-3) from the mandible to the clavicle and removal of the sternocleidomastoid muscle, internal jugular vein, and spinal accessory muscle on one side of the neck. The associated complications include shoulder drop and poor cosmesis (visible neck depression). Modified radical neck dissection, which preserves one or more of the nonlymphatic structures, is used more often (Fonseca, 2009). A selective neck dissection (in comparison to a radical dissection) preserves one or more of the lymph node groups, the internal jugular vein, the sternocleidomastoid muscle, and the spinal accessory nerve (Fig. 46-4).

Reconstructive techniques may be performed with a variety of grafts. A cutaneous flap (skin and subcutaneous tissue), such as the deltopectoral flap, may be used. A myocutaneous flap (subcutaneous tissue, muscle, and skin) is a more frequently used graft; the pectoralis major muscle is usually used. For large grafts, a microvascular free flap may be used. This involves the transfer of muscle, skin, or bone with an artery and vein to the area of reconstruction, using microinstrumentation. Areas used for a free flap include the scapula, the radial area of the forearm, or the fibula. The fibula, which provides a larger bone area, may be used if mandibular reconstruction is involved (Fonseca, 2009).

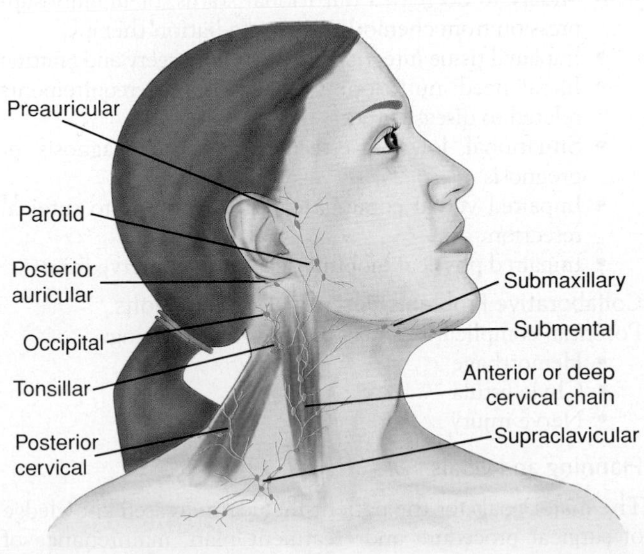
Preauricular
Parotid
Posterior auricular
Occipital
Tonsillar
Posterior cervical
Submaxillary
Submental
Anterior or deep cervical chain
Supraclavicular

FIGURE 46-3 • Lymphatic drainage of the head and neck.

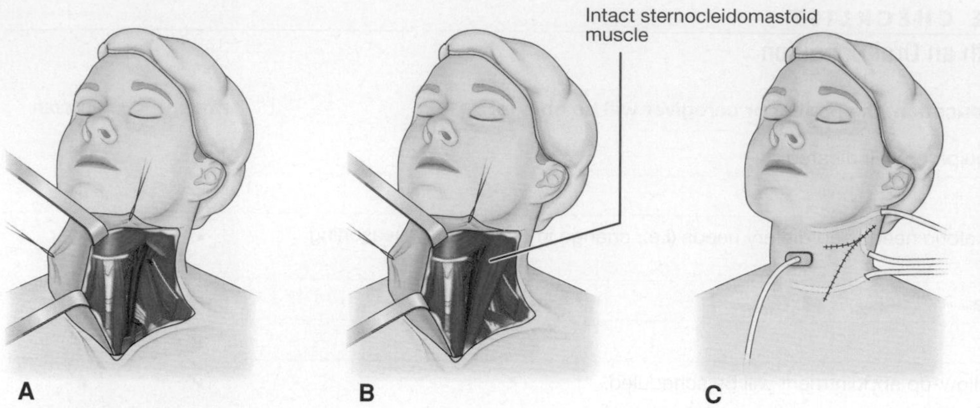
Intact sternocleidomastoid muscle
A
B
C

FIGURE 46-4 • **A.** A classic radical neck dissection in which the sternocleidomastoid and smaller muscles are removed. All tissue is removed, from the ramus of the jaw to the clavicle. The jugular vein has also been removed. **B.** The selective neck dissection is similar but preserves the sternocleidomastoid muscle, internal jugular vein, and spinal accessory nerve. **C.** The wound is closed, and portable suction drainage tubes are in place.

NURSING PROCESS

The Patient Undergoing a Neck Dissection

Assessment

Preoperatively, the patient's physical and psychological preparation for major surgery is assessed, along with his or her knowledge of the preoperative and postoperative procedures. Postoperatively, the patient is assessed for complications such as altered respiratory status, wound infection, and hemorrhage. As healing occurs, neck range of motion is assessed to determine whether there has been a decrease in range of motion due to nerve or muscle damage.

Diagnosis

Nursing Diagnoses

Based on the assessment data, major nursing diagnoses may include the following:

- Deficient knowledge about preoperative and postoperative procedures
- Ineffective airway clearance related to obstruction by mucus, hemorrhage, or edema
- Acute pain related to surgical incision
- Risk for infection related to surgical intervention secondary to decreased nutritional status, or immunosuppression from chemotherapy or radiation therapy
- Impaired tissue integrity secondary to surgery and grafting
- Imbalanced nutrition: less than body requirements related to disease process or treatment
- Situational low self-esteem related to diagnosis or prognosis
- Impaired verbal communication secondary to surgical resection
- Impaired physical mobility secondary to nerve injury

Collaborative Problems/Potential Complications

Potential complications may include the following:

- Hemorrhage
- Chyle fistula
- Nerve injury

Planning and Goals

The major goals for the patient include increased knowledge of surgical procedure and treatment plan, maintenance of respiratory status, decreased pain, absence of infection, viability of the graft, maintenance of adequate intake of food and fluids, effective coping strategies, effective communication, maintenance of shoulder and neck motion, and absence of complications.

Nursing Interventions

Providing Preoperative Patient Education

Before surgery, the patient should be informed about the nature and extent of the surgery and what the postoperative period will be like. The patient is encouraged to ask questions and to express concerns about the upcoming surgery and the expected results. During this exchange, the nurse has an opportunity to assess the patient's coping abilities, answer questions, and develop a plan for offering assistance. A sense of mutual understanding and rapport make the postoperative experience less traumatic for the patient. The patient's expressions of concern, anxieties, and fears guide the nurse in providing support postoperatively.

Providing General Postoperative Care

The general postoperative nursing interventions are similar to those presented in Chapter 19 and are directed toward the identified nursing diagnoses and goals.

Maintaining the Airway

After the endotracheal tube or airway has been removed and the effects of the anesthesia have worn off, the patient may be placed in the Fowler's position to facilitate breathing and promote comfort. This position also increases lymphatic and venous drainage, facilitates swallowing, decreases venous pressure on the skin flaps, and prevents regurgitation and aspiration of stomach contents. Signs of respiratory distress, such as dyspnea, cyanosis, changes in mental status, and changes in vital signs, are assessed because they may suggest edema, hemorrhage, inadequate oxygenation, or inadequate drainage.

Quality and Safety Nursing Alert

In the immediate postoperative period, the nurse assesses for stridor (coarse, high-pitched sound on inspiration) by listening frequently over the trachea with a stethoscope. This finding must be reported immediately because it indicates obstruction of the airway.

Pneumonia may occur in the postoperative phase if pulmonary secretions are not removed. To aid in the removal of secretions, coughing and deep breathing are encouraged. With the nurse supporting the neck, the patient should assume a sitting position so that excessive secretions can be coughed up and expectorated. If this is ineffective, the patient's respiratory tract may have to be suctioned. Care is taken to protect the suture lines during suctioning. If a tracheostomy tube is in place, suctioning is performed through the tube. The patient may also be instructed on use of Yankauer suction (tonsil tip suction) to remove oral secretions. Humidified air or oxygen is provided through the tracheostomy to keep secretions thin. Temperature should not be taken orally.

Relieving Pain

Pain and the patient's fear of pain are assessed and managed. Patients with head and neck cancer often report less pain than patients with other types of cancer; however, the nurse needs to be aware that each person's pain experience is individual (see Chapter 12). Pain management is monitored on a continual basis by the nursing staff and adjusted on an individual basis. Patient-controlled analgesia may be prescribed for postoperative pain management, thereby reducing the wait time for pain relief.

Providing Wound Care

Wound drainage tubes are usually inserted during surgery to prevent the collection of fluid subcutaneously. The drainage tubes are connected to a portable suction device (e.g., Jackson-Pratt), and the container is emptied periodically. Between 80 and 120 mL of serosanguineous secretions may drain over the first 24 hours. Excessive drainage may be indicative of a chyle fistula or hemorrhage (see later discussion). Dressings are reinforced as needed and are observed for evidence of hemorrhage and constriction, which impair respiration and perfusion of the graft. A graft, if present, is assessed for color and temperature and for the presence of a pulse, if applicable, to determine viability. The graft should be pale pink and warm to the touch. The surgical incisions are also assessed for signs of infection (purulent, malodorous drainage), which are reported immediately. Prophylactic antibiotics may be prescribed in the early postoperative period. Aseptic technique is used when cleansing skin around the drains; dressings are changed as prescribed by the surgeon, usually on the second through the fifth postoperative days.

Maintaining Adequate Nutrition

Nutritional status is assessed preoperatively; early intervention to correct nutritional imbalances may decrease the risk of postoperative complications. Frequently, nutrition is less than optimal because of inadequate intake, and the patient often requires enteral or parenteral supplements preoperatively and postoperatively to attain and maintain a positive nitrogen balance. Supplements (e.g., Ensure, Sustacal, Boost) that are nutritionally dense may help reestablish a positive nitrogen balance (Garg, Yoo, & Winquist, 2010). They may be taken enterally by mouth, by a nasogastric feeding tube, or by a gastrostomy feeding tube (see Chapter 45).

The patient who can chew may take food by mouth; the patient's chewing ability determines whether some diet modification (e.g., soft, puréed, or liquid foods) is necessary. Food preferences should also be discussed with the patient. Oral care before eating may enhance the patient's appetite, and oral care after eating is important to prevent infection and dental caries. Most patients can maintain and gain weight.

Supporting Coping Measures

Preoperatively, information about the planned surgery is given to the patient and family. Any questions are answered as accurately as possible. The nurse must pay attention to nonverbal behavior that may indicate something different from what the patient is able to articulate. Postoperatively, psychological nursing interventions are aimed at supporting the patient who has had a change in body image or who has major concerns related to the prognosis. The patient may have difficulty communicating and may be concerned about his or her ability to breathe and swallow normally. The nurse supports the patient's family in encouraging and reassuring the patient that adjusting to the results of this surgery will take time.

The person who has had extensive neck surgery often is sensitive about his or her appearance. This can occur when the operative area is covered by bulky dressings, when the incision line is visible, or later after healing has occurred and the appearance of the neck and possibly the lower face has been significantly altered. If the nurse accepts the patient's appearance and expresses a positive, optimistic attitude, the patient is more likely to be encouraged. The patient also needs an opportunity to express fears and concerns regarding the success of the surgery and the prognosis. The ACS may be a resource to provide a volunteer who meets with the patient either preoperatively or postoperatively and shares his or her own experience about the diagnosis, treatment, and recovery. The Look Good Feel Better programs of the ACS also are a source of information about clothing and cosmetics that can be used to improve body image and self-esteem (see the Resources section at the end of this chapter).

People with cancer of the head and neck frequently have used alcohol or tobacco before surgery; postoperatively, they are encouraged to abstain from these substances. Alternative methods of coping need to be explored. A referral to Alcoholics Anonymous, a smoking cessation program, and family counseling may be appropriate.

Promoting Effective Communication

Communication begins preoperatively, when the patient and family determine which method of communication will be the best postoperatively. Useful communication methods for the patient who has undergone a laryngectomy include Magic Slates, writing materials, pictorial guides, computer aids, and hand signals. During the postoperative period, the call bell must be readily accessible to the patient at all times.

The nurse obtains a consultation with a speech or language therapist. Alternative speech techniques, such as an electrolarynx (a mechanical device held against the neck) or esophageal speech, may be taught by a speech or language therapist. The most widely used technique for creating laryngeal speech is tracheoesophageal puncture. A surgically created fistula extends from the superior wall of the tracheal stoma into the proximal esophageal wall. A voice prosthesis is then inserted into the fistula to assist with speech (see Chapter 22).

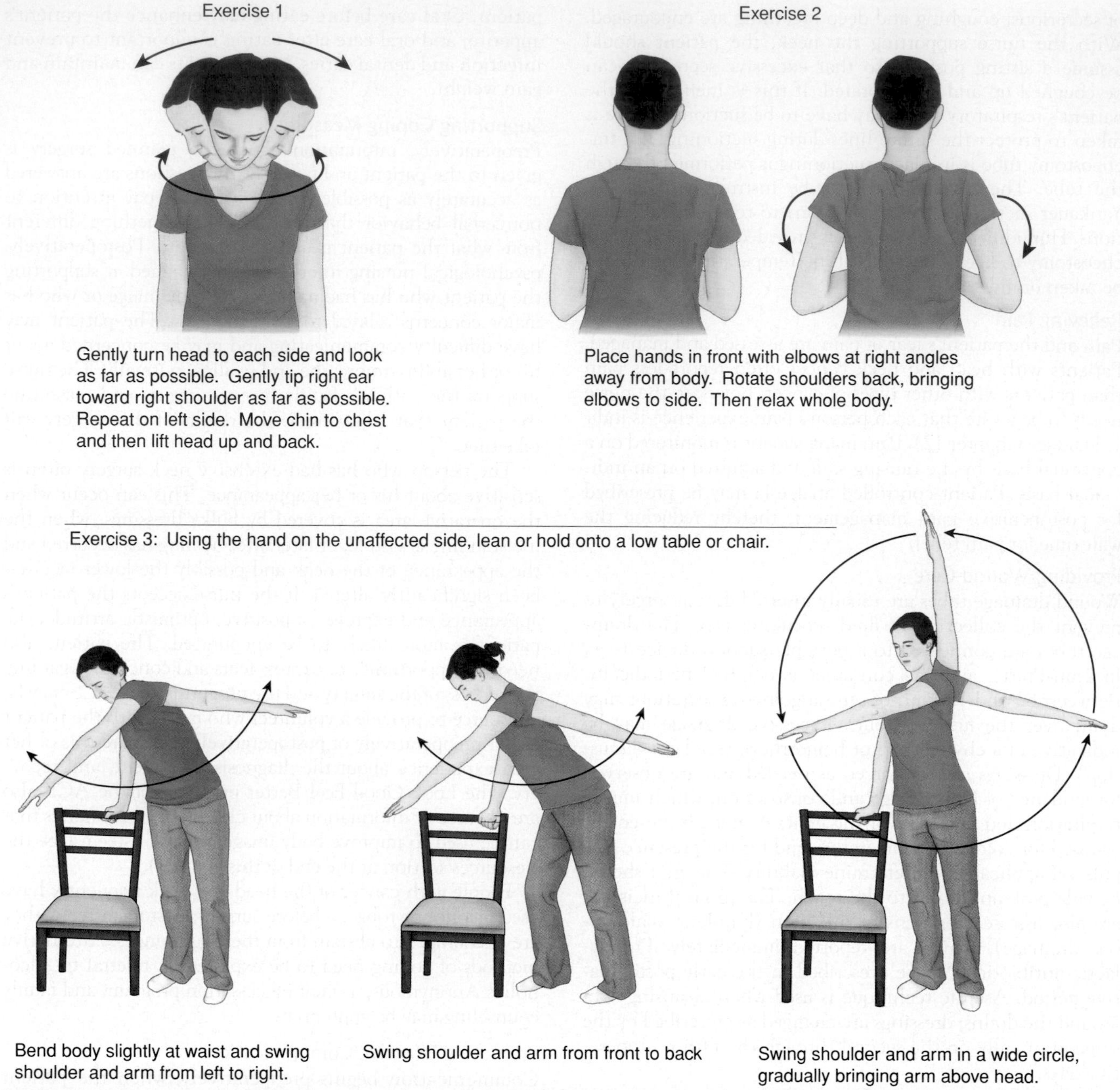
Exercise 1
Gently turn head to each side and look as far as possible. Gently tip right ear toward right shoulder as far as possible. Repeat on left side. Move chin to chest and then lift head up and back.
Exercise 2
Place hands in front with elbows at right angles away from body. Rotate shoulders back, bringing elbows to side. Then relax whole body.
Exercise 3: Using the hand on the unaffected side, lean or hold onto a low table or chair.
Bend body slightly at waist and swing shoulder and arm from left to right.
Swing shoulder and arm from front to back
Swing shoulder and arm in a wide circle, gradually bringing arm above head.

FIGURE 46-5 • Three rehabilitation exercises after head and neck surgery. The objective is to regain maximum shoulder function and neck motion after neck surgery. From *Exercise for radical neck surgery patients.* Head and Neck Service, Department of Surgery, Memorial Hospital, New York, NY.

Maintaining Physical Mobility

Excision of muscles and nerves results in weakness at the shoulder that can cause shoulder drop, which is a forward curvature of the shoulder. Many problems can be avoided with a conscientious exercise program. These exercises are usually started after the drains have been removed and the neck incision is sufficiently healed. The purpose of the exercises depicted in Figure 46-5 is to promote maximal shoulder function and neck motion after surgery. Physical therapists and occupational therapists can assist patients in performing these exercises.

Monitoring and Managing Potential Complications

Hemorrhage. Hemorrhage may occur from carotid artery rupture as a result of necrosis of the graft or damage to the artery itself from tumor or infection. The following measures are indicated:

- Vital signs are assessed frequently (every 1 to 2 hours or every 15 minutes if the patient is critical). Once the patient is stabilized, assessment is increased to every 4 hours. Tachycardia, tachypnea, and hypotension may indicate hemorrhage and impending hypovolemic shock.
- The patient is instructed to avoid the Valsalva maneuver to prevent stress on the graft and carotid artery.
- Signs of impending rupture, such as high epigastric pain or discomfort, are reported.
- Dressings and wound drainage are observed for excessive bleeding.

Chart 46-3 HOME CARE CHECKLIST
Recovering From Neck Surgery

At the completion of home care education, the patient or caregiver will be able to:	PATIENT	CAREGIVER
• Demonstrate the use of suction equipment.	✔	✔
• State rationale for humidification.	✔	✔
• State dietary modifications needed to meet caloric needs.	✔	✔
• Demonstrate enteral or parenteral feeding techniques.	✔	✔
• Demonstrate care of incision and drains.	✔	✔
• Identify signs and symptoms (e.g., bleeding, respiratory distress, drainage) to be reported to primary provider.	✔	✔
• State when next checkup is needed.	✔	✔
• Demonstrate exercises.	✔	
• Identify available support groups.	✔	✔

- If hemorrhage occurs, assistance is summoned immediately.
- Hemorrhage requires the continuous firm application of pressure to the bleeding site or major associated vessel.
- The head of the patient's bed should be elevated to maintain airway patency and prevent aspiration.
- A controlled, calm manner allays the patient's anxiety.
- The surgeon is notified immediately, because a vascular or ligature tear requires surgical intervention.

Chyle Fistula. A chyle fistula (milklike drainage from the thoracic duct into the thoracic cavity) may develop as a result of damage to the thoracic duct during surgery. The diagnosis is made if there is excess drainage present in the chest tube drainage that has a 3% fat content and a specific gravity of 1.012 or greater. Treatment of a small leak (500 mL or less) requires a diet of medium-chain fatty acids or parenteral nutrition. Surgical intervention may be needed to repair the damaged duct (Tessier, 2011).

Nerve Injury. Nerve injury can occur if the cervical plexus or spinal accessory nerves are severed during surgery. Because lower facial paralysis may occur as a result of injury to the facial nerve, this complication is observed for and reported. Likewise, if the superior laryngeal nerve is damaged, the patient may have difficulty swallowing liquids and food because of the partial lack of sensation of the glottis. Speech therapy may be indicated to assist with the problems related to nerve injury.

Promoting Home and Community-Based Care

Educating Patients About Self-Care. The patient and caregiver require instructions about management of the wound, the dressing, and any drains that remain in place. Patients who require oral suctioning or who have a tracheostomy may be very anxious about their care at home; the transition to home can be eased if the caregiver is given several opportunities to demonstrate the ability to meet the patient's needs (Chart 46-3). The patient and caregiver are also instructed about possible complications such as bleeding and respiratory distress and when to notify the health care provider of signs and symptoms of these complications.

If the patient cannot take food by mouth, detailed instructions and demonstration of enteral or parenteral feedings will be required. Education in techniques of effective oral hygiene is also important.

Continuing Care. A referral for home care nursing may be necessary in the early period after discharge. The nurse assesses healing, ensures that feedings are being administered properly, and monitors for any complications. The nurse also assesses the patient's adjustment to changes in physical appearance and status and ability to communicate and eat normally. Physical and speech therapy also are likely to be continued at home.

The patient is given information regarding local support groups such as "I Can Cope" or "New Voice Club," if indicated. The local chapter of the ACS may be contacted for information and equipment needed for the patient (see the Resources section).

Evaluation

Expected patient outcomes may include:

1. Exhibits increased knowledge of course of treatment
2. Demonstrates good respiratory exchange
   a. Lungs are clear to auscultation
   b. Breathes easily with no shortness of breath
   c. Demonstrates ability to use suction effectively
3. Remains free of infection
   a. Maintains normal laboratory values
   b. Is afebrile
4. Graft is pink and warm to touch.
5. Maintains adequate intake of foods and fluids
   a. Accepts altered route of feeding
   b. Is well hydrated
   c. Maintains or gains weight
6. Demonstrates ability to cope
   a. Discusses emotional responses to the diagnosis
   b. Attends support group meetings
7. Verbalizes comfort and relief of pain
8. Attains maximal mobility
   a. Adheres to physical therapy exercises
   b. Attains maximal range of motion
9. Exhibits no complications
   a. Vital signs stable
   b. No excessive bleeding or discharge
   c. Able to move muscles of lower face

DISORDERS OF THE ESOPHAGUS

The esophagus is a mucus-lined, muscular tube that carries food from the mouth to the stomach. It begins at the base of the pharynx and ends about 4 cm below the diaphragm. Its ability to transport food and fluid is facilitated by two sphincters. The upper esophageal sphincter, also called the *hypopharyngeal sphincter,* is located at the junction of the pharynx and the esophagus. The lower esophageal sphincter, also called the *gastroesophageal sphincter* or *cardiac sphincter,* is located at the junction of the esophagus and the stomach. An incompetent lower esophageal sphincter allows reflux (backward flow) of gastric contents. There is no serosal layer of the esophagus; therefore, if surgery is necessary, it is more difficult to perform suturing or anastomosis.

Disorders of the esophagus include motility disorders (achalasia, diffuse spasm), hiatal hernias, diverticula, perforation, foreign bodies, chemical burns, gastroesophageal reflux disease (GERD), Barrett's esophagus, benign tumors, and carcinoma. **Dysphagia** (difficulty swallowing), the most common symptom of esophageal disease, may vary from an uncomfortable feeling that a bolus of food is caught in the upper esophagus to acute **odynophagia** (pain on swallowing). Obstruction of food (solid and soft) and even liquids may occur anywhere along the esophagus. Often, the patient can indicate that the problem is located in the upper, middle, or lower third of the esophagus.

Achalasia

Achalasia is absent or ineffective peristalsis of the distal esophagus accompanied by failure of the esophageal sphincter to relax in response to swallowing. Narrowing of the esophagus just above the stomach results in a gradually increasing dilation of the esophagus in the upper chest. Achalasia may progress slowly and occurs most often in people 40 years or older.

Clinical Manifestations

The main symptom is difficulty in swallowing both liquids and solids. The patient has a sensation of food sticking in the lower portion of the esophagus. As the condition progresses, food is commonly regurgitated either spontaneously or intentionally by the patient to relieve the discomfort produced by prolonged distention of the esophagus by food that will not pass into the stomach. The patient may also report chest pain and **pyrosis** (heartburn) that may or may not be associated with eating. Secondary pulmonary complications may result from aspiration of gastric contents.

Assessment and Diagnostic Findings

X-ray studies show esophageal dilation above the narrowing at the gastroesophageal junction. Barium swallow, computed tomography (CT) scan of the chest, and endoscopy may be used for diagnosis; however, manometry, a process in which the esophageal pressure is measured by a radiologist or gastroenterologist, confirms the diagnosis (Pohl & Tutian, 2007).

Management

The patient is instructed to eat slowly and to drink fluids with meals. As a temporary measure, calcium channel blockers and nitrates have been used to decrease esophageal pressure and improve swallowing. Injection of botulinum toxin (Botox) into quadrants of the esophagus via endoscopy has been helpful because it inhibits the contraction of smooth muscle. Periodic injections are required to maintain remission (Dughera, Chiaverina, Cacciotella, et al., 2011).

Achalasia may be treated conservatively by pneumatic dilation to stretch the narrowed area of the esophagus (Fig. 46-6). Pneumatic dilation has a high success rate. Although perforation is a potential complication, its incidence is low. The procedure can be painful; therefore, moderate sedation in the form of an analgesic or tranquilizer, or both, is administered for the treatment. The patient is monitored for perforation. Abdominal tenderness and fever may indicate perforation (Dughera et al., 2011) (see later discussion).

Achalasia may be treated surgically by esophagomyotomy. The procedure is usually performed laparoscopically, either with a complete lower esophageal sphincter myotomy and an antireflux procedure or without an antireflux procedure (Cowgill, Villadoid, Boyle, et al., 2009). The esophageal muscle fibers are separated to relieve the lower esophageal stricture.

Diffuse Esophageal Spasm (Nutcracker Esophagus)

Diffuse esophageal spasm is a motor disorder of the esophagus. The cause is unknown, but stress may be a factor. It is more common in women and usually manifests in middle age (Agrawal, Hila, Tutuian, et al., 2006).

Clinical Manifestations

Diffuse esophageal spasm is characterized by difficulty (dysphagia) or pain (odynophagia) on swallowing and by chest pain similar to that of coronary artery spasm.

Assessment and Diagnostic Findings

Esophageal manometry, which measures the motility of the esophagus and the pressure within the esophagus, indicates that simultaneous contractions of the esophagus occur irregularly. Diagnostic x-ray studies after ingestion of barium show separate areas of spasm.

Management

Conservative therapy includes administration of sedative agents and long-acting nitrates to relieve pain. Calcium channel blockers (e.g., nifedipine [Procardia], verapamil [Calan]) have also been used to manage diffuse spasm. Small, frequent feedings and a soft diet are usually recommended to decrease the esophageal pressure and irritation that lead to spasm. Dilation performed by bougienage (the use of progressively sized flexible dilators; see later discussion in the Chemical Burns section), pneumatic dilation, or esophagomyotomy may be necessary if the pain becomes intolerable.

If none of the conservative approaches is successful in managing symptoms, surgery may be considered. The laparoscopic

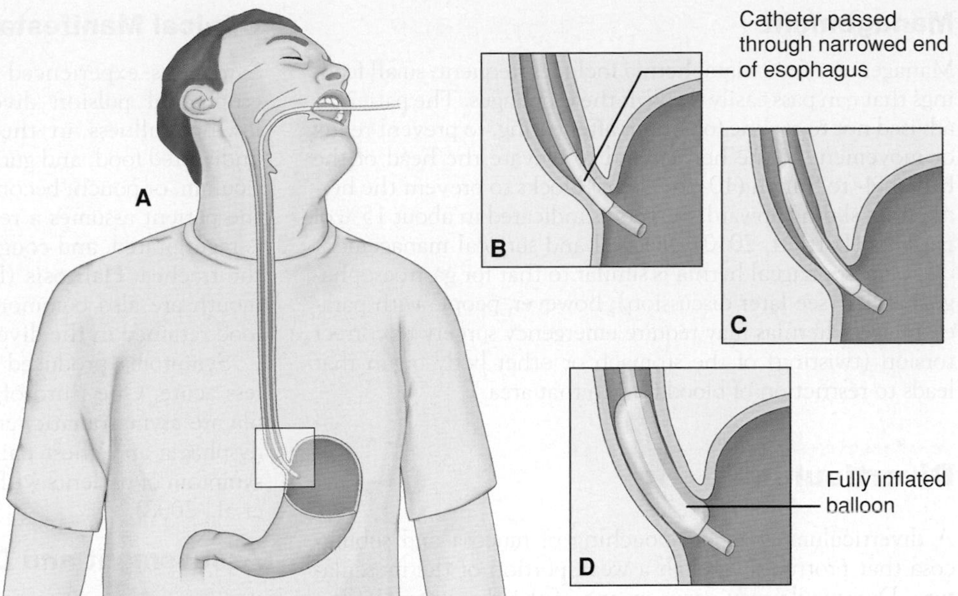

A
Catheter passed through narrowed end of esophagus
B
C
Fully inflated balloon
D

FIGURE 46-6 • Treatment of achalasia by pneumatic dilation. **A–C.** The dilator is passed, guided by a previously inserted guide wire. **D.** When the balloon is in proper position, it is distended by pressure sufficient to dilate the narrowed area of the esophagus.

modified Heller myotomy is preferred because it has been shown to reduce reflux better than the open surgical approach. Additionally, it is associated with reduced lengths of hospital stay and pain as compared to open procedures such as transhiatal esophagectomies or Nissen fundoplications (Cowgill et al., 2009) (see Hiatal Hernia and Cancer of the Esophagus sections).

Hiatal Hernia

In the condition known as hiatus (or hiatal) **hernia**, the opening in the diaphragm through which the esophagus passes becomes enlarged, and part of the upper stomach tends to move up into the lower portion of the thorax. Hiatal hernia occurs more often in women than in men. There are two types of hiatal hernias: sliding and paraesophageal. Sliding, or type I, hiatal hernia occurs when the upper stomach and the gastroesophageal junction are displaced upward and slide in and out of the thorax (Fig. 46-7A). About 90% of patients with esophageal hiatal hernia have a sliding hernia (National Library of Medicine and National Institutes of Health [NLM/NIH], 2012). A paraesophageal hernia occurs when all or part of the stomach pushes through the diaphragm beside the esophagus (see Fig. 46-7B). Paraesophageal hernias are further classified as types II, III, or IV, depending on the extent of herniation, with type IV having the greatest herniation.

Clinical Manifestations

The patient with a sliding hernia may have heartburn, regurgitation, and dysphagia, but at least 50% of patients are asymptomatic (NLM/NIH, 2012). Sliding hiatal hernia is often implicated in reflux. The patient with a paraesophageal hernia usually feels a sense of fullness or chest pain after eating, or there may be no symptoms. Reflux usually does not occur, because the gastroesophageal sphincter is intact. Hemorrhage, obstruction, and strangulation can occur with any type of hernia.

Assessment and Diagnostic Findings

Diagnosis is confirmed by x-ray studies, barium swallow, and chest CT.

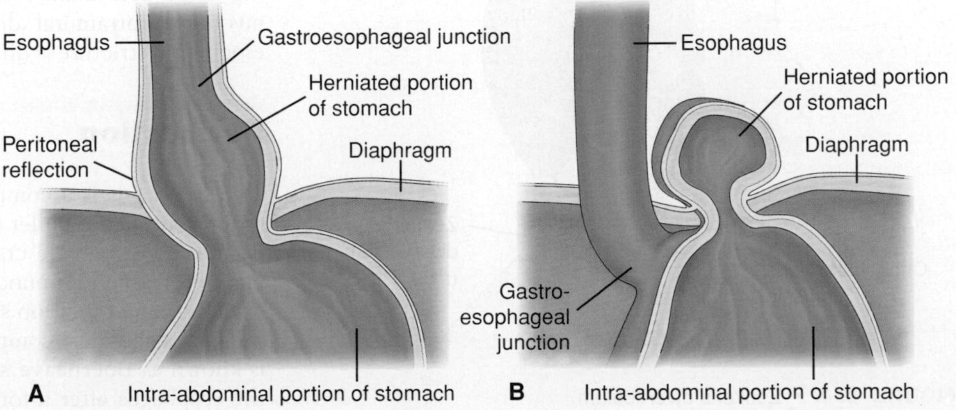

Esophagus
Gastroesophageal junction
Herniated portion of stomach
Peritoneal reflection
Diaphragm
A
Intra-abdominal portion of stomach
Esophagus
Herniated portion of stomach
Diaphragm
Gastro-esophageal junction
B
Intra-abdominal portion of stomach

FIGURE 46-7 • **A.** Sliding esophageal hernia. The upper stomach and gastroesophageal junction have moved upward and slide in and out of the thorax. **B.** Paraesophageal hernia. All or part of the stomach pushes through the diaphragm next to the gastroesophageal junction.

Management

Management for a hiatal hernia includes frequent, small feedings that can pass easily through the esophagus. The patient is advised not to recline for 1 hour after eating, to prevent reflux or movement of the hernia, and to elevate the head of the bed on 4- to 8-inch (10- to 20-cm) blocks to prevent the hernia from sliding upward. Surgery is indicated in about 15% of patients (Bartlett, 2010). Medical and surgical management of a paraesophageal hernia is similar to that for gastroesophageal reflux (see later discussion); however, people with paraesophageal hernias may require emergency surgery to correct torsion (twisting) of the stomach or other body organ that leads to restriction of blood flow to that area.

Diverticulum

A diverticulum is an out-pouching of mucosa and submucosa that protrudes through a weak portion of the musculature. Diverticula may occur in one of the three areas of the esophagus—the pharyngoesophageal or upper area of the esophagus, the midesophageal area, or the epiphrenic or lower area of the esophagus—or they may occur along the border of the esophagus intramurally.

The most common type of diverticulum, which is found three times more frequently in men than in women, is Zenker's diverticulum (also known as pharyngoesophageal pulsion diverticulum or a pharyngeal pouch) (Fig. 46-8). It occurs posteriorly through the cricopharyngeal muscle in the midline of the neck. It is usually seen in people older than 60 years (Ferreira, Simmons, & Baron, 2008). Other types of diverticula include midesophageal, epiphrenic, and intramural diverticula.

Midesophageal diverticula are uncommon. Symptoms are less acute, and usually the condition does not require surgery. Epiphrenic diverticula are usually larger diverticula in the lower esophagus just above the diaphragm. They may be related to the improper functioning of the lower esophageal sphincter or to motor disorders of the esophagus. Intramural diverticulosis is the occurrence of numerous small diverticula associated with a stricture in the upper esophagus.

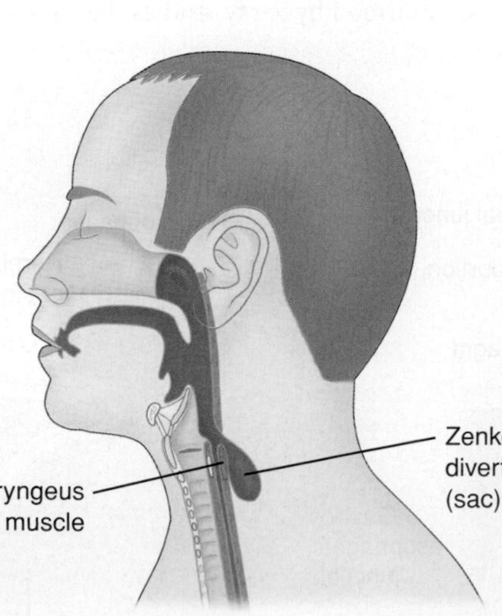
Zenker's diverticulum (sac)
Cricopharyngeus muscle

FIGURE 46-8 • Zenker's diverticulum.

Clinical Manifestations

Symptoms experienced by the patient with a pharyngoesophageal pulsion diverticulum include difficulty swallowing, fullness in the neck, belching, regurgitation of undigested food, and gurgling noises after eating. The diverticulum, or pouch, becomes filled with food or liquid. When the patient assumes a recumbent position, undigested food is regurgitated, and coughing may be caused by irritation of the trachea. **Halitosis** (bad breath) and a sour taste in the mouth are also common because of the decomposition of food retained in the diverticulum.

Symptoms produced by midesophageal diverticula are less acute. One third of patients with epiphrenic diverticula are asymptomatic, and the remaining two thirds report dysphagia and chest pain. Dysphagia is the most common symptom of patients with intramural diverticulosis (Ferreira et al., 2008).

Assessment and Diagnostic Findings

A barium swallow may determine the exact nature and location of a diverticulum. Manometric studies are often performed for patients with epiphrenic diverticula to rule out a motor disorder. Esophagoscopy usually is contraindicated because of the danger of perforation of the diverticulum, with resulting mediastinitis (inflammation of the organs and tissues that separate the lungs). Blind insertion of a nasogastric tube should be avoided.

Management

Because Zenker's diverticulum is progressive, the only means of cure is surgical removal of the diverticulum. During surgery, care is taken to avoid trauma to the common carotid artery and internal jugular veins. The sac is dissected free and amputated flush with the esophageal wall. In addition to a diverticulectomy, a myotomy of the cricopharyngeal muscle is often performed to relieve spasticity of the musculature, which otherwise seems to contribute to a continuation of the previous symptoms. A nasogastric tube may be inserted at the time of surgery. Postoperatively, the surgical incision must be observed for evidence of leakage from the esophagus and a developing fistula. Food and fluids are withheld until x-ray studies show no leakage at the surgical site. The diet begins with liquids and is progressed as tolerated.

Surgery is indicated for epiphrenic and midesophageal diverticula only if the symptoms are troublesome and becoming worse. Treatment consists of a diverticulectomy and long myotomy. Intramural diverticula usually regress after the esophageal stricture is dilated.

Perforation

The esophagus is a common site of injury. Perforation may result from stab or bullet wounds of the neck or chest, trauma from a motor vehicle crash, caustic injury from a chemical burn, or inadvertent puncture by a surgical instrument during examination or dilation such as endoscopy.

One of the most common types of esophageal perforation is known as **Boerhaave syndrome**, a spontaneous rupture of the esophagus after a forceful vomiting episode (may occur after eating a large meal). Boerhaave syndrome is further characterized by a transmural laceration of the lower esophagus, causing gastric contents to empty into the mediastinum, causing mediastinal sepsis, which is a life-threatening emergency (Roy, 2011).

Clinical Manifestations

The patient has excruciating retrosternal pain followed by dysphagia. Infection, fever, leukocytosis, and severe hypotension may be noted. Additionally, mediastinal sepsis can occur with Boerhaave syndrome, which may be also accompanied by pneumothorax and subcutaneous emphysema.

Assessment and Diagnostic Findings

X-ray studies, fluoroscopy by either a barium swallow or esophagram, or a chest CT scan may be used to identify the site and scope of the injury.

Management

Because of the high risk of infection, broad-spectrum antibiotic therapy is initiated. If the perforation is small enough and without symptoms, surgical intervention may not be necessary. The patient may not consume anything orally (i.e., NPO). Nutritional needs are met by parenteral or enteral nutrition (see Chapter 45). The type of nutritional support depends on the location of the injury. Enteral or parenteral nutrition is provided for at least 1 month to give the esophagus a chance to heal. A repeat barium swallow study is performed after 1 month, and the involved area is reevaluated. If there is no evidence of perforation, foods are reintroduced, beginning with liquids and then slowly progressing to solids as tolerated.

Surgery is performed if the esophageal perforation is large or if mediastinitis or infection of the thoracic cavity is a threat. In a cervical esophagostomy, the upper portion of the esophagus is attached to an opening made in the neck; this "spit" fistula allows for the drainage of saliva. The lower portion of the esophagus remaining within the chest is closed. After 6 months, during which the patient is allowed to heal and recover from possible infection, surgery is again performed to reconnect the two parts of the esophagus.

Postoperative nutritional status is a major concern. The patient is not allowed any oral nourishment for 6 months. Enteral or parenteral support is maintained (see Chapter 45). Water to moisten the patient's mouth is allowed for comfort measures only. The postoperative nursing management is similar to that for patients who have had thoracic or abdominal surgery.

Foreign Bodies

Many swallowed foreign bodies pass through the GI tract without the need for medical intervention. However, some swallowed foreign bodies (e.g., dentures, fish bones, pins, small batteries, items containing mercury or lead) may injure the esophagus or obstruct its lumen and must be removed. Pain and dysphagia may be present, and dyspnea may occur as a result of pressure on the trachea. The foreign body may be identified by x-ray. Perforation may have occurred (see earlier discussion).

Glucagon, because of its relaxing effect on the esophageal muscle, may be injected intramuscularly. An endoscope may be used to remove the impacted food or object from the esophagus. A mixture consisting of sodium bicarbonate and tartaric acid may be prescribed to increase intraluminal pressure by the formation of a gas. Caution must be used with this treatment because of the risk of perforation.

Chemical Burns

Chemical burns of the esophagus occur most often when a patient, either intentionally or unintentionally, swallows a strong acid or base (e.g., lye). This patient is emotionally distraught as well as in acute physical pain. Chemical burns of the esophagus may also be caused by undissolved medications in the esophagus. This occurs more frequently in older adults than it does among the general adult population (Kardon, 2011). A chemical burn may also occur after swallowing of a battery, which may release a caustic alkaline. An acute chemical burn of the esophagus may be accompanied by severe burns of the lips, mouth, and pharynx, with pain on swallowing. There may be difficulty in breathing due to either edema of the throat or a collection of mucus in the pharynx.

The patient, who may be profoundly toxic, febrile, and in shock, is treated immediately for shock, pain, and respiratory distress. Esophagoscopy and barium swallow are performed as soon as possible to determine the extent and severity of damage. The patient is given nothing by mouth, and IV fluids are administered. Vomiting and gastric lavage are avoided to prevent further exposure of the esophagus to the caustic agent. The use of corticosteroids to reduce inflammation and minimize subsequent scarring and stricture formation is of questionable value. Antibiotics are prescribed if there is documented infection.

After the acute phase has subsided, the patient may need nutritional support via enteral or parenteral feedings. The patient may require further treatment to prevent or manage strictures of the esophagus. Dilation by bougienage may be sufficient but may need to be repeated periodically. (In bougienage, cylindrical rubber tubes of different sizes, called *bougies*, are advanced into the esophagus via the oral cavity. Progressively larger bougies are used to dilate the esophagus. The procedure usually is performed in the endoscopy suite or clinic by the gastroenterologist.) Some strictures require rigid dilators, such as Savary dilators. These dilators are used in the same fashion as bougies but may be more successful for opening difficult strictures. For strictures that do not respond to either method of dilation, surgical management may be necessary. Reconstruction may be accomplished by esophagectomy and colon interposition to replace the portion of esophagus removed. This surgery is quite complex and should be considered only when other options have failed.

Gastroesophageal Reflux Disease

Some degree of **gastroesophageal reflux (GERD)** (backflow of gastric or duodenal contents into the esophagus) is normal. Excessive reflux may occur because of an incompetent lower esophageal sphincter, pyloric stenosis, hiatal hernia, or a motility disorder. The incidence of GERD seems to increase with aging. GERD can be related to Barrett's esophagus (Souza, 2012) (see later discussion).

Clinical Manifestations

Symptoms may include pyrosis (burning sensation in the esophagus), dyspepsia (indigestion), regurgitation, dysphagia or odynophagia (pain on swallowing), hypersalivation, and esophagitis. The symptoms may mimic those of a heart attack. The patient's history aids in obtaining an accurate diagnosis (Bickley, 2009).

Assessment and Diagnostic Findings

Diagnostic testing may include an endoscopy or barium swallow to evaluate damage to the esophageal mucosa. Ambulatory 12- to 36-hour esophageal pH monitoring is used to evaluate the degree of acid reflux. Esophageal pH monitoring was historically an uncomfortable procedure, but the advent of wireless capsule pH monitoring is better tolerated and quite accurate (Gawron & Hirano, 2010).

Management

Management begins with educating the patient to avoid situations that decrease lower esophageal sphincter pressure or cause esophageal irritation. The patient is instructed to eat a low-fat diet; to avoid caffeine, tobacco, beer, milk, foods containing peppermint or spearmint, and carbonated beverages; to avoid eating or drinking 2 hours before bedtime; to maintain normal body weight; to avoid tight-fitting clothes; to elevate the head of the bed on 6- to 8-inch (15- to 20-cm) blocks; and to elevate the upper body on pillows (Johnson, 2012). If reflux persists, antacids or histamine-2 (H_2) receptor antagonists, such as famotidine (Pepcid), nizatidine (Axid), or ranitidine (Zantac), may be prescribed. Proton pump inhibitors (medications that decrease the release of gastric acid, such as lansoprazole [Prevacid], rabeprazole [AcipHex], esomeprazole [Nexium], omeprazole [Prilosec], and pantoprazole [Protonix]) may be used; however, these products may increase intragastric bacterial growth and the risk of infection (Dent, 2006). In addition, the patient may receive prokinetic agents, which accelerate gastric emptying. These agents include bethanechol (Urecholine), domperidone (Motilium), and metoclopramide (Reglan). Because metoclopramide can have extrapyramidal side effects that are increased in certain neuromuscular disorders, such as Parkinson's disease, it should be used only if no other option exists, and the patient should be monitored closely. On occasion, when metoclopramide cannot be used, low doses of erythromycin (E-Mycin) may be successful (Gastroparesis and Dysmotilities Association, 2011).

If medical management is unsuccessful, surgical intervention may be necessary. Surgical management involves a Nissen fundoplication (wrapping of a portion of the gastric fundus around the sphincter area of the esophagus). A Nissen fundoplication can be performed by the open method or by laparoscopy.

Barrett's Esophagus

Barrett's esophagus is a condition in which the lining of the esophageal mucosa is altered. It typically occurs in association with GERD; indeed, long-standing untreated GERD may lead to Barrett's esophagus. Reflux eventually causes changes in the cells lining the lower esophagus. The cells that are laid to cover the exposed area are no longer squamous in origin. These precancerous cells initiate the healing process and can be a precursor to esophageal cancer.

Clinical Manifestations

The patient complains of symptoms of GERD, notably frequent heartburn. The patient may also complain of symptoms related to peptic ulcers or esophageal stricture, or both.

Assessment and Diagnostic Findings

An **esophagogastroduodenoscopy (EGD)** is performed. This usually reveals an esophageal lining that is red rather than pink. Biopsies are performed, and high-grade **dysplasia** (HGD) is evidenced by the squamous mucosa of the esophagus replaced by columnar epithelium that resembles that of the stomach or intestines. HGD has been found to be associated with a 30% increased risk of development of cancer (Wang & Sampliner, 2008).

Management

Monitoring varies depending on the extent of cell changes. Follow-up endoscopy is performed within 6 months if there are minor cell changes. Treatment is individualized for each patient. The options include intensive surveillance with biopsies, endoscopic ablation therapy (e.g., photodynamic therapy [PDT]), and esophagectomy. PDT has proven to be efficacious in the eradication of Barrett's esophagus with HGD and in the prevention of cancer. Lastly, it has shown a reduced mortality and morbidity when compared to esophageal reconstructive surgery (Comay, Blackhouse, Goeree, et al., 2007).

Benign Tumors of the Esophagus

Benign tumors can arise anywhere along the esophagus. The most common lesion is a leiomyoma (tumor of the smooth muscle), which can occlude the lumen of the esophagus. Most benign tumors are asymptomatic and are distinguished from cancerous lesions by a biopsy. Small lesions are excised during esophagoscopy; lesions that occur within the wall of the esophagus may require treatment via a thoracotomy (Castell & Richter, 2003).

NURSING PROCESS

The Patient With a Noncancerous Condition of the Esophagus

Assessment

Emergency conditions of the esophagus (perforation, chemical burns) usually occur in the home or away from medical help and require emergency medical care. The patient is treated for shock and respiratory distress and transported as quickly as possible to a health care facility. Foreign bodies in the esophagus do not pose an immediate threat to life unless pressure is exerted on the trachea, resulting in dyspnea or interfering with respiration, or unless there is leakage of caustic alkali from a battery or exposure to another corrosive agent. Educating the public to prevent inadvertent swallowing of foreign bodies or corrosive agents is a major health goal.

For nonemergency symptoms, a complete health history may reveal the nature of the esophageal disorder. The nurse asks about the patient's appetite. Has it remained the same, increased, or decreased? Is there any discomfort with swallowing? If so, does it occur only with certain foods? Is it associated with pain? Does a change in position affect the discomfort? The patient is asked to describe the pain. Does anything aggravate it? Are there any other symptoms that occur regularly, such as regurgitation, nocturnal regurgitation, eructation (belching), heartburn, substernal pressure, a sensation that food is sticking in the throat, a feeling of fullness after eating a small amount of food, nausea, vomiting, or weight loss? Are the symptoms aggravated by emotional upset? If the patient reports any of these symptoms, the nurse asks about when they occur, their relationship to eating, and factors that relieve or aggravate them (e.g., position change, belching, antacids, vomiting) (Bickley, 2009).

This history also includes questions about past or present causative factors, such as infections and chemical, mechanical, or physical irritants; alcohol and tobacco use; and the amount of daily food intake. The nurse determines whether the patient appears emaciated and auscultates the patient's chest to assess for pulmonary complications (Bickley, 2009).

Nursing Diagnosis

Based on the assessment data, nursing diagnoses may include the following:

- Imbalanced nutrition: less than body requirements related to difficulty swallowing
- Risk for aspiration related to difficulty swallowing or to tube feeding
- Acute pain related to difficulty swallowing, ingestion of an abrasive agent, tumor, or frequent episodes of gastric reflux
- Deficient knowledge about the esophageal disorder, diagnostic studies, medical management, surgical intervention, and rehabilitation

Planning and Goals

The major goals for the patient may include attainment of adequate nutritional intake, avoidance of respiratory compromise from aspiration, relief of pain, and increased knowledge level.

Nursing Interventions

Encouraging Adequate Nutritional Intake

The patient is encouraged to eat slowly and to chew all food thoroughly so that it can pass easily into the stomach. Small, frequent feedings of nonirritating foods are recommended to promote digestion and to prevent tissue irritation. Sometimes liquid swallowed with food helps the food pass through the esophagus, but usually liquids should be consumed between meals. Food should be prepared in an appealing manner to help stimulate the appetite. Irritants such as tobacco and alcohol should be avoided. A baseline weight is obtained, and daily weights are recorded. The patient's intake of nutrients is assessed.

Decreasing Risk of Aspiration

The patient who has difficulty swallowing or difficulty handling secretions should be kept in at least a semi-Fowler's position to decrease the risk of aspiration. The patient is instructed in the use of oral suction to decrease the risk of aspiration further.

Relieving Pain

Small, frequent feedings (six to eight per day) are recommended because large quantities of food overload the stomach and promote gastric reflux. The patient is advised to avoid any activities that increase pain and to remain upright for 1 to 4 hours after each meal to prevent reflux. The head of the bed should be placed on 4- to 8-inch (10- to 20-cm) blocks. Eating before bedtime is discouraged.

The patient is advised that excessive use of over-the-counter antacids can cause rebound acidity. Antacid use should be directed by the primary provider, who can recommend the daily, safe dose needed to neutralize gastric juices and prevent esophageal irritation. H_2 antagonists are administered as prescribed to decrease gastric acid irritation.

Providing Patient Education

The patient is prepared physically and psychologically for diagnostic tests, treatments, and possible surgery. Nursing interventions include reassuring the patient and explaining the procedures and their purposes. Some disorders of the esophagus evolve over time, whereas others are the result of trauma (e.g., chemical burns, perforation). In instances of trauma, the emotional and physical preparation for treatment is more difficult because of the short time available and the circumstances of the injury. Treatment interventions must be evaluated continually, and the patient is given sufficient information to participate in care and diagnostic tests. If endoscopic diagnostic methods are used, the patient is instructed regarding the moderate sedation that will be used during the procedure. If outpatient procedures are performed with the use of moderate sedation, someone must be available to drive the patient home after the procedure. If surgery is required, immediate and long-term evaluation is similar to that for a patient undergoing thoracic surgery.

Promoting Home and Community-Based Care

Educating Patients About Self-Care. The self-care required of the patient depends on the nature of the disorder and on the surgery or treatment measures used (e.g., diet, positioning, medications). If an ongoing condition exists, the nurse helps the patient plan for needed physical and psychological adjustments and for follow-up care (Chart 46-4).

Special equipment, such as suction or enteral or parenteral feeding devices, may be required. The patient may need assistance in planning meals, using medications as prescribed, and resuming activities. Education about nutritional requirements and how to measure the adequacy of nutrition is important. Older adults and patients who are debilitated in particular often need assistance and education about ways they can adjust to their limitations and resume activities that are important to them.

Continuing Care. Patients with chronic esophageal conditions require an individualized approach to their management at home. Foods may need to be prepared in a special way (blenderized foods, soft foods), and the patient may need to

Chart 46-4 HOME CARE CHECKLIST
The Patient With an Esophageal Condition

At the completion of home care education, the patient or caregiver will be able to:	PATIENT	CAREGIVER
• Demonstrate the use of suction equipment.	✔	✔
• State dietary modifications needed to meet caloric needs.	✔	✔
• Demonstrate enteral or parenteral feeding techniques.	✔	✔
• Demonstrate care of incision, if indicated.	✔	✔
• Identify signs and symptoms (e.g., difficulty swallowing, pain, respiratory distress) to be reported to primary provider.	✔	✔
• State when next checkup is needed.	✔	✔
• Identify available support groups (other patients, social networks, healthcare staff, etc.).	✔	✔

eat more frequently (e.g., six to eight small servings per day). The medication schedule is adjusted to the patient's daily activities as much as possible. Analgesic medications and antacids can usually be taken as needed every 3 to 4 hours.

Postoperative home health care focuses on nutritional support, management of pain, and respiratory function. Some patients are discharged from the hospital with enteral feeding by means of a gastrostomy or jejunostomy tube or parenteral nutrition. The patient and caregiver need specific instructions regarding management of the equipment and treatments. Home care visits by a nurse may be necessary to assess the patient and the caregiver's ability to provide the necessary care. (See Chapter 45 for more information about parenteral nutrition and management of the patient with a gastrostomy.) A multidisciplinary team that includes a dietitian, a social worker, and family members is helpful. Hospice care and consideration of end-of-life issues are appropriate for some patients.

Evaluation

Expected patient outcomes may include:

1. Achieves an adequate nutritional intake
   a. Eats small, frequent meals
   b. Drinks small sips of water with small servings of food
   c. Avoids irritants (alcohol, tobacco, very hot beverages)
   d. Maintains desired weight
2. Does not aspirate or develop pneumonia
   a. Maintains upright position during feeding
   b. Uses oral suction equipment effectively
3. Is free of pain or able to control pain within a tolerable level
   a. Avoids large meals and irritating foods
   b. Takes medications as prescribed and with adequate fluids (at least 4 oz), and remains upright for at least 10 minutes after taking medications
   c. Maintains an upright position after meals for 1 to 4 hours
   d. Reports that there is less eructation (belching) and chest pain
4. Increases knowledge level of esophageal condition, diagnostic tests, treatment, and prognosis
   a. States cause of condition
   b. Discusses rationale for medical or surgical management and diet or medication regimen
   c. Describes treatment program
   d. Practices preventive measures so that injuries are avoided

Cancer of the Esophagus

In the United States, there are about 17,460 newly diagnosed cases of carcinoma of the esophagus annually; of these, 13,950 are men and 3,510 are women (ACS, 2012). It is seen more frequently in African Americans than in Caucasians and usually occurs in the fifth or sixth decade of life. Cancer of the esophagus has a much higher incidence (10 to 100 times higher) in other parts of the world, including China and northern Iran. About 15,070 Americans die of esophageal cancer annually (ACS, 2012).

Pathophysiology

Esophageal cancer can be of two cell types: adenocarcinoma and squamous cell carcinoma. The rate of adenocarcinoma is rapidly increasing in the United States as well as in other Western countries. It is found primarily in the distal esophagus and gastroesophageal junction (ACS, 2012).

Risk factors for esophageal cancer include chronic esophageal irritation or GERD. In the United States, cancer of the esophagus has been associated with ingestion of alcohol and the use of tobacco. There is an apparent association between GERD and adenocarcinoma of the esophagus. People with Barrett's esophagus (which is caused by chronic irritation of the mucous membranes due to reflux of gastric and duodenal contents) have a higher incidence of esophageal cancer (ACS, 2012). Risk factors for squamous cell carcinoma of the esophagus include chronic ingestion of hot liquids or foods, nutritional deficiencies, poor oral hygiene, exposure to nitrosamines in the environment or food, cigarette smoking or chronic alcohol exposure (especially in Western cultures), and some esophageal medical conditions such as caustic injury.

Tumor cells of adenocarcinoma and of squamous cell carcinoma may spread beneath the esophageal mucosa or directly into, through, and beyond the muscle layers into the lymphatics. In the latter stages, obstruction of the esophagus is noted, with possible perforation into the mediastinum and erosion into the great vessels (Castell & Richter, 2003).

Clinical Manifestations

Many patients have an advanced ulcerated lesion of the esophagus before symptoms are manifested. Symptoms include dysphagia, initially with solid foods and eventually with liquids; a sensation of a mass in the throat; painful swallowing; substernal pain or fullness; and, later, regurgitation of undigested food with halitosis and hiccups. The patient first becomes aware of intermittent and increasing difficulty in swallowing. As the tumor grows and the obstruction becomes nearly complete, even liquids cannot pass into the stomach. Regurgitation of food and saliva occurs, hemorrhage may take place, and progressive loss of weight and strength occurs from inadequate nutrition. Later symptoms include substernal pain, persistent hiccup, respiratory difficulty, and halitosis.

The delay between the onset of early symptoms and the patient seeking medical advice is often 12 to 18 months. Any person having swallowing difficulties should be encouraged to consult a physician immediately.

Assessment and Diagnostic Findings

Currently, diagnosis is confirmed most often by EGD with biopsy and brushings. The biopsy can be used to determine the presence of disease and cell differentiation. At presentation, most patients have moderately differentiated tumors.

Several imaging techniques may provide useful diagnostic information. A CT scan of the chest and abdomen is beneficial for detecting any anatomic evidence of metastatic disease, especially of the lungs, liver, and kidney. A positron emission tomography scan may also help detect metastasis. Endoscopic ultrasound is used to determine whether the cancer has spread to the lymph nodes and other mediastinal structures; it can also determine the size and invasiveness of the tumor. Exploratory laparoscopy is the best method for finding positive lymph nodes in patients with distal lesions.

Future diagnostic techniques that may serve as predictors for dysplastic progression in patients with Barrett's esophagus involve molecular markers. Some data has shown that a small percentage of people may have a genetic predisposition to esophageal cancer. Researchers have identified at least 11 genotypes that may increase esophageal cancer risk (American Association for Cancer Research [AACR], 2012). The usefulness of molecular markers in treating esophageal cancer is being researched.

Medical Management

If esophageal cancer is detected at an early stage, treatment goals may be directed toward cure; however, it is often detected in late stages, making relief of symptoms the only reasonable goal of therapy. Treatment may include surgery, radiation, chemotherapy, or a combination of these modalities, depending on the type of cancer cell, the extent of the disease, and the patient's condition. A standard treatment plan for a person who is newly diagnosed with esophageal cancer includes the following: preoperative combination chemotherapy and radiation therapy for 4 to 6 weeks, followed by a period of no medical intervention for 4 weeks, and, lastly, surgical resection of the esophagus (NCI, 2012b).

Standard surgical management includes a total resection of the esophagus (esophagectomy) with removal of the tumor plus a wide tumor-free margin of the esophagus and the lymph nodes in the area (Cox, 1999; Forastiere, Heitmiller, & Kleinberg, 1997). The surgical approach may be through the thorax or the abdomen, depending on the location of the tumor. When tumors occur in the cervical or upper thoracic area, esophageal continuity may be maintained by a free jejunal graft transfer, in which the tumor is removed and the area is replaced with a portion of the jejunum (Fig. 46-9). A segment of the colon may be used, or the stomach can be elevated into the chest and the proximal section of the esophagus anastomosed to the stomach (Ott, Bader, Lordick, et al., 2009).

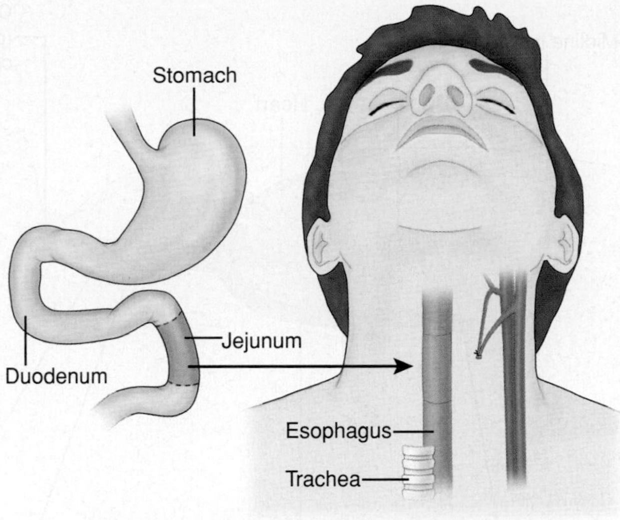

Stomach
Jejunum
Duodenum
Esophagus
Trachea

FIGURE 46-9 • Esophageal reconstruction with free jejunal transfer. A portion of the jejunum is grafted between the esophagus and pharynx to replace the abnormal portion of the esophagus. The vascular structures are also anastomosed.

Tumors of the lower thoracic esophagus are more amenable to surgery than are tumors located higher in the esophagus. GI tract integrity is maintained by anastomosing the lower esophagus to the stomach (Fig. 46-10).

Surgical resection of the esophagus has a relatively high mortality rate because of infection, pulmonary complications, or leakage through the anastomosis. Postoperatively, the patient has a nasogastric tube in place that should not be manipulated. The patient is given nothing by mouth until x-ray studies confirm that the anastomosis is free from an esophageal leak, there is no obstruction, and there is no evidence of pulmonary aspiration.

Palliative treatment may be necessary to keep the esophagus open, to assist with nutrition, and to control saliva. Palliation may be accomplished with dilation of the esophagus, laser therapy, placement of an endoprosthesis (stent) via EGD, radiation, or chemotherapy.

Nursing Management

Preoperative nursing management is directed toward improving the patient's nutritional and physical status in preparation for surgery, radiation therapy, or chemotherapy. A program to promote weight gain based on a high-calorie and high-protein diet, in liquid or soft form, is provided if adequate food can be taken by mouth. If this is not possible, parenteral or enteral nutrition is initiated. Nutritional status is monitored throughout treatment. The patient is

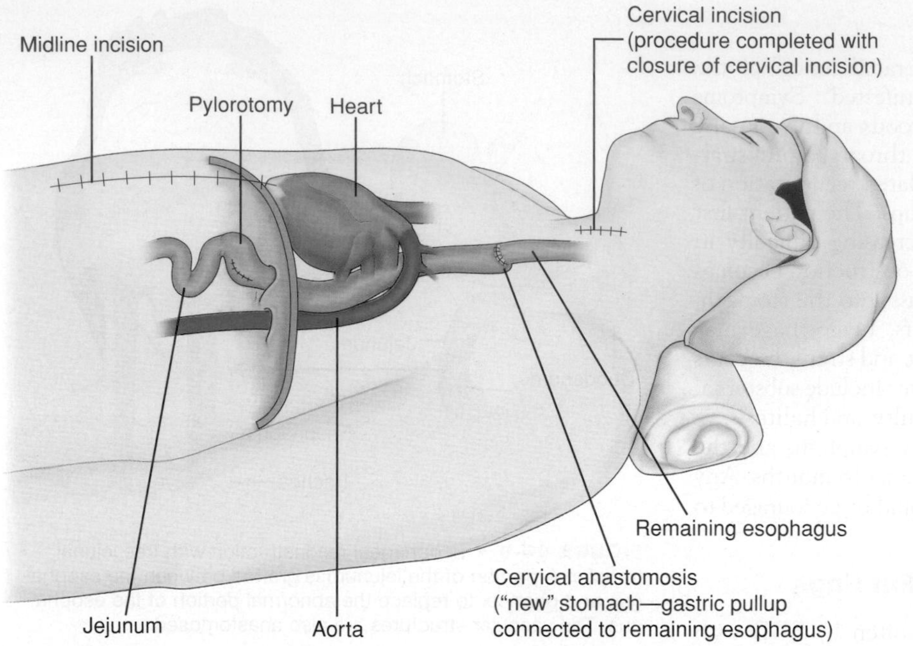
Midline incision
Pylorotomy
Heart
Cervical incision (procedure completed with closure of cervical incision)
Remaining esophagus
Cervical anastomosis ("new" stomach—gastric pullup connected to remaining esophagus)
Jejunum
Aorta

FIGURE 46-10 • Transhiatal esophagectomy. Surgical removal of tumor of the lower esophagus with anastomosis of the remaining esophagus to the stomach. Redrawn with permission from Heitmiller, R. F. Closed chest esophageal resection. *Operative Techniques in Thoracic and Cardiovascular Surgery, 4*(3), 263 © 1999 Elsevier Inc.

informed about the nature of the postoperative equipment that will be used, including that required for closed chest drainage, nasogastric suction, parenteral fluid therapy, and gastric intubation.

Immediate postoperative care is similar to that provided for patients undergoing thoracic surgery. It is not uncommon for patients to have a tracheostomy and be placed in an intensive care unit or step-down unit. After recovering from the effects of anesthesia, the patient is placed in a low Fowler's position, and later in a Fowler's position, to help prevent reflux of gastric secretions. The patient is observed carefully for regurgitation and dyspnea. A common postoperative complication is aspiration pneumonia. Therefore, the patient is placed on a vigorous pulmonary plan of care that includes incentive spirometry, sitting up in a chair, and, if necessary, nebulizer treatments. Chest physiotherapy is avoided due to the risk of aspiration. The patient's temperature is monitored to detect any elevation that may indicate aspiration or seepage of fluid through the operative site into the mediastinum, which would indicate an esophageal leak. Drainage from the cervical neck wound, usually saliva, is evidence of an early esophageal leak. Typically, no treatment other than nothing by mouth and parenteral or enteral support is warranted.

Cardiac complications include atrial fibrillation, which occurs due to irritation of the vagus nerve at the time of surgery. Typical medical management includes digitalization or the use of beta-blockers, depending on the patient's response. Rarely, cardioversion may be used.

During surgery, a nasogastric tube is inserted and taped in place. It is connected to low intermittent suction. The nasogastric tube is not manipulated; if displacement occurs, it is not replaced because damage to the anastomosis may occur. The nasogastric tube is typically removed 5 days after surgery; before the patient is allowed to eat, a barium swallow is performed to assess for any anastomotic leak.

Once feeding begins, the nurse encourages the patient to swallow small sips of water. Eventually, the diet is advanced as tolerated to a soft, mechanical diet. When the patient can increase his or her food and fluid intake to an adequate amount, parenteral fluids are discontinued. After each meal, the patient remains upright for at least 2 hours to allow the food to move through the GI tract. It is a challenge to encourage the patient to eat because the appetite is usually poor. Family involvement and home-cooked favorite foods may help the patient to eat. Antacids may help patients with gastric distress. Metoclopramide is useful in promoting gastric motility.

If chemotherapy and radiation are part of the therapy, the patient's appetite will be further depressed, and esophagitis may occur, causing pain when food is eaten. Liquid supplements may be more easily tolerated. However, supplements such as Boost and Ensure should be avoided because they promote **vagotomy syndrome** (dumping syndrome), which can occur with each meal or approximately 20 minutes to 2 hours after eating. The vagotomy syndrome occurs due to interruption of vagal nerve fibers, which in turn causes an alteration in the storage function of the stomach and the pyloric emptying mechanism. As a result, large amounts of solids and liquids rapidly "dump" into the duodenum. The patient experiences severe abdominal cramping, followed by a liquid bowel movement that may or may not be associated with diaphoresis, rapid heart rate or rapid respirations, or both. It can be quite disabling but typically resolves without incident, and the patient is left feeling extremely tired. The vagotomy syndrome is common following esophageal surgery, but as the patient's recovery progresses and the patient begins to eat soft foods and remains in an upright position for 2 hours after eating, the frequency and severity of episodes decrease.

Often, in either the preoperative or the postoperative period, an obstructed or nearly obstructed esophagus causes difficulty with excess saliva, and drooling becomes a problem. Oral suction may be used if the patient cannot handle oral secretions, or a wick-type gauze may be placed at the corner of the mouth to direct secretions to a dressing or emesis basin. The possibility that the patient may aspirate saliva into the tracheobronchial tree and develop pneumonia is a concern.

When the patient is ready to go home, the family is instructed about how to promote nutrition, what observations to make, what measures to take if complications occur, how to keep the patient comfortable, and how to obtain needed physical and emotional support.

Critical Thinking Exercises

1 ebp A 35-year-old white man presents to the medical clinic with a white spot on the inside of his mouth. He states that he has been using smokeless tobacco for 15 years. He does not like to smoke cigarettes because he is afraid of contracting lung cancer. How would you educate this patient about the similarities between smokeless tobacco and cigarettes as they relate to cancer? What other questions you would ask this patient? What would be the next step that this patient should take? What is the strength of the evidence that guides your planned intervention with this patient?

2 pq You are a nurse in the intensive care unit (ICU) and have been assigned a new postoperative patient who has had a neck resection due to cancer. The patient returns to the ICU with neck drains that are set to bulb suction. You note that the patient has drained 100 mL of bright red blood within the first 2 hours postoperatively. Two hours later, the patient has another 150 mL of bloody drainage. What is your first nursing action? Why?

3 pq A 70-year-old woman presents to the emergency department with complaints of chest pain. Upon further questioning, this patient reports she had an endoscopic procedure for evaluation of a hiatal hernia earlier today. She now reports excruciating chest pain and shortness of breath. Chest x-ray reveals a widened mediastinum. Why might this patient be experiencing chest pain? Twenty minutes later, the patient continues to be uncomfortable, a surgery consult is obtained, and the patient is noted to have enlarging tissue of the neck and face and a stridorous voice. It is identified that this patient has increasing subcutaneous emphysema. What would be the most appropriate plan of action for this patient? What would the nurse anticipate and why?

Brunner Suite Resources

Explore these additional resources to enhance learning for this chapter:

- NCLEX-Style Questions and Other Resources on thePoint, http://thePoint.lww.com/Brunner13e
- Study Guide
- PrepU
- Clinical Handbook
- Handbook of Laboratory and Diagnostic Tests

References

*Asterisk indicates classic reference.

Books

American Cancer Society. (2012). *Cancer facts and figures*. Atlanta, GA: Author.

Bickley, L. S. (2009). *Bates' guide to physical examination and history taking* (10th ed.). Philadelphia: Lippincott Williams & Wilkins.

*Castell, D. O., & Richter, J. E. (2003). *The esophagus* (4th ed.). Philadelphia: Lippincott Williams & Wilkins.

*Cox, J. L. (1999). Operative techniques in thoracic and cardiovascular surgery: A comparative atlas. Philadelphia: W. B. Saunders.

Fonseca, R. J. (Ed.). (2009). *Oral and maxillofacial surgery* (2nd ed.). Philadelphia: W. B. Saunders.

Mendenhall, W. M., Werning, J. W., & Pfister, D. G. (2011). Treatment of head and neck cancer. In V.T. DeVita, Jr., T. S. Lawrence TS, & S. A. Rosenberg (Eds.). *Cancer: Principles and practice of oncology* (9th ed., pp. 729–780). Philadelphia: Lippincott Williams & Wilkins.

Perry, M. C. (2008). *The chemotherapy source book* (4th ed.). Philadelphia: Lippincott Williams & Wilkins.

U.S. Department of Health and Human Services. (2010). *Oral health in America: A report of the surgeon general. Executive summary*. Rockville, MD: National Institutes of Dental and Craniofacial Research, National Institutes of Health.

Journals and Electronic Documents

Adelstein, D. J., & Rodriguez, C. P. (2010). Human papillomavirus: Changing paradigms in oropharyngeal cancer. *Current Oncology Report, 12*(2), 115–120.

Agrawal, A., Hila, A., Tutuian, R., et al. (2006). Clinical relevance of the nutcracker esophagus: Suggested revision of criteria for diagnosis. *Journal of Clinical Gastroenterology, 40*(6), 504.

American Academy of Periodontology. (2008). *Mouth-Body connection*. Available at: www.perio.org

American Association for Cancer Research. (2012). *Genetic predictors of esophageal cancer identified*. Available at: bethesdatrials.cancer.gov

Bartlett, D. (2010). Nissen fundoplication for repair of hiatal hernia. *OR Nurse 2010, 4*(2), 18–24.

Buescher, J. J. (2007). Temporomandibular joint disorders. *American Family Physician, 76*(10), 1477–1484.

Centers for Disease Control and Prevention. (2010a). *Healthy people 2010: Oral health objectives*. Available at: www.cdc.gov/oralhealth/topics/healthy_people.htm#2HealthyPeople2010

Centers for Disease Control and Prevention. (2010b). *Oral health: Preventing cavities, gum disease, tooth loss, and oral cancers: At a glance 2010*. Available at: www.cdc.gov/OralHealth/topics

Comay, D., Blackhouse, G., Goeree, R., et al. (2007). Photodynamic therapy for Barrett's esophagus with high-grade dysplasia: A cost-effectiveness analysis. *Canadian Journal of Gastroenterology, 21*(4), 217–222.

Cowgill, S. M., Villadoid, D., Boyle, R., et al. (2009). Laparoscopic Heller myotomy for achalasia: Results after 10 years. *Surgical Endoscopy, 23*(12), 2644–2649.

De Bree, R., van der Waal, I., & Leemans, C. R. (2008). Management of Frey syndrome. *Head and Neck, 29*(8), 773–778.

*Dent J. (2006). From 1906 to 2006—a century of major evolution of understanding of gastro-oesophageal reflux. *Alimentary Pharmacology and Therapeutics, 24*(9), 1269–1281.

Dughera, L., Chiaverina, M., Cacciotella, L., et al. (2011). Management of achalasia. *Clinical and Experimental Gastroenterology, 25*(4), 33–41.

Ferreira, L. E., Simmons, D. T., & Baron T. H. (2008). Zencker's diverticula: Pathophysiology, clinical presentation, and flexible endoscopic management. *Disease of the Esophagus, 21*(1), 1–8.

*Forastiere, A. A., Heitmiller, R. F., & Kleinberg, L. (1997). Multimodal therapy for esophageal cancer. *Chest, 112*(4), 195S–200S.

Garg, S., Yoo, J., & Winquist, E. (2010). Nutritional support for head and neck cancer patients receiving radiotherapy: A systematic review. *Supportive Care in Cancer, 18*(6), 667–677.

Gastroparesis and Dysmotilities Association. (2011). *Prokinetic drugs (motility Rx)*. Available at: www.digestivedistress.com/motility-rx

Gawron, A. J., & Hirano, I. (2010). Advances in diagnostic testing for gastroesophageal reflux disease. *World Journal of Gastroenterology, 16*(30), 3750–3756.

Gillison, M. L. (2007). Current topics in the epidemiology of oral cavity and oropharyngeal cancers. *Head Neck, 29*(8), 779–792.

Holmlund, A., Holm, G., & Lind, L. (2010). Number of teeth as a predictor of cardiovascular mortality in a cohort of 7,674 subjects followed for 12 years. *Journal of Periodontology, 81*(6), 870–876.

Ingawale, S., & Goswami, T. (2009). Temporomandibular joint: Disorders, treatments, and biomechanics. *Annals of Biomedical Engineering, 37*(5), 976–996.

Johnson, K. (2012). *Lifestyle changes to manage heartburn*. Available at: www.webmd.com/heartburn-gerd/guide/lifestyle-changes-heartburn

Kardon, E. M. (2011). *Caustic ingestions*. Available at: emedicine.medscape.com/article/813772-overview#a0199

National Cancer Institute. (2012a). *Salivary gland cancer treatment*. Available at: www.cancer.gov/cancertopics/pdq/treatment/salivarygland/HealthProfessional

National Cancer Institute. (2012b). *What You Need to Know About™ cancer of the esophagus*. Available at: www.cancer.gov/cancertopics/wyntk/esophagus/page9

National Institute of Dental and Craniofacial Research. (2011). *TMJ (temporomandibular joint and muscle disorders)*. Available at: www.nidcr.nih.gov/OralHealth/topics/TMJ

National Library of Medicine and National Institutes of Health. (2012). *Hiatal hernia*. Available at: www.nlm.nih.gov/medlineplus/hiatalhernia.html

Ott, K., Bader, F., Lordick, F., et al. (2009). Surgical factors influence the outcome after Ivor Lewis esophagectomy with intrathoracic anastomosis for adenocarcinoma of esophagogastric junction: A consecutive series of 240 patients at an experienced center. *Annals of Surgical Oncology, 16*(4), 1017–1025.

Pohl, D., & Tutian, R. (2007). Achalasia: An overview of diagnosis and treatment. *Journal of Gastrointestinal and Liver Diseases*, 6(3), 297–303.

Robertson, D., & Smith, A. J. (2012). The microbiology of the acute dental abscess. *Journal of Medical Microbiology, 58*(2), 155–162.

Roy, P. K. (2011). *Boerhaave syndrome*. Available at: emedicine.medscape.com/article/171683-overview

Scrivani, S. J., Keith, D. A., & Kaban, L. B. (2008). Temporomandibular disorders. *New England Journal of Medicine*, 359, 2693–2705.

Sidebottom, A. (2009). Current thinking in temporomandibular joint management. *British Journal of Oral and Maxillofacial Surgery, 47*(2), 91–94.

Souza R. (2012). The pathogenesis of gastroesophageal reflux disease. *Gastroenterology and Hepatology*, 8(4) 263–265.

Tessier, D. J. (2011). *Chyle fistula*. Available at: emedicine.medscape.com/article/190025-overview

U.S. Department of Health and Human Services. (2012). *Healthy people 2020: Oral health*. Available at: www.healthypeople.gov/2020/topicsobjectives2020/overview.aspx?topicid=32

Van Lierop, A. C., Basson, O., & Fagan, J. J. (2008). Is total glossectomy for advanced carcinoma of the tongue justified? *South African Journal of Surgery*, 46(1), 22–25.

Wang, K. K., & Sampliner, R. E. (2008). Updated guidelines 2008 for the diagnosis, surveillance and therapy of Barrett's esophagus. *American Journal of Gastroenterology, 103*, 788–797.

Williams, R. C. (2008). Understanding and managing periodontal disease: A notable past, a promising future. *Journal of Periodontology, 79*(8), 1552–1559.

World Health Organization. (2012). *The objectives of the WHO Global Oral Health Programme (ORH)*. Available at: www.who.int/oral_health/objectives/en/index.html

Resources

Academy of General Dentistry (AGD), agd.org

American Cancer Society, www.cancer.org; www.cancer.org/Treatment/SupportProgramsServices/i-can-cope

American Dental Association (ADA), www.ada.org

Centers for Disease Control and Prevention (CDC), www.cdc.gov

Look Good Feel Better, lookgoodfeelbetter.org/

National Institute of Dental and Craniofacial Research (NIDCR), National Institutes of Health, www.nidr.nih.gov

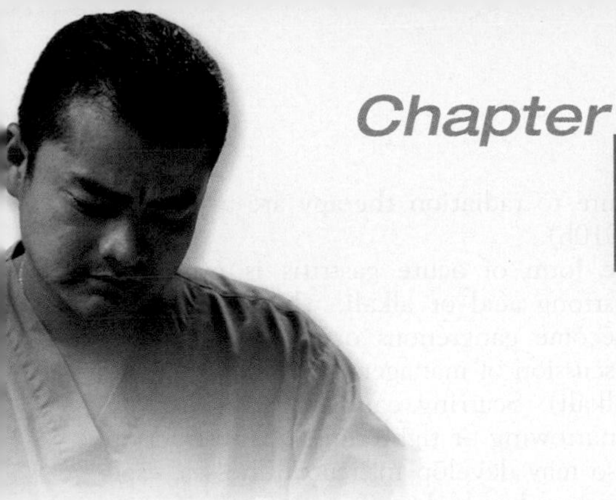

Chapter 47

Management of Patients With Gastric and Duodenal Disorders

Learning Objectives

On completion of this chapter, the learner will be able to:

1 Compare the etiology, clinical manifestations, and management of acute gastritis, chronic gastritis, and peptic ulcer.
2 Use the nursing process as a framework for care of patients with peptic ulcer.
3 Describe the pharmacologic, dietary, and surgical treatment of peptic ulcer.
4 Describe the causes, classifications, and morbid complications associated with obesity and management strategies for treating obesity.
5 Use the nursing process as a framework for care of patients who undergo bariatric surgical procedures.
6 Use the nursing process as a framework for care of patients with gastric cancer.
7 Identify the complications of bariatric or other gastric surgical procedures and their prevention and management.
8 Describe the home health care needs of the patient who has had bariatric or other gastric surgical procedures.
9 Discuss the etiology, clinical manifestations, and management of tumors of the small intestine.

Glossary

achlorhydria: lack of hydrochloric acid in digestive secretions of the stomach

antrectomy: removal of the pyloric (antrum) portion of the stomach with anastomosis (surgical connection) to the duodenum (gastroduodenostomy or Billroth I) or anastomosis to the jejunum (gastrojejunostomy or Billroth II)

bariatric: relating to obesity; term derives from two Greek words meaning “weight” and “treatment”

dumping syndrome: physiologic response to rapid emptying of gastric contents into the jejunum, manifested by nausea, weakness, sweating, palpitations, syncope, and possibly diarrhea; occurs in patients who have had partial gastrectomy and gastrojejunostomy

duodenum: first portion of the small intestine, between the stomach and the jejunum

dysphagia: difficulty swallowing

enteroclysis: fluoroscopic x-ray of the small intestine; a tube is placed from the nose or mouth through the esophagus and the stomach to the duodenum, a barium-based liquid contrast material is infused through the tube, and x-rays are taken as it travels through the duodenum

gastric: refers to the stomach

gastric outlet obstruction: any condition that mechanically impedes normal gastric emptying; there is obstruction of the channel of the pylorus and duodenum through which the stomach empties

gastritis: inflammation of the stomach

Helicobacter pylori: a spiral-shaped gram-negative bacterium that colonizes the gastric mucosa; is involved in most cases of peptic ulcer disease

hematemesis: vomiting of blood

hematochezia: bright red, bloody stools

melena: tarry or black stools; indicative of occult blood in stools

morbidly obese: body mass index exceeding 40 kg/m^2

obese: body mass index exceeding 30 kg/m^2

omentum: fold of the peritoneum that surrounds the stomach and other organs of the abdomen

overweight: body mass index exceeding 25 kg/m^2

peritoneum: thin membrane that lines the inside of the wall of the abdomen and covers all of the abdominal organs

pyloroplasty: surgical procedure to increase the opening of the pyloric orifice

pylorus: opening between the stomach and the duodenum

pyrosis: a burning sensation in the stomach and esophagus that moves up to the mouth; commonly called heartburn

serosa: thin membrane that covers the outer surface of the stomach; visceral peritoneum covering the outer surface of the stomach

steatorrhea: fatty stool; typically malodorous with an oily appearance and floats in water

stenosis: narrowing or tightening of an opening or passage in the body

A person’s nutritional status depends not only on the type and amount of intake but also on the functioning of the gastric and intestinal portions of the gastrointestinal (GI) system. The scope of disorders that may affect a person’s nutritional status is of particular note—according to the National Institute of Diabetes and Digestive and Kidney Diseases (NIDDK; 2010a), as many as 70 million Americans are diagnosed with digestive diseases. This statistic does not include the rates of obesity in the United States, which has reached epidemic proportions. Indeed, the Centers for Disease Control and Prevention (CDC; 2009) claims that the United States has become an "obesogenic" nation. Twice as many Americans are **obese** today than in 1980; furthermore, 32.6% are **overweight**, 34.3% are obese, and 5.9% are **morbidly obese** (Marzen-Groller & Cheever, 2010). Given the high prevalence of Americans who have digestive diseases or who are obese, nurses will encounter adults with these disorders in virtually every inpatient and outpatient clinical setting. This chapter describes disorders of the stomach and small intestine and the disease of obesity, as well as their treatment and related nursing care.

Gastritis

Gastritis (inflammation of the **gastric** or stomach mucosa) is a common GI problem, accounting for approximately 2 million visits to outpatient clinics annually in the United States. It affects women and men equally and is more common in older adults (Wehbi, Griglione, Snyder, et al., 2012). Gastritis may be acute, lasting several hours to a few days, or chronic, resulting from repeated exposure to irritating agents or recurring episodes of acute gastritis. Gastritis may also be classified as erosive or nonerosive, based upon pathologic manifestations present in the stomach wall (Wehbi et al., 2012).

Nonerosive gastritis (acute and chronic) is most often caused by infection with ***Helicobacter pylori*** (*H. pylori*) (Wehbi et al., 2012). It is estimated that between 20% and 50% of all Americans are infected with *H. pylori* (NIDDK, 2010b). Erosive gastritis (acute and chronic) is most often caused by long-term use of nonsteroidal anti-inflammatory drugs (NSAIDs) (e.g., ibuprofen [Motrin]); alcohol abuse and recent exposure to radiation therapy are also implicated (NIDDK, 2010b).

A more severe form of acute gastritis is caused by the ingestion of strong acid or alkali, which may cause the mucosa to become gangrenous or to perforate (see Chapter 72 for discussion of management of patients who ingest acids or alkali). Scarring can occur, resulting in pyloric **stenosis** (narrowing or tightening) or obstruction. Acute gastritis also may develop in acute illnesses, especially when the patient has had major traumatic injuries; burns; severe infection; hepatic, kidney, or respiratory failure; or major surgery. Gastritis may be the first sign of an acute systemic infection (Ruggiero, 2012).

Chronic gastritis is sometimes associated with autoimmune diseases such as pernicious anemia (see Chapter 33). Chronic gastritis caused by *H. pylori* infection is implicated in the development of peptic ulcers, gastric cancer, and mucosa-associated lymphoid tissue lymphoma (Mukherjee & Sepulveda, 2012).

Pathophysiology

In gastritis, the gastric mucous membrane becomes edematous and hyperemic (congested with fluid and blood) and undergoes superficial erosion (Fig. 47-1). It secretes a scanty amount of gastric juice, containing very little acid but much mucus. Superficial ulceration may occur as a result of erosive disease and may lead to hemorrhage (Porth & Matfin, 2009).

Clinical Manifestations

The patient with acute gastritis may have a rapid onset of symptoms, such as abdominal discomfort, headache, lassitude, nausea, anorexia, vomiting, and hiccupping, which can last from a few hours to a few days. Erosive gastritis may cause

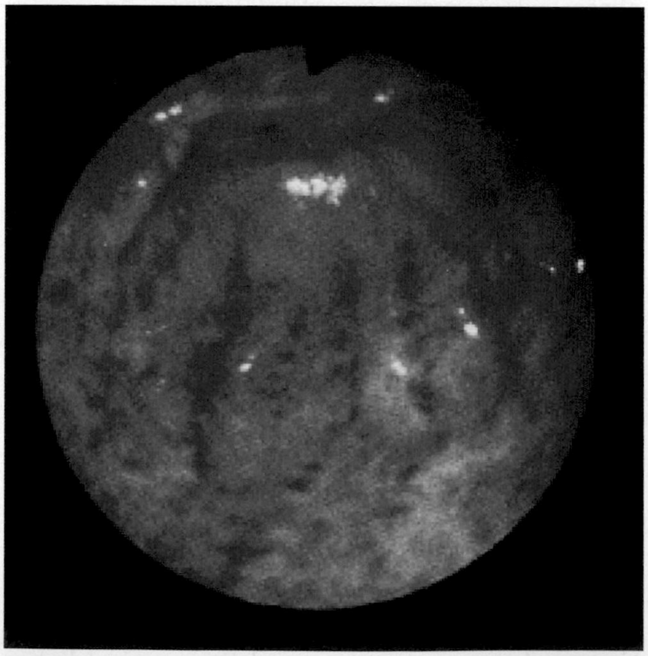

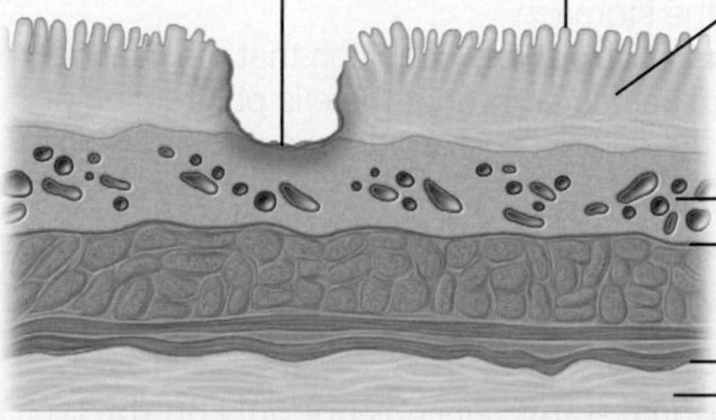
Cells secrete hydrochloric (HCl) acid or pepsinogen and intrinsic factor
Damaged mucosa
Mucosa (provides a barrier to protect stomach from HCl acid)
Submucosa
Muscularis
Serosa

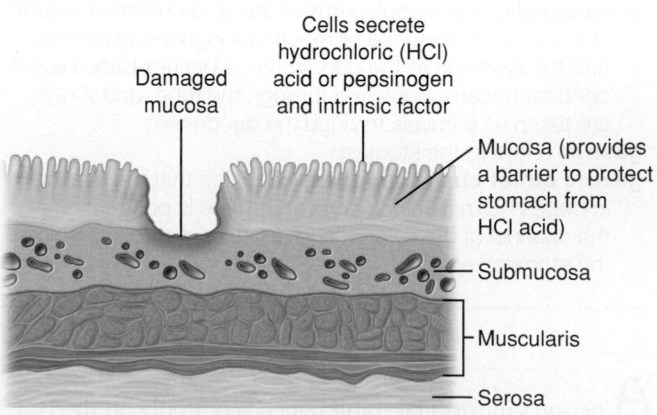

FIGURE 47-1 • Endoscopic view of erosive gastritis (*left*). Damage from irritants (*right*) results in increased intracellular pH, impaired enzyme function, disrupted cellular structures, ischemia, vascular stasis, and tissue death. Reproduced with permission from Porth, C. M., & Matfin, G. (2009). *Pathophysiology: Concepts of altered health states* (8th ed.). Philadelphia: Lippincott Williams & Wilkins.

bleeding, which may manifest as blood in vomit or as black, tarry stools (i.e., **melena**) or bright red, bloody stools (i.e., **hematochezia**) (NIDDK, 2010b).

The patient with chronic gastritis may complain of anorexia, heartburn after eating, belching, a sour taste in the mouth, or nausea and vomiting. Some patients may have only mild epigastric discomfort or report intolerance to spicy or fatty foods or slight pain that is relieved by eating. Patients with chronic gastritis may not be able to absorb vitamin B_{12} because of diminished production of intrinsic factor by the stomach's parietal cells, which may lead to pernicious anemia (see Chapter 33 for further discussion). However, some patients with chronic gastritis have no symptoms (Fujinami, Kodo, Hosokawsa, et al., 2012; Wataru, Yasuhiko, Katsunori, et al., 2012).

Assessment and Diagnostic Findings

Gastritis is sometimes associated with **achlorhydria** (lack of hydrochloric acid [HCl], hypochlorhydria (low levels of HCl), or hyperchlorhydria (high levels of HCl). Diagnosis is determined by an endoscopy and histologic examination of a tissue specimen obtained by biopsy (Greenberger, Blumberg, & Burakoff, 2012). Diagnostic measures for detecting *H. pylori* infection may be used and are discussed later in this chapter in the Peptic Ulcer Disease section.

Medical Management

The gastric mucosa is capable of repairing itself after an episode of acute gastritis. As a rule, the patient recovers in about 1 day, although the appetite may be diminished for an additional 2 or 3 days. Acute gastritis is also managed by instructing the patient to refrain from alcohol and food until symptoms subside. When the patient can take nourishment by mouth, a nonirritating diet is recommended. If the symptoms persist, intravenous (IV) fluids may need to be administered. If bleeding is present, management is similar to the procedures used to control upper GI tract hemorrhage discussed later in this chapter.

Therapy is supportive and may include nasogastric (NG) intubation, antacids, histamine-2 receptor antagonists (H_2 blockers) (e.g., famotidine [Pepcid], ranitidine [Zantac]), proton pump inhibitors (e.g., omeprazole [Prilosec], lansoprazole [Prevacid]), and IV fluids (NIDDK, 2010b). Fiberoptic endoscopy may be necessary. In extreme cases, emergency surgery may be required to remove gangrenous or perforated tissue. A gastric resection or a gastrojejunostomy (anastomosis of jejunum to stomach to detour around the pylorus) may be necessary to treat **gastric outlet obstruction**, also called *pyloric obstruction*, a narrowing of the pyloric orifice, which cannot be relieved by medical management.

Chronic gastritis is managed by modifying the patient's diet, promoting rest, reducing stress, recommending avoidance of alcohol and NSAIDs, and initiating medications that may include antacids, H_2 blockers, or proton pump inhibitors (NIDDK, 2010b). *H. pylori* may be treated with selected drug combinations (Table 47-1).

Nursing Management

Reducing Anxiety

If the patient has ingested acids or alkalis, emergency measures may be necessary (see Chapter 72 for further discussion). The nurse offers supportive therapy to the patient and family during treatment and after the ingested acid or alkali has been neutralized or diluted. In some cases, the nurse may need to prepare the patient for additional diagnostic studies (endoscopies) or surgery. The patient may be anxious because of pain and planned treatment modalities. The nurse uses a calm approach to assess the patient and to answer all questions as completely as possible.

Promoting Optimal Nutrition

For acute gastritis, the nurse provides physical and emotional support and helps the patient manage the symptoms, which may include nausea, vomiting, heartburn, and fatigue. The patient should take no foods or fluids by mouth—possibly for a few days—until the acute symptoms subside, thus allowing the gastric mucosa to heal. If IV therapy is necessary, the nurse monitors fluid intake and output along with serum electrolyte values. After the symptoms subside, the nurse may offer the patient ice chips followed by clear liquids. Introducing solid food as soon as possible may provide adequate oral nutrition, decrease the need for IV therapy, and minimize irritation to the gastric mucosa. As food is introduced, the nurse evaluates and reports any symptoms that suggest a repeat episode of gastritis.

The nurse discourages the intake of caffeinated beverages, because caffeine is a central nervous system stimulant that increases gastric activity and pepsin secretion. The nurse also discourages alcohol use. Discouraging cigarette smoking is important because nicotine reduces the secretion of pancreatic bicarbonate, which inhibits the neutralization of gastric acid in the duodenum (Alexandre, Pandol, Gorelick, et al., 2011). When appropriate, the nurse initiates and refers the patient for alcohol counseling and smoking cessation programs.

Promoting Fluid Balance

Daily fluid intake and output are monitored to detect early signs of dehydration (minimal fluid intake of 1.5 L/day, minimal output of 0.5 mL/kg/h). If food and oral fluids are withheld, IV fluids (3 L/day) usually are prescribed and a record of fluid intake plus caloric value (1 L of 5% dextrose in water = 170 calories of carbohydrate) needs to be maintained. Electrolyte values (sodium, potassium, chloride) are assessed every 24 hours to detect any imbalance.

The nurse must always be alert to any indicators of hemorrhagic gastritis, which include **hematemesis** (vomiting of blood), tachycardia, and hypotension. All stools should be examined for the presence of frank or occult bleeding. If these occur, the primary provider is notified and the patient's vital signs are monitored as the patient's condition warrants. Guidelines for managing upper GI tract bleeding are discussed later in this chapter.

Relieving Pain

Measures to help relieve pain include instructing the patient to avoid foods and beverages that may irritate the gastric mucosa as well as the correct use of medications to relieve chronic gastritis. The nurse must regularly assess the patient's level of pain and the extent of comfort achieved through the use of medications and avoidance of irritating substances.

TABLE 47-1 Selected Pharmacotherapy for Peptic Ulcer Disease and Gastritis

Pharmacologic Agent	Major Action	Key Nursing Considerations
Antibiotics		
Amoxicillin (Amoxil)	A bactericidal antibiotic that assists with eradicating *H. pylori* bacteria in the gastric mucosa	• May cause diarrhea • Should not be used in patients allergic to penicillin
Clarithromycin (Biaxin)	Exerts bactericidal effects to eradicate *H. pylori* bacteria in the gastric mucosa	• May cause GI upset, headache, altered taste • Many drug–drug interactions (e.g., colchicine [Colcrys], lovastatin [Mevacor], warfarin [Coumadin])
Metronidazole (Flagyl)	A synthetic antibacterial and antiprotozoal agent that assists with eradicating *H. pylori* bacteria in the gastric mucosa when administered with other antibiotics and proton pump inhibitors	• Should be administered with meals to decrease GI upset; may cause anorexia and metallic taste • Patient should avoid alcohol; increases blood-thinning effects of warfarin
Tetracycline	Exerts bacteriostatic effects to eradicate *H. pylori* bacteria in the gastric mucosa	• May cause photosensitivity reaction; advise patient to use sunscreen. • May cause GI upset • Must be used with caution in patients with renal or hepatic impairment • Milk or dairy products may reduce effectiveness
Antidiarrheal		
Bismuth subsalicylate (Pepto-Bismol)	Suppresses *H. pylori* bacteria in the gastric mucosa and assists with healing of mucosal ulcers	• Given concurrently with antibiotics to eradicate *H. pylori* infection • Should be taken on empty stomach
H_2 Receptor Antagonists		
Cimetidine (Tagamet)	Decreases amount of HCl produced by stomach by blocking action of histamine on histamine receptors of parietal cells in the stomach	• Least expensive of H_2 receptor antagonists • May cause confusion, agitation, or coma in older adults or those with renal or hepatic insufficiency • Long-term use may cause diarrhea, dizziness, gynecomastia. • Many drug–drug interactions (e.g., amiodarone [Cordarone], amitriptyline [Elavil], benzodiazepines, metoprolol [Lopressor], nifedipine [Procardia], phenytoin [Dilantin], warfarin)
Famotidine (Pepcid)	Same as for cimetidine	• Best choice for critically ill patient because it is known to have the least risk of drug–drug interactions; does not alter liver metabolism • Prolonged half-life in patients with renal insufficiency • Short-term relief for GERD
Nizatidine (Axid)	Same as for cimetidine	• Used for treatment of ulcers and GERD • Prolonged half-life in patients with renal insufficiency • May cause headache, dizziness, diarrhea, nausea/vomiting, GI upset, and urticaria
Ranitidine (Zantac)	Same as for cimetidine	• Prolonged half-life in patients with renal and hepatic insufficiency • Causes fewer side effects than cimetidine • May cause headache, dizziness, constipation, nausea and vomiting, or abdominal discomfort
Proton Pump Inhibitors of Gastric Acid		
Esomeprazole (Nexium)	Decreases gastric acid secretion by slowing the H^+, K^+-ATPase pump on the surface of the parietal cells of the stomach	• Used mainly for treatment of duodenal ulcer disease and *H. pylori* infection • A delayed-release capsule that is to be swallowed whole and taken before meals
Lansoprazole (Prevacid)	Decreases gastric acid secretion by slowing the H^+, K^+-ATPase pump on the surface of the parietal cells	• A delayed-release capsule that is to be swallowed whole and taken before meals
Omeprazole (Prilosec)	Same as for lansoprazole	• A delayed-release capsule that is to be swallowed whole and taken before meals • May cause diarrhea, nausea, constipation, abdominal pain, vomiting, headache, or dizziness
Pantoprazole (Protonix)	Same as for lansoprazole	• A delayed-release tablet that is to be swallowed whole and taken before meals • May cause diarrhea and hyperglycemia, headache, abdominal pain, and abnormal liver function tests
Rabeprazole (AcipHex)	Same as for lansoprazole	• A delayed-release tablet to be swallowed whole • May cause abdominal pain, diarrhea, nausea, and headache • Drug–drug interactions with digoxin, iron, and warfarin
Prostaglandin E_1 Analogue		
Misoprostol (Cytotec)	Synthetic prostaglandin; protects the gastric mucosa from agents that cause ulcers; also increases mucus production and bicarbonate levels	• Used to prevent ulceration in patients using NSAIDs • Administer with food. • May cause diarrhea and cramping (including uterine cramping)
Sucralfate (Carafate)	Creates a viscous substance in the presence of gastric acid that forms a protective barrier, binding to the surface of the ulcer, and prevents digestion by pepsin	• Used mainly for the treatment of duodenal ulcers • Should be taken without food but with water • Other medications should be taken 2 hours before or after this medication. • May cause constipation or nausea

H. pylori, *Helicobacter pylori*; GI, gastrointestinal; H_2, histamine-2; HCl, hydrochloric acid; GERD, gastroesophageal reflux disease; H^+, K^+-ATPase, hydrogen-potassium adenosine triphosphatase; NSAIDs, nonsteroidal anti-inflammatory drugs.

Chart 47-1 HOME CARE CHECKLIST
The Patient With Gastritis

At the completion of home care education, the patient or caregiver will be able to:	PATIENT	CAREGIVER
• Identify foods and other substances that may cause gastritis.	✔	✔
• Report inability to ingest adequate solids and liquids.	✔	✔
• Describe medication regimen.	✔	✔
• State need for vitamin B_{12} injections if patient has pernicious anemia.	✔	✔
• State schedule of follow-up appointments with health care provider.	✔	✔

Promoting Home and Community-Based Care

Educating Patients About Self-Care

The nurse evaluates the patient's knowledge about gastritis and develops an individualized education plan that includes information about stress management, diet, and medications (Chart 47-1). Dietary instructions take into account the patient's daily caloric needs as well as cultural aspects of food preferences and patterns of eating. The nurse and patient review foods and other substances to be avoided (e.g., spicy, irritating, or highly seasoned foods; caffeine; nicotine; alcohol). Consultation with a dietitian may be recommended.

Providing information about prescribed medications, which may include antacids, H_2 blockers, or proton pump inhibitors, may help the patient to better understand why these medications assist in recovery and prevent recurrence. The importance of completing the medication regimen as prescribed to eradicate *H. pylori* infection must be reinforced to the patient and caregiver (see later discussion of use these medications in the Peptic Ulcer Disease section).

Continuing Care

The nurse reinforces previous instruction and conducts ongoing assessment of the patient's symptoms and progress. Patients with malabsorption of vitamin B_{12} need information about lifelong vitamin B_{12} injections; the nurse may instruct a family member or caregiver how to administer the injections or make arrangements for the patient to receive the injections from a health care provider. Finally, the nurse emphasizes the importance of keeping follow-up appointments with health care providers.

Peptic Ulcer Disease

Peptic ulcer disease affects 14.5 million Americans and is responsible for approximately 1.4 million outpatient clinic visits and 489,000 inpatient hospitalizations annually (NIDDK, 2010a). A peptic ulcer may be referred to as a gastric, duodenal, or esophageal ulcer, depending on its location. A peptic ulcer is an excavation (hollowed-out area) that forms in the mucosal wall of the stomach, in the **pylorus** (the opening between the stomach and duodenum), in the **duodenum** (the first part of the small intestine), or in the esophagus. Erosion of a circumscribed area of mucous membrane is the cause (Fig. 47-2). This erosion may extend as deeply as the muscle layers or through the muscle to the **peritoneum** (thin membrane that lines the inside of the wall of the abdomen).

Peptic ulcers are more likely to occur in the duodenum than in the stomach. As a rule they occur alone, but they may occur in multiples. Chronic gastric ulcers tend to occur in the lesser curvature of the stomach, near the pylorus. Esophageal ulcers occur as a result of the backward flow of HCl from the stomach into the esophagus (gastroesophageal reflux disease [GERD]).

The rates of peptic ulcer disease among middle-aged adults have diminished over the past three decades, whereas the rates among older adults have increased (Anand, Bank, Qureshi, et al., 2012). Those who are 65 years and older present to both outpatient and inpatient settings for treatment of peptic ulcers more than any other age group (Everhart, 2009). Women and men have equivalent risks of developing peptic ulcers (Anand et al., 2012).

In the past, stress and anxiety were thought to be causes of ulcers, but research has documented that most peptic ulcers result from infection with the gram-negative bacteria *H. pylori*, which may be acquired through ingestion of food and water. Person-to-person transmission of the bacteria also occurs through close contact and exposure to emesis. Although *H. pylori* infection is common in the United States, most infected people do not develop ulcers. It is not known why *H. pylori* infection does not cause ulcers in all people, but most likely the predisposition to ulcer formation depends

FIGURE 47-2 • Deep peptic ulcer. From Rubin, R., Strayer, D. S., & Rubin, E. (2012). *Rubin's pathology: Clinicopathologic foundations of medicine* (6th ed.). Philadelphia: Lippincott Williams & Wilkins.

on certain factors, such as the type of *H. pylori* and other as yet unknown factors (Alakkari, Zullo, & O'Connor, 2011; Zelickson, Bronder, Johnson, et al., 2011).

The use of NSAIDs such as ibuprofen and aspirin is also a major risk factor for peptic ulcers. Furthermore, infection with *H. pylori* and concomitant use of NSAIDs are synergistic risks (Anand et al., 2012). Excessive secretion of HCl in the stomach may contribute to the formation of peptic ulcers, and stress may be associated with its increased secretion. It is believed that smoking and alcohol consumption may be risks, although the evidence is inconclusive. There is no evidence that the ingestion of milk, caffeinated beverages, and spicy foods are associated with the development of peptic ulcers (Anand et al., 2012).

Familial tendency also may be a significant predisposing factor. People with blood type O are more susceptible to peptic ulcers than are those with blood type A, B, or AB. There also is an association between peptic ulcers and chronic pulmonary disease or chronic kidney disease (Anand et al., 2012).

Peptic ulcers are found in rare cases of patients with tumors that cause secretion of excessive amounts of the hormone gastrin. The Zollinger-Ellison syndrome (ZES) consists of severe peptic ulcers, extreme gastric hyperacidity, and gastrin-secreting benign or malignant tumors of the pancreas.

Pathophysiology

Peptic ulcers occur mainly in the gastroduodenal mucosa because this tissue cannot withstand the digestive action of gastric acid (HCl) and pepsin. The erosion is caused by the increased concentration or activity of acid–pepsin or by decreased resistance of the mucosa. A damaged mucosa cannot secrete enough mucus to act as a barrier against HCl. The use of NSAIDs inhibits the secretion of mucus that protects the mucosa. Patients with duodenal ulcers secrete more acid than normal, whereas patients with gastric ulcers tend to secrete normal or decreased levels of acid. Damage to the gastroduodenal mucosa results in decreased resistance to bacteria, and thus infection from *H. pylori* bacteria may occur (Ruggiero, 2010, 2012).

ZES is suspected when a patient has several peptic ulcers or an ulcer that is resistant to standard medical therapy. It is identified by the following: hypersecretion of gastric juice, duodenal ulcers, and gastrinomas (islet cell tumors) in the pancreas. Seventy-five percent of tumors are found in the "gastric triangle," which encompasses the cystic and common bile ducts, the second and third portions of the duodenum, and the junction of the head and body of the pancreas (Metz, 2012; Ozlem, Yazici, Ince, et al., 2011). Most gastrinomas grow slowly; 60% to 90% of these are malignant (Lopez, Falconi, Waldmann, et al., 2012; Norton, Fraker, Alexander, et al., 2012). Diarrhea and **steatorrhea** (i.e., fat in the stool) may be evident. The patient may have coexisting parathyroid adenomas or hyperplasia and may therefore exhibit signs of hypercalcemia. The most common symptom is epigastric pain.

Stress ulcer is the term given to the acute mucosal ulceration of the duodenal or gastric area that occurs after physiologically stressful events, such as burns, shock, severe sepsis, and multiple organ traumas. These ulcers, which are clinically different from peptic ulcers, are most common in ventilator-dependent patients after trauma or surgery. Fiberoptic endoscopy within 24 hours of trauma or surgery reveals shallow erosions of the stomach wall; by 72 hours, multiple gastric erosions are observed. As the stressful condition continues, the ulcers spread. When the patient recovers, the lesions are reversed. This pattern is typical of stress ulceration.

Differences of opinion exist as to the actual cause of mucosal ulceration in stress ulcers. Usually, the ulceration is preceded by shock; this leads to decreased gastric mucosal blood flow and to reflux of duodenal contents into the stomach. In addition, large quantities of pepsin are released. The combination of ischemia, acid, and pepsin creates an ideal climate for ulceration (Ruggiero, 2010).

Specific types of ulcers that result from stressful conditions include Curling's ulcers and Cushing's ulcers. Curling's ulcer is frequently observed about 72 hours after extensive burns and involves the antrum of the stomach or the duodenum. Cushing's ulcers are common in patients with head injury and brain trauma. They may occur in the esophagus, stomach, or duodenum and are usually deeper and more penetrating than typical stress ulcers (Hoshino, Satoh, Narita, et al., 2011; Wijdicks, 2011).

Clinical Manifestations

Symptoms of an ulcer may last for a few days, weeks, or months and may disappear only to reappear, often without an identifiable cause. Many people with ulcers have no symptoms, and perforation or hemorrhage may occur in 20% to 30% of patients who had no preceding manifestations (Greenberger et al., 2012).

As a rule, the patient with an ulcer complains of dull, gnawing pain or a burning sensation in the midepigastrium or the back. There are few clinical manifestations that differentiate gastric ulcers from duodenal ulcers; however, classically, the pain associated with gastric ulcers most commonly occurs immediately after eating, whereas the pain associated with duodenal ulcers most commonly occurs 2 to 3 hours after meals. In addition, approximately 50% to 80% of patients with duodenal ulcers awake with pain during the night, whereas 30% to 40% of patients with gastric ulcers voice this type of complaint. Patients with duodenal ulcers are more likely to express relief of pain after eating or after taking an antacid than patients with gastric ulcers (Anand et al., 2012).

Other nonspecific symptoms of either gastric ulcers or duodenal ulcers may include **pyrosis** (heartburn), vomiting, constipation or diarrhea, and bleeding. Pyrosis is a burning sensation in the stomach and esophagus that moves up to the mouth. It is often accompanied by sour eructation (burping), which is common when the patient's stomach is empty.

Although vomiting is rare in uncomplicated peptic ulcer, it may be a symptom of a complication of an ulcer. It results from gastric outlet obstruction, caused by either muscular spasm of the pylorus or mechanical obstruction from scarring or acute swelling of the inflamed mucous membrane adjacent to the ulcer. Vomiting may or may not be preceded by nausea; usually, it follows a bout of severe pain and bloating, which is relieved by vomiting. Emesis may contain undigested food eaten many hours earlier. Constipation or diarrhea may occur, probably as a result of diet and medications.

Fifteen percent of patients with peptic ulcer experience bleeding (Greenberger et al., 2012; Milosavljevic, Kostic-Milosavljevic, Jovanovic, et al., 2011). Patients may present with GI bleeding as evidenced by hematemesis or the passage of melena (Anand et al., 2012; Greenberger et al., 2012).

Assessment and Diagnostic Findings

A physical examination may reveal pain, epigastric tenderness, or abdominal distention. Upper endoscopy is the preferred diagnostic procedure because it allows direct visualization of inflammatory changes, ulcers, and lesions. Through endoscopy, a biopsy of the gastric mucosa and any suspicious lesions can be obtained. Endoscopy may reveal lesions that, because of their size or location, are not evident on x-ray studies. *H. pylori* infection may be determined by endoscopy and histologic examination of a tissue specimen obtained by biopsy, or a rapid urease test of the biopsy specimen. Other less invasive diagnostic measures for detecting *H. pylori* include serologic testing for antibodies against the *H. pylori* antigen, stool antigen test, and urea breath test (Anand et al., 2012).

The patient who has a bleeding peptic ulcer may require periodic complete blood counts (CBCs) to determine the extent of blood loss and whether or not blood transfusions are advisable (see Chapter 32). Stools may be tested periodically until they are negative for occult blood. Gastric secretory studies are of value in diagnosing achlorhydria and ZES.

Medical Management

Once the diagnosis is established, the patient is informed that the condition can be managed. Recurrence may develop; however, peptic ulcers treated with antibiotics to eradicate *H. pylori* have a lower recurrence rate than those not treated with antibiotics. The goals are to eradicate *H. pylori* as indicated and to manage gastric acidity. Methods used include medications, lifestyle changes, and surgical intervention.

Pharmacologic Therapy

Currently, the most commonly used therapy for peptic ulcers is a combination of antibiotics, proton pump inhibitors, and bismuth salts that suppress or eradicate *H. pylori*. Recommended therapy for 10 to 14 days includes triple therapy with two antibiotics (e.g., metronidazole [Flagyl] or amoxicillin [Amoxil] and clarithromycin [Biaxin]) plus a proton pump inhibitor (e.g., lansoprazole [Prevacid], omeprazole [Prilosec], or rabeprazole [AcipHex]), or quadruple therapy with two antibiotics (metronidazole and tetracycline) plus a proton pump inhibitor and bismuth salts (Pepto-Bismol) (Anand et al., 2012). Research is being conducted to develop a vaccine against *H. pylori* (Czinn & Blanchard, 2011).

H_2 blockers and proton pump inhibitors that reduce gastric acid secretion are used to treat ulcers not associated with *H. pylori* infection. Table 47-2 provides information about the medication regimens for peptic ulcer disease. (See Table 47-1 for details about specific medications.)

The patient is advised to adhere to and complete the medication regimen to ensure complete healing of the ulcer. The patient is advised to avoid the use of NSAIDs. Because most patients become symptom free within a week, the nurse stresses to the patient the importance of following the prescribed regimen so that the healing process can continue uninterrupted and the return of chronic ulcer symptoms can be prevented. Maintenance dosages of H_2 blockers are usually recommended for 1 year.

For patients with ZES, hypersecretion of acid may be controlled with high doses of H_2 blockers. These patients

TABLE 47-2 Drug Regimens for Peptic Ulcer Disease

Indications	Drug Regimen	Nursing Considerations
Ulcer healing	**H_2 receptor antagonists:** Ranitidine 150 mg bid or 300 mg at bedtime Cimetidine 400 mg bid or 800 mg at bedtime Famotidine 20 mg bid or 40 mg at bedtime Nizatidine 150 mg bid or 300 mg at bedtime	Should be used for 6 weeks for duodenal ulcer and 8 weeks for gastric ulcer
	PPIs: Omeprazole 20 mg daily Lansoprazole 30 mg daily Rabeprazole 20 mg daily Pantoprazole 40 mg daily Esomeprazole 40 mg daily	Should be used for 4 weeks for duodenal ulcer and 6 weeks for gastric ulcer Healing occurs in 90% of patients who adhere to therapy
Initial *H. pylori* therapy	**First line:** PPI bid plus clarithromycin 500 mg bid plus amoxicillin 1,000 mg bid *or* metronidazole 500 mg bid for 10–14 days	Efficacy of therapy is approximately 85%
	Second line: Bismuth subsalicylate 2 tabs qid plus tetracycline 250 mg qid plus metronidazole 250 mg qid (optional: add PPI daily) for 14 days	Qid dosing may decrease adherence to the regimen
Therapy for retreatment of *H. pylori* therapy failure	Repeat first-line therapy, substitute metronidazole for amoxicillin (or vice versa) for 14 days; may add Bismuth subsalicylate. Add second-line *H. pylori* therapy.	Efficacy of retreatment not known; success of more than 2 courses of treatment is very low
Prophylactic therapy for NSAID ulcers	Peptic ulcer healing doses of PPIs (above) Misoprostol 200 mcg bid	Prevents recurrent ulceration in approximately 80%–90% of patients

H_2, histamine-2; bid, two times a day; PPIs, proton pump inhibitors; qid, four times a day; NSAID, nonsteroidal anti-inflammatory drug.
Adapted from Anand, B. S., Bank, S., Qureshi, W. A., et al. (2012). Peptic ulcer disease. *Medscape*. Available at: emedicine.medscape.com/article/181753

may require twice the normal dose, and dosages usually need to be increased with prolonged use. Octreotide (Sandostatin), a medication that suppresses gastrin levels, also may be prescribed.

Patients at risk for stress ulcers (e.g., patients with head injury or extensive burns) may be treated prophylactically with IV H_2 blockers and cytoprotective agents (e.g., misoprostol, sucralfate) because of the risk of upper GI tract hemorrhage (Clarke, Ferraro, Gbadehan, et al., 2011).

Smoking Cessation

Smoking decreases the secretion of bicarbonate from the pancreas into the duodenum, resulting in increased acidity of the duodenum. Research indicates that continued smoking may significantly inhibit ulcer repair (Zelickson et al., 2011). Therefore, the patient is encouraged to stop smoking. (Refer to Chapter 27 for information on how the nurse may promote cessation of tobacco use.)

Dietary Modification

The intent of dietary modification for patients with peptic ulcers is to avoid oversecretion of acid and hypermotility in the GI tract. These can be minimized by avoiding extremes of temperature in food and beverages and overstimulation from the consumption of alcohol, coffee (including decaffeinated coffee, which also stimulates acid secretion), and other caffeinated beverages. In addition, an effort is made to neutralize acid by eating three regular meals a day. Small, frequent feedings are not necessary as long as an antacid or a H_2 blocker is taken. Diet compatibility becomes an individual matter: The patient eats foods that are tolerated and avoids those that produce pain.

Surgical Management

The introduction of antibiotics to eradicate *H. pylori* and of H_2 blockers as treatment for ulcers has greatly reduced the need for surgical intervention (Wilkins, Khan, Nabh, et al., 2012). However, surgery is usually recommended for patients with intractable ulcers (those failing to heal after 12 to 16 weeks of medical treatment), life-threatening hemorrhage, perforation, or obstruction and for those with ZES that is unresponsive to medications (Zelickson et al., 2011). Surgical procedures include vagotomy, with or without **pyloroplasty** (transecting nerves that stimulate acid secretion and opening the pylorus), and **antrectomy**, which is removal of the pyloric (antrum) portion of the stomach with anastomosis (surgical connection) to either the duodenum (gastroduodenostomy or Billroth I) or jejunum (gastrojejunostomy or Billroth II) (Table 47-3; see also the Gastric Surgery section later in this chapter).

Follow-Up Care

Recurrence of peptic ulcer disease within 1 year may be prevented with the prophylactic use of H_2 blockers taken at a reduced dose. Not all patients require maintenance therapy; it may be prescribed only for those with two or three recurrences per year, those who have had a complication such as bleeding or gastric outlet obstruction, or those for whom gastric surgery poses too high a risk. The likelihood of recurrence is reduced if the patient avoids smoking, coffee (including decaffeinated coffee) and other caffeinated beverages, alcohol, and ulcerogenic medications (e.g., NSAIDs).

NURSING PROCESS

The Patient With Peptic Ulcer Disease

Assessment

The nurse asks the patient to describe the pain, its pattern and whether or not it occurs predictably (e.g., after meals, during the night), and strategies used to relieve it (e.g., food, antacids). If the patient reports a recent history of vomiting, the nurse determines how often emesis has occurred and notes important characteristics of the vomitus: Is it bright red, does it resemble coffee grounds, or is there undigested food from previous meals? Has the patient noted any bloody or tarry stools?

The nurse also asks the patient to list his or her usual food intake for a 72-hour period. Lifestyle and other habits are a concern as well. For example, does he or she smoke cigarettes? If yes, how many? Does the patient ingest alcohol? If yes, how much and how often? Are NSAIDs used? Is there a family history of ulcer disease?

The nurse assesses the patient's vital signs and reports tachycardia and hypotension, which may indicate anemia from GI bleeding. The stool is tested for occult blood, and a physical examination, including palpation of the abdomen for localized tenderness, is performed.

Diagnosis

Nursing Diagnoses

Based on the assessment data, nursing diagnoses may include the following:

- Acute pain related to the effect of gastric acid secretion on damaged tissue
- Anxiety related to an acute illness
- Imbalanced nutrition: less than body requirements related to changes in diet

Collaborative Problems/Potential Complications

Potential complications may include the following:

- Hemorrhage
- Perforation
- Penetration
- Gastric outlet obstruction

Planning and Goals

The goals for the patient may include relief of pain, reduced anxiety, maintenance of nutritional requirements, knowledge about the management and prevention of ulcer recurrence, and absence of complications.

Nursing Interventions

Relieving Pain

Pain relief can be achieved with prescribed medications. The patient should avoid aspirin and other NSAIDs and alcohol. In addition, meals should be eaten at regularly paced intervals in a relaxed setting. Medications prescribed to treat the peptic ulcer should provide relief of ulcer-associated pain. Some patients benefit from learning relaxation techniques to help manage stress and pain.

Reducing Anxiety

The nurse assesses the patient's level of anxiety. Explaining diagnostic tests and administering medications as scheduled

TABLE 47-3 Surgical Procedures for Peptic Ulcer Disease

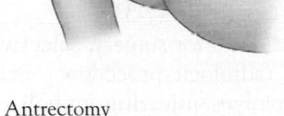

Operation	Description	Adverse Effects
Vagotomy 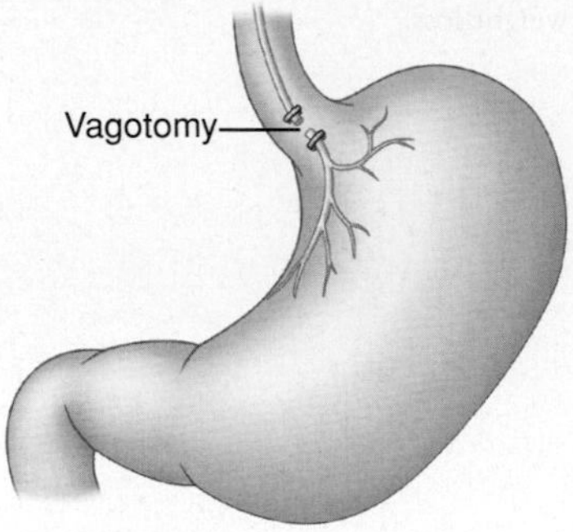Vagotomy	Severing of the vagus nerve. Decreases gastric acid by diminishing cholinergic stimulation to the parietal cells, making them less responsive to gastrin. May be performed via open surgical approach, laparoscopy, or thoracoscopy. May be performed to reduce gastric acid secretion. A drainage type of procedure (see pyloroplasty) is usually performed to assist with gastric emptying (because there is total denervation of the stomach).	Some patients experience problems with feeling of fullness, dumping syndrome, diarrhea, and gastritis.
Truncal vagotomy	Severs the right and left vagus nerves as they enter the stomach at the distal part of the esophagus; most commonly used to decrease acid secretions	Some patients experience problems with feeling of fullness, dumping syndrome, diarrhea, or constipation.
Selective vagotomy	Severs vagal innervation to the stomach but maintains innervation to the rest of the abdominal organs	Fewer associated adverse effects than with truncal vagotomy
Proximal (parietal cell) gastric vagotomy without drainage	Denervates acid-secreting parietal cells but preserves vagal innervation to the gastric antrum and pylorus	No associated dumping syndrome
Pyloroplasty 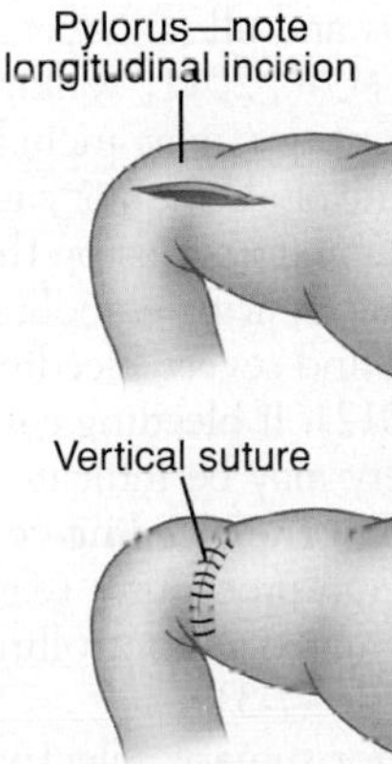Pylorus—note longitudinal incicion Vertical suture	Longitudinal incision is made into the pylorus and transversely sutured closed to enlarge the outlet and relax the muscle; usually accompanies truncal and selective vagotomies.	See adverse effects associated with truncal and selective vagotomies, as appropriate.
Antrectomy Billroth I (gastroduodenostomy) 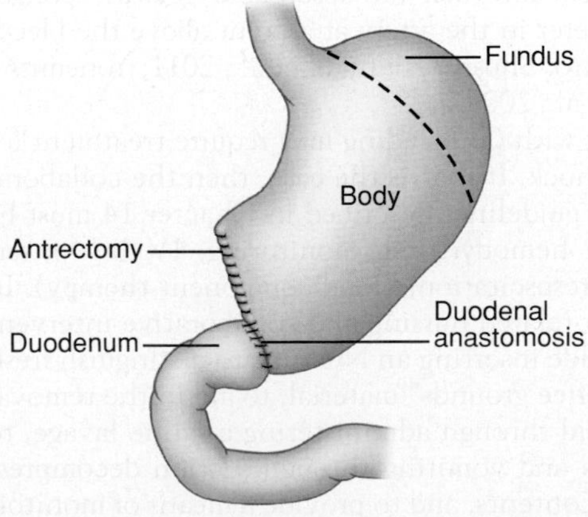Fundus Body Antrectomy Duodenal anastomosis Duodenum	Removal of the lower portion of the antrum of the stomach (which contains the cells that secrete gastrin) as well as a small portion of the duodenum and pylorus. The remaining segment is anastomosed to the duodenum. May be performed in conjunction with a truncal vagotomy.	Patients may have problems with feeling of fullness, dumping syndrome, and diarrhea.

(continues on page 1270)

TABLE 47-3 Surgical Procedures for Peptic Ulcer Disease (continued)

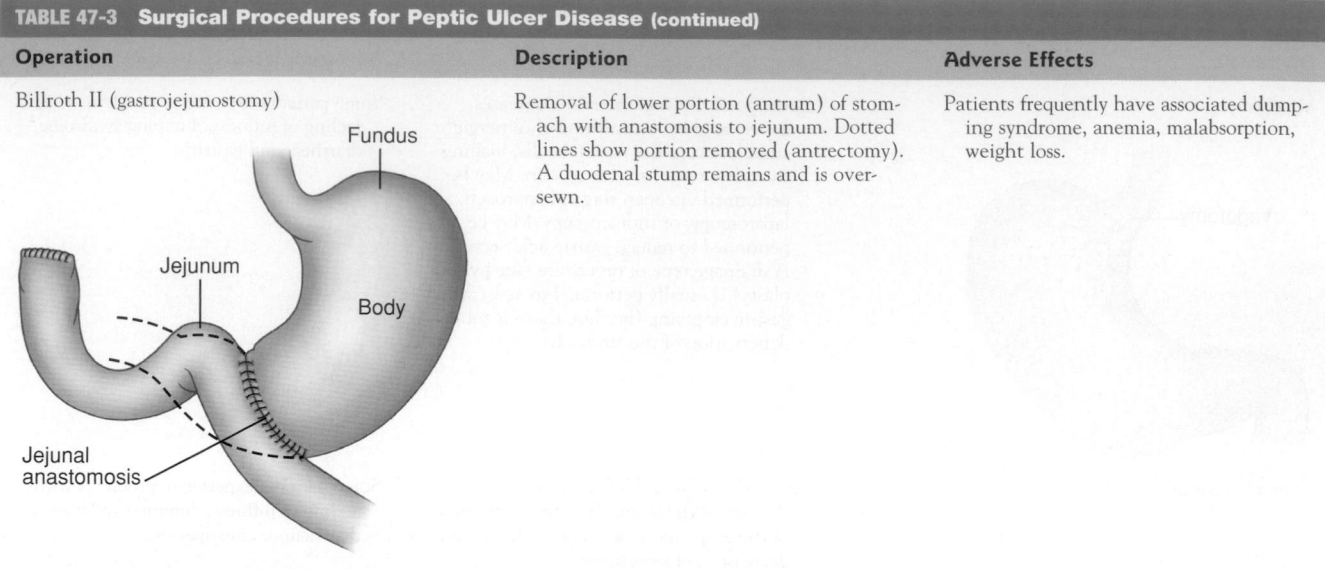

Operation	Description	Adverse Effects
Billroth II (gastrojejunostomy) 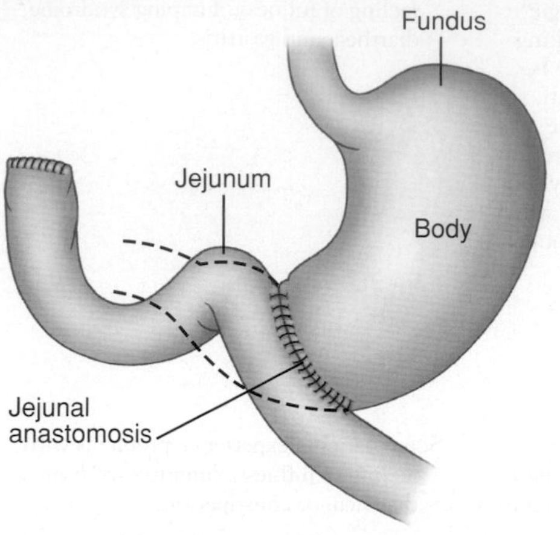 Fundus Jejunum Body Jejunal anastomosis	Removal of lower portion (antrum) of stomach with anastomosis to jejunum. Dotted lines show portion removed (antrectomy). A duodenal stump remains and is oversewn.	Patients frequently have associated dumping syndrome, anemia, malabsorption, weight loss.

help reduce anxiety. The nurse interacts with the patient in a relaxed manner; helps identify stressors; and explains various coping techniques and relaxation methods, such as biofeedback, hypnosis, or behavior modification. The patient's family is also encouraged to participate in care and to provide emotional support.

Maintaining Optimal Nutritional Status

The nurse assesses the patient for malnutrition and weight loss. After recovery from an acute phase of peptic ulcer disease, the patient is advised about the importance of adhering to the medication regimen and dietary restrictions.

Monitoring and Managing Potential Complications

Hemorrhage. Gastritis and hemorrhage from peptic ulcer are the two most common causes of upper GI tract bleeding (which may also occur with esophageal varices, as discussed in Chapter 49). Hemorrhage, the most common complication, occurs in 15% of patients with peptic ulcers (Greenberger et al., 2012). Bleeding may be manifested by hematemesis or melena (Kumar & Mills, 2011). The vomited blood can be bright red, or it can have a dark coffee grounds appearance from the oxidation of hemoglobin to methemoglobin. When the hemorrhage is large (2,000 to 3,000 mL), most of the blood is vomited. Because large quantities of blood may be lost quickly, immediate correction of blood loss may be required to prevent hemorrhagic shock. When the hemorrhage is small, much or all of the blood is passed in the stools, which appear tarry black because of the digested hemoglobin. Management depends on the amount of blood lost and the rate of bleeding.

The nurse assesses the patient for faintness or dizziness and nausea, which may precede or accompany bleeding. The nurse must monitor vital signs frequently and evaluate the patient for tachycardia, hypotension, and tachypnea. Other nursing interventions include monitoring the hemoglobin and hematocrit, testing the stool for gross or occult blood, and recording hourly urinary output to detect anuria or oliguria (absence of or decreased urine production).

Many times, the bleeding from a peptic ulcer stops spontaneously; however, the incidence of recurrent bleeding is high. Because bleeding can be fatal, the cause and severity of the hemorrhage must be identified quickly and the blood loss treated to prevent hemorrhagic shock. The nurse monitors the patient carefully so that bleeding can be detected quickly. Patients suspected of having an ulcer who present with symptoms of GI bleeding should undergo evaluation with endoscopy within 24 hours to confirm the diagnosis and allow targeted endoscopic interventions (Milosavljevic et al., 2011; Wilkins et al., 2012). These endoscopic interventions may include injecting the bleeding site with epinephrine or alcohol, or cauterizing the site, or clipping the ulcer, all in efforts to stop the bleeding (Anand et al., 2012). Arteriography with embolization may be needed if there is persistent and severe bleeding (Kumar & Mills, 2011; Wilkins et al., 2012). If bleeding cannot be managed by these methods, surgery may be indicated, in which the area of the ulcer is removed or the bleeding vessels are ligated. Many patients also undergo procedures (e.g., vagotomy and pyloroplasty, gastrectomy) aimed at controlling the underlying cause of the ulcers (see Table 47-3).

For patients who are not candidates for surgery, selective embolization—an interventional radiologic procedure—may be indicated. This procedure involves injecting emboli of autologous blood clots (i.e., the patient's own clots), sometimes with Gelfoam (i.e., an absorbable gelatin sponge) through a catheter in the artery at a point above the bleeding lesion (Ichiro, Shushi, Akihiko, et al., 2011; Yonemitsu, Kawai, Sato, et al., 2009).

The patient with GI bleeding may require treatment for hemorrhagic shock. If that is the case, then the collaborative treatment guidelines described in Chapter 14 must be followed (e.g., hemodynamic monitoring, IV line insertion and fluid resuscitation, blood component therapy). In addition, other related nursing and collaborative interventions may include inserting an NG tube to distinguish fresh blood from "coffee grounds" material, to aid in the removal of clots and acid through administering a saline lavage, to prevent nausea and vomiting through suction decompression of gastric contents, and to provide a means of monitoring further bleeding.

Perforation and Penetration. Perforation is the erosion of the ulcer through the gastric **serosa** into the peritoneal cavity

Chart 47-2 HOME CARE CHECKLIST
The Patient With Peptic Ulcer Disease

At the completion of home care education, the patient or caregiver will be able to:	PATIENT	CAREGIVER
• State the medication regimen and importance of complying with medication schedule.	✔	✔
• State dietary restrictions and foods that may exacerbate condition.	✔	✔
• Identify smoking cessation groups.	✔	
• Identify methods to reduce stress.	✔	✔
• State signs and symptoms of complications:		
• Hemorrhage—cool skin, confusion, increased heart rate, labored breathing, blood in stool	✔	✔
• Penetration and perforation—severe abdominal pain, rigid and tender abdomen, vomiting, elevated temperature, increased heart rate	✔	✔
• Gastric outlet obstruction—nausea and vomiting, distended abdomen, abdominal pain	✔	✔
• State need for follow-up medical care.	✔	✔

without warning. It is an abdominal emergency and requires immediate surgery. Penetration is erosion of the ulcer through the gastric serosa into adjacent structures such as the pancreas, biliary tract, or gastrohepatic **omentum** (membranous fold of the peritoneum). Symptoms of penetration include back and epigastric pain not relieved by medications that were effective in the past. Like perforation, penetration usually requires surgical intervention.

Signs and symptoms of perforation include the following:

- Sudden, severe upper abdominal pain (persisting and increasing in intensity); pain may be referred to the shoulders, especially the right shoulder, because of irritation of the phrenic nerve in the diaphragm
- Vomiting
- Collapse (fainting)
- Extremely tender and rigid (boardlike) abdomen
- Hypotension and tachycardia, indicating shock

Because chemical peritonitis develops within a few hours of perforation and is followed by bacterial peritonitis, the perforation must be closed as quickly as possible and the abdominal cavity lavaged of stomach or intestinal contents. In some patients, it may be safe and advisable to perform surgery to treat the ulcer disease in addition to suturing the perforation.

During surgery and postoperatively, the stomach contents are drained by means of an NG tube. The nurse monitors fluid and electrolyte balance and assesses the patient for localized infection or peritonitis (increased temperature, abdominal pain, paralytic ileus, increased or absent bowel sounds, abdominal distention). Antibiotic therapy is administered parenterally as prescribed.

Gastric Outlet Obstruction. Gastric outlet obstruction occurs when the area distal to the pyloric sphincter becomes scarred and stenosed from spasm or edema or from scar tissue that forms when an ulcer alternately heals and breaks down. The patient may have nausea and vomiting, constipation, epigastric fullness, anorexia, and, later, weight loss.

In treating the patient with gastric outlet obstruction, the first consideration is to insert an NG tube to decompress the stomach. Confirmation that obstruction is the cause of the discomfort is accomplished by assessing the amount of fluid aspirated from the NG tube. A residual of more than 400 mL suggests obstruction. Usually, an upper GI study or endoscopy is performed to confirm gastric outlet obstruction. Decompression of the stomach and management of extracellular fluid volume and electrolyte balances may improve the patient's condition and avert the need for surgical intervention. Balloon dilation of the pylorus via endoscopy may be beneficial. If the obstruction is unrelieved by medical management, surgery (in the form of a vagotomy and antrectomy or gastrojejunostomy and vagotomy) may be required.

Promoting Home and Community-Based Care

Educating Patients About Self-Care. The nurse instructs the patient about the factors that relieve and those that aggravate the condition. The nurse reviews information about medications to be taken at home, including name, dosage, frequency, and possible side effects, stressing the importance of continuing to take medications even after signs and symptoms have decreased or subsided (Chart 47-2). The nurse instructs the patient to avoid certain medications and foods that exacerbate symptoms (e.g., NSAIDs, alcohol). If relevant, the nurse also informs the patient about the irritant effects of smoking on the ulcer and provides information about smoking cessation programs.

Continuing Care. The nurse reinforces the importance of follow-up care for approximately 1 year; the need to report recurrence of symptoms; and the need for treating possible problems that occur after surgery, such as intolerance to specific foods. The nurse also reminds the patient and family of the importance of participating in health promotion activities and recommended health screening.

Evaluation

Expected patient outcomes may include:

1. Reports freedom from pain between meals and at night
2. Reports feeling less anxiety
3. Maintains weight
4. Demonstrates knowledge of self-care activities
   a. Avoids irritating foods and beverages (alcohol) and medications (NSAIDs)
   b. Takes medications as prescribed
5. No evidence of complications (e.g., hemorrhage, perforation or penetration, gastric outlet obstruction)

Obesity

Obesity is not merely a condition; rather, it is a metabolic disease that is characterized by fat that accumulates to the extent that health is impaired (American Society for Metabolic and Bariatric Surgery [ASMBS], 2012). Patients identified as overweight have a body mass index (BMI) of 25 to 29.9 kg/m^2, and those considered obese have a BMI that exceeds 30 kg/m^2 (Artham, Lavie, Milani, et al., 2010). In the United States, obesity is a rapidly growing problem; approximately 66% of all adults are overweight or obese (Echols, 2010).

Traditionally, the term *morbid obesity* applies to adults whose BMI exceeds 40 kg/m^2. This term is beginning to be used less consistently by some clinicians, however, because morbidity that is associated with excess body weight may occur in some persons well before their BMIs reach 40 kg/m^2. Obesity-related mortality rates are 30% greater for every gain of 5 kg/m^2 of body mass beyond a BMI of 25 kg/m^2. Furthermore, the average lifespan for persons with BMIs in excess of 30 kg/m^2 is 3 years less than for persons with BMIs of 25 kg/m^2 or less (ASMBS, 2012). Many diseases and disorders are associated with obesity; these are displayed in Chart 47-3. Because obesity is associated with significant morbidity and mortality with BMIs that exceed 30 kg/m^2, and because morbidity and mortality rates increase steadily with increases in BMI, the ASMBS (2012) endorses the use of a classification system for obesity that is linked to health risks (Table 47-4).

Patients who are obese are not only at risk for a host of diseases and disorders; they also frequently suffer from low self-esteem, impaired body image, depression, and diminished quality of life. Obese adults are frequently stigmatized as being lazy, overindulgent, and lacking self-control (Marzen-Groller & Cheever, 2010). Studies suggest that health care providers, including nurses, also harbor negative attitudes toward patients who are obese (Poon & Tarrant, 2009) (Chart 47-4). Although behaviors that include poor eating habits and sedentary lifestyles may cause weight gain, environmental, genetic, metabolic, cultural, and socioeconomic factors are all believed to interconnect in a complex and still poorly understood relationship that then leads to obesity (Marzen-Groller & Cheever, 2010).

Chart 47-3 Diseases and Disorders Associated With Obesity

- Asthma
- Cancers (breast, endometrial, prostate, kidney, colon, gall bladder)
- Cerebrovascular accident (stroke)
- Cholecystitis
- Cholelithiasis
- Chronic back pain
- Coronary artery disease
- Diabetes (type 2)
- Heart failure
- Hypercholesterolemia
- Hypertension
- Nonalcoholic fatty liver disease
- Obstructive sleep apnea
- Osteoarthritis
- Pulmonary embolism

Adapted from American Society for Metabolic and Bariatric Surgery. (2012). *ASMBS position statement: Bariatric surgery in class I obesity (BMI 30–35 kg/m^2)*. Available at: www.asmbs.org/2012/09/bariatric-surgery-in-class-1-obesity-bmi-30-35-kgm2/; and Marzen-Groller, K., & Cheever, K. H. (2010). Facilitating students' competence in caring for the bariatric surgical patient: The case study approach. *Bariatric Nursing and Surgical Patient Care Journal*, *5*(2), 117–125.

TABLE 47-4 Classification of Obesity

Classification	BMI Range	Health Risk
Overweight	25–30 kg/m^2	Mild
Class I	30–35 kg/m^2	Moderate
Class II	35–40 kg/m^2	Severe
Class III	>40 kg/m^2	Very severe

BMI, body mass index.
Adapted from American Society for Metabolic and Bariatric Surgery. (2012). *ASMBS position statement: Bariatric surgery in class I obesity (BMI 30–35 kg/m^2)*. Available at: www.asmbs.org/2012/09/bariatric-surgery-in-class-1-obesity-bmi-30–35-kgm2/

Medical Management

There are three general approaches to treating obesity: lifestyle modifications, pharmacotherapy, and bariatric surgery.

Lifestyle Modification

The lifestyle modification approach to treating obesity consists of placing a person on a weight loss diet in conjunction with behavioral modification and exercise. Diet therapy is the most commonly prescribed therapy; it is generally safe and can be highly effective (see Chapter 5). The involvement of dietitians in meal planning can be effective in promoting dietary management. Unfortunately, a very small number of individuals who are prescribed weight loss diets adhere to dietary guidelines. Acupuncture in combination with diet restrictions has been found to be effective in enhancing weight loss and improving dyslipidemia (Abdi, Abbasi-Parizad, Zhao, et al., 2012; Abdi, Zhao, Darbandi, et al., 2012). Hypnosis in combination with diet therapy for stress reduction and energy intake reduction also has been shown to increase weight loss (Steyer & Ables, 2009).

Pharmacologic Management

Patients who are not successful at meeting weight loss goals from lifestyle modification may be prescribed antiobesity medications. Orlistat (Xenical), which is available both by prescription and over the counter as Alli, reduces caloric intake by binding to gastric and pancreatic lipase to prevent digestion of fats. Side effects of orlistat include increased frequency of bowel movements, gas with oily discharge, decreased food absorption, decreased bile flow, and decreased absorption of some vitamins. A multivitamin is usually recommended. Orlistat should not be taken by pregnant or nursing women or transplant recipients. Lorcaserin (Belviq), a new antiobesity drug, acts by stimulating serotonin receptors in the satiety and appetite centers of the hypothalamus in the brain, thus curbing appetite (Caveney, Caveney, Somaratne, et al., 2011). Associated side effects

Chart 47-4

NURSING RESEARCH PROFILE

Attitudes Toward Patients Who Are Obese

Poon, M. Y., & Tarrant, M. (2009). Obesity: Attitudes of undergraduate student nurses and registered nurses. *Journal of Clinical Nursing*, *18*(16), 2355–2365.

Purpose

Obesity is a global public health concern. Rising rates of obesity throughout the world suggest that it is poised to soon overtake smoking as the primary preventable cause of death in developed nations. Yet, it is believed that many persons who are obese are reluctant to engage in preventive health care practices because they feel that health care providers have negative attitudes toward them. The rates of obesity are steadily rising in China; however, few studies have examined the attitudes of Chinese nurses toward this population. Therefore, the purpose of this study was to ascertain the attitudes that Chinese nursing students and nurses have toward patients who are obese.

Design

This was a cross-sectional study that enrolled 352 prelicensure baccalaureate nursing students and 198 registered nurses enrolled in a baccalaureate degree completion program in Hong Kong. Each participant completed a survey that included the Fat Phobia Scale, the Attitudes Toward Obese Adult Patients Scale, and demographic information.

Findings

Both student nurse participants and registered nurse participants had average scores on the Fat Phobia Scale and neutral attitude scores toward caring for patients who are obese. However, the registered nurse participants had generally higher levels of fat phobia and more negative attitudes toward caring for patients who are obese than the student nurses.

Nursing Implications

These findings suggest that although Chinese nurses tend to have fat phobia, they do not let these feelings affect their attitudes toward caring for patients who are obese. However, registered nurses' attitudes toward this population was generally more negative than student nurses' attitudes in this study. These differences in attitudes may suggest that more negative attitudes develop either with age or with years of practice and experience. Nursing students and experienced nurses should be socialized to examine their attitudes toward patients who are obese so that they can deliver sensitive, optimal care to these patients.

are few and tend to diminish with continued use; these include headaches, dry mouth, fatigue, and nausea. Sibutramine HCl (Meridia) is an antiobesity drug that had been pulled from the U.S. market by the U.S. Food and Drug Administration (FDA) when its use was linked with higher risk of adverse cardiovascular events, including myocardial infarctions and strokes. Rimonabant (Acomplia) is an antiobesity drug that was approved for use in some European nations and was being clinically trialed in the United States when it was pulled from the market for being linked with risks of suicidality in patients.

Depression may contribute to weight gain, and treatment of the depression with an antidepressant may be helpful (Billes & Greenway, 2011; Ye, Chen, & Yang, 2011). Overall, although antiobesity drugs help some patients lose weight, their use rarely results in loss of more than 10% of total body weight. Furthermore, studies are needed to evaluate their long-term efficacy and risks (Caveney et al., 2011).

Surgical Management

Bariatric surgery, or surgery for obesity, is performed only after other nonsurgical attempts at weight control have failed. Most insurance companies will only authorize bariatric surgery after a patient who is obese tries 6 to 18 months of a medically supervised diet that fails to reach its goal of weight loss (ASMBS, 2011a).

The number of bariatric procedures in the United States dramatically increased over the past decade, from 13,386 in 1998 to 220,000 in 2008 (Dumon & Murayama, 2011). Bariatric surgical procedures work by restricting a patient's ability to eat (restrictive procedure), interfering with ingested nutrient absorption (malabsorptive procedures), or both. Different bariatric surgical procedures entail different lifestyle modifications, and patients must be well informed about the specific lifestyle changes, eating habits, and bowel habits that may result from a particular procedure.

Studies have shown that the average weight loss after bariatric surgery in the majority of patients is approximately 25% to 35% of previous body weight within the first 18 to 24 months; comorbid conditions such as diabetes, hypertension, and sleep apnea may resolve; and dyslipidemia improves (Gianos, Abdemur, Fendrich, et al., 2011). Bariatric surgery has been extended to carefully selected adolescents because of the positive results it has achieved in adults (Barnett, 2011; Michalsky, Kramer, Fullmer, et al., 2011).

Patient selection is critical, and the preliminary process may necessitate months of counseling, education, and evaluation by a multidisciplinary team, including social workers, dietitians, a nurse counselor, a psychologist or psychiatrist, and a surgeon. The selection criteria for patients has changed considerably since the advent of bariatric surgery, with patients with BMIs as low as 30 kg/m^2 now considered candidates for surgical intervention if they have comorbid conditions and fail to lose weight after supervised dieting (ASMBS, 2012) (Chart 47-5).

Because bariatric surgery involves a drastic change in the functioning of the digestive system, patients need counseling before and after the surgery. Guidelines have been developed to assist in the care of patients having bariatric surgery (Aills, Blankenship, Buffington, et al., 2008; Heber, Greenway, Kaplan, et al., 2010; LeMont, Moorehead, Parish, et al., 2004).

Roux-en-Y gastric bypass, gastric banding, vertical-banded gastroplasty, sleeve gastrectomy, and biliopancreatic diversion with duodenal switch are the current bariatric procedures of choice. These procedures may be performed by laparoscopy or by an open surgical technique. The Roux-en-Y gastric bypass and the sleeve gastrectomy are the two most commonly used procedures (ASMBS, 2012; Mechanick, Kushner, Sugerman, et al., 2008). The Roux-en-Y gastric bypass is a combined restrictive and malabsorptive procedure. The sleeve gastrectomy, gastric banding, and vertical-banded gastroplasty are

Chart 47-5 Selection Criteria for Bariatric Surgery

Body mass index (BMI) ≥40 kg/m^2 without excessive surgical risk
BMI ≥30 kg/m^2 with one or more obesity-associated comorbid conditions (e.g., coronary artery disease, type 2 diabetes, obstructive sleep apnea, hypertension, asthma, debilitating arthritis, or impaired quality of life)
Failure of previous nonsurgical attempts at weight loss, including nonprofessional programs
Expectation that patient will adhere to postoperative care, follow-up visits, and recommended medical management, including the use of dietary supplements

Exclusions

Reversible endocrine or other disorders that can cause obesity
Current drug or alcohol abuse
Uncontrolled, severe psychiatric illness
Lack of comprehension of risks, benefits, expected outcomes, alternatives, and lifestyle changes required with bariatric surgery

Adapted from American Society for Metabolic and Bariatric Surgery. (2012). *ASMBS position statement: Bariatric surgery in class I obesity (BMI 30–35 kg/m^2)*. Available at: www.asmbs.org/2012/09/bariatric-surgery-in-class-1-obesity-bmi-30-35-kgm2/; and Mechanick, J. I., Kushner, R. F., Sugerman, H. J., et al. (2008). American Association of Clinical Endocrinologists, the Obesity Society, and American Society for Metabolic and Bariatric Surgery: Medical guidelines for clinical practice for the perioperative nutritional, metabolic, and nonsurgical support of the bariatric surgery patient. *Surgery for Obesity and Related Diseases*, *4*, S109–S184.

restrictive procedures, and biliopancreatic diversion with duodenal switch combines gastric restriction with intestinal malabsorption. Figure 47-3A–E provides additional details about these procedures.

NURSING PROCESS

The Patient Undergoing Bariatric Surgery

Assessment

Preoperatively, the nurse assesses for contraindications to major abdominal surgery. Has the patient attempted to lose weight by nonoperative means such as nutritional counseling, dieting, or an exercise program? Has the patient received counseling regarding possible risks and benefits of surgery, complications, postsurgical outcomes, dietary changes, and the need for lifelong follow-up? The nurse ensures that the patient has been screened for mental or behavioral disorders that may interfere with postsurgical outcomes. Dietary counseling is initiated preoperatively to prepare for postoperative dietary changes.

The nurse ensures that preoperative screening tests are obtained and scrutinizes the results. Typical laboratory tests include a CBC, electrolytes, blood urea nitrogen (BUN), and creatinine. Patients who are obese may have sleep apnea, GERD, heart disease, nonalcoholic fatty liver disease, diabetes (or prediabetes), and vitamin and mineral deficiencies; thus, other screening tests that may be obtained include a sleep study, upper endoscopy, electrocardiogram, lipid panel, liver function tests, glucose, and hemoglobin A1c, as well as iron, vitamin B_{12}, thiamine, folate, vitamin D, and calcium levels.

Postoperatively, the nurse assesses the patient to ensure that goals for recovery are met and that the patient exhibits absence of complications secondary to the surgical intervention. (See Chapter 19 for general assessment of the postoperative patient.)

Diagnosis

Nursing Diagnoses

Based on the assessment data, major nursing diagnoses may include the following:

- Deficient knowledge about the dietary limitations during the immediate preoperative and postoperative phases
- Anxiety related to impending surgery
- Acute pain related to surgical procedure
- Risk for deficient fluid volume related to nausea, gastric irritation, and pain
- Risk for infection related to anastomotic leak
- Imbalanced nutrition: less than body requirements related to dietary restrictions
- Disturbed body image related to body changes from bariatric surgery
- Risk for constipation and/or diarrhea related to gastric irritation and surgical changes in anatomic structures from bariatric surgery

Collaborative Problems/Potential Complications

Potential complications may include the following:

- Hemorrhage
- Bile reflux
- Dumping syndrome
- Dysphagia
- Bowel or gastric outlet obstruction

Planning and Goals

Preoperative goals include that the patient will become knowledgeable about the preoperative and postoperative dietary routine/restrictions and will have decreased anxiety about the surgery. Postoperative goals include relief of pain, maintenance of homeostatic fluid balance and asepsis, adherence to detailed diet instructions to include progression of food intake as well as fluid intake (to prevent dehydration), knowledge about vitamin supplements and the need for lifelong follow-up, achievement of a positive body image, and maintenance of normal bowel habits (ASMBS, 2011b; Compher, Hanlon, Kang, et al., 2012).

Nursing Interventions

Ensuring Dietary Restrictions

The nurse counsels the patient anticipating bariatric surgery to ingest nothing but clear liquids for a specified period of time preoperatively (typically about 48 hours). The patient's diet will be quite limited postoperatively; because of this, patients scheduled for bariatric surgery are given guidelines on which foods and liquids they may consume postoperatively prior to surgery so that they may stock up on these items at home before they are admitted to the hospital. These typically include sugar-free drinks, gelatins and puddings, flavored electrolyte drinks, fat-free milk, protein drinks, sugar-free applesauce, and low-fat soups.

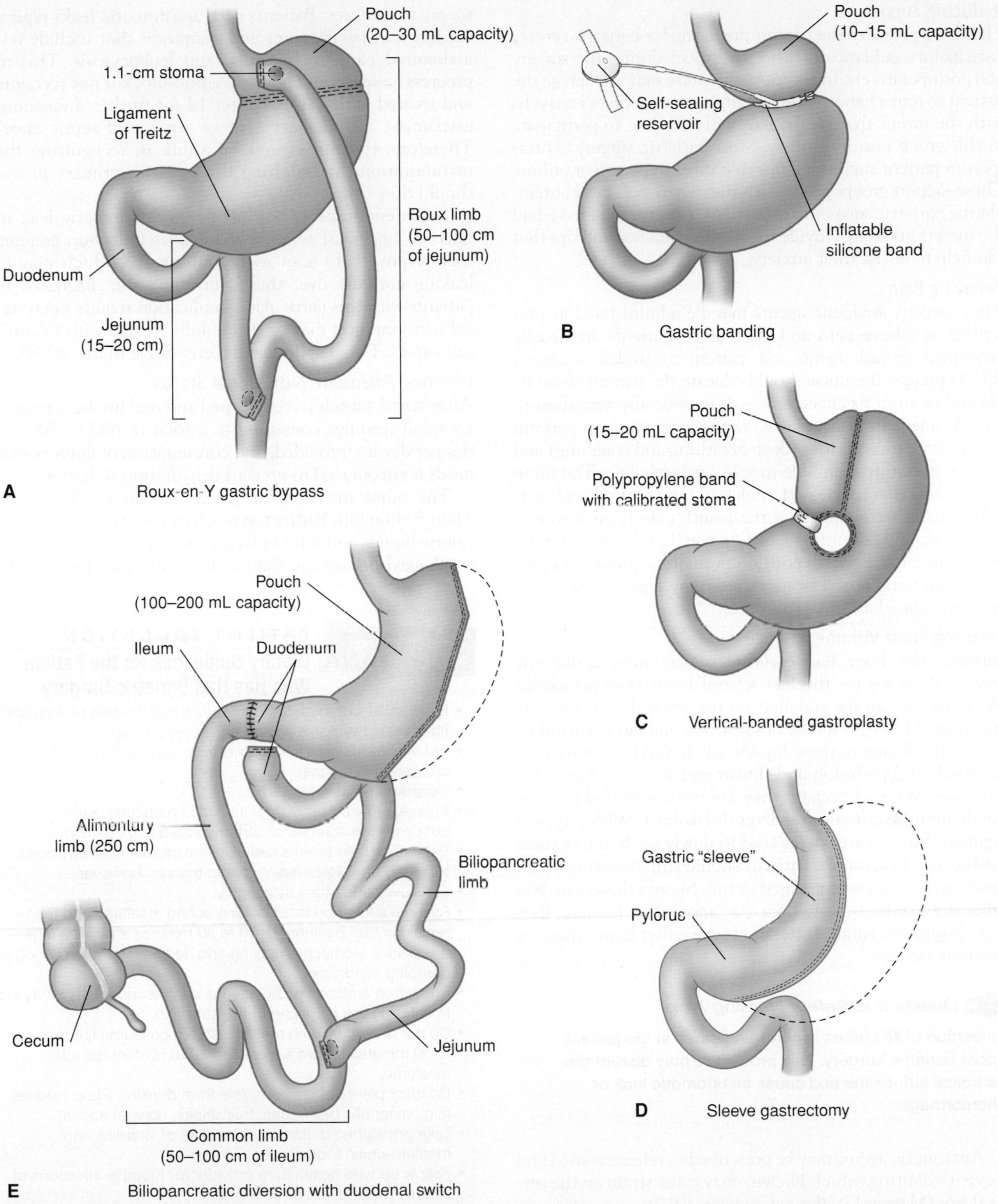
Pouch
(20–30 mL capacity)
1.1-cm stoma
Ligament
of Treitz
Roux limb
(50–100 cm
of jejunum)
Duodenum
Jejunum
(15–20 cm)
A
Roux-en-Y gastric bypass
Pouch
(10–15 mL capacity)
Self-sealing
reservoir
Inflatable
silicone band
B
Gastric banding
Pouch
(15–20 mL capacity)
Polypropylene band
with calibrated stoma
C
Vertical-banded gastroplasty
Pouch
(100–200 mL capacity)
Ileum
Duodenum
Alimontary
limb (250 cm)
Biliopancreatic
limb
Cecum
Jejunum
Common limb
(50–100 cm of ileum)
E
Biliopancreatic diversion with duodenal switch
Gastric "sleeve"
Pylorus
D
Sleeve gastrectomy

FIGURE 47-3 • Surgical procedures for morbid obesity. **A.** Roux-en-Y gastric bypass. A horizontal row of staples across the fundus of the stomach creates a pouch with a capacity of 20 to 30 mL. The jejunum is divided distal to the ligament of Treitz, and the distal end is anastomosed to the new pouch. The proximal segment is anastomosed to the jejunum. **B.** Gastric banding. A prosthetic device is used to restrict oral intake by creating a small pouch of 10 to 15 mL that empties through the narrow outlet into the remainder of the stomach. **C.** Vertical-banded gastroplasty. A vertical row of staples along the lesser curvature of the stomach creates a new, smaller stomach pouch of 10 to 15 mL. **D.** Sleeve gastrectomy. The stomach is incised vertically and up to 85% of the stomach is surgically removed, leaving a "sleeve"-shaped tube that retains intact nervous innervation and does not obstruct or decrease the size of the gastric outlet. **E.** Biliopancreatic diversion with duodenal switch (also called *sleeve gastrectomy with duodenal switch*). Half of the stomach is removed, leaving a small area that holds about 60 mL. The entire jejunum is excluded from the rest of the gastrointestinal tract. The duodenum is disconnected and sealed off. The ileum is divided above the ileocecal junction, and the distal end of the jejunum is anastomosed to the first portion of the duodenum. The distal end of the biliopancreatic limb is anastomosed to the ileum.

Reducing Anxiety

The nurse provides the patient preparing for bariatric surgery anticipatory guidance of what to expect during the surgery and postoperatively. In addition, the nurse may encourage the patient to join a bariatric surgery support group preoperatively, with the intent that the patient will continue to participate in this group postoperatively. Most bariatric surgery centers sponsor patient support groups that meet in person or online. These support groups provide a forum where patients contemplating bariatric surgery may talk with patients who have had the surgery and may provide them with guidance and tips that can help to lessen their anxiety.

Relieving Pain

After surgery, analgesic agents may be administered as prescribed to relieve pain and discomfort. Patients are usually prescribed opioid agents via patient-controlled analgesia (PCA) pumps; the nurse should educate the patient about its use and monitor its effectiveness. It is especially important to provide adequate pain relief so that the patient can perform pulmonary care activities (deep breathing and coughing) and leg exercises, turn from side to side, and ambulate. The nurse assesses the effectiveness of analgesic intervention and consults with other members of the health care team if pain is not adequately controlled (see Chapter 12). Positioning the patient in a low Fowler's position promotes comfort and promotes emptying of the stomach after any type of gastric surgery, including bariatric procedures.

Ensuring Fluid Volume Balance

Patients who have had bariatric surgery most commonly receive IV fluids for the first several hours postoperatively. Once they are awake and alert on the surgical unit, they are encouraged to begin intake of sugar-free oral fluids. Introducing small volumes of these liquids is believed to stimulate GI peristalsis and perfusion and thwart gastric reflux. Sugar-free fluids are preferred because they are not implicated in causing dumping syndrome (see later discussion). With a typical regimen, patients are encouraged to slowly sip 30 mL of these fluids every 15 minutes. Patients should stop ingesting fluids, however, if they feel nauseated or full. Should this occur, NG tubes are contraindicated for decompression because they may disrupt the surgical site and cause either hemorrhage or anastomotic leak.

> ***Quality and Safety Nursing Alert***
>
> **Insertion of NG tubes is contraindicated in the patient post bariatric surgery. This procedure may disrupt the surgical suture line and cause anastomotic leak or hemorrhage.**

Antiemetic agents may be prescribed to relieve nausea and prevent vomiting, which likewise may cause strain on the surgical site (Marzen-Groller & Cheever, 2010).

Preventing Infection/Anastomotic Leak

Disruption at the site of anastomosis (i.e., surgically resected site) may cause leakage of gastric contents into the peritoneal cavity, causing infection and possible sepsis. Patients at risk for this particular complication tend to be older, male, and with greater body mass. In addition, anastomotic leak is more commonly associated with open rather than laparoscopic procedures. Patients with anastomotic leaks typically exhibit nonspecific signs and symptoms that include fever, abdominal pain, tachycardia, and leukocytosis. This may progress to sepsis and possibly septic shock if not recognized and treated early (see Chapter 14 for further discussion of assessment and management of sepsis and septic shock). Therefore, the nurse must be astute in recognizing these manifestations and alerting the patient's primary provider should they occur.

A patient suspected of having an anastomotic leak may have an upper GI series that includes follow-up computed tomography (CT) scan with contrast dye, which may find leaking contrast dye, thus confirming the diagnosis. Few patients with this particular complication require open surgical revision; most may be successfully treated with IV antibiotics and CT-guided drainage (Perez & De Maria, 2008).

Ensuring Adequate Nutritional Status

After bowel sounds have returned and oral intake is resumed, six small feedings consisting of a total of 600 to 800 calories per day are provided, and consumption of fluids between meals is encouraged to prevent dehydration (Chart 47-6).

The nurse instructs the patient to eat slowly and stop when feeling full. Eating too much or too fast or eating high-calorie liquids and soft foods can result in vomiting or painful esophageal distention. Gastric retention may be evidenced

Chart 47-6 **PATIENT EDUCATION**

Dietary Guidelines for the Patient Who Has Had Bariatric Surgery

- Eat smaller but more frequent meals that contain protein and fiber; each meal size should not exceed 1 cup.
- Eat only foods high in nutrients (e.g., peanut butter, cheese, chicken, fish, beans).
- Consume fat as tolerated.
- Ensure a low carbohydrate intake; in particular, avoid concentrated sources of carbohydrates (e.g., candy).
- Eat two protein snacks daily; animal protein may be poorly tolerated after Roux-en-Y gastric bypass, however.
- Eat slowly and chew thoroughly.
- Assume a low Fowler's position during mealtime and then remain in that position for 20 to 30 minutes after mealtime—this delays stomach emptying and decreases the likelihood of dumping syndrome.
- Know that antispasmodic agents, as prescribed, also may aid in delaying the emptying of the stomach.
- Do not drink fluid with meals; instead, consume fluids up to 30 minutes before a meal and 30 to 60 minutes after mealtime.
- Do drink plenty of water; refrain from drinking liquid calories (e.g., alcoholic beverages, fruit drinks, nondiet sodas).
- Take prescribed dietary supplements of vitamins and medium-chain triglycerides.
- Follow up with health care provider for monthly injections of vitamin B_{12} and iron as prescribed.
- Walk for at least 30 minutes daily.

Adapted from Aills, L., Blankenship, J., Buffington, C., et al. (2008). ASMBS guidelines: ASMBS allied health nutritional guidelines for the surgical weight loss patient. *Surgery for Obesity and Related Diseases*, *4*, S73–S108; Manchester, S., & Roye, G. (2011). Bariatric surgery: An overview for dietetics professionals. *Nutrition Today*, *46*(6), 264–273; and Muoneke, P. U., & Reedy, S. (2011). Nutritional resources for post-bariatric surgery patients and healthcare workers. *Bariatric Nursing and Surgical Patient Care*, *6*(2), 99–102.

by abdominal distention, nausea, and vomiting. A nutritionist is typically consulted to assist with diet restrictions and diet progression.

Common dietary deficiencies include malabsorption of organic iron, which may require supplementation with oral or parenteral iron, and a low serum level of vitamin B_{12}; the patient may be prescribed monthly vitamin B_{12} intramuscular injections to prevent pernicious anemia (see Chapter 33 for further discussion).

Supporting Body Image Changes

Within the first 18 to 24 months postprocedure, the average patient post bariatric surgery loses between 25% and 35% of presurgical body weight, with the majority of weight lost within the first 6 months. Most patients report greatly improved perceptions of their body image, as well as improved quality of life (Sarwer, Wadden, Moore, et al., 2010; Teufel, Rieber, Meile, et al., 2012). However, some patients report lingering dissatisfaction with their body images. In particular, some patients may report dissatisfaction related to loose skin folds and may eventually seek elective body-contouring surgical options (e.g., breast reductions, breast lifts, abdominoplasty). Others report distorted perceptions of themselves as still carrying excessive weight despite losing substantial weight. This phenomenon is referred to as phantom fat (Stenson, 2009). The nurse may support the patient who reports dissatisfaction with body image post weight loss by acknowledging the patient's feelings as real, sharing that these perceptions are not unusual, and providing links to supports groups or counselors, as necessary.

Ensuring Maintenance of Bowel Habits

Patients may complain of either diarrhea or constipation postprocedure. Diarrhea is more common an occurrence post bariatric surgery, particularly after malabsorptive procedures (Mechanick et al., 2008). Both may be prevented if the patient consumes a nutritious diet that is high in fiber. Steatorrhea also may occur as a result of rapid gastric emptying, which prevents adequate mixing with pancreatic and biliary secretions. In mild cases, reducing the intake of fat and administering an antimotility medication (e.g., loperamide [Imodium]) may control symptoms. Persistent diarrhea or steatorrhea may warrant further diagnostic testing, such as an upper endoscopy or colonoscopy with biopsies to rule out the presence of additional pathology, such as celiac diseases (Mechanick et al., 2008) (see Chapter 48).

Monitoring and Managing Potential Complications

After surgery, the nurse assesses the patient for complications from the bariatric surgery, such as hemorrhage, bile reflux, dumping syndrome, dysphagia, and bowel or gastric outlet obstruction.

Hemorrhage. Postoperative hemorrhage may be a complication following bariatric surgery. Bleeding occurs in 1% to 5% of cases after open Roux-en-Y gastric bypass surgery (Ferreira, Song, & Baron, 2011). The nurse should be alert to the typical signs and symptoms of rapid blood loss and hemorrhagic shock; in addition, the patient may vomit considerable amounts of bright red blood. (See Chapter 14 for further discussion of assessment and management of hemorrhagic shock.)

Bile Reflux. Bile reflux gastritis and esophagitis may occur with procedures that manipulate or remove the pylorus, which acts as a barrier to the reflux of duodenal contents. Burning epigastric pain and vomiting of bilious material manifest this condition. Eating or vomiting does not relieve the situation. Agents that bind with bile acid, such as cholestyramine (Questran), may be helpful in mitigating symptoms (Menon & Jones, 2011).

Dumping Syndrome. **Dumping syndrome** is an unpleasant set of vasomotor and GI symptoms that occur in up to 76% of patients who have had bariatric surgery. For many years, it had been theorized that the hypertonic gastric food boluses that quickly transit into the intestines drew extracellular fluid from the circulating blood volume into the small intestines to dilute the high concentration of electrolytes and sugars, resulting in symptoms. Now, it is thought that this rapid transit of the food bolus from the stomach into the small intestines instead causes a rapid and exuberant release of metabolic peptides that are responsible for the symptoms of dumping syndrome (Mechanick et al., 2008; Rohof, Bisschops, Tack, et al. 2011; Sweed, Edmonson, & Cohen, 2009).

Early symptoms include a sensation of fullness, weakness, faintness, dizziness, palpitations, diaphoresis, cramping pains, and diarrhea. These symptoms resolve once the intestine has been evacuated (i.e., with defecation). Later, there is a rapid elevation of blood glucose, followed by increased insulin secretion. This results in a reactive hypoglycemia, which also is unpleasant for the patient. Vasomotor symptoms that occur 10 to 90 minutes after eating are pallor, perspiration, palpitations, headache, and feelings of warmth, dizziness, and even drowsiness. Anorexia may also be a result of the dumping syndrome, because the patient may be reluctant to eat. This is thought to be a preferred symptom among some patients post bariatric surgery, however, because they are intent on losing weight (Mechanick et al., 2008).

Dysphagia. **Dysphagia**, or difficulty swallowing, may occur in patients who have had any type of restrictive bariatric procedure. If it occurs, it tends to be most severe 4 to 6 weeks postoperatively and may persist for up to 6 months after surgery. Dysphagia may be prevented by educating patients to eat slowly, to chew food thoroughly, and to avoid eating tough foods such as steak or dry chicken or doughy bread. Patients with severe dysphagia who have had gastric banding may benefit from having their bands adjusted. Patients who have had other restrictive procedures may experience relief of symptoms after having stomal strictures relieved endoscopically (ASMBS, 2008).

Bowel and Gastric Outlet Obstruction. Bowel or gastric outlet obstruction may occur as a complication of bariatric surgery. The typical manifestations of gastric outlet obstruction were noted previously (see discussion in the Peptic Ulcer Disease section); however, there is one key difference in treating the patient post bariatric surgery with a gastric outlet obstruction: Insertion of NG tubes is contraindicated. Alternative treatment options may include endoscopic procedures aimed at relieving the obstruction, such as balloon dilation, or surgical revisions.

Promoting Home and Community-Based Care

The patient is usually discharged from the hospital in 4 days (this may be within 24 to 72 hours for patients who have had laparoscopic procedures) with detailed dietary instructions (see Chart 47-7 later in this chapter).

Educating Patients About Self-Care. The nurse provides information about nutrition, nutritional supplements, pain management, the importance of physical activity, and the symptoms of dumping syndrome and measures to prevent or minimize these symptoms. Patients who undergo laparoscopic or open Roux-en-Y procedures may have one or more Jackson-Pratt drains, which may remain in place after discharge. The nurse educates the patient or caregiver about how to empty, measure, and record the amount of drainage. The nurse must emphasize the continued need for follow-up (even after weight loss goals are met) and continued support group participation (Compher et al., 2012).

Continuing Care. After bariatric surgery, all patients require lifelong monitoring of weight loss, comorbidities, metabolic and nutritional status, and dietary and activity behaviors because they are at risk for developing malnutrition or weight gain (Heber et al., 2010; Tariq & Chand, 2011). Women of childbearing age who have bariatric surgery are advised to use contraceptives for at least 18 months after surgery to avoid pregnancy until their weight stabilizes (Mechanick et al., 2008). After weight loss, the patient may elect additional surgical interventions for body contouring. These may include breast reductions, lipoplasty to remove fat deposits, or a panniculectomy or abdominoplasty to remove excess abdominal skin folds.

Evaluation

Expected patient outcomes may include the following:

1. Relief of pain
   a. Reports relief of pain
   b. Engages in early mobilization activities as prescribed
2. Maintenance of fluid balance
   a. Able to tolerate progressive fluid intake without complaints of nausea or gastric reflux
   b. Voids 0.5 mL/kg/h
3. Maintenance of asepsis
   a. No evidence of infection (e.g., no fever, no leukocytosis, no complaints of abdominal pain)
4. Achievement of nutritional balance
   a. Able to consume small, frequent meals as prescribed
   b. Adheres to prescribed intake of vitamins and supplements
   c. Achieves and maintains weight reduction goals
5. Promotion of positive body image
   a. Verbalizes continued satisfaction with weight reduction plan and its effect on body image
6. Maintenance of usual bowel habits
   a. No evidence of diarrhea
   b. No evidence of constipation
7. Has no complications (e.g., no bleeding, bile reflux, dumping syndrome, dysphagia, or bowel or gastric outlet obstruction)

Gastric Cancer

Although the incidence of gastric or stomach cancer continues to decrease in the United States, it still accounts for more than 10,000 deaths annually (American Cancer Society, 2012). Gastric cancer is a more common diagnosis among older adults, with the median age at diagnosis of 70 years in men and 74 years in women (Cabebe, Mehta, & Fischer, 2012). Men have a higher incidence of gastric cancer than women. Asian Americans, Native Americans, Hispanic Americans, and African Americans are twice as likely as Caucasian Americans to develop gastric cancer. The incidence of gastric cancer is much greater in Japan, which has instituted mass screening programs for earlier diagnosis. Diet appears to be a significant factor: A diet high in smoked, salted, or pickled foods and low in fruits and vegetables may increase the risk of gastric cancer. Other factors related to the incidence of gastric cancer include chronic inflammation of the stomach, *H. pylori* infection, pernicious anemia, smoking, achlorhydria, gastric ulcers, previous subtotal gastrectomy (more than 20 years ago), and genetics. The prognosis is generally poor; the diagnosis is usually made late because most patients are asymptomatic during the early stages of the disease. Most cases of gastric cancer are discovered only after local invasion has advanced or metastases are present (Bang, Cutsem, Feyereislova, et al., 2010).

Pathophysiology

Most gastric cancers are adenocarcinomas. They can occur anywhere in the stomach, although up to 40% develop in the lower part, 40% in the middle part, and 15% in the upper part; 10% involve more than one of these parts of the stomach (Cabebe et al., 2012). The tumor infiltrates the surrounding mucosa, penetrating the wall of the stomach and adjacent organs and structures. The liver, pancreas, esophagus, and duodenum are often already affected at the time of diagnosis. Metastasis through lymph to the peritoneal cavity occurs later in the disease.

Clinical Manifestations

Symptoms of early disease, such as pain relieved by antacids, resemble those of benign ulcers and are seldom definitive. Symptoms of progressive disease include dyspepsia (indigestion), early satiety, weight loss, abdominal pain just above the umbilicus, loss or decrease in appetite, bloating after meals, nausea and vomiting, and symptoms similar to those of peptic ulcer disease.

Assessment and Diagnostic Findings

The physical examination is usually not helpful in detecting the cancer because most early gastric tumors are not palpable. Advanced gastric cancer may be palpable as a mass. Ascites and hepatomegaly (enlarged liver) may be apparent if the cancer cells have metastasized to the liver. Palpable nodules around the umbilicus, called *Sister Mary Joseph's nodules*, are a sign of a GI malignancy, usually a gastric cancer. Esophagogastroduodenoscopy for biopsy and cytologic washings is the diagnostic study of choice, and a barium x-ray examination of the upper GI tract may also be performed (Kuo, Holloway, & Nguyen, 2012; Lucenco, Marincas, Cirimbei, et al., 2012). Endoscopic ultrasound is an important tool to assess tumor depth and any lymph node involvement. CT scanning completes the diagnostic studies, particularly to assess for surgical resectability of the tumor before surgery is scheduled. CT scans of the chest, abdomen, and pelvis are valuable in staging gastric cancer.

Medical Management

A diagnostic laparoscopy may be the initial surgical approach to evaluate the gastric tumor, obtain tissue for pathologic diagnosis, and detect metastasis. The patient with a tumor that is deemed resectable undergoes an open surgical procedure to resect the tumor and appropriate lymph nodes. If the tumor can be removed while it is still localized to the stomach, the patient may be cured. The patient with an unresectable tumor and advanced disease undergoes chemotherapy, and cure is less likely.

A total gastrectomy may be performed for a resectable cancer in the midportion or body of the stomach. The entire stomach is removed along with the duodenum, the lower portion of the esophagus, supporting mesentery, and lymph nodes. Reconstruction of the GI tract is performed by anastomosing the end of the jejunum to the end of the esophagus, a procedure called an *esophagojejunostomy*. A radical subtotal gastrectomy is performed for a resectable tumor in the middle and distal portions of the stomach. A Billroth I or a Billroth II operation (see Table 47-3) is performed. The Billroth I involves a limited resection and offers a lower cure rate than the Billroth II. The Billroth II procedure is a wider resection that involves removing approximately 75% of the stomach and decreases the possibility of lymph node spread or metastatic recurrence. A proximal subtotal gastrectomy may be performed for a resectable tumor located in the proximal portion of the stomach or cardia. A total gastrectomy or an esophagogastrectomy is usually performed in place of this procedure to achieve a more extensive resection (Jang, Park, & Kim, 2010; Lee & Kim, 2010).

Common complications of advanced gastric cancer that often require surgery include gastric outlet obstruction, bleeding, and severe pain. Gastric perforation is an emergency situation requiring surgical intervention. A gastric resection may be the most effective palliative procedure for advanced gastric cancer. Palliative procedures such as gastric or esophageal bypass, gastrostomy, or jejunostomy may temporarily alleviate symptoms such as nausea and vomiting. Palliative rather than radical surgery may be performed if there is metastasis to other vital organs, such as the liver, or to achieve a better quality of life.

In instances where the tumor is not completely resectable, treatment with chemotherapy may offer further control of the disease or palliation. Commonly used single-agent chemotherapeutic medications include 5-fluorouracil (5-FU), cisplatin (Platinol), doxorubicin (Adriamycin), etoposide (Etopophos), and mitomycin C (Mutamycin). For improved response rates, it is more common to administer combination therapy, primarily 5-FU–based therapy, with other agents. Studies are being conducted to assess the use of chemotherapy before surgery. Trastuzumab (a recombinant humanized anti–HER-2 monoclonal antibody) prescribed in combination with cisplatin has shown survival improvement by nearly 3 months in patients with advanced gastric cancer who are HER-2 positive (McRee, Cowherd, Wang, et al., 2011; Meza-Junco, Au, Sawyer, 2011). Drugs that target angiogenesis pathways are also under investigation (Meza-Junco & Sawyer, 2012). Antiangiogenesis agents such as bevacizumab inhibit vascular endothelial growth factor activity, with the goal of regressing tumors by starvation (Goel, Duda, Xu, et al., 2011). Radiation therapy is mainly used for palliation in patients with obstruction, GI bleeding secondary to tumor, and significant pain. Assessment of tumor markers (blood analysis for antigens indicative of cancer) such as carcinoembryonic antigen (CEA), carbohydrate antigen (CA 19–9), and CA 50 may help determine the effectiveness of treatment. If these values were elevated before treatment, they should decrease if the tumor is responding to the treatment (Del Monte, Ranieri, Mazzetta, et al., 2012; Kim, Lee, Kim, et al., 2009).

Gerontologic Considerations

The number of older patients (75 years and older) with gastric cancer is increasing (Krejs, 2010). Sixty percent of cancer-related deaths occur in people 65 years and older (Krejs, 2010). Confusion, agitation, and restlessness may be the only symptoms seen in older adult patients, who may have no gastric symptoms until their tumors are well advanced. At this time, they present with reduced functional ability and other signs and symptoms of malignancy.

Surgery is more hazardous for the older adult, and the risk increases proportionately with increasing age. Nonetheless, gastric cancer should be treated with traditional surgery in older patients. Patient education is important to prepare older patients with cancer for treatment, to help them manage adverse effects, and to face the challenges that cancer and aging present.

NURSING PROCESS

The Patient With Gastric Cancer

Assessment

The nurse obtains a dietary history from the patient, focusing on recent nutritional intake and status. Has the patient lost weight? If so, how much and over what period of time? Can the patient tolerate a full diet? If not, what foods can he or she eat? What other changes in eating habits have occurred? Does the patient have an appetite? Does the patient feel full after eating a small amount of food? Is the patient in pain? Do foods, antacids, or medications relieve the pain, make no difference, or worsen the pain? Is there a history of infection with *H. pylori*? Other health information to obtain includes the patient's smoking and alcohol history and family history (e.g., any first- or second-degree relatives with gastric or other cancer). A psychosocial assessment, including questions about social support, individual and family coping skills, and financial resources, helps the nurse plan for care in acute and community settings.

After the interview, the nurse performs a complete physical examination, carefully assesses the patient's abdomen for tenderness or masses, and palpates and percusses the abdomen to detect ascites.

Nursing Diagnosis

Based on the assessment data, major nursing diagnoses may include the following:

- Anxiety related to the disease and anticipated treatment
- Imbalanced nutrition: less than body requirements related to early satiety or anorexia

- Acute pain related to tumor mass
- Grieving related to the diagnosis of cancer
- Deficient knowledge regarding self-care activities

Planning and Goals

The major goals for the patient may include reduced anxiety, optimal nutrition, relief of pain, and adjustment to the diagnosis and anticipated lifestyle changes.

Nursing Interventions

Reducing Anxiety

A relaxed, nonthreatening atmosphere is provided so the patient can express fears, concerns, and possibly anger about the diagnosis and prognosis. The nurse encourages the family or significant other to support the patient, offering reassurance and supporting positive coping measures. The nurse advises the patient about any procedures and treatments so that the patient knows what to expect.

Promoting Optimal Nutrition

The nurse encourages the patient to eat small, frequent portions of nonirritating foods to decrease gastric irritation. Food supplements should be high in calories, as well as vitamins A and C and iron, to enhance tissue repair. If the patient is unable to eat adequately prior to surgery to meet nutritional requirements, parenteral nutrition may be necessary. Because the patient may develop dumping syndrome (see previous discussion in this chapter) when enteral feeding resumes after gastric resection, the nurse explains ways to prevent and manage it (six small feedings daily that are low in carbohydrates and sugar; fluids between meals rather than with meals) and informs the patient that symptoms often resolve after several months. If a total gastrectomy is performed, injection of vitamin B_{12} will be required for life, because intrinsic factor, secreted by parietal cells in the stomach, binds to vitamin B_{12} so that it may be absorbed in the ileum. This deficiency in vitamin B_{12} metabolism can result in decreased production of red blood cells, or pernicious anemia. The nurse monitors the IV therapy and nutritional status and records intake, output, and daily weights to ensure that the patient is maintaining or gaining weight. The nurse assesses for signs of dehydration (thirst, dry mucous membranes, poor skin turgor, tachycardia, decreased urine output) and reviews the results of daily laboratory studies to note any metabolic abnormalities (sodium, potassium, glucose, BUN). Antiemetic agents are administered as prescribed.

Relieving Pain

The nurse administers analgesic agents as prescribed. A continuous IV infusion of an opioid or a PCA pump set to infuse an opioid may be necessary to mitigate postoperative pain. The nurse routinely assesses the frequency, intensity, and duration of the pain to determine the effectiveness of the analgesic agent. The nurse works with the patient to help manage pain by suggesting nonpharmacologic methods for pain relief, such as position changes, imagery, distraction, relaxation exercises (using relaxation audiotapes), backrubs, massage, and periods of rest and relaxation (see Chapter 12 for further discussion of pain management).

Providing Psychosocial Support

The nurse helps the patient express fears, concerns, and grief about the diagnosis. The nurse answers the patient's questions honestly and encourages the patient to participate in treatment decisions. Some patients mourn the loss of a body part and perceive their surgery as a type of mutilation. Some express disbelief and need time and support to accept the diagnosis.

The nurse offers emotional support and involves family members and significant others whenever possible. This includes recognizing mood swings and defense mechanisms (e.g., denial, rationalization, displacement, regression) and reassuring the patient, family members, and significant others that emotional responses are normal and expected. The services of clergy, psychiatric clinical nurse specialists, psychologists, social workers, and psychiatrists are made available, if needed. The nurse projects an empathetic attitude and spends time with the patient. Many patients may begin to participate in self-care activities after they have acknowledged their loss.

Promoting Home and Community-Based Care

Educating Patients About Self-Care. Self-care activities depend on the type of treatments used—surgery, chemotherapy, radiation, or palliative care. Patient and family education include information about diet and nutrition, treatment regimens, activity and lifestyle changes, pain management, and possible complications (Chart 47-7). Consultation with a dietitian is essential to determine how the patient's nutritional needs can best be met at home. The nurse instructs the patient or caregiver about administration of enteral or parenteral nutrition. If chemotherapy or radiation is prescribed, the nurse provides explanations to the patient and family about what to expect, including the length of treatments, the expected side effects (e.g., nausea, vomiting, anorexia, fatigue, neutropenia), and the need for transportation to appointments for treatment. Psychological counseling may also be helpful (see Chapter 15).

Continuing Care. The need for ongoing care in the home depends on the patient's condition and treatment. The home care nurse reinforces nutritional counseling and supervises the administration of any enteral or parenteral feedings; the patient or caregiver must become skillful in administering the feedings and in detecting and preventing untoward effects or complications related to the feedings (see Chapter 45 to review management of enteral and parenteral feedings). The nurse instructs the patient or caregiver to record the patient's daily intake, output, and weight and explains strategies to manage pain, nausea, vomiting, or other symptoms. The nurse also educates the patient or caregiver to recognize and report signs and symptoms of complications that require immediate attention, such as bleeding, obstruction, perforation, or any symptoms that become progressively worse. The nurse must explain the chemotherapy or radiation therapy regimen and ensure that the patient and family or significant other understand the care that will be needed during and after treatments (see Chapter 15). Because the prognosis for gastric cancer is poor, the nurse may need to assist the patient, family, or significant other with decisions regarding end-of-life care and make referrals as warranted.

Evaluation

Expected patient outcomes may include the following:

1. Reports less anxiety
   a. Expresses fears and concerns about surgery
   b. Seeks emotional support

Chart 47-7 HOME CARE CHECKLIST
The Patient With Gastric Cancer

At the completion of home care education, the patient or caregiver will be able to:	PATIENT	CAREGIVER
• Demonstrate safe management of enteral or parenteral feedings, if applicable.	✔	✔
• Describe dietary restrictions.	✔	✔
• Identify potential side effects of chemotherapy or radiation therapy, if applicable.	✔	✔
• Identify signs and symptoms of wound infection.	✔	✔
• State signs and symptoms of obstruction or perforation.	✔	✔
• Describe follow-up needs.	✔	✔
• Make decisions about end-of-life care as appropriate.	✔	✔

2. Attains optimal nutrition
   a. Eats small, frequent meals high in calories, iron, and vitamins A and C
   b. Complies with enteral or parenteral nutrition as needed
3. Has decreased pain
4. Performs self-care activities and adjusts to lifestyle changes
   a. Resumes typical activities within 3 months
   b. Alternates periods of rest and activity
   c. Manages enteral feedings
5. Verbalizes knowledge of disease management
   a. Acknowledges disease process
   b. Reports control of symptoms
   c. Verbalizes fears and concerns about dying; involves family/caregiver in discussions
   d. Completes advance directives and other appropriate documents

Gastric Surgery

Gastric surgery may be performed on patients with peptic ulcers who have life-threatening hemorrhage, obstruction, perforation, or penetration or whose condition does not respond to medication. It also may be indicated for patients with gastric cancer or trauma. Surgical procedures include a vagotomy and pyloroplasty, a partial gastrectomy, or a total gastrectomy (see Table 47-3).

Nursing Management

Before surgery, the nurse assesses the patient's and family's knowledge of preoperative and postoperative surgical routines and the rationale for surgery. The nurse also assesses the patient's nutritional status: Has the patient lost weight? How much? Over how much time? Does the patient have nausea and vomiting? Has the patient had hematemesis? The nurse assesses for the presence of bowel sounds and palpates the abdomen to detect masses or tenderness.

After surgery, the nurse assesses the patient for complications secondary to the surgical intervention, such as hemorrhage, infection, abdominal distention, atelectasis, or impaired nutritional status (see Chapter 19).

In addition to the complications to which all postoperative patients are subject, the patient undergoing gastric surgery is at increased risk for many of the same complications that are risky for patients who have had bariatric surgery, including hemorrhage, nutritional deficiencies, bile reflux, dumping syndrome, and dysphagia (see previous discussions).

Tumors of the Small Intestine

Tumors of the small intestine are uncommon; of these, approximately 64% are malignant (Espat, Somasundar, Fisichella, et al., 2011). Malignant tumors of the small intestine account for only about 2% of all GI cancers (Espat et al., 2011); annually, there are approximately 5,000 new diagnoses of cancer of the small intestine in the United States (Everhart, 2009). Rates are higher among older adults (median age at diagnosis of 67 years) and are also higher among African Americans and men (Everhart, 2009). Malignant tumors are often not discovered until they have metastasized to distant sites. Benign tumors may place patients at an increased risk for malignancy. The relative rarity of tumors of the small intestine, the diversity of tumor types (that may include adenocarcinomas, carcinoid tumors, lymphomas, or sarcomas), and the nonspecific nature of their manifestations complicate their diagnosis and treatment (Espat et al., 2011).

Clinical Manifestations

Tumors of the small intestine often present insidiously with vague, nonspecific symptoms. Most benign tumors are discovered incidentally on an x-ray study, during surgery, or at autopsy. When the patient is symptomatic, benign tumors often present with intermittent pain. The next most common presentation is occult bleeding. Malignant tumors often result in symptoms that lead to their diagnosis, although these symptoms may reflect advanced disease. Most patients have sustained weight loss and may be malnourished at diagnosis. Occult GI bleeding is less common than is found in patients with benign tumors, and complaints of pain are common. The patient also frequently presents with complaints of nausea, vomiting, and intestinal obstruction (Espat et al., 2011). Intestinal perforation is rare and associated with a poorer overall prognosis (Tan, Bang, & Ho, 2012).

Assessment and Diagnostic Findings

A CBC may reveal low hemoglobin levels and hematocrits that are consistent with anemia if the patient has an occult source of GI bleeding. The bilirubin may also be elevated if tumor mass has caused biliary obstruction. CEA levels may also be elevated, consistent with a malignant mass.

An upper GI x-ray series with small bowel follow-through using oral water-insoluble contrast with frequent and detailed x-rays to follow the contrast through the small bowel is the traditional approach to diagnosis. A more sensitive examination is an **enteroclysis**, in which an NG tube is advanced into the small bowel to a position above the area in question; the area is then studied by single- and double-contrast techniques. Abdominal CT is used to determine the extent of disease (Espat et al., 2011).

Management

Benign tumors of the small intestine include adenomas, lipomas, hemangiomas, and hamartomas (a focal malformation that resembles a neoplasm, but unlike a neoplasm does not result in compression of adjacent tissue). These tumors may be treated endoscopically by excision/resection or electrocautery if the patient is symptomatic. Routine monitoring is recommended to assess for malignant transformation.

The most common primary malignant tumor of the small intestine is adenocarcinoma; the second and third portions of the duodenum are most often involved. These tumors may present with obstruction. If the tumor is located at the ampulla of Vater, obstructive jaundice is likely. Other rare malignant tumors of the small intestine include carcinoid tumors, lymphoma, and GI stromal tumors. Abdominal surgery may be required to remove these rare tumors. Chemotherapy is commonly part of the treatment regimen.

The nursing process related to the care of the patient with a tumor of the small intestine is similar to that of the patient with gastric cancer. Each patient requires specialized care, astute assessment for complications, prompt interventions, and individualized education for self-care.

Critical Thinking Exercises

1 A 32-year-old patient is returning to your unit post laparoscopic sleeve gastrectomy. She is alert and oriented when aroused. Her pain level is 2 (on a 0 to 10 numeric pain scale). She denies nausea. Her blood pressure is 128/78 mm Hg, heart rate is 128 bpm, respiratory rate is 18 breaths/min and regular, and temperature is 36.7°C (98.1°F). Describe your priorities for care of this patient.

2 You are caring for a 48-year-old man with osteoarthritis of both knees who was recently diagnosed with peptic ulcer disease. He wants to know if he can continue to take his ibuprofen for pain. What would you tell him? Give the rationale for your answer.

3 Your aunt is 7 months post bariatric surgery. She tells you that she is doing well and has been successful in reaching her weight loss goal and does not have to see her primary provider anymore. What would you tell her? On what evidence do you base your answer?

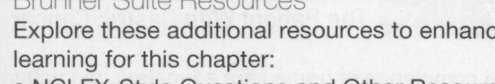

Brunner Suite Resources

Explore these additional resources to enhance learning for this chapter:

- NCLEX-Style Questions and Other Resources on thePoint, http://thePoint.lww.com/Brunner13e
- Study Guide
- PrepU
- Clinical Handbook
- Handbook of Laboratory and Diagnostic Tests

References

*Asterisk indicates nursing research.

Books

Greenberger, N., Blumberg, R., & Burakoff, R. (2012). *Current diagnosis & treatment: Gastroenterology, hepatology, & endoscopy* (2nd ed). New York: McGraw-Hill.

National Institute of Diabetes and Digestive and Kidney Diseases. (2010a). *Digestive diseases statistics for the United States*. NIH Publication No. 10–3873. Bethesda, MD: U.S. Department of Health and Human Services; National Institutes of Health.

Perez, A., & De Maria, E. J. (2008). Anastomotic leaks after laparoscopic gastric bypass. In N. T. Nygen, E. J. De Maria, S. Ikramaddin, et al. (Eds.). *The SAGES manual: A practical guide to bariatric surgery*. New York: Springer.

Journals and Electronic Documents

Gastric Cancer

American Cancer Society. (2012). *Cancer facts & figures 2012*. Available at: www.cancer.org

Bang, Y., Cutsem, E., Feyereislova, A., et al. (2010). Trastuzumab in combination with chemotherapy versus chemotherapy alone for treatment of HER2-positive advanced gastric or gastric or gastro-oesophageal junction cancer (ToGA): A phase 3, open-label, randomized controlled trial. *Lancet, 376*(9742), 687–697.

Cabebe, E. W., Mehta, V. K., & Fisher, G. (2012). Gastric cancer. *Medscape*. Available at: www.emedicine.medscape.com/article/278744-overview

Del Monte, S., Ranieri, D., Mazzetta, F., et al. (2012). Free peritoneal tumor cells detection in gastric and colorectal cancer patients. *Journal of Surgical Oncology, 106*(1), 17–23.

Ferreira, L., Song, L., & Baron, T. (2011). Management of acute postoperative hemorrhage in the bariatric patient. *Gastrointestinal Endoscopic Clinics of North America, 21*(2), 287–294.

Goel, S., Duda, D., Xu, L., et al. (2011). Normalization of vasculature for treatment of cancer and other diseases. *Physiological Reviews, 91*(3), 1071–1121.

Jang, Y., Park, M., & Kim, J. (2010). Advanced gastric cancer in the middle one-third of the stomach: Should surgeons perform total gastrectomy? *Journal of Surgical Oncology, 101*(6), 451–456.

Kim, H., Lee, K., Kim, Y., et al. (2009). Chemotherapy-induced transient CEA and CA 19–9 surges in patients with metastatic or recurrent gastric cancer. *Acta Oncologica, 48*(3), 385–390.

Krejs, G. (2010). Gastric cancer: Epidemiology and risk factors, *Digestive Diseases, 28*(4–5), 355–358.

Kuo, P., Holloway, R., & Nguyen, N. (2012). Current and future techniques in the evaluation of dysphagia. *Journal of Gastroenterology and Hepatology, 27*(5), 873–881.

Lee, J., & Kim, Y. (2010). Which is the optimal extent of resection in middle third gastric cancer between total gastrectomy and subtotal gastrectomy? *Journal of Gastric Cancer, 10*(4), 226–233.

Lucenco, L., Marincas, M., Cirimbei, C., et al. (2012). The 10 years' experience in the laparoscopic treatment of benign pathology of the ESO gastric function. *Journal of Medicine and Life, 5*(2), 179–184.

McRee, A., Cowherd, S., Wang, A., et al. (2011). Chemoradiation therapy in the management of gastrointestinal malignancies. *Future Oncology, 7*(3), 409–426.

Menon, S., & Jones, B. (2011). Postinfective bile acid malabsorption: Is this a long-term condition? *European Journal of Gastroenterology & Hepatology, 23*(4), 308–310.

Meza-Junco, J., Au, H., & Sawyer, M. (2011). Critical appraisal of trastuzumab in treatment of advanced stomach cancer. *Cancer Management Research, 3*, 57–64.

Meza-Junco, J., & Sawyer, M. (2012). Metastatic gastric cancer—focus on targeted therapies. *Biologics Targets and Therapy, 6*, 137–146.

Sweed, M., Edmonson, D., & Cohen, S. (2009). Tumors of the esophagus, gastroesophageal junction, and stomach. *Seminars in Oncology Nursing, 25*(1), 61–75.

Obesity

Abdi, H., Abbasi-Parizad, P., Zhao, B., et al. (2012). Effects of auricular acupuncture on anthropometric, lipid profile, inflammatory, and immunologic markers: A randomized controlled trial study. *Journal of Complementary Medicine, 18*(7), 668–677.

Abdi, H., Zhao, B., Darbandi, M., et al. (2012). The effects of body acupuncture on obesity: Anthropometric parameters, lipid profile, and inflammatory and immunologic markers. *Scientific World Journal*. Epub 2012 April 29.

Aills, L., Blankenship, J., Buffington, C., et al. (2008). ASMBS guidelines: ASMBS allied health nutritional guidelines for the surgical weight loss patient. *Surgery for Obesity and Related Diseases, 4*, S73–S108.

American Society for Metabolic and Bariatric Surgery. (2008). *Bariatric surgery: Postoperative concerns*. Available at: www.asmbs.org/2012/01/bariatric-surgery-postoperative-concerns/

American Society for Metabolic and Bariatric Surgery. (2011a). *Preoperative supervised weight loss requirements*. Available at: www.asmbs.org/2012/01/preoperative-supervised-weight-loss-requirements/

American Society for Metabolic and Bariatric Surgery. (2011b). *Sleeve gastrectomy as a bariatric procedure (update)*. Available at: www.asmbs.org/2012/06/sleeve-gastrectomy-as-a-bariatric-procedure-update/

American Society for Metabolic and Bariatric Surgery. (2012). *ASMBS position statement: Bariatric surgery in class I obesity (BMI 30–35 kg/m^2)*. Available at: www.asmbs.org/2012/09/bariatric-surgery-in-class-1-obesity-bmi-30-35-kgm2/

Artham, S., Lavie, C., Milani, R., et al. (2010). Value of weight reduction in patients with cardiovascular disease. *Current Treatment Options in Cardiovascular Medicine, 12*(1), 21–35.

Barnett, S. (2011). Contemporary surgical management of the obese adolescent. *Current Opinion in Pediatrics, 23*(3), 351–355.

Billes, S., & Greenway, F. (2011). Combination therapy with naltrexone and bupropion for obesity. *Expert Opinion Pharmacotherapy, 12*(11), 1813–1826.

Caveney, E., Caveney, B., Somaratne, R., et al. (2011). Pharmaceutical interventions for obesity: A public health perspective. *Diabetes, Obesity and Metabolism, 13*(6), 490–497.

Centers for Disease Control and Prevention. (2009). *Adult obesity. Obesity rises among adults*. Available at: www.cdc.gov/obesity/data/index.html

Compher, C., Hanlon, A., Kang, Y., et al. (2012). Attendance at clinical visits predicts weight loss after gastric bypass surgery. *Obesity Surgery, 22*(6), 927–834.

Dumon, K., & Murayama, K. (2011). Bariatric surgery outcomes. *Surgical Clinics of North America, 91*(6), 1313–1338.

Echols, J. (2010). Obesity weight management and bariatric surgery case management programs. *Professional Case Management, 15*(1), 17–26.

Gianos, M., Abdemur, A., Fendrich I., et al. (2011). Outcomes of bariatric surgery in patients with body mass index <35 kg/m2. *Surgery for Obesity and Related Diseases, 8*(1), 25–30.

Heber, D., Greenway, F., Kaplan, L., et al. (2010). Endocrine and nutritional management of the post-bariatric patient: An Endocrine Society clinical practice guideline. *Journal of Clinical Endocrinology Metabolism*, 95(11), 4823–4843.

LeMont, D., Moorehead, M. K., Parish, M. S., et al. (2004). *Suggestions for the pre-surgical psychological assessment of bariatric surgery candidates*. Available at: www.asmbs.org/

Marzen-Groller, K., & Cheever, K. H. (2010). Facilitating students' competence in caring for the bariatric surgical patient: The case study approach. *Bariatric Nursing and Surgical Patient Care Journal, 5*(2), 117–125.

Mechanick, J. I., Kushner, R. F., Sugerman, H. J., et al. (2008). American Association of Clinical Endocrinologists, the Obesity Society, and American Society for Metabolic and Bariatric Surgery: Medical guidelines for clinical practice for the perioperative nutritional, metabolic, and nonsurgical support of the bariatric surgery patient. *Surgery for Obesity and Related Diseases, 4*, S109–S184.

Michalsky, M., Kramer, R., Fullmer, M., et al. (2011). Developing criteria for pediatric/adolescent bariatric surgery programs. *Pediatrics, 128*(2), S65–S70.

*Poon, M. Y., & Tarrant, M. (2009). Obesity: Attitudes of undergraduate student nurses and registered nurses. *Journal of Clinical Nursing, 18*(16), 2355–2365.

Rohof, W., Bisschops, R., Tack, J., et al. (2011). Postoperative problems 2011: Fundoplication and obesity surgery. *Gastroenterology Clinics of North America, 40*(4), 809–821.

Sarwer, D. B., Wadden, T. A., Moore, R. H., et al. (2010). Changes in quality of life and body image following gastric bypass surgery. *Surgery in Obesity and Related Diseases, 6*(6), 608–614.

Stenson, J. (2009). 'Phantom fat' can linger after weight loss. *NBCNews.com*. Available at: www.msnbc.msn.com/id/31489881/ns/health-womens/he

Steyer, T., & Ables, A. (2009). Complementary and alternative therapies for weight loss. *Primary Care, 36*(2), 395–406.

Tariq, N., & Chand, B. (2011). Presurgical evaluation and postoperative care for the bariatric patient. *Gastrointestinal Endoscopy Clinics of North America, 21*(2), 229–240.

Teufel, M., Rieber, N., Meile, T., et al. (2012). Body image after sleeve gastrectomy: Reduced dissatisfaction and increased dynamics. *Obesity Surgery, 22*(8), 1232–1237.

Ye, Z., Chen, L., Yang, Z., et al. (2011). Metabolic effects of fluoxetine in adults with type 2 diabetes mellitus: A meta-analysis of randomized placebo-controlled trials. *PLoS One, 6*(7), e21551.

Peptic Ulcers and Gastritis

Alexandre, M., Pandol, S., Gorelick, F., et al. (2011). The emerging role of smoking in the development of pancreatitis. *Pancreatology, 11*(5) 469–474.

Alakkari, A., Zullo, A., & O'Connor, H. (2011). *Helicobacter pylori* and nonmalignant diseases. *Helicobacter, 16*(Suppl. 1), 33–37.

Anand, B. S., Bank, S., Qureshi, W. A., et al. (2012). Peptic ulcer disease. *Medscape*. Available at: emedicine.medscape.com/article/181753

Clarke, R. C., Ferraro, R. M., Gbadehan, E., et al. (2011). Stress-induced gastritis. *Medscape*. Available at: emedicine.medscape.com/article/176319

Czinn, S., & Blanchard, T. (2011). Vaccinating against *Helicobacter pylori* infection. *Nature Reviews. Gastroenterology & Hepatology, 8*(3), 133–140.

Fujinami, H., Kudo, T., Hosokawsa, A., et al. (2012). A study of the cause of peptic ulcer bleeding. *World Journal of Gastrointestinal Endoscopy, 4*(7), 323–327.

Hoshino, C., Satoh, N., Narita, M., et al. (2011). Another 'Cushing ulcer'. *BMJ Case Report*, doi:10.1136/bcr.02.2011.3888.

Ichiro, I., Shushi, H., Akihiko, I., et al. (2011). Empiric transcatheter arterial embolization for massive bleeding from duodenal ulcers: Efficacy and complications. *Journal of Vascular Interventional Radiology, 22*(7), 911–916.

Kumar, R., & Mills, A. (2011). Gastrointestinal bleeding. *Emergency Medical Clinics of North America, 29*(2), 239–252.

Lopez, C., Falconi, M., Waldmann, J., et al. (2012). Partial pancreaticoduodenectomy can provide cure for duodenal gastrinoma associated with multiple endocrine neoplasia type 1. *Annals of Surgery*, doi:10.1097/sla0b013e3182536339.

Metz, D. (2012). Diagnosis of the Zollinger-Ellis syndrome. *Clinical Gastroenterology and Hepatology, 10*, 126–130.

Milosavljevic, T., Kostic-Milosavljevic, M., Jovanovic, I., et al. (2011). Complications of peptic ulcer disease. *Digestive Diseases, 29*(5), 491–493.

Mukherjee, S., & Sepulveda, A. R. (2012). Chronic gastritis. *Medscape*. Available at: emedicine.medscape.com/article/176156

National Institute of Diabetes and Digestive and Kidney Diseases. (2010b). Gastritis. *National Digestive Diseases Information Clearinghouse*. Available at: digestive.niddk.nih.gov/ddiseases/pubs/gastritis/index.aspx

Norton, J., Fraker, D., Alexander, H., et al. (2012). Value of surgery in patients with negative imaging and sporadic Zollinger-Ellison syndrome. *Annals of Surgery, 256*(3), 509–517.

Ozlem, T., Yazici, D., Ince, U., et al. (2011). Bulky gastrinoma of the common bile duct: Unusual localization of extrapancreatic gastrinoma case report. *Turkish Journal of Gastroenterology, 22*(2), 219–223.

Ruggiero, P. (2010). *Helicobacter pylori* and inflammation. *Current Pharmaceutical Design, 16*(38), 4225–4236.

Ruggiero, P. (2012). *Helicobacter pylori* infection: What's new. *Current Opinion in Infectious Disease, 25*(3), 337–344.

Wataru, I., Yasuhiko, A., Katsunori, I., et al. (2012). Gastric hypochlorhydria is associated with an exacerbation of dyspeptic symptoms in female patients. *Journal of Gastroenterology*, doi: 10.1007/s00535-012-0634-8.

Wehbi, M., Griglione, N. M., Snyder, R. H., et al. (2012). Acute gastritis. *Medscape*. Available at: emedicine.medscape.com/article/175909

Wijdicks, E. (2011). Cushing's ulcer: The eponym and his own. *Neurosurgery, 68*(6), 1695–1698.

Wilkins, T., Khan, N., Nabh, A., et al. (2012). Diagnosis and management of upper gastrointestinal bleeding. *American Family Physician, 85*(5), 469–476.

Yonemitsu, T., Kawai, N., Sato, M., et al. (2009). Evaluation of transcatheter arterial embolization with gelatin sponge particles, microcoils, and N-butyl

cyanoacrylate for acute bleeding in a coagulopathic condition. *Journal of Vascular Interventional Radiology*, 20(9), 1176–1187.
Zelickson, M., Bronder, C., Johnson, J., et al. (2011). *Helicobacter pylori* is not predominant etiology for peptic ulcers requiring operation. *American Surgeon*, 77(8), 1054–1060.

Tumors of the Small Intestine
Espat, N. J., Somasundar, P. S., Fisichella, P. M., et al. (2011). Malignant neoplasms of the small intestine. *Medscape*. Available at: emedicine.medscape.com/article/282684
Everhart, J. E. (2009). *The burden of digestive diseases in the United States*. Available at: www2.niddk.nih.gov/AboutNIDDK/ReportsAndStrategicPlanning/BurdenOfDisease/DigestiveDiseases/
Tan, K., Bang, S., & Ho, C. (2012). Surgery for perforated small bowel malignancy: A single institution's experience over 4 years. *Surgeon*, 10(1), 6–8.

Resources

Agency for Healthcare Research and Quality (AHRQ), www.ahrq.gov
American Cancer Society, www.cancer.org
American Gastroenterological Association (AGA), www.gastro.org
American Society for Metabolic and Bariatric Surgery (ASMBS), www.asmbs.org
Centers for Disease Control and Prevention (CDC), www.cdc.gov/nccdphp/dnpa/obesity/index.htm
Lindstrom Obesity Advocacy, wlsappeals.com
National Comprehensive Cancer Network (NCCN) Clinical Practice Guidelines, www.nccn.org
National Digestive Diseases Information Clearinghouse (NDDIC), digestive.niddk.nih.gov

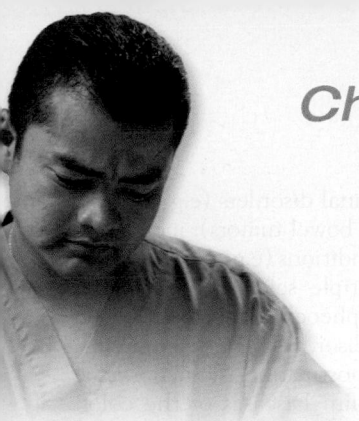

Chapter

48

Management of Patients With Intestinal and Rectal Disorders

Learning Objectives

On completion of this chapter, the learner will be able to:

1. Identify the health care learning needs of patients with constipation or diarrhea.
2. Compare the conditions of malabsorption with regard to their pathophysiology, clinical manifestations, and management.
3. Use the nursing process as a framework for care of patients with diverticular disease.
4. Compare Crohn's disease (regional enteritis) and ulcerative colitis with regard to their pathophysiology; clinical manifestations; diagnostic evaluation; and medical, surgical, and nursing management.
5. Use the nursing process as a framework for care of the patient with inflammatory bowel disease.
6. Describe the responsibilities of the nurse in meeting the needs of the patient with an intestinal diversion.
7. Describe the various types of intestinal obstructions and their management.
8. Use the nursing process as a framework for care of the patient with colorectal cancer.
9. Describe the nursing management of the patient with an anorectal condition.

Glossary

abscess: localized collection of purulent material surrounded by inflamed tissues

borborygmus: rumbling noise caused by the movement of gas through the intestines

colostomy: surgical opening into the colon by means of a stoma to allow drainage of bowel contents; one type of fecal diversion

constipation: an abnormal infrequency or irregularity of defecation, abnormal hardening of stools that makes their passage difficult and sometimes painful, a decrease in stool volume, retention of stool in the rectum for a prolonged period often with a sense of incomplete evacuation after defecation, or a persistent sensation of abdominal fullness

diarrhea: an increased frequency of bowel movements (more than three per day), an increased amount of stool (more than 200 g/day), and altered consistency (i.e., increased liquidity) of stool

diverticulitis: inflammation of a diverticulum from obstruction by fecal matter resulting in abscess formation

diverticulosis: presence of several diverticula in the intestine

diverticulum: saclike out-pouching of the lining of the bowel protruding through the muscle of the intestinal wall

fecal incontinence: involuntary passage of feces

fissure: normal or abnormal fold, groove, or crack in body tissue

fistula: anatomically abnormal tract that arises between two internal organs or between an internal organ and the body surface

gastrocolic reflex: peristaltic movements of the large bowel occurring five to six times daily that are triggered by distention of the stomach

hemorrhoids: dilated portions of the anal veins

ileostomy: surgical opening into the ileum by means of a stoma to allow drainage of bowel contents; one type of fecal diversion

inflammatory bowel disease (IBD): group of chronic disorders (ulcerative colitis and Crohn's disease) that result in inflammation or ulceration (or both) of the bowel lining

irritable bowel syndrome (IBS): chronic functional disorder that affects frequency of defecation and consistency of stool; is associated with no specific structural or biochemical alterations

Kock pouch: type of continent ileal reservoir created surgically by making an internal pouch with a portion of the ileum and placing a nipple valve flush with the stoma

malabsorption: impaired transport across the mucosa

peritonitis: inflammation of the lining of the abdominal cavity

steatorrhea: excess of fatty wastes in the feces

tenesmus: ineffective and sometimes painful straining and urge to eliminate either feces or urine

Valsalva maneuver: forcible exhalation against a closed glottis followed by a rise in intrathoracic pressure and subsequent possible dramatic rise in arterial pressure; may occur during straining at stool

wound-ostomy-continence (WOC) nurse: nurse specially educated in the appropriate management of fecal and urinary diversions; guides patients, their families, surgeons, and nurses by recommending appropriate use of skin, wound, ostomy, and continence products; formerly called *enterostomal therapist*

Between 60 and 70 million people in the United States are diagnosed with some type of disease of the gastrointestinal (GI) tract. These diseases account for more than 104 million office visits to health care facilities and clinics and approximately 14 million hospital admissions annually. GI diseases cost the American public more than $140 billion each year and account for approximately 240,000 deaths each year (National Digestive Diseases Information Clearinghouse [NDDIC], 2012a). The types of diseases and disorders that affect the lower GI tract are many and varied, including constipation, diarrhea, diverticulitis, and inflammatory bowel disease (IBD), to name a few.

In all age groups, a fast-paced lifestyle, high levels of stress, irregular eating habits, insufficient intake of fiber and water, and lack of daily exercise contribute to GI disorders. There is a growing understanding of the biopsychosocial implications of GI disease. That is, the mind and emotions can have a profound impact on the GI system. Nurses can have an impact on these GI disorders by identifying behavior patterns that put patients at risk, by educating the public about prevention and management, and by helping those affected to improve their condition and prevent complications.

ABNORMALITIES OF FECAL ELIMINATION

Changes in patterns of fecal elimination are symptoms of functional disorders or diseases of the GI tract. The most common changes seen are constipation, diarrhea, and fecal incontinence. The nurse should be aware of the causes and therapeutic management of these disorders and nursing management techniques. Education is important for patients with these conditions.

Constipation

Constipation is an abnormal infrequency or irregularity of defecation, abnormal hardening of stools that makes their passage difficult and sometimes painful, a decrease in stool volume, retention of stool in the rectum for a prolonged period often with a sense of incomplete evacuation after defecation, or a persistent sensation of abdominal fullness (Apau, 2010a). Any variation from normal habits may be considered a problem. It is estimated that 6.3 million Americans are clinically constipated at any time, between 10% and 19% of the American population (more than 63 million people) may be affected periodically, and women and adults older than 65 years are disproportionately constipated (Belsey, Greenfield, Candy, et al., 2010; Bouras & Tangalos, 2009; Ostaszkiewicz, Hornby, Millar, et al., 2010). Notably, constipation is a symptom and not a disease; however, constipation can indicate an underlying disease or motility disorder of the GI tract.

Constipation can be caused by certain medications (i.e., tranquilizers, anticholinergic agents, antidepressants, antihypertensive agents, bile acid sequestrants, diuretic agents, opioids, aluminum-based antacids, iron preparations, selected antibiotics, muscle relaxants, and cytotoxic drugs (e.g., vincristine [Oncovin]); rectal or anal disorders (e.g., hemorrhoids, fissures); obstruction (e.g., bowel tumors); metabolic, neurologic, and neuromuscular conditions (e.g., Hirschsprung disease, Parkinson's disease, multiple sclerosis); endocrine disorders (e.g., hypothyroidism, pheochromocytoma); lead poisoning; and connective tissue disorders (e.g., scleroderma, systemic lupus erythematosus). Constipation is a major issue for patients taking opioids for pain. Diseases of the colon commonly associated with constipation include irritable bowel syndrome (IBS) and diverticular disease. Constipation can also occur with an acute disease process in the abdomen (e.g., appendicitis) (Apau, 2010a; Bouras & Tangalos, 2009; Spinzi, Amato, Imperiali, et al., 2009).

Other causes of constipation may include weakness, immobility, debility, fatigue, and an inability to increase intra-abdominal pressure to facilitate the passage of stools, as may occur in patients with emphysema or spinal cord injury, for instance. Many people develop constipation because they do not take the time to defecate or ignore the urge to defecate. Constipation is also a result of dietary habits (i.e., low consumption of fiber and inadequate fluid intake), lack of regular exercise, and a stress-filled life. Fiber is particularly important to bowel health because it increases the bulk of stool, generally easing the passage. More importantly, it also promotes optimal fermentation, providing good bowel wall health (Emmanuel, 2011).

Perceived constipation can also be an issue. This subjective problem occurs when a person's bowel elimination pattern is not consistent with what he or she considers normal. Chronic laxative use may contribute to this problem and is a major health issue in the United States. Laxative abuse, particularly the anthracene derivatives such as senna and cascara, can lead to destruction of the nerves of the colon that are essential for normal peristalsis (Apau, 2010a). "Normal" bowel function varies substantially from three bowel movements a day to three times per week (Apau, 2010a).

Pathophysiology

The pathophysiology of constipation is poorly understood, but it is thought to include interference with one of three major functions of the colon: mucosal transport (i.e., mucosal secretions facilitate the movement of colon contents), myoelectric activity (i.e., mixing of the rectal mass and propulsive actions), or the processes of defecation (e.g., pelvic floor dysfunction). Any of the causative factors identified previously can interfere with any of these three processes.

The urge to defecate is stimulated normally by rectal distention that initiates a series of four actions: stimulation of the inhibitory rectoanal reflex, relaxation of the internal sphincter muscle, relaxation of the external sphincter muscle and muscles in the pelvic region, and increased intra-abdominal pressure. Interference with any of these processes can lead to constipation.

If all organic causes are eliminated, idiopathic or functional constipation is diagnosed. When the urge to defecate is ignored, the rectal mucous membrane and musculature become insensitive to the presence of fecal masses, and consequently a stronger stimulus is required to produce the necessary peristaltic rush for defecation. The initial effect of fecal retention is to produce irritability of the colon, which at this stage frequently goes into spasm, especially after meals, giving rise to colicky midabdominal or low abdominal pains. After several years of this process, the colon loses muscular tone and becomes essentially unresponsive to normal stimuli (similar to an overstretched balloon). Atony or decreased muscle tone occurs with aging. This may lead to constipation because the stool is retained for longer periods.

Clinical Manifestations

Clinical manifestations of constipation include fewer than three bowel movements per week; abdominal distention; pain and pressure; decreased appetite; headache; fatigue; indigestion; a sensation of incomplete evacuation; straining at stool; and the elimination of small-volume, lumpy, hard, dry stools. To make diagnosis more efficient, an international committee developed the Rome criteria (in the city of Rome, Italy). To be considered true chronic constipation, the previously mentioned manifestations must be present for at least 12 weeks of the preceding 12 months (Belsey et al., 2010; Muller-Lissner & Ward, 2011).

Assessment and Diagnostic Findings

Chronic constipation is usually considered idiopathic, but secondary causes should be excluded. In patients with severe, intractable constipation, further diagnostic testing is needed (Apau, 2010a). The diagnosis of constipation is based on the patient's history, physical examination, possibly the results of a barium enema or sigmoidoscopy, and stool testing for occult blood. These tests are used to determine whether this symptom results from spasm or narrowing of the bowel. Anorectal manometry (i.e., pressure studies such as a balloon expulsion test) may be performed to assess malfunction of the sphincter. Defecography and colonic transit studies can also assist in the diagnosis because they permit assessment of active anorectal function. Tests such as pelvic floor magnetic resonance imaging (MRI) may identify occult pelvic floor defects (Apau, 2010a; Bouras & Tangalos, 2009).

Complications

Complications of constipation include hypertension, fecal impaction, **hemorrhoids** (dilated portions of anal veins), **fissures** (normal or abnormal fold, groove, or crack in body tissue), and megacolon. Increased arterial pressure can occur with defecation. Straining at stool, which results in the **Valsalva maneuver** (i.e., forcibly exhaling with the glottis closed), has a striking effect on arterial blood pressure. During active straining, the flow of venous blood in the chest is temporarily impeded because of increased intrathoracic pressure. This pressure tends to collapse the large veins in the chest. The atria and the ventricles receive less blood, and consequently less blood is ejected by the left ventricle. Cardiac output is decreased, and there is a transient drop in arterial pressure. Almost immediately after this period of hypotension, an increase in arterial pressure occurs; the pressure is elevated momentarily to a point far exceeding the original level (i.e., rebound phenomenon). In patients with hypertension, this compensatory reaction may be exaggerated greatly, and the peak pressure attained may be dangerously high—sufficient to rupture a major artery in the brain or elsewhere.

Fecal impaction occurs when an accumulated mass of dry feces cannot be expelled. The mass may be palpable on digital examination, may produce pressure on the colonic mucosa that results in ulcer formation, and frequently causes seepage of liquid stools. Treatment can be embarrassing and also painful, because impaction removal usually involves digital dislodgement and enema administration.

Hemorrhoids and anal fissures can develop as a result of constipation. Hemorrhoids develop as a result of perianal vascular congestion caused by straining. Anal fissures may result from the passage of the hard stool through the anus, tearing the lining of the anal canal.

Megacolon is a dilated and atonic colon caused by a fecal mass that obstructs the passage of colon contents. Symptoms include constipation, liquid fecal incontinence, and abdominal distention. Megacolon can lead to perforation of the bowel.

Gerontologic Considerations

Visits to primary providers for treatment of constipation are common in people 65 years and older. The most common complaint they voice is the need to strain at stool. The aging process inevitably generates changes in the colon; but the extent and physiologic implications for defecation remain unclear. The clinical situation is made more complex by ubiquitous factors among the aged (Bouras & Tangalos, 2009) (Chart 48-1).

Older adults who have loose-fitting dentures or have lost their teeth have difficulty chewing and frequently choose soft, processed foods that are low in fiber. Older adults tend to have decreased food intake, reduced mobility, and weak abdominal and pelvic muscles, and they are more likely to have multiple chronic illnesses requiring multiple medications (polypharmacy) that often cause constipation. Low-fiber convenience foods are widely used by people who have lost interest in eating. Some older people reduce their fluid

Chart 48-1 RISK FACTORS
Constipation in the Older Adult

The "Ten Ds of Constipation":

- side effects of *d*rugs
- defecatory *d*ysfunction
- *d*egenerative disease
- decreased *d*ietary intake
- *d*ementia
- *d*ecreased mobility
- *d*ependence on others for assistance
- *d*ecreased privacy
- *d*ehydration
- *d*epression

Others:

- Female gender (3 times more common)
- Low socioeconomic status
- Living in rural areas

Adapted from Bouras, E. P., & Tangalos, E. G. (2009). Chronic constipation in the elderly. *Gastroenterology Clinics of North America*, *38*(3), 463–480; and Johanson, J. F. (2011). Current topics: Clinical factors and complications of constipation in elderly. *E-IMPACCT*. Available at: www.cmecorner.com/portal/topic_detail.asp?1D=elderlyconstipation.org

intake if they are not eating regular meals. Depression, weakness, and prolonged bed rest also contribute to constipation by decreasing intestinal motility and anal sphincter tone. Nerve impulses are dulled, and there is a decreased urge to defecate. Many older people overuse laxatives in an attempt to have a daily bowel movement and become dependent on them. Chronic constipation profoundly impairs quality of life comparable to other conditions such as diabetes, rheumatoid arthritis, and osteoarthritis (Belsey et al., 2010).

Medical Management

Treatment targets the underlying cause of constipation and prevention of recurrence. It includes education, exercise, bowel habit training, increased fiber and fluid intake, and judicious use of laxatives. Management may also include discontinuing laxative abuse. Patients can be educated to sit on the toilet with legs supported and to utilize the **gastrocolic reflex** (peristaltic movements of the large bowel occurring five to six times daily that are triggered by distention of the stomach) by attempting to defecate following a meal and a warm drink (Ostaszkiewicz et al., 2010). Routine exercise to strengthen abdominal muscles is encouraged. Biofeedback is a technique that can be used to help patients learn to relax the sphincter mechanism to expel stool. It is considered first-line therapy once anorectal structural lesions have been excluded as the cause for constipation (Emmanuel, 2011). Daily dietary intake of 25 to 30 g/day of fiber (soluble and bulk forming) is recommended, especially for the treatment of constipation in the older adult. If laxative use is necessary, one of the following may be prescribed: bulk-forming agents (fiber laxatives), saline and osmotic agents, lubricants, stimulants, or fecal softeners. The physiologic action and patient education information related to these laxatives are presented in Table 48-1. Enemas and rectal suppositories are generally not recommended for treating constipation unless rectal evacuation is a

TABLE 48-1 Laxative Medications

Medications (Classification and Selected Drugs)	Action	Patient Education
Bulk Forming		
psyllium hydrophilic mucilloid (Metamucil), methylcellulose (Citrucel)	Polysaccharides and cellulose derivatives mix with intestinal fluids, swell, and stimulate peristalsis.	Take with 8 oz water and follow with 8 oz water; do not take dry. Report abdominal distention or unusual amount of flatulence.
Saline Agent		
magnesium hydroxide (Milk of Magnesia)	Nonabsorbable magnesium ions alter stool consistency by drawing water into the intestines by osmosis; peristalsis is stimulated. Action occurs within 2 hours.	The liquid preparation is more effective than the tablet form. Only short-term use is recommended because of toxicity (central nervous system or neuromuscular depression, electrolyte imbalance). Magnesium laxatives should not be taken by patients with renal insufficiency.
Lubricant		
Mineral oil glycerin suppository	Nonabsorbable hydrocarbons soften fecal matter by lubricating the intestinal mucosa; the passage of stool is facilitated. Action occurs within 6–8 hours for mineral oil and within 30 minutes for glycerin suppository.	Do not take mineral oil with meals, because it can impair the absorption of fat-soluble vitamins and delay gastric emptying. Swallow carefully, because drops of oil that gain access to the pharynx can produce a lipid pneumonia. Glycerin suppositories must be inserted fully and retained.
Stimulant		
bisacodyl (Dulcolax), senna (Senokot)	Irritates the colonic epithelium by stimulating sensory nerve endings and increasing mucosal secretions and decreasing large intestinal water absorption. Action occurs within 6–8 hours.	Catharsis may cause fluid and electrolyte imbalance, especially in the older adult. Tablets should be swallowed, not crushed or chewed. Avoid milk or antacids within 1 hour of taking medication, because the enteric coating may dissolve prematurely. Stimulant laxatives are *not* indicated for long-term use.
Fecal Softener		
docusate (Colace)	Hydrates the stool by its surfactant action on the colonic epithelium (increases the wetting efficiency of intestinal water); aqueous and fatty substances are mixed. Does *not* exert a laxative action.	Can be used safely by patients who should avoid straining (cardiac patients, patients with anorectal disorders). Will not evacuate hard stool because it is not a true laxative.
Osmotic Agent		
polyethylene glycol and electrolytes (CoLyte)	Cleanses colon rapidly and induces diarrhea	This is a large-volume product. It takes time to consume it safely. It can cause considerable nausea and bloating. Has high level of research evidence support.

Adapted from American Society for Gastrointestinal Endoscopy. (2012). *Understanding bowel preparation*. Available at: asge.org/patients/patients problem. Then, glycerin suppositories are considered first-line therapy, followed by bisacodyl suppositories or mini-enemas (Emmanuel, 2011). If long-term laxative use is necessary, a bulk-forming agent may be prescribed in combination with an osmotic laxative.

Specific medications may be prescribed to enhance colonic transit by increasing propulsive motor activity. These may include cholinergic agents (e.g., bethanechol [Urecholine]), cholinesterase inhibitors (e.g., neostigmine [Prostigmin]), or prokinetic agents (e.g., metoclopramide [Reglan]). Prokinetic agents including serotonin (5-HT_4) receptor agonists such as prucalopride (Resolor) and prostones such as lubiprostone (Amitiza) stimulate chloride channels in the gut. Medical probiotics (i.e., ingested live organisms) may help some constipated individuals by creating improved bacterial balance. The use of sacral neuromodulation by sacral stimulation has been tried and is being analyzed for efficacy. Alternative therapies are gaining popularity, including abdominal massage, aromatherapy, acupuncture, and the use of Chinese herbal medications (Kyle, 2010; Lin, Fu, Dunning, et al., 2009). They should be used only for patients with unremitting constipation (Emmanuel, 2011; Kyle, 2010; Singh & Rao, 2010).

Nursing Management

The nurse elicits information about the onset and duration of constipation, current and past elimination patterns, the patient's expectation of normal bowel elimination, and lifestyle information (e.g., exercise and activity level, occupation, food and fluid intake, and stress level) during the health history interview. Past medical and surgical history, current medications, and laxative and enema use are important, as is information about the sensation of rectal pressure or fullness, abdominal pain, excessive straining at defecation, and flatulence.

Patient education and health promotion are important functions of the nurse (Chart 48-2). After the health history is obtained, the nurse sets specific goals for education. Goals for the patient include restoring or maintaining a regular pattern of elimination by responding to the urge to defecate, ensuring adequate intake of fluids and high-fiber foods, learning about methods to avoid constipation, relieving anxiety about bowel elimination patterns, and avoiding complications.

Chart 48-2 **HEALTH PROMOTION**

Preventing Constipation

- Describe the physiology of defecation.
- Emphasize the importance of responding to the urge to defecate.
- Discuss normal variations in patterns of defecation.
- Describe how to establish a bowel routine, and explain that having a regular time for defecation (e.g., best time is after a meal) may aid in initiating the reflex.
- Provide dietary information: Suggest eating high-residue, high-fiber foods (e.g., fruits, vegetables); adding bran daily (must be introduced gradually); and increasing fluid intake (unless contraindicated).
- Explain how an exercise regimen, increased ambulation, and abdominal muscle toning will increase muscle strength and help propel colon contents.
- Describe abdominal toning exercises (contracting abdominal muscles 4 times daily and leg-to-chest lifts 10 to 20 times each day).
- Explain that the normal position (semisquatting) maximizes the use of abdominal muscles and force of gravity.
- Avoid overuse or long-term use of stimulant laxatives.

Diarrhea

Diarrhea is an increased frequency of bowel movements (more than three per day), an increased amount of stool (more than 200 g/day), and altered consistency (i.e., increased liquidity) of stool. It is usually associated with urgency, perianal discomfort, incontinence, or a combination of these factors. Any condition that causes increased intestinal secretions, decreased mucosal absorption, or altered motility can produce diarrhea. IBS, IBD, and lactose intolerance are frequently the underlying disease processes that cause diarrhea (Avadhani & Miley, 2011; Hall, 2010; Kent & Banks, 2010; Surawicz, 2010).

Diarrhea can be acute or chronic. Acute diarrhea is most often associated with infection and is usually self-limiting, lasting up to 7 to 14 days; chronic diarrhea persists for more than 2 to 3 weeks and may return sporadically. Diarrhea can be caused by certain medications (e.g., thyroid hormone replacement, stool softeners and laxatives, prokinetic agents, antibiotics, chemotherapy, antiarrhythmic agents, antihypertensive agents, magnesium-based antacids), certain tube-feeding formulas, metabolic and endocrine disorders (e.g., diabetes, Addison's disease, thyrotoxicosis), and viral or bacterial infectious processes (e.g., *Clostridium difficile*–associated disease), dysentery, shigellosis, food poisoning, Norwalk virus). Other disease processes associated with diarrhea include nutritional and malabsorptive disorders (e.g., celiac disease), anal sphincter defect, Zollinger-Ellison syndrome, paralytic ileus, intestinal obstruction, and acquired immunodeficiency syndrome (AIDS) (Avadhani & Miley, 2011; Hall, 2010; Kent & Banks, 2010).

Pathophysiology

Types of diarrhea include secretory, osmotic, malabsorptive, infectious, and exudative. Secretory diarrhea is usually high-volume diarrhea. Often associated with bacterial toxins and neoplasms, it is caused by increased production and secretion of water and electrolytes by the intestinal mucosa into the intestinal lumen. Osmotic diarrhea occurs when water is pulled into the intestines by the osmotic pressure of unabsorbed particles, slowing the reabsorption of water. It can be caused by lactase deficiency, pancreatic dysfunction, or intestinal hemorrhage. Malabsorptive diarrhea combines mechanical and biochemical actions, inhibiting effective absorption of nutrients. Low serum albumin levels lead to intestinal mucosa swelling and liquid stool. Infectious diarrhea results from infectious agents invading the intestinal mucosa. *C. difficile* is the most commonly identified agent in antibiotic-associated diarrhea in the hospital (Avadhani & Miley, 2011). Exudative diarrhea is caused by changes in mucosal integrity, epithelial loss, or tissue destruction by radiation or chemotherapy (Hall, 2010; Surawicz, 2010). Diarrhea may also be caused by laxative misuse.

Clinical Manifestations

In addition to the increased frequency and fluid content of stools, the patient usually has abdominal cramps, distention, **borborygmus** (i.e., a rumbling noise caused by the movement of gas through the intestines), anorexia, and thirst. Painful spasmodic contractions of the anus and **tenesmus** (i.e., ineffective, sometimes painful straining with a strong urge) may occur with defecation. Other symptoms depend on the cause and severity of the diarrhea but are related to dehydration and to fluid and electrolyte imbalances.

Watery stools are characteristic of disorders of the small bowel, whereas loose, semisolid stools are associated more often with disorders of the large bowel. Voluminous, greasy stools suggest intestinal **malabsorption** (impaired transport across the mucosa), and the presence of blood, mucus, and pus in the stools suggests inflammatory enteritis or colitis. Oil droplets on the toilet water are almost always diagnostic of pancreatic insufficiency. Nocturnal diarrhea may be a manifestation of diabetic neuropathy. The possibility of C. *difficile* infection should be considered in all patients with unexplained diarrhea who are taking or have recently taken antibiotics (Avadhani & Miley, 2011; Hall, 2010; Surawicz, 2010).

Assessment and Diagnostic Findings

When the cause of the diarrhea is not obvious, the following diagnostic tests may be performed: complete blood cell count; serum chemistries; urinalysis; routine stool examination; and stool examinations for infectious or parasitic organisms, bacterial toxins, blood, fat, electrolytes, and white blood cells. Endoscopy or barium enema may assist in identifying the cause.

Complications

Complications of diarrhea include the potential for cardiac dysrhythmias because of significant fluid and electrolyte loss (especially loss of potassium). Loss of bicarbonate with diarrhea can also lead to metabolic acidosis. Urinary output less than 0.5 mL/kg/h for 2 to 3 consecutive hours, muscle weakness, paresthesia, hypotension, anorexia, and drowsiness with a potassium level less than 3.5 mEq/L (3.5 mmol/L) must be reported. Chronic diarrhea can also result in skin care issues related to irritant dermatitis. Cleansing with a wet wipe and applying barrier cream can prevent dermatitis (Hall, 2010; Kent & Banks, 2010).

Gerontologic Considerations

Older patients can become dehydrated quickly and develop low potassium levels (i.e., hypokalemia) as a result of diarrhea. The nurse observes for clinical manifestations of muscle weakness, dysrhythmias, or decreased peristaltic motility that may lead to paralytic ileus. The older patient taking digitalis (e.g., digoxin [Lanoxin]) must be aware of how quickly dehydration and hypokalemia can occur with diarrhea. The nurse educates the patient to recognize the symptoms of hypokalemia, because low levels of potassium potentiate the action of digitalis, leading to digitalis toxicity.

Medical Management

Management is directed at controlling symptoms, preventing complications, and eliminating or treating the underlying disease. Until the definitive cause is discovered, infection control measures that restrict the transmission of infectious organisms (e.g., C. *difficile*–associated diarrhea) are warranted (Simor, 2010). Certain medications (e.g., antibiotics, anti-inflammatory agents) and antidiarrheal agents (e.g., loperamide [Imodium], diphenoxylate with atropine [Lomotil]) may be used to reduce the severity of the diarrhea and treat the underlying disease. In most cases, loperamide is the medication of choice because it has fewer side effects than diphenoxylate with atropine (Hall, 2010; Kent & Banks, 2010; Simor, 2010). Notably, antidiarrheal agents should not be used until C. *difficile* infection is ruled out (Simor, 2010). Some research supports the use of probiotics (live organisms administered to a host) in some forms of diarrhea. These organisms consist of *Saccharomyces boulardii* (yeast) or lactic acid bacteria such as *Lactobacillus* species (Williams, 2010).

Nursing Management

The nurse's role includes assessing and monitoring the characteristics and pattern of diarrhea. A health history should address the patient's medication therapy, medical and surgical history, and dietary patterns and intake. Reports of recent acute illness or recent travel to another geographic area are important. Assessment includes abdominal auscultation and palpation for tenderness. Inspection of the abdomen, mucous membranes, and skin is important to determine hydration status. Stool samples are obtained for testing. The perianal area should also be assessed for skin excoriation.

During an episode of acute diarrhea, the nurse encourages bed rest and intake of liquids and foods low in bulk until the acute attack subsides. When the patient is able to tolerate food intake, the nurse recommends a bland diet of semisolid and solid foods. The patient should avoid caffeine, carbonated beverages, and very hot and very cold foods, because they stimulate intestinal motility. It may be necessary to restrict milk products, fat, whole-grain products, fresh fruits, and vegetables for several days. The nurse administers antidiarrheal medications such as diphenoxylate with atropine or loperamide as prescribed. Oral rehydration therapy alone will not resolve symptoms of diarrhea. Intravenous (IV) fluid therapy may be necessary for rapid rehydration in some patients, especially in older patients and in patients with preexisting GI conditions (e.g., IBD). It is important to monitor serum electrolyte levels closely. The nurse immediately reports evidence of dysrhythmias or a change in a patient's level of consciousness.

The perianal area may become excoriated because diarrheal stool contains digestive enzymes that can irritate the skin. The patient should follow a perianal skin care routine to decrease irritation and excoriation (see Chapter 61). The skin of an older person is very sensitive because of decreased turgor and reduced subcutaneous fat layers. Gentle cleansing with a perineal cleansing solution (i.e., wet wiping method) and the use of a barrier cream or a liquid skin sealant will prevent and/or treat the excoriation (Bryant, 2012).

Fecal Incontinence

Fecal incontinence describes the involuntary passage of stool from the rectum. Factors that influence this disorder include the ability of the rectum to sense and accommodate stool, the amount and consistency of stool, the integrity of the anal sphincters and musculature, and rectal motility. Fecal incontinence can have a substantially negative impact on quality of life (Landerfeld, Bowen, Feld, et al., 2008; Malmstrom, Anderson, Wolinksi, et al., 2010).

Pathophysiology

Fecal incontinence has many causes and risk factors and may be a symptom of an underlying condition. In general, it results from conditions that interrupt or disrupt the structure or function of the anorectal unit (Bryant, 2012). Causes include trauma (e.g., after surgical procedures involving the rectum), neurologic disorders (e.g., stroke, multiple sclerosis, diabetic neuropathy, dementia), inflammation, infection, chemotherapy, radiation treatment, fecal impaction, pelvic floor relaxation, laxative abuse, medications, or advancing age (i.e., weakness or loss of anal or rectal muscle tone). Fecal incontinence in younger women is most often the result of childbirth-related injuries (Ahmad, McCallum, & Mercer-Jones, 2010; Apau, 2010b; Bliss & Norton, 2010; Hurnauth, 2011; Nazarko, 2011).

Clinical Manifestations

Patients may have minor soiling, occasional urgency and loss of control, or complete incontinence. Patients may also experience poor control of flatus, diarrhea, or constipation. Passive incontinence occurs without warning.

Assessment and Diagnostic Findings

Assessing the patient's medical history is helpful in identifying the most likely etiology. Diagnostic studies are necessary because the treatment of fecal incontinence depends on the cause. A rectal examination and endoscopic examinations such as a flexible sigmoidoscopy are performed to rule out tumors, inflammation, fissures, or impaction. X-ray studies such as barium enema, computed tomography (CT) scans, anorectal manometry, and transit studies may be helpful in identifying alterations in intestinal mucosa and muscle tone or in detecting other structural or functional problems.

Medical Management

Although there is no known cure for idiopathic fecal incontinence (no abnormality in sphincters, rectal compliance, intact sensation), specific management techniques can help the patient achieve a better quality of life. If fecal incontinence is related to diarrhea, the incontinence may disappear when diarrhea is successfully treated. Fecal incontinence is frequently a symptom of a fecal impaction (Bliss & Norton, 2010). After the impaction is removed and the rectum is cleansed, normal functioning of the anorectal area can resume. If the fecal incontinence is related to the use of contributory drugs (e.g., calcium channel blockers, laxatives, metformin [Glucophage], antacids containing magnesium, sildenafil [Viagra]), the incontinence may improve or cease when the drug regimen is altered (Ahmad et al., 2010; Apau, 2010b). When fecal incontinence is related to a more permanent condition, other treatments are initiated. Biofeedback therapy with pelvic floor muscle training can be of assistance if the problem is decreased sensory awareness or sphincter control. Bowel training programs can also be effective. Surgical procedures include surgical reconstruction or repair of anal sphincter, artificial sphincter implantation, anal sphincter bulking by injection of synthetic agents, sacral nerve stimulation, or fecal diversion (Ahmad et al., 2010; Sharpe & Read, 2010).

Nursing Management

The nurse obtains a thorough health history, including information about previous surgical procedures, chronic illnesses, dietary patterns, bowel habits and problems, current medication regimen, and, in women, number of pregnancies and types of deliveries (e.g., vaginal delivery with forceps). Stool charts (e.g., Bristol Stool Chart) may help with identifying frequency, volume, and consistency of the feces (Nazarko, 2011). The nurse also completes an examination of the rectal area. If a fecal impaction is noted, it must be removed before instituting any preventive therapies (Bryant, 2012).

The nurse initiates a bowel training program that involves setting a schedule to establish bowel regularity. The goal is to help the patient achieve fecal continence. If this is not possible, the goal should be to manage the problem so the patient can have predictable, planned elimination (Manthey, Bliss, Savik, et al., 2010). Sometimes it is necessary to use suppositories to stimulate the anal reflex. After the patient has achieved a regular schedule, the suppository can be discontinued. Biofeedback in conjunction with pelvic floor exercises can be used to help the patient improve sphincter contractility and rectal sensitivity. Bowel regulation also involves the therapeutic use of diet and fiber. Foods that thicken stool (e.g., applesauce) and fiber products (e.g., psyllium) help improve continence. Conversely, foods that loosen stool (e.g., rhubarb, figs, prunes, plums) should be avoided. Some patients with fecal incontinence may benefit from the use of antidiarrheal medications. Loperamide, diphenoxylate with atropine, and difenoxin can be used; loperamide is the preferred medication because it does not cause central nervous system adverse effects (Chien & Bradway, 2010).

Fecal incontinence can disrupt perineal skin integrity. Maintaining skin integrity is a priority, especially in the debilitated or older adult patient. Incontinence briefs or adult diapers, although helpful in containing the fecal material, permit increased skin contact with feces and may cause skin excoriation. In general, incontinence briefs are to be used only for brief periods of time. The nurse encourages and instructs about meticulous skin hygiene and uses perineal skin cleansers and skin protection products to protect perineal skin (Bryant, 2012; Nix & Haugen, 2010). Some patients may benefit from occasional use of anal plugs. However, research suggests that most people find them unacceptable (Ahmad et al., 2010).

Continence sometimes cannot be achieved, and the nurse assists the patient and family to accept and cope with this chronic situation. Patients with dementia may benefit from toileting assistance, including prompted or timed voiding and habit training, which is the setting of a regular time to go to the bathroom (e.g., after breakfast to have a bowel movement) (Hagglund, 2010; Schnelle, Simmons,

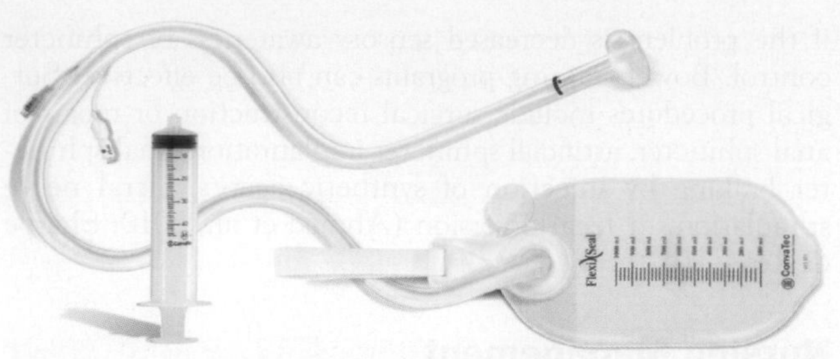

FIGURE 48-1 • Flexi-Seal Fecal Management System. Reprinted with permission from ConvaTec, Inc.

Beuscher, et al., 2009). The patient can use fecal incontinence devices, which include external collection devices and internal drainage systems. External devices are special rectal pouches (called *fecal incontinence collectors*) that are drainable. They are attached to a synthetic adhesive skin barrier specially designed to conform to the buttocks. Designed for more acutely ill patients, internal drainage systems can be used to eliminate fecal skin contact and are especially useful when there is extensive excoriation or skin breakdown. New fecal and bowel management systems (e.g., Flexi-Seal Fecal Management System) are available. These systems, which consist of a tube with a low-pressure balloon that conforms to the internal rectal area, may be used for up to 29 consecutive days (Fig. 48-1). Given the better safety data of the newer bowel management systems versus rectal catheters, the latter are now generally contraindicated (Hurnauth, 2011; Kyle, 2010).

Irritable Bowel Syndrome

Irritable bowel syndrome (IBS) is a chronic functional bowel disorder that affects frequency of defecation and consistency of stool and is one of the most common GI conditions. Approximately 12% to 14% of adults in the United States report classic symptoms of IBS (Banerjee, 2010). IBS accounts for millions of office visits to health care facilities and clinics and is a leading cause of workforce absenteeism (Banerjee, 2010). Sometimes called *spastic colon*, it occurs more commonly in women than in men, and the cause remains unknown (Banerjee, 2010; Smith, 2010). Although no anatomic or biochemical abnormalities have been found that account for its common symptoms, various factors are associated with the syndrome: heredity, psychological stress or conditions such as depression and anxiety, a diet high in fat and stimulating or irritating foods, alcohol consumption, and smoking. The diagnosis is made only after tests confirm the absence of structural or other disorders (Banerjee, 2010; Smith, 2010).

Pathophysiology

IBS results from a functional disorder of intestinal motility. The change in motility may be related to neuroendocrine dysregulation, especially changes in serotonin signaling, infection, irritation, or a vascular or metabolic disturbance. The peristaltic waves are affected at specific segments of the intestine and in the intensity with which they propel the fecal matter forward. There is no evidence of inflammation or tissue changes in the intestinal mucosa (Banerjee, 2010; Reed, 2010; Smith, 2010).

Clinical Manifestations

Symptoms can vary widely, ranging in intensity and duration from mild and infrequent to severe and continuous. The main symptom is an alteration in bowel patterns: constipation (classified as IBS-C), diarrhea (classified as IBS-D), or a combination of both (classified as IBS-A for "alternating") (Banerjee, 2010). Pain, bloating, and abdominal distention often accompany changes in bowel pattern. The abdominal pain is sometimes precipitated by eating and is frequently relieved by defecation. The abdominal pain can be accompanied by a variety of non-GI functional disorders, including fibromyalgia, backache, lethargy, and urinary frequency (Smith, 2010).

Assessment and Diagnostic Findings

International consensus meetings have yielded specific diagnostic criteria for IBS. These include recurrent abdominal pain or discomfort at least 3 days a month in the past 3 months with two or more of the following: (1) improvement with defecation, (2) onset associated with change in frequency of stool, and (3) onset associated with change in appearance (form) of stool (Banerjee, 2010; Reed, 2010). A definite diagnosis of IBS also necessitates testing to confirm the absence of structural or other disorders. Stool studies, contrast x-ray studies, and proctoscopy may be performed to rule out other colon diseases. Barium enema and colonoscopy may reveal spasm, distention, or mucus accumulation in the intestine (Fig. 48-2). Manometry and electromyography are used to study intraluminal pressure changes generated by spasticity.

Medical Management

The goals of treatment are relieving abdominal pain, controlling the diarrhea or constipation, and reducing stress. Restriction and then gradual reintroduction of foods that are possibly irritating may help determine what types of food are acting as irritants (e.g., beans, caffeinated products, corn, wheat, dairy lactose, fried foods, alcohol, spicy foods, aspartame). A high-fiber diet is prescribed to help control the diarrhea

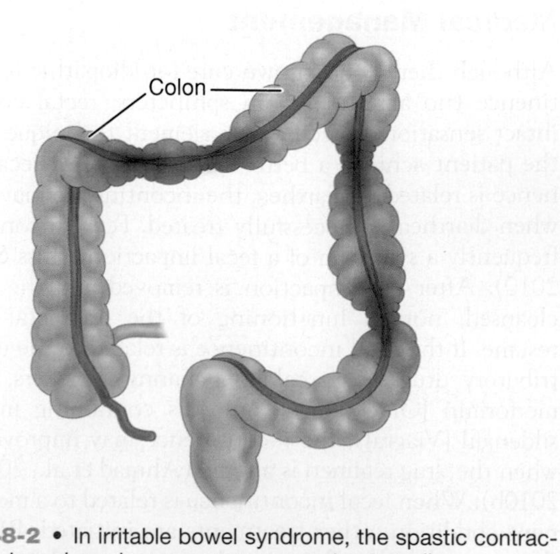
Colon

FIGURE 48-2 • In irritable bowel syndrome, the spastic contractions of the bowel can be seen on x-ray contrast studies.

and constipation. Exercise can assist in reducing anxiety and increasing intestinal motility. Patients often find it helpful to participate in a stress reduction or behavior modification program. Hydrophilic colloids (i.e., bulk) and antidiarrheal agents (e.g., loperamide) may be given to control the diarrhea and fecal urgency. Antidepressants can assist in treating underlying anxiety and depression but also have secondary benefits. Antidepressants may affect serotonin levels, thus slowing intestinal transit time and improving diarrhea and abdominal comfort. Anticholinergic or antispasmodic agents (e.g., propantheline [Pro-Banthine]) may be prescribed to decrease smooth muscle spasm, decreasing cramping and constipation (Reed, 2010).

Lubiprostone, a chloride channel regulator in the gut, is now being used for treating persons with IBS-C (constipation dominant) (Amitiza, 2011). However, alosetron (Lotronex) has been approved to treat those with IBS-D (diarrhea dominant) (NDDIC, 2012a). Other alternatives for IBS management include probiotics and alternative medicines. Probiotics are bacteria that include *Lactobacillus* and *Bifidobacterium* that can be administered to help decrease abdominal bloating and gas. In addition, an over-the-counter combination probiotic (VSL #3) that contains eight bacterial organisms has also had good results (Williams, 2010). Complementary medicine approaches to treatment of IBS include artichoke leaf extract, peppermint oil, and caraway oil. They reputedly diminish IBS symptoms; however, formal studies are needed to examine their effectiveness (Banerjee, 2010).

Nursing Management

The nurse's role is to provide patient and family education. The nurse emphasizes and reinforces good dietary habits (e.g., avoidance of food triggers). A good way to identify problem foods is to keep a 1- to 2-week symptom and food diary. Patients are encouraged to eat at regular times and to chew food slowly and thoroughly. They should understand that although adequate fluid intake is necessary, fluid should not be taken with meals because this results in abdominal distention. Alcohol use and cigarette smoking are discouraged. Stress management via relaxation techniques, cognitive-behavioral therapy, yoga, and exercise can be recommended (Banerjee, 2010; Blesse, 2010; Lahmann, Rohricht, Sauer, et al., 2010; Reed, 2010).

Conditions of Malabsorption

The inability of the digestive system to absorb one or more of the major vitamins (especially A and B_{12}), minerals (i.e., iron and calcium), and nutrients (i.e., carbohydrates, fats, and proteins) occurs in conditions of malabsorption. Interruptions in the complex digestive process may occur anywhere in the digestive system and cause decreased absorption. Diseases of the small intestine are the most common cause of malabsorption (Huether & McCance, 2012).

Pathophysiology

The conditions that cause malabsorption can be grouped into the following categories (Klapparoth, 2012):

- Mucosal (transport) disorders causing generalized malabsorption (e.g., celiac disease, regional enteritis, radiation enteritis)
- Infectious diseases causing generalized malabsorption (e.g., small bowel bacterial overgrowth, tropical sprue, Whipple's disease)
- Luminal disorders causing malabsorption (e.g., bile acid deficiency, Zollinger-Ellison syndrome, pancreatic insufficiency, or chronic pancreatitis)
- Postoperative malabsorption (e.g., after gastric or intestinal resection)
- Disorders that cause malabsorption of specific nutrients (e.g., disaccharidase deficiency leading to lactose intolerance; food allergies)

Table 48-2 lists the clinical and pathologic aspects of malabsorptive diseases.

Clinical Manifestations

The hallmarks of malabsorption syndrome from any cause are diarrhea or frequent, loose, bulky, foul-smelling stools that have increased fat content and are often grayish (steatorrhea). Patients often have associated abdominal distention, pain, increased flatus, weakness, weight loss, and a decreased sense of well-being. The symptoms can overlap those of IBS (Food allergies and food intolerances, 2011). The chief result of malabsorption is malnutrition, manifested by weight loss and other signs of vitamin and mineral deficiency (e.g., easy bruising, osteoporosis, anemia). Patients with a malabsorption syndrome, if untreated, become weak and emaciated because of starvation and dehydration. Failure to absorb the fat-soluble vitamins A, D, and K causes a corresponding avitaminosis.

Assessment and Diagnostic Findings

Several diagnostic tests may be prescribed, including stool studies for quantitative and qualitative fat analysis, lactose tolerance tests, D-xylose absorption tests, and Schilling tests. The hydrogen breath test that is used to evaluate carbohydrate absorption (see Chapter 44) is performed if carbohydrate malabsorption is suspected. Fat malabsorption may occur before other clinical signs and symptoms appear (Lewis, 2011), thus early stool fat testing is recommended. Endoscopy with biopsy of the mucosa is the best diagnostic tool. Biopsy of the small intestine is performed to assay enzyme activity or to identify infection or destruction of mucosa. Ultrasound studies, CT scans, and x-ray findings can reveal pancreatic or intestinal tumors that may be the cause. A complete blood cell count is used to detect anemia. Pancreatic function tests can assist in the diagnosis of specific disorders.

Medical Management

Intervention is aimed at avoiding dietary substances that aggravate malabsorption and at supplementing nutrients that have been lost. Common supplements are water-soluble vitamins (e.g., B_{12}, folic acid), fat-soluble vitamins (i.e., A, D, and K), and minerals (e.g., calcium, iron). Causative disease states may be managed surgically or nonsurgically. For example, the best treatment for chronic pancreatitis due to excessive alcohol consumption is alcohol cessation (Lewis, 2011). Dietary therapy is aimed at reducing gluten intake in patients with celiac disease. Folic acid supplements are prescribed for

TABLE 48-2 Characteristics of Diseases of Malabsorption

Diseases/Disorders	Pathophysiology	Clinical Features
Gastric resection with gastrojejunostomy	Decreased pancreatic stimulation because of duodenal bypass; poor mixing of food, bile, pancreatic enzymes; decreased intrinsic factor	Weight loss, moderate steatorrhea, anemia (combination of iron deficiency, vitamin B_{12} malabsorption, folate deficiency)
Pancreatic insufficiency (chronic pancreatitis, pancreatic carcinoma, pancreatic resection, cystic fibrosis)	Reduced intraluminal pancreatic enzyme activity, with maldigestion of lipids and proteins	History of abdominal pain followed by weight loss; marked steatorrhea, azotorrhea (excess of nitrogenous matter in the feces or urine); also frequent glucose intolerance (70% in pancreatic insufficiency)
Ileal dysfunction (resection or disease)	Loss of ileal absorbing surface leads to reduced bile salt pool size and reduced vitamin B_{12} absorption; bile in colon inhibits fluid absorption.	Diarrhea, weight loss with steatorrhea, especially when >100-cm resection, decreased vitamin B_{12} absorption
Stasis syndromes (surgical strictures, blind loops, enteric fistulas, multiple jejunal diverticula, scleroderma)	Overgrowth of intraluminal intestinal bacteria, especially anaerobic organisms, to $>10^6$/mL results in deconjugation of bile salts, leading to decreased effective bile salt pool size, also bacterial utilization of vitamin B_{12}	Weight loss, steatorrhea; low vitamin B_{12} absorption; may have low D-xylose absorption
Zollinger-Ellison syndrome	Hyperacidity in duodenum inactivates pancreatic enzymes.	Ulcer diathesis, steatorrhea
Lactose intolerance	Deficiency of intestinal lactase results in high concentration of intraluminal lactose with osmotic diarrhea.	Varied degrees of diarrhea and cramps after ingestion of lactose-containing foods; positive lactose intolerance test, decreased intestinal lactase
Celiac disease (gluten-sensitive enteropathy)	Toxic response to a gluten fraction gliadin by surface epithelium results in destruction of absorbing surface of intestine.	Weight loss, diarrhea, bloating, anemia (low iron, folate), osteomalacia, steatorrhea, azotorrhea, low D-xylose absorption; folate and iron malabsorption
Tropical sprue	Unknown toxic factor results in mucosal inflammation, partial villous atrophy	Weight loss, diarrhea, anemia (low folate, vitamin B_{12}); steatorrhea; low D-xylose absorption, low vitamin B_{12} absorption
Whipple's disease	Bacterial invasion of intestinal mucosa	Arthritis, hyperpigmentation, lymphadenopathy, serous effusions, fever, weight loss, steatorrhea, azotorrhea
Certain parasitic diseases (giardiasis, strongyloidiasis, coccidiosis, capillariasis)	Damage to or invasion of surface mucosa	Diarrhea, weight loss; steatorrhea; organism may be seen on jejunal biopsy or recovered in stool.
Immunoglobulinopathy	Decreased local intestinal defenses, lymphoid hyperplasia, lymphopenia	Frequent association with *Giardia:* hypogammaglobulinemia or isolated immunoglobulin A deficiency

Adapted from Klapparoth, J. M. (2012). Malabsorption. *Medscape*. Available at: www.emedicine.medscape.com/article/180785-overview;Mason, J. (2011). Mechanisms of nutrient absorption and malabsorption. *UpToDate*. Available at: www.uptodate.com; and Mason, J. B., & Milovic, V. (2012). Clinical features and diagnosis of malabsorption. *UpToDate*. Available at: www.uptodate.com

patients with tropical sprue. Antibiotics (e.g., tetracycline [Tetracyn], ampicillin [Polycillin], rifaximin [Xifaxan]) are sometimes needed in the treatment of tropical sprue and bacterial overgrowth syndromes. Antidiarrheal agents may be used to decrease intestinal spasms. Probiotics (e.g., *Lactobacillus acidophilus*) and other "friendly" bacteria may be recommended to help improve microbial balance (Food allergies and food intolerances, 2011). Parenteral fluids may be necessary to treat dehydration.

Gerontologic Considerations

The older patient may have more subtle symptoms of malabsorption that may be extraintestinal, including fatigue and confusion. Medical management may include the administration of corticosteroids, which may cause a host of adverse effects such as hypertension, hypokalemia, and confusion. Antibiotics may reduce vitamin K–producing intestinal flora, resulting in a prolonged prothrombin time and international normalized ratio if the patient is concurrently taking warfarin (Coumadin). Urinary retention, altered mental status, or glaucoma may occur as adverse effects of anticholinergic drug therapy in older people (Bolin, 2010).

Nursing Management

The nurse provides patient and family education regarding diet, medications, and the use of nutritional supplements (Chart 48-3 provides a discussion of the management of lactose intolerance). Patients with diarrhea must be monitored for fluid and electrolyte imbalances, particularly metabolic acidosis. The nurse conducts ongoing assessments to determine whether the clinical manifestations related to the nutritional deficits have abated. Patient education includes information about the risk of osteoporosis related to malabsorption of calcium.

ACUTE INFLAMMATORY INTESTINAL DISORDERS

Any part of the lower GI tract is susceptible to acute inflammation caused by bacterial, viral, or fungal infection. Two such conditions are appendicitis and diverticulitis, both of which may lead to **peritonitis**, which is an inflammation of the lining of the abdominal cavity.

Chart 48-3 PATIENT EDUCATION
Managing Lactose Intolerance

- Deficiency of lactase, a digestive enzyme essential for the digestion and absorption of lactose ("milk sugar") from the intestines, results in an intolerance to milk.
- Elimination of milk and milk substances can prevent symptoms.
- Many processed foods have fillers, such as dried milk, added to them; recognizing this can help in determining which foods may need to be eliminated.
- Pretreatment of foods with lactase preparations (e.g., Lactaid drops) before ingestion can reduce symptoms.
- Ingestion of lactase enzyme tablets with the first bite of food can reduce symptoms.
- Most people can tolerate 1 to 2 cups of milk or milk products daily without major problems; they are best tolerated if ingested in small amounts during the day.
- Lactase activity of yogurt with "active cultures" helps the digestion of lactose within the intestine better than lactase preparations.
- Milk and milk products are rich sources of calcium and vitamin D; elimination of milk from the diet may result in calcium and vitamin D deficiencies; decreased intake without supplements can lead to osteoporosis.

Appendicitis

The appendix is a small, fingerlike appendage about 10 cm (4 in) long that is attached to the cecum just below the ileocecal valve. The appendix fills with food and empties regularly into the cecum. Because it empties inefficiently and its lumen is small, the appendix is prone to obstruction and is particularly vulnerable to infection (i.e., appendicitis).

Appendicitis, the most common cause of acute surgical abdomen in the United States, is the most common reason for emergency abdominal surgery. Although it can occur at any age, it more commonly occurs between the ages of 10 and 30 years (Hennelly & Bachur, 2011; Spirt, 2010; Vissers & Lennarz, 2010).

Pathophysiology

The appendix becomes inflamed and edematous as a result of becoming kinked or occluded by a fecalith (i.e., hardened mass of stool), tumor, lymphoid hyperplasia, or foreign body. The inflammatory process increases intraluminal pressure, initiating a progressively severe, generalized, or periumbilical pain that becomes localized to the right lower quadrant of the abdomen within a few hours. Eventually, the inflamed appendix fills with pus (Spirt, 2010). Once obstructed, the appendix becomes ischemic, bacterial overgrowth occurs, and eventually gangrene occurs (Vissers & Lennarz, 2010).

Clinical Manifestations

Vague epigastric or periumbilical pain (i.e., visceral pain that is dull and poorly localized) progresses to right lower quadrant pain (i.e., parietal pain that is sharp, discrete, and well localized) and is usually accompanied by a low-grade fever and nausea and sometimes by vomiting. Loss of appetite is common. In up to 50% of presenting cases, local tenderness is elicited at McBurney's point when pressure is applied (Black & Martin, 2012) (Fig. 48-3). Rebound tenderness (i.e., production or intensification of pain when pressure is released) may be present. The extent of tenderness and muscle spasm and the existence of constipation or diarrhea depend not so much on the severity of the appendiceal infection as on the location of the appendix. If the appendix curls around behind the cecum, pain and tenderness may be felt in the lumbar region. If its tip is in the pelvis, these signs may be elicited only on rectal examination. Pain on defecation suggests that the tip of the appendix is resting against the rectum; pain on urination suggests that the tip is near the bladder or impinges on the ureter. Some rigidity of the lower portion of the right rectus muscle may occur. Rovsing's sign may be elicited by palpating the left lower quadrant; this paradoxically causes pain to be felt in the right lower quadrant (see Fig. 48-3). Two other physical examination findings may include the psoas sign (i.e., pain that occurs upon slow extension of the right thigh with the patient lying on the left side) and the obturator sign (i.e., pain that occurs with passive internal rotation of the flexed right thigh with the patient supine (Spirt, 2010). If the appendix has ruptured, the pain becomes more diffuse; abdominal distention develops as a result of paralytic ileus, and the patient's condition worsens.

Constipation can also occur with appendicitis. Laxatives administered in this instance may result in perforation of the inflamed appendix. In general, a laxative or cathartic should not be administered when a person has fever, nausea, and abdominal pain.

Assessment and Diagnostic Findings

Diagnosis is based on results of a complete history and physical examination and on laboratory findings and imaging studies. The patient is commonly younger; age is a crucial differential finding (Spirt, 2010). The complete blood cell count demonstrates an elevated white blood cell count with an elevation

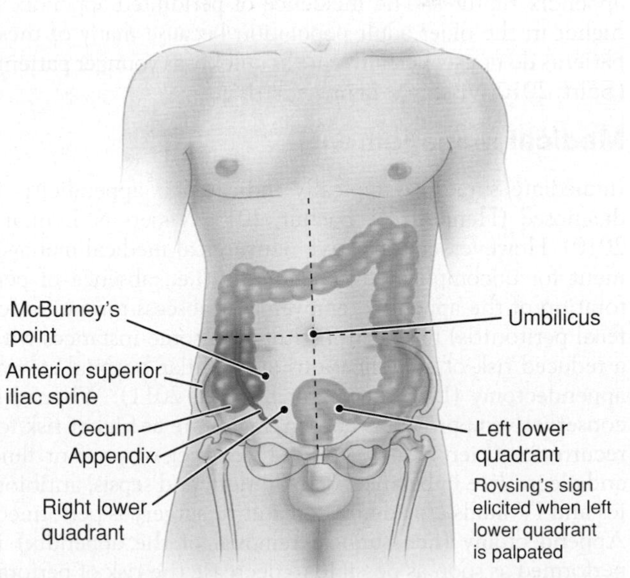
McBurney's point
Umbilicus
Anterior superior iliac spine
Cecum
Appendix
Left lower quadrant
Rovsing's sign elicited when left lower quadrant is palpated
Right lower quadrant

FIGURE 48-3 • When the appendix is inflamed, tenderness can be noted in the right lower quadrant at McBurney's point, which is between the umbilicus and the anterior superior iliac spine. Rovsing's sign is pain felt in the right lower quadrant after the left lower quadrant has been palpated.

of the neutrophils. Abdominal x-ray films, ultrasound studies, and CT scans may reveal a right lower quadrant density or localized distention of the bowel. CT scanning is particularly helpful when the diagnosis is uncertain, especially if nonsurgical diagnoses (e.g., pancreatitis) are under consideration (Spirt, 2010). A pregnancy test may be ordered for women of childbearing age to rule out ectopic pregnancy and before radiologic studies are done. A urinalysis is usually obtained to rule out urinary tract infection (Vissers & Lennarz, 2010). A diagnostic laparoscopy may be used to rule out acute appendicitis in equivocal cases.

Complications

The major complication of appendicitis is perforation of the appendix, which can lead to peritonitis, **abscess** (localized collection of purulent material surrounded by inflamed tissues) formation, or portal pylephlebitis, which is septic thrombosis of the portal vein caused by vegetative emboli that arise from septic intestines. Perforation generally occurs 24 hours after the onset of pain. Symptoms include a fever of 37.7°C (100°F) or greater, a toxic appearance, and continued abdominal pain or tenderness. Patients with peritonitis are often found to be supine and motionless (Spirt, 2010).

Gerontologic Considerations

Acute appendicitis is uncommon in the older adult population. Older adult patients should have an electrocardiogram and chest x-ray to rule out cardiac ischemia and pneumonia, respectively (Spirt, 2010). When appendicitis does occur, classic signs and symptoms are altered and may vary greatly. Pain may be absent or minimal. Symptoms may be vague, suggesting bowel obstruction or another process. Fever and leukocytosis may not be present. As a result, diagnosis and prompt treatment may be delayed, causing complications and mortality. The patient may have no symptoms until the appendix ruptures. The incidence of perforated appendix is higher in the older adult population because many of these patients do not seek health care as quickly as younger patients (Spirt, 2010; Vissers & Lennarz, 2010).

Medical Management

Immediate surgery is typically indicated if appendicitis is diagnosed (Hennelly & Bachur, 2011; Vissers & Lennarz, 2010). However, conservative nonsurgical medical management for uncomplicated appendicitis (i.e., absence of perforation of the appendix, empyema or abscess formation, or fecal peritonitis) has been instituted in some instances with a reduced risk of complications and similar hospital stay as appendectomy (Lien, Lee, Wang, et al., 2011). When this conservative approach is chosen, males are at higher risk for recurrence (Lien et al., 2011). To correct or prevent fluid and electrolyte imbalance, dehydration, and sepsis, antibiotics and IV fluids are administered until surgery is performed. Appendectomy (i.e., surgical removal of the appendix) is performed as soon as possible to decrease the risk of perforation. It may be performed using general or spinal anesthesia with a low abdominal incision (laparotomy) or by laparoscopy. The laparoscopic approach is the preferred approach (Nakhamiyayev, Galldin, Chiarello, et al., 2010). Both laparotomy and laparoscopy are safe and effective in the treatment of appendicitis with perforation. However, recovery after laparoscopic surgery is generally quicker. Consequently, laparoscopic appendectomy has become the more common approach (Hennelly & Bachur, 2011).

When perforation of the appendix occurs, an abscess may form. If this occurs, the patient may be initially treated with antibiotics, and the surgeon may place a drain in the abscess. After the abscess is drained and there is no further evidence of infection, an appendectomy is then typically performed (Hennelly & Bachur, 2011).

Nursing Management

Goals include relieving pain, preventing fluid volume deficit, reducing anxiety, eliminating infection due to the potential or actual disruption of the GI tract, maintaining skin integrity, and attaining optimal nutrition.

The nurse prepares the patient for surgery, which includes an IV infusion to replace fluid loss and promote adequate renal function and antibiotic therapy to prevent infection. If there is evidence or likelihood of paralytic ileus, a nasogastric tube is inserted. An enema is not administered because it can lead to perforation.

After surgery, the nurse places the patient in a high Fowler's position. This position reduces the tension on the incision and abdominal organs, helping to reduce pain. An opioid, usually morphine sulfate, is prescribed to relieve pain. When tolerated, oral fluids are administered. Any patient who was dehydrated before surgery receives IV fluids. Food is provided as desired and tolerated on the day of surgery when normal bowel sounds are present.

The patient may be discharged on the day of surgery if the temperature is within normal limits, there is no undue discomfort in the operative area, and the appendectomy was uncomplicated. Discharge instruction for the patient and family is imperative. The nurse instructs the patient to make an appointment to have the surgeon remove any sutures and inspect the wound between the fifth and seventh days after surgery. Incision care and activity guidelines are discussed; heavy lifting is to be avoided postoperatively, although normal activity can usually be resumed within 2 to 4 weeks.

If there is a possibility of peritonitis, a drain is left in place at the area of the incision. Patients at risk for this complication may be kept in the hospital for several days and are monitored carefully for signs of intestinal obstruction or secondary hemorrhage. Secondary abscesses may form in the pelvis, under the diaphragm, or in the liver, causing elevation of the temperature, pulse rate, and white blood cell count.

When the patient is ready for discharge, the patient and family are educated about how to care for the incision and perform dressing changes and irrigations as prescribed. A home care nurse may be needed to assist with this care and to monitor the patient for complications and wound healing. Other complications of appendectomy are listed in Table 48-3.

Diverticular Disease

A **diverticulum** is a saclike herniation of the lining of the bowel that extends through a defect in the muscle layer. Diverticula may occur anywhere in the small intestine or

TABLE 48-3 Potential Complications and Nursing Interventions After Appendectomy

Complication	Nursing Interventions
Peritonitis	Monitor for abdominal tenderness, fever, vomiting, abdominal rigidity, and tachycardia. Employ constant nasogastric suction. Correct dehydration as prescribed.
Pelvic abscess	Administer antibiotic agents as prescribed. Evaluate for anorexia, chills, fever, and diaphoresis. Observe for diarrhea, which may indicate pelvic abscess. Prepare patient for rectal examination. Prepare patient for surgical drainage procedure.
Subphrenic abscess (abscess under the diaphragm)	Assess patient for chills, fever, and diaphoresis. Prepare for x-ray examination. Prepare for surgical drainage of abscess.
Ileus (paralytic and mechanical)	Assess for bowel sounds. Employ nasogastric intubation and suction. Replace fluids and electrolytes by IV route as prescribed. Prepare for surgery, if diagnosis of mechanical ileus is established.

colon but most commonly occur in the sigmoid colon (at least 95%) (Spirt, 2010; Stocchi, 2010). However, persons of Asian heritage tend to develop diverticula in the right colon, probably because of genetic differences (Carloni, Sage, Roodie, et al., 2010).

Diverticulosis is when multiple diverticula are present without inflammation or symptoms. Diverticular disease of the colon is very common in developed countries, and its prevalence increases with age: As many as 65% of Americans older than 80 years have diverticulosis (Beitz, 2008; Shahedi & Katz, 2013; Spirt, 2010; Young-Fadok & Pemberton, 2011). A low intake of dietary fiber is considered a predisposing factor, but the exact cause has not been identified. Most patients with diverticular disease are asymptomatic, so its exact prevalence is unknown. Clinical data suggest that its incidence may be increasing in the United States (Manwaring & Champagne, 2011).

Diverticulitis results when food and bacteria retained in a diverticulum produce infection and inflammation that can impede drainage and lead to perforation or abscess formation. At least 10% to 25% of patients with diverticulosis have diverticulitis at some point (Spirt, 2010). When the disorder occurs in those younger than 50 years, obesity may play a role (Bianco, Loyson, McNinch, et al., 2010; Stocchi, 2010). Diverticulitis may occur as an acute attack or may persist as a continuing, smoldering infection. The symptoms manifested generally result from complications: abscess, **fistula** (abnormal tract) formation, obstruction, perforation, peritonitis, and hemorrhage.

Pathophysiology

Diverticula form when the mucosa and submucosal layers of the colon herniate through the muscular wall because of high intraluminal pressure, low volume in the colon (i.e., fiber-deficient contents), and decreased muscle strength in the colon wall (i.e., muscular hypertrophy from hardened fecal masses). Bowel contents can accumulate in the diverticulum and decompose, causing inflammation and infection. The diverticulum can also become obstructed and then inflamed if the obstruction continues. The inflammation of the weakened colonic wall of the diverticulum can cause it to perforate, giving rise to irritability and spasticity of the colon (i.e., diverticulitis). In addition, abscesses develop and may eventually perforate, leading to peritonitis and erosion of the arterial blood vessels, resulting in bleeding. When a patient develops symptoms of diverticulitis, microperforation of the colon has occurred (Beitz, 2008; Spirt, 2010).

Clinical Manifestations

Chronic constipation often precedes the development of diverticulosis by many years. Frequently, no problematic symptoms occur with diverticulosis. Signs and symptoms of diverticulosis are relatively mild and include bowel irregularity with intervals of diarrhea, nausea and anorexia, and bloating or abdominal distention (Tursi, 2010). With repeated local inflammation of the diverticula, the large bowel may narrow with fibrotic strictures, leading to cramps, narrow stools, and increased constipation or at times intestinal obstruction. Weakness, fatigue, and anorexia are common symptoms. With diverticulitis, the patient reports an acute onset of mild to severe pain in the left lower quadrant, accompanied by nausea, vomiting, fever, chills, and leukocytosis. The condition, if untreated, can lead to peritonitis and septicemia.

Assessment and Diagnostic Findings

Diverticulosis is typically diagnosed by colonoscopy, which permits visualization of the extent of diverticular disease and biopsy of tissue to rule out other diseases. In the past, barium enema was the preferred diagnostic test, but it is now used less frequently than colonoscopy. If there are symptoms of peritoneal irritation when the diagnosis is diverticulitis, barium enema is contraindicated because of the potential for perforation (Spirt, 2010).

CT with contrast agent is the diagnostic test of choice if the suspected diagnosis is diverticulitis; it can also reveal abscesses. Abdominal x-rays may demonstrate free air under the diaphragm if a perforation has occurred from the diverticulitis. Laboratory tests that assist in diagnosis include a complete blood cell count, revealing an elevated white blood cell count and elevated erythrocyte sedimentation rate (ESR). Colonoscopy is contraindicated in acute diverticulitis because the risk of perforation in the presence of local infection may result in sepsis (Spirt, 2010).

Complications

Complications of diverticulitis include peritonitis, abscess formation, fistulas, and bleeding. If an abscess develops, the associated findings are tenderness, a palpable mass, fever, and leukocytosis. An inflamed diverticulum that perforates results in abdominal pain localized over the involved segment, usually the sigmoid; local abscess or peritonitis follows. Abdominal pain, a rigid boardlike abdomen, loss of bowel sounds, and signs and symptoms of shock occur with peritonitis. On occasion, the diverticular inflammation creates an abnormal passage between body structures such as a colovesical fistula (between bowel and bladder). Noninflamed or slightly inflamed diverticula may erode areas adjacent to arterial branches, causing massive rectal bleeding.

Gerontologic Considerations

The incidence of diverticular disease increases with age because of degeneration and structural changes in the circular muscle layers of the colon and because of cellular hypertrophy. The symptoms are less pronounced in the older adult than in other adults. The older adult may not have abdominal pain until infection occurs. They may delay reporting symptoms because they fear surgery or are afraid that they may have cancer. Blood in the stool is overlooked frequently, especially in the older adult, because of a failure to examine the stool or the inability to see changes if vision is impaired (Shahedi & Katz, 2013; Young-Fadok & Pemberton, 2011).

Medical Management

Dietary and Pharmacologic Management

Diverticulitis can usually be treated on an outpatient basis with diet and medication. When symptoms occur, rest, analgesic medications, and antispasmodic agents are recommended. Initially, a clear liquid diet is consumed until the inflammation subsides; then a high-fiber, low-fat diet is recommended. This type of diet helps increase stool volume, decrease colonic transit time, and reduce intraluminal pressure. Antibiotics are prescribed for 7 to 10 days. A bulk-forming laxative is also prescribed.

In acute cases of diverticulitis with significant symptoms, hospitalization is required. Hospitalization is often indicated for those who are older, immunocompromised, or taking corticosteroids. Withholding oral intake, administering IV fluids, and instituting nasogastric suctioning if vomiting or distention occurs are used to rest the bowel. Broad-spectrum antibiotics (e.g., ampicillin/sulbactam [Unasyn], ticarcillin/clavulanate [Timentin], ertapenem [Invanz]) (Spirt, 2010) are prescribed for 7 to 10 days. An opioid (e.g., oxycodone) or other analgesic agents may be prescribed for pain relief. If surgery is necessary, pain management will include parenteral opioids such as hydromorphone (Dilaudid), morphine, or fentanyl with progression to oral analgesic agents (e.g., oxycodone with acetaminophen [Percocet]) (Kodali & Oberoi, 2011). Oral intake is increased as symptoms subside. A low-fiber diet may be necessary until signs of infection decrease.

Antispasmodic agents such as propantheline bromide and oxyphencyclimine (Daricon) may be prescribed. Often, it is not possible for patients to consume the 20 to 30 g of daily fiber that is recommended. Normal stools can be achieved by supplementing dietary fiber with bulk preparations (psyllium) or stool softeners (docusate), by instilling warm oil into the rectum, or by inserting an evacuant suppository (bisacodyl). Such a prophylactic plan can reduce the bacterial flora of the bowel, diminish the bulk of the stool, and soften the fecal mass so that it moves more easily through the area of inflammatory obstruction. Probiotics have been suggested as a way to promote prevention of relapse in that the healthy bacteria may promote a better balance of microbes in the intestine and augment immune competence (Tursi, 2010).

Surgical Management

Although acute diverticulitis usually subsides with medical management, immediate surgical intervention is necessary if complications (e.g., perforation, peritonitis, hemorrhage, obstruction) occur (Martel, Bouchard, Soto, et al., 2010; Stocchi, 2010). In cases of abscess formation without peritonitis, hemorrhage, or obstruction, CT-guided percutaneous drainage may be performed to drain the abscess, and IV antibiotics are administered. After the abscess is drained and the acute episode of inflammation has subsided (after approximately 6 weeks), surgery may be recommended to prevent repeated episodes. Two types of surgery are typically considered either to treat acute complications or prevent further episodes of inflammation:

- One-stage resection, in which the inflamed area is removed and a primary end-to-end anastomosis is completed
- Multiple-stage procedures for complications such as obstruction or perforation (Fig. 48-4)

The type of surgery performed depends on the extent of complications found during surgery. When possible, the area of diverticulitis is resected and the remaining bowel is joined end to end (i.e., primary resection and end-to-end anastomosis). This is performed using traditional surgical or laparoscopically assisted colectomy. More primary anastomoses are completed because surgeons are now able to perform intraoperative colonic lavage to decrease intestinal bacterial load and the risk of anastomotic failure from sepsis (Beitz, 2008). Laparoscopic lavage has been suggested and tested, demonstrating that in selected patients (e.g., those that are younger and without comorbid conditions), it saves the patient from bowel resection and creation of a stoma (Stocchi, 2010). A two-stage resection may be performed in which the diseased colon is resected (as in a one-stage procedure) but no anastomosis is performed. In this procedure,

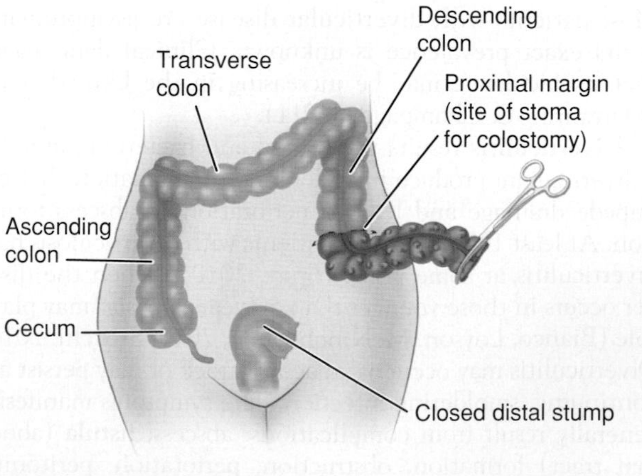
Descending colon
Transverse colon
Proximal margin (site of stoma for colostomy)
Ascending colon
Cecum
Closed distal stump

FIGURE 48-4 • Hartmann's procedure for diverticulitis: primary resection for diverticulitis of the colon. The affected segment (*clamp attached*) has been divided at its distal end. In a primary anastomosis, the proximal margin (*dotted line*) is transected and the bowel attached end to end. In a two-stage procedure, a colostomy is constructed at the proximal margin with the distal stump oversewn (Hartmann's procedure, as shown) and the stump is left in the pelvis. The distal stump may be brought to the surface as a mucous fistula if there is concern about blood supply. The second stage consists of colostomy takedown and anastomosis.

one end of the bowel is brought out to the abdominal wall and the distal end is closed over and left in the abdomen (Hartmann's procedure), or if the blood supply to the distal colon is questionable, both ends of the bowel are brought out to the abdominal wall (double-barrel) (Beitz, 2008; Stocchi, 2010). Both Hartmann's and double-barrel colostomies are usually reanastomosed at a later time.

NURSING PROCESS

The Patient With Diverticulitis

Assessment

During the health history, the nurse asks the patient about the onset and duration of pain and about past and present elimination patterns. The nurse reviews dietary habits to determine fiber intake and asks the patient about straining at stool, history of constipation with periods of diarrhea, tenesmus, abdominal bloating, and distention.

Assessment includes auscultation for the presence and character of bowel sounds and palpation for left lower quadrant pain, tenderness, or firm mass. The stool is inspected for pus, mucus, or blood. Temperature, pulse, and blood pressure are monitored for abnormal variations.

Diagnosis

Nursing Diagnoses

Based on the assessment data, nursing diagnoses may include the following:

- Constipation related to narrowing of the colon from thickened muscular segments and strictures
- Acute pain related to inflammation and infection

Collaborative Problems/Potential Complications

Potential complications may include the following:

- Peritonitis
- Abscess formation
- Bleeding

Planning and Goals

The major goals for the patient may include attainment and maintenance of normal elimination patterns, pain relief, and absence of complications.

Nursing Interventions

Maintaining Normal Elimination Patterns

The nurse recommends a fluid intake of 2 L/day (within limits of the patient's cardiac and renal reserve) and suggests foods that are soft but have increased fiber, such as prepared cereals or soft-cooked vegetables, to increase the bulk of the stool and facilitate peristalsis, thereby promoting defecation. An individualized exercise program is encouraged to improve abdominal muscle tone. It is important to review the patient's daily routine to establish a schedule for meals and a set time for defecation and to assist in identifying habits that may have suppressed the urge to defecate. The nurse encourages daily intake of bulk laxatives such as psyllium, which helps propel feces through the colon. Stool softeners are administered as prescribed to decrease straining at stool, which decreases intestinal pressure. Oil retention enemas may be prescribed to soften the stool, making it easier to pass. Some people with diverticulosis may have food triggers such as nuts and popcorn that bring on a diverticulitis attack, whereas others may not report food triggers. If triggers are identified, patients should be urged to avoid them.

Relieving Pain

Opioid analgesics (e.g., hydromorphone, morphine) to relieve the pain of diverticulitis and antispasmodic agents to decrease intestinal spasm are administered as prescribed (Kodali & Oberoi, 2011). The nurse records the intensity, duration, and location of pain to determine whether the inflammatory process worsens or subsides.

Monitoring and Managing Potential Complications

The major nursing focus is to prevent complications by identifying patients at risk and managing their symptoms as needed. The nurse assesses for the following signs and symptoms of perforation: increased abdominal pain and tenderness accompanied by abdominal rigidity, elevated white blood cell count, elevated ESR, increased temperature, tachycardia, and hypotension. Perforation is a surgical emergency. The clinical manifestations of perforation and peritonitis and the care of the patient with peritonitis are presented in the next section. The nurse monitors vital signs and urine output and administers IV fluids to replace volume loss as needed.

Promoting Home and Community-Based Care

Because patients and their family members and health care providers tend to focus on the most obvious needs and issues, the nurse reminds the patient and family about the importance of continuing health promotion and screening practices. The nurse educates patients who have not been involved in these practices in the past about their importance and refers the patients to appropriate health care providers.

Evaluation

Expected patient outcomes may include:

1. Attains a normal pattern of elimination
   a. Reports less abdominal cramping and pain
   b. Reports the passage of soft, formed stool without pain
   c. Adds unprocessed bran to foods
   d. Drinks at least 10 glasses of fluid each day (if fluid intake is tolerated)
   e. Exercises daily
2. Reports decreased pain
   a. Requests analgesic agents as needed
   b. Adheres to a low-fiber diet during acute episodes
3. Recovers without complications
   a. Is afebrile
   b. Has normal blood pressure
   c. Has a soft, nontender abdomen with normal bowel sounds
   d. Maintains adequate urine output
   e. Has no blood in the stool

Peritonitis

Peritonitis is inflammation of the peritoneum, which is the serous membrane lining the abdominal cavity and covering the viscera. Usually, it is a result of bacterial infection but

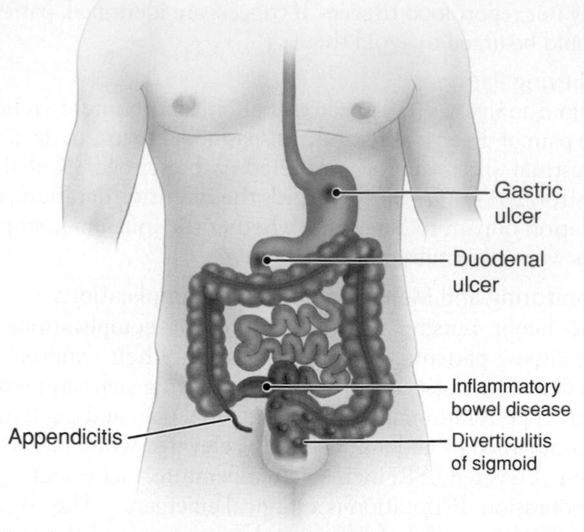
Gastric ulcer
Duodenal ulcer
Inflammatory bowel disease
Appendicitis
Diverticulitis of sigmoid

FIGURE 48-5 • Common gastrointestinal causes of peritonitis.

may occur secondary to a fungal or mycobacterial infection (Craig & Infante, 2011); the organisms come from diseases of the GI tract or, in women, from the internal reproductive organs (e.g., fallopian tube). The most common bacteria implicated are *Escherichia coli* and *Klebsiella, Proteus, Pseudomonas,* and *Streptococcus* species. Inflammation and paralytic ileus are the direct effects of the infection. Peritonitis can also result from external sources such as injury or trauma (e.g., gunshot wound, stab wound) or an inflammation that extends from an organ outside the peritoneal area, such as the kidney. Other common causes of peritonitis are appendicitis, perforated ulcer, diverticulitis, and bowel perforation (Fig. 48-5). Peritonitis may also be associated with abdominal surgical procedures and peritoneal dialysis (Neumann & Moran, 2010). Rarely, an aseptic peritonitis occurs, likely caused by reactions of the peritoneum to intra-abdominal foreign material or medications (e.g., associated with peritoneal dialysis) (Craig & Infante, 2011).

Pathophysiology

Peritonitis is caused by leakage of contents from abdominal organs into the abdominal cavity, usually as a result of inflammation, infection, ischemia, trauma, or tumor perforation. Bacterial proliferation occurs. Edema of the tissues results, and exudation of fluid develops in a short time. Fluid in the peritoneal cavity becomes turbid with increasing amounts of protein, white blood cells, cellular debris, and blood. The immediate response of the intestinal tract is hypermotility, soon followed by paralytic ileus with an accumulation of air and fluid in the bowel.

Clinical Manifestations

Symptoms depend on the location and extent of the inflammation. The early clinical manifestations of peritonitis frequently are the signs and symptoms of the disorder causing the condition (e.g., manifestations of infection). At first, a diffuse pain is felt. The pain tends to become constant, localized, and more intense over the site of the pathologic process (site of maximal peritoneal irritation). Movement usually aggravates it. The affected area of the abdomen becomes extremely tender and distended, and the muscles become rigid. Rebound tenderness and paralytic ileus may be present. Diminished perception of pain in peritonitis can occur in people receiving corticosteroids or analgesic agents. Patients with diabetes who have advanced neuropathy and patients with cirrhosis who have ascites may not experience pain during an acute bacterial infectious process. Usually, anorexia, nausea, and vomiting occur and peristalsis is diminished. A temperature of 37.8°C to 38.3°C (100°F to 101°F) can be expected, along with an increased pulse rate. With progression of the condition, patients may become hypotensive.

Assessment and Diagnostic Findings

The white blood cell count is almost always elevated. The hemoglobin and hematocrit levels may be low if blood loss has occurred. Serum electrolyte studies may reveal altered levels of potassium, sodium, and chloride.

An abdominal x-ray may show air and fluid levels as well as distended bowel loops. Abdominal ultrasound may reveal abscesses and fluid collections, and ultrasound-guided aspiration may assist in easier placement of drains. A CT scan of the abdomen may show abscess formation. Peritoneal aspiration and culture and sensitivity studies of the aspirated fluid may reveal infection and identify the causative organisms. MRI may be used for diagnosis of intra-abdominal abscesses (Taheri, Singh, & Duerksen, 2011).

Complications

Frequently, the inflammation is not localized, and the entire abdominal cavity shows evidence of widespread infection. Sepsis is the major cause of death from peritonitis. Shock may result from septicemia or hypovolemia (see Chapter 14 for further discussion). The inflammatory process may cause intestinal obstruction, primarily from the development of bowel adhesions.

Medical Management

Fluid, colloid, and electrolyte replacement is the major focus of medical management. The administration of several liters of an isotonic solution is prescribed. Hypovolemia occurs because massive amounts of fluid and electrolytes move from the intestinal lumen into the peritoneal cavity and deplete the fluid in the vascular space.

Analgesic medications are prescribed for pain. Antiemetic agents are administered as prescribed for nausea and vomiting. Intestinal intubation and suction assist in relieving abdominal distention and in promoting intestinal function. Fluid in the abdominal cavity can cause pressure that restricts expansion of the lungs and causes respiratory distress. Oxygen therapy by nasal cannula or mask generally promotes adequate oxygenation, but airway intubation and ventilatory assistance occasionally are required (refer to Chapter 14 for further discussion).

Antibiotic therapy is initiated early in the treatment of peritonitis. Large doses of a broad-spectrum antibiotic are administered IV until the specific organism causing the infection is identified and appropriate antibiotic therapy can be initiated. If peritonitis is caused by peritoneal dialysis, prompt antibiotic therapy is crucial. If the peritonitis does not respond to the therapy in 5 days, the peritoneal dialysis catheter should be removed (Neumann & Moran, 2010).

Surgical objectives include removing the infected material and correcting the cause. Surgical treatment is directed toward excision (e.g., appendix), resection with or without anastomosis (e.g., intestine), repair (e.g., perforation), and drainage (e.g., abscess). With extensive sepsis, a fecal diversion may need to be created. In selected instances, ultrasound-guided and CT-guided peritoneal drainage of abdominal and extraperitoneal abscesses has allowed for avoidance or delay of surgical therapy until the acute septic process has subsided (Craig & Infante, 2011; Taheri et al., 2011).

The two most common postoperative complications are wound evisceration and abscess formation. Any suggestion from the patient that an area of the abdomen is tender or painful or "feels as if something just gave way" must be reported. The sudden occurrence of serosanguineous wound drainage strongly suggests wound dehiscence (see Chapter 19).

Nursing Management

Intensive care is often needed. The patient's blood pressure is monitored by arterial line if shock is present. The central venous pressure or pulmonary artery wedge pressure and urine output are monitored frequently. In addition, ongoing assessment of pain, GI function, and fluid and electrolyte balance is important. The nurse reports the nature of the pain, its location in the abdomen, and any changes in location. Administering analgesic medication and positioning the patient for comfort are helpful in decreasing pain. The patient is placed on the side with knees flexed; this position decreases tension on the abdominal organs. Accurate recording of all intake and output (I&O) and central venous pressures and pulmonary artery pressures assist in calculating fluid replacement. The nurse administers and closely monitors IV fluids. Nasogastric intubation may be necessary.

Signs that indicate that peritonitis is subsiding include a decrease in temperature and pulse rate, softening of the abdomen, return of peristaltic sounds, passing of flatus, and bowel movements. The nurse increases fluid and food intake gradually and reduces parenteral fluids as prescribed. A worsening clinical condition may indicate a complication, and the nurse must prepare the patient for emergency surgery.

Drains are frequently inserted during the surgical procedure, and the nurse must monitor and record the character of the drainage postoperatively. Care must be taken when moving and turning the patient to prevent the drains from being dislodged. The nurse prepares the patient and family for discharge by educating the patient about how to care for the incision and drains if the patient will be sent home with the drains still in place. Referral for home care may be indicated for further monitoring and patient and family education.

INFLAMMATORY BOWEL DISEASE

Inflammatory bowel disease (IBD) is a group of chronic disorders: Crohn's disease (i.e., regional enteritis) and ulcerative colitis that result in inflammation or ulceration (or both) of the bowel lining. Both disorders have striking similarities but also several differences. Table 48-4 compares Crohn's disease and ulcerative colitis.

The incidence of IBD in the United States has increased in the past century, and more than 30,000 new cases occur annually. People between 15 and 30 years of age are at the greatest risk of developing IBD, followed by those between 50 and 70 years of age. Women and men tend to be equally affected, and family history appears to predispose people to IBD, particularly if a first-degree relative has the disease. Although it was previously believed that psychological conditions such as anxiety or depression predisposed certain people to IBD, this theory has been discarded (NDDIC, 2012b). Crohn's disease and ulcerative colitis are more prevalent in Jewish people than in any other ethnic group, especially North American Jews of eastern European descent. Since familial clusters also occur a genetic linkage is evident. A positive family history is the most important independent risk factor for IBD (Peppercorn, 2012c).

Despite extensive research, the cause of IBD is still unknown. Researchers theorize that it is triggered by environmental agents such as pesticides, food additives, tobacco, and radiation (NDDIC, 2012b). Nonsteroidal anti-inflammatory drugs have been found to exacerbate IBD. Allergies and immune disorders have also been suggested as causes. Abnormal response to dietary or bacterial antigens has been studied extensively, and genetic factors also are being studied. There is a high prevalence of coexistent IBS, which complicates the overall symptom presentation.

Crohn's Disease (Regional Enteritis)

Crohn's disease is usually first diagnosed in adolescents or young adults but can appear at any time of life. The incidence of Crohn's disease has risen over the past 30 years (NDDIC, 2012b). Crohn's disease is seen more often in smokers than in nonsmokers (Peppercorn, 2012c).

Pathophysiology

Crohn's disease is a subacute and chronic inflammation of the GI tract wall that extends through all layers (i.e., transmural lesion). Although its characteristic histopathologic changes can occur anywhere in the GI tract, it most commonly occurs in the distal ileum and, to a lesser degree, the ascending colon. It is characterized by periods of remission and exacerbation. It is posited that defects in the immune system in genetically predisposed individuals allow bacteria to invade the gut mucosa, resulting in an overactive adaptive immune response (Etchevers, Ordas, & Ricart, 2010). The disease process begins with edema and thickening of the mucosa. Ulcers begin to appear on the inflamed mucosa. These lesions are not in continuous contact with one another and are separated by normal tissue. Hence, these clusters of ulcers tend to take on a classic "cobblestone" appearance. Fistulas, fissures, and abscesses form as the inflammation extends into the peritoneum. Granulomas occur in 50% of patients (Hanauer & Norton, 2011). As the disease advances, the bowel wall thickens and becomes fibrotic, and the intestinal lumen narrows. Diseased bowel loops sometimes adhere to other loops surrounding them (Peppercorn, 2012c).

TABLE 48-4 Comparison of Crohn's Disease and Ulcerative Colitis

Factor	Crohn's Disease	Ulcerative Colitis
Course	Prolonged, variable	Exacerbations, remissions
Pathology		
Early	Transmural thickening	Mucosal ulceration
Late	Deep, penetrating granulomas	Minute, mucosal ulcerations
Clinical Manifestations		
Location	Ileum, ascending colon (usually)	Rectum, descending colon
Bleeding	Usually not, but if it occurs, it tends to be mild	Common—severe
Perianal involvement	Common	Rare—mild
Fistulas	Common	Rare
Rectal involvement	About 20%	Almost 100%
Diarrhea	Less severe	Severe
Abdominal mass	Common	Rare
Diagnostic Study Findings		
Barium studies	Regional, discontinuous skip lesions Narrowing of colon Thickening of bowel wall Mucosal edema Stenosis, fistulas	Diffuse involvement No narrowing of colon No mucosal edema Stenosis rare Shortening of colon
Sigmoidoscopy	May be unremarkable unless accompanied by perianal fistulas	Abnormal inflamed mucosa
Colonoscopy	Distinct ulcerations separated by relatively normal mucosa in ascending colon	Friable mucosa with pseudopolyps or ulcers in descending colon
Therapeutic Management	Corticosteroids, aminosalicylates (sulfasalazine [Azulfidine]) Antibiotics Parenteral nutrition Partial or complete colectomy, with ileostomy or anastomosis Rectum can be preserved in some patients. Recurrence common	Corticosteroids, aminosalicylates (sulfasalazine) useful in preventing recurrence Bulk hydrophilic agents Antibiotics Proctocolectomy, with ileostomy Rectum can be preserved in only a few patients "cured" by colectomy.
Systemic Complications	Small bowel obstruction Right-sided hydronephrosis Nephrolithiasis Cholelithiasis Arthritis Retinitis, iritis Erythema nodosum	Toxic megacolon Perforation Hemorrhage Malignant neoplasms Pyelonephritis Nephrolithiasis Cholangiocarcinoma Arthritis Retinitis, iritis Erythema nodosum

Adapted from Peppercorn, M. (2012a). Clinical manifestations, diagnosis and prognosis of Crohn's disease in adults. *UpToDate*. Available at: www.uptodate.com; and Peppercorn, M. (2012b). Clinical manifestations, diagnosis and prognosis of ulcerative colitis in adults. *UpToDate*. Available at: www.uptodate.com

Clinical Manifestations

The onset of symptoms is usually insidious in Crohn's disease, with prominent right lower quadrant abdominal pain and diarrhea unrelieved by defecation. Scar tissue and the formation of granulomas interfere with the ability of the intestine to transport products of upper intestinal digestion through the constricted lumen, resulting in crampy abdominal pains. There is abdominal tenderness and spasm. Because eating stimulates intestinal peristalsis, the crampy pains occur after meals. To avoid these bouts of crampy pain, the patient tends to limit food intake, reducing the amounts and types of food to such a degree that normal nutritional requirements are often not met. As a result, weight loss, malnutrition, and secondary anemia occur (Cronin, 2010). Ulcers in the membranous lining of the intestine and other inflammatory changes result in a weeping, edematous intestine that continually empties an irritating discharge into the colon. Disrupted absorption causes chronic diarrhea and nutritional deficits. The result is a person who is thin and emaciated from inadequate food intake and constant fluid loss. In some patients, the inflamed intestine may perforate, leading to intra-abdominal and anal abscesses. Fever and leukocytosis occur. Chronic symptoms include diarrhea, abdominal pain, **steatorrhea** (i.e., excessive fat in the feces), anorexia, weight loss, and nutritional deficiencies.

Abscesses, fistulas, and fissures are common. Manifestations may extend beyond the GI tract and can include joint disorders (e.g., arthritis), skin lesions (e.g., erythema nodosum), ocular disorders (e.g., conjunctivitis), and oral ulcers. The clinical course and symptoms can vary; in some patients, periods of remission and exacerbation occur, but in others, the disease follows a fulminating course. When intestinal symptoms worsen, extraintestinal manifestations often worsen as well. Both usually improve simultaneously. Mucosal healing is being suggested as an end point in IBD therapy. As mucosal lesions heal, extraintestinal manifestations subside (Devlin & Panaccione, 2010).

Assessment and Diagnostic Findings

A proctosigmoidoscopy is usually performed initially to determine whether the rectosigmoid area is inflamed. A stool examination is also performed; the result may be positive for occult blood and steatorrhea. The most conclusive diagnostic aid for Crohn's disease has classically been a barium study of the upper GI tract that shows a "string sign" on an x-ray film of the terminal ileum, indicating the constriction of a segment of intestine. Newer studies, such as video capsule endoscopy, are gaining favor because they can provide extensive evaluation of the small bowel (Hanauer & Norton, 2011). Endoscopy, colonoscopy, and intestinal biopsies may be used to confirm the diagnosis. A barium enema may show ulcerations (the cobblestone appearance described earlier), fissures, and fistulas. A CT scan may show bowel wall thickening and fistula formation.

A complete blood cell count is performed to assess hematocrit and hemoglobin levels (usually decreased) as well as the white blood cell count (may be elevated). The ESR is usually elevated. Albumin and protein levels may be decreased, indicating malnutrition.

Complications

Complications of Crohn's disease include intestinal obstruction or stricture formation, perianal disease, fluid and electrolyte imbalances, malnutrition from malabsorption, and fistula and abscess formation. The most common type of small bowel fistula caused by Crohn's disease is the enterocutaneous fistula (i.e., an abnormal opening between the small bowel and the skin). Abscesses can be the result of an internal fistula that results in fluid accumulation and infection. Patients with colonic Crohn's disease are also at increased risk of colon cancer (Hanauer & Norton, 2011).

Ulcerative Colitis

Ulcerative colitis is a recurrent ulcerative and inflammatory disease of the mucosal and submucosal layers of the colon and rectum. The prevalence of ulcerative colitis is highest in Caucasians and people of Jewish heritage. It is typically accompanied by systemic complications and a high mortality rate. Approximately 5% of patients with ulcerative colitis develop colon cancer (NDDIC, 2012b).

Pathophysiology

Ulcerative colitis affects the superficial mucosa of the colon and is characterized by multiple ulcerations, diffuse inflammations, and desquamation or shedding of the colonic epithelium. Bleeding occurs as a result of the ulcerations. The mucosa becomes edematous and inflamed. The lesions are contiguous, occurring one after the other. Abscesses form, and infiltrate is seen in the mucosa and submucosa, with clumps of neutrophils found in the lumens of the crypts (i.e., crypt abscesses) that line the intestinal mucosa. The disease process usually begins in the rectum and spreads proximally to involve the entire colon. Eventually, the bowel narrows, shortens, and thickens because of muscular hypertrophy and fat deposits. Because the inflammatory process is not transmural (i.e., it affects the inner lining only), fistulas, obstruction, and fissures are uncommon in ulcerative colitis (NDDIC, 2012b).

Clinical Manifestations

The clinical course is usually one of exacerbations and remissions. The predominant symptoms of ulcerative colitis include diarrhea, passage of mucus and pus, left lower quadrant abdominal pain, intermittent tenesmus, and rectal bleeding. The bleeding may be mild or severe, and pallor, anemia, and fatigue result. The patient may have anorexia, weight loss, fever, vomiting, and dehydration, as well as cramping, tenesmus, and the passage of 10 to 20 liquid stools each day. The disease is classified as mild, severe, or fulminant, depending on the severity of the symptoms. Hypocalcemia and anemia frequently develop. Rebound tenderness may occur in the right lower quadrant. Extraintestinal manifestations include skin lesions (e.g., erythema nodosum), eye lesions (e.g., uveitis), joint abnormalities (e.g., arthritis), and liver disease (Huether & McCance, 2012).

Assessment and Diagnostic Findings

The patient should be assessed for tachycardia, hypotension, tachypnea, fever, and pallor. Other assessments address the level of hydration and nutritional status. The abdomen is examined for bowel sounds, distention, and tenderness. These findings assist in determining the severity of the disease.

The stool is positive for blood, and laboratory test results reveal low hematocrit and hemoglobin levels in addition to an elevated white blood cell count, low albumin levels, and an electrolyte imbalance. Elevated antineutrophil cytoplasmic antibody levels are common (Peppercorn, 2012c). Abdominal x-ray studies are useful for determining the cause of symptoms. Free air in the peritoneum and bowel dilation or obstruction should be excluded as a source of the presenting symptoms. Sigmoidoscopy or colonoscopy and barium enema are valuable in distinguishing ulcerative colitis from other diseases of the colon with similar symptoms. A barium enema may show mucosal irregularities, focal strictures or fistulas, shortening of the colon, and dilation of bowel loops. Colonoscopy may reveal friable, inflamed mucosa with exudate and ulcerations, and it assists in defining the extent and severity of the disease. CT scanning, MRI, and ultrasound studies can identify abscesses and perirectal involvement. Leukocyte tagging (see Chapter 44) is useful when severe colitis prohibits the use of colonoscopy to determine the extent of inflammation.

Careful stool examination for parasites and other microbes is performed to rule out dysentery caused by common intestinal organisms, especially *Entamoeba histolytica*, *C. difficile* and *Campylobacter*, *Salmonella*, *Shigella*, and *Cryptospora* species.

Complications

Complications of ulcerative colitis include toxic megacolon, perforation, and bleeding as a result of ulceration, vascular engorgement, and highly vascular granulation tissue. In toxic megacolon, the inflammatory process extends into the muscularis, inhibiting its ability to contract and resulting in colonic distention. Symptoms include fever, abdominal pain and distention, vomiting, and fatigue. If the patient with toxic megacolon does not respond within 24 to 72 hours to medical management with nasogastric suction, IV fluids with electrolytes, corticosteroids, and antibiotics, surgery is required. A subtotal colectomy may be performed if bowel perforation has not occurred; otherwise, colectomy is indicated (Sheth & Lamont, 2012). For many patients, surgery becomes necessary to relieve the effects of the disease and to treat these serious complications; an ileostomy usually is performed. The surgical procedures involved and the care of patients with this type of fecal diversion are discussed later in this chapter.

Patients with IBD also have a significantly increased risk of osteoporotic fractures due to decreased bone mineral density. Corticosteroid therapy may also contribute to the diminished bone density.

Management of Chronic Inflammatory Bowel Disease

Medical treatment for both Crohn's disease and ulcerative colitis is aimed at reducing inflammation, suppressing inappropriate immune responses, providing rest for a diseased bowel so that healing may take place, improving quality of life, and preventing or minimizing complications. Most patients have long periods of well-being interspersed with short intervals of illness. Management depends on the disease location, severity, and complications (Blonski, Buchner, & Lichtenstein, 2011; Farrell & Savage, 2010; Grimpen & Pavli, 2010).

Nutritional Therapy

Oral fluids and a low-residue, high-protein, high-calorie diet with supplemental vitamin therapy and iron replacement are prescribed to meet nutritional needs, reduce inflammation, and control pain and diarrhea. Fluid and electrolyte imbalances from dehydration caused by diarrhea are corrected by IV therapy as necessary if the patient is hospitalized or by oral fluids if the patient is managed at home. Any foods that exacerbate diarrhea are avoided. Milk may contribute to diarrhea in those with lactose intolerance. Cold foods and smoking are avoided because both increase intestinal motility. Parenteral nutrition may be indicated (see Chapter 45).

Pharmacologic Therapy

Sedatives and antidiarrheal and antiperistaltic medications are used to minimize peristalsis in order to rest the inflamed bowel. They are continued until the patient's stools approach normal frequency and consistency.

Aminosalicylates such as sulfasalazine (Azulfidine) are often effective for mild or moderate inflammation and are used to prevent or reduce recurrences in long-term maintenance regimens. Sulfa-free aminosalicylates (e.g., mesalamine [Asacol, Pentasa]) are effective in preventing and treating recurrence of inflammation. Antibiotics (e.g., metronidazole [Flagyl]) are used for secondary infections, particularly for purulent complications such as abscesses, perforation, and peritonitis.

Corticosteroids are used to treat severe and fulminant disease and can be administered orally (e.g., prednisone) in outpatient treatment or parenterally (e.g., hydrocortisone [Solu-Cortef]) in hospitalized patients. Topical (i.e., rectal administration) corticosteroids (e.g., budesonide [Entocort]) are also widely used in the treatment of distal colon disease. When the dosage of corticosteroids is reduced or stopped, the symptoms of disease may return. If corticosteroids are continued, numerous adverse sequelae may ensue; these are discussed in Chapter 52 and summarized in Table 52-5 (Blonski et al., 2011; Cronin, 2010; Dahan, Amidon & Zimmerman, 2010; Grimpen & Pavli, 2010).

Immunomodulators (e.g., azathioprine [Imuran], mercaptopurine [6-MP], methotrexate [MTX], cyclosporine [Neoral]) have been used to alter the immune response. The exact mechanism of action of these medications in treating IBD is unknown. They are used in patients with severe disease who have not responded favorably to other therapies. These medications are useful in maintenance regimens to prevent relapses (Stansfield, 2010). Newer biologic therapies incorporate monoclonal antibodies, including infliximab (Remicade), adalimumab (Humira), certolizumab pegol (Cimzia), and natalizumab (Tysabri) for treating Crohn's disease (Blonski et al., 2011; Devlin & Panaccione, 2010) and infliximab for treating ulcerative colitis (Blonski et al., 2011). Clinical outcomes for the biologic therapies are promising, although adverse effects may seriously limit their usefulness. Other biologic therapies recently tested in Crohn's disease include anticytokine therapy using anti-interleukin (IL)-type drugs (e.g., anti-IL-12). To date, the safety profile is satisfactory (Grimpen & Pavli, 2010). Another intervention being tried for IBD is gene therapy (Dahan et al., 2010). A recent study suggests that the newer biologics and azathioprine are more effective for mucosal healing in Crohn's disease versus traditional treatment with methotrexate (Laharie, Reffet, Belleannée, et al., 2011). A major issue associated with appropriate pharmacologic treatment of IBD is nonadherance. Patients who are nonadherent may suffer a greater chance of disease relapse with severe associated symptoms.

Surgical Management

When nonsurgical measures fail to relieve the severe symptoms of IBD, surgery may be necessary. Ultimately, 75% of patients with Crohn's disease undergo surgery within 10 years of diagnosis, and between 25% and 60% will require further repeat surgery within the same time frame (Hanauer & Norton, 2011). The most common indications for surgery are medically intractable disease, poor quality of life, or complications from the disease or its treatment. Recurrence of inflammation and disease after surgery in Crohn's disease is inevitable (Umanskiy & Fichera, 2010).

A common procedure performed for strictures of the small intestines is laparoscope-guided strictureplasty, in which the blocked or narrowed sections of the intestines are widened, leaving the intestines intact. In some cases, a small bowel resection is performed; diseased segments of the small intestines are resected, and the remaining portions of the intestines are anastomosed. Surgical removal of up to 50% of the small bowel usually can be tolerated. In cases of severe Crohn's disease of the colon, a total colectomy and ileostomy may be needed.

A newer surgical procedure developed for patients with severe Crohn's disease is intestinal transplant. This technique is now available to children and to young and middle-aged adults who have lost intestinal function from disease. It may provide improvement in quality of life for some patients. The associated technical and immunologic problems remain formidable, and the costs and mortality rates continue to be high. None of the surgical procedures for Crohn's disease are curative. Ultimately, achievement of remission depends on medical therapy (Hanauer & Norton, 2011).

At least 25% of patients with ulcerative colitis eventually have total colectomies (NDDIC, 2012b). When the colon is surgically removed, the patient is considered "cured" in that extraintestinal manifestations subside and the disease process is otherwise limited to the colon. Indications for surgery include lack of improvement and continued deterioration, profuse bleeding, perforation, continued stricture formation, and cancer. Surgical excision usually improves quality of life. Proctocolectomy with ileostomy (i.e., complete excision of colon, rectum, and anus) is recommended when the rectum is severely diseased. If the rectum can be preserved, restorative proctocolectomy with ileal pouch anal anastomosis (IPAA) is the procedure of choice for ulcerative colitis. Although complications following IPAA may occur, quality of life is generally good to excellent (Beliard & Prudhomme, 2010; Buckman & Heise, 2010). IPAA is also the procedure of choice in familial adenomatous polyposis (Carmichael & Mills, 2011).

Other types of surgical procedures, known as fecal diversions, are discussed later in this chapter.

Total Colectomy With Ileostomy

An **ileostomy**—the surgical creation of an opening into the ileum or small intestine (usually by means of an ileal stoma on the abdominal wall)—is commonly performed after a total colectomy (i.e., excision of the entire colon). It allows for drainage of fecal matter (i.e., effluent) from the ileum to the outside of the body. The drainage is liquid to unformed and occurs at frequent intervals. Nursing management of the patient with an ileostomy is discussed later in this chapter.

Continent Ileostomy

Another procedure involves the creation of a continent ileal reservoir (i.e., **Kock pouch**) by diverting a portion of the distal ileum to the abdominal wall and creating a stoma (see Fig. 55-8 in Chapter 55). This procedure eliminates the need for an external fecal collection bag. Approximately 30 cm (11 in) of the distal ileum is reconstructed to form a reservoir with a nipple valve that is created by pulling a portion of the terminal ileal loop back into the ileum. GI effluent can accumulate in the pouch for several hours and then be removed by means of a catheter inserted through the nipple valve. In many patients, a total colectomy is also performed with the Kock pouch. Possible indications for a total colectomy with Kock pouch placement (rather than a restorative proctocolectomy with IPAA) include a badly diseased rectum, lack of rectal sphincter tone, or inability to achieve fecal continence post IPAA.

The major challenge with the Kock pouch is malfunction of the nipple valve, which often requires additional corrective surgery. Kock pouches for both urinary and GI disorders are used less commonly because they have greater complication rates than do newer procedures (e.g., neobladder and ileoanal reservoir) (Doughty & Landmann, 2012; Shariat & Bochner, 2012).

Restorative Proctocolectomy With Ileal Pouch Anal Anastomosis

A restorative proctocolectomy with IPAA is the surgical procedure of choice in cases where the rectum can be preserved in that it eliminates the need for a permanent ileostomy. It establishes an ileal reservoir that functions as a "new" rectum, and anal sphincter control of elimination is retained. The procedure involves connecting the ileum to the anal pouch (made from a small intestine segment), and the surgeon connects the pouch to the anus in conjunction with removing the colon and the rectal mucosa (i.e., total abdominal colectomy and mucosal proctectomy) (Fig. 48-6). A temporary

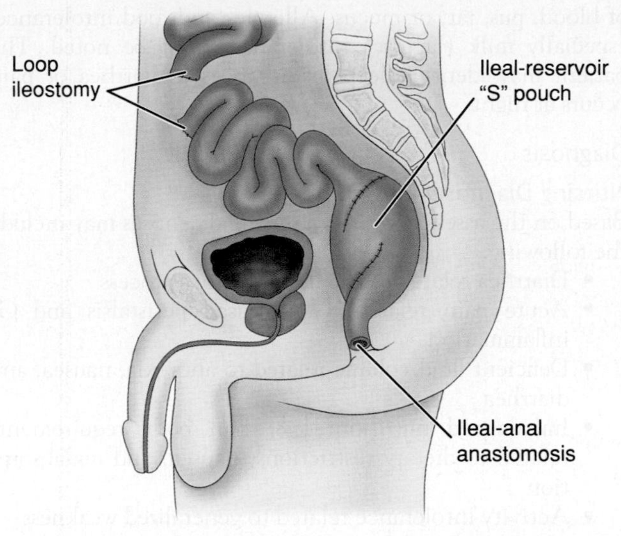
Loop ileostomy
Ileal-reservoir "S" pouch
Ileal-anal anastomosis

FIGURE 48-6 • A mucosal proctectomy precedes anastomosis of the ileal reservoir. A temporary loop ileostomy diverts effluent for several months to allow healing.

diverting loop ileostomy that promotes healing of the surgical anastomoses is constructed at the time of surgery and closed about 3 months later (Psillos & Catanzaro, 2011).

With IPAA or restorative proctocolectomy, the diseased colon and rectum are removed, voluntary defecation is maintained, and anal continence is preserved. The ileal reservoir decreases the number of bowel movements by 50%, from approximately 14 to 20 per day to 7 to 10 per day (Psillos & Catanzaro, 2011). Nighttime elimination is gradually reduced to one bowel movement. Complications of ileoanal anastomosis include irritation of the perianal skin from leakage of fecal contents, stricture formation at the anastomosis site, small bowel obstruction, and "pouchitis" (inflammation of the ileoanal pouch due to altered microbial levels). For patients with IBD history, dietary intolerances may persist after the IPAA is formed. Increased stool output, flatulence, and perineal irritation are associated with consumption of nuts, corn, chocolate, spicy foods, onions, and citrus fruits. Consequently, some patients with a J-pouch (i.e., a continent coloanal reservoir; see later discussion under Surgical Management in the Colorectal Cancer section) may need to alter their diet to avoid complications and perineal breakdown (Buckman & Heise, 2010).

NURSING PROCESS

Management of the Patient With Chronic Inflammatory Bowel Disease

PRACTICE & LEARN

Assessment

The nurse obtains a health history to identify the onset, duration, and characteristics of abdominal pain; the presence of diarrhea or fecal urgency, straining at stool (tenesmus), nausea, anorexia, or weight loss; and family history of IBD. It is important to discuss dietary patterns, including the amounts of alcohol, caffeine, and nicotine-containing products used daily and weekly. The nurse asks about patterns of bowel elimination, including character, frequency, and presence of blood, pus, fat, or mucus. Allergies and food intolerance, especially milk (lactose) intolerance, must be noted. The patient may identify sleep disturbances if diarrhea or pain occurs at night.

Diagnosis

Nursing Diagnoses

Based on the assessment data, nursing diagnoses may include the following:

- Diarrhea related to the inflammatory process
- Acute pain related to increased peristalsis and GI inflammation
- Deficient fluid volume related to anorexia, nausea, and diarrhea
- Imbalanced nutrition: less than body requirements related to dietary restrictions, nausea, and malabsorption
- Activity intolerance related to generalized weakness
- Anxiety related to impending surgery
- Ineffective coping related to repeated episodes of diarrhea
- Risk for impaired skin integrity related to malnutrition and diarrhea
- Risk for ineffective self-health management related to insufficient knowledge concerning the process and management of the disease

Collaborative Problems/Potential Complications

Potential complications may include the following:

- Electrolyte imbalance
- Cardiac dysrhythmias related to electrolyte imbalances
- GI bleeding with fluid volume loss
- Perforation of the bowel

Planning and Goals

The major goals for the patient include attainment of normal bowel elimination patterns, relief of abdominal pain and cramping, prevention of fluid volume deficit, maintenance of optimal nutrition and weight, avoidance of fatigue, reduction of anxiety, promotion of effective coping, absence of skin breakdown, increased knowledge about the disease process and self-health management, and avoidance of complications.

Nursing Interventions

Maintaining Normal Elimination Patterns

The nurse assists the patient in determining if there is a relationship between diarrhea and certain foods, activities, or emotional stressors. Identifying precipitating factors, the frequency of bowel movements, and the character, consistency, and amount of stool passed is important. The nurse provides ready access to a bathroom, commode, or bedpan and keeps the environment clean and odor free. It is important to administer antidiarrheal medications as prescribed, to record the frequency and consistency of stools after therapy is initiated, and to encourage bed rest to decrease peristalsis.

Relieving Pain

The character of the pain is described as dull, burning, or crampy. It is important to ask about its onset. Does it occur before or after meals, during the night, or before elimination? Is the pattern constant or intermittent? Is it relieved with medications? The nurse administers anticholinergic medications 30 minutes before a meal as prescribed to decrease intestinal motility and administers analgesic agents as prescribed for pain. Position changes, local application of heat (as prescribed), diversional activities, and prevention of fatigue also are helpful for reducing pain.

Maintaining Fluid Intake

To detect fluid volume deficit, the nurse keeps an accurate record of I&O. The nurse monitors daily weights for fluid gains or losses and assesses the patient for signs of fluid volume deficit (i.e., dry skin and mucous membranes, decreased skin turgor, oliguria, fatigue, decreased temperature, increased hematocrit, elevated urine specific gravity, and hypotension). It is important to encourage oral intake of fluids and to monitor the flow rate of any IV fluids. The nurse initiates measures to decrease diarrhea (e.g., dietary restrictions, stress reduction, antidiarrheal agents).

Maintaining Optimal Nutrition

Parenteral nutrition is indicated in patients with IBD with severe malnutrition and intolerance to enteral nutrition and are expected to likely remain intolerant to enteral nutrition for more than 1 to 2 weeks (DeLegge, 2011; Seres, 2012). With parenteral nutrition, the nurse maintains an accurate record of fluid I&O as well as the daily weight. The patient should gain 0.5 kg (1.1 lb) daily during parenteral nutrition therapy. Because parenteral nutrition is very high in glucose and can cause hyperglycemia, blood glucose levels are monitored every 6 hours. Once the symptoms of IBD exacerbation have diminished and the patient has gained or stabilized weight, parenteral nutrition is stopped and the patient is advanced on oral elemental feedings. Elemental feedings are high in protein and low in fat and residue. They are digested primarily in the jejunum, do not stimulate intestinal secretions, and allow the bowel to continue to rest. The nurse notes intolerance if the patient exhibits nausea, vomiting, diarrhea, or abdominal distention.

If oral foods are tolerated, small, frequent, low-residue feedings are given to avoid overdistending the stomach and stimulating peristalsis. The patient must restrict activity to conserve energy, reduce peristalsis, and reduce caloric requirements.

Promoting Rest

The nurse recommends intermittent rest periods during the day and schedules or restricts activities to conserve energy and reduce the metabolic rate. It is important to encourage activity within the limits of the patient's capacity. The nurse suggests bed rest for a patient who is febrile, has frequent diarrheal stools, or is bleeding. However, the patient on bed rest should perform active exercises to maintain muscle tone and prevent thromboembolic complications. If the patient cannot perform these active exercises, the nurse performs passive exercises and joint range of motion. Activity restrictions are modified as needed on a day-to-day basis.

Reducing Anxiety

Rapport can be established by being attentive and displaying a calm, confident manner. The nurse allows time for the patient to ask questions and express feelings. Careful listening and sensitivity to nonverbal indicators of anxiety (e.g., restlessness, tense facial expressions) are helpful. The patient may be emotionally labile because of the consequences of the disease and the uncertainty of exacerbations with complications. The nurse tailors information about possible impending surgery to the patient's level of understanding and desire for detail. If surgery is planned, pictures and illustrations help explain the surgical procedure and help the patient visualize what a stoma looks like.

Enhancing Coping Measures

Because the patient may feel isolated, helpless, and out of control, understanding and emotional support are essential. The patient may respond to stress in a variety of ways that may alienate others (e.g., anger, denial, social self-isolation).

The nurse needs to recognize that the patient's behavior may be affected by a number of factors. Any patient suffering the discomforts of frequent bowel movements and rectal soreness is anxious, discouraged, and unhappy. It is important to develop a relationship with the patient that supports all attempts to cope with these stressors. It is also important to communicate that the patient's feelings are understood by encouraging the patient to talk and express his or her feelings and to discuss any concerns. Stress reduction measures that may be used include relaxation techniques, visualization, breathing exercises, and biofeedback. Professional counseling may be needed to help the patient and family manage issues associated with chronic illness and resulting disability.

Preventing Skin Breakdown

The nurse examines the patient's skin frequently, especially the perianal skin. Perianal care, including the use of a skin barrier (e.g., petroleum ointment [Vaseline]), is important after each bowel movement. The nurse gives immediate attention to reddened or irritated areas over bony prominences and uses pressure-relieving devices to prevent skin breakdown. Consultation with a **wound-ostomy-continence (WOC) nurse** (or WOCN; a nurse specially educated in the management of a variety of fecal and urinary diversions) is often helpful.

Monitoring and Managing Potential Complications

Serum electrolyte levels are monitored daily, and electrolyte replacements are administered as prescribed. Evidence of dysrhythmias or changes in level of consciousness must be reported immediately.

The nurse closely monitors rectal bleeding and administers blood component therapy and volume expanders as prescribed to prevent hypovolemia. It is important to monitor the blood pressure for hypotension and to obtain coagulation profiles and hemoglobin and hematocrit levels frequently. Vitamin K may be prescribed to increase clotting factors.

The nurse closely monitors the patient for indications of perforation (i.e., acute increase in abdominal pain, rigid abdomen, vomiting, or hypotension) and obstruction and toxic megacolon (i.e., abdominal distention, decreased or absent bowel sounds, change in mental status, fever, tachycardia, hypotension, dehydration, and electrolyte imbalances).

Promoting Home and Community-Based Care

Educating Patients About Self-Care. The nurse assesses the patient's understanding of the disease process and his or her need for additional information about medical management (e.g., medications, diet) and surgical interventions. The nurse provides information about nutritional management; a bland, low-residue, high-protein, high-calorie, and high-vitamin diet relieves symptoms and decreases diarrhea. It is important to explain the rationale for the use of corticosteroids and anti-inflammatory, antibacterial, antidiarrheal, and antispasmodic medications. The nurse emphasizes the importance of taking medications as prescribed and not abruptly discontinuing them (especially corticosteroids) to avoid development of serious medical problems (Chart 48-4). The nurse reviews ileostomy care as necessary (see Nursing Management of the Patient Requiring an Ileostomy section). Patient education information can be obtained from the Crohn's and Colitis Foundation of America (CCFA) and from a patient skills education program developed by the American College of Surgeons (see Resources section).

Continuing Care. Patients with chronic IBD are managed at home with follow-up care by their primary provider or through an outpatient clinic. Those whose nutritional status is compromised and who are receiving parenteral nutrition need home care nursing to ensure that their nutritional requirements are being met and that they or their caregivers can follow through with the instructions for parenteral nutrition. Patients who are undergoing medical treatment need to understand that their disease can be controlled and that they can lead a healthy life between exacerbations. Control implies management based on an understanding of the

Chart 48-4 HOME CARE CHECKLIST
Inflammatory Bowel Disease

At the completion of home care education, the patient or caregiver will be able to:	PATIENT	CAREGIVER
• Verbalize an understanding of the disease process.	✔	✔
• Discuss nutritional management: bland, low-residue, high-protein, high-vitamin diet; identify foods to include and foods to be avoided.	✔	✔
• Describe medication regimen; identify medications by name, use, route, and frequency.	✔	✔
• Identify measures to be used to treat exacerbation of symptoms, to include rest, dietary modifications, and medications.	✔	✔
• Identify measures to be used to promote fluid and electrolyte balance during acute exacerbations.	✔	✔
• Demonstrate management of parenteral nutrition therapy, if applicable; identify possible complications and interventions.	✔	✔
• Incorporate stress reduction measures into lifestyle.	✔	

disease and its treatment. Patients in the home setting need information about their medications (i.e., name, dose, side effects, and frequency of administration) and need to take medications on schedule. Medication reminders such as containers that separate pills according to day and time or daily checklists are helpful.

During a flare-up, the nurse encourages the patient to rest as needed and to modify activities according to his or her energy level. Patients should limit tasks that impose strain on the lower abdominal muscles. They should sleep in a room close to the bathroom because of the frequent diarrhea (10 to 20 times per day); quick access to a toilet helps alleviate worry about having an "accident." Room deodorizers help control odors.

Dietary modifications can control but do not cure the disease; the nurse recommends a low-residue, high-protein, high-calorie diet, especially during an acute phase. It is important to encourage the patient to keep a record of the foods that irritate the bowel and to avoid them and to drink at least eight glasses of water each day.

The prolonged nature of the disease has an impact on the patient and often strains his or her family life and financial resources. Family support is vital; however, some family members may be resentful or feel guilty, tired, or unable to cope with the emotional demands of the illness and the physical demands of providing care. Some patients with IBD do not socialize for fear of being embarrassed. Many prefer to eat alone. Because they have lost control over elimination, they may fear losing control over other aspects of their lives. They need time to express their fears and frustrations. Individual and family counseling may be helpful.

Evaluation

Expected patient outcomes may include:

1. Reports a decrease in the frequency of diarrheal stools
   a. Complies with dietary restrictions; maintains bed rest
   b. Takes medications as prescribed
2. Has reduced pain
3. Maintains fluid volume balance
   a. Drinks 1 to 2 L of oral fluids daily
   b. Has normal body temperature
   c. Displays adequate skin turgor and moist mucous membranes
4. Attains optimal nutrition; tolerates small, frequent feedings without diarrhea
5. Avoids fatigue
   a. Rests periodically during the day
   b. Adheres to activity restrictions
6. Is less anxious
   a. Seeks emotional support as appropriate
   b. Verbalizes less feelings of anxiety and concern
7. Copes successfully with diagnosis
   a. Verbalizes feelings freely
   b. Uses appropriate stress reduction behaviors
8. Maintains skin integrity
   a. Cleans perianal skin after defecation
   b. Uses appropriate skin barrier
9. Acquires an understanding of the disease process
   a. Modifies diet appropriately to decrease diarrhea
   b. Adheres to medication regimen as prescribed
10. Recovers without complications
    a. Electrolytes within normal ranges
    b. Normal sinus or baseline cardiac rhythm
    c. Maintains fluid balance
    d. Experiences no perforation or rectal bleeding

Nursing Management of the Patient Requiring an Ileostomy

Some patients with IBD eventually require a permanent fecal diversion with creation of an ileostomy to manage symptoms and to treat or prevent complications. The Plan of Nursing Care summarizes care for the patient requiring an ostomy (Chart 48-5).

Providing Preoperative Care

A period of preparation with intensive replacement of fluid, blood, and protein is necessary prior to surgery. Antibiotics may be prescribed. If the patient has been taking corticosteroids, they will be continued during the surgical phase to prevent steroid-induced adrenal insufficiency. Usually, the patient is given a low-residue diet, provided in frequent, small feedings. All other preoperative measures are similar to those for general abdominal surgery. The abdomen is marked for the proper placement of the stoma by the surgeon or the WOC nurse. Care is taken to ensure that the stoma is conveniently placed—usually in the right lower quadrant about 5 cm (2 in) below the waist, in an area away from previous scars, bony prominences, skin folds, or fistulas. The stoma site must be visible to the patient.

The patient must have a thorough understanding of the surgery to be performed and what to expect after surgery. Information about an ileostomy is presented to the patient by means of written materials, models, and discussion. Preoperative education includes management of drainage from the stoma; the nature of drainage; and the need for nasogastric intubation, parenteral fluids, and possibly perineal packing.

Providing Postoperative Care

General abdominal surgery care is required. As with other patients undergoing abdominal surgery, the nurse encourages those with an ileostomy to engage in early ambulation. It is important to administer prescribed pain medications as required. The nurse observes the stoma for color and size. It should be pink to bright red and shiny. Typically, a temporary clear or transparent plastic bag (i.e., appliance or pouch) with an adhesive facing is placed over the ileostomy in the operating room and firmly pressed onto the surrounding skin. The nurse monitors the ileostomy for fecal drainage, which should begin about 24 to 48 hours after surgery. The drainage is a continuous liquid from the small intestine because the stoma does not have a controlling sphincter. The contents drain into the pouch and are thus kept from coming into contact with the skin. They are collected, measured, and discarded when the pouch becomes full. If a continent ileal reservoir was created, as described for the Kock pouch, continuous drainage is provided by an indwelling reservoir catheter for 2 to 3 weeks after surgery. This allows the suture lines to heal.

Because these patients lose large fluid volumes in the early postoperative period, an accurate record of fluid I&O,

(text continues on page 1311)

Chart 48-5

PLAN OF NURSING CARE

The Patient Undergoing Ostomy Surgery

NURSING DIAGNOSIS: Deficient knowledge about the surgical procedure and preoperative preparation
GOAL: Understands the surgical process and the necessary preoperative preparations

Nursing Interventions	Rationale	Expected Outcomes
Preoperative Care		
1. Ascertain whether the patient has had a previous surgical experience, and ask for recollections of positive and negative impressions.	**1.** Fear of a repeated negative experience increases anxiety. Talking about the experience with a nurse helps clarify misconceptions and helps the patient ventilate any repressed emotions. Positive experiences are reinforced.	• Expresses anxieties and fears about the surgical process • Projects a positive attitude toward the surgical procedure • Repeats in own words information given by the surgeon • Identifies normal anatomy and physiology of gastrointestinal (GI) tract and how it will be altered; can point to expected location of abdominal wound and stoma; describes stoma appearance and size • Adheres to "bowel prep" regimen of antimicrobial agents or mechanical cleansing • Tolerates the presence of nasogastric/nasoenteric tube
2. Determine what information the surgeon gave the patient and family and whether it was understood. Clarify and elaborate as necessary. Determine whether the stoma is permanent or temporary. Be aware of the patient's prognosis if carcinoma exists.	**2.** Clarification prevents misunderstandings and alleviates anxiety.	
3. Use pictures or drawings to illustrate the location and appearance of the surgical wounds (abdominal, perineal) and the stoma if the patient is receptive.	**3.** Knowledge, for some, alleviates anxiety because fear of the unknown is decreased. Others choose not to know because it makes them more anxious.	
4. Explain that oral/parenteral antimicrobial agents will be administered to cleanse the bowel preoperatively. Mechanical cleansing may also be required.	**4.** Antimicrobial agents and mechanical cleansing (e.g., laxatives, enemas) reduce intestinal bacterial flora.	
5. Assist the patient during nasogastric/nasoenteric intubation. Measure drainage from the tube.	**5.** Nasoenteral intubation is used for decompression and drainage of GI contents before surgery.	

NURSING DIAGNOSIS: Disturbed body image
GOAL: Attainment of a positive self-concept

Nursing Interventions	Rationale	Expected Outcomes
1. Encourage the patient to verbalize feelings about the stoma. Offer to be present when the stoma is first viewed and touched.	**1.** Free expression of feelings allows the patient the opportunity to verbalize and identify concerns. Expressed concerns can be therapeutically addressed by health care team members.	• Freely expresses concerns and fears • Accepts support • Seeks help as needed • States is willing to talk with another patient with a stoma
2. Suggest that the spouse or significant other view the stoma.	**2.** Helps patient to overcome fears about partner's response.	
3. Offer counseling, if desired.	**3.** Provides opportunity for additional support.	
4. Arrange for a visit or a phone call with another patient with a stoma.	**4.** People with stomas can offer support and share mutual feelings and experiences.	

NURSING DIAGNOSIS: Anxiety related to the loss of bowel control
GOAL: Reduction of anxiety

Nursing Interventions	Rationale	Expected Outcomes
Postoperative Care		
1. Provide information about expected bowel function: **a.** Characteristics of effluent **b.** Frequency of discharge	**1.** Emotional adjustment is facilitated if adequate information is provided at the level of the learner.	• Expresses interest in learning about altered bowel function • Handles equipment correctly • Changes the appliance unassisted • Irrigates colostomy successfully • Progresses toward a regular schedule of elimination

(continues on page 1310)

Chart 48-5

PLAN OF NURSING CARE

The Patient Undergoing Ostomy Surgery (continued)

Nursing Interventions	Rationale	Expected Outcomes
2. Explain how to prepare the appliance for an adequate fit. **a.** Choose the drainage appliance that will provide a secure fit around the stoma. Measure the stoma size with a measuring guide provided by the ostomy equipment manufacturer and compare with the opening on the pouch. The barrier opening should be sized to "hug" the stoma and cover the peristomal skin. *Note:* Newer wafer barriers can be pulled or molded to the size of the stoma. **b.** Remove any plastic covering that protects the appliance adhesive. *Note:* The pouch is applied by pressing the adhesive for 30 seconds to the skin or skin barrier.	**2.** Adequate fit is necessary for successful use of the appliance. **a.** The appliance opening should be larger than the stoma for an adequate fit. Available brands come in different sizes to fit the stoma. Adjustments are made as necessary. **b.** The appliance is ready to apply directly to the skin or skin protector.	
3. Demonstrate how to change the appliance or empty the pouch before leakage occurs. Be aware that the older adult may have diminished vision and difficulty handling equipment.	**3.** Manipulation of the appliance is a learned motor skill that requires practice and positive reinforcement.	
4. When appropriate, demonstrate how to irrigate the colostomy (usually on the 4th or 5th day). Recommend that irrigation be performed at a consistent time, depending on the type of colostomy.	**4.** Colostomy irrigation is used to regulate the passage of fecal material; alternatively, the bowel can be allowed to evacuate naturally. Irrigation is not routinely indicated.	

NURSING DIAGNOSIS: Risk for impaired skin integrity related to irritation of the peristomal skin by the effluent
GOAL: Maintenance of skin integrity

Nursing Interventions	Rationale	Expected Outcomes
1. Provide information about signs and symptoms of irritated or inflamed skin. Use pictures if possible.	**1.** Peristomal skin should be slightly pink without abrasions and similar to that of the entire abdomen.	• Describes appearance of healthy skin • Correctly cleanses the skin • Successfully applies a skin barrier • Gently removes the drainage appliance without skin damage • Demonstrates intact skin around the colostomy stoma
2. Instruct patient how to cleanse the peristomal skin gently.	**2.** Mild friction with warm water and a gentle soap cleanses the skin and minimizes irritation and possible abrasions. After rinsing the soap, patting the skin dry prevents tissue trauma.	
3. Demonstrate how to apply a skin barrier (powder, gel, paste, wafer).	**3.** Skin barriers protect the peristomal skin from enzymes and bacteria.	
4. Demonstrate how to remove the pouch.	**4.** Gently separate adhesive from the skin to avoid irritation. Never pull!	

NURSING DIAGNOSIS: Imbalanced nutrition: less than body requirements related to avoidance of foods that may cause GI discomfort
GOAL: Achievement of an optimal nutritional intake

Nursing Interventions	Rationale	Expected Outcomes
1. Conduct a complete nutritional assessment to identify any foods that may increase peristalsis by irritating the bowel.	**1.** Patients react differently to certain foods because of individual sensitivity.	• Modifies diet to avoid offensive foods yet maintains adequate nutritional intake • Avoids cellulose-based foods, such as peanuts • Modifies intake of certain fruits
2. Advise the patient to avoid food products with a cellulose or hemicellulose base (nuts, seeds).	**2.** Cellulose food products are the nondigestible residue of plant foods. They hold water, provide bulk, and stimulate elimination.	
3. Recommend moderation in intake of certain irritating fruits such as prunes, grapes, and bananas.	**3.** These fruits tend to increase the quantity of effluent.	

Chart 48-5

PLAN OF NURSING CARE

The Patient Undergoing Ostomy Surgery (continued)

NURSING DIAGNOSIS: Sexual dysfunction related to altered body image
GOAL: Attainment of satisfactory sexual performance

Nursing Interventions	Rationale	Expected Outcomes
1. Encourage the patient to verbalize concerns and fears. The sexual partner is welcomed to participate in the discussion.	1. Expressed needs help the therapist develop a plan of care.	• Expresses fears and concerns • Discusses alternative sexual positions • Accepts services of a professional counselor
2. Recommend alternative sexual positions.	2. Avoid patient embarrassment with the visual appearance of the stoma. Avoid peristomal skin irritation or stomal trauma secondary to friction.	
3. Seek assistance from a sexual therapist or wound-ostomy-continence nurse.	3. Some patients need professional sexual counseling.	

NURSING DIAGNOSIS: Risk for deficient fluid volume related to anorexia and vomiting and increased loss of fluids and electrolytes from GI tract
GOAL: Attainment of fluid balance

Nursing Interventions	Rationale	Expected Outcomes
1. Estimate fluid intake and output (I&O): a. Strict I&O	1. Provides indication of fluid balance a. An early indicator of fluid imbalance is a daily, significant difference between I&O. The average person ingests (food, fluids) and loses (from urine, feces, lungs) about 2 L of fluid every 24 hours.	• Maintains fluid balance • Maintains normal serum and urinary values for sodium and potassium • Normal skin turgor • Surface of tongue is pink, with a moist mucous membrane.
b. Daily weights	b. A gain/loss of 1 L of fluid is reflected in a body weight change of 1 kg (2.2 lbs).	
2. Assess serum and urinary values of sodium and potassium.	2. Sodium is the major electrolyte regulating water balance. Vomiting results in decreased urinary and serum sodium levels. Urinary sodium values, in contrast to serum values, reflect early, sensitive changes in sodium balance. Sodium works in conjunction with potassium, which is also decreased with vomiting. A significant deficiency in potassium is associated with a decrease in intracellular potassium bicarbonate, which leads to acidosis and compensatory hyperventilation.	
3. Observe and record skin turgor and the appearance of the tongue.	3. Adequate hydration is reflected by the skin's ability to return to its normal shape after being grasped between the fingers. *Note:* In the older person, it is normal for the return to be delayed. Changes in the mucous membrane covering the tongue are accurate and early indicators of hydration status.	

including fecal discharge, is necessary to help gauge each patient's fluid needs. There may be 1,000 to 2,000 mL of fluid lost each day in addition to expected fluid loss through urine, perspiration, respiration, and other sources. With this loss, sodium and potassium are depleted. The nurse monitors laboratory values and administers electrolyte replacements as prescribed. Fluids are administered IV for 4 to 5 days to replace lost fluids.

Nasogastric suction may be a part of the immediate postoperative care, with the tube requiring frequent irrigation, as prescribed. The purpose of nasogastric suction is to prevent a buildup of gastric contents while the intestines are not functioning. After the tube is removed, the nurse offers sips of clear liquids and gradually progresses the diet. Nausea and abdominal distention, which may indicate intestinal obstruction, must be reported immediately.

If rectal packing has been used, it is removed by the end of the first week. Because this procedure may be uncomfortable, the nurse may administer an analgesic agent an hour before the removal. After the packing is removed, the perineum is irrigated two or three times daily until full healing takes place.

Providing Emotional Support

The patient may think that everyone is aware of the ileostomy and may view the stoma as a mutilation compared with other abdominal incisions that heal and are hidden. Because there is loss of a body part and a major change in anatomy and function, the patient often goes through the phases of grief—denial, anger, bargaining, depression, and acceptance. Nursing support through these phases is important, and understanding of the patient's emotional state should determine the approach taken. For example, education may be ineffective until the patient is ready to learn. Concern about body image may lead to questions related to family relationships, sexual function, and, for women, the ability to become pregnant and deliver a baby normally. Patients need to know that someone understands and cares about them. A calm, nonjudgmental attitude exhibited by the nurse aids in gaining the patient's confidence. It is important to recognize the dependency needs of these patients. Their prolonged illness can make them irritable, anxious, and unhappy. The nurse can coordinate patient care through meetings attended by consultants such as the physician, psychologist, psychiatrist, social worker, WOC nurse, and dietitian. The team approach is important in facilitating the often complex care of the patient (Boyles, 2010a, 2010b). Research studies suggest that gender may affect the nature of postprocedure ostomy concerns and adaptation. Both men and women have reported concerns with dietary management, physical activity, social support, and sexuality. However, women report more specific social and psychological issues than men (Grant, McMullen, Altschuler, et al., 2011).

Conversely, a surgical procedure to create an ileostomy can produce dramatic positive changes in patients who have suffered from IBD for several years. After the discomfort of the disease has decreased and the patient learns how to take care of the ileostomy, he or she often develops a more positive outlook. Until the patient progresses to this phase, an empathetic and tolerant approach by the nurse plays an important part in recovery. The sooner the patient masters the physical care of the ileostomy, the sooner he or she will psychologically accept it. Strong self-care skills are associated with better outcomes and adjustment for patients (Boyles, 2010a, 2010b).

Support from other people with ostomies is also helpful. The United Ostomy Associations of America (UOAA) is dedicated to the rehabilitation of people with ostomies. This organization gives patients useful information about living with an ostomy through an educational program of literature, lectures, and exhibits. Local associations offer visiting services by qualified members who provide hope and rehabilitation services to patients with new ostomies. Hospitals and other health care agencies may have a WOC nurse on staff who can serve as a valuable resource person for the patient with an ileostomy.

Managing Skin and Stoma Care

The patient with a traditional ileostomy cannot establish regular bowel habits because the contents of the ileum are fluid and are discharged continuously. The patient must wear a pouch at all times. Stomal size and pouch size vary initially; the stoma should be rechecked 3 weeks after surgery, when the edema has subsided. The final size and type of appliance is selected in 3 months, after the patient's weight has stabilized and the stoma shrinks to a stable shape.

The location and length of the stoma are significant to the patient's management of the ileostomy. The surgeon positions the stoma as close to the midline as possible and at a location where even a patient who is obese with a protruding abdomen can see it and care for it easily. Usually, the ileostomy stoma is about 2.5 cm (1 in) long, which makes it convenient for the attachment of an appliance.

Skin excoriation around the stoma can be a persistent problem. Peristomal skin integrity may be compromised by several factors, such as an allergic reaction to the ostomy appliance, skin barrier, or paste; chemical irritation from the effluent; mechanical injury from the removal of the appliance; and infection. If irritation and yeast growth occur, nystatin powder (Mycostatin) is dusted lightly on the peristomal skin and a pouch with skin barrier is applied over the affected area.

Changing an Appliance

A regular schedule for changing the pouch before leakage occurs must be established for those with a traditional ileostomy. The patient can be taught to change the pouch in a manner similar to that described in Chart 48-6.

The amount of time a person can keep the appliance sealed to the body surface depends on the location of the stoma and on body structure. The usual wearing time, which also depends on the type of skin barrier, is 5 to 10 days. The appliance is emptied every 4 to 6 hours, or at the same time the patient empties the bladder. An emptying spout at the bottom of the appliance is closed with a special clip or Velcro closure made for this purpose.

Most pouches are disposable and odor proof. Foods such as spinach and parsley act as deodorizers in the intestinal tract; foods that cause odors include asparagus, cabbage, onions, and fish. Bismuth subcarbonate tablets, which may be prescribed and taken orally three or four times each day, are effective in reducing odor. Oral diphenoxylate with atropine can be prescribed to diminish intestinal motility, thereby thickening the stool and assisting in odor control. Foods such as rice, mashed potatoes, and applesauce may also thicken stool.

Irrigating a Continent Ileostomy

For a continent ileostomy (i.e., Kock pouch), the nurse instructs the patient to drain the pouch, as described in Chart 48-7. A catheter is inserted into the reservoir to drain the fluid. The length of time between drainage periods is gradually increased until the reservoir needs to be drained only every 4 to 6 hours and irrigated once each day. A pouch is not necessary; instead, most patients wear a small dressing over the opening.

When the fecal discharge is thick, water can be injected through the catheter to loosen and soften it. The consistency of the effluent is affected by food intake. At first, drainage is only 60 to 80 mL, but as time goes on, the amount increases significantly. The internal Kock pouch stretches, eventually accommodating 500 to 1,000 mL. The patient learns to use the sensation of pressure in the pouch as a gauge to determine how often the pouch should be drained.

Managing Dietary and Fluid Needs

A low-residue diet is followed for the first 6 to 8 weeks. Strained fruits and vegetables are given. These foods are

(text continues on page 1315)

Chart 48-6

GUIDELINES
Changing an Ostomy Appliance

Equipment Needed
• Mild soap • Clean cloths or towels • Skin barrier (e.g., Stomahesive) • Cutting guide • Appliance pouch

Optional Equipment
• Barrier powder • Antifungal spray or powder • Barrier washer

Implementation

Nursing Action	Rationale
1. Promote patient comfort and involvement in the procedure. **a.** Have the patient assume a relaxed position. **b.** Provide privacy. **c.** Explain details of the procedure. **d.** Expose the ileostomy area; remove the ileostomy belt (if worn).	**1.** Providing a relaxed atmosphere and adequate explanations help the patient to become an active participant in the procedure.
2. Remove the appliance. **a.** Have the patient sit on the toilet or on a chair facing the toilet. A patient who prefers to stand should face the toilet. **b.** The appliance (pouch) can be removed by gently pushing the skin away from the adhesive.	**2.** These positions facilitate disposal or drainage.
3. Cleanse the skin. **a.** Wash the skin gently with a soft cloth moistened with tepid water and mild soap; the patient may prefer to bathe before putting on a clean appliance.	**3.** The patient may shower with or without the pouch. **a.** Micropore or waterproof tape applied to the sides of the faceplate keeps it secure during bathing.
b. Rinse the soap off and dry the skin thoroughly after cleansing.	**b.** Moisture or soap residue interferes with appliance adhesion.

Pouching options

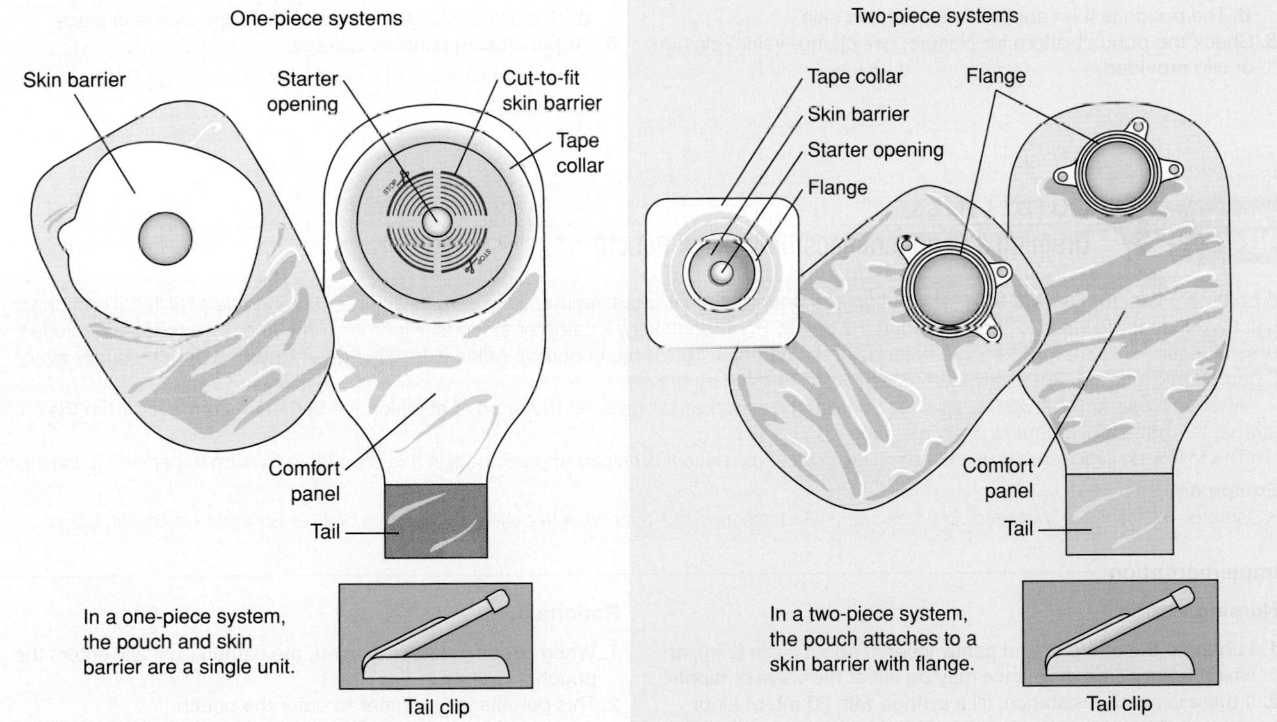
One-piece systems
Skin barrier
Starter opening
Cut-to-fit skin barrier
Tape collar
Comfort panel
Tail
Tail clip
In a one-piece system, the pouch and skin barrier are a single unit.
Two-piece systems
Tape collar
Flange
Skin barrier
Starter opening
Flange
Comfort panel
Tail
Tail clip
In a two-piece system, the pouch attaches to a skin barrier with flange.

(continues on page 1314)

Chart 48-6

GUIDELINES

Changing an Ostomy Appliance (continued)

Nursing Action	Rationale
4. Apply appliance. When there is *no* skin irritation: **a.** Apply an appropriate skin barrier to the peristomal skin before the appliance is applied. **b.** Remove cover from adherent surface of disk of disposable plastic appliance, and apply directly to the skin. **c.** Press firmly in place for 30 seconds to ensure adherence.	**4.** Many appliances have a built-in skin barrier. The skin should be thoroughly dried before applying the appliance.
When there is skin irritation: **a.** Cleanse the skin thoroughly but gently; pat dry. Apply barrier powder.	**a.** Cleansing removes debris and protects irritated skin under wafer.
b. If more irritation develops, apply triamcinolone aerosol (Kenalog spray); blot excess moisture with a cotton pledget and dust lightly with nystatin (Mycostatin) powder.	**b.** The corticosteroid preparation (triamcinolone) helps decrease inflammation. The antifungal agent (nystatin) treats those types of infections that are common around stomas. A prescription is required for either medication. A skin barrier is a substance that facilitates healing of excoriated skin. It adheres well even to moist, irritated skin.
c. Apply a wafer or barrier (e.g., Stomahesive), which is commercially available. The opening in the wafer should be the same size as the stoma; use a cutting guide (supplied with appliance as indicated). The wafer is applied directly to the skin. If using a moldable barrier, stretch it to the size of the stoma. Barrier powder can be used to dry irritated skin before barrier application. OR Another alternative is to apply a special barrier washer (e.g., Eakin Seals). The special barrier adheres well to irritated skin.	**c.** A special barrier protects skin from effluent, promotes healing and helps with adherence.
d. The pouch is then applied to the treated skin.	**d.** This allows skin to heal while the appliance is in place.
5. Check the pouch bottom for closure; use clamp, Velcro closure, or clip provided.	**5.** Proper closure controls leakage.

Chart 48-7

GUIDELINES

Draining a Continent Ileostomy (Kock Pouch)

A continent ileostomy is the surgical creation of a pouch of small intestine that can serve as an internal receptacle for fecal and urinary discharge; a nipple valve is constructed at the outlet. Postoperatively, a catheter extends from the stoma and is attached to a closed drainage suction system. To ensure patency of the catheter, 10–20 mL of normal saline is instilled gently into the pouch usually every 3 hours; return flow is not aspirated but is allowed to drain by gravity.

After approximately 2 weeks, when the healing process has progressed to the point at which the catheter is removed from the stoma, the patient is taught to drain the pouch.

The following procedure is used to drain the pouch; the patient is helped to participate in this procedure to learn to perform it unassisted.

Equipment

• Catheter • Tissues • Water-soluble lubricant • Gauze squares • Syringe • Irrigating solution in a bowl • Emesis or receiving basin

Implementation

Nursing Action	Rationale
1. Lubricate the catheter and gently insert it about 5 cm (2 in), at which point some resistance may be felt at the valve or nipple.	**1.** When gentle pressure is used, the catheter usually enters the pouch.
2. If there is much resistance, fill a syringe with 20 mL of air or water and inject it through the catheter, while still exerting some pressure on the catheter.	**2.** This permits the catheter to enter the pouch.
3. Place the other end of the catheter in a drainage basin held below the level of the stoma. Later, this process can be carried out at the toilet with drainage delivered into the toilet bowl.	**3.** Gravity facilitates drainage. Drainage may include flatus as well as effluent.
4. After drainage, the catheter is removed and the area around the stoma is gently washed with warm water. Pat dry and apply an absorbent pad over the stoma. Fasten the pad with hypoallergenic tape.	**4.** The entire procedure requires about 5–10 minutes; at first, it is performed every 3 hours. The time between procedures is gradually lengthened to 3 times daily.

important sources of vitamins A and C. Later, there are few dietary restrictions, except for avoiding foods that are high in fiber or hard-to-digest kernels, such as celery, popcorn, corn, poppy seeds, caraway seeds, and coconut, which may result in a stomal obstruction (food blockage) for the person with an ileostomy. Foods are reintroduced one at a time. The nurse assesses the patient's tolerance for these foods and reminds him or her to chew food thoroughly.

Fluids may be a problem during the summer, when fluid lost through perspiration adds to the fluid loss through the ileostomy. Fluids such as sports drinks (Gatorade) are helpful in maintaining electrolyte balance. If the fecal discharge is too watery, fibrous foods (e.g., whole-grain cereals, fresh fruit skins, beans, corn, nuts) are restricted. If the effluent is excessively dry, salt intake is increased. Increased intake of water or fluid does not increase the effluent, because excess water is excreted in the urine.

Preventing Complications

Monitoring for complications is an ongoing activity for the patient with an ileostomy. Peristomal skin irritation, which results from leakage of effluent, is the most common complication of an ileostomy. A drainable pouching system that does not fit well is often the cause. Components of the drainable pouching system include the pouch, a solid skin barrier, and adhesive. The WOC nurse typically recommends the appropriate drainable pouching system. The solid skin barrier is the component of this system that is most important in ensuring healthy peristomal skin. Solid skin barriers are typically shaped as rectangular or elliptical wafers and are composed of polymers and hydrocolloids. They protect the skin around the stoma from effluent from the stoma and provide a stable interface between the stoma and the pouch. It is critical that the barrier be sized appropriately to "hug" the stoma (up to the stoma but not touching) and not expose peristomal skin.

Other common complications include diarrhea, stomal stenosis, urinary calculi, and cholelithiasis. Even in the presence of a properly fitted drainable pouching system, diarrhea can be problematic. Diarrhea, manifested by very irritating effluent that rapidly fills the pouch (every hour or sooner), can quickly lead to dehydration and electrolyte losses. Supplemental water, sodium, and potassium are administered to prevent hypovolemia and hypokalemia. Antidiarrheal agents are administered. Stenosis is caused by circular scar tissue that forms at the stoma site. The scar tissue must be surgically released. Urinary calculi may occur in patients with ileostomies and are at least partly attributed to dehydration from decreased fluid intake. Intense lower abdominal pain that radiates to the legs, hematuria, and signs of dehydration indicate that the urine should be strained. Fluid intake is encouraged. Sometimes, small stones are passed during urination; treatment that crushes or removes larger stones is necessary (see Chapter 55).

Permanent ileostomy is more commonly indicated in patients with Crohn's disease. Crohn's disease is a risk factor for cholelithiasis (i.e., gallstones) due to altered absorption of bile acids (Afdhal, 2011). Spasm of the gallbladder causes severe right upper quadrant abdominal pain that can radiate to the back and right shoulder (see Chapter 50).

Promoting Home and Community-Based Care

Educating Patients About Self-Care. The spouse and family should be familiar with the adjustments that will be necessary when the patient returns home. They need to know why it is necessary for the patient to occupy the bathroom for 10 minutes or more at certain times of the day and why certain equipment is needed. Their understanding is necessary to reduce tension; a relaxed patient tends to have fewer problems. Visits from a WOC nurse may be arranged to ensure that the patient is progressing as expected and to provide additional guidance and education as needed.

Continuing Care. The patient needs to know the commercial name of the drainable pouching system to be used so that he or she can obtain a ready supply; the patient should also know how to obtain other supplies. The names and contact information of a local WOC nurse and local self-help groups are often helpful. Any restrictions on driving or working also need to be reviewed. The nurse instructs the patient about common postoperative complications and how to recognize and report them (Chart 48-8).

Concept Mastery Alert

Visit thePoint to view an Interactive Tutorial on IBD and related fundamental concepts.

Chart 48-8 HOME CARE CHECKLIST

Managing Ostomy Care

At the completion of home care education, the patient or caregiver will be able to:	PATIENT	CAREGIVER
• Demonstrate ostomy care, including wound cleansing, irrigation, and appliance changing.	✔	✔
• Describe the importance of maintaining peristomal skin integrity.	✔	✔
• Identify sources for obtaining additional dressing and appliance supplies.	✔	✔
• Identify dietary restrictions (foods that can cause diarrhea and constipation).	✔	✔
• Identify measures to be used to promote fluid and electrolyte balance.	✔	✔
• Describe medication regimen: Identify medications by name, use, route, and frequency.	✔	✔
• Describe potential complications and necessary actions to be taken if complications occur.	✔	✔
• Identify how to contact wound-ostomy-continence or home health nurse.	✔	✔

INTESTINAL OBSTRUCTION

Intestinal obstruction exists when blockage prevents the normal flow of intestinal contents through the intestinal tract. Two types of processes can impede this flow:

- *Mechanical obstruction:* An intraluminal obstruction or a mural obstruction from pressure on the intestinal wall occurs. Examples are intussusception, polypoid tumors and neoplasms, stenosis, strictures, adhesions, hernias, abscesses, and bezoars (i.e., foreign objects created by ingesting unusual substances) (Hodin & Bordeianou, 2011).
- *Functional obstruction:* The intestinal musculature cannot propel the contents along the bowel. Examples are amyloidosis, muscular dystrophy, endocrine disorders such as diabetes, or neurologic disorders such as Parkinson's disease. The blockage also can be temporary and the result of the manipulation of the bowel during surgery.

The obstruction can be partial or complete. Severity depends on the region of bowel affected, the degree to which the lumen is occluded, and especially the degree to which the vascular supply to the bowel wall is disturbed.

Most bowel obstructions occur in the small intestine. Small bowel obstructions are one of the most challenging clinical situations facing surgeons and are responsible for nearly 300,000 hospital admissions in North America every year (Zielinski & Bannon, 2011). Adhesions are the most common cause of small bowel obstruction, followed by hernias and neoplasms (Zielinski & Bannon, 2011). Other causes include intussusception, volvulus (i.e., twisting of the bowel), and paralytic ileus. Most obstructions in the large bowel occur in the sigmoid colon. The most common causes are carcinoma, diverticulitis, inflammatory bowel disorders, and benign tumors. Table 48-5 and Figure 48-7 list mechanical causes of obstruction and describe how they occur.

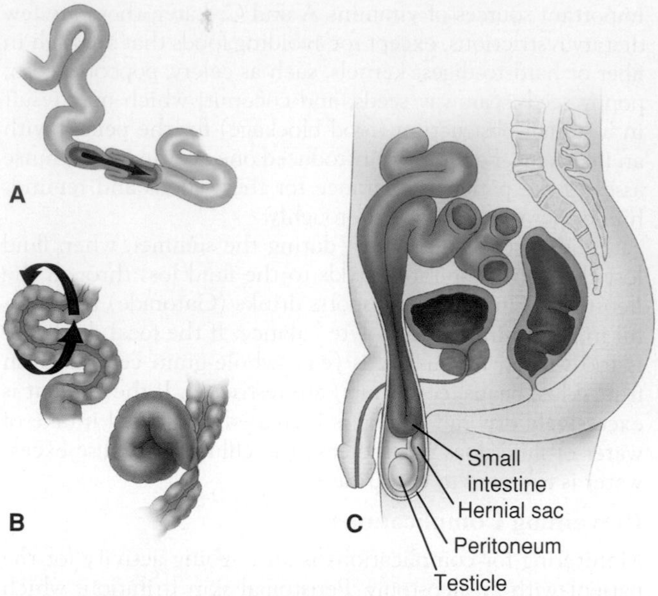
A
B
C
Small intestine
Hernial sac
Peritoneum
Testicle

FIGURE 48-7 • Three causes of intestinal obstruction. **A.** Intussusception; invagination or shortening of the colon caused by the movement of one segment of bowel into another. **B.** Volvulus of the sigmoid colon; the twist is counterclockwise in most cases. Note the edematous bowel. **C.** Hernia (inguinal). The sac of the hernia is a continuation of the peritoneum of the abdomen. The hernial contents are intestine, omentum, or other abdominal contents that pass through the hernial opening into the hernial sac.

Small Bowel Obstruction

Pathophysiology

Intestinal contents, fluid, and gas accumulate above the intestinal obstruction. The abdominal distention and retention of fluid reduce the absorption of fluids and stimulate more gastric secretion. With increasing distention, pressure within the intestinal lumen increases, causing a decrease in venous and arteriolar capillary pressure. This causes edema, congestion, necrosis, and eventual rupture or perforation of the intestinal wall, with resultant peritonitis.

Clinical Manifestations

The initial symptom is usually crampy pain that is wavelike and colicky due to persistent peristalsis both above and below the blockage. The patient may pass blood and mucus but no fecal matter and no flatus. Vomiting occurs. If the obstruction is complete, the peristaltic waves initially become extremely vigorous and eventually assume a reverse direction, with the intestinal contents propelled toward the mouth instead of toward the rectum. If the obstruction is in the ileum, fecal vomiting takes place. First, the patient vomits the stomach

TABLE 48-5 Mechanical Causes of Intestinal Obstruction

Cause	Course of Events	Result
Adhesions	Loops of intestine become adherent to areas that heal slowly or scar after abdominal surgery; occurs most commonly in small intestine	After surgery, adhesions produce a kinking of an intestinal loop.
Intussusception	One part of the intestine slips into another part located below it (like a telescope shortening); occurs more commonly in infants than adults	The intestinal lumen becomes narrowed, and blood supply becomes strangulated.
Volvulus	Bowel twists and turns on itself and occludes the blood supply	Intestinal lumen becomes obstructed. Gas and fluid accumulate in the trapped bowel.
Hernia	Protrusion of intestine through a weakened area in the abdominal muscle wall	Intestinal flow may be completely obstructed. Blood flow to the area may be obstructed as well.
Tumor	A tumor that exists within the wall of the intestine extends into the intestinal lumen, or a tumor outside the intestine causes pressure on the wall of the intestine. Most common type is colorectal adenocarcinoma.	Intestinal lumen becomes partially obstructed; if the tumor is not removed, complete obstruction results.

contents, then the bile-stained contents of the duodenum and the jejunum, and finally, with each paroxysm of pain, the darker, fecal-like contents of the ileum. The signs of dehydration become evident: intense thirst, drowsiness, generalized malaise, aching, and a parched tongue and mucous membranes. The patient may continue to have flatus and stool early in the process due to distal peristalsis. The abdomen becomes distended. The lower the obstruction in the GI tract, the more marked the abdominal distention; this may cause reflux vomiting. Vomiting results in loss of hydrogen ions and potassium from the stomach, leading to reduction of chlorides and potassium in the blood and to metabolic alkalosis. Dehydration and acidosis develop from loss of water and sodium. With acute fluid losses, hypovolemic shock may occur; septic shock may also occur (see Chapter 14 for discussion of septic shock).

Assessment and Diagnostic Findings

Diagnosis is based on the symptoms described previously and on imaging studies. Abdominal x-ray and CT findings include abnormal quantities of gas, fluid, or both in the intestines and sometimes collapsed distal bowel. Laboratory studies (i.e., electrolyte studies and a complete blood cell count) reveal a picture of dehydration, loss of plasma volume, and possible infection. The approach to small bowel obstruction focuses on (1) confirming the diagnosis, (2) identifying the etiology, and (3) determining the likelihood of strangulation (Zielinski & Bannon, 2011).

Medical Management

Decompression of the bowel through a nasogastric tube (see Chapter 45) is necessary for all patients with small bowel obstruction; this intervention improves mortality, improves symptoms, and minimizes the risk of aspiration (Lauder, Garcea, Strickland, et al., 2010; Zielinski & Bannon, 2011). Sometimes a nonoperative clinical trial is tried. Monitoring for bowel ischemia is mandatory. When the bowel is completely obstructed, the possibility of strangulation and tissue necrosis (i.e., tissue death) warrants surgical intervention. Before surgery, IV fluids are necessary to replace the depleted water, sodium, chloride, and potassium.

The surgical treatment of intestinal obstruction depends on the cause of the obstruction. For the most common causes of obstruction, such as hernia and adhesions, the surgical procedure involves repairing the hernia or dividing the adhesion to which the intestine is attached. In some instances, the portion of affected bowel may be removed and an anastomosis performed. The complexity of the surgical procedure depends on the duration of the intestinal obstruction and the condition of the intestine. Laparoscopy has become increasingly common because it can facilitate diagnosis and is easily converted to open laparotomy if warranted (Lauder et al., 2010; Zielinski & Bannon, 2011).

Nursing Management

Nursing management of the nonsurgical patient with a small bowel obstruction includes maintaining the function of the nasogastric tube, assessing and measuring the nasogastric output, assessing for fluid and electrolyte imbalance, monitoring nutritional status, and assessing improvement (e.g., return of normal bowel sounds, decreased abdominal distention, subjective improvement in abdominal pain and tenderness, passage of flatus or stool).

> ***Quality and Safety Nursing Alert***
>
> **Maintaining fluid and electrolyte balance is a priority area to monitor in the patient with a small bowel obstruction. The presence of the nasogastric tube in conjunction with the patient's nothing-by-mouth (NPO) status place the patient at significant risk of fluid imbalance. Thus, measures to promote fluid balance are critically important.**

The nurse reports discrepancies in I&O, worsening of pain or abdominal distention, and increased nasogastric output. If the patient's condition does not improve, the nurse prepares him or her for surgery. Nursing care of the patient after surgical repair of a small bowel obstruction is similar to that for other abdominal surgeries (see Chapter 19).

Large Bowel Obstruction

Pathophysiology

As in small bowel obstruction, large bowel obstruction results in an accumulation of intestinal contents, fluid, and gas proximal to the obstruction. It can lead to severe distention and perforation unless some gas and fluid can flow back through the ileal valve. Large bowel obstruction, even if complete, may be undramatic if the blood supply to the colon is not disturbed. However, if the blood supply is cut off, intestinal strangulation and necrosis occur; this condition is life threatening. In the large intestine, dehydration occurs more slowly than in the small intestine because the colon can absorb its fluid contents and can distend to a size considerably beyond its normal full capacity.

Adenocarcinoid tumors account for the majority of large bowel obstructions (Dalal, Gollub, Miner, et al., 2011; White, Abdool, Frenkiel, et al., 2011). Most tumors occur beyond the splenic flexure, making them accessible with a flexible sigmoidoscope.

Clinical Manifestations

Large bowel obstruction differs clinically from small bowel obstruction in that the symptoms develop and progress relatively slowly. In patients with obstruction in the sigmoid colon or the rectum, constipation may be the only symptom for months. The shape of the stool is altered as it passes the obstruction that is gradually increasing in size. Blood loss in the stool may result in iron deficiency anemia. The patient may experience weakness, weight loss, and anorexia. Eventually, the abdomen becomes markedly distended, loops of large bowel become visibly outlined through the abdominal wall, and the patient has crampy lower abdominal pain. Finally, fecal vomiting develops. Symptoms of shock may occur.

Assessment and Diagnostic Findings

Diagnosis is based on symptoms and on imaging studies. Abdominal x-ray and abdominal CT or MRI findings reveal a distended colon and pinpoint the site of the obstruction. Barium studies are contraindicated. Occasionally, flexible sigmoidoscopy is used to confirm the diagnosis (White et al., 2011).

Medical Management

Restoration of intravascular volume, correction of electrolyte abnormalities, and nasogastric aspiration and decompression are instituted immediately. A colonoscopy may be performed to untwist and decompress the bowel. A cecostomy, in which a surgical opening is made into the cecum, may be performed in patients who are poor surgical risks and urgently need relief from the obstruction. The procedure provides an outlet for releasing gas and a small amount of drainage. A rectal tube may be used to decompress an area that is lower in the bowel. As an alternative, a metal colonic stent may be used as either a palliative intervention or as a bridge to definitive surgery. The colonic stent is placed endoscopically with the assistance of an image intensifier, which creates a fluoroscopic image (Dalal et al., 2011; Ronnekleiv-Kelly & Kennedy, 2011; White et al., 2011). The usual treatment is surgical resection to remove the obstructing lesion. A temporary or permanent colostomy may be necessary. An ileoanal anastomosis may be performed if removal of the entire large bowel is necessary.

Nursing Management

The nurse's role is to monitor the patient for symptoms indicating that the intestinal obstruction is worsening and to provide emotional support and comfort. The nurse administers IV fluids and electrolytes as prescribed. If the patient's condition does not respond to nonsurgical treatment, the nurse prepares the patient for surgery. This preparation includes preoperative education as the patient's condition indicates. After surgery, routine postoperative nursing care is provided, including abdominal wound care.

Colorectal Cancer

Tumors of the colon and rectum are relatively common; the colorectal area (the colon and rectum combined) is now the third most common site of new cancer cases in the United States. Colorectal cancer is a disease of Western cultures. In the United States, almost 145,000 new cases and 50,000 deaths from colorectal cancer occur annually (American Cancer Society [ACS], 2012; National Cancer Institute [NCI], 2012). Colorectal cancer is the third leading cause of cancer death in both men and women in the United States (NCI, 2012; Popek & Tsikitis, 2011). The lifetime risk of developing colorectal cancer is 1 in 20 (NCI, 2012).

The incidence of colorectal cancer increases with age (the incidence is highest in people older than 85 years) and is higher in people with a family history of colon cancer and those with IBD or polyps (the mean age at diagnosis for cancer of the colon and rectum is 70 years) (NCI, 2012). The exact cause of colon and rectal cancer is still unknown, but risk factors have been identified (Chart 48-9). A specific form of hereditary colorectal cancer is *Lynch syndrome,* or hereditary non-polyposis colorectal cancer (HNPCC) (Kalady, 2011). HNPCC-defining cancers include those of the colorectum, uterus, stomach, ovaries, urinary epithelium, and small bowel. HNPCC is characterized by early age of onset. Another disorder with high risk of colorectal cancer is familial adenomatous polyposis, in which patients can develop hundreds of colonic polyps that can become malignant (Carmichael & Mills, 2011).

Chart 48-9 RISK FACTORS
Colorectal Cancer

- Increasing age
- Family history of colon cancer (Lynch syndrome) or polyps (familial adenomatous polyposis)
- Previous colon cancer or adenomatous polyps
- High consumption of alcohol
- Cigarette smoking
- Obesity
- History of gastrectomy
- History of inflammatory bowel disease
- High-fat, high-protein (with high intake of beef), low-fiber diet
- Genital cancer (e.g., endometrial cancer, ovarian cancer) or breast cancer (in women)

Adapted from John, E., & Khachemoune, A. (2010). Gardner syndrome: Skin manifestations, differential diagnosis, and management. *American Journal of Clinical Dermatology, 11*(2), 117–122; and Popek, S., & Tsikitis, V. L. (2011). Epidemiology of inherited colon cancer. *Seminars in Colon & Rectal Surgery, 22*(2), 77–81.

Improved screening strategies have helped reduce the number of deaths from colon cancer in recent years. Early diagnosis and prompt treatment could save almost three of every four people. The stage at presentation is the single consistent variable affecting prognosis in colon cancer (Aldoss & Iqbal, 2011). If the disease is detected and treated at an early stage before it spreads, the 5-year survival rate is 90%; however, only 39% of colorectal cancers are detected at an early stage (NCI, 2012). Survival rates after late diagnosis are very low. Most people are asymptomatic for long periods and seek health care only when they notice a change in bowel habits or rectal bleeding. Prevention and early screening are key to detection and reduction of mortality rates.

Pathophysiology

Cancer of the colon and rectum is predominantly (95%) adenocarcinoma (i.e., arising from the epithelial lining of the intestine) (ACS, 2012). It may start as a benign polyp but may become malignant, invade and destroy normal tissues, and extend into surrounding structures. Cancer cells may migrate away from the primary tumor and spread to other parts of the body (most often to the liver, peritoneum, and lungs) (NCI, 2012).

Clinical Manifestations

The symptoms are greatly determined by the location of the tumor, the stage of the disease, and the function of the affected intestinal segment. The most common presenting symptom is a change in bowel habits. The passage of blood in or on the stools is the second most common symptom. Symptoms may also include unexplained anemia, anorexia, weight loss, and fatigue.

The symptoms most commonly associated with right-sided lesions are dull abdominal pain and melena (i.e., black, tarry stools). The symptoms most commonly associated with left-sided lesions are those associated with obstruction (i.e., abdominal pain and cramping, narrowing stools, constipation, distention), as well as bright red blood in the stool. Symptoms associated with rectal lesions are tenesmus (i.e., ineffective, painful straining at stool with urge), rectal pain, the feeling of incomplete evacuation after a bowel movement, alternating constipation and diarrhea, and bloody stool.

Assessment and Diagnostic Findings

Along with an abdominal and rectal examination, the most important diagnostic procedures for cancer of the colon are fecal occult blood testing, double-contrast barium enema, proctosigmoidoscopy, and colonoscopy (NCI, 2012) (see Chapter 44). The majority of colorectal cancer cases can be identified by colonoscopy with biopsy or cytology smears.

Carcinoembryonic antigen (CEA) studies may also be performed. CEA is a tumor marker that is useful in assessing progression or recurrence of cancers, especially GI cancers (Carcinoembryonic antigen, 2012). With complete excision of the tumor, the elevated levels of CEA should return to normal within 48 hours. Elevations of CEA at a later date suggest recurrence.

Complications

Tumor growth may cause partial or complete bowel obstruction. Extension of the tumor and ulceration into the surrounding blood vessels result in hemorrhage. Perforation, abscess formation, peritonitis, sepsis, and shock may occur.

Gerontologic Considerations

Carcinomas of the colon and rectum are common malignancies in advanced age. In men, only the incidence of prostate cancer and lung cancer exceeds that of colorectal cancer. In women, only the incidence of breast cancer exceeds that of colorectal cancer (NCI, 2012). Symptoms are often insidious. Patients with colorectal cancer usually report fatigue, which is caused primarily by iron deficiency anemia. In early stages, minor changes in bowel patterns and occasional bleeding may occur. The later symptoms most commonly reported by the older adult are abdominal pain, obstruction, tenesmus, and rectal bleeding.

Colon cancer in the older adult has been closely associated with dietary carcinogens. Lack of fiber is a major causative factor because the passage of feces through the intestinal tract is prolonged, which extends exposure to possible carcinogens. Excess dietary fat, high alcohol consumption, and smoking all increase the incidence of colorectal tumors. Physical activity and dietary folate have protective effects (ACS, 2012; NCI, 2012).

Medical Management

Treatment for colorectal cancer depends on the stage of the disease (Chart 48-10) and consists of surgery to remove the tumor, supportive therapy, and adjuvant therapy. Patients who receive some form of adjuvant therapy, which may include chemotherapy, radiation therapy, immunotherapy, or multimodality therapy, typically demonstrate delays in tumor recurrence and increases in survival time (Aldoss & Iqbal, 2011; Masi, Fornaro, Caparello, et al., 2011; Popek & Tsikitis, 2011; Van Loon & Venook, 2011). For the patient with isolated metastases, newer, less invasive approaches have increased life expectancy. Management includes hepatic arterial infusion chemotherapy, radiofrequency ablation, and selective internal radiation therapy (Masi et al., 2011; Ronnekleiv-Kelly, & Kennedy, 2011; Van Loon & Venook, 2011).

The patient with symptoms of intestinal obstruction is treated with IV fluids and nasogastric suction. If there has been significant bleeding, blood component therapy may be required.

Adjuvant Therapy

The standard adjuvant therapy administered to patients with Dukes' class C or TNM (tumor, nodes, metastasis) system (T1–4, N1–2, M0) colon cancer is the 5-fluorouracil (5-FU [Adrucil]) plus leucovorin calcium (Wellcovorin) plus oxaliplatin (Eloxatin) approach (Aldoss & Iqbal, 2011) (see Chart 48-10 for Dukes' classification and Chapter 15 for further discussion of TNM system). Other agents include irinotecan (Camptosar) and capecitabine (Xeloda). Patients with Dukes' class B or C rectal cancer are given 5-FU and high doses of pelvic irradiation. Molecular markers (i.e., fragments of deoxyribonucleic acid [DNA]) are being identified and used for prognostic value. In the future, chemotherapy will likely be individualized to target these markers (Aldoss & Iqbal, 2011). Radiation therapy is used before, during, and after surgery to shrink the tumor; to achieve better results from surgery; and to reduce the risk of recurrence. For inoperative or unresectable tumors, radiation is used to provide significant relief from symptoms. Intracavitary and implantable devices are used to deliver radiation to the site. The response to adjuvant therapy varies. Patients at risk for poor outcomes include those with higher Dukes' or TNM stage (see Chapter 15), elevated CEA levels, insufficient lymph node sampling, and presentation with colonic perforation or obstruction (Aldoss & Iqbal, 2011).

Surgical Management

Surgery is the main treatment for most colon and rectal cancers. It may be curative or palliative. Advances in

Chart 48-10 Staging of Colorectal Cancer: Dukes' Classification—Modified Staging System

Class A: Tumor limited to muscular mucosa and submucosa
Class B_1: Tumor extends into mucosa
Class B_2: Tumor extends through entire bowel wall into serosa or pericolic fat, no nodal involvement
Class C_1: Positive nodes, tumor is limited to bowel wall
Class C_2: Positive nodes, tumor extends through entire bowel wall
Class D: Advanced and metastasis to liver, lung, or bone.

Another staging system, the TNM (tumor, nodal involvement, metastasis) classification, may be used to describe the anatomic extent of the primary tumor, depending on:

- Size, invasion depth, and surface spread
- Extent of nodal involvement
- Presence or absence of metastasis

The higher the score in each category, the worse the disease and prognosis.

surgical techniques can enable the patient with cancer to have sphincter-sparing devices that restore continuity of the GI tract. The type of surgery recommended depends on the location and size of the tumor. Cancers limited to one site can be removed through the colonoscope. Laparoscopic colotomy with polypectomy minimizes the extent of surgery needed in some cases. A laparoscope is used as a guide in making an incision into the colon; the tumor mass is then excised. Laparoscopic colectomy has also been shown to have equivalent surgical outcomes to open colectomy and is associated with decreased hospital length of stay, decreased use of pain medications, and improved quality of life (Carmichael & Mills, 2011). The use of the neodymium/yttrium-aluminum-garnet (Nd:YAG) laser is effective with some lesions as well. Bowel resection is indicated for most class A lesions and all class B and C lesions. Surgery is sometimes recommended for class D colon cancer, but the goal of surgery in this instance is palliative; if the tumor has spread and involves surrounding vital structures, it is considered nonresectable. Insertion of a stent (made of self-expanding metal mesh) for acute malignant colorectal cancer obstruction is an option for patients who require decompression in order to permit elective surgical intervention (Rodriguez-Bigas, 2011).

Possible surgical procedures include the following:

- Segmental resection with anastomosis (i.e., removal of the tumor and portions of the bowel on either side of the growth, as well as the blood vessels and lymphatic nodes) (Fig. 48-8)
- Abdominoperineal resection with permanent sigmoid colostomy (i.e., removal of the tumor and a portion of the sigmoid and all of the rectum and anal sphincter, also called *Miles resection*) (Fig. 48-9)
- Temporary colostomy followed by segmental resection and anastomosis and subsequent reanastomosis of the colostomy, allowing initial bowel decompression and bowel preparation before resection
- Permanent colostomy or ileostomy for palliation of unresectable obstructing lesions
- Construction of a coloanal reservoir called a *colonic J-pouch*, which is performed in two steps. A temporary loop ileostomy is constructed to divert intestinal flow, and the newly constructed J-pouch (made from 6 to 10 cm of colon) is reattached to the anal stump. About 3 months after the initial stage, the ileostomy is reversed and intestinal continuity is restored. The anal sphincter and therefore continence are preserved.

A **colostomy** is the surgical opening into the colon by means of a stoma to allow drainage of bowel contents; it is one type of fecal diversion. The colostomy can be created as a temporary or permanent fecal diversion. It allows the drainage or evacuation of colon contents to the outside of the body. The consistency of the drainage is related to the placement of the colostomy, which is dictated by the location of the tumor and the extent of invasion into surrounding tissues (Fig. 48-10). With improved surgical techniques that allow salvage of the anal sphincter, colostomies are performed in fewer than one third of patients with colorectal cancer (Masi et al., 2011; Ronnekleiv-Kelly, & Kennedy, 2011; Van Loon & Venook, 2011).

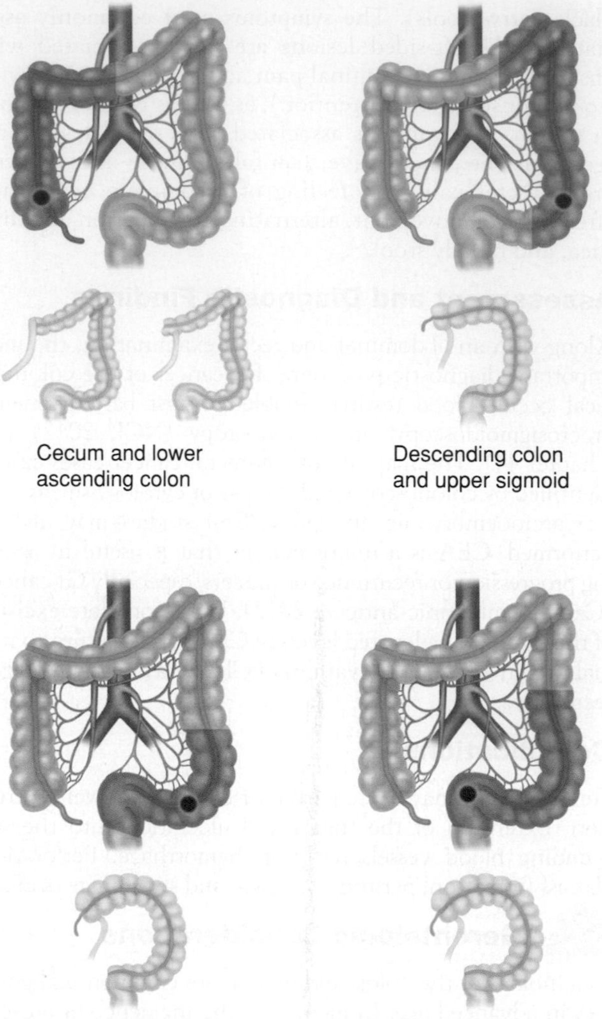
Cecum and lower ascending colon
Descending colon and upper sigmoid
Low sigmoid and upper rectum
Rectal sigmoid resection

FIGURE 48-8 • Examples of areas where cancer can occur, the area that is removed, and how the anastomosis is performed (*small diagrams*).

Gerontologic Considerations

Older adults are at increased risk for complications after surgery and may have difficulty managing colostomy care. Some older adult patients may have decreased vision, impaired hearing, and difficulty with fine motor coordination. It may be helpful for patients to handle ostomy equipment and simulate cleaning the peristomal skin and irrigating the stoma before surgery. Skin care is a major concern in older patients with a colostomy because of the skin changes that occur with aging—the epithelial and subcutaneous fatty layers become thin, and the skin is irritated easily. To prevent skin breakdown, special attention is paid to skin cleansing and the proper fit of an appliance. Newer stoma skin barriers do not have to be cut but can be molded into shape around the stoma (e.g., ConvaTec; Fig. 48-11). Arteriosclerosis may also be an issue; it causes decreased blood flow to the wound and stoma site. As a result, transport of nutrients is delayed and healing time may be prolonged. Some patients have delayed elimination after irrigation because of decreased peristalsis and mucus production. Most patients require 6 months before they feel comfortable with their ostomy care.

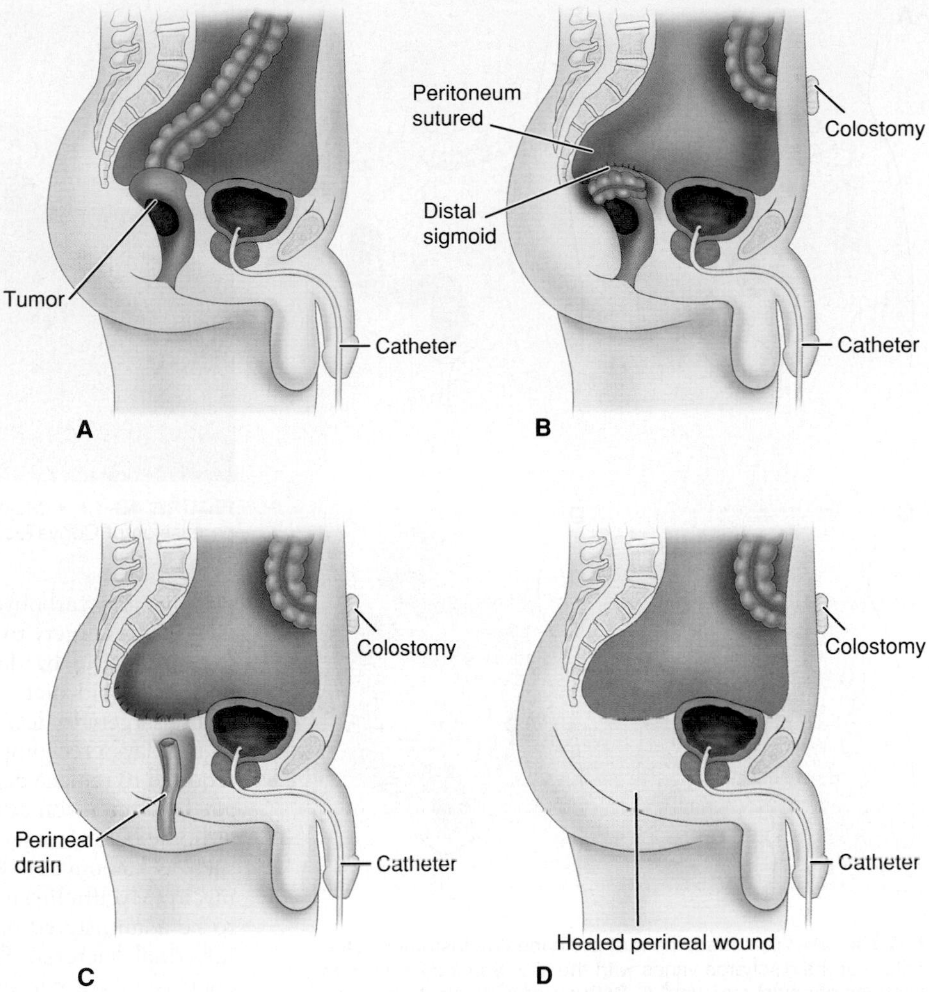
Tumor
Catheter
A
Peritoneum sutured
Colostomy
Distal sigmoid
Catheter
B
Colostomy
Perineal drain
Catheter
C
Colostomy
Catheter
Healed perineal wound
D

FIGURE 48-9 • Abdominoperineal resection for carcinoma of the rectum. **A.** Prior to surgery. Note tumor in rectum. **B.** During surgery, the sigmoid is removed and the colostomy is established. The distal bowel is dissected free to a point below the pelvic peritoneum, which is sutured over the closed end of the distal sigmoid and rectum. **C.** Perineal resection includes removal of the rectum and free portion of the sigmoid from below. A perineal drain is inserted. **D.** The final result after healing. Note the healed perineal wound and the permanent colostomy.

NURSING PROCESS

The Patient With Colorectal Cancer

Assessment

The nurse obtains a health history about the presence of fatigue, abdominal or rectal pain (e.g., location, frequency, duration, association with eating or defecation), past and present elimination patterns, and characteristics of stool (e.g., color, odor, consistency, presence of blood or mucus). Additional information includes a history of IBD or colorectal polyps, a family history of colorectal disease, and current medication therapy. The nurse assesses dietary patterns, including fat and fiber intake, as well as amounts of alcohol consumed and history of smoking. The nurse describes and documents a history of weight loss and feelings of weakness and fatigue.

Assessment includes auscultation of the abdomen for bowel sounds and palpation of the abdomen for areas of tenderness, distention, and solid masses. Stool specimens are inspected for character and presence of blood.

Diagnosis

Nursing Diagnoses

Based on the assessment data, major nursing diagnoses may include the following:

- Imbalanced nutrition: less than body requirements related to nausea and anorexia
- Risk for deficient fluid volume related to vomiting and dehydration
- Anxiety related to impending surgery and the diagnosis of cancer
- Risk for ineffective self-health management related to knowledge deficit concerning the diagnosis, the surgical procedure, and self-care after discharge
- Impaired skin integrity related to the surgical incisions (abdominal and perianal), the formation of a stoma, and frequent fecal contamination of peristomal skin
- Disturbed body image related to colostomy
- Ineffective sexuality patterns related to presence of ostomy and changes in body image and self-concept

Collaborative Problems/Potential Complications

Potential complications may include the following:

- Intraperitoneal infection
- Complete large bowel obstruction
- GI bleeding
- Bowel perforation
- Peritonitis, abscess, and sepsis

Planning and Goals

The major goals for the patient may include attainment of optimal level of nutrition; maintenance of fluid balance; reduction of anxiety; learning about the diagnosis, surgical procedure, and self-care after discharge; maintenance of

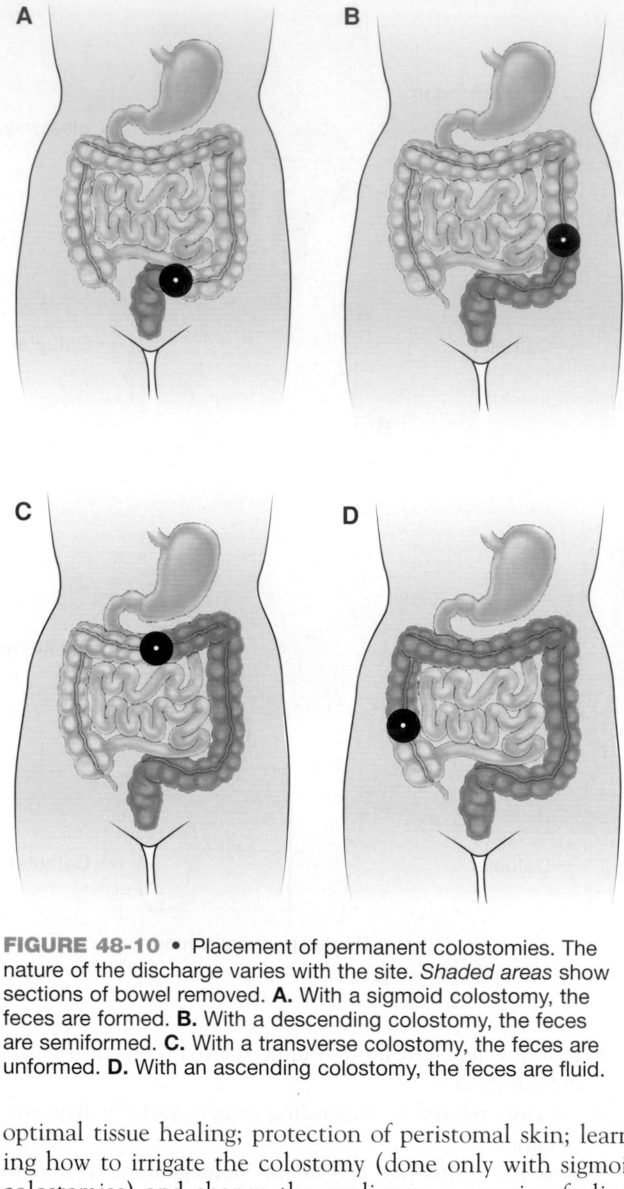
A
B
C
D

FIGURE 48-10 • Placement of permanent colostomies. The nature of the discharge varies with the site. *Shaded areas* show sections of bowel removed. **A.** With a sigmoid colostomy, the feces are formed. **B.** With a descending colostomy, the feces are semiformed. **C.** With a transverse colostomy, the feces are unformed. **D.** With an ascending colostomy, the feces are fluid.

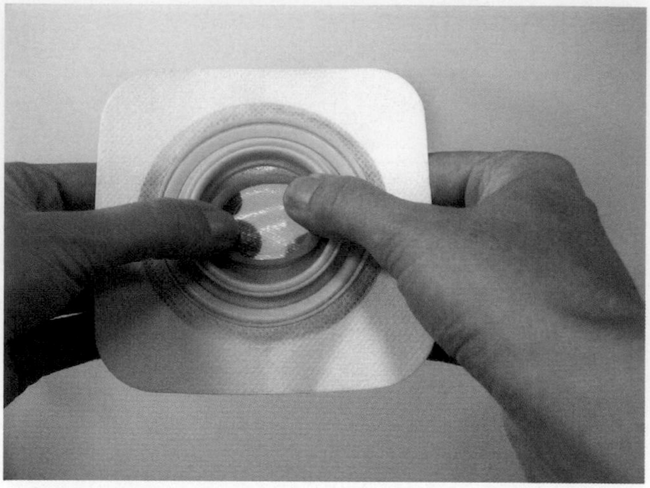

FIGURE 48-11 • Moldable stomal skin barrier. Reprinted with permission of ConvaTec, Inc.

optimal tissue healing; protection of peristomal skin; learning how to irrigate the colostomy (done only with sigmoid colostomies) and change the appliance; expressing feelings and concerns about the colostomy and the impact on self; and avoidance of complications.

Nursing Interventions

Preparing the Patient for Surgery

The patient awaiting surgery for colorectal cancer has many concerns, needs, and fears. He or she may be physically debilitated and emotionally distraught with concerns about lifestyle changes after surgery, prognosis, ability to perform in established roles, and finances. Priorities for nursing care include preparing the patient physically for surgery; providing information about postoperative care, including stoma care if a colostomy is to be created; and supporting the patient and family emotionally. Ideally, a WOC nurse or surgeon should identify the stoma site preoperatively to ensure that its placement is visible and accessible to the patient.

Physical preparation for surgery involves building the patient's stamina in the days preceding surgery and cleansing the bowel the day before surgery. If the patient's condition permits, the nurse recommends a diet high in calories, protein, and carbohydrates and low in residue for several days before surgery to provide adequate nutrition and minimize cramping by decreasing excessive peristalsis. A full or clear liquid diet may be prescribed for 24 to 48 hours before surgery to decrease bulk. If the patient is hospitalized in the days preceding surgery, parenteral nutrition may be required to replace depleted nutrients, vitamins, and minerals. In some instances, parenteral nutrition is administered at home before surgery. Some surgeons prescribe antibiotics such as kanamycin (Kantrex), ciprofloxacin (Cipro), neomycin (Mycifradin) metronidazole, and cephalexin (Keflex) to be administered orally the day before surgery to reduce intestinal bacteria. The bowel is cleansed with laxatives, enemas, or colonic irrigations the evening before and the morning of surgery. IV antibiotics such as cefazolin (Ancef) and metronidazole are administered immediately before surgery (Rodriguez-Bigas, 2011).

For the patient who is very ill and hospitalized, the nurse measures and records I&O, including vomitus, to provide an accurate record of fluid balance. The patient's intake of oral food and fluids may be restricted to prevent vomiting. The nurse administers antiemetic agents as prescribed. Full or clear liquids may be tolerated, or the patient may be NPO. A nasogastric tube may be inserted to drain accumulated fluids and prevent abdominal distention. The nurse monitors the abdomen for increasing distention, loss of bowel sounds, and pain or rigidity, which may indicate obstruction or perforation. It also is important to monitor IV fluids and electrolytes. Monitoring serum electrolyte levels can detect the hypokalemia and hyponatremia that occur with GI fluid loss. The nurse observes for signs of hypovolemia (e.g., tachycardia, hypotension, decreased pulse volume); assesses hydration status; and reports decreased skin turgor, dry mucous membranes, and concentrated urine.

The nurse assesses the patient's knowledge about the diagnosis, prognosis, surgical procedure, and expected level of functioning after surgery. Education is provided about the physical preparation for surgery, the expected appearance and care of the wound, the technique of ostomy care (if applicable), dietary restrictions, pain control, and medication management (see Chart 48-5). If the patient is admitted the day of surgery, the physician's office may arrange for the patient to be seen by a WOC nurse in the days preceding surgery. The WOC nurse helps determine the optimal site for the stoma and provides education about care. If the patient is hospitalized before the day of surgery, the WOC nurse is involved in the preoperative education. All procedures are explained in language the patient understands.

Providing Emotional Support

Patients anticipating bowel surgery for colorectal cancer may be very anxious. They may grieve about the diagnosis, the impending surgery, and possible permanent colostomy. Patients undergoing surgery for a temporary colostomy may express fears and concerns similar to those of a person with a permanent stoma. All members of the health care team, including the WOC nurse, should be available for assistance and support. The nurse's role is to assess the patient's anxiety level and coping mechanisms and suggest methods for reducing anxiety, such as deep-breathing exercises and visualizing a successful recovery from surgery and cancer. The nurse can arrange a meeting with a spiritual advisor if the patient desires or with the physician if the patient wishes to discuss the treatment or prognosis. To promote patient comfort, the nurse projects a relaxed, professional, and empathetic attitude.

The patient undergoing a colostomy may find the anticipated changes in body image and lifestyle profoundly disturbing. Because the stoma is located on the abdomen, the patient may think that everyone will be aware of the ostomy. The nurse helps reduce this fear by presenting facts about the surgical procedure and the creation and management of the ostomy. If the patient is receptive, the nurse can use diagrams, photographs, and appliances to explain and clarify. Because the patient is experiencing emotional stress, the nurse will likely need to repeat some of the information. The nurse provides time for the patient and family to ask questions; the nurse's acceptance and understanding of the patient's concerns and feelings convey a caring, competent attitude that promotes confidence and cooperation. Consultation with a WOC nurse during the preoperative period can be extremely helpful, as can speaking with a person who is successfully managing a colostomy. The UOAA provides useful information about living with an ostomy through literature, lectures, and exhibits (see the Resources section). Visiting services by qualified members and rehabilitation services for patients with new ostomies are provided.

Providing Postoperative Care

Postoperative nursing care for patients undergoing colon resection or colostomy is similar to nursing care for any abdominal surgery patient (see Chapter 19), including pain management during the immediate postoperative period. The nurse also monitors the patient for complications. The nurse assesses the abdomen for returning peristalsis and assesses the initial stool characteristics. It is important to help patients with a colostomy out of bed on the first postoperative day and to encourage them to begin participating in managing the colostomy.

Maintaining Optimal Nutrition

The nurse educates all patients undergoing surgery for colorectal cancer about the health benefits to be derived from consuming a healthy diet. The diet is individualized as long as it is nutritionally sound and does not cause diarrhea or constipation. The return to normal diet is rapid.

A complete nutritional assessment is important for the patient with a colostomy. The patient avoids foods that cause excessive odor and gas, including foods in the cabbage family, eggs, asparagus, fish, beans, and high-cellulose products such as peanuts. It is important to determine whether the elimination of specific foods is causing any nutritional deficiency. Nonirritating foods are substituted for those that are restricted so that deficiencies are corrected. The nurse advises the patient to experiment with an irritating food several times before restricting it, because an initial sensitivity may decrease with time. The nurse can help the patient identify any foods or fluids that may be causing diarrhea, such as fruits, high-fiber foods, soda, coffee, tea, or carbonated beverages. Diphenoxylate with atropine may be prescribed as needed to control the diarrhea. For constipation, prune or apple juice or a mild laxative is effective. The nurse suggests fluid intake of at least 2 L/day.

Providing Wound Care

The nurse frequently examines the abdominal dressing during the first 24 hours after surgery to detect signs of hemorrhage. It is important to help the patient splint the abdominal incision during coughing and deep breathing to lessen tension on the edges of the incision. The nurse monitors temperature, pulse, and respiratory rate for elevations that may indicate an infectious process. If the patient has a colostomy, the stoma is examined for swelling (slight edema from surgical manipulation is normal), color (a healthy stoma is pink or red), discharge (a small amount of oozing is normal), and bleeding (an abnormal sign if bright red or beyond trace amounts).

If the malignancy has been removed using the perineal route, the perineal wound is observed for signs of hemorrhage. This wound may contain a drain or packing that is removed gradually. Bits of tissue may slough off for a week. This process is hastened by mechanical irrigation of the wound or with sitz baths performed two or three times each day initially. The condition of the perineal wound and any bleeding, infection, or necrosis is documented.

Monitoring and Managing Complications

The patient is observed for signs and symptoms of complications. It is important to frequently assess the abdomen, including bowel sounds and abdominal girth, to detect bowel obstruction. The nurse monitors vital signs for increased temperature, pulse, and respirations and for decreased blood pressure that may indicate an intra-abdominal infectious process. Rectal bleeding must be reported immediately because it indicates hemorrhage. The nurse monitors hemoglobin and hematocrit levels and administers blood component therapy as prescribed. Any abrupt change in abdominal pain is reported promptly. Elevated white blood cell counts and temperature or symptoms of shock are reported because they may indicate sepsis. The nurse administers antibiotics as prescribed.

Pulmonary complications are always a concern with abdominal surgery; patients older than 50 years are at risk, especially if they are or have been receiving sedative agents or are being maintained on bed rest for a prolonged period. Pneumonia and atelectasis are two major pulmonary complications. Frequent activity (e.g., turning the patient from side to side every 2 hours), deep breathing, coughing, and early

TABLE 48-6 Potential Complications and Nursing Interventions After Intestinal Surgery

Complication	Nursing Interventions
General Complications	
Paralytic ileus	Initiate or continue nasogastric intubation as prescribed. Prepare patient for x-ray study. Ensure adequate fluid and electrolyte replacement. Administer prescribed antibiotics if patient has symptoms of peritonitis.
Mechanical obstruction	Assess patient for intermittent colicky pain, nausea, and vomiting.
Intra-abdominal Septic Conditions	
Peritonitis	Evaluate patient for nausea, hiccups, chills, spiking fever, tachycardia, boardlike abdomen. Administer antibiotics as prescribed. Prepare patient for drainage procedure. Administer parenteral fluid and electrolyte therapy as prescribed. Prepare patient for surgery if condition deteriorates.
Abscess formation	Administer antibiotics as prescribed. Apply warm compresses as prescribed. Prepare for surgical drainage.
Surgical Wound Complications	
Infection	Monitor temperature; report temperature elevation. Observe for redness, tenderness, hardening (induration), and pain around the surgical wound. Assist in establishing local drainage. Obtain specimen of drainage material for culture and sensitivity studies.
Wound disruption	Observe for sudden drainage of profuse serous fluid from wound. Cover wound area with sterile moist dressings supported with binder or similar method. Prepare patient immediately for surgery.
Intraperitoneal infection and abdominal wound infection	Monitor for evidence of constant or generalized abdominal pain, rapid pulse, and elevation of temperature. Prepare for tube decompression of bowel. Administer fluids and electrolytes by IV route as prescribed. Administer antibiotics as prescribed.
Anastomotic Complications	
Dehiscence of anastomosis	Prepare patient for surgery.
Fistulas	Prepare for tube decompression of bowel. Administer parenteral fluids as prescribed to correct fluid and electrolyte deficits.

ambulation can reduce the risk of these complications. Table 48-6 lists additional potential postoperative complications.

The incidence of complications related to the colostomy is usually less than that of an ileostomy. Some common complications are prolapse of the stoma, parastomal hernia, perforation (from improper stoma irrigation), stoma retraction, mucocutaneous separation, and skin irritation (Boyles, 2010b; Kalashnikova, Achkasov, Fadeeva, et al., 2011; Sung, Kwon, Jo, et al., 2010). Leakage from an anastomotic site can occur if the remaining bowel segments are diseased or weakened. Leakage from an intestinal anastomosis causes peritonitis with abdominal distention and rigidity, temperature elevation, and signs of shock. Surgical repair is necessary.

Removing and Applying the Colostomy Appliance

The colostomy begins to function 3 to 6 days after surgery. The nurse manages the colostomy and educates the patient about its care until the patient can take over its management. The nurse educates about skin care and how to apply, empty, and remove the drainage pouch. Care of the peristomal skin is an ongoing concern because excoriation or ulceration can develop quickly. The presence of such irritation makes adhering the ostomy appliance difficult, and adhering the ostomy appliance to irritated skin can worsen the skin condition. The effluent discharge and the degree to which it is irritating vary with the type of ostomy. With a transverse colostomy, the stool is soft and unformed and irritating to the skin. With a descending or sigmoid colostomy, the stool is fairly solid and less irritating to the skin. Ileostomies have the most irritating effluent. Other skin problems include yeast infections and allergic dermatitis.

If the patient wants to bathe or shower before putting on a clean appliance, micropore tape applied to the sides of the pouch keeps it secure during bathing. (Refer to Chart 48-6 for guidelines on changing an ostomy appliance.)

As soon as the patient has learned a routine for evacuation, pouches may be dispensed with, and a closed ostomy appliance or a stoma cap is used to cover the stoma. Except for gas and a slight amount of mucus, nothing escapes from the colostomy opening between irrigations. New assistive devices are available to help nurses learn ostomy assessment and ostomy product selection. An ostomy care algorithm is available and is being turned into a digital format with Internet-based access (Beitz, Gerlach, Ginsburg, et al., 2010) (Chart 48-11 and Fig. 48-12). Other algorithms are available to assist with identification of ostomy complications (Kalashnikova et al., 2011).

Irrigating the Colostomy

The purpose of irrigating a colostomy is to empty the colon of gas, mucus, and feces so that the patient can go about social and business activities without fear of fecal drainage. A stoma does not have voluntary muscular control and may empty at irregular intervals. Regulating the passage of fecal material is achieved by irrigating the colostomy or allowing the bowel to evacuate naturally without irrigations. This choice depends on the person and the type of the colostomy (i.e., descending or sigmoid colostomies). By irrigating the stoma at a regular time, there is less gas and retention of the irrigant. The time for irrigating the colostomy should be consistent with the schedule that the person will follow after leaving the hospital. Chart 48-12 describes the irrigating procedure. Colostomy irrigation is not recommended for persons with extensive pelvic irradiation because it carries a risk of perforation.

Supporting a Positive Body Image

The patient is encouraged to verbalize feelings and concerns about altered body image and to discuss the surgery and the stoma (if one was created). A supportive environment and attitude on the nurse's part are crucial in promoting the patient's adaptation to the change brought about by the surgery. If applicable, the patient must learn colostomy care and begin to plan for incorporating stoma care into daily life. The nurse helps the patient overcome aversion to the stoma or fear of self-injury by providing care and education in an open, accepting manner and by encouraging the patient to talk about his or her feelings about the stoma. The nurse's positive, supportive facial expression and other nonverbal cues help the patient develop a positive attitude toward independent stoma care.

(text continues on page 1327)

Chart 48-11

NURSING RESEARCH PROFILE

Content Validation of Stomal and Peristomal Complications Algorithm

Beitz, J., Gerlach, M., Ginsburg, P., et al. (2010). Content validation of a standardized algorithm for ostomy care. *Ostomy Wound Management*, *56*(10), 22–38.

Purpose

The number of ostomy care nurse clinicians is limited; hence, most ostomy care is provided by nonspecialized, nonexpert clinicians or unskilled caregivers and family. The purpose of this study was to obtain content validation data for a new nurse-designed standardized algorithm for ostomy care developed by expert wound-ostomy-continence (WOC) nurses.

Design

Using a cross-sectional, mixed methods (quantitative and qualitative) study design and a 30-item instrument with a four-point Likert-type scale, participants were asked to quantify the degree of validity of the ostomy algorithm's decisions and components. Participants included 166 WOC nurses who self-identified as having expertise in ostomy care. They were surveyed online for 6 weeks in the summer and fall of 2009. In addition to quantitative ratings, participants' open-ended comments were thematically analyzed.

Findings

WOC nurses rated the ostomy algorithm as having strong content validity. Using a scale of 1 to 4, the mean score of the entire algorithm was 3.8 (4 = very relevant/relevant). The ostomy algorithm's Content Validity Index (CVI) was 0.95 (out of 1.0). Algorithm components CVI's ranged from 0.90 to 0.98. Qualitative data analysis revealed various themes that were both positive and negative; for example, the inability to capture all aspects and patient attributes affecting patient care.

Nursing Implications

Findings from this study support the strong content validity for the ostomy algorithm. Future nursing research to ascertain its construct validity (the ability of the ostomy algorithm to affect ostomy patients care positively) with nonexpert users is warranted. Continued validation and use of this algorithm may facilitate expeditious and evidence-based decisions for management of patients with ostomies.

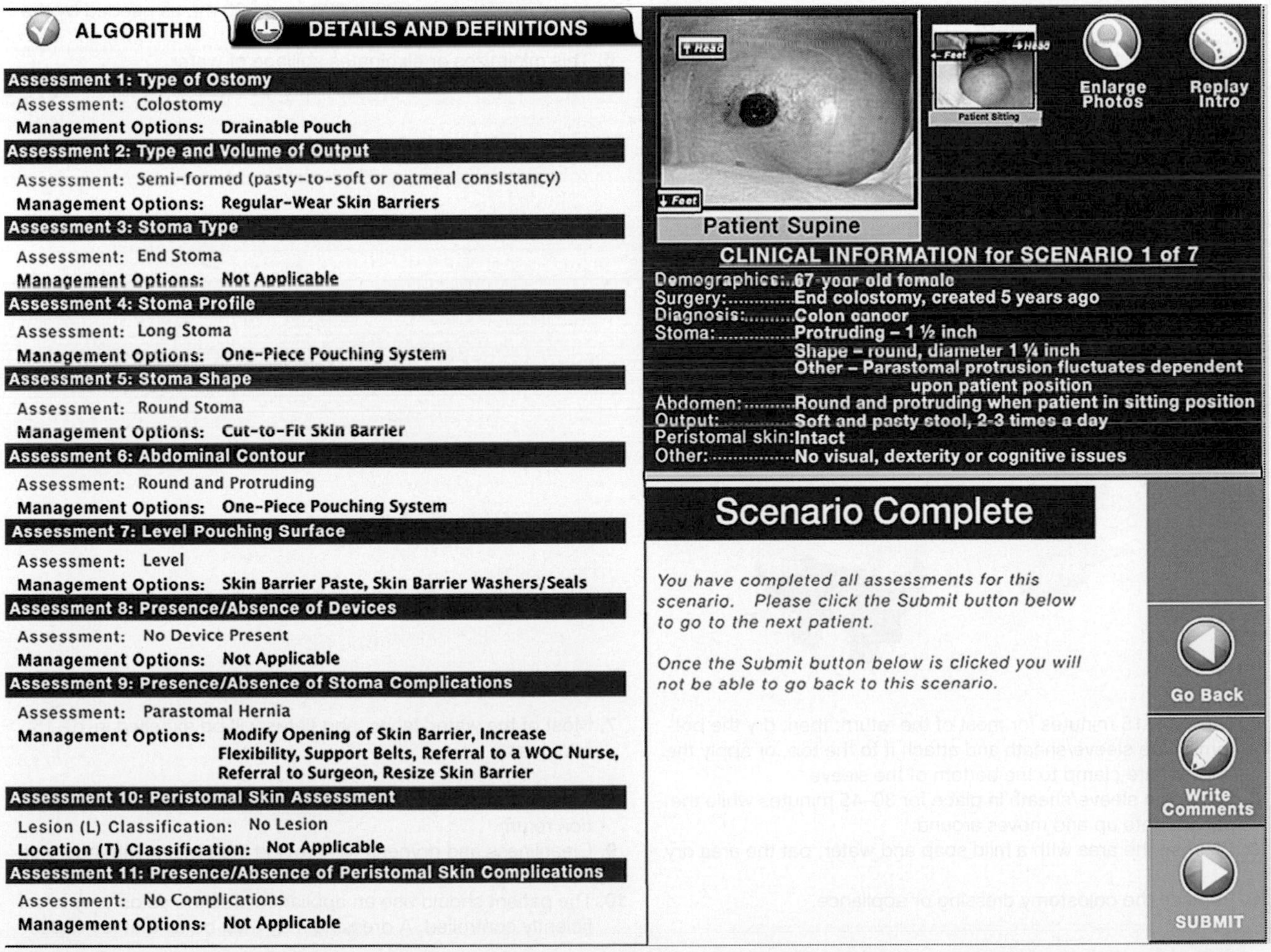
ALGORITHM
DETAILS AND DEFINITIONS
Assessment 1: Type of Ostomy
Assessment: Colostomy
Management Options: Drainable Pouch
Assessment 2: Type and Volume of Output
Assessment: Semi-formed (pasty-to-soft or oatmeal consistancy)
Management Options: Regular-Wear Skin Barriers
Assessment 3: Stoma Type
Assessment: End Stoma
Management Options: Not Applicable
Assessment 4: Stoma Profile
Assessment: Long Stoma
Management Options: One-Piece Pouching System
Assessment 5: Stoma Shape
Assessment: Round Stoma
Management Options: Cut-to-Fit Skin Barrier
Assessment 6: Abdominal Contour
Assessment: Round and Protruding
Management Options: One-Piece Pouching System
Assessment 7: Level Pouching Surface
Assessment: Level
Management Options: Skin Barrier Paste, Skin Barrier Washers/Seals
Assessment 8: Presence/Absence of Devices
Assessment: No Device Present
Management Options: Not Applicable
Assessment 9: Presence/Absence of Stoma Complications
Assessment: Parastomal Hernia
Management Options: Modify Opening of Skin Barrier, Increase Flexibility, Support Belts, Referral to a WOC Nurse, Referral to Surgeon, Resize Skin Barrier
Assessment 10: Peristomal Skin Assessment
Lesion (L) Classification: No Lesion
Location (T) Classification: Not Applicable
Assessment 11: Presence/Absence of Peristomal Skin Complications
Assessment: No Complications
Management Options: Not Applicable
Head
Feet
Patient Sitting
Enlarge Photos
Replay Intro
Patient Supine
CLINICAL INFORMATION for SCENARIO 1 of 7
Demographics: 67-year old female
Surgery: End colostomy, created 5 years ago
Diagnosis: Colon cancer
Stoma: Protruding – 1 ½ inch
Shape – round, diameter 1 ¼ inch
Other – Parastomal protrusion fluctuates dependent upon patient position
Abdomen: Round and protruding when patient in sitting position
Output: Soft and pasty stool, 2-3 times a day
Peristomal skin: Intact
Other: No visual, dexterity or cognitive issues
Scenario Complete
You have completed all assessments for this scenario. Please click the Submit button below to go to the next patient.
Once the Submit button below is clicked you will not be able to go back to this scenario.
Go Back
Write Comments
SUBMIT

FIGURE 48-12 • Screenshot of the Internet-based ostomy care algorithm designed to assist with identification of ostomy care needs and product selection. The algorithm consists of 11 assessments, beginning with the type of ostomy, intended to provide a pathway to help nonspecialist clinicians optimize care for their patients with an ostomy. © ConvaTec, Inc. Reprinted with permission.

Chart 48-12

GUIDELINES

Irrigating a Colostomy

A colostomy is irrigated to empty the colon of feces, gas, or mucus, cleanse the lower intestinal tract, and establish a regular pattern of evacuation so that normal life activities may be pursued. A suitable time for the irrigation is selected that is compatible with the patient's posthospital pattern of activity (preferably after a meal). Irrigation should be performed at the same time each day.

Before the procedure, the patient sits on a chair in front of the toilet or on the toilet itself. An irrigating reservoir containing 500–1,500 mL of lukewarm tap water is hung 45–50 cm (18–20 in) above the stoma (shoulder height when the patient is seated). The dressing or pouch is removed. The following procedure is used; the patient is helped to participate in the procedure so that he or she can learn to perform it unassisted.

Equipment

- Irrigating sleeve or sheath • Irrigating catheter or cone and tubing • Lubricant • Clamp • Mild soap • Cloth or towel
- New colostomy dressing or appliance

Implementation

Nursing Action	Rationale
1. Apply an irrigating sleeve or sheath to the stoma. Place the end in the commode.	1. This helps control odor and splashing and allows feces and water to flow directly into the commode.
2. Allow some of the solution to flow through the tubing and catheter/cone.	2. Air bubbles in the setup are released so that air is not introduced into the colon, which would cause crampy pain.
3. Lubricate the irrigating cone and gently insert it into the stoma (Fig. A). Insert the cone into the stoma and hold it gently, but firmly, against the stoma to prevent backflow of water.	3. Lubrication permits ease of insertion of the cone. A cone is used to prevent internal damage if a catheter is used.
4. Allow water to flow slowly while advancing catheter (Fig. B).	4. A slow rate of flow helps to relax the bowel.
5. Allow tepid fluid to enter the colon slowly. If cramping occurs, clamp off the tubing and allow the patient to rest before progressing. Water should flow in over a 5- to 10-minute period.	5. Painful cramps usually are caused by too rapid a flow or by too much solution; 300 mL of fluid may be all that is needed to stimulate evacuation. Volume may be increased with subsequent irrigations to 500, 1,000, or 1,500 mL as needed by the patient for effective results.
6. Hold the cone in place 10 seconds after the water has been instilled, then gently remove it (Fig. C).	6. This minimizes or eliminates spillage of water.

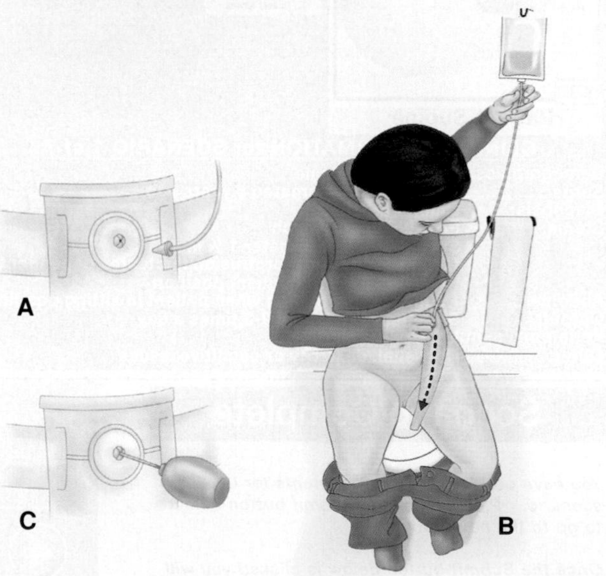
A
B
C

Nursing Action	Rationale
7. Allow 10–15 minutes for most of the return; then, dry the bottom of the sleeve/sheath and attach it to the top, or apply the appropriate clamp to the bottom of the sleeve.	7. Most of the water, feces, and flatus will be expelled in 10–15 minutes.
8. Leave the sleeve/sheath in place for 30–45 minutes while the patient gets up and moves around.	8. Ambulation stimulates peristalsis and completion of the irrigation return.
9. Cleanse the area with a mild soap and water; pat the area dry.	9. Cleanliness and dryness will provide the patient with hours of comfort.
10. Replace the colostomy dressing or appliance.	10. The patient should use an appliance until the colostomy is sufficiently controlled. A dressing may then be all that is needed.

Discussing Sexuality Issues

The nurse encourages the patient to discuss feelings about sexuality and sexual function. The nurse can say, "Many people after surgery wonder how this will affect their role as a sexual partner. Do you have concerns about that?" Some patients may initiate questions about sexual activity directly or give indirect clues about their fears. Some may view the surgery as mutilating and a threat to their sexuality; some fear impotence. Others may express worry about odor or leakage from the pouch during sexual activity. Although the appliance presents no deterrent to sexual activity, some patients wear silk or cotton covers and smaller pouches during sex. Alternative sexual positions are recommended, as well as alternative methods of stimulation to satisfy sexual drives. The nurse assesses the patient's needs and attempts to identify specific concerns. If the nurse is uncomfortable with this or if the patient's concerns seem complex, the nurse may seek assistance from a WOC nurse, sex counselor or therapist, or advanced practice nurse.

Promoting Home and Community-Based Care

Educating Patients About Self-Care. Patient education and discharge planning require the combined efforts of the physician, nurse, WOC nurse, social worker, and dietitian. Patients are given specific information about ostomy care and signs and symptoms of potential complications that is individualized to their needs. Dietary instructions are essential to help patients identify and eliminate irritating foods that can cause diarrhea or constipation. It is important to educate patients about their prescribed medications (i.e., action, purpose, and possible side and toxic effects).

The nurse reviews treatments (e.g., irrigations, wound cleansing) and dressing changes and encourages the family to participate. Because the length of hospital stay is short, the patient may not become proficient in all stoma care techniques before discharge home. However, it is critical that the patient or family member learn how to empty and change the pouch before leaving the hospital. Many patients need referral to a home care agency and the telephone number of the local chapter of the ACS. The home care nurse provides further care and education and assesses the patient's and family's adjustment to the colostomy. The home environment is assessed for adequacy of resources that allow the patient to manage self-care activities. A family member may assume responsibility for purchasing the equipment and supplies needed at home.

Patients need very specific directions about when to call their primary provider. They need to know which complications require prompt attention (i.e., bleeding, abdominal distention and rigidity, diarrhea, fever, wound drainage, and disruption of suture line). If radiation therapy is planned, the possible side effects (i.e., anorexia, vomiting, diarrhea, fatigue) are reviewed.

Continuing Care. Ongoing care of the patient with cancer and a colostomy often extends well beyond the initial hospital stay. Home care nurses manage ostomy follow-up care, manage the assessment and care of the debilitated patient, and coordinate adjuvant therapy. The home visits also provide an opportunity to assess the patient's physical and emotional status and the patient's and family's ability to carry out recommended care strategies. Visits from a WOC nurse are available to the patient and family as they learn to care for the ostomy and work through their feelings about it, the diagnosis of cancer, and the future. Some patients are interested in and can benefit from involvement in an ostomy support group (e.g., the UOAA).

Evaluation

Expected patient outcomes may include:

1. Consumes a healthy diet
   a. Avoids foods and fluids that cause diarrhea, constipation, and obstruction
   b. Substitutes nonirritating foods and fluids for those that are restricted
2. Maintains fluid balance
   a. Experiences no vomiting or diarrhea
   b. Experiences no signs or symptoms of dehydration
3. Feels less anxious
   a. Expresses concerns and fears freely
   b. Uses coping measures to manage stress
4. Acquires information about diagnosis, surgical procedure, preoperative preparation, and self-care after discharge
   a. Discusses the diagnosis, surgical procedure, and postoperative self-care
   b. Demonstrates techniques of ostomy care
5. Maintains clean incision, stoma, and perineal wound
6. Expresses feelings and concerns about self
   a. Gradually increases participation in stoma and peristomal skin care
   b. Discusses feelings related to changed appearance
7. Discusses sexuality in relation to ostomy and to changes in body image
8. Recovers without complications
   a. Is afebrile
   b. Regains normal bowel activity
   c. Exhibits no signs and symptoms of perforation or bleeding
   d. Identifies signs and symptoms that should be reported to the health care provider

Polyps of the Colon and Rectum

A polyp is a mass of tissue that protrudes into the lumen of the bowel. Polyps can occur anywhere in the intestinal tract and rectum. They can be classified as neoplastic (i.e., adenomas and carcinomas) or non-neoplastic (i.e., mucosal and hyperplastic). Non-neoplastic polyps, which are benign epithelial growths, are common in the Western world. They occur more commonly in the large intestine than in the small intestine. Although most polyps do not develop into invasive neoplasms, they must be identified and followed closely (National Institute of Diabetes and Digestive and Kidney Diseases [NIDDK], 2011). Adenomatous polyps are more common in men. The proportion of these polyps arising in the proximal part of the colon increases with age (after 50 years of age). Prevalence rates vary from 25% to 60%, depending on age. Non-neoplastic polyps occur in 80% of the population, and their frequency increases with age. Up to two thirds of people older than 65 years are at risk for colonic adenomas (ACS, 2012).

Clinical manifestations depend on the size of the polyp and the amount of pressure it exerts on intestinal tissue. The most common symptom is rectal bleeding. Lower abdominal pain may also occur. If the polyp is large enough, symptoms of obstruction occur. The diagnosis is based on history and digital rectal examination, double-contrast barium enema studies, sigmoidoscopy, or colonoscopy.

After a polyp is identified, it should be removed. Several methods are used: colonoscopy with the use of special equipment (i.e., biopsy forceps and snares), laparoscopy, or colonoscopic excision with laparoscopic visualization. The latter technique enables immediate detection of potential problems and allows laparoscopic resection and repair of the major complications of perforation and bleeding that may occur with polypectomy. Microscopic examination of the polyp then identifies the type of polyp and indicates what further surgery is required, if any.

DISEASES OF THE ANORECTUM

Anorectal disorders are common, and the majority of the population will experience one at some time during their lives. Patients with anorectal disorders seek medical care primarily because of pain, rectal bleeding, or change in bowel habits. Other common complaints are protrusion of hemorrhoids, anal discharge, perianal itching, swelling, anal tenderness, stenosis, and ulceration. Constipation results from delaying defecation because of anorectal pain.

There has been a steady increase in the prevalence of sexually transmitted infections (STIs; also called *sexually transmitted diseases*, or STDs) in recent decades, leading to the identification of new anorectal syndromes. These syndromes include STIs such as syphilis, gonorrhea, herpes, chlamydia, and candidiasis, and they are most commonly seen in male homosexuals who practice anorectal intercourse.

Anorectal Abscess

An anorectal abscess is caused by obstruction of an anal gland with dried debris, resulting in retrograde infection. People with Crohn's disease or immunosuppressive conditions such as AIDS are particularly susceptible to these infections. Many of these abscesses result in fistulas (Lewis & Maron, 2010; Rizzo, Naig, & Johnson, 2010).

An abscess may occur in a variety of spaces in and around the rectum, usually in the path of least resistance, where anatomic structures are in close proximity without hard or thick structures to separate them. It often contains a quantity of foul-smelling pus and is painful. If the abscess is superficial, swelling, redness, and tenderness are observed. A deeper abscess may result in severe lower abdominal pain and fever.

Palliative therapy consists of sitz baths and analgesic agents. However, prompt surgical treatment to incise and drain the abscess is the treatment of choice (Yano, Asano, Matsuda, et al., 2010). When a deeper infection exists with the possibility of a fistula, the fistulous tract must be excised. If possible, the fistula is excised when the abscess is incised and drained, or a second procedure may be necessary to do so. The wound may be packed with an absorptive dressing (e.g., calcium alginate or hydrofiber) and allowed to heal by granulation.

Anal Fistula

An anal fistula is a tiny, tubular, fibrous tract that extends into the anal canal from an opening located beside the anus in the perianal skin (Fig. 48-13A). Fistulas usually result from an infection. They may also develop from trauma, fissures, or Crohn's disease (up to 25% of patients with Crohn's disease develop a perianal fistula). Purulent drainage or stool may leak constantly from the cutaneous opening. Other symptoms may be the passage of flatus or feces from the vagina or bladder, depending on the location of the fistula tract. Untreated fistulas may cause systemic infection with related symptoms (Michalopoulos, Papadopoulos, Tziris, et al., 2010).

Medical therapy includes antibiotics or anti-inflammatory type agents (e.g., azathioprine, infliximab, cyclosporine). Fistula recurrence is common (Lewis & Maron, 2010). Surgery is recommended because few fistulas heal spontaneously. A fistulectomy (i.e., excision of the fistulous tract) is the recommended surgical procedure. The lower bowel is evacuated thoroughly with several prescribed enemas. The fistula is dissected out or laid open by an incision from its rectal opening to its outlet. The wound is packed with gauze.

Anal Fissure

An anal fissure is a longitudinal tear or ulceration in the lining of the anal canal usually just distal to the dentate line (see Fig. 48-13B). Fissures are usually caused by the trauma of

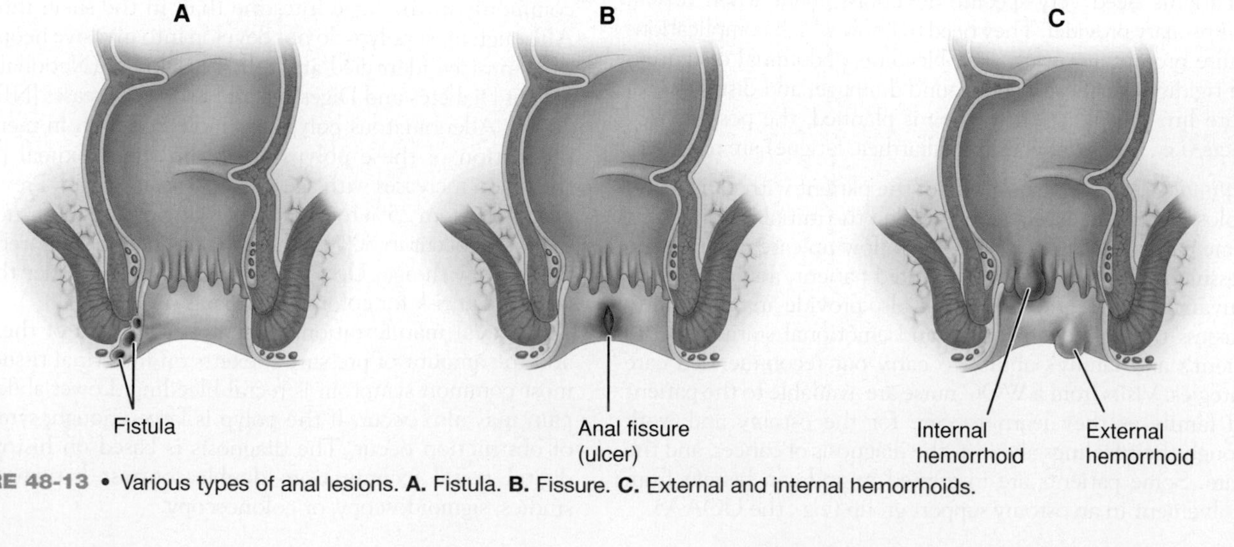
A
B
C
Fistula
Anal fissure
(ulcer)
Internal
hemorrhoid
External
hemorrhoid

FIGURE 48-13 • Various types of anal lesions. **A.** Fistula. **B.** Fissure. **C.** External and internal hemorrhoids.

passing a large, firm stool or from persistent tightening of the anal canal because of stress and anxiety (leading to constipation). Other causes include childbirth, trauma, and the overuse of laxatives (Etzioni, 2011).

Extremely painful defecation, burning, and bleeding characterize fissures. Bright red blood may be seen on the toilet tissue after a bowel movement. Most of these fissures heal if treated by conservative measures that include dietary modification with addition of fiber supplements, stool softeners and bulk agents, an increase in water intake, sitz baths, and emollient suppositories. A suppository combining an anesthetic with a corticosteroid helps relieve the discomfort. Anal dilation under anesthesia may be required. Novel therapies such as perianal or intra-anal application of nitroglycerin ointment, calcium channel blockers, minoxidil, or botulinum toxin (Botox) injections have increased the rate of healing and lowered pain levels in chronic anal fissures. Some literature suggests that these therapies be tried before surgery. They are theorized to work by increasing blood supply to the region and relaxing the anal sphincter (Etzioni, 2011; Madalinski, 2011).

If fissures do not respond to conservative treatment, surgery is indicated. Most surgeons consider the procedure of choice to be the lateral internal sphincterotomy with excision of the fissure (Madalinski, 2011).

Hemorrhoids

Hemorrhoids are dilated portions of veins in the anal canal. They are very common, affecting 1 million Americans each year (Sneider & Maykel, 2010). By 50 years of age, about 50% of people have hemorrhoids (NDDIC, 2012c). Shearing of the mucosa during defecation results in the sliding of the structures in the wall of the anal canal, including the hemorrhoidal and vascular tissues. Increased pressure in the hemorrhoidal tissue due to pregnancy may initiate hemorrhoids or aggravate existing ones. Hemorrhoids are classified into two types: those above the internal sphincter are called *internal hemorrhoids*, and those appearing outside the external sphincter are called *external hemorrhoids* (Sneider & Maykel, 2010) (see Fig. 48-13C). They are also classified by degree of prolapse (Pramateftakis, 2010; Sneider & Maykel, 2010):

- First degree—do not prolapse and protrude into anal canal
- Second degree—prolapse outside the anal canal during defecation but reduce spontaneously
- Third degree—prolapsed to the extent that they require manual reduction
- Fourth degree—prolapsed to the extent that they may not be reduced

Hemorrhoids cause itching and pain and are the most common cause of bright red bleeding with defecation. External hemorrhoids are associated with severe pain from the inflammation and edema caused by thrombosis (i.e., clotting of blood within the hemorrhoid). This may lead to ischemia of the area and eventual necrosis. Internal hemorrhoids are not usually painful until they bleed or prolapse when they become enlarged.

Hemorrhoid symptoms and discomfort can be relieved by good personal hygiene and by avoiding excessive straining during defecation. A high-residue diet that contains fruit and bran along with an increased fluid intake may be all the treatment that is necessary to promote the passage of soft, bulky stools to prevent straining. If this treatment is not successful, the addition of hydrophilic bulk-forming agents such as psyllium may help. Warm compresses, sitz baths, analgesic ointments and suppositories, astringents (e.g., witch hazel), and bed rest reduce engorgement (Sneider & Maykel, 2010).

There are several types of nonsurgical treatments for hemorrhoids. Infrared photocoagulation, bipolar diathermy, and laser therapy are used to affix the mucosa to the underlying muscle. Injection of sclerosing agents is also effective for small, bleeding hemorrhoids. Sclerotherapy involves injecting a sclerosing agent (5% phenol in saline) into the base of the hemorrhoid to cause blood vessel thrombosis (Sneider & Maykel, 2010). These procedures help prevent prolapse.

A conservative surgical treatment of internal hemorrhoids is the rubber band ligation procedure. The hemorrhoid is visualized through the anoscope, and its proximal portion above the mucocutaneous lines is grasped with an instrument. A small rubber band is then slipped over the hemorrhoid. Tissue distal to the rubber band becomes necrotic after several days and sloughs off. Fibrosis occurs; the result is that the lower anal mucosa is drawn up and adheres to the underlying muscle. Although this treatment has been satisfactory for some patients, it has proven painful for others and may cause secondary hemorrhage. It has also been known to cause perianal infection (Sneider & Maykel, 2010).

Cryosurgical hemorrhoidectomy, another method for removing hemorrhoids, involves freezing the hemorrhoid for a sufficient time to cause necrosis. Although it is relatively painless, this procedure is not widely used because the discharge is foul smelling and wound healing is prolonged. The Nd:YAG laser is useful in excising hemorrhoids, particularly external hemorrhoidal tags. The treatment is quick and relatively painless. Hemorrhage and abscess are rare postoperative complications. The previously described methods for treating hemorrhoids are not effective for advanced thrombosed veins—that is, third to fourth degree prolapsed internal hemorrhoids, which must be treated by more extensive surgery (Lin, Liu, & Chen, 2010). Stapled hemorrhoidopexy, a newer procedure, uses surgical staples to treat prolapsing hemorrhoids and is associated with less postoperative pain and fewer complications. If it is not successful, hemorrhoidectomy, or surgical excision, may be performed to remove all of the redundant tissue involved in the process. During surgery, the rectal sphincter is usually dilated digitally, and the hemorrhoids are removed with a clamp and cautery or are ligated and then excised. After the surgical procedures are completed, a small tube may be inserted through the sphincter to permit the escape of flatus and blood; pieces of absorbable gelatin sponge (Gelfoam) or oxidized cellulose (Oxycel) gauze may be placed over the anal wounds.

Sexually Transmitted Anorectal Diseases

Three infectious syndromes that are related to STIs have been identified: proctitis, proctocolitis, and enteritis. Proctitis involves the rectum. It is commonly associated with recent anal-receptive intercourse with an infected partner. Symptoms include mucopurulent discharge or bleeding, rectal pain, and diarrhea. The pathogens most frequently involved are *Neisseria gonorrhoeae*, *Chlamydia trachomatis*, herpes simplex virus, and *Treponema pallidum*. Proctocolitis involves the rectum and lowest portion of the descending colon. Symptoms are similar to proctitis but may also include watery or bloody diarrhea, cramps, pain, and bloating. Enteritis involves more of the descending colon, and symptoms include watery, bloody diarrhea; abdominal pain; and weight loss. The most common pathogens causing enteritis are *E. histolytica*, *Giardia lamblia*, and *Shigella*, and *Campylobacter* species (Barry, Kent, Philip, et al., 2010; Lee & Wilkins, 2010; Raychaudhuri & Birley, 2010).

Sigmoidoscopy is performed to identify portions of the anorectum involved. Samples are taken with rectal swabs, and cultures are obtained to identify the pathogens involved. Antibiotics (i.e., ceftriaxone [Rocephin] or cefixime [Suprax], doxycycline [Vibramycin], and penicillin G) are the treatment of choice for bacterial infections (Centers for Disease Control and Prevention [CDC], 2012). Acyclovir [Zovirax] is given to patients with viral infections. Antiamebic therapy (i.e., metronidazole) is appropriate for infections with *E. histolytica* and *G. lamblia*. Ciprofloxacin (Cipro) is effective for *Shigella*. The antibiotics erythromycin (E-Mycin) and ciprofloxacin are the treatment of choice for *Campylobacter* infection (CDC, 2012).

Pilonidal Sinus or Cyst

A pilonidal sinus or cyst is found in the intergluteal cleft on the posterior surface of the lower sacrum (Fig. 48-14). Current theories suggest that it results from local trauma, causing penetration of hairs into the epithelium and subcutaneous tissue. It may also be formed congenitally by an infolding of epithelial tissue beneath the skin, which may communicate with the skin surface through one or several small sinus openings. Hair frequently is seen protruding from these openings, and this gives the cyst its name, *pilonidal* (i.e., a nest of hair) (Humphries & Duncan, 2010). The cysts rarely cause symptoms until adolescence or early adult life, when infection

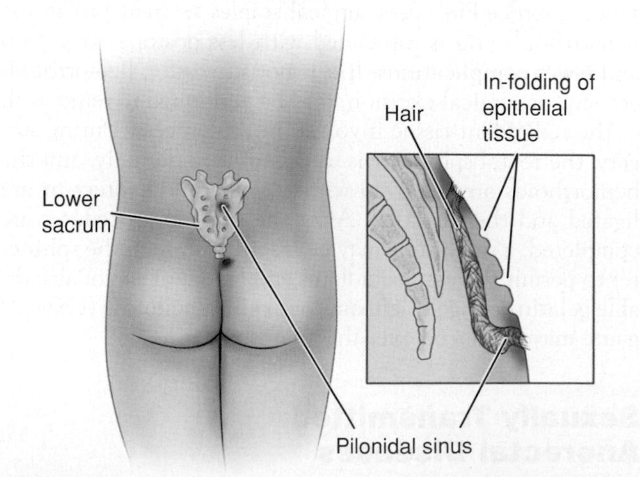
In-folding of epithelial tissue
Hair
Lower sacrum
Pilonidal sinus

FIGURE 48-14 • (*Left*) Pilonidal sinus on lower sacrum about 5 cm (2 in) above the anus in the intergluteal cleft. (*Right*) Hair particles emerge from the sinus tract, and localized indentations (pits) can appear on the skin near the sinus openings.

produces an irritating drainage or an abscess. Perspiration and friction easily irritate this area.

In the early stages of the inflammation, the infection may be controlled by antibiotic therapy, but after an abscess has formed, surgery is indicated. The abscess is incised and drained under local anesthesia. After the acute process resolves, further surgery is performed to excise the cyst and the secondary sinus tracts. The wound is allowed to heal by granulation. Absorptive dressings are placed in the wound to keep its edges separated while healing occurs (Al-Khamis, McCallum, Kino, et al., 2010).

Nursing Management of Patients With Anorectal Conditions

Promoting Home and Community-Based Care

Educating Patients About Self-Care

Most patients with anorectal conditions are not hospitalized. Those who undergo surgical procedures to correct the condition often are discharged directly from the outpatient surgical center. If they are hospitalized, it is for a short time, usually only 24 hours. Patient education is essential to facilitate recovery at home.

The nurse instructs the patient to keep the perianal area as clean as possible by gently cleansing with warm water and then drying with absorbent cotton wipes. The patient should avoid rubbing the area with toilet tissue. Instructions are provided about how to take a sitz bath and how to test the temperature of the water.

During the first 24 hours after rectal surgery, painful spasms of the sphincter and perineal muscles may occur. The nurse instructs the patient that ice and analgesic ointments may decrease the pain. Warm compresses may promote circulation and soothe irritated tissues. Sitz baths taken three or four times each day can relieve soreness and pain by relaxing sphincter spasm. Twenty-four hours after surgery, topical anesthetic agents may be beneficial in relieving local irritation and soreness. Medications may include topical anesthetics (i.e., suppositories), astringents, antiseptics, tranquilizers, and antiemetic agents. Patients are more adherent and less apprehensive if they are free of pain.

Wet dressings saturated with equal parts of cold water and witch hazel help relieve edema. When wet compresses are being used continuously, petrolatum is applied around the anal area to prevent skin maceration. The patient is instructed to assume a prone position at intervals because this position reduces edema of the tissue.

Continuing Care

Sitz baths may be given in the bathtub or plastic sitz bath unit three or four times each day. Sitz baths should follow each bowel movement for 1 to 2 weeks after surgery. The nurse encourages intake of at least 2 L of water daily to provide adequate hydration and recommends high-fiber foods to promote bulk in the stool and to make it easier to pass fecal matter through the rectum. Bulk laxatives such as psyllium may be recommended, and stool softeners (e.g., docusate) may be prescribed. The patient is advised to set aside a time for bowel movements and to heed the urge to defecate as promptly as possible. The nurse encourages the patient to respond quickly to the urge to defecate in order to prevent constipation. The diet is modified to increase fluids and fiber. Moderate exercise is encouraged, and the patient is taught about the prescribed diet, the significance of proper eating habits and exercise, and the laxatives that can be taken safely.

Critical Thinking Exercises

1 ebp A fellow college student in your dormitory is newly diagnosed with ulcerative colitis. Because you are a nursing student, she asks you to help her investigate this diagnosis. How would you look up this information? How would you find the "evidence" for best practices in care? What databases and search terms would you use? What kind of research evidence is stronger than others?

2 pq You work on a surgical unit and are assigned to care for a 68-year-old patient originally admitted for medical therapy for an attack of diverticulitis. The surgeon now identifies the need for emergency surgery. What patient signs and symptoms would the patient have manifested that swayed the surgeon's decision to operate? What are the priorities of nursing care as you prepare to send this patient to the operating room within 1 hour?

3 ebp You work in a long-term care facility, and a female resident asks how to prevent constipation. What will you tell her about nutrition, diet, drug therapy, and activities? What sources could you use to help you search for "best practice" answers?

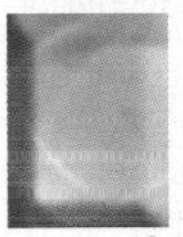

Brunner Suite Resources

Explore these additional resources to enhance learning for this chapter:

• NCLEX-Style Questions and Other Resources on thePoint, http://thePoint.lww.com/Brunner13e
• Study Guide
• PrepU
• Clinical Handbook
• Handbook of Laboratory and Diagnostic Tests

References

*Asterisk indicates nursing research.

Books

Huether, S. E., & McCance, K. L. (2012). *Understanding pathophysiology* (5th ed). St. Louis: Mosby.

Journals and Electronic Documents

Constipation

American Society for Gastrointestinal Endoscopy. (2012). *Understanding bowel preparation*. Available at: asge.org/patients/patients

Apau, D. (2010a). Assessing the cause of constipation and appropriate interventions. *Gastrointestinal Nursing*, 8(6), 24–27.

Belsey, J., Greenfield, S., Candy, D., et al. (2010). Systematic review: Impact of constipation on quality of life in adults and children. *Alimentary Pharmacology Therapeutics*, 31(9), 938–949.

Bouras, E. P., & Tangalos, E. G. (2009). Chronic constipation in the elderly. *Gastroenterology Clinics of North America*, 38(3), 463–480.

Emmanuel, A. (2011). Current management strategies and therapeutic targets in chronic constipation. *Therapeutic Advances in Gastroenterology*, 4(5), 37–48.

Kyle, G. (2010). Considering the options for treating constipation. *Practice Nursing*, 21(3), 124–130.

Lin, L. W., Fu, Y., Dunning, T., et al. (2009). Efficacy of the traditional Chinese medicines for the management of constipation: A systematic review. *Journal of Alternative and Complementary Medicine*, 15(12), 1335–1346.

Muller-Lissner, S., & Ward, A. (2011). Constipation in adults. *American Family Physician*, 83(8), 904–905.

National Digestive Diseases Information Clearinghouse. (2012a). *Digestive diseases statistics for the United States*. Available at: digestive.niddk.nih.gov/statistics/statistics.aspx

*Ostaszkiewicz, J., Hornby, L., Millar, L., et al. (2010). The effects of conservative treatment for constipation on symptom severity and quality of life in community-dwelling adults. *Journal of WOCN*, 37(2), 193–198.

Singh, S., & Rao, S. C. (2010). Pharmacologic management of chronic constipation. *Gastroenterology Clinics of North America*, 39, 509–527.

Spinzi, G., Amato, A., Imperiali, G., et al. (2009). Constipation in the elderly: Management strategies. *Drugs and Aging*, 26(6), 469–474.

Diarrhea

*Avadhani, A., & Miley, H. (2011). Probiotics for prevention of antibiotic-associated diarrhea and clostridium difficile-associated disease in hospitalized adults—a meta-analysis. *Journal of the American Academy of Nurse Practitioners*, 23(6), 269–274.

Bryant, R. (2012). Types of skin damage and differential diagnosis. (2012). In R. Bryant & D. Nix. *Acute and chronic wounds: Current management concepts* (4th ed., pp. 83–106). St. Louis: Elsevier-Mosby.

Hall, V. (2010). Acute uncomplicated diarrhea management. *Practice Nursing*, 21(3), 118–122.

Kent, A. J., & Banks, M. R. (2010). Pharmacological management of diarrhea. *Gastroenterology Clinics of North America*, 39(3), 495–507.

Simor, A. E. (2010). Diagnosis, management, and prevention of clostridium difficile infection in long-term care facilities: A review. *Journal of the American Geriatric Society*, 58(8), 1556–1564.

Surawicz, C. M. (2010). Mechanisms of diarrhea. *Current Gastroenterology Reports*, 12(4), 236–241.

Williams, N. T. (2010). Probiotics. *American Journal of Health System Pharmacists*, 67(6), 449–458.

Fecal Incontinence

Ahmad, M., McCallum, I., & Mercer-Jones, M. (2010). Management of fecal incontinence in adults. *British Medical Journal*, 340(7608), C2964–C3005.

Apau, D. (2010b). Fecal incontinence: Assessment and conservative management. *Gastrointestinal Nursing*, 8(8), 18–22.

Bliss, D. Z., & Norton, C. (2010). Conservative management of fecal incontinence. *American Journal of Nursing*, 110(9), 30–38.

Chien, D., & Bradway, C. (2010). Acquired fecal incontinence in community-dwelling adults. *Nurse Practitioner*, 35(1), 15–22.

*Hagglund, D. (2010). A systematic literature review of incontinence care for persons with dementia: The research evidence. *Journal of Clinical Nursing*, 19(3–4), 303–312.

Hurnauth, C. (2011). Management of fecal incontinence in acutely ill patients. *Nursing Standard*, 25(22), 48–56.

Kyle, G. (2010). Bowel and bladder care at the end of life. *British Journal of Nursing*, 19(7), 408–414.

Landerfeld, C. S., Bowen, B. J., Feld, A. D., et al. (2008). National Institutes of Health state-of-the-science conference statement: Prevention of fecal and urinary incontinence in adults. *Annals of Internal Medicine*, 148(6), 449–460.

Malmstrom, T. K., Anderson, E. M., Wolinski, F. D., et al. (2010). Urinary and fecal incontinence and quality of life in African-Americans. *Journal of the American Geriatric Society*, 58(10), 1941–1945.

Manthey, A., Bliss, D. Z., Savik, K., et al. (2010). Goals of fecal incontinence management identified by community-living incontinent adults. *Western Journal of Nursing Research*, 32(5), 644–661.

Nazarko, L. (2011). Fecal incontinence: Diagnosis, treatment, and management. *Nursing & Residential Care*, 13(6), 270–275.

Nix, D., & Haugen, V. (2010). Prevention and management of incontinence-associated dermatitis. *Drugs and Aging*, 27(6), 491–496.

*Schnelle, J. F., Simmons, S. F., Beuscher, L., et al. (2009). Prevalence of constipation symptoms in fecally incontinent nursing home residents. *Journal of the American Geriatric Society*, 57(4), 647–652.

Sharpe, A., & Read, A. (2010). Sacral nerve stimulation for the management of fecal incontinence. *British Journal of Nursing*, 19(7), 415–419.

Irritable Bowel Syndrome

Amitiza (lubiprostone). Available at: www.amitizahcp.com

Banerjee, S. (2010). Irritable bowel syndrome. *Journal of Pain and Palliative Care Pharmacotherapy, 24*(3), 271–274.

Blesse, L. C. (2010). Irritable bowel syndrome: Relieving the symptoms, and the frustration. *Journal of the American Academy of Physician Assistants, 23*(11), 46–52.

Lahmann, C., Rohricht, F., Sauer, N., et al. (2010). Functional relaxation as complementary therapy in irritable bowel syndrome: A randomized controlled clinical trial. *Journal of Alternative and Complementary Medicine, 16*(1), 47–52.

Reed, L. (2010). Irritable bowel syndrome. *Practice Nurse, 40*(1), 15–20.

*Smith, G. D. (2010). Health related quality of life and symptom classification in patients with irritable bowel syndrome. *Journal of Nursing and Healthcare in Chronic Illness, 2*(11), 4–12.

Malabsorption

Bolin, T. (2010). Malabsorption may contribute to malnutrition in the elderly. *Nutrition, 26*(7–8), 852–853.

Food allergies and food intolerances. Both are on the rise—and it's important to know the difference. (2011). *Harvard Women's Health Watch, 18*(9), 4–6.

Klapparoth, J. M. (2012). Malabsorption. *Medscape,* Available at: www.emedicine.medscape.com/article/180785-overview

Lewis, S. (2011). Nutrient deficiency and supplementation in chronic pancreatitis. *Topics in Clinical Nutrition, 26*(2), 126–137.

Appendicitis

Black, C. E., & Martin, R. F. (2012). Acute appendicitis in adults: Clinical manifestations and diagnosis. *UpToDate.* Available at: www.uptodate.com/contents/acute-appendicitis-in-adults-clinical-manifestations

Hennelly, K. E., & Bachur, R. (2011). Appendicitis update. *Current Opinion in Pediatrics, 23*(3), 281–285.

Lien, W. C., Lee, W. C., Wang, H. P., et al. (2011). Male gender is a risk factor for recurrent appendicitis following nonoperative treatment. *World Journal of Surgery, 35*(7), 1636–1642.

Nakhamiyayev, V., Galldin, L., Chiarello, M., et al. (2010). Laparoscopic appendectomy is the preferred approach for appendicitis: A retrospective review of two practice patterns. *Surgical Endoscopy, 24*(4), 859–864.

Spirt, M. (2010). Complicated abdominal infections: A focus on appendicitis and diverticulitis. *Postgraduate Medicine, 122*(1), 39–51.

Vissers, R. J., & Lennarz, W. B. (2010). Pitfalls in appendicitis. *Emergency Medicine Clinics of North America, 28*(1), 103–118.

Diverticular Disease

Beitz, J. (2008). Ruptured diverticulum: Issues in management. *OR Nurse, 2*(1), 32–40.

Bianco, M., Loyson, M., McNinch, D., et al. (2010). Obesity and acute diverticular disease: Is there an association? (Research abstract). *Annals of Emergency Medicine, 56*(3), S31.

Carloni, A., Sage, E., Roodie, J., et al. (2010). Right colonic diverticulitis: An uncommon disease in Western countries. *Acta Chirurgia Belgica, 110,* 57–89.

Kodali, B., & Oberoi, J. (2011). Management of postoperative pain. *UpToDate.* Available at: www.uptodate.com

Manwaring, M., & Champagne, B. (2011). Diverticular disease: Genetic, geographic, and environmental aspects. *Seminars in Colon & Rectal Surgery, 22*(3), 148–153.

Martel, G., Bouchard, A., Soto, C. M., et al. (2010). Laparoscopic colectomy for complex diverticular disease: A justifiable choice? *Surgical Endoscopy, 24*(9), 2273–2280.

Shahedi, K., & Katz, J. (2013). Diverticulitis. *Medscape.* Available at: www.emedicine.medscape.com/article/173388-overview

Stocchi, L. (2010). Current indications and role of surgery in the management of sigmoid diverticulitis. *World Journal of Gastroenterology, 16*(7), 804–817.

Tursi, A. (2010). Diverticular disease: What is the best long-term treatment? *Nature Reviews. Gastroenterology & Hepatology, 7*(2), 77–78.

Young-Fadok, T., & Pemberton, J. H. (2011). Epidemiology and pathophysiology of colonic diverticular disease. *UpToDate.* Available at: www.uptodate.com

Peritonitis

Craig, M., & Infante, S. (2011). Clinical consult: Abdominal mysteries: Pain, peritonitis, pancreatitis, pseudocyst. *Nephrology Nursing Journal, 38*(2), 173–186.

Neumann, J. L., & Moran, J. (2010). Peritonitis due to a peritoneal vaginal fistula. *Nephrology Nursing Journal, 37*(2), 177–181.

Taheri, M. R., Singh, H., & Duerksen, D. R. (2011). Peritonitis after gastrostomy tube replacement: Case series and review of literature. *Journal of Parenteral and Enteral Nutrition, 35*(1), 56–60.

Inflammatory Bowel Disease (Crohn's/Ulcerative Colitis)

Afdhal, N. H. (2011). Epidemiology and risk factors for gallstones. *UpToDate.* Available at: www.uptodate.com

Blonski, W., Buchner, A. M., & Lichtenstein, G. R. (2011). Inflammatory bowel disease. *Current Opinion in Gastroenterology, 27*(4), 346–357.

Cronin, E. (2010). Prednisolone in the management of patients with Crohn's disease. *British Journal of Nursing, 19*(21), 1333–1336.

Dahan, A., Amidon, G. L., & Zimmerman, E. M. (2010). Drug targeting strategies for the treatment of inflammatory bowel disease: A mechanistic update. *Expert Review in Clinical Immunology, 6*(4), 543–550.

DeLegge, M. H. (2011). Nutrition and dietary interventions in adults with inflammatory bowel disease. *UpToDate.* Available at: www.uptodate.com

Devlin, S. M., & Panaccione, R. (2010). Evolving inflammatory bowel disease treatment paradigms: Top-down versus step-up. *Medical Clinics of North America, 94*(1), 1–18.

Etchevers, M. J., Ordas, I., & Ricart, E. (2010). Optimizing the use of tumour necrosis factor inhibitors in Crohn's disease. *Drugs 2010, 70*(2), 109–120.

Farrell, D., & Savage, E. (2010). Symptom burden in inflammatory bowel disease. Rethinking conceptual and theoretical underpinnings. *International Journal of Nursing Practice, 16*(5), 437–442.

Grimpen, F., & Pavli, P. (2010). Advances in the management of inflammatory bowel disease. *Internal Medicine Journal, 40*(4), 258–264.

Hanauer, S., & Norton, B. (2011). PCE updates on Crohn's disease: Surgical indications and postoperative care. *PCE Updates, 1*(3), 1–14. Available at: practicingclinicians.com/crohns3/

Laharie, D., Reffet, A., Belleannée, G., et al. (2011). Mucosal healing with methotrexate in Crohn's disease: A prospective comparative study with azathioprine and infliximab. *Alimentary Pharmacology and Therapeutics, 33*(6), 714–721.

National Digestive Diseases Information Clearinghouse. (2012b). *Ulcerative colitis.* Available at: digestive.niddk.nih.gov/ddiseases/pubs/colitiis/index.aspx#what

Peppercorn, M. (2012c). Definition of and risk factors for inflammatory bowel disease. *UpToDate.* Available at: www.uptodate.com

Seres, D. (2012). Nutrition support in critically ill patients: An overview. *UpToDate.* Available at: www.uptodate.com

Stansfield, C. (2010). Risk assessment for new therapies in inflammatory bowel disease. *Nurse Prescribing, 8*(11), 551–555.

Umanskiy, K., & Fichera, A. (2010). Health-related quality of life in inflammatory bowel disease: The impact of surgical therapy. *World Journal of Gastroenterology, 16*(40), 5024–5034.

Types of Ostomies: Ileostomy, Colostomy

*Beitz, J., Gerlach, M., Ginsburg, P., et al. (2010). Content validation of a standardized algorithm for ostomy care. *Ostomy Wound Management, 56*(10), 22–38.

Boyles, A. (2010a). Patient outcomes and quality of life following stoma-forming surgery. *Gastrointestinal Nursing, 8*(8), 30–35.

Boyles, A. (2010b). Stomal and peristomal complications: Predisposing factors and management. *Gastrointestinal Nursing, 8*(7), 26–36.

Cronin, E. (2010). An overview of stoma bridges and a case study on their management. *British Journal of Nursing (Stoma Care Supplement), 19*(17), S16–S20.

Doughty, P., & Landmann, R. G. (2012). Management of patients with a colostomy or ileostomy. *UpToDate.* Available at: www.uptodate.com

*Grant, M., McMullen, C. K., Altschuler, A., et al. (2011). Gender differences in quality of life among long-term colorectal cancer survivors with ostomies. *Oncology Nursing Forum, 38*(5), 587–596.

Kalashnikova, I., Achkasov, S., Fadeeva, S., et al. (2011). The development and use of algorithms for diagnosing and choosing treatment of ostomy complications: Results of a prospective evaluation. *Ostomy Wound Management, 57*(1), 20–27.

Shariat, S. F., & Bochner, B. (2012). Urinary diversions and reconstruction following cystectomy. *UpToDate.* Available at: www.uptodate.com

Sheth, S. G., & Lamont, J. T. (2012). Toxic megacolon. *UpToDate.* Available at: www.uptodate.com

*Sung, Y. H., Kwon, I., Jo, S., et al. (2010). Factors affecting ostomy-related complications in Korea. *Journal of WOCN, 37*(2), 166–172.

Ileoanal Pouch

Beliard, A., & Prudhomme, M. (2010). Ileal reservoir with ileo-anal anastomosis: Long-term complications. *Journal of Visceral Surgery, 147,* E137–E144.

Buckman, S. A., & Heise, C.P. (2010). Nutrition considerations surrounding restorative proctocolectomy. *Nutrition in Clinical Practice, 25*(3), 250–256.

Psillos, A., & Catanzaro, J. (2011). Ileal pouch anal anastomosis: An overview of surgery, recovery, and achieving post surgical continence. *Ostomy Wound Management, 57*(12), 22–28.

Small Bowel Obstruction

Hodin, R. A., & Bordeianou, L. (2011). Small bowel obstruction. Causes and management. *UpToDate*. Available at: www.uptodate.com

Lauder, C. I., Garcea, G., Strickland, A., et al. (2010). Abdominal adhesion prevention: Still a sticky subject? *Digestive Surgery, 27*(2), 347–358.

Zielinski, M. D., & Bannon, M. P. (2011). Current management of small bowel obstruction. *Advances in Surgery, 45*(1), 1–29.

Large Bowel Obstruction

Dalal, K. M., Gollub, M. J., Miner, T. J., et al. (2011). Management of patients with malignant bowel obstruction and stage IV colorectal cancer. *Journal of Palliative Medicine, 14*(7), 822–828.

Ronnekleiv-Kelly, S. M., & Kennedy, G. D. (2011). Management of stage IV rectal cancer: Palliative options. *World Journal of Gastroenterology, 17*(7), 835–847.

White, S. I., Abdool, I., Frenkiel, B., et al. (2011). Management of malignant left-sided large bowel obstruction: A comparison between colonic stents and surgery. *ANZ Journal of Surgery, 81*(4), 257–260.

Colorectal Cancer/Polyps

Aldoss, I., & Iqbal, S. (2011). Adjuvant treatment and predictors of response in colon cancer. *Seminars in Colon & Rectal Surgery, 22*(2), 131–136.

American Cancer Society. (2012). *Cancer facts & figures 2012*. Available at: www.cancer.org/acs/groups/content/@epidemiologysurveillance/documents/document/acspc-031941.pdf

Carcinoembryonic antigen. (2012). *MedicineNet*. Available at: www.medicinenet.com/carcinoembryonic-antigen/article.html

Carmichael, J. C., & Mills, S. (2011). Surgical management of familial adenomatous polyposis. *Seminars in Colon & Rectal Surgery, 22*(2), 108–111.

Kalady, M. F. (2011). Surgical management of hereditary nonpolyposis colorectal cancer. *Advances in Surgery, 45*, 265–274.

Masi, G., Fornaro, L., Caparello, C., et al. (2011). Liver metastases from colorectal cancer: How to best complement medical treatment with surgical approaches. *Future Oncology, 7*(11), 1299–1323.

National Cancer Institute. (2012). *Colon and rectal cancer*. Available at: www.cancer.gov/cancertopics/types/colon-and-rectal

National Institute of Diabetes and Digestive and Kidney Diseases. (2011). Colonic polyps. Available at: www.niddk.nih.gov

Popek, S., & Tsikitis, V. L. (2011). Epidemiology of inherited colon cancer. *Seminars in Colon & Rectal Surgery, 22*(2), 77–81.

Rodriguez-Bigas, M. A. (2011). Surgical management of primary colon cancer. *UpToDate*. Available at: www.uptodate.com

Van Loon, K., & Venook, A. P. (2011). Adjuvant treatment of colon cancer: What is next? *Current Opinion in Oncology, 23*(4), 403–409.

Anorectal Abscess

Rizzo, J. A., Naig, A. L., & Johnson, E. K. (2010). Anorectal abscess and fistula-in-ano: Evidence-based management. *Surgical Clinics of North America, 90*(1), 45–68.

Yano, T., Asano, M., Matsuda, Y., et al. (2010). Prognostic factors for recurrence following the initial drainage of an anorectal abscess. *International Journal of Colorectal Diseases, 25*(12), 1495–1498.

Anal Fistula/Fissure

Etzioni, D. A. (2011). Current management of anal fissure. *Seminars in Colon & Rectal Surgery, 22*(1), 2–8.

Lewis, R. T., & Maron, D. J. (2010). Anorectal Crohn's disease. *Surgical Clinics of North America, 90*(1), 83–97.

Madalinski, M. H. (2011). Identifying the best therapy for chronic anal fissure. *World Journal of Gastrointestinal Pharmacology and Therapeutics, 2*(2), 9–16.

Michalopoulos, A., Papadopoulos, V., Tziris, N., et al. (2010). Perianal fistulas. *Techniques in Coloproctology, 14*(Suppl. 1), S15–S17.

Hemorrhoids

National Digestive Diseases Information Clearinghouse. (2012c). *Hemorrhoids*. Available at: digestive.niddk.nih.gov/ddiseases/pubs/hemorrhoids/index.aspx

*Lin, Y. H., Liu, K. W., & Chen, H. P. (2010). Hemorrhoidectomy: Prevalence and risk factors of urine retention among post recipients. *Journal of Clinical Nursing, 19*(19–20), 2771–2776.

Pramateftakis, M. G. (2010). The role of hemorrhoidopexy in the management of 3rd degree hemorrhoids. *Techniques in Coloproctology, 14*(Suppl. 1), S5–S7).

Sneider, E. B., & Maykel, J. A. (2010). Diagnosis and management of symptomatic hemorrhoids. *Surgical Clinics of North America, 90*(1), 17–32.

Sexually Transmitted Infections of the Anorectum

Barry, P. M., Kent, C. K., Philip, S. S., et al. (2010). Results of a program to test women for rectal chlamydia and gonorrhea. *Obstetrics & Gynecology, 115*(4), 753–759.

Centers for Disease Control and Prevention. (2012). *Campylobacter*. Available at: www.cdc.gov/nczved/divisions/dfbmd/diseases/campylobacter/#treat

Lee, P. K., & Wilkins, K. B. (2010). Condyloma and other infections including human immunodeficiency virus. *Surgical Clinics of North America, 90*(1), 99–112.

Raychaudhuri, M., & Birley, H. D. L. (2010). Audit of routine rectal swabs for gonorrhoea culture in women. *International Journal of STDs & AIDS, 21*(2), 143–144.

Pilonidal Disease

Al-Khamis, A., McCallum, I., Kino, P. M., et al. (2010). Healing by primary vs. secondary intention after surgical treatment for pilonidal sinus. *Cochrane Database of Systematic Reviews*, (1), CD006213. DOI: 10.1002/14651858.CD006213.pub3.

Humphries, A. E., & Duncan, J. E. (2010). Evaluation and management of pilonidal disease. *Surgical Clinics of North America, 90*(1), 113–124.

Resources

American Cancer Society, www.cancer.org

American College of Surgeons, Ostomy Home Skills Program, www.facs.org/patienteducation/skills/ostomy.html

American Society of Colon and Rectal Surgeons (ASCRS), www.fascrs.org

Centers for Disease Control and Prevention, www.cdc.gov

Colon Cancer Alliance, www.ccalliance.org

Crohn's and Colitis Foundation of America (CCFA), www.ccfa.org

Healthy New Jersey Information for Healthy Living, University of Medicine and Dentistry of New Jersey, Constipation, www.healthynj.org/diseases/constipation.html

International Foundation for Functional Gastrointestinal Disorders (IFFGD), www.iffgd.org

J-Pouch Group (source for J-Pouch surgery support), www.j-pouch.org

Medicine Online, Colon Cancer Information Library, www.medicineonline.com

Merck Manual, Colorectal Cancer, merckmanuals.com/professional/gastrointestinal_disorders/tumors_of_the_gi_tract/colorectal_cancer.html

National Association for Continence, www.nafc.org

National Cancer Institute, National Institutes of Health, www.cancer.gov

National Digestive Diseases Information Clearinghouse (NDDIC), digestive.niddk.nih.gov

United Ostomy Associations of America (UOAA), www.ostomy.org

Wound Ostomy and Continence Nurses Society, www.wocn.org

Unit 11

Metabolic and Endocrine Function

Case Study

A PATIENT WITH A NEW DIAGNOSIS OF DIABETES

Mr. Johansen is a 78-year-old retired farmer who is newly diagnosed with type 2 diabetes. He is also 50 pounds (23 kilograms) over his ideal body weight. He was recently prescribed an oral antidiabetic agent that he takes twice daily along with lovastatin for elevated cholesterol levels. Mr. Johansen lives in a rural area with harsh, snowy winters. His home is more than 35 miles from his primary provider's office. Through a state-funded grant, Mr. Johansen is invited to enroll in a tele-health program. Mr. Johansen will participate in diabetes education webinars aimed at reducing weight, maintaining a healthy glycemic index, and improving his nutritional status. In addition, he will be able to provide periodic data to his primary provider on his blood glucose, vital signs, and weight.

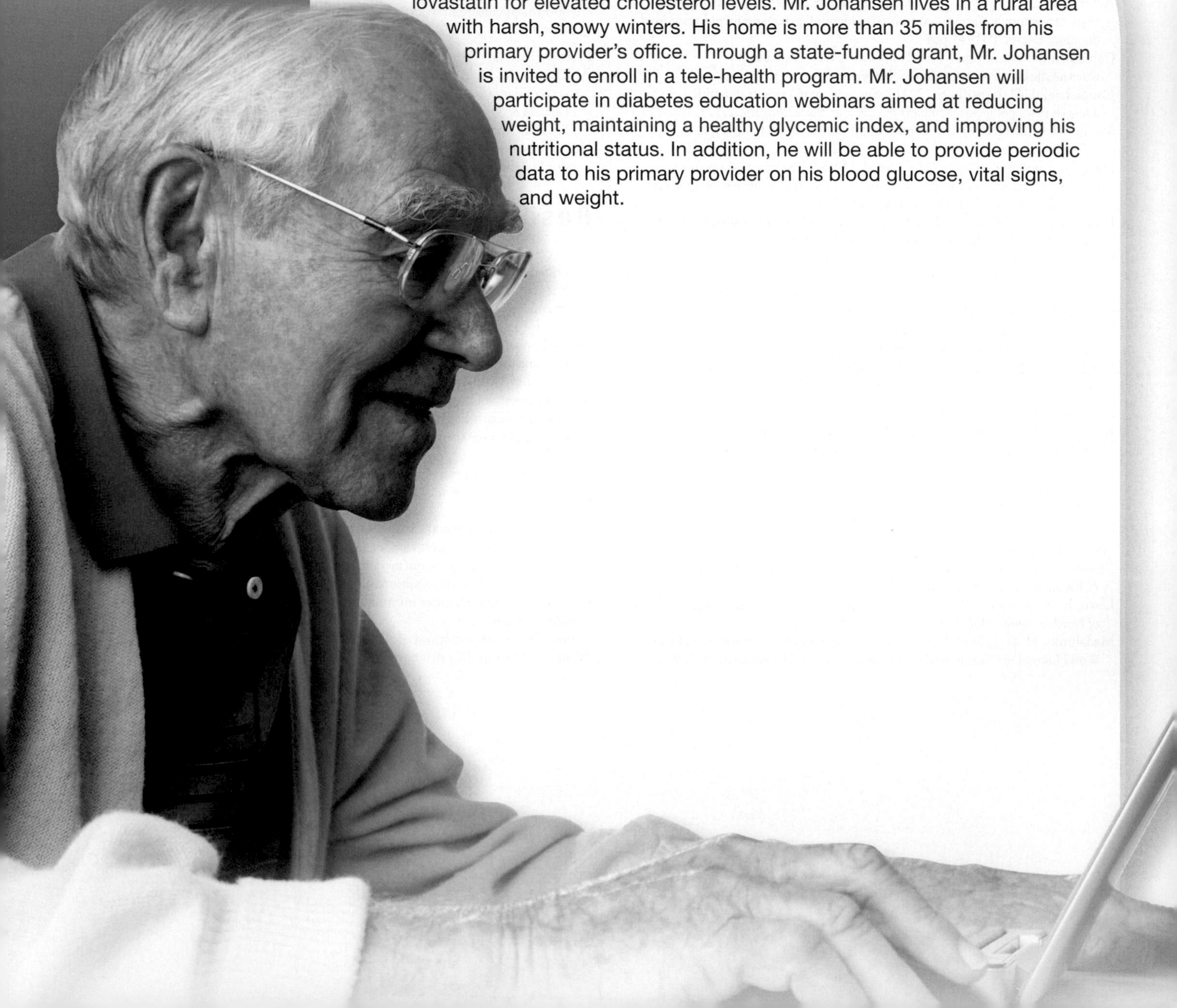

QSEN Competency Focus: *Informatics*

The complexities inherent in today's health care system challenge nurses to demonstrate integration of specific interdisciplinary core competencies. These competencies are aimed at ensuring the delivery of safe, quality patient care (Institute of Medicine, 2003). The concepts from the Quality and Safety Education for Nurses (QSEN) Institute (2012) provide a framework for the knowledge, skills, and attitudes (KSAs) required for nurses to demonstrate competency in these key areas, which include *patient-centered care, interdisciplinary teamwork and collaboration, evidence-based practice, quality improvement, safety,* and *informatics.*

Informatics Definition: Use information and technology to communicate, manage knowledge, mitigate error, and support decision making.

RELEVANT PRE-LICENSURE KSAs	APPLICATION AND REFLECTION
Knowledge	
Explain why information and technology skills are essential for safe patient care.	Describe the differences in Mr. Johansen's ability to self-manage his therapeutic regimen with and without the tele-health system. Given the distance he must travel to reach his primary provider, and the additional threats of bad weather during the winter months, how could his rural status hamper his ability to best manage his therapeutic regimen? How can tele-health improve the quality of care that he receives?
Skills	
Apply technology and information management tools to support safe processes of care.	Identify the skill set required by both Mr. Johansen and his health care providers if the tele-health system is to be effective in improving Mr. Johansen's health.
Attitudes	
Appreciate the necessity for all health care professionals to seek lifelong, continuous learning of information technology skills.	Reflect on your attitudes toward the use of technology by patients such as Mr. Johansen. Do you feel threatened by monitoring patients from a distance and prefer instead to have face-to-face encounters? If so, how might this attitude create barriers to effective patient care?

Cronenwett, L., Sherwood, G., Barnsteiner, J., et al. (2007). Quality and safety education for nurses. *Nursing Outlook, 55*(3), 122–131.
Institute of Medicine. (2003). *Health professions education: A bridge to quality.* Washington, DC: National Academies Press.
QSEN Institute. (2012). *Competencies: Prelicensure KSAs.* Available at: qsen.org/competencies/pre-licensure-ksas

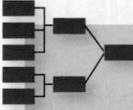

Read More About This Case

More information about this case study and the relationships between nursing diagnoses, interventions, and expected outcomes is available online. Visit thePoint for Applying Concepts From NANDA-I, NIC, and NOC.

Chapter

49

Assessment and Management of Patients With Hepatic Disorders

Learning Objectives

On completion of this chapter, the learner will be able to:

1. Identify the metabolic functions of the liver and the alterations that occur with hepatic disorders.
2. Explain liver function tests and the clinical manifestations of liver dysfunction in relation to pathophysiologic alterations of the liver.
3. Relate jaundice, portal hypertension, ascites, varices, nutritional deficiencies, and hepatic coma to pathophysiologic alterations of the liver.
4. Describe the medical, surgical, and nursing management of patients with esophageal varices.
5. Compare the various types of hepatitis and their causes, prevention, clinical manifestations, management, prognosis, and home health care needs.
6. Use the nursing process as a framework for care of the patient with cirrhosis of the liver.
7. Compare the nonsurgical and surgical management of patients with cancer of the liver.
8. Describe the postoperative nursing care of the patient undergoing liver transplantation.

Glossary

ascites: an albumin-rich fluid accumulation in the peritoneal cavity

asterixis: involuntary flapping movements of the hands

cirrhosis: a chronic liver disease characterized by fibrotic changes, the formation of dense connective tissue within the liver, subsequent degenerative changes, and loss of functioning cells

constructional apraxia: inability to draw figures in two or three dimensions

endoscopic variceal ligation (EVL): procedure that uses a modified endoscope loaded with an elastic rubber band passed through an overtube directly onto the varix (or varices) to be banded to ligate the area and stop bleeding

fetor hepaticus: sweet, slightly fecal odor to the breath, presumed to be of intestinal origin; prevalent with the extensive collateral portal circulation in chronic liver disease

fulminant hepatic failure: sudden, severe onset of acute liver failure that occurs within 8 weeks after the first symptoms of jaundice

hepatic encephalopathy: central nervous system dysfunction frequently associated with elevated ammonia levels that produce changes in mental status, altered level of consciousness, and coma

jaundice: condition where the body tissues, including the sclerae and the skin, become tinged yellow or greenish-yellow, due to high bilirubin levels

orthotopic liver transplantation (OLT): grafting of a donor liver into the normal anatomic location, with removal of the diseased native liver

portal hypertension: elevated pressure in the portal circulation resulting from obstruction of venous flow into and through the liver

sclerotherapy: the injection of substances into or around esophagogastric varices to cause constriction, thickening, and hardening of the vessel and stop bleeding

xenograft: transplantation of organs from one species to another

Liver function is complex, and hepatic dysfunction affects all body systems. For this reason, the nurse must understand how the liver functions and must have expert clinical assessment and management skills to care for patients undergoing complex diagnostic and treatment procedures. The nurse also must have an understanding of technologic advances in the management of hepatic disorders. Liver disorders are common and may result from a virus, obesity, and insulin resistance, or exposure to toxic substances, such as alcohol, or tumors.

ASSESSMENT OF THE LIVER

Anatomic and Physiologic Overview

The liver—the largest gland of the body—can be considered a chemical factory that manufactures, stores, alters, and excretes a large number of substances involved in metabolism (Hall, 2011). The location of the liver is essential because it receives nutrient-rich blood directly from the

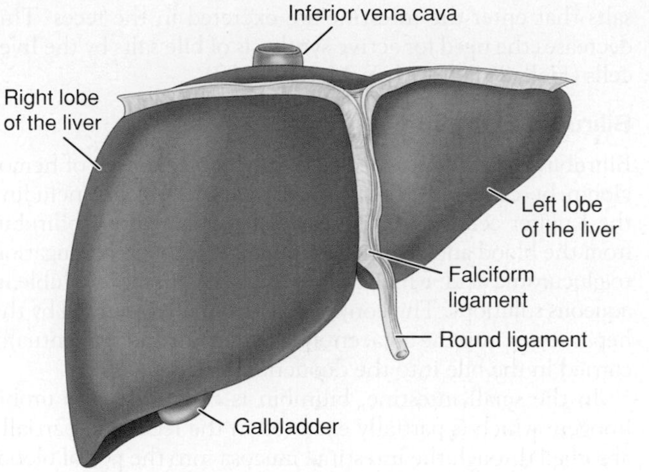
Inferior vena cava
Right lobe of the liver
Left lobe of the liver
Falciform ligament
Round ligament
Galbladder

FIGURE 49-1 • The liver and biliary system.

gastrointestinal (GI) tract and then either stores or transforms these nutrients into chemicals that are used elsewhere in the body for metabolic needs. The liver is especially important in the regulation of glucose and protein metabolism. The liver manufactures and secretes bile, which has a major role in the digestion and absorption of fats in the GI tract. The liver removes waste products from the bloodstream and secretes them into the bile. The bile produced by the liver is stored temporarily in the gallbladder until it is needed for digestion, at which time the gallbladder empties and bile enters the intestine (Fig. 49-1).

Anatomy of the Liver

The liver is a large, highly vascular organ located behind the ribs in the upper right portion of the abdominal cavity. It weighs between 1,200 and 1,500 g in the average adult and is divided into four lobes. A thin layer of connective tissue surrounds each lobe, extending into the lobe itself and dividing the liver mass into small, functional units called *lobules* (Hall, 2011).

The circulation of the blood into and out of the liver is of major importance to liver function. The blood that perfuses the liver comes from two sources. Approximately 80% of the blood supply comes from the portal vein, which drains the GI tract and is rich in nutrients but lacks oxygen. The remainder of the blood supply enters by way of the hepatic artery and is rich in oxygen. Terminal branches of these two blood vessels join to form common capillary beds, which constitute the sinusoids of the liver (Fig. 49-2). Thus, a mixture of venous and arterial blood bathes the liver cells (hepatocytes). The sinusoids empty into venules that occupy the center of each liver lobule and are called the *central veins*. The central veins join to form the hepatic vein, which constitutes the venous drainage from the liver and empties into the inferior vena cava, close to the diaphragm (Hall, 2011).

In addition to hepatocytes, phagocytic cells belonging to the reticuloendothelial system are present in the liver. Other organs that contain reticuloendothelial cells are the spleen, bone marrow, lymph nodes, and lungs. In the liver, these cells are called *Kupffer cells* (Hall, 2011). As the most common phagocyte in the human body, their main function is to

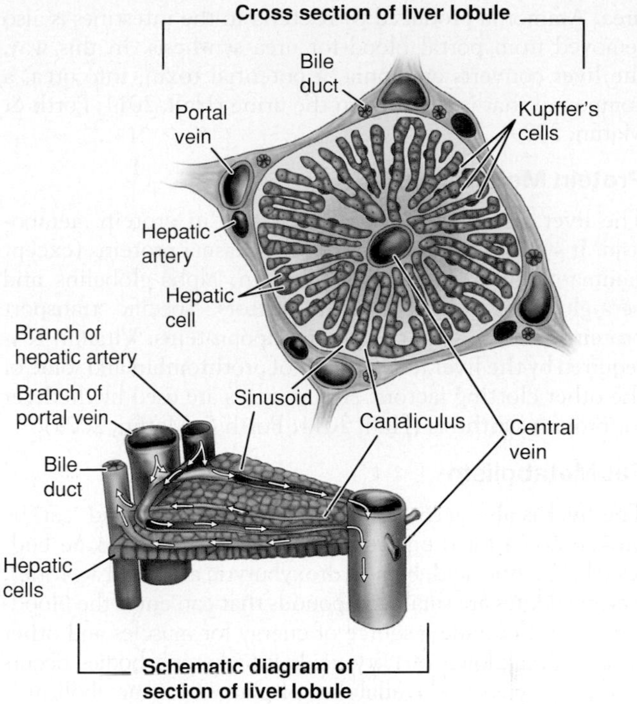
Cross section of liver lobule
Bile duct
Portal vein
Kupffer's cells
Hepatic artery
Hepatic cell
Branch of hepatic artery
Branch of portal vein
Sinusoid
Canaliculus
Central vein
Bile duct
Hepatic cells
Schematic diagram of section of liver lobule

FIGURE 49-2 • A section of liver lobule showing the location of hepatic veins, hepatic cells, liver sinusoids, and branches of the portal vein and hepatic artery.

engulf particulate matter (e.g., bacteria) that enters the liver through the portal blood.

The smallest bile ducts, called *canaliculi*, are located between the lobules of the liver. The canaliculi receive secretions from the hepatocytes and carry them to larger bile ducts, which eventually form the hepatic duct. The hepatic duct from the liver and the cystic duct from the gallbladder join to form the common bile duct, which empties into the small intestine. The sphincter of Oddi, located at the junction where the common bile duct enters the duodenum, controls the flow of bile into the intestine.

Functions of the Liver

Glucose Metabolism

The liver plays a major role in the metabolism of glucose and the regulation of blood glucose concentration. After a meal, glucose is taken up from the portal venous blood by the liver and converted into glycogen, which is stored in the hepatocytes. Subsequently, the glycogen is converted back to glucose (glycogenolysis) and released as needed into the bloodstream to maintain normal levels of blood glucose. However, this process provides a limited amount of glucose. Additional glucose can be synthesized by the liver through a process called *gluconeogenesis*. For this process, the liver uses amino acids from protein breakdown or lactate produced by exercising muscles. This process occurs in response to hypoglycemia (Hall, 2011).

Ammonia Conversion

The use of amino acids from protein for gluconeogenesis results in the formation of ammonia as a by-product. The liver converts this metabolically generated ammonia into urea. Ammonia produced by bacteria in the intestines is also removed from portal blood for urea synthesis. In this way, the liver converts ammonia, a potential toxin, into urea, a compound that is excreted in the urine (Hall, 2011; Porth & Matfin, 2009).

Protein Metabolism

The liver also plays an important role in protein metabolism. It synthesizes almost all of the plasma proteins (except gamma-globulin), including albumin, alpha-globulins and beta-globulins, blood clotting factors, specific transport proteins, and most of the plasma lipoproteins. Vitamin K is required by the liver for synthesis of prothrombin and some of the other clotting factors. Amino acids are used by the liver for protein synthesis (Hall, 2011; Porth & Matfin, 2009).

Fat Metabolism

The liver is also active in fat metabolism. Fatty acids can be broken down for the production of energy and ketone bodies (acetoacetic acid, beta-hydroxybutyric acid, and acetone). Ketone bodies are small compounds that can enter the bloodstream and provide a source of energy for muscles and other tissues. Breakdown of fatty acids into ketone bodies occurs primarily when the availability of glucose for metabolism is limited, as in starvation or in uncontrolled diabetes. Fatty acids and their metabolic products are also used for the synthesis of cholesterol, lecithin, lipoproteins, and other complex lipids (Hall, 2011; Porth & Matfin, 2009).

Vitamin and Iron Storage

Vitamins A, B, and D and several of the B-complex vitamins are stored in large amounts in the liver. Certain substances, such as iron and copper, are also stored in the liver. Because the liver is rich in these substances, liver extracts have been used for therapy for more than a century for a wide range of nutritional disorders; however, caution must be employed regarding the use of any animal organ extract because of possible risk of exposure to pathogenic organisms.

Bile Formation

Bile is continuously formed by the hepatocytes and collected in the canaliculi and bile ducts. It is composed mainly of water and electrolytes such as sodium, potassium, calcium, chloride, and bicarbonate, and it also contains significant amounts of lecithin, fatty acids, cholesterol, bilirubin, and bile salts. Bile is collected and stored in the gallbladder and is emptied into the intestine when needed for digestion. The functions of bile are excretory, as in the excretion of bilirubin; bile also serves as an aid to digestion through the emulsification of fats by bile salts.

Bile salts are synthesized by the hepatocytes from cholesterol. After conjugation or binding with amino acids (taurine and glycine), bile salts are excreted into the bile. The bile salts, together with cholesterol and lecithin, are required for emulsification of fats in the intestine, which is necessary for efficient digestion and absorption. Bile salts are then reabsorbed, primarily in the distal ileum, into portal blood for return to the liver and are again excreted into the bile. This pathway from hepatocytes to bile to intestine and back to the hepatocytes is called the *enterohepatic circulation*. Because of the enterohepatic circulation, only a small fraction of the bile salts that enter the intestine are excreted in the feces. This decreases the need for active synthesis of bile salts by the liver cells (Hall, 2011; Porth & Matfin, 2009).

Bilirubin Excretion

Bilirubin is a pigment derived from the breakdown of hemoglobin by cells of the reticuloendothelial system, including the Kupffer cells of the liver. Hepatocytes remove bilirubin from the blood and chemically modify it through conjugation to glucuronic acid, which makes the bilirubin more soluble in aqueous solutions. The conjugated bilirubin is secreted by the hepatocytes into the adjacent bile canaliculi and is eventually carried in the bile into the duodenum.

In the small intestine, bilirubin is converted into urobilinogen, which is partially excreted in the feces and partially absorbed through the intestinal mucosa into the portal blood. Much of this reabsorbed urobilinogen is removed by the hepatocytes and secreted into the bile once again (enterohepatic circulation). Some of the urobilinogen enters the systemic circulation and is excreted by the kidneys in the urine. Elimination of bilirubin in the bile represents the major route of its excretion.

Drug Metabolism

The liver metabolizes many medications, such as barbiturates, opioids, sedatives, anesthetics, and amphetamines (Karch, 2012). Metabolism generally results in drug inactivation, although activation may also occur. One of the important pathways for medication metabolism involves conjugation (binding) of the medication with a variety of compounds, such as glucuronic acid or acetic acid, to form more soluble substances. These substances may be excreted in the feces or urine, similar to bilirubin excretion. Bioavailability is the fraction of the administered medication that actually reaches the systemic circulation. The bioavailability of an oral medication (absorbed from the GI tract) can be decreased if the medication is metabolized to a great extent by the liver before it reaches the systemic circulation; this is known as first-pass effect. Some medications have such a large first-pass effect that their use is essentially limited to the parenteral route, or oral doses must be substantially larger than parenteral doses to achieve the same effect.

Gerontologic Considerations

Chart 49-1 summarizes age-related changes in the liver. In the older adult, the most common change in the liver is a decrease in size and weight, accompanied by a decrease in total hepatic blood flow. However, in general, these decreases are proportional to the decreases in body size and weight seen in normal aging. Results of liver function tests do not normally change with age; abnormal results in older patients indicate abnormal liver function and are not a result of the aging process itself.

Metabolism of medications by the liver decreases in the older adult, but such changes are usually accompanied by changes in intestinal absorption, renal excretion, and altered body distribution of some medications secondary to changes in fat deposition. These alterations necessitate careful medication administration and monitoring; if appropriate, reduced dosages may be needed to prevent medication toxicity.

Chart 49-1 Age-Related Changes of the Hepatobiliary System

- Steady decrease in size and weight of the liver, particularly in women
- Decrease in blood flow
- Decrease in replacement/repair of liver cells after injury
- Reduced drug metabolism
- Slow clearance of hepatitis B surface antigen
- More rapid progression of hepatitis C infection and lower response rate to therapy
- Decline in drug clearance capability
- Increased prevalence of gallstones due to the increase in cholesterol secretion in bile
- Decreased gallbladder contraction after a meal
- Atypical clinical presentation of biliary disease
- More severe complications of biliary tract disease

Adapted from Townsend, C. M., Beauchamp, R. D., Evers, B. M., et al. (2012). *Sabiston's textbook of surgery. The biological basis of modern surgical practice.* Philadelphia: Elsevier.

Assessment

Health History

If liver function test results are abnormal, the patient is evaluated for liver disease. In such cases, the health history focuses on previous exposure of the patient to hepatotoxic substances or infectious agents. The patient's occupational, recreational, and travel history may assist in identifying exposure to hepatotoxins (e.g., industrial chemicals, other toxins). The patient's history of alcohol and drug use, including but not limited to the use of intravenous (IV) or injection drugs, provides additional information about exposure to toxins and infectious agents. Many medications (including acetaminophen [Tylenol], ketoconazole [Nizoral], and valproic acid [Depakene]) are responsible for hepatic dysfunction and disease (Karch, 2012). A thorough medication history should address all current and past prescription medications, over-the-counter medications, herbal remedies, and dietary supplements.

Lifestyle behaviors that increase the risk of exposure to infectious agents are identified. IV or injection drug use, sexual practices, and foreign travel are all potential risk factors for liver disease. The amount and type of alcohol consumption are identified using screening tools (questionnaires) that have been developed for this purpose (see Chapter 5). The amount of alcohol required to produce chronic liver disease varies widely, but men who consume 60 to 80 g/day of alcohol (approximately four glasses of beer, wine, or mixed drinks) and women whose alcohol intake is 40 to 60 g/day are considered at high risk for cirrhosis. **Cirrhosis** is a chronic liver disorder characterized by fibrotic changes, the formation of dense connective tissue within the liver, subsequent degenerative changes, and loss of functional liver tissue (Porth & Matfin, 2009) (see later discussion).

The history also includes an evaluation of the patient's past medical history to identify risk factors for the development of liver disease. Current and past medical conditions, including those of a psychological or psychiatric nature, are identified. The family history includes questions about familial liver disorders that may have their origin in alcohol abuse or gallstone disease, as well as other familial or genetic disorders (Chart 49-2).

Chart 49-2 GENETICS IN NURSING PRACTICE
Hepatic Disorders

A number of hepatic disorders have an underlying genetic cause. Some examples of hepatic disorders caused by genetic abnormalities include:

- $Alpha_1$-antitrypsin deficiency
- Alagille syndrome
- Budd-Chiari syndrome
- Congenital hepatic fibrosis
- Hemochromatosis
- Gilbert's syndrome
- Wilson's disease

Nursing Assessments

Family History Assessment

- Collect family history for three generations of maternal and paternal relatives in the patient's family.
- Assess family history for relatives with early-onset hepatic disease

Patient Assessment

- Assess for physical signs of indigestion, reflux, hemorrhoids, gallstones, intolerance to fatty foods, intolerance to alcohol, nausea and vomiting attacks, abdominal bloating, and constipation.
- Assess for associated nervous system disorders such as depression and mood changes, especially anger and irritability.
- Assess for associated blood sugar problems such as hypoglycemia.

Management Issues Specific to Genetics

- Inquire whether any affected family members have had genetic testing.
- If indicated, refer for further genetic counseling and evaluation so that family members can discuss inheritance, risk to other family members, and availability of genetic testing and gene-based interventions.
- Offer appropriate genetic information and resources.
- Provide support to families with newly diagnosed hepatic disorders.
- Participate in the management and coordination of care for patients with genetics-related hepatic conditions and other genetic conditions as well as those patients who are predisposed to develop or pass on a genetics-related hepatic condition or other genetic condition.

Genetics Resources

Merck Manual for Health Care Professionals: provides up-to-date information about hepatic and biliary disorders, www.merckmanuals.com/professional/hepatic_and_biliary_disorders.html

See Chapter 8, Chart 8-6 for additional genetics resources.

The history also addresses symptoms that suggest liver disease. Symptoms that may have their origin in liver disease but are not specific to hepatic dysfunction include jaundice, malaise, weakness, fatigue, pruritus, abdominal pain, fever, anorexia, weight gain, edema, increasing abdominal girth, hematemesis, melena, hematochezia (passage of bloody stools), easy bruising, changes in mental acuity, personality changes, sleep disturbances, and decreased libido in men and secondary amenorrhea in women.

Physical Assessment

The nurse assesses the patient for physical signs that may occur with liver dysfunction, including the pallor often seen with chronic illness and jaundice. The skin, mucosa, and sclerae are inspected for jaundice, and the extremities are assessed for muscle atrophy, edema, and skin excoriation secondary to scratching. The nurse observes the skin for petechiae or ecchymotic areas (bruises), spider angiomas (Fig. 49-3), and palmar erythema. The male patient is assessed for unilateral or bilateral gynecomastia and testicular atrophy due to hormonal changes. The patient's cognitive status (recall, memory, abstract thinking) and neurologic status are assessed. The nurse observes for general tremor, asterixis, weakness, and slurred speech. These symptoms are discussed later.

Nonalcoholic fatty liver disease (NAFLD) and nonalcoholic steatohepatitis (NASH) are two diseases within the spectrum of fatty liver to fibrosis and cirrhosis that are strongly associated with obesity (McDonald, Burroughs, Feagan, et al., 2010; Schattenberg & Schuppan, 2011). Two preliminary studies further suggest that being overweight and drinking too much alcohol can cause severe harm to the liver (European Association for the Study of the Liver, 2013). One study found that overweight and obese women who were heavy drinkers had a significantly increased risk of developing and dying of chronic liver disease. The other study found an increased risk of liver cancer in people with alcoholic cirrhosis who also have fatty liver disease, type 2 diabetes, and are overweight or obese. In patients who are overweight, obese, or have high alcohol intake, the nurse observes for signs of associated liver dysfunction.

The nurse assesses for the presence of an abdominal fluid wave (discussed later). The abdomen is palpated to assess liver size and to detect any tenderness over the liver. The liver may be palpable in the right upper quadrant. A palpable liver presents as a firm, sharp ridge with a smooth surface (Fig. 49-4). The nurse estimates the size of the liver by percussing its upper and lower borders. If the liver is not palpable but tenderness is suspected, tapping the lower right thorax briskly may elicit tenderness. For comparison, the nurse then performs a similar maneuver on the left lower thorax (Bickley, 2009).

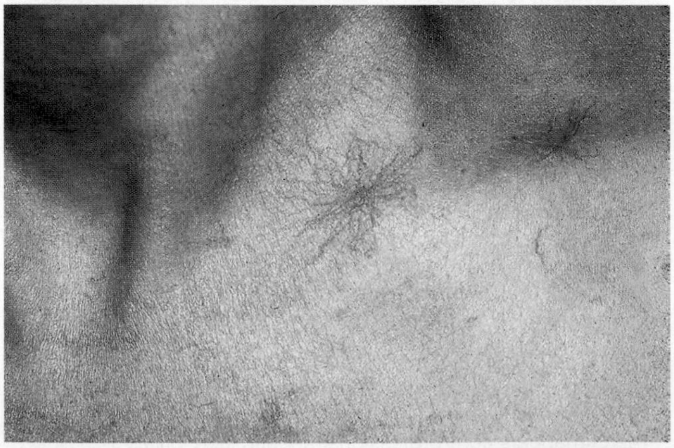

FIGURE 49-3 • Spider angioma. This vascular (arterial) spider appears on the skin. Beneath the elevated center and radiating branches, the blood vessels are looped and tortuous.

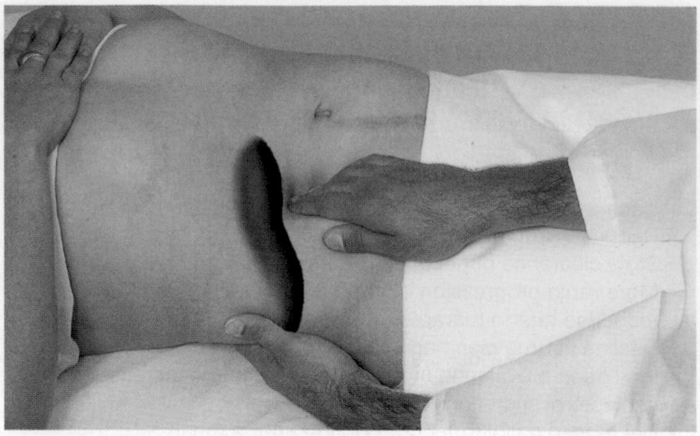

FIGURE 49-4 • Technique for palpating the liver. The examiner places one hand under the right lower rib cage and presses downward during inspiration with light pressure with the other hand. From Bickley, L. S. (2009). *Bates' guide to physical examination and history taking* (10th ed.). Philadelphia: Lippincott Williams & Wilkins.

If the liver is palpable, the examiner notes and records its size, its consistency, any tenderness, and whether its outline is regular or irregular. If the liver is enlarged, the degree to which it descends below the right costal margin is recorded to provide some indication of its size. The examiner determines whether the liver's edge is sharp and smooth or blunt and whether the enlarged liver is nodular or smooth. The liver of a patient with cirrhosis is small and hard in late-stage cirrhosis, whereas the liver of a patient with acute hepatitis is soft and the hand easily moves the edge.

Tenderness of the liver indicates recent acute enlargement with consequent stretching of the liver capsule. The absence of tenderness may imply that the enlargement is of long-standing duration. The liver of a patient with viral hepatitis is tender, whereas that of a patient with alcoholic hepatitis is not. Enlargement of the liver is an abnormal finding that requires evaluation (Bickley, 2009).

Diagnostic Evaluation

A wide range of diagnostic studies may be performed in patients with hepatic disorders. The nurse should educate the patient about the purpose, what to expect, and any possible side effects related to these examinations prior to testing. The nurse should note trends in results because they provide information about disease progression as well as the patient's response to therapy.

Liver Function Tests

More than 70% of the parenchyma of the liver may be damaged before liver function test results become abnormal.

Function is generally measured in terms of serum enzyme activity (i.e., serum aminotransferases, alkaline phosphatase, lactic dehydrogenase) and serum concentrations of proteins (albumin and globulins), bilirubin, ammonia, clotting factors, and lipids (Fischbach & Dunning, 2009). Several of these tests may be helpful for assessing patients with liver disease. However, the nature and extent of hepatic dysfunction cannot be determined by these tests alone, because other disorders can affect test results.

Serum aminotransferases (previously called *transaminases*) are sensitive indicators of injury to the liver cells and are useful in detecting acute liver disease such as hepatitis. Alanine aminotransferase (ALT), aspartate aminotransferase (AST), and gamma-glutamyl transferase (GGT) (also called *gamma-glutamyl transpeptidase* [GGTP]) are the most frequently used tests of liver damage (Fischbach & Dunning, 2009). ALT levels increase primarily in liver disorders and may be used to monitor the course of hepatitis or cirrhosis or the effects of treatments that may be toxic to the liver. AST is present in tissues that have high metabolic activity; therefore, the level may be increased if there is damage to or death of tissues of organs such as the heart, liver, skeletal muscle, and kidney. Although not specific to liver disease, levels of AST may be increased in cirrhosis, hepatitis, and liver cancer. Increased GGT levels are associated with cholestasis but can also be due to alcoholic liver disease. Although the kidney has the highest level of the enzyme, the liver is considered the source of normal serum activity. The test determines liver cell dysfunction and is a sensitive indicator of cholestasis. Its main value in liver disease is confirming the hepatic origin of an elevated alkaline phosphatase level. Common liver function tests are summarized in Table 49-1.

TABLE 49-1 Common Laboratory Tests to Assess Liver Function

Test	Normal	Clinical Functions
Pigment Studies		
Serum bilirubin, direct	0.1–0.4 mg/dL (1.7–3.7 mcmol/L)	These studies measure the ability of the liver to conjugate and excrete bilirubin. Results are abnormal in liver and biliary tract disease and are associated with jaundice clinically.
Serum bilirubin, total	0.3–1 mg/dL (5–17 mcmol/L)	
Urine bilirubin	<0.25 mg/24 h (< .42 mcmol/24 h)	
Urine urobilinogen	(Urine urobilinogen) 0.05–2.5 mg/24 h (0.5–4 Ehrlich U/24 h)	
Fecal urobilinogen (infrequently used)	(Fecal urobilinogen) 50–300 mg/24 h (100–400 Ehrlich U/100 g)	
Protein Studies		
Total serum protein	7–7.5 g/dL (70–75 g/L)	Proteins are manufactured by the liver. Their levels may be affected in a variety of liver impairments: albumin is affected in cirrhosis, chronic hepatitis, edema; and ascites; globulins are affected in cirrhosis, liver disease, chronic obstructive jaundice, and viral hepatitis.
Serum albumin	3.5–5.5 g/dL (40–55 g/L)	
Serum globulin	2.3–3.5 g/dL (23–35 g/L)	
Serum protein electrophoresis		
Albumin	4–6 g/dL (40–60 g/L)	
α_1-Globulin	0.15–0.25 g/dL (1.5–2.5 g/L)	
α_2-Globulin	0.43–0.75 g/dL (4.3–7.5 g/L)	
β-Globulin	0.5–1 g/dL (5–10 g/L)	
γ-Globulin	0.6–1.3 g/dL (6–13 g/L)	
A/G ratio	A > G or 1.5:1–2.5:1	A/G ratio is reversed in chronic liver disease (decreased albumin and increased globulin).
Prothrombin Time	100% or 12–16 seconds	Prothrombin time may be prolonged in liver disease. It will not return to normal with vitamin K in severe liver cell damage.
Serum Alkaline Phosphatase	Varies with method: *Adults:* 30–120 U/L	Serum alkaline phosphatase is manufactured in bones, liver, kidneys, and intestine and excreted through biliary tract. In the absence of bone disease, it is a sensitive measure of biliary tract obstruction. Results may vary because this test is temperature and lab method dependent.
Serum Aminotransferase Studies		
AST	10–40 U/mL (0.34–0.68 U/L)	The studies are based on release of enzymes from damaged liver cells. These enzymes are elevated in liver cell damage. Normal values differ in men and women.
ALT	8–40 U/mL (0.14–0.68 U/L)	
GGT, GGTP	0–30 U/L IU/L	Values are elevated in alcohol abuse and markers for biliary cholestasis.
LDH	100–200 units (100–225 U/L)	
Ammonia (plasma)	15–45 mcg/dL (11–32 mcmol/L)	Liver converts ammonia to urea. Ammonia level rises in liver failure.
Cholesterol		
Ester	60%–70% of total cholesterol, fraction of total cholesterol .60–.70	Cholesterol levels are elevated in biliary obstruction and decreased in parenchymal liver disease.
HDL	*Male:* 35–70 mg/dL; *Female:* 35–85 mg/dL	
LDL	<130 mcg/dL	

A/G, albumin/globulin; AST, aspartate aminotransferase; ALT, alanine aminotransferase; GGT, gamma-glutamyl transferase; GGTP, gamma-glutamyl transpeptidase; LDH, lactate dehydrogenase; HDL, high-density lipoprotein; LDL, low-density lipoprotein.

Adapted from Koda-Kimble, M. A., Young, L. Y., Krodian, W. A., et al. (Eds.). (2008). *Applied therapeutics: The clinical use of drugs*. Philadelphia: Lippincott Williams & Wilkins.

Liver Biopsy

Liver biopsy is the removal of a small amount of liver tissue, usually through needle aspiration. It permits examination of liver cells. The most common indication is to evaluate diffuse disorders of the parenchyma and to diagnose space-occupying lesions. Liver biopsy is especially useful when clinical findings and laboratory tests are not diagnostic. Peritonitis caused by blood or bile after liver biopsy are the major complications; therefore, coagulation studies are obtained, their values are noted, and abnormal results are treated before liver biopsy is performed. Other techniques for liver biopsy are preferred if **ascites** (an accumulation of albumin-rich fluid in the peritoneal cavity) or coagulation abnormalities exist. A liver biopsy can be performed percutaneously with ultrasound guidance or transvenously through the right internal jugular vein to right hepatic vein under fluoroscopic control. Liver biopsy can also be performed laparoscopically. Nursing interventions related to percutaneous liver biopsy are summarized in Chart 49-3.

Other Diagnostic Tests

Ultrasonography, computed tomography (CT) scans, and magnetic resonance imaging (MRI) are used to identify normal structures and abnormalities of the liver and biliary tree. A radioisotope liver scan may be performed to assess liver size, blood flow, and obstruction.

Laparoscopy (insertion of a fiberoptic endoscope through a small abdominal incision) is used to examine the liver and other pelvic structures. It is also used to perform guided liver biopsy, to determine the cause of ascites, and to diagnose and stage tumors of the liver and other abdominal organs.

MANIFESTATIONS OF HEPATIC DYSFUNCTION

Hepatic dysfunction results from damage to the liver's parenchymal cells, directly from primary liver diseases, or indirectly from either obstruction of bile flow or derangements of hepatic circulation. Liver dysfunction may be acute or chronic; the latter is far more common.

Chronic liver disease, including cirrhosis, is the 12th leading cause of death in the United States among young and middle-aged adults (Xu, Kochanek, Murphy, et al., 2010). At least 40% of those deaths are associated with alcohol use. The rate of chronic liver disease for men is twice that for women, and chronic liver disease is more common in Asian and African countries than it is in Europe and the United States. Compensated cirrhosis, in which the damaged liver is still able to perform normal functions, often goes undetected for extended periods, and as many as 1% of people may have subclinical or compensated cirrhosis (Bope & Kellerman, 2011). Approximately 80% of patients diagnosed with cirrhosis compensate and remain asymptomatic for the next 10 years (Hansen, Sasaki, & Zucker, 2010).

Disease processes that lead to hepatocellular dysfunction may be caused by infectious agents such as bacteria and viruses and by anoxia, metabolic disorders, toxins and medications, nutritional deficiencies, and hypersensitivity states. The most common cause of parenchymal damage is malnutrition, especially that related to alcoholism.

The parenchymal cells respond to most noxious agents by replacing glycogen with lipids, producing fatty infiltration with or without cell death or necrosis. This is commonly associated with inflammatory cell infiltration and growth of fibrous tissue. Cell regeneration can occur if the disease process is not too toxic to the cells. The result of chronic parenchymal disease is the shrunken, fibrotic liver seen in cirrhosis.

In some conditions, lipids may accumulate in the hepatocytes, resulting in the abnormal condition called *fatty liver disease*. If unrelated to alcohol, this disease is referred to as NAFLD. A condition known as NASH represents a more serious condition within the broad spectrum of NAFLDs and may result in damage, fibrotic changes in the liver, and cirrhosis (McDonald et al., 2010; Schiff, 2013).

The consequences of liver disease are numerous and varied. Their ultimate effects are often incapacitating or life threatening, and their presence is ominous. Among the most common and significant manifestations of liver disease are jaundice, portal hypertension, ascites and varices, nutritional deficiencies (resulting from the inability of damaged liver cells to metabolize certain vitamins), and hepatic encephalopathy or coma.

Jaundice

The bilirubin concentration in the blood may be increased in the presence of liver disease, if the flow of bile is impeded (e.g., by gallstones in the bile ducts), or if there is excessive destruction of red blood cells. With bile duct obstruction, bilirubin does not enter the intestine; as a consequence, urobilinogen is absent from the urine and decreased in the stool (Hall, 2011; Porth & Matfin, 2009).

When the bilirubin concentration in the blood is abnormally elevated, all of the body tissues, including the sclerae and the skin, become tinged yellow or greenish-yellow, a condition known as **jaundice**. Jaundice becomes clinically evident when the serum bilirubin level exceeds 2.5 mg/dL (43 fmol/L) (Fischbach & Dunning, 2009). Increased serum bilirubin levels and jaundice may result from impairment of hepatic uptake, conjugation of bilirubin, or excretion of bilirubin into the biliary system. There are several types of jaundice: hemolytic, hepatocellular, and obstructive jaundice, and jaundice due to hereditary hyperbilirubinemia. Hepatocellular and obstructive jaundice are the two types commonly associated with liver disease.

Hemolytic Jaundice

Hemolytic jaundice is the result of an increased destruction of the red blood cells; the effect is that the plasma is rapidly flooded with bilirubin so that the liver, although functioning normally, cannot excrete the bilirubin as quickly as it is formed. This type of jaundice is encountered in patients with hemolytic transfusion reactions and other hemolytic disorders. In these patients, the bilirubin in the blood is predominantly unconjugated or free. Fecal and urine urobilinogen levels are increased, but the urine is free of bilirubin. Patients with this type of jaundice, unless their hyperbilirubinemia is extreme, do not experience symptoms or complications as a

Chart 49-3

GUIDELINES
Assisting With Percutaneous Liver Biopsy

Equipment

- Liver biopsy tray (contains needles, scalpel, specimen tubes, etc.) • Sterile gloves • Antiseptic solution • Local anesthetic
- Sterile dressing • Sphygmomanometer to monitor blood pressure

Implementation

Nursing Interventions	Rationale
Preprocedure	
1. Ascertain that results of coagulation tests (prothrombin time, partial thromboplastin time, and platelet count) are available and that compatible donor blood is available.	1. Many patients with liver disease have clotting defects and are at risk for bleeding.
2. Check for signed consent; confirm that informed consent has been provided. Confirm patient identity using two identifiers.	2. Ensures that the patient consents to this invasive procedure. Correctly identifies the patient.
3. Measure and record the patient's pulse, respirations, and blood pressure immediately before biopsy.	3. Prebiopsy values provide a basis on which to compare the patient's vital signs and evaluate status after the procedure.
4. Describe the following to the patient in advance: steps of the procedure, sensations expected, after-effects anticipated, restrictions of activity and monitoring procedures to follow.	4. Explanations allay fears and ensure cooperation.
During Procedure	
1. Support the patient during the procedure.	1. Encouragement and support of the nurse enhance comfort and promote a sense of security.
2. Expose the right side of the patient's upper abdomen (right hypochondriac).	2. The skin at the site of penetration will be cleansed, and a local anesthetic will be infiltrated.
3. Instruct the patient to inhale and exhale deeply several times, finally to exhale, and to hold breath at the end of expiration. The physician promptly introduces the biopsy needle by way of the transthoracic (intercostal) or transabdominal (subcostal) route, penetrates the liver, aspirates, and withdraws.	3. Holding the breath immobilizes the chest wall and the diaphragm; penetration of the diaphragm thereby is avoided, and the risk of lacerating the liver is minimized.
4. Instruct the patient to resume breathing.	4. The patient often continues holding his or her breath because of anxiety.

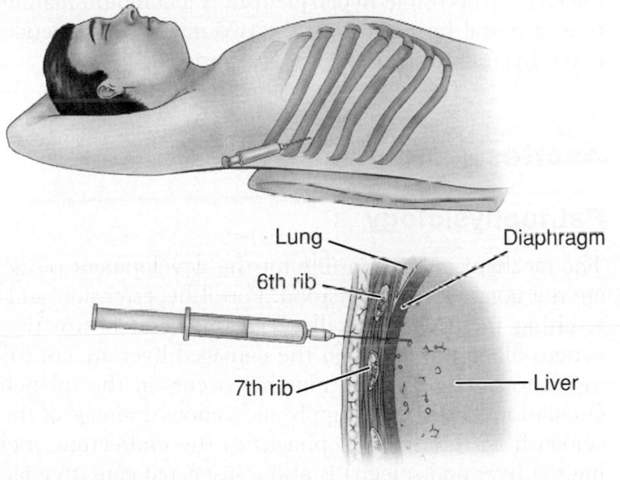
Lung
Diaphragm
6th rib
7th rib
Liver

Nursing Interventions	Rationale
Postprocedure	
1. Immediately after the biopsy, assist the patient to turn on to the right side; place a pillow under the costal margin, and caution the patient to remain in this position, recumbent and immobile, for several hours. Instruct the patient to avoid coughing or straining.	1. In this position, the liver capsule at the site of penetration is compressed against the chest wall, and the escape of blood or bile through the perforation is prevented.
2. Measure and record the patient's pulse, respiratory rate, and blood pressure at 10–15-minute intervals for the first hour, then every 30 minutes for the next 1–2 hours or until the patient's condition stabilizes.	2. Changes in vital signs may indicate bleeding, severe hemorrhage, or bile peritonitis, which as the most frequent complications of liver biopsy.
3. If the patient is discharged after the procedure, instruct the patient to avoid heavy lifting and strenuous activity for 1 week.	3. Activity restriction reduces the risk of bleeding at the biopsy puncture site.

result of the jaundice per se. However, prolonged jaundice, even if mild, predisposes to the formation of pigment stones in the gallbladder, and extremely severe jaundice (levels of free bilirubin exceeding 20 to 25 mg/dL) poses a risk for central nervous system effects (Goldman & Schafer, 2012).

Hepatocellular Jaundice

Hepatocellular jaundice is caused by the inability of damaged liver cells to clear normal amounts of bilirubin from the blood. The cellular damage may be caused by hepatitis viruses, other viruses that affect the liver (e.g., yellow fever virus, Epstein-Barr virus), chemical toxins (e.g., carbon tetrachloride, chloroform, phosphorus, arsenicals, certain medications), or alcohol. Cirrhosis of the liver is a form of hepatocellular disease that may produce jaundice. It is usually associated with excessive alcohol intake, but it may also be a late result of liver cell necrosis caused by viral infection. In prolonged obstructive jaundice, cell damage eventually develops, and both types of jaundice (i.e., obstructive and hepatocellular jaundice) appear together.

Patients with hepatocellular jaundice may be mildly or severely ill, with lack of appetite, nausea, malaise, fatigue, weakness, and possible weight loss. In some cases of hepatocellular disease, jaundice may not be obvious. The serum bilirubin concentration and the urine urobilinogen level may be elevated. In addition, AST and ALT levels may be increased, indicating cellular necrosis. The patient may report headache, chills, and fever if the cause is infectious. Depending on the cause and extent of the liver cell damage, hepatocellular jaundice may be completely reversible.

Obstructive Jaundice

Obstructive jaundice resulting from extrahepatic obstruction may be caused by occlusion of the bile duct from a gallstone, an inflammatory process, a tumor, or pressure from an enlarged organ (e.g., liver, gallbladder). The obstruction may also involve the small bile ducts within the liver (i.e., intrahepatic obstruction); this may be caused, for example, by pressure on these channels from inflammatory swelling of the liver or by an inflammatory exudate within the ducts themselves. Intrahepatic obstruction resulting from stasis and inspissation (thickening) of bile within the canaliculi may occur after the ingestion of certain medications, which are referred to as cholestatic agents. These include phenothiazines, antithyroid medications, sulfonylureas, tricyclic antidepressant agents, nitrofurantoin, androgens and estrogens, and some antibiotics.

Regardless of whether the obstruction is intrahepatic or extrahepatic, and regardless of its cause, bile cannot flow normally into the intestine and becomes backed up into the liver. It is then reabsorbed into the blood and carried throughout the entire body, staining the skin, mucous membranes, and sclerae. It is excreted in the urine, which becomes deep orange and foamy. Because of the decreased amount of bile in the intestinal tract, the stools become light or clay colored. The skin may itch intensely, requiring repeated soothing baths. Dyspepsia and intolerance to fatty foods may develop because of impaired fat digestion in the absence of intestinal bile. In general, AST, ALT, and GGT levels rise only moderately, but bilirubin and alkaline phosphatase levels are elevated.

Hereditary Hyperbilirubinemia

Increased serum bilirubin levels (hyperbilirubinemia), resulting from any of several inherited disorders, can also produce jaundice. Gilbert's syndrome is a familial disorder characterized by an increased level of unconjugated bilirubin that causes jaundice. Although serum bilirubin levels are increased, liver histology and liver function test results are normal, and there is no hemolysis. This syndrome affects 3% to 8% of the population, predominantly males (Bope & Kellerman, 2011).

Other conditions that are probably caused by inborn errors of biliary metabolism include Dubin-Johnson syndrome (chronic idiopathic jaundice, with pigment in the liver) and Rotor's syndrome (chronic familial conjugated hyperbilirubinemia, without pigment in the liver); the "benign" cholestatic jaundice of pregnancy, with retention of conjugated bilirubin, probably secondary to unusual sensitivity to the hormones of pregnancy; and benign recurrent intrahepatic cholestasis.

Portal Hypertension

Portal hypertension is the increased pressure throughout the portal venous system that results from obstruction of blood flow into and through the damaged liver. Commonly associated with hepatic cirrhosis, it can also occur with noncirrhotic liver disease. Although splenomegaly (enlarged spleen) with possible hypersplenism is a common manifestation of portal hypertension, the two major consequences of portal hypertension are ascites and varices.

Ascites

Pathophysiology

The mechanisms responsible for the development of ascites are not completely understood. Portal hypertension and the resulting increase in capillary pressure and obstruction of venous blood flow through the damaged liver are contributing factors. The vasodilation that occurs in the splanchnic circulation (the arterial supply and venous drainage of the GI system from the distal esophagus to the midrectum, including the liver and spleen) is also a suspected causative factor. The failure of the liver to metabolize aldosterone increases sodium and water retention by the kidney. Sodium and water retention, increased intravascular fluid volume, increased lymphatic flow, and decreased synthesis of albumin by the damaged liver all contribute to the movement of fluid from the vascular system into the peritoneal space. The process becomes self-perpetuating; loss of fluid into the peritoneal space causes further sodium and water retention by the kidney in an effort to maintain the vascular fluid volume.

As a result of liver damage, large amounts of albumin-rich fluid, 20 L or more, may accumulate in the peritoneal cavity as ascites (Hall, 2011). (Ascites may also occur with disorders such as cancer, kidney disease, and heart failure.) With the movement of albumin from the serum to the peritoneal cavity, the osmotic pressure of the serum decreases. This, combined

Physiology ⋯ Pathophysiology

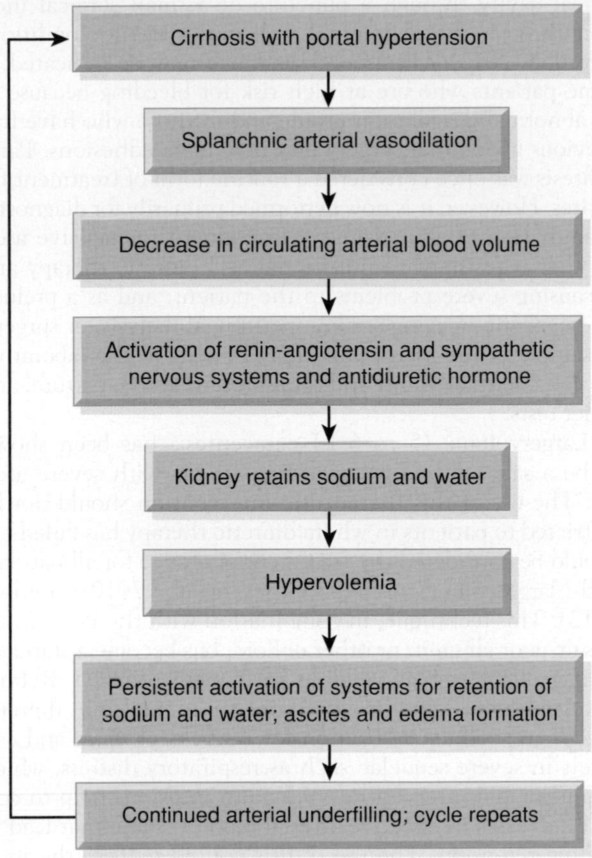
Cirrhosis with portal hypertension
Splanchnic arterial vasodilation
Decrease in circulating arterial blood volume
Activation of renin-angiotensin and sympathetic nervous systems and antidiuretic hormone
Kidney retains sodium and water
Hypervolemia
Persistent activation of systems for retention of sodium and water; ascites and edema formation
Continued arterial underfilling; cycle repeats

FIGURE 49-5 • Pathogenesis of ascites (arterial vasodilation theory).

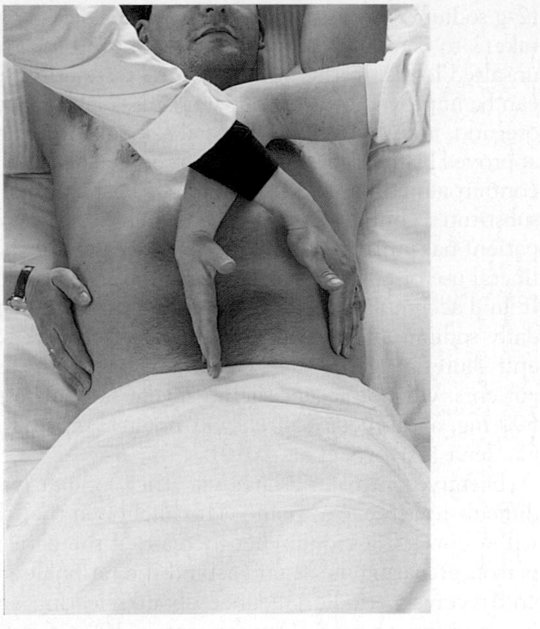

FIGURE 49-6 • Assessing for abdominal fluid wave. The examiner places the hands along the sides of the patient's flanks, then strikes one flank sharply, detecting any fluid wave with the other hand. An assistant's hand is placed (ulnar side down) along the patient's midline to prevent the fluid wave from being transmitted through the tissues of the abdominal wall.

with increased portal pressure, results in movement of fluid into the peritoneal cavity (Fig. 49-5).

Clinical Manifestations

Increased abdominal girth and rapid weight gain are common presenting symptoms of ascites. The patient may be short of breath and uncomfortable from the enlarged abdomen, and striae and distended veins may be visible over the abdominal wall. Umbilical hernias also occur frequently in those patients with cirrhosis. Fluid and electrolyte imbalances are common.

Assessment and Diagnostic Findings

The presence and extent of ascites are assessed by percussion of the abdomen. When fluid has accumulated in the peritoneal cavity, the flanks bulge when the patient assumes a supine position. The presence of fluid can be confirmed either by percussing for shifting dullness, by detecting a fluid wave (Fig. 49-6), or by performing ballottement technique (Fig. 49-7). A fluid wave is likely to be found only if a large amount of fluid is present (Weber & Kelley, 2009). The ballottement technique is a palpation technique performed to identify a mass or enlarged organ within an abdomen with ascites (Weber & Kelley, 2009). Ballottement can be performed in two different ways: single handed or bimanually (see Fig. 49-7). Daily measurement and recording of abdominal girth and body weight are essential to assess the progression of ascites and its response to treatment.

Medical Management

The medical management of the patient with ascites includes dietary modifications, pharmacologic therapy, bed rest, paracentesis, the use of shunts, and other therapies.

Nutritional Therapy

The goal of treatment for the patient with ascites is a negative sodium balance to reduce fluid retention. Table salt, salty foods, salted butter and margarine, and all canned and frozen foods that are not specifically prepared for low-sodium

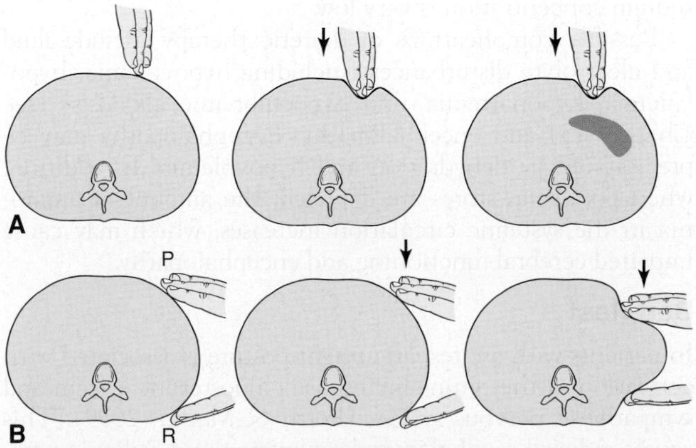
A
P
B
R

FIGURE 49-7 • Performing ballottement with one hand (**A**) and bimanually (**B**). From Weber, J., & Kelley, J. (2009). *Health assessment in nursing* (4th ed.). Philadelphia: Lippincott Williams & Wilkins.

(2-g sodium) diets should be avoided (Dudek, 2010). It may take 2 to 3 months for the patient's taste buds to adjust to unsalted foods. In the meantime, the taste of unsalted foods can be improved by using salt substitutes such as lemon juice, oregano, and thyme. Commercial salt substitutes need to be approved by the patient's primary provider, because those that contain ammonia could precipitate hepatic coma. Most salt substitutes contain potassium and should be avoided if the patient has impaired renal function. The patient should make liberal use of powdered, low-sodium milk and milk products. If fluid accumulation is not controlled with this regimen, the daily sodium allowance may be reduced further to 500 mg, and diuretic agents may be administered. However, most patients will not accept such a severe sodium restriction as 500 mg, so clinicians often will not recommend it (Gines, Cardenas, Arroyo, et al., 2010).

Dietary control of ascites via strict sodium restriction is difficult to achieve at home. The likelihood that the patient will follow a 2-g sodium diet increases if the patient and the person preparing meals understand the rationale for the diet and receive periodic guidance about selecting and preparing appropriate foods. Approximately 10% of patients with ascites respond to these measures alone. Patients who do not respond and those who find sodium restriction difficult require diuretic therapy (Gordon, 2012).

Pharmacologic Therapy

The use of diuretic agents along with sodium restriction is successful in 90% of patients with ascites (Goldman & Schafer, 2012). Spironolactone (Aldactone), an aldosterone-blocking agent, is most often the first-line therapy in patients with ascites from cirrhosis. When used with other diuretic agents, spironolactone helps prevent potassium loss. Oral diuretic agents such as furosemide (Lasix) may be added but should be used cautiously, because long-term use may induce severe sodium depletion (hyponatremia).

Ammonium chloride and acetazolamide (Diamox) are contraindicated because of the possibility of precipitating hepatic coma. Daily weight loss should not exceed 1 to 2 kg (2.2 to 4.4 lb) in patients with ascites and peripheral edema or 0.5 to 0.75 kg (1.1 to 1.65 lb) in patients without edema (Bope & Kellerman, 2011; Feldman, Friedman, & Brandt, 2010). Fluid restriction is not attempted unless the serum sodium concentration is very low.

Possible complications of diuretic therapy include fluid and electrolyte disturbances (including hypovolemia, hypokalemia, hyponatremia, and hypochloremic alkalosis) (see Chapter 13) and encephalopathy. Encephalopathy may be precipitated by dehydration and hypovolemia. In addition, when potassium stores are depleted, the amount of ammonia in the systemic circulation increases, which may cause impaired cerebral functioning and encephalopathy.

Bed Rest

In patients with ascites, an upright posture is associated with activation of the renin–angiotensin–aldosterone system and sympathetic nervous system (Porth & Matfin, 2009). This causes reduced renal glomerular filtration and sodium excretion and a decreased response to loop diuretics. Therefore, bed rest may be a useful therapy, especially for patients whose condition is refractory to diuretic agents.

Paracentesis

Paracentesis is the removal of fluid (ascites) from the peritoneal cavity through a puncture or a small surgical incision through the abdominal wall under sterile conditions (Gordon, 2012). Ultrasound guidance may be indicated in some patients who are at high risk for bleeding because of an abnormal coagulation profile and in those who have had previous abdominal surgery and may have adhesions. Paracentesis was once considered a routine form of treatment for ascites. However, it is now performed primarily for diagnostic examination of ascitic fluid; in treatment for massive ascites that is resistant to nutritional and diuretic therapy and is causing severe problems to the patient; and as a prelude to diagnostic imaging studies, peritoneal dialysis, or surgery. A sample of the ascitic fluid may be sent to the laboratory for cell count, albumin and total protein levels, culture, and other tests.

Large-volume (5 to 6 L) paracentesis has been shown to be a safe method for treating patients with severe ascites. The use of this therapeutic intervention should not be restricted to patients in whom diuretic therapy has failed but should be considered the treatment of choice for all patients with large-volume ascites (Gines et al., 2010; Gordon, 2012). This technique, in combination with the IV infusion of salt-poor albumin or other colloid, has become a standard management strategy yielding an immediate effect. Refractive, massive ascites is unresponsive to multiple diuretic agents and sodium restriction for 2 weeks or more and can result in severe sequelae such as respiratory distress, which requires rapid intervention. Albumin infusions help to correct decreases in effective arterial blood volume that lead to sodium retention. The use of this colloid reduces the incidence of postparacentesis circulatory dysfunction with renal dysfunction, hyponatremia, and rapid reaccumulation of ascites associated with decreased effective arterial volume (Gines et al., 2010; Gordon, 2012). The beneficial effects of albumin administration on hemodynamic stability and renal functional status may be related to an improvement in cardiac function as well as a decrease in the degree of arterial vasodilation. Although the patient with cirrhosis has a greatly increased extracellular blood volume, the kidney incorrectly senses that the effective volume has decreased. The renin–angiotensin–aldosterone axis is stimulated, and sodium is reabsorbed (Gines et al., 2010; Gordon, 2012). In addition, antidiuretic hormone secretion increases, which leads to increased retention of free water and sometimes to the development of dilutional hyponatremia. Therapeutic paracentesis provides only temporary removal of fluid; ascites rapidly recurs, necessitating repeated fluid removal. Nursing care of the patient undergoing paracentesis is presented in Chart 49-4.

Transjugular Intrahepatic Portosystemic Shunt

Transjugular intrahepatic portosystemic shunt (TIPS) is a method of treating ascites in which a cannula is threaded into the portal vein by the transjugular route (Fig. 49-8). To reduce portal hypertension, an expandable stent is inserted to serve as an intrahepatic shunt between the portal circulation and the hepatic vein. This is extremely effective in decreasing sodium retention, improving the renal response to

Chart 49-4

GUIDELINES

Assisting With a Paracentesis

Equipment

- Paracentesis tray (contains trocar, syringe, needles, drainage tube) • Sterile gloves • Antiseptic solution • Local anesthetic
- Sterile dressing • Drainage collection bottles, receptacles • Sphygmomanometer to monitor blood pressure

Implementation

Nursing Interventions	Rationale
Preprocedure	
1. Check for signed consent form, and identify the patient using two identifiers.	**1.** Ensures that patient has agreed to procedure. Correctly identifies the patient.
2. Prepare the patient by providing the necessary information and education and by offering reassurance.	**2.** Education increases the patient's understanding of the procedure and the reason for it.
3. Instruct the patient to void.	**3.** An empty bladder minimizes the risk of inadvertent puncture of the bladder and minimizes discomfort from a full bladder.
4. Gather appropriate sterile equipment and collection receptacles.	**4.** Sterility of equipment is essential to minimize risk of infection; having equipment available enables the procedure to be performed smoothly.
5. Place the patient in upright position on the edge of the bed or in a chair with feet supported on a stool. Fowler's position should be used by the patient confined to bed.	**5.** An upright position results in movement of the peritoneal fluid close to the abdominal wall and promotes easier puncture and removal of fluid.
6. Place the sphygmomanometer cuff around patient's arm.	**6.** This allows the nurse to monitor the patient's blood pressure during procedure.
Procedure	
1. The primary provider, using aseptic technique, inserts the trocar through a puncture below the umbilicus. The trocar or needle is connected to a drainage tube, the end of which is inserted into a collecting receptacle.	**1.** Sterile technique minimizes the risk of infection. Bleeding at the puncture site is minimal at this location. The fluid drains by gravity or mild siphon into the container.
2. Help the patient maintain position throughout the procedure.	**2.** The patient who is fatigued or weak may have difficulty maintaining an optimal position for drainage of fluid.
3. Measure and record blood pressure at frequent intervals throughout the procedure.	**3.** Decreased blood pressure may occur with vascular collapse, which can result from removal of the fluid from the peritoneal cavity and fluid shifts.
4. Monitor the patient closely for signs of vascular collapse: pallor, increased pulse rate, or decreased blood pressure.	**4.** Vascular collapse (hypovolemia) may occur as fluid moves from the vascular system to replace fluid drained from peritoneal cavity.

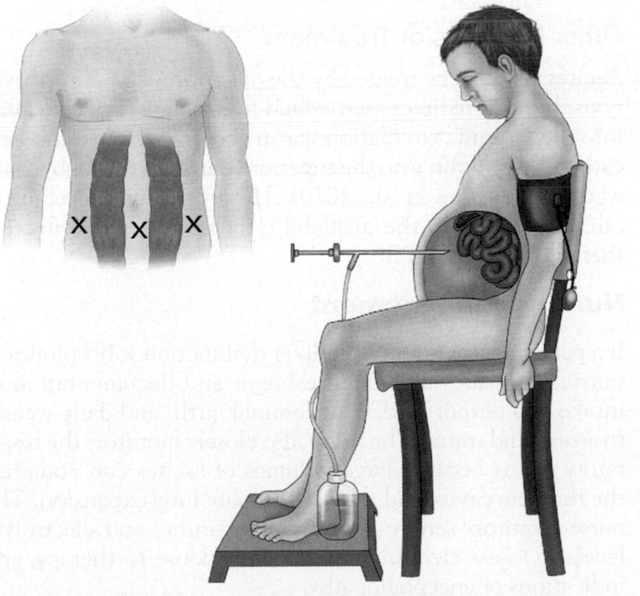

Figure on left shows possible sites for insertion of trocar.

(continues on page 1348)

Chart 49-4

GUIDELINES

Assisting With a Paracentesis (continued)

Nursing Action	Rationale
Postprocedure	
1. Return the patient to bed or to a comfortable sitting position.	**1.** The weak or fatigued patient may have difficulty resuming a comfortable position without assistance.
2. Measure, describe, and record the fluid collected.	**2.** The volume of fluid removed may range from small to very large, and its removal may affect fluid and vascular status; volume should be included in input and output records. The characteristics of the fluid (clear vs. cloudy, red vs. colorless) may be helpful in diagnostic evaluation.
3. Label samples of fluid and send to laboratory.	**3.** Peritoneal fluid is analyzed as part of the diagnostic workup.
4. Monitor vital signs every 15 minutes for 1 hour, every 30 minutes for 2 hours, every hour for 2 hours, and then every 4 hours.	**4.** Vital signs (blood pressure, pulse rate) may change as fluid shifts occur after removal of fluid, especially if a large volume of fluid has been removed.
5. Measure the patient's temperature.	**5.** An elevated temperature is a sign of infection and should be reported to the patient's primary provider.
6. Assess for hypovolemia, electrolyte shifts, changes in mental status, and encephalopathy.	**6.** Changes in fluid and electrolyte states and mental and cognitive status may occur with removal of fluid and fluid shifts, and should be reported.
7. When taking vital signs, check puncture site for leakage or bleeding.	**7.** Leakage of fluid may occur because of changes in abdominal pressure and may contribute to further loss of fluid if undetected. Leakage suggests a possible site for infection, and bleeding may occur in patients with altered clotting secondary to liver disease.
8. Provide patient education regarding need to monitor for bleeding or excessive drainage from puncture site, importance of avoiding heavy lifting or straining, the need to change position slowly, and frequency of monitoring for fever.	**8.** The patient (or family members) needs to monitor the puncture site for bleeding and excessive drainage if the patient is discharged home after the procedure. Heavy lifting or straining is avoided to enable the puncture site to close. Slow changes in position are recommended because of the risk of hypovolemia related to fluid removal. Monitoring for fever is needed to detect infection.

diuretic therapy, and preventing recurrence of fluid accumulation (Gines et al., 2010). TIPS is an effective management strategy for refractive ascites. However, due to a higher risk of encephalopathy and higher cost of TIPS compared with large-volume paracentesis plus albumin, many consider TIPS a second-line therapy for refractive ascites (Gines et al., 2010; Gordon, 2012).

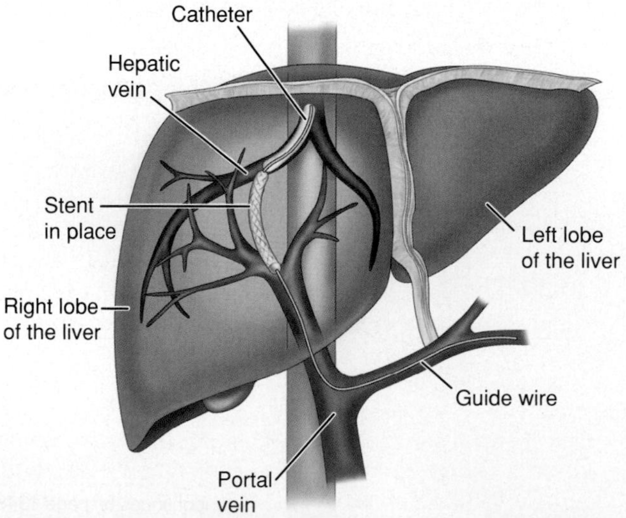
Catheter
Hepatic vein
Stent in place
Left lobe of the liver
Right lobe of the liver
Guide wire
Portal vein

FIGURE 49-8 • Transjugular intrahepatic portosystemic shunt. A stent is inserted via catheter to the portal vein to divert blood flow and reduce portal hypertension.

Because the development of ascites in patients with cirrhosis is associated with a 50% mortality rate, patients considered candidates for liver transplantation may be referred for TIPS if paracentesis is contraindicated.

Other Methods of Treatment

Ascites can also be treated by the insertion of a peritoneovenous shunt to redirect ascitic fluid from the peritoneal cavity into the systemic circulation via an abdominal and a thoracic catheter that drain into the superior vena cava through a one-way valve (Gines et al., 2010). However, this procedure is rarely used due to the availability of newer, more effective therapies such as TIPS.

Nursing Management

If a patient with ascites from liver dysfunction is hospitalized, nursing measures include assessment and documentation of intake and output (I&O), abdominal girth, and daily weight to assess fluid status. The nurse also closely monitors the respiratory status because large volumes of ascites can compress the thoracic cavity and inhibit adequate lung expansion. The nurse monitors serum ammonia, creatinine, and electrolyte levels to assess electrolyte balance, response to therapy, and indications of encephalopathy.

Promoting Home and Community-Based Care

Educating Patients About Self-Care

The patient treated for ascites is likely to be discharged with some ascites still present. Before hospital discharge, the nurse

Chart 49-5 HOME CARE CHECKLIST
Management of Ascites

At the completion of home care education, the patient or caregiver will be able to:	PATIENT	CAREGIVER
• Make appropriate dietary choices consistent with dietary prescription and recommendations.	✔	✔
• State the importance of weighing self daily and keeping a daily record of weight.	✔	✔
• Maintain record of daily weight, and identify daily weight-loss goals.	✔	✔
• List weight changes (loss or gain) that should be reported to the primary provider.	✔	✔
• Explain the rationale for monitoring and recording daily intake and output.	✔	✔
• Identify changes in output that should be reported to primary provider (e.g., decreasing urine output).	✔	✔
• Identify rationale for fluid restrictions (if needed), and comply with fluid restriction.	✔	✔
• Discuss importance of avoiding nonsteroidal anti-inflammatory agents, medications (e.g., cough mixtures) containing alcohol, antibiotics, or antacids containing salt.	✔	✔
• Describe effects, side effects, and monitoring parameters for diuretic therapy.	✔	✔
• Identify need to stop all alcohol intake as critical to well-being.	✔	✔
• Explain how to contact Alcoholics Anonymous or alcohol counselors in related organizations if indicated.	✔	✔
• Demonstrate how to care for skin, alleviate pressure over bony prominences by turning when in bed or chair, and decrease edema by position changes.	✔	✔
• Identify early signs and symptoms of complications (encephalopathy, spontaneous bacterial peritonitis, dehydration, electrolyte abnormalities, azotemia).	✔	✔

educates the patient and family about the treatment plan, including the need to avoid all alcohol intake, adhere to a low-sodium diet, take medications as prescribed, and check with the primary provider before taking any new medications. Additional patient and family education is summarized in Chart 49-5.

Continuing Care

A referral for home care may be warranted, especially if the patient lives alone or cannot provide self-care. The home visit enables the nurse to assess changes in the patient's condition and weight, abdominal girth, skin, and cognitive and emotional status. The home care nurse assesses the home environment and the availability of resources needed to adhere to the treatment plan (e.g., a scale to obtain daily weights, facilities to prepare and store appropriate foods, resources to purchase needed medications). The nurse also assesses the patient's adherence to the treatment plan and the ability to buy, prepare, and eat appropriate foods. The nurse reinforces previous education and emphasizes the need for regular follow-up and the importance of keeping scheduled health care appointments.

Esophageal Varices

Esophageal varices are present in 30% of patients with compensated cirrhosis and 60% of patients with decompensated cirrhosis at the time of diagnosis (Triantos, Goulis, & Burroughs, 2010) (see Clinical Manifestations in the Hepatic Cirrhosis section for further discussion). Varices are varicosities that develop from elevated pressure in the veins that drain into the portal system. They are prone to rupture and often are the source of massive hemorrhages from the upper GI tract and the rectum. In addition, abnormalities in blood clotting, often seen in patients with severe liver disease, increase the likelihood of bleeding and significant blood loss.

Once esophageal varices form, they increase in size over time and may eventually bleed (Triantos et al., 2010). In cirrhosis, they are the most significant source of bleeding. The first bleeding episode has a mortality rate of 10% to 30% depending on the severity of the liver disease and is one of the major causes of death in patients with cirrhosis. Overall mortality associated with acute variceal bleeding ranges from 10% to 40%. The mortality rate is related to failure to control a bleeding episode and the occurrence of early rebleeding (Triantos et al., 2010). Patients surviving the first episode of variceal bleeding are at very high risk for recurrent bleeding (approximately 70%) and death (30% to 50%) (Triantos et al., 2010).

Pathophysiology

Esophageal varices are dilated, tortuous veins that are usually found in the submucosa of the lower esophagus but may develop higher in the esophagus or extend into the stomach. This condition is almost always caused by portal hypertension, which results from obstruction of the portal venous circulation within the damaged liver.

Because of increased obstruction of the portal vein, venous blood from the intestinal tract and spleen seeks an outlet through collateral circulation (new pathways for return of blood to the right atrium). The effect is increased pressure, particularly in the vessels in the submucosal layer of the lower esophagus and upper part of the stomach. These collateral

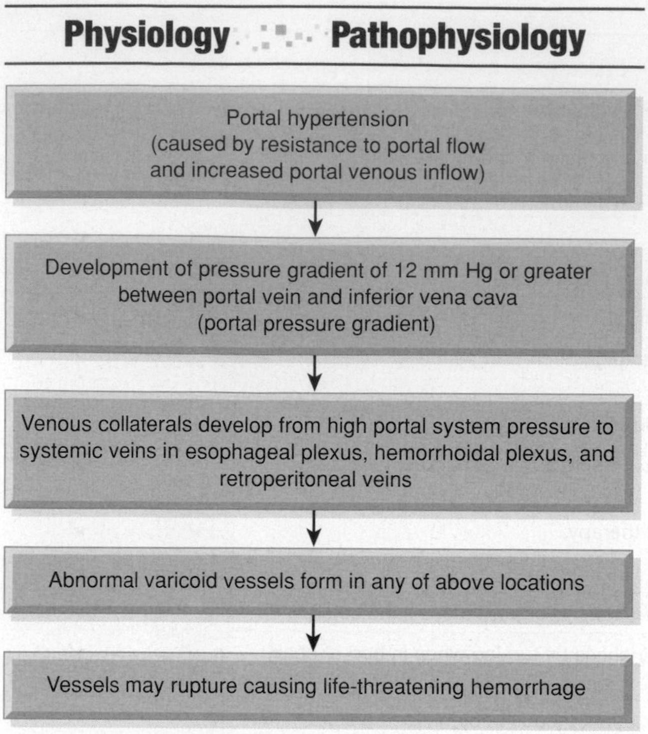
Physiology Pathophysiology
Portal hypertension
(caused by resistance to portal flow
and increased portal venous inflow)
Development of pressure gradient of 12 mm Hg or greater
between portal vein and inferior vena cava
(portal pressure gradient)
Venous collaterals develop from high portal system pressure to
systemic veins in esophageal plexus, hemorrhoidal plexus, and
retroperitoneal veins
Abnormal varicoid vessels form in any of above locations
Vessels may rupture causing life-threatening hemorrhage

FIGURE 49-9 • Pathogenesis of bleeding esophageal varices.

vessels are not very elastic; rather, they are tortuous and fragile, and they bleed easily (Fig. 49-9). Less common causes of varices are abnormalities of the circulation in the splenic vein or superior vena cava and hepatic venothrombosis.

Bleeding esophageal varices are life threatening and can result in hemorrhagic shock that produces decreased cerebral, hepatic, and renal perfusion. In turn, there is an increased nitrogen load from bleeding into the GI tract and an increased serum ammonia level, increasing the risk of encephalopathy. Usually, the dilated veins cause no symptoms. However, if the portal pressure increases sharply and the mucosa or supporting structures become thin, massive hemorrhaging occurs.

Factors that contribute to hemorrhage are muscular exertion from lifting heavy objects; straining at stool; sneezing, coughing, or vomiting; esophagitis; irritation of vessels by poorly chewed foods or irritating fluids; and reflux of stomach contents (especially alcohol). Salicylates and any medication that erodes the esophageal mucosa or interferes with cell replication also may contribute to bleeding.

Clinical Manifestations

The patient with bleeding esophageal varices may present with hematemesis, melena, or general deterioration in mental or physical status and often has a history of alcohol abuse. Signs and symptoms of shock (cool clammy skin, hypotension, tachycardia) may be present (see Chapter 14).

Assessment and Diagnostic Findings

Endoscopy is used to identify the bleeding site, along with ultrasonography, CT scanning, and angiography. A newer diagnostic tool, the endoscopic video capsule, can detect esophageal varices but does not substitute for endoscopy unless this test cannot be performed (Goldman & Schafer, 2012; Triantos et al., 2010). Because varices are present in 50% of patients with cirrhosis, it is recommended that patients who have been diagnosed with cirrhosis undergo screening endoscopy. If no varices are detected on initial endoscopy, the test should be repeated in 2 to 3 years in an effort to identify and treat large varices, which are the ones most likely to bleed. If small varices are identified on initial endoscopy, the test should be repeated in 1 to 2 years (Feldman et al., 2010).

Endoscopy

Immediate endoscopy (see Chapter 44) is indicated to identify the cause and the site of bleeding; approximately 40% of patients with suspected bleeding from esophageal varices are actually bleeding from another source (gastritis, ulcer) (Sostres & Lanas, 2011). Nursing support is essential during this often stressful experience. Careful monitoring can detect early signs of cardiac dysrhythmias, perforation, and hemorrhage.

After the examination, fluids are not given until the patient's gag reflex returns. Lozenges and gargles may be used to relieve throat discomfort if the patient's physical condition and mental status permit. If the patient is actively bleeding, oral intake will not be permitted, and the patient will be prepared for further diagnostic and therapeutic procedures.

Portal Hypertension Measurements

Portal hypertension may be suspected if dilated abdominal veins and hemorrhoids are detected. A palpable enlarged spleen (splenomegaly) and ascites may also be present. Portal venous pressure can be measured directly or indirectly. Indirect measurement of the hepatic vein pressure gradient is the most common procedure. The measurement requires insertion of a catheter with a balloon into the antecubital or femoral vein. The catheter is advanced under fluoroscopy to a hepatic vein. Fluid is infused once the catheter is in position to inflate the balloon. A "wedged" pressure (similar to pulmonary artery wedge pressure) is obtained by occluding the blood flow in the blood vessel; pressure in the unoccluded vessel is also measured. Although the values obtained may underestimate portal pressure, this measurement may be taken several times to evaluate the results of therapy.

Direct measurement of portal vein pressure can be obtained by several methods. During laparotomy, a needle may be introduced into the spleen; a manometer reading of more than 20 mL saline is abnormal. Another direct measurement requires insertion of a catheter into the portal vein or one of its branches. Endoscopic measurement of pressure within varices is used only in conjunction with endoscopic sclerotherapy (see later discussion).

Laboratory Tests

Laboratory studies may include various liver function tests, such as serum aminotransferases, bilirubin, alkaline phosphatase, and serum proteins. Splenoportography, which involves serial or segmental x-rays, is used to detect extensive collateral circulation in esophageal vessels, which would indicate varices. Other tests are hepatoportography and celiac angiography. These are usually performed in the operating room or x-ray department.

Medical Management

Bleeding from esophageal varices is an emergency that can quickly lead to hemorrhagic shock. The patient is critically ill, requiring aggressive medical care and expert nursing care, and is usually transferred to the intensive care unit (ICU) for close monitoring and management. (See Chapter 14 for a discussion of care of the patient in shock.) The extent of bleeding is evaluated, and vital signs are monitored continuously if hematemesis and melena are present.

Because patients with bleeding esophageal varices have intravascular volume depletion and are subject to electrolyte imbalance, IV fluids, electrolytes, and volume expanders are provided to restore fluid volume and replace electrolytes. Transfusion of blood components also may be required.

Caution must be taken with volume resuscitation so that overhydration does not occur, because this would raise portal pressure and increase bleeding. An indwelling urinary catheter is usually inserted to permit frequent monitoring of urine output.

Although a variety of pharmacologic, endoscopic, and surgical approaches are used to treat bleeding esophageal varices, none is ideal, and most are associated with considerable risk to the patient. Nonsurgical treatment of bleeding esophageal varices is preferable because of the high mortality rate of emergency surgery to control bleeding esophageal varices and because of the poor physical condition that is typical of the patient with severe liver dysfunction.

Pharmacologic Therapy

In suspected variceal bleeding, vasoactive drugs need to be given as soon as possible and before endoscopy (Cardenas, 2010). In an actively bleeding patient, medications are administered initially because they can be obtained and administered quicker than other therapies. Octreotide (Sandostatin), a synthetic analogue of the hormone somatostatin, is effective in decreasing bleeding from esophageal varices, and lacks the vasoconstrictive effects of vasopressin. Because of this safety and efficacy profile, octreotide is considered the preferred treatment regimen for immediate control of variceal bleeding. These medications cause selective splanchnic vasoconstriction by inhibiting glucagon release and are used mainly in the management of active hemorrhage (Cat & Liu-DeRyke, 2010).

Vasopressin (Pitressin) may be the initial mode of therapy in urgent situations because it produces constriction of the splanchnic arterial bed and decreases portal pressure. As described previously, splanchnic circulation comprises the arterial blood supply and venous drainage of the entire GI tract from the distal esophagus to the midrectum, including the liver and spleen. Vasopressin constricts distal esophageal and proximal gastric veins, thus reducing the inflow into the portal system and therefore the portal pressure. Vital signs and the presence or absence of blood in the gastric aspirate indicate the effectiveness of vasopressin. Monitoring of I&O and electrolyte levels is necessary because hyponatremia may develop, and vasopressin may have an antidiuretic effect.

Coronary artery disease is a contraindication to the use of vasopressin because coronary vasoconstriction is a side effect that may precipitate myocardial infarction. The combination of vasopressin with nitroglycerin (administered by the IV, sublingual, or transdermal route) has been effective in reducing or preventing the side effects (constriction of coronary vessels and angina) caused by vasopressin alone. Side effects of vasopressin include myocardial and extremity ischemia as well as cardiac dysrhythmias; therefore, vasopressin is used only in urgent situations or when other agents such as octreotide are not available. Vasopressin must be administered with close monitoring (Opio & Garcia-Tsao, 2011; Triantos et al., 2010).

Beta-blocking agents such as propranolol or nadolol that decrease portal pressure are the most common medications used both to prevent a first bleeding episode in patients with known varices and to prevent rebleeding (Opio & Garcia-Tsao, 2011; Triantos et al., 2010). Beta-blockers should not be used in acute variceal hemorrhage, but they are effective prophylaxis against such an episode. Nitrates such as isosorbide (Isordil) lower portal pressure by venodilation and decreased cardiac output and may be used in combination with beta-blockers (Opio & Garcia-Tsao, 2011). Further studies of these and other medications are necessary to evaluate their use in the treatment and prevention of bleeding episodes.

Balloon Tamponade

There is infrequent use of balloon tamponade therapy today; it may be used to temporarily control hemorrhage and to stabilize a patient with massive bleeding prior to other definitive management (Feldman et al., 2010).

When indicated, balloon tamponade can be successful; however, there are risks. Displacement of the tube and the inflated balloon into the oropharynx can cause life-threatening obstruction of the airway and asphyxiation. This may occur if the patient pulls on the tube because of confusion or discomfort. It may also result from rupture of the gastric balloon, which causes the esophageal balloon to move into the oropharynx. Sudden rupture of the balloon causes airway obstruction and aspiration of gastric contents into the lungs. Therefore, the tube must be tested before insertion to minimize this risk by ensuring that the balloons can attain and maintain inflation. Aspiration of blood and secretions into the lungs is frequently associated with balloon tamponade, especially in the stuporous or comatose patient. Endotracheal intubation before insertion of the tube protects the airway and minimizes the risk of aspiration. Ulceration and necrosis of the nose, the mucosa of the stomach, or the esophagus may occur if the tube is left in place too long, inflated too long, or inflated at too high a pressure. The therapy is used for as short a time as possible to control bleeding while emergency treatment is completed and definitive therapies are instituted (no longer than 12 hours, preferably less) (Wiegand, 2011).

Quality and Safety Nursing Alert

The patient being treated with balloon tamponade must remain under close observation in the ICU because of the risk of serious complications. The patient must be monitored closely and continuously. Precautions must be taken to ensure that the patient does not pull on or inadvertently displace the tube.

Nursing measures include frequent mouth and nasal care. For secretions that accumulate in the mouth, tissues should be within easy reach of the patient. Oral suction may be necessary to remove oral secretions.

Although balloon tamponade stops the bleeding in 90% of patients, bleeding recurs in 60% to 70%, necessitating other treatment modalities, such as endoscopic therapies (see later discussion) (D'Amico, Berzigotti & Garcia-Pagan, 2010; Feldman et al., 2010). Once the tube is removed, the patient must be assessed frequently because of the high risk of recurrent bleeding.

Endoscopic Sclerotherapy

In endoscopic **sclerotherapy** (Fig. 49-10), also referred to as injection sclerotherapy, a sclerosing agent (i.e., sodium morrhuate, ethanolamine oleate, sodium tetradecyl sulfate, or ethanol) is injected through a fiberoptic endoscope into or adjacent to the bleeding esophageal varices to promote thrombosis and eventual sclerosis (Feldman et al., 2010). The process of sclerotherapy causes inflammation of the involved vein with eventual thrombosis and loss of the lumen of the vessel. The procedure has been used successfully to treat acute GI hemorrhage but is not recommended for prevention of first and subsequent variceal bleeding episodes where endoscopic variceal ligation (EVL), also known as esophageal banding therapy (discussed later), is the first-line treatment (Triantos et al., 2010).

After treatment for acute hemorrhage, the patient must be observed for bleeding, perforation of the esophagus, aspiration pneumonia, and esophageal stricture. Antacids, histamine-2 (H_2) antagonists such as cimetidine (Tagamet), or proton pump inhibitors such as pantoprazole (Protonix) may be administered after the procedure to counteract the chemical effects of the sclerosing agent on the esophagus and the acid reflux associated with the therapy.

Endoscopic Variceal Ligation (Esophageal Banding Therapy)

In variceal banding (Fig. 49-11), also referred to as **endoscopic variceal ligation (EVL)**, a modified endoscope loaded with an elastic rubber band is passed through an overtube directly onto the varix (or varices) to be banded. After the bleeding varix is suctioned into the tip of the endoscope, the rubber band is slipped over the tissue, causing necrosis, ulceration, and eventual sloughing of the varix.

Variceal banding is comparable to endoscopic sclerotherapy in its effectiveness in controlling acute bleeding. Compared with sclerotherapy, variceal banding also significantly reduces the rebleeding rate, mortality, procedure-related complications, and number of sessions needed to eradicate varices. Esophageal band ligation has replaced sclerotherapy as the treatment of choice in the management of esophageal varices. Complications include superficial ulceration and dysphagia, transient chest discomfort, and, rarely, esophageal strictures. Band ligation in combination with pharmacologic therapy may be more effective than monotherapy (i.e., a single mode of therapy) in the treatment of acute hemorrhage. EVL is recommended for patients who have experienced variceal bleeding while receiving beta-blocker therapy and for those who cannot tolerate beta-blocking agents (Triantos et al., 2010).

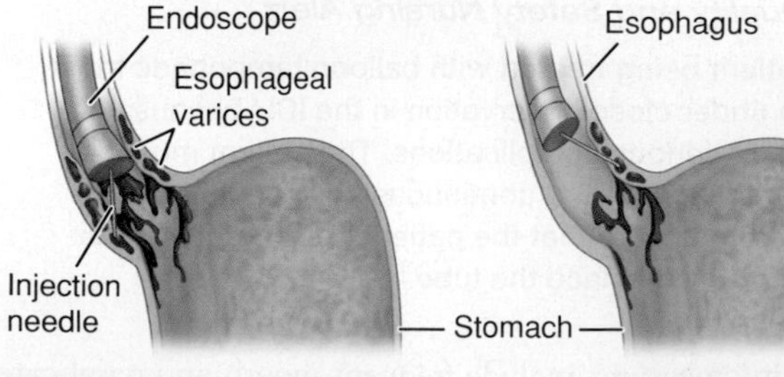
Endoscope
Esophageal varices
Injection needle
Esophagus
Stomach

FIGURE 49-10 • Endoscopic or injection sclerotherapy. Injection of sclerosing agent into esophageal varices through an endoscope promotes thrombosis and eventual sclerosis, thereby obliterating the varices.

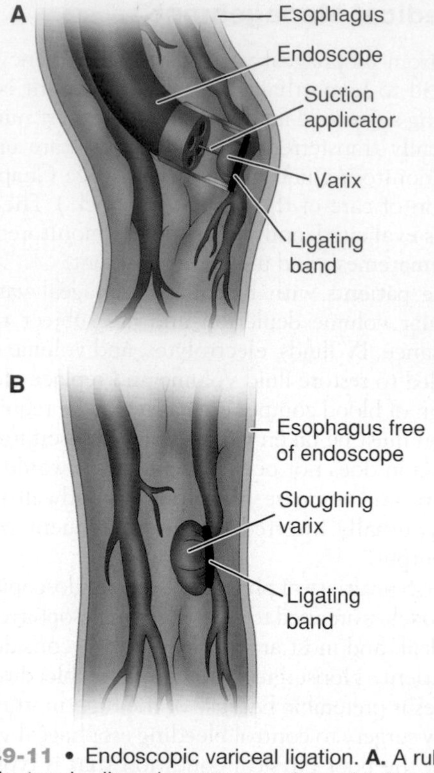
A
Esophagus
Endoscope
Suction applicator
Varix
Ligating band
B
Esophagus free of endoscope
Sloughing varix
Ligating band

FIGURE 49-11 • Endoscopic variceal ligation. **A.** A rubber band–like ligature is slipped over an esophageal varix via an endoscope. **B.** Necrosis results, and the varix eventually sloughs off.

Transjugular Intrahepatic Portosystemic Shunt

A TIPS procedure (see Fig. 49-8) is indicated for the treatment of an acute episode of uncontrolled variceal bleeding refractory to pharmacologic or endoscopic therapy. In 10% to 20% of patients for whom urgent band ligation or sclerotherapy and medications are not successful in eradicating bleeding, a TIPS procedure can effectively control acute variceal hemorrhage by rapidly lowering portal pressure. TIPS is not recommended for the secondary prevention of variceal bleeding because the costs of treatment do not offset its potential benefits when compared with other treatments. This recommendation may change with the advent of new covered stents when compared to uncovered stents (Triantos et al., 2010). Potential complications of TIPS include bleeding, sepsis, heart failure, organ perforation, shunt thrombosis, and progressive liver failure.

Additional Therapies

The use of endoscopically placed tissue adhesives and fibrin glue have been successful in the treatment of esophageal varices. Coated expandable stents (placed via endoscope) have also been used effectively for the same purpose. These developing technologies require further study to determine their role in the management of bleeding esophageal varices (Triantos et al., 2010).

Surgical Management

Several surgical procedures have been developed to treat esophageal varices and to minimize rebleeding, but these procedures are often accompanied by significant risk. Procedures that may be used for esophageal varices are direct surgical ligation of varices; splenorenal, mesocaval, and portacaval venous shunts to relieve portal pressure; and esophageal transection with devascularization. The use of these procedures is controversial, and studies regarding their effectiveness and outcomes continue. What is known thus far is that these procedures are very effective in controlling variceal bleeding. They may be considered as second-line management (rescue therapy) in those patients for whom all other treatments have failed, those who are not candidates for liver transplantation, and those who require a bridge to transplantation. There is a high incidence of encephalopathy after the surgical shunting procedures, and morbidity and mortality statistics remain high (Triantos et al., 2010). The TIPS procedure has largely replaced the use of surgical decompression shunts and ligation procedures.

Surgical Bypass Procedures

Surgical decompression (shunt surgery) of the portal circulation may be used with the advent of a variceal bleeding episode. Although effective in eradicating bleeding, survival statistics and encephalopathy are worse than other preventative measures such as a TIPS procedure when this method is employed for prophylaxis, and shunt surgery for this purpose has largely been abandoned worldwide (Triantos et al., 2010).

One of the various surgical shunting procedures (Fig. 49-12) is the distal splenorenal shunt, which is made between the splenic vein and the left renal vein after splenectomy. A mesocaval shunt is created by anastomosing the superior mesenteric vein to the proximal end of the vena cava or to the side of the vena cava using grafting material. The goal of distal splenorenal and mesocaval shunts is to decrease portal pressure by draining only a portion of venous blood from the portal bed; therefore, they are considered selective shunts. The liver continues to receive some portal flow, and the incidence of encephalopathy may be reduced. Portacaval shunts are considered nonselective shunts because they divert all portal flow to the vena cava via end-to-side or side-to-side approaches.

These procedures are extensive and are not always successful because of secondary thrombosis in the veins used for the shunt and because of complications (e.g., encephalopathy, accelerated liver failure). The effectiveness of these procedures has been studied extensively. All shunt procedures are equally effective in preventing recurrent variceal bleeding but may cause further impairment of liver function and encephalopathy. Partial portacaval shunts with interposition grafts are as effective as other shunts but are associated with a lower rate of encephalopathy (D'Amico et al., 2010). The severity of the disease (by a classification such as the Child-Pugh system, discussed later) and the potential for future liver transplantation guide the treatment decision. If the cause of

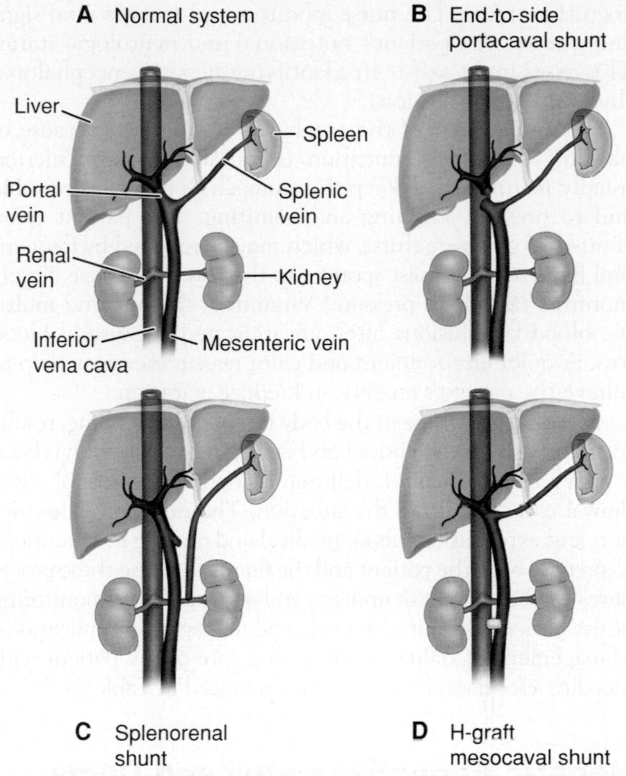
A Normal system
B End-to-side portacaval shunt
Liver
Spleen
Portal vein
Splenic vein
Renal vein
Kidney
Inferior vena cava
Mesenteric vein
C Splenorenal shunt
D H-graft mesocaval shunt

FIGURE 49-12 • Portosystemic shunts. **A.** Normal portal system. **B–D.** Examples of portal shunts to reduce portal pressure.

portal hypertension is the rare Budd-Chiari syndrome (which is manifested by non-cirrhotic portal hypertension caused by hepatic vein thrombosis) or other venous obstructive disease, a portacaval or a mesoatrial shunt may be performed (see Fig. 49-10). The mesoatrial shunt is required when the infrahepatic vena cava is thrombosed and must be bypassed.

Devascularization and Transection

Devascularization and staple-gun transection procedures to separate the bleeding site from the high-pressure portal system have been used in the emergency management of variceal bleeding. The lower end of the esophagus is reached through a small gastrostomy incision; a staple gun permits anastomosis of the transected ends of the esophagus. Rebleeding is a risk, and the outcomes of these procedures vary among patient populations.

Quality and Safety Nursing Alert

The surgical procedures used to treat esophageal varices do not alter the course of the progressive liver disease, and bleeding may recur as new collateral vessels develop. The risk of complications (hypovolemic or hemorrhagic shock, hepatic encephalopathy, electrolyte imbalance, metabolic and respiratory alkalosis, alcohol withdrawal syndrome, and seizures) is high.

Nursing Management

Nursing assessment includes monitoring the patient's physical condition and evaluating emotional responses and cognitive status. The nurse monitors and records vital signs and assesses the patient's nutritional and neurologic status. This assessment assists in identifying hepatic encephalopathy (see later discussion).

If complete rest of the esophagus is indicated because of bleeding, parenteral nutrition is initiated. Gastric suction usually is initiated to keep the stomach as empty as possible and to prevent straining and vomiting. The patient often complains of severe thirst, which may be relieved by frequent oral hygiene and moist sponges to the lips. The nurse closely monitors the blood pressure. Vitamin K therapy and multiple blood transfusions often are indicated because of blood loss. A quiet environment and calm reassurance may help to relieve the patient's anxiety and reduce agitation.

Bleeding anywhere in the body is anxiety provoking, resulting in a crisis for the patient and family. If the patient has been a heavy user of alcohol, delirium secondary to alcohol withdrawal can complicate the situation. The nurse provides support and explanations about medical and nursing interventions to prepare both the patient and the family, because these procedures can be difficult to undergo and observe. Close monitoring of the patient helps in detecting and managing complications. Management modalities and nursing care of the patient with bleeding esophageal varices are summarized in Table 49-2.

Hepatic Encephalopathy and Coma

Hepatic encephalopathy, or portosystemic encephalopathy, is a life-threatening complication of liver disease that occurs with profound liver failure. Patients with this condition may have no overt signs of the illness but have abnormalities on neuropsychologic testing (Khungar & Poordad, 2012a; Sundaram & Shaikh, 2009). Hepatic encephalopathy is the neuropsychiatric manifestation of hepatic failure associated with portal hypertension and the shunting of blood from the portal venous system into the systemic circulation (Sundaram & Shaikh, 2011). This reversible metabolic form of encephalopathy can improve with recovery of liver function. The onset is often insidious and subtle, and initially the disease is termed *subclinical* or *minimal hepatic encephalopathy*.

Pathophysiology

Despite the frequency with which hepatic encephalopathy occurs, the precise pathophysiology is not fully understood (Sundarum & Shaikh, 2009). Two major alterations underlie its development in acute and chronic liver disease. First, hepatic insufficiency may result in encephalopathy because of the inability of the liver to detoxify toxic by-products of metabolism. Second, portosystemic shunting, in which collateral vessels develop as a result of portal hypertension, allows elements of the portal blood (laden with potentially toxic substances usually extracted by the liver) to enter the systemic circulation (Khungar & Poordad, 2012a). Ammonia is considered the major etiologic factor in the development of encephalopathy. Ammonia enters the brain and excites peripheral benzodiazepine-type receptors on astrocyte cells, increasing neurosteroid synthesis, and stimulating gamma-aminobutyric acid (GABA) neurotransmission. GABA

TABLE 49-2 Select Modalities and Nursing Care for the Patient With Bleeding Esophageal Varices

Treatment Modality*	Action	Nursing Priorities
Nonsurgical Modalities		
Pharmacologic agents		Observe response to therapy.
propranolol (Inderal)/ nadolol (Corgard)	Reduces portal pressure by β-adrenergic blocking action	Monitor for side effects: *propranolol* and *nadolol*—decreased pulse pressure, impaired cardiovascular response to hemorrhage;
vasopressin (Pitressin)	Reduces portal pressure by constricting splanchnic arteries	*vasopressin*—angina (nitroglycerin may be prescribed to prevent or treat angina).
octreotide (Sandostatin)	Reduces portal pressure by selective vasodilation of portal system	Support patient during treatment.
Injection sclerotherapy	Promotes thrombosis and sclerosing of bleeding sites by injection of sclerosing agent into the esophageal varices	Observe for aspiration, perforation of the esophagus, and recurrence of bleeding after treatment.
Endoscopic variceal ligation	Provides thrombosis and mucosal necrosis of bleeding sites by band ligation	Observe for recurrence of bleeding, esophageal perforation.
Transjugular intrahepatic portosystemic shunt	Reduces portal pressure by creating a shunt within the liver between the portal and systemic venous systems	Observe for rebleeding and signs of infection.
Balloon tamponade	Exerts pressure directly to bleeding sites in esophagus and stomach	Explain procedure to patient briefly to obtain cooperation with insertion and maintenance of esophageal/gastric tamponade tube and reduce patient's fear of the procedure. Monitor closely to prevent inadvertent removal or displacement of tube, subsequent airway obstruction, and aspiration. Provide frequent oral hygiene.
Surgical Modalities		
Portal-systemic shunt	Reduces portal hypertension by diverting blood flow away from obstructed portal system	Observe for development of portal-systemic encephalopathy (altered mental status, neurologic dysfunction), hepatic failure, and rebleeding. Requires intensive, expert nursing care for prolonged period.
Surgical ligation of varices	Ties off blood vessels at the site of bleeding	Observe for rebleeding.
Esophageal transection and devascularization	Separates bleeding site from portal system	Observe for rebleeding. Provide postthoracotomy care.

*Several modalities may be used concurrently or in sequence.

causes depression of the central nervous system that inhibits neurotransmission and synaptic regulation (Khungar & Poordad, 2012a), producing sleep and behavior patterns associated with hepatic encephalopathy.

Circumstances that increase serum ammonia levels tend to aggravate or precipitate hepatic encephalopathy. The largest source of ammonia is the enzymatic and bacterial digestion of dietary and blood proteins in the GI tract. Ammonia from these sources increases as a result of GI bleeding (i.e., bleeding esophageal varices, chronic GI bleeding), a high-protein diet, bacterial infection, or uremia. The ingestion of ammonium salts also increases the blood ammonia level. In the presence of alkalosis or hypokalemia, increased amounts of ammonia are absorbed from the GI tract and from the renal tubular fluid. Conversely, serum ammonia is decreased by elimination of protein from the diet and by the administration of antibiotic agents, such as neomycin sulfate (Mycifradin, Neo-Fradin), which reduce the number of intestinal bacteria capable of converting urea to ammonia (Dudek, 2010; Khungar & Poordad, 2012b).

Other factors unrelated to increased serum ammonia levels that can cause hepatic encephalopathy in susceptible patients include excessive diuresis, dehydration, infections, surgery, fever, and some medications (sedatives, tranquilizers, analgesics, and diuretics that cause potassium loss). Additional causes include elevated levels of serum manganese (Sundarum & Shaikh, 2009), as well as changes in the types of circulating amino acids, mercaptans, and levels of dopamine and other neurotransmitters in the central nervous system (Feldman et al., 2010). Mercaptans are toxic metabolites of sulfur-containing compounds that are excreted by the liver under normal conditions. Mercaptans and these other so-called "false" neurotransmitters may be generated from an intestinal source or from metabolism of protein by the liver and, with defective hepatic clearance, may precipitate encephalopathy.

Clinical Manifestations

The earliest symptoms of hepatic encephalopathy include mental status changes and motor disturbances. The patient appears confused and unkempt and has alterations in mood and sleep patterns. The patient tends to sleep during the day and has restlessness and insomnia at night. As hepatic encephalopathy progresses, the patient may become difficult to awaken and completely disoriented with respect to time and place. With further progression, the patient lapses into frank coma and may have seizures.

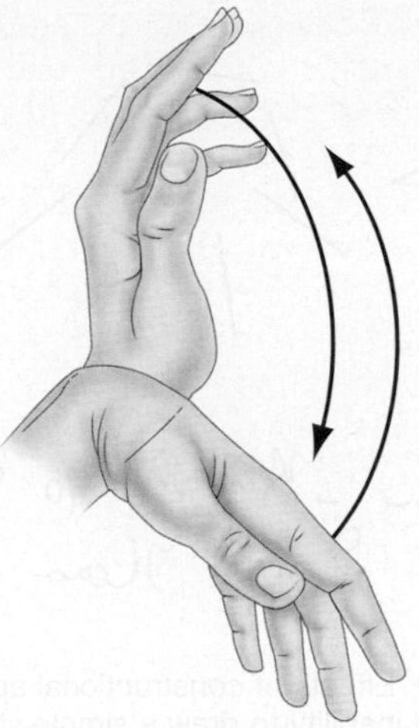

FIGURE 49-13 • Asterixis ("liver flap") may occur in hepatic encephalopathy. The patient is asked to hold the arm out with the hand held upward (dorsiflexed). Within a few seconds, the hand falls forward involuntarily and then quickly returns to the dorsiflexed position.

Concept Mastery Alert

It is vital for the nurse to understand the four stages of hepatic encephalopathy and common signs and symptoms. These key points, as well as selected nursing diagnoses, are summarized in Table 49-3.

Asterixis, an involuntary flapping of the hands, may be seen in stage II encephalopathy (Fig. 49-13). Simple tasks, such as handwriting, become difficult. A handwriting or drawing sample (e.g., star figure), taken daily, may provide graphic evidence of progression or reversal of hepatic encephalopathy. Inability to reproduce a simple figure in two or three dimensions (Fig. 49-14) is referred to as **constructional apraxia**. In the early stages of hepatic encephalopathy, the deep tendon reflexes are hyperactive; with worsening of the encephalopathy, these reflexes disappear and the extremities may become flaccid.

TABLE 49-3 Stages of Hepatic Encephalopathy and Applicable Nursing Diagnoses*

Stage	Clinical Symptoms	Clinical Signs and EEG Changes	Selected Potential Nursing Diagnoses
1	Normal level of consciousness with periods of lethargy and euphoria; reversal of day–night sleep patterns	Asterixis; impaired writing and ability to draw line figures. Normal EEG.	Activity intolerance Self-care deficit Disturbed sleep pattern
2	Increased drowsiness; disorientation; inappropriate behavior; mood swings; agitation	Asterixis; fetor hepaticus. Abnormal EEG with generalized slowing.	Impaired social interaction Ineffective role performance Risk for injury Confusion
3	Stuporous; difficult to rouse; sleeps most of time; marked confusion; incoherent speech	Asterixis; increased deep tendon reflexes; rigidity of extremities. EEG markedly abnormal.	Imbalanced nutrition Impaired mobility Impaired verbal communication
4	Comatose; may not respond to painful stimuli	Absence of asterixis; absence of deep tendon reflexes; flaccidity of extremities. EEG markedly abnormal.	Risk for aspiration Impaired gas exchange Impaired tissue integrity

EEG, electroencephalogram.
*Nursing diagnoses are likely to progress; thus, most nursing diagnoses present at earlier stages will occur during later stages as well.

FIGURE 49-14 • Effects of constructional apraxia. Deterioration of handwriting and inability to draw a simple star figure occurs with progressive hepatic encephalopathy. With permission from Sherlock, S., & Dooley, J. (2002). *Diseases of the liver and biliary system* (11th ed.). Oxford, UK: Blackwell Scientific.

Occasionally, **fetor hepaticus**, a sweet, slightly fecal odor to the breath that is presumed to be of intestinal origin, may be noticed. The odor has also been described as similar to that of freshly mowed grass, acetone, or old wine. Fetor hepaticus is prevalent with extensive collateral portal circulation in chronic liver disease.

Assessment and Diagnostic Findings

Several diagnostic algorithms and a variety of psychometric tests are used in determining the presence and severity of hepatic encephalopathy. The electroencephalogram shows generalized slowing, an increase in the amplitude of brain waves, and characteristic triphasic waves. The survival rate after a first episode of overt hepatic encephalopathy in patients with cirrhosis is approximately 40% at 1 year. Patients should be referred for liver transplantation after this initial episode (Khungar & Poordad, 2012b; Sundarum & Shaikh, 2011).

Medical Management

Medical management focuses on identifying and eliminating the precipitating cause, if possible, initiating ammonia-lowering therapy, minimizing potential medical complications of cirrhosis and depressed consciousness, and reversing the underlying liver disease, if possible. Correction of the possible reasons for the deterioration such as bleeding, electrolyte abnormalities, sedation, or azotemia is essential (Feldman et al., 2010; Khungar & Poordad, 2012b). Lactulose (Cephulac) is administered to reduce serum ammonia levels. It acts by trapping and expelling the ammonia in the feces (Karch, 2012). Two or three soft stools per day are desirable; this indicates that lactulose is performing as intended.

Quality and Safety Nursing Alert

The patient receiving lactulose is monitored closely for the development of watery diarrhea stools, because they indicate a medication overdose. Serum ammonia levels are closely monitored as well.

Possible side effects of lactulose include intestinal bloating and cramps, which usually disappear within a week. To mask the sweet taste, which some patients dislike, it can be diluted with fruit juice. The patient is closely monitored for hypokalemia and dehydration. Other laxatives are not prescribed during lactulose administration because their effects disturb dosage regulation. Lactulose may be administered by nasogastric tube or enema for patients who are comatose or for those in whom oral administration is contraindicated or impossible (Karch, 2012).

Other management strategies include IV administration of glucose to minimize protein breakdown, administration of vitamins to correct deficiencies, and correction of electrolyte imbalances (especially potassium). Antibiotics may also be added to the treatment regimen. Neomycin, metronidazole (Flagyl), and rifaximin (Xifaxan) have been used to reduce levels of ammonia-forming bacteria in the colon. However, no benefit has been shown for long-term treatment with these antibiotics (Khungar & Poordad, 2012b). Additional principles of management of hepatic encephalopathy include the following:

- Neurologic status is assessed frequently.
- Mental status is monitored by keeping a daily record of handwriting and arithmetic performance.
- I&O and body weight are recorded each day.
- Vital signs are measured and recorded every 4 hours.
- Potential sites of infection (peritoneum, lungs) are assessed frequently, and abnormal findings are reported promptly.
- Serum ammonia level is monitored daily.
- Protein intake is moderately restricted only in patients who are comatose or who have encephalopathy that is refractory to lactulose and antibiotic therapy (Chart 49-6). Long-term restriction of dietary protein to less than 1.2 g/kg daily should be avoided. If animal protein precipitates encephalopathy, vegetable or dairy proteins may be used because most patients can better tolerate a diet of vegetable protein (McDonald et al., 2010).
- Patients and families are advised about foods that are high in protein (e.g., meat, eggs), which may need to be limited in the diet for the short term to reduce production of ammonia.
- Enteral feeding is provided for patients whose encephalopathic state persists.

Chart 49-6 Nutritional Management of Hepatic Encephalopathy

- Minimize the formation and absorption of toxins, principally ammonia, from the intestine.
- Keep daily protein intake between 1.2 and 1.5 g/kg body weight per day.
- Avoid protein restriction if possible, even in those with encephalopathy.
- For patients who are truly protein intolerant, provide additional nitrogen in the form of an amino acid supplement. The use of branched-chain amino acids should be a consideration in patients with cirrhosis. It has improved outcomes in varied populations with the disease.
- Provide small, frequent meals and 3 small snacks per day in addition to a late-night snack before bed.

Adapted from Mueller, C. M. (2012). *The A.S.P.E.N. adult nutrition support core curriculum* (2nd ed.). Silver Spring, MD: American Society for Parenteral and Enteral Nutrition.

- Reduction in the absorption of ammonia from the GI tract is accomplished by the use of gastric suction, enemas, or oral antibiotics.
- Electrolyte status is monitored and corrected if abnormal.
- Sedatives, tranquilizers, and analgesic medications are discontinued.
- Benzodiazepine antagonists such as flumazenil (Romazicon) may be administered to improve encephalopathy, whether or not the patient has previously taken benzodiazepines. This action may have short-term efficacy because patients with hepatic encephalopathy have an increased concentration of benzodiazepine receptors.

Nursing Management

Table 49-3 presents the stages of hepatic encephalopathy, common signs and symptoms, and potential nursing diagnoses for each stage. The nurse is responsible for maintaining a safe environment to prevent injury, bleeding, and infection. The nurse administers the prescribed treatments and monitors the patient for the numerous potential complications. The potential for respiratory compromise is great given the patient's depressed neurologic status. The nurse encourages deep breathing and position changes to prevent the development of atelectasis, pneumonia, and other respiratory complications. Despite aggressive pulmonary care, patients may develop respiratory compromise. They may require intubation and mechanical ventilation to protect the airway, and they are frequently admitted to the ICU.

The nurse communicates with the patient's family to inform them about the patient's status and supports them by explaining the procedures and treatments that are part of the patient's care. If the patient recovers from hepatic encephalopathy and coma, rehabilitation is likely to be prolonged. Therefore, the patient and family will require assistance to understand the causes of this severe complication and to recognize that it may recur.

Promoting Home and Community-Based Care

Educating Patients About Self-Care

If the patient has recovered from hepatic encephalopathy and is to be discharged home, the nurse educates the family about subtle signs of recurrent encephalopathy. The goals for caloric intake and protein intake should be 35 to 40 kcal/kg body weight per day and 1 to 1.5 g/kg body weight per day (Mueller, 2012). Protein intake should not be limited too severely, because it has been shown that doing so worsens nutritional status and increases mortality (Kerwin & Nussbaum, 2011). Continued use of lactulose after discharge is not uncommon, and the patient and family should closely monitor its efficacy and side effects. They should also be cautioned that constipation can precipitate encephalopathy and should be prevented through the prescribed use of lactulose, which is crucial in preventing constipation.

Continuing Care

Referral for home care is warranted for the patient who returns home after recovery from hepatic encephalopathy. The home care nurse assesses the patient's physical and mental status and collaborates closely with the primary provider. The home visit provides an opportunity for the nurse to assess the home environment and the ability of the patient and family to monitor signs and symptoms and follow the treatment regimen. The nurse must evaluate the patient's fluid volume status and be alert for changes indicative of hypovolemia due to decreased intake and for decreased urine output associated with hepatorenal syndrome (see later discussion). Monitoring of laboratory values continues to be important, and the home care nurse must obtain prescriptions to correct abnormalities, especially electrolyte imbalances, which also can worsen encephalopathy.

The safety of the home environment is assessed closely to identify areas of risk for falls and other injuries. Home care visits are especially important if the patient lives alone because encephalopathy may affect the patient's ability to remember or follow the treatment regimen. The nurse reinforces previous education and reminds the patient and family about the importance of dietary restrictions, close monitoring, and follow-up. In addition, the nurse must observe the patient for subtle behavior changes of worsening hepatic encephalopathy. Patients with all types and stages of hepatic encephalopathy should have periodic neurologic evaluations to determine their cognitive function so that they do not engage in potentially harmful activities. Even subtle neuropsychiatric abnormalities may preclude patients from driving, operating machinery, or participating in other activities that require psychomotor coordination.

Patients and families may need additional support during those times that the patient exhibits mood disturbances and sleep disorders. Patients should be as active as possible during the day and develop a normal sleep–wake pattern. Sedating medications should be avoided because they may precipitate encephalopathy. Patients and families may require assistance in developing plans to cope with changes in mood and mental status changes. This plan should identify support persons to attend to the patient in the home situation if needed. Social workers and case managers may make appropriate referrals for assistance with physical and psychosocial support and care. Referrals to psychologists, psychiatric liaison nurses, case managers, social workers, or therapists may assist family members with coping. Spiritual advisors may also provide another outlet for communication and guidance. If alcohol played a role in the development of the liver disease and encephalopathy, referral to Alcoholics Anonymous or Al-Anon may provide needed support and education.

Other Manifestations of Hepatic Dysfunction

Edema and Bleeding

Many patients with liver dysfunction develop generalized edema caused by hypoalbuminemia due to decreased hepatic production of albumin. The production of blood clotting factors by the liver is also reduced, leading to an increased incidence of bruising, epistaxis, bleeding from wounds, and, as described previously, GI bleeding. Abnormalities in the number and effectiveness of platelets also contribute to the bleeding in liver dysfunction. Congestion of the spleen secondary to portal hypertension causes increased pooling of platelets in the organ (hypersplenism). The resultant thrombocytopenia generally correlates with spleen size. In patients who are alcoholics, suppression of bone marrow by the acute toxic effects of alcohol or folate deficiency may contribute to the thrombocytopenia (Goldman & Shafer, 2012). These factors predispose patients to easy bruising, petechiae formation, and bleeding from a variety of sources such as the GI or genitourinary tract (Goldman & Shafer, 2012).

Vitamin Deficiency

Decreased production of several clotting factors may be partially due to deficient absorption of vitamin K from the GI tract. This probably is caused by the inability of liver cells to use vitamin K to make prothrombin (Hall, 2011). Absorption of the other fat-soluble vitamins (vitamins A, D, and E) as well as dietary fats may also be impaired because of decreased secretion of bile salts into the intestine.

Another group of problems common to patients with severe chronic liver dysfunction results from inadequate intake of sufficient vitamins. These include the following:

- Vitamin A deficiency, resulting in night blindness and eye and skin changes
- Thiamine deficiency, leading to beriberi, polyneuritis, and Wernicke-Korsakoff psychosis
- Riboflavin deficiency, resulting in characteristic skin and mucous membrane lesions
- Pyridoxine deficiency, resulting in skin and mucous membrane lesions and neurologic changes
- Vitamin C deficiency, resulting in the hemorrhagic lesions of scurvy
- Vitamin K deficiency, resulting in hypoprothrombinemia, characterized by spontaneous bleeding and ecchymoses
- Folic acid deficiency, resulting in macrocytic anemia

Because of these avitaminoses, the diet of every patient with chronic liver disease (especially if alcohol related) is supplemented with vitamins A, B complex, C, K, and folic acid.

Metabolic Abnormalities

Abnormalities of glucose metabolism also occur; the blood glucose level may be abnormally high shortly after a meal (similar to that when diabetes is present), but hypoglycemia may occur during fasting because of decreased hepatic glycogen reserves and decreased gluconeogenesis. Medications must be used cautiously and in reduced dosages because the ability to metabolize medications is decreased in the patient with liver failure.

Many endocrine abnormalities also occur with liver dysfunction because the liver cannot properly metabolize hormones, including androgens and sex hormones. Failure of the damaged liver to inactivate estrogens normally can cause gynecomastia, amenorrhea, testicular atrophy, loss of pubic hair in the male, menstrual irregularities in the female, and other disturbances of sexual function and sex characteristics.

Pruritus and Other Skin Changes

Patients with liver dysfunction resulting from biliary obstruction commonly develop severe pruritus due to retention of bile salts. Patients may develop vascular (or arterial) spider angiomas on the skin, usually above the waistline. These are numerous small vessels resembling a spider's legs. They are most often associated with cirrhosis, especially in alcoholic liver disease. Patients may also develop reddened palms ("liver palms" or palmar erythema).

VIRAL HEPATITIS

Viral hepatitis is a systemic, viral infection in which necrosis and inflammation of liver cells produce a characteristic cluster of clinical, biochemical, and cellular changes. To date, five definitive types of viral hepatitis that cause liver disease have been identified: hepatitis A, B, C, D, and E. Hepatitis A and E are similar in mode of transmission (fecal–oral route), whereas hepatitis B, C, and D share many other characteristics.

Hepatitis is easily transmitted and causes high morbidity and prolonged loss of time from school or employment. Acute viral hepatitis affects 0.5% to 1% of people in the United States each year. Hepatitis A was responsible for 1,670 cases in the United States in 2010, with an overall incidence of 0.5 cases per 100,000 cases. During the same year, hepatitis B was the offending agent in a total of 3,350 cases of acute viral hepatitis nationwide, with an incidence of 1.1 cases per 100,000 population. The occurrence rate of viral hepatitis C in 2010 was 850 cases, with an incidence rate of 0.3 cases per 100,000 population, which represents an increase of approximately 6% since 2006 (Centers for Disease Control & Prevention [CDC], 2010). The occurrence rate for hepatitis A and B has been decreasing steadily since 1990, largely because of the use of hepatitis A and B vaccines as well as public health education regarding high-risk behaviors (Goldman & Schafer, 2012). It is estimated that 60% to 90% of viral hepatitis cases go unreported. The occurrence of subclinical cases, failure to recognize mild cases, and misdiagnosis are thought to contribute to the underreporting. Table 49-4 compares the major forms of viral hepatitis.

Hepatitis A Virus

Hepatitis A virus (HAV) accounts for 20% to 25% of cases of clinical hepatitis in the United States (CDC, 2010). Hepatitis A, formerly called *infectious hepatitis*, is caused by an RNA virus of the enterovirus family. In the United States, the disease is seen mainly in the adult population. Fewer than 25% of children have antibodies to HAV. This form of hepatitis is transmitted primarily through the fecal–oral route, by the ingestion of food or liquids infected with the virus. It is more prevalent in countries with overcrowding and poor sanitation. The virus has been found in the stool of infected patients before the onset of symptoms and during the first few days of illness.

Typically, a child or a young adult acquires the infection at school through poor hygiene, hand-to-mouth contact, or other close contact. The virus is carried home, where haphazard sanitary habits spread it through the family. An infected food handler can spread the disease, and people can contract it by consuming water or shellfish from sewage-contaminated waters. Outbreaks have occurred in day care centers and institutions as a result of poor hygiene among

TABLE 49-4 Comparison of Major Forms of Viral Hepatitis

	Hepatitis A	Hepatitis B	Hepatitis C	Hepatitis D	Hepatitis E
Previous names	Infectious hepatitis	Serum hepatitis	Non-A, non-B hepatitis		
Epidemiology					
Cause	Hepatitis A virus (HAV)	Hepatitis B virus (HBV)	Hepatitis C virus (HCV)	Hepatitis D virus (HDV)	Hepatitis E virus (HEV)
Mode of transmission	Fecal–oral route; poor sanitation. Person-to-person contact. Waterborne; foodborne. Transmission possible with oral–anal contact during sex.	Parenterally; by intimate contact with carriers or those with acute disease; sexual and oral–oral contact. Perinatal transmission from mothers to infants. An important occupational hazard for health care personnel.	Transfusion of blood and blood products; exposure to contaminated blood through equipment or drug paraphernalia. Transmission possible with sex with infected partner; risk increased with sexually transmitted infection.	Same as HBV. HBV surface antigen necessary for replication; pattern similar to that of hepatitis B.	Fecal–oral route; person to person contact may be possible, although risk appears low.
Incubation	15–50 d *Average*: 30 d	28–160 d *Average*: 70–80 d	15–160 d *Average*: 50 d	Same as hepatitis B *Average*: 35 d	28–35 d *Average*: 31 d
Immunity	Homologous	Homologous	Second attack may indicate weak immunity or infection with another agent.	Homologous	Unknown
Nature of Illness					
Signs and symptoms	May occur with or without symptoms; flulike illness *Preicteric phase*: Headache, malaise, fatigue, anorexia, fever *Icteric phase*: Dark urine, jaundice of sclera and skin, tender liver	May occur without symptoms May develop arthralgias, rash	Similar to HBV; less severe and anicteric	Similar to HBV	Similar to HAV; very severe in pregnant women
Outcome	Usually mild with recovery. Fatality rate <1%. No carrier state or increased risk of chronic hepatitis, cirrhosis, or hepatic cancer.	May be severe. Fatality rate 1%–10%. Carrier state possible. Increased risk of chronic hepatitis, cirrhosis, and hepatic cancer.	Frequent occurrence of chronic carrier state and chronic liver disease. Increased risk of hepatic cancer.	Similar to HBV but greater likelihood of carrier state, chronic active hepatitis, and cirrhosis	Similar to HAV except very severe in pregnant women

Adapted from Kumar, V., Abbas, A. K., Fausto, N., et al. (2010). *Robbins and Cotran pathologic basis of disease* (8th ed.). Philadelphia: Saunders Elsevier; and Greenberger, N. J., Blumberg, R. S., & Burakoff, R. (Eds.). (2012). *Current diagnosis and treatment: Gastroenterology, hepatology & endoscopy*. New York: McGraw-Hill.

people with developmental disabilities. Hepatitis A can be transmitted during sexual activity; this is more likely with oral–anal contact or anal intercourse and with multiple sex partners (Goldman & Schafer, 2012). It is rarely, if ever, transmitted by blood transfusions.

The incubation period is estimated to be between 2 and 6 weeks, with a mean of approximately 4 weeks (CDC, 2010). The illness may be prolonged, lasting 4 to 8 weeks. It usually lasts longer and is more severe in those older than 40 years. Most patients recover from hepatitis A; it rarely progresses to acute liver necrosis or fulminant hepatic failure resulting in cirrhosis of the liver or death. The mortality rate of hepatitis A is approximately 0.5% for those younger than 40 years and 1% to 2% for older people. In patients with underlying chronic liver disease, morbidity and mortality are increased in the presence of an acute hepatitis A infection. No carrier state exists, and no chronic hepatitis is associated with hepatitis A. The virus is present only briefly in the serum; by the time jaundice occurs, the patient is likely to be noninfectious. Although hepatitis A confers immunity against itself, the person may contract other forms of hepatitis.

Clinical Manifestations

Many patients are anicteric (without jaundice) and symptomless. When symptoms appear, they resemble those of a mild, flulike upper respiratory tract infection, with low-grade fever. Anorexia, an early symptom, is often severe. It is thought to result from release of a toxin by the damaged liver or from failure of the damaged liver cells to detoxify an abnormal product. Later, jaundice and dark urine may become apparent. Indigestion is present in varying degrees, marked by vague epigastric distress, nausea, heartburn, and flatulence. The patient may also develop a strong aversion to the taste of cigarettes or the presence of cigarette smoke and other strong odors (Papadakis & McPhee, 2013). These symptoms tend to clear as soon as the jaundice reaches its peak, perhaps 10 days after its initial appearance. Symptoms may be mild in children; in adults, they may be more severe and the course of the disease prolonged.

Assessment and Diagnostic Findings

The liver and spleen are often moderately enlarged for a few days after onset; other than jaundice, there are few other physical signs. Hepatitis A antigen may be found in the stool 7 to 10 days before illness and for 2 to 3 weeks after symptoms appear. HAV antibodies are detectable in the serum, although usually not until symptoms appear. Analysis of subclasses of immunoglobulins can help determine whether the antibody represents acute or past infection.

Prevention

A number of strategies exist to prevent transmission of HAV. Patients and their families are encouraged to follow general precautions that can prevent transmission of the virus. Scrupulous hand hygiene, safe water supplies, and proper control of sewage disposal are just a few of these prevention strategies.

Effective (95% to 100% after two to three doses) and safe HAV vaccines include Havrix and Vaqta (Mandell, Bennett & Dolin, 2010). It is recommended that the two-dose vaccine be given to adults 18 years of age or older, with the second dose given 6 to 12 months after the first. Protection against hepatitis A develops within several weeks after the first dose of the vaccine. Children and adolescents 2 to 18 years of age receive three doses; the second dose is given 1 month after the first, and the third dose is given 6 to 12 months later. Hepatitis A routine immunization of young children has proved to be effective in reducing disease incidence and maintaining very low incidence levels among vaccine recipients and across all age groups in many settings (Goldman & Schafer, 2012). Hepatitis A vaccine is recommended for people traveling to locations where sanitation and hygiene are unsatisfactory. Vaccination is also recommended for those from high-risk groups, such as homosexual men, IV or injection drug users, staff of day care centers, and health care personnel (Mandell et al., 2010). The vaccine has also been used to interrupt community-wide outbreaks. A combined hepatitis A and B vaccine (Twinrix) is available for vaccination of people 18 years of age and older with indications for both hepatitis A and B vaccination. Vaccination consists of three doses, given on the same schedule as that used for single-antigen hepatitis B vaccine.

For people who have not been previously vaccinated, hepatitis A can be prevented by intramuscular administration of globulin during the incubation period, if given within 2 weeks of exposure. This bolsters the person's antibody production and provides 6 to 8 weeks of passive immunity. Immune globulin may suppress overt symptoms of the disease; the resulting subclinical case of hepatitis A would produce immunity to subsequent episodes of the virus.

Immune globulin is also recommended for household members and sexual contacts of people with hepatitis A. Susceptible people in the same household as the patient are usually also infected by the time the diagnosis is made and should receive immune globulin. Institutional contacts of patients with hepatitis A should also receive postexposure prophylaxis with immune globulin. Prophylaxis is not necessary for casual contacts of an infected person, such as classmates, coworkers, or hospital employees (Mandell et al., 2010). Although rare, systemic reactions to immune globulin do occur. Caution is required when anyone who has previously had angioedema, hives, or other allergic reactions is treated with any human immune globulin. Epinephrine should be available in case of systemic, anaphylactic reaction.

Pre-exposure prophylaxis is recommended for those traveling to developing countries or settings with poor or uncertain sanitation conditions who do not have sufficient time to acquire protection by administration of hepatitis A vaccine (Mandell et al., 2010). Prevention strategies for hepatitis A are outlined in Chart 49-7.

Medical Management

Bed rest during the acute stage and a nutritious diet are part of the treatment and nursing care. During the period of anorexia, the patient should receive frequent small feedings, supplemented if necessary by IV fluids with glucose. Because the patient often has an aversion to food, gentle persistence and creativity may be required to stimulate appetite. Optimal food and fluid levels are necessary to counteract weight loss and to speed recovery. Even before the icteric phase, however, many patients recover their appetites (Chart 49-8).

The patient's sense of well-being and laboratory test results are generally appropriate guides to bed rest and restriction of physical activity. Gradual but progressive ambulation hastens recovery, provided the patient rests after activity and does not participate in activities to the point of fatigue (Wu, Wu, Lien, et al., 2012).

Nursing Management

Management usually occurs in the home unless symptoms are severe. Therefore, the nurse assists the patient and family in coping with the temporary disability and fatigue that are common in hepatitis and educates them to seek additional health care if the symptoms persist or worsen. The patient and family also need specific guidelines about diet, rest, follow-up blood work, and the importance of avoiding alcohol, as well as sanitation and hygiene measures (particularly hand hygiene) to prevent spread of the disease to other family members.

Specific education for patients and families about reducing the risk of contracting hepatitis A includes good personal hygiene, stressing careful hand hygiene (after bowel movements and before eating) and environmental sanitation (safe food and water supply, effective sewage disposal).

Hepatitis B Virus

Unlike HAV, the hepatitis B virus (HBV) is transmitted primarily through blood (percutaneous and permucosal routes). HBV can be found in blood, saliva, semen, and vaginal secretions and can be transmitted through mucous membranes and breaks in the skin. HBV is also transferred from carrier mothers to their infants, especially in areas with a high incidence (e.g., Southeast Asia). The infection usually is not transmitted via the umbilical vein but from the mother at the time of birth and during close contact afterward.

HBV has a long incubation period. It replicates in the liver and remains in the serum for relatively long periods, allowing transmission of the virus. Risk factors for HBV infection are summarized in Chart 49-9. Screening of blood donors has greatly reduced the occurrence of hepatitis B after blood transfusion.

Chart 49-7 HEALTH PROMOTION

Prevention of Hepatitis

Hepatitis A

- Encourage proper community and home sanitation.
- Encourage conscientious individual hygiene.
- Educate patients regarding safe practices for preparing and dispensing food.
- Support effective health supervision of schools, dormitories, extended care facilities, barracks, and camps.
- Promote community health education programs.
- Facilitate mandatory reporting of viral hepatitis to local health departments.
- Recommend vaccination for travelers to developing countries, illegal drug users (injection and noninjection drug users), men who have sex with men, people with chronic liver disease, and recipients (e.g., hemophiliacs) of pooled plasma products.
- Recommend pre-exposure vaccination for all children 12–23 months of age. Continue existing immunization programs for children 2–18 years of age
- Promote vaccination to interrupt community-wide outbreaks.

Hepatitis B

- Recommend vaccination for persons at risk for infection by sexual exposure, by percutaneous or mucosal exposure to blood.
- Recommend vaccination for international travelers to regions with high or intermediate levels of endemic hepatitis B virus infection and for persons with chronic liver disease or with human immunodeficiency virus infection.
- Recommend vaccination of all infants in the United States regardless of the mother's hepatitis B.
- Advise avoidance of high-risk behaviors.
- Use standard precautions in clinical care.
- Use needleless IV and injection systems in health care.
- Use barrier precautions in situations of contact with blood or body fluids.
- Monitor cleaning, disinfection, and sterilization of reusable devices in patient care settings.
- Avoid multidose vials in patient care settings.

Hepatitis C

- Advise avoidance of high risk behaviors such as IV drug use.
- Use needleless IV and injection systems in health care.
- Use barrier precautions in situations of contact with blood or body fluids.
- Monitor cleaning, disinfection, and sterilization of reusable devices in patient care settings.
- Avoid multidose vials in patient care settings.
- Use standard precautions in clinical care.

Adapted from Ferri, F. F. (Ed.). (2011). *Practical guide to the care of the medical patient* (8th ed.). Philadelphia: Mosby Elsevier; Michelin, A., & Henderson, D. K. (2010). Infection control guidelines for prevention of health-care associated transmission of hepatitis B and C viruses. *Clinics in Liver Disease, 14*(1), 119–136; and MacCannell, T., Laramie, A. K., Gomaa, A., et al. (2010). Occupational exposure of health care personnel to hepatitis B and hepatitis C: Prevention and surveillance strategies. *Clinics in Liver Disease, 14*(1), 23–36.

Most people (more than 90%) who contract HBV infection develop antibodies and recover spontaneously in 6 months. The mortality rate from acute hepatitis B has been reported to be as high as 1%. Another 10% of patients who have hepatitis B progress to a carrier state or develop chronic hepatitis with persistent HBV infection and hepatocellular injury and inflammation. It remains a major worldwide cause of cirrhosis and hepatocellular carcinoma (HCC) with higher mortality rates (Papadakis & McPhee, 2013). In fact, approximately 15% of those who develop chronic hepatitis B during adulthood die of cirrhosis or liver cancer. Mortality rates are even higher (25%) for those whose chronic infection occurs during childhood (Wasley, Kruszon-Moran, Kuhnert, et al., 2010). The numbers of chronically infected children in the United States has been estimated to have declined since the incorporation of domestic and worldwide routine infant and childhood vaccination programs. Although an estimated 730,000 adult residents of the United States are afflicted with chronic hepatitis B infection, there has been a small but significant decrease in the prevalence among U.S.-born adults who are 20 to 49 years of age (Wasley et al., 2010).

Chart 49-8 **Dietary Management of Hepatitis**

- Recommend small, frequent meals; minimize periods without food intake.
- Provide intake of 25–30 kcal/day.
- Provide protein intake of 1–1.5 g/kg/day.
- Carefully monitor fluid balance.
- Be aware that that enteral feedings may be necessary if anorexia, nausea, and vomiting persist.
- Instruct patient to abstain from alcohol during acute illness and for at least 6 months after recovery.
- Advise patient to avoid substances (medications, herbs, illicit drugs, and toxins) that may affect liver function, such as St. John's wort in patients taking hepatitis C virus protease inhibitors.

Adapted from Mueller, C. M. (2012). *The A.S.P.E.N. adult nutrition support core curriculum* (2nd ed.). Silver Spring, MD: American Society for Parenteral and Enteral Nutrition.

Chart 49-9 RISK FACTORS

Hepatitis B

- Frequent exposure to blood, blood products, or other body fluids
- Health care workers: hemodialysis staff, oncology and chemotherapy nurses, personnel at risk for needlesticks, operating room staff, respiratory therapists, surgeons, dentists
- Hemodialysis
- Male homosexual and bisexual activity
- IV/injection drug use
- Close contact with carrier of hepatitis B virus
- Mother-to-child transmission
- Travel to or residence in area with uncertain sanitary conditions
- Multiple sexual partners
- Recent history of sexually transmitted infection
- Receipt of blood or blood products (e.g., clotting factor concentrate)
- Tattooing

Adapted from Bope, E. T., & Kellerman, R. D. (Eds.). (2011). *Conn's current therapy.* Philadelphia: Saunders.

Gerontologic Considerations

The immune system is altered in the aged. A less responsive immune system may be responsible for the increased incidence and severity of hepatitis B among older adults and the increased incidence of liver abscesses secondary to decreased phagocytosis by the Kupffer cells. The older patient with hepatitis B has a serious risk of severe liver cell necrosis or fulminant hepatic failure, particularly if other illnesses are present. With the advent of hepatitis B vaccine as the standard for prevention, the incidence of hepatic diseases may decrease in the future.

Clinical Manifestations

Clinically, HBV closely resembles hepatitis A, but the incubation period is much longer (1 to 6 months). Signs and symptoms of hepatitis B may be insidious and variable. Fever and respiratory symptoms are rare; some patients have arthralgias and rashes. The patient may have loss of appetite, dyspepsia, abdominal pain, generalized aching, malaise, and weakness. Jaundice may or may not be evident. If jaundice occurs, light-colored stools and dark urine accompany it. The liver may be tender and enlarged to 12 to 14 cm vertically. The spleen is enlarged and palpable in a few patients; the posterior cervical lymph nodes may also be enlarged. Subclinical episodes also occur frequently.

Assessment and Diagnostic Findings

HBV is a deoxyribonucleic acid (DNA) virus composed of the following antigenic particles:

- HBcAg—hepatitis B core antigen (antigenic material in an inner core)
- HBsAg—hepatitis B surface antigen (antigenic material on the viral surface, a marker of active replication and infection)
- HBeAg—an independent protein circulating in the blood
- HBxAg—gene product of X gene of HBV DNA

Each antigen elicits its specific antibody and is a marker for different stages of the disease process:

- anti-HBc—antibody to core antigen of HBV; persists during the acute phase of illness; may indicate continuing HBV in the liver
- anti-HBs—antibody to surface determinants on HBV; detected during late convalescence; usually indicates recovery and development of immunity
- anti-HBe—antibody to hepatitis B e-antigen; usually signifies reduced infectivity
- anti-HBxAg—antibody to the hepatitis B x-antigen; may indicate ongoing replication of HBV

HBsAg appears in the circulation in 80% to 90% of infected patients 1 to 10 weeks after exposure to HBV and 2 to 8 weeks before the onset of symptoms or an increase in transferase levels. Patients with HBsAg that persists for 6 months or longer after acute infection are considered to be HBsAg carriers (Gluud & Gluud, 2009). HBeAg is the next antigen of HBV to appear in the serum. It usually appears within 1 week of the appearance of HBsAg but before changes in aminotransferase levels; it disappears from the serum within 2 weeks. HBV DNA, detected by polymerase chain reaction testing, appears in the serum at about the same time as HBeAg. HBcAg is not always detected in the serum in HBV infection.

The prevalence of chronic hepatitis B infection in the United States was 0.27% during the years 1999 to 2008 in persons 6 years or older. This is a slightly lower percentage of the population than past national estimates during the years 1976 to 1980. The prevalence of exposure to the virus (past or chronic infection) was 4.6% during the years 1999 to 2008—also a lower national estimate than reported previously (Ioannou, 2011).

Prevention

Preventing Transmission

Continued screening of blood donors for the presence of hepatitis B antigen (HBAg) further decreases the risk of transmission by blood transfusion. The use of disposable syringes, needles, and lancets and the introduction of needleless IV administration systems have reduced the risk of spreading this infection from one patient to another or to health care personnel during the collection of blood samples or the administration of parenteral therapy. In the clinical laboratory and the hemodialysis unit, work areas are disinfected daily. Gloves are worn when handling all blood and body fluids, as well as HBAg–positive specimens, or when there is potential exposure to blood (e.g., blood drawing) or to patients' secretions. Eating and smoking are prohibited in the laboratory and in other areas exposed to secretions, blood, or blood products. Patient education regarding the nature of the disease, its infectiousness, and prognosis is a critical factor in preventing transmission and protecting contacts (see Chart 49-7).

Active Immunization: Hepatitis B Vaccine

Active immunization is recommended for people who are at high risk for hepatitis B (e.g., health care personnel, hemodialysis patients). In addition, people with hepatitis C and other chronic liver diseases should receive the vaccine. A yeast-recombinant hepatitis B vaccine (Recombivax HB) is used to provide active immunity and has shown rates of protection greater than 90% in healthy people (Mandell et al., 2010). Although antibody levels may become low or undetectable, immunologic memory may remain intact for at least 5 to 10 years. Measurable levels of antibodies may not be essential for protection. In general, in those with normal immune systems, booster doses are not required, and no data support the use of booster doses of hepatitis B vaccine among immunocompetent people who have responded to the vaccination series. However, booster doses are recommended for people who are immunocompromised (Mandell et al., 2010). Additional information is required to determine if booster injections are needed for adults 15 years or more after initial vaccination as well as those at high risk for HBV infection.

A hepatitis B vaccine prepared from plasma of humans chronically infected with HBV is used only rarely in patients who are immunodeficient or allergic to recombinant yeast-derived vaccines.

Both forms of the hepatitis B vaccine are administered intramuscularly in three doses; the second and third doses are given 1 and 6 months, respectively, after the first dose. The third dose is very important in producing prolonged immunity. Hepatitis B vaccination should be administered to adults in the deltoid muscle. Antibody response may be measured by anti-HBs levels 1 to 3 months after completion of the basic course of vaccine, but this testing is not routine and is not currently recommended. People who do not respond may benefit from one to three additional doses (Mandell et al., 2010).

People at high risk, including nurses and other health care personnel exposed to blood or blood products, should receive active immunization. Health care workers who have had frequent contact with blood are screened for anti-HBs to determine whether immunity is already present from previous exposure. The vaccine produces active immunity to HBV in 90% of healthy people (MacCannell, Laramie, Gomaa, et al., 2010). It does not provide protection to those already exposed to HBV, and it provides no protection against other types of viral hepatitis.

Because hepatitis B infection is frequently transmitted sexually, hepatitis B vaccination is recommended for all unvaccinated people being evaluated for a sexually transmitted infection (STI). It is also recommended for those with a history of an STI, people with multiple sex partners, people who have sex with IV or injection drug users, and sexually active men who have sex with other men (Feldman et al., 2010; Mandell et al., 2010).

Universal childhood vaccination for hepatitis B prevention has been instituted in the United States, and universal vaccination of all infants is encouraged. Catch-up vaccination is recommended for all children and prepubertal adolescents up to the age of 19 years who have not been previously immunized (Feldman et al., 2010). Development of chronic carrier states has not been reported in adult responders to the vaccine.

Passive Immunity: Hepatitis B Immune Globulin

Hepatitis B immune globulin (HBIG) provides passive immunity to hepatitis B and is indicated for people exposed to HBV who have never had hepatitis B and have never received hepatitis B vaccine. Specific indications for postexposure vaccine with HBIG include (1) inadvertent exposure to HBAg-positive blood through percutaneous (needlestick) or transmucosal (splashes in contact with mucous membrane) routes, (2) sexual contact with people positive for HBAg, and (3) perinatal exposure (infants born to HBV-infected mothers should receive HBIG within 12 hours after delivery). HBIG is prepared from plasma selected for high titers of anti-HBs. Prompt immunization with HBIG (within hours to a few days after exposure to hepatitis B) increases the likelihood of protection. Both active and passive immunization are recommended for people who have been exposed to hepatitis B through sexual contact or through the percutaneous or transmucosal routes. If HBIG and hepatitis B vaccine are administered at the same time, separate sites and separate syringes should be used. HBIG is considered very safe, and there has been no evidence that infectious diseases have been transmitted due its administration (Roberts & Hedges, 2010).

Medical Management

Goals are to minimize infectivity and liver inflammation and decrease symptoms. Of all the agents that have been used to treat chronic type B viral hepatitis, alpha-interferon is the single modality of therapy that offers the most promise. A regimen of 5 million U daily or 10 million U three times weekly for 16 to 24 weeks results in remission of disease in approximately one third of patients (Ferri, 2011). A prolonged course of treatment may also have additional benefits and is under study. Interferon must be administered by injection and has significant side effects, including fever, chills, anorexia, nausea, myalgias, and fatigue. Delayed side effects are more serious and may necessitate dosage reduction or discontinuation. These include bone marrow suppression, thyroid dysfunction, alopecia, and bacterial infections. Several recombinant forms of alpha-interferon are also available, including the pegylated form (peginterferon alfa-2a [Pegasys]), with once-weekly dosing. Pegylated interferon, also referred to as peginterferon, has largely replaced standard interferon due to its dosing schedule (Ferri, 2011).

Two antiviral agents, lamivudine (Epivir) and adefovir (Hepsera), which are oral nucleoside analogues, have been approved for use in chronic hepatitis B in the United States. Studies have revealed improved seroconversion rates, loss of detectable virus, improved liver function, and reduced progression to cirrhosis with lamivudine. It can be used for patients with decompensated cirrhosis who are awaiting liver transplantation (Ferri, 2011). Patients with decompensated cirrhosis have such severely damaged liver parenchyma that normal liver function severely deteriorates, resulting in life-threatening ascites, encephalopathy, or variceal hemorrhage (Feldman et al., 2010). Adefovir may be effective in people who are resistant to lamivudine.

Bed rest may be recommended until the symptoms of hepatitis have subsided. Activities are restricted until the hepatic enlargement and levels of serum bilirubin and liver enzymes have decreased. Gradually increased activity is then allowed.

Adequate nutrition should be maintained. Proteins are restricted if symptoms indicate that the liver's ability to metabolize protein by-products is impaired. Measures to control the dyspeptic symptoms and general malaise include the use of antacids and antiemetic agents, but all medications should be avoided if vomiting occurs. If vomiting persists, the patient may require hospitalization and fluid therapy. Because of the mode of transmission, the patient is evaluated for other bloodborne diseases (e.g., human immunodeficiency virus infection).

Nursing Management

Convalescence may be prolonged, with complete symptomatic recovery sometimes requiring 3 to 4 months or longer (Papadakis & McPhee, 2013). During this stage, gradual resumption of physical activity is encouraged after the jaundice has resolved.

The nurse identifies psychosocial issues and concerns, particularly the effects of separation from family and friends if the patient is hospitalized during the acute and infective stages. Even if not hospitalized, the patient will be unable to work and must avoid sexual contact. Planning is required to minimize social isolation. Planning that includes the family helps to reduce their fears and anxieties about the spread of the disease.

Promoting Home and Community-Based Care

Educating Patients About Self-Care

Because of the prolonged period of convalescence, the patient and family must be prepared for home care. Provision for adequate rest and nutrition must be ensured. The nurse educates family members and friends who have had intimate contact with the patient about the risks of contracting hepatitis B and makes arrangements for them to receive hepatitis B vaccine or HBIG as prescribed. Those at risk must be made aware of the early signs of hepatitis B and of ways to reduce risk by avoiding all modes of transmission. Patients with all forms of hepatitis should avoid drinking alcohol (Feldman et al., 2010).

Continuing Care

Follow-up visits by a home care nurse may be needed to assess the patient's progress and answer family members' questions about disease transmission. During a home visit, the nurse assesses the patient's physical and psychological status and confirms that the patient and family understand the importance of adequate rest and nutrition. The nurse also reinforces previous education. Because of the risk of transmission through sexual intercourse, strategies to prevent exchange of body fluids are recommended, such as abstinence or the use of condoms. The nurse emphasizes the importance of keeping follow-up appointments and participating in other health promotion activities and recommended health screenings.

Hepatitis C Virus

Blood transfusions and sexual contact once accounted for most cases of hepatitis C in the United States, but other parenteral means, such as sharing of contaminated needles by IV or injection drug users and unintentional needlesticks and other injuries in health care workers now account for a significant number of cases. Approximately 35,000 new cases of hepatitis C are reported in the United States each year. About 4 million people (1.8% of the U.S. population) have been infected with the hepatitis C virus (HCV), making it the most common chronic bloodborne infection nationally. A fourfold increase in the number of adults diagnosed with HCV infection is projected from 1990 to 2015. The highest prevalence of hepatitis C is in adults 40 to 59 years of age; in this age group, its prevalence is highest among African Americans. There are 10,000 to 12,000 deaths each year in the United States due to hepatitis C, and it has been suggested that deaths from this cause are underestimated. HCV is the underlying cause of about one third of cases of HCC, and it is the most common reason for liver transplantation (Gluud & Gluud, 2009; Fortune, Rosen & Burton, 2010).

People who are at particular risk for hepatitis C include IV or injection drug users, sexually active people with multiple partners, patients receiving frequent transfusions, those who require large volumes of blood, and health care personnel (Chart 49-10). The incubation period is variable and may range from 15 to 160 days. The clinical course of acute hepatitis C is similar to that of hepatitis B; symptoms are usually mild. However, a chronic carrier state occurs frequently, and there is an increased risk of chronic liver disease, including cirrhosis or liver cancer, after hepatitis C. Small amounts of alcohol taken regularly appear to cause progression of the disease. Therefore, alcohol and medications that may affect the liver should be avoided.

Chart 49-10 **RISK FACTORS**

Hepatitis C

- Recipient of blood products or organ transplant before 1992 or clotting factor concentrates before 1987
- Health care and public safety workers after needlestick injuries or mucosal exposure to blood
- Children born to women infected with hepatitis C virus
- Past/current illicit IV/injection drug use
- Multiple contacts with a hepatitis C virus–infected person
- Multiple sex partners, history of sexually transmitted infection, unprotected sex

Adapted from Kumar, V., Abbas, A.cK., Fausto, N., et al. (2010). *Robbins and Cotran pathologic basis of disease* (8th ed.). Philadelphia: Saunders Elsevier.

There is no benefit from rest, diet, or vitamin supplements. Studies have demonstrated that a combination of two antiviral agents, peginterferon and ribavirin (Rebetol), is effective in producing improvement in patients with hepatitis C and in treating relapses. Some patients experience complete remission with combination therapy (McDonald et al., 2010). Peginterferon (Pegasys) has become the preferred agent for treatment, largely replacing standard interferon in hepatitis C. A polyethylene glycol (PEG) moiety (portion of a molecule) is added to the interferon to keep it in the body longer without reducing its efficacy; this extends the dosing interval to once a week. Some studies have shown it to have a somewhat improved virologic response rate compared with interferon (McDonald et al., 2010). Hemolytic anemia, the most frequent side effect, may be severe enough to require discontinuation of treatment. Ribavirin must be used with caution in women of childbearing age.

The U.S. Food and Drug Administration (FDA) has approved the protease inhibitors telaprevir (Incivek) and boceprevir (Victrelis) for the treatment of hepatitis C genotype 1 in combination with peginterferon and ribavirin. Neither protease inhibitor may be used as monotherapy for the disease (FDA, 2011a, 2011b). The triple therapy of protease inhibitor, peginterferon, and ribavirin is recommended as standard treatment for hepatitis C by the American Association for the Study of Liver Diseases (Ghany, Nelson, Strader, et al., 2011).

Screening of blood has reduced the incidence of hepatitis C associated with blood transfusion, and public health programs are helping to reduce the number of cases associated with shared needles in IV or injection drug use (see Chart 49-7).

Hepatitis D Virus

Hepatitis D virus (delta agent) infection occurs in some cases of hepatitis B. Because the virus requires HBsAg for its replication, only people with hepatitis B are at risk for hepatitis D. Anti-delta antibodies in the presence of HBAg on testing confirm the diagnosis. Hepatitis D is common among IV or injection drug users, hemodialysis patients, and recipients of multiple blood transfusions. Sexual contact with those who have hepatitis B is considered to be an important mode of transmission of hepatitis B and D. The incubation period varies between 30 and 150 days (Feldman et al., 2010).

The symptoms of hepatitis D are similar to those of hepatitis B, except that patients are more likely to develop fulminant hepatic failure and to progress to chronic active hepatitis and cirrhosis. Treatment is similar to that of other forms of hepatitis. Currently, interferon alfa is the only licensed drug available in the treatment for hepatitis D viral infection. The rate of recurrence is high, and the efficacy of interferon is related to the dose and duration of treatment. High-dose, long-duration therapy for at least a year is recommended (Greenberger, Blumberg, & Burakoff, 2012).

Hepatitis E Virus

It is believed that hepatitis E virus (HEV) is transmitted by the fecal–oral route, principally through contaminated water in areas with poor sanitation. The incubation period is variable, estimated to range between 15 and 65 days. In general, hepatitis E resembles hepatitis A. It has a self-limited course with an abrupt onset. Jaundice is almost always present. Chronic forms do not develop.

Avoiding contact with the virus through good hygiene, including handwashing, is the major method of prevention of hepatitis E. The effectiveness of immune globulin in protecting against HEV is uncertain.

Hepatitis G Virus and GB Virus-C

It has long been believed that there is another non-A–E agent causing hepatitis in humans. The incubation period for posttransfusion hepatitis is 14 to 145 days—too long for hepatitis B or C. In the United States, about 5% of chronic liver disease remains cryptogenic (i.e., does not appear to be autoimmune or viral in origin), and 50% of these patients have received blood transfusions before developing disease. Therefore, another form of hepatitis, referred to as hepatitis G virus (HGV) or GB virus-C (GBV-C), has been described; these are thought to be two different isolates of the same virus, which are percutaneously transmitted. Autoantibodies are absent.

The clinical significance of this virus remains uncertain. Risk factors are similar to those for hepatitis C. There is no clear relationship between HGV/GBV-C infection and progressive liver disease. Persistent infection does occur but does not affect the clinical course (Papadakis & McPhee, 2013).

NONVIRAL HEPATITIS

Certain chemicals have toxic effects on the liver and produce acute liver cell necrosis or toxic hepatitis when inhaled, injected parenterally, or taken by mouth. The chemicals most commonly implicated in this disease are carbon tetrachloride, phosphorus, chloroform, and gold compounds. These substances are true hepatotoxins. Many medications can induce hepatitis but are only sensitizing rather than toxic. Drug-induced hepatitis is similar to acute viral hepatitis, but parenchymal destruction tends to be more extensive. Medications that can lead to hepatitis include isoniazid (Nydrazid); halothane (Fluothane); acetaminophen; methyldopa (Aldomet); and certain antibiotics, antimetabolites, and anesthetic agents.

Toxic Hepatitis

At the onset of disease, toxic hepatitis resembles viral hepatitis. Obtaining a history of exposure to hepatotoxic chemicals, medications, botanical agents, or other toxic agents assists in early treatment and removal of the causative agent. Anorexia, nausea, and vomiting are the usual symptoms; jaundice and hepatomegaly are noted on physical assessment. Symptoms are more intense for the more severely toxic patient.

Recovery from acute toxic hepatitis is rapid if the hepatotoxin is identified early and removed or if exposure to the agent has been limited. Recovery is unlikely if there is a prolonged period between exposure and onset of symptoms. There are no effective antidotes. The fever rises; the patient becomes toxic and prostrated. Vomiting may be persistent, with the emesis containing blood. Clotting abnormalities may be severe, and hemorrhages may appear under the skin. The severe GI symptoms may lead to vascular collapse. Delirium, coma, and seizures develop, and within a few days the patient may die of fulminant hepatic failure (discussed later) unless he or she receives a liver transplant.

Short of liver transplantation, few treatment options are available. Therapy is directed toward restoring and maintaining fluid and electrolyte balance, blood replacement, and comfort and supportive measures. A few patients recover from acute toxic hepatitis only to develop chronic liver disease. If the liver heals, there may be scarring, followed by postnecrotic cirrhosis.

Drug-Induced Hepatitis

Drug-induced liver disease is the most common cause of acute liver failure, accounting for more than 50% of all cases in the United States (Davern, 2012). Manifestations of sensitivity to a medication may occur on the first day of its use or not until several months later. Usually, the onset is abrupt, with chills, fever, rash, pruritus, arthralgia, anorexia, and nausea. Later, there may be jaundice, dark urine, and an enlarged and tender liver. After the offending medication is withdrawn, symptoms may gradually subside. However, reactions can be severe, or even fatal, even if the medication is stopped. If fever, rash, or pruritus occurs from any medication, its use should be stopped immediately.

Although any medication can affect liver function, the use of acetaminophen (found in many over-the-counter medications used to treat fever and pain) has been identified as the leading cause of acute liver failure (Davern, 2012). Other causes commonly associated with liver injury include many anesthetic agents, medications used to treat rheumatic and musculoskeletal disease, antidepressants, psychotropic medications, anticonvulsants, and antituberculosis agents.

A short course of high-dose corticosteroids may be used in patients with severe hypersensitivity reactions, although its efficacy is uncertain. Liver transplantation is an option for drug-induced hepatitis, but outcomes may not be as successful as with other causes of liver failure.

FULMINANT HEPATIC FAILURE

Fulminant hepatic failure is the clinical syndrome of sudden and severely impaired liver function in a previously healthy person. The generally accepted definition is that fulminant hepatic failure develops within 8 weeks after the first symptoms of jaundice (Feldman et al., 2010).

Patterns of the progression from jaundice to encephalopathy have been identified and have led to proposals of time-based classifications. However, no agreement as to these classifications has been reached. Three categories are frequently cited: hyperacute, acute, and subacute liver failure. In hyperacute liver failure, the duration of jaundice before the onset of encephalopathy is 0 to 7 days; in acute liver failure, it is 8 to 28 days; and in subacute liver failure, it is 28 to 72 days. The prognosis for fulminant hepatic failure is much worse than for chronic liver failure. However, in fulminant failure, the hepatic lesion is potentially reversible, and survival rates are approximately 20% to 50%, depending greatly on the cause. Those who do not survive die of massive hepatocellular injury and necrosis (Feldman et al., 2010).

Viral hepatitis is a common cause of fulminant hepatic failure; other causes include toxic medications (e.g., acetaminophen) and chemicals (e.g., carbon tetrachloride), metabolic disturbances (e.g., Wilson's disease, a hereditary syndrome with deposition of copper in the liver), and structural changes (e.g., Budd-Chiari syndrome, an obstruction to outflow in major hepatic veins).

Jaundice and profound anorexia may be the initial reasons the patient seeks health care. Fulminant hepatic failure is often accompanied by coagulation defects, renal failure and electrolyte disturbances, cardiovascular abnormalities, infection, hypoglycemia, encephalopathy, and cerebral edema.

The key to optimized treatment is rapid recognition of acute liver failure and intensive intervention. Supporting the patient in the ICU and assessing the indications for and feasibility of liver transplantation are hallmarks of management. The use of antidotes for certain conditions may be indicated such as *N*-acetylcysteine for acetaminophen toxicity and penicillin for mushroom poisoning. Treatment modalities may include plasma exchanges (plasmapheresis) to correct coagulopathy, to reduce serum ammonia levels, and to stabilize the patient awaiting liver transplantation, and prostaglandin therapy to enhance hepatic blood flow. Although these treatment modalities may be implemented, no evidence exists indicating any clinical improvement with their use (Feldman et al., 2010). Hepatocytes within synthetic fiber columns have been tested as liver support systems (liver assist devices) to provide a bridge to transplantation.

Research into interventions for acute liver failure has begun to focus on techniques that combine the efficacy of a whole liver with the convenience and biocompatibility of hemodialysis. The acronyms ELAD (*e*xtracorporeal *l*iver *a*ssist *d*evices) and BAL (*b*io*a*rtificial *l*iver) have been used to describe these hybrid devices. These short-term devices, which remain experimental, may help patients survive until transplantation is possible. The BAL device exposes separated plasma to a cartridge containing porcine liver cells after the plasma has flowed through a charcoal column that removes substances toxic to hepatocytes. The ELAD exposes whole blood to cartridges containing human hepatoblastoma cells, resulting in removal of toxic substances. In the near future, similar extracorporeal circuits using **xenografts** (transplantation of organs from one species to another) may be studied as a bridge to liver transplantation. These approaches appear promising and have had success in animal studies. In human clinical application, the use of various BAL systems has resulted in improved neurologic and biochemical parameters. Adding albumin to extracorporeal dialysis in a process known as molecular adsorbent recirculating system (MARS) has been used to remove protein-bound toxins and is potentially useful in unstable patients with fulminant hepatic failure (O'Beirne, Murphy & Wendon, 2010; Papadakis & McPhee, 2013).

In patients who have fulminant hepatic failure with stage 4 encephalopathy (see Table 49-3), there is a high risk of cerebral edema, a life-threatening complication. The cause is not fully understood, although disruption of the blood–brain barrier and plasma leakage into the cerebrospinal fluid may be one cause. An increase in the intracellular osmolarity within cerebral astrocyte cells, possibly related to increased sodium and glutamine in these cells, may be another (O'Beirne et al., 2010). These patients require intracranial pressure monitoring. Measures to promote adequate cerebral perfusion include careful fluid balance and hemodynamic assessments, a quiet environment, and diuresis with mannitol (Osmitrol), an osmotic diuretic.

The use of barbiturate anesthesia or pharmacologic paralysis and sedation is indicated to prevent surges in intracranial pressure related to agitation. Other support measures include monitoring for and treating hypoglycemia, coagulopathies, and infection. Despite these treatment modalities, the mortality rate remains high. Consequently, liver transplantation (discussed later) is the treatment of choice for fulminant hepatic failure.

HEPATIC CIRRHOSIS

Cirrhosis is a chronic disease characterized by replacement of normal liver tissue with diffuse fibrosis that disrupts the structure and function of the liver. There are three types of cirrhosis or scarring of the liver:

- Alcoholic cirrhosis, in which the scar tissue characteristically surrounds the portal areas. This is most frequently caused by chronic alcoholism and is the most common type of cirrhosis.
- Postnecrotic cirrhosis, in which there are broad bands of scar tissue. This is a late result of a previous bout of acute viral hepatitis.
- Biliary cirrhosis, in which scarring occurs in the liver around the bile ducts. This type of cirrhosis usually results from chronic biliary obstruction and infection (cholangitis); it is much less common.

The portion of the liver chiefly involved in cirrhosis consists of the portal and the periportal spaces, where the bile canaliculi of each lobule communicate to form the liver bile ducts. These areas become the sites of inflammation, and the bile ducts become occluded with inspissated (thickened) bile and pus. The liver attempts to form new bile channels; hence, there is an overgrowth of tissue made up largely of disconnected, newly formed bile ducts and surrounded by scar tissue.

Pathophysiology

Several factors have been implicated in the etiology of cirrhosis. Nutritional deficiency with reduced protein intake contributes to liver destruction in cirrhosis, but excessive alcohol intake is the major causative factor in fatty liver and its consequences. However, cirrhosis can occur in people who do not consume alcohol and in those who consume a normal diet and have a high alcohol intake.

Some people appear to be more susceptible than others to this disease, whether or not they have alcoholism or are malnourished. Other factors may play a role, including exposure to certain chemicals (carbon tetrachloride, chlorinated naphthalene, arsenic, or phosphorus) or infectious schistosomiasis. Twice as many men as women are affected, although, for unknown reasons, women are at greater risk for development of alcohol-induced liver disease. Most patients are between 40 and 60 years of age. Each year, more than 27,000 people die of chronic liver diseases and cirrhosis in the United States (Mehta & Rothstein, 2009).

Alcoholic cirrhosis is characterized by episodes of necrosis involving the liver cells, which sometimes occur repeatedly throughout the course of the disease. The destroyed liver cells are gradually replaced by scar tissue. Eventually, the amount of scar tissue exceeds that of the functioning liver tissue. Islands of residual normal tissue and regenerating liver tissue may project from the constricted areas, giving the cirrhotic liver its characteristic hobnail appearance. The disease usually has an insidious onset and a protracted course, occasionally proceeding over a period of 30 or more years.

The prognoses for different forms of cirrhosis caused by various liver diseases have been investigated in several studies. Of the many prognostic indicators, the Child-Pugh classification seems most useful in predicting the outcome of patients with liver disease (Table 49-5). It is also used in choosing management approaches.

Clinical Manifestations

Signs and symptoms of cirrhosis increase in severity as the disease progresses, and severity is used to categorize the disorder as compensated or decompensated cirrhosis (Chart 49-11). Compensated cirrhosis, with its less severe, often vague symptoms, may be discovered secondarily at a routine physical examination. The hallmarks of decompensated cirrhosis result from failure of the liver to synthesize proteins, clotting factors, and other substances and manifestations of portal hypertension (see earlier sections of this chapter for clinical manifestations and management of portal hypertension, ascites, varices, and hepatic encephalopathy).

TABLE 49-5 Modified Child-Pugh Classification of the Severity of Liver Disease*

	Points Assigned		
Parameter	**1**	**2**	**3**
Ascites	Absent	Slight	Moderate
Bilirubin (mg/dL)	≤2	2–3	>3
Albumin (g/dL)	>3.5	2.8–3.5	<2.8
Prothrombin time (seconds over control)	1–3	4–6	>6
Encephalopathy	None	Grade 1–2	Grade 3–4

*Total score of 1–6, grade A; 7–9, grade B; 10–15, grade C.

Chart 49-11 ASSESSMENT

Assessing for Cirrhosis

Be alert to the following signs and symptoms:

Compensated

- Intermittent mild fever
- Vascular spiders
- Palmar erythema (reddened palms)
- Unexplained epistaxis
- Ankle edema
- Vague morning indigestion
- Flatulent dyspepsia
- Abdominal pain
- Firm, enlarged liver
- Splenomegaly

Decompensated

- Ascites
- Jaundice
- Weakness
- Muscle wasting
- Weight loss
- Continuous mild fever
- Clubbing of fingers
- Purpura (due to decreased platelet count)
- Spontaneous bruising
- Epistaxis
- Hypotension
- Sparse body hair
- White nails
- Gonadal atrophy

Liver Enlargement

Early in the course of cirrhosis, the liver tends to be large, and the cells are loaded with fat. The liver is firm and has a sharp edge that is noticeable on palpation. Abdominal pain may be present because of recent, rapid enlargement of the liver, which produces tension on the fibrous covering of the liver (Glisson's capsule). Later in the disease, the liver decreases in size as scar tissue contracts the liver tissue. The liver edge, if palpable, is nodular.

Portal Obstruction and Ascites

Portal obstruction and ascites—late manifestations of cirrhosis—are caused partly by chronic failure of liver function and partly by obstruction of the portal circulation. Almost all of the blood from the digestive organs is collected in the portal veins and carried to the liver. Because a cirrhotic liver does not allow free blood passage, blood backs up into the spleen and the GI tract, and these organs become the seat of chronic passive congestion—that is, they are stagnant with blood and therefore cannot function properly. Indigestion and altered bowel function result. Fluid rich in protein may accumulate in the peritoneal cavity, producing ascites. This can be detected through percussion for shifting dullness or a fluid wave (see Fig. 49-5).

Infection and Peritonitis

Bacterial peritonitis may develop in patients with cirrhosis and ascites in the absence of an intra-abdominal source of infection or an abscess. This condition is referred to as spontaneous bacterial peritonitis (SBP). Bacteremia due to translocation of intestinal flora is believed to be the most likely route of infection. Clinical signs may be absent, necessitating paracentesis for diagnosis. Antibiotic therapy is effective in the treatment and prevention of recurrent episodes of SBP. The most severe complication of SBP is hepatorenal syndrome, a form of renal failure unresponsive to administration of fluid or diuretic agents. This type of renal failure is characterized by a lack of pathologic changes in the kidney; there is no evidence of dehydration or obstruction of the urinary tract or any other renal disorder.

Gastrointestinal Varices

The obstruction to blood flow through the liver caused by fibrotic changes also results in the formation of collateral blood vessels in the GI system and shunting of blood from the portal vessels into blood vessels with lower pressures. As a result, the patient with cirrhosis often has prominent, distended abdominal blood vessels, which are visible on abdominal inspection (caput medusae) and distended blood vessels throughout the GI tract. The esophagus, stomach, and lower rectum are common sites of collateral blood vessels. These distended blood vessels form varices or hemorrhoids, depending on their location (see Fig. 49-7).

Because these vessels were not intended to carry the high pressure and volume of blood imposed by cirrhosis, they may rupture and bleed. Therefore, assessment must include observation for occult and frank bleeding from the GI tract.

Edema

Another late symptom of cirrhosis is edema, which is attributed to chronic liver failure. A reduced plasma albumin concentration predisposes the patient to the formation of edema. Although edema is generalized, it often affects the lower extremities, the upper extremities, and the presacral area. Facial edema is not typical. Overproduction of aldosterone occurs, causing sodium and water retention and potassium excretion.

Vitamin Deficiency and Anemia

Because of inadequate formation, use, and storage of certain vitamins (notably vitamins A, C, and K), signs of deficiency are common, particularly hemorrhagic phenomena associated with vitamin K deficiency. Chronic gastritis and impaired GI function, together with inadequate dietary intake and impaired liver function, account for the anemia that is often associated with cirrhosis. The patient's anemia, poor nutritional status, and poor state of health result in severe fatigue, which interferes with the ability to carry out routine activities of daily living.

Mental Deterioration

Additional clinical manifestations include deterioration of mental and cognitive function with impending hepatic encephalopathy and hepatic coma, as described previously. Neurologic assessment is indicated, including assessment of the patient's general behavior, cognitive abilities, orientation to time and place, and speech patterns.

Assessment and Diagnostic Findings

The extent of liver disease and the type of treatment are determined after review of the laboratory findings. The functions of the liver are complex, and many diagnostic tests provide information about liver function (see Table 49-1). The patient needs to know why these tests are being performed and how to cooperate.

In severe parenchymal liver dysfunction, the serum albumin level tends to decrease, and the serum globulin level rises. Enzyme tests indicate liver cell damage: serum alkaline phosphatase, AST, ALT, and GGT levels increase, and the serum cholinesterase level may decrease. Bilirubin tests are performed to measure bile excretion or retention; increased levels of bilirubin can occur with cirrhosis and other liver disorders. Prothrombin time is prolonged.

Ultrasound scanning is used to measure the difference in density of parenchymal cells and scar tissue. CT, MRI, and radioisotope liver scans give information about liver size and hepatic blood flow and obstruction. Diagnosis is confirmed by liver biopsy. Arterial blood gas analysis may reveal a ventilation–perfusion imbalance and hypoxia.

Medical Management

The management of the patient with cirrhosis is usually based on the presenting symptoms. For example, antacids or H_2 antagonists are prescribed to decrease gastric distress and minimize the possibility of GI bleeding. Vitamins and nutritional supplements promote healing of damaged liver cells and improve the patient's general nutritional status. Potassium-sparing diuretic agents such as spironolactone or triamterene (Dyrenium) may be indicated to decrease ascites, if present; these diuretics are preferred because they minimize the fluid and electrolyte changes commonly seen with other agents. An adequate diet and avoidance of alcohol are essential. Although the fibrosis of the cirrhotic liver cannot be reversed, its progression may be halted or slowed by such measures.

Many medications have been shown to possess antifibrotic activity for the treatment of cirrhosis. Some of these medications include colchicine, angiotensin system inhibitors, statins, diuretics including spironolactone (Aldactone), immunosuppressants, and glitazones such as pioglitazone (Acto) or rosiglitazone (Avandia). These medications have reasonable safety profiles, but their long-term safety and efficacy in patients with cirrhosis has yet to be demonstrated (Leung, 2011).

Many patients who have end-stage liver disease (ESLD) with cirrhosis use the herb milk thistle (*Silybum marianum*) to treat jaundice and other symptoms. This herb has been used for centuries because of its healing and regenerative properties for liver disease. Silymarin from milk thistle has anti-inflammatory and antioxidant properties that may have beneficial effects, especially in hepatitis (Rakel, 2012). The natural compound SAM-e (S-adenosylmethionine) may improve outcomes in liver disease by improving liver function, possibly through enhancing antioxidant function. Primary biliary cirrhosis has been treated with ursodeoxycholic acid (Actigall, Urso) to improve liver function.

Nursing Management

Nursing management for the patient with cirrhosis of the liver is described in detail in Chart 49-12. Nursing interventions

(*text continues on page 1376*)

Chart 49-12

PLAN OF NURSING CARE
The Patient With Impaired Liver Function

NURSING DIAGNOSIS: Activity intolerance related to fatigue, lethargy, and malaise
GOAL: Patient reports decrease in fatigue and reports increased ability to participate in activities

Nursing Interventions	Rationale	Expected Outcomes
1. Assess level of activity tolerance and degree of fatigue, lethargy, and malaise when performing routine activities of daily living.	**1.** Provides baseline for further assessment and criteria for assessment of effectiveness of interventions.	• Exhibits increased interest in activities and events • Participates in activities and gradually increases exercise within physical limits • Reports increased strength and well-being • Reports absence of abdominal pain and discomfort • Plans activities to allow ample periods of rest • Takes vitamins as prescribed
2. Assist with activities and hygiene when fatigued.	**2.** Promotes exercise and hygiene within patient's level of tolerance.	
3. Encourage rest when fatigued or when abdominal pain or discomfort occurs.	**3.** Conserves energy and protects the liver.	
4. Assist with selection and pacing of desired activities and exercise.	**4.** Stimulates patient's interest in selected activities.	
5. Provide diet high in carbohydrates with protein intake consistent with liver function.	**5.** Provides calories for energy and protein for healing.	
6. Administer supplemental vitamins (A, B complex, C, and K).	**6.** Provides additional nutrients.	

NURSING DIAGNOSIS: Imbalanced nutrition: less than body requirements related to abdominal distention and discomfort and anorexia
GOAL: Positive nitrogen balance, no further loss of muscle mass; meets nutritional requirements

Nursing Interventions	Rationale	Expected Outcomes
1. Assess dietary intake and nutritional status through diet history and diary, daily weight measurements, and laboratory data.	**1.** Identifies deficits in nutritional intake and adequacy of nutritional state.	• Exhibits improved nutritional status by increased weight (without fluid retention) and improved laboratory data • States rationale for dietary modifications • Identifies foods high in carbohydrates and within protein requirements (moderate to high protein in cirrhosis and hepatitis, low protein in hepatic failure) • Reports improved appetite • Participates in oral hygiene measures • Reports increased appetite; identifies rationale for smaller, frequent meals • Demonstrates intake of high-calorie diet; adheres to protein restriction • Identifies foods and fluids that are nutritious and permitted on diet • Gains weight without increased edema or ascites formation • Reports increased appetite and well-being • Excludes alcohol from diet • Takes medications for gastrointestinal disorders as prescribed • Reports normal gastrointestinal function with regular bowel function
2. Provide diet high in carbohydrates with protein intake consistent with liver function.	**2.** Provides calories for energy, sparing protein for healing.	
3. Assist patient in identifying low-sodium foods.	**3.** Reduces edema and ascites formation.	
4. Elevate the head of the bed during meals.	**4.** Reduces discomfort from abdominal distention and decreases sense of fullness produced by pressure of abdominal contents and ascites on the stomach.	
5. Provide oral hygiene before meals and pleasant environment for meals at mealtime.	**5.** Promotes positive environment and increased appetite; reduces unpleasant taste.	
6. Offer smaller, more frequent meals (6/day).	**6.** Decreases feeling of fullness, bloating.	
7. Encourage patient to eat meals and supplementary feedings.	**7.** Encouragement is essential for the patient with anorexia and gastrointestinal discomfort.	
8. Provide attractive meals and an aesthetically pleasing setting at mealtime.	**8.** Promotes appetite and sense of well-being.	
9. Eliminate alcohol.	**9.** Eliminates "empty calories" and further damage from alcohol.	
10. Apply an ice collar for nausea.	**10.** May reduce incidence of nausea.	
11. Administer medications prescribed for nausea, vomiting, diarrhea, or constipation.	**11.** Reduces gastrointestinal symptoms and discomforts that decrease the appetite and interest in food.	
12. Encourage increased fluid intake and exercise if the patient reports constipation.	**12.** Promotes normal bowel pattern and reduces abdominal discomfort and distention.	

(continues on page 1370)

Chart 49-12

PLAN OF NURSING CARE

The Patient With Impaired Liver Function (continued)

NURSING DIAGNOSIS: Impaired skin integrity related to pruritus from jaundice and edema
GOAL: Decrease potential for pressure ulcer development; breaks in skin integrity

Nursing Interventions	Rationale	Expected Outcomes
1. Assess degree of discomfort related to pruritus and edema. 2. Note and record degree of jaundice and extent of edema. 3. Keep patient's fingernails short and smooth. 4. Provide frequent skin care; avoid the use of soaps and alcohol-based lotions. 5. Massage every 2 hours with emollients; turn every 2 hours. 6. Initiate use of alternating-pressure mattress or low air loss bed. 7. Recommend avoiding the use of harsh detergents. 8. Assess skin integrity every 4–8 hours. Instruct patient and family in this activity. 9. Restrict sodium as prescribed. 10. Perform range-of-motion exercises every 4 hours; elevate edematous extremities whenever possible.	1. Assists in determining appropriate interventions. 2. Provides baseline for detecting changes and evaluating effectiveness of interventions. 3. Prevents skin excoriation and infection from scratching. 4. Removes waste products from skin while preventing dryness of skin. 5. Promotes mobilization of edema. 6. Minimizes prolonged pressure on bony prominences susceptible to breakdown. 7. May decrease skin irritation and need for scratching. 8. Edematous skin and tissue have compromised nutrient supply and are vulnerable to pressure and trauma. 9. Minimizes edema formation. 10. Promotes mobilization of edema.	• Exhibits intact skin without redness, excoriation, or breakdown • Reports relief from pruritus • Exhibits no skin excoriation from scratching • Uses nondrying soaps and lotions; states rationale for the use of nondrying soaps and lotions • Turns self periodically; exhibits reduced edema of dependent parts of the body • Exhibits no areas of skin breakdown • Exhibits decreased edema; normal skin turgor

NURSING DIAGNOSIS: Risk for injury related to altered clotting mechanisms and altered level of consciousness
GOAL: Reduced risk of injury

Nursing Interventions	Rationale	Expected Outcomes
1. Assess level of consciousness and cognitive level. 2. Provide safe environment (pad side rails, remove obstacles in room, prevent falls). 3. Provide frequent surveillance to orient patient, and avoid the use of restraints. 4. Replace sharp objects (razors) with safer items. 5. Observe each stool for color, consistency, and amount. 6. Be alert to symptoms of anxiety, epigastric fullness, weakness, and restlessness. 7. Test each stool and emesis for occult blood. 8. Observe for hemorrhagic manifestations: ecchymosis, epistaxis, petechiae, and bleeding gums. 9. Record vital signs at frequent intervals, depending on patient acuity (every 1–4 hours). 10. Keep patient quiet, and limit activity.	1. Assists in determining patient's ability to protect self and comply with required self-protective actions; may detect deterioration of hepatic function. 2. Minimizes falls and injury if falls occur. 3. Protects patient from harm while stimulating and orienting patient; the use of restraints may disturb patient further. 4. Avoids cuts and bleeding. 5. Permits detection of bleeding in gastrointestinal tract. 6. May indicate early signs of bleeding and shock. 7. Detects early evidence of bleeding. 8. Indicates altered clotting mechanisms. 9. Provides baseline and evidence of hypovolemia and hemorrhagic shock. 10. Minimizes risk of bleeding and straining.	• Is oriented to time, place, and person • Exhibits no hallucinations and demonstrates no efforts to get up unassisted or to leave hospital • Exhibits no ecchymoses (bruises), cuts, or hematoma • Uses electric razor rather than sharp-edged razor • Exhibits absence of frank bleeding from gastrointestinal tract • Exhibits absence of restlessness, epigastric fullness, and other indicators of hemorrhage and shock • Exhibits negative results of test for occult gastrointestinal bleeding • Is free of ecchymotic areas or hematoma formation • Exhibits normal vital signs • Maintains rest and remains quiet if active bleeding occurs • Identifies rationale for blood transfusions and measures to treat bleeding • Uses measures to prevent trauma (e.g., uses soft toothbrush, blows nose gently, avoids bumps and falls, avoids straining during defecation) • Experiences no side effects of medications • Takes all medications as prescribed • Identifies rationale for precautions with the use of all medications • Cooperates with treatment modalities

Chart 49-12

PLAN OF NURSING CARE

The Patient With Impaired Liver Function (continued)

Nursing Interventions	Rationale	Expected Outcomes
11. Assist physician in passage of tube for esophageal balloon tamponade, if its insertion is indicated.	**11.** Promotes nontraumatic insertion of tube in anxious and combative patient for immediate treatment of bleeding.	
12. Observe during blood transfusions.	**12.** Permits detection of transfusion reactions (risk increased with multiple blood transfusions needed for active bleeding from esophageal varices).	
13. Measure and record nature, time, and amount of vomitus.	**13.** Assists in evaluating extent of bleeding and blood loss.	
14. Maintain patient in fasting state, if indicated.	**14.** Reduces risk of aspiration of gastric contents and minimizes risk of further trauma to esophagus and stomach by preventing vomiting.	
15. Administer vitamin K as prescribed.	**15.** Promotes clotting by providing fat-soluble vitamin necessary for clotting.	
16. Remain with patient during episodes of bleeding.	**16.** Reassures anxious patient and permits monitoring and detection of further needs of the patient.	
17. Offer cold liquids by mouth when bleeding stops (if prescribed).	**17.** Minimizes risk of further bleeding by promoting vasoconstriction of esophageal and gastric blood vessels.	
18. Institute measures to prevent trauma.	**18.** Promotes safety of patient.	
a. Maintain safe environment.	**a.** Minimizes risk of trauma and bleeding by avoiding falls and cuts, etc.	
b. Encourage *gentle* blowing of nose.	**b.** Reduces risk of nosebleed (epistaxis) secondary to trauma and decreased clotting.	
c. Provide soft toothbrush, and avoid the use of toothpicks.	**c.** Prevents trauma to oral mucosa while promoting good oral hygiene.	
d. Encourage intake of foods with high content of vitamin C.	**d.** Promotes healing.	
e. Apply cold compresses where indicated.	**e.** Minimizes bleeding into tissues by promoting local vasoconstriction.	
f. Record location of bleeding sites.	**f.** Permits detection of new bleeding sites and monitoring of previous sites of bleeding.	
g. Use small-gauge needles for injections.	**g.** Minimizes oozing and blood loss from repeated injections.	
19. Administer medications carefully; monitor for side effects.	**19.** Reduces risk of side effects secondary to damaged liver's inability to detoxify (metabolize) medications normally.	

NURSING DIAGNOSIS: Disturbed body image related to changes in appearance, sexual dysfunction, and role function
GOAL: Patient verbalizes feelings consistent with improvement of body image and self-esteem

Nursing Interventions	Rationale	Expected Outcomes
1. Assess changes in appearance and the meaning these changes have for patient and family.	**1.** Provides information for assessing impact of changes in appearance, sexual function, and role on the patient and family.	• Verbalizes concerns related to changes in appearance, life, and lifestyle • Shares concerns with significant others • Identifies past coping strategies that have been effective • Uses past effective coping strategies to deal with changes in appearance, life, and lifestyle • Maintains good grooming and hygiene • Identifies short-term goals and strategies to achieve them • Takes an active role in decision making about self and care • Identifies resources that are not harmful • Verbalizes that some of previous lifestyle practices have been harmful
2. Encourage patient to verbalize reactions and feelings about these changes.	**2.** Enables patient to identify and express concerns; encourages patient and significant others to share these concerns.	
3. Assess patient's and family's previous coping strategies.	**3.** Permits encouragement of those coping strategies that are familiar to patient and have been effective in the past.	

(continues on page 1372)

Chart 49-12

PLAN OF NURSING CARE
The Patient With Impaired Liver Function (continued)

Nursing Interventions	Rationale	Expected Outcomes
4. Assist and encourage patient to maximize appearance (such as strategies to limit the appearance of jaundice and ascites through careful selection of colors and type of clothing) and explore alternatives to previous sexual and role functions.	4. Encourages patient to continue safe roles and functions while encouraging exploration of alternatives.	• Uses healthy expressions of frustration, anger, anxiety
5. Assist patient in identifying short-term goals.	5. Accomplishing these goals serves as positive reinforcement and increases self-esteem.	
6. Encourage and assist patient in decision making about care.	6. Promotes patient's control of life and improves sense of well-being and self-esteem.	
7. Identify with patient resources to provide additional support (counselor, spiritual advisor).	7. Assists patient in identifying resources and accepting assistance from others when indicated.	
8. Assist patient in identifying previous practices that may have been harmful to self (alcohol and drug abuse). Involve patient in goal setting, and provide positive feedback for accomplishments.	8. Recognition and acknowledgment of the harmful effects of these practices are necessary for identifying a healthier lifestyle.	

NURSING DIAGNOSIS: Chronic pain and discomfort related to enlarged tender liver and ascites
GOAL: Increased level of comfort

Nursing Interventions	Rationale	Expected Outcomes
1. Maintain bed rest when patient experiences abdominal discomfort.	1. Reduces metabolic demands and protects the liver.	• Reports pain and discomfort if present • Maintains bed rest and decreases activity in presence of pain
2. Administer antispasmodic and analgesic agents as prescribed.	2. Reduces irritability of the gastrointestinal tract and decreases abdominal pain and discomfort.	• Takes antispasmodic and analgesic agents as indicated and as prescribed
3. Observe, record, and report presence and character of pain and discomfort.	3. Provides baseline to detect further deterioration of status and to evaluate interventions.	• Reports decreased pain and abdominal discomfort
4. Reduce sodium and fluid intake if prescribed.	4. Minimizes further formation of ascites.	• Reduces sodium and fluid intake to prescribed levels if indicated to treat ascites
5. Prepare patient and assist with paracentesis.	5. Removal of ascites fluid may decrease abdominal discomfort.	• Exhibits decreased abdominal girth and appropriate weight changes
6. Encourage the use of distracting activities such as music, reading, or meditation.	6. Distraction may limit the perception of pain.	• Reports decreased discomfort after paracentesis

NURSING DIAGNOSIS: Fluid volume excess related to ascites and edema formation
GOAL: Restoration of normal fluid volume

Nursing Interventions	Rationale	Expected Outcomes
1. Restrict sodium and fluid intake if prescribed.	1. Minimizes formation of ascites and edema.	• Consumes diet low in sodium and within prescribed fluid restriction
2. Administer diuretic agents, potassium, and protein supplements as prescribed.	2. Promotes excretion of fluid through the kidneys and maintenance of normal fluid and electrolyte balance.	• Takes diuretic agents, potassium, and protein supplements as indicated without experiencing side effects
3. Record intake and output every 1 to 8 hours depending on response to interventions and on patient acuity.	3. Indicates effectiveness of treatment and adequacy of fluid intake.	• Exhibits increased urine output • Exhibits decreasing abdominal girth • Exhibits no rapid increase in weight
4. Measure and record abdominal girth and weight daily.	4. Monitors changes in ascites formation and fluid accumulation.	• Identifies rationale for sodium and fluid restriction
5. Explain rationale for sodium and fluid restriction.	5. Promotes patient's understanding of restriction and cooperation with it.	• Shows a decrease in ascites with decreased weight
6. Prepare patient and assist with paracentesis.	6. Paracentesis will temporarily decrease amount of ascites present.	

Chart 49-12

PLAN OF NURSING CARE

The Patient With Impaired Liver Function (continued)

NURSING DIAGNOSIS: Confusion related to abnormal liver function and increased serum ammonia level
GOAL: Improved mental status; safety maintained; ability to cope with cognitive and behavioral changes

Nursing Interventions	Rationale	Expected Outcomes
1. Restrict dietary protein as prescribed for transient period.	1. Reduces source of ammonia (protein foods).	• Adheres to protein restriction • Demonstrates an interest in events and activities in environment
2. Give frequent, small feedings of carbohydrates.	2. Promotes consumption of adequate carbohydrates for energy requirements and spares protein from breakdown for energy.	• Demonstrates normal attention span • Follows and participates in conversation appropriately
3. Protect from infection.	3. Minimizes risk of further increase in metabolic requirements.	• Is oriented to person, place, and time • Remains in bed when indicated
4. Keep environment warm and draft free.	4. Minimizes shivering, which would increase metabolic requirements.	• Reports no urinary or fecal incontinence • Experiences no seizures
5. Pad the side rails of the bed.	5. Provides protection for the patient should hepatic coma and seizure activity occur.	• No neurologic or respiratory depression • Patient develops no cognitive impairments, but if they develop, they are quickly identified and treated, enhancing the potential of recovery.
6. Limit visitors.	6. Minimizes patient's activity and metabolic requirements.	
7. Provide careful nursing surveillance to ensure patient's safety.	7. Provides close monitoring of new symptoms and minimizes trauma to the confused patient.	• Patient and family describe adequate feelings of coping and lowered anxiety. They demonstrate ability to listen and to make decisions as able.
8. Avoid opioids and barbiturates.	8. Prevents masking of symptoms of hepatic coma and prevents drug overdose secondary to reduced ability of the damaged liver to metabolize opioids and barbiturates; prevents respiratory depression.	• Patient and family communicate their feelings and their needs in a secure and caring environment.
9. Awaken at intervals (every 2–4 hours) to assess cognitive status.	9. Provides stimulation to the patient and opportunity for observing patient's level of consciousness.	
10. Identify subtle changes in behavior or sleep–wake pattern (consistent staff caring for the patient enhances this assessment as they become familiar with patient's baseline).	10/11. These changes may herald worsening of encephalopathy, which requires rapid intervention, including medication.	
11. Assess handwriting or drawing skill daily as indication of cognitive ability.		
12. Encourage patient and family to participate in therapeutic strategies to enhance coping with episodes of mental deterioration.	12. Promoting activities such as listening to music, relaxation techniques, or preillness coping strategies can reduce anxiety.	
13. Encourage patient and family to discuss feeling of fear, powerlessness, or emotional distress related to patient's mental deterioration.	13. Actively listening demonstrates caring and concern.	

NURSING DIAGNOSIS: Risk for imbalanced body temperature: hyperthermia related to inflammatory process of cirrhosis or hepatitis
GOAL: Maintenance of normal body temperature, free from infection

Nursing Interventions	Rationale	Expected Outcomes
1. Record temperature regularly (every 4 hours).	1. Provides baseline to detect fever and to evaluate interventions.	• Exhibits normal temperature and reports absence of chills or sweating
2. Encourage fluid intake.	2. Corrects fluid loss from perspiration and fever and increases patient's level of comfort.	• Demonstrates adequate intake of fluids • Exhibits no evidence of local or systemic infection
3. Apply cool sponges or ice bag for elevated temperature.	3. Promotes reduction of fever and increases patient's comfort.	• Develops no nosocomial infections related to invasive procedures/lines
4. Administer antibiotics as prescribed.	4. Ensures appropriate serum concentration of antibiotics to treat infection.	

(continues on page 1374)

Chart 49-12

PLAN OF NURSING CARE

The Patient With Impaired Liver Function (continued)

Nursing Interventions	Rationale	Expected Outcomes
5. Avoid exposure to infections.	**5.** Minimizes risk of further infection and further increases in body temperature and metabolic rate.	
6. Keep patient at rest while temperature is elevated.	**6.** Reduces metabolic rate.	
7. Assess for abdominal pain, tenderness.	**7.** May occur with bacterial peritonitis.	
8. Use sterile technique for all invasive procedures.	**8.** Many evidence-based practice guidelines (e.g., central venous catheter care) recommend the use of sterile technique to prevent nosocomial infections.	

NURSING DIAGNOSIS: Ineffective breathing pattern related to ascites and restriction of thoracic excursion secondary to ascites, abdominal distention, and fluid in the thoracic cavity
GOAL: Improved respiratory status

Nursing Interventions	Rationale	Expected Outcomes
1. Elevate head of bed to at least 30 degrees.	**1.** Reduces abdominal pressure on the diaphragm and permits fuller thoracic excursion and lung expansion.	• Experiences improved respiratory status • Reports decreased shortness of breath • Reports increased strength and sense of well-being • Exhibits normal respiratory rate (12–18 breaths/min) with no adventitious sounds • Exhibits full thoracic excursion without shallow respirations • Exhibits normal arterial blood gases • Exhibits adequate oxygen saturation by pulse oximetry • Experiences absence of confusion or cyanosis
2. Conserve patient's strength by providing rest periods and assisting with activities.	**2.** Reduces metabolic and oxygen requirements.	
3. Change position every 2 hours.	**3.** Promotes expansion and oxygenation of all areas of the lungs.	
4. Assist with paracentesis or thoracentesis.	**4.** Paracentesis and thoracentesis (performed to remove fluid from the abdominal and thoracic cavities, respectively) may be frightening to the patient.	
a. Explain procedure and its purpose to patient.	**a.** Helps obtain patient's cooperation with procedures.	
b. Have patient void before paracentesis.	**b.** Prevents inadvertent bladder injury.	
c. Support and maintain position during procedure.	**c.** Prevents inadvertent organ or tissue injury.	
d. Record both the amount and the character of fluid aspirated.	**d.** Provides record of fluid removed and indication of severity of limitation of lung expansion by fluid.	
e. Observe for evidence of coughing, increasing dyspnea, or pulse rate.	**e.** Indicates irritation of the pleural space and evidence of pneumothorax or hemothorax.	

COLLABORATIVE PROBLEM: Gastrointestinal bleeding and hemorrhage
GOAL: Absence of episodes of gastrointestinal bleeding and hemorrhage

Nursing Interventions	Rationale	Expected Outcomes
1. Assess patient for evidence of gastrointestinal bleeding or hemorrhage. If bleeding does occur: **a.** Monitor vital signs (blood pressure, pulse, respiratory rate) every 4 hours or more frequently, depending on acuity. **b.** Assess skin temperature, level of consciousness every 4 hours or more frequently, depending on acuity. **c.** Monitor gastrointestinal secretions and output (emesis, stool for occult or obvious bleeding). Test emesis for blood once per shift and with any color change. Hematest each stool. **d.** Monitor hematocrit and hemoglobin for trends and changes.	**1.** Allows early detection of signs and symptoms of bleeding and hemorrhage.	• Experiences no episodes of bleeding and hemorrhage • Vital signs are within acceptable range for patient. • No evidence of bleeding from gastrointestinal tract • Hematocrit and hemoglobin levels within acceptable limits • Turns and moves without straining and increasing intra-abdominal pressure • No straining with bowel movements • No further bleeding episodes if aggressive treatment of bleeding and hemorrhage was needed • Patient and family state rationale for treatments.

Chart 49-12

PLAN OF NURSING CARE

The Patient With Impaired Liver Function (continued)

Nursing Interventions	Rationale	Expected Outcomes
2. Avoid activities that increase intra-abdominal pressure (straining, turning). **a.** Avoid coughing/sneezing. **b.** Assist patient to turn. **c.** Keep all needed items within easy reach. **d.** Use measures to prevent constipation such as adequate fluid intake, stool softeners. **e.** Ensure small meals.	**2.** Minimizes increases in intra-abdominal pressure that could lead to rupture and bleeding of esophageal or gastric varices.	• Patient and family identify supports available to them. • Patient and family describe signs and symptoms of a recurrent bleeding episode and identify needed action.
3. Have equipment (Sengstaken-Blakemore tube™, medications, IV fluids) available if indicated.	**3.** Equipment, medications, and supplies will be readily available if patient experiences bleeding from ruptured esophageal or gastric varices.	
4. Assist with procedures and therapy needed to treat gastrointestinal bleeding and hemorrhage.	**4.** Gastrointestinal bleeding and hemorrhage require emergency measures (e.g., insertion of Sengstaken-Blakemore tube™, administration of fluids and medications).	
5. Monitor respiratory status every hour, and minimize risk of respiratory complications if balloon tamponade is needed.	**5.** The patient is at high risk for respiratory complications, including asphyxiation if gastric balloon of tamponade tube ruptures or migrates upward.	
6. Prepare patient physically and psychologically for other treatment modalities if needed.	**6.** The patient who experiences hemorrhage is very anxious and fearful; minimizing anxiety assists in control of hemorrhage.	
7. Monitor patient for recurrence of bleeding and hemorrhage.	**7.** Risk of rebleeding is high with all treatment modalities used to halt gastrointestinal bleeding.	
8. Keep family informed of patient's status.	**8.** Family members are likely to be anxious about the patient's status; providing information will reduce their anxiety level and promote more effective coping.	
9. Once recovered from bleeding episode, provide patient and family with information regarding signs and symptoms of gastrointestinal bleeding.	**9.** Risk of rebleeding is high. Subtle signs may be more quickly identified.	

COLLABORATIVE PROBLEM: Hepatic encephalopathy
GOAL: Absence of changes in cognitive status and of injury

Nursing Interventions	Rationale	Expected Outcomes
1. Assess cognitive status every 4–8 hours. **a.** Assess patient's orientation to person, place, and time. **b.** Monitor patient's level of activity, restlessness, and agitation. Assess for presence of flapping hand tremors (asterixis). **c.** Obtain and record daily sample of patient's handwriting or ability to construct a simple figure (e.g., star). **d.** Assess neurologic signs (deep tendon reflexes, ability to follow instructions).	**1.** Data will provide baseline of patient's cognitive status and enable detection of changes.	• Remains awake, alert, and aware of surroundings • Is oriented to time, place, and person • Deep tendon reflexes remain within normal limits • Exhibits no restlessness or agitation • Record of handwriting demonstrates no deterioration in cognitive function. • States rationale for treatment used to prevent or treat hepatic encephalopathy • Demonstrates stable serum ammonia level within acceptable limits • Consumes adequate caloric intake and adheres to protein restriction
2. Monitor medications to prevent administration of those that may precipitate hepatic encephalopathy (sedative, hypnotic, analgesic agents).	**2.** Medications are a common precipitating factor in development of hepatic encephalopathy in patients at risk.	• Takes medications as prescribed • Breath sounds normal without adventitious sounds

(continues on page 1376)

Chart 49-12

PLAN OF NURSING CARE

The Patient With Impaired Liver Function (continued)

Nursing Interventions	Rationale	Expected Outcomes
3. Monitor laboratory data, especially serum ammonia level.	**3.** Increases in serum ammonia level are associated with hepatic encephalopathy and coma.	• Skin and tissue intact without evidence of pressure or breaks in integrity • Verbalizes understanding of need for treatments and procedures to promote recovery
4. Notify physician of even subtle changes in patient's neurologic assessment, cognitive function, sleep pattern, or mood.	**4.** Allows early initiation of treatment of hepatic encephalopathy and prevention of hepatic coma.	
5. Limit sources of protein from diet if indicated.	**5.** Reduces breakdown and conversion of protein to ammonia.	
6. Administer medications prescribed to reduce serum ammonia level (e.g., lactulose, antibiotics, glucose, benzodiazepine antagonist [flumazenil] if indicated).	**6.** Reduces serum ammonia level.	
7. Assess respiratory status, and initiate measures to prevent complications.	**7.** The patient who develops hepatic coma is at risk for respiratory complications (i.e., pneumonia, atelectasis, infection).	
8. Protect patient's skin and tissue from pressure and breakdown.	**8.** The patient in coma is at risk for skin breakdown and pressure ulcer formation.	
9. Provide support and active listening for patient and family as patient's mental status deteriorates.	**9.** The patient with hepatic encephalopathy can experience episodes of mental deterioration due to liver failure. This can produce feelings of fear and anxiety.	

are directed toward promoting patient's rest, improving nutritional status, providing skin care, reducing risk of injury, and monitoring and managing potential complications.

Promoting Rest

The patient with cirrhosis requires rest and other supportive measures to permit the liver to reestablish its functional ability. If the patient is hospitalized, weight and I&O are measured and recorded daily. The nurse adjusts the patient's position in bed for maximal respiratory efficiency, which is especially important if ascites is marked, because it interferes with adequate thoracic excursion. Oxygen therapy may be required in liver failure to oxygenate the damaged cells and prevent further cell destruction.

Rest reduces the demands on the liver and increases the liver's blood supply. Because the patient is susceptible to the hazards of immobility, efforts to prevent respiratory, circulatory, and vascular disturbances are initiated. These measures may help prevent such problems as pneumonia, thrombophlebitis, and pressure ulcers. After nutritional status improves and strength increases, the nurse encourages the patient to increase activity gradually. Activity and mild exercise, as well as rest, are planned.

Improving Nutritional Status

The patient with cirrhosis without ascites, edema, or signs of impending hepatic coma should receive a nutritious, high-protein diet, if tolerated, supplemented by vitamins of the B complex, as well as A, C, and K. The nurse encourages the patient to eat. If ascites is present, small, frequent meals may be better tolerated than three large meals because of the abdominal pressure exerted by ascites.

Imbalance of the intestinal flora is not uncommon. Research suggests that the oral ingestion of 1 cup of probiotic yogurt three times a day reduces intestinal flora imbalance by decreasing *Escherichia coli* counts (Liu, Zhang, Zhang, et al., 2010).

Patients with fatty stools (steatorrhea) should receive water-soluble forms of fat-soluble vitamins A, D, and E (Aquasol A, D, and E). Folic acid and iron are prescribed to prevent anemia. If the patient shows signs of impending or advancing coma despite medical interventions, the amount of protein in the diet may be decreased temporarily. Protein is restricted if encephalopathy develops that cannot be effectively managed with lactulose and other medical treatment strategies. Incorporating vegetable protein to meet protein needs may decrease the risk for encephalopathy. Sodium restriction is also indicated to prevent ascites. Patients with prolonged or severe anorexia and those who are vomiting or eating poorly for any reason may receive nutrients by the enteral or parenteral route.

Providing Skin Care

Providing careful skin care is important because of subcutaneous edema, the patient's immobility, jaundice, and increased susceptibility to skin breakdown and infection. Frequent changes in position are necessary to prevent pressure ulcers. Irritating soaps and the use of adhesive tape are avoided to prevent trauma to the skin. Lotion may be soothing to irritated skin; the nurse takes measures to minimize scratching by the patient.

Reducing Risk of Injury

The nurse protects the patient with cirrhosis from falls and other injuries. The side rails should be in place and pads used in case the patient becomes agitated or restless. To minimize agitation, the nurse orients the patient to time and place and explains all procedures. The nurse instructs the patient to ask for assistance to get out of bed. The nurse carefully evaluates any injury because of the possibility of internal bleeding.

Because of the risk for bleeding from abnormal clotting, the patient should use an electric razor rather than a safety razor. A soft-bristled toothbrush helps minimize bleeding gums, and pressure applied to all venipuncture sites helps minimize bleeding.

Monitoring and Managing Potential Complications

A major role of the nurse is monitoring of the patient with cirrhosis for complications.

Bleeding and Hemorrhage

The patient is at increased risk for bleeding and hemorrhage because of decreased production of prothrombin and decreased ability of the diseased liver to synthesize the necessary substances for blood coagulation (see the Esophageal Varices section).

Hepatic Encephalopathy

Hepatic encephalopathy and coma, which are complications of cirrhosis, may manifest as deteriorating mental status and dementia or as physical signs such as abnormal voluntary and involuntary movements. Hepatic encephalopathy was discussed earlier in the chapter in detail.

Monitoring is essential to identify early deterioration in mental status. The nurse monitors the patient's mental status closely and reports changes so that treatment for encephalopathy can be initiated promptly. An extensive baseline and ongoing neurologic evaluation is key to identify progression through the four stages of encephalopathy (see Table 49-3).

Each advancing stage demands more intensive nursing interventions aimed at providing for patient safety and prevention and early identification of life-threatening complications such as respiratory failure and cerebral edema, which would necessitate interventions in an ICU. Because electrolyte disturbances can contribute to encephalopathy, serum electrolyte levels are carefully monitored and corrected if abnormal. Oxygen is administered if oxygen desaturation occurs. The nurse monitors for fever or abdominal pain, which may signal the onset of bacterial peritonitis or other infection (see earlier discussion of hepatic encephalopathy).

Fluid Volume Excess

Patients with advanced chronic liver disease develop cardiovascular abnormalities. These occur due to an increased cardiac output and decreased peripheral vascular resistance, possibly resulting from the release of vasodilators. A hyperdynamic circulatory state develops in patients with cirrhosis, and plasma volume increases. This increase in circulating plasma volume is probably multifactorial, but some studies have implicated excess production of nitrous oxide, such as that seen in sepsis, as one causative factor (Gordon, 2012). The greater the degree of hepatic decompensation, the more severe the hyperdynamic state. Close assessment of cardiovascular and respiratory status is of key importance for the care of patients with this disorder. Pulmonary compromise is always a potential complication of ESLD because of plasma volume excess; consequently, the nurse has an important role in preventing pulmonary complications. Administering diuretic agents, implementing fluid restrictions, and enhancing patient positioning can optimize pulmonary function. Fluid retention may be noted in the development of ascites, lower extremity swelling, and dyspnea. Monitoring of I&O, daily weight changes, changes in abdominal girth, and edema formation is part of nursing assessment in the hospital or in the home setting. Patients are also monitored for nocturia and, later, for oliguria, because these states indicate increasing severity of liver dysfunction (Gordon, 2012).

Promoting Home and Community-Based Care

Educating Patients About Self-Care

During the hospital stay, the nurse and other health care providers prepare the patient with cirrhosis for discharge, focusing on dietary education. Of greatest importance is the exclusion of alcohol from the diet. The patient may need referral to Alcoholics Anonymous, psychiatric care, or counseling or may benefit from support from a spiritual advisor. The patient should also avoid the consumption of raw shellfish.

Sodium restriction will continue for a considerable time, if not permanently. The patient will require written education, reinforcement, and support from the staff as well as family members.

Successful treatment depends on convincing the patient of the need to adhere completely to the therapeutic plan. This includes rest, lifestyle changes, adequate dietary intake, and the elimination of alcohol. The nurse also instructs the patient and family about symptoms of impending encephalopathy, possible bleeding tendencies, and susceptibility to infection.

Recovery is neither rapid nor easy; there are frequent setbacks and apparent lack of improvement. Many patients find it difficult to refrain from using alcohol for comfort or escape. The nurse has a significant role in offering support and encouragement to the patient and in providing positive feedback when the patient experiences success.

Continuing Care

Referral for home care may assist the patient in dealing with the transition from hospital to home. The use of alcohol may have been an important part of normal home and social life in the past. The home care nurse assesses the patient's progress at home and the manner in which the patient and family are coping with the elimination of alcohol and the dietary restrictions. The nurse also reinforces previous education and answers questions that may not have occurred to the patient or family until the patient is back home and trying to establish new patterns of eating, drinking, and lifestyle.

CANCER OF THE LIVER

Hepatic tumors may be malignant or benign. Benign liver tumors were uncommon until oral contraceptives were in widespread use. Now, benign liver tumors such as hepatic adenomas occur most frequently in women in their reproductive years who are taking oral contraceptives (Greenberger et al., 2012).

Primary Liver Tumors

Few cancers originate in the liver. Primary liver tumors usually are associated with chronic liver disease, hepatitis B and C infections, and cirrhosis. HCC is the most common type of primary liver cancer, with more than half a million cases diagnosed each year on a worldwide basis. HCC is the third leading cause of cancer-related mortality worldwide. It is rare in the United States and northern Europe, accounting for fewer than five cases per 100,000 inhabitants (McGlynn & London, 2011). Other types of primary liver cancer include cholangiocellular carcinoma and combined hepatocellular and cholangiocellular carcinoma. HCC is usually nonresectable because of rapid growth and metastasis. If found early, resection of primary liver cancer may be possible; however, early detection is not common.

Cirrhosis, chronic infection with hepatitis B and C, and exposure to certain chemical toxins (e.g., vinyl chloride, arsenic) have been implicated as causes of HCC. Cigarette smoking has also been identified as a risk factor, especially when combined with alcohol use. Some evidence suggests that aflatoxin, a metabolite of the fungus *Aspergillus flavus*, may be a risk factor for HCC. This is especially true in areas where HCC is endemic (i.e., Asia and Africa). Aflatoxin and other similar toxic molds can contaminate food such as ground nuts and grains and may act as co-carcinogens with hepatitis B (Greenberger et al., 2012; McMahon, 2010). The risk of contamination is greatest when these foods are stored unrefrigerated in tropical or subtropical climates.

Liver Metastases

Metastases from other primary sites, particularly the digestive system, breast, and lung, are found in the liver 2.5 times more frequently than tumors due to primary liver cancers (Goldman & Schafer, 2012; Tsim, Frampton, Haby, et al., 2010). Malignant tumors are likely to reach the liver eventually, by way of the portal system or lymphatic channels, or by direct extension from an abdominal tumor. Moreover, the liver apparently is an ideal place for these malignant cells to thrive. Often, the first evidence of cancer in an abdominal organ is the appearance of liver metastases; unless exploratory surgery or an autopsy is performed, the primary tumor may never be identified.

Clinical Manifestations

The early manifestations of malignancy of the liver include pain—a continuous dull ache in the right upper quadrant, epigastrium, or back. Weight loss, loss of strength, anorexia, and anemia may also occur. The liver may be enlarged and irregular on palpation. Jaundice is present only if the larger bile ducts are occluded by the pressure of malignant nodules in the hilum of the liver. Ascites develops if such nodules obstruct the portal veins or if tumor tissue is seeded in the peritoneal cavity.

Assessment and Diagnostic Findings

The diagnosis of liver cancer is based on clinical signs and symptoms, the history and physical examination, and the results of laboratory and x-ray studies. Increased serum levels of bilirubin, alkaline phosphatase, AST, GGT, and lactic dehydrogenase may occur. Leukocytosis (increased white blood cells), erythrocytosis (increased red blood cells), hypercalcemia, hypoglycemia, and hypocholesterolemia may also be seen on laboratory assessment.

The serum level of alpha-fetoprotein, which serves as a tumor marker, is elevated in 30% to 40% of patients with primary liver cancer, commonly to levels greater than 200 ng/mL (Feldman et al., 2010) The level of carcinoembryonic antigen, a marker of advanced cancer of the digestive tract, may be elevated. These two markers together are useful to distinguish between metastatic liver disease and primary liver cancer.

Many patients have metastases from the primary liver tumor to other sites by the time the diagnosis is made; metastases occur primarily to the lung but may also occur to regional lymph nodes, adrenals, bone, kidneys, heart, pancreas, or stomach.

X-rays, liver scans, CT scans, ultrasound studies, MRI, arteriography, and laparoscopy may be part of the diagnostic workup and may be performed to determine the extent of the cancer. Positron emission tomography (PET) scans are used to evaluate a wide range of metastatic tumors of the liver.

Confirmation of a tumor's histology can be made by biopsy under imaging guidance (CT scan or ultrasound) or laparoscopically. Local or systemic dissemination of the tumor by needle biopsy or fine-needle biopsy can occur. Because of the small but real risks of tumor seeding (0.5% to 2%), hemorrhage, and false-negative results from biopsy, many centers avoid biopsy, particularly in patients who may be candidates for liver resection or liver transplantation. Assessment of the imaging characteristics for the diagnosis of HCC is preferred in these instances, and a confirmed diagnosis of HCC is made by frozen section at the time of surgery (Goldman & Schafer, 2012).

Medical Management

Although surgical resection of the liver tumor is possible in some patients, the underlying cirrhosis is so prevalent in cancer of the liver that it increases the risks associated with surgery. Radiation therapy and chemotherapy have been used to treat cancer of the liver with varying degrees of success. Although these therapies may prolong survival and improve quality of life by reducing pain and discomfort, their major effect is palliative.

Radiation Therapy

The use of external-beam radiation for the treatment of liver tumors has been limited by the radiosensitivity of normal hepatocytes and the risk of destruction of normal liver parenchyma. More effective methods of delivering radiation to tumors of the liver include (1) IV or intra-arterial injection of antibodies tagged with radioactive isotopes that specifically attack tumor-associated antigens and (2) percutaneous placement of a high-intensity source for interstitial radiation therapy (delivery of radiation directly to the tumor cells). Internal radiotherapy can result in reduction in tumor size, but its effect on survival is yet to be determined.

Chemotherapy

Typically, studies of patients with advanced cases of liver cancer have shown that the use of systemic chemotherapeutic agents leads to poor outcomes. For patients with stable hepatic function (Child class A), new anticancer drugs (targeted molecular therapies), one known as sorafenib (Nexavar), have been developed and approved for use and have shown some degree of improvement in treating patients with HCC (DiBisceglie & Befeler, 2010). Systemic chemotherapy may be used to treat metastatic liver lesions. Embolization of tumor vessels with chemotherapy (a process known as transarterial chemoembolization) produces anoxic necrosis with high concentrations of trapped chemotherapeutic agents. This therapy has begun to show some promising results. An implantable pump has been used to deliver a high concentration of chemotherapy by constant infusion to the liver through the hepatic artery in cases of metastatic disease. This method has shown a moderate response rate (Feldman et al., 2010).

Percutaneous Biliary Drainage

Percutaneous biliary or transhepatic drainage is used to bypass biliary ducts obstructed by liver, pancreatic, or bile duct tumors in patients who have inoperable tumors or are considered poor surgical risks. Under fluoroscopy, a catheter is inserted through the abdominal wall and past the obstruction into the duodenum. Such procedures are used to reestablish biliary drainage, relieve pressure and pain from the buildup of bile behind the obstruction, and decrease pruritus and jaundice. As a result, the patient is made more comfortable, and quality of life and survival are improved.

For several days after its insertion, the catheter is opened to external drainage. The bile is observed closely for amount, color, and presence of blood and debris. Complications of percutaneous biliary drainage include sepsis, leakage of bile, hemorrhage, and reobstruction of the biliary system by debris in the catheter or by encroaching tumor. Therefore, the patient is observed for fever and chills, bile drainage around the catheter, changes in vital signs, and evidence of biliary obstruction, including increased pain or pressure, pruritus, and recurrence of jaundice.

Other Nonsurgical Treatments

Laser hyperthermia has been used to treat hepatic metastases. Heat has been directed to tumors through several methods to cause necrosis of the tumor cells while sparing normal tissue. In radiofrequency thermal ablation, a needle electrode is inserted into the liver tumor under imaging guidance. Radiofrequency energy passes through to the noninsulated needle tip, causing heat and tumor cell death from coagulation necrosis.

Immunotherapy is another treatment modality under investigation. In this therapy, lymphocytes with antitumor reactivity are administered to the patient with hepatic cancer. Tumor regression has been demonstrated in patients with metastatic cancer for whom standard treatment has failed.

Transcatheter arterial embolization interrupts the arterial blood flow to small tumors by injecting small particulate embolic or chemotherapeutic agents (as described previously) into the artery supplying the tumor. As a result, ischemia and necrosis of the tumor occur.

For multiple small lesions, ultrasound-guided injection of alcohol promotes dehydration of tumor cells and tumor necrosis (Feldman et al., 2010).

Surgical Management

Surgical resection is the treatment of choice when HCC is confined to one lobe of the liver and the function of the remaining liver is considered adequate for postoperative recovery. In the case of metastasis, hepatic resection can be performed if the primary site can be completely excised and the metastasis is limited. However, metastases to the liver are rarely limited or solitary. Capitalizing on the regenerative capacity of the liver cells, some surgeons have successfully removed 90% of the liver. However, the presence of cirrhosis limits the ability of the liver to regenerate. Laparoscopic liver resection for malignant tumors has also been described. Staging of liver tumors aids in predicting the likelihood of surgical cure.

In preparation for surgery, the patient's nutritional, fluid, and general physical status are assessed, and efforts are undertaken to ensure the best physical condition possible. Extensive diagnostic studies may be performed. Specific studies may include liver scan, liver biopsy, cholangiography, selective hepatic angiography, percutaneous needle biopsy, peritoneoscopy, laparoscopy, ultrasound, CT scan, PET scan, MRI, and blood tests, particularly determinations of serum alkaline phosphatase, AST, and GGT and its isoenzymes.

Lobectomy

Removal of a lobe of the liver is the most common surgical procedure for excising a liver tumor. If it is necessary to restrict blood flow from the hepatic artery and portal vein for longer than 15 minutes, it is likely that hypothermia will be used. For a right-liver lobectomy or an extended right lobectomy (including the medial left lobe), a thoracoabdominal incision is used. An extensive abdominal incision is made for a left lobectomy.

Local Ablation

In patients who are not candidates for resection or transplantation, ablation of HCC may be accomplished by chemicals such as ethanol or by physical means such as radiofrequency ablation or microwave coagulation. These techniques may be performed under ultrasound or CT guidance laparoscopically or percutaneously. Radiofrequency ablation is becoming a standard mode of treatment; a tumor up to 5 cm in size can be destroyed in one session. The most common complications following ablation are local pain or bleeding. Serious complications are rare (Feldman et al., 2010).

Immunotherapy with interferon may be used after surgical resection for HCC to prevent recurrence of the lesion in those patients who have developed the lesion related to hepatitis B or C.

Liver Transplantation

Liver transplantation offers good patient outcomes. Candidates with liver cancer meet stringent selection criteria, including having small, early-stage lesions (Earl & Chapman, 2011). This treatment involves removing the liver and replacing it with a healthy donor organ. Studies report decreased recurrence rates of the primary liver malignancy after transplantation, with improvement in 5-year survival rates to consistently greater than 70% (Tsim et al., 2010). Metastasis and recurrence may be enhanced by the immunosuppressive therapy that is needed to prevent rejection of the transplanted liver. In patients with small (less than 5 cm), single lesions, liver transplantation has been shown to be beneficial, but its use is limited by organ shortages. The increasing use of living donor transplantation may improve this situation and decrease the waiting time and tumor proliferation that is characteristic of patients with liver cancer (see later discussion).

Nursing Management

For patients with liver cancer anticipating surgery, support, education, and encouragement are provided to help them prepare psychologically for the surgery. After surgery, potential problems related to cardiopulmonary involvement may include vascular complications and respiratory and liver dysfunction. Metabolic abnormalities require careful attention. Because extensive blood loss may occur as well, the patient receives infusions of blood and IV fluids. The patient requires constant, close monitoring and care for the first 2 or 3 days, similar to postsurgical abdominal and thoracic nursing care.

If the patient is to receive chemotherapy or radiation therapy in an effort to relieve symptoms, he or she may be discharged home while still receiving one or both of these therapies. The patient may also go home with a biliary drainage system or hepatic artery catheter in place. In most cases, the hepatic artery catheter has been inserted surgically and has a prefilled infusion pump implanted subcutaneously that delivers a continuous chemotherapeutic dose until completed (Yarbro, Wujcik, & Gobel, 2011). A hepatic artery port may also be inserted to provide access for intermittent chemotherapy infusion. This port dwells under the skin, but because it provides direct arterial access, it is not used for continuous infusion therapy in the home environment; the access line is discontinued once the chemotherapeutic agent has infused. The patient and family require education about care of the biliary catheter and the effects and side effects of hepatic artery chemotherapy. This education is necessary because of participation of the patient and family in patient care in the home setting.

Promoting Home and Community-Based Care

Educating Patients About Self-Care

The nurse educates the patient to recognize and report the potential complications and side effects of the chemotherapy and the desirable and undesirable effects of the specific chemotherapy regimen. The nurse also emphasizes the importance of follow-up visits to assess the response to chemotherapy and radiation therapy. In addition, if the patient is receiving chemotherapy on an outpatient basis, the nurse explains the patient's and family's role in managing the chemotherapy infusion and in assessing the infusion or insertion site. The nurse encourages the patient to resume routine activities as soon as possible while cautioning about falls and activities that may damage the infusion pump or site.

Patients at home with a biliary drainage system in place typically fear that the catheter will become dislodged; this fear is often shared by the patient's family. Reassurance and instruction can help reduce their fear that the catheter will fall out easily. The patient and family also require education on catheter care, including instruction on how to keep the catheter site clean and dry and how to assess the catheter and its insertion site. Irrigation of the catheter with sterile normal saline solution or water may be prescribed to keep the catheter patent and free of debris. The patient and caregivers are taught proper technique to avoid introducing bacteria into the biliary system or catheter during irrigation. They are instructed not to aspirate or draw back on the syringe during irrigation in order to prevent entry of irritating duodenal contents into the biliary tree or catheter. The patient and caregivers are also instructed about the signs of complications and are encouraged to notify the nurse or primary provider if problems or questions arise.

Patients with implantable ports are instructed about the chemotherapy regimen, types of medications, effects and side effects that may occur, and appropriate management strategies if problems occur. If a hepatic artery port is inserted for intermittent chemotherapy, patients and their families are provided the same educational content. Such a port has an internal one-way valve; therefore, it is not aspirated for a blood return before the infusion is initiated. The patient is instructed to assess the port site between infusions and to note and report any sign of infection or inflammation.

Continuing Care

In many cases, referral for home care enables the patient with liver cancer to be at home in a familiar environment with family and friends. Because of the poor prognosis associated with liver cancer, the home care nurse serves a vital role in assisting the patient and family to cope with the symptoms that may occur and the prognosis. The home care nurse assesses the patient's physical and psychological status, adequacy of pain relief, nutritional status, and presence of symptoms indicating complications of treatment or progression of disease. During home visits, the nurse assesses the function of the chemotherapy pump, the infusion site, and the biliary drainage system, if indicated. The nurse collaborates with the other members of the health care team, the patient, and the family to ensure effective pain management and to manage potential problems, which include weakness, pruritus, inadequate dietary intake, jaundice, and symptoms associated with metastasis to other sites. The home care nurse also assists the patient and family in making decisions about hospice care and assists with initiation of referrals. The patient is encouraged to discuss preferences for end-of-life care with family members and health care providers (see Chapter 16).

Liver Transplantation

Liver transplantation is used to treat life-threatening ESLD for which no other form of treatment is available. The transplantation procedure involves total removal of the diseased liver and replacement with a healthy liver from a cadaver donor or with the right lobe from a live donor in the same anatomic location (**orthotopic liver transplantation [OLT]**). Removal of the liver creates a space for the new liver and permits anatomic reconstruction of the hepatic vasculature and biliary tract as close to normal as possible.

The success of liver transplantation depends on successful immunosuppression. Immunosuppressant agents currently used include cyclosporine (Neoral), tacrolimus (Prograf), corticosteroids, azathioprine (Imuran), mycophenolate mofetil (CellCept), sirolimus (formerly known as rapamycin [Rapamune]), antithymocyte globulin (Thymoglobulin), basiliximab (Simulect), and daclizumab (Zenapax). There is no one, accepted, optimal immunosuppressive regimen. Most centers have developed their own therapeutic practices, largely based on experience. Multiple immunosuppressive strategies can be used to prevent transplanted organ rejection. Most strategies involve the use of more than one agent. Using multiple immunosuppressive agents has the effect of blocking multiple targets in the immune response cascade. This allows the use of lower doses of each drug, thus avoiding toxicity associated with high doses of these powerful drugs. Many patients are treated with "triple therapy" using corticosteroids, a calcineurin inhibitor such as tacrolimus or cyclosporine, and either an antiproliferative agent (mycophenolate mofetil) or a TOR (target of rapamycin) inhibitor such as sirolimus (Doherty, 2010). Some transplant centers also prescribe steroid-free immunosuppressant regimens after liver transplantation. This regimen has been found to be a safe alternative. Still other transplant centers advocate monotherapy with a calcineurin inhibitor alone to provide long-term immunosuppression (Doherty, 2010; Townsend, Beauchamp, Evers, et al., 2012).

Despite the success of immunosuppression in reducing the incidence of rejection of transplanted organs, liver transplantation is not routine and may be accompanied by complications related to the lengthy surgical procedure, immunosuppressive therapy, infection, and technical difficulties encountered in reconstructing the blood vessels and biliary tract. Long-standing systemic problems resulting from the primary liver disease may complicate the pre- and postoperative course. Previous surgery of the abdomen, including procedures to treat complications of advanced liver disease (i.e., shunt procedures used to treat portal hypertension and esophageal varices) increase the complexity of the transplantation procedure.

General indications for liver transplantation include irreversible advanced chronic liver disease, fulminant hepatic failure, metabolic liver diseases, and some hepatic malignancies. Examples of disorders that are indications for liver transplantation include hepatocellular liver diseases (e.g., viral hepatitis, drug- or alcohol-induced liver disease, Wilson's disease) and cholestatic diseases (primary biliary cirrhosis, sclerosing cholangitis, NASH, and biliary atresia).

The patient being considered for liver transplantation frequently has many systemic problems that influence pre- and postoperative care. Because transplantation is more difficult if the patient has developed severe GI bleeding and hepatic coma, efforts are made to perform the procedure before the disease progresses to this stage. The patient must undergo a thorough evaluation of hepatic reserve and general health. Part of this evaluation includes classification of the degree of medical need, an objective determination known as the Model for End-Stage Liver Disease (MELD) classification, which stratifies the level of illness of those awaiting a liver transplant. The MELD score is derived from a complex formula incorporating bilirubin levels, prothrombin time (reported as international normalized ratio), creatinine, and the cause of the liver disease (i.e., cholestatic, alcoholic, or other). This system has replaced the Child-Pugh and Child-Pugh-Turcotte classifications and other related scoring systems for prioritizing patients on the liver transplantation list (Fox & Brown, 2012; Hansen et al., 2010). Although the Child-Pugh score classifies the severity of liver disease and stratifies patients into levels for varied treatment regimens, the MELD score is an indicator of short-term mortality for those with ESLD. Organs are allocated using the MELD score in an effort to provide transplants to the most severely ill patients.

Liver transplantation is an established therapeutic modality, and the number of liver transplant centers is increasing. Patients requiring transplantation are often referred from distant hospitals to these centers. To prepare the patient and family for liver transplantation, nurses in all settings must understand the processes and procedures of liver transplantation.

Many ethical issues arise concerning liver transplantation, particularly concerning the allocation of organs. The way in which some persons contracted liver disease (e.g., alcohol use, hepatitis) leads others to question allocation of organs to them, and some believe that preference should be given to people who need liver transplants but do not have a history of socially unacceptable behavior. More issues arise when a patient requires a second transplant operation because of a return to alcohol or drug use or failure to follow immunosuppressive regimens (Chart 49-13). Transplant recipients must go through a rigorous selection and preparation process that includes counseling and education to aid them in making critical choices for their improved health. Nurses and other health care providers need to be aware of and confront their own biases and work toward improved understanding and acceptance.

Surgical Procedure

During the procedure, the donor liver is freed from other structures, the bile is flushed from the gallbladder to prevent damage to the walls of the biliary tract, and the liver is perfused with a preservative and cooled. Before the donor liver is placed in the recipient, it is flushed with cold lactated Ringer's solution to remove potassium and air bubbles. The presence of portal hypertension increases the difficulty of the procedure. To minimize this problem, many centers use venovenous bypass, which decompresses the venous system below the diaphragm by temporarily shunting blood to the superior vena cava via the axillary vein (Townsend et al., 2012).

Anastomoses (connections) of the blood vessels and bile duct are performed between the donor liver and the recipient liver. There are two types of biliary anastomoses. Biliary reconstruction is performed with an end-to-end anastomosis of the donor and recipient common bile ducts; a stented T-tube may be inserted for external drainage of bile. In patients with biliary disease such as primary sclerosing cholangitis or if the recipient's bile duct is not suitable for anastomosis for other reasons, a biliary-enteric end-to-side anastomosis with a 40- to 50-cm Roux-en-Y loop of jejunum is created for biliary drainage (known as a Roux-en-Y procedure) (Fig. 49-15A); in this case, bile drainage is internal, and a T-tube is not inserted (Townsend

Chart 49-13 ETHICAL DILEMMA

What Ethical Principles Apply When a Candidate for a Second Liver Transplantation Abuses Alcohol?

Case Scenario

You are a staff nurse working in a university health science center critical care unit that has an affiliated solid organ transplant service. A 38-year-old man is emergently admitted to the unit with liver failure and hemorrhagic shock from bleeding esophageal varices. He received a liver transplant 3 years ago for end-stage liver disease due to alcohol abuse. He had reputedly not drank alcohol for the better part of the past 4 years and had meticulously adhered to his regimen of antirejection medications until 6 months ago, when his wife died in a motor vehicle crash. Since that time, according to family members, he sank into depression, resumed drinking alcohol, and stopped taking his medications. The medical social worker and the transplant coordinator note that he is the sole caretaker for his two young sons, ages 8 and 10 years, and his 68-year-old father, who is disabled with emphysema. As you prepare to enter his room, his family arrives. The oldest son is pushing the patient's father in his wheelchair; you note that he is also receiving portable oxygen by nasal cannula. The youngest son is crying and rubbing his eyes. The oldest son grabs your hand and anxiously asks you "Is my Dad dying?"

Discussion

This patient has already received one liver transplant, and he will not survive this acute episode unless he receives another one. Yet, given his unstable condition, he is not a good candidate for a successful transplant—he may not survive the surgery or the immediate postoperative period, and the transplanted liver could be given to a patient who is not as unstable. Moreover, should he receive a second transplanted organ and survive this hospitalization, if he abuses alcohol or does not adhere to his medication regimen, he will very likely reject his second transplanted liver. All of these issues are considerations when the supply of viable organs is less than the number of patients on organ transplant wait lists. On any given day in the United States, it is estimated that 18 people die waiting for organs (U.S. Department of Health and Human Services [HHS], 2013).

Analysis

- Describe the ethical principles that are in conflict in this case (see Chart 3-3 in Chapter 3). Which principle has preeminence when making decisions about allocation of scarce resources, such as organs?
- Do you believe that this patient should be restricted from receiving a second liver because of his lapse in the therapeutic regimen? Is his lapse "excusable"? Should the fact that he is the sole caretaker of two young children and a disabled parent have any consideration in whether or not he receives a second transplant? Should he be excluded from receiving a second transplant because of his critical illness? Who should receive organs first—patients who are the sickest, or those most likely to receive benefit?

Reference

U.S. Department of Health and Human Services. (2013). *Donate the gift of life.* Available at: www.organdonor.gov/index.html

Resources

See Chapter 3, Chart 3-6 for ethics resources.

et al., 2012). Figure 49-15B and C illustrates the final appearance of the grafted liver and final closure and drain placement.

Several additional techniques have been developed to expand the donor pool for liver transplantation. In a split-liver transplant, a single organ is used to provide grafts for two individuals with ESLD, with the smaller patient receiving the smaller left lobe. This procedure has resulted in a higher complication rate and lower survival rate than traditional liver transplantation. Living donor transplantation is being increasingly performed from adult to adult using full right lobes, although it is controversial because it is a major surgical procedure for the donor, and some donor deaths have occurred.

Living donor liver transplantation (LDLT) is considered for patients who have a high potential for mortality while awaiting a cadaveric liver donor, such as those patients with HCC

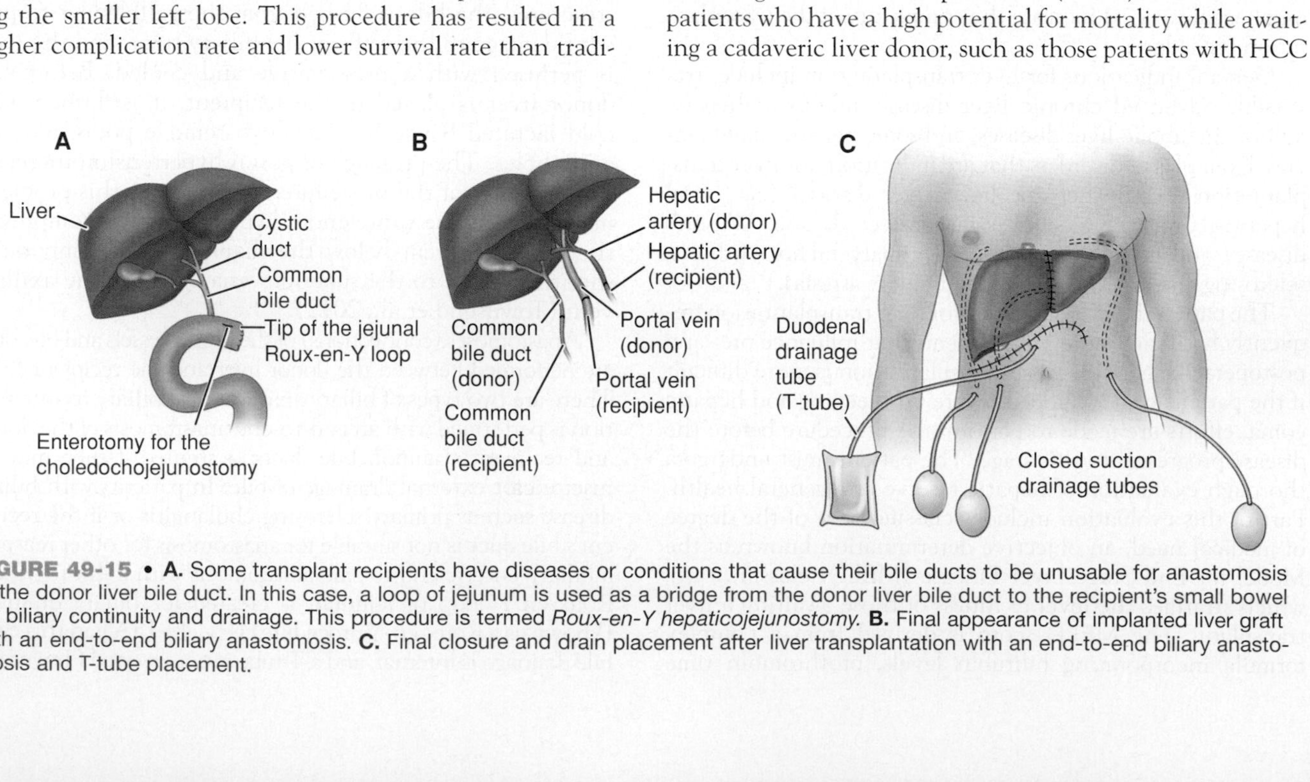
A
Liver
Cystic duct
Common bile duct
Tip of the jejunal Roux-en-Y loop
Enterotomy for the choledochojejunostomy
B
Hepatic artery (donor)
Hepatic artery (recipient)
Common bile duct (donor)
Portal vein (donor)
Portal vein (recipient)
Common bile duct (recipient)
C
Duodenal drainage tube (T-tube)
Closed suction drainage tubes

FIGURE 49-15 • **A.** Some transplant recipients have diseases or conditions that cause their bile ducts to be unusable for anastomosis to the donor liver bile duct. In this case, a loop of jejunum is used as a bridge from the donor liver bile duct to the recipient's small bowel for biliary continuity and drainage. This procedure is termed *Roux-en-Y hepaticojejunostomy.* **B.** Final appearance of implanted liver graft with an end-to-end biliary anastomosis. **C.** Final closure and drain placement after liver transplantation with an end-to-end biliary anastomosis and T-tube placement.

or those with severe complications of cirrhosis including GI bleeding or hepatic encephalopathy (Koffron & Stein, 2008). The results thus far have indicated that this procedure is most successful when donor and recipient are appropriately selected using careful screening criteria (Townsend et al., 2012). The option of LDLT decreases the waiting list mortality and produces positive recipient outcomes with a low risk of morbidity and mortality for the donor. The LDLT procedure involves transplantation of the right hepatic lobe from an adult donor to the recipient. Potential donors are evaluated by a donor advocate team. Donors must be completely healthy and have hepatic size and anatomy compatible with right lobe transplantation (Koffron & Stein, 2008). There is an extensive informed consent process for live liver donors. The donor advocate ensures that the concern for donor safety is paramount, especially in the intra- and postoperative period when complications can occur. The clear separation of donor and recipient teams ensures that the donor is treated without ulterior motives, which might occur if the same team cared for both the donor and recipient (Townsend et al., 2012).

In the LDLT procedure, the surgeon performs a formal right hepatic lobectomy. The right lobe is then flushed with preservative solution and vascular reconstruction is completed to prepare for implantation. The recipient operation involves an inferior vena cava–sparing hepatectomy with anastomosis of the donor right-sided vascular and biliary structures to the corresponding recipient structures (Koffron & Stein, 2008; Townsend et al., 2012).

Liver transplantation is a long surgical procedure, partly because the patient with liver failure often has portal hypertension, requiring ligation of many venous collateral vessels. Blood loss during the surgical procedure may be extensive. If the patient has adhesions from previous abdominal surgery, lysis of adhesions is often necessary. If a shunt procedure was performed previously, it must be surgically reversed to permit adequate portal venous blood supply to the new liver. During the lengthy surgery, it is important to provide regular updates to the family about the progress of the operation and the patient's status.

Complications

The postoperative complication rate is high, primarily because of technical complications or infection. Immediate postoperative complications may include bleeding, infection, and rejection. Disruption, infection, obstruction of the biliary anastomosis, and impaired biliary drainage may occur. Vascular thrombosis and stenosis are other potential complications. These complications occur in patients receiving either a cadaver or live donor organ. Despite the development of some complications, the 1-year patient survival rate approaches 90% and the 5-year patient survival rate is approximately 80% (Townsend et al., 2012).

Bleeding

Bleeding is common in the postoperative period and may result from coagulopathy, portal hypertension, and fibrinolysis caused by ischemic injury to the donor liver. Hemodynamic instability and transient hypotension may occur in this phase, secondary to blood loss, loss of vasomotor tone, and vasodilatation due to rewarming the hypothermic patient or due to pre-existing cardiac conditions such as cardiomyopathy (Grogan, 2011; Niemann & Kramer, 2011). Administration of platelets, fresh-frozen plasma, or other blood products may be necessary. Hypertension may co-occur postoperatively but is more common later in the postoperative phase, although its cause is uncertain. Currently available hypertension guidelines do not have specific recommendations for management in the liver transplant population (Najeed, Saghir, Hein, et al., 2011). However, calcium channel blockers such as nifedipine or amlodipine are frequently used for their vasodilatory effects. These drugs also are preferable due to their low interaction level with the cytochrome P450 enzyme system with resultant minimal risk of disruption of immunosuppressant levels. Angiotensin-converting enzyme inhibitors and angiotensin receptor blockers are not first-line drugs for the treatment of hypertension during the first year after transplant due to low levels of renin during this time period. Thiazide diuretics are reserved for patients requiring more than one medication for blood pressure control (Najeed et al., 2011). Blood pressure elevation that is significant or sustained is also managed with lifestyle modifications, a low-sodium diet, and an exercise regimen.

Infection

Infection is the leading cause of death after liver transplantation. Pulmonary and fungal infections are common; susceptibility to infection is increased by the immunosuppressive therapy that is needed to prevent rejection (McDonald et al., 2010). Therefore, precautions must be taken to prevent health care–associated infections. The nurse uses strict asepsis when manipulating central venous catheters, arterial lines, and urine, bile, and other drainage systems; obtaining specimens; and changing dressings. Meticulous hand hygiene is crucial. In the ICU, the nurse vigilantly monitors for early clinical manifestations of sepsis (see Charts 14-5 and 14-6 in Chapter 14) and uses evidence-based practice guidelines (or bundles) developed by the Institute for Healthcare Improvement (IHI) in the care of the postoperative liver transplant patient. Some of these care guidelines include prevention of sepsis through prevention of central line–associated bloodstream infections and its rapid treatment (see Chart 14-2) and prevention of ventilator-associated pneumonia (VAP) (see Chart 21-13 in Chapter 21).

Rejection

Rejection is a major concern. A transplanted liver is perceived by the immune system as a foreign antigen. This triggers an immune response, leading to the activation of T lymphocytes that attack and destroy the transplanted liver. Immunosuppressive agents are used as long-term therapy to prevent this response and rejection of the transplanted liver. These agents inhibit the activation of immunocompetent T lymphocytes to prevent the production of effector T cells.

Although the 1- and 5-year survival rates have increased dramatically with the use of new immunosuppressive therapies, these advances are not without major side effects. A major side effect of cyclosporine, which has been widely used in transplantation, is nephrotoxicity; this problem seems to be dose related. Cyclosporine-related side effects have caused many centers to use tacrolimus instead as first-line therapy because of its efficacy and lower side-effect profile.

Corticosteroids, azathioprine, mycophenolate mofetil, sirolimus, antithymocyte globulin, basiliximab, and daclizumab are also used in various regimens of immunosuppression. These agents may be used as the initial therapy to prevent rejection or used later to treat rejection. Liver biopsy and ultrasound may be required to evaluate suspected episodes of rejection.

Retransplantation is usually attempted if the transplanted liver fails, but the success rate of retransplantation does not approach that of initial transplantation. The greater rates of organ dysfunction and loss after a second or third liver transplant is related to technical intraoperative difficulties and higher bleeding risk (Razonable, Findlay, O'Riordan, et al., 2011).

Complications of the LDLT Donor

Improved surgical techniques have made the LDLT procedure an increasingly safe one; however, complications do occur for donors as well. The most frequently occurring complications include pulmonary emboli, portal vein thrombosis, bile duct injury, and liver insufficiency secondary to a resection that is too extensive (Townsend et al., 2012).

Nursing Management

The patient considering transplantation, together with the family, must make difficult choices about treatment, the use of financial resources, and relocation to another area to be closer to the medical center. They must also be aware of the risks and benefits of the procedure and its consequences. In addition, they must also cope with the patient's long-standing health problems and any social and family problems associated with behaviors that may have caused the patient's liver failure. As a result, considerable emotional stress occurs while the patient and family consider liver transplantation and wait for an available liver. The nurse must be aware of these issues and attuned to the emotional and psychological status of the patient and family. Referral to a psychiatric liaison nurse, psychologist, psychiatrist, or spiritual advisor may help them cope with the stressors associated with ESLD and liver transplantation.

If the patient and family are considering undergoing a LDLT, they are subject to additional stressors. Both the patient and the potential donor must undergo a thorough and exhaustive physical and psychological workup to ensure that all involved parties are physically and emotionally prepared. Often, but not always, the donor is a close family member. Coercion must be excluded as influencing the decision to donate a portion of one's liver to another. The potential donor must be aware of the risks associated with the procedure.

If the patient and family believe that liver transplantation may be appropriate, the nurse, surgeon, hepatologist, and other health care team members provide the patient and family with full explanations about the procedure, the chances of success, and the risks (for the donor—bleeding and venous thromboembolism), including the side effects of long-term immunosuppression and postoperative complications in the recipient as well as bleeding and biliary abnormalities (Townsend et al., 2012). The need for close follow-up and lifelong compliance with the therapeutic regimen, including immunosuppression, is emphasized to the patient and family.

Preoperative Nursing Interventions

Once the patient has been accepted as a candidate, he or she is placed on a waiting list at the transplant center, and patient information is entered into the United Network for Organ Sharing (UNOS) computer system. The UNOS system uses the MELD score to determine organ allocation priorities so that the patient with the highest MELD score will receive the first available organ. Candidates may be matched with appropriate organs as they become available. MELD scores provide the necessary information regarding medical need.

Except in the case of LDLT, a liver becomes available for transplantation only with the death of another person, usually someone who had been healthy except for severe brain injury and brain death. Therefore, the patient and family undergo a stressful waiting period, and the nurse is often their major source of support. The patient must be accessible at all times in case an appropriate liver becomes available. During this time, liver function may deteriorate further, and the patient may experience complications from the progressing disease. Because of the shortage of donor organs, many patients die awaiting transplantation.

Malnutrition, massive ascites, and fluid and electrolyte disturbances are treated before surgery to increase the likelihood of a successful outcome. If the patient's liver dysfunction has a very rapid onset, as in fulminant hepatic failure, there is little time or opportunity for the patient to consider and weigh options and their consequences; the patient may be in a coma and the decision to proceed with transplantation made by the family.

The nurse coordinator is an integral member of the transplant team and plays an important role in preparing the patient for liver transplantation. The nurse serves as an advocate for the patient and family and assumes the important role of liaison between the patient and the other members of the transplant team. The nurse also serves as a resource to other nurses and health care team members involved in evaluating and caring for the patient.

Postoperative Nursing Interventions

The organ recipient is maintained in an environment as free from bacteria, viruses, and fungi as possible, because immunosuppressive medications reduce the body's natural defenses. In the immediate postoperative period, cardiovascular, pulmonary, renal, neurologic, and metabolic functions are monitored continuously. Mean arterial and pulmonary artery pressures are also monitored continuously. Cardiac output, central venous pressure, pulmonary capillary wedge pressure, arterial and mixed venous blood gases, oxygen saturation, oxygen demand and delivery, urine output, heart rate, and blood pressure are used to evaluate the patient's hemodynamic status and intravascular fluid volume. Liver function tests, electrolyte levels, the coagulation profile, chest x-ray, electrocardiogram, and fluid output (including urine, bile from the T-tube, and drainage from Jackson-Pratt tubes) are monitored closely. Because the liver is responsible for the storage of glycogen and the synthesis of protein and clotting factors, these substances need to be monitored and replaced in the immediate postoperative period.

There is a high risk of atelectasis and an altered ventilation–perfusion ratio caused by insult to the diaphragm during the surgical procedure, prolonged anesthesia, immobility, and postoperative pain. The patient will have an endotracheal

Chart 49-14

NURSING RESEARCH PROFILE

Caregiver Stress After Liver Transplantation

Weng, L. C., Huang, H. L., Wang, Y. W., et al. (2011). Primary caregiver stress in caring for a living-related liver transplant recipient during the postoperative stage. *Journal of Advanced Nursing, 67*(8), 1749–1757.

Purpose

The purpose of this study was to examine the level of stress experienced by the primary caretaker of patients who are recipients of a living-related liver transplantation during the postoperative stage.

Design

The researchers carried out a qualitative study in which they performed face-to-face semistructured interviews to identify the subjective experience of the study participants. The study participants were identified from contact through a tertiary medical center in Taiwan. The sample size was 6 out of a possible 12 caregivers. All of the participants were women, and 5 of the 6 were the wives of a transplant recipient.

Findings

The core theme that arose from this qualitative design included the finding that the stress of the primary caregivers of living-related liver transplant recipients is related to the gap between expectations and the actual primary caregiving experience. The specific themes about the stress the participants experienced were (1) unstable sentiment toward liver transplantation, (2) entanglement of burden, (3) nonsynchronized family interaction, (4) distance from health care professionals, and (5) concern about the protector role function.

Nursing Implications

Nursing professionals who work with and prepare families for a loved one's liver transplantation should work with the primary caregivers throughout the process, beginning with the pretransplant phase and especially into the early postoperative phase. In addition, nurses should provide sufficient resources and supportive services to the recipient and family members. The lines of communication with all involved in the transplant process must be kept continually open with health care professionals, ensuring their availability. Prior to discharge, there should be education and training on home care skills for caregivers, and home care consults and regular periodic follow-up should be planned.

tube in place and will require mechanical ventilation during the initial postoperative period. Suctioning is performed as required, and sterile humidification is provided. Evidence-based practice guidelines are implemented to prevent the development of VAP in the postoperative liver transplant recipient (see Chart 21-13 in Chapter 21).

As the patient's condition stabilizes, efforts are made to promote recovery from the trauma of this complex surgery. After removal of the endotracheal tube, the nurse encourages the patient to use an incentive spirometer to decrease the risk of atelectasis. Following extubation, the patient is assisted to get out of bed, to ambulate as tolerated, and to participate in self-care to prevent the complications associated with immobility. Close monitoring for signs and symptoms of liver dysfunction and rejection continue throughout the hospital stay. Plans are made for close follow-up after discharge as well. Education is initiated during the preoperative period and continues after surgery.

The live donor is frequently admitted to an ICU setting along with the recipient. The donor also requires close monitoring for cardiovascular, hemodynamic, and pulmonary stability. The nurse closely assesses the donor for signs of hemorrhage, biliary complication, respiratory decompensation, and infection. The donor is mobilized early in the postoperative phase to prevent the development of complications such as pulmonary embolism. Studies suggest that the donor may experience more pain than the recipient, possibly requiring more analgesia for pain control (Lentine & Patel, 2012). Patient education focuses on prevention and recognition of complications as well as activity progression and pain management.

Promoting Home and Community-Based Care

Educating Patients About Self-Care

Educating the patient, family, and caregivers about long-term measures to promote health is crucial for the success of transplantation and is an important role of the nurse (Chart 49-14). The patient and family must understand why they need to adhere closely to the therapeutic regimen, with special emphasis on the methods of administration, rationale, and side effects of the prescribed immunosuppressive agents. The nurse provides written and verbal education about how and when to take the medications. To avoid running out of medication or skipping a dose, the patient must make sure that an adequate supply of medication is available. Education is provided about the signs and symptoms that indicate problems necessitating consultation with the transplant team. The patient with a T-tube in place must be educated about how to manage the tube, drainage, and skin care.

Continuing Care

The nurse emphasizes the importance of follow-up blood tests and appointments with the transplant team. Trough blood levels of immunosuppressive agents are obtained, along with other blood tests that assess the function of the liver and kidneys. During the first months, the patient is likely to require blood tests two or three times a week. As the patient's condition stabilizes, blood studies and visits to the transplant team are less frequent. The importance of routine ophthalmologic examinations is emphasized because of the increased incidence of cataracts and glaucoma associated with the long-term corticosteroid therapy used with transplantation. Regular oral hygiene and follow-up dental care, with administration of prophylactic antibiotics before dental examinations and treatments, are recommended because of the immunosuppression.

The nurse reminds the patient that preventing rejection and infection is essential and increases the chances for survival and a more normal life than before transplantation. Many patients live successful and productive lives after receiving a liver transplant. In fact, pregnancy can be considered 2 years after transplantation. Although successful outcomes have been reported, these pregnancies are considered high risk for both mother and infant. Transplant recipients should be advised about birth control. The 2-year waiting period allows time to establish good health, stable liver function, and lower maintenance levels of immunosuppressive therapy (Carey, 2010; Feldman et al., 2010).

Liver Abscesses

Two categories of liver abscess have been identified: amebic and pyogenic. Amebic liver abscesses are most commonly caused by *Entamoeba histolytica*. Most amebic liver abscesses occur in the developing countries of the tropics and subtropics because of poor sanitation and hygiene. Pyogenic liver abscesses are much less common, but they are more common in developed countries than the amebic type (Townsend et al., 2012).

Pathophysiology

Whenever an infection develops anywhere along the biliary or GI tract, infecting organisms may reach the liver through the biliary system, portal venous system, or hepatic arterial or lymphatic system. Most bacteria are destroyed promptly, but occasionally some gain a foothold. The bacterial toxins destroy the neighboring liver cells, and the resulting necrotic tissue serves as a protective wall for the organisms.

Meanwhile, leukocytes migrate into the infected area. The result is an abscess cavity full of a liquid containing living and dead leukocytes, liquefied liver cells, and bacteria. Pyogenic abscesses of this type may be either single or multiple and small. Examples of causes of pyogenic liver abscess include cholangitis (usually related to benign or malignant obstruction of the biliary tree) and abdominal trauma.

Clinical Manifestations

The clinical picture is one of sepsis with few or no localizing signs. Fever with chills and diaphoresis, malaise, anorexia, nausea, vomiting, and weight loss may occur. The patient may complain of dull abdominal pain and tenderness in the right upper quadrant of the abdomen. Hepatomegaly, jaundice, anemia, and pleural effusion may develop. Sepsis and shock may be severe and life threatening. In the past, the mortality rate was 100% because of the vague clinical symptoms, inadequate diagnostic tools, and inadequate surgical drainage of the abscess. With the aid of ultrasound, CT, MRI, and liver scans, early diagnosis and surgical drainage of abscesses have greatly reduced the mortality rate.

Assessment and Diagnostic Findings

Although blood cultures are obtained, the organism may not be identified. Aspiration of the liver abscess, guided by ultrasound, CT, or MRI, may be performed to assist in diagnosis and to obtain cultures of the organism. Percutaneous drainage of pyogenic abscesses is carried out to evacuate the abscess material and promote healing. A catheter may be left in place for continuous drainage; the patient must be instructed about its management.

Medical Management

Treatment includes IV antibiotic therapy; the specific antibiotic used in treatment depends on the organism identified. Continuous supportive care is indicated because of the serious condition of the patient. Open surgical drainage may be required if antibiotic therapy and percutaneous drainage are ineffective.

Nursing Management

Although the manifestations of liver abscess vary with the type of abscess, most patients appear acutely ill. Others appear to be chronically ill and debilitated. The nursing management depends on the patient's physical status and the medical management that is indicated. For patients who undergo evacuation and drainage of an abscess, monitoring of the drainage and skin care are imperative. Strategies must be implemented to contain the drainage and to protect the patient from other sources of infection. Vital signs are monitored to detect changes in the patient's physical status. Deterioration in vital signs or the onset of new symptoms such as increasing pain, which may indicate rupture or extension of the abscess, is reported promptly. The nurse administers IV antibiotic therapy as prescribed. The white blood cell count and other laboratory test results are monitored closely for changes consistent with worsening infection. The nurse prepares the patient for discharge by providing instruction about symptom management, signs and symptoms that should be reported to the physician, management of drainage, and the importance of taking antibiotics as prescribed.

Critical Thinking Exercises

1 A 49-year-old man with obesity and diabetes underwent a physical with routine blood work due to his complaint of fatigue and occasional right upper quadrant abdominal discomfort. His liver transaminases are found to be elevated. Identify what further diagnostic studies the nurse should anticipate. What nursing interventions and patient education should the nurse undertake to prepare him to undergo these studies?

2 A 56-year-old woman is being treated for ESLD with cirrhosis related to hepatitis B. Describe the nursing plan of care. What specific interventions and nursing management strategies are indicated? Develop a plan for patient education for this patient, and identify outcomes.

3 **ebp** A 68-year-old man is admitted to the hospital with a diagnosis of metastatic liver cancer. What is the evidence base for available treatment options for this patient? Identify the criteria used to evaluate the strength of the evidence for the treatment practices you have identified.

4 **pq** A 36-year-old man is admitted for liver transplantation. Identify the priorities, approach, and techniques you would use to provide postoperative care for this patient following a cadaver transplant. How would your priorities, approach, and techniques differ if the patient receives an organ from a living, related donor?

Brunner Suite Resources
Explore these additional resources to enhance learning for this chapter:
• NCLEX-Style Questions and Other Resources on thePoint, http://thePoint.lww.com/Brunner13e
• Study Guide
• PrepU
• Clinical Handbook
• Handbook of Laboratory and Diagnostic Tests

References

**Asterisk indicates nursing research.*

Books

Bickley, L. S. (2009). *Bates' guide to physical examination and history taking* (10th ed.). Philadelphia: Wolters Kluwer Health/Lippincott Williams & Wilkins.
Bope, E. T., & Kellerman, R. D. (Eds.). (2011). *Conn's current therapy*. Philadelphia: Saunders.
Carey, W. D. (2010). *Current clinical medicine*. Philadelphia: Saunders.
DiBisceglie, A. M., & Befeler, A. S. (2010). Tumors & cysts of the liver. In M. Feldman, L. S. Friedman, & L. J. Brandt (Eds.). *Sleisinger & Fordtran's gastrointestinal & liver diseases* (9th ed.). Philadelphia: Saunders Elsevier
Doherty, G. (2010). *Current diagnosis and treatment: Surgery* (13th ed.). New York: McGraw-Hill.
Dudek, S. G. (2010). *Nutrition essentials for nursing practice* (6th ed.). Philadelphia: Lippincott Williams & Wilkins.
Feldman, M., Friedman, L. S., & Brandt, L. J. (2010). *Sleisinger & Fordtran's gastrointestinal & liver disease* (9th ed.). Philadelphia: Saunders Elsevier.
Fischbach, F., & Dunning, M. B. (2009). *A manual of laboratory and diagnostic tests* (8th ed.). Philadelphia: Lippincott Williams & Wilkins.
Ferri, F. F. (Ed.). (2011). *Practical guide to the care of the medical patient* (8th ed.). Philadelphia: Mosby Elsevier.
Fortune, B. E., Rosen, H. R., & Burton, J. R. (2010). Management of HCV infections and liver transplantation. In J. McDonald, A. Burroughs, B. Feagan, et al. (Eds.). *Evidence-based gastroenterology & hepatology* (3rd ed.). Malden, MA: Blackwell.
Gines, P., Cardenas, A., Arroyo, V., & Rodes, J. (2010). Ascites, hepatorenal syndrome and spontaneous bacterial peritonitis. In J. McDonald, A. Burroughs, B. Feagan, et al. (Eds.). *Evidence-based gastroenterology & hepatology* (3rd ed.). Malden, MA: Blackwell.
Goldman, L., & Schafer, A. I. (2012). *Goldman's Cecil medicine* (24th ed.). Philadelphia: Saunders Elsevier.
Greenberger, N. J., Blumberg, R. S., & Burakoff, R. (Eds.). (2012). *Current diagnosis and treatment: Gastroenterology, hepatology & endoscopy*. New York: McGraw-Hill.
Hall, J. E. (2011). *Guyton & Hall textbook of medical physiology* (12th ed.). Philadelphia: Saunders Elsevier.
Karch, A. M. (2012). *Lippincott's nursing drug guide*. Philadelphia: Lippincott Williams & Wilkins.
Mandell, G. L., Bennett, J. E., & Dolin, R. (2010). *Mandell, Douglas and Bennett's principles and practice of infectious diseases* (7th ed.). Philadelphia: Elsevier.
McDonald, J., Burroughs, A., Feagan, B., et al. (Eds.). (2010). *Evidence-based gastroenterology & hepatology* (3rd ed.). Malden, MA: Blackwell.
Mueller, C. M. (2012). *The A.S.P.E.N. adult nutrition support core curriculum* (2nd ed.). Silver Spring, MD: American Society for Parenteral and Enteral Nutrition.
O'Beirne, J., Murphy, N., & Wendon, J. (2010). Fulminant hepatic failure: Treatment. In J. McDonald, A. Burroughs, B. Feagan, et al. (Eds.). *Evidence-based gastroenterology & hepatology* (3rd ed.). Malden, MA: Blackwell.
Papadakis, M. A., & McPhee, S. J. (Eds.). (2013). *Current medical diagnosis and treatment* (52nd ed.). New York: McGraw-Hill.
Porth, C. M., & Matfin, G. (2009). *Pathophysiology: Concepts of altered health states* (8th ed.). Philadelphia: Lippincott Williams & Wilkins.
Rakel, D. (2012). *Integrative medicine* (3rd ed.). Philadelphia: Elsevier Saunders.
Roberts, J. R., & Hedges, J. R. (Eds.). (2010). *Clinical procedures in emergency medicine*. Philadelphia: Saunders Elsevier.
Schiff, E. R., Maddrey, W. C., & Sorrell, M. F. (2012). *Schiff's diseases of the liver* (11th ed.). Chichester, West Sussex: John Wiley & Sons.
Townsend, C. M., Beauchamp, R. D., Evers, B. M., et al. (2012). *Sabiston's textbook of surgery: The biological basis of modern surgical practice*. Philadelphia: Elsevier.
Triantos, C., Goulis, J., & Burroughs, A. K. (2010). Portal hypertensive bleeding. In J. McDonald, A. Burroughs, B. Feagan, et al. (Eds.). *Evidence-based gastroenterology & hepatology* (3rd ed.). Malden, MA: Blackwell.
Weber, J., & Kelley, J. (2009). *Health assessment in nursing* (4th ed.). Philadelphia: Lippincott Williams & Wilkins.
Wiegand, D. L. M. (Ed.). (2011). *AACN procedure manual for critical care* (6th ed.). Philadelphia: Elsevier.
Yarbro, C. H., Wujcik, D., & Gobel, B. H. (2011). *Cancer nursing: Principles & practice* (7th ed.). Burlington, MA: Jones & Bartlett.

Journals and Electronic Documents

General

Davern, T. J. (2012). Drug-induced liver disease. *Clinics in Liver Disease, 16*(2), 231–245.
European Association for the Study of the Liver. (2013). Too much drinking, weight may harm liver. *News release*, April 25, 2013.
Gordon, F. D. (2012). Ascites. *Clinics in Liver Diseases, 16*(2), 285–299.
Hansen, L., Sasaki, A., & Zucker, B. (2010). End-stage disease: Challenges and practice implications. *Nursing Clinics of North America, 45*(3), 411–426.
Kerwin, A. J., & Nussbaum, M. S. (2011). Adjuvant nutrition management of patients with liver failure, including transplant. *Surgical Clinics of North America, 91*(3), 564–578.
Khungar, V., & Poordad, F. (2012a). Hepatic encephalopathy. *Clinics in Liver Disease, 16*(2), 301–320.
Khungar, V., & Poordad, F. (2012b). Management of overt hepatic encephalopathy. *Clinics in Liver Disease, 16*(1), 73–89.
*Liu, J., Zhang, Y, Zhang, J., et al. (2010). Probiotic yogurt effects on intestinal flora of patients with chronic liver disease. *Nursing Research, 59*(6), 426–432.
Schattenberg, J. M., & Schuppan, D. (2011). Nonalcoholic steatohepatitis: The therapeutic challenge of a global epidemic. *Current Opinion in Lipidology, 22*(6), 479–488.
Sundarum, V., & Shaikh, O. S. (2009). Hepatic encephalopathy: Pathophysiology and emerging therapies. *Medical Clinics of North America, 93*(4), 819–836.
Sundarum, V., & Shaikh, O. S. (2011). Acute liver failure: Current practice and recent advances. *Gastroenterology Clinics of North America, 40*(3), 523–539.
Xu, J., Kochanek, K. D., Murphy, S. L., et al. (2010). Deaths: Final data for 2007. *National Vital Statistics Report, 58*(19), 1–135.

Cirrhosis and Esophageal Varices

Cardenas, A. (2010). Management of acute variceal bleeding: Emphasis on endoscopic therapy.*Clinics in Liver Disease, 14*(2), 251–262.
Cat, T. B., & Liu-DeRyke, X. (2010). Medical management of variceal hemorrhage. *Critical Care Nursing Clinics of North America, 22*(3), 281–293.
D'Amico, M., Berzigotti, A., & Garcia-Pagan, J. C. (2010). Refractory acute variceal bleeding: What to do next? *Clinics in Liver Disease, 14*(2), 297–305.
Leung, J. (2011). Colchicine or methotrexate with ursodiol are effective after 20 years in a subset of patients with primary biliary cirrhosis. *Clinical Gastroenterology and Hepatology, 9*(9), 776–780.
Mehta, G., & Rothstein, K. D. (2009). Health maintenance issues in cirrhosis. *Medical Clinics of North America, 93*(4), 901–915.
Opio, C. K., & Garcia-Tsao, G. (2011). Managing varices: Drugs, bands and shunts. *Gastroenterology Clinics of North America, 40*(3), 561–579.
Sostres, C., & Lanas, A. (2011). Epidemiology and demographics of upper gastrointestinal bleeding: Prevalence, incidence & mortality. *Gastrointestinal Endoscopy Clinics of North America, 21*(4), 567–581.
*Wu, L. J., Wu, M. S., Lien, G. S., et al. (2012). Fatigue and physical activity levels in patients with liver cirrhosis. *Journal of Clinical Nursing, 21*(1), 129–138.

Hepatitis

Centers for Disease Control and Prevention, Division of Viral Hepatitis. (2010). *Viral hepatitis surveillance—United States, 2010*. Available at: www.cdc.gov/hepatitis/Statistics/2010Surveillance/index.htm
Ghany, M. G., Nelson, D. R., Strader, D. B., et al. (2011). An update on treatment of genotype 1 chronic hepatitis C virus infection: 2011 practice guideline by the American Association for the Study of Liver Diseases. *Hepatology, 54*(4), 1433–1444.

Gluud, L. L., & Gluud, C. (2009). Meta-analyses on viral hepatitis. *Infectious Disease Clinics of North America, 23*(2), 315–330.

Ioannou, G. N. (2011). Hepatitis B virus in the United States: Infection, exposure and immunity rates in a nationally representative survey. *Annals of Internal Medicine, 154*(5), 319–328.

MacCannell, T., Laramie, A. K., Gomaa, A., et al. (2010). Occupational exposure of health care personnel to hepatitis B and hepatitis C: Prevention and surveillance strategies. *Clinics in Liver Disease, 14*(1), 23–36

McMahon, B. J. (2010). Natural history of chronic hepatitis B. *Clinics in Liver Disease, 14*(3), 381–396.

U.S. Food and Drug Administration. (2011a). FDA approves Incivek for hepatitis C. *FDA news release*, May 23, 2011.

U.S. Food and Drug Administration. (2011b). FDA approves Victrelis for hepatitis C. *FDA news release*, May 13, 2011.

Liver Cancer

Earl, T. M., & Chapman, W. C. (2011). Conventional surgical treatment of hepatocellular carcinoma. *Clinics in Liver Disease, 15*(2), 353–370.

McGlynn, K. A., & London, W. T. (2011). The global epidemiology of hepatocellular cancer: Present and future. *Clinics in Liver Disease, 15*(2), 223–243.

Tsim, N. C., Frampton, A. E., Haby, N. A., et al. (2010). Surgical treatment for liver cancer. *World Journal of Gastroenterology, 16*(8), 927–933.

Liver Transplantation

Fox, A. N., & Brown, R. S. (2012). Is the patient a candidate for liver transplantation? *Clinics in Liver Disease, 16*(2), 435–448.

Grogan, T. A. (2011). Liver transplantation: Issues and nursing care requirements. *Critical Care Nursing Clinics of North America, 23*(3), 443–456.

Koffron, A., & Stein, J. A. (2008). Liver transplantation: Indications, pretransplant evaluation, surgery and post transplant complications. *Medical Clinics of North America, 92*(4), 861–888.

*Lentine, K. L., & Patel, A. (2012). Risks and outcomes of living donation. *Advances in Chronic Kidney Disease, 19*(4), 220–228.

Najeed, S. A., Saghir, S., Hein, B., et al. (2011). Management of hypertension in liver transplant patients. *International Journal of Cardiology, 152*(1), 4–6.

Niemann, C. U., & Kramer, D. J. (2011). Transplant critical care: Standards for intensive care of the patient with liver failure before and after transplantation. *Liver Transplantation, 17*(4), 485–487.

Razonable, R. R., Findlay, J. Y., O'Riordan, A., et al. (2011). Critical care issues in patients after liver transplantation. *Liver Transplantation, 17*(5), 511–527.

Wasley, A., Kruszon-Moran, D., Kuhnert, W., et al. (2010). The prevalence of hepatitis B virus infection in the United States in the era of vaccination. *Journal of Infectious Diseases, 202*(2), 192–201.

*Weng, L. C., Huang, H. L., Wang, Y. W., et al. (2011). Primary caregiver stress in caring for a living-related liver transplantation recipient during the postoperative stage. *Journal of Advanced Nursing, 67*(8), 1749–1757.

Resources

Al-Anon Family Groups Headquarters, www.al-anon.alateen.org

Alcoholics Anonymous World Services, aa.org

American Association for the Study of Liver Diseases (ASSLD), www.aasld.org

American College of Gastroenterology (ACG), www.acg.gi.org

American Liver Foundation, www.liverfoundation.org

Hepatitis Foundation International, www.hepfi.org

National Council on Alcoholism and Drug Dependence (NCADD), www.ncadd.org

National Digestive Diseases Information Clearinghouse (NDDIC), digestive.niddk.nih.gov

National Institute on Alcohol Abuse and Alcoholism, www.niaaa.nih.gov

United Network for Organ Sharing (UNOS), www.unos.org

Chapter 50

Assessment and Management of Patients With Biliary Disorders

Learning Objectives

On completion of this chapter, the learner will be able to:

1. Identify the structure and function of the biliary tract and pancreas.
2. Compare approaches to management of cholelithiasis.
3. Use the nursing process as a framework for care of patients with cholelithiasis and those undergoing laparoscopic or open cholecystectomy.
4. Differentiate between acute and chronic pancreatitis.
5. Use the nursing process as a framework for care of patients with acute pancreatitis.
6. Describe the nutritional and metabolic effects of surgical treatment of tumors of the pancreas.

Glossary

amylase: pancreatic enzyme; aids in the digestion of carbohydrates

cholecystectomy: removal of the gallbladder

cholecystitis: inflammation of the gallbladder which can be acute or chronic

cholecystojejunostomy: anastomosis of the jejunum to the gallbladder to divert bile flow

cholecystokinin (CCK): hormone; major stimulus for digestive enzyme secretion; stimulates contraction of the gallbladder

cholecystostomy: surgical opening and drainage of the gallbladder

choledocholithiasis: stones in the common bile duct

choledochostomy: opening into the common bile duct

cholelithiasis: calculi in the gallbladder

dissolution therapy: the use of medications to break up/dissolve gallstones

endocrine: secreting internally; hormonal secretion of a ductless gland

endoscopic retrograde cholangiopancreatography (ERCP): procedure using fiberoptic technology to visualize the biliary system

exocrine: secreting externally; hormonal secretion from excretory ducts

laparoscopic cholecystectomy: removal of gallbladder through an endoscopic procedure

lipase: pancreatic enzyme; aids in the digestion of fats

lithotripsy: disintegration of gallstones by shock waves

pancreaticojejunostomy: joining of the pancreatic duct to the jejunum by side-to-side anastomosis; allows drainage of the pancreatic secretions into the jejunum

pancreatitis: inflammation of the pancreas; may be acute or chronic

secretin: hormone responsible for stimulating bicarbonate secretion from the pancreas; also used as an aid in diagnosing pancreatic exocrine disease

steatorrhea: frothy, foul-smelling stools with a high fat content, results from impaired digestion of proteins and fats due to a lack of pancreatic juice in the intestine

trypsin: pancreatic enzyme; aids in the digestion of proteins

wound-ostomy-continence (WOC) nurse: nurse specially educated in appropriate skin, wound, ostomy, and continence care; also referred to as wound care specialist or enterostomal therapist

Zollinger-Ellison syndrome: hypersecretion of gastric acid that produces peptic ulcers as a result of a non–beta-cell tumor of the pancreatic islets

Disorders of the biliary tract and pancreas are common and include gallbladder stones and pancreatic dysfunction. An understanding of the structure and function of the biliary tract and pancreas is essential, along with an understanding of how biliary tract disorders are closely linked with liver disease. Patients with acute or chronic biliary tract or pancreatic disease require care from nurses who are knowledgeable about the diagnostic procedures and interventions that are used in the management of gallbladder and pancreatic disorders.

ANATOMIC AND PHYSIOLOGIC OVERVIEW

The Gallbladder

The gallbladder, a pear-shaped, hollow, saclike organ that is 7.5 to 10 cm (3 to 4 in) long, lies in a shallow depression on the inferior surface of the liver, to which it is attached by loose connective tissue. The capacity of the gallbladder is 30 to 50 mL of bile. Its wall is composed largely of smooth

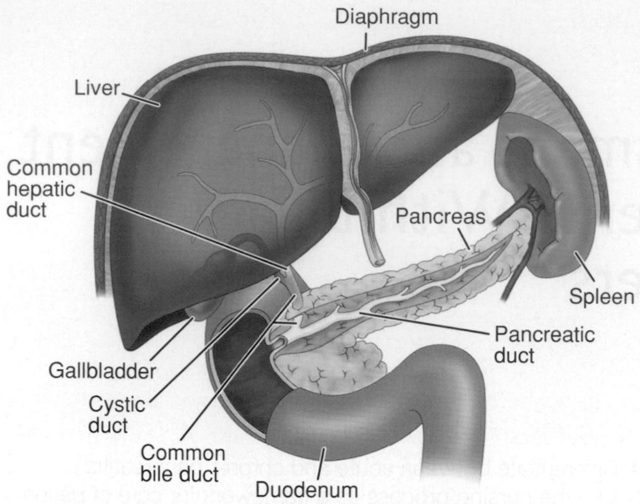
Diaphragm
Liver
Common hepatic duct
Pancreas
Spleen
Pancreatic duct
Gallbladder
Cystic duct
Common bile duct
Duodenum

FIGURE 50-1 • The liver, biliary system, and pancreas.

muscle. The gallbladder is connected to the common bile duct by the cystic duct (Fig. 50-1).

The gallbladder functions as a storage depot for bile. Between meals, when the sphincter of Oddi is closed, bile produced by the hepatocytes enters the gallbladder. During storage, a large portion of the water in bile is absorbed through the walls of the gallbladder; thus, bile in the gallbladder is 5 to 10 times more concentrated than that originally secreted by the liver. When food enters the duodenum, the gallbladder contracts and the sphincter of Oddi (located at the junction of the common bile duct with the duodenum) relaxes. Relaxation of this sphincter allows the bile to enter the intestine. This response is mediated by secretion of the hormone **cholecystokinin (CCK)** from the intestinal wall (Ferri, 2012; Marx, 2009). CCK is the major stimulus for digestive enzyme secretion and acts by stimulating the gallbladder to contract.

Bile is composed of water and electrolytes (sodium, potassium, calcium, chloride, and bicarbonate) along with significant amounts of lecithin, fatty acids, cholesterol, bilirubin, and bile salts. The bile salts, together with cholesterol, assist in emulsification of fats in the distal ileum. They are then reabsorbed into the portal blood for return to the liver, after which they are once again excreted into the bile. This pathway from hepatocytes to bile to intestine and back to the hepatocytes is called the *enterohepatic circulation*. Because of this circulation, only a small fraction of the bile salts that enter the intestine are excreted in the feces. This decreases the need for active synthesis of bile salts by the liver cells.

Approximately half of the bilirubin (a pigment derived from the breakdown of red blood cells) is a component of bile. It is converted by the intestinal flora into urobilinogen, which is a highly soluble substance. Urobilinogen is either excreted in the feces or returned to the portal circulation, where it is re-excreted into the bile. About 5% is normally absorbed into the general circulation and then excreted by the kidneys (Porth & Matfin, 2009).

If the flow of bile is impeded (e.g., by gallstones in the bile ducts), bilirubin does not enter the intestine. As a result, blood levels of bilirubin increase. This causes increased renal excretion of urobilinogen, which results from conversion of bilirubin in the small intestine, and decreased excretion in the stool. These changes produce many of the signs and symptoms seen in gallbladder disorders.

The Pancreas

The pancreas is located in the upper abdomen (see Fig. 50-1). It has both **exocrine** (secreting externally; hormonal secretion from excretory ducts) and **endocrine** (secreting internally; hormonal secretion of a ductless gland) functions. The exocrine functions include secretion of pancreatic enzymes into the gastrointestinal (GI) tract through the pancreatic duct. The endocrine functions include secretion of insulin, glucagon, and somatostatin directly into the bloodstream.

The Exocrine Pancreas

The secretions of the exocrine portion of the pancreas are collected in the pancreatic duct, which joins the common bile duct and enters the duodenum at the ampulla of Vater. Surrounding the ampulla is the sphincter of Oddi, which partially controls the rate at which secretions from the pancreas and the gallbladder enter the duodenum.

The secretions of the exocrine pancreas are digestive enzymes high in protein content and an electrolyte-rich fluid. The secretions, which are very alkaline because of their high concentration of sodium bicarbonate, are capable of neutralizing the highly acid gastric juice that enters the duodenum. Pancreatic enzymes include **amylase**, which aids in the digestion of carbohydrates; **trypsin**, which aids in the digestion of proteins; and **lipase**, which aids in the digestion of fats. Other enzymes that promote the breakdown of more complex foodstuffs are also secreted.

Hormones originating in the GI tract stimulate the secretion of these exocrine pancreatic juices. The hormone **secretin** is the major stimulus for increased bicarbonate secretion from the pancreas, and the major stimulus for digestive enzyme secretion is the hormone CCK. The vagus nerve also influences exocrine pancreatic secretion.

The Endocrine Pancreas

The islets of Langerhans, the endocrine part of the pancreas, are collections of cells embedded in the pancreatic tissue. They are composed of alpha, beta, and delta cells. The hormone produced by the beta cells is called *insulin*, the alpha cells secrete glucagon, and the delta cells secrete somatostatin.

Insulin

A major action of insulin is to lower blood glucose by permitting entry of glucose into the cells of the liver, muscle, and other tissues, where it is either stored as glycogen or used for energy. Insulin also promotes the storage of fat in adipose tissue and the synthesis of proteins in various body tissues. In the absence of insulin, glucose cannot enter the cells and is excreted in the urine. This condition, called *diabetes*, can be diagnosed by high levels of glucose in the blood. In diabetes, stored fats and protein are used for energy instead of glucose, causing loss of body mass. (Diabetes is discussed in detail in Chapter 51.) The level of glucose in the blood normally regulates the rate of insulin secretion from the pancreas (Porth & Matfin, 2009).

Glucagon

The effect of glucagon (opposite to that of insulin) is chiefly to raise the blood glucose by converting glycogen to glucose in the liver. Glucagon is secreted by the pancreas in response to a decrease in the level of blood glucose.

Somatostatin

Somatostatin exerts a hypoglycemic effect by interfering with release of growth hormone from the pituitary and glucagon from the pancreas, both of which tend to raise blood glucose levels.

Endocrine Control of Carbohydrate Metabolism

Glucose required for energy is derived by metabolism of ingested carbohydrates and also from proteins by the process of gluconeogenesis. Glucose can be stored temporarily in the form of glycogen in the liver, muscles, and other tissues. The endocrine system controls the level of blood glucose by regulating the rate at which glucose is synthesized, stored, and moved to and from the bloodstream. Through the action of hormones, blood glucose is normally maintained at less than 100 mg/dL (5.6 mmol/L) (McPherson & Pincus, 2011). Insulin is the primary hormone that lowers the blood glucose level. Hormones that raise the blood glucose level are glucagon, epinephrine, adrenocorticosteroids, growth hormone, and thyroid hormone.

The endocrine and exocrine functions of the pancreas are interrelated. The major exocrine function is to facilitate digestion through secretion of enzymes into the proximal duodenum. Secretin and CCK are hormones from the GI tract that aid in the digestion of food substances by controlling the secretions of the pancreas. Neural factors also influence pancreatic enzyme secretion. Considerable dysfunction of the pancreas must occur before enzyme secretion decreases and protein and fat digestion becomes impaired. Pancreatic enzyme secretion is normally 1,500 to 3,000 mL/day (Mulholland, Lillemoe, Doherty, et al., 2011).

Gerontologic Considerations

There is little change in the size of the pancreas with age. However, there is an increase in fibrous material and some fatty deposition in the normal pancreas in people older than 70 years. Some localized arteriosclerotic changes occur with age. There is also a decreased rate of pancreatic secretion (decreased lipase, amylase, and trypsin) and decreased bicarbonate output in older people. Some impairment of normal fat absorption occurs with increasing age, possibly because of delayed gastric emptying and pancreatic insufficiency (Eliopoulos, 2010). Decreased calcium absorption may also occur. These changes require care in interpreting diagnostic test results in the normal older patient and in providing dietary counseling.

DISORDERS OF THE GALLBLADDER

Several disorders affect the biliary system and interfere with normal drainage of bile into the duodenum. These disorders include inflammation of the biliary system and carcinoma that obstructs the biliary tree. Gallbladder disease with stones is the most common disorder of the biliary system. Although not all occurrences of gallbladder inflammation (cholecystitis) are related to stones in the gallbladder (**cholelithiasis**) or stones in the common bile duct (**choledocholithiasis**), more than 90% of patients with acute cholecystitis have gallstones. However, most of the 15 million Americans with gallstones have no pain and are unaware of the presence of stones (Bope & Kellerman, 2011).

Cholecystitis

Cholecystitis (inflammation of the gallbladder which can be acute or chronic) causes pain, tenderness, and rigidity of the upper right abdomen that may radiate to the midsternal area or right shoulder and is associated with nausea, vomiting, and the usual signs of an acute inflammation. An empyema of the gallbladder develops if the gallbladder becomes filled with purulent fluid (pus).

Calculous cholecystitis is the cause of more than 90% of cases of acute cholecystitis (Feldman, Friedman, & Brandt, 2010; Rakel & Rakel, 2011). In calculous cholecystitis, a gallbladder stone obstructs bile outflow. Bile remaining in the gallbladder initiates a chemical reaction; autolysis and edema occur; and the blood vessels in the gallbladder are compressed, compromising its vascular supply. Gangrene of the gallbladder with perforation may result. Bacteria play a minor role in acute cholecystitis; however, secondary infection of bile occurs in approximately 50% of cases. The organisms involved are generally enteric (normally live in the GI tract) and include *Escherichia coli*, *Klebsiella* species, and *Streptococcus*. Bacterial contamination is not believed to stimulate the actual onset of acute cholecystitis (Feldman et al., 2010).

Acalculous cholecystitis describes acute gallbladder inflammation in the absence of obstruction by gallstones. Acalculous cholecystitis occurs after major surgical procedures, severe trauma, or burns. Other factors associated with this type of cholecystitis include torsion, cystic duct obstruction, primary bacterial infections of the gallbladder, and multiple blood transfusions. It is speculated that acalculous cholecystitis is caused by alterations in fluids and electrolytes and alterations in regional blood flow in the visceral circulation. Bile stasis (lack of gallbladder contraction) and increased viscosity of the bile are also thought to play a role. The occurrence of acalculous cholecystitis with major surgical procedures or trauma makes its diagnosis difficult.

Cholelithiasis

Calculi, or gallstones, usually form in the gallbladder from the solid constituents of bile; they vary greatly in size, shape, and composition (Fig. 50-2). They are uncommon in children and young adults but become more prevalent with increasing age. It is estimated that the prevalence of gallstones ranges from 5% to 20% in women between the ages of 20 and 55 years and from 25% to 30% in women older than 50 years. Cholelithiasis affects approximately 50% of women by age 70 years. The prevalence in men is approximately one third to one half the rates of occurrence in women (Feldman et al., 2010).

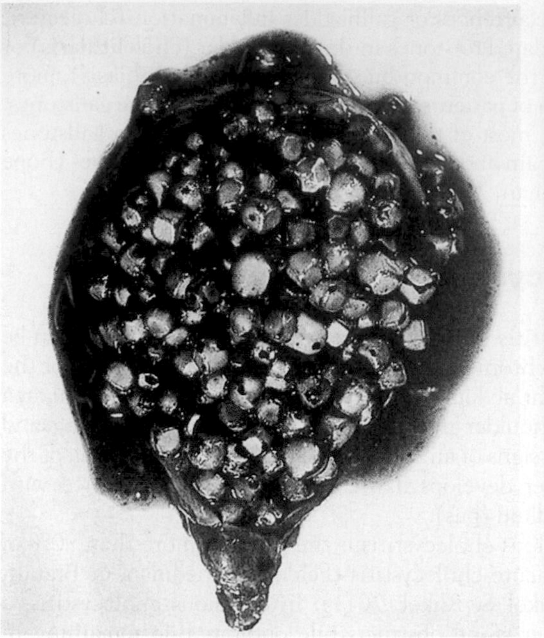
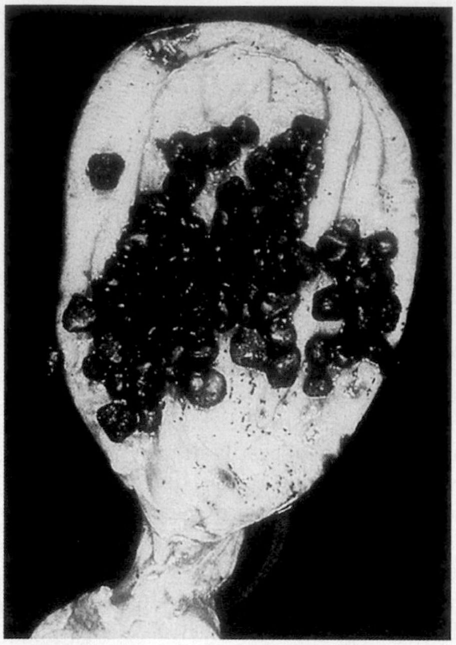

FIGURE 50-2 • Examples of cholesterol gallstones (*left*) made up of a coalescence of multiple small stones and pigment gallstones (*right*) composed of calcium bilirubinate. From Rubin, R., Strayer, D. S., & Rubin, E. (2012). *Rubin's pathology* (6th ed.). Philadelphia: Lippincott Williams & Wilkins.

Pathophysiology

There are two major types of gallstones: those composed predominantly of pigment and those composed primarily of cholesterol. Pigment stones probably form when unconjugated pigments in the bile precipitate to form stones; these stones account for about 10% to 25% of cases in the United States (Feldman et al., 2010). The risk of developing such stones is increased in patients with cirrhosis, hemolysis, and infections of the biliary tract. Pigment stones cannot be dissolved and must be removed surgically.

Cholesterol stones account for most of the remaining 75% of cases of gallbladder disease in the United States. Cholesterol, which is a normal constituent of bile, is insoluble in water. Its solubility depends on bile acids and lecithin (phospholipids) in bile. In gallstone-prone patients, there is decreased bile acid synthesis and increased cholesterol synthesis in the liver, resulting in bile supersaturated with cholesterol, which precipitates out of the bile to form stones. The cholesterol-saturated bile predisposes to the formation of gallstones and acts as an irritant that produces inflammatory changes in the mucosa of the gallbladder (Feldman et al., 2010).

Two to three times more women than men develop cholesterol stones and gallbladder disease; affected women are usually older than 40 years, multiparous, and obese (Feldman et al., 2010; Goldman & Schafer, 2012). Stone formation is more frequent in people who use oral contraceptives, estrogens, or clofibrate; these medications are known to increase biliary cholesterol saturation (Feldman et al., 2010). The incidence of stone formation increases with age as a result of increased hepatic secretion of cholesterol and decreased bile acid synthesis. In addition, there is an increased risk because of malabsorption of bile salts in patients with GI disease or T-tube fistula and in those who have undergone ileal resection or bypass. The incidence is also greater in people with diabetes (Chart 50-1).

Clinical Manifestations

Gallstones may be silent, producing no pain and only mild GI symptoms. Such stones may be detected incidentally during surgery or evaluation for unrelated problems.

The patient with gallbladder disease resulting from gallstones may develop two types of symptoms: those due to disease of the gallbladder itself and those due to obstruction of the bile passages by a gallstone. The symptoms may be acute or chronic. Epigastric distress, such as fullness, abdominal distention, and vague pain in the right upper quadrant of the abdomen, may occur. This distress may follow a meal rich in fried or fatty foods.

Pain and Biliary Colic

If a gallstone obstructs the cystic duct, the gallbladder becomes distended, inflamed, and eventually infected (acute cholecystitis). The patient develops a fever and may have a palpable abdominal mass. The patient may have biliary colic with excruciating upper right abdominal pain that radiates to

Chart 50-1 **RISK FACTORS**
Cholelithiasis

- Obesity
- Women, especially those who have had multiple pregnancies or who are of Native American or U.S. southwestern Hispanic ethnicity
- Frequent changes in weight
- Rapid weight loss (leads to rapid development of gallstones and high risk of symptomatic disease)
- Treatment with high-dose estrogen (e.g., in prostate cancer)
- Low-dose estrogen therapy—carries a small increase in the risk of gallstones
- Ileal resection or disease
- Cystic fibrosis
- Diabetes

the back or right shoulder. Biliary colic is usually associated with nausea and vomiting, and it is noticeable several hours after a heavy meal. The patient moves about restlessly, unable to find a comfortable position. In some patients, the pain is constant rather than colicky.

Such a bout of biliary colic is caused by contraction of the gallbladder, which cannot release bile because of obstruction by the stone. When distended, the fundus of the gallbladder comes in contact with the abdominal wall in the region of the right 9th and 10th costal cartilages. This produces marked tenderness in the right upper quadrant on deep inspiration and prevents full inspiratory excursion.

The pain of acute cholecystitis may be so severe that analgesic medications are required. The use of morphine has traditionally been avoided because of concern that it could cause spasm of the sphincter of Oddi, and meperidine (Demerol) has been used instead. This is controversial, because morphine is the preferred analgesic agent for management of acute pain, and some metabolites of meperidine are toxic to the central nervous system (CNS). Furthermore, all opioids stimulate the sphincter of Oddi to some degree (Porth & Matfin, 2009).

If the gallstone is dislodged and no longer obstructs the cystic duct, the gallbladder drains and the inflammatory process subsides after a relatively short time. If the gallstone continues to obstruct the duct, abscess, necrosis, and perforation with generalized peritonitis may result.

Jaundice

Jaundice occurs in a few patients with gallbladder disease, usually with obstruction of the common bile duct. The bile, which is no longer carried to the duodenum, is absorbed by the blood and gives the skin and mucous membranes a yellow color. This is frequently accompanied by marked pruritus (itching) of the skin.

Changes in Urine and Stool Color

The excretion of the bile pigments by the kidneys gives the urine a very dark color. The feces, no longer colored with bile pigments, are grayish (like putty) or clay colored.

Vitamin Deficiency

Obstruction of bile flow interferes with absorption of the fat-soluble vitamins A, D, E, and K. Patients may exhibit deficiencies of these vitamins if biliary obstruction has been prolonged. For example, a patient may have bleeding caused by vitamin K deficiency (vitamin K is necessary for normal blood clotting).

Assessment and Diagnostic Findings

A wide range of diagnostic studies may be performed in patients with biliary disorders. Table 50-1 identifies various procedures and their diagnostic uses. The nurse should educate the patient about the purpose, what to expect, and any possible side effects related to these examinations prior to testing. The nurse should note trends in results because they provide information about disease progression as well as the patient's response to therapy.

Abdominal X-Ray

If gallbladder disease is suspected, an abdominal x-ray may be obtained to exclude other causes of symptoms. However, only 10% to 15% of gallstones are calcified sufficiently to be visible on such x-ray studies (Rakel & Rakel, 2011).

TABLE 50-1 Studies Used in the Diagnosis of Biliary Tract and Pancreatic Disease

Studies	Diagnostic Uses
Magnetic resonance cholangiopancreatography (MRCP)	Visualizes the biliary tree and capable of detecting biliary tract obstruction
Cholecystogram, cholangiogram	Visualize gallbladder and bile duct
Celiac axis arteriography	Visualize liver and pancreas
Laparoscopy	Visualize anterior surface of liver, gallbladder, and mesentery through a trocar
Ultrasonography	Show size of abdominal organs and presence of masses
Helical computed tomography and magnetic resonance imaging	Detect neoplasms; diagnose cysts, pseudocysts, abscess, and hematomas
Endoscopic retrograde cholangiopancreatography	Visualize biliary structures and pancreas via endoscopy
Endoscopic ultrasound	Identify small tumors and facilitate fine-needle aspiration biopsy of tumors or lymph nodes for diagnosis
Serum alkaline phosphatase	In absence of bone disease, to measure biliary tract obstruction
Gamma-glutamyl, gamma-glutamyl transpeptidase, lactate dehydrogenase	Markers for biliary stasis; also elevated in alcohol abuse
Cholesterol levels	Elevated in biliary obstruction; decreased in parenchymal liver disease

Ultrasonography

Ultrasonography has replaced cholecystography (discussed later) as the diagnostic procedure of choice because it is rapid and accurate and can be used in patients with liver dysfunction and jaundice. It does not expose patients to ionizing radiation. The procedure is most accurate if the patient fasts overnight so that the gallbladder is distended. Ultrasonography can detect calculi in the gallbladder or a dilated common bile duct with 90% accuracy (Rakel & Rakel, 2011).

Radionuclide Imaging or Cholescintigraphy

Cholescintigraphy is used successfully in the diagnosis of acute cholecystitis or blockage of a bile duct (Rakel & Rakel, 2011; Privette, Carlisle, & Palma, 2011). In this procedure, a radioactive agent is administered intravenously (IV). It is taken up by the hepatocytes and excreted rapidly through the biliary tract. The biliary tract is then scanned, and images of the gallbladder and biliary tract are obtained. This test is more expensive than ultrasonography, takes longer to perform, and exposes the patient to radiation. It is often used when ultrasonography is not conclusive, such as in acalculous cholecystitis (Feldman et al., 2010; Privette et al., 2011).

Cholecystography

Although cholecystography has been replaced by ultrasonography as the test of choice, it is still used if ultrasound equipment is not available or if the ultrasound results are inconclusive. Oral cholangiography may be performed to detect gallstones and to assess the ability of the gallbladder to fill, concentrate its contents, contract, and empty. If the patient is not allergic to iodine or seafood, an iodide-containing contrast agent that is excreted by the liver and concentrated in the gallbladder is administered 10 to 12 hours before the x-ray study (Feldman et al., 2010; Marx, 2009). The normal gallbladder fills with this radiopaque substance. If gallstones are present, they appear as shadows on the x-ray film.

Oral cholecystography is likely to continue to be used as part of the evaluation of the few patients who have been treated with gallstone **dissolution therapy** (the use of medications to break up/dissolve gallstones) or **lithotripsy** (disintegration of gallstones by shock waves).

Endoscopic Retrograde Cholangiopancreatography

Endoscopic retrograde cholangiopancreatography (ERCP) permits direct visualization of structures that previously could be seen only during laparotomy. This procedure examines the hepatobiliary system via a side-viewing flexible fiberoptic endoscope inserted through the esophagus to the descending duodenum (Fig. 50-3). Multiple position changes are required to pass the endoscope during the procedure, beginning in the left semiprone position.

Fluoroscopy and multiple x-rays are used during ERCP to evaluate the presence and location of ductal stones. Careful insertion of a catheter through the endoscope into the common bile duct is the most important step in sphincterotomy (division of the muscles of the biliary sphincter) for gallstone extraction via this technique (see later discussion).

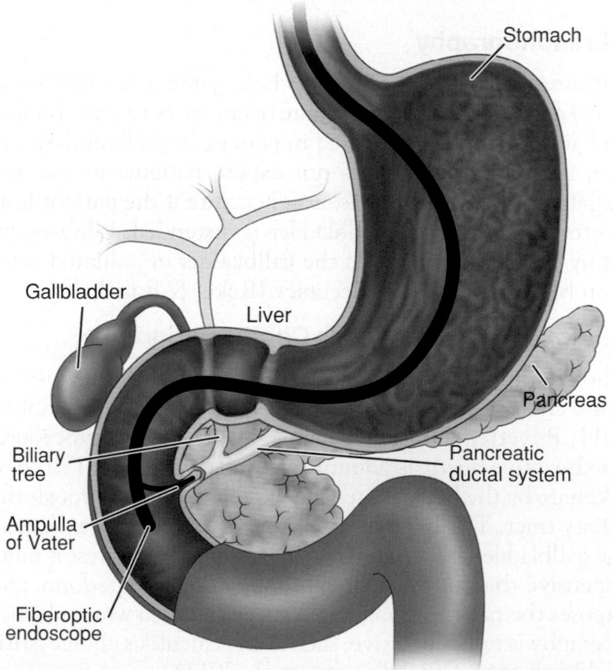
Stomach
Gallbladder
Liver
Pancreas
Biliary tree
Pancreatic ductal system
Ampulla of Vater
Fiberoptic endoscope

FIGURE 50-3 • Endoscopic retrograde cholangiopancreatography. A fiberoptic duodenoscope, with side-viewing apparatus, is inserted into the duodenum. The ampulla of Vater is catheterized, and the biliary tree is injected with contrast agent. The pancreatic ductal system is also assessed, if indicated. This procedure is of special value in visualizing neoplasms of the ampulla area and extracting a biopsy specimen.

Nursing Implications

The procedure requires a cooperative patient to permit insertion of the endoscope without damage to the GI tract structures, including the biliary tree. Before the procedure, the patient is educated about the procedure and his or her role in it. The patient takes nothing by mouth for several hours before the procedure. Moderate sedation is used, and the sedated patient must be monitored closely. It may be necessary to administer medications, such as glucagon or anticholinergic agents, to make cannulation easier by decreasing duodenal peristalsis. The nurse observes closely for signs of respiratory and CNS depression, hypotension, oversedation, and vomiting (if glucagon is administered). During ERCP, the nurse monitors IV fluids, administers medications, and positions the patient.

After the procedure, the nurse monitors the patient's condition, observing vital signs and assessing for signs of perforation or infection. The nurse also monitors the patient for side effects of any medications received during the procedure and for return of the gag and cough reflexes after the use of local anesthetic agents.

Percutaneous Transhepatic Cholangiography

Percutaneous transhepatic cholangiography (PTC) is rarely used for diagnostic purposes alone due to the multitude of other less invasive and reliable imaging studies. PTC is reserved for those patients in whom an ERCP may be unsafe due to previous surgery involving the biliary tract (Feldman et al., 2010). The use of PTC has mainly been replaced by ERCP and magnetic resonance cholangiopancreatography (MRCP). PTC involves the injection of dye directly into the biliary tract. Because of the relatively large concentration of dye that is introduced into the biliary system, including the hepatic ducts within the liver, the entire length of the common bile duct, the cystic duct, and the gallbladder is outlined clearly.

This procedure can be carried out even in the presence of liver dysfunction and jaundice. It is useful for (1) distinguishing jaundice caused by liver disease (hepatocellular jaundice) from that caused by biliary obstruction, (2) investigating the GI symptoms of a patient whose gallbladder has been removed, (3) locating stones within the bile ducts, and (4) diagnosing cancer involving the biliary system (Miller, Eriksson, Fleisher, et al., 2009).

This sterile procedure is performed under moderate sedation on a patient who has been fasting; the patient also receives local anesthesia. Coagulation parameters and platelet count should be normal to minimize the risk of bleeding. Broad-spectrum antibiotics are administered during the procedure because of the high prevalence of bacterial colonization from obstructed biliary systems (Feldman et al., 2010; Miller et al., 2009). After infiltration with a local anesthetic agent has occurred, a flexible needle is inserted into the liver from the right side in the midclavicular line immediately beneath the right costal margin. Successful entry of a duct is noted when bile is aspirated or on injection of a contrast agent. Ultrasound can be used to guide puncture of the duct. Bile is aspirated, and samples are sent for bacteriology and cytology (Feldman et al., 2010; Rakel & Rakel, 2011). A water-soluble contrast agent is injected to fill the biliary system. The fluoroscopy table is tilted and the patient is repositioned to allow x-rays to be taken in multiple projections. Delayed x-ray views can identify abnormalities of more distant ducts and determine the length of a stricture or multiple strictures. Before the needle is removed, as much dye and bile as possible are aspirated to forestall subsequent leakage into the needle tract and eventually into the peritoneal cavity, thus minimizing the risk of bile peritonitis.

Although the complication rate after this procedure is low, the nurse must closely observe the patient for symptoms of bleeding, peritonitis, and septicemia. The nurse assesses the patient for pain and indications of these complications and reports them promptly to the primary provider, takes measures to reassure the patient, and ensures patient comfort. Antibiotic agents are often prescribed to minimize the risk of sepsis and septic shock.

Medical Management

The major objectives of medical therapy are to reduce the incidence of acute episodes of gallbladder pain and cholecystitis by supportive and dietary management and, if possible, to remove the cause of cholecystitis by pharmacologic therapy, endoscopic procedures, or surgical intervention. Although nonsurgical procedures eliminate risks associated with surgery, these approaches are associated with persistent symptoms or recurrent stone formation. Most of the nonsurgical approaches, including lithotripsy and dissolution of gallstones, provide only temporary solutions to gallstone problems and are infrequently used in the United States. In some instances, other treatment approaches may be indicated; these are described later.

Removal of the gallbladder (**cholecystectomy**) through traditional surgical approaches has largely been replaced by **laparoscopic cholecystectomy** (removal of the gallbladder through a small incision through the umbilicus). As a result, surgical risks have decreased, along with the length of hospital stay and the long recovery period required after standard surgical cholecystectomy. In relatively rare instances, a standard surgical procedure may be necessary.

Nutritional and Supportive Therapy

Approximately 80% of the patients with acute gallbladder inflammation achieve remission with rest, IV fluids, nasogastric suction, analgesia, and antibiotic agents. Unless the patient's condition deteriorates, surgical intervention is delayed just until the acute symptoms subside (usually within a few days). At this time, the patient should undergo a laparoscopic cholecystectomy (Goldman & Schafer, 2012).

The diet immediately after an episode is usually low-fat liquids. These can include powdered supplements high in protein and carbohydrate stirred into skim milk. Cooked fruits, rice or tapioca, lean meats, mashed potatoes, non–gas-forming vegetables, bread, coffee, or tea may be added as tolerated. The patient should avoid eggs, cream, pork, fried foods, cheese, rich dressings, gas-forming vegetables, and alcohol. It is important to remind the patient that fatty foods may induce an episode of cholecystitis. Dietary management may be the major mode of therapy in patients who have had only dietary intolerance to fatty foods and vague GI symptoms (Dudek, 2010; Rakel & Rakel, 2011).

Pharmacologic Therapy

Ursodeoxycholic acid (UDCA [Urso, Actigall]) and chenodeoxycholic acid (chenodiol or CDCA [Chenix]) have been used to dissolve small, radiolucent gallstones composed primarily of cholesterol (Karch, 2012). UDCA has fewer side effects than chenodiol and can be administered in smaller doses to achieve the same effect. It acts by inhibiting the synthesis and secretion of cholesterol, thereby desaturating bile. Treatment with UDCA can reduce the size of existing stones, dissolve small stones, and prevent new stones from forming. Six to 12 months of therapy is required in many patients to dissolve stones, and monitoring of the patient for recurrence of symptoms or the occurrence of side effects (e.g., GI symptoms, pruritus, headache) is required during this time. The effective dose of medication depends on body weight. This method of treatment is generally indicated for patients who refuse surgery or for whom surgery is contraindicated (DiCiaula, Wang, Wang, et al., 2010).

Patients with significant, frequent symptoms, cystic duct occlusion, or pigment stones are not candidates for pharmacologic therapy. Laparoscopic or open cholecystectomy is more appropriate for symptomatic patients with acceptable operative risk (DiCiaula et al., 2010).

Nonsurgical Removal of Gallstones

Dissolving Gallstones

Several methods have been used to dissolve gallstones by infusion of a solvent (mono-octanoin or methyl tertiary butyl ether [MTBE]) into the gallbladder. The solvent can be infused through the following routes: through a tube or catheter inserted percutaneously directly into the gallbladder, through a tube or drain inserted through a T-tube tract to dissolve stones not removed at the time of surgery, endoscopically with ERCP; or via a transnasal biliary catheter.

In the latter procedure, the catheter is introduced through the mouth and inserted into the common bile duct. The upper end of the tube is then rerouted from the mouth to the nose and left in place. This enables the patient to eat and drink normally while passage of stones is monitored or chemical solvents are infused to dissolve the stones. This method of dissolution of stones is rarely used due to its lack of success, potential side effects, and rates of recurrence rate of up to 50% (DiCiaula et al., 2010; Townsend, Beauchamp, Evers, et al., 2012). Laparoscopic cholecystectomy is the standard for management. Dissolution therapies are used for those patients who may not be candidates for the procedure due to safety concerns regarding general anesthesia (DiCiaula et al., 2010; Townsend et al., 2012).

Stone Removal by Instrumentation

Several nonsurgical methods are used to remove stones that were not removed at the time of cholecystectomy or have become lodged in the common bile duct (Fig. 50-4A,B). A catheter and instrument with a basket attached are threaded through the T-tube tract or fistula formed at the time of T-tube insertion; the basket is used to retrieve and remove the stones lodged in the common bile duct.

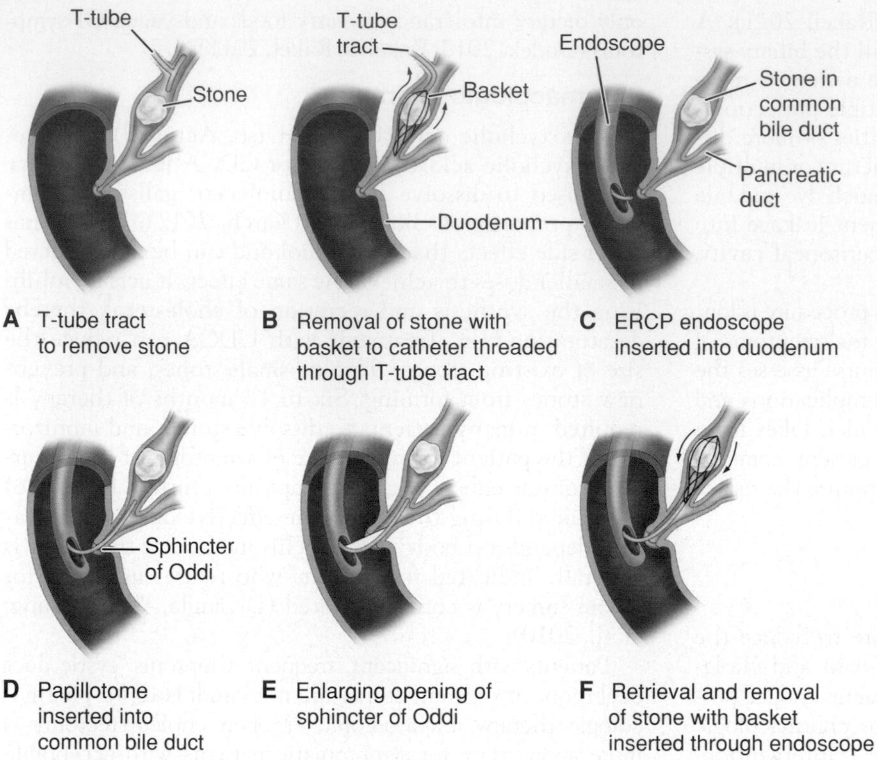
T-tube
Stone
T-tube tract
Basket
Endoscope
Stone in common bile duct
Pancreatic duct
Duodenum
Sphincter of Oddi
A T-tube tract to remove stone
B Removal of stone with basket to catheter threaded through T-tube tract
C ERCP endoscope inserted into duodenum
D Papillotome inserted into common bile duct
E Enlarging opening of sphincter of Oddi
F Retrieval and removal of stone with basket inserted through endoscope

FIGURE 50-4 • **A–F.** Nonsurgical techniques for removing gallstones.

A second procedure involves the use of the ERCP endoscope (Fig. 50-4C). After the endoscope is inserted, a cutting instrument is passed through the endoscope into the ampulla of Vater of the common bile duct. It may be used to cut the submucosal fibers, or papilla, of the sphincter of Oddi, enlarging the opening, which may allow the lodged stones to pass spontaneously into the duodenum. Another instrument with a small basket or balloon at its tip may be inserted through the endoscope to retrieve the stones (Fig. 50-4D–F). The patient is observed closely for bleeding, perforation, and the development of pancreatitis or sepsis.

The ERCP procedure is particularly useful in diagnosis and treatment of patients who have symptoms after biliary tract surgery, patients with intact gallbladders, and patients for whom surgery is particularly hazardous.

Intracorporeal Lithotripsy

Stones in the gallbladder or common bile duct may be fragmented by means of laser pulse technology. A laser pulse is directed under fluoroscopic guidance with the use of devices that can distinguish between stones and tissue. The laser pulse produces rapid expansion and disintegration of plasma on the stone surface, resulting in a mechanical shock wave. Electrohydraulic lithotripsy uses a probe with two electrodes that deliver electric sparks in rapid pulses, creating expansion of the liquid environment surrounding the gallstones. This results in pressure waves that cause stones to fragment. This technique can be used percutaneously with a basket or balloon catheter system or by direct visualization through an endoscope. Repeated procedures may be necessary because of stone size, local anatomy, bleeding, or technical difficulty. A nasobiliary tube can be inserted to allow for biliary decompression and to prevent stone impaction in the common bile duct. This approach allows time for improvement in the patient's clinical condition until gallstones are cleared endoscopically, percutaneously, or surgically.

Extracorporeal Shock Wave Lithotripsy

Extracorporeal shock wave therapy (lithotripsy or ESWL) has been used for nonsurgical fragmentation of gallstones. Lithotripsy, which is a noninvasive procedure, uses repeated shock waves directed at the gallstones in the gallbladder or common bile duct to fragment the stones. The waves are transmitted to the body through a fluid-filled bag or by immersing the patient in a water bath. After the stones are gradually broken up, the stone fragments can be spontaneously passed from the gallbladder or common bile duct, removed by endoscopy, or dissolved with oral bile acid or solvents. Because the procedure requires no incision and no hospitalization, patients are usually treated as outpatients, but usually several sessions are necessary. This procedure has largely been replaced by laparoscopic cholecystectomy. ESWL is used in some centers for a small percentage of suitable patients (those with common bile duct stones who may not be surgical candidates), sometimes in combination with dissolution therapy (Feldman et al., 2010; Rakel & Rakel, 2011).

Surgical Management

Surgical treatment of gallbladder disease and gallstones is carried out to relieve persistent symptoms, to remove the cause of biliary colic, and to treat acute cholecystitis. Surgery may be delayed until the patient's symptoms have subsided, or it may be performed as an emergency procedure, if necessitated by the patient's condition.

Preoperative Measures

Chest x-ray, electrocardiogram, and liver function tests may be performed in addition to x-ray studies of the gallbladder. Vitamin K may be administered if the prothrombin level

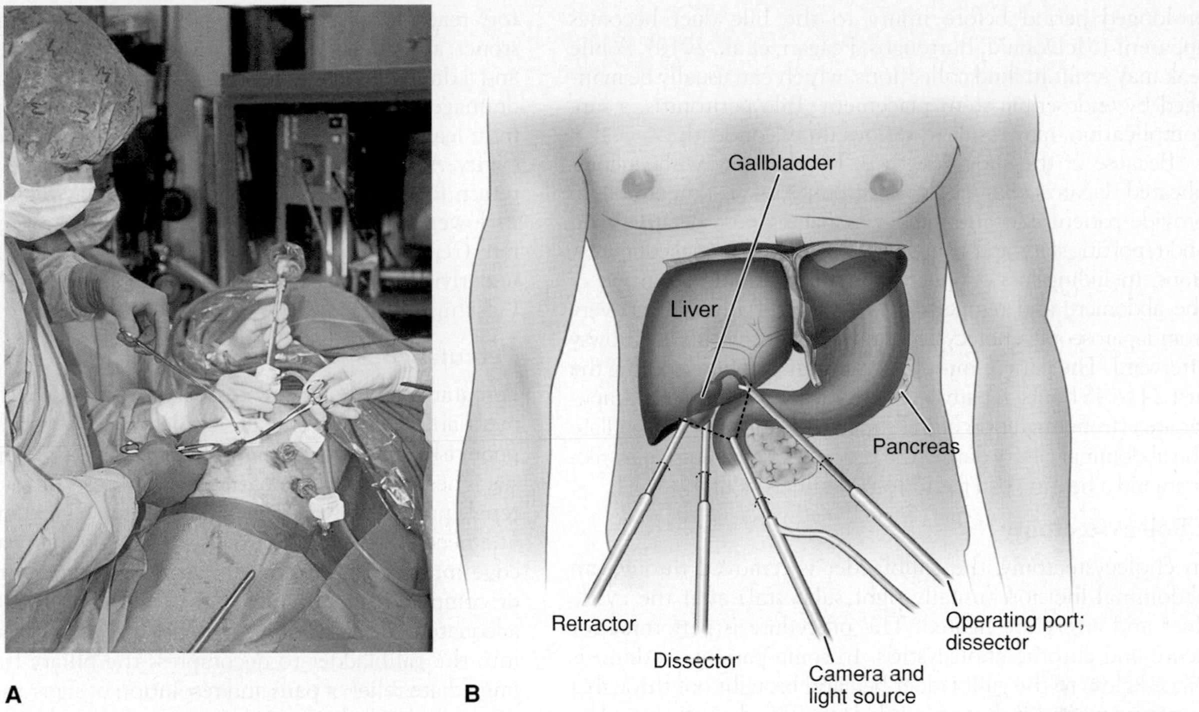
Gallbladder
Liver
Pancreas
Retractor
Operating port; dissector
Dissector
Camera and light source
A
B

FIGURE 50-5 • In laparoscopic cholecystectomy (**A**), the surgeon makes four small incisions (less than one half inch each) in the abdomen and inserts a laparoscope with a miniature camera through the umbilical incision (**B**). The camera apparatus displays the gallbladder and adjacent tissues on a screen, allowing the surgeon to visualize the sections of the organ for removal.

is low. Nutritional requirements are considered, and, if the nutritional status is suboptimal, it may be necessary to provide IV glucose with protein supplements to aid wound healing and help prevent liver damage.

Patient education for gallbladder surgery is similar to that for any upper abdominal laparotomy or laparoscopy. Instructions and explanations are given before surgery about turning and deep breathing. Postoperative pneumonia and atelectasis can be avoided by deep-breathing exercises and frequent turning. The patient should be informed that drainage tubes and a nasogastric tube and suction might be required during the immediate postoperative period if an open cholecystectomy is performed.

Laparoscopic Cholecystectomy

Laparoscopic cholecystectomy (Fig. 50-5) is the standard of therapy for symptomatic gallstones. Approximately 700,000 patients in the United States require surgery each year for removal of the gallbladder, and 80% to 90% of them are candidates for laparoscopic cholecystectomy (Feldman et al., 2010). If the common bile duct is thought to be obstructed by a gallstone, an ERCP with sphincterotomy may be performed to explore the duct before laparoscopy.

Before the procedure, the patient is educated that an open abdominal procedure may be necessary, and general anesthesia is administered. Laparoscopic cholecystectomy is performed through a small incision or puncture made through the abdominal wall at the umbilicus. The abdominal cavity is insufflated with carbon dioxide (pneumoperitoneum) to assist in inserting the laparoscope and to aid in visualizing the abdominal structures. The fiberoptic scope is inserted through the small umbilical incision. Several additional punctures or small incisions are made in the abdominal wall to introduce other surgical instruments into the operative field. A camera attached to the laparoscope permits the surgeon to view the intra-abdominal field and biliary system on a television monitor. After the cystic duct is dissected, the common bile duct can be visualized by ultrasound or cholangiography to evaluate the anatomy and identify stones. The cystic artery is dissected free and clipped. The gallbladder is separated from the hepatic bed and removed from the abdominal cavity after bile and small stones are aspirated. Stone forceps also can be used to remove or crush larger stones.

With the laparoscopic procedure, the patient does not experience the paralytic ileus that occurs with open abdominal surgery and has less postoperative abdominal pain. The patient is often discharged from the hospital on the same day of surgery or within 1 or 2 days and resumes full activity and employment within 1 week after the procedure.

Conversion to a traditional abdominal surgical procedure occurs in 2.2% of cases in the United States and 3.6% to 8.2% of cases internationally. Conversion to an open procedure occurs if there is inflammation in and around the gallbladder, making safe dissection of the porta hepatis difficult (Feldman et al., 2010). (The porta hepatis is the fissure of the liver where the portal vein and the hepatic artery enter and the hepatic ducts exit the liver.) Careful screening of patients and identification of those at low risk for complications limit the frequency of conversion to an open abdominal procedure. However, with increasing use of laparoscopic procedures, the number of such conversions may increase.

The most serious complication after laparoscopic cholecystectomy is a bile duct injury, which may be identified and corrected at the time of the procedure. Patients with a postoperative bile leak may not develop symptoms until several days after the procedure, and some have an even more prolonged period before injury to the bile duct becomes apparent (McDonald, Burroughs, Feagan, et al., 2010). A bile leak may result in fluid collections, which can usually be managed by endoscopic stent placement. Bile peritonitis, a rare complication, may result in serious illness or death.

Because of the short length of hospital stay with uncomplicated laparoscopic cholecystectomies, it is important to provide patient education about managing postoperative pain and reporting signs and symptoms of intra-abdominal complications, including loss of appetite, vomiting, pain, distention of the abdomen, and temperature elevation. Although recovery from laparoscopic cholecystectomy is rapid, patients are drowsy afterward. The patient must have assistance at home during the first 24 to 48 hours. If pain occurs in the right shoulder or scapular area (from migration of the carbon dioxide used to insufflate the abdominal cavity during the procedure), the nurse may recommend a heating pad for 15 to 20 minutes hourly.

Cholecystectomy

In cholecystectomy, the gallbladder is removed through an abdominal incision (usually right subcostal) after the cystic duct and artery are ligated. The procedure is performed for acute and chronic cholecystitis. In some patients, a drain is placed close to the gallbladder bed and brought out through a puncture wound if there is a bile leak. The drain type is chosen based on the physician's preference. A small leak should close spontaneously in a few days, with the drain preventing accumulation of bile. Usually, only a small amount of serosanguineous fluid drains in the initial 24 hours after surgery; afterward, the drain is removed. The drain is typically maintained if there is excess oozing or bile leakage. Insertion of a T-tube into the common bile duct during the open procedure is now uncommon; it is used only in the setting of a complication (i.e., retained common bile duct stone). Bile duct injury is a serious complication of cholecystectomy, but it occurs less frequently than with the laparoscopic approach, which has largely replaced traditional surgical cholecystectomy.

Small-Incision Cholecystectomy

Small-incision cholecystectomy is a surgical procedure in which the gallbladder is removed through a small abdominal incision, as the name implies. If needed, the surgical incision is extended to remove larger gallbladder stones. Drains may or may not be used. The short length hospital stay has been identified as a major advantage of this type of procedure (Goldman & Schafer, 2012). The procedure is controversial because it limits exposure to all involved biliary structures.

Choledochostomy

Choledochostomy is reserved for the patient with acute cholecystitis who may be too ill to undergo a surgical procedure. It involves making an incision in the common duct, usually for removal of stones. After the stones have been evacuated, a tube is usually inserted into the duct for drainage of bile until edema subsides. This tube is connected to gravity drainage tubing; the patient is monitored closely, and a laparoscopic cholecystectomy is planned for a future date after acute inflammation has resolved.

Surgical Cholecystostomy

Cholecystostomy is performed when the patient's condition precludes more extensive surgery or when an acute inflammatory reaction is severe. The gallbladder is surgically opened, stones and the bile or the purulent drainage are removed, and a drainage tube is secured with a purse-string suture. The drainage tube is connected to a drainage system to prevent bile from leaking around the tube or escaping into the peritoneal cavity. After recovery from the acute episode, the patient may return for subsequent laparoscopic cholecystectomy. Despite its lower risk, surgical cholecystostomy has a high mortality rate (reported to be as high as 10% to 30%) because of the underlying disease process (Barie, 2010; McKay, Abulfaraj, & Lipschitz, 2012).

Percutaneous Cholecystostomy

Percutaneous cholecystostomy has been used in the treatment and diagnosis of acute cholecystitis in patients who are poor risks for any surgical procedure or for general anesthesia. These may include patients with sepsis or severe cardiac, renal, pulmonary, or liver failure. Under local anesthesia, a fine needle is inserted through the abdominal wall and liver edge into the gallbladder under the guidance of ultrasound or computed tomography (CT). Bile is aspirated to ensure adequate placement of the needle, and a catheter is inserted into the gallbladder to decompress the biliary tract. Almost immediate relief of pain and resolution of signs and symptoms of sepsis and cholecystitis have been reported with this procedure. Antibiotic agents are administered before, during, and after the procedure.

Gerontologic Considerations

Surgical intervention for disease of the biliary tract is the most common operative procedure performed in the older adult. Cholesterol saturation of bile increases with age because of increased hepatic secretion of cholesterol and decreased bile acid synthesis.

Although the incidence of gallstones increases with age, the older patient may not exhibit the typical symptoms of fever, pain, chills, and jaundice. Symptoms of biliary tract disease in the older adult may be accompanied or preceded by those of septic shock, which include oliguria, hypotension, changes in mental status, tachycardia, and tachypnea.

Although surgery in the older adult presents a risk because of preexisting associated diseases, the mortality rate from serious complications of biliary tract disease itself is also high. The risk of death and complications is increased in the older patient who undergoes emergency surgery for life-threatening disease of the biliary tract. Despite chronic illness in many older patients, elective cholecystectomy is usually well tolerated and can be carried out with low risk if expert assessment and care are provided before, during, and after the surgical procedure.

Because of changes in reimbursement for health care expenses, there are fewer elective surgical procedures performed, including cholecystectomies. As a result, patients requiring the procedure are seen in later stages of disease. At the same time, patients undergoing surgery are increasingly older than 60 years and may have complicated acute cholecystitis. The higher risk of complications and shorter length of hospital stay make it essential that older patients and their family members receive specific information about signs and symptoms of complications and measures to prevent them.

NURSING PROCESS

The Patient Undergoing Surgery for Gallbladder Disease

Assessment

The patient who is to undergo surgical treatment of gallbladder disease is often admitted to the hospital or same-day surgery unit on the morning of surgery. Preadmission testing is often completed a week or longer before admission. At that time, the nurse educates the patient about the need to avoid smoking, to enhance pulmonary recovery postoperatively, and to avoid respiratory complications. The need to avoid aspirin, nonsteroidal medications, and other agents (over-the-counter medications and herbal remedies) that can alter coagulation and other biochemical processes is also emphasized.

Assessment should focus on the patient's respiratory status. If a traditional surgical approach is planned, the high abdominal incision required during surgery may interfere with full respiratory excursion. The nurse notes a history of smoking, previous respiratory problems, shallow respirations, a persistent or ineffective cough, and the presence of adventitious breath sounds. Nutritional status is evaluated through a dietary history and a general examination performed at the time of preadmission testing. The nurse also reviews previously obtained laboratory results to obtain information about the patient's nutritional status.

Diagnosis

Nursing Diagnoses

Based on the assessment data, major postoperative nursing diagnoses may include the following:

- Acute pain and discomfort related to surgical incision
- Impaired gas exchange related to the high abdominal surgical incision (if traditional surgical cholecystectomy was performed)
- Impaired skin integrity related to altered biliary drainage after surgical intervention (if a T-tube was inserted because of retained stones in the common bile duct or another drainage device was employed)
- Imbalanced nutrition: less than body requirements related to inadequate bile secretion
- Deficient knowledge about self-care activities related to incision care, dietary modifications (if needed), medications, and reportable signs or symptoms (e.g., fever, bleeding, vomiting)

Collaborative Problems/Potential Complications

Potential complications may include the following:

- Bleeding
- GI symptoms (may be related to biliary leak or injury to the bowel)

Planning and Goals

The goals for the patient include relief of pain, adequate ventilation, intact skin and improved biliary drainage, optimal nutritional intake, absence of complications, and understanding of self-care routines.

Nursing Interventions

After recovery from anesthesia, the patient is placed in the low Fowler's position. Fluids may be administered IV, and nasogastric suction (a nasogastric tube was probably inserted immediately before surgery for a nonlaparoscopic procedure) may be instituted to relieve abdominal distention. Water and other fluids are administered within hours after laparoscopic procedures. A soft diet is started after bowel sounds return, which is usually the next day if the laparoscopic approach is used.

Relieving Pain

The location of the subcostal incision in nonlaparoscopic gallbladder surgery often causes the patient to avoid turning and moving, to splint the affected site, and to take shallow breaths to prevent pain. Because full expansion of the lungs and gradually increased activity are necessary to prevent postoperative complications, the nurse administers analgesic agents as prescribed to relieve the pain and to help the patient turn, cough, breathe deeply, and ambulate as indicated. The use of a pillow or binder over the incision may reduce pain during these maneuvers.

Improving Respiratory Status

Patients undergoing biliary tract surgery are especially prone to pulmonary complications, as are all patients with upper abdominal incisions. Therefore, the nurse reminds the patient to take deep breaths and cough every hour to expand the lungs fully and prevent atelectasis. The early and consistent use of incentive spirometry also helps improve respiratory function. Early ambulation prevents pulmonary complications as well as other complications, such as thrombophlebitis. Pulmonary complications are more likely to occur in patients who are older, those who are obese, and those with preexisting pulmonary disease.

Maintaining Skin Integrity and Promoting Biliary Drainage

In patients who have undergone a cholecystostomy or choledochostomy, the drainage tube must be connected immediately to a drainage receptacle. The nurse should fasten the tubing to the dressings or to the patient's gown, with enough leeway for the patient to move without dislodging or kinking the tube. Because a drainage system remains attached when the patient is ambulating, the drainage bag may be placed in a bathrobe pocket or fastened so that it is below the waist or common duct level. If a Penrose drain is used, the nurse changes the dressings as required.

After these surgical procedures, the patient is observed for indications of infection, leakage of bile into the peritoneal cavity, and obstruction of bile drainage. If bile is not draining properly, an obstruction is probably causing bile to be forced back into the liver and bloodstream. Because jaundice may result, the nurse should assess the color of the sclerae. The nurse should note and report right upper quadrant abdominal pain, nausea and vomiting, bile drainage around any drainage tube, clay-colored stools, and a change in vital signs.

Bile may continue to drain from the drainage tract in considerable quantities for some time, necessitating frequent changes of the outer dressings and protection of the skin from irritation (bile is corrosive to the skin).

To prevent total loss of bile, the physician may want the drainage tube (T-tube) or collection receptacle elevated above the level of the abdomen so that the bile drains externally only if pressure develops in the duct system. Every 24 hours, the nurse measures the bile collected and records the amount, color, and character of the drainage. After several days of drainage, the T-tube may be clamped for 1 hour before and after each meal to deliver bile to the duodenum to aid in digestion (Ren, Zhang, & Shang, 2008; Townsend et al., 2012). Within 7 days to 3 weeks, the drainage tube is removed (Chohan & Munden, 2008). The patient who goes home with a drainage tube in place requires instruction and reassurance about the function and care of the T-tube (Chohan & Munden, 2008).

In all patients with biliary drainage, the nurse (or the patient, if at home) observes the color of stools daily. Urine and stool specimens may be sent to the laboratory for examination for bile pigments. In this way, it is possible to determine whether the bile pigment is disappearing from the blood and is draining again into the duodenum. Maintaining a careful record of fluid intake and output is important.

Improving Nutritional Status

The nurse encourages the patient to eat a diet that is low in fats and high in carbohydrates and proteins immediately after surgery. At the time of hospital discharge, there are usually no special dietary instructions other than to maintain a healthy diet and avoid excessive fats. Fat restriction usually is lifted in 4 to 6 weeks, when the biliary ducts dilate to accommodate the volume of bile once held by the gallbladder and when the ampulla of Vater again functions effectively. After this time, when the patient eats fat, adequate bile will be released into the GI tract to emulsify the fats and allow their digestion. This is in contrast to the condition before surgery, when fats may not have been digested completely or adequately and flatulence may have occurred. One purpose of gallbladder surgery is to allow a normal diet.

Monitoring and Managing Potential Complications

Bleeding may occur as a result of inadvertent puncture or injury to a major blood vessel. Postoperatively, the nurse closely monitors vital signs and inspects the surgical incisions and any drains for bleeding. The nurse also assesses the patient for increased tenderness and rigidity of the abdomen. If these signs and symptoms occur, they are reported to the surgeon. The nurse instructs the patient and family to report any change in the color of stools, because this may indicate complications. GI symptoms, although not common, may occur with manipulation of the intestines during surgery.

After laparoscopic cholecystectomy, the nurse assesses the patient for anorexia, vomiting, pain, abdominal distension, and temperature elevation. These may indicate infection or disruption of the GI tract and should be reported to the surgeon promptly. Because the patient is discharged soon after laparoscopic surgery, the patient and family are instructed verbally and in writing about the importance of reporting these symptoms promptly.

Promoting Home and Community-Based Care

Educating Patients About Self-Care. The nurse educates the patient about the medications that are prescribed (vitamins, anticholinergic and antispasmodic agents) and their actions. The nurse also informs the patient and family about symptoms that should be reported to the primary provider, including jaundice, dark urine, pale-colored stools, pruritus, and signs of inflammation and infection such as pain or fever.

Some patients report one to three bowel movements a day, which is a result of a continual trickle of bile through the choledochoduodenal junction after cholecystectomy. Usually, such frequency diminishes over a period of a few weeks to several months.

If a patient is discharged from the hospital with a drainage tube still in place, the patient and family need education about its management. The nurse educates them in proper care of the drainage tube and the importance of reporting promptly any changes in the amount or characteristics of drainage. Assistance in securing the appropriate dressings reduces the patient's anxiety about going home with the drain or tube still in place. Chart 50-2 provides additional details.

Continuing Care. With sufficient support at home, most patients recover quickly from a cholecystectomy. However, older or frail patients and those who live alone may require a referral for home care. During home visits, the nurse assesses the patient's physical status, especially wound healing, and progress toward recovery. Assessing the patient for adequacy of pain relief and pulmonary exercises is also important. If the patient has a drainage system in place, the nurse assesses it for patency and appropriate management by the patient

Chart 50-2 PATIENT EDUCATION

Managing Self-Care After Laparoscopic Cholecystectomy

Managing Pain

- You may experience pain or discomfort in your right shoulder from the gas used to inflate your abdominal area during surgery. Sitting upright in bed or a chair, walking, or using a heating pad may ease the discomfort.
- Take analgesic medications as needed and as prescribed. Report to your surgeon if pain is unrelieved even with analgesic use.

Resuming Activity

- Begin light exercise (walking) immediately.
- Take a shower or bath after 1 or 2 days.
- Drive a car after 3 or 4 days.
- Avoid lifting objects exceeding 5 pounds after surgery, usually for 1 week.
- Resume sexual activity when desired.

Caring for the Wound

- Check puncture site daily for signs of infection.
- Wash puncture site with mild soap and water.
- Allow special adhesive strips on the puncture site to fall off. Do not pull them off.

Resuming Eating

- Resume your normal diet.
- If you had fat intolerance before surgery, gradually add fat back into your diet in small increments.

Managing Follow-Up Care

- Make an appointment with your surgeon for 7 to 10 days after discharge.
- Call your surgeon if you experience any signs or symptoms of infection at or around the puncture site: redness, tenderness, swelling, heat, or drainage.
- Call your surgeon if you experience a fever of 37.7°C (100°F) or more for 2 consecutive days.
- Call your surgeon if you develop nausea, vomiting, or abdominal pain.

and family. Assessing for signs of infection and educating the patient about the signs and symptoms of infection are also important nursing interventions. The patient's understanding of the therapeutic regimen (medications, gradual return to normal activities) is assessed, and previous education is reinforced. The nurse emphasizes the importance of keeping follow-up appointments and reminds the patient and family of the importance of participating in health promotion activities and recommended health screening.

Evaluation

Expected patient outcomes may include:

1. Reports decrease in pain
   a. Splints abdominal incision to decrease pain
   b. Avoids foods that cause pain
   c. Uses postoperative analgesia as prescribed
2. Demonstrates appropriate respiratory function
   a. Achieves full respiratory excursion, with deep inspiration and expiration
   b. Coughs effectively, using pillow to splint abdominal incision
   c. Uses postoperative analgesia as prescribed
   d. Exercises as prescribed (e.g., turns, ambulates)
3. Exhibits normal skin integrity around biliary drainage site (if applicable)
   a. Is free of fever; abdominal pain; change in vital signs; and presence of bile, foul-smelling drainage, or pus around drainage tube
   b. Demonstrates correct management of drainage tube (if applicable)
   c. Identifies signs and symptoms of biliary obstruction to be noted and reported
   d. Has serum bilirubin level within normal range
4. Obtains relief from dietary intolerance
   a. Maintains adequate dietary intake and avoids foods that cause GI symptoms
   b. Reports decreased or absent nausea, vomiting, diarrhea, flatulence, and abdominal discomfort
5. Absence of complications
   a. Has normal vital signs (blood pressure, pulse, respiratory rate and pattern, and temperature)
   b. Reports absence of bleeding from GI tract and from biliary drainage tube or catheter (if present) and no evidence of bleeding in stool
   c. Reports return of appetite and no evidence of vomiting, abdominal distention, or pain
   d. Lists symptoms that should be reported to surgeon promptly and demonstrates an understanding of self-care, including wound care

DISORDERS OF THE PANCREAS

Pancreatitis (inflammation of the pancreas) is a serious disorder. The most basic classification system used to describe or categorize the various stages and forms of pancreatitis divides the disorder into acute and chronic forms. Acute pancreatitis can be a medical emergency associated with a high risk of life-threatening complications and mortality, whereas chronic pancreatitis often goes undetected because classic clinical and diagnostic findings are not always present in the early stages of the disease (Affronti, 2011). By the time symptoms occur in chronic pancreatitis, approximately 90% of normal acinar cell function (exocrine function) has been lost (Mulholland et al., 2011). Acute pancreatitis does not usually lead to chronic pancreatitis unless complications develop. However, chronic pancreatitis can be characterized by acute episodes.

Although the mechanisms causing pancreatic inflammation are unknown, pancreatitis is commonly described as autodigestion of the pancreas. It is believed that the pancreatic duct becomes temporarily obstructed, accompanied by hypersecretion of the exocrine enzymes of the pancreas. These enzymes enter the bile duct, where they are activated and, together with bile, back up (reflux) into the pancreatic duct, causing pancreatitis.

Acute Pancreatitis

Acute pancreatitis ranges from a mild, self-limited disorder to a severe, rapidly fatal disease that does not respond to any treatment. Approximately 200,000 cases of acute pancreatitis occur in the United States each year, of which 80% are the result of cholelithiasis or sustained alcohol abuse (Feldman et al., 2010; McDonald et al., 2010; Wu & Conwell, 2010). Mild acute pancreatitis is characterized by edema and inflammation confined to the pancreas. Minimal organ dysfunction is present, and return to normal function usually occurs within 6 months. Although this is considered the milder form of pancreatitis, the patient is acutely ill and at risk for hypovolemic shock, fluid and electrolyte disturbances, and sepsis. A more widespread and complete enzymatic digestion of the gland characterizes severe acute pancreatitis. Enzymes damage the local blood vessels, and bleeding and thrombosis can occur. The tissue may become necrotic, with damage extending into the retroperitoneal tissues. Local complications include pancreatic cysts or abscesses and acute fluid collections in or near the pancreas. Patients who develop systemic complications with organ failure, such as pulmonary insufficiency with hypoxia, shock, renal failure, and GI bleeding, are also characterized as having severe acute pancreatitis.

Gerontologic Considerations

Acute pancreatitis affects people of all ages, but the mortality rate associated with acute pancreatitis increases with advancing age (Anand, Park, & Wu, 2012). In addition, the pattern of complications changes with age. Younger patients tend to develop local complications; the incidence of multiple organ failure increases with age, possibly as a result of progressive decreases in physiologic function of major organs with increasing age. Close monitoring of major organ function (i.e., lungs, kidneys) is essential, and aggressive treatment is necessary to reduce mortality from acute pancreatitis in the older adult patient.

Pathophysiology

Self-digestion of the pancreas by its own proteolytic enzymes, principally trypsin, causes acute pancreatitis. Eighty percent of patients with acute pancreatitis have biliary tract disease such as gallstones or a history of long-term alcohol abuse (Feldman et al., 2010; Privette et al., 2011). These patients usually have had undiagnosed chronic pancreatitis before their first episode of acute pancreatitis. Gallstones enter the common bile duct and lodge at the ampulla of Vater, obstructing the flow of pancreatic juice or causing a reflux of bile from the common bile duct into the pancreatic duct, thus activating the powerful enzymes within the pancreas. Normally, these remain in an inactive form until the pancreatic secretions reach the lumen of the duodenum. Activation of the enzymes can lead to vasodilation, increased vascular permeability, necrosis, erosion, and hemorrhage (McDonald et al., 2010; Townsend et al., 2012).

Other less common causes of pancreatitis include bacterial or viral infection, with pancreatitis occasionally developing as a complication of mumps viral infection. Spasm and edema of the ampulla of Vater, caused by duodenitis, can probably produce pancreatitis. Blunt abdominal trauma, peptic ulcer disease, ischemic vascular disease, hyperlipidemia, hypercalcemia, and the use of corticosteroids, thiazide diuretics, oral contraceptives, and other medications have also been associated with an increased incidence of pancreatitis. Acute pancreatitis may develop after surgery on or near the pancreas or after instrumentation of the pancreatic duct. Acute idiopathic pancreatitis accounts for up to 10% of the cases of acute pancreatitis. Some postulate that these cases may be related to occult microlithiasis (small stones in the bile) (Townsend et al., 2012). In addition, there is a small incidence of hereditary pancreatitis.

The mortality rate of patients with acute pancreatitis is 2% to 10% because of shock, anoxia, hypotension, or fluid and electrolyte imbalances. This mortality rate may also be related to the 10% to 30% of patients with severe acute disease characterized by pancreatic and peripancreatic necrosis (Talukdar, Clemens, & Vege, 2012; Townsend et al., 2012). Pancreatitis may result in complete recovery, may recur without permanent damage, or may progress to chronic pancreatitis. The patient who is admitted to the hospital with a diagnosis of pancreatitis is acutely ill and needs expert nursing and medical care.

The severity of acute pancreatitis and its outcomes can be predicted based on clinical and laboratory data (Chart 50-3).

Chart 50-3 Criteria for Predicting Severity of Pancreatitis*

Criteria on Admission to Hospital

Age >55 years
White blood cells (WBCs) >16,000 mm^3
Serum glucose >200 mg/dL (>11.1 mmol/L)
Serum lactose dehydrogenase (LDH) >350 IU/L (>350 U/L)
AST >250 IU/L

Criteria Within 48 Hours of Hospital Admission

Fall in hematocrit >10% (>0.10)
Blood urea nitrogen (BUN) increase >5 mg/dL (>1.7 mmol/L)
Serum calcium <8 mg/dL (<2 mmol/L)
Base deficit >4 mEq/L (>4 mmol/L)
Fluid retention or sequestration >6 L
Partial pressure of oxygen (PO_2) <60 mm Hg

Two or fewer signs, 1% mortality; 3 or 4 signs, 15% mortality; 5 or 6 signs, 40% mortality; >6 signs, 100% mortality.

**Note:* The more risk factors a patient has, the greater the severity and likelihood of complications or death.

Adapted from Ranson, J. H., Rifkind, K. M., Roses, D. F., et al. (1974). Prognostic signs and the role of operative management in acute pancreatitis. *Surgery, Gynecology & Obstetrics*, *139*(1), 69–81.

Clinical Manifestations

Severe abdominal pain is the major symptom of pancreatitis that causes the patient to seek medical care. Abdominal pain and tenderness and back pain result from irritation and edema of the inflamed pancreas. Increased tension on the pancreatic capsule and obstruction of the pancreatic ducts also contribute to the pain. Typically, the pain occurs in the midepigastrium. Pain is frequently acute in onset, occurring 24 to 48 hours after a very heavy meal or alcohol ingestion, and it may be diffuse and difficult to localize. It is generally more severe after meals and is unrelieved by antacids. Pain may be accompanied by abdominal distention; a poorly defined, palpable abdominal mass; decreased peristalsis; and vomiting that fails to relieve the pain or nausea.

The patient appears acutely ill. Abdominal guarding is present. A rigid or boardlike abdomen may develop and is generally an ominous sign, usually indicating peritonitis (Privette et al., 2011). Ecchymosis (bruising) in the flank or around the umbilicus may indicate severe pancreatitis. Nausea and vomiting are common in acute pancreatitis. The emesis is usually gastric in origin but may also be bile stained. Fever, jaundice, mental confusion, and agitation may also occur.

Hypotension is typical and reflects hypovolemia and shock caused by the loss of large amounts of protein-rich fluid into the tissues and peritoneal cavity. In addition to hypotension, the patient may develop tachycardia; cyanosis; and cold, clammy skin. Acute renal failure is common.

Respiratory distress and hypoxia are common, and the patient may develop diffuse pulmonary infiltrates, dyspnea, tachypnea, and abnormal blood gas values. Myocardial depression, hypocalcemia, hyperglycemia, and disseminated intravascular coagulation may also occur with acute pancreatitis.

Assessment and Diagnostic Findings

The diagnosis of acute pancreatitis is based on a history of abdominal pain, the presence of known risk factors, physical examination findings, and diagnostic findings. Serum amylase and lipase levels are used in making the diagnosis of acute pancreatitis, although their elevation can be attributed to many other causes (Feldman et al., 2010). In most cases, serum amylase and lipase levels are elevated within 24 hours of the onset of the symptoms. Serum amylase usually returns to normal within 48 to 72 hours, but serum lipase levels may remain elevated for a longer period, often days longer than amylase. Urinary amylase levels also become elevated and remain elevated longer than serum amylase levels. The white blood cell count is usually elevated; hypocalcemia is present in many patients and correlates well with the severity of pancreatitis. Transient hyperglycemia and glucosuria and elevated serum bilirubin levels occur in some patients with acute pancreatitis.

X-ray studies of the abdomen and chest may be obtained to differentiate pancreatitis from other disorders that can cause similar symptoms and to detect pleural effusions. Ultrasound studies, contrast-enhanced CT scans, and magnetic resonance imaging (MRI) scans are used to identify an increase in the diameter of the pancreas and to detect pancreatic cysts, abscesses, or pseudocysts.

Hematocrit and hemoglobin levels are used to monitor the patient for bleeding. Peritoneal fluid, obtained through paracentesis or peritoneal lavage, may contain increased levels of pancreatic enzymes. ERCP is rarely used in the diagnostic evaluation of acute pancreatitis, because the patient is acutely ill; however, it may be valuable in the treatment of gallstone pancreatitis.

Medical Management

Management of acute pancreatitis is directed toward relieving symptoms and preventing or treating complications. All oral intake is withheld to inhibit stimulation of the pancreas and its secretion of enzymes. Parenteral nutrition plays an important role in the nutritional support of patients with severe acute pancreatitis, particularly in those who are debilitated and those with a prolonged paralytic ileus (more than 48 to 72 hours) (Townsend et al., 2012; Wu & Conwell, 2010). Ongoing research has shown positive outcomes with the use of enteral feedings. The current recommendation is that, whenever possible, the enteral route should be used to meet nutritional needs in patients with pancreatitis. This strategy also has been found to prevent infectious complications safely and cost-effectively (Townsend et al., 2012; Wu & Conwell, 2010). Enteral feedings should be started early in the course of acute pancreatitis. Patients who do not tolerate enteral feeding require parenteral nutrition. Nasogastric suction may be used to relieve nausea and vomiting and to decrease painful abdominal distention and paralytic ileus. Research data do not support the routine use of nasogastric tubes to remove gastric secretions in an effort to limit pancreatic secretion. Histamine-2 (H_2) antagonists such as cimetidine (Tagamet) and ranitidine (Zantac) may be prescribed to decrease pancreatic activity by inhibiting secretion of gastric acid. Proton pump inhibitors such as pantoprazole (Protonix) may be used for patients who do not tolerate H_2 antagonists or for whom this therapy is ineffective (Karch, 2012).

Pain Management

Adequate administration of analgesia is essential during the course of acute pancreatitis to provide sufficient pain relief and to minimize restlessness, which may stimulate pancreatic secretion further. Pain relief may require parenteral opioids such as morphine, fentanyl (Sublimaze), or hydromorphone (Dilaudid) (Rakel & Rakel, 2011). There is no clinical evidence to support the use of meperidine for pain relief in pancreatitis; in fact, accumulation of its metabolites can cause CNS irritability and possibly seizures. The current recommendation for pain management is the use of opioids, with assessment for their effectiveness and altering therapy if pain is not controlled or increased (Marx, 2009). More research is needed to identify the best option for pain management in the patient with acute pancreatitis (Marx, 2009). Antiemetic agents may be prescribed to prevent vomiting.

Intensive Care

Correction of fluid and blood loss and low albumin levels is necessary to maintain fluid volume and prevent renal failure. The patient is usually acutely ill and is monitored in the intensive care unit, where hemodynamic monitoring and arterial blood gas monitoring are initiated. Antibiotic agents may be prescribed if infection is present. The role of prophylactic antibiotics is controversial and still under study. Insulin may be required if hyperglycemia occurs. Intensive insulin therapy (continuous infusion) in the critically ill patient has undergone much study and has shown promise in terms of positive patient outcomes when compared with intermittent insulin dosing. Glycemic control with normal or near-normal blood glucose levels improves patient outcomes (Griesdale, 2009).

Respiratory Care

Aggressive respiratory care is indicated because of the high risk of elevation of the diaphragm, pulmonary infiltrates and effusion, and atelectasis. Hypoxemia occurs in a significant number of patients with acute pancreatitis, even with normal x-ray findings. Respiratory care may range from close monitoring of arterial blood gases to the use of humidified oxygen to intubation and mechanical ventilation (see Chapter 21 for further discussion).

Biliary Drainage

Placement of biliary drains (for external drainage) and stents (indwelling tubes) in the pancreatic duct through endoscopy has been performed to reestablish drainage of the pancreas. This has resulted in decreased pain and increased weight gain.

Surgical Intervention

Although the acutely ill patient is at high risk for surgical complications, surgery may be performed to assist in the diagnosis of pancreatitis (diagnostic laparotomy); to establish pancreatic drainage; or to resect or débride an infected, necrotic pancreas. The patient who undergoes pancreatic surgery may have multiple drains in place postoperatively, as well as a surgical incision that is left open for irrigation and repacking every 2 to 3 days to remove necrotic debris (Fig. 50-6).

Postacute Management

Oral feedings that are low in fat and protein are initiated gradually. Caffeine and alcohol are eliminated from the diet. If the episode of pancreatitis occurred during treatment with thiazide diuretics, corticosteroids, or oral contraceptives, these medications are discontinued. Follow-up may include ultrasound, x-ray studies, or ERCP to determine whether the pancreatitis is resolving and to assess for abscesses and pseudocysts. ERCP may also be used to identify the cause of acute pancreatitis if it is in question and for endoscopic sphincterotomy and removal of gallstones from the common bile duct.

Nursing Management

Relieving Pain and Discomfort

Because the pathologic process responsible for pain is autodigestion of the pancreas, the objectives of therapy are to relieve pain and decrease secretion of pancreatic enzymes. The pain of acute pancreatitis is often very severe, necessitating the liberal use of analgesic agents. The current recommendation

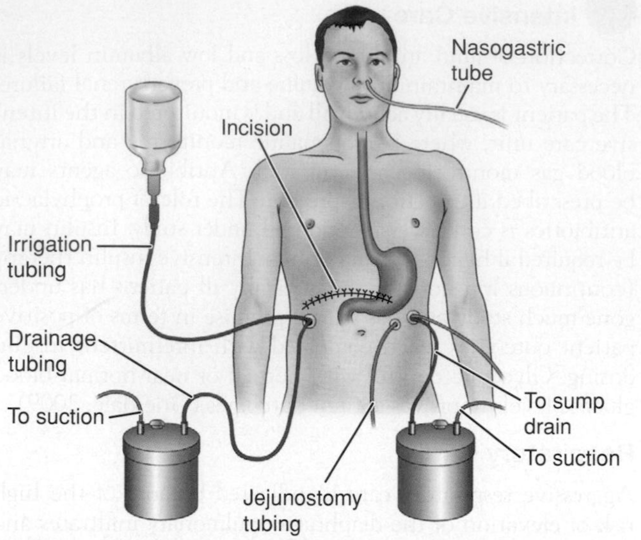
Nasogastric tube
Incision
Irrigation tubing
Drainage tubing
To suction
To sump drain
To suction
Jejunostomy tubing

FIGURE 50-6 • Multiple sump tubes are used after pancreatic surgery. Triple-lumen tubes consist of ports that provide tubing for irrigation, air venting, and drainage.

for pain management in this population is parenteral opioids, including morphine, hydromorphone, or fentanyl via patient-controlled analgesia or bolus (Marx, 2009). In critically ill patients, a continuous infusion may be needed. Because most opioids stimulate spasm of the sphincter of Oddi to some degree, consensus has not been reached on the most effective agent. Ensuring patient comfort, regardless of the opioid prescribed, is the most essential aspect of care. The nurse frequently assesses the pain and the effectiveness of the pharmacologic (and nonpharmacologic) interventions. Changes may be needed in the regimen for pain management based on the achievement of pain control. Pain assessment tools (see Chapter 12) are available for the nurse to ensure an accurate rating of pain. Nonpharmacologic interventions such as proper positioning, music, distraction, and imagery may be effective in reducing pain when used along with medications.

In addition, oral feedings are withheld to decrease the secretion of secretin. Parenteral fluids and electrolytes are prescribed to restore and maintain fluid balance. Nasogastric suction may be used to relieve nausea and vomiting or to treat abdominal distention and paralytic ileus. The nurse provides frequent oral hygiene and care to decrease discomfort from the nasogastric tube and relieve dryness of the mouth.

The acutely ill patient is maintained on bed rest to decrease the metabolic rate and reduce the secretion of pancreatic and gastric enzymes. If the patient experiences increasing severity of pain, the nurse reports this to the physician because the patient may be experiencing hemorrhage of the pancreas or the dose of analgesic medication may be inadequate.

The patient with acute pancreatitis often has a clouded sensorium because of severe pain, fluid and electrolyte disturbances, and hypoxia. Therefore, the nurse provides frequent and repeated but simple explanations about the need for withholding fluids, maintenance of gastric suction, and bed rest.

Improving Breathing Pattern

The nurse maintains the patient in a semi-Fowler's position to decrease pressure on the diaphragm by a distended abdomen and to increase respiratory expansion. Frequent changes of position are necessary to prevent atelectasis and pooling of respiratory secretions. Pulmonary assessment, including monitoring of pulse oximetry or arterial blood gases, is essential to detect changes in respiratory status so that early treatment can be initiated. The nurse instructs the patient in techniques of coughing and deep breathing and in the use of incentive spirometry to improve respiratory function and assists the patient to perform these activities every hour.

Improving Nutritional Status

It is important to assess the patient's nutritional status and to note factors that alter the patient's nutritional requirements (e.g., temperature elevation, surgery, drainage). Laboratory test results and daily weights are useful to monitor the nutritional status.

Enteral or parenteral nutrition may be prescribed. In addition to administering enteral or parenteral nutrition, the nurse monitors serum glucose levels every 4 to 6 hours. As the acute symptoms subside, oral feedings are gradually reintroduced. Between acute attacks, the patient receives a diet that is high in protein and low in fat (Dudek, 2010). The patient should avoid heavy meals and alcoholic beverages.

Maintaining Skin Integrity

The patient is at risk for skin breakdown because of poor nutritional status, enforced bed rest, and restlessness, which may result in pressure ulcers and breaks in tissue integrity. In addition, the patient who has undergone surgery may have multiple drains or an open surgical incision and is at risk for skin breakdown and infection. The nurse carefully assesses the wound, drainage sites, and skin for signs of infection, inflammation, and breakdown. The nurse carries out wound care as prescribed and takes precautions to protect intact skin from contact with drainage. Consultation with a **wound-ostomy-continence (WOC) nurse**, a nurse specially educated in appropriate skin, wound, ostomy, and continence care, also referred to as a wound care specialist or enterostomal therapist, is often helpful in identifying appropriate skin care devices and protocols. The patient must be turned every 2 hours; the use of specialty beds may be indicated to prevent skin breakdown.

Monitoring and Managing Potential Complications

Fluid and electrolyte disturbances are common complications because of nausea, vomiting, movement of fluid from the vascular compartment to the peritoneal cavity, diaphoresis, fever, and the use of gastric suction. The nurse assesses the patient's fluid and electrolyte status by noting skin turgor and moistness of mucous membranes. The nurse weighs the patient daily and carefully measures fluid intake and output, including urine output, nasogastric secretions, and diarrhea. In addition, it is important to assess for other factors that may affect fluid and electrolyte status, including increased body temperature and wound drainage. The nurse assesses the patient for ascites and measures abdominal girth daily if ascites is suspected.

Fluids are administered IV and may be accompanied by infusion of blood or blood products to maintain the blood volume and to prevent or treat hypovolemic shock. Emergency medications must be readily available because of the risk of circulatory collapse and shock. The nurse promptly reports decreased blood pressure and reduced urine output, which indicate hypovolemia and shock or renal failure. Low serum calcium and magnesium levels may occur and require prompt treatment.

Pancreatic necrosis is a major cause of morbidity and mortality in patients with acute pancreatitis because of resulting hemorrhage, septic shock, and multiple organ dysfunction syndrome (MODS). The patient may undergo diagnostic procedures for confirmation of pancreatic necrosis. If the patient is found to have pancreatic necrosis with infection, this may require surgical débridement and/or insertion of multiple drains. The patient with pancreatic necrosis with or without infection is usually critically ill and requires expert medical and nursing management, including hemodynamic monitoring in the intensive care unit.

In addition to carefully monitoring vital signs and other signs and symptoms, the nurse is responsible for administering prescribed fluids, medications, and blood products; assisting with supportive management, such as the use of a ventilator; preventing additional complications; and providing physical and psychological care.

Shock and MODS may occur with acute pancreatitis. Hypovolemic shock may occur as a result of hypovolemia and sequestering of fluid in the peritoneal cavity. Hemorrhagic shock may occur with hemorrhagic pancreatitis. Septic shock may occur with bacterial infection of the pancreas. Cardiac dysfunction may occur as a result of fluid and electrolyte disturbances, acid–base imbalances, and release of toxic substances into the circulation.

The nurse closely monitors the patient for early signs of neurologic, cardiovascular, renal, and respiratory dysfunction. The nurse must be prepared to respond quickly to rapid changes in the patient's status, treatments, and therapies. In addition, it is important to inform the family about the status and progress of the patient and to allow them to spend time with the patient. (Management of shock and MODS is discussed in detail in Chapter 14.)

Promoting Home and Community-Based Care

Educating Patients About Self-Care

After an episode of acute pancreatitis, the patient is often still weak and debilitated for weeks or months. A prolonged period may be needed to regain strength and return to the previous level of activity. Because of the severity of the acute illness, the patient may not recall much of the education given during the acute phase. Patient education often needs to be repeated and reinforced. The nurse educates the patient about the factors implicated in the onset of acute pancreatitis and about the need to avoid high-fat foods, heavy meals, and alcohol. The patient and family should receive verbal and written instructions about signs and symptoms of acute pancreatitis and possible complications that should be reported promptly to the primary provider.

If acute pancreatitis is a result of biliary tract disease, such as gallstones and gallbladder disease, additional explanations are needed about required dietary modifications. If the pancreatitis is a result of alcohol abuse, the nurse reinforces the need to avoid all alcohol.

Continuing Care

A referral for home care is often indicated. This enables the nurse to assess the patient's physical and psychological status and adherence to the therapeutic regimen. The nurse also assesses the home situation and reinforces instructions about fluid and nutrition intake and avoidance of alcohol. After the acute attack has subsided, some patients may be inclined to return to their previous drinking habits. The nurse provides specific information about resources and support groups that may be of assistance in avoiding alcohol in the future. Referral to Alcoholics Anonymous as appropriate or other support groups is essential. See the accompanying plan of nursing care in Chart 50-4 for the patient with acute pancreatitis.

Chronic Pancreatitis

Chronic pancreatitis is an inflammatory disorder characterized by progressive destruction of the pancreas. As cells are replaced by fibrous tissue with repeated attacks of pancreatitis, pressure within the pancreas increases. The result is obstruction of the pancreatic and common bile ducts and the duodenum. Additionally, there is atrophy of the epithelium of the ducts, inflammation, and destruction of the secreting cells of the pancreas.

Alcohol consumption in Western societies and malnutrition worldwide are the major causes of chronic pancreatitis. The median age of patients diagnosed with chronic pancreatitis has been reported to be 58 years (Yadav, Timmons, Benson, et al., 2011). Frequently, at that age, patients already report a long history of alcohol abuse. Excessive and prolonged consumption of alcohol accounts for approximately 70% to 80% of all cases of chronic pancreatitis (Townsend et al., 2012). The incidence of pancreatitis is 50 times greater in people with alcoholism than in those who do not abuse alcohol. Long-term alcohol consumption causes hypersecretion of protein in pancreatic secretions, resulting in protein plugs and calculi within the pancreatic ducts. Alcohol also has a direct toxic effect on the cells of the pancreas. Damage to these cells is more likely to occur and to be more severe in patients whose diets are poor in protein content and either very high or very low in fat.

Smoking is another factor in the development of chronic pancreatitis. Because they are often associated, it is difficult to separate the effects of the alcohol abuse and smoking (Townsend et al., 2012).

Clinical Manifestations

Chronic pancreatitis is characterized by recurring attacks of severe upper abdominal and back pain, accompanied by vomiting. Attacks are often so painful that opioids, even in large doses, do not provide relief. The risk of opioid dependence is increased in pancreatitis because of the chronic nature and severity of the pain. As the disease progresses, recurring attacks of pain are more severe, more frequent, and of longer duration. Some patients experience continuous severe pain, and others have dull, nagging constant pain. Periods of well-being sometimes follow the episodes of pain (Trikudanathan, Navaneethan, & Vege, 2012). In some patients, chronic pancreatitis is painless. The natural history of abdominal pain (character, timing, severity) is variable, and many studies have documented a decrease in pain ("burnout") over time in a majority of patients (Trikudanathan et al., 2012).

(text continues on page 1408)

Chart 50-4

PLAN OF NURSING CARE

Care of the Patient With Acute Pancreatitis

NURSING DIAGNOSIS: Acute pain and discomfort related to edema, distention of the pancreas, peritoneal irritation, and excess stimulation of pancreatic secretions
GOAL: Relief of pain and discomfort

Nursing Interventions	Rationale	Expected Outcomes
1. Using a pain scale, assess pain level at baseline, before and after administration of anticholinergic and analgesic medications.	**1.** Baseline assessment and control of pain are important because restlessness increases the body's metabolism, which stimulates the secretion of pancreatic and gastric enzymes.	• Patient rates pain using pain scale • Reports relief of pain, discomfort, and abdominal cramping • Moves and turns without increasing pain and discomfort • Rests comfortably and sleeps for increasing periods • Reports increased feelings of well-being and security with the health care team • Identifies rationale for fluid and dietary restrictions and the use of nasogastric drainage
2. Administer anticholinergic medications as prescribed.	**2.** Anticholinergic medications reduce gastric and pancreatic secretion.	
3. Administer morphine, fentanyl, or hydromorphone frequently, as prescribed, to achieve level of pain acceptable to patient.	**3.** Morphine, fentanyl, and hydromorphone act by depressing the central nervous system and thereby increasing the patient's pain threshold. Meperidine (Demerol) is avoided because it has failed acute pain studies and possesses toxic metabolites.	
4. Maintain the patient NPO (nothing by mouth) as prescribed.	**4.** Pancreatic secretion is increased by food and fluid intake.	
5. Maintain the patient on bed rest.	**5.** Bed rest decreases body metabolism and thus reduces pancreatic and gastric secretions.	
6. Maintain continuous nasogastric drainage if paralytic ileus or nausea and vomiting, abdominal distention are present. **a.** Measure gastric secretions at specified intervals. **b.** Observe and record color and viscosity of gastric secretions. **c.** Ensure that the nasogastric tube is patent to permit free drainage.	**6.** Nasogastric suction relieves nausea, vomiting, and abdominal distention. Decompression of the intestines (if intestinal intubation is used) also assists in relieving respiratory distress.	
7. Report unrelieved pain or increasing intensity of pain.	**7.** Pain may increase pancreatic enzymes and may also indicate pancreatic hemorrhage.	
8. Assist patient to assume positions of comfort; turn and reposition every 2 hours.	**8.** Frequent turning relieves pressure and assists in preventing pulmonary and vascular complications.	
9. Use nonpharmacologic interventions for relieving pain (e.g., relaxation, focused breathing, diversion).	**9.** The use of nonpharmacologic methods will enhance the effects of analgesic medications.	
10. Listen to patient's expression of pain experience.	**10.** Demonstration of caring can help to decrease anxiety.	

NURSING DIAGNOSIS: Impaired comfort related to nasogastric tube
GOAL: Relief of discomfort associated with nasogastric intubation used to treat ileus, vomiting, distention

Nursing Interventions	Rationale	Expected Outcomes
1. Use water-soluble lubricant around external nares.	**1.** Prevents irritation of nares.	• Exhibits intact skin and tissue of nares at site of nasogastric tube insertion • Reports no pain or irritation of nares or oropharynx • Exhibits moist, clean mucous membranes of mouth and nasopharynx • States that thirst is relieved by oral hygiene • Identifies rationale for nasogastric tube and suction
2. Turn patient at intervals; avoid pressure or tension on nasogastric tube.	**2.** Relieves pressure of tube on esophageal and gastric mucosa.	
3. Provide oral hygiene and gargling solutions without alcohol.	**3.** Relieves dryness and irritation of oropharynx.	
4. Explain rationale for the use of nasogastric drainage.	**4.** Assists patient to cooperate with the drainage, nasogastric tube, and suction.	

Chart 50-4

PLAN OF NURSING CARE

Care of the Patient With Acute Pancreatitis (continued)

NURSING DIAGNOSIS: Imbalanced nutrition: less than body requirements related to inadequate dietary intake, impaired pancreatic secretions, increased nutritional needs secondary to acute illness, and increased body temperature
GOAL: Improvement in nutritional status

Nursing Interventions	Rationale	Expected Outcomes
1. Assess current nutritional status and increased metabolic requirements.	**1.** Alteration in pancreatic secretions interferes with normal digestive processes. Acute illness, infection, and fever increase metabolic needs.	• Maintains normal body weight • Demonstrates no additional weight loss • Maintains normal serum glucose levels • Reports decreasing episodes of vomiting and diarrhea • Reports return of normal stool characteristics and bowel pattern • Consumes foods high in carbohydrates, low in fat and protein • Explains rationale for high-carbohydrate, low-fat, low-protein diet • Eliminates alcohol from diet • Explains rationale for limiting coffee intake and avoiding spicy foods • Participates in Alcoholics Anonymous as appropriate or other counseling approach • Returns to and maintains desirable weight
2. Monitor serum glucose levels and administer insulin as prescribed.	**2.** Impairment of endocrine function of the pancreas leads to increased serum glucose levels.	
3. Administer IV fluid and electrolytes, enteral or parenteral nutrition as prescribed.	**3.** Parenteral administration of fluids and electrolytes, and enteral or parenteral nutrients are essential to provide fluids, calories, electrolytes, and nutrients when oral intake is prohibited.	
4. Provide high-carbohydrate, low-protein, low-fat diet when tolerated.	**4.** These foods increase caloric intake without stimulating pancreatic secretions beyond the ability of the pancreas to respond.	
5. Instruct patient to eliminate alcohol, and refer to Alcoholics Anonymous, if indicated.	**5.** Alcohol intake produces further damage to pancreas and precipitates attacks of acute pancreatitis.	
6. Counsel patient to avoid excessive use of coffee and spicy foods.	**6.** Coffee and spicy foods increase pancreatic and gastric secretions.	
7. Monitor daily weights.	**7.** This provides a baseline and a means to measure weight gain or weight loss.	

NURSING DIAGNOSIS: Ineffective breathing pattern related to splinting from severe pain, pulmonary infiltrates, pleural effusion, and atelectasis
GOAL: Improvement in respiratory function

Nursing Interventions	Rationale	Expected Outcomes
1. Assess respiratory status (rate, pattern, breath sounds), pulse oximetry, and arterial blood gases.	**1.** Acute pancreatitis produces retroperitoneal edema, elevation of the diaphragm, pleural effusion, and inadequate lung ventilation. Intra-abdominal infection and labored breathing increase the body's metabolic demands, which further decreases pulmonary reserve and leads to respiratory failure.	• Demonstrates normal respiratory rate and pattern and full lung expansion • Demonstrates normal breath sounds and absence of adventitious breath sounds • Demonstrates normal arterial blood gases and pulse oximetry • Maintains semi-Fowler's position when in bed • Changes position in bed frequently • Coughs and takes deep breaths at least every hour • Demonstrates normal body temperature • Exhibits no signs or symptoms of respiratory infection or impairment • Is alert and responsive to environment
2. Maintain semi-Fowler's position.	**2.** Decreases pressure on diaphragm and allows greater lung expansion.	
3. Instruct and encourage patient to take deep breaths and to cough every hour.	**3.** Taking deep breaths and coughing will clear the airways and reduce atelectasis.	
4. Assist patient to turn and change position every 2 hours.	**4.** Changing position frequently assists aeration and drainage of all lobes of the lungs.	
5. Reduce the excessive metabolism of the body. **a.** Administer antibiotics as prescribed. **b.** Place patient in an air-conditioned room. **c.** Administer nasal oxygen as required for hypoxia. **d.** Use a hypothermia blanket if necessary.	**5.** Pancreatitis produces a severe peritoneal and retroperitoneal reaction that causes fever, tachycardia, and accelerated respirations. Placing the patient in an air-conditioned room and supporting the patient with oxygen therapy decrease the workload of the respiratory system and the tissue utilization of oxygen. Reduction of fever and pulse rate decreases the metabolic demands on the body.	

(continues on page 1408)

Chart 50-4

PLAN OF NURSING CARE

Care of the Patient With Acute Pancreatitis (continued)

COLLABORATIVE PROBLEM: Fluid and electrolyte disturbances, hypovolemia, shock
GOAL: Improvement in fluid and electrolyte status, prevention of hypovolemia and shock

Nursing Interventions	Rationale	Expected Outcomes
1. Assess fluid and electrolyte status (skin turgor, mucous membranes, urine output, vital signs, hemodynamic parameters).	**1.** The amount and type of fluid and electrolyte replacement are determined by the status of the blood pressure, the laboratory evaluations of serum electrolyte and blood urea nitrogen levels, the urinary volume, and the assessment of the patient's condition.	• Exhibits moist mucous membranes and normal skin turgor • Exhibits normal blood pressure without evidence of postural (orthostatic) hypotension • Excretes adequate urine volume • Exhibits normal, not excessive, thirst • Maintains normal pulse and respiratory rate • Remains alert and responsive • Exhibits normal arterial pressures and blood gases • Exhibits normal electrolyte levels • Exhibits no signs or symptoms of calcium deficit (e.g., tetany, carpopedal spasm) • Exhibits no additional losses of fluids and electrolytes through vomiting, diarrhea, or diaphoresis • Reports stabilization of weight • Demonstrates no increase in abdominal girth • Demonstrates no fluid wave on palpation of the abdomen • Demonstrates stable organ function without manifestations of failure
2. Assess sources of fluid and electrolyte loss (vomiting, diarrhea, nasogastric drainage, excessive diaphoresis).	**2.** Electrolyte losses occur from nasogastric suctioning, severe diaphoresis, and emesis, and as a result of the patient being in a fasting state.	
3. Combat shock if present. **a.** Administer corticosteroids as prescribed if patient does not respond to conventional treatment. **b.** Evaluate the amount of urinary output. Attempt to maintain this at 0.5 mL/kg/h.	**3.** Extensive acute pancreatitis may cause peripheral vascular collapse and shock. Blood and plasma may be lost into the abdominal cavity; therefore, there is a decreased blood and plasma volume. The toxins from the bacteria of a necrotic pancreas may cause shock.	
4. Administer blood products, fluids, and electrolytes (sodium, potassium, chloride) as prescribed.	**4.** Patients with hemorrhagic pancreatitis lose large amounts of blood and plasma, which decreases effective circulation and blood volume.	
5. Administer plasma and blood products as prescribed.	**5.** Replacement with blood, plasma, or albumin assists in ensuring effective circulating blood volume.	
6. Keep a supply of IV calcium gluconate readily available.	**6.** Calcium may be prescribed to prevent or treat tetany, which may result from calcium losses into retroperitoneal (peripancreatic) exudate.	
7. Assess abdomen for ascites formation. **a.** Measure abdominal girth daily. **b.** Weigh patient daily. **c.** Assess abdomen for ascites. (See Figs. 49-6 and 49-7.)	**7.** During acute pancreatitis, plasma may be lost into the abdominal cavity, which diminishes the blood volume.	
8. Monitor for manifestations of multiple organ dysfunction syndrome: neurologic, cardiovascular, renal, and respiratory dysfunction.	**8.** All body systems may fail if pancreatitis is severe and treatment is ineffective.	

Weight loss is a major problem in chronic pancreatitis. More than 80% of patients experience significant weight loss, which is usually caused by decreased dietary intake secondary to anorexia or fear that eating will precipitate another attack (Bope & Kellerman, 2011). Malabsorption occurs late in the disease, when as little as 10% of pancreatic function remains (Bope & Kellerman, 2011). As a result, digestion, especially of proteins and fats, is impaired. The stools become frequent, frothy, and foul smelling because of impaired fat digestion, which results in stools with a high fat content referred to as **steatorrhea**. As the disease progresses, calcification of the gland may occur, and calcium stones may form within the ducts.

Assessment and Diagnostic Findings

ERCP is the most useful study in the diagnosis of chronic pancreatitis. It provides details about the anatomy of the pancreas and the pancreatic and biliary ducts. It is also helpful in obtaining tissue for analysis and differentiating pancreatitis from other conditions, such as carcinoma. Various imaging procedures, including MRI, CT scans, and ultrasound, are used in the diagnostic evaluation of patients with suspected pancreatic disorders. A CT scan or ultrasound study is also helpful to detect pancreatic cysts.

A glucose tolerance test evaluates pancreatic islet cell function and provides necessary information for making decisions about surgical resection of the pancreas. An abnormal glucose tolerance test may indicate the presence of diabetes associated with pancreatitis. Acute exacerbations of chronic pancreatitis may result in increased serum amylase levels. Steatorrhea is best confirmed by laboratory analysis of fecal fat content (Bope & Kellerman, 2011).

Medical Management

The management of chronic pancreatitis depends on its probable cause in each patient. Treatment is directed toward preventing and managing acute attacks, relieving pain and discomfort, and managing exocrine and endocrine insufficiency of pancreatitis.

Nonsurgical Management

Nonsurgical approaches may be indicated for the patient who refuses surgery, is a poor surgical risk, or when the disease and symptoms do not warrant surgical intervention. Endoscopy to remove pancreatic duct stones, correct strictures, and drain cysts may be effective in selected patients to manage pain and relieve obstruction via ERCP (Trikudanathan et al., 2012).

Management of abdominal pain and discomfort is similar to that of acute pancreatitis; however, the focus is usually on the use of nonopioid methods to manage pain and the implementation of the World Health Organization's (WHO) three-step ladder for the treatment of chronic pain. This involves initiating monotherapy and, if ineffective, instituting combination therapy with peripherally acting and centrally acting medications. Antioxidants have shown effect in the relief of pain and in improving quality of life and are often administered to patients with chronic pancreatitis (Trikudanathan et al., 2012). Researchers have proposed that yoga may be an effective nonpharmacologic method for pain reduction and for relief of other coexisting symptoms of chronic pancreatitis (Trikudanathan et al., 2012). Persistent, unrelieved pain is often the most difficult aspect of management (Trikudanathan et al., 2012). The primary provider, nurse, and dietitian emphasize to the patient and family the importance of avoiding alcohol and foods that have produced abdominal pain and discomfort in the past. The health care team stresses to the patient that no other treatment is likely to relieve pain if the patient continues to consume alcohol.

Diabetes resulting from dysfunction of the pancreatic islet cells is treated with diet, insulin, or oral antidiabetic agents. The hazard of severe hypoglycemia with alcohol consumption is stressed to the patient and family. Pancreatic enzyme replacement is indicated for the patient with malabsorption and steatorrhea.

Surgical Management

Chronic pancreatitis is not often managed by surgery. However, surgery may be indicated to relieve persistent abdominal pain and discomfort, restore drainage of pancreatic secretions, and reduce the frequency of acute attacks of pancreatitis and hospitalization (Trikudanathan et al., 2012). The type of surgery performed depends on the anatomic and functional abnormalities of the pancreas, including the location of disease within the pancreas, the presence of diabetes, exocrine insufficiency, biliary stenosis, and pseudocysts of the pancreas. Other considerations for surgery selection include the patient's likelihood for continued use of alcohol and the likelihood that the patient will be able to manage the endocrine or exocrine changes that are expected after surgery.

Pancreaticojejunostomy (also referred to as Roux-en-Y), with a side-to-side anastomosis or joining of the pancreatic duct to the jejunum, allows drainage of the pancreatic secretions into the jejunum. Pain relief occurs within 6 months in more than 85% of the patients who undergo this procedure, but pain returns in a substantial number of patients as the disease progresses (Trikudanathan et al., 2012).

Other surgical procedures may be performed for different degrees and types of underlying disorders. These procedures include revision of the sphincter of the ampulla of Vater, internal drainage of a pancreatic cyst into the stomach (see later discussion), insertion of a stent, and wide resection or removal of the pancreas. A Whipple resection (pancreaticoduodenectomy) can be carried out to relieve the pain of chronic pancreatitis (see later discussion under Tumors of the Head of the Pancreas). In an effort to provide permanent pain relief and avoid endocrine and exocrine insufficiency that ensue with major resections of the pancreas, surgeons have designed new procedures that combine limited resection of the head of the pancreas with a pancreaticojejunostomy. These procedures, known as the Beger or Frey operations, remove most of the head of the pancreas except for a shell of pancreatic tissue posteriorly (Trikudanathan et al., 2012).

When chronic pancreatitis develops as a result of gallbladder disease, surgery is performed to explore the common duct and remove the stones; usually, the gallbladder is removed at the same time. In addition, an attempt is made to improve the drainage of the common bile duct and the pancreatic duct by dividing the sphincter of Oddi, a muscle that is located at the ampulla of Vater (this surgical procedure is known as a sphincterotomy). A T-tube usually is placed in the common bile duct, requiring a drainage system to collect the bile postoperatively. Nursing care after such surgery is similar to that indicated after other biliary tract surgery.

Approximately two thirds of all patients with chronic pancreatitis can be managed with endoscopic or laparoscopic intervention (Yamada, 2009). Endoscopic and laparoscopic procedures such as distal pancreatectomy, longitudinal decompression of the pancreatic duct, nerve denervation, and stenting have been performed in patients with jaundice or recurrent inflammation and are being refined. Minimally invasive procedures to treat chronic pancreatitis may prove to be successful adjuncts in the management of this complex disorder (Trikudanathan et al., 2012).

Patients who undergo surgery for chronic pancreatitis may experience weight gain and improved nutritional status; this may result from reduction in pain associated with eating rather than from correction of malabsorption. However, morbidity and mortality after these surgical procedures are high because of the poor physical condition of the patient before surgery and the concomitant presence of cirrhosis. Even after undergoing these surgical procedures, the patient is likely to continue to have pain and impaired digestion secondary to pancreatitis.

Concept Mastery Alert

Visit thePoint to view an Interactive Tutorial on pancreatitis and related fundamental concepts.

Pancreatic Cysts

As a result of the local necrosis that occurs at the time of acute pancreatitis, collections of fluid may form close to the pancreas. These fluid collections become walled off by fibrous tissue and are called *pancreatic pseudocysts*. Pseudocysts are amylase-rich fluid collections contained within a wall of fibrous granulation tissue that occur within 4 to 6 weeks after an episode of acute pancreatitis. They are a result of pancreatic necrosis, which produces a pancreatic ductal leak *into* pancreatic tissue weakened by extravasating enzymes (Feldman et al., 2010; Mulholland et al., 2011). Pseudocysts are distinguished from true cysts by the characteristics of the lining of the walls of these anomalies. The lining of pseudocysts consists of fibrous granulation tissue, whereas true cysts have epithelium-lined walls (Mulholland et al., 2011). Pseudocysts are the most common type of pancreatic "cyst." Less common cysts occur as a result of congenital anomalies or secondary to chronic pancreatitis or trauma to the pancreas.

Diagnosis of pancreatic cysts and pseudocysts is made by ultrasound, CT scan, and ERCP. ERCP may be used to define the anatomy of the pancreas and evaluate the patency of pancreatic drainage. Pancreatic pseudocysts may be of considerable size. When pancreatic pseudocysts enlarge, they impinge on and displace the adjacent stomach or the colon because of the location of pseudocysts behind the posterior peritoneum. Eventually, through pressure or secondary infection, they produce symptoms and require drainage.

Drainage into the GI tract or through the skin and abdominal wall may be established. In the latter instance, the drainage is likely to be profuse and destructive to tissue because of the enzyme contents. Hence, steps (including application of skin ointment) must be taken to protect the skin near the drainage site from excoriation. A suction apparatus may be used to continuously aspirate digestive secretions from the drainage tract so that skin contact with the digestive enzymes is avoided. Expert nursing attention is required to ensure that the suction tube does not become dislodged and suction is not interrupted. Consultation with a WOC nurse is indicated to identify appropriate strategies for maintaining drainage and protecting the skin.

Cancer of the Pancreas

Pancreatic cancer is the fourth leading cause of cancer death in men in the United States and the fifth leading cause of cancer death in women. It is very rare before the age of 45 years, and the majority of patients present in or beyond the sixth decade of life (Feldman et al., 2010). The incidence of pancreatic cancer increases with age, peaking in the seventh and eighth decades for both men and women (American Cancer Society [ACS], 2012). The frequency of pancreatic cancer has decreased slightly over the past 25 years among non-Caucasian men. There is a slight male preponderance, and in the United States, incidence is highest in African American males (Feldman et al., 2010; Othman & Wallace, 2012). Cigarette smoking, exposure to industrial chemicals or toxins in the environment, and a diet high in fat, meat, or both are associated risk factors (ACS, 2012).

The risk of pancreatic cancer increases as the extent of cigarette smoking increases. Diabetes, chronic pancreatitis, and hereditary pancreatitis are also associated with pancreatic cancer. The pancreas can also be the site of metastasis from other tumors.

Cancer may develop in the head, body, or tail of the pancreas; clinical manifestations vary depending on the site and whether functioning insulin-secreting pancreatic islet cells are involved. Approximately 70% of pancreatic cancers originate in the head of the pancreas and give rise to a distinctive clinical picture (Feldman et al., 2010). Functioning islet cell tumors, whether benign (adenoma) or malignant (carcinoma), are responsible for the syndrome of hyperinsulinism. The symptoms are typically nonspecific, and patients usually do not seek medical attention until late in the disease. Only about 7% of cases are diagnosed in early stages; 80% to 85% of patients have advanced, unresectable tumor when first detected. As a result, pancreatic carcinoma has only a 5% survival rate at 5 years regardless of the stage of disease at diagnosis or treatment (ACS, 2012).

Clinical Manifestations

Pain, jaundice, or both are present in more than 80% of patients and, along with weight loss, are considered classic signs of pancreatic carcinoma (Feldman et al., 2010). However, they often do not appear until the disease is far advanced. Other signs include rapid, profound, and progressive weight loss as well as vague upper or midabdominal pain or discomfort that is unrelated to any GI function and is often difficult to describe. Such discomfort radiates as a boring pain in the midback and is unrelated to posture or activity. It is often progressive and severe, requiring the use of opioids. It is often more severe at night and is accentuated when lying supine. Relief may be obtained by sitting up and leaning forward.

Malignant cells from pancreatic cancer are often shed into the peritoneal cavity, increasing the likelihood of metastasis. The formation of ascites is common. An important sign, if present, is the onset of symptoms of insulin deficiency: glucosuria, hyperglycemia, and abnormal glucose tolerance. Therefore, diabetes may be an early sign of carcinoma of the pancreas. Meals often aggravate epigastric pain, which usually occurs before the appearance of jaundice and pruritus.

Assessment and Diagnostic Findings

Spiral (helical) CT is more than 85% to 90% accurate in the diagnosis and staging of pancreatic cancer and currently is the most useful preoperative imaging technique. MRI may also be used. ERCP is also used in the diagnosis of pancreatic carcinoma. Endoscopic ultrasound is useful in identifying small tumors and in performing fine-needle aspiration biopsy of the primary tumor or lymph nodes (Goldman & Schafer, 2012; Othman & Wallace, 2012). Cells obtained during ERCP are sent to the laboratory for analysis. GI x-ray findings may demonstrate deformities in adjacent organs caused by the impinging pancreatic mass.

A histologic diagnosis is not usually required in patients who are candidates for surgery. The tissue diagnosis is made at the time of the surgical procedure. Percutaneous fine-needle aspiration biopsy of the pancreas, which is used to diagnose pancreatic tumors, is also used to confirm the diagnosis in patients whose tumors are not resectable so that a palliative plan of care can be determined. This may eliminate the stress and postoperative pain of ineffective surgery. In this procedure, a needle is inserted through the anterior abdominal wall into the pancreatic mass, guided by CT, ultrasound, ERCP, or other imaging techniques. The aspirated material is examined for malignant cells. Although percutaneous biopsy is a valuable diagnostic tool, it has some potential drawbacks: a false-negative result if small tumors are missed and the risk of seeding of cancer cells along the needle track. Low-dose radiation to the site may be used before the biopsy to reduce this risk.

PTC is another procedure that may be performed to identify obstructions of the biliary tract by a pancreatic tumor. Several tumor markers (e.g., cancer antigen 19–9, carcinoembryonic antigen, DU-PAN-2) may be used in the diagnostic workup, but they are nonspecific for pancreatic carcinoma. These tumor markers are useful as indicators of disease progression.

Angiography, CT scans, and laparoscopy may be performed to determine whether the tumor can be removed surgically. Intraoperative ultrasonography has been used to determine whether there is metastatic disease to other organs.

Medical Management

If the tumor is resectable and localized (typically tumors in the head of the pancreas), the surgical procedure to remove it is usually extensive (see later discussion). However, total excision of the lesion often is not possible for two reasons: (1) extensive growth of tumor before diagnosis and (2) probable widespread metastases (especially to the liver, lungs, and bones). More often, treatment is limited to palliative measures.

Although pancreatic tumors may be resistant to standard radiation therapy, the patient may be treated with radiation and chemotherapy (5-fluorouracil [5-FU, Adrucil], leucovorin [Wellcovorin], and gemcitabine [Gemzar]). Currently, gemcitabine is the standard of care for patients with metastatic pancreatic cancer (Feldman et al., 2010). At present, newer biologic agents, including farnesyltransferase inhibitors and monoclonal antibodies, are under study for the treatment of metastatic pancreatic cancer (Feldman et al., 2010). If the patient undergoes surgery, intraoperative radiation therapy may be used to deliver a high dose of radiation to the tumor with minimal injury to other tissues; this may also be helpful in relief of pain. Interstitial implantation of radioactive sources has also been used, although the rate of complications is high. A large biliary stent inserted percutaneously or by endoscopy may be used to relieve jaundice.

Nursing Management

Pain management and attention to nutritional requirements are important nursing measures that improve the level of patient comfort. Skin care and nursing measures are directed toward relief of pain and discomfort associated with jaundice, anorexia, and profound weight loss. Specialty mattresses are beneficial and protect bony prominences from pressure. Pain associated with pancreatic cancer may be severe and may require liberal use of opioids; patient-controlled analgesia should be considered for the patient with severe, escalating pain.

Because of the poor prognosis and likelihood of short survival, end-of-life preferences are discussed and honored. If appropriate, the nurse refers the patient to hospice care. (See Chapters 15 and 16 for care of the patient with cancer and end-of-life care, respectively.)

Promoting Home and Community-Based Care

Educating Patients About Self-Care

Specific education for the patient and family varies with the stage of disease and the treatment choices made by the patient. If the patient elects to receive chemotherapy, the nurse focuses on prevention of side effects and complications of the agents used. If surgery is performed to relieve obstruction and establish biliary drainage, education addresses management of the drainage system and monitoring for complications. The nurse educates the family about changes in the patient's status that should be reported to the primary provider.

Continuing Care

A referral for home care is indicated to help the patient and family deal with the physical problems and discomforts associated with pancreatic cancer and the psychological impact of the disease. The home care nurse assesses the patient's physical status, fluid and nutritional status, skin integrity, and the adequacy of pain management. The nurse educates the patient and family on strategies to prevent skin breakdown and relieve pain, pruritus, and anorexia. It is important to discuss and arrange palliative care (hospice services) as indicated in an effort to relieve patient discomfort, assist with care, and comply with the patient's end-of-life decisions and wishes.

Tumors of the Head of the Pancreas

Tumors of the head of the pancreas comprise 60% to 80% of all pancreatic tumors (Goldman & Schafer, 2012). Tumors in this region of the pancreas obstruct the common bile duct where the duct passes through the head of the pancreas to join the pancreatic duct and empty at the ampulla of Vater into the duodenum. The tumors producing the obstruction may arise from the pancreas, the common bile duct, or the ampulla of Vater.

Clinical Manifestations

The obstructed flow of bile produces jaundice, clay-colored stools, and dark urine. Malabsorption of nutrients and fat-soluble vitamins may result if the tumor obstructs the entry of bile to the GI tract. Abdominal discomfort or pain and pruritus may be noted, along with anorexia, weight loss, and malaise. If these signs and symptoms are present, cancer of the head of the pancreas is suspected.

The jaundice of this disease must be differentiated from that due to a biliary obstruction caused by a gallstone in the common duct. Jaundice caused by a gallstone is usually intermittent and occurs more commonly in women and in people who are obese and who have had previous symptoms of gallbladder disease.

Assessment and Diagnostic Findings

Diagnostic studies may include duodenography, angiography by hepatic or celiac artery catheterization, pancreatic scanning, PTC, ERCP, and percutaneous needle biopsy of the pancreas. Results of a biopsy of the pancreas may aid in the diagnosis.

Medical Management

Before extensive surgery can be performed, a period of preparation is necessary because the patient's nutritional status and physical condition are often quite compromised. Various liver and pancreatic function studies are performed. A diet high in protein along with pancreatic enzymes is often prescribed. Preoperative preparation includes adequate hydration, correction of prothrombin deficiency with vitamin K, and treatment of anemia to minimize postoperative complications. Enteral or parenteral nutrition and blood component therapy are frequently required.

A biliary drainage procedure may be performed, usually with a catheter via percutaneous access, to relieve the jaundice and, perhaps, to provide time for a thorough diagnostic evaluation. Total pancreatectomy (removal of the pancreas) may be performed if there is no evidence of direct extension of the tumor to adjacent tissues or regional lymph nodes. A pancreaticoduodenectomy (Whipple procedure or resection) is used for potentially resectable cancer of the head of the pancreas (Fig. 50-7). This procedure involves removal of the gallbladder, a portion of the stomach, duodenum, proximal jejunum, head of the pancreas, and distal common bile duct.

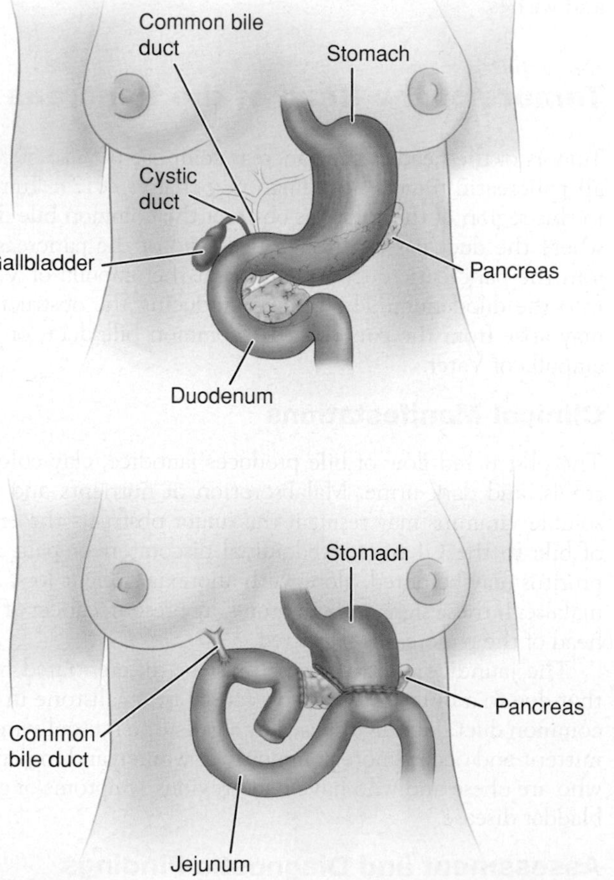
Common bile duct
Stomach
Cystic duct
Gallbladder
Pancreas
Duodenum
Stomach
Pancreas
Common bile duct
Jejunum

FIGURE 50-7 • Pancreatoduodenectomy (Whipple procedure or resection). End result of resection of carcinoma of the head of the pancreas or the ampulla of Vater. The common duct is sutured to the side of the jejunum (choledochojejunostomy), and the remaining portion of the pancreas and the end of the stomach are sutured to the side of the jejunum.

Reconstruction involves anastomosis of the remaining pancreas and stomach to the jejunum (Townsend et al., 2012). The result is removal of the tumor, allowing flow of bile into the jejunum. If the tumor cannot be excised, the jaundice may be relieved by diverting the bile flow into the jejunum by anastomosing the jejunum to the gallbladder, a procedure known as **cholecystojejunostomy**.

The postoperative management of patients who have undergone a pancreatectomy or a pancreaticoduodenectomy is similar to the management of patients after extensive GI or biliary surgery. The patient's physical status is often suboptimal, increasing the risk of postoperative complications. Hemorrhage, vascular collapse, and hepatorenal failure remain the major postoperative complications. The mortality rate associated with these procedures has decreased because of advances in nutritional support and improved surgical techniques. A nasogastric tube with suction and parenteral nutrition allow the GI tract to rest while promoting adequate nutrition.

Nursing Management

Preoperatively and postoperatively, nursing care is directed toward promoting patient comfort, preventing complications, and assisting the patient to return to and maintain as normal and comfortable a life as possible. The nurse closely monitors the patient in the intensive care unit after surgery; in the immediate postoperative period, multiple IV and arterial lines are used for fluid and blood replacement and hemodynamic monitoring, and a mechanical ventilator may be used. It is important to note and report changes in vital signs, arterial blood gases and pressures, pulse oximetry, laboratory values, and urine output. The nurse must also consider the patient's compromised nutritional status and risk of bleeding. Depending on the type of surgical procedure performed, malabsorption syndrome and diabetes are likely; the nurse must address these issues during acute and long-term patient care.

Although the patient's physiologic status is the focus of the health care team in the immediate postoperative period, the patient's psychological and emotional states must be considered, along with those of the family. The patient has undergone a major high-risk surgery and is critically ill; anxiety and depression may affect recovery. The immediate and long-term outcomes of this extensive surgical resection are uncertain, and the patient and family require emotional support and understanding in the critical and stressful preoperative and postoperative periods.

Promoting Home and Community-Based Care

Educating Patients About Self-Care

The patient who has undergone this extensive surgery requires careful and thorough preparation for self-care at home. The nurse educates the patient and family about strategies to relieve pain and discomfort, along with strategies to manage drains, if present, and to care for the surgical incision. The patient and family members may require education about the use of appropriate analgesic medications, parenteral nutrition, wound care, skin care, and management of drainage.

The nurse may need to educate the patient and family about the need for modifications in the diet because of malabsorption and hyperglycemia resulting from the surgery. It is important to educate the patient and family about the continuing need for pancreatic enzyme replacement, a low-fat diet, and vitamin supplementation. The nurse describes—verbally and in writing—the signs and symptoms of complications and educates the patient and family about indicators of complications that should be reported promptly.

Discharge of the patient to a long-term care or rehabilitation facility may be warranted after surgery as extensive as pancreatectomy or pancreaticoduodenectomy, particularly if the patient's preoperative status was not optimal. Information about the education that has been provided is shared with the long-term care staff so that instructions can be clarified and reinforced. During the recovery or long-term phase of care, the patient and family receive further education about self-care in the home.

Continuing Care

A referral for home care may be indicated when the patient returns home. The home care nurse assesses the patient's physical and psychological status and the ability of the patient and family to manage needed care. The home care nurse provides needed physical care and monitors the adequacy of pain management. In addition, it is important to assess the patient's nutritional status and monitor the use of enteral or parenteral nutrition, if used. The nurse discusses the use of hospice services with the patient and family and makes a referral if indicated.

Pancreatic Islet Tumors

The pancreas contains the islets (islands) of Langerhans, which are small nests of cells that secrete hormones directly into the bloodstream and therefore are part of the endocrine system. The hormone insulin is essential for the metabolism of glucose. Diabetes (see Chapter 51) is the result of deficient insulin secretion. At least two types of tumors of the pancreatic islet cells are known: those that secrete insulin (insulinoma) and those in which insulin secretion is not increased (nonfunctioning islet cell cancer). All of these types of tumors combined are termed *neuroendocrine tumors* (NETs). Insulinomas produce hypersecretion of insulin and cause an excessive rate of glucose metabolism. The resulting hypoglycemia may produce symptoms of weakness, mental confusion, and seizures. These symptoms may be relieved almost immediately by oral or IV administration of glucose. The 5-hour glucose tolerance test is helpful to diagnose insulinoma and to distinguish a diagnosis of NET from other causes of hypoglycemia.

Surgical Management

If a tumor of the islet cells (a type of NET) has been diagnosed, surgical treatment with removal of the tumor is usually recommended (Joseph, Wang, Boudreaux, et al., 2010). The tumors may be benign adenomas, or they may be malignant. Complete removal usually results in almost immediate relief of symptoms. In some patients, symptoms may be produced by simple hypertrophy of this tissue rather than a tumor of the islet cells. In such cases, a partial pancreatectomy (removal of the tail and part of the body of the pancreas) is performed.

Nursing Management

In preparing the patient for surgery, the nurse must be alert for symptoms of hypoglycemia and be ready to administer glucose as prescribed if symptoms occur. Postoperatively, the nursing management is the same as after other upper abdominal surgical procedures, with special emphasis on monitoring serum glucose levels. Patient education is determined by the extent of surgery and alterations in pancreatic function.

Hyperinsulinism

Hyperinsulinism is caused by overproduction of insulin by the pancreatic islets. Symptoms resemble those of excessive doses of insulin and are attributable to the same mechanism: an abnormal reduction in blood glucose levels. Clinically, it is characterized by episodes during which the patient experiences unusual hunger, nervousness, sweating, headache, and faintness; in severe cases, seizures and episodes of unconsciousness may occur. The findings at the time of surgery or at autopsy may indicate hyperplasia (overgrowth) of the islets of Langerhans or a benign or malignant tumor involving the islets that is capable of producing large amounts of insulin (see preceding discussion). Occasionally, tumors of nonpancreatic origin produce an insulinlike material that can cause severe hypoglycemia and may be responsible for seizures coinciding with blood glucose levels that are too low to sustain normal brain function (i.e., lower than 30 mg/dL [1.6 mmol/L] (Goldman & Schafer, 2012; McPherson & Pincus, 2011).

All of the symptoms that accompany spontaneous hypoglycemia are relieved by the oral or parenteral administration of glucose. Surgical removal of the hyperplastic or neoplastic tissue from the pancreas is the only successful method of treatment. About 15% of patients with spontaneous or functional hypoglycemia eventually develop diabetes.

Ulcerogenic Tumors

Some tumors of the islets of Langerhans are associated with hypersecretion of gastric acid that produces ulcers in the stomach, duodenum, and jejunum. This is referred to as **Zollinger-Ellison syndrome**. The hypersecretion is so excessive that even after partial gastric resection, enough acid is produced to cause further ulceration. If a marked tendency to develop gastric and duodenal ulcers is noted, an ulcerogenic tumor of the islets of Langerhans is considered.

These tumors, which may be benign or malignant, are treated by excision, if possible. Frequently, however, removal is not possible because of extension beyond the pancreas. In many patients, a total gastrectomy may be necessary to reduce the secretion of gastric acid sufficiently to prevent further ulceration. This procedure is also indicated to treat gastric carcinoid tumors that may arise from the effect of prolonged hypersecretion of gastric acids (Joseph et al., 2010).

The many disorders of the biliary and pancreatic systems result in a host of clinical and biochemical abnormalities presented in detail in this chapter. The recognition and management of these illnesses pose a challenge to nurses. There is now much evidence about the prevention, diagnosis, and treatment of these disorders. However, there is still much to learn as the medical and nursing communities strive to provide optimum care to patients afflicted with biliary and pancreatic disorders.

Critical Thinking Exercises

1 A 75-year-old man has confirmed cholelithiasis. He also has many other chronic medical conditions, including coronary artery disease, heart failure, and chronic obstructive pulmonary disease. He has been evaluated for treatment by an internist and a surgeon. Surgical treatment of cholelithiasis is not appropriate at this time. What options exist for this patient if he is symptomatic? What medication regimen might be used, especially if the patient is asymptomatic? How would you educate and prepare the patient for nonsurgical interventions?

2 ebp A 68-year-old woman has undergone Whipple procedure as a potentially curative intervention for pancreatic cancer. What factors in her past medical history might you investigate to determine her risk factors for developing this disease? Once she has had the surgery, to what immediate postoperative complications must you be alert? What assessment strategies are used to monitor for the development of complications? Describe two evidence-based preventive interventions for postoperative care.

3 ebp A 35-year-old woman has been experiencing more frequent, recurrent episodes of midepigastric pain. She visits her physician, who states that her symptoms suggest possible gallstones. What tests do you expect her to undergo to evaluate for this problem? If she is found to have gallstones, what intervention do you expect her to undergo? If she undergoes a surgical intervention, would you instruct her about an open laparotomy, a laparoscopic procedure, or both? What information should you collect regarding the patient's home situation before she is discharged? What information should you provide about expectations for postoperative pain and other complications, and what instructions should you provide to the patient about pharmacologic and nonpharmacologic pain management strategies? What is the evidence base for the pain management strategies you provide, and what is the strength of that evidence?

4 pq A 52-year-old man has been hospitalized following a motor vehicle crash with intestinal and liver injuries requiring surgery. He underwent a bowel resection and liver repair that has been complicated by postoperative and GI bleeding requiring multiple blood transfusions. His hospitalization has also been complicated by sepsis. He presents today with fever, nausea, vomiting, and right upper quadrant pain. What tests might you expect to be ordered? If the patient undergoes an abdominal ultrasound revealing gallbladder wall abnormalities (thickening) but no gallstones, what diagnosis might be causative of his symptoms? What is the next most likely test that could be done to determine his diagnosis? What are your priority nursing diagnoses for this patient? What nursing interventions carry the greatest priority for this patient?

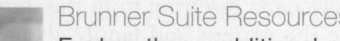

Brunner Suite Resources

Explore these additional resources to enhance learning for this chapter:

- NCLEX-Style Questions and Other Resources on thePoint, http://thePoint.lww.com/Brunner13e
- Study Guide
- PrepU
- Clinical Handbook
- Handbook of Laboratory and Diagnostic Tests

References

*Asterisk indicates nursing research.
**Double asterisk indicates classic reference.

Books

Bope, E. T., & Kellerman, R. D (Eds.). (2011). *Conn's current therapy*. Philadelphia: Saunders.

Chohan, N. D., & Munden, D. H. (Eds.). (2008). *Nurse's 5-minute clinical consult: Procedures*. Philadelphia: Lippincott Williams & Wilkins.

Dudek, S. G. (2010). *Nutrition essentials for nursing practice* (6th ed.). Philadelphia: Lippincott Williams & Wilkins.

Eliopoulos, C. (2010). *Gerontological nursing* (7th ed.). Philadelphia: Lippincott Williams & Wilkins.

Feldman, M., Friedman, L. S., & Brandt, L. J. (Eds.). (2010). *Sleisenger and Fordtran's gastrointestinal & liver disease* (9th ed.). Philadelphia: Saunders Elsevier.

Ferri, F. F. (2012). *Ferri's clinical advisor*. Philadelphia: Mosby Elsevier.

Goldman, L., & Schafer, A. I. (2012). *Goldmans's Cecil medicine* (24th ed.). Philadelphia: Saunders Elsevier.

Karch, A. M. (2012). *2012 Lippincott's nursing drug guide*. Philadelphia: Lippincott Williams & Wilkins.

Marx, J. A. (2009). *Rosen's emergency medicine: Concepts and clinical practice*. Philadelphia: Mosby Elsevier.

McDonald, J., Burroughs, A., Feagan, B., et al. (Eds.). (2010). *Evidence-based gastroenterology & hepatology* (3rd ed.). Malden, MA: Blackwell.

McPherson, R. A., & Pincus, M. R. (2011). *Henry's clinical diagnosis and management by laboratory methods* (22nd ed.). Philadelphia: Elsevier.

Miller, R. D., Eriksson, L. I., Fleisher, L. E., et al. (2009). *Miller's anesthesia*. Edinburgh, UK: Churchill Livingstone.

Mulholland, D. M., Lillemoe, K., Doherty, G., et al. (2011). *Greenfield's surgery* (5th ed.). Philadelphia: Lippincott Williams & Wilkins.

Porth, C. M., & Matfin, G. (2009). *Pathophysiology: Concepts of altered health states* (8th ed.). Philadelphia: Lippincott Williams & Wilkins.

Rakel, R. E., & Rakel, D. P. (Eds.). (2011). *Textbook of family medicine* (8th ed.). Philadelphia: Saunders Elsevier.

Townsend, C. M., Beauchamp, D., Evers, M., et al. (2012). *Sabiston's textbook of surgery: The biologic basis of modern surgical practice*. Philadelphia: Saunders Elsevier.

Yamada, T. (2009). *Textbook of gastroenterology* (5th ed.). Hoboken, NJ: Wiley-Blackwell.

Journals and Electronic Documents

Gallbladder Disease

Barie, P. S. (2010). Acute acalculous cholecystitis. *Gastroenterology Clinics of North America*, 39(2), 343–357.

DiCiaula, A., Wang, D. Q., Wang, H. H., et al. (2010). Targets for current pharmacologic therapy in cholesterol gallstone disease. *Gastroenterology Clinics of North America*, 39(2), 245–264.

McKay, A., Abulfaraj, M., & Lipschitz, J. (2012). Short and long-term outcomes following percutaneous cholecystostomy for acute cholecystitis in high-risk patients. *Surgical Endoscopy*, 26(5), 1343–1351.

Privette, T. W., Carlisle, M. C., & Palma, J. K. (2011). Emergencies of the liver, gallbladder and pancreas. *Emergency Medical Clinics of North America*, 29(2), 293–317.

*Ren, R., Zhang, C., & Shang, X. (2008). Influence of early clamping of T-tube on biliary tract function of patients. *Chinese Nursing Research*, 22(2A), 294–296.

Pancreatic Disorders

Affronti, J. (2011). Chronic pancreatitis and exocrine insufficiency. *Primary care: Clinics in Office Practice*, *38*(3), 515–537.

American Cancer Society. (2012). *Cancer facts & figures 2012*. Available at: www.cancer.org/Research/CancerFactsFigures

Anand, N., Park, J. H., & Wu, B. U. (2012). Modern management of acute pancreatitis. *Gastroenterology Clinics of North America*, *41*(1), 1–8.

Griesdale, D. E. (2009). Intensive insulin therapy and mortality among critically ill patients: A meta-analysis including NICE-SUGAR. *Canadian Medical Association Journal*, *180*(8), 821–827.

Joseph, S., Wang, Y. Z., Boudreaux, J. P., et al. (2010). Neuroendocrine tumors: Current recommendations for diagnosis and surgical management. *Endocrinology and Metabolism Clinics of North America*, *40*(1), 205–231.

Othman, M. O., & Wallace, M. B. (2012). The role of endoscopic ultrasonography in the diagnosis and management of pancreatic cancer. *Gastroenterology Clinics of North America*, *41*(1), 179–188.

**Ranson, J. H., Rifkind, K. M., Roses, D. F., et al. (1974). Prognostic signs and the role of operative management in acute pancreatitis. *Surgery, Gynecology & Obstetrics*, *139*(1), 69–81.

Talukdar, R., Clemens, M., & Vege, S. S. (2012). Moderately severe acute pancreatitis. *Pancreas*, *41*(2), 306–309.

Trikudanathan, G., Navaneethan, U., & Vege, S. S. (2012). Modern treatment of patients with chronic pancreatitis. *Gastroenterology Clinics of North America*, *41*(1), 63–76.

Wu, B., & Conwell, D. L. (2010). Update in acute pancreatitis. *Current Gastroenterology Report*, *12*(2), 83–90.

Yadav, D., Timmons, L., Benson, J. T., et al. (2011). Incidence, prevalence, and survival of chronic pancreatitis: A population-based study. *American Journal of Gastroenterology*, *106*(12), 2192–2199.

Resources

American Gastroenterological Association (AGA), www.gastro.org

Endocrine Society, www.endo-society.org

National Digestive Diseases Information Clearinghouse (NDDIC), digestive.niddk.nih.gov

National Pancreas Foundation (NPF), www.pancreasfoundation.org

Chapter **51**

Assessment and Management of Patients With Diabetes

Learning Objectives

On completion of this chapter, the learner will be able to:

1. Differentiate between type 1 and type 2 diabetes.
2. Describe etiologic factors associated with diabetes.
3. Relate the clinical manifestations of diabetes to the associated pathophysiologic alterations.
4. Identify the diagnostic and clinical significance of blood glucose test results.
5. Explain the dietary modifications used for management of people with diabetes.
6. Describe the relationships among diet, exercise, and medication (i.e., insulin or oral antidiabetic agents) for people with diabetes.
7. Develop an education plan for insulin self-management.
8. Identify the role of oral antidiabetic agents in therapy for patients with diabetes.
9. Use the nursing process as a framework for care of patients who have hyperglycemia with diabetic ketoacidosis or hyperglycemic hyperosmolar syndrome.
10. Describe management strategies for a person with diabetes to use during "sick days."
11. Describe the major macrovascular, microvascular, and neuropathic complications of diabetes and the self-care behaviors that are important in their prevention.
12. Identify the programs and community support groups available for people with diabetes.

Glossary

continuous glucose monitoring system (CGMS): a device used to continuously monitor blood glucose levels

continuous subcutaneous insulin infusion, insulin pump: a small device that delivers insulin on a 24-hour basis as basal insulin

diabetes: a group of metabolic diseases characterized by hyperglycemia resulting from defects in insulin secretion, insulin action, or both

diabetic ketoacidosis (DKA): a metabolic derangement in type 1 diabetes that results from a deficiency of insulin; highly acidic ketone bodies are formed, resulting in acidosis

fasting plasma glucose (FPG): blood glucose determination obtained in the laboratory after fasting for at least 8 hours

gestational diabetes: any degree of glucose intolerance with its onset during pregnancy

glycated hemoglobin (glycosylated hemoglobin, Hgb A_{1C} or A1C): a measure of glucose control that is a result of glucose molecule attaching to hemoglobin for the life of the red blood cell (120 days)

glycemic index: the amount a given food increases the blood glucose level compared with an equivalent amount of glucose

hyperglycemia: elevated blood glucose level

hyperglycemic hyperosmolar syndrome (HHS): a metabolic disorder of type 2 diabetes resulting from a relative insulin deficiency initiated by an illness that raises the demand for insulin

hypoglycemia: low blood glucose level

impaired fasting glucose (IFG), impaired glucose tolerance (IGT): a metabolic stage intermediate between normal glucose homeostasis and diabetes; referred to as prediabetes

insulin: a hormone secreted by the beta cells of the islets of Langerhans of the pancreas that is necessary for the metabolism of carbohydrates, proteins, and fats; a deficiency of insulin results in diabetes

ketone: a highly acidic substance formed when the liver breaks down free fatty acids in the absence of insulin

medical nutrition therapy (MNT): nutritional therapy prescribed for management of diabetes that usually is administered by a registered dietician

nephropathy: a long-term complication of diabetes in which the kidney cells are damaged; characterized by microalbuminuria in early stages and progressing to end-stage kidney disease

neuropathy: a long-term complication of diabetes resulting from damage to the nerve cell

prediabetes: impaired glucose metabolism in which blood glucose concentrations fall between normal levels and those considered diagnostic for diabetes; includes impaired fasting glucose and impaired glucose tolerance, not clinical entities in their own right but risk factors for future diabetes and cardiovascular disease

retinopathy: a complication of diabetes in which the small blood vessels that nourish the retina in the eye are damaged

self-monitoring of blood glucose (SMBG): a method of capillary blood glucose testing

sulfonylureas: a class of oral antidiabetic medication for treating type 2 diabetes; stimulates insulin secretion and insulin action

thiazolidinediones: a class of oral antidiabetic medications that reduces insulin resistance in target tissues, enhancing insulin action without directly stimulating insulin secretion

type 1 diabetes: a metabolic disorder characterized by an absence of insulin production and secretion from autoimmune destruction of the beta cells of the islets of Langerhans in the pancreas; formerly called *insulin-dependent diabetes, juvenile diabetes,* or *type I diabetes*

type 2 diabetes: a metabolic disorder characterized by the relative deficiency of insulin production and a decreased insulin action and increased insulin resistance; formerly called *non–insulin-dependent diabetes, adult-onset diabetes,* or *type II diabetes*

Diabetes is a group of metabolic diseases characterized by increased levels of glucose in the blood **(hyperglycemia)** resulting from defects in insulin secretion, insulin action, or both (American Diabetes Association [ADA], 2013). Care of the patient with diabetes, formerly known as diabetes mellitus but now more commonly referred to as diabetes, requires an understanding of the epidemiology, pathophysiology, diagnostic testing, medical and nursing care, and rehabilitation of patients with diabetes. Nurses care for patients with diabetes in all settings. This chapter focuses on the nursing management of patients with diabetes.

DIABETES

Epidemiology

It is estimated that more than 25.8 million people in the United States have diabetes, although almost one third of these cases are undiagnosed. In 2010, the number of people older than 20 years who were newly diagnosed with diabetes increased by about 1.9 million (Centers for Disease Control and Prevention [CDC], 2011). By 2030, the number of cases is expected to exceed 30 million. In 2000, the worldwide estimate of the prevalence of diabetes was 171 million people, and by 2030, this is expected to increase to more than 360 million (World Health Organization [WHO], 2012). Diabetes is especially prevalent in older adults.

Minority populations are disproportionately affected by diabetes. The age-adjusted prevalence of diabetes is increasing among all gender and race groups, but compared to Caucasians, African Americans and members of other racial and ethnic groups (Native Americans and persons of Hispanic origin) are more likely to develop diabetes, are at greater risk for many of the complications, and have higher death rates due to diabetes (CDC, 2011). Chart 51-1 summarizes risk factors for diabetes.

Diabetes can have far-reaching and devastating physical, social, and economic consequences, including the following (CDC, 2011):

- In the United States, diabetes is the leading cause of nontraumatic amputations, blindness in working-age adults, and end-stage kidney disease (ESKD).
- Diabetes is a leading cause of death from disease, primarily because of the high rate of cardiovascular disease (myocardial infarction [MI], stroke, and peripheral vascular disease) among people with diabetes.
- Hospitalization rates for people with diabetes are 2.4 times greater for adults and 5.3 times greater for children than for the general population.

The economic cost of diabetes continues to increase because of increasing health care costs and an aging population. The U.S. Department of Health and Human Services (HHS) has identified diabetes as an important topic area. Nationwide efforts are needed to decrease its occurrence and to increase the quality of life of those people who have the disease (HHS, 2010).

Classification

The major classifications of diabetes are type 1 diabetes, type 2 diabetes, gestational diabetes, and diabetes associated with other conditions or syndromes (ADA, 2009). The different types of diabetes vary in cause, clinical course, and treatment (Table 51-1). The classification system is dynamic in two ways. First, research findings suggest many differences among individuals within each category. Second, except for people with type 1 diabetes, patients may move from one category to another. For example, a woman with gestational diabetes may, after delivery, move into the type 2 category. **Prediabetes** is classified as **impaired glucose tolerance (IGT)** or **impaired fasting glucose (IFG)** and refers to a condition in which blood glucose concentrations fall between normal levels and those considered diagnostic for diabetes (CDC, 2011).

Pathophysiology

CONCEPTS IN ACTION

Insulin is a hormone secreted by beta cells, which are one of four types of cells in the islets of Langerhans in the pancreas (Porth & Matfin, 2009). Insulin is an anabolic, or storage, hormone. When a person eats a meal, insulin secretion increases and moves glucose from the blood into muscle, liver, and fat cells. In those cells, insulin has the following actions:

- Transports and metabolizes glucose for energy
- Stimulates storage of glucose in the liver and muscle (in the form of glycogen)
- Signals the liver to stop the release of glucose

Chart 51-1 RISK FACTORS
Diabetes

- Family history of diabetes (e.g., parents or siblings with diabetes)
- Obesity (i.e., ≥20% over desired body weight or body mass index ≥30 kg/m²)
- Race/ethnicity (e.g., African Americans, Hispanic Americans, Native Americans, Asian Americans, Pacific Islanders)
- Age ≥45 years
- Previously identified impaired fasting glucose or impaired glucose tolerance
- Hypertension (≥140/90 mm Hg)
- High-density lipoprotein (HDL) cholesterol level ≤35 mg/dL (0.90 mmol/L) and/or triglyceride level ≥250 mg/dL (2.8 mmol/L)
- History of gestational diabetes or delivery of a baby over 9 lb

Used with permission of American Diabetes Association. (2009). Diagnosis and classification of diabetes mellitus. *Diabetes Care, 32*(Suppl. 1), S62–S67.

TABLE 51-1 Classification of Diabetes and Related Glucose Intolerances

Current Classification	Clinical Characteristics and Clinical Implications
Type 1 (5%–10% of all diabetes; previously classified as juvenile diabetes, juvenile-onset diabetes, ketosis-prone diabetes, brittle diabetes, and insulin-dependent diabetes mellitus [IDDM])	Onset any age, but usually young (<30 y) Usually thin at diagnosis; recent weight loss Etiology includes genetic, immunologic, and environmental factors (e.g., virus) Often have islet cell antibodies Often have antibodies to insulin even before insulin treatment Little or no endogenous insulin Need exogenous insulin to preserve life Ketosis prone when insulin absent Acute complication of hyperglycemia: diabetic ketoacidosis
Type 2 (90%–95% of all diabetes: obese—80% of type 2, nonobese—20% of type 2; previously classified as adult-onset diabetes, maturity-onset diabetes, ketosis-resistant diabetes, stable diabetes, and non–insulin-dependent diabetes mellitus [NIDDM])	Onset any age, usually >30 y Usually obese at diagnosis Causes include obesity, heredity, and environmental factors No islet cell antibodies Decrease in endogenous insulin, or increased with insulin resistance Most patients can control blood glucose through weight loss if obese Oral antidiabetic agents may improve blood glucose levels if dietary modification and exercise are unsuccessful May need insulin on a short- or long-term basis to prevent hyperglycemia Ketosis uncommon, except in stress or infection Acute complication: hyperglycemic hyperosmolar syndrome
Diabetes associated with other conditions or syndromes (previously classified as secondary diabetes)	Accompanied by conditions known or suspected to cause the disease: pancreatic diseases, hormonal abnormalities, medications such as corticosteroids and estrogen-containing preparations Depending on the ability of the pancreas to produce insulin, the patient may require treatment with oral antidiabetic agents or insulin.
Gestational diabetes	Onset during pregnancy, usually in the 2nd or 3rd trimester Due to hormones secreted by the placenta, which inhibit the action of insulin Above-normal risk for perinatal complications, especially macrosomia (abnormally large babies) Treated with diet and, if needed, insulin to strictly maintain normal blood glucose levels Occurs in about 2%–5% of all pregnancies Glucose intolerance transitory but may recur: • In subsequent pregnancies • 30%–40% will develop overt diabetes (usually type 2) within 10 years (especially if obese) Risk factors include obesity, age >30 y, family history of diabetes, previous large babies (>9 lb) Screening tests (glucose challenge test) should be performed on all pregnant women between 24 and 28 weeks of gestation Should be screened for diabetes periodically
Prediabetes (previously classified as previous abnormality of glucose tolerance [PrevAGT])	Previous history of hyperglycemia (e.g., during pregnancy or illness) Current normal glucose metabolism Impaired glucose tolerance or impaired fasting glucose screening after age 40 years if there is a family history of diabetes or if symptomatic Encourage ideal body weight, because loss of 10–15 lb may improve glycemic control

- Enhances storage of dietary fat in adipose tissue
- Accelerates transport of amino acids (derived from dietary protein) into cells
- Insulin also inhibits the breakdown of stored glucose, protein, and fat

During fasting periods (between meals and overnight), the pancreas continuously releases a small amount of insulin (basal insulin); another pancreatic hormone called *glucagon* (secreted by the alpha cells of the islets of Langerhans) is released when blood glucose levels decrease, which stimulates the liver to release stored glucose. The insulin and the glucagon together maintain a constant level of glucose in the blood by stimulating the release of glucose from the liver.

Initially, the liver produces glucose through the breakdown of glycogen (glycogenolysis). After 8 to 12 hours without food, the liver forms glucose from the breakdown of noncarbohydrate substances, including amino acids (gluconeogenesis).

Type 1 Diabetes

Type 1 diabetes affects approximately 5% to 10% of people with the disease (CDC, 2011). It is characterized by destruction of the pancreatic beta cells (Porth & Matfin, 2009). Combined genetic, immunologic, and possibly environmental (e.g., viral) factors are thought to contribute to beta-cell destruction. Although the events that lead to beta-cell destruction are not fully understood, it is generally accepted that a genetic susceptibility is a common underlying factor in the development of type 1 diabetes. People do not inherit type 1 diabetes itself but rather a genetic predisposition, or tendency, toward development of type 1 diabetes. This genetic tendency has been found in people with certain human leukocyte antigen types. There is also evidence of an autoimmune response in type 1 diabetes. This is an abnormal response in which antibodies are directed against normal tissues of the body, responding to these tissues as if they were foreign. Autoantibodies against islet cells and against endogenous (internal) insulin have been detected in people at the time of diagnosis and even several years before the development of clinical signs of type 1 diabetes. In addition to genetic and immunologic components, environmental factors, such as viruses or toxins, that may initiate destruction of the beta cell are being investigated.

Regardless of the specific cause, the destruction of the beta cells results in decreased insulin production, unchecked glucose production by the liver, and fasting hyperglycemia. In addition, glucose derived from food cannot be stored in the liver but instead remains in the bloodstream and contributes to postprandial (after meals) hyperglycemia. If the concentration of glucose in the blood exceeds the renal threshold for glucose, usually 180 to 200 mg/dL (9.9 to 11.1 mmol/L), the kidneys may not reabsorb all of the filtered glucose; the glucose then appears in the urine (glycosuria). When excess glucose is excreted in the urine, it is accompanied by excessive loss of fluids and electrolytes. This is called *osmotic diuresis*.

Because insulin normally inhibits glycogenolysis (breakdown of stored glucose) and gluconeogenesis (production of new glucose from amino acids and other substrates), these processes occur in an unrestrained fashion in people with insulin deficiency and contribute further to hyperglycemia. In addition, fat breakdown occurs, resulting in an increased production of **ketone** bodies, a highly acidic substance formed when the liver breaks down free fatty acids in the absence of insulin.

Diabetic ketoacidosis (DKA) is a metabolic derangement that occurs most commonly in persons with type 1 diabetes and results from a deficiency of insulin; highly acidic ketone bodies are formed, and metabolic acidosis occurs. The three major metabolic derangements are hyperglycemia, ketosis, and metabolic acidosis (Porth & Matfin, 2009). DKA is commonly preceded by a day or more of polyuria, polydipsia, nausea, vomiting, and fatigue with eventual stupor and coma if not treated. The breath has a characteristic fruity odor due to the presence of ketoacids.

Type 2 Diabetes

Type 2 diabetes affects approximately 90% to 95% of people with the disease (CDC, 2011). It occurs more commonly among people who are older than 30 years and obese, although its incidence is rapidly increasing in younger people because of the growing epidemic of obesity in children, adolescents, and young adults (CDC, 2011).

The two main problems related to insulin in type 2 diabetes are insulin resistance and impaired insulin secretion. Insulin resistance refers to a decreased tissue sensitivity to insulin. Normally, insulin binds to special receptors on cell surfaces and initiates a series of reactions involved in glucose metabolism. In type 2 diabetes, these intracellular reactions are diminished, making insulin less effective at stimulating glucose uptake by the tissues and at regulating glucose release by the liver (Fig. 51-1). The exact mechanisms that lead to insulin resistance and impaired insulin secretion in type 2 diabetes are unknown, although genetic factors are thought to play a role.

To overcome insulin resistance and to prevent the buildup of glucose in the blood, increased amounts of insulin must be secreted to maintain the glucose level at a normal or slightly elevated level. If the beta cells cannot keep up with the increased demand for insulin, the glucose level rises and type 2 diabetes develops. Insulin resistance may also lead to metabolic syndrome, which is a constellation of symptoms including hypertension, hypercholesterolemia, abdominal obesity, and other abnormities (Porth & Matfin, 2009).

Despite the impaired insulin secretion that is characteristic of type 2 diabetes, there is enough insulin present to prevent the breakdown of fat and the accompanying production of ketone bodies. Therefore, DKA does not typically occur in type 2 diabetes. However, uncontrolled type 2 diabetes may lead to another acute problem—hyperglycemic hyperosmolar syndrome (HHS) (see later discussion).

Physiology ⟷ Pathophysiology

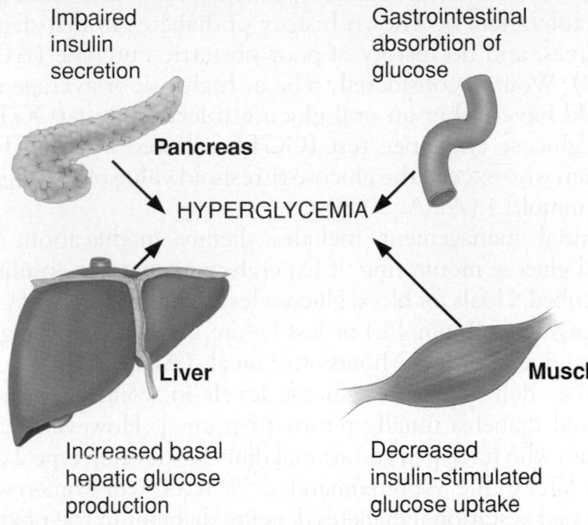
Impaired insulin secretion
Gastrointestinal absorbtion of glucose
Pancreas
HYPERGLYCEMIA
Liver
Muscle
Increased basal hepatic glucose production
Decreased insulin-stimulated glucose uptake

FIGURE 51-1 • Pathogenesis of type 2 diabetes.

Because type 2 diabetes is associated with a slow, progressive glucose intolerance, its onset may go undetected for many years. If the patient experiences symptoms, they are frequently mild and may include fatigue, irritability, polyuria, polydipsia, poorly healing skin wounds, vaginal infections, or blurred vision (if glucose levels are very high).

For most patients (approximately 75%), type 2 diabetes is detected incidentally (e.g., when routine laboratory tests or ophthalmoscopic examinations are performed). One consequence of undetected diabetes is that long-term diabetes complications (e.g., eye disease, peripheral neuropathy, peripheral vascular disease) may have developed before the actual diagnosis of diabetes is made (ADA, 2009), signifying that the blood glucose has been elevated for a time before diagnosis.

Gestational Diabetes

Gestational diabetes is any degree of glucose intolerance with its onset during pregnancy. Hyperglycemia develops during pregnancy because of the secretion of placental hormones, which causes insulin resistance. Gestational diabetes occurs in as many as 18% of pregnant women and increases their risk for hypertensive disorders during pregnancy (CDC, 2011).

Women who are considered to be at high risk for gestational diabetes and should be screened by blood glucose testing at their first prenatal visit are those with marked obesity, a personal history of gestational diabetes, glycosuria, or a strong family history of diabetes. High-risk ethnic groups include Hispanic Americans, Native Americans, Asian Americans, African Americans, and Pacific Islanders. If these high-risk women do not have gestational diabetes at initial screening, they should be retested between 24 and 28 weeks of gestation. All women of average risk should be tested at 24 to 28 weeks of gestation. Testing is not specifically recommended for women identified as being at low risk. Low-risk women are those who meet all of the following criteria: age younger than 25 years, normal weight before pregnancy, member of an ethnic group with low prevalence of gestational diabetes, no history of abnormal glucose tolerance, no known history of diabetes in first-degree relatives, and no history of poor obstetric outcome (ADA, 2009). Women considered to be at high risk or average risk should have either an oral glucose tolerance test (OGTT) or a glucose challenge test (GCT) followed by OGTT in women who exceed the glucose threshold value of 140 mg/dL (7.8 mmol/L) (ADA, 2009).

Initial management includes dietary modification and blood glucose monitoring. If hyperglycemia persists, insulin is prescribed. Goals for blood glucose levels during pregnancy are 105 mg/dL (5.8 mmol/L) or less before meals and 130 mg/dL (7.2 mmol/L) or less 2 hours after meals (ADA, 2009).

After delivery, blood glucose levels in women with gestational diabetes usually return to normal. However, many women who have had gestational diabetes develop type 2 diabetes later in life. Approximately 35% to 60% of women who have had gestational diabetes develop diabetes in the next 10 to 20 years (CDC, 2011).

Prevention

The Diabetes Prevention Program Research Group (2002) reported that type 2 diabetes can be prevented with appropriate changes in lifestyle. Persons at high risk for type 2 diabetes (body mass index [BMI] greater than 24, fasting and postprandial plasma glucose levels elevated but not to levels diagnostic of diabetes) received standard lifestyle recommendations plus metformin (Glucophage), an oral antidiabetic agent, standard lifestyle recommendations plus placebo, or an intensive program of lifestyle modifications. The 16-lesson curriculum of the intensive program of lifestyle modifications focused on weight reduction of greater than 7% of initial body weight and physical activity of moderate intensity. It also included behavior modification strategies designed to help patients achieve the goals of weight reduction and participation in exercise. Compared to the placebo group, the lifestyle intervention group had a 58% lower incidence of diabetes and the metformin group had a 31% lower incidence of diabetes. These findings were found in both genders and all racial and ethnic groups. These findings demonstrate that type 2 diabetes can be prevented or delayed in persons at high risk for the disease (Diabetes Prevention Program Research Group, 2002).

Clinical Manifestations

Clinical manifestations depend on the patient's level of hyperglycemia. Classic clinical manifestations of diabetes include the "three Ps": polyuria, polydipsia, and polyphagia. Polyuria (increased urination) and polydipsia (increased thirst) occur as a result of the excess loss of fluid associated with osmotic diuresis. Patients also experience polyphagia (increased appetite) that results from the catabolic state induced by insulin deficiency and the breakdown of proteins and fats (Porth & Matfin, 2009). Other symptoms include fatigue and weakness, sudden vision changes, tingling or numbness in hands or feet, dry skin, skin lesions or wounds that are slow to heal, and recurrent infections. The onset of type 1 diabetes may also be associated with sudden weight loss or nausea, vomiting, or abdominal pains, if DKA has developed.

Chart 51-2 Criteria for the Diagnosis of Diabetes

1. Symptoms of diabetes plus casual plasma glucose concentration equal to or greater than 200 mg/dL (11.1 mmol/L). Casual is defined as any time of day without regard to time since last meal. The classic symptoms of diabetes include polyuria, polydipsia, and unexplained weight loss.

Or

2. Fasting plasma glucose greater than or equal to 126 mg/dL (7.0 mmol/L). Fasting is defined as no caloric intake for at least 8 hours.

Or

3. Two-hour postload glucose equal to or greater than 200 mg/dL (11.1 mmol/L) during an oral glucose tolerance test. The test should be performed as described by the World Health Organization, using a glucose load containing the equivalent of 75 g anhydrous glucose dissolved in water.

In the absence of unequivocal hyperglycemia with acute metabolic decompensation, these criteria should be confirmed by repeat testing on a different day. The third measure is not recommended for routine clinical use.

Used with permission of American Diabetes Association. (2009). Report of the expert committee on the diagnosis and classification of diabetes mellitus. *Diabetes Care, 32*(Suppl. 1), S62–S67.

Assessment and Diagnostic Findings

An abnormally high blood glucose level is the basic criterion for the diagnosis of diabetes. **Fasting plasma glucose (FPG)** (blood glucose determination obtained in the laboratory after fasting for at least 8 hours), random plasma glucose, and glucose level 2 hours after receiving glucose (2-hour postload) may be used (Fischbach & Dunning, 2009). The OGTT and the intravenous (IV) glucose tolerance test are no longer recommended for routine clinical use. See Chart 51-2 for the ADA's diagnostic criteria for diabetes (ADA, 2013).

In addition to the assessment and diagnostic evaluation performed to diagnose diabetes, ongoing specialized assessment of patients with known diabetes and evaluation for complications in patients with newly diagnosed diabetes are important components of care. Parameters that should be regularly assessed are discussed in Chart 51-3.

Gerontologic Considerations

Type 2 diabetes is the seventh leading cause of death and affects approximately 20% of older adults (Eliopoulos, 2010). There is a high prevalence among African Americans and those who are 65 to 74 years of age.

Early detection is important but may be challenging because symptoms may be absent or nonspecific. A glucose tolerance test is more effective in diagnosis than urine testing for glucose in older patients due to the higher renal threshold for glucose (Eliopoulos, 2010).

Medical Management

The main goal of diabetes treatment is to normalize insulin activity and blood glucose levels to reduce the development of vascular and neuropathic complications. The Diabetes Control and Complications Trial (DCCT), a 10-year prospective clinical trial conducted from 1983 to 1993, demonstrated the importance of achieving blood glucose control

Chart 51-3 ASSESSMENT

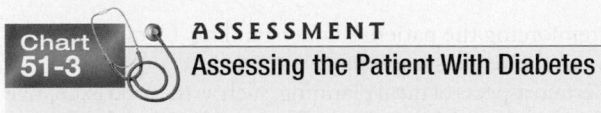

Assessing the Patient With Diabetes

History

- Symptoms related to the diagnosis of diabetes:
  - Symptoms of hyperglycemia
  - Symptoms of hypoglycemia
    - Frequency, timing, severity, and resolution
- Results of blood glucose monitoring
- Status, symptoms, and management of chronic complications of diabetes:
  - Eye; kidney; nerve; genitourinary and sexual, bladder, and gastrointestinal
  - Cardiac; peripheral vascular; foot complications associated with diabetes
- Adherence to/ability to follow prescribed dietary management plan
- Adherence to prescribed exercise regimen
- Adherence to/ability to follow prescribed pharmacologic treatment (insulin or oral antidiabetic agents)
- Use of tobacco, alcohol, and prescribed and over-the-counter medications/drugs
- Lifestyle, cultural, psychosocial, and economic factors that may affect diabetes treatment
- Effects of diabetes or its complications on functional status (e.g., mobility, vision)

Physical Examination

- Blood pressure (sitting and standing to detect orthostatic changes)
- Body mass index (height and weight)
- Funduscopic examination and visual acuity
- Foot examination (lesions, signs of infection, pulses)
- Skin examination (lesions and insulin injection sites)
- Neurologic examination
  - Vibratory and sensory examination using monofilament
  - Deep tendon reflexes
- Oral examination

Laboratory Examination

- Hgb A_{1C} (A1C)
- Fasting lipid profile
- Test for microalbuminuria
- Serum creatinine level
- Urinalysis
- Electrocardiogram

Need for Referrals

- Ophthalmologist
- Podiatrist
- Dietitian
- Diabetes educator
- Others if indicated

in the normal, nondiabetic range. This landmark trial demonstrated that intensive glucose control dramatically reduced the development and progression of complications such as **retinopathy** (small blood vessels that nourish the retina in the eye are damaged), **nephropathy** (the kidney cells are damaged), and **neuropathy** (nerve cells are damaged). Intensive treatment is defined as three or four insulin injections per day or **continuous subcutaneous insulin infusion, insulin pump** therapy plus frequent blood glucose monitoring and weekly contacts with diabetes educators. The ADA now recommends that all patients with diabetes strive for glucose control (Hgb A1c less than 7%) to reduce their risk of complications (ADA, 2013).

Intensive therapy must be initiated with caution and must be accompanied by thorough education of the patient and family and by responsible behavior of the patient. Careful screening of patients for capability and responsibility is a key step in initiating intensive therapy.

The therapeutic goal for diabetes management is to achieve normal blood glucose levels (euglycemia) without hypoglycemia while maintaining a high quality of life. Diabetes management has five components: nutritional therapy, exercise, monitoring, pharmacologic therapy, and education. Diabetes management involves constant assessment and modification of the treatment plan by health professionals and daily adjustments in therapy by the patient. Although the health care team directs the treatment, it is the individual patient who must manage the complex therapeutic regimen. For this reason, patient and family education is an essential component of diabetes treatment and is as important as all other components of the regimen.

Nutritional Therapy

Nutrition, meal planning, weight control, and increased activity are the foundation of diabetes management (ADA, 2013). The most important objectives in the dietary and nutritional management of diabetes are control of total caloric intake to attain or maintain a reasonable body weight, control of blood glucose levels, and normalization of lipids and blood pressure to prevent heart disease. Success in this area alone is often associated with reversal of hyperglycemia in type 2 diabetes. However, achieving these goals is not always easy. Because **medical nutrition therapy (MNT)**—nutritional therapy prescribed for management of diabetes usually administered by a registered dietician—is complex, a registered dietitian who understands the therapy has the major responsibility for designing and educating about this aspect of the therapeutic plan. Nurses and all other members of the health care team must be knowledgeable about nutritional therapy and supportive of patients who need to implement nutritional and lifestyle changes. Nutritional management of diabetes includes the following goals:

1. To achieve and maintain:
   a. Blood glucose levels in the normal range or as close to normal as is safely possible
   b. A lipid and lipoprotein profile that reduces the risk for vascular disease
   c. Blood pressure levels in the normal range or as close to normal as is safely possible
2. To prevent, or at least slow, the rate of development of the chronic complications of diabetes by modifying nutrient intake and lifestyle
3. To address individual nutrition needs, taking into account personal and cultural preferences and willingness to change
4. To maintain the pleasure of eating by only limiting food choices when indicated by scientific evidence

For patients who are obese and have diabetes (especially those with type 2 diabetes), weight loss is the key to treatment. (It is also a major factor in preventing diabetes.)

In general, overweight is considered to be a BMI of 25 to 29; obesity is defined as 20% above ideal body weight or a BMI equal to or greater than 30 (WHO, 2011). Calculation of BMI is discussed in Chapter 5. Patients who are obese, have type 2 diabetes, and require insulin or oral agents to control blood glucose levels may be able to reduce or eliminate the need for medication through weight loss. A weight loss as small as 5% to 10% of total weight may significantly improve blood glucose levels. For patients who are obese, have diabetes, and do not take insulin or **sulfonylureas** (a class of oral antidiabetic medication for treating type 2 diabetes that stimulates insulin secretion and insulin action), consistent meal content or timing is important but not as critical. Rather, decreasing the overall caloric intake assumes more importance. However, meals should not be skipped. Pacing food intake throughout the day places more manageable demands on the pancreas.

Consistently following a meal plan is one of the most challenging aspects of diabetes management. It may be more realistic to restrict calories only moderately. For patients who have lost weight incorporating new dietary habits into their lifestyles, diet education, behavioral therapy, group support, and ongoing nutrition counseling are encouraged to maintain weight loss.

Meal Planning and Related Education

The meal plan must consider the patient's food preferences, lifestyle, usual eating times, and ethnic and cultural background. For patients who require insulin to help control blood glucose levels, maintaining as much consistency as possible in the amount of calories and carbohydrates ingested at each meal is essential. In addition, consistency in the approximate time intervals between meals, with the addition of snacks if necessary, helps prevent hypoglycemic reactions and maintain overall blood glucose control. For patients who can master the insulin-to-carbohydrate calculations, lifestyle can be more flexible and diabetes control more predictable. For those using intensive insulin therapy, there may be greater flexibility in the timing and content of meals by allowing adjustments in insulin dosage for changes in eating and exercise habits. Advances in insulin management (new insulin analogues, insulin algorithms, insulin pumps) permit greater flexibility of schedules than was previously possible. This contrasts with the concept of maintaining a constant dose of insulin, requiring strict scheduling of meals to match the onset and duration of the insulin.

The first step in preparing a meal plan is a thorough review of the patient's diet history to identify eating habits and lifestyle and cultural eating patterns (Welch, Allen, Zagarins, et al., 2011). This includes a thorough assessment of the patient's need for weight loss, gain, or maintenance. In most instances, people with type 2 diabetes require weight reduction.

In educating about meal planning, clinical dietitians use various tools, materials, and approaches. Initial education addresses the importance of consistent eating habits, the relationship of food and insulin, and the provision of an individualized meal plan. In-depth follow-up education then focuses on management skills, such as eating at restaurants, reading food labels, and adjusting the meal plan for exercise, illness, and special occasions. The nurse plays an important role in communicating pertinent information to the dietitian and reinforcing the patient's understanding. Communication between the team is important.

Certain aspects of meal planning, such as the food exchange system, may be difficult to learn. This may be related to limitations in the patient's intellectual level or to emotional issues, such as difficulty accepting the diagnosis of diabetes or feelings of deprivation and undue restriction in eating. In any case, it helps to emphasize that using the exchange system (or any food classification system) provides a new way of thinking about food rather than a new way of eating. It is also important to simplify information as much as possible and to provide opportunities for the patient to practice and repeat activities and information.

Caloric Requirements. Calorie-controlled diets are planned by first calculating a person's energy needs and caloric requirements based on age, gender, height, and weight. An activity element is then factored in to provide the actual number of calories required for weight maintenance. To promote a 1- to 2-pound weight loss per week, 500 to 1,000 calories are subtracted from the daily total. The calories are distributed into carbohydrates, proteins, and fats, and a meal plan is then developed, taking into account the patient's lifestyle and food preferences.

Patients may be underweight at the onset of type 1 diabetes because of rapid weight loss from severe hyperglycemia. The goal initially may be to provide a higher-calorie diet to regain lost weight and blood glucose control.

Caloric Distribution. A meal plan for diabetes focuses on the percentages of calories that come from carbohydrates, proteins, and fats.

Carbohydrates. The caloric distribution currently recommended is higher in carbohydrates than in fat and protein. In general, carbohydrate foods have the greatest effect on blood glucose levels because they are more quickly digested than other foods and are converted into glucose rapidly. However, research into the appropriateness of a higher-carbohydrate diet in patients with decreased glucose tolerance is ongoing, and recommendations may change accordingly. Currently, the ADA and the Academy of Nutrition and Dietetics (formerly the American Dietetic Association) recommend that for all levels of caloric intake, 50% to 60% of calories should be derived from carbohydrates, 20% to 30% from fat, and the remaining 10% to 20% from protein (Dudek, 2010). The majority of the selections for carbohydrates should come from whole grains. These recommendations are also consistent with those of the American Heart Association and American Cancer Society.

Carbohydrates consist of sugars (e.g., sucrose) and starches (e.g., rice, pasta, bread). Low glycemic index diets (described later) may reduce postprandial glucose levels. Therefore, the nutrition guidelines recommend that all carbohydrates should be eaten in moderation to avoid high postprandial blood glucose levels (Dudek, 2010).

Foods high in carbohydrates, such as sucrose (concentrated sweets), are not totally eliminated from the diet but should be eaten in moderation (up to 10% of total calories), because they are typically high in fat and lack vitamins, minerals, and fiber.

Fats. The recommendations regarding fat content of the diabetic diet include both reducing the total percentage of calories from fat sources to less than 30% of total calories and limiting the amount of saturated fats to 10% of total calories. Additional recommendations include limiting the total intake of dietary cholesterol to less than 300 mg/day. This approach may help reduce risk factors such as increased serum cholesterol levels, which are associated with the development of coronary artery disease—the leading cause of death and disability among people with diabetes (ADA, 2008a).

Protein. The meal plan may include the use of some nonanimal sources of protein (e.g., legumes, whole grains) to help reduce saturated fat and cholesterol intake. In addition, the amount of protein intake may be reduced in patients with early signs of kidney disease.

Fiber. Increased fiber in the diet may improve blood glucose levels, decrease the need for exogenous insulin, and lower total cholesterol and low-density lipoprotein levels in the blood (ADA, 2008b; Geil, 2008).

There are two types of dietary fibers: soluble and insoluble. Soluble fiber—in foods such as legumes, oats, and some fruits—plays more of a role in lowering blood glucose and lipid levels than does insoluble fiber, although the clinical significance of this effect is probably small (ADA, 2008b; Geil, 2008). Soluble fiber slows stomach emptying and the movement of food through the upper digestive tract. The potential glucose-lowering effect of fiber may be caused by the slower rate of glucose absorption from foods that contain soluble fiber. Insoluble fiber is found in whole-grain breads and cereals and in some vegetables. This type of fiber along with soluble fiber increases satiety, which is helpful for weight loss. At least 25 g of fiber should be ingested daily.

One risk involved in suddenly increasing fiber intake is that it may require adjusting the dosage of insulin or oral agents to prevent hypoglycemia. Other problems may include abdominal fullness, nausea, diarrhea, increased flatulence, and constipation if fluid intake is inadequate. If fiber is added to or increased in the meal plan, it should be done gradually and in consultation with a dietitian. Exchange lists (ADA, 2008b; Geil, 2008) serve as an excellent guide for increasing fiber intake. Fiber-rich food choices within the vegetable, fruit, and starch/bread exchanges are highlighted in the lists.

Food Classification Systems. To educate about diet principles and help in meal planning, several systems have been developed in which foods are organized into groups with common characteristics, such as number of calories, composition of foods (i.e., amount of protein, fat, or carbohydrate in the food), or effect on blood glucose levels. Several of these are listed next.

Exchange Lists. A commonly used tool for nutritional management is the exchange lists for meal planning (ADA, 2008b; Geil, 2008). There are six main exchange lists: bread/starch, vegetable, milk, meat, fruit, and fat. Foods within one group (in the portion amounts specified) contain equal numbers of calories and are approximately equal in grams of protein, fat, and carbohydrate. Meal plans can be based on a recommended number of choices from each exchange list. Foods on one list may be interchanged with one another, allowing for variety while maintaining as much consistency as possible in the nutrient content of foods eaten. Table 51-2 presents three sample lunch menus that are interchangeable in terms of carbohydrate, protein, and fat content.

Exchange list information on combination foods such as pizza, chili, and casseroles, as well as convenience foods, desserts, snack foods, and fast foods, is available from the ADA (see the Resources section). Some food manufacturers and restaurants publish exchange lists that describe their products.

Nutrition Labels. Food manufacturers are required to have the nutrition content of foods listed on their packaging, and reading food labels is an important skill for patients to learn and use when food shopping. The label includes information about how many grams of carbohydrate are in a serving of food. This information can be used to determine how much medication is needed. For example, a patient who takes premeal insulin may use the algorithm of 1 unit of insulin for 15 g of carbohydrate. Patients can also be educated to have a "carbohydrate budget" per meal (e.g., 45 to 60 g).

Carbohydrate counting is a nutritional tool used for blood glucose management because carbohydrates are the main nutrients in food that influence blood glucose levels. This method provides flexibility in food choices, can be less complicated to understand than the diabetic food exchange list, and allows more accurate management with multiple daily injections (insulin before each meal). However, if carbohydrate counting is not used with other meal-planning techniques, weight gain can result. A variety of methods are used to count carbohydrates. When developing a diabetic meal plan using carbohydrate counting, all food sources should be considered.

Once digested, 100% of carbohydrates are converted to glucose. Approximately 50% of protein foods (meat, fish, and poultry) are also converted to glucose, and this has minimal effect on blood glucose levels.

TABLE 51-2 Selected Sample Menus From Exchange Lists

Exchanges	Sample Lunch #1	Sample Lunch #2	Sample Lunch #3
2 starch	2 slices bread	Hamburger bun	1 cup cooked pasta
3 meat	2 oz sliced turkey and 1 oz low-fat cheese	3 oz lean beef patty	3 oz boiled shrimp
1 vegetable	Lettuce, tomato, onion	Green salad	½ cup plum tomatoes
1 fat	1 tsp mayonnaise	1 tbsp salad dressing	1 tsp olive oil
1 fruit	1 medium apple	1¼ cup watermelon	1¼ cup fresh strawberries
"Free" items (optional)	Unsweetened iced tea Mustard, pickle, hot pepper	Diet soda 1 tbsp catsup, pickle, onions	Ice water with lemon Garlic, basil

Carbohydrate counting consists of counting grams of carbohydrates as noted by food labels or other sources of information such as the CalorieKing books. If target goals are not reached by counting carbohydrates alone, protein is factored into the calculations. This is especially true if the meal consists only of meat, fish, and no starchy vegetables.

Although carbohydrate counting is now commonly used for blood glucose management with type 1 and type 2 diabetes, it is not a perfect system. All carbohydrates affect the blood glucose level to different degrees, regardless of equivalent serving size (i.e., the glycemic index—see later discussion). When carbohydrate counting is used, reading labels on food items is the key to success. Knowing what the "carbohydrate budget" for the meal is and knowing how many grams of carbohydrate are in a serving of a food, the patient can calculate the amount in one serving.

Healthy Food Choices. An alternative to counting grams of carbohydrate is measuring servings or choices. This method is used more often by people with type 2 diabetes. It is similar to the food exchange list and emphasizes portion control of total servings of carbohydrate at meals and snacks. One carbohydrate serving is equivalent to 15 g of carbohydrate. Examples of one serving are an apple 2 inches in diameter and one slice of bread. Vegetables and meat are counted as one third of a carbohydrate serving. This system works well for those who have difficulty with more complicated systems.

MyPlate Food Guide. The Food Guide (i.e., MyPlate) is another tool used to develop meal plans. It is commonly used for patients with type 2 diabetes who have a difficult time following a calorie-controlled diet. Foods are categorized into five major groups (grains, vegetables, fruits, dairy, and protein), plus fats and oils (see Chapter 5). Foods (grains, fruits, and vegetables) that are lowest in calories and fat and highest in fiber should make up the basis of the diet. For those with diabetes, as well as for the general population, 50% to 60% of the daily caloric intake should be from these three groups. Foods higher in fat (particularly saturated fat) should account for a smaller percentage of the daily caloric intake. Fats, oils, and sweets should be used sparingly to obtain weight and blood glucose control and to reduce the risk for cardiovascular disease. Reliance on MyPlate may result in fluctuations in blood glucose levels, however, because high-carbohydrate foods may be grouped with low-carbohydrate foods. The guide is appropriately used only as a first-step educational tool for patients who are learning how to control food portions and how to identify which foods contain carbohydrate, protein, and fat.

Glycemic Index. One of the main goals of diet therapy in diabetes is to avoid sharp, rapid increases in blood glucose levels after food is eaten. The term **glycemic index** is used to describe how much a given food increases the blood glucose level compared with an equivalent amount of glucose. The effects of the use of the glycemic index on blood glucose levels and on long-term patient outcomes are unclear, but it may be beneficial (ADA, 2013). Although more research is necessary, the following guidelines may be helpful when making dietary recommendations:

- Combining starchy foods with protein- and fat-containing foods tends to slow their absorption and lower the glycemic index.
- In general, eating foods that are raw and whole results in a lower glycemic index than eating chopped, puréed, or cooked foods (except meat).
- Eating whole fruit instead of drinking juice decreases the glycemic index, because fiber in the fruit slows absorption.
- Adding foods with sugars to the diet may result in a lower glycemic index if these foods are eaten with foods that are more slowly absorbed.

Patients can create their own glycemic index by monitoring their blood glucose level after ingestion of a particular food. This can help improve blood glucose control through individualized manipulation of the diet. Many patients who use frequent monitoring of blood glucose levels can use this information to adjust their insulin doses in accordance with variations in food intake.

Other Dietary Concerns

Alcohol Consumption. Patients with diabetes do not need to give up alcoholic beverages entirely, but patients and primary providers must be aware of the potential adverse effects of alcohol specific to diabetes. Alcohol is absorbed before other nutrients and does not require insulin for absorption. Large amounts can be converted to fats, increasing the risk for DKA. In general, the same precautions regarding the use of alcohol by people without diabetes should be applied to patients with diabetes. Moderation is recommended. A major danger of alcohol consumption by the patient with diabetes is hypoglycemia, especially for patients who take insulin or insulin secretagogues (medications that increase the secretion of insulin by the pancreas). Alcohol may decrease the normal physiologic reactions in the body that produce glucose (gluconeogenesis). Therefore, if a patient with diabetes consumes alcohol on an empty stomach, there is an increased likelihood of hypoglycemia. In addition, excessive alcohol intake may impair the patient's ability to recognize and treat hypoglycemia or to follow a prescribed meal plan to prevent hypoglycemia. To reduce the risk of hypoglycemia, the patient should be cautioned to consume food along with the alcohol; however, carbohydrate consumed with alcohol may raise blood glucose.

Alcohol consumption may lead to excessive weight gain (from the high caloric content of alcohol), hyperlipidemia, and elevated glucose levels (especially with mixed drinks and liqueurs). Patient education regarding alcohol intake must emphasize moderation in the amount of alcohol consumed. Moderate intake is considered to be one alcoholic beverage per day for women and two per day for men. Lower-calorie or less-sweet drinks (e.g., light beer, dry wine) and food intake along with alcohol consumption are advised (ADA, 2013). Patients with type 2 diabetes who wish to control their weight should incorporate the calories from alcohol into the overall meal plan.

Sweeteners. The use of artificial sweeteners is acceptable, especially if it assists in overall dietary adherence. Moderation in the amount of sweetener used is encouraged to avoid potential adverse effects. There are two main types of sweeteners: nutritive and nonnutritive. The nutritive sweeteners contain calories, and the nonnutritive sweeteners have few or no calories in the amounts normally used.

Nutritive sweeteners include fructose (fruit sugar), sorbitol, and xylitol, all of which provide calories in amounts similar to those in sucrose (table sugar). They cause less elevation in blood sugar levels than sucrose does and are often used in sugar-free foods. Sweeteners containing sorbitol may have a laxative effect.

Nonnutritive sweeteners have minimal or no calories. They are used in food products and are also available for table use. They produce minimal or no elevation in blood glucose levels and have been approved by the U.S. Food and Drug Administration (FDA) as safe for people with diabetes. Nonnutritive sweeteners include saccharin, aspartame (NutraSweet), acesulfame-K (Sunett), and sucralose (Splenda).

Misleading Food Labels. Foods labeled "sugarless" or "sugar-free" may still provide calories equal to those of the equivalent sugar-containing products if they are made with nutritive sweeteners. Therefore, these foods should not be considered "free" foods to be eaten in unlimited quantity, because they can elevate blood glucose levels. Foods labeled "dietetic" are not necessarily reduced-calorie foods. Patients are advised that foods labeled dietetic may still contain significant amounts of sugar or fat.

Patients must read the labels of "health foods"—especially snacks—because they often contain carbohydrates (e.g., honey, brown sugar, corn syrup, flour) and saturated vegetable fats (e.g., coconut or palm oil), hydrogenated vegetable fats, or animal fats, which may be contraindicated in people with elevated blood lipid levels.

Exercise

Exercise is extremely important in diabetes management because of its effects on lowering blood glucose and reducing cardiovascular risk factors (ADA, 2013). Exercise lowers blood glucose levels by increasing the uptake of glucose by body muscles and by improving insulin utilization. It also improves circulation and muscle tone. Resistance (strength) training, such as weight lifting, can increase lean muscle mass, thereby increasing the resting metabolic rate. These effects are useful in diabetes in relation to losing weight, easing stress, and maintaining a feeling of well-being. Exercise also alters blood lipid concentrations, increasing levels of high-density lipoproteins and decreasing total cholesterol and triglyceride levels. This is especially important for people with diabetes because of their increased risk of cardiovascular disease.

Exercise Recommendations

Ideally, a person with diabetes should engage in regular exercise. General considerations for exercise in patients with diabetes are presented in Chart 51-4. Exercise recommendations must be altered as necessary for patients with diabetic complications such as retinopathy, autonomic neuropathy, sensorimotor neuropathy, and cardiovascular disease (ADA, 2013). Increased blood pressure associated with exercise may aggravate diabetic retinopathy and increase the risk of a hemorrhage into the vitreous or retina.

In general, a slow, gradual increase in the exercise period is encouraged. For many patients, walking is a safe and beneficial form of exercise that requires no special equipment (except for proper shoes) and can be performed anywhere. People with diabetes should discuss an exercise program with their primary provider and undergo a careful medical evaluation with appropriate diagnostic studies before beginning a program (ADA, 2013).

Chart 51-4 PATIENT EDUCATION
General Considerations for Exercise in People With Diabetes

- Exercise 3 times each week with no more than 2 consecutive days without exercise.
- Perform resistance training twice a week (for people with type 2 diabetes).
- Exercise at the same time of day (preferably when blood glucose levels are at their peak) and for the same duration each session.
- Use proper footwear and, if appropriate, other protective equipment (i.e., helmets for cycling).
- Avoid trauma to the lower extremities, especially in patients with numbness due to peripheral neuropathy.
- Inspect feet daily after exercise.
- Avoid exercise in extreme heat or cold.
- Avoid exercise during periods of poor metabolic control.

Adapted from American Diabetes Association. (2013). Executive summary: Standards of medical care in diabetes—2012. *Diabetes Care, 36*(Suppl. 1), S11–S66.

For patients who are older than 30 years and who have two or more risk factors for heart disease, an exercise stress test is recommended. Risk factors for heart disease include hypertension, obesity, high cholesterol levels, abnormal resting electrocardiogram (ECG), sedentary lifestyle, smoking, male gender, and a family history of heart disease. An abnormal stress test may indicate cardiac ischemia. Typically, an abnormal stress test is followed by a cardiac catheterization and, in some cases, with an intervention such as angioplasty, stent placement, or cardiac surgery.

Exercise Precautions

Patients who have blood glucose levels exceeding 250 mg/dL (14 mmol/L) and who have ketones in their urine should not begin exercising until the urine test results are negative for ketones and the blood glucose level is closer to normal. Exercising with elevated blood glucose levels increases the secretion of glucagon, growth hormone, and catecholamines. The liver then releases more glucose, and the result is an increase in the blood glucose level (ADA, 2013).

The physiologic decrease in circulating insulin that normally occurs with exercise cannot occur in patients treated with insulin. Initially, patients who require insulin should be taught to eat a 15-g carbohydrate snack (a fruit exchange) or a snack of complex carbohydrates with a protein before engaging in moderate exercise to prevent unexpected hypoglycemia. The exact amount of food needed varies from person to person and should be determined by blood glucose monitoring.

Another potential concern for patients who take insulin is hypoglycemia that occurs many hours after exercise. To avoid postexercise hypoglycemia, especially after strenuous or prolonged exercise, the patient may need to eat a snack at the end of the exercise session and at bedtime and monitor the blood glucose level more frequently. Patients who are capable, knowledgeable, and responsible can learn to adjust their own insulin doses by working closely with a diabetes educator. Others need specific instructions on what to do when they exercise.

Patients taking insulin and participating in extended periods of exercise should test their blood glucose levels before, during, and after the exercise period, and they should snack on carbohydrates as needed to maintain blood glucose levels. Other participants or observers should be aware that the person exercising has diabetes, and they should know what assistance to give if severe hypoglycemia occurs.

In people with type 2 diabetes who are overweight or obese, exercise in addition to dietary management both improves glucose metabolism and enhances loss of body fat. Exercise coupled with weight loss improves insulin sensitivity and may decrease the need for insulin or oral antidiabetic agents (ADA, 2013). Eventually, the patient's glucose tolerance may return to normal. Patients with type 2 diabetes who are not taking insulin or an oral agent may not need extra food before exercise.

Gerontologic Considerations

Physical activity that is consistent and realistic is beneficial to older adults with diabetes. Physical fitness in the older adult population with diabetes may lead to improved glycemic control, decreased risk for chronic vascular disease, and an improved quality of life (Eliopoulos, 2010). Advantages of exercise in this population include a decrease in hyperglycemia, a general sense of well-being, and better use of ingested calories, resulting in weight reduction. Because there is an increased incidence of cardiovascular problems in older adults, a physical examination and exercise stress test may be warranted before an exercise program is initiated. A pattern of gradual, consistent exercise, including a combination of aerobic exercise and resistance training, should be planned that does not exceed the patient's physical capacity. Physical impairment due to other chronic diseases must also be considered. In some cases, a physical therapy evaluation may be indicated, with the goal of determining exercises specific to the patient's needs and abilities. Tools such as the *Armchair Fitness* video may be helpful.

Monitoring Glucose Levels and Ketones

Blood glucose monitoring is a cornerstone of diabetes management, and **self-monitoring of blood glucose (SMBG)** levels has dramatically altered diabetes care. SMBG is a method of capillary blood glucose testing in which the patient pricks his or her finger and applies a drop of blood to a test strip that is read by a meter. It is recommended that SMBG occurs when circumstances call for it (e.g., before meals and snacks) for patients prescribed frequent insulin injections or an insulin pump (ADA, 2013).

Self-Monitoring of Blood Glucose

Using SMBG and learning how to respond to the results enable people with diabetes to individualize their treatment regimen to obtain optimal blood glucose control. This allows for detection and prevention of hypoglycemia and hyperglycemia and plays a crucial role in normalizing blood glucose levels, which in turn may reduce the risk of long-term diabetic complications.

Various methods for SMBG are available. Most involve obtaining a drop of blood from the fingertip, applying the blood to a special reagent strip, and allowing the blood to stay on the strip for the amount of time specified by the manufacturer (usually 5 to 30 seconds). The meter gives a digital readout of the blood glucose value. The meters available for SMBG offer various features and benefits such as monthly averages, tracking of events such as exercise and food consumption, and downloading capacity. Most meters are biosensors that can use blood obtained from alternative test sites, such as the forearm. They have a special lancing device that is useful for patients who have painful fingertips or experience pain with fingersticks.

Because laboratory methods measure plasma glucose, most blood glucose monitors approved for patient use in the home and some test strips calibrate blood glucose readings to plasma values. Plasma glucose values are 10% to 15% higher than whole blood glucose values, and it is crucial for patients with diabetes to know whether their monitor and strips provide whole blood or plasma results.

Methods for SMBG must match the skill level and physical capabilities of patients. Factors affecting SMBG performance include visual acuity, fine motor coordination, cognitive ability, comfort with technology and willingness to use it, and cost (Eliopoulos, 2010). Some meters can be used by patients with visual impairments; these meters have audio components to assist in performing the test and obtaining the result. In addition, meters are available to check both blood glucose and blood ketone levels by those who are particularly susceptible to DKA. Most insurance companies, and programs such as Medicare and Medicaid, cover some or all of the costs of meters and strips.

All methods of SMBG carry the risk that patients may obtain and report erroneous blood glucose values as a result of incorrect techniques. Some common sources of error include improper application of blood (e.g., drop too small), damage to the reagent strips caused by heat or humidity, the use of outdated strips, and improper meter cleaning and maintenance.

Nurses play an important role in providing initial education about SMBG techniques. Equally important is evaluating the techniques of patients who are experienced in self-monitoring. Every 6 to 12 months, patients should conduct a comparison of their meter result with a simultaneous laboratory-measured blood glucose level in their provider's office and have their technique observed (Funnell, Brown, Childs, et al., 2009). The accuracy of the meter and strips can also be assessed with control solutions specific to that meter whenever a new vial of strips is used and whenever the validity of the reading is in doubt.

Candidates for Self-Monitoring of Blood Glucose. SMBG is a useful tool for managing self-care for everyone with diabetes. It is a key component of treatment for any intensive insulin therapy regimen (i.e., two to four injections per day or the use of an insulin pump) and for diabetes management during pregnancy. It is also recommended for patients with the following conditions:

- Unstable diabetes (severe swings from very high to very low blood glucose levels within a 24-hour day)
- A tendency to develop severe ketosis or hypoglycemia
- Hypoglycemia without warning symptoms

For patients not taking insulin, SMBG is helpful for monitoring the effectiveness of exercise, diet, and oral antidiabetic agents. It can also help motivate patients to continue with treatment. For patients with type 2 diabetes, SMBG is recommended during periods of suspected hyperglycemia (e.g., illness) or hypoglycemia (e.g., unusual increased activity levels) and when the medication or dosage of medication is modified (ADA, 2013).

Frequency of Self-Monitoring of Blood Glucose. For most patients who require insulin, SMBG is recommended two to four times daily (usually before meals and at bedtime). For patients who take insulin before each meal, SMBG is required at least three times daily before meals to determine each dose (ADA, 2013). Those not receiving insulin may be instructed to assess their blood glucose levels at least two or three times per week, including a 2-hour postprandial test. For all patients, testing is recommended whenever hypoglycemia or hyperglycemia is suspected; with changes in medications, activity, or diet; and with stress or illness.

Responding to Self-Monitoring of Blood Glucose Results. Patients are asked to keep a record or logbook of blood glucose levels so that they can detect patterns. Testing is done at the peak action time of the medication to evaluate the need for dosage adjustments. To evaluate basal insulin and determine bolus insulin doses, testing is performed before meals. To determine bolus doses of regular or rapid-acting insulin (lispro [Humalog], aspart [NovoLog], or glulisine [Apidra]), testing is done 2 hours after meals. Patients with type 2 diabetes are encouraged to test daily before and 2 hours after the largest meal of the day until stabilized. Thereafter, testing should be done periodically before and after meals. Patients who take insulin at bedtime or who use an insulin infusion pump should also test at 3 AM once a week to document that the blood glucose level is not decreasing during the night. If the patient is unwilling or cannot afford to test frequently, then once or twice a day may be sufficient if the time of testing is varied (e.g., before breakfast one day and before lunch the next day).

A tendency to discontinue SMBG is more likely to occur if the patient does not receive instruction about using the results to alter the treatment regimen, if positive reinforcement is not given, and if costs of testing increase. At the very least, the patient should be given parameters for contacting the primary provider. Patients using intensive insulin therapy regimens may be instructed in the use of algorithms (rules or decision trees) for changing the insulin doses based on patterns of values greater or less than the target range and the amount of carbohydrate to be consumed. Baseline patterns should be established by SMBG for 1 to 2 weeks.

Using a Continuous Glucose Monitoring System

A **continuous glucose monitoring system (CGMS)** can be used to continuously monitor blood glucose levels (Fig. 51-2). A sensor attached to an infusion set, which is similar to an insulin pump infusion set, is inserted subcutaneously in the abdomen and connected to the device worn on the patient's clothing or placed in a pocket. After 72 hours, the data from the device are downloaded, and blood glucose readings are analyzed. Although the CGMS cannot be used for making decisions about specific insulin doses, it can be used to determine whether treatment is adequate over a 24-hour period. This device is most useful in patients with type 1 diabetes (ADA, 2013).

Testing for Glycated Hemoglobin

Glycated hemoglobin (also referred to as **glycosylated hemoglobin, $HgbA_{1C}$, or A1C**) is a measure of glucose control that is a result of the glucose molecule attaching to hemoglobin for the life of the red blood cell (ADA, 2013). When blood glucose levels are elevated, glucose molecules attach to hemoglobin in red blood cells. The longer the amount of

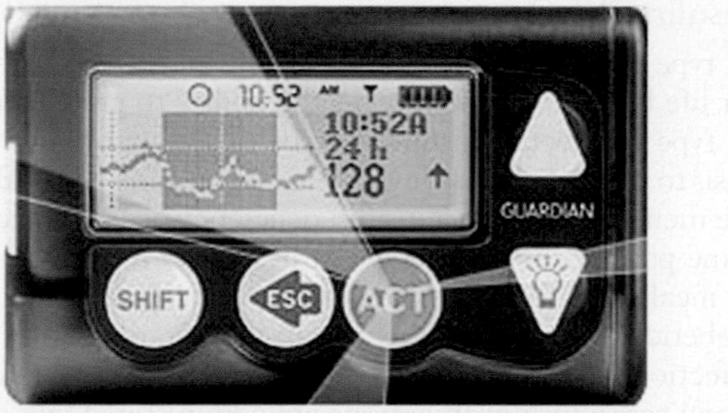
10:52
10:52A
24 h
128
GUARDIAN
SHIFT
ESC
ACT

FIGURE 51-2 • MiniMed Guardian® REAL-Time CGM System. Manufactured by the diabetes division of Medtronic, Inc. Used with permission.

glucose in the blood remains above normal, the more glucose binds to hemoglobin and the higher the glycated hemoglobin level becomes. This complex (hemoglobin attached to the glucose) is permanent and lasts for the life of an individual red blood cell, approximately 120 days. If near-normal blood glucose levels are maintained, with only occasional increases, the overall value will not be greatly elevated. However, if the blood glucose values are consistently high, then the test result is also elevated. If the patient reports mostly normal SMBG results but the glycated hemoglobin is high, there may be errors in the methods used for glucose monitoring, errors in recording results, or frequent elevations in glucose levels at times during the day when the patient is not usually monitoring blood sugar levels. Normal values typically range from 4% to 6% and indicate consistently near-normal blood glucose concentrations. The target range for people with diabetes is less than 7% (ADA, 2013).

Testing for Ketones

Ketones (or ketone bodies) are by-products of fat breakdown, and they accumulate in the blood and urine. Ketones in the urine signal that there is a deficiency of insulin and control of type 1 diabetes is deteriorating. When there is almost no effective insulin available, the body starts to break down stored fat for energy.

Most commonly, the patient uses a urine dipstick (Ketostix or Chemstrip uK) to detect ketonuria. The reagent pad on the strip turns purple when ketones are present. (One of the ketone bodies is called *acetone,* and this term is frequently used interchangeably with the term *ketones.*) Other strips are available for measuring both urine glucose and ketones (Keto-Diastix or Chemstrip uGK). Large amounts of ketones may depress the color response of the glucose test area. A meter that enables testing of blood for ketones is available.

Urine ketone testing should be performed whenever patients with type 1 diabetes have glycosuria or persistently elevated blood glucose levels (more than 240 mg/dL or 13.2 mmol/L for two testing periods in a row) and during illness, in pregnancy with preexisting diabetes, and in gestational diabetes (ADA, 2013).

Pharmacologic Therapy

Insulin is secreted by the beta cells of the islets of Langerhans and lowers the blood glucose level after meals by facilitating the uptake and utilization of glucose by muscle, fat, and liver cells. In the absence of adequate insulin, pharmacologic therapy is essential.

Insulin Therapy

In type 1 diabetes, exogenous insulin must be administered for life because the body loses the ability to produce insulin. In type 2 diabetes, insulin may be necessary on a long-term basis to control glucose levels if meal planning and oral agents are ineffective or when insulin deficiency occurs. In addition, some patients in whom type 2 diabetes is usually controlled by meal planning alone or by meal planning and an oral antidiabetic agent may require insulin temporarily during illness, infection, pregnancy, surgery, or some other stressful event. In many cases, insulin injections are administered two or more times daily to control the blood glucose level. Because the insulin dose required by the individual patient is determined by the level of glucose in the blood, accurate monitoring of blood glucose levels is essential; thus, SMBG is a cornerstone of insulin therapy.

Preparations. A number of insulin preparations are available. They vary according to three main characteristics: time course of action, species (source), and manufacturer. Human insulins are produced by recombinant deoxyribonucleic acid (DNA) technology and are the only type of insulin available in the United States.

Time Course of Action. Insulins may be grouped into several categories based on the onset, peak, and duration of action (Table 51-3).

Rapid-acting insulins produce a more rapid effect that is of shorter duration than regular insulin. Because of their rapid onset, the patient should be instructed to eat no more than 5 to 15 minutes after injection. Because of the short duration of action of these insulin analogues, patients with type 1 diabetes and some patients with type 2 or gestational diabetes also require a long-acting insulin (basal insulin) to maintain glucose control. Basal insulin is necessary to maintain blood glucose levels irrespective of meals. A constant level of insulin is required at all times. Intermediate-acting insulins function as basal insulins but may have to be split into two injections to achieve 24-hour coverage.

Short-acting insulins are called *regular insulin* (marked R on the bottle). Regular insulin is a clear solution and is usually administered 20 to 30 minutes before a meal, either alone or in combination with a longer-acting insulin. Regular insulin is the only insulin approved for IV use.

Intermediate-acting insulins are called *NPH insulin* (neutral protamine Hagedorn) or *Lente insulin*. Intermediate-acting insulins, which are similar in their time course of action, appear white and cloudy. If NPH or Lente insulin is taken alone, it is not crucial that it be taken 30 minutes before the meal. However, patients should eat some food around the time of the onset and peak of these insulins.

"Peakless" basal or very long acting insulins are approved by the FDA for use as a basal insulin—that is, the insulin is absorbed very slowly over 24 hours and can be given once a day. Because the insulin is in a suspension with a pH of 4, it cannot be mixed with other insulins because this would cause precipitation. It was originally approved to be given once a day at bedtime; however, it has now been approved to be given once a day at any time of the day but must be given at the same time each day to prevent overlap of action. Many patients fall asleep, forgetting to take their bedtime insulin, or may be wary of taking insulin before going to sleep. Having these patients take their insulin in the morning ensures that the dose is taken.

Quality and Safety Nursing Alert

When administering insulin, it is very important to read the label carefully and to be sure that the correct type of insulin is administered. It is also important to avoid mistaking Lantus insulin for Lente insulin and vice versa.

The nurse should emphasize which meals—and snacks—are being "covered" by which insulin doses. In general, the rapid- and short-acting insulins are expected to cover the increase in glucose levels after meals, immediately after the injection; the intermediate-acting insulins are expected to cover subsequent meals; and the long-acting insulins provide a relatively constant level of insulin and act as a basal insulin.

Insulin Regimens. Insulin regimens vary from one to four injections per day. Usually, there is a combination of a short-acting insulin and a longer-acting insulin. The normally functioning pancreas continuously secretes small amounts of insulin during the day and night. In addition, whenever blood glucose increases after ingestion of food, there is a rapid burst of insulin secretion in proportion to the glucose-raising effect of the food. The goal of all but the simplest, one-injection insulin regimens is to mimic this normal pattern of insulin secretion in response to food intake and activity patterns. Table 51-4 describes several insulin regimens and the advantages and disadvantages of each.

TABLE 51-3 Categories of Insulin

Time Course	Agent	Onset	Peak	Duration	Indications
Rapid acting	lispro (Humalog) aspart (NovoLog) glulisine (Apidra)	10–15 min 5–15 min 5–15 min	1 h 40–50 min 30–60 min	2–4 h 2–4 h 2 h	Used for rapid reduction of glucose level, to treat postprandial hyperglycemia, and/or to prevent nocturnal hypoglycemia
Short acting	regular (Humalog R, Novolin R, Iletin II Regular)	1½–1 h	2–3 h	4–6 h	Usually administered 20–30 min before a meal; may be taken alone or in combination with longer-acting insulin
Intermediate acting	NPH (neutral protamine Hagedorn) (Humulin N, Iletin II Lente, Iletin II NPH, Novolin N [NPH])	2–4 h 3–4 h	4–12 h 4–12 h	16–20 h 16–20 h	Usually taken after food
Very long acting	glargine (Lantus) detemir (Levemir)	1 h	Continuous (no peak)	24 h	Used for basal dose

TABLE 51-4 Insulin Regimens

Schematic Representation	Description	Advantages	Disadvantages
Normal pancreas 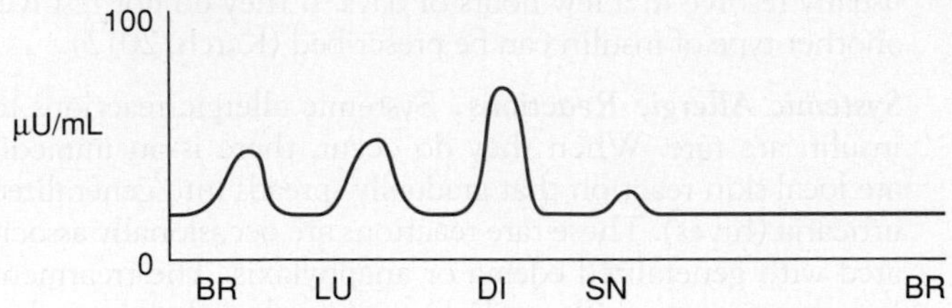 100 µU/mL 0 BR LU DI SN BR	Insulin release increases when blood glucose levels rise and continues at a low steady rate between meals.		
One injection per day 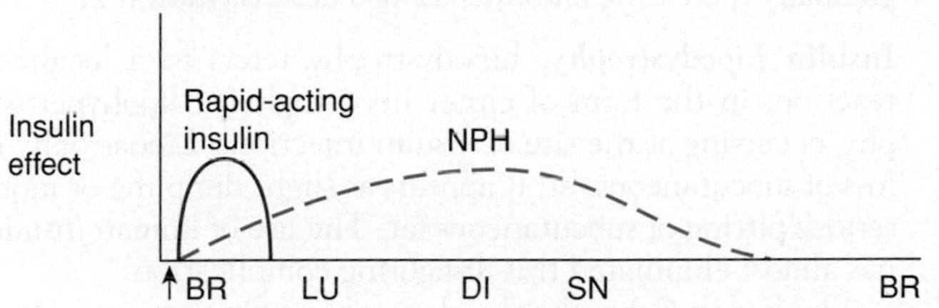 Insulin effect Rapid-acting insulin NPH BR LU DI SN BR	Before breakfast: • NPH or • NPH with rapid-acting insulin	Simple regimen	Difficult to control fasting blood glucose if effects of NPH do not last Afternoon hypoglycemia may result from attempts to control fasting glucose level by increasing NPH dose.
Two injections per day–mixed 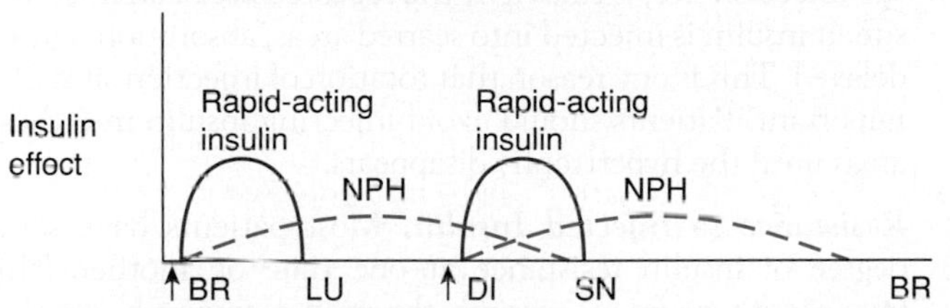 Insulin effect Rapid-acting insulin Rapid-acting insulin NPH NPH BR LU DI SN BR	Before breakfast and dinner: • NPH or • NPH with rapid-acting insulin or • Premixed (rapid-acting insulin) insulin	Simplest regimen that attempts to mimic normal pancreas	Need relatively fixed schedule of meals and exercise Cannot independently adjust NPH or regular if premixed insulin is used
Three or four injections per day 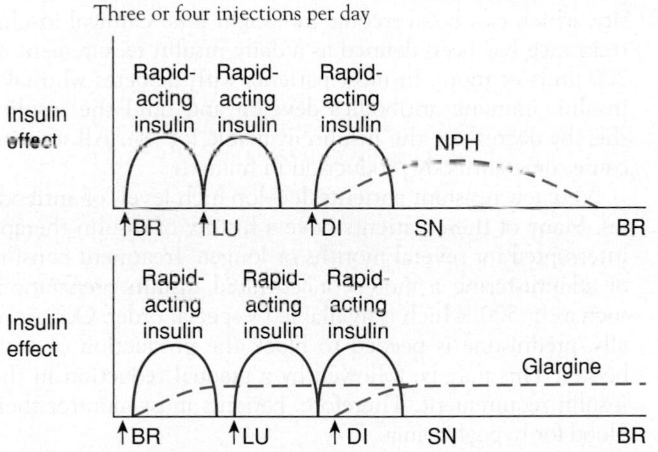 Insulin effect Rapid-acting insulin Rapid-acting insulin Rapid-acting insulin NPH BR LU DI SN BR Insulin effect Rapid-acting insulin Rapid-acting insulin Rapid-acting insulin Glargine BR LU DI SN BR	Rapid-acting insulin before each meal with: • NPH at dinner or • NPH at bedtime or • Glargine 1 or 2 times/d	More closely mimics normal pancreas than 3-injection regimen Each premeal dose of regular insulin decided independently More flexibility with meals and exercise	Requires more injections than other regimens Requires multiple blood glucose tests on a daily basis Requires intensive education and follow-up
Insulin pump 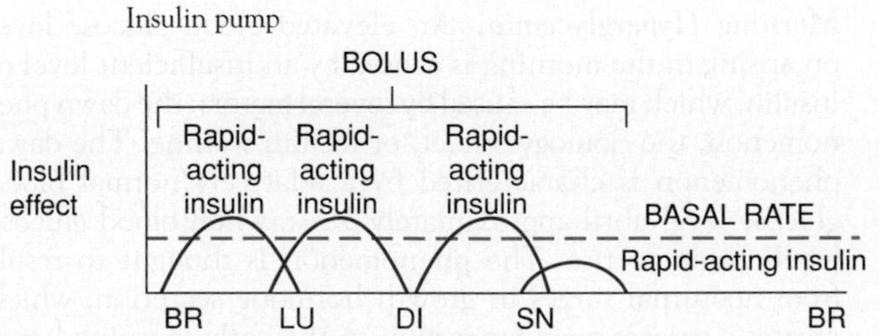 BOLUS Insulin effect Rapid-acting insulin Rapid-acting insulin Rapid-acting insulin BASAL RATE Rapid-acting insulin BR LU DI SN BR	Uses ONLY rapid-acting insulin infused at continuous, low rate called *basal rate* (commonly 0.5–1.5 units/h) and premeal *bolus doses* activated by pump wearer	Most closely mimics normal pancreas Decreases unpredictable peaks of intermediate- and long-acting insulins Increases meal and exercise flexibility	Requires intensive training and frequent follow-up Potential for mechanical problems Requires multiple blood glucose tests on a daily basis Potential increase in expenses (depending on insurance coverage)

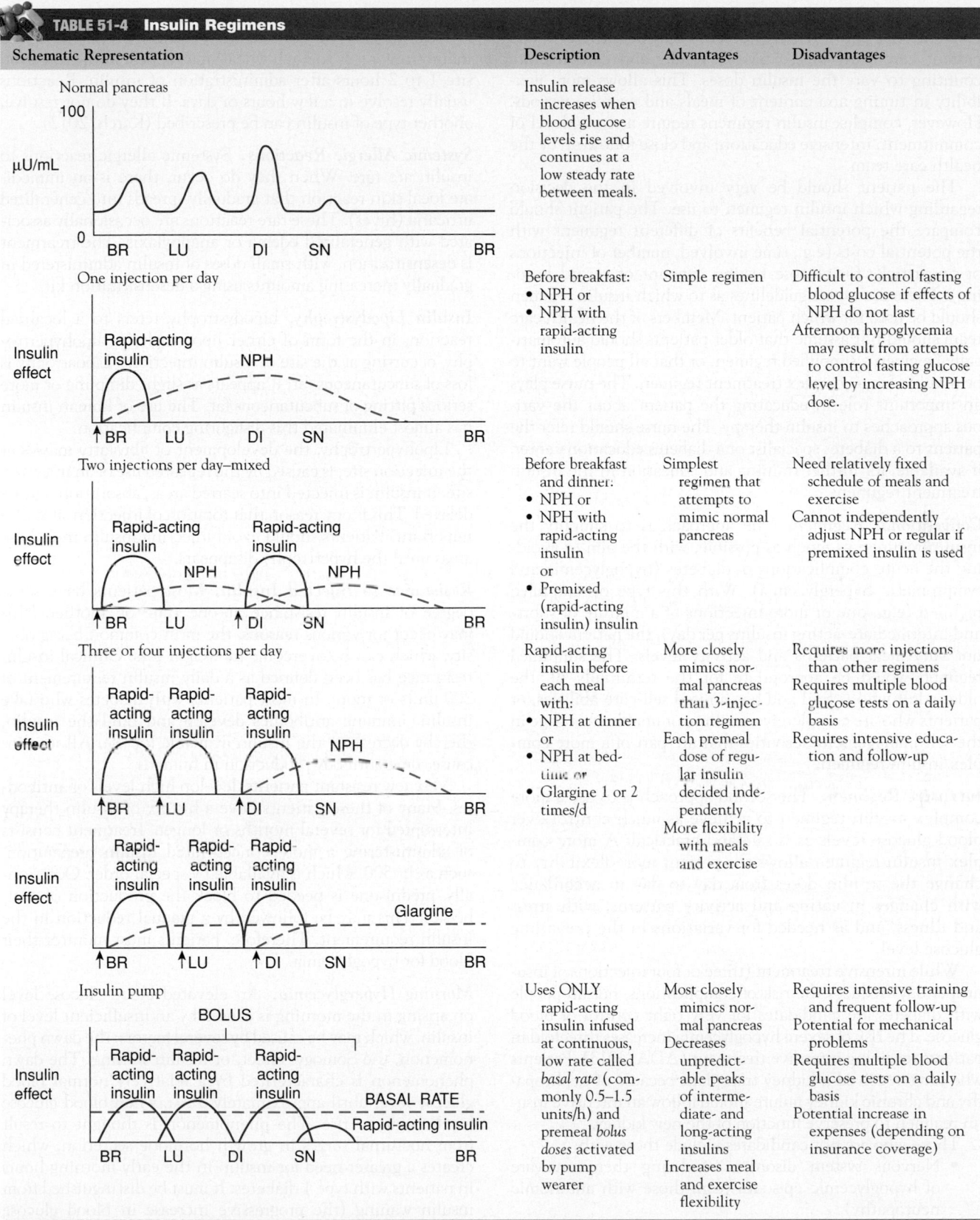

BR, breakfast; LU, lunch; DI, dinner; SN, snack; NPH, neutral protamine Hagedorn; regular; ↑, insulin injections.
Note: Rapid-acting insulin—lispro, aspart, or glulisine [Apidra].

There are two general approaches to insulin therapy: conventional and intensive (described in detail later). The patient can learn to use SMBG results and carbohydrate counting to vary the insulin doses. This allows more flexibility in timing and content of meals and exercise periods. However, complex insulin regimens require a strong level of commitment, intensive education, and close follow-up by the health care team.

The patient should be very involved in the decision regarding which insulin regimen to use. The patient should compare the potential benefits of different regimens with the potential costs (e.g., time involved, number of injections or fingersticks for glucose testing, amount of record keeping). There are no set guidelines as to which insulin regimen should be used for which patient. Members of the health care team should not assume that older patients should automatically be given a simplified regimen, or that all people want to be involved in a complex treatment regimen. The nurse plays an important role in educating the patient about the various approaches to insulin therapy. The nurse should refer the patient to a diabetes specialist or a diabetes education center, if available, for further training and education in the insulin treatment regimens.

Conventional Regimen. One approach is to simplify the insulin regimen as much as possible, with the aim of avoiding the acute complications of diabetes (hypoglycemia and symptomatic hyperglycemia). With this type of simplified regimen (e.g., one or more injections of a mixture of short- and intermediate-acting insulins per day), the patient should not vary meal patterns and activity levels. The simplified regimen would be appropriate for the terminally ill, the older adult who is frail and has limited self-care abilities, or patients who are completely unwilling or unable to engage in the self-management activities that are part of a more complex insulin regimen.

Intensive Regimen. The second approach is to use a more complex insulin regimen to achieve as much control over blood glucose levels as is safe and practical. A more complex insulin regimen allows the patient more flexibility to change the insulin doses from day to day in accordance with changes in eating and activity patterns, with stress and illness, and as needed for variations in the prevailing glucose level.

While intensive treatment (three or four injections of insulin per day) reduces the risk of complications, not all people with diabetes are candidates for very tight control of blood glucose. The risk of severe hypoglycemia increases threefold in patients receiving intensive treatment (ADA, 2013). Patients who have received a kidney transplant because of nephropathy and chronic kidney failure should follow an intensive insulin regimen to preserve function of the new kidney.

Those who are not candidates include those with:

- Nervous system disorders rendering them unaware of hypoglycemic episodes (e.g., those with autonomic neuropathy)
- Recurring severe hypoglycemia
- Irreversible diabetic complications, such as blindness or ESKD
- Cerebrovascular or cardiovascular disease
- Ineffective self-care skills

Complications of Insulin Therapy. Local Allergic Reactions A local allergic reaction (redness, swelling, tenderness, and induration or a 2- to 4-cm wheal) may appear at the injection site 1 to 2 hours after administration of insulin. Reactions usually resolve in a few hours or days. If they do not resolve, another type of insulin can be prescribed (Karch, 2012).

Systemic Allergic Reactions. Systemic allergic reactions to insulin are rare. When they do occur, there is an immediate local skin reaction that gradually spreads into generalized urticaria (hives). These rare reactions are occasionally associated with generalized edema or anaphylaxis. The treatment is desensitization, with small doses of insulin administered in gradually increasing amounts using a desensitization kit.

Insulin Lipodystrophy. Lipodystrophy refers to a localized reaction, in the form of either lipoatrophy or lipohypertrophy, occurring at the site of insulin injections. Lipoatrophy is loss of subcutaneous fat; it appears as slight dimpling or more serious pitting of subcutaneous fat. The use of human insulin has almost eliminated this disfiguring complication.

Lipohypertrophy, the development of fibrofatty masses at the injection site, is caused by the repeated use of an injection site. If insulin is injected into scarred areas, absorption may be delayed. This is one reason that rotation of injection sites is so important. Patients should avoid injecting insulin into these areas until the hypertrophy disappears.

Resistance to Injected Insulin. Most patients have some degree of insulin resistance at one time or another. This may occur for various reasons, the most common being obesity, which can be overcome by weight loss. Clinical insulin resistance has been defined as a daily insulin requirement of 200 units or more. In most patients with diabetes who take insulin, immune antibodies develop and bind the insulin, thereby decreasing the insulin available for use. All insulins cause some antibody production in humans.

Very few resistant patients develop high levels of antibodies. Many of these patients have a history of insulin therapy interrupted for several months or longer. Treatment consists of administering a more concentrated insulin preparation, such as U-500, which is available by special order. Occasionally, prednisone is needed to block the production of antibodies. This may be followed by a gradual reduction in the insulin requirement. Therefore, patients must monitor their blood for hypoglycemia.

Morning Hyperglycemia. An elevated blood glucose level on arising in the morning is caused by an insufficient level of insulin, which may be caused by several factors: the dawn phenomenon, the Somogyi effect, or insulin waning. The dawn phenomenon is characterized by a relatively normal blood glucose level until approximately 3 AM, when blood glucose levels begin to rise. The phenomenon is thought to result from nocturnal surges in growth hormone secretion, which creates a greater need for insulin in the early morning hours in patients with type 1 diabetes. It must be distinguished from insulin waning (the progressive increase in blood glucose from bedtime to morning) and from the Somogyi effect (nocturnal hypoglycemia followed by rebound hyperglycemia). Insulin waning is frequently seen if the evening NPH dose is administered before dinner; it is prevented by moving the evening dose of NPH insulin to bedtime.

TABLE 51-5 Causes of Morning Hyperglycemia

Characteristic	Treatment
Insulin Waning	
Progressive rise in blood glucose from bedtime to morning	Increase evening (predinner or bedtime) dose of intermediate- or long-acting insulin, or institute a dose of insulin before the evening meal if one is not already part of the treatment regimen.
Dawn Phenomenon	
Relatively normal blood glucose until about 3 AM, when the level begins to rise	Change time of injection of evening intermediate-acting insulin from dinnertime to bedtime.
Somogyi Effect	
Normal or elevated blood glucose at bedtime, a decrease at 2–3 AM to hypoglycemic levels, and a subsequent increase caused by the production of counter-regulatory hormones	Decrease evening (predinner or bedtime) dose of intermediate-acting insulin, or increase bedtime snack.

It may be difficult to tell from a patient's history what the cause is for morning hyperglycemia. To determine the cause, the patient must be awakened once or twice during the night to test blood glucose levels. Testing at bedtime, at 3 AM, and on awakening provides information that can be used to make adjustments in insulin to avoid morning hyperglycemia.

Concept Mastery Alert

Causes of morning hyperglycemia may be easily confused. Table 51-5 summarizes the differences among insulin waning, the dawn phenomenon, and the Somogyi effect.

Methods of Insulin Delivery. Methods of insulin delivery include traditional subcutaneous injections, insulin pens, jet injectors, and insulin pumps. (See Nursing Management later in this section for discussion of traditional subcutaneous injections.)

Insulin Pens. Insulin pens use small (150- to 300-unit) prefilled insulin cartridges that are loaded into a penlike holder. A disposable needle is attached to the device for insulin injection. Insulin is delivered by dialing in a dose or pushing a button for every 1- or 2-unit increment administered. People using these devices still need to insert the needle for each injection (Fig. 51-3); however, they do not need to carry insulin bottles or draw up insulin before each injection. These devices are most useful for patients who need to inject only one type of insulin at a time (e.g., premeal rapid-acting insulin three times a day and bedtime NPH insulin) or who can use the premixed insulins. These pens are convenient for those who administer insulin before dinner if eating out or traveling. They are also useful for patients with impaired manual dexterity, vision, or cognitive function, which makes the use of traditional syringes difficult.

Jet Injectors. As an alternative to needle injections, jet injection devices deliver insulin through the skin under pressure in an extremely fine stream. These devices are more expensive and require thorough training and supervision when first used. In addition, patients should be cautioned that absorption rates, peak insulin activity, and insulin levels may be different when changing to a jet injector. (Insulin administered by jet injector is usually absorbed faster.) The use of jet injectors has been associated with bruising in some patients.

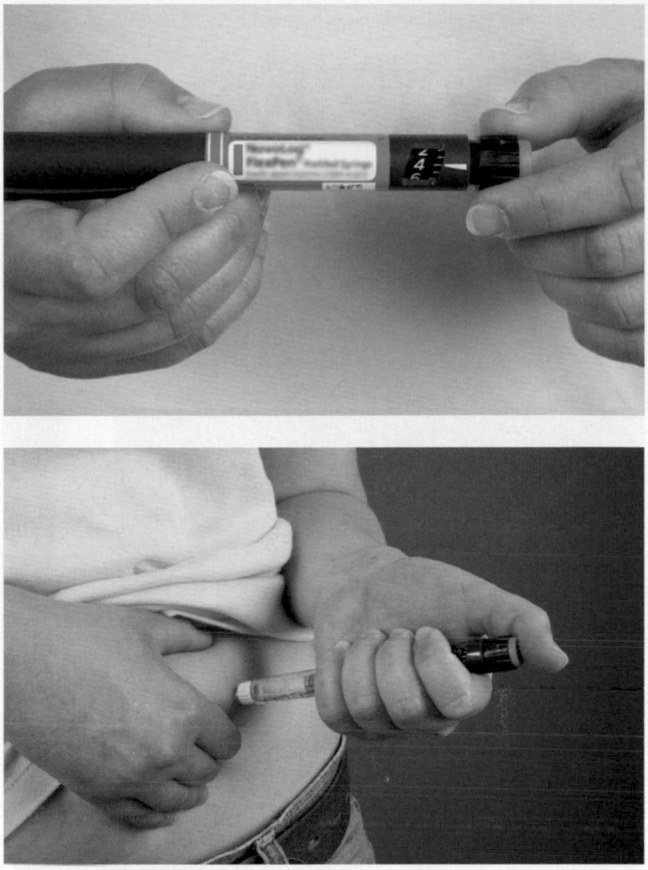

FIGURE 51-3 • Prefilled insulin syringe.

Insulin Pumps. Continuous subcutaneous insulin infusion involves the use of small, externally worn devices (insulin pumps) that closely mimic the functioning of the normal pancreas (ADA, 2013). Insulin pumps contain a 3-mL syringe attached to a long (24- to 42-in), thin, narrow-lumen tube with a needle or Teflon catheter attached to the end (Fig. 51-4). The patient inserts the needle or catheter into subcutaneous tissue (usually on the abdomen) and secures it with tape or a transparent dressing. The needle or catheter is changed at least every 3 days. The pump is then worn either on the patient's clothing or in a pocket. Some women keep the pump tucked into the front or side of the bra.

When an insulin pump is used, insulin is delivered by subcutaneous infusion at a basal rate (e.g., 0.5 to 2 units per hour). When a meal is consumed, the patient calculates a dose of insulin to metabolize the meal by counting the total amount of carbohydrate for the meal using a predetermined insulin-to-carbohydrate ratio; for example, a ratio of 1 unit of insulin for every 15 g of carbohydrate would require 3 units

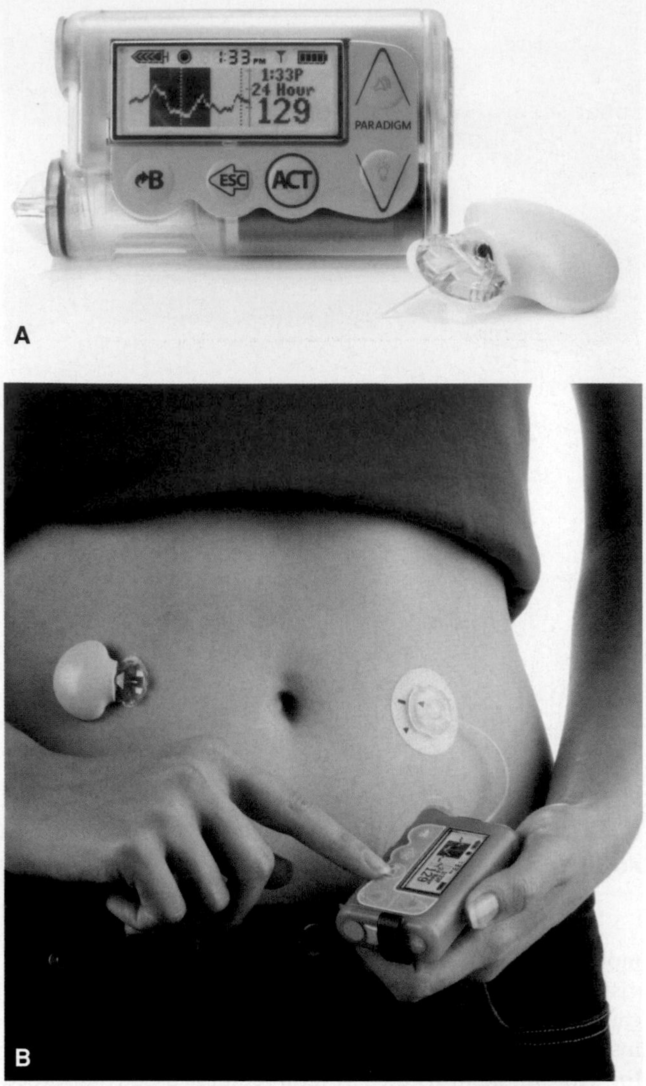
1:33P
24 Hour
129
PARADIGM
ESC
ACT
A
B

FIGURE 51-4 • **A.** MiniMed Paradigm® REAL-Time Revel™ Insulin Pump. **B.** Patient using this insulin pump for self-management of blood glucose and insulin doses. Manufactured by the diabetes division of Medtronic, Inc. Used with permission.

for a meal with 45 g of carbohydrate. This allows flexibility of meal timing and content.

Possible disadvantages of insulin pumps are unexpected disruptions in the flow of insulin from the pump that may occur if the tubing or needle becomes occluded, if the supply of insulin runs out, or if the battery is depleted, increasing the risk of DKA. Effective education to produce knowledgeable patients minimizes this risk. There is the potential for infection at needle insertion sites. Hypoglycemia may occur with insulin pump therapy; however, this is usually related to the lowered blood glucose levels that many patients achieve rather than to a specific problem with the pump itself. The tight diabetes control associated with the use of an insulin pump may increase the incidence of hypoglycemia unawareness because of the very gradual decline in serum glucose level, from more than 70 mg/dL (3.9 mmol/L) to less than 60 mg/dL (3.3 mmol/L).

Some patients find that wearing the pump for 24 hours each day is inconvenient. However, the pump can easily be disconnected, per patient preference, for limited periods, such as for showering, exercise, or sexual activity.

Candidates for the insulin pump must be willing to assess their blood glucose level several times daily. In addition, they must be psychologically stable and open about having diabetes, because the insulin pump is often a visible sign to others and a constant reminder to patients that they have diabetes. Most important, patients using insulin pumps must have extensive education in the use of the pump and in self-management of blood glucose and insulin doses. They must work closely with a team of health care professionals who are experienced in insulin pump therapy—specifically, a diabetologist/endocrinologist, a dietitian, and a certified diabetes educator.

The most common risk of insulin pump therapy is ketoacidosis, which can occur if there is an occlusion in the infusion set or tubing. Because only rapid-acting insulin is used in the pump, any interruption in the flow of insulin may rapidly cause the patient to be without insulin. The patient should be taught to administer insulin by manual injection if an insulin interruption is suspected (e.g., no response in blood glucose level after a meal bolus).

Many insurance companies cover the cost of pump therapy. If not, the extra expense of the pump and associated supplies may be a deterrent for some patients. Medicare covers insulin pump therapy for patients with type 1 diabetes.

Insulin pumps have been used in patients with type 2 diabetes whose beta-cell function has diminished and who require insulin. Patients with a hectic lifestyle often do well with an insulin pump. There is no risk of DKA when there is an interruption of the flow of insulin in people with type 2 diabetes wearing an insulin pump.

Future Insulin Delivery. Research into mechanical delivery of insulin has involved implantable insulin pumps that can be externally programmed according to blood glucose test results. Clinical trials with these devices are continuing. In addition, there is research into the development of implantable devices that both measure the blood glucose level and deliver insulin as needed. Methods of administering insulin by the oral route (oral spray or capsule) and skin patch are undergoing intensive study.

Transplantation of Pancreatic Cells. Transplantation of the whole pancreas or a segment of the pancreas is being performed on a limited population (mostly patients with diabetes who are receiving a kidney transplantation simultaneously). One main issue is weighing the risks of antirejection medications against the advantages of pancreas transplantation. Implantation of insulin-producing pancreatic islet cells is another approach. This latter approach involves a less extensive surgical procedure and a potentially lower incidence of immunogenic problems. However, thus far, independence from exogenous insulin has been limited to 2 years after transplantation of islet cells. Results of studies of patients with islet cell transplants using less toxic antirejection drugs have shown some promise. These procedures are not widely available due to a critical shortage of organs for transplantation (National Institute of Diabetes and Digestive and Kidney Diseases [NIDDK], 2012a).

Oral Antidiabetic Agents

Oral antidiabetic agents may be effective for patients who have type 2 diabetes that cannot be treated effectively with MNT and exercise alone. In the United States, oral antidiabetic agents include first- and second-generation sulfonylureas, biguanides, alpha-glucosidase inhibitors, non-sulfonylurea insulin secretogogues (meglitinides and phenylalanine derivatives), thiazolidinediones (glitazones), and dipeptide peptidase-4 (DPP-4) inhibitors (Table 51-6). The **thiazolidinediones** are a class a class of oral antidiabetic medications that reduces insulin resistance in target tissues, enhancing insulin action without directly stimulating insulin secretion. Sulfonylureas and meglitinides are considered insulin secretagogues because their action increases the secretion of insulin by the pancreatic beta cells.

Patients must understand that oral agents are prescribed as an addition to (not as a substitute for) other treatment modalities, such as MNT and exercise. The use of oral antidiabetic medications may need to be halted temporarily and insulin prescribed if hyperglycemia develops that is attributable to infection, trauma, or surgery. (See later section on glycemic control in the hospitalized patient.)

Because mechanisms of action vary (Fig. 51-5), effects may be enhanced with the use of a multidose, or more than one medication (ADA, 2013). The use of multiple medications with different mechanisms of action is very common today. A combination of oral agents with insulin, usually glargine at bedtime, has also been used as a treatment for some patients with type 2 diabetes. Insulin therapy may be used from the onset for newly diagnosed patients with type 2 diabetes who are symptomatic and have high blood glucose and A1C levels (ADA, 2013).

Other Pharmacologic Therapy

Additional medications are available for use in the pharmacologic management of diabetes. These are adjunct therapies, not a substitute for insulin if insulin is required to control diabetes.

Pramlintide (Symlin), a synthetic analogue of human amylin, a hormone that is secreted by the beta cells of the pancreas, is approved for treatment of both type 1 and type 2 diabetes (Karch, 2012). It is used to control hyperglycemia in adults who have not achieved acceptable levels of glucose control despite the use of insulin at mealtimes. It is used with insulin, not in place of insulin, but the two cannot be combined in the same syringe. The goal of therapy is to minimize fluctuations in daily glucose levels and provide better glucose control. There is a high risk of hypoglycemia; therefore, a source of glucose must be available if hypoglycemia occurs. Pramlintide must be injected subcutaneously 2 inches from an insulin injection site (Karch, 2012). Patients are instructed to monitor their blood glucose levels closely during the initial period of use of pramlintide.

Exenatide (Byetta) is used as an adjunct therapy or monotherapy for patient with type 2 diabetes (Karch, 2012). It is derived from a hormone that is produced in the small intestine and has been found to be deficient in type 2 diabetes. It is normally released after food is ingested to delay gastric emptying and enhance insulin secretion, resulting in dampening of the rise in blood glucose levels after meals and a feeling of satiety. The return of the blood glucose level to normal results in decreased production of the hormone. Hypoglycemia is not a side effect of exenatide if adjustments are made in the sulfonylurea dose. Exenatide is associated with appetite suppression and weight loss. Exenatide must be injected twice a day within 1 hour before breakfast and dinner (Karch, 2012). It is not a substitute for insulin in patients who require insulin to control their diabetes.

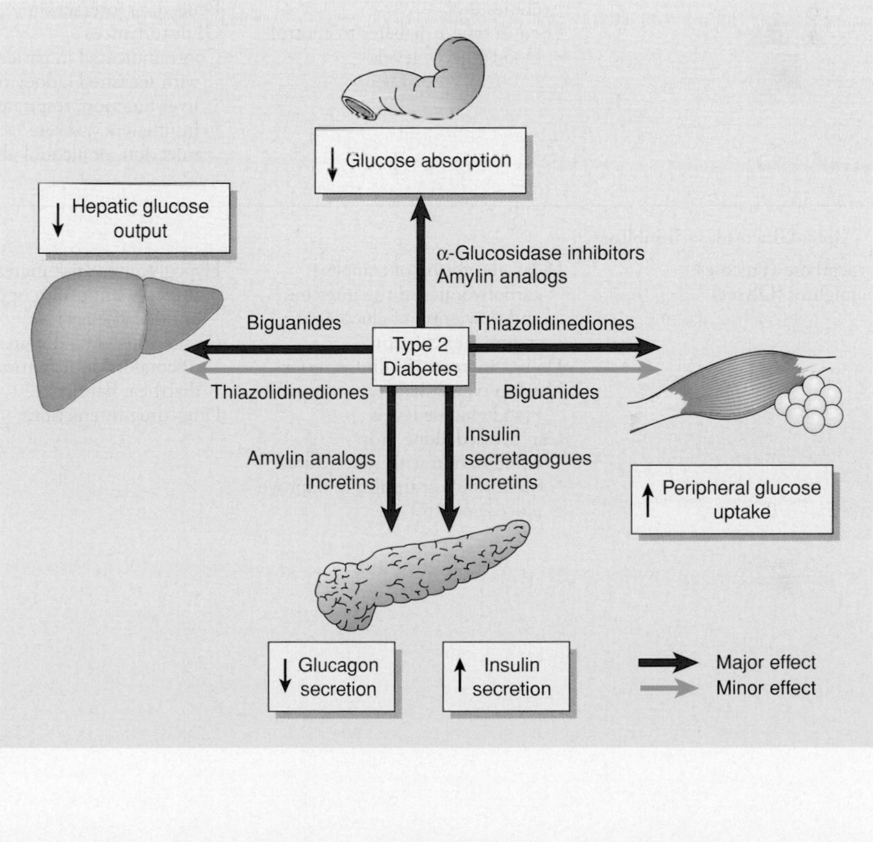
↓ Glucose absorption
↓ Hepatic glucose output
α-Glucosidase inhibitors
Amylin analogs
Biguanides
Thiazolidinediones
Type 2 Diabetes
Thiazolidinediones
Biguanides
Amylin analogs
Incretins
Insulin secretagogues
Incretins
↑ Peripheral glucose uptake
↓ Glucagon secretion
↑ Insulin secretion
Major effect
Minor effect

FIGURE 51-5 • Action sites of hypoglycemic agents and mechanisms of lowering blood glucose in type 2 diabetes. The incretins are the dipeptidyl peptidase-4 inhibitors and glucagonlike peptide-1 agonists.

TABLE 51-6 Oral Antidiabetic Agents

Generic (Trade) Name	Action/Indications	Side Effects	Implications
First-Generation Sulfonylureas			
chlorpropamide (Diabinese) tolazamide (Tolinase) tolbutamide (Orinase)	Less commonly used Used in type 2 diabetes to control blood glucose levels Stimulate beta cells of the pancreas to secrete insulin; may improve binding between insulin and insulin receptors or increase the number of insulin receptors	Hypoglycemia Mild GI symptoms Weight gain Drug–drug interactions (NSAIDs, warfarin [Coumadin], sulfonamides) Skin reactions	Monitor patient for hypoglycemia. Monitor blood glucose and urine ketone levels to assess effectiveness of therapy. Patients at high risk for hypoglycemia: advanced age, renal insufficiency. When taken with beta-adrenergic blocking agents, may mask usual warning signs and symptoms of hypoglycemia. Instruct patients to avoid the use of alcohol. Check for interactions with other medications. Hypersensitivity in those with sulfa allergy.
Second-Generation Sulfonylureas			
glipizide (Glucotrol, Glucotrol XL) glyburide (Micronase, Glynase, DiaBeta) glimepiride (Amaryl)	Stimulate beta cells of the pancreas to secrete insulin; may improve binding between insulin and insulin receptors or increase the number of insulin receptors Used in type 2 diabetes to control blood glucose levels Have more potent effects than 1st-generation sulfonylureas May be used in combination with metformin or insulin to improve glucose control	Hypoglycemia Mild GI symptoms Weight gain Drug–drug interactions (NSAIDs, warfarin, sulfonamides)	Monitor patient for hypoglycemia. Monitor blood glucose and urine ketone levels to assess effectiveness of therapy. Patients at high risk for hypoglycemia: advanced age, renal insufficiency. When taken with beta-adrenergic blocking agents, may mask usual warning signs and symptoms of hypoglycemia. Instruct patients to avoid the use of alcohol. Contraindicated with sulfa allergy.
Biguanides			
metformin (Glucophage, Glucophage XL, Fortamet) metformin with glyburide (Glucovance)	Inhibit production of glucose by the liver Increase body tissue sensitivity to insulin Decrease hepatic synthesis of cholesterol Used in type 2 diabetes to control blood glucose levels	Lactic acidosis Hypoglycemia if metformin is used in combination with insulin or other antidiabetic agents Drug–drug interaction GI disturbances Contraindicated in patients with impaired kidney or liver function, respiratory insufficiency, severe infection, or alcohol abuse	Monitor for lactic acidosis and hypoglycemia. Monitor kidney function. Patients taking metformin are at increased risk for acute kidney failure and lactic acidosis with the use of iodinated contrast material for diagnostic studies; metformin should be stopped 48 hours prior to and for 48 hours after the use of contrast agent or until kidney function is evaluated and normal. Check for interactions with other medications.
Alpha-Glucosidase Inhibitors			
acarbose (Precose) miglitol (Glyset)	Delay absorption of complex carbohydrates in the intestine and slow entry of glucose into systemic circulation Do not increase insulin secretion Used in type 2 diabetes to control blood glucose levels Can be used alone or in combination with sulfonylureas, metformin, or insulin to improve glucose control	Hypoglycemia (risk increased if used with insulin or other antidiabetic agents) GI side effects (abdominal discomfort or distention, diarrhea, flatulence) Drug–drug interactions	Must be taken with first bite of food to be effective Monitor for GI side effects (diarrhea, abdominal distention). Monitor for blood glucose levels to assess effectiveness of therapy. Monitor liver function studies every 3 months for 1 year, then periodically. Contraindicated in patients with GI or kidney dysfunction, or cirrhosis ***Alert*: Hypoglycemia must be treated with glucose, not sucrose.**

Generic (Trade) Name	Action/Indications	Side Effects	Implications
Non-Sulfonylurea Insulin Secretagogues			
repaglinide (Prandin) categorized as a meglitinide nateglinide (Starlix) categorized as a D-phenylalanine derivative	Stimulate pancreas to secrete insulin Used in type 2 diabetes to control blood glucose levels Can be used alone or in combination with metformin or thiazolidinediones to improve glucose control	Hypoglycemia/weight gain less likely than sulfonylureas Drug–drug interactions (with ketoconazole [Nizoral], fluconazole [Diflucan], erythromycin [E-Mycin], rifampin [Rifadin], isoniazid [INH])	Monitor blood glucose levels to assess effectiveness of therapy. Has rapid action and short half-life Should be taken only if able to eat a meal immediately Educate patients about symptoms of hypoglycemia. Monitor patients with impaired liver function and renal impairment. Has no effect on plasma lipids Is taken before each meal Check for interactions with other medications.
Thiazolidinediones (or Glitazones)			
pioglitazone (Actos) rosiglitazone (Avandia)	Sensitize body tissue to insulin; stimulate insulin receptor sites to lower blood glucose and improve action of insulin May be used alone or in combination with sulfonylurea, metformin, or insulin	Hypoglycemia (risk increased with the use of insulin or other antidiabetic agents) Anemia Weight gain, edema Decrease effectiveness of oral contraceptives Possible liver dysfunction Drug–drug interactions Hyperlipidemia (has variable effect on lipids; pioglitazone may be preferred choice in patients with lipid abnormalities) Impaired platelet function	Monitor blood glucose levels to assess effectiveness of therapy. Monitor liver function tests. Arrange dietary education to establish weight control program. Instruct patient taking oral contraceptives about increased risk of pregnancy.
Dipeptidyl Peptidase-4 (DPP-4) Inhibitors			
sitagliptin (Januvia) vildagliptin (Galvus)	Increase and prolong the action of incretin, a hormone that increases insulin release and decreases glucagon levels, with the result of improved glucose control	Upper respiratory infection Stuffy or runny nose and sore throat Headache Stomach discomfort and diarrhea Hypoglycemia, if used with sulfonylurea	Usually administered once a day Used alone or with other oral antidiabetic agents Instruct patient about signs and symptoms of hypoglycemia and other adverse effects to report. Monitor kidney function.

GI, gastrointestinal; NSAIDs, nonsteroidal anti-inflammatory drugs.

Nursing Management

Nursing management of patients with diabetes can involve treatment of a wide variety of physiologic disorders, depending on the patient's health status and whether the patient is newly diagnosed or seeking care for an unrelated health problem. Glucose control in patients diagnosed with diabetes as well as those who have not been diagnosed is an important consideration in the hospital setting. Nursing management of patients with DKA and HHS and of those with diabetes as a secondary diagnosis is discussed in subsequent sections of this chapter.

Because all patients with diabetes must master the concepts and skills necessary for long-term management and avoidance of potential complications of diabetes, a solid educational foundation is necessary for competent self-care and is an ongoing focus of nursing care.

Managing Glucose Control in the Hospital Setting

Hyperglycemia can prolong lengths of stay and increase infection rates and mortality; thus, nurses need to address glucose management in all hospital patients (Seggelke & Everhart, 2012). Hyperglycemia occurs most often in patients with known diabetes (i.e., type 1, type 2, gestational) and in those newly diagnosed with diabetes or stress hyperglycemia. Nursing management of hyperglycemia in the hospital uses the following principles:

- Blood glucose targets are 140 to 180 mg/dL.
- Insulin (subcutaneous or IV) is preferred to oral antidiabetic agents to manage hyperglycemia.
- Hospital insulin protocols or order sets should minimize complexity, ensure adequate staff training, include standardized hypoglycemic treatment, and make guidelines available for glycemic goals and insulin dosing.
- Appropriate timing of blood glucose checks, meal consumption, and insulin dose are all crucial for glucose control and to avoid hypoglycemia.

Providing Patient Education

Diabetes is a chronic illness that requires a lifetime of special self-management behaviors. Because MNT, physical activity, medication, and physical and emotional stress affect diabetic control, patients must learn to balance a multitude of factors.

Developing a Diabetes Education Plan

Changes in the health care system as a whole have had a major impact on diabetes education and training. Patients with new-onset type 1 diabetes are hospitalized for short periods or may be managed completely on an outpatient basis. Patients with new-onset type 2 diabetes are rarely hospitalized for initial care. Outpatient diabetes education and training programs have proliferated with increasing support of third-party reimbursement. All encounters with patients with diabetes are opportunities for reinforcement of self-management skills, regardless of the setting.

Many hospitals employ nurses who specialize in diabetes education and management and who are certified by the National Certification Board for Diabetes Educators as certified diabetes educators. However, because of the large number of patients with diabetes who are admitted to every unit of a hospital for reasons other than diabetes or its complications, all nurses play a vital role in identifying patients with diabetes, assessing self-care skills, providing basic education, reinforcing the education provided by the specialist, and referring patients for follow-up care after discharge. Diabetes patient education programs that have been peer-reviewed by the ADA as meeting National Standards for Diabetes Self-Management Education can be reimbursed for education.

Organizing Information. There are various strategies for organizing and prioritizing the vast amount of information that must be taught to patients with diabetes. In addition, many hospitals and outpatient diabetes centers have devised written guidelines, care plans, and documentation forms (often based on ADA guidelines) that may be used to document and evaluate education. One approach is to organize education using the seven tips for managing diabetes identified and developed by the American Association of Diabetes Educators (AADE, 2007): healthy eating, being active, monitoring, taking medication, problem solving, healthy coping, and reducing risks. The AADE can be contacted for additional information about assessment and documentation of outcomes of this approach to education.

Another general approach is to organize information and skills into two main categories: basic, initial, or "survival" skills and information, and in-depth (advanced) or continuing education.

Educating Patients About Survival Skills. Survival skills must be learned by all patients with newly diagnosed type 1 or type 2 diabetes and all patients receiving insulin for the first time. Basic information that patients must know to survive is included in Chart 51-5.

For patients with newly diagnosed type 2 diabetes, emphasis is initially placed on meal planning and exercise. Those who are starting to take oral sulfonylureas or insulin secretagogues need to know about detecting, preventing, and treating hypoglycemia. If diabetes has gone undetected for many years, the patient may already be experiencing some chronic diabetic complications. Therefore, for some patients with newly diagnosed type 2 diabetes, basic diabetes education must include information on preventive skills, such as foot care and eye care (e.g., planning yearly or more frequent complete [dilated eye] examinations by an ophthalmologist, understanding that retinopathy is largely asymptomatic until advanced stages).

Chart 51-5 PATIENT EDUCATION

Basic Survival Skills for People With Diabetes

Basic survival information to be included in education:

1. Pathophysiology
   a. Basic definition of diabetes (having a high blood glucose level)
   b. Normal blood glucose ranges and target blood glucose levels
   c. Effect of insulin and exercise (decrease glucose)
   d. Effect of food and stress, including illness and infections (increase glucose)
   e. Basic treatment approaches
2. Treatment modalities
   a. Administration of insulin and oral antidiabetes medications
   b. Meal planning (food groups, timing of meals)
   c. Monitoring of blood glucose and urine ketones
3. Recognition, treatment, and prevention of acute complications
   a. Hypoglycemia
   b. Hyperglycemia
4. Pragmatic information
   a. Where to buy and store insulin, syringes, and glucose monitoring supplies
   b. When and how to contact the primary provider

Patients also need to realize that once they master the basic skills and information, further diabetes education must be pursued. Acquiring in-depth and advanced diabetes knowledge occurs throughout the patient's lifetime, both formally through programs of continuing education and informally through experience and sharing of information with other people with diabetes.

Planning In-Depth and Continuing Education. This education involves more details related to survival skills (e.g., learning to vary food choices [carbohydrate counting] and insulin, preparing for travel) as well as learning preventive measures for avoiding long-term diabetic complications. Preventive measures include foot care, eye care, general hygiene (e.g., skin care and oral hygiene), and risk factor management (e.g., blood pressure control and blood glucose normalization (cholesterol/lipid control) (AADE, 2005).

More advanced continuing education may include alternative methods for insulin delivery, such as the insulin pump, and algorithms or rules for evaluating and adjusting insulin doses. The degree of advanced diabetes education to be provided depends on the patient's interest and ability. However, learning preventive measures (especially foot and eye care) is mandatory for early detection and treatment to reduce the occurrence of amputations and blindness in patients with diabetes.

Assessing Readiness to Learn

Before initiating diabetes education, the nurse assesses the patient's (and family's) readiness to learn. When patients are first diagnosed with diabetes (or first told of their need for insulin), they often go through stages of the grieving process. These stages may include shock and denial, anger, depression, negotiation, and acceptance. The amount of time it takes for the patient and family members to work through the grieving process varies from patient to patient. They may experience helplessness, guilt, altered body image, loss of self-esteem, and concern about the future. The nurse must assess the patient's coping strategies and reassure the patient and family that feelings of depression and shock are normal.

Asking the patient and family about their major concerns or fears is an important way to learn about any misinformation that may be contributing to anxiety. Simple, direct information should be provided to dispel misconceptions. Once the patient masters survival skills, more information is provided.

Patients who are in the hospital rarely have the luxury of waiting until they feel ready to learn; short lengths of hospital stay necessitate initiation of survival skill education as early as possible. This gives the patient the opportunity to practice skills with supervision by the nurse before discharge. Follow-up by home health nurses is often necessary for reinforcement of survival skills.

The nurse evaluates the patient's social situation for factors that may influence the diabetes treatment and education plan, such as:

- Low literacy level (may be evaluated while assessing for visual deficits by having the patient read from educational materials)
- Limited financial resources or lack of health insurance
- Presence or absence of family support
- Typical daily schedule (the patient is asked about timing and number of usual daily meals, work and exercise schedule, plans for travel)
- Neurologic deficits caused by stroke, other neurologic disorders, or other disabling conditions, obtained from the patient's health history and physical assessment (the patient is assessed for aphasia or decreased ability to follow simple commands)
- Cultural beliefs that may impact adherence to a regimen (Chart 51-6)

Educating Experienced Patients

Nurses need to continue to assess the skills and self-care behaviors of patients who have had diabetes for many years. Assessment of these patients must include direct observation of skills, not just the patient's self-report of self-care behaviors. In addition, these patients must be fully aware of preventive measures related to foot care, eye care, and risk factor management. Those experiencing long-term diabetic complications for the first time may go through the grieving process again. Some patients may have a renewed interest in diabetes self-care in the hope of delaying further complications. Others may be overwhelmed by feelings of guilt and depression. The patient is encouraged to discuss feelings and fears related to complications. Meanwhile, the nurse provides appropriate information regarding diabetic complications.

Determining Education Methods

Maintaining flexibility with regard to education approaches is important. Providing education on skills and information in a logical sequence is not always the most helpful method for patients. For example, many patients fear self-injection. Before they learn how to prepare, purchase, store, and mix insulins, they should be taught to insert the needle and inject insulin (or practice with saline solution).

Various tools can be used to complement education. Many of the companies that manufacture products for diabetes self-care also provide booklets and videotapes to assist in patient education. Educational materials are also available from the AADE and the ADA. It is important to use a variety of written handouts that match the patient's learning needs (including different languages, low-literacy information, and large print) and reading level and to ensure that these materials are technically accurate. Patients can continue learning about

Chart 51-6 NURSING RESEARCH PROFILE
Impact of a Comprehensive Diabetes Management Program

Welch, G., Allen, N. A., Zagarins, A. E., et al. (2011). Comprehensive diabetes management program for poorly controlled Hispanic type 2 patients at a community health center. *Diabetes Educator, 37*(5), 680–688.

Purpose

The purpose of this study was to determine the usefulness of the Comprehensive Diabetes Management Program (CDMP) in an urban Hispanic population with type 2 diabetes. CDMP is an interactive, Web-based diabetes management tool that focuses on clinical management, lifestyle modification, and psychological health.

Design

This was a randomized controlled trial in which patients were assigned to one of two groups: Attention Control or Intervention. The Attention Control group received some information through booklets and attended seven 1-hour group sessions regarding diabetes management conducted by bilingual/bicultural staff. The Intervention group received seven 1-hour sessions conducted by bilingual/bicultural instructors in addition to the CDMP interactive computer program. A one-page summary of self-management behaviors and identification of barriers to self-management was generated and used in the plan of care. Cultural sensitivity was an important element of the intervention. Clinical data were measured at baseline as well as diabetes distress and self-management behaviors. Measures were repeated at 1 year.

Findings

Attendance in the group sessions was high in both groups. The physical measures for blood glucose and blood pressure controls were significantly improved in the Intervention group. Diabetes distress increased in the Attention Control group but decreased in the Intervention group. More participation in eye and foot examinations were found in the Intervention group.

Nursing Implications

Diabetes education in self-management is vital to achieving the metabolic control needed to prevent complications. Cultural sensitivity is a key component of education in particular groups. In this study, participants who were of Hispanic origin needed materials translated into Spanish and bilingual instructors. Nutrition education requires consideration of ethnic differences in food components. This helps to minimize or eliminate barriers as well as facilitates adherence over the long time. Nurses need to be sensitive to cultural differences and incorporate them into a plan of care as much as possible so that adherence to that plan of care may be feasible.

diabetes care by participating in activities sponsored by local hospitals and diabetes organizations. In addition, magazines and Web sites with information on diabetes management are available (see the Resources section).

Educating Patients to Self-Administer Insulin

Insulin injections are self-administered into the subcutaneous tissue with the use of special insulin syringes. Basic information includes explanations of the equipment, insulins, and syringes and how to mix insulin, if necessary.

Storing Insulin. Whether insulin is the short- or the long-acting preparation, vials not in use, including spare vials or pens, should be refrigerated. Extremes of temperature should be avoided; insulin should not be allowed to freeze and should not be kept in direct sunlight or in a hot car. The insulin vial in use should be kept at room temperature to reduce local irritation at the injection site, which may occur if cold insulin is injected. If a vial of insulin will be used up within 1 month, it may be kept at room temperature. The patient should be instructed to always have a spare vial of the type or types of insulin that he or she uses (ADA, 2013). Cloudy insulins should be thoroughly mixed by gently inverting the vial or rolling it between the hands before drawing the solution into a syringe or a pen (Karch, 2012). The patient needs to be educated to pay attention to the expiration date on any type of insulin.

Bottles of intermediate-acting insulin should also be inspected for flocculation, which is a frosted, whitish coating inside the bottle. This occurs most commonly with insulins that are exposed to extremes of temperature. If a frosted, adherent coating is present, some of the insulin is bound, inactive, and should not be used.

Selecting Syringes. Syringes must be matched with the insulin concentration (e.g., U-100). Currently, three sizes of U-100 insulin syringes are available:

- 1-mL syringe, 100-unit capacity
- 0.5-mL syringe, 50-unit capacity
- 0.3-mL syringe, 30-unit capacity

The concentration of insulin used in the United States is U-100; that is, there are 100 units per milliliter (or cubic centimeter). Small syringes allow patients who require small amounts of insulin to measure and draw up the amount of insulin accurately. There is a U-500 (500 units/mL) concentration of insulin available by special order for patients who have severe insulin resistance and require massive doses of insulin.

Most insulin syringes have a disposable 27- to 29-gauge needle that is approximately 0.5 inch long. The smaller syringes are marked in 1-unit increments and may be easier to use for patients with visual deficits and those taking very small doses of insulin. The 1-mL syringes are marked in 1- and 2-unit increments. A small disposable insulin needle (31 gauge, 8 mm long) is available for very thin patients and children.

Mixing Insulins. When rapid- or short-acting insulins are to be given simultaneously with longer-acting insulins, they are usually mixed together in the same syringe; the longer-acting insulins must be mixed thoroughly before drawing into the syringe. The most important issue is that patients be consistent in how they prepare their insulin injections from day to day.

There are varying opinions regarding which type of insulin (short-acting or longer-acting) should be drawn up into the syringe first when they are going to be mixed, but the ADA recommends that the regular insulin be drawn up first. The most important issues are (1) that patients are consistent in technique, so as not to draw up the wrong dose in error or the wrong type of insulin, and (2) that patients not inject one type of insulin into the bottle containing a different type of insulin. Injecting cloudy insulin into a vial of clear insulin contaminates the entire vial of clear insulin and alters its action.

For patients who have difficulty mixing insulins, several options are available. They may use a premixed insulin, they may have prefilled syringes prepared (see Fig. 51-3), or they may take two injections. Premixed insulins are available in several different ratios of NPH insulin to regular insulin (Karch, 2012). The ratio of 70/30 (70% NPH and 30% regular insulin in one bottle) is most common; this combination is available as Novolin 70/30 (Novo Nordisk) and Humulin 70/30 (Lilly). Combinations with a ratio of 75% NPL (neutral protamine lispro) and 25% insulin lispro are also available. NPL is used only in the mix with Humalog; its action is the same as NPH. The appropriate initial dosage of premixed insulin must be calculated so that the ratio of NPH to regular insulin most closely approximates the separate doses needed.

For patients who can inject insulin but who have difficulty drawing up a single or mixed dose, syringes may be prefilled with the help of home care nurses or family and friends. A 3-week supply of insulin syringes may be prepared and kept in the refrigerator but warmed to room temperature before administration. The prefilled syringes should be stored with the needle in an upright position to avoid clogging of the needle; they should be mixed thoroughly by inverting syringe several times before the insulin is injected.

Withdrawing Insulin. Most (if not all) of the printed materials available on insulin dose preparation instruct patients to inject air into the bottle of insulin equivalent to the number of units of insulin to be withdrawn. The rationale for this is to prevent the formation of a vacuum inside the bottle, which would make it difficult to withdraw the proper amount of insulin.

Selecting and Rotating the Injection Site. The four main areas for injection are the abdomen, upper arms (posterior surface), thighs (anterior surface), and hips (Fig. 51-6). Insulin is absorbed faster in some areas of the body than others. The speed of absorption is greatest in the abdomen and decreases progressively in the arm, thigh, and hip, respectively.

Systematic rotation of injection sites within an anatomic area is recommended to prevent localized changes in fatty tissue (lipodystrophy). In addition, to promote consistency in insulin absorption, the patient should be encouraged to use all available injection sites within one area rather than randomly rotating sites from area to area. For example, some patients almost exclusively use the abdominal area, administering each injection 0.5 to 1 inch away from the previous injection. Another approach to rotation is always to use the same area at the same time of day. For example, patients may inject morning doses into the abdomen and evening doses into the arms or legs.

A few general principles apply to all rotation patterns. First, the patient should try not to use the exact same site

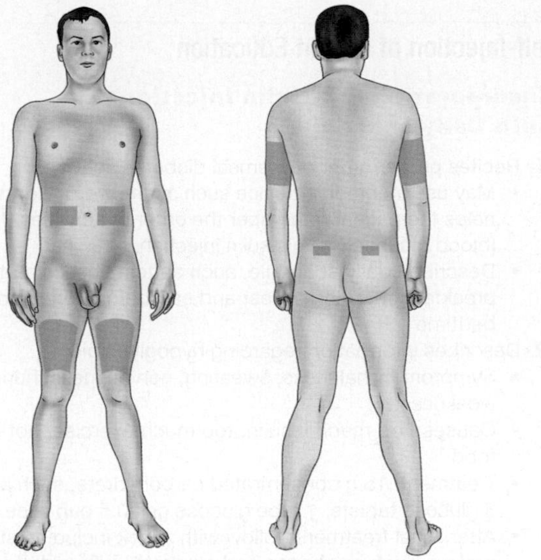

FIGURE 51-6 • Suggested areas for insulin injection.

more than once in 2 to 3 weeks. In addition, if the patient is planning to exercise, insulin should not be injected into the limb that will be exercised because this will cause the drug to be absorbed faster, which may result in hypoglycemia.

Preparing the Skin. The use of alcohol to cleanse the skin is not necessary, but patients who have learned this technique often continue to use it. They should be cautioned to allow the skin to dry after cleansing with alcohol. If the skin is not allowed to dry before the injection, the alcohol may be carried into the tissues, resulting in a localized reddened area and a burning sensation.

Inserting the Needle. There are varying approaches to inserting the needle for insulin injections. The correct technique is based on the need for the insulin to be injected into the subcutaneous tissue (Chart 51-7). Injection that is too deep (e.g., intramuscular) or too shallow (intradermal) may affect the rate of absorption of the insulin. For a normal or overweight person, a 90-degree angle is the best insertion angle. Aspiration (inserting the needle and then

Chart 51-7 **PATIENT EDUCATION**

Self-Injection of Insulin

1. With one hand, stabilize the skin by spreading it or pinching up a large area.

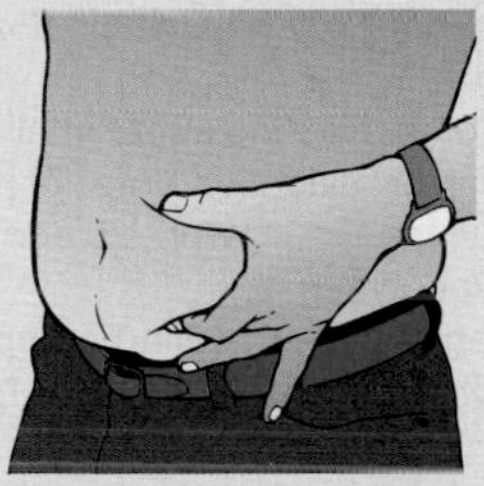

Pinching the skin

2. Pick up syringe with the other hand, and hold it as you would a pencil. Insert needle straight into the skin.*

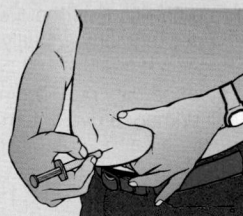

Inserting the needle into the skin

3. To inject the insulin, push the plunger all the way in.

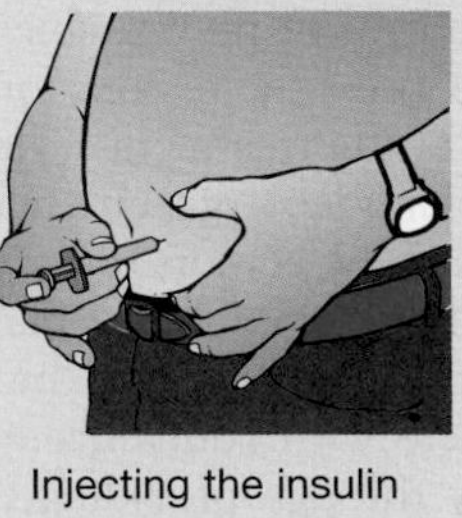

Injecting the insulin

4. Pull needle straight out of skin. Press cotton ball over injection site for several seconds.

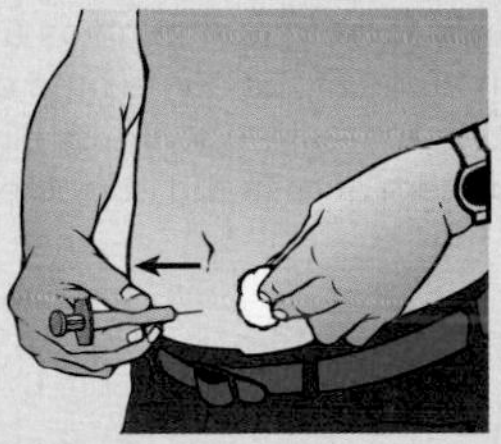

Removing the needle and holding cotton ball over site

5. Use disposable syringe *only once* and discard into hard plastic container (with a tight-fitting top) such as an empty bleach or detergent container.† Follow state regulations for disposal of syringes and needles.

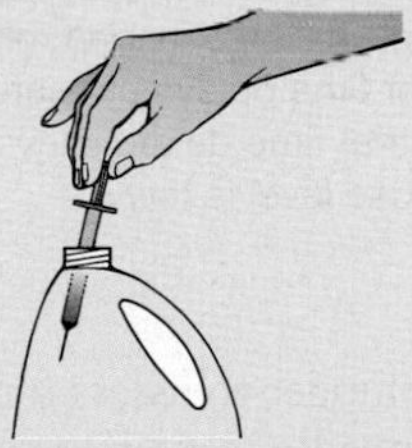

Disposing of syringe

*Some patients may be taught to insert the needle at a 45-degree angle.

†Although some studies suggest that reusing disposable syringes may be safe, it is recommended that this be done only in the absence of poor personal hygiene, an acute concurrent illness, open wounds on the hands, or decreased resistance to infection.

Chart 51-8 Outcome Criteria for Determining Effectiveness of Self-Injection of Insulin Education

Equipment

Insulin

1. Identifies information on label of insulin bottle:
   - Type (e.g., NPH, regular, 70/30)
   - Manufacturer (Lilly, Novo Nordisk)
   - Concentration (e.g., U-100)
   - Expiration date
2. Checks appearance of insulin:
   - Clear or milky white
   - Checks for flocculation (clumping, frosted appearance)
3. Identifies where to purchase and store insulin:
   - Indicates approximately how long bottle will last (1,000 units per bottle U-100 insulin)
   - Indicates how long opened bottles can be used

Syringes

1. Identifies concentration (U-100) marking on syringe
2. Identifies size of syringe (e.g., 100 unit, 50 unit, 30 unit)
3. Describes appropriate disposal of used syringe

Preparation and Administration of Insulin Injection

1. Draws up correct amount and type of insulin
2. Properly mixes 2 insulins if necessary
3. Inserts needle and injects insulin
4. Describes site rotation:
   - Demonstrates injection with all anatomic areas to be used
   - Describes pattern for rotation, such as using abdomen only or using certain areas at the same time of day
   - Describes system for remembering site locations, such as horizontal pattern across the abdomen as if drawing a dotted line

Knowledge of Insulin Action

1. Lists prescription:
   - Type and dosage of insulin
   - Timing of insulin injections
2. Describes approximate time course of insulin action:
   - Identifies long- and short-acting insulins by name
   - States approximate time delay until onset of insulin action
   - Identifies need to delay food until 5–15 minutes after injection of rapid-acting insulin (lispro, aspart, glulisine [Apidra])
   - Knows that longer time delays are safe when blood glucose level is high and that time delays may need to be shortened when blood glucose level is low

Incorporation of Insulin Injections Into Daily Schedule

1. Recites proper order of premeal diabetes activities:
   - May use mnemonic device such as the word "tie," which helps the patient remember the order of activities ("t" = test [blood glucose], "i" = insulin injection, "e" = eat)
   - Describes daily schedule, such as test, insulin, eat before breakfast and dinner; test and eat, before lunch and bedtime
2. Describes information regarding hypoglycemia:
   - Symptoms: shakiness, sweating, nervousness, hunger, weakness
   - Causes: too much insulin, too much exercise, not enough food
   - Treatment: 15 g concentrated carbohydrate, such as 2 or 3 glucose tablets, 1 tube glucose gel, 0.5 cup juice
   - After initial treatment, follow with snack including starch and protein, such as cheese and crackers, milk and crackers, half sandwich.
3. Describes information regarding prevention of hypoglycemia:
   - Avoids delays in meal timing
   - Eats a meal or snack approximately every 4–5 hours (while awake)
   - Does not skip meals
   - Increases food intake before exercise if blood glucose level is <100 mg/dL
   - Checks blood glucose regularly
   - Identifies safe modification of insulin doses consistent with management plan
   - Carries a form of fast-acting sugar at all times
   - Wears a medical identification bracelet
   - Educates family, friends, coworkers about signs and treatment of hypoglycemia
   - Has family, roommates, traveling companions learn to use injectable glucagon for severe hypoglycemic reactions
4. Maintains regular follow-up for evaluation of diabetes control:
   - Keeps written record of blood glucose, insulin doses, hypoglycemic reactions, variations in diet
   - Keeps all appointments with health professionals
   - Sees primary provider regularly (usually 2–4 times per year)
   - States how to contact primary provider in case of emergency
   - States when to call primary provider to report variations in blood glucose levels

pulling back on the plunger to assess for blood being drawn into the syringe and needle in vein) is not necessary. Many patients who have been using insulin for an extended period have eliminated this step from their insulin injection routine with no apparent adverse effects. Chart 51-8 details how to evaluate the effectiveness of self-injection of insulin education.

Disposing of Syringes and Needles. Insulin syringes and pens, needles, and lancets should be disposed of according to local regulations. If community disposal programs are unavailable, used sharps should be placed in a puncture-resistant container. The patient should contact local trash authorities for instructions about proper disposal of filled containers, which should not be mixed with containers to be recycled.

Promoting Home and Community-Based Care

Educating the Patient About Self-Care

If poor glucose control or preventable complications occur, the nurse needs to assess the reasons for the patient's ineffective management of the treatment regimen. It should not be assumed that problems with diabetes management are related to the patient's willful decision to ignore self-management. The patient may have forgotten or may have never learned certain information, or there may be cultural or religious beliefs that interfere with adherence. The problem may be correctable simply through providing complete information and ensuring that the patient understands the information. The focus of diabetes education should be patient empowerment. Patient education must address behavior change, self-efficacy, and health beliefs.

If knowledge deficit is not the issue, physical or emotional factors may be impairing the patient's ability to perform self-care skills. For example, decreased visual acuity may impair the patient's ability to administer insulin accurately, measure the blood glucose level, or inspect the skin and feet. In addition, decreased joint mobility (especially in older adults) or preexisting disability may impair the patient's ability to inspect the bottom of the feet. Denial of the diagnosis or depression may impair the patient's ability to carry out multiple daily self-care measures. The patient whose family, personal, or work problems may be of higher priority may benefit from assistance in establishing priorities. The nurse must also assess the patient for infection or emotional stress, which may lead to elevated blood glucose levels despite adherence to the treatment regimen.

The following approaches are helpful for promoting self-care management skills:

- Address any underlying factors (e.g., knowledge deficit, self-care deficit, illness) that may affect diabetic control.
- Simplify the treatment regimen if it is too difficult for the patient to follow.
- Adjust the treatment regimen to meet patient requests (e.g., adjust diet or insulin schedule to allow increased flexibility in meal content or timing).
- Establish a specific plan or contract with each patient with simple, measurable goals.
- Provide positive reinforcement of self-care behaviors performed instead of focusing on behaviors that were neglected (e.g., positively reinforce blood glucose tests that were performed instead of focusing on the number of missed tests).
- Help the patient identify personal motivating factors rather than focusing on wanting to please primary providers.
- Encourage the patient to pursue life goals and interests, and discourage an undue focus on diabetes.

Continuing Care

The degree to which patients interact with primary providers to obtain ongoing care depends on many factors. Age, socioeconomic level, existing complications, type of diabetes, and comorbid conditions may dictate the frequency of follow-up visits. Many patients with diabetes are seen by home health nurses for diabetes education, wound care, insulin preparation, or assistance with glucose monitoring. Even patients who achieve excellent glucose control and have no complications can expect to see their primary provider at least twice a year for ongoing evaluation and should receive routine nutrition updates. In addition, the nurse should remind the patient to participate in recommended health promotion activities (e.g., immunizations) and age-appropriate health screenings (e.g., pelvic examinations, mammograms).

Participation in support groups is encouraged for patients who have had diabetes for many years as well as for those who are newly diagnosed. Such participation may help the patient and family cope with changes in lifestyle that occur with the onset of diabetes and its complications. People who participate in support groups often share valuable information and experiences and learn from others. Support groups provide an opportunity for discussion of strategies to deal with diabetes and its management and to clarify and verify information with nurses or other health care professionals. Participation in support groups may also promote healthy activities.

Concept Mastery Alert

Visit thePoint to view an Interactive Tutorial on diabetes and related fundamental concepts.

ACUTE COMPLICATIONS OF DIABETES

There are three major acute complications of diabetes related to short-term imbalances in blood glucose levels: hypoglycemia, DKA, and HHS.

Hypoglycemia (Insulin Reactions)

Hypoglycemia means low (hypo) sugar in the blood (glycemia) and occurs when the blood glucose falls to less than 70 mg/dL (3.7 mmol/L). Severe hypoglycemia is when glucose levels are less the 40 mg/dL (2.5 mmol/L) (Seggelke & Everhart, 2012). It can occur when there is too much insulin or oral hypoglycemic agents, too little food, or excessive physical activity. Hypoglycemia may occur at any time of the day or night. It often occurs before meals, especially if meals are delayed or snacks are omitted. For example, midmorning hypoglycemia may occur when the morning insulin is peaking, whereas hypoglycemia that occurs in the late afternoon coincides with the peak of the morning NPH insulin. Middle-of-the-night hypoglycemia may occur because of peaking evening or predinner NPH insulins, especially in patients who have not eaten a bedtime snack.

Gerontologic Considerations

In older patients with diabetes, hypoglycemia is a particular concern for many reasons:

- Older adults frequently live alone and may not recognize the symptoms of hypoglycemia.
- With decreasing kidney function, it takes longer for oral hypoglycemic agents to be excreted by the kidneys.
- Skipping meals may occur because of decreased appetite or financial limitations.
- Decreased visual acuity may lead to errors in insulin administration.

Clinical Manifestations

The clinical manifestations of hypoglycemia may be grouped into two categories: adrenergic symptoms and central nervous system (CNS) symptoms.

In mild hypoglycemia, as the blood glucose level falls, the sympathetic nervous system is stimulated, resulting in a surge of epinephrine and norepinephrine. This causes symptoms such as sweating, tremor, tachycardia, palpitation, nervousness, and hunger.

In moderate hypoglycemia, the drop in blood glucose level deprives the brain cells of needed fuel for functioning. Signs of impaired function of the CNS may include inability to concentrate, headache, lightheadedness, confusion, memory lapses, numbness of the lips and tongue, slurred speech, impaired coordination, emotional changes, irrational or combative behavior, double vision, and drowsiness. Any combination of these symptoms (in addition to adrenergic symptoms) may occur with moderate hypoglycemia.

In severe hypoglycemia, CNS function is so impaired that the patient needs the assistance of another person for treatment of hypoglycemia. Symptoms may include disoriented behavior, seizures, difficulty arousing from sleep, or loss of consciousness.

Concept Mastery Alert

It is important to check the patient's blood glucose level and correlate it with the patient's symptoms. If the patient's blood glucose level is low, but he or she is not exhibiting any symptoms, the nurse should double-check the glucose level to ensure that it is correct.

Assessment and Diagnostic Findings

Symptoms of hypoglycemia may occur suddenly and vary considerably from person to person. Decreased hormonal (adrenergic) response to hypoglycemia may contribute to lack of symptoms of hypoglycemia. This occurs in some patients who have had diabetes for many years. It may be related to autonomic neuropathy, which is a chronic diabetic complication (see later discussion). As the blood glucose level falls, the normal surge in adrenalin does not occur, and the usual adrenergic symptoms, such as sweating and shakiness, do not take place. The hypoglycemia may not be detected until moderate or severe CNS impairment occurs. Affected patients must perform SMBG on a frequent regular basis, especially before driving or engaging in other potentially dangerous activities.

Management

Treating With Carbohydrates

Immediate treatment must be given when hypoglycemia occurs (Seggelke & Everhart, 2012). The usual recommendation is for 15 g of a fast-acting concentrated source of carbohydrate (see Chart 51-8). It is not necessary to add sugar to juice, even if it is labeled as unsweetened juice, because the fruit sugar in juice contains enough carbohydrate to raise the blood glucose level. Adding table sugar to juice may cause a sharp increase in the blood glucose level, and patients may experience hyperglycemia for hours after treatment.

Initiating Emergency Measures

In emergency situations, for adults who are unconscious and cannot swallow, an injection of glucagon 1 mg can be administered either subcutaneously or intramuscularly (Seggelke & Everhart, 2012). Glucagon is a hormone produced by the alpha cells of the pancreas that stimulates the liver to breakdown glycogen, the stored glucose. Injectable glucagon is packaged as a powder in 1-mg vials and must be mixed with a diluent immediately before being injected. After injection of glucagon, the patient may take as long as 20 minutes to regain consciousness. A concentrated source of carbohydrate followed by a snack should be given to the patient on awakening to prevent recurrence of hypoglycemia (because the duration of the action of 1 mg of glucagon is brief—its onset is 8 to 10 minutes, and its action lasts 12 to 27 minutes) and to replenish liver stores of glucose. Some patients experience nausea after the administration of glucagon. If this occurs, the patient should be turned to the side to prevent aspiration in case the patient vomits.

Glucagon is sold by prescription only and should be part of the emergency supplies available to patients with diabetes who require insulin. Family members, friends, neighbors, and coworkers should be instructed in the use of glucagon, especially for patients who have little or no warning of hypoglycemic episodes. Patients should be instructed to notify their primary provider after severe hypoglycemia has occurred and been treated. Close monitoring for 24 hours following a hypoglycemic episode is indicated because the patient is at increased risk of another episode (Seggelke & Everhart, 2012).

In hospitals and emergency departments, for patients who are unconscious or cannot swallow, 25 to 50 mL of dextrose 50% in water ($D_{50}W$) may be administered IV. The effect is usually seen within minutes. The patient may complain of a headache and of pain at the injection site. Ensuring patency of the IV line used for injection of 50% dextrose is essential because hypertonic solutions such as 50% dextrose are very irritating to veins.

Providing Patient Education

Hypoglycemia is prevented by a consistent pattern of eating, administering insulin, and exercising. Between-meal and bedtime snacks may be needed to counteract the maximum insulin effect. In general, the patient should cover the time of peak activity of insulin by eating a snack and by taking additional food when physical activity is increased. Routine blood glucose tests are performed so that changing insulin requirements may be anticipated and the dosage adjusted. Because unexpected hypoglycemia can occur, all patients treated with insulin should wear an identification bracelet or tag stating that they have diabetes.

Patients, family members, and coworkers must be instructed to recognize the symptoms of hypoglycemia. Family members in particular must be made aware that any subtle (but unusual) change in behavior may be an indication of hypoglycemia. They should be taught to encourage and even insist that the person with diabetes assess blood glucose levels if hypoglycemia is suspected. Some patients become very resistant to testing or eating and become angry at family members who are trying to treat the hypoglycemia. Family members must be taught to persevere and to understand that the hypoglycemia can cause irrational behavior, due to low supply glucose to the brain.

Autonomic neuropathy or beta-blockers such as propranolol (Inderal) to treat hypertension or cardiac dysrhythmias may mask the typical symptoms of hypoglycemia. It is very important that these patients perform blood glucose tests on a frequent and regular basis. Patients who have type 2 diabetes and who take oral sulfonylurea agents may also develop hypoglycemia, which can be prolonged and severe; this is a particular risk for older adult patients.

It is important that patients with diabetes, especially those receiving insulin, learn to carry some form of simple sugar with them at all times (ADA, 2013). There are commercially prepared glucose tablets and gels that the patient may find convenient to carry. If the patient has a hypoglycemic reaction and does not have any of the recommended emergency foods available, he or she should eat any available food (preferably a carbohydrate food).

Patients are advised to refrain from eating high-calorie, high-fat dessert foods (e.g., cookies, cakes, doughnuts, ice cream) to treat hypoglycemia because their high fat content may slow the absorption of the glucose and resolution of the hypoglycemic symptoms. The patient may subsequently eat more of the foods when symptoms do not resolve rapidly, which may cause very high blood glucose levels for several hours and may contribute to weight gain.

Patients who feel unduly restricted by their meal plan may view hypoglycemic episodes as a time to reward themselves with desserts. Instructing these patients to incorporate occasional desserts into the meal plan may be more effective, because this may make it easier for them to limit their treatment of hypoglycemic episodes to simple (low-calorie) carbohydrates such as juice or glucose tablets. Patients should be instructed to report all severe hypoglycemic episodes in addition to any increase in incidence, frequency, and severity to the primary provider.

Diabetic Ketoacidosis

DKA is caused by an absence or markedly inadequate amount of insulin. This deficit in available insulin results in disorders in the metabolism of carbohydrate, protein, and fat. The three main clinical features of DKA are as follows:

- Hyperglycemia
- Dehydration and electrolyte loss
- Acidosis

Pathophysiology

Without insulin, the amount of glucose entering the cells is reduced, and production and release of glucose by the liver (gluconeogenesis) is increased, leading to hyperglycemia (Fig. 51-7). In an attempt to rid the body of the excess glucose, the kidneys excrete the glucose along with water and electrolytes (e.g., sodium, potassium). This osmotic diuresis, which is characterized by excessive urination (polyuria), leads to dehydration and marked electrolyte loss (Porth & Matfin, 2009). Patients with severe DKA may lose up to 6.5 L of water and up to 400 to 500 mEq each of sodium, potassium, and chloride over a 24-hour period.

Another effect of insulin deficiency or deficit is the breakdown of fat (lipolysis) into free fatty acids and glycerol. The free fatty acids are converted into ketone bodies by the liver. Ketone bodies are acids; their accumulation in the circulation due to lack of insulin leads to metabolic acidosis.

Three main causes of DKA are decreased or missed dose of insulin, illness or infection, and undiagnosed and untreated diabetes (DKA may be the initial manifestation of type 1 diabetes). An insulin deficiency may result from an insufficient dosage of insulin prescribed or from insufficient insulin being administered by the patient. Errors in insulin dosage may be made by patients who are ill and who assume that if they are eating less or if they are vomiting, they must decrease their insulin doses. (Because illness, especially

Physiology ⋯ Pathophysiology

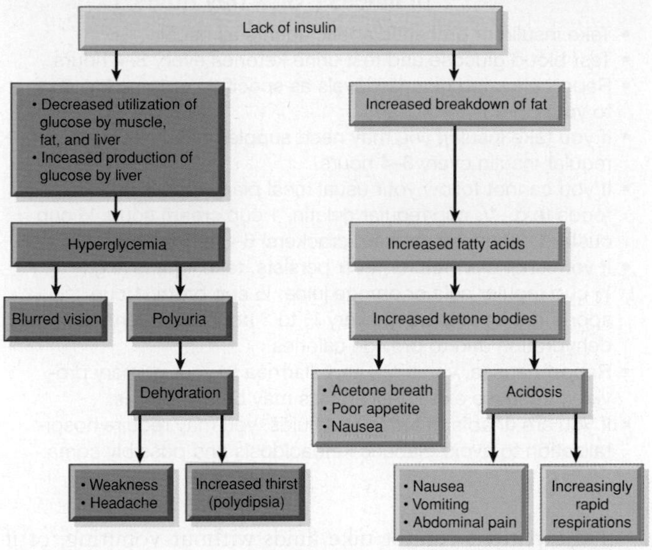

Lack of insulin
• Decreased utilization of glucose by muscle, fat, and liver
• Inceased production of glucose by liver
Increased breakdown of fat
Hyperglycemia
Increased fatty acids
Blurred vision
Polyuria
Increased ketone bodies
Dehydration
• Acetone breath
• Poor appetite
• Nausea
Acidosis
• Weakness
• Headache
Increased thirst (polydipsia)
• Nausea
• Vomiting
• Abdominal pain
Increasingly rapid respirations

FIGURE 51-7 • Abnormal metabolism that causes signs and symptoms of diabetic ketoacidosis. Redrawn from Pearce, M. A., Rosenberg, C. S., & Davidson, M. D. (2003). Patient education. In Davidson, M. B. (Ed.). *Diabetes mellitus: Diagnosis and treatment*. New York: Churchill Livingstone.

infections, can cause increased blood glucose levels, the patient does not need to decrease the insulin dose to compensate for decreased food intake when ill and may even need to increase the insulin dose.)

Other potential causes of decreased insulin include patient error in drawing up or injecting insulin (especially in patients with visual impairments), intentional skipping of insulin doses (especially in adolescents with diabetes who are having difficulty coping with diabetes or other aspects of their lives), or equipment problems (e.g., occlusion of insulin pump tubing). Illness and infections are associated with insulin resistance. In response to physical (and emotional) stressors, there is an increase in the level of "stress" hormones—glucagon, epinephrine, norepinephrine, cortisol, and growth hormone. These hormones promote glucose production by the liver and interfere with glucose utilization by muscle and fat tissue, counteracting the effect of insulin. If insulin levels are not increased during times of illness and infection, hyperglycemia may progress to DKA (ADA, 2013).

Prevention

For prevention of DKA related to illness, "sick day rules" for managing diabetes when ill (Chart 51-9) should be reviewed with patients. The most important concept in this is to never eliminate insulin doses when nausea and vomiting occur. Instead, the patient should take the usual insulin dose (or previously prescribed special sick day doses) and then attempt to consume frequent small portions of carbohydrates (including foods usually avoided, such as juices, regular sodas, and gelatin). Drinking fluids every hour is important to prevent dehydration. Blood glucose and urine ketones must be assessed every 3 to 4 hours.

Chart 51-9 PATIENT EDUCATION

Guidelines to Follow During Periods of Illness ("Sick Day Rules")

- Take insulin or oral antidiabetic agents as usual.
- Test blood glucose and test urine ketones every 3–4 hours.
- Report elevated glucose levels as specified or urine ketones to your primary provider.
- If you take insulin, you may need supplemental doses of regular insulin every 3–4 hours.
- If you cannot follow your usual meal plan, substitute soft foods (e.g., 1/3 cup regular gelatin, 1 cup cream soup, ½ cup custard, 3 squares graham crackers) 6–8 times/day.
- If vomiting, diarrhea, or fever persists, take liquids (e.g., ½ cup regular cola or orange juice, ½ cup broth, 1 cup sports drink [Gatorade]) every ½ to 1 hour to prevent dehydration and to provide calories.
- Report nausea, vomiting, and diarrhea to your primary provider, because extreme fluid loss may be dangerous.
- If you are unable to retain oral fluids, you may require hospitalization to avoid diabetic ketoacidosis and possibly coma.

If the patient cannot take fluids without vomiting, or if elevated glucose or ketone levels persist, the provider must be contacted. Patients are taught to have foods available for use on sick days. In addition, a supply of urine test strips (for ketone testing) and blood glucose test strips should be available. The patient must know how to contact his or her primary provider 24 hours a day. These materials should be assembled in a "sick day" kit.

After the acute phase of DKA has resolved, the nurse should assess for underlying causes. If there are psychological reasons for the patient missing insulin doses, the patient and family may be referred for evaluation and counseling or therapy.

Clinical Manifestations

The hyperglycemia of DKA leads to polyuria, polydipsia (increased thirst), and marked fatigue. In addition, the patient may experience blurred vision, weakness, and headache. Patients with marked intravascular volume depletion may have orthostatic hypotension (drop in systolic blood pressure of 20 mm Hg or more on changing from a reclining to a standing position). Volume depletion may also lead to frank hypotension with a weak, rapid pulse.

The ketosis and acidosis of DKA lead to gastrointestinal symptoms such as anorexia, nausea, vomiting, and abdominal pain. The patient may have acetone breath (a fruity odor), which occurs with elevated ketone levels. In addition, hyperventilation (with very deep, but not labored, respirations) may occur. These Kussmaul respirations represent the body's attempt to decrease the acidosis, counteracting the effect of the ketone buildup (Porth & Matfin, 2009). In addition, mental status in DKA varies widely. The patient may be alert, lethargic, or comatose.

Assessment and Diagnostic Findings

Blood glucose levels may vary between 300 and 800 mg/dL (16.6 and 44.4 mmol/L). Some patients have lower glucose values, and others have values of 1,000 mg/dL (55.5 mmol/L) or higher (usually depending on the degree of dehydration). The severity of DKA is not necessarily related to the blood glucose level. Evidence of ketoacidosis is reflected in low serum bicarbonate (0 to 15 mEq/L) and low pH (6.8 to 7.3) values. A low partial pressure of carbon dioxide (PCO_2 10 to 30 mm Hg) reflects respiratory compensation (Kussmaul respirations) for the metabolic acidosis. Accumulation of ketone bodies (which precipitates the acidosis) is reflected in blood and urine ketone measurements.

Sodium and potassium concentrations may be low, normal, or high, depending on the amount of water loss (dehydration). Despite the plasma concentration, there has been a marked total body depletion of these (and other) electrolytes, and they will need to be replaced.

Increased levels of creatinine, blood urea nitrogen (BUN), and hematocrit may also be seen with dehydration. After rehydration, continued elevation in the serum creatinine and BUN levels suggests underlying renal insufficiency.

Management

In addition to treating hyperglycemia, management of DKA is aimed at correcting dehydration, electrolyte loss, and acidosis before correcting the hyperglycemia with insulin.

Rehydration

In dehydrated patients, rehydration is important for maintaining tissue perfusion. In addition, fluid replacement enhances the excretion of excessive glucose by the kidneys. The patient may need as much as 6 to 10 L of IV fluid to replace fluid losses caused by polyuria, hyperventilation, diarrhea, and vomiting.

Initially, 0.9% sodium chloride (normal saline [NS]) solution is administered at a rapid rate, usually 0.5 to 1 L per hour for 2 to 3 hours. Half-strength NS (0.45%) solution (also known as hypotonic saline solution) may be used for patients with hypertension or hypernatremia and those at risk for heart failure. After the first few hours, half-strength NS solution is the fluid of choice for continued rehydration, provided the blood pressure is stable and the sodium level is not low. Moderate to high rates of infusion (200 to 500 mL per hour) may be needed for several more hours. When the blood glucose level reaches 300 mg/dL (16.6 mmol/L) or less, the IV solution may be changed to dextrose 5% in water (D_5W) to prevent a precipitous decline in the blood glucose level (Fowler, 2009).

Monitoring of fluid volume status involves frequent measurements of vital signs (including monitoring for orthostatic changes in blood pressure and heart rate), lung assessment, and monitoring of intake and output. Initial urine output lags behind IV fluid intake as dehydration is corrected. Plasma expanders may be necessary to correct severe hypotension that does not respond to IV fluid treatment. Monitoring for signs of fluid overload is especially important for patients who are older, have renal impairment, or are at risk for heart failure.

Restoring Electrolytes

The major electrolyte of concern during treatment of DKA is potassium. The initial plasma concentration of potassium may be low, normal, or high, but more often than not, tends to be high (hyperkalemia) from disruption of the cellular sodium-potassium pump (in the face of acidosis). Therefore, the serum potassium level must be monitored frequently. Some of the factors related to treating DKA that affect potassium concentration include rehydration, which leads to increased plasma volume and subsequent decreases in the concentration of serum potassium. Rehydration also leads to increased urinary excretion of potassium. Insulin administration enhances the movement of potassium from the extracellular fluid into the cells.

Cautious but timely potassium replacement is vital to avoid dysrhythmias that may occur with hypokalemia. As much as 40 mEq per hour may be needed for several hours. Because extracellular potassium levels decrease during DKA treatment, potassium must be infused even if the plasma potassium level is normal.

Frequent (every 2 to 4 hours initially) ECGs and laboratory measurements of potassium are necessary during the first 8 hours of treatment. Potassium replacement is withheld only if hyperkalemia is present or if the patient is not urinating.

Quality and Safety Nursing Alert

Because a patient's serum potassium level may drop quickly as a result of rehydration and insulin treatment, potassium replacement must begin once potassium levels drop to normal in the patient with DKA.

Reversing Acidosis

Ketone bodies (acids) accumulate as a result of fat breakdown. The acidosis that occurs in DKA is reversed with insulin, which inhibits fat breakdown, thereby ending ketone production and acid buildup. Insulin is usually infused IV at a slow, continuous rate (e.g., 5 units per hour). Hourly blood glucose values must be measured. IV fluid solutions with higher concentrations of glucose, such as NS solution (e.g., D_5NS, D_5.45NS), are administered when blood glucose levels reach 250 to 300 mg/dL (13.8 to 16.6 mmol/L) to avoid too rapid a drop in the blood glucose level (i.e., hypoglycemia) during treatment.

Regular insulin, the only type of insulin approved for IV use, may be added to IV solutions. The nurse must convert hourly rates of insulin infusion (frequently prescribed as units per hour) to IV drip rates. For example, if 100 units of regular insulin are mixed into 500 mL of 0.9% NS, then 1 unit of insulin equals 5 mL; therefore, an initial insulin infusion rate of 5 units per hour would equal 25 mL per hour. The insulin is often infused separately from the rehydration solutions to allow frequent changes in the rate and content of the latter.

Insulin must be infused continuously until subcutaneous administration of insulin can be resumed. Any interruption in administration may result in the reaccumulation of ketone bodies and worsening acidosis. Even if blood glucose levels are decreasing and returning to normal, the insulin drip must not be stopped until subcutaneous insulin therapy has been started. Rather, the rate or concentration of the dextrose infusion may be increased to prevent hypoglycemia. Blood glucose levels are usually corrected before the acidosis is corrected. Therefore, IV insulin may be continued for 12 to 24 hours, until the serum bicarbonate level increases (to at least 15 to 18 mEq/L) and until the patient can eat. In general, bicarbonate infusion to correct severe acidosis is avoided during treatment of DKA because it precipitates further, sudden (and potentially fatal) decreases in serum potassium levels. Continuous insulin infusion is usually sufficient for reversal of DKA.

Quality and Safety Nursing Alert

When hanging the insulin drip, the nurse must flush the insulin solution through the entire IV infusion set and discard the first 50 mL of fluid. Insulin molecules adhere to the inner surface of plastic IV infusion sets; therefore, the initial fluid may contain a decreased concentration of insulin.

Hyperglycemic Hyperosmolar Syndrome

Hyperglycemic hyperosmolar syndrome (HHS) is a metabolic disorder of type 2 diabetes resulting from a relative insulin deficiency initiated by an illness that raises the demand for insulin. This is a serious condition in which hyperosmolarity and hyperglycemia predominate, with alterations of the sensorium (sense of awareness). At the same time, ketosis is usually minimal or absent. The basic biochemical defect is lack of effective insulin (i.e., insulin resistance). Persistent hyperglycemia causes osmotic diuresis, which results in losses of water and electrolytes. To maintain osmotic equilibrium, water shifts from the intracellular fluid space to the extracellular fluid space. With glycosuria and dehydration, hypernatremia and increased osmolarity occur. Table 51-7 compares DKA and HHS.

HHS occurs most often in older people (50 to 70 years of age) who have no known history of diabetes or who have type 2 diabetes (Reynolds, 2012). HHS often can be traced to an infection or a precipitating event such as an acute illness (e.g., cerebrovascular accident [CVA]), medications that exacerbate hyperglycemia (e.g., thiazides), or treatments such as dialysis. The history includes days to weeks of polyuria with adequate fluid intake. What distinguishes HHS from DKA is that ketosis and acidosis generally do not occur in HHS, partly because of differences in insulin levels. In DKA, no insulin is present, and this promotes the breakdown of stored glucose, protein, and fat, which leads to the production of ketone bodies and ketoacidosis. In HHS, the insulin level is too low to prevent hyperglycemia (and subsequent osmotic diuresis), but it is high enough to prevent fat breakdown. Patients with HHS do not have the ketosis-related gastrointestinal symptoms that lead them to seek medical attention. Instead, they may tolerate polyuria and polydipsia until neurologic changes or an underlying illness (or family members or others) prompts them to seek treatment.

Clinical Manifestations

The clinical picture of HHS is one of hypotension, profound dehydration (dry mucous membranes, poor skin turgor), tachycardia, and variable neurologic signs (e.g., alteration of consciousness, seizures, hemiparesis) (see Table 51-7).

TABLE 51-7 Comparison of Diabetic Ketoacidosis and Hyperglycemic Hyperosmolar Syndrome

Characteristics	DKA	HHS
Patients most commonly affected	Can occur in type 1 or type 2 diabetes; more common in type 1 diabetes	Can occur in type 1 or type 2 diabetes; more common in type 2 diabetes, especially older patients with type 2 diabetes
Precipitating event	Omission of insulin; physiologic stress (infection, surgery, CVA, MI)	Physiologic stress (infection, surgery, CVA, MI)
Onset	Rapid (<24 h)	Slower (over several days)
Blood glucose levels	Usually >250 mg/dL (>13.9 mmol/L)	Usually >600 mg/dL (>33.3 mmol/L)
Arterial pH level	<7.3	Normal
Serum and urine ketones	Present	Absent
Serum osmolality	300–350 mOsm/L	>350 mOsm/L
Plasma bicarbonate level	<15 mEq/L	Normal
BUN and creatinine levels	Elevated	Elevated
Mortality rate	1%–5%	10%–20%

DKA, diabetic ketoacidosis; HHS, hyperglycemic hyperosmolar syndrome; CVA, cerebrovascular accident; MI, myocardial infarction; BUN, blood urea nitrogen.
Adapted from Reynolds, I. G. (2012). How to recognize and intervene for hyperosmolar hyperglycemic syndrome. *American Nursing Today*, 7(7), 12–15.

Assessment and Diagnostic Findings

Diagnostic assessment includes a range of laboratory tests, including blood glucose, electrolytes, BUN, complete blood count, serum osmolality, and arterial blood gas analysis. The blood glucose level is usually 600 to 1,200 mg/dL, and the osmolality exceeds 320 mOsm/kg (Reynolds, 2012). Electrolyte and BUN levels are consistent with the clinical picture of severe dehydration (see Chapter 14). Mental status changes, focal neurologic deficits, and hallucinations are common secondary to the cerebral dehydration that results from extreme hyperosmolality. Postural hypotension accompanies the dehydration.

Management

The overall approach to the treatment of HHS is similar to that of DKA: fluid replacement, correction of electrolyte imbalances, and insulin administration. Because patients with HHS are typically older, close monitoring of volume and electrolyte status is important for prevention of fluid overload, heart failure, and cardiac dysrhythmias. Fluid treatment is started with 0.9% or 0.45% NS, depending on the patient's sodium level and the severity of volume depletion. Central venous or hemodynamic pressure monitoring guides fluid replacement. Potassium is added to IV fluids when urinary output is adequate and is guided by continuous ECG monitoring and frequent laboratory determinations of potassium (Reynolds, 2012).

Extremely elevated blood glucose concentrations decrease as the patient is rehydrated. Insulin plays a less important role in the treatment of HHS because it is not needed for reversal of acidosis, as in DKA. Nevertheless, insulin is usually administered at a continuous low rate to treat hyperglycemia, and replacement IV fluids with dextrose are administered (as in DKA) after the glucose level has decreased to the range of 250 to 300 mg/dL (13.8 to 16.6 mmol/L) (Fowler, 2009; Reynolds, 2012).

Other therapeutic modalities are determined by the underlying illness and the results of continuing clinical and laboratory evaluation. It may take 3 to 5 days for neurologic symptoms to clear, and treatment of HHS usually continues well after metabolic abnormalities have resolved. After recovery from HHS, many patients can control their diabetes with MNT alone or with MNT and oral antidiabetic medications. Insulin may not be needed once the acute hyperglycemic complication is resolved. Frequent SBGM is important in prevention of recurrence of HHS.

NURSING PROCESS

The Patient With Diabetic Ketoacidosis or Hyperglycemic Hyperosmolar Syndrome

Assessment

For the patient with DKA, the nurse monitors the ECG for dysrhythmias indicating abnormal potassium levels. Vital signs (especially blood pressure and pulse), arterial blood gases, breath sounds, and mental status are assessed every hour and recorded on a flow sheet. Neurologic status checks are included as part of the hourly assessment because cerebral edema can be a severe and sometimes fatal outcome. Blood glucoses are checked every hour (Fowler, 2009).

For the patient with HHS, the nurse assesses vital signs, fluid status, and laboratory values. Fluid status and urine output are closely monitored because of the high risk of kidney failure secondary to severe dehydration. Because HHS tends to occur in older patients, the physiologic changes that occur with aging should be considered. Careful assessment of cardiovascular, pulmonary, and kidney function throughout the acute and recovery phases of HHS is important (Reynolds, 2012).

Diagnosis

Nursing Diagnoses

Based on the assessment data, major nursing diagnoses may include the following:

- Risk for deficient fluid volume related to polyuria and dehydration
- Risk for electrolyte imbalance related to fluid loss or shifts
- Deficient knowledge about diabetes self-care skills or information
- Anxiety related to loss of control, fear of inability to manage diabetes, misinformation related to diabetes, fear of diabetes complications

Collaborative Problems/Potential Complications

Potential complications may include the following:

- Fluid overload, pulmonary edema, and heart failure
- Hypokalemia
- Hyperglycemia and ketoacidosis
- Hypoglycemia
- Cerebral edema

Planning and Goals

The major goals for the patient may include maintenance of fluid and electrolyte balance, increased knowledge about diabetes survival skills and self-care, decreased anxiety, and absence of complications.

Nursing Interventions

Maintaining Fluid and Electrolyte Balance

Intake and output are measured. IV fluids and electrolytes are administered as prescribed, and oral fluid intake is encouraged when it is permitted. Laboratory values of serum electrolytes (especially sodium and potassium) are monitored. Vital signs are monitored hourly for signs of dehydration (tachycardia, orthostatic hypotension) along with assessment of breath sounds, level of consciousness, presence of edema, and cardiac status (ECG rhythm strips).

Increasing Knowledge About Diabetes Management

The development of DKA or HHS suggests the need for the nurse to carefully assess the patient's understanding of and adherence to the diabetes management plan. Further, factors that may have led to the development of DKA or HHS are explored with the patient and family. If the patient's blood glucose monitoring, dietary intake, use of antidiabetes (insulin or oral agents) medications, and exercise patterns differ from those identified in the diabetes management plan, their relationship to the development of DKS or HHS is discussed, along with early manifestations of DKA or HHS. If other factors, such as trauma, illness, surgery, or stress, are implicated, appropriate strategies to respond to these and similar situations in the future are described so that the patient can respond in the future without developing life-threatening complications. The nurse may need to provide education about survival skills again to patients who may not be able to recall the instructions. If the patient has omitted insulin or oral antidiabetes agents that have been prescribed, the nurse explores the reasons for doing so and addresses those issues to prevent future recurrence and readmissions for treatment of these complications.

If the patient has not previously been diagnosed with diabetes, the opportunity is used to educate the patient about the need for maintaining blood glucose at a normal level and learning about diabetes management and survival skills.

Decreasing Anxiety

Educating the patient about cognitive strategies may be useful for relieving tension, overcoming anxiety, decreasing fear, and achieving relaxation (see Chapter 5). Examples include:

- *Imagery:* The patient concentrates on a pleasant experience or restful scene.
- *Distraction:* The patient thinks of an enjoyable story or recites a favorite poem or song.
- *Optimistic self-recitation:* The patient recites optimistic thoughts ("I know all will go well").
- *Music:* The patient listens to soothing music (an easy-to-administer, inexpensive, noninvasive intervention).

Monitoring and Managing Potential Complications

Fluid Overload. Fluid overload can occur because of the administration of a large volume of fluid at a rapid rate, which is often required to treat patients with DKA or HHS. This risk is increased in older patients and in those with preexisting cardiac or kidney disease. To avoid fluid overload and resulting heart failure and pulmonary edema, the nurse monitors the patient closely during treatment by measuring vital signs and intake and output at frequent intervals. Central venous pressure monitoring and hemodynamic monitoring may be initiated to provide additional measures of fluid status. Physical examination focuses on assessment of cardiac rate and rhythm, breath sounds, venous distention, skin turgor, and urine output. The nurse monitors fluid intake and keeps careful records of IV and other fluid intake, along with urine output measurements.

Hypokalemia. Hypokalemia is a potential complication during the treatment of DKA. Low serum potassium levels may result from rehydration, increased urinary excretion of potassium, movement of potassium from the extracellular fluid into the cells with insulin administration, and restoration of the cellular sodium–potassium pump. Prevention of hypokalemia includes cautious replacement of potassium; however, before its administration, it is important to ensure that a patient's kidneys are functioning. Because of the adverse effects of hypokalemia on cardiac function, monitoring of the cardiac rate, cardiac rhythm, ECG, and serum potassium levels is essential.

Cerebral Edema. Although the exact cause of cerebral edema is unknown, rapid correction of hyperglycemia, resulting in fluid shifts, is thought to be the cause. Cerebral edema, which occurs more often in children than in adults, can be prevented by gradual reduction in the blood glucose level. An hourly flow sheet is used to enable close monitoring of the blood glucose level, serum electrolyte levels, fluid intake, urine output, mental status, and neurologic signs. Precautions are taken to minimize activities that could increase intracranial pressure.

Educating Patients About Self-Care

The patient is educated about survival skills, including treatment modalities (diet, insulin administration, monitoring of blood glucose, and, for type 1 diabetes, monitoring of urine ketones); the patient is also educated about recognition, treatment, and prevention of DKA and HHS (Reynolds, 2012). Education addresses those factors leading to DKA or HHS. Follow-up education is arranged with a home care nurse and dietitian or an outpatient diabetes education center. This is particularly important for patients who have experienced DKA or HHS because of the need to address factors that led to its occurrence (i.e., dehydration). For patients who have had HHS, avoiding dehydration and paying attention to increased urination or thirst is even more important than insulin administration. The importance of self-monitoring and of monitoring and follow-up by primary providers is reinforced, and the patient is reminded about the importance of keeping follow-up appointments.

Evaluation

Expected patient outcomes may include:

1. Achieves fluid and electrolyte balance
   a. Demonstrates intake and output balance
   b. Exhibits electrolyte values within normal limits
   c. Exhibits vital signs that remain stable, with resolution of orthostatic hypotension and tachycardia
2. Demonstrates knowledge about DKA and HHS
   a. Identifies factors leading to DKA and HHS
   b. Describes signs and symptoms of DKA and HHS
   c. Describes short- and long-term consequences of DKA and HHS
   d. Identifies strategies to prevent the development of DKA and HHS
   e. States when contact with primary provider is needed to treat early signs of DKS and HHS
3. Decreased anxiety
   a. Identifies strategies to decrease anxiety and fear
4. Absence of complications
   a. Exhibits normal cardiac rate and rhythm and normal breath sounds
   b. Exhibits no jugular venous distention
   c. Exhibits blood glucose and urine ketone levels within target range
   d. Exhibits no manifestations of hypoglycemia or hyperglycemia
   e. Shows improved mental status without signs of cerebral edema

LONG-TERM COMPLICATIONS OF DIABETES

The number of deaths attributable to ketoacidosis and infection in patients with diabetes has steadily declined, but diabetes related complications have increased. Long-term complications are becoming more common as more people live longer with diabetes; these complications can affect almost every organ system of the body and are a major cause of disability. The general categories of long-term diabetic complications are macrovascular disease, microvascular disease, and neuropathy.

The causes and pathogenesis of each type of complication are still being investigated. However, it appears that increased levels of blood glucose play a role in neuropathic disease, microvascular complications, and risk factors contributing to macrovascular complications. Hypertension may also be a major contributing factor, especially in macrovascular and microvascular diseases (ADA, 2013).

Long-term complications are seen in both type 1 and type 2 diabetes but usually do not occur within the first 5 to 10 years after diagnosis. However, evidence of these complications may be present at the time of diagnosis of type 2 diabetes, because patients may have had undiagnosed diabetes for many years. Kidney (microvascular) disease is more prevalent in patients with type 1 diabetes, and cardiovascular (macrovascular) complications are more prevalent in older patients with type 2 diabetes.

Macrovascular Complications

Diabetic macrovascular complications result from changes in the medium to large blood vessels. Blood vessel walls thicken, sclerose, and become occluded by plaque that adheres to the vessel walls. Eventually, blood flow is blocked. These atherosclerotic changes tend to occur more often and at an earlier age in patients with diabetes. Coronary artery disease, cerebrovascular disease, and peripheral vascular disease are the three main types of macrovascular complications that occur frequently in patients with diabetes.

MI is twice as common in men with diabetes and three times as common in women with diabetes, compared to people without diabetes. There is also an increased risk of complications resulting from MI and an increased likelihood of a second MI. Coronary artery disease accounts for an increased incidence of death among patients with diabetes. The typical ischemic symptoms may be absent in patients with diabetes. Therefore, the patient may not experience the early warning signs of decreased coronary blood flow and may have "silent" MIs, which may be discovered only as changes on the ECG. In some cases, ECG changes may not be apparent. This lack of ischemic symptoms may be secondary to autonomic neuropathy (see later discussion). (See Chapter 27 for a detailed discussion of coronary vascular disorders.)

Cerebral blood vessels are similarly affected by accelerated atherosclerosis. Occlusive changes or the formation of an embolus elsewhere in the vasculature that lodges in a cerebral blood vessel can lead to transient ischemic attacks and strokes. People with diabetes have twice the risk of developing cerebrovascular disease and an increased risk of death from CVA (Bader & Littlejohns, 2010). In addition, recovery from a stroke may be impaired in patients who have elevated blood glucose levels at the time of and immediately after a stroke. Because symptoms of CVA may be similar to symptoms of acute diabetic complications (HHS or hypoglycemia), it is very important to assess the blood glucose level (and treat abnormal levels) rapidly in patients with these symptoms so that testing and treatment of CVA (stroke) can be initiated promptly if indicated.

Atherosclerotic changes in the large blood vessels of the lower extremities are responsible for the increased incidence (two to three times higher than in nondiabetic people) of occlusive peripheral arterial disease in patients with diabetes (ADA, 2013). Signs and symptoms of peripheral vascular disease include diminished peripheral pulses and intermittent claudication (pain in the buttock, thigh, or calf during walking). The severe form of arterial occlusive disease in the lower extremities is largely responsible for the increased incidence of gangrene and subsequent amputation in patients with diabetes. Neuropathy and impairments in wound healing also play a role in diabetic foot disease (see later discussion).

Role of Diabetes in Macrovascular Diseases

Researchers continue to investigate the relationship between diabetes and macrovascular diseases. The main feature unique to diabetes is elevated blood glucose; however, a direct link has not been found between hyperglycemia and atherosclerosis. Although it may be tempting to attribute the increased prevalence of macrovascular diseases to the increased prevalence of certain risk factors (e.g., obesity, increased triglyceride levels, hypertension) in patients with diabetes, there is a higher-than-expected rate of macrovascular diseases among patients with diabetes compared with patients without diabetes who have the same risk factors (ADA, 2013). Therefore, diabetes itself is seen as an independent risk factor for accelerated atherosclerosis. Other potential factors that may play a role in diabetes-related atherosclerosis include platelet and clotting factor abnormalities, decreased flexibility of red blood cells, decreased oxygen release, changes in the arterial wall related to hyperglycemia, and possibly hyperinsulinemia.

Management

The focus of management is aggressive modification and reduction of risk factors. This involves prevention and treatment of the commonly accepted risk factors for atherosclerosis. MNT and exercise are important in managing obesity, hypertension, and hyperlipidemia. In addition, the use of medications to control hypertension and hyperlipidemia is indicated. Smoking cessation is essential. Control of blood glucose levels may reduce triglyceride concentrations and can significantly reduce the incidence of complications (ADA, 2013).

Microvascular Complications

Diabetic microvascular disease (or microangiopathy) is characterized by capillary basement membrane thickening. The basement membrane surrounds the endothelial cells of the capillary. Researchers believe that increased blood glucose levels react through a series of biochemical responses to thicken the basement membrane to several times its normal thickness. Two areas affected by these changes are the retina and the kidneys (Porth & Matfin, 2009).

Diabetic Retinopathy

Diabetic retinopathy is the leading cause of blindness among people between 20 and 74 years of age in the United States; it occurs in both type 1 and type 2 diabetes (ADA, 2013).

People with diabetes are subject to many visual complications (Table 51-8). The pathology referred to as diabetic retinopathy is caused by changes in the small blood vessels in the retina, which is the area of the eye that receives images and sends information about the images to the brain (Fig. 51-8). The retina is richly supplied with blood vessels of all kinds: small arteries and veins, arterioles, venules, and capillaries. Retinopathy has three main stages: nonproliferative (background), preproliferative, and proliferative.

Almost all patients with type 1 diabetes and the majority of patients with type 2 diabetes have some degree of retinopathy after 20 years (ADA, 2013). Changes in the microvasculature include microaneurysms, intraretinal hemorrhage, hard exudates, and focal capillary closure. Although most patients do not develop visual impairment, it can be devastating if

TABLE 51-8 Ocular Complications of Diabetes

Eye Disorder	Characteristics
Retinopathy	Damage to the small blood vessels that nourish the retina
Background	Early stage, asymptomatic retinopathy. Blood vessels within the retina develop microaneurysms that leak fluid, causing swelling and forming deposits (exudates). In some cases, macular edema causes distorted vision.
Preproliferative	Represents increased destruction of retinal blood vessels
Proliferative	Abnormal growth of new blood vessels on the retina. New vessels rupture, bleeding into the vitreous and blocking light. Ruptured blood vessels in the vitreous form scar tissue, which can pull on and detach the retina.
Cataracts	Opacity of the lens of the eye; cataracts occur at an earlier age in patients with diabetes.
Lens changes	The lens of the eye can swell when blood glucose levels are elevated. For some patients, visual changes related to lens swelling may be the first symptoms of diabetes. It may take up to 2 months of improved blood glucose control before hyperglycemic swelling subsides and vision stabilizes. Therefore, patients are advised not to change eyeglass prescriptions during the 2 months after discovery of hyperglycemia.
Extraocular muscle palsy	This may occur as a result of diabetic neuropathy. The involvement of various cranial nerves responsible for ocular movements may lead to double vision. This usually resolves spontaneously.
Glaucoma	Results from occlusion of the outflow channels by new blood vessels. Glaucoma may occur with slightly higher frequency in the diabetic population.

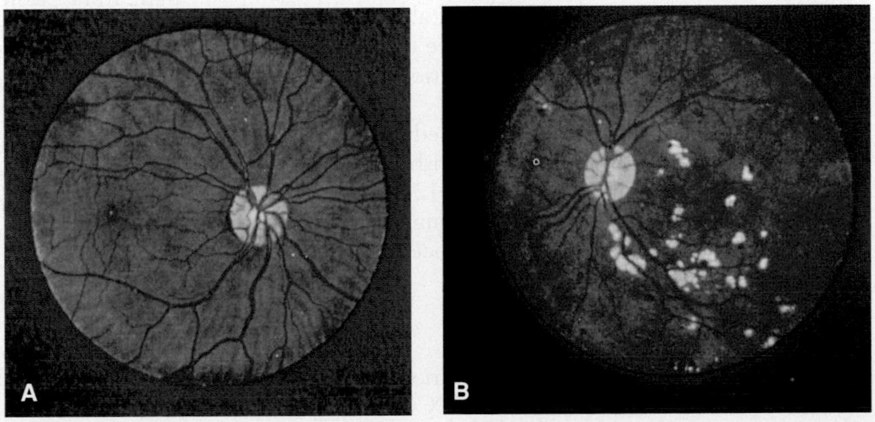
A
B

FIGURE 51-8 • Diabetic retinopathy. **A.** In the fundus photograph of a normal eye, the light circular area over which a number of blood vessels converge is the optic disc, where the optic nerve meets the back of the eye. **B.** The fundus photograph of a patient with diabetic retinopathy shows characteristic waxy-looking retinal lesions, microaneurysms of the vessels, and hemorrhages. Courtesy of American Optometric Association.

it occurs. A complication of nonproliferative retinopathy—macular edema—occurs in approximately 10% of people with type 1 or type 2 diabetes and may lead to visual distortion and loss of central vision (ADA, 2013).

An advanced form of background retinopathy—preproliferative retinopathy—is considered to be a precursor to the more serious proliferative retinopathy. In preproliferative retinopathy, there are more widespread vascular changes and loss of nerve fibers. Epidemiologic evidence suggests that 10% to 50% of patients with preproliferative retinopathy will develop proliferative retinopathy within a short time (possibly as little as 1 year). As with background retinopathy, if visual changes occur during the preproliferative stage, they are usually caused by macular edema.

Proliferative retinopathy represents the greatest threat to vision and is characterized by the proliferation of new blood vessels growing from the retina into the vitreous. These new vessels are prone to bleeding. The visual loss associated with proliferative retinopathy is caused by this vitreous hemorrhage, retinal detachment, or both. The vitreous is normally clear, allowing light to be transmitted to the retina. When there is a hemorrhage, the vitreous becomes clouded and cannot transmit light, resulting in loss of vision. Another consequence of vitreous hemorrhage is that resorption of the blood in the vitreous leads to the formation of fibrous scar tissue. This scar tissue may place traction on the retina, resulting in retinal detachment and subsequent visual loss.

Clinical Manifestations

Retinopathy is a painless process. In nonproliferative and preproliferative retinopathy, blurry vision secondary to macular edema occurs in some patients, although many patients are asymptomatic. Even patients with a significant degree of proliferative retinopathy and some hemorrhaging may not experience major visual changes. However, symptoms indicative of hemorrhaging include floaters or cobwebs in the visual field, sudden visual changes including spotty or hazy vision, or complete loss of vision.

Assessment and Diagnostic Findings

Diagnosis is by direct visualization of the retina through dilated pupils with an ophthalmoscope or with a technique known as fluorescein angiography. Fluorescein angiography can document the type and activity of the retinopathy. Dye is injected into an arm vein and is carried to various parts of the body through the blood, but especially through the vessels of the retina of the eye. This technique allows an ophthalmologist, using special instruments, to see the retinal vessels in bright detail and gives useful information that cannot be obtained with just an ophthalmoscope.

Side effects of this diagnostic procedure may include nausea during the dye injection; yellowish, fluorescent discoloration of the skin and urine lasting 12 to 24 hours; and occasionally allergic reactions, usually manifested by hives or itching. However, the diagnostic procedure is generally safe.

Medical Management

The first focus of management of retinopathy is on primary and secondary prevention. The DCCT study (1993) demonstrated that in patients without preexisting retinopathy, maintenance of blood glucose to a normal or near-normal level in type 1 diabetes through intensive insulin therapy and patient education decreased the risk of retinopathy by 76%, compared with conventional therapy. The progression of retinopathy was decreased by 54% in patients with very mild to moderate nonproliferative retinopathy at the time of initiation of treatment. Better control of blood glucose levels in patients with type 2 diabetes reduces the risk of retinopathy as well.

Other strategies that may slow the progression of diabetic retinopathy include control of hypertension, control of blood glucose, and cessation of smoking.

For advanced cases of diabetic retinopathy, the main treatment is argon laser photocoagulation. The laser treatment destroys leaking blood vessels and areas of neovascularization. For patients who are at increased risk for hemorrhage, panretinal photocoagulation may significantly reduce the rate of progression to blindness. Panretinal photocoagulation involves the systematic application of multiple (more than 1,000) laser burns throughout the retina (except in the macular region). This stops the widespread growth of new vessels and hemorrhaging of damaged vessels. The role of "mild" panretinal photocoagulation (with only one third to one half as many laser burns) in the early stages of proliferative retinopathy or in patients with preproliferative changes is being investigated. For patients with macular edema, focal photocoagulation is used to apply smaller laser burns to specific areas of microaneurysms in the macular region. This may reduce the rate of visual loss from macular edema (ADA, 2013).

Photocoagulation treatments are usually performed on an outpatient basis, and most patients can return to their usual activities by the next day. Limitations may be placed on activities involving weight bearing or bearing down. In most cases, the treatment does not cause intense pain, although patients may report varying degrees of discomfort such as a headache. Usually, an anesthetic eye drop is all that is needed during the treatment. A few patients may experience slight visual loss, loss of peripheral vision, or impairments in adaptation to the dark. However, the risk of slight visual changes from the laser treatment itself is much less than the potential for loss of vision from progression of retinopathy.

A major hemorrhage into the vitreous may occur, with the vitreous fluid becoming mixed with blood, preventing light from passing through the eye; this can cause blindness. A vitrectomy is a surgical procedure in which vitreous humor filled with blood or fibrous tissue is removed with a special drill-like instrument and replaced with saline or another liquid. A vitrectomy is performed for patients who already have visual loss and in whom the vitreous hemorrhage has not cleared on its own after 6 months. The purpose is to restore useful vision; recovery to near-normal vision is not usually expected.

Nursing Management

Nursing management of patients with diabetic retinopathy or other eye disorders involves implementing the individual plan of care and providing patient education. Education focuses on prevention through regular ophthalmologic examinations, blood glucose control, and self-management of eye care regimens. The effectiveness of early diagnosis and prompt treatment is emphasized in educating the patient and family. Research suggests that the use of a tailored telephone intervention can increase the rates of annual screening for diabetic retinopathy, especially in an underserved ethnically diverse urban community (Jones, Walker, Schechter, et al., 2010).

If vision loss occurs, nursing care must also address the patient's adjustment to impaired vision and the use of adaptive devices for diabetes self-care as well as activities of daily living. (See Chapter 63 for discussion of nursing care for patients with low vision and blindness.)

Promoting Home and Community-Based Care

Educating Patients About Self-Care

Because the course of the retinopathy may be long and stressful, patient education is essential. In educating and counseling patients, it is important to stress the following:

- Retinopathy may appear after many years of diabetes, and its appearance does not necessarily mean that the diabetes is on a downhill course.
- The odds for maintaining vision are in the patient's favor, especially with adequate control of glucose levels and blood pressure.
- Frequent eye examinations allow for the detection and prompt treatment of retinopathy.

A patient's response to vision loss depends on personality, self-concept, and coping mechanisms. Acceptance of blindness occurs in stages; some patients may learn to accept blindness in a rather short period, and others may never do so. An important issue in educating patients is that several complications of diabetes may occur simultaneously. For example, a patient who is blind due to diabetic retinopathy may also have peripheral neuropathy and may experience impairment of manual dexterity and tactile sensation, or kidney failure. This can be devastating to the patient and family. Psychological counseling may be warranted. To prevent further losses, glycemic control remains a priority.

Continuing Care

The importance of careful diabetes management is emphasized as one means of slowing the progression of visual changes. The patient is reminded of the need to see an ophthalmologist regularly. If eye changes are progressive and unrelenting, the patient should be prepared for inevitable blindness. Therefore, consideration is given to making referrals for educating the patient in Braille and for training him or her with guide (i.e., service) dogs. Referral to state agencies should be made to ensure that the patient receives services for the blind. Family members are also taught how to assist the patient to remain as independent as possible despite decreasing visual acuity.

Referral for home care may be indicated for some patients, particularly those who live alone, those who are not coping well, and those who have other health problems or complications of diabetes that may interfere with their ability to perform self-care. During home visits, the nurse can assess the patient's home environment and his or her ability to manage diabetes despite visual impairments. (See Chapter 63 for a detailed discussion of medical management and nursing care for patients with visual disturbances.)

Nephropathy

Nephropathy, or kidney disease secondary to diabetic microvascular changes in the kidney, is a common complication of diabetes (ADA, 2013). In the United States each year, people with diabetes account for almost 50% of new cases of ESKD, and about 25% of those require dialysis or transplantation. About 20% to 30% of people with type 1 or types 2 diabetes develop nephropathy, but fewer of those with type 2 diabetes progress to ESKD. Native American, Latino, African American, Asian American, and Pacific Island people with type 2 diabetes are at greater risk for ESKD than non-Latino whites (ADA, 2013).

Patients with type 1 diabetes frequently show initial signs of kidney disease after 10 to 15 years; while patients with type 2 diabetes tend to develop kidney disease within 10 years after the diagnosis of diabetes. Many patients with type 2 diabetes have had diabetes for many years before the diabetes is diagnosed and treated. Therefore, they may have evidence of nephropathy at the time of diagnosis (ADA, 2013). If blood glucose levels are elevated consistently for a significant period of time, the kidney's filtration mechanism is stressed, allowing blood proteins to leak into the urine. As a result, the pressure in the blood vessels of the kidney increases. It is thought that this elevated pressure serves as the stimulus for the development of nephropathy. Various medications and diets are being tested to prevent these complications.

The DCCT (1993) results showed that intensive treatment for type 1 diabetes with a goal of achieving a glycolated hemoglobin level as close to the nondiabetic range as possible reduced the occurrence of early signs of nephropathy. Similarly, the United Kingdom Prospective Diabetes Study Group (UKPDS, 1998) demonstrated a reduced incidence of overt nephropathy in patients with type 2 diabetes who controlled their blood glucose levels.

Clinical Manifestations

Most of the signs and symptoms of kidney dysfunction in patients with diabetes are similar to those seen in patients without diabetes (see Chapter 54). In addition, as kidney failure progresses, the catabolism (breakdown) of both exogenous and endogenous insulin decreases, and frequent hypoglycemic episodes may result. Insulin needs change as a result of changes in the catabolism of insulin, changes in diet related to the treatment of nephropathy, and changes in insulin clearance that occur with decreased kidney function.

Assessment and Diagnostic Findings

Albumin is one of the most important blood proteins that leak into the urine. Although small amounts may leak undetected for years, its leakage into the urine is among the earliest signs that can be detected. Clinical nephropathy eventually develops in more than 85% of people with microalbuminuria but in fewer than 5% of people without microalbuminuria (Chart 51-10). The urine should be checked annually for the presence of microalbumin. If the microalbuminuria exceeds 30 mg/24 hours on two consecutive random urine tests, a

Chart 51-10 NURSING RESEARCH PROFILE

Diabetes Complications

Calvin, D., Quinn, L., Dancy, B., et al. (2011). African Americans' perception of risk for diabetes complications. *Diabetes Educator*, *37*(5), 689–698.

Purpose

Diabetes and its complications affect African Americans more than other ethnic groups. The purpose of this study was to assess the perception of risk for diabetic complications in an African American population and how it compares with the actual risk in physical measurements such as hemoglobin AIC, microalbuminuria, and blood pressure control.

Design

This was a descriptive, exploratory, correlational study of 143 patients with type 2 diabetes who attended 1 of 3 diabetes clinics in an urban center. Participants were administered 3 instruments of risk perception: the Risk Perception Survey–Diabetes, the Revised Illness Perception Questionnaire, and the Well-Being Questionnaire. Physiologic measures included hemoglobin A1C, urine microalbumin, and blood pressure.

Findings

The physiologic measurements showed that participants were at increased risk for developing complications. Microalbumin was present in 30% of the cohort, hemoglobin A1C was out of range for 66.7% (mean A1C was 8.7%) of the cohort, and blood pressure was out of range (>130/80 mm Hg) in 61.3% of the cohort. However, the perception for risk was low. Amputation, blindness, and kidney disease were rated as the lowest perceived risk to one's personal health.

Nursing Implications

Diabetes is a self-managed disease that requires patients to be active participants in care in order to achieve good glycemic control and prevent complications such as retinopathy, nephropathy, and neuropathy. If patients do not perceive that they are at risk for developing complications, they may not take action to prevent them. In addition, if patients do not have any symptoms, they may not see the benefit of treatment. Nurses who are providing education about diabetes self-management should assess each patient's perception of risk. If the perception of risk is low, education concerning the reality of risk needs to be done. Individualized treatment plans identified to target and minimize risk should be developed for each patient. Primary providers need to continuously be aware of the patient's perception of risk of complications, especially if the treatment plan is not being followed.

24-hour urine sample should be obtained and tested. If results are positive, treatment is indicated (see later discussion).

In addition, tests for serum creatinine and BUN levels should be conducted annually. Diagnostic testing for cardiac or other systemic disorders may also be required with progression of other complications, and caution is indicated if contrast agents are used with these tests. Contrast agents and dyes used for some diagnostic tests may not be easily cleared by the damaged kidney, and the potential benefits of these diagnostic tests must be weighed against their potential risks.

Hypertension often develops in patients (with and without diabetes) who are in the early stages of kidney disease. However, hypertension is the most common complication in all people with diabetes (ADA, 2013). Therefore, this symptom may or may not be due to kidney disease; other diagnostic criteria must also be present.

Management

In addition to achieving and maintaining near-normal blood glucose levels, management for all patients with diabetes should include careful attention to the following:

- Control of hypertension (the use of angiotensin-converting enzyme [ACE] inhibitors, such as captopril [Capoten]), because control of hypertension may also decrease or delay the onset of early proteinuria
- Prevention or vigorous treatment of urinary tract infections
- Avoidance of nephrotoxic medications and contrast dye
- Adjustment of medications as kidney function changes
- Low-sodium diet
- Low-protein diet

If the patient has already developed microalbuminuria with levels that exceed 30 mg/24 hours on two consecutive tests, an ACE inhibitor should be prescribed. ACE inhibitors lower blood pressure and reduce microalbuminuria, thereby protecting the kidney. Alternatively, angiotensin receptor blocking agents may be prescribed. This preventive strategy should be part of the standard of care for all people with diabetes. Carefully designed low-protein diets also appear to reverse early leakage of small amounts of protein from the kidney.

In chronic or ESKD, two types of treatment are available: dialysis (hemodialysis or peritoneal dialysis) and transplantation from a relative or a cadaver. Hemodialysis for patients with diabetes is similar to that for patients without the disease (see Chapter 54). Because hemodialysis creates additional stress on patients with cardiovascular disease, it may not be appropriate for some patients.

Continuous ambulatory peritoneal dialysis is being used by patients with diabetes, mainly because of the independence it allows. In addition, insulin can be mixed into the dialysate, which may result in better blood glucose control and end the need for insulin injections. In some cases, patients may require higher doses of insulin because the dialysate contains glucose. Major risks of peritoneal dialysis are infection and peritonitis. The mortality rate for patients with diabetes undergoing dialysis is higher than that for patients without diabetes undergoing dialysis and is closely related to the severity of cardiovascular problems.

Kidney disease is frequently accompanied by advancing retinopathy that may require laser treatments and surgery. Severe hypertension also worsens eye disease because of the additional stress it places on the blood vessels. Patients being treated with hemodialysis who require eye surgery may be changed to peritoneal dialysis and have their hypertension aggressively controlled for several weeks before surgery to prevent bleeding and damage to the retina. The rationale for this change is that hemodialysis requires anticoagulant agents that can increase the risk of bleeding after the surgery, and peritoneal dialysis minimizes pressure changes in the eyes.

In medical centers performing large numbers of transplantations, the chances are 75% to 80% that the transplanted kidney will continue to function in patients with diabetes for at least 5 years. Like the original kidneys, transplanted kidneys can eventually be damaged if blood glucose levels are consistently high after the transplantation. Therefore, monitoring blood glucose levels frequently and adjusting insulin levels in patients with diabetes are essential for long-term success of kidney transplantation.

Diabetic Neuropathies

Diabetic neuropathy refers to a group of diseases that affect all types of nerves, including peripheral (sensorimotor), autonomic, and spinal nerves. The disorders appear to be clinically diverse and depend on the location of the affected nerve cells. The prevalence increases with the age of the patient and the duration of the disease (NIDDK, 2012b).

The etiology of neuropathy may involve elevated blood glucose levels over a period of years. Control of blood glucose levels to normal or near-normal levels decreases the incidence of neuropathy. The pathogenesis of neuropathy may be attributed to either a vascular or metabolic mechanism or both. Capillary basement membrane thickening and capillary closure may be present. In addition, there may be demyelinization of the nerves, which is thought to be related to hyperglycemia. Nerve conduction is disrupted when there are aberrations of the myelin sheaths.

The two most common types of diabetic neuropathy are sensorimotor polyneuropathy and autonomic neuropathy. Sensorimotor polyneuropathy is also called *peripheral neuropathy*. Cranial mononeuropathies—those affecting the oculomotor nerve—also occur in diabetes, especially in older adults.

Peripheral Neuropathy

Peripheral neuropathy most commonly affects the distal portions of the nerves, especially the nerves of the lower extremities; it affects both sides of the body symmetrically and may spread in a proximal direction.

Clinical Manifestations

Although approximately half of patients with diabetic neuropathy do not have symptoms, initial symptoms may include paresthesias (prickling, tingling, or heightened sensation) and burning sensations (especially at night). As the neuropathy progresses, the feet become numb. In addition, a decrease in proprioception (awareness of posture and movement of the body and of position and weight of objects in relation to the body) and a decreased sensation of light touch may lead to an unsteady gait. Decreased sensations of pain and temperature place patients with neuropathy at increased risk for injury and undetected foot infections. Deformities of the foot may also occur; neuropathy-related joint changes are sometimes referred to as Charcot joints. These joint deformities result from the abnormal weight distribution on joints resulting from lack of proprioception.

On physical examination, a decrease in deep tendon reflexes and vibratory sensation is found. For patients who have few or no symptoms of neuropathy, these physical findings may be the only indication of neuropathic changes. For patients with signs or symptoms of neuropathy, it is important to rule out other possible causes, including alcohol-induced and vitamin-deficiency neuropathies.

Management

Intensive insulin therapy and control of blood glucose levels delay the onset and slow the progression of neuropathy. Pain, particularly of the lower extremities, is a disturbing symptom in some people with neuropathy secondary to diabetes. In some cases, neuropathic pain spontaneously resolves within 6 months; for others, pain persists for many years. Various approaches to pain management can be tried. These include analgesic agents (preferably nonopioid); tricyclic antidepressants and other antidepressant medications (duloxetine [Cymbalta]); antiseizure medications (pregabalin [Lyrica], carbamazepine [Tegretol], or gabapentin [Neurontin]); mexiletine (Mexitil, an antiarrhythmic agent); and transcutaneous electrical nerve stimulation.

Duloxetine and pregabalin are approved specifically for treating painful diabetic peripheral neuropathy (NIDDK, 2012b).

Autonomic Neuropathies

Neuropathy of the autonomic nervous system results in a broad range of dysfunctions affecting almost every organ system of the body (NIDDK, 2012b).

Clinical Manifestations

Three manifestations of autonomic neuropathy are related to the cardiac, gastrointestinal, and renal systems. Cardiovascular symptoms range from a fixed, slightly tachycardic heart rate and orthostatic hypotension to silent, or painless, myocardial ischemia and infarction. Delayed gastric emptying may occur with the typical gastrointestinal symptoms of early satiety, bloating, nausea, and vomiting. "Diabetic" constipation or diarrhea (especially nocturnal diarrhea) may occur as a result. In addition, there may be unexplained wide swings in blood glucose levels related to inconsistent absorption of the glucose from ingested foods secondary to the inconsistent gastric emptying.

Urinary retention, a decreased sensation of bladder fullness, and other urinary symptoms of neurogenic bladder result from autonomic neuropathy. The patient with a neurogenic bladder is predisposed to development of urinary tract infections because of the inability to empty the bladder completely. This is especially true of patients with poorly controlled diabetes because hyperglycemia impairs resistance to infection.

Hypoglycemic Unawareness

Autonomic neuropathy affecting the adrenal medulla is responsible for diminished or absent adrenergic symptoms of hypoglycemia. Patients may report that they no longer feel the typical shakiness, sweating, nervousness, and palpitations associated with hypoglycemia. Frequent blood glucose monitoring is recommended for these patients. The inability to detect and treat these warning signs of hypoglycemia puts patients at risk for development of dangerously low blood glucose levels. Therefore, goals for blood glucose levels may need to be adjusted to reduce the risk for hypoglycemia. Patients and families need to be taught to recognize subtle and atypical symptoms of hypoglycemia, such as numbness around the mouth and impaired ability to concentrate.

Sudomotor Neuropathy

The neuropathic condition called *sudomotor neuropathy* refers to a decrease or absence of sweating (anhidrosis) of the extremities, with a compensatory increase in upper body sweating. Dryness of the feet increases the risk for the development of foot ulcers.

Sexual Dysfunction

Sexual dysfunction, especially erectile dysfunction in men, is a complication of diabetes. The effects of autonomic neuropathy on female sexual functioning are not well documented. Reduced vaginal lubrication has been mentioned as a possible neuropathic effect. Other possible changes in sexual function in women with diabetes include decreased libido and lack of orgasm. Vaginal infection, which increases in incidence in women with diabetes, may be associated with decreased lubrication and vaginal pruritus (itching) and tenderness. Urinary tract infections and vaginitis may also affect sexual function.

Impotence (inability of the penis to become rigid and sustain an erection adequate for penetration) occurs with greater frequency in men with diabetes than in other men of the same age. Some men with autonomic neuropathy have normal erectile function and can experience orgasm but do not ejaculate normally. Retrograde ejaculation occurs; seminal fluid is propelled backward through the posterior urethra and into the urinary bladder. Examination of the urine confirms the diagnosis because of the large number of active sperm present. Fertility counseling may be necessary for couples attempting conception.

Diabetic neuropathy is not the only cause of impotence in men with diabetes. Medications such as antihypertensive agents, psychological factors, and other medical conditions (e.g., vascular insufficiency) that may affect other men also play a role in impotence in men with diabetes (see Chapter 59).

Management

Management strategies for autonomic neuropathy focus on alleviating symptoms and on modification and management of risk factors. Detection of painless cardiac ischemia is important so that education about avoiding strenuous exercise can be provided. Orthostatic hypotension may respond to a diet high in sodium, discontinuation of medications that impede autonomic nervous system responses, the use of sympathomimetic and other agents (e.g., caffeine) that stimulate an autonomic response, mineralocorticoid therapy, and the use of lower-body elastic garments that maximize venous return and prevent pooling of blood in the extremities.

Treatment of delayed gastric emptying includes a low-fat diet, frequent small meals, frequent blood glucose monitoring, and the use of agents that increase gastric motility (e.g., metoclopramide [Reglan], bethanechol [Myotonachol]). Treatment of diabetic diarrhea may include bulk-forming laxatives or antidiarrheal agents. Constipation is treated with a high-fiber diet and adequate hydration; medications, laxatives, and enemas may be necessary if constipation is severe. Management of sexual dysfunction in women and men is discussed in Chapters 57 and 59, respectively. Intermittent straight catheterization may be necessary to prevent urinary tract infections in patients with neurogenic bladders.

Treatment of sudomotor dysfunction focuses on education about skin care and heat intolerance.

Foot and Leg Problems

Between 50% and 75% of lower extremity amputations are performed on people with diabetes. More than 50% of these amputations are thought to be preventable, provided patients are taught foot care measures and practice them on a daily basis (ADA, 2013). Complications of diabetes that contribute to the increased risk of foot problems and infections include the following:

- *Neuropathy:* Sensory neuropathy leads to loss of pain and pressure sensation, and autonomic neuropathy leads to increased dryness and fissuring of the skin (secondary to decreased sweating). Motor neuropathy results in muscular atrophy, which may lead to changes in the shape of the foot.
- *Peripheral vascular disease:* Poor circulation of the lower extremities contributes to poor wound healing and the development of gangrene.
- *Immunocompromise:* Hyperglycemia impairs the ability of specialized leukocytes to destroy bacteria. Therefore, in poorly controlled diabetes, there is a lowered resistance to certain infections.

The typical sequence of events in the development of a diabetic foot ulcer begins with a soft tissue injury of the foot, formation of a fissure between the toes or in an area of dry skin, or formation of a callus (Fig. 51-9). Patients with an insensitive foot do not feel injuries, which may be thermal (e.g., from using heating pads, walking barefoot on hot concrete, testing bathwater with the foot), chemical (e.g., burning the foot while using caustic agents on calluses, corns, or bunions), or traumatic (e.g., injuring skin while cutting nails, walking

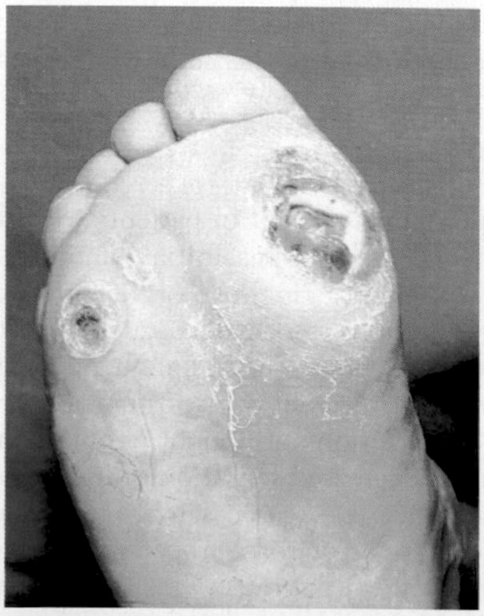

FIGURE 51-9 • Neuropathic ulcers occur on pressure points in areas with diminished sensation in diabetic polyneuropathy. Because pain is absent, the ulcer may go unnoticed.

with an undetected foreign object in the shoe, or wearing ill-fitting shoes and socks).

If the patient is not in the habit of thoroughly inspecting both feet on a daily basis, the injury or fissure may go unnoticed until a serious infection has developed. Drainage, swelling, redness of the leg (from cellulitis), or gangrene may be the first sign of foot problems that the patient notices. Treatment of foot ulcers involves bed rest, antibiotics, and débridement. In addition, controlling glucose levels, which tend to increase when infections occur, is important for promoting wound healing. When peripheral vascular disease is present, foot ulcers may not heal because of the decreased ability of oxygen, nutrients, and antibiotics to reach the injured tissue. Amputation (see Chapter 43) may be necessary to prevent the spread of infection, particularly if it involves the bone (osteomyelitis) (see Chapter 42).

Foot assessment and foot care instructions are most important when caring for patients who are at high risk for foot infections. Some of the high-risk characteristics include:

- Duration of diabetes more than 10 years
- Age greater than 40 years
- Current smoker and history of smoking
- Decreased peripheral pulses
- Decreased sensation
- Anatomic deformities or pressure areas (e.g., bunions, calluses, hammer toes)
- History of previous foot ulcers or amputation

Management

Educating patients about proper foot care is a nursing intervention that can prevent costly and painful complications that result in disability (Chart 51-11).

In addition to the daily visual and manual inspection of the feet, the feet should be examined during every health care visit or at least once per year (more often if there is an increase in risk) by a podiatrist, physician, or nurse (ADA, 2013; NIDDK, 2008). All patients should be assessed for neuropathy and undergo evaluation of neurologic status by an experienced examiner using a monofilament device (Boulton, Armstrong, Albert, et al., 2008) (Fig. 51-10). Pressure areas, such as calluses, or thick toenails should be treated by a podiatrist in addition to routine trimming of nails.

Additional aspects of preventive foot care include the following:

- Properly bathing, drying, and lubricating the feet, taking care not to allow moisture (water or lotion) to accumulate between the toes
- Wearing closed-toed shoes that fit well. A podiatrist can provide the patient with inserts (orthotics) to remove pressure from pressure points on the foot. New shoes should be broken in slowly (i.e., worn for 1 to 2 hours initially, with gradual increases in the length of time worn) to avoid blister formation. Patients with bony deformities may need custom-made shoes with extra width or depth. High-risk behaviors, such as walking barefoot, using heating pads on the feet, wearing open-toed shoes, soaking the feet, and shaving calluses, should be avoided.
- Trimming toenails straight across and filing sharp corners to follow the contour of the toe. If the patient has visual deficits, is unable to reach the feet because of disability, or has thickened toenails, a podiatrist should cut the nails.

Chart 51-11 PATIENT EDUCATION
Foot Care Tips

1. Take care of your diabetes.
   - Work with your health care team to keep your blood glucose level within a normal range.
2. Inspect your feet every day.
   - Look at your bare feet every day for cuts, blisters, red spots, and swelling.
   - Use a mirror to check the bottoms of your feet, or ask a family member for help if you have trouble seeing.
   - Check for changes in temperature.
3. Wash your feet every day.
   - Wash your feet in warm, not hot, water.
   - Dry your feet well. Be sure to dry between the toes.
   - Do not soak your feet.
   - Do not check water temperature with your feet; use a thermometer or elbow.
4. Keep the skin soft and smooth.
   - Rub a thin coat of skin lotion over the tops and bottoms of your feet, but not between your toes.
5. Smooth corns and calluses gently.
   - Use a pumice stone to smooth corns and calluses.
6. Trim your toenails each week or when needed.
   - Trim your toenails straight across, and file the edges with an emery board or nail file.
7. Wear shoes and socks at all times.
   - Never walk barefoot.
   - Wear comfortable shoes that fit well and protect your feet.
   - Feel inside your shoes before putting them on each time to make sure the lining is smooth and there are no objects inside.
8. Protect your feet from hot and cold.
   - Wear shoes at the beach or on hot pavement.
   - Wear socks at night if your feet get cold.
9. Keep the blood flowing to your feet.
   - Put your feet up when sitting.
   - Wiggle your toes and move your ankles up and down for 5 minutes, 2 or 3 times/day.
   - Do not cross your legs for long periods of time.
   - Do not smoke.
10. Check with your primary provider.
    - Have your primary provider check your bare feet and find out whether you are likely to have serious foot problems. Remember that you may not feel the pain of an injury.
    - Call your primary provider right away if a cut, sore, blister, or bruise on your foot does not begin to heal after 1 day.
    - Follow your primary provider's advice about foot care.
    - Do not self-medicate or use home remedies or over-the-counter agents to treat foot problems.

Adapted from National Institute of Diabetes and Digestive and Kidney Diseases. (2008). *Prevent diabetes problems: Keep your feet and skin healthy.* NIH Publication No. 08–4282. Bethesda, MD: U.S. Department of Health and Human Services, National Institutes of Health. Available at: diabetes.niddk.nih.gov/dm/pubs/complications_feet

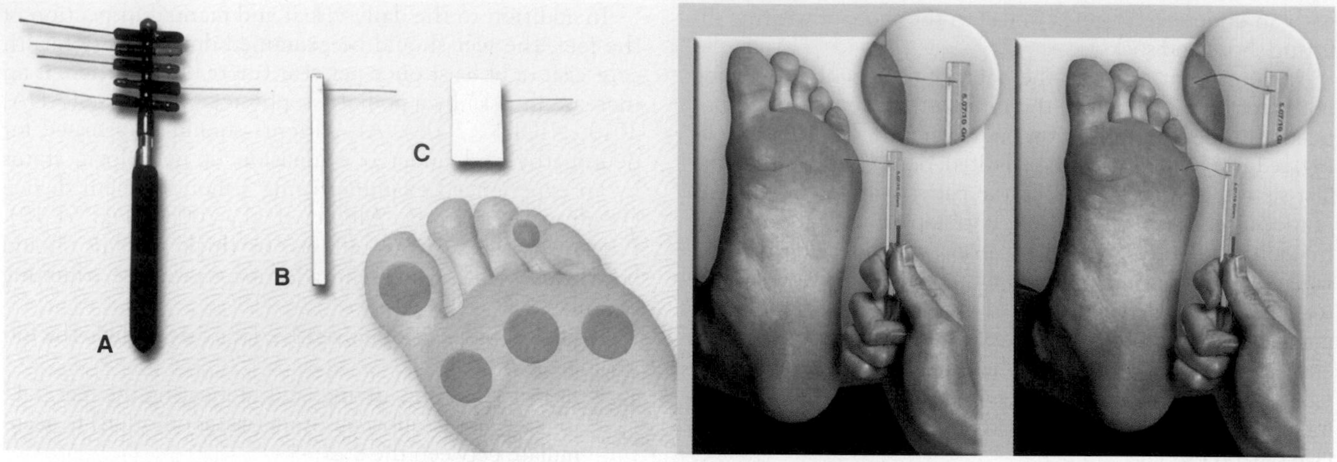
A
B
C

FIGURE 51-10 • The monofilament test is used to assess the sensory threshold in patients with diabetes. The test instrument—a monofilament—is gently applied to about five pressure points on the foot (as shown in image on *left*). **A.** Example of a monofilament used for advanced quantitative assessment. **B.** Semmes-Weinstein monofilament used by clinicians. **C.** Disposable monofilament used by patients. The examiner applies the monofilament to the test area to determine whether the patient feels the device. Adapted with permission from Cameron, B. L. (2002). Making diabetes management routine. *American Journal of Nursing*, *102*(2), 26–32.

- Reducing risk factors, such as smoking and elevated blood lipids, that contribute to peripheral vascular disease
- Avoiding home remedies, over-the-counter agents, and self-medicating to treat foot problems

Blood glucose control is important for avoiding decreased resistance to infections and for preventing diabetic neuropathy. Maintaining adequate nutrition and blood glucose control promote wound healing.

SPECIAL ISSUES IN DIABETES CARE

Patients With Diabetes Who Are Undergoing Surgery

During periods of physiologic stress, such as surgery, blood glucose levels tend to increase, because levels of stress hormones (epinephrine, norepinephrine, glucagon, cortisol, and growth hormone) increase. If hyperglycemia is not controlled during surgery, the resulting osmotic diuresis may lead to excessive loss of fluids and electrolytes. Patients with type 1 diabetes also risk developing ketoacidosis during periods of stress.

Hypoglycemia is also a concern in patients with diabetes who are undergoing surgery. This is a special concern during the preoperative period if surgery is delayed beyond the morning in a patient who received a morning injection of intermediate-acting insulin.

There are various approaches to managing glucose control during the perioperative period. Frequent blood glucose monitoring is essential throughout the pre- and postoperative periods, regardless of the method used for glucose control. Examples of these approaches are described in Chart 51-12. The use of IV insulin and dextrose has become widespread with the increased availability of meters for intraoperative glucose monitoring.

During the postoperative period, patients with diabetes must also be closely monitored for cardiovascular complications because of the increased prevalence of atherosclerosis, wound infections, and skin breakdown (especially in patients with decreased sensation in the extremities due to neuropathy).

Management of Hospitalized Patients With Diabetes

At any one time, as many as 25% of hospitalized general medical-surgical patients have diabetes (Seggelke & Everhart, 2012). Often, diabetes is not the primary medical diagnosis, yet problems with control of diabetes frequently result from changes in the patient's normal routine or from surgery or illness. Patients who are hospitalized and have a diagnosis of diabetes should have this clearly indicated on their medical record, and glucose monitoring needs to be prescribed (ADA, 2013). During the course of treatment, blood glucose control may worsen. Control of blood glucose levels is important because hyperglycemia in patients who are hospitalized can increase the length of hospital stay, the risk of infection, and mortality (Seggelke & Everhart, 2012). The principles of glucose control are discussed earlier in this chapter.

In addition, this is an opportunity for patients with diabetes to update their knowledge about diabetes self-care and prevention of complications. Nurses caring for patients with diabetes should focus attention on the diabetes as well as the primary health issue.

Self-Care Issues

For patients who are actively involved in diabetes self-management (especially insulin dose adjustment), relinquishing control over meal timing, insulin timing, and insulin dosage can be particularly difficult and anxiety provoking. The patient may fear hypoglycemia and express much concern over possible delays in receiving attention from the nurse if hypoglycemic symptoms occur or may disagree with a planned dose of insulin.

The nurse acknowledges the patient's concerns and involves the patient in the plan of care as much as possible. If

Chart 51-12 Approaches to Management of Glucose Control During the Perioperative Period for Those With a Diagnosis of Diabetes

- Monitor blood glucose levels frequently (every 1–2 hours).
- For patients taking insulin:
  1. The morning of surgery, all subcutaneous insulin doses are withheld, unless the blood glucose level is elevated (e.g., >200 mg/dL [11.1 mmol/L]), in which case a small dose of subcutaneous regular insulin may be prescribed. The blood glucose level is controlled during surgery with the IV infusion of regular insulin, which is balanced by an infusion of dextrose. The insulin and dextrose infusion rates are adjusted according to frequent (hourly) capillary glucose determinations. After surgery, the insulin infusion may be continued until the patient can eat. If IV insulin is discontinued, subcutaneous regular insulin may be administered at set intervals (every 4–6 hours), or intermediate-acting insulin may be administered every 12 hours with supplemental regular insulin as necessary until the patient is eating and the usual pattern of insulin dosing is resumed.
     - Carefully monitor the insulin infusion rate and blood glucose levels in a patient with diabetes who is receiving IV insulin. IV insulin has a much shorter duration of action than subcutaneous insulin. If the infusion is interrupted or discontinued, hyperglycemia will develop rapidly (within 1 hour in type 1 diabetes and within a few hours in type 2 diabetes).
     - Ensure that subcutaneous insulin is administered 30 minutes before the IV insulin infusion is discontinued.
  2. One half to two thirds of the patient's usual morning dose of insulin (either intermediate-acting insulin alone or both short- and intermediate-acting insulins) is administered subcutaneously in the morning before surgery. The remainder is then administered after surgery.
  3. The patient's usual daily dose of subcutaneous insulin is divided into 4 equal doses of regular insulin. These are then administered at 6-hour intervals. The last 2 approaches do not provide the control achieved by IV administration of insulin and dextrose.
- Patients with type 2 diabetes who do not usually take insulin may require insulin during the perioperative period to control blood glucose elevations. Patients who are taking metformin may be instructed to discontinue the oral agent 24–48 hours before surgery, if possible. Some of these patients may resume their usual regimen of diet and oral agent during the recovery period. Other patients (whose diabetes is probably not well controlled with diet and an oral antidiabetic agent before surgery) need to continue with insulin injections after discharge.
- For patients with type 2 diabetes who are undergoing minor surgery but who do not normally take insulin, glucose levels may remain stable provided no dextrose is infused during the surgery. After surgery, these patients may require small doses of regular insulin until the usual diet and oral agent are resumed.

the patient disagrees with certain aspects of the care related to diabetes, the nurse must communicate this to other members of the health care team. Nurses and other health care providers must pay particular attention to patients who are successful in managing self-care; they should assess these patients' self-care management skills and encourage them to continue if their performance is correct and effective.

Hospitalization of a patient with diabetes should be considered an opportunity to evaluate the patient's self-care skills and to reinforce and/or deliver education that might be needed. The nurse observes the patient preparing and injecting the insulin, monitoring blood glucose, and performing foot care. (Simply questioning the patient about these skills without actually observing performance of the skills is not sufficient.) The patient's knowledge about diet can be assessed with the help of a dietitian through direct questioning and review of the patient's menu choices. The patient's understanding about signs and symptoms, treatment, and prevention of hypoglycemia and hyperglycemia is assessed, along with knowledge of risk factors for macrovascular disease, including hypertension, increased lipids, and smoking. In addition, the patient is asked the date of his or her last eye examination (including dilation of the pupils). Education about these issues is critical.

Hyperglycemia During Hospitalization

Hyperglycemia may occur in patients who are hospitalized as a result of the original illness that led to the need for hospitalization. A number of other factors may contribute to hyperglycemia; examples include:

- Changes in the usual treatment regimen (e.g., increased food, decreased insulin, decreased activity)
- Medications (e.g., corticosteroids such as prednisone, which are used in the treatment of a variety of inflammatory disorders)
- IV dextrose, which may be part of the maintenance fluids or may be used for the administration of antibiotics and other medications, without adequate insulin therapy
- Overly vigorous treatment of hypoglycemia
- Inappropriate withholding of insulin or inappropriate use of "sliding scales"
- Mismatched timing of meals and insulin (e.g., postmeal hyperglycemia may occur if short-acting insulin is administered immediately before or even after a meal)

Nursing actions to correct some of these factors are important for avoiding hyperglycemia. Assessment of the patient's usual home routine is important. The nurse should try to approximate as much as possible the home schedule of insulin, meals, and activities. Monitoring blood glucose levels has been identified by the ADA as an additional "vital sign" essential in assessment of patients (ADA, 2013). The results of blood glucose monitoring provide information needed to obtain orders for extra doses of insulin (at times when insulin is usually taken), which is an important nursing function. Insulin doses must not be withheld when blood glucose levels are normal. It is very important to test blood glucose before a meal and administer insulin at that time, not on a rigid set time schedule as other medications are administered. Insulin should be administered when the meal is served to prevent hypoglycemia and elicit a physiologic response.

Short-acting insulin is usually needed to avoid postprandial hyperglycemia (even in patients with normal premeal glucose levels), and NPH insulin does not peak until many hours after the dose is given. IV antibiotics should be mixed in normal saline (if possible) to avoid excess infusion of dextrose (especially in patients who are eating). It is important to avoid overly vigorous treatment of hypoglycemia, which may lead to hyperglycemia.

Hypoglycemia During Hospitalization

Hypoglycemia in patients who are hospitalized is usually the result of too much insulin or delays in eating. Specific examples include:

- Overuse of sliding-scale regular insulin, particularly as a supplement to regularly scheduled, twice-daily short- and intermediate-acting insulins
- Lack of change in insulin dosage when dietary intake is changed (e.g., in the patient taking nothing by mouth [NPO])
- Overly vigorous treatment of hyperglycemia (e.g., giving too-frequent successive doses of regular insulin before the time of peak insulin activity is reached), resulting in a cumulative effect
- Delayed meal after administration of lispro, aspart, or glulisine [Apidra] insulin (patient should eat within 5 to 15 minutes after insulin administration)

Treatment of hypoglycemia should be based on the established hospital protocol (Seggelke & Everhart, 2012). If the initial treatment does not increase the glucose level adequately, the same treatment may be repeated after 15 minutes. The nurse must assess the pattern of glucose values and avoid giving doses of insulin that repeatedly lead to hypoglycemia. Successive doses of subcutaneous regular insulin should be administered no more frequently than every 3 to 4 hours. For patients receiving intermediate insulin before breakfast and dinner, the nurse must use caution in administering supplemental doses of regular insulin at lunch and bedtime. Hypoglycemia may occur when two insulins peak at similar times (e.g., morning NPH peaks with lunchtime regular insulin and may lead to late-afternoon hypoglycemia; dinnertime NPH peaks with bedtime regular insulin and may lead to nocturnal hypoglycemia). To avoid hypoglycemic reactions caused by delayed food intake, the nurse should arrange for snacks to be given to the patient if meals are going to be delayed because of procedures, physical therapy, or other activities.

Common Alterations in Diet

Dietary modifications commonly prescribed during hospitalization require special consideration for patients who have diabetes (ADA, 2013).

Nothing by Mouth

For patients who must be NPO in preparation for diagnostic or surgical procedures, the nurse must ensure that the usual insulin dosage has been changed. These changes may include eliminating the rapid-acting insulin and giving a decreased amount (e.g., half the usual dose) of intermediate-acting insulin. Another approach is to use frequent (every 3 to 4 hours) dosing of rapid-acting insulin only. IV dextrose may be administered to provide calories and to avoid hypoglycemia.

Even without food, glucose levels may increase as a result of hepatic glucose production, especially in patients with type 1 diabetes and lean patients with type 2 diabetes. Furthermore, in type 1 diabetes, elimination of the insulin dose may lead to the development of DKA. Administration of basal insulin to patients with type 1 diabetes who are NPO is an important nursing action.

For patients with type 2 diabetes who are taking insulin, DKA does not usually develop when insulin doses are eliminated because the patient's pancreas produces some insulin. Therefore, skipping the insulin dose altogether (when the patient is receiving IV dextrose) may be safe; however, close monitoring of blood glucose levels is essential.

For patients who are NPO for extended periods (24 hours), glucose testing and insulin administration should be performed at regular intervals, usually four times per day. Insulin regimens for the patient who is NPO for an extended period may include NPH insulin every 12 hours, rapid-acting insulin only every 4 to 6 hours, or an IV insulin drip. These patients should receive dextrose infusions to provide some calories and limit ketosis.

To prevent the problems that result from the need to withhold food, diagnostic tests and procedures and surgery should be scheduled early in the morning when possible.

Clear Liquid Diet

When the diet is advanced to include clear liquids, patients with diabetes receive more simple carbohydrate foods, such as juice and gelatin desserts, than are usually included in the diabetic diet. Because patients who are hospitalized should maintain their nutritional status as much as possible to promote healing, the use of reduced-calorie substitutes such as diet soda or diet gelatin desserts would not be appropriate when the only source of calories is clear liquids. Simple carbohydrates, if eaten alone, cause a rapid rise in blood glucose levels; therefore, it is important to try to match peak times of insulin effect with peaks in the blood glucose concentration. If the patient receives insulin at regular intervals while NPO, the scheduled times for glucose tests and insulin injections should match mealtimes.

Enteral Tube Feedings

Tube feeding formulas contain more simple carbohydrates and less protein and fat than the typical meal plan for diabetes. This results in increased levels of glucose in patients with diabetes who are receiving tube feedings. Insulin doses must be administered at regular intervals (e.g., NPH every 12 hours or regular insulin every 4 to 6 hours) when continuous tube feedings are administered. If insulin is administered at routine (prebreakfast and predinner) times, hypoglycemia during the day may result (because the patient receives more insulin without more calories); hyperglycemia may occur during the night if feedings continue but insulin action decreases.

A common cause of hypoglycemia in patients receiving both continuous tube feedings and insulin is inadvertent or purposeful discontinuation of the feeding. The nurse must discuss with the medical team any plans for temporarily discontinuing the tube feeding (e.g., when the patient is away from the unit). Planning ahead may allow for alterations to be made in the insulin dose or for administration of IV dextrose. In addition, if problems with the tube feeding develop unexpectedly (e.g., the patient pulls out the tube, the tube clogs, the feeding is discontinued when residual gastric contents are found), the nurse must notify the primary provider, assess blood glucose levels more frequently, and administer IV dextrose if indicated.

Parenteral Nutrition

Patients receiving parenteral nutrition may receive both IV insulin (added to the parenteral nutrition container) and subcutaneous intermediate- or short-acting insulins. If the patient is receiving continuous parenteral nutrition, the blood glucose level should be monitored and insulin administered at regular intervals. If the parenteral nutrition is infused over a limited number of hours, subcutaneous insulin should be administered so that peak times of insulin action coincide with times of parenteral nutrition infusion.

Hygiene

Nurses caring for hospitalized patients with diabetes must focus attention on oral hygiene and skin care. Because these patients are at increased risk for periodontal disease, the nurse assists with daily dental care. The patient may also require assistance in keeping the skin clean and dry, especially in areas of contact between two skin surfaces (e.g., groin, axilla, under the breasts), where chafing and fungal infections tend to occur.

Careful assessments of the oral cavity and the skin are important. The skin is assessed for dryness, cracks, breakdown, and redness, especially at pressure points and on the lower extremities. The patient is asked about symptoms of neuropathy, such as tingling and pain or numbness of the feet. Deep tendon reflexes are assessed.

As with any patient confined to bed, nursing care must emphasize the prevention of skin breakdown at pressure points. The heels are particularly susceptible to breakdown because of loss of sensation of pain and pressure associated with sensory neuropathy.

Feet should be cleaned, dried, lubricated with lotion (but not between the toes), and inspected frequently. If the patient is in the supine position, pressure on the heels can be alleviated by elevating the lower legs on a pillow, with the heels positioned over the edge of the pillow. When the patient is seated in a chair, the feet should be positioned so that pressure is not placed on the heels. If the patient has an ulcer on one foot, the nurse provides preventive care to the unaffected foot as well as special care of the affected foot.

As always, every opportunity should be taken to educate the patient about diabetes self-management, including daily oral, skin, and foot care. Female patients should also be instructed about measures for the avoidance of vaginal infections, which occur more frequently when blood glucose levels are elevated. Patients often take their cues from nurses and realize the importance of daily personal hygiene if this is emphasized during their hospitalization.

Stress

Physiologic stress, such as infections and surgery, contributes to hyperglycemia and may precipitate DKA or HHS. Emotional stress related to hospitalization for any reason can also have a negative impact on diabetic control. An increase in stress hormones leads to an increase in glucose levels, especially if intake of food and insulin remains unchanged. In addition, during periods of emotional stress, people with diabetes may alter their usual pattern of meals, exercise, and medication. This can contribute to hyperglycemia or even hypoglycemia (e.g., in the patient taking insulin or oral antidiabetic agents who stops eating in response to stress).

People with diabetes must be made aware of the potential deterioration in diabetic control that can accompany emotional stress. They must be encouraged to follow the diabetes treatment plan as much as possible during times of stress. In addition, learning strategies for minimizing stress and coping with stress when it does occur are important aspects of diabetes education. Healthy coping is one of the seven steps to managing diabetes identified by the AADE (2007).

Gerontologic Considerations

Because people with diabetes are living longer, both type 1 and type 2 diabetes are being seen more frequently in older patients hospitalized for various reasons. Regardless of the type or duration of diabetes, the goals of diabetes treatment may need to be altered when caring for older adults who are hospitalized. The focus is on quality-of-life issues, such as maintaining independent functioning and promoting general well-being.

Some of the barriers to learning and self-care during hospital stays and in preparing patients for discharge include decreased vision, hearing loss, memory deficits, decreased mobility and fine motor coordination, increased tremors, depression and isolation, decreased financial resources, and limitations related to disabilities and other medical disorders. Assessing these barriers is important in planning diabetes treatment and educational activities. Presenting brief, simplified instructions with ample opportunity for practice of skills is important. The use of special devices such as a magnifier for the insulin syringe, an insulin pen, or a mirror for foot inspection is helpful. Frequent evaluation of self-care skills (insulin administration, blood glucose monitoring, foot care, diet planning) is essential, especially in patients with deteriorating vision and memory.

If appropriate, family members may be called on to assist with diabetes survival skills, and referral to community resources may be made. It is preferable to educate the patient or family members to test blood glucose at home; the choice of meter should be tailored to the patient's visual and cognitive status and dexterity.

Quality and Safety Nursing Alert

Careful monitoring for diabetes complications must not be neglected in older adults. Hypoglycemia is especially dangerous, because it may go undetected and result in falls. Dehydration is a concern in patients who have chronically elevated blood glucose levels. Assessment for long-term complications, especially eye and foot problems, is important. Avoiding blindness and amputation through early detection and treatment of retinopathy and foot ulcers may mean the difference between placement in a long-term care facility and continued independent living for the older adult with diabetes.

Nursing Management

Monitoring and Managing Potential Complications

Assessment for hypoglycemia and hyperglycemia involves frequent blood glucose monitoring (usually prescribed before meals and at bedtime) and monitoring for signs and symptoms of hypoglycemia or prolonged hyperglycemia (including DKA or HHS), as described previously. Inadequate control of blood glucose levels may hinder recovery from the primary health problem. Blood glucose levels are monitored, and insulin is administered as prescribed. The nurse must ensure that prescribed insulin dosage is modified as needed to compensate for changes in the patient's schedule or eating pattern. Treatment is given for hypoglycemia (with oral glucose) or hyperglycemia (with supplemental regular insulin no more often than every 3 to 4 hours). Blood glucose records are assessed for patterns of hypoglycemia and hyperglycemia at the same time of day, and findings are reported to the primary provider for modification in insulin orders. In the patient with prolonged elevations in blood glucose, laboratory values and the patient's physical condition are monitored for signs and symptoms of DKA or HHS.

Promoting Home and Community-Based Case

Educating Patients About Self-Care

Even if patients have had diabetes for many years, their knowledge and adherence to the plan of care must be assessed. A new plan of care may need to be devised using concepts mentioned earlier. The nurse also reminds the patient and family about the importance of health promotion activities and recommended health screening.

Continuing Care

A patient who is hospitalized may require referral for home care. The home care nurse can use this opportunity to assess the patient's knowledge about diabetes management and the patient's and family's ability to carry out that management. The nurse reinforces the education provided in the hospital, clinic, office, or diabetes education center and assesses the home care environment to determine its adequacy for self-care and safety.

Critical Thinking Exercises

1 ebp A 25-year-old pregnant woman with gestational diabetes is admitted to the hospital for an elective cesarean section. What is the evidence base for glucose management before, during, and after her surgery? Identify the criteria used to evaluate the strength of the evidence for the practices you have identified.

2 A 45-year-old patient is newly diagnosed with type 2 diabetes. His current diet consists of high-fat and high-carbohydrate foods, high sodium intake, and minimal vegetable consumption. He does not exercise and has smoked one pack of cigarettes a day since his early 20s. Generate an education plan for this patient. Identify immediate and long-term goals of management for his diabetes.

3 pq A 68-year-old man who was found at home unresponsive by his neighbors is admitted to the emergency department with possible DKA. Identify the pathophysiology and signs and symptoms of DKA. What are your priorities for the assessment, medical management, and nursing care for this patient with DKA?

4 You are providing education to a 55-year-old woman with type 1 diabetes who is being discharged from the hospital after cardiac artery bypass surgery. She indicates that she has had diabetes for 10 years. Identify the long-term complication that you would consider for a patient with type 1 diabetes. What factors do you need to assess to determine her risks for long-term complications?

Brunner Suite Resources

Explore these additional resources to enhance learning for this chapter:

- NCLEX-Style Questions and Other Resources on thePoint, http://thePoint.lww.com/Brunner13e
- Study Guide
- PrepU
- Clinical Handbook
- Handbook of Laboratory and Diagnostic Tests

References

*Asterisk indicates nursing research.

**Double asterisk indicates classic reference.

Books

Bader, M., & Littlejohns, L. R. (2010). *AANN core curriculum for neuroscience nursing* (5th ed.). Glenview, IL: American Association of Neuroscience Nurses.

Dudek, S. G. (2010). *Nutrition essentials for nursing practice* (6th ed.). Philadelphia: Lippincott Williams & Wilkins.

Eliopoulos, C. (2010). *Gerontological nursing* (7th ed.). Philadelphia: Lippincott Williams & Wilkins.

Fischbach, F. T., & Dunning, M. B. (2009). *A manual of laboratory and diagnostic tests* (8th ed.). Philadelphia: Lippincott Williams & Wilkins.

Karch, A. M. (2012). *2012 Lippincott's nursing drug guide*. Philadelphia: Lippincott Williams & Wilkins.

National Institute of Diabetes and Digestive and Kidney Diseases. (2012a). *Pancreatic islet transplantation*. NIH Publication No. 07–4693. Bethesda, MD: U.S. Department of Health and Human Services, National Institutes of Health. Available at: diabetes.niddk.nih.gov/dm/pubs/pancreaticislet/Pancreatic_Islet_Transplantation_508.pdf

National Institute of Diabetes and Digestive and Kidney Diseases. (2012b). *Diabetic neuropathies: The nerve damage of diabetes*. NIH Publication No. 08–3185. Bethesda, MD: U.S. Department of Health and Human Services, National Institutes of Health. Available at: diabetes.niddk.nih.gov/dm/pubs/neuropathies/

National Institute of Diabetes and Digestive and Kidney Diseases. (2008). *Prevent diabetes problems: Keep your feet and skin healthy*. NIH Publication No. 08–4282. Bethesda, MD: U.S. Department of Health and Human Services, National Institutes of Health. Available at: diabetes.niddk.nih.gov/dm/pubs/complications_feet

Porth, C. M., & Matfin, G. (2009). *Pathophysiology: Concepts of altered health states* (8th ed.). Philadelphia: Lippincott Williams & Wilkins.

U.S. Department of Health and Human Services. (2010). *Healthy people 2020*. Washington, DC: Author.

Journals and Electronic Documents

American Association of Diabetes Educators. (2005). The scope of practice, standards of practice, and standards of professional performance for diabetes educators. *Diabetes Educators, 31*(4), 487–513.

American Association of Diabetes Educators. (2007). AADE position statement: Individualization of diabetes self-management education. *Diabetes Educator, 33*(1), 45–49.

American Diabetes Association (ADA). (2008a). Lipoprotein management in patients with cardiometabolic risk. Consensus statement from ADA and American College of Cardiology Foundation [consensus statement]. *Diabetes Care, 31*(4), 811–822.

American Diabetes Association (ADA). (2008b). Nutrition recommendations and interventions for diabetes [position statement]. *Diabetes Care, 31*(Suppl 1), S61–S78.
American Diabetes Association. (2009). Diagnosis and classification of diabetes mellitus. *Diabetes Care, 32*(Suppl. 1), S62–S67.
American Diabetes Association. (2013). Standards of medical care in diabetes—2013. *Diabetes Care, 36*(Suppl. 1), S11–S66.
Boulton, A., Armstrong, D., Albert, S., et al. (2008) Comprehensive foot examination and risk assessment. *Diabetes Care, 31*(8), 1679–1685.
*Calvin, D., Quinn, L., Dancy, B., et al. (2011). African Americans' perception of risk for diabetes complications. *Diabetes Educator, 37*(5), 689–698.
Centers for Disease Control and Prevention. (2011). *National diabetes fact sheet, 2011*. Available at: www.cdc.gov/diabetes/pubs/pdf/ndfs_2011.pdf
**Diabetes Control and Complications Trial Research Group. (1993). The effect of intensive treatment of diabetes on the development and progression of long-term complications in insulin-dependent diabetes mellitus. *New England Journal of Medicine, 329*(14), 977–986.
**Diabetes Prevention Program Research Group. (2002). Reduction in the incidence of type 2 diabetes with lifestyle intervention or metformin. *New England Journal of Medicine, 346*(6), 393–403.
Fowler, M. (2009). Hyperglycemic crisis in adults: Pathophysiology, presentation, pitfalls, and prevention. *Clinical Diabetes, 27*(1), 19–23.
Funnell, M. M., Brown, T. L., Childs, B. P., et al. (2009). National standards for diabetes self-management education. *Diabetes Care, 32*(Suppl. 1), S87–S94.
Geil, P. B. (2008). Choose your foods: Exchange lists for diabetes: The 2008 revision of exchange lists for meal planning. *Diabetes Spectrum, 21*(4), 281–283.
*Jones, H. L., Walker, E. A., Schechter, C. B., & Blanco, E. (2010). Vision is precious: A successful behavioral intervention to increase the rate of screening for diabetic retinopathy in inner-city adults. *Diabetes Educator, 36*(1), 118–126.
*Penckofer, S., Ferrans, C., & Velsor-Friedrich, B. (2007). The psychological impact of living with diabetes: Women's day-to-day experiences. *Diabetes Educator, 33*(4), 680–690.
Reynolds, I. G. (2012). How to recognize and interviene for hyperosmolar hyperglycemic syndrome. *American Nursing Today, 7*(7), 12–15.
Seggelke, S. A., & Everhart, B. (2012). Managing glucose levels in hospital patients. *American Nurse Today, 7*(9), 27–32.
**United Kingdom Prospective Diabetes Study Group. (1998). Intensive blood glucose control with sulfonylureas or insulin compared with conventional treatment and risk of complications with type 2 diabetes. *Lancet, 352*(9131), 837–853.
*Welch, G., Allen, N. A., Zagarins, A. E., et al. (2011). Comprehensive diabetes management program for poorly controlled Hispanic type 2 patients at a community health center. *Diabetes Educator, 37*(5), 680–688.
World Health Organization. (2011). *Obesity and overweight*. Available at: www.who.int/mediacentre/factsheets/fs311/en/
World Health Organization. (2012). *Diabetes*. Fact Sheet No. 312. www.who.int/mediacentre/factsheets/fs312/en/index.html

Resources

Academy of Nutrition and Dietetics (formerly the American Dietetic Association), www.eatright.org
American Association of Diabetes Educators (AADE), www.aadenet.org/
American Diabetes Association, www.diabetes.org
American Foundation for the Blind (AFB), www.afb.org
Armchair Fitness Series, CC-M Productions, 7755 16th Street, N.W., Washington, DC 20012, armchairfitness.stores.yahoo.net/index.html
Centers for Disease Control and Prevention (CDC), www.cdc.gov/diabetes/pubs/factsheet.htm
JDRF (formerly the Juvenile Diabetes Research Foundation), www.jdrf.org
MedicAlert Foundation, www.medicalert.org
National Diabetes Information Clearinghouse, digestive.niddk.nih.gov
National Library Services for the Blind and Physically Handicapped (NLS), www.loc.gov/nls/

Chapter 52

Assessment and Management of Patients With Endocrine Disorders

Learning Objectives

On completion of this chapter, the learner will be able to:

1. Describe the functions of each of the endocrine glands and their hormones.
2. Identify the diagnostic tests used to determine alterations in function of each of the endocrine glands.
3. Use the nursing process as a framework for care of patients with hyperthyroidism.
4. Develop a plan of nursing care for the patient undergoing thyroidectomy.
5. Compare hyperparathyroidism and hypoparathyroidism: their causes, clinical manifestations, management, and nursing interventions.
6. Compare Addison's disease with Cushing syndrome: their causes, clinical manifestations, management, and nursing interventions.
7. Describe nursing management of patients with adrenal insufficiency.
8. Use the nursing process as a framework for care of patients with Cushing syndrome.
9. Identify the education needs of patients requiring corticosteroid therapy.

Glossary

acromegaly: progressive enlargement of peripheral body parts resulting from excessive secretion of growth hormone

addisonian crisis: acute adrenocortical insufficiency; characterized by hypotension, cyanosis, fever, nausea/vomiting, and classic signs of shock

Addison's disease: chronic adrenocortical insufficiency due to inadequate adrenal cortex function

adrenalectomy: surgical removal of one or both adrenal glands

adrenocorticotropic hormone (ACTH): hormone secreted by the anterior pituitary, essential for growth and development

adrenogenital syndrome: masculinization in women, feminization in men, or premature sexual development in children; result of abnormal secretion of adrenocortical hormones, especially androgens

androgens: male sex hormones

basal metabolic rate: chemical reactions occurring when the body is at rest

calcitonin: hormone secreted by the thyroid gland; participates in calcium regulation

Chvostek's sign: spasm of the facial muscles produced by sharply tapping over the facial nerve in front of the parotid gland and anterior to the ear; suggestive of latent tetany in patients with hypocalcemia

corticosteroids: hormones produced by the adrenal cortex or their synthetic equivalents; also referred to as adrenal-cortical hormone and adrenocorticosteroid; consist of glucocorticoids, mineralocorticoids, and androgens

Cushing syndrome: group of symptoms produced by an oversecretion of adrenocorticotropic hormone; characterized by truncal obesity, "moon face," acne, abdominal striae, and hypertension

diabetes insipidus: condition in which abnormally large volumes of dilute urine are excreted as a result of deficient production of vasopressin

dwarfism: generalized limited growth resulting from insufficient secretion of growth hormone during childhood

endocrine: secreting internally; hormonal secretion of a ductless gland

euthyroid: state of normal thyroid hormone production

exocrine: secreting externally; hormonal secretion from excretory ducts

exophthalmos: abnormal protrusion of one or both eyeballs

glucocorticoids: steroid hormones secreted by the adrenal cortex in response to adrenocorticotropic hormone; produce a rise of liver glycogen and blood glucose

goiter: enlargement of the thyroid gland

Graves' disease: a form of hyperthyroidism; characterized by a diffuse goiter and exophthalmos

hormones: chemical transmitter substances produced in one organ or part of the body and carried by the bloodstream to other cells or organs on which they have a specific regulatory effect; produced mainly by endocrine glands

hypophysectomy: removal or destruction of all or part of the pituitary gland

mineralocorticoids: steroid hormones secreted by the adrenal cortex

myxedema: severe hypothyroidism; can be with or without coma

negative feedback: regulating mechanism in which an increase or decrease in the level of a substance decreases or increases the function of the organ producing the substance

pheochromocytoma: adrenal medulla tumor

syndrome of inappropriate antidiuretic hormone (SIADH) secretion: excessive secretion of antidiuretic hormone from the pituitary gland despite low serum osmolality level

thyroidectomy: surgical removal of all or part of the thyroid gland

thyroiditis: inflammation of the thyroid gland; may lead to chronic hypothyroidism or may resolve spontaneously

thyroid-stimulating hormone (TSH): released from the pituitary gland; causes stimulation of the thyroid, resulting in release of T_3 and T_4

thyroid storm: severe life-threatening hyperthyroidism precipitated by stress; characterized by high fever, extreme tachycardia, and altered mental state

thyrotoxicosis: condition produced by excessive endogenous or exogenous thyroid hormone

thyroxine (T_4): thyroid hormone; active iodine compound formed and stored in the thyroid; deiodinated in peripheral tissues to form triiodothyronine; maintains body metabolism in a steady state

triiodothyronine (T_3): thyroid hormone; formed and stored in the thyroid; released in smaller quantities, biologically more active, and with faster onset of action than T_4; widespread effect on cellular metabolism

Trousseau's sign: carpopedal spasm induced when blood flow to the arm is occluded using a blood pressure cuff or tourniquet, causing ischemia to the distal nerves; suggestive sign for latent tetany in hypocalcemia

vasopressin: antidiuretic hormone secreted by the posterior pituitary

---

The endocrine system plays a vital role in orchestrating cellular interactions, metabolism, growth, and senescence (Barzilai & Gabriely, 2010). This interconnected network of glands is closely linked with the nervous and immune systems, regulating the functions of multiple body organs. Disorders of the endocrine system are common and are manifested as hyperfunction and hypofunction.

Nursing interventions are essential in the management of patients with endocrine disorders. This chapter focuses on the anatomy and physiology of the endocrine system; the most common endocrine disorders of the pituitary, thyroid, parathyroid, and adrenal glands; clinical manifestations; diagnostic studies; medical management; and nursing interventions. The unique endocrine and exocrine functions of the pancreas, pancreatic function, and associated pancreatic disorders are discussed in Chapters 50 and 51; reproductive structures, including the ovaries and testes, are discussed in Chapters 56 and 59, respectively.

ASSESSMENT OF THE ENDOCRINE SYSTEM

Anatomic and Physiologic Overview

The endocrine system involves the release of chemical transmitter substances known as **hormones**. These substances regulate and integrate body functions by acting on local or distant target sites. Hormones are generally produced by the endocrine glands but may also be produced by specialized tissues such as those found in the gastrointestinal (GI) system, the kidney, and white blood cells. The GI mucosa produces hormones (e.g., gastrin, enterogastrone, secretin, cholecystokinin) that are important in the digestive process; the kidneys produce erythropoietin, a hormone that stimulates the bone marrow to produce red blood cells; and the white blood cells produce cytokines (hormonelike proteins) that actively participate in inflammatory and immune responses.

The endocrine system has a unique relationship with the immune and the nervous systems. Chemicals such as neurotransmitters (e.g., epinephrine) released by the nervous system can also function as hormones when needed. The immune system responds to the introduction of foreign agents by means of chemical messengers (cytokines), which are hormonelike proteins, and is also subject to regulation by adrenal corticosteroid hormones (Porth, 2011).

Glands of the Endocrine System

The endocrine system is composed of the pituitary gland, thyroid gland, parathyroid glands, adrenal glands, pancreatic islets, ovaries, and testes (Fig. 52-1). Most hormones secreted from **endocrine** glands are released directly into the bloodstream. However, **exocrine** glands, such as sweat glands, secrete their products through ducts onto epithelial surfaces or into the GI tract.

Function and Regulation of Hormones

Hormones help regulate organ function in concert with the nervous system. The rapid action by the nervous system is balanced by slower hormonal action. This dual regulatory system permits precise control of organ functions in response to changes within and outside the body. Table 52-1 lists the major hormones, their target tissues, and some of their properties.

The endocrine glands are composed of secretory cells arranged in minute clusters known as acini. A rich blood supply provides a vehicle for the hormones produced by the endocrine glands to enter the bloodstream rapidly. The amount of circulating hormones depends on their unique function and the body's needs. In the healthy physiologic state, hormone concentration in the bloodstream is maintained at a relatively constant level. To prevent accumulation, these hormones must be inactivated continuously by a **negative feedback** system so that when the hormone concentration increases, further production of that hormone is inhibited. Conversely, when the hormone concentration decreases, the rate of production of that hormone increases.

Classification and Action of Hormones

Hormones are classified into four categories according to their structure: (1) amines and amino acids (e.g., epinephrine, norepinephrine, and thyroid hormones); (2) peptides, polypeptides, proteins, and glycoproteins (e.g., thyrotropin-releasing

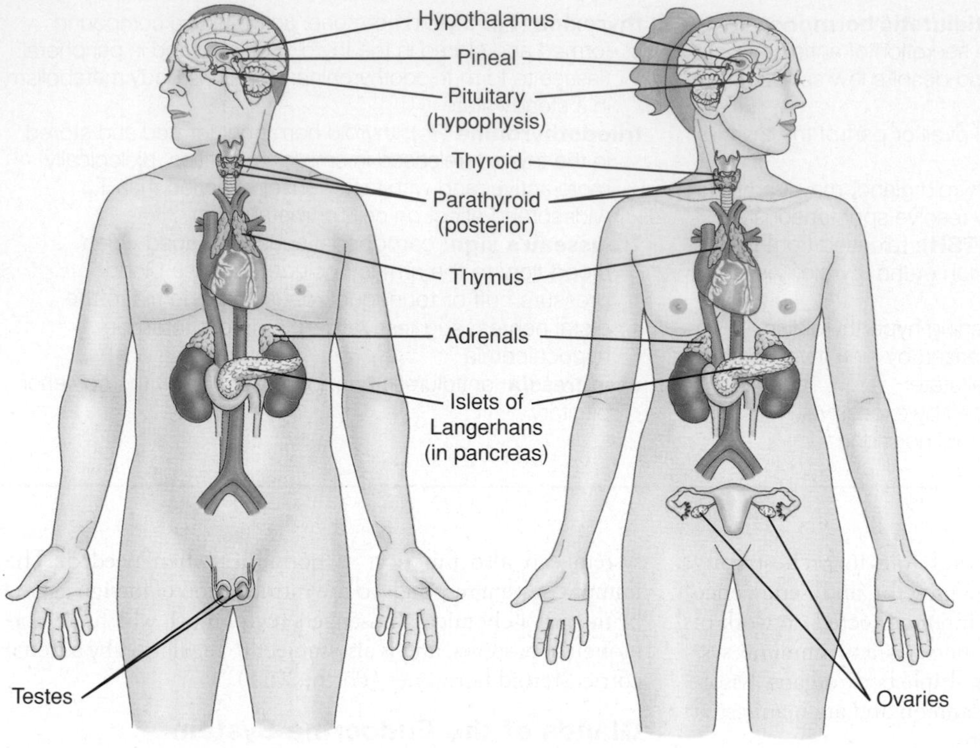
Hypothalamus
Pineal
Pituitary
(hypophysis)
Thyroid
Parathyroid
(posterior)
Thymus
Adrenals
Islets of
Langerhans
(in pancreas)
Testes
Ovaries

FIGURE 52-1 • Major hormone-secreting glands of the endocrine system.

hormone [TRH], follicle-stimulating hormone [FSH], and growth hormone [GH]); (3) steroids (e.g., **corticosteroids**, which are hormones produced by the adrenal cortex or their synthetic equivalents); and (4) fatty acid derivatives (e.g., eicosanoid, retinoids) (Porth, 2011). Although most hormones released by endocrine glands can be transported to distant target sites for action, some hormones and hormone-like substances never enter the bloodstream. Some hormones act locally in the area where they are released; this is called *paracrine action* (e.g., the effect of sex hormones on the ovaries). Others may act on the actual cells from which they were released; this is called *autocrine action* (e.g., the effect of insulin from pancreatic beta cells on those cells) (Porth, 2011).

Hormones can alter the function of the target tissue by interacting with chemical receptors located either on the cell membrane or in the interior of the cell. For example, *peptide* and *protein hormones* interact with receptor sites on the cell surface, resulting in stimulation of the intracellular enzyme adenyl cyclase. This causes increased production of cyclic 3′,5′-adenosine monophosphate (AMP). The cyclic AMP inside the cell alters enzyme activity. Thus, cyclic AMP is the "second messenger" that links the peptide hormone at the cell surface to a change in the intracellular environment. Some protein and peptide hormones also act by changing membrane permeability and act within seconds or minutes. The mechanism of action for *amine hormones* is similar to that for peptide hormones.

Steroid hormones, because of their smaller size and higher lipid solubility, penetrate cell membranes and interact with intracellular receptors. The steroid–receptor complex modifies cell metabolism and the formation of messenger ribonucleic acid (mRNA) from deoxyribonucleic acid (DNA). The mRNA then stimulates protein synthesis within the cell. Steroid hormones require several hours to exert their effects, because they exert their action by the modification of protein synthesis.

Assessment

Health History

Although specific endocrine disorders are often accompanied by specific clinical symptoms, more general manifestations may also occur. A thorough health history and review of systems are necessary for diagnosis and management of these disorders. Patients should be asked if they have experienced changes in energy level, tolerance to heat or cold, weight, thirst, frequency of urination, fat and fluid distribution, secondary sexual characteristics such as loss or growth of hair, menstrual cycle, memory, concentration, sleep patterns, and mood, as well as vision changes, joint pain, and sexual dysfunction. It is important to document (1) the severity of these changes, (2) the length of time the patient has experienced these changes, (3) the way in which these changes have affected the patient's ability to carry out activities of daily living, (4) the effect of the changes on the patient's self-perception, and (5) family history.

Physical Assessment

The physical examination should include vital signs; head-to-toe inspection; and palpation of skin, hair, and thyroid. Findings should be compared with previous findings, if available. Physical, psychological, and behavioral changes should be noted. Examples of changes in physical characteristics on examination may include appearance of facial hair in women, "moon face," "buffalo hump," exophthalmos, vision changes,

TABLE 52-1 Major Action and Source of Selected Hormones

Source	Hormone	Major Action
Hypothalamus	Releasing and inhibiting hormones Corticotropin-releasing hormone (CRH) Thyrotropin-releasing hormone (TRH) Growth hormone–releasing hormone (GHRH) Gonadotropin-releasing hormone (GnRH)	Controls the release of pituitary hormones
	Somatostatin	Inhibits growth hormone and thyroid-stimulating hormone
Anterior pituitary	Growth hormone (GH)	Stimulates growth of bone and muscle, promotes protein synthesis and fat metabolism, decreases carbohydrate metabolism
	Adrenocorticotropic hormone (ACTH)	Stimulates synthesis and secretion of adrenocortical hormones
	Thyroid-stimulating hormone (TSH)	Stimulates synthesis and secretion of thyroid hormone
	Follicle-stimulating hormone (FSH)	*Female:* Stimulates growth of ovarian follicle, ovulation *Male:* Stimulates sperm production
	Luteinizing hormone (LH)	*Female:* Stimulates development of corpus luteum, release of oocyte, production of estrogen and progesterone *Male:* Stimulates secretion of testosterone, development of interstitial tissue of testes
	Prolactin	Prepares female breast for breast-feeding
Posterior pituitary	Antidiuretic hormone (ADH)	Increases water reabsorption by kidney
	Oxytocin	Stimulates contraction of pregnant uterus, milk ejection from breasts after childbirth
Adrenal cortex	Mineralocorticosteroids, mainly aldosterone	Increase sodium absorption, potassium loss by kidney
	Glucocorticoids, mainly cortisol	Affect metabolism of all nutrients; regulates blood glucose levels, affects growth, has anti-inflammatory action, and decreases effects of stress
	Adrenal androgens, mainly dehydroepiandrosterone (DHEA) and androstenedione	Have minimal intrinsic androgenic activity; they are converted to testosterone and dihydrotestosterone in the periphery
Adrenal medulla	Epinephrine Norepinephrine	Serve as neurotransmitters for the sympathetic nervous system
Thyroid (follicular cells)	Thyroid hormones Triiodothyronine (T_3) Thyroxine (T_4)	Increase the metabolic rate; increase protein and bone turnover; increase responsiveness to catecholamines; necessary for fetal and infant growth and development
Thyroid C cells	Calcitonin	Lowers blood calcium and phosphate levels
Parathyroid glands	Parathormone (parathyroid hormone [PTH])	Regulates serum calcium
Pancreatic islet cells	Insulin	Lowers blood glucose by facilitating glucose transport across cell membranes of muscle, liver, and adipose tissue
	Glucagon	Increases blood glucose concentration by stimulation of glycogenolysis and glyconeogenesis
	Somatostatin	Delays intestinal absorption of glucose
Kidney	1,25-Dihydroxyvitamin D	Stimulates calcium absorption from the intestine
	Renin	Activates renin–angiotensin–aldosterone system
	Erythropoietin	Increases red blood cell production
Ovaries	Estrogen	Affects development of female sex organs and secondary sex characteristics
	Progesterone	Influences menstrual cycle; stimulates growth of uterine wall; maintains pregnancy
Testes	Androgens, mainly testosterone	Affect development of male sex organs and secondary sex characteristics; aid in sperm production

Reproduced with permission from Porth, C. M., & Matfin, G. (2009). *Pathophysiology: Concepts of altered health states* (8th ed.). Philadelphia: Lippincott Williams & Wilkins.

edema, thinning of the skin, obesity of the trunk, thinness of the extremities, increased size of the feet and hands, edema, and hypo- or hyperreflexia. The patient may also exhibit changes in mood and behavior such as nervousness, lethargy, and fatigue (Hogan-Quigley, Palm, & Bickley, 2012).

Diagnostic Evaluation

A variety of diagnostic studies are used to evaluate the endocrine system. The nurse should educate the patient about the purpose of the prescribed studies, what to expect, and any possible side effects related to these examinations prior to testing. The nurse should note trends in results because they provide information about disease progression as well as the patient's response to therapy.

Blood Tests

Blood tests are used to determine the levels of circulating hormones, the presence of autoantibodies, and the effect of a specific hormone on other substances (e.g., the effect of insulin on blood glucose levels). The serum levels of a specific hormone may provide information to determine the presence of hypofunction or hyperfunction of the endocrine system and the site of dysfunction. Radioimmunoassays are radioisotope-labeled antigen tests that are commonly indicated blood tests used to measure the levels of hormones or other substances.

Urine Tests

Urine tests are used to measure the amount of hormones or the end products of hormones excreted by the kidneys. One-time specimens or, in some disorders, 24-hour urine specimens are collected to measure hormones or their metabolites. For example, urinary levels of free catecholamines (norepinephrine, epinephrine, and dopamine) may be measured in patients with suspected tumors of the adrenal medulla (**pheochromocytoma**). Several disadvantages related to urine tests that must be considered are that patients may be unable to urinate at scheduled intervals and that some medications or disease states may affect the test results (Porth, 2011).

Additional Diagnostic Studies

Stimulation tests are used to confirm hypofunction of an endocrine organ. The tests determine how an endocrine gland responds to the administration of stimulating hormones that are normally produced or released by the hypothalamus or pituitary gland. If the endocrine gland responds to this stimulation, the specific disorder may be in the hypothalamus or pituitary. Failure of the endocrine gland to respond to this stimulation helps identify the problem as being in the endocrine gland itself.

Suppression tests are used to detect hyperfunction of an endocrine organ. They determine if the organ is not responding to the negative feedback mechanisms that normally control secretion of hormones from the hypothalamus or pituitary gland. Suppression tests measure the effect of an administered exogenous dose of the hormone on the endogenous secretion of the hormone or on the secretion of stimulation hormones from the hypothalamus or pituitary gland.

Imaging studies include radioactive scanning, magnetic resonance imaging (MRI), computed tomography (CT), ultrasonography, positron emission tomography (PET), and dual-energy x-ray absorptiometry (DEXA) (Porth, 2011).

Genetic screening is increasingly becoming more routine in the assessment of endocrine disorders (Chart 52-1). DNA testing is expected to lead to the identification of specific genes associated with endocrine disorders, selective targeting for drug development, and increased understanding of the function of the endocrine system. Genetic screening is used to determine the presence of a gene mutation that may predispose an individual to a certain condition (Greco & Mahon, 2012). The pros and cons of genetic screening must be considered carefully by the primary provider and patient.

THE PITUITARY GLAND

Anatomic and Physiologic Overview

The pituitary gland, or hypophysis, is commonly referred to as the master gland because of the influence it has on secretion of hormones by other endocrine glands (Porth, 2011) (Fig. 52-2). The round structure, about 1.27 cm (1/2 in) in diameter, is located on the inferior aspect of the brain and is divided into anterior and posterior lobes. It is controlled by

Chart 52-1 GENETICS IN NURSING PRACTICE

Metabolic and Endocrine Disorders

Some examples of metabolic and endocrine disorders influenced by genetic factors include the following:

- Alpha$_1$-antitrypsin deficiency
- Cystic fibrosis
- Diabetes, both type 1 and type 2
- Hereditary hemochromatosis
- Multiple endocrine neoplasia type I and type II
- Von Hippel-Lindau syndrome

Nursing Assessments

Family History Assessment

- Assess family history for relatives with early-onset hepatic, pancreatic, or endocrine disease.
- Inquire about family members with diabetes and their ages at onset.
- Assess family history of other related genetic conditions such as cystic fibrosis, alpha$_1$-antitrypsin deficiency, and hereditary hemochromatosis.

Patient Assessment

- Assess for physical symptoms such as mucosal neuromas, hypertrophied lips, skeletal abnormalities, and marfanoid appearance.
- Assess for signs of arthritis and bronze pigmentation of the skin (hereditary hemochromatosis).

Management Issues Specific to Genetics

- Inquire whether deoxyribonucleic acid (DNA) mutation testing has been performed on any affected family member.
- If indicated, refer for further genetic counseling and evaluation so that family members can discuss inheritance, risk to other family members, and availability of genetic testing and gene-based interventions.
- Offer appropriate genetics information and resources.
- Assess patient's understanding of genetics information.
- Provide support to families with newly diagnosed genetics-related metabolic and endocrine conditions.
- Participate in management and coordination of care of patients with genetics-related metabolic or endocrine disorders and other genetic conditions and people predisposed to develop or pass on a genetics-related metabolic or endocrine disorder or other genetic condition.

Genetics Resources

See Chapter 8, Chart 8-6 for genetics resources.

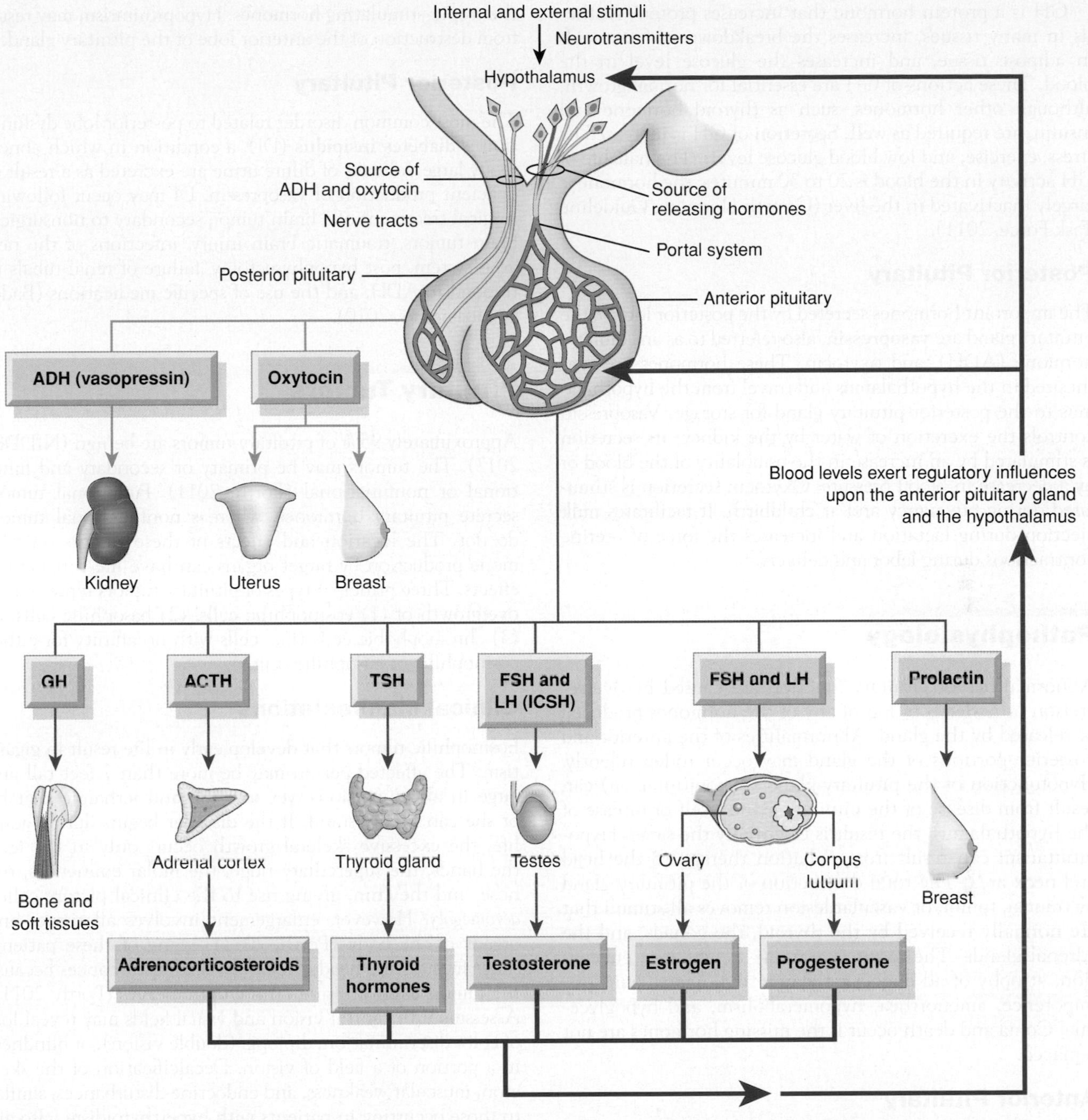
Internal and external stimuli
Neurotransmitters
Hypothalamus
Source of
ADH and oxytocin
Source of
releasing hormones
Nerve tracts
Portal system
Posterior pituitary
Anterior pituitary
ADH (vasopressin)
Oxytocin
Kidney
Uterus
Breast
Blood levels exert regulatory influences
upon the anterior pituitary gland
and the hypothalamus
GH
ACTH
TSH
FSH and
LH (ICSH)
FSH and LH
Prolactin
Bone and
soft tissues
Adrenal cortex
Thyroid gland
Testes
Ovary
Corpus
luteum
Breast
Adrenocorticosteroids
Thyroid
hormones
Testosterone
Estrogen
Progesterone

FIGURE 52-2 • The pituitary gland, the relationship of the brain to pituitary action, and the hormones secreted by the anterior and posterior pituitary lobes. ADH, antidiuretic hormone; GH, growth hormone; ACTH, adrenocorticotropic hormone; TSH, thyroid-stimulating hormone; FSH, follicle-stimulating hormone; LH, luteinizing hormone; ICSH, interstitial cell-stimulating hormone.

the hypothalamus, which is an adjacent area of the brain that is connected to the pituitary by the pituitary stalk.

Anterior Pituitary

The major hormones of the anterior pituitary gland are FSH, luteinizing hormone (LH), prolactin, **adrenocorticotropic hormone (ACTH)**, **thyroid-stimulating hormone (TSH)**, and GH (also referred to as somatotropin). The secretion of these major hormones is controlled by releasing factors secreted by the hypothalamus. These releasing factors reach the anterior pituitary by way of the bloodstream in a special circulation called the *pituitary portal blood system*. Other hormones include melanocyte-stimulating hormone and beta-lipotropin; the function of lipotropin is poorly understood.

The hormones released by the anterior pituitary enter the general circulation and are transported to their target organs. The main function of TSH, ACTH, FSH, and LH is the release of hormones from other endocrine glands. Prolactin acts on the breast to stimulate milk production. Hormones that stimulate other organs and tissues are discussed in conjunction with their target organs.

GH is a protein hormone that increases protein synthesis in many tissues, increases the breakdown of fatty acids in adipose tissue, and increases the glucose level in the blood. These actions of GH are essential for normal growth, although other hormones, such as thyroid hormone and insulin, are required as well. Secretion of GH is increased by stress, exercise, and low blood glucose levels. The half-life of GH activity in the blood is 20 to 30 minutes; the hormone is largely inactivated in the liver (Growth Hormone Guideline Task Force, 2011).

Posterior Pituitary

The important hormones secreted by the posterior lobe of the pituitary gland are **vasopressin**, also referred to as antidiuretic hormone (ADH), and oxytocin. These hormones are synthesized in the hypothalamus and travel from the hypothalamus to the posterior pituitary gland for storage. Vasopressin controls the excretion of water by the kidney; its secretion is stimulated by an increase in the osmolality of the blood or by a decrease in blood pressure. Oxytocin secretion is stimulated during pregnancy and at childbirth. It facilitates milk ejection during lactation and increases the force of uterine contractions during labor and delivery.

Pathophysiology

Abnormalities of pituitary function are caused by oversecretion or undersecretion of any of the hormones produced or released by the gland. Abnormalities of the anterior and posterior portions of the gland may occur independently. Hypofunction of the pituitary gland (hypopituitarism) can result from disease of the pituitary gland itself or disease of the hypothalamus; the result is essentially the same. Hypopituitarism can result from radiation therapy to the head and neck area. The total destruction of the pituitary gland by trauma, tumor, or vascular lesion removes all stimuli that are normally received by the thyroid, the gonads, and the adrenal glands. The result is extreme weight loss, emaciation, atrophy of all endocrine glands and organs, hair loss, impotence, amenorrhea, hypometabolism, and hypoglycemia. Coma and death occur if the missing hormones are not replaced.

Anterior Pituitary

Oversecretion (hypersecretion) of the anterior pituitary gland most commonly involves ACTH or GH and results in **Cushing syndrome** or **acromegaly**, respectively. Acromegaly, an excess of GH in adults, results in enlargement of peripheral body parts without an increase in height. It occurs in approximately four cases per 1 million people per year (National Institute of Diabetes and Digestive and Kidney Diseases [NIDDK], 2012). Oversecretion of GH results in gigantism in children; a person may be 7 or even 8 feet tall. Conversely, insufficient secretion of GH during childhood results in generalized limited growth and **dwarfism** (Porth, 2011). Undersecretion (hyposecretion) commonly involves all of the anterior pituitary hormones and is termed *panhypopituitarism*. In this condition, the thyroid gland, the adrenal cortex, and the gonads atrophy (shrink) because of loss of the tropic-stimulating hormones. Hypopituitarism may result from destruction of the anterior lobe of the pituitary gland.

Posterior Pituitary

The most common disorder related to posterior lobe dysfunction is **diabetes insipidus** (DI), a condition in which abnormally large volumes of dilute urine are excreted as a result of deficient production of vasopressin. DI may occur following surgical treatment of a brain tumor, secondary to nonsurgical brain tumors, traumatic brain injury, infections of the nervous system, post hypophysectomy, failure of renal tubals to respond to ADH, and the use of specific medications (Bader & Littlejohns, 2010).

Pituitary Tumors

Approximately 95% of pituitary tumors are benign (NIDDK, 2012). The tumors may be primary or secondary and functional or nonfunctional (Porth, 2011). Functional tumors secrete pituitary hormones, whereas nonfunctional tumors do not. The location and effects of these tumors on hormone production by target organs can have life-threatening effects. Three principal types of pituitary tumors represent an overgrowth of (1) eosinophilic cells, (2) basophilic cells, or (3) chromophobic cells (i.e., cells with no affinity for either eosinophilic or basophilic stains).

Clinical Manifestations

Eosinophilic tumors that develop early in life result in gigantism. The affected person may be more than 7 feet tall and large in all proportions, yet so weak and lethargic that he or she can hardly stand. If the disorder begins during adult life, the excessive skeletal growth occurs only in the feet, the hands, the superciliary ridge, the molar eminences, the nose, and the chin, giving rise to the clinical picture called *acromegaly*. However, enlargement involves all tissues and organs of the body (Porth, 2011). Many of these patients suffer from severe headaches and visual disturbances because the tumors exert pressure on the optic nerves (Porth, 2011). Assessment of central vision and visual fields may reveal loss of color discrimination, diplopia (double vision), or blindness in a portion of a field of vision. Decalcification of the skeleton, muscular weakness, and endocrine disturbances, similar to those occurring in patients with hyperthyroidism, also are associated with this type of tumor.

Basophilic tumors give rise to Cushing syndrome with features largely attributable to hyperadrenalism, including masculinization and amenorrhea in females, truncal obesity, hypertension, osteoporosis, and polycythemia.

Chromophobic tumors represent 90% of pituitary tumors. These tumors usually produce no hormones but destroy the rest of the pituitary gland, causing hypopituitarism. People with this disease are often obese and somnolent and exhibit fine, scanty hair; dry, soft skin; a pasty complexion; and small bones. They also experience headaches, loss of libido, and visual defects progressing to blindness. Other signs and symptoms include polyuria, polyphagia, a lowering of the **basal metabolic rate** (chemical reactions occurring when the body is at rest), and a subnormal body temperature.

Assessment and Diagnostic Findings

Diagnostic evaluation requires a careful history and physical examination, including assessment of visual acuity and visual fields. CT and MRI scans are used to diagnose the presence and extent of pituitary tumors. Serum levels of pituitary hormones may be obtained along with measurements of hormones of target organs (e.g., thyroid, adrenal) to assist in diagnosis.

Medical Management

Surgical removal of the pituitary gland (**hypophysectomy**) through a transsphenoidal approach is the usual treatment. Stereotactic radiation therapy, which requires the use of a neurosurgery-type stereotactic frame, may be used to deliver external-beam radiation therapy precisely to the pituitary tumor with minimal effect on normal tissue (see Chapter 15). Other treatments include conventional radiation therapy, bromocriptine (Parlodel, a dopamine antagonist), and octreotide (Sandostatin, a synthetic analogue of GH). These medications inhibit the production or release of GH and may bring about marked improvement of symptoms. Octreotide and lanreotide (Somatuline Depot, a somatostatin analogue) may also be used preoperatively to improve the patient's clinical condition and to shrink the tumor (Karch, 2012).

Surgical Management

Hypophysectomy is the treatment of choice in patients with Cushing syndrome resulting from excessive production of ACTH by a pituitary tumor. Hypophysectomy may also be performed on occasion as a palliative measure to relieve bone pain secondary to metastasis of malignant lesions of the breast and prostate.

Several approaches are used to remove or destroy the pituitary gland, including surgical removal by transfrontal, subcranial, or oronasal–transsphenoidal approaches; irradiation; and cryosurgery. The transsphenoidal approach and the nursing management of a patient undergoing cranial surgery are discussed in Chapter 66. Features or symptoms of acromegaly are unaffected by surgical removal of the tumor.

The absence of the pituitary gland alters the function of many body systems. Menstruation ceases and infertility occurs after total or near-total ablation of the pituitary gland. Replacement therapy with corticosteroids and thyroid hormone is necessary.

Diabetes Insipidus

DI is the most common disorder of the posterior lobe of the pituitary gland and is characterized by a deficiency of ADH (vasopressin). Excessive thirst (polydipsia) and large volumes of dilute urine are manifestations of the disorder. It may occur secondary to head trauma, brain tumor, or surgical ablation or irradiation of the pituitary gland (Bader & Littlejohns, 2010). It may also occur with infections of the central nervous system (meningitis, encephalitis, tuberculosis) or with tumors (e.g., metastatic disease, lymphoma of the breast or lung). Another cause of DI is failure of the renal tubules to respond to ADH; this nephrogenic form may be related to hypokalemia, hypercalcemia, and a variety of medications (e.g., lithium, demeclocycline [Declomycin]).

> **Concept Mastery Alert**
>
> **It is important to distinguish between DI, a disorder of insufficient output of pituitary ADH, and diabetes, a metabolic disorder characterized by hyperglycemia caused by deficient insulin secretion or action (or both).**

Clinical Manifestations

Without the action of ADH on the distal nephron of the kidney, an enormous daily output (greater than 250 mL per hour) of very dilute urine with a specific gravity of 1.001 to 1.005 occurs (Bader & Littlejohns, 2010; John & Day, 2012). The urine contains no abnormal substances such as glucose or albumin. Because of the intense thirst, the patient tends to drink 2 to 20 L of fluid daily and craves cold water. In adults, the onset of DI may be insidious or abrupt.

The disease cannot be controlled by limiting fluid intake, because the high-volume loss of urine continues even without fluid replacement. Attempts to restrict fluids cause the patient to experience an insatiable craving for fluid and to develop hypernatremia and severe dehydration.

Assessment and Diagnostic Findings

The fluid deprivation test is carried out by withholding fluids for 8 to 12 hours or until 3% to 5% of the body weight is lost. The patient is weighed frequently during the test. Plasma and urine osmolality studies are performed at the beginning and end of the test. The inability to increase the specific gravity and osmolality of the urine is characteristic of DI. The patient continues to excrete large volumes of urine with low specific gravity and experiences weight loss, increasing serum osmolality, and elevated serum sodium levels. The patient's condition needs to be monitored frequently during the test, and the test is terminated if tachycardia, excessive weight loss, or hypotension develops.

Other diagnostic procedures include concurrent measurements of plasma levels of ADH and plasma and urine osmolality as well as a trial of desmopressin (synthetic vasopressin) therapy and intravenous (IV) infusion of hypertonic saline solution. If the diagnosis is confirmed and the cause (e.g., head injury) is not obvious, the patient is carefully assessed for tumors that may be causing the disorder.

Medical Management

The objectives of therapy are (1) to replace ADH (which is usually a long-term therapeutic program), (2) to ensure adequate fluid replacement, and (3) to identify and correct the underlying intracranial pathology. Nephrogenic causes require different management approaches.

Pharmacologic Therapy

Desmopressin (DDAVP), a synthetic vasopressin without the vascular effects of natural ADH, is particularly valuable because it has a longer duration of action and fewer adverse effects than other preparations previously used to treat the disease. It is administered intranasally; the patient sprays the solution into the nose through a flexible calibrated plastic tube. One or two administrations daily (i.e., every 12 to 24 hours) usually control the symptoms (Papadakis, McPhee, & Rabow,

2013). Vasopressin causes vasoconstriction; thus, it must be used cautiously in patients with coronary artery disease.

Chlorpropamide (Diabinese) and thiazide diuretics are also used in mild forms of the disease because they potentiate the action of vasopressin (Karch, 2012). Hyperglycemia is possible.

If the DI is renal in origin, the previously described treatments are ineffective. Thiazide diuretics, mild salt depletion, and prostaglandin inhibitors (ibuprofen [Advil, Motrin], indomethacin [Indocin], and aspirin) are used to treat the nephrogenic form of DI.

Nursing Management

Physical assessment and patient education are the pillars of skilled nursing management of the patient with a diagnosis of DI. Initially, the nurse reviews the patient history and physical assessment. The nurse is responsible to educate the patient, family, and other caregivers about follow-up care, prevention of complications, and emergency measures. Specific verbal and written instructions should include the dose, actions, side effects, and administration of all medications and the signs and symptoms of hyponatremia. The nurse should demonstrate and observe a return demonstration of medication administration to ensure that the patient received the prescribed dosage. The patient should be advised to wear a medical identification bracelet and carry required medication and information about DI at all times.

Syndrome of Inappropriate Antidiuretic Hormone Secretion

The **syndrome of inappropriate antidiuretic hormone (SIADH) secretion** includes excessive ADH secretion from the pituitary gland even in the face of subnormal serum osmolality. Patients with SIADH cannot excrete a dilute urine, retain fluids, and develop a sodium deficiency known as dilutional hyponatremia. SIADH is often of nonendocrine origin; for instance, the syndrome may occur in patients with bronchogenic carcinoma in which malignant lung cells synthesize and release ADH. SIADH has also occurred in patients with severe pneumonia, pneumothorax, and other disorders of the lungs, as well as malignant tumors that affect other organs (Porth, 2011).

Disorders of the central nervous system, such as head injury, brain surgery or tumor, and infection, are thought to produce SIADH by direct stimulation of the pituitary gland (John & Day, 2012). Some medications (e.g., vincristine [Oncovin], phenothiazines, tricyclic antidepressants, thiazide diuretics) and nicotine have been implicated in SIADH; they either directly stimulate the pituitary gland or increase the sensitivity of renal tubules to circulating ADH.

Medical Management

Interventions include eliminating the underlying cause, if possible, and restricting fluid intake (John & Day, 2012). Because retained water is excreted slowly through the kidneys, the extracellular fluid volume contracts and the serum sodium concentration gradually increases toward normal. Diuretic agents such as furosemide (Lasix) may be used along with fluid restriction if severe hyponatremia is present.

Nursing Management

Close monitoring of fluid intake and output, daily weight, urine and blood chemistries, and neurologic status is indicated for the patient at risk for SIADH. Supportive measures and explanations of procedures and treatments assist the patient in managing this disorder.

THE THYROID GLAND

The thyroid gland—the largest endocrine gland—is a butterfly-shaped organ located in the lower neck, anterior to the trachea (Fig 52-3). It consists of two lateral lobes connected by an isthmus. The gland is about 5 cm long and 3 cm wide and weighs about 30 g. The blood flow to the thyroid is very high (about 5 mL/min per gram of thyroid tissue), approximately five times the blood flow to the liver. The thyroid gland produces three hormones: **thyroxine (T_4)**, **triiodothyronine (T_3)**, and **calcitonin**.

Anatomic and Physiologic Overview

Various hormones and chemicals are responsible for normal thyroid function. Key among them are thyroid hormone, calcitonin, and iodine.

Thyroid Hormone

Thyroid hormone is comprised of T_4 and T_3, two separate hormones produced by the thyroid gland. Both are amino acids that contain iodine molecules bound to the amino acid structure; T_4 contains four iodine atoms in each molecule, and T_3 contains three. These hormones are synthesized and stored bound to proteins in the cells of the thyroid gland until needed for release into the bloodstream. Three thyroid-binding hormones—thyroxine-binding globulin (TBG), transthyretin (formerly known as thyroid-binding prealbumin), and albumin—bind and transport T3 and T4 (Porth & Matfin, 2009).

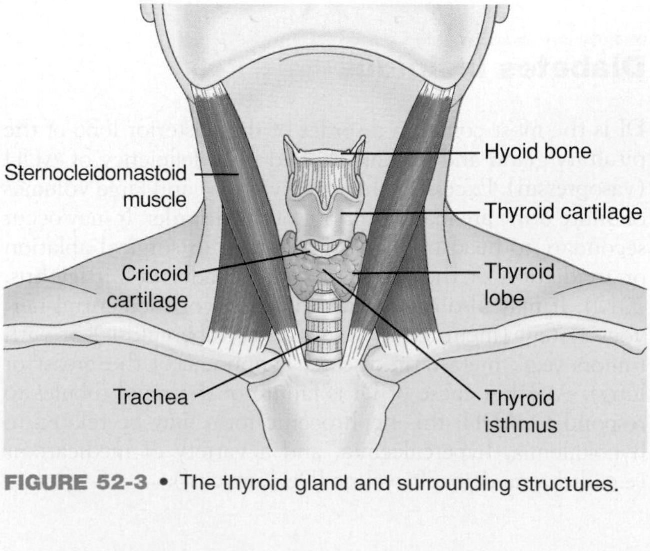
Hyoid bone
Sternocleidomastoid muscle
Thyroid cartilage
Cricoid cartilage
Thyroid lobe
Trachea
Thyroid isthmus

FIGURE 52-3 • The thyroid gland and surrounding structures.

Synthesis of Thyroid Hormone

Iodine is essential to the thyroid gland for synthesis of its hormones. The major use of iodine in the body is by the thyroid, and the major derangement in iodine deficiency is alteration of thyroid function. Iodide is ingested in the diet and absorbed into the blood in the GI tract. The thyroid gland is extremely efficient at taking up iodide from the blood and concentrating it within the cells, where iodide ions are converted to iodine molecules, which react with tyrosine (an amino acid) to form the thyroid hormones.

Regulation of Thyroid Hormone

The secretion of T_3 and T_4 by the thyroid gland is controlled by TSH (also called *thyrotropin*) from the anterior pituitary gland. TSH controls the rate of thyroid hormone release through a negative feedback mechanism. In turn, the level of thyroid hormone in the blood determines the release of TSH. If the thyroid hormone concentration in the blood decreases, the release of TSH increases, which causes increased output of T_3 and T_4. The term **euthyroid** refers to thyroid hormone production that is normal.

TRH, secreted by the hypothalamus, exerts a modulating influence on the release of TSH from the pituitary. Environmental factors, such as a decrease in temperature, may lead to increased secretion of TRH, resulting in elevated secretion of thyroid hormones. Figure 52-4 shows the hypothalamic–pituitary–thyroid axis, which regulates thyroid hormone production.

Function of Thyroid Hormone

The main function of thyroid hormone is to control cellular metabolic activity. T_4, a relatively weak hormone, maintains body metabolism in a steady state. T_3 is about five times as potent as T_4 and has a more rapid metabolic action. These hormones accelerate metabolic processes by increasing the level of specific enzymes that contribute to oxygen consumption and altering the responsiveness of tissues to other hormones. The thyroid hormones influence cell replication and are important in brain development. Thyroid hormone is also necessary for normal growth. Thyroid hormones affect virtually every major organ system and tissue function, including the basal metabolic rate, tissue thermogenesis, serum cholesterol levels, and vascular resistance (Porth, 2011).

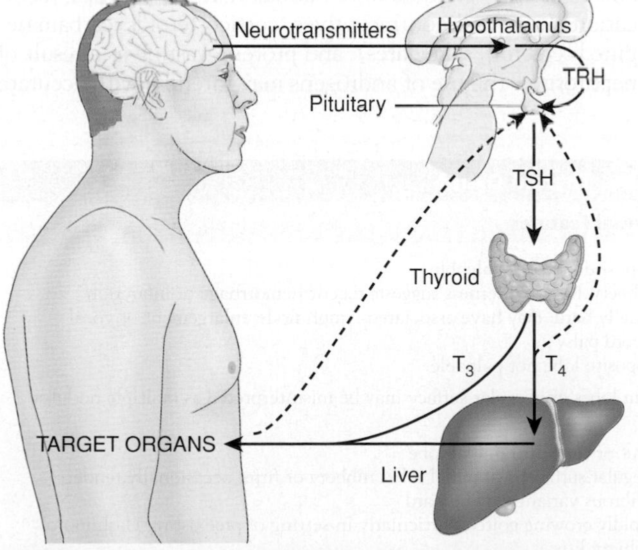
Neurotransmitters
Hypothalamus
TRH
Pituitary
TSH
Thyroid
T3
T4
TARGET ORGANS
Liver

FIGURE 52-4 • The hypothalamic–pituitary–thyroid axis. Thyroid-releasing hormone (TRH) from the hypothalamus stimulates the pituitary gland to secrete thyroid-stimulating hormone (TSH). TSH stimulates the thyroid to produce thyroid hormone (triiodothyronine [T_3] and thyroxine [T_4]). High circulating levels of T_3 and T_4 inhibit further TSH secretion and thyroid hormone production through a negative feedback mechanism (*dashed lines*).

Calcitonin

Calcitonin, or thyrocalcitonin, is another important hormone secreted by the thyroid gland. It is secreted in response to high plasma levels of calcium, and it reduces the plasma level of calcium by increasing its deposition in bone.

Pathophysiology

Inadequate secretion of thyroid hormone during fetal and neonatal development results in stunted physical and mental growth (neonatal hypothyroidism) because of general depression of metabolic activity (Mitchell & Hsu, 2011). In adults, hypothyroidism manifests as lethargy, slow mentation, and generalized slowing of body functions.

Oversecretion of thyroid hormones (hyperthyroidism) is manifested by a greatly increased metabolic rate. Many of the other characteristics of hyperthyroidism result from the increased response to circulating catecholamines (epinephrine and norepinephrine). Oversecretion of thyroid hormones is usually associated with an enlarged thyroid gland known as a **goiter**. Goiter also commonly occurs with iodine deficiency. In this latter condition, lack of iodine results in low levels of circulating thyroid hormones, which causes increased release of TSH; the elevated TSH causes overproduction of thyroglobulin (a precursor of T_3 and T_4) and hypertrophy of the thyroid gland.

Assessment

Physical Examination

The thyroid gland is inspected and palpated routinely in all patients. Inspection begins with identification of landmarks. The lower neck region between the sternocleidomastoid muscles is inspected for swelling or asymmetry. The patient is instructed to extend the neck slightly and swallow. Thyroid tissue rises normally with swallowing. The thyroid is then palpated for size, shape, consistency, symmetry, and the presence of tenderness (Hogan-Quigley et al., 2012).

The clinician may examine the thyroid from an anterior or a posterior position. In the posterior position, both hands encircle the patient's neck. The thumbs rest on the nape of the neck, while the index and middle fingers palpate for the thyroid isthmus and the anterior surfaces of the lateral lobes. When palpable, the isthmus is perceived as firm and of a rubber-band consistency.

The left lobe is examined by positioning the patient so that the neck flexes slightly forward and to the left. The thyroid cartilage is then displaced to the left with the fingers of the right hand. This maneuver displaces the left lobe deep into the sternocleidomastoid muscle, where it can be more easily palpated. The left lobe is then palpated by placing the left thumb deep into the posterior area of the sternocleidomastoid muscle, while the index and middle fingers exert opposite pressure in the anterior portion of the muscle. Having the patient swallow during the maneuver may assist the examiner to locate the thyroid as it ascends in the neck. The procedure is reversed to examine the right lobe. The isthmus is the only portion of the thyroid that is normally palpable. If a patient has a very thin neck, two thin, smooth, nontender lobes may also be palpable.

If palpation discloses an enlarged thyroid gland, both lobes are auscultated using the diaphragm of the stethoscope. Auscultation identifies the localized audible vibration of a bruit. This is indicative of increased blood flow through the thyroid gland associated with hyperthyroidism and necessitates referral to a physician. Other abnormal findings that require referral for further evaluation may include a soft texture (Graves' disease), firmness (Hashimoto's thyroiditis or malignancy), and tenderness (thyroiditis) (Table 52-2) (Hogan-Quigley et al., 2012).

Diagnostic Evaluation

Assessment measures in addition to palpation and auscultation include thyroid function tests, such as laboratory measurement of thyroid hormones, thyroid scanning, biopsy, and ultrasonography. The most widely used tests are serum immunoassay for TSH and free T_4 (Papadakis et al., 2013). Measurement of TSH has a high sensitivity and specificity (Fischbach & Dunning, 2009). Free T_4 levels correlate with metabolic status; they are elevated in hyperthyroidism and decreased in hypothyroidism. Ultrasound, CT, and MRI may be used to clarify or confirm the results of other diagnostic studies.

Thyroid Tests

Serum Thyroid-Stimulating Hormone

Measurement of the serum TSH concentration is the single best screening test of thyroid function in outpatients because of its high sensitivity (Fischbach & Dunning, 2009). The ability to detect minute changes in serum TSH makes it possible to distinguish subclinical thyroid disease from euthyroid states in patients with low or high normal values. Measurement of TSH is also used for monitoring thyroid hormone replacement therapy and for differentiating between disorders of the thyroid gland itself and disorders of the pituitary or hypothalamus.

There are conflicting recommendations regarding routine screening for thyroid disease. The American Academy of Family Physicians (AAFP) and the American Association of Clinical Endocrinologists (AACE) recommend periodic screening of older women. The American Thyroid Association (ATA) recommends TSH screening for all adults beginning at 35 years of age and every 5 years thereafter (U.S. Department of Health and Human Services [HHS], 2006). The U.S. Preventive Services Task Force (USPSTF) could not make a recommendation due to insufficient evidence (HHS, 2012). Risk assessment is appropriate for those at higher risk for thyroid dysfunction, including the older adult, postpartum women, persons with high levels of radiation exposure, and patients with Down syndrome (HHS, 2012).

Serum Free T_4

Serum free T_4 is a direct measurement of free (unbound) thyroxine, the only metabolically active fraction of T_4. The range of free T_4 in serum is normally 0.9 to 1.7 ng/dL (11.5 to 21.8 pmol/L). When measured by the dialysis method, free T_4 is not affected by variations in protein binding and is the procedure of choice for monitoring the changes in T_4 secretion during treatment for hyperthyroidism.

Serum T_3 and T_4

Measurement of total T_3 or T_4 includes protein-bound and free hormone levels that occur in response to TSH secretion. T_4 is 70% bound to TBG; T_3 is bound less firmly. Only 0.03% of T_4 and 0.3% of T_3 are unbound. Serious systemic illnesses, medications (e.g., oral contraceptives, corticosteroids, carbamazepine [Tegretol], salicylates), and protein wasting as a result of nephrosis or the use of androgens may interfere with accurate

TABLE 52-2 Summary of Findings on Physical Examination of the Thyroid Gland

Physical Finding	Differential Diagnosis	Special Features
Single nodule	Autonomously functioning adenoma	Opposite lobe not palpable
	Adenoma or adenomatous nodule	Rubbery, firm; tenderness suggests recent hemorrhage or infarction.
	Cancer	Usually hard; may have associated lymph node enlargement or vocal cord palsy
	Hyperplasia secondary to unilobar agenesis	Opposite lobe not palpable
Multiple nodules	Multinodular goiter Hashimoto's thyroiditis	Firm lobes or irregular surface may be misinterpreted as multiple nodules
Diffuse goiter	Graves' disease	Bruit or thrill; pyramidal lobe
	Hashimoto's thyroiditis	Irregular surface; pyramidal lobe; rubbery or firm; occasionally tender; fibrous variant may be hard
	Thyroid lymphoma	Rapidly growing goiter, particularly in setting of preexisting Hashimoto's thyroiditis
	Multinodular goiter	Nodules may be hidden within gland and may become apparent with thyroid hormone suppression
Tenderness	Subacute thyroiditis	Unilateral or bilateral; tenderness often severe
	Hemorrhagic or infarcted adenoma	Discrete nodule with tenderness
	Hashimoto's thyroiditis	Mild tenderness
	Cancer	Irregular, firm thyroid nodule with chronic tenderness

test results (Fischbach & Dunning, 2009). Normal range for T_4 is 5.4 to 11.5 mcg/dL (57 to 148 nmol/L) (Fischbach & Dunning, 2009). Although serum T_3 and T_4 levels generally increase or decrease together, the T_3 level appears to be a more accurate indicator of hyperthyroidism, which causes a greater increase in T_3 than in T_4 levels. The normal range for serum T_3 is 80 to 200 ng/dL (1.2 to 3.1 nmol/L) (Fischbach & Dunning, 2009).

T_3 Resin Uptake Test

The T_3 resin uptake test is an indirect measure of unsaturated TBG. Its purpose is to determine the amount of thyroid hormone bound to TBG and the number of available binding sites. This provides an index of the amount of thyroid hormone already present in the circulation. Normally, TBG is not fully saturated with thyroid hormone, and additional binding sites are available to combine with radioiodine-labeled T_3 added to the blood specimen. The normal T_3 uptake value is 25% to 35% (relative uptake fraction, 0.25 to 0.35), which indicates that about one third of the available sites of TBG are occupied by thyroid hormone (Fischbach & Dunning, 2009). If the number of free or unoccupied binding sites is low, as in hyperthyroidism, the T_3 uptake is greater than 35% (0.35). If the number of available sites is high, as occurs in hypothyroidism, the test result is less than 25% (0.25).

T_3 uptake is useful in evaluating thyroid hormone levels in patients who have received diagnostic or therapeutic doses of iodine. The test results may be altered by the use of estrogens, androgens, salicylates, phenytoin, anticoagulants, or corticosteroids.

Thyroid Antibodies

Autoimmune thyroid diseases include both hypothyroid and hyperthyroid conditions. Results of testing by immunoassay techniques for antithyroid antibodies are positive in chronic autoimmune thyroid disease (90%), Hashimoto's thyroiditis (100%), Graves' disease (80%), and other organ-specific autoimmune diseases, such as systemic lupus erythematosus (SLE) and rheumatoid arthritis. Antithyroid antibody titers are normally present in 5% to 10% of the population and increase with age.

Radioactive Iodine Uptake

The radioactive iodine uptake test measures the rate of iodine uptake by the thyroid gland. The patient is administered a tracer dose of iodine 123 (^{123}I) or another radionuclide, and a count is made over the thyroid gland with a scintillation counter, which detects and counts the gamma rays released from the breakdown of ^{123}I in the thyroid. It measures the proportion of the administered dose that is present in the thyroid gland at a specific time after its administration. It is a simple test and provides reliable results. It is affected by the patient's intake of iodide or thyroid hormone; therefore, a careful preliminary clinical history is essential in evaluating results. Normal values vary from one geographic region to another and with the intake of iodine. Patients with hyperthyroidism exhibit a high uptake of the ^{123}I (in some patients, as high as 90%), whereas patients with hypothyroidism exhibit a very low uptake.

Fine-Needle Aspiration Biopsy

The use of a small-gauge needle to sample the thyroid tissue for biopsy is a safe and accurate method of detecting malignancy. It is often the initial test for evaluation of thyroid masses. Results are reported as (1) negative (benign), (2) positive (malignant), (3) indeterminate (suspicious), and (4) inadequate (nondiagnostic).

Thyroid Scan, Radioscan, or Scintiscan

In a thyroid scan, a scintillation detector or gamma camera moves back and forth across the area to be studied in a series of parallel tracks, and a visual image is made of the distribution of radioactivity in the area being scanned. The most commonly used isotopes of iodine are ^{123}I and iodine 131 (^{131}I) (Fischbach & Dunning, 2009).

Scans are helpful in determining the location, size, shape, and anatomic function of the thyroid gland, particularly when thyroid tissue is substernal or large (Fischbach & Dunning, 2009). Identifying areas of increased function ("hot" areas) or decreased function ("cold" areas) can assist in diagnosis. Although most areas of decreased function do not represent malignancies, lack of function increases the likelihood of malignancy, particularly if only one nonfunctioning area is present. Scanning of the entire body, to obtain the total body profile, may be carried out in a search for a functioning thyroid metastasis (i.e., a lesion that produces thyroid hormones).

Serum Thyroglobulin

Thyroglobulin (Tg) can be measured reliably in the serum by radioimmunoassay. Clinically, it is used to detect persistence or recurrence of thyroid carcinoma.

Nursing Implications

When thyroid tests are scheduled, it is necessary to determine if the patient is allergic to iodine (shellfish) and whether the patient has taken medications or agents that contain iodine, because these may alter the test results (Fischbach & Dunning, 2009). Obvious sources of iodine-containing medications include contrast agents and those used to treat thyroid disorders such as radioactive iodine. Less obvious sources of iodine are topical antiseptics; multivitamin preparations; food supplements that may contain kelp and seaweed (frequently found in health food stores); and amiodarone (Cordarone), an antiarrhythmic agent (Surks, 2012a). Other medications that may affect test results are estrogens, salicylates, amphetamines, chemotherapeutic agents, antibiotics, corticosteroids, and mercurial diuretics. This information should be documented in the patient's medical record and the laboratory requisition. Chart 52-2 gives a partial list

Chart 52-2 PHARMACOLOGY

Selected Medications That May Alter Thyroid Test Results

amiodarone (Cordarone)
aspirin
cimetidine (Tagamet)
diazepam (Valium)
furosemide (Lasix)
heparin
lithium (Lithotabs)
phenytoin (Dilantin) and other anticonvulsants
propranolol (Inderal)

Adapted from Fischbach, F., & Dunning, M. B. (2009). A manual of laboratory and diagnostic tests (8th ed.). Philadelphia: Lippincott Williams & Wilkins.

of medications that may interfere with accurate testing of thyroid gland function.

Hypothyroidism

Hypothyroidism results from suboptimal levels of thyroid hormone. Thyroid deficiency can affect all body functions and can range from mild, subclinical forms to **myxedema** (severe deficiency discussed later), an advanced life-threatening form (Ross, 2012a). The most common cause of hypothyroidism in adults is autoimmune thyroiditis (Hashimoto's disease), in which the immune system attacks the thyroid gland (Porth & Matfin, 2009). Symptoms of hyperthyroidism may later be followed by those of hypothyroidism and myxedema. Hypothyroidism also commonly occurs in patients with previous hyperthyroidism that has been treated with radioiodine or antithyroid medications or **thyroidectomy** (surgical removal of all or part of the thyroid gland). The condition occurs most frequently in older women. In addition, there is an increased incidence of thyroid cancer in men who have undergone radiation therapy for head and neck cancer. Therefore, testing of thyroid function is recommended for all patients who receive such treatment. Other causes of hypothyroidism are presented in Chart 52-3.

More than 95% of patients with hypothyroidism have primary or thyroidal hypothyroidism, which refers to dysfunction of the thyroid gland itself. If the cause of the thyroid dysfunction is failure of the pituitary gland, the hypothalamus, or both, the hypothyroidism is known as central hypothyroidism. If the cause is entirely a pituitary disorder, it may be referred to as pituitary or secondary hypothyroidism. If the cause is a disorder of the hypothalamus resulting in inadequate secretion of TSH due to decreased stimulation of TRH, it is referred to as hypothalamic or tertiary hypothyroidism. If thyroid deficiency is present at birth, it is referred to as neonatal hypothyroidism. In such instances, the mother may also have thyroid deficiency. The term *myxedema* refers to the accumulation of mucopolysaccharides in subcutaneous and other interstitial tissues. Although myxedema occurs in longstanding hypothyroidism, the term is used appropriately only to describe the extreme symptoms of severe hypothyroidism (Ross, 2012a).

Chart 52-3 Causes of Hypothyroidism

- Autoimmune disease (Hashimoto's thyroiditis, post-Graves' disease)
- Atrophy of thyroid gland with aging
- Therapy for hyperthyroidism
  - Radioactive iodine (^{131}I)
  - Thyroidectomy
- Medications
  - Lithium
  - Iodine compounds
  - Antithyroid medications
- Radiation to head and neck in treatment for head and neck cancers, lymphoma
- Infiltrative diseases of the thyroid (amyloidosis, scleroderma, lymphoma)
- Iodine deficiency and iodine excess

Clinical Manifestations

Extreme fatigue makes it difficult for the person to complete a full day's work or participate in usual activities. Reports of hair loss, brittle nails, and dry skin are common, and numbness and tingling of the fingers may occur. On occasion, the voice may become husky, and the patient may complain of hoarseness. Menstrual disturbances such as menorrhagia or amenorrhea occur, in addition to loss of libido. Hypothyroidism affects women more frequently than men and occurs most often between 40 and 70 years of age. The prevalence of the disease increases with increasing age (Eliopoulos, 2010).

Severe hypothyroidism results in a subnormal body temperature and pulse rate. The patient usually begins to gain weight even without an increase in food intake, although he or she may be cachectic. The skin becomes thickened because of an accumulation of mucopolysaccharides in the subcutaneous tissues. The hair thins and falls out, and the face becomes expressionless and masklike. The patient often complains of being cold even in a warm environment.

At first, the patient may be irritable and may complain of fatigue, but as the condition progresses, the emotional responses are subdued. The mental processes become dulled, and the patient appears apathetic. Speech is slow, the tongue enlarges, the hands and feet increase in size, and deafness may occur. The patient frequently complains of constipation.

Advanced hypothyroidism may produce personality and cognitive changes characteristic of dementia. Inadequate ventilation and sleep apnea can occur with severe hypothyroidism. Pleural effusion, pericardial effusion, and respiratory muscle weakness may also occur.

Severe hypothyroidism is associated with an elevated serum cholesterol level, atherosclerosis, coronary artery disease, and poor left ventricular function. The patient with advanced hypothyroidism is hypothermic and abnormally sensitive to sedative, opioid, and anesthetic agents, which must be administered with extreme caution.

Patients with unrecognized hypothyroidism who are undergoing surgery are at increased risk for intraoperative hypotension, postoperative heart failure, and altered mental status.

Myxedema coma is a rare life-threatening condition. It is the decompensated state of severe hypothyroidism in which the patient is hypothermic and unconscious (Ross, 2012a). This condition may develop with undiagnosed hypothyroidism and may be precipitated by infection or other systemic disease or by use of sedatives or opioid analgesic agents. Patients may also experience myxedema coma if they forget to take their thyroid replacement medication. The condition occurs most often among older women in the winter months and appears to be precipitated by cold. However, the disorder can affect any age group.

In myxedema coma, the patient may initially show signs of depression, diminished cognitive status, lethargy, and somnolence (Ross, 2012a). Increasing lethargy may progress to stupor. The patient's respiratory drive is depressed, resulting in alveolar hypoventilation, progressive carbon dioxide retention, narcosis, and coma. In addition patients with myxedema coma can also exhibit hyponatremia, hypoglycemia, hypoventilation, hypotension, bradycardia, and hypothermia. These symptoms, along with cardiovascular collapse and shock, require aggressive and intensive supportive and hemodynamic therapy if the patient is to survive. Although there has been a decline in mortality rates over the past two decades due to early intervention and improved therapies, the mortality rate remains high at 30% to 40%. Patients at greatest risk are older adults, individuals with cardiac complications, reduced consciousness, persistent hypothermia, and sepsis (Ross, 2012a).

Medical Management

The objectives in the management of hypothyroidism are to restore a normal metabolic state by replacing the missing hormone, as well as prevention of disease progression and complications.

Pharmacologic Therapy

Synthetic levothyroxine (Synthroid or Levothroid) is commonly prescribed for treating hypothyroidism and suppressing nontoxic goiters (Karch, 2012). Its dosage is based on the patient's serum TSH concentration. If replacement therapy is adequate, the symptoms of myxedema disappear and normal metabolic activity is resumed. Although rare, if the disease progresses to myxedema coma, IV administration of T_4 and T_3 are recommended rather than T_4 alone. T_4 is administered IV until the patient is able to safely take T_4 in oral form. T_3 is administered IV until the patient is stable (Ross, 2012a).

Administration of high-dose glucocorticoids (hydrocortisone) every 8 to 12 hours for 24 hours followed by low-dose therapy is recommended until coexisting adrenal insufficiency is ruled out (Ross, 2012a). Serum levels of T_4, TSH, and cortisol should be obtained before and after administration of ACTH and before administration of glucocorticoid and thyroid hormone therapies. Associated conditions such as infection, GI bleeding, hyponatremia, hypotension, bradycardia, hypoglycemia, and hypothermia will also require appropriate pharmacologic management (Ross, 2012).

Prevention of Cardiac Dysfunction

Any patient who has had hypothyroidism for a long period usually has associated elevated serum cholesterol, atherosclerosis, and coronary artery disease. As long as metabolism is subnormal and the tissues (including the myocardium) require relatively little oxygen, a reduction in blood supply is tolerated without overt symptoms of coronary artery disease. When thyroid hormone is administered, the oxygen demand increases, but oxygen delivery cannot be increased unless, or until, the atherosclerosis improves. This occurs very slowly, if at all. The occurrence of angina and acute coronary syndrome (ACS; see Chapter 27) is the signal that the oxygen needs of the myocardium exceed its blood supply. Angina or dysrhythmias can occur when thyroid replacement is initiated because thyroid hormones enhance the cardiovascular effects of catecholamines.

> ***Quality and Safety Nursing Alert***
>
> **The nurse must monitor for signs and symptoms of ACS, which can occur in response to therapy in patients with severe, long-standing hypothyroidism or myxedema coma, especially during the early phase of treatment. ACS must be aggressively treated at once to avoid morbid complications (e.g., myocardial infarction).**

Obviously, if angina or dysrhythmias occur, thyroid hormone administration must be discontinued immediately. Later, when it can be resumed safely, it should be prescribed cautiously at a lower dosage and with close monitoring by the primary provider and the nurse.

Prevention of Medication Interactions

Oral thyroid hormones interact with many other medications. There is a decrease in thyroid hormone absorption when patients are also taking magnesium-containing antacids. Thyroid hormones may also decrease the pharmacologic effects of digitalis glycosides. Doses of anticoagulant agents need to be decreased when beginning thyroid replacement because of the increased risk of bleeding (Karch, 2012).

Even in small IV doses, hypnotic and sedative agents may induce profound somnolence, lasting far longer than anticipated and leading to narcosis (stuporlike condition). Furthermore, they are likely to cause respiratory depression, which can easily be fatal because of decreased respiratory reserve and alveolar hypoventilation. The dose of these medications should be one half or one third of that typically prescribed for patients of similar age and weight with normal thyroid function.

Supportive Therapy

Severe hypothyroidism and myxedema coma require prompt, aggressive management to maintain vital functions. Arterial blood gases may be measured to determine carbon dioxide retention and to guide the use of assisted ventilation to combat hypoventilation. Oxygen saturation levels should be monitored using pulse oximetry. Fluids are administered cautiously because of the danger of water intoxication. Passive rewarming with a blanket is recommended versus active rewarming such as application of external heat (e.g., heating pads) (Ross, 2012a). The latter should be avoided to prevent increased oxygen demands and hypotension. If myxedema has progressed to myxedema coma, refer to the current recommendations stated in the previous Pharmacologic Therapy section.

Nursing Management

Nursing care of the patient with hypothyroidism and myxedema is summarized in the plan of nursing care in Chart 52-4. In patients with hypothyroidism, the effects of analgesic, sedative, and anesthetic agents are prolonged. The nurse should carefully monitor patients who are prescribed these agents for adverse effects. Older patients are at increased risk because of concurrent changes in liver and renal function.

> ***Quality and Safety Nursing Alert***
>
> **Medications are administered to the patient with hypothyroidism with extreme caution because of the potential for altered metabolism and excretion, as well as depressed metabolic rate and respiratory status.**

Promoting Home and Community-Based Care

Educating Patients About Self-Care

The patient and family require education and support to manage this complex disorder at home. Oral and written instructions should be provided regarding the following:

- Desired actions and side effects of medications
- Correct medication administration ("Take first thing in the morning with a full glass of water.")

(text continues on page 1478)

Chart 52-4

PLAN OF NURSING CARE
Care of the Patient With Hypothyroidism

NURSING DIAGNOSIS: Activity intolerance related to fatigue and depressed cognitive process
GOAL: Increased participation in activities and increased independence

Nursing Interventions	Rationale	Expected Outcomes
1. Promote independence in self-care activities. **a.** Space activities to promote rest and exercise as tolerated. **b.** Assist with self-care activities when patient is fatigued. **c.** Provide stimulation through conversation and nonstressful activities. **d.** Monitor patient's response to increasing activities.	**1.** Encouragement needed in fatigued, often depressed patient **a.** Encourages activities while allowing time for adequate rest **b.** Permits patient to participate to the extent possible in self-care activities **c.** Promotes interest without stressing the patient **d.** Guards against over- and underexertion by the patient	• Participates in self-care activities • Reports decreased level of fatigue • Displays interest and awareness in environment • Participates in activities and events in environment • Participates in family events and activities • Reports no chest pain, increased fatigue, or breathlessness with increased level of activity

NURSING DIAGNOSIS: Risk for imbalanced body temperature
GOAL: Maintenance of normal body temperature

Nursing Interventions	Rationale	Expected Outcomes
1. Provide extra layer of clothing or extra blanket. **2.** Avoid and discourage the use of external heat source (e.g., heating pads, electric or warming blankets). **3.** Monitor patient's body temperature and report decreases from patient's baseline value. **4.** Protect from exposure to cold and drafts.	**1.** Minimizes heat loss **2.** Reduces risk of peripheral vasodilation and vascular collapse **3.** Detects decreased body temperature and onset of myxedema coma **4.** Increases patient's level of comfort and decreases further heat loss	• Experiences relief of discomfort and cold intolerance • Maintains baseline body temperature • Reports adequate feeling of warmth and lack of chilling • Uses extra layer of clothing or extra blanket • Explains rationale for avoiding external heat source

NURSING DIAGNOSIS: Constipation related to depressed gastrointestinal function
GOAL: Return of normal bowel function

Nursing Interventions	Rationale	Expected Outcomes
1. Encourage increased fluid intake within limits of fluid restriction. **2.** Provide foods high in fiber. **3.** Instruct patient about foods with high water content. **4.** Monitor bowel function. **5.** Encourage increased mobility within patient's exercise tolerance. **6.** Encourage patient to use laxatives and enemas sparingly.	**1.** Promotes passage of soft stools **2.** Increases bulk of stools and more frequent bowel movements **3.** Provides rationale for patient to increase fluid intake **4.** Permits detection of constipation and return to normal bowel pattern **5.** Promotes evacuation of the bowel **6.** Minimizes patient's dependence on laxatives and enemas and encourages normal pattern of bowel evacuation	• Reports normal bowel function • Identifies and consumes foods high in fiber • Drinks recommended amount of fluid each day • Participates in gradually increasing exercises • Uses laxatives as prescribed and avoids excessive dependence on laxatives and enemas

NURSING DIAGNOSIS: Deficient knowledge about the therapeutic regimen for lifelong thyroid replacement therapy
GOAL: Knowledge and acceptance of the prescribed therapeutic regimen

Nursing Interventions	Rationale	Expected Outcomes
1. Explain rationale for thyroid hormone replacement. **2.** Describe desired effects of medication to patient. **3.** Assist patient to develop schedule and checklist to ensure self-administration of thyroid replacement. **4.** Describe signs and symptoms of over- and underdose of medication.	**1.** Provides rationale for patient to use thyroid hormone replacement as prescribed **2.** Provides encouragement to patient by identifying improved physical status and well-being that will occur with thyroid hormone therapy and return to a euthyroid state **3.** Increases chances that medication will be taken as prescribed **4.** Serves as check for patient to determine if therapeutic goals are met	• Describes therapeutic regimen correctly • Explains rationale for thyroid hormone replacement • Identifies positive outcomes of thyroid hormone replacement • Administers medication to self as prescribed • Identifies adverse side effects that should be reported promptly to primary provider: recurrence of symptoms of hypothyroidism and occurrence of symptoms of hyperthyroidism

Chart 52-4

PLAN OF NURSING CARE

Care of the Patient With Hypothyroidism (continued)

Nursing Interventions	Rationale	Expected Outcomes
5. Explain the necessity for long-term follow-up to patient and family.	**5.** Increases likelihood that hypo- or hyperthyroidism will be detected and treated	• Restates need for periodic/long-term follow-up visits to primary provider

NURSING DIAGNOSIS: Ineffective breathing pattern related to depressed ventilation
GOAL: Improved respiratory status and maintenance of normal breathing pattern

Nursing Interventions	Rationale	Expected Outcomes
1. Monitor respiratory rate, depth, pattern, pulse oximetry, and arterial blood gases.	**1.** Identifies patient's baseline to monitor further changes and evaluate effectiveness of interventions	• Shows improved respiratory status and maintenance of normal breathing pattern • Demonstrates normal respiratory rate, depth, and pattern • Takes deep breaths, coughs, and uses incentive spirometry when encouraged • Demonstrates normal breath sounds without adventitious sounds on auscultation • Explains rationale for cautious use of medications • Cooperates with suction procedure and ventilator support when necessary
2. Encourage deep breathing, coughing, and the use of incentive spirometry.	**2.** Prevents atelectasis and promotes adequate ventilation	
3. Administer medications (hypnotics and sedatives) with caution.	**3.** Patients with hypothyroidism are very susceptible to respiratory depression with the use of hypnotics and sedatives	
4. Maintain patent airway through suction and ventilatory support if indicated (see Chapter 21 for care of patients requiring mechanical ventilation).	**4.** The use of an artificial airway and ventilatory support may be necessary with respiratory depression	

NURSING DIAGNOSIS: Acute confusion related to depressed metabolism and altered cardiovascular and respiratory status
GOAL: Improved thought processes

Nursing Interventions	Rationale	Expected Outcomes
1. Orient patient to time, place, date, and events around him or her.	**1.** Provides reality orientation to patient	• Shows improved cognitive functioning • Identifies time, place, date, and events correctly • Responds appropriately when stimulated • Responds spontaneously as treatment becomes effective • Interacts spontaneously with family and environment • Explains that change in mental and cognitive processes is a result of disease processes • Takes medications as prescribed to prevent decrease in cognitive processes
2. Provide stimulation through conversation and nonthreatening activities.	**2.** Provides stimulation within patient's level of tolerance for stress	
3. Explain to patient and family that change in cognitive and mental functioning is a result of disease process.	**3.** Reassures patient and family about the cause of the cognitive changes and that a positive outcome is possible with appropriate treatment	
4. Monitor cognitive and mental processes and response of these to medication and other therapy.	**4.** Permits evaluation of the effectiveness of treatment	

COLLABORATIVE PROBLEM: Myxedema and myxedema coma
GOAL: Evidence of progression to pre-coma baseline without incurring additional complications

Nursing Interventions	Rationale	Expected Outcomes
1. Monitor patient for increasing severity of signs and symptoms of hypothyroidism: **a.** Decreased level of consciousness **b.** Decreased vital signs (blood pressure, respiratory rate, temperature, pulse rate) **c.** Increasing difficulty in awakening or arousing patient	**1.** Extreme hypothyroidism may lead to myxedema, myxedema coma, and slowing of all body systems if untreated	• Exhibits reversal of myxedema and myxedema coma • Responds appropriately to questions and surroundings • Vital signs return to normal or near-normal ranges • Respiratory status improves with adequate spontaneous ventilatory effort • Reports no episodes of angina or other indicators of cardiac insufficiency • Experiences minimal or no complications caused by immobility
2. Assist in ventilatory support if respiratory depression and failure occur.	**2.** Ventilatory support is necessary to maintain adequate oxygenation and maintenance of airway	
3. Administer prescribed medications (e.g., thyroxine) with extreme caution.	**3.** The slow metabolism and atherosclerosis of myxedema may result in angina with administration of thyroxine	
4. Turn and reposition patient at intervals.	**4.** Minimizes risks associated with immobility	
5. Avoid the use of hypnotic, sedative, and analgesic agents.	**5.** Altered metabolism of these agents greatly increases the risks of their use in myxedema	

Chart 52-5 HOME CARE CHECKLIST
The Patient With Hypothyroidism (Myxedema)

At the completion of home care education, the patient or caregiver will be able to:	PATIENT	CAREGIVER
• State present and potential effects of hypothyroidism on the body.	✔	✔
• State precipitating factors and interventions for complications (hyperthyroidism, myxedema coma).	✔	✔
• Explain the purpose, dose, route, schedule, side effects, and precautions of prescribed medication (synthetic thyroid hormone).	✔	✔
• State that compliance with medical regimen is lifelong.	✔	✔
• State the need to avoid extreme cold temperature until condition is stable.	✔	✔
• State importance of regular follow-up visits with primary provider.	✔	✔
• Identify dietary strategies to promote weight reduction and prevent constipation (high fiber, low calorie, adequate fluid intake).	✔	✔
• State potential for menstrual irregularities and potential for pregnancy for women.	✔	✔
• State the importance of avoiding infection.	✔	✔
• Identify changes in personality as related to hypothyroidism.	✔	✔
• Identify areas of activity limitations and impact on lifestyle.	✔	✔

- Importance of continuing to take the medications as prescribed even after symptoms improve
- When to seek medical attention
- Importance of nutrition and diet to promote weight loss and normal bowel patterns
- Importance of periodic follow-up testing

The patient and family should be educated that the symptoms observed during the course of the disorder will disappear with effective treatment (Chart 52-5).

Continuing Care

If indicated, a referral is made for home care. The home care nurse monitors the patient's recovery and ability to cope with the recent changes, and assesses the patient's physical and cognitive status and the patient's and family's understanding of the education provided before hospital discharge. The home care nurse documents and reports to the patient's primary provider subtle signs and symptoms that may indicate either inadequate or excessive thyroid hormone.

Gerontologic Considerations

The prevalence of hypothyroidism increases with age, most often among women (Eliopoulos, 2010). The higher prevalence of hypothyroidism among older people may be related to alterations in immune function with age and complicated by multiple comorbidities.

Most patients with primary hypothyroidism present with long-standing mild to moderate hypothyroidism. Subclinical disease is common among older women and can be asymptomatic or mistaken for other medical conditions. Subtle symptoms of hypothyroidism, such as fatigue, muscle aches, and mental confusion, may be attributed to the normal aging process by patients, families, and health care providers; therefore, these symptoms require close attention (Dominguez, Bevilacqua, DiBella, et al., 2008). In addition, signs and symptoms of hypothyroidism in older people are often atypical, and manifestations of hypothyroidism and hyperthyroidism may blur. Patients may have few or no symptoms until dysfunction is severe. Depression, apathy, and decreased mobility or activity may be the major initial symptoms and may be accompanied by significant weight loss. Constipation affects one fourth of older patients.

In older patients with mild to moderate hypothyroidism, thyroid hormone replacement is individually tailored and must be started with low dosages and increased gradually to prevent serious cardiovascular side effects (Eliopoulos, 2010). Angina, for example, may occur with rapid thyroid replacement in the presence of coronary artery disease secondary to the hypothyroid state. Heart failure and tachydysrhythmias may worsen during the transition from the hypothyroid state to the normal metabolic state. Dementia may become more apparent during early thyroid hormone replacement in older patients.

Older patients with severe hypothyroidism and atherosclerosis may become confused and agitated if their metabolic rate is increased too quickly. Marked clinical improvement follows the administration of hormone replacement; such medication must be continued for life, even though signs of hypothyroidism disappear within 3 to 12 weeks.

Myxedema and myxedema coma usually occur exclusively in patients older than 50 years (Ross, 2012a). The high mortality rate of myxedema coma mandates immediate IV administration of thyroid hormone and supportive management of other clinical manifestations such as hyponatremia, hypoglycemia, hypothermia, hypoventilation, and hypotension.

Older patients require periodic follow-up monitoring of serum TSH levels, because poor adherence with therapy may occur or the patient may take the medications erratically. A careful history can identify the need for further education about the importance of the medication.

Hyperthyroidism

Hyperthyroidism, a common endocrine disorder, is a form of **thyrotoxicosis** resulting from an excessive synthesis and secretion of endogenous or exogenous thyroid hormones by the thyroid (Bahn, Burch, Cooper, et al., 2011). The most common causes are Graves' disease, toxic multinodular goiter, and toxic adenoma. Other causes include **thyroiditis** (inflammation of the thyroid gland) and excessive ingestion of thyroid hormone.

Graves' disease, the most common cause of hyperthyroidism, is an autoimmune disorder that results from an excessive output of thyroid hormones caused by abnormal stimulation of the thyroid gland by circulating immunoglobulins (Bahn et al., 2011; Porth, 2011). This disease affects women eight times more frequently than men, with onset usually between the second and fourth decades (Papadakis et al., 2013). The disorder may appear after an emotional shock, stress, or an infection, but the exact significance of these relationships is not understood.

Clinical Manifestations

Patients with hyperthyroidism exhibit a characteristic group of signs and symptoms. The presenting symptom is often nervousness. These patients are often emotionally hyperexcitable, irritable, and apprehensive; they cannot sit quietly; they suffer from palpitations; and their pulse is abnormally rapid at rest as well as on exertion. They tolerate heat poorly and perspire unusually freely. The skin is flushed continuously, with a characteristic salmon color in Caucasians, and is likely to be warm, soft, and moist. However, patients may report dry skin and diffuse pruritus. A fine tremor of the hands may be observed. Patients may exhibit ophthalmopathy, such as **exophthalmos** (abnormal protrusion of one or both eyeballs), which produces a startled facial expression. Despite treatment, these ocular changes are not always reversible (Hogan-Quigley et al., 2012).

Other manifestations include an increased appetite and dietary intake, weight loss, fatigability and weakness (difficulty in climbing stairs and rising from a chair), amenorrhea, and changes in bowel function. Atrial fibrillation occurs in 15% of in older adult patients with new-onset hyperthyroidism (Porth & Matfin, 2009).

Cardiac effects may include sinus tachycardia or dysrhythmias, increased pulse pressure, and palpitations; these changes may be related to increased sensitivity to catecholamines or to changes in neurotransmitter turnover. Myocardial hypertrophy and heart failure may occur if the hyperthyroidism is severe and untreated.

The course of the disease may be mild, characterized by remissions and exacerbations, and terminate with spontaneous recovery in a few months or years. Conversely, it may progress relentlessly, with the untreated person becoming emaciated, intensely nervous, delirious, and even disoriented; eventually, the heart fails.

Symptoms of hyperthyroidism may occur with the release of excessive amounts of thyroid hormone as a result of inflammation after irradiation of the thyroid or destruction of thyroid tissue by tumor. Such symptoms may also occur with excessive administration of thyroid hormone for treatment of hypothyroidism. Long-standing use of thyroid hormone in the absence of close monitoring may be a cause of symptoms of hyperthyroidism. It is also likely to result in premature osteoporosis, particularly in women.

Assessment and Diagnostic Findings

The thyroid gland invariably is enlarged to some extent. It is soft and may pulsate; a thrill often can be palpated, and a bruit is heard over the thyroid arteries (Hogan-Quigley et al., 2012). These are signs of greatly increased blood flow through the thyroid gland. In advanced cases, the diagnosis is made on the basis of the symptoms, a decrease in serum TSH, increased free T_4, and an increase in radioactive iodine uptake.

Medical Management

Appropriate treatment of hyperthyroidism depends on the underlying cause and often consists of a combination of therapies, including antithyroid agents, radioactive iodine, and surgery. Treatment of hyperthyroidism is directed toward reducing thyroid hyperactivity to relieve symptoms and preventing complications. The use of radioactive iodine is the most common form of treatment for Graves' disease in North America. Beta-adrenergic blocking agents (e.g., propranolol [Inderal], atenolol [Tenormin], metoprolol [Lopressor]) are used as adjunctive therapy for symptomatic relief, particularly in transient thyroiditis (Porth & Matfin, 2009; Ross, 2012c). Surgical removal of the entire thyroid gland (or part of it) is a nonpharmacologic alternative (Chiu, Yao, Chan, et al., 2012).

The three treatments (radioactive iodine therapy, antithyroid medications [e.g., thionamides], and surgery) share the same complications: relapse or recurrent hyperthyroidism and permanent hypothyroidism. The rate of relapse increases in patients who have had very severe disease, a long history of dysfunction, ocular and cardiac symptoms, large goiter, or relapse after previous treatment. The relapse rate after radioactive iodine therapy depends on the dose used in treatment. Patients receiving a lower dose of radioactive iodine are more likely to require subsequent treatment than those treated with a higher dose. The remission rate achieved with a single dose of radioactive iodine is 80% (Brent, 2008).

Pharmacologic Therapy

Two forms of pharmacotherapy are available for treating hyperthyroidism and controlling excessive thyroid activity: (1) the use of irradiation by administration of the radioisotope ^{131}I for destructive effects on the thyroid gland and (2) antithyroid medications that interfere with the synthesis of thyroid hormones and other agents that control manifestations of hyperthyroidism.

Radioactive Iodine Therapy

Radioactive iodine has been used to treat toxic adenomas, toxic multinodular goiter, and most varieties of thyrotoxicosis. Radioactive iodine is contraindicated during pregnancy because it crosses the placenta. Women of childbearing age should be given a pregnancy test 48 hours before administration of radioactive iodine. They should also be instructed to not conceive for at least 6 months following treatment. To ensure that radioactivity is no longer actively concentrated in breast tissue, radioactive iodine should not be administered until at least 6 weeks after lactation stops (Bahn et al., 2011).

The goal of radioactive iodine therapy (^{131}I) is to eliminate the hyperthyroid state with the administration of sufficient radiation in a single dose (Bahn et al., 2011). Almost all of the iodine that enters and is retained in the body becomes concentrated in the thyroid gland. Therefore, the radioactive isotope of iodine is concentrated in the thyroid gland, where it destroys thyroid cells without jeopardizing other radiosensitive tissues. Over a period of several weeks, thyroid cells exposed to the radioactive iodine are destroyed, resulting in reduction of the hyperthyroid state and inevitably hypothyroidism.

Some patients at high risk for complications of hyperthyroidism (e.g., older adults, patients with cardiovascular disease) may need pretreatment with antithyroid medications. Methimazole (MMI) is given 4 to 6 weeks prior to administration of radioactive iodine. The antithyroid medications are stopped 3 days before and restarted 3 days after administering radioactive iodine and then tapered over 4 to 6 weeks (Ross, 2012g).

The use of an ablative dose of radioactive iodine initially causes an acute release of thyroid hormone from the thyroid gland and may cause increased symptoms. The patient is observed for signs of **thyroid storm** (Chart 52-6), a life-threatening condition manifested by cardiac dysrhythmias, fever, and neurologic impairment (Ross, 2012b). Beta-blockers are used to control these symptoms.

Thyroid hormone replacement is started 4 to 18 weeks after the antithyroid medications have been stopped based on the results of thyroid function tests. TSH measurements can be misleading in the early months following treatment with radioactive iodine. Therefore, serum free T_4 is the principal test (Bahn et al., 2011; Ross, 2012e) measured at 3 to 6 weeks following administration of radioactive iodine and then every 1 to 2 months until normal thyroid function is established. If TSH and free T_4 are both persistently low, the total T_3 then must be measured to differentiate between persistent hyperthyroidism (T_3 elevated) or transient hypothyroidism (T_3 normal or low) (Ross, 2012e). Once a normal thyroid state has been established, TSH should be measured every 6 to 12 months for life (Ross, 2012e).

A major advantage of treatment with radioactive iodine is that it avoids many of the side effects associated with antithyroid medications. However, some patients may elect to be treated with antithyroid medications rather than radioactive iodine for a variety of reasons, including fear of radiation.

Patients who receive radioactive iodine should be informed that they can contaminate their household and other persons through saliva, urine, or radiation emitting from their body. They should avoid sexual contact, sleeping in the same bed with other persons, having close contact with children and pregnant women, and sharing utensils and cups. The patient should follow the instructions provided by the provider regarding the time restrictions for these cautions because they are dose related (Ross, 2012e).

Antithyroid Medications

Antithyroid medications (thionamides) are summarized in Table 52-3. The objective of pharmacotherapy is to inhibit one or more stages in thyroid hormone synthesis or hormone release. Antithyroid agents block the utilization of iodine by interfering with the iodination of tyrosine and the coupling of iodotyrosines in the synthesis of thyroid hormones. This prevents the synthesis of thyroid hormone. The most commonly used antithyroid drugs in the United States are methimazole (MMI, Tapazole) or propylthiouracil (PTU).

Chart 52-6 Thyroid Storm (Thyrotoxic Crisis, Thyrotoxicosis)

Thyroid storm (thyrotoxic crisis) is a form of severe hyperthyroidism, usually of abrupt onset. Untreated, it is almost always fatal, but with proper treatment the mortality rate is reduced substantially. The patient with thyroid storm or crisis is critically ill and requires astute observation and aggressive and supportive nursing care during and after the acute stage of illness.

Clinical Manifestations

Thyroid storm is characterized by:

- High fever (hyperpyrexia), >38.5°C (>101.3°F)
- Extreme tachycardia (>130 bpm)
- Exaggerated symptoms of hyperthyroidism with disturbances of a major system—for example, gastrointestinal (weight loss, diarrhea, abdominal pain) or cardiovascular (edema, chest pain, dyspnea, palpitations)
- Altered neurologic or mental state, which frequently appears as delirium psychosis, somnolence, or coma

Life-threatening thyroid storm is usually precipitated by stress, such as injury, infection, thyroid and nonthyroid surgery, tooth extraction, insulin reaction, diabetic ketoacidosis, pregnancy, digitalis intoxication, abrupt withdrawal of antithyroid medications, extreme emotional stress, or vigorous palpation of the thyroid. These factors can precipitate thyroid storm in the partially controlled or completely untreated patient with hyperthyroidism. Current methods of diagnosis and treatment for hyperthyroidism have greatly decreased the incidence of thyroid storm, making it uncommon today.

Management

Immediate objectives are reduction of body temperature and heart rate and prevention of vascular collapse. Measures to accomplish these objectives include:

- A hypothermia mattress or blanket, ice packs, a cool environment, hydrocortisone, and acetaminophen (Tylenol). Salicylates (e.g., aspirin) are not used because they displace thyroid hormone from binding proteins and worsen the hypermetabolism.
- Humidified oxygen is administered to improve tissue oxygenation and meet the high metabolic demands. Arterial blood gas levels or pulse oximetry may be used to monitor respiratory status.
- IV fluids containing dextrose are administered to replace liver glycogen stores that have been decreased in the patient who is hyperthyroid.
- Propylthiouracil (PTU) or methimazole is administered to impede formation of thyroid hormone and block conversion of T_4 to T_3, the more active form of thyroid hormone.
- Hydrocortisone is prescribed to treat shock or adrenal insufficiency.
- Iodine is administered to decrease output of T_4 from the thyroid gland. For cardiac problems such as atrial fibrillation, dysrhythmias, and heart failure, sympatholytic agents may be administered. Propranolol, combined with digitalis, has been effective in reducing severe cardiac symptoms.

TABLE 52-3 Pharmacologic Agents Used to Treat Hyperthyroidism

Agent	Action	Nursing Considerations
Propylthiouracil (PTU)	Blocks synthesis of hormones (conversion of T_4 to T_3)	Monitor cardiac parameters. Observe for conversion to hypothyroidism. Must be given by mouth Watch for rash, nausea, vomiting, agranulocytosis, SLE.
Methimazole (Tapazole)	Inhibits synthesis of thyroid hormone	More toxic than PTU Watch for rash and other symptoms as for PTU.
Sodium iodide	Suppresses release of thyroid hormone	Given 1 hour after PTU or methimazole Watch for edema, hemorrhage, gastrointestinal upset.
Potassium iodide	Suppresses release of thyroid hormone	Discontinue for rash. Watch for signs of toxic iodinism.
Saturated solution of potassium iodide (SSKI)	Suppresses release of thyroid hormone	Mix with juice or milk. Give by straw to prevent staining of teeth.
Dexamethasone (Decadron)	Suppresses release of thyroid hormone	Monitor input and output. Monitor glucose. May cause hypertension, nausea, vomiting, anorexia, infection
Beta-blocker (e.g., propranolol [Inderal])	Beta-adrenergic blocking agent	Monitor cardiac status. Hold for bradycardia or decreased cardiac output. Use with caution in patients with heart failure.

SLE, systemic lupus erythematosus
Adapted from Morton, P. G., & Fontaine, D. K. (2013). *Critical care nursing: A holistic approach*. Philadelphia: Lippincott Williams & Wilkins.

The medications are used until the patient is euthyroid (i.e., neither hyperthyroid nor hypothyroid). These medications block extrathyroidal conversion of T_4 to T_3 (Bahn et al., 2011).

Prior to initiating therapy with these drugs, baseline blood tests are performed, including complete blood count (white blood cell [WBC] count with differential) and liver profile (transaminases and bilirubin) (Bahn et al., 2011; Ross, 2012d). The therapeutic dose is determined on the basis of clinical criteria, including changes in pulse rate, pulse pressure, body weight, size of the goiter, and results of laboratory studies. The patient should be instructed to take the medication in the morning on an empty stomach 30 minutes before eating to avoid decrease in absorption associated with some foods such as walnuts, soybean flour, cottonseed meal, and dietary fiber (Lexicomp, 2012). Because antithyroid medications do not interfere with release or activity of previously formed thyroid hormones, it may take several weeks until relief of symptoms occurs. At that time, the maintenance dose is established, and the medication is gradually tapered over several months.

Toxic complications of antithyroid medications are relatively uncommon; nevertheless, the importance of periodic follow-up is emphasized, because medication sensitization, fever, rash, urticaria, or even agranulocytosis and thrombocytopenia (decrease in granulocytes and platelets) may develop (Bahn et al., 2011). With any sign of infection, especially pharyngitis and fever or the occurrence of mouth ulcers, the patient is advised to stop the medication, notify the primary provider immediately, and undergo hematologic studies (Bahn et al., 2011). PTU is recommended during the first trimester of pregnancy rather than MMI due to the teratogenic effects of MMI (Ross, 2012f). Due to risk of hepatotoxicity, PTU should be discontinued after the first trimester and the patient should be switched to MMI for the remainder of the pregnancy and when nursing (Ross, 2012b).

Discontinuation of antithyroid medications before therapy is complete usually results in relapse within 6 months. It is important that the possibility of relapse be discussed so that a treatment strategy will be in place if relapse occurs.

Adjunctive Therapy

Iodine or iodide is necessary for thyroid function, and deficiency or excess can lead to thyroid dysfunction. Iodine or iodide solutions were once the only therapy available for patients with hyperthyroidism but are no longer the sole method of treatment. Antithyroid medications (MMI and PTU) and radioactive iodine are not the mainstays of pharmacologic treatment for hyperthyroidism. In the short term, iodine solutions are considered effective. When treating patients with thyroid storm and those requiring surgical intervention for hyperthyroidism, iodine solutions are considered highly effective. Iodine solutions inhibit the release of T_4 and T_3 within hours of administration and inhibit thyroid hormone synthesis. The maximum inhibitory effect on serum thyroid hormone production is only 10 days (Ross, 2012g). Solutions such as potassium iodide (KI), Lugol's solution, and saturated solution of potassium iodide (SSKI) may be used in combination with antithyroid agents or beta-adrenergic blockers to prepare the patient with hyperthyroidism for surgery. Solutions of iodine and iodide compounds are more palatable in milk or fruit juice and are administered through a straw to prevent staining of the teeth. The nurse should be aware that there have been reported cases of local esophageal or duodenal mucosal injury and hemorrhage as a result of administration of Lugol's solution (960 mg/day) to manage thyroid storm (Bahn et al., 2011).

Beta-adrenergic blocking agents (e.g., propranolol [Inderal], atenolol [Tenormin], metoprolol [Lopressor]) are important in decreasing heart rate, systolic blood pressure, muscle weakness, nervousness, tremor, anxiety, and heat intolerance. The patient continues taking the beta-blocker until the free T_4 is within the normal range and the TSH level approaches normal.

Surgical Management

Surgery to remove thyroid tissue was once the main method of treating hyperthyroidism. Today, surgery is reserved for special circumstances—for example, in pregnant women who are allergic to antithyroid medications, in patients with large goiters, or in patients who are unable to take antithyroid agents. Surgery for treatment of hyperthyroidism is performed soon after the thyroid function has returned to normal (4 to 6 weeks).

The surgical removal of about five sixths of the thyroid tissue (subtotal thyroidectomy) reliably results in a prolonged remission in most patients with exophthalmic goiter. Its use today is reserved for patients with obstructive symptoms, for pregnant women in the second trimester, and for patients with a need for rapid normalization of thyroid function. Before surgery, an antithyroid medication is given until signs of hyperthyroidism have disappeared. A beta-adrenergic blocking agent (e.g., propranolol) may be used to reduce the heart rate and other signs and symptoms of hyperthyroidism. Medications that may prolong clotting (e.g., aspirin) are stopped several weeks before surgery to reduce the risk of postoperative bleeding. Patients receiving iodine medication must be monitored for evidence of iodine toxicity (iodinism), which requires immediate withdrawal of the medication. Symptoms of iodinism include swelling of the buccal mucosa, excessive salivation, cold symptoms, and skin eruptions. The incidence of relapse with a total thyroidectomy is nearly 0%, whereas recurrence following a subtotal thyroidectomy is 8% at 5 years (Bahn et al., 2011).

Gerontologic Considerations

Although hyperthyroidism is much less common in older adults than hypothyroidism, patients 65 years and older need careful assessment to avoid missing subtle signs and symptoms. Subclinical hyperthyroidism has been reported in approximately 8% of this age group (Gesing, Lewinski, & Karbownik-Lewinska, 2012). This age group may present with atypical vague and nonspecific signs and symptoms of thyroid disease such as anorexia and weight loss, absence of ocular signs, or isolated atrial fibrillation (Dominguez et al., 2008). New or worsening heart failure or angina is more likely to occur in older patients rather than in younger patients. Symptoms such as tachycardia, fatigue, mental confusion, weight loss, change in bowel habits, and depression can be attributed to age and other illnesses that are common in older people. The older patient may complain of difficulty climbing stairs or rising from a chair because of muscle weakness (Mitrou, Raptis, & Dimitreadis, 2011).

Evaluation for thyroid disease with a serum TSH measurement is indicated in older patients who have unexplained physical or mental deterioration (Bahn et al., 2011). Free T_4 and T_3 should be included in the initial screening when hyperthyroidism is highly suspected. Once thyrotoxicosis is confirmed, additional tests such as radioactive iodine uptake and thyroid scan are ordered to differentiate between causes such as Graves' disease, toxic nodular goiter, acute thyroiditis, and other disorders. Toxic nodular goiter is the most common cause of thyrotoxicosis in older patients. Patients have the option of treatment using antithyroid medications, radioactive iodine, and surgery. Radioactive iodine is generally recommended for treatment of thyrotoxicosis caused by toxic nodular goiter in older patients unless an enlarged thyroid gland is pressing on the airway. Prior to administration of radioactive iodine (Bahn et al., 2011) in this at-risk group, pretreatment with beta-blockade is indicated if the patient's resting heart rate is 90 bpm or if the patient has coexistent cardiovascular disease and symptomatic hyperthyroidism. MMI should be administered prior to treatment with radioactive iodine to prevent complications due to worsening of hyperthyroidism. Patients who elect to have surgery to treat toxic nodular goiter must be pretreated with an antithyroid medication (methimazole) to achieve a euthyroid state. However, preoperative iodine should not be used to prevent exacerbating the hyperthyroidism (Bahn et al., 2011). Long-term use of certain antithyroid medications such as PTU is not recommended for treatment of toxic nodular goiter in older patients due to risk of side effects. Although rare, there is evidence that PTU can result in agranulocytosis and hepatic injury. However, the use of antithyroid medications versus radioactive iodine or surgery may be the patient's preferred choice or the option for some older patients and other ill persons with "limited longevity" who can be monitored at least every 3 months (Bahn et al., 2011).

The use of beta-adrenergic blocking agents (e.g., propranolol [Inderal] and atenolol [Tenormin]) may be indicated to decrease the cardiovascular and neurologic signs and symptoms of thyrotoxicosis. These agents must be used with extreme caution in older patients to minimize adverse effects on cardiac function that may produce heart failure. The dosage of other medications used to treat other chronic illnesses in older patients may also need to be modified because of the altered rate of metabolism associated with hyperthyroidism.

NURSING PROCESS

The Patient With Hyperthyroidism

Assessment

The health history and examination focus on symptoms related to accelerated or exaggerated metabolism. These include the patient's and family's reports of irritability and increased emotional reaction and the impact that these changes have had on the patient's interactions with family, friends, and coworkers. The history includes other stressors and the patient's ability to cope with stress.

The nurse initially and periodically assesses the patient's nutritional status and the presence of symptoms related to the hypermetabolic state. This hypermetabolic state may affect the cardiovascular system including heart rate and rhythm, blood pressure, heart sounds, and peripheral pulses. Other specific changes may also include alteration in vision and appearance of the external eye. Because emotional changes are associated with hyperthyroidism, the patient's emotional state and psychological status are evaluated, as well as such symptoms as irritability, anxiety, sleep disturbances, apathy, and lethargy, all of which may occur with hyperthyroidism. The family may also provide information about recent changes in the patient's emotional status.

Diagnosis

Nursing Diagnoses

Based on the assessment data, major nursing diagnoses may include the following:

- Imbalanced nutrition: less than body requirements related to exaggerated metabolic rate, excessive appetite, and increased GI activity

- Ineffective coping related to irritability, hyperexcitability, apprehension, and emotional instability
- Situational low self-esteem related to changes in appearance, excessive appetite, and weight loss
- Risk for imbalanced body temperature

Collaborative Problems/Potential Complications

Potential complications may include the following:

- Thyrotoxicosis or thyroid storm
- Hypothyroidism

Planning and Goals

The goals for the patient may be improved nutritional status, improved coping ability, improved self-esteem, maintenance of normal body temperature, and absence of complications.

Nursing Interventions

Improving Nutritional Status

Hyperthyroidism affects all body systems, including the GI system. The appetite is increased but may be satisfied by several well-balanced meals of small size, even up to six meals a day. Foods and fluids are selected to replace fluid lost through diarrhea and diaphoresis and to control the diarrhea that results from increased peristalsis. Rapid movement of food through the GI tract may result in nutritional imbalance and further weight loss. To reduce diarrhea, highly seasoned foods and stimulants such as coffee, tea, cola, and alcohol are discouraged. High-calorie, high-protein foods are encouraged. A quiet atmosphere during mealtime may aid digestion. Weight and dietary intake are recorded to monitor nutritional status.

Enhancing Coping Measures

The patient with hyperthyroidism needs reassurance that the emotional reactions being experienced are a result of the disorder and that with effective treatment those symptoms will be controlled. Because of the negative effect that these symptoms have on family and friends, they too need reassurance that the symptoms are expected to disappear with treatment.

It is important to use a calm, unhurried approach with the patient. Stressful experiences should be minimized and maintain a quiet uncluttered environment. The nurse encourages relaxing activities that will not overstimulate the patient. It is important to balance periods of activity with rest.

If thyroidectomy is planned, the patient needs to know that pharmacologic therapy is necessary to prepare the thyroid gland for surgical treatment. The nurse provides education and reminds the patient to take the medications as prescribed. Because of hyperexcitability and shortened attention span, the patient may require repetition of this education and written instructions.

Improving Self-Esteem

The patient with hyperthyroidism is likely to experience changes in appearance, appetite, and weight. These factors, along with the patient's inability to cope well with family and the illness, may result in loss of self-esteem. The nurse conveys an understanding of the patient's concern about these problems and promotes the use of effective coping strategies. The patient and family should be reassured that these changes are a result of the thyroid dysfunction and are, in fact, out of the patient's control. The nurse refers the patient to professional counseling as necessary if the patient is unable to cope with physical appearance.

If the patient experiences ocular changes secondary to hyperthyroidism, eye care and protection may be necessary. The nurse educates the patient about instillation of eye drops or ointment prescribed to soothe the eyes and protect the exposed cornea. Smoking should be highly discouraged, and smoking cessation strategies are recommended. The patient may be embarrassed by the need to eat large meals. Caregivers and family should avoid commenting on the patient's large dietary intake while making sure that the patient receives sufficient nutritious food.

Maintaining Normal Body Temperature

The patient with hyperthyroidism frequently finds a normal room temperature too warm because of an exaggerated metabolic rate and increased heat production. If the patient is hospitalized, the environment should be maintained at a cool, comfortable temperature, and the bedding and clothing should be changed as needed. Cool baths and cool or cold fluids may also provide relief.

Monitoring and Managing Potential Complications

The nurse closely monitors the patient with hyperthyroidism for signs and symptoms that may be indicative of thyroid storm. Cardiac and respiratory function are assessed by measuring vital signs and cardiac output, electrocardiographic (ECG) monitoring, arterial blood gases, and pulse oximetry. Assessment continues after treatment is initiated because of the potential effects of treatment on cardiac function. Oxygen is administered to prevent hypoxia, to improve tissue oxygenation, and to meet the high metabolic demands. IV fluids may be necessary to maintain blood glucose levels and to replace lost fluids. Antithyroid medications (MMI or PTU) may be prescribed to reduce thyroid hormone levels. In addition, beta-blockers and digitalis may be prescribed to treat cardiac symptoms. If shock develops, treatment strategies must be implemented (see Chapter 14).

Hypothyroidism is likely to occur with any of the treatments used for hyperthyroidism. Therefore, the nurse periodically monitors the patient. Most patients report a greatly improved sense of well-being after treatment of hyperthyroidism, and some fail to continue to take prescribed thyroid replacement therapy. Therefore, part of patient and family education is instruction about the importance of continuing therapy indefinitely after discharge and a discussion of the consequences of failing to take medication.

Promoting Home and Community-Based Care

Educating Patients About Self-Care. The nurse educates the patient with hyperthyroidism about how and when to take prescribed medication and provides instruction about the essential role of the medication in the broader therapeutic plan. Because of the hyperexcitability and decreased attention span associated with hyperthyroidism, the nurse provides a written plan for the patient to use at home. The type and amount of information given depend on the patient's stress and anxiety levels. The patient and family members receive verbal and written education about the actions and possible side effects of the medications as well as adverse effects that should be reported if they occur (Chart 52-7).

If a total or subtotal thyroidectomy is anticipated, the patient needs to be educated about what to expect. Information is repeated as the time of surgery approaches. The nurse

Chart 52-7 HOME CARE CHECKLIST
The Patient With Hyperthyroidism

At the completion of home care education, the patient or caregiver will be able to:	PATIENT	CAREGIVER
• State present and potential effects of hyperthyroidism on the body.	✔	✔
• State precipitating factors and interventions for complications (hypothyroidism, thyroid storm).	✔	✔
• State the purpose, dose, route, schedule, side effects, and precautions of prescribed medications (propylthiouracil, radioactive iodine).	✔	✔
• State the need to contact primary provider before taking over-the-counter medications.	✔	✔
• State the need for regular follow-up visits with primary provider.	✔	✔
• Identify the need for planned rest periods and methods to improve sleep patterns.	✔	✔
• Identify the need for increased dietary intake until weight stabilizes.	✔	✔
• Identify areas of physical and emotional stress.	✔	✔
• State that emotional lability is part of disease process.	✔	✔
• Describe the potential benefits and risks of surgical intervention or radioactive iodine therapy.	✔	✔
• Identify potential for menstrual irregularities, increased risk for osteoporosis, and potential for pregnancy for women.	✔	✔
• State the need to wear medical identification, and carry medical information card.	✔	✔
• Identify rationale for smoking cessation, and take steps to stop smoking.	✔	

also advises the patient to avoid stressful situations that may precipitate thyroid storm.

Continuing Care. Referral for home care, if indicated, allows the home care nurse to assess the home and family environment, as well as the patient's and family's understanding of the importance of adhering to the therapeutic regimen and the recommended follow-up monitoring. The nurse reinforces to the patient and family the importance of long-term follow-up because of the risk of hypothyroidism after thyroidectomy or treatment with antithyroid medications or radioactive iodine. The nurse also assesses the patient for changes indicating return to normal thyroid function and signs and symptoms of hyperthyroidism and hypothyroidism. Furthermore, the nurse reminds the patient and family about the importance of health promotion activities and recommended health screening.

Evaluation

Expected patient outcomes may include:

1. Improves nutritional status
   a. Reports adequate dietary intake and decreased hunger
   b. Identifies high-calorie, high-protein foods; identifies foods to be avoided
   c. Avoids the use of alcohol and other stimulants
   d. Stops smoking
   e. Reports decreased episodes of diarrhea
2. Demonstrates effective coping methods in dealing with family, friends, and coworkers
   a. Explains reasons for irritability and emotional instability
   b. Avoids stressful situations, events, and people
   c. Participates in relaxing, nonstressful activities
3. Achieves increased self-esteem
   a. Verbalizes feelings about self and illness
   b. Describes feelings of frustration and loss of control
   c. Describes reasons for increased appetite
4. Maintains normal body temperature
5. Absence of complications
   a. Serum thyroid hormone and TSH levels within normal limits
   b. Identifies signs and symptoms of thyroid storm and hypothyroidism
   c. Vital signs and results of ECG, arterial blood gases, and pulse oximetry within normal limits
   d. States importance of regular follow-up and lifelong maintenance of prescribed therapy

Thyroid Tumors

Tumors of the thyroid gland are classified on the basis of being benign or malignant, the presence or absence of associated thyrotoxicosis, and the diffuse or irregular quality of the glandular enlargement. If the enlargement is sufficient to cause a visible swelling in the neck, the tumor is referred to as a goiter.

All grades of goiter are encountered, from those that are barely visible to those producing disfigurement. Some are symmetric and diffuse; others are nodular. Some are accompanied by hyperthyroidism, in which case they are described as toxic; others are associated with a euthyroid state and are referred to as nontoxic goiters.

Endemic (Iodine-Deficient) Goiter

The most common type of goiter that occurs when iodine intake is deficient is the simple or colloid goiter. In addition to being caused by an iodine deficiency, simple goiter may be caused by an intake of large quantities of goitrogenic substances in patients with unusually susceptible glands. These substances include excessive amounts of iodine. Lithium prescribed for the treatment of bipolar disorder has also been found to also have antithyroid actions (Janicak, 2012; Surks, 2012b).

Simple goiter represents a compensatory hypertrophy of the thyroid gland, caused by stimulation by the pituitary gland. The pituitary gland produces thyrotropin or TSH, a hormone that controls the release of thyroid hormone from the thyroid gland. Its production increases if there is subnormal thyroid activity, as when insufficient iodine is available for production of the thyroid hormone. Such goiters usually cause no symptoms, except for the swelling in the neck, which may result in tracheal compression when excessive swelling is present.

Many goiters of this type recede after the iodine imbalance is corrected. Supplementary iodine, such as SSKI, is prescribed to suppress the pituitary's thyroid-stimulating activity. When surgery is recommended, the risk of postoperative complications is minimized by ensuring a preoperative euthyroid state through treatment with antithyroid medications and iodide to reduce the size and vascularity of the goiter. The introduction of iodized salt has been the single most effective means of preventing goiter in at-risk populations.

Nodular Goiter

Some thyroid glands are nodular because of areas of hyperplasia (overgrowth). No symptoms may arise as a result of this condition, but not uncommonly these nodules slowly increase in size, with some descending into the thorax, where they cause local pressure symptoms. Some nodules become malignant, and some are associated with a hyperthyroid state. Therefore, the patient with many thyroid nodules may eventually require surgery.

Thyroid Cancer

Cancer of the thyroid is much less prevalent than other forms of cancer, but the incidence has been increasing steadily since the 1990s. It accounts for 90% of endocrine malignancies. Although it is the fastest-growing cancer rate among men and women, three of four cases occur in women. Unlike other cancers, 80% of new cases are in patients under the age of 65 years (American Cancer Society [ACS], 2012).

External radiation of the head, neck, or chest in infancy and childhood increases the risk of thyroid carcinoma. The incidence of thyroid cancer appears to increase 5 to 40 years after irradiation. Consequently, people who underwent radiation treatment or were otherwise exposed to radiation as children should consult their primary provider, request an isotope thyroid scan as part of the evaluation, follow recommended treatment of abnormalities of the gland, and continue with annual checkups.

Assessment and Diagnostic Findings

Lesions that are single, hard, and fixed on palpation or associated with cervical lymphadenopathy suggest malignancy. Thyroid function tests may be helpful in evaluating thyroid nodules and masses; however, results are rarely conclusive. Needle biopsy of the thyroid gland is used as an outpatient procedure to make a diagnosis of thyroid cancer, to differentiate cancerous thyroid nodules from noncancerous nodules, and to stage the cancer if detected. The procedure is safe and usually requires only a local anesthetic agent. However, patients who undergo the procedure are monitored closely, because cancerous tissues may be missed during the procedure. A second type of aspiration or biopsy uses a large-bore needle rather than the fine needle used in standard biopsy; it may be used when the results of the standard biopsy are inconclusive or with rapidly growing tumors. Additional diagnostic studies include ultrasound, MRI, CT, thyroid scans, radioactive iodine uptake studies, and thyroid suppression tests.

Medical Management

The treatment of choice for thyroid carcinoma is surgical removal. Total or near-total thyroidectomy is performed if possible. Modified neck dissection or more extensive radical neck dissection is performed if there is lymph node involvement.

Efforts are made to spare parathyroid tissue to reduce the risk of postoperative hypocalcemia and tetany. After surgery, ablation procedures are carried out with radioactive iodine to eradicate residual thyroid tissue (ACS, 2011). Radioactive iodine is also used for thyroid cancers with metastasis (ACS, 2011).

After surgery, thyroid hormone is administered in suppressive doses to lower the levels of TSH to a euthyroid state (Bahn et al., 2011). If the remaining thyroid tissue is inadequate to produce sufficient thyroid hormone, thyroxine is required permanently.

Several routes are available for administering radiation to the thyroid or tissues of the neck, including oral administration of radioactive iodine (Bahn et al., 2011) and external administration of radiation therapy. Short-term side effects of radioactive iodine treatment may include neck soreness, nausea, and upset stomach; salivary glands being tender and swollen; dry mouth; changes in taste; and, rarely, pain (Bahn et al., 2011). The patient who receives external sources of radiation therapy is at risk for mucositis, dryness of the mouth, dysphagia, redness of the skin, anorexia, and fatigue (see Chapter 15). Chemotherapy is infrequently used to treat thyroid cancer.

Patients whose thyroid cancer is detected early and who are appropriately treated usually do very well. Patients who have had papillary cancer—the most common and least aggressive tumor—have a 10-year survival rate greater than 90%. Long-term survival is also common in follicular cancer, which is a more aggressive form of thyroid cancer (Papadakis et al., 2013). However, continued thyroid hormone therapy and periodic follow-up and diagnostic testing are important to ensure the patient's well-being.

Later follow-up includes clinical assessment for recurrence of nodules or masses in the neck and signs of hoarseness, dysphagia, or dyspnea. The recommendations for long-term follow-up of patients with differentiated thyroid cancer are based on the stage of cancer and results of the follow-up examination 1 year following the initial treatment. The first year evaluation includes clinical examination, TSH and free thyroxine, and measurement of serum thyroglobulin within 6 months following the initial treatment, and a routine neck ultrasound with the first 6 to 12 months following initial treatment. Tests used to confirm sites of metastasis if there is clinical evidence of recurrence include radioiodine imaging, CT, MRI, skeletal x-rays, and skeletal radionucleotide imaging. Fluorodeoxyglucose (FGDA) PET is useful to establish prognosis if there is evidence of distant metastases (Tuttle, 2012). Free T_4, TSH, and serum calcium and phosphorus levels are monitored to determine whether the thyroid hormone supplementation is adequate and to note whether calcium balance is maintained.

Although local and systemic reactions to radiation may occur and may include neutropenia or thrombocytopenia, these complications are rare when radioactive iodine is used. Patients who undergo surgery that is combined with radioactive iodine have a higher survival rate than those who undergo surgery alone. Patient education emphasizes the importance of taking prescribed medications and following recommendations for follow-up monitoring. The patient who is undergoing radiation therapy is also instructed in how to assess and manage side effects of treatment (see Chapter 15).

Nursing Management

Important preoperative goals are to prepare the patient for surgery and reduce anxiety. Often, the patient's home life has become tense because of his or her restlessness, irritability, and nervousness secondary to hyperthyroidism. Efforts are necessary to protect the patient from tension and stress to avoid precipitating thyroid storm. If the patient reports increased stress when with family or friends, suggestions are made to limit contact with them. Quiet and relaxing activities are encouraged.

Providing Preoperative Care

The nurse instructs the patient about the importance of eating a diet high in carbohydrates and proteins. A high daily caloric intake is necessary because of the increased metabolic activity and rapid depletion of glycogen reserves. Supplementary vitamins, particularly thiamine and ascorbic acid, may be prescribed. The patient is reminded to avoid tea, coffee, cola, and other stimulants.

The nurse also informs the patient about the purpose of preoperative tests, if they are to be performed, and explains what preoperative preparations to expect. This information should help to reduce the patient's anxiety about the surgery. In addition, special efforts are made to ensure a good night's rest before surgery, although many patients are admitted to the hospital on the day of surgery.

Preoperative education includes demonstrating to the patient how to support the neck with the hands after surgery to prevent stress on the incision. This involves raising the elbows and placing the hands behind the neck to provide support and reduce strain and tension on the neck muscles and the surgical incision.

Providing Postoperative Care

The nurse periodically assesses the surgical dressings and reinforces them if necessary. When the patient is in a recumbent position, the nurse observes the sides and the back of the neck as well as the anterior dressing for bleeding. In addition to monitoring the pulse and blood pressure for any indication of internal bleeding, the nurse must be alert for complaints of a sensation of pressure or fullness at the incision site. Such symptoms may indicate subcutaneous hemorrhage and hematoma formation and should be reported.

Difficulty in respiration can occur as a result of edema of the glottis, hematoma formation, or injury to the recurrent laryngeal nerve. This complication requires that an airway be inserted. Therefore, a tracheostomy set is kept at the bedside at all times, and the surgeon is summoned at the first indication of respiratory distress. If the respiratory distress is caused by hematoma, surgical evacuation is required.

The intensity of pain is assessed, and analgesic agents are administered as prescribed for pain. The nurse should anticipate apprehension in the patient and should inform the patient that oxygen will assist breathing. When moving and turning the patient, the nurse carefully supports the patient's head and avoids tension on the sutures. The most comfortable position is the semi-Fowler's position, with the head elevated and supported by pillows.

IV fluids are administered during the immediate postoperative period. Water may be given by mouth as soon as nausea subsides and bowel sounds are present. Usually, there is a little difficulty in swallowing; initially, cold fluids and ice may be taken better than other fluids. Often, patients prefer a soft diet to a liquid diet in the immediate postoperative period.

The patient is advised to talk as little as possible to reduce edema to the vocal cords; however, when the patient does speak, any voice changes are noted, indicating possible injury to the recurrent laryngeal nerve, which lies just behind the thyroid next to the trachea. An overbed table is provided for access to frequently used items so that the patient avoids turning his or her head. The table can also be used to support a humidifier when vapor-mist inhalations are prescribed for the relief of excessive mucus accumulation.

The patient is encouraged to be out of bed as soon as possible and to eat foods that are easily swallowed. A high-calorie diet may be prescribed to promote weight gain. The incision may be closed using absorbable sutures, nonabsorbable sutures, and adhesive strips. Absorbable sutures dissolve within the body. If nonabsorbable sutures are used, the timeline for removal may vary; however, these types of sutures are usually removed 5 to 7 days following surgery. Adhesives will peel off spontaneously. The patient is usually discharged from the hospital on the day of surgery or soon afterward if the postoperative course is uncomplicated.

Monitoring and Managing Potential Complications

Hemorrhage, hematoma formation, edema of the glottis, and injury to the recurrent laryngeal nerve are complications reviewed previously in this chapter. Occasionally in thyroid surgery, the parathyroid glands are injured or removed, producing a disturbance in calcium metabolism. As the blood calcium level falls, hyperirritability of the nerves occurs, with spasms of the hands and feet and muscle twitching (see Chapter 13). This group of symptoms is termed *tetany*, and the nurse must immediately report its appearance because laryngospasm, although rare, may occur and obstruct the airway. Tetany of this type is usually treated with IV calcium gluconate. This calcium abnormality is usually temporary after thyroidectomy unless all parathyroid tissue was removed.

Quality and Safety Nursing Alert

Following thyroid surgery, the patient should be monitored closely for signs of tetany, including hyperirritability of the nerves, with spasms of the hands and feet and muscle twitching. Laryngospasm, although rare, may occur and obstruct the airway.

Promoting Home and Community-Based Care

Predischarge education is essential because the patient is usually discharged within 1 or 2 days. The patient, family, and caregivers need to be knowledgeable about the signs and symptoms that should be reported. Discharge education includes strategies for managing postoperative pain at home and for increasing humidification. The nurse explains to the patient and family the need for rest, relaxation, and adequate nutrition and to avoid putting strain on the incision and sutures. The patient is permitted to resume his or her former activities and responsibilities completely once recovered from surgery.

Family responsibilities and factors relating to the home environment that produce emotional tension have often been implicated as precipitating causes of thyrotoxicosis. A home visit provides an opportunity to evaluate these factors and to suggest ways to improve the home and family environment. If indicated, a referral to home care is made. The home care nurse reviews the history; performs a physical assessment; assesses the surgical incision; develops a plan of care with the patient and family; and educates the patient, family, and caregivers about wound care, signs and symptoms to report, stress reduction, and the importance of keeping appointments with the primary provider.

THE PARATHYROID GLANDS

Anatomic and Physiologic Overview

The parathyroid glands (normally four) are situated in the neck and embedded in the posterior aspect of the thyroid gland (Fig. 52-5). Parathormone (parathyroid hormone)—the protein hormone produced by the parathyroid glands—regulates calcium and phosphorus metabolism. Increased secretion of parathormone results in increased calcium absorption from the kidney, intestine, and bones, which raises the serum calcium level (Porth & Matfin, 2009). Some actions of this hormone are increased by the presence of vitamin D. Parathormone also tends to lower the blood phosphorus level. The serum level of ionized calcium regulates the output of parathormone. Increased serum calcium results in decreased parathormone secretion, creating a negative feedback system.

Pathophysiology

Excess parathormone can result in markedly increased levels of serum calcium, which is a potentially life-threatening situation. When the product of serum calcium and serum phosphorus (calcium × phosphorus) rises, calcium phosphate may precipitate in various organs of the body (e.g., the kidneys) and cause tissue calcification.

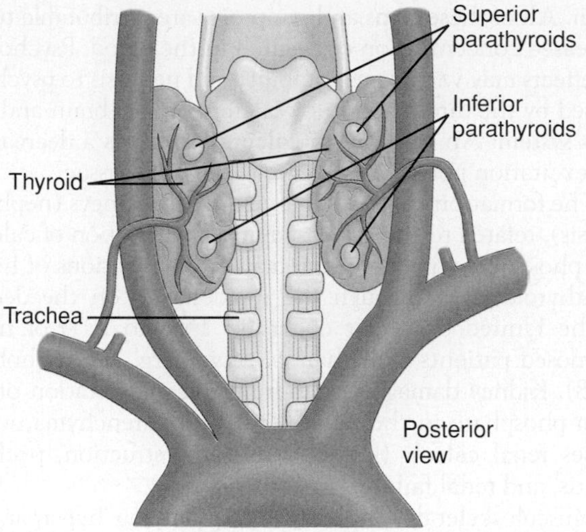

Superior parathyroids
Inferior parathyroids
Thyroid
Trachea
Posterior view

FIGURE 52-5 • The parathyroid glands are located behind the thyroid gland. The parathyroids may be embedded in the thyroid tissue.

Hyperparathyroidism

Hyperparathyroidism is caused by overproduction of parathormone by the parathyroid glands and is characterized by bone decalcification and the development of renal calculi (kidney stones) containing calcium.

Primary hyperparathyroidism occurs two to four times more often in women than in men and is most common in people between 60 and 70 years of age. Its incidence is approximately 25 cases per 100,000 people. The disorder is rare in children younger than 15 years, but its incidence increases 10-fold between the ages of 15 and 65 years. Half of the people diagnosed with hyperparathyroidism do not have symptoms (Fuleihan & Silverberg, 2012).

Secondary hyperparathyroidism, with manifestations similar to those of primary hyperparathyroidism, occurs in patients who have chronic kidney failure and so-called renal rickets as a result of phosphorus retention, increased stimulation of the parathyroid glands, and increased parathormone secretion.

Concept Mastery Alert

Conditions that involve lowered, *not* elevated, serum calcium levels may contribute to hyperparathyroidism because as calcium levels decrease, parathyroid hormone levels increase.

Clinical Manifestations

The patient may have no symptoms or may experience signs and symptoms resulting from involvement of several body systems. Apathy, fatigue, muscle weakness, nausea, vomiting, constipation, hypertension, and cardiac dysrhythmias may occur. All of these signs and symptoms are attributable to the increased concentration of calcium in the blood. Psychological effects may vary from irritability and neurosis to psychoses caused by the direct action of calcium on the brain and nervous system. An increase in calcium produces a decrease in the excitation potential of nerve and muscle tissue.

The formation of stones in one or both kidneys (nephrolithiasis), related to the increased urinary excretion of calcium and phosphorus, is one of the major complications of hyperparathyroidism. Although the incidence is on the decline in the United States, it occurs in 15% to 20% of newly diagnosed patients (Fuleihan & Silverberg, 2012; Shoback, 2008). Kidney damage results from the precipitation of calcium phosphate in the renal pelvis and parenchyma, which causes renal calculi (kidney stones), obstruction, pyelonephritis, and renal failure.

Musculoskeletal symptoms accompanying hyperparathyroidism may be caused by demineralization of the bones or by bone tumors composed of benign giant cells resulting from overgrowth of osteoclasts. The patient may develop skeletal pain and tenderness, especially of the back and joints; pain on weight bearing; pathologic fractures; deformities; and shortening of body stature. Bone loss attributable to hyperparathyroidism increases the risk of fracture.

The incidence of peptic ulcer and pancreatitis is increased with hyperparathyroidism and may be responsible for many of the GI symptoms that occur.

Assessment and Diagnostic Findings

Primary hyperparathyroidism is diagnosed by persistent elevation of serum calcium levels and an elevated concentration of parathormone. Radioimmunoassays for parathormone are sensitive and differentiate primary hyperparathyroidism from other causes of hypercalcemia in more than 80% of patients with elevated serum calcium levels (Fuleihan & Silverberg, 2012). An elevated serum calcium level alone is a nonspecific finding, because serum levels may be altered by diet, medications, and kidney and bone changes. Bone changes may be detected on x-ray or bone scans in advanced disease. The double-antibody parathyroid hormone test is used to distinguish between primary hyperparathyroidism and malignancy as a cause of hypercalcemia. Ultrasound, MRI, thallium scan, and fine-needle biopsy have been used to evaluate the function of the parathyroids and to localize parathyroid cysts, adenomas, or hyperplasia.

Medical Management

Surgical Management

The recommended treatment for primary hyperparathyroidism is the surgical removal of abnormal parathyroid tissue (parathyroidectomy) (Rodgers, Lew, & Solorzano, 2008). In the past, the standard parathyroidectomy involved a bilateral neck exploration under general anesthesia. Today, minimally invasive parathyroidectomy techniques allow for unilateral neck exploration using local anesthesia; these are performed on an outpatient basis. In some cases, only the removal of a single diseased gland is necessary, reducing morbidity rates associated with surgery. For asymptomatic patients who have only mildly elevated serum calcium concentrations and normal kidney function, surgery may be delayed and the patient monitored closely for worsening of hypercalcemia, bone deterioration, renal impairment, or the development of kidney stones.

Surgery is recommended for asymptomatic patients who meet the following criteria: (1) younger than 50 years, (2) unable or unlikely to participate in follow-up care, (3) serum calcium level more than 1 mg/dL (0.25 mmol/L) above normal reference range, (4) urinary calcium level greater than 400 mg/day (10 mmol/day), (5) a 30% or greater decrease in kidney function, or (6) with complaints of primary hyperparathyroidism, including nephrocalcinosis, osteoporosis, or a severe psychoneurologic disorder (AACE/AAES Task Force on Primary Hyperparathyroidism, 2005).

Hydration Therapy

Patients with hyperparathyroidism are at risk for renal calculi. Therefore, a daily fluid intake of 2,000 mL or more is encouraged to help prevent calculus formation. The patient is instructed to report other manifestations of renal calculi, such as abdominal pain and hematuria. Thiazide diuretics are avoided, because they decrease the renal excretion of calcium and further elevate serum calcium levels. Because of the risk of hypercalcemic crisis (see later discussion), the patient is instructed to avoid dehydration and to seek immediate health care if conditions that commonly produce dehydration (e.g., vomiting, diarrhea) occur.

Mobility

The nurse encourages the patient to be mobile. The patient with limited mobility is encouraged to walk. Bones subjected to the normal stress of walking give up less calcium. Bed rest increases calcium excretion and the risk of renal calculi. Oral phosphates lower the serum calcium level in some patients; long-term use is not recommended because of the risk of ectopic calcium phosphate deposition in soft tissues.

Diet and Medications

Nutritional needs are met, but the patient is advised to avoid a diet with restricted or excess calcium. If the patient has a coexisting peptic ulcer, prescribed antacids and protein feedings are necessary. Because anorexia is common, efforts are made to improve the appetite. Prune juice, stool softeners, and physical activity, along with increased fluid intake, help offset constipation, which is common postoperatively.

Nursing Management

The insidious onset and chronic nature of hyperparathyroidism along with its diverse and commonly vague symptoms may result in depression and frustration. The family may have considered the patient's illness to be psychosomatic. An awareness of the course of the disorder and an understanding approach by the nurse may help the patient and family deal with their reactions and feelings.

The nursing management of the patient undergoing parathyroidectomy is essentially the same as that of a patient undergoing thyroidectomy. However, the previously described precautions about airway patency, dehydration, immobility, and diet are particularly important in the patient who is awaiting or recovering from parathyroidectomy. Although not all parathyroid tissue is removed during surgery in an effort to control the calcium–phosphorus balance,

Chart 52-8 HOME CARE CHECKLIST
The Patient With Hyperparathyroidism

At the completion of home care education, the patient or caregiver will be able to:	PATIENT	CAREGIVER
• State present and potential effects of hyperparathyroidism on the body.	✔	✔
• State precipitating factors and interventions for complications.	✔	✔
• State importance of regular follow-up visits with primary provider.	✔	✔
• Describe potential benefits and risks of parathyroidectomy.	✔	✔
• State the purpose, dose, route, schedule, side effects, and precautions of prescribed medications (loop diuretics, phosphate, calcitonin).	✔	✔
• State the need to contact primary provider before taking over-the-counter medication containing calcium.	✔	✔
• State the need to take pain medications on a scheduled basis.	✔	✔
• Describe nonpharmacologic methods of pain management.	✔	✔
• Identify safety hazards and methods of injury prevention.	✔	✔
• Identify areas of activity limitations and impact on lifestyle.	✔	✔
• State the need for increased fluid intake and diet low in calcium and vitamin D.	✔	✔

the nurse closely monitors the patient to detect symptoms of tetany (which may be an early postoperative complication). Most patients quickly regain function of the remaining parathyroid tissue and experience only mild, transient postoperative hypocalcemia. In patients with significant bone disease or bone changes, a more prolonged period of hypocalcemia should be anticipated. The nurse educates the patient and family about the importance of follow-up laboratory testing to ensure return of serum calcium levels to normal (Chart 52-8).

Complications: Hypercalcemic Crisis

Acute hypercalcemic crisis can occur with extreme elevation of serum calcium levels. Serum calcium levels greater than 13 mg/dL (3.25 mmol/L) result in neurologic, cardiovascular, and kidney symptoms that can be life threatening (Fischbach & Dunning, 2009). Rapid rehydration with large volumes of IV isotonic saline fluids to maintain urine output of 100 to 150 mL per hour is combined with administration of calcitonin (Shane & Berenson, 2012). Calcitonin promotes renal excretion of excess calcium and reduces bone resorption. The saline infusion should be stopped and a loop diuretic may be needed if the patient develops edema. Dosage and rates of infusion depend on the patient profile. The patient should be monitored carefully for fluid overload. Loop diuretics are not recommended as initial therapy in the absence of heart failure and kidney insufficiency. Bisphosphonates are added to promote a sustained decrease in serum calcium levels by promoting calcium deposition in bone and reducing the GI absorption of calcium. Cytotoxic agents (e.g., mithramycin), calcitonin, and dialysis may be used in emergency situations to decrease serum calcium levels quickly.

Quality and Safety Nursing Alert

The patient in acute hypercalcemic crisis requires close monitoring for life-threatening complications (e.g., airway obstruction) and prompt treatment to reduce serum calcium levels.

A combination of calcitonin and corticosteroids has been administered in emergencies to reduce the serum calcium level by increasing calcium deposition in bone. Other agents that may be administered to decrease serum calcium levels include bisphosphonates (e.g., etidronate [Didronel], pamidronate [Aredia]) (Shane & Berenson, 2012).

Expert assessment and care are required to minimize complications and reverse the life-threatening hypercalcemia. Medications are administered with care, and attention is given to fluid balance to promote return of normal fluid and electrolyte balance. Supportive measures are necessary for the patient and family. (See Chapters 13 and 15 for further discussion of hypercalcemic crisis.)

Hypoparathyroidism

Hypoparathyroidism is caused by abnormal parathyroid development, destruction of the parathyroid glands (surgical removal or autoimmune response), and vitamin D deficiency. The most common cause is the near-total removal of the thyroid gland. The result is inadequate secretion of parathormone (Goltzman, 2012).

Deficiency of parathormone results in increased blood phosphate (hyperphosphatemia) and decreased blood calcium (hypocalcemia) levels. In the absence of parathormone, there is decreased intestinal absorption of dietary calcium and decreased resorption of calcium from bone and through the renal tubules. Decreased renal excretion of phosphate causes hypophosphaturia, and low serum calcium levels result in hypocalciuria.

Clinical Manifestations

Hypocalcemia causes irritability of the neuromuscular system and contributes to the chief symptom of hypoparathyroidism—tetany. Tetany is a general muscle hypertonia, with tremor and spasmodic or uncoordinated contractions occurring with or without efforts to make voluntary movements. Symptoms of latent tetany are numbness, tingling, and cramps in the extremities, and the patient complains of stiffness in the hands and feet. In overt tetany, the signs include bronchospasm, laryngeal spasm, carpopedal spasm (flexion of the elbows and wrists and extension of the carpophalangeal joints and dorsiflexion of the feet), dysphagia, photophobia, cardiac dysrhythmias, and seizures. Other symptoms include anxiety, irritability, depression, and even delirium. ECG changes and hypotension also may occur.

Assessment and Diagnostic Findings

A positive Chvostek's sign or a positive Trousseau's sign suggests latent tetany. **Chvostek's sign** is positive when a sharp tapping over the facial nerve just in front of the parotid gland and anterior to the ear causes spasm or twitching of the mouth, nose, and eye (see Fig. 13-6A in Chapter 13). **Trousseau's sign** is positive when carpopedal spasm is induced by occluding the blood flow to the arm for 3 minutes with a blood pressure cuff (see Fig. 13-6B in Chapter 13). The diagnosis of hypoparathyroidism often is difficult because of the vague symptoms, such as aches and pains. Therefore, laboratory studies are especially helpful. Tetany develops at very low serum calcium levels. Serum phosphate levels are increased, and x-rays of bone show increased density. Calcification is detected on x-rays of the subcutaneous or paraspinal basal ganglia of the brain.

Medical Management

The goal of therapy is to increase the serum calcium level to 9 to 10 mg/dL (2.2 to 2.5 mmol/L) and to eliminate the symptoms of hypoparathyroidism and hypocalcemia. Management is determined by the underlying cause and patient profile. Treatment may include combinations of Calcitrol, calcium, magnesium, and vitamin D_2 (ergocalciferol) or vitamin D_3 (calcitriol), the latter being preferred. A thiazide diuretic (e.g., hydrochlorothiazide) may be administered to help decrease urinary calcium excretion (Goltzman, 2012; Rosen, 2012). Recombinant parathyroid hormone has been approved for treatment of osteoporosis but not for hypoparathyroidism at this time (Goltzman, 2012; Rosen, 2012).

When hypocalcemia and tetany occur after a thyroidectomy, the immediate treatment is administration of IV calcium gluconate. If this does not decrease neuromuscular irritability and seizure activity immediately, sedative agents such as pentobarbital may be administered.

Because of neuromuscular irritability, the patient with hypocalcemia and tetany requires an environment that is free of noise, drafts, bright lights, or sudden movement. Tracheostomy or mechanical ventilation may become necessary, along with bronchodilating medications, if the patient develops respiratory distress.

Therapy for chronic hypoparathyroidism is determined after serum calcium levels are obtained. A diet high in calcium and low in phosphorus is prescribed. Although milk, milk products, and egg yolk are high in calcium, they are restricted because they also contain high levels of phosphorus. Spinach also is avoided because it contains oxalate, which would form insoluble calcium substances. Oral tablets of calcium salts, such as calcium gluconate, may be used to supplement the diet. Aluminum hydroxide gel or aluminum carbonate (Gelusil, Amphojel) also is administered after meals to bind phosphate and promote its excretion through the GI tract.

Nursing Management

Nursing management of the patient with possible acute hypoparathyroidism includes the following:

- Care of postoperative patients who have undergone thyroidectomy, parathyroidectomy, or radical neck dissection is directed toward detecting early signs of hypocalcemia and anticipating signs of tetany, seizures, and respiratory difficulties.
- Calcium gluconate should be available for emergency IV administration. If the patient requiring administration of calcium gluconate has a cardiac disorder, is subject to dysrhythmias, or is receiving digitalis, the calcium gluconate is administered slowly and cautiously.
- Calcium and digitalis increase systolic contraction and also potentiate each other; this can produce potentially fatal dysrhythmias. Consequently, the cardiac patient requires continuous cardiac monitoring and careful assessment.

An important aspect of nursing care is patient education about medications and diet therapy. The patient needs to know the reason for high calcium and low phosphate intake and the symptoms of hypocalcemia and hypercalcemia; he or she should know to contact the primary provider immediately if these symptoms occur (Chart 52-9).

THE ADRENAL GLANDS

Anatomic and Physiologic Overview

Each person has two adrenal glands, one attached to the upper portion of each kidney (Porth, 2011). Each adrenal gland is, in reality, two endocrine glands with separate, independent functions. The adrenal medulla at the center of the gland secretes catecholamines, and the outer portion of the gland, the adrenal cortex, secretes steroid hormones (Fig. 52-6). The secretion of hormones from the adrenal cortex is regulated by the hypothalamic–pituitary–adrenal axis. The hypothalamus secretes corticotropin-releasing hormone (CRH), which stimulates the pituitary gland to secrete ACTH, which in turn stimulates the adrenal cortex to secrete glucocorticoid hormone (cortisol). Increased levels of the adrenal hormone then inhibit the production or secretion of CRH and ACTH. This system is an example of a negative feedback mechanism.

Adrenal Medulla

The adrenal medulla functions as part of the autonomic nervous system. Stimulation of preganglionic sympathetic nerve fibers, which travel directly to the cells of the adrenal medulla, causes release of the catecholamine hormones epinephrine and norepinephrine. About 90% of the secretion of the human adrenal medulla is epinephrine (also called *adrenaline*). Catecholamines regulate metabolic pathways to promote catabolism of stored fuels to meet caloric needs from

Chart 52-9 HOME CARE CHECKLIST
The Patient With Hypoparathyroidism

At the completion of home care education, the patient or caregiver will be able to:	PATIENT	CAREGIVER
• State present and potential effects of hypoparathyroidism on the body.	✔	✔
• State precipitating factors and interventions for complications (seizure, cardiac dysrhythmias, cardiac arrest).	✔	✔
• State necessary actions for seizure activity.		✔
• State importance of regular follow-up visits with primary provider.	✔	✔
• State purpose, dose, route, schedule, side effects, and precautions of prescribed medications (calcium, phosphate binders).	✔	✔
• State the need to alternate activity and rest periods.	✔	✔
• Identify areas of activity limitations and impact on lifestyle.	✔	✔
• Identify foods high in calcium and vitamin D, low in phosphorus.	✔	✔

endogenous sources. The major effects of epinephrine release are to prepare to meet a challenge (fight-or-flight response). Secretion of epinephrine causes decreased blood flow to tissues that are not needed in emergency situations, such as the GI tract, and increased blood flow to tissues that are important for effective fight or flight, such as cardiac and skeletal muscle. Catecholamines also induce the release of free fatty acids, increase the basal metabolic rate, and elevate the blood glucose level.

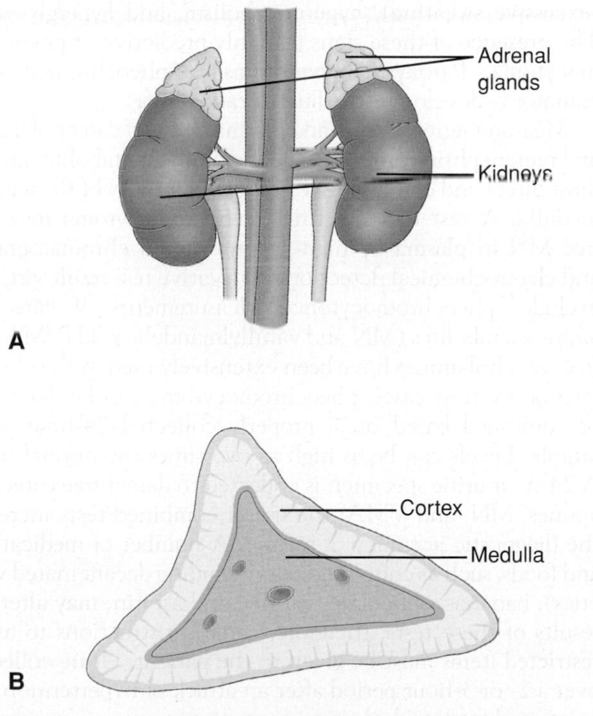
Adrenal glands
Kidneys
A
Cortex
Medulla
B

FIGURE 52-6 • **A.** The adrenal glands sit on top of the kidneys. **B.** Each gland is composed of an outer cortex and an inner medulla. Each area secretes specific hormones. The adrenal medulla secretes catecholamines—epinephrine and norepinephrine; the adrenal cortex secretes glucocorticoids, mineralocorticoids, and sex hormones. Adapted from Porth, C. (2006). *Essentials of pathophysiology: Concepts of altered health states* (2nd ed.). Philadelphia: Lippincott Williams & Wilkins.

Adrenal Cortex

A functioning adrenal cortex is necessary for life; adrenocortical secretions make it possible for the body to adapt to stress of all kinds. The three types of steroid hormones produced by the adrenal cortex are **glucocorticoids**, mainly cortisol; **mineralocorticoids**, mainly aldosterone; and sex hormones, mainly **androgens** (male sex hormones) (Porth & Matfin, 2009). Without the adrenal cortex, severe stress would cause peripheral circulatory failure, circulatory shock, and prostration. Survival in the absence of a functioning adrenal cortex is possible only with nutritional, electrolyte, and fluid replacement and appropriate replacement with exogenous adrenocortical hormones.

Glucocorticoids

The glucocorticoids are so named because they have an important influence on glucose metabolism: Increased cortisol secretion results in elevated blood glucose levels. However, the glucocorticoids have major effects on the metabolism of almost all organs of the body. Glucocorticoids are secreted from the adrenal cortex in response to the release of ACTH from the anterior lobe of the pituitary gland. This system represents an example of negative feedback. The presence of glucocorticoids in the blood inhibits the release of CRH from the hypothalamus and also inhibits ACTH secretion from the pituitary. The resultant decrease in ACTH secretion causes diminished release of glucocorticoids from the adrenal cortex.

Corticosteroids are the classification of drugs that include glucocorticoids. These drugs are administered to inhibit the inflammatory response to tissue injury and to suppress allergic manifestations. Their side effects include the development of diabetes, osteoporosis, peptic ulcer, increased protein breakdown resulting in muscle wasting and poor wound healing, and redistribution of body fat. When large doses of exogenous glucocorticoids are administered, the release of ACTH and endogenous glucocorticoids are inhibited. This can cause the adrenal cortex to atrophy. If exogenous glucocorticoid administration is discontinued suddenly, adrenal insufficiency results because of the inability of the atrophied cortex to respond adequately.

Mineralocorticoids

Mineralocorticoids exert their major effects on electrolyte metabolism. They act principally on the renal tubular and GI epithelium to cause increased sodium ion absorption in exchange for excretion of potassium or hydrogen ions. ACTH only minimally influences aldosterone secretion. It is primarily secreted in response to the presence of angiotensin II in the bloodstream. Angiotensin II is a substance that elevates the blood pressure by constricting arterioles. Its concentration is increased when renin is released from the kidney in response to decreased perfusion pressure. The resultant increased aldosterone levels promote sodium reabsorption by the kidney and the GI tract, which tends to restore blood pressure to normal. The release of aldosterone is also increased by hyperkalemia. Aldosterone is the main hormone for the long-term regulation of sodium balance.

Adrenal Sex Hormones (Androgens)

Androgens, the third major type of steroid hormones produced by the adrenal cortex, exert effects similar to those of male sex hormones. The adrenal gland may also secrete small amounts of some estrogens, or female sex hormones. ACTH controls the secretion of adrenal androgens. When secreted in normal amounts, the adrenal androgens have little effect, but when secreted in excess, they produce masculinization in women, feminization in men, or premature sexual development in children. This is called the **adrenogenital syndrome**.

Pheochromocytoma

Pheochromocytoma is a tumor that is usually benign and originates from the chromaffin cells of the adrenal medulla. This tumor is the cause of high blood pressure in 0.1% of patients with hypertension and is usually fatal if undetected and untreated. Although uncommon, it is one form of hypertension that is usually cured by surgery. In 90% of patients (Porth, 2011), the tumor arises in the medulla; in the remaining patients, it occurs in the extra-adrenal chromaffin tissue located in or near the aorta, ovaries, spleen, or other organs. Pheochromocytoma may occur at any age, but its peak incidence is between 40 and 50 years of age and affects men and women equally (Young & Kaplan, 2012). Ten percent of the tumors are bilateral, and 10% are malignant. Because of the high incidence of pheochromocytoma in family members of affected people, the patient's family members should be alerted and screened for this tumor. Pheochromocytoma may occur in the familial form as part of multiple endocrine neoplasia type 2; therefore, it should be considered a possibility in patients who have medullary thyroid carcinoma and parathyroid hyperplasia or tumor.

Clinical Manifestations

The nature and severity of symptoms of functioning tumors of the adrenal medulla depend on the relative proportions of epinephrine and norepinephrine secretion. The typical triad of symptoms is headache, diaphoresis, and palpitations in the patient with hypertension. The tumor is discovered incidentally in approximately 10% of patients (Young, Kaplan, & Kebebew, 2012). Hypertension and other cardiovascular disturbances are common. The hypertension may be intermittent or persistent. If the hypertension is sustained, it may be difficult to distinguish from other causes of hypertension. Other symptoms may include tremor, headache, flushing, and anxiety. Hyperglycemia may result from conversion of liver and muscle glycogen to glucose due to epinephrine secretion; insulin may be required to maintain normal blood glucose levels.

The clinical picture in the paroxysmal form of pheochromocytoma is usually characterized by acute, unpredictable attacks lasting seconds or several hours. Symptoms usually begin abruptly and subside slowly. During these attacks, the patient is extremely anxious, tremulous, and weak. The patient may experience headache, vertigo, blurring of vision, tinnitus, air hunger, and dyspnea. Other symptoms include polyuria, nausea, vomiting, diarrhea, abdominal pain, and a feeling of impending doom. Palpitations and tachycardia are common (Porth, 2011). Blood pressures exceeding 250/150 mm Hg have been recorded. Such blood pressure elevations are life threatening and can cause severe complications, such as cardiac dysrhythmias, dissecting aneurysm, stroke, and acute kidney failure. Postural hypotension (decrease in systolic blood pressure, lightheadedness, dizziness on standing) occurs in 70% of patients with untreated pheochromocytoma.

Assessment and Diagnostic Findings

Pheochromocytoma is suspected if signs of sympathetic nervous system overactivity occur in association with marked elevation of blood pressure. These signs can be associated with the "five Hs": *h*ypertension, *h*eadache, *h*yperhidrosis (excessive sweating), *h*ypermetabolism, and *h*yperglycemia. The presence of these signs is highly predictive of pheochromocytoma. Paroxysmal symptoms of pheochromocytoma commonly develop in the fifth decade of life.

Measurements of urine and plasma levels of catecholamines and metanephrine (MN), a catecholamine metabolite, are the most direct and conclusive tests for overactivity of the adrenal medulla. A test for detecting pheochromocytoma measures free MN in plasma by high-pressure liquid chromatography and electrochemical detection. A negative test result virtually excludes pheochromocytoma. Measurements of catecholamine metabolites (MN and vanillylmandelic acid [VMA]) or free catecholamines have been extensively used in the clinical setting. In most cases, pheochromocytoma can be diagnosed or confirmed based on a properly collected 24-hour urine sample. Levels can be as high as two times the normal limit. A 24-hour urine specimen is collected to detect free catecholamines, MN, and VMA; the use of combined tests increases the diagnostic accuracy of testing. A number of medications and foods, such as coffee and tea (including decaffeinated varieties), bananas, chocolate, vanilla, and aspirin, may alter the results of these tests; therefore, careful instructions to avoid restricted items must be given to the patient. Urine collected over a 2- or 3-hour period after an attack of hypertension can be assayed for catecholamine content.

The total plasma catecholamine (epinephrine and norepinephrine) concentration is measured with the patient supine and at rest for 30 minutes. To prevent elevation of catecholamine levels resulting from the stress of venipuncture, a butterfly needle, scalp vein needle, or venous catheter may be inserted 30 minutes before the blood specimen is obtained.

Factors that may elevate catecholamine concentrations must be controlled to obtain valid results; these factors include consumption of coffee or tea (including decaffeinated varieties), the use of tobacco, emotional and physical stress, and the use of many prescription and over-the-counter medications (e.g., amphetamines, nose drops or sprays, decongestant agents, bronchodilators).

Normal plasma values of epinephrine are 100 pg/mL (590 pmol/L); normal values of norepinephrine are generally less than 100 to 550 pg/mL (590 to 3,240 pmol/L). Values of epinephrine greater than 400 pg/mL (2,180 pmol/L) or norepinephrine values greater than 2,000 pg/mL (11,800 pmol/L) are considered diagnostic of pheochromocytoma. Values that fall between normal levels and those diagnostic of pheochromocytoma indicate the need for further testing.

A clonidine suppression test may be performed if the results of plasma and urine tests of catecholamines are inconclusive. Clonidine (Catapres) is a centrally acting antiadrenergic medication that suppresses the release of neurogenically mediated catecholamines. The suppression test is based on the principle that catecholamine levels are normally increased through the activity of the sympathetic nervous system. In pheochromocytoma, increased catecholamine levels result from the diffusion of excess catecholamines into the circulation, bypassing normal storage and release mechanisms. Therefore, in patients with pheochromocytoma, clonidine does not suppress the release of catecholamines (Young & Kaplan, 2012).

Imaging studies, such as CT, MRI, and ultrasonography, may also be carried out to localize the pheochromocytoma and to determine whether more than one tumor is present. The use of ^{131}I-metaiodobenzylguanidine (MIBG) scintigraphy may be required to determine the location of the pheochromocytoma and to detect metastatic sites outside the adrenal gland. MIBG is a specific isotope for catecholamine-producing tissue. It has been helpful in identifying tumors not detected by other tests or procedures. MIBG scintigraphy is a noninvasive, safe procedure that has increased the accuracy of diagnosis of adrenal tumors (Fischbach & Dunning, 2009).

Other diagnostic studies may focus on evaluating the function of other endocrine glands because of the association of pheochromocytoma in some patients with other endocrine tumors.

Medical Management

During an episode or attack of hypertension, tachycardia, anxiety, and the other symptoms of pheochromocytoma, bed rest with the head of the bed elevated is prescribed to promote an orthostatic decrease in blood pressure.

Pharmacologic Therapy

The patient may be treated preoperatively on an inpatient or outpatient basis. Regardless of the setting, monitoring of blood pressure and cardiac function is essential. The goals are to control hypertension before and during surgery and volume expansion. Although there are no randomized controlled studies that have compared the efficacy of different treatments, the use of an alpha-adrenergic blocker, sometimes in tandem with a beta-adrenergic blocker, has led to positive outcomes. In some instances, calcium channel blockers and metyrosine have been also been used successfully.

Preoperatively, the patient may begin treatment with a low dose of an alpha-adrenergic blocker (phenoxybenzamine [Dibenzyline]) 10 to 14 days or longer prior to surgery (Young, Kaplan, & Kebebew, 2012). The patient should be informed about the potential for adverse effects of these medications, which include orthostasis, nasal stuffiness, increased fatigue, and retrograde ejaculation in men. The dosage is increased every 2 to 3 days as needed to control blood pressure, with the final dose usually ranging between 20 to 100 mg daily (Young, Kaplan, & Kebebew, 2012). Some practitioners may prefer the use of a selective alpha$_1$-adrenergic blocker with fewer side effects, such as prazosin (Minipress), terazosin (Hytrin), or doxazosin (Cardura), when long-term therapy is indicated. Unless contraindicated, patients are encouraged to begin a high-sodium diet on the second or third day after the introduction of alpha-adrenergic blockade.

After administration of the alpha-blockers, the blood pressure should be monitored closely in both inpatient and outpatient settings. In an outpatient setting, the blood pressure should be taken twice daily in a sitting and standing position. The targets are 120/80 mm Hg (seated) with a standing systolic pressure greater than 90 mm Hg. Age and comorbid disease should be taken into consideration when establishing and evaluating targets. Cautious low-dose beta-adrenergic blockade (e.g., propranolol [Inderal]) is only initiated once adequate alpha-adrenergic control has been achieved (Young, Kaplan, & Kebebew, 2012). The dosage is adjusted according to the patient response and ability to tolerate the drug. The dosage may be initiated in daily divided doses and on day 2 adjusted to a single long-acting dose. On subsequent days, the dose may be increased to control tachycardia as needed with a target heart rate of 60 to 80 bpm.

Calcium channel blockers such as nifedipine (Procardia) are sometimes used as an alternative or supplement to preoperative alpha- and beta-blockers, when blood pressure control is inadequate or the patient is unable to tolerate the side effects. Nifedipine and nicardipine (Cardene) may be used safely without causing undue hypotension. For episodes of severe hypertension, nifedipine is a fast and effective treatment, because the capsules can be pierced and chewed. The patient needs to be well hydrated before, during, and after surgery to prevent hypotension. Additional medications that may be used preoperatively include catecholamine synthesis inhibitors, such as alpha-methyl-*p*-tyrosine (metyrosine [Demser]). These are occasionally used if adrenergic blocking agents (i.e., alpha- and beta-blockers) are not effective. Long-term use of metyrosine may result in many adverse effects, including sedation, depression, diarrhea, anxiety, nightmares, dysuria, impotence, elevated aspartate aminotransferase, anemia, thrombocytopenia, crystalluria, galactorrhea (breast discharge), and extrapyramidal signs (e.g., drooling, speech impairment, tremors).

Surgical Management

The definitive treatment of pheochromocytoma is surgical removal of the tumor, usually with **adrenalectomy** (removal of one or both adrenal glands). Surgery may be performed using a laparoscopic approach or an open operation. The laparoscopic approach is the preferred method for patients with solitary intra-adrenal pheochromocytomas less than 8 cm in diameter without malignancy (Young, Kaplan, & Kebebew, 2012). Bilateral adrenalectomy may be necessary if tumors are present in both adrenal glands. Patient preparation includes control of blood pressure and blood volumes; usually, this is carried out over 10 to 14 days, as described previously. A calcium channel blocker (nicardipine [Cardene]) can be given intraoperatively exclusively or in combination with the alpha- and beta-blockers to control blood pressure.

A hypertensive crisis, however, can still arise as a result of manipulation of the tumor during surgical excision, causing a release of stored epinephrine and norepinephrine, with marked increases in blood pressure and changes in heart rate. In response to this crisis, sodium nitroprusside (Nitropress), phentolamine (OraVerse), or nicardipine can be administered. Lidocaine (Xylocaine) or esmolol (Brevibloc) is used to control cardiac arrhythmias (Young, Kaplan, & Kebebew, 2012). Exploration of other possible tumor sites is frequently undertaken to ensure removal of all tumor tissue. As a result, the patient is subject to the stress and effects of a long surgical procedure, which may increase the risk of hypertension postoperatively.

Corticosteroid replacement is required if bilateral adrenalectomy has been necessary. Corticosteroids may also be required for the first few days or weeks after removal of a single adrenal gland. IV administration of corticosteroids (methylprednisolone [Solu-Medrol]) may begin on the evening before surgery and continue during the early postoperative period to prevent adrenal insufficiency. Oral preparations of corticosteroids (prednisone) are prescribed after the acute stress of surgery diminishes.

Hypotension and hypoglycemia may occur in the postoperative period because of the sudden withdrawal of excessive amounts of catecholamines. Therefore, careful attention is directed toward monitoring and treating these changes. Hypertension may continue if not all pheochromocytoma tissue was removed, if pheochromocytoma recurs, or if the blood vessels were damaged by severe and prolonged hypertension. Several days after surgery, urine and plasma levels of catecholamines and their metabolites are measured to determine whether the surgery was successful.

Nursing Management

The patient who has undergone surgery to treat pheochromocytoma has experienced a stressful preoperative and postoperative course and may remain fearful of repeated attacks. Although it is usually expected that all pheochromocytoma tissue has been removed, there is a possibility that other sites were undetected and that attacks may recur. The patient is monitored until stable with special attention given to ECG changes, arterial pressures, fluid and electrolyte balance, and blood glucose levels. IV access will be required for administration of fluids and medications.

Promoting Home and Community-Based Care

Educating Patients About Self-Care

During the pre- and postoperative phases of care, the nurse educates the patient about the importance of follow-up monitoring to ensure that pheochromocytoma does not recur undetected. After adrenalectomy, the use of corticosteroids may be needed. Therefore, the nurse educates the patient about their purpose, the medication schedule, and the risks of skipping doses or stopping their administration abruptly.

The nurse educates the patient and family about how to measure the patient's blood pressure and when to notify the primary provider about changes in blood pressure. In addition, the nurse provides verbal and written instructions about the procedure for collecting 24-hour urine specimens to monitor urine catecholamine levels.

Continuing Care

A follow-up visit from a home care nurse may be indicated to assess the patient's postoperative recovery, surgical incision, knowledge regarding medication, and adherence to the medication schedule. This may help reinforce previous education about management and monitoring. The home care nurse also obtains blood pressure measurements and assists the patient in preventing or dealing with problems that may result from long-term use of corticosteroids.

Because of the risk of recurrence of hypertension, periodic checkups are required, especially in young patients and in those whose families have a history of pheochromocytoma. The patient is scheduled for periodic follow-up appointments to observe for return of normal blood pressure and plasma and urine levels of catecholamines.

Adrenocortical Insufficiency (Addison's Disease)

Addison's disease, or adrenocortical insufficiency, occurs when adrenal cortex function is inadequate to meet the patient's need for cortical hormones. Autoimmune or idiopathic atrophy of the adrenal glands is responsible for the vast majority of cases. Other causes include surgical removal of both adrenal glands and infection of the adrenal glands. Tuberculosis and histoplasmosis are the most common infections that destroy adrenal gland tissue. Although autoimmune destruction has replaced tuberculosis as the principal cause of Addison's disease, tuberculosis should be considered in the diagnostic workup because of its increasing incidence. Inadequate secretion of ACTH from the pituitary gland also results in adrenal insufficiency because of decreased stimulation of the adrenal cortex.

Therapeutic use of corticosteroids is the most common cause of adrenocortical insufficiency (Porth, 2011). Symptoms of adrenocortical insufficiency may result from the sudden cessation of exogenous adrenocortical hormonal therapy, which suppresses the body's normal response to stress and interferes with normal feedback mechanisms. Treatment with daily administration of corticosteroids for 2 to 4 weeks may suppress function of the adrenal cortex; therefore, adrenal insufficiency should be considered in any patient who has been treated with corticosteroids.

Clinical Manifestations

Addison's disease is characterized by muscle weakness; anorexia; GI symptoms; fatigue; emaciation; dark pigmentation of the mucous membranes and the skin, especially of the knuckles, knees, and elbows; hypotension; and low blood glucose, low serum sodium, and high serum potassium levels. Depression, emotional lability, apathy, and confusion are present in 20% to 40% of patients (Nieman, 2012a). In severe cases, the disturbance of sodium and potassium metabolism may be marked by depletion of sodium and water and severe, chronic dehydration.

With disease progression, **addisonian crisis** develops. This condition is characterized by hypotension, cyanosis, fever, nausea, vomiting, and classic signs of shock. In addition, the patient may have pallor; complain of headache, abdominal pain, and diarrhea; and may show signs of confusion and restlessness. Even slight overexertion, exposure to cold, acute infection, or a decrease in salt intake may lead to circulatory collapse, shock, and death, if untreated. The stress of surgery or dehydration resulting from preparation for diagnostic tests or surgery may precipitate an addisonian or hypotensive crisis.

Assessment and Diagnostic Findings

Although the clinical manifestations presented appear specific, the onset of Addison's disease usually occurs with nonspecific symptoms. The diagnosis is confirmed by laboratory test results. Combined measurements of early-morning serum cortisol and plasma ACTH are performed to differentiate primary adrenal insufficiency from secondary adrenal insufficiency and from normal adrenal function. Patients with primary insufficiency have a greatly increased plasma ACTH level and a serum cortisol concentration lower than the normal range or in the low-normal range (Nieman, 2012b). Other laboratory findings include decreased levels of blood glucose (hypoglycemia) and sodium (hyponatremia), an increased serum potassium concentration (hyperkalemia), and an increased white blood cell count (leukocytosis).

Medical Management

Immediate treatment is directed toward combating circulatory shock: restoring blood circulation, administering fluids and corticosteroids, monitoring vital signs, and placing the patient in a recumbent position with the legs elevated. Hydrocortisone (Solu-Cortef) is administered by IV, followed by 5% dextrose in normal saline. Vasopressors may be required if hypotension persists.

Antibiotics may be administered if infection has precipitated adrenal crisis in a patient with chronic adrenal insufficiency. In addition, the patient is assessed closely to identify other factors, stressors, or illnesses that led to the acute episode.

Oral intake may be initiated as soon as tolerated. IV fluids are gradually decreased after oral fluid intake is adequate to prevent hypovolemia. If the adrenal gland does not regain function, the patient needs lifelong replacement of corticosteroids and mineralocorticoids to prevent recurrence of adrenal insufficiency (Porth & Matfin, 2009). During stressful procedures or significant illnesses, additional supplementary therapy with glucocorticoids is required to prevent addisonian crisis. In addition, the patient may need to supplement dietary intake with salt during GI losses of fluids through vomiting and diarrhea.

Nursing Management

Assessing the Patient

The health history and examination focus on the presence of symptoms of fluid imbalance and the patient's level of stress. The nurse should monitor the blood pressure and pulse rate as the patient moves from a lying, sitting, and standing position to assess for inadequate fluid volume. A decrease in systolic pressure (20 mm Hg or more) may indicate depletion of fluid volume, especially if accompanied by symptoms. The skin should be assessed for changes in color and turgor, which could indicate chronic adrenal insufficiency and hypovolemia. The patient is assessed for change in weight, muscle weakness, fatigue, and any illness or stress that may have precipitated the acute crisis.

Monitoring and Managing Addisonian Crisis

The patient at risk is monitored for signs and symptoms indicative of addisonian crisis, which can include shock; hypotension; rapid, weak pulse; rapid respiratory rate; pallor; and extreme weakness (see Chapter 14). Physical and psychological stressors such as cold exposure, overexertion, infection, and emotional distress should be avoided.

The patient with addisonian crisis requires immediate treatment with IV administration of fluid, glucose, and electrolytes, especially sodium; replacement of missing steroid hormones; and vasopressors. The nurse anticipates and meets the patient's needs to promote return to a precrisis state.

Restoring Fluid Balance

The nurse encourages the patient to consume foods and fluids that assist in restoring and maintaining fluid and electrolyte balance. Along with the dietitian, the nurse helps the patient select foods high in sodium during GI disturbances and in very hot weather.

The nurse educates the patient and family to administer hormone replacement as prescribed and to modify the dosage during illness and other stressful situations. Written and verbal instructions are provided about the administration of corticosteroids (hydrocortisone, cortisone, and prednisone) and mineralocorticoids (fludrocortisone [Florinef]) as prescribed.

Improving Activity Tolerance

Until the patient's condition is stabilized, the nurse takes precautions to avoid unnecessary activity and stress that could precipitate another hypotensive episode. Efforts are made to detect signs of infection or the presence of other stressors. Explaining the rationale for minimizing stress during the acute crisis assists the patient to increase activity gradually.

Promoting Home and Community-Based Care

Educating Patients About Self-Care

Because of the need for lifelong replacement of adrenal cortex hormones to prevent addisonian crises, the patient and family members receive explicit education about the rationale for replacement therapy and proper dosage. In addition, the patient, family, and caregivers are educated about the signs of excessive or insufficient hormone replacement. The patient should have an emergency kit in the form of corticosteroid in prefilled syringes, 100-mg vials of hydrocortisone [Solu-Cortef], or 4 mg of dexamethasone and 0.9% sterile saline to reconstitute the corticosteroid and syringes (Nieman, 2012c). Specific verbal and written instructions about how and when to use the injection are also provided to the patient and family or caregivers. The entire dose of 100-mg hydrocortisone or 4 mg of dexamethasone should be administered and medical

Chart 52-10 HOME CARE CHECKLIST
The Patient With Adrenal Insufficiency (Addison's Disease)

At the completion of home care education, the patient or caregiver will be able to:	PATIENT	CAREGIVER
• State present and potential effects of adrenal insufficiency on the body.	✔	✔
• State warning signs of adrenal crisis and the need for emergency care.	✔	✔
• Explain components of an emergency kit and indications for their use; demonstrate how to use them.	✔	✔
• State strategies for dealing with stress and avoiding adrenal crisis.	✔	✔
• State the purpose, dose, route, schedule, side effects, and precautions of prescribed medications (corticosteroid replacement).	✔	✔
• State that compliance with medical regimen is lifelong.	✔	✔
• State importance of regular follow-up visits with primary provider.	✔	✔
• Recognize the need for dosage adjustment during times of stress.	✔	✔
• State the need to wear medical alert identification, and carry medical information card.	✔	✔
• State the need to notify health care providers about disease before treatment or procedure.	✔	✔
• State the need to avoid strenuous activity in hot, humid weather.	✔	✔
• State the need for increased fluid intake and salt with excessive perspiration.	✔	✔
• State the need for high-carbohydrate, high-protein diet with adequate sodium intake.	✔	✔
• Identify needed activity limitations and impact on lifestyle.	✔	✔

attention sought immediately after administering the drug. Chart 52-10 summarizes education for patients with Addison's disease and their caregivers.

Continuing Care

Although most patients can return to their job and family responsibilities soon after hospital discharge, others cannot do so because of concurrent illnesses or incomplete recovery from the episode of adrenal insufficiency. In these circumstances, a referral for home care enables the home care nurse to assess the patient's recovery, monitor hormone replacement, and evaluate stress in the home. The nurse assesses the patient's and family's knowledge about medication therapy and dietary modifications and provides education as needed. The home care nurse educates the patient and family on the importance of keeping follow-up visits with the health care provider and participating in health promotion activities and health screening.

Cushing Syndrome

Cushing syndrome results from excessive, rather than deficient, adrenocortical activity (Porth, 2011). Cushing syndrome is commonly caused by the use of corticosteroid medications and is infrequently the result of excessive corticosteroid production secondary to hyperplasia of the adrenal cortex. However, overproduction of endogenous corticosteroids may be caused by several mechanisms, including a tumor of the pituitary gland that produces ACTH and stimulates the adrenal cortex to increase its hormone secretion despite production of adequate amounts. Primary hyperplasia of the adrenal glands in the absence of a pituitary tumor is less common. Another less common cause of Cushing syndrome is the ectopic production of ACTH by malignancies; bronchogenic carcinoma is the most common type. Regardless of the cause, the normal feedback mechanisms that control the function of the adrenal cortex become ineffective, and the usual diurnal pattern of cortisol is lost. The signs and symptoms of Cushing syndrome are primarily a result of oversecretion of glucocorticoids and androgens, although mineralocorticoid secretion may be affected as well (Porth, 2011).

Concept Mastery Alert

Nurses must recognize that imbalanced adrenocortical hormone secretion characterizes both Addison's disease (hypoproduction) and Cushing syndrome (hyperproduction).

Clinical Manifestations

When overproduction of the adrenocortical hormone occurs, arrest of growth, obesity, and musculoskeletal changes occur along with glucose intolerance. The classic picture of Cushing syndrome in the adult is that of central-type obesity, with a fatty "buffalo hump" in the neck and supraclavicular areas, a heavy trunk, and relatively thin extremities. The skin is thin, fragile, and easily traumatized; ecchymoses (bruises) and striae develop. The patient complains of weakness and lassitude. Sleep is disturbed because of altered diurnal secretion of cortisol.

Excessive protein catabolism occurs, producing muscle wasting and osteoporosis. Kyphosis, backache, and compression fractures of the vertebrae may result. Retention of sodium and water occurs as a result of increased mineralocorticoid activity, producing hypertension and heart failure.

The patient develops a "moon-faced" appearance and may experience increased oiliness of the skin and acne. Hyperglycemia or overt diabetes may develop. The patient may also report weight gain, slow healing of minor cuts, and bruises.

Women between the ages of 20 and 40 years are five times more likely than men to develop Cushing syndrome. In females of all ages, virilization may occur as a result of excess androgens. Virilization is characterized by the appearance of masculine traits and the recession of feminine traits. There is an excessive growth of hair on the face (hirsutism), the breasts atrophy, menses cease, the clitoris enlarges, and the voice deepens. Libido is lost in men and women. Distress and depression are common and are increased by the severity of the physical changes that occur with this syndrome. If Cushing syndrome is a consequence of pituitary tumor, visual disturbances may occur because of pressure of the growing tumor on the optic chiasm. Chart 52-11 summarizes the clinical manifestations of Cushing syndrome.

Assessment and Diagnostic Findings

The three tests used to diagnose Cushing syndrome are serum cortisol, urinary cortisol, and low-dose dexamethasone (Decadron) suppression tests (Nieman, 2012d). Two of these three tests need to be unequivocally abnormal to diagnose Cushing syndrome. If the results of all three tests are normal, the patient likely does not have Cushing syndrome (but may have a mild case, or the manifestations may be cyclic). For these patients, further testing is not recommended unless symptoms progress. If test results are either slightly abnormal or discordant, further testing is recommended.

Serum cortisol levels are usually higher in the early morning (6 to 8 AM) and lower in the evening (4 to 6 PM). This variation is lost in patients with Cushing syndrome (Fischbach & Dunning, 2009).

A urinary cortisol test requires a 24-hour urine collection. The nurse instructs the patient how to collect and store the specimen. If the results of the urinary cortisol test are three times the upper limit of the normal range and one other test is abnormal, Cushing syndrome can be assumed.

An overnight dexamethasone suppression test is used to diagnosis pituitary and adrenal causes of Cushing syndrome. It can be performed on an outpatient basis. Dexamethasone (1 mg or 8 mg) is administered orally late in the evening or at bedtime, and a plasma cortisol level is obtained at 8 AM the next morning. Suppression of cortisol to less than 5 mg/dL indicates that the hypothalamic–pituitary–adrenal axis is functioning properly (Fischbach & Dunning, 2009). Stress, obesity, depression, and medications such as antiseizure agents, estrogen (during pregnancy or as oral medications), and rifampin (Rifadin) can falsely elevate cortisol levels.

Indicators of Cushing syndrome include an increase in serum sodium and blood glucose levels and a decrease in serum potassium, a reduction in the number of blood eosinophils, and disappearance of lymphoid tissue. Measurements of plasma and urinary cortisol levels are obtained. Several blood samples may be collected to determine whether the normal diurnal variation in plasma levels is present; this variation is frequently absent in adrenal dysfunction. If several blood samples are required, they must be collected at the times specified, and the time of collection must be noted on the requisition slip.

Chart 52-11 Clinical Manifestations of Cushing Syndrome

Ophthalmic

Cataracts
Glaucoma

Cardiovascular

Hypertension
Heart failure

Endocrine/Metabolic

Truncal obesity
Moon face
Buffalo hump
Sodium retention
Hypokalemia
Metabolic alkalosis
Hyperglycemia
Menstrual irregularities
Impotence
Negative nitrogen balance
Altered calcium metabolism
Adrenal suppression

Immune Function

Decreased inflammatory responses
Impaired wound healing
Increased susceptibility to infections

Skeletal

Osteoporosis
Spontaneous fractures
Aseptic necrosis of femur
Vertebral compression fractures

Gastrointestinal

Peptic ulcer
Pancreatitis

Muscular

Myopathy
Muscle weakness

Dermatologic

Thinning of skin
Petechiae
Ecchymoses
Striae
Acne

Psychiatric

Mood alterations
Psychoses

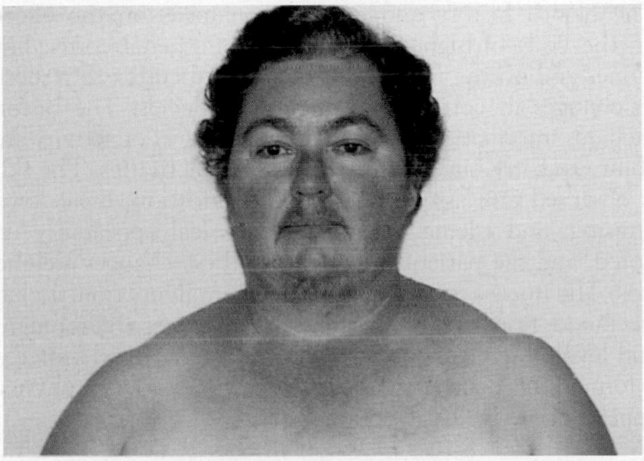

This woman with Cushing syndrome has several classic signs, including facial hair, buffalo hump, and moon face. From Rubin, R., Strayer, D. S., & Rubin, E. (2012). *Rubin's pathology* (6th ed.). Philadelphia: Lippincott Williams & Wilkins.

Medical Management

If Cushing syndrome is caused by pituitary tumors rather than tumors of the adrenal cortex, treatment is directed at the pituitary gland. Surgical removal of the tumor by transsphenoidal hypophysectomy (see Chapter 66) is the treatment of choice and has an 80% success rate. Radiation of the pituitary gland also has been successful, although it may take several months for control of symptoms. Adrenalectomy is the treatment of choice in patients with primary adrenal hypertrophy.

Postoperatively, symptoms of adrenal insufficiency may begin to appear 12 to 48 hours after surgery because of reduction of the high levels of circulating adrenal hormones. Temporary replacement therapy with hydrocortisone may be necessary for several months, until the adrenal glands begin to respond normally to the body's needs. If both adrenal glands have been removed (bilateral adrenalectomy), lifetime replacement of adrenal cortex hormones is necessary.

Adrenal enzyme inhibitors (e.g., metyrapone [Metopirone], aminoglutethimide [Cytadren], mitotane [Lysodren], and ketoconazole [Nizoral]) may be used to reduce hyperadrenalism if the syndrome is caused by ectopic ACTH secretion by a tumor that cannot be eradicated. Close monitoring is necessary, because symptoms of inadequate adrenal function may result and side effects of the medications may occur.

If Cushing syndrome is a result of the administration of corticosteroids, an attempt is made to reduce or taper the medication to the minimum dosage needed to treat the underlying disease process (e.g., autoimmune or allergic disease, rejection of a transplanted organ). Frequently, alternate-day therapy decreases the symptoms of Cushing syndrome and allows recovery of the adrenal glands' responsiveness to ACTH.

NURSING PROCESS

The Patient With Cushing Syndrome

Assessment

The health history and examination focus on the effects on the body of high concentrations of adrenal cortex hormones and on the inability of the adrenal cortex to respond to changes in cortisol and aldosterone levels. The history includes information about the patient's level of activity and ability to carry out routine and self-care activities. The skin is observed and assessed for trauma, infection, breakdown, bruising, and edema. Changes in physical appearance are noted, and the patient's responses to these changes are elicited. The nurse assesses the patient's mental function, including mood, responses to questions, awareness of environment, and level of depression. The family is often a good source of information about gradual changes in the patient's physical appearance as well as emotional status.

Diagnosis

Nursing Diagnoses

Based on the assessment data, major nursing diagnoses include the following:

- Risk for injury related to weakness
- Risk for infection related to altered protein metabolism and inflammatory response
- Self-care deficit related to weakness, fatigue, muscle wasting, and altered sleep patterns
- Impaired skin integrity related to edema, impaired healing, and thin and fragile skin
- Disturbed body image related to altered physical appearance, impaired sexual functioning, and decreased activity level
- Ineffective coping related to mood swings, irritability, and depression

Collaborative Problems/Potential Complications

Potential complications may include the following:

- Addisonian crisis
- Adverse effects of adrenocortical activity

Planning and Goals

The major goals for the patient include decreased risk of injury, decreased risk of infection, increased ability to carry out self-care activities, improved skin integrity, improved body image, improved mental function, and absence of complications.

Nursing Interventions

Decreasing Risk of Injury

Establishing a protective environment helps prevent falls, fractures, and other injuries to bones and soft tissues. The patient who is very weak may require assistance from the nurse in ambulating to avoid falling or bumping into sharp corners of furniture. Foods high in protein, calcium, and vitamin D are recommended to minimize muscle wasting and osteoporosis. Referral to a dietitian may assist the patient in selecting appropriate foods that are also low in sodium and calories.

Decreasing Risk of Infection

The patient should avoid unnecessary exposure to others with infections. The nurse frequently assesses the patient for subtle signs of infection, because the anti-inflammatory effects of corticosteroids may mask the common signs of inflammation and infection.

Preparing the Patient for Surgery

The patient is prepared for adrenalectomy, if indicated, and the postoperative course (see later discussion). If Cushing syndrome is a result of a pituitary tumor, a transsphenoidal hypophysectomy may be performed (see Chapter 66). Diabetes and peptic ulcer are common in patients with Cushing syndrome. Therefore, insulin therapy and medication to prevent or treat peptic ulcer are initiated if needed. Before, during, and after surgery, blood glucose monitoring and assessment of stools for blood are carried out to monitor for these complications. If the patient has other symptoms of Cushing syndrome, these are considered in the preoperative preparation. For example, if the patient has experienced weight gain, special instruction is given about postoperative breathing exercises.

Encouraging Rest and Activity

Although the patient with Cushing syndrome experiences insomnia, weakness, fatigue, and muscle wasting, the nurse should encourage moderate activity to prevent complications of immobility and promote increased self-esteem. It is important to help the patient plan and space rest periods throughout the day and promote a relaxing, quiet environment for rest and sleep.

Promoting Skin Integrity

Meticulous skin care is necessary to avoid traumatizing the patient's fragile skin. The use of adhesive tape is avoided, because it can irritate the skin and tear the fragile tissue when the tape is removed. The nurse frequently assesses the skin and bony prominences and encourages and assists the patient to change positions frequently to prevent skin breakdown.

Improving Body Image

If treated successfully, the major physical changes associated with Cushing syndrome disappear in time. The patient may

Chart 52-12 HOME CARE CHECKLIST
The Patient With Cushing Syndrome

At the completion of home care education, the patient or caregiver will be able to:	PATIENT	CAREGIVER
• State present and potential effects of Cushing syndrome on the body.	✔	✔
• Identify signs and symptoms of excessive and insufficient adrenal hormone.	✔	✔
• State the relationship between adrenal hormones, emotional state, and stress.	✔	✔
• Identify methods for managing labile emotions.	✔	✔
• Describe protective skin care measures and the use of protective devices and practices.	✔	✔
• State the importance of regular follow-up visits with primary provider.	✔	✔
• State the purpose, dose, route, schedule, side effects, and precautions for prescribed medications (adrenocortical inhibitors).	✔	✔
• Identify the need to wear medical alert identification, and carry medical information card.	✔	✔
• State importance of compliance with medical regimen.	✔	✔
• State the need to contact primary provider before taking over-the-counter medications.	✔	✔
• Identify foods high in potassium and low in sodium, calories, and carbohydrates.	✔	✔
• Identify areas of activity limitations and impact on lifestyle.	✔	✔

benefit from discussion of the effect the changes have had on his or her self-concept and relationships with others. Weight gain and edema may be modified by a low-carbohydrate, low-sodium diet, and a high protein intake may reduce some of the other bothersome symptoms.

Improving Coping

Explanations to the patient and family members about the cause of emotional instability are important in helping them cope with the mood swings, irritability, and depression that may occur. Psychotic behavior may occur in a few patients and should be reported. The nurse encourages the patient and family members to verbalize their feelings and concerns.

Monitoring and Managing Potential Complications

Addisonian Crisis. The patient with Cushing syndrome whose symptoms are treated by withdrawal of corticosteroids, by adrenalectomy, or by removal of a pituitary tumor is at risk for adrenal hypofunction and addisonian crisis. If high levels of circulating adrenal hormones have suppressed the function of the adrenal cortex, atrophy of the adrenal cortex is likely. If the circulating hormone level is decreased rapidly because of surgery or abrupt cessation of corticosteroid agents, manifestations of adrenal hypofunction and addisonian crisis may develop. Therefore, the patient with Cushing syndrome should be assessed for signs and symptoms of addisonian crisis as discussed previously. If addisonian crisis occurs, the patient is treated for circulatory collapse and shock (see Chapter 14).

Adverse Effects of Adrenocortical Activity. The nurse assesses fluid and electrolyte status by monitoring laboratory values and daily weights. Because of the increased risk of glucose intolerance and hyperglycemia, blood glucose monitoring is initiated. The nurse reports elevated blood glucose levels to the primary provider so that treatment can be prescribed if indicated.

Promoting Home and Community-Based Care

Educating Patients About Self-Care. The patient, family, and caregivers should be educated that acute adrenal insufficiency and underlying symptoms will recur if medication is stopped abruptly without medical supervision. The nurse stresses the need for dietary modifications to ensure adequate calcium intake without increasing the risks for hypertension, hyperglycemia, and weight gain. The nurse educates the patient and family about how to monitor blood pressure, blood glucose levels, and weight. Patients should be advised to wear a medical alert bracelet and to notify other health care providers (e.g., dentist) about their condition (Chart 52-12).

Continuing Care. The need for follow-up depends on the origin and duration of the disease and its management. The patient who has been treated by adrenalectomy or removal of a pituitary tumor requires close monitoring to ensure that adrenal function has returned to normal and adequacy of circulating adrenal hormones. Home care referral may be indicated to ensure a safe environment that minimizes stress and risk of falls and other side effects. The home care nurse assesses the patient's physical and psychological status and reports changes to the primary provider. The nurse also assesses the patient's understanding of the medication regimen and his or her compliance with the regimen and reinforces previous education about the medications and the importance of taking them as prescribed. The nurse emphasizes the importance of regular medical follow-up, the side effects and toxic effects of medications, and the need to wear medical identification with Addison's and Cushing diseases. In addition, the nurse reminds the patient and family about the importance of health promotion activities and recommended health screening, including bone mineral density testing.

Evaluation

Expected patient outcomes may include:

1. Decreases risk of injury
   a. Is free of fractures or soft tissue injuries
   b. Is free of ecchymotic areas

2. Decreases risk of infection
   a. Experiences no temperature elevation, redness, pain, or other signs of infection or inflammation
   b. Avoids contact with others who have infections
3. Increases participation in self-care activities
   a. Plans activities and exercises to allow alternating periods of rest and activity
   b. Reports improved well-being
   c. Is free of complications of immobility
4. Attains/maintains skin integrity
   a. Has intact skin, without evidence of breakdown or infection
   b. Exhibits decreased edema in extremities and trunk
   c. Changes position frequently and inspects bony prominences daily
5. Achieves improved body image
   a. Verbalizes feelings about changes in appearance, sexual function, and activity level
   b. States that physical changes are a result of excessive corticosteroids
6. Exhibits improved thought processes
7. Exhibits absence of complications
   a. Exhibits normal vital signs and weight and is free of symptoms of addisonian crisis
   b. Identifies signs and symptoms of adrenocortical hypofunction that should be reported and measures to take in case of severe illness and stress
   c. Identifies strategies to minimize complications of Cushing syndrome
   d. Complies with recommendations for follow-up appointments and health screening

Primary Aldosteronism

The principal action of aldosterone is to conserve body sodium. Under the influence of this hormone, the kidneys excrete less sodium and more potassium and hydrogen. Excessive production of aldosterone, which occurs in some patients with functioning tumors of the adrenal gland, causes a distinctive pattern of biochemical changes and a corresponding set of clinical manifestations that are diagnostic of this condition.

Clinical Manifestations

Patients with aldosteronism exhibit a profound decline in the serum levels of potassium (hypokalemia) and hydrogen ions (alkalosis), as demonstrated by an increase in pH and serum bicarbonate concentration. The serum sodium level is normal or elevated, depending on the amount of water reabsorbed with the sodium. Hypertension is the most prominent and almost universal sign of aldosteronism (Stewart, 2009).

Hypokalemia is responsible for the variable muscle weakness, cramping, and fatigue in patients with aldosteronism, as well as the kidneys' inability to acidify or concentrate the urine. Accordingly, the urine volume is excessive, leading to polyuria. Serum, by contrast, becomes abnormally concentrated, contributing to excessive thirst (polydipsia) and arterial hypertension. A secondary increase in blood volume and possible direct effects of aldosterone on nerve receptors, such as the carotid sinus, are other factors that result in hypertension.

Hypokalemic alkalosis may decrease the ionized serum calcium level and predispose the patient to tetany and paresthesias. Chvostek's and Trousseau's signs may be used to assess neuromuscular irritability before overt paresthesia and tetany occur. Glucose intolerance may occur, because hypokalemia interferes with insulin secretion from the pancreas.

Assessment and Diagnostic Findings

In addition to a high or normal serum sodium level and a low serum potassium level, diagnostic studies indicate high serum aldosterone and low serum renin levels. Measurement of the aldosterone excretion rate after salt loading is a useful diagnostic test for primary aldosteronism. The renin–aldosterone stimulation test and bilateral adrenal venous sampling are useful in differentiating the cause of primary aldosteronism. Antihypertensive medication may be discontinued up to 2 weeks before testing.

Medical Management

Treatment of primary aldosteronism usually involves surgical removal of the adrenal tumor through adrenalectomy. Hypokalemia resolves for all patients after surgery, but hypertension may persist. Spironolactone (Aldactone) has been the drug of choice to control hypertension. Eplerenone (Inspra) is a newer, more expensive drug available with few side effects to control hypertension (Young, Kaplan, & Rose, 2012). Serum potassium and creatinine should be monitored frequently during the first 4 to 6 weeks of taking spironolactone. Ongoing monitoring will be determined by the clinical course. The half-life of digoxin may be increased when taken with spironolactone, and its dosage may need to be adjusted. The effectiveness of spironolactone in controlling blood pressure may be affected if taken with salicylates or nonsteroidal anti-inflammatory drugs (Young, Kaplan, & Rose, 2012). Adrenalectomy is performed through an incision in the flank or the abdomen. In general, the postoperative care resembles that for other abdominal surgery. However, the patient is susceptible to fluctuations in adrenocortical hormones and requires administration of corticosteroids, fluids, and other agents to maintain blood pressure and prevent acute complications. If the adrenalectomy is bilateral, replacement of corticosteroids will be lifelong; if one adrenal gland is removed, replacement therapy may be needed temporarily because of suppression of the remaining adrenal gland by high levels of adrenal hormones. A normal serum glucose level is maintained with insulin, appropriate IV fluids, and dietary modifications.

Nursing Management

Nursing management in the postoperative period includes frequent assessment of vital signs to detect early signs and symptoms of adrenal insufficiency and crisis or hemorrhage. Explaining all treatments and procedures, providing comfort measures, and providing rest periods can reduce the patient's stress and anxiety level.

Corticosteroid Therapy

Corticosteroids are used extensively for adrenal insufficiency and are also widely used in suppressing inflammation and

TABLE 52-4 Commonly Used Corticosteroid Preparations

Generic Names	Trade Names
beclomethasone	Beconase, Beclovent, Vanceril, Vancenase, Propaderm
betamethasone	Celestone, Betameth, Betnesol, Betnelan
cortisone	Cortone, Cortate, Cortogen
dexamethasone	Decadron, Dexameth, Deronil, Dexalone, Dexasone, Dexone, Hexadrol
hydrocortisone	Cortisol, Cortef, Hydrocortone, Solu-Cortef
methylprednisolone	Medrol, Solu-Medrol, Meprolone
prednisone	Meticorten, Deltasone, Orasone, Panasol, Novo-prednisone
prednisolone	Meticortelone, Delta-Cortef, Prelone, Predalone
triamcinolone	Aristocort, Kenacort, Kenalog, Cenocort, Azmacort, Aristospan

autoimmune reactions, controlling allergic reactions, and reducing the rejection process in transplantation. Commonly used corticosteroid preparations are listed in Table 52-4. Their anti-inflammatory and antiallergy actions make corticosteroids effective in treating rheumatic or connective tissue diseases, such as rheumatoid arthritis and SLE. They are also frequently used in the treatment of asthma, multiple sclerosis, and other autoimmune disorders.

High doses appear to allow patients to tolerate high degrees of stress. Such antistress action may be caused by the ability of corticosteroids to aid circulating vasopressor substances in keeping the blood pressure elevated; other effects, such as maintenance of the serum glucose level, also may keep blood pressure elevated.

Side Effects

Although the synthetic corticosteroids are safer for some patients because of relative freedom from mineralocorticoid activity, most natural and synthetic corticosteroids produce similar kinds of side effects. The dose required for anti-inflammatory and antiallergy effects also produces metabolic effects, pituitary and adrenal gland suppression, and changes in the function of the central nervous system. Therefore, although corticosteroids are highly effective therapeutically, they may also be very dangerous. Dosages of these medications are frequently altered to allow high concentrations when necessary and then tapered in an attempt to avoid undesirable effects. This requires that patients be observed closely for side effects and that the dose be reduced when high doses are no longer required. Suppression of the adrenal cortex may persist up to 1 year after a course of corticosteroids.

Therapeutic Uses of Corticosteroids

The dosage of corticosteroids is determined by the nature and chronicity of the illness as well as the patient's other medical conditions. Rheumatoid arthritis, bronchial asthma, and multiple sclerosis are chronic disorders that corticosteroids do not cure; however, these medications may be useful when other measures do not provide adequate control of symptoms. In addition, corticosteroids may be used to treat acute exacerbations of these disorders.

In such situations, the adverse effects of corticosteroids are weighed against the patient's current condition. These medications may be used for a period but then are gradually reduced or tapered as the symptoms subside. The nurse plays an important role in providing encouragement and understanding during times when the patient is experiencing (or is apprehensive about experiencing) recurrence of symptoms while taking smaller doses.

Treatment of Acute Conditions

Acute flare-ups and crises are treated with large doses of corticosteroids. Examples include emergency treatment for bronchial obstruction in status asthmaticus and for septic shock from septicemia caused by gram-negative bacteria. Other measures, such as anti-infective agents or medications, are also used with corticosteroids to treat shock and other major symptoms. At times, corticosteroids are continued past the acute flare-up stage to prevent serious complications.

Ophthalmologic Treatment

Outer eye infections can be treated by topical application of corticosteroid eye drops, because the agents do not cause systemic toxicity. However, long-term application can cause an increase in intraocular pressure, which leads to glaucoma in some patients. In addition, prolonged use of corticosteroids can sometimes lead to cataract formation.

Dermatologic Disorders

Topical administration of corticosteroids in the form of creams, ointments, lotions, and aerosols is especially effective in many dermatologic disorders. It may be more effective in some conditions to use occlusive dressings around the affected part to achieve maximum absorption of the medication. Penetration and absorption are also increased if the medication is applied when the skin is hydrated or moist (e.g., immediately after bathing).

Absorption of topical agents varies with body location. For example, absorption is greater through the layers of skin on the scalp, face, and genital area than on the forearm; as a result, the use of topical agents on these sites increases the risk of side effects. The availability of over-the-counter topical corticosteroids increases the risk of side effects in patients who are unaware of their potential risks. Excessive use of these agents, especially on large surface areas of inflamed skin, can lead to decreased therapeutic effects and increased side effects.

Dosage

Attempts have been made to determine the best time to administer pharmacologic doses of steroids. If symptoms have been controlled on a 6- or 8-hour program, a once-daily or every-other-day schedule may be implemented. In keeping with the natural secretion of cortisol, the best time of day for the total corticosteroid dose is in the early morning, between 7 and 8 AM. Large-dose therapy at 8 AM, when the adrenal gland is most active, produces maximal suppression of the gland. A large 8 AM dose is more physiologic because it allows the body to escape effects of the steroids from 4 PM to 6 AM, when serum levels are normally low, hence minimizing cushingoid effects. If symptoms of the disorder being treated are suppressed, alternate-day therapy is helpful in reducing pituitary–adrenal suppression in patients requiring prolonged therapy. Some patients report discomfort associated with symptoms of their primary illness on the second day; therefore, the nurse must explain to patients that this regimen is necessary to minimize side effects and suppression of adrenal function.

Tapering

Corticosteroid dosages are reduced gradually (tapered) to allow normal adrenal function to return and to prevent steroid-induced adrenal insufficiency. Up to 1 year or longer after the use of corticosteroids, the patient is still at risk for adrenal insufficiency in times of stress. For example, if surgery for any reason is necessary, the patient is likely to require IV corticosteroids during and after surgery to reduce the risk of acute adrenal crisis.

Nursing Management

Nursing management of corticosteroid therapy includes many important interventions. Table 52-5 provides an overview of the effects of corticosteroid therapy and the nursing implications.

TABLE 52-5 Corticosteroid Therapy and Implications for Nursing Practice

Side Effects	Nursing Interventions
Cardiovascular Effects	
Hypertension	Monitor for elevated blood pressure.
Thrombophlebitis	Assess for signs and symptoms of deep venous thrombosis: redness, warmth, tenderness, and edema of an extremity.
Thromboembolism	Remind patient to avoid positions and situations that restrict blood flow (e.g., crossing legs, prolonged sitting in same position).
Accelerated atherosclerosis	Encourage foot and leg exercises when recumbent. Encourage low-sodium diet. Encourage limited intake of fat.
Immunologic Effects	
Increased risk of infection and masking of signs of infection	Assess for subtle signs of infection and inflammation. Encourage patient to avoid exposure to others with upper respiratory infection. Monitor patient for fungal infections. Encourage handwashing.
Ophthalmologic Changes	
Glaucoma	Encourage yearly eye examinations.
Corneal lesions	Refer patient to ophthalmologist if changes in visual acuity are detected.
Musculoskeletal Effects	
Muscle wasting	Encourage high protein intake.
Poor wound healing	Encourage high protein intake and vitamin C supplementation.
Osteoporosis with vertebral compression fractures, pathologic fractures of long bones, aseptic necrosis of head of the femur	Encourage diet high in calcium and vitamin D or calcium and vitamin D supplementation if indicated. Take measures to avoid falls and other trauma. Use caution in moving and turning patient. Encourage postmenopausal women on corticosteroids to consider bone mineral density testing and treatment, if indicated. Instruct patient to rise slowly from bed or chair to avoid falling due to postural hypotension.
Metabolic Effects	
Alterations in glucose metabolism	Monitor blood glucose levels at periodic intervals. Instruct patient about medications, diet, and exercise prescribed to control blood glucose level.
Steroid withdrawal syndrome	Report signs of adrenal insufficiency. Administer corticosteroids and mineralocorticoids as prescribed. Instruct patient about importance of taking corticosteroids as prescribed without abruptly stopping therapy. Encourage patient to obtain and wear a medical identification bracelet. Advise patient to notify all health care providers (e.g., dentist) about the need for corticosteroid therapy.
Changes in Appearance	
Moon face Weight gain	Encourage low-calorie, low-sodium diet.
Acne	Assure patient that most changes in appearance are temporary and will disappear if and when corticosteroid therapy is no longer necessary.
Fluid and Electrolyte Imbalances	Monitor intake and output and electrolytes. Administer fluids and electrolytes as prescribed.

Promoting Home and Community-Based Care

Educating Patients About Self-Care

The patient, family, and caregivers should be educated that acute adrenal insufficiency and underlying symptoms will recur if corticosteroid therapy is stopped abruptly without medical supervision. The patient should be instructed to always have an adequate supply of the corticosteroid medication to avoid running out.

Continuing Care

The patient who requires continued corticosteroid therapy is monitored to ensure understanding of the medications and the need for a dosage that treats the underlying disorder while minimizing the side effects. Home care referral may be indicated to ensure a safe environment that minimizes stress, risk of falls, and other side effects. The home care nurse assesses the patient's physical and psychological status and reports changes to the primary provider. The nurse also assesses the patient's understanding of the medication regimen and his or her compliance with the regimen and reinforces previous education about the medications and the importance of taking them as prescribed. The nurse emphasizes the importance of regular medical follow-up, the side effects, and effects of abruptly discontinuing corticosteroids. In addition, the nurse reminds the patient and family about the importance of health promotion activities and recommended health screening, including bone mineral density testing.

Critical Thinking Exercises

1 A 22-year-old man has been recently diagnosed with hyperthyroidism. Identify the nursing plan of care you would use to manage the hyperthyroidism. How would you explain this disorder to the patient? Describe your primary and secondary objectives for educating this patient about hyperthyroidism.

2 ebp A 40-year-old woman is seen in the emergency department for a decreased level of consciousness and possible myxedema coma. Describe the assessment techniques appropriate to evaluate the patient. Review the possible causes, describe the actions you would take and the rationale for each action, and identify the evidence base that supports the actions. What criteria would you use to evaluate the strength of the evidence?

3 pq Identify the priorities, approach, and techniques you would use to perform a comprehensive assessment on a 75-year-old woman with Addison's disease. How will your priorities, approach, and techniques differ if the patient is disoriented? If the patient has a visual impairment or is hard of hearing? If the patient is from a culture with very different values from your own?

4 A 50-year-old man has a new prescription for corticosteroid therapy. Describe the assessment techniques appropriate to evaluate the patient and the use of the medication. What instructions should you provide regarding precautions to take related to corticosteroid therapy?

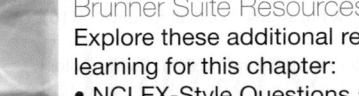

Brunner Suite Resources

Explore these additional resources to enhance learning for this chapter:

- NCLEX-Style Questions and Other Resources on thePoint, http://thePoint.lww.com/Brunner13e
- Study Guide
- PrepU
- Clinical Handbook
- Handbook of Laboratory and Diagnostic Tests

References

Books

American Cancer Society. (2012). *Cancer facts & figures 2012*. Atlanta, GA: Author.

Bader, M. K., & Littlejohns, L. (2010). *AANN core curriculum for neuroscience nursing*, Glenview, IL: American Association of Neuroscience Nurses.

Eliopoulos, C. (2010). *Gerontological nursing* (7th ed.). Philadelphia: Lippincott Williams & Wilkins.

Fischbach, F., & Dunning, M. B. (2009). *A manual of laboratory and diagnostic tests* (8th ed.). Philadelphia: Lippincott Williams & Wilkins.

Hogan-Quigley, B., Palm, M. L., & Bickley, L. S. (2012). *Bates' nursing guide to physical examination and history taking*. Philadelphia: Lippincott Williams & Wilkins.

Karch, A. M. (2012). *Lippincott's nursing drug guide*. Philadelphia: Lippincott Williams & Wilkins.

Papadakis, M. A., McPhee, S., & Rabow, M. W. (Eds.). (2013). *Current medical diagnosis and treatment*. New York: Lange Medical Books/McGraw-Hill.

Porth, C. M. (2011). *Essentials of pathophysiology: Concepts for altered health states* (3rd ed.). Philadelphia: Lippincott Williams & Wilkins.

Porth, C. M., & Matfin, G. (2009). *Pathophysiology: Concepts of altered health states* (8th ed.). Philadelphia: Lippincott Williams & Wilkins.

Journals and Electronic Documents

General

Barzilai, N., & Gabriely, L. (2010). Genetic studies reveal the role of the endocrine and metabolic systems in aging. *Journal of Clinical Endocrinology and Metabolism*, 95(10), 4493–4500.

Greco, K., & Mahon, S. (2012). Genomic health care has arrived but are nurses competent to deliver it? *American Nurse Today*, 7(11), 37–38.

Pituitary Gland

Growth Hormone Guideline Task Force. (2011). Evaluation and treatment of adult growth hormone deficiency: An Endocrine Society clinical practice guideline. *Journal of Clinical Endocrinology and Metabolism*, 96(6), 1587–1609.

John, C. A., & Day, M. W. (2012). Central neurogenic diabetes insipidus, syndrome of inappropriate secretion of antidiuretic hormone, and cerebral salt-wasting syndrome in traumatic brain injury. *Critical Care Nurse*, 32(2), e1–e7.

National Institute of Diabetes and Digestive and Kidney Diseases. (2012). *Acromegaly*. NIH Publication No. 08–3924. Bethesda, MD: U.S. Department of Health and Human Services, National Institutes of Health. Available at: endocrine.niddk.nih.gov/pubs/acro/acro.aspx

Thyroid Gland

American Cancer Society. (2011). *Radioactive iodine treatment for thyroid cancer*. Available at: www.cancer.org/cancer/thyroidcancer/overviewguide/thyroid-cancer-overview-treating-radioactive-iodine

Bahn, R. S., Burch, H. B., Cooper, D. S., et al. (2011). Hyperthyroidism and other causes of thyrotoxicosis: Management guidelines of the American Thyroid Association and American Association of Clinical Endocrinologists. *Thyroid, 21*(6), 593–646.

Brent, G. A. (2008). Graves' disease. *New England Journal of Medicine*, 358(24), 2594–2605.

Chiu, C. G., Yao, R., Chan, S. K., et al. (2012). Hemithyroidectomy is the preferred initial operative approach for an intermediate fine needle aspiration. *Canadian Journal of Surgery*, 55(3), 191–198.

Dominguez, L. J., Bevilacqua, M., DiBella, G., et al. (2008). Diagnosing and managing thyroid disease in the nursing home. *Journal of the American Medical Directors Association*, 9(1), 9–17.

TABLE 53-4 Changes in Urine Color and Possible Causes

Urine Color	Possible Cause
Colorless to pale yellow	Dilute urine due to diuretic agents, alcohol consumption, diabetes insipidus, glycosuria, excess fluid intake, renal disease
Yellow to milky white	Pyuria, infection, vaginal cream
Bright yellow	Multiple vitamin preparations
Pink to red	Hemoglobin breakdown, red blood cells, gross blood, menses, bladder or prostate surgery, beets, blackberries, medications (phenytoin, [Dilantin], rifampin [Rifadin], thioridazine [Mellaril], cascara sagrada, senna products)
Blue, blue green	Dyes, methylene blue, *Pseudomonas* species organisms, medications (amitriptyline HCL [Amitriptyline], triamterine [Dyrenium])
Orange to amber	Concentrated urine due to dehydration, fever, bile, excess bilirubin or carotene, medications (phenazopyridine hydrochloride [Pyridium], nitrofurantoin [Furadantin])
Brown to black	Old red blood cells, urobilinogen, bilirubin, melanin, porphyrin, extremely concentrated urine due to dehydration, medications (cascara sagrada, metronidazole [Flagyl], iron preparations, quinine sulfate [Qualaquin], senna products, methyldopa [Aldomet], nitrofurantoin)

concentrations of less than 30 mg/dL, the test cannot be used for early detection of diabetic nephropathy. Microalbuminuria (excretion of 20 to 200 mg/dL of protein in the urine) is an early sign of diabetic nephropathy. Common benign causes of transient proteinuria are fever, strenuous exercise, and prolonged standing.

Causes of persistent proteinuria include glomerular diseases, malignancies, collagen diseases, diabetes, preeclampsia, hypothyroidism, heart failure, exposure to heavy metals, and the use of medications, such as nonsteroidal anti-inflammatory drugs and angiotensin-converting enzyme inhibitors (Karch, 2012).

Specific Gravity

Specific gravity is an expression of the degree of concentration of the urine that measures the density of a solution compared to the density of water, which is 1.000. Specific gravity is altered by the presence of blood, protein, and casts in the urine. The normal range of urine specific gravity is 1.010 to 1.025 (Porth & Matfin, 2009).

Methods for determination of specific gravity include the following:

- Multiple-test dipstick (most common method), with a specific reagent area for specific gravity
- Urinometer (least accurate method), in which urine is placed in a small cylinder and the urinometer is floated in the urine; a specific gravity reading is obtained at the meniscus level of the urine
- Refractometer, an instrument used in a laboratory setting, which measures differences in the speed of light passing through air and the urine sample

Urine specific gravity depends largely on hydration status. When fluid intake decreases, specific gravity normally increases. With high fluid intake, specific gravity decreases. In patients with kidney disease, urine specific gravity does not vary with fluid intake, and the patient's urine is said to have a fixed specific gravity. Disorders or conditions that cause decreased urine specific gravity include diabetes insipidus, glomerulonephritis, and severe renal damage. Those that can cause increased specific gravity include diabetes, nephritis, and fluid deficit.

Osmolality

Osmolality is the most accurate measurement of the kidney's ability to dilute and concentrate urine. It measures the number of solute particles in a kilogram of water. Serum and urine osmolality are measured simultaneously to assess the body's fluid status. In healthy adults, serum osmolality is 280 to 300 mOsm/kg, and normal urine osmolality is 200 to 800 mOsm/kg. For a 24-hour urine sample, the normal value is 300 to 900 mOsm/kg (Fischbach & Dunning, 2009).

Renal Function Tests

Renal function tests are used to evaluate the severity of kidney disease and to assess the status of the patient's kidney function. These tests also provide information about the effectiveness of the kidney in carrying out its excretory function. Renal function test results may be within normal limits until the GFR is reduced to less than 50% of normal. Renal function can be assessed most accurately if several tests are performed and their results are analyzed together. Common tests of renal function include renal concentration tests, creatinine clearance, and serum creatinine and blood **urea nitrogen** (nitrogenous end product of protein metabolism) levels.

Concept Mastery Alert

Understanding common tests of renal function, the purpose of each test, and the range of normal values is important to ensure accurate assessment of kidney function. This key information is summarized in Table 53-5.

Other tests for evaluating renal function that may be helpful include serum electrolyte levels (see Chapter 13).

Diagnostic Imaging

Kidney, Ureter, and Bladder Studies

An x-ray study of the abdomen or kidneys, ureters, and bladder (KUB) may be performed to delineate the size, shape, and position of the kidneys and to reveal urinary system abnormalities (Fischbach & Dunning, 2009).

General Ultrasonography

Ultrasonography is a noninvasive procedure that uses sound waves passed into the body through a transducer to detect abnormalities of internal tissues and organs. Abnormalities such as fluid accumulation, masses, congenital malformations, changes in organ size, and obstructions can be identified. During the test, the lower abdomen and genitalia may need to be exposed. Ultrasonography requires a full bladder; therefore, fluid intake is encouraged before the procedure.

TABLE 53-5 Renal Function Tests

Test	Purpose	Normal Values
Renal Concentration		
Specific gravity	A measure of the degree of concentration of the urine.	1.010–1.025
Urine osmolality	Concentrating ability is lost early in kidney disease; hence, these test findings may disclose early defects in renal function.	250–900 mO sm/kg/24 h, 50–1,200 mOsm/kg random sample
24-Hour Urine		
Creatinine clearance	Detects and evaluates progression of renal disease. Test measures volume of blood cleared of endogenous creatinine in 1 minute, which provides an approximation of the glomerular filtration rate. Sensitive indicator of renal disease used to follow progression of renal disease.	Measured in mL/min/1.73 m^2 Age (y) / Male / Female <30 / 88–146 / 81–134 30–40 / 82–140 / 75–128 40–50 / 75–133 / 69–122 50–60 / 68–126 / 64–116 60–70 / 61–120 / 58–110 70–80 / 55–113 / 52–105
Serum		
Creatinine level	Measures effectiveness of renal function. Creatinine is the end product of muscle energy metabolism. In normal function, the level of creatinine, which is regulated and excreted by the kidneys, remains fairly constant in the body.	0.6–1.2 mg/dL (50–110 mmol/L)
Urea nitrogen (BUN)	Serves as index of renal function. Urea is the nitrogenous end product of protein metabolism. Test values are affected by protein intake, tissue breakdown, and fluid volume changes.	7–18 mg/dL Patients >60 y: 8–20 mg/dL
BUN-to-creatinine ratio	Evaluates hydration status. An elevated ratio is seen in hypovolemia; a normal ratio with an elevated BUN and creatinine is seen with intrinsic renal disease.	About 10:1

BUN, blood urea nitrogen.
Adapted from Ali, B., & Gray-Vickrey, P. (2011). Limiting the damage from acute kidney injury. *Nursing 2011, 41*(3), 22–31; and Fischbach, F., & Dunning, M. B. (2009). *A manual of laboratory and diagnostic tests* (8th ed.). Philadelphia: Lippincott Williams & Wilkins.

Because of its sensitivity, renal ultrasonography has replaced many other tests as the initial diagnostic procedure (Ali & Gray-Vickrey, 2011).

Bladder Ultrasonography

Bladder ultrasonography is a noninvasive method of measuring urine volume in the bladder. It may be indicated for urinary frequency, inability to void after removal of an indwelling urinary catheter, measurement of postvoiding residual urine volume, inability to void postoperatively, or assessment of the need for catheterization during the initial stages of an intermittent catheterization training program. Portable, battery-operated devices are available for bedside use. The scan head is placed on the patient's abdomen and directed toward the bladder (Fig. 53-8). The device automatically calculates and displays urine volume.

Computed Tomography and Magnetic Resonance Imaging

Computed tomography (CT) and magnetic resonance imaging (MRI) are noninvasive techniques that provide excellent cross-sectional views of the anatomy of the kidney and urinary tract (Ali & Gray-Vickrey, 2011). They are used to evaluate genitourinary masses, nephrolithiasis, chronic renal infections, renal or urinary tract trauma, metastatic disease, and soft tissue abnormalities. Occasionally, an oral or intravenous (IV) radiopaque contrast agent is used in CT scanning to enhance visualization.

Nursing Interventions

Preparation should include educating the patient about relaxation techniques and explaining that he or she will be able to communicate with the staff by means of a microphone located inside the scanner. Many MRI suites provide headphones so that patients can listen to the music of their choice during the procedure. Nursing care guidelines for patient preparation and precautions for any imaging procedure that requires a contrast agent (contrast medium) are explained in Chart 53-5.

Before the patient enters the room where the MRI is to be performed, all metal objects and credit cards (the magnetic field can erase them) are removed. This includes medication patches (e.g., nicotine and nitroglycerin) that have a metal backing, which can cause burns if they are not removed. No metal objects (e.g., oxygen tanks, ventilators, stethoscopes) may be brought into the MRI room. The magnetic field is so strong that any metal-containing items will be pulled toward the magnet, causing severe injury and possible death. A patient history is obtained to determine the presence of any metal objects (e.g., aneurysm clips, orthopedic hardware, pacemakers, artificial heart valves, intrauterine devices). These objects could malfunction, be dislodged, or heat up as they absorb energy. Cochlear implants are inactivated by MRI; therefore, other imaging procedures are considered. A sedative agent may be prescribed, because claustrophobia is a problem for some patients.

Prior to MRI of the urinary system, the patient needs to be informed to avoid alcohol, caffeine-containing beverages, and smoking for at least 2 hours and food for at least 1 hour prior to the scan. Patients should continue taking their usual medication, except for iron supplements, which can interfere with the imaging (Fischbach & Dunning, 2009).

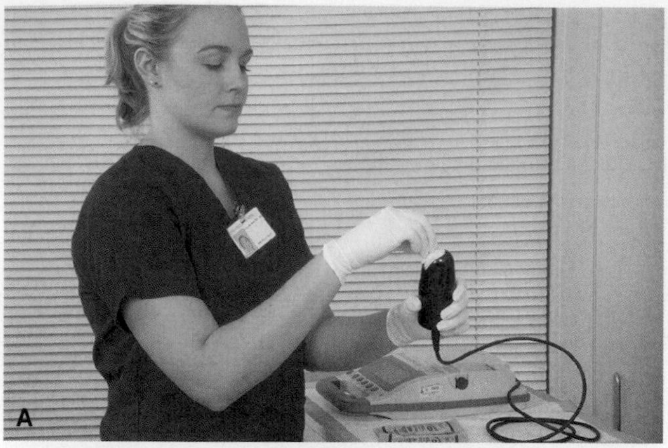
A

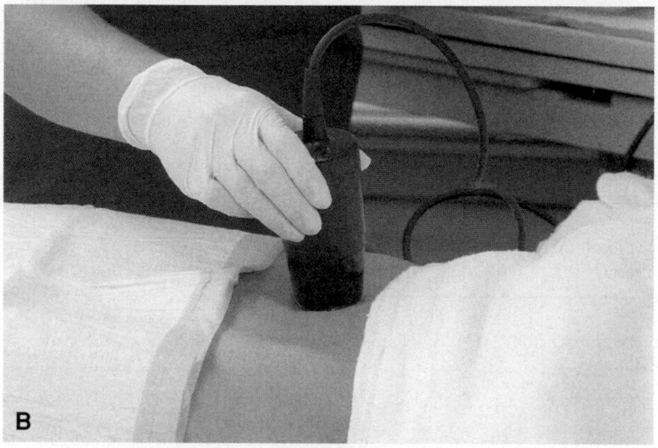
B

FIGURE 53-8 • Bladder ultrasonography. **A.** The nurse puts on gloves and cleans the rounded end of the scan head with an alcohol pad. **B.** After locating the symphysis pubis, the nurse places the scan head about 2.5 cm (1 in) superior to the symphysis pubis. From Springhouse (2008). *Lippincott's visual encyclopedia of clinical skills*. Philadelphia: Lippincott Williams & Wilkins.

Chart 53-5 Patient Care During Urologic Testing With Contrast Agents

For some patients, contrast agents are nephrotoxic and allergenic. Emergency equipment and medications should be available in case of an anaphylactic reaction to the contrast agent. Emergency supplies include epinephrine, corticosteroids, vasopressors, oxygen, and airway and suction equipment.

The following guidelines can help the nurse respond quickly in the event of a problem.

Nursing Actions for Patient Preparation

- Obtain the patient's allergy history with emphasis on allergy to iodine, shellfish, and other seafood, because many contrast agents contain iodine.
- Notify primary provider and radiologist if the patient is allergic or suspected to be allergic to iodine.
- Obtain health history. Contrast agents should be used with caution in older patients and in patients who have diabetes, multiple myeloma, renal insufficiency, or volume depletion.
- Obtain medication history. Nephrotoxic medications include antimicrobial agents, angiotensin-converting enzyme inhibitors, and nonsteroidal anti-inflammatory drugs, as well as others.
- Check kidney function in at-risk patients.
- Inform the patient that he or she may experience a temporary feeling of warmth, flushing of the face, and an unusual flavor (similar to that of seafood) in the mouth when the contrast agent is infused.

Nursing Actions During and Postprocedure

- Monitor patient closely for allergic reaction, and monitor urine output.
- Maintain hydration status.

Adapted from Rank, W. (2013). Preventing contrast media–induced nephrotoxicity. *Nursing 2013*, *43*(4), 48–51.

Nuclear Scans

Nuclear scans require injection of a radioisotope (a technetium 99m–labeled compound or iodine 123 [^{123}I] hippurate) into the circulatory system; the isotope is then monitored as it moves through the blood vessels of the kidneys. A scintillation camera is placed behind the kidney with the patient in a supine, prone, or seated position. Hypersensitivity to the radioisotope is rare. The technetium scan provides information about kidney perfusion. The ^{123}I-hippurate renal scan provides information about kidney function, such as GFR.

Nuclear scans are used to evaluate acute and chronic renal failure, renal masses, and blood flow before and after kidney transplantation. The radioisotope is injected at a specified time to achieve the proper concentration in the kidneys. After the procedure is completed, the patient is encouraged to drink fluids to promote excretion of the radioisotope by the kidneys.

Intravenous Urography

IV urography includes various tests such as excretory urography, intravenous pyelography (IVP), and infusion drip pyelography. A radiopaque contrast agent is administered IV. An IVP shows the kidneys, ureter, and bladder via x-ray imaging as the dye moves through the upper and then the lower urinary system. A nephrotomogram may be carried out as part of the study to visualize different layers of the kidney and the diffuse structures within each layer and to differentiate solid masses or lesions from cysts in the kidneys or urinary tract.

IV urography may be used as the initial assessment of many suspected urologic conditions, especially lesions in the kidneys and ureters. It also provides an approximate estimate of renal function. After the contrast agent (sodium diatrizoate or meglumine diatrizoate) is administered IV, multiple x-rays are obtained to visualize drainage structures in the upper and lower urinary systems.

Infusion drip pyelography requires IV infusion of a large volume of a dilute contrast agent to opacify the renal parenchyma and fill the urinary tract. This examination method is useful when prolonged opacification of the drainage structures is desired so that tomograms (body-section radiography) can be made. Images are obtained at specified intervals after the start of the infusion. These images show the filled and distended collecting system. The patient preparation is the same as for excretory urography, except fluids are not restricted.

Retrograde Pyelography

In retrograde pyelography, catheters are advanced through the ureters into the renal pelvis by means of cystoscopy. A contrast agent is then injected. Retrograde pyelography is usually performed if IV urography provides inadequate visualization of the collecting systems. It may also be used before extracorporeal shock wave lithotripsy and in patients with urologic cancer who need follow-up and have an allergy to IV contrast agents. Possible complications include infection, hematuria, and perforation of the ureter. Retrograde pyelography is used infrequently because of improved techniques in excretory urography.

Cystography

Cystography aids in evaluating vesicoureteral reflux (backflow of urine from the bladder into one or both ureters) and in assessing for bladder injury. A catheter is inserted into the bladder, and a contrast agent is instilled to outline the bladder wall. The contrast agent may leak through a small bladder perforation stemming from bladder injury, but such leakage is usually harmless. Cystography can also be performed with simultaneous pressure recordings inside the bladder.

Voiding Cystourethrography

Voiding cystourethrography uses fluoroscopy to visualize the lower urinary tract and assess urine storage in the bladder. It is commonly used as a diagnostic tool to identify vesicoureteral reflux. A urethral catheter is inserted, and a contrast agent is instilled into the bladder. When the bladder is full and the patient feels the urge to void, the catheter is removed, and the patient voids.

Renal Angiography

A renal angiogram, or renal arteriogram, provides an image of the renal arteries. The femoral (or axillary) artery is pierced with a needle, and a catheter is threaded up through the femoral and iliac arteries into the aorta or renal artery. A contrast agent is injected to opacify the renal arterial supply. Angiography is used to evaluate renal blood flow in suspected renal trauma, to differentiate renal cysts from tumors, and to evaluate hypertension. It is used preoperatively for renal transplantation.

Nursing Interventions

Before the procedure, a laxative may be prescribed to evacuate the colon so that unobstructed x-rays can be obtained. Injection sites (groin for femoral approach or axilla for axillary approach) may be shaved. The peripheral pulse sites (radial, femoral, and dorsalis pedis) are marked for easy access during postprocedural assessment (Rank, 2013). (See Chart 53-5 for considerations for the patient receiving a contrast agent.)

After the procedure, vital signs are monitored until stable. If the axillary artery was the injection site, blood pressure measurements are taken on the opposite arm. The injection site is examined for swelling and hematoma. Peripheral pulses are palpated, and the color and temperature of the involved extremity are noted and compared with those of the uninvolved extremity. Cold compresses may be applied to the injection site to decrease edema and pain. Possible complications include hematoma formation, arterial thrombosis or dissection, false aneurysm formation, and altered renal function.

MAG3 Renogram

This relatively new scan is used to further evaluate kidney function in some centers. The patient is given an injection containing a small amount of radioactive material, which will show how the kidneys are functioning. The patient needs to lie still for about 35 minutes while special cameras take images (Albala, Gomelia, Morey, et al., 2010).

Urologic Endoscopic Procedures

Endourology, or urologic endoscopic procedures, can be performed in one of two ways: using a cystoscope inserted into the urethra, or percutaneously, through a small incision.

The cystoscopic examination is used to directly visualize the urethra and bladder. The cystoscope, which is inserted through the urethra into the bladder, has an optical lens system that provides a magnified, illuminated view of the bladder (Fig. 53-9). The use of a high-intensity light and interchangeable lenses allows excellent visualization and permits still and motion pictures to be taken. The cystoscope is manipulated to allow complete visualization of the urethra and bladder as well as the ureteral orifices and prostatic urethra. Small ureteral catheters can be passed through the cystoscope for assessment of the ureters and the pelvis of each kidney.

The cystoscope also allows the urologist to obtain a urine specimen from each kidney to evaluate its function. Cup forceps can be inserted through the cystoscope for biopsy. Calculi may be removed from the urethra, bladder, and ureter using cystoscopy. If a lower tract cystoscopy is performed, the patient is usually conscious, and the procedure is usually no more uncomfortable than a catheterization. To minimize posttest urethral discomfort, viscous lidocaine is administered several minutes before the study. If the cystoscopy includes examination of the upper tracts, a sedative agent may be administered before the procedure. General anesthesia is usually administered to ensure that there are no involuntary muscle spasms when the scope is being passed through the ureters or kidney.

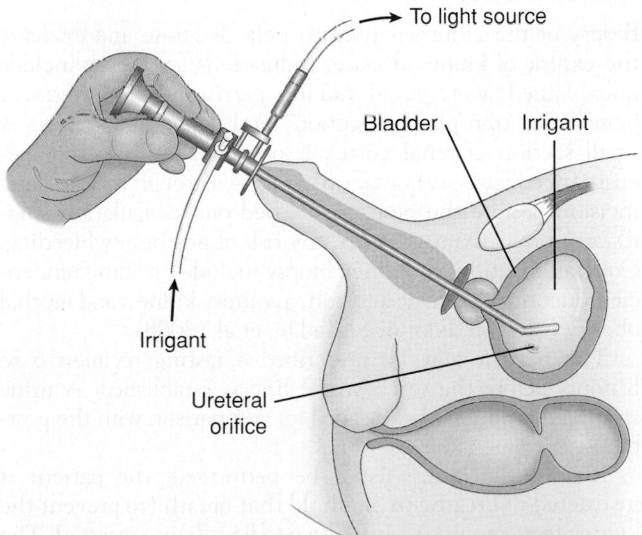

To light source
Bladder
Irrigant
Irrigant
Ureteral orifice

FIGURE 53-9 • Cystoscopic examination. A rigid or semirigid cystoscope is introduced into the bladder. The upper cord is an electric line for the light at the distal end of the cystoscope. The lower tubing leads from a reservoir of sterile irrigant that is used to inflate the bladder.

Nursing Interventions

The nurse describes the procedure to the patient and family to prepare them and to allay their fears. If an upper cystoscopy is to be performed, the patient is usually restricted to nothing by mouth (NPO) for several hours beforehand.

Postprocedural management is directed at relieving any discomfort resulting from the examination. Some burning on voiding, blood-tinged urine, and urinary frequency from trauma to the mucous membranes can be expected. Moist heat to the lower abdomen and warm sitz baths are helpful in relieving pain and relaxing the muscles.

After a cystoscopic examination, the patient with obstructive pathology may experience urine retention if the instruments used during the examination caused edema. The nurse carefully monitors the patient with prostatic hyperplasia for urine retention. Warm sitz baths and antispasmodic medication, such as flavoxate (Urispas), may be prescribed to relieve temporary urine retention caused by poor relaxation of the urinary sphincter; however, intermittent catheterization may be necessary for a few hours after the examination. The nurse monitors the patient for signs and symptoms of urinary tract infection. Because edema of the urethra secondary to local trauma may obstruct urine flow, the patient is also monitored for signs and symptoms of obstruction.

Biopsy

Renal and Ureteral Brush Biopsy

Brush biopsy techniques provide specific information when abnormal x-ray findings of the ureter or renal pelvis raise questions about whether a defect is a tumor, a stone, a blood clot, or an artifact. First, a cystoscopic examination is conducted. Then, a ureteral catheter is introduced, followed by a biopsy brush that is passed through the catheter. The suspected lesion is brushed back and forth to obtain cells and surface tissue fragments for histologic analysis.

Kidney Biopsy

Biopsy of the kidney is used to help diagnose and evaluate the extent of kidney disease. Indications for biopsy include unexplained acute renal failure, persistent proteinuria or hematuria, transplant rejection, and glomerulopathies. A small section of renal cortex is obtained either percutaneously (needle biopsy) or by open biopsy through a small flank incision. Before the biopsy is carried out, coagulation studies are conducted to identify any risk of postbiopsy bleeding. Contraindications to kidney biopsy include bleeding tendencies, uncontrolled hypertension, a solitary kidney, and morbid obesity (Morton, Fontaine, Hudak, et al., 2008).

The patient may be prescribed a fasting regimen 6 to 8 hours before the test. An IV line is established. A urine specimen is obtained and saved for comparison with the postbiopsy specimen.

If a needle biopsy is to be performed, the patient is instructed to breathe in and hold that breath (to prevent the kidney from moving) while the needle is being inserted. The sedated patient is placed in a prone position with a sandbag under the abdomen. The skin at the biopsy site is infiltrated with a local anesthetic agent. The biopsy needle is introduced just inside the renal capsule of the outer quadrant of the kidney. The location of the needle may be confirmed by fluoroscopy or by ultrasound, in which case a special probe is used.

With open biopsy, a small incision is made over the kidney, allowing direct visualization. Preparation for an open biopsy is similar to that for any major abdominal surgery.

Nursing Interventions

After a biopsy procedure, IV fluids may be administered to help clear the kidneys and prevent clot formation. Urine may contain blood (usually clearing in 24 to 48 hours) from oozing at the site. Postoperative renal colic occasionally occurs and responds to analgesic agents.

Critical Thinking Exercises

1 ebp Two days after a cystoscopic examination, your patient complains of lower abdominal pain and burning on urination. Describe the assessment techniques appropriate to evaluate the patient. Review the possible causes, describe the actions you would take and the rationale for each action, and identify the evidence base that supports the actions. What criteria would you use to evaluate the strength of the evidence?

2 pq A 46-year-old mother of two toddlers is admitted to the emergency room for evaluation of urinary dysfunction and is scheduled for a kidney biopsy. The patient is now highly anxious. Describe your priorities for educating this patient about a kidney biopsy. What are your priority interventions for decreasing the patient's anxiety?

3 ebp You make a home visit to a 50 year old who has stress incontinence. Identify assessments and possible interventions that you would use to evaluate and manage the incontinence. Identify the evidence for the assessments and nursing interventions you chose and the strength of that evidence.

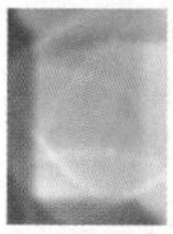

Brunner Suite Resources

Explore these additional resources to enhance learning for this chapter:

- NCLEX-Style Questions and Other Resources on thePoint, http://thePoint.lww.com/Brunner13e
- Study Guide
- PrepU
- Clinical Handbook
- Handbook of Laboratory and Diagnostic Tests

References

*Asterisk indicates nursing research.

Books

Albala, D. M., Gomelia, L. G., Morey, A. F., & Stein, J. P. (2010). *Oxford American handbook of urology*. New York: Oxford University Press.

Eliopoulos, C. (2010). *Gerontological nursing* (7th ed.). Philadelphia: Lippincott Williams & Wilkins.

Fischbach, F., & Dunning, M. B. (2009). *A manual of laboratory and diagnostic tests* (8th ed.). Philadelphia: Lippincott Williams & Wilkins.

Hall, J. (2011). *Guyton and Hall textbook of medical physiology* (12th ed.). St. Louis: Elsevier Saunders.
Karch, A. M. (2012). *Lippincott's nursing drug guide*. Philadelphia: Lippincott Williams & Wilkins.
Morton, P. G., Fontaine, D. K., Hudak, C. M., et al. (2008). *Critical care nursing: A holistic approach*. Philadelphia: Lippincott Williams & Wilkins.
Porth, C. M., & Matfin, G. (2009). *Pathophysiology: Concepts of altered health states* (8th ed.). Philadelphia: Lippincott Williams & Wilkins.
Weber, J., & Kelley, J. (2010). *Health assessment in nursing* (4th ed.). Philadelphia: Lippincott Williams & Wilkins.

Journals and Electronic Documents

Ali, B., & Gray-Vickrey, P. (2011). Limiting the damage from acute kidney injury. *Nursing 2011, 41*(3), 22–31.
Casadio, V. (2009). Accuracy of urine telomerase activity to detect bladder cancer in symptomatic patients. *International Journal of Biological Markers, 24*(4), 253–257.
Collins, M., & Claros, E. (2011). Recognizing the face of dehydration. *Nursing 2011, 41*(8), 26–31.
Crawford, A., & Harris, H. (2011). Balancing act: Sodium (NA+) and potassium (K+). *Nursing 2011, 41*(7), 44–50.
*Elstad, E. A., Taubenberger, S. T., Botelho, E. M., et al. (2010). Beyond incontinence: The stigma of other uninary symptoms. *Journal of Advanced Nursing, 66*(11), 2460–2470.
Gray-Vickrey, P. (2010). Gathering "pearls" of knowledge for assessing older adults. *Nursing 2010, 40*(3), 34–42.
Rank, W. (2013). Preventing contrast media–induced nephrotoxicity. *Nursing 2013, 43*(4), 48–51.
Vanesse, G. J., & Berliner, N. (2010). Anemia in elderly patients: An emerging problem for the 21st century. *Hematology, 2010*, 271–275.
Yaklin, K. M. (2011). Acute kidney injury: An overview of pathophysiology and treatments. *American Journal of Nephrology Nursing, 38*(1), 13–18.

Resources

American Association of Kidney Patients (AAKP), www.aakp.org
American Urological Association, www.auafoundation.org
National Kidney Foundation, www.kidney.org
National Institute of Diabetes and Digestive and Kidney Diseases, National Institutes of Health, www.niddk.nih.gov

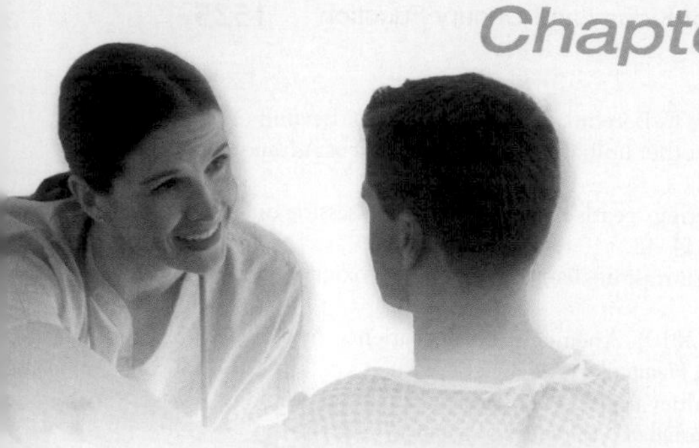

Chapter 54

Management of Patients With Kidney Disorders

Learning Objectives

On completion of this chapter, the learner will be able to:

1. Describe the key factors associated with the development of kidney disorders.
2. Differentiate between the causes of chronic kidney disease and acute kidney injury.
3. Compare and contrast the pathophysiology, clinical manifestations, medical management, and nursing management for patients with kidney disorders.
4. Describe the nursing management of patients with chronic kidney disease and acute kidney injury.
5. Compare and contrast the renal replacement therapies, including hemodialysis, peritoneal dialysis, continuous renal replacement therapies, and kidney transplantation.
6. Describe the nursing management of the patient on dialysis who is hospitalized.
7. Develop a postoperative plan of nursing care for the patient undergoing kidney surgery and transplantation.

Glossary

acute kidney injury (AKI): rapid loss of renal function due to damage to the kidneys; formerly called *acute renal failure*
acute nephritic syndrome: type of renal failure with glomerular inflammation
acute tubular necrosis (ATN): type of acute kidney injury in which there is damage to the kidney tubules
anuria: total urine output less than 50 mL in 24 hours
arteriovenous fistula: type of vascular access for dialysis; created by surgically connecting an artery to a vein
arteriovenous graft: type of surgically created vascular access for dialysis by which a piece of biologic, semibiologic, or synthetic graft material connects the patient's artery to a vein
azotemia: abnormal concentration of nitrogenous wastes in the blood
chronic kidney disease: kidney damage or a decrease in the glomerular filtration rate lasting for 3 or more months
continuous ambulatory peritoneal dialysis (CAPD): method of peritoneal dialysis whereby a patient manually performs exchanges or cycles throughout the day
continuous cyclic peritoneal dialysis (CCPD): method of peritoneal dialysis in which a peritoneal dialysis machine (cycler) automatically performs exchanges, usually while the patient sleeps
continuous renal replacement therapy (CRRT): method used to replace normal kidney function by circulating the patient's blood through a hemofilter and returning it to the patient
dialysate: the electrolyte solution that circulates through the dialyzer in hemodialysis and through the peritoneal membrane in peritoneal dialysis
dialyzer: artificial kidney; contains a semipermeable membrane through which particles of a certain size can pass
diffusion: movement of solutes (waste products) from an area of higher concentration to an area of lower concentration
effluent: term used to describe the drained fluid from a peritoneal dialysis exchange
end-stage kidney disease (ESKD): final stage of chronic kidney disease that results in retention of uremic waste products and the need for renal replacement therapies; also called chronic renal failure
exchange: denotes a complete cycle including fill, dwell, and drain phases of peritoneal dialysis
glomerular filtration rate (GFR): amount of plasma filtered through the glomeruli per unit of time
glomerulonephritis: inflammation of the glomerular capillaries
hemodialysis: procedure during which a patient's blood is circulated through a dialyzer to remove waste products and excess fluid
interstitial nephritis: inflammation within the renal tissue
nephrosclerosis: hardening of the renal arteries
nephrotic syndrome: type of renal failure with increased glomerular permeability and massive proteinuria
nephrotoxic: any substance, medication, or action that destroys kidney tissue
oliguria: urine output less than 0.5 mL/kg/h
osmosis: movement of water through a semipermeable membrane from an area of lower solute concentration to an area of higher solute concentration
peritoneal dialysis: procedure that uses the lining of the patient's peritoneal cavity as the semipermeable membrane for exchange of fluid and solutes
peritonitis: inflammation of the peritoneal membrane (lining of the peritoneal cavity)
polyuria: large amounts of urine
pyelonephritis: inflammation of the renal pelvis
ultrafiltration: process whereby water is removed from the blood by means of a pressure gradient between the patient's blood and the dialysate
uremia: an excess of urea and other nitrogenous wastes in the blood
urinary casts: proteins secreted by damaged kidney tubules

The kidneys and urinary system helps regulate the body's internal environment and is essential for the maintenance of life. Nurses working in any clinical setting may encounter patients with various kidney injuries and diseases, and thus need to be knowledgeable about these disorders. This chapter provides an overview of electrolyte imbalances and systemic manifestations that are common in patients with kidney disorders. The main causes are discussed, together with management strategies to prevent damage and preserve renal function. Chronic kidney disease (CKD) and acute kidney injury (AKI) are discussed, as is the care of patients with other renal conditions requiring dialysis, continuous renal replacement therapies (CRRTs), transplantation, and kidney surgery.

FLUID AND ELECTROLYTE IMBALANCES IN KIDNEY DISORDERS

Patients with kidney disorders commonly experience fluid and electrolyte imbalances and require careful assessment and close monitoring for signs of potential problems. The patient whose fluid intake exceeds the ability of the kidneys to excrete fluid is said to have fluid overload. If fluid intake is inadequate, the patient is said to be volume depleted and may show signs and symptoms of fluid volume deficit. The fluid intake and output (I&O) record, a key monitoring tool, is used to document important fluid parameters, including the amount of fluid taken in (orally or parenterally), the volume of urine excreted, and other fluid losses (diarrhea, vomiting, diaphoresis). Patient weight is also important, and documenting trends in weight is a key assessment strategy essential for determining the daily fluid allowance and indicating signs of fluid volume excess or deficit.

Quality and Safety Nursing Alert

The most accurate indicator of fluid loss or gain in an acutely ill patient is weight. An accurate daily weight must be obtained and recorded. A 1-kg weight gain is equal to 1,000 mL of retained fluid.

Clinical Manifestations

The signs and symptoms of common fluid and electrolyte disturbances that can occur in patients with kidney disorders and their general management strategies are listed in Table 54-1. The

TABLE 54-1 Common Fluid and Electrolyte Disturbances in Kidney Disorders

Disturbance	Manifestations	General Management Strategies
Fluid volume deficit	Acute weight loss ≥5%, decreased skin turgor, dry mucous membranes, oliguria or anuria, increased hematocrit, BUN level increased out of proportion to creatinine level, hypothermia	Fluid challenge, fluid replacement orally or parenterally
Fluid volume excess	Acute weight gain ≥5%, edema, crackles, shortness of breath, decreased BUN, decreased hematocrit, distended neck veins	Fluid and sodium restriction, diuretic agents, dialysis
Sodium deficit	Nausea, malaise, lethargy, headache, abdominal cramps, apprehension, seizures	Diet, normal saline or hypertonic saline solutions
Sodium excess	Dry, sticky mucous membranes, thirst, rough dry tongue, fever, restlessness, weakness, disorientation	Fluids, diuretic agents, dietary restriction
Potassium deficit	Anorexia, abdominal distention, paralytic ileus, muscle weakness, ECG changes, dysrhythmias	Diet, oral or parenteral potassium replacement therapy
Potassium excess	Diarrhea, colic, nausea, irritability, muscle weakness, ECG changes	Dietary restriction, diuretics, IV glucose, insulin and sodium bicarbonate, cation-exchange resin, calcium gluconate, dialysis
Calcium deficit	Abdominal and muscle cramps, stridor, carpopedal spasm, hyperactive reflexes, tetany, positive Chvostek's or Trousseau's sign, tingling of fingers and around mouth, ECG changes	Diet, oral or parenteral calcium salt replacement
Calcium excess	Deep bone pain, flank pain, muscle weakness, depressed deep tendon reflexes, constipation, nausea and vomiting, confusion, impaired memory, polyuria, polydipsia, ECG changes	Fluid replacement, etidronate, pamidronate, mithramycin, calcitonin, glucocorticoids, phosphate salts
Bicarbonate deficit	Headache, confusion, drowsiness, increased respiratory rate and depth, nausea and vomiting, warm flushed skin	Bicarbonate replacement, dialysis
Bicarbonate excess	Depressed respirations, muscle hypertonicity, dizziness, tingling of fingers and toes	Fluid replacement if volume depleted; ensure adequate chloride
Protein deficit	Chronic weight loss, emotional depression, pallor, fatigue, soft flabby muscles	Diet, dietary supplements, hyperalimentation, albumin
Magnesium deficit	Dysphagia, muscle cramps, hyperactive reflexes, tetany, positive Chvostek's or Trousseau's sign, tingling of fingers, dysrhythmias, vertigo	Diet, oral or parenteral magnesium replacement therapy
Magnesium excess	Facial flushing, nausea and vomiting, sensation of warmth, drowsiness, depressed deep tendon reflexes, muscle weakness, respiratory depression, cardiac arrest	Calcium gluconate, mechanical ventilation, dialysis
Phosphorus deficit	Deep bone pain, flank pain, muscle weakness and pain, paresthesia, apprehension, confusion, seizures	Diet, oral or parenteral phosphorus supplementation therapy

BUN, blood urea nitrogen; ECG, electrocardiographic; IV, intravenous.

nurse continually assesses, monitors, and informs appropriate members of the health care team if the patient exhibits any of these signs. Management strategies for fluid and electrolyte disturbances in kidney disease are discussed in greater depth later in this chapter (see also Chapter 13).

Gerontologic Considerations

With aging, the kidney is less able to respond to acute fluid and electrolyte changes. Older adult patients may develop atypical and nonspecific signs and symptoms of altered renal function and fluid and electrolyte imbalances. A fluid balance deficit in older adults can lead to constipation, falls, medication toxicity, urinary tract and respiratory tract infections, delirium, seizures, electrolyte imbalances, hyperthermia, and delayed wound healing (Thomas, 2010). Recognition of acute changes in fluid and electrolytes is further hampered by their association with preexisting disorders and the misconception that they are normal changes of aging.

KIDNEY DISORDERS

Chronic Kidney Disease

Chronic kidney disease is an umbrella term that describes kidney damage or a decrease in the glomerular filtration rate (GFR) lasting for 3 or more months. CKD is associated with decreased quality of life, increased health care expenditures, and premature death (Neyhart, McCoy, Rodegast, et al., 2010). Untreated CKD can result in **end-stage kidney disease (ESKD)**, which is the final stage of renal failure. ESKD results in retention of uremic waste products and the need for renal replacement therapies, dialysis, or kidney transplantation. Risk factors include cardiovascular disease, diabetes, hypertension, and obesity. Recent research reported that 10% of the U.S. population aged 20 years and older has CKD (Centers for Disease Control and Prevention [CDC], 2010).

Diabetes is the primary cause of CKD. More than 35% of the U.S. population aged 20 years and older with diabetes have CKD (CDC, 2010). Diabetes is the leading cause of renal failure in patients starting renal replacement therapy. The second leading cause is hypertension, followed by glomerulonephritis and **pyelonephritis**; polycystic, hereditary, or congenital disorders; and renal cancers (U.S. Renal Data System [USRDS], 2011). More than 20% of the U.S. population aged 20 years and older with hypertension have CKD (CDC, 2010).

Pathophysiology

In the early stages of CKD, there can be significant damage to the kidneys without signs or symptoms. The pathophysiology of CKD is not yet clearly understood, but the damage to the kidneys is thought to be caused by prolonged acute inflammation that is not organ specific and thus has subtle systemic manifestations.

Stages of Chronic Kidney Disease

CKD has been classified into five stages by the National Kidney Foundation (NKF) (Chart 54-1). Stage 5 results when the kidneys cannot remove the body's metabolic wastes or perform their regulatory functions; thus, renal replacement therapies are required to sustain life. Screening and early intervention are important, because not all patients progress to stage 5 CKD. Patients with CKD are at increased risk for cardiovascular disease, which is the leading cause of morbidity and mortality (Neyhart et al., 2010). Treatment of hypertension, anemia, and hyperglycemia and detection of proteinuria all help to slow disease progression and improve patient outcomes (Eskridge, 2010).

Chart 54-1 Stages of Chronic Kidney Disease

Stages are based on the GFR. The normal GFR is 125 mL/min/1.73 m².

Stage 1

GFR ≥90 mL/min/1.73 m²
Kidney damage with normal or increased GFR

Stage 2

GFR = 60–89 mL/min/1.73 m²
Mild decrease in GFR

Stage 3

GFR = 30–59 mL/min/1.73 m²
Moderate decrease in GFR

Stage 4

GFR = 15–29 mL/min/1.73 m²
Severe decrease in GFR

Stage 5

GFR <15 mL/min/1.73 m²
End-stage kidney disease or chronic renal failure

GFR, glomerular filtration rate.
Adapted from Porth, C. M., & Matfin, G. (2009). *Pathophysiology: Concepts of altered health states* (8th ed.). Philadelphia: Lippincott Williams & Wilkins.

Concept Mastery Alert

The ability to identify the five stages of CKD is a vital skill for the nurse. Chart 54-1 lists each of these stages and the related GFR for each.

Clinical Manifestations

Elevated serum creatinine levels indicate underlying kidney disease; as the creatinine level increases, symptoms of CKD begin. Anemia, due to decreased erythropoietin production by the kidney, metabolic acidosis, and abnormalities in calcium and phosphorus herald the development of CKD (Chamney, Pugh-Clarke, Kafkia, et al., 2010). Fluid retention, evidenced by both edema and congestive heart failure, develops. As the disease progresses, abnormalities in electrolytes occur, heart failure worsens, and hypertension becomes more difficult to control.

Assessment and Diagnostic Findings

The **glomerular filtration rate (GFR)** is the amount of plasma filtered through the glomeruli per unit of time. Creatinine clearance is a measure of the amount of creatinine the kidneys are able to clear in a 24-hour period. Normal values differ in men and women. Calculation of GFR, an important assessment parameter in CKD, is discussed in Chapter 53.

Medical Management

The management of patients with CKD includes treatment of the underlying causes. Regular clinical and laboratory assessment is important to keep the blood pressure below 130/80 mm Hg. Medical management also includes early referral for initiation of renal replacement therapies as indicated by the patient's renal status. Prevention of complications is accomplished by controlling cardiovascular risk factors; treating hyperglycemia; managing anemia; smoking cessation, weight loss, and exercise programs as needed; and reduction in salt and alcohol intake.

Gerontologic Considerations

Changes in kidney function with normal aging increase the susceptibility of older patients to kidney dysfunction and renal failure (Miller, 2012). In addition, the incidence of systemic diseases, such as atherosclerosis, hypertension, heart failure, diabetes, and cancer, increases with advancing age, predisposing older adults to kidney disease associated with these disorders. Therefore, acute problems need to be prevented if possible or recognized and treated quickly to avoid kidney damage. Thus, nurses in all settings need to be alert to signs and symptoms of kidney dysfunction in older patients.

Older patients frequently take multiple prescription and over-the-counter medications. Because alterations in renal blood flow, glomerular filtration, and renal clearance increase the risk of medication-associated changes in renal function, precautions are indicated with all medications. When older patients undergo extensive diagnostic tests or when new medications (e.g., diuretic agents) are added, precautions must be taken to prevent dehydration, which can compromise marginal renal function and lead to renal failure (Stilos, 2010).

Nephrosclerosis

Nephrosclerosis (hardening of the renal arteries) is most often due to prolonged hypertension and diabetes. Nephrosclerosis is a major cause of CKD and ESKD secondary to many disorders.

Pathophysiology

There are two forms of nephrosclerosis: malignant (accelerated) and benign. Malignant nephrosclerosis is often associated with significant hypertension (diastolic blood pressure higher than 130 mm Hg). It usually occurs in young adults and twice as often in men compared to women (CDC, 2010). Damage is caused by decreased blood flow to the kidney resulting in patchy necrosis of the renal parenchyma. Over time, fibrosis occurs and glomeruli are destroyed.

The disease process progresses rapidly. Without dialysis, more than half of patients die of **uremia** (an excess of urea and other nitrogenous waste products in the blood) in a few years. Benign nephrosclerosis can be found in older adults, associated with atherosclerosis and hypertension.

Assessment and Diagnostic Findings

Symptoms are rare early in the disease, even though the urine usually contains protein and occasional casts. Renal insufficiency and associated signs and symptoms occur late in the disease.

Medical Management

Treatment of nephrosclerosis is aggressive antihypertensive therapy. An angiotensin-converting enzyme (ACE) inhibitor, alone or in combination with other antihypertensive medications, significantly reduces its incidence. (See Chapter 31 for additional information on hypertension.)

Primary Glomerular Diseases

Diseases that destroy the glomerulus of the kidney are the third most common cause of stage 5 CKD. In these disorders, the glomerular capillaries are primarily involved. Antigen–antibody complexes form in the blood and become trapped in the glomerular capillaries (the filtering portion of the kidney), inducing an inflammatory response. Immunoglobulin G (IgG)—the major immunoglobulin (antibody) found in the blood—can be detected in the glomerular capillary walls. The major clinical manifestations of glomerular injury include proteinuria, hematuria, decreased GFR, decreased excretion of sodium, edema, and hypertension (Chart 54-2).

Acute Nephritic Syndrome

Acute nephritic syndrome is a type of renal failure with glomerular inflammation (Porth & Matfin, 2009). **Glomerulonephritis** is an inflammation of the glomerular capillaries that can occur in acute and chronic forms.

Chart 54-2 Terms Typically Used When Describing Glomerular Disease

Primary: Disease is mainly in glomeruli
Secondary: Glomerular diseases that are the consequence of systemic disease
Idiopathic: Cause is unknown
Acute: Occurs over days or weeks
Chronic: Occurs over months or years
Rapidly progressing: Constant loss of renal function with minimal chance of recovery
Diffuse: Involves all glomeruli
Focal: Involves some glomeruli
Segmental: Involves portions of individual glomeruli
Membranous: Evidence of thickened glomerular capillary walls
Proliferative: Number of glomerular cells involved is increasing

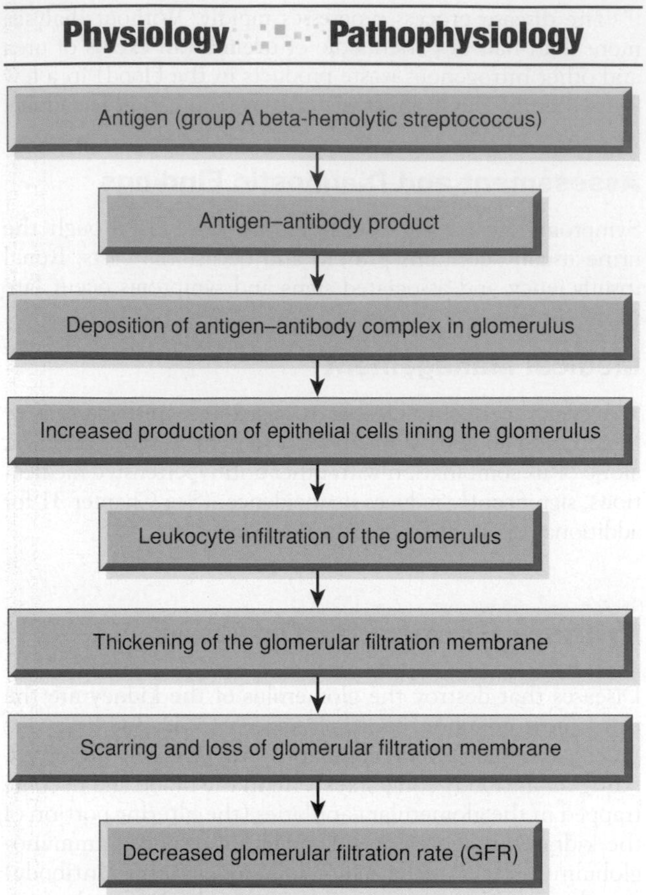
Physiology Pathophysiology
Antigen (group A beta-hemolytic streptococcus)
Antigen–antibody product
Deposition of antigen–antibody complex in glomerulus
Increased production of epithelial cells lining the glomerulus
Leukocyte infiltration of the glomerulus
Thickening of the glomerular filtration membrane
Scarring and loss of glomerular filtration membrane
Decreased glomerular filtration rate (GFR)

FIGURE 54-1 • Sequence of events in acute nephritic syndrome.

Pathophysiology

Primary glomerular diseases include postinfectious glomerulonephritis, rapidly progressive glomerulonephritis, membrane proliferative glomerulonephritis, and membranous glomerulonephritis. Postinfectious causes are group A beta-hemolytic streptococcal infection of the throat that precedes the onset of glomerulonephritis by 2 to 3 weeks (Fig. 54-1). It may also follow impetigo (infection of the skin) and acute viral infections (upper respiratory tract infections, mumps, varicella zoster virus, Epstein-Barr virus, hepatitis B, and human immunodeficiency virus [HIV] infection). In some patients, antigens outside the body (e.g., medications, foreign serum) initiate the process, resulting in antigen–antibody complexes being deposited in the glomeruli. In other patients, the kidney tissue itself serves as the inciting antigen.

Clinical Manifestations

The primary presenting features of an acute glomerular inflammation are hematuria, edema, **azotemia** (an abnormal concentration of nitrogenous wastes in the blood), and proteinuria (excess protein in the urine) (Porth & Matfin, 2009). The hematuria may be microscopic (identifiable only through microscopic examination) or macroscopic (visible to the eye). The urine may appear cola colored because of red blood cells (RBCs) and protein plugs or casts; RBC casts indicate glomerular injury. Glomerulonephritis may be mild and the hematuria discovered incidentally through a routine urinalysis, or the disease may be severe, with AKI and oliguria.

Some degree of edema and hypertension is present in most patients. Marked proteinuria due to the increased permeability of the glomerular membrane may also occur, with associated pitting edema, hypoalbuminemia, hyperlipidemia, and fatty casts in the urine. Blood urea nitrogen (BUN) and serum creatinine levels may increase as urine output decreases. In addition, anemia may be present.

In the more severe form of the disease, patients also complain of headache, malaise, and flank pain. Older patients may experience circulatory overload with dyspnea, engorged neck veins, cardiomegaly, and pulmonary edema. Atypical symptoms include confusion, somnolence, and seizures, which are often confused with the symptoms of a primary neurologic disorder.

Assessment and Diagnostic Findings

In acute nephritic syndrome, the kidneys become large, edematous, and congested. All renal tissues, including the glomeruli, tubules, and blood vessels, are affected to varying degrees. Patients with an immunoglobulin A (IgA) nephropathy have an elevated serum IgA and low to normal complement levels. Electron microscopy and immunofluorescent analysis help identify the nature of the lesion; however, a kidney biopsy may be needed for definitive diagnosis. (See Chapter 53 for discussion of kidney biopsy.)

If the patient improves, the amount of urine increases and the urinary protein and sediment diminish. The percentage of adults who recover is unknown. Some patients develop severe uremia (an excess of urea and other nitrogenous wastes in the blood) within weeks and require dialysis for survival. Others, after a period of apparent recovery, insidiously develop chronic glomerulonephritis.

Complications

Complications of acute glomerulonephritis include hypertensive encephalopathy, heart failure, and pulmonary edema. Hypertensive encephalopathy is a medical emergency, and therapy is directed toward reducing the blood pressure without impairing renal function. This can occur in acute nephritic syndrome or preeclampsia with chronic hypertension of greater than 140/90 mm Hg. Rapidly progressive glomerulonephritis is characterized by a rapid decline in renal function. Without treatment, ESKD develops in a matter of weeks or months. Signs and symptoms are similar to those of acute glomerulonephritis (hematuria and proteinuria), but the course of the disease is more severe and rapid. Crescent-shaped cells accumulate in Bowman's space, disrupting the filtering surface. Plasma exchange (plasmapheresis) and treatment with high-dose corticosteroids and cytotoxic agents have been used to reduce the inflammatory response. Dialysis is initiated in acute glomerulonephritis if signs and symptoms of uremia are severe. The prognosis for patients with acute nephritic syndrome is excellent and rarely causes CKD (Porth & Matfin, 2009).

Medical Management

Management consists primarily of treating symptoms, attempting to preserve kidney function, and treating complications promptly. Treatment may include prescribing corticosteroids, managing hypertension, and controlling proteinuria. Pharmacologic therapy depends on the cause of acute glomerulonephritis. If residual streptococcal infection is suspected, penicillin is the agent of choice; however, other antibiotic agents may be prescribed. Dietary protein is restricted when renal insufficiency and nitrogen retention (elevated BUN) develop. Sodium is restricted when the patient has hypertension, edema, and heart failure.

Nursing Management

Although most patients with acute uncomplicated glomerulonephritis are cared for as outpatients, nursing care is important in every setting.

Providing Care in the Hospital

In a hospital setting, carbohydrates are given liberally to provide energy and reduce the catabolism of protein. I&O is carefully measured and recorded. Fluids are given based on the patient's fluid losses and daily body weight. Insensible fluid loss through the lungs (300 mL) and skin (600 mL) is considered when estimating fluid loss (see Chapter 13). If treatment is effective, diuresis will begin, resulting in decreased edema and blood pressure. Proteinuria and microscopic hematuria may persist for many months; in fact, 20% of patients have some degree of persistent proteinuria or decreased GFR 1 year after presentation (Porth & Matfin, 2009). Other nursing interventions focus on patient education about the disease process, explanations of laboratory and other diagnostic tests, and preparation for safe and effective self-care at home.

Promoting Home and Community-Based Care

Educating Patients About Self-Care

Patient education is directed toward managing symptoms and monitoring for complications. Fluid and diet restrictions must be reviewed with the patient to avoid worsening of edema and hypertension. The patient is instructed verbally and in writing to notify the primary provider if symptoms of renal failure occur (e.g., fatigue, nausea, vomiting, diminishing urine output) or at the first sign of any infection.

Continuing Care

The importance of follow-up evaluations of blood pressure, urinalysis for protein, and BUN and serum creatinine levels to determine if the disease has progressed is stressed to the patient. A referral for home care may be indicated; a visit from a home care nurse provides an opportunity for careful assessment of the patient's progress and detection of early signs and symptoms of renal insufficiency. If corticosteroids, immunosuppressant agents, or antibiotic medications are prescribed, the home care nurse or nurse in the outpatient setting uses the opportunity to review the dosage, desired actions, and adverse effects of medications and the precautions to be taken.

Chronic Glomerulonephritis

Chronic glomerulonephritis may be due to repeated episodes of acute nephritic syndrome, hypertensive nephrosclerosis, hyperlipidemia, chronic tubulointerstitial injury, or hemodynamically mediated glomerular sclerosis. Secondary glomerular diseases that can have systemic effects include systemic lupus erythematosus, Goodpasture syndrome (caused by antibodies to the glomerular basement membrane), diabetic glomerulosclerosis, and amyloidosis.

Pathophysiology

The kidneys are reduced to as little as one fifth their normal size (consisting largely of fibrous tissue). The cortex layer shrinks to 1 to 2 mm in thickness or less. Bands of scar tissue distort the remaining cortex, making the surface of the kidney rough and irregular. Numerous glomeruli and their tubules become scarred, and the branches of the renal artery are thickened. The resulting severe glomerular damage can progress to stage 5 CKD and require a renal replacement therapy.

Clinical Manifestations

The symptoms of chronic glomerulonephritis vary. Some patients with severe disease have no symptoms at all for many years (Porth & Matfin, 2009). The condition may be discovered when hypertension or elevated BUN and serum creatinine levels are detected. Most patients report general symptoms, such as loss of weight and strength, increasing irritability, and an increased need to urinate at night (nocturia). Headaches, dizziness, and digestive disturbances are also common.

As chronic glomerulonephritis progresses, signs and symptoms of CKD may develop. The patient appears poorly nourished, with a yellow-gray pigmentation of the skin and periorbital and peripheral (dependent) edema. Blood pressure may be normal or severely elevated. Retinal findings include hemorrhage, exudate, narrowed tortuous arterioles, and papilledema. Anemia causes pale mucous membranes. Cardiomegaly, a gallop rhythm, distended neck veins, and other signs and symptoms of heart failure may be present. Crackles can be heard in the bases of the lungs.

Peripheral neuropathy with diminished deep tendon reflexes and neurosensory changes occur late in the disease. The patient becomes confused and demonstrates a limited attention span. An additional late finding includes evidence of pericarditis with a pericardial friction rub and pulsus paradoxus (difference in blood pressure during inspiration and expiration of greater than 10 mm Hg).

Assessment and Diagnostic Findings

A number of laboratory abnormalities occur. Urinalysis reveals a fixed specific gravity of about 1.010, variable proteinuria, and **urinary casts** (proteins secreted by damaged kidney tubules). As renal failure progresses and the GFR falls below 50 mL/min, the following changes occur:

- Hyperkalemia due to decreased potassium excretion, acidosis, catabolism, and excessive potassium intake from food and medications
- Metabolic acidosis from decreased acid secretion by the kidney and inability to regenerate bicarbonate
- Anemia secondary to decreased erythropoiesis (production of RBCs)
- Hypoalbuminemia with edema secondary to protein loss through the damaged glomerular membrane
- Increased serum phosphorus level due to decreased renal excretion of phosphorus
- Decreased serum calcium level (calcium binds to phosphorus to compensate for elevated serum phosphorus levels)

- Mental status changes
- Impaired nerve conduction due to electrolyte abnormalities and uremia

Chest x-rays may show cardiac enlargement and pulmonary edema. The electrocardiogram (ECG) may be normal or may indicate left ventricular hypertrophy associated with hypertension and signs of electrolyte disturbances, such as tall, tented (or peaked) T waves associated with hyperkalemia. Computed tomography (CT) and magnetic resonance imaging (MRI) scans show a decrease in the size of the renal cortex.

Medical Management

Management of symptoms guides the treatment. If the patient has hypertension, efforts are made to reduce the blood pressure with sodium and water restriction, antihypertensive agents, or both. Weight is monitored daily, and diuretic medications are prescribed to treat fluid overload. Proteins of high biologic value (dairy products, eggs, meats) are provided to promote good nutritional status. Adequate calories are provided to spare protein for tissue growth and repair. Urinary tract infections (UTIs) must be treated promptly to prevent further kidney damage.

Dialysis is initiated early in the course of the disease to keep the patient in optimal physical condition, prevent fluid and electrolyte imbalances, and minimize the risk of complications of renal failure. The course of dialysis is smoother if treatment begins before the patient develops complications.

Nursing Management

Whether the patient is hospitalized or cared for in the home, the nurse observes the patient for common fluid and electrolyte disturbances in kidney disease (see Table 54-1). Changes in fluid and electrolyte status and in cardiac and neurologic status are reported promptly to the primary provider. Throughout the course of the disease and treatment, the nurse gives emotional support by providing opportunities for the patient and family to verbalize their concerns, have their questions answered, and explore their options.

Promoting Home and Community-Based Care

Educating Patients About Self-Care

The nurse has a major role in educating the patient and family about the prescribed treatment plan and the risks associated with noncompliance. Instructions to the patient include explanations and scheduling for follow-up evaluations: blood pressure, urinalysis for protein and casts, and laboratory studies of BUN and serum creatinine levels. If long-term dialysis is needed, the nurse educates the patient and family about the procedure, how to care for the access site, dietary restrictions, and other necessary lifestyle modifications. These topics are discussed later in this chapter.

Periodic hospitalization, visits to the outpatient clinic or office, and home care referrals provide the nurse in each setting with the opportunity for careful assessment of the patient's progress and continued education about changes to report to the primary provider (worsening signs and symptoms of renal failure, such as nausea, vomiting, and diminished urine output). Specific education may include explanations about recommended diet and fluid modifications and medications (purpose, desired effects, adverse effects, dosage, and administration schedule).

Continuing Care

Periodic laboratory evaluations of creatinine clearance and BUN and serum creatinine levels are carried out to assess residual renal function and the need for dialysis or transplantation. If dialysis is initiated, the patient and family require considerable assistance and support in dealing with therapy and its long-term implications. The patient and family are reminded of the importance of participation in health promotion activities, including health screening. The patient is instructed to inform all health care providers about the diagnosis of glomerulonephritis so that all medical management, including pharmacologic therapy, is based on altered renal function.

Nephrotic Syndrome

Nephrotic syndrome is a type of renal failure characterized by increased glomerular permeability and is manifested by massive proteinuria (Porth & Matfin, 2009). Clinical findings include a marked increase in protein (particularly albumin) in the urine (proteinuria), a decrease in albumin in the blood (hypoalbuminemia), diffuse edema, high serum cholesterol, and low-density lipoproteins (hyperlipidemia).

The syndrome is apparent in any condition that seriously damages the glomerular capillary membrane and results in increased glomerular permeability to plasma proteins. Although the liver is capable of increasing the production of albumin, it cannot keep up with the daily loss of albumin through the kidneys. Thus, hypoalbuminemia results (Fig. 54-2).

Physiology Pathophysiology

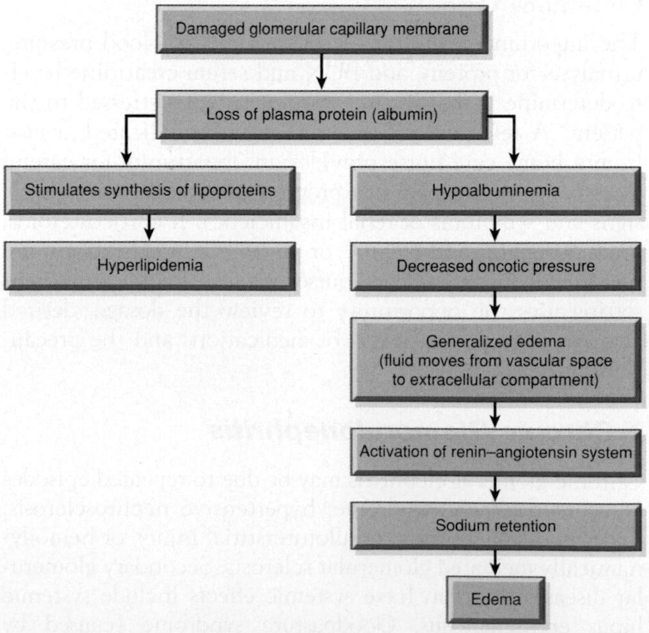
Damaged glomerular capillary membrane
Loss of plasma protein (albumin)
Stimulates synthesis of lipoproteins
Hypoalbuminemia
Hyperlipidemia
Decreased oncotic pressure
Generalized edema (fluid moves from vascular space to extracellular compartment)
Activation of renin–angiotensin system
Sodium retention
Edema

FIGURE 54-2 • Sequence of events in nephrotic syndrome.

Pathophysiology

Nephrotic syndrome occurs with many intrinsic kidney diseases and systemic diseases that cause glomerular damage. It is not a specific glomerular disease but a constellation of clinical findings that result from the glomerular damage (Porth & Matfin, 2009).

Clinical Manifestations

The major manifestation of nephrotic syndrome is edema. It is usually soft and pitting and commonly occurs around the eyes (periorbital), in dependent areas (sacrum, ankles, and hands), and in the abdomen (ascites). Patients may also exhibit irritability, headache, and malaise.

Assessment and Diagnostic Findings

Proteinuria (predominately albumin) exceeding 3.5 g/day is the hallmark of the diagnosis of nephrotic syndrome (Porth & Matfin, 2009). Protein electrophoresis and immunoelectrophoresis may be performed on the urine to categorize the type of proteinuria. The urine may also contain increased white blood cells (WBCs) as well as granular and epithelial casts. A needle biopsy of the kidney may be performed for histologic examination of renal tissue to confirm the diagnosis.

Complications

Complications of nephrotic syndrome include infection (due to a deficient immune response), thromboembolism (especially of the renal vein), pulmonary embolism, AKI (due to hypovolemia), and accelerated atherosclerosis (due to hyperlipidemia).

Medical Management

Treatment is focused on addressing the underlying disease state causing proteinuria, slowing progression of CKD, and relieving symptoms. Typical treatment includes diuretic agents for edema, ACE inhibitors to reduce proteinuria, and lipid-lowering agents for hyperlipidemia.

Nursing Management

In the early stages of nephrotic syndrome, nursing management is similar to that of the patient with acute glomerulonephritis, but as the condition worsens, management is similar to that of the patient with ESKD (see the following section).

Patients with nephrotic syndrome need adequate instruction about the importance of following all medication and dietary regimens so that their condition can remain stable as long as possible. Patients must be made aware of the importance of communicating any health-related change to their health care providers as soon as possible so that appropriate medication and dietary changes can be made before further changes occur within the glomeruli.

Polycystic Kidney Disease

Pathophysiology

Polycystic kidney disease (PKD) is a genetic disorder characterized by the growth of numerous cysts in the kidneys. When cysts form in the kidneys, they are filled with fluid, destroying the nephrons. PKD cysts can profoundly enlarge the kidneys while replacing much of the normal structure, resulting in reduced kidney function and leading to kidney failure.

PKD can also cause cysts in the liver and problems in other areas, such as blood vessels in the brain and heart. The number of cysts and the resulting complications help distinguish PKD from the usually harmless cysts that can form in the kidneys in later years of life. In the United States, PKD and cystic diseases are a leading cause of kidney failure. Two major inherited forms of PKD exist:

- *Autosomal dominant PKD* is the most common inherited form. Symptoms usually develop between 30 and 40 years of age, but they can begin earlier, even in childhood. About 90% of all PKD cases are autosomal dominant PKD.
- *Autosomal recessive PKD* is a rare inherited form. Symptoms of autosomal recessive PKD begin in the earliest months of life or in utero.

When autosomal dominant PKD causes kidneys to fail, which usually happens after many years, the patient requires dialysis or kidney transplantation. Approximately one half of individuals with autosomal dominant PKD progress to stage 5 CKD, requiring renal replacement therapy.

Clinical Manifestations

Signs and symptoms of PKD result from loss of renal function and the increasing size of the kidneys as the cysts grow. Kidney damage can result in hematuria, **polyuria** (large amounts of urine), hypertension, development of renal calculi and associated UTIs, and proteinuria. The growing cysts are noted with reports of abdominal fullness and flank pain (back and lower sides).

Assessment and Diagnostic Findings

PKD is a genetic disease; therefore, careful evaluation of family history is necessary. Palpation of the abdomen will often reveal enlarged cystic kidneys. Diagnosis is usually made with ultrasound imaging of the kidneys (Porth & Matfin, 2009).

Medical Management

PKD has no cure, and treatment is largely supportive and includes blood pressure control, pain control, and antibiotic agents to resolve infections. Once the kidneys fail, renal replacement therapy is indicated (see later discussion in chapter). Genetic testing and counseling may be indicated.

RENAL CANCER

Renal cancer accounts for about 5% of all cancers in the United States (American Cancer Society [ACS], 2012), where the incidence of renal cancer at all stages has increased in the past two decades. The incidence of renal cell carcinoma is higher in both men and women with an increased body mass index. Tobacco use continues to be a significant risk factor (Chart 54-3). American Indians and Alaskans have a higher mortality rate from renal carcinoma compared to other race and population groups (Baldwin, 2010).

Chart 54-3 RISK FACTORS
Renal Cancer

- Gender—affects men more than women
- Tobacco use
- Occupational exposure to industrial chemicals, such as petroleum products, heavy metals, and asbestos
- Obesity
- Unopposed estrogen therapy
- Polycystic kidney disease

Adapted from Porth, C. M., & Matfin, G. (2009). *Pathophysiology: Concepts of altered health states* (8th ed.). Philadelphia: Lippincott Williams & Wilkins; and Wood, L. S. (2009). Renal cell carcinoma: Screening, diagnosis, and prognosis. *Clinical Journal of Oncology Nursing, 13*(6), 3–7.

The most common type of renal carcinoma arises from the renal epithelium and accounts for more than 85% of all kidney tumors. These tumors may metastasize early to the lungs, bone, liver, brain, and contralateral kidney. One quarter of patients have metastatic disease at the time of diagnosis. Although enhanced imaging techniques account for improved detection of early-stage kidney cancer, it is unknown why the rate of late-stage kidney cancers is high (Marchese, 2010).

Clinical Manifestations

Many renal tumors produce no symptoms and are discovered on a routine physical examination as a palpable abdominal mass. Signs and symptoms, which occur in only 10% of patients, include hematuria, pain, and a mass in the flank (Wood, 2009). The usual sign that first calls attention to the tumor is painless hematuria, which may be either intermittent and microscopic or continuous and gross. There may be a dull pain in the back from the pressure produced by compression of the ureter, extension of the tumor into the perirenal area, or hemorrhage into the kidney tissue. Colicky pains occur if a clot or mass of tumor cells passes down the ureter. Symptoms from metastasis may be the first manifestations of renal tumor and may include unexplained weight loss, increasing weakness, and anemia.

Assessment and Diagnostic Findings

The diagnosis of a renal tumor may require intravenous (IV) urography, cystoscopic examination, renal angiograms, ultrasonography, or a CT scan (see Chapter 53). These tests may be exhausting for patients already debilitated by the systemic effects of a tumor as well as for older patients and those who are anxious about the diagnosis and outcome. The nurse assists the patient to prepare physically and psychologically for these procedures and monitors carefully for signs and symptoms of dehydration and exhaustion.

Medical Management

The goal of medical management is to detect the tumor early and to eradicate slow-growing tumors before metastasis occurs. Treatment most often includes a combination of surgery and pharmacologic management. Radiation therapy may be used for palliation in patients who are not candidates for surgery or other treatments.

Surgical Management

Nephrectomy

A radical nephrectomy is the preferred treatment if the tumor can be removed. This includes removal of the kidney (and tumor), adrenal gland, surrounding perinephric fat and Gerota's fascia, and lymph nodes. Laparoscopic nephrectomy can be performed for removal of the kidney with a small tumor. This procedure incurs less morbidity and a shorter recovery time. Radiation therapy, hormonal therapy, or chemotherapy may be used along with surgery. Immunotherapy may also be helpful. For patients with bilateral tumors or cancer of a functional single kidney, nephron-sparing surgery (partial nephrectomy) may be considered. Favorable results have been achieved in patients with small local tumors and a normal contralateral kidney (Wood, 2010).

Nephron-sparing surgery is increasingly being used to treat patients with solid renal lesions. The technical success rate of nephron-sparing surgery is excellent, and operative morbidity and mortality are low.

Patients with upper tract transitional cell carcinoma may benefit from laparoscopic nephroureterectomy. Although it is a lengthier surgical procedure, it has the same efficacy and is better tolerated by patients than open nephroureterectomy.

Renal Artery Embolization

In patients with metastatic renal carcinoma, the renal artery may be occluded to impede the blood supply to the tumor and thus kill the tumor cells. After angiographic studies are completed, a catheter is advanced into the renal artery, and embolizing materials (e.g., Gelfoam, autologous blood clot, steel coils) are injected into the artery and carried with the arterial blood flow to occlude the tumor vessels mechanically. This decreases the local blood supply, making removal of the kidney (nephrectomy) easier. It also stimulates an immune response because infarction of the renal cell carcinoma releases tumor-associated antigens that enhance the patient's response to metastatic lesions. The procedure may also reduce the number of tumor cells entering the venous circulation during surgical manipulation.

After renal artery embolization and tumor infarction, a characteristic symptom complex called *postinfarction syndrome* occurs, lasting 2 to 3 days (Yaklin, 2011). The patient has pain localized to the flank and abdomen, elevated temperature, and gastrointestinal (GI) symptoms. Pain is treated with parenteral analgesic agents, and acetaminophen (Tylenol) is administered to control fever. Antiemetic medications, restriction of oral intake, and IV fluids are used to treat the GI symptoms.

Pharmacologic Therapy

Depending on the stage of the tumor, percutaneous partial or radical nephrectomy may be followed by treatment with chemotherapeutic agents. Treatment with biologic response modifiers such as interleukin 2 (IL-2) is effective. IL-2, a protein that regulates cell growth, is used alone or in combination with lymphokine-activated killer cells (WBCs that have been stimulated by IL-2 to increase their ability to kill cancer cells). Interferon, another biologic response modifier, appears to have a direct antiproliferative effect on renal tumors.

Treatment options for renal cell carcinoma have changed dramatically since 2005, when the U.S. Food and Drug Administration (FDA) approved several new therapies (Wood, 2010). These agents inhibit pathways relevant in the pathogenesis of renal cell carcinoma, interfering with tumor angiogenesis, cell progression, and metastasis.

Another promising experimental approach to renal cell carcinoma is a vaccination to stimulate immune response, with autologous tumor cells with IL-2–, granulocyte-macrophage stimulating factor–, and dendritic cell–type vaccines. If patients with renal cancer do not respond to immunotherapy, allogeneic stem cell transplantation may be indicated (Wood, 2010).

Nursing Management

The patient with a renal tumor usually undergoes extensive diagnostic and therapeutic procedures (Wood, 2009). Treatment includes surgery, radiation therapy, and medications. After surgery, the patient usually has catheters and drains in place to maintain a patent urinary tract, to remove drainage, and to permit accurate measurement of urine output. Because of the location of the surgical incision, the patient's position during surgery, and the nature of the surgical procedure, pain and muscle soreness are common. Pharmacologic management often includes immunosuppressant agents; therefore, patients are monitored for infection.

The patient requires frequent analgesia during the postoperative period and assistance with turning, coughing, the use of incentive spirometry, and deep breathing to prevent atelectasis and other pulmonary complications. The patient and family require assistance and support to cope with the diagnosis and uncertain prognosis. (See discussion later in this chapter of postoperative care of the patient undergoing kidney surgery and Chapter 15 for discussion of care of the patient with cancer.)

Promoting Home and Community-Based Care

Educating Patients About Self-Care

The nurse educates the patient about how to inspect and care for the incision and to perform other general postoperative care including activity and lifting restrictions, driving, and pain management. Instructions are provided about when to notify the primary provider about problems (e.g., fever, respiratory difficulty, wound drainage, blood in the urine, pain or swelling of the legs).

The nurse encourages the patient to eat a healthy diet and to drink adequate liquids to avoid constipation and to maintain an adequate urine volume. Education and emotional support are provided related to the diagnosis, treatment, and continuing care because many patients are concerned about the loss of the other kidney, the possible need for dialysis, or the recurrence of cancer.

Continuing Care

Follow-up care is essential to detect signs of metastases and to reassure the patient and family about the patient's status and well-being. The patient who has had surgery for renal carcinoma should have a yearly physical examination and chest x-ray, because late metastases are not uncommon. All subsequent symptoms should be evaluated with possible metastases in mind.

If follow-up chemotherapy is necessary, the patient and family are informed about the treatment plan or chemotherapy protocol, what to expect with each visit, and when to notify the primary provider. Evaluation of remaining renal function (creatinine clearance, BUN and serum creatinine levels) may also be carried out periodically. A home care nurse may monitor the patient's physical status and psychological well-being and coordinate other indicated services and resources.

RENAL FAILURE

Renal failure results when the kidneys cannot remove the body's metabolic wastes or perform their regulatory functions. The substances normally eliminated in the urine accumulate in the body fluids as a result of impaired renal excretion, affecting endocrine and metabolic functions as well as fluid, electrolyte, and acid–base disturbances. Renal failure is a systemic disease and a final common pathway of many different kidney and urinary tract diseases. Each year, the number of deaths from irreversible renal failure increases (USRDS, 2011).

Acute Kidney Injury

Acute kidney injury (AKI) is a rapid loss of renal function due to damage to the kidneys. Depending on the duration and severity of AKI, a wide range of potentially life-threatening metabolic complications can occur, including metabolic acidosis as well as fluid and electrolyte imbalances. Treatment is aimed at replacing renal function temporarily to minimize potentially lethal complications and reduce potential causes of increased kidney injury with the goal of minimizing long-term loss of renal function. AKI is a problem seen in patients who are hospitalized and those in outpatient settings. A widely accepted criterion for AKI is a 50% or greater increase in serum creatinine above baseline (normal creatinine is less than 1 mg/dL) (Dirkes, 2011). Urine volume may be normal, or changes may occur. Possible changes include **oliguria** (less than 0.5 mL/kg/h, nonoliguria (greater than 800 mL/day), or **anuria** (less than 50 mL/day) (Yaklin, 2011).

Pathophysiology

Although the pathogenesis of AKI and oliguria is not always known, many times there is a specific underlying cause. Some of the factors may be reversible if identified and treated promptly, before kidney function is impaired. This is true of the following conditions that reduce blood flow to the kidney and impair kidney function: (1) hypovolemia; (2) hypotension; (3) reduced cardiac output and heart failure; (4) obstruction of the kidney or lower urinary tract by tumor, blood clot, or kidney stone; and (5) bilateral obstruction of the renal arteries or veins. If these conditions are treated and corrected before the kidneys are permanently damaged, the increased BUN and creatinine levels, oliguria, and other signs may be reversed.

Although renal stones are not a common cause of AKI, some types may increase the risk of AKI. Some hereditary stone diseases (see Chapter 55), primary struvite stones, and infection-related urolithiasis associated with anatomic and functional urinary tract anomalies and spinal cord injury may cause recurrent bouts of obstruction as well as crystal-specific damage to tubular epithelial cells and interstitial renal cells.

Classifications of Acute Kidney Injury

The term *acute kidney injury* has replaced the term *acute renal failure* because it better describes this syndrome in patients, not only simply those who require renal replacement therapies but also those patients who experience minor changes in renal function. The Acute Dialysis Quality Initiative Group developed this classification system to enable a more accurate and descriptive diagnosis of kidney injury (Dirkes, 2011). Classification criteria for AKI include assessment of three grades of severity and two outcome-level classifications. This five-point system is known as the RIFLE classification system (Dirkes, 2011). RIFLE stands for *r*isk, *i*njury, *f*ailure, *l*oss and *e*nd-stage kidney disease (Yaklin, 2011). Risk, injury, and failure are considered grades of AKI severity, whereas loss and ESKD are considered outcomes of loss that requires some form of renal replacement therapy, at least temporarily (Dirkes, 2011). Table 54-2 lists the classification criteria for the RIFLE system for AKI (Murphy & Byrne, 2010). This classification system is used by health care professionals to identify kidney injury and improve patient outcomes.

Categories of Acute Kidney Injury

The major categories of AKI are prerenal (hypoperfusion of kidney), intrarenal (actual damage to kidney tissue), and postrenal (obstruction to urine flow). Prerenal AKI, which occurs in 60% to 70% of cases, is the result of impaired blood flow that leads to hypoperfusion of the kidney commonly caused by volume depletion (burns, hemorrhage, GI losses), hypotension (sepsis, shock), and renal artery stenosis, ultimately leading to

TABLE 54-2 The RIFLE Classification for Acute Kidney Injury

Class	GFR Criteria	Urinary Output Criteria
R (Risk)	Increased serum creatinine 1.5 × baseline OR GFR decreased ≥25%	0.5 mL/kg/h for 6 h
I (Injury)	Increased serum creatinine 2 × baseline OR GFR decreased ≥50%	0.5 mL/kg/h for 12 h
F (Failure)	Increased serum creatinine 3 × baseline OR GFR decreased ≥75% OR Serum creatinine ≥354 mmol/L with an acute rise of at least 44 mmol/L	<0.3 mL/kg/h for 24 h OR Anuria for 12 h
L (Loss)	Persistent acute renal failure = complete loss of kidney function >4 wk	
E (ESKD)	ESKD >3 mo	

GFR, glomerular filtration rate; ESKD, end-stage kidney disease.
Adapted from Murphy, F., & Byrne, G. (2010). The role of the nurse in the management of acute kidney injury. *British Journal of Nursing, 19*(3), 146–152.

a decrease in the GFR (Yaklin, 2011). Intrarenal AKI is the result of actual parenchymal damage to the glomeruli or kidney tubules. **Acute tubular necrosis (ATN)**, or AKI in which there is damage to the kidney tubules, is the most common type of intrinsic AKI. Characteristics of ATN are intratubular obstruction, tubular back leak (abnormal reabsorption of filtrate and decreased urine flow through the tubule), vasoconstriction, and changes in glomerular permeability. These processes result in a decrease of GFR, progressive azotemia, and fluid and electrolyte imbalances. CKD, diabetes, heart failure, hypertension, and cirrhosis can lead to ATN (Yaklin, 2011). Postrenal AKI usually results from obstruction distal to the kidney by conditions such as renal calculi, strictures, blood clots, benign prostatic hypertrophy, malignancies, and pregnancy. Pressure rises in the kidney tubules, and eventually the GFR decreases. Common causes of each type of AKI are further summarized in Chart 54-4.

Chart 54-4 Causes of Acute Kidney Injury

Prerenal Failure

- *Volume depletion resulting from:*
  - Hemorrhage
  - Renal losses (diuretic agents, osmotic diuresis)
  - Gastrointestinal losses (vomiting, diarrhea, nasogastric suction)
- *Impaired cardiac efficiency resulting from:*
  - Myocardial infarction
  - Heart failure
  - Dysrhythmias
  - Cardiogenic shock
- *Vasodilation resulting from:*
  - Sepsis
  - Anaphylaxis
  - Antihypertensive medications or other medications that cause vasodilation

Intrarenal Failure

- *Prolonged renal ischemia resulting from:*
  - Pigment nephropathy (associated with the breakdown of blood cells containing pigments that in turn occlude kidney structures)
  - Myoglobinuria (trauma, crush injuries, burns)
  - Hemoglobinuria (transfusion reaction, hemolytic anemia)
- *Nephrotoxic agents such as:*
  - Aminoglycoside antibiotics (gentamicin, tobramycin)
  - Radiopaque contrast agents
  - Heavy metals (lead, mercury)
  - Solvents and chemicals (ethylene glycol, carbon tetrachloride, arsenic)
  - Nonsteroidal anti-inflammatory drugs
  - Angiotensin-converting enzyme inhibitors
- *Infectious processes such as:*
  - Acute pyelonephritis
  - Acute glomerulonephritis

Postrenal Failure

- *Urinary tract obstruction, including:*
  - Calculi (stones)
  - Tumors
  - Benign prostatic hyperplasia
  - Strictures
  - Blood clots

TABLE 54-3 Comparing Clinical Characteristics of Acute Kidney Injury

	Categories		
Characteristics	**Prerenal**	**Intrarenal**	**Postrenal**
Etiology	Hypoperfusion	Parenchymal damage	Obstruction
Blood urea nitrogen value	↑ (out of normal 20:1 proportion to creatinine)	↑	↑
Creatinine	↑	↑	↑
Urine output	↓	Varies, often ↓	Varies, may be ↓, or sudden anuria
Urine sodium	↓ to <20 mEq/L	↑ to >40 mEq/L	Varies, often ↓ to ≤20 mEq/L
Urinary sediment	Normal, few hyaline casts	Abnormal casts and debris	Usually normal
Urine osmolality	↑ to 500 mOsm	~350 mOsm, similar to serum	Varies, ↑ or equal to serum
Urine specific gravity	↑	Low normal	Varies

Phases of Acute Kidney Injury

There are four phases of AKI: initiation, oliguria, diuresis, and recovery.

- The initiation period begins with the initial insult and ends when oliguria develops.
- The oliguria period is accompanied by an increase in the serum concentration of substances usually excreted by the kidneys (urea, creatinine, uric acid, organic acids, and the intracellular cations [potassium and magnesium]). The minimum amount of urine needed to rid the body of normal metabolic waste products is 400 mL. In this phase, uremic symptoms first appear and life-threatening conditions such as hyperkalemia develop.

  Some patients have decreased renal function with increasing nitrogen retention but actually excrete normal amounts of urine (1 to 2 L/day). This is the nonoliguric form of renal failure and occurs predominantly after exposure of the patient to nephrotoxic agents, burns, traumatic injury, and the use of halogenated anesthetic agents.
- The diuresis period is marked by a gradual increase in urine output, which signals that glomerular filtration has started to recover. Laboratory values stabilize and eventually decrease. Although the volume of urinary output may reach normal or elevated levels, renal function may still be markedly abnormal. Because uremic symptoms may still be present, the need for expert medical and nursing management continues. The patient must be observed closely for dehydration during this phase; if dehydration occurs, the uremic symptoms are likely to increase.
- The recovery period signals the improvement of renal function and may take 3 to 12 months. Laboratory values return to the patient's normal level. Although a permanent 1% to 3% reduction in the GFR may occur, it is not clinically significant (Ali & Gray-Vickrey, 2011).

Clinical Manifestations

Almost every system of the body is affected with failure of the normal renal regulatory mechanisms. The patient may appear critically ill and lethargic. The skin and mucous membranes are dry from dehydration. Central nervous system signs and symptoms include drowsiness, headache, muscle twitching, and seizures. Table 54-3 summarizes common clinical characteristics in all three categories of AKI.

Assessment and Diagnostic Findings

Assessment of the patient with AKI includes evaluation for changes in the urine, diagnostic tests that evaluate the kidney contour, and a variety of laboratory values. (See Chapter 53 for information about the normal characteristics of urine, diagnostic findings, and laboratory values in the renal system.)

In AKI, urine output varies from scanty to a normal volume, hematuria may be present, and the urine has a low specific gravity (compared with a normal value of 1.010 to 1.025). One of the earliest manifestations of tubular damage is the inability to concentrate the urine (Porth & Matfin, 2009). Patients with prerenal azotemia have a decreased amount of sodium in the urine (less than 20 mEq/L) and normal urinary sediment. Patients with intrarenal azotemia usually have urinary sodium levels greater than 40 mEq/L with urinary casts and other cellular debris.

Ultrasonography is a critical component of the evaluation of patients with renal failure. A renal sonogram or a CT or MRI scan may show evidence of anatomic changes.

The BUN level increases steadily at a rate that depends on the degree of catabolism (breakdown of protein), renal perfusion, and protein intake. Serum creatinine levels are useful in monitoring kidney function and disease progression and increase with glomerular damage.

With a decline in the GFR, oliguria, and anuria, patients are at high risk for hyperkalemia. Protein catabolism results in the release of cellular potassium into the body fluids, causing severe hyperkalemia (high serum potassium levels). Hyperkalemia may lead to dysrhythmias, such as ventricular tachycardia and cardiac arrest. Sources of potassium include normal tissue catabolism, dietary intake, blood in the GI tract, or blood transfusion and other sources (e.g., IV infusions, potassium penicillin, and extracellular shift in response to metabolic acidosis).

Progressive metabolic acidosis occurs in renal failure because patients cannot eliminate the daily metabolic load of acid-type substances produced by the normal metabolic processes. In addition, normal renal buffering mechanisms fail. This is reflected by a decrease in the serum carbon dioxide (CO_2)-combining power and blood pH.

There may be an increase in blood phosphate concentrations; calcium levels may be low due to decreased absorption of calcium from the intestine and as a compensatory mechanism for the elevated blood phosphate levels. Anemia is another common laboratory finding in AKI, as a result of reduced erythropoietin production, uremic GI lesions, reduced RBC lifespan, and blood loss from the GI tract.

Prevention

AKI has a high mortality rate that ranges from 40% to 90%. Factors that influence mortality include increased age, comorbid conditions, and preexisting kidney and vascular diseases and respiratory failure (Dirkes, 2011). Therefore, prevention of AKI is essential (Chart 54-5).

A careful history is obtained to identify exposure to nephrotoxic agents or environmental toxins. The kidneys are susceptible to the adverse effects of medications because they are repeatedly exposed to substances in the blood. Patients taking nephrotoxic medications (e.g., aminoglycosides, gentamicin [Garamycin], tobramycin, colistimethate [Coly-Mycin], polymyxin B, amphotericin B, vancomycin, amikacin [Amikin], cyclosporine [Neoral]) should be monitored closely for changes in renal function. Kidney function needs to be monitored prior to initiation of these medications and during therapy (Karch, 2012).

Any agent that reduces renal blood flow (e.g., long-term analgesic use) may cause renal insufficiency. Chronic use of analgesic agents, particularly nonsteroidal anti-inflammatory drugs (NSAIDs), may cause **interstitial nephritis** (inflammation within the renal tissue) and papillary necrosis. Patients with heart failure or cirrhosis with ascites are at particular risk for NSAID-induced renal failure. Increased age, preexisting kidney disease, and the simultaneous administration of several nephrotoxic agents increase the risk of kidney damage.

Chart 54-5 Preventing Acute Kidney Injury

1. Provide adequate hydration to patients at risk for dehydration, including:
   - Before, during, and after surgery
   - Patients undergoing intensive diagnostic studies requiring fluid restriction and contrast agents (e.g., barium enema, IV pyelograms), especially older patients who may have marginal renal reserve
   - Patients with neoplastic disorders or disorders of metabolism (e.g., gout) and those receiving chemotherapy
2. Prevent and treat shock promptly with blood and fluid replacement.
3. Monitor central venous and arterial pressures and hourly urine output of critically ill patients to detect the onset of renal failure as early as possible.
4. Treat hypotension promptly.
5. Continually assess renal function (urine output, laboratory values) when appropriate.
6. Take precautions to ensure that the appropriate blood is administered to the correct patient in order to avoid severe transfusion reactions, which can precipitate renal failure.
7. Prevent and treat infections promptly. Infections can produce progressive kidney damage.
8. Pay special attention to wounds, burns, and other precursors of sepsis.
9. To prevent infections from ascending in the urinary tract, give meticulous care to patients with indwelling catheters. Remove catheters as soon as possible.
10. To prevent toxic drug effects, closely monitor dosage, duration of use, and blood levels of all medications metabolized or excreted by the kidneys.

Radiocontrast-induced nephropathy (CIN) is a major cause of hospital-acquired AKI. Patients undergo more than 1 million radiocontrast studies in the United States annually; of these, approximately 150,000 patients will experience CIN, and at least 1% of them will require dialysis and experience a prolonged length of hospital stay (Murphy & Byrne, 2010). This is a potentially preventable condition. Baseline levels of creatinine greater than 2 mg/dL identify patients at high risk. Limiting the patient's exposure to contrast agents and nephrotoxic medications will reduce the risk of CIN (Murphy & Byrne, 2010; Rank, 2013). Administration of *N*-acetylcysteine and sodium bicarbonate before and during procedures reduces risk, but prehydration with saline is considered the most effective method to prevent CIN (Murphy & Byrne, 2010; Rank, 2013).

Gerontologic Considerations

About half of all patients who develop AKI during hospitalization are older than 60 years. The etiology of AKI in older adults includes prerenal causes such as dehydration, intrarenal causes such as **nephrotoxic** agents (e.g., medications, contrast agents), and complications of major surgery (Dirkes, 2011). Suppression of thirst, enforced bed rest, lack of access to drinking water, and confusion all contribute to the older patient's failure to consume adequate fluids and may lead to dehydration, further compromising already decreased renal function.

AKI in older adults is also often seen in the community setting. Nurses in the ambulatory setting need to be aware of the risk. All medications need to be monitored for potential side effects that could result in damage to the kidney either through reduced circulation or nephrotoxicity. Outpatient procedures that require fasting or a bowel preparation may cause dehydration and therefore require careful monitoring.

Medical Management

The kidneys have a remarkable ability to recover from insult. The objectives of treatment for AKI are to restore normal chemical balance and prevent complications until repair of renal tissue and restoration of renal function can occur. Management includes eliminating the underlying cause; maintaining fluid balance; avoiding fluid excesses; and, when indicated, providing renal replacement therapy. Prerenal azotemia is treated by optimizing renal perfusion, whereas postrenal failure is treated by relieving the obstruction. Intrarenal azotemia is treated with supportive therapy, with removal of causative agents, aggressive management of prerenal and postrenal failure, and avoidance of associated risk factors. Shock and infection, if present, are treated promptly (see Chapter 14).

Maintenance of fluid balance is based on daily body weight, serial measurements of central venous pressure, serum and urine concentrations, fluid losses, blood pressure, and the clinical status of the patient. The parenteral and oral intake and the output of urine, gastric drainage, stools, wound drainage, and perspiration are calculated and are used as the basis for fluid replacement. The insensible fluid produced through the normal metabolic processes and lost through the skin and lungs is also considered in fluid management.

Fluid excesses can be detected by the clinical findings of dyspnea, tachycardia, and distended neck veins. The patient's lungs are auscultated for moist crackles. Because pulmonary edema may be caused by excessive administration of parenteral fluids, extreme caution must be used to prevent fluid overload. The development of generalized edema is assessed by examining the presacral and pretibial areas several times daily. Mannitol (Osmitrol), furosemide (Lasix), or ethacrynic acid (Edecrin) may be prescribed to initiate diuresis (Warise, 2010).

Adequate renal blood flow in patients with prerenal causes of AKI may be restored by IV fluids or transfusions of blood products. If AKI is caused by hypovolemia secondary to hypoproteinemia, an infusion of albumin may be prescribed. Dialysis may be initiated to prevent complications of AKI, such as hyperkalemia, metabolic acidosis, pericarditis, and pulmonary edema. Dialysis corrects many biochemical abnormalities; allows for liberalization of fluid, protein, and sodium intake; diminishes bleeding tendencies; and promotes wound healing. **Hemodialysis** (a procedure that circulates the patient's blood through an artificial kidney [dialyzer] to remove waste products and excess fluid), **peritoneal dialysis** (PD; a procedure that uses the patient's peritoneal membrane [the lining of the peritoneal cavity] as the semipermeable membrane to exchange fluid and solutes), or a variety of **continuous renal replacement therapies (CRRTs)** (methods used to replace normal kidney function by circulating the patient's blood through a hemofilter) may be performed. These and other treatment modalities for patients with renal dysfunction are discussed later in this chapter.

Pharmacologic Therapy

Hyperkalemia is the most life threatening of the fluid and electrolyte changes that occur in patients with kidney disorders. Therefore, the patient is monitored for hyperkalemia through serial serum electrolyte levels (potassium value greater than 5.0 mEq/L [5 mmol/L]), ECG changes (tall, tented, or peaked T waves), and changes in clinical status (see Chapter 13). Other symptoms of hyperkalemia include irritability, abdominal cramping, diarrhea, paresthesia, and generalized muscle weakness. Muscle weakness may present as slurred speech, difficulty breathing, paresthesia, and paralysis. As the potassium level increases, both cardiac and other muscular function declines, making this a true medical emergency (Murphy & Byrne, 2010).

The elevated potassium levels may be reduced by administering cation-exchange resins (sodium polystyrene sulfonate [Kayexalate]) orally or by retention enema. Kayexalate works by exchanging sodium ions for potassium ions in the intestinal tract. Sorbitol may be administered in combination with Kayexalate to induce a diarrhea-type effect (it induces water loss in the GI tract). If a Kayexalate retention enema is administered (the colon is the major site of potassium exchange), a rectal catheter with a balloon may be used to facilitate retention if necessary. The patient should retain the Kayexalate for at least 30 minutes (preferable several hours) to promote potassium removal (Karch, 2012). Afterward, a cleansing enema may be prescribed to remove remaining medication as a precaution against fecal impaction.

If the patient is hemodynamically unstable (low blood pressure, changes in mental status, dysrhythmia), IV dextrose 50%, insulin, and calcium replacement may be administered to shift potassium back into the cells. The shift of potassium into the intracellular space is temporary, so arrangements for dialysis need to be made on an emergent basis.

Many medications are eliminated through the kidneys; therefore, dosages must be reduced when a patient has AKI. Examples of commonly used agents that require adjustment are antibiotic medications (especially aminoglycosides), digoxin (Lanoxin), phenytoin (Dilantin), ACE inhibitors, and magnesium-containing agents.

In addition, many medications have been used in patients with AKI in an attempt to improve patient outcomes. Diuretic agents are often used to control fluid volume, but they have not been shown to improve recovery from AKI (Warise, 2010).

In patients with severe acidosis, the arterial blood gases and serum bicarbonate levels (CO_2-combining power) must be monitored because the patient may require sodium bicarbonate therapy or dialysis. If respiratory problems develop, appropriate ventilatory measures must be instituted. The elevated serum phosphate level may be controlled with phosphate-binding agents (e.g., calcium or lanthanum carbonate) that help prevent a continuing rise in serum phosphate levels by decreasing the absorption of phosphate from the intestinal tract.

Nutritional Therapy

AKI causes severe nutritional imbalances (because nausea and vomiting contribute to inadequate dietary intake), impaired glucose use and protein synthesis, and increased tissue catabolism. The patient is weighed daily and loses 0.2 to 0.5 kg (0.5 to 1 lb) daily if the nitrogen balance is negative (i.e., caloric intake falls below caloric requirements). If the patient gains or does not lose weight or develops hypertension, fluid retention should be suspected.

Nutritional support is based on the underlying cause of AKI, the catabolic response, the type and frequency of renal replacement therapy, comorbidities, and nutritional status. Replacement of dietary proteins is individualized to provide the maximum benefit and minimize uremic symptoms. Caloric requirements are met with high-carbohydrate meals, because carbohydrates have a protein-sparing effect (i.e., in a high-carbohydrate diet, protein is not used for meeting energy requirements but is "spared" for growth and tissue healing). Foods and fluids containing potassium or phosphorus (e.g., bananas, citrus fruits and juices, coffee) are restricted.

The oliguric phase of AKI may last 10 to 14 days and is followed by the diuretic phase, at which time urine output begins to increase, signaling the patient is in the recovery phase (Ali & Gray-Vickrey, 2011). Results of blood chemistry tests are used to determine the amounts of sodium, potassium, and water needed for replacement, along with assessment for over- or underhydration. Following the diuretic phase, the patient is placed on a high-protein, high-calorie diet and is encouraged to resume activities gradually.

Nursing Management

The nurse has an important role in caring for the patient with AKI. The nurse monitors for complications, participates in emergency treatment of fluid and electrolyte imbalances, assesses the patient's progress and response to treatment, and provides physical and emotional support. In addition, the nurse keeps family members informed about the patient's condition, helps them understand the treatments, and provides psychological support. Although the development of AKI may be the most serious problem, the nurse continues to provide nursing care indicated for the primary disorder (e.g., burns, shock, trauma, obstruction of the urinary tract).

Monitoring Fluid and Electrolyte Balance

Because of the serious fluid and electrolyte imbalances that can occur with AKI, the nurse monitors the patient's serum electrolyte levels and physical indicators of these complications during all phases of the disorder. IV solutions must be carefully selected based on the patient's fluid and electrolyte status. The patient's cardiac function and musculoskeletal status are monitored closely for signs of hyperkalemia.

> ***Quality and Safety Nursing Alert***
>
> **Hyperkalemia is the most immediate life-threatening imbalance seen in AKI. Parenteral fluids, all oral intake, and all medications are screened carefully to ensure that sources of potassium are not inadvertently administered or consumed.**

The nurse monitors fluid status by paying careful attention to fluid intake (IV medications should be administered in the smallest volume possible), urine output, apparent edema, distention of the jugular veins, alterations in heart sounds and breath sounds, and increasing difficulty in breathing. Accurate daily weights, as well as I&O records, are essential. Indicators of deteriorating fluid and electrolyte status are reported immediately to the primary provider, and preparation is made for emergency treatment. Severe fluid and electrolyte disturbances may be treated with hemodialysis, PD, or CRRT.

Reducing Metabolic Rate

The nurse takes steps to reduce the patient's metabolic rate. Bed rest may be indicated to reduce exertion and the metabolic rate during the most acute stage of the disorder. Fever and infection, both of which increase the metabolic rate and catabolism, are prevented or treated promptly.

Promoting Pulmonary Function

Attention is given to pulmonary function, and the patient is assisted to turn, cough, and take deep breaths frequently to prevent atelectasis and respiratory tract infection. Drowsiness and lethargy may prevent the patient from moving and turning without encouragement and assistance.

Preventing Infection

Asepsis is essential with invasive lines and catheters to minimize the risk of infection and increased metabolism. An indwelling urinary catheter is avoided whenever possible due to the high risk of UTI associated with its use but may be required to provide ongoing data required to monitor fluid I&O.

Providing Skin Care

The skin may be dry or susceptible to breakdown as a result of edema; therefore, meticulous skin care is important. Additionally, excoriation and itching of the skin may result from the deposit of irritating toxins in the patient's tissues. Bathing the patient with cool water, frequent turning, and keeping the skin clean and well moisturized and the fingernails trimmed to avoid excoriation are often comforting and prevent skin breakdown.

Providing Psychosocial Support

The patient with AKI may require treatment with hemodialysis, PD, or CRRT. The length of time that these treatments are necessary varies with the cause and extent of damage to the kidneys. The patient and family need assistance, explanation, and support during this period. The purpose of the treatment is explained to the patient and family by the primary provider. However, high levels of anxiety and fear may necessitate repeated explanation and clarification by the nurse. The family members may initially be afraid to touch and talk to the patient during these procedures but should be encouraged and assisted to do so.

In an intensive care setting, many of the nurse's functions are devoted to the technical aspects of patient care; however, it is essential that the psychological needs and other concerns of the patient and family be addressed. Continued assessment of the patient for complications of AKI and precipitating causes is essential.

End-Stage Kidney Disease or Chronic Renal Failure

When a patient has sustained enough kidney damage to require renal replacement therapy on a permanent basis, the patient has moved into the fifth or final stage of CKD, also referred to as ESKD or chronic renal failure.

Pathophysiology

As renal function declines, the end products of protein metabolism (normally excreted in urine) accumulate in the blood. Uremia develops and adversely affects every system in the body. The greater the buildup of waste products, the more pronounced the symptoms.

The rate of decline in renal function and progression of ESKD is related to the underlying disorder, the urinary excretion of protein, and the presence of hypertension. The disease tends to progress more rapidly in patients who excrete significant amounts of protein or have elevated blood pressure than in those without these conditions.

Clinical Manifestations

Because virtually every body system is affected in ESKD, patients exhibit a number of signs and symptoms. The severity of these signs and symptoms depends in part on the degree of renal impairment, other underlying conditions, and the patient's age. Cardiovascular disease is the predominant

Chart 54-6 ASSESSMENT
Assessing for End-Stage Kidney Disease

Be alert to the following signs and symptoms:

Neurologic
- Weakness and fatigue
- Confusion
- Inability to concentrate
- Disorientation
- Tremors
- Seizures
- Asterixis
- Restlessness of legs
- Burning of soles of feet
- Behavior changes

Integumentary
- Gray-bronze skin color
- Dry, flaky skin
- Pruritus
- Ecchymosis
- Purpura
- Thin, brittle nails
- Coarse, thinning hair

Cardiovascular
- Hypertension
- Pitting edema (feet, hands, sacrum)
- Periorbital edema
- Pericardial friction rub
- Engorged neck veins
- Pericarditis
- Pericardial effusion
- Pericardial tamponade
- Hyperkalemia
- Hyperlipidemia

Pulmonary
- Crackles
- Thick, tenacious sputum
- Depressed cough reflex
- Pleuritic pain
- Shortness of breath
- Tachypnea
- Kussmaul-type respirations
- Uremic pneumonitis

Gastrointestinal
- Ammonia odor to breath ("uremic fetor")
- Metallic taste
- Mouth ulcerations and bleeding
- Anorexia, nausea, and vomiting
- Hiccups
- Constipation or diarrhea
- Bleeding from gastrointestinal tract

Hematologic
- Anemia
- Thrombocytopenia

Reproductive
- Amenorrhea
- Testicular atrophy
- Infertility
- Decreased libido

Musculoskeletal
- Muscle cramps
- Loss of muscle strength
- Renal osteodystrophy
- Bone pain
- Bone fractures
- Footdrop

cause of death in patients with ESKD. Peripheral neuropathy, a disorder of the peripheral nervous system, is present in some patients. Patients complain of severe pain and discomfort. Restless leg syndrome and burning feet can occur in the early stage of uremic peripheral neuropathy (Williams & Manias, 2009). The precise mechanisms for many of these systemic signs and symptoms have not been identified. However, it is generally thought that the accumulation of uremic waste products is the probable cause. Chart 54-6 summarizes the systemic signs and symptoms.

Assessment and Diagnostic Findings

Glomerular Filtration Rate

As the GFR decreases (due to nonfunctioning glomeruli), the creatinine clearance decreases, whereas the serum creatinine and BUN levels increase. Serum creatinine is a more sensitive indicator of renal function than BUN. The BUN is affected not only by kidney disease but also by protein intake in the diet, catabolism (tissue and RBC breakdown), parenteral nutrition, and medications such as corticosteroids.

Sodium and Water Retention

The kidney cannot concentrate or dilute the urine normally in ESKD. Appropriate responses by the kidney to changes in the daily intake of water and electrolytes, therefore, do not occur. Some patients retain sodium and water, increasing the risk for edema, heart failure, and hypertension. Hypertension may also result from activation of the renin–angiotensin–aldosterone axis and the concomitant increased aldosterone secretion. Other patients have a tendency to lose sodium and run the risk of developing hypotension and hypovolemia. Vomiting and diarrhea may cause sodium and water depletion, which worsens the uremic state.

Acidosis

Metabolic acidosis occurs in ESKD because the kidneys are unable to excrete increased loads of acid. Decreased acid secretion results from the inability of the kidney tubules to excrete ammonia (NH_3^-) and to reabsorb sodium bicarbonate (HCO_3^-). There is also decreased excretion of phosphates and other organic acids.

Chart 54-7

PLAN OF NURSING CARE

The Patient With End-Stage Kidney Disease (continued)

Nursing Interventions	Rationale	Expected Outcomes
2. Promote independence in self-care activities as tolerated; assist if fatigued.	2. Promotes improved self-esteem.	
3. Encourage alternating activity with rest.	3. Promotes activity and exercise within limits and adequate rest.	
4. Encourage patient to rest after dialysis treatments.	4. Adequate rest is encouraged after dialysis treatments, which are exhausting to many patients.	

NURSING DIAGNOSIS: Risk for situational low self-esteem related to dependency, role changes, change in body image, and change in sexual function
GOAL: Improved self-esteem

Nursing Interventions	Rationale	Expected Outcomes
1. Assess patient's and family's responses and reactions to illness and treatment.	1. Provides data about problems encountered by patient and family in coping with changes in life.	• Identifies previously used coping styles that have been effective and those no longer possible due to disease and treatment (alcohol or drug use; extreme physical exertion) • Patient and family identify and verbalize feelings and reactions to disease and necessary changes in their lives • Seeks professional counseling, if necessary, to cope with changes resulting from renal failure • Reports satisfaction with method of sexual expression
2. Assess relationship of patient and significant family members.	2. Identifies strengths and supports of patient and family.	
3. Assess usual coping patterns of patient and family members.	3. Coping patterns that may have been effective in past may be harmful in view of restrictions imposed by disease and treatment.	
4. Encourage open discussion of concerns about changes produced by disease and treatment: a. Role changes b. Changes in lifestyle c. Changes in occupation d. Sexual changes e. Dependence on health care team	4. Encourages patient to identify concerns and steps necessary to deal with them.	
5. Explore alternate ways of sexual expression other than sexual intercourse.	5. Alternative forms of sexual expression may be acceptable.	
6. Discuss role of giving and receiving love, warmth, and affection.	6. Sexuality means different things to different people, depending on stage of maturity.	

COLLABORATIVE PROBLEMS: Hyperkalemia; pericarditis, pericardial effusion, and pericardial tamponade; hypertension; anemia; bone disease and metastatic calcifications
GOAL: Absence of complications

Nursing Interventions	Rationale	Expected Outcomes
Hyperkalemia		
1. Monitor serum potassium levels. Notify primary provider if level >5.5 mEq/L, and prepare to treat hyperkalemia.	1. Hyperkalemia causes potentially life-threatening changes in the body.	• Has normal potassium level • Experiences no muscle weakness or diarrhea • Exhibits normal ECG pattern • Vital signs are within normal limits
2. Assess patient for muscle weakness, diarrhea, electrocardiographic (ECG) changes (tall-tented T waves and widened QRS).	2. Cardiovascular signs and symptoms are characteristic of hyperkalemia.	
Pericarditis, Pericardial Effusion, and Pericardial Tamponade		
1. Assess patient for fever, chest pain, and a pericardial friction rub (signs of pericarditis); if present, notify primary provider.	1. About 30%–50% of patients with chronic renal failure develop pericarditis due to uremia; fever, chest pain, and a pericardial friction rub are classic signs.	• Has strong and equal peripheral pulses • Absence of a paradoxical pulse • Absence of pericardial effusion or tamponade on cardiac ultrasound • Has normal heart sounds

Chart 54-7

PLAN OF NURSING CARE

The Patient With End-Stage Kidney Disease (continued)

Nursing Interventions	Rationale	Expected Outcomes
2. If patient has pericarditis, assess for the following every 4 hours: a. Paradoxical pulse, >10 mm Hg b. Extreme hypotension c. Weak or absent peripheral pulses d. Altered level of consciousness e. Bulging neck veins	2. Pericardial effusion is a common fatal sequela of pericarditis. Signs of an effusion include a paradoxical pulse (>10 mm Hg drop in blood pressure during inspiration) and signs of shock due to compression of the heart by a large effusion. Cardiac tamponade exists when the patient is severely compromised hemodynamically.	
3. Prepare patient for cardiac ultrasound to aid in diagnosis of pericardial effusion and cardiac tamponade.	3. Cardiac ultrasound is useful in visualizing pericardial effusions and cardiac tamponade.	
4. If cardiac tamponade develops, prepare patient for emergency pericardiocentesis.	4. Cardiac tamponade is a life-threatening condition, with a high mortality rate. Immediate aspiration of fluid from the pericardial space is essential.	
Hypertension		
1. Monitor and record blood pressure as indicated.	1. Provides objective data for monitoring. Elevated levels may indicate nonadherence to the treatment regimen.	• Blood pressure within normal limits • Reports no headaches, visual problems, or seizures • Edema is absent • Demonstrates compliance with dietary and fluid restrictions
2. Administer antihypertensive medications as prescribed.	2. Antihypertensive medications play a key role in treatment of hypertension associated with chronic renal failure.	
3. Encourage compliance with dietary and fluid restriction therapy.	3. Adherence to diet and fluid restrictions and dialysis schedule prevents excess fluid and sodium accumulation.	
4. Instruct patient to report signs of fluid overload, vision changes, headaches, edema, or seizures.	4. These are indications of inadequate control of hypertension and the need to alter therapy.	
Anemia		
1. Monitor red blood cell (RBC) count and hemoglobin and hematocrit levels as indicated.	1. Provides assessment of degree of anemia.	• Patient has a normal skin color without pallor • Exhibits hematology values within acceptable limits • Experiences no bleeding from any site
2. Administer medications as prescribed, including iron and folic acid supplements, an erythrocyte stimulating agent, and multivitamins.	2. RBCs need iron, folic acid, and vitamins to be produced. An erythrocyte-stimulating agent stimulates the bone marrow to produce RBCs.	
3. Avoid drawing unnecessary blood specimens.	3. Anemia is worsened by drawing numerous specimens.	
4. Educate patient to prevent bleeding: Avoid vigorous nose blowing and contact sports, and use a soft toothbrush.	4. Bleeding from anywhere in the body worsens anemia.	
5. Administer blood component therapy as indicated.	5. Blood component therapy may be needed if the patient has symptoms.	
Bone Disease and Metastatic Calcifications		
1. Administer the following medications as prescribed: phosphate binders, calcium supplements, vitamin D supplements.	1. Chronic renal failure causes numerous physiologic changes affecting calcium, phosphorus, and vitamin D metabolism.	• Exhibits serum calcium, phosphorus, and aluminum levels within acceptable ranges • Exhibits no symptoms of hypocalcemia • Has no bone demineralization on bone scan • Discusses importance of maintaining activity level and exercise program
2. Monitor serum lab values as indicated (calcium, phosphorus, aluminum levels), and report abnormal findings to primary provider.	2. Hyperphosphatemia, hypocalcemia, and excess aluminum accumulation are common in chronic renal failure.	
3. Assist patient with an exercise program.	3. Bone demineralization increases with immobility.	

Gerontologic Considerations

Diabetes, hypertension, chronic glomerulonephritis, interstitial nephritis, and urinary tract obstruction are among the causes for ESKD in older adults. The signs and symptoms of kidney disease in older adults are often nonspecific. The occurrence of symptoms of other disorders (heart failure, dementia) can mask the symptoms of kidney disease and delay or prevent diagnosis and treatment.

Hemodialysis and PD are used effectively in treating older patients with ESKD. Initiation of dialysis among older patients has dramatically increased in the past decade. Implementation of palliative care has also increased among patients who choose not to start dialysis or who decide to stop dialysis. Although there is no specific age limitation for kidney transplantation, concomitant disorders (e.g., coronary artery disease, peripheral vascular disease) have made it a less common treatment for older adults. However, the outcome is comparable to that in younger patients. Some older patients elect not to undergo dialysis or transplantation. Conservative management and palliative care, including nutritional therapy, fluid control, and medications such as phosphate binders, may be considered in patients who are not suitable for or elect not to have dialysis or transplantation (Hopkins, Kott, Rose, et al., 2011; Young, 2009). Palliative care for the patient with ESKD focuses on relieving suffering, promoting health-related quality of life, and facilitating dignity at the end of life (Harrison & Watson, 2011; Young, 2009) (see Chapter 16).

Concept Mastery Alert

Visit thePoint to view an Interactive Tutorial on renal failure and associated fundamental concepts.

RENAL REPLACEMENT THERAPIES

The use of renal replacement therapies becomes necessary when the kidneys can no longer remove wastes, maintain electrolytes, and regulate fluid balance. This can occur rapidly or over a long period of time, and the need for replacement therapy can be acute (short term) or chronic (long term). The main renal replacement therapies include the various types of dialysis and kidney transplantation.

Dialysis

Types of dialysis include hemodialysis, CRRT, and PD. Acute or urgent dialysis is indicated when there is a high and increasing level of serum potassium, fluid overload, or impending pulmonary edema; increasing acidosis; pericarditis; and advanced uremia (Porth & Matfin, 2009). It may also be used to remove medications or toxins (poisoning or medication overdose) from the blood or for edema that does not respond to other treatment, hepatic coma, hyperkalemia, hypercalcemia, hypertension, and uremia.

Chronic or maintenance dialysis is indicated in advanced CKD and ESKD in the following instances: the presence of uremic signs and symptoms affecting all body systems (nausea and vomiting, severe anorexia, increasing lethargy, mental confusion), hyperkalemia, fluid overload not responsive to diuretics and fluid restriction, and a general lack of well-being. An urgent indication for dialysis in patients with renal failure is pericardial friction rub, which is indicative of uremic pericarditis.

The decision to initiate dialysis should be reached only after thoughtful discussion among the patient, family, primary provider, and other health care team members. The nurse can assist the patient and family by answering their questions, clarifying the information provided, and supporting their decision.

Successful kidney transplantation eliminates the need for dialysis. Not only is the quality of life much improved in patients with ESKD who undergo transplantation, but physiologic function is improved as well. Patients who undergo kidney transplantation from living donors before dialysis is initiated generally have longer survival of the transplanted kidney than patients who receive transplantation after dialysis treatment is initiated (Serur & Charlton, 2011).

Hemodialysis

Hemodialysis is used for patients who are acutely ill and require short-term dialysis for days to weeks until kidney function resumes and for patients with advanced CKD and ESKD who require long-term or permanent renal replacement therapy. Hemodialysis prevents death but does not cure kidney disease and does not compensate for the loss of endocrine or metabolic activities of the kidneys. More than 90% of patients requiring long-term renal replacement therapy are on chronic hemodialysis (USRDS, 2011). Most patients receive intermittent hemodialysis that involves treatments three times a week with an average treatment duration of 3 to 5 hours in an outpatient setting. Hemodialysis can also be performed at home by the patient and a caregiver. With home dialysis, treatment time and frequency can be adjusted to meet optimal patient needs.

The objectives of hemodialysis are to extract toxic nitrogenous substances from the blood and to remove excess fluid. A **dialyzer** (also referred to as an artificial kidney) is a synthetic semipermeable membrane through which blood is filtered to remove uremic toxins and a desired amount of fluid. In hemodialysis, the blood, laden with toxins and nitrogenous wastes, is diverted from the patient to a machine via the use of a blood pump to the dialyzer, where toxins are filtered from the blood and the blood is returned to the patient.

Diffusion, osmosis, and ultrafiltration are the principles on which hemodialysis is based (see Chapter 13). The toxins and wastes in the blood are removed by **diffusion**—that is, they move from an area of higher concentration in the blood to an area of lower concentration in the dialysate. The **dialysate** is a solution that circulates through the dialyzer, made up of all the electrolytes in their ideal extracellular concentrations. The electrolyte level in the patient's blood can be brought under control by properly adjusting the electrolytes in the dialysate solution. The semipermeable membrane impedes the diffusion of large molecules, such as RBCs and proteins.

Excess fluid is removed from the blood by **osmosis**, in which water moves from an area of low concentration potential (the blood) to an area of high concentration potential (the dialysate bath). In **ultrafiltration**, fluid moves under high pressure to an area of lower pressure. This process is much more efficient than osmosis for fluid removal and is accomplished by applying negative pressure or a suctioning force to the dialysis membrane. Because patients with disease requiring dialysis usually cannot excrete water, this force is necessary to remove fluid to achieve fluid balance.

The body's buffer system is maintained using a dialysate bath made up of bicarbonate (most common) or acetate, which is metabolized to form bicarbonate. The anticoagulant heparin is administered to keep blood from clotting in the extracorporeal dialysis circuit. Cleansed blood is returned to the body with the goal of removing fluid, balancing electrolytes, and managing acidosis.

Dialyzers

Dialyzers are hollow-fiber devices containing thousands of tiny capillary tubes that carry the blood through the dialyzer. The tubes are porous and act as a semipermeable membrane, allowing toxins, fluid, and electrolytes to pass across the membrane. The constant flow of the solution maintains the concentration gradient to facilitate the exchange of wastes from the blood across the semipermeable membrane into the dialysate solution, where they are removed and discarded (Fig. 54-3).

Dialyzers have undergone many technologic changes in performance and biocompatibility. High-flux dialysis uses highly permeable membranes to increase the clearance of low- and mid-molecular-weight molecules. These special membranes are used with higher than traditional rates of flow for the blood entering and exiting the dialyzer (500 to 550 mL/min). High-flux dialysis increases the efficiency of treatments while shortening their duration and reducing the need for heparin.

Vascular Access

Access to the patient's vascular system must be established to allow blood to be removed, cleansed, and returned to the patient's vascular system at the rapid rates of 300 and 800 mL/min. Several types of access can be surgically created or placed during procedures performed in interventional radiology suites or at the bedside.

Vascular Access Devices

Immediate access to the patient's circulation for acute hemodialysis is achieved by inserting a double-lumen, noncuffed, large-bore catheter into the subclavian, internal jugular, or femoral vein by the physician (Fig. 54-4). This method of vascular access involves some risk (e.g., hematoma, pneumothorax,

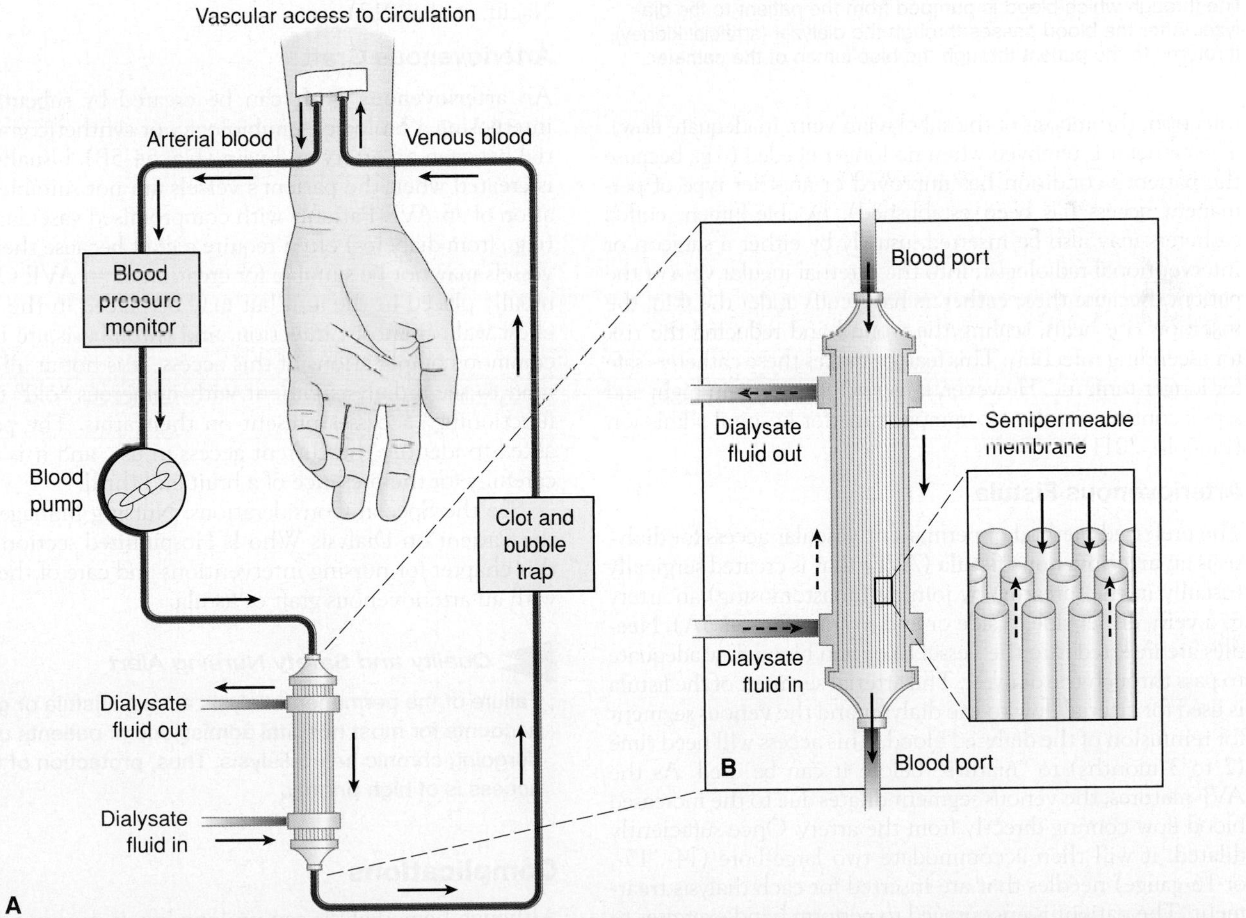

Vascular access to circulation
Arterial blood
Venous blood
Blood pressure monitor
Blood pump
Clot and bubble trap
Dialysate fluid out
Dialysate fluid in
A
Blood port
Dialysate fluid out
Semipermeable membrane
Dialysate fluid in
B
Blood port

FIGURE 54-3 • Hemodialysis system. Blood from an artery is pumped (**A**) into a dialyzer, where it flows through the synthetic capillary tubes (**B**), which act as the semipermeable membrane (*inset*). The dialysate, which has a particular chemical composition, flows into the dialyzer around the capillary tubes that the blood flows through. The waste products in the blood diffuse across the semipermeable membrane into the dialysate solution.

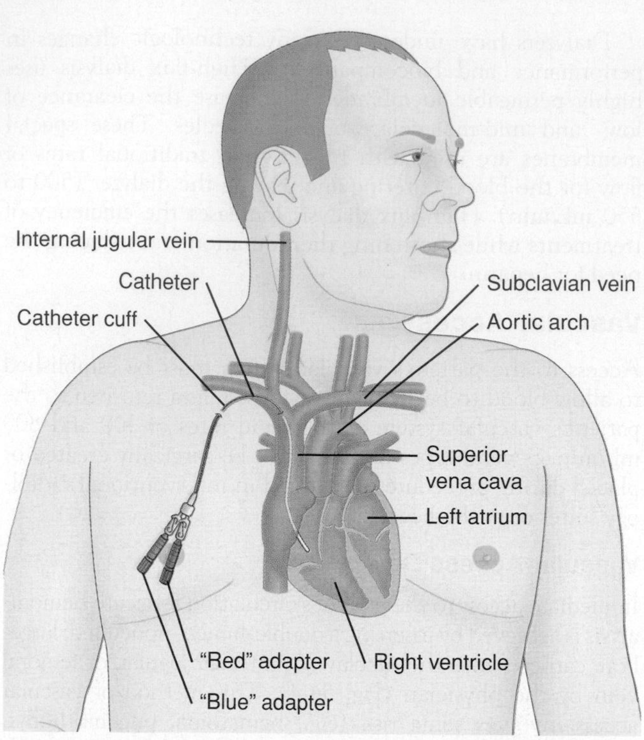
Internal jugular vein
Catheter
Catheter cuff
Subclavian vein
Aortic arch
Superior vena cava
Left atrium
"Red" adapter
Right ventricle
"Blue" adapter

FIGURE 54-4 • Double-lumen, cuffed hemodialysis catheter used in acute hemodialysis. The red lumen is attached to a blood line through which blood is pumped from the patient to the dialyzer. After the blood passes through the dialyzer (artificial kidney), it returns to the patient through the blue lumen of the catheter.

infection, thrombosis of the subclavian vein, inadequate flow). The catheter is removed when no longer needed (e.g., because the patient's condition has improved or another type of permanent access has been established). Double-lumen, cuffed catheters may also be inserted, usually by either a surgeon or interventional radiologist, into the internal jugular vein of the patient. Because these catheters have cuffs under the skin, the insertion site heals, sealing the wound and reducing the risk for ascending infection. This feature makes these catheters safe for longer-term use. However, infection rates remain high, and sepsis continues to be a common cause for hospital admission (Lincoln, 2011).

Arteriovenous Fistula

The preferred method of permanent vascular access for dialysis is an **arteriovenous fistula** (AVF) that is created surgically (usually in the forearm) by joining (anastomosing) an artery to a vein, either side to side or end to side (Fig. 54-5A). Needles are inserted into the vessel to obtain blood flow adequate to pass through the dialyzer. The arterial segment of the fistula is used for arterial flow to the dialyzer and the venous segment for reinfusion of the dialyzed blood. This access will need time (2 to 3 months) to "mature" before it can be used. As the AVF matures, the venous segment dilates due to the increased blood flow coming directly from the artery. Once sufficiently dilated, it will then accommodate two large-bore (14-, 15-, or 16-gauge) needles that are inserted for each dialysis treatment. The patient is encouraged to perform hand exercises to increase the size of these vessels (i.e., squeezing a rubber ball for forearm fistulas) to accommodate the large-bore needles. Once established, this access has the longest useful life and

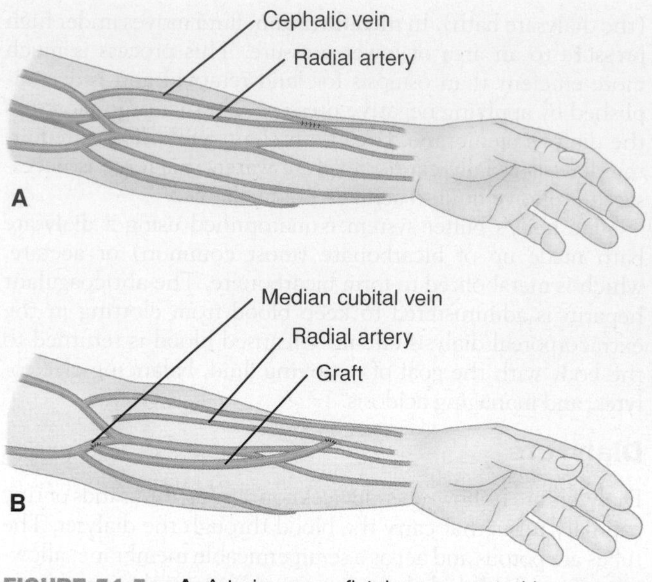
Cephalic vein
Radial artery
A
Median cubital vein
Radial artery
Graft
B

FIGURE 54-5 • **A.** Arteriovenous fistulas are created by anastomosing a patient's vein to an artery. This illustrates a side-to-side anastomosis. **B.** Arteriovenous grafts are established by connecting the artery and vein using synthetic tubing.

thus is the best option for vascular access for the chronic hemodialysis patient (Roy-Chaudhury, El-Khatib, Campos-Naciff, et al., 2012).

Arteriovenous Graft

An **arteriovenous graft** can be created by subcutaneously interposing a biologic, semibiologic, or synthetic graft material between an artery and vein (Fig. 54-5B). Usually, a graft is created when the patient's vessels are not suitable for creation of an AVF. Patients with compromised vascular systems (e.g., from diabetes) often require a graft because their native vessels may not be suitable for creation of an AVF. Grafts are usually placed in the arm but may be placed in the thigh or chest wall. Stenosis, infection, and thrombosis are the most common complications of this access. It is not at all uncommon to see a dialysis patient with numerous "old" or "nonfunctioning" accesses present on their arms. The patient is asked to identify the current access in use, and it is checked carefully for the presence of a bruit and thrill.

See the Special Considerations: Nursing Management of the Patient on Dialysis Who Is Hospitalized section later in this chapter for nursing interventions and care of the patient with an arteriovenous graft or fistula.

Quality and Safety Nursing Alert

Failure of the permanent dialysis access (fistula or graft) accounts for most hospital admissions of patients undergoing chronic hemodialysis. Thus, protection of the access is of high priority.

Complications

Although hemodialysis can prolong life, it does not alter the natural course of the underlying CKD, nor does it completely replace kidney function. The CKD complications previously discussed will continue to worsen and require treatment.

With the initiation of dialysis, disturbances of lipid metabolism (hypertriglyceridemia) are accentuated and contribute to cardiovascular complications. Heart failure, coronary artery disease, angina, stroke, and peripheral vascular disease may occur and can incapacitate the patient. Cardiovascular disease remains the leading cause of death in patients receiving dialysis (Danquah, Zimmerman, Diamond, et al., 2010).

Anemia is compounded by blood lost during hemodialysis. Gastric ulcers may result from the physiologic stress of chronic illness, medication, and preexisting medical conditions (e.g., diabetes). Patients with uremia report a metallic taste and nausea when they require dialysis. Vomiting may occur during the hemodialysis treatment when rapid fluid shifts and hypotension occur. These contribute to the malnutrition seen in patients on dialysis. Poor calcium metabolism and renal osteodystrophy can result in bone pain and fractures, interfering with mobility. As time on dialysis continues, calcification of major blood vessels has been reported and linked to hypertension and other vascular complications. Phosphorus deposits in the skin can occur and cause itching.

Many people undergoing hemodialysis experience major sleep problems that further complicate their overall health status (Danquah et al., 2010). Early-morning or late-afternoon dialysis may be a risk factor for developing sleep disturbances.

Other complications of dialysis treatment may include the following:

- Episodes of shortness of breath often occur as fluid accumulates between dialysis treatments.
- Hypotension may occur during the treatment as fluid is removed. Nausea and vomiting, diaphoresis, tachycardia, and dizziness are common signs of hypotension.
- Painful muscle cramping may occur, usually late in dialysis as fluid and electrolytes rapidly leave the extracellular space.
- Exsanguination may occur if blood lines separate or dialysis needles become dislodged.
- Dysrhythmias may result from electrolyte and pH changes or from removal of antiarrhythmic medications during dialysis.
- Air embolism is rare but can occur if air enters the vascular system.
- Chest pain may occur in patients with anemia or arteriosclerotic heart disease.
- Dialysis disequilibrium results from cerebral fluid shifts. Signs and symptoms include headache, nausea and vomiting, restlessness, decreased level of consciousness, and seizures. It is rare and more likely to occur in AKI or when BUN levels are very high (exceeding 150 mg/dL).

Nursing Management

The nurse in the dialysis unit has an important role in monitoring, supporting, assessing, and educating the patient. During dialysis, the patient, the dialyzer, and the dialysate bath require constant monitoring because numerous complications are possible, including clotting of the dialysis tubing or dialyzer, air embolism, inadequate or excessive fluid removal, hypotension, cramping, vomiting, blood leaks, contamination, and access complications. Nursing care of the patient and maintenance of the vascular access device are especially important and are discussed in the Special Considerations: Nursing Management of the Patient on Dialysis Who Is Hospitalized section.

Promoting Pharmacologic Therapy

Many medications are removed from the blood during hemodialysis; therefore, dosage or timing of the medication administration may require adjustment. Medications that are water soluble are readily removed during hemodialysis treatment, and those that are fat soluble or adhere to other substances (like albumin) are not dialyzed out very well. This is the reason some drug overdoses are treated with emergency hemodialysis and others are not.

Patients undergoing hemodialysis who require medications (e.g., cardiac glycosides, antibiotic agents, antiarrhythmic medications, antihypertensive agents) are monitored closely to ensure that blood and tissue levels of these medications are maintained without toxic accumulation. Antihypertensive therapy, often part of the regimen of patients on dialysis, is one example when communication, education, and evaluation can make a difference in patient outcomes. The patient must know when—and when not—to take the medication. For example, if an antihypertensive agent is taken on a dialysis day, hypotension may occur during dialysis, causing dangerously low blood pressure. Many medications that are taken once daily can be held until after the dialysis treatment.

Promoting Nutritional and Fluid Therapy

Diet is important for patients on hemodialysis because of the effects of uremia. Goals of nutritional therapy are to minimize uremic symptoms and fluid and electrolyte imbalances; to maintain good nutritional status through adequate protein, calorie, vitamin, and mineral intake; and to enable the patient to eat a palatable and enjoyable diet. Restricting dietary protein decreases the accumulation of nitrogenous wastes, reduces uremic symptoms, and may even postpone the initiation of dialysis for a few months. Restriction of fluid is also part of the dietary prescription because fluid accumulation may occur, leading to weight gain, heart failure, and pulmonary edema.

With the initiation of hemodialysis, the patient usually requires some restriction of dietary protein, sodium, potassium, phosphorus, and fluid intake. Protein intake is restricted to about 1.2 to 1.3 g/kg ideal body weight per day; therefore, protein must be of high biologic quality. Sodium is usually restricted to 2 to 3 g/day; fluids are restricted to an amount equal to the daily urine output plus 500 mL/day. The goal for patients on hemodialysis is to keep their interdialytic (between dialysis treatments) weight gain under 1.5 kg. Potassium restriction depends on the amount of residual renal function and the frequency of dialysis. Dietary restriction is an unwelcome change in lifestyle for many patients with ESKD. Patients can feel stigmatized in social situations because there may be few food choices available for their diet. If the restrictions are ignored, life-threatening complications, such as hyperkalemia and pulmonary edema, may result. Thus, the patient may feel punished for responding to basic human drives to eat and drink. The nurse who cares for a patient with symptoms or complications resulting

Chart 54-8

NURSING RESEARCH PROFILE

The Benefits of Home Blood Pressure Monitoring in Hemodialysis Patients

Lingerfelt, K., & Hodnicki, D. (2012). Hypertension management in patients receiving hemodialysis: The benefits of home blood pressure monitoring. *Nephrology Nursing Journal*, *39*(1), 31–36.

Purpose

Patients with end-stage kidney disease on hemodialysis are at risk for cardiovascular events due to hypertension that can be difficult to assess due to unreliable blood pressure (BP) measurements at dialysis centers. The purpose of this pilot project was to describe BP evaluation and prevent overtreatment of elevated BPs.

Design

This descriptive study surveyed 36 patients on hemodialysis in a single hemodialysis center. Hemodialysis patients whose in-center blood pressure measurements did not meet the recommended goal of 130/80 mm Hg after consecutive dialysis treatments over a 1 to 2-week period of time were candidates for the study. Participants were followed with biweekly BP monitoring over a 6-week period. Data related to fluid gain between dialysis treatments and self-reported sodium intake was collected.

Findings

After 6 weeks of biweekly reviews and treatment adjustments when indicated, home BP measurements were evaluated to determine if the BP goal in hypertension management was met. Not all participants with BP measurements above 130/80 using peridialysis measurements needed medication adjustments. Home BP measurements taken after hemodialysis and prior to fluid and uremic toxin level accumulation may provide a more accurate depiction of hypertension control compared to predialysis BP measurements.

Nursing Implications

Adjustment of antihypertensive medications takes place in the dialysis center, hospital, community setting, and the outpatient office. The findings from this study indicate that home BP monitoring provides a thorough assessment of the patient's BP and may prevent overtreatment in some patients. The risk for symptomatic hypotension for patients on hemodialysis may be reduced if home BP monitoring is available to this patient population.

from dietary indiscretion must avoid harsh, judgmental, or punitive tones when communicating with him or her. Regular education with reinforcement is needed to achieve this difficult change in lifestyle (Lingerfelt & Thornton, 2011) (Chart 54-8).

Meeting Psychosocial Needs

Patients requiring long-term hemodialysis are often concerned about the unpredictability of the illness and the disruption of their lives. They often have financial problems, difficulty holding a job, waning sexual desire and impotence, clinical depression, and fear of dying. Younger patients worry about marriage, having children, and the burden that they bring to their families. The regimented lifestyle that frequent dialysis treatments and restrictions in food and fluid intake impose can be demoralizing to the patient and family.

Dialysis alters the lifestyle of the patient and family. The amount of time required for dialysis and primary provider visits and being chronically ill can create conflict, frustration, guilt, and depression. It may be difficult for the patient, spouse, and family to express anger and negative feelings.

The nurse needs to give the patient and family the opportunity to express feelings of anger and concern about the limitations that the disease and treatment impose, possible financial problems, and job insecurity. If anger is not expressed, it may be directed inward and lead to depression, despair, and attempts at suicide (suicide is more prevalent in patients on dialysis); however, if anger is projected outward to other people, it may destroy already threatened family relationships (Makaroff, 2012).

Although these feelings are normal in this situation, they are often profound and overwhelming. Counseling and psychotherapy may be necessary. Depression may require treatment with antidepressant agents. Referring the patient and family to a mental health provider with expertise in the care of patients receiving dialysis may also be helpful. Clinical nurse specialists, psychologists, and social workers may be helpful in assisting the patient and family to cope with the changes brought about by renal failure and its treatment.

The sense of loss that the patient experiences cannot be underestimated because every aspect of a "normal life" is disrupted. Some patients use denial to deal with the overwhelming array of medical problems (e.g., infections, hypertension, anemia, neuropathy). Staff who are tempted to label the patient as noncompliant must consider the impact of renal failure and its treatment on the patient and family and the coping strategies that they may use.

Palliative care principles that focus on symptom control are becoming increasingly important as greater attention is focused on quality-of-life issues (Bayoumi & El-Fouly, 2010). Patients and their families are encouraged to discuss end-of-life options and to develop advanced directives or living wills.

Promoting Home and Community-Based Care

Educating Patients About Self-Care

Preparing a patient for hemodialysis is essential. Assessment helps identify the learning needs of the patient and family members. In many cases, the patient is discharged home before learning needs and readiness to learn can be thoroughly evaluated; therefore, hospital-based nurses, dialysis staff, and home care nurses must work together to provide appropriate education that meets the patient's and family's changing needs and readiness to learn.

The diagnosis of ESKD and the need for dialysis can overwhelm the patient and family. In addition, many patients with ESKD have clinical depression, a shortened attention span, a decreased level of concentration, and altered perception. Therefore, education must occur in brief, 10- to 15-minute sessions, with time added for clarification, repetition, reinforcement, and questions from the patient and family. The nurse needs to convey a nonjudgmental attitude to enable the patient and family to discuss options and their feelings about those options. Team conferences are helpful for sharing information and providing every team member the opportunity to discuss the needs of the patient and family.

Home Hemodialysis

Most patients who undergo hemodialysis do so in an outpatient setting, but home hemodialysis is an option for some. Home hemodialysis requires a highly motivated patient who is willing to take responsibility for the procedure and is able to adjust each treatment to meet the body's changing needs. It also requires the commitment and cooperation of a caregiver to assist the patient. However, many patients are not comfortable imposing on others this way and do not wish to subject family members to the feeling that their home is being turned into a clinic. The health care team never forces a patient to use home hemodialysis, because this treatment requires changes in the home and family. Home hemodialysis must be the patient's and family's decision (Loiselle, O'Connor, & Michaud, 2011).

The patient undergoing home hemodialysis and the caregiver assisting that patient must be trained to prepare, operate, and disassemble the dialysis machine; maintain and clean the equipment; administer medications (e.g., heparin) into the machine lines; and handle emergency problems (hemodialysis dialyzer rupture, electrical or mechanical problems, hypotension, shock, and seizures). Because home hemodialysis places primary responsibility for the treatment on the patient and the family member, they must understand and be capable of performing all aspects of the hemodialysis procedure (Chart 54-9).

Before home hemodialysis is initiated, the home environment, household and community resources, and ability and willingness of the patient and family to carry out this treatment are assessed. The home is surveyed to see if electrical outlets, plumbing facilities, and storage space are adequate. Modifications may be needed to enable the patient and assistant to perform dialysis safely and to deal with emergencies.

Once home hemodialysis is initiated, the home care nurse must visit periodically to evaluate compliance with the recommended techniques, to assess the patient for complications, to reinforce previous instruction, and to provide reassurance.

Continuing Care

The health care team's goal in treating patients with chronic renal failure is to maximize their vocational potential, functional status, and quality of life. To facilitate renal rehabilitation, appropriate follow-up and monitoring by members of the health care team (physicians, dialysis nurses, social worker, psychologist, home care nurses, and others as appropriate) are essential to identify and resolve problems early on. Many patients with chronic renal failure can resume relatively normal lives, doing the things that are important to them: traveling, exercising, working, or actively participating in family activities. If appropriate interventions are available early in the course of dialysis, the potential for better health improves, and the patient can remain active in family and community life. Outcome goals for renal rehabilitation include employment for those able to work, improved physical functioning of all patients, improved understanding about adaptation and options for living well, increased control over the effects of kidney disease and dialysis, and resumption of activities enjoyed before dialysis.

Continuous Renal Replacement Therapies

CRRTs may be indicated for patients with acute or chronic renal failure who are too clinically unstable for traditional

Chart 54-9 HOME CARE CHECKLIST
Hemodialysis

At the completion of home care education, the patient or caregiver will be able to:	PATIENT	CAREGIVER
• Discuss renal failure and its effects on the body	✔	✔
• Describe the cause of renal failure and why hemodialysis is necessary.	✔	✔
• Describe the basic principles of hemodialysis.	✔	✔
• Discuss common problems that may occur during hemodialysis and their prevention and management.	✔	✔
• Demonstrate knowledge about prescribed medications and the reason for their use, potential side effects, guidelines on when to notify the primary provider, and the schedule of medications on dialysis and nondialysis days.	✔	✔
• Acknowledge dietary and fluid restrictions, rationale, and consequences of noncompliance.	✔	✔
• Describe commonly measured laboratory values, results, and implications.	✔	✔
• List guidelines for prevention and detection of fluid overload, meaning of "dry" weight, and how to weigh self.	✔	✔
• Demonstrate vascular access care, how to check patency, signs and symptoms of infection, and prevention of complications.	✔	✔
• Discuss strategies for detection, management, and relief of pruritus, neuropathy, and other complications of renal failure.	✔	✔
• Develop strategies to manage or reduce anxiety and maintain independence.	✔	✔
• Coordinate financial arrangements for dialysis and strategies to identify and obtain resources.	✔	✔

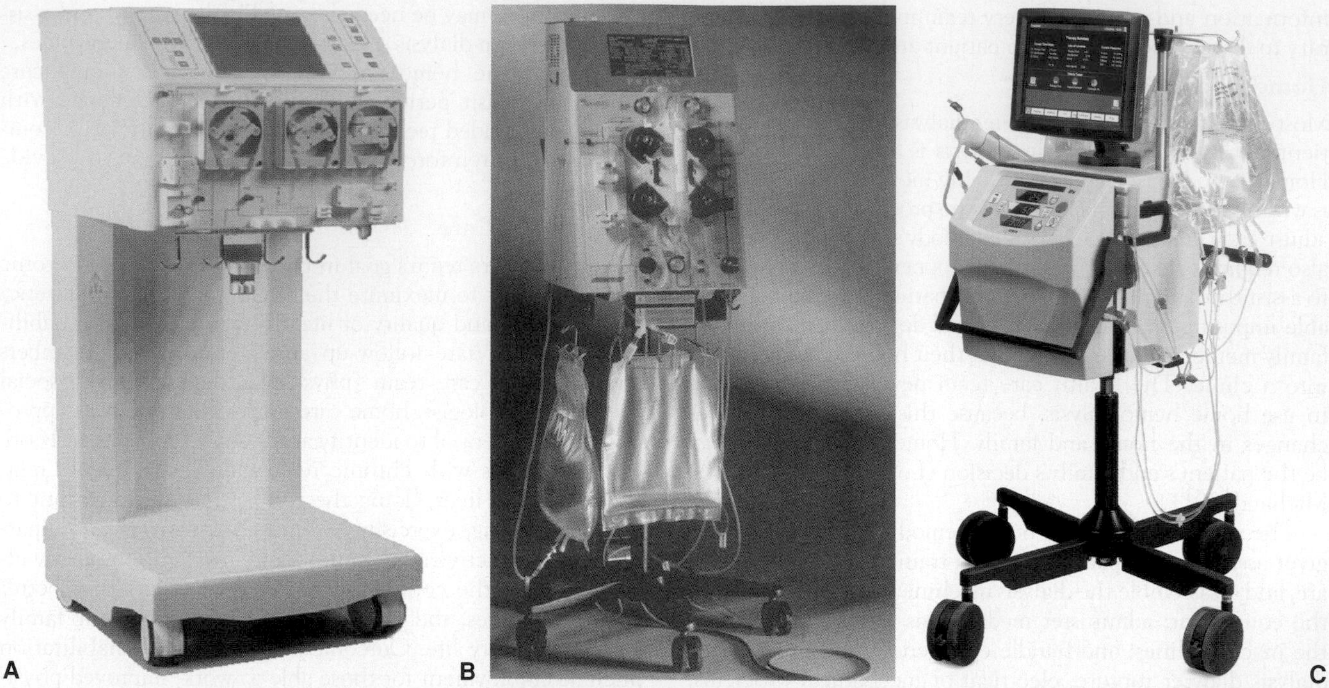

A B C

FIGURE 54-6 • Devices for administering continuous renal replacement therapy (CRRT) offer an integrated fluid warmer for the heating of infusion and dialysate fluids, a weighing system to reduce the possibility of error in assessing fluid balance, and a battery backup that allows treatments to continue when the patient is moved. **A.** Diapact CRRT System, B-Braun Medical, Inc., Bethlehem, PA. **B.** PRISMA, Gambro Corporation, Lakewood, CO. **C.** System One, NxStage Medical Inc., Lawrence, MA.

hemodialysis, for patients with fluid overload secondary to oliguric (low urine output) renal failure, and for patients whose kidneys cannot handle their acutely high metabolic or nutritional needs. Some forms of CRRT may not require dialysis machines or dialysis personnel to carry out the procedures and can be initiated quickly. Several types of CRRT are available and widely used in critical care units (Fig. 54-6). The methods are similar, as they require access to the circulation and blood to pass through an artificial filter. A hemofilter (an extremely porous blood filter containing a semipermeable membrane) is used in all types.

Continuous Venovenous Hemofiltration

Continuous venovenous hemofiltration (CVVH) is often used to manage AKI. Blood from a double-lumen venous catheter is pumped (using a small blood pump) through a hemofilter and then returned to the patient through the same catheter. CVVH provides continuous slow fluid removal (ultrafiltration); therefore, hemodynamic effects are mild and better tolerated by patients with unstable conditions. CVVH requires a dual-lumen venous catheter and critical care nurses trained in management of the therapy who can set up, initiate, maintain, and terminate the system. Many hospitals have developed a collaborative approach to managing the CVVH therapy between the critical care and nephrology nursing staff.

Continuous Venovenous Hemodialysis

Continuous venovenous hemodialysis (CVVHD) is similar to CVVH. Blood is pumped from a double-lumen venous catheter through a hemofilter and returned to the patient through the same catheter. In addition to the benefits of ultrafiltration, CVVHD uses a concentration gradient to facilitate the removal of uremic toxins and fluid by adding a dialysate solution into the circuit. A dual-lumen venous catheter is required; hemodynamic effects are usually mild; and critical care nurses can set up, initiate, maintain, and terminate the system with the support of the nephrology nursing staff.

Peritoneal Dialysis

The goals of PD are to remove toxic substances and metabolic wastes and to reestablish normal fluid and electrolyte balance. PD may be the treatment of choice for patients with renal failure who are unable or unwilling to undergo hemodialysis or kidney transplantation. Patients who are susceptible to the rapid fluid, electrolyte, and metabolic changes that occur during hemodialysis experience fewer of these problems with the slower rate of PD. Therefore, patients with diabetes or cardiovascular disease, many older patients, and those who may be at risk for adverse effects of systemic heparin are likely candidates for PD. Additionally, severe hypertension, heart failure, and pulmonary edema not responsive to usual treatment regimens have been successfully treated with PD. Fewer than 8% of patients with ESKD receive PD as their treatment modality (USRDS, 2011).

In PD, the peritoneal membrane that covers the abdominal organs and lines the abdominal wall serves as the semipermeable membrane. A sterile dextrose dialysate fluid is introduced into the peritoneal cavity through an abdominal catheter at established intervals (Fig. 54-7). Once the sterile solution is in the peritoneal cavity, uremic toxins such as urea and creatinine begin to be cleared from the blood. Diffusion and osmosis occur as waste products move from an area of higher concentration (the bloodstream) to an area of lesser

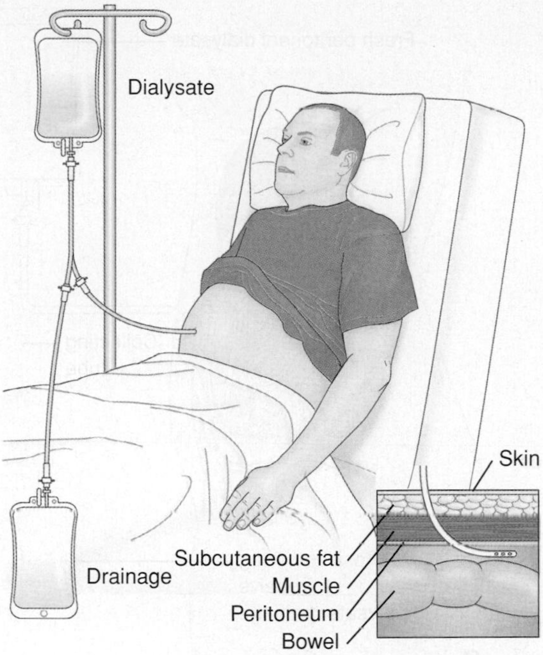
Dialysate
Skin
Subcutaneous fat
Drainage
Muscle
Peritoneum
Bowel

FIGURE 54-7 • In peritoneal dialysis and in acute intermittent peritoneal dialysis, dialysate is infused into the peritoneal cavity by gravity, after which the clamp on the infusion line is closed. After a dwell time (when the dialysate is in the peritoneal cavity), the drainage tube is unclamped and the fluid drains from the peritoneal cavity, again by gravity. A new container of dialysate is infused as soon as drainage is complete. The duration of the dwell time depends on the type of peritoneal dialysis.

concentration (the dialysate fluid) through a semipermeable membrane (the peritoneum) (Prowant, Moore, Satalowich, et al., 2010). This movement of solute from the blood into the dialysate fluid is called *clearance*. Because substances cross the peritoneal membrane at different rates, adjustments in solution dwell time in the peritoneal cavity and amount of fluid used are made to facilitate the process. Ultrafiltration (water removal) occurs in PD through an osmotic gradient created by using a dialysate fluid with a higher glucose concentration than the blood.

Procedure

As with other forms of treatment, the decision to begin PD is made by the patient and family in consultation with the physician. The patient may be acutely ill, thus requiring short-term treatment to correct severe disturbances in fluid and electrolyte status, or may have ESKD and need to receive ongoing treatments.

Preparing the Patient

The nurse's preparation of the patient and family for PD depends on the patient's physical and psychological status, level of alertness, previous experience with dialysis, and understanding of and familiarity with the procedure.

The nurse explains the procedure to the patient and assists in obtaining signed consent for insertion of the catheter. Baseline vital signs, weight, and serum electrolyte levels are recorded. Evaluation of the abdomen for placement of the catheter is done to facilitate self-care. Typically, the catheter is placed on the nondominant side to allow the patient easier access to the catheter connection site when exchanges are done. The patient is encouraged to empty the bladder and bowel to reduce the risk of puncture of internal organs during the insertion procedure. Broad-spectrum antibiotic agents may be administered to prevent infection. The peritoneal catheter can be inserted in interventional radiology, in the operating room, or at the bedside. Depending on the situation, this will need to be explained to the patient and family.

Preparing the Equipment

In addition to assembling the equipment for PD, the nurse consults with the physician to determine the concentration of dialysate to be used and the medications to be added. Heparin may be added to prevent fibrin formation and resultant occlusion of the peritoneal catheter. Potassium chloride may be prescribed to prevent hypokalemia. Antibiotic agents may be added to treat **peritonitis** (inflammation of the peritoneal membrane) caused by infection. Regular insulin may be added for patients with diabetes due to the high dextrose concentration of the PD solution. Aseptic technique is imperative whenever medications are added to the PD solution.

Before medications are added, the dialysate is warmed to body temperature to prevent patient discomfort and abdominal pain and to dilate the vessels of the peritoneum to increase urea clearance. Solutions that are too cold cause pain, cramping, and vasoconstriction and reduce clearance. Dry heating (heating cabinet, incubator, or heating pad) is recommended. Methods not recommended include soaking the bags of solution in warm water (can introduce bacteria to the exterior of the bags of solution and increase the chance of peritonitis) and the use of a microwave oven to heat the fluid (increases the danger of burning the peritoneum).

Immediately before initiating dialysis, using aseptic technique, the nurse assembles the administration set and tubing. The tubing is filled with the prepared dialysate to reduce the amount of air entering the catheter and peritoneal cavity, which could increase abdominal discomfort and interfere with instillation and drainage of the fluid.

Inserting the Catheter

Ideally, the peritoneal catheter is inserted in the operating room or radiology suite to maintain surgical asepsis and minimize the risk of contamination. Catheters for long-term use are usually soft and flexible and made of silicone with a radiopaque strip to permit visualization on x-ray. These catheters have three sections: (1) an intraperitoneal section, with numerous openings and an open tip to allow dialysate to flow freely; (2) a subcutaneous section that passes from the peritoneal membrane and tunnels through muscle and subcutaneous fat to the skin; and (3) an external section for connection to the dialysate and tubing system. Most of these catheters have two cuffs made of Dacron polyester. The cuffs stabilize the catheter, limit movement, prevent leaks, and provide a barrier against microorganisms. One cuff is placed just distal to the peritoneum, and the other cuff is placed subcutaneously. The subcutaneous tunnel (5 to 10 cm long) further protects against bacterial infection (Fig. 54-8).

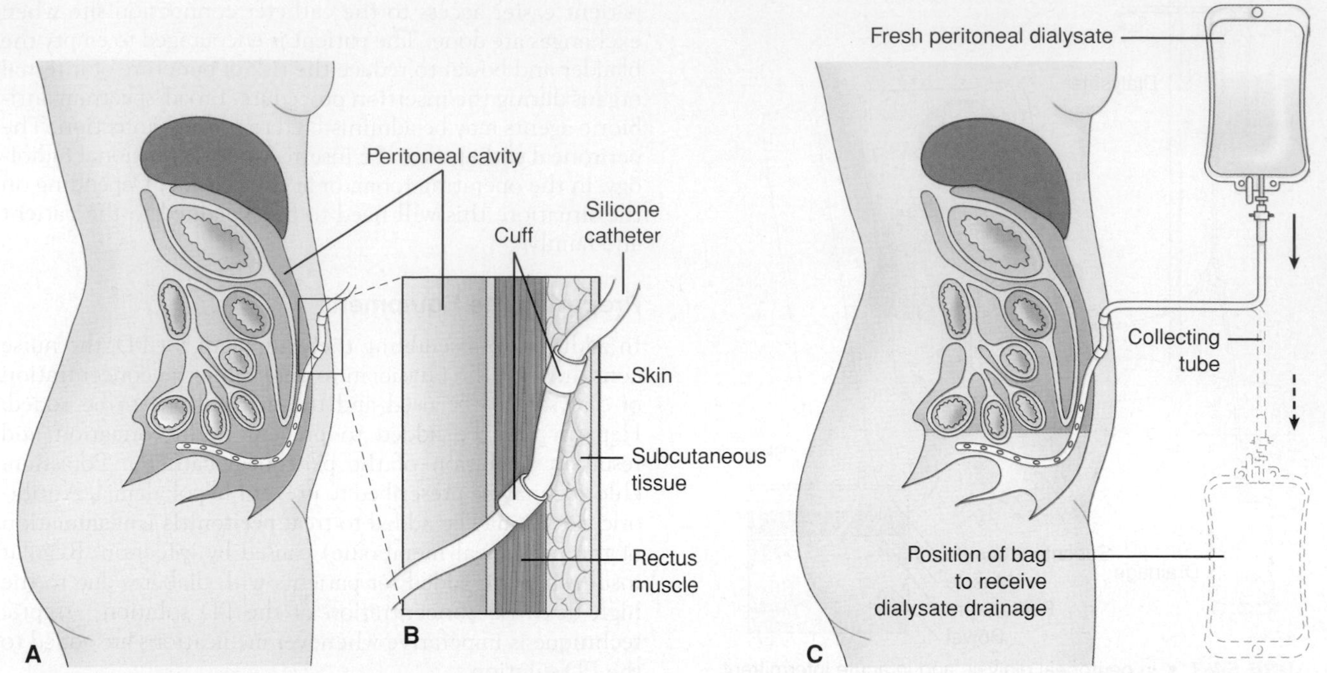
Peritoneal cavity
Cuff
Silicone catheter
Skin
Subcutaneous tissue
Rectus muscle
A
B
Fresh peritoneal dialysate
Collecting tube
Position of bag to receive dialysate drainage
C

FIGURE 54-8 • Continuous ambulatory peritoneal dialysis. **A.** The peritoneal catheter is implanted through the abdominal wall. **B.** Dacron cuffs and a subcutaneous tunnel provide protection against bacterial infection. **C.** Dialysate flows by gravity through the peritoneal catheter into the peritoneal cavity. After a prescribed period of time, the fluid is drained by gravity and discarded. New solution is then infused into the peritoneal cavity until the next drainage period. Dialysis thus continues on a 24-hour-a-day basis, during which the patient is free to move around and engage in his or her usual activities.

Performing the Exchange

PD involves a series of exchanges or cycles. An **exchange** is the entire cycle including the infusion (fill), dwell, and drainage of the dialysate. This cycle is repeated throughout the course of the dialysis. The dialysate is infused by gravity into the peritoneal cavity. A period of about 5 to 10 minutes is usually required to infuse 2 to 3 L of fluid. The prescribed dwell, or equilibration, time allows diffusion and osmosis to occur. At the end of the dwell time, the drainage portion of the exchange begins. The tube is unclamped, and the solution drains from the peritoneal cavity by gravity through a closed system. Drainage is usually completed in 10 to 20 minutes. The drainage fluid is normally colorless or straw colored and should not be cloudy. Bloody drainage may be seen in the first few exchanges after insertion of a new catheter but should not occur after that time. The number of cycles or exchanges and their frequency are prescribed based on monthly laboratory values and the presence of uremic symptoms. The exchanges can be performed manually during the waking hours by the patient (**continuous ambulatory peritoneal dialysis [CAPD]**) or via the use of a PD machine (cycler) that automatically performs exchanges, usually while the patient is sleeping at night (**continuous cycling peritoneal dialysis [CCPD]**).

The removal of excess water during PD occurs because dialysate has a high dextrose concentration, making it hypertonic. An osmotic gradient is created between the blood and the dialysate solution. Dextrose solutions of 1.5%, 2.5%, and 4.25% are available in several volumes, from 1,000 mL to 3,000 mL. The higher the dextrose concentration, the greater the osmotic gradient and the more water will be removed. Selection of the appropriate solution is based on the patient's fluid status (Prowant et al., 2010).

Complications

Most complications of PD are minor; however, several, if unattended, can have serious consequences.

Acute Complications

Peritonitis

Peritonitis is the most common and serious complication of PD. The first sign of peritonitis is cloudy dialysate drainage fluid. Diffuse abdominal pain and rebound tenderness occur much later. Hypotension and other signs of shock may also occur with advancing infection. The patient with peritonitis may be treated as an inpatient or outpatient (most common), depending on the severity of the infection and the patient's clinical status. Drainage fluid is examined for cell count; Gram stain and culture are used to identify the organism and guide treatment. Antibiotic agents (aminoglycosides or cephalosporins) are usually added to subsequent exchanges until Gram stain or culture results are available for appropriate antibiotic determination. Intraperitoneal administration of antibiotic agents is as effective as IV administration and therefore most often used. Antibiotic therapy continues for 10 to 14 days (White & Vinet, 2010). Careful selection and calculation of the antibiotic dosage are needed to prevent nephrotoxicity and further compromise of residual renal function.

Regardless of the organism causing peritonitis, the patient loses large amounts of protein through the peritoneum. Acute malnutrition and delayed healing may result. Therefore, attention must be given to detecting and promptly treating peritonitis.

Leakage

Leakage of dialysate through the catheter site may occur immediately after the catheter is inserted. Usually, the leak stops spontaneously if dialysis is withheld for several days, giving the tissue surrounding the cuffs located on the abdominal catheter a chance to infiltrate the Dacron and seal the insertion tunnel. It also allows the exit site time to heal. During this time, it is important to reduce factors that might delay healing, such as undue abdominal muscle activity and straining during bowel movement (Yang, Wang, Yeh, et al., 2011). In many cases, leakage can be avoided by using small volumes (500 mL) of dialysate, gradually increasing the volume up to 2,000 to 3,000 mL.

Bleeding

A bloody **effluent** (drainage) may be observed occasionally, especially in young, menstruating women. (The hypertonic fluid pulls blood from the uterus, through the opening in the fallopian tubes, and into the peritoneal cavity.) Bleeding is also common during the first few exchanges after a new catheter insertion because some blood enters the abdominal cavity following insertion. In many cases, no cause can be found for the bleeding, although catheter displacement from the pelvis has occasionally been associated with bleeding. Some patients have had bloody effluent after an enema or from minor trauma. Most often, bleeding stops in 1 to 2 days and requires no specific intervention. More frequent exchanges and the addition of heparin to the dialysate during this time may be necessary to prevent blood clots from obstructing the catheter.

Long-Term Complications

Hypertriglyceridemia is common in patients undergoing long-term PD, suggesting that the therapy may accelerate atherogenesis. Cardiovascular disease is the leading cause of morbidity and mortality in patients with kidney failure, and many patients have suboptimal blood pressure control (Eskridge, 2010). Beta-blockers and ACE inhibitors should be used to control hypertension or protect the heart, and the use of aspirin and statins should be considered.

Other complications that may occur with long-term PD include abdominal hernias (incisional, inguinal, diaphragmatic, and umbilical), probably resulting from continuously increased intra-abdominal pressure. The persistently elevated intra-abdominal pressure also aggravates symptoms of hiatal hernia and hemorrhoids. Low back pain and anorexia from fluid in the abdomen and a constant sweet taste related to glucose absorption may also occur.

Mechanical problems occasionally occur and may interfere with instillation or drainage of the dialysate. Formation of clots in the peritoneal catheter and constipation are factors that may contribute to these problems.

Approaches

PD can be performed using several different approaches: acute intermittent peritoneal dialysis, CAPD, and CCPD.

Acute Intermittent Peritoneal Dialysis

Indications for acute intermittent PD, a variation of PD, include uremic signs and symptoms (nausea, vomiting, fatigue, altered mental status), fluid overload, acidosis, and hyperkalemia. Although PD is not as efficient as hemodialysis in removing solute and fluid, it permits a more gradual change in the patient's fluid volume status and in waste product removal. Therefore, it may be the treatment of choice for the hemodynamically unstable patient. It can be carried out manually (the nurse warms, spikes, and hangs each container of dialysate) or by a cycler machine. Exchange times range from 30 minutes to 2 hours. A common routine is hourly exchanges consisting of a 10-minute infusion, a 30-minute dwell time, and a 20-minute drain time (Prowant et al., 2010).

Maintaining the PD cycle is a nursing responsibility. Strict aseptic technique is maintained when changing solution containers and emptying drainage containers. Vital signs, weight, I&O, laboratory values, and patient status are frequently monitored. The nurse uses a flow sheet or the electronic medical record to document each exchange and records vital signs, dialysate concentration, medications added, exchange volume, dwell time, dialysate fluid balance for each exchange (fluid lost or gained), and cumulative fluid balance. The nurse also carefully assesses skin turgor and mucous membranes to evaluate fluid status and monitor the patient for edema.

> ***Quality and Safety Nursing Alert***
>
> **If the peritoneal fluid does not drain properly, the nurse can facilitate drainage by turning the patient from side to side or raising the head of the bed. The catheter should never be pushed further into the peritoneal cavity.**

Other measures to promote drainage include checking the patency of the catheter by inspecting for kinks, closed clamps, or an air lock. The nurse monitors for complications, including peritonitis, bleeding, respiratory difficulty, and leakage of peritoneal fluid. Abdominal girth may be measured periodically to determine if the patient is retaining large amounts of dialysis solution. In addition, the nurse must ensure that the PD catheter remains secure and that the dressing remains dry. Physical comfort measures, frequent turning, and skin care are provided. The patient and family are educated about the procedure and are kept informed about progress (fluid loss, weight loss, laboratory values). Emotional support and encouragement are given to the patient and family during this stressful and uncertain time.

Continuous Ambulatory Peritoneal Dialysis

CAPD is the second most common form of dialysis for patients with ESKD (USRDS, 2011). CAPD is performed at home by the patient or a trained caregiver who is usually a family member. The procedure allows the patient reasonable freedom and control of daily activities but requires a serious commitment to be successful. Chart 54-10 discusses suitability for CAPD.

CAPD works on the same principles as other forms of PD: diffusion and osmosis. Less extreme fluctuations in the patient's laboratory values occur with CAPD than with intermittent PD or hemodialysis because the dialysis is constantly in progress. The serum electrolyte levels usually remain in the normal range.

Chart 54-10 Considerations in Continuous Ambulatory Peritoneal Dialysis

Although continuous ambulatory peritoneal dialysis (CAPD) is not suitable for all patients with end-stage kidney disease (ESKD), it is a viable therapy for those who can perform self-care and fluid exchanges and fit therapy into their own routines. Often, patients report having more energy and feeling healthier once they begin CAPD. Nurses can be instrumental in helping patients with ESKD find the dialysis therapy that best suits their lifestyle. Those considering CAPD need to understand the advantages and disadvantages along with the indications and contraindications for this form of therapy.

Advantages

- Freedom from a hemodialysis machine
- Control over daily activities
- Opportunities to eat a more liberal diet than allowed with hemodialysis, increase fluid intake, raise serum hematocrit values, improve blood pressure control, avoid venipuncture, and gain a sense of well-being

Disadvantages

- Continuous dialysis 24 hours a day, 7 days a week
- Dietary alterations related to protein and potassium losses. Patients may be encouraged to increase the intake of protein and potassium in the diet due to these losses with peritoneal dialysis fluid exchanges.

Indications

- Patient's willingness, motivation, and ability to perform dialysis at home
- Strong family or community support system (essential for success), particularly if the patient is an older adult
- Special problems with long-term hemodialysis, such as dysfunctional or failing vascular access devices, excessive thirst, severe hypertension, postdialysis headaches, and severe anemia requiring frequent transfusion
- Interim therapy while awaiting kidney transplantation
- ESKD secondary to diabetes because hypertension, uremia, and hyperglycemia are easier to manage with CAPD than with hemodialysis

Contraindications

- Adhesions from previous surgery (adhesions reduce clearance of solutes) or systemic inflammatory disease
- Chronic backache and preexisting disk disease, which could be aggravated by the continuous pressure of dialysis fluid in the abdomen
- Severe arthritis or poor hand strength necessitating assistance in performing the exchange. However, patients who are blind or partially blind and those with other physical limitations can learn to perform CAPD.

Procedure

The patient performs exchanges four or five times a day, 24 hours a day, 7 days a week, at intervals scheduled throughout the day. Different manufacturers supply different equipment. A Y-shaped system is most commonly used, in which a bag containing dialysate solution comes connected to one branch of the "Y" and a sterile empty bag is connected to the second branch. This leaves the third part of the "Y" open and available for connection to the transfer set on the PD catheter. To perform an exchange, the patient (or person doing the exchange) washes his or her hands, dons a mask, and then removes the cap from the transfer set while maintaining sterility. The open end of the "Y" set is connected to the end of the transfer set and the dialysate infused where it will dwell. After the dialysate is infused, the patient clamps off the transfer set and the tubing set, disconnects the tubing set, and applies a new cap to the transfer set, making it a closed system. The patient drains the fluid (effluent) from the peritoneal cavity through the catheter (over about 20 to 30 minutes) into an empty bag. Once the effluent has been fully drained, fresh fluid is instilled into the peritoneal cavity.

The longer the dwell time, the better the clearance of uremic toxins. If dwell time is excessive, the patient will absorb some of the effluent back into the body simply because the osmotic gradient is lost. Once equilibrium is reached, the movement of fluid and toxins stops.

Complications

To reduce the risk of peritonitis, the patient (and all caregivers) must use meticulous care to avoid contaminating the catheter, fluid, or tubing and to avoid accidentally disconnecting the catheter from the tubing. Whenever a connection or disconnection is made, hand hygiene must be used and a mask worn by anyone within 6 feet of the area to avoid contamination with airborne bacteria. Excess manipulation should be avoided, and meticulous care of the catheter entry site is provided using a standardized protocol.

Continuous Cyclic Peritoneal Dialysis

CCPD uses a machine called a *cycler* to provide the fluid exchanges. It is programmed to deliver an established amount of PD solution that will dwell in the peritoneal cavity for a programmed period of time before it drains from the peritoneal cavity via gravity. The cycler is also set to deliver a specific number of fluid changes in a designated period of time. Because it is programmed, it also keeps track of the total amounts removed and will sound an alarm if limits are not met. It requires that a person set up and break down the system for use, which typically takes about 15 minutes.

CCPD combines overnight intermittent PD with a prolonged dwell time during the day. The peritoneal catheter is connected to a cycler machine every evening, usually just before the patient goes to sleep for the night. Because the machine is very quiet, the patient can sleep, and the extra-long tubing allows the patient to move and turn normally during sleep.

In the morning, the patient disconnects from the cycler. Sometimes, dialysate is left in the abdominal cavity for a longer day dwell cycle. This day exchange is drained during the day either by using a "Y" set or reattaching to the cycler. This process is done every day to achieve the effects of dialysis required.

CCPD has a lower infection rate than other forms of PD because there are fewer opportunities for contamination with bag changes and tubing disconnections (Kelman, 2012). It also allows the patient to be free from exchanges throughout the day, making it possible to engage in work and activities of daily living more freely.

Nursing Management

Meeting Psychosocial Needs

In addition to the complications of PD described previously, patients who elect to do PD may experience altered body image because of the presence of the abdominal catheter, bag, tubing, and cycler. Waist size increases from 1 to 2 inches (or more) with fluid in the abdomen. This affects clothing selection and may make the patient feel "fat." Body image may be so altered that patients do not want to look at or care for the catheter for days or weeks. The nurse may arrange for the patient to talk with other patients who have adapted well to PD. Although some patients have no psychological problems with the catheter—they think of it as their lifeline and as a life-sustaining device—other patients feel they are doing exchanges all day long and have no free time, particularly in the beginning. They may experience depression because they feel overwhelmed with the responsibility of self-care.

Patients undergoing PD may also experience altered sexuality patterns and sexual dysfunction. The patient and partner may be reluctant to engage in sexual activities, partly because of the catheter being psychologically "in the way" of sexual performance. The peritoneal catheter, drainage bag, and about 2 L of dialysate may interfere with the patient's sexual function and body image as well. In patients on CCPD, the presence of the dialysis cycler in the bedroom and the continual connection during the sleeping hours can also cause interference with intimacy. Although these problems may resolve with time, some problems may warrant special counseling. Questions by the nurse about concerns related to sexuality and sexual function often provide the patient with a welcome opportunity to discuss these issues and a first step toward their resolution.

Promoting Home and Community-Based Care

Educating Patients About Self-Care

Patients are educated as inpatients or outpatients to perform PD once their condition is medically stable. Education usually takes 5 days to 2 weeks. Patients are taught according to their own learning ability and knowledge level and only as much at one time as they can handle without feeling uncomfortable or becoming overwhelmed. Education topics for the patient and family who will be performing PD at home are described in Chart 54-11. The use of an adult learning theory–based curriculum may decrease peritonitis and exit site infection rates.

Because of protein loss with continuous PD, the patient is instructed to eat a high-protein, nutritious diet. The patient is also encouraged to increase his or her daily fiber intake to help prevent constipation, which can impede the flow of

Chart 54-11 HOME CARE CHECKLIST

Peritoneal Dialysis, Continuous Ambulatory Peritoneal Dialysis, or Continuous Ambulatory Cyclic Dialysis

At the completion of home care education, the patient or caregiver will be able to:	PATIENT	CAREGIVER
• Discuss basic information about normal kidney function.	✔	✔
• Discuss basic information about the disease process.	✔	✔
• Discuss the basic principles of peritoneal dialysis.	✔	✔
• Demonstrate catheter and exit site care.	✔	✔
• Demonstrate measurement of vital signs and weight measurement.	✔	✔
• Discuss monitoring and management of fluid balance.	✔	✔
• Discuss basic principles of aseptic technique.	✔	✔
• Demonstrate the continuous ambulatory peritoneal dialysis (CAPD) exchange procedure using aseptic technique (continuous cyclic peritoneal dialysis [CCPD] patients should also demonstrate exchange procedure in case of failure or unavailability of cycling machine).	✔	✔
• Demonstrate cycler setup procedure and maintenance if on CCPD.	✔	✔
• Discuss complications of peritoneal dialysis; prevention, recognition, and management of complications.	✔	✔
• Demonstrate procedure for adding medications to the dialysis solution.	✔	✔
• Demonstrate procedure for obtaining sterile dialysis fluid samples.	✔	✔
• Discuss routine laboratory tests needed and implications of results.	✔	✔
• Discuss dietary restrictions.		
• Discuss medications: name of medications, their actions, potential side effects, and when to contact primary provider.	✔	✔
• Discuss ordering, storage, and inventory of dialysis supplies.	✔	✔
• Describe plan for follow-up care.	✔	✔
• Demonstrate maintenance of home dialysis records.	✔	✔
• Describe actions in case of emergency.	✔	✔

dialysate into or out of the peritoneal cavity. Many patients gain 3 to 5 pounds within a month of initiating PD, so they may be asked to limit their carbohydrate intake to avoid excessive weight gain. Potassium, sodium, and fluid restrictions are not usually needed. Patients commonly lose about 2 to 3 L of fluid over and above the volume of dialysate infused into the abdomen during a 24-hour period, permitting a normal fluid intake even in a patient who is anephric (a patient without kidneys).

Continuing Care

Follow-up care through phone calls, visits to the dialysis clinic, outpatient department, and continuing home care assists patients in the transition to home and promotes their active participation in their own health care. Patients often check with the nurse to see if they are making the correct choices about dialysate or control of blood pressure, or simply to discuss a problem.

Patients may be seen by the PD team as outpatients once a month or more often if needed. The exchange procedure is evaluated at that time to see that strict aseptic technique is being used. Blood chemistry values are followed closely to make certain the therapy is adequate for the patient.

If a referral is made for home care, the home care nurse assesses the home environment and suggests modifications to accommodate the equipment and facilities needed to carry out PD. In addition, the nurse assesses the patient's and family's understanding of PD and evaluates their technique in performing PD. Assessments include checking for changes related to kidney disease; complications such as peritonitis; and treatment-related problems such as heart failure, inadequate drainage, and weight gain or loss. The nurse continues to reinforce and clarify education about PD and ESKD and assesses the patient's and family's progress in coping with the procedure. This is also an opportunity to remind patients about the need to participate in appropriate health promotion activities and health screening (e.g., gynecologic examinations, colonoscopy).

Special Considerations: Nursing Management of the Patient on Dialysis Who Is Hospitalized

Whether undergoing hemodialysis or PD, the patient may be hospitalized for treatment of complications related to the dialysis treatment, the underlying kidney disorder, or health problems not related to renal dysfunction or its treatment.

Protecting Vascular Access

When the patient undergoing hemodialysis is hospitalized for any reason, care must be taken to protect the vascular access. The nurse assesses the vascular access for patency and takes precautions to ensure that the extremity with the vascular access is not used for measuring blood pressure or for obtaining blood specimens; tight dressings, restraints, or jewelry over the vascular access must be avoided as well.

The bruit, or "thrill," over the venous access site must be evaluated at least every shift. Absence of a palpable thrill or audible bruit may indicate blockage or clotting in the vascular access. Clotting can occur if the patient has an infection anywhere in the body (serum viscosity increases) or if the blood pressure has dropped. When blood flow is reduced through the access for any reason (hypotension, application of blood pressure cuff or tourniquet), the access can clot. If a patient has a hemodialysis catheter or implanted hemodialysis access device, the nurse must observe for signs and symptoms of infection such as redness, swelling, drainage from the exit site, fever, and chills. The nurse must assess the integrity of the dressing and change it as needed. Patients with kidney disease are more prone to infection; therefore, appropriate infection control measures must be used for all procedures. The patient's vascular access device should not be used for any purpose other than dialysis, unless it is an emergency situation and no other access is available. In this situation, a dialysis nurse or physician should cannulate the vascular access.

Taking Precautions During Intravenous Therapy

When the patient needs IV therapy, the rate of administration must be as slow as possible. Accurate I&O records are essential.

> ***Quality and Safety Nursing Alert***
>
> **Because patients on dialysis cannot excrete water, rapid administration of IV fluid can result in pulmonary edema.**

Monitoring Symptoms of Uremia

As metabolic end products accumulate, symptoms of uremia worsen. Patients whose metabolic rate accelerates (those receiving corticosteroid medications or parenteral nutrition, those with infections or bleeding disorders, those undergoing surgery) accumulate waste products more quickly and may require daily dialysis. These same patients are more likely than other patients receiving dialysis to experience complications.

Detecting Cardiac and Respiratory Complications

Cardiac and respiratory assessment must be conducted frequently. As fluid builds up, fluid overload, heart failure, and pulmonary edema develop. Crackles in the bases of the lungs may indicate pulmonary edema.

Pericarditis may result from the accumulation of uremic toxins. If not detected and treated promptly, this serious complication may progress to pericardial effusion and cardiac tamponade. Pericarditis is detected by the patient's report of substernal chest pain (if the patient can communicate), low-grade fever (often overlooked), and pericardial friction rub. A paradoxical pulse (a decrease in blood pressure of more than 10 mm Hg during inspiration) is often present. When pericarditis progresses to effusion, the friction rub disappears, heart sounds become distant and muffled, ECG waves show very low voltage, and the pulsus paradoxus worsens (Park, Lee, Kim, et al., 2011).

The effusion may progress to life-threatening cardiac tamponade, noted by narrowing of the pulse pressure in addition to muffled or inaudible heart sounds, crushing chest pain, dyspnea, and hypotension.

Quality and Safety Nursing Alert

Although pericarditis, pericardial effusion, and cardiac tamponade can be detected by chest x-ray, they should also be detected through astute nursing assessment. Because of their clinical significance, assessment of the patient for these complications is a priority.

Controlling Electrolyte Levels and Diet

Electrolyte alterations are common, and potassium changes can be life threatening. All IV solutions and medications to be administered are evaluated for their electrolyte content. Serum laboratory values are assessed daily. If blood transfusions are required, they may be administered during hemodialysis, if possible, so that excess potassium can be removed. Dietary intake must also be monitored. The patient's frustrations related to dietary restrictions typically increase if the food is unappetizing. The nurse needs to recognize that this may lead to dietary indiscretion and hyperkalemia.

Hypoalbuminemia is an indicator of malnutrition in patients undergoing long-term or maintenance dialysis. Although some patients can be treated with adequate nutrition alone, some patients remain hypoalbuminemic for reasons that are poorly understood.

Managing Discomfort and Pain

Complications such as pruritus and pain secondary to neuropathy must be managed. Antihistamine agents, such as diphenhydramine hydrochloride (Benadryl), are commonly used, and analgesic medications may be prescribed. However, because elimination of the metabolites of medications occurs through dialysis rather than through renal excretion, medication dosages may need to be adjusted. Keeping the skin clean and well moisturized using bath oils, superfatted soap, and creams or lotions helps promote comfort and reduce itching. Instructing the patient to keep the nails trimmed to avoid scratching and excoriation also promotes comfort.

Monitoring Blood Pressure

Hypertension in renal failure is common. It is usually the result of fluid overload and, in part, oversecretion of renin. Many patients undergoing dialysis receive some form of antihypertensive therapy. Patients require detailed education and reinforcement of information regarding their antihypertensive regimen, because it is not uncommon for patients to need more than one antihypertensive agent (Eskridge, 2010). Rapid fluid fluctuations in patients receiving dialysis also create challenges to maintaining blood pressure control. Antihypertensive agents must be withheld before dialysis to avoid hypotension due to the combined effect of fluid removal with the dialysis treatment and the medication.

Typically, these patients require single or multiple antihypertensive agents to achieve normal blood pressure, thus adding to the total number of medications needed on an ongoing basis.

Preventing Infection

Patients with ESKD commonly have low WBC counts (and decreased phagocytic ability), low RBC counts (anemia), and impaired platelet function. Together, these pose a high risk of infection and potential for bleeding after even minor trauma. Preventing and controlling infection are essential because the incidence of infection is high. Infection of the vascular access site and pneumonia are common.

Caring for the Catheter Site

Patients receiving CAPD usually know how to care for the catheter site; however, the hospital stay is an opportunity to assess catheter care technique and correct misperceptions or deviations from recommended technique. Recommended daily or three-or-four-times-weekly routine catheter site care is typically performed during showering or bathing. The exit site should not be submerged in bathwater. The most common cleaning method is soap and water; liquid soap is recommended. During care, the nurse and patient need to make sure that the catheter remains secure to avoid tension and trauma. The patient may wear a gauze or semitransparent dressing over the exit site.

Administering Medications

All medications and the dosage prescribed for any patient on dialysis must be closely monitored to avoid those that are toxic to the kidneys and may threaten remaining renal function. Medications are also scrutinized for potassium and magnesium content; those medications that contain them are avoided. Care must be taken to evaluate all problems and symptoms that the patient reports without automatically attributing them to renal failure or to dialysis therapy.

Providing Psychological Support

Patients undergoing dialysis for a while may begin to reevaluate their status, the treatment modality, their satisfaction with life, and the impact of these factors on their families and support systems. Nurses must provide opportunities for these patients to express their feelings and reactions and to explore options. The decision to begin dialysis does not require that dialysis be continued indefinitely, and it is not uncommon for patients to consider discontinuing treatment. These feelings and reactions must be taken seriously, and the patient should have the opportunity to discuss them with the dialysis team as well as with a psychologist, psychiatrist, psychiatric nurse, trusted friend, or spiritual advisor. The patient's informed decision about discontinuing treatment, after thoughtful deliberation, should be respected.

KIDNEY SURGERY

A patient may undergo surgery to remove obstructions that affect the kidney (tumors or calculi), to insert a tube for draining the kidney (nephrostomy, ureterostomy), or to remove the kidney involved in unilateral kidney disease, renal carcinoma, or kidney transplantation.

Management of Patients Undergoing Kidney Surgery

Preoperative Considerations

Surgery is performed only after a thorough evaluation of renal function. Patient preparation to ensure that optimal renal function is maintained is essential. Fluids are encouraged to promote increased excretion of waste products before surgery unless contraindicated because of preexisting renal or cardiac dysfunction. If kidney infection is present preoperatively, broad-spectrum antimicrobial agents may be prescribed to prevent bacteremia. Antibiotic agents must be given with extreme care because many are toxic to the kidneys. Coagulation studies (prothrombin time, partial thromboplastin time, platelet count) may be indicated if the patient has a history of bruising and bleeding. The preoperative preparation is similar to that described in Chapter 17.

Because many patients facing kidney surgery are apprehensive, the nurse encourages the patient to recognize and verbalize concerns. Confidence is reinforced by establishing a relationship of trust and by providing expert care. Patients faced with the prospect of losing a kidney may think that they will have to depend on dialysis for the rest of their lives. The nurse reassures the patient and family that normal function may be maintained by a single healthy kidney.

Perioperative Concerns

Kidney surgery requires various patient positions to expose the surgical site adequately. Three surgical approaches are common: flank, lumbar, and thoracoabdominal (Fig. 54-9). During surgery, plans are carried out for managing altered urinary drainage. These may include inserting a nephrostomy or other drainage tube.

Postoperative Management

Because the kidney is a highly vascular organ, hemorrhage and shock are the chief potential complications of kidney surgery. Fluid and blood component replacement is frequently necessary in the immediate postoperative period to treat intraoperative blood loss.

Abdominal distention and paralytic ileus are fairly common after renal and ureteral surgery and are thought to be due to a reflex paralysis of intestinal peristalsis and manipulation of the colon or duodenum during surgery. Abdominal distention is relieved by decompression through a nasogastric tube (see Chapter 48 for treatment of paralytic ileus). Oral fluids are permitted when the passage of flatus is noted.

If infection occurs, antibiotics are prescribed after a culture reveals the causative organism. The toxic effects that antibiotic agents have on the kidneys (nephrotoxicity) must be kept in mind when assessing the patient. Low-dose heparin therapy may be initiated postoperatively to prevent thromboembolism in patients who have had any type of urologic surgery.

Nursing Management

In addition to those interventions listed in this section, Chart 54-12 provides a plan of nursing care for the patient undergoing kidney surgery.

Providing Immediate Postoperative Care

Immediate postoperative care of the patient who has undergone surgery of the kidney includes assessment of all body systems. Respiratory and circulatory status, pain level, fluid and electrolyte status, and patency and adequacy of urinary drainage systems are assessed.

Respiratory Status

As with any surgery, the use of anesthesia increases the risk of respiratory complications. Noting the location of the surgical incision assists the nurse in anticipating respiratory problems and pain. Respiratory status is assessed by monitoring the rate, depth, and pattern of respirations. The location of the incision frequently causes pain on inspiration and coughing; therefore, the patient tends to splint the chest wall and take shallow respirations. Auscultation is performed to assess normal and adventitious breath sounds.

Circulatory Status and Blood Loss

The patient's vital signs and arterial or central venous pressure are monitored. Skin color and temperature and urine output provide information about circulatory status. The surgical incision and drainage tubes are observed frequently to help detect unexpected blood loss and hemorrhage.

Pain

Postoperative pain is a major problem for the patient because of the location of the surgical incision and patient's position on the operating table to permit access to the kidney. The location and severity of pain are assessed before and after

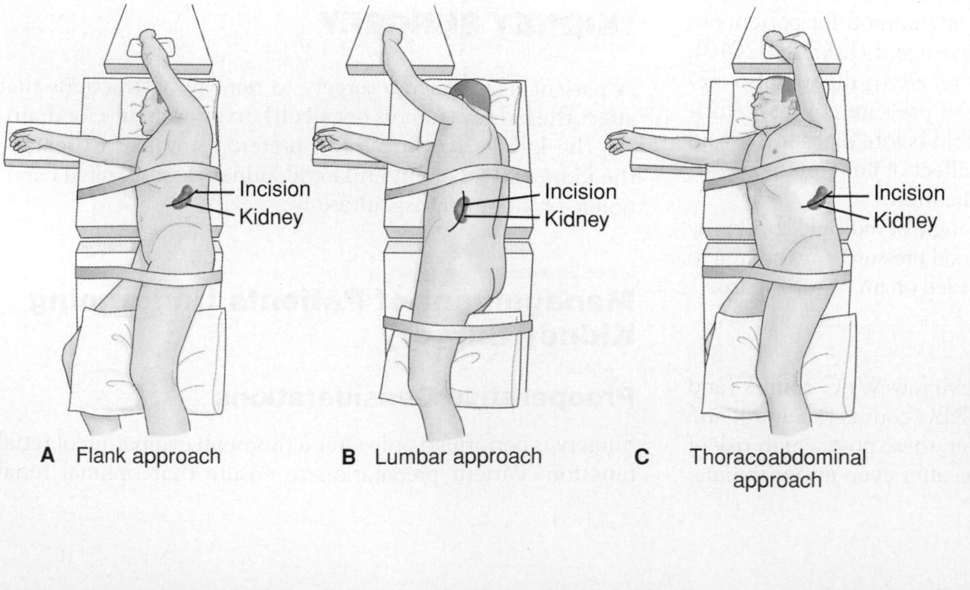
Incision
Kidney
Incision
Kidney
Incision
Kidney

A Flank approach **B** Lumbar approach **C** Thoracoabdominal approach

FIGURE 54-9 • Patient positioning and incisional approaches—flank (**A**), lumbar (**B**), thoracoabdominal (**C**)—for kidney surgery are associated with significant postoperative discomfort.

Chart 54-12 PLAN OF NURSING CARE

Care of the Patient Undergoing Kidney Surgery

NURSING DIAGNOSIS: Ineffective airway clearance related to pain of high abdominal or flank incision, abdominal discomfort, and immobility; risk for ineffective breathing pattern related to high abdominal incision
GOAL: Improved airway clearance

Nursing Interventions	Rationale	Expected Outcomes
1. Administer analgesic agent as prescribed. **2.** Splint incision with hands or pillow to assist patient in coughing. **3.** Assist patient to change positions frequently. **4.** Encourage the use of incentive spirometer if indicated or prescribed. **5.** Assist with and encourage early ambulation.	**1.** Enables patient to take deep breaths and cough. **2.** Splints incision and promotes adequate cough and prevention of atelectasis. **3.** Promotes drainage and inflation of all lobes of the lungs. **4.** Encourages adequate deep breaths. **5.** Mobilizes pulmonary secretions.	• Takes deep breaths and coughs adequately when encouraged and assisted • Exhibits respiratory rate of 12–18 breaths/min • Exhibits normal breath sounds without adventitious sounds • Exhibits full thoracic excursion without shallow respirations • Uses incentive spirometer with encouragement • Splints incision while taking deep breaths and coughing • Reports progressively less pain and discomfort with coughing and deep breaths • Exhibits normal blood gas levels and chest x-ray • Exhibits normal body temperature with no signs of atelectasis or pneumonia on assessment

NURSING DIAGNOSIS: Acute pain and discomfort related to surgical incision, positioning, and stretching of muscles during kidney surgery
GOAL: Relief of pain and discomfort

Nursing Interventions	Rationale	Expected Outcomes
1. Assess level of pain. **2.** Administer analgesic agents as prescribed. **3.** Splint incision with hands or pillow during movement or deep breathing and coughing exercises. **4.** Assist and encourage early ambulation.	**1.** Provides baseline for later evaluation of pain relief strategies. **2.** Promotes pain relief. **3.** Minimizes sensation of pulling or tension on incision and provides sense of support to the patient. **4.** Promotes resumption of muscle activity exercise.	• Reports relief of severe pain and discomfort • Takes analgesia as prescribed • Exercises aching muscles within recommendations • Uses distraction, relaxation exercises, and imagery to relieve pain • Exhibits no behavioral manifestations of pain and discomfort (e.g., restlessness, perspiration, verbal expressions of pain) • Participates in deep-breathing and coughing exercises • Gradually increases physical activity and exercise

NURSING DIAGNOSIS: Fear and anxiety related to diagnosis, outcome of surgery, and alteration in urinary function
GOAL: Reduction of fear and anxiety

Nursing Interventions	Rationale	Expected Outcomes
1. Assess patient's anxiety and fear before surgery if possible. **2.** Assess patient's knowledge about procedure and expected surgical outcome preoperatively. **3.** Evaluate the meaning of alterations resulting from surgical procedure for the patient and family or partner. **4.** Encourage patient to verbalize reactions, feelings, and fears. **5.** Encourage patient to share feelings with spouse or partner.	**1.** Provides a baseline for postoperative assessment. **2.** Provides a basis for further education. **3.** Enables understanding of patient's reactions and responses to expected and unexpected results of surgery. **4.** Affirms patient's understanding of and ultimate resolution of feelings and fears. **5.** Enables patient and partner to receive mutual support and reduces sense of isolation from each other.	• Verbalizes reactions and feelings to staff • Shares reactions and feelings with family or partner • Grieves appropriately for self and for changes in role and function • Identifies information needed to promote own adaptation and coping • Participates in activities and events in immediate environment • Accepts visit from support group if indicated • Identifies support person or support group

(continues on page 1564)

Chart 54-12

PLAN OF NURSING CARE

Care of the Patient Undergoing Kidney Surgery (continued)

Nursing Interventions	Rationale	Expected Outcomes
6. Offer and arrange for visit from member of support group (e.g., ostomy group, transplant group, if indicated).	6. Provides support from another person who has encountered the same or a similar surgical procedure and an example of how others have coped with the alteration.	

NURSING DIAGNOSIS: Impaired urinary elimination related to urinary drainage; risk for infection related to altered urinary drainage
GOAL: Maintenance of urinary elimination; infection-free urinary tract

Nursing Interventions	Rationale	Expected Outcomes
1. Assess urinary drainage system immediately.	1. Provides basis for further assessment and action	• Exhibits adequate urinary output and patent drainage system • Exhibits urinary output consistent with fluid intake • Demonstrates normal laboratory values: blood urea nitrogen, serum creatinine levels, urine specific gravity, and osmolality • Exhibits sterile urine on urine culture • Exhibits clear, dilute urine without debris or encrustation in the drainage system • States rationale for avoiding manipulation of catheter, drainage, or irrigation system • Exhibits normal placement of urinary stent or ureteral catheters until removed by physician • Maintains closed urinary drainage system • Exhibits normal body temperature without signs or symptoms of urinary tract infection • Cleans catheter with soap and water • Consumes adequate fluid intake (6–8 glasses of water or more per day, unless contraindicated) • Urinary drainage system remains in place until physician removes or discontinues it • Maintains urinary drainage system without infection or obstruction • Maintains urinary diversion as instructed • Maintains self-care so that environment is odor free • States rationale for close follow-up and maintains recommended schedule of appointments with health care providers
2. Assess adequacy of urinary output and patency of drainage system.	2. Provides baseline.	
3. Use asepsis and hand hygiene when providing care and manipulating drainage system.	3. Prevents or reduces risk of contamination of urinary drainage system.	
4. Maintain closed urinary drainage system.	4. Reduces risk of bacterial contamination and infection.	
5. If irrigation of the drainage system is necessary, use sterile gloves and sterile irrigating solution and a closed drainage and irrigation system.	5. Permits irrigation when necessary while maintaining closed drainage system, minimizing risk of infection.	
6. If irrigation is necessary and prescribed, perform it gently with sterile saline and the prescribed amount of irrigating fluid.	6. Maintains patency of the catheter or drainage system and prevents sudden increases in pressure in the urinary tract that may cause trauma, pressure on sutures or urinary tract structures, and pain.	
7. Assist patient in turning and moving in bed and when ambulating to prevent displacement or inadvertent removal of urinary stent or ureteral catheters if in place.	7. Prevents trauma from accidental displacement of urinary stent or ureteral catheter necessitating repeated instrumentation of the urinary tract (e.g., cystoscopy) to replace them.	
8. Observe urine color, volume, odor, and components.	8. Provides information about adequacy of urine output, condition and patency of drainage system, and debris in urine.	
9. Minimize trauma and manipulation of catheter, drainage system, and urethra.	9. Reduces risk of contamination of drainage system and eliminates site of bacterial invasion.	
10. Clean catheter gently with soap during bathing, avoiding any to-and-fro movement of catheter.	10. Removes debris and encrustations without causing trauma to or contamination of urethra.	
11. Anchor drainage tube.	11. Prevents movement or slipping of drainage tube, minimizing trauma to and contamination of urethra or catheter.	
12. Maintain adequate fluid intake.	12. Promotes adequate urine output and prevents urinary stasis.	
13. Assist with and encourage early ambulation while ensuring placement of urinary drainage system.	13. Minimizes cardiovascular and pulmonary complications while preventing loss, dislodging, or disruption of drainage system.	
14. If patient is to be discharged with urinary drainage system (catheter) in place or a urinary diversion, instruct patient and family member in care.	14. Knowledge and understanding of the drainage system or urinary diversion are essential to prevent infection and other complications.	

Chart 54-12

PLAN OF NURSING CARE

Care of the Patient Undergoing Kidney Surgery (continued)

NURSING DIAGNOSIS: Risk for imbalanced fluid volume related to surgical fluid loss, altered urinary output, parenteral fluid administration
GOAL: Normal fluid balance will be maintained

Nursing Interventions	Rationale	Expected Outcomes
1. Weigh patient daily.	1. Daily weight is the most sensitive indicator of fluid loss or gain.	• Patient's weight will be within 2–3 lb of patient's baseline.
2. Take accurate intake and output measurements.	2. Detects fluid retention due to poor cardiac or renal output.	• Intake that exceeds output will be detected early.
3. Place all parenteral therapy on an infusion pump.	3. Ensures that the patient does not receive excess or insufficient IV fluids.	• The exact amount of solution is infused with no adverse effects resulting from over- or underinfusion.
4. Monitor amount and characteristics of urine.	4. Assists in early detection of possible complications of surgery or tube insertion.	• Urine is clear and absent of blood, pus, or any foreign substances.
5. Monitor vital signs: temperature, pulse, respirations, and blood pressure.	5. When fluid volume or cardiac output is altered, vital signs are affected.	• Temperature, pulse, respiration, and blood pressure are normal.
6. Auscultate heart and lungs every shift.	6. When fluid volume is increased because of poor cardiac or renal output, fluid accumulates in the lungs. In addition, heart sounds change as heart failure develops; frequent auscultation ensures early detection.	• Normal heart and lung sounds are present.

analgesic medications are administered. Abdominal distention, which increases discomfort, is also noted.

Urinary Drainage

Urine output and drainage from tubes inserted during surgery are monitored for amount, color, and type or characteristics. Decreased or absent drainage is promptly reported to the primary provider because it may indicate obstruction that could cause pain, infection, and disruption of the suture lines.

Monitoring and Managing Potential Complications

Bleeding is a major complication of kidney surgery. If undetected and untreated, it can result in hypovolemia and hemorrhagic shock. The nurse's role is to observe for these complications, to report their signs and symptoms, and to administer prescribed parenteral fluids and blood and blood components. Monitoring of vital signs, skin condition, the urinary drainage system, the surgical incision, and the level of consciousness is necessary to detect evidence of bleeding, decreased circulating blood, and fluid volume and cardiac output. Frequent monitoring of vital signs (initially monitored at least at hourly intervals) and urinary output is necessary for early detection of these complications.

If bleeding goes undetected or is not detected promptly, the patient may lose significant amounts of blood and may experience hypoxemia. In addition to hypovolemic shock due to hemorrhage, this type of blood loss may precipitate a myocardial infarction or transient ischemic attack. Bleeding may be suspected when the patient experiences fatigue and when urine output is less than 0.5 mL/kg/hr. As bleeding persists, late signs of hypovolemia occur, such as cool skin, flat neck veins, and change in level of consciousness or responsiveness. Transfusions of blood components are indicated, along with surgical repair of the bleeding vessel.

Pneumonia may be prevented through the use of an incentive spirometer, adequate pain control, and early ambulation. Early signs of pneumonia include fever, increased heart and respiratory rates, and adventitious breath sounds.

Preventing infection involves using asepsis when changing dressings and handling and preparing catheters, other drainage tubes, central venous catheters, and IV catheters for administration of fluids. Insertion sites are monitored closely for signs and symptoms of inflammation: redness, drainage, heat, and pain. Special care must be taken to prevent UTI, which is associated with the use of indwelling urinary catheters. Catheters and other invasive tubes are removed as soon as they are no longer needed.

Antibiotics are commonly administered postoperatively to prevent infection. If antibiotic agents are prescribed, serum creatinine and BUN values must be monitored closely because many antibiotic agents are toxic to the kidney or can accumulate to toxic levels if renal function is decreased.

Preventing fluid imbalance is critical when caring for a patient undergoing kidney surgery, because both fluid loss and fluid excess are possible adverse effects of the surgery. Fluid loss may occur during surgery as a result of excessive urinary drainage when the obstruction is removed, or it may occur if diuretic agents are used. Such loss may also occur with GI losses, with diarrhea resulting from antibiotic use, or with nasogastric drainage. When postoperative IV therapy is inadequate to match the output or fluids lost, a fluid deficit results. Fluid excess, or overload, may result from cardiac effects of anesthesia, administration of excessive amounts of fluids, or the patient's inability to excrete fluid because of changes in renal function. Decreased urine output may be an indication of fluid excess.

Astute assessment skills are needed to detect early signs of fluid excess (such as weight gain, pedal edema, urine output below 0.5 mL/kg/hr, and slightly elevated pulmonary artery wedge pressure if available) before they become severe (appearance of adventitious breath sounds, shortness of breath).

Fluid excess may be treated with fluid restriction and administration of furosemide (Lasix) or other diuretic agents.

If renal insufficiency is present, these medications may prove ineffective; therefore, dialysis may be necessary to prevent heart failure and pulmonary edema.

Deep vein thrombosis (DVT) may occur postoperatively because of surgical manipulation of the iliac vessels during surgery or prolonged immobility. Anti-embolism stockings are applied, and the patient is monitored closely for signs and symptoms of thrombosis and encouraged to exercise the legs. Heparin may be administered postoperatively to reduce the risk of thrombosis.

Promoting Home and Community-Based Care

Educating Patients About Self-Care

If the patient has a drainage system in place, measures are taken to ensure that both the patient and family understand the importance of maintaining the system correctly at home and preventing infection. Verbal and written instructions and guidelines are provided to the patient and family at the time of hospital discharge. The patient may be asked to demonstrate management of the drainage system to ensure understanding. The importance of strategies to prevent postoperative complications (urinary tract obstruction and infection, DVT, atelectasis, and pneumonia) is stressed to the patient and family. The signs, symptoms, problems, and questions that should be referred to the physician or other primary provider are reviewed by the nurse with the patient and family.

Continuing Care

The need for postoperative assessment and care after kidney surgery continues regardless of the setting: the home, subacute care unit, outpatient clinic or office, or rehabilitation facility. Referral for home care is indicated for the patient going home with a urinary drainage system in place. During the home visit, the home care nurse reviews the instructions and guidelines given to the patient at hospital discharge. The nurse assesses the patient's ability to carry out the instructions in the home and answers questions that the patient or family has about management of the drainage system and the surgical incision.

In addition, the home care nurse obtains vital signs and assesses the patient for signs and symptoms of urinary tract obstruction and infection. The nurse also ensures that pain is adequately controlled and that the patient is complying with recommendations. The home care nurse encourages adequate fluid intake and increased levels of activity. Together, the nurse, patient, and family review the signs, symptoms, problems, and questions that should be referred to the primary provider. If the patient has a drainage tube in place, the nurse assesses the site and the patency of the system and monitors the patient for complications, such as DVT, bleeding, or pneumonia.

Because it is easy for the patient, family, and health care team to focus on the patient's immediate disorder to the exclusion of other health issues, the nurse reminds the patient and family about the importance of participating in health promotion activities, including appropriate health screenings.

Kidney Transplantation

Kidney transplantation has become the treatment of choice for most patients with ESKD. In the United States and globally, there are many more patients on the waiting list for kidney transplantation than there are organ donors. More than 93,000 Americans are on the waiting list to receive a kidney (Organ Procurement and Transplantation Network [OPTN], 2012). Patients choose kidney transplantation for various reasons, such as the desire to avoid dialysis or to improve their sense of well-being and the wish to lead a more normal life. In addition, the cost of maintaining a successful transplantation is one third the cost of dialysis treatment (Wiederhold, Langer, & Landenberger, 2011) Kidney transplantation is an elective procedure, not an emergency lifesaving procedure. Therefore, patients should be in the best possible condition prior to transplantation (Weng, Dai, Huang, et al., 2010).

Kidney transplantation involves transplanting a kidney from a living or deceased donor to a recipient who no longer has renal function (Chart 54-13). A living donor is a person who is alive at the time of donation and may or may not be related to the recipient. A deceased or cadaveric transplant comes from someone who has died and donated his or her organs. Transplantation from well-matched living donors who are related to the patient (those with compatible ABO and human leukocyte antigens) is slightly more successful than from cadaver donors (Ding, 2010). A contemporary development in kidney transplantation is paired donor exchange. In paired donor exchange, recipients essentially swap willing donors (Serur & Charlton, 2011). Although medically eligible to donate a kidney, a willing donor may be incompatible with the intended recipient due to blood type or antigens. The donor then agrees to donate the kidney to a compatible and unknown recipient, with the intention that the donor's originally intended organ recipient will be part of the donation chain and be the recipient of a donated kidney through organized donor and recipient matches.

Prior to either receiving or donating an organ, an extensive medical evaluation is performed. Not everyone is suitable for kidney transplantation. Contraindications include recent malignancy, active or chronic infection, severe irreversible extrarenal disease (e.g., inoperable cardiac disease, chronic lung disease, severe peripheral vascular disease), active infection (e.g., HIV, hepatitis B and C), morbid obesity (body mass index greater than 35), current substance abuse, inability to give informed consent, and history

Chart 54-13 Kidney Donation

An inadequate number of available kidneys remains the greatest limitation to treating patients with end-stage kidney disease successfully. For those interested in donating a kidney, the National Kidney Foundation provides written information describing the organ donation program and a card specifying the organs to be donated in the event of death.

The organ donation card is signed by the donor and two witnesses and should be carried by the donor at all times. Procurement of an adequate number of kidneys for potential recipients is still a major problem, despite national legislation that requires relatives of deceased patients or patients declared brain dead to be asked if they would consider organ donation.

In some states in the United States, drivers can indicate their desire to be organ donors on their driver's license application or renewal; however, this decision should be discussed with the family because the organ procurement agency will approach the family to discuss this option.

of nonadherence to treatment regimens (Counts, 2008; Weng et al., 2010). Donors may be rejected for the same reasons or any condition that is determined to have an impact on the remaining kidney. Examples include hypertension and diabetes because both are known causes of kidney disease. It is imperative when donors are evaluated that serious consideration be given to the overall long-term health of the donor. Every precaution must be taken to ensure that the remaining kidney in the donor will remain healthy. If these conditions are met, the donor should remain healthy after donation and have a normal lifespan. Because one kidney can easily handle the body's needs, no long-term adjustments will need to be made.

The recipient's native kidneys are not usually removed. The transplanted kidney is placed in the patient's iliac fossa anterior to the iliac crest because it allows for easier access to the blood supply needed to perfuse the kidney. The ureter of the newly transplanted kidney is transplanted into the bladder or anastomosed to the ureter of the recipient (Fig. 54-10). Once the blood supply has been reestablished to the transplanted kidney in the operating room, urine should begin to flow. The production of urine at this stage is an important indicator of the overall success of the procedure and ultimate long-term outcome.

Preoperative Management

Preoperative management goals include bringing the patient's metabolic state to a level as close to normal as possible through diet, possibly dialysis and medical management, making sure that the patient is free of infection, and preparing the patient for surgery and the postoperative course.

Medical Management

A complete physical examination is performed on the donor and the recipient to detect and treat any conditions that could cause complications after the donor and transplantation procedure. Tissue typing, blood typing, and antibody screening are performed to determine compatibility of the tissues and cells of the donor and recipient. Other diagnostic tests must be completed to identify conditions requiring treatment before transplantation for either individual. The lower urinary tract is studied to assess bladder neck function and to detect ureteral reflux.

Both patients must be free of infection at the time of kidney transplantation. After surgery, medications to prevent transplant rejection will be prescribed to the transplant recipient. These medications suppress the immune response, leaving the patient immunosuppressed and at risk for infection. Therefore, both the donor and recipient are evaluated and treated for any infections, including gingival (gum) disease and dental caries.

A psychosocial evaluation is conducted to assess the organ recipient's ability to adjust to the transplant, coping styles, social history, social support available, and financial resources. A history of psychiatric illness is important to obtain because psychiatric conditions are often aggravated by the corticosteroids needed for immunosuppression after transplantation (Weng et al., 2010). A psychosocial evaluation is also conducted to assess the organ donor's motive for giving the organ. The donor should not be coerced to donate this organ; it should be an altruistic act (Chart 54-14). If a dialysis routine has been established, hemodialysis is often performed the day before the scheduled transplantation procedure to optimize the physical status of the patient receiving the transplant.

Nursing Management

The nursing aspects of preoperative care for the patient undergoing kidney transplant and donation are similar to those for patients undergoing other types of kidney or elective abdominal surgery. Preoperative education can be conducted in a variety of settings, including the outpatient preadmission area, the hospital, or the transplantation clinic during the preliminary workup phase. Patient education directed at the kidney transplant donor and recipient addresses postoperative pulmonary

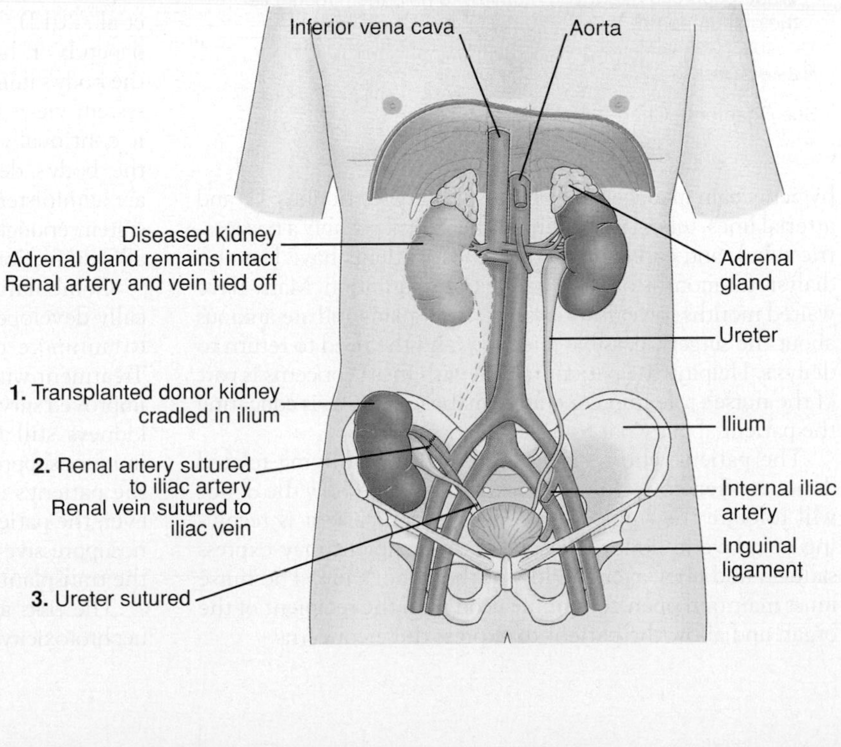
Inferior vena cava
Aorta
Diseased kidney
Adrenal gland remains intact
Renal artery and vein tied off
Adrenal gland
Ureter
1. Transplanted donor kidney cradled in ilium
Ilium
2. Renal artery sutured to iliac artery
Renal vein sutured to iliac vein
Internal iliac artery
Inguinal ligament
3. Ureter sutured

FIGURE 54-10 • Kidney transplantation. **1.** The transplanted kidney is placed in the iliac fossa. **2.** The renal artery of the donated kidney is sutured to the iliac artery, and the renal vein is sutured to the iliac vein. **3.** The ureter of the donated kidney is sutured to the bladder or to the patient's ureter.

Chart 54-14

ETHICAL DILEMMA

The Gift of Life for Whom? Whose Life Is It?

Case Scenario

You are a nurse working in a solid organ transplant clinic that is affiliated with a university health science center. A 42-year-old man with a history of polycystic kidney disease (PKD) and each of his three biological siblings, none of whom have PKD, have been worked up to see if one may be a match for a living transplant. The youngest sibling, who is the only sister in the family, is a perfect donor match. The brother with PKD is jubilant. However, the sister confides in you when other family members are not present, "You know, my parents and my brothers really guilted me into this. I'm not really certain I want to do this. What if I need this kidney one day?"

Discussion

Living donor transplants offer the best possibility for successful organ engraftment. Biologic siblings are the best candidates for finding good organ matches; there is a greater likelihood that siblings will share all human leukocyte antigens. Yet, whereas some families rally together and bond during times of illness and stress, others can become fragmented. Old resentments, grudges, and dysfunctional family patterns may surface during major life event crises such as these. Organ transplant nurses are frequent witnesses to these types of dysfunctional family processes.

Analysis

- Describe the ethical principles that are in conflict in this case (see Chart 3-3 in Chapter 3). Which principle should have preeminence as you proceed to work with this family?
- There are two patients in this case—one is the brother with PKD who is hoping to receive a donated kidney from a sibling, and the other is the sister with healthy kidneys who is reluctant to donate one to her brother. Should the autonomous right of one patient have preeminence over the autonomous right of the other? How would you respond to the sister when she confides in you?
- For whom is there an obligation to advocate for in this situation—the brother, the sister, or the extended family? What resources might you mobilize to help them reconcile their differences?

Resources

See Chapter 3, Chart 3-6 for ethics resources.

hygiene, pain management options, dietary restrictions, IV and arterial lines, tubes (indwelling catheter and possibly a nasogastric tube), and early ambulation. Most patients have been on dialysis for months or years before transplantation. Many have waited months to years for a kidney transplant and are anxious about the surgery, possible rejection, and the need to return to dialysis. Helping the patient to deal with these concerns is part of the nurse's role in preoperative management, as is educating the patient about what to expect after surgery.

The patient who receives a kidney from a living related donor is often concerned about the donor and how the donor will tolerate the surgical procedure. If the patient is receiving a cadaveric donor transplant, the recipient may express sadness and grief over the loss of the donor's life. The nurse must maintain open communication with the recipient of the organ and allow the patient to express these concerns.

The nurse working in an intensive care setting may provide care to the organ donor who is brain dead prior to organ removal. The overall goal is to preserve the function of the organs through maintaining hemodynamic stability, decreasing the risk for infection, and monitoring laboratory values while providing dignified care to the donor and family members (Flodén & Forsberg, 2009). Continuing care for the donor can be complex and last several hours. Care is often provided in collaboration with the organ procurement coordinator and transplant coordinator.

Postoperative Management

The goal of postoperative care is to maintain homeostasis until the transplanted kidney is functioning well. The patient whose kidney functions immediately has a more favorable prognosis than the patient whose kidney does not.

Often, the organ donor will be on the same unit as the transplant recipient. The donor will require the same level of care provided to the recipient, including follow-up after the procedure and lifelong. Studies have shown that the organ donor may experience more pain than the recipient, possibly requiring more analgesia for pain control (Lentine & Patel, 2012). Fluid, electrolyte, and hemodynamic status will also be closely monitored in the organ donor (Tong, Chapman, Wong, et al., 2012).

Medical Management

After a kidney transplantation, rejection and failure can occur within 24 hours (hyperacute), within 3 to 14 days (acute), or after many years. A hyperacute rejection is caused by an immediate antibody-mediated reaction that leads to generalized glomerular capillary thrombosis and necrosis. An acute rejection typically occurs within a few days to weeks of the transplant surgery, and the patient experiences tenderness at the transplant site, a decrease in serum creatinine values, fever, malaise, and oliguria (Ding, 2010; Wiederhold et al., 2011). An acute rejection requires early recognition and treatment with immunosuppressant therapy, whereas a hyperacute reaction requires immediate removal of the transplanted organ (Tong et al., 2012). The long-term survival of a transplanted kidney depends on how well it matches the recipient and how well the body's immune response is controlled. The body's immune system views the transplanted kidney as "foreign"; therefore, it continually works to reject it. To overcome or minimize the body's defense mechanisms, immunosuppressive agents are administered. Optimally, medications modify the immune system enough to prevent rejection, although not enough to allow infections or malignancies to occur (Table 54-4).

Combinations of corticosteroids and medications specifically developed to affect the action of lymphocytes are used to minimize the body's reaction to the transplanted organ. Treatment with combinations of new agents has dramatically improved survival rates, and now 90% to 95% of transplanted kidneys still function after 1 year (Ding, 2010). Doses of immunosuppressive agents are often adjusted depending on the patient's immunologic response to the transplant. However, the patient will be required to take some form of immunosuppressive therapy for the entire time that he or she has the transplanted kidney.

The risks associated with taking these medications include nephrotoxicity, hypertension, hyperlipidemia, hirsutism, tremors,

TABLE 54-4 Posttransplant Immunosuppressant Agents

Agent	Action	Nursing Implications
azathioprine (Azasan, Imuran)	Antagonizes purine metabolism and appears to inhibit DNA, RNA, and normal protein synthesis in rapidly growing cells; suppresses T-cell effects	• Give oral drug in divided doses with food. • Monitor leukocyte and platelet counts and bleeding.
belatacept (Nulojix)	Inhibition of T-lymphocyte proliferation and cytokine production	• Contraindicated in patients with EBV seronegativity or unknown EBV serostatus, liver transplantation, breast-feeding • Monitor for symptoms of infection, hypertension, progressive multifocal leukoencephalopathy. • Administer IV.
cyclosporine (Gengraf, Neoral, Sandimmune, Restasis)	Selective and reversible inhibition of 1st-phase of T-cell activation with T lymphocytes	• Do not dilute oral solution with grapefruit juice (use orange juice, regular or chocolate milk, or apple juice and administer immediately after mixing). • Observe for adverse reactions for 30 minutes after initiation of the IV infusion. • Be aware that nephrotoxicity is reported in up to 1/3 of transplant recipients. • Give medication with food to reduce gastrointestinal upset. • Administer medication at the same time each day.
everolimus (Afinitor, Zortess)	Protein–tyrosine kinase inhibitor	• Monitor for hypersensitivity reaction. • Watch for changes in pulmonary status and cough. • Avoid administration of live vaccines. • Administer at the same time each day with food; do not crush or allow patient to chew tablet.
mycophenolate mofetil (CellCept) mycophenolic acid (Myfortic)	Inhibition of T- and B- lymphocyte responses, thus inhibiting antibody formation and generation of cytotoxic T cells	• Administer on an empty stomach. • Do not crush or open capsules. • Avoid contact with powder in capsules; wash thoroughly with soap and water if contact occurs. • Obtain baseline complete blood count with differential prior to initiating therapy. • Instruct the patient to avoid over-the-counter antacids.
prednisone (Apo-Prednisone, Deltasone, Orasone, Panasol, Winpred)	Immediate-acting synthetic analogue of hydrocortisone that has anti-inflammatory and immunosuppressant properties	• Give with meals or a snack. • Crush tablet and give with fluid if patient is unable to swallow it whole. • Note that the drug should not be stopped abruptly; the dosage is reduced gradually. • Monitor weight, blood pressure, blood glucose levels, and sleep patterns. • Monitor for hypocalcemia. • Instruct patient to avoid or minimize alcohol intake.
sirolimus (Rapamune)	Inhibits the response of helper T- and B lymphocytes	• Administer 4 hours after oral cyclosporine. • Instruct patient to swallow tablets whole and to avoid chewing or crushing tablets. • Refrigerate drug and protect from light. • Instruct patient to avoid grapefruit juice within 2 hours of taking drug and to mix the drug only with orange juice or water. • Instruct the patient to limit exposure to sunlight.
tacrolimus (Prograf, Protopic)	Inhibits helper T lymphocytes	• Monitor for neurotoxicity (tremors and changes in mental status). • Assess for hypertension. • Monitor tacrolimus levels.

DNA, deoxyribonucleic acid; RNA, ribonucleic acid; EBV, Epstein-Barr virus; IV, intravenous.
Adapted from Karch, A. (2012). *2012 Lippincott's nursing drug guide*. Philadelphia: Lippincott Williams & Wilkins.

blood dyscrasias, cataracts, gingival hyperplasia, and several types of cancer (Ding, 2010).

Nursing Management

Assessing the Patient for Transplant Rejection

After kidney transplantation, the nurse assesses the patient for signs and symptoms of transplant rejection: oliguria, edema, fever, increasing blood pressure, weight gain, and swelling or tenderness over the transplanted kidney or graft. Patients receiving cyclosporine (Neoral) may not exhibit the usual signs and symptoms of acute rejection. In these patients, the only sign may be an asymptomatic rise in the serum creatinine level (more than a 20% rise is considered acute rejection) (Ding, 2010; Danovitch, 2010).

Preventing Infection

The results of blood chemistry tests and leukocyte and platelet counts are monitored closely because immunosuppression depresses the formation of leukocytes and platelets. The patient is closely monitored for infection because of susceptibility to impaired healing and infection related to immunosuppressive therapy and complications of renal failure. Clinical manifestations of infection include shaking chills, fever, rapid heartbeat (tachycardia), and respirations (tachypnea), as well as either an increase or a decrease in WBCs (leukocytosis or leukopenia).

Infection may be introduced through many sources. Urine cultures are performed frequently because of the high incidence of bacteriuria during early and late stages of transplantation. Any type of wound drainage should be viewed as a potential source of infection because drainage is an excellent culture medium for bacteria. Catheter and drain tips may be cultured when removed by cutting off the tip of the catheter or drain (using aseptic technique) and placing the tip in a sterile container to be taken to the laboratory for culture (Chart 54-15).

The nurse ensures that the patient is protected from exposure to infection by hospital staff, visitors, and other patients with active infections. Attention to hand hygiene by all who come in contact with the patient is imperative.

Monitoring Urinary Function

A kidney from a living donor who is related to the patient usually begins to function immediately after surgery and may produce large quantities of dilute urine. A kidney from a cadaver donor may undergo ATN and therefore may not function for 2 or 3 weeks, during which time anuria, oliguria, or polyuria may be present. During this stage, the patient may experience significant changes in fluid and electrolyte status. Therefore, careful monitoring is indicated. The output from the urinary catheter (connected to a closed drainage system) is measured every hour. IV fluids are administered on the basis of urine volume and serum electrolyte levels and as prescribed by the physician. Hemodialysis may be necessary postoperatively to maintain homeostasis until the transplanted kidney is functioning well. It also may be required if fluid overload and hyperkalemia occur. After successful kidney transplantation, the vascular access device may clot, possibly from improved coagulation with the return of renal function. The vascular access for hemodialysis is monitored to ensure patency and to evaluate for evidence of infection.

Addressing Psychological Concerns

The rejection of a transplanted kidney is of great concern to the patient, the family, and the health care team for many months. The fear of kidney rejection and the complications of immunosuppressive therapy (Cushing syndrome, diabetes, capillary fragility, osteoporosis, glaucoma, cataracts, acne, nephrotoxicity) place tremendous psychological stress on the patient. Anxiety and uncertainty about the future and difficult posttransplantation adjustment are often sources of stress for the patient and family.

An important nursing function is the assessment of the patient's stress and coping. The nurse uses each visit with the patient to determine if the patient and family are coping effectively and the patient is adhering to the prescribed medication regimen. If indicated or requested, the nurse refers the patient for counseling.

Monitoring and Managing Potential Complications

The patient undergoing kidney transplantation is at risk for the postoperative complications that are associated with any surgical procedure. In addition, the patient's physical condition may be compromised because of the effects of longstanding renal failure and its treatment. Therefore, careful assessment of the complications related to renal failure and those associated with a major surgery are important aspects of nursing care. Breathing exercises, early ambulation, and care of the surgical incision are important aspects of postoperative care.

GI ulceration and corticosteroid-induced bleeding may occur. Fungal colonization of the GI tract (especially the mouth) and urinary bladder may occur secondary to corticosteroid and antibiotic therapy. Closely monitoring the patient and notifying the physician about the occurrence of these complications are important nursing interventions. In addition, the patient is monitored closely for signs and symptoms of adrenal insufficiency if the treatment has included the use of corticosteroids.

Promoting Home and Community-Based Care

Educating Patients About Self-Care. The nurse works closely with the patient and family to be sure that they understand the need for continuing immunosuppressive therapy as prescribed. In addition, the patient and family are instructed to assess for and report signs and symptoms of transplant rejection, infection, or significant adverse effects of the immunosuppressive regimen. These include decreased urine output; weight gain; malaise; fever; respiratory distress; tenderness over the transplanted kidney; anxiety; depression; changes in eating, drinking, or other habits; and changes in

Chart 54-15 Kidney Transplant Rejection and Infection

Renal graft rejection and failure may occur within 24 hours (hyperacute), within 3 to 14 days (most commonly acute), or after many years (most commonly chronic). It is not uncommon for rejection to occur during the first year after transplantation.

Detecting Rejection

Ultrasonography may be used to detect enlargement of the kidney; percutaneous renal biopsy (most reliable) and x-ray techniques are used to evaluate transplant rejection. If the body rejects the transplanted kidney, the patient needs to return to dialysis. The rejected kidney may or may not be removed, depending on when the rejection occurs (acute vs. chronic) and the risk for infection if the kidney is left in place.

Potential Infection

About 75% of kidney transplant recipients have at least one episode of infection in the first year after transplantation because of immunosuppressant therapy. Immunosuppressants of the past made the transplant recipient more vulnerable to opportunistic infections (candidiasis, cytomegalovirus, *Pneumocystis* pneumonia) and infection with other relatively nonpathogenic viruses, fungi, and protozoa, which can be a major hazard. Immunosuppressive therapy, such as cyclosporine (Neoral) has reduced the incidence of opportunistic infections because it selectively exerts its effect, sparing T cells that protect the patient from life-threatening infections. In addition, combination immunosuppressant therapy and improved clinical care have produced 1-year patient survival rates approaching 100% and graft survival exceeding 90%. Infections, however, remain a major cause of death at all points in time for kidney transplant recipients (Ding, 2010).

blood pressure. The patient is instructed to inform other health care providers (e.g., dentist) about the kidney transplant and the use of immunosuppressive agents.

Continuing Care. The patient needs to know that follow-up care after transplantation is a lifelong necessity. Individual verbal and written instructions are provided concerning diet, medication, fluids, daily weight, daily measurement of urine, management of I&O, prevention of infection and chronic rejection, resumption of activity, and avoidance of contact sports in which the transplanted kidney may be injured. Because of the risk of other potential complications, the patient is followed closely by a health care team that includes the nephrologist, transplant surgeon, transplant coordinator or nurse, social worker, and dietician. Medications are often obtained at one pharmacy or through the pharmacy at the hospital where the transplant surgery was performed for the purpose of quality control. Follow-up with providers from the transplant team will initially occur once a week upon discharge from the hospital and taper with time. Laboratory studies will also be obtained and followed on an ongoing basis to monitor the function of the kidney.

Cardiovascular disease is the major cause of morbidity and mortality after transplantation, due in part to the increasing age of patients with transplants. An additional problem is possible malignancy; patients receiving long-term immunosuppressive therapy are at higher risk for cancers than the general population. The nurse reminds the patient of the importance of health promotion and health screening and provides information on local transplantation support groups at the transplant hospital or through the procurement organization.

The American Association of Kidney Patients and the NKF (listed in the Resources section of this chapter) are nonprofit organizations that serve the needs of those with kidney disease. These groups can provide many helpful suggestions for patients and family members learning to cope with dialysis and transplantation.

RENAL TRAUMA

The kidneys are protected by the rib cage and musculature of the back posteriorly and by a cushion of abdominal wall and viscera anteriorly. They are highly mobile and are fixed only at the renal pedicle (stem of renal blood vessels and the ureter). With traumatic injury, the kidneys can be thrust against the lower ribs, resulting in contusion and rupture. Rib fractures or fractures of the transverse process of the upper lumbar vertebrae may be associated with renal contusion or laceration. Failure to wear seat belts contributes to the incidence of renal trauma in motor vehicle crashes. Up to 80% of patients with renal trauma have associated injuries of other internal organs (Miller-Graziano, 2011).

Injuries may be blunt (automobile and motorcycle crashes, falls, athletic injuries, assaults) or penetrating (gunshot wounds, stabbings). Blunt renal trauma accounts for 80% to 90% of all renal injuries; penetrating renal trauma accounts for the remaining 10% to 20% (Counts, 2008; USRDS, 2011).

Blunt renal trauma is classified into one of four groups, as follows:

- *Contusion:* Bruises or hemorrhages under the renal capsule; capsule and collecting system intact
- *Minor laceration:* Superficial disruption of the cortex; renal medulla and collecting system are not involved
- *Major laceration:* Parenchymal disruption extending into cortex and medulla, possibly involving the collecting system
- *Vascular injury:* Tears of renal artery or vein

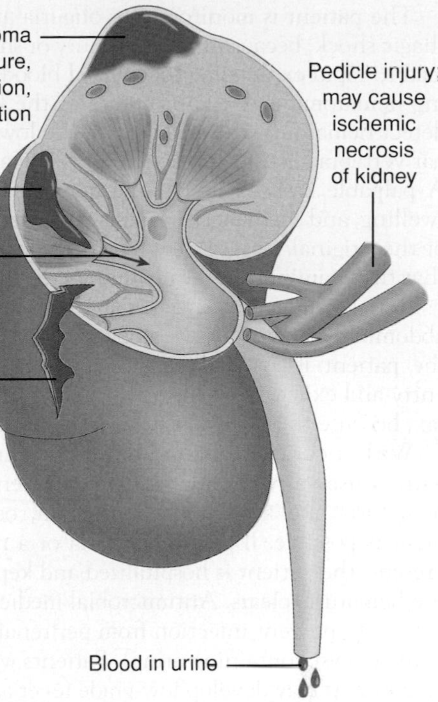
Expanding hematoma may cause rupture, extravasation, exsanguination
Pedicle injury; may cause ischemic necrosis of kidney
Contusion
Bleeding into collecting system, shock
Laceration
Blood in urine

FIGURE 54-11 • Types and pathophysiologic effects of kidney injuries: contusions, lacerations, rupture, and pedicle injury.

The most common renal injuries are contusions, lacerations, ruptures, and renal pedicle injuries or small internal lacerations of the kidney (Fig. 54-11). The kidneys receive half of the blood flow from the abdominal aorta; therefore, even a fairly small renal laceration can produce massive bleeding. About 70% of patients are in shock when admitted to the hospital (Miller-Graziano, 2011). In some cases, there is an isolated renal artery thrombosis.

Clinical manifestations include pain, renal colic (due to blood clots or fragments obstructing the collecting system), hematuria, mass or swelling in the flank, ecchymoses, and lacerations or wounds of the lateral abdomen and flank. Hematuria is the most common manifestation of renal trauma; its presence after trauma suggests kidney injury. There is no relationship between the degree of hematuria and the degree of injury. Hematuria may not occur, or it may be detectable only on microscopic examination. Signs and symptoms of hypovolemia and shock (see Chapter 14) are likely with significant hemorrhage.

Medical Management

The goals of management in patients with renal trauma are to control hemorrhage, pain, and infection as well as to preserve and restore renal function. All urine is saved and sent to the laboratory for analysis to detect RBCs and to evaluate the course of bleeding. Hematocrit and hemoglobin levels are monitored closely; decreasing values indicate hemorrhage.

The patient is monitored for oliguria and signs of hemorrhagic shock, because a pedicle injury or shattered kidney can lead to rapid exsanguination (lethal blood loss). An expanding hematoma may cause rupture of the kidney capsule. To detect hematoma, the area around the lower ribs, upper lumbar vertebrae, flank, and abdomen is palpated for tenderness. A palpable flank or abdominal mass with local tenderness, swelling, and ecchymosis suggests renal hemorrhage. The area of the original mass can be outlined with a marking pen so that the examiner can evaluate the area for change.

Renal trauma is often associated with other injuries to the abdominal organs (liver, colon, small intestines); therefore, the patient is assessed for skin abrasions, lacerations, and entry and exit wounds of the upper abdomen and lower thorax, because these may be associated with kidney injury.

With a contusion of the kidney, healing may take place with conservative measures. If the patient has microscopic hematuria and a normal IV urogram, outpatient management is possible. If gross hematuria or a minor laceration is present, the patient is hospitalized and kept on bed rest until the hematuria clears. Antimicrobial medications may be prescribed to prevent infection from perirenal hematoma or urinoma (a cyst containing urine). Patients with retroperitoneal hematomas may develop low-grade fever as absorption of the clot takes place.

Surgical Management

In renal trauma, any sudden change in the patient's condition suggests hemorrhage and requires rapid surgical intervention. Depending on the patient's condition and the nature of the injury, major lacerations may be treated through surgical intervention or conservatively (bed rest, no surgery). Vascular injuries require immediate exploratory surgery because of the high incidence of involvement of other organ systems and the serious complications that may result if these injuries are untreated. The patient is often in shock and requires aggressive fluid resuscitation. The damaged kidney may have to be removed (nephrectomy).

Early postoperative complications (within 6 months) include rebleeding, perinephritic abscess formation, sepsis, urine extravasation, and fistula formation. Other complications include stone formation, infection, cysts, vascular aneurysms, and loss of renal function. Hypertension can be a complication of any surgery but usually is a late complication of kidney injury.

Nursing Management

The patient with renal trauma must be assessed frequently during the first few days after injury to detect flank and abdominal pain, muscle spasm, and swelling over the flank. During this time, the patient who has undergone surgery is educated about care of the incision and the importance of an adequate fluid intake. In addition, instructions about changes that should be reported to the physician, such as fever, hematuria, flank pain, or any signs and symptoms of decreasing kidney function, are provided. Guidelines for gradually increasing activity, lifting, and driving are also provided in accordance with the physician's prescription.

Follow-up nursing care includes monitoring the blood pressure to detect hypertension and advising the patient to restrict activities for about 1 month after trauma to minimize the incidence of delayed or secondary bleeding. The patient should be advised to schedule periodic follow-up assessments of renal function (creatinine clearance, BUN, and serum creatinine analyses). If a nephrectomy was necessary, the patient is advised to wear medical identification.

Critical Thinking Exercises

1 You are a staff nurse in an outpatient dialysis facility. A 50-year-old man with ESKD is seen in the clinic for the first time and states that he wants to begin home hemodialysis. The patient lives alone and is employed full-time. Identify what further assessments are needed at this time. Develop an education plan to explain the different types of dialysis, goals, and level of involvement on the part of the patient.

2 ebp A 45-year-old married woman visits the nephrology department to discuss options for dealing with her ESKD. Identify the criteria used to evaluate the strength of the evidence for dialysis compared to kidney transplantation.

3 ebp You are caring for a 35-year-old woman who has been recently diagnosed with renal cancer. What is the evidence base for her treatment options? Identify the criteria used to evaluate the strength of the evidence.

4 pq A 23-year-old man is admitted to the surgical intensive care unit after a motor vehicle crash. The patient sustained multiple abdominal injuries that resulted in intra-abdominal hemorrhage requiring emergent surgery. The patient's blood pressure was 58/42 upon arrival to the emergency department, and the patient was not responsive. Due to broken ribs and a pneumothorax, the patient is also being mechanically ventilated. The nurse notes that the patient's urine output is less than 0.5 mL/kg/h and that the patient remains hypotensive. Based on the assessment information, what type of AKI is this patient most likely experiencing? Describe the priorities in caring for the patient as he moves through the four phases of AKI.

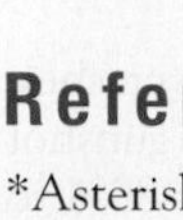

Brunner Suite Resources

Explore these additional resources to enhance learning for this chapter:

- NCLEX-Style Questions and Other Resources on thePoint, http://thePoint.lww.com/Brunner13e
- Study Guide
- PrepU
- Clinical Handbook
- Handbook of Laboratory and Diagnostic Tests

References

*Asterisk indicates nursing research.

Books

Baldwin, C. M. (2010). Cultural differences in cancer care. In J. Eggert (Ed.). *Cancer basics*. Pittsburgh: Oncology Nursing Society.

Counts, C. S. (2008). *ANNA core curriculum for nephrology nursing* (5th ed.). Pitman, NJ: American Nephrology Nurses Association.

Danovitch, G. M. (2010). *Handbook of kidney transplantation* (5th ed.). Philadelphia: Lippincott Williams & Wilkins.

Karch, A. (2012). *2012 Lippincott's nursing drug guide*. Philadelphia: Lippincott Williams & Wilkins.

Miller, C. A. (2012). *Nursing for wellness in older adults* (6th ed.). Philadelphia: Lippincott Williams & Wilkins.

Porth, C. M., & Matfin, G. (2009). *Pathophysiology: Concepts of altered health states* (8th ed.). Philadelphia: Lippincott Williams & Wilkins.

U.S. Renal Data System. (2011). *Annual data report: Atlas of chronic kidney disease and end-stage renal disease in the United States*. Bethesda, MD: National Institutes of Health, National Institute of Diabetes and Digestive and Kidney Diseases.

Journals and Electronic Documents

General

American Cancer Society. (2012). *Cancer facts & figures 2012*. Atlanta, GA: Author.

Centers for Disease Control and Prevention. (2010). *National chronic kidney disease fact sheet 2010*. Available at: www.cdc.gov/diabetes/pubs/factsheets/kidney.htm

*Hopkins, D. J., Kott, M., Rose, P., et al. (2011). End-of-life issues and the patient with renal disease: An evidence-based practice project. *Nephrology Nursing Journal, 38*(1), 79–83.

Miller-Graziano, C. (2011). What's new in shock? *Shock, 35*(6), 539–541.

Rank, W. (2013). Preventing contrast media-induced nephrotoxicity. *Nursing 2013, 43*(4), 48–51.

Young, S. (2009). Rethinking and integrating nephrology palliative care: A nephrology nursing perspective. *Canadian Association of Nephrology Nurses and Technologists Journal, 19*(1), 36–44.

Chronic Kidney Disease

Chamney, M., Pugh-Clarke, K., Kafkia, T., et al. (2010). Continuing education article: Management of anaemia in chronic kidney disease. *Journal of Renal Care, 36*(2), 102–111.

*Eskridge, M. S. (2010). Hypertension and chronic kidney disease: The role of lifestyle modification and medication management. *Nephrology Nursing Journal, 37*(1), 55.

*Harrison, K., & Watson, S. (2011). Palliative care in advanced kidney disease: A nurse-led joint renal and specialist palliative care clinic. *International Journal of Palliative Nursing, 17*(1), 42–46.

Murphy, F., Bennett, L., & Jenkins, K. (2010). Managing anaemia of chronic kidney disease. *British Journal of Nursing, 19*(20), 1281–1286.

*Neyhart, C. D., McCoy, L., Rodegast, B., et al. (2010). A new nursing model for the care of patients with chronic kidney disease: The UNC Kidney Center Nephrology Nursing Initiative. *Nephrology Nursing Journal, 37*(2), 121–131.

Thomas, N. (2010). Recognising and managing chronic kidney disease. *Practice Nurse, 39*(10), 19–22.

Acute Kidney Injury

Ali, B., & Gray-Vickrey, P. (2011). Limiting the damage from acute kidney injury. *Nursing 2011, 41*(3), 22–31.

Dirkes, S. (2011). Acute kidney injury: Not just acute kidney injury anymore? *Critical Care Nurse, 31*(1), 37–50.

Murphy, F., & Byrne, G. (2010). The role of the nurse in the management of acute kidney injury. *British Journal of Nursing, 19*(3), 146–152.

Warise, L. (2010). Update: Diuretic therapy in acute kidney injury—a clinical case study. *MEDSURG Nursing, 19*(3), 149–152.

Yaklin, K. M. (2011). Acute kidney injury: An overview of pathophysiology and treatments. *Nephrology Nursing Journal, 38*(1), 13–19.

Renal Carcinoma

Marchese, K. (2010). Getting ready for certification: Renal cell carcinoma. *Urologic Nursing, 30*(6), 353–354.

Wood, L. S. (2009). Renal cell carcinoma: Screening, diagnosis, and prognosis. *Clinical Journal of Oncology Nursing, 13*(6), 3–7.

Wood, L. S. (2010). New therapeutic strategies for renal cell carcinoma. *Urologic Nursing, 30*(1), 40–54.

Chronic Renal Failure

*Bayoumi, M., & El-Fouly, Y. (2010). Effects of teaching programme on quality of life for patients with end-stage renal disease. *Journal of Renal Care, 36*(2), 96–101.

Kelman, E. (2012). Applying International Society of Peritoneal Dialysis guidelines/recommendations (2010) for management of peritonitis. *Canadian Association of Nephrology Nurses and Technologists Journal, 22*(2), 29–29.

Lincoln, M. (2011). Preventing catheter-associated bloodstream infections in hemodialysis centers: The facility perspective. *Nephrology Nursing Journal, 38*(5), 411–415.

*Makaroff, K. L. (2012). Experiences of kidney failure: A qualitative meta-synthesis. *Nephrology Nursing Journal, 39*(1), 21–29.

Organ Procurement and Transplantation Network (OPTN). (2012). *Waiting list candidates as of today*. Available at: optn.transplant.hrsa.gov/data/

Park, E. A., Lee, W., Kim, K. H., et al. (2011). Rapid progression of pericardial calcification containing a "calcium paste" in a patient with end-stage renal disease. *Circulation, 123*(9), 262–264.

Roy-Chaudhury, P., El-Khatib, M., Campos-Naciff, B., et al. (2012). Back to the future: How biology and technology could change the role of PTFE grafts in vascular access management. *Seminars in Dialysis, 25*(5), 495–504.

Stilos, K. (2010). Conservative management of end-stage renal disease and withdrawal of dialysis. *Canadian Association of Nephrology Nurses and Technologists Journal, 20*(2), 36–37.

Williams, A. F., & Manias, E. (2009). Perceptions of pain control by consumers with chronic kidney disease. *Journal of Nursing & Healthcare of Chronic Illnesses, 1*(3), 199-209.

Yang, Y., Wang, H., Yeh, C., et al. (2011), Early initiation of continuous ambulatory peritoneal dialysis in patients undergoing surgical implantation of Tenckhoff catheters. *Peritoneal Dialysis International, 31*(5), 551–555.

Renal Replacement Therapies

*Danquah, F., Zimmerman, L., Diamond, P. M., et al. (2010). Frequency, severity, and distress of dialysis-related symptoms reported by patients on hemodialysis. *Nephrology Nursing Journal, 37*(6), 627–638.

*Lingerfelt, K., & Hodnicki, D. (2012). Hypertension management in patients receiving hemodialysis: The benefits of home blood pressure monitoring. *Nephrology Nursing Journal, 39*(1), 31–36.

*Lingerfelt, K. L., & Thornton, K. (2011). An educational project for patients on hemodialysis to promote self-management behaviors of end stage renal disease education. *Nephrology Nursing Journal, 38*(6), 483–489.

*Loiselle, M., O'Connor, A. M., & Michaud, C. (2011). Developing a decision support intervention regarding choice of dialysis modality. *Canadian Association of Nephrology Nurses and Technologists Journal, 21*(3), 13–18.

*Prowant, B. F., Moore, H., Satalowich, R., et al. (2010). Peritoneal dialysis survival in relation to patient body size and peritoneal transport characteristics. *Nephrology Nursing Journal, 37*(6), 641–646.

White, S., & Vinet, A. (2010). Partnering with patients to improve peritonitis rates. *Canadian Association of Nephrology Nurses and Technologists Journal, 20*(1), 38–41.

Kidney Transplantation

Ding, D. (2010). Post-kidney transplant rejection and infection complications. *Nephrology Nursing Journal, 37*(4), 419–426.

*Flodén, A., & Forsberg, A. (2009). A phenomenographic study of ICU-nurses' perceptions of and attitudes to organ donation and care of potential donors. *Intensive & Critical Care Nursing, 25*(6), 306–313.

*Lentine, K. L., & Patel, A. (2012). Risks and outcomes of living donation. *Advances in Chronic Kidney Disease, 19*(4), 220–228.

Segur, D., & Charlton, M. (2011). Kidney paired donation 2011. *Progress in Transplantation, 21*(3), 215–218.

*Tong, A., Chapman, J. R., Wong, G., et al. (2012). The motivations and experiences of living kidney donors: A thematic synthesis. *American Journal of Kidney Diseases, 60*(1), 15–26.

*Weng, L., Dai, Y., Huang, H., et al. (2010). Self-efficacy, self-care behaviours and quality of life of kidney transplant recipients. *Journal of Advanced Nursing, 66*(4), 828–838.

*Wiederhold, D., Langer, G., & Landenberger, M. (2011). Ambivalent lived experiences and instruction need of patients in the early period after kidney transplantation: A phenomenological study. *Nephrology Nursing Journal, 38*(5), 417–424.

Resources

American Association of Kidney Patients (AAKP), www.aakp.org
American Kidney Fund, www.kidneyfund.org
American Nephrology Nurses' Association (ANNA), www.annanurse.org
American Urological Association (AUA), www.auanet.org
Arteriovenous Fistula First, www.fistulafirst.org
National Institute of Diabetes and Digestive and Kidney Diseases (NIDDK), www.niddk.nih.gov
National Kidney and Urologic Diseases Information Clearinghouse (NKUDIC), digestive.niddk.nih.gov
National Kidney Foundation, www.kidney.org
United Network for Organ Sharing (UNOS), www.unos.org

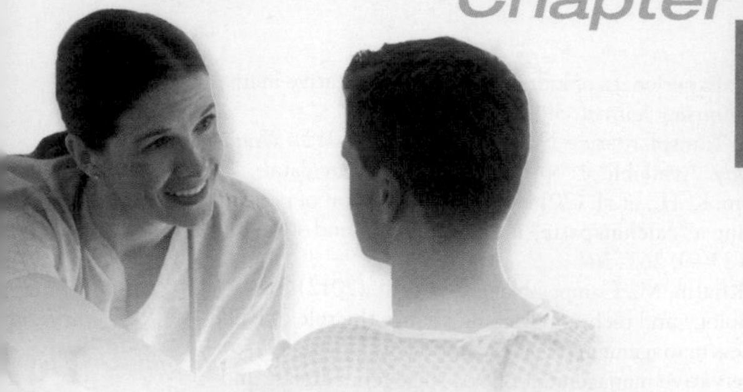

Chapter 55

Management of Patients With Urinary Disorders

Learning Objectives

On completion of this chapter, the learner will be able to:

1. Identify factors contributing to upper and lower urinary tract infections (UTIs).
2. Use the nursing process as a framework for care of the patient with a lower UTI.
3. Differentiate between the various adult dysfunctional voiding patterns.
4. Develop a patient education plan for a patient who has mixed (stress and urge) urinary incontinence.
5. Identify potential causes of an obstruction of the urinary tract along with the medical and surgical management of the patient with this condition.
6. Use the nursing process as a framework for care of the patient with kidney stones.
7. Describe the pathophysiology, clinical manifestations, medical management, and nursing management for patients with genitourinary trauma and urinary tract cancers.
8. Use the nursing process as a framework for care of the patient undergoing urinary diversion surgery.

Glossary

bacteriuria: bacteria in the urine
cystectomy: surgical removal of the urinary bladder
cystitis: inflammation of the urinary bladder
ileal conduit: transplantation of the ureters to an isolated section of the terminal ileum, with one end of the ureters brought to the abdominal wall
interstitial cystitis: inflammation of the bladder wall that eventually causes disintegration of the lining and loss of bladder elasticity
micturition: voiding or urination
neurogenic bladder: bladder dysfunction that results from a disorder or dysfunction of the nervous system and leads to urinary incontinence
nocturia: awakening at night to urinate
overflow incontinence: involuntary urine loss associated with overdistention of the bladder
prostatitis: inflammation of the prostate gland
pyelonephritis: inflammation of the renal pelvis
pyuria: white blood cells in the urine
residual urine: urine that remains in the bladder after voiding
suprapubic catheter: a urinary catheter that is inserted through a suprapubic incision into the bladder
ureterosigmoidostomy: transplantation of the ureters into the sigmoid colon, allowing urine to flow through the colon and out the rectum
ureterovesical or vesicoureteral reflux: backward flow of urine from the bladder into one or both ureters
urethritis: inflammation of the urethra
urethrovesical reflux: An obstruction to free-flowing urine leading to the reflux of urine from the urethra into the bladder
urinary frequency: voiding more often than every 3 hours
urinary incontinence: involuntary or uncontrolled loss of urine from the bladder
urosepsis: spread of infection from the urinary tract to the bloodstream that results in a systemic infection

The urinary system is responsible for providing the route for drainage of urine formed by the kidneys. Care of the patient with disorders of the urinary tract requires an understanding of the anatomy, physiology, diagnostic testing, nursing care, and rehabilitation of patients with the multiple processes that affect the urinary system. Nurses care for patients with urologic disorders in all settings. This chapter focuses on the nursing management of patients with common urinary dysfunctions, including infections, dysfunctional voiding patterns, urolithiasis, genitourinary trauma, cancer of the urinary tract, and urinary diversions.

INFECTIONS OF THE URINARY TRACT

Urinary tract infections (UTIs) are caused by pathogenic microorganisms in the urinary tract (the normal urinary tract is sterile above the urethra). UTIs are generally classified as infections involving the upper or lower urinary tract and further classified as uncomplicated or complicated, depending on other patient-related conditions (Chart 55-1).

Lower UTIs include bacterial **cystitis** (inflammation of the urinary bladder), bacterial **prostatitis** (inflammation of the prostate gland), and bacterial **urethritis** (inflammation of the urethra). There can be acute or chronic nonbacterial causes of inflammation in any of these areas that can be misdiagnosed as bacterial infections. Upper UTIs are much less common and include acute or chronic **pyelonephritis** (inflammation of the renal pelvis), interstitial nephritis (inflammation of the kidney), and renal abscesses. Upper and lower UTIs are further classified as uncomplicated or complicated, depending on whether the UTI is recurrent and the duration of the infection. Most uncomplicated UTIs

Chart 55-1 Classifying Urinary Tract Infections

Urinary tract infections (UTIs) are classified by location: the lower urinary tract (which includes the bladder and structures below the bladder) or the upper urinary tract (which includes the kidneys and ureters). They can also be classified as uncomplicated or complicated UTIs.

Lower UTIs

Cystitis, prostatitis, urethritis

Upper UTIs

Acute pyelonephritis, chronic pyelonephritis, renal abscess, interstitial nephritis, perirenal abscess

Uncomplicated Lower or Upper UTIs

Community-acquired infection; common in young women and not usually recurrent

Complicated Lower or Upper UTIs

Often acquired in the hospital and related to catheterization; occur in patients with urologic abnormalities, pregnancy, immunosuppression, diabetes, and obstructions and are often recurrent

are community acquired. Complicated UTIs usually occur in people with urologic abnormalities or recent catheterization and are often acquired during hospitalization.

A UTI is the second most common infection in the body. Most cases occur in women; one out of every five women in the United States will develop a UTI during her lifetime. The urinary tract is the most common site of nosocomial infection, accounting for greater than 40% of the total number reported by hospitals and affecting about 600,000 patients each year. In most of these hospital-acquired UTIs, instrumentation of the urinary tract or catheterization is the precipitating cause. More than 250,000 cases of acute pyelonephritis occur in the United States each year, with 100,000 patients requiring hospitalization. Approximately 11.3 million women are diagnosed with UTIs in the United States annually, representing an expenditure of about $3.5 billion in direct health care costs (Litwin & Saigal, 2012). This amount does not include the indirect costs associated with time lost from work and the negative impact on the person's lifestyle.

Lower Urinary Tract Infections

Several mechanisms maintain the sterility of the bladder: the physical barrier of the urethra, urine flow, ureterovesical junction competence, various antibacterial enzymes and antibodies, and antiadherent effects mediated by the mucosal cells of the bladder. Abnormalities or dysfunctions of these mechanisms are contributing risk factors for lower UTIs (Chart 55-2).

Pathophysiology

For infection to occur, bacteria must gain access to the bladder, attach to and colonize the epithelium of the urinary tract

Chart 55-2 RISK FACTORS Urinary Tract Infection

- Inability or failure to empty the bladder completely
- Obstructed urinary flow caused by:
  - Congenital abnormalities
  - Urethral strictures
  - Contracture of the bladder neck
  - Bladder tumors
  - Calculi (stones) in the ureters or kidneys
  - Compression of the ureters
- Decreased natural host defenses or immunosuppression
- Instrumentation of the urinary tract (e.g., catheterization, cystoscopic procedures)
- Inflammation or abrasion of the urethral mucosa
- Contributing conditions such as:
  - Female gender
  - Diabetes
  - Pregnancy
  - Neurologic disorders
  - Gout
  - Altered states caused by incomplete emptying of the bladder and urinary stasis

to avoid being washed out with voiding, evade host defense mechanisms, and initiate inflammation. Many UTIs result from fecal organisms ascending from the perineum to the urethra and the bladder and then adhering to the mucosal surfaces.

Bacterial Invasion of the Urinary Tract

By increasing the normal slow shedding of bladder epithelial cells (resulting in bacteria removal), the bladder can clear large numbers of bacteria. Glycosaminoglycan (GAG), a hydrophilic protein, normally exerts a nonadherent protective effect against various bacteria. The GAG molecule attracts water molecules, forming a water barrier that serves as a defensive layer between the bladder and the urine. GAG may be impaired by certain agents (cyclamate, saccharin, aspartame, and tryptophan metabolites). The normal bacterial flora of the vagina and urethral area also interfere with adherence of *Escherichia coli*. Urinary immunoglobulin A (IgA) in the urethra may also provide a barrier to bacteria.

Reflux

An obstruction to free-flowing urine is a condition known as **urethrovesical reflux**, which is the reflux (backward flow) of urine from the urethra into the bladder (Fig. 55-1). With coughing, sneezing, or straining, the bladder pressure increases, which may force urine from the bladder into the urethra. When the pressure returns to normal, the urine flows back into the bladder, bringing into the bladder bacteria from the anterior portions of the urethra. Urethrovesical reflux is also caused by dysfunction of the bladder neck or urethra. The urethrovesical angle and urethral closure pressure may be altered with menopause, increasing the incidence of infection in postmenopausal women. Reflux is most often noted in young children, and treatment is based on its severity.

Ureterovesical or **vesicoureteral reflux** refers to the backward flow of urine from the bladder into one or both ureters

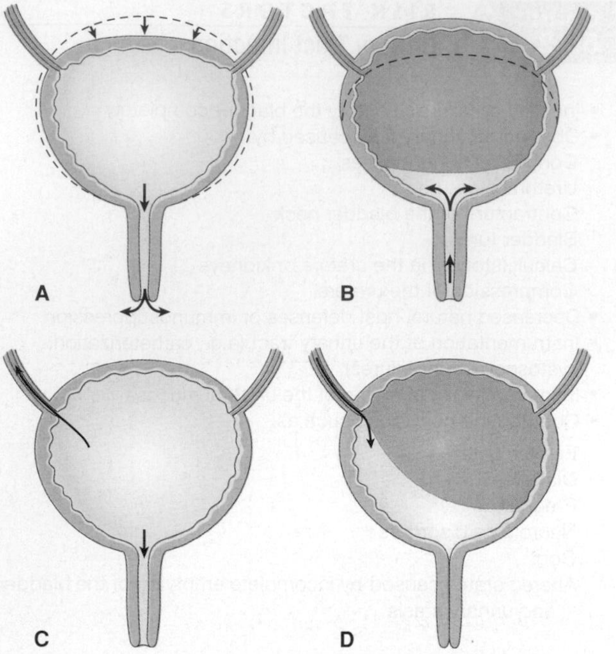
A
B
C
D

FIGURE 55-1 • Mechanisms of urethrovesical and ureterovesical reflux may cause urinary tract infection. *Urethrovesical reflux:* With coughing and straining, bladder pressure rises, which may force urine from the bladder into the urethra. **A.** When bladder pressure returns to normal, the urine flows back to the bladder (**B**), which introduces bacteria from the urethra to the bladder. *Ureterovesical reflux:* With failure of the ureterovesical valve, urine moves up the ureters during voiding (**C**) and flows into the bladder when voiding stops (**D**). This prevents complete emptying of the bladder. It also leads to urinary stasis and contamination of the ureters with bacteria-laden urine.

(see Fig. 55-1). Normally, the ureterovesical junction prevents urine from traveling back into the ureter. The ureters tunnel into the bladder wall so that the bladder musculature compresses a small portion of the ureter during normal voiding. When the ureterovesical valve is impaired by congenital causes or ureteral abnormalities, the bacteria may reach the kidneys and eventually destroy them.

Uropathogenic Bacteria

Bacteriuria is bacteria in the urine. Because urine samples (especially in women) are commonly contaminated by the bacteria normally present in the urethral area, a bacterial count exceeding 10^5 colonies/mL of clean-catch midstream urine is the measure that distinguishes true bacteriuria from contamination. In men, contamination of the collected urine sample occurs less frequently; hence, bacteriuria is defined as 10^4 colonies/mL urine (Hogan, Young, Gabbert, et al., 2011). Community-acquired UTIs are among the most common bacterial infections in women (Porth & Matfin, 2009).

The organisms most frequently responsible for UTIs are those normally found in the lower gastrointestinal (GI) tract, usually *E. coli*. However, isolation of *E. coli* is decreasing compared with previous observations, especially in males and in patients with indwelling bladder catheters, who instead have higher rates of *Pseudomonas* and *Enterococcus* organisms than females and noncatheterized patients (Ling Man & Le Low, 2010).

Routes of Infection

Bacteria enter the urinary tract in three ways: by the transurethral route (ascending infection), through the bloodstream (hematogenous spread), or by means of a fistula from the intestine (direct extension).

The most common route of infection is transurethral, in which bacteria (often from fecal contamination) colonize the periurethral area and subsequently enter the bladder by means of the urethra. In women, the short urethra offers little resistance to the movement of uropathogenic bacteria. Sexual intercourse forces the bacteria from the urethra into the bladder. This accounts for the increased incidence of UTIs in sexually active women. Bacteria may also enter the urinary tract by means of the blood from a distant site of infection or through direct extension by way of a fistula from the intestinal tract.

Clinical Manifestations

A variety of signs and symptoms are associated with UTI. About half of all patients with bacteriuria have no symptoms. Signs and symptoms of an uncomplicated lower UTI (cystitis) include burning on urination, **urinary frequency** (voiding more than every 3 hours), urgency, **nocturia** (awakening at night to urinate), incontinence, and suprapubic or pelvic pain. Hematuria and back pain may also be present. In older people, these symptoms are less common (see Gerontologic Considerations section).

In patients with complicated UTIs, manifestations can range from asymptomatic bacteriuria to gram-negative sepsis with shock. Complicated UTIs often are caused by a broader spectrum of organisms, have a lower response rate to treatment, and tend to recur. Many patients with catheter-associated UTIs are asymptomatic; however, any patient with a catheter who suddenly develops signs and symptoms of septic shock should be evaluated for **urosepsis** (the spread of infection from the urinary tract to the bloodstream that results in a systemic infection).

Gerontologic Considerations

The incidence of bacteriuria in older adults differs from that in younger adults. Bacteriuria increases with age and disability, and women are affected more frequently than men. UTI is the most common cause of acute bacterial sepsis in patients older than 65 years, in whom gram-negative sepsis carries a mortality rate exceeding 50% (Nicolle, 2009). Urologists see many asymptomatic patients with bacteriuria, and 20% are women older than 65 years. In older patients who reside in nursing homes, 25% to 50% of women and 15% to 40% of men have chronic bacteriuria (Das, Perrelli, Towle, et al., 2009).

In the older adult population at large, structural abnormalities secondary to decreased bladder tone and neurogenic bladder (dysfunctional bladder) secondary to stroke or autonomic neuropathy of diabetes may prevent complete emptying of the bladder and increase the risk of UTI (Nicolle, 2009). When indwelling catheters are used, the risk of UTI increases dramatically (Muzzi-Bjornson & Macera, 2011). Older women often have incomplete emptying of the bladder and urinary stasis. In the absence of estrogen, postmenopausal women are susceptible to colonization and increased adherence of bacteria to the vagina and urethra. Oral or topical estrogen has been used to restore the glycogen content of vaginal epithelial cells and an acidic pH for some postmenopausal women with recurrent cystitis.

The antibacterial activity of prostatic secretions that protect men from bacterial colonization of the urethra and bladder decreases with aging. Although UTIs are less common in men, the prevalence of infection in men older than 50 years approaches that of women in the same age group. The increase of UTIs in men as they age is largely a result of prostatic hyperplasia or carcinoma, strictures of the urethra, and neuropathic bladder. The use of catheterization or cystoscopy in evaluation or treatment may also contribute to the higher incidence of UTIs in this group. The incidence of bacteriuria increases in men with confusion, dementia, or bowel or bladder incontinence. The most common cause of recurrent UTIs in older males is chronic bacterial prostatitis. Resection of the prostate gland may help reduce its incidence (see Chapter 59).

In institutionalized older patients, such as those in long-term care facilities, infecting pathogens are often resistant to many antibiotics. Chart 55-3 lists other factors that may contribute to UTI in older patients in these environments. Diligent hand hygiene, careful perineal care, and frequent toileting may decrease the incidence of UTIs.

The organisms responsible for UTIs in the institutionalized older adult may differ from those found in patients residing in the community; this is thought to result in part from the frequent use of antibiotic agents by patients in long-term care facilities. *E. coli* is the most common organism seen in older patients in the community or hospital. However, patients with indwelling catheters are more likely to be infected with organisms such as *Proteus*, *Klebsiella*, *Pseudomonas*, or *Staphylococcus*. Patients who have been previously treated with antibiotics may be infected with *Enterococcus* species. Frequent reinfections are common in older adults.

The most common subjective presenting symptom of UTI in older adults is generalized fatigue. The most common objective finding is a change in cognitive functioning, especially in those with dementia, because these patients usually exhibit even more profound cognitive changes with the onset of a UTI.

Controversy continues about the necessity of treatment for asymptomatic bacteriuria in institutionalized older patients, because resulting antibiotic-resistant organisms and sepsis may be greater threats to patients. It is recommended that antibiotics be prescribed only when symptoms develop (Hart, 2011). However, treatment regimens are generally the same as those for younger adults, although age-related changes in the intestinal absorption of medications and decreased renal function and hepatic flow may necessitate alterations in the antimicrobial regimen. Renal function must be monitored, and medication dosages should be altered accordingly.

Chart 55-3 Factors That Contribute to Urinary Tract Infection in Older Adults

- High incidence of multiple chronic medical conditions
- Frequent use of antimicrobial agents
- Presence of infected pressure ulcers
- Immunocompromise
- Cognitive impairment
- Immobility and incomplete emptying of bladder

Adapted from Nicolle, L. E. (2009). Urinary tract infections in the elderly. *Clinics in Geriatric Medicine*, *25*(3), 423–436.

Quality and Safety Nursing Alert

Older patients often lack the typical symptoms of UTI and sepsis. Although frequency and urgency may occur, nonspecific symptoms, such as altered sensorium, lethargy, anorexia, new incontinence, hyperventilation, and low-grade fever, may be the only clues.

Assessment and Diagnostic Findings

Results of various tests, such as bacterial colony counts, cellular studies, and urine cultures, help confirm the diagnosis of UTI. In an uncomplicated UTI, the strain of bacteria determines the antibiotic of choice (Hart, 2011).

Urine Cultures

Urine cultures are useful for documenting a UTI and identifying the specific organism present. UTI is diagnosed by bacteria in the urine culture. A colony count of at least 10^5 colony-forming units (CFU) per milliliter of urine on a clean-catch midstream or catheterized specimen is a major criterion for infection (O'Shea, 2010). However, UTI and subsequent sepsis have occurred with lower bacterial colony counts. About one third of women with symptoms of acute infections have negative midstream urine culture results and may go untreated if 10^5 CFU/mL is used as the criterion for infection (O'Shea, 2010). The presence of any bacteria in specimens obtained by suprapubic needle aspiration of the urinary bladder or catheterization (insertion of a tube into the urinary bladder) is considered indicative of infection.

The following groups of patients should have urine cultures obtained when bacteriuria is present:

- All men (because of the likelihood of structural or functional abnormalities)
- All children
- Women with a history of compromised immune function or renal problems
- Patients with diabetes
- Patients who have undergone recent instrumentation (including catheterization) of the urinary tract
- Patients who have been recently hospitalized or who live in long-term care facilities
- Patients with prolonged or persistent symptoms
- Patients with three or more UTIs in the previous year
- Pregnant women
- Postmenopausal women
- Women who are sexually active
- Women who have new sexual partners

Cellular Studies

Microscopic hematuria is present in about half of patients with an acute UTI (see Chapter 53). **Pyuria** (white blood cells [WBCs] in the urine) occurs in all patients with UTI;

however, it is not specific for bacterial infection. Pyuria can also be seen with kidney stones, interstitial nephritis, and renal tuberculosis.

Other Studies

A multiple-test dipstick often includes testing for WBCs, known as the leukocyte esterase test, and nitrite testing. Tests for sexually transmitted infections may be performed because acute urethritis caused by sexually transmitted organisms (i.e., *Chlamydia trachomatis*, *Neisseria gonorrhoeae*, herpes simplex) or acute vaginitis infections (caused by *Trichomonas* or *Candida* species) may be responsible for symptoms similar to those of UTIs.

Diagnostic studies such as computed tomography (CT) and ultrasonography are useful diagnostic tools. A CT scan may detect pyelonephritis or abscesses, and ultrasonography is extremely sensitive for detecting obstruction, abscesses, tumors, and cysts. Transrectal ultrasonography (to assess the prostate and bladder) is the procedure of choice for men with recurrent or complicated UTIs. A cystourethroscopy may be indicated to visualize the ureters or to detect strictures, calculi, or tumors (Lane & Takhar, 2011).

Medical Management

Management of UTIs typically involves pharmacologic therapy and patient education. The nurse educates the patient about prescribed medication regimens and infection prevention measures.

Acute Pharmacologic Therapy

The ideal medication for treatment of UTI is an antibacterial agent that eradicates bacteria from the urinary tract with minimal effects on fecal and vaginal flora, thereby minimizing the incidence of vaginal yeast infections. (Yeast vaginitis occurs in as many as 25% of patients treated with antimicrobial agents that affect vaginal flora. Yeast vaginitis can cause more symptoms and be more difficult and costly to treat than the original UTI.) Additionally, the antibacterial agent should be affordable and should have few adverse effects and low resistance. Because the organism in initial, uncomplicated UTIs in women is most likely *E. coli* or other fecal flora, the agent should be effective against these organisms. Various treatment regimens have been successful in treating uncomplicated lower UTIs in women: single-dose administration, short-course (3 to 4 days) regimens, or 7- to 10-day regimens (Lovatt, 2010). The trend is toward a shortened course of antibiotic therapy for uncomplicated UTIs, because most cases are cured after 3 days of treatment. Patients in institutional settings may require 7 to 10 days of medication for the treatment to be effective. Medications commonly used to treat UTIs are listed in Table 55-1. Regardless of the regimen prescribed, the patient is instructed to take all doses prescribed, even if relief of symptoms occurs promptly. Longer medication courses are indicated for men, pregnant women, and women with pyelonephritis and other types of complicated UTIs. Hospitalization and intravenous (IV) antibiotics are occasionally necessary (Karch, 2012).

Long-Term Pharmacologic Therapy

Although brief pharmacologic treatment of UTIs for 3 days is usually adequate in women, infection recurs in about 20% of women treated for uncomplicated UTIs. Infections that recur within 2 weeks of therapy do so because organisms of the original offending strain remain. Relapses suggest that the source of bacteriuria may be the upper urinary tract or that initial treatment was inadequate or administered for too short a time. Recurrent infections in men are usually caused by persistence of the same organism; further evaluation and treatment are indicated (Lovatt, 2010; Karch 2012).

Reinfection with new bacteria is the reason for more than 90% of recurrent UTIs in women (Colgan, Williams, & Johnson, 2011). If the diagnostic evaluation reveals no structural abnormalities in the urinary tract, the woman with recurrent UTIs may be instructed to begin treatment on her own whenever symptoms occur and to contact her primary provider only when symptoms persist, fever occurs, or the number of treatment episodes exceeds four in a 6-month period. The patient may be taught to use dip-slide culture devices to detect bacteria.

TABLE 55-1 Examples of Medications Used to Treat Urinary Tract Infections and Pyelonephritis

Drug Classes	Generic (Brand) Name	Major Indications
Anti-infective, urinary tract	nitrofurantoin (Macrodantin, Furadantin)	UTI
Bactericidal	cephalexin (Keflex)	Genitourinary infection
Cephalosporin	cefadroxil (Duricef, Ultracef)	UTI
Fluoroquinolone	ciprofloxacin (Cipro) ofloxacin (Floxin) norfloxacin (Noroxin) gatifloxacin (Zymar)	UTI Pyelonephritis
Fluoroquinolone	levofloxacin (Levaquin)	Uncomplicated UTI
Penicillin	ampicillin (Principen, Omnipen) amoxicillin (Amoxil)	UTI—not commonly used alone due to *Escherichia coli* resistance Pyelonephritis UTI—not commonly used alone due to *E. coli* resistance
Trimethoprim–sulfamethoxazole combination	Co-trimoxazole (Bactrim, Septra)	UTI Pyelonephritis
Urinary analgesic agent	Phenazopyridine (Pyridium)	For relief of burning, pain, and other symptoms associated with UTI

UTI, urinary tract infection.
Adapted from Karch, A. (2012). *2012 Lippincott's nursing drug guide*. Philadelphia: Lippincott Williams & Wilkins.

If infection recurs after completing antimicrobial therapy, another short course (3 to 4 days) of full-dose antimicrobial therapy followed by a regular bedtime dose of an antimicrobial agent may be prescribed. If there is no recurrence, medication is taken every other night for 6 to 7 months. Long-term use of antimicrobial agents decreases the risk of reinfection and may be indicated in patients with recurrent infections.

If recurrence is caused by persistent bacteria from preceding infections, the cause (i.e., kidney stone, abscess), if known, must be treated. After treatment and sterilization of the urine, low-dose preventive therapy (trimethoprim with or without sulfamethoxazole) each night at bedtime is often prescribed.

Current evidence about the effectiveness of daily intake of cranberry juice to prevent UTIs is inconclusive (Stapleton, Dziura, Hooton, et al., 2012). Patients who like cranberry juice can be encouraged to include it in their increased fluid intake due to the potential protective effect (Dudek, 2010).

NURSING PROCESS

The Patient With a Lower Urinary Tract Infection

Nursing care of the patient with a lower UTI focuses on treating the underlying infection and preventing its recurrence.

Assessment

A history of pertinent signs and symptoms is obtained from the patient with a suspected UTI. The presence of pain, frequency, urgency, hesitancy, and changes in urine are assessed, documented, and reported. The patient's usual pattern of voiding is assessed to detect factors that may predispose him or her to UTI. Infrequent emptying of the bladder, the association of symptoms of UTI with sexual intercourse, contraceptive practices, and personal hygiene are assessed. The patient's knowledge about prescribed antimicrobial medications and preventive health care measures is also assessed. Additionally, the urine is assessed for volume, color, concentration, cloudiness, and odor—all of which are altered by bacteria in the urinary tract.

Diagnosis

Nursing Diagnoses

Based on the assessment data, nursing diagnoses may include the following:

- Acute pain related to infection within the urinary tract
- Deficient knowledge about factors predisposing the patient to infection and recurrence, detection and prevention of recurrence, and pharmacologic therapy

Collaborative Problems/Potential Complications

Potential complications may include the following:

- Sepsis (urosepsis)
- Renal failure, which may occur as the long-term result of either an extensive infective or inflammatory process

Planning and Goals

Major goals for the patient may include relief of pain and discomfort, increased knowledge of preventive measures and treatment modalities, and absence of complications.

Nursing Interventions

Relieving Pain

The pain associated with a UTI is quickly relieved once effective antimicrobial therapy is initiated. Antispasmodic agents may also be useful in relieving bladder irritability and pain. Analgesic agents and the application of heat to the perineum help relieve pain and spasm. The patient is encouraged to drink liberal amounts of fluids (water is the best choice) to promote renal blood flow and to flush the bacteria from the urinary tract. Urinary tract irritants (e.g., coffee, tea, citrus, spices, colas, alcohol) are avoided. Frequent voiding (every 2 to 3 hours) is encouraged to empty the bladder completely, because doing so can significantly lower urine bacterial counts, reduce urinary stasis, and prevent reinfection (Prynn, 2012).

Monitoring and Managing Potential Complications

Early recognition of UTI and prompt treatment are essential to prevent recurrent infection and the possibility of complications, such as renal failure, sepsis (urosepsis), strictures, and obstructions. The goal of treatment is to prevent infection from progressing and causing permanent renal damage and failure. Thus, the patient must be educated to recognize early signs and symptoms, to test for bacteriuria, and to initiate treatment as prescribed. Appropriate antimicrobial therapy, liberal fluid intake, frequent voiding, and hygienic measures are commonly prescribed for managing UTIs. The patient is instructed to notify the primary provider if fatigue, nausea, vomiting, fever, or pruritus occurs. Periodic monitoring of renal function and evaluation for strictures, obstructions, or stones may be indicated for patients with recurrent UTIs.

Patients with UTIs are at increased risk for gram-negative sepsis. For each day a urinary catheter is in place, the risk of catheter-associated urinary tract infection (CAUTI) development increases. CAUTIs are undesirable from the patient perspective due to unpleasant symptoms, additional pharmacologic therapy, serious complications, and delayed discharge (Bernard, Hunter, & Moore, 2012; Kahnen, Flanders, & Magalong, 2011). Considering these deleterious patient consequences, and the financial implications, the National Database of Nurse Quality Indicators (2010) has designated CAUTIs as a nurse-sensitive indicator (see the Resources section later in the chapter).

However, if an indwelling catheter is necessary, the following specific nursing interventions are initiated to prevent infection and urosepsis:

- Using strict aseptic technique during insertion of the smallest catheter possible
- Securing the catheter to prevent movement
- Frequently inspecting urine color, odor, and consistency
- Performing meticulous daily perineal care with soap and water
- Maintaining a closed system
- Following the manufacturer's instructions when using the catheter port to obtain urine specimens

Careful assessment of vital signs and level of consciousness may alert the nurse to kidney involvement or impending sepsis. Positive blood cultures and elevated WBC counts must be reported immediately. At the same time, appropriate antibiotic therapy and increased fluid intake are prescribed (IV antibiotic therapy and fluids may be required). Aggressive early treatment is the key to reducing the mortality rate associated

Chart 55-4

NURSING RESEARCH PROFILE

Decreasing the Incidence of Catheter-Associated Urinary Tract Infections

Elpern, E. H., Killeen, K., Ketchem, A., et al. (2009). Reducing use of indwelling urinary catheters and associated urinary tract infections. *American Journal of Critical Care, 18*(6), 535–542.

Purpose

The use of indwelling urinary catheters can lead to complications, most commonly catheter-associated urinary tract infections. These infections can result in sepsis, prolonged hospitalization, additional hospital costs, and mortality. The purpose of this study was to implement and evaluate an intervention to reduce catheter-associated associated urinary tract infections in a medical intensive care unit (MICU) by decreasing the use of urinary catheters.

Design

This quantitative study used a quasi-experimental design and enrolled 337 patients in an MICU. Indications for continuing urinary catheterization with indwelling devices were developed by unit clinicians. For a 6-month intervention period, participants who had indwelling catheters were evaluated daily by using criteria for appropriate catheter continuance. Recommendations were made to discontinue indwelling urinary catheters in participants who did not meet the criteria. Days of use of a urinary catheter and rates of catheter-associated urinary tract infections during the intervention were compared with those of the preceding 11 months.

Findings

During the study period, 337 participants had a total of 1,432 days of urinary catheterization. With the use of guidelines, duration of use was significantly reduced to a mean of 238.6 device days per month from 311.7 device days per month. The number of catheter-associated urinary tract infections per 1,000 days of use was a mean of 4.7 each month before the intervention and zero during the 6-month intervention period.

Nursing Implications

This small study demonstrates that implementation of an intervention to judge the appropriateness of indwelling urinary catheters may result in significant reductions in duration of catheterization and occurrences of catheter-associated urinary tract infections.

with gram-negative sepsis, especially in older patients (Elpern, Killeen, Ketchem, et al., 2009; Muzzi-Bjornson & Macera, 2011). See Chart 55-4 for more information on decreasing the incidence of CAUTIs.

Promoting Home and Community-Based Care

Educating Patients About Self-Care. In helping patients learn about and prevent or manage a recurrent UTI, the nurse implements education that meets the patient's needs. Health-related behaviors that help prevent recurrent UTIs include practicing careful personal hygiene, increasing fluid intake to promote voiding and dilution of urine, urinating regularly and more frequently, and adhering to the therapeutic regimen. For a detailed discussion of patient education, see Chart 55-5.

Evaluation

Expected patient outcomes may include:

1. Experiences relief of pain
   a. Reports absence of pain, urgency, frequency, nocturia, or hesitancy on voiding
   b. Takes analgesic, antispasmodic, and antibiotic agents as prescribed
2. Explains UTIs and their treatment
   a. Demonstrates knowledge of preventive measures and prescribed treatments
   b. Drinks 8 to 10 glasses of fluids daily
   c. Voids every 2 to 3 hours
   d. Produces urine that is clear and odorless
3. Experiences no complications
   a. Reports no symptoms of infection (fever, frequency)
   b. Has normal renal function, negative urine and blood cultures
   c. Exhibits normal vital signs and temperature; no signs or symptoms of sepsis (urosepsis)
   d. Maintains adequate urine output more than 0.5 mL/kg/hr

Chart 55-5

PATIENT EDUCATION

Preventing Recurrent Urinary Tract Infections

Hygiene

- Shower rather than bathe in the tub because bacteria in the bathwater may enter the urethra.
- After each bowel movement, clean the perineum and urethral meatus from front to back. This will help reduce concentrations of pathogens at the urethral opening and, in women, the vaginal opening.

Fluid Intake

- Drink liberal amounts of fluids daily to flush out bacteria.
- Avoid coffee, tea, colas, alcohol, and other fluids that are urinary tract irritants.

Voiding Habits

- Void every 2–3 hours during the day, and completely empty the bladder. This prevents overdistention of the bladder and compromised blood supply to the bladder wall. Both predispose the patient to urinary tract infection. Precautions expressly for women include voiding immediately after sexual intercourse.

Interventions

- Take medication *exactly* as prescribed. Special timing of administration may be required.
- If bacteria continue to appear in the urine, long-term antimicrobial therapy may be required to prevent colonization of the periurethral area and recurrence of infection.
- For recurrent infection, consider acidification of the urine through ascorbic acid (vitamin C), 1,000 mg daily, or cranberry juice.
- If prescribed, test urine for presence of bacteria following manufacturer's and health care provider's instructions.
- Notify the primary provider if fever occurs or if signs and symptoms persist.
- Consult the primary provider regularly for follow-up.

Upper Urinary Tract Infections

Pyelonephritis is a bacterial infection of the renal pelvis, tubules, and interstitial tissue of one or both kidneys. Causes involve either the upward spread of bacteria from the bladder or spread from systemic sources reaching the kidney via the bloodstream. Pathogenic bacteria from a bladder infection can ascend into the kidney, resulting in pyelonephritis. An incompetent ureterovesical valve or obstruction occurring in the urinary tract increases the susceptibility of the kidneys to infection (see Fig. 55-1), because static urine provides a good medium for bacterial growth. Bladder or prostate tumors, strictures, benign prostatic hyperplasia, and urinary stones are some potential causes of obstruction that can lead to infections. Systemic infections (such as tuberculosis) can spread to the kidneys and result in abscesses.

Pyelonephritis may be acute or chronic. Acute pyelonephritis usually leads to enlargement of the kidneys with interstitial infiltrations of inflammatory cells (Lane & Takhar, 2011). Abscesses may be noted on or within the renal capsule and at the corticomedullary junction. Eventually, atrophy and destruction of tubules and the glomeruli may result. When pyelonephritis becomes chronic, the kidneys become scarred, contracted, and nonfunctioning. Chronic pyelonephritis is a cause of chronic kidney disease that can result in the need for renal replacement therapies such as transplantation or dialysis.

Acute Pyelonephritis

Clinical Manifestations

The patient with acute pyelonephritis has chills, fever, leukocytosis, bacteriuria, and pyuria. Low back pain, flank pain, nausea and vomiting, headache, malaise, and painful urination are common findings. Physical examination reveals pain and tenderness in the area of the costovertebral angle (see Fig. 53-6 in Chapter 53). In addition, symptoms of lower urinary tract involvement, such as urgency and frequency, are common.

Assessment and Diagnostic Findings

An ultrasound study or a CT scan may be performed to locate an obstruction in the urinary tract. Relief of obstruction is essential to prevent complications and eventual kidney damage. An IV pyelogram may be indicated with pyelonephritis if functional and structural renal abnormalities are suspected (Colgan et al., 2011). Radionuclide imaging with gallium citrate and indium-111 (^{111}In)–labeled WBCs may be useful to identify sites of infection that may not be visualized on CT scan or ultrasound. Urine culture and sensitivity tests are performed to determine the causative organism so that appropriate antimicrobial agents can be prescribed.

Medical Management

Patients with acute uncomplicated pyelonephritis are most often treated on an outpatient basis if they are not exhibiting acute symptoms of sepsis, dehydration, nausea, or vomiting. In addition, they must be responsible and reliable to ensure that all medications will be taken as prescribed. For outpatients, a 2-week course of antibiotic agents is recommended because renal parenchymal disease is more difficult to eradicate than mucosal bladder infections. Commonly prescribed agents include some of the same medications prescribed for the treatment of UTIs (see Table 55-1).

Pregnant women may be hospitalized for 2 or 3 days of parenteral antibiotic therapy (Colgan et al., 2011). Oral antibiotic agents may be prescribed once the patient is afebrile and showing clinical improvement.

Following acute pyelonephritis treatment, the patient may develop a chronic or recurring symptomless infection persisting for months or years. After the initial antibiotic regimen, the patient may need antibiotic therapy for up to 6 weeks if a relapse occurs. A follow-up urine culture is obtained 2 weeks after completion of antibiotic therapy to document clearing of the infection.

Hydration with oral or parenteral fluids is essential in all patients with UTIs when there is adequate kidney function. Hydration helps facilitate "flushing" of the urinary tract and reduces pain and discomfort.

Chronic Pyelonephritis

Repeated bouts of acute pyelonephritis may lead to chronic pyelonephritis.

Clinical Manifestations

The patient with chronic pyelonephritis usually has no symptoms of infection unless an acute exacerbation occurs. Noticeable signs and symptoms may include fatigue, headache, poor appetite, polyuria, excessive thirst, and weight loss. Persistent and recurring infection may produce progressive scarring of the kidney, resulting in renal failure (see Chapter 54).

Assessment and Diagnostic Findings

The extent of the disease is assessed by an IV urogram and measurements of creatinine clearance, blood urea nitrogen, and creatinine levels.

Complications

Complications of chronic pyelonephritis include end-stage kidney disease (from progressive loss of nephrons secondary to chronic inflammation and scarring), hypertension, and formation of kidney stones (from chronic infection with urea-splitting organisms).

Medical Management

Bacteria, if detected in the urine, are eradicated if possible. Long-term use of prophylactic antimicrobial therapy may help limit recurrence of infections and renal scarring. Impaired renal function alters the excretion of antimicrobial agents and necessitates careful monitoring of renal function, especially if the medications are potentially toxic to the kidneys.

Nursing Management

The patient may require hospitalization or may be treated as an outpatient. When the patient requires hospitalization, fluid intake and output are carefully measured and recorded. Unless contraindicated, 3 to 4 L of fluids per day is encouraged to dilute the urine, decrease burning on urination, and prevent dehydration. The nurse assesses the patient's temperature every 4 hours and administers antipyretic and antibiotic agents as prescribed. Symptomatic patients are often more comfortable on bed rest.

Patient education focuses on prevention of further infection by consuming adequate fluids, emptying the bladder regularly, and performing recommended perineal hygiene. The importance of taking antimicrobial medications exactly as prescribed is stressed, as is the need for keeping follow-up appointments.

ADULT VOIDING DYSFUNCTION

Both neurogenic and nonneurogenic disorders can cause adult voiding dysfunction (Table 55-2). The **micturition** (voiding or urination) process involves several highly coordinated neurologic responses that mediate bladder function. A functional urinary system allows for appropriate bladder filling and complete bladder emptying (see Chapter 53). If voiding dysfunction goes undetected and untreated, the upper urinary system may be compromised. Chronic incomplete bladder emptying from poor detrusor pressure results in recurrent bladder infection. Incomplete bladder emptying due to bladder outlet obstruction (such as benign prostatic hyperplasia), causing high-pressure detrusor contractions, can result in hydronephrosis from the high detrusor pressure that radiates up the ureters to the renal pelvis.

Urinary Incontinence

More than 25 million adults in the United States are estimated to have **urinary incontinence** (involuntary loss of urine from the bladder), with most of them experiencing overactive bladder syndrome, making this disorder more prevalent than diabetes or ulcer disease (Meiner, 2011; Miller, 2012). Despite widespread media coverage, urinary incontinence remains underdiagnosed and underreported. Patients may be too embarrassed to seek help, causing them to ignore or conceal symptoms. Many patients use absorbent pads or other devices without having their condition properly diagnosed and treated. Health care providers must be alert to subtle cues of urinary incontinence and stay informed about current management strategies.

The cost of urologic care in the United States is more than $11 billion (Omli, Skotnes, Romild, et al., 2010). The costs of care for patients with urinary incontinence include the expenses of absorbent products, medications, and surgical or nonsurgical treatment modalities, as well as psychosocial costs (i.e., embarrassment, loss of self-esteem, and social isolation).

Although urinary incontinence is commonly regarded as a condition that occurs in older multiparous women, it can occur in young nulliparous women, especially during vigorous high-impact activity. Age, gender, and number of vaginal deliveries are established risk factors that explain, in part, the increased incidence in women (Chart 55-6). Urinary incontinence is a symptom of many possible disorders.

Types of Urinary Incontinence

This section provides common terms used to describe the many types of urinary incontinence.

Stress incontinence is the involuntary loss of urine through an intact urethra as a result of sneezing, coughing, or changing position (Meiner, 2011; Miller, 2012). It predominantly affects women who have had vaginal deliveries and is thought

TABLE 55-2 Conditions Causing Adult Voiding Dysfunction

Condition	Voiding Dysfunction	Treatment
Neurogenic Disorders		
Cerebellar ataxia	Incontinence or dyssynergia	Timed voiding; anticholinergic agents
Cerebrovascular accident	Retention or incontinence	Anticholinergic agents; bladder retraining
Dementia	Incontinence	Prompted voiding; anticholinergic agents
Diabetes	Incontinence and/or incomplete bladder emptying	Timed voiding; EMG/biofeedback; pelvic floor nerve stimulation; anticholinergic/antispasmodic agents; well-controlled blood glucose levels
Multiple sclerosis	Incontinence or incomplete bladder emptying	Timed voiding; EMG/biofeedback to learn pelvic muscle exercises and urge inhibition; pelvic floor nerve stimulation; antispasmodic agents
Parkinson's disease	Incontinence	Anticholinergic/antispasmodic agents
Spinal Cord Dysfunction		
Acute injury	Urinary retention	Indwelling catheter
Degenerative disease	Incontinence and/or incomplete bladder emptying	EMG/biofeedback; pelvic floor nerve stimulation; anticholinergic agents
Nonneurogenic Disorder		
"Bashful bladder"	Inability to initiate voiding in public bathrooms	Relaxation therapy; EMG/biofeedback
Overactive bladder	Urgency, frequency, and/or urge incontinence	EMG/biofeedback; pelvic floor nerve stimulation; bladder drill (see Chart 55-8); anticholinergic agents
Post general surgery	Acute urine retention	Catheterization
Post prostatectomy	Incontinence	*Mild:* Biofeedback; bladder drill (see Chart 55-8); pelvic floor nerve stimulation *Moderate/severe:* Surgery—artificial sphincter
Stress incontinence	Incontinence with cough, laugh, sneeze, position change	*Mild:* Biofeedback; bladder drill (see Chart 55-8); periurethral bulking with collagen *Moderate/severe:* Surgery

EMG, electromyogram.

Chart 55-6 RISK FACTORS
Urinary Incontinence

- Pregnancy—vaginal delivery, episiotomy
- Menopause
- Genitourinary surgery
- Pelvic muscle weakness
- Incompetent urethra due to trauma or sphincter relaxation
- Immobility
- High-impact exercise
- Diabetes
- Stroke
- Age-related changes in the urinary tract
- Morbid obesity
- Cognitive disturbances—dementia, Parkinson's disease
- Medications—diuretic, sedative, hypnotic, and opioid agents
- Caregiver or toilet unavailable

Adapted from Omli, R., Skotnes, L. H., Romild, U., et al. (2010). Pad per day usage, urinary incontinence and urinary tract infections in nursing home residents. *Age and Aging*, *39*, 549–554.

to be the result of decreasing ligament and pelvic floor support of the urethra and decreasing or absent estrogen levels within the urethral walls and bladder base. In men, stress incontinence is often experienced after a radical prostatectomy for prostate cancer because of the loss of urethral compression that the prostate had supplied before the surgery, and possibly bladder wall irritability.

Urge incontinence is the involuntary loss of urine associated with a strong urge to void that cannot be suppressed (Meiner, 2011; Miller, 2012). The patient is aware of the need to void but is unable to reach a toilet in time. An uninhibited detrusor contraction is the precipitating factor. This can occur in a patient with neurologic dysfunction that impairs inhibition of bladder contraction or in a patient without overt neurologic dysfunction.

Functional incontinence refers to those instances in which lower urinary tract function is intact but other factors, such as severe cognitive impairment (e.g., Alzheimer's dementia), make it difficult for the patient to identify the need to void or physical impairments make it difficult or impossible for the patient to reach the toilet in time for voiding (Specht, 2011).

Iatrogenic incontinence refers to the involuntary loss of urine due to extrinsic medical factors, predominantly medications. One such example is the use of alpha-adrenergic agents to decrease blood pressure. In some people with an intact urinary system, these agents adversely affect the alpha receptors responsible for bladder neck closing pressure; the bladder neck relaxes to the point of incontinence with a minimal increase in intra-abdominal pressure, thus mimicking stress incontinence. As soon as the medication is discontinued, the apparent incontinence resolves.

Mixed urinary incontinence, which encompasses several types of urinary incontinence, is involuntary leakage associated with urgency and also with exertion, effort, sneezing, or coughing (Miller, 2012).

Only with appropriate recognition of the problem, assessment, and referral for diagnostic evaluation and treatment can the outcome of incontinence be determined. All people with incontinence should be considered for evaluation and treatment.

Gerontologic Considerations

Although urinary incontinence is not a normal consequence of aging, age-related changes in the urinary tract do predispose the older person to incontinence. However, if nurses and other health care providers accept incontinence as an inevitable part of illness or aging or consider it irreversible and untreatable, it cannot be treated successfully. Collaborative, interdisciplinary efforts are essential in assessing and effectively treating urinary incontinence. Urinary incontinence can decrease an older person's ability to maintain an independent lifestyle, which increases dependence on caregivers and may lead to institutionalization. Between 25% and 50% of all nursing home residents have urinary incontinence (Omli et al., 2010).

Many older people experience transient episodes of incontinence that tend to be abrupt in onset. When this occurs, the nurse should question the patient, as well as the family if possible, about the onset of symptoms and any signs or symptoms of a change in other organ systems. Acute UTI, infection elsewhere in the body, constipation, decreased fluid intake, and a change in a chronic disease pattern, such as elevated blood glucose levels in patients with diabetes or decreased estrogen levels in menopausal women, can provoke the onset of urinary incontinence. If the cause is identified and modified or eliminated early at the onset of incontinence, the incontinence itself may be eliminated. Although the bladder of the older person is more vulnerable to altered detrusor activity, age alone is not a risk factor for urinary incontinence (Ling Man & Le Low, 2010; Stewart, 2012).

Decreased bladder muscle tone is a normal age-related change found in older adults. This leads to decreased bladder capacity, increased residual urine, and an increase in urgency (Ling Man & Le Low, 2010).

Many medications affect urinary continence in addition to causing other unwanted or unexpected effects. All medications need to be assessed for potential interactions.

Assessment and Diagnostic Findings

Once incontinence is recognized, a thorough history is necessary. This includes a detailed description of the problem and a history of medication use. The patient's voiding history, a diary of fluid intake and output, and bedside tests (e.g., residual urine, stress maneuvers) may be used to help determine the type of urinary incontinence involved. Extensive urodynamic tests may be performed (see Chapter 53). Urinalysis and urine culture are performed to identify infection.

Urinary incontinence may be transient or reversible if the underlying cause is successfully treated and the voiding pattern reverts to normal. Chart 55-7 provides causes of transient incontinence.

Medical Management

Management depends on the type of urinary incontinence and its causes. Management of urinary incontinence may be behavioral, pharmacologic, or surgical.

Behavioral Therapy

Behavioral therapies are the first choice to decrease or eliminate urinary incontinence (Chart 55-8). In using these techniques, health care professionals help patients avoid

Chart 55-7 Causes of Transient Incontinence

- Delirium or confusion
- Urinary tract infections
- Atrophic vaginitis, urethritis, prostatitis
- Pharmacologic agents (anticholinergic agents, sedatives, alcohol, analgesic agents, diuretics, muscle relaxants, adrenergic agents)
- Psychological factors (depression, regression)
- Excessive urine production (increased intake, diabetes insipidus, diabetic ketoacidosis)
- Limited or restricted activity
- Stool impaction or constipation

potential adverse effects of pharmacologic or surgical interventions. Pelvic floor muscle exercises (sometimes referred to as Kegel exercises) represent the cornerstone of behavioral intervention for addressing symptoms of stress, urge, and mixed incontinence (Agency for Healthcare Research and Quality [AHRQ], 2012). Other behavioral treatments include the use of a voiding diary, biofeedback, verbal instruction (prompted voiding), and physical therapy (Ling Man & Le Low, 2010).

Pharmacologic Therapy

Pharmacologic therapy works best when used as an adjunct to behavioral interventions. Anticholinergic agents inhibit bladder contraction and are considered first-line medications for urge incontinence. Several tricyclic antidepressant medications (e.g., amitriptyline [Endep], amoxapine [Asendin]) can also decrease bladder contractions as well as increase bladder neck resistance (Karch, 2012). Pseudoephedrine sulfate (Sudafed), which acts on alpha-adrenergic receptors, causing urinary retention, may be used to treat stress incontinence; it needs to be used with caution in men with prostatic hyperplasia and patients with hypertension. Hormone therapy (e.g., estrogen) taken orally, transdermally, or topically was once the treatment of choice for urinary incontinence in postmenopausal women because it restores the mucosal, vascular, and muscular integrity of the urethra. However, research suggests incontinence increases in women taking estrogen alone compared to placebo (Weinstein, 2012).

Surgical Management

Surgical correction may be indicated in patients who have not achieved continence using behavioral and pharmacologic therapy. Surgical options vary according to the underlying anatomy and the physiologic problem. Most procedures involve lifting and stabilizing the bladder or urethra to restore the normal urethrovesical angle or to lengthen the urethra.

Women with stress incontinence may undergo an anterior vaginal repair, retropubic suspension, or needle suspension to reposition the urethra. Procedures to compress the urethra and increase resistance to urine flow include sling procedures and placement of periurethral bulking agents such as artificial collagen.

Periurethral bulking is a semipermanent procedure in which small amounts of artificial collagen are placed within the walls of the urethra to enhance the closing pressure of the urethra. This procedure takes only 10 to 20 minutes and may be performed under local anesthesia or moderate sedation. A cystoscope is inserted into the urethra. An instrument is inserted through the cystoscope to deliver a small amount of collagen into the urethral wall at locations selected by the urologist. The patient is usually discharged home after voiding. There are no restrictions following the procedure, although occasionally more than one collagen bulking session may be necessary if the initial procedure did not halt stress incontinence. Collagen placement anywhere in the body is considered semipermanent because its durability averages between 12 and 24 months, until the body absorbs the material. Periurethral bulking with collagen offers an alternative to surgery, such as one for an older person who is frail (Keegan, Atiemo, Cody, et al., 2012). It is also an option for people who are seeking help with stress incontinence who prefer to avoid surgery and who do not have access to behavioral therapies.

An artificial urinary sphincter can be used to close the urethra and promote continence. Two types of artificial sphincters are a periurethral cuff and a cuff inflation pump.

Men with overflow and stress incontinence may undergo a transurethral resection to relieve symptoms of prostatic enlargement. An artificial sphincter can be used after prostatic surgery for sphincter incompetence (Fig. 55-2). After surgery, periurethral bulking agents can be injected into the periurethral area to increase compression of the urethra.

Nursing Management

Nursing management is based on the premise that incontinence is not inevitable with illness or aging and that it is often reversible and treatable. The nursing interventions are determined in part by the type of treatment that is undertaken. For behavioral therapy to be effective, the nurse must provide support and encouragement, because it is easy for the patient to become discouraged if therapy does not quickly improve the level of continence. Patient education is important and should be provided verbally

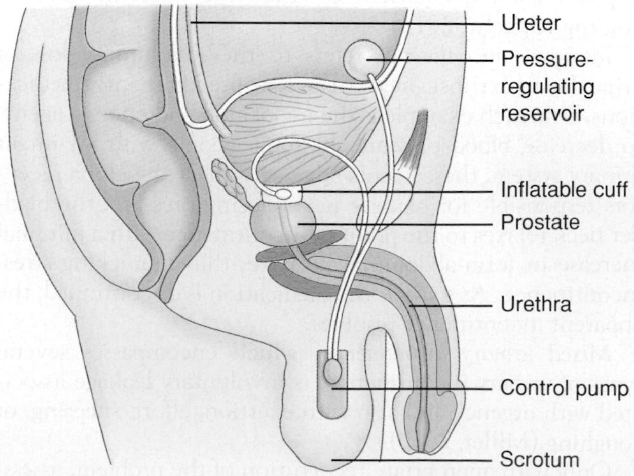

Ureter
Pressure-regulating reservoir
Inflatable cuff
Prostate
Urethra
Control pump
Scrotum

FIGURE 55-2 • Male artificial urinary sphincter. An inflatable cuff is inserted surgically around the urethra or neck of the bladder. To empty the bladder, the cuff is deflated by squeezing the control pump located in the scrotum.

Chart 55-8 HEALTH PROMOTION

Interventions for Urinary Incontinence

Behavioral strategies are largely carried out, coordinated, and monitored by the nurse. These interventions may or may not be augmented by the use of medications.

Fluid Management

An adequate daily fluid intake of approximately 50–60 ounces (1,500–1,600 mL), taken as small increments between breakfast and the evening meal, helps to reduce urinary urgency related to concentrated urine production, decreases the risk of urinary tract infection, and maintains bowel functioning. (Constipation, resulting from inadequate daily fluid intake, can increase urinary urgency and urine retention.) The best fluid is water. Fluids containing caffeine, carbonation, alcohol, or artificial sweetener should be avoided because they irritate the bladder wall, thus resulting in urinary urgency. Some patients who have heart failure or end-stage kidney disease need to discuss their daily fluid limit with their primary provider.

Standardized Voiding Frequency

After establishing a patient's natural voiding and urinary incontinence tendencies, voiding on a schedule can be very effective in those with and without cognitive impairment, although patients with cognitive impairment may require assistance with this technique from nursing personnel or family members. The object is to purposely empty the bladder before the bladder reaches the critical volume that would cause an urge or stress incontinence episode. This approach involves the following:

- **Timed voiding** involves establishing a set voiding frequency (such as every 2 hours if incontinent episodes tend to occur 2 or more hours after voiding). The individual chooses to "void by the clock" at the given interval while awake rather than wait until a voiding urge occurs.
- **Prompted voiding** is timed voiding that is carried out by staff or family members when the individual has cognitive difficulties that make it difficult to remember to void at set intervals. The caregiver checks the patient to assess if he or she has remained dry and, if so, assists the patient to use the bathroom while providing positive reinforcement for remaining dry.
- **Habit retraining** is timed voiding at an interval that is more frequent than the individual would usually choose. This technique helps to restore the sensation of the need to void in individuals who are experiencing diminished sensation of bladder filling due to various medical conditions such as a cerebrovascular accident.
- **Bladder retraining**, also known as "bladder drill," incorporates a timed voiding schedule and urinary urge inhibition exercises to inhibit voiding, or leaking urine, in an attempt to remain dry for a set time. When the first timing interval is easily reached on a consistent basis without urinary urgency or incontinence, a new voiding interval, usually 10–15 minutes beyond the last, is established. Again, the individual practices urge inhibition exercises to delay voiding or avoid incontinence until the next preset interval arrives. When an acceptable voiding interval is reached, the patient continues that timed voiding sequence throughout the day.

Pelvic Muscle Exercise

Also known as Kegel exercises, pelvic muscle exercise (PME) aims to strengthen the voluntary muscles that assist in bladder and bowel continence in both men and women. Research shows that written or verbal instruction alone is usually inadequate to educate an individual about how to identify and strengthen the pelvic floor for sufficient bladder and bowel control. Biofeedback-assisted PME uses either electromyography or manometry to help the individual identify the pelvic muscles as he or she attempts to learn which muscle group is involved when performing PME. The biofeedback method also allows assessment of the strength of this muscle area.

PME involves gently tightening the same muscles used to stop flatus or the stream of urine for 5–10-second increments, followed by 10-second resting phases. To be effective, these exercises need to be performed 2 or 3 times a day for at least 6 weeks. Depending on the strength of the pelvic musculature when initially evaluated, anywhere from 10–30 repetitions of PME are prescribed at each session. Older patients may need to exercise for an even longer time to strengthen the pelvic floor muscles. Pelvic muscle exercises are helpful for women with stress, urge, or mixed incontinence and for men who have undergone prostate surgery.

Vaginal Cone Retention Exercises

Vaginal cone retention exercises are an adjunct to the Kegel exercises. Vaginal cones of varying weight are inserted intravaginally twice a day. The patient tries to retain the cone for 15 minutes by contracting the pelvic muscles.

Transvaginal or Transrectal Electrical Stimulation

Commonly used to treat urinary incontinence, electrical stimulation is known to elicit a passive contraction of the pelvic floor musculature, thus re-educating these muscles to provide enhanced levels of continence. This modality is often used with biofeedback-assisted pelvic muscle exercise training and voiding schedules. At high frequencies, it is effective for stress incontinence. At low frequencies, electrical stimulation can also relieve symptoms of urinary urgency, frequency, and urge incontinence. Intermediate ranges are used for mixed incontinence.

Neuromodulation

Neuromodulation via transvaginal or transrectal nerve stimulation of the pelvic floor inhibits detrusor overactivity and hypersensory bladder signals and strengthens weak sphincter muscles.

Adapted from Agency for Healthcare Research and Quality. (2012). Experts seek better diagnosis and treatment for women's urinary incontinence and chronic pelvic pain. *Research Activities*, (383), 1–4; and Sangsawang, B., & Serisathien, Y. (2012). Effect of pelvic floor muscle exercise programme on stress urinary incontinence among pregnant women. *Journal of Advanced Nursing*, *68*(9), 1997–2007.

and in writing (Chart 55-9). The patient should be educated to develop and use a log or diary to record timing of pelvic floor muscle exercises, frequency of voiding, any changes in bladder function, and any episodes of incontinence (Miller, 2012).

If pharmacologic treatment is used, the patient and family are educated about its purpose. Patients with mixed incontinence must be informed that anticholinergic and antispasmodic agents can help decrease urinary urgency and frequency and urge incontinence but do not decrease the urinary incontinence related to stress incontinence. If surgical correction is undertaken, the procedure and its desired outcomes are described to the patient and family. Follow-up contact with the patient enables the nurse to

Chart 55-9 PATIENT EDUCATION

Strategies for Promoting Urinary Continence

- Increase your awareness of the amount and timing of all fluid intake.
- Avoid taking diuretic agents after 4 PM.
- Avoid bladder irritants, such as caffeine, alcohol, and aspartame (NutraSweet).
- Take steps to avoid constipation: Drink adequate fluids, eat a well-balanced diet high in fiber, exercise regularly, and take stool softeners if recommended.
- Void regularly, 5–8 times a day (about every 2–3 hours):
  First thing in the morning
  Before each meal
  Before retiring to bed
  Once during the night if necessary
- Perform all pelvic floor muscle exercises as prescribed, every day.
- Stop smoking (smokers usually cough frequently, which increases incontinence).

answer the patient's questions and to provide reinforcement and encouragement.

Urinary Retention

Urinary retention is the inability to empty the bladder completely during attempts to void. Chronic urine retention often leads to **overflow incontinence** (involuntary urine loss associated with overdistention of the bladder). **Residual urine** is urine that remains in the bladder after voiding. In a healthy adult younger than 60 years, complete bladder emptying should occur with each voiding. In adults older than 60 years, 50 to 100 mL of residual urine may remain after each voiding because of the decreased contractility of the detrusor muscle.

Urinary retention can occur postoperatively in any patient, particularly if the surgery affected the perineal or anal regions and resulted in reflex spasm of the sphincters. General anesthesia reduces bladder muscle innervation and suppresses the urge to void, impeding bladder emptying.

Pathophysiology

Urinary retention may result from diabetes, prostatic enlargement, urethral pathology (infection, tumor, calculus), trauma (pelvic injuries), pregnancy, or neurologic disorders (e.g., stroke, spinal cord injury, multiple sclerosis, or Parkinson's disease). Some medications cause urinary retention either by inhibiting bladder contractility or by increasing bladder outlet resistance (Karch, 2012).

Assessment and Diagnostic Findings

The assessment of a patient for urinary retention is multifaceted because the signs and symptoms are challenging to detect. The following questions serve as a guide in assessment:

- What was the time of the last voiding, and how much urine was voided?
- Is the patient voiding small amounts of urine frequently?
- Is the patient dribbling urine?
- Does the patient complain of pain or discomfort in the lower abdomen? (Discomfort may be relatively mild if the bladder distends slowly.)
- Is the pelvic area rounded and swollen (could indicate urine retention and a distended bladder)?
- Does percussion of the suprapubic region elicit dullness (possibly indicating urine retention and a distended bladder)?
- Are other indicators of urinary retention present, such as restlessness and agitation?
- Does a postvoid bladder ultrasound test reveal residual urine?

The patient may verbalize an awareness of bladder fullness and a sensation of incomplete bladder emptying. Signs and symptoms of UTI (hematuria, urgency, frequency, and nocturia) may be present. A series of urodynamic studies (described in Chapter 53) may be performed to identify the type of bladder dysfunction and to aid in determining appropriate treatment. A voiding diary can be used to provide a written record of the amount of urine voided and the frequency of voiding. Postvoid residual urine may be assessed by using either straight catheterization or an ultrasound bladder scanner and is considered diagnostic of urinary retention. Normally, residual urine amounts to no more than 50 mL in the middle-aged adult and less than 50 to 100 mL in the older adult (Hogan-Quigley, Palm, & Bickley, 2012; Weber & Kelley, 2010).

Complications

The retention of urine can lead to chronic infections that if unresolved predispose the patient to renal calculi (urolithiasis or nephrolithiasis), pyelonephritis, sepsis, or hydronephrosis. In addition, urine leakage can lead to perineal skin breakdown, especially if regular hygiene measures are neglected.

Nursing Management

Strategies are instituted to prevent overdistention of the bladder and to treat infection or correct obstruction. However, many complications can be prevented with careful assessment and appropriate nursing interventions. The nurse explains to the patient why normal voiding is not occurring and monitors urine output closely. The nurse also provides reassurance about the temporary nature of retention and successful management strategies.

Promoting Urinary Elimination

Nursing measures to encourage normal voiding patterns include providing privacy, ensuring an environment and body position conducive to voiding, and assisting the patient with the use of the bathroom or bedside commode, rather than a bedpan, to provide a more natural setting for voiding. If his condition allows, the male patient may stand beside the bed to use the urinal; most men find this position more comfortable and natural.

Additional measures include applying warmth to relax the sphincters (i.e., sitz baths, warm compresses to the perineum, showers), giving the patient hot tea, and offering encouragement and reassurance. Simple trigger techniques, such as turning on the water faucet while the patient is trying to void, may also be used. Other examples of trigger techniques are stroking the abdomen or inner thighs, tapping above the pubic area, and dipping the patient's hands in warm water. After surgery or childbirth, prescribed analgesic agents should be administered because pain in the perineal area can make voiding difficult. A combination of techniques may be necessary to initiate voiding.

When the patient cannot void, catheterization is used to prevent overdistention of the bladder (see later discussion of neurogenic bladder and catheterization). In the case of prostatic obstruction, attempts at catheterization (by the urologist) may not be successful, requiring insertion of a **suprapubic catheter** (catheter inserted through a small abdominal incision into the bladder). After urinary drainage is restored, bladder retraining is initiated for the patient who cannot void spontaneously.

Promoting Home and Community-Based Care

In addition to the strategies listed for promoting urinary continence found in Chart 55-9, modifications to the home environment can provide simple and effective ways to assist in treating urinary incontinence and retention. For example, the patient may need to remove obstacles, such as throw rugs or other objects, to provide easy, safe access to the bathroom. Other modifications that the nurse may recommend include installing support bars in the bathroom; placing a bedside commode, bedpan, or urinal within easy reach; leaving lights on in the bedroom and bathroom; and wearing clothing that is easy to remove quickly.

Neurogenic Bladder

Neurogenic bladder is a dysfunction that results from a disorder or dysfunction of the nervous system and leads to urinary incontinence. It may be caused by spinal cord injury, spinal tumor, herniated vertebral disk, multiple sclerosis, congenital disorders (spina bifida or myelomeningocele), infection, or complications of diabetes (Klausner & Steers, 2011) (see Chapters 51, 68, and 69).

Pathophysiology

The two types of neurogenic bladder are spastic (or reflex) bladder and flaccid bladder. Spastic bladder is the more common type and is caused by any spinal cord lesion above the voiding reflex arc (upper motor neuron lesion). The result is a loss of conscious sensation and cerebral motor control. A spastic bladder empties on reflex, with minimal or no controlling influence to regulate its activity.

Flaccid bladder is caused by a lower motor neuron lesion, commonly resulting from trauma. This form of neurogenic bladder is also increasingly being recognized in patients with diabetes. The bladder continues to fill and becomes greatly distended, and overflow incontinence occurs. The bladder muscle does not contract forcefully at any time. Because sensory loss may accompany a flaccid bladder, the patient feels no discomfort.

Assessment and Diagnostic Findings

Evaluation for neurogenic bladder involves measurement of fluid intake, urine output, and residual urine volume; urinalysis; and assessment of sensory awareness of bladder fullness and degree of motor control. Comprehensive urodynamic studies are also performed.

Complications

The most common complication of neurogenic bladder is infection resulting from urinary stasis and catheterization. Long-term complications include urolithiasis (stones in the urinary tract), vesicoureteral reflux, and hydronephrosis—all of which can lead to destruction of the kidney (Klausner & Steers, 2011).

Medical Management

The problems resulting from neurogenic bladder disorders vary considerably from patient to patient and are a major challenge to the health care team. Several long-term objectives appropriate for all types of neurogenic bladders include preventing overdistention of the bladder, emptying the bladder regularly and completely, maintaining urine sterility with no stone formation, and maintaining adequate bladder capacity with no reflux.

Specific interventions include continuous, intermittent, or self-catheterization (discussed later in this chapter); the use of an external condom-type catheter; a diet low in calcium (to prevent calculi); and encouragement of mobility and ambulation. A liberal fluid intake is encouraged to reduce the urinary bacterial count, reduce stasis, decrease the concentration of calcium in the urine, and minimize the precipitation of urinary crystals and subsequent stone formation.

A bladder retraining program may be effective in treating a spastic bladder or urine retention. The use of a timed, or habit, voiding schedule may be established. To further enhance emptying of a flaccid bladder, the patient may be taught to "double void." After each voiding, the patient is instructed to remain on the toilet, relax for 1 to 2 minutes, and then attempt to void again in an effort to further empty the bladder.

Pharmacologic Therapy

Parasympathomimetic medications, such as bethanechol (Urecholine), may help to increase the contraction of the detrusor muscle.

Surgical Management

In some cases, surgery may be carried out to correct bladder neck contractures or vesicoureteral reflux or to perform some type of urinary diversion procedure.

Catheterization

In patients with a urologic disorder or with marginal kidney function, care must be taken to ensure that urinary drainage is adequate and that kidney function is preserved. When urine cannot be eliminated naturally and must be drained artificially, catheters may be inserted directly into the bladder, the ureter, or the renal pelvis (Myles, 2011). Catheters vary in size, shape, length, material, and configuration. The type of catheter used depends on its purpose.

Catheterization is performed to achieve the following:

- Relieve urinary tract obstruction
- Assist with postoperative drainage in urologic and other surgeries
- Provide a means to monitor accurate urine output in critically ill patients

- Promote urinary drainage in patients with neurogenic bladder dysfunction or urine retention
- Prevent urinary leakage in patients with stage III to IV pressure ulcers (see Chapter 10)

A patient should be catheterized only if necessary, because catheterization commonly leads to UTI. Catheters impede most of the natural defenses of the lower urinary tract by obstructing the periurethral ducts, irritating the bladder mucosa, and providing an artificial route for organisms to enter the bladder. Organisms may be introduced from the urethra into the bladder during catheterization, or they may migrate along the epithelial surface of the urethra or external surface of the catheter. In addition, urinary catheters have been associated with other complications, such as bladder spasms, urethral strictures, and pressure necrosis.

Indwelling Catheters

When an indwelling catheter cannot be avoided, a closed drainage system is essential. This drainage system is designed to prevent any disconnections, thereby reducing the risk of contamination. Triple-lumen catheters are commonly used after transurethral prostate surgery (see Chapter 59). This system has a triple-lumen indwelling urethral catheter attached to a closed sterile drainage system. With the triple-lumen catheter, urinary drainage occurs through one channel. The retention balloon of the catheter is inflated with water or air through the second channel, and the bladder is continuously irrigated with sterile irrigating solution through the third channel.

The spout (or drainage port) of any urinary drainage bag can become contaminated when opened to drain the bag. Bacteria enter the urinary drainage bag, multiply rapidly, and then migrate to the drainage tubing, catheter, and bladder. By keeping the drainage bag lower than the patient's bladder and not allowing urine to flow back into the bladder, this risk is reduced.

Suprapubic Catheters

Suprapubic catheterization allows bladder drainage by inserting a catheter or tube into the bladder through a suprapubic (above the pubis) incision or puncture (Fig. 55-3). The catheter or suprapubic drainage tube is then threaded into the bladder and secured with sutures or tape, and the area around the catheter is covered with a sterile dressing. The catheter is connected to a sterile closed drainage system, and the tubing is secured to prevent tension on the catheter. This may be a temporary measure to divert the flow of urine from the urethra when the urethral route is impassable (because of injuries, strictures, prostatic obstruction), after gynecologic or other abdominal surgery when bladder dysfunction is likely to occur, and occasionally after pelvic fractures.

Suprapubic bladder drainage may be maintained continuously for several weeks. When the patient's ability to void is to be tested, the catheter is clamped for 4 hours, during which time the patient attempts to void. After the patient voids, the catheter is unclamped, and the residual urine is measured. If the amount of residual urine is less than 100 mL on two separate occasions (morning and evening), the catheter is usually removed. However, if the patient complains of pain or discomfort, the suprapubic catheter is usually left in place until the patient can void successfully.

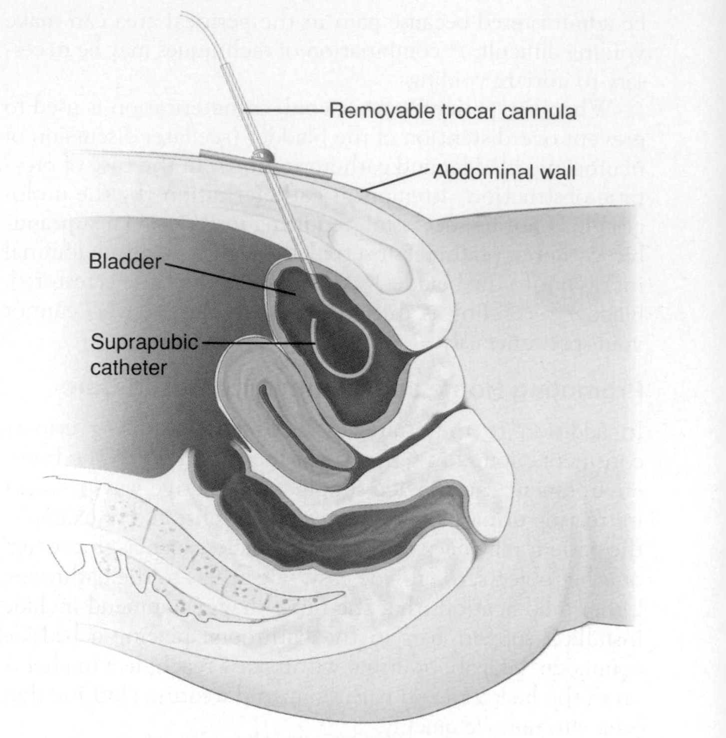
Removable trocar cannula
Abdominal wall
Bladder
Suprapubic
catheter

FIGURE 55-3 • Suprapubic bladder drainage. A trocar cannula is used to puncture the abdominal and bladder walls. The catheter is threaded through the trocar cannula, which is then removed, leaving the catheter in place. The catheter is secured by tape or sutures to prevent unintentional removal.

Suprapubic drainage offers certain advantages. Patients can usually void sooner after surgery than those with urethral catheters, and they may be more comfortable. The catheter allows greater mobility, permits measurement of residual urine without urethral instrumentation, and presents less risk of bladder infection. The suprapubic catheter is removed when it is no longer required, and a sterile dressing is placed over the site.

The patient requires liberal amounts of fluid to prevent encrustation around the catheter. Other potential problems include the formation of bladder stones, acute and chronic infections, and problems collecting urine. A wound-ostomy-continence (WOC) nurse may be consulted to assist the patient and family in selecting the most suitable urine collection system and to educate them about its use and care (Gray, 2009).

Nursing Management

Assessing the Patient and the System

For patients with indwelling catheters, the nurse assesses the drainage system to ensure that it provides adequate urinary drainage. The color, odor, and volume of urine are also monitored. An accurate record of fluid intake and urine output provides essential information about the adequacy of renal function and urinary drainage.

Patients at high risk for UTI from catheterization need to be identified and monitored carefully. These include women; older adults; and patients who are debilitated, malnourished, chronically ill, immunosuppressed, or have diabetes (Wilde, Brasch, Getliffe, et al., 2010). They are observed for signs and symptoms of UTI: cloudy malodorous urine, hematuria, fever, chills, anorexia, and malaise. Any drainage and excoriation in the area around the urethral orifice is noted. Urine cultures provide the most accurate means of assessing a patient for infection.

Gerontologic Considerations

The older patient with an indwelling catheter may not exhibit the typical signs and symptoms of infection. Therefore, any subtle change in physical condition or mental status must be considered a possible indication of infection and promptly investigated because sepsis may occur before the infection is diagnosed. Figure 55-4 summarizes the sequence of events leading to infection and leakage of urine that often follow long-term use of an indwelling catheter in an older patient (Muzzi-Bjornson & Macera, 2011).

Preventing Infection

Certain principles of care are essential to prevent infection in patients with a closed urinary drainage system (Chart 55-10). The catheter is an object foreign to the body and produces a reaction in the urethral mucosa with some urethral discharge. Vigorous cleansing of the meatus while the catheter is in place is discouraged because the cleansing action can move the catheter back and forth, increasing the risk of infection. To remove obvious encrustations from the external catheter surface, the area is washed gently with soap during the daily bath. The catheter is anchored as securely as possible to prevent it from moving in the urethra (Gray, 2010). Encrustations arising from urinary salts may serve as a nucleus for stone formation; however, using silicone catheters results in significantly less crust formation.

A liberal fluid intake, within the limits of the patient's cardiac and renal reserve, and an increased urine output must be

Physiology Pathophysiology

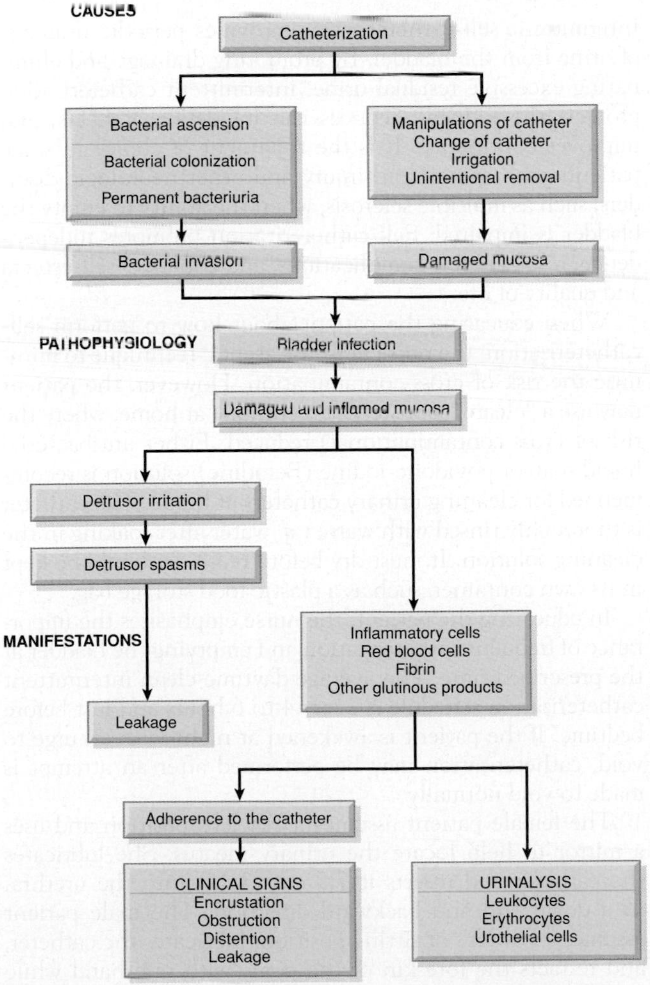
CAUSES
Catheterization
Bacterial ascension
Bacterial colonization
Permanent bacteriuria
Manipulations of catheter
Change of catheter
Irrigation
Unintentional removal
Bacterial invasion
Damaged mucosa
PATHOPHYSIOLOGY
Bladder infection
Damaged and inflamed mucosa
Detrusor irritation
Detrusor spasms
MANIFESTATIONS
Inflammatory cells
Red blood cells
Fibrin
Other glutinous products
Leakage
Adherence to the catheter
CLINICAL SIGNS
Encrustation
Obstruction
Distention
Leakage
URINALYSIS
Leukocytes
Erythrocytes
Urothelial cells

FIGURE 55-4 • Pathophysiology and manifestations of bladder infection with long-term catheterization in older patients.

Chart 55-10 Preventing Infection in the Patient With an Indwelling Urinary Catheter

- Evaluate the benefit of placing an indwelling urinary catheter versus the risk the patient developing a catheter-associated urinary tract infection.
- Use scrupulous aseptic technique during insertion of the catheter. Use a preassembled, sterile, closed urinary drainage system.
- To prevent contamination of the closed system, *never* disconnect the tubing. The drainage bag must *never* touch the floor. The bag and collecting tubing are changed if contamination occurs, if urine flow becomes obstructed, or if tubing junctions start to leak at the connections.
- If the collection bag *must* be raised above the level of the patient's bladder, clamp the drainage tube. This prevents backflow of contaminated urine into the patient's bladder from the bag.
- Ensure a free flow of urine to prevent infection. Improper drainage occurs when the tubing is kinked or twisted, allowing pools of urine to collect in the tubing loops.
- To reduce the risk of bacterial proliferation, empty the collection bag at least every 8 hours through the drainage spout—more frequently if there is a large volume of urine.
- Avoid contamination of the drainage spout. A receptacle in which to empty the bag is provided for each patient.
- Never irrigate the catheter routinely. If the patient is prone to obstruction from clots or large amounts of sediment, use a three-way system with continuous irrigation.
- Never disconnect the tubing to obtain urine samples, to irrigate the catheter, or to ambulate or transport the patient.
- Never leave the catheter in place longer than is necessary to decrease the risk of a catheter-acquired urinary tract infection.
- Avoid routine catheter changes. The catheter is changed only to correct problems such as leakage, blockage, or encrustations.
- Avoid unnecessary handling or manipulation of the catheter by the patient or staff.
- Carry out hand hygiene before and after handling the catheter, tubing, or drainage bag.
- Wash the perineal area with soap and water at least twice a day; avoid a to-and-fro motion of the catheter. Dry the area well, but avoid applying powder because it may irritate the perineum.
- Monitor the patient's voiding when the catheter is removed. The patient must void within 8 hours; if unable to void, the patient may require catheterization with a straight catheter.
- Obtain a urine specimen for culture at the first sign of infection.

ensured to flush the catheter and to dilute urinary substances that might form encrustations (Kahnen et al., 2011).

Urine cultures are obtained as prescribed or indicated when monitoring the patient for infection; many catheters have an aspiration (puncture) port from which a specimen can be obtained.

Bacteriuria is considered inevitable in patients with indwelling catheters; therefore, controversy remains about the usefulness of taking cultures and treating asymptomatic bacteriuria, because overtreatment may lead to resistant strains of bacteria. Continual observation for fever, chills, and other signs and symptoms of systemic infection is necessary. Infections are treated aggressively.

Minimizing Trauma

Trauma to the urethra can be minimized by:

- Using an appropriately sized catheter
- Lubricating the catheter adequately with a water-soluble lubricant during insertion
- Inserting the catheter far enough into the bladder to prevent trauma to the urethral tissues when the retention balloon of the catheter is inflated

Manipulation of the catheter is the most common cause of trauma to the bladder mucosa in the catheterized patient. Infection can occur when urine invades the damaged mucosa.

The catheter is secured properly to prevent it from moving, causing traction on the urethra, or being unintentionally removed, and care is taken to ensure that the catheter position permits leg movement. In male patients, the drainage tube (not the catheter) is taped laterally to the thigh to prevent pressure on the urethra at the penoscrotal junction, which can eventually lead to formation of an urethrocutaneous fistula. In female patients, the drainage tubing attached to the catheter is taped to the thigh to prevent tension and traction on the bladder.

Special care should be taken to ensure that any patient who is confused does not remove the catheter with the retention balloon still inflated, because this could cause bleeding and considerable injury to the urethra.

Retraining the Bladder

When an indwelling urinary catheter is in place, the detrusor muscle does not actively contract the bladder wall to stimulate emptying because urine is continuously draining from the bladder. As a result, the detrusor may not immediately respond to bladder filling when the catheter is removed, resulting in either urine retention or urinary incontinence. This condition, known as postcatheterization detrusor instability, can be managed with bladder retraining (Chart 55-11).

Immediately after the indwelling catheter is removed, the patient is placed on a timed voiding schedule, usually every 2 to 3 hours. At the given time interval, the patient is instructed to void. The bladder is then scanned using a portable ultrasonic bladder scanner, and if the bladder has not emptied completely, straight catheterization may be performed (Ling Man & Le Low, 2010). After a few days, as the nerve endings in the bladder wall become resensitized to the bladder filling and emptying, bladder function usually returns to normal. If the person has had an indwelling catheter in place for an extended period (e.g., greater than 1 month), bladder retraining will take longer; in some cases, function may never return to normal, and long-term intermittent catheterization may become necessary.

Chart 55-11 Bladder Retraining After Indwelling Catheterization

- Instruct the patient to drink a measured amount of fluid from 8 AM to 10 PM to avoid bladder overdistention. Offer no fluids (except sips) after 10 PM.
- At specific times, ask the patient to void by applying pressure over the bladder, tapping the abdomen, or running water to trigger the bladder.
- Immediately after the voiding attempt, catheterize the patient to determine the amount of residual urine.
- Measure the volumes of urine voided and obtained by catheterization.
- Palpate the bladder at repeated intervals to assess for distention.
- Instruct the patient who has no voiding sensation to be alert to any signs that indicate a full bladder, such as perspiration, cold hands or feet, or feelings of anxiety.
- Lengthen the intervals between catheterizations as the volume of residual urine decreases. Catheterization is usually discontinued when the volume of residual urine is <100 mL.

Assisting With Intermittent Self-Catheterization

Intermittent self-catheterization provides periodic drainage of urine from the bladder. By promoting drainage and eliminating excessive residual urine, intermittent catheterization protects the kidneys, reduces the incidence of UTIs, and improves continence. It is the treatment of choice in some patients with spinal cord injury and other neurologic disorders, such as multiple sclerosis, when the ability to empty the bladder is impaired. Self-catheterization promotes independence, results in few complications, and enhances self-esteem and quality of life.

When educating the patient about how to perform self-catheterization, the nurse must use aseptic technique to minimize the risk of cross-contamination. However, the patient may use a "clean" (nonsterile) technique at home, where the risk of cross-contamination is reduced. Either antibacterial liquid soap or povidone-iodine (Betadine) solution is recommended for cleaning urinary catheters at home. The catheter is thoroughly rinsed with warm tap water after soaking in the cleaning solution. It must dry before reuse. It should be kept in its own container, such as a plastic food storage bag.

In educating the patient, the nurse emphasizes the importance of frequent catheterization and emptying the bladder at the prescribed time. The average daytime clean intermittent catheterization schedule is every 4 to 6 hours and just before bedtime. If the patient is awakened at night with an urge to void, catheterization may be performed after an attempt is made to void normally.

The female patient assumes a Fowler's position and uses a mirror to help locate the urinary meatus. She lubricates the catheter and inserts it 7.5 cm (3 in) into the urethra, in a downward and backward direction. The male patient assumes a Fowler's or sitting position, lubricates the catheter, and retracts the foreskin of the penis with one hand while grasping the penis and holding it at a right angle to the body. (This maneuver straightens the urethra and makes it easier to insert the catheter.) He inserts the catheter 15 to 25 cm (6 to 10 in) until urine begins to flow. After removal, the catheter is cleaned, rinsed, dried, and placed in a plastic bag or case. Patients who follow an intermittent catheterization routine should consult a primary provider at regular intervals to assess urinary function and to detect complications. If the patient cannot perform intermittent self-catheterization, a family member or caregiver may be taught to carry out the procedure at regular intervals during the day.

An alternative to self-catheterization is creation of the Mitrofanoff umbilical appendicovesicostomy, which provides easy access to the bladder but requires an extensive surgical procedure (Wille, Zagaja, Shalhav, et al., 2011). In this procedure, the bladder neck is closed and the appendix is used to create access to the bladder from the skin surface through a submucosal tunnel created with the appendix. One end of the appendix is brought to the skin surface and used as a stoma, and the other end is tunneled into the bladder. The appendix serves as an artificial urinary sphincter when an alternative is necessary to empty the bladder. A surgically prepared continent urine reservoir with a sphincter mechanism is required in cases of bladder cancer and severe **interstitial cystitis** (inflammation of the bladder wall). Various types of urinary diversions may be used when a radical **cystectomy** (surgical removal of the bladder) is necessary (see discussion later in chapter).

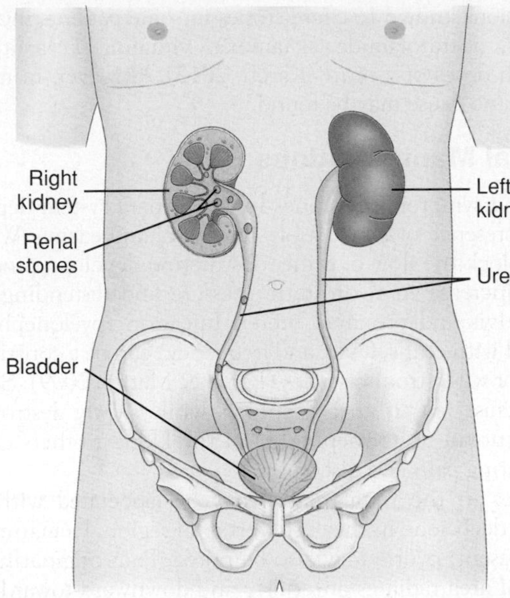
Right
kidney
Renal
stones
Bladder
Left
kidney
Ureters

FIGURE 55-5 • Examples of potential sites of calculi formation (urolithiasis) in the urinary tract.

UROLITHIASIS AND NEPHROLITHIASIS

Urolithiasis and nephrolithiasis refer to stones (calculi) in the urinary tract and kidney, respectively. Urinary stones account for more than 320,000 hospital admissions each year (Frassetto & Kohlstadt, 2011). The occurrence of urinary stones occurs predominantly in the third to fifth decades of life and affects men more than women. About half of patients with a single renal stone have another episode within 5 years (Meschi, Nouvenne, & Borghi, 2011).

Pathophysiology

Stones are formed in the urinary tract when urinary concentrations of substances such as calcium oxalate, calcium phosphate, and uric acid increase. Referred to as supersaturation, this depends on the amount of the substance, ionic strength, and pH of the urine. Stones may be found anywhere from the kidney to the bladder and may vary in size from minute granular deposits, called *sand* or *gravel,* to bladder stones as large as an orange. The different sites of calculi formation in the urinary tract are shown in Figure 55-5.

Stone formation is not clearly understood, and there are a number of theories about their causes. One theory is that there is a deficiency of substances that normally prevent crystallization in the urine, such as citrate, magnesium, nephrocalcin, and uropontin (Jeong, Kang, Bang, et al., 2011). Another theory relates to fluid volume status of the patient (stones tend to occur more often in dehydrated patients). Certain factors favor the formation of stones, including infection, urinary stasis, and periods of immobility, all of which slow renal drainage and alter calcium metabolism. In addition, increased calcium concentrations in the blood and urine promote precipitation of calcium and formation of stones (about 75% of all renal stones are calcium based) (Worcester & Coe, 2010). Causes of hypercalcemia (high serum calcium) and hypercalciuria (high urine calcium) may include the following:

- Hyperparathyroidism
- Renal tubular acidosis
- Cancers (e.g., leukemia, multiple myeloma)
- Granulomatous diseases (e.g., sarcoidosis, tuberculosis), which may cause increased vitamin D production by the granulomatous tissue
- Excessive intake of vitamin D
- Excessive intake of milk and alkali
- Myeloproliferative diseases such as polycythemia vera, which produce an unusual proliferation of blood cells from the bone marrow

For patients with stones containing uric acid, struvite, or cystine, a thorough physical examination and metabolic workup are indicated because of associated disturbances contributing to the stone formation. Uric acid stones (5% to 10% of all stones) may be seen in patients with gout or myeloproliferative disorders. Struvite stones account for 15% of urinary calculi and form in persistently alkaline, ammonia-rich urine caused by the presence of urease-splitting bacteria such as *Proteus*, *Pseudomonas*, *Klebsiella*, *Staphylococcus*, or *Mycoplasma* species. Predisposing factors for struvite stones include neurogenic bladder, foreign bodies, and recurrent UTIs. Cystine stones (1% to 2% of all stones) occur exclusively in patients with a rare inherited defect in renal absorption of cystine (an amino acid) (Porth & Matfin, 2009).

Several conditions, as well as certain metabolic risk factors, predispose patients to stone formation. These include anatomic derangements such as polycystic kidney disease, horseshoe kidneys, chronic strictures, and medullary sponge disease. Urinary stone formation can occur in patients with inflammatory bowel disease and in those with an ileostomy or bowel resection because these patients absorb more oxalate.

Medications known to cause stones in some patients include antacids, acetazolamide (Diamox), vitamin D, laxatives, and high doses of aspirin (Karch, 2012). However, in many patients, no cause may be found.

Clinical Manifestations

Signs and symptoms of stones in the urinary system depend on the presence of obstruction, infection, and edema. When stones block the flow of urine, obstruction develops, producing an increase in hydrostatic pressure and distending the renal pelvis and proximal ureter. Infection (pyelonephritis and UTI with chills, fever, and frequency) can be a contributing factor with struvite stones (Porth & Matfin, 2009). Some stones cause few, if any, symptoms while slowly destroying the functional units (nephrons) of the kidney; others cause excruciating pain and discomfort.

Stones in the renal pelvis may be associated with an intense, deep ache in the costovertebral region. Hematuria is often present; pyuria may also be noted. Pain originating in the renal area radiates anteriorly and downward toward the bladder in the female and toward the testis in the male. If the pain suddenly becomes acute, with tenderness over the costovertebral area, and nausea and vomiting occur, the patient is having an episode of renal colic. Diarrhea and abdominal discomfort are due to renointestinal reflexes and the anatomic proximity of the kidneys to the stomach, pancreas, and large intestine.

Stones lodged in the ureter (ureteral obstruction) cause acute, excruciating, colicky, wavelike pain that radiates down the thigh and to the genitalia. Often, the patient has a desire to void, but little urine is passed, and it usually contains blood because of the abrasive action of the stone. This group of symptoms is called *ureteral colic*. Colic is mediated by prostaglandin E, a substance that increases ureteral contractility and renal blood flow and that leads to increased intraureteral pressure and pain. In general, the patient is able to pass stones 0.5 to 1 cm in diameter. Stones larger than 1 cm in diameter usually must be removed or fragmented (broken up by lithotripsy) so that they can be removed or passed spontaneously.

Stones lodged in the bladder usually produce symptoms of irritation and may be associated with UTI and hematuria. If the stone obstructs the bladder neck, urinary retention occurs. If infection is associated with a stone, the condition is far more serious, with urosepsis threatening the patient's life.

Assessment and Diagnostic Findings

The diagnosis is confirmed by x-rays of the kidneys, ureters, and bladder (KUB) or by ultrasonography, IV urography, or retrograde pyelography. Blood chemistries and a 24-hour urine test for measurement of calcium, uric acid, creatinine, sodium, pH, and total volume are part of the diagnostic workup. Dietary and medication histories and family history of renal stones are obtained to identify factors predisposing the patient to the formation of stones.

When stones are recovered (whether freely passed by the patient or removed through special procedures), chemical analysis is carried out to determine their composition. Stone analysis can provide a clear indication of the underlying disorder. For example, calcium oxalate or calcium phosphate stones usually indicate disorders of oxalate or calcium metabolism, whereas urate stones suggest a disturbance in uric acid metabolism (Worcester & Coe, 2010).

Medical Management

The goals of management are to eradicate the stone, determine the stone type, prevent nephron destruction, control infection, and relieve any obstruction that may be present. The immediate objective of treatment of renal or ureteral colic is to relieve the pain until its cause can be eliminated. Opioid analgesic agents are administered to prevent shock and syncope that may result from the excruciating pain. Nonsteroidal anti-inflammatory drugs (NSAIDs) are effective in treating renal stone pain because they provide specific pain relief. They also inhibit the synthesis of prostaglandin E, reducing swelling and facilitating passage of the stone. Generally, once the stone has passed, the pain is relieved. Hot baths or moist heat to the flank area may also be helpful. Unless the patient is vomiting or has heart failure or any other condition requiring fluid restriction, fluids are encouraged. This increases the hydrostatic pressure behind the stone, assisting it in its downward passage. A high, around-the-clock fluid intake reduces the concentration of urinary crystalloids, dilutes the urine, and ensures a high urine output.

Nutritional Therapy

Nutritional therapy plays an important role in preventing renal stones (Dudek, 2010; Meschi et al., 2011) (Chart 55-12). Fluid intake is the mainstay of most medical therapy for renal stones. Unless fluids are contraindicated, patients with renal stones should drink eight to ten 8-ounce glasses of water daily or have IV fluids prescribed to keep the urine dilute. A urine output exceeding 2 L/day is advisable.

Calcium Stones

Historically, patients with calcium-based renal stones were recommended to restrict calcium in their diet. However,

Chart 55-12 PATIENT EDUCATION

Preventing Kidney Stones

- Avoid protein intake; usually protein is restricted to 60 g/day to decrease urinary excretion of calcium and uric acid.
- A sodium intake of 3–4 g/day is recommended. Table salt and high-sodium foods should be reduced, because sodium competes with calcium for reabsorption in the kidneys.
- Low-calcium diets are not generally recommended, except for true absorptive hypercalciuria. Evidence shows that limiting calcium, especially in women, can lead to osteoporosis and does not prevent renal stones.
- Avoid intake of oxalate-containing foods (e.g., spinach, strawberries, rhubarb, tea, peanuts, wheat bran).
- During the day, drink fluids (ideally water) every 1–2 hours.
- Drink two glasses of water at bedtime and an additional glass at each nighttime awakening to prevent urine from becoming too concentrated during the night.
- Avoid activities leading to sudden increases in environmental temperatures that may cause excessive sweating and dehydration.
- Contact your primary provider at the first sign of a urinary tract infection.

evidence has questioned this practice, except for patients with type II absorptive hypercalciuria (half of all patients with calcium stones), in whom stones are clearly the result of excess dietary calcium. Liberal fluid intake is encouraged along with dietary restriction of protein and sodium; however, dietary changes cannot be recommended with confidence because of insufficient evidence. It was once thought that a high-protein diet was associated with increased urinary excretion of calcium and uric acid, thereby causing a supersaturation of these substances in the urine. Similarly, a high sodium intake was thought to increase the amount of calcium in the urine. These beliefs have not been supported by research. Medications such as ammonium chloride may be used, and if increased parathormone production (resulting in increased serum calcium levels in blood and urine) is a factor in the formation of stones, therapy with thiazide diuretics may be beneficial in reducing the calcium loss in the urine and lowering the elevated parathormone levels (Porth & Matfin, 2009).

Uric Acid Stones

For uric acid stones, the patient is placed on a low-purine diet to reduce the excretion of uric acid in the urine. Foods high in purine (shellfish, anchovies, asparagus, mushrooms, and organ meats) are avoided, and other proteins may be limited. Allopurinol (Zyloprim) may be prescribed to reduce serum uric acid levels and urinary uric acid excretion.

Cystine Stones

A low-protein diet is prescribed, the urine is alkalinized, and fluid intake is increased.

Oxalate Stones

A dilute urine is maintained, and the intake of oxalate is limited. Many foods contain oxalate; however, only certain foods increase the urinary excretion of oxalate. These include spinach, strawberries, rhubarb, chocolate, tea, peanuts, and wheat bran.

Interventional Procedures

If the stone does not pass spontaneously or if complications occur, common interventions include endoscopic or other procedures. For example, ureteroscopy, extracorporeal shock wave lithotripsy (ESWL), or endourologic (percutaneous) stone removal may be necessary.

Ureteroscopy (Fig. 55-6A) involves first visualizing the stone and then destroying it. Access to the stone is accomplished by inserting a ureteroscope into the ureter and then inserting a laser, electrohydraulic lithotriptor, or ultrasound device through the ureteroscope to fragment and remove the stones. A stent may be inserted and left in place for 48 hours or more after the procedure to keep the ureter patent. Length of hospital stay is generally brief, and some patients can be treated as outpatients.

ESWL is a noninvasive procedure used to break up stones in the calyx of the kidney (Fig. 55-6B). After the stones are fragmented to the size of grains of sand, the remnants of the stones are spontaneously voided. In ESWL, a high-energy amplitude of pressure, or shock wave, is generated by the abrupt release of energy and transmitted through water and soft tissues. When the shock wave encounters a substance of different intensity (a renal stone), a compression wave causes the surface of the stone to fragment. Repeated shock waves focused on the stone eventually reduce it to many small pieces that are excreted in the urine.

Discomfort from the multiple shocks may occur, although the shock waves usually do not cause damage to other tissue. The patient is observed for obstruction and infection resulting from blockage of the urinary tract by stone fragments. All urine is strained after the procedure; voided gravel or sand is sent to the laboratory for chemical analysis. Several treatments may be necessary to ensure disintegration of stones. Although lithotripsy is a costly treatment, the high charges are offset by a decrease in the length of hospital stay and avoidance of a surgical procedure (Frassetto & Kohlstadt, 2011).

Endourologic methods of stone removal (Fig. 55-6C) may be used to extract renal calculi that cannot be removed by other procedures. A percutaneous nephrostomy or a percutaneous nephrolithotomy (which are similar procedures) may be performed. A nephroscope is introduced through a percutaneous route into the renal parenchyma. Depending on its size, the stone may be extracted with forceps or by a stone retrieval basket. If the stone is too large to initially be removed, an ultrasound probe inserted through a nephrostomy tube is used to pulverize the stone. Small stone fragments and stone dust are then removed.

Electrohydraulic lithotripsy is a similar method in which an electrical discharge is used to create a hydraulic shock wave to break up the stone. A probe is passed through the cystoscope, and the tip of the lithotriptor is placed near the stone. The strength of the discharge and pulse frequency can be varied. This procedure is performed under topical anesthesia. After the stone is extracted, the percutaneous nephrostomy tube is left in place for a time to ensure that the ureter is not obstructed by edema or blood clots. The most common complications are hemorrhage, infection, and urinary extravasation. After the tube is removed, the nephrostomy tract usually closes spontaneously.

Chemolysis, stone dissolution using infusions of chemical solutions (e.g., alkylating agents, acidifying agents) for the purpose of dissolving the stone, is an alternative treatment sometimes used in patients who are at risk for complications with other types of therapy, who refuse to undergo other methods, or who have stones (struvite) that dissolve easily. A percutaneous nephrostomy is performed, and the warm chemical solution is allowed to flow continuously onto the stone. The solution exits the renal collecting system by means of the ureter or the nephrostomy tube. The pressure inside the renal pelvis is monitored during the procedure.

Several of these treatment modalities may be used in combination to ensure removal of the stones.

Surgical Management

Surgical removal was the major mode of therapy before the advent of lithotripsy. However, today, surgery is performed in only 1% to 2% of patients (Frassetto & Kohlstadt, 2011). Surgical intervention is indicated if the stone does not respond to other forms of treatment. It may also be performed to correct anatomic abnormalities within the kidney to improve urinary drainage. If the stone is in the kidney, the surgery performed may be a nephrolithotomy (incision into the kidney with removal of the stone) or a nephrectomy, if the kidney

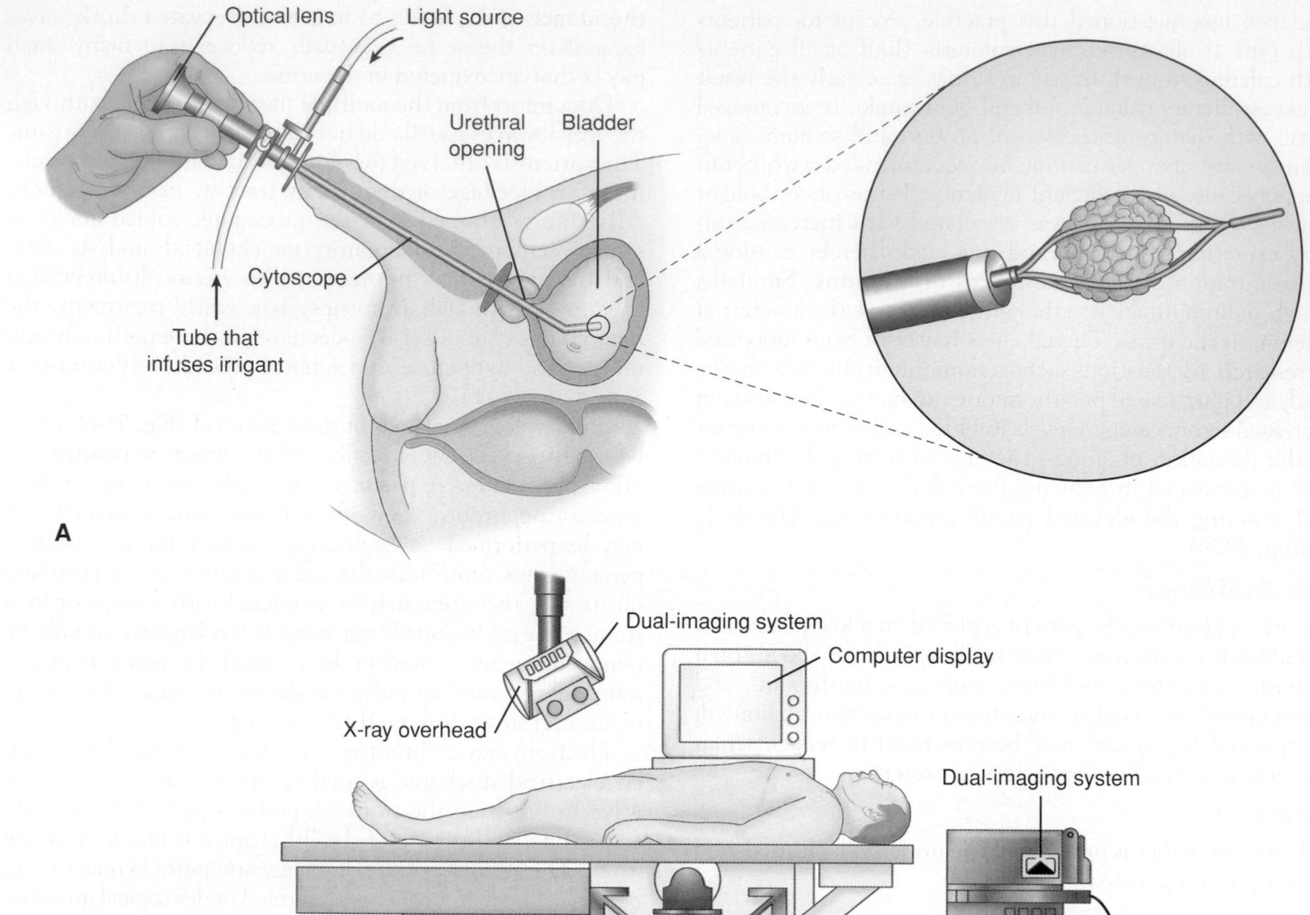
Optical lens
Light source
Urethral opening
Bladder
Cytoscope
Tube that infuses irrigant
A
Dual-imaging system
Computer display
X-ray overhead
Dual-imaging system
X-ray film cassette
Shock wave generating system
Ultrasound system
B
Kidney stone fragments
Alligator forceps
C

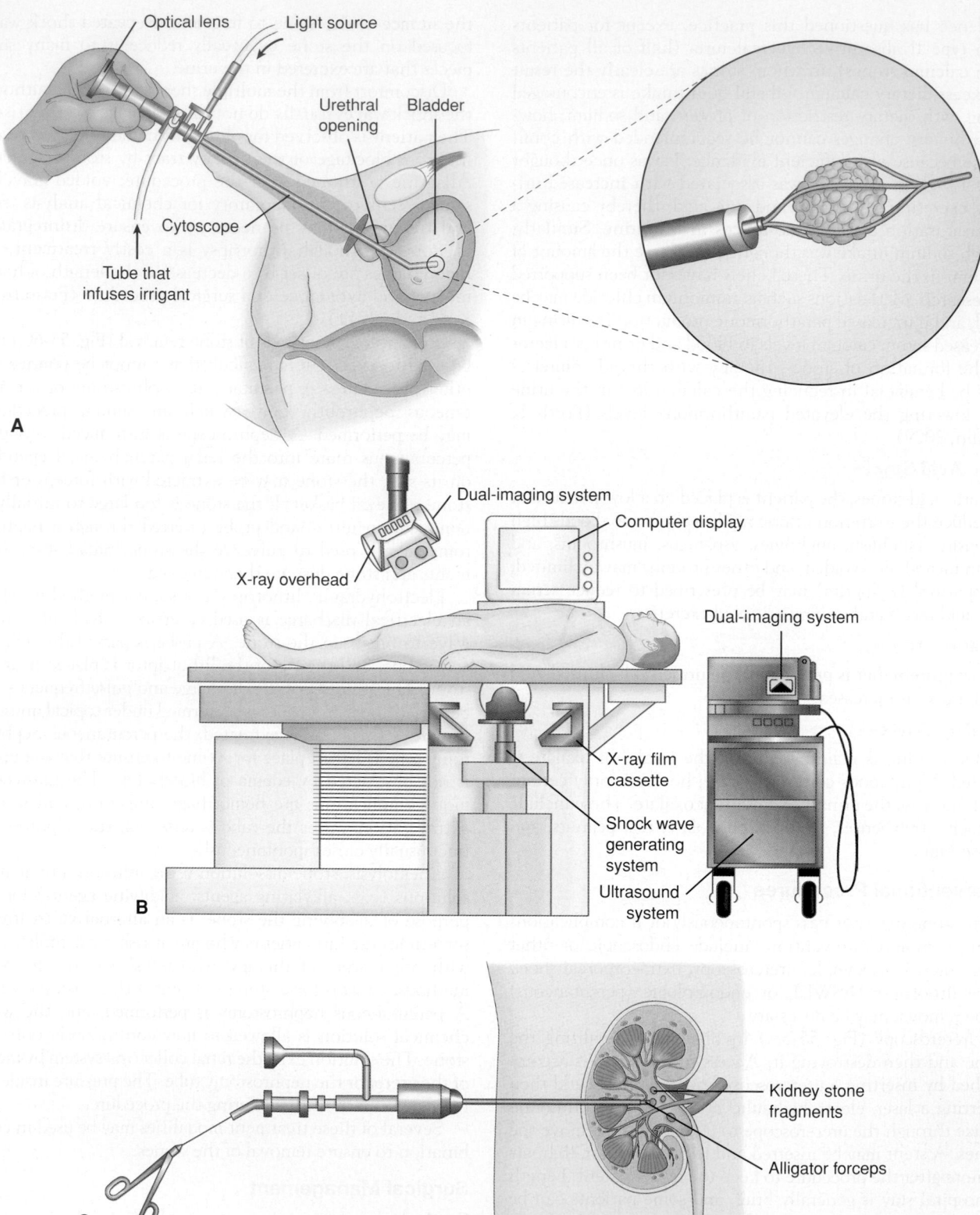

FIGURE 55-6 • Methods of treating renal stones. **A.** During ureteroscopy, which is used for removing small stones located in the ureter close to the bladder, a ureteroscope is inserted into the ureter to visualize the stone. The stone is then fragmented or captured and removed. **B.** Extracorporeal shock water lithotripsy is used for most symptomatic, nonpassable upper urinary stones. Electromagnetically generated shock waves are focused over the area of the renal stone. The high-energy dry shock waves pass through the skin and fragment the stone. **C.** Percutaneous nephrolithotomy is used to treat larger stones. A percutaneous tract is formed, and a nephroscope is inserted through it. Then, the stone is extracted or pulverized.

is nonfunctional secondary to infection or hydronephrosis. Stones in the kidney pelvis are removed by a pyelolithotomy, those in the ureter by ureterolithotomy, and those in the bladder by cystotomy. If the stone is in the bladder, an instrument may be inserted through the urethra into the bladder, and the stone crushed. Such a procedure is called a *cystolitholapaxy*. Nursing management following kidney surgery is discussed in Chapter 54.

NURSING PROCESS

The Patient With Kidney Stones

Assessment

The patient with suspected renal stones is assessed for pain and discomfort as well as associated symptoms, such as nausea, vomiting, diarrhea, and abdominal distention. The severity and location of pain are determined, along with any radiation of the pain. Nursing assessment also includes observing for signs and symptoms of UTI (chills, fever, frequency, and hesitancy) and obstruction (frequent urination of small amounts, oliguria, or anuria). The urine is inspected for blood and is strained for stones or gravel.

The history focuses on factors that predispose the patient to urinary tract stones or that may have precipitated the current episode of renal or ureteral colic. The patient's knowledge about renal stones and measures to prevent their occurrence or recurrence is also assessed.

Diagnosis

Nursing Diagnoses

Based on the assessment data, nursing diagnoses may include the following:

- Acute pain related to inflammation, obstruction, and abrasion of the urinary tract
- Deficient knowledge regarding prevention of recurrence of renal stones

Collaborative Problems/Potential Complications

Potential complications may include the following:

- Infection and urosepsis (from UTI and pyelonephritis)
- Obstruction of the urinary tract by a stone or edema with subsequent acute renal failure

Planning and Goals

The major goals for the patient may include relief of pain and discomfort, prevention of recurrence of renal stones, and absence of complications.

Nursing Interventions

Relieving Pain

Severe acute pain is often the presenting symptom of a patient with renal and urinary calculi and requires immediate attention. Opioid analgesic agents (IV or intramuscular) may be prescribed and administered to provide rapid relief along with an IV NSAID. The patient is encouraged and assisted to assume a position of comfort. If activity brings pain relief, the patient is assisted to ambulate. The pain level is monitored closely, and an increase in severity is reported promptly to the physician so that relief can be provided and additional treatment initiated.

Monitoring and Managing Potential Complications

Increased fluid intake is encouraged to prevent dehydration and increase hydrostatic pressure within the urinary tract to promote passage of the stone. If the patient cannot take adequate fluids orally, IV fluids are prescribed. The total urine output and patterns of voiding are monitored. Ambulation is encouraged as a means of moving the stone through the urinary tract.

All urine is strained through gauze because uric acid stones may crumble. Any blood clots passed in the urine should be crushed and the sides of the urinal and bedpan inspected for clinging stones. Because renal stones increase the risk of infection, sepsis, and obstruction of the urinary tract, the patient is instructed to report decreased urine volume, bloody or cloudy urine, fever, and pain.

Patients with calculi require frequent nursing observation to detect the spontaneous passage of a stone. The patient is instructed to immediately report any sudden increases in pain intensity because of the possibility of a stone fragment obstructing a ureter. Vital signs, including temperature, are monitored closely to detect early signs of infection. UTIs may be associated with renal stones due to an obstruction from the stone or from the stone itself. All infections should be treated with the appropriate antibiotic agent before efforts are made to dissolve the stone.

Promoting Home and Community-Based Care

Educating Patients About Self-Care. Because the risk of recurring renal stones is high, the nurse provides education about the causes of kidney stones and recommendations to prevent their recurrence (see Chart 55-12). The patient is encouraged to follow a regimen to avoid further stone formation, including maintaining a high fluid intake because stones form more readily in concentrated urine. A patient who has shown a tendency to form stones should drink enough fluid to excrete greater than 2,000 mL (preferably 3,000 to 4,000 mL) of urine every 24 hours (Meschi et al., 2011).

Urine cultures may be performed every 1 to 2 months in the first year and periodically thereafter. Recurrent UTI is treated vigorously. Because prolonged immobilization slows renal drainage and alters calcium metabolism, increased mobility is encouraged whenever possible. In addition, excessive ingestion of vitamins (especially vitamin D) and minerals is discouraged.

If lithotripsy, percutaneous stone removal, ureteroscopy, or other surgical procedures for stone removal have been performed, the nurse instructs the patient about the signs and symptoms of complications (e.g., urinary retention, infection) that need to be reported to the physician. The importance of follow-up to assess kidney function and to ensure the eradication or removal of all kidney stones is emphasized to the patient and family.

If ESWL has been performed, the nurse must provide instructions for home care and necessary follow-up. The patient is encouraged to increase fluid intake to assist in the passage of stone fragments, which may occur for 6 weeks to several months after the procedure. The patient and family are instructed about signs and symptoms of complications. It is also important to inform the patient to expect hematuria (it is anticipated in all patients), but it should disappear within 4 to 5 days. If the patient has a stent in the ureter, hematuria may be expected until the stent is removed. The patient is instructed to check his or her temperature daily and notify the physician if the temperature is greater than 38°C (about 101°F) or the pain is unrelieved by the prescribed medication. The patient is also informed that a bruise may be observed on the treated side of the back.

Continuing Care. Close monitoring of the patient in follow-up care is essential to ensure that treatment has been effective and that no complications develop. The nurse has the opportunity to assess the patient's understanding of ESWL and possible complications. In addition, the nurse has the opportunity to assess the patient's understanding of factors that increase the risk of recurrence of renal calculi and strategies to reduce those risks.

The nurse must assess the patient's ability to monitor urinary pH and interpret the results during follow-up visits. Because of the high risk of recurrence, the patient with renal stones needs to understand the signs and symptoms of stone formation, obstruction, and infection and the importance of reporting these signs promptly. If medications are prescribed for the prevention of stone formation, the nurse explains their actions, importance, and side effects to the patient.

Evaluation

Expected patient outcomes may include:

1. Reports relief of pain
2. States increased knowledge of health-seeking behaviors to prevent recurrence
   a. Consumes increased fluid intake (at least eight 8-ounce glasses of fluid per day)
   b. Participates in appropriate activity
   c. Consumes diet prescribed to reduce dietary factors predisposing to stone formation
   d. Recognizes symptoms (fever, chills, flank pain, hematuria) to be reported to primary provider
   e. Monitors urinary pH as directed
   f. Takes prescribed medication as directed to reduce stone formation
3. Experiences no complications
   a. Reports no signs or symptoms of infection or urosepsis
   b. Voids 200 to 400 mL per voiding of clear urine without evidence of bleeding
   c. Experiences absence of urgency, frequency, and hesitancy
   d. Maintains normal body temperature

GENITOURINARY TRAUMA

Various types of injuries to the flank, back, or upper abdomen may result in trauma to the ureters, bladder, or urethra. Approximately 10% of all injuries seen in the emergency department involve the genitourinary system (Blair, 2011). (Renal trauma is discussed in Chapter 54.)

Specific Injuries

Ureteral Trauma

Penetrating trauma and unintentional injury during surgery are the major causes of trauma to the ureters. Gunshot wounds account for 95% of ureteral injuries, which may range from contusions to complete transection (Blair, 2012). Unintentional injury to the ureter may occur during gynecologic or urologic surgery. There are no specific signs or symptoms of ureteral injury; many traumatic injuries are discovered during exploratory surgery. If the ureteral trauma is not detected and urine leakage continues, fistulas can develop.

IV urography detects 90% of ureteral injuries and can be performed on the operating table in patients undergoing emergency surgery (Blair, 2012). Surgical repair with placement of stents (to divert urine away from an anastomosis) is usually necessary.

Bladder Trauma

Injury to the bladder may occur with pelvic fractures and multiple trauma or from a blow to the lower abdomen when the bladder is full (Kong, Bultitude, Royce, et al., 2011). Blunt trauma may result in contusion evident as an ecchymosis—a large bruise resulting from escape of blood into the tissues and involving a segment of the bladder wall—or in rupture of the bladder extraperitoneally, intraperitoneally, or both. Complications from these injuries include hemorrhage, shock, sepsis, and extravasation of blood into the tissues, which must be treated promptly.

Urethral Trauma

Urethral injuries usually occur with blunt trauma to the lower abdomen or pelvic region. Many patients also have associated pelvic fractures. The classic triad of symptoms comprises blood at the urinary meatus, inability to void, and a distended bladder.

Medical Management

The goals of management in patients with genitourinary trauma are to control hemorrhage, pain, and infection and to maintain urinary drainage. Genitourinary trauma is frequently associated with renal trauma (see Chapter 54). Hematocrit and hemoglobin levels are monitored closely; decreasing values can indicate hemorrhage within the genitourinary system. The patient is also monitored for oliguria, signs of hemorrhagic shock, and signs and symptoms of acute peritonitis.

Surgical Management

In urethral trauma, unstable patients who need monitoring of urine output may need a suprapubic catheter inserted. The patient is catheterized after urethrography has been performed to minimize the risk of urethral disruption and extensive, long-term complications, such as stricture, incontinence, and impotence. Surgical repair may be performed immediately or at a later time. Delayed surgical repair tends to be the favored procedure because it is associated with fewer long-term complications, such as impotence, strictures, and incontinence (Koraitim, 2012). After surgery, an indwelling urinary catheter may remain in place for up to 1 month.

Nursing Management

The patient with genitourinary trauma should be assessed frequently during the first few days after injury to detect flank and abdominal pain, muscle spasm, and swelling over the flank (Blair, 2011).

During this time, patients can be instructed about care of the incision and the importance of an adequate fluid intake. In addition, instructions about changes that should be reported to the physician, such as fever, hematuria, flank pain, or any signs and symptoms of decreasing kidney function, are provided. The patient with a ruptured bladder may have gross bleeding for several days after repair. Guidelines for increasing activity gradually, lifting, and driving are also provided.

Follow-up nursing care includes monitoring the blood pressure to detect hypertension and advising the patient to restrict activities for about 1 month after trauma to minimize the incidence of delayed or secondary bleeding.

URINARY TRACT CANCERS

Urinary tract cancers include those of the urinary bladder; kidney and renal pelvis; ureters; and other urinary structures, such as the prostate. (Renal cancer is discussed in Chapter 54, and prostate cancer is discussed in Chapter 59.)

Cancer of the Bladder

Cancer of the urinary bladder is more common in people older than 55 years. It affects more men than women (4:1) and is more common in Caucasians than in African Americans. Bladder cancer is a leading cause of cancer, accounting for more than 15,000 deaths in the United States annually, and has a high incidence worldwide (National Cancer Institute [NCI], 2011).

Bladder cancer, combined with prostatic cancer, is the most common urologic malignancy, accounting for 90% of all tumors seen (Centers for Disease Control and Prevention [CDC], 2010). Cancers arising from the prostate, colon, and rectum in males and from the lower gynecologic tract in females may metastasize to the bladder.

Tobacco use continues to be a leading risk factor for all urinary tract cancers. People who smoke develop bladder cancer twice as often as those who do not smoke (NCI, 2011) (Chart 55-13).

Chart 55-13 RISK FACTORS
Bladder Cancer

- Cigarette smoking—risk proportional to pack-years of smoking
- Exposure to environmental carcinogens—dyes, rubber, leather, ink, or paint
- Recurrent or chronic bacterial infection of the urinary tract
- Bladder stones
- High urinary pH
- High cholesterol intake
- Pelvic radiation therapy
- Cancers arising from the prostate, colon, and rectum in males

Adapted from Brausi, M., Witjes, J. A., Lamm, D., et al. (2011). A review of current guidelines and best practice recommendations for the management of nonmuscle invasive bladder cancer by the International Bladder Cancer Group. *Journal of Urology, 186*(6), 2158–2167.

Clinical Manifestations

Bladder tumors usually arise at the base of the bladder and involve the ureteral orifices and bladder neck. Visible, painless hematuria is the most common symptom of bladder cancer. Infection of the urinary tract is a common complication, producing frequency and urgency. However, any alteration in voiding or change in the urine may indicate cancer of the bladder. Pelvic or back pain may occur with metastasis.

Assessment and Diagnostic Findings

The diagnostic evaluation includes ureteroscopy (the mainstay of diagnosis), excretory urography, CT, ultrasonography, and bimanual examination with the patient anesthetized (Chatterton, Bageja, Challacombe, et al., 2011). Biopsies of the tumor and adjacent mucosa are the definitive diagnostic procedures. Transitional cell carcinomas and carcinomas in situ shed recognizable cancer cells. Cytologic examination of fresh urine and saline bladder washings provide information about the prognosis and staging, especially for patients at high risk for recurrence of primary bladder tumors. (See Chapter 15 for more information on cancer grading and staging.)

Although the mainstay diagnostic tools such as cytology and CT have a high detection rate, they are costly. Diagnostic tools such as bladder tumor antigens, nuclear matrix proteins, adhesion molecules, cytoskeletal proteins, and growth factors are being investigated and implemented to support the early detection and diagnosis of bladder cancer, as well as to monitor for recurrence of bladder cancer (Tilki, Burger, Dalbagni, et al., 2011).

Medical Management

Treatment of bladder cancer depends on the grade of the tumor (the degree of cellular differentiation), the stage of tumor growth (the degree of local invasion and the presence or absence of metastasis), and the multicentricity (having many centers) of the tumor (Chatterton et al., 2011). The patient's age and physical, mental, and emotional status are considered when determining treatment modalities.

Surgical Management

Transurethral resection or fulguration (cauterization) may be performed for simple papillomas (benign epithelial tumors). These procedures (described in more detail in Chapter 59) eradicate the tumors through surgical incision or electrical current with the use of instruments inserted through the urethra. After this bladder-sparing surgery, intravesical administration of bacille Calmette-Guérin (BCG) is the treatment of choice. BCG is an attenuated live strain of *Mycobacterium bovis*, the causative agent in tuberculosis. The exact action of BCG is unknown, but it is thought to produce a local inflammatory and a systemic immunologic response (Ward-Smith, Miller, & Caughron, 2010).

Management of superficial bladder cancers presents a challenge because there are usually widespread abnormalities in the bladder mucosa. The entire lining of the urinary tract, or urothelium, is at risk because carcinomatous changes can occur in the mucosa of the bladder, renal pelvis, ureter, and urethra. About 25% to 40% of superficial tumors recur after transurethral resection or fulguration (Brausi, Witjes, Lamm, et al., 2011). Patients with benign papillomas should undergo cytology and cystoscopy periodically for the rest of their lives because aggressive malignancies may develop from these tumors.

A simple cystectomy or a radical cystectomy is performed for invasive or multifocal bladder cancer. Radical cystectomy in men involves removal of the bladder, prostate, and seminal vesicles and immediate adjacent perivesical tissues. In women, radical cystectomy involves removal of the bladder, lower ureter, uterus, fallopian tubes, ovaries, anterior vagina, and urethra. It may include removal of pelvic lymph nodes. Removal of the bladder requires a urinary diversion procedure, which is described later in this chapter.

Although radical cystectomy remains the standard of care for invasive bladder cancer in the United States, researchers are exploring trimodality therapy—transurethral resection of the bladder tumor, radiation, and chemotherapy—in an effort to spare patients the need for cystectomy (Brausi et al., 2011). This approach to transitional cell bladder cancer mandates lifelong surveillance with periodic cystoscopy. Although most patients respond completely and their bladders remain free from invasive relapse, one fourth develop a relapse of noninvasive disease. This may be managed with transurethral resection of the bladder tumor and intravesical therapies but carries an additional risk that a late cystectomy may be required.

Pharmacologic Therapy

Chemotherapy with a combination of methotrexate, 5-fluorouracil, vinblastine (Velban), doxorubicin (Adriamycin), and cisplatin (Platinol) has been effective in producing partial remission of transitional cell carcinoma of the bladder in some patients. IV chemotherapy may be accompanied by radiation therapy. Topical chemotherapy (intravesical chemotherapy or instillation of antineoplastic agents into the bladder, resulting in contact of the agent with the bladder wall) is considered when there is a high risk of recurrence, when cancer in situ is present, or when tumor resection has been incomplete. Topical chemotherapy delivers a high concentration of medication (thiotepa [Thioplex], doxorubicin, mitomycin [Mutamycin], and BCG Live [TheraCys]) to the tumor to promote tumor destruction. Bladder cancer may also be treated by direct infusion of the cytotoxic agent through the bladder's arterial blood supply to achieve a higher concentration of the chemotherapeutic agent with fewer systemic toxic effects (Vasdev, Shaw, & Thorpe, 2011).

BCG Live is now considered the most predominant and conservative intravesical agent for recurrent bladder cancer, especially superficial transitional cell carcinoma, because it is an immunotherapeutic agent that enhances the body's immune response to cancer. BCG Live has a 43% advantage in preventing tumor recurrence, a significantly better rate than the 16% to 21% advantage of intravesical chemotherapy. In addition, BCG Live is particularly effective in the treatment of carcinoma in situ, eradicating it in more than 80% of cases. In contrast to intravesical chemotherapy, BCG Live has also been shown to decrease the risk of tumor progression. Although BCG Live treatment is the current standard of care, this treatment fails or is not tolerated in a significant proportion of patients (Askeland, Newton, O'Donnell, et al., 2012).

The optimal course of BCG Live appears to be a 6-week course of weekly instillations, followed by a 3-week course at 3 months for tumors that do not respond. In high-risk cancers, maintenance BCG Live administered in a 3-week course at 6, 12, 18, and 24 months may limit recurrence and prevent progression. However, the adverse effects associated with this prolonged therapy may limit its widespread applicability.

The patient is allowed to eat and drink before the instillation procedure. Once the bladder is full, the patient must retain the intravesical solution for 2 hours before voiding. At the end of the procedure, the patient is encouraged to void and to drink liberal amounts of fluid to flush the medication from the bladder.

Radiation Therapy

Radiation of the tumor may be performed preoperatively to reduce microextension of the neoplasm and viability of tumor cells, thus decreasing the chances that the cancer may recur in the immediate area or spread through the circulatory or lymphatic systems. Radiation therapy is also used in combination with surgery or to control the disease in patients with inoperable tumors.

For more advanced bladder cancer or for patients with intractable hematuria (especially after radiation therapy), a large, water-filled balloon placed in the bladder produces tumor necrosis by reducing the blood supply of the bladder wall (hydrostatic therapy). The instillation of formalin, phenol, or silver nitrate relieves hematuria and strangury (slow and painful discharge of urine) in some patients.

Investigational Therapy

The use of photodynamic techniques in treating superficial bladder cancer is under investigation. This procedure involves systemic injection of a photosensitizing material (hematoporphyrin), which the cancer cell picks up. A laser-generated light then changes the hematoporphyrin in the cancer cell into a toxic agent. This process has received renewed interest with regulatory approval of several photosensitizing medications and light applicators as potential palliative and curative treatments (Brausi et al., 2011). Chemoprevention is the use of drugs, vitamins, or other substances to reduce the risk of developing cancer or to reduce the risk of it returning. Chemoprevention and photodynamic therapy are being tested in clinical trials (NCI, 2011).

URINARY DIVERSIONS

Urinary diversion procedures are performed to divert urine from the bladder to a new exit site, usually through a surgically created opening (stoma) in the skin. These procedures are primarily performed when a bladder tumor necessitates cystectomy. Urinary diversion has also been used in managing pelvic malignancy, birth defects, strictures, trauma to the ureters and urethra, neurogenic bladder, chronic infection causing severe ureteral and renal damage, and intractable interstitial cystitis. It may also be used as a last resort in managing incontinence.

Controversy exists about the best method of establishing permanent diversion of the urinary tract. New techniques are frequently introduced in an effort to improve patient outcomes and quality of life. The age of the patient, condition of the bladder, body build, degree of obesity, degree of ureteral dilation, status of renal function, and the patient's

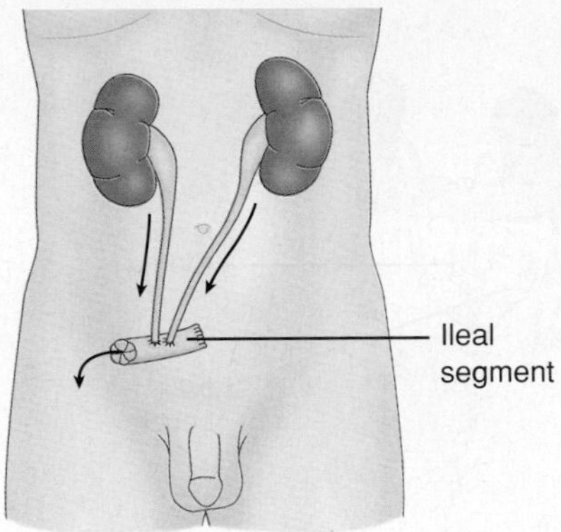
Ileal
segment

A Conventional ileal conduit.
The surgeon transplants the ureters to an isolated section of the terminal ileum (ileal conduit), bringing one end to the abdominal wall. The ureter may also be transplanted into the transverse sigmoid colon (colon conduit) or proximal jejunum (jejunal conduit).

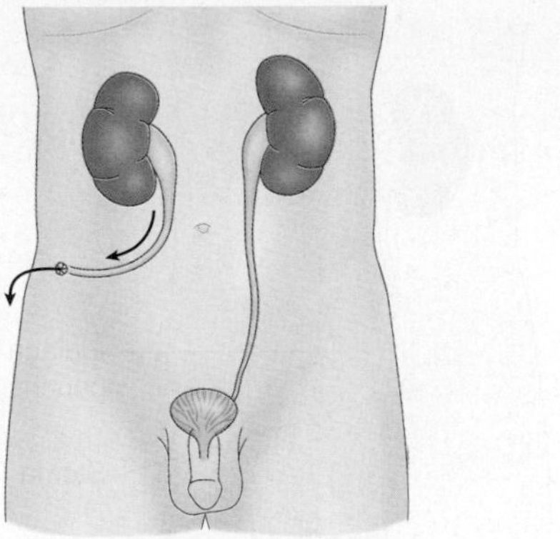

B Cutaneous ureterostomy.
The surgeon brings the detached ureter through the abdominal wall and attaches it to an opening in the skin.

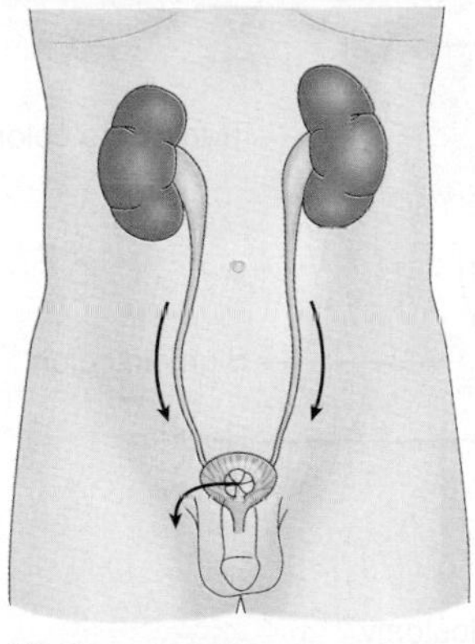

C Vesicostomy.
The surgeon sutures the bladder to the abdominal wall and creates an opening (stoma) through the abdominal and bladder walls for urinary drainage.

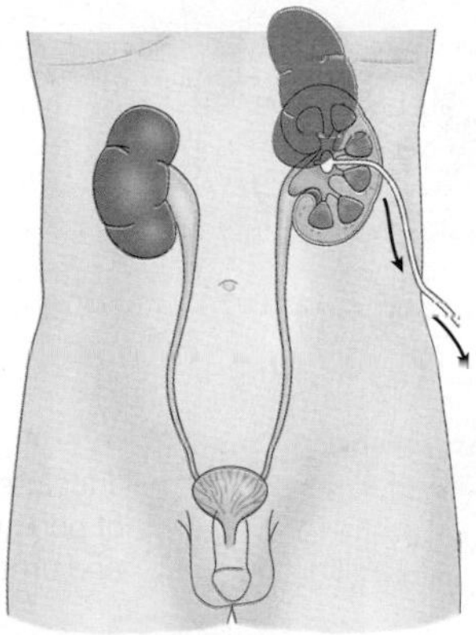

D Nephrostomy.
The surgeon inserts a catheter into the renal pelvis via an incision in the flank or by percutaneous catheter placement into the kidney.

FIGURE 55-7 • Types of cutaneous diversions include the conventional ileal conduit (**A**), cutaneous ureterostomy (**B**), vesicostomy (**C**), and nephrostomy (**D**).

learning ability and willingness to participate in postoperative care are all taken into consideration when determining the appropriate surgical procedure.

The extent to which the patient accepts urinary diversion depends to a large degree on the location or position of the stoma, whether the drainage device (pouch or bag) establishes a watertight seal to the skin, and the patient's ability to manage the pouch and drainage apparatus (Myles, 2011).

There are two types of urinary diversion. In a cutaneous urinary diversion, urine drains through an opening created in the abdominal wall and skin (Fig. 55-7). In a continent urinary diversion, a portion of the intestine is used to create a new reservoir for urine (Fig. 55-8).

Cutaneous Urinary Diversions

Ileal Conduit

The **ileal conduit** (ileal loop) is the oldest and most common of the urinary diversion procedures in use because of the low number of complications and surgeons' familiarity with the procedure. In an ileal conduit, the urine is diverted by implanting the ureter into a 12-cm loop of ileum that is led out through the abdominal wall. This loop of ileum is a simple conduit (passageway) for urine from the ureters to the surface. A loop of the sigmoid colon may also be used. An ileostomy bag is used to collect the urine. The resected (cut) ends of the remaining intestine are anastomosed (connected) to provide an intact bowel (Myles, 2011).

Stents, usually made of thin, pliable tubing, are placed in the ureters to prevent occlusion secondary to postsurgical edema. The bilateral ureteral stents allow urine to drain from the kidney to the stoma and provide a method for accurate measurement of urine output. They may be left in place 10 to 21 days postoperatively. Jackson-Pratt drains or other types of drains are inserted to prevent the accumulation of fluid in the space created by removal of the bladder.

After surgery, a skin barrier and a transparent, disposable urinary drainage bag are applied around the conduit and connected to drainage. A custom-cut appliance is used until the edema subsides and the stoma shrinks to normal size. The

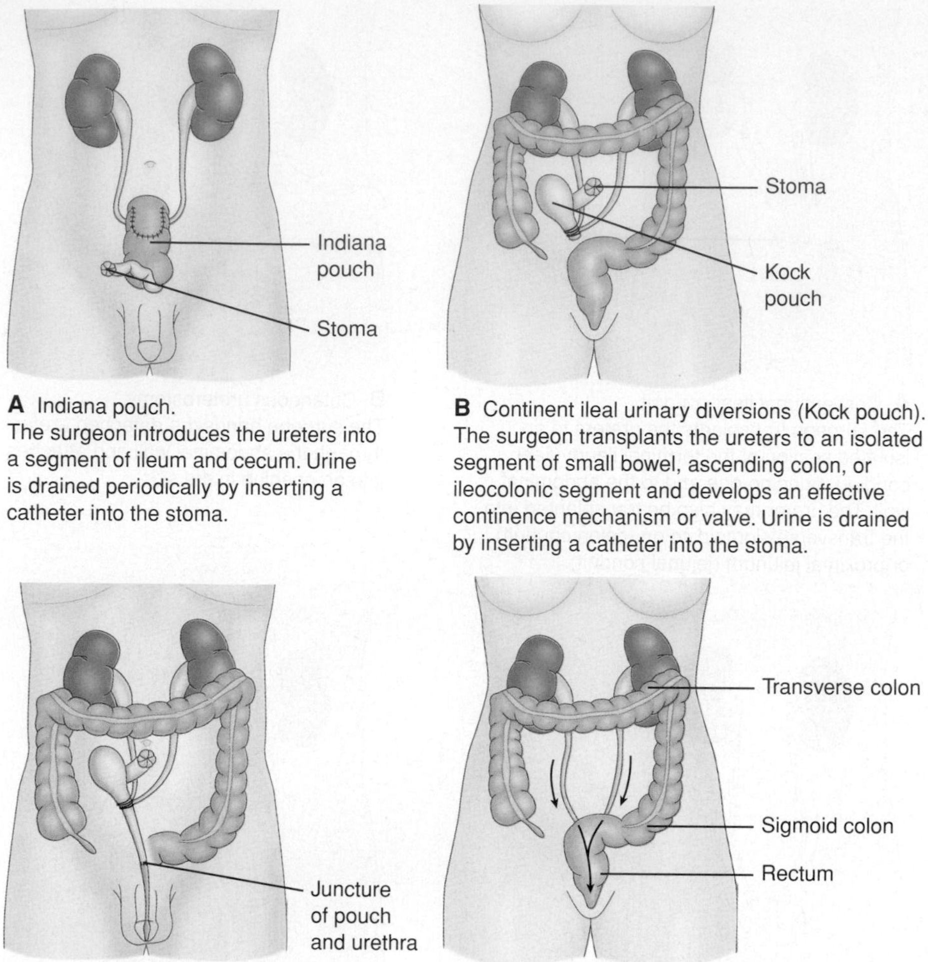
Indiana
pouch
Stoma
Stoma
Kock
pouch
Juncture
of pouch
and urethra
Transverse colon
Sigmoid colon
Rectum

A Indiana pouch.
The surgeon introduces the ureters into a segment of ileum and cecum. Urine is drained periodically by inserting a catheter into the stoma.

B Continent ileal urinary diversions (Kock pouch).
The surgeon transplants the ureters to an isolated segment of small bowel, ascending colon, or ileocolonic segment and develops an effective continence mechanism or valve. Urine is drained by inserting a catheter into the stoma.

C
In male patients, the Kock pouch can be modified by attaching one end of the pouch to the urethra, allowing more normal voiding. The female urethra is too short for this modification.

D Ureterosigmoidostomy.
The surgeon introduces the ureters into the sigmoid colon, thereby allowing urine to flow through the colon and out of the rectum.

FIGURE 55-8 • Types of continent urinary diversions include the Indiana pouch (**A**); the Kock pouch (**B, C**), also called a *continent ileal diversion;* and ureterosigmoidostomy (**D**).

clear bag allows the stoma to be inspected and the patency of the stent and the urine output to be monitored. The ileal bag drains urine (not feces) continuously. The appliance (bag) usually remains in place as long as it is watertight; it is changed when necessary to prevent leakage of urine.

Complications

Complications that may follow placement of an ileal conduit include wound infection or wound dehiscence, urinary leakage, ureteral obstruction, hyperchloremic acidosis, small bowel obstruction, ileus, and gangrene of the stoma (Gray, 2009). Delayed complications include ureteral obstruction, contraction or narrowing of the stoma (stenosis), renal deterioration due to chronic reflux, pyelonephritis, and renal calculi.

Nursing Management

In the immediate postoperative period, urine volumes are monitored hourly. Throughout the patient's hospitalization, the nurse monitors closely for complications, reports signs and symptoms of them promptly, and intervenes quickly to prevent their progression.

A urine output below 0.5 mL/kg/hr may indicate dehydration or an obstruction in the ileal conduit, with possible backflow or leakage from the ureteroileal anastomosis. A catheter may be inserted through the urinary conduit to monitor the patient for possible stasis or residual urine from a constricted stoma. Urine may drain through the bilateral ureteral stents as well as around the stents. If the ureteral stents are not draining, the nurse may be instructed to carefully irrigate with 5 to 10 mL sterile normal saline solution, being careful not to exert tension that could dislodge the stent. Hematuria may be noted in the first 48 hours after surgery but usually resolves spontaneously.

Providing Stoma and Skin Care

Because the patient requires specialized care, a consultation is initiated with a WOC nurse. The stoma is inspected frequently for color and viability. A healthy stoma is pink or red. A change from this normal color to purple, brown, or black suggests that the vascular supply may be compromised. If cyanosis and a compromised blood supply persist, surgical intervention may be necessary. The stoma is not sensitive to touch, but the skin around the stoma becomes sensitive if urine or the appliance irritates it (Black, 2011). The skin is inspected for (1) signs of irritation and bleeding of the stoma mucosa, (2) encrustation and skin irritation around the stoma (from alkaline urine coming in contact with exposed skin), and (3) wound infections.

Testing Urine and Caring for the Ostomy

Moisture in bed linens or clothing or the odor of urine around the patient should alert the nurse to the possibility of leakage from the appliance, potential infection, or a problem in hygienic management. Because severe alkaline encrustation can accumulate rapidly around the stoma, the urine pH is kept below 6.5 by administration of ascorbic acid by mouth. Urine pH is determined by testing the urine draining from the stoma, not from the collecting appliance. A properly fitted appliance is essential to prevent exposure of the skin around the stoma to urine. If the urine is foul smelling, the stoma is catheterized, if prescribed, to obtain a urine specimen for culture and sensitivity testing.

Encouraging Fluids and Relieving Anxiety

Because mucous membrane is used in forming the conduit, the patient may excrete a large amount of mucus mixed with urine. This causes anxiety in many patients. To help relieve this anxiety, the nurse reassures the patient that this is a normal occurrence after an ileal conduit procedure. The nurse encourages adequate fluid intake to flush the ileal conduit and decrease the accumulation of mucus.

Selecting the Ostomy Appliance

Various urine collection appliances are available, and the nurse is instrumental (often with consultation with a WOC nurse) in selecting an appropriate one (Black, 2011). The urinary appliance may consist of one or two pieces and may be disposable (usually used once and discarded) or reusable. The choice of appliance is determined by the location of the stoma and by the patient's normal activity, manual dexterity, visual function, body build, economic resources, and preference.

Promoting Home and Community-Based Care

Educating Patients About Self-Care

Patient education begins in the hospital but continues in the home setting because patients are usually discharged within days of surgery. The nurse instructs the patient how to assess and manage the urinary diversion as well as how to deal with changes in body image. A WOC nurse is invaluable in consulting with the nurse on various aspects of care and patient education.

Changing the Appliance. The patient and family are educated about how to apply and change the appliance so that they are comfortable carrying out the procedure and can do so proficiently. Ideally, the appliance system is changed before the system leaks and at a time that is convenient for the patient. Many patients find that early morning is most convenient because the urine output is reduced. A variety of appliances are available; an average collecting appliance lasts 3 to 7 days before leakage occurs.

Regardless of the type of appliance used, a skin barrier is essential to protect the skin from irritation and excoriation. To maintain skin integrity, a skin barrier or leaking pouch is never patched with tape to prevent accumulation of urine under the skin barrier or faceplate. The patient is instructed to avoid moisturizing soaps and body washes when cleaning the area because they interfere with the adhesion of the pouch. The degree to which the stoma protrudes is not the same in all patients; thus, there are various accessories and custom-made appliances to solve individual problems. Patient guidelines for applying reusable and disposable systems are presented in Chart 55-14.

Chart 55-14 PATIENT EDUCATION

Using Urinary Diversion Collection Appliances

Applying a Reusable Pouch System

1. Gather all necessary supplies.
2. Prepare new appliance according to the manufacturer's directions:
   - Apply double-faced adhesive disk that has been properly sized to fit the reusable pouch faceplate.
   - Remove paper backing and set pouch aside, or apply thin layer of contact cement to one side of the reusable pouch faceplate.
   - Set pouch aside.
3. Remove soiled pouch gently. Lay aside to clean later.
4. Clean peristomal skin (skin around stoma) with small amount of soap and water. Rinse thoroughly and dry. If a film of soap remains on the skin and the site does not dry, the appliance will not adhere adequately.
5. Use a wick (rolled gauze pad or tampon) over the stoma to absorb urine and keep the skin dry throughout the appliance change.
6. Inspect peristomal skin for irritation.
7. Note that a skin protector wipe or barrier ring may be applied before centering the faceplate opening directly over the stoma.
8. Position appliance over stoma, and press gently into place.
9. If desired, use a pouch cover or apply cornstarch under the pouch to prevent perspiration and skin irritation.
10. Clean soiled pouch, and prepare for reuse.

Applying a Disposable Pouch System

1. Gather all necessary supplies.
2. Measure stoma, and prepare an opening in the skin barrier about 1/8-inch larger than the stoma and the same shape as the stoma.
3. Remove paper backing from skin barrier, and set aside.
4. Gently remove old appliance, and set aside.
5. Clean peristomal skin with warm water, and dry thoroughly.
6. Inspect peristomal skin (skin around stoma) for irritation.
7. Use a wick (rolled gauze pad or tampon) over the stoma to absorb urine, and keep the skin dry during the appliance change.
8. Center opening of skin barrier over stoma, and apply with firm, gentle pressure to attain a watertight seal.
9. If using a two-piece system, snap pouch onto the flanged wafer that adheres to skin.
10. Close drainage tap or spout at bottom of pouch.
11. Note that a pouch cover can be used or cornstarch applied under pouch to prevent perspiration and skin irritation.
12. Apply hypoallergenic tape around the skin barrier in a picture-frame manner.
13. Dispose of soiled appliance.

Controlling Odor. The patient is instructed to avoid foods that give the urine a strong odor (e.g., asparagus, cheese, eggs). Most appliances contain odor barriers, but, if needed, a few drops of liquid deodorizer or diluted white vinegar may be introduced through the drain spout into the bottom of the pouch with a syringe or eyedropper to reduce odors. Ascorbic acid by mouth helps acidify the urine and suppress urine odor. Patients should be cautioned not to put aspirin tablets in the pouch to control odor, because they may ulcerate the stoma. In addition, the patient is reminded that odor will develop if the pouch is worn longer than recommended and not cared for properly.

Managing the Ostomy Appliance. The patient is instructed to empty the pouch by means of a drain valve when it is one third full because the weight of more urine will cause the pouch to separate from the skin. Some patients prefer wearing a leg bag attached with an adapter to the drainage apparatus. To promote uninterrupted sleep, a collecting bottle and tubing (one unit) are snapped onto an adapter that connects to the ileal appliance. A small amount of urine is left in the bag when the adapter is attached to prevent the bag from collapsing against itself. The tubing may be threaded down the pajama or pants leg to prevent kinking. The collecting bottle and tubing are rinsed daily with cool water and once a week with a 3:1 solution of water and white vinegar.

Cleaning and Deodorizing the Appliance. Usually, the reusable appliance is rinsed in warm water and soaked in a 3:1 solution of water and white vinegar or a commercial deodorizing solution for 30 minutes (Myles, 2011). It is rinsed with tepid water and air-dried away from direct sunlight. (Hot water and exposure to direct sunlight dry the pouch and increase the incidence of cracking.) After drying, the appliance may be powdered with cornstarch and stored. Two appliances are necessary—one to be worn while the other is air-drying.

Continuing Care

Follow-up care is essential to determine how the patient has adapted to the altered body image and lifestyle changes. Referral for home care is indicated to determine how well the patient and family are coping with the urinary drainage diversion. The home care nurse assesses the patient's physical status and emotional response. In addition, the nurse assesses the ability of the patient and family to manage the urinary diversion and appliance, reinforces previous education, and provides additional information (e.g., community resources, sources of ostomy supplies, insurance coverage for supplies).

As the postoperative edema subsides, the home care nurse assists in determining the appropriate changes needed in the ostomy appliance. The size of the stoma is measured every 3 to 6 weeks for the first few months postoperatively. The correct appliance size is determined by measuring the widest part of the stoma with a ruler. The permanent appliance should be no more than 1.6 mm (1/8 inch) larger than the diameter of the stoma and the same shape as the stoma to prevent contact of the skin with drainage.

The nurse educates the patient and family about resources (see the Resources section at the end of this chapter). Local chapters of the American Cancer Society (ACS) can provide medical equipment and supplies and other resources for the patient who has undergone ostomy surgery for cancer.

The home care nurse assesses the patient for potential long-term complications such as ureteral obstruction, stenosis, hernias, or deterioration of renal function (Borwell, 2011). The nurse also reinforces previous education about these complications.

Cutaneous Ureterostomy

A cutaneous ureterostomy (see Fig. 55-7), in which the ureters are directed through the abdominal wall and attached to an opening in the skin, is used for selected patients with ureteral obstruction (i.e., advanced pelvic cancer) because it requires less extensive surgery than other urinary diversion procedures. It is also an appropriate procedure for patients who have had previous abdominal irradiation.

A urinary appliance is fitted immediately after surgery. The management of the patient with a cutaneous ureterostomy is similar to the care of the patient with an ileal conduit, although the stomas are usually flush with the skin or retracted.

Continent Urinary Diversions

Continent Ileal Urinary Reservoir (Indiana Pouch)

The most common continent urinary diversion is the Indiana pouch, created for the patient whose bladder is removed or no longer functions. The Indiana pouch uses a segment of the ileum and cecum to form the reservoir for urine (see Fig. 55-8A). The ureters are tunneled through the muscular bands of the intestinal pouch and anastomosed. The reservoir is made continent by narrowing the efferent portion of the ileum and sewing the terminal ileum to the subcutaneous tissue, forming a continent stoma flush with the skin. The pouch is sewn to the anterior abdominal wall around a cecostomy tube. Urine collects in the pouch until a catheter is inserted and the urine is drained (Black, 2011).

The pouch must be drained at regular intervals by a catheter to prevent absorption of metabolic waste products from the urine, reflux of urine to the ureters, and UTI. Postoperative nursing care of the patient with a continent ileal urinary pouch is similar to nursing care of the patient with an ileal conduit. However, these patients usually have additional drainage tubes (cecostomy catheter from the pouch, stoma catheter exiting from the stoma, ureteral stents, and Penrose drain, as well as a urethral catheter). All drainage tubes must be carefully monitored for patency and amount and type of drainage. In the immediate postoperative period, the cecostomy tube is irrigated two or three times daily to remove mucus and prevent blockage.

Other variations of continent urinary reservoirs include the Kock pouch (U-shaped pouch constructed of ileum, with a nipplelike one-way valve; see Fig. 55-8B, C) and the Charleston pouch (uses the ileum and ascending colon as the pouch, with the appendix and colon junction serving as the one-way valve mechanism). With both of these methods, the pouch must be drained at regular intervals by inserting a catheter. (See Chart 48-7 in Chapter 48 for guidelines for draining a Kock pouch.)

Ureterosigmoidostomy

Ureterosigmoidostomy, another form of continent urinary diversion, is a transplantation of the ureters into the sigmoid colon, allowing urine to flow through the colon and out the rectum (see Fig. 55-8D). It is usually performed in patients who have had extensive pelvic irradiation, previous small bowel resection, or coexisting small bowel disease.

After surgery, voiding occurs from the rectum (for life), and an adjustment in lifestyle will be necessary because of urinary frequency. Drainage has a consistency equivalent to watery diarrhea, and the patient has some degree of nocturia. Patients usually need to plan activities around the frequent need to urinate, which in turn may affect the patient's social life. However, patients have the advantage of urinary control without having to wear an external appliance.

Nursing Management

In addition to the usual preoperative regimen, the patient may be placed on a liquid diet for several days preoperatively to reduce residue in the colon. Antibiotic agents (neomycin, kanamycin) are administered to disinfect the bowel. Ureterosigmoidostomy requires a competent anal sphincter, adequate renal function, and active renal peristalsis. The degree of anal sphincter control may be determined by assessing the patient's ability to retain enemas.

The postoperative regimen initially includes placing a catheter in the rectum to drain the urine and prevent reflux of urine into the ureters and kidneys. The tube is taped to the buttocks, and special skin care is given around the anus to prevent excoriation. Irrigations of the rectal tube may be prescribed, but force is never used because of the danger of introducing bacteria into the newly implanted ureters.

Monitoring Fluid and Electrolytes

In ureterosigmoidostomy, larger areas of the bowel mucosa are exposed to urine and electrolyte reabsorption. As a result, electrolyte imbalance and acidosis may occur. Potassium and magnesium in the urine may cause diarrhea. Fluid and electrolyte balance is maintained in the immediate postoperative period by closely monitoring the serum electrolyte levels and administering appropriate IV fluids. Acidosis may be prevented by placing the patient on a low-chloride diet supplemented with sodium potassium citrate.

The patient should be instructed never to wait longer than 2 to 3 hours before emptying urine from the intestine. This keeps rectal pressure low and minimizes the absorption of urinary constituents from the colon. It is essential to educate the patient about the symptoms of UTI: fever, flank pain, urgency, and frequency.

Retraining the Anal Sphincter

After the rectal catheter is removed, the patient learns to control the anal sphincter through special sphincter exercises. At first, urination is frequent. With reassurance and encouragement and the passage of time, the patient gains greater control and learns to differentiate between the need to void and the need to defecate.

Promoting Dietary Measures

Specific dietary instructions include avoidance of gas-forming foods (flatus can cause stress incontinence and offensive odors). Other ways to avoid gas are to avoid chewing gum, smoking, and any other activity that involves swallowing air. Salt intake may be restricted to prevent hyperchloremic acidosis. Potassium intake is increased through foods and medication because potassium may be lost in acidosis.

Monitoring and Managing Potential Complications

Pyelonephritis (upper UTI) due to reflux of bacteria from the colon is fairly common. Long-term antibiotic therapy may be prescribed to prevent infection. A late complication is adenocarcinoma of the sigmoid colon, possibly from cellular changes due to exposure of the colonic mucosa to urine. Urinary carcinogens promote late malignant transformation of the colon after a ureterosigmoidostomy, warranting lifelong medical follow-up.

Other Urinary Diversion Procedures

Variations in urinary diversion surgical procedures are devised frequently in an effort to identify and perfect procedures that will improve patient outcomes and reduce the incidence of postoperative problems (Myles, 2011). These include cecal, patched cecal, and Mainz reservoirs. These techniques involve isolating a part of the large intestine to form a reservoir for urine and creating an abdominal stoma. Another surgical procedure, the Camey procedure, uses a portion of the ileum as a bladder substitute. In this procedure, the isolated ileum serves as the reservoir for urine; it is anastomosed directly to the portion of the remaining urethra after cystectomy. This procedure permits emptying of the bladder through the urethra. However, the Camey procedure applies only to men because the entire urethra is removed when a cystectomy is performed in women.

NURSING PROCESS

The Patient Undergoing Urinary Diversion Surgery

Preoperative Assessment

The following are key preoperative nursing assessment concerns:

- Cardiopulmonary function assessments are performed because patients undergoing cystectomy are often older people who may be at greater risk for cardiac and respiratory complications.
- A nutritional status assessment is important because of possible poor nutritional intake related to underlying health problems.
- Learning needs are assessed in consultation with a WOC nurse to evaluate the patient's and the family's understanding of the procedure as well as the changes in physical structure and function that result from the surgery. The patient's self-concept and self-esteem are assessed in addition to methods for coping with stress and loss. The patient's mental status, manual dexterity and coordination, vision, and preferred method of

learning are noted because they affect postoperative self-care.

Preoperative Nursing Diagnoses

Based on the assessment data, preoperative nursing diagnoses may include the following:

- Anxiety related to anticipated losses associated with the surgical procedure
- Imbalanced nutrition: less than body requirements related to inadequate nutritional intake
- Deficient knowledge about the surgical procedure and postoperative care

Preoperative Planning and Goals

The major goals for the patient may include relief of anxiety; improved preoperative nutritional status; and increased knowledge about the surgical procedure, expected outcomes, and postoperative care.

Preoperative Nursing Interventions

Relieving Anxiety

The threat of cancer and removal of the bladder create anxiety related to changes in body image. Patients may face problems adapting to an external appliance, a stoma, a surgical incision, and altered toileting habits. Men must also adapt to sexual impotency; a penile implant is considered if the patient is a candidate for the procedure. Women also have anxiety related to altered appearance, body image, and self-esteem. A supportive approach, both physical and psychosocial, is needed and includes assessing the patient's self-concept and manner of coping with stress and loss; helping the patient to identify ways to maintain his or her lifestyle and independence with as few changes as possible; and encouraging the patient to express fears and anxieties about the ramifications of the upcoming surgery. A visitor from the Ostomy Visitation Program of the ACS can provide emotional support and make adaptation easier both before and after surgery.

Ensuring Adequate Nutrition

A low-residue diet is prescribed to cleanse the bowel to minimize fecal stasis, decompress the bowel, and minimize postoperative ileus. In addition, antibiotic medications are administered to reduce pathogenic flora in the bowel and to reduce the risk of infection. Because the patient undergoing a urinary diversion procedure for cancer may be severely malnourished due to the tumor, previous treatments, and anorexia, enteral or parenteral nutrition may be prescribed to promote healing. Adequate preoperative hydration is imperative to ensure urine flow during surgery and to prevent hypovolemia during the prolonged surgical procedure.

Explaining Surgery and Its Effects

Participation of a WOC nurse is invaluable for informed preoperative education and postoperative care planning. Explanations of the surgical procedure, the appearance of the stoma, the rationale for preoperative bowel preparation, the reasons for wearing a collection device, and the anticipated effects of the surgery on sexual functioning are part of patient education. The placement of the stoma site is planned preoperatively with the patient standing, sitting, and lying down to locate the stoma away from bony prominences, skin creases, and folds. The stoma should also be placed away from old scars, the umbilicus, and the belt line.

For ease of self-care, the patient must be able to see and reach the site comfortably. The site is marked with indelible ink so that it can be located easily during surgery. The patient is assessed for allergies or sensitivity to tape or adhesives. Patch testing of certain appliances may be necessary before the ostomy equipment is selected. This is particularly important if the patient is or may be allergic to latex (see Chapter 17).

Preoperative Evaluation

To measure the effectiveness of care, the nurse evaluates the patient's preoperative anxiety level and nutritional status as well as preexisting knowledge and expectations of surgery.

Expected patient outcomes may include:

1. Exhibits reduced anxiety about surgery and expected losses
   a. Verbalizes fears with health care team and family
   b. Expresses positive attitude about outcome of surgery
2. Exhibits adequate nutritional status
   a. Maintains adequate intake before surgery
   b. Maintains body weight
   c. States rationale for enteral or parenteral nutrition if needed
   d. Exhibits normal skin turgor, moist mucous membranes, adequate urine output, and absence of excessive thirst
3. Demonstrates knowledge about the surgical procedure and postoperative course
   a. Identifies limitations expected after surgery
   b. Discusses expected immediate postoperative environment (tubes, equipment, nursing surveillance)
   c. Practices deep-breathing, coughing, and foot exercises

Postoperative Assessment

The role of the nurse in the immediate postoperative period is to prevent complications and to assess the patient carefully for any signs and symptoms of complications. The catheters and any drainage devices are monitored closely. Urine volume, patency of the drainage system, and color of the drainage are assessed. A sudden decrease in urine volume or increase in drainage is reported promptly to the primary provider, because these may indicate obstruction of the urinary tract, inadequate blood volume, or bleeding. In addition, the patient's need for pain control is assessed regularly, as with all postoperative patients.

Postoperative Diagnosis

Nursing Diagnoses

Based on the assessment data, major postoperative nursing diagnoses may include the following:

- Risk for impaired skin integrity related to problems in managing the urine collection appliance
- Acute pain related to surgical incision
- Disturbed body image related to urinary diversion
- Ineffective sexuality pattern related to structural and physiologic alterations
- Deficient knowledge about management of urinary function

Collaborative Problems/Potential Complications

Potential complications may include the following:

- Peritonitis due to disruption of anastomosis
- Stoma ischemia and necrosis due to compromised blood supply to stoma
- Stoma retraction and separation of mucocutaneous border due to tension or trauma

Postoperative Planning and Goals

The major goals for the patient may include maintaining skin integrity, relieving pain, increasing self-esteem, developing appropriate coping mechanisms to accept and deal with altered urinary function and sexuality, increasing knowledge about management of urinary function, and preventing potential complications.

Postoperative Nursing Interventions

Postoperative management focuses on monitoring urinary function, preventing postoperative complications (infection and sepsis, respiratory complications, fluid and electrolyte imbalances, fistula formation, and urine leakage), and promoting patient comfort. Catheters or drainage systems are monitored, and urine output is monitored carefully. A nasogastric tube is inserted during surgery to decompress the GI tract and to relieve pressure on the intestinal anastomosis. It is usually kept in place for several days after surgery. As soon as bowel function resumes—as indicated by bowel sounds, the passage of flatus, and a soft abdomen—oral fluids are permitted. Until that time, IV fluids and electrolytes are administered. The patient is assisted to ambulate as soon as possible to prevent complications of immobility.

Maintaining Skin Integrity

Strategies to promote skin integrity begin with reducing and controlling those factors that increase the patient's risk of poor nutrition and poor healing. Meticulous skin care and management of the drainage system are provided by the nurse until the patient can manage them and is comfortable doing so. Care is taken to keep the system intact to protect the skin from exposure to drainage. Supplies must be readily available to manage the drainage in the immediate postoperative period. Consistency in implementing the skin care program throughout the postoperative period results in maintenance of skin integrity and patient comfort. In addition, maintenance of skin integrity around the stoma enables the patient and family to adjust more easily to the alterations in urinary function and helps them learn skin care techniques.

Relieving Pain

Analgesic medications are administered liberally postoperatively to relieve pain and promote comfort, thereby allowing the patient to turn, cough, and perform deep-breathing exercises. Patient-controlled analgesia and regular administration of analgesic agents around the clock are two options that may be used to ensure adequate pain relief. A pain intensity scale is used to evaluate the adequacy of the medication and the approach to pain management (see Chapter 12 for further discussion of pain management).

Improving Body Image

The patient's ability to cope with the changes associated with the surgery depends to some degree on his or her body image and self-esteem before the surgery and the support and reaction of others. Allowing the patient to express concerns and anxious feelings can help, especially in adjusting to the changes in toileting habits. The nurse can also help improve the patient's self-concept by educating about skills needed to be independent in managing the urinary drainage devices. Education about ostomy care is conducted in a private setting to encourage the patient to ask questions without fear of embarrassment. Explaining why the nurse must wear gloves when performing ostomy care can prevent the patient from misinterpreting the use of gloves as a sign of aversion to the stoma.

Exploring Sexuality Issues

Patients who experience altered sexual function as a result of the surgical procedure may mourn this loss. Encouraging the patient and partner to share their feelings about this loss with each other and acknowledging the importance of sexual function and expression may encourage the patient and partner to seek sexual counseling and to explore alternative ways of expressing sexuality. A visit from another "ostomate" who is functioning fully in society and family life may also assist the patient and family in recognizing that full recovery is possible.

Monitoring and Managing Potential Complications

Complications are not unusual because of the complexity of the surgery, the underlying reason (cancer, trauma) for the urinary diversion procedure, and the patient's frequently less-than-optimal nutritional status. Complications may include respiratory disorders (e.g., atelectasis, pneumonia), fluid and electrolyte imbalances, breakdown of any anastomosis, sepsis, fistula formation, fecal or urine leakage, and skin irritation. If these occur, the patient will remain hospitalized for an extended length of time and will probably require parenteral nutrition, GI decompression by means of nasogastric suction, and further surgery. The goals of management are to establish drainage, provide adequate nutrition for healing to occur, and prevent sepsis.

Peritonitis. Peritonitis can occur postoperatively if urine leaks at the anastomosis. Signs and symptoms include abdominal pain and distention, muscle rigidity with guarding, nausea and vomiting, paralytic ileus (absence of bowel sounds), fever, and leukocytosis.

Urine output must be monitored closely, because a sudden decrease in output with a corresponding increase in drainage from the incision or drains may indicate urine leakage. In addition, the urine drainage device is observed for leakage. The pouch is changed if a leak is observed. Small leaks in the anastomosis may seal themselves, but surgery may be needed for larger leaks.

Vital signs (blood pressure, pulse and respiratory rates, temperature) are monitored. Changes in vital signs, as well as increasing pain, nausea and vomiting, and abdominal distention, are reported and may indicate peritonitis.

Stoma Ischemia and Necrosis. The stoma is monitored because ischemia and necrosis of the stoma can result from tension on the mesentery blood vessels, twisting of the bowel segment (conduit) during surgery, or arterial insufficiency. The new stoma must be inspected at least every 4 hours to assess the adequacy of its blood supply. The stoma should be red or pink.

If the blood supply to the stoma is compromised, the color changes to purple, brown, or black. These changes are reported immediately. The physician or WOC nurse may insert a small, lubricated tube into the stoma and shine a flashlight into the lumen of the tube to assess for superficial ischemia or necrosis. A necrotic stoma requires surgical intervention. If the ischemia is superficial, the dusky stoma is observed and may slough its outer layer in several days.

Stoma Retraction and Separation. Stoma retraction and separation of the mucocutaneous border can occur as a result of trauma or tension on the internal bowel segment used for creation of the stoma. In addition, mucocutaneous separation can occur if the stoma does not heal as a result of accumulation of urine on the stoma and mucocutaneous border. Using a collection drainage pouch with an antireflux valve is helpful because the valve prevents urine from pooling on the stoma and mucocutaneous border. Meticulous skin care to keep the area around the stoma clean and dry promotes healing. If a separation of the mucocutaneous border occurs, surgery is not usually needed. The separated area is protected by applying karaya powder, stoma adhesive paste, and a properly fitted skin barrier and pouch. By protecting the separation, healing is promoted. If the stoma retracts into the peritoneum, surgical intervention is mandatory.

If surgery is needed to manage these complications, the nurse provides explanations to the patient and family. The need for additional surgery is usually perceived as a setback by the patient and family. Emotional support of the patient and family is provided along with physical preparation of the patient for surgery.

Promoting Home and Community-Based Care

Educating Patients About Self-Care. A major postoperative objective is to assist the patient to achieve the highest level of independence and self-care possible. The nurse and WOC nurse work closely with the patient and family to instruct and assist them in all phases of managing the ostomy. Adequate supplies and complete instruction are necessary to enable the patient and a family member to develop competence and confidence in their skills. Written and verbal instructions are provided, and the patient is encouraged to contact the nurse or primary provider with follow-up questions. Follow-up telephone calls from the nurse to the patient and family after discharge may provide added support and provide another opportunity to answer their questions. Follow-up visits and reinforcement of correct skin care and appliance management techniques also promote skin integrity. Specific techniques for managing the appliance are described in Chart 55-14.

The patient is encouraged to participate in decisions regarding the type of collecting appliance and the time of day to change the appliance. The patient is assisted and encouraged to look at and touch the stoma early to overcome any fears. The patient and family need to know the characteristics of a normal stoma:

- Pink or red and moist, like the inside of the mouth
- Insensitive to pain because it has no nerve endings
- Vascular, which means it may bleed when cleaned

In addition, if a segment of the GI tract was used to create the urinary diversion, mucus may be visible in the urine. By learning what is normal, the patient and family become familiar with what signs and symptoms they should report to the physician or nurse and what problems they can handle themselves.

Information provided to the patient and the extent of involvement in self-care are determined by the patient's physical recovery and ability to accept and acquire the knowledge and skill needed for independence. Verbal and written instructions are provided, and the patient is given the opportunity to practice and demonstrate the knowledge and skills needed to manage urinary drainage.

Continuing Care. Follow-up care is essential to determine how the patient has adapted to the changes in body image and lifestyle adjustments. Visits from a home care nurse are important to assess the patient's adaptation to the home setting and management of the ostomy. Education and reinforcement may assist the patient and family to cope with altered urinary function. It is important to assess for long-term complications that may occur, such as pouch leakage or rupture, stone formation, stenosis of the stoma, deterioration in renal function, or incontinence.

Long-term monitoring for anemia is performed to identify vitamin B12 deficiency, which may occur when a significant portion of the terminal ileum is removed. This may take several years to develop and can be treated with vitamin B12 injections. The patient and family are informed about the United Ostomy Associations of America (UOAA) and any local ostomy support groups to provide ongoing support, assistance, and education.

Postoperative Evaluation

Expected patient outcomes may include:

1. Maintains skin integrity
   a. Maintains intact skin and demonstrates skill in managing drainage system and appliance
   b. States actions to take if skin excoriation occurs
2. Reports relief of pain
3. Exhibits improved body image as evidenced by the following:
   a. Voices acceptance of urinary diversion, stoma, and appliance
   b. Demonstrates increasingly independent self-care, including hygiene and grooming
   c. States acceptance of support and assistance from family members, health care providers, and other ostomates
4. Copes with sexuality issues
   a. Verbalizes concern about possible alterations in sexuality and sexual function
   b. Reports discussion of sexual concerns with partner and appropriate counselor
5. Demonstrates knowledge needed for self-care
   a. Performs self-care and proficient management of urinary diversion and appliance
   b. Asks questions relevant to self-management and prevention of complications
   c. Identifies signs and symptoms needing care from physician, nurse, or other health care providers
6. Absence of complications as evidenced by the following:
   a. Reports absence of pain or tenderness in abdomen
   b. Has temperature within normal range
   c. Reports no urine leakage from incision or drains
   d. Has urine output within desired volume limits

e. Maintains stoma that is red or pink, moist, and appropriate in size without edema
f. Has intact and healed border of the stoma

Critical Thinking Exercises

1 ebp As a community health nurse, you are caring for a patient who was recently discharged from a rehabilitation unit. During a routine visit, you are approached by the adult daughter, who is caring for her mother. She requests that her mother, who can ambulate with assistance, have an indwelling urinary catheter inserted "for convenience sake." What is the evidence base that determines your response? Identify the criteria used to evaluate the strength of the evidence.

2 ebp You are a nurse in a busy urology practice who often cares for women of varying ages with incontinence. Describe the major types of incontinence, and compare and contrast each. What are the evidence-based management techniques used in treating the different types of incontinence? Identify the criteria used to evaluate the strength of the evidence for these practices.

3 pq A 35-year-old man is admitted to a medical-surgical nursing unit with a suspected renal stone. Describe the pathophysiology of renal stone formation. What diagnostic tests should be performed to confirm the diagnosis? What is the most common priority nursing diagnosis in the patient admitted with a renal stone? What history and physical findings are common in the patient who is trying to pass the stone? If the patient does not pass the stone or develops complications, what are the interventional procedures available for stone removal?

4 pq You are caring for a 56-year-old woman who was admitted to an oncology unit with hematuria and is undergoing an evaluation for bladder cancer. Identify the priorities, approach, and techniques you would use to provide care for this patient. How will your priorities, approach, and techniques differ if the patient is diagnosed with bladder cancer?

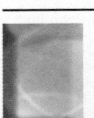

Brunner Suite Resources
Explore these additional resources to enhance learning for this chapter:
- NCLEX-Style Questions and Other Resources on thePoint, http://thePoint.lww.com/Brunner13e
- Study Guide
- PrepU
- Clinical Handbook
- Handbook of Laboratory and Diagnostic Tests

References

*Asterisk indicates nursing research.

Books

Dudek, S. G. (2010). *Nutrition essentials for nursing practice* (6th ed.). Philadelphia: Lippincott Williams & Wilkins.
Hogan-Quigley, B., Palm, M. J., & Bickley, L. S. (2012). *Bates' nursing guide to physical examination and history taking*. Philadelphia: Lippincott Williams & Wilkins.
Karch, A. (2012). *2012 Lippincott's nursing drug guide*. Philadelphia: Lippincott Williams & Wilkins.
Litwin, M. S., & Saigal, C. S. (2112). *Urologic diseases in America. NIH Publication No. 12–7865. U.S. Department of Health and Human Services, Public Health Service, National Institutes of Health, National Institute of Diabetes and Digestive and Kidney Diseases*. Washington, DC: U.S. Government Printing Office.
Meiner, S. E. (2011). *Gerontologic nursing* (4th ed.). St. Louis: Elsevier Mosby.
Miller, C. A. (2012). *Nursing for wellness in older adults* (6th ed.). Philadelphia: Lippincott Williams & Wilkins.
Porth, C. M., & Matfin, G. (2009). *Pathophysiology: Concepts of altered health states* (8th ed.). Philadelphia: Lippincott Williams & Wilkins.
Weber, J., & Kelley, J. (2010). *Health assessment in nursing* (4th ed.). Philadelphia: Lippincott Williams & Wilkins.

Journals and Electronic Documents

Adult Voiding Disorders

Agency for Healthcare Research and Quality. (2012). Experts seek better diagnosis and treatment for women's urinary incontinence and chronic pelvic pain. *Research Activities*, (383), 1–4
Keegan, P. E., Atiemo, K., Cody, J. D., et al. (2012). Urethral injection therapy for urinary incontinence in women. *Cochrane Database of Systematic Reviews*, (2), CD003881.
Klausner, A. P., & Steers, W. D. (2011). The neurogenic bladder: An update with management strategies for primary care physicians. *Medical Clinics of North America*, 95(1), 111–120.
Ling Man, E. S., & Le Low, L. P. (2010). Nurses' experiences of caring for non-catheterized older infirmed patients: A descriptive study of what nurses actually do? *Journal of Clinical Nursing*, 19(9–10), 1387–1395.
Prynn, P. (2012). Continence issues in the elderly. *Practice Nurse*, 42(1), 20–22.
Specht, J. K. (2011). Promoting continence in individuals with dementia. *Journal of Gerontological Nursing*, 37(2), 17–21.
Stewart, E. (2012). Treating urinary incontinence in older women. *British Journal of Community Nursing*, 15(11), 526–532.
*Weinstein, M. (2012). Hormone therapy and urinary incontinence. *Menopause*, 19(3), 255–256.

Genitourinary Trauma

Blair, M. (2011). CNE SERIES. Overview of genitourinary trauma. *Urologic Nursing*, 31(3), 139–146.
Kong, J. P., Bultitude, M. F., Royce, P., et al. (2011). Lower urinary tract injuries following blunt trauma: A review of contemporary management. *Reviews in Urology*, 13(3), 119–130.
Koraitim, M. (2012). Effect of early realignment on length and delayed repair of postpelvic fracture urethral injury. *Urology*, 79, 912–916.

Urinary Cancers

Askeland, E. J., Newton, M. R., O'Donnell, M. A., et al. (2012). Bladder cancer immunotherapy: BCG and beyond. *Advances in Urology*, *2012*, 181987. Available at: www.ncbi.nlm.nih.gov/pmc/articles/PMC3388311/?tool=pubmed
Brausi, M., Witjes, J. A., Lamm, D., et al. (2011). A review of current guidelines and best practice recommendations for the management of nonmuscle invasive bladder cancer by the International Bladder Cancer Group. *Journal of Urology*, 186(6), 2158–2167.
Centers for Disease Control and Prevention. (2010). *United States cancer statistics (USCS)*. Available at: www.cdc.gov/Features/CancerStatistics/
*Chatterton, K., Bugeja, P., Challacombe, B., et al. (2011). Nurses' experience establishing a nurse-led bladder cancer surveillance flexible cystoscopy service. *Australian Journal of Advanced Nursing*, 28(4), 53–59.
*Gemmill, R., Sun, V., Ferrell, B., et al. (2010). Going with the flow: Quality-of-life outcomes of cancer survivors with urinary diversion. *Journal of Wound, Ostomy & Continence Nursing*, 37(1), 65–72.
National Cancer Institute. (2011). *SEER stat fact sheets: Bladder*. Available at: seer.cancer.gov/statfacts/html/urinb.html
Tilki, D., Burger, M., Dalbagni, G., et al. (2011). Urine markers for detection and surveillance of non–muscle-invasive bladder cancer. *European Urology*, 60(3), 484–492.
Vasdev, N., Shaw, M. B., & Thorpe, A. C. (2011). Neoadjuvant chemotherapy for muscle invasive bladder cancer. *Current Urology*, 5(2) 57–61.
*Ward-Smith, P., Miller, B., & Caughron, M. (2010). Applicability of the FISH test for bladder cancer. *Urologic Nursing*, 30(4), 218–221, 251.

Urinary Diversion
Black, P. (2011). Choosing the correct stoma appliance. *Journal of Community Nursing, 25*(6), 44–49.
Borwell, B. (2011). Stoma management and palliative care. *Journal of Community Nursing, 25*(4), 4–10.
Gray, M. (2009). Context for WOC practice: Unavoidable pressure ulcers, wound swabs, bladder cancer, and pelvic floor muscle training: Time to call for a WOC nurse consult! *Journal of Wound, Ostomy & Continence Nursing, 36*(4), 360–363.
Myles, M. L. (2011). Urinary diversions. *MEDSURG Nursing, 20*(2), 94–95.
*Wilde, M. H., Brasch, J., Getliffe, K., et al. (2010). Study on the use of long-term urinary catheters in community-dwelling individuals. *Journal of Wound, Ostomy & Continence Nursing, 37*(3), 301–310.
Wille, M. A., Zagaja, G. P., Shalhav, A. L., & Gundeti, M. S. (2011). Continence outcomes in patients undergoing robotic assisted laparoscopic Mitrofanoff appendicovesicostomy. *Journal of Urology, 185*(4), 1438–1443.

Urolithiasis/Nephrolithiasis
Frassetto, L., & Kohlstadt, I. (2011). Treatment and prevention of kidney stones: An update. *American Family Physician, 84*(11), 1234–1242.
Jeong, I. G., Kang, T., Bang, J. K., et al. (2011). Association between metabolic syndrome and the presence of kidney stones in a screened population. *American Journal of Kidney Diseases, 58*(3), 383–388.
Meschi, T., Nouvenne, A., & Borghi, L. (2011). Lifestyle recommendations to reduce the risk of kidney stones. *Urologic Clinics of North America, 38*(3), 313–320.
Worcester, E. M., & Coe, F. L. (2010). Calcium kidney stones. *New England Journal of Medicine, 363*(10), 954–963.

Infection/Inflammation
Bernard, M. S., Hunter, K. F., & Moore, K. N. (2012). A review of strategies to decrease the duration of indwelling urethral catheters and potentially reduce the incidence of catheter- associated urinary tract infections. *Urologic Nursing, 32*(1), 29–37.
Colgan, R., Williams, M., & Johnson, J. R. (2011). Diagnosis and treatment of acute pyelonephritis in women. *American Family Physician, 84*(5), 519–526.
Das, R., Perrelli, E., Towle, V., et al. (2009). Antimicrobial susceptibility of bacteria isolated from urine samples obtained from nursing home residents. *Infection Control & Hospital Epidemiology, 30*(11), 1116–1119.
*Elpern, E. H., Killeen, K., Ketchem, A., et al. (2009). Reducing use of indwelling urinary catheters and associated urinary tract infections. *American Journal of Critical Care, 18*(6), 535–542.
Gray, M. (2010). Reducing catheter-associated urinary tract infection in the critical care unit. *AACN Advanced Critical Care, 21*(3), 247–257.
Hart, A. M. (2011). Antibiotic stewardship to preserve benefits. *Clinical Advisor for Nurse Practitioners, 14*(11), 46–55.
Hogan, L., Young, C., Gabbert, W., et al. (2011). Unique predisposing factors for male urinary tract infections. *Journal of the American Academy of Nurse Practitioners, 23*(10), 525–529.
Kahnen, D. A., Flanders, S., & Magalong, T. (2011). Catheter-associated urinary tract infections: Making them matter! *Med-Surg Matters, 20*(6), 4–7.
Lane, D. R., & Takhar, S. S. (2011). Diagnosis and management of urinary tract infection and pyelonephritis. *Emergency Medicine Clinics of North America, 29*(3), 539–552.
Lovatt, P. (2010). Drug profile: Trimethoprim for uncomplicated UTIs. *Nurse Prescribing, 8*(7), 330–332.
*Muzzi-Bjornson, L., & Macera, L. (2011). Preventing infection in elders with long-term indwelling urinary catheters. *Journal of the American Academy of Nurse Practitioners, 23*(3), 127–134.
Nicolle, L. E. (2009). Urinary tract infections in the elderly. *Clinics in Geriatric Medicine, 25*(3), 423–436.
*Omli, R., Skotnes, L. H., Romild, U., et al. (2010). Pad per day usage, urinary incontinence and urinary tract infections in nursing home residents. *Age and Aging, 39*, 549–554.
O'Shea, L. (2010). Diagnosing urinary tract infections. *Practice Nurse, 40*(9), 20–25.
Stapleton, A. E., Dziura, J., Hooton, T. M., et al. (2012). Recurrent urinary tract infection and urinary Escherichia coli in women ingesting cranberry juice daily: A randomized controlled trial. *Mayo Clinic Proceedings, 87*(2), 143–150.

Resources

American Cancer Society (ACS), www.cancer.org
American Urological Association, www.auanet.org
Centers for Disease Control and Prevention (CDC), www.cdc.gov
Institute for Healthcare Improvement (IHI), www.ihi.org/Pages/default.aspx
National Association for Continence (NAFC), www.nafc.org
National Cancer Institute (NCI), www.cancer.gov
National Database of Nurse Quality Indicators (NDNQI), www.nursingquality.org
National Institute of Diabetes and Digestive and Kidney Diseases (NIDDK), National Institutes of Health, www.niddk.nih.gov
National Kidney and Urologic Diseases Information Clearinghouse (NKUDIC), kidney.niddk.nih.gov/
National Kidney Foundation, www.kidney.org
Wound Ostomy and Continence Nurses Society, www.wocn.org

Unit 13

Reproductive Function

Case Study

A PATIENT WITH A DIFFICULT HEALTH CARE CHOICE INVOLVING LOSSES

Mrs. Cole is a 49-year-old woman who has been undergoing cancer staging after positive breast biopsy results. The surgeon has informed her that she has stage IIB infiltrating ductal carcinoma. The surgeon has discussed with her two different surgical approaches—breast conserving or modified radical mastectomy (MRM). If she chooses MRM, Mrs. Cole must decide whether she will undergo breast reconstruction or use a breast prosthesis. The nurse notes that Mrs. Cole is trembling and near tears. Mrs. Cole tells the nurse that she is uncertain about how her husband will respond if she chooses MRM. She is also concerned that she will not feel feminine after MRM but states that she is very frightened about anything less because a friend died of metastatic breast cancer.

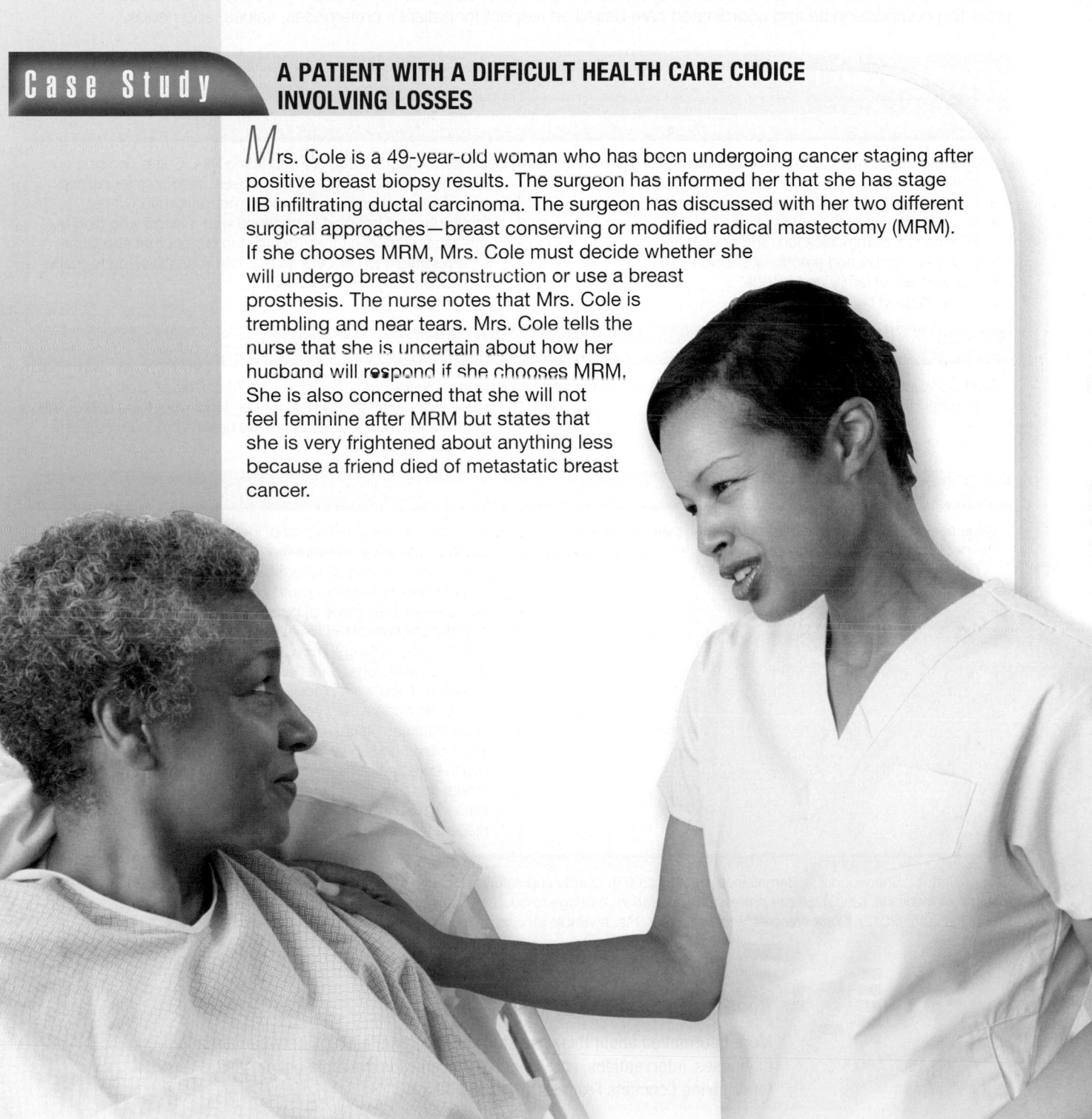

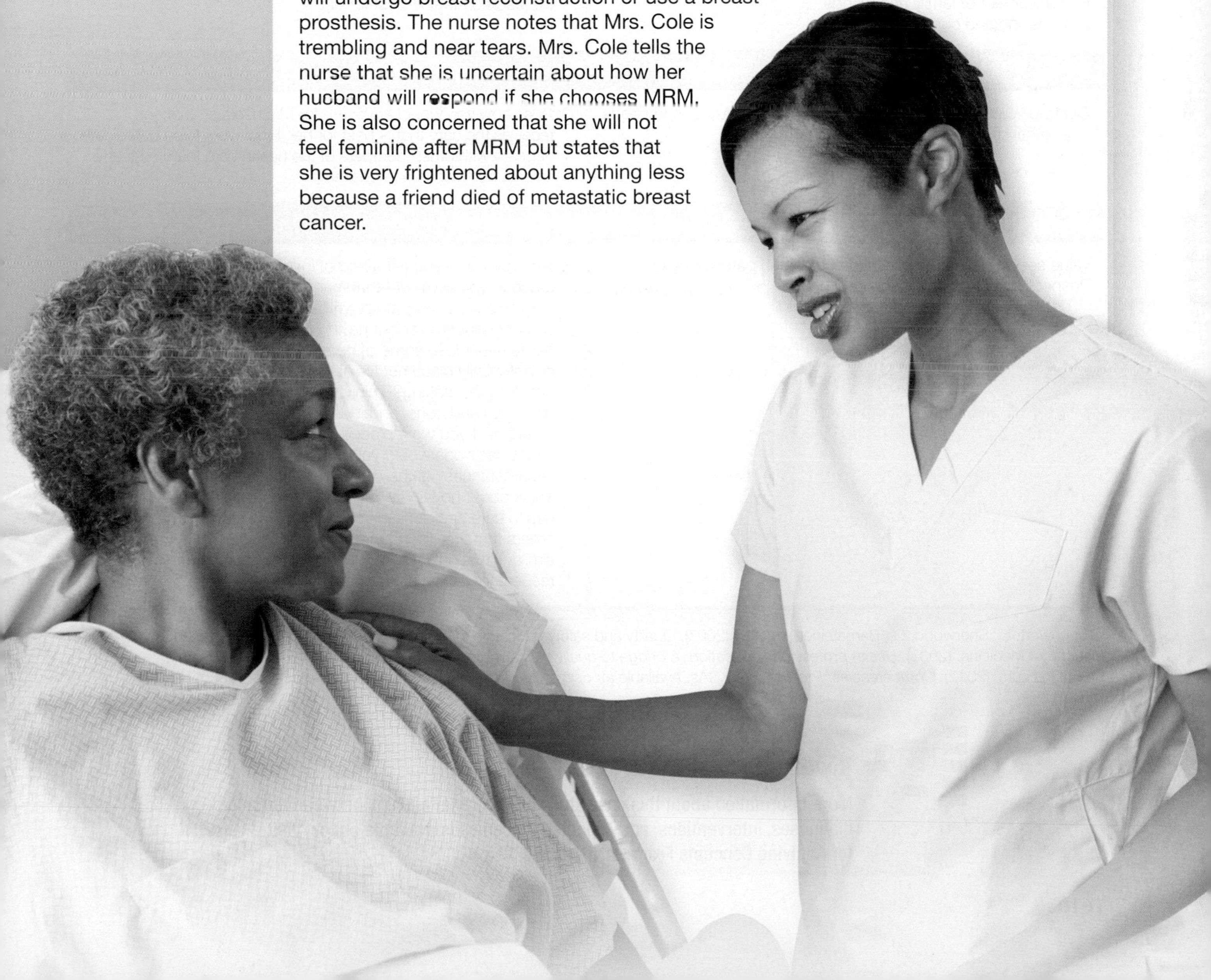

QSEN Competency Focus: *Patient-Centered Care*

The complexities inherent in today's health care system challenge nurses to demonstrate integration of specific interdisciplinary core competencies. These competencies are aimed at ensuring the delivery of safe, quality patient care (Institute of Medicine, 2003). The concepts from the Quality and Safety Education for Nurses (QSEN) Institute (2012) provide a framework for the knowledge, skills, and attitudes (KSAs) required for nurses to demonstrate competency in these key areas, which include *patient-centered care, interdisciplinary teamwork and collaboration, evidence-based practice, quality improvement, safety,* and *informatics.*

Patient-Centered Care Definition: Recognize the patient or designee as the source of control and full partner in providing compassionate and coordinated care based on respect for patient's preferences, values, and needs.

RELEVANT PRE-LICENSURE KSAs	APPLICATION AND REFLECTION
Knowledge	
Integrate understanding of multiple dimensions of patient-centered care: • Patient/family/community preferences, values • Coordination and integration of care • Information, communication, and education • Physical comfort and emotional support • Involvement of family and friends • Transition and continuity	Describe the factors that are impacting Mrs. Cole's decision on which surgical approach she chooses, including her perception of her husband's preferences, her perception of her femininity, and her past experience with a friend who died of metastatic breast cancer. Identify the factors that should be incorporated into her care so that she is empowered to make the decision that is best for her.
Skills	
Communicate patient values, preferences, and expressed needs to other members of health care team.	Discuss how you would advocate for Mrs. Cole so that she receives the care that is best for her. How would you coordinate her care with other members of the health care team so that it reflects her wishes and desires?
Attitudes	
Value seeing health care situations "through patient's eyes." Respect and encourage individual expression of patient values, preferences, and expressed needs.	Reflect on a time you felt afraid of losing something of importance to you. How did that fear affect your decision making? How might you empathize with Mrs. Cole, who verbalizes to you her concerns about having to make a choice that forces her to either lose some of her self-perceived feminine identity or potentially place her life at risk? How might this impact her decision making? Think about how you might weigh your own pros and cons, and identify what you think your decision would be if you were placed in a similar situation. What different decisions would you and others make, given the same circumstances? Is there one "right" or "wrong" decision? Think about how individual preferences and values affect how health care decisions are made. Can you respect decisions of others that are different from decisions you would make in a similar situation? Why is it important for nurses to demonstrate respect for patients' autonomous decision making?

Cronenwett, L., Sherwood, G., Barnsteiner, J., et al. (2007). Quality and safety education for nurses. *Nursing Outlook*, *55*(3), 122–131.
Institute of Medicine. (2003). *Health professions education: A bridge to quality*. Washington, DC: National Academies Press.
QSEN Institute. (2012). *Competencies: Prelicensure KSAs*. Available at: qsen.org/competencies/pre-licensure-ksas

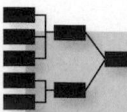

Read More About This Case

More information about this case study and the relationships between nursing diagnoses, interventions, and expected outcomes is available online. Visit thePoint for Applying Concepts From NANDA-I, NIC, and NOC.

Chapter 56

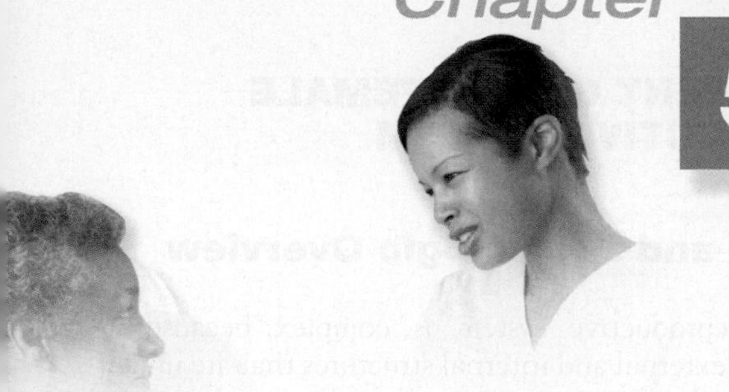

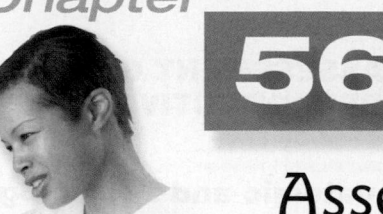

Assessment and Management of Female Physiologic Processes

Learning Objectives

On completion of this chapter, the learner will be able to:

1 Describe the anatomy and physiology of the female reproductive system.
2 Describe approaches to effective assessment of female physiologic processes.
3 Identify the diagnostic examinations and tests used to determine alterations in female reproductive function, and describe the nurse's role before, during, and after these examinations and procedures.
4 Identify common female physiologic processes and related nursing implications.
5 Describe methods of contraception and implications for health care and education.
6 Describe the causes and management of infertility.
7 Use the nursing process to plan for care of the patient with an ectopic pregnancy.
8 Develop an education plan for women who are approaching or have completed menopause.

Glossary

adnexa: the fallopian tubes and ovaries
amenorrhea: absence of menstrual flow
androgens: hormones produced by the ovaries and adrenals that affect many aspects of female health, including follicle development, libido, oiliness of hair and skin, and hair growth
cervix: bottom (inferior) part of the uterus that is located in the vagina
corpus luteum: site within a follicle that changes after ovulation to produce progesterone
cystocele: weakness of the anterior vaginal wall that allows the bladder to protrude into the vagina
dysmenorrhea: painful menstruation
dyspareunia: difficult or painful sexual intercourse
endometrial ablation: procedure performed through a hysteroscope in which the lining of the uterus is burned away or ablated to treat abnormal uterine bleeding
endometrium: mucous membrane lining of the uterus
estrogens: several hormones produced in the ovaries that develop and maintain the female reproductive system
follicle-stimulating hormone (FSH): hormone released by the pituitary gland to stimulate estrogen production and ovulation
fornix: upper part of the vagina
fundus: body of the uterus
graafian follicle: cystic structure that develops on the ovary as ovulation begins
hymen: tissue that covers the vaginal opening partially or completely before vaginal penetration
hysteroscopy: an endoscopic procedure performed using a long telescopelike instrument inserted through the cervix to diagnose uterine problems
introitus: perineal opening to the vagina
luteal phase: stage in the menstrual cycle in which the endometrium becomes thicker and more vascular
luteinizing hormone (LH): hormone released by the pituitary gland that stimulates progesterone production
menarche: beginning of menstrual function
menopause: permanent cessation of menstruation resulting from the loss of ovarian follicular activity
menstruation: sloughing and discharge of the lining of the uterus if conception does not take place
ovaries: almond-shaped reproductive organs that produce eggs at ovulation and play a major role in hormone production
ovulation: discharge of a mature ovum from the ovary
perimenopause: the period immediately prior to menopause and the first year after menopause
progesterone: hormone produced by the corpus luteum that prepares the uterus for receiving the fertilized ovum
proliferative phase: stage in the menstrual cycle before ovulation when the endometrium increases
rectocele: weakness of the posterior vaginal wall that allows the rectal cavity to protrude into the submucosa of the vagina
secretory phase: stage of the menstrual cycle in which the endometrium becomes thickened, becomes more vascular, and ovulation occurs
uterine prolapse: the cervix and uterus descend into the lower vagina

Nurses who work with women need an understanding of the physical, developmental, psychological, and sociocultural influences on women's health, as well as health practices. It is necessary to consider how medications and diseases specifically affect women. In addition, women's sexuality is complex and often affected by many factors, and related issues need careful evaluation and treatment.

ROLE OF NURSES IN WOMEN'S HEALTH

As their presence in the labor market continues to increase, women face challenges in their roles, lifestyles, and family patterns. Furthermore, they encounter environmental hazards and stress, prompting greater attention on health and health-promoting practices. As a result, many women are taking a greater interest in and responsibility for their own health and health care. Because nurses in all settings encounter women with health care needs, they need a solid understanding of the unique issues related to women's health to provide optimal care.

The Affordable Care Act will require private insurers to cover preventative health care services for women without cost sharing such as copayments, deductibles, or co-insurance (Kaiser Family Foundation, 2012). Evidence-based screening and counseling for depression, diabetes, cholesterol, obesity, various cancers, human immunodeficiency virus (HIV), and sexually transmitted infections (STIs) are also covered (Kaiser Family Foundation, 2012). A range of women's services will also be covered without cost sharing. These will include annual well woman visits, support for breast-feeding, counseling and screening for intimate partner violence, all U.S. Food and Drug Administration (FDA)-approved contraception methods, and mammograms (Kaiser Family Foundation, 2012).

In recent years, research has focused on the disparity in health care between men and women. Walsh and colleagues (2009) identified areas that would benefit with improvement in women's health—for example, fewer women than men are treated with lipid-lowering medications such as 3-hydroxy-3-methylglutaryl coenzyme A (HMG-CoA) reductase inhibitors (i.e., statins) to prevent cardiovascular disease (CVD). One study reported that only 24% of eligible women reported current statin use (Spatz, Canavan, & Desai, 2009). More aggressive management of CVD risk factors to actualize treatment goals is thus warranted.

Women have other unique health care needs. Women with diabetes have a higher mortality rate than men with diabetes and are at much higher risk for CVD (Walsh et al., 2009). A meta-analysis looked at the duration of vasomotor symptoms during menopause, finding that vasomotor symptoms peaked 1 year after the last menses, lasted 4 years for most women, and were often treated with hormonal therapy to relieve symptoms; however, the safe duration of this therapy is unknown (Politi, Shlentz, & Col, 2008).

ASSESSMENT OF THE FEMALE REPRODUCTIVE SYSTEM

Anatomic and Physiologic Overview

The female reproductive system is complex because it involves many external and internal structures that are under hormonal control.

Anatomy of the Female Reproductive System

The female reproductive system consists of external and internal pelvic structures. Other anatomic structures that affect the female reproductive system include the hypothalamus and pituitary gland of the endocrine system. The female breast is included in discussions of reproductive health and is covered in Chapter 57.

External Genitalia

The female external genitalia are composed of a variety of tissue types starting with the mons pubis, which is a thick pad of adipose tissue that covers the symphysis pubis and cushions it during intercourse (Fig. 56-1). Moving downward are two thick folds of connective tissue covered with pubic hair known as the labia majora, which extend from the mons pubis to the perineum. The labia majora cover the oval-shaped area known as the vestibule, where the labia minora originate. The labia minora are two narrow folds of hairless skin, beginning at the clitoris and extending to the fourchette. This area is highly vascular and rich in nerve supply and glands that lubricate the vulva, the collective name for the external genitalia. The labia minora join at the top to form the prepuce, a hoodlike structure that partially covers the clitoris. The clitoris, an erectile organ located beneath the pubic arch, consists of a shaft and glans. It secretes smegma, a pheromone (olfactory erotic stimulant), and is sensitive to touch and temperature. Below the clitoris is the urinary meatus, the external opening of the female urethra that has a slit appearance. Below

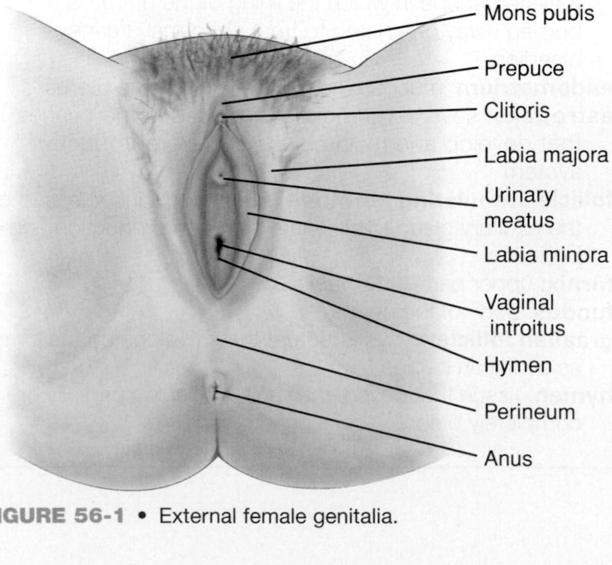
Mons pubis
Prepuce
Clitoris
Labia majora
Urinary meatus
Labia minora
Vaginal introitus
Hymen
Perineum
Anus

FIGURE 56-1 • External female genitalia.

the meatus is the introitus (vaginal opening). On each side of the introitus are Bartholin glands, which secrete mucus through tiny ducts that lie within the labia minora and are external to the hymen (membrane that encircles the introitus). The vestibule is bounded by the clitoris, fourchette, and labia minora and contains the urethral meatus. Skene's gland, located within the urethral meatus, produce mucus for lubrication (Link, 2011).

The hymen opening varies widely among women, and the size of the opening is an unreliable indicator of sexual experience (Link, 2011). The fourchette is located in the midline below the vaginal opening where the labia majora and labia minor merge. The perineum is the area between the vagina and the rectum or anus; it is skin-covered muscular tissue (Link, 2011).

A number of muscles support the external genitalia. Support for the pelvic organs is from the deep muscle layer known as the levator ani. This muscle, which forms most of the pelvic diaphragm, is composed of the iliococcygeus, pubococcygeus, and the puborectalis muscles. It supports the organs of reproduction and provides elasticity of the pelvic floor. When viewed from above, it looks like joined cupped hands. Its main function is to provide support to the organs of the pelvis when increased pressure from coughing and sneezing occurs. When this muscle group is contracted, the pelvic floor is lifted upward, supporting continence (Marecki & Seo, 2010). Muscles that contribute to the strength of the pelvic floor may be damaged during childbirth (Marecki & Seo, 2010). The bulbocavernosus, ischiocavernosus, and transverse perineal muscles encircle and support the vagina and urethra as well as the anal sphincter.

Internal Reproductive Structures

The internal structures consist of the vagina, uterus, ovaries, and fallopian or uterine tubes (Fig. 56-2).

Vagina

The vagina, a tubular-shaped canal lined with glandular mucous membrane, is 7.5 to 10 cm (3 to 4 in) long and extends upward and backward from the vulva to the cervix. It is thin walled and can be distended during birth. It is highly vascular and has little sensation. Anterior to it are the bladder and the urethra, and posterior to it lies the rectum. The anterior and posterior walls of the vagina normally touch each other. The **fornix** (the upper part of the vagina) surrounds the **cervix** (the inferior part of the uterus) (Link, 2011).

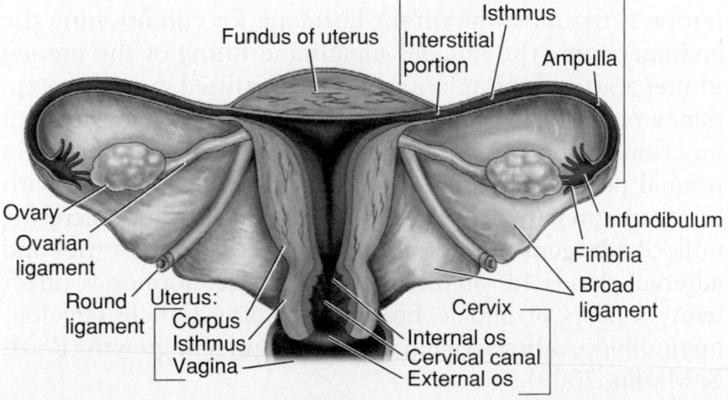
Fallopian tube
Isthmus
Fundus of uterus
Interstitial portion
Ampulla
Ovary
Ovarian ligament
Infundibulum
Fimbria
Broad ligament
Round ligament
Uterus:
Corpus
Isthmus
Vagina
Cervix
Internal os
Cervical canal
External os

FIGURE 56-2 • Internal female reproductive structures.

Uterus

The uterus, a pear-shaped, muscular organ, is about 7.5 cm (3 in) long and 5 cm (2 in) wide at its upper part. Its walls are about 1.25 cm (0.5 in) thick. The size of the uterus varies, depending on parity (number of pregnancies), size of the infants, and uterine abnormalities (e.g., fibroids, which are a type of tumor that may distort the uterus). A nulliparous woman (one who has not completed a pregnancy to the stage of fetal viability) usually has a smaller uterus than a multiparous woman (one who has completed two or more pregnancies to the stage of fetal viability). The uterus lies posterior to the bladder and is held in position by several ligaments. The round ligaments extend anteriorly and laterally to the internal inguinal ring and down the inguinal canal, where they blend with the tissues of the labia majora. The broad ligaments are folds of peritoneum extending from the lateral pelvic walls and enveloping the fallopian tubes. The uterosacral ligaments extend posterior to the sacrum.

The uterus has four parts: the cervix, fundus, corpus, and isthmus. The cervix is the opening to the uterus and projects into the vagina. A larger upper part, the **fundus**, is the rounded portion above the insertion of the fallopian tubes. The corpus or body is the main portion of the uterus located between the fundus and the isthmus. The isthmus is referred to as the lower uterine segment during pregnancy and joins the corpus to the cervix.

The cervix is divided into two portions. The portion above the site of attachment of the cervix to the vaginal vault is known as the supravaginal portion; the portion below the attachment site that protrudes into the vagina is known as the vaginal portion. The cervix is composed of fibrous connective tissue. The diameter varies from 2 to 5 cm depending on the childbearing history. The length is usually 2.5 to 3 cm in the nonpregnant woman. The vaginal portion is smooth, firm, and doughnut shaped with a visible central opening referred to as the external os. The external os is round before the first birth and is often slitlike in shape after childbirth. The internal os is the opening in the cervix to the uterine cavity. In response to cyclic hormones, the cervix produces mucus, which is an important factor in fertility awareness. The vaginal surface of the cervix is covered with squamous epithelium, a rapid cellular growth site of cervical cancer and precancerous changes.

The uterine wall has three layers. The endometrium, the innermost layer, is highly vascular and responds to hormone stimulation to prepare to receive the developing ovum. It sloughs if pregnancy does not occur, resulting in menstruation; if pregnancy occurs, it sloughs after delivery. The myometrium, the middle layer, is made up of several smooth muscle layers. The outer layer of the myometrium is composed of longitudinal fibers, chiefly in the fundus, which provide the power to expel the fetus. The middle layer of the myometrium is composed of fibers interlaced with blood vessels in a figure-eight pattern called a *living ligature* because it contracts after childbirth to help control blood loss (Smith & Chelnow, 2012). The inner layer of the myometrium is composed of circular fibers that are concentrated around the internal cervical os to help keep the cervix closed during pregnancy. The other layer of the uterus is composed of parietal peritoneum, which covers most of the uterus. From here, the oviducts or fallopian (or uterine) tubes extend outward, and their lumina are internally continuous with the uterine cavity (Porth & Matfin, 2009). The fallopian tubes or oviducts are the passageway for eggs from the ovary to the uterus. They curve around each ovary and are attached to the uterine fundus. Fallopian tubes are about 10 cm in length and are comprised of four parts. The infundibulum, the most distal portion, is covered in fimbriae, whose wave-like motion helps to pull the egg into the tube. The ampulla is usually the site of fertilization of the egg or ovum. The fallopian tube then narrows from its 0.6 cm diameter in the isthmus, ending in the narrowest part of the interstitial portion, which opens into the uterine cavity. The fallopian tube also secretes nutrients for growth and development of the ovum after fertilization while it passes down the tube to the uterus.

Ovaries

The **ovaries** lie behind the broad ligaments and behind and below the fallopian tubes. They are almond-shaped bodies about 3 cm (1.2 in) long. At birth, they contain thousands of tiny egg cells, or ova. The ovaries and the fallopian tubes together are referred to as the **adnexa**.

Function of the Female Reproductive System

Ovulation

At puberty (usually between 11 and 13 years of age), the ova begin to mature and menstrual cycles begin (Link, 2011). In the follicular phase, an ovum enlarges into a cystic structure called a **graafian follicle** until it reaches the surface of the ovary, where transport occurs. The ovum (or oocyte) is discharged into the peritoneal cavity. This periodic discharge of matured ovum is referred to as **ovulation**. The ovum usually finds its way into the fallopian tube, where it is carried to the uterus. If it is penetrated by a spermatozoon, the male reproductive cell, a union occurs and conception takes place. After the discharge of the ovum, the cells of the graafian follicle undergo a rapid change. Gradually, they become yellow and produce **progesterone**, a hormone that prepares the uterus for receiving the fertilized ovum. Ovulation usually occurs 2 weeks prior to the next menstrual period.

Menstrual Cycle

The menstrual cycle is a complex process involving the reproductive and endocrine systems. The ovaries produce steroid hormones, predominantly estrogens and progesterone. Several different **estrogens** are produced by the ovarian follicle, which consists of the developing ovum and its surrounding cells. The most potent of the ovarian estrogens is estradiol. Estrogens are responsible for developing and maintaining the female reproductive organs and the secondary sex characteristics associated with the adult female. Estrogens play an important role in breast development and in monthly cyclic changes in the uterus (Porth & Matfin, 2009).

Progesterone is also important in regulating the changes that occur in the uterus during the menstrual cycle. It is

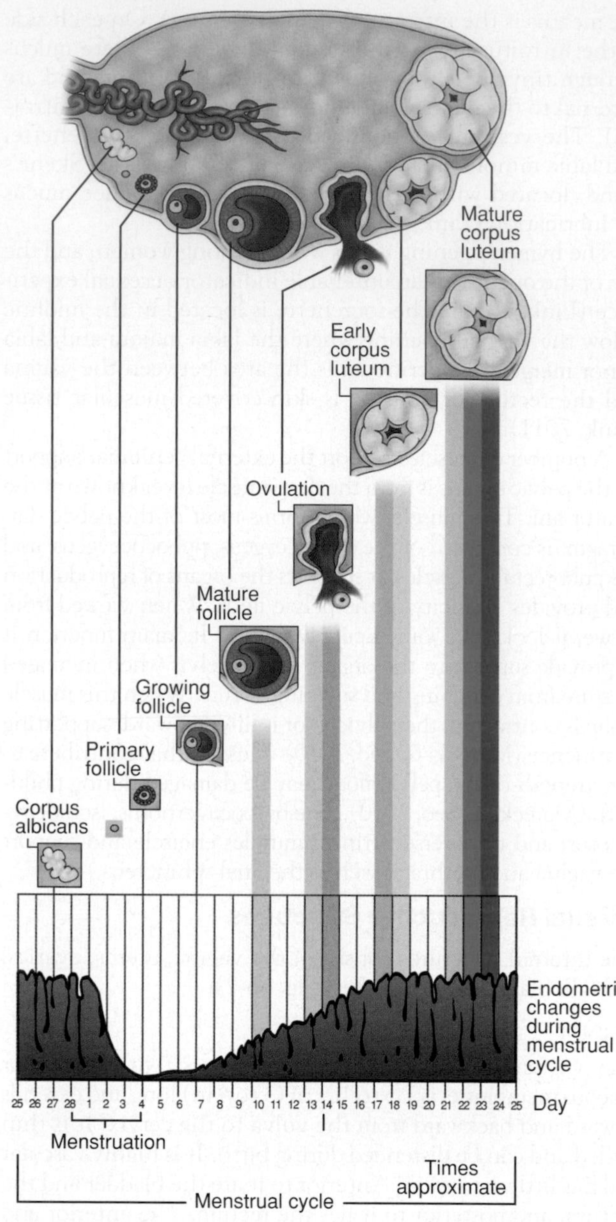

Mature corpus luteum
Early corpus luteum
Ovulation
Mature follicle
Growing follicle
Primary follicle
Corpus albicans
Endometrial changes during menstrual cycle
25 26 27 28 1 2 3 4 5 6 7 8 9 10 11 12 13 14 15 16 17 18 19 20 21 22 23 24 25 Day
Menstruation
Times approximate
Menstrual cycle

FIGURE 56-3 • The menstrual cycle and corresponding changes in the endometrium.

secreted by the **corpus luteum** (site within a follicle) or the ovarian follicle after the ovum has been released. Progesterone is the most important hormone for conditioning the **endometrium** (the mucous membrane lining of the uterus) in preparation for implantation of a fertilized ovum. If pregnancy occurs, the progesterone secretion becomes largely a function of the placenta and is essential for maintaining a normal pregnancy. In addition, progesterone, working with estrogen, prepares the breast for producing and secreting milk. **Androgens** are hormones produced by the ovaries and adrenal glands in small amounts. These hormones affect many aspects of female health, including follicle development, libido, oiliness of hair and skin, and hair growth (Porth & Matfin, 2009).

Two gonadotropic hormones are released by the pituitary gland: **follicle-stimulating hormone (FSH)** and **luteinizing**

TABLE 56-1 Hormonal Changes During the Menstrual Cycle

(Times approximate) Phase	Menstrual	Follicular	Ovulation	Luteal	Premenstrual
Days	1 2 3 4 5	6 7 8 9 10 11 12 13	14 15 16	17 18 19 20 21 22 23 24 25	26 27 28 1 2
Ovary	Degenerating corpus luteum; beginning follicular development	Growth and maturation of follicle	Ovulation	Active corpus luteum	Degenerating corpus luteum
Estrogen Production	Low	Increasing	High	Declining, then a secondary rise	Decreasing
Progesterone Production	None	Low	Low	Increasing	Decreasing
FSH Production	Increasing	High, then declining	Low	Low	Increasing
LH Production	Low	Low, then increasing	High	High	Decreasing
Endometrium	Degeneration and shedding of superficial layer. Coiled arteries dilate, then constrict again.	Reorganization and proliferation of superficial layer	Continued growth	Active secretion and glandular dilation; highly vascular; edematous	Vasoconstriction of coiled arteries; beginning degeneration

FSH, follicle-stimulating hormone; LH, luteinizing hormone.

hormone (LH). FSH is primarily responsible for stimulating the ovaries to secrete estrogen. LH is primarily responsible for stimulating progesterone production. Feedback mechanisms, in part, regulate FSH and LH secretion. For example, elevated estrogen levels in the blood inhibit FSH secretion but promote LH secretion, whereas elevated progesterone levels inhibit LH secretion. In addition, gonadotropin-releasing hormone (GnRH) from the hypothalamus affects the rate of FSH and LH release.

The secretion of ovarian hormones follows a cyclic pattern that results in changes in the uterine endometrium and in menstruation (Fig. 56-3, Table 56-1). This cycle is typically 28 days in length, but there are many normal variations (from 21 to 42 days). In the **proliferative phase** at the beginning of the cycle (just after menstruation), FSH output increases, and estrogen secretion is stimulated. This causes the endometrium to thicken and become more vascular. In the **secretory phase** near the middle portion of the cycle (day 14 in a 28-day cycle), LH output increases, and ovulation occurs. Under the combined stimulus of estrogen and progesterone, the endometrium reaches the peak **luteal phase**, in which the endometrium is thick and highly vascular. In the luteal phase, which begins after ovulation, progesterone is secreted by the corpus luteum.

If the ovum is fertilized, estrogen and progesterone levels remain high, and the complex hormonal changes of pregnancy follow. If the ovum has not been fertilized, FSH and LH output diminishes; estrogen and progesterone secretion falls; the ovum disintegrates; and the endometrium, which has become thick and congested, becomes hemorrhagic. The product, menstrual flow, consisting of old blood, mucus, and endometrial tissue, is discharged through the cervix and into the vagina. After the menstrual flow stops, the cycle begins again; the endometrium proliferates and thickens from estrogenic stimulation, and ovulation recurs (Porth & Matfin, 2009).

Menopausal Period

The menopausal period marks the end of a woman's reproductive capacity. It usually occurs between 45 and 52 years of age but may occur as early as 42 or as late as 55; the median age is 51 years. Perimenopause precedes this and can begin as early as 35 years of age. Physical, emotional, and menstrual changes may occur, and this transition offers another opportunity for health promotion and disease prevention education and counseling. Menopause is not a pathologic phenomenon but a normal part of aging and maturation. Menstruation ceases, and because the ovaries are no longer active, the reproductive organs become smaller. No more ova mature; therefore, no ovarian hormones are produced. (An earlier menopause may occur if the ovaries are surgically removed or are destroyed by radiation or chemotherapy or because of an unknown etiology.) Multifaceted changes also occur throughout the woman's body. These changes are neuroendocrine, biochemical, and metabolic and are related to normal maturation or aging (Table 56-2).

Assessment

A nurse who is obtaining information from a woman for the health history and performing physical assessment is in an ideal position to discuss the woman's general health issues, health promotion, and health-related concerns. Relevant topics include fitness, nutrition, cardiovascular risks, health screening, sexuality, menopause, abuse, health risk behaviors, emotional well-being, and immunizations. Selected

TABLE 56-2 Age-Related Changes in the Female Reproductive System

Changes	Physiologic Effects	Signs and Symptoms
Cessation of ovarian function and decreased estrogen production	Decreased ovulation	Decreased/loss of ability to conceive; increased infertility
	Onset of menopause	Irregular menses with eventual cessation of menses
	Vasomotor instability and hormonal fluctuations	Hot flashes or flushing; night sweats, sleep disturbances; mood swings; fatigue
	Decreased bone formation	Bone loss and increased risk for osteoporosis and osteoporotic fractures; loss of height
	Decreased vaginal lubrication	Dyspareunia, resulting in lack of interest in sex
	Thinning of urinary and genital tracts	Increased risk for urinary tract infection
	Increased pH of vagina	Increased incidence of inflammation (atrophic vaginitis) with discharge, itching, and vulvar burning
	Thinning of pubic hair and shrinking of labia	
Relaxation of pelvic musculature	Prolapse of uterus, cystocele, rectocele	Dyspareunia, incontinence, feelings of perineal pressure

health screening and counseling issues are summarized in Chart 56-1.

Health History

In addition to the general health history, the nurse asks about past illnesses and experiences specific to a woman's health. Data should be collected about the following:

- Menstrual history (including menarche, length of cycles, duration and amount of flow, presence of cramps or pain, bleeding between periods or after intercourse, bleeding after menopause)
- Pregnancies (number of pregnancies, outcomes of pregnancies)
- Exposure to medications (diethylstilbestrol [DES], immunosuppressive agents, others)

Chart 56-1 Selected Health Screening and Counseling Issues for Women

Ages 19–39 Years

Sexuality and Reproductive Issues

Annual pelvic examination to begin at 21 years of age
Annual clinical breast examination
Contraceptive options
High-risk sexual behaviors

Health and Risk Behaviors

Hygiene
Injury prevention
Nutrition
Exercise patterns
Risk for abuse, maltreatment, and neglect
Use of tobacco, drugs, and alcohol
Life stresses
Immunizations

Diagnostic Testing*

Pap smear every 3 years after onset of sexual intercourse
Sexually transmitted infection screening as indicated

Ages 40–64 Years

Sexuality and Reproductive Issues

Annual pelvic examination
Annual clinical breast examination
Contraceptive options
High-risk sexual behaviors
Menopausal concerns

Health and Risk Behaviors

Hygiene
Bone loss and injury prevention
Nutrition
Exercise patterns
Risk for abuse, maltreatment, and neglect
Use of tobacco, drugs, and alcohol
Life stresses
Immunizations

Diagnostic Testing*

Pap smear every 2 to 3 years after 3 consecutive negative tests if no history of cervical abnormalities, human immunodeficiency virus infection, or diethylstilbestrol exposure
Annual mammography
Cholesterol and lipid profile
Colorectal cancer screening beginning at age 50 years
Bone mineral density testing
Thyroid-stimulating hormone testing
Hearing and eye examinations

Age 65 Years and Older

Sexuality and Reproductive Issues

Annual pelvic examination
Annual clinical breast examination
High-risk sexual behaviors

Health and Risk Behaviors

Hygiene
Injury prevention, falls
Nutrition
Exercise patterns
Risk for abuse, maltreatment, and neglect
Use of tobacco, drugs, and alcohol
Life stresses
Immunizations

Diagnostic Testing*

Mammography
Cholesterol and lipid profile
Colorectal cancer screening
Bone mineral density testing
Thyroid-stimulating hormone testing
Hearing and eye examinations

*Each individual's risks (family history, personal history) influence the need for specific assessments and their frequency.
Adapted from Smith, R. A, Cokkinides, V., & Brawley, O. W. (2012). Cancer screening in the United States, 2012. *CA: A Cancer Journal for Clinicians*, *62*(2), 129–142; and Eliopoulos, C. (2010). *Gerontological nursing* (7th ed.). Philadelphia: Lippincott Williams & Wilkins.

Chart 56-2

GENETICS IN NURSING PRACTICE

Reproductive Processes

Various female reproductive disorders are influenced by genetic factors. Some examples are:

- Hereditary breast or ovarian cancer syndromes
- Hereditary nonpolyposis colon cancer syndrome (risk for uterine cancer)
- Müllerian aplasia
- 21-Hydroxylase deficiency (female masculinization)
- Turner syndrome (45,XO)

Nursing Assessments

Family History Assessment

- Assess family history for other family members with similar reproductive problems/abnormalities.
- Inquire about ethnic background (e.g., Ashkenazi Jewish populations and hereditary breast/ovarian cancer mutations).
- Inquire about relatives with other cancers, including early-onset ovarian, uterine, renal, prostate cancers.

Patient Assessment

- In females with delayed puberty or primary amenorrhea, assess for clinical features of Turner syndrome (short stature, webbing of the neck, widely spaced nipples).
- Assess for other congenital anomalies in females with Müllerian defect, including renal and vertebral anomalies.

Management Issues Specific to Genetics

- Inquire whether genetic testing (deoxyribonucleic acid [DNA] chromosomal, metabolic) has been carried out on affected family member(s).
- If indicated, refer for further genetic counseling and evaluation so that family members can discuss inheritance, risk to other family members and availability of genetic testing, and gene-based interventions.
- Offer appropriate genetic information and resources.
- Assess patient's understanding of genetic information.
- Provide support to families with newly diagnosed gene-related reproductive disorders.
- Participate in management and coordination of care of patients with genetic conditions and individuals predisposed to develop or pass on a genetic condition.

Genetics Resources

American Cancer Society: offers general information about cancer and support resources for families, www.cancer.org

Gene Clinics: a listing of common genetic disorders with up-to-date clinical summaries, genetic counseling, and testing information, www.geneclinics.org

National Cancer Institute: current information about cancer research, treatment, and resources for health care providers, individuals, and families, www.nci.nih.gov

See Chapter 8, Chart 8-6 for additional genetics resources.

- **Dysmenorrhea** (pain with menses), **dyspareunia** (pain with intercourse), pelvic pain
- Symptoms of vaginitis (i.e., odor or itching)
- Problems with urinary function, including frequency, urgency, and incontinence
- Bowel problems
- Sexual history
- STIs and methods of treatment
- Current or previous sexual abuse or physical abuse
- Past surgery or other procedures on reproductive tract structures (including female genital mutilation [FGM] or female circumcision)
- Chronic illness or disability that may affect health status, reproductive health, need for health screening, or access to health care
- Presence or family history of a genetic disorder (Chart 56-2 presents information about genetic reproductive disorders.)

In collecting data related to reproductive health, the nurse can educate the patient about normal physiologic processes, such as menstruation and menopause, and assess possible abnormalities. Many problems experienced by young or middle-age women can be corrected easily. However, if they are not treated, they may result in anxiety and health problems. Issues related to sexuality and sexual function are typically more often brought to the attention of the gynecologic or women's health care provider than other health care providers; however, nurses caring for all women should consider these issues part of routine health assessment.

Sexual History

A sexual assessment includes both subjective and objective data. Health and sexual histories, physical examination findings, and laboratory results are all part of the database. The purpose of a sexual history is to obtain information that provides a picture of a woman's sexuality and sexual practices and to promote sexual health. The sexual history may enable a patient to discuss sexual matters openly and to discuss sexual concerns with an informed health professional. This information can be obtained after the gynecologic–obstetric or genitourinary history is completed. By incorporating the sexual history into the general health history, the nurse can move from areas of lesser sensitivity to areas of greater sensitivity after establishing initial rapport.

Taking the sexual history becomes a dynamic process reflecting an exchange of information between the patient and the nurse and provides the opportunity to clarify myths and explore areas of concern that the patient may not have felt comfortable discussing in the past. In obtaining a sexual history, the nurse must not assume the patient's sexual preference until clarified. When asking about sexual health, the nurse also cannot assume that the patient is married or unmarried. Asking a patient to label herself as single, married, widowed, or divorced may be considered by some women as inappropriate. Asking about a partner or about current meaningful relationships may be a less offensive way to initiate a sexual history.

The PLISSIT (*p*ermission, *l*imited *i*nformation, *s*pecific *s*uggestions, *i*ntensive *t*herapy) model of sexual assessment and intervention may be used to provide a framework for nursing interventions (Annon, 1976). The assessment begins by introducing the topic and asking the woman for permission to discuss issues related to sexuality with her.

The nurse can begin by explaining the purpose of obtaining a sexual history (e.g., "I ask all my patients about their sexual health. May I ask you some questions about this?"). History taking continues by inquiring about present sexual activity and sexual orientation (e.g., "Are you currently having sex? With a man, a woman, or both?"). Inquiries about possible sexual dysfunction may include, "Are you having any problems related to your current sexual performance?" Such problems may be related to medication, life changes, disability, or the onset of physical or emotional illness. A patient can be asked about her thoughts on what is causing the current problem (Weber & Kelley, 2010).

Information about sexual function can be introduced during the health history (see Chapter 5). By initiating an assessment about sexual concerns, the nurse communicates to the patient that issues about changes or problems in sexual functioning are valid health issues, which provides a safe environment for discussing these sensitive topics. Young women may be apprehensive about having irregular periods, may be concerned about STIs, or may need contraception. They may want information about using tampons, emergency contraception, or issues related to pregnancy. Perimenopausal women may have concerns about irregular menses. Menopausal women may be concerned about vaginal dryness and discomfort with intercourse. Women of any age may have concerns about relationships, sexual satisfaction, orgasm, or masturbation.

Risk of STIs can be assessed by asking about the number of sexual partners in the past year or in the patient's lifetime. An open-ended question related to the patient's need for further information should be included (e.g., "Do you have any questions or concerns about your sexual health?"). Women can be advised that intercourse should never be painful; pain should be investigated by a care provider. They should also be encouraged to talk openly about their sexual feelings with their partner; in an intimate relationship, feelings are facts.

Female Genital Mutilation or Cutting

FGM, also known as female genital cutting, refers to the partial or total removal of the external female genitalia or other injury to female organs. FGM is illegal in the United States and is condemned by some organizations (e.g., World Health Organization [WHO], Amnesty International). An increasing number of women entering the American health care system underwent FGM before coming to the United States (Turner, 2007), and others have undergone FGM since their arrival in this country. Because FGM can affect sexual function, menstrual hygiene, and bladder function, the possibility of FGM must be considered in the sexual history, particularly in women from cultures and countries where the practice is more common (i.e., some sub-Saharan African nations).

Nurses caring for patients who have undergone FGM need to be sensitive, empathetic, knowledgeable, culturally competent, and nonjudgmental (Turner, 2007). Respect for others' health beliefs, practices, and behaviors, as well as recognition of the complexity of issues involved, is crucial. The nurse should use terminology with which the woman is familiar; *cutting* is usually a more acceptable term than *mutilation*. Speculums are not used in some developing countries; the function of this instrument should be explained and an appropriately sized speculum used to examine women who have experienced FGM.

Family Violence

Family violence is a broad term that includes child abuse, elder abuse, and intimate partner violence or abuse of women and men. Approximately 95% to 99% of intimate partner violence is committed by men against women and can be emotional, physical, sexual, or economic in nature (Wong, Tiwari, Fong, et al., 2011). These women are encountered daily in nursing practice. The lifetime prevalence of intimate partner violence is between 10% and 69% (Wong et al., 2011).

Intimate partner violence involves fear of one partner by another and repeated physical or sexual assault in a context of coercive control and, more broadly, emotional degradation, threats, and intimidation. Violence is rarely a one-time occurrence in a relationship; it usually continues and escalates in severity. This is an important point to emphasize when a woman states that her partner has hurt her but has promised to change. Perpetrators can change their behavior, although not without extensive counseling and motivation. If a woman states that she is being hurt, sensitive care is required (Chart 56-3).

By knowing about this major public health problem, being alert to abuse-related problems, and learning how to elicit information from women about abuse, maltreatment, and neglect in their lives, nurses can intervene in a problem that might otherwise go undetected and thus save lives by making women safer through education and support. Asking each woman about violence in her life in a safe environment (i.e., a private room with the door closed) is part of a comprehensive assessment (More information on family violence, abuse, and neglect, including sexual assault and rape, can be found in Chapter 72; see Chart 72-8 for information on questions to ask when assessing for abuse, maltreatment, and neglect.).

No specific signs or symptoms are diagnostic of abuse; however, nurses may see an injury that does not fit the account of how it happened (e.g., a bruise on the side of the upper arm after "I walked into a door"). Manifestations of abuse, maltreatment, and neglect may involve suicide attempts, drug and alcohol abuse, frequent emergency department visits, vague pelvic pain, somatic complaints, and depression. However, there may be no obvious signs or symptoms. Women in abusive situations have higher levels of depression and often report that they "do not feel well," possibly due to the stress or fear and anticipation of impending abuse (Chart 56-4).

Incest and Childhood Sexual Abuse

More than one in five women have experienced incest or childhood sexual abuse, and nurses frequently encounter women who have been sexually traumatized. It has been reported that female survivors of sexual abuse have more health problems and undergo more surgery than women who were not victims of abuse. Victims of childhood sexual abuse are reported to experience more chronic depression, post-traumatic stress disorder, morbid obesity, marital instability, gastrointestinal problems, and headaches, as well as use health care services more frequently than people who were not victims. In women, chronic pelvic pain is often associated with physical violence, emotional neglect, and sexual abuse in childhood (Schuiling & Likis, 2013). Women who have experienced rape or sexual abuse may be very anxious about

Chart 56-3 **Managing Reported Abuse, Maltreatment, and Neglect**

1. Reassure the woman that she is not alone. *Rationale: Women often believe that they are alone in experiencing abuse, maltreatment, and neglect at the hands of their partners.*
2. Express your belief that no one should be hurt, that abuse is the fault of the batterer and is against the law. *Rationale: Doing so lets the woman know that no one deserves to be abused and that she has not caused the abuse.*
3. Assure the woman that her information is confidential, although it does become part of her medical record. *If children are suspected of being abused or are being abused, the law requires that this be reported to the authorities.* Some states require reporting of spousal or partner abuse. Domestic violence agencies and medical and nursing groups disagree with this policy and are trying to have it changed. Serious opposition is based on the fact that reporting does not and cannot currently guarantee a woman's safety and may place her in more danger. It may also interfere with a patient's willingness to discuss her personal life and concerns with health care providers. This places a serious barrier in the way of comprehensive nursing care. If nurses are in doubt about laws on reporting abuse, they need to check with their local or state domestic violence agency. *Rationale: Women are often afraid that their information will be reported to the police or protective services and their children may be taken away.*
4. Document the woman's statement of abuse and take photographs of any visible injuries if written formal consent has been obtained. (Emergency departments usually have a camera available if one is not on the nursing unit.) *Rationale: Doing so provides documentation of injuries that may be needed for later legal or criminal proceedings.*
5. Provide education. *Rationale: The following options may be lifesaving for the woman and her children:*
   - Inform the woman that shelters are available to ensure safety for her and her children. (Lengths of stay in shelters vary by state but are often up to 2 months. Staff often assist with housing, jobs, and the emotional distress that accompanies the breakup of the family.) Provide list of shelters.
   - Inform the woman that violence gets worse, not better.
   - If the woman chooses to go to a shelter, let her make the call.
   - If the woman chooses to return to the abuser, remain nonjudgmental and provide information that will make her safer than she was before disclosing her situation.
   - Make sure that the woman has a 24-hour hotline telephone number that provides information and support (Spanish translation and a device for the deaf are also available), police number, and 911.
   - Assist her to set up a safety plan in case she decides to return home. (A safety plan is an organized plan for departure with packed bags and important papers hidden in a safe spot.)

pelvic examinations, labor, pelvic or breast irradiation, or any treatment or examination that involves hands-on treatment or requires removal of clothing. Nurses should be prepared to offer support and referral to psychologists, community resources, and self-help groups.

Health Issues in Women With Disabilities

Approximately 20% of women have disabilities and encounter physical, architectural, and attitudinal barriers that may limit their full participation in society. Women with disabilities may experience stereotyping and increased risk of abuse, maltreatment, and neglect. They have reported that others, including health care providers, often equate them with their disability. Studies have shown that women with disabilities receive less primary health care and preventive health screening than other women, often because of access problems and health care providers who focus on the causes of disability rather than on health issues that are of concern

Chart 56-4 **NURSING RESEARCH PROFILE**

Depression and Intimate Partner Violence

Wong, J. Y., Tiwari, A., Fong, S. Y., et al. (2011). Depression among women experiencing intimate partner violence in a Chinese community. *Nursing Research*, *60*(1), 58–65.

Purpose

Depression is a significant mental health impact of abuse, but there has been little research on the empirical evidence of the factors associated with depression among Chinese women who have been abused. The purpose of this study was to identify the factors associated with a higher level of depression among Chinese women suffering abuse.

Design

A total of 200 women with a mean age of 38 years were included in this cross-sectional study. Demographic data were collected and the tools used included the Chinese Abuse Screen, Chinese Beck Depression Inventory Version II, Revised Conflict Tactics Scale, and Interpersonal Support Evaluation List 12. Interviews were audiotaped and then transcribed verbatim. Data were analyzed using structured multiphase regression analysis.

Results

Approximately 11% of the women reported intimate partner violence, and 75% reported severe levels of depression. Factors associated with higher rates of depression were low educational level ($p = .038$), immigration ($p = .0250$), financial support from friends and relatives ($p = .006$), and chronic psychological abuse ($p < .001$). One protective factor against depression was the perception of social support ($p < .001$).

Nursing Implications

Nurses need to be aware that an overwhelming number of women who have been abused have moderate or severe levels of depression. Furthermore, there is a need to increase the awareness of the detrimental mental health impact that abuse has on women. Screening for depression needs to take place with women who have suffered abuse and provisions made for social support, particularly among women who have immigrated, to minimize depression.

to all women (Sudduth & Linton, 2011). To address these issues, the health history must include questions about barriers to health care encountered by women with disabilities and the effect of their disability on their health status and health care.

Other issues to be addressed are identified in Chart 56-5. If a patient has hearing loss, vision loss, or another disability that affects communication, it may be necessary to obtain the assistance of an interpreter or to establish another method of communication. Nurses assessing women with disabilities may require additional time and the assistance of others to be certain that accurate information is obtained in a sensitive and unhurried manner. Women with disabilities may have had previous negative experiences with health care providers (Sharts-Hopko, Smeltzer, Ott, et al., 2010; Smeltzer, Sharts-Hopko, Ott, et al., 2007), and it is important that nurses provide them with knowledgeable and sensitive care. Nurses should take every opportunity to discuss health-related issues and concerns of women with disabilities. (See Chapter 9 for further discussion of health care of patients with disabilities.)

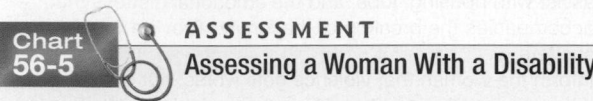

Chart 56-5 ASSESSMENT

Assessing a Woman With a Disability

Health History

Address questions directly to the woman herself rather than to people accompanying her. Ask about:

- Self-care limitations resulting from her disability (ability to feed and dress self, the use of assistive devices, transportation requirements, other assistance needed)
- Sensory limitations (lack of sensation, low vision, deaf or hard of hearing)
- Accessibility issues (ability to get to health care provider, transfer to examination table, accessibility of office/clinic of health care provider, previous experiences with health care providers, health screening practices; her understanding of physical examination)
- Cognitive or developmental changes that affect understanding
- Limitations secondary to disability that affect general health issues and reproductive health and health care
- Sexual function and concerns (those of all women and those that may be affected by the presence of a disabling condition)
- Menstrual history and menstrual hygiene practices
- Physical, sexual, or psychological abuse (including abuse by care providers; abuse by neglect, withholding or withdrawing assistive devices or personal or health care)
- Presence of secondary disabilities (i.e., those resulting from the patient's primary disability: pressure ulcers, spasticity, osteoporosis, etc.)
- Health concerns related to aging with a disability

Physical Assessment

Provide instructions directly to the woman herself rather than to people accompanying her; provide written or audiotaped instructions.

Ask the woman what assistance she needs for the physical examination and provide assistance if needed:

- Undressing and dressing
- Providing a urine specimen
- Standing on scale to be weighed (provide alternative means of obtaining weight if she is unable to stand on scale)
- Moving on and off the examination table
- Assuming, changing, and maintaining positions

Consider the fatigue experienced by the woman during a lengthy examination and allow rest.

Provide assistive devices and other aids/methods needed to allow adequate communication with the patient (interpreters, signers, large-print written materials).

Complete examination that would be indicated for any other woman; having a disability is *never* justification for omitting parts of the physical examination, including the pelvic examination.

Lesbians and Bisexual Women

Women who identify themselves as lesbian or bisexual may have concerns about disclosure and confidentiality, discriminatory attitudes, and treatment (ACOG, 2012b) (Chart 56-6). There are no known cases of woman-to-woman transmission of HIV, but lesbian and bisexual women are at higher risks of other STIs (ACOG, 2012b). Research on breast cancer and ovarian cancer is also lacking in this population. Current recommendations for cancer screening for all women should be followed, including mammography, colorectal cancer

Chart 56-6

Health Care for Women Who Are Lesbians

Lesbians can generally be defined as women who have sex with or primary emotional partnerships with women, but there is no universally accepted definition; variability exists in relationships and sexual preferences. Lesbians are found in every ethnic group and socioeconomic class. They can be single, celibate, or divorced and are seen in all age groups, including teens and seniors.

Lesbians have often encountered insensitivity in health care encounters. When they are asked if they are sexually active and respond affirmatively, contraception is immediately urged as health care providers may assume incorrectly that they practice heterosexual intercourse. Similar to many other marginalized groups of women, they often feel invisible and underuse health care. Whether heterosexual or homosexual, nurses need to consider lesbianism within the continuum of human sexual behavior and need to use gender-neutral questions and terms that are nonjudgmental and accepting. Lesbian teens are at risk for suicide and sexually transmitted infections (STIs). Many lesbians do participate in heterosexual activity and often consider themselves at low risk for STIs. Because human papillomavirus, herpes infections, and other organisms implicated in STIs are transmitted by secretions and contact, lesbians may need information on STIs and contraception. If sex toys are used and not cleaned, pelvic infections can occur.

Lesbians have lower health screening rates than other women. They are at high risk for cancer, heart disease, depression, and alcohol abuse. They may have a higher body mass index, may bear fewer or no children, and often have fewer health preventive screenings than heterosexual women. These factors may increase the risk of colon, endometrial, ovarian, and breast cancer, as well as cardiovascular disease and diabetes. Adolescent lesbians are at risk for smoking and suicide/depression. Nurses need to understand the unique needs of this population and provide appropriate and sensitive care.

Adapted from American College of Obstetricians and Gynecologists. (2012b). Committee Opinion No. 525: Health care for lesbians and bisexual women. *Obstetrics and Gynecology*, *119*(1), 1077–1084.

screening, hormone therapy, and osteoporosis screening (ACOG, 2012b). Some research has reported that lesbians abuse alcohol and drugs to a greater degree than heterosexual women because their social venues may contribute to alcohol and drug use. In addition, youth who self-identify as lesbian, gay, or bisexual or who lack support from parents and families may experience increased mental health issues (e.g., depression and substance abuse) (ACOG, 2012b).

Gerontologic Considerations

Older women are functioning at various levels across the health spectrum; some function at a high level in their jobs or families, whereas others may be very ill. Nurses need to be prepared to care for older women who may be bright, energetic, and ambitious or who are coping with multiple family crises, including their own health issues, as well as for those who are experiencing a life-altering or life-threatening health problem. Older women are at risk for several conditions, including diabetes, dyslipidemia, hypertension, and thyroid disease, all of which have symptoms that may be dismissed as typical aging. Nurses can help prevent morbidity and mortality from these conditions by encouraging women to obtain regular health screenings (Thakur & Supiano, 2007). In addition, knowledge related to heart disease prevention, pharmacology, diet, signs of dementia or cognitive decline, fall prevention, osteoporosis prevention, gynecologic and breast cancers, and sexuality are important for providing high-level nursing care. Health disparities, cultural competency, and end-of-life issues also need to be considered.

Physical Assessment

Periodic examinations and routine cancer screening are important for all women. Annual breast and pelvic examinations are important for all women 21 years or older and for those who are sexually active, regardless of age (ACOG, 2012a; Hawkins, Roberto-Nichols, & Stanley-Haney, 2012). Patients deserve understanding and support because of the emotional and physical considerations associated with gynecologic examinations. Women may be embarrassed by the usual questions asked by a gynecologist or women's health care provider. Because gynecologic conditions are of a personal and private nature to most women, such information is shared only with those directly involved in patient care.

The approach to the gynecologic examination needs to be systematic and thorough (Carcio & Secor, 2010). The nurse can alleviate feelings of anxiety with explanations and education (Chart 56-7). It may be helpful to emphasize that a pelvic examination should not usually be uncomfortable. Before the examination begins, the patient is asked to empty her bladder and to provide a urine specimen if urine tests are part of the total assessment. Voiding ensures patient comfort and eases the examination because a full bladder can make palpation of pelvic organs uncomfortable for the patient and difficult for the examiner.

Positioning

The supine lithotomy position is used most commonly (Weber & Kelley, 2010). This position offers several advantages:

- It is more comfortable for some women.
- It allows better eye contact between patient and examiner.

Chart 56-7 PATIENT EDUCATION
The Pelvic Examination

A pelvic examination includes assessment of the appearance of the vulva, vagina, and cervix and the size and shape of the uterus and ovaries to ensure reproductive health and absence of illness. Telling the patient the following information should make the examination proceed more smoothly:

- You may have a feeling of fullness or pressure during the examination, but you should not feel pain. It is important to relax, because if you are very tense, you may feel discomfort.
- It is normal to feel uncomfortable and apprehensive.
- A narrow, warmed speculum will be inserted to visualize the cervix.
- A Papanicolaou (Pap) smear will be obtained, if indicated, and should not be uncomfortable.
- You may watch the examination with a mirror if you choose.
- The examination usually takes no longer than 5 minutes.
- Draping will be used to minimize exposure and reduce embarrassment.

Adapted from Weber, J., & Kelley, J. (2010). *Health assessment in nursing* (4th ed.). Philadelphia: Lippincott Williams & Wilkins.

- It may provide an easier means for the examiner to carry out the bimanual examination.
- It enables the woman to use a mirror to see her anatomy (if she chooses) to visualize any conditions that require treatment or to learn about using certain contraceptive methods.

Inspection

After the patient is prepared, the examiner inspects the external genitalia by looking at the labia majora and minora, noting the epidermal tissue of the labia majora; the skin fades to the pink mucous membrane of the vaginal introitus. Lesions of any type (e.g., venereal warts, pigmented lesions [melanoma]) are evaluated. In the nulliparous woman, the labia minora come together at the opening of the vagina. In a woman who has delivered children vaginally, the labia minora may gape and vaginal tissue may protrude.

Trauma to the anterior vaginal wall during childbirth may have resulted in incompetency of the musculature, and a bulge caused by the bladder protruding into the submucosa of the anterior vaginal wall (**cystocele**) may be seen. Childbirth trauma may also have affected the posterior vaginal wall, producing a bulge caused by rectal cavity protrusion (**rectocele**). The cervix and uterus may descend under pressure through the vaginal canal and be seen at the introitus (**uterine prolapse**). To identify such protrusions, the examiner asks the patient to "bear down."

The introitus should be free of superficial mucosal lesions. The labia minora may be separated by the fingers of the gloved hand and the lower part of the vagina palpated. In women who have not had vaginal intercourse, a hymen of variable thickness may be felt circumferentially within the vaginal opening. The hymeneal ring usually permits the insertion of one finger. Rarely, the hymen totally occludes the vaginal entrance (imperforate hymen).

Speculum Examination

The bivalve speculum, either metal or plastic, is available in many sizes (Weber & Kelley, 2010). Metal specula are cleaned and sterilized between patients; plastic specula are for one-time use. The speculum can be warmed with a heating pad or warm water to make insertion more comfortable for the patient.

The speculum is gently inserted into the posterior portion of the introitus and slowly advanced to the top of the vagina; this should not be painful or uncomfortable for the woman. The speculum is then slowly opened. In the metal types, a set-screw of the thumb rest is tightened; in the plastic types, a clip is locked to hold the blades in place (Weber & Kelley, 2010).

Inspecting the Cervix

The cervix is inspected. In nulliparous women, the cervix usually is 2 to 3 cm wide and smooth. In women who have borne children, the cervix may have a laceration, usually transverse, giving the cervical os a "fishmouth" appearance. Epithelium from the endocervical canal may have grown onto the surface of the cervix, appearing as beefy-red surface epithelium circumferentially around the os. Occasionally, the cervix of a woman whose mother took DES during pregnancy has a hooded appearance (a peaked aspect superiorly or a ridge of tissue surrounding it); this is evaluated by colposcopy when identified.

Malignant changes may not be obviously differentiated from the rest of the cervical mucosa. Small, benign cysts may appear on the cervical surface. Called *nabothian (retention) cysts*, these are usually bluish or white and are a normal finding after childbirth (Weber & Kelley, 2010). A polyp of endocervical mucosa may protrude through the os and usually is dark red. Polyps can cause irregular bleeding; they are rarely malignant and usually are removed easily in an office or clinic setting. A carcinoma may appear as a cauliflowerlike growth (Weber & Kelley, 2010). Bluish coloration of the cervix is a sign of early pregnancy (Chadwick sign).

Obtaining Pap Smears and Other Samples

A Pap smear is obtained by rotating a small spatula at the os, followed by a cervical brush rotated in the os. The material obtained is spread on a glass slide and sprayed/fixed immediately, or inserted into liquid (ThinPrep) (Chart 56-8). A small broomlike device can also be used to obtain specimens for the Pap smear.

A specimen of any purulent material appearing at the cervical os is obtained for culture. A sterile applicator is used to obtain the specimen, which is immediately placed in an appropriate medium for transfer to a laboratory. In a patient who has a high risk for infection, routine cultures for gonococcal and chlamydial organisms are recommended because of the high incidence of both diseases and the complications of pelvic infection, fallopian tube damage, and subsequent infertility.

Vaginal discharge, which may be normal or may result from vaginitis, may be present. Table 56-3 summarizes the characteristics of vaginal discharge found in different conditions.

Inspecting the Vagina

The vagina is inspected as the examiner withdraws the speculum. It is smooth in young girls and thickens after puberty,

Chart 56-8 GUIDELINES

Obtaining an Optimal Pap Smear

Equipment Needed

• Speculum • Gloves • Slide, spatula, and cytobrush or thin prep kit

Implementation

Nursing Action	Rationale
1. Do not obtain a Papanicolaou (Pap) smear if the woman is menstruating or has other frank bleeding (*Exception:* High suspicion of neoplasia).	1. Blood obscures a proper reading of cells.
2. If performing more than one test, obtain the Pap smear first.	2. By performing the Pap smear first, the chance of a bloody smear is avoided.
3. Label the frosted end of the slide with the patient's name in pencil, or label ThinPrep Pap bottle.	3. Ink may rub off or blur. Labeling with a pencil prevents improper identification.
4. Put on gloves before gently inserting the unlubricated speculum. (Speculum may be moistened with warm water.)	4. Gloves provide protection and warm water prevents discomfort. Lubricants may obscure cells on Pap smear.
5. Place the longer end of the Ayre spatula in the cervical canal, and rotate it in a full circle to obtain a sample from the exocervix. Spread the material obtained onto the Pap smear slide.	5. This technique obtains a sampling of the exocervix and squamocolumnar junction.
6. Insert a cytobrush 2 cm into the cervical canal and rotate 180 degrees. Roll the brush onto the Pap smear slide. (With ThinPrep Pap smears, the brushings are not spread onto a slide. The spatula and brush are placed in a bottle of fixative and swirled.)	6. This technique obtains a sampling of the endocervical cells and may sample cells from the squamocolumnar junction if it is high in the canal.
7. In women who have had a hysterectomy for a gynecologic cancer, use a cotton applicator moistened with normal saline solution to obtain a sampling of cells from the vaginal cuff or posterior vagina. Women who have had a hysterectomy for benign conditions do not require frequent Pap smears.	7. Normal saline solution prevents drying, which makes interpretation difficult for the cytologist and prevents absorption of cells into the cotton, increasing the yield on the slide.
8. Immediately spray the slide, or, if a ThinPrep, swirl the brush and spatula in the solution.	8. Exposure to light or air causes distortion of cells.

TABLE 56-3 Characteristics of Vaginal Discharge

Cause of Discharge	Symptoms	Odor	Consistency/Color
Physiologic	None	None	Mucus/white
Candida species infection	Itching, irritation	Yeast odor or none	Thin to thick, curdlike/white
Bacterial vaginosis	Odor	Fishy, often noticed after intercourse	Thin/grayish or yellow
Trichomonas species infection	Irritation, odor	Malodorous	Copious, often frothy/yellow-green
Atrophic	Vulvar or vaginal dryness	Occasional mild malodor	Usually scant and mucoid/may be blood tinged

with many rugae (folds) and redundancy in the epithelium. In menopausal women, the vagina thins and has fewer rugae because of decreased estrogen.

Bimanual Palpation

To complete the pelvic examination, the examiner performs a bimanual examination (ACOG, 2012a). The examiner should avoid startling the patient by telling her that she will be touched in private areas (Carcio & Secor, 2010). The fingers are advanced vertically along the vaginal canal, and the vaginal wall is palpated. Any firm part of the vaginal wall may represent old scar tissue from childbirth trauma but may also require further evaluation.

Cervical Palpation

The cervix is palpated and assessed for its consistency, mobility, size, and position. The normal cervix is uniformly firm but not hard. Softening of the cervix is a finding in early pregnancy. Hardness and immobility of the cervix may reflect invasion by a neoplasm. Pain on gentle movement of the cervix is called a *positive chandelier sign* (positive cervical motion tenderness; recorded as +CMT) and usually indicates a pelvic infection.

Uterine Palpation

To palpate the uterus, the examiner places the opposite hand on the abdominal wall halfway between the umbilicus and the pubis and presses firmly toward the vagina. Movement of the abdominal wall causes the body of the uterus to descend, and the organ becomes freely movable between the hand used to examine the abdomen and the fingers of the hand used to examine the pelvis. Uterine size, mobility, and contour can be estimated through palpation. Fixation of the uterus in the pelvis may be a sign of endometriosis or malignancy.

The body of the uterus is normally twice the diameter and twice the length of the cervix, curving anteriorly toward the abdominal wall. Some women have a retroverted or retroflexed uterus, which tips posteriorly toward the sacrum, whereas others have a uterus that is neither anterior nor posterior and is described as midline.

Adnexal Palpation

The right and left adnexal areas are palpated to evaluate the fallopian tubes and ovaries. The fingers of the hand examining the pelvis are moved first to one side, then to the other, while the hand palpating the abdominal area is moved correspondingly to either side of the abdomen and downward. The adnexa (ovaries and fallopian tubes) are trapped between the two hands and palpated for an obvious mass, tenderness, and mobility. Commonly, the ovaries are slightly tender, and the patient is informed that slight discomfort on palpation is normal.

Vaginal and Rectal Palpation

Bimanual palpation of the vagina and cul-de-sac is accomplished by placing the index finger in the vagina and the middle finger in the rectum. To prevent cross-contamination between the vaginal and rectal orifices, the examiner puts on new gloves. A gentle movement of these fingers toward each other compresses the posterior vaginal wall and the anterior rectal wall and assists the examiner in identifying the integrity of these structures. During this procedure, the patient may sense an urge to defecate. The nurse assures the patient that this is unlikely to occur. Ongoing explanations are provided to reassure and educate the patient about the procedure.

Gerontologic Considerations

Yearly examinations aid early identification of reproductive system disorders in aging women (Eliopoulos, 2010). Nurses play an important role in encouraging all women to have an annual gynecologic examination. Women older than 65 years may choose to stop cervical cancer screening if they have had a hysterectomy or have had three normal cytology tests and no abnormal test in the past 10 years (Smith, Cokkinides, & Brawley, 2012). (See Table 15-3 in Chapter 15).

Perineal pruritus is abnormal in older women and should be evaluated because it may indicate a disease process (diabetes or malignancy). Vulvar dystrophy (a thickened or whitish discoloration of tissue) may be visible, and biopsy is needed to rule out abnormal cells. Topical cortisone and hormone creams may be prescribed for symptomatic relief.

With relaxing pelvic musculature, uterine prolapse and relaxation of the vaginal walls can occur. Appropriate evaluation and surgical repair can provide relief if the patient is a candidate for surgery. After surgery, the patient should know that tissue repair and healing may require more time with aging. Pessaries (latex devices that provide support) are often used if surgery is contraindicated or before surgery to see if surgery can be avoided. They are fitted by a health care provider and may reduce discomfort and pressure. The use of a pessary requires the patient to have routine gynecologic examinations to monitor for irritation or infection (Eliopoulos, 2010). The patient must be assessed for allergy prior to insertion of a latex pessary (see Chapter 57 for details about pessaries).

Diagnostic Evaluation

A wide range of diagnostic studies may be performed in the management of female physiologic processes. The nurse should educate the patient about the purpose, what to expect, and any possible side effects related to these examinations prior to testing. The nurse should be aware of contraindications, potential complications, and trends in results. Trends provide information about disease progression as well as the patient's response to therapy.

Cytologic Test for Cancer (Pap Smear)

The Papanicolaou (Pap) smear is used to detect cervical cancer. Cervical secretions are gently removed from the cervical os and may be transferred to a glass slide and fixed immediately by spraying with a fixative or immersed in solution. If the Pap smear reveals atypical cells, the liquid method allows for human papillomavirus (HPV) testing (see Chapter 57 for further discussion of HPV, a common STI that can cause venereal warts or cervical cancer). Guidelines for obtaining an optimal Pap smear are described in Chart 56-8.

Terminology used to describe findings includes the following categories:

- No abnormal or atypical cells
- Atypical squamous cells of undetermined significance
- Inflammatory reactions and microbes identified
- Positive deoxyribonucleic acid (DNA) test for HPV
- Precancerous and cancerous lesions of the cervix identified

The patient may incorrectly assume that an abnormal Pap smear signifies cancer. If the Pap smear (liquid immersion method) shows atypical cells and no high-risk HPV types, the next Pap smear is performed in 1 year. If a specific infection is causing inflammation, it is treated appropriately, and the Pap smear is repeated. If the repeat Pap smear reveals atypical squamous cells with high-risk HPV types, colposcopy may be indicated. Pap smears that indicate precancerous lesions should be repeated in 4 to 6 months and colposcopy performed if the lesion has not resolved. Patients with Pap smears that indicate cancerous lesions require prompt colposcopy (Fischbach & Dunning, 2009).

If the Pap smear results are abnormal, prompt notification, evaluation, and treatment are crucial. Notification of patients is often the responsibility of nurses in a women's health care practice or clinic. Pap smear follow-up is essential because it can prevent cervical cancer. Interventions are tailored to meet the needs and health beliefs of the particular patient (Ochoa-Frongia, Thompson, Lewis-Kelly, et al., 2012). Intensive telephone counseling, tracking systems, brochures, videos, and financial incentives have all been used to encourage follow-up. The nurse provides clear explanations and emotional support along with a carefully designed setting-specific follow-up protocol designed to meet the needs of the patient.

Colposcopy and Cervical Biopsy

If the cervical cytology screening result requires evaluation, a colposcopy is performed. The colposcope is an instrument with a magnifying lens that allows the examiner to visualize the cervix and obtain a sample of abnormal tissue for analysis (Fischbach & Dunning, 2009). Nurse practitioners and gynecologists require special training in this diagnostic technique.

After inserting a speculum and visualizing the cervix and vaginal walls, the examiner applies acetic acid to the cervix. Subsequent abnormal findings that indicate the need for biopsy include leukoplakia (white plaque visible before applying acetic acid), acetowhite tissue (white epithelium after applying acetic acid), punctation (dilated capillaries occurring in a dotted or stippled pattern), mosaicism (a tile-like pattern), and atypical vascular patterns. If biopsy specimens show precancerous cells, the patient usually requires cryotherapy, laser therapy, or a cone biopsy (excision of an inverted tissue cone from the cervix).

Cryotherapy and Laser Therapy

Cryotherapy (freezing cervical tissue with nitrous oxide) and laser treatment are used in the outpatient setting. Cryotherapy may result in cramping and occasional feelings of faintness (vasovagal response). A watery discharge is normal for a few weeks after the procedure as the cervix heals; however, excessive bleeding, pain, or fever should be reported to the provider (Fischbach & Dunning, 2009).

Cone Biopsy and Loop Electrosurgical Excision Procedure

If endocervical curettage findings indicate abnormal changes or if the lesion extends into the canal, the patient may undergo a cone biopsy. This can be performed surgically or with a procedure called *loop electrosurgical excision procedure* (LEEP), which uses a laser beam (Fischbach & Dunning, 2009).

Usually performed in the outpatient setting, LEEP is associated with a high success rate in removal of abnormal cervical tissue. The gynecologist excises a small amount of cervical tissue, and the pathologist examines the borders of the specimen to determine if disease is present. A patient who has received anesthesia for a surgical cone biopsy is advised to rest for 24 hours after the procedure and to leave any vaginal packing in place until it is removed (usually the next day). The patient is instructed to report any excessive bleeding.

The nurse or primary provider provides guidelines regarding postoperative sexual activity, bathing, and other activities. Because open tissue may be potentially exposed to HIV and other pathogens, the patient is cautioned to avoid intercourse until healing is complete and verified at follow-up.

Endometrial (Aspiration) Biopsy

Endometrial biopsy, a method of obtaining endometrial tissue, is performed as an outpatient procedure. This procedure is usually indicated in cases of midlife irregular bleeding, postmenopausal bleeding, and irregular bleeding while taking hormone therapy or tamoxifen. A tissue sample obtained through biopsy permits diagnosis of cellular changes in the endometrium.

Women who undergo endometrial biopsy may experience slight discomfort. The examiner may apply a tenaculum (a clamplike instrument that stabilizes the uterus) after the pelvic examination and then inserts a thin, hollow, flexible suction tube (Pipelle or sampler) through the cervix into the uterus.

Findings on aspiration may include normal endometrial tissue, hyperplasia, or endometrial cancer. Simple hyperplasia is an overgrowth of the uterine lining and is usually treated with progesterone. Complex hyperplasia, which refers to overgrowth of cells with abnormal features, is a risk factor for uterine cancer and is treated with progesterone and careful follow-up. Women who are overweight, who are older than 45 years, who have a history of nulliparity and infertility, or who have a family history of colon cancer seem to be at higher risk for hyperplasia. (See Chapter 57 for discussion of endometrial cancer.)

Dilation and Curettage

Dilation and curettage (D&C) may be diagnostic (identifies the cause of irregular bleeding) or therapeutic (often temporarily stops irregular bleeding). The cervical canal is widened with a dilator, and the uterine endometrium is scraped with a curette. The purpose of the procedure is to secure endometrial or endocervical tissue for cytologic examination, to control abnormal uterine bleeding, and as a therapeutic measure for incomplete abortion.

Because D&C is usually carried out under anesthesia and requires surgical asepsis, it is usually performed in the operating room. However, it may take place in the outpatient setting with the patient receiving a local anesthetic supplemented with diazepam (Valium) or midazolam (Versed).

The nurse explains the procedure, preparation, and expectations regarding postoperative discomfort and bleeding. The patient is instructed to void before the procedure. The patient is placed in the lithotomy position, the cervix is dilated with a dilating instrument, and endometrial scrapings are obtained by a curette. A perineal pad is placed over the perineum after the procedure, and excessive bleeding is reported. No restrictions are placed on dietary intake. If pelvic discomfort or low back pain occurs, mild analgesic medications usually provide relief. The primary provider indicates when sexual intercourse may be safely resumed. To reduce the risk of infection and bleeding, most physicians advise no vaginal penetration or use of tampons for 2 weeks.

Endoscopic Examinations

Laparoscopy (Pelvic Peritoneoscopy)

A laparoscopy involves inserting a laparoscope (a tube about 10 mm wide and similar to a small periscope) into the peritoneal cavity through a 2-cm (0.75-in) incision below the umbilicus to allow visualization of the pelvic structures (Fig. 56-4). Laparoscopy may be used for diagnostic purposes (e.g., in cases of pelvic pain when no cause can be found) or treatment. Laparoscopy facilitates many surgical procedures, such as tubal ligation, ovarian biopsy, myomectomy, hysterectomy, and lysis of adhesions (scar tissue that can cause pelvic discomfort). A surgical instrument (intrauterine sound or cannula) may be positioned inside the uterus to permit manipulation or movement during laparoscopy, affording better visualization. The pelvic organs can be visualized after the injection of carbon dioxide intraperitoneally into the cavity. Called *insufflation,* this technique separates the intestines from the pelvic organs. If a patient is undergoing sterilization, the fallopian or uterine tubes may be electrocoagulated, sutured, or ligated and a segment removed for histologic verification (clips are an alternative device for occluding the tubes).

After the laparoscopy is completed, the laparoscope is withdrawn, carbon dioxide is allowed to escape through the

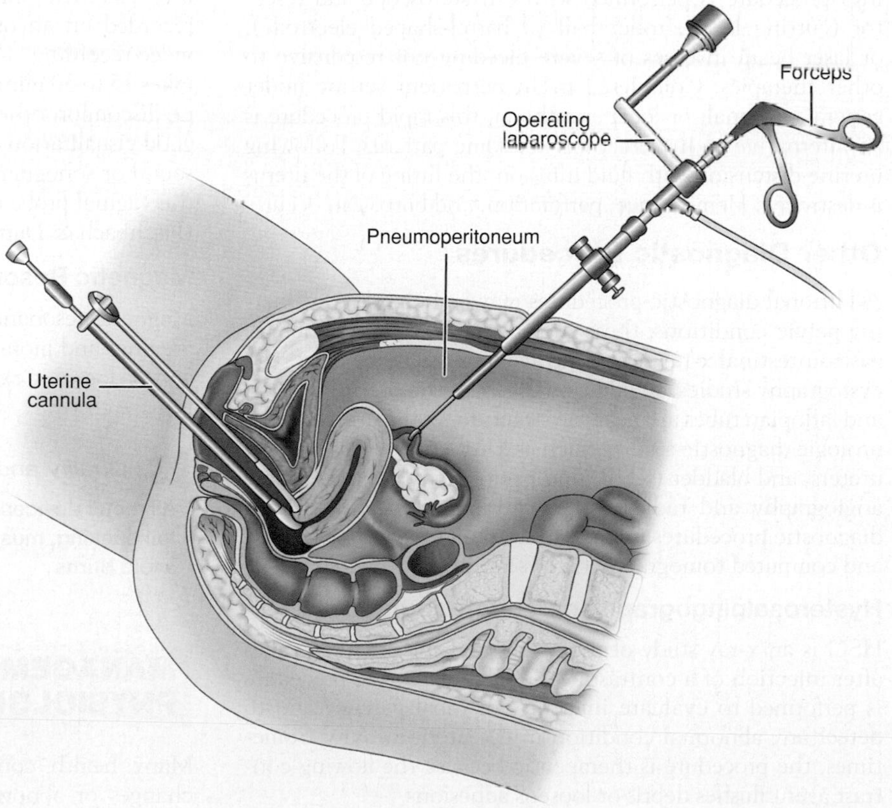
Forceps
Operating laparoscope
Pneumoperitoneum
Uterine cannula

FIGURE 56-4 • Laparoscopy. The laparoscope (*right*) is inserted through a small incision in the abdomen. A forceps is inserted through the scope to grasp the fallopian tube. To improve the view, a uterine cannula (*left*) is inserted into the vagina to push the uterus upward. Insufflation of gas creates an air pocket (pneumoperitoneum), and the pelvis is elevated (note the angle), which forces the intestines higher in the abdomen.

outer cannula, the small skin incision is closed with sutures or a clip, and the incision is covered with an adhesive bandage. The patient is carefully monitored for several hours to detect any untoward signs indicating bleeding (most commonly from vascular injury to the hypogastric vessels), bowel or bladder injury, or burns from the coagulator. These complications are rare, making laparoscopy a cost-effective and safe short-stay procedure. The patient may experience abdominal or shoulder pain related to the use of carbon dioxide gas.

Hysteroscopy

Hysteroscopy (transcervical intrauterine endoscopy) allows direct visualization of all parts of the uterine cavity by means of a lighted optical instrument. The procedure is best performed about 5 days after menstruation ceases, in the estrogenic phase of the menstrual cycle. The vagina and vulva are cleansed, and a paracervical anesthetic block is performed or lidocaine spray is used. The instrument used for the procedure, a hysteroscope, is passed into the cervical canal and advanced 1 or 2 cm under direct vision. Uterine-distending fluid (normal saline solution or dextrose 5% in water) is infused through the instrument to dilate the uterine cavity and enhance visibility. Hysteroscopy, which has few complications, is useful for evaluating endometrial pathology.

Hysteroscopy may be indicated as an adjunct to a D&C and laparoscopy in cases of infertility, unexplained bleeding, retained intrauterine device (IUD), and recurrent early pregnancy loss. Treatment for some conditions (e.g., fibroid tumors) can be accomplished during this procedure, and sterilization may also be performed. Hysteroscopy is contraindicated in patients with cervical or endometrial carcinoma or acute pelvic inflammation.

An **endometrial ablation** (destruction of the uterine lining) procedure is performed with a hysteroscope and resector (cutting loop), roller ball (a barrel-shaped electrode), or laser beam in cases of severe bleeding not responsive to other therapies. Completed in an outpatient setting under general, regional, or local anesthesia, this rapid procedure is an alternative to hysterectomy for some patients. Following uterine distension with fluid infusion, the lining of the uterus is destroyed. Hemorrhage, perforation, and burns can occur.

Other Diagnostic Procedures

Additional diagnostic procedures may be helpful in evaluating pelvic conditions; these include x-rays, barium enemas, gastrointestinal x-ray series, intravenous (IV) urography, and cystography studies. In addition, because the uterus, ovaries, and fallopian tubes are near the structures of the urinary tract, urologic diagnostic studies, such as x-ray study of the kidney, ureters, and bladder (KUB) and pyelography are used, as are angiography and radioisotope scanning, if needed. Other diagnostic procedures include hysterosalpingography (HSG) and computed tomography (CT) scanning.

Hysterosalpingography or Uterotubography

HSG is an x-ray study of the uterus and the fallopian tubes after injection of a contrast agent. The diagnostic procedure is performed to evaluate infertility or tubal patency and to detect any abnormal condition in the uterine cavity. Sometimes, the procedure is therapeutic because the flowing contrast agent flushes debris or loosens adhesions.

Prior to HSG, laxatives and an enema may be administered to evacuate the intestinal tract so that gas shadows do not distort the x-ray findings. A mild sedative or an analgesic agent, such as ibuprofen (Advil, Motrin) may be prescribed. The patient is placed in the lithotomy position and the cervix is exposed with a bivalve speculum. A cannula is inserted into the cervix, and the contrast agent is injected into the uterine cavity and the fallopian tubes. X-rays are taken to show the path and the distribution of the contrast agent.

Some patients experience nausea, vomiting, cramps, and faintness. After the test, the patient is advised to wear a perineal pad for several hours, because the radiopaque contrast agent may stain clothing.

Computed Tomography

CT scans have several advantages over ultrasonography, but they involve radiation exposure and are more costly. They are more effective than ultrasonography for patients who are obese or for patients with a distended bowel. CT scans can also demonstrate a tumor and any extension into the retroperitoneal lymph nodes and skeletal tissue, although they have limited value in diagnosing other gynecologic abnormalities.

Ultrasonography

Ultrasonography (or ultrasound) is a useful adjunct to the physical examination, particularly in obstetric patients or in patients with abnormal pelvic examination findings. It is a simple procedure based on sound wave transmission that uses pulsed ultrasonic waves at frequencies exceeding 20,000 Hz (formerly cycles per second) by way of a transducer placed in contact with the abdomen (abdominal scan) or a vaginal probe (vaginal ultrasound). Mechanical energy is converted into electrical impulses, which in turn are amplified and recorded on an oscilloscope screen while a photograph or video recording of the patterns is taken. The entire procedure takes 15 to 30 minutes and involves no ionizing radiation and no discomfort other than a full bladder, which is necessary for good visualization during an abdominal scan. A vaginal ultrasound or sonogram does not require a full bladder; however, the vaginal probe can cause mild discomfort in some women (Fischbach & Dunning, 2009).

Magnetic Resonance Imaging

Magnetic resonance imaging (MRI) produces patterns that are finer and more definitive than other imaging procedures, and it does not expose patients to radiation. However, MRI is more costly.

Quality and Safety Nursing Alert

All metal devices, including medication skin patches with foil backing, must be removed before MRI is performed to avoid burns.

MANAGEMENT OF FEMALE PHYSIOLOGIC PROCESSES

Many health concerns of women are related to normal changes or abnormalities of the menstrual cycle and may result from women's lack of understanding of the menstrual cycle, developmental changes, and factors that may affect the pattern of the menstrual cycle. Educating women about the menstrual cycle and changes over time is an important aspect of the nurse's role in providing quality care to women. Education should begin early so that menstruation and the lifelong changes in the menstrual cycle can be anticipated and accepted as a normal part of life.

Menstruation

Menstruation, the sloughing and discharge of the lining of the uterus that takes place if conception does not occur, happens about every 28 days during the reproductive years, although normal cycles can vary from 25 to 35 days (Porth & Matfin, 2009) (see Fig. 56-3). The flow usually lasts 4 to 5 days, during which time 50 to 60 mL of blood is lost.

A perineal pad or tampon is generally used to absorb menstrual discharge. Tampons are used extensively. There is no significant evidence of untoward effects from their use, provided that there is no difficulty in inserting them. However, a tampon should not be used for more than 4 to 6 hours, and the lowest absorbency should be used to prevent toxic shock syndrome (Mayo Clinic, 2011). If a tampon is difficult to remove or shreds when removed, less absorbent tampons should be used. If the string breaks or retracts, the woman should squat in a comfortable position, insert one finger into the vagina, try to locate the tampon, and remove it. If she feels uncomfortable attempting this maneuver or cannot remove the tampon, she should consult a gynecologic health care provider promptly.

Psychosocial Considerations

Girls who are approaching **menarche** (the onset of menstruation) should be educated about the normal process of the menstrual cycle before it occurs. Psychologically, it is much healthier and appropriate to refer to this event as a "period" rather than as "being sick." With adequate nutrition, rest, and exercise, most women feel little discomfort, although some report breast tenderness and a feeling of fullness 1 or 2 days before menstruation begins. Others report fatigue and some discomfort in the lower back, legs, and pelvis on the first day and temperament or mood changes. Slight deviations from a usual pattern of daily living are considered normal, but excessive deviation may require evaluation. Regular exercise and a healthy diet have been found to decrease discomfort for some women. Heating pads or nonsteroidal anti-inflammatory drugs (NSAIDs) may be very effective for cramps. For women with excessive cramping or dysmenorrhea, referral to a women's health care provider is appropriate; following evaluation, practitioners may prescribe oral contraceptive agents.

Cultural Considerations

Culture refers to knowledge, beliefs, customs, and values acquired as members of a racial, ethnic, religious, or social group. The United States is becoming more culturally diverse. Various aspects of culture affect many health care encounters, and these encounters can be positive if nurses understand the various cultures of their patients.

Cultural views and beliefs about menstruation differ. Some women believe that it is detrimental to change a pad or tampon too frequently; they think that allowing the discharge to accumulate increases the flow, which is considered desirable. Some women believe they are vulnerable to illness during menstruation. Others believe it is harmful to swim, shower, have their hair "permed," have their teeth filled, or eat certain foods during menstruation. They may also avoid using contraception during menstruation.

In such situations, nurses are in a position to provide women with facts in an accepting and culturally sensitive manner. The objective is to be mindful of these unexpressed, deep-rooted beliefs and to provide the facts with care. Aspects of gynecologic problems cannot always be expressed easily. The nurse needs to convey confidence and openness and to offer facts to facilitate communication. Suggestions to improve care include overcoming language barriers, providing appropriate materials in the patient's language, asking about traditional beliefs and dietary practices, and asking about fears regarding care. Patience, sensitivity, and a desire to learn about other cultures and groups will enhance the nursing care of all women.

Menstrual Disorders

Menstrual disorders may include premenstrual syndrome (PMS); dysmenorrhea; **amenorrhea** (absence of menstrual flow); and excessive bleeding, irregular bleeding, or bleeding between cycles or unrelated to cycles. These disorders need to be discussed with a health care provider and managed individually.

Premenstrual Syndrome

PMS is a cluster of physical, emotional, and behavioral symptoms that are usually related to the luteal phase of the menstrual cycle. PMS is very common, affecting many women at some time in their lives (Tharpe, Farley, & Jordan, 2013) (Chart 56-9).

Clinical Manifestations

Major symptoms of PMS include physical symptoms such as headache, fatigue, low back pain, painful breasts, and a feeling of abdominal fullness. Behavioral and emotional symptoms may include general irritability, mood swings, fear of losing control, binge eating, and crying spells. Symptoms vary widely from one woman to another and from one cycle to the next in the same woman. Great variability is found in the degree of symptoms. Many women are affected to some degree, but some are severely affected. Premenstrual dysphoric disorder (PMDD) is a severe form of PMS with significant severity of symptoms (Hawkins et al., 2012).

Medical Management

Because there is no single treatment or known cure for PMS, it is helpful for women to keep a record of their symptoms so they can anticipate and therefore cope with them. Regular exercise may be helpful. Although women have been advised to avoid caffeine, high-fat foods, and refined sugars, little

Chart 56-9 Causes, Manifestations, and Treatment of Premenstrual Syndrome

Cause

- Unknown; may be related to hormonal changes combined with other factors (diet, stress, and lack of exercise)
- Many women have some symptoms related to menses, but premenstrual syndrome affects 75%–95% of women at some point and is a complex of symptoms that result in dysfunction.

Physical Symptoms

- Fluid retention (e.g., bloating, breast tenderness)
- Headache
- Low back pain

Affective Symptoms

- Depression
- Anger
- Irritability
- Anxiety
- Confusion
- Withdrawal
- Symptoms begin in the 5 days preceding menses, and relief occurs within 4 days of onset of menses. Dysfunction usually occurs in relationships, parenting, work, or school.

Treatment

- The use of social support and family resources
- Nutritious diet consisting of whole grains, fruits, and vegetables; increased water intake may help.
- Serotonin reuptake inhibitors
- Alprazolam (Xanax) has been effective, but risk of physical and psychological dependence is high.
- Spironolactone, a diuretic agent, may be effective in treating fluid retention.
- Initiation/maintenance of exercise program
- Stress reduction techniques

Adapted from Hawkins, J. W., Roberto-Nichols, D. M., & Stanley-Haney, J. L. (2012). *Guidelines for nurse practitioners in gynecologic settings* (10th ed.). New York: Springer; and Tharpe, N., Farley, C., & Jordan, R. (2013). *Clinical practice guidelines for midwifery & women's health*. Boston: Jones & Bartlett.

research demonstrates the efficacy of dietary changes. Alternative therapies that have been used include vitamins B_6 (pyridoxine) and E, calcium, magnesium, and oil of primrose capsules (Hawkins et al., 2012).

Pharmacologic treatments include selective serotonin reuptake inhibitors (e.g., fluoxetine [Prozac, Sarafem]), prostaglandin inhibitors (e.g., ibuprofen and naproxen [Anaprox, Aleve]), diuretic agents, antianxiety agents, and calcium supplements. Oral contraceptives containing drospirenone (a synthetic progestin) and extended regimens also may be effective (Lopez, Kaptein, & Helmerhorst, 2008).

Nursing Management

The nurse obtains a health history, noting the time when symptoms began and their nature and intensity. The nurse then determines whether symptoms occur before or shortly after the menstrual flow begins. In addition, the nurse can show the patient how to record the timing and intensity of symptoms. A nutritional history is also elicited to determine if the diet is high in salt, caffeine, or alcohol or low in essential nutrients.

The patient's goals may include reduction of anxiety, mood swings, crying, binge eating, fear of losing control, improved coping with day-to-day stressors, improved relationships with family and coworkers, and increased knowledge about PMS. Positive coping measures are promoted. This may involve encouraging the woman's partner to offer support and assistance with child care. The patient can try to plan her working time to accommodate the days she is less productive because of PMS. The nurse encourages the patient to use exercise, meditation, imagery, and creative activities to reduce stress. The nurse also encourages the patient to take medications as prescribed and provides instructions about the desired effects of the medications. Enrolling in a PMS group may help the patient learn to recognize and cope with this condition.

If the patient has severe symptoms of PMS or PMDD, the nurse assesses her for suicidal, uncontrollable, and violent behavior. An immediate psychiatric evaluation is necessary for women with any suggestions of suicidal tendencies. In rare cases, uncontrollable behavior may lead to violence toward family members. If abuse, maltreatment, and neglect of children or other members of a patient's family are suspected, it is important to implement and follow reporting protocols.

Dysmenorrhea

Primary dysmenorrhea is painful menstruation, with no identifiable pelvic pathology. It occurs at the time of menarche or shortly thereafter. It is characterized by crampy pain that begins before or shortly after the onset of menstrual flow and continues for 48 to 72 hours. Pelvic examination findings are normal. Dysmenorrhea is thought to result from excessive production of prostaglandins, which causes painful contraction of the uterus. In secondary dysmenorrhea, pelvic pathology such as endometriosis, tumors such as leiomyomata or malignancies, polyps, or pelvic inflammatory disease (PID) contributes to symptoms. Patients frequently have pain that occurs several days before menses, with ovulation, and occasionally with intercourse. It may be accompanied by nausea, diarrhea, dizziness, and backache (Hawkins et al., 2012).

Assessment and Diagnostic Findings

A pelvic examination is performed to rule out possible disorders, such as endometriosis, PID, adenomyosis, and uterine fibroids. A laparoscopy may be performed to identify organic causes (see Fig. 56-4).

Management

In primary dysmenorrhea, the reason for the discomfort is explained, and the patient is assured that menstruation is a normal function of the reproductive system. If the patient is young and accompanied by her mother, the mother may also need reassurance. Many young women expect to have painful periods if their mothers did. The discomfort of cramps can be treated once anxiety and concern about its cause are dispelled by adequate explanation. Symptoms usually subside with appropriate medication. Useful medications include prostaglandin antagonists such as NSAIDs (e.g., ibuprofen [Motrin], naproxen [Anaprox], and mefenamic acid [Ponstel] or aspirin. If one medication does not provide relief, another may be recommended. Usually, these medications are well tolerated, but some women experience gastrointestinal side effects. Contraindications include allergy, peptic ulcer history, sensitivity to aspirin-containing medications, asthma, and pregnancy. Low-dose oral contraceptives provide relief in more than 90% of patients and may be prescribed for women with dysmenorrhea who are sexually active but do not desire pregnancy (Hawkins et al., 2012).

Continuous low-level local heat may also be effective in relieving primary dysmenorrhea. Heat therapy and medication have been found to work well in combination. The patient is encouraged to continue her usual activities and to increase physical exercise if possible because this relieves discomfort for some women. Taking analgesic agents before cramps start, in anticipation of discomfort, is advised.

Management of secondary dysmenorrhea is directed at diagnosis of and treatment for the underlying cause (e.g., endometriosis or PID) (see Chapter 57.)

Amenorrhea

Amenorrhea, or the absence of menstrual flow, is a symptom of a variety of disorders and dysfunctions. Primary amenorrhea (delayed menarche) refers to the situation in which a young woman who by age 14 years has not begun developing secondary sex characteristics or who by age 16 years or older has developed secondary sex characteristics but has not started menstruation. There are many reasons for primary amenorrhea, including genetic and congenital disorders, malnutrition, or hyperthyroidism (Fritz & Speroff, 2011).

The nurse encourages the patient to express her concerns and anxiety about this problem because the patient may feel that she is different from her peers. A complete physical examination, careful health history, and simple laboratory tests help rule out possible causes, such as metabolic or endocrine disorders and systemic diseases. Treatment is directed toward correcting any abnormalities.

Secondary amenorrhea (an absence of menses for three cycles or 6 months after a normal menarche) may be caused by pregnancy, breast-feeding, menopause, too little body fat (about 22% required for menses), eating disorder, thyroid disease, polycystic ovary syndrome, Asherman's syndrome, cervical stenosis, excessive exercise, or medications (Hawkins et al., 2012). In adolescents, secondary amenorrhea can be caused by minor emotional upset related to being away from home, attending college, tension due to schoolwork, or interpersonal problems.

Secondary nutritional disturbances may also be factors. Obesity can result in anovulation and subsequent amenorrhea. Eating disorders, such as anorexia and bulimia, often result in lack of menses because the decrease in body fat and caloric intake affects hormonal function. Intense exercise can induce menstrual disturbances. Competitive female athletes often experience amenorrhea. On occasion, a pituitary or thyroid dysfunction may cause amenorrhea. These dysfunctions can be treated successfully by treatment of the underlying endocrine disorder. Infrequent periods (oligomenorrhea) may be related to thyroid disorders, polycystic ovarian syndrome, or premature ovarian failure. Women who are HIV positive are apt to miss menstrual periods and need to be evaluated for pregnancy, thyroid disorders, hyperprolactinemia, and menopause.

Abnormal Uterine Bleeding

Dysfunctional uterine bleeding is defined as irregular, painless bleeding of endometrial origin that may be excessive, prolonged, or without pattern. Dysfunctional uterine bleeding can occur at any age but is most common at opposite ends of the reproductive lifespan. It is usually secondary to anovulation (lack of ovulation) and is common in adolescents and women approaching menopause.

Adolescents account for many cases of abnormal uterine bleeding; they often do not ovulate regularly as the pituitary–ovarian axis matures. Perimenopausal women also experience this condition because of irregular ovulation secondary to decreasing ovarian hormone production. Other causes may include fibroids, obesity, and hypothalamic dysfunction.

Abnormal or unusual vaginal bleeding that is atypical in time or amount must be evaluated because it could possibly be a manifestation of a major disorder. A physical examination is performed, and the patient is evaluated for conditions such as pregnancy, neoplasm, infection, anatomic abnormalities, endocrine disorders, trauma, blood dyscrasias, platelet dysfunction, and hypothalamic disorders.

Menorrhagia

Menorrhagia is prolonged or excessive bleeding at the time of the regular menstrual flow. In young women, the cause is usually related to endocrine disturbance; in later life, it usually results from inflammatory disturbances, tumors of the uterus, or hormonal imbalance.

Women with menorrhagia are urged to see a primary provider and to describe the amount of bleeding by pad count and saturation (i.e., absorbency of perineal pad or tampon and number saturated hourly). Persistent heavy bleeding can result in anemia. It can also be a sign of a bleeding disorder or a result of anticoagulant therapy. Treatment may involve endometrial ablation or hysterectomy.

Metrorrhagia

Metrorrhagia (vaginal bleeding between regular menstrual periods) is probably the most significant form of menstrual dysfunction because it may signal cancer, benign tumors of the uterus, or other gynecologic problems. This condition warrants prompt evaluation and treatment. Although bleeding between menstrual periods by women taking oral contraceptive agents is usually not serious, irregular bleeding by women taking hormone therapy (HT) should be evaluated.

Menometrorrhagia is heavy vaginal bleeding between and during periods. It, too, requires evaluation.

Dyspareunia

Dyspareunia (difficult or painful intercourse) can be superficial, deep, primary, or secondary and may occur at the beginning of, during, or after intercourse. Dyspareunia may be related to many factors, including injury during childbirth; lack of vaginal lubrication; a history of incest, sexual abuse, or assault; endometriosis; pelvic or vaginal infection; vaginal dryness due to breast-feeding or menopause; gastrointestinal disorders; fibroids; urinary tract infection; STIs; or vulvodynia (vulvar pain that affects women of all ages without any discernible physical cause). Depending on the cause of dyspareunia, counseling, extra lubrication, or antidepressant medications may be prescribed (Schuiling & Likis, 2013). Women's health issues related to sexuality may be affected by many factors. Thus, these issues need to be taken seriously, carefully assessed, and treated.

Contraception

There are approximately 62 million women in the United States in their childbearing years (i.e., between the ages of 15 and 44 years). Those who are sexually active and do not want to become pregnant, but could become pregnant if they and their partners fail to use a contraceptive method, are at risk of unintended pregnancy. Thus, approximately 43 million women of childbearing age are at risk for unintended pregnancy. Ninety-three percent of at-risk mothers with two children and 86% of at-risk women with no children use contraceptives (Alan Guttmacher Institute, 2012). Family planning benefits mothers, newborns, families, and communities.

Under the Affordable Care Act, a designated list of preventive services must be covered, without out-of-pocket costs to the consumer, by all private health plans written on or after August 1, 2012. Those services include provision of all FDA-approved contraceptive methods, along with sterilization procedures and contraceptive counseling for all women (Alan Guttmacher Institute, 2012).

Nurses can provide information and support by asking women directly when they plan to have their next pregnancy and about their need for contraception. Many women who are sexually active or who are considering becoming sexually active can benefit from learning about contraception. Nurses who are involved in helping patients make contraceptive choices need to listen, take time to answer questions, and educate and assist patients in choosing the method they prefer. It is important for women to receive unbiased and nonjudgmental information, understand the benefits and risks of each method, learn about alternatives and how to use them, and receive positive reinforcement and acceptance of their choice. Some women have received misinformation (i.e., contraception causes cancer or weight gain). Adolescents who worry that they may be unable to become pregnant later in life may not be willing to use contraception.

Abstinence

Abstinence, or celibacy, is the only completely effective means of preventing pregnancy. Abstinence may not be a desired or available option for many women because of cultural expectations and their own and their partner's values and sexual needs.

Sterilization

After abstinence, sterilization by bilateral tubal occlusion or vasectomy is the most effective means of contraception. Both procedures must be considered permanent because neither is easily reversible. Women and men who choose these methods should be certain that they no longer wish to have children, no matter how the circumstances in their life may change. Vasectomy (male sterilization) and tubal ligation (female sterilization) are compared in Table 56-4. (See Chapter 59 for discussion of vasectomy.)

Hormonal Contraception

Oral contraceptives block ovarian stimulation by preventing the release of FSH from the anterior pituitary gland. In the absence of FSH, a follicle does not ripen, and ovulation does not occur. Progestins (synthetic forms of progesterone) suppress the LH surge, prevent ovulation, and also render the cervical mucus impenetrable to sperm. Hormonal contraceptive agents may be oral, transdermal, vaginal, or injectable. Combined oral contraceptive agents that contain both estrogens and progestins are currently used by many women to prevent pregnancy (Hatcher, Trussell, Nelson, et al., 2011).

TABLE 56-4 Comparison of Sterilization Methods

Sterilization Method	Advantages	Disadvantages
Vasectomy	• Highly effective • Relieves female of contraceptive burden • Inexpensive in long run • Permanent • Highly acceptable procedure to most patients • Very safe • Quickly performed	• Expensive in short term • Serious long-term effects suggested (although currently unproved) • Permanent (Although reversal is possible, it is expensive and requires highly technical and major surgery, and results cannot be guaranteed.) • Regret in 5%–10% of patients • No protection against STIs, including HIV • Not effective until sperm remaining in reproductive system are ejaculated
Hysteroscopic and laparoscopic tubal sterilization	• Low incidence of complications • Short recovery • Leaves small or no scar • Quickly performed	• Permanent • Reversal difficult and expensive • Sterilization procedures technically difficult • Requires surgeon, operating room (aseptic conditions), trained assistants, medications, surgical equipment (Essure [insertion of coil or spring in fallopian tubes] requires hysteroscopy rather than surgery) • Expensive at the time performed • If failure, high probability of ectopic pregnancy • No protection against STIs, including HIV

STIs, sexually transmitted infections; HIV, human immunodeficiency virus.

Chart 56-10 PHARMACOLOGY
Comparison of Hormonal Contraceptive Regimens

There are two kinds of hormonal contraceptives: combined (consisting of an estrogen and a progestin) and progestin only.

Combined Preparations (pills, transdermal patches, vaginal rings)

- Monophasic preparations supply the same dose of estrogen and progestin for 21 days.
- Biphasic preparations and triphasic pills vary the amount of hormonal components during the cycle.
- Usually leads to a lighter-than-normal menstrual flow, which results from withdrawal.

Progestin-Only "Mini" Preparations

- Preparations provide less protection against conception than combined preparations.
- About 40% of women taking progestin-only preparations have ovulatory cycles.
- Progestin-only preparations are useful for women who have had estrogen-related side effects (e.g., headaches, hypertension, leg pain, chloasma or skin discoloration, weight gain, or nausea) on combination pills.
- Progestin-only preparations are useful for lactating women who need a hormonal contraceptive method.
- Depo-Provera, a progestin-only injection, lasts for 3 months.
- Implanon, a subdermal implant lasts for 3 years.

Chart 56-10 compares different oral contraceptive regimens, and Chart 56-11 describes the benefits and risks of oral contraceptive use.

Contraindications

Absolute contraindications to hormonal contraceptives include current or past thromboembolic disorder, cerebrovascular disease, or artery disease; migraine headaches with visual auras; known or suspected breast cancer; known or suspected current or past estrogen-dependent neoplasia; pregnancy; current or past benign or malignant liver tumors; liver dysfunction; clotting disorders; congenital hyperlipidemia; and abnormal vaginal bleeding (ACOG, 2006).

Relative contraindications include hypertension, bile-induced jaundice, acute phase of mononucleosis, and sickle cell disease. Controlled hypertension in otherwise healthy young nonsmokers is generally not a contraindication to the use of combination agents but does require a low dose and careful blood pressure monitoring. Women older than 35 years who smoke are at risk for cardiac problems and should not use hormonal contraceptives. Occasionally, neuro-ocular complications arise, but a cause-and-effect relationship has not been established. If visual disturbances occur, hormonal contraceptives should be discontinued. Chart 56-12 summarizes patient education guidelines that are important for women using combination hormonal contraceptives.

Coexisting medical disorders may make contraception a complex issue. These disorders include chronic hypertension, lipid disorders, diabetes, migraines, fibroids, obesity, lupus, depression, seizures, and HIV infection or acquired immunodeficiency syndrome. Contraception needs to be addressed individually in women with these conditions, and primary providers may prescribe injectable hormonal contraceptives (e.g., Depo-Provera) or IUDs (ACOG, 2006).

Chart 56-11 PHARMACOLOGY
Benefits and Risks of Combination Hormonal Contraceptives

Benefits

- Prevents unintended pregnancy
- Decreased cramps and bleeding
- Regular bleeding cycle
- Decreased incidence of anemia
- Decrease in acne with some formulations
- Protection from uterine and ovarian cancer
- Decreased incidence of ectopic pregnancy
- Protection from benign breast disease
- Decreased incidence of pelvic infection

Risks

- Rare in healthy women
- Bothersome side effects (e.g., breakthrough bleeding, breast tenderness)
- Nausea, weight gain, mood changes
- Small increased risk of developing blood clots, stroke, or heart attack, related more to smoking than to oral contraceptive use alone
- Possible increased incidence of benign liver tumors and gallbladder disorders
- No protection from sexually transmitted infections (possible increased risk with unsafe sex)

Methods of Hormonal Contraception

Various hormonal methods of birth control are approved by the FDA. Combination methods include the combination of oral contraceptive pills, vaginal ring (NuvaRing), and transdermal patch (Ortho Evra). Progestin-only methods include the progestin-only pills or minipills, progestin-only emergency contraception (Plan B), once-every 3 month injection (Depo-Provera), levonorgestrel-releasing intrauterine system (Mirena), and single-rod subdermal implant (Implanon).

Quality and Safety Nursing Alert

Patients need to be aware that hormonal contraceptives protect them from pregnancy but not from STIs or HIV infection. In addition, sex with multiple partners or sex without a condom may also result in chlamydial and other infections, including HIV infection.

Chart 56-12 PATIENT EDUCATION
Using Combination Contraceptives

- Use condoms to protect against sexually transmitted infections.
- Take pill at exactly the same time every day *or* put the patch on once a week or remove the vaginal ring after 3 weeks.
- Stop smoking or cut down on smoking.
- Report the following symptoms immediately:
  A abdominal pains
  C chest pains
  H headaches
  E eye problems (blurred vision or spots)
  S severe leg pains

Oral Contraceptives

Many women use oral contraceptive preparations of synthetic estrogens and progestins. A variety of formulations are available. Extended regimens of oral hormonal contraceptive agents are an option for women who have heavy or uncomfortable menstrual bleeding or who wish to have fewer periods. With the use of these regimens, women may have an increased occurrence of breakthrough bleeding; the blood may be dark brown rather than red. It may be more difficult to tell if a pregnancy occurs with this method, although pregnancy is unlikely if pills are taken as prescribed. Studies are ongoing to assess the risks of exposure to increased estrogen resulting from this method.

Transdermal Contraceptives

Ortho Evra is a thin, beige, matchbook-size skin patch that releases an estrogen and a progestin continuously. It is changed every week for 3 weeks, and no patch is used during the fourth week, resulting in withdrawal bleeding. The effectiveness of Ortho Evra is comparable to that of oral contraceptives. Its risks are similar to those of oral contraceptives and include an increased risk of venous thromboemboli formation. The patch may be applied to the torso, chest, arms, or thighs; it should not be applied to the breasts. The patch is convenient and more easily remembered than a daily pill but is not as effective for women who weigh more than 90 kg (198 lb). One additional side effect with the patch includes possible skin reaction such as irritation, redness, pigment changes, or rash at the site of the patch (Hatcher et al., 2011).

Vaginal Contraceptives

NuvaRing (etonogestrel/ethinyl estradiol vaginal ring) is a combination hormonal contraceptive that releases estrogen and progestin. It is inserted in the vagina for 3 weeks and then removed, resulting in withdrawal bleeding. It is as effective as oral contraceptive agents and results in lower hormone blood levels than oral contraceptives. NuvaRing is flexible, does not require sizing or fitting, and is effective when placed anywhere in the vagina. Patients are occasionally reluctant to consider vaginal methods of contraception unless discussed openly and as a convenient alternative to other routes of administration. Some women are uncomfortable with this method and may fear that the ring may migrate or be uncomfortable or be noticed by a partner. The nurse can be helpful in dispelling misconceptions. The patient can be informed that although some women notice a slight increase in vaginal discharge, this effective method of contraception has been found to increase vaginal health-promoting lactobacillus (ACOG, 2006). NuvaRing is usually more expensive than oral contraceptives.

Injectable Contraceptives

An intramuscular injection of Depo-Provera (a long-acting progestin) every 3 months inhibits ovulation and provides a reliable, private, and convenient contraceptive method (Karch, 2012). A subcutaneous formulation is also available. It can be used by lactating women and those with hypertension, liver disease, migraine headaches, heart disease, and hemoglobinopathies. With continued use, women must be prepared for irregular bleeding episodes and spotting decrease, or amenorrhea.

Advantages of Depo-Provera include reduction of menorrhagia, dysmenorrhea, and anemia due to heavy menstrual bleeding. It may reduce the risk of pelvic infection, has been associated with improvement in hematologic status in women with sickle cell disease, and does not interfere with the efficacy of seizure agents. It decreases the risk of endometrial cancer, PID, endometriosis, and uterine fibroids (Hatcher et al., 2011).

Possible side effects of Depo-Provera include irregular menstrual bleeding, bloating, headaches, hair loss, decreased sex drive, bone loss, and weight loss or weight gain. The contraceptive does not protect against STIs. Fertility may be delayed when women discontinue this method; therefore, other methods of contraception may be more appropriate for the woman who wishes to conceive within a year of discontinuing contraception.

Depo-Provera is contraindicated in women who are pregnant and those who have abnormal vaginal bleeding of unknown cause, breast or pelvic cancer, or sensitivity to synthetic progestin. The long-term effects on infants of nursing mothers who use Depo-Provera are unknown but are thought to be negligible.

Implants

Implanon is a single-rod subdermal implant that is usually placed inside the upper arm via a small incision (Karch, 2012). It is effective for 3 years. Implanon may cause irregular bleeding but may improve dysmenorrhea, and it does not affect bone density. This contraceptive can be used by women who are lactating.

Intrauterine Device

An IUD is a small plastic device, usually T shaped, that is inserted into the uterine cavity to prevent pregnancy. A string attached to the IUD is visible and palpable at the cervical os. An IUD prevents conception by causing a local inflammatory reaction that prevents fertilization.

Advantages include effectiveness over a long period of time, few if any systemic effects, and reduction of patient error (ACOG, 2011). This reversible method of birth control is as effective as sterilization and more effective than barrier methods.

Disadvantages include possible excessive bleeding, cramps, and backaches; a slight risk of tubal pregnancy; slight risk of pelvic infection on insertion; displacement of the device; and, rarely, perforation of the cervix and uterus. If a pregnancy occurs with an IUD in place, the device is removed immediately to avoid infection. Spontaneous abortion (miscarriage) may occur on removal. An IUD is not usually used in women who have not had children, because a small nulliparous uterus may not tolerate it. Women with multiple partners, women with heavy or crampy periods, or those with a history of ectopic pregnancy or pelvic infection should be encouraged to use other methods of contraception.

Mechanical Barriers

Diaphragm

The diaphragm is an effective contraceptive device that consists of a round, flexible spring (50 to 90 mm wide) covered with a domelike latex rubber cup. A spermicidal (contraceptive) jelly or cream is used to coat the concave side of the diaphragm before it is inserted deep into the vagina, covering the cervix completely. The spermicide inhibits spermatozoa from entering the cervical canal. The diaphragm is not felt by the user or her partner when properly fitted and inserted. Because women vary in size, the diaphragm must be sized and fitted by an experienced clinician. The woman is instructed in using and caring for the device. A return demonstration ensures that the woman can insert the diaphragm correctly and that it covers the cervix.

Each time the woman uses the diaphragm, she should examine it carefully. By holding it up to a bright light, she should ensure that it has no pinpoint holes, cracks, or tears. She then applies spermicidal jelly or cream and inserts the diaphragm. The diaphragm should remain in place at least 6 hours after coitus (no more than 12 hours). Additional spermicide is necessary if more than 6 hours have passed before intercourse occurs and before each act of repeated intercourse. On removal, the diaphragm should be cleansed thoroughly with mild soap and water, rinsed, and dried before being stored in its original container.

Disadvantages include allergic reactions in those who are sensitive to latex and an increased incidence of urinary tract infections. Toxic shock syndrome has been reported in some diaphragm users but is rare.

Quality and Safety Nursing Alert

The nurse must assess the woman for possible latex allergy because the use of latex barrier methods (e.g., diaphragm, cervical cap, male condoms) may cause severe allergic reactions, including anaphylaxis, in patients with latex allergy.

Cervical Cap

The cervical cap is much smaller (22 to 35 mm) than the diaphragm and covers only the cervix. If a woman can feel her cervix, she can usually learn to use a cervical cap. The chief advantage is that the cap may be left in place for 2 days after coitus. Although convenient to use, the cervical cap may cause cervical irritation; therefore, before fitting a cap, most primary providers obtain a Pap smear and repeat the smear after 3 months. The cap is used with a spermicide and does not require additional spermicide for repeated intercourse.

Female Condom

The female condom was developed to give control of barrier protection to women—to provide them with protection from STIs and HIV as well as pregnancy. The female condom (Reality) consists of a cylinder of polyurethane enclosed at one end by a closed ring that covers the cervix and at the other end by an open ring that covers the perineum (Fig. 56-5). Advantages include some degree of protection from STIs (i.e., HPV, herpes simplex virus, and HIV) (Hatcher et al., 2011). Disadvantages include the inability to use the female condom with some positions (i.e., standing). Women have found that it can be noisy and slippery.

Spermicides

Spermicides are made from nonoxynol-9 or octoxynol and are available over the counter as foams, gels, films, and suppositories and also on condoms. Spermicides do not protect women from HIV or other STIs (Alan Guttmacher Institute, 2012). Advantages of spermicides include they are nonhormonal, are user controlled, do not cause systemic side effects, and are immediately effective (Hatcher et al., 2011).

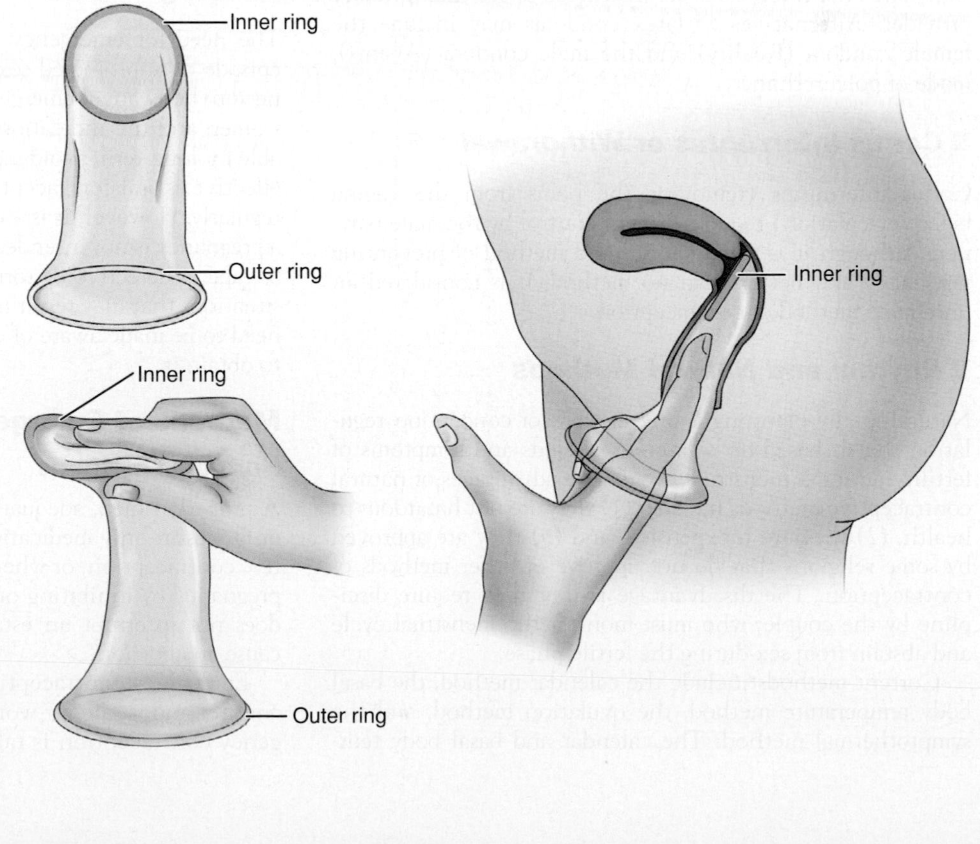
Inner ring
Outer ring
Inner ring
Outer ring
Inner ring

FIGURE 56-5 • Female condom. To insert the female condom, hold the inner ring between the thumb and middle finger. Put the index finger on the pouch between the thumb and other fingers, and squeeze the ring. Slide the condom into the vagina as far as it will go. The inner ring keeps the condom in place.

Male Condom

The male condom is an impermeable, snug-fitting cover applied to the erect penis before it enters the vaginal canal. The tip of the condom is pinched while being applied to leave space for ejaculate. If no space is left, ejaculation may cause a tear or hole in the condom and reduce its effectiveness. The penis, with the condom held in place, is removed from the vagina while still erect to prevent the ejaculate from leaking. Condoms are now available in large and small sizes.

The latex condom also creates a barrier against transmission of STIs (gonorrhea, chlamydial infection, and HIV) by body fluids and may reduce the risk of herpes virus transmission. However, natural condoms (those made from animal tissue) do not protect against HIV infection. Nurses need to reassure women that they have a right to insist that their male partners use condoms and a right to refuse sex without condoms, although women in abusive relationships may increase their risk of abuse, maltreatment, and neglect by doing so. Some women carry condoms with them to be certain that one is available. Nurses should be familiar and comfortable with instructions about using condoms because many women need to know about this way of protecting themselves from HIV and other STIs. Condoms do not provide complete protection from STIs, however, because HPV may be transmitted by skin-to-skin contact. Other STIs may be transmitted if any abraded skin is exposed to body fluids. This information should be included in patient education.

The nurse needs to consider the possibility of latex allergy. Swelling and itching can also occur. Possible warning signs of latex allergy include oral itching after blowing up a balloon or eating kiwis, bananas, pineapples, passion fruit, avocados, or chestnuts. Because many contraceptives are made of latex, patients who experience burning or itching while using latex contraceptives are instructed to see their primary provider. Alternatives to latex condoms may include the female condom (Reality) and the male condom (Avanti), made of polyurethane.

Coitus Interruptus or Withdrawal

Coitus interruptus (removing the penis from the vagina before ejaculation) requires careful control by the male partner. Although it is a frequently used method of preventing pregnancy and better than no method, it is considered an unreliable method of contraception.

Rhythm and Natural Methods

Natural family planning is any method of conception regulation that is based on awareness of signs and symptoms of fertility during a menstrual cycle. The advantages of natural contraceptive methods include: (1) they are not hazardous to health, (2) they are inexpensive, and (3) they are approved by some religions that do not approve of other methods of contraception. The disadvantage is that they require discipline by the couple, who must monitor the menstrual cycle and abstain from sex during the fertile phase.

Current methods include the calendar method, the basal body temperature method, the ovulation method, and the symptothermal method. The calendar and basal body temperature methods are older than the ovulation method and the symptothermal method. Combinations of these methods are often used (Fehring, Schneider, Raviele, et al., 2007). The fertile phase (in which sexual abstinence is required) is estimated to occur about 14 days before menstruation, although it may occur between the 10th and 17th days. Spermatozoa can fertilize an ovum up to 72 hours after intercourse, and the ovum can be fertilized for 24 hours after leaving the ovary. The pregnancy rate with the rhythm (i.e., calendar) method is about 40% yearly.

Women who carefully determine their "safe period," based on a precise recording of menstrual dates for at least 1 year, and who follow a carefully worked out formula may achieve very effective protection. A long abstinence period during each cycle is required. These prerequisites require more time and control than many couples have. Changes in cervical mucus and basal body temperature due to hormonal changes related to ovulation form the scientific basis for the symptothermal method of ovulatory timing. Courses in natural family planning are offered at many Catholic hospitals and some family planning clinics.

Ovulation detection methods (e.g., Clearblue Fertility Monitor) are available in most pharmacies. The presence of the enzyme guaiacol peroxidase in cervical mucus signals ovulation 6 days beforehand and also affects mucosal viscosity. Over-the-counter test kits are easy to use and reliable but can be expensive. Ovulation prediction kits are more effective for planning conception than for avoiding it. But if they are used in combination with cervical mucus changes and the calendar method, they may be effective; further research is needed (Carcio & Secor, 2010).

Douching is not a contraceptive method and may enhance rather than decrease the chances of conception.

Emergency Contraception

The need for emergency contraception may arise after an episode of unprotected sexual intercourse. Therefore, nurses need to be aware of emergency contraception as an option for women and the indications for its use. It is clearly not suitable for long-term avoidance of pregnancy because it is not as effective as oral contraceptives or other reliable methods used regularly. However, it is valuable following intercourse when a pregnancy is not intended and in emergency situations such as rape, a defective or torn condom or diaphragm, or other situations that may result in unintended conception. Women need to be made aware of emergency contraception and how to obtain it.

Methods of Emergency Contraception

Hormonal Methods

A properly timed, adequate dose of estrogen and a progestin or progestin-only medication after intercourse without effective contraception, or when a method has failed, can prevent pregnancy by inhibiting or delaying ovulation. This method does not interrupt an established pregnancy and does not cause an abortion.

Emergency contraception is currently available over the counter and is safe for women of all ages. The sooner emergency contraception is taken, the more effective it is. It is considered safe and effective by the FDA and can be purchased as Plan B (progestin only) packages of emergency contraception with patient literature (ACOG, 2013a).

This method must be used not more than 5 days following intercourse. Nausea, a common side effect, can be minimized by taking the medication with meals and with an antiemetic agent. Other side effects, such as breast soreness and irregular bleeding, may occur but are transient. Patients who use this method should be advised of the potential failure rate and also counseled about other contraceptive methods. There are no known contraindications to the use of this method, except an established pregnancy (Allen & Goldberg, 2007).

The nurse reviews with the patient instructions for emergency contraception based on the medication regimen prescribed. If the woman is breast-feeding, a progestin-only formulation is prescribed. To avoid exposing infants to synthetic hormones through breast milk, the patient can manually express milk and bottle-feed for 24 hours after treatment. The patient should be informed that her next menstrual period may begin a few days earlier or a few days later than expected. She is instructed to return for a pregnancy test if she has not had a menstrual period in 3 weeks and should be offered another visit to provide a regular method of contraception if she does not have one currently.

Postcoital Intrauterine Device Insertion

Postcoital IUD insertion, another form of emergency contraception, involves insertion of a copper-bearing IUD within 5 days of coitus in women who want this method of contraception; however, it may be inappropriate for some women or if contraindications exist. The mechanism of action is unknown, but it is thought that the IUD interferes with fertilization (Allen & Goldberg, 2007). The patient may experience discomfort on insertion and may have heavier menstrual periods and increased cramping. Contraindications include a confirmed or suspected pregnancy or any contraindication to regular copper IUD use. The patient must be informed that there is a risk that insertion of an IUD may disrupt a pregnancy that is already present.

Nursing Management

Patients who use emergency contraception may be anxious, embarrassed, and lacking information about birth control. The nurse must be supportive and nonjudgmental and provide facts and appropriate patient education. If the patient repeatedly uses this method of birth control, she should be informed that the failure rate with this method is higher than with a regularly used method. A toll-free telephone information service (1-888-NOT-2-LATE) operates 24 hours a day in English and Spanish and provides information and referrals to health care providers. Nurses can educate and inform women about emergency contraception options to reduce unintended pregnancies and abortions. (See the Resources section at the end of this chapter for more information.)

Abortion

Interruption of pregnancy or expulsion of the product of conception before the fetus is viable is called *abortion*. The fetus is generally considered to be viable any time after the fifth to sixth month of gestation.

Spontaneous Abortion

It is estimated that 1 of every 5 to 10 conceptions ends in spontaneous abortion. Most of these occur because an abnormality in the fetus makes survival impossible. Other causes may include systemic diseases, hormonal imbalance, or anatomic abnormalities. If a pregnant woman experiences bleeding and cramping, a threatened abortion is diagnosed because an actual abortion is usually imminent. Spontaneous abortion occurs most commonly in the second or third month of gestation.

There are various types of spontaneous abortion, depending on the nature of the process (threatened, inevitable, incomplete, or complete). In a threatened abortion, the cervix does not dilate. With bed rest and conservative treatment, the abortion may be prevented. If not, an abortion is imminent. If only some of the tissue is passed, the abortion is referred to as incomplete. An emptying or evacuation procedure (D&C, or dilation and evacuation [D&E]) or administration of oral misoprostol (Cytotec) is usually required to remove the remaining tissue. If the fetus and all related tissue are spontaneously evacuated, the abortion is termed *complete*, and no further treatment is required.

Habitual Abortion

Habitual or recurrent abortion is defined as successive, repeated, spontaneous abortions of unknown cause. As many as 60% of abortions may result from chromosomal anomalies. After two consecutive abortions, the patient is referred for genetic counseling and testing, and other possible causes are explored.

If bleeding occurs in a pregnant woman with a past history of habitual abortion, conservative measures, such as bed rest and administration of progesterone to support the endometrium, are attempted to save the pregnancy. Supportive counseling is crucial in this stressful condition. Bed rest, sexual abstinence, a light diet, and no straining on defecation may be recommended in an effort to prevent spontaneous abortion. If infection is suspected, antibiotic agents may be prescribed.

In the condition known as incompetent or dysfunctional cervix, the cervix dilates painlessly in the second trimester of pregnancy, often resulting in a spontaneous abortion. In such cases, a surgical procedure called *cervical cerclage* may be used to prevent the cervix from dilating prematurely, although its effectiveness is unclear. It involves placing a purse-string suture around the cervix at the level of the internal os. Bed rest is usually advised to keep the weight of the uterus off the cervix. About 2 to 3 weeks before term or at the onset of labor, the suture is cut. Delivery is usually by cesarean section.

Medical Management

After a spontaneous abortion, all tissue passed vaginally is saved for examination, if possible. The patient and all personnel who care for her are alerted to save any discharged material. In the rare case of heavy bleeding, the patient may require blood component transfusions and fluid replacement. An estimate of the bleeding volume can be determined by recording the number of perineal pads and the degree of saturation over 24 hours. When an incomplete abortion occurs, oxytocin may be prescribed to cause uterine contractions before D&E or uterine suctioning.

Nursing Management

Because patients experience loss and anxiety, emotional support and understanding are important aspects of nursing care. Women may be grieving or relieved, depending on their feelings about the pregnancy. Providing opportunities for the patient to talk and express her emotions is helpful and also provides clues for the nurse in planning more specific care.

Elective Abortion

A voluntary induced termination of pregnancy is called an *elective abortion* and is usually performed by skilled health care providers. In 1973, the U.S. Supreme Court in *Roe v. Wade* ruled that decisions about abortion reside with a woman and her physician in the first trimester. During the second trimester, the state may regulate practice in the interest of a woman's health and during the final weeks of pregnancy may choose to protect the life of the fetus, except when necessary to preserve the life or health of the woman.

The U.S. rate of abortion is among the highest in the industrialized Western world. These numbers indicate the need for effective contraceptive education, information about emergency contraception, and counseling.

Medical Management

Before the abortion procedure is performed (Chart 56-13), a nurse or counselor trained in pregnancy counseling should talk with the patient and explore her fears, feelings, and options. The nurse then identifies the patient's choice (i.e., continuing pregnancy and parenthood, continuing pregnancy followed by adoption, or terminating pregnancy by abortion). If abortion is chosen, the patient has a pelvic examination to determine uterine size. A pelvic ultrasound may also be performed. Laboratory studies before an abortion must include a pregnancy test to confirm the pregnancy, hematocrit to rule out anemia, and Rh determination. Patients with anemia will require an iron supplement, and patients who are Rh negative may require Rho(D) immune globulin (RhoGAM) to prevent isoimmunization. Before the procedure, all patients should be screened for STIs to prevent introducing pathogens upward through the cervix during the procedure.

> *Quality and Safety Nursing Alert*
>
> **Women who have resorted to unskilled attempts to end a pregnancy may become critically ill because of infection, hemorrhage, or uterine rupture. If a woman has undergone such efforts to end a pregnancy, prompt medical attention, broad-spectrum antibiotics, and replacement of fluids and blood components may be required before careful attempts are made to evacuate the uterus.**

Surgical terminations include D & C or vacuum aspiration of uterine contents. Medications can also be used. Mifepristone (Mifeprex), formerly known as RU-486, is used only in early pregnancy (up to 49 days from the last menstrual period). It works by blocking progesterone. Cramping and bleeding similar to a heavy menstrual period occur. After counseling and consent and often a sonogram to confirm the pregnancy, mifepristone is administered. This is followed by a dose of misoprostol orally or vaginally. If the pregnancy persists, a suction aspiration is performed. Contraindications include ectopic pregnancy, adrenal failure, allergy to the medications, bleeding disorder, irritable bowel syndrome, or uncontrolled seizure disorders. Several deaths from sepsis have occurred following medical abortion; researchers and the FDA are closely monitoring the morbidity and mortality associated with medical abortion. Currently, there is no evidence that a previous medical or surgical abortion increases the risk of adverse future pregnancy outcomes (Virk, Zhang, & Olsen, 2007).

Nursing Management

Patient education is an important aspect of care for women who elect to terminate a pregnancy. A patient undergoing elective abortion is informed about what the procedure entails and the expected course after the procedure. The patient is scheduled for a follow-up appointment 2 weeks after the procedure and is instructed about signs and symptoms (i.e., fever, heavy bleeding, or pain) that should be reported.

Available contraceptive methods are reviewed with the patient at this time. Effectiveness depends on the method used and the extent to which the woman and her partner follow the instructions for use. A woman who has used any method of birth control should be assessed for her understanding of the method and its potential side effects as well as her satisfaction with the method. If the woman has not been using contraception, the nurse explains all methods and their benefits and risks and helps the patient make a contraceptive choice for use after abortion. Related education issues, such as the need to use barrier contraceptive devices (i.e., condoms) for protection against transmission of STIs and HIV infection and the availability of emergency contraception, are becoming increasingly important.

Psychological support is another important aspect of nursing care. The nurse needs to be aware that women terminate pregnancies for many reasons. Some women terminate pregnancies because of severe genetic defects. Women who have been raped or impregnated in incestuous relationships or by an abusive partner may elect to terminate their pregnancies. The care of a woman undergoing termination of pregnancy is stressful, and assistance needs to be provided in a safe and nonjudgmental way. Nurses have the right to refuse to participate in a procedure that is against their religious beliefs but are professionally obligated not to impose their beliefs or judgments on their patients.

Infertility

Infertility is defined as a couple's inability to achieve pregnancy after 1 year of unprotected intercourse (Hatcher et al., 2011). Primary infertility refers to a couple who has never had a child. Secondary infertility means that at least one

Chart 56-13 Types of Elective Abortions

Vacuum Aspiration

- The cervix is dilated manually with instrumentation or by laminaria (small suppositories made of seaweed that swells as it absorbs water).
- A uterine aspirator is introduced.
- Suction is applied, and tissue is removed from the uterus.

This is the most common type of termination procedure and is used early in pregnancy, up to 14 weeks. Laminaria may be used to soften and dilate the cervix prior to the procedure.

Dilation and Evacuation

Cervical dilation with laminaria followed by vacuum aspiration

Labor Induction

These procedures account for fewer than 1% of all terminations and generally take place in an inpatient setting.

1. Installation of normal saline or urea results in uterine contractions.
   - Although rare, serious complications can occur, including cardiovascular collapse, cerebral edema, pulmonary edema, renal failure, and disseminated intravascular coagulopathy.
2. Prostaglandins
   - Prostaglandins are introduced into the amniotic fluid or by vaginal suppository or intramuscular injection in later pregnancy.
   - Strong uterine contractions begin within 4 hours and usually result in abortion.
   - Gastrointestinal side effects (e.g., nausea, vomiting, diarrhea, and abdominal cramping) and fever can occur.
3. IV oxytocin
   - Used for later abortions for genetic indications. Requires patient to go through labor.

Medical Abortion

Mifepristone

- Mifepristone (formerly known as RU-486) is a progesterone antagonist that prevents implantation of the ovum.
- Administered orally within 10 days of an expected menstrual period, mifepristone produces a medical abortion in most patients.
- Combined with a prostaglandin suppository, mifepristone causes abortion in up to 95% of patients.
- Prolonged bleeding may occur. Other side effects may include abdominal pain, nausea, vomiting, and diarrhea. This method may not be used in women with adrenal failure, asthma, long-term corticosteroid therapy, an intrauterine device in place, porphyria, or a history of allergy to mifepristone or other prostaglandins. It is less effective when used in pregnancies more than 49 days from the beginning of the last menstrual period.

Methotrexate

- Methotrexate has also been used to terminate pregnancy because it is lethal to the fetus. It has been found to have minimal risk and few side effects in the woman. Its low cost may provide an alternative for some women.

Misoprostol

- Misoprostol is a synthetic prostaglandin analogue that produces cervical effacement and uterine contractions.
- Inserted vaginally, misoprostol is effective in terminating a pregnancy in about 75% of cases.
- When combined with methotrexate or mifepristone, misoprostol's effectiveness rate is high.

conception has occurred, but currently the couple cannot achieve a pregnancy. It is often a complex physical problem, and its causes are usually related to azoospermia, anovulation, or tubal obstruction.

Diagnostic Findings

Ovarian and Ovulation Factors

Diagnostic studies performed to determine if ovulation is regular and whether the progestational endometrium is adequate for implantation may include a serum progesterone level and an ovulation index. The ovulation index involves a urine dipstick test to determine whether the surge in LH that precedes follicular rupture has occurred.

Tubal and Uterine Factors

HSG is used to rule out uterine or tubal abnormalities. A contrast agent injected into the uterus through the cervix produces an outline of the shape of the uterine cavity and the patency of the tubes. This process sometimes removes mucus or tissue that is lodged in the tubes. Laparoscopy permits direct visualization of the tubes and other pelvic structures and can assist in identifying conditions that may interfere with fertility (e.g., endometriosis).

Fibroids, polyps, and congenital malformations are possible causative factors affecting the uterus. Their presence may be determined by pelvic examination, hysteroscopy, saline sonogram (a variation of a sonogram), and HSG. Endometriosis, even if mild, is associated with reduced fertility (Schuiling & Likis, 2013).

Male Factors

An analysis of semen provides information about the number of sperm (density), percentage of moving forms, quality of forward movement (forward progression), and morphology (shape and form). From 2 to 6 mL of watery alkaline semen is normal. A normal count has 60 to 100 million sperm/mL. However, the incidence of impregnation is lessened only when the count decreases to fewer than 20 million sperm/mL.

Men may also be affected by varicoceles (varicose veins around the testicle), which decrease semen quality by increasing testicular temperature. Retrograde ejaculation or ejaculation into the bladder is assessed by urinalysis after ejaculation. Blood tests for male partners may include measuring testosterone, FSH and LH (both of which are involved in maintaining testicular function), and prolactin levels.

Medical Management

The treatment of infertility is complex and often requires advanced technology. The specific type of treatment depends on the cause of the problem, if it can be identified. Many infertile couples have normal test results for ovulation, sperm production, and fallopian tube patency.

Ovulatory dysfunction is complex, but many women with ovulation disorders have polycystic ovary syndrome (see Chapter 57) and may be treated with clomiphene (Clomid) to induce ovulation or insulin-sensitizing agents. Once insulin levels are normalized, ovulation often occurs. Some women have high prolactin levels, which inhibit ovulation, and they are treated with dopaminergic drugs after a pituitary adenoma is ruled out by MRI. If a woman has premature ovarian failure, oocyte donation may be considered.

Pharmacologic Therapy

Pharmacologically induced ovulation is undertaken when women do not ovulate on their own or ovulate irregularly. These couples are often treated with clomiphene to stimulate ovulation. Gonadotropin treatment may also be used if conception does not occur. Various other medications are used, depending on the main cause of infertility (Chart 56-14).

Blood tests and ultrasounds are used to monitor ovulation. Multiple pregnancies (i.e., twins, triplets, or more) may occur with the use of these medications. Ovarian hyperstimulation syndrome (OHSS) may also occur. This condition is characterized by enlarged multicystic ovaries and is complicated by a shift of fluid from the intravascular space into the abdominal cavity. The fluid shift can result in ascites, pleural effusion, and edema; hypovolemia may also occur. Risk factors include younger age, history of polycystic ovarian syndrome, high serum estradiol levels, a larger number of follicles, and pregnancy.

Artificial Insemination

Artificial insemination is the deposit of semen into the female genital tract by artificial means. If the sperm cannot penetrate the cervical canal normally, artificial insemination using a partner's or husband's semen or that of a donor may be considered. When the sperm of the woman's partner is defective or absent (azoospermia) or when there is a risk of transmitting a genetic disease, donor sperm may be used. Safeguards are put in place to address legal, ethical, emotional, and religious issues. Written consent is obtained to protect all parties involved, including the woman, the donor, and the resulting child. The donor's semen is frozen, and the donor is evaluated to ensure that he is free of genetic disorders and STIs, including HIV infection.

Certain conditions must be met before semen is transferred to the vagina or uterus. The woman must have no abnormalities of the genital system, the fallopian tubes must be patent, and ova must be available. In the male, sperm need to be normal in shape, amount, motility, and endurance. The time of ovulation should be determined as accurately as possible so that the 2 or 3 days during which fertilization is possible each month can be targeted for treatment.

Ultrasonography and blood studies of varying hormone levels are used to pinpoint the best time for insemination and to monitor for OHSS. Fertilization seldom occurs from a single insemination. Usually, insemination is attempted between days 10 and 17 of the cycle; three different attempts may be made during one cycle. The woman may have received clomiphene or other medications to stimulate ovulation before insemination. The recipient is placed in the lithotomy position on the examination table, a speculum is inserted, and the vagina and cervix are swabbed with a cotton-tipped applicator to remove any excess secretions. The sperm are washed before insertion to remove biochemicals and to select the most active sperm. Semen is drawn into a sterile syringe, and a cannula is attached. The semen is then directed to the external os. In intrauterine insemination, semen is placed into the uterine cavity.

Chart 56-14 PHARMACOLOGY
Medications That Induce Ovulation

- Clomiphene citrate (Clomid, Serophene) is an estrogen antagonist that increases gonadotropin release, resulting in follicular rupture or ovulation. Clomiphene is used when the hypothalamus is not stimulating the pituitary gland to release follicle-stimulating hormone (FSH) and luteinizing hormone (LH). This medication stimulates follicles in the ovary. It is usually taken for 5 days beginning on the 5th day of the menstrual cycle. Ovulation should occur 4–8 days after the last dose. Patients receive instructions about timing intercourse to facilitate fertilization.
- Menotropins (Repronex, Pergonal), a combination of FSH and LH, may be used to stimulate the ovaries to produce eggs. These agents are used for women with deficiencies in FSH and LH. When followed by administration of human chorionic gonadotropin, menotropins stimulates the ovaries, so monitoring by ultrasound and hormone levels is essential because overstimulation may occur.
- Follitropin alfa (Gonal-F), follitropin beta (Follistim), and urofollitropin (Bravelle) may be used to treat ovulation disorders or to stimulate a follicle and egg production for intrauterine insemination or in vitro fertilization or other assisted reproductive technologies.
- Gonadotropin-releasing hormone agonists (leuprolide [Lupron], nafarelin acetate [Synarel]) suppress FSH, prevent premature egg release, and shrink fibroids.
- Bromocriptine (Parlodel) may be used in treatment for infertility due to elevated prolactin levels.
- Progesterone (Prometrium Crinone, progesterone in oil) vaginal suppositories help improve the uterine lining after ovulation.
- Urofollitropin (Metrodin, Bravelle), which contains FSH with a small amount of LH, is used in some disorders (e.g., polycystic ovarian syndrome) to stimulate follicle growth. Clomiphene is then used to stimulate ovulation.
- Chorionic gonadotropin (Ovidrel, Novarel, Pregnyl), which mimics LH, releases an egg after hyperstimulation and supports the corpus luteum.
- Metformin (Glucophage, Fortamet) may be used in polycystic ovarian syndrome to induce regular ovulation.
- Aspirin and heparin may be used to prevent recurrent pregnancy loss in patients with elevated antiphospholipid antibodies.

In Vitro Fertilization

In vitro fertilization (IVF) involves ovarian stimulation, egg retrieval, fertilization, and embryo transfer. This procedure is accomplished by first stimulating the ovary to produce multiple eggs or ova, usually with medications, because success rates are greater with more than one embryo. Many different protocols exist for inducing ovulation with one or more agents. Patients are carefully selected and evaluated, and cycles are carefully monitored using ultrasound and monitoring hormone levels. At the appropriate time, the ova are recovered by transvaginal ultrasound retrieval. Sperm and eggs are coincubated for up to 36 hours, and the embryos are transferred about 48 hours after retrieval. Implantation should occur in 3 to 5 days.

Gamete intrafallopian transfer (GIFT), a variation of IVF, is the treatment of choice for patients with ovarian failure. GIFT is considered in unexplained infertility and when there is religion-based discomfort with IVF. The most common indications for IVF and GIFT are irreparable tubal damage, endometriosis (see Chapter 57), unexplained infertility, inadequate sperm, and exposure to DES. Success rates for GIFT vary from 20% to 30%.

Other Assisted Reproductive Technologies

In intracytoplasmic sperm injection (ICSI), an ovum is retrieved as described previously, and a single sperm is injected through the zona pellucida, through the egg membrane, and into the cytoplasm of the oocyte. The fertilized egg is then transferred back to the donor. ICSI is the treatment of choice in severe male factor infertility.

Women who cannot produce their own eggs (i.e., premature ovarian failure) have the option of using the eggs of a donor after stimulation of the donor's ovaries. The recipient also receives hormones in preparation for these procedures. Couples may also choose this modality if the female partner has a genetic disorder that may be passed on to children.

Nursing Management

Nursing interventions that are appropriate when working with couples during infertility evaluations include assisting in reducing stress in the relationship, encouraging cooperation, protecting privacy, fostering understanding, and referring the couple to appropriate resources when necessary. Because infertility evaluations and treatments are expensive, time-consuming, invasive, stressful, and not always successful, couples need support in working together to deal with this process.

Smoking is strongly discouraged because it has an adverse effect on the success of assisted reproduction. Diet, exercise, stress reduction techniques, folic acid supplementation, health maintenance, and disease prevention are emphasized in many infertility programs. Couples may also consider adoption, child-free living, and gestational carriers (the use of a surrogate to carry the fetus for the infertile couple). Nurses can be helpful listeners and information resources in these deliberations.

RESOLVE: The National Infertility Association, a nonprofit self-help group that provides information and support for patients who are infertile, was founded by a nurse who experienced difficulty conceiving. The literature on infertility produced by this group and others is an important resource for patients and professionals. (See the Resources section at the end of this chapter.)

Preconception/Periconception Health Care

Nurses can be instrumental in encouraging all women of childbearing age, including those with chronic illness or disabilities, to consider issues that may affect health during pregnancy (Smeltzer, 2007; Smeltzer & Wetzel-Effinger, 2009). Women who plan their pregnancies and are healthy and well informed tend to have better outcomes. This is an important issue because half of all pregnancies in the United States are unintended.

Nurses can make a difference through education and counseling; preconception counseling can decrease the incidence of birth defects. Women who smoke should be encouraged to stop smoking, and it may help to offer smoking cessation classes. Women should take folic acid supplements to prevent neural tube defects. Women with diabetes should have good glycemic control prior to conception. It is necessary to assess rubella immunity and other immunizations as well as a family history of genetic defects; genetic counseling may be appropriate. Women taking teratogenic medications and women concerned about genetic disorders should be encouraged to discuss effective contraception and childbearing plans with their primary provider (see Chart 56-2).

Ectopic Pregnancy

The incidence of ectopic pregnancy and the risk of death due to ectopic pregnancy are decreasing. However, ectopic pregnancy remains the leading cause of pregnancy-related death in the first trimester. Ectopic pregnancy occurs when a fertilized ovum (a blastocyst) becomes implanted on any tissue other than the uterine lining (e.g., the fallopian tube, ovary, abdomen, cervix, or scar tissue from previous caesarean section). The most common site of ectopic implantation is the fallopian tube (ACOG, 2010) (Fig. 56-6).

Possible causes of ectopic pregnancy include salpingitis, peritubal adhesions (after pelvic infection, endometriosis, appendicitis), structural abnormalities of the fallopian tube (rare and usually related to DES exposure), previous ectopic pregnancy, previous tubal surgery, multiple previous induced abortions (particularly if followed by infection), tumors that distort the tube, and IUD and progestin-only contraceptive agents. PID appears to be the major risk factor. Improved antibiotic therapy for PID usually prevents total tubal closure but may leave a stricture or narrowing, predisposing to ectopic implantation. The odds of recurrent ectopic pregnancy are three times higher if an infectious pathology caused the first ectopic pregnancy. After a second ectopic pregnancy occurs, assisted reproduction is considered.

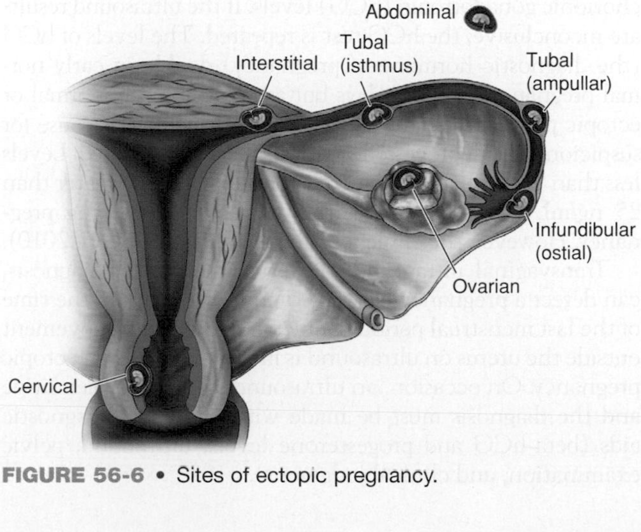
Abdominal
Tubal
(isthmus)
Interstitial
Tubal
(ampullar)
Infundibular
(ostial)
Ovarian
Cervical

FIGURE 56-6 • Sites of ectopic pregnancy.

Risk factors are important, but all women need to be educated about early treatment and have a high index of suspicion in the case of a period that does not seem normal, the presence of pain, or pain with a suspected pregnancy. Women may have fatal hemorrhage with ruptured ectopic pregnancies if they delay seeking attention or if their primary providers are not alert to the possibility of this diagnosis.

Clinical Manifestations

Signs and symptoms vary depending on whether tubal rupture has occurred. Delay in menstruation from 1 to 2 weeks followed by slight bleeding (spotting) or a report of a slightly abnormal period suggests the possibility of an ectopic pregnancy. Symptoms may begin late, with vague soreness on the affected side (probably due to uterine contractions and distention of the tube), and may proceed to sharp, colicky pain. Most patients experience some pelvic or abdominal pain and some spotting or bleeding. Gastrointestinal symptoms, dizziness, or lightheadedness may occur. Patients may think the abnormal bleeding is a menstrual period, especially if a recent period occurred and was normal.

If implantation occurs in the fallopian tube, the tube becomes more and more distended and can rupture if the ectopic pregnancy remains undetected for 4 to 6 weeks or longer after conception. When the tube ruptures, the ovum is discharged into the abdominal cavity, and the woman experiences agonizing pain, dizziness, faintness, and nausea and vomiting due to the peritoneal reaction to blood escaping from the tube. Air hunger and symptoms of shock may occur, and the signs of hemorrhage—rapid and thready pulse, decreased blood pressure, subnormal temperature, restlessness, pallor, and sweating—are evident. Later, the pain becomes generalized in the abdomen and radiates to the shoulder and neck because of accumulating intraperitoneal blood that irritates the diaphragm.

Assessment and Diagnostic Findings

Ectopic pregnancies must be diagnosed promptly to prevent life-threatening hemorrhage, which is the major complication of rupture. During vaginal examination, a large mass of clotted blood that has collected in the pelvis behind the uterus or a tender adnexal mass may be palpable, although there are often no abnormal findings. If an ectopic pregnancy is suspected, the patient is evaluated by sonography and human chorionic gonadotropin (hCG) levels. If the ultrasound results are inconclusive, the hCG test is repeated. The levels of hCG (the diagnostic hormone of pregnancy) double in early normal pregnancies every 3 days but are reduced in abnormal or ectopic pregnancies. A less-than-normal increase is cause for suspicion. Serum progesterone levels are also measured. Levels less than 5 ng/mL are considered abnormal; levels greater than 25 ng/mL are associated with a normally developing pregnancy. However, the clinical utility is limited (ACOG, 2010).

Transvaginal ultrasound, the usual method of diagnosis, can detect a pregnancy between 5 and 6 weeks from the time of the last menstrual period. Detectable fetal heart movement outside the uterus on ultrasound is firm evidence of an ectopic pregnancy. On occasion, an ultrasound study is not definitive and the diagnosis must be made with combined diagnostic aids (beta-hCG and progesterone levels, ultrasound, pelvic examination, and clinical judgment).

Occasionally, the clinical picture makes the diagnosis relatively easy. However, when the clinical signs and symptoms are inconclusive, which is often the case, other procedures may be needed. Laparoscopy can be used because the physician can visually detect an unruptured tubal pregnancy and thereby circumvent the risk of its rupture.

Medical Management

Surgical Management

When surgery is performed early, almost all patients recover rapidly; if tubal rupture occurs, mortality increases. The type of surgery is determined by the size and extent of local tubal damage. Conservative surgery includes "milking" an ectopic pregnancy from the tube. Resection of the involved fallopian tube with end-to-end anastomosis may be effective. Some surgeons attempt to salvage the tube with a salpingotomy, which involves opening and evacuating the tube and controlling bleeding. More extensive surgery includes removing the tube alone (salpingectomy) or with the ovary (salpingo-oophorectomy). Depending on the amount of blood lost, blood component therapy and treatment for hemorrhagic shock may be necessary before and during surgery. Surgery may also be indicated in women unlikely to comply with close monitoring or those who live too far away from a health care facility to obtain the monitoring needed with nonsurgical management.

Pharmacologic Therapy

Another option is the use of methotrexate without surgery. Because methotrexate stops the pregnancy from progressing by interfering with DNA synthesis and the multiplication of cells, it interrupts early, small, unruptured ectopic pregnancies. The patient must be hemodynamically stable; have no active renal or hepatic disease; have no evidence of thrombocytopenia or leukopenia; and have a very small, unruptured ectopic pregnancy on ultrasound. Other indications may include no fetal cardiac activity and no active abdominal bleeding. In most cases, an intramuscular methotrexate dose of 50 mg/m^2 is used. Oral use of methotrexate is considered less effective, and direct injection into the ectopic mass is infrequently administered (ACOG, 2010).

Side effects of methotrexate include abdominal cramping, mucositis, and renal and hepatic damage. Allergic reactions have occurred in patients receiving high doses. NSAIDs may enhance methotrexate toxicity, and folic acid may lower its efficacy (ACOG, 2010).

NURSING PROCESS

The Patient With an Ectopic Pregnancy

Assessment

The health history includes the menstrual pattern and any (even slight) bleeding since the last menstrual period. The nurse elicits the patient's description of pain and its location. The nurse asks the patient whether any sharp, colicky pains have occurred. Then, the nurse notes whether pain radiates to the shoulder and neck (possibly caused by rupture and pressure on the diaphragm).

In addition, the nurse monitors vital signs, level of consciousness, and the nature and amount of vaginal bleeding.

If possible, the nurse assesses how the patient is coping with the abnormal pregnancy and likely loss.

Diagnosis

Nursing Diagnoses

Based on the assessment data, major nursing diagnoses may include the following:

- Acute pain related to the progression of the tubal pregnancy
- Grieving related to the loss of pregnancy and anticipatory effect on future pregnancies
- Deficient knowledge related to the treatment and effect on future pregnancies

Collaborative Problems/Potential Complications

Potential complications may include:

- Hemorrhage
- Hemorrhagic shock

Planning and Goals

The major goals may include relief of pain; acceptance and resolution of grief and pregnancy loss; increased knowledge about ectopic pregnancy, its treatment, and its outcome; and absence of complications.

Nursing Interventions

Relieving Pain

The abdominal pain associated with ectopic pregnancy may be described as cramping or severe continuous pain. If the patient is to have surgery, preanesthetic medications may provide pain relief. Postoperatively, analgesic agents are administered liberally; this promotes early ambulation and enables the patient to cough and take deep breaths.

Supporting the Grieving Process

Patients' distress levels vary. If the pregnancy was desired, loss may or may not be expressed verbally by the patient and her partner. The impact may not be fully realized until much later. The nurse should be available to listen and provide support. The patient's partner, if appropriate, should participate in this process. Even if the pregnancy was unintended, a loss has been experienced, and a grief reaction may occur.

Monitoring and Managing Potential Complications

Potential complications of ectopic pregnancy are hemorrhage and shock. Careful assessment is essential to detect the development of these complications. Continuous monitoring of vital signs, level of consciousness, amount of bleeding, and intake and output provide information about the possibility of hemorrhage and the need to prepare for IV therapy. Bed rest is indicated. Hematocrit, hemoglobin, and blood gases are monitored to assess hematologic status and adequacy of tissue perfusion. Significant deviations in these laboratory values are reported immediately, and the patient is prepared for possible surgery. Blood component therapy may be required if blood loss has been rapid and extensive. If hypovolemic shock occurs, the treatment is directed toward reestablishing tissue perfusion and adequate blood volume. (See Chapter 14 for a discussion of the IV fluids and medications used in treating shock.)

The nurse has an important role in prevention by being alert to patients with abnormal bleeding who may be at risk for an ectopic pregnancy and referring them immediately for care. It is necessary to keep a high index of suspicion in daily practice when a woman of childbearing age, particularly one who is not using an effective method of contraception consistently, reports abdominal discomfort or abnormal bleeding.

Promoting Home and Community-Based Care

Educating Patients About Self-Care. If the patient has experienced life-threatening hemorrhage and shock, these complications are addressed and treated before any in-depth education can begin. At this time, the patient's and the nurse's attention is focused on the crisis, not on learning. At a later time, the patient begins to ask questions about what happened and why certain procedures were performed. Procedures are explained in terms that the distressed and apprehensive patient can understand. The patient's partner is included in education when possible. After the patient recovers from postoperative discomfort, it may be more appropriate to address any questions and concerns that she and her partner have, including the effect of this pregnancy or its treatment on future pregnancies. The patient should be advised that ectopic pregnancies may recur. The patient is educated about possible complications and instructed to report early signs and symptoms. It is important to review signs and symptoms with the patient and instruct her to report an abnormal menstrual period promptly.

Continuing Care. Because of the risk of subsequent ectopic pregnancies, the patient is advised to seek preconception counseling before considering future pregnancies and to seek early prenatal care. Follow-up contact allows the nurse to answer questions and clarify information for the patient and her partner.

Evaluation

Expected patient outcomes may include:

1. Experiences relief of pain
   a. Reports a decrease in pain and discomfort
   b. Ambulates as prescribed; performs coughing and deep breathing
2. Begins to accept loss of pregnancy and expresses grief by verbalizing feelings and reactions to loss
3. Verbalizes an understanding of the causes of ectopic pregnancy
4. Experiences no complications
   a. Exhibits no signs of bleeding, hemorrhage, or shock
   b. Has decreased amounts of discharge (on perineal pad)
   c. Has normal skin color and turgor
   d. Exhibits stable vital signs and adequate urine output
   e. Levels of beta-hCG return to normal

Perimenopause

Perimenopause is the period extending from the first signs of menopause—usually hot flashes, vaginal dryness, or irregular menses—to beyond the complete cessation of menses. It has also been defined as the period around menopause, lasting to 1 year after the last menstrual period. Women often have varied beliefs about aging, and these must be considered when caring for or educating perimenopausal patients.

Nursing Management

Perimenopausal women often benefit from information about the subtle physiologic changes they are experiencing. Perimenopause has been described as an opportune time for educating women about health promotion and disease prevention strategies. When discussing health-related concerns with midlife women, nurses should consider the following issues:

- Sexuality, fertility, contraception, and STIs
- Unintended pregnancy (if contraception is not used correctly and consistently)
- Oral contraceptive use. Oral contraceptives provide perimenopausal women with protection against uterine cancer, ovarian cancer, anemia, pregnancy, and fibrocystic breast changes as well as relief from perimenopausal symptoms (Alan Guttmacher Institute, 2012). This option should be discussed with perimenopausal women. Women who smoke and are 35 years or older should not take oral contraceptive agents because of an increased risk of CVD. Contraception is discussed in detail earlier in this chapter.
- Breast health. About 16% of cases of breast cancer occur in perimenopausal women, so breast self-examination, routine physical examinations, and mammograms are essential.

Menopause

Menopause is the permanent physiologic cessation of menses associated with declining ovarian function (Porth & Matfin, 2009). Most women stop menstruating between 48 and 55 years of age. Postmenopause is the period beginning from about 1 year after menses cease. Menopause may be associated with some atrophy of breast tissue and genital organs, loss in bone density, and vascular changes.

Menopause starts gradually and is usually signaled by changes in menstruation. The monthly flow may increase or decrease, become irregular, and finally cease. Often, the interval between periods is longer; a lapse of several months between periods is not uncommon. Changes signaling menopause begin to occur as early as the late 30s, when ovulation occurs less frequently, estrogen levels fluctuate, and FSH levels increase in an attempt to stimulate estrogen production (Porth & Matfin, 2009).

Postmenopausal Bleeding

Bleeding 1 year after menses cease at menopause must be investigated, and a malignant condition must be considered until proven otherwise. A transvaginal ultrasound can be used to measure the thickness of the endometrial lining (Fischbach & Dunning, 2009). The uterine lining in postmenopausal women should be thin because of low estrogen levels. A thicker lining warrants further evaluation by endometrial biopsy or a D & C.

Clinical Manifestations

Because of these hormonal changes, some women notice irregular menses, breast tenderness, and mood changes long before menopause occurs. The hot or warm flashes and night sweats reported by some women are thought to be caused by hormonal changes and denote vasomotor instability. They may vary in intensity from a barely perceptible warm feeling to a sensation of extreme warmth accompanied by profuse sweating, causing discomfort, sleep disturbances, and subsequent fatigue. Other physical changes may include increased bone loss (see Chapter 42).

The entire genitourinary system is affected by the reduced estrogen level. Changes in the vulvovaginal area may include a gradual thinning of pubic hair and a gradual shrinkage of the labia. Vaginal secretions decrease, and women may report dyspareunia (discomfort during intercourse). The vaginal pH increases during menopause, predisposing women to bacterial infections and atrophic vaginitis. Discharge, itching, and vulvar burning may result.

Some women report fatigue, forgetfulness, weight gain, irritability, trouble sleeping, feeling "blue," and feelings of panic. Menopausal complaints need to be evaluated carefully because they may indicate other disorders. Most women have few problems and are relieved to be free from menstrual periods.

Psychological Considerations

Women's reactions and feelings related to loss of reproductive capacity may vary. Some women may experience role confusion, whereas others experience a sense of sexual and personal freedom. Women may be relieved that the childbearing phase of their lives is over. Each woman's personal views about menopause and circumstances affect her response and must be considered on an individual basis. Nurses need to be sensitive to all possibilities and take their cues from the patient.

Medical Management

Women approaching menopause often have many concerns about their health. Some have concerns based on a family history of heart disease, osteoporosis, or cancer. Each woman needs to be as knowledgeable as possible about her health options and should be encouraged to discuss her concerns with her primary provider so that she can make an informed decision about managing menopausal symptoms and maintaining her health.

Hormone Therapy

HT or menopausal hormonal therapy (previously referred to as hormone replacement therapy [HRT]) has been found to increase some health disorders and to be less effective in preventing others than previously believed. Although HT decreases hot flashes and reduces the risk of osteoporotic fractures as well as colorectal cancer, studies have shown that it increases the risk of breast cancer, heart attack, stroke, and blood clots (Heiss, Wallace, Anderson, et al., 2008). Thus, the benefits of HT are inadequate given the increased risk of these other disorders. Because of these findings, many women have discontinued HT or are reluctant to begin HT. Although some women have elected to use HT in low doses on the advice of their health care providers, the effect of these low doses has not been studied. The current recommendation for treatment of hot flashes with HT is to use the lowest dose possible for the shortest time possible.

Nurses need to be knowledgeable about HT-related issues to be able to respond to women's questions about HT use.

Methods of Administration

Both estrogen and progestin are prescribed for women who have not had a hysterectomy; progestin prevents proliferation of the uterine lining and hyperplasia. Women who no longer have a uterus because of hysterectomy can take estrogen without progestin (i.e., unopposed estrogen) because there is no longer a risk of estrogen-induced hyperplasia of the uterine lining. Although there is a slight increase of risk of stroke in women taking estrogen alone following hysterectomy, the risk of breast cancer is unchanged (Heiss et al., 2008).

Some women take both estrogen and progestin daily; others take estrogen for 25 consecutive days each month, with progestin taken in cycles (e.g., 10 to 14 days of the month). Women who take HT for 25 days often experience bleeding after completing the progestin. Other women take estrogen and progestin every day and usually experience no bleeding. They occasionally have irregular spotting, which should be evaluated by their primary provider. Progestin administration may be oral, transdermal, vaginal, or intrauterine.

Estrogen patches, which are replaced once or twice weekly, are another option but require a progestin along with them if the woman still has a uterus. Another type of patch provides estrogen and progestin treatment. Skin should be dry at the area of application, and cleansing the site with alcohol may improve adhesiveness. Vaginal treatment with an estrogen cream, suppository, or an estradiol ring (Estring) may be used for vasomotor symptoms, vaginal dryness, or atrophy (Karch, 2012). The estradiol ring is a small, flexible vaginal ring that slowly releases estrogen in small doses over 3 months.

Risks and Benefits

HT is contraindicated in women with a history of breast cancer, vascular thrombosis, impaired liver function, uterine cancer, and undiagnosed abnormal vaginal bleeding. The risk of venous thromboembolism is increased with HT (ACOG, 2013b). Women who elect to take HT should be educated about the signs and symptoms of deep vein thrombosis (DVT) and pulmonary embolism (PE) and instructed to report these signs and symptoms immediately (see Chapter 30 for discussion of DVT and Chapter 23 for discussion of PE). Women who take HT should be assessed for leg redness, tenderness, chest pain, and shortness of breath. Furthermore, they need to be informed about the importance of regular follow-up care, including a yearly physical examination and mammogram. An endometrial biopsy is indicated for any irregular bleeding. Because the risk of complications increases the longer HT is used, HT should be used for the shortest time possible (Karch, 2012). Estrogen alone or in combination with a progestin does not reduce the risk of dementia or cognitive impairment.

Alternative Therapy for Hot Flashes

Because women often seek information about alternatives to the use of HT, nurses must be knowledgeable about other approaches that women can use to promote their health in the peri- and postmenopausal periods. Problematic hot flashes have been treated with low-dose venlafaxine (Effexor) and other medications. Similarly, vitamin B_6 and vitamin E may be effective. Some women have expressed interest in other alternative treatments (e.g., natural estrogens and progestins, black cohosh, ginseng, dongquai, soy products, and several other herbal preparations); however, few data exist about their safety or effectiveness. Therefore, assessment of menopausal women should address their use of complementary and alternative therapies and supplements.

Maintaining Bone Health

Acceleration of bone loss resulting in osteoporosis and microarchitectural deterioration of bone tissue occurs at menopause and leads to increased bone fragility and risk of fracture. (Osteoporosis and its treatment are described in detail in Chapter 42.)

Maintaining Cardiovascular Health

A variety of strategies can help to lower the risk of heart disease in women, including lifestyle changes and behavioral strategies. (Prevention and treatment of CVD are discussed in detail in Chapter 25.)

Behavioral Strategies

As stated previously, regular physical exercise is beneficial. It may also reduce stress, enhance well-being, and improve self-image. In addition, weight-bearing exercise may prevent loss of muscle tissue and bone tissue.

Women are also encouraged to participate in other health-promoting activities. These include regular health screening recommended for women at the time of menopause: gynecologic examinations, mammograms, colonoscopy, fecal occult blood testing, and bone mineral density testing if at risk for osteoporosis.

Nutritional Therapy

Women are encouraged to decrease their fat and caloric intake and increase their intake of whole grains, fiber, fruit, and vegetables.

Nursing Management

Nurses can encourage women to view menopause as a natural change resulting in freedom from symptoms related to menses. No relationship exists between menopause and mental health problems; however, social circumstances (e.g., adolescent children, ill partners, and dependent or ill parents) that may coincide with menopause can be stressful.

Measures should be taken to promote general health (ACOG, 2012a). The nurse explains to the patient that cessation of menses is a normal occurrence that is rarely accompanied by nervous symptoms or illness. The current expected lifespan after menopause for the average woman is 30 to 35 years, which may encompass as many years as the childbearing phase of her life. Normal sexual urges continue, and women retain their usual response to sex long after menopause. Many women enjoy better health after menopause than before, especially those who have experienced dysmenorrhea. The individual woman's evaluation of herself and her worth, now and in the future, is likely to affect her emotional reaction to menopause. Patient education and counseling regarding healthy lifestyles, health promotion, and health screening are of paramount importance (ACOG, 2012a) (Chart 56-15).

Chart 56-15 HOME CARE CHECKLIST

The Woman Approaching Menopause

At the completion of home care education, the patient or caregiver will be able to:	PATIENT	CAREGIVER
• Describe menopause as a normal period in a woman's life.	✔	✔
• State that fatigue and stress may worsen hot flashes.	✔	✔
• State that a nutritious diet and weight control will enhance physical and emotional well-being.	✔	✔
• State the importance of exercising for at least 30 minutes 3 or 4 times a week to maintain good health.	✔	✔
• Describe involvement in outside activities as beneficial in reducing anxiety and tension.	✔	✔
• Identify the following as changes that often occur in midlife: departure of children, aging, dependence of parents, possible loss of loved ones.	✔	✔
• Describe this phase of life as having the potential for intellectual growth, personal accomplishment, and initiation of new activities.	✔	✔
• State the following points about sexual activity:		
• Frequent sexual activity helps to maintain the elasticity of the vagina.	✔	
• Contraception is advised until 1 year passes without menses.	✔	
• Safer sex is important at any age.	✔	
• Sexual functioning may be enhanced at midlife.	✔	
• Identify the importance of an annual physical examination to screen for problems and to promote general health.	✔	
• Identify strategies and methods to prevent or manage the following problems:		
• *Itching or burning of vulvar areas:* See primary provider to rule out dermatologic abnormalities and, if appropriate, to obtain a prescription for a lubricating or hormonal cream.	✔	
• *Dyspareunia (painful intercourse) due to vaginal dryness:* Use a water-soluble lubricant (e.g., K-Y Jelly, Astroglide, Replens), hormone cream, or contraceptive foam.	✔	
• *Decreased perineal muscle tone and bladder control:* Practice Kegel exercises daily (contract the perineal muscles as though stopping urination; hold for 5–10 seconds and release; repeat frequently during the day).	✔	
• *Dry skin:* Use mild emollient skin cream and lotions to prevent dry skin.	✔	
• *Weight control:* Join a weight reduction support group such as Weight Watchers or a similar group if appropriate, or consult a registered dietitian for guidance about the tendency to gain weight, particularly around the hips, thighs, and abdomen.	✔	
• *Osteoporosis:* Observe recommended calcium and vitamin D intake, including calcium supplements, if indicated, to slow the process of osteoporosis; avoid smoking, alcohol, and excessive caffeine, all of which increase bone loss. Perform weight-bearing exercises. Undergo bone density testing when appropriate.	✔	
• *Risk for urinary tract infection (UTI):* Drink 6–8 glasses of water daily as a possible way to reduce the incidence of UTI related to atrophic changes of the urethra.	✔	
• *Vaginal bleeding:* Report any bleeding after 1 year of no menses to the primary provider *immediately, no matter how minimal.*	✔	

Critical Thinking Exercises

1 **ebp** A 20-year-old female college student comes to the student health clinic for a gynecologic examination because she anticipates having sex with her new boyfriend. She asks you for advice about avoiding pregnancy and STIs. What is the evidence base for contraceptive methods for this college student? Specify the criteria used to evaluate the strength of the evidence for the practices that you identify.

2 **pq** A 28-year-old patient accompanied by her partner is returning to your unit post surgery following an ectopic pregnancy. She is alert and oriented when aroused. Her pain level is 4 (on a 0 to 10 numeric pain scale), and she is crying. Her blood pressure is 128/78 mm Hg, heart rate is 128 bpm, respiratory rate is 18 breaths/min and regular, and temperature is 36.7°C (98.1°F). Describe your priorities for assessment and nursing care for this patient and her partner.

3 During her annual physical examination, a 45-year-old woman states that she is concerned about approaching menopause. Identify what further assessments are needed at this time. Develop an education plan that will include critical elements the woman will need as she approaches menopause. What resources would you recommend to her?

Brunner Suite Resources

Explore these additional resources to enhance learning for this chapter:

- **NCLEX-Style Questions and Other Resources on thePoint, http://thePoint.lww.com/Brunner13e**
- Study Guide
- PrepU
- Clinical Handbook
- Handbook of Laboratory and Diagnostic Tests

References

*Asterisk indicates nursing research.

**Double asterisk indicates classic reference.

Books

**Annon, J. S. (1976). *The behavioral treatment of sexual problems*. Honolulu, HI: Enabling Systems.

Carcio, H., & Secor, M. C. (2010). *Advanced health assessment of women* (2nd ed.). New York: Springer.

Eliopoulos, C. (2010). *Gerontological nursing* (7th ed.). Philadelphia: Lippincott Williams & Wilkins.

Fischbach, F. T., & Dunning, M. B. (2009). *A manual of laboratory and diagnostic tests* (8th ed.). Philadelphia: Lippincott Williams & Wilkins.

Fritz, M. A., & Speroff, L. (2011). *Clinical gynecologic endocrinology and infertility*. Philadelphia: Lippincott Williams & Wilkins.

Hatcher, R., Trussell, J., Nelson, A., et al. (2011). *Contraceptive technology* (20th ed.). New Tork: Ardent Media.

Hawkins, J. W., Roberto-Nichols, D. M., & Stanley-Haney, J. L. (2012). *Guidelines for nurse practitioners in gynecologic settings* (10th ed.). New York: Springer.

Karch, A. (2012). *2012 Lippincott's nursing drug guide*. Philadelphia: Lippincott Williams & Wilkins.

Link, D. G. (2011). Reproductive anatomy, physiology, and the menstrual cycle. In S. Mattson & J. E. Smith (Eds.). *Core curriculum for maternal-newborn nursing* (4th ed.). St. Louis: Saunders.

Porth, C. M., & Matfin, G. (2009). *Pathophysiology: Concepts of altered health states* (8th ed.). Philadelphia: Lippincott Williams & Wilkins.

Schuiling, K., & Likis, F. (2013). *Women's gynecologic health*. Boston: Jones & Bartlett.

Tharpe, N., Farley, C., & Jordan, R. (2013). *Clinical practice guidelines for midwifery & women's health*. Boston: Jones & Bartlett.

Weber, J., & Kelley, J. (2010). *Health assessment in nursing* (4th ed.). Philadelphia: Lippincott Williams & Wilkins.

Journals and Electronic Documents

Alan Guttmacher Institute. (2012). *Contraceptive use in the United States*. Available at: www.guttmacher.org/pubs/fb_contr_use.pdf

Allen, R., & Goldberg, A. (2007). Emergency contraception: A clinical review. *Clinical Obstetrics and Gynecology, 50*(4), 927–936.

American College of Obstetricians and Gynecologists. (2006). ACOG Practice Bulletin No. 73: Use of hormonal contraception in women with coexisting medical conditions.. *Obstetrics and Gynecology, 107*(6), 1453–1472.

American College of Obstetricians and Gynecologists. (2010). ACOG Practice Bulletin No. 94: Medical management of ectopic pregnancy. *Obstetrics and Gynecology, 111*(6), 1479–1485.

American College of Obstetricians and Gynecologists. (2011). ACOG Practice Bulletin No. 121: Long-acting reversible contraception: Implants and intrauterine devices. *Obstetrics and Gynecology, 118*(1), 184–196.

American College of Obstetricians and Gynecologists. (2012a). Committee Opinion No. 534: Well-woman visit. *Obstetrics and Gynecology, 120*(8), 421–424.

American College of Obstetricians and Gynecologists. (2012b). Committee Opinion No. 525: Health care for lesbians and bisexual women. *Obstetrics and Gynecology, 119*(1), 1077–1084.

American College of Obstetricians and Gynecologists. (2013a). Committee Opinion No. 556: Postmenopausal estrogen therapy: Administration and risk of venous thromboembolism. Available at: www.acog.org/~/media/Committee%20Opinions/Committee%20on%20Gynecologic%20Practice/co556.pdf?dmc=1&ts=20130406T1350390473

American College of Obstetricians and Gynecologists. (2013b). Statement on FDA approval of OTC emergency contraception. Available at: www.acog.org/About%20ACOG/News%20Room/News%20Releases/2013/Statement%20on%20FDA%20Approval%20of%20OTC%20Emergency%20Contraception.aspx

*Fehring, R., Schneider, M., Raviele, K., et al. (2007). Efficacy of cervical mucus observations plus electronic hormonal fertility monitoring as a method of natural family planning. *Journal of Obstetrical, Gynecologic, and Neonatal Nursing, 36*(2), 152–160.

Heiss, G., Wallace, R., Anderson, G. L., et al. (2008). Health risks and benefits 3 years after stopping randomized treatment with estrogen and progestin. *Journal of American Medical Association, 299*(9), 1036–1045.

Kaiser Family Foundation. (2012). *Women's health insurance coverage. Women's health insurance coverage: Fact sheet*. Available at: www.kff.org/womenshealth/upload/6000–10.pdf

Lopez, L. M., Kaptein, A., & Helmerhorst, F. M. (2008). Oral contraceptives containing drospirenone for premenstrual syndrome. *Cochrane Database of Systematic Reviews*, (1), CD006586.

Marecki, M., & Seo, J. Y. (2010). Perinatal urinary and fecal incontinence suffering in silence. *Journal of Perinatal Neonatal Nursing, 24*(4), 330–340.

Mayo Clinic. (2011). *Toxic shock syndrome: Prevention*. Available at: www.mayoclinic.com/health/toxic-shock-syndrome/DS00221/DSECTION=prevention

*Ochoa-Frongia, L., Thompson, H. S., Lewis-Kelly, Y., et al. (2012). Breast and cervical cancer screening and health beliefs among African American women attending educational programs. *Health Promotion Practice, 13*(4), 447–453.

Politi, M. C., Shlentz, M. D., Col, N. F., et al. (2008). Revisiting the duration of vasomotor symptoms of menopause: A meta-analysis. *Journal of Internal General Medicine, 23*(9), 1507–1513.

*Sharts-Hopko, N. C., Smeltzer, S., Ott, B. B., et al. (2010). Healthcare experiences of women with visual impairment. *Clinical Nurse Specialist, 24*(3), 149–153.

Smeltzer, S. C. (2007). Pregnancy in women with physical disabilities *Journal of Obstetrical, Gynecologic, and Neonatal Nursing, 36*(1), 88–96.

Smeltzer, S. C., & Wetzel-Effinger, L. (2009). Pregnancy in women with spinal cord injury. *Topics in Spinal Cord Injury Rehabilitation, 15*(1), 29–42.

*Smeltzer, S. C., Sharts-Hopko, N. C., Ott, B., et al. (2007). Perspectives of women with disabilities on reaching those who are hard to reach. *Journal of Neuroscience Nursing, 39*(3), 163–171.

Smith, J. R., & Chelnow, D. (2012). Management of the third stage of labor. *Medscape*. Available at: emedicine.medscape.com/article/275304-overview

Smith, R. A., Cokkinides, V., & Brawley, O. W. (2012). Cancer screening in the United States, 2012. *CA: A Cancer Journal for Clinicians, 62*(2), 129–142.

Spatz, E. S., Canavan, M. E., & Desai, M. M. (2009). From here to JUPITER: Identifying new patients for statin therapy using data from the 1999–2004 National Health and Nutrition Examination Survey. *Circulation Cardiovascular Quality and Outcomes, 2*(10), 41–48.

Sudduth, A., &. Linton, D. (2011). Gynecologic care of women with disabilities: Implications for nurses. *Nursing for Women's Health, 15*(2), 138–147.

Thakur, S., & Supiano, M. (2007). Screening for common clinical conditions in older women. *Clinical Obstetrics and Gynecology, 50*(3), 767–775.

Turner, D. (2007). Female genital cutting. *Nursing for Women's Health, 11*(4), 366–372.

Disorders of the female reproductive system can be minor or serious but are often anxiety producing and distressing. Some disorders are self-limited and cause only minor inconvenience to the woman; others are life threatening and require immediate attention and long-term therapy. Many disorders are managed by the patient at home, whereas others require hospitalization and surgical intervention. Nurses not only need to be knowledgeable about these disorders but also need to be sensitive to patient concerns and possible discomfort in discussing and dealing with these disorders.

VULVOVAGINAL INFECTIONS

Vulvovaginal infections are common, and nurses have an important role in providing information that may prevent their occurrence. To help prevent these infections, women need to understand their anatomy and normal vulvovaginal health.

The vagina is protected against infection by its normally low pH (3.5 to 4.5), which is maintained in part by the actions of *Lactobacillus acidophilus*, the dominant bacteria in a healthy vaginal ecosystem. These bacteria suppress the growth of anaerobes and produce lactic acid, which maintains normal pH. They also produce hydrogen peroxide, which is toxic to anaerobes. The risk of infection increases if a woman's resistance is reduced by stress or illness, if the pH is altered, or if a pathogen is introduced. Continued research into causes and treatments is needed, along with better ways to encourage growth of lactobacilli.

The epithelium of the vagina is highly responsive to estrogen, which induces glycogen formation. The subsequent breakdown of glycogen into lactic acid assists in producing a low vaginal pH. When estrogen decreases during lactation and menopause, glycogen also decreases. With reduced glycogen formation, infections may occur. In addition, as estrogen production ceases during the peri- and postmenopausal periods, the vagina and labia may atrophy (thin), making the vaginal area more susceptible to infection. When patients are treated with antibiotic agents, the normal vaginal flora are reduced. This results in altered pH and growth of fungal organisms. Other factors that may initiate or predispose to infections include contact with an infected partner and wearing tight, nonabsorbent, and heat- and moisture-retaining clothing.

Concept Mastery Alert

Nurses must be able to identify the risk factors for vulvovaginal infections in order to effectively educate patients about preventing these common disorders. Chart 57-1 summarizes this key information.

Vaginitis (inflammation of the vagina) is a group of conditions that cause vulvovaginal symptoms such as itching, irritation, burning, and abnormal discharge. Bacterial vaginitis is the most common cause, followed by vulvovaginal candidiasis and trichomoniasis (Centers for Disease Control and Prevention [CDC], 2010a) (Table 57-1). Other types include desquamative vaginitis, atrophic vaginitis, various vulvar dermatologic conditions, and vulvodynia. Normal vaginal discharge, which may occur in slight amounts during ovulation or just before the onset of menstruation, is clear to white, odorless, and viscous. It becomes more profuse when vaginitis occurs. Urethritis may accompany vaginitis because of the proximity of the urethra to the vagina. Discharge that occurs with vaginitis may produce itching, odor, redness, burning, or edema, which may be aggravated by voiding and defecation. After the causative organism has been identified, appropriate treatment (discussed later) is prescribed. This may include an oral medication or a local medication that is inserted into the vagina using an applicator.

Chart 57-1 RISK FACTORS
Vulvovaginal Infections

- Premenarche
- Pregnancy
- Perimenopause/Menopause
- Poor personal hygiene
- Tight undergarments
- Synthetic clothing
- Frequent douching
- Allergies
- Use of oral contraceptives
- Long-term or repeated use of broad-spectrum antibiotics
- Diabetes
- Low estrogen levels
- Intercourse with infected partner
- Oral–genital contact (yeast can inhabit the mouth and intestinal tract)
- Human immunodeficiency virus infection

Adapted from Schuiling, K. D., & Likis, F. E. (2013). *Women's gynecologic health* (2nd ed.). Burlington, MA: Jones & Bartlett.

Candidiasis

Vulvovaginal **candidiasis** is a fungal or yeast infection caused by strains of *Candida* (see Table 57-1). *Candida albicans* accounts for most cases, but other strains, such as *Candida glabrata*, may also be implicated. Many women with a healthy vaginal ecosystem harbor *Candida* but are asymptomatic. Certain conditions favor the change from an asymptomatic state to colonization with symptoms. For example, the use of antibiotic agents decreases bacteria, thereby altering the natural protective organisms usually present in the vagina. Although infections can occur at any time, they occur more commonly in pregnancy or with a systemic condition such as diabetes or human immunodeficiency virus (HIV) infection, or when patients are taking medications such as corticosteroids or oral contraceptive agents (Schuiling & Likis, 2013). Information about incidence is incomplete because there is no mechanism for the reporting of cases.

Clinical Manifestations

Clinical manifestations include a vaginal discharge that causes pruritus (itching) and subsequent irritation. The discharge may be watery or thick but usually has a white, cottage cheese–like appearance. Symptoms are usually more severe just before menstruation and may be less responsive to treatment during pregnancy. Diagnosis is made

TABLE 57-1 Vaginal Infections and Vaginitis

Infection	Cause	Clinical Manifestations	Management Strategies
Candidiasis	*Candida albicans*, *glabrata*, or *tropicalis*	Inflammation of vaginal epithelium, producing itching, reddish irritation White, cheeselike discharge clinging to epithelium	Eradicate the fungus by administering an antifungal agent. Some more frequently used vaginal creams and suppositories are miconazole (Monistat) and clotrimazole (Gyne-Lotrimin). Review other causative factors (e.g., antibiotic therapy, nylon underwear, tight clothing, pregnancy, oral contraceptive agents). Assess for diabetes and human immunodeficiency virus infection in patients with recurrent monilia.
Gardnerella-associated bacterial vaginosis	*Gardnerella vaginalis* and vaginal anaerobes	Usually no edema or erythema of vulva or vagina Gray-white to yellow-white discharge clinging to external vulva and vaginal walls	Administer metronidazole (Flagyl), with instructions about avoiding alcohol while taking this medication. If infection is recurrent, may treat partner
Trichomonas vaginalis vaginitis	*Trichomonas vaginalis*	Inflammation of vaginal epithelium, producing burning and itching Frothy yellow-white or yellow-green vaginal discharge	Relieve inflammation, restore acidity, and reestablish normal bacterial flora; provide oral metronidazole for patient and partner.
Bartholinitis (infection of greater vestibular gland)	*Escherichia coli* *T. vaginalis* Staphylococcus Streptococcus Gonococcus	Erythema around vestibular gland Swelling and edema Abscessed vestibular gland	Drain the abscess; provide antibiotic therapy; excise gland of patients with chronic bartholinitis.
Cervicitis—acute and chronic	Chlamydia Gonococcus Streptococcus Many pathogenic bacteria	Profuse purulent discharge Backache Urinary frequency and urgency	Determine the cause—perform cytologic examination of cervical smear and appropriate cultures. Eradicate the gonococcal organism, if present: penicillin (as directed) or spectinomycin or tetracycline, if patient is allergic to penicillin. Tetracycline, doxycycline (Vibramycin) to eradicate chlamydia Eradicate other causes.
Atrophic vaginitis	Lack of estrogen; glycogen deficiency	Discharge and irritation from alkaline pH of vaginal secretions	Provide topical vaginal estrogen therapy; improve nutrition if necessary; relieve dryness through the use of moisturizing medications.

Adapted from Karch, A. M. (2012). *2012 Lippincott's nursing drug guide*. Philadelphia: Lippincott Williams & Wilkins.

by microscopic identification of spores and **hyphae** (long, branching filamentous structures) on a glass slide prepared from a discharge specimen mixed with potassium hydroxide. With candidiasis, the pH is 4.5 or less (Schuiling & Likis, 2013). Manifestations may be fairly uncomplicated, occurring sporadically in healthy women, or recurrent and complicated in women who have diabetes or are pregnant, immunocompromised, or obese.

Medical Management

The goal of management is to eliminate symptoms. Treatments include antifungal agents such as miconazole (Monistat), nystatin (Mycostatin), clotrimazole (Gyne-Lotrimin), and terconazole (Terazol) cream. These agents are inserted into the vagina with an applicator at bedtime. There are 1-night, 3-night, and 7-night treatment courses available (Karch, 2012). Oral medication (fluconazole [Diflucan]) is also available in a one-pill dose. Relief should be noted within 3 days.

Some vaginal creams are available without a prescription; however, patients are cautioned to use these creams only if they are certain that they have a yeast or monilial infection. Patients often use these remedies for problems other than yeast infections. If a woman is uncertain about the cause of her symptoms or if relief has not been obtained after using these creams, she should be instructed to seek health care promptly. Yeast infections can become recurrent or complicated. Women may have more than four infections in a year and severe symptoms due to preexisting conditions such as diabetes or immunosuppression. Cell-mediated immunity may be a factor. Women with recurrent yeast infections benefit from a comprehensive gynecologic assessment.

Bacterial Vaginosis

Bacterial vaginosis is caused by an overgrowth of anaerobic bacteria and *Gardnerella vaginalis* normally found in the vagina and an absence of lactobacilli (see Table 57-1). Risk factors include douching after menses, smoking, multiple sex partners, and other sexually transmitted infections (STIs) (also referred to as sexually transmitted diseases [STDs]). Bacterial vaginosis is not considered an STI but is associated with sexual activity, and incidence is increased in female same-sex partners (Schuiling & Likis, 2013).

Clinical Manifestations

Bacterial vaginosis can occur throughout the menstrual cycle and does not produce local discomfort or pain. More than half of patients with bacterial vaginosis do not notice any symptoms. Discharge, if noticed, is heavier than normal and gray to yellowish white in color. It is characterized by a fishlike odor that is particularly noticeable after sexual intercourse or during menstruation as a result of an increase in vaginal pH. The pH of the discharge is usually greater than 4.7 because of the amines that result from enzymes from anaerobes. The fishlike odor can be detected readily by adding a drop of potassium hydroxide to a glass slide with a sample of vaginal discharge, which releases amines; this is referred to as a positive "whiff" test. Under the microscope, vaginal cells are coated with bacteria and are described as "clue cells." Lactobacilli, which serve as a natural host defense, are usually absent. Bacterial vaginosis is not usually considered a serious condition, although it can be associated with premature labor, premature rupture of membranes, endometritis, and pelvic infection (Schuiling & Likis, 2013).

Medical Management

Metronidazole (Flagyl), administered orally twice a day for 1 week, is effective; a vaginal gel is also available. Clindamycin (Cleocin) vaginal cream or ovules (oval suppositories) are also effective. Treatment of patients' partners does not seem to be effective, but the use of condoms may be helpful. Research is exploring the link between vitamin D deficiency and bacterial vaginosis (Harris, 2011).

Trichomoniasis

Trichomonas vaginalis is a flagellated protozoan that causes a common STI often called *trich*. About 3.7 million cases occur each year in the United States (CDC, 2010b). Trichomoniasis may be transmitted by an asymptomatic carrier who harbors the organism in the urogenital tract (see Table 57-1). It may increase the risk of contracting HIV from an infected partner and may play a role in development of cervical neoplasia, postoperative infections, adverse pregnancy outcomes, pelvic inflammatory disease (PID), and infertility. Incidence is 10 times higher in African American women, and approximately 85% of women are asymptomatic (Schuiling & Likis, 2013).

Clinical Manifestations

Clinical manifestations include a vaginal discharge that is thin (sometimes frothy), yellow to yellow-green, malodorous, and very irritating. An accompanying vulvitis may result, with vulvovaginal burning and itching. Diagnosis is made most often by microscopic detection of the motile causative organisms or less frequently by culture. Inspection with a speculum often reveals vaginal and cervical erythema (redness) with multiple small petechiae ("strawberry spots") (Schuiling & Likis, 2013). Testing of a trichomonal discharge demonstrates a pH greater than 4.5.

Medical Management

The most effective treatment for trichomoniasis is metronidazole or tinidazole (Tindamax). Both partners receive a one-time loading dose or a smaller dose three times a day for 1 week (CDC, 2010a). The one-time dose is more convenient; consequently, adherence tends to be greater. The week-long treatment has occasionally been noted to be more effective. Some patients complain of an unpleasant but transient metallic taste when taking metronidazole. Nausea and vomiting, as well as a hot, flushed feeling (disulfiram-like reaction), occur when this medication is taken with an alcoholic beverage. Patients are strongly advised to abstain from alcohol during treatment and for 24 hours after taking metronidazole or 72 hours after completion of a course of tinidazole (CDC, 2010a).

Gerontologic Considerations

After menopause, the vaginal mucosa becomes thinner and may atrophy. This condition can be complicated by infection from phylogenic bacteria, resulting in atrophic vaginitis (see Table 57-1). Leukorrhea (vaginal discharge) may cause itching and burning. Management is similar to that for bacterial vaginosis if bacteria are present. Estrogenic hormones, either taken orally or inserted into the vagina in a cream form, can also be effective in restoring the epithelium.

Desquamative inflammatory vaginitis is an uncommon but severe purulent form of vaginal infection that occurs mostly in perimenopausal Caucasian women. It results in vaginal inflammation, discharge, and dyspareunia. Anti-inflammatory and antibiotic local treatment is usually effective (Sobel, Reichman, Misra, et al., 2011).

NURSING PROCESS

The Patient With a Vulvovaginal Infection

Assessment

The woman with vulvovaginal symptoms should be examined as soon as possible after the onset of symptoms. She should be instructed not to **douche** (rinse the vaginal canal), because doing so removes the discharge needed to make the diagnosis. The area is observed for erythema, edema, excoriation, and discharge. Each of the infection-producing organisms produces its own characteristic discharge and effect (see Table 57-1). The patient is asked to describe any discharge and other symptoms, such as odor, itching, or burning. Dysuria often occurs as a result of local irritation of the urinary meatus. A urinary tract infection may need to be ruled out by obtaining a urine specimen for culture and sensitivity testing.

The patient is asked about the occurrence of factors that may contribute to vulvovaginal infection:

- Physical and chemical factors, such as constant moisture from tight or synthetic clothing, perfumes and powders, soaps, bubble bath, poor hygiene, and the use of feminine hygiene products
- Psychogenic factors (e.g., stress, fear of STIs, abuse)
- Medical conditions or endocrine factors, such as a predisposition to *Monilia* in a patient who has diabetes
- The use of medications such as antibiotics, which may alter the vaginal flora and allow an overgrowth of monilial organisms
- New sex partner, multiple sex partners, previous vaginal infection

The patient is also asked about factors that could contribute to infection, including hygiene practices (douching) and the use or non-use of condoms. The nurse needs to assess for the use of complementary or alternative therapies; yogurt, acidophilus pills, garlic, acupuncture, glucosamine, and dietary changes such as low-carbohydrate or low-oxalate diets are some of the options reported in one survey (Nyirjesy, Robinson, Mathew, et al., 2011).

The nurse may prepare a vaginal smear (wet mount) to assist in diagnosing the infection. A common method for preparing the smear is to collect vaginal secretions with an applicator and place the secretions on two separate glass slides. A drop of saline solution is added to one slide and a drop of 10% potassium hydroxide is added to another slide for examination under a microscope. If bacterial vaginosis is present, the slide with normal saline solution added shows epithelial cells dotted with bacteria (clue cells). If *Trichomonas* species is present, small motile cells are seen. In the presence of yeast, the potassium hydroxide slide reveals typical characteristics. Discharge associated with bacterial vaginosis produces a strong odor when mixed with potassium hydroxide. Testing the pH of the discharge with Nitrazine paper assists in proper diagnosis (Schuiling & Likis, 2013).

Diagnosis

Nursing Diagnoses

Based upon the assessment data, major nursing diagnoses may include the following:

- Impaired comfort related to burning, odor, or itching from the infectious process
- Anxiety related to stressful symptoms
- Risk for infection or spread of infection
- Deficient knowledge about proper hygiene and preventive measures

Planning and Goals

Major goals may include relief of impaired comfort, reduction of anxiety related to symptoms, prevention of reinfection or infection of sexual partner, and acquisition of knowledge about methods for preventing vulvovaginal infections and managing self-care.

Nursing Interventions

Relieving Impaired Comfort

Treatment with the appropriate medication usually relieves discomfort. Sitz baths may be occasionally recommended and may provide temporary relief of symptoms.

Reducing Anxiety

Vulvovaginal infections are upsetting and require treatment. The patient who experiences such an infection may be very anxious about the significance of the symptoms and possible causes. Explaining the cause of symptoms may reduce anxiety related to fear of a more serious illness. Discussing ways to help prevent vulvovaginal infections may help patients adopt specific strategies to decrease infection and the related symptoms.

Preventing Reinfection or Spread of Infection

Patient education should include the fact that vulvovaginal candidiasis is not an STI and incidence can be decreased by completing treatment, avoiding unnecessary antibiotic agents, wearing cotton underwear, and not douching. Prevention strategies for vaginal infections include promotion of adequate rest, reduction of life stress, and a healthy diet low in refined sugars (Schuiling & Likis, 2013).

The patient needs to be informed about the importance of adequate treatment for herself and her partner, if indicated. Other strategies to prevent persistence or spread of infection include abstaining from sexual intercourse when infected, treatment for sexual partners, and minimizing irritation of the affected area. When medications such as antibiotic agents are prescribed for any infection, the nurse instructs the patient about the usual precautions related to using these agents. If vaginal itching occurs several days after use, the patient can be reassured that this is usually not an allergic reaction but may be a yeast or monilial infection resulting from altered vaginal bacteria. Treatment for monilial infection is prescribed if indicated.

Another goal of treatment is to reduce tissue irritation caused by scratching or wearing tight clothing. The area needs to be kept clean by daily bathing and adequate hygiene after voiding and defecation. The use of a hair dryer on a cool setting will dry the area, and application of topical corticosteroids may decrease irritation.

When educating the patient about medications such as suppositories and devices such as applicators to dispense cream or ointment, the nurse may demonstrate the procedure by using a plastic model of the pelvis and vagina. The nurse should also stress the importance of hand hygiene before and after each administration of medication. To prevent the medication from escaping from the vagina, the patient should recline for 30 minutes after it is inserted, if possible. The patient is informed that seepage of medication may occur, and the use of a perineal pad may be helpful.

Promoting Home and Community-Based Care

Educating Patients About Self-Care. Vulvovaginal conditions are treated on an outpatient basis unless a patient has other medical problems. Patient education, tact, and reassurance are important aspects of nursing care. Women may express embarrassment, guilt, or anger and may be concerned that the infection could be serious or that it may have been acquired from a sex partner. (In some instances, treatment plans include the partner.)

The nurse assesses the patient's learning needs about the immediate problem. The patient needs to know the characteristics of normal as opposed to abnormal discharge. Questions often arise about douching. Normally, douching and the use of feminine hygiene sprays are unnecessary because daily baths or showers and proper hygiene after voiding and defecating keep the perineal area clean. Douching tends to eliminate normal flora, reducing the body's ability to ward off infection. In addition, repeated douching may result in vaginal epithelial breakdown and chemical irritation and has been associated with other pelvic disorders. No data support any benefit with douching (CDC, 2010a). In the case of recurrent yeast infections, the perineum should be kept as dry as possible. Loose-fitting cotton instead of tight-fitting synthetic, nonabsorbent, heat-retaining underwear is recommended.

Vulvar self-examination is a good health practice for all women. Becoming familiar with one's own anatomy and reporting anything that seems new or different may result in early detection and treatment of any new disorders. Nurses can also play a role in educating women about the risks of unprotected intercourse, particularly with partners who have had sex with others.

Evaluation

Expected patient outcomes may include:

1. Experiences increased comfort
   a. Cleans the perineum as instructed
   b. Reports that itching is relieved
   c. Maintains urine output within normal limits and without dysuria
2. Experiences relief of anxiety
3. Remains free from infection
   a. Has no signs of inflammation, pruritus, odor, or dysuria
   b. Notes that vaginal discharge appears normal (thin, clear, not frothy)
4. Participates in self-care
   a. Takes medication as prescribed
   b. Wears absorbent underwear
   c. Avoids unprotected sexual intercourse
   d. Douches only as prescribed
   e. Performs vulvar self-examination regularly and reports any new findings to primary provider

Human Papillomavirus

Human papillomavirus (HPV) is the most common STI in the United States, acquired by an estimated 6.2 million patients every year and the majority occurring in adolescents and young adults (CDC, 2012a). Most infections are self-limiting and without symptoms, and others can cause cervical and anogenital cancers. Infections can be latent (asymptomatic and detected only by deoxyribonucleic acid [DNA] hybridization tests for HPV), subclinical (visualized only after application of acetic acid followed by inspection under magnification), or clinical (visible condylomata acuminata).

Pathophysiology

HPV can be found in lesions of the skin, cervix, vagina, anus, penis, and oral cavity. More than 100 types of HPV exist. Some are low risk in that they are unlikely to cause cancerous changes. These include types 6, 11, 42, 43, 44, 54, 61, 70, and 72. The most common strains of HPV, 6 and 11, usually cause **condylomata** (warty growths) on the vulva. These are often visible or may be palpable by patients. Condylomata are rarely premalignant but are an outward manifestation of the virus. Strains 6 and 11 are associated with a low risk for cervical cancer. High-risk oncogenic types, including 16, 18, 31, and 45, affect the cervix, causing cell changes or dysplasia (found on a Papanicolaou [Pap] smear). The effects of these strains are usually invisible on examination but may be seen on colposcopy (Castle, 2011). High-risk HPV types cause almost all cases of cervical cancer (Pruitt, 2012).

The incidence of HPV in young, sexually active women is high. The infection often disappears as the result of an effective immune system response. It is thought that two proteins produced by high-risk types of HPV interfere with tumor suppression by normal cells. Risk factors include being young, being sexually active, having multiple sex partners, and having sex with a partner who has or has had multiple partners. It can be transmitted by other means, however, as it has been found in young girls who have not been sexually active. Perinatal transmission is being researched, as is autoinoculation.

Medical Management

Options for treatment of external genital warts by a primary provider include topical application of trichloroacetic acid, podophyllin (Podofin, Podocon), cryotherapy, as well as surgical removal. Topical agents that can be applied by patients to external lesions include podofilox (Condylox) and imiquimod (Aldara). Because the safety of podophyllin, imiquimod, and podofilox during pregnancy has not been determined, these agents should not be used during pregnancy. Electrocautery and laser therapy are alternative therapies that may be indicated for patients with a large number or area of genital warts.

Treatment usually eradicates perineal warts or condylomata. However, they may resolve spontaneously without treatment and may also recur even with treatment.

If the treatment includes application of a topical agent by the patient, she needs to be carefully instructed in the use of the agent prescribed and must be able to identify the warts and be able to apply the medication to them. The patient is instructed to anticipate mild pain or local irritation with the use of these agents.

Women with HPV should have annual Pap smears because of the potential of HPV to cause **dysplasia** (abnormal changes in cells). Much remains unknown about subclinical and latent HPV disease. Women are often exposed to HPV by partners who are unknowing carriers. The use of condoms can reduce the likelihood of transmission, but transmission can also occur during skin-to-skin contact in areas not covered by condoms.

In many cases, patients are angry about having warts or HPV and do not know who infected them because the incubation period can be long and partners may have no symptoms. Acknowledging the emotional distress that occurs when an STI is diagnosed and providing support and facts are important nursing actions.

Prevention

The best strategy is prevention of HPV. The Advisory Committee on Immunization Practices (ACIP) of the CDC recommends routine vaccination of boys and girls 11 to 12 years of age, before they become sexually active. The vaccination is administered in three intramuscular doses, with the initial dose followed by a second dose in 2 months and a third dose 6 months after the first dose. Completion of all three doses of the vaccine is important for immunity to develop (Teitelman, Stringer, Nguyen, et al., 2011).

Although this vaccine is considered an important medical breakthrough with the potential to decrease the impact of HPV-related disease in men and women, it does not replace other strategies important in prevention of HPV. Women still need cervical cancer screening as recommended (Pruitt, 2012).

Herpes virus Type 2 Infection (Herpes Genitalis, Herpes Simplex Virus)

Herpes genitalis is a recurrent, lifelong viral infection that causes herpetic lesions (blisters) on the external genitalia and occasionally the vagina and cervix. It is an STI but possibly may also be transmitted asexually from wet surfaces or by self-transmission (i.e., touching a cold sore and then touching the genital area). The initial infection is usually very painful and lasts about 1 week, but it can also be asymptomatic. Recurrences are less painful and usually produce less severe symptoms (Schuiling & Likis, 2013). Some patients have few or no recurrences, whereas others have frequent bouts. Recurrences can be associated with stress, sunburn, dental work, or inadequate rest or poor nutrition, or any situations that tax the immune system.

At least 50 million people in the United States have genital herpes infection, and most have not been diagnosed (CDC, 2012b). The prevalence of other STIs has decreased slightly, possibly because of increased condom use, but herpes can be transmitted by contact with skin that is not covered by a condom. Transmission is possible even when a carrier does not have symptoms (subclinical shedding). Lesions increase vulnerability to HIV infection and other STIs. Vaccines for herpes genitalis are in clinical trials; however, at this time, there is no commercially available vaccine (CDC, 2012b).

Pathophysiology

There are nine types of herpes viruses belonging to three different groups that cause infections in humans (Porth & Matfin, 2009). These include herpes simplex type 1 (HSV-1), usually associated with cold sores of the lips; herpes simplex type 2 (HSV-2), usually associated with genital herpes; varicella zoster or shingles; Epstein-Barr virus; cytomegalovirus; human B-lymphotrophic virus; and others.

There is considerable overlap between HSV-1 and HSV-2, which are clinically indistinguishable (Porth & Matfin, 2009). Close human contact by the mouth, oropharynx, mucosal surface, vagina, or cervix appears necessary to acquire the infection. Other susceptible sites are skin lacerations and conjunctivae. Usually, the virus is killed at room temperature by drying. When viral replication diminishes, the virus ascends the peripheral sensory nerves and remains inactive in the nerve ganglia. Another outbreak may occur when the host is subjected to stress. In pregnant women with active herpes, infants delivered vaginally may become infected with the virus. There is a risk of fetal morbidity and mortality if this occurs; therefore, a cesarean delivery may be performed if the virus recurs near the time of delivery.

Clinical Manifestations

Itching and pain occur as the infected area becomes red and edematous. Infection may begin with macules and papules and progress to vesicles and ulcers. The vesicular state often appears as a blister, which later coalesces, ulcerates, and encrusts. In women, the labia are the usual primary site, although the cervix, vagina, and perianal skin may be affected. In men, the glans penis, foreskin, or penile shaft is typically affected. Influenzalike symptoms may occur 3 or 4 days after the lesions appear. Inguinal lymphadenopathy (enlarged lymph nodes in the groin), minor temperature elevation, malaise, headache, myalgia (aching muscles), and dysuria (pain on urination) are often noted. Pain is evident during the first week and then decreases. The lesions last 4 to 15 days before crusting over (Schuiling & Likis, 2013).

Rarely, complications may arise from extragenital spread, such as to the buttocks, upper thighs, or even the eyes, as a result of touching lesions and then touching other areas. Patients should be advised to wash their hands after contact with lesions. Other potential problems are aseptic meningitis, neonatal transmission, and severe emotional stress related to the diagnosis.

Medical Management

Currently, there is no cure for HSV-2 infection, but treatment is aimed at relieving the symptoms. Management goals include preventing the spread of infection, making patients comfortable, decreasing potential health risks, and initiating a counseling and education program. Three oral antiviral agents—acyclovir (Zovirax), valacyclovir (Valtrex), and famciclovir (Famvir)—can suppress symptoms and shorten the course of the infection (Schuiling & Likis, 2013). These agents are effective at reducing the duration of lesions and preventing recurrences. Resistance and long-term side effects do not appear to be major problems. Recurrent episodes are often milder than the initial episode.

NURSING PROCESS

The Patient With a Genital Herpes Infection

Assessment

The health history and a physical and pelvic examination are important in establishing the nature of the infectious condition. In addition, patients are assessed for risk of STIs. The perineum is inspected for painful lesions. Inguinal nodes are assessed and are often enlarged and tender during an occurrence of HSV.

Diagnosis

Nursing Diagnoses

Based on the assessment data, major nursing diagnoses may include the following:

- Acute pain related to the genital lesions
- Risk for infection or spread of infection
- Anxiety related to the diagnosis
- Deficient knowledge about the disease and its management

Planning and Goals

Major goals may include relief of pain and discomfort, control of infection and its spread, relief of anxiety, knowledge of and adherence to the treatment regimen and self-care, and knowledge about implications for the future.

Nursing Interventions

Relieving Pain

The lesions should be kept clean, and proper hygiene practices are advocated. Sitz baths may ease discomfort. Clothing should be clean, loose, soft, and absorbent. Aspirin and other analgesic agents are usually effective in controlling pain. Occlusive ointments and powders are avoided because they prevent the lesions from drying.

The patient is encouraged to increase fluid intake, to be alert for possible bladder distention, and to contact her primary provider immediately if she cannot void because of discomfort. Painful voiding may occur if urine comes in contact with the herpes lesions. Discomfort with urination can be reduced by pouring warm water over the vulva during voiding or by sitz baths. When oral acyclovir or other antiviral agents are prescribed, the patient is instructed about when to take the medication and what side effects to note, such as rash and headache. Rest, fluids, and a nutritious diet are recommended to promote recovery.

Preventing Infection and Its Spread

The risk of reinfection and spread of infection to others or to other structures of the body can be reduced by proper hand hygiene, the use of barrier methods with sexual contact, and adherence to prescribed medication regimens. Avoidance of contact when obvious lesions are present does not eliminate the risk because the virus can be shed in the absence of symptoms, and lesions may not be visible.

Relieving Anxiety

Concern about the presence of herpes infection, future occurrences of lesions, and the impact of the infection on future relationships and childbearing may cause considerable patient anxiety. Nurses serve as important sources of support by listening to patients' concerns and providing information and instruction. The patient may be angry with her partner if the partner is the probable source of the infection. The patient may need assistance in discussing the infection and its implications with her current sexual partners and in future sexual relationships. The nurse can refer the patient to a support group to assist in coping with the diagnosis (see the Resources section at the end of the chapter).

Increasing Knowledge About the Disease and Its Treatment

Patient education is an essential part of nursing care of the patient with a genital herpes infection. This includes an adequate explanation about the infection and how it is transmitted, management and treatment strategies, strategies to minimize spread of infection, the importance of adherence to the treatment regimen, and self-care strategies. Because of the increased risk of HIV and other STIs in the presence of skin lesions, an important part of patient education involves instructing the patient to protect herself from exposure to HIV and other STIs (Chart 57-2).

Promoting Home and Community-Based Care

Educating Patients About Self-Care. Self-care measures for people with genital herpes appear in Chart 57-2.

Chart 57-2 HOME CARE CHECKLIST
The Patient With Genital Herpes

At the completion of home care education, the patient or caregiver will be able to:	PATIENT	CAREGIVER
• State that herpes is transmitted mainly by direct contact.	✔	✔
• State that abstinence from sex is required for a brief period (intercourse is avoided during treatment, but other options such as hand holding and kissing are acceptable).	✔	✔
• State that intercourse during a herpes outbreak not only increases the risk of transmission but also increases the likelihood of contracting human immunodeficiency virus and other sexually transmitted infections.	✔	✔
• State that transmission is possible even in the absence of active lesions.	✔	✔
• State that condoms may provide some protection against viral transmission.	✔	✔
• Explain that obstetric care provider should be informed about the history of herpes. In cases of recurrence at time of delivery, cesarean section may be considered.	✔	✔
• Describe appropriate hygiene practices (hand hygiene, perineal cleanliness, gentle washing of lesions with mild soap and running water and lightly drying lesions) and importance of avoiding occlusive ointments, strong perfumed soaps, or bubble bath.	✔	✔
• State that control of the condition may require changes in sexual behavior and the use of medications.	✔	✔
• Describe strategies to avoid self-infection (e.g., avoid touching lesions during an outbreak).	✔	✔
• Explain rationale for avoiding self-infection (i.e., lesions can become infected from germs on the hand, and the virus from the lesion can be transmitted from the hand to another area of the body or another person).	✔	✔
• Describe health promotion strategies: Wear loose, comfortable clothing; eat a balanced diet; get adequate rest and relaxation.	✔	✔
• State rationale for avoiding exposure to the sun, which can cause recurrences (and skin cancer).	✔	✔
• Identify importance of taking medications as prescribed, keeping follow-up appointments with health care provider, and reporting repeated recurrences (may not be as severe as the initial episode).	✔	✔
• Describe possible benefits of joining a group to share solutions and experiences and hear about newer treatments (see Resources section).	✔	✔

Evaluation

Expected patient outcomes may include:

1. Experiences a reduction in pain and discomfort
2. Keeps infection under control
   a. Demonstrates proper hygiene techniques
   b. Takes medication as prescribed
   c. Consumes adequate fluids
   d. Assesses own current lifestyle (diet, adequate fluid intake, safer sex practices, stress management)
3. Uses strategies to reduce anxiety
   a. Verbalizes issues and concerns related to genital herpes infection
   b. Discusses strategies to deal with issues and concerns with current and future sexual partners
   c. Initiates contact with support group if indicated
4. Demonstrates knowledge about genital herpes and strategies to control and minimize recurrences
   a. Identifies methods of transmission of herpes infection and strategies to prevent transmission to others
   b. Discusses strategies to reduce recurrence of lesions
   c. Takes medications as prescribed
   d. Reports no recurrence of lesions

Endocervicitis and Cervicitis

Endocervicitis is an inflammation of the mucosa and the glands of the cervix that may occur when organisms gain access to the cervical glands after intercourse and, less often, after procedures such as abortion, intrauterine manipulation, or vaginal delivery. If untreated, the infection may extend into the uterus, fallopian tubes, and pelvic cavity. Inflammation can irritate the cervical tissue, resulting in spotting or bleeding and **mucopurulent cervicitis** (inflammation of the cervix with exudate).

Chlamydia and Gonorrhea

Chlamydia and gonorrhea are the most common causes of endocervicitis, although *Mycoplasma* may also be involved. Chlamydia causes about 2.5 million infections every year in the United States; it is most commonly found in young, sexually active people with more than one partner and is transmitted through sexual intercourse (CDC, 2012c). It needs to be reported to the CDC and can result in serious complications, including pelvic infection, an increased risk of ectopic pregnancy, and infertility (CDC, 2012c). Chlamydial infections of the cervix often produce no symptoms, but cervical discharge, dyspareunia, dysuria, and bleeding may occur. Other complications include conjunctivitis and perihepatitis. If pregnant women are infected, stillbirth, neonatal death, and premature labor may occur.

Chlamydial infection and gonorrhea often coexist (Schuiling & Likis, 2013). The inflamed cervix that results from this infection may leave a woman more vulnerable to HIV transmission from an infected partner. Gonorrhea is also a major cause of PID, tubal infertility, ectopic pregnancy, and chronic pelvic pain (CDC, 2012d).

Most women with gonorrhea have no symptoms, but without treatment, many may develop PID (CDC, 2012d). Diagnosis can be confirmed by urine culture or other methods such as using a swab to obtain a sample of cervical or penile discharge from the patient's partner (CDC, 2012c).

Medical Management

The CDC recommends treating chlamydia with doxycycline (Vibramycin) for 1 week or with a single dose of azithromycin (Zithromax). Because of the high incidence of coinfection with chlamydia and gonorrhea, treatment for gonorrhea should include treatment for chlamydia as well (CDC, 2012c). Partners must also be treated. Pregnant women are cautioned not to take tetracycline because of potential adverse effects on the fetus. In these cases, erythromycin may be prescribed. Results are usually good if treatment begins early. Possible complications from delayed or no treatment are tubal disease, ectopic pregnancy, PID, and infertility.

Cultures for chlamydia and other STIs should be obtained from all patients who have been sexually assaulted when they first seek medical attention; patients are treated prophylactically. Cultures should then be repeated in 2 weeks. Annual screening for chlamydia is recommended for all sexually active young women and older women with new sex partners or multiple partners (CDC, 2012c).

Nursing Management

All sexually active women may be at risk for chlamydia, gonorrhea, and other STIs, including HIV. Nurses can assist patients in assessing their own risk. Recognition of risk is a first step before changes in behavior occur. Patients should be discouraged from assuming that a partner is "safe" without open, honest discussion. Nonjudgmental attitudes, educational counseling, and role-playing may be helpful.

Because chlamydia, gonorrhea, and other STIs may have a serious effect on future health and fertility, and because many of these disorders can be prevented by the use of condoms and spermicides and careful choice of partners, nurses can play a major role in counseling patients about safer sex practices. Exploring options with patients, addressing knowledge deficits, and correcting misinformation may reduce morbidity and mortality.

Patients should be advised to refer partners for evaluation and treatment. All sexually active women aged 25 and younger should be screened annually. Those older than 25 years should be screened if risk factors are present. Repeat testing should occur 3 months after treatment (CDC, 2012c).

Promoting Home and Community-Based Care

Educating Patients About Self-Care

Nurses can educate women and help them improve communication skills and initiate discussions about sex with their partners. Communicating with partners about sex, risk, postponing intercourse, and using safer sex behaviors, including the use of condoms, may be lifesaving. Some young women report having sex but not being comfortable enough to discuss sexual risk issues. Nurses can help women to advocate for their own health by discussing safety with partners prior to sexual activity.

Reinforcing the need for annual screening for chlamydia and other STIs is an important part of patient education. Instructions also include the need for the patient to abstain from sexual intercourse until all of her sex partners are treated (CDC, 2012c).

Pelvic Inflammatory Disease

Pelvic inflammatory disease (PID) is an inflammatory condition of the pelvic cavity that may begin with cervicitis and involve the uterus (endometritis), fallopian tubes (salpingitis), ovaries (oophoritis), pelvic peritoneum, or pelvic vascular system. Infection, which may be acute, subacute, recurrent, or chronic and localized or widespread, is usually caused by bacteria but may be attributed to a virus, fungus, or parasite. Gonorrheal and chlamydial organisms are common causes, but most cases are associated with more than one organism. Each year, there are an estimated 750,000 cases of PID (CDC, 2011a).

Short- and long-term consequences can occur. The fallopian tubes become narrow and scarred, which increases the risk of ectopic pregnancy (fertilized eggs trapped in the tube), infertility, recurrent pelvic pain, tubo-ovarian **abscess** (a collection of purulent material), and recurrent disease. The true incidence of PID is unknown because some cases are asymptomatic and others present atypically (CDC, 2011a; Turner, 2012).

Pathophysiology

The exact pathogenesis of PID has not been determined, but it is presumed that organisms usually enter the body through the vagina, pass through the cervical canal, colonize the endocervix, and move upward into the uterus. Under various conditions, the organisms may proceed to one or both fallopian tubes and ovaries and into the pelvis. In bacterial infections that occur after childbirth or abortion, pathogens are disseminated directly through the tissues that support the uterus by way of the lymphatics and blood vessels (Fig. 57-1A). In pregnancy, the increased blood supply required by the placenta provides a wider pathway for infection. These postpartum and postabortion infections tend to be unilateral. Infections can cause perihepatic inflammation when the organism invades the peritoneum.

In gonorrheal infections, the gonococci pass through the cervical canal and into the uterus, where the environment, especially during menstruation, allows them to multiply rapidly and spread to the fallopian tubes and into the pelvis (see Fig. 57-1B). The infection is usually bilateral.

In rare instances, organisms (e.g., tuberculosis) gain access to the reproductive organs by way of the bloodstream from the lungs (see Fig. 57-1C). One of the most common causes of **salpingitis** (inflammation of the fallopian tube) is chlamydia, possibly accompanied by gonorrhea.

Pelvic infection is most often sexually transmitted but can also occur with invasive procedures such as endometrial biopsy, abortion, hysteroscopy, or insertion of an intrauterine device. Bacterial vaginosis (a vaginal infection) may predispose women to pelvic infection. Risk factors include early age at first intercourse, multiple sexual partners, frequent intercourse, intercourse without condoms, sex with a partner with an STI, and a history of STIs or previous pelvic infection.

Clinical Manifestations

Symptoms of pelvic infection usually begin with vaginal discharge, dyspareunia, lower abdominal pelvic pain, and tenderness that occurs after menses. Pain may increase with voiding or with defecation. Other symptoms include fever, general malaise, anorexia, nausea, headache, and possibly vomiting. On pelvic examination, intense tenderness may be noted on palpation of the uterus or movement of the cervix (cervical motion tenderness). Symptoms may be acute and severe or low grade and subtle (CDC, 2011a).

Complications

Pelvic or generalized peritonitis, abscesses, strictures, and fallopian tube obstruction may develop. Obstruction may cause an ectopic pregnancy in the future if a fertilized egg cannot pass a tubal stricture, or scar tissue may occlude the tubes, resulting in sterility. Adhesions are common and often result in chronic pelvic pain; they eventually may require removal of the uterus, fallopian tubes, and ovaries. Other complications include bacteremia with septic shock, chronic pelvic and abdominal pain, and recurring PID (Schuiling & Likis, 2013).

Medical Management

Broad-spectrum antibiotic therapy is prescribed, usually a combination of ceftriaxone (Rocephin), azithromycin, and doxycycline. Women are most often treated as outpatients

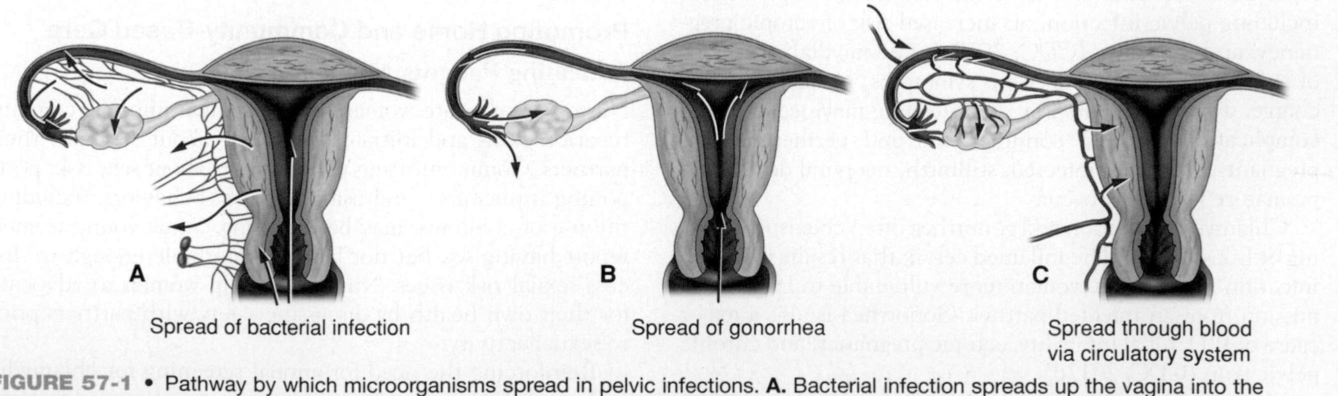
A
Spread of bacterial infection
B
Spread of gonorrhea
C
Spread through blood
via circulatory system

FIGURE 57-1 • Pathway by which microorganisms spread in pelvic infections. **A.** Bacterial infection spreads up the vagina into the uterus and through the lymphatics. **B.** Gonorrhea spreads up the vagina into the uterus and then to the tubes and ovaries. **C.** Bacterial infection can reach the reproductive organs through the bloodstream (hematogenous spread).

and monitored carefully. Indications for hospitalization include surgical emergencies, pregnancy, no clinical response to oral antimicrobial therapy, inability to follow or tolerate an outpatient oral regimen, severe illness (i.e., nausea, vomiting, or high fever), and tubo-ovarian abscess (Schuiling & Likis, 2013). Treatment of sexual partners is necessary to prevent reinfection.

Nursing Management

The nurse assesses for both the physical and emotional effects of PID. The patient may feel well one day and experience vague symptoms and discomfort the next. She may also suffer from constipation and menstrual difficulties.

If the patient is hospitalized, the nurse prepares the patient for further diagnostic evaluation and surgical intervention as prescribed. Accurate recording of vital signs, intake and output, and the characteristics and amount of vaginal discharge is necessary as a guide to therapy.

The nurse administers analgesic agents as prescribed for pain relief. Adequate rest and a healthy diet are encouraged. In addition, the nurse minimizes the transmission of infection by adhering to appropriate infection control practices and performing meticulous hand hygiene.

Promoting Home and Community-Based Care

Educating Patients About Self-Care

The patient must be informed of the need for precautions and must be encouraged to take part in procedures to prevent infecting others and protect herself from reinfection. The use of condoms is essential to prevent infection and sequelae. If reinfection occurs or if the infection spreads, symptoms may include abdominal pain, nausea and vomiting, fever, malaise, malodorous purulent vaginal discharge, and leukocytosis. Patient education consists of explaining how pelvic infections occur, how they can be controlled and avoided, and the associated signs and symptoms. Guidelines and instructions provided to the patient are summarized in Chart 57-3.

All patients who have had PID need to be informed of the signs and symptoms of ectopic pregnancy (pain, abnormal bleeding, delayed menses, faintness, dizziness, and shoulder pain), because they are prone to this complication. (See Chapter 56 for a discussion of ectopic pregnancy.)

Human Immunodeficiency Virus Infection and Acquired Immunodeficiency Syndrome

Any discussion of vulvovaginal infections and STIs must include the topic of HIV and acquired immunodeficiency syndrome (AIDS), which is described in Chapter 37. An estimated 25% of those living with HIV are women (CDC, 2011b). Because HIV infection may be detected during prenatal testing and screening for STIs, nurses and other women's health care clinicians are often the first professionals to provide care for a woman with HIV infection. Thus, clinicians need to be knowledgeable about this disorder and sensitive to women's issues and concerns.

After informed consent is obtained, women who are at risk for HIV are offered testing by a nurse or counselor. Because patients may be reluctant to discuss risk-taking behavior, routine screening should be offered to all women between the ages of 13 and 64 years in all health care settings (Shuiling & Likis, 2013). Early detection permits early treatment to delay progression of the disease. The nurse also needs to remember that many women do not see themselves as at risk for acquiring HIV infection. (See Chapter 37 for further discussion of HIV infection and AIDS.)

Women with HIV and women with partners who have HIV must be counseled about safer sex and informed about the dangers of unprotected sex. Inconsistent use of condoms results in a higher seroconversion rate. Because there is a risk of perinatal transmission, decisions to conceive or to use contraception must be based on education, accurate information, and care. Pregnant women are advised to have an HIV test. The use of antiretroviral agents by pregnant women significantly decreases perinatal transmission of HIV infection. Therefore, the use of these agents during pregnancy is critical

Chart 57-3 HOME CARE CHECKLIST

The Patient With Pelvic Inflammatory Disease

At the completion of home care education, the patient or caregiver will be able to:	PATIENT	CAREGIVER
• State that any pelvic pain or abnormal discharge, particularly after sexual exposure, childbirth, or pelvic surgery, should be evaluated as soon as possible.	✔	✔
• State that antibiotic agents may be prescribed after insertion of intrauterine devices.	✔	✔
• Describe proper perineal care procedures (wiping from front to back after defecation or urination).	✔	✔
• State that douching reduces the natural flora that combat infecting organisms and may introduce bacteria upward.	✔	✔
• Identify the importance of consulting a health care provider if unusual vaginal discharge or odor is noted.	✔	✔
• Discuss the importance of following health practices (i.e., proper nutrition, exercise, and weight control), and safer sex practices (i.e., using condoms, avoiding multiple sexual partners).	✔	✔
• Explain the importance of consistent use of condoms before intercourse or any penile–vaginal contact if there is any chance of transmitting infection.	✔	✔
• State that a gynecologic examination should be performed at least once a year.	✔	✔

and must also be discussed. For women who choose to avoid conception, the use of condoms alone and with oral contraceptives are possible choices.

STRUCTURAL DISORDERS

Fistulas of the Vagina

A **fistula** is an abnormal opening between two internal hollow organs or between an internal hollow organ and the exterior of the body. The name of the fistula indicates the two areas that are connected abnormally—for example, a vesicovaginal fistula is an opening between the bladder and the vagina, and a rectovaginal fistula is an opening between the rectum and the vagina (Fig. 57-2). Fistulas may be congenital in origin but are most common in developing countries due to labor complications in women. In the United States, they occur most often due to damage resulting from injury sustained during surgery, vaginal delivery, irritable bowel disease, radiation therapy, or disease processes such as carcinoma (Wong, Wong, Rezvan, et al., 2012).

Clinical Manifestations

Symptoms depend on the specific defect. For example, in a patient with a vesicovaginal fistula, urine escapes continuously into the vagina. With a rectovaginal fistula, there is fecal incontinence, and flatus is discharged through the vagina. The combination of fecal discharge with leukorrhea results in malodor that is difficult to control.

Assessment and Diagnostic Findings

A history of the symptoms experienced by the patient is important to identify the structural alterations and to assess the impact of the symptoms on the patient's quality of life. Although there is no reported specificity for its use, methylene blue dye is commonly used to help delineate the course of the fistula (Wong et al., 2012). In a vesicovaginal fistula, the dye is instilled into the bladder and appears in the vagina. After a negative methylene blue test result, indigo carmine is injected intravenously (IV); the appearance of the dye in the vagina indicates ureterovaginal fistula. Cystoscopy or IV pyelography may then be used to determine the exact location.

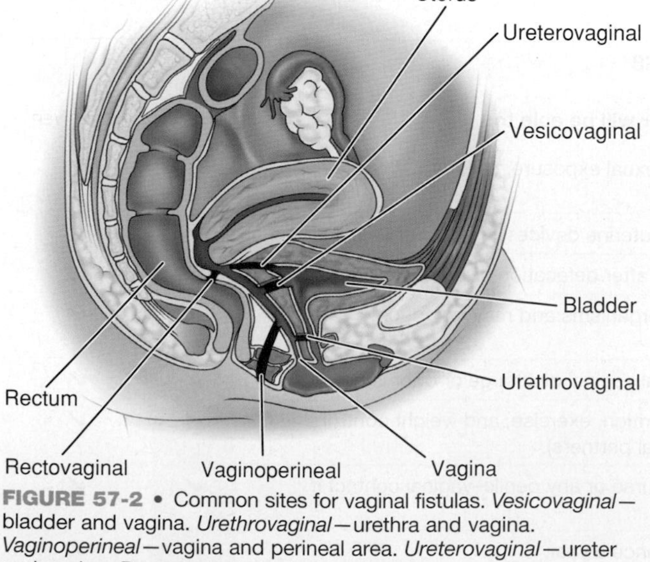
Uterus
Ureterovaginal
Vesicovaginal
Bladder
Urethrovaginal
Rectum
Rectovaginal
Vaginoperineal
Vagina

FIGURE 57-2 • Common sites for vaginal fistulas. *Vesicovaginal*—bladder and vagina. *Urethrovaginal*—urethra and vagina. *Vaginoperineal*—vagina and perineal area. *Ureterovaginal*—ureter and vagina. *Rectovaginal*—rectum and vagina.

Medical Management

The goal is to eliminate the fistula and to treat infection and excoriation. A fistula may heal without surgical intervention, but surgery is often required. If the primary provider determines that a fistula will heal without surgical intervention, care is planned to relieve discomfort, prevent infection, and improve the patient's self-concept and self-care abilities. Measures to promote healing include proper nutrition, cleansing douches and enemas, rest, and administration of prescribed intestinal antibiotic agents. A rectovaginal fistula heals faster when the patient eats a low-residue diet and when the affected tissue drains properly. Warm perineal irrigations promote healing.

Sometimes, a fistula does not heal and cannot be surgically repaired. In this situation, care must be planned and implemented on an individual basis. Cleanliness, frequent sitz baths, and deodorizing douches are required, as are perineal pads and protective undergarments. Meticulous skin care is necessary to prevent excoriation. Applying bland creams or lightly dusting with cornstarch may be soothing. In addition, attending to the patient's social and psychological needs is an essential aspect of care.

If the patient is to have a fistula repaired surgically, preoperative treatment of any existing vaginitis is important to ensure success. Usually, the vaginal approach is used to repair vesicovaginal and urethrovaginal fistulas; the abdominal approach is used to repair fistulas that are large or complex. Fistulas that are difficult to repair or very large may require surgical repair with a urinary or fecal diversion. Tissue transfer techniques (skin or tissue grafting) may be used (Wong et al., 2012).

Because fistulas usually are related to obstetric, surgical, or radiation trauma, occurrence in a patient without previous vaginal delivery or a history of surgery must be evaluated carefully. Crohn's disease or lymphogranuloma venereum are other possible causes.

Despite the best surgical intervention, fistulas may recur. After surgery, medical follow-up continues for at least 2 years to monitor for a possible recurrence.

Pelvic Organ Prolapse: Cystocele, Rectocele, Enterocele

Age and parity can put strain on the ligaments and structures that make up the female pelvis and pelvic floor. Childbirth can result in tears of the levator sling musculature, resulting in structural weakness. Hormone deficiency also may play a role. Some degree of prolapse (weakening of the vaginal walls allowing the pelvic organs to descend and protrude into the vaginal canal) may be found in older women. Risk factors

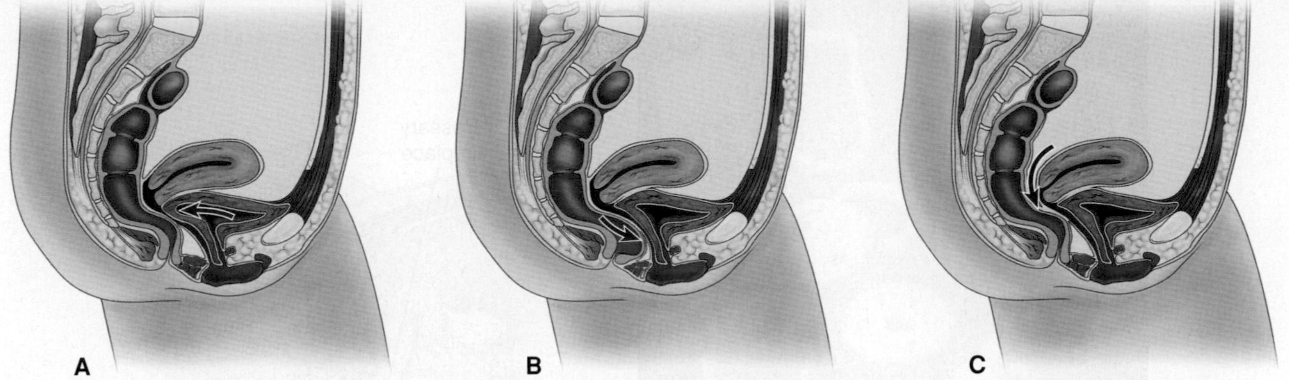
A
B
C

FIGURE 57-3 • Diagrammatic representation of the three most common types of pelvic floor relaxation. **A.** Cystocele. **B.** Rectocele. **C.** Enterocele. *Arrows* depict sites of maximum protrusion.

include age, parity, and vaginal delivery, and there may also be a familial predisposition.

Cystocele is a downward displacement of the bladder toward the vaginal orifice (Fig. 57-3) from damage to the anterior vaginal support structures. It usually results from injury and strain during childbirth. The condition usually appears years later when genital atrophy associated with aging occurs, but younger, multiparous, premenopausal women may also be affected.

Rectocele is an upward pouching of the rectum that pushes the posterior wall of the vagina forward. Both rectoceles and perineal lacerations, which occur because of muscle tears below the vagina, may affect the muscles and tissues of the pelvic floor and may occur during childbirth. Sometimes, the lacerations may completely sever the fibers of the anal sphincter (complete tear). An **enterocele** is a protrusion of the intestinal wall into the vagina. Prolapse results from a weakening of the support structures of the uterus itself; the cervix drops and may protrude from the vagina. If complete prolapse occurs, it may also be referred to as procidentia.

Clinical Manifestations

Because a cystocele causes the anterior vaginal wall to bulge downward, the patient may report a sense of pelvic pressure and urinary problems such as incontinence, frequency, and urgency. Back pain and pelvic pain may occur as well. The symptoms of rectocele resemble those of cystocele, with one exception: Instead of urinary symptoms, patients may experience rectal pressure. Constipation, uncontrollable gas, and fecal incontinence may occur in patients with complete tears. Prolapse can result in feelings of pressure and ulcerations and bleeding. Dyspareunia may occur with these disorders.

Medical Management

Kegel exercises, which involve contracting or tightening the vaginal muscles, are prescribed to help strengthen these weakened muscles (Schuiling & Likis, 2013). The exercises are more effective in the early stages of a cystocele. Kegel exercises are easy to perform and are recommended for all women, including those with strong pelvic floor muscles (Chart 57-4).

A pessary can be used to avoid surgery (Lone, Thakar, Sultan, et al., 2011). This device is inserted into the vagina and positioned to keep an organ, such as the bladder, uterus, or intestine, properly aligned when a cystocele, rectocele, or prolapse has occurred. Pessaries are usually ring- or doughnut shaped and are made of various materials, such as rubber or plastic (Fig. 57-4). Rubber pessaries must be avoided in women with latex allergy. The size and type of pessary are selected and fitted by a gynecologic health care provider. The patient should have the pessary removed, examined, and cleaned by her health care provider at prescribed intervals. At these checkups, vaginal walls should be examined for pressure points or signs of irritation. Normally, the patient experiences no pain, discomfort, or discharge with a pessary, but if chronic irritation occurs, alternative measures may be needed.

A Colpexin Sphere is another nonsurgical device used to treat pelvic organ prolapse. This intravaginal device is similar to a pessary, but it supports the pelvic floor muscles and facilitates exercise of these muscles. It is removed daily for cleaning.

Surgical Management

In many cases, surgery helps correct structural abnormalities. The procedure to repair the anterior vaginal wall is called anterior **colporrhaphy**, repair of a rectocele is referred to as

Chart 57-4 PATIENT EDUCATION

Performing Kegel (Pelvic Muscle) Exercises

Purposes: To strengthen and maintain the tone of the pubococcygeal muscle, which supports the pelvic organs; reduce or prevent stress incontinence and uterine prolapse; enhance sensation during sexual intercourse; and hasten postpartum healing

1. Become aware of pelvic muscle function by “drawing in” the perivaginal muscles and anal sphincter as if to control urine or defecation, but not contracting the abdominal, buttock, or inner thigh muscles.
2. Sustain contraction of the muscles for up to 10 seconds, followed by at least 10 seconds of relaxation.
3. Perform these exercises 30–80 times a day.

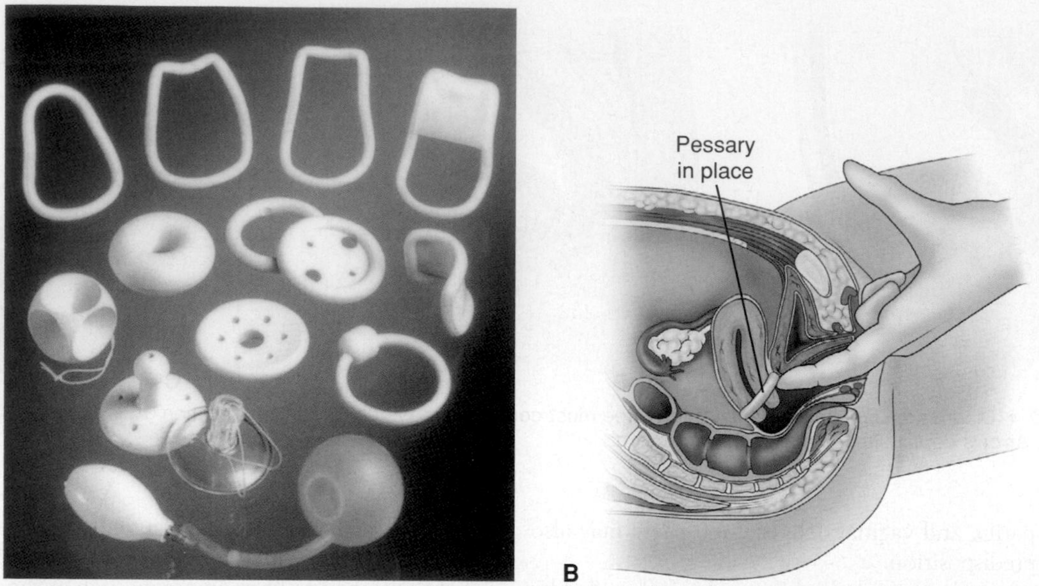
Pessary
in place
A
B

FIGURE 57-4 • Examples of pessaries. **A.** Various shapes and sizes of pessaries available. **B.** Insertion of one type of pessary.

a posterior colporrhaphy, and repair of perineal lacerations is called a **perineorrhaphy**. These repairs are frequently performed laparoscopically, resulting in short hospital lengths of stay and good outcomes. A laparoscope is inserted through a small abdominal incision, the pelvis is visualized, and surgical repairs are performed. Surgical mesh is sometimes used but has been associated with the complication of vaginal erosion (American College of Obstetricians and Gynecologists [ACOG], 2011a). The U.S. Food and Drug Administration (FDA, 2012) has recommended additional postmarketing studies comparing surgical mesh and tissue-derived products in pelvic organ prolapse surgeries.

Uterine Prolapse

Usually, the uterus and the cervix lie at right angles to the long axis of the vagina with the body of the uterus inclined slightly forward. The uterus is normally freely movable on examination. Individual variations may result in an anterior, middle, or posterior uterine position. A backward positioning of the uterus, known as retroversion and retroflexion, is not uncommon (Fig. 57-5).

If the structures that support the uterus weaken (typically from childbirth), the uterus may work its way down the vaginal canal (prolapse) and even appear outside the vaginal orifice (procidentia) (Fig. 57-6). As the uterus descends, it may pull the vaginal walls and even the bladder and rectum with it. Symptoms include pressure and urinary problems (incontinence or retention) from displacement of the bladder. The symptoms are aggravated when a woman coughs, lifts a heavy object, or stands for a long time. Normal activities, even walking up stairs, may aggravate the symptoms.

Medical Management

There are surgical and nonsurgical options for treatment. With surgery, the uterus is sutured back into place and repaired to strengthen and tighten the muscle bands. In postmenopausal women, the uterus may be removed (**hysterectomy**) or repaired by colpopexy. Colpocleisis, or vaginal closure, may be an option for women who do not wish to have sexual intercourse or to bear children. Conservative treatment including pessaries, pelvic floor muscle training, or both can usually result in symptomatic improvement (Culligan, 2012). Pessaries may be the treatment of choice in older women or those unable to tolerate surgery (Lone et al., 2011).

Nursing Management

Implementing Preventive Measures

Some disorders related to "relaxed" pelvic muscles (cystocele, rectocele, and uterine prolapse) may be prevented. During pregnancy, early visits to the primary provider permit early detection of problems. During the postpartum period, the woman can be educated to perform pelvic muscle exercises, commonly known as Kegel exercises, to increase muscle mass and strengthen the muscles that support the uterus and then to continue them as a preventive action (Schuiling & Likis, 2013).

Delays in obtaining evaluation and treatment may result in complications such as infection, cervical ulceration, cystitis, and hemorrhoids. The nurse encourages the patient to obtain prompt treatment for these structural disorders.

Implementing Preoperative Nursing Care

Before surgery, the patient needs to know the extent of the proposed surgery, the expectations for the postoperative period, and the effect of surgery on future sexual function. In addition, the patient having a rectocele repair needs to know that before surgery, a laxative and a cleansing enema may be prescribed. She may be asked to administer these at home the day before surgery. The patient is usually placed in a lithotomy position for surgery, with special attention given to moving both legs in and out of the stirrups simultaneously to prevent muscle strain and excess pressure on the legs and thighs. Other preoperative interventions are similar to those described in Chapter 17.

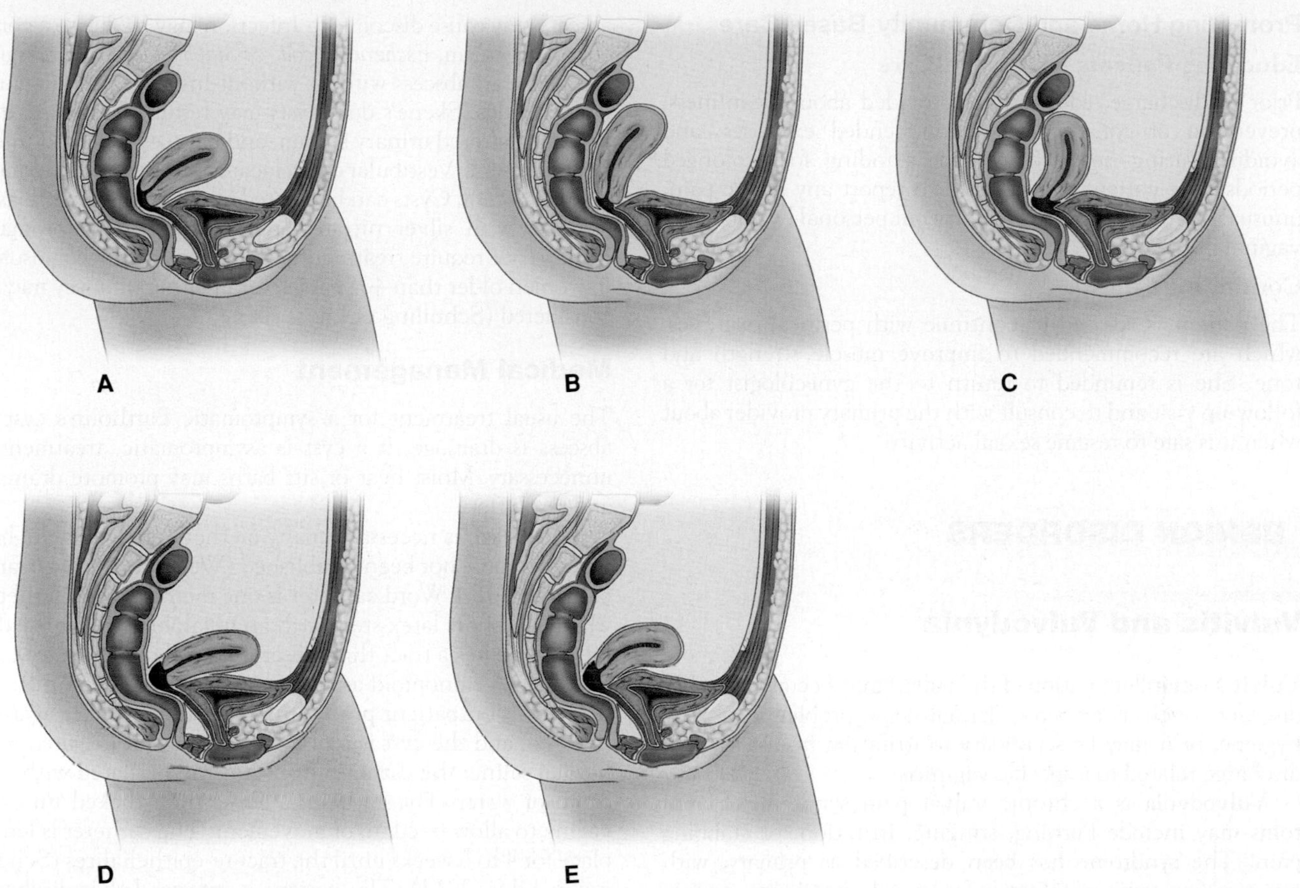
A
B
C
D
E

FIGURE 57-5 • Positions of the uterus. **A.** The most common position of the uterus detected on palpation. **B.** In *retroversion,* the uterus turns posteriorly as a whole unit. **C.** In *retroflexion,* the fundus bends posteriorly. **D.** In *anteversion,* the uterus tilts forward as a whole unit. **E.** In *anteflexion,* the uterus bends anteriorly.

Initiating Postoperative Nursing Care

Immediate postoperative goals include preventing infection and pressure on any existing suture line. This may require perineal care and may preclude using dressings. The patient is encouraged to void within a few hours after surgery for cystocele and complete tear. If the patient does not void within this period and reports discomfort or pain in the bladder region after 6 hours, she needs to be catheterized. An indwelling catheter may be indicated for 2 to 4 days, so some women may return home with a catheter in place. Various other bladder care methods are described in Chapter 55. After each voiding or bowel movement, the perineum may be cleaned with warm, sterile saline solution and dried with sterile absorbent material if a perineal incision has been made.

After an external perineal repair, the perineum is kept as clean as possible. Commercially available sprays containing combined antiseptic and anesthetic solutions are soothing and effective, and an ice pack applied locally may relieve discomfort. However, the weight of the ice bag must rest on the bed, not on the patient.

Routine postoperative care is similar to that given after abdominal surgery. The patient is positioned in bed with her head and knees elevated slightly. The patient may go home the day of or the day after surgery; the length of hospital stay depends on the surgical approach used.

After surgery for a complete perineal laceration (through the rectal sphincter), special care and attention are required. The bladder is drained through the catheter to prevent strain on the sutures. Throughout recovery, stool-softening agents are administered nightly after the patient begins a soft diet.

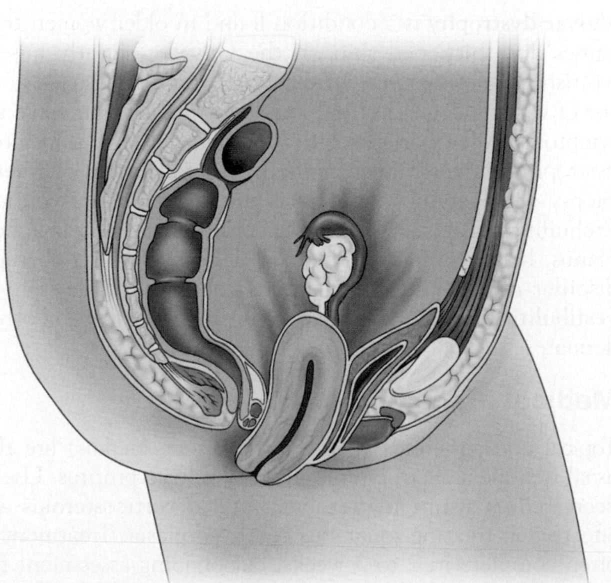

FIGURE 57-6 • Complete prolapse of the uterus through the introitus.

Promoting Home and Community-Based Care

Educating Patients About Self-Care

Prior to discharge, education is provided about cleanliness, prevention of constipation, recommended exercises, and avoiding lifting heavy objects or standing for prolonged periods. The patient is instructed to report any pelvic pain, unusual discharge, inability to carry out personal hygiene, and vaginal bleeding.

Continuing Care

The patient is advised to continue with perineal exercises, which are recommended to improve muscle strength and tone. She is reminded to return to the gynecologist for a follow-up visit and to consult with the primary provider about when it is safe to resume sexual activity.

BENIGN DISORDERS

Vulvitis and Vulvodynia

Vulvitis (an inflammation of the vulva) may occur with other disorders, such as diabetes, dermatologic problems, or poor hygiene, or it may be secondary to irritation from a vaginal discharge related to a specific vaginitis.

Vulvodynia is a chronic vulvar pain syndrome. Symptoms may include burning, stinging, irritation, or stabbing pain. The syndrome has been described as primary, with onset at first tampon insertion or sexual experience, or secondary, beginning months or years after first tampon insertion or sexual experience. Women affected by this are usually between 18 and 25 years of age. Vulvodynia may be classified as organic if it has a known cause (infection, trauma, or irritants) or idiopathic if no cause is known. It can be chronic or unremitting, intermittent or episodic, or may occur only in response to contact (Schuiling & Likis, 2013). The pathophysiology is unknown. **Vestibulodynia** is the most frequent type of vulvodynia, producing sharp pain on pressure on the vestibule or posterior aspect of the vaginal opening.

Medical Management

Treatment methods for vulvodynia vary and depend on cause. Topical treatments (i.e., estrogens, corticosteroids, trichloroacetic acid), surgery, and interferon, as well as biofeedback and dietary changes, have been used. Some cases seem to be similar to peripheral neuralgia and may respond to treatment with tricyclic antidepressant agents. Patients with dyspareunia or painful intercourse may benefit from referral to a therapist with expertise in sexual dysfunction (Schuiling & Likis, 2013).

Vulvar Cysts

Bartholin's cyst results from the obstruction of a duct in one of the paired Bartholin's or vestibular mucous-secreting glands located in the posterior third of the vulva, near the vestibule. This cyst is the most common vulvar disorder. A simple cyst may be asymptomatic, but an infected cyst or abscess may cause discomfort. Infection may be due to a gonococcal organism, *Escherichia coli*, or *Staphylococcus aureus* and can cause an abscess with or without involving the inguinal lymph nodes. Skene's duct cysts may result in pressure, dyspareunia, altered urinary stream, and pain, especially if infection is present. Vestibular cysts, located inferior to the hymen, may also occur. Cysts can be treated by resection or with laser, ablation with silver nitrate, and puncture. Asymptomatic cysts do not require treatment. Malignancy can occur, usually in women older than 40 years, so drainage and biopsy may be considered (Schuiling & Likis, 2013).

Medical Management

The usual treatment for a symptomatic Bartholin's cyst or abscess is drainage. If a cyst is asymptomatic, treatment is unnecessary. Moist heat or sitz baths may promote drainage and resolution.

If drainage is necessary, many methods are available; best practices have not been established (Wechter, Wu, Marzano, et al., 2009). A Word catheter is one method. This catheter, which is a short latex stem with an inflatable bulb at the distal end, creates a tract that preserves the gland and allows for drainage. A nonopioid analgesic agent may be administered before this outpatient procedure. A local anesthetic agent is injected, and the cyst is incised or lanced and irrigated with normal saline; the catheter is inserted and inflated with 2 to 3 mL of water. The catheter stem is then tucked into the vagina to allow freedom of movement. The catheter is left in place for 4 to 6 weeks until the tract re-epithelializes (Schuiling & Likis, 2013). The patient is informed that discharge should be expected because the catheter allows drainage of the cyst. She is instructed to contact her primary provider if pain occurs, in which case the bulb may be too large for the cavity and fluid may need to be removed. Routine hygiene is encouraged.

Vulvar Dystrophy

Vulvar dystrophy is a condition found in older women that causes dry, thickened skin on the vulva or slightly raised, whitish papules, fissures, or macules. Symptoms usually consist of varying degrees of itching, but some patients have no symptoms. A few patients with vulvar cancer have associated dystrophy (vulvar cancer is discussed later in this chapter). Biopsy with careful follow-up is the standard intervention (Schuiling & Likis, 2013). Benign dystrophies include lichen planus, lichen simplex chronicus, **lichen sclerosus** (benign disorder of the vulva), squamous cell hyperplasia, vulvar **vestibulitis** (inflammation of the vulvar vestibule), and other dermatoses.

Medical Management

Topical corticosteroids (i.e., hydrocortisone creams) are the usual treatment. Petrolatum jelly may relieve pruritus. Use is decreased as symptoms resolve. Topical corticosteroids are effective in treating squamous cell hyperplasia. Treatment is often complete in 2 to 3 weeks, but ongoing assessment for vulvar atrophy should occur at least annually (Schuiling & Likis, 2013).

If malignant cells are detected on biopsy, local excision, laser therapy, local chemotherapy, and immunologic treatment are used. Vulvectomy is avoided, if possible, to spare the patient from the stress of disfigurement and possible sexual dysfunction.

Nursing Management

Key nursing responsibilities for patients with vulvar dystrophies focus on education. Important topics include hygiene and self-monitoring for signs and symptoms of complications.

Promoting Home and Community-Based Care

Educating Patients About Self-Care

Instructions for patients with benign vulvar dystrophies include the importance of maintaining good personal hygiene and keeping the vulva dry. Lanolin or hydrogenated vegetable oil is recommended for relief of dryness. Sitz baths may help but should not be overused because dryness may result or increase. The patient is instructed to notify her primary provider about any change or ulceration because biopsy may be necessary to rule out squamous cell carcinoma.

By encouraging all patients to perform genital self-examinations regularly and have any itching, lesions, or unusual symptoms assessed by a primary provider, nurses can help prevent complications and progression of vulvar lesions. Ongoing assessment should occur at least annually.

Ovarian Cysts

The ovary is a common site for cysts, which may be simple enlargements of normal ovarian constituents, the graafian follicle, or the corpus luteum, or they may arise from abnormal growth of the ovarian epithelium. Ovarian cysts are often detected on routine pelvic examination. Although these cysts are typically benign, they nevertheless should be evaluated to exclude ovarian cancer, particularly in postmenopausal women (ACOG, 2007, reaffirmed 2009).

The patient may or may not report acute or chronic abdominal pain. Symptoms of a ruptured cyst mimic various acute abdominal emergencies, such as appendicitis or ectopic pregnancy. Larger cysts may produce abdominal swelling and exert pressure on adjacent abdominal organs.

Postoperative nursing care after surgery to remove an ovarian cyst is similar to that after abdominal surgery, with one exception. The marked decrease in intra-abdominal pressure resulting from removal of a very large cyst usually leads to considerable abdominal distention. This may be prevented to some extent by applying a snug-fitting abdominal binder.

Some surgeons discuss the option of a hysterectomy when a woman is undergoing bilateral ovary removal because of a suspicious mass; it may increase life expectancy and avoid a later second surgery. Patient preference is a priority in determining its appropriateness.

Polycystic ovary syndrome (PCOS) is a type of hormonal imbalance or cystic disorder that affects the ovaries. This complex endocrine condition involves a disorder in the hypothalamic–pituitary and ovarian network or axis, resulting in chronic anovulation and clinical androgen excess, often along with multiple small ovarian cysts. It is common and occurs in approximately 6% of women of childbearing age (Hall, 2011). Features can include obesity, insulin resistance, impaired glucose tolerance, dyslipidemia, sleep apnea, and infertility. Symptoms are related to androgen excess. Irregular menstrual periods, resulting from lack of regular ovulation, infertility, obesity, and hirsutism, may be a presenting complaint. Cysts form in the ovaries because the hormonal milieu cannot cause ovulation on a regular basis.

Diagnosis is based on clinical criteria, including hyperandrogenism, menstrual dysfunction, and polycystic ovaries on ultrasound examination. Two out of three of these criteria must be present to make the diagnosis (Schuiling & Likis, 2013). Women with PCOS are at increased risk for physical conditions including diabetes, increased blood lipids, and cardiovascular disease, as well as a host of psychosocial issues including but not limited to anger, frustration, and anxiety (Hall, 2011; Weiss & Bulmer, 2011) (Chart 57-5).

Chart 57-5 NURSING RESEARCH PROFILE

Life With Polycystic Ovary Syndrome

Weiss, T. R., & Bulmer, S. M. (2011). Young women's experiences living with polycystic ovary syndrome. *Journal of Obstetric, Gynecologic, and Neonatal Nursing, 40*(6), 709–718.

Purpose

The purpose of this study was to explore the psychosocial experience of living with polycystic ovary syndrome in women in late adolescence and their early 20s. This syndrome occurs in 5%–10% of all women of reproductive age, yet few studies have addressed the experiences of young women.

Design

This study used a qualitative phenomenologic methodology. In-depth interviews were conducted with 12 young women 18–23 years of age. The women were recruited from college campuses.

Findings

Several themes emerged from the responses of participants. The themes included concerns for older self (i.e., future health issues including cancer, diabetes, and heart disease), feeling physically inferior, coping with symptoms, patient–provider relationship, seeking usable information and support, and coming to grips with a chronic condition.

Nursing Implications

Nurses should work with providers and patients to develop a comprehensive holistic plan of care that includes psychosocial issues. Patients may need education in developing effective communication skills with their provider. Support groups may be helpful, if available. Encouraging regular exercise and good nutrition is essential in helping these women optimize their health during adolescence and throughout life.

Medical Management

The treatment of large ovarian cysts is usually surgical removal. However, oral contraceptives may be used in young, healthy patients to suppress ovarian activity and resolve small cysts that appear to be fluid filled or physiologic.

Oral contraceptive agents are also usually prescribed to treat PCOS (Weis, 2011). When pregnancy is desired, medications to stimulate ovulation (clomiphene [Clomid]) are often effective. Lifestyle modification is critical, and weight management is part of the treatment plan.

Weight loss as little as 5% of total body weight can help with hormone imbalance and infertility (Schuiling & Likis, 2013). Metformin (Glucophage) often regulates periods and can help with weight loss. Women with this diagnosis are at increased risk for endometrial cancer due to anovulation.

Benign Tumors of the Uterus: Fibroids (Leiomyomas, Myomas)

Myomatous or **fibroid tumors** of the uterus are estimated to occur in 25% to 40% of women during their reproductive years (Schuiling & Likis, 2013). It is thought that women are genetically predisposed to develop this condition, which is almost always benign. Fibroids arise from the muscle tissue of the uterus and can be solitary or multiple, in the lining (intracavitary), muscle wall (intramural), and outside surface (serosal) of the uterus. They usually develop slowly in women between 25 and 40 years of age and may become quite large. A growth spurt with enlargement of the fibroid tumor may occur in the decade before menopause, possibly related to anovulatory cycles and high levels of unopposed estrogen. Fibroids are a common reason for hysterectomy because they often result in menorrhagia, which can be difficult to control.

Clinical Manifestations

Fibroids may cause no symptoms, or they may produce abnormal vaginal bleeding. Other symptoms result from pressure on the surrounding organs and include pain, backache, pressure, bloating, constipation, and urinary problems. Menorrhagia (excessive bleeding) and metrorrhagia (irregular bleeding) may occur because fibroids may distort the uterine lining (Fig. 57-7). Fibroids may interfere with fertility.

Medical Management

Treatment of uterine fibroids may include medical or surgical intervention and depends to a large extent on the size, symptoms, and location, as well as the woman's age and her reproductive plans. Fibroids usually shrink and disappear during menopause, when estrogen is no longer produced. Simple observation and follow-up may be all the management that is necessary. The patient with minor symptoms is closely monitored. If she plans to have children, treatment is as conservative as possible. As a rule, large tumors that produce pressure symptoms are surgically removed (**myomectomy**). A hysterectomy may be performed if symptoms are severe and childbearing is completed (see later discussion of nursing care for a patient having a hysterectomy).

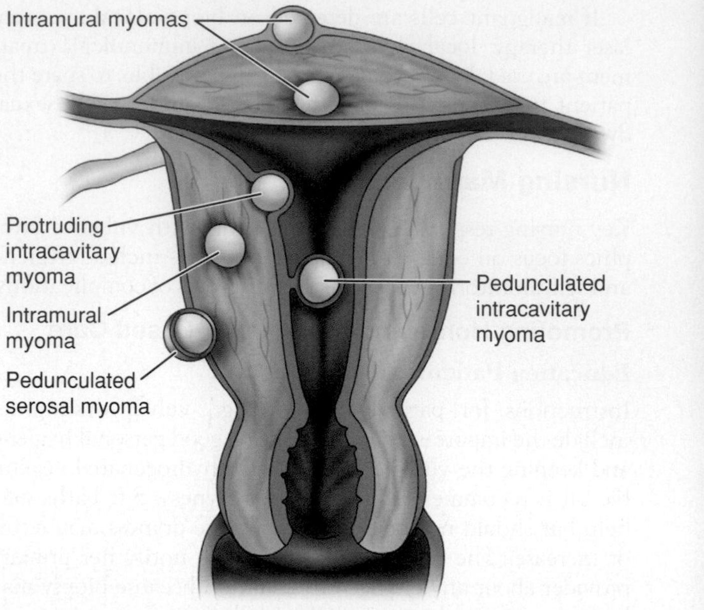
Intramural myomas
Protruding intracavitary myoma
Intramural myoma
Pedunculated serosal myoma
Pedunculated intracavitary myoma

FIGURE 57-7 • Myomas (fibroids). Those that impinge on the uterine cavity are called *intracavitary myomas*.

Several alternatives to hysterectomy have been developed for the treatment of excessive bleeding due to fibroids (Schuiling & Likis, 2013). These include the following:

- *Hysteroscopic resection of myomas:* A laser is used through a hysteroscope passed through the cervix; no incision or overnight stay is needed.
- *Laparoscopic myomectomy:* Removal of a fibroid through a laparoscope inserted through a small abdominal incision
- *Laparoscopic myolysis:* The use of a laser or electrical needles to cauterize and shrink the fibroid
- *Laparoscopic cryomyolysis:* Electric current is used to coagulate the fibroid.
- *Uterine artery embolization (UAE):* Polyvinyl alcohol or gelatin particles are injected into the blood vessels that supply the fibroid via the femoral artery, resulting in infarction and resultant shrinkage. This percutaneous image-guided therapy offers an alternative to hormone therapy or surgery. UAE may result in infrequent but serious complications such as pain, infection, amenorrhea, necrosis, and bleeding. Although rare, deaths and ovarian failure may occur. Women need to weigh the risks and benefits carefully, especially if they have not completed childbearing. This procedure has been found to cause fewer complications than hysterectomy, but women may need further treatment in the future (Hickey, Marino, & Brownfoot, 2012).
- *Magnetic resonance–guided focused ultrasound surgery (MRgFUS):* Ultrasonic energy is passed through the abdominal wall to target and destroy the fibroid. This noninvasive procedure is approved by the FDA for premenopausal women with bothersome symptoms due to fibroids and who do not want more children. It is an outpatient treatment.

Medications (e.g., leuprolide [Lupron]) or other gonadotropin-releasing hormone (GnRH) analogues, which induce a temporary menopauselike environment, may be prescribed to shrink the fibroids. This treatment consists of monthly injections, which may cause hot flashes and vaginal dryness. Treatment is usually short term (i.e., before surgery) to shrink the fibroids, allowing easier surgery, and to alleviate anemia, which may occur as a result of heavy menstrual flow. This treatment is used on a temporary basis because it leads to vasomotor symptoms and loss of bone density.

Antifibrotic agents are under investigation for long-term treatment of fibroids. Mifepristone (Mifeprex), a progesterone antagonist is also utilized.

Endometriosis

Endometriosis is a chronic disease affecting between 6% and 10% of women of reproductive age (Falcone & Lebovic, 2011) and consisting of a benign lesion or lesions that contain endometrial tissue (similar to that lining the uterus) found in the pelvic cavity outside the uterus. Extensive endometriosis may cause few symptoms, or an isolated lesion may produce severe symptoms. It is a major cause of chronic pelvic pain and infertility.

Endometriosis has been diagnosed more frequently as a result of the increased use of laparoscopy. There is a high incidence among patients who bear children late and among those who have fewer children. In countries where tradition favors early marriage and early childbearing, endometriosis is rare. There also appears to be a familial predisposition to endometriosis; it is more common in women whose close female relatives are affected. Other factors that may suggest increased risk include a shorter menstrual cycle (less than every 27 days), flow longer than 7 days, outflow obstruction, and younger age at menarche. Characteristically, endometriosis is found in young, nulliparous women between 25 and 35 years of age and in adolescents, particularly those with dysmenorrhea that does not respond to nonsteroidal anti-inflammatory drugs (NSAIDs) or oral contraceptive agents (Falcone & Lebovic, 2011).

Pathophysiology

Misplaced endometrial tissue responds to and depends on ovarian hormonal stimulation. During menstruation, this ectopic tissue bleeds, mostly into areas having no outlet, which causes pain and adhesions. The lesions are typically small and puckered, with a blue/brown/gray powder-burn appearance and brown or blue-black appearance, indicating concealed bleeding.

Endometrial tissue contained within an ovarian cyst has no outlet for the bleeding; this formation is referred to as a pseudocyst or chocolate cyst. Adhesions, cysts, and scar tissue may result, causing pain and infertility (Hall, 2011). Endometriosis may increase the risk of ovarian cancer.

Currently, the best-accepted theory regarding the origin of endometrial lesions is the transplantation theory, which suggests that a backflow of menses (retrograde menstruation) transports endometrial tissue to ectopic sites through the fallopian tubes. Why some women with retrograde menstruation develop endometriosis and others do not is unknown. Endometrial tissue can also be spread by lymphatic or venous channels.

Clinical Manifestations

Symptoms vary but include dysmenorrhea, dyspareunia, and pelvic discomfort or pain. Dyschezia (pain with bowel movements) and radiation of pain to the back or leg may occur. Depression, loss of work due to pain, and relationship difficulties may result. Infertility may occur because of fibrosis and adhesions or because of a variety of substances (prostaglandins, cytokines, other factors) produced by the implants of endometriosis and scar tissue on anatomical sites.

Assessment and Diagnostic Findings

A health history, including an account of the menstrual pattern, is necessary to elicit specific symptoms. On bimanual pelvic examination, fixed tender nodules are sometimes palpated, and uterine mobility may be limited, indicating adhesions. Laparoscopic examination confirms the diagnosis and helps stage the disease. In stage 1, patients have superficial or minimal lesions; stage 2, mild involvement; stage 3, moderate involvement; and stage 4, extensive involvement and dense adhesions, with obliteration of the cul-de-sac.

Medical Management

Treatment depends on the symptoms, the patient's desire for pregnancy, and the extent of the disease. If the woman does not have symptoms, routine examination may be all that is required. Other therapy for varying degrees of symptoms may be NSAIDs, oral contraceptive agents, GnRH agonists, or surgery. Pregnancy often alleviates symptoms, because neither ovulation nor menstruation occurs.

Pharmacologic Therapy

Palliative measures include the use of medications, such as analgesic agents and prostaglandin inhibitors, for pain. Hormonal therapy is effective in suppressing endometriosis and relieving dysmenorrhea (menstrual pain). Oral contraceptive agents provide effective pain relief and may prevent disease progression (Schuiling & Likis, 2013). Infrequently, side effects may occur with oral contraceptives, such as fluid retention, weight gain, and nausea. These can usually be managed by changing brands or formulations.

Several types of hormonal therapy are also available in addition to oral contraceptive agents. A synthetic androgen, danazol (Danocrine), causes atrophy of the endometrium and subsequent amenorrhea. The medication inhibits the release of gonadotropin with minimal overt sex hormone stimulation. The drawbacks of this medication are that it is expensive and may cause troublesome side effects such as fatigue, depression, weight gain, oily skin, decreased breast size, mild acne, hot flashes, and vaginal atrophy (Schuiling & Likis, 2013). GnRH agonists decrease estrogen production and cause subsequent amenorrhea. Side effects are related to low estrogen levels (e.g., hot flashes and vaginal dryness). Loss of bone density is often offset by concurrent use of estrogen. Leuprolide, a GnRH agonist, is injected monthly to suppress hormones and induce an artificial menopause, thus avoiding menstrual effects and relieving endometriosis. A combination of medical therapies is often used (Schuiling & Likis, 2013). Most women continue treatment despite side effects, and symptoms diminish for 80% to 90% of women with mild to moderate endometriosis. Hormonal medications are not used in patients with a history of abnormal vaginal bleeding or liver, heart, or kidney disease. Bone density is followed carefully because of the risk of bone loss; hormone therapy is usually short term.

Surgical Management

If conservative measures are not helpful, surgery may be necessary to relieve pain and improve the possibility of pregnancy. Surgery may be combined with the use of medical therapy. The procedure selected depends on the patient. Laparoscopy may be used to fulgurate (cut with high-frequency current) endometrial implants and to release adhesions. Laser surgery is another option made possible by laparoscopy. Laser therapy vaporizes or coagulates the endometrial implants, thereby destroying this tissue. Other surgical options include endocoagulation and electrocoagulation, laparotomy, abdominal hysterectomy, **oophorectomy** (removal of the ovary), bilateral **salpingo-oophorectomy** (removal of the ovary and its fallopian tube), and appendectomy. Many women need further intervention following conservative surgeries; therefore, total hysterectomy is the definitive procedure (Schuiling & Likis, 2013).

Nursing Management

The health history and physical examination focus on specific symptoms (e.g., pelvic pain), the effect of prescribed medications, and the woman's reproductive plans. This information helps in determining the treatment plan. Explaining the various diagnostic procedures may help to alleviate the patient's anxiety. Patient goals include relief of pain, dysmenorrhea, dyspareunia, and avoidance of infertility.

As the treatment progresses, the woman with endometriosis and her partner may find that pregnancy is not easily possible, and the psychosocial impact of this realization must be recognized and addressed. Alternatives, such as in vitro fertilization or adoption, may be discussed at an appropriate time and referrals offered.

The nurse's role in patient education is to dispel myths and encourage the patient to seek care if dysmenorrhea or dyspareunia occurs. The Endometriosis Association (see the Resources section at the end of this chapter) is a helpful resource for patients seeking further information and support for this condition, which can cause disabling pain and severe emotional distress.

Chronic Pelvic Pain

Chronic pelvic pain is a common disorder of women that may be related to several of the previously discussed gynecologic disorders. Chronic pelvic pain—that is, pelvic pain that persists for more than 6 months—may be cyclic or intermittent and noncyclic (Baron, Florendo, Sandbo, et al., 2011). Causes may be of reproductive, genitourinary, or gastrointestinal origin. A history of sexual and physical abuse, PID, endometriosis, interstitial cystitis, musculoskeletal disorders, irritable bowel syndrome, and previous surgery resulting in abdominal adhesions may be associated with chronic pelvic pain. Dysmenorrhea, dyspareunia, and lower abdominal pain may also be associated with sexual and physical abuse.

Chronic pelvic pain is often difficult to treat. Treatment depends on physical and diagnostic test results and may include antidepressant, analgesic, and oral contraceptive agents; GnRH agonists; exercise; and various surgical procedures (Schuiling & Likis, 2013). (See Chapter 72 for further discussion of family violence, abuse, and neglect.)

Adenomyosis

In adenomyosis, the tissue that lines the endometrium invades the uterine wall. The incidence is highest in multiparous women (Schuiling & Likis, 2013). Symptoms include hypermenorrhea (excessive and prolonged bleeding), dysmenorrhea, polymenorrhea (abnormally frequent bleeding), and pelvic pain. Physical examination findings on palpation include an enlarged, firm, and tender uterus. Treatment depends on the severity of bleeding and pain. Hysterectomy may be the best option in providing relief of symptoms.

Endometrial Hyperplasia

Endometrial hyperplasia (a buildup of endometrial tissue) can be a precursor to endometrial cancer and often results from unopposed estrogen from any source. Estrogen therapy alone without progesterone in a woman with a uterus can cause this condition. Women with anovulatory cycles, PCOS, or obesity may have high circulating levels of estrogen. Tamoxifen (Nolvadex) may also be a causative factor. Diagnosis is by biopsy or ultrasound findings of thickness of the endometrium. Hyperplasia with atypia on a pathology or biopsy report indicates risk of progression. Progestin treatment may be effective, but hysterectomy may be advised if pathology from an endometrial biopsy shows atypia. Abnormal bleeding is the most common symptom.

MALIGNANT CONDITIONS

Gynecologic cancer is any cancer that starts in a woman's reproductive organs (CDC, 2010c). According to the American Cancer Society (ACS, 2012), the projected incidence for female reproductive cancers in the United States includes about 12,170 new cases of cervical cancer; 47,130 new cases of uterine cancer; 22,280 new cases of ovarian cancer; 2,680 new cases of vaginal cancer; and 4,490 new cases of vulvar cancer. Cervical cancer is the third most common cause of cancer death among women worldwide, and the incidence is declining.

Although some cancers are difficult to detect or prevent, annual pelvic examination with a Pap smear is a painless and relatively inexpensive method of early detection (CDC, 2010c). Primary providers can encourage women to follow this health practice by providing examinations that are educational and supportive and offer women an opportunity to ask questions and clarify misinformation.

Women diagnosed with gynecologic malignancies experience anxiety related to their prognosis. The occurrence of physical symptoms may cause more psychological distress.

Intervention directed toward physical and psychological symptoms requires a multidisciplinary approach.

Nurses should be aware of ongoing clinical trials that are being conducted to identify effective treatments for many conditions. They are often in a position to answer questions about clinical trials and to encourage patients to consider participation if appropriate. Women's participation in cancer research may not occur in part because women are unaware of ongoing relevant research (see ClinicalTrials.gov in the Resources section).

Cancer of the Cervix

Death from cervical cancer is less common due to early detection of cell changes by Pap smear (CDC, 2010c). However, it is still the third most common female reproductive cancer and is estimated to affect more than 12,000 women in the United States every year (CDC, 2010c). Risk factors are presented in Chart 57-6.

Preventive measures include regular pelvic examinations and Pap tests for all women, especially older women past childbearing age. Preventive counseling should encourage delaying first intercourse, avoiding HPV infection, engaging only in safer sex, ceasing smoking, and receiving HPV immunization.

There are several different types of cervical cancer. Most of these cancers are squamous cell carcinomas, and the rest are adenocarcinomas or mixed adenosquamous carcinomas. Adenocarcinomas begin in mucus-producing glands and are often due to HPV infection. Most cervical cancers, if not detected and treated, spread to regional pelvic lymph nodes, and local recurrence is not uncommon.

Chart 57-6 RISK FACTORS
Cervical Cancer

- Sexual activity:
  - Multiple sex partners
  - Early age (<20 years) at first coitus (exposes the vulnerable young cervix to potential viruses from a partner)
- Sex with uncircumcised men
- Sexual contact with men whose partners have had cervical cancer
- Early childbearing
- Exposure to human papillomavirus, types 16 and 18
- Human immunodeficiency virus infection and other causes of immunodeficiency
- Smoking and exposure to secondhand smoke
- Exposure to diethylstilbestrol in utero
- Family history of cervical cancer
- Low socioeconomic status (may be related to early marriage and early childbearing)
- Nutritional deficiencies (folate, beta-carotene, and vitamin C levels are lower in women with cervical cancer than in women without it)
- Chronic cervical infection
- Overweight status

Adapted from Eggert, J. (2010). *Cancer basics.* Pittsburgh: Oncology Nursing Society; and Schuiling, K. D., & Likis, F. E. (2013). *Women's gynecologic health* (2nd ed.). Burlington, MA: Jones & Bartlett.

Clinical Manifestations

Early cervical cancer rarely produces symptoms. If symptoms are present, they may go unnoticed as a thin, watery vaginal discharge often noticed after intercourse or douching. When symptoms such as discharge, irregular bleeding, or pain or bleeding after sexual intercourse occur, the disease may be advanced. Advanced disease should not occur if all women have access to gynecologic care and avail themselves to it. The nurse's role in access to care and its utilization is crucial.

In advanced cervical cancer, the vaginal discharge gradually increases and becomes watery and, finally, dark and foul smelling from necrosis and infection. The bleeding, which occurs at irregular intervals between periods (metrorrhagia) or after menopause, may be slight (just enough to spot the undergarments) and occurs usually after mild trauma or pressure (e.g., intercourse, douching, or bearing down during defecation). As the disease continues, the bleeding may persist and increase. Leg pain, dysuria, rectal bleeding, and edema of the extremities signal advanced disease.

As the cancer advances, it may invade the tissues outside the cervix, including the lymph glands anterior to the sacrum. In one third of patients with invasive cervical cancer, the disease involves the fundus. The nerves in this region may be affected, producing excruciating pain in the back and the legs that is relieved only by large doses of opioid analgesic agents. If the disease progresses, it often produces extreme emaciation and anemia that usually is accompanied by fever (due to secondary infection and abscesses in the ulcerating mass) and by fistula formation. Because the survival rate for in situ cancer is 100% and the rate for women with more advanced stages of cervical cancer decreases dramatically, early detection is essential.

Assessment and Diagnostic Findings

Diagnosis may be made on the basis of abnormal Pap smear results, followed by biopsy results identifying severe dysplasia (cervical intraepithelial neoplasia type III [CIN III], high-grade squamous intraepithelial lesions [HGSILs, also referred to as HSILs], or carcinoma in situ). HPV infections are usually implicated in these conditions. Screening should begin within 3 years of the initiation of sexual intercourse or at 21 years of age (Smith, Cokkinides, & Brawley, 2012) (see Table 15-3 in Chapter 15). Liquid-based cytology allows for co-testing for HPV. Carcinoma in situ is technically classified as severe dysplasia and is defined as cancer that has extended through the full thickness of the epithelium of the cervix, but not beyond. This is often referred to as preinvasive cancer.

In its very early stages, cervical cancer is found microscopically by Pap smear. In later stages, pelvic examination may reveal a large, reddish growth or a deep, ulcerating lesion. The patient may report spotting or bloody discharge.

When the patient has been diagnosed with invasive cervical cancer, clinical staging estimates the extent of the disease so that treatment can be planned more specifically and prognosis reasonably predicted. The tumor, nodes, and metastases (TNM) system is the most widely used staging system. In this system, "T" describes the primary tumor size and location, "N" describes lymph node involvement, and

"M" describes metastasis, or spread of the disease (Schuiling & Likis, 2013). (See Chapter 15 for a detailed discussion of cancer staging.)

Signs and symptoms are evaluated, and x-rays, laboratory tests, and special examinations, such as biopsy and colposcopy, are performed (Schuiling & Likis, 2013). Depending on the stage of the cancer, other tests and procedures may be performed to determine the extent of disease and appropriate treatment. These tests may include dilation and curettage (D&C), computed tomography (CT), magnetic resonance imaging (MRI), IV urography, cystography, positron emission tomography, and barium x-ray studies. Treatment depends on the stage of the disease.

Medical Management

Precursor or Preinvasive Lesions

When precursor lesions, such as low-grade squamous intraepithelial lesions (LGSILs, also referred to as LSILs; CIN I and II or mild to moderate dysplasia), are found by colposcopy and biopsy, careful monitoring by frequent Pap smears or conservative treatment is possible. Conservative treatment may consist of monitoring, **cryotherapy** (freezing with liquid nitrogen), or laser therapy. A **loop electrocautery excision procedure (LEEP)** may also be used to remove abnormal cells. In this procedure, a thin wire loop with laser is used to cut away a thin layer of cervical tissue. LEEP is an outpatient procedure that usually is performed in a gynecologist's office; it takes only a few minutes. Analgesia is given before the procedure, and a local anesthetic agent is injected into the area. This procedure allows the pathologist to examine the removed tissue sample to determine if the borders of the tissue are disease free. Another procedure referred to as a cone biopsy or **conization** (removing a cone-shaped portion of the cervix) is performed when biopsy findings demonstrate CIN III or HGSIL (equivalent to severe dysplasia) and carcinoma in situ.

If preinvasive cervical cancer (carcinoma in situ) occurs when a woman has completed childbearing, a simple hysterectomy (removal of the uterus only) is usually recommended (Schuiling & Likis, 2013). If a woman is pregnant or wishes to have children and invasion is less than 1 mm, conization may be sufficient. Frequent follow-up examinations are necessary to monitor for recurrence (Schuiling & Likis, 2013).

Patients who have precursor or premalignant lesions need reassurance that they do not have invasive cancer. However, the importance of close follow-up is emphasized because the condition, if untreated for a long time, may progress to cancer. Patients with cervical cancer in situ also need to know that this is usually a slow-growing and nonaggressive type of cancer that is not expected to recur after appropriate treatment.

Invasive Cancer

Treatment of invasive cervical cancer depends on the stage of the lesion, the patient's age and general health, and the judgment and experience of the provider. Surgery and radiation treatment (intracavitary and external) are most often used. Surgical procedures that may be used to treat cervical cancer are summarized in Chart 57-7. Ethical issues often arise in the treatment of invasive cancers. These issues are summarized in Chart 57-8.

Frequent follow-up after surgery is imperative because the risk of recurrence is high and usually occurs within the first 2 years. Recurrences are often in the upper quarter of the vagina, and ureteral obstruction may be a sign. Weight loss, leg edema, and pelvic pain may be signs of lymphatic obstruction and metastasis.

Radiation, which is often part of treatment to reduce recurrent disease, may be delivered by an external beam or by **brachytherapy** (method by which the radiation source is placed near the tumor in a sealed source) or both. The field to be irradiated as well as the dose and method of radiation are determined by stage, volume of tumor, and lymph node involvement (Eggert, 2010).

A variety of chemotherapeutic approaches are used to treat advanced cervical cancer. They are often used in combination with surgery and radiation. Studies are ongoing to find the best approach to treat advanced cervical cancer (Eggert, 2010). Vaginal stenosis is a frequent side effect of radiation. Preventive therapy such as the use of a vaginal dilator to avoid severe permanent vaginal stenosis.

Some patients with recurrences of cervical cancer are considered for **pelvic exenteration**, in which several pelvic organs are removed. This is a complex, extensive surgical procedure that is reserved for women with a high likelihood of cure. Unilateral leg edema, sciatica, and ureteral obstruction indicate likely disease progression. Patients with these symptoms have advanced disease and are not considered candidates for this major surgical procedure. Surgery is complex because it is performed close to the bowel, bladder, ureters, and great vessels. Possible complications include pulmonary embolism (PE), pulmonary edema, myocardial infarction, cerebrovascular accident, hemorrhage, sepsis, small bowel obstruction, fistula formation, obstruction of the ileal conduit, bladder dysfunction, and pyelonephritis, most often in the first 18 months postoperatively. Vein

Chart 57-7 Surgical Procedures for Cervical Cancer

- Total hysterectomy—removal of the uterus, cervix, and ovaries
- Radical hysterectomy—removal of the uterus, ovaries, fallopian tubes, proximal vagina, and bilateral lymph nodes through an abdominal incision (*Note:* "Radical" indicates that an extensive area of the paravaginal, paracervical, parametrial, and uterosacral tissues is removed with the uterus.)
- Radical vaginal hysterectomy—vaginal removal of the uterus, ovaries, fallopian tubes, and proximal vagina
- Bilateral pelvic lymphadenectomy—removal of the common iliac, external iliac, hypogastric, and obturator lymphatic vessels and nodes
- Pelvic exenteration—removal of the pelvic organs, including the bladder or rectum and pelvic lymph nodes, and construction of diversional conduit, colostomy, and vagina
- Radical trachelectomy—removal of the cervix and selected nodes to preserve childbearing capacity in a woman of reproductive age with cervical cancer

Adapted from Eggert, J. (2010). *Cancer basics.* Pittsburgh: Oncology Nursing Society; and Schuiling, K. D., & Likis, F. E. (2013). *Women's gynecologic health* (2nd ed.). Burlington, MA: Jones & Bartlett.

Chart 57-8 ETHICAL DILEMMA

What If the Patient Refuses to Autonomously Make Decisions?

Case Scenario

You work on a gynecology-oncology unit of a university health science center. Your patient is a 24-year-old woman from a sub-Saharan African nation who is a graduate student in engineering enrolled in the university. She is admitted to the unit with a new diagnosis of advanced cervical cancer that was discovered when she presented to the emergency department with intractable pelvic pain and bilateral leg edema. You are present with the surgeon when he tells the patient and her husband that he thinks she should have a pelvic exenteration and explains what this surgical procedure entails. The patient's husband asks if there is a different procedure that may be performed that could preserve the patient's ability to eventually have children. The surgeon notes that this is possible but not advisable, given the extent and spread of the tumor mass, and that the alternative type of organ preservation surgery is associated with a worst prognosis for the patient. The husband insists that the surgeon not perform the pelvic exenteration. He notes that his wife's life would not be worth living in their native country if she were incapable of bearing children. Throughout this discussion, the wife has remained silent. When you ask her what her wishes might be, she tells you that she wants whatever treatment her husband chooses for her.

Discussion

There are some cultures that adhere to paternalistic tenets; the male "head of household" makes all major decisions, including health care decisions. The patient may be adhering to her cultural norms when she tells you that she will acquiesce to whatever decision her husband makes for her. Arguably, however, she may feel coerced or obligated to acquiesce to his decisions.

Analysis

- Describe the ethical principles that are in conflict in this case (see Chart 3-3 in Chapter 3). Which principle should have preeminence as you proceed with your assessment of this patient?
- Describe any concerns you may have regarding preservation of the patient's autonomy. What steps might you take to ensure that her autonomy is preserved?
- Assume that you have spoken in private with the patient and have verified that she understands that by not having a pelvic exenteration, she may be putting her life in greater jeopardy while not necessarily guaranteeing that she will ever be able to have children. How can you professionally reconcile her decision to put her life at greater risk so that she may adhere to her cultural norms (i.e., being obedient to her husband)? Does this type of decision truly preserve her autonomy?

Resources

See Chapter 3, Chart 3-6 for ethics resources.

constriction must be avoided postoperatively. Patients with varicose veins or a history of thromboembolic disease may be treated prophylactically with heparin. Anti-embolism stockings are prescribed to reduce the risk of deep vein thrombosis (DVT). Nursing care of these patients is complex and requires coordination and care by experienced health care professionals. Pelvic exenteration is discussed in further detail later in this chapter.

Cancer of the Uterus (Endometrium)

Although the incidence of cancer of the uterine endometrium (fundus or corpus) has stabilized, its death rate has increased, possibly because of increased lifespan and coexisting comorbidities. This cancer is the most frequently occurring gynecologic cancer in the United States. After breast, colorectal, and lung cancer, endometrial cancer is the fourth most common cancer in women. More than 42,100 new cases of uterine cancer occur each year, with more than 7,800 deaths (ACS, 2012). Most women are diagnosed between 55 and 64 years of age (Grube, Ammon, & Killen, 2011). Many women with endometrial cancer are obese; obesity increases the risk of morbidity and mortality from this disease. This disease occurs twice as often in Caucasian women as in African American women, who have a less favorable prognosis, and research is needed to explain this disparity (Almadrones-Cassidy, 2010). Cumulative exposure to estrogen is considered the major risk factor (Chart 57-9). This exposure occurs with the use of estrogen therapy without the use of progestin, early menarche, late menopause, nulliparity, and anovulation. Other risk factors include infertility, diabetes, and the use of tamoxifen. Tamoxifen, which is taken for treatment or prevention of breast cancer, may cause proliferation of the uterine lining (ACOG, 2011b). Women who take tamoxifen should be monitored by their oncologists and/or gynecologic health care providers.

Pathophysiology

Most uterine cancers are endometrioid (i.e., originating in the lining of the uterus). There are three types. Type 1, which accounts for about 80% of cases, is estrogen related and occurs in women who are younger, obese, and perimenopausal. It is usually low grade with a favorable prognosis. Type 2, which occurs in about 10% of cases, is high grade and usually serous cell or clear cell. Older and African

Chart 57-9 RISK FACTORS

Uterine Cancer

- Age—usually >50 years; median age, 64 years
- Obesity that results in increased estrone levels (related to excess weight) resulting from conversion of androstenedione to estrone in body fat, which exposes the uterus to unopposed estrogen
- Unopposed estrogen therapy (estrogen used without progesterone, which offsets the risk of unopposed estrogen)
- Other—nulliparity, truncal obesity, late menopause (after 52 years of age) and the use of tamoxifen

Adapted from Eggert, J. (2010). *Cancer basics.* Pittsburgh: Oncology Nursing Society; and Schuiling, K. D., & Likis, F. E. (2013). *Women's gynecologic health* (2nd ed.). Burlington, MA: Jones & Bartlett.

American women are at higher risk for type 2. Type 3, which also occurs in about 10% of cases, is the hereditary and genetic types, some of which are related to Lynch II syndrome, also known as hereditary nonpolyposis colorectal cancer (Schuiling & Likis, 2013).

Assessment and Diagnostic Findings

All women should be encouraged to have annual check-ups, including a gynecologic examination. Any woman who is experiencing irregular bleeding should be evaluated promptly. If a menopausal woman experiences bleeding, an endometrial aspiration or biopsy is performed to rule out hyperplasia, which is a possible precursor of endometrial cancer. The procedure is quick and usually not painful. Transvaginal ultrasound can also be used to measure the thickness of the endometrium (Grube et al., 2011). (Postmenopausal women should have a very thin endometrium due to low levels of estrogen; a thicker lining warrants further investigation.) A biopsy or aspiration for tissue pathology is diagnostic.

Medical Management

Treatment for endometrial cancer consists of surgical staging, total or radical hysterectomy (discussed later in this chapter), and bilateral salpingo-oophorectomy and lymph node sampling (Galic & Gupta, 2012). Laparoscopy or robotic-assisted laparoscopic surgery is less invasive than abdominal surgery. Lymph node sampling and visualization of the peritoneum can be accomplished in many women in this manner. Cancer antigen 125 (CA-125) levels must be monitored, because elevated levels are a significant predictor of extrauterine disease or metastasis. Depending on the stage, the therapeutic approach is individualized and is based on stage, type, differentiation, degree of invasion, and node involvement. Radiation may be used in the form of external-beam radiation or vaginal brachytherapy (Schuiling & Likis, 2013). Whole pelvis radiotherapy may be used if there is any spread beyond the uterus. Recurrent cancer usually occurs inside the vaginal vault or in the upper vagina, and metastasis usually occurs in lymph nodes or the ovary. Recurrent lesions in the vagina are treated with surgery and radiation. Recurrent lesions beyond the vagina are treated with hormonal therapy or chemotherapy. Progestin therapy is used frequently. Patients should be prepared for such side effects as nausea, depression, rash, or mild fluid retention with progestin therapy.

Cancer of the Vulva

Primary cancer of the vulva represents 5% of all gynecologic malignancies and is seen mostly in postmenopausal women, although its incidence in younger women is increasing. The median age for cancer limited to the vulva is 50 years, whereas the median age for invasive vulvar cancer is 70 years (Schuiling & Likis, 2013). Possible risk factors include smoking, HPV infection, HIV infection, and immunosuppression. Squamous cell carcinoma accounts for most primary vulvar tumors. Less common are Bartholin's gland cancer, vulvar sarcoma, and malignant melanoma. Little is known about what causes this disease; however, increased risk may be related to chronic vulvar irritation. In younger women, HPV infection may be implicated, especially types 16, 18, and 31. Prevention includes delaying onset of sexual activity to avoid early exposure to HPV, administration of the HPV vaccine, and avoidance of smoking. Regular pelvic examinations, Pap smears, and vulvar self-examination are helpful in early detection. Women with persistent irritation or itching should be encouraged to seek evaluation.

Clinical Manifestations

Long-standing pruritus and soreness are the most common symptoms of vulvar cancer. Itching occurs in half of all patients with vulvar malignancy. Bleeding, foul-smelling discharge, and pain may also be present and are usually signs of advanced disease. Cancerous lesions of the vulva are visible and accessible and grow relatively slowly. Early lesions appear as a chronic dermatitis; later, patients may note a lump that continues to grow and becomes a hard, ulcerated, cauliflower-like growth. Biopsy should be performed on any vulvar lesion that persists, ulcerates, or fails to heal quickly with proper therapy. Vulvar malignancies may appear as a lump or mass, redness, or a lesion that fails to heal.

Medical Management

Vulvar intraepithelial lesions are preinvasive and are also called *vulvar carcinoma in situ*. They may be treated by local excision, laser ablation, application of chemotherapeutic creams, or cryosurgery.

When invasive vulvar carcinoma exists, treatment may include wide excision or removal of the vulva (**vulvectomy**). An effort is made to individualize treatment, depending on the extent of the disease. A wide excision is performed only if lymph nodes are normal. More pervasive lesions require vulvectomy with deep pelvic node dissection. Vulvectomy is very effective at prolonging life but is frequently followed by complications (i.e., scarring, wound breakdown, leg swelling, vaginal stenosis, or rectocele). To reduce complications, only necessary tissue is removed. External-beam radiation may be used, resulting in sunburnlike irritation that usually resolves in 6 to 12 months. Laser therapy and chemotherapy are other possible treatment options.

If a widespread area is involved or the disease is advanced, a radical vulvectomy with bilateral groin dissection may be performed. Antibiotic and heparin prophylaxis may be prescribed preoperatively and continued postoperatively to prevent infection, DVT, and PE. Sequential compression devices (SCDs) are applied to reduce the risk of venous thromboembolism (VTE).

Although the role of systemic chemotherapy in the treatment of vulvar cancer remains to be determined, chemotherapy may be helpful when used in combination with radiation therapy in treatment for advanced disease. The combination of radiation and chemotherapy may reduce the size of the cancer, resulting in less extensive subsequent surgery (Eggert, 2010).

Clinical trials to determine the most effective treatment are difficult to conduct because there are few patients with this condition. Morbidity with recurrence of the disease is high, and patterns of recurrence vary. Reconstruction after vulvectomy is performed by plastic surgeons when appropriate and desired.

Nursing Management

Obtaining the Health History

The health history is a valuable tool for establishing rapport with the patient. The reason the patient is seeking health care is apparent. What the nurse can tactfully elicit is the reason a delay, if any, occurred in seeking health care—for example, because of modesty, economics, denial, neglect, or fear (abusive partners sometimes prevent women from seeking health care). Factors involved in any delay in seeking health care and treatment may also affect recovery. The patient's health habits and lifestyle are assessed, and her receptivity to education is evaluated. Psychosocial factors are also assessed. Preoperative preparation and psychological support begin at this time.

Providing Preoperative Care

Relieving Anxiety

Prior to surgery, the patient must be allowed time to talk and ask questions. Fear often decreases when a woman who is to undergo wide excision of the vulva or vulvectomy learns that the possibility for subsequent sexual relations is good. The nurse reinforces the information the surgeon has given to the patient and addresses the patient's questions and concerns.

Preparing Skin for Surgery

Skin preparation may include cleansing the lower abdomen, inguinal areas, upper thighs, and vulva with a detergent germicide for several days before the surgical procedure. The patient may be instructed to do this at home.

Providing Postoperative Care

Relieving Pain

Because of the wide excision, the patient may experience severe pain and discomfort even with minimal movement. Therefore, analgesic agents are administered preventively (i.e., around the clock at designated times) to relieve pain, increase the patient's comfort level, and allow mobility. Patient-controlled analgesia (see Chapter 12) may be used to relieve pain and promote patient comfort. Careful positioning using pillows usually increases comfort, as do soothing back rubs. A low Fowler's position or, occasionally, a pillow placed under the knees reduces pain by relieving tension on the incision; however, efforts must be made to avoid pressure behind the knees, which increases the risk of DVT. Positioning the patient on her side, with pillows between her legs and against the lumbar region, provides comfort and reduces tension on the surgical wound.

Improving Skin Integrity

A pressure-reducing mattress may be used to prevent pressure ulcers. Moving from one position to another requires time and effort; the use of an overbed trapeze bar may help the patient move herself more easily. The extent of the surgical incision and the type of dressing are considered when choosing strategies to promote skin integrity. Intact skin needs to be protected from drainage and moisture. Dressings are changed as needed to ensure patient comfort, to perform wound care and irrigation (if prescribed), and to permit observation of the surgical site.

The wound is usually cleansed daily with warm, normal saline irrigations or other antiseptic solutions as prescribed, or a transparent dressing may be in place over the wound to minimize exposure to the air and subsequent pain. The appearance of the surgical site and the characteristics of drainage are assessed and documented. After the dressings are removed, a bed cradle may be used to keep the bed linens away from the surgical site. The patient is always protected from exposure when visitors arrive or someone else enters the room.

Supporting Positive Sexuality and Sexual Function

The patient who undergoes vulvar surgery usually experiences concerns about body image, sexual attractiveness, and functioning. Establishing a trusting nurse–patient relationship is important for the patient to feel comfortable with expressing her concerns and fears. The patient is encouraged to discuss her concerns with her sexual partner as well.

Because alterations in sexual sensation and functioning depend on the extent of surgery, the nurse needs to know about any structural and functional changes resulting from the surgery. Referral of the patient and her partner to a sex counselor may help them address these changes and resume satisfying sexual activity.

Monitoring and Managing Potential Complications

Location, extent, and exposure of the surgical site and incision put the patient at risk for contamination of the site and infection and sepsis. The patient is monitored closely for local and systemic signs and symptoms of infection: purulent drainage, redness, increased pain, fever, and increased white blood cell count. The nurse assists in obtaining specimens for culture if infection is suspected and administers antibiotic agents as prescribed. Hand hygiene—always a crucial infection-preventing measure—is of particular importance along with wearing masks whenever there is an extensive area of exposed tissue. Catheters, drains, and dressings are handled carefully with gloves on to avoid cross-contamination. A low-residue diet prevents straining on defecation and wound contamination.

The patient is at risk for complications of VTE, which include DVT and PE, because of the positioning required during surgery, postoperative edema, and the immobility needed to promote healing. SCDs are applied, and other prophylactic measures may be prescribed for patients at high risk. The patient is encouraged and reminded to perform ankle exercises to minimize venous pooling, which leads to VTE. The patient is encouraged and assisted in changing positions by using the overbed trapeze bar. Pressure behind the knees is avoided when positioning the patient, because this may increase venous pooling. The patient is assessed for signs and symptoms of DVT (leg pain, redness, warmth, edema) and PE (chest pain, tachycardia, dyspnea). Fluid intake is encouraged to prevent dehydration, which also increases the risk of DVT.

The extent of the surgical incision and possibly wide excision of tissue increase the risk of postoperative bleeding and hemorrhage. Pressure dressings are applied after surgery to minimize this risk.

Quality and Safety Nursing Alert

The patient must be monitored closely for signs of hemorrhage and resulting hypovolemic shock. These signs may include decreased blood pressure; increased pulse rate; decreased urine output; decreased mental status; and cold, clammy skin.

If hemorrhage and shock occur, interventions include fluid replacement, blood component therapy, and vasopressor medications. Laboratory results (e.g., hematocrit and hemoglobin levels) and hemodynamic monitoring are used to assess the patient's response to treatment. Depending on the specific cause of hemorrhage, the patient may be returned to the operating room. (See Chapter 14 for a detailed discussion of shock.)

Promoting Home and Community-Based Care

Educating Patients About Self-Care

Preparing the patient for hospital discharge begins before hospital admission. The patient and family are informed about what to expect during the immediate postoperative and recovery periods. Depending on the changes resulting from the surgery, the patient and her family may need instruction about wound care, urinary catheterization, and possible complications. The patient is encouraged to share her concerns and to assume increasing responsibility for her own care. She is encouraged and assisted in learning to care for the surgical site. A referral for home care is made as indicated.

Nurses are in an ideal position to educate women about performing regular vulvar self-examinations. Using a mirror, patients can see what constitutes normal female anatomy and learn about changes that should be reported (e.g., lesions, ulcers, masses, and persistent itching). Nurses must urge women to seek health care if they notice anything abnormal, because vulvar cancer is one of the most curable of all malignant conditions.

Continuing Care

Patients may be discharged early in their postoperative recovery to home or a subacute facility. During this phase, the patient's physical status and psychological responses to the surgery are assessed. In addition, the patient is assessed for complications and healing of the surgical site. During home visits, the nurse assesses the home to determine if modifications are needed to facilitate care. The home visit is used to reinforce previous education and to assess the patient's and the family's understanding of and adherence to the prescribed treatment strategies. Follow-up phone calls by the nurse to the patient between home visits are usually reassuring to the patient and family, who may be responsible for performing complex care procedures. Attention to the patient's psychological responses is important because the patient may become discouraged and depressed because of alterations in body image and a slow recovery. Communication between the nurse involved in the patient's immediate postoperative care and the home care nurse is essential to ensure continuity of care.

Cancer of the Vagina

Cancer of the vagina is rare and usually takes years to develop (ACS, 2012). Primary cancer of the vagina is uncommon and usually squamous in origin. Malignant melanoma and sarcomas can occur. Most vaginal cancers are secondary and invasive at the time of diagnosis. Risk factors include previous cervical cancer, in utero exposure to diethylstilbestrol (DES), previous vaginal or vulvar cancer, previous radiation therapy, history of HPV, or pessary use. Any patient with previous cervical cancer should be examined regularly for vaginal lesions.

Vaginal pessaries, which are used to support prolapsed tissues, can be a source of chronic irritation. As such, they have been associated with vaginal cancer, but only when the devices were not cared for properly (i.e., the device was not cleaned regularly or the patient did not return to the primary provider regularly for vaginal examinations).

Patients often do not have symptoms but may report slight bleeding after intercourse, spontaneous bleeding, vaginal discharge, pain, and urinary or rectal symptoms (or both). Diagnosis is often by Pap smear. Encouraging close follow-up is the focus of nursing interventions with women who were exposed to DES in utero. Emotional support for mothers who received DES before its risks were discovered and their daughters who were exposed to DES in utero is essential.

Medical Management

Treatment of early lesions may include local excision, topical chemotherapy, or laser. Laser therapy is a common treatment option in early vaginal and vulvar cancer. Surgery for more advanced lesions depends on the size and the stage of the cancer. If radical vaginectomy is required, a vagina can be reconstructed with tissue from the intestine, muscle, or skin grafts. After vaginal reconstructive surgery and radiation, regular intercourse may be helpful in preventing vaginal stenosis. Water-soluble lubricants are helpful in reducing pain with intercourse (dyspareunia).

Following surgery, radiation therapy may be administered by a variety of methods, including external-beam radiation, which is usually an outpatient procedure, or brachytherapy, which is internal radiation therapy. Internal radiation may be given with intracavitary radioactive material contained in a seed, wire, needle, or tube, which is placed into a cavity such as the uterus or vagina. Interstitial radiation is another type of internal radiation treatment in which the radioactive material is placed in or near the cancer but not into a body cavity and is used in cervical and ovarian malignancies. These treatments may be high dose for a short period or low dose, which may take longer. Treatment during hospitalization or during outpatient therapy depends on several factors, including the status of the patient and the mode of delivery.

Cancer of the Fallopian Tubes

Malignancies of the fallopian tube are the least common type of genital cancer (ACS, 2012). Although this type of cancer can occur at any age, the average age at diagnosis is 55 years. Symptoms include abdominal pain, abnormal bleeding, and vaginal discharge. An enlarged fallopian tube may be found on sonogram if dilated and fluid filled, or it may appear or be palpated as a mass. Surgery followed by radiation therapy is the usual treatment.

Cancer of the Ovary

Ovarian cancer is the leading cause of gynecologic cancer deaths in the United States (Blewitt, 2010). Despite careful physical examination, ovarian tumors are often difficult to detect because they are usually deep in the pelvis. No early screening mechanism exists at present. Tumor-associated antigens are helpful in determining follow-up care after diagnosis and treatment but not in early general screening (Li, 2012).

Epidemiology

One in 70 women will develop ovarian cancer in her lifetime. Family history is the most significant risk factor (Hunn & Rodriguez, 2012). The incidence of this type of cancer increases after 40 years of age, and half of women at diagnosis are older than 60 years (Blewitt, 2010). The frequency of ovarian cancer is highest in industrialized countries and affects women of all races and ethnic backgrounds.

Most cases are random, but 5% to 10% of ovarian cancers are familial (Blewitt, 2010). In most cases, the mutations are in the *BRCA1* gene and sometimes in the *BRCA2* gene. A family history in a first-degree relative (mother, daughter, or sister), older age, early menarche, late menopause, and obesity may increase the risk of ovarian cancer. However, most women who develop ovarian cancer have no known risk factors, and no definitive causative factors have been determined.

Patients with concerns about their family history should be referred to a cancer genetics center to obtain information and testing, if indicated (see Chapter 8). Women with inherited types of ovarian cancer tend to be younger when the diagnosis is made than the average age at the time of diagnosis. Prophylactic oophorectomy in women with genetic mutations has been found to be associated with decreased risk of ovarian and other gynecologic cancers, as well as breast cancer, and is an option for women who have completed childbearing (Schuiling & Likis, 2013). Hereditary nonpolyposis colon cancer increases the risk of uterine cancer and slightly increases the risk of ovarian cancer.

Pathophysiology

Types of tumors include germ cell tumors, which arise from the cells that produce eggs; stromal cell tumors, which arise in connective tissue cells that produce hormones; and epithelial tumors, which originate from the outer surface of the ovary. Most ovarian cancers are epithelial in origin. Of the many different cell types in ovarian cancer, epithelial tumors constitute 90%. Germ cell and stromal tumors make up the other 10% (Schuiling & Likis, 2013).

Primary peritoneal carcinoma is closely related to ovarian cancer. Extraovarian primary peritoneal carcinoma (EOPPC) resembles ovarian cancer histologically and can occur in women with and without ovaries. Symptoms and treatment are similar. Because of the possibility of EOPPC, oophorectomy lessens the chance, but does not guarantee, that the patient will not develop carcinoma.

Clinical Manifestations

Symptoms of ovarian cancer are nonspecific and may include increased abdominal girth, pelvic pressure, bloating, back pain, constipation, abdominal pain, urinary urgency, indigestion, flatulence, increased waist size, leg pain, and pelvic pain. Symptoms are often vague, so many women tend to ignore them. Ovarian cancer is often silent, but enlargement of the abdomen from an accumulation of fluid is a common sign. All women with gastrointestinal symptoms without a known cause must be evaluated for potential ovarian cancer. Vague, undiagnosed, persistent gastrointestinal symptoms should alert the nurse to the possibility of an early ovarian malignancy. A palpable ovary in a woman who has gone through menopause is investigated immediately, because ovaries normally become smaller and less palpable after menopause.

Assessment and Diagnostic Findings

Any enlarged ovary must be investigated. Pelvic examination often does not detect early ovarian cancer, and pelvic imaging techniques are not always definitive. Ovarian tumors are classified as benign if there is no proliferation or invasion, borderline if there is proliferation but no invasion, and malignant if there is invasion. Of all new cases of ovarian tumors, 15% are classified as borderline and have low malignancy potential. However, by the time of diagnosis, most ovarian cancers are advanced (Blewitt, 2010; Gershenson, 2012).

Diagnostic test may include an MRI scan, transvaginal and pelvic ultrasound, chest x-rays, and a blood test for CA-125. An abdominal CT scan with and without contrast may be used to rule out metastasis (Blewitt, 2010).

Medical Management

Surgical Management

Surgical staging, exploration, and reduction of tumor mass are the basics of treatment. Surgical removal is the treatment of choice. Staging the tumor by the TNM system is performed to guide treatment (Chart 57-10). Likely treatment involves a total abdominal hysterectomy with removal of the fallopian tubes and ovaries and possibly the omentum (bilateral salpingo oophorectomy and omentectomy), tumor debulking, para-aortic and pelvic lymph node sampling, diaphragmatic biopsies, random peritoneal biopsies, and cytologic washings. Postoperative management may include taxanes or platinum-based chemotherapy (discussed in the next section).

Chart 57-10 Main Stages of Ovarian Cancer

- **I.** Cancer is contained within the ovary (or ovaries).
- **II.** Cancer is in one or both ovaries and has involved other organs (i.e., uterus, fallopian tubes, bladder, the sigmoid colon, or the rectum) within the pelvis.
- **III.** Cancer involves one or both ovaries, and one or both of the following are present: (1) cancer has spread beyond the pelvis to the lining of the abdomen; (2) cancer has spread to lymph nodes.
- **IV.** The most advanced stage of ovarian cancer. Cancer is in one or both ovaries. There is distant metastasis to the liver, lungs, or other organs outside the peritoneal cavity; ovarian cancer cells in the pleural cavity are evidence of stage IV disease.

Adapted from Blewitt, K. (2010). Ovarian cancer: Listen for the disease that whispers. *Nursing, 40*(11), 24–31.

Borderline tumors resemble ovarian cancer but have much more favorable outcomes. Women diagnosed with this type of cancer tend to be younger (early 40s). A conservative surgical approach is used. The affected ovary is removed, but the uterus and the contralateral ovary may remain in place. Adjuvant therapy may not be warranted.

Pharmacologic Therapy

Chemotherapy is usually administered IV on an outpatient basis using a combination of platinum and taxane agents. Paclitaxel (Taxol) plus carboplatin (Paraplatin) are most often used because of their excellent clinical benefits and manageable toxicity. Leukopenia, neurotoxicity, and fever may occur.

Because paclitaxel often causes leukopenia, patients may need to take granulocyte colony-stimulating factor as well. Paclitaxel is contraindicated in patients with hypersensitivity to medications formulated in polyoxyethylated castor oil and in patients with baseline neutropenia. Because of possible adverse cardiac effects, paclitaxel is not used in patients with cardiac disorders. Hypotension, dyspnea, angioedema, and urticaria indicate severe reactions that usually occur soon after the first and second doses are administered. Nurses who administer chemotherapy are prepared to assist in treating anaphylaxis. Patients should be prepared for inevitable hair loss.

Carboplatin may be used in the initial treatment and in patients with recurrence. It is used with caution in patients with renal impairment. Usually, six cycles are given. A positive clinical response is normalization of the tumor marker CA-125, negative CT results, and a normal physical and gynecologic examination.

Liposomal therapy (delivery of chemotherapy in a liposome) allows the highest possible dose of chemotherapy to the tumor target with a reduction in adverse effects. Liposomes are used as drug carriers because they are nontoxic, biodegradable, easily available, and relatively inexpensive. This encapsulated chemotherapy allows increased duration of action and better targeting. The encapsulation of doxorubicin (Doxil) lessens the incidence of nausea, vomiting, and alopecia. Patients must be monitored for bone marrow suppression and gastrointestinal and cardiac effects.

Combination IV and intraperitoneal chemotherapy is an option for some patients. However, this treatment may result in pain, fatigue, and hematologic, gastrointestinal, metabolic, and neurologic toxicities, thus decreasing the quality of life (Eggert, 2010). Because of these effects, the decision to use intraperitoneal chemotherapy is individualized.

Genetic engineering and identification of cancer genes may make gene therapy a future possibility; gene therapy is under investigation. Emerging proteomic technologies (tissue-based protein analysis) look promising; they may allow earlier diagnosis and treatment decision making. New biomarkers need further validation, but protein signature patterns are now being tested. These technologies may result in individualized treatment strategies for epithelial ovarian cancer.

Recurrence of ovarian cancer is common, and many patients may require treatment with multiple agents. Treatment is directed toward control of the cancer, maintenance of quality of life, and palliation. Liposomal preparations, intraperitoneal drug administration, anti-cancer vaccines, monoclonal antibodies directed against cancer antigens, gene therapy, and antiangiogenic treatments (to prevent formation of new blood vessels in an effort to halt growth of ovarian cancer) may be used in the treatment for recurrence. Many agents are being studied in clinical trials in advanced ovarian cancer (Blewitt, 2010; Eggert, 2010).

Nursing Management

Nursing measures involve those related to the patient's treatment plan, which may include surgery, chemotherapy, palliative care, or a combination of these. Emotional support, comfort measures, and information, plus attentiveness and caring, are important components of nursing care for the patient and her family (Blewitt, 2010).

Nursing interventions after pelvic surgery to remove the tumor are similar to those after other abdominal surgeries. If ovarian cancer occurs in a young woman and the tumor is unilateral, it is removed. Childbearing, if desired, is encouraged in the near future. After childbirth, surgical reexploration may be performed, and the remaining ovary may be removed. If both ovaries are involved, bilateral oophorectomy is performed and chemotherapy follows.

Patients with advanced ovarian cancer may develop ascites and pleural effusion. Nursing care may include administering IV fluids prescribed to alleviate fluid and electrolyte imbalances, administering parenteral nutrition to provide adequate nutrition, providing postoperative care after intestinal bypass to alleviate any obstruction, controlling pain, and managing drainage tubes. Comfort measures for women with ascites may include providing small frequent meals, decreasing fluid intake, administering diuretic agents, and providing rest. Patients with pleural effusion may experience shortness of breath, hypoxia, pleuritic chest pain, and cough. Thoracentesis is usually performed to relieve these symptoms. The patient with ovarian cancer often has complex needs and benefits from the assistance and support of an oncology clinical nurse specialist.

Hysterectomy

Hysterectomy is the surgical removal of the uterus to treat cancer, dysfunctional uterine bleeding, endometriosis, nonmalignant growths, persistent pain, pelvic relaxation and prolapse, and previous injury to the uterus. Hysterectomies are decreasing as the number of other therapeutic options (i.e., laser therapy, endometrial ablation, UAE, and medications to shrink fibroid tumors) has increased (Roberts, Tsourapas, Middleton, et al., 2011).

A total hysterectomy involves removal of the uterus and the cervix. Hysterectomy can be supracervical or subtotal, in which the uterus is removed but the cervix is spared. Radical hysterectomy involves removal of the uterus as well as the surrounding tissue, including the upper third of the vagina and pelvic lymph nodes. The procedure can be performed through the vagina, through an abdominal incision, or laparoscopically (in which the uterus is removed in sections through small incisions using a laparoscope). Malignant conditions usually require a total abdominal hysterectomy and bilateral salpingo-oophorectomy.

A laparoscopically assisted approach can also be used for vaginal hysterectomy. This procedure is performed as a short-stay procedure or ambulatory surgery in carefully selected patients. Robotic-assisted hysterectomies are performed in approximately 25% of cases, and research is comparing outcomes (e.g., length of hospital stay, blood loss) in robotic-assisted procedures to laparoscopy or open hysterectomies (Saver, 2010).

Preoperative Management

Patients are advised to discontinue anticoagulant medications, NSAIDs such as aspirin, and vitamin E prior to surgery to reduce the risk of bleeding. Pregnancy is ruled out on the day of surgery. Prophylactic antibiotic agents may be administered prior to surgery and discontinued the next day. Prevention of thromboembolic events is critical, and methods depend on the risk profile of the patient.

Postoperative Management

The principles of general postoperative care for abdominal surgery apply. Major risks are infection and hemorrhage. In addition, because the surgical site is close to the bladder, voiding problems may occur, particularly after a vaginal hysterectomy. Edema or nerve trauma may cause temporary loss of bladder tone (bladder atony), and an indwelling catheter may be inserted.

NURSING PROCESS

The Patient Undergoing a Hysterectomy

Assessment

The health history and the physical and pelvic examination are completed, and laboratory tests are performed. Additional assessment data include the patient's psychosocial responses, because the need for a hysterectomy may elicit strong emotional reactions. If the hysterectomy is performed to remove a malignant tumor, anxiety related to fear of cancer and its consequences adds to the stress of the patient and her family. Women who have had a hysterectomy may be at risk for psychological and physical symptoms. Alternatively, women may note improved physical and mental health after hysterectomy as troublesome symptoms may be alleviated.

Diagnosis

Nursing Diagnoses

Based on the assessment data, major nursing diagnoses may include the following:

- Anxiety related to the diagnosis of cancer, fear of pain, possible perception of loss of femininity or childbearing potential
- Disturbed body image related to altered fertility and fears about sexuality and relationships with partner and family
- Acute pain related to surgery and other adjuvant therapy
- Deficient knowledge of the perioperative aspects of hysterectomy and postoperative self-care

Collaborative Problems/Potential Complications

Potential complications may include the following:

- Hemorrhage
- DVT
- Bladder dysfunction
- Infection

Planning and Goals

The major goals may include relief of anxiety, acceptance of loss of the uterus, absence of pain or discomfort, increased knowledge of self-care requirements, and absence of complications.

Nursing Interventions

Relieving Anxiety

Anxiety stems from several factors: unfamiliar environment; the effects of surgery on body image and reproductive ability; fear of pain and other discomfort; and, possibly, feelings of embarrassment about exposure in the perioperative period. The nurse determines what the experience means to the patient and encourages her to verbalize her concerns. Throughout the preoperative, postoperative, and recovery periods, explanations are given about physical preparations and procedures that are performed.

Patient education addresses the outcomes of surgery, possible feelings of loss, and options for management of any symptoms that occur. Women vary in their preferences for information and participation in decision making, including choice of treatment options, accurate and useful information at the appropriate time, support from their health care providers, and access to professional and lay support systems (Blewitt, 2010).

Improving Body Image

The patient may have strong emotional reactions to having a hysterectomy and personal feelings related to the diagnosis, views of significant others who may be involved (family, partner), religious beliefs, and fears about prognosis. Concerns such as the inability to have children and the effect on femininity may surface, as may questions about the effects of surgery on sexual relationships, function, and satisfaction. The patient needs reassurance that she will still have a vagina and that she can experience sexual activity after temporary postoperative abstinence while tissues heal. Information that sexual satisfaction and orgasm arise from clitoral stimulation rather than from the uterus reassures many women. Most women note some change in sexual feelings after hysterectomy, but they vary in intensity. In some cases, the vagina is shortened by surgery, and this may affect sensitivity or comfort.

When hormonal balance is upset, as often occurs with reproductive system disorders, the patient may experience depressed mood and heightened emotional sensitivity to people and situations. The nurse needs to approach and evaluate each patient individually in light of these factors. A nurse who exhibits interest, concern, and willingness to listen to the patient's fears will help the patient progress through the surgical experience.

Relieving Pain

Postoperative pain and discomfort are common. Therefore, the nurse assesses the intensity of the patient's pain and assists the patient with analgesia as prescribed. If the patient has abdominal distention or flatus, a rectal tube and application of heat to the abdomen may be prescribed. When abdominal auscultation reveals return of bowel sounds and peristalsis, additional fluids and a soft diet are permitted. Early ambulation facilitates the return of normal peristalsis.

Monitoring and Managing Potential Complications

Hemorrhage. Vaginal bleeding and hemorrhage may occur after hysterectomy. To detect these complications early, the nurse counts the perineal pads used or checks the incision site, assesses the extent of saturation with blood, and monitors vital signs. Abdominal dressings are monitored for drainage if an abdominal surgical approach has been used. In preparation for hospital discharge, the nurse gives prescribed guidelines for activity restrictions to promote healing and to prevent postoperative bleeding. Because many women may go home the day of surgery or within a day or two, they are instructed to contact the nurse or surgeon if bleeding is beyond what is expected, which should be minimal.

Venous Thromboembolism. Because of positioning during surgery, postoperative edema, and decreased activity postoperatively, the patient is at risk for DVT and PE. To minimize the risk, anti-embolism stockings are applied. In addition, the patient is encouraged and assisted to change positions frequently, although pressure under the knees is avoided, and to exercise her legs and feet while in bed. The nurse helps the patient ambulate early in the postoperative period. The nurse also assesses for DVT (leg pain, redness, warmth, edema) and PE (chest pain, tachycardia, dyspnea). If the patient is being discharged home soon after surgery, she is instructed to avoid prolonged sitting in a chair with pressure at the knees, sitting with crossed legs, and inactivity. Furthermore, she is instructed to contact her primary provider if symptoms of DVT or PE occur.

Bladder Dysfunction. Because of possible difficulty in voiding postoperatively, occasionally an indwelling catheter may be inserted before or during surgery and is left in place in the immediate postoperative period. If a catheter is in place, it is usually removed shortly after the patient begins to ambulate. After the catheter is removed, urinary output is monitored; additionally, the abdomen is assessed for distention. If the patient does not void within a prescribed time, measures are initiated to encourage voiding (e.g., assisting the patient to the bathroom, pouring warm water over the perineum). If the patient cannot void, catheterization may be necessary. On rare occasions, the patient may be discharged home with the catheter in place and is instructed in its management.

Promoting Home and Community-Based Care

Educating Patients About Self-Care. The information provided to the patient is tailored to her needs. She must know what limitations or restrictions, if any, to expect. She is instructed to check the surgical incision daily and to contact her primary provider if redness or purulent drainage or discharge occurs. She is informed that her periods are now over but that she may have a slightly bloody discharge for a few days; if bleeding recurs after this time, it should be reported immediately. The patient is instructed about the importance of an adequate oral intake and of maintaining bowel and urinary tract function. The patient is informed that she is likely to recover quickly but that postoperative fatigue is not unusual.

The patient should resume activities gradually. This does not mean sitting for long periods, because doing so may cause blood to pool in the pelvis, increasing the risk of VTE. The nurse explains that showers are preferable to tub baths to reduce the possibility of infection and to avoid the dangers of injury that may occur when getting in and out of the bathtub. The patient is instructed to avoid straining, lifting, having sexual intercourse, or driving until permitted. Vaginal discharge, foul odor, excessive bleeding, any leg redness or pain, or an elevated temperature should be reported, and the nurse reinforces education regarding activities and restrictions.

Continuing Care. Follow-up telephone contact provides the nurse with the opportunity to determine whether the patient is recovering without problems and to answer any questions that may have arisen. The patient is reminded about postoperative follow-up appointments. If the patient's ovaries were removed and she finds vasomotor symptoms troublesome, hormone therapy may be considered at a low dose for a short amount of time. Providing information regarding the findings of the Women's Health Initiative (2002) study about the benefits and risks of hormone therapy promotes informed decision making about its use. The patient is reminded to discuss risks and benefits of hormone therapy and alternative therapies with her primary provider and gynecologic care provider. Decisions about use of hormone therapy need to be made individually in consultation with these providers.

Evaluation

Expected patient outcomes may include:

1. Experiences decreased anxiety
2. Has improved body image
   a. Discusses changes resulting from surgery with her partner
   b. Verbalizes understanding of her disorder and the treatment plan
   c. Displays minimal depression or anxiety
3. Experiences minimal pain and discomfort
   a. Reports relief of abdominal pain and discomfort
   b. Ambulates without pain
4. Verbalizes knowledge and understanding of self-care
   a. Practices deep-breathing, turning, and leg exercises as instructed
   b. Increases activity and ambulation daily
   c. Reports adequate fluid intake and adequate urinary output
   d. Identifies reportable symptoms
   e. Schedules and keeps follow-up appointments
5. Absence of complications
   a. Has minimal vaginal bleeding and exhibits normal vital signs
   b. Ambulates early
   c. Notes no chest or calf pain and no redness, tenderness, or swelling in the extremities
   d. Reports no urinary problems or abdominal distention

Radiation Therapy

Radiation may be used in the treatment of cervical, uterine, and less frequently in ovarian cancers either alone or in combination with surgery and chemotherapy. Several approaches are used to deliver radiation to the female reproductive system: external radiation, intraoperative radiation therapy (IORT), and internal (intracavitary) irradiation or brachytherapy. The cervix and uterus can serve as a receptacle for radioactive sources for internal radiation therapy.

Methods of Radiation Therapy

External Radiation Therapy

This method of delivering radiation destroys cancerous cells at the skin surface or deeper in the body. Other methods of delivering radiation therapy are more commonly used to treat cancer of the female reproductive system than this method.

Intraoperative Radiation Therapy

IORT allows radiation to be applied directly to the affected area during surgery. An electron beam is directed at the disease site. This direct-view irradiation may be used when para-aortic nodes are involved or for unresectable (inoperable) or partially resectable neoplasms. Benefits include accurate beam direction (which precisely limits the radiation to the tumor) and the ability during treatment to block sensitive organs from radiation. IORT is usually combined with external-beam irradiation pre- or postoperatively.

Internal (Intracavitary) Irradiation

After the patient receives an anesthetic agent and an examination, specially prepared applicators are inserted into the endometrial cavity and vagina. These devices are not loaded with radioactive material until the patient returns to her room. X-rays are obtained to verify the precise relationship of the applicator to the normal pelvic anatomy and to the tumor. When this step is completed, the radiation oncologist loads the applicators with predetermined amounts of radioactive material. This procedure, called *afterloading*, allows for precise control of the radiation exposure received by the patient, with minimal exposure of physicians, nurses, and other health care personnel. A patient undergoing internal radiation treatment remains isolated in a private room until the application is completed. Adjacent rooms may need to be evacuated and a lead shield placed at the doorway to the patient's room.

Of the various applicators developed for intracavitary treatment, some are inserted into the endometrial cavity and endocervical canal as multiple small irradiators (e.g., Heyman capsules). Others consist of a central tube (a tandem or intrauterine "stem") placed through the dilated endocervical canal into the uterine cavity, which remains in a fixed relationship with the irradiators placed in the upper vagina on each side of the cervix (vaginal ovoids) (Fig. 57-8).

When the applicator is inserted, an indwelling urinary catheter is also inserted. Vaginal packing is inserted to keep the applicator in place and to keep other organs, such as the bladder and rectum, as far from the radioactive source as possible. The objective of the internal treatment is to

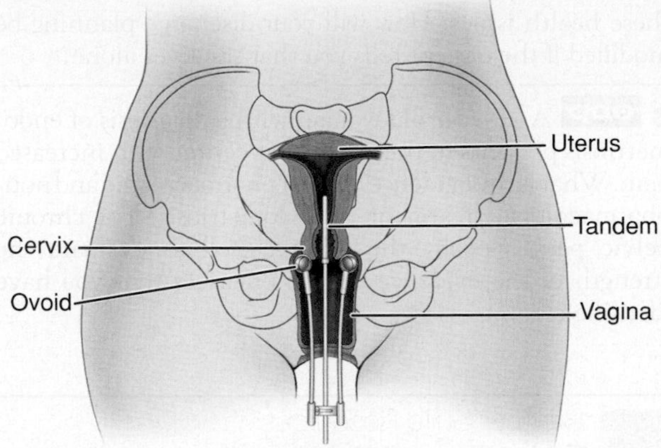
Uterus
Tandem
Cervix
Ovoid
Vagina

FIGURE 57-8 • Placement of tandem and ovoids for internal radiation therapy.

maintain the distribution of internal radiation at a fixed dosage throughout the application, which may last 24 to 72 hours, depending on dose calculations made by the radiation physicist.

Automated high dose rate intracavitary brachytherapy systems have been developed that allow outpatient radiation therapy. Treatment time is shorter, thereby decreasing patient discomfort. Staff exposure to radiation is also avoided. Isotopes of radium and cesium are used for intracavitary irradiation.

Nursing Considerations for Radiation Safety

Special precautions for the safety of the patient and the nurse are important considerations when the patient is receiving radiation therapy. The radiation safety department will identify specific safety precautions to those people who will be in contact with the patient, including health care providers and family. Nursing concerns include providing the patient with emotional support and physical comfort. Further details about nursing management are provided in Chapter 15.

Critical Thinking Exercises

1 pq Identify the priorities, approach, and techniques you would use to provide care for an 18-year-old woman who presents to your clinic with a vulvovaginal infection. How will your priorities, approach, and techniques differ if the patient has a visual impairment or is hard of hearing? If the patient is from a culture with very different values from your own? Describe your priorities, approach, and techniques to prevent reinfection or spread of the infection.

2 A 70-year-old woman with hypertension who has had surgery for breast cancer is scheduled to undergo a total hysterectomy. She reports that she is very anxious because of her previous surgery. Describe the preoperative education for her and the postoperative care that can be anticipated. How will her history of breast cancer affect her care? What modifications in care, postoperative education, and discharge planning may be necessary because of these health issues? How will your discharge planning be modified if the patient tells you that she lives alone?

3 ebp A 28-year-old woman with a diagnosis of endometriosis presents to the outpatient center with increased pain. What is the evidence base for pharmacologic and nonpharmacologic treatment of endometriosis? For chronic pelvic pain? Specify the criteria used to evaluate the strength of the evidence for the practices that you have identified.

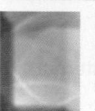

Brunner Suite Resources

Explore these additional resources to enhance learning for this chapter:

- NCLEX-Style Questions and Other Resources on thePoint, http://thePoint.lww.com/Brunner13e
- Study Guide
- PrepU
- Clinical Handbook
- Handbook of Laboratory and Diagnostic Tests

References

*Asterisk indicates nursing research.

**Double asterisk indicates classic reference.

Books

Almadrones-Cassidy, L. (Ed.). (2010). *Gynecologic cancers (site-specific cancer series)*. Pittsburgh: Oncology Nursing Society.

American Cancer Society. (2012). *Cancer facts and figures 2012*. Atlanta, GA: Author.

American College of Obstetricians and Gynecologists. (2011b). *Compendium of selected publications*. Washington, DC: Author.

Eggert, J. (2010). *Cancer basics*. Pittsburgh: Oncology Nursing Society.

Hall, J. (2011). *Guyton and Hall textbook of medical physiology* (12th ed.). St. Louis: Elsevier Saunders.

Karch, A. M. (2012). *2012 Lippincott's nursing drug guide*. Philadelphia: Lippincott Williams & Wilkins.

Porth, C. M., & Matfin, G. (2009). *Pathophysiology: Concepts of altered health states* (8th ed.). Philadelphia: Lippincott Williams & Wilkins.

Schuiling, K. D., & Likis, F. E. (2013). *Women's gynecologic health* (2nd ed.). Burlington, MA: Jones & Bartlett.

Journals and Electronic Documents

American College of Obstetricians and Gynecologists. (2007, reaffirmed 2009). Practice Bulletin 83: Management of adnexal masses. *Obstetrics and Gynecology, 110*(1), 201–214.

American College of Obstetricians and Gynecologists. (2011a). Committee Opinion #513: Vaginal placement of synthetic mesh for pelvic organ prolapse *Obstetrics and Gynecology, 118*(12), 1459–1464.

Baron, S., Florendo, J., Sandbo, S., et al. (2011). Sexual pain disorders in women: Evaluation and treatment. *Clinical Reviews, 21*(5), 32–41.

Blewitt, K. (2010). Ovarian cancer: Listen for the disease that whispers. *Nursing, 40*(11), 24–31.

Castle, P. (2011). Abuses in human papillomavirus DNA testing. *Obstetrics and Gynecology, 118*(1), 1–3.

Centers for Disease Control and Prevention. (2010a). *2010 STD treatment guidelines*. Available at: www.cdc.gov/std/treatment/2010/

Centers for Disease Control and Prevention. (2010b). *Trichomoniasis—CDC fact sheet*. Available at: www.cdc.gov/std/trichomonas/stdfact-trichomoniasis.htm

Centers for Disease Control and Prevention. (2010c). *Basic information about gynecologic cancers*. Available at: www.cdc.gov/cancer/gynecologic/basic_info/index.htm

Centers for Disease Control and Prevention. (2011a). *Pelvic inflammatory disease (PID)—CDC fact sheet*. Available at: www.cdc.gov/std/PID/STDFact-PID.htm

Centers for Disease Control and Prevention. (2011b). *HIV among women*. Available at: www.cdc.gov/hiv/topics/women/index.htm

Centers for Disease Control and Prevention. (2012a). *Genital HPV infection—fact sheet*. Available at: www.cdc.gov/std/HPV/STDFact-HPV.htm

Centers for Disease Control and Prevention. (2012b). *Genital herpes—CDC fact sheet*. Available at: www.cdc.gov/std/herpes/stdfact-herpes.htm

Centers for Disease Control and Prevention. (2012c). *Chlamydia—CDC fact sheet*. Available at: www.cdc.gov/std/chlamydia/STDFact-Chlamydia.htm

Centers for Disease Control and Prevention. (2012d). *Gonorrhea—CDC fact sheet*. Available at: www.cdc.gov/std/gonorrhea/STDFact-gonorrhea.htm

Culligan, P. J. (2012). Nonsurgical management of pelvic organ prolapse. *Obstetrics and Gynecology, 119*(4), 852–860.

Falcone, T., & Lebovic, D. (2011). Clinical management of endometriosis. *Obstetrics and Gynecology, 118*(3), 691–705.

Galic, V., & Gupta, D. (2012). Fertility-sparing management of endometrial carcinoma. *Female Patient, 37*(6), 39–47.

Gershenson, D. (2012). Treatment of ovarian cancer in young women. *Clinical Journal of Obstetrics and Gynecology, 55*(1), 65–74.

Grube, W., Ammon, T., & Killen, M. (2011). The role of ultrasound imaging in detection endometrial cancer in postmenopausal women with vaginal bleeding. *Journal of Obstetric, Gynecologic, and Neonatal Nursing, 40*(5), 632–637.

*Harris, A. (2011). Vitamin D deficiency and bacterial vaginosis in pregnancy. *Nursing for Women's Health, 15*(5), 423–430.

Hickey, M., Marino, J., & Brownfoot, F. (2012) Uterine artery embolization associated with greater need for reintervention than surgical treatment for symptomatic uterine fibroids: Quality of life similar though study underpowered. *Evidence Based Medicine, 17*(3), 87–88.

Hunn, J., & Rodriguez, G. (2012). Ovarian cancer: Etiology, risk factors and epidemiology. *Clinical Gynecology and Obstetrics, 55*(1), 3–23.

Li, A. (2012). New biomarkers for ovarian cancer. *Contemporary Obstetrics and Gynecology, 57*(4), 28–34.

Lone, F., Thakar, R., Sultan, A., et al. (2011). A 5 year prospective study of vaginal pessary use for pelvic organ prolapse. *International Journal of Gynecology and Obstetrics, 114*(1), 56–59.

Nyirjesy, P., Robinson, J., Mathew, L., et al. (2011). Alternative therapies in women with chronic vaginitis. *Obstetrics and Gynecology, 117*(4), 856–861.

Pruitt, B. (2012). For all the right reasons: The HPV vaccine to prevent cervical cancer. *American Journal for Nurse Practitioners, 16*(5–6), 31–35.

Roberts, T., Tsourapas, A., Middleton, L., et al. (2011). Hysterectomy, endometrial ablation and levonorgestrel releasing intrauterine system (Mirena) for treatment of heavy menstrual bleeding: Cost effectiveness analysis. *BMJ Online*. Available at: www.bmj.com/content/342/bmj.d2202.pdf%2Bhtml

Saver, C. (2010). Robotics: Little data much debate. *OR Manager, 26*(8), 16–18.

Smith, R. A, Cokkinides, V., & Brawley, O. W. (2012). Cancer screening in the United States, 2012. *CA: A Cancer Journal for Clinicians, 62*(2), 129–142.

Sobel, J. Reichman O., Misra, D., et al. (2011). Prognosis and treatment of desquamative inflammatory vaginitis. *Obstetrics and Gynecology, 117*(4), 850–855.

*Teitelman, A., Stringer, M., Nguyen, G., et al. (2011). Social cognitive and clinical factors associated with HPV vaccine initiation among urban, economically disadvantaged women. *Journal of Obstetric, Gynecologic, and Neonatal Nursing, 40*(6), 691–701.

Turner, D. (2012). Pelvic inflammatory disease: A continuing challenge. *American Journal of Nurse Practitioners, 16*(1–2), 20–23.

U.S. Food and Drug Administration. (2012). *Medical devices: Pelvic organ prolapse (POP)*. Available at: www.fda.gov/MedicalDevices/ProductsandMedicalProcedures/ImplantsandProsthetics/UroGynSurgicalMesh/ucm262299.htm

Wechter, M., Wu, J., Marzano, D., et al. (2009). Management of Bartholin duct cysts and abscesses: A systematic review. *Obstetrics and Gynecology Survey, 64*(6), 395–404.

**Women's Health Initiative. (2002). Risks and benefits of estrogen plus progestin in healthy postmenopausal women: Principal results from the Women's Health Initiative randomized controlled trial. *Journal of the American Medical Association, 288*(3), 321–333.

Wong, M. J., Wong, K., Rezvan, A., et al. (2012). Urogenital fistula. *Female Pelvic and Reconstructive Surgery, 18*(2), 71–78.
*Weiss, T., & Bulmer, S. (2011). Young women's experiences living with polycystic ovary syndrome. *Journal of Obstetric, Gynecologic, and Neonatal Nursing, 40*(6), 709–718.

Resources

American Cancer Society, www.cancer.org
American Sexual Health Association (ASHA; formerly the American Social Health Organization), www.ashasexualhealth.org/std-sti/hpv.html
Association of Reproductive Health Professionals (ARHP), www.arhp.org
Association of Women's Health, Obstetric and Neonatal Nurses (AWHONN), www.awhonn.org
California STD/HIV Prevention Training Center, www.stdhivtraining.org
Centers for Disease Control and Prevention (CDC), Office of Women's Health, www.cdc.gov/women/
ClinicalTrials.gov, National Institutes of Health, www.clinicaltrials.gov
Effective Interventions, www.effectiveinterventions.org
Endo-Online, Endometriosis Association, www.endometriosisassn.org
Foundation for Women's Cancer (formerly the Gynecologic Cancer Foundation), Foundationforwomen'scancer.org
Gay and Lesbian Medical Association (GLMA), www.glma.org
Herpes Hotline, 1–919–361–8488
National Ovarian Cancer Coalition (NOCC), www.ovarian.org
National STD Hotline, 1–800–227–8922 or 1–800–342–2437
Oncology Nursing Society (ONS), www.ons.org
Ovarian Cancer National Alliance, www.ovariancancer.org
Planned Parenthood Federation of America, Plannedparenthood.org
RESOLVE: The National Infertility Association, www.resolve.org

Chapter 58

Assessment and Management of Patients With Breast Disorders

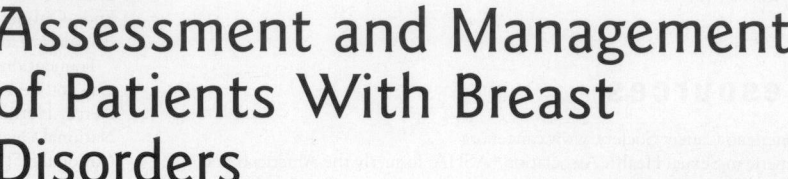

Learning Objectives

On completion of this chapter, the learner will be able to:

1 Describe the anatomy and physiology of the breast.
2 Identify the assessment and diagnostic studies used to diagnose breast disorders.
3 Identify and describe the pathophysiology of breast disorders, both benign and malignant.
4 Summarize evidence-based guidelines for the early detection of breast cancer.
5 Develop a plan for educating patients and consumer groups about breast self-awareness.
6 Describe the different modalities used to treat breast cancer.
7 Use the nursing process as a framework for care of the patient undergoing surgery for the treatment of breast cancer.
8 Describe the physical, psychosocial, and rehabilitative needs of the patient who has had breast surgery for the treatment of breast cancer.

Glossary

adjuvant chemotherapy: the use of anticancer medications in addition to other treatments to delay or prevent a recurrence of the disease

adjuvant hormonal therapy: the use of synthetic hormones or other medications given after primary treatment to increase the chances of a cure by stopping or slowing the growth of certain cancers that are affected by hormone stimulation (sometimes called *endocrine* or *antiestrogen therapy*)

aromatase inhibitors: medications that block the production of estrogens by the adrenal glands

atypical hyperplasia: abnormal increase in the number of cells in a specific area within the ductal or lobular areas of the breast; this abnormal proliferation increases the risk for cancer

benign proliferative breast disease: various types of atypical, yet noncancerous, breast tissue that increase the risk of breast cancer

brachytherapy: delivery of radiation therapy through internal implants called *seeds* to a localized area of tissue

***BRCA1* and *BRCA2*:** genes on chromosome 17 that, when damaged or mutated, increase a woman's risk for breast and/or ovarian cancer compared with women without the mutation

breast conservation treatment: surgery to remove a breast tumor and a margin of tissue around the tumor without removing any other part of the breast; may or may not include lymph node removal and radiation therapy

dose-dense chemotherapy: administration of chemotherapeutic agents at standard doses with shorter time intervals between each cycle of treatment

ductal carcinoma in situ (DCIS): cancer cells starting in the ductal system of the breast but not penetrating surrounding tissue

estrogen and progesterone receptor assay: test to determine whether the breast tumor is nourished by hormones; this information helps to determine prognosis and treatment

fibrocystic breast changes: term used to describe certain benign changes in the breast, typically palpable nodularity, lumpiness, swelling, or pain

fine-needle aspiration (FNA): removal of fluid for diagnostic analysis from a cyst or cells from a mass using a needle and syringe

galactography: the use of mammography after an injection of radiopaque dye to diagnose problems in the ductal system of the breast

gynecomastia: firm, overdeveloped breast tissue typically seen in adolescent boys

lobular carcinoma in situ (LCIS): atypical change and proliferation of the lobular cells of the breast

lymphedema: chronic swelling of an extremity due to interrupted lymphatic circulation, typically from an axillary lymph node dissection

mammoplasty: surgery to reconstruct or change the size or shape of the breast; can be performed for reduction or augmentation

mastalgia: breast pain, usually related to hormonal fluctuations or irritation of a nerve

mastitis: inflammation or infection of the breast

modified radical mastectomy: removal of the breast tissue, nipple–areola complex, and a portion of the axillary lymph nodes

Paget's disease: form of breast cancer that begins in the ductal system and involves the nipple, areola, and surrounding skin

prophylactic mastectomy: removal of the breast to reduce the risk of breast cancer in women considered to be at high risk

sentinel lymph node: first lymph node(s) in the lymphatic basin that receives drainage from the primary tumor in the breast; identified by a radioisotope and/or blue dye

stereotactic core biopsy: computer-guided method of core needle biopsy that is useful when masses in the breast cannot be felt but can be visualized using mammography

surgical biopsy: surgical removal of all or a portion of a mass for microscopic examination by a pathologist

tissue expander followed by permanent implant: series of breast reconstruction surgeries after a mastectomy; involves stretching the skin and muscle before inserting the permanent implant

total mastectomy: removal of the breast tissue and nipple–areola complex

transverse rectus abdominal myocutaneous (TRAM) flap: method of breast reconstruction in which a flap of skin, fat, and muscle from the lower abdomen, with its attached blood supply, is rotated to the mastectomy site

ultrasonography: imaging method using high-frequency sound waves to diagnose whether masses are solid or fluid filled

Nurses care for patients with breast disorders in many settings. To care for these patients effectively, nurses require an understanding of the assessment, diagnostic testing, nursing management, and rehabilitation needs of patients with multiple processes that affect the breasts. A breast disorder, whether benign or malignant, can cause great anxiety and fear of potential disfigurement, loss of sexual attractiveness, and even death. Nurses, therefore, must have expertise in the assessment and management of not only the physical symptoms but also the psychosocial symptoms associated with various breast disorders.

BREAST ASSESSMENT

Anatomic and Physiologic Overview

Male and female breasts mature comparably until puberty, when estrogen and other hormones initiate breast development in females. This development usually occurs from 10 to 16 years of age, although the range can vary from 9 to 18 years. Stages of breast development are described as Tanner stages 1 through 5:

- Stage 1 describes a prepubertal breast.
- Stage 2 is breast budding, the first sign of puberty in a female.
- Stage 3 involves further enlargement of breast tissue and the areola (a darker tissue ring around the nipple).
- Stage 4 occurs when the nipple and areola form a secondary mound on top of the breast tissue.
- Stage 5 is the continued development of a larger breast with a single contour.

The breasts are located between the second and sixth ribs over the pectoralis muscle from the sternum to the midaxillary line. An area of breast tissue, called the *tail of Spence*, extends into the axilla (Weber & Kelley, 2010). Fascial bands, called *Cooper's ligaments*, support the breast on the chest wall. The inframammary fold (or crease) is a ridge of fat at the bottom of the breast.

Each breast contains 12 to 20 cone-shaped lobes, which are made up of glandular elements (lobules and ducts) and separated by fat and fibrous tissue that binds the lobes together. Milk is produced in the lobules and then carried through the ducts to the nipple. Figure 58-1 shows the anatomy of the fully developed breast.

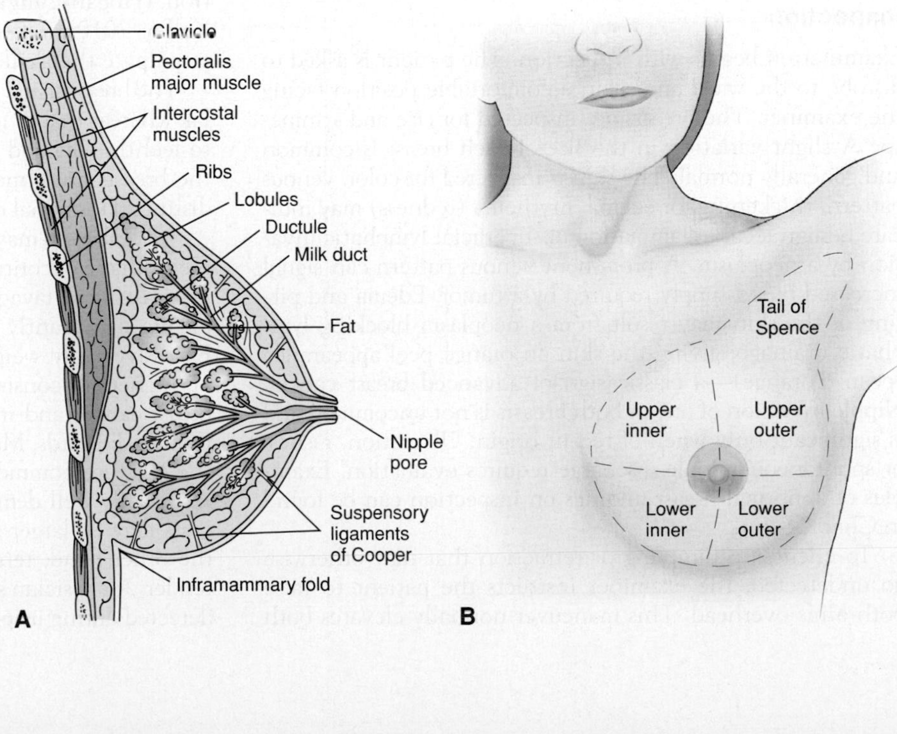
Clavicle
Pectoralis major muscle
Intercostal muscles
Ribs
Lobules
Ductule
Milk duct
Fat
Nipple pore
Suspensory ligaments of Cooper
Inframammary fold
A
Tail of Spence
Upper inner
Upper outer
Lower inner
Lower outer
B

FIGURE 58-1 • A. Anatomy of the breast. **B.** Areas of breast, including the tail of Spence.

Assessment

Health History

When a patient presents with a breast problem, the nurse conducts a general health assessment, including history of medical disorders and previous surgery; family history of diseases, particularly cancer; gynecologic and obstetric history; present medications (including prescriptions, vitamins, and herbals); past and present use of hormonal contraceptives, hormone therapy (HT) (formerly referred to as hormone replacement therapy [HRT]), or fertility treatments; and social habits (e.g., smoking, drinking alcohol). Psychosocial information, such as the patient's marital status, occupation, and availability of resources and support people, is obtained. Any recent x-rays or other diagnostic tests are noted. Focused questions pertaining to the breast disorder are asked concerning the onset of the disorder and the length of time it has been present. In addition, the patient is asked if any masses are palpable and if there is any associated pain, swelling, redness, nipple discharge, or change in the skin. Knowledge and comfort in practicing breast self-examination (BSE) should also be ascertained from the patient.

Physical Assessment: Female Breast

A female breast examination can be conducted during any general physical or gynecologic examination or whenever the patient reports an abnormality. The American College of Obstetricians and Gynecologists (ACOG, 2011), the American Cancer Society (ACS, 2010), and the National Comprehensive Cancer Network (NCCN, 2012) recommend annual clinical breast examination for women 40 years and older. They continue to recommend clinical breast examination every 1 to 3 years for women between the ages of 20 and 39 years, although its value is not clear in women who are at low risk for breast cancer (ACOG, 2011). A thorough breast examination, including instruction in BSE, takes at least 10 minutes.

Inspection

Examination begins with inspection. The patient is asked to disrobe to the waist and sit in a comfortable position facing the examiner. The breasts are inspected for size and symmetry. A slight variation in the size of each breast is common and generally normal. The skin is inspected for color, venous pattern, thickening, or edema. Erythema (redness) may indicate benign local inflammation or superficial lymphatic invasion by a neoplasm. A prominent venous pattern can signal increased blood supply required by a tumor. Edema and pitting of the skin may result from a neoplasm blocking lymphatic drainage, giving the skin an orange peel appearance (peau d'orange)—a classic sign of advanced breast cancer. Nipple inversion of one or both breasts is not uncommon and is significant only when of recent origin. Ulceration, rashes, or spontaneous nipple discharge requires evaluation. Examples of abnormal breast findings on inspection can be found in Chart 58-1.

To elicit skin dimpling or retraction that may otherwise go undetected, the examiner instructs the patient to raise both arms overhead. This maneuver normally elevates both breasts equally. The patient is then instructed to place her hands on her waist and push in. These movements, which cause contraction of the pectoral muscles, do not normally alter the breast contour or nipple direction. Any dimpling or retraction during these position changes suggests an underlying mass. The clavicular and axillary regions are inspected for swelling, discoloration, lesions, or enlarged lymph nodes (Weber & Kelley, 2010).

Palpation

The breasts are palpated with the patient sitting up (upright) and lying down (supine). In the supine position, the patient's shoulder is first elevated with a small pillow to help balance the breast on the chest wall. Failure to do this allows the breast tissue to slip laterally, and a breast mass may be missed. The entire surface of the breast and the axillary tail is systematically palpated using the flat part (pads) of the second, third, and fourth fingertips, held together, making dime-size circles. The examiner may choose to proceed in a clockwise direction, following imaginary concentric circles from the outer limits of the breast toward the nipple. Other acceptable methods are to palpate from each number on the face of the clock toward the nipple in a clockwise fashion or along imaginary vertical lines on the breast (Fig. 58-2).

Palpation of the axillary and clavicular areas is easily performed with the patient seated (Fig. 58-3). To examine the axillary lymph nodes, the examiner gently abducts the patient's arm from the thorax. With the left hand, the patient's right forearm is grasped and supported. The right hand is then free to palpate the axilla. Any lymph nodes that may be lying against the thoracic wall are noted. Normally, these lymph nodes are not palpable, but if they are enlarged, their location, size, mobility, and consistency are noted. During palpation, the examiner notes any patient-reported tenderness or masses. If a mass is detected, it is described by its location (e.g., right breast, 2 cm from the nipple at 2 o'clock position). Size, shape, consistency, border delineation, and mobility are included in the description (Hogan-Quigley, Palm, & Bickley, 2012; Weber & Kelley, 2010). The examiner then repeats these same steps to palpate the axillary nodes in the other breast.

The breast tissue of the adolescent is usually firm and lobular, whereas that of the postmenopausal woman is more likely to feel thinner and fattier. During pregnancy and lactation, the breasts are firmer and larger with lobules that are more distinct. Hormonal changes cause the areola to darken.

Obesity may have a proinflammatory effect on the breast that can contribute to increased rates of atypia. Atypia in breast ductal lavage and C-reactive protein levels in the nipple are significantly correlated with body mass index (BMI). Excessive body weight, as reflected by a BMI of 25 kg/m^2 or higher, is consistently associated with postmenopausal breast cancer and increases the risk of dying of this disease (Djuric, Edwards, Madan, et al., 2009).

Cysts are commonly found in menstruating women and are usually well defined and freely movable. Premenstrually, cysts may be larger and more tender. Malignant tumors, on the other hand, tend to be hard, poorly defined, and nontender. A physician should further evaluate any abnormalities detected during inspection and palpation.

Chart 58-1 Abnormal Findings During Inspection of the Breasts

Retraction Signs

- Signs include skin dimpling, creasing, or changes in the contour of the breast or nipple.
- They may be secondary to contraction of fibrotic tissue that can occur with underlying malignancy.
- They may be secondary to scar tissue formation after breast surgery.
- Retraction signs may appear only with position changes.

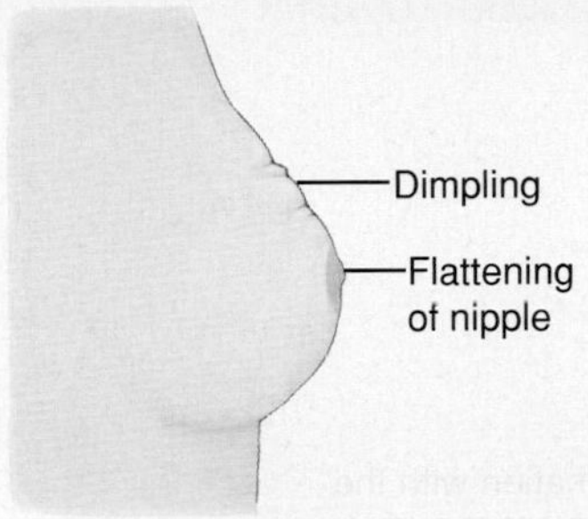
Dimpling
Flattening of nipple
Retraction signs

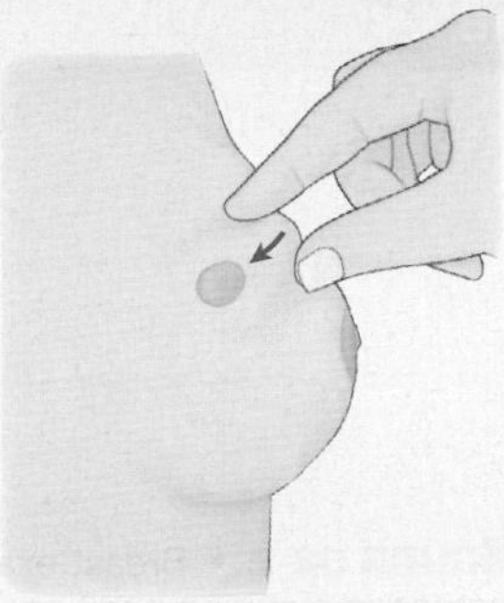
Retraction with compression

Increased Venous Prominence

- Unilateral localized increase in venous pattern associated with malignant tumors
- Normal with bilateral and symmetrical breast enlargement associated with pregnancy and lactation

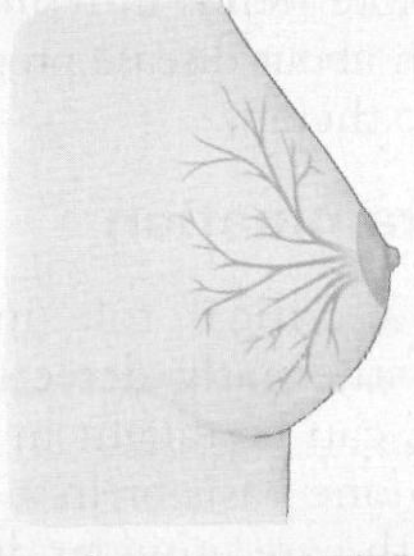
Increased venous prominence

Peau d'Orange (Edema)

- Associated with inflammatory breast cancer
- Caused by interference with lymphatic drainage
- Breast skin has orange peel appearance.
- Skin pores enlarge.
- May be noted on the areola
- Skin becomes thick, hard, and immobile.

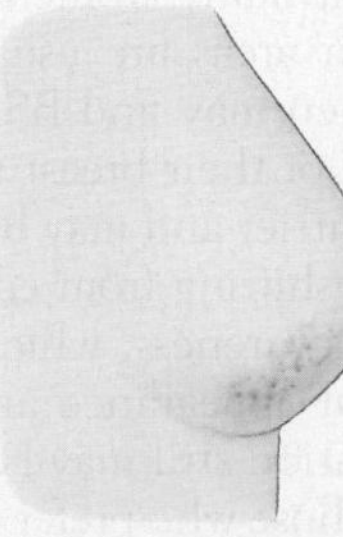
Peau d'orange

Nipple Inversion

- Considered normal if long-standing
- Associated with fibrosis and malignancy if recent development

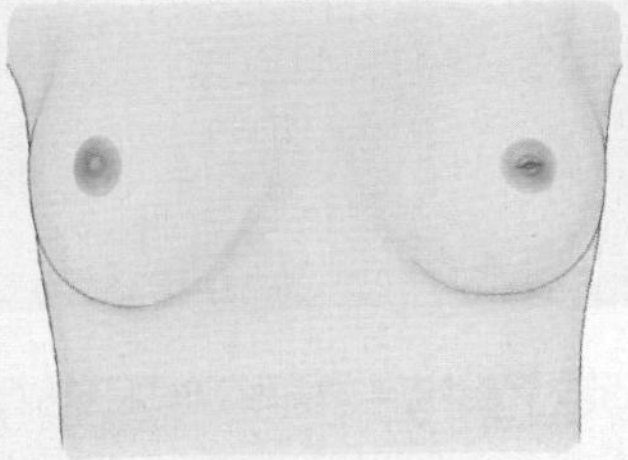
Nipple inversion

Acute Mastitis (Inflammation of the Breasts)

- Associated with lactation but may occur at any age
- Nipple cracks or abrasions noted
- Breast skin reddened and warm to touch
- Tenderness
- Systemic signs include fever and increased pulse.

Paget's Disease (Malignancy of Mammary Ducts)

- Early signs—erythema of nipple and areola
- Late signs—thickening, scaling, and erosion of the nipple and areola

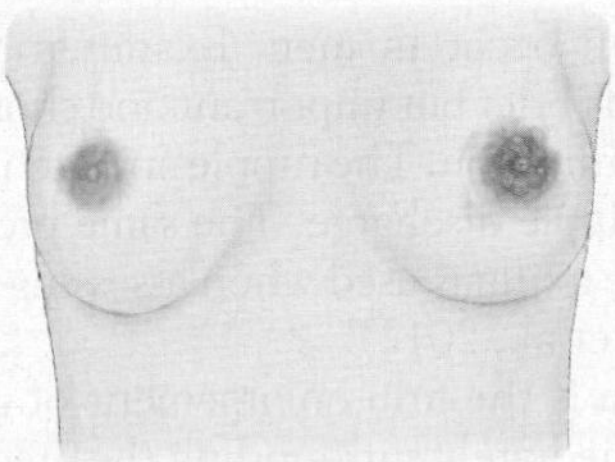
Paget's disease

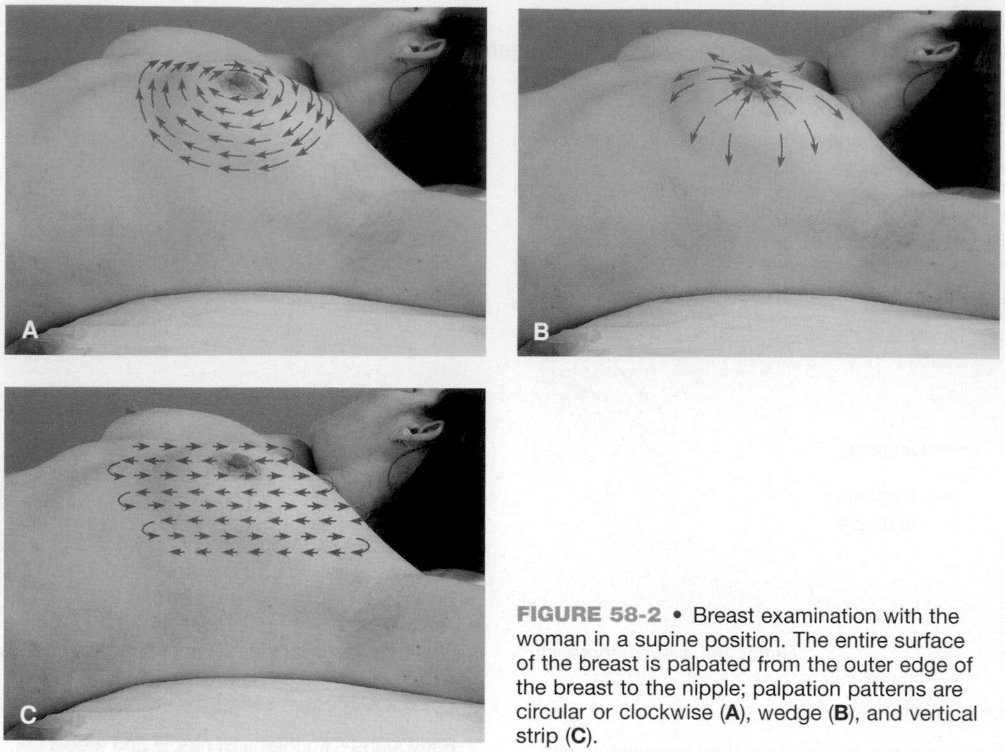
A
B
C

FIGURE 58-2 • Breast examination with the woman in a supine position. The entire surface of the breast is palpated from the outer edge of the breast to the nipple; palpation patterns are circular or clockwise (**A**), wedge (**B**), and vertical strip (**C**).

Physical Assessment: Male Breast

Breast cancer can occur in men. Examination of the male breast and axilla is brief but important and should be included in a physical examination. The nipple and areola are inspected for masses and nipple discharge. The same procedure for palpating the female axilla is used when assessing the male axilla (Hogan-Quigley et al., 2012).

Gynecomastia is the firm enlargement of glandular tissue beneath and immediately surrounding the areola of the male. This is different from the enlargement of soft, fatty tissue, which is caused by obesity.

Diagnostic Evaluation

A wide range of diagnostic studies may be performed in patients with breast conditions. The nurse should educate the patient about the purpose, what to expect, and any possible side effects related to these examinations prior to testing. The nurse should note trends in results, because they often provide information about disease progression as well as the patient's response to therapy.

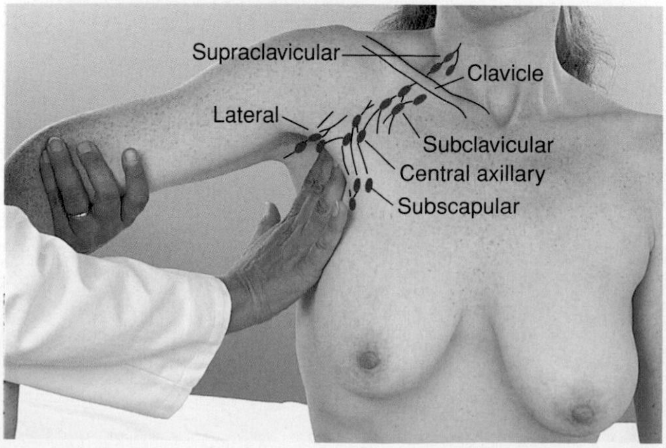
Supraclavicular
Clavicle
Lateral
Subclavicular
Central axillary
Subscapular

FIGURE 58-3 • Palpating axillary nodes in breast examination.

Breast Self-Examination

The nurse plays a critical role in BSE education—a modality used for the early detection of breast cancer (Chart 58-2). BSE can be taught in a variety of settings, either on a one-to-one basis or in a group. It can also be initiated by a health care provider during a patient's routine physical examination.

Variations in breast tissue occur during the menstrual cycle, pregnancy, and the onset of menopause. Women on HT can also experience fluctuations. Normal changes must be distinguished from those that may signal disease. Most women notice increased tenderness and lumpiness before their menstrual periods; therefore, BSE is best performed after menses (day 5 to day 7, counting the first day of menses as day 1). In addition, many women have grainy-textured breast tissue, but such areas are usually less nodular after menses. Younger women may find BSE particularly difficult because of the density of their breast tissue. As women age, their breasts become fattier and may be easier to examine.

Current practice is shifting from educating about BSE to promoting breast self-awareness, which is a woman's attentiveness to the normal appearance and feel of her breasts. However, self-examination still may be appropriate for some high-risk women and those who prefer it. The ACOG (2011) endorses breast self-awareness, which can include self-examination. For every woman, knowing how her breasts normally feel helps detect any changes or signs of a problem. BSE may play an important role in screening, especially for women who develop cancer in the interval after a negative result on

Chart 58-2

NURSING RESEARCH PROFILE

Health Beliefs of African American Women on Breast Self-Exam

Registe, M., & Porterfield, S. P. (2012). Health beliefs of African American women on breast self-exam. *Journal for Nurse Practitioners, 8*(6), 446–451.

Purpose

Screening is the key to finding breast cancer in its early and treatable stage. A woman's decision to participate in cancer prevention, such as routine breast self- examination (BSE) and mammogram, is influenced by cultural, ethnic, and economic differences. This study investigated African American women's health beliefs with regard to breast cancer screening behaviors.

Design

This was a descriptive, correlational study designed to examine the relationship between health beliefs, knowledge, attitude, and the performance of BSE among 131 African American women between the ages of 20 and 65 years living in the southeastern United States with no history of breast cancer. The women were recruited to participate in the study through advertisements, flyers, church newsletters, and social clubs, as well as through friends and family of women self-identified as African American. A demographic questionnaire was used to determine personal history related to race (ethnicity), age, marital status, education, employment, health insurance, socioeconomic status, and BSE practice. The Champion's Breast Cancer Screening Instrument Scale was used to examine participants' perception of susceptibility to breast cancer, belief in the seriousness of the threat of breast cancer to themselves, benefits of BSE, barriers to BSE, and confidence in performing BSE, as well as health motivation.

Findings

Of the participants, 109 women reported practicing BSE within the past 12 months. Those who practiced BSE were more likely to have better knowledge of breast cancer, more confidence in their ability to perform BSE, lower perceived barriers to BSE, and a higher value of their health than those who did not practice BSE. It was found that 21 participants had never practiced BSE; fear of not being able to perform BSE correctly was the main barrier.

Nursing Implications

One role of the nurse is to identify sociocultural factors that may influence screening for breast cancer and incorporate them into practice. African American women, although having a lower incidence of breast cancer, are more likely to die of the disease than Caucasian women. An increased awareness and understanding of facilitators and barriers to breast cancer screening can assist providers in addressing these issues and in the development of culturally sensitive and appropriate educational interventions tailored to African American women.

mammography or clinical breast examination or who have a false-negative imaging or clinical examination result. It can also promote detection in unscreened women (ACOG, 2011). The goal, with or without BSE, is to report any breast changes to a primary provider.

Family history can increase the risk of breast cancer in men, particularly if other men in the family have had breast cancer. The risk is also higher if there is a proven breast cancer gene abnormality in the family. An abnormal *BRCA2* gene accounts for up to 40% of male breast cancers (Williams, 2012). Instructions about BSE should be provided to men if they have a family history of breast cancer.

Patients who elect to perform BSE should receive proper instruction on technique (Chart 58-3). They should be informed that routine BSE will help them become familiar with their "normal abnormalities." If a change is detected, they should seek medical attention.

Patients should be instructed about optimal timing for BSE (5 to 7 days after menses begin for premenopausal women and once monthly for postmenopausal women). When demonstrating examination techniques, the feel of normal breast tissue should be reviewed and ways to identify breast changes discussed. Patients should then perform a BSE demonstration on themselves or on a breast model. Patients who have had breast cancer surgery should be instructed to examine their breast or chest wall for any new changes or nodules that may indicate a recurrence of the disease.

BSE videos, shower cards, and pamphlets can be obtained from local chapters of the ACS Cancer Resource Center (see the Resources section at the end of this chapter).

Mammography

Mammography is a breast imaging technique used to visualize the breast to detect small abnormalities that could warn of cancer (Fischbach & Dunning, 2009). The procedure takes about 15 minutes and can be performed in a hospital radiology department or independent imaging center. Two views are taken of each breast. The breast is mechanically compressed from top to bottom (craniocaudal view) and side to side (mediolateral oblique view) (Fig. 58-4). Women may experience some discomfort because maximum compression is necessary for proper visualization. The new mammogram is compared with previous mammograms, and any changes may indicate a need for further investigation. Mammography may detect a breast tumor before it is clinically palpable (i.e., smaller than 1 cm); however, it has limitations (Fischbach & Dunning, 2009). The false-negative rate ranges between 5% and 10%. Younger women, or those taking HTs, may have dense breast tissue, making it more difficult to detect lesions with mammography.

Patients scheduled for a mammogram may voice concern about exposure to radiation. The radiation exposure is equivalent to about 1 hour of exposure to sunlight, so patients would have to have many mammograms in a year to increase their cancer risk. The benefits of this test outweigh the risks (Heywang-Kobrunner, Hacker, & Sedlacek, 2011). To ensure that a mammogram is reliable, it is important that a woman find a reputable facility. A facility that is certified by the Mammography Quality Standards Act must meet stringent quality standards, be accredited by the U.S. Food and Drug Administration (FDA), and be inspected annually.

In 2009, the U.S. Preventive Services Task Force (USPSTF), a group of health experts, changed the mammography recommendations to state that healthy women should have mammography every 2 years beginning at age 50 years. This change was based on calculations that starting annual screening mammography later and getting it less often would cause less harm and be as safe as starting it earlier and

Chart 58-3

PATIENT EDUCATION

Breast Self-Examination

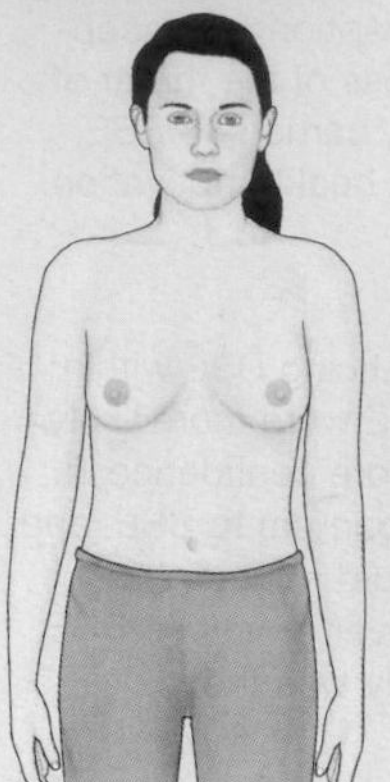

Step 1

1. Stand in front of a mirror.
2. Check both breasts for anything unusual.
3. Look for discharge from the nipple, puckering, dimpling, or scaling of the skin.

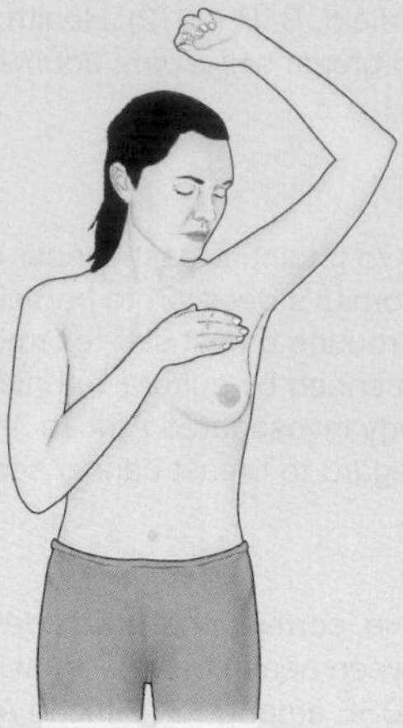

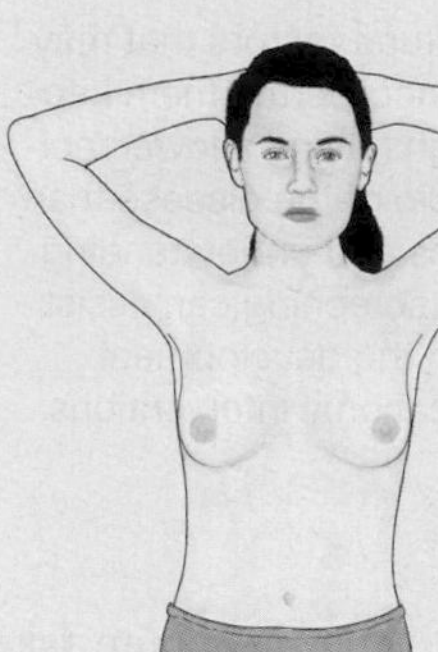

Step 2

Steps 2 and 3 are done to check for any changes in the contour of your breasts. As you do them, you should be able to feel your muscles tighten.

1. Watch closely in the mirror as you clasp your hands behind your head and press your hands forward.
2. Note any change in the contour of your breasts.

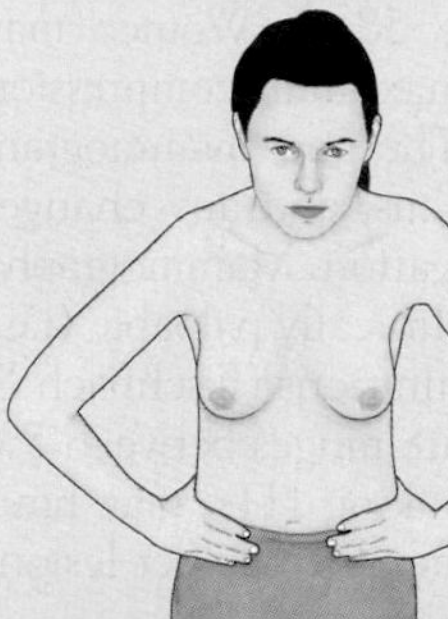

Step 3

1. Next, press your hands firmly on your hips and bow slightly toward the mirror as you pull your shoulders and elbows forward.
2. Note any change in the contour of your breasts.

Step 4

Some women do step 4 of the examination in the shower. Your fingers will glide easily over soapy skin, so you can concentrate on feeling for changes inside the breast.

1. Raise your left arm.
2. Use three or four fingers of your right hand to feel your left breast firmly, carefully, and thoroughly.
3. Beginning at the outer edge, press the flat part of your fingers in small circles, moving the circles slowly around the breast.
4. Gradually work toward the nipple.
5. Be sure to cover the whole breast.
6. Pay special attention to the area between the breast and the underarm, including the underarm itself.
7. Feel for any unusual lumps or masses under the skin.
8. If you have any spontaneous discharge during the month—whether or not it is during your breast self-examination—see your primary provider.
9. Repeat the examination on your right breast.

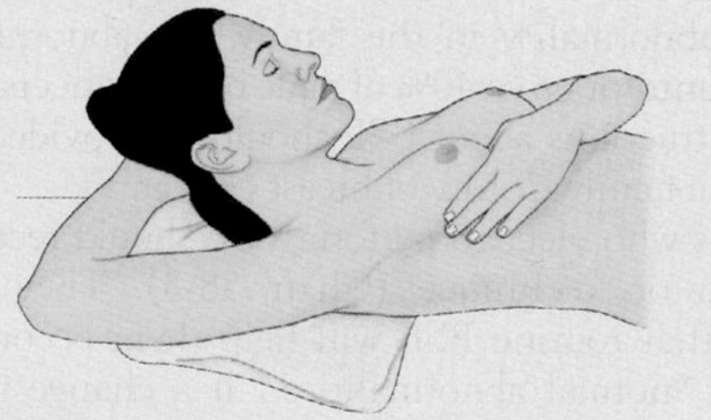

Step 5

1. Step 5 should be repeated lying down.
2. Lie flat on your back with your left arm over your head and a pillow or folded towel under your left shoulder. (This position flattens your breast and makes it easier to check.)
3. Use the same circular motion described earlier.
4. Repeat on your right breast.

Adapted from U.S. Department of Health and Human Services, Public Health Service. (2007). *What you need to know about breast cancer.* Bethesda, MD: National Institutes of Health.

getting it more often. Findings support the recommendation that mammography every other year will reduce a woman's probability of being called back for normal imaging or normal breast biopsy. Screening every other year may, however, may be associated with a small increase in the probability of being diagnosed with later-stage cancer (Hubbard, Kerlikowske, Flowers, et al., 2011).

The ACS, however, continues to recommend that women 40 years and older have a mammogram every year and that they continue to do so for as long as they do not have serious,

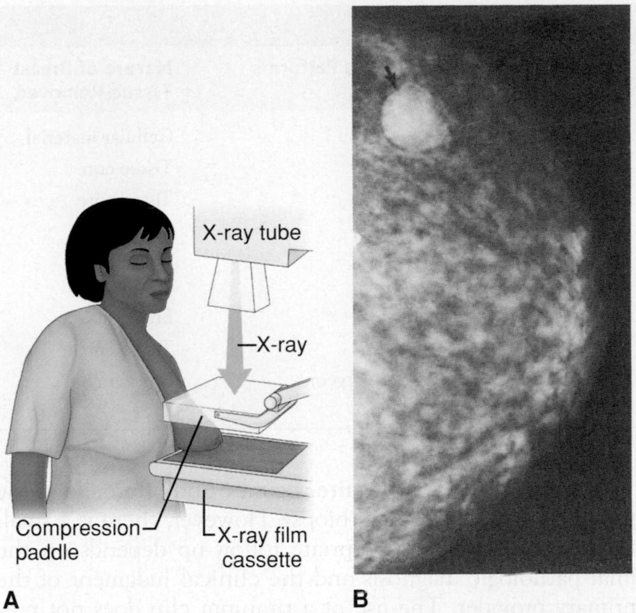
X-ray tube
X-ray
Compression paddle
X-ray film cassette
A
B

FIGURE 58-4 • The mammography procedure (**A**) relies on x-ray imaging to produce the mammogram (**B**), which in this case reveals a breast lump.

chronic health problems such as congestive heart failure, end-stage kidney disease, chronic obstructive pulmonary disease, and moderate to severe dementia. Age alone should not be the reason to stop having regular mammograms (ACS, 2010).

Newer techniques for breast screening include digital mammography and computer-assisted detection (CAD) programs. Digital mammography records x-ray images on a computer instead of on film, thus allowing radiologists to adjust the contrast and focus on an image without having to take additional x-rays. Although the accuracy of both film and digital screening mammography is similar for most women, digital mammography has been shown to be better at detecting estrogen receptor–negative tumors and cancer in extremely dense breasts. Both of these subgroups are more common in younger women, who may therefore choose digital mammography if they wish to have screening mammography (Kerlikowske, Hubbard, Miglioretti, et al., 2011). CAD can be helpful when used by a skilled provider to find abnormal areas that should be checked more closely. Refinements and improvements have been made to CAD software with a focus on increasing sensitivity for masses and reduced false-positive marks rates (Kim, Moon, Kim, et al., 2010).

Galactography

Galactography is a diagnostic procedure that involves injection of less than 1 mL of radiopaque material through a cannula inserted into a ductal opening on the areola, which is followed by a mammogram. It is performed to evaluate an abnormality within the duct when the patient has bloody nipple discharge on expression, spontaneous nipple discharge, or a solitary dilated duct noted on mammography.

Ultrasonography

Ultrasonography (ultrasound) is used as a diagnostic adjunct to mammography to help distinguish fluid-filled cysts from other lesions. A thin coating of lubricating jelly is spread over the area to be imaged. A transducer is then placed on the breast. The transducer transmits high-frequency sound waves through the skin toward the area of concern. The sound waves that are reflected back form a two-dimensional image, which is then displayed on a computer screen. No radiation is emitted during the procedure.

Ultrasonography has advantages and disadvantages. Although it can diagnose cysts with great accuracy, it cannot definitively rule out malignant lesions. Microcalcifications, which are detectable on mammography, cannot be identified on ultrasonography. Finally, examination techniques and interpretation criteria are not standardized.

Magnetic Resonance Imaging

Magnetic resonance imaging (MRI) of the breast is a highly sensitive test that has become a useful diagnostic adjunct to mammography. A magnet is linked to a computer that creates detailed images of the breast without exposure to radiation. An intravenous (IV) injection of gadolinium, a contrast dye, is given to improve visibility. The patient lies face down, and the breast is placed through a depression in the table. A coil is placed around the breast, and the patient is placed inside the MRI machine. The entire procedure takes about 30 to 40 minutes.

Breast MRI is useful for evaluation of contralateral disease, invasive lobular carcinoma, and assessment of chemotherapeutic response (Teller, Jefford, Gabram, et al., 2010). The ACS recommends an annual MRI scan in addition to mammography in women at high risk for breast cancer (i.e., those with greater than 20% lifetime risk). Candidates include women who have a *BRCA1* or *BRCA2* mutation, a first-degree relative with either of these mutations, certain rare genetic syndromes, or radiation to the chest between 10 and 30 years of age (ACS, 2010). MRI should be used in addition to mammography, not instead of it.

Some disadvantages of MRI include high cost, variations in technique and interpretation, and the potential for patient claustrophobia. The procedure cannot always accurately distinguish between malignant and benign breast conditions. MRI is contraindicated in patients with implantable metal devices (e.g., aneurysm clips, pacemakers, ports of tissue expanders) because of the metallic force. Foil-backed medication patches (e.g., nicotine, nitroglycerin, fentanyl) must be removed prior to MRI to avoid burns to the skin.

Procedures for Tissue Analysis

Percutaneous Biopsy

Percutaneous biopsy is performed on an outpatient basis to sample palpable and nonpalpable lesions. Less invasive than a surgical biopsy, percutaneous biopsy is a needle or core biopsy that obtains tissue by making a small puncture in the skin. Table 58-1 outlines the different types of biopsies that can be performed to obtain a tissue diagnosis.

Concept Mastery Alert

It is important to understand the differences between common procedures used for patients with breast disorders. Mammography is used to detect breast abnormalities, whereas biopsy is performed to confirm a diagnosis of breast cancer.

TABLE 58-1 Types of Breast Biopsies

Procedure	Palpable Mass	Health Professional Who Performs Procedure	Nature of Breast Tissue Removed
Fine-needle aspiration	Yes	Surgeon	Cellular material
Core needle biopsy	Yes	Surgeon	Tissue core
Stereotactic core biopsy	No	Radiologist	Tissue core
Ultrasound-guided core biopsy	No	Radiologist	Tissue core
Magnetic resonance imaging (MRI)-guided core biopsy	No	Radiologist	Tissue core
Excisional biopsy	Yes	Surgeon	Entire mass
Incisional biopsy	Yes	Surgeon	Tissue core
Wire needle localization biopsy; may be guided by mammogram, ultrasound, or MRI	No	Radiologist inserts wire; surgeon performs biopsy	Entire mass

Fine-Needle Aspiration

Fine-needle aspiration (FNA) is a noninvasive biopsy technique that is generally well tolerated by most women. A local anesthetic may or may not be used. A small-gauge needle (25- or 22 gauge) attached to a syringe is inserted into the mass or area of nodularity. Suction is applied to the syringe, and multiple passes are made through the mass. A simple cyst often disappears on aspiration, and the fluid is usually discarded. If no fluid is obtained, any cellular material obtained in the hub of the needle is spread on a glass slide or placed in a preservative and sent to the laboratory for analysis (CancerQuest, 2011). For nonpalpable masses, the same procedure can be performed by a radiologist using ultrasound guidance (ultrasound-guided FNA).

FNA is less expensive than other diagnostic methods, and results are usually available quickly. However, false-negative or false-positive results are possible, and appropriate follow-up depends on the clinical judgment of the treating physician.

Core Needle Biopsy

Core needle biopsy is similar to FNA, except a larger-gauge needle is used (usually 14 gauge). A local anesthetic is applied, and tissue cores are removed via a spring-loaded device. This procedure allows for a more definitive diagnosis than FNA, because actual tissue, not just cells, is removed. It is often performed for relatively large tumors that are close to the skin surface.

Stereotactic Core Biopsy

Stereotactic core biopsy is performed on nonpalpable lesions detected by mammography. The patient lies prone on the stereotactic table. The breast is suspended through an opening in the table and compressed between two x-ray plates. Images are then obtained using digital mammography. The exact coordinates of the lesion to be sampled are located with the aid of a computer. Next, a local anesthetic is injected into the entry site on the breast. A small nick is made in the skin, a core needle is inserted, and samples of the tissue are taken for pathologic examination. Often, several passes are taken to ensure that the lesion is well sampled. Postbiopsy films are then taken to check that sampling has been adequate. A small titanium clip is often placed at the biopsy site so that the site can easily be located if further treatment is indicated.

Stereotactic biopsy is quite accurate and often allows the patient to avoid a surgical biopsy. However, there is a small false-negative rate. Appropriate follow-up depends on the final pathologic diagnosis and the clinical judgment of the primary provider. The use of a titanium clip does not preclude subsequent MRIs.

Ultrasound-Guided Core Biopsy

The principles for ultrasound-guided core biopsy are similar to those of stereotactic core biopsy, but by using ultrasound guidance, computer coordination and mammographic compression are not necessary. An ultrasound-guided core biopsy does not use radiation and is also faster and less expensive than stereotactic core biopsy.

Magnetic Resonance Imaging–Guided Core Biopsy

MRI-guided core biopsy can be performed by a technologist when the abnormal area in the breast is too small to be felt and to guide the instruments to the site of the abnormal growth. The number of facilities that are equipped to perform this procedure is increasing.

Surgical Biopsy

Surgical biopsy is usually performed using local anesthesia and IV sedation. After an incision is made, the lesion is excised and sent to a laboratory for pathologic examination.

Types of Surgical Breast Biopsy

Excisional Biopsy. Excisional biopsy is the standard procedure for complete pathologic assessment of a palpable breast mass. The entire mass, plus a margin of surrounding tissue, is removed. This type of biopsy may also be referred to as a lumpectomy. Depending on the clinical situation, a frozen section analysis of the specimen may be performed at the time of the biopsy by the pathologist, who does an immediate reading intraoperatively and provides a provisional diagnosis. This can help confirm a diagnosis in a patient who has had no previous tissue analysis performed.

Incisional Biopsy. Incisional biopsy surgically removes a portion of a mass. This is performed to confirm a diagnosis and to conduct special studies (e.g., ER/PR, HER-2/neu [also referred to as ERBB2]; see later discussion for explanation of these terms) that will aid in determining treatment, which are discussed later in this chapter. Complete excision of the area may not be possible or immediately beneficial to the patient, depending on the clinical situation. This procedure is often performed on women with locally advanced breast cancer or on women with suspected cancer recurrence, whose treatment may depend on the results of these special studies. However, pathologic information may be easily obtained from core needle biopsy, and incisional biopsy is becoming less common.

Wire Needle Localization. Wire needle localization is a technique used to locate nonpalpable masses or suspicious calcium deposits detected on a mammogram, ultrasound, or MRI that require an excisional biopsy. The radiologist inserts a long, thin wire through a needle, which is then inserted into the area of abnormality using x-ray or ultrasound guidance (whichever imaging technique originally identified the abnormality). The wire remains in place after the needle is withdrawn to ensure the precise location. The patient is then taken to the operating room, where the surgeon follows the wire to the tip and excises the area.

Nursing Management

During the preoperative visit, the nurse assesses the patient for any specific educational, physical, or psychosocial needs that the patient may have. This can be accomplished by reviewing the medical and psychosocial history and encouraging the patient to verbalize fears, concerns, and questions. Patients are often worried not only about the procedure but also about the potential implications of the pathology results. Providing a thorough explanation about what to expect in a supportive manner can help alleviate anxiety. Patients often have difficulty absorbing all the information given to them; therefore, written materials should be provided to reinforce education.

The nurse instructs the patient to discontinue any agents that can increase the risk of bleeding, including products containing aspirin, nonsteroidal anti-inflammatory drugs, vitamin E supplements, herbal substances (such as ginkgo biloba and garlic supplements), and warfarin (Coumadin) (Harmer, 2011). The patient may be instructed not to eat or drink for several hours prior to the procedure or after midnight the night before the procedure, depending on the type of biopsy planned. Most breast biopsy procedures today are performed with the use of moderate sedation and local anesthesia.

Immediate postoperative assessment includes monitoring the effects of the anesthesia and inspecting the surgical dressing for any signs of bleeding. Once the sedation has worn off, the nurse reviews the care of the biopsy site, pain management, and activity restrictions with the patient. Prior to discharge from the ambulatory surgical center or the office, the patient must be able to tolerate fluids, ambulate, and void. The patient must have somebody to accompany her home. The dressing covering the incision is usually removed after 48 hours, but the Steri-Strips, which are applied directly over the incision, should remain in place for approximately 7 to 10 days. The use of a supportive bra following surgery is encouraged to limit movement of the breast and reduce discomfort. A follow-up telephone call from the nurse 24 to 48 hours after the procedure can provide the patient with the opportunity to ask any questions and can be a source of great comfort and reassurance.

Most women return to their usual activities the day after the procedure but are encouraged to avoid jarring or high-impact activities for 1 week to promote healing of the biopsy site. Discomfort is usually minimal, and most women find acetaminophen (Tylenol) sufficient for pain relief, although a mild opioid analgesic agent may be prescribed if needed.

Follow-up after the biopsy includes a return visit to the surgeon for discussion of the final pathology report and assessment of the healing of the biopsy site. Depending on the results of the biopsy, the nurse's role varies. If the pathology report is benign, the nurse reviews incision care and explains what the patient should expect as the biopsy site heals (i.e., changes in sensation may occur weeks or months after the biopsy due to nerve injury within the breast tissue). If a diagnosis of cancer is made, the nurse's role changes dramatically. This is discussed in depth later in this chapter.

CONDITIONS AFFECTING THE NIPPLE

Nipple Discharge

Nipple discharge in a woman who is not lactating may be related to many causes, such as carcinoma, papilloma, pituitary adenoma, cystic breasts, and various medications. Oral contraceptives, pregnancy, HT, chlorpromazine (Thorazine)-type medications, and frequent breast stimulation may be contributing factors. In some athletic women, nipple discharge may occur during running or aerobic exercises. Nipple discharge should be evaluated by a health care provider, but it is not often a cause for alarm. One in three women has clear discharge on expression, which is usually normal. A green discharge could indicate an infection. Any discharge that is spontaneous, persistent, or unilateral is of concern. Although bloody discharge can indicate a malignancy, it is often caused by a benign wartlike growth on the lining of the duct called an *intraductal papilloma*.

Nipple discharge should be evaluated for the presence of occult (hidden) blood by performing a guaiac test. A galactogram can also be performed to detect abnormalities within the duct that may be causing the discharge. If there is a high level of suspicion, an excisional biopsy may be indicated. (See procedures for tissue analysis earlier in this chapter.)

Fissure

A fissure is a longitudinal ulcer that may develop in breast-feeding women. If the nipple becomes irritated, a painful, raw area may form and become a site of infection. Daily washing with water, massage with breast milk or lanolin, and exposure to air are helpful. Breast-feeding can continue with the use of a nipple shield. However, if the fissure is severe or extremely painful, the woman is advised to stop breast-feeding. A breast pump can be used until breast-feeding can be resumed. Persistent ulceration requires further diagnosis and therapy. Guidance from a nurse or lactation consultant may be helpful, because nipple irritation can result from improper positioning (i.e., the infant has not grasped the areola fully) during breast-feeding.

BREAST INFECTIONS

Mastitis

Mastitis, an inflammation or infection of breast tissue, occurs most commonly in breast-feeding women, although it may also occur in nonlactating women. The infection may result from a transfer of microorganisms to the breast by the patient's hands or from a breast-fed infant with an oral, eye, or skin infection. Mastitis may also be caused by bloodborne organisms. As inflammation progresses, the breast texture becomes tough or doughy, and the patient complains of dull to severe pain in the infected region. A nipple that is discharging purulent material, serum, or blood should be investigated.

Treatment consists of antibiotics and local application of cold compresses to relieve discomfort. A broad-spectrum antibiotic agent may be prescribed for 7 to 10 days. The patient should wear a snug bra and perform personal hygiene carefully. Adequate rest and hydration are important aspects of management.

Lactational Abscess

A breast abscess may develop as a consequence of acute mastitis. The area affected becomes tender and red. Purulent matter can usually be aspirated with a needle, but incision and drainage may be required. Specimens of the aspirated material are obtained for culture so that an organism-specific antibiotic agent can be prescribed.

BENIGN CONDITIONS OF THE BREAST

Breast Pain

Breast pain (**mastalgia**) may be cyclical or noncyclical. Cyclical pain is usually related to hormonal fluctuations and accounts for nearly 75% of all complaints. Noncyclical pain is far less common and does not vary with the menstrual cycle. Women who experience injury or trauma to the breast or those who have had a breast biopsy may experience noncyclical pain. Patients should be reassured that breast pain is rarely indicative of cancer. However, if the pain persists after menses begin, the patient should see her primary provider.

Nursing Management

The nurse may recommend that the patient wear a supportive bra both day and night for a week, decrease her salt and caffeine intake, and take ibuprofen (Advil) as needed for its anti-inflammatory actions. Vitamin E supplements may also be helpful.

Cysts

Cysts are fluid-filled sacs that develop as breast ducts dilate. Cysts occur most commonly in women 30 to 55 years of age and may be exacerbated during perimenopause. Although their cause is unknown, cysts usually disappear after menopause, suggesting that estrogen is a factor. Cystic areas often fluctuate in size and are usually larger premenstrually. They may be painless or may become very tender premenstrually. Occasionally, a patient may report an intermittent shooting sensation or a dull ache. Various breast masses are compared in Table 58-2. Cysts that are confirmed on an ultrasound and are not bothersome can often be left alone. To confirm a diagnosis or to relieve pain, FNA can be performed. Cysts do not increase the risk of breast cancer (ACS, 2011).

Fibrocystic breast changes, often called *fibrocystic breast disease*, is a nonspecific term used to describe an array of benign findings including palpable nodularity, lumpiness, swelling, or pain. The changes do not necessarily indicate a cystic process.

Fibroadenomas

Fibroadenomas are firm, round, movable, benign tumors. They can occur from puberty to menopause with a peak incidence at 30 years of age. These masses are nontender and are sometimes removed for definitive diagnosis.

Benign Proliferative Breast Disease

The two most common types of **benign proliferative breast disease** (atypical, yet noncancerous, breast tissue) found on biopsy are atypical hyperplasia and lobular carcinoma in situ (LCIS). These diagnoses increase a woman's risk of breast cancer.

Atypical Hyperplasia

Atypical hyperplasia is a premalignant lesion of the breast and is recognized as a precursor lesion to both noninvasive and invasive breast cancer. Imbalance in the normal regulation of cell proliferation is a defining feature of the cancer type. Women with atypical hyperplasia have a fourfold increased risk of breast cancer compared to women in the general population, with a cumulative incidence approaching 30% at 25 years (Santisteban, Reynolds, Barr Fritcher, et al., 2010).

Lobular Carcinoma in Situ

Lobular carcinoma in situ (LCIS) is an incidental microscopic finding of abnormal tissue growth in the lobules of the breast. LCIS increases the risk of subsequent invasive breast cancer in either breast by approximately 7% over 10 years (Maughan, Lutterbie, & Ham, 2010). Affected women

TABLE 58-2 Comparison of Various Breast Masses

The most common breast masses are due to cysts, fibroadenomas, or malignancy. Biopsy is usually needed for confirmation, but the following characteristics are diagnostic clues:

Characteristics	Cysts	Fibroadenomas	Malignancy
	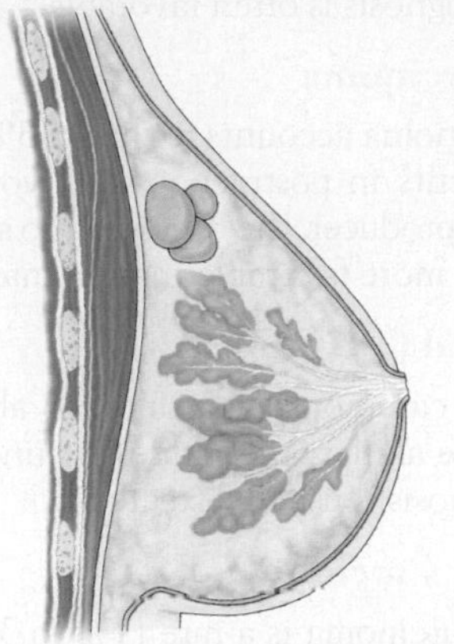	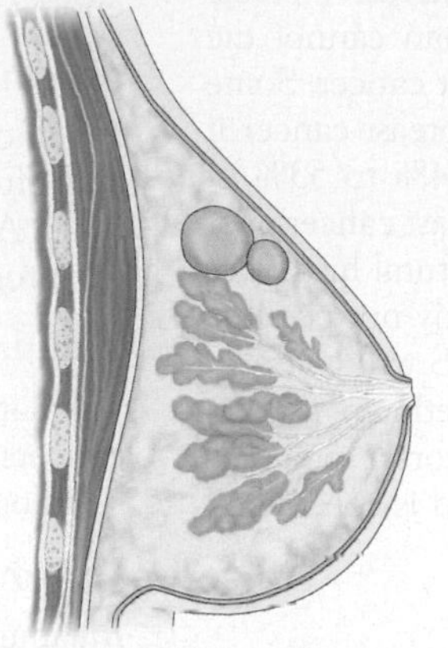	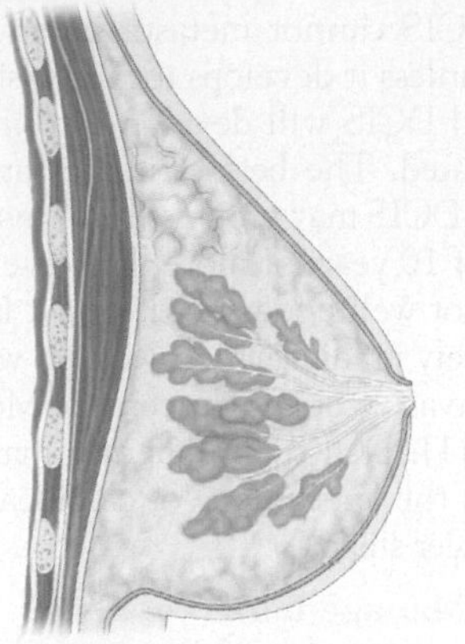
Age	30–55 y, regress after menopause except with use of estrogen therapy	Puberty to menopause	30–90 y; most common, 40–80 y
Number	Single or multiple	Usually single	Usually single
Shape	Round	Round, disk, or lobular	Irregular or stellate
Consistency	Soft to firm, usually elastic	Usually firm	Firm or hard
Mobility	Mobile	Mobile	May be fixed to skin or underlying tissues
Tenderness	Usually tender	Usually nontender	Usually nontender
Retraction signs	Absent	Absent	May be present

should undergo rigorous breast cancer surveillance that consists of annual mammography and clinical breast examination every 6 months (NCCN, 2012). Patients should be offered information about chemoprevention with selective estrogen receptor modulators (SERMs), such as tamoxifen (Soltamox) (Karch, 2012; Maughan et al., 2010). See the discussion of chemoprevention later in this chapter.

Other Benign Conditions

Cystosarcoma phyllodes is a rare fibroepithelial tumor that tends to grow rapidly. It is rarely malignant and is treated with surgical excision. If it is malignant, mastectomy may follow. Lymph node removal is usually not performed, because metastasis is rare.

Fat necrosis is a condition of the breast that is often associated with a history of trauma. Surgical procedures such as a breast biopsy can cause fat necrosis. It may be indistinguishable from carcinoma, and the entire mass is usually excised.

Intraductal papilloma is a wartlike growth that often involves the large milk ducts near the nipple, causing bloody nipple discharge. Surgery usually involves removal of the papilloma and a segment of the duct where the papilloma is found.

Superficial thrombophlebitis of the breast (Mondor disease) is an uncommon condition that is usually associated with pregnancy, trauma, or breast surgery. Pain and redness occur as a result of a superficial thrombophlebitis in the vein that drains the outer part of the breast. The mass is usually linear, tender, and erythematous. Treatment consists of analgesic agents and heat.

MALIGNANT CONDITIONS OF THE BREAST

Breast cancer is a major health problem in the United States. At present, there is no cure. Current statistics indicate that over a lifetime (birth to death), a woman's risk of developing breast cancer is about 12%, or one in eight. Currently, about 230,480 new cases of invasive breast cancer are diagnosed in women each year. Risk of developing breast cancer also increases with increasing age. About two of three invasive breast cancers are found in women 55 years or older. However, after increasing for more than two decades, female breast cancer incidence rates decreased by about 2% per year from 1999 to 2005. The decrease was seen only in women 50 years or older and may be due, at least in part, to the decline in the use of HT. About 5% to 10% of breast cancer cases are thought to be hereditary, resulting directly from gene defects (call mutations) inherited from a biologic parent (ACS, 2011).

Female breast cancer incidence rates vary substantially by race and ethnicity. Higher death rates in African Americans have been attributed to later stage at diagnosis and poorer stage-specific survival. Research suggests that racial disparities in cancer mortality are driven in large part by differences in socioeconomic status. Enhanced efforts are needed to ensure that all women have access to high-quality prevention, detection, and treatment services (DeSantis, Siegel, Bandi, et al., 2011).

Types of Breast Cancer

Ductal Carcinoma in Situ

Ductal carcinoma in situ (DCIS) is characterized by the proliferation of malignant cells inside the milk ducts without invasion into the surrounding tissue. Unlike invasive breast cancer, DCIS cannot metastasize and a woman cannot die of DCIS unless it develops into invasive breast cancer. Some but not all DCIS will develop into invasive breast cancer if left untreated. The best estimates are that 14% to 53% of untreated DCIS may progress into invasive breast cancer over a period of 10 years or more. However, the natural history of DCIS is not well understood, and it is currently not possible to accurately predict which women with DCIS will go on to develop invasive breast cancer (De Morgan, Redman, D'Este, et al., 2011). DCIS is frequently manifested on a mammogram with the appearance of calcifications and is considered breast cancer stage 0.

Medical Management

Current management takes into account (1) assurance of an accurate diagnosis, (2) assessment of DCIS size and grade, and (3) careful margin evaluation. The pathologist analyzes the piece of breast tissue to determine the type and grade of the DCIS or how abnormal the cells look when compared with normal breast cells and how fast they are growing. Grade III (high-grade DCIS) cells tend to grow more quickly than grade I (low-grade) and grade II (moderate-grade) cells and look much different from normal breast cells. Accurate grading of DCIS is critical, because high nuclear grade and the presence of necrosis (the premature death of cells in living tissue) are highly predictive of the inability to achieve adequate margins or borders of healthy tissue around the cancer, of local recurrence, and of the probability of missed areas of invasion. The pros and cons of irradiating conservatively treated patients with DCIS should be carefully weighed on a case-by-case basis, considering recent trials have shown that radiation has a beneficial effect on distant recurrence, breast cancer–specific mortality, and overall survival. In the premammographic era, DCIS was treated as one disease, resulting in mastectomy. Breast conservation (treatment of a breast cancer without the loss of the breast) is now seen curative for well-defined subsets of women (Sanders & Simpson, 2011).

Invasive Cancer

The NCCN (2012), a nonprofit group of the world's 21 leading cancer centers, disseminates estimates for various types of cancer.

Infiltrating Ductal Carcinoma

Infiltrating ductal carcinoma—the most common histologic type of breast cancer—accounts for 80% of all cases. The tumors arise from the duct system and invade the surrounding tissues. They often form a solid irregular mass in the breast.

Infiltrating Lobular Carcinoma

Infiltrating lobular carcinoma accounts for 10% to 15% of breast cancers. The tumors arise from the lobular epithelium and typically occur as an area of ill-defined thickening in the breast. They are often multicentric and can be bilateral.

Medullary Carcinoma

Medullary carcinoma accounts for about 5% of breast cancers, and it tends to be diagnosed more often in women younger than 50 years. The tumors grow in a capsule inside a duct. They can become large and may be mistaken for a fibroadenoma. The prognosis is often favorable.

Mucinous Carcinoma

Mucinous carcinoma accounts for about 3% of breast cancers and often presents in postmenopausal women 75 years and older. A mucin producer, the tumor is also slow growing; thus, the prognosis is more favorable than in many other types.

Tubular Ductal Carcinoma

Tubular ductal carcinoma accounts for about 2% of breast cancers. Because axillary metastases are uncommon with this histology, prognosis is usually excellent.

Inflammatory Carcinoma

Inflammatory carcinoma is a rare (1% to 3%) and aggressive type of breast cancer that has unique symptoms. The cancer is characterized by diffuse edema and brawny erythema of the skin, often referred to as peau d'orange (resembling an orange peel). This is caused by malignant cells blocking the lymph channels in the skin. An associated mass may or may not be present; if there is a mass, it is often a large area of indiscrete thickening. Inflammatory carcinoma can be confused with an infection because of its presentation. The disease can spread to other parts of the body rapidly. Chemotherapy often plays an initial role in controlling disease progression, but radiation and surgery may also be useful.

Paget's Disease

Paget's disease of the breast accounts for 2% of diagnosed cases of breast cancer. Symptoms typically include a scaly, erythematous, pruritic lesion of the nipple (Harmer, 2011). Paget's disease often represents DCIS of the nipple but may have an invasive component. If no lump can be felt in the breast tissue and the biopsy shows DCIS without invasion, the prognosis is very favorable.

Risk Factors

There is no single, specific cause of breast cancer. A combination of genetic, hormonal, and possibly environmental factors may increase the risk of its development (Table 58-3). More than 80% of all cases of breast cancer are sporadic, meaning that patients have no known family history of the disease. The remaining cases are either familial (there is a family history of breast cancer, but it is not passed on genetically) or genetically acquired. There is no evidence that smoking, silicone breast implants, the use of antiperspirants, underwire bras, or abortion (induced or spontaneous) increases the risk of the disease.

As stated previously, breast cancer can be genetically inherited, resulting in significant risk. Approximately 5% to 10% of breast cancer cases develop as a result of genetic mutations. Factors that may indicate a genetic link include multiple first-degree relatives with early-onset breast cancer, breast and ovarian cancer in the same family, male breast cancer, and Ashkenazi Jewish background. ***BRCA1*** and ***BRCA2*** are tumor suppressor genes that normally function to identify damaged deoxyribonucleic

TABLE 58-3 Risk Factors for Breast Cancer

Risk Factor	Comments
Female gender	99% of cases occur in women.
Increasing age	Increasing age is associated with an increased risk.
Personal history of breast cancer	Once treated for breast cancer, the risk of developing breast cancer in same or opposite breast is significantly increased.
Family history of breast cancer	Having first-degree relative with breast cancer (mother, sister, daughter) increases the risk twofold; having two first-degree relatives increases the risk fivefold. The risk is higher if the relative was premenopausal at the time of diagnosis. The risk is increased if a father or brother had breast cancer (exact risk is unknown).
Genetic mutation	*BRCA1* and *BRCA2* mutations account for majority of inherited cases of breast cancer (see additional information in text).
Hormonal Factors	
• Early menarche	Before 12 years of age
• Late menopause	After 55 years of age
• Nulliparity	No full-term pregnancies
• Late age at first full-term pregnancy	After 30 years of age
• Hormone therapy (formerly referred to as hormone replacement therapy)	Current or recent use of combined postmenopausal hormone therapy (estrogen and progesterone) Long-term use (several years or more)
Exposure to ionizing radiation during adolescence and early adulthood	The risk is highest if breast tissue was exposed while still developing (during adolescence), such as women who received mantle radiation (to the chest area) for treatment of Hodgkin lymphoma in their younger years.
History of benign proliferative breast disease	Having had atypical ductal or lobular hyperplasia or lobular carcinoma in situ increases the risk.
Obesity	Obesity and weight gain during adulthood increases the risk of postmenopausal breast cancer. During menopause, estrogen is primarily produced in fat tissue. More fat tissue can increase estrogen levels, thereby increasing breast cancer risk.
High-fat diet	More research is needed.
Alcohol intake (beer, wine, or liquor)	Two to five drinks daily increases the risk about one and a half times.

Adapted from National Comprehensive Cancer Network. (2012). *NCCN clinical practice guidelines in oncology: Breast cancer.* Available at: www.nccn.org

acid (DNA) and thereby restrain abnormal cell growth (Yarbo, Wujcik, & Gobel, 2010). Mutations in these genes on chromosome 17 are responsible for the majority of hereditary breast cancer in the United States. *BRCA* mutations in women have been associated with an overall risk of breast cancer between 56% and 84% (Odle, 2011). Currently, *BRCA*-positive women are counseled to start screening, typically using mammography, once a year and then MRI 6 months after the yearly mammography (Salhab, Bismohun, & Mokbel, 2010) by 25 years of age, or 5 to 10 years earlier than their youngest affected family member (Litton, Ready, Chen, et al., 2012). Male *BRCA2* mutation carriers may have a lifetime risk of 6% to 7% of developing breast cancer. Male carriers should be offered support in how to communicate risk information to their family (Stromsvik, Raheim, Oyen, et al., 2009).

Protective Factors

Certain factors may be protective against the development of breast cancer. For example, research suggests the efficacy of physical activity in the prevention of breast cancer. In physically active subjects, there was an average overall risk reduction of 25% to 30%. Thirty to 60 minutes of exercise per day at a moderate intensity is regarded as optimal (Graf & Wessely, 2010). Breast-feeding is thought to decrease risk because it prevents the return of menstruation, thereby decreasing exposure to endogenous estrogen (Morris, 2009). Management of stress, through the use of such activities as meditation, prayer, or involvement in support groups, may also be protective (Love & Lindsey, 2010).

Breast Cancer Prevention Strategies in the High-Risk Patient

Patients often over- or underestimate their risk of developing breast cancer. A consultation with a breast specialist is of paramount importance prior to embarking on any of the prevention strategies that follow. Once patients have an accurate assessment of their risk, along with the knowledge of the pros and cons of each prevention strategy, they can make a decision that is most appropriate for their situation.

Long-Term Surveillance

Long-term surveillance focuses on early detection. As recommended by the ACS (2011), women at high risk for breast cancer benefit from additional screening using MRI along with a yearly mammogram. Clinical breast examinations may be performed twice a year starting as early as 25 years of age. Mammograms may also be performed as early as 25 years of age. Data concerning the effectiveness of BSE are limited. In addition to yearly mammography and MRI, other screening tests, including ultrasonography, may be useful.

Chemoprevention

Chemoprevention is the main modality that aims to prevent the disease. Several national, randomized clinical trials in the past two decades have led to FDA approval of tamoxifen and raloxifene (Evista) as effective chemopreventive agents for use in high-risk women (Cummings, Eckert, Krueger, et al., 1999; Fisher, Constantino, Wickerham, et al., 1998; Vogel, Costantino, Wickerham, et al., 2006).

Nurses can help women who are considering chemoprevention by providing them with information about the benefits, risks, and possible side effects of both tamoxifen and raloxifene.

Prophylactic Mastectomy

Prophylactic mastectomy is another primary prevention modality that can reduce the risk of breast cancer by 90% (National Cancer Institute [NCI], 2011) and is sometimes referred to as a "risk-reducing" mastectomy. The procedure consists of a total mastectomy (removal of breast tissue only) and is usually accompanied by immediate breast reconstruction. Possible candidates include women with a strong family history of breast cancer, a diagnosis of LCIS or atypical hyperplasia, a mutation in a *BRCA* gene, and previous cancer in one breast. Because of physical and psychological ramifications including anxiety, depression, and altered body image, this procedure should be undertaken only after extensive counseling related to its risks and benefits. The procedure does not confer 100% protection against the development of breast cancer (Tirona, Sehgal, & Ballester, 2010).

A multidisciplinary approach should be used to help the patient arrive at a decision that is best for her. Consultation with a genetic counselor, plastic surgeon, medical oncologist, and psychiatrist can be invaluable. The patient needs to understand that this surgery is elective and not emergent. The nurse can play a valuable role in providing the patient with information, clarification, and support during the decision-making process.

Clinical Manifestations

Breast cancers can occur anywhere in the breast but are usually found in the upper outer quadrant, where the most breast tissue is located. Generally, the lesions are nontender, fixed rather than mobile, and hard with irregular borders. Complaints of diffuse breast pain and tenderness with menstruation are usually associated with benign breast disease.

With the increased use of mammography, more women are seeking treatment at earlier stages of the disease. These women often have no signs or symptoms other than a mammographic abnormality. Some women with advanced disease seek initial treatment after ignoring symptoms. Advanced signs may include skin dimpling, nipple retraction, or skin ulceration.

Assessment and Diagnostic Findings

Techniques to determine the diagnosis of breast cancer include various types of biopsy, which have been described previously. Tumor staging and analysis of additional prognostic factors are used to determine the prognosis and optimal treatment regimen (see below).

Staging

Staging involves classifying the cancer by the extent of the disease in the body. It is based on whether the cancer is invasive or noninvasive, the size of the tumor, how many lymph nodes are involved, and if it has spread to other parts of the body. The stage of a cancer is one of the most important factors in determining prognosis and treatment options. The most common system used to describe the stages of breast cancer is the American Joint Committee on Cancer (AJCC) TNM (tumor, nodes, metastasis) system (Edge, Byrd, Compton, et al., 2010) (see Chart 15-3 in Chapter 15).

Other diagnostic tests may be performed before or after the surgery to help in the staging of the disease. The extent of testing often depends on the clinical presentation of the disease and may include chest x-rays, computed tomography (CT), MRI, positron emission tomography (PET), bone scans, and blood work (complete blood count, comprehensive metabolic panel, tumor markers [i.e., carcinoembryonic antigen, cancer antigen 15-3]).

Prognosis

Several different factors must be taken into consideration when determining the prognosis of a patient with breast cancer. Two of the most important factors are tumor size and whether the tumor has spread to the lymph nodes under the arm (axilla).

Generally, the smaller the tumor appears, the better the prognosis. A tumor starts with a genetic alteration in a single cell and takes time to divide and double in size. A carcinoma may double in size 30 times to become 1 cm or larger, at which point it becomes clinically apparent. Doubling time varies, but breast tumors are often present for several years before they become palpable. Nurses can reassure patients that once breast cancer is diagnosed, they have a safe period of several weeks to make decisions regarding treatment.

Prognosis also depends on the extent of spread of the breast cancer. The 5-year survival rate is approximately 88% for a stage I breast cancer and 15% for a stage IV breast cancer (ACS, 2011). The most common route of regional spread is to the axillary lymph nodes. Other sites of lymphatic spread include the internal mammary and supraclavicular nodes (Fig. 58-5). Distant metastasis can affect any organ, but the most common sites are bone, lung, liver, pleura, adrenals, skin, and brain (ACS, 2011).

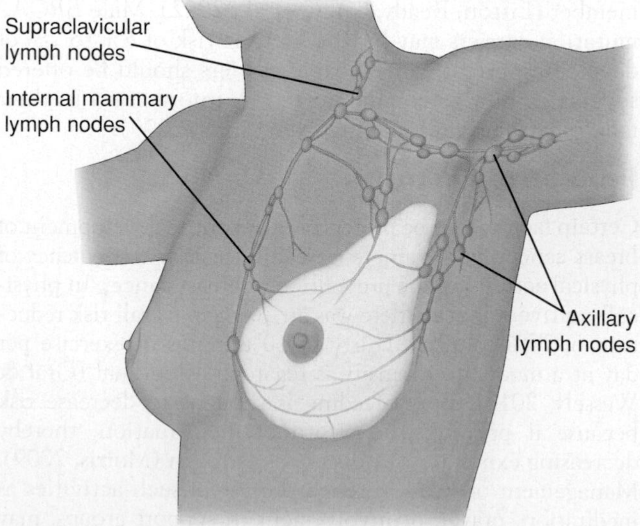
Supraclavicular lymph nodes
Internal mammary lymph nodes
Axillary lymph nodes

FIGURE 58-5 • Lymphatic drainage of the breast.

Chart 58-4 Pathologic Factors Associated With Favorable Prognosis for Breast Cancer

- Noninvasive tumors or invasive tumors <1 cm
- Negative axillary lymph nodes
- Estrogen receptor (ER) and progesterone receptor (PR) proteins
- Well-differentiated tumors
- Low expression of HER-2/neu oncogene (also known as ERBB2)
- No vascular or lymphatic invasion
- Diploid tumors with low S-phase fraction

In addition to the type of breast cancer and the stage, other factors may help determine prognosis (Chart 58-4). Excessive number of copies of certain genes (amplification) or excessive amounts of their protein product (overexpression) may represent a poorer prognosis. The HER-2/neu (also known as ERBB2) oncogene is the classic example; approximately 25% of invasive breast cancers, which typically involve the more aggressive tumors, have amplification or overexpression of the HER-2/neu gene (Harbeck, Pegram, Ruschoff, et al., 2010). The proliferative rate or rapidity in growth rate (S-phase fraction) and DNA content (ploidy) of a tumor is associated with overall survival rate (Pazaiti & Fentiman, 2011).

Concept Mastery Alert

Nurses must be able to differentiate between risk factors and prognostic factors for breast cancer. Risk factors may increase the risk of developing the disease. Prognostic factors are important in verifying the diagnosis of breast cancer.

Surgical Management

The main goal of surgery is to gain local control of the disease. With breast cancer being diagnosed today at earlier stages, options for less invasive surgical procedures are available (Eggert, 2010). Surgical treatment options for noninvasive and invasive breast cancer are summarized in Table 58-4.

Modified Radical Mastectomy

Modified radical mastectomy is performed to treat invasive breast cancer. The procedure involves removal of breast tissue, including the nipple–areola complex. In addition, a portion of the axillary lymph nodes are also removed in axillary lymph node dissection (ALND). If immediate breast reconstruction is desired, the patient is referred to a plastic surgeon prior to the mastectomy so that she has the opportunity to explore all available options. In modified radical mastectomy, the pectoralis major and pectoralis minor muscles are left intact, unlike in radical mastectomy, in which the muscles are removed.

TABLE 58-4 Surgical Treatment Options for Noninvasive and Invasive Breast Cancer

Noninvasive Breast Cancer	Invasive Breast Cancer
Breast conservation* alone	Breast conservation* with one of the following: Sentinel lymph node biopsy Axillary lymph node dissection
Total mastectomy alone	Total mastectomy with sentinel lymph node biopsy or Modified radical mastectomy

*Breast conservation treatment includes lumpectomy, wide excision, partial or segmental mastectomy, and quadrantectomy. These are relatively synonymous terms that describe removal of varying amounts of breast tissue.

Total Mastectomy

Like modified radical mastectomy, **total mastectomy** (i.e., simple mastectomy) also involves removal of the breast and nipple–areola complex but does not include ALND. Total mastectomy may be performed in patients with noninvasive breast cancer (e.g., DCIS), which does not have a tendency to spread to the lymph nodes. It may also be performed prophylactically in patients who are at high risk for breast cancer (e.g., LCIS, *BRCA* mutation). A total mastectomy may also be performed in conjunction with sentinel lymph node biopsy (SLNB) for patients with invasive breast cancer.

Breast Conservation Treatment

The goal of **breast conservation treatment** (i.e., lumpectomy, wide excision, partial or segmental mastectomy, quadrantectomy) is to excise the tumor in the breast completely and obtain clear margins while achieving an acceptable cosmetic result. If the procedure is being performed to treat a noninvasive breast cancer, lymph node removal is not necessary. For an invasive breast cancer, lymph node removal (SLNB or ALND) is indicated. The lymph nodes are removed through a separate semicircular incision in the axilla.

Sentinel Lymph Node Biopsy

The status of the lymph nodes is the most important prognostic factor in breast cancer. Approximately two thirds of women with early-stage breast cancer who have ALND have negative nodes. In the mid-1990s, SLNB emerged as a less invasive alternative to ALND and is now considered a standard of care for the treatment of early-stage breast cancer. ALND is associated with potential morbidity, including lymphedema, cellulitis, decreased arm mobility, and sensory changes. Studies have shown that SLNB is highly accurate and is associated with a local recurrence rate similar to that of ALND (Aslani, Swanson, Kennecke, et al., 2011; Krag, Anderson, Julian, et al., 2010). Table 58-5 compares SLNB and ALND.

The **sentinel lymph node**, which is the first node (or nodes) in the lymphatic basin that receives drainage from the primary tumor in the breast, is identified by injecting a radioisotope and/or blue dye into the breast; the radioisotope or dye then travels via the lymphatic pathways to the node. In SLNB, the surgeon uses a handheld probe to locate the sentinel lymph node, excises it, and sends it for pathologic analysis, which is often performed immediately during the surgery using frozen section analysis. If the sentinel lymph node is positive, the surgeon can proceed with an immediate ALND, thus sparing the patient a return trip to the operating room and additional anesthesia. (The patient could also opt

TABLE 58-5 Comparison of Sentinel Lymph Node Biopsy and Axillary Lymph Node Dissection

Sentinel Lymph Node Biopsy	Axillary Lymph Node Dissection
Shorter operating room time (~15–30 minutes)	Longer operating room time (~60–90 minutes)
No surgical drain	Surgical drain
Local anesthesia with IV moderate sedation as outpatient surgery (unless being performed in conjunction with total mastectomy)	General anesthesia; usually overnight admission (sometimes done as outpatient surgery)
Lymphedema incidence approximately 0%–8%	Lymphedema incidence approximately 10%–30%
Presence of neuropathic sensations postoperatively (prevalence lower than after axillary lymph node dissection)	Presence of neuropathic sensations postoperatively
Decreased range of motion in affected arm unlikely postoperatively but may occur	Decreased range of motion likely postoperatively
Seroma (collection of serous fluid in the axilla) may occur postoperatively.	Seroma may occur postoperatively.

to return for additional surgery at a later time.) If the sentinel lymph node is negative, a standard ALND is not needed, thus sparing the patient the sequelae of the procedure. After the procedure is complete, all specimens are sent to pathology for more thorough analysis.

Concept Mastery Alert

ALND is a prognostic procedure; it is performed to determine if chemotherapy is indicated. Although ALND may help in preventing metastasis, that is not the reason it is performed.

Nursing Management

Patients who undergo SLNB in conjunction with breast conservation treatments are generally discharged the same day. Patients who undergo SLNB with total mastectomy usually stay in the hospital overnight, possibly longer if breast reconstruction is being performed. The patient must be informed that although frozen section analysis is highly accurate, false-negative results can occur. A negative sentinel lymph node on frozen section analysis may show metastatic disease on subsequent analysis, indicating that ALND is still necessary. The patient should also be reassured that the radioisotope and blue dye are generally safe. The nurse informs patients that they may notice a blue-green discoloration in the urine or stool for the first 24 hours as the blue dye is excreted. The incidence of lymphedema, decreased arm mobility, and seroma formation (collection of serous fluid) in the axilla is generally low, but the patient should be prepared for these possibilities. Women who have SLNB alone have neuropathic sensations similar to those who undergo ALND, although the prevalence and severity of these sensations and the resulting distress are lower with SLNB (Baron, Fey, Borgen, et al., 2007).

The nurse must not overlook the psychosocial needs of the patient who has undergone SLNB. Although SLNB is a less invasive procedure than ALND and results in a shorter recovery period, a patient who has undergone SLNB also has many difficult issues surrounding her breast cancer diagnosis and treatment. The nurse must listen, provide emotional support, and refer the patient to appropriate specialists when indicated.

NURSING PROCESS

The Patient Undergoing Surgery for Breast Cancer

Assessment

The health history is a valuable tool to assess the patient's reaction to the diagnosis and her ability to cope with it. Pertinent questions include the following:

- How is the patient responding to the diagnosis?
- What coping mechanisms does she find most helpful?
- What psychological or emotional supports does she have and use?
- Is there a partner, family member, or friend available to assist her in making treatment choices?
- What are her educational needs?
- Is she experiencing any discomfort?

Diagnosis

Preoperative Nursing Diagnoses

Based on the assessment data, major preoperative nursing diagnoses may include the following:

- Deficient knowledge about the planned surgical treatments
- Anxiety related to the diagnosis of cancer
- Fear related to specific treatments and body image changes
- Risk for defensive or ineffective coping related to the diagnosis of breast cancer and related treatment options
- Decisional conflict related to treatment options

Postoperative Nursing Diagnoses

Based on the assessment data, major postoperative nursing diagnoses may include the following:

- Acute pain and discomfort related to surgical procedure
- Peripheral neurovascular dysfunction related to nerve irritation in affected arm, breast, or chest wall
- Disturbed body image related to loss or alteration of the breast
- Risk for impaired coping related to the diagnosis of cancer and surgical treatment
- Self-care deficit related to partial immobility of upper extremity on operative side
- Risk for sexual dysfunction related to loss of body part, change in self-image, and fear of partner's responses

- Deficient knowledge: drain management after breast surgery, arm exercises to regain mobility of affected extremity, hand and arm care after ALND

Collaborative Problems/Potential Complications

Potential complications may include the following:

- Lymphedema
- Hematoma/seroma formation
- Infection

Planning and Goals

The major goals may include increased knowledge about the disease and its treatment; reduction of preoperative and postoperative fear, anxiety, and emotional stress; improvement of decision-making ability; pain management; improvement in coping abilities; improvement in sexual function; and the absence of complications.

Preoperative Nursing Interventions

Providing Education and Preparation About Surgical Treatments

Patients with newly diagnosed breast cancer are expected to absorb an abundance of new information during a very emotionally difficult time, and this may lead to difficulty in making treatment decisions. The nurse plays a key role in reviewing treatment options by reinforcing information provided to the patient and answering any questions. The nurse fully prepares the patient for what to expect before, during, and after surgery. Patients undergoing breast conservation with ALND, or a total or modified radical mastectomy, generally remain in the hospital overnight (or longer if they have immediate reconstruction). Surgical drains will be inserted in the mastectomy incision and in the axilla if the patient undergoes ALND. A surgical drain is generally not needed after SLNB. The patient should be informed that she will go home with the drain(s) and that complete instructions about drain care will be provided prior to discharge. In addition, the patient should be informed that she will often have decreased arm and shoulder mobility after ALND and that she will be shown range-of-motion exercises prior to discharge. The patient should also be reassured that appropriate analgesia and comfort measures will be provided to alleviate any postoperative discomfort.

Reducing Fear and Anxiety and Improving Coping Ability

The nurse must help the patient cope with the physical and emotional effects of surgery. Many fears may emerge during the preoperative phase. These can include fear of pain, mutilation (after mastectomy), and loss of sexual attractiveness; concern about inability to care for oneself and one's family; concern about taking time off from work; and coping with an uncertain future. Providing the patient with realistic expectations about the healing process and expected recovery can help alleviate fears. Maintaining open communication and assuring the patient that she can contact the nurse at any time with questions or concerns can be a source of comfort. The patient should also be made aware of available resources at the treatment facility as well as in the breast cancer community such as social workers, psychiatrists, and support groups. Some women find it helpful and reassuring to talk to a breast cancer survivor who has undergone similar treatments.

Promoting Decision-Making Ability

The patient may be eligible for more than one therapeutic approach; she may be presented with treatment options and then asked to make a choice. This can be very frightening for some patients, and they may prefer to have someone else make the decision for them (e.g., surgeon, family member). The nurse can be instrumental in ensuring that the patient and family members truly understand their options. The nurse can then help the patient weigh the risks and benefits of each option. The patient may be presented with the option of having breast conservation treatment followed by radiation or a mastectomy. The nurse can explore the issues with the patient by asking questions such as the following:

- How would you feel about losing your breast?
- Are you considering breast reconstruction?
- If you choose to retain your breast, would you consider undergoing radiation treatments 5 days a week for 5 to 6 weeks?

Questions such as these can help the patient focus. Once the patient's decision is made, it is very important to support it.

Postoperative Nursing Interventions

Relieving Pain and Discomfort

Many patients tolerate breast surgery quite well and have minimal pain during the postoperative period. This is particularly true of less invasive procedures such as breast conservation treatment with SLNB. However, all patients must be carefully assessed, because individual patients can have varying degrees of pain. Patients who have had more invasive procedures, such as a modified radical mastectomy with immediate reconstruction, may have considerably more pain. All patients are discharged home with analgesic medication (e.g., oxycodone and acetaminophen [Percocet]) and are encouraged to take it if needed. An over-the-counter analgesic agent such as acetaminophen may provide sufficient relief. Patients sometimes complain of a slight increase in pain after the first few days of surgery; this may occur as patients regain sensation around the surgical site and become more active. However, patients who report excruciating pain must be evaluated to rule out any potential complications such as infection or hematoma. Alternative methods of pain management, such as taking warm showers and using distraction methods (e.g., guided imagery), may also be helpful. (See Chapter 12 for further discussion of methods that relieve pain.)

Managing Postoperative Sensations

Because nerves in the skin and axilla are often cut or injured during breast surgery, patients experience a variety of sensations. Common sensations include tenderness, soreness, numbness, tightness, pulling, and twinges. These sensations may occur along the chest wall, in the axilla, and along the inside aspect of the upper arm. After mastectomy, some patients experience phantom sensations and report a feeling that the breast or nipple is still present. Overall, patients do not find these sensations severe or distressing (Baron et al., 2007). Sensations usually persist for several months and then begin to diminish, although some may persist for as long as 5 years and possibly longer. Patients should be reassured that this is a normal part of healing and that these sensations are not indicative of a problem.

Promoting Positive Body Image

Patients who have undergone mastectomy often find it very difficult to view the surgical site for the first time. No matter how prepared the patient may think she is, the appearance of an absent breast can be very emotionally distressing. Ideally, the patient sees the incision for the first time when she is with the nurse or another health care provider who is available for support.

The nurse first assesses the patient's readiness and provides gentle encouragement. It is important to maintain the patient's privacy while assisting her as she views the incision; this allows her to express feelings safely to the nurse. Asking the patient what she perceives, acknowledging her feelings, and allowing her to express her emotions are important nursing actions. Reassuring the patient that her feelings are a normal response to breast cancer surgery may be comforting. If the patient has not had immediate reconstruction, providing her with a temporary breast form to place in her bra on discharge can help alleviate feelings of embarrassment or self-consciousness.

Promoting Positive Adjustment and Coping

Providing ongoing assessment of how the patient is coping with her diagnosis of breast cancer and her surgical treatment is important in determining her overall adjustment. Assisting the patient in identifying and mobilizing her support systems can be beneficial to her well-being. The patient's spouse or partner may also need guidance, support, and education. The patient and partner may benefit from a wide network of available community resources, including the Reach to Recovery program of the ACS, advocacy groups, or a spiritual advisor. Encouraging the patient to discuss issues and concerns with other patients who have had breast cancer may help her to understand that her feelings are normal and that other women who have had breast cancer can provide invaluable support and understanding.

The patient may also have considerable anxiety about the treatments that will follow surgery (i.e., chemotherapy and radiation) and their implications. Providing her with information about the plan of care and referring her to the appropriate members of the health care team also promote coping during recovery. Some women require additional support to adjust to their diagnosis and the changes that it brings. If a woman displays ineffective coping, consultation with a mental health provider may be indicated.

Improving Sexual Function

Once discharged from the hospital, most patients are physically allowed to engage in sexual activity. However, any change in the patient's body image, self-esteem, or the response of her partner may increase her anxiety level and affect sexual function. Some partners may have difficulty looking at the incision, whereas others may be completely unaffected. Encouraging the patient to openly discuss how she feels about herself and about possible reasons for a decrease in libido (e.g., fatigue, anxiety, self-consciousness) may help clarify issues for her. Helpful suggestions for the patient may include varying the time of day for sexual activity (when the patient is less tired), assuming positions that are more comfortable, and expressing affection using alternative measures (e.g., hugging, kissing, manual stimulation).

Most patients and their partners adjust with minimal difficulty if they openly discuss their concerns. However, if issues cannot be resolved, a referral for counseling (e.g., psychologist, psychiatrist, psychiatric clinical nurse specialist, social worker, sex therapist) may be helpful. The ambulatory care nurse in the outpatient clinic or hospital should inquire whether the patient is having difficulty with sexuality issues, because many patients are reluctant or embarrassed to bring it up themselves.

Monitoring and Managing Potential Complications

Lymphedema. **Lymphedema** is a complication in which there is a chronic swelling of an extremity due to interrupted lymphatic circulation. The swelling is due to the accumulation of protein-rich fluid in the interstial space and is a common postoperative complication after ALND (Cheifetz & Haley, 2010). Lymphedema occurs after treatment for breast cancer, often affecting both the breast and ipsilateral limb. It is associated with a painful swelling of the arm as well as weakness, shoulder pain, and tingling sensations in the arm and shoulder. Research results suggest that within 5 years of ALND, 30.3% of the women developed lymphedema (Tuma, 2011). Because sentinel lymph node dissection (SLND) involves more focused surgery and less disruption of the axilla, only 13% of patients post SLND experience some degree of lymphedema (Helyer, Varnic, Le, et al., 2010). Risk factors for lymphedema in mixed age groups include ALND, concomitant radiation therapy, increased age, presence of a concomitant infection, preexisting cardiovascular conditions, and obesity (Clough-Gorr, Ganz, & Silliman, 2010).

Lymphedema results if functioning lymphatic channels are inadequate to ensure a return flow of lymph fluid to the general circulation. After axillary lymph nodes are removed, collateral circulation must assume this function. Transient edema in the postoperative period occurs until collateral circulation has completely taken over this function, which generally occurs within a month. Performing prescribed exercises, elevating the arm above the heart several times a day, and gentle muscle pumping (making a fist and releasing) can help reduce the transient edema. The patient needs reassurance that this transient swelling is not lymphedema.

Once lymphedema develops, it tends to be chronic, so preventive strategies are vital. After ALND, the patient is taught hand and arm care to prevent injury or trauma to the affected extremity, thus decreasing the likelihood for development of lymphedema (Chart 58-5). The patient is instructed to follow these guidelines for the rest of her life. She is also instructed to contact her primary provider immediately if she suspects that she has lymphedema, because early intervention provides the best chance for control. If allowed to progress without treatment, the swelling can become more difficult to manage. Treatment may consist of a course of antibiotic agents if an infection is present. A referral to a rehabilitation specialist (e.g., occupational or physical therapist) may be necessary for a compression sleeve or glove, exercises, manual lymph drainage, and a discussion of ways to modify daily activities to avoid worsening lymphedema. Ongoing research is seeking to identifying which lymph nodes drain the arm before surgery so that they can be preserved when possible, helping to prevent the development of lymphedema.

Chart 58-5 PATIENT EDUCATION

Hand and Arm Care After Axillary Lymph Node Dissection

- Avoid blood pressures, injections, and blood draws in affected extremity.
- Use sunscreen (higher than 15 SPF) for extended exposure to sun.
- Apply insect repellent to avoid insect bites.
- Wear gloves for gardening.
- Use cooking mitt for removing objects from oven.
- Avoid cutting cuticles; push them back during manicures.
- Use electric razor for shaving armpit.
- Avoid lifting objects heavier than 5–10 pounds.
- If a trauma or break in the skin occurs, wash the area with soap and water, and apply an over-the-counter antibacterial ointment (Bacitracin or Neosporin). Observe the area and extremity for 24 hours; if redness, swelling, or a fever occurs, call the surgeon or nurse.

Hematoma or Seroma Formation. Hematoma formation (collection of blood inside a cavity) may occur after either mastectomy or breast conservation and usually develops within the first 12 hours after surgery. The nurse assesses for signs and symptoms of hematoma at the surgical site, which may include swelling, tightness, pain, and bruising of the skin. The surgeon should be notified immediately if there is gross swelling or increased bloody output from the drain. Depending on the surgeon's assessment, a compression wrap may be applied to the incision for approximately 12 hours, or the patient may be returned to the operating room so that the incision may be reopened to identify the source of bleeding. Some hematomas are small, and the body absorbs the blood naturally. The patient may take warm showers or apply warm compresses to help increase the absorption. A hematoma usually resolves in 4 to 5 weeks.

A seroma, a collection of serous fluid, may accumulate under the breast incision after mastectomy or breast conservation or in the axilla. Signs and symptoms may include swelling, heaviness, discomfort, and a sloshing of fluid. Seromas may develop temporarily after the drain is removed or if the drain is in place and becomes obstructed. Seromas rarely pose a threat and may be treated by unclogging the drain or manually aspirating the fluid with a needle and syringe. Large, long-standing seromas that have not been aspirated may lead to infection. Small seromas that are not bothersome to the patient usually resolve on their own.

Infection. Although infection is rare, it is a risk after any surgical procedure. This risk may be higher in patients with conditions such as diabetes, immune disorders, and advanced age, as well as in those with poor hygiene. Patients are taught to monitor for signs and symptoms of infection (redness, warmth around incision, tenderness, foul-smelling drainage, temperature greater than 40°C [100.4°F], chills) and to contact the surgeon or nurse for evaluation. Treatment consists of oral or IV antibiotics (for more severe infections) for 1 or 2 weeks. Cultures are taken of any foul-smelling discharge.

Promoting Home and Community-Based Care

Educating Patients About Self-Care. Patients who undergo breast cancer surgery receive a tremendous amount of information both pre- and postoperatively. It is often difficult for the patient to absorb all of the information, partly because of the emotional distress that often accompanies the diagnosis and treatment. Prior to discharge, the nurse must assess the patient's readiness to assume self-care responsibilities and identify any gaps in knowledge. A review of education, with reinforcement, may be required to ensure that the patient and family are prepared to manage the necessary care at home. The nurse reiterates symptoms that the patient should report, such as infection, seroma, hematoma, or arm swelling. All instruction should be reinforced during office visits and by telephone.

Most patients are discharged 1 or 2 days after ALND or mastectomy (possibly later if they have had immediate reconstruction) with surgical drains in place. Initially, the drainage fluid appears bloody, but it gradually changes to a serosanguineous and then a serous fluid over the next several days. The patient is given instructions about drainage management at home (Chart 58-6). If the patient lives alone and drainage management is difficult, a referral for a home care nurse should be made. The drains are usually removed when the output is less than 30 mL in a 24-hour period (approximately 7 to 10 days). The home care nurse also reviews pain management and incision care.

Generally, the patient may shower on the second postoperative day and wash the incision and drain site with soap and water to prevent infection. If immediate reconstruction has been performed, showering may be contraindicated until the drain is removed. A dry dressing may be applied to the incision each day for 7 days. The patient should realize that sensation may be decreased in the operative area because the nerves were disrupted during surgery, and she should be informed that gentle care is needed to avoid injury. After the incision has completely healed (usually after 4 to 6 weeks),

Chart 58-6 HOME CARE CHECKLIST

Surgical Breast Cancer Patient With a Drainage Device

At the completion of home care education, the patient or caregiver will be able to:	PATIENT	CAREGIVER
• Demonstrate how to empty and measure fluid from the drainage device.	✔	✔
• Demonstrate how to milk clots through the tubing of the drainage device.	✔	✔
• State observations that require contacting the physician or nurse (e.g., sudden change in color of drainage, sudden cessation of drainage, signs or symptoms of an infection).	✔	✔
• Care for the drain site as per surgeon's recommendation.	✔	✔
• Identify when the drain is ready for removal (usually when draining <30 mL for a 24-hour period).	✔	✔

Chart 58-7 **PATIENT EDUCATION**

Exercise After Breast Surgery

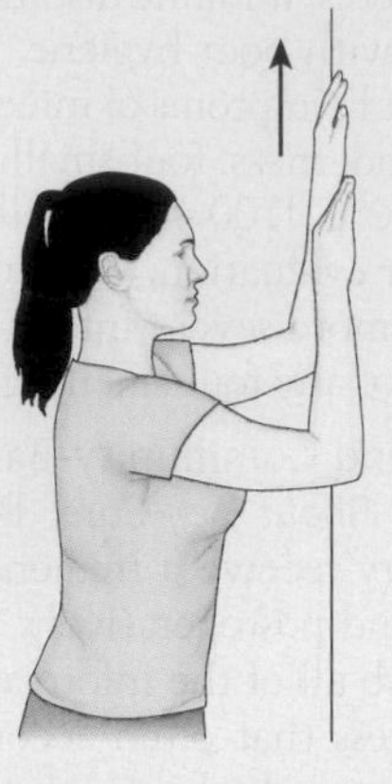

1. *Wall handclimbing.* Stand facing the wall with feet apart and toes as close to the wall as possible. With elbows slightly bent, place the palms of the hand on the wall at shoulder level. By flexing the fingers, work the hands up the wall until arms are fully extended. Then reverse the process, working the hands down to the starting point.

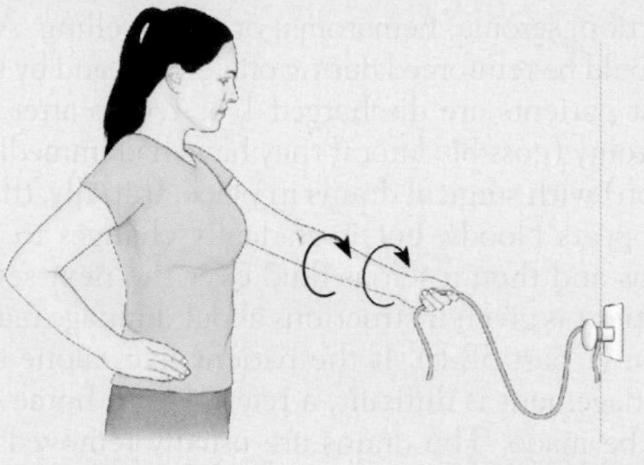

2. *Rope turning.* Tie a light rope to a doorknob. Stand facing the door. Take the free end of the rope in the hand on the side of surgery. Place the other hand on the hip. With the rope-holding arm extended and held away from the body (nearly parallel with the floor), turn the rope, making as wide swings as possible. Begin slowly at first; speed up later.

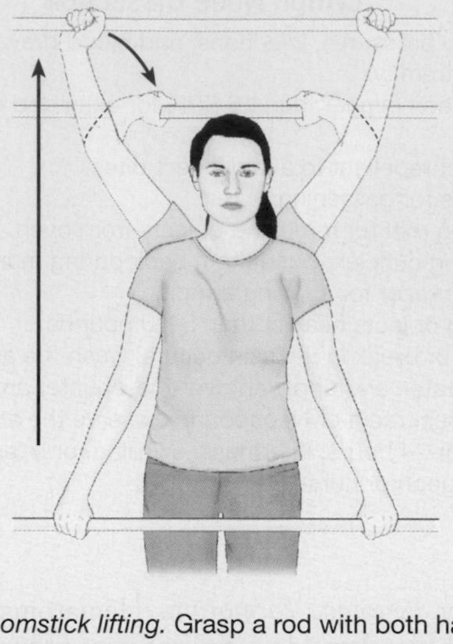

3. *Rod or broomstick lifting.* Grasp a rod with both hands, held about 2 feet apart. Keeping the arms straight, raise the rod over the head. Bend elbows to lower the rod behind the head. Reverse maneuver, raising the rod above the head, then return to the starting position.

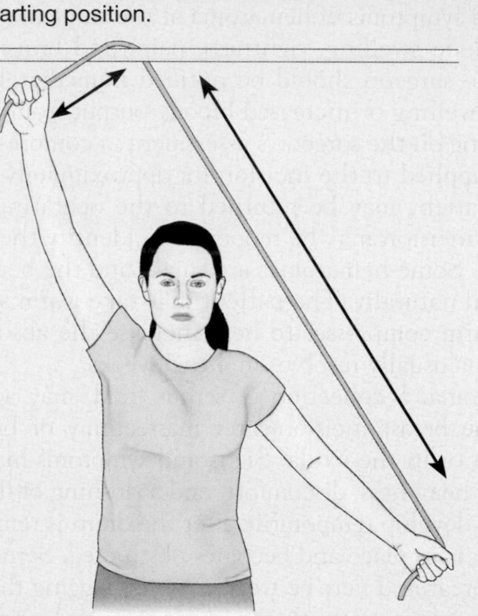

4. *Pulley tugging.* Toss a light rope over a shower curtain rod or doorway curtain rod. Stand as close to the rope as possible. Grasp an end in each hand. Extend the arms straight and away from the body. Pull the left arm up by tugging down with the right arm, then the right arm up and the left down in a seesawing motion.

lotions or creams may be applied to the area to increase skin elasticity. The patient can begin to use deodorant on the affected side, although many women note that they no longer perspire as much as before the surgery.

After ALND, patients are taught arm exercises on the affected side to restore range of motion (Chart 58-7). After SLNB, patients may also benefit from these exercises, although they are less likely to have decreased range of motion than those who have undergone ALND. Range-of-motion exercises are initiated on the second postoperative day; however, instruction often occurs on the first postoperative day. The goals of the exercise regimen are to increase circulation and muscle strength, prevent joint stiffness and contractures, and restore full range of motion. The patient is instructed to perform range-of-motion exercises at home three times a day for 20 minutes at a time until full range of motion is restored (generally 4 to 6 weeks). Most patients find that after the drain is removed, range of motion returns quickly if they have adhered to their exercise program.

If the patient is having any discomfort, taking an analgesic agent 30 minutes before beginning the exercises can be helpful. Taking a warm shower before exercising can also loosen stiff muscles and provide comfort. When exercising, the patient is encouraged to use the muscles in both arms and to maintain proper posture. Specific exercises may need to be prescribed and introduced gradually if the patient has had skin grafts; has a tense, tight surgical incision; or has had immediate reconstruction. Self-care activities, such as brushing the teeth, washing the face, and brushing the hair, are physically and emotionally therapeutic because they aid in restoring arm function and provide a sense of normalcy for the patient.

The patient is instructed about postoperative activity limitation. Generally, heavy lifting (more than 5 to 10 lb) is avoided for about 4 to 6 weeks, although normal household and work-related activities are promoted to maintain muscle tone. Brisk walking, the use of stationary bikes and stepping machines, and stretching exercises may begin as soon as the patient feels comfortable (Graf & Wessely, 2010). Once the drain is removed, the patient may begin to drive if she has full arm range of motion and is no longer taking opioid analgesic agents. General guidelines for activity focus on the gradual introduction of previous activities (e.g., bowling, weight training) once fully healed, although checking with the physician or nurse beforehand is recommended.

Continuing Care. Patients who have difficulty managing their postoperative care at home may benefit from a home health care referral. The home care nurse assesses the patient's incision and surgical drain(s), adequacy of pain management, adherence to the exercise plan, and overall physical and psychological functioning. In addition, the home care nurse reinforces previous education and communicates important physiologic findings and psychosocial issues to the patient's primary provider, nurse, or surgeon.

The frequency of follow-up visits after surgery may vary but generally should occur every 3 to 6 months for the first several years. The patient may alternate visits with the surgeon, medical oncologist, or radiation oncologist, depending on the treatment regimen. The ambulatory care nurse can also be a great source of comfort and security for the patient and family and should encourage them to telephone if they have any questions or concerns. It is common for people to ignore routine health care when a major health issue arises, so women who have been treated for breast cancer should be reminded of the importance of participating in routine health screening.

Evaluation

Expected preoperative patient outcomes may include:

1. Exhibits knowledge about diagnosis and surgical treatment options
   a. Asks relevant questions about diagnosis and available surgical treatments
   b. States rationale for surgery
   c. Describes advantages and disadvantages of treatment options
2. Verbalizes willingness to deal with anxiety and fears related to the diagnosis and the effects of surgery on self-image and sexual functioning
3. Demonstrates ability to cope with diagnosis and treatment
   a. Verbalizes feelings appropriately and recognizes normalcy of mood lability
   b. Proceeds with treatment in timely fashion
   c. Discusses impact of diagnosis and treatment on family and work
4. Makes decisions regarding treatment options in timely fashion

Expected postoperative patient outcomes may include:

1. Reports that pain has decreased and states pain and discomfort management strategies are effective
2. Identifies postoperative sensations and recognizes that they are a normal part of healing
3. Exhibits clean, dry, and intact surgical incisions without signs of inflammation or infection
4. Lists the signs and symptoms of infection to be reported to the nurse or surgeon
5. Verbalizes feelings regarding change in body image
6. Discusses meaning of the diagnosis, surgical treatment, and fears appropriately
7. Participates actively in self-care measures
   a. Performs exercises as prescribed
   b. Participates in self-care measures as prescribed
8. Discusses issues of sexuality and resumption of sexual relations
9. Demonstrates knowledge of postdischarge recommendations and restrictions
   a. Describes follow-up care and activities
   b. Demonstrates appropriate care of incisions and drainage system
   c. Demonstrates arm exercises, and describes exercise regimen and activity limitations during postoperative period
   d. Describes care of affected arm and hand, and lists indications to contact the surgeon or nurse
10. Experiences no complications
    a. Identifies signs and symptoms of reportable complications (e.g., fever, redness, heat, pain, edema)
    b. Explains how to contact appropriate providers in case of complications

Radiation Therapy

Radiation therapy is used to decrease the chance of a local recurrence in the breast by eradicating residual microscopic cancer cells. Breast conservation treatment followed by radiation therapy for stages I and II breast cancer results in a survival rate equal to that of a modified radical mastectomy (NCCN, 2012). If radiation therapy, which is part of breast conservation treatment (Chart 58-8), is contraindicated, a mastectomy would then be indicated.

External-beam radiation (the most common type) typically begins about 6 weeks after breast conservation to allow the surgical site to heal. If systemic chemotherapy is indicated, radiation therapy usually begins after its completion. Before radiation begins, the patient undergoes a planning session called a *simulation,* in which the anatomic areas to be treated are mapped out and then identified with small permanent ink markings. External-beam radiation, which delivers high-energy photons from a linear accelerator, is administered to the entire breast region (whole breast radiation). Each treatment lasts only a few minutes and is generally given 5 days a week for 5 to 6 weeks. After completion of radiation to

Chart 58-8 Contraindications to Breast Conservation Treatment

Note: Breast-conservation treatment includes both surgery and radiation.

Absolute Contraindications

- First or second trimester of pregnancy
- Presence of multicentric disease in the breast
- Prior radiation to the breast or chest region

Relative Contraindications

- History of collagen vascular disease
- Large tumor-to-breast ratio
- Tumor beneath nipple

the entire breast, many patients receive a "boost"—a dose of radiation to the lumpectomy site where the cancer cells were located. The boost consists of the same dose of radiation but is less penetrating and directed to a smaller area. The treatments are not painful.

Because most breast cancer recurrences appear at or near the lumpectomy site, the need for whole breast radiation has been questioned. Partial breast radiation (radiation to the lumpectomy site alone) continues be evaluated at some institutions in carefully selected patients. One approach is **brachytherapy**, in which radiation is delivered by an internal device that is placed close to the tumor within the breast. This technique can lead to an improved quality of life because the treatments are administered over 4 to 5 days instead of 5 to 6 weeks. Another approach is intraoperative radiation therapy, in which a single intense dose of radiation is delivered to the surgical site in the operating room immediately following the lumpectomy (Ivanov, Dickler, Lum, et al., 2011). Many questions remain unanswered, and longer follow-up with larger studies is needed to document the long-term effectiveness and potential side effects of these techniques. In the meantime, whole breast radiation remains the standard treatment of choice.

After mastectomy, postoperative radiation may be indicated for women at high risk for cancer recurrence (i.e., chest wall involvement, four or more positive lymph nodes, tumors larger than 5 cm, positive surgical margins).

Side Effects

Generally, radiation therapy is well tolerated. Acute side effects consist of mild to moderate erythema, breast edema, and fatigue. Occasionally, skin breakdown may occur in the inframammary fold or near the axilla toward the end of treatment. Fatigue can be depressing, as can the frequent trips to the radiation oncology unit for treatment. The patient needs to be reassured that the fatigue is normal and not a sign of recurrence. Side effects usually resolve within a few weeks to a few months after treatment is completed. Rare long-term effects of radiation therapy include pneumonitis, rib fracture, and breast fibrosis or necrosis (Eggert, 2010).

Nursing Management

Nurses play a significant role in supporting patients throughout their treatment with radiation therapy. (See Chapter 15 for discussion of radiation therapy.)

Self-care instructions for patients receiving radiation are provided to assist in the maintenance of skin integrity during the treatments and for several weeks after completion. They pertain only to the area being treated and not to the rest of the body. Instructions include:

- Use mild soap with minimal rubbing.
- Avoid perfumed soaps or deodorants.
- Use hydrophilic lotions (Lubriderm, Eucerin, Aquaphor) for dryness.
- Use a nondrying, antipruritic soap (Aveeno) if pruritus occurs.
- Avoid tight clothes, underwire bras, excessive temperatures, and ultraviolet light.

Follow-up care includes educating the patient to minimize sun exposure to the treated area (i.e., using sunblock with sun protection factor [SPF] of 15 or higher) and reassuring the patient that short-term minor twinges and pain in the breast are normal after radiation treatment.

Systemic Treatments

Chemotherapy

Adjuvant chemotherapy involves the use of anticancer agents in addition to other treatments (i.e., surgery, radiation) to delay or prevent a recurrence of breast cancer. It is recommended for patients who have positive lymph nodes or who have invasive tumors greater than 1 cm in size, regardless of nodal status. It is considered in patients with tumors that are 0.6 to 1 cm, are moderately to poorly differentiated, or have unfavorable features (NCCN, 2012). Table 58-6 outlines general indications for adjuvant chemotherapy. A survival benefit has been shown in both pre- and postmenopausal women who have received chemotherapy, although data are limited in women older than 70 years. Chemotherapy is most commonly initiated after breast surgery and before radiation.

TABLE 58-6 General Indications for Adjuvant Chemotherapy for Breast Cancer

Nodal Status, Tumor Size	Adjuvant Chemotherapy
Node negative, ≤0.5 cm	None
Node negative 0.6–1 cm (well differentiated)	None
Node negative, 0.6–1 cm (moderately or poorly differentiated and/or unfavorable features)	Consider chemotherapy
Node negative, >1 cm	Chemotherapy
Node positive, any tumor size	Chemotherapy

- In addition to chemotherapy, patients with HER-2/neu-positive tumors will receive trastuzumab if they have node-positive disease; or node-negative disease with a tumor >1 cm. Trastuzumab is a monoclonal antibody that targets and inactivates the HER-2/neu protein. HER-2/neu is overproduced in 25%–30% of tumors and is associated with rapid growth and poor prognosis.
- Following chemotherapy, patients with hormone receptor positive (ER+/PR+) tumors will receive hormonal therapy (tamoxifen or aromatase inhibitor) if they have either node-positive disease; node-negative disease with a tumor >1 cm; or node-negative disease with a tumor 0.6–1 cm and moderately or poorly differentiated and/or unfavorable features.

Note: These are only general guidelines. Recommendations may vary depending on factors such as prognostic variables, patient age, and comorbid conditions.
Adapted from National Comprehensive Cancer Network. (2012). *NCCN clinical practice guidelines in oncology: Breast cancer.* Available at: www.nccn.org.

Chemotherapy regimens for breast cancer combine several agents (polychemotherapy), generally administered over a period of 3 to 6 months. Decisions regarding the optimal regimen are based on a variety of factors, including tumor characteristics (i.e., tumor size, lymph node status, hormone receptor status, HER-2/neu status) and the patient's age, physical status, and existing comorbid conditions. A regimen that includes cyclophosphamide (Cytoxan), methotrexate (Trexall), and fluorouracil (Fluoroplex) (collectively referred to as CMF) has been the most widely used adjuvant therapy. It is usually well tolerated and may be considered for patients with a low risk of recurrence. CMF also may be considered for use in patients who have a high risk of cardiac toxicity or who have other limiting comorbidities. Anthracycline-based regimens (e.g., doxorubicin [Adriamycin], epirubicin [Ellence]) have shown longer survival in patients. However, the benefit relative to CMF is modest and is accompanied by increased toxicity (Eggert, 2010). Selection of patients most likely to benefit from anthracycline therapy would allow better use of current cytotoxic agents and reduce the risk of patients receiving toxicity with little or no effect. Identifying biomarkers that can accurately predict benefit from anthracyclines will also highlight key resistance/susceptibility pathways that can then be exploited clinically to further increase efficacy (Munro, Cameron, & Bartlett, 2010). Cyclophosphamide, doxorubicin, and fluorouracil (CAF) and doxorubicin and cyclophosphamide (AC) are examples of combination regimens often administered to higher-risk patients.

The taxanes (paclitaxel [Taxol], docetaxel [Taxotere]) are generally incorporated into treatment regimens for patients with larger, node-negative cancers and for those with positive axillary lymph nodes. The addition of four cycles of paclitaxel after a standard course of AC (regimen known as ACT) has been found to increase the disease-free period and improve overall survival in patients with operable breast cancer and positive lymph nodes (DeLaurentis, Cancello, D'Agostino, et al., 2008).

Much attention has been focused on **dose-dense chemotherapy**, which is the administration of chemotherapeutic agents at standard doses with shorter time intervals between each cycle of treatment. A systematic review and meta-analysis of existing data from randomized controlled trials that compared dose-dense chemotherapy with a standard chemotherapy schedule in women with nonmetastatic breast cancer demonstrated that dose-dense chemotherapy results in better overall and disease-free survival, particularly in women with hormone receptor–negative breast cancer. However, additional data from randomized controlled trials are needed before dose-dense chemotherapy can be considered the standard of care (Bonilla, Ben-Aharon, Vidal, et al., 2010).

Side Effects

Many of the side effects of adjuvant chemotherapy can be managed well, allowing patients to maintain their daily routines and work schedules. In large part, this is the a result of the meticulous educational and psychological preparation provided to patients and their families by oncology nurses, oncologists, social workers, and other members of the health care team. In addition, strides have been made in the effectiveness of antiemetic agents used to alleviate nausea and vomiting and the use of hematopoietic growth factors to treat neutropenia and anemia.

Common physical side effects of chemotherapy for breast cancer may include nausea, vomiting, bone marrow suppression, taste changes, alopecia (hair loss), mucositis, neuropathy, skin changes, and fatigue. A weight gain of more than 10 pounds occurs in about half of all patients; the cause is unknown. Premenopausal women may also experience temporary or permanent amenorrhea.

Specific side effects vary with the type of chemotherapeutic agent used. In general, CMF and the taxanes are better tolerated than the anthracyclines. However, the taxanes can cause peripheral neuropathy, arthralgias, and myalgias, particularly at high doses. During taxane administration, hypersensitivity reactions may occur; therefore, the patient must be premedicated. Alopecia is also common. The side effects of the anthracyclines may be severe and include cardiotoxicity in addition to nausea and vomiting, bone marrow suppression, and alopecia. Their vesicant properties can lead to tissue necrosis if infiltration of the medication infusion occurs.

Nursing Management

Nurses play an important role in helping patients manage the physical and psychosocial sequelae of chemotherapy. (Chapter 15 provides an in-depth discussion of side effect management.) Instructing the patient about the use of antiemetic agents and reviewing the optimal dosage schedule can help minimize nausea and vomiting. The different classes of antiemetic agents include serotonin (5-HT-3) receptor antagonists (palonosetron [Aloxi], granisetron [Kytril], ondansetron [Zofran]); neurokinin-1 receptor antagonists (aprepitant [Emend]); dopamine receptor antagonists (prochlorperazine [Compazine], metoclopramide [Reglan]); benzodiazepines (lorazepam [Ativan]); and corticosteroids (dexamethasone [Decadron]). Measures to ease the symptoms of mucositis may include rinsing with normal saline or sodium bicarbonate solution, avoiding hot and spicy foods, and using a soft toothbrush.

Some patients may require hematopoietic growth factors to minimize the effects of chemotherapy-induced neutropenia and anemia. Granulocyte colony-stimulating factors boost the white blood cell count, helping to reduce the incidence of neutropenic fever and infection. The short-acting form, filgrastim (Neupogen), is injected subcutaneously or IV for 7 to 10 days after chemotherapy administration. The long-acting form, pegfilgrastim (Neulasta), is injected once, no earlier than 24 hours after chemotherapy (Karch, 2012). Erythropoietin growth factor increases the production of red blood cells, thus decreasing the symptoms of anemia. The short-acting form, epoetin alfa (Epogen) is usually administered weekly. The long-acting form, darbepoetin alfa (Aranesp), can be administered every 2 to 3 weeks. The nurse instructs the patient and family on proper injection technique of hematopoietic growth factors and about symptoms that require follow-up with a physician (Chart 58-9).

To prevent some of the emotional trauma associated with alopecia, it often helps to have a patient obtain a wig before hair loss begins to occur. The nurse may provide a list of

Chart 58-9 HOME CARE CHECKLIST

Self-Administration of Hematopoietic Growth Factors

At the completion of home care education, the patient or caregiver will be able to:	PATIENT	CAREGIVER
• State the purpose for the injections.	✔	✔
• Identify the equipment necessary for self-injection.	✔	✔
• Identify appropriate body sites for self-injection.	✔	✔
• Demonstrate how to draw up the solution in a syringe if indicated. (*Note:* Darbepoetin and pegfilgrastim come in prefilled syringes.)	✔	✔
• Demonstrate how to give an injection properly.	✔	✔
• State possible side effects of medication.	✔	✔
• Demonstrate correct disposal of sharps.	✔	✔
• Describe proper storage of supplies.	✔	✔
• State reasons for contacting the primary provider or nurse (e.g., excessive pain, fever).	✔	✔

wig suppliers in the patient's geographic region. Familiarity with creative ways to use scarves and turbans may also help minimize the patient's distress. The patient needs reassurance that new hair will grow back when treatment is completed, although the color and texture may be different. The ACS offers the Look Good Feel Better program, which provides useful tips for applying cosmetics during the period a patient is receiving chemotherapy (see the Resources section at the end of the chapter).

Chemotherapy may negatively affect the patient's self-esteem, sexuality, and sense of well-being. This, combined with the stress of a potentially life-threatening disease, can be overwhelming. Providing support and promoting open communication are important aspects of nursing care. Referring the patient to the dietitian, social worker, psychiatrist, or spiritual advisor can provide additional support. Numerous community support and advocacy groups are available for patients and their families. Complementary therapies, such as guided imagery, meditation, and relaxation exercises, can also be used in conjunction with conventional treatments.

Hormonal Therapy

The use of **adjuvant hormonal therapy**, with or without the addition of chemotherapy, is considered in women who have hormone receptor–positive tumors. Its use can be determined by the results of an **estrogen and progesterone receptor assay** (a test to determine whether the breast tumor is nourished by hormones). About two thirds of breast cancers depend on estrogen for growth and express a nuclear receptor that binds to the estrogen; thus, they are estrogen receptor positive (ER+). Similarly, tumors that express the progesterone receptor are progesterone receptor positive (PR+). Hormonal therapy involves the use of synthetic hormones or other medications that compete with estrogen by binding to the receptor sites (SERMs), or by blocking estrogen production by the adrenal glands (**aromatase inhibitors**). Generally, tumors that are ER+/PR+ have the greatest likelihood of responding to hormonal therapy and have a more favorable prognosis than those that are ER–/PR–. Both pre- and perimenopausal women are more likely to have non–hormone-dependent lesions, whereas postmenopausal women are more likely to have hormone-dependent lesions.

Traditionally, the SERM tamoxifen has been the main hormonal agent used in treatment of pre- and postmenopausal breast cancer and remains the mainstay in premenopausal women. As a SERM, tamoxifen has estrogen antagonistic (estrogen-blocking) and agonistic (estrogenlike) effects on certain tissues. Its antagonistic effects in the breast prevent estrogen from binding to the receptor sites, thus preventing tumor growth. Tamoxifen has positive agonistic effects on blood lipid profiles and bone mineral density in postmenopausal women. It also has agonistic effects on endometrial tissue and blood coagulation processes, leading to an increased incidence of endometrial cancer and thromboembolic events (e.g., deep vein thrombosis, superficial phlebitis, pulmonary embolism). Nevertheless, the benefits in most women with breast cancer outweigh the risks.

The aromatase inhibitors anastrozole (Arimidex), letrozole (Femara), and exemestane (Aromasin) are important components in the hormonal management of postmenopausal women. Most of the circulating estrogens in postmenopausal women are derived from the conversion of the adrenal androgen androstenedione to estrone and the conversion of testosterone to estradiol. Aromatase inhibitors work by blocking the enzyme aromatase from performing the conversion, thereby decreasing the level of circulating estrogen in peripheral tissues. Clinical trials have demonstrated that the aromatase inhibitors are superior to tamoxifen in terms of overall response rate and clinical benefit and that inhibitors appear to be effective and feasible compared with tamoxifen as first-line hormonal therapy in postmenopausal women with advanced breast cancer (Xu, Liu, & Li, 2011). These data ensure that aromatase inhibitors will play an increasingly central role in the long-term management of breast cancer. Trials are ongoing to determine the optimal treatment regimen and the timing of the treatment. Table 58-7 outlines the adverse effects of adjuvant hormonal therapy. Chart 58-10 outlines appropriate patient education to manage the adverse effects.

Targeted Therapy

An exciting area of research in the systemic treatment of breast cancer involves the use of targeted therapies. Trastuzumab (Herceptin) is a monoclonal antibody that

TABLE 58-7 Adverse Reactions Associated With Adjuvant Hormonal Therapy Used to Treat Breast Cancer

Therapeutic Agent	Adverse Reactions/Side Effects
Selective Estrogen Receptor Modulator	
tamoxifen (Soltamox)	Hot flashes, vaginal dryness/discharge/bleeding, irregular menses, nausea, mood disturbances, rashes; increased risk for endometrial cancer; increased risk for thromboembolic events (deep vein thrombosis, pulmonary embolism, superficial phlebitis)
Aromatase Inhibitors	
anastrozole (Arimidex) letrozole (Femara) exemestane (Aromasin)	Musculoskeletal symptoms (arthritis, arthralgia, myalgia), increased risk of osteoporosis/fractures, nausea/vomiting, hot flashes, fatigue, mood disturbances, rashes

Adapted from Karch, A. M. (2012). *2012 Lippincott's nursing drug guide*. Philadelphia: Lippincott Williams & Wilkins.

binds specifically to the HER-2/neu protein. This protein, which regulates cell growth, is present in small amounts on the surface of normal breast cells and in most breast cancers. Approximately 25% to 30% of tumors overexpress (overproduce) the HER-2/neu protein and are associated with rapid growth and poor prognosis. Trastuzumab targets and inactivates the HER-2/neu protein, thus slowing tumor growth.

Unlike chemotherapy, trastuzumab spares the normal cells and has limited adverse reactions, which may include fever, chills, nausea, vomiting, diarrhea, and headache. However, when trastuzumab is administered to patients who have previously been treated with an anthracycline, the risk of cardiac toxicity is increased. The medication has been shown to improve survival rates in women with HER-2/neu–positive metastatic breast cancer and is now regarded as standard therapy. It may be administered as a single agent or in combination with chemotherapy. More recently, trastuzumab has been shown to be effective in treating early-stage breast cancer that is HER-2/neu positive. Analysis of disease-free survival after treatment with trastuzumab for 1 year following adjuvant chemotherapy in patients with HER2-positive early breast cancer showed a significant benefit in favor of patients in the 1-year trastuzumab group (4-year disease-free survival 78.6%) compared with the observation group (4-year disease-free survival 72.2%) (Gianni, Dafni, Gelber, et al., 2011).

Treatment of Recurrent and Metastatic Breast Cancer

Despite the advances made in the treatment of breast cancer, it may recur locally (on the chest wall or in the conserved breast), regionally (in the remaining lymph nodes), or systemically (in distant organs). In metastatic disease, the bone, usually the hips, spine, ribs, skull, or pelvis, is the most common site of spread. Other sites of metastasis include the lungs, liver, pleura, and brain.

The overall prognosis and optimal treatment are determined by a variety of factors such as the site and extent of recurrence, the time to recurrence from the original diagnosis, history of prior treatments, the patient's performance status, and any existing comorbid conditions. Patients with bone metastases generally have a longer overall survival compared with metastases in visceral organs.

Local recurrence in the absence of systemic disease is treated aggressively with surgery, radiation, and hormonal therapy. Chemotherapy may also be used for tumors that are not hormonally sensitive. Local recurrence may be an indicator that systemic disease will develop in the

Chart 58-10 PATIENT EDUCATION

Managing Side Effects of Adjuvant Hormonal Therapy in Breast Cancer

Hot Flashes

- Wear breathable, layered clothing.
- Avoid caffeine and spicy foods.
- Perform breathing exercises (paced respirations).
- Consider medications (vitamin E, antidepressants) or acupuncture.

Vaginal Dryness

- Use vaginal moisturizers for everyday dryness (e.g., Replens, vitamin E suppository).
- Apply vaginal lubrication during intercourse (e.g., Astroglide, K-Y Jelly).

Nausea and Vomiting

- Consume a bland diet.
- Try to take medication in the evening.

Musculoskeletal Symptoms

- Take nonsteroidal analgesic agents as recommended.
- Take warm baths.

Risk of Endometrial Cancer

- Report any irregular bleeding to a gynecologist for evaluation.

Risk for Thromboembolic Events

- Report any redness, swelling, or tenderness in the lower extremities, or any unexplained shortness of breath.

Risk for Osteoporosis or Fractures

- Undergo a baseline bone density scan.
- Perform regular weight-bearing exercises.
- Take calcium supplements with vitamin D.
- Take bisphosphonates (e.g., alendronate) or calcitonin as prescribed.

future, particularly if it occurs within 2 years of the original diagnosis.

Metastatic breast cancer involves control of the disease rather than cure. Treatment includes hormonal therapy, chemotherapy, and targeted therapy. Surgery or radiation may be indicated in select situations. Premenopausal women who have hormonally dependent tumors may eliminate the production of estrogen by the ovaries through oophorectomy (removal of the ovaries) or suppression of estrogen production by medications such as leuprolide (Lupron) or goserelin (Zoladex).

Patients with advanced breast cancer are monitored closely for signs of disease progression. Baseline studies are obtained at the time of recurrence. These may include complete blood count; comprehensive metabolic panel; tumor markers (i.e., carcinoembryonic antigen, cancer antigen 15–3); bone scan; CT of the chest, abdomen, and pelvis; and MRI of symptomatic areas. Additional x-rays may be performed to evaluate areas of pain or abnormal areas seen on bone scan (e.g., long bones, pelvis). These studies are repeated at regular intervals to assess for effectiveness of treatment and to monitor progression of disease.

Nursing Management

Nurses play an important role in not only educating patients and managing their symptoms but also in providing emotional support. Many patients find that recurrence of the disease is more distressing than the initial cancer diagnosis. They not only have to contend with another round of treatments but are also faced with a greater uncertainty about their future and long-term survival. The nurse can help the patient identify coping strategies and set priorities to optimize quality of life. Family members and significant others should be included in the treatment plan and follow-up care. Referrals to support groups, as well as psychiatry or psychiatric clinical nurse specialist, social work, and complementary medicine programs (e.g., guided imagery, meditation, yoga), should be made as indicated.

Nurses also play important roles in providing palliative care, if indicated. The highest priorities should include alleviating pain and providing comfort measures. A frank discussion with the patient and family regarding their preferences for end-of-life care should occur before the need arises to ensure a smooth transition without disruption of care. Referrals to hospice and home health care should be initiated as necessary. (See Chapter 15 for more information on the general care of the patient with advanced cancer and Chapter 16 for discussion of end-of-life care.)

Reconstructive Procedures After Mastectomy

Breast reconstruction can provide a significant psychological benefit for women who are already struggling with the emotional distress of losing a breast. A consultation with a plastic surgeon can help the patient understand procedures for which she is a candidate and the pros and cons of each. Factors to consider include body size and shape, comorbid conditions (e.g., hypertension, diabetes, obesity), personal habits such as smoking, and patient preference. The patient must be informed that although breast reconstruction can provide a good cosmetic result, it will never precisely duplicate the natural breast. Realistic preparation can help the patient avoid unrealistic expectations. Once reconstruction is complete, the opposite breast may require augmentation, reduction, or mastopexy to achieve symmetry on both sides. The patient must also be informed that breast reconstruction neither will interfere with breast cancer treatments nor affect the risk of cancer recurrence. Reconstruction is considered an integral component in the surgical treatment of breast cancer and is usually covered by insurance companies.

Many women elect immediate reconstruction at the time of the mastectomy operation. This can be beneficial in that it saves the woman from undergoing general anesthesia a second time and saves the cost and stress of future hospitalizations. However, it does increase the length of the surgical procedure. Delayed reconstruction is preferable in women who are having a difficult time deciding on the type of reconstruction that they desire. It may also be preferable in patients with advanced disease such as inflammatory breast cancer, where the breast cancer treatments should begin without delay. Any delays in healing after immediate reconstruction may interfere with the initiation of treatment.

Tissue Expander Followed by Permanent Implant

Breast reconstruction using a **tissue expander followed by a permanent implant** is the simplest and most common method used today (Fig. 58-6). To accommodate an implant, the skin remaining after a mastectomy and the underlying muscle must gradually be stretched by a process called *tissue expansion*. The surgeon places a tissue expander (a balloonlike device) through the mastectomy incision underneath the pectoralis muscle. A small amount of saline is injected through a metal port intraoperatively to partially inflate the expander. Then, for about 6 to 8 weeks, at weekly intervals, the patient receives additional saline injections through the port until the expander is fully inflated. It remains fully expanded for about 6 weeks to allow the skin to loosen. The expander is then exchanged for a permanent implant. This is usually performed as an outpatient surgical procedure.

Advantages of this expansion procedure are a shorter operating time and a shorter recuperation period than for autologous reconstruction (see the Tissue Transfer Procedures section). A disadvantage is a tendency for the implant to feel firm and round, with little natural ptosis (sag). Women with a small to medium opposite breast with little ptosis are good candidates for this procedure. Women who have had radiation or who have connective tissue disease are not good candidates because of the decreased elasticity of the skin.

Quality and Safety Nursing Alert

The patient must be cautioned not to have an MRI while the tissue expander is in place because the port contains metal. This is not an issue once the permanent implant is in place because it does not contain any metal.

The patient should be informed that for the rest of her life she should not engage in any exercises that will develop the pectoralis muscle, because this can result in distortion of the reconstructed breast.

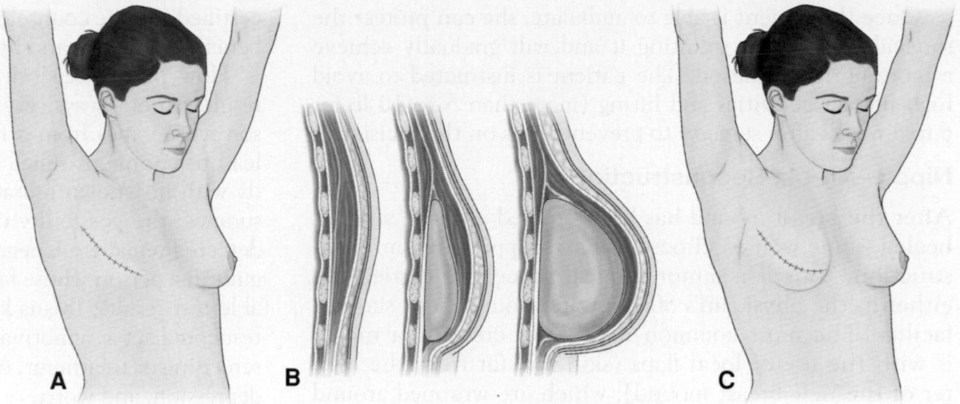
A
B
C

FIGURE 58-6 • Breast reconstruction with tissue expander. **A.** Mastectomy incision line prior to tissue expansion. **B.** The expander is placed under the pectoralis muscle and is gradually filled with saline solution through a port to stretch the skin enough to accept a permanent implant. **C.** The breast mound is restored. Although permanent, scars will fade with time. The nipple and areola are reconstructed later. Adapted from American Society of Plastic Surgeons.

Tissue Transfer Procedures

Autologous reconstruction is the use of the patient's own tissue to create a breast mound. A flap of skin, fat, and muscle with its attached blood supply is rotated to the mastectomy site to create a mound that simulates the breast. Donor sites may include the **transverse rectus abdominal myocutaneous (TRAM) flap** (abdominal muscle) (Fig. 58-7), gluteal flap (buttock muscle), or the latissimus dorsi flap (back muscle) (Fig. 58-8). The results more closely resemble a real breast because the skin and fat from the donor sites are similar in consistency to a natural breast. These procedures avoid the use of synthetic material. However, they involve longer recuperation than a tissue expander procedure. The risk of potential complications (e.g., infection, bleeding, flap necrosis) is also greater. Therefore, patients must be in relatively good health, and those with medical conditions (e.g., atherosclerosis, pulmonary disease, heart failure) that affect circulation or compromise oxygen delivery are not good candidates. Other poor candidates include those with poorly controlled diabetes or morbid obesity and heavy smokers.

The TRAM flap is the most commonly performed tissue transfer procedure. A free TRAM procedure may also be performed; in this case, the skin, fat, muscle, and blood supply are completely detached from the body and then transplanted to the mastectomy site using microvascular surgery (the use of a microscope to reconnect the vessels). Postoperatively, patients who have undergone TRAM procedures often face a lengthy recovery (often 6 to 8 weeks) and have incisions both at the mastectomy site and at the donor site in the abdomen.

> ***Quality and Safety Nursing Alert***
>
> **The nurse must assess the newly constructed breast site for changes in color, circulation, and temperature because flap loss is a potential complication. Mottling or an obvious decrease in skin temperature is reported to the surgeon immediately.**

Breathing and leg exercises are essential because the patient is more limited in her activity and is at greater risk for respiratory complications and deep vein thrombosis. Measures to help the patient reduce tension on the abdominal incision during the first postoperative week include elevating the head of the bed 45 degrees and flexing the patient's knees.

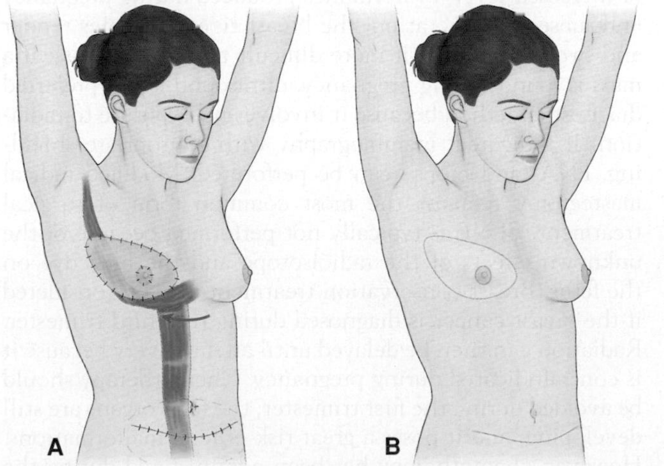
A
B

FIGURE 58-7 • Breast reconstruction: transverse rectus abdominal myocutaneous flap. **A.** A breast mound is created by tunneling abdominal skin, fat, and muscle to the mastectomy site. **B.** Final location of scars. Adapted from American Society of Plastic Surgeons.

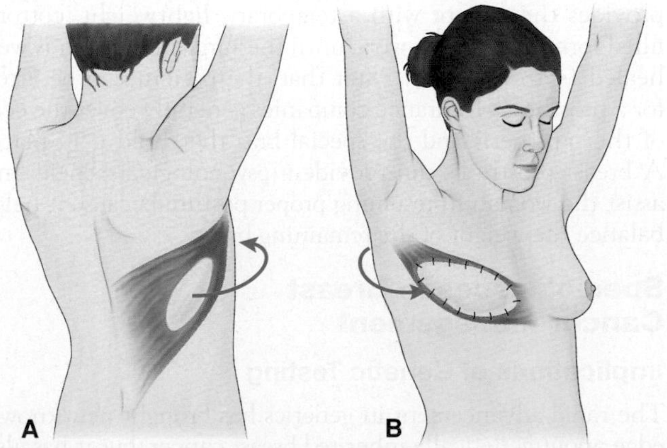
A
B

FIGURE 58-8 • Breast reconstruction: latissimus dorsi flap. **A.** The latissimus muscle with an ellipse of skin is rotated from the back to the mastectomy site. **B.** Because the flap is usually not bulky enough to provide an adequate breast mound, an implant is often also required. Adapted from American Society of Plastic Surgeons.

Once the patient is able to ambulate, she can protect the surgical incision by splinting it and will gradually achieve a more upright position. The patient is instructed to avoid high-impact activities and lifting (more than 5 to 10 lb for 6 to 8 weeks after surgery) to prevent stress on the incision.

Nipple–Areola Reconstruction

After the breast mound has been created and the site has healed, some women choose to have nipple–areola reconstruction. This is a minor surgical procedure carried out either in the physician's office or at an outpatient surgical facility. The most common method of creating a nipple is with the use of local flaps (skin and fat from the center of the new breast mound), which are wrapped around each other to create a projecting nipple. The areola is created using a skin graft. The most common donor site is the upper inner thigh, because this skin has darker pigmentation than the skin on the reconstructed breast. After the nipple graft has healed, micropigmentation (tattooing) can be performed to achieve a more natural color. The surgeon can usually match the reconstructed nipple–areola complex with that of the contralateral breast for an acceptable cosmetic result.

Prosthetics

Not all patients desire or are candidates for reconstructive surgery. A breast prosthesis—an external form that simulates the breast—is another option. Prostheses are available in different shapes, sizes, colors, and materials, although they are most often made of silicone. They can be placed inside a pocket in a bra or can adhere directly to the chest wall. The nurse can provide the patient with the names of shops where she can be fitted for a prosthesis, or the patient can call the Reach to Recovery program of the ACS for appropriate referrals. The patient should be encouraged to find a shop with a comfortable, supportive atmosphere that employs a certified prosthetics consultant. Generally, medical supply shops are not recommended because often they do not have the appropriate resources to ensure the proper fitting of a prosthesis.

Prior to discharge from the hospital, the nurse usually provides the patient with a temporary, lightweight, cotton-filled form that can be worn until the surgical incision is well healed (4 to 6 weeks). After that, the patient can be fitted for a prosthesis. Insurance companies generally cover the cost of the prosthesis and the special bras that hold it in place. A breast prosthesis can provide a psychological benefit and assist the woman in resuming proper posture because it helps balance the weight of the remaining breast.

Special Issues in Breast Cancer Management

Implications of Genetic Testing

The rapid advancement in genetics has brought new knowledge about genetically inherited breast cancer, but it has also raised potential ethical and psychosocial issues. Although the actual testing for the *BRCA1* and *BRCA2* genes involves a simple blood test, it is these issues that must be addressed. Before undergoing genetic testing, a person should meet either with a clinician who has expertise in this area or with a certified genetic counselor to discuss risk factors as well as the benefits, sequelae, and limitations of testing.

How people react when they receive their actual test results is not always easy to predict. A negative test in a person who comes from a family with a known mutation may lead to enormous relief. However, a negative test in a family with no known mutation may be a source of undue reassurance; the possibility of existing genes that cannot yet be detected remains. A negative test may also lead to feelings of guilt in a person whose family members did not receive favorable test results; this is known as survivor's guilt. A positive test could act as a motivator in a person to pursue appropriate screening or treatment, or it could cause tremendous anxiety, depression, and worry.

In addition, test results may be ambiguous, leading to feelings of confusion and uncertainty. People must be informed that not all gene carriers develop breast cancer (incomplete penetrance) and that not all noncarriers are protected.

Other issues include those of cost: Who should pay for genetic testing and the services that relate to it? Difficult ethical questions arise concerning whether the person who is tested should disclose the test results. Is it ethical to withhold results from family members who may be at risk? If they are told, what effect will it have on them? People considering testing must be informed that there is no guarantee that test results will remain confidential. Once confidentiality is breached, it could unleash potential discrimination in employment and insurability. There are federal and state laws to protect the individual if this should happen. People must be well informed of all issues and potential implications prior to undergoing genetic testing (see Chapter 8). Nurses play a role in educating and counseling patients and their family members about the implications of genetic testing. Nurses provide support and clarification and make referrals to appropriate specialists when indicated.

Pregnancy and Breast Cancer

Breast cancer during pregnancy is defined as breast cancer diagnosed during gestation or within 1 year of childbirth and occurs in 1 in 3,000 women (NCCN, 2012). Because of increased levels of hormones produced during pregnancy and subsequent lactation, the breast tissue becomes tender and swollen, making it more difficult to detect a mass. If a mass is found during pregnancy, ultrasound is the preferred diagnostic method because it involves no exposure to radiation. If indicated, mammography with appropriate shielding, FNA, and biopsy can be performed. Modified radical mastectomy remains the most common form of surgical treatment. SLNB is typically not performed because of the unknown effects of the radioisotope and the blue dye on the fetus. Breast conservation treatment may be considered if the breast cancer is diagnosed during the third trimester. Radiation can then be delayed until after delivery because it is contraindicated during pregnancy. Chemotherapy should be avoided during the first trimester; the fetal organs are still developing, and it poses a great risk of fetal malformations. However, chemotherapy has been administered during the second and third trimesters with few reported abnormalities. Long-term effects on the fetus are still being studied. If a woman is close to term, a cesarean section may be performed as soon as maturation of the fetus allows and then treatment is initiated. If aggressive disease is detected early in pregnancy and chemotherapy is advised, termination of the pregnancy may be considered. If a mass is found while a woman is breast-feeding, she is urged to stop to allow the breast to involute (return to its baseline state) before any type of surgery is performed.

Fertility issues and the future desire for children are major concerns of young breast cancer survivors. Most cancer therapies have a substantial morbidity on reproductive function, not only because they increase the risk of early menopause but also because they are associated with a decreased ovarian reserve and a loss of fertility. It is estimated that physiologic age of the ovaries in a cancer survivor may be 10 years older than the actual chronologic age.

Chemotherapy causes a progressive dose-related depletion of ovarian follicles and granulosa cells that translates into oligomenorrhea and subsequent premature ovarian failure, ultimately leading to what is known as chemotherapy-induced amenorrhea (CIA). Regardless of the beneficial effects that hormonal changes can have as part of the adjuvant endocrine strategy, CIA is an adverse event to take into account for the selection of the best adjuvant treatment. Physicians should inform patients with cancer about their reproductive future and options to preserve fertility before treatment (Haba-Rodriquez & Calderay, 2010). Fertile Hope, a national nonprofit organization, can also provide updated information on reproduction (see the Resources section at the end of the chapter).

Quality of Life and Survivorship

With increased early detection and improved treatment modalities, women with breast cancer have become the largest group of cancer survivors. However, the treatment or simply the diagnosis of breast cancer may have long-term effects that negatively affect the patient and her family. The patient should be prepared early on for the potential long-term effects of the disease so that she has realistic expectations and can make informed decisions.

Breast cancer survivors may experience a variety of issues as a result of their diagnosis and treatment (Shockney, 2011). Estrogen withdrawal from chemotherapy-induced menopause and hormonal treatments can lead to a variety of symptoms, including hot flashes, vaginal dryness, urinary tract infections, weight gain, decreased sex drive, and increased risk of osteoporosis. HT to alleviate symptoms is contraindicated in women with breast cancer. Certain chemotherapeutic agents can cause long-term cardiac effects and neuropathy. In addition, patients may experience impaired cognitive functioning, such as difficulty concentrating (often referred to as "chemo brain"). Rare long-term effects of radiation can include pneumonitis and rib fractures. Long-term sequelae after breast surgery may include lymphedema (mainly after ALND), pain, and sensory disturbances. Once lymphedema develops, it tends to be a chronic problem, so prevention strategies (discussed earlier) are vital.

Long-term psychosocial sequelae may include fears of recurrence, mood changes (e.g., worry, sadness, anger, frustration), an increased sense of vulnerability, uncertainty, feelings of loss (e.g., fertility), concerns about body image, self-concept, and sexuality; emotional distress related to role adjustments and family response; and concerns about finances and employment. Depression and anxiety have been documented in 20% to 30% of women with breast cancer. Interventions should be targeted to meet informational needs, manage uncertainty, control symptoms, address cultural differences, and enhance social and emotional support (Knobf, 2011).

Gerontologic Considerations

Until recently, breast reconstruction has not been offered to older women undergoing mastectomies as a result of bias that older women may not want reconstructions or that their comorbidities preclude reconstructive surgery. Breast cancer reconstruction in older women is a feasible option that should be offered to patients. Most women tolerate the procedure well and have good cosmetic outcomes. Patients can be offered both implant-based reconstruction and autologous tissue transfer with minimal complications as long as appropriate preoperative selection criteria are used. The safety of reconstruction, together with increase in life expectancy and healthier lifestyles, makes breast reconstruction after mastectomy desirable at any age (Howard-McNatt, Forsberg, Levine, et al., 2011).

A thorough assessment must be performed before any treatment is initiated, and careful monitoring must occur throughout the course of treatment to avoid complications. The physical and psychosocial assessment of the older woman should include general health, currently existing comorbidities, performance status, cognitive status, current medications, available resources, and support systems.

Breast Health of Women With Disabilities

Disparities in obtaining a mammogram at recommended screening intervals persist for women with disabilities. Prevalence of self-reported use of mammography is lower for women with a disability (72.2% for women 40 years and older and 78.1% for women 50 to 74 years) than women without a disability (77.8% and 82.6%, respectively). As many as 30% of women in the United States have a disability; therefore, efforts to reduce disparities in breast cancer screening might be more effective if they target all segments of the population and explicitly include women with disabilities. Barriers to the use of mammography in women with disabilities include physical inaccessibility of office space and medical equipment; limited transportation and parking options; provider discomfort in providing care to these women or recommending a mammogram less frequently than to women without disabilities; and time and assistance constraints associated with undressing, transferring, and positioning for medical examinations. To promote health and wellness, health agencies, providers, and health care plans must promote cancer prevention and educational programs that are inclusive and responsive to the special needs of women with disabilities (Courtney-Long, Armour, Frammartino, et al., 2011). Women with disabilities also identified a lack of health promotion messages and materials that reflect their unique needs as problematic (Centers for Disease Control and Prevention [CDC], 2012).

An essential role of the nurse is to assist women with disabilities to identify accessible health screening and to advocate for greater accessibility of imaging centers and other health care facilities. Reminding women of the need for recommended clinical breast examinations and mammograms is an important part of nursing care.

RECONSTRUCTIVE BREAST SURGERY

Breast reconstruction is elective surgery that can enhance a woman's self-image and sense of well-being. Women desire reconstruction for a variety of physical and psychological reasons. Therefore, it is important for the health care team to conduct a thorough assessment prior to reconstructive surgery to evaluate the woman's underlying desire, motivation, and expectations. Preparing a woman realistically could help her to avoid potential disappointment. A variety of reconstructive options are available today for women who desire a correction in the size or the shape of the breast, including reduction **mammoplasty** (breast reduction), augmentation mammoplasty (breast enlargement), and mastopexy (breast lift). Several options are also available to reconstruct the breast after a mastectomy.

Reduction Mammoplasty

Reduction mammoplasty is usually performed on women who have breast hypertrophy (excessively large breasts). The weight of the enlarged breasts can cause discomfort, fatigue, embarrassment, and poor posture.

Reduction mammoplasty is an outpatient procedure that is performed under general anesthesia. Most commonly, an anchor-shaped incision that circles the areola is made, extending downward and following the natural curve of the crease beneath the breast (inframammary fold). Depending on the size of the breast, the nipple may be moved up to a higher position while still attached to the breast tissue, or it may be separated and transplanted to a new location. Drains are placed in the incision and remain for 2 to 5 days.

During the preoperative consultation, the patient should be informed that there is a possibility that sensory changes of the nipple (such as numbness) may occur. These sensations are normal and usually resolve after several months but can sometimes persist. The procedure may also make breastfeeding impossible, although some women have breast-fed successfully. The patient must also be aware that if she gains weight (usually more than 10 lb), her breasts may also enlarge.

After reduction mammoplasty, many women verbalize feelings of extreme satisfaction, possibly because of the relief they experience. The patient is instructed to wear a supportive bra 24 hours a day for 2 weeks to prevent tension on the swollen breast and incision line. Vigorous exercise (e.g., jumping, jogging) should be avoided for about 6 weeks after surgery.

Augmentation Mammoplasty

Augmentation mammoplasty is requested by women who desire larger or fuller breasts. The procedure is performed by placing a breast implant either under the pectoralis muscle (subpectoral) or under the breast tissue (subglandular). The subpectoral approach is preferred because it interferes less with clinical breast examinations and mammograms. The incision line can be placed in the inframammary fold, in the axilla, or around the areola. The procedure is performed as an outpatient procedure under general anesthesia. A drain is not necessary. Postoperative instructions are the same as for reduction mammoplasty.

Saline implants are typically used for augmentation mammoplasty. Because of concerns that silicone implants could cause autoimmune diseases, they were removed from the market in 1992. The FDA has approved the use of silicone gel-filled implants as long as they have been manufactured by three specific companies. This approval covers women of all ages for breast reconstruction and women 22 years and older for breast augmentation (FDA, 2013). Women with breast implants should be aware that mammograms may be more difficult to read, so they should seek experienced breast radiologists.

Mastopexy

Mastopexy is performed when the patient is happy with the size of her breasts but wishes to have the shape improved and a lift performed. This is also an outpatient surgical procedure, and postoperative instructions are the same as for reduction mammoplasty.

DISEASES OF THE MALE BREAST

Gynecomastia

Gynecomastia, or firm overdeveloped breast tissue, is the most common breast condition in the male. Adolescent boys can be affected because of hormones secreted by the testes. This type of gynecomastia is virtually always benign and resolves spontaneously in 1 to 2 years. Gynecomastia can also occur in older men and usually presents as a firm, tender mass underneath the areola. In these patients, gynecomastia may be diffuse and related to the use of certain medications (e.g., digitalis, ranitidine [Zantac]). It may also be associated with certain conditions, including feminizing testicular tumors, infection in the testes, and liver disease resulting from factors such as alcohol abuse or a parasitic infection.

Patients in their late teens to late 40s presenting with idiopathic (unknown cause) gynecomastia should have a testicular examination and possibly a testicular ultrasound. Treatment of the enlarged breast tissue is based on patient preference and is usually reserved for those men who cannot tolerate the cosmetic appearance of the breast or who have severe pain associated with the condition. Observation is acceptable in most cases because gynecomastia may resolve on its own. Surgical removal of the tissue through a small incision around the areola is the best treatment option. Liposuction performed by a plastic surgeon is another possibility, although this does not allow for pathologic examination of the tissue.

Male Breast Cancer

The lifetime risk of breast cancer in men is about 1 in 1,000. The number of breast cancer cases in men relative to the population has been fairly stable over the past 30 years. Nearly

2,140 new cases of invasive breast cancer were diagnosed in men in 2011 (ACS, 2011). Although breast carcinoma in both genders shares certain characteristics, notable differences have emerged. Familial cases in men usually have *BRCA2* rather than *BRCA1* mutations. Klinefelter syndrome, a chromosomal condition reflecting decreased testosterone levels, is the strongest risk factor for developing male breast carcinoma. Presentation is usually a painless lump, but is often late, with more than 40% of individuals having stage III or IV disease. When survival is adjusted for age at diagnosis and stage of disease, outcomes for male and female patients with breast cancer is similar (Gomez-Raposo, Zambrana, Sereno, et al., 2010).

Early detection is uncommon in male breast cancer because of the rare nature of the disease. Neither patient nor provider suspects male breast cancer early in its development. Treatment generally consists of a total mastectomy with either SLNB or ALND. As in women with breast cancer, prognosis depends on the stage of disease at presentation. Involvement of the axillary lymph nodes is the most important prognostic indicator. Male breast cancers are very likely to be ER+, and tamoxifen, although it has several side effects, is a mainstay of treatment.

Because breast cancer is primarily a disease of women, men may feel that a certain stigma is attached to their diagnosis. Health care professionals must be sensitive to their needs and provide information and support.

Critical Thinking Exercises

1 A 37-year-old woman with one child has had annual Papanicolaou (Pap) smears and gynecologic examinations. She has had no past medical issues but has a positive family history for breast cancer. Her mother, sister, and aunt (her mother's sister) had the disease. How would you assess her risk of breast cancer? What preventive measures are indicated?

2 ebp A 54-year-old woman with breast cancer has completed surgery, chemotherapy, and radiation treatment. She tolerated the treatments very well both physically and emotionally. She is now taking tamoxifen, which she started 6 months ago, and she feels depressed and anxious. What evidence-based recommendations are indicated for this woman? Identify the criteria to evaluate the strength of the evidence for these recommendations.

3 A 24-year-old African American woman with no family history of breast cancer has been advised by her primary provider of the risks of breast cancer screening (mammography) at her age. What other recommendations and their frequency are appropriate for this woman? What resources are appropriate for her?

4 You are providing postoperative care for a 44-year-old mother of two small children after a mastectomy. Describe the nursing plan of care for this patient. What issues would be most important to include in patient education? Describe at least three topics in the postoperative phase of care that need to be addressed. Explain how the patient education will differ if home care is planned.

5 pq Identify the priorities, approach, and techniques that you would use to perform a comprehensive admission assessment on a 60-year-old patient with a breast disorder. How would your priorities, approach, and techniques differ for a man admitted with a breast disorder? For a patient whose primary language is Spanish?

Brunner Suite Resources

Explore these additional resources to enhance learning for this chapter:

- NCLEX-Style Questions and Other Resources on thePoint, http://thePoint.lww.com/Brunner13e
- Study Guide
- PrepU
- Clinical Handbook
- Handbook of Laboratory and Diagnostic Tests

References

*Asterisk indicates nursing research.
**Double asterisk indicates classic reference.

Books

American Cancer Society. (2011). *Breast cancer facts and figures 2011*. Atlanta: Author.

Edge, S. B., Byrd, D. R., Compton, C. C., et al. (Eds.). (2010). *AJCC cancer staging manual* (7th ed.). New York: Springer.

Eggert, J. (2010). *Cancer basics*. Pittsburgh: Oncology Nursing Society.

Fischbach, F., & Dunning, M. B. (2009). *A manual of laboratory and diagnostic tests* (8th ed.). Philadelphia: Lippincott Williams & Wilkins.

Harmer, V. (2011). *Breast cancer nursing care and management* (2nd ed.). Somerset: Wiley-Blackwell.

Hogan-Quigley, B., Palm, M. J., & Bickley, L. S. (2012). *Bates' nursing guide to physical examination and history taking*. Philadelphia: Lippincott Williams & Wilkins.

Karch, A. M. (2012). *2012 Lippincott's nursing drug guide*. Philadelphia: Lippincott Williams & Wilkins.

Love, S. M., & Lindsey, K. (2010). *Dr. Susan Love's breast book* (5th ed.). Cambridge, MA: Da Capo Press.

Shockney, L. (2011). *Breast cancer survivorship care: A resource for nurses*. Sudbury, MA: Jones & Bartlett.

Weber, J., & Kelley, J. (2010). *Health assessment in nursing* (4th ed.). Philadelphia: Lippincott Williams & Wilkins.

Yarbo, C. H., Wujcik, D., & Gobel, B. H. (Eds.). (2010). *Cancer nursing: Principles and practices* (7th ed.). Sudbury, MA: Jones & Bartlett.

Journals and Electronic Documents

American Cancer Society. (2010). *Detailed guide: Breast cancer. 2010*. Available at: www.cancer.org

American College of Obstetricians and Gynecologists. (2011). Practice Bulletin No. 122: Breast cancer screening. *Obstetrics and Gynecology, 118*(2 Pt. 1), 372–382.

Aslani, N., Swanson, T., Kennecke, H., et al. (2011). Factors that determine whether a patient receives completion axillary lymph node dissection after a positive sentinel lymph node biopsy for breast cancer in British Columbia. *Canadian Journal of Surgery, 54*(4), 237–242.

*Baron, R. H., Fey, J. V., Borgen, P. I., et al. (2007). Eighteen sensations after breast cancer surgery: A five-year comparison of sentinel lymph node biopsy and axillary lymph node dissection. *Annals of Surgical Oncology, 14*(5), 1653–1661.

Bonilla, L., Ben-Aharon, I., Vidal, L., et al. (2010). Dose-dense chemotherapy in nonmetastatic breast cancer: A systematic review and meta-analysis of

randomized controlled trials. *Journal of the National Cancer Institute, 102*(24), 1845–1854.
CancerQuest. (2011). *Introduction to patient information, detection and diagnosis, fine needle aspiration*. Available at: www.cancerquest.org
Centers for Disease Control and Prevention. (2012). *National Center on Birth Defects and Developmental Disabilities: Right to know campaign breast cancer screening program*. Available at: www.cdc.gov/ncbddd/disabilityandhealth/righttoknow/freematerials.html
Cheifetz, O., & Haley, L. (2010). Management of secondary lymphedema related to breast cancer. *Canadian Family Physician, 56*(12), 1277–1284.
Clough-Gorr, K. M., Ganz, P. A., & Silliman, R. A. (2010). Older breast cancer survivors: Factors associated with self-reported symptoms of persistent lymphedema over 7 years of follow-up. *Breast Journal, 16*(2), 147–155.
Courtney-Long, E., Armour, B., Frammartino, B., et al. (2011). Factors associated with self-reported mammography use for women with and women without a disability. *Journal of Women's Health, 20*(9), 1279–1286.
**Cummings, S. R., Eckert, S., Krueger, K. A., et al. (1999). The effect of raloxifene on risk of breast cancer in postmenopausal women: Results from the MORE randomized trial. *Journal of the American Medical Association, 281*(23), 2189–2197.
DeLaurentis, M., Cancello, G., D'Agostino, D., et al. (2008). Taxane-based combinations as adjuvant chemotherapy of early breast cancer: A meta-analysis of randomized trials. *Journal of Clinical Oncology, 26*(1), 44–53.
De Morgan, S., Redman, S., D'Este, C., et al. (2011). Knowledge, satisfaction with information, decisional conflict and psychological morbidity amongst women diagnosed with ductal carcinoma in situ (DCIS). *Patient Education and Counseling, 84*(1), 62–68.
DeSantis, C., Siegel, R., Bandi, P., et al. (2011). Breast cancer statistics, 2011. *CA: A Cancer Journal for Clinicians, 61*(6), 409–418.
Djuric, Z., Edwards, A., Madan, S., et al. (2009). Obesity is associated with atypia in breast ductal lavage of women with proliferative breast disease. *Cancer Epidemiology, 33*(3-4), 242–248.
**Fisher, B., Costantino, J. P., Wickerham, D. L., et al. (1998). Tamoxifen for prevention of breast cancer. Report of the National Surgical Adjuvant Breast and Bowel Project P-1 study. *Journal of the National Cancer Institute*, 90(18), 1371–1388.
Gianni, L., Dafni, U., Gelber, R. D., et al. (2011). Treatment with trastuzumab for 1 year after adjuvant chemotherapy in patients with HER2-positive early breast cancer: A 4-year follow-up of a randomized controlled trial. *Lancet Oncology, 12*(3), 236–244.
Gomez-Raposo, C., Zambrana, T. F., Sereno, M. M., et al. (2010). Male breast cancer. *Cancer Treatment Reviews, 36*(6), 451–457.
Graf, C., & Wessely, N. (2010). Physical activity in the prevention and therapy of breast cancer. *Breast Care 2010, 5*, 389–394.
Haba-Rodriguez, J., & Calderay, M. (2010). Impact of breast cancer treatment on fertility. *Breast Cancer Research & Treatment, 123*, 59–63.
Harbeck, A., Pegram, M. D., Ruschoff, J., et al. (2010). Targeted therapy in metastatic breast cancer: The HER2/neu oncogene. *Breast Care 2010, 5*, 3–7.
Helyer, L. K., Varnic, M., Le, L. W., et al. (2010). Obesity is a risk factor for developing postoperative lymphedema in breast cancer patients. *Breast Journal, 16*(1), 48–54.
Heywang-Kobrunner, S. H., Hacker, A., & Sedlacek, S. (2011). Advantages and disadvantages of mammography screening. *Breast Care, 6*(3), 199–207.
Howard-McNatt, M., Forsberg, C., Levine, E. A., et al. (2011). Breast cancer reconstruction in the elderly. *American Surgeon, 77*(2), 1640–1643.
Hubbard, R. A., Kerlikowske, C. I., Flowers, B. C., et al. (2011). The benefits and harms of more or less frequent screening mammography. *Annuals of Internal Medicine, 155*(8), 1–14.
Ivanov, O., Dickler, A., Lum, B. Y. F., et al. (2011). Twelve-month follow-up results of a trial utilizing Axxent electronic brachytherapy to deliver intraoperative radiation therapy for early-stage breast cancer. *Annals of Surgical Oncology, 18*(2), 453–458.
Kerlikowske, K., Hubbard, R. A., Miglioretti, D. L., et al. (2011). Comparative effectiveness of digital versus film screen mammography in community practice in the United States: A cohort study. *Annals of Internal Medicine, 155*(8), 493–502.
Kim, S. J., Moon, W. K., Kim, S. Y., et al. (2010). Comparison of two software versions of a commercially available computer-aided detection (CAD) system for detecting breast cancer. *Acta Radiologica, 51*(5), 482–490.
Knobf, M. T. (2011). Clinical update: Psychosocial responses in breast cancer survivors. *Seminars in Oncology Nursing, 27*(3), e1–e14.
Krag, D. N., Anderson, S. J., Julian, T. B., et al. (2010). Sentinel-lymph-node resection compared with conventional axillary-lymph-node dissection in clinically node-negative patients with breast cancer: Overall survival findings from the NSABP B-32 randomized phase 3 trial. *Lancet Oncology, 11*(10), 927–933.
Litton, J. K., Ready, K., Chen, H., et al. (2012). Earlier age of onset of BRCA mutation-related cancers in subsequent generations. *Cancer, 118*(2), 321–325.
Maughan, K. L., Lutterbie, M. A., & Ham, P. S. (2010). Treatment of breast cancer. *American Family Physician, 81*(11), 1339–1346.
Morris, G. J. (2009). Breastfeeding, parity, and reduction of breast cancer risk. *Breast Journal, 15*(5), 562–563.
Munro, A. F., Cameron, D. A., & Bartlett, J. M. S. (2010). Targeting anthracyclines in early breast cancer: New candidate predictive biomarkers emerge. *Oncogene, 29*(38), 5231–5240.
National Cancer Institute. (2011). *Preventive mastectomy: Fact sheet*. Available at: www.cancer.gov/cancertopics/factsheet/Therapy/preventive-mastectomy
National Comprehensive Cancer Network. (2012). *NCCN clinical practice guidelines in oncology: Breast cancer*. Available at: www.nccn.org
Odle, T. G. (2011). Breast cancer survivorship and surveillance. *Radiologic Technology, 83*(1), 63M–87M.
Pazaiti, A., & Fentiman, I. S. (2011). Basal phenotype breast cancer: Implications for treatment and prognosis. *Women's Health, 7*(2), 181–202.
Salhab, M., Bismohun, S., & Mokbel, K. (2010). Risk-reducing strategies for women carrying BRCA1/2 mutations with a focus on prophylactic surgery. *BMC Women's Health*. Available at: www.ncbi.nlm.nih.gov/pmc/articles/PMC2987888/pdf/1472-6874-10-28.pdf
Sanders, M. E., & Simpson, J. F. (2011). Can we know what to do when DCIS is diagnosed? *Oncology, 25*(9), 852—856.
Santisteban, M., Reynolds, C., Barr Fritcher, E. G., et al. (2010). Ki67: A time-varying biomarker of risk of breast cancer in atypical hyperplasia. *Breast Cancer Research & Treatment, 121*, 431–437.
Stromsvik, N., Raheim, M., Oyen, N., et al. (2009). Men in the women's world of hereditary breast and ovarian cancer—a systematic review. *Familial Cancer, 8*(3), 221–229.
Teller, P., Jefford, V., Gabram, S., et al. (2010). The utility of breast MRI in the management of breast cancer. *Breast Journal, 16*(4), 394–403.
Tirona, M. T., Sehgal, R., & Ballester, O. (2010). Prevention of breast cancer (part II): Risk reduction strategies. *Cancer Investigations, 28*(10), 1070–1077.
Tuma, R. (2011). Nomogram predicts lymphedema risk in breast cancer patients.*Oncology Times, 33*(18), 24–25.
U.S. Food and Drug Administration. (2013). *FDA approves new silicone gel-filled breast implant*. Available at: www.fda.gov/NewsEvents/Newsroom/PressAnnouncements/ucm340447.htm
**Vogel, V. G., Costantino, J. P., Wickerham, D. L., et al. (2006). Effects of tamoxifen vs. raloxifene on the risk of developing invasive breast cancer and other disease outcomes. The NSABP Study of Tamoxifen and Raloxifene (STAR) P-2 trial. *Journal of the American Medical Association, 295*(23), 2727–2741.
Williams, L. (2010). Male breast cancer in family leads to high perception of risk, low likelihood of genetic counseling. *Medical News Today*. Available at: www.medicalnewstoday.com/releases/196366.php
Xu, H. B., Liu, Y. J., & Li, L. (2011). Aromatase inhibitor versus tamoxifen in postmenopausal woman with advanced breast cancer: A literature-based meta-analysis. *Clinical Breast Cancer, 11*(4), 246–251.

Resources

ABCD: After Breast Cancer Diagnosis, abcdbreastcancersupport.org
American Cancer Society, www.cancer.org
American Society of Plastic Surgeons (ASPS), www.plasticsurgery.org
Cancer Care, Inc., www.cancercare.org
Fertile Hope, www.fertilehope.org
National Breast Cancer Coalition, www.breastcancerdeadline2020.org/homepage.html
National Cancer Institute (NCI), www.cancer.gov/cancertopics/types/breast
National Lymphedema Network (NLN), www.lymphnet.org
Oncology Nursing Society (ONS), www.ons.org
Reach to Recovery Program—I Can Cope Program, www.cancer.org/cancer/breastcancer/moreinformation/breastreconstructionaftermastectomy/breast-reconstruction-after-mastectomy-reach-to-recovery
Susan G. Komen for the Cure, ww5.komen.org
Young Survival Coalition (YSC), www.youngsurvival.org

Chapter

59

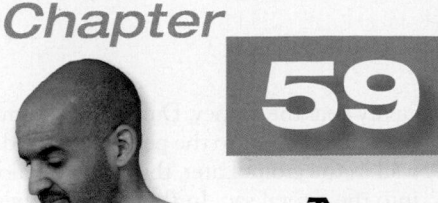

Assessment and Management of Problems Related to Male Reproductive Processes

Learning Objectives

On completion of this chapter, the learner will be able to:

1. Describe structures and function of the male reproductive system.
2. Discuss nursing assessment of the male reproductive system and identify diagnostic tests that complement assessment.
3. Discuss the causes and management of male sexual dysfunction.
4. Compare the types of prostatectomy with regard to advantages and disadvantages.
5. Use the nursing process as a framework for care of the patient with prostate cancer or undergoing prostatectomy.
6. Describe the nursing management of patients with testicular cancer.
7. Describe the various disorders of the penis, including pathophysiology, clinical manifestations, and management.

Glossary

androgen deprivation therapy (ADT): surgical (orchiectomy) or medical castration (e.g., with luteinizing hormone–releasing hormone agonists)

benign prostatic hyperplasia (BPH): noncancerous enlargement or hypertrophy of the prostate; the most common pathologic condition in older men

brachytherapy: delivery of radiation therapy through internal implants called *seeds* to a localized area of tissue

circumcision: excision of the foreskin, or prepuce, of the glans penis

cystostomy: surgical creation of an opening into the urinary bladder

cryptorchidism: most common congenital defect in males; characterized by failure of one or both of the testes to descend into the scrotum

epididymitis: infection of the epididymis that usually descends from an infected prostate or urinary tract; also may develop as a complication of gonorrhea, chlamydia, or *Escherichia coli*

erectile dysfunction: the inability to either achieve or maintain an erection sufficient to accomplish sexual intercourse; also called *impotence*

hydrocele: a collection of fluid, generally in the tunica vaginalis of the testis, although it also may collect within the spermatic cord

orchiectomy: surgical removal of one or both of the testes

orchitis: acute inflammation of the testes (testicular congestion) caused by pyogenic, viral, spirochetal, parasitic, traumatic, chemical, or unknown factors

Peyronie's disease: buildup of fibrous plaques in the sheath of the corpus cavernosum, causing curvature of the penis when it is erect

phimosis: condition in which the foreskin is constricted so that it cannot be retracted over the glans; can occur congenitally or from inflammation and edema

priapism: an uncontrolled, persistent erection of the penis from either neural or vascular causes, including medications, sickle cell thrombosis, leukemic cell infiltration, spinal cord tumors, and tumor invasion of the penis or its vessels

prostatectomy: open or laparoscopic surgical removal of the entire prostate, the prostate urethra, and the attached seminal vesicles plus the ampulla of the vas deferens

prostate-specific antigen (PSA): substance that is produced by the prostate gland; is used in combination with digital rectal examination to screen for prostate cancer

prostatism: obstructive and irritative symptom complex that includes increased frequency and hesitancy in starting urination, a decrease in the volume and force of the urinary stream, acute urinary retention, and recurrent urinary tract infections

prostatitis: inflammation of the prostate gland caused by infectious agents (bacteria, fungi, mycoplasma) or various other problems (e.g., urethral stricture, prostatic hyperplasia)

retrograde ejaculation: during ejaculation, semen travels to the urinary bladder instead of exiting through the penis

spermatogenesis: production of sperm in the testes

testosterone: male sex hormone secreted by the testes; induces and preserves the male sex characteristics

transurethral resection of the prostate (TURP): resection of the prostate through endoscopy; the surgical and optical instrument is introduced directly through the urethra to the prostate, and the gland is then removed in small chips with an electrical cutting loop

varicocele: an abnormal dilation of the veins of the pampiniform venous plexus in the scrotum (the network of veins from the testis and the epididymis, which constitute part of the spermatic cord)

vasectomy: ligation and transection of part of the vas deferens, with or without removal of a segment of the vas, to prevent the passage of the sperm from the testes; also called *male sterilization*

Disorders of the male reproductive system include a wide variety of conditions that usually affect both urinary and reproductive systems. Because these disorders involve the genitalia and often affect sexuality, the patient may experience anxiety and embarrassment. The nurse must be aware of the patient's need for privacy as well as his need for education and support. This requires an openness to discuss critical and sensitive issues with the patient, including his partner when appropriate, as well as effective assessment, management, and communication. Nurses must be comfortable when examining male genitalia and must recognize their own attitudes and perceptions about male reproductive problems. Education of the patient and partner about treatment and self-care strategies is essential (Tabloski, 2009).

ASSESSMENT OF THE MALE REPRODUCTIVE SYSTEM

Anatomic and Physiologic Overview

In the male, several organs serve as parts of both the urinary tract and the reproductive system. Disorders in the male reproductive organs may interfere with the functions of one or both of these systems. As a result, diseases of the male reproductive system are usually treated by a urologist. The structures in the male reproductive system include the (1) external male genitalia, consisting of the testes, epididymides, scrotum, and penis, and the (2) internal male genitalia, consisting of the vas deferens (ductus deferens), ejaculatory duct, and prostatic and membranous sections of the urethra, seminal vesicles, and certain accessory glands, such as the prostate gland and Cowper glands (bulbourethral glands) (Fig. 59-1).

The testes have a dual function: **spermatogenesis** (production of sperm) and secretion of the male sex hormone **testosterone**, which induces and preserves the male sex characteristics. The testes are formed in the embryo, within the abdominal cavity near the kidney. During the last month of fetal life, they descend posterior to the peritoneum and pierce the abdominal wall in the groin. Later, they progress along the inguinal canal into the scrotal sac. In this descent, they are accompanied by blood vessels, lymphatics, nerves, and ducts, which support the tissue and make up the spermatic cord. This cord extends from the internal inguinal ring through the abdominal wall and the inguinal canal to the scrotum. As the testes descend into the scrotum in the final 2 to 3 months of gestation, a tubular extension of peritoneum accompanies them (Hall, 2011). Normally, this tissue is obliterated during fetal development; only the tunica vaginalis, which covers the testes, remains. If the peritoneal process remains open into the abdominal cavity, a potential sac remains into which abdominal contents may enter to form an indirect inguinal hernia.

The testes, or ovoid sex glands, are encased in the scrotum, which keeps them at a slightly lower temperature than the rest of the body to facilitate spermatogenesis (Hall, 2011). The testes consist of numerous seminiferous tubules in which the spermatozoa form. Collecting tubules transmit the spermatozoa into the epididymis, a hoodlike structure lying on the testes and containing winding ducts that lead into the vas deferens. This firm, tubular structure passes upward through the inguinal canal to enter the abdominal cavity behind the peritoneum. It then extends downward toward the base of the bladder. An out-pouching from this structure is the seminal vesicle, which acts as a reservoir for testicular secretions. The tract is continued as the ejaculatory duct, which passes through the prostate gland to enter the urethra. Testicular secretions take this pathway when they exit the penis during ejaculation.

The penis is the organ for both copulation and urination. It consists of the glans penis, the body, and the root. The glans penis is the soft, rounded portion at the distal end of the penis. The urethra (the tube that carries urine) opens at the tip of the glans. The glans is naturally covered by elongated penile skin—the foreskin—which may be retracted to expose the glans. However, many men as newborns have undergone **circumcision**, which is a procedure to remove the

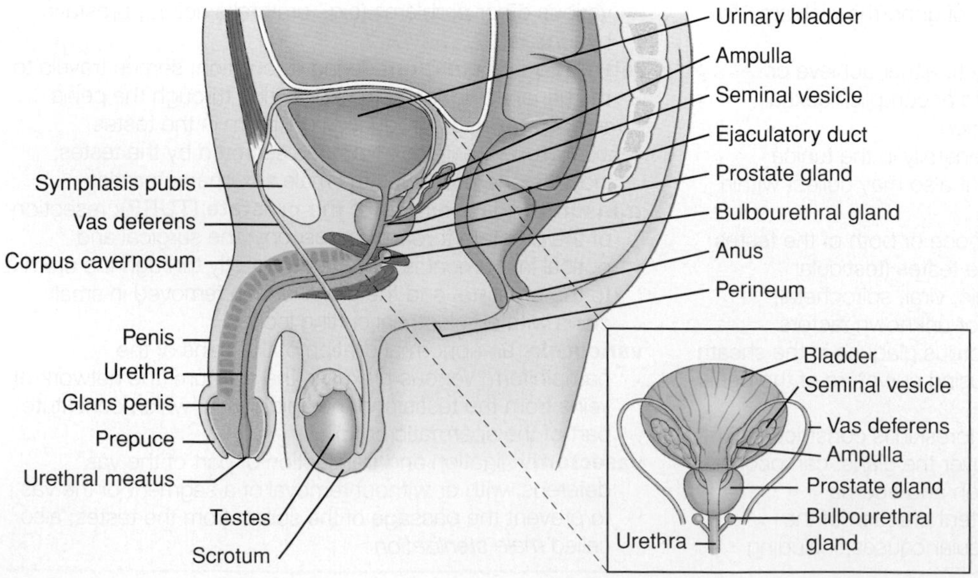

Urinary bladder
Ampulla
Seminal vesicle
Ejaculatory duct
Prostate gland
Bulbourethral gland
Anus
Perineum
Symphasis pubis
Vas deferens
Corpus cavernosum
Penis
Urethra
Glans penis
Prepuce
Urethral meatus
Testes
Scrotum
Bladder
Seminal vesicle
Vas deferens
Ampulla
Prostate gland
Bulbourethral gland
Urethra

FIGURE 59-1 • Structures of the male reproductive system.

foreskin. The body of the penis is composed of erectile tissues containing numerous blood vessels that become dilated, leading to an erection during sexual excitement. The urethra, which passes through the penis, extends from the bladder through the prostate to the distal end of the penis.

The prostate gland, lying just below the neck of the bladder, is composed of four zones and four lobes. It surrounds the urethra and is traversed by the ejaculatory duct, a continuation of the vas deferens. This gland produces a secretion that is chemically and physiologically suitable to the needs of the spermatozoa in their passage from the testes. Cowper glands lie below the prostate, within the posterior aspect of the urethra. This gland empties its secretions into the urethra during ejaculation, providing lubrication.

Gerontologic Considerations

As men age, the prostate gland enlarges; prostate secretion decreases; the scrotum hangs lower; the testes decrease in weight, atrophy, and become softer; and pubic hair becomes sparser and stiffer. Changes in gonadal function include a decline in plasma testosterone levels and reduced production of progesterone (Table 59-1). Other changes include decreasing sexual function, decreased libido (sexual desire), slower sexual response, longer time before sexual arousal can occur again, and urinary incontinence. Libido and potency often decrease in as many as two thirds of men older than 70 years (Tabloski, 2009). Vascular problems cause about half of the cases of impotence in men older than 50 years.

However, male reproductive capability is maintained with advancing age. Although degenerative changes occur in the seminiferous tubules and sperm production decreases, spermatogenesis continues, allowing men to produce viable sperm throughout their lives (McCance & Huether, 2009).

Male hypogonadism (decreased function of the testes) starts gradually at approximately 50 years of age, resulting in decreased testosterone production. The older man notices that the sexual response slows, erection takes longer, full erections may not be attained, and ejaculation takes longer to occur and control or resolution may occur without orgasm. Sexual function can be affected by psychological problems, illnesses, and medications (Althof & Needle, 2011). In general, the entire sexual act takes longer. Sexual activity is closely correlated with the man's sexual activity in his earlier years; if he was more active than average as a young man, he will most likely continue to be more active than average in his later years.

Men older than 50 years are at increased risk for genitourinary tract cancers, including those of the kidney, bladder, prostate, and penis. The digital rectal examination (DRE), prostate-specific antigen (PSA) test, and urinalysis, which screens for hematuria, may uncover a higher percentage of malignancies at earlier stages and lead to lower treatment-associated morbidity as well as a lower mortality.

Urinary incontinence occurs in one fifth of community-dwelling older men and rises to nearly 50% in men in long-term care settings (Mazur, Helffand, & McVary, 2012; Tabloski, 2009). Older adults admitted to acute care settings should be screened for this problem. Urinary incontinence may have many causes, including medications, neurologic disease, or benign prostatic hyperplasia (BPH). Diagnostic tests are performed to exclude reversible causes. New-onset urinary incontinence is a nursing priority that requires evaluation.

Assessment

Health History

Male sexuality is a complex phenomenon that is strongly influenced by personal, cultural, religious, and social factors. Sexuality and male reproductive function become concerns in the presence of illness and disability (Mulhall, Incrocci, Goldstein, et al., 2011). Throughout the assessment process, the nurse must recognize the importance of sexuality to the patient. Assessment of male reproductive function begins with an evaluation of urinary function and symptoms. The patient is asked about his usual state of health and any recent change in general physical and sexual activity. Any symptoms or changes in function are explored fully and described in detail. Symptoms related to bladder function and urination, collectively referred to as **prostatism**, are explored further. They may occur with an obstruction caused by an enlarged prostate gland: increased urinary frequency, decreased force of urine stream, and "double" or "triple" voiding (the patient needs to urinate two or three times over a period of several minutes to completely empty his bladder). The patient is also assessed for dysuria (painful urination), hematuria (blood in the urine), nocturia (urination during the night), and hematospermia (blood in the ejaculate).

Assessment also involves addressing sexual function, including manifestations of sexual dysfunction. The extent

TABLE 59-1 Age-Related Changes in the Male Reproductive System

Age-Related Changes	Physiologic Changes	Manifestations
Decrease in sex hormone secretion, especially testosterone	Decreased muscle strength and sexual energy	Changes in sexual response—prolonged time to reach full erection, rapid penile detumescence, and prolonged refractory period Decrease in number of viable sperm
	Shrinkage and loss of firmness of testes; thickening of seminiferous tubules	Smaller testes
	Fibrotic changes of corpora cavernosa	Erectile dysfunction
	Enlargement of prostate gland	Weakening of prostatic contractions Hyperplasia of prostate gland Signs and symptoms of obstruction of lower urinary tract (urgency, frequency, nocturia)

of the history depends on the patient's presenting symptoms and the presence of factors that may affect sexual function such as chronic illnesses or disability (e.g., diabetes, multiple sclerosis, stroke, cardiac disease), the use of medications that affect sexual function (e.g., antihypertensive and anticholesterolemic medications, psychotropic agents), stress, the use of alcohol, and the patient's willingness to discuss sexual issues.

By initiating an assessment about sexual concerns, the nurse conveys the message that changes in sexual functioning are valid topics and provides a safe environment for discussing these sensitive topics. A number of models are available to assist in assessing patient's problems and concerns. The PLISSIT (*p*ermission, *l*imited *i*nformation, *s*pecific *s*uggestions, *i*ntensive *t*herapy) model of sexual assessment and intervention may be used to provide a framework for nursing interventions (Rowland, 2012). It provides a graded counseling approach that allows health care professionals to deal with sexual issues with a level of comfort and expertise. The model begins by asking the patient's permission (P) to discuss sexual functioning. Limited information (LI) about sexual function may then be provided to the patient. As the discussion progresses, the nurse may offer specific suggestions (SS) for interventions. A professional who specializes in sex therapy may provide more intensive therapy (IT) as needed. The BETTER (*b*ringing up the topic, *e*xplaining, *t*elling, *t*iming, *e*ducate about treatment-related sexual side effects, *r*ecording) model was developed more recently to assist health care professionals to include sexuality in the assessment of patients with cancer (Mulhall et al., 2011).

Patients may find it difficult to express their feelings and concerns regarding their sexuality, especially after a body image change (e.g., after major surgery such as amputation). Discussing sexuality with patients who have an illness or disability can be uncomfortable for nurses and other health care providers; this, in turn, makes discussion of these issues more difficult and uncomfortable for patients. Health care professionals may unconsciously have stereotypes about the sexuality of people who are ill or have a disability (e.g., the belief that people with disabilities are asexual or should be sexually inactive). In addition, patients are often embarrassed to initiate a discussion about sexual issues with their health care providers (Miner, 2012).

Physical Assessment

In addition to the usual aspects of the physical examination, two essential components address disorders of the male genital or reproductive system: the DRE and the testicular examination.

Digital Rectal Examination

The DRE is used to screen for prostate cancer and is recommended annually for every man older than 50 years (45 years for men at high risk [African American men and men with a strong family history of prostate cancer]) (American Cancer Society [ACS], 2012). The DRE enables the skilled examiner, using a lubricated, gloved finger placed in the rectum, to assess the size, symmetry, shape, and consistency of the posterior surface of the prostate gland (Fig. 59-2). The clinician

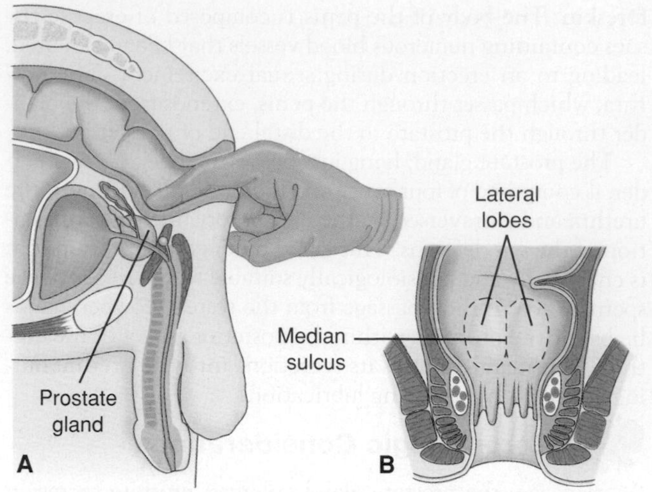
Lateral lobes
Median sulcus
Prostate gland
A
B

FIGURE 59-2 • A. Palpation of the prostate gland during digital rectal examination enables the examiner to assess the size, shape, and texture of the gland. **B.** The prostate is round, with a palpable median sulcus or groove separating the lateral lobes. It should feel rubbery and free of nodules and masses.

assesses for tenderness of the prostate gland on palpation and for the presence and consistency of any nodules. The DRE may be performed with the patient leaning over an examination table or positioning the man in a side-lying position with legs flexed toward the abdomen or supine with legs resting in stirrups. To minimize discomfort and relax the anal sphincter during the rectal examination, the patient is instructed to take a deep breath and exhale slowly as the practitioner inserts a finger. If possible, he should turn his feet inward so his toes are touching. Although this examination may be uncomfortable and embarrassing for the patient, it is an important screening tool.

Testicular Examination

The male genitalia are inspected for abnormalities and palpated for masses. The scrotum is palpated carefully for nodules, masses, or inflammation. Examination of the scrotum can reveal such disorders as hydrocele, inguinal hernia, testicular torsion, orchitis, epididymitis, or a tumor of the testis. The penis is inspected and palpated for ulcerations, nodules, inflammation, discharge, and curvature. If the patient is uncircumcised, the foreskin should be retracted for visualization of the glans penis. The testicular examination provides an excellent opportunity to instruct the patient on how to perform a testicular self-examination (TSE) and its importance in early detection of testicular cancer. TSE should begin during adolescence. (For more details on TSE, see later discussion in the chapter and Chart 59-7.)

Diagnostic Evaluation

A wide range of diagnostic studies may be performed in men with reproductive disorders. The nurse should educate the patient about the purpose, what to expect, and any possible side effects related to these tests and examinations prior to testing. The nurse should note trends in results because they provide information about disease progression as well as the patient's response to therapy.

Prostate-Specific Antigen Test

The cells within the prostate gland produce a protein that can be measured in the blood called the **prostate-specific antigen (PSA)**. It is a sensitive but not specific test for prostate cancer. In the absence of prostate cancer, serum PSA levels vary with age, race, and prostate volume. Increased levels may indicate prostate cancer. However, a number of other conditions such as BPH, acute urinary retention, and acute prostatitis may cause high PSA levels. Values of PSA may also increase after ejaculation. PSA levels are measured in nanograms per milliliter (ng/mL). In most laboratories, values less than 4 ng/mL are generally considered normal, and values greater than 4 ng/mL are considered elevated (Fischbach & Dunning, 2009). The use of age-specific reference ranges is encouraged to help minimize unnecessary biopsies.

A serum PSA level and a DRE, which are recommended by the ACS (2012), are used to screen for prostate cancer for men with at least a 10-year life expectancy and for men at high risk, including those with a strong family history of prostate cancer and of African American ethnicity. The PSA test is also used to monitor patients for recurrence after treatment for cancer of the prostate, based on evidence-based guidelines (National Comprehensive Cancer Network [NCCN], 2012a). Neither the DRE nor PSA is 100% accurate, but when used together, their accuracy increases.

Ultrasonography

Transrectal ultrasound (TRUS) may be performed in patients with abnormalities detected by DRE and in those with elevated PSA levels. After DRE has been completed, a lubricated, condom-covered, rectal probe transducer is inserted into the rectum (Fischbach & Dunning, 2009). Water may be introduced into the condom to help transmit sound waves to the prostate. TRUS may be used in detecting nonpalpable prostate cancers and in staging localized prostate cancer. Needle biopsies of the prostate are commonly guided by TRUS.

Prostate Fluid or Tissue Analysis

Specimens of prostate fluid or tissue may be obtained for culture if disease or inflammation of the prostate gland is suspected. A biopsy of the prostate gland may be necessary to obtain tissue for histologic examination. This may be performed at the time of prostatectomy or by means of a perineal or transrectal needle biopsy. Six to 12 biopsies from all four prostate zones may be obtained during a TRUS-guided biopsy.

Tests of Male Sexual Function

If the patient cannot engage in sexual intercourse to his satisfaction, a detailed history is obtained. Nocturnal erections occur in healthy males of all ages. Nocturnal penile tumescence tests may be conducted in a sleep laboratory to monitor changes in penile circumference during sleep using various methods to determine number, duration, rigidity, and circumference of penile erections; the results help identify whether the erectile dysfunction is caused by physiologic or psychological factors. Additional tests, including psychological evaluations, are also part of the diagnostic workup and are usually conducted by a specialized team of health care providers.

DISORDERS OF MALE SEXUAL FUNCTION

Erectile Dysfunction

Erectile dysfunction, also called *impotence*, is the inability to achieve or maintain an erect penis (Galloway, 2011). The man may report decreased frequency of erections, inability to achieve a firm erection, or rapid detumescence (subsiding of erection). In the United States, 30 million men experience erectile dysfunction; more than half of men 40 to 70 years of age are unable to attain or maintain an erection sufficient for satisfactory sexual performance (McAninch & Lue, 2012). The physiology of erection and ejaculation is complex and involves parasympathetic and sympathetic components. Erection involves the release of nitric oxide into the corpus cavernosum during sexual stimulation. Its release activates cyclic guanosine monophosphate (cGMP), causing smooth muscle relaxation. This allows flow of blood into the corpus cavernosum, resulting in erection (Porth & Matfin, 2009; Wein, Kavoussi, Partin, et al., 2012).

Erectile dysfunction has both psychogenic and organic causes. Psychogenic causes include anxiety, fatigue, depression, pressure to perform sexually, negative body image, absence of desire, and privacy, as well as trust and relationship issues. Organic causes include cardiovascular disease, endocrine disease (diabetes, pituitary tumors, testosterone deficiency, hyperthyroidism, and hypothyroidism), cirrhosis, chronic kidney failure, genitourinary conditions (radical pelvic surgery), hematologic conditions (Hodgkin lymphoma, leukemia), neurologic disorders (neuropathies, parkinsonism, spinal cord injury [SCI], multiple sclerosis), trauma to the pelvic or genital area, alcohol, smoking, medications (Chart 59-1), and drug abuse.

Assessment and Diagnostic Findings

The diagnosis of erectile dysfunction requires a sexual and medical history; an analysis of presenting symptoms; a physical examination, including a neurologic examination; a detailed assessment of all medications, alcohol, and drugs used; and various laboratory studies. Nocturnal penile tumescence tests are conducted to monitor changes in penile circumference. This test can help to determine if erectile impotence has an organic or a psychological cause. In healthy men, nocturnal penile erections closely parallel rapid eye movement (REM) sleep in occurrence and duration. Organically impotent men show inadequate sleep-related erections that correspond to their waking performance. Arterial blood flow to the penis is measured using a Doppler probe. In addition, nerve conduction tests and extensive psychological evaluations may be carried out. Figure 59-3 describes the evaluation and treatment of erectile dysfunction.

Medical Management

Treatment can be medical, surgical, or both, depending on the cause. Treatment of erectile dysfunction includes therapy for associated disorders (e.g., alcoholism, diabetes) or adjustment of medications (Hackett, 2009). Endocrine

Chart 59-1 PHARMACOLOGY

Medications Associated With Erectile Dysfunction

- *Antiadrenergics and antihypertensives:* guanethidine (Ismelin), clonidine (Catapres), hydralazine (Apresoline), metoprolol (Lopressor)
- *Anticholinergics and phenothiazines:* prochlorperazine (Compazine), trihexyphenidyl (Artane)
- *Antiseizure agents:* carbamazepine (Tegretol)
- *Antifungals:* ketoconazole (Nizoral)
- *Antihormone (prostate cancer treatment):* flutamide (Eulexin), leuprolide (Lupron)
- *Antipsychotics:* haloperidol (Haldol), chlorpromazine (Thorazine)
- *Antispasmodics:* oxybutynin (Ditropan)
- *Anxiolytics, sedative–hypnotics, tranquilizers:* lorazepam (Ativan), triazolam (Halcion)
- *Beta-blockers:* nadolol (Corgard)
- *Calcium channel blockers:* nifedipine (Adalat, Procardia)
- *Carbonic anhydrase inhibitors:* acetazolamide (Diamox)
- *Chemotherapeutic agents:* busulfan (Myleran), cyclophosphamide (Cytoxan)
- *Histamine-2 antagonists:* nizatidine (Axid), ranitidine (Zantac)
- *Nonsteroidal anti-inflammatory drugs:* naproxen (Naprosyn), indomethacin (Indocin)
- *Diuretics:* hydrochlorothiazide (HydroDIURIL), furosemide (Lasix), spironolactone (Aldactone), verapamil (Calan)
- *Antidepressants:* tricyclic antidepressants: amitriptyline (Elavil), desipramine (Norpramin); selective serotonin reuptake inhibitors: fluoxetine (Prozac), sertraline (Zoloft)
- *Parkinson's disease medications:* carbidopa/levodopa (Sinemet), benztropine (Cogentin)
- *Antihistamines:* diphenhydramine (Benadryl), dimenhydrinate (Dramamine)
- *Other substances:* alcohol, amphetamines, barbiturates, cocaine, marijuana, methadone, nicotine, opioids

Adapted from Karch, A. M. (2012). *2012 Lippincott's nursing drug guide*. Philadelphia: Lippincott Williams & Wilkins.

therapy instituted to treat erectile dysfunction secondary to hypothalamic–pituitary–gonadal dysfunction may reverse the condition. Insufficient penile blood flow may be treated with vascular surgery. Patients with erectile dysfunction from psychogenic causes are referred to a health care provider or therapist who specializes in sexual dysfunction. Patients with erectile dysfunction secondary to organic causes may be candidates for penile implants.

Currently available therapies for the treatment of erectile dysfunction include pharmacologic therapy (including urethral suppositories), penile implants, and vacuum constriction devices (Table 59-2). These options should be considered in a stepwise fashion, with increasing invasiveness and risk balanced against the likelihood of efficacy. The patient and, if possible, his partner, should be informed of the relevant treatment options and their associated risks and benefits. The choice of treatment is made jointly by the primary provider, patient, and partner, taking into consideration patient preferences and expectations.

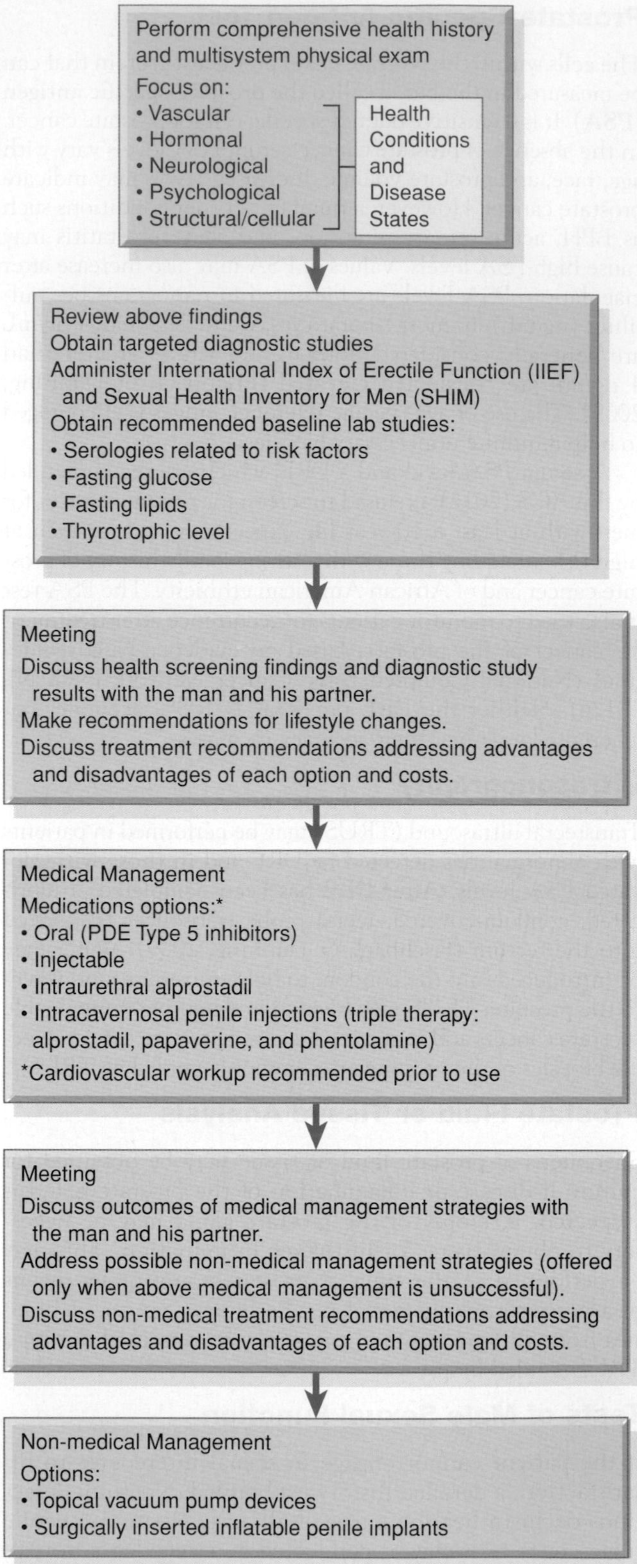
Perform comprehensive health history and multisystem physical exam
Focus on:
• Vascular
• Hormonal
• Neurological
• Psychological
• Structural/cellular
Health Conditions and Disease States
Review above findings
Obtain targeted diagnostic studies
Administer International Index of Erectile Function (IIEF) and Sexual Health Inventory for Men (SHIM)
Obtain recommended baseline lab studies:
• Serologies related to risk factors
• Fasting glucose
• Fasting lipids
• Thyrotrophic level
Meeting
Discuss health screening findings and diagnostic study results with the man and his partner.
Make recommendations for lifestyle changes.
Discuss treatment recommendations addressing advantages and disadvantages of each option and costs.
Medical Management
Medications options:*
• Oral (PDE Type 5 inhibitors)
• Injectable
• Intraurethral alprostadil
• Intracavernosal penile injections (triple therapy: alprostadil, papaverine, and phentolamine)
*Cardiovascular workup recommended prior to use
Meeting
Discuss outcomes of medical management strategies with the man and his partner.
Address possible non-medical management strategies (offered only when above medical management is unsuccessful).
Discuss non-medical treatment recommendations addressing advantages and disadvantages of each option and costs.
Non-medical Management
Options:
• Topical vacuum pump devices
• Surgically inserted inflatable penile implants

FIGURE 59-3 • Evaluation and treatment of erectile dysfunction.

Pharmacologic Therapy

Phosphodiesterase type 5 (PDE-5) inhibitors (oral medications that are used to treat erectile dysfunction) are first-line therapy (Heidelbaugh, 2010; Lee, 2011). Currently available PDE-5 inhibitors include sildenafil (Viagra), vardenafil (Levitra), and tadalafil (Cialis). Each of these agents has a similar mechanism of action but a different pharmacologic action and clinical use. Erection involves the release of nitric oxide in the vasculature of the corpus cavernosum as a result of sexual stimulation. This subsequently leads to smooth

TABLE 59-2 Treatments for Erectile Dysfunction

Method	Description	Advantages and Disadvantages	Duration
Pharmacologic Therapy • Oral medications (sildenafil [Viagra]; vardenafil [Levitra]; tadalafil [Cialis]) 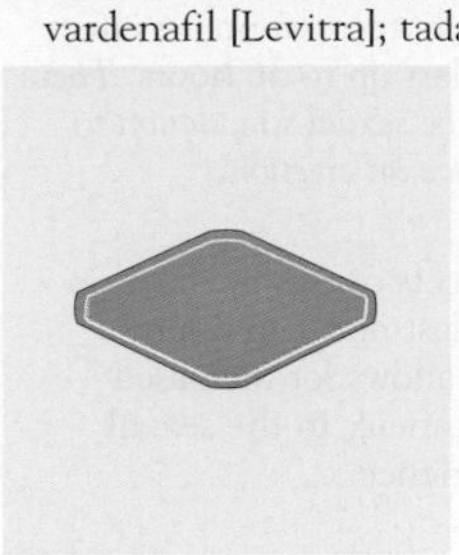 Oral medication	Smooth muscle relaxant causing blood to flow into penis	Can cause headache and diarrhea Contraindicated for men taking nitrate medications Used with caution in patients with retinopathy, especially diabetic retinopathy	Taken orally 1 hour before intercourse Stimulation is required to achieve erection. Erection can last 1 hour.
• Injection (alprostadil, papaverine, phentolamine) 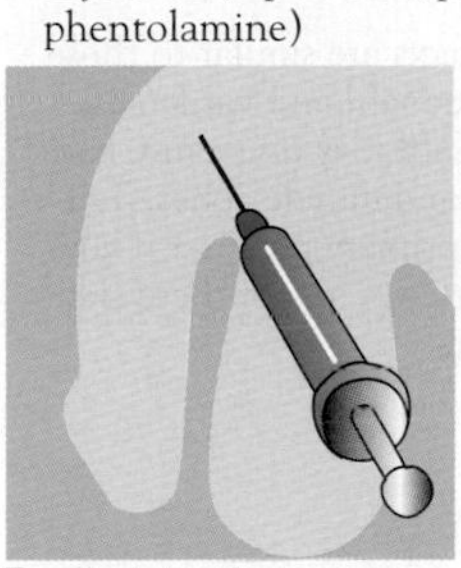 Penile injection	Smooth muscle relaxant causing blood to flow into penis	Firm erections are achievable in >50% of cases. Pain at injection site; plaque formation, risk of priapism	Injection 20 minutes before intercourse Erection can last up to 1 hour.
• Urethral suppository (alprostadil)  Penile suppository	Smooth muscle relaxant causing blood to flow into penis	May be used twice a day Urethral and genital pain; risk of hypertension and syncope Not recommended with pregnant partners	Inserted 10 minutes before intercourse Erection can last up to 1 hour.
Penile Implants • Semirigid rod • Inflatable • Soft silicone  Penile implant	Surgically implanted into corpus cavernosum	Reliable Requires surgery Healing takes up to 3 weeks. Subsequent cystoscopic surgery is difficult. Semirigid rod results in permanent semierection.	Indefinite Inflatable prosthesis—saline returns from penile receptacle to reservoir.
Negative-Pressure (Vacuum) Devices  Penile vacuum pump	Induction of erection with vacuum; maintained with constriction band around base of penis	Few side effects Cumbersome to use before intercourse Vasocongestion of penis can cause pain or numbness.	To prevent penile injury, constriction band must not be left in place >1 hour.

TABLE 59-3 Pharmacologic Treatment of Erectile Dysfunction

	Sildenafil (Viagra)	Vardenafil (Levitra)	Tadalafil (Cialis)
When to take	Take the medication 30 minutes to 4 hours before intercourse. *There must be sexual stimulation to produce an erection.*	Follow the same directions as with sildenafil; take the medication 1 hour before intercourse. The peak action occurs in 30–120 minutes. *There must be sexual stimulation to produce an erection.*	Take the medication before sexual activity. Effect peaks at 30 minutes to 6 hours; effect may last up to 36 hours. *There must be sexual stimulation to produce an erection.*
Frequency of use	If you take this medication more than once a day, it will not have an increased effect. You may take it 7 days per week if you wish, but only once in 24 hours. It does not build up in your bloodstream. Remember to take it only when you want to have intercourse.	The recommended frequency for this medication is 10 mg in 24 hours.	The effects of this medication may last up to 36 hours. This allows for increased spontaneity in the sexual experience.
Side effects	Side effects include headache, flushing, indigestion, nasal congestion, abnormal vision, diarrhea, dizziness, and rash. You may also have low blood sugar and abnormal liver function tests; your primary provider can determine this.	Side effects include headache, flushing, runny nose, indigestion, sinusitis, flulike syndrome, dizziness, nausea, back pain, and joint pain. Tell your primary provider if you experience any of these effects. You may also have abnormally elevated liver enzymes; your primary provider can determine this.	Side effects are similar to those of sildenafil and vardenafil. Tadalafil may also cause back pain and muscle aches. Tell your primary provider if you experience any of these side effects.
Contraindications	Do not take if you are taking nitrate medications such as nitroglycerin (e.g., Nitro-Bid) or isosorbide mononitrate (e.g., Imdur). Do not take if you have high uncontrolled blood pressure, coronary artery disease, or have had a heart attack within the past 6 months. Do not take if you have been diagnosed with a cardiac dysrhythmia or kidney or liver dysfunction.		
Drug interactions	This medication can react with other medications that you may be taking. Provide your primary provider and pharmacist with a complete list of all prescribed and over-the-counter medications that you are using.		
Use of PDE-5 inhibitors with penile injections or urethral suppositories	The use of PDE-5 inhibitors with other forms of therapy for erectile dysfunction has not been tested and should be avoided.		

PDE-5, phosphodiesterase type 5.

muscle relaxation in blood vessels supplying the corpus cavernosum, resulting in increased blood flow and an erection. During sexual stimulation, PDE-5 inhibitors increase blood flow to the penis (Lee, 2011; Porth & Matfin, 2009).

When PDE-5 inhibitors are taken about 1 hour before sexual activity, they are effective in producing an erection with sexual stimulation; the erection can last about 1 to 2 hours. The most common side effects of these medications include headache, flushing, dyspepsia, diarrhea, nasal congestion, and lightheadedness. These agents are contraindicated in men who take organic nitrates (e.g., isosorbide [Isordil], nitroglycerin), because taken together, these medications can cause side effects such as severe hypotension (Lee, 2011; Porth & Matfin, 2009; Wein et al., 2012). In addition, PDE-5 inhibitors must be used with caution in patients with retinopathy, especially in those with diabetic retinopathy. Patient education about the use of these medications and their side effects is summarized in Table 59-3.

For patients in whom PDE-5 inhibitors are contraindicated or ineffective, other pharmacologic measures to induce erections include injecting vasoactive agents, such as alprostadil, papaverine, and phentolamine, directly into the penis. Complications include **priapism** (a persistent abnormal erection) and development of fibrotic plaques at the injection sites. Alprostadil is also formulated in a gel pellet that can be inserted into the tip of the urethra with an applicator to create an erection.

Penile Implants

Two general types of penile implants are available: the malleable, noninflatable, nonhydraulic prosthesis (also called the *semirigid rod*) and the inflatable, hydraulic prostheses (McAninch & Lue, 2012; Montague, 2011). The semirigid rod (e.g., the Small-Carrion prosthesis) results in a permanent semierection but can be bent into an unnoticeable position when appropriate. The inflatable prosthesis simulates natural erections and natural flaccidity. Complications after implantation include infection, erosion of the prosthesis through the skin (more common with the semirigid rod than with the inflatable prosthesis), and persistent pain, which may require removal of the implant. Subsequent cystoscopic surgery, such as transurethral resection of the prostate (TURP), is more difficult with a semirigid rod than with the inflatable prosthesis.

Factors to consider in choosing a penile prosthesis are the patient's activities of daily living, social activities, and the expectations of the patient and his partner. Ongoing counseling for the patient and his partner is usually necessary to help them adapt to the prosthesis.

Negative-Pressure Devices

Negative-pressure (vacuum) devices may also be used to induce an erection. A plastic cylinder is placed over the flaccid penis, and negative pressure is applied. When an erection is attained, a constriction band is placed around the base of the penis to maintain the erection. To avoid penile injury, the patient is instructed not to leave the constricting band in place for longer than 1 hour. Only devices with a vacuum limiter are recommended for use (Heidelbaugh, 2010). Although many men find this method satisfactory, others experience premature loss of penile rigidity or pain when applying suction or during intercourse.

Nursing Management

Personal satisfaction and the ability to sexually satisfy a partner are common concerns of patients. Men with illnesses and disabilities may need the assistance of a sex therapist to identify, implement, and integrate their sexual beliefs and behaviors into a healthy and satisfying lifestyle. The nurse can inform patients about support groups for men with erectile dysfunction and their partners.

Disorders of Ejaculation

Premature ejaculation (PE) is defined as the occurrence of ejaculation sooner than desired, either before or shortly after penetration, causing distress to either one or both partners. It is one of the most common complaints of men or couples, affecting 20% to 30% of men (McAninch & Lue, 2012; Wein et al., 2012). The spectrum of responses ranges from occasional ejaculation with intercourse or self-stimulation to complete inability to ejaculate under any circumstances. Various forms of PE have been identified: (1) lifelong PE caused by neurobiologic or genetic conditions, (2) acquired PE (medical or psychological), (3) natural variable PE (normal variation), and (4) prematurelike ejaculatory dysfunction (psychological). Other ejaculatory problems may include inhibited (delayed or retarded) ejaculation, which is the involuntary inhibition of the ejaculatory reflex (Chart 59-2). **Retrograde ejaculation** occurs when semen travels toward the bladder instead of exiting through the penis, resulting in infertility. This form of PE may occur after prior prostate or urethral surgery, with diabetes, or with the use of medications such as antihypertensive agents.

Evaluation of PE involves a thorough sexual history focusing on the duration of symptoms, time to ejaculation, degree of voluntary control over ejaculation, frequency of occurrence, and course of the problem since the first sexual encounter (McAninch & Lue, 2012; Wein et al., 2012). Treatment, which depends on the nature and severity of PE and perceived distress that it causes, includes behavioral and psychological approaches, as well as pharmacologic therapy that attempts to alter the sensory input or retard the ejaculatory response. Behavioral therapy (e.g., counseling, sex therapy, psychoeducation, and couples therapy) often involves both the man and his sexual partner. The couple is encouraged to identify their sexual needs and to communicate those needs to each other. Pharmacologic management involves selective serotonin reuptake inhibitors, alpha$_1$ adrenoceptor antagonists, the tricyclic antidepressant clomipramine (Anafranil), and topical anesthetic agents. In some cases, a combination of pharmacologic and behavioral therapy may be effective.

Inhibited ejaculation is most often caused by psychological factors, neurologic disorders (e.g., SCI, multiple sclerosis,

Chart 59-2 GENETICS IN NURSING PRACTICE

Male Reproductive Disorders

Various male reproductive disorders are influenced by genetic factors. Some examples are:

- Klinefelter syndrome (47, XXY)
- Kallmann syndrome
- 21-Hydroxylase deficiency
- Congenital absence of the vas deferens
- Y chromosome deletions

Nursing Assessments

Family History Assessment

- Collect a three-generation family history on both maternal and paternal sides of the family.
- Assess family history for other family members with similar reproductive problems/abnormalities.

Patient Assessment

- In males with delayed puberty or infertility, assess for clinical features of Klinefelter syndrome (tall stature, gynecomastia, learning disabilities).
- Assess males with delayed or absent puberty for clinical features of Kallmann syndrome (cleft lip with or without cleft palate, abnormal eye movements, hearing loss, and abnormalities of tooth development).
- Assess males for history of early growth spurt, which is a symptom of 21 hydroxylase deficiency.

Management Issues Specific to Genetics

- Inquire whether genetic testing (deoxyribonucleic acid [DNA] chromosomal, metabolic) has been carried out on affected family member(s).
- If indicated, refer for further genetic counseling and evaluation so that family members can discuss inheritance, risk to other family members, and availability of genetic testing, and gene-based interventions.
- Offer appropriate genetic information and resources.
- Assess patient's understanding of genetic information.
- Provide support to families with newly diagnosed gene-related reproductive disorders.
- Participate in management and coordination of care of male patients with reproductive disorders, patients with genetic conditions, and individuals predisposed to develop or pass on a genetic condition.

Genetics Resources

See Chapter 8, Chart 8-6 for genetics resources.

neuropathy secondary to diabetes), surgery (prostatectomy), and medications. Chemical, vibratory, and electrical methods of stimulation have been used with some success. Treatment usually addresses the physical and psychological factors involved in inhibited ejaculation (Rowland, 2012). Although outpatient therapy may involve numerous sessions (12 to 18), it often results in a success rate of 70% to 80%. The outcome depends on a previous satisfying sexual experience history, a short duration of the ejaculatory problem, feelings of sexual desire, feelings of attraction to one's sexual partner, motivation for treatment, and absence of serious psychological problems.

For men with retrograde ejaculation, the urine may be collected shortly after ejaculation, revealing a large amount of sperm in the urine. This urine may also be collected to obtain adequate viable sperm for use in artificial insemination. In men with SCI, techniques that may be used to obtain sperm for artificial insemination include self-stimulation, vibratory stimulation, or electroejaculation. Electroejaculation involves the use of a specially designed probe that is inserted into the rectum next to the prostate. The probe delivers a current that stimulates the nerves and produces contraction of the pelvic muscles and ejaculation. However, spontaneous or stimulated ejaculation may cause autonomic dysreflexia (overstimulation of the autonomic nervous system) in patients with SCI at T6 or higher, creating a life-threatening situation (see Chapter 68). If this disorder is not treated promptly, it may lead to seizures, stroke, and even death.

INFECTIONS OF THE MALE GENITOURINARY TRACT

Acute uncomplicated cystitis in adult men is uncommon but occasionally occurs. Asymptomatic bacteriuria may also result from genitourinary manipulation, catheterization, or instrumentation. Urinary tract infections (UTIs) are discussed in Chapter 55.

According to the Centers for Disease Control and Prevention (CDC, 2010a), more than 19 million people develop sexually transmitted infections (STIs) annually in the United States; almost half of all STIs occur in people 18 to 24 years of age. The incidence of STIs has declined over the past several years, except in specific populations, including men who have sex with men. STIs affect people from all walks of life—from all social, educational, economic, and racial backgrounds. The single greatest risk factor for contracting an STI is the number of sexual partners. As the number of partners increases, so does the risk of exposure to a person infected with an STI. For men who have sex with men, the CDC recommends annual testing for human immunodeficiency virus (HIV), syphilis, *Chlamydia*, and gonorrhea (CDC, 2010b).

There are many sources of urethritis (gonococcal and nongonococcal), genital ulcers (genital herpes infections, primary syphilis, chancroid, granuloma inguinale, and lymphogranuloma venereum), genital warts (human papillomavirus [HPV]), scabies, pediculosis pubis, molluscum contagiosum, hepatitis and enteric infections, proctitis, and acquired immunodeficiency syndrome (AIDS). Trichomoniasis and STIs characterized by genital ulcers are thought to increase susceptibility to HIV infection. Trichomoniasis is associated with nonchlamydial, nongonococcal urethritis.

Current treatment guidelines for STIs are available from the CDC (2010a, 2012; Champion & Collins, 2012). Treatment must target the patient as well as his sexual partners and sometimes an unborn child. A thorough history, including a sexual history, is crucial to identify patients at risk and to direct care and education. Partners of men with STIs must also be examined, treated, and counseled to prevent reinfection and complications in both partners and to limit the spread of the disease. Sexual abstinence during treatment and recovery is advised to prevent the transmission of STIs. The use of synthetic condoms for at least 6 months after completion of treatment is recommended to decrease transmission of HPV infection as well as other STIs. It is important to assess and test for other STIs because patients who have one STI may also have another. The use of spermicides with nonoxynol-9 (known as N-9) is discouraged; these agents do not protect against HIV infection and may increase the risk of transmission of the virus. (See Chapters 37 and 71 for more detailed discussions of HIV infection, AIDS, and other STIs.)

PROSTATIC DISORDERS

Prostatitis

Prostatitis is an inflammation of the prostate gland that is often associated with lower urinary tract symptoms and symptoms of sexual discomfort and dysfunction. The condition affects 5% to 10% of men. It is the most common urologic diagnosis in men younger than 50 years and the third most common such diagnosis in men older than 50 years (Touma & Nickel, 2011; Wein et al., 2012). Prostatitis may be caused by infectious agents (bacteria, fungi, mycoplasma) or other conditions (e.g., urethral stricture, BPH). *Escherichia coli* is the most commonly isolated organism, although *Klebsiella* and *Proteus* species are also found (Touma & Nickel, 2011). The microorganisms colonize the urinary tract and ascend to the prostate, ultimately causing infection. The causal pathogen is usually the same in recurrent infections.

There are four types of prostatitis: acute bacterial prostatitis (type I), chronic bacterial prostatitis (type II), chronic prostatitis/chronic pelvic pain syndrome (CP/CPPS) (type III), and asymptomatic inflammatory prostatitis (type IV). Type III, which occurs in more than 90% of cases, is further classified as type IIIA or type IIIB, depending on the presence (type IIIA) or absence (type IIIB) of white blood cells in semen after prostate massage (Sharp, Takacs, & Powell, 2010).

Clinical Manifestations

Acute prostatitis is characterized by the sudden onset of fever, dysuria, perineal prostatic pain, and severe lower urinary tract symptoms: dysuria, frequency, urgency, hesitancy, and nocturia. Approximately 5% of cases of type I prostatitis (acute bacterial prostatitis) progress to type II prostatitis (chronic bacterial prostatitis) (Wein et al., 2012). Patients with type II disease are typically asymptomatic between episodes. Patients with type III prostatitis often have no bacteria in the urine in the presence of genitourinary pain. Patients with type IV prostatitis are usually diagnosed incidentally during a workup for infertility, an elevated PSA test, or other disorders.

Medical Management

The goal of treatment is to eradicate the causal organisms. Hospital admission may be necessary for patients with unstable vital signs, sepsis, or intractable pelvic pain; those who are frail or immunosuppressed; or those who have diabetes or renal insufficiency. Specific treatment is based on the type of prostatitis and on the results of culture and sensitivity testing of the urine (Sharp et al., 2010). If bacteria are cultured from the urine, antibiotic agents, including trimethoprim-sulfamethoxazole (Bactrim) or a fluoroquinolone (e.g., ciprofloxacin [Cipro]), may be prescribed, and continuous therapy with low-dose antibiotic agents may be used to suppress the infection. If the patient is afebrile and has a normal urinalysis, anti-inflammatory agents may be used. Alpha-adrenergic blocker therapy (e.g., tamsulosin [Flomax]), may be prescribed to promote bladder and prostate relaxation.

Factors contributing to prostatitis, including stress, neuromuscular factors, and myofascial pain, are also addressed. Supportive, nonpharmacologic therapies may be prescribed. These include biofeedback, pelvic floor training, physical therapy, reduction of prostatic fluid retention by ejaculation through sexual intercourse or masturbation, sitz baths, stool softeners, and evaluation of sexual partners to reduce the possibility of cross-infection.

Nursing Management

If the patient experiences symptoms of acute prostatitis (fever, severe pain and discomfort, inability to urinate, malaise), he may be hospitalized for intravenous (IV) antibiotic therapy. Nursing management includes administration of prescribed antibiotic agents and provision of comfort measures, including prescribed analgesic agents and sitz baths.

The patient with chronic prostatitis is usually treated on an outpatient basis and needs to be educated about the importance of continuing antibiotic therapy and recognizing recurrent signs and symptoms of prostatitis.

Promoting Home and Community-Based Care

Educating Patients About Self-Care

The nurse educates the patient about the importance of completing the prescribed course of antibiotic therapy. If IV antibiotic agents are to be administered at home, the nurse educates the patient and family about correct and safe administration. Arrangements for a home care nurse to oversee administration may be needed. Warm sitz baths (10 to 20 minutes) may be taken several times daily. Fluids are encouraged to satisfy thirst but are not "forced," because an effective medication level must be maintained in the urine. Foods and liquids with diuretic action or that increase prostatic secretions, such as alcohol, coffee, tea, chocolate, cola, and spices, should be avoided. A suprapubic catheter may be necessary for severe urinary retention. During periods of acute inflammation, sexual arousal and intercourse should be avoided. To minimize discomfort, the patient should avoid sitting for long periods. Medical follow-up is necessary for at least 6 months to 1 year, because prostatitis caused by the same or different organisms can recur. The patient is advised that the UTI may recur and is educated to recognize its symptoms.

Benign Prostatic Hyperplasia (Enlarged Prostate)

Benign prostatic hyperplasia (BPH), a noncancerous enlargement or hypertrophy of the prostate, is one of the most common diseases in aging men. It can cause bothersome lower urinary tract symptoms that affect quality of life by interfering with normal daily activities and sleep patterns (Palone, 2010). BPH typically occurs in men older than 40 years. By the time they reach 60 years, 50% of men have BPH. It affects as many as 90% of men by 85 years of age. BPH is the second most common cause of surgical intervention in men older than 60 years.

Pathophysiology

The cause of BPH is not well understood, but testicular androgens have been implicated. Dihydrotestosterone (DHT), a metabolite of testosterone, is a critical mediator of prostatic growth. Estrogens may also play a role in the cause of BPH; BPH generally occurs when men have elevated estrogen levels and when prostate tissue becomes more sensitive to estrogens and less responsive to DHT. Smoking, heavy alcohol consumption, obesity, reduced activity level, hypertension, heart disease, diabetes, and a Western diet (high in animal fat and protein and refined carbohydrates, low in fiber) are risk factors for BPH (Zarowitz, 2010).

BPH develops over a prolonged period; changes in the urinary tract are slow and insidious. BPH is a result of complex interactions involving resistance in the prostatic urethra to mechanical and spastic effects, bladder pressure during voiding, detrusor muscle strength, neurologic functioning, and general physical health (McCance & Huether, 2009). The hypertrophied lobes of the prostate may obstruct the bladder neck or urethra, causing incomplete emptying of the bladder and urinary retention. As a result, a gradual dilation of the ureters (hydroureter) and kidneys (hydronephrosis) can occur. Urinary retention may result in UTIs because urine that remains in the urinary tract serves as a medium for infective organisms.

Clinical Manifestations

BPH may or may not lead to lower urinary tract symptoms; if symptoms occur, they may range from mild to severe. Severity of symptoms increases with age, and half of men with BPH report having moderate to severe symptoms. Obstructive and irritative symptoms may include urinary frequency, urgency, nocturia, hesitancy in starting urination, decreased and intermittent force of stream and the sensation of incomplete bladder emptying, abdominal straining with urination, a decrease in the volume and force of the urinary stream, dribbling (urine dribbles out after urination), and complications of acute urinary retention and recurrent UTIs. Normally, residual urine amounts to no more than 50 mL in the middle-aged adult and less than 50 to 100 mL in the older adult (Weber & Kelley, 2010). Ultimately, chronic urinary retention and large residual volumes can lead to azotemia (accumulation of nitrogenous waste products) and kidney failure.

Generalized symptoms may also be noted, including fatigue, anorexia, nausea, vomiting, and pelvic discomfort. Other disorders that produce similar symptoms include urethral stricture, prostate cancer, neurogenic bladder, and urinary bladder stones.

Assessment and Diagnostic Findings

The health history focuses on the urinary tract, previous surgical procedures, general health issues, family history of prostate disease, and fitness for possible surgery (Palone, 2010). A patient voiding diary is used to record voiding frequency and urine volume. A DRE often reveals a large, rubbery, and nontender prostate gland. A urinalysis to screen for hematuria and UTI is recommended. A PSA level is obtained if the patient is without a terminal disease and for whom knowledge of the presence of prostate cancer would change management. The American Urological Association (AUA) Symptom Index or International Prostate Symptom Score (IPSS) can be used to assess the severity of symptoms (McAninch & Lue, 2012).

Other diagnostic tests may include recording urinary flow rate and the measurement of postvoid residual urine. If invasive therapy is considered, urodynamic studies, urethrocystoscopy, and ultrasound may be performed. Complete blood studies are performed. Cardiac status and respiratory function are assessed because a high percentage of patients with BPH have cardiac or respiratory disorders due to their age.

Medical Management

The goals of medical management of BPH are to improve quality of life, improve urine flow, relieve obstruction, prevent disease progression, and minimize complications. Treatment depends on the severity of symptoms, the cause of disease, the severity of the obstruction, and the patient's condition.

If a patient is admitted on an emergency basis because he is unable to void, he is immediately catheterized. The ordinary catheter may be too soft and pliable to advance through the urethra into the bladder. In such cases, a thin wire (stylet) is introduced (by a urologist) into the catheter to prevent the catheter from collapsing when it encounters resistance. A metal catheter with a pronounced prostatic curve may be used if obstruction is severe. A **cystostomy** (incision into the bladder) may be needed to provide urinary drainage.

Discussion of all treatment options by the primary provider enables the patient to make an informed decision based on symptom severity, the effect of BPH on his quality of life, and preference. Patients with mild symptoms and patients with moderate or severe symptoms who are not bothered by them and have not developed complications may be managed with "watchful waiting." With this approach, the patient is monitored and reexamined annually but receives no active intervention (Zarowitz, 2010). Other therapeutic choices include pharmacologic treatment, minimally invasive procedures, and surgery.

Pharmacologic Therapy

Pharmacologic treatment for BPH includes the use of alpha-adrenergic blockers and 5-alpha-reductase inhibitors (Wein et al., 2012). Alpha-adrenergic blockers, which include alfuzosin (Uroxatral), terazosin (Hytrin), doxazosin (Cardura), and tamsulosin, relax the smooth muscle of the bladder neck and prostate. This improves urine flow and relieves symptoms of BPH. Side effects include dizziness, headache, asthenia/fatigue, postural hypotension, rhinitis, and sexual dysfunction (McAninch & Lue, 2012; Wein et al., 2012).

Another method of treatment involves hormonal manipulation with antiandrogen agents. The 5-alpha-reductase inhibitors finasteride (Proscar) and dutasteride (Avodart) are used to prevent the conversion of testosterone to DHT and decrease prostate size. Side effects include decreased libido, ejaculatory dysfunction, erectile dysfunction, gynecomastia (breast enlargement), and flushing. Combination therapy (doxazosin and finasteride) has decreased symptoms and reduced clinical progression of BPH (McAninch & Lue, 2012; Wein et al., 2012).

The use of phytotherapeutic agents and other dietary supplements (*Serenoa repens* [saw palmetto berry] and *Pygeum africanum* [African plum]) are not recommended, although they are commonly used. They may function by interfering with the conversion of testosterone to DHT. In addition, *S. repens* may directly block the ability of DHT to stimulate prostate cell growth. These agents should not be used with finasteride, dutasteride, or estrogen-containing medications.

Surgical Treatment

Other treatment options include minimally invasive procedures and resection of the prostate gland.

Minimally Invasive Therapy

Several forms of minimally invasive therapy may be used to treat BPH. Transurethral microwave thermotherapy (TUMT) involves the application of heat to prostatic tissue. High-energy TUMT devices (CoreTherm, Prostatron, Targis) and low-energy devices (TherMatrx) are available (Wein et al., 2012). A transurethral probe is inserted into the urethra, and microwaves are directed to the prostate tissue. The targeted tissue becomes necrotic and sloughs. To minimize damage to the urethra and decrease the discomfort from the procedure, some systems have a water-cooling apparatus.

Other minimally invasive treatment options include (transurethral needle ablation [TUNA]) by radiofrequency energy and the UroLume stent. TUNA uses low-level radiofrequencies delivered by thin needles placed in the prostate gland to produce localized heat that destroys prostate tissue while sparing other tissues. The body then reabsorbs the dead tissue. Prostatic stents are associated with significant complications (e.g., encrustation, infection, chronic pain); therefore, they are used only for patients with urinary retention and in patients who are poor surgical risks (Wein et al., 2012).

Surgical Resection

Surgical resection of the prostate gland is another option for patients with moderate to severe lower urinary tract symptoms of BPH and for those with acute urinary retention or other complications. The specific surgical approach (open or endoscopic) and the energy source (electrocautery vs. laser) are based on the surgeon's experience, the size of the prostate gland, the presence of other medical disorders, and the patient's preference. If surgery is to be performed, all clotting defects must be corrected and medications for anticoagulation withheld because bleeding is a complication of prostate surgery.

Transurethral resection of the prostate (TURP) remains the benchmark for surgical treatment for BPH. It involves the surgical removal of the inner portion of the prostate through an endoscope inserted through the urethra; no external skin incision is made. It can be performed with ultrasound guidance. The treated tissue either vaporizes or becomes necrotic and sloughs. The procedure is performed in the outpatient setting and usually results in less postoperative bleeding than a traditional surgical prostatectomy.

Other surgical options for BPH include transurethral incision of the prostate (TUIP), transurethral electrovaporization, laser therapy, and open prostatectomy (McAninch & Lue, 2012; Wein et al., 2012). TUIP is an outpatient procedure used to treat smaller prostates. One or two cuts are made in the prostate and prostate capsule to reduce constriction of the urethra and decrease resistance to flow of urine out of the bladder; no tissue is removed. Open prostatectomy involves the surgical removal of the inner portion of the prostate via a suprapubic, retropubic, or perineal (rare) approach for large prostate glands. Prostatectomy may also be performed laparoscopically or by robotic-assisted laparoscopy.

Nursing management of patients undergoing these procedures is described later in this chapter.

Cancer of the Prostate

Prostate cancer is the most common cancer in men other than nonmelanoma skin cancer. It is the second most common cause of cancer death in American men, exceeded only by lung cancer, and is responsible for 10% of cancer-related deaths in men. Among men diagnosed with prostate cancer, 98% survive at least 5 years, 84% survive at least 10 years, and 56% survive 15 years (ACS, 2012).

Prostate cancer is common in the United States and northwestern Europe but is rare in Africa, Central America, South America, China, and other parts of Asia. African American men have a high risk of prostate cancer; furthermore, they are more than twice as likely to die of prostate cancer as men of other racial or ethnic groups. There is a need for education about prostate cancer and screening in African American men (Ross, Dark, Orom, et al., 2011). Health care providers should ensure the delivery of culturally sensitive education programs and counseling about prostate cancer screening to not only the African American patient at risk for prostate cancer but also to his friends and family (Jones, Steeves, & Williams, 2010) (Chart 59-3).

Chart 59-3 NURSING RESEARCH PROFILE
Influences on African American Men's Decisions Regarding Prostate Cancer Screening

Jones, R. A., Steeves, R., & Williams, I. (2010). Family, friend interactions among African-American men deciding whether or not to have a prostate cancer screening. *Urologic Nursing, 30*(3), 189–194.

Purpose

African American men are more than 2.4 times more likely to die of prostate cancer than men in other cultural groups. It is essential for this group to be aware of the increased risk and the need for screening for early detection of prostate cancer. The purpose of this study was to examine how rural African American men decide to have prostate cancer screening.

Design

A hermeneutic, phenomenologic, qualitative research design was used to interview 17 rural African American men to understand their experiences in making the decision to obtain a prostate cancer screening. Specifically, the researchers asked participants about their interactions with family members and friends and how these interactions affected their decision-making process. The men shared their experiences about how their conversations influenced their decision making about prostate cancer screening. The participants were recruited from rural Virginia through the use of radio public service announcements, flyers in African American barbershops, and by word of mouth. The inclusion criteria specified that the participant: (1) self-identified as a person of Black/African American origin, (2) was 40 years or older, (3) was able to read and understand English, (4) was not diagnosed as having prostate cancer, and (5) was willing to provide informed consent. All participants were interviewed at a place and time of convenience for them, either at home or at a public library conference room. Each person received $30.00 as compensation for participation in the study.

Findings

The study participants were 40 to 71 years of age, with a mean age of 52 years. Twelve men reported an annual income between $1,000 and $30,000. The majority of the men (i.e., 12 men [71%]) were employed, and 8 men (47%) had private health insurance. Five of the men (29%) received Medicare/Medicaid. Only 35% were high school graduates. Seven of the men (41%) were married, another 7 men (41%) were single, and 3 (18%) were divorced. Each man verbalized that he had male friends with whom he had frequent contact. Three themes emerged from the interview data: (1) family and friend involvement is important, (2) trust in the primary provider is necessary, and (3) knowing a friend or family member with prostate cancer influenced the men's decision making.

Nursing Implications

It is important for nurses and other health care providers to understand the support networks that African American men use to make decisions about obtaining prostate cancer screenings. The informal networks of family and friends and their opinions are highly valued and have a positive impact on health screening decision making. Health care providers must work to create rapport and establish an open and trusting clinical environment in order to form a strong patient–provider relationship that will establish a foundation so that education, screening, and informed decision making can occur.

Other risk factors for prostate cancer include increasing age; the incidence of prostate cancer increases rapidly after the age of 50 years. More than 70% of cases occur in men older than 65 years. A familial predisposition may occur in men who have a father or brother previously diagnosed with prostate cancer, especially if their relatives were diagnosed at a young age. Genes that may be associated with increased risk of prostate cancer include hereditary prostate cancer 1 (*HPC1*) and *BRCA1* and *BRCA2* mutations (Blaine, Honeywell, Allanson, et al., 2009; Wein et al., 2012). The risk of prostate cancer is also greater in men whose diet contains excessive amounts of red meat or dairy products that are high in fat (ACS, 2012). Endogenous hormones, such as androgens and estrogens, also may be associated with the development of prostate cancer.

Clinical Manifestations

Cancer of the prostate in its early stages rarely produces symptoms. Usually, symptoms that develop from urinary obstruction occur in advanced disease. Prostate cancer tends to vary in its course. If the cancer is large enough to encroach on the bladder neck, signs and symptoms of urinary obstruction occur (difficulty and frequency of urination, urinary retention, and decreased size and force of the urinary stream). Other symptoms may include blood in the urine or semen and painful ejaculation. Hematuria may occur if the cancer invades the urethra or bladder. Sexual dysfunction is common before the diagnosis is made.

Prostate cancer can spread to lymph nodes and bone. Symptoms of metastases include backache, hip pain, perineal and rectal discomfort, anemia, weight loss, weakness, nausea, oliguria (decreased urine output), and spontaneous pathologic fractures. These symptoms may be the first indications of prostate cancer.

Assessment and Diagnostic Findings

If prostate cancer is detected early, the likelihood of cure is high (Eggert, 2010). It can be diagnosed through an abnormal finding with the DRE, serum PSA, and ultrasound-guided TRUS with biopsy. Detection is more likely with the use of combined diagnostic procedures. Routine repeated DRE (preferably by the same examiner) is important because early cancer may be detected as a nodule within the gland or as an extensive hardening in the posterior lobe. The more advanced lesion is "stony hard" and fixed. DRE also provides useful clinical information about the rectum, anal sphincter, and quality of stool.

The diagnosis of prostate cancer is confirmed by a histologic examination of tissue removed surgically by TURP, open prostatectomy, or ultrasound-guided transrectal needle biopsy. Fine-needle aspiration is a quick, painless method of obtaining prostate cells for cytologic examination and determining the stage of disease.

Most prostate cancers are detected when a man seeks medical attention for symptoms of urinary obstruction or are found by routine DRE and PSA testing. Cancer detected incidentally when TURP is performed for clinically benign disease and lower urinary tract symptoms occurs in about 1 of 10 cases.

Concept Mastery Alert

DRE and PSA testing are important screening procedures, because abnormal DRE and elevated levels of PSA may raise suspicion of prostate cancer. However, a diagnosis of cancer requires confirmation with a prostate biopsy.

TRUS helps detect nonpalpable prostate cancers and assists with staging of localized prostate cancer. Needle biopsies of the prostate are commonly guided by TRUS. The biopsies are examined by a pathologist to both determine if cancer is present and to grade the tumor. The most commonly used tumor grading system is the Gleason score. This system assigns a grade of 1 to 5 for the most predominant architectural pattern of the glands of the prostate and a secondary grade of 1 to 5 to the second most predominant pattern. The Gleason score is then reported as, for example, 2 + 4; the combined value can range from 2 to 10. With each increase in Gleason score, there is an increase in tumor aggressiveness. Lower Gleason scores indicate well-differentiated and less aggressive tumor cells; higher Gleason scores indicate undifferentiated cells and more aggressive cancer. A total score of 8 to 10 indicates a high-grade cancer (McAninch & Lue, 2012).

Categorization of low-, intermediate-, and high-risk prostate cancer is determined by the extent of cancer in the prostate gland, whether or not the cancer is localized to the prostate, the aggressiveness of the cells, and the spread to the lymph nodes and beyond. Level of risk, in turn, is used to determine treatment options.

Bone scans, skeletal x-rays, and magnetic resonance imaging (MRI) may be used to identify metastatic bone disease. Pelvic computed tomography (CT) scans may be performed to determine if the cancer has spread to the lymph nodes. The radiolabeled monoclonal antibody capromab pendetide with indium 111 (ProstaScint) is an antibody that can be used to detect either recurrent prostate cancer at low PSA levels or metastatic disease (NCCN, 2012b).

Medical Management

Treatment is based on the patient's life expectancy, symptoms, risk of recurrence after definitive treatment, size of the tumor, Gleason score, PSA level, likelihood of complications, and patient preference. Therapy is often guided by the use of a nomogram or risk stratification scheme suggested by the NCCN (2012b) clinical practice guidelines. A multidisciplinary team approach is essential for the development of appropriate treatment. Management may be nonsurgical and involve watchful waiting or be surgical and entail **prostatectomy**. Nursing care of the patient with cancer of the prostate is summarized in Chart 59-4.

For patients with prostate cancer who choose nonsurgical watchful waiting, this approach involves actively monitoring the course of disease and intervening only if the cancer progresses or if symptoms warrant other intervention. It is an option for patients with life expectancy of less than 5 years and low-risk cancers. Advantages include absence of side effects of more aggressive treatment, improved quality of life, avoidance of unnecessary treatment, and decreased initial costs. Disadvantages include

(text continues on page 1730)

Chart 59-4

PLAN OF NURSING CARE

The Patient With Prostate Cancer

NURSING DIAGNOSIS: Anxiety related to concern and lack of knowledge about the diagnosis, treatment plan, and prognosis
GOAL: Reduced stress and improved ability to cope

Nursing Interventions	Rationale	Expected Outcomes
1. Obtain health history to determine the following: **a.** Patient's concerns **b.** His level of understanding of his health problem **c.** His past experience with cancer **d.** Whether he knows his diagnosis of malignancy and its prognosis **e.** His support systems and coping methods	**1.** Nurse clarifies information and facilitates patient's understanding and coping.	• Appears relaxed • States that anxiety has been reduced or relieved • Demonstrates understanding of illness, diagnostic tests, and treatment when questioned • Verbalizes adequate coping ability • Engages in open communication with others
2. Provide education about diagnosis and treatment plan. **a.** Explain in simple terms what diagnostic tests to expect, how long they will take, and what will be experienced during each test. **b.** Review treatment plan, and encourage patient to ask questions.	**2.** Helping the patient to understand the diagnostic tests and treatment plan will help decrease his anxiety and promote cooperation.	
3. Assess his psychological reaction to his diagnosis/prognosis and how he has coped with past stresses.	**3.** This information provides clues in determining appropriate measures to facilitate coping.	
4. Provide information about institutional and community resources for coping with prostate cancer: social services, support groups, community agencies.	**4.** Institutional and community resources can help the patient and family cope with the illness and treatment on an ongoing basis.	

NURSING DIAGNOSIS: Urinary retention related to urethral obstruction secondary to prostatic enlargement or tumor and loss of bladder tone due to prolonged distention/retention
GOAL: Improved pattern of urinary elimination

Nursing Interventions	Rationale	Expected Outcomes
1. Determine patient's usual pattern of urinary function.	**1.** Provides a baseline for comparison and goal to work toward.	• Voids at normal intervals • Reports absence of frequency, urgency, or bladder fullness • Displays no palpable suprapubic distention after voiding • Maintains balanced intake and output
2. Assess for signs and symptoms of urinary retention: amount and frequency of urination, suprapubic distention, complaints of urgency and discomfort.	**2.** Voiding 20–30 mL frequently and output less than intake suggest retention.	
3. Catheterize patient to determine amount of residual urine.	**3.** Determines amount of urine remaining in bladder after voiding.	
4. Initiate measures to treat retention. **a.** Encourage assuming normal position for voiding. **b.** Recommend using Valsalva maneuver preoperatively, if not contraindicated. **c.** Administer prescribed cholinergic agent. **d.** Monitor effects of medication.	**4.** Promotes voiding: **a.** Usual position provides relaxed conditions conducive to voiding. **b.** Valsalva maneuver exerts pressure to force urine out of bladder. **c.** Stimulates bladder contraction. **d.** If unsuccessful, another measure may be required.	
5. Consult with primary provider regarding intermittent or indwelling catheterization; assist with procedure as required.	**5.** Catheterization will relieve urinary retention until the specific cause is determined; it may be an obstruction that can be corrected only surgically.	
6. Monitor catheter function; maintain sterility of closed system; irrigate as required.	**6.** Adequate functioning of catheter is to be ensured to empty bladder and to prevent infection.	
7. Prepare patient for surgery if indicated.	**7.** Surgical removal of obstruction may be necessary.	

(continues on page 1728)

Assessing Skin Color

The color gradations that occur in people with dark skin are largely determined by genetics; they may be described as light, medium, or dark. In people with dark skin, melanin is produced at a faster rate and in larger quantities than in people with light skin. Healthy dark skin has a reddish base or undertone. The buccal mucosa, tongue, lips, and nails normally are pink. The skin of exposed portions of the body, especially in sunny, warm climates, tends to be more pigmented than the rest of the body. Almost every process that occurs on the skin causes some color change. For example, **hypopigmentation** (i.e., loss of pigmentation) may be caused by a fungal infection, eczema, or **vitiligo** (i.e., white patches); **hyperpigmentation** (i.e., increase in pigmentation) can occur after sun injury or as a result of aging. Dark pigment responds with discoloration after injury or inflammation, and patients with dark skin more often experience postinflammatory hyperpigmentation than those with lighter skin. The hyperpigmentation eventually fades but may require months to a year to do so.

Changes in skin color in people with dark skin are more noticeable and may cause more concern because the discoloration is more readily visible. Because of the increased number of melanocytes in darker skin, pigment changes can become quite obvious and cause great psychological discomfort. Some variation in skin pigment levels is considered normal. Examples include the pigmented crease across the bridge of the nose, pigmented streaks in the nails, and pigmented spots on the sclera of the eye. Women of color often develop a dark line along the midline of the lower abdomen during pregnancy.

Table 60-1 provides an overview of color changes in light- and dark-skinned people.

Cyanosis

Cyanosis is the bluish discoloration that results from a lack of oxygen in the blood (Fig. 60-3). It appears with shock or with respiratory or circulatory compromise. In people with light skin, cyanosis manifests as a bluish hue to the lips, fingertips, and nail beds. Other indications of decreased tissue perfusion include cold, clammy skin; a rapid, thready pulse; and rapid, shallow respirations. The conjunctivae of the eyelids are examined for pallor and **petechiae** (i.e., pinpoint red spots that result from blood leakage into skin).

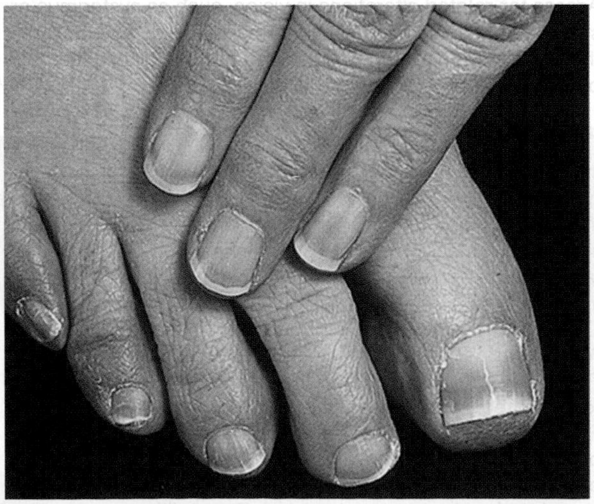

FIGURE 60-3 • Examples of skin color changes: the bluish tint of cyanosis (*left*) and the yellow hue of jaundice (*right*). From Bickley, L. S., & Szilagyi, P. G. (2013). *Bates' guide to physical examination and history taking* (11th ed.). Philadelphia: Lippincott Williams & Wilkins.

In a person with dark skin, the skin usually assumes a grayish cast. To detect cyanosis, the areas around the mouth and lips and over the cheekbones and earlobes should be observed.

Erythema

Erythema is redness of the skin caused by the congestion of capillaries. In light-skinned people, it is easily observable. To determine possible inflammation, the skin is palpated for increased warmth and for smoothness (i.e., edema) or hardness (i.e., intracellular infiltration). Because dark skin tends to assume a purple-gray cast when an inflammatory process is present, it may be difficult to detect erythema.

Jaundice

Jaundice, a yellowing of the skin, is directly related to elevations in serum bilirubin and is often first observed in the sclerae and mucous membranes (see Fig. 60-3).

Assessing Rash

In instances of pruritus (i.e., itching), the patient is asked to indicate which areas of the body are involved. The skin is then stretched gently to decrease the reddish tone and make the rash more visible. Pointing a penlight laterally across the skin may highlight the rash, making it easier to observe. The differences in skin texture are then assessed by running the tips of the fingers lightly over the skin. The borders of the rash may be palpable. The patient's mouth and ears are included in the examination (rubeola, or measles, causes a red cast to appear on the ears, and skin cancers are quite common on the crest of the ears). The patient's temperature is assessed, and the lymph nodes are palpated especially in the axilla, inguinal fold, and behind the knees (popliteal area).

Assessing Skin Lesions

Skin lesions are the most prominent characteristics of dermatologic conditions. They vary in size, shape, and cause and are classified according to their appearance and origin. Skin lesions can be described as primary or secondary. Primary lesions are the initial lesions and are characteristic of the disease itself. Secondary lesions result from changes in primary lesions resulting from external causes, such as scratching, trauma, infections, or changes caused by wound healing. Depending on the stage of development, skin lesions are further categorized by type and appearance (Table 60-2).

A preliminary assessment of the eruption or lesion helps identify the type of **dermatosis** (i.e., abnormal skin condition) and indicates whether the lesion is primary or secondary. At the same time, the anatomic distribution of the eruption or lesion should be observed because certain diseases affect certain sites of the body and are distributed in characteristic patterns and shapes (Figs. 60-4 and 60-5). To determine the extent of the regional distribution, the left and right sides of the body should be compared while the color and shape of

TABLE 60-1 Color Changes in Light and Dark Skin

Etiology	Light Skin	Dark Skin
Pallor		
Anemia—decreased hematocrit Shock—decreased perfusion, vasoconstriction	Generalized pallor	Brown skin appears yellow-brown, dull; black skin appears ashen gray, dull. (Observe areas with least pigmentation: conjunctivae, mucous membranes.)
Local arterial insufficiency	Marked localized pallor (lower extremities, especially when elevated)	Ashen gray, dull; cool to palpation
Albinism—total absence of pigment melanin	Whitish pink	Tan, cream, white
Vitiligo—a condition characterized by destruction of the melanocytes in circumscribed areas of the skin (may be localized or widespread)	Patchy, milky white spots, often symmetric bilaterally	Same
Cyanosis		
Increased amount of unoxygenated hemoglobin:	Dusky blue	Dark but dull, lifeless; only severe cyanosis is apparent in skin. (Observe conjunctivae, oral mucosa, nail beds.)
Central—chronicheart and lung diseases cause arterial desaturation	Nail beds dusky	
Peripheral—exposure to cold, anxiety		
Erythema		
Hyperemia—increased blood flow through engorged arterial vessels, as in inflammation, fever, alcohol intake, blushing	Red, bright pink	Purplish tinge, but difficult to see. (Palpate for increased warmth with inflammation, taut skin, and hardening of deep tissues.)
Polycythemia—increased red blood cells, capillary stasis	Ruddy blue in face, oral mucosa, conjunctivae, hands and feet	Well concealed by pigment. (Observe for redness in lips.)
Carbon monoxide poisoning	Bright, cherry red in face and upper torso	Cherry red nail beds, lips, and oral mucosa
Venous stasis—decreased blood flow from area, engorged venules	Dusky rubor of dependent extremities (a prelude to necrosis with pressure ulcer)	Easily masked. (Use palpation to identify warmth or edema.)
Jaundice		
Increased serum bilirubin concentration (>2–3 mg/100 mL) due to liver dysfunction or hemolysis, as after severe burns or some infections	Yellow first in sclerae, hard palate, and mucous membranes; then over skin	Check sclerae for yellow near limbus; do not mistake normal yellowish fatty deposits in the periphery under eyelids for jaundice. (Jaundice is best noted at junction of hard and soft palate, on palms.)
Carotenemia—increased level of serum carotene from ingestion of large amounts of carotene-rich foods	Yellow-orange tinge in forehead, palms and soles, and nasolabial folds, but no yellowing in sclerae or mucous membranes	Yellow-orange tinge in palms and soles
Uremia—renal failure causes retained urochrome pigments in the blood	Orange-green or gray overlying pallor of anemia; may also have ecchymoses and purpura	Easily masked. (Rely on laboratory and clinical findings.)
Brown-Tan		
Addison's disease—cortisol deficiency stimulates increased melanin production	Bronzed appearance, an "external tan"; most apparent around nipples, perineum, genitalia, and pressure points (inner thighs, buttocks, elbows, axillae)	Easily masked. (Rely on laboratory and clinical findings.)
Café-au-lait spots—caused by increased melanin pigment in basal cell layer	Tan to light brown, irregularly shaped, oval patch with well-defined borders often not visible in the very dark skinned person	

Adapted from Kelly, A. P., & Taylor, S. C. (2009). *Dermatology for skin of color*. New York: McGraw-Hill Medical.

the lesions are assessed. The degree of pigmentation of the patient's skin may affect the appearance of a lesion. Lesions may be black, purple, or gray on dark skin and tan or red in patients with light skin. A metric ruler is used to measure the size of the lesions so that any further extension can be compared with this baseline measurement. After observation, the lesions are palpated to determine their texture, shape, and border and to see if they are soft and filled with fluid or hard and fixed to the surrounding tissue.

Skin lesions are described clearly and in detail on the patient's health record, using precise terminology:

- Color of the lesion
- Any redness, heat, pain, or swelling
- Size and location of the involved area
- Pattern of eruption (e.g., macular, papular, scaling, oozing, discrete, confluent)
- Distribution of the lesion (e.g., bilateral, symmetric, linear, circular)

If acute open wounds or lesions are found on inspection of the skin, a comprehensive assessment should be made and documented. This assessment should address the following issues:

- *Wound bed:* Inspect for necrotic and granulation tissue, epithelium, exudate, color, and odor.
- *Wound edges and margins:* Observe for undermining (i.e., extension of the wound under the surface skin), and evaluate for condition of skin (i.e., necrotic).

(text continues on page 1762)

TABLE 60-2 Primary and Secondary Skin Lesions

Lesion	Description	Examples
Primary Lesions		
MACULE, PATCH 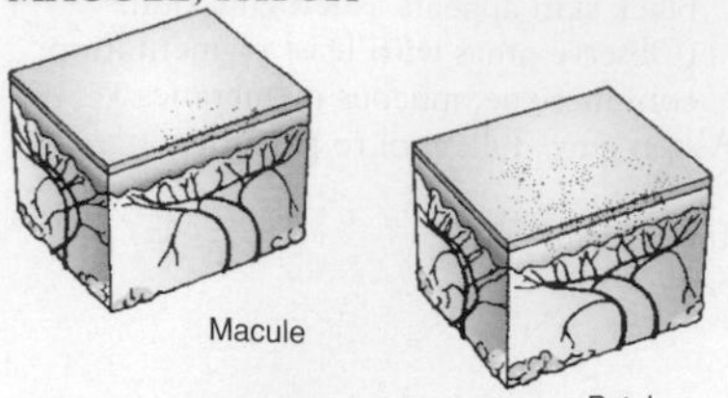 Macule Patch	Flat, nonpalpable skin color change (color may be brown, white, tan, purple, red) • *Macule:* <1 cm; circumscribed border • *Patch:* >1 cm; may have irregular border	Freckles, flat moles, petechia, rubella, vitiligo, port wine stains, ecchymosis
PAPULE, PLAQUE 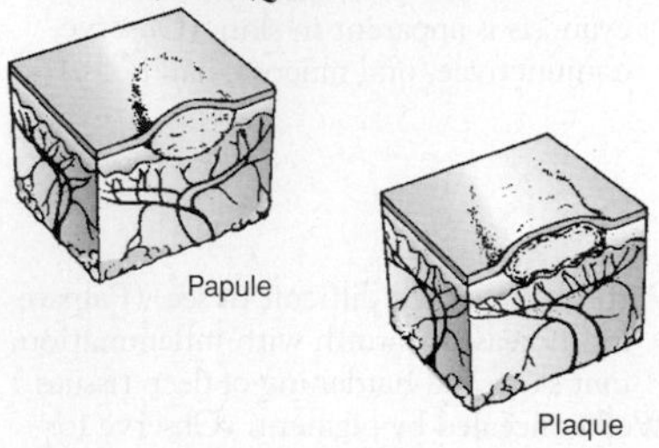 Papule Plaque	Elevated, palpable, solid mass with a circumscribed border Plaque may be coalesced papules with flat top • *Papule:* <0.5 cm • *Plaque:* >0.5 cm	*Papules:* Elevated nevi, warts, lichen planus *Plaques:* Psoriasis, actinic keratosis
NODULE, TUMOR 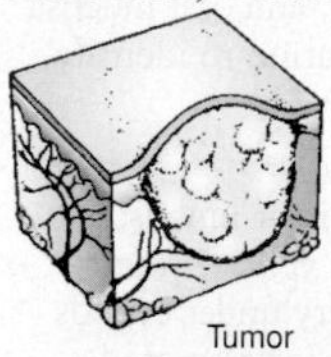 Tumor	Elevated, palpable, solid mass that extends deeper into the dermis than a papule • *Nodule:* 0.5–2 cm; circumscribed • *Tumor:* >1–2 cm; tumors do not always have sharp borders	*Nodules:* Lipoma, squamous cell carcinoma, poorly absorbed injection, dermatofibroma *Tumors:* Larger lipoma, carcinoma
VESICLE, BULLA 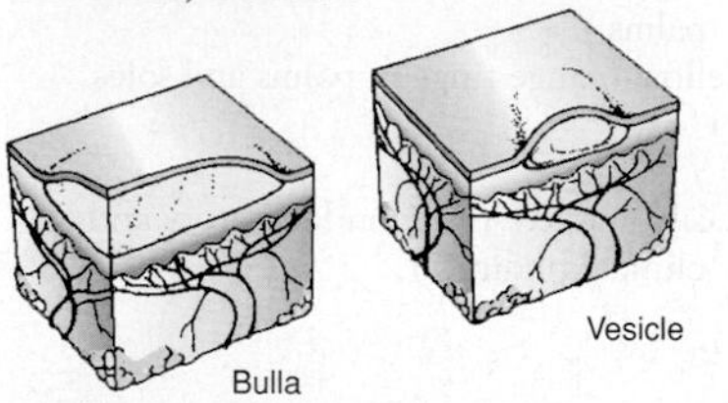 Vesicle Bulla	Circumscribed, elevated, palpable mass containing serous fluid • *Vesicle:* <0.5 cm • *Bulla:* >0.5 cm	*Vesicles:* Herpes simplex/zoster, varicella, poison ivy, 2nd-degree burn (blister) *Bulla:* Pemphigus, contact dermatitis, large burn blisters, poison ivy, bullous impetigo
WHEAL 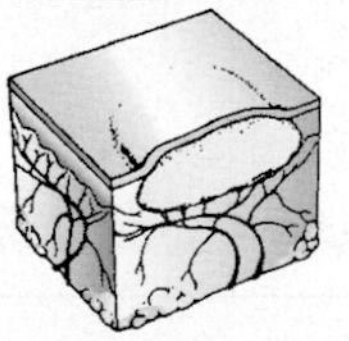	Elevated mass with transient borders; often irregular; size and color vary Caused by movement of serous fluid into the dermis; does not contain free fluid in a cavity (e.g., as a vesicle does)	Urticaria (hives), insect bites
PUSTULE 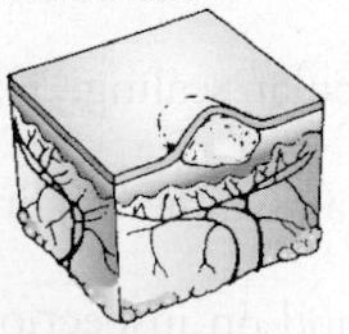	Pus-filled vesicle or bulla	Acne, impetigo, furuncles, carbuncles
CYST 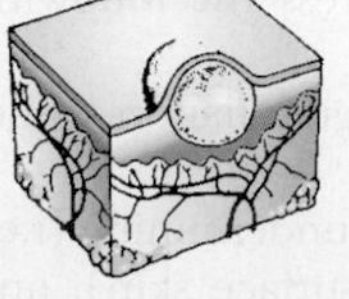	Encapsulated fluid-filled or semisolid mass in the subcutaneous tissue or dermis	Sebaceous cyst, epidermoid cysts

Lesion	Description	Examples
Secondary Lesions		
EROSION 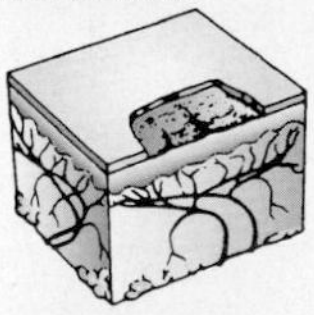	Loss of superficial epidermis that does not extend to dermis; depressed, moist area	Ruptured vesicles, scratch marks
ULCER 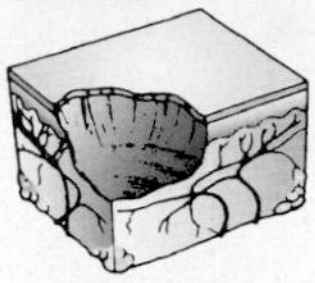	Skin loss extending past epidermis; necrotic tissue loss; bleeding and scarring possible	Stasis ulcer of venous insufficiency, pressure ulcer
FISSURE 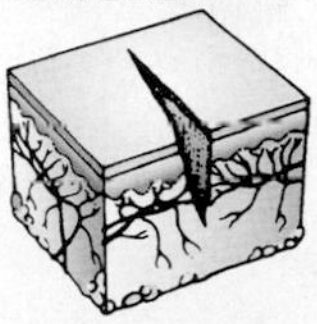	Linear crack in the skin that may extend to dermis	Chapped lips or hands, tinea pedis
SCALES 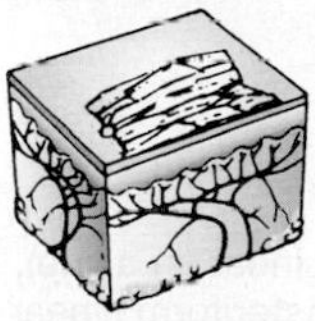	Flakes secondary to desquamated, dead epithelium that may adhere to skin surface; color varies (silvery, white); texture varies (thick, fine)	Dandruff, psoriasis, dry skin, pityriasis rosea
CRUST 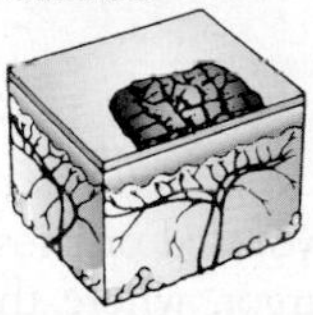	Dried residue of serum, blood, or pus on skin surface Large, adherent crust is a scab.	Residue left after vesicle rupture: impetigo, herpes, eczema
SCAR (CICATRIX) 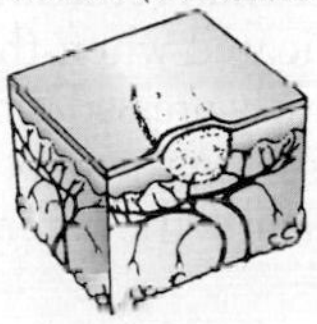	Skin mark left after healing of a wound or lesion; represents replacement by connective tissue of the injured tissue • *Young scars:* Red or purple • *Mature scars:* White or glistening	Healed wound or surgical incision
KELOID 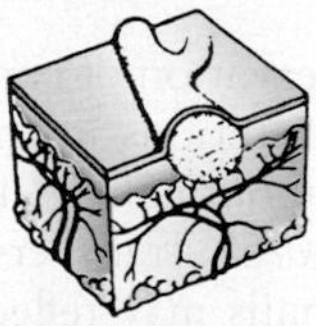	Hypertrophied scar tissue secondary to excessive collagen formation during healing; elevated, irregular, red Greater incidence among African Americans	Keloid of ear piercing or surgical incision
ATROPHY 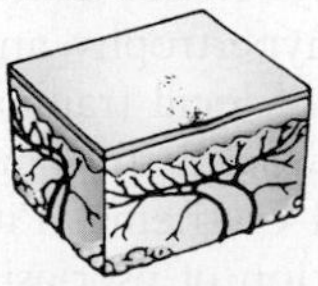	Thin, dry, transparent appearance of epidermis; loss of surface markings; secondary to loss of collagen and elastin; underlying vessels may be visible	Aged skin, arterial insufficiency
LICHENIFICATION 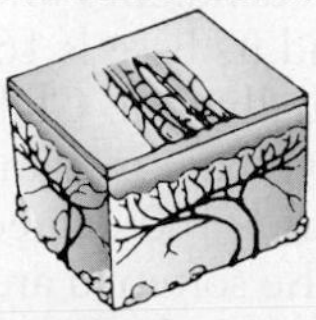	Thickening and roughening of the skin or accentuated skin markings that may be secondary to repeated rubbing, irritation, scratching	Contact dermatitis

Adapted from Bickley, L. S. (2010). *Bates' guide to physical examination and history taking* (11th ed.). Philadelphia: Lippincott Williams & Wilkins and Weber, J. W., & Kelley, J. (2009). *Health assessment in nursing* (4th ed.). Philadelphia: Lippincott Williams & Wilkins.

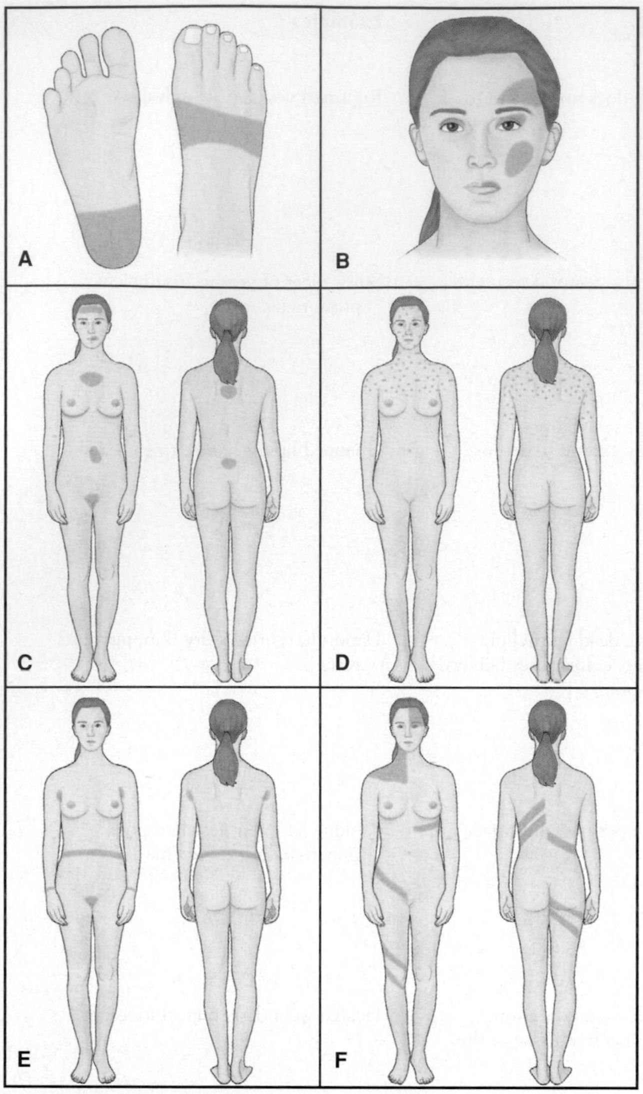
A
B
C
D
E
F

FIGURE 60-4 • Anatomic distribution of common skin disorders. **A.** Contact dermatitis (shoes). **B.** Contact dermatitis (cosmetics, perfumes, earrings). **C.** Seborrheic dermatitis. **D.** Acne. **E.** Scabies. **F.** Herpes zoster (shingles).

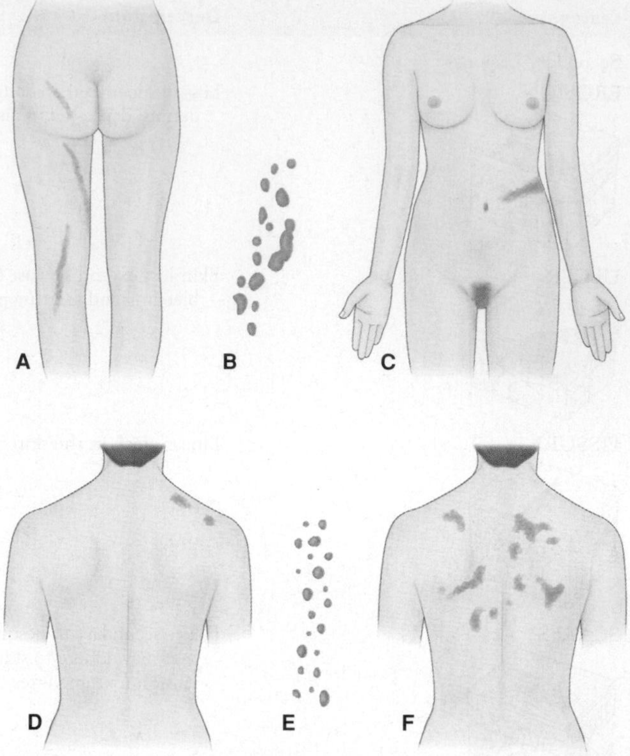
A
B
C
D
E
F

FIGURE 60-5 • Skin lesion configurations. **A.** Linear (in a line). **B.** Annular and arciform (circular or arcing). **C.** Zosteriform (linear along a nerve route). **D.** Grouped (clustered). **E.** Discrete (separate and distinct). **F.** Confluent (merged).

- *Wound size:* Measure in millimeters or centimeters, as appropriate, to determine diameter and depth of the wound and surrounding erythema.
- *Surrounding skin:* Assess for color, suppleness and moisture, irritation, and scaling.

Assessing Vascularity and Hydration

After the color of the skin has been evaluated and lesions have been inspected, an assessment of vascular changes in the skin is performed. A description of vascular changes includes location, distribution, color, size, and the presence of pulsations. Common vascular changes include petechiae, ecchymoses, **telangiectasias** (venous stars), and angiomas (Table 60-3).

Skin moisture, temperature, and texture are assessed primarily by palpation. The turgor (i.e., elasticity) of the skin, which decreases in normal aging, may be a factor in assessing the hydration status of a patient. To assess skin turgor, the skin should be gently pinched between the thumb and forefinger. The skin is observed to see how long it takes to return to base location. People who are dehydrated or those with dry skin will display decreased skin turgor, where the skin remains tented after being pinched rather than returning to normal almost immediately. Edema is indicated when the skin appears tense and shiny, when a finger gently pressed into the skin leaves an indentation or "pit." Measuring the depth of the pit and length of time to resolution indicates the extent of edema (DeGowan, 2009).

Assessing the Nails

A brief inspection of the nails includes observation of configuration, color, and consistency. Many alterations in the nail or nail bed reflect local or systemic abnormalities in progress or resulting from past events (Fig. 60-6). Transverse depressions known as Beau's lines in the nails may reflect retarded growth of the nail matrix because of severe illness or, more commonly, local trauma. Ridging, hypertrophy, and other changes may also be visible because of local trauma. Paronychia, an inflammation of the skin around the nail, is usually accompanied by tenderness and erythema. Pitted surface of the nails is a definite indication of psoriasis. Spoon-shape nails can indicate a severe iron deficiency anemia. The angle between the normal nail and its base is 160 degrees. When palpated, the nail base is usually firm. Clubbing of the nails, which can occur from hypoxia, is manifested by a straightening of the normal angle (180 degrees or greater) and softening of the nail base. The softened area feels spongelike when palpated (Bickley, 2010; Weber & Kelley, 2009).

TABLE 60-3 Vascular Lesions

PETECHIA (*PL.* PETECHIAE) 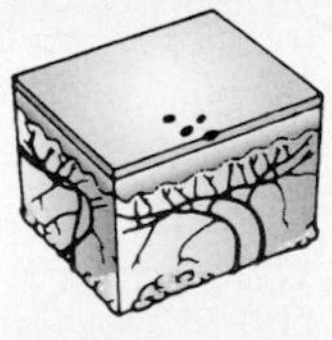	Round red or purple macule Small (1–2 mm) Secondary to blood extravasation Associated with bleeding tendencies or emboli to skin
ECCHYMOSIS (*PL.* ECCHYMOSES) 	Round or irregular macular lesion Larger than petechia Color varies and changes—black, yellow, and green hues Secondary to blood extravasation Associated with trauma, bleeding tendencies
CHERRY ANGIOMA 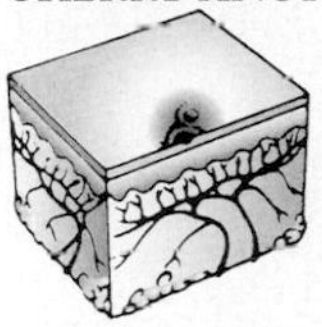	Papular and round Red or purple Noted on trunk, extremities May blanch with pressure Normal age-related skin alteration Usually not clinically significant
SPIDER ANGIOMA 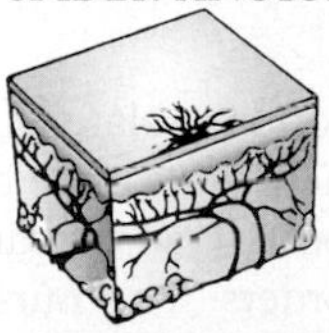	Red, arteriole lesion Central body with radiating branches Noted on face, neck, arms, trunk Rare below the waist May blanch with pressure Associated with liver disease, pregnancy, vitamin B deficiency
TELANGIECTASIA (VENOUS STAR) 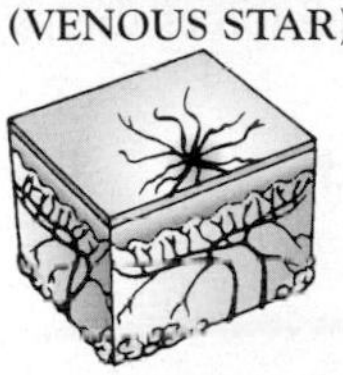	Shape varies—spiderlike or linear Color bluish or red Does not blanch when pressure is applied Noted on legs, anterior chest Secondary to superficial dilation of venous vessels and capillaries Associated with increased venous pressure states (varicosities)

Adapted from Bickley, L. S. (2010). *Bates' guide to physical examination and history taking* (11th ed.). Philadelphia: Lippincott Williams & Wilkins and Weber, J. W., & Kelley, J. (2009). *Health assessment in nursing* (4th ed.). Philadelphia: Lippincott Williams & Wilkins.

Assessing the Hair

The hair assessment is carried out by inspection and palpation. Gloves are worn by the examiner, and the examination room should be well lighted. The hair is separated so that the condition of the skin underneath can be easily seen. The examiner notes the color, texture, and distribution of hair shafts. The wooden end of a cotton swab can be used to make small parts in the hair so that the scalp can be inspected. Any abnormal lesions, evidence of itching, inflammation, scaling, or signs of infestation (i.e., lice or mites) are documented.

Color and Texture

Natural hair color ranges from white to black. Hair begins to turn gray with age, initially during the third decade of life, when the loss of melanin begins to become apparent. However, it is not unusual for the hair of younger people to turn gray as a result of hereditary traits. The person with albinism (i.e., partial or complete absence of pigmentation) has a genetic predisposition to white hair from birth. The natural state of the hair can be altered by using hair dyes, bleaches, and curling or relaxing products. The use of these products has varying impact on hair, depending on its natural characteristics. For example, the use of straightening chemicals on the hair of most people of African descent can cause extensive breakage and hair loss. The types of products used are identified in the assessment (Kelly & Taylor, 2009).

The texture of scalp hair ranges from fine to coarse, silky to brittle, oily to dry, and shiny to dull, and hair can be straight, curly, or kinky. Dry, brittle hair may result from the overuse of hair dyes, hair dryers, and curling irons or from endocrine disorders, such as thyroid dysfunction. Oily hair is usually caused by increased secretion from the sebaceous glands close to the scalp. If the patient reports a recent change in hair texture, the underlying reason is pursued; the alteration may arise simply from the overuse of commercial hair products or from changing to a new shampoo.

Distribution

Body hair distribution varies with location. Hair over most of the body is fine, except in the axillae and pubic areas, where it is coarse. Pubic hair, which develops at puberty, forms a diamond shape extending up to the umbilicus in boys and men. Female pubic hair resembles an inverted triangle. If the pattern found is more characteristic of the opposite gender, it may indicate an endocrine disorder and further investigation is in order. Racial differences in hair are expected, such as straight hair in Asians and curly, coarser hair in people of African descent.

Men tend to have more body and facial hair than women. Alopecia can occur over the entire body or be confined to a specific area. Scalp hair loss may be localized to patchy areas or may range from generalized thinning to total baldness. When assessing scalp hair loss, it is important to investigate the underlying cause with the patient. Patchy hair loss may be from habitual hair pulling or twisting; excessive traction on the hair (e.g., braiding too tightly); excessive use of dyes, straighteners, and oils; chemotherapeutic agents (e.g., doxorubicin [Adriamycin], cyclophosphamide [Cytoxan]); bacterial or fungal infection; or moles or lesions on the scalp. Well-defined patches of hair loss generally indicate the condition called *alopecia areata*. Regrowth may be erratic, and distribution may never attain the previous thickness (Wolff, 2009).

Hair Loss

The most common cause of hair loss is male pattern baldness (i.e., androgenic alopecia), which affects more than half of the male population and is believed to be related to heredity, aging, and androgen (male hormone) levels. Androgen is necessary for male pattern baldness to develop. The pattern of hair loss begins with receding of the hairline in the frontal temporal area and progresses to gradual thinning and complete loss of hair over the top of the scalp and crown. The typical pattern of male hair loss is illustrated in Figure 60-7. Although androgenic alopecia is considered a male disorder, millions of women also experience it. Women tend to retain some of the hair on the crown of the scalp and do not become completely bald (Wolff, 2009).

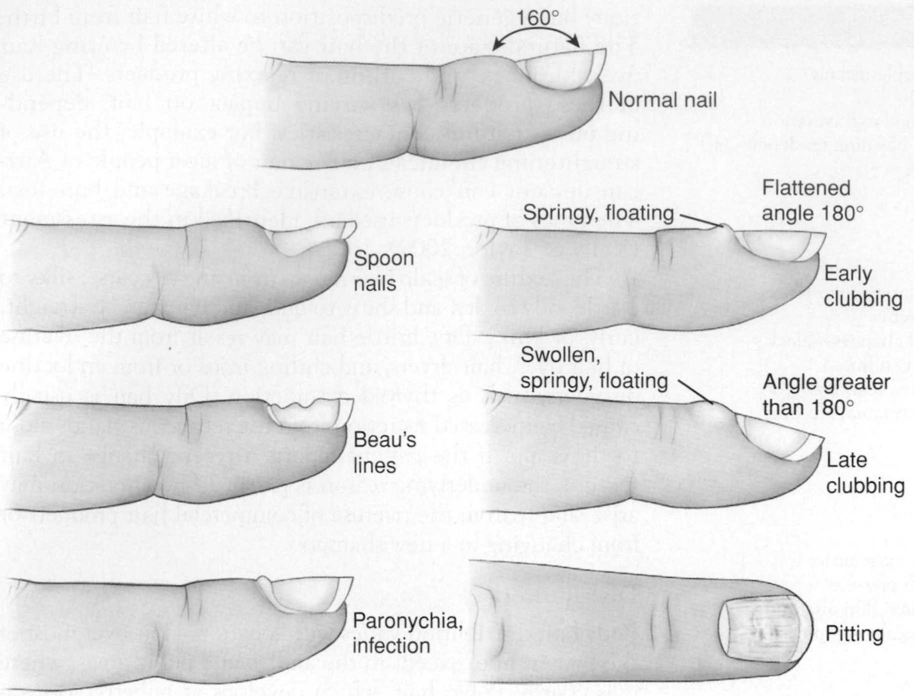
160°
Normal nail
Springy, floating
Flattened angle 180°
Spoon nails
Early clubbing
Swollen, springy, floating
Angle greater than 180°
Beau's lines
Late clubbing
Paronychia, infection
Pitting

FIGURE 60-6 • Common nail disorders.

Other Changes

Male pattern hair distribution may be seen in some women at the time of menopause, when the hormone estrogen is no longer produced by the ovaries. In women with hirsutism, excessive hair may grow on the face, chest, shoulders, and pubic area. If menopause is ruled out as the underlying cause, other hormonal changes related to pituitary or adrenal dysfunction must be investigated.

Because patients with skin conditions may be viewed negatively by others, these patients may become distraught and avoid interaction with people. Skin conditions can lead to isolation, job loss, and economic hardship as well as poor self-esteem.

Some conditions may lead to feelings of depression, frustration, self-consciousness, poor self-image, and rejection. Itching and skin irritation (features of many skin diseases) may be constant annoyances. These discomforts may result in loss of sleep, anxiety, and depressive symptoms, all of which reinforce the general distress and fatigue that frequently accompany skin disorders.

For patients experiencing physical and psychological discomforts, the nurse needs to provide understanding, explanations of the problem, appropriate instructions related to treatment, nursing support, and encouragement. It is imperative to overcome any aversion that may be felt when caring for patients with unattractive skin disorders. The nurse should show no sign of hesitancy when approaching patients with skin disorders. Such hesitancy only reinforces the psychological trauma of the disorder.

Skin Consequences of Selected Systemic Diseases

Diabetes

Because diabetes causes changes in circulation and cell nutrition, it can have a great impact on skin status. Some of the more common skin conditions encountered in diabetes are discussed in this section. Further information can be found in Chapter 51.

Diabetic Dermopathy

Diabetic dermopathy (shin spots) is a frequent occurrence in people with diabetes. These lesions are found on the lower anterior legs, forearms, and thighs and over other bony

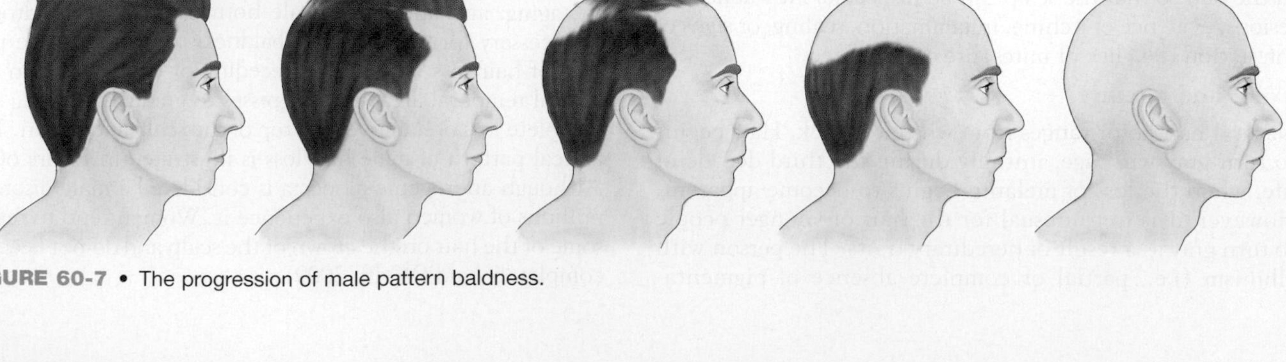

FIGURE 60-7 • The progression of male pattern baldness.

prominences. They are caused by breakdown of the small vessels that supply the skin. Each spot starts as a dull red bump, smaller than a pencil eraser. It slowly spreads to about 1 inch (the size of a quarter), becomes increasingly scaly, and eventually leaves a brownish scar on the skin. The lesions are usually bilateral and occur in linear clusters (Goldsmith, Katz, Gilchrest, et al., 2012).

Stasis Dermatitis

Stasis dermatitis is not unique to diabetes, but because of the blood vessel damage that results from diabetes, it is very common in patients with diabetes. Large vessels are damaged, compromising circulation to the lower arms and legs. The skin suffers from lack of nutrients, becoming very dry and fragile. Minor injuries heal slowly, and ulcers form easily. The skin takes on a thick leathery texture and a yellowish, waxy hue.

Skin Infections

The skin of patients with diabetes is prone to bacterial and fungal infections. Bacterial infections appear as small pimples around hair follicles. The most frequently affected sites include the lower legs, lower abdomen, and buttocks. Sometimes, these lesions enlarge to become furuncles or carbuncles. If the blood glucose level is not well controlled, these infections may be very slow to heal.

Fungal infections are quite common in areas that remain moist all the time (under breasts, upper thighs, in axilla). *Candida* (i.e., yeast) infections appear beefy red and have small pustules around the border of the area, with the skin appearing moist and raw.

Dermatophyte infections are dry and only minimally red, with more scale. Common sites are the toenails and feet.

Nurses must be alert to the signs of these common infections. If necessary, they should bring them to the attention of the patient's primary provider and help the patient or family learn basic skin maintenance techniques.

Leg and Foot Ulcers

Because of changes in peripheral nerves, patients with diabetes do not always sense minor injuries to the lower legs and feet. Infections begin and if left untreated may lead to ulcerations. Ulcerations are often not noticed and become quite large before being treated. Ulcerations unresponsive to treatment are a leading cause of diabetic foot and leg amputations.

Human Immunodeficiency Virus Disease

Cutaneous signs may be the first manifestation of human immunodeficiency virus (HIV), appearing in more than 90% of HIV-infected people as immune function deteriorates. These skin signs correlate with low CD4 counts and may become very atypical in immunocompromised people. Some disorders such as Kaposi's sarcoma, oral hairy leukoplakia, facial molluscum contagiosum, and oral candidiasis may suggest that CD4 counts are less than 200 to 300 cells/mcL. Skin infections, both bacterial and viral, are common and will appear more severe than expected. Acute flare of chronic conditions such as seborrhea or acne may indicate a new infection. Being sensitive to these changes can alert the nurse so that early interventions can be initiated (Schwartz, 2011).

Diagnostic Evaluation

A wide range of diagnostic studies may be performed in patients with altered integumentary function. The nurse should educate the patient about the purpose, what to expect, and any possible side effects related to these examinations prior to testing. The nurse should note trends in results because they provide information about whether lesions are primary or secondary, disease progression, and the patient's response to therapy.

Skin Biopsy

Performed to obtain tissue for microscopic examination, a skin biopsy may be obtained by scalpel excision or by a skin punch instrument that removes a small core of tissue. Biopsies are performed on skin nodules, plaques, blisters, and other lesions to rule out malignancy and to establish an exact diagnosis.

Immunofluorescence

Designed to identify the site of an immune reaction, immunofluorescence testing combines an antigen or antibody with a fluorochrome dye. Antibodies can be made fluorescent by attaching them to a dye. Direct immunofluorescence tests on skin are techniques to detect autoantibodies directed against portions of the skin. The indirect immunofluorescence test detects specific antibodies in the patient's serum.

Patch Testing

Performed to identify substances to which the patient has developed an allergy, patch testing involves applying the suspected allergens, such as nickel or fragrances, to normal skin under occlusive patches. Patients wear these occluded strips on their backs for 48 hours, and the area is assessed after 72 hours. The development of redness, fine elevations, or itching is considered a weak positive reaction; fine blisters, papules, and severe itching indicate a moderately positive reaction; and blisters, pain, and ulceration indicate a strong positive reaction. The nurse educates the patient with a positive reaction to avoid the allergen, which is often quite difficult, because of the common ambience of some topical allergens.

Skin Scrapings

Tissue samples are scraped from suspected fungal lesions with a scalpel blade moistened with oil so that the scraped skin adheres to the blade. The scraped material is transferred to a glass slide, covered with a coverslip, and examined microscopically. The spores and hyphae of dermatophyte infections, as well as infestations such as scabies, can be visualized.

Tzanck Smear

The Tzanck smear is a test used to examine cells from blistering skin conditions, such as herpes zoster, varicella, herpes simplex, and all forms of pemphigus. The secretions from a suspected lesion are applied to a glass slide, stained, and examined.

Wood's Light Examination

Wood's light is a special lamp that produces long-wave ultraviolet rays, which result in a characteristic blue to dark purple fluorescence. The color of the fluorescent light is best seen in a darkened room, where it is possible to differentiate epidermal from dermal lesions and hypo- and hyperpigmented lesions from normal skin. The patient is reassured that the light is not harmful to skin or eyes. Lesions that still contain melanin almost disappear under ultraviolet light, whereas lesions that are devoid of melanin increase in whiteness with ultraviolet light.

Clinical Photographs

Photographs are taken to document the nature and extent of the skin condition and are used to determine progress or improvement resulting from treatment. They are sometimes used to track the status of moles to document if the characteristics of the mole are changing.

Nursing Implications

The nurse may be responsible to ensure that consent forms are completed for surgical procedures and for clinical photography, that all specimens collected are managed according to protocol, that a log is maintained tracking specimens to and from the laboratory, and that results are received in a timely manner. The nurse educates the patient regarding appropriate care of surgical sites and implication of test results (Uhlenhake & Feldman, 2010).

Critical Thinking Exercises

1 ebp You are working in a dermatology clinic. A 44-year-old woman presents for treatment of squamous cell carcinoma that is present on her upper chest. She notes that she plays tennis and rarely uses sunscreen because she likes having a "healthy-looking tan." You note on her health history that she also has three young children at home. You wish to provide education about the appropriate use of sunscreens for her and her family. Identify the evidence to support the use of protection from the sun, including sunscreens, to prevent further occurrence of skin cancer. Discuss the strength of the evidence that supports the use of sunscreens. Identify the criteria used to evaluate the strength of the evidence for this practice.

2 As a home health nurse, you are making a first-time visit to a new patient, an 80-year-old woman who was discharged from an acute care rehabilitation facility after receiving a total hip replacement for a hip fracture 2 weeks ago. The patient shows you areas on her forearm where she had intravenous lines placed while she was in the hospital that are still visibly bruised. She asks you why it seems to take longer that it used to for any type of wound to heal and why her skin seems so much more fragile than it did than when she was younger. How do you respond?

3 pq You work in an internal medicine clinic. A woman who has just seen her primary provider for a routine visit reports to you that she forgot to ask something. Her sister was recently diagnosed with melanoma, and she is wondering about the implications for herself and her children. How would you explain this condition to her? List the priority precautions that would you advise for this woman and her children.

Brunner Suite Resources

Explore these additional resources to enhance learning for this chapter:

- NCLEX-Style Questions and Other Resources on thePoint, http://thePoint.lww.com/Brunner13e
- Study Guide
- PrepU
- Clinical Handbook
- Handbook of Laboratory and Diagnostic Tests

References

Books

Arenson, C., Busby-Whitehead, J., O'Brien, J. G., et al. (2009). *Reichel's care of the elderly* (6th ed.). New York: Cambridge University Press.

Bickley, L. S. (2010). *Bates' guide to physical examination and history taking* (11th ed.). Philadelphia: Lippincott Williams & Wilkins.

DeGowin R. L. (2009). *DeGowin's diagnostic examination* (9th ed.). New York: McGraw-Hill.

Goldsmith, L., Katz, S. I., Gilchrest, B., et al. (2012). *Fitzpatrick's dermatology in general medicine* (7th ed.). New York: McGraw-Hill.

James, W. D., Elston, D., & Berger, T. G. (2011). *Andrews' diseases of the skin* (11th ed.). Philadelphia: W. B. Saunders.

Kelly, A. P., & Taylor, S. C. (2009). *Dermatology for skin of color*. New York: McGraw-Hill Medical.

Porth, C. M., & Matfin, G. (2009). *Pathophysiology: Concepts of altered health states* (8th ed.). Philadelphia: Lippincott Williams & Wilkins.

Weber, J. W., & Kelley, J. (2009). *Health assessment in nursing* (4th ed.). Philadelphia: Lippincott Williams & Wilkins.

Wolff, K. (2009). *Color atlas & synopsis of clinical dermatology* (6th ed.). New York: McGraw-Hill.

Wolff, K., Goldsmith, L., Katz, S. I., et al. (2008). *Fitzpatrick's dermatology in general medicine* (7th ed.). New York: McGraw-Hill.

Journals and Electronic Documents

Schwartz, R. A. (2011). Cutaneous manifestations of HIV. *Medscape*. Available at: emedicine.medscape.com/article/1133746-overview

Terushkin, V., Bender, A., Psaty, E. L., et al. (2010). Estimated equivalent of vitamin D production. *Journal of the American Academy of Dermatology, 62*(6), 929.

Uhlenhake, E., & Feldman, S. R. (2010). Dermatology patient safety: Problems and solutions. *Journal of Dermatological Treatment, 21*(2), 86–92.

Wilhelmi, B. J., & Molnar, J. A. (2012). Finger nail and tip injuries. *Medscape*. Available at: emedicine.medscape.com/article/1285680-overview

Resources

American Academy of Dermatology (AAD), www.aad.org

Dermatology Atlas (DermIS), a cooperation between the Department of Clinical Social Medicine (University of Heidelberg) and the Department of Dermatology (University of Erlangen), www.dermis.net

New Zealand Dermatology Society (DermNet NZ), www.dermnetnz.org

Skin Cancer Foundation (lists approved sunscreens and other sun protection products), www.skincancer.org

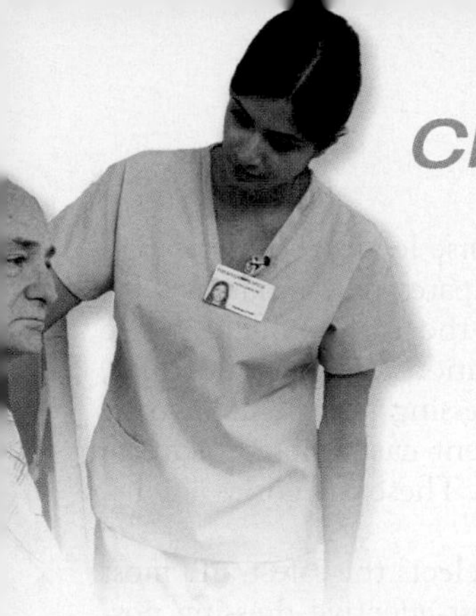

Chapter 61

Management of Patients With Dermatologic Problems

Learning Objectives

On completion of this chapter, the learner will be able to:

1. Describe the management of the patient with a wound, pruritus, or a dermatologic secretory disorder.
2. Describe the management and nursing care of the patient with infections of the skin and parasitic skin diseases.
3. Describe the management and nursing care of the patient with noninfectious inflammatory dermatoses, including contact dermatitis or psoriasis.
4. Use the nursing process as a framework for care of patients with blistering disorders, including toxic epidermal necrolysis and Stevens-Johnson syndrome.
5. Describe the management and nursing care of the patient with skin tumors (benign, malignant, and metastatic).
6. Use the nursing process as a framework for care of the patient with melanoma.
7. Compare the various types of dermatologic and plastic reconstructive procedures.
8. Describe the management and nursing care of patients undergoing plastic and cosmetic procedures.

Glossary

acantholysis: separation of epidermal cells from each other due to damage or abnormality of the intracellular substance
balneotherapy: a bath with therapeutic additives
bullae: large, fluid-filled blisters
carbuncle: localized skin infection involving several hair follicles
cheilitis: inflammation of the lips (when dry, cracking, inflamed skin occurs at the corners of the mouth, it is called *angular cheilitis;* when caused by sun exposure, it is called *solar cheilitis*)
comedones: the primary lesions of acne, caused by sebum blockage in the hair follicle
cytotoxic: destructive of cells
débridement: removal of necrotic or dead tissue by mechanical, surgical, chemical, or autolytic means
dermatitis: any inflammation of the skin
dermatosis: any abnormal skin lesion
epidermopoiesis: development of epidermal cells
furuncle: localized skin infection of a single hair follicle; also known as a boil
hydrophilic: a material that absorbs moisture
hydrophobic: a material that repels moisture
hygroscopic: a material that absorbs moisture from the air
lichenification: thickening of the horny layer of the skin; also known as scaling
liniments: lotions with added oil for increased softening of the skin
mitogenic: a substance that stimulates mitosis or cell division and reproduction
pruritus: itching
pyodermas: bacterial skin infections
striae: bandlike streaks on the skin, distinguished by color, texture, depression, or elevation from the tissue in which they are found; usually purplish or white
suspensions: liquid preparations in which powder is suspended, requiring shaking before use
tinea: a superficial fungal infection on the skin or scalp
xerosis: overly dry skin

Dermatologic disorders are encountered frequently by nurses across many practice settings. Nursing management of patients with dermatologic problems includes administering topical and systemic medications and managing dressings. The objectives of therapy are to prevent additional damage, prevent secondary infection, reverse the inflammatory process, and relieve the symptoms.

SKIN CARE FOR PATIENTS WITH SKIN CONDITIONS

Some skin problems are markedly aggravated by soap and water; therefore, bathing routines are modified according to the condition. Denuded skin, whether the area of desquamation is large or small, is excessively prone to damage by chemicals and trauma. The friction of a towel, if applied with vigor, is sufficient to produce a brisk inflammatory response that causes any existing lesion to flare up and extend.

Protecting the Skin

Basic skin care in bathing a patient with skin problems is as follows:

- A mild, lipid-free soap or soap substitute is used (e.g., Dove, Cetaphil, Aquanil).
- The area is rinsed completely and blotted dry with a soft cloth.
- Deodorant soaps and laundry detergents are avoided.

Special care is necessary when changing dressings. The use of pledgets saturated with oil, sterile saline, or another prescribed solution helps loosen crusts, remove exudates, or free an adherent dry dressing (Dale & Wright, 2011).

Preventing Secondary Infection

Skin lesions should be regarded as potentially infectious, and proper precautions should be observed until the diagnosis is established. Most lesions with purulent drainage contain infectious material. The nurse and primary provider must adhere to standard precautions and wear gloves when inspecting the skin or changing a dressing. The use of standard precautions and proper disposal of any contaminated dressing is carried out according to Occupational Safety and Health Administration (OSHA) regulations (OSHA, 2012).

Reversing the Inflammatory Process

The type of skin lesion (e.g., oozing, infected, or dry) usually determines the type of local medication or treatment that is prescribed. As a rule, if the skin is acutely inflamed (i.e., hot, red, and swollen) and oozing, it is best to apply wet dressings and soothing lotions. For chronic conditions in which the skin surface is dry and scaly, water-soluble emulsions, creams, ointments, and pastes are used (Goldsmith, Katz, Gilchrist, et al., 2012). The therapy is modified as the responses of the skin indicate. The patient and the nurse note whether the medication or dressings irritate the skin.

WOUND CARE FOR SKIN CONDITIONS

There are three types of wound dressings: passive, interactive, and active. *Passive dressings* have only a protective function and maintain a moist environment for natural healing. They include those that just cover the area (e.g., DuoDERM, Tegaderm) and may remain in place for several days. *Interactive dressings* are capable of absorbing wound exudate while (1) maintaining a moist environment in the area of the wound and (2) allowing the surrounding skin to remain dry. They include hydrocolloids, alginates, and hydrogels. It is thought that interactive dressings are able to modify the physiology of the wound environment by modulating and stimulating cellular activity and by releasing growth factor (Fonder, Lazarus, Cowan, et al., 2008). *Active dressings* improve the healing process and decrease healing time. They include skin grafts and biologic skin substitutes. Both interactive and active dressings create a moist environment at the interface of the wound with the dressing.

Because so many wound care products are available, it is often difficult to select the most appropriate product for a specific wound. Selection of products should be made carefully because of their expense. Both clinical efficacy and health-related outcomes (e.g., decreased pain, increased mobility) should be used to measure the success of a product for a wound. Even with the availability of a large variety of dressings, an appropriate selection can be made if certain principles are maintained. These principles are referred to as the five rules of wound care (Krasner, Rodeheaver, Sibbald, et al., 2012):

- *Rule 1: Categorization*. The nurse learns about dressings by generic category and compares new products with those that already make up the category. The nurse becomes familiar with indications, contraindications, and side effects. The best dressing may be created by combining products in different categories to achieve several goals at the same time. These categories are discussed in subsequent sections.
- *Rule 2: Selection*. The nurse selects the safest and most effective, easy-to-use, and cost-effective dressing possible. In many cases, nurses carry out the primary provider's prescriptions for dressings, but they must be prepared to give the primary provider feedback about the dressing's effect on the wound, ease of use for the patient, and other considerations when applicable.
- *Rule 3: Change*. The nurse changes dressings based on patient, wound, and dressing assessments, not on standardized routines.

Quality and Safety Nursing Alert

The natural wound-healing process should not be disrupted. Unless the wound is infected or has a heavy discharge, it is common to leave chronic wounds covered for 48 to 72 hours and acute wounds for 24 hours.

- *Rule 4: Evolution*. As the wound progresses through the phases of wound healing, the dressing protocol is altered to optimize healing. It is rare, especially in cases of chronic wounds, that the same dressing material is appropriate throughout the healing process. The nurse educates the patient or family caregiver about wound care and ensures that the family has access to appropriate dressing choices.
- *Rule 5: Practice*. Practice with dressing material is required for the nurse to learn the performance parameters of the particular dressing. Refining the skills of applying appropriate dressings correctly and learning about new dressing products are essential nursing responsibilities. Dressing changes should not be delegated to unlicensed personnel; these techniques require the knowledge base and assessment skills of professional nurses.

Autolytic Débridement

Autolytic **débridement** is a process that uses the body's own digestive enzymes to break down necrotic tissue. The wound is kept moist with occlusive dressings. Eschar and necrotic debris are softened, liquefied, and separated from the bed of the wound.

Several commercially available products contain the same enzymes that the body produces naturally and are referred to as enzymatic débriding agents; an example is collagenase (Santyl). Application of these products speeds the rate at which necrotic tissue is removed. This method, although slower than surgical débridement, is more discriminating for tissue removal and does not damage healthy tissue surrounding the wound. When enzymatic débridement is being used under an occlusive dressing, a foul odor is produced by the breakdown of cellular debris. This odor does not indicate that the wound is infected. The nurse should expect this reaction

TABLE 61-1 Quick Guide to Function and Action of Wound Dressings

Function	Action	Example
Absorption	Absorbs exudate	Alginates, composite dressings, foams, gauze, hydrocolloids, hydrogels
Cleansing	Removes purulent drainage, foreign debris, and devitalized tissue	Wound cleansers
Débridement	*Autolytic*—covers a wound and allows enzymes to self-digest sloughed skin	Absorption beads, pastes, powders; alginates; composite dressings; foams; hydrate gauze; hydrogels; hydrocolloids; transparent films; wound care systems
	Chemical or enzymatic—applied topically to break down devitalized tissue	Enzymatic débridement agents
	Mechanical—removes devitalized tissue with mechanical force	Wound cleansers, gauze (wet to dry), whirlpool
Diathermy	Produces electrical current to promote warmth and new tissue growth	
Hydration	Adds moisture to a wound	Gauze (saturated with saline) solution, hydrogels, wound care systems
Maintain moist environment	Manages moisture levels in a wound and maintains a moist environment	Composites, contact layers, foams, gauze (impregnated or saturated), hydrogels, hydrocolloids, transparent films, wound care systems
Manage high-output wounds	Manages excessive quantities of exudate	Pouching systems
Pack or fill dead space	Prevents premature wound closure or fills shallow areas and provides absorption	Absorbent beads, powders, pastes; alginates; composites, foams; gauze (impregnated and nonimpregnated)
Protect and cover wound	Provides protection from the external environment	Composites, compression bandages/wraps, foams, gauze dressings, hydrogels, hydrocolloids, transparent film dressings
Protect periwound skin	Prevents moisture and mechanical trauma from damaging delicate tissue around wound	Composites, foams, hydrocolloids, pouching systems, skin sealants, transparent film dressings
Provide therapeutic compression	Provides appropriate levels of support to the lower extremities in venous stasis disease	Compression bandages, wraps, graduated compression stockings

Adapted from Krasner, D., Rodeheaver, G., Sibbald, G., et al. (2012). *Chronic wound care: A clinical source book for healthcare professionals* (5th ed.). Malvern, PA: HMP Communications.

and help the patient and family understand the reason for the odor (Ramundo & Gray, 2008).

Categories of Dressings

Table 61-1 provides a guide to the function and action of wound dressings.

Occlusive Dressings

Occlusive dressings may be commercially produced or made inexpensively from sterile or nonsterile gauze squares or wrap. Occlusive dressings cover topical medication that is applied to a skin lesion. The area is kept airtight by using plastic film (e.g., plastic wrap). Plastic film is thin and readily adapts to all sizes, body shapes, and skin surfaces. Generally, plastic wrap should be used no more than 12 hours each day. Plastic surgical tape containing a corticosteroid in the adhesive layer can be cut to size and applied to individual lesions.

Wet Dressings

Wet dressings (wet compresses applied to the skin) were traditionally used for acute, weeping, inflammatory lesions. They have become almost obsolete because of the many newer products available for wound care.

Moisture-Retentive Dressings

Commercially produced moisture-retentive dressings can perform the same functions as wet dressings but are more efficient at removing exudate because of their higher moisture-vapor transmission rate; some have reservoirs that can hold excessive exudate. A number of moisture-retentive dressings are already impregnated with saline solution, petrolatum, zinc-saline solution, hydrogel, or antimicrobial agents, thereby eliminating the need to coat the skin to avoid maceration. The main advantages of moisture-retentive dressings over wet dressings are improved fibrinolysis, accelerated epidermal resurfacing, reduced pain, fewer infections, less scar tissue, gentle autolytic débridement, and decreased frequency of dressing changes. Depending on the product used and the type of dermatologic conditions encountered, most moisture-retentive dressings may remain in place from 12 to 24 hours; some can remain in place as long as a week (Baranoski & Ayello, 2012).

Hydrogels

Hydrogels are polymers with 90% to 95% water content. They are available in impregnated sheets or as gel in a tube. Their high moisture content makes them ideal for autolytic débridement of wounds. They are semitransparent, allowing for wound inspection without dressing removal. They are comfortable and soothing for the painful wound. They require a secondary dressing to keep them in place. Hydrogels are appropriate for superficial wounds with high serous output, such as abrasions, skin graft sites, and draining venous ulcers (Baranoski & Ayello, 2012).

Hydrocolloids

Hydrocolloids are composed of a water-impermeable, polyurethane outer covering separated from the wound by a hydrocolloid material. They are adherent and nonpermeable to water vapor and oxygen. As water evaporates over the wound, it is absorbed into the dressing, which softens and discolors with the increased water content. The dressing can be removed without damage to the wound. As the dressing absorbs water, it produces a foul-smelling, yellowish covering over the wound. This is a normal chemical interaction between the dressing and wound exudate and should not be confused with purulent drainage from the wound. Unfortunately, most of the hydrocolloid dressings are opaque, preventing inspection of the wound without removal of the dressing.

Available in sheets and in gels, hydrocolloids are a good choice for both exudative and acute wounds. Easy to use and comfortable, hydrocolloid dressings promote débridement and formation of granulation tissue. Most can be left in place for as long as 7 days and can be submerged in water for bathing or showering. Hydrocolloid dressings are more effective than saline gauze or paraffin gauze dressings in completely healing chronic wounds (Baranoski & Ayello, 2012).

Foam Dressings

Foam dressings consist of microporous polyurethane with an absorptive **hydrophilic** (water-absorbing) surface that covers the wound and a **hydrophobic** (water-resistant) backing to block leakage of exudate. They are nonadherent and require a secondary dressing to keep them in place. Moisture is absorbed into the foam layer, decreasing maceration of surrounding tissue. A moist environment is maintained, and removal of the dressing does not damage the wound. The foams are opaque and must be removed for wound inspection. Foams are a good choice for exudative wounds. They are especially helpful over bony prominences because they provide contoured cushioning (Baranoski & Ayello, 2012).

Calcium Alginates

Calcium alginates are derived from seaweed and consist of very absorbent calcium alginate fibers. They are hemostatic and bioabsorbable and can be used as sheets or mats of absorbent material. As the exudate is absorbed, the fibers turn into a viscous hydrogel. They are useful in areas where the tissue is more irritated or macerated. The alginate dressing forms a moist pocket over the wound while the surrounding skin stays dry. The dressing also reacts with wound fluid, which forms a foul-smelling coating. Alginates work well when packed into a deep cavity, wound, or sinus tract with heavy drainage (Krasner et al., 2012). They are nonadherent and require a secondary dressing. A consensus statement of wound experts suggests that alginates are superior to other modern dressings for débriding necrotic wounds (Vaneau, Chaby, Guillot, et al., 2007).

Advances in Wound Treatment

Increasing understanding of how skin heals has led to several advances in therapy. Growth factors are cytokines or proteins that have potent **mitogenic** (increase in cell division and reproduction) activity. Low levels of cytokines circulate in the blood continuously and control key factors involved in wound healing. Many investigators believe that a deficit of cytokines contributes to poor wound healing. Becaplermin (Regranex Gel), which contains a recombinant human platelet-derived growth factor, promotes chemotactic recruitment and proliferation of the cells involved in wound healing (Fonder et al., 2008).

Bioengineered skin substitutes have emerged in the past 25 years as the most effective method for management of chronic wounds. Most of these skin substitutes are cultures of keratinocytes delivered on a petrolatum gauze. They work by maintaining wound moisture, providing a structure for regeneration of cells, and supplying beneficial cytokines. These substitutes include AlloDerm, Apligraf, Dermagraft, Epicel, and Laserskin (Bryant & Nix, 2011).

Some oral medications may accelerate healing chronic venous ulcers of the lower legs. Pentoxifylline (Trental) increases peripheral blood flow by decreasing the viscosity of blood. It has some fibrinolytic action and decreases leukocyte adhesion to the wall of the blood vessels. Enteric-coated aspirin was traditionally thought to be therapeutic, but research findings suggest that aspirin inhibits the tensile strength of skin and prolongs the healing time (Fonder et al., 2008). Micronized purified flavonoid fraction (Daflon), although not yet approved for use in the United States by the U.S. Food and Drug Administration (FDA), is widely used in Europe with good results. These medications are most effective when the venous ulcers are concomitantly treated with pressure dressings (Bryant & Nix, 2011).

Medical Management

Medical management of dermatologic problems includes a host of prescribed and over-the-counter therapies. Occasionally, therapeutic baths are used.

Pharmacologic Therapy

Because skin is easily accessible, topical medications are often used. High concentrations of some medications can be applied directly to the affected site with little systemic absorption and therefore with few systemic side effects. However, some medications are readily absorbed through the skin and can produce systemic effects. Because topical preparations may induce allergic contact **dermatitis** (skin inflammation) in sensitive patients, any untoward response should be reported immediately and the medication discontinued.

Medicated lotions, creams, ointments, and powders are frequently used to treat skin lesions. In general, moisture-retentive dressings, with or without medication, are used in the acute stage; lotions and creams are reserved for the subacute stage; and ointments are used when inflammation has become chronic and the skin is dry with scaling or **lichenification** (thickening of the horny layer of the skin).

With all types of topical medication, the patient is educated to apply the medication gently but thoroughly and, when necessary, to cover the medication with a dressing to protect clothing. Table 61-2 lists commonly used topical preparations and medications.

Lotions

Lotions are frequently used to replenish lost skin oils or to relieve pruritus. They must be applied every 3 or 4 hours for sustained therapeutic effect. They are usually applied directly to the skin, but a dressing soaked in the lotion can be placed on the affected area. However, if left in place for a longer period, it may crust and cake on the skin.

Lotions are of two types: suspensions and liniments. **Suspensions** consist of either a powder in water that requires shaking before application, or clear solutions, which contain

TABLE 61-2 Common Topical Preparations and Medications

Preparation	Product Name
Moisturizer creams	AcidMantle Cream, Curel Cream, Dermasil, Eucerin, Lubriderm, Noxzema Skin Cream
Moisturizer ointments	Aquaphor Ointment, Eutra Swiss Skin Cream, Vaseline Ointment
Topical anesthetic agents	lidocaine (Xylocaine) of various strengths in the form of spray, ointment, gel; lidocaine 2.5% and prilocaine 2.5% (EMLA cream)
Topical antibiotic agents	bacitracin, bacitracin and polymixin B (Polysporin), mupirocin 2% (Bactroban ointment or cream), erythromycin 2% (Emgel, Eryderm Solution), clindamycin phosphate 1% (Cleocin cream, gel, solution), gentamicin sulfate 1% (Garamycin cream or ointment), 1% silver sulfadiazine cream (Silvadene)

Adapted from Goldsmith, L., Katz, S. I., Gilchrist B. A., et al. (2012). *Fitzpatrick's dermatology in general medicine* (8th ed.). New York: McGraw-Hill.

completely dissolved active ingredients. A suspension such as calamine lotion provides a rapid cooling and drying effect as it evaporates, leaving a thin, medicinal layer of powder on the affected skin. **Liniments** are lotions with oil added to prevent crusting. Because lotions are easy to use, therapeutic compliance is generally high.

Powders

Powders usually have a talc, zinc oxide, bentonite, or cornstarch base and are dusted on the skin with a shaker or with cotton sponges. Although their therapeutic action is brief, powders act as **hygroscopic** agents that absorb and retain moisture from the air and reduce friction between skin surfaces and clothing or bedding.

Creams

Creams may be suspensions of oil in water or emulsions of water in oil, with additional ingredients to prevent bacterial and fungal growth. Both may cause an allergic reaction such as contact dermatitis. Oil-in-water creams are easily applied and usually are the most cosmetically acceptable to the patient. Although they can be used on the face, they tend to have a drying effect. Water-in-oil emulsions are greasier and are preferred for drying and flaking dermatoses. Creams usually are rubbed into the skin by hand. They are used for their moisturizing and emollient effects.

Gels

Gels are semisolid emulsions that become liquid when applied to the skin or scalp. They are cosmetically acceptable to the patient because they are not visible after application, and they are greaseless and nonstaining. The newer water-based gels appear to penetrate the skin more effectively and cause less stinging on application. They are especially useful for acute dermatitis in which there is weeping exudate (e.g., poison ivy) and are applied the same way as creams.

Pastes

Pastes are mixtures of powders and ointments and are used in inflammatory blistering conditions. They adhere to the skin and may be difficult to remove without using an oil (e.g., olive oil, mineral oil). Pastes are applied with a wooden tongue depressor or gloved hand.

Ointments

Ointments retard water loss and lubricate and protect the skin. They are the preferred vehicle for delivering medication to chronic or localized dry skin conditions, such as eczema or psoriasis. Ointments are applied with a wooden tongue depressor or gloved hand.

Sprays and Aerosols

Spray and aerosol preparations may be used on any widespread dermatologic condition. They evaporate on contact and are used infrequently.

Topical Corticosteroids

Corticosteroids are widely used in treating dermatologic conditions to provide anti-inflammatory, antipruritic, and vasoconstrictive effects. The patient is educated to apply this medication according to strict guidelines, using it sparingly but rubbing it into the prescribed area thoroughly. Absorption of topical corticosteroids is enhanced when the skin is hydrated or the affected area is covered by an occlusive or moisture-retentive dressing (Karch, 2013). Inappropriate use of topical corticosteroids can result in local and systemic side effects, especially when the medication is absorbed through inflamed and excoriated skin, is used under occlusive dressings, or is used for long periods on sensitive areas. Local side effects may include skin atrophy and thinning, **striae** (band-like streaks), and telangiectasias (small, red lesions caused by dilation of blood vessels). Thinning of the skin results from the ability of corticosteroids to inhibit skin collagen synthesis. The thinning process can be reversed by discontinuing the medication, but striae and telangiectasia are permanent. Systemic side effects may include hyperglycemia and symptoms of Cushing syndrome (see Chapter 52). Caution is required when applying corticosteroids around the eyes because long-term use may cause glaucoma or cataracts, and the anti-inflammatory effect of corticosteroids may mask existing viral or fungal infections.

Concentrated (fluorinated) corticosteroids are never applied on the face or intertriginous areas (i.e., axilla and groin) because these areas have a thinner stratum corneum and absorb the medication much more quickly than areas such as the forearm or legs. Persistent use of concentrated topical corticosteroids in any location may produce acnelike dermatitis, known as steroid-induced acne, and hypertrichosis (excessive hair growth). Because some topical corticosteroid preparations are available without prescription, patients should be cautioned about prolonged and inappropriate use. Table 61-3 lists topical corticosteroid preparations according to potency.

Intralesional Therapy

Intralesional therapy consists of injecting a sterile suspension of medication (usually a corticosteroid) into or just below a lesion. Although this treatment may have an anti-inflammatory effect, local atrophy may result if the medication is injected into subcutaneous fat. Skin lesions treated

TABLE 61-3 Potency: Topical Corticosteroids

Potency	Topical Corticosteroid	Preparations
OTC	0.5–1% hydrocortisone	Cream, lotion, or ointment
Lowest	dexamethasone 0.1% (Decaderm)	Cream, ointment, aerosol, gel
	alclometasone 0.05% (Aclovate)	Cream, ointment
	hydrocortisone 2.5% (Hytone)	Cream, lotion, ointment
Low–medium	desonide 0.05% (DesOwen, Tridesilon)	Cream, lotion, ointment
	fluocinolone acetonide 0.025% (Synalar)	Cream, solution
	hydrocortisone valerate 0.2% (Westcort)	Cream, solution
	betamethasone valerate 0.1% (Valisone)	Cream, ointment
	fluticasone propionate 0.05% (Cutivate)	Cream, ointment
Medium–high	triamcinolone acetonide 0.1–0.5% (Aristocort)	Cream, ointment, lotion
	fluocinonide 0.05% (Lidex)	Cream, ointment, gel
	desoximetasone 0.05–0.25% (Topicort)	Cream, ointment, gel
	fluocinolone 0.2% (Synalar)	Cream, ointment
	diflorasone diacetate 0.05% (Psorcon)	Cream, ointment
Very high	clobetasol propionate 0.05% (Temovate)	Cream, ointment, gel
	betamethasone dipropionate 0.05% (Diprolene)	Cream, ointment, gel
	halobetasol propionate 0.05% (Ultravate)	Cream, ointment

OTC, over the counter.
Adapted from Karch, A. M. (2013). *2013 Lippincott's nursing drug guide*. Philadelphia: Lippincott Williams & Wilkins.

with intralesional therapy include psoriasis, keloids, and cystic acne. Occasionally, immunotherapeutic and antifungal agents are administered as intralesional therapy.

Systemic Medications

Systemic medications are also prescribed for skin conditions. These include corticosteroids for short-term therapy for contact dermatitis or for long-term treatment of a chronic **dermatosis** (skin lesion), such as pemphigus vulgaris. Other frequently used systemic medications include antibiotic, antifungal, antihistamine, sedative, analgesic, tranquilizing, **cytotoxic** (destructive of cells), and immunosuppressive agents.

Therapeutic Baths (Balneotherapy)

Baths or soaks, known as **balneotherapy**, are traditional treatments that are useful when large areas of skin are diseased or inflamed. Baths may remove crusts, scales, and previously applied topical medications and relieve the inflammation and pruritus (itching) that accompany acute dermatoses. Although some patients still find these treatments effective, the time and effort required have limited patient acceptance of this form of treatment. In addition, there have been many advances in oral and topical treatment options that are easier to use and take less time for the patient. Therefore, while balneotherapy remains a viable option, most nurses, especially those in hospital settings, will not find themselves administering this type of treatment as often as in the past.

Nursing Management

Management begins with a health history, direct observation, and a complete physical examination. (Chapter 60 provides a description of integumentary assessment.) Because of its visibility, a skin condition is usually difficult to ignore or conceal from others and may therefore cause the patient emotional distress. The major goals for the patient may include maintenance of skin integrity, relief of discomfort, promotion of restful sleep, self-acceptance, knowledge about skin care, and avoidance of complications.

Nursing management for patients who must perform self-care for skin problems, such as applying medications and dressings, focuses mainly on educating the patient about how to wash the affected area and pat it dry, apply medication to the lesion while the skin is moist, cover the area with plastic (e.g., Telfa pads, plastic wrap, vinyl gloves, plastic bag) if recommended, and cover it with an elastic bandage, dressing, or paper tape to seal the edges. Dressings that contain or cover a topical corticosteroid should be removed for 12 of every 24 hours to prevent skin thinning, striae, and telangiectasia.

Other forms of dressings, such as those used to cover topical medications, include soft cotton cloth and stretchable cotton dressings (e.g., Surgitube, Tubegauz) that can be used for fingers, toes, hands, and feet. The hands can be covered with disposable polyethylene or vinyl gloves sealed at the wrists; the feet can be wrapped in plastic bags covered by cotton socks. Gloves and socks that are already impregnated with emollients, making application to the hands and feet more convenient, are also available. When large areas of the body must be covered, cotton cloth topped by an expandable stockinette can be used. Disposable diapers or cloths folded in diaper fashion are useful for dressing the groin and the perineal areas. Axillary dressings can be made of cotton cloth, or a commercially prepared dressing may be used and taped in place or held by dress shields. A turban or plastic shower cap is useful for holding dressings on the scalp. A face mask, made from gauze with holes cut out for the eyes, nose, and mouth, may be held in place with gauze ties looped through holes cut in the four corners of the mask.

PRURITUS

General Pruritus

Pruritus (itching) is the most common symptom of patients with dermatologic disorders (Garcia-Albea & Limaye, 2012). Itch receptors are unmyelinated, penicillate (brushlike)

Chart 61-1 Systemic Disorders Associated With Generalized Pruritus

- Chronic kidney disease
- Obstructive biliary disease (primary biliary cirrhosis, extrahepatic biliary obstruction, drug-induced cholestasis)
- Endocrine disease (thyrotoxicosis, hypothyroidism, diabetes)
- Psychiatric disorders (emotional stress, anxiety, neurosis, phobias)
- Malignancies (polycythemia vera, Hodgkin lymphoma, lymphoma, leukemia, multiple myeloma, mycosis fungoides, and cancers of the lung, breast, central nervous system, and gastrointestinal tract)
- Neurologic disorders (multiple sclerosis, brain abscess, brain tumor)
- Hematologic disorders (iron deficiency anemia)
- Infestations (scabies, lice, other insects)
- Pruritus of pregnancy (pruritic urticarial papules of pregnancy, cholestasis of pregnancy, pemphigoid of pregnancy)
- Folliculitis (bacterial, candidiasis, dermatophyte)
- Skin conditions (seborrheic dermatitis, folliculitis, atopic dermatitis)

Adapted from Goldsmith, L., Katz, S. I., Gilchrist B. A., et al. (2012). *Fitzpatrick's dermatology in general medicine* (8th ed.). New York: McGraw-Hill.

nerve endings that are found exclusively in the skin, mucous membranes, and cornea. Although pruritus is usually caused by primary skin disease with resultant rash or lesions, it may occur without a rash or lesion. This is referred to as essential pruritus, which generally has a rapid onset, may be severe, and interferes with normal daily activities.

Pruritus may be the first indication of a systemic internal disease such as diabetes, blood disorders, or cancer (occult malignancy of the breast or colon, lymphoma). It may also accompany kidney, hepatic, and thyroid diseases (Chart 61-1). Some common oral medications such as aspirin, antibiotics, hormones (i.e., estrogens, testosterone, or oral contraceptives), and opioids (i.e., morphine or cocaine) may cause pruritus directly or by increasing sensitivity to ultraviolet light. Certain soaps and chemicals, radiation therapy, prickly heat (miliaria), and contact with woolen garments are associated with pruritus as well. Pruritus may also be caused by psychological factors, such as excessive stress in family or work situations, and is called *psychodermatosis* (Garcia-Albea & Limaye, 2012).

Gerontologic Considerations

Pruritus occurs frequently in older adults as a result of dry skin. Older adults are also more likely to have a systemic illness that triggers pruritus, are at higher risk for occult malignancy, and are more likely to be taking multiple medications than younger people. All of these factors increase the incidence of pruritus in older adults.

Pathophysiology

Scratching the pruritic area causes the inflamed cells and nerve endings to release histamine, which produces more pruritus, generating a vicious itch–scratch cycle. If the patient responds to an itch by scratching, the integrity of the skin may be altered, and excoriation, redness, raised areas (i.e., wheals), infection, or changes in pigmentation may result. Pruritus usually is more severe at night and is less frequently reported during waking hours, probably because the person is distracted by daily activities. At night, when there are fewer distractions, the slightest pruritus cannot be easily ignored. Severe itching can be debilitating.

Medical Management

A thorough history and physical examination usually provide clues to the underlying cause of the pruritus, such as hay fever, allergy, recent administration of a new medication, or a change of cosmetics or soaps. After the cause has been identified, treatment of the condition should relieve the pruritus. Signs of infection and environmental clues, such as warm, dry air or irritating bed linens, should be identified. In general, washing with soap and hot water is avoided. Bath oils containing a surfactant that allows the oil to mix with bathwater (e.g., Lubath, Alpha Keri) may be sufficient for cleaning. However, an older adult patient or a patient with unsteady balance should avoid adding oil because it increases the danger of slipping in the bathtub. A warm bath with a mild soap followed by application of a bland emollient to moist skin can control **xerosis** (overly dry skin). Applying a cold compress, ice cube, or cool agents that contain menthol and camphor (which constrict blood vessels) may also help relieve pruritus (Goldsmith et al., 2012).

Pharmacologic Therapy

Topical antipruritic agents (e.g., lidocaine, prilocaine) or capsaicin cream (Capzasin) may be useful in providing relief from localized pruritus. Topical corticosteroids are effective when used to diminish pruritus that occurs secondary to inflammatory conditions because of their anti-inflammatory effects. Oral antihistamines are frequently prescribed and may be effective when the pruritus is nocturnal, particularly agents such as diphenhydramine (Benadryl) or hydroxyzine (Atarax), which also cause somnolence, resulting in a restful and comfortable sleep. Other nonsedating antihistamines are not beneficial in relieving pruritus. Doxepin (Sinequan), a tricyclic antidepressant, at doses of 10 to 25 mg four times daily, seems to be effective at relieving pruritus. Selective serotonin reuptake inhibitor antidepressants (e.g., fluoxetine [Prozac], sertraline [Zoloft]) may be effective, but results from studies to date are inconclusive with regard to their efficacy (Garcia-Albea & Limaye, 2012).

Nursing Management

The nurse reinforces the reasons for the prescribed therapeutic regimen and educates the patient about specific points of care. If baths have been prescribed, the patient is reminded to use tepid (not hot) water and to shake off the excess water and blot between intertriginous areas (body folds) with a towel. Rubbing vigorously with the towel is avoided because this overstimulates the skin and causes more itching. It also removes water from the stratum corneum. Immediately after bathing, the skin should be lubricated with an emollient to trap moisture.

The patient is instructed to avoid situations that cause vasodilation. Examples include exposure to an overly warm environment and ingestion of alcohol or hot foods and liquids. All can induce or intensify pruritus. Using a humidifier is helpful if environmental air is dry. Activities that result in perspiration should be limited because perspiration may irritate and promote pruritus. If the patient is troubled at night with itching that interferes with sleep, the nurse can advise wearing cotton clothing next to the skin rather than synthetic materials. The room should be kept cool and humidified. Vigorous scratching should be avoided and nails kept trimmed to prevent skin damage and infection. When the underlying cause of pruritus is unknown and further testing is required, the nurse explains each test and the expected outcome.

Perineal and Perianal Pruritus

Pruritus of the genital and anal regions may be caused by small particles of fecal material lodged in the perianal crevices or attached to anal hairs. Alternatively, it may result from perianal skin damage caused by scratching, moisture, and decreased skin resistance as a result of corticosteroid or antibiotic therapy. Other possible causes of perianal itching include local lesions such as hemorrhoids, fungal or yeast infections, and pinworm infestation. Conditions reviewed in Chart 61-1 may also result in pruritus. Occasionally, no cause can be identified.

Management

The patient is instructed to follow proper hygiene measures and to discontinue home and over-the-counter remedies. The perineal or anal area should be rinsed with lukewarm water and blotted dry with cotton balls. Premoistened tissues may be used after defecation. Cornstarch can be applied in the skinfold areas to absorb perspiration.

As part of health education, the nurse instructs the patient to avoid bathing in water that is too hot and to avoid using bubble baths, sodium bicarbonate, and detergent soaps, all of which aggravate dryness. To keep the perineal or perianal skin as dry as possible, patients should avoid wearing underwear made of synthetic fabrics. The patient should also avoid vasodilating agents or stimulants (e.g., alcohol, caffeine) and mechanical irritants such as rough or woolen clothing. A diet that includes adequate fiber may help maintain soft stools and prevent minor trauma to the anal mucosa.

SECRETORY DISORDERS

The main secretory function of the skin is performed by the sweat glands, which help regulate body temperature. These glands excrete perspiration that evaporates, thereby cooling the body. The sweat glands are located in various parts of the body and respond to different stimuli. Those on the trunk generally respond to thermal stimulation; those on the palms and soles respond to nervous stimulation; and those in the axillae and on the forehead respond to both kinds of stimulation. Normal perspiration has no odor. Body odor is produced by the increase in bacteria on the skin and the interaction of bacterial waste products with the chemicals of perspiration. As a rule, moist skin is warm, and dry skin is cool, but this is not always true. It is not unusual to observe warm, dry skin in a patient who is dehydrated and very hot, dry skin in some febrile states.

Normally, sweat can be controlled with the use of antiperspirants and deodorants. Most antiperspirants are aluminum salts that block the opening to the sweat duct. Pure deodorants inhibit bacterial growth and block the metabolism of sweat; they have no antiperspirant effect. Fragrance-free deodorants are available for those with sensitive skin.

Hidradenitis Suppurativa

Hidradenitis suppurativa (HS) is a chronic suppurative folliculitis of the perianal, axillary, and genital areas or under the breasts. It can produce abscesses or sinuses with scarring. It develops after puberty and diminishes in incidence after 50 years of age. African Americans are at greater risk for HS. In addition, men are at greater risk for anogenital HS, whereas women are at greater risk for axillary HS. The cause is unknown, but it appears to have a genetic basis (Beshara, 2012).

Pathophysiology

It had been assumed for many years that HS was caused by an abnormal blockage and infection of the sweat glands. However, more recent evidence suggests that it is a primary disorder of follicular occlusion that causes eventual hypertrophic formation of scar tissue in the area of the sweat glands (Beshara, 2012).

Clinical Manifestations

HS occurs more frequently in the axillae but also appears in inguinal folds, on the mons pubis, around the buttocks, areolae of the breasts, submammary fold, nape of the neck, and shoulders. The patient may present with a firm, pea-sized nodule that causes discomfort, or with a history of this type of nodule that then ruptures and discharges purulent drainage. The nodule then propagates, and multiple similar nodules will form adjacent to the initial nodule. These nodules become deep seated and, as they rupture, form scars. The nodules may coalesce or form "bridges," become infected, and result in abscesses. As they coalesce, the patient will present with complaints of persistent pain (Beshara, 2012).

Management

The patient is educated to use warm compresses and wear loose-fitting clothes over the nodules or lesions. Oral antibiotic agents such as erythromycin (E-Mycin), tetracycline, minocycline (Minocin), and doxycycline (Vibramycin) are frequently prescribed. Nonsteroidal anti-inflammatory drugs (NSAIDs) may be indicated to relieve the pain. Silver-impregnated alginate dressings may be useful with some lesions. Incision and drainage of large suppurating areas with gauze packs inserted to facilitate drainage are often necessary. Rarely, the entire area is excised, removing the scar tissue and any infection. This surgery is drastic in that it may require the use of skin grafts (see later discussion) and is performed only as a last resort. Carbon dioxide laser surgery (see later discussion) may be more effective than this type of excisional surgery (Beshara, 2012).

Seborrheic Dermatoses

Seborrhea is excessive production of sebum (secretion of sebaceous glands) in areas where sebaceous glands are normally found in large numbers, such as the face, scalp, eyebrows, eyelids, sides of the nose and upper lip, malar regions (cheeks), ears, axillae, under the breasts, groin, and gluteal crease of the buttocks. Seborrheic dermatitis is a chronic inflammatory disease of the skin with a predilection for areas that are well supplied with sebaceous glands or lie between skin folds, where the bacteria count is high (Schmidt, 2011).

Clinical Manifestations

Two forms of seborrheic dermatoses can occur: an oily form and a dry form. Either form may start in childhood and continue throughout life. The oily form appears moist or greasy. There may be patches of sallow, greasy skin, with or without scaling, and slight erythema, predominantly on the forehead, nasolabial fold, beard area, scalp, and between adjacent skin surfaces in the regions of the axillae, groin, and breasts. Small pustules or papulopustules resembling acne may appear on the trunk. The dry form, consisting of flaky desquamation of the scalp with a profuse amount of fine, powdery scales, is commonly called *dandruff*. The mild forms of the disease are asymptomatic. When scaling occurs, it is often accompanied by pruritus, which may lead to scratching and secondary infections and excoriation.

Seborrheic dermatitis has a genetic predisposition. Hormones, nutritional status, infection, and emotional stress influence its course. The remissions and exacerbations of this condition should be explained to the patient. If a person has not previously been diagnosed with this condition and suddenly appears with a severe outbreak, a complete history and physical examination should be conducted.

Medical Management

Because there is no known cure for seborrhea, the objectives of therapy are to control the disorder and allow the skin to repair itself. Seborrheic dermatitis of the body and face may respond to a topically applied corticosteroid cream, which allays the secondary inflammatory response. However, this medication should be used with caution near the eyelids because it can lead to glaucoma and cataracts. Patients with seborrheic dermatitis may develop a secondary candidal (yeast) infection in body creases or folds. To avoid this, patients should be advised to ensure maximum aeration of the skin and to clean carefully areas where there are creases or folds in the skin (Schmidt, 2011). Patients with persistent candidiasis should be evaluated for diabetes.

The mainstay of dandruff treatment is proper, frequent shampooing (at least three times weekly) with medicated shampoos (Schmidt, 2011). Two or three different types of shampoo should be used in rotation to prevent the seborrhea from becoming resistant to a particular shampoo. The shampoo is left on at least 5 to 10 minutes. As the condition of the scalp improves, the treatment can be less frequent. Antiseborrheic shampoos include those containing selenium sulfide suspension, zinc pyrithione, salicylic acid, or sulfur compounds, and tar shampoo that contains sulfur or salicylic acid.

Nursing Management

A person with seborrheic dermatitis is advised to avoid external irritants, excessive heat, and perspiration; rubbing and scratching prolong the disorder. To avoid secondary infection, the patient should air the skin and keep skin folds clean and dry.

Instructions for using medicated shampoos are reinforced for people with dandruff who require treatment. Frequent shampooing is contrary to some cultural practices; the nurse should be sensitive to these differences when educating the patient about home care.

The patient is cautioned that seborrheic dermatitis is a chronic condition that tends to reappear. The goal is to keep it under control (Schmidt, 2011). Patients need to be encouraged to adhere to the treatment program. Those who become discouraged and disheartened by the effect on body image should be treated with sensitivity and encouraged to express their feelings.

Acne Vulgaris

Acne vulgaris is a common disorder affecting susceptible hair follicles, most commonly on the face, neck, and upper trunk. It is characterized by **comedones** (primary acne lesions), both closed and open, and by papules, pustules, nodules, and cysts (see Table 60–2 for more information).

Acne is the most commonly encountered skin condition that affects up to 80% of Americans at some time during their lives. Acne is most prevalent during adolescence among males but is more prevalent in adulthood among females. Acne is traditionally considered a skin disorder of adolescence; however, by age 45 years, up to 5% of adults report having acne. Acne appears to stem from an interplay of genetic, hormonal, and bacterial factors (Fulton, 2012).

Pathophysiology

During puberty, androgens stimulate the sebaceous glands, causing them to enlarge and secrete a natural oil (sebum) that rises to the top of the hair follicle and flows out onto the skin surface. In adolescents who develop acne, androgenic stimulation produces a heightened response in the sebaceous glands so that acne occurs when accumulated sebum plugs the pilosebaceous ducts. Sebaceous plugging then causes a localized inflammatory response (Fulton, 2012).

Clinical Manifestations

The main lesions of acne are comedones. Closed comedones (whiteheads) form from impacted lipids or oils and keratin that plug the dilated follicle. Closed comedones may evolve into open comedones (blackheads), in which the contents of the ducts are in open communication with the external environment. The color of open comedones results from an accumulation of lipid, bacterial, and epithelial debris. Some closed comedones may rupture, resulting in an inflammatory reaction caused by leakage of follicular contents (e.g., sebum, keratin, bacteria) into the dermis. The resultant inflammation is seen clinically as erythematous papules, inflammatory pustules, and inflammatory cysts. Mild papules and cysts drain and heal without treatment. Deeper papules and cysts cause scarring of the skin. Acne is usually graded as mild, moderate, or severe based on the number and type of lesions (Fulton, 2012).

Assessment and Diagnostic Findings

The diagnosis of acne is based on the history and physical examination, evidence of lesions characteristic of acne, and age. Women may report a history of flare-ups a few days before menses. The presence of the typical comedones along with excessively oily skin is characteristic. Oiliness is more prominent in the midfacial area; other parts of the face may appear dry. When there are numerous lesions, some of which are open, the person may exude a distinct sebaceous odor. Biopsy of lesions is seldom necessary for a definitive diagnosis.

Medical Management

The goals of management are to reduce bacterial colonies, decrease sebaceous gland activity, prevent the follicles from becoming plugged, reduce inflammation, combat secondary infection, minimize scarring, and eliminate factors that predispose the person to acne. The therapeutic regimen depends on the type of lesion (e.g., comedones, papule, pustule, cyst). The duration of treatment depends on the extent and severity of the acne. In severe cases, treatment may extend over years.

There is no predictable cure for the acne, but combinations of therapies are available that can effectively control its activity. Table 61-4 summarizes the treatment modalities for acne vulgaris.

Nutrition and Hygiene Therapy

Diet is not believed to play a major role in therapy. However, the elimination of a specific food or food product associated with a flare-up of acne, such as chocolate, cola, fried foods, or milk products, should be promoted. Maintenance of good nutrition equips the immune system for effective action against bacteria and infection.

For mild cases of acne, washing twice each day with a cleansing soap may be all that is required. Oil-free cosmetics and creams should be chosen. These products are usually designated as useful for acne-prone skin.

TABLE 61-4 Type of Treatments Commonly Used for Acne Vulgaris

Type of Therapy	Prescribed Treatment Agent
Topical	Benzoyl peroxide wash, gel
	Benzoyl peroxide and erythromycin (Benzamycin gel)
	Resorcinol (as ingredient in other preparations)
	Salicylic acid (as ingredient in other preparations)
	Sulfur (as ingredient in other preparations)
	Tretinoin (Retin A, Avita)
	Other comedogenics (adapalene [Differin], azelaic acid [Azelex], tazarotene [Tazorac])
	Topical antibiotic agents (clindamycin [Cleocin T], erythromycin [Akne-mycin])
Systemic	Oral antibiotic agents (tetracycline, doxycycline [Vibramycin], minocycline [Minocin], trimethoprim [Proloprim], sulfamethoxazole and trimethoprim [Bactrim])
	Oral contraceptive agents (women only)
	Spironolactone (Aldactone)
Surgical	Extraction of comedones
	Superficial peels (glycolic acid, salicylic acid)
	Intralesional corticosteroids for anti-inflammatory action
	Phototherapy
	Fractional laser treatments

Note: Treatments listed are common but do not include all available forms of therapy. Adapted from Fulton, J. (2012). Acne vulgaris. *Medscape*. Available at: emedicine.medscape.com/article/1069804-overview

Pharmacologic Therapy

Topical Therapy

Over-the-counter acne medications contain either salicylic acid or benzoyl peroxide, both of which are very effective at removing the sebaceous follicular plugs. However, the skin of some people is sensitive to these products, which can cause irritation or excessive dryness, especially when used with some prescribed topical medications. The patient should be instructed to discontinue use of the product if severe irritation occurs.

Benzoyl peroxide preparations are widely used because they produce a rapid and sustained reduction of inflammatory lesions. They depress sebum production and promote breakdown of comedo plugs and have an antibacterial effect. Initially, benzoyl peroxide causes redness and scaling, but the skin usually adjusts quickly to its use. Typically, the patient applies a gel of benzoyl peroxide once daily. In many instances, this is the only treatment needed. Benzoyl peroxide, benzoyl erythromycin (Benzamycin), and benzoyl sulfur (Sulfoxyl) are available over the counter and by prescription.

Vitamin A acid (tretinoin) applied topically is used to clear the keratin plugs from the pilosebaceous ducts. The patient should be informed that symptoms may worsen during early weeks of therapy because inflammation, erythema, and peeling may occur. The patient is cautioned against sun exposure while using this topical medication because it may cause an exaggerated sunburn. Package insert directions should be followed carefully. Improvement may take 8 to 12 weeks.

Topical antibiotic treatment for acne is common. Topical antibiotic agents suppress bacterial growth; reduce superficial free fatty acid levels; decrease comedones, papules, and pustules; and produce no systemic side effects (Fulton, 2012; Karch, 2013).

Systemic Therapy

Oral antibiotic agents administered in small doses over a long period are very effective in treating moderate and severe acne, especially when the acne is inflammatory and results in pustules, abscesses, and scarring. Therapy may continue for months to years. The tetracycline family of antibiotics is contraindicated in children younger than 12 years and in pregnant women. Administration during pregnancy can affect the development of teeth, causing enamel hypoplasia and permanent discoloration of teeth in infants. Side effects of tetracyclines include photosensitivity, nausea, diarrhea, cutaneous infection in either gender, and vaginitis in women. In some women, broad-spectrum antibiotics may suppress normal vaginal bacteria and predispose the patient to candidiasis, which is a fungal infection.

Synthetic vitamin A compounds (i.e., retinoids) are used with dramatic results in patients with nodular cystic acne unresponsive to conventional therapy. One compound is isotretinoin, which is used for active inflammatory papular pustular acne that has a tendency to scar. Isotretinoin reduces sebaceous gland size and inhibits sebum production. It also causes the epidermis to shed (epidermal desquamation), thereby unseating and expelling existing comedones.

The most common side effect, experienced by almost all patients, is **cheilitis** (inflammation of the lips). Dry and chafed skin and mucous membranes are frequent side effects. These changes are reversible with the withdrawal of the medication. Most important, isotretinoin, like other vitamin A metabolites, is teratogenic in humans, meaning that it can have an adverse effect on the normal development of a fetus, causing defects. Effective contraceptive measures for women of childbearing age are mandatory during treatment and for about 4 to 8 weeks thereafter. To avoid additive toxic effects, patients are cautioned not to take vitamin A supplements while taking isotretinoin.

Estrogen therapy (including progesterone–estrogen preparations) suppresses sebum production and reduces skin oiliness. It is usually reserved for young women when the acne begins somewhat later than usual and tends to flare up at certain times in the menstrual cycle. Estrogen-dominant oral contraceptive compounds may be administered on a prescribed cyclic regimen. Estrogen is not administered to male patients because of undesirable side effects such as enlargement of the breasts and decrease in body hair (Fulton, 2012).

Surgical Management

Treatment includes comedo extraction; injections of corticosteroids into the inflamed lesions; and incision and drainage of large, fluctuant (moving in palpable waves), nodular cystic lesions. Phototherapy (using either red or blue light) is showing promise in treating acne (Fulton, 2012). Patients with deep scars may be treated with deep abrasive therapy (dermabrasion), in which the epidermis and some superficial dermis are removed down to the level of the scars.

Comedones may be removed with a comedo extractor. The site is first cleaned with alcohol. The opening of the extractor is then placed over the lesion, and direct pressure is applied to cause extrusion of the plug through the extractor. Removal of comedones leads to erythema, which may take several weeks to subside. Recurrence of comedones after extraction is common.

Nursing Management

Nursing care of patients with acne consists largely of monitoring and managing potential complications of skin treatments. Major nursing activities include patient education, particularly in proper skin care techniques, and managing potential problems related to the skin disorder or therapy. Providing positive reassurance, listening attentively, and being sensitive to the feelings of the patient with acne are essential to the patient's psychological well-being and understanding of the disease and treatment plan.

Preventing Scarring

Prevention of scarring is the ultimate goal of therapy. The chance of scarring increases with the severity of the grade of acne. Severe acne (25 to more than 50 comedones, papules, or pustules) usually requires longer-term therapy with systemic antibiotic agents or isotretinoin. Patients should be warned that discontinuing these medications may lead to more flare-ups and increase the chance of deep scarring. Furthermore, manipulation of the comedones, papules, and pustules increases the potential for scarring.

When acne surgery is prescribed to extract deep-seated comedones or inflamed lesions or to incise and drain cystic lesions, the intervention itself may result in further scarring. Dermabrasion, which levels existing scar tissue, can also increase scar formation. Hyper- or hypopigmentation also may affect the tissue involved. The patient should be informed of these potential outcomes before choosing surgical intervention for acne.

Promoting Home and Community-Based Care

Educating Patients About Self-Care

In addition to providing instructions for taking prescribed medications, the nurse informs patients to wash the face and other affected areas with mild soap and water twice each day to remove surface oils and prevent obstruction of the oil glands. Mild abrasive soaps and drying agents are prescribed to eliminate the oily feeling that troubles many patients. At the same time, patients are cautioned to avoid excessive abrasion, because it makes acne worse.

All forms of friction and trauma are avoided, including propping the hands against the face, rubbing the face, and wearing tight collars and helmets. Patients are instructed to avoid manipulation of pimples or blackheads. Squeezing merely worsens the problem, because a portion of the blackhead is pushed down into the skin, which may cause the follicle to rupture. Because cosmetics, shaving creams, and lotions can aggravate acne, these substances are best avoided unless the patient is advised otherwise.

INFECTIOUS DERMATOSES

Bacterial Skin Infections

Also called **pyodermas**, pus-forming bacterial infections of the skin may be primary or secondary. Primary skin infections originate in previously normal-appearing skin and are usually caused by a single organism. Secondary skin infections arise from a preexisting skin disorder or from disruption of the skin integrity from injury or surgery. In either case, several microorganisms may be implicated (e.g., *Staphylococcus aureus*, group A streptococci). Common primary bacterial skin infections are impetigo and folliculitis. Folliculitis may lead to furuncles or carbuncles.

Impetigo

Impetigo is a superficial infection of the skin caused by staphylococci, streptococci, or multiple bacteria. Bullous impetigo, a more deep-seated infection of the skin caused by *S. aureus*, is characterized by the formation of **bullae** (i.e., large, fluid-filled blisters) from original vesicles. The bullae rupture, leaving raw, red areas. Nonbullous impetigo accounts for approximately 70% of cases. This type of impetigo tends to affect skin that has already been disrupted by cuts, abrasions, bites, or other types of trauma. *S. aureus* is also commonly implicated,

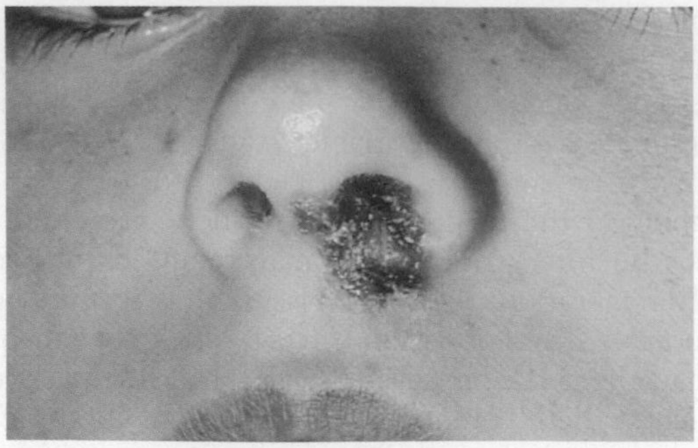

FIGURE 61-1 • Impetigo of the nostril.

including methicillin-resistant *Staphylococcus aureus* (MRSA), as well as *Streptococcus pyogenes* (Lewis & Friedman, 2013).

The exposed areas of the body, face, hands, neck, and extremities are most frequently involved. Impetigo is contagious and may spread to other parts of the patient's skin or to other members of the family who touch the patient or use towels or combs that are soiled with the exudate of the lesions.

Impetigo is seen in people of all races and ages. It is particularly common in children living in poor hygienic conditions. Chronic health problems, poor hygiene, and malnutrition may predispose an adult to impetigo. It is more prevalent in warm, humid climates and is therefore more commonly seen in the southeastern United States than in northern climates (Lewis & Friedman, 2013).

Clinical Manifestations

The lesions of impetigo are most commonly seen on the face or extremities. They begin as small, red macules, which quickly become discrete, thin-walled vesicles that rupture and become covered with a loosely adherent honey-yellow crust (Fig. 61-1). These crusts are easily removed to reveal smooth, red, moist surfaces on which new crusts soon develop (Lewis & Friedman, 2013).

Medical Management

Topical antibacterial therapy (e.g., mupirocin [Bactroban], retapamulin [Altabax]) is typically prescribed when the disease is limited to a small area. The medication must be applied to the lesions several times daily for 5 to 7 days. Lesions are first soaked or washed with soap solution to remove the central site of bacterial growth, giving the topical antibiotic an opportunity to reach the infected site. After the crusts are removed, the prescribed topical antibiotic cream is applied. Gloves are worn when providing patient care (Lewis & Friedman, 2013).

Systemic antibiotic agents may be prescribed to treat infections that are widespread or in cases where there are systemic manifestations (e.g., a fever is present). These antibiotics are effective in reducing contagious spread, treating deep infections, and preventing acute glomerulonephritis (kidney infection), which may occur as a consequence of streptococcal skin diseases. Amoxicillin–clavulanate (Augmentin), cloxacillin (Cloxapen), or dicloxacillin (Dynapen) may be prescribed. In cases where MRSA is present, antibiotics prescribed may include clindamycin (Cleocin), trimethoprim–sulfamethoxazole (Bactrim), or vancomycin (Vancocin) (Lewis & Friedman, 2013).

Nursing Management

The nurse educates the patient and family members to bathe at least once daily with bactericidal soap. Cleanliness and good hygiene practices help to prevent the spread of the lesions from one skin area to another and from one person to another. In particular, patients and family members must be educated to practice hand hygiene every time after a lesion is touched. Each person should have a separate towel and washcloth. Because impetigo is a contagious disorder, infected people should avoid contact with other people until the lesions heal (Lewis & Friedman, 2013).

Folliculitis, Furuncles, and Carbuncles

Folliculitis is an inflammatory condition of the cells within the wall and ostia of the hair follicles that is typically caused by a bacterial or fungal infection. Lesions may be superficial or deep. Single or multiple papules or pustules appear close to the hair follicles. Folliculitis commonly affects the beard area of men who shave, as well as women's legs, if they shave. Other areas include the axillae, trunk, and buttocks. Follicular disorders are usually caused by staphylococci, although if the immune system is impaired, the causative organisms may be gram-negative bacilli (Satter, 2012).

Pseudofolliculitis barbae (shaving bumps) occur predominately on the faces of African Americans and other curly-haired men as a result of shaving. The sharp ingrowing hairs have a curved root that grows at a more acute angle and pierces the skin, provoking an irritative reaction. The only entirely effective treatment is to avoid shaving. Other treatments include using special lotions or antibiotics or using a hand brush to dislodge the hairs mechanically. If the patient must remove facial hair, a depilatory cream or electric razor may be used.

A **furuncle** (boil) is an acute inflammation arising deep in one or more hair follicles and spreading into the surrounding dermis (Fig. 61-2). This inflammation is a deep form of folliculitis. Furunculosis refers to multiple or recurrent lesions. Furuncles may occur anywhere on the body but are more prevalent in areas subjected to irritation, pressure, friction, and excessive perspiration, such as the back of the neck, the axillae, and the buttocks.

A furuncle may start as a small, red, raised, painful pimple. Frequently, the infection progresses and involves the skin and

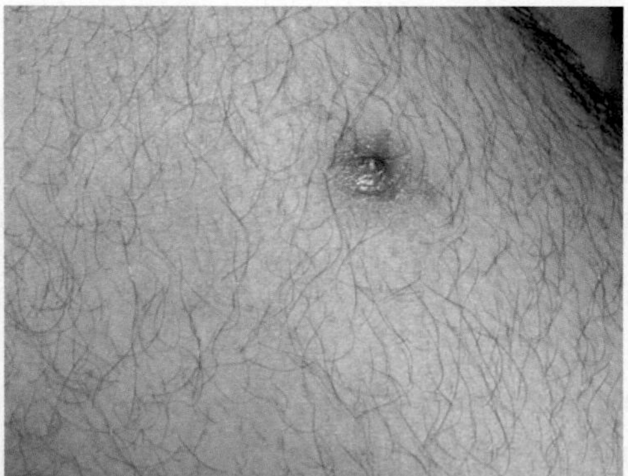

FIGURE 61-2 • Painful furuncle on the thigh. From Goodheart H. P. (2003). *Goodheart's photoguide of common skin disorders* (2nd ed.). Philadelphia: Lippincott Williams & Wilkins.

subcutaneous fatty tissue, causing tenderness, pain, and surrounding cellulitis. The area of redness and induration represents an effort of the body to keep the infection localized. The bacteria (usually staphylococci) produce necrosis of the invaded tissue. The characteristic pointing of a boil follows in a few days. When this occurs, the center becomes yellow or black, and the boil is said to have "come to a head."

A **carbuncle** is an abscess of the skin and subcutaneous tissue that represents an extension of a furuncle that has invaded several follicles and is large and deep seated. It is usually caused by a staphylococcal infection. Carbuncles appear most commonly in areas where the skin is thick and inelastic; the back of the neck and the buttocks are common sites. The extensive inflammation frequently prevents a complete walling off of the infection; purulent material may be absorbed, resulting in high fever, pain, leukocytosis, and even extension of the infection to the bloodstream.

Furuncles and carbuncles are more likely to occur in patients with underlying systemic diseases, such as diabetes or hematologic malignancies, and in those receiving immunosuppressive therapy for other diseases. Both are more prevalent in hot climates, especially on skin beneath occlusive clothing.

Medical Management

In treating staphylococcal infections, it is important not to rupture or destroy the protective wall of induration that localizes the infection. The boil or pimple should never be squeezed. Systemic antibiotic therapy, selected by culture and sensitivity study, is generally indicated. Oral dicloxacillin and cephalosporins are first-line medications. If MRSA is suspected, antibiotic agents selected may include clindamycin, trimethoprim–sulfamethoxazole, or minocycline (Satter, 2012). To promote comfort, bed rest is advised for patients who have boils on the perineum or in the anal region.

When the pus has localized and is fluctuant, a small incision with a scalpel can speed resolution by relieving the tension and ensuring direct evacuation of the pus and debris. The patient is instructed to keep the draining lesion covered with a dressing.

Nursing Management

Intravenous (IV) fluids, fever reduction, and other supportive treatments are indicated for patients who are acutely ill from infection. Warm, moist compresses hasten resolution of the furuncle or carbuncle. The surrounding skin may be cleaned gently with antibacterial soap, and an antibacterial ointment may be applied. Soiled dressings are handled according to standard precautions. Nursing personnel should carefully follow standard precautions to avoid becoming carriers of staphylococci. Disposable gloves are worn when caring for these patients.

Quality and Safety Nursing Alert

Nurses must take special precautions in caring for boils on the face because the skin area drains directly into the cranial venous sinuses. Sinus thrombosis with fatal pyemia can develop after manipulating a boil in this location. The infection can travel through the sinus tract and penetrate the brain cavity, causing a brain abscess.

Promoting Home and Community-Based Care

Educating Patients About Self-Care

To prevent and control staphylococcal skin infections such as boils and carbuncles, the staphylococcal pathogen must be eliminated from the skin and environment. Efforts must be made to increase the patient's resistance and provide a hygienic environment. If lesions are actively draining, the mattress and pillow should be covered with plastic material and wiped with disinfectant daily; the bed linens, towels, and clothing should be laundered after each use; and the patient should use an antibacterial soap and shampoo for an indefinite period, often several months.

Viral Skin Infections

Herpes Zoster

Herpes zoster, also called *shingles*, is an infection caused by the varicella-zoster viruses (VZVs), which are members of a group of deoxyribonucleic acid (DNA) viruses. The viruses that cause chickenpox (varicella) and herpes zoster are indistinguishable, hence the two-part name. The disease is characterized by a painful vesicular eruption along the area of distribution of the sensory nerves from one or more posterior ganglia. After a case of chickenpox runs its course, the VZV responsible for the outbreak lies dormant inside nerve cells near the brain and spinal cord. Later, when these latent viruses are reactivated because of declining cellular immunity, they travel by way of the peripheral nerves to the skin, where the viruses multiply and create a red rash of small, fluid-filled blisters.

It is thought that during the aging process, natural immunity to the varicella virus wanes, allowing the virus to reactivate. Herpes zoster develops over the lifetime of about 10% to 20% of all adults who had chickenpox earlier in life, usually after 50 years of age. The rates of occurrence tend to be the same in both men and women but are slightly higher in Caucasians compared to African Americans. There is an increased frequency of herpes zoster infections in patients with weakened immune systems, including those with human immunodeficiency virus (HIV) infection and in those with cancer. In these patients, the infection can become widespread and cause significant complications (Janninger, Eastern, Hospenthal, et al., 2013).

Clinical Manifestations

Manifestations typically occur in three phases, including the preeruptive, acute eruptive, and postherpetic neuralgia (PHN) phases. During the preeruptive phase, the previously dormant VZV becomes reactivated within the dorsal root ganglia of the spinal cord. Manifestations that follow, therefore, tend to follow the dermatome that corresponds with the ganglion or ganglia that are affected. The patient will typically complain of pain, or sometimes pruritus or paresthesias, over the sensory region that follows that dermatome. This phase lasts from 1 to 10 days, with 48 hours being typical (Janninger et al., 2013).

The acute eruptive phase is heralded by the appearance of unilateral patchy erythematous areas in the dermatomal

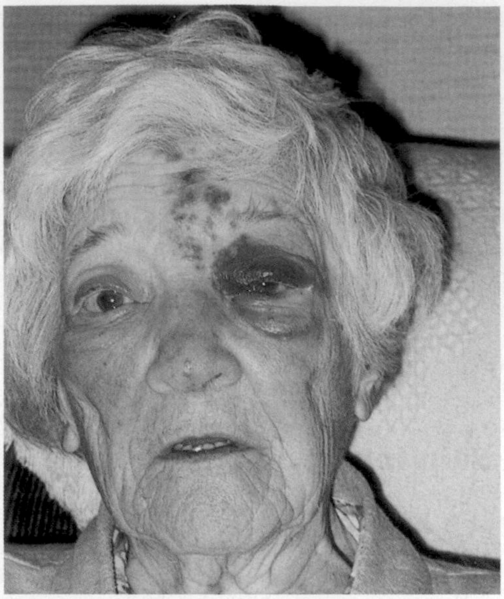

FIGURE 61-3 • Herpes zoster (shingles).

area that is affected. Vesicles develop that appear initially clear, then become cloudy, and eventually rupture and crust (Fig. 61-3). The pain that accompanies this stage is typically described as severe and unrelenting. This phase typically lasts between 10 to 15 days (Janninger et al., 2013).

The last phase—PHN—is variable in terms of both duration and manifestations. As many as 45% of all patients have PHN pain that may be severe for 30 or more days after the lesions have healed. The pain is typically localized to the dermatomal area that was affected. The pain tends to last longer in older adults. Approximately 50% of adults older than 60 years with herpes zoster experience PHN pain for longer than 60 days (Janninger et al., 2013).

Herpes zoster ophthalmicus (HZO) is a rare subtype of herpes zoster that causes severe consequences. Typically in HZO, a branch of the trigeminal nerve is affected that innervates the ocular and periocular structures. This may cause significant pain and morbid ocular complications, including blindness (Janninger et al., 2013).

Medical Management

Herpes zoster infection can be arrested if oral antiviral agents such as acyclovir (Zovirax), valacyclovir (Valtrex), or famciclovir (Famvir) are administered within 24 hours of the initial eruption. IV acyclovir, if started early, is effective in significantly reducing the pain and halting the progression of the disease (Janninger et al., 2013).

The goals of herpes zoster management are to relieve the pain and to reduce or avoid complications, which include infection, scarring, and PHN and eye complications. Pain is controlled with analgesic agents because adequate pain control during the acute phase helps prevent persistent pain patterns. Systemic corticosteroids may be prescribed to reduce the incidence and duration of PHN. Healing usually occurs more quickly in those who have been treated with corticosteroids. Triamcinolone (Aristocort, Kenacort, Kenalog) injected subcutaneously under painful areas is effective as an anti-inflammatory agent. Patients with HZO require emergent treatment by an ophthalmologist (Janninger et al., 2013).

People who have been exposed to varicella by primary infection or by vaccination are not at risk for infection with VZV after exposure to patients with herpes zoster.

A vaccination for childhood varicella developed in the 1970s has been used successfully to decrease the incidence of childhood disease. A more potent formulation of the vaccine, the varicella-zoster vaccine (Zostavax) was developed to boost VZV cellular immunity in people older than 60 years. This vaccination is now recommended as part of prevention strategies in older adults who are not immunocompromised, including those with a history of herpes zoster, because it may recur (Janninger et al., 2013).

Nursing Management

The patient and family members are instructed about the importance of taking antiviral agents as prescribed and in keeping follow-up appointments with the primary provider. The nurse assesses the patient's discomfort and response to medication and collaborates with the primary provider to make necessary adjustments to the treatment regimen. The patient is educated about how to apply wet dressings or medication to the lesions and to follow proper hand hygiene techniques to avoid spreading the virus.

Diversionary activities and relaxation techniques are encouraged to ensure restful sleep and to alleviate discomfort. A caregiver may be required to assist with dressings, particularly if the patient is an older adult and unable to apply them. Food preparation for patients who cannot care for themselves or prepare nourishing meals must be arranged.

Herpes Simplex

Herpes simplex is a common skin infection. There are two types of the causative virus, which are identified by viral typing. Generally, herpes simplex type 1 occurs on the skin of the lips, mouth, gums, or tongue (or on the skin around the mouth) and type 2 occurs in the genital area, but both viral types can be found in both locations. (See Chapters 22 and 57 for discussion of herpes simplex type 1 and type 2, respectively.)

Fungal (Mycotic) Skin Infections

Fungi, tiny members of a subdivision of the plant kingdom that thrive on organic matter, cause various common skin infections. In some cases, they affect only the skin and its appendages (hair and nails). In other cases, internal organs are involved, and the diseases may be life threatening. However, superficial infections rarely cause even temporary disability and respond readily to treatment. Secondary infection with bacteria, *Candida*, or both organisms may occur.

The most common fungal skin infection is **tinea,** which is also called *ringworm* because of its characteristic appearance of a ring or rounded tunnel under the skin. Tinea infections affect the head, body, groin, feet, and nails (Miller, Rashid, & Silverberg, 2011). Table 61-5 summarizes the tinea infections and treatments.

To obtain a specimen for diagnosis, the lesion is cleaned and a scalpel or glass slide is used to remove scales from the

TABLE 61-5 Tinea (Ringworm) Infections

Type and Location	Clinical Manifestations	Treatment
Tinea capitis (head; contagious fungal infection of the hair shaft)	• Most common type seen in children • Oval, scaling, erythematous patches • Small papules or pustules on the scalp • Brittle hair that breaks easily	• Griseofulvin for 4–6 weeks or itraconazole (Sporanox) for 4–6 weeks or terbinafine (Lamisil) for 2–4 weeks • Shampoo hair 2 or 3 times with ketoconazole (Nizoral) or selenium sulfide shampoo.
Tinea corporis (body)	• Found in all age groups; however, most frequently found in adolescents and pregnant women • Begins with red macule, which spreads to a ring of papules or vesicles with central clearing • Lesions found in clusters; many spread to the hair, scalp, or nails • Very pruritic • Infected pet may be source	• Local infections—topical antifungal creams once or twice daily (e.g., clotrimazole [Lotrimin], econazole [Spectazole], ketoconazole [Nizoral]) • Extensive infections or concomitant tinea capitis or immunosuppressive conditions (e.g., active neoplasms)—oral antifungal medications (e.g., fluconazole [Diflucan] for 2–4 weeks, itraconazole for 1 week, terbinafine for 2 weeks)
Tinea cruris (groin area; "jock itch")	• Begins with small, red scaling patches, which spread to form circular elevated plaques • Very pruritic • Clusters of pustules may be seen around borders.	• Local infections—see treatment for tinea corporis. • Extensive infections or concomitant tinea pedis or immunosuppressive conditions (e.g., active neoplasms)—see treatment for tinea corporis. • Educate patients to pat dry skin folds thoroughly (avoid rubbing) after bathing and to use separate towels for groin and other body parts.
Tinea pedis (foot; "athlete's foot")	• Most common type found in adults • Soles of one or both feet have scaling and mild redness with maceration in the toe webs. • More acute infections may have clusters of clear vesicles on dusky base.	• Local infections—see treatment for tinea corporis. • Extensive infections or concomitant tinea pedis or immunosuppressive conditions (e.g., active neoplasms)—see treatment for tinea corporis. • Educate patients to put on socks before undershorts to avoid cross-contamination to groin.
Tinea unguium (toenails; onychomycosis)	• Prevalence in adult males is 3%, in adult females is 1.4%; • Nails thicken, crumble easily, and lack luster. • Whole nail may be destroyed.	• Oral antifungal medications for 12 weeks (e.g., itraconazole, terbinafine) with or without concomitant topical ciclopirox olamine nail lacquer (Penlac) • Nail avulsion may be indicated, either surgically or chemically using a 40%–50% urea compound

Adapted from Kao, G. F. (2013). Tinea capitis. *Medscape*. Available at: emedicine.medscape.com/article/1091351-overview; Lesher, J. L. (2012). Tinea corporis. *Medscape*. Available at: emedicine.medscape.com/article/1091473-overview; Miller, A. C., Rashid, R. M., & Silverberg, M. A. (2011). Tinea in emergency medicine. *Medscape*. Available at: emedicine.medscape.com/article/787217-overview; Robbins, C. M., & Elewski, B. E. (2012). Tinea pedis. *Medscape*. Available at: emedicine.medscape.com/article/1091684-overview; Tosti, A. (2013). Onychomycosis. *Medscape*. Available at: emedicine.medscape.com/article/1105828-overview; and Weiderkehr, M., & Schwartz, R. A. (2012). Tinea cruris. *Medscape*. Available at: emedicine.medscape.com/article/1091806-overview

margin of the lesion. The scales are dropped onto a slide to which potassium hydroxide has been added. The diagnosis is made by examination of the infected scales microscopically for spores and hyphae or by isolating the organism in culture. Under Wood's light, a specimen of infected hair appears fluorescent; this may be helpful in diagnosing some cases of tinea capitis (Miller et al., 2011).

Parasitic Skin Infestations

Parasitic skin infestations include those of the skin by lice (pediculosis) and the itch mite (scabies).

Pediculosis: Lice Infestation

Lice infestation affects people of all ages. Three varieties of lice infest humans: *Pediculus humanus capitis* (head louse), *Pediculus humanus corporis* (body louse), and *Phthirus pubis* (pubic louse or "crab"). Lice are called *ectoparasites* because they live on the outside of the host's body. They depend on the host for their nourishment, feeding on human blood approximately five times each day. They inject their digestive juices and excrement into the skin, which causes severe itching (Guenther, Maguiness, & Austin, 2012).

Types of Pediculosis

Pediculosis Capitis

Pediculosis capitis is an infestation of the scalp by the head louse. The female louse lays her eggs (nits) close to the scalp. The nits become firmly attached to the hair shafts with a tenacious substance. The young lice hatch in about 10 days and reach maturity in 2 weeks. Head lice may be transmitted directly by physical contact or indirectly by infested combs, brushes, wigs, hats, and bedding (Guenther et al., 2012).

Pediculosis Corporis and Pubis

Pediculosis corporis is an infestation of the body by the body louse. This is a disease of those who live in close quarters. Pediculosis pubis is extremely common. The infestation is generally localized in the genital region and is transmitted chiefly by sexual contact (Guenther et al., 2012).

Clinical Manifestations

Head lice are found most commonly along the back of the head and behind the ears. To the naked eye, the eggs look like silvery, glistening oval bodies. The bite of the insect causes intense pruritus, and the resultant scratching often leads to secondary bacterial infection, such as impetigo or furunculosis. The infestation is more common in children and people with long hair (Guenther et al., 2012).

With body lice, the areas of the skin that come in closest contact with the underclothing (i.e., neck, trunk, and thighs) are chiefly involved. The body louse lives primarily in the seams of underwear and clothing, to which it clings as it pierces the skin with its proboscis. Its bites cause characteristic minute hemorrhagic points. Widespread excoriation may appear as a result of intense pruritus and scratching, especially on the trunk and neck. Among the secondary lesions produced are parallel linear scratches and a slight degree of eczema. In long-standing cases, the skin may become thick, dry, and scaly, with dark pigmented areas (Guenther et al., 2012).

Pruritus, particularly at night, is the most common symptom of pediculosis pubis. Reddish-brown dust (i.e., excretions of the insects) may be found in the patient's underclothing. The pubic area should be examined with a magnifying glass for lice crawling down a hair shaft or nits cemented to the hair or at the junction with the skin. Infestation by pubic lice may coexist with sexually transmitted infections (STI) such as gonorrhea, herpes, or syphilis. There may also be infestation of the hairs of the chest, axillae, beard, and eyelashes. Gray-blue macules may sometimes be seen on the trunk, thighs, and axillae as a result of either the reaction of the insects' saliva with bilirubin (converting it to biliverdin) or an excretion produced by the salivary glands of the louse (Guenther et al., 2012).

Medical Management

Treatment of head and pubic lice involves washing the hair with a shampoo containing pyrethrin compounds with piperonyl butoxide (RID or R&C Shampoo) or rinsing with permethrin (Nix). Although it had been first-line treatment for many years, lindane (Kwell) is no longer recommended because of its neurotoxic adverse effects (Guenther et al., 2012). The patient is instructed to shampoo the scalp and hair according to the product directions. After the hair is rinsed thoroughly, it is combed with a fine-toothed comb dipped in vinegar to remove any remaining nits or nit shells freed from the hair shafts. They are extremely difficult to remove and may have to be picked off one by one.

The patient with body lice is instructed to bathe with soap and water. Typically, no medications are indicated because the lice live on the patient's clothing. Topical medications used to treat head and pubic lice may be applied to the clothing, however, particularly in the seams of garments (see following discussion about general hygiene measures). If the eyelashes are involved, petrolatum may be thickly applied twice daily for 8 days, followed by mechanical removal of any remaining nits (Guenther et al., 2012).

All articles of clothing, towels, and bedding that may have lice or nits should be washed in hot water—at least 54°C (130°F)—or dry-cleaned to prevent reinfestation. Upholstered furniture, rugs, and floors should be vacuumed frequently. Combs and brushes are also disinfected with the shampoo or discarded. All family members and close contacts are treated (Guenther et al., 2012).

Complications, such as severe pruritus, pyoderma, and dermatitis, are treated with antipruritics, systemic antibiotics, and topical corticosteroids. Body lice can transmit epidemic rickettsial disease (e.g., epidemic typhus, relapsing fever, and trench fever) to humans (Guenther et al., 2012). The causative organism may be in the gastrointestinal tract of the insect and may be excreted on the skin surface of the infested person.

Nursing Management

The nurse informs the patient that head lice may infest anyone and are not a sign of uncleanliness. Because the condition spreads rapidly, treatment must be started immediately. Epidemics among those living in close quarters (e.g., dormitories, military barracks) may be managed by having everyone shampoo their hair on the same night. Cohabitants and family members should be warned not to share combs, brushes, and hats; they should be inspected for head lice daily for at least 2 weeks.

Treatment is necessary for all family members and sexual contacts of patients with body and/or pubic lice. The nurse educates them about personal hygiene and methods to prevent or control infestation. The patient and partner must also be scheduled for a diagnostic workup for coexisting STIs.

Scabies

Scabies is an infestation of the skin by the itch mite *Sarcoptes scabiei*. The disease is most commonly found in people living in substandard hygienic conditions and in people who are sexually active. The mites frequently involve the fingers, and hand contact may produce infection (Cordoro, Wilson, & Kauffman, 2012).

Clinical Manifestations

It takes approximately 4 weeks from the time of contact for the patient's symptoms to appear. The patient complains of severe itching caused by a delayed type of immunologic reaction to the mite or its fecal pellets. During examination, the patient is asked where the pruritus is most severe. A magnifying glass and a penlight are held at an oblique angle to the skin while a search is made for the small, raised burrows created by the mites. The burrows may be multiple, straight or wavy, brown or black, threadlike lesions, most commonly observed between the fingers and on the wrists. Other sites are the extensor surfaces of the elbows, the knees, the edges of the feet, the points of the elbows, around the nipples, in the axillary folds, under pendulous breasts, and in or near the groin or gluteal fold, penis, or scrotum. Red, pruritic eruptions usually appear between adjacent skin areas. However, the burrow is not always visible (Cordoro et al., 2012).

One classic sign of scabies is the increased itching that occurs during the overnight hours, perhaps because the increased warmth of the skin has a stimulating effect on the parasite. Hypersensitivity to the organism and its products of excretion also may contribute to the pruritus. If the infection has spread, other members of the family and close friends also complain of pruritus about 1 month later (Cordoro et al., 2012).

Secondary lesions are quite common and include vesicles, papules, excoriations, and crusts. Bacterial superinfection may result from constant excoriation of the burrows and papules (Cordoro et al., 2012).

Gerontologic Considerations

Older adult patients living in long-term care facilities are susceptible to outbreaks of scabies because of close living quarters, poor hygiene due to limited physical ability, and the potential for incidental spread of the organisms by staff members. The vivid inflammatory reaction seen in younger people seldom occurs; rather, the older adult may have peripheral sensory deficits and be less prone to scratch or may be physically unable to scratch. Scratching is an effective mechanism that partially eradicates mite infestation; thus, this results in a more severe subtype. The lesions crust over (causing "crusted scabies") and, in time, may become hyperkeratotic (Cordoro et al., 2012).

Health care personnel in extended care facilities should wear gloves when providing hands-on care to a patient suspected of having scabies until the diagnosis is confirmed and treatment completed. It is advisable to treat all residents, staff, and families of patients at the same time to prevent reinfection. The scales that are present with crusted scabies must be removed so that the antiscabicidal medication may be effective. Crusts may be removed with warm water soaks followed by application of 5% salicylic acid in petrolatum cream (Cordoro et al., 2012).

Assessment and Diagnostic Findings

The diagnosis is confirmed by recovering *S. scabiei* or the mites' by-products from the skin. A sample of superficial epidermis is scraped from the top of the burrows or papules with a small scalpel blade. The scrapings are placed on a microscope slide and examined through a microscope at low power to demonstrate evidence of the mite (Cordoro et al., 2012).

Medical Management

The patient is instructed to take a warm, soapy bath or shower to remove the scaling debris from the crusts and then to pat the skin dry thoroughly and allow it to cool. A prescription scabicide, 5% permethrin, is considered the medication of choice. It is applied thinly to the entire skin from the neck down, sparing only the face and scalp (which are not affected in scabies). The medication is left on for 12 to 24 hours, after which the patient is instructed to wash thoroughly. One application may be curative, but it is advisable to repeat the treatment in 1 week (Cordoro et al., 2012).

Nursing Management

The patient should wear clean clothing and sleep between freshly laundered bed linens. All bedding and clothing should be washed in hot water and dried on the hot dryer cycle. If bed linens or clothing cannot be washed in hot water, dry cleaning is advised.

After treatment is completed, the patient may apply an ointment, such as a topical corticosteroid, to skin lesions because the scabicide may irritate the skin. The patient's hypersensitivity does not cease on destruction of the mites. Pruritus may continue for several weeks as a manifestation of hypersensitivity, particularly in people who are atopic (allergic). This is not a sign that the treatment has failed. The patient is instructed not to apply more scabicide, because it will cause more irritation and increased itching, and not to take frequent hot showers, because they can dry the skin and produce pruritus. Oral antihistamines such as diphenhydramine or hydroxyzine can help control the pruritus. If a secondary infection is present, treatment with oral antibiotic agents may be indicated (Cordoro et al., 2012).

All family members and close contacts should be treated simultaneously to eliminate the mites. Some scabicides are approved for use in infants and pregnant women. If scabies is sexually transmitted, the patient may require treatment for coexisting STI. Scabies may also coexist with pediculosis.

NONINFECTIOUS INFLAMMATORY DERMATOSES

Irritant Contact Dermatitis

Contact dermatitis (also called *eczema*) is an inflammatory reaction of the skin to physical, chemical, or biologic agents. The epidermis is damaged by repeated physical and chemical irritations. Contact dermatitis may be of the primary irritant type, in which a nonallergic reaction results from exposure to an irritating substance, or it may be an allergic reaction resulting from exposure of sensitized people to contact allergens (allergic dermatoses are discussed in Chapter 38).

Common causes of irritant dermatitis are soaps, detergents, scouring compounds, and industrial chemicals. Predisposing factors include extremes of heat and cold, frequent contact with soap and water, and a preexisting skin disease. Persons at risk include those whose occupations require repeated handwashing (e.g., nurses) or repeated exposure to food or other irritants (e.g., cleaners, hairdressers, food preparation workers). Women tend to be affected more commonly than men (Hogan, 2011).

Clinical Manifestations

The eruptions begin when the causative agent contacts the skin. The first reactions include pruritus, burning, and erythema, followed closely by edema, papules, vesicles, and oozing or weeping. In the subacute phase, these vesicular changes are less marked, and they alternate with crusting, drying, fissuring, and peeling. If repeated reactions occur or if the patient continually scratches the skin, lichenification and pigmentation occur. Secondary bacterial invasion may follow (Hogan, 2011).

Medical Management

The objectives of management are to soothe and heal the involved skin and protect it from further damage. The distribution pattern of the reaction is identified to differentiate between allergic and irritant contact dermatitis. A detailed history is obtained. If possible, the offending irritant is removed. Local irritation should be avoided, and soap is not generally used until healing occurs.

Many preparations are advocated for relieving dermatitis. In general, a barrier cream that contains ceramide (e.g., Impruv, CeraVe) or dimethicone (e.g., Cetaphil) is used for small patches of erythema. Cool, wet dressings also are applied over small areas of vesicular dermatitis. Wet dressings may help clear the oozing eczematous lesions. A thin layer of cream or ointment containing a corticosteroid is then commonly used, although the efficacy of corticosteroids has not

Chart 61-2 PATIENT EDUCATION

Strategies for Avoiding Contact Dermatitis

The following precautions may help prevent repeated cases of contact dermatitis. Follow these instructions for at least 4 months after your skin appears to be completely healed.

- Study the pattern and location of your dermatitis and think about which things have touched your skin and which things may have caused the problem; try to avoid contact with these materials.
- Avoid heat, soap, and rubbing, all of which are external irritants.
- Choose bath soaps, laundry detergents, and cosmetics that do not contain fragrance.
- Avoid using a fabric softener dryer sheet (Bounce, Cling Free). Fabric softeners that are added to the washer may be used.
- Avoid topical medications, lotions, or ointments, except those specifically prescribed for your condition.
- Wash your skin thoroughly immediately after exposure to possible irritants.
- When wearing gloves (such as for washing dishes or general cleaning), be sure they are cotton lined. Do not wear them more than 15 or 20 minutes at a time.

Adapted from Hogan, D. J. (2011). Irritant contact dermatitis. *Medscape*. Available at: emedicine.medscape.com/article/1049353-overview

been demonstrated in research (Hogan, 2011). The patient is educated about how to treat and prevent future bouts of irritant dermatitis (Chart 61-2).

Psoriasis

Considered one of the most common chronic noncommunicable skin diseases, psoriasis is typically characterized by the appearance of silvery plaques that most commonly appear on the skin over the elbows, knees, scalp, lower back, and buttocks, although lesions may appear anywhere, including the oral cavity, eyes (including the lids, conjunctivae, and corneas), and joints (Meffert, Arffa, Gordon, et al., 2013). Psoriasis affects approximately 7.5 million Americans, or 2% to 3% of the world's population (Dowling, 2010). Onset may occur at any age, with a median onset at 28 years. It is more prevalent among women and Caucasians and among persons who are obese. It is thought that most patients with psoriasis have a genetic predisposition to develop the disease (Meffert et al., 2013). Psoriasis has a tendency to improve and then recur periodically throughout life (Porth & Matfin, 2009).

Pathophysiology

Current evidence supports an autoimmune basis for psoriasis (Porth & Matfin, 2009). Periods of emotional stress and anxiety aggravate the condition, and trauma, infections, and seasonal and hormonal changes may also serve as triggers.

In this disease, the epidermis becomes infiltrated by activated T cells and cytokines, resulting in both vascular engorgement and proliferation of keratinocytes. Epidermal hyperplasia results. These epidermal cells tend to improperly retain their nuclei, crippling their ability to release lipids that encourage cellular adhesion. This results in rapid turnover of poorly matured cells that do not adhere well to each other, resulting in the classic presentation of plaquelike lesions that have a silvery, scaly, flaky appearance (Meffert et al., 2013).

Clinical Manifestations

Psoriasis may range in severity from a cosmetic source of annoyance to a physically disabling and disfiguring disorder. Lesions appear as red, raised patches of skin covered with silvery scales. The scaly patches are formed by the buildup of living and dead skin (Fig. 61-4). If the scales are scraped away, the dark red base of the lesion is exposed, producing multiple bleeding points. The patches are not moist and may be pruritic. In many cases, the nails are also involved, with pitting, discoloration, crumbling beneath the free edges, and separation of the nail plate (Weber & Kelley, 2010).

Complications

Asymmetric rheumatoid factor–negative arthritis of multiple joints occurs in up to 30% of people with psoriasis, most typically after the skin lesions appear. The most typical joints affected include those in the hands or feet, although sometimes larger joints such as the elbows, knees, or hips may be affected (Meffert et al., 2013). It is recommended that a rheumatologist be consulted to assist in the diagnosis and long-term treatment of this disorder. (See Chapter 39 for further discussion of spondyloarthropathies, including psoriatic arthritis.)

Generalized exfoliative dermatitis, also called *erythroderma*, may also result from psoriasis (see discussion later in this chapter).

Assessment and Diagnostic Findings

The presence of the classic plaque-type lesions generally confirms the diagnosis of psoriasis. If in doubt, the health care provider should assess for signs of nail and scalp involvement and for a positive family history. Biopsy of the skin is of little diagnostic value.

Medical Management

The goals of management are to slow the rapid turnover of epidermis, to promote resolution of the psoriatic lesions, and to control the natural cycles of the disease. There is no known cure.

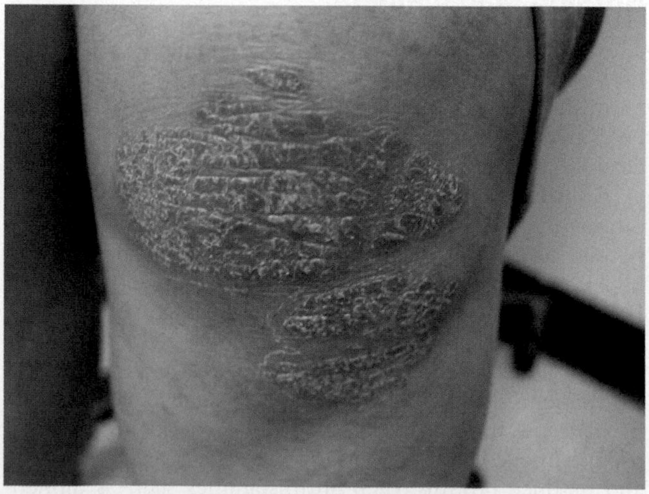

FIGURE 61-4 • Psoriasis.

The therapeutic approach should be one that the patient understands; it should be cosmetically acceptable and minimally disruptive of lifestyle. Treatment involves the commitment of time and effort by the patient and possibly the family. Any precipitating or aggravating factors are addressed. An assessment is made of lifestyle because psoriasis is significantly affected by stress. Management of emotional factors should be addressed as part of the overall treatment of psoriasis. The patient is informed that treatment of severe psoriasis can be time-consuming, expensive, and aesthetically unappealing at times. Many patients report difficulty adhering to treatment plans, either for time reasons or lack of response to the treatment (National Institute for Health and Clinical Excellence [NICE], 2012).

Gentle removal of scales is an important principle of psoriasis treatment. This can be accomplished by taking baths with added oils (e.g., olive oil, mineral oil), colloidal oatmeal preparations or coal tar preparations. A soft brush may be used to gently scrub the psoriatic plaques. After bathing, the application of emollient creams containing alpha-hydroxy acids (e.g., Lac-Hydrin, Penederm) or salicylic acid can soften thick scales. The patient and family should be encouraged to establish a regular skin care routine that can be maintained even when the psoriasis is not in an acute stage (NICE, 2012).

Pharmacologic Therapy

Three types of therapy are commonly indicted: topical, phototherapy, and systemic.

Topical Agents

Topically applied agents are used to slow the overactive epidermis. Topical corticosteroids may be applied for their anti-inflammatory effects (see Table 61-3). Choosing the correct strength of corticosteroid for the involved site and choosing the most effective vehicle base are important aspects of topical treatment. In general, high-potency topical corticosteroids should not be used on the face and intertriginous areas, and their use on other areas should be limited to a 4-week course of twice-daily applications. A 4-week break should be taken before repeating treatment with the high-potency corticosteroids. For long-term therapy, moderate-potency corticosteroids are used. On the face and intertriginous areas, only low-potency corticosteroids are appropriate for long-term use (NICE, 2012) (see Table 61-3).

Occlusive dressings may be applied to increase the effectiveness of the corticosteroid. Large plastic bags may be used—one for the upper body with openings cut for the head and arms and one for the lower body with openings for the legs. Large rolls of tubular plastic can be used to cover the arms and legs. Another option is a vinyl jogging suit. The medication is applied, and the suit is put on over it. The hands can be wrapped in gloves, the feet in plastic bags, and the head in a shower cap. Occlusive dressings should not remain in place longer than 8 hours. The skin should be inspected carefully for the appearance of atrophy, hypopigmentation, striae, and telangiectasias—all of which are side effects of corticosteroids.

When psoriasis involves large areas of the body, topical corticosteroid treatment can be expensive and involve some systemic risk. The more potent corticosteroids, when applied to large areas of the body, have the potential to cause adrenal suppression through percutaneous absorption of the medication. In this event, other treatment modalities (e.g., nonsteroidal topical medications, ultraviolet light) may be used instead or in combination to decrease the need for corticosteroids (NICE, 2012).

Two relatively new topical nonsteroidal treatments are calcipotriene (Dovonex) and tazarotene (Tazorac). Treatment with these agents tends to suppress **epidermopoiesis** (i.e., development of epidermal cells) and cause sloughing of the rapidly growing epidermal cells. Calcipotriene 0.05% is a derivative of vitamin D_2. It works by decreasing the mitotic turnover of the psoriatic plaques. Its most common side effect is local irritation. The intertriginous areas and face should be avoided when using this medication. The patient should be monitored for symptoms of hypercalcemia. Calcipotriene is available as a cream for use on the body and a solution for the scalp. It is not recommended for use by older adult patients because of their more fragile skin or by pregnant or lactating women (Meffert et al., 2013).

Tazarotene, a retinoid, causes sloughing of the scales covering psoriatic plaques. As with other retinoids, it causes increased sensitivity to sunlight by loss of the outermost layer of skin, so the patient should be cautioned to use an effective sunscreen and avoid other photosensitizers (e.g., tetracycline, antihistamines). Tazarotene is listed as a category X drug in pregnancy; reports indicate evidence of fetal risk, and the risk of use in pregnant women clearly outweighs any possible benefits. A negative result on a pregnancy test should be obtained before initiating this medication in women of childbearing age, and an effective contraceptive should be continued during treatment. Side effects include burning, erythema, or irritation at the site of application, and worsening of psoriasis (Meffert et al., 2013).

Intralesional injections of the corticosteroid triamcinolone acetonide (Aristocort, Kenalog-10, Trymex) can be administered directly into highly visible or isolated patches of psoriasis that are resistant to other forms of therapy. Care must be taken to ensure that the medication is not injected into normal skin (Meffert et al., 2013).

Phototherapy

For patients who do not respond well to topical treatments, phototherapy using narrow-band ultraviolet-B (UVB) therapy may be effective as a single-therapy modality. However, phototherapy is generally more effective when it is administered as ultraviolet-A (UVA) in conjunction with a photosensitizing oral medication (a combination referred to as PUVA). Here, the patient takes a photosensitizing medication (i.e., psoralen) in a standard dose and is subsequently exposed to long-wave ultraviolet light as the medication plasma levels peak. It is thought that when psoralen-treated skin is exposed to UVA light, the psoralen binds with DNA and decreases epidermal cellular proliferation. PUVA has been associated with long-term risks of skin cancer, cataracts, and premature aging of the skin (Porth & Matfin, 2009; Meffert et al., 2013; NICE, 2012).

The patient is usually treated two or three times each week until the psoriasis clears. An interim period of 48 hours between treatments is necessary to allow any burns resulting from PUVA therapy to become evident. After the psoriasis clears, the patient begins a maintenance program. Once little or no disease is active, less potent therapies are used to keep minor flare-ups under control (NICE, 2012).

Systemic Agents

Although systemic corticosteroids may cause rapid improvement of psoriasis, the usual risks and the possibility of triggering a severe flare-up on withdrawal limit their use; therefore, they are not indicated for treatment of psoriasis.

Systemic cytotoxic preparations, such as methotrexate, have long been used successfully in treating extensive psoriasis that fails to respond to other forms of therapy. Methotrexate appears to inhibit DNA synthesis in epidermal cells, thereby reducing the turnover time of the psoriatic epidermis. However, the medication can be toxic, especially to the liver, kidneys, and bone marrow. Laboratory studies must be monitored to ensure that the hepatic, hematopoietic, and renal systems are functioning adequately. The patient should avoid drinking alcohol while taking methotrexate because alcohol ingestion increases the possibility of liver damage. The medication is teratogenic and thus should not be administered to pregnant women.

Cyclosporine (Neoral), a cyclic peptide used to prevent rejection of transplanted organs, has shown some success in treatment of severe, therapy-resistant cases of psoriasis. However, its use is limited by side effects such as hypertension and nephrotoxicity (Meffert et al., 2013).

The newest line of treatments for psoriasis includes a group called *biologic agents* because of their derivation from immunomodulators and bioengineered proteins (such as antibodies or recombinant cytokines) and their targeted action directly on the T cells. These agents act by inhibiting activation and migration, eliminating the T cells completely, slowing postsecretory cytokines or inducing immune deviation.

Infliximab (Remicade) is a monoclonal antibody that binds to tumor necrosis factor-alpha (TNF-α) and can only be administered by IV infusion. Ustekinumab (Stelara) is also a monoclonal antibody that specifically interferes with the effect of interleukins (ILs), particularly IL-12 and IL-23. Etanercept (Enbrel) is a fusion protein that binds to soluble TNF-α and blocks its interaction with cell surface receptors. Alefacept (Amevive) is a fusion protein that inhibits T-cell proliferation. Adalimumab (Humira) is a recombinant human immunoglobulin G1 (IgG1) monoclonal antibody against TNF-α. These biologic agents have significant side effects, making close monitoring essential (Meffert et al., 2013).

Nursing Management

Psoriasis may cause despair and frustration for the patient; observers may stare, comment, ask embarrassing questions, or even avoid the person. The disease can eventually exhaust the patient's resources, interfere with his or her job, and negatively affect many aspects of life. Teenagers are especially vulnerable to the psychological effects of this disorder.

The nurse assesses the impact of the disease on the patient and the coping strategies used for conducting normal activities and interactions with family and friends (Dowling, 2010). Many patients need reassurance that the condition is not infectious, not a reflection of poor personal hygiene, and not skin cancer. The nurse can create an environment in which the patient feels comfortable discussing important quality-of-life issues related to his or her psychosocial and physical response to this chronic illness.

The nurse explains with sensitivity that although there is no cure for psoriasis and lifetime management is necessary, the condition can usually be controlled. The pathophysiology of psoriasis is reviewed, as are the factors that provoke it—irritation or injury to the skin (e.g., cut, abrasion, sunburn), current illness (e.g., pharyngeal streptococcal infection), and emotional stress. It is emphasized that repeated trauma to the skin and an unfavorable environment (e.g., cold) may exacerbate psoriasis. The patient is cautioned about taking any nonprescription medications because some may aggravate mild psoriasis. As well, the patient is advised to seek treatment from the same primary provider for any acute illnesses or chronic conditions to minimize chances of receiving prescriptions that may interfere with each other (NICE, 2012).

Reviewing and explaining the treatment regimen are essential to ensure adherence to the therapeutic regimen. For example, if the patient has a mild condition confined to localized areas, such as the elbows or knees, application of an emollient to maintain softness and minimize scaling may be all that is required. Most patients need a comprehensive plan of care that ranges from using topical medications and shampoos to more complex and lengthy treatment with systemic medications and photochemotherapy, such as PUVA therapy. Patient education materials that include a description of the therapy and specific guidelines are helpful but cannot replace face-to-face discussions of the treatment plan.

To avoid injuring the skin, the patient is advised not to pick at or scratch the affected areas. Measures to prevent dry skin are encouraged because dry skin worsens psoriasis. Too-frequent washing produces more soreness and scaling. Water should be warm, not hot, and the skin should be dried by patting with a towel rather than by rubbing. Emollients have a moisturizing effect, providing an occlusive film on the skin surface so that normal water loss through the skin is halted and allowing the trapped water to hydrate the stratum corneum. A bath oil or emollient cleansing agent can comfort sore and scaling skin. Softening the skin can prevent fissures.

A therapeutic relationship between health care professionals and the patient with psoriasis includes education and support. Introducing the patient to successful coping strategies used by others with psoriasis and making suggestions for reducing or coping with stressful situations at home, school, and work can facilitate a more positive outlook and acceptance of the chronicity of the disease (Dowling, 2010).

Promoting Home and Community-Based Care

Educating Patients About Self-Care

Printed patient education materials may be provided to reinforce face-to-face discussions about treatment guidelines and other considerations. Patients using topical corticosteroid preparations repeatedly on the face and around the eyes should be aware that cataract development is possible. Strict guidelines for applying these medications should be emphasized, because overuse can result in skin atrophy, striae, and medication resistance (NICE, 2012).

PUVA, which is reserved for moderate to severe psoriasis, produces photosensitization. If exposure to the sun is unavoidable, the skin must be protected with sunscreen and clothing. Gray- or green-tinted wraparound sunglasses should be worn to protect the eyes during and after treatment,

Chart 61-3 HOME CARE CHECKLIST
The Patient With Psoriasis

At the completion of home care education, the patient or caregiver will be able to:	PATIENT	CAREGIVER
• Describe the etiology of psoriasis.	✔	✔
• Describe optimal skin maintenance practices to maintain moisture of skin and prevent infection.	✔	✔
• Demonstrate correct application of prescribed topical medications.	✔	✔
• Describe common side effects of oral medication, if prescribed.	✔	✔
• Demonstrate appropriate therapeutic bath technique, if prescribed.	✔	✔
• Verbalize optimism about condition.	✔	
• Identify a support person with whom to discuss feelings and concerns.	✔	

and ophthalmologic examinations should be performed on a regular basis. Contraceptives should be used by sexually active women of reproductive age because the teratogenic effect of PUVA has not been determined.

If indicated, referral may be made to a mental health professional who can help to ease emotional strain and give support. Belonging to a support group may also help patients recognize that they are not alone in experiencing life adjustments in response to a visible, chronic disease. The National Psoriasis Foundation publishes periodic bulletins and reports about new and relevant developments in this condition (see Resources section at the end of this chapter).

Chart 61-3 provides a Home Care Checklist for the patient with psoriasis.

Generalized Exfoliative Dermatitis

Generalized exfoliative dermatitis, also called *erythroderma,* is characterized by a scaling erythematous dermatitis that may involve more than 90% of the skin (Umar & Kelly, 2012). It may be associated with chills, fever, prostration, and severe pruritus. There is a profound loss of stratum corneum (i.e., outermost layer of the skin), which causes capillary leakage, hypoalbuminemia, and negative nitrogen balance. Because of widespread dilation of cutaneous vessels, large amounts of body heat can be lost.

Generalized exfoliative dermatitis has a variety of causes. It may occur as a result of a reactive process (e.g., drug allergy) or may be secondary to an underlying skin disease (e.g., psoriasis, contact dermatitis, atopic dermatitis) or a systemic disease (e.g., lymphoma, leukemia). The cause is idiopathic (i.e., unknown) in approximately 30% of cases; idiopathic generalized exfoliative dermatitis is also called *red man syndrome* (Umar & Kelly, 2012). Although generalized exfoliative dermatitis may occur at any age, it more commonly appears in those older than 40 years. It occurs two to four times more commonly in men than in women (Umar & Kelly, 2012).

Clinical Manifestations

This condition starts as a patchy or generalized erythematous eruption accompanied by fever, malaise, and chills. The skin color changes from pink to dark red. Afterward, the characteristic exfoliation (i.e., scaling) begins, usually in the form of thin flakes that leave the underlying skin smooth and red, with new scales forming as the older ones come off. Hair loss may accompany this disorder. The progression of these clinical manifestations varies, depending on the underlying cause. For instance, the progression from fever and scaling may be acute and progress over hours or a couple of days when generalized exfoliative dermatitis results from a drug reaction, or may be insidious and progress over weeks when it is secondary to a skin disease, such as psoriasis (Umar & Kelly, 2012).

Assessment and Diagnostic Findings

The presence of scaling erythematous dermatitis, particularly when it occurs in tandem with a known skin disease or a new prescription medication, increases the suspicion that generalized exfoliative dermatitis may be diagnosed. The patient frequently also presents with hypoalbuminemia and a negative nitrogen balance, as described previously, as well as an increased erythrocyte sedimentation rate that is consistent with an underlying acute inflammatory process. A skin biopsy is indicated, because it may confirm the underlying cause and diagnosis (Umar & Kelly, 2012).

Medical Management

The objectives of management are to discern and treat any underlying disorder, to maintain fluid and electrolyte balance, and to prevent infection. The treatment is individualized and supportive and should be initiated as soon as the condition is diagnosed.

The patient may be hospitalized. All medications that may be implicated are discontinued. A comfortable room temperature should be maintained because the patient does not have normal thermoregulatory control as a result of temperature fluctuations caused by vasodilation and evaporative water loss. Fluid and electrolyte balance must be maintained because there is considerable water and protein loss from the skin surface. Administration of plasma volume expanders may be indicated. Patients in a negative nitrogen balance may be prescribed enteral or parenteral therapy (Umar & Kelly, 2012) (see Chapter 45 for further discussion).

Nursing Management

Continual nursing assessment is carried out to detect infection. The disrupted, erythematous, moist skin is susceptible to infection and becomes colonized with pathogenic organisms, which produce more inflammation. Antibiotic agents, which are prescribed if infection is present, are selected on the basis of culture and sensitivity.

Hypothermia may occur because increased blood flow in the skin, coupled with increased water loss through the skin, leads to heat loss by radiation, conduction, and evaporation. Changes in vital signs are closely monitored and reported.

Topical therapy is used to provide symptomatic relief. Soothing baths, compresses, and lubrication with emollients are used to treat the extensive dermatitis. Topical corticosteroids are the mainstay of treatment (see Table 61-3). The patient is likely to be extremely irritable because of the severe pruritus; sedating antihistamines to be administered before bedtime may be prescribed to relieve the itching and promote sleep (e.g., hydroxyzine [Atarax, Vistaril]).

The use of oral or parenteral corticosteroids may be prescribed; however, their use is controversial. They are contraindicated when the cause of the condition is either not known or is suspected to be secondary to an underlying skin disease, such as psoriasis. When generalized exfoliative dermatitis occurs as a complication of psoriasis, systemic corticosteroids can exacerbate the condition (Umar & Kelly, 2012). When a specific cause is known, more specific therapy may be used. The patient is advised to avoid all irritants in the future, particularly medications that are known to cause the condition.

BLISTERING DISEASES

Blisters of the skin have many origins, including bacterial, fungal, or viral infections; allergic contact reactions; burns; metabolic disorders; and immunologically mediated (i.e., autoimmune) reactions. Some of these have been discussed previously (e.g., herpes simplex and zoster infections, contact dermatitis).

In the past, blistering diseases that resulted from immunoglobulin G (IgG)-mediated autoimmune reactions were generally classified as *pemphigus*. Today, there are three recognized subsets of these types of IgG-mediated skin disorders, including pemphigus vulgaris, pemphigus foliaceus, and paraneoplastic pemphigus. Pemphigus vulgaris accounts for 70% of these blistering diseases, whereas the others are much less common (Zeina, Sakka, & Mansoor, 2011). Bullous pemphigoid, another type of IgG-mediated blistering disorder, has distinct characteristics that make this disease different from the pemphigus disorders. Blistering diseases that result from other autoimmune responses may also occur. For instance, dermatitis herpetiform occurs as a consequence of an immunoglobulin A (IgA)-mediated response from gluten sensitivity (Miller & Zaman, 2011). The diagnosis for these types of disorders is made by histologic examination of a biopsy specimen, usually by a dermatopathologist.

Pemphigus Vulgaris

Pemphigus vulgaris is characterized by the appearance of bullae (blisters) of various sizes on apparently normal skin and mucous membranes. Pemphigus vulgaris is an autoimmune disease where the IgG antibody is directed against a specific cell surface antigen in epidermal cells. A blister forms from the antigen–antibody reaction. The level of serum antibody is predictive of disease severity. Genetic factors may also have a role in its development, with the highest incidence in people of Jewish or Mediterranean descent. This disorder usually occurs in men and women in middle and late adulthood. The condition may be associated with use of penicillins and captopril (Capoten) and with other concomitant autoimmune diseases, most commonly myasthenia gravis (Zeina et al., 2011).

Assessment and Diagnostic Findings

Most patients present with oral lesions appearing as irregularly shaped erosions that are painful, bleed easily, and heal slowly. The skin bullae enlarge, rupture, and leave large, painful eroded areas that are accompanied by crusting and oozing. A characteristic odor emanates from the bullae and the exuding serum. There is blistering or sloughing of uninvolved skin when minimal pressure is applied (Nikolsky's sign). The eroded skin heals slowly, and large areas of the body eventually are involved.

Specimens from the blister and surrounding skin demonstrate **acantholysis** (separation of epidermal cells from each other because of damage to or an abnormality of the intracellular substance), and immunofluorescent studies show intraepidermal presence of IgG (Zeina et al., 2011).

The most common complications arise when the disease process is widespread. Before the advent of corticosteroid and immunosuppressive therapy, patients were very susceptible to secondary bacterial infection. Skin bacteria have relatively easy access to the bullae as they ooze, rupture, and leave denuded areas exposed to the environment. Fluid and electrolyte imbalance results from fluid and protein loss as the bullae rupture.

Medical Management

The goals of therapy are to bring the disease under control as rapidly as possible, to prevent loss of serum and the development of secondary infection, and to promote re-epithelization (i.e., renewal of epithelial tissue).

Corticosteroids are administered in high doses to control the disease and keep the skin free of blisters. The high dosage level is maintained until remission is apparent. In some cases, corticosteroid therapy must be maintained for life. High-dose corticosteroid therapy has serious side effects (see Chapter 52).

Immunosuppressive agents (e.g., azathioprine [Imuran], mycophenolate mofetil [CellCept]) may be prescribed early in the course of the disease to help control the disease and reduce the corticosteroid dose. The immunosuppressant agent cyclophosphamide (Cytoxan) may be tried when other medications fail to induce remission. The monoclonal antibody rituximab (Rituxan) is demonstrating promise as an effective therapeutic agent in clinical trials (Zeina et al., 2011).

Bullous Pemphigoid

Bullous pemphigoid is a chronic disease that is characterized by periodic flare-up and remission. If untreated, it may be fatal. It is most commonly seen in the older adult, with a peak incidence at about 65 years of age. There is no gender or racial predilection, and the disease can be found throughout the world (Chan, 2013).

Assessment and Diagnostic Findings

Bullous pemphigoid is characterized most commonly by the general appearance of tense bullae that have a particular tendency to appear on the flexor surfaces of the arms. When the blisters break, the skin has shallow erosions that heal fairly quickly. Pruritus can be intense, even before the appearance of the blisters (Chan, 2013).

Immunofluorescent studies of skin biopsy specimens from patients with bullous pemphigoid reveal depositions of IgG and complement C3 at the junction of the dermis and epidermis (Chan, 2013).

Medical Management

Medical treatment includes topical corticosteroids for localized eruptions and systemic anti-inflammatory or immunosuppressant medications for widespread involvement. Systemic corticosteroids (e.g., prednisone) may be continued for months, in alternate-day doses. The patient needs to understand the implications of long-term corticosteroid therapy (see Chapter 52). Tetracycline may be prescribed, although not for its antimicrobial effectiveness but because its anti-inflammatory properties are believed to be particularly efficacious in treating this disorder. Alternative medications may include immunosuppressive agents (e.g., azathioprine) or monoclonal antibodies (e.g., rituximab). Most patients will achieve remission, although this may require from 6 to 60 months of treatment (Chan, 2013).

Dermatitis Herpetiformis

Dermatitis herpetiformis is an intensely pruritic, chronic disease that manifests with small, tense blisters that are distributed over the extensor surfaces of the elbows and knees, as well as the buttocks and back. It most commonly occurs between 20 and 40 years of age but can appear at any age. It is more common in people of northern European heritage and is slightly more common in men. Patients with dermatitis herpetiformis have a defect in gluten metabolism; many have a concomitant diagnosis of celiac disease (Miller & Zaman, 2011).

Assessment and Diagnostic Findings

Patients typically present with erythematous papules with small, clustered (i.e., herpetiform) vesicles that tend to have a symmetrical distribution on affected extensor surfaces of the skin. Erosions and crusts may also be present, which may result from excoriation and scratching as a reaction to the intense pruritus (Miller & Zaman, 2011).

Immunofluorescent studies of skin biopsy specimens from patients with dermatitis herpetiformis reveal granular patterns of IgA deposits in the papillary dermis (Miller & Zaman, 2011).

Medical Management

Most patients respond to dapsone (Aczone) and to a gluten-free diet. All patients should be screened for glucose-6-phosphate dehydrogenase deficiency because dapsone can induce severe hemolysis in those with this deficiency. Patients benefit from dietary counseling because the dietary restrictions are lifelong, and a gluten-free diet is often difficult to follow (Miller & Zaman, 2011). Patients need emotional support as they deal with the process of learning new habits and accepting major changes in their lives.

NURSING PROCESS

Care of the Patient With Blistering Diseases

Assessment

Patients with blistering disorders may experience significant disability. There is constant itching and possible pain in the denuded areas of skin. There may be drainage from the denuded areas, which may be malodorous. Effective assessment and nursing management become a challenge.

Disease activity is monitored clinically by examining the skin for the appearance of new blisters. Particular attention is given to assessing for signs and symptoms of infection. Hyperpigmentation may be seen in areas of resolving blisters.

Diagnosis

Nursing Diagnoses

Based on the assessment data, major nursing diagnoses may include the following:

- Acute pain of skin and oral cavity related to blistering and erosions
- Impaired skin integrity related to ruptured bullae and denuded areas of the skin
- Disturbed body image related to the appearance of the skin
- Risk for infection related to loss of protective barrier of skin and mucous membranes
- Deficient fluid volume related to loss of tissue fluids

Planning and Goals

The major goals for the patient may include relief of discomfort from lesions, skin healing, improved body image, absence of infection, and achievement of fluid and electrolyte balance.

Nursing Interventions

Relieving Oral Discomfort

Depending on the skin disorder, the patient's entire oral cavity may be affected with erosions and denuded surfaces. Necrotic tissue may develop over these areas, adding to the patient's discomfort and interfering with eating. Weight loss and hypoproteinemia may result. Meticulous oral hygiene is important to keep the oral mucosa clean and allow the epithelium to regenerate. Frequent rinsing of the mouth with chlorhexidine solution is prescribed to rid the mouth of debris and to soothe ulcerated areas. Commercial mouthwashes are avoided. The lips are kept moist with petrolatum. Cool mist therapy helps humidify environmental air.

Enhancing Skin Integrity and Relieving Discomfort

Cool, wet dressings or baths are protective and soothing. The patient with painful and extensive lesions should be premedicated with analgesic agents before skin care is initiated. Patients with large areas of blistering have a characteristic odor that decreases when secondary infection is controlled. After the patient's skin is bathed, it is dried carefully and dusted liberally with nonirritating powder (e.g., cornstarch), which enables the patient to move about freely in bed. Fairly large amounts are necessary to keep the patient's skin from adhering to the sheets. Tape should never be used because it may produce more blisters. Hypothermia is common, and measures to keep the patient warm and comfortable are priority nursing activities. The nursing management of patients with bullous skin conditions is similar to that for patients with extensive burns (see Chapter 62).

Promoting a Positive Body Image

Attention to the psychological needs of the patient requires listening to the patient, being available, providing expert nursing care, and educating the patient and the family. The patient is encouraged to express anxieties, discomfort, and feelings of hopelessness. Arranging for a family member or a close friend to spend more time with the patient can be supportive. When patients receive information about the disease and its treatment, uncertainty and anxiety often decrease, and the patient's capacity to act on his or her own behalf is enhanced. Psychological counseling may assist the patient in dealing with fears, anxiety, and promote positive self-esteem.

Preventing Infection

The patient is susceptible to infection because the barrier function of the skin is compromised. Bullae are also susceptible to infection, and sepsis may follow (see Chapter 14). The skin is cleaned to remove debris and dead skin and to prevent infection.

Secondary infection may be accompanied by an unpleasant odor from skin or oral lesions. *Candida albicans* of the mouth (i.e., thrush) commonly affects patients receiving high-dose corticosteroid therapy. The oral cavity is inspected daily, and any changes are reported. Oral lesions are slow to heal.

Quality and Safety Nursing Alert

Because infection is the leading cause of death in patients with blistering diseases, meticulous assessment for signs and symptoms of local and systemic infection is required. Seemingly trivial complaints or minimal changes are investigated because corticosteroids can mask or alter typical signs and symptoms of infection.

The patient's vital signs are monitored, and temperature fluctuations are documented. The patient is observed for chills, and all secretions and excretions are monitored for changes suggesting infection. Results of culture and sensitivity tests are monitored. Antimicrobial agents are administered as prescribed, and response to treatment is assessed. Health care personnel must perform effective hand hygiene and wear gloves.

In patients who are hospitalized, environmental contamination is reduced as much as possible. Protective isolation measures and standard precautions are warranted.

Promoting Fluid Balance

Extensive denudation of the skin leads to fluid and electrolyte imbalance because of significant loss of fluids and sodium chloride from the skin. This sodium chloride loss is responsible for many of the systemic symptoms associated with the disease and is treated by IV administration of saline solution.

A large amount of protein and blood is also lost from the denuded skin areas. Blood component therapy may be prescribed to maintain the blood volume, hemoglobin level, and plasma protein concentration. Serum albumin, protein, hemoglobin, and hematocrit values are monitored.

The patient is encouraged to maintain adequate oral fluid intake. Cool, nonirritating fluids are encouraged to maintain hydration. Small, frequent meals or snacks of high-protein, high-calorie foods (e.g., oral nutritional supplements, eggnog, milk shakes) help maintain nutritional status. Parenteral nutrition is considered if the patient cannot eat an adequate diet.

Evaluation

Expected patient outcomes may include:

1. Reports relief from pain of oral lesions
   a. Identifies therapies that reduce pain
   b. Uses mouthwashes and anesthetic or antiseptic aerosol mouth spray
   c. Drinks chilled fluids at 2-hour intervals
2. Achieves skin healing
   a. States purpose of therapeutic regimen
   b. Cooperates with soaks and bath regimen
   c. Reminds caregivers to use liberal amounts of nonirritating powder on bed linens
3. Reports body image has improved
   a. Verbalizes concerns about condition, self, and relationships with others
   b. Participates in self-care
4. Remains free of infections and sepsis
   a. Has cultures from bullae, skin, and orifices that are negative for pathogenic organisms
   b. Has no purulent drainage
   c. Shows signs that skin is clearing
   d. Has normal body temperature
5. Maintains fluid and electrolyte balance
   a. Keeps intake record to ensure adequate fluid intake and normal fluid and electrolyte balance
   b. Verbalizes the rationale for IV infusion therapy
   c. Has urine output that is greater than 0.5 mL/kg/h
   d. Has serum chemistry and hemoglobin and hematocrit values within normal limits

Toxic Epidermal Necrolysis and Stevens-Johnson Syndrome

Toxic epidermal necrolysis (TEN) and Stevens-Johnson syndrome (SJS) are potentially fatal acute skin disorders characterized by widespread erythema and macule formation with blistering, resulting in epidermal detachment or sloughing and erosion formation. These diseases are believed to be one and the same but manifest along a spectrum of reactions, with TEN being the most severe. The mortality rate from TEN is 40% and from SJS is 5%. TEN and SJS are most commonly triggered by a reaction to medications in adults, whereas infections are the most common precipitants in children. Antibiotics, especially sulfonamides, antiseizure agents, NSAIDs, and allopurinol (Zyloprim) are the most frequent medications implicated (Cohen, Jellinek, & Schwartz, 2011; Klein, 2013).

TEN and SJS occur in all ages and both genders, with a slight predilection for women. The incidence is increased in older people because they more commonly take many medications. People who are immunosuppressed, including those with HIV infection and acquired immunodeficiency syndrome (AIDS), have a high risk of TEN and SJS. Although the incidence of these diseases in the general population is about 2 to 12 cases per 1 million people in the United States, the risk associated with sulfonamides in HIV-positive people may approach 1 case per 1,000 (Cohen et al., 2011; Porth & Maffin, 2009). The mechanism leading to TEN seems to be a cell-mediated cytotoxic reaction (Cohen et al., 2011).

Clinical Manifestations

TEN and SJS are characterized initially by conjunctival burning or itching, cutaneous tenderness, fever, cough, sore throat, headache, extreme malaise, and myalgias (i.e., aches and pains). These signs are followed by a rapid onset of erythema involving much of the skin surface and mucous membranes, including the oral mucosa, conjunctiva, and genitalia. In severe cases of mucosal involvement, there may be danger of damage to the larynx, bronchi, and esophagus from ulcerations. Large, flaccid bullae develop in some areas; in other areas, large sheets of epidermis are shed, exposing the underlying dermis. Fingernails, toenails, eyebrows, and eyelashes may be shed along with the surrounding epidermis. The skin is excruciatingly tender, and the loss of skin leaves a weeping surface similar to that of a total-body, partial-thickness burn; hence, the condition was previously referred to as scalded skin syndrome (Cohen et al., 2011; Klein, 2013).

Complications

Keratoconjunctivitis, sepsis, and multiple organ dysfunction syndrome (MODS) are potential complications of TEN and SJS. Keratoconjunctivitis can impair vision and result in conjunctival retraction, scarring, and corneal lesions. Sepsis and MODS can be life threatening (Cohen et al., 2011) (see Chapter 14).

Assessment and Diagnostic Findings

Histologic studies of frozen skin cells from a fresh lesion and cytodiagnosis of collections of cellular material from a freshly denuded area are conducted. A history of the use of medications known to precipitate TEN or SJS may confirm medication reaction as the underlying cause, especially if the medications were prescribed within 4 weeks prior to the onset of illness (Cohen et al., 2011).

Results from a complete blood count (CBC) may show leukopenia and a normochromic normocytic anemia. Skin biopsy results confirm the diagnosis, showing necrotic keratinocytes with full-thickness epithelial necrosis and detachment (Cohen et al., 2011).

Medical Management

The goals of treatment include control of fluid and electrolyte balance, prevention of sepsis, and prevention of ophthalmic complications. Supportive care is the mainstay of treatment.

Any medications that may be implicated as precipitating TEN or SJS are discontinued immediately. If possible, the patient is treated in a regional burn center because aggressive treatment similar to that for severe burns is required. Patients hospitalized in regional burn centers have more favorable prognoses than those who do not benefit from this specialized treatment (Cohen et al., 2011). Surgical débridement or hydrotherapy in a Hubbard tank (large steel tub) may be performed to remove involved skin.

Tissue samples from the nasopharynx, eyes, ears, blood, urine, skin, and unruptured blisters are obtained for culture to identify pathogenic organisms. IV crystalloid fluids are prescribed to maintain fluid and electrolyte balance, using parameters similar to those used to guide care of patients with burns. Similarly, thermoregulation, wound care, and pain management guidelines used to treat patients with burns are also implemented (Cohen et al., 2011) (see Chapter 62). Patients frequently require nutritional and metabolic support with total parenteral nutrition (Cohen et al., 2011) (see Chapter 45).

Initial treatment with systemic corticosteroids (e.g., methylprednisolone [Solu-Medrol]) is controversial. In most cases, the risk of infection, fluid and electrolyte imbalance, delayed healing, and difficulty in initiating oral corticosteroids early in the course of the disease outweigh its benefits.

Administration of intravenous immunoglobulin (IVIG) may provide rapid improvement and skin healing at dosages of 1 g/kg/day for 4 days (Klein, 2013). Other medications that may be effective include the immunosuppressive agents cyclosporine (Neoral) or cyclophosphamide (Cytoxan) (Klein, 2013).

Protecting the skin with topical agents is crucial. Various topical antibacterial and anesthetic agents are used to prevent wound sepsis and to assist with pain management. Systemic antibiotic therapy is used with extreme caution. Temporary biologic dressings (e.g., pigskin, amniotic membrane) or plastic semipermeable dressings (e.g., Vigilon) may be used to reduce pain, decrease evaporation, and prevent secondary infection until the epithelium regenerates. Meticulous oropharyngeal and eye care is essential when there is involvement of the mucous membranes and the eyes.

NURSING PROCESS

Care of the Patient With Toxic Epidermal Necrolysis or Stevens-Johnson Syndrome

Assessment

A careful inspection of the skin is made, including its appearance and the extent of involvement. The normal skin is closely observed to determine if new areas of blisters are developing. Drainage from blisters is monitored for amount, color, and odor. The oral cavity is inspected daily for blistering and erosive lesions; the patient is assessed daily for itching, burning, and dryness of the eyes. The patient's ability to swallow and drink fluids, as well as speak normally, is determined.

The patient's vital signs are monitored, and special attention is given to the presence and character of fever and the respiratory rate, depth, rhythm, and cough. The characteristics and amount of respiratory secretions are observed. Assessment for high fever, tachycardia, and extreme weakness and fatigue is essential because these factors indicate the process of epidermal necrosis, increased metabolic needs, and possible gastrointestinal and respiratory mucosal sloughing. Urine volume, specific gravity, and color are monitored. The insertion sites of IV lines are inspected for signs of local infection. Body weight is recorded daily.

The patient is asked to describe fatigue and pain levels. An attempt is made to evaluate the patient's level of anxiety. The patient's basic coping mechanisms are assessed, and effective coping strategies are identified.

Diagnosis

Nursing Diagnoses

Based on the assessment data, major nursing diagnoses may include the following:

- Impaired tissue integrity (i.e., oral, eye, and skin) related to epidermal shedding
- Deficient fluid volume and electrolyte losses related to loss of fluids from denuded skin
- Risk for imbalanced body temperature (i.e., hypothermia) related to heat loss secondary to skin loss
- Acute pain related to denuded skin and oral lesions
- Anxiety related to the physical appearance of the skin and prognosis

Collaborative Problems/Potential Complications

Potential complications may include the following:

- Sepsis
- Conjunctival retraction, scars, and corneal lesions

Planning and Goals

The major goals for the patient may include skin and oral tissue healing, fluid balance, prevention of heat loss, relief of pain, reduced anxiety, and absence of complications.

Nursing Interventions

Maintaining Skin and Mucous Membrane Integrity

The local care of the skin is an important area of nursing management. The skin denudes easily, especially when the patient is lifted and turned. The nursing staff must take special care to avoid friction involving the skin when moving the patient in bed. The skin should be checked after each position change to ensure that no new denuded areas have appeared. The nurse applies the prescribed topical agents to reduce the bacterial population of the wound surface. Warm compresses, if prescribed, should be applied gently to denuded areas. The topical antibacterial agent may be used in conjunction with hydrotherapy in a tank, bathtub, or shower. The nurse monitors the patient's condition during the treatment and encourages the patient to exercise the extremities during hydrotherapy.

The painful oral lesions make oral hygiene difficult. Careful oral hygiene is performed to keep the oral mucosa clean. Prescribed chlorhexidine mouthwashes, anesthetics, or coating agents are used frequently to rid the mouth of debris, soothe ulcerative areas, and control foul mouth odor. The oral cavity is inspected several times each day, and any changes are documented and reported. Petrolatum or a prescribed ointment is applied to the lips.

Attaining Fluid Balance

The patient's vital signs, urine output, and mental status are assessed for indications of hypovolemia. Mental changes from fluid and electrolyte imbalance, sensory overload, or sensory deprivation may occur. Laboratory test results are evaluated, and abnormal results are reported. The patient is weighed daily (with a bed scale if necessary).

Oral lesions may result in dysphagia, making tube feeding or parenteral nutrition necessary until oral ingestion can be tolerated. A daily calorie count and accurate recording of all intake and output are essential.

Preventing Hypothermia

The patient with TEN is prone to chilling. Dehydration may be made worse by exposing the denuded skin to a continuous current of warm air. The patient is usually sensitive to changes in room temperature. Measures similar to those implemented for a burn patient, such as cotton blankets, ceiling-mounted heat lamps, and heat shields, are useful in maintaining body temperature. To minimize shivering and heat loss, the nurse should work rapidly and efficiently when large wounds are exposed for wound care. The patient's temperature is monitored frequently.

Relieving Pain

The nurse assesses the patient's pain, its characteristics, factors that influence the pain, and the patient's behavioral responses. Prescribed analgesic agents are administered on a regular schedule, and the nurse documents pain relief and any side effects. Analgesics are administered before painful treatments are performed. Providing thorough explanations and speaking calmly to the patient during treatments can allay the anxiety that may intensify pain. Offering emotional support and reassurance and implementing measures that promote rest and sleep are basic in achieving pain control. As the pain diminishes and the patient has more physical and emotional energy, self-management techniques for pain relief, such as progressive muscle relaxation and imagery, may be shared (see Chapter 12).

Reducing Anxiety

Because the lifestyle of the patient with TEN or SJS has been abruptly changed to one of complete dependence, an assessment of his or her emotional state may reveal anxiety, depression, and fear of dying. The patient can be reassured that these reactions are normal. The patient also needs nursing support, honest communication, and hope that the situation can improve. The patient is encouraged to express his or her feelings. Listening to the patient's concerns and being readily available with skillful and compassionate care are important anxiety-relieving interventions. Emotional support by a psychiatric nurse, spiritual advisor, psychologist, or psychiatrist may be helpful to promote coping during the long recovery period.

Monitoring and Managing Potential Complications

Sepsis. The major cause of death from TEN is from severe sepsis. The organisms most often involved are *S. aureus*, *Pseudomonas aeruginosa*, *C. albicans*, and gram-negative species (Cohen et al., 2011). Monitoring vital signs closely and noticing changes in respiratory, kidney, and gastrointestinal function may quickly detect the beginning of an infection. Strict asepsis is always maintained during routine skin care measures. Hand hygiene and wearing sterile gloves when carrying out procedures are essential. When the condition involves a large portion of the body, the patient should be in reverse isolation to prevent possible cross-infection from other patients. Visitors should wear protective garments and wash their hands before and after coming into contact with the patient. People with any infections or infectious disease should not visit the patient until they are no longer a danger to the patient. The nurse is critical in identifying early signs and symptoms of infection and notifying the primary provider. Antibiotic agents are not generally begun until there is indication for their use (Cohen et al., 2011).

Conjunctival Retraction, Scars, and Corneal Lesions. The eyes are inspected daily for signs of pruritus, burning, and dryness, which may indicate progression to keratoconjunctivitis—the principal eye complication. Applying a cool, damp cloth over the eyes may relieve burning sensations. The eyes are kept clean and observed for signs of discharge or discomfort, and the progression of symptoms is documented and reported. Administering an eye lubricant, when prescribed, may alleviate dryness and prevent corneal abrasion. Using eye patches or reminding the patient to blink periodically may also counteract dryness. The patient is instructed to avoid rubbing the eyes or putting any medication into the eyes that has not been prescribed or recommended by the primary provider.

Promoting Home and Community-Based Care

Educating Patients About Self-Care. Patients with TEN or SJS with involvement of large areas of the skin require care that is similar to that of patients with thermal burns. As the patient completes the acute inpatient stage of illness, the focus is directed toward rehabilitation and outpatient care or care in a rehabilitation center. Throughout this care, the patient and family members are involved in the care and are instructed in the procedures, such as wound care and dressing changes, that will need to be continued at home. The patient and family members are assisted in acquiring dressing supplies that will be needed at home.

The patient and family members are also provided with education about pain management; nutrition; measures to increase mobility; and prevention of complications, including prevention of infection. They are educated about the signs and symptoms of complications and instructed when to notify the health care provider. When appropriate, instructions are provided in writing to the patient and family so that they can refer to these instructions when necessary at later times.

Continuing Care. Interdisciplinary follow-up care is imperative to ensure that the patient's progress continues. Some patients will require care in a rehabilitation center before returning home. Others will require outpatient physical and occupational therapy for an extended period. When the patient returns home, the home care nurse coordinates the care provided by the various members of the health care team (e.g., physician, physical therapist, occupational therapist, dietician). The nurse also monitors the patient's progress, provides ongoing assessment to identify complications, and monitors the patient's adherence to the plan of care. The patient's adaptation to the home care environment and the patient's and family's needs for support and assistance are also assessed. Referrals to community agencies are made as appropriate.

Evaluation

Expected patient outcomes may include:

1. Achieves increasing skin and oral tissue healing
   a. Demonstrates areas of healing skin
   b. Swallows fluids and speaks clearly
2. Attains fluid balance
   a. Demonstrates laboratory values within normal ranges
   b. Maintains urine volume and specific gravity within acceptable range
   c. Shows stable vital signs
   d. Increases intake of oral fluids without discomfort
   e. Maintains weight or gains weight, if appropriate
3. Attains thermoregulation
   a. Registers body temperature within normal range
   b. Reports no chills
4. Achieves pain relief
   a. Uses analgesic agents as prescribed
   b. Uses self-management techniques for relief of pain
5. Reports less anxiety
   a. Discusses concerns freely
   b. Sleeps for progressively longer periods
6. Absence of complications, such as sepsis and impaired vision
   a. Has body temperature within normal range
   b. Laboratory values within normal ranges
   c. Has no abnormal discharges or signs of infection
   d. Continues to see objects at baseline acuity level
   e. Shows no signs of keratoconjunctivitis

SKIN TUMORS

Tumors of the skin are common and occur along a spectrum from those that are benign to those that are highly malignant.

Benign Skin Tumors

Cysts

Cysts of the skin are epithelium-lined cavities that contain fluid or solid material. Epidermal cysts (epidermoid cysts) occur frequently and may be described as slow-growing, firm, elevated tumors found most frequently on the face, neck, upper chest, and back. Removal of the cysts provides a cure (Weber & Kelley, 2010).

Pilar cysts (trichilemmal cysts), formerly called *sebaceous cysts*, are most frequently found on the scalp. They originate from the middle portion of the hair follicle and from the cells of the outer hair root sheath. Treatment is surgical removal.

Seborrheic and Actinic Keratoses

Seborrheic keratoses are benign, wartlike lesions of various sizes and colors, ranging from light tan to black. They are usually located on the face, shoulders, chest, and back and are the most common skin tumors seen in middle-aged and older adults. They may be cosmetically unacceptable to the patient. A black keratosis may be erroneously diagnosed as melanoma. Treatment is removal of the tumor tissue by excision, electrodesiccation (destruction of the skin lesions by monopolar high-frequency electric current) and curettage, or application of carbon dioxide or liquid nitrogen. However, there is no harm in allowing these growths to remain because there is no medical significance to their presence (Porth & Matfin, 2009).

Actinic keratoses are premalignant skin lesions that develop in chronically sun-exposed areas of the body. They appear as rough, scaly patches with underlying erythema. These lesions may gradually transform into squamous cell carcinoma (SCC) (see later discussion); they are usually removed by cryotherapy, electrodesiccation, or lasers, or they may be treated with topical chemotherapeutic creams (e.g., 5-fluorouracil cream) (Porth & Matfin, 2009).

Verrucae: Warts

Warts are common, benign skin tumors caused by infection with the human papillomavirus, which belongs to the DNA virus group. People of all ages may be affected, but the warts occur most frequently between the ages of 12 and 16 years. There are many types of warts.

As a rule, warts are asymptomatic, except when they occur on weight-bearing areas, such as the soles of the feet. They may be treated with locally applied laser therapy, liquid nitrogen, salicylic acid plasters, or electrodesiccation (Porth & Matfin, 2009).

Warts occurring on the genitalia and perianal areas are known as condylomata acuminata. They may be transmitted sexually and are treated with liquid nitrogen, cryosurgery, electrosurgery, topically applied trichloroacetic acid, and curettage. Condylomata that affect the uterine cervix predispose the patient to cervical cancer (see Chapter 57).

Angiomas

Angiomas are benign vascular tumors that involve the skin and the subcutaneous tissues. They are present at birth and may occur as flat, violet-red patches (port-wine angiomas) or as raised, bright red, nodular lesions (i.e., hemangiomas of infancy, formerly known as strawberry angiomas). The latter tend to involute spontaneously within the first few years of life, but port-wine angiomas usually persist indefinitely. Most patients use masking cosmetics to camouflage the lesions. The argon laser is being used on various angiomas with some success (Porth & Matfin, 2009).

Pigmented Nevi: Moles

Moles are common skin tumors of various sizes and shades, ranging from yellowish brown to black. They may be flat, macular lesions or elevated papules or nodules that occasionally contain hair. Most pigmented nevi are harmless lesions. However, in rare cases, malignant changes occur, and a melanoma develops at the site of the nevus. Nevi that show a change in color or size, become symptomatic (e.g., itch), or develop irregular borders should be removed to determine if malignant changes have occurred. Moles that occur in unusual places should be examined carefully for any irregularity and for notching of the border and variation in color. Early melanomas may display some redness and irritation and areas of bluish pigmentation where the pigment-containing cells have spread deeper into the skin. Late melanomas have areas of paler color, where pigment cells have stopped producing melanin. Nevi larger than 5 mm should be examined carefully (Porth & Matfin, 2009). Excised nevi should be examined histologically.

Keloids

Keloids are benign overgrowths of fibrous tissue at the site of a scar or trauma. They appear to be more common among dark-skinned people. Keloids are asymptomatic but may cause disfigurement and cosmetic concern (Porth & Matfin, 2009). The treatment, which is not always satisfactory, consists of surgical excision, intralesional corticosteroid therapy, and radiation.

Dermatofibroma

A dermatofibroma is a common, benign tumor of connective tissue that occurs predominantly on the extremities. It is a firm, dome-shaped nodule that may be skin colored or pinkish brown (Weber & Kelley, 2010). Excisional biopsy is the recommended method of treatment.

Neurofibromatosis: Von Recklinghausen's Disease

Neurofibromatosis is a hereditary condition manifested by pigmented patches (*café-au-lait* macules), axillary freckling, and cutaneous neurofibromas that vary in size. Developmental changes may occur in the nervous system, muscles, and bone. Malignant degeneration of the neurofibromas occurs in some patients (Porth & Matfin, 2009).

Malignant Skin Tumors

Skin cancer is the most common cancer in the United States. If the incidence continues at the present rate, an estimated 20% of all Americans will eventually develop skin cancer, especially basal cell carcinoma (BCC) (Shelestak & Lindow, 2011) (Chart 61-4). Because the skin is easily inspected, skin cancer is readily seen and detected and is therefore believed to be amenable to early intervention. However, the U.S. Preventive Services Task Force (USPSTF, 2009) notes that there

Chart 61-4 RISK FACTORS
Skin Cancer

- Fair-skinned, fair-haired, blue-eyed people, particularly those of Celtic origin, with insufficient skin pigmentation to protect underlying tissues
- People who sustain sunburn and who do not tan
- Chronic sun exposure (certain occupations, such as farming, construction work)
- Exposure to chemical pollutants (industrial workers in arsenic, nitrates, coal, tar and pitch, oils and paraffins)
- Adults >50 years of age
- Male gender (twice the rate of basal cell carcinoma as seen in women)
- History of x-ray therapy for acne or benign lesions
- Scars from severe burns
- Chronic skin irritations
- Immunosuppression
- Genetic factors

Adapted from Bader, R. S., Santacroce, L., Diomede, L., et al. (2013). Basal cell carcinoma. *Medscape*. Available at: emedicine.medscape.com/article/276624-overview; and Porth, C. M., & Matfin, G. (2009). *Pathophysiology: Concepts of altered health states* (8th ed.). Philadelphia: Lippincott Williams & Wilkins.

is a gap in the evidence-based literature to support the assertion that early detection reduces skin cancer rates of morbidity and mortality; clearly, more studies are indicated.

Exposure to the sun is the leading cause of skin cancer; incidence is related to the total amount of exposure to the sun. Sun damage is cumulative, and harmful effects may be severe by 20 years of age. The increase in skin cancer probably reflects changing lifestyles and the emphasis on sunbathing and related activities in light of changes in the environment, such as holes in the earth's ozone layer (Porth & Matfin, 2009). Skin cancers tend to be categorized as non-melanoma types and melanomas.

Basal Cell and Squamous Cell Carcinoma

The most common types of non-melanoma skin cancers are BCC and SCC (Shulstad & Proper, 2010).

Clinical Manifestations

BCC is the most common type of skin cancer but is rarely fatal, accounting for less than 0.1% of all cancer-related deaths. It generally appears on sun-exposed areas of the body, such as the face, neck, hands, and scalp. Ultraviolet damage, from sunlight or the use of indoor tanning booths, is most frequently associated with BCC, with a latency period from time of exposure that may span 20 to 50 years. Those patients who sunburn before the age of 25 years are at particular risk to develop BCC 20 to 50 years after the fact (Bader, Santacroce, Diomede, et al., 2013).

BCC usually begins as a small, waxy nodule with rolled, translucent, pearly borders; telangiectatic vessels may be present. As it grows, it undergoes central ulceration and sometimes crusting (Fig. 61-5). The tumors appear most frequently on the face. BCC is characterized by invasion and erosion of contiguous (adjoining) tissues. It rarely metastasizes, but recurrence is common. However, a neglected lesion can result in the loss of a nose, an ear, or a lip. Other variants of BCC may appear as shiny, flat, gray or yellowish plaques (Bader et al., 2013).

SCC is a malignant proliferation arising from the epidermis. Its precursor is typically actinic keratosis (see previous discussion). Although it usually appears on sun-damaged skin, it may arise from normal skin or from preexisting skin lesions. It is of greater concern than BCC because it is a truly invasive carcinoma, metastasizing by the blood or lymphatic system (Shulstad & Proper, 2010).

SCC may metastasize in 10% of all cases, most commonly to regional lymph nodes and sometimes the lungs and liver. Lesions larger than 2 cm are associated with a doubled rate of recurrence post tumor excision and a tripling of the rate of associated metastases (i.e., 30%) (Shulstad & Proper, 2010). SCC appears as a rough, thickened, scaly tumor that may be asymptomatic or may involve bleeding (see Fig. 61-5B). The border of an SCC lesion may be wider, more infiltrated, and more inflammatory than that of a BCC lesion. Secondary infection can occur. Exposed areas, especially of the upper extremities and of the face, lower lip, ears, nose, and forehead, are common sites.

The incidence of BCC and SCC is increased in all people who are immunocompromised, including those infected with HIV. Clinically, the tumors have the same appearance as in people who are not infected with HIV; however, in people who are HIV positive, the tumors may grow more rapidly and recur more frequently. These tumors are managed the same as those for the general population. Frequent follow-up (every 4 to 6 months) is recommended to monitor for recurrence (Porth & Matfin, 2009).

The prognosis for BCC is usually good because tumors remain localized. Although some require wide excision with resultant disfigurement, the risk of death from BCC is low. The prognosis for SCC depends on the incidence of metastases, which is related to the histologic type and the level or depth of invasion. Regional lymph nodes should be evaluated for metastases.

Medical Management

The goal of treatment is to eradicate the tumor. The treatment method depends on the tumor location; the cell type, location, and depth; the cosmetic desires of the patient; the history of previous treatment; whether the tumor is invasive; and whether metastatic nodes are present. The management of BCC and SCC includes surgical excision, which may include Mohs micrographic surgery, electrosurgery, or cryosurgery. In those patients who are not surgical candidates, alternatives such as radiation therapy, photodynamic therapy

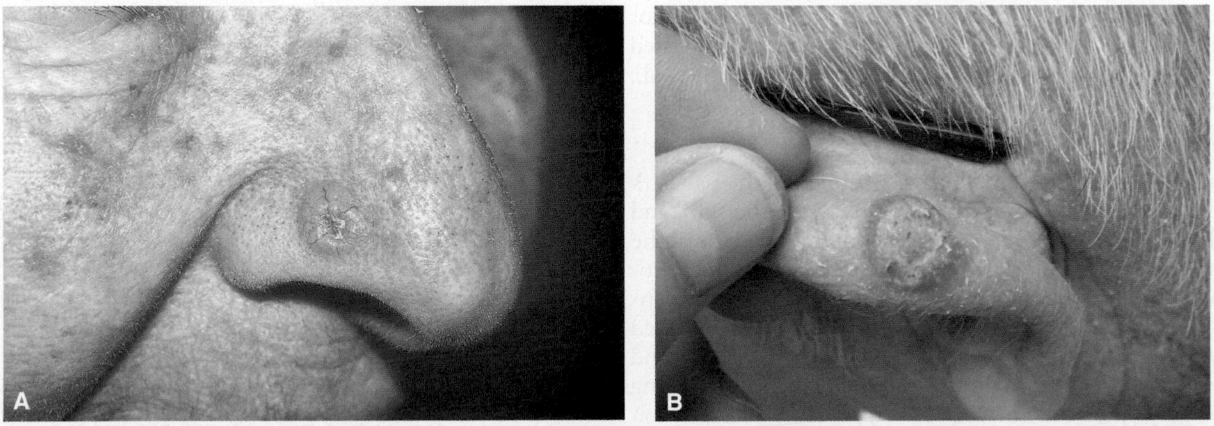
A
B

FIGURE 61-5 • Basal cell carcinoma (**A**) and squamous cell carcinoma (**B**). From Goodheart, H. P. (2011). *Goodheart's same-site differential diagnosis: A rapid method of diagnosing and treating common skin disorders*. Philadelphia: Lippincott Williams & Wilkins.

(PDT), or topical chemotherapeutic creams may be viable options (Bader et al., 2013).

Surgical Management

The main goal is to remove the tumor entirely. The best way to maintain cosmetic appearance is to place the incision properly along natural skin tension lines and natural anatomic body lines. In this way, scars are less noticeable. The size of the incision depends on the tumor size and location but usually involves a length-to-width ratio of 3:1.

The adequacy of the surgical excision is verified by microscopic evaluation of sections of the specimen. When the tumor is large, reconstructive surgery with the use of a skin flap or skin grafting may be required. The incision is closed in layers to enhance cosmetic effect. A pressure dressing applied over the wound provides support. Infection after a simple excision is uncommon if proper surgical asepsis is maintained.

Mohs Micrographic Surgery

This technique is the most accurate surgical technique and best conserves normal tissue. The procedure removes the tumor layer by layer. The first layer excised includes all evident tumor and a small margin of normal-appearing tissue. The specimen is frozen and analyzed by section to determine if all of the tumor has been removed. If not, additional layers of tissue are shaved and examined until all tissue margins are tumor free. In this manner, only the tumor and a safe, normal tissue margin are removed. Mohs surgery is the recommended tissue-sparing procedure, with extremely high cure rates for BCC and SCC. It is the treatment of choice and the most effective for tumors around the eyes, nose, upper lip, and auricular and periauricular areas (Mooney & Parry, 2011).

Electrosurgery

Electrosurgery is the destruction or removal of tissue by electrical energy. The current is converted to heat, which then passes to the tissue from a cold electrode. Electrosurgery may be preceded by curettage (excising the skin tumor by scraping its surface with a curette). Electrodesiccation is then implemented to achieve hemostasis and to destroy any viable malignant cells at the base of the wound or along its edges. Electrodesiccation and curettage is useful for lesions smaller than 1 to 2 cm (0.4 to 0.8 in) in diameter.

This method takes advantage of the fact that the tumor is softer than surrounding skin and therefore can be outlined by a curette, which "feels" the extent of the tumor. The tumor is removed and the base cauterized. The process is repeated twice. Usually, healing occurs within 1 month (Bader et al., 2013).

Cryosurgery

Cryosurgery destroys the tumor by deep-freezing the tissue. A thermocouple needle apparatus is inserted into the skin, and liquid nitrogen is directed to the center of the tumor until the tumor base is –40°C to –60°C (–40°F to –76°F). Liquid nitrogen has the lowest boiling point of all cryogens, is inexpensive, and is easy to obtain. The tumor tissue is frozen, allowed to thaw, and then refrozen. The site thaws naturally and then becomes gelatinous and heals spontaneously. Swelling and edema follow the freezing. The appearance of the lesion varies. Normal healing, which may take 4 to 6 weeks, occurs faster in areas with a good blood supply (Bader et al., 2013).

Nonsurgical Alternative Therapies

Some older adult patients may defer surgical treatment options. Furthermore, in some cases, lesions may be extensive or are located on sites where wide surgical excision is not practical to achieve (e.g., cancer of the eyelid, tip of the nose). In these cases, local radiation therapy (see Chapter 15) or PDT may prove reasonable alternatives. PDT involves application of 5-aminolevulinic acid (ALA-PDT) to the lesion, which is then followed by photoactivation with directed blue light for approximately 1 hour. This has the effect of locally destroying the neoplastic cells, with good cosmetic results. Topical application of 5-flourouracil cream (a chemotherapeutic agent) may be tried as another alternative to managing superficial BCC (Bader et al., 2013).

The patient should be informed that the skin may become red and blistered after any of these therapies. A bland skin ointment prescribed by the primary provider may be applied to relieve discomfort. The patient should also be cautioned to avoid exposure to the sun.

Nursing Management

Because many skin cancers are removed by excision, patients are usually treated in outpatient surgical units. The role of the nurse is to educate the patient about prevention of skin cancer (Chart 61-5) and self-care after treatment.

Promoting Home and Community-Based Care

Educating Patients About Self-Care

The wound is usually covered with a dressing to protect the site from physical trauma, external irritants, and contaminants.

Chart 61-5 HEALTH PROMOTION

Preventing Skin Cancer

Minimize sun exposure:

- To the extent possible, avoid sun between the hours of 10 AM and 4 PM.
- Wear protective clothing (e.g., long-sleeved clothing, broad-brimmed hats).
- Seek shady areas when outdoors.
- Use caution around snow and water because of reflective sun rays.

Use sunscreen:

- Use a sunscreen with a sun protection factor (SPF) of 15 or higher that protects against both ultraviolet-A (UVA) and ultraviolet-B (UVB) rays.
- Apply generously 20 minutes prior to sun exposure (e.g., going outdoors).
- Reapply every 2 hours, or immediately after swimming.
- Use lip balm with SPF of 15 or higher.

Do not use artificial ultraviolet sources (e.g., tanning beds and booths).

Check your skin regularly:

- Perform self-examination monthly.
- Schedule an examination by primary provider yearly, if over the age of 50 years.

Adapted from Barrow, M. M. (2010). Approaching skin cancer education with a clear message: "Be safe. Be SunAWARE." *Journal of the Dermatology Nurses' Association*, *2*(5), 209–213; and Shelestak, D., & Lindow, K. (2011). Beliefs and practices regarding skin cancer prevention. *Journal of the Dermatology Nurses' Association*, *3*(3), 150–155.

The patient is advised when to report for a dressing change or is given written and verbal information on how to change dressings, including the type of dressing to purchase, how to remove dressings and apply fresh ones, and the importance of hand hygiene before and after the procedure.

The patient is advised to watch for excessive bleeding and tight dressings that compromise circulation. If the lesion is in the perioral area, the patient is instructed to drink liquids through a straw and limit talking and facial movement. Dental work should be avoided until the area is completely healed.

After the sutures are removed, an emollient cream may be used to help reduce dryness. Applying a sunscreen over the wound is advised to prevent postoperative hyperpigmentation if the patient spends time outdoors.

Follow-up examinations should be at regular intervals, usually every 3 months for a year, and should include palpation of the adjacent lymph nodes. The patient should also be instructed to seek treatment for any moles that are subject to repeated friction and irritation and to watch for indications of potential malignancy in moles as described previously. The importance of lifelong follow-up evaluations is emphasized.

Melanoma

A melanoma is a cancerous neoplasm in which neoplastic melanocytes are present in the epidermis and the dermis (and sometimes the subcutaneous cells). It is the most lethal of all skin cancers. The incidence of melanoma in the United States continues to rise faster than any other malignancy among men and is second only to the rapidly climbing incidence of lung cancer among women. The estimated lifetime risk of developing melanoma in Caucasians is 1 in 50 (Rubin, 2009; Tan, Schulman, & Talavera, 2013). Risk factors for melanoma are noted in Chart 61-6.

Melanoma can occur in one of several forms: superficial spreading melanoma (SSM), lentigo maligna melanoma (LMM), nodular melanoma (NM), acral lentiginous melanoma (ALM), and mucosal lentiginous melanoma (MLM). These types have specific clinical and histologic features as well as different biologic behaviors. Most melanomas arise from cutaneous epidermal melanocytes, but some appear in preexisting nevi (i.e., moles) in the skin or develop in the uveal tract of the eye or from the mucosal lining of the gastrointestinal or genitourinary tract (Tan et al., 2013).

Melanomas spread in two growth phases: radial and vertical. During the first growth phase—the radial phase—cutaneous melanomas tend to spread radially within the layer of the epidermis. It is during this earlier phase of radial growth that the tumor is most amenable to treatment. The second growth phase—the vertical phase—is characterized by vertical tumor growth into the dermal layer and eventual metastasis. Those melanomas that progress more rapidly from the radial to the vertical growth phase are considered more aggressive types and have a poorer prognosis (Tan et al., 2013).

Chart 61-6 RISK FACTORS
Melanoma

- Caucasians, particularly fairer-skinned or freckled, blue-eyed, blond or red-haired people of Celtic or Scandinavian origin
- History of sunburns, particularly in childhood
- Previous history of melanoma (Multiple primary melanomas are not uncommon.)
- Family history of melanoma (Of patients diagnosed with melanoma, 10% have a positive family history of melanoma.)
- Personal or family history of multiple atypical nevi (formerly called *dysplastic nevi*)
- Family history of astrocytoma or pancreatic cancer

Adapted from Rubin, K. M. (2009). Dysplastic nevi and the risk of melanoma. *Journal of the Dermatology Nurses' Association*, 1(4), 228–235; and Tan, W. W., Schulman, P., & Talavera, F. (2013). Malignant melanoma. *Medscape*. Available at: emedicine.medscape.com/article/280245-overview

Clinical Manifestations

Superficial Spreading Melanoma

SSM occurs anywhere on the body and is the most common form of melanoma, accounting for 70% of all cases. It usually affects middle-aged people and occurs most frequently on the head, neck, or trunk in men and in the lower extremities in women. These types of melanomas typically develop over time from a long-present stable and benign nevus (Tan et al., 2013). The lesion tends to be circular, with irregular outer portions. The margins of the lesion may be flat or elevated and palpable (Fig. 61-6). This type of melanoma may appear in a combination of colors, with hues of tan, brown, and black mixed with gray, blue-black, or white. Sometimes, a dull pink rose color can be seen in a small area within the lesion.

Lentigo Maligna Melanoma

LMM is a slowly evolving, pigmented lesion that occurs on sun-exposed skin areas, especially the dorsum of the hand, the head, and the neck in older adults. Often, the lesion is present for many years before it is examined by a primary provider. It first appears as a tan, flat lesion, but in time undergoes changes in size and color. LMMs account for approximately 10% to 15% of all melanomas (Tan et al., 2013).

Nodular Melanoma

NM is a spherical, blueberrylike nodule with a relatively smooth surface and a relatively uniform blue-black color (see Fig. 61-6). It may be dome shaped with a smooth surface. It may have other shadings of red, gray, or purple. NMs sometimes appear as irregularly shaped plaques. The patient may describe this as a blood blister that fails to resolve. NM is

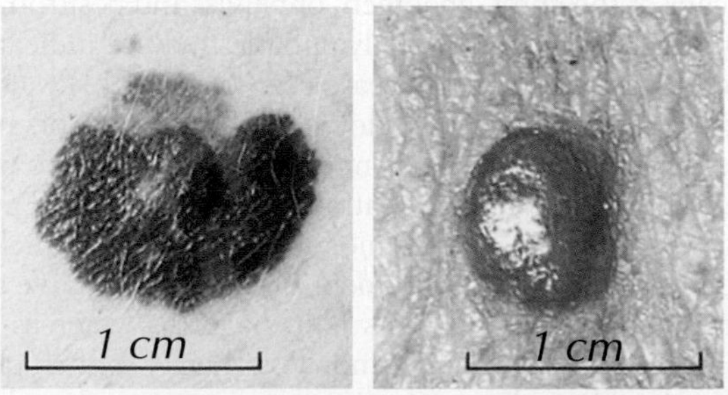
1 cm
1 cm

FIGURE 61-6 • Two forms of melanoma: superficial spreading (*left*) and nodular (*right*). From Bickley, L. S. (2009). *Bates' guide to physical examination* (10th ed.). Philadelphia: Lippincott Williams & Wilkins.

associated with rapid vertical invasion into the adjacent dermis and therefore has a poorer prognosis than most other types of melanomas. Like LMMs, they account for approximately 10% to 15% of all cases of melanomas (Tan et al., 2013).

Acral Lentiginous Melanoma

ALM occurs in areas not excessively exposed to sunlight and where hair follicles are absent. It is found on the palms of the hands, on the soles, in the nail beds, and in the mucous membranes in both light- and dark-skinned people. These melanomas appear as irregular, pigmented macules that develop nodules. Like NMs, they may become invasive early and are also associated with a poor prognosis (Tan et al., 2013).

Mucosal Lentiginous Melanoma

MLM is a noncutaneous melanoma arising from the mucosal epithelium that lines the respiratory, gastrointestinal, and genitourinary tracts, as well as the conjunctiva. MLMs account for approximately 3% of melanoma and tend to be seen in the very old adult. They tend to be associated with high mortality rates (Tan et al., 2013).

Assessment and Diagnostic Findings

Biopsy results confirm the diagnosis of melanoma. An excisional biopsy specimen provides information on the type, level of invasion, and thickness of the lesion. An excisional biopsy specimen that includes a 1- to 2-cm margin of normal tissue and a portion of underlying subcutaneous fatty tissue is sufficient for staging a melanoma in situ or an early, noninvasive melanoma. Incisional biopsy should be performed when the suspicious lesion is too large to be removed safely without extensive scarring. Biopsy specimens obtained by shaving, curettage, or needle aspiration are not considered reliable histologic proof of disease (Tan et al., 2013).

A thorough history and physical examination should include a meticulous skin examination and palpation of regional lymph nodes that drain the lesional area. Because melanoma occurs in families, a positive family history of melanoma is investigated so that first-degree relatives, who may be at high risk for melanoma, can be evaluated for atypical lesions. After the diagnosis of melanoma has been confirmed, a chest x-ray, CBC, complete chemistry panel with creatinine, liver function tests, and lactate dehydrogenase (LDH) are usually performed. The LDH may be elevated in the presence of metastatic disease. Depending on the results of these tests, magnetic resonance imaging of the brain, computed tomography scans of the chest, abdomen, or pelvis, and positron emission tomography scans of the lymphatics may be indicated to further stage the extent of disease (Tan et al., 2013).

Staging of the tumors follows the TNM (tumor, nodes, metastasis) classification system (see Chapter 15) and is used to determine appropriate treatment and prognosis. Patients with stage I disease (i.e., tumor less than 2 mm thick and without lymph node involvement or metastases) have a better than 90% survival 5 years post diagnosis, whereas those with stage II disease (i.e., tumor less than 4 mm thick and without lymph node involvement or metastases) and stage III disease (i.e., tumor of any thickness with lymph node involvement but no metastasis) have a likelihood of surviving 5 years post diagnosis between 45% and 77% and between 27% and 70%, respectively. The prognosis is grim for patients who present with stage IV disease (i.e., tumors with associated distant metastases), with less than 20% survival after 5 years (Tan et al., 2013).

Medical Management

Treatment depends on the stage of the tumor and the tumor type. Surgical excision is the treatment of choice for small, superficial lesions. Deeper lesions require wide local excision, after which skin grafting may be necessary. Sentinel lymph node biopsy is commonly performed to examine the nodes nearest the tumor and to spare the patient the long-term sequelae of extensive removal of lymph nodes if the sample nodes are negative. If these are positive, lymph node dissection may be indicated (Tan et al., 2013).

Despite many years of investigation, no dependable systemic treatment for advanced melanoma has been identified. Recently, greater understanding of the stem cells of melanoma indicates that effective therapies will have to be individualized for each patient. Clinical observations suggest that very few single-agent therapies provide significant clinical benefits because of the complex molecular nature of melanoma and the need to further define individual cell variations (Sekulic, Haluska, Miller, et al., 2008).

The observation that between 40% and 60% of all patients with melanoma have a *BRAF* genetic mutation is useful in guiding targeted therapy. In particular, treatment with the monoclonal antibodies vemurafenib (Zelboraf) and dabrafenib in separate clinical trials were each associated with improved survival in patients with stage IV melanoma who were positive for this genetic mutation (Tan et al., 2103).

Current treatments for metastatic melanoma rarely, if ever, produce a satisfactory outcome. Further surgical intervention may be performed to debulk the tumor or to remove part of the organ involved (e.g., lung, liver, or colon). However, the rationale for more extensive surgery is for relief of symptoms, not for cure. Chemotherapy for metastatic melanoma may be used; however, only dacarbazine (DTIC-Dome) has demonstrated effectiveness in controlling the disease. Other therapies that are demonstrating promise in some patients with advanced disease include the biologic agents interferon alfa, IL-2, and granulocyte-macrophage colony-stimulating factor (Tan et al., 2013).

NURSING PROCESS

Care of the Patient With Melanoma

Assessment

Assessment of the patient with melanoma is based on the patient's history and symptoms. The patient is asked specifically about pruritus, tenderness, and pain, which are not features of a benign nevus. The patient is also questioned about changes in preexisting moles or the development of new, pigmented lesions. People at risk are assessed carefully.

A magnifying lens and good lighting are needed for inspecting the skin for irregularity and changes in the mole. Signs that suggest malignant changes are referred to as the ABCDs of moles (Chart 61-7).

Chart 61-7

ASSESSMENT

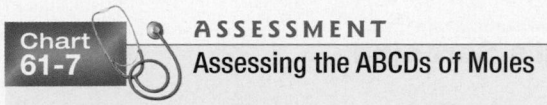

Assessing the ABCDs of Moles

Melanomas may be distinguished from benign nevi, using the following characteristics:

A for Asymmetry

- The lesion does not appear balanced on both sides. If an imaginary line were drawn down the middle, the two halves would not look alike.
- The lesion has an irregular surface with uneven elevations (irregular topography) either palpable or visible. A change in the surface may be noted from smooth to scaly.

B for Irregular Border

- Angular indentations or multiple notches appear in the border.
- The border is fuzzy or indistinct, as if rubbed with an eraser.

C for Variegated Color

- Normal moles are usually a uniform light to medium brown. Darker coloration indicates that the melanocytes have penetrated to a deeper layer of the dermis.
- Colors that may indicate malignancy if found together within a single lesion are shades of red, white, and blue; shades of blue are ominous.
- White areas within a pigmented lesion are suspicious.
- Some melanomas, however, are not variegated but are uniformly colored (bluish-black, bluish-gray, bluish-red).

D for Diameter

- A diameter >6 mm (about the size of a pencil eraser) is considered more suspicious, although this finding without other signs is not significant. Many benign skin growths are larger than 6 mm, whereas some early melanomas may be smaller.

Adapted from Rubin, K. M. (2009). Dysplastic nevi and the risk of melanoma. *Journal of the Dermatology Nurses' Association*, 1(4), 228–235; Tan, W. W., Schulman, P., & Talavera, F. (2013). Malignant melanoma. *Medscape*. Available at: emedicine.medscape.com/article/280245-overview

Common sites of melanomas are the skin of the back; the legs (especially in women); between the toes; and on the feet, face, scalp, fingernails, and backs of hands. In dark-skinned people, melanomas are most likely to occur in less pigmented sites: palms, soles, subungual areas, and mucous membranes. Satellite lesions (i.e., those situated near the mole) are inspected.

Diagnosis

Nursing Diagnoses

Based on the assessment data, major nursing diagnoses may include the following:

- Acute pain related to surgical excision and grafting
- Anxiety related to possible life-threatening consequences of melanoma and disfigurement
- Deficient knowledge about early signs of melanoma

Collaborative Problems/Potential Complications

Potential complications may include the following:

- Metastasis

Planning and Goals

The major goals for the patient may include relief of pain and discomfort, reduced anxiety, increased knowledge of early signs of melanoma, and absence of complications.

Nursing Interventions

Relieving Pain and Discomfort

Surgical removal of melanoma in different locations presents different challenges, taking into consideration the removal of the primary melanoma, the intervening lymphatic vessels, and the lymph nodes to which metastases may spread. Nursing management of the patient having surgery in these regions is discussed in the appropriate chapters.

Nursing interventions after surgery for a melanoma center on promoting comfort, because wide excision surgery may be necessary. A split- or full-thickness skin graft may be necessary when large defects are created by surgical removal of a melanoma. Anticipating the need for and administration of appropriate analgesic medications is important.

Reducing Anxiety

Psychological support is essential when disfiguring surgery is performed. Support includes encouraging the patient to express anxieties and feelings about the seriousness of the neoplasm and conveying understanding of these feelings. During the diagnostic workup and staging of the depth, type, and extent of the tumor, the nurse answers questions, provides information, and helps clarify misconceptions. Learning that he or she has a melanoma can cause the patient considerable fear and anguish. Pointing out the patient's resources, past effective coping mechanisms, and social support systems helps the patient cope with the diagnosis and need for treatment and continuing follow-up. Family members should be included in all discussions to enable them to clarify information, ask questions that the patient might be reluctant to ask, and provide emotional support to the patient.

Monitoring and Managing Potential Complications

Metastasis of melanoma is closely related to prognosis: the deeper and thicker (more than 4 mm) the melanoma, the greater the likelihood of metastasis. The role of the nurse in caring for the patient with metastatic disease is to provide holistic care. The nurse must be knowledgeable about the most effective current therapies and must deliver supportive care, provide and clarify information about the therapy and the rationale for its use, identify potential side effects of therapy and ways to manage them, and instruct the patient and family about the expected outcomes of treatment. The nurse monitors and documents symptoms that may indicate metastasis: lung (e.g., difficulty breathing, shortness of breath, increasing cough), bone (e.g., pain, decreased mobility and function, pathologic fractures), and liver (e.g., change in liver enzyme levels, pain, jaundice). Nursing care is based on the patient's symptoms and emotional needs.

Although the chance of a cure for melanoma that has metastasized is poor, the nurse encourages hope while maintaining a realistic perspective about the disease and ultimate outcome. Furthermore, the nurse provides time for the patient to express fears and concerns regarding future activities and relationships, offers information about support groups and contact people, and arranges palliative and hospice care if appropriate (see Chapter 16).

Promoting Home and Community-Based Care

Educating Patients About Self-Care. The best hope of decreasing the incidence of skin cancer lies in educating patients about the early signs. Patients at risk are educated to examine their skin and scalp monthly in a systematic manner and to seek prompt medical attention if changes are detected. The American Academy of Dermatology (AAD) provides multimedia resources on performing skin self-examination (see the Resources section at the end of this chapter). The nurse also points out that a key factor in the development of melanoma is exposure to sunlight. Because melanoma is thought to be genetically linked, the family and the patient should be educated about sun-avoiding measures and the importance of annual assessment by a primary provider.

Evaluation

Expected patient outcomes may include:

1. Experiences relief of pain and discomfort
   a. States pain is diminishing
   b. Exhibits healing of surgical scar without heat, redness, or swelling
2. Is less anxious
   a. Expresses fears and perceptions
   b. Asks questions about medical condition
   c. Requests facts about melanoma
   d. Identifies support and comfort provided by family member or significant other
3. Demonstrates understanding of the means for detecting and preventing melanoma
   a. Demonstrates how to conduct self-examination of skin on a monthly basis
   b. Verbalizes the following danger signals of melanoma: change in size, color, shape, or outline of mole, mole surface, or skin around mole
   c. Identifies measures to protect self from exposure to sunlight
4. Experiences absence of complications
   a. Recognizes abnormal signs and symptoms that should be reported to the primary provider
   b. Adheres to recommended follow-up procedures and prevention strategies

Metastatic Skin Tumors

The skin is an important yet uncommon site of metastatic cancer. All types of cancer may metastasize to the skin. Cancers with a predilection to metastasize to the skin include melanomas and cancers of the breast, nasal sinuses, larynx, and oral cavity. Of these, skin metastases from carcinoma of the breast is most frequently seen, accounting for 30% of all cases. The clinical appearance of metastatic skin lesions is not distinctive, except perhaps in some cases of breast cancer in which diffuse, brawny hardening of the skin of the involved breast is seen. In most instances, metastatic lesions occur as multiple cutaneous or subcutaneous nodules of various sizes that may be skin colored or different shades of red (Helm & Lee, 2012).

Kaposi's Sarcoma

Kaposi's sarcoma (KS) is a malignancy of endothelial cells that line the small blood vessels. KS is manifested clinically by lesions of the skin, oral cavity, gastrointestinal tract, and lungs. The skin lesions consist of reddish-purple to dark blue macules, plaques, or nodules (Nolen, Beebe, King, et al., 2011). KS is subdivided into four categories:

- *Classic KS* occurs predominantly in older adult men of Mediterranean or Jewish ancestry. Most patients have nodules or plaques on the lower extremities that rarely metastasize beyond this area. Classic KS is chronic, relatively benign, and rarely fatal.
- *Endemic (African) KS* affects people predominantly in the eastern half of Africa near the equator. Men are affected more often than women, and children can be affected as well. The disease may resemble classic KS, or it may infiltrate and progress to lymphadenopathic forms.
- *Iatrogenic/organ transplant–associated KS* occurs in transplant recipients and in patients receiving long-term immunosuppressants, such as azathioprine, cyclosporine, or corticosteroids (e.g., prednisone)
- *AIDS-related or epidemic KS* occurs in people with AIDS. This form of KS is characterized by local skin lesions and disseminated visceral and mucocutaneous diseases. This is a more aggressive tumor type than other forms of KS (Nolen et al., 2011). More information on AIDS-related KS can be found in Chapter 37.

PLASTIC RECONSTRUCTIVE AND COSMETIC PROCEDURES

The word *plastic* comes from a Greek word meaning "to form." Plastic or reconstructive procedures are performed to reconstruct or alter congenital or acquired defects to restore or improve the body's form and function. Often the terms *plastic* and *reconstructive* are used interchangeably. This type of surgery includes closure of wounds, removal of skin tumors, repair of soft tissue injuries or burns, correction of deformities, and repair of cosmetic defects. Plastic surgery can be used to repair many parts of the body and numerous structures, such as bone, cartilage, fat, fascia, mucous membrane, muscle, nerve, and cutaneous structures. Bone inlays and transplants for deformities and nonunion can be performed, muscle can be transferred, nerves can be reconstructed and spliced, and cartilage can be replaced. As important as any of these measures is the reconstruction of the cutaneous tissues around the neck and the face; this is usually referred to as aesthetic or cosmetic surgery.

Cosmetic procedures are generally considered to be ones that correct defects that are not life threatening or caused by disease. An example would be removal of a benign mole or sebaceous cyst from the face. Most health insurance plans do not cover procedures deemed to be cosmetic, and these procedures can be expensive. Procedures that are performed to correct a surgical defect, such as removal of a skin cancer or correction of a significant congenital defect such as a cleft lip, are generally covered by insurance.

Wound Coverage: Grafts and Flaps

Various surgical techniques, including skin grafts and flaps, are used to cover skin wounds.

Skin Grafts

Skin grafting is a technique in which a section of skin is detached from its own blood supply and transferred as free tissue to a distant (recipient) site. Skin grafting can be used to repair almost any type of wound and is the most common form of reconstructive surgery.

Skin grafts are commonly used to repair surgical defects such as those that result from excision of skin tumors, to cover areas denuded of skin (e.g., burns), and to cover wounds in which insufficient skin is available to permit wound closure. They are also used when primary closure of the wound increases the risk of complications or when primary wound closure would interfere with function.

Skin grafts may be classified as autografts, homografts, or xenografts. An autograft is tissue obtained from the patient's own skin. A homograft is tissue obtained from a donor of the same species. These grafts are also referred to as allogeneic or allografts. A xenograft or heterograft is tissue obtained from another species. A common xenograft for human skin is the pig.

Grafts are also referred to by their thickness. A skin graft may be a split-thickness (i.e., thin, intermediate, or thick) or a full-thickness graft, depending on the amount of dermis included in the specimen. A split-thickness graft can be cut at various thicknesses and is commonly used to cover large wounds or defects for which a full-thickness graft or flap is impractical (Fig. 61-7). A full-thickness graft consists of epidermis and the entire dermis without the underlying fat. It is used to cover wounds that are too large to be closed directly (Grande & Mezebish, 2011).

Site Selection

The site where the intact skin is harvested is called the *donor site*. Selection of the donor site is made to match the color and texture of skin at the surgical site and to leave as little scarring as possible.

Graft Application

The skin graft is taken from the donor or host site and applied to the desired site, called the *recipient site bed* or *recipient graft bed*.

For a graft to survive and be effective, certain conditions must be met:

- The recipient site must have an adequate blood supply so that normal physiologic function can resume.
- The graft must be in close contact with its bed to avoid accumulation of blood or fluid between the graft and the recipient site.
- The graft must be fixed firmly (immobilized) so that it remains in place on the recipient site.
- The area must be free of infection.

The graft, when applied to the recipient site, may be sutured in place; alternatively, it may be slit and spread apart to cover a greater area. The process of revascularization (establishing the blood supply) and reattachment of a skin graft to a recipient site bed is referred to as a "take." After a skin graft is put in place, it may be left exposed (in areas that are impossible to immobilize) or covered with a light dressing or a pressure dressing, depending on the area of the body (Grande & Mezebish, 2011).

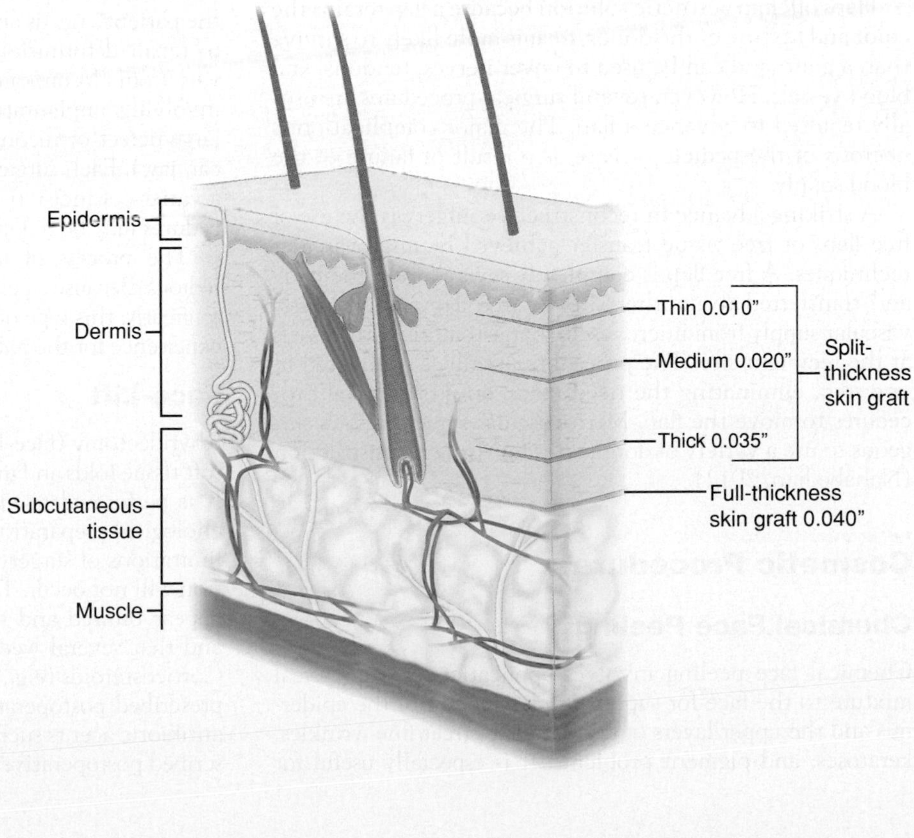
Epidermis
Dermis
Subcutaneous tissue
Muscle
Thin 0.010"
Medium 0.020"
Thick 0.035"
Split-thickness skin graft
Full-thickness skin graft 0.040"

FIGURE 61-7 • Layers of skin appropriate for split- and full-thickness graft.

Nursing Interventions

The nurse must ensure that both the surgical and the donor sites receive proper postoperative care. The surgical site is covered by the harvested skin, and the donor site heals by re-epithelization of the raw, exposed dermis. Both sites are protected by dressings as they heal. Prevention of infection is essential, as it is with all surgical sites. Both sites can be kept soft and pliable with cream (e.g., lanolin).

> ***Quality and Safety Nursing Alert***
>
> **Both the donor site and the grafted area must be protected from exposure to extremes in temperature, external trauma, and sunlight because these areas are sensitive, especially to thermal injuries.**

Flaps

Another form of wound coverage is provided by flaps. A flap is a segment of tissue that remains attached at one end (i.e., a base or pedicle) while the other end is moved to a recipient area. Its survival depends on functioning arterial and venous blood supplies and lymphatic drainage in its pedicle or base (Clark, Wang, & Downs, 2011). A flap differs from a graft in that a portion of the tissue is attached to its original site and retains its blood supply. An exception is the free flap, which is described later.

Flaps may consist of skin, mucosa, muscle, adipose tissue, omentum, and bone. They are used for wound coverage and provide bulk, especially when bone, tendon, blood vessels, or nerve tissue is exposed. Flaps are used to repair defects caused by congenital deformity, trauma, or tumor ablation (removal, usually by excision) in an adjacent part of the body (Clark et al., 2011).

Flaps offer an aesthetic solution because a flap retains the color and texture of the donor area; is more likely to survive than a graft; and can be used to cover nerves, tendons, and blood vessels. However, several surgical procedures are usually required to advance a flap. The major complication is necrosis of the pedicle or base as a result of failure of the blood supply.

A striking advance in reconstructive surgery is the use of free flaps or free tissue transfer achieved by microvascular techniques. A free flap is completely severed from the body and transferred to another site. A free flap receives early vascular supply from microvascular anastomosis with vessels at the recipient site. The procedure usually is completed in one step, eliminating the need for a series of surgical procedures to move the flap. Microvascular surgery allows surgeons to use a variety of donor sites for tissue reconstruction (Nahabedian, 2012).

Cosmetic Procedures

Chemical Face Peeling

Chemical face peeling involves application of a chemical mixture to the face for superficial destruction of the epidermis and the upper layers of the dermis to treat fine wrinkles, keratoses, and pigment problems. It is especially useful for wrinkles at the upper and lower lip, forehead, and periorbital areas. The type of chemical used depends on the planned depth of the peel. The patient who is conscious feels a burning sensation that continues for 12 to 24 hours. Frequent small doses of analgesic and tranquilizing agents are prescribed to keep the patient comfortable. The most common complications include discoloration of the skin, infection of the burned area, persistent sensory changes or itching, and occasionally permanent scarring of the skin (Fabbrocini, 2012).

Dermabrasion

Dermabrasion is a form of skin abrasion used to treat acne scarring, aging, and sun-damaged skin. A special instrument (e.g., motor-driven wire brush, diamond-impregnated disk) is used. The epidermis and some superficial dermis are removed by a sanding-type action, and enough of the dermis is preserved to allow re-epithelization of the treated areas. Results are best in the face because it is rich in intradermal epithelial elements (Harmon & Prather, 2011).

Patients with a history of herpes simplex viral infection are typically prescribed prophylactic antiviral medications (e.g., valacyclovir [Valtrex]) preprocedurally so that the physiological stress of the procedure is less likely to cause a cutaneous herpes eruption. Tretinoin cream (Renova) may be prescribed with instructions to apply it 2 to 3 weeks preoperatively; this is associated with accelerating re-epithelialization post dermabrasion. Patients must be educated preprocedurally about the postprocedural dressing regimen and when to return to the health care provider to have dressing changes performed (Harmon & Prather, 2011).

Facial Reconstructive Surgery

Reconstructive procedures on the face are individualized to the patient's needs and desired outcomes. They are performed to repair deformities or restore normal function. They may vary from closure of small defects to complicated procedures involving implantation of prosthetic devices to conceal a large defect or reconstruct a lost part of the face (e.g., nose, ear, jaw). Each surgical procedure is customized and involves a variety of incisions, flaps, and grafts. Multiple surgical procedures may be required.

The process of facial reconstruction is often slow and tedious. Because a person's facial appearance affects self-esteem so greatly, this type of reconstruction is often a very emotional experience for the patient.

Face-Lift

Rhytidectomy (face-lift) is a surgical procedure that removes soft tissue folds and minimizes cutaneous wrinkles on the face. It is performed to create a more youthful appearance. Psychological preparation requires that the patient recognize the limitations of surgery and the fact that miraculous rejuvenation will not occur. The patient is informed that the face may appear bruised and swollen after the dressings are removed and that several weeks may pass before the edema subsides. Corticosteroids (e.g., methylprednisolone) and vitamin C are prescribed postoperatively to minimize edema. Prophylactic antibiotic agents such as cephalexin (Keflex) may also be prescribed postoperatively (Jacono, 2011).

TABLE 61-6 Nursing Considerations in Cosmetic Procedures

Nursing Consideration	Interventions and Patient Education
Maintaining airway and pulmonary function	Cosmetic surgeries involving the face and neck can cause considerable swelling; bandages can restrict breathing or eating. Check dressings frequently, and ensure that no constriction occurs as swelling develops.
Relieving pain and achieving comfort	Procedures that involve a large surface area will cause considerable pain. Cool compresses or ice packs will relieve the burning of dermabrasion or chemical peels. Oral analgesic agents should be administered regularly to control pain.
Maintaining adequate nutrition	When the face is involved, the patient may be unable to fully open the mouth, and chewing may be painful. Provide soft or liquid diet that is high in protein to assist with healing.
Enhancing communication	Depending on the type of cosmetic procedure, a nonverbal method of communication might be necessary until pain and swelling have subsided.
Improving self-concept	Recovery time from cosmetic procedures is slow. Expected results will take weeks to become apparent. Persons of color will experience increased pigmentation long after the initial wounds have healed. Helping patients to understand postoperative expectations will allow them to feel more comfortable with the healing process.
Promoting family coping	Most cosmetic procedures are performed in an outpatient facility; therefore, family members are integral to postoperative care. They should understand what to expect as the patient emerges from the procedure room: the type of dressings that will be in place, the skin care plan that is prescribed, and how to cope with the patient's pain.
Monitoring and managing potential complications	Infection is the most common complication, but excessive pain, nerve damage, and emotional distress about appearance are also common. If opioid analgesic agents are used, there may be gastrointestinal upset, mental status changes, or allergic reaction to the medication. Alert the caregiver to signs of these complications and how and when to report changes in status.

Laser Treatment of Cutaneous Lesions

Lasers are devices that amplify or generate highly specialized light energy. They can mobilize immense heat and power when focused at close range and are valuable tools in providing dermatologic abrasion therapy. The laser modalities used for this purpose today include scanned carbon dioxide laser, pulsed carbon dioxide laser, pulsed erbium/yttrium-aluminum-garnet (Er:YAG) laser, fractional Er:YAG laser resurfacing, combination carbon dioxide and Er:YAG lasers, and fractionated photothermolysis (Sandhu & Goldberg, 2012).

Each of these lasers is a precise surgical instrument that vaporizes and excises water-containing tissues with minimal damage. Because the beams used can seal blood and lymphatic vessels, they create a dry surgical field that makes many procedures easier and quicker. Therefore, these lasers are generally safe to use on patients with bleeding disorders or those receiving anticoagulant therapy. They are primarily used to improve the appearance of facial wrinkles, although they are also useful in removing epidermal nevi, tattoos, certain warts, skin cancer, ingrown toenails, and keloids. Incisions made with the laser beam heal and scar much like those made by a scalpel. Patients with a history of herpes simplex viral infection typically receive preprocedural antiviral prophylaxis (Sandhu & Goldberg, 2012).

Nursing Management

The majority of dermatologic and reconstructive procedures are performed in the physician's office or in an outpatient surgical department; therefore, most care takes place in the home. Most procedures, except very extensive reconstruction, are performed under local anesthesia or moderate sedation, therefore requiring a very short recovery time. Unless there are complications, the patient does not need hospitalization. The nurse must prepare both the patient and family for what to expect during the postoperative recovery time. Table 61-6 lists a few of the nursing considerations that must be reviewed in educating the patient and family.

Critical Thinking Exercises

1 ebp You are caring for a middle-aged man who has congestive heart failure. He has developed a venous stasis ulcer on his lower leg just above the ankle. A moisture-retentive dressing that is impregnated with hydrogel has been prescribed with dressing changes every 3 days. Identify the evidence that supports the use of moisture-retentive dressings for venous ulcers. Discuss the strength of the evidence regarding their effectiveness in the promotion of wound healing

2 You are providing care for a 35-year-old patient with SJS. What should be included in the admission and ongoing nursing assessment? Describe the nursing diagnoses, goals, nursing interventions, and expected outcomes for this patient.

3 pq A 50-year-old professional golfer presenting to the clinic states that his mother had melanoma and that now he is concerned about several lesions on his arms and neck. Describe your focused priority assessments and interventions. What are your priorities for education that you should provide to this patient about prevention of melanoma and self-examination of the skin?

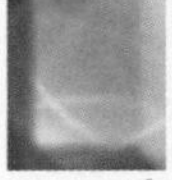

Brunner Suite Resources

Explore these additional resources to enhance learning for this chapter:

- NCLEX-Style Questions and Other Resources on thePoint, http://thePoint.lww.com/Brunner13e
- Study Guide
- PrepU
- Clinical Handbook
- Handbook of Laboratory and Diagnostic Tests

References

*Asterisk indicates nursing research.

Books

Baranoski, S., & Ayello, E. (2012). *Wound care essentials: Practice principles* (3rd ed.). Philadelphia: Lippincott Williams & Wilkins.

Bryant, R. A., & Nix, D. P. (2011). *Acute and chronic wounds* (4th ed.). St. Louis: Mosby.

Goldsmith, L., Katz, S. I., Gilchrist B. A., et al. (2012). *Fitzpatrick's dermatology in general medicine* (8th ed.). New York: McGraw-Hill.

Karch, A. M. (2013). *2013 Lippincott's nursing drug guide*. Philadelphia: Lippincott Williams & Wilkins.

Krasner, D., Rodeheaver, G., Sibbald, G., et al. (2012). *Chronic wound care: A clinical source book for healthcare professionals* (5th ed.). Malvern, PA: HMP Communications.

Porth, C. M., & Matfin, G. (2009). *Pathophysiology: Concepts of altered health states* (8th ed.). Philadelphia: Lippincott Williams & Wilkins.

Weber, J., & Kelley, J. (2010). *Health assessment in nursing* (4th ed.). Philadelphia: Lippincott Williams & Wilkins.

Journals and Electronic Documents

Bader, R. S., Santacroce, L., Diomede, L., et al. (2013). Basal cell carcinoma. *Medscape*. Available at: emedicine.medscape.com/article/276624-overview

Beshara, M. A. (2012). Hidradenitis suppurativa: A clinician's tool for early diagnosis and treatment. *Journal of the Dermatology Nurses' Association*, 4(1), 50–53.

Chan, L. S. (2013). Bullous pemphigoid. *Medscape*. Available at: emedicine.medscape.com/article/1062391-overview

Clark, J. M., Wang, T. D., & Downs, B. W. (2011). Skin flap design. *Medscape*. Available at: emedicine.medscape.com/article/875968-overview

Cohen, V., Jellinek, S. P., & Schwartz, R. A. (2011). Toxic epidermal necrolysis. *Medscape*. Available at: emedicine.medscape.com/article/229698-overview

Cordoro, K. M., Wilson, B. B., & Kauffman, C. L. (2012). Dermatologic manifestations of scabies. *Medscape*. Available at: emedicine.medscape.com/article/1109204-overview

Dale, B. A., & Wright, D. H. (2011). Say goodbye to wet-to-dry wound care dressings: Changing the culture of wound care management within your agency. *Home Healthcare Nurse*, 29(7), 429–440.

Dowling, V. L., (2010). The psychological impact of psoriasis: A review of short-term psychotherapy group participation for psoriasis patients. *Journal of the Dermatology Nurses' Association*, 2(4), 163–167.

Fabbrocini, B. (2012). Chemical peels. *Medscape*. Available at: emedicine.medscape.com/article/1829120-overview

Fonder, M. A., Lazarus, G. S., Cowan, D., et al. (2008). Treating the chronic wound: A practical approach to the care of nonhealing wounds and wound care dressings. *Journal of American Academy of Dermatology*, 58(2), 185–206.

Fulton, J. (2012). Acne vulgaris. *Medscape*. Available at: emedicine.medscape.com/article/1069804-overview

Garcia-Albea, V., & Limaye, K. (2012). The clinical conundrum of pruritus. *Journal of the Dermatology Nurses' Association*, 4(2), 97–105.

Grande, D. J., & Mezebish, D. S. (2011). Skin grafting. *Medscape*. Available at: emedicine.medscape.com/article/1129479-overview

Guenther, L., Maguiness, S., & Austin, T. W. (2012). Pediculosis (lice). *Medscape*. Available at: emedicine.medscape.com/article/225013-overview

Harmon, C. B., & Prather, C. L. (2011). Dermabrasion procedures. *Medscape*. Available at: emedicine.medscape.com/article/1126231-overview

Helm, T. N., & Lee, T. C. (2012). Dermatologic manifestations of metastatic carcinomas. *Medscape*. Available at: emedicine.medscape.com/article/1101058-overview

Hogan, D. J. (2011). Irritant contact dermatitis. *Medscape*. Available at: emedicine.medscape.com/article/1049353-overview

Jacono, A. (2011). SMAS facelift rhytidectomy. *Medscape*. Available at: emedicine.medscape.com/article/841892-overview

Janninger, C. K., Eastern, J. S., Hospenthal, D. R., et al. (2013). Herpes zoster. *Medscape*. Available at: emedicine.medscape.com/article/1132465-overview

Klein, P. A. (2013). Dermatologic manifestations of Stevens-Johnson syndrome and toxic epidermal necrolysis. *Medscape*. Available at: emedicine.medscape.com/article/1124127-overview

Lewis, L. S., & Friedman, A. D. (2013). Impetigo. *Medscape*. Available at: emedicine.medscape.com/article/965254-overview

Meffert, J., Arffa, R., Gordon, R., et al. (2013). Psoriasis. *Medscape*. Available at: emedicine.medscape.com/article/1943419-overview

Miller, A. C., Rashid, R. M., & Silverberg, M. A. (2011). Tinea in emergency medicine. *Medscape*. Available at: emedicine.medscape.com/article/787217-overview

Miller, J. L., & Zaman, S. A. (2011). Dermatitis herpetiformis. *Medscape*. Available at: emedicine.medscape.com/article/1062640-overview

Mooney, M. A., & Parry, E. (2011). Mohs microscopic surgery. *Medscape*. Available at: emedicine.medscape.com/article/1125510-overview

Nahabedian, M. Y. (2012). Free tissue transfer flaps. *Medscape*. Available at: emedicine.medscape.com/article/1284841-overview

National Institute for Health and Clinical Excellence. (2012). *Psoriasis: The assessment and management of psoriasis*. London, UK: Author. Available at: www.guideline.gov/popups/printView.aspx?id=38575

Nolen, M. E., Beebe, V. R., King, J. M., et al. (2011). Nonmelanoma skin cancer: Part 2. *Journal of the Dermatology Nurses' Association*, 3(6), 326–341.

Occupational Safety and Health Administration. (2012). *Toxic and hazardous substances: Bloodborne pathogens*. Available at: www.osha.gov/pls/oshaweb/owadisp.show_document?p_table=STANDARDS&p_id=10051

Ramundo, J., & Gray, M. (2008). Enzymatic wound debridement. *Journal of Wound, Ostomy, and Continence Nursing*, 35(3), 273–280.

Rubin, K. M. (2009). Dysplastic nevi and the risk of melanoma. *Journal of the Dermatology Nurses' Association*, 1(4), 228–235.

Sandhu, N., & Goldberg, D. J. (2012). Carbon dioxide cutaneous laser resurfacing. *Medscape*. Available at: emedicine.medscape.com/article/1120283-overview

Satter, E. K. (2012). Folliculitis. *Medscape*. Available at: emedicine.medscape.com/article/1070456-overview

Schmidt, J. A. (2011). Seborrheic dermatitis: A clinical practice snapshot. *Journal of the Dermatology Nurses' Association*, 3(5), 294–299.

Sekulic, A., Haluska, P., Miller, A. J., et al.; Melanoma Study Group of the Mayo Clinic Cancer Center. (2008). Malignant melanoma in the 21st century: The emerging molecular landscape. *Mayo Clinic Proceedings*, 83(7), 825–846.

*Shelestak, D., & Lindow, K. (2011). Beliefs and practices regarding skin cancer prevention. *Journal of the Dermatology Nurses' Association*, 3(3), 150–155.

Shulstad, R. M., & Proper, S. (2010). Squamous cell carcinoma: A review of etiology, pathogenesis, treatment, and variants. *Journal of the Dermatology Nurses' Association*, 2(1), 12–16.

Tan, W. W., Schulman, P., & Talavera, F. (2013). Malignant melanoma. *Medscape*. Available at: emedicine.medscape.com/article/280245-overview

Umar, S. H., & Kelly, A. P. (2012). Erythroderma (generalized exfoliative dermatitis). *Medscape*. Available at: emedicine.medscape.com/article/1106906-overview

U.S. Preventive Services Task Force. (2009). Screening for skin cancer: U.S. Preventive Services Task Force recommendation statement. *Annals of Internal Medicine*, 150(3), 188–193.

Vaneau, M., Chaby, G., Guillot, B., et al. (2007). Consensus panel recommendations for chronic and acute wound dressings. *Archives of Dermatology*, 143(10), 1291–1294.

Zeina, B., Sakka, N., & Mansoor, S. (2011). Pemphigus vulgaris. *Medscape*. Available at: emedicine.medscape.com/article/1064187-overview

Resources

American Academy of Dermatology (AAD), www.aad.org

American Melanoma Foundation (AMF), www.melanomafoundation.org

Dermatology Atlas (DermIS), a cooperation between the Department of Clinical Social Medicine (University of Heidelberg) and the Department of Dermatology (University of Erlangen), www.dermis.net

Foundation for Ichthyosis and Related Skin Types (FIRST), www.firstskinfoundation.org

National Eczema Association, www.nationaleczema.org

National Psoriasis Foundation, www.psoriasis.org

New Zealand Dermatology Society (DermNET NZ), www.dermnetnz.org

Skin Cancer Foundation, www.skincancer.org

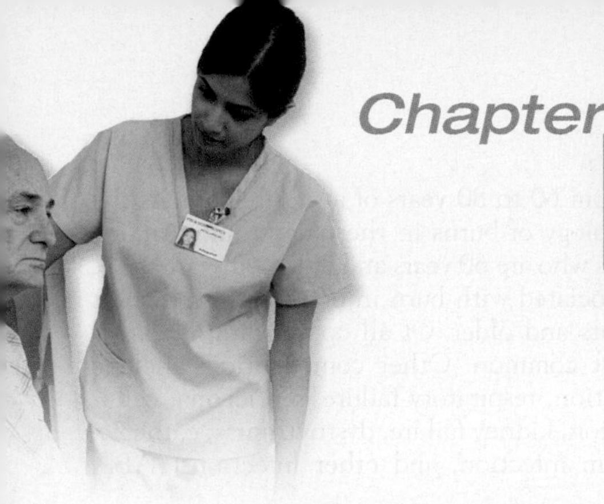

Chapter 62

Management of Patients With Burn Injury

Learning Objectives

On completion of this chapter, the learner will be able to:

1. Discuss the incidence of burn injury in the United States.
2. Describe the factors that affect the severity of burn injury.
3. Describe the local and systemic effects of a major burn injury.
4. Identify priorities of care and potential complications for each phase of burns.
5. Identify fluid replacement requirements during the emergent/resuscitative and acute phases of a burn injury.
6. Discuss the nurse's role in burn wound management during the acute/intermediate phase of burn care.
7. Use the nursing process as a framework of care for the patient with burns during the emergent/resuscitative and rehabilitation phases of burn care.
8. Describe the psychosocial challenges associated with burn injuries and identify strategies for intervention.

Glossary

autograft: a graft derived from one part of a patient's body and used on another part of that same patient's body

carboxyhemoglobin: a compound of carbon monoxide and hemoglobin, formed in the blood with exposure to carbon monoxide

collagen: a protein present in skin, tendon, bone, cartilage, and connective tissue

contracture: shrinkage of burn scar through collagen maturation

débridement: removal of foreign material and devitalized tissue until surrounding healthy tissue is exposed

donor site: the area from which skin is taken to provide a skin graft for another part of the body

eschar: devitalized tissue resulting from a burn or wound

escharotomy: a linear excision made through eschar to release constriction of underlying tissue

excision: surgical removal of tissue

fasciotomy: an incision made through the fascia to release constriction of underlying muscle

homograft: a graft transferred from one human (living or cadaveric) to another human; also called *allograft*

hydrotherapy: cleansing of wounds through the use of bath, shower, shower cart table, or immersion

Nikolsky's sign: rubbing of the epidermis, resulting in exfoliation or separation from the dermal layer

xenograft: a graft obtained from an animal of a species other than that of the recipient (e.g., pigskin); also called *heterograft*

Nearly all adults will experience a burn injury at some point in their lives (Purdue, Arnoldo, & Hunt, 2011). Burn injuries are painful, costly, disfiguring, require intensive and extensive rehabilitation therapy, and may be associated with long-term disability. Furthermore, "larger burns are associated with morbidity and mortality disproportionate to their initial appearance" (Purdue et al., 2011, p. 201).

Advances in burn care over the past 80 years have contributed to significant improvements in morbidity and mortality of patients with burns. These advances include the introduction of systemic antibiotic and topical antimicrobial agents, advances in fluid resuscitation, aggressive nutrition, early excision and wound closure, the introduction of engineered tissue therapies, advances in critical care therapies, and the advent of specialized burn centers (Kasten, Mackley, & Kagan, 2011). The role of the nurse in the interdisciplinary treatment team includes provision of holistic evidence-based care during all phases of the injury to optimize patient outcomes.

Overview of Burn Injury

Incidence

A burn injury can affect people of all ages and socioeconomic groups. An estimated 450,000 people are treated for burns annually. Approximately 45,000 patients are hospitalized each year with burn injuries, with 25,000 admitted to hospitals with burn centers (American Burn Association [ABA], 2011b). Of those people admitted to burn centers, 68% had injuries that occurred at home, 10% had industry-related injuries, 5% had recreationally related injuries, and the remaining 17% had injuries from other sources. Of all these injuries, 44% were flame related, 33% were scald injuries, 9% were from direct source contact, 4% were electrical, 3% were chemical, 1% were inhalation only, and the remaining were from unspecified or miscellaneous categories (American Burn Association National Burn Repository [ABA NBR], 2011).

Patients with burns have particularly prolonged lengths of hospital stay. Many factors contribute to a lengthy length of stay. For example, many of these patients require surgical interventions, extensive pain control, immobilization and rehabilitation, and prolonged intravenous (IV) medication regimens, especially with antibiotics and opioids. In addition, patients with smoke inhalation and electrical injuries require particularly lengthy care regimens (Peck, 2011a).

Men have more than twice the incidence of burn injury than women; for both men and women, the most frequent age group for burns is between 20 and 30 years (ABA NBR, 2011). In the United States, the race and ethnicity of patients treated in burn centers is as follows: 60% Caucasian, 19% African American, 15% Hispanic, 2% Asian American, 0.7% Native American, and 3% identified as "other" (ABA NBR, 2011). One study noted an association between the severity of burns and socioeconomic groups, with severe injuries more common in lower socioeconomic groups (Park, Shin, Kim, et al., 2009). The socioeconomic maldistribution of burns has been referred to as "dramatic examples of the inequity of injury" (Peck, 2011a, p. 1090).

The National Fire Protection Association reported 3,120 fire fatalities in 2010. This translates to one fire death every 169 minutes and fire injury every 30 minutes in 2010 (Karter, 2011). The overall mortality rate for all total body surface area (TBSA) burns is 3.9%, with the incidence of fatality increasing as burn size increases (ABA NBR, 2011). One systematic review (Colohan, 2010), which examined the predictors of mortality in adult burn patients, reported an overall mortality rate of 13.9%. The strongest predictors for mortality in burn injuries included increased percent of TBSA burned, presence of inhalation injury, and increased age. Similar results were reported in a European systematic review on severe burn injuries. Risk factors for death included increased age, greater TBSA, and presence of concomitant chronic diseases. This same study cited multiple organ failure and sepsis as major causes of death (Brusselaers, Monstrey, Vogelaers, et al., 2010).

Gerontologic Considerations

Common age-related changes, such as diminished mobility, postural stability, strength, coordination, sensation, visual acuity, as well as declining memory, predispose older adults to burn injury. The majority of burn injuries in the older adult occur in the home and as the result of carelessness, such as from smoking, cooking, or tub scald injuries (Campbell, DeGolia, Fallon, et al., 2009).

One study that compared burn-related outcomes in patients younger than 65 years to those older than 65 years found that the older group of patients had 1.9 times greater probability of death. Furthermore, no patients older than 85 years survived in this study, regardless of TBSA burned (Albornoz, Villegas, Sylvester, et al., 2011).

Registry data from 163,000 hospitalized burn patients over a 9-year period suggest that 12% of burn injuries requiring hospitalization occur in patients 60 years and older. Mortality associated with burns are greater in older adult patients than in younger patients when comparing injuries with similar severity. The overall mortality from burns in the older adult is approximately 15%, and the mortality more than doubles as age increases from 60 to 80 years of age. Fire/flame is the most common etiology of burns in the older adult. Among patients with burns who are 60 years and older, 60% are male. Complications associated with burn injuries are also highest in patients 60 years and older. Of all complications, pneumonia is the most common. Other complications include urinary tract infection, respiratory failure, septicemia, cellulitis, wound infection, kidney failure, dysrhythmias, catheter-related bloodstream infection, and other infections (ABA NBR, 2011).

The skin of the older adult is thinner and less elastic, which affects the depth of injury and its ability to heal. Pulmonary function becomes impaired with age; therefore, airway exchange, lung elasticity, and ventilation can be altered. This can be exacerbated if the older adult has a history of smoking. Decreased cardiac output, coronary artery disease, and decreased cardiovascular compensatory response may increase the risk of complications in older adult patients with burn injuries. There may be a fine line between adequate fluid resuscitation and fluid overload in this population. Decreased kidney and hepatic function can impact medication dosing due to altered medication clearance. Malnutrition may impact morbidity and mortality in older adults, especially those who are institutionalized. The presence of malnutrition pre-injury poses increased risk because a patient who has been critically injured may be hypermetabolic and exhibit accelerated catabolism, resulting in rapid malnutrition (Campbell et al. 2009; Keck, Lumenta, Andel, et al., 2009). Older patients may have varying degrees of mental capacity on admission or through the course of care, making assessment of pain, anxiety, and delirium a challenge for the burn team.

Prevention

An important goal of nurses in community and home settings is to provide education about the prevention of burn injury, especially among older adults (Chart 62-1). According to one group of researchers (Keck, Lumenta, et al., 2009), disease prevention is often perceived as a higher priority compared to injury prevention, which results in lower government funding for injury prevention programs. Nurses need to assess an older adult patient's ability to safely perform activities of daily living, assist older adult patients and families to modify their environment to ensure safety, and make referrals as needed.

An important area of education and awareness is the role of alcohol and cigarettes in fires among adults of all ages. Alcohol significantly impairs judgment and ability. Smoking in bed while intoxicated is a common cause of fires. As best stated by Peck (2011b), "recognizing the presence of smoke is not the same as being able to escape from the fire" (p. 516).

Outlook for Survival and Recovery

World fire statistics rank the United States 14th in terms of fire deaths and 2nd in terms of fire costs (adjusted direct fire losses, as a percentage of gross domestic product) (Geneva Association, 2012). The National Center for Injury Prevention and Control of the Centers for Disease Control and

Chart 62-1

HEALTH PROMOTION

Burn Prevention

- Advise that matches and lighters be kept out of the reach of children.
- Emphasize the importance of never leaving children unattended around fire or in bathroom/bathtub.
- Educate about the installation and maintenance of smoke and carbon monoxide detectors on every level of the home and changing batteries annually on birthday.
- Recommend the development and practice of a home exit fire drill with all members of the household.
- Advocate setting the water heater temperature no higher than 48.9°C (120°F).
- Educate about the perils of smoking in bed, smoking while using home oxygen, or falling asleep while smoking.
- Caution against using flammable liquids to start fires and/or throwing flammable liquids onto an already burning fire.
- Warn of the danger of removing the radiator cap from a hot car engine.
- Recommend avoidance of overhead electrical wires and underground wires when working outside.
- Advise that hot irons and curling irons be kept out of the reach of children.
- Discourage running electric cords under carpets or rugs.
- Recommend storage of flammable liquids well away from a fire source, such as a pilot light.
- Educate about the importance of being aware of loose clothing when cooking over a stovetop or flame.
- Recommend having a working fire extinguisher in the home and knowing how to use it.

Prevention (CDC) identifies unintentional injury as the 5th leading cause of death in the United States for both sexes, all races, and all ages; of these, fire or burn is the 7th most common cause of death (CDC, 2011).

A number of factors have been instrumental in reducing mortality from fire, including smoke alarms, carbon monoxide detectors, safer appliances for heating and cooking, flame-resistant materials, child-resistant lighters, and sprinkler systems (Peck, 2011b).

Great strides in research on treatment have helped to increase the survival rate of patients with burn injuries. Long-term outcomes can now be evaluated because patients with very large burns are surviving their injuries. Research in areas of fluid resuscitation, emergency burn treatment, inhalation injury and management, nutritional requirements, early **excision** (surgical removal of tissue), skin grafting, and the use of skin substitutes have contributed greatly to the decrease in burn deaths (Branski, Herndon, & Barrow, 2012; Palmieri, 2009). Continued research and advances in the areas of critical care, rehabilitation, psychosocial, and scar management are essential for continued progress in burn care.

Severity

The severity of each burn injury is determined by multiple factors. These factors include age of the patient; depth of the burn; amount of surface area of the body that is burned; the presence of inhalation injury; presence of other injuries; location of the injury in areas such as the face, the perineum, hands, or feet; and the presence of a past medical history. Careful assessment enables the burn team to estimate the likelihood of survival and develop an individualized plan of care for each patient.

Age

Young children and older adults continue to have increased morbidity and mortality when compared to other age groups with similar injuries and present a challenge for the burn team. This is an important factor when determining the severity of injury and possible outcome for the patient.

Burn Depth

Burns are classified according to the depth of tissue destruction as depicted in Table 62-1. First-degree burns are superficial injuries that involve only the outermost layer of skin. These burns are erythematous, but the epidermis is intact; if rubbed, the burned tissue does not separate from the underlying dermis. This is known as a negative **Nikolsky's sign**. A typical first-degree burn is a sunburn or superficial scald. This type of injury does not contribute to morbidity and is therefore not included in calculations of TBSA (Purdue et al., 2011) (see later discussion).

Second-degree burns involve the entire epidermis and varying portions of the dermis. They are painful and are typically associated with blister formation. Healing time depends on the depth of dermal injury and typically ranges from 2 to 3 weeks. Hair follicles and skin appendages remain intact. If these injuries take more than 3 weeks to heal, they may require grafting due to propensity for scarring (Purdue et al., 2011).

Third-degree (full-thickness) burns involve total destruction of the epidermis and dermis and, in some cases, destruction of underlying tissue. Wound color ranges widely from pale white to red, brown, or charred. The burned area lacks sensation because nerve fibers are damaged. The wound appears leathery; hair follicles and sweat glands are destroyed. The severity of this burn is often deceiving to patients because they have no pain in the injury area (Fig. 62-1).

Fourth-degree burns (deep burn necrosis) are those injuries that extend into deep tissue, muscle, or bone (Purdue et al., 2011) (Fig. 62-2).

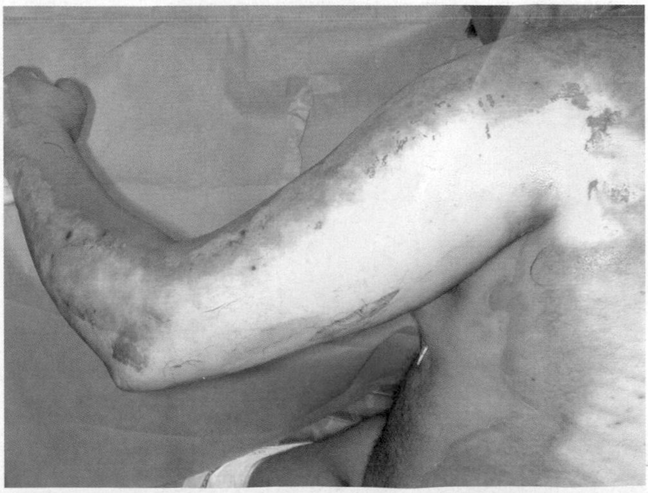

FIGURE 62-1 • Third-degree (full-thickness) burn to arm and upper back with surrounding second-degree (partial-thickness) burn. Used with permission from University of Texas Medical Branch, Galveston, TX.

TABLE 62-1 Characteristics of Burns According to Depth

Causes	Skin Involvement	Clinical Manifestations	Wound Appearance	Recuperative Course and Treatment
First Degree (Superficial)				
Sunburn Low-intensity flash Superficial scald	Epidermis; possibly a portion of dermis	Tingling Hyperesthesia (hypersensitivity) Pain that is soothed by cooling Peeling Itching	Reddened; blanches with pressure; dry Minimal or no edema Possible blisters	Complete recovery within a few days Oral pain medications, cool compresses, skin lubricants (e.g., ointments, emollients); topical antimicrobial agents not indicated
Second Degree (Partial Thickness)				
Scalds Flash flame Contact	Epidermis, portion of dermis	Pain Hyperesthesia Sensitive to air currents	Blistered, mottled red base; disrupted epidermis; weeping surface Edema	Recovery in 2–3 weeks Some scarring and depigmentation possible; may require grafting
Third Degree (Full Thickness)				
Flame Prolonged exposure to hot liquids Electric current Chemical Contact	Epidermis, dermis, and sometimes subcutaneous tissue; may involve connective tissue, and muscle	Insensate Shock Myoglobinuria (red pigment in urine) and possible hemolysis (blood cell destruction) Possible contact points (entrance or exit wounds in electrical burns)	Dry; pale white, red brown, leathery, or charred Coagulated vessels may be visible. Edema	Eschar may slough Grafting necessary Scarring and loss of contour and function
Fourth Degree (Full Thickness That Includes Fat, Fascia, Muscle, and/or Bone)				
Prolonged exposure or high voltage electrical injury	Deep tissue, muscle and bone	Shock Myoglobinuria (red pigment in urine) and possible hemolysis (blood cell destruction)	Charred	Amputations likely Grafting of no benefit, given depth and severity of wound(s)

Adapted from American Burn Association. (2011a). *Advanced burn life support (ABLS) course provider manual 2011*. Chicago: Author; and Lewis, G. M., Heimbach, D. M., & Gibran, N. S. (2012). Evaluation of the burn wound: Management decisions. In D. Herndon (Ed.). *Total burn care* (4th ed., pp. 125–130). Edinburgh, UK: Saunders Elsevier.

Burn depth determines whether re-epithelialization will occur. Determining burn depth can be difficult even for the experienced burn care provider. The following factors are considered in determining the depth of a burn: how the injury occurred, causative agent (such as flame or scalding liquid), temperature and duration of contact with the causative agent, and thickness of the skin.

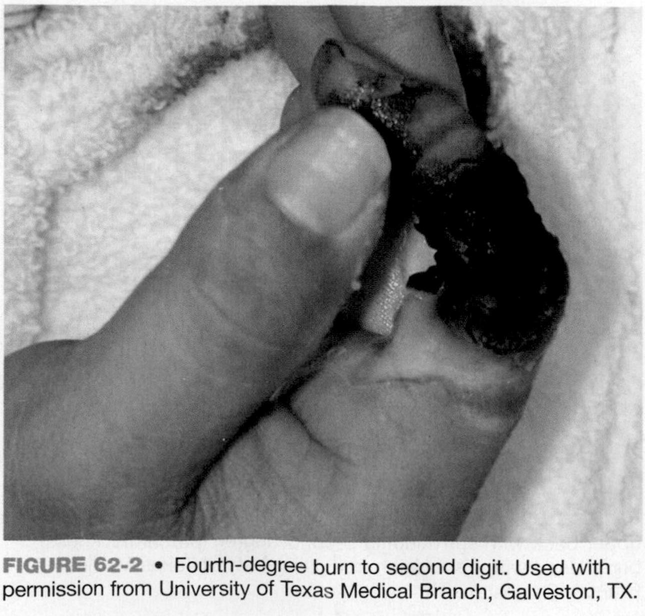

FIGURE 62-2 • Fourth-degree burn to second digit. Used with permission from University of Texas Medical Branch, Galveston, TX.

Extent of Body Surface Area Injured

Various methods are used to estimate the TBSA affected by burns; among them are the rule of nines, the Lund and Browder method, and the palmer method. These tools assist the treatment team in making decisions about the plan of care, which may include transfer of the patient to a hospital with a burn center. Burn centers are specially equipped with the resources and personnel to treat patients with burns from the time of injury through their rehabilitation. Burn center designation is jointly conferred by the ABA and the American College of Surgeons (ABA, 2005). Chart 62-2 provides the ABA criteria for referral to a burn center.

Rule of Nines

The most commonly used method to estimate the extent of burns in adults is the rule of nines (Fig. 62-3). This system is based on dividing anatomic regions, each representing approximately 9% of the TBSA, allowing clinicians to quickly obtain an estimate. If a portion of an anatomic area is burned, the TBSA is calculated accordingly—for example, if approximately half of the anterior arm is burned, the TBSA burned would be 4.5% (ABA, 2011a).

Lund and Browder Method

A more precise method of estimating the extent of a burn is the Lund and Browder method (ABA, 2011a), which recognizes the percentage of surface area of various anatomic parts, especially the head and legs, as it relates to the age of the patient. By dividing the body into very small areas and providing an

Chart 62-2 **American Burn Association Criteria for Referral to a Burn Center**

- Partial-thickness burns covering 10% of total body surface area or greater
- Burns involving the face, hands, feet, genitalia, perineum, or major joints
- Third-degree burns
- Electrical burns, including lightning injury
- Chemical burns
- Inhalation injury
- Burn injury in patients with preexisting medical disorders
- Any patients with burns and concomitant trauma
- Children with burn injuries in facilities that do not specialize in pediatric care
- Patients who will require special social, emotional, or long-term rehabilitation

Adapted from American Burn Association. (2011a). *Advanced burn life support (ABLS) course provider manual 2011*. Chicago: Author.

estimate of the proportion of TBSA accounted for by each body part, clinicians can obtain a reliable estimate of TBSA burned. The initial evaluation is made on arrival of the patient to the hospital and should be revised within the first 72 hours, because demarcation of the wound and its depth present themselves more clearly by this time. The Lund and Browder chart is readily available in both printed and electronic sources.

Palmer Method

In patients with scattered burns, the palmer method may be used to estimate the extent of the burns. The size of the patient's hand, including the fingers, is approximately 1% of that patient's TBSA (ABA, 2011a).

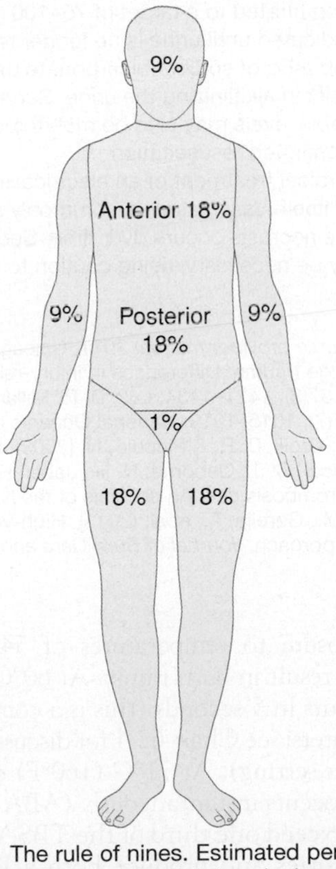
9%
Anterior 18%
9%
Posterior
18%
9%
1%
18%
18%

FIGURE 62-3 • The rule of nines. Estimated percentage of total body surface area (TBSA) in the adult is calculated by sectioning the body surface into areas with a numerical value related to nine. (*Note:* The anterior and posterior head total 9% of TBSA.)

Pathophysiology

Burn injuries are some of the worst traumatic injuries sustained because the initial injury can evolve and worsen over time. The clinical course of burns depends on the initial insult and the subsequent environmental and treatment factors (Williams, 2009). Burn injury is the result of a chemical injury or heat transfer from one site to another, causing tissue destruction through coagulation, protein denaturation, or ionization of cellular contents. The burn wound is not homogenous; rather, tissue necrosis occurs at the center of the injury with regions of tissue viability toward the periphery. The central area of the wound is termed the *zone of coagulation* due to the characteristic coagulation necrosis of cells that occurs (Fig. 62-4). The surrounding zone, the *zone of stasis,* describes an area of injured cells that may remain viable but, with persistent decreased blood flow, will undergo necrosis within 24 to 48 hours. The outermost zone, the *zone of hyperemia,* sustains minimal injury and may fully recover over time (ABA, 2011a; Evers, Bhavsar, & Mailänder, 2010).

The skin and the mucosa of the upper airways are the most common sites of tissue destruction, although deep tissues, including the viscera, can be damaged by electrical burns (Chart 62-3) or by prolonged contact with a heat or chemical source. The release of local mediators and changes in blood flow, tissue edema, and infection can cause progression of the burn injury.

Another potential mechanism of burn injury is radiation exposure. This has received increased attention in recent years due to threats of terrorism and recent world events. Radiation injuries produce two detrimental effects. The first

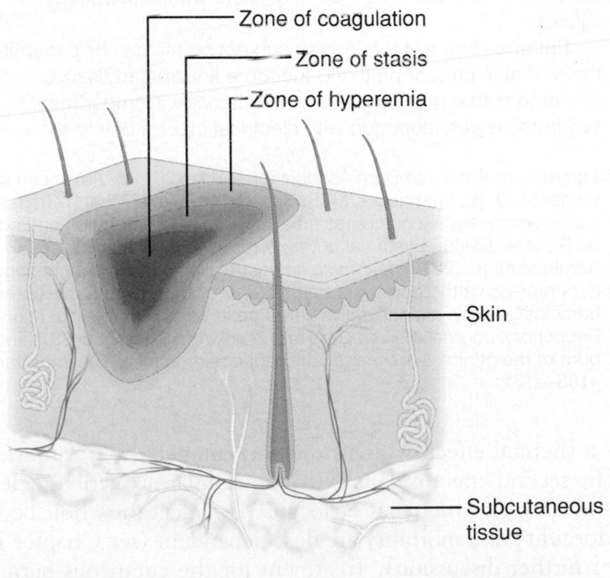
Zone of coagulation
Zone of stasis
Zone of hyperemia
Skin
Subcutaneous tissue

FIGURE 62-4 • Zones of burn injury. Each burned area has three zones of injury. The zone of coagulation (the innermost area, where cellular death occurs) sustains the most damage. The zone of stasis (the middle area) has a compromised blood supply, inflammation, and tissue injury. The zone of hyperemia (the outermost area) sustains the least damage.

Chart 62-3 Electrical Burns

Electrical injuries are some of the most devastating and complex of all burn injuries. Heat generated by electricity is directly responsible for tissue damage, but unlike most thermal burns, visual examination is not predictive of burn size and severity. It is helpful to know the circumstances of the injury to anticipate potential tissue damage and complications. Superficial injuries of an electrical burn may present themselves as contact points on physical examination. Deep tissue injuries may not be visible on initial clinical presentation but in many circumstances should be assumed as present and timely intervention initiated. Mechanisms of injury include flash, conductive, and lightning injury.

Flash Injury

An electrical flash generates light and heat without current flow and usually causes a thermal burn of exposed areas due to the heat generated, or a flame burn from ignition of clothing. Flash burns have fewer complications, and patients generally require shorter lengths of hospital stay.

Conductive Injury

Conductive electrical injuries occur when the current overcomes the skin's resistance and travels through the body. The amount and severity of damage is directly proportional to the strength of the current (voltage), duration of contact with the source, which organs lay along the pathway of current, and whether the current is direct or alternating. Direct current (DC) travels in one direction, often with an entrance and exit wound, and is associated with an explosion and likely concomitant trauma. Alternating current (AC) passes back and forth from the point of contact, through the body and back to the source many times per second and can hold the victim to it, increasing contact time. Conduction of electricity through the nerves and vessels and along the outside of bones generates heat, causing damage to adjacent tissues. Deep muscle injury may be present without injury to superficial muscles, masking the true extent of the injury. In addition, electrical current immediately contracts muscles as it travels through the body, and skeletal injuries may occur with high-voltage injuries.

Entrance and exit wounds or contact points can help identify the probable current path and therefore anticipated tissues and organs that may have sustained damage. Compartment syndrome is very common with electrical injuries due to the edema that results from injured tissue compounded by the large fluid volumes required for resuscitation to prevent kidney failure. As a result, invasive decompressive therapies such as fasciotomies, nerve releases, ocular releases, and laparotomies may be required.

Lightning Injury

Lightning injuries, the third mechanism, can result from a direct strike, a high-voltage DC injury, or a side flash wherein the current discharges from an object nearby through the air to an adjacent object or person. Side flashes are the most common and result in immediate deep polarization of the entire myocardium with possible cardiac arrest. Respiratory arrest is also common because electric current can temporarily inactivate the brain's respiratory center. In the United States, 80–100 people are killed each year and an additional 300 injured by lightning (ABA, 2011a). People struck by lightning may suffer from various long-term symptoms that can be debilitating. These include memory loss, attention deficits, sleep disorders, chronic pain, numbness, dizziness, stiffness in joints, irritability, fatigue, weakness, muscle spasms, and depression.

Management

Resuscitation fluid calculations based on total body surface area are inaccurate in conductive electrical injuries, including some lightning injuries. It is difficult to quantify the extent of tissue injury without surgical exploration, because the damage may not be visible on physical examination. Serum creatine kinase levels may be useful to assist in determining the degree of muscle injury in the early phases of care. Myoglobinuria is common with muscle damage and may cause kidney failure if not treated. IV fluid administration titrated to a target of 75–100 mL of urine output per hour is indicated until urine is no longer red. It is common practice to add 50 mEq of sodium bicarbonate per liter of IV fluid in an effort to assist in alkalinizing the urine. Serum myoglobin and urine myoglobin levels may also be monitored as indicators of the need for continued resuscitation.

Finally, the surgical treatment of an electrical injury is as complex as the injury itself. Vasculature is commonly affected, thus progressive tissue necrosis occurs over time. Sequential surgical débridement may be necessary, using caution to preserve viable tissue.

Adapted from American Burn Association (ABA). (2011a). *Advanced burn life support (ABLS) course provider manual 2011*. Chicago: Author; Lumenta, D. B., Vierhapper, M. F., Kamolz, L. P., et al. (2011). Train surfing and other high voltage trauma: Differences in injury-related mechanisms and operative outcomes after fasciotomy, amputation and soft-tissue coverage. *Burns, 37*(8), 1427–1434; Luz, D. P., Millan, L. S., Alessi, M. S., et al. (2009). Electrical burns: A retrospective analysis across a 5-year period. *Burns, 35*(7), 1015–1019; National Oceanic and Atmospheric Administration. (2013). *Lightning safety*. Available at: www.lightningsafety.noaa.gov/index.htm; Orgill, D. P., & Piccolo, N. (2009). Escharotomy and decompressive therapies in burns. *Journal of Burn Care and Research, 30*(5), 759–768; Pannucci, C. J., Osborne, N. H., Jaber, R. M., et al. (2010). Early fasciotomy in electrically injured patients as a marker for injury severity and deep vein thrombosis risk: An analysis of the National Burn Repository. *Journal of Burn Care and Research, 31*(6), 882–887; and Tiengo, C., Castagnetti, M., Garolla, A., et al. (2011). High-voltage electrical burn of the genitalia, perineum, and upper extremities: The importance of a multidisciplinary approach. *Journal of Burn Care and Research, 32*(6), e168–e171.

is a thermal effect, which results in cutaneous burn injuries. The second effect is damage to the cellular deoxyribonucleic acid (DNA), which may be localized or affect the whole body. Morbidity and mortality are dose dependent (see Chapter 73 for further discussion). Treatment for the cutaneous burn is the same as other burns discussed in this chapter. Fluid resuscitation needs may be higher due to tissue injury from radiation exposure (Milner & Feldman, 2012).

The depth of the injury depends on the temperature of the burning agent and the duration of contact with the agent. In adults, exposure to temperatures of 54°C (130°F) for 30 seconds will result in burn injury. At 60°C (140°F), tissue destruction occurs in 5 seconds (this is a common setting for home water heaters; see Chart 62-1 for discussion of appropriate water heater setting). At 71°C (160°F) or higher, a full-thickness burn occurs instantaneously (ABA, 2011a).

Burns that exceed one third of the TBSA are considered major burn injuries and produce both a local and a systemic inflammatory response (Kramer, 2012). In fact, the inflammatory response exceeds the inflammatory responses

TABLE 62-2 Pathophysiologic Changes With Severe Burns

Body System	Physiologic Changes
Cardiovascular	Cardiac depression, edema, hypovolemia
Pulmonary	Vasoconstriction, edema
Gastrointestinal	Impaired motility and absorption, vasoconstriction, loss of mucosal barrier function with bacterial translocation, increased pH
Kidney	Vasoconstriction
Other	Anemia, immunodepression

Adapted from Evers, L. H., Bhavsar, D., & Mailänder, P. (2010). The biology of burn injury. *Experimental Dermatology, 19*, 777–783.

that are seen in trauma or sepsis (Endorf & Dries, 2011). Once a burn reaches 30% of TBSA, the localized release of cytokines and other inflammatory mediators produce a systemic effect (Keck, Herndon, Kamolz, et al., 2009). Major burn injuries are characterized by burn wound edema, generalized edema formation in noninjured tissue, altered cardiovascular function, and impaired organ perfusion (Kramer, 2012).

The initial systemic event after a major burn injury is hemodynamic instability, which results from loss of capillary integrity and a subsequent shift of fluid, sodium, and protein from the intravascular space into the interstitial space, producing hypovolemic shock (Keck, Herndon, et al., 2009). Severe thermal injuries ultimately encompass changes in pathophysiology of all body systems as presented in Table 62-2. These changes precipitate conditions such as shock, sepsis, acute respiratory distress syndrome (ARDS), ileus, and kidney failure (Evers et al., 2010).

Cardiovascular Alterations

In burns greater than 45% TBSA, intrinsic contractile deficits of the myocardium occur. Even in a burn as small as 2%, a case has been reported of transient acute cardiomyopathy (Wikiel, Gemma, Yowler, et al., 2011). Upon injury, there is an immediate decrease in cardiac output that precedes the loss of plasma volume, although the exact mechanism why this occurs is unknown. It has been suggested that it is likely a neurogenic response to the release of inflammatory mediators (Keck, Herndon, et al., 2009). Due to vasoconstrictive compensatory responses, the workload of the heart increases. Cardiac output can be maintained by the body's stress response, but this may be at the cost of increasing myocardial oxygen demands (Kramer, 2012).

Hypovolemia is the immediate consequence of ensuing fluid loss and results in decreased perfusion and oxygen delivery. As fluid loss continues and vascular volume decreases, cardiac output continues to decrease and the blood pressure drops. This is the onset of burn shock. Burn shock results from the combination of direct cutaneous injury, intravascular volume loss, and systemic inflammatory mediators affecting multiple physiologic systems (Kramer, 2012). This systemic inflammation causes the release of free oxygen radicals that increase vascular permeability, causing subsequent peripheral edema. It has been proposed that the use of antioxidants during resuscitation may decrease this process of increased vascular permeability (Endorf & Dries, 2011). As a compensatory response to intravascular fluid loss, the sympathetic nervous system releases catecholamines, resulting in an increase in peripheral resistance (vasoconstriction) and an increase in pulse rate that further decreases tissue perfusion.

Prompt fluid resuscitation maintains the blood pressure in the low to normal range and improves cardiac output (see later discussion). Despite adequate fluid resuscitation, cardiac filling pressures (central venous pressure, pulmonary artery pressure, and pulmonary artery wedge pressure) remain low during the burn shock period. If adequate fluids are not administered, distributive shock occurs (see Chapter 14).

Generally, the greatest volume of fluid leak occurs in the first 24 to 36 hours after the burn, peaking by 6 to 8 hours. As the capillaries begin to regain their integrity, burn shock resolves and fluid returns to the vascular compartment. Diuresis will begin and continues for several days to 2 weeks.

At the time of burn injury, some red blood cells may be destroyed and others damaged, resulting in anemia. Despite this, the hematocrit may be elevated due to plasma loss. Abnormalities in coagulation, including a decrease in platelets (thrombocytopenia) and prolonged clotting and prothrombin times, also occur with burn injury.

Fluid and Electrolyte Alterations

Edema forms rapidly after a burn injury. A superficial burn will cause edema to form within 4 hours after injury, whereas a deeper burn will continue to form edema over a longer period of time, up to 18 hours postinjury. This is caused by increased perfusion to the injured area and reflects the amount of microvascular and lymphatic damage to the tissue. There is loss of capillary integrity, and fluid is localized to the burn itself, resulting in blister formation and edema in the area of injury. As described previously, in larger burns (20% to 30% TBSA), inflammatory mediators stimulate local and systemic reactions resulting in extensive shift of intravascular fluid into the surrounding interstitium (Kramer, 2012).

As the taut, burned tissue becomes unyielding to the edema underneath its surface, it begins to act like a tourniquet, especially if the burn is circumferential. As edema increases, pressure on small blood vessels in the distal extremities causes an obstruction of blood flow and consequent ischemia, causing compartment syndrome (see Chapter 43 for discussion of compartment syndrome and Fig. 43-5 for a picture of the Wick catheter monitoring system). Treatments may include decompression via **escharotomy,** a surgical incision into the **eschar** (devitalized tissue resulting from a burn), or **fasciotomy** (surgical opening of the full length of fascial compartments) (Kasten et al., 2011; Orgill & Piccolo, 2009) (see later discussion). (See Figs. 62-5 and 62-6.)

Reabsorption of edema begins at about 4 hours postinjury and is complete by 4 days post burn injury. However, the reabsorption depends on the depth of injury to the tissue. Although adequate fluid resuscitation is paramount to maintaining tissue perfusion, excessive fluid administration increases edema formation in both burned and nonburned tissue. A recent systematic review that addressed treatments for acute edema with burn injuries (Edgar, Hons, Fish, et al., 2011) identified bias in randomized control trials that have been published to evaluate treatments for edema, and an

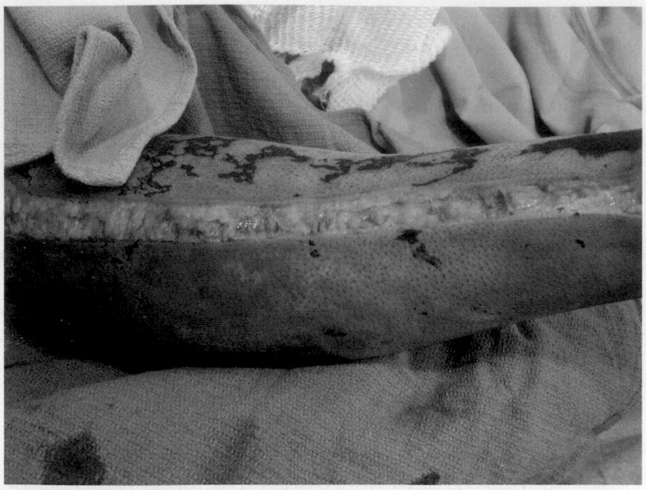

FIGURE 62-5 • Escharotomy of forearm. Used with permission from University of Texas Medical Branch, Galveston, TX.

overall paucity of research, resulting in an inability to pool the data and make strong evidence-based recommendations for treatment. Nonetheless, findings to date do support the importance of preventive interventions designed to reduce ischemia associated with edema, such as elevation of affected areas, escharotomies, and fasciotomies.

Immediately after burn injury, hyperkalemia (excessive potassium) may result from massive cell destruction. Hypokalemia (potassium depletion) may occur later with fluid shifts and inadequate potassium replacement. During burn shock, serum sodium levels vary in response to fluid resuscitation. Hyponatremia (serum sodium depletion) may be present as a result of plasma loss. Hyponatremia may also occur during the first week of the acute phase, as water shifts from the interstitial space and returns to the vascular space.

Pulmonary Alterations

Inhalation injury is a general term for injury caused by inhalation of thermal and/or chemical irritants (Woodson, 2009).

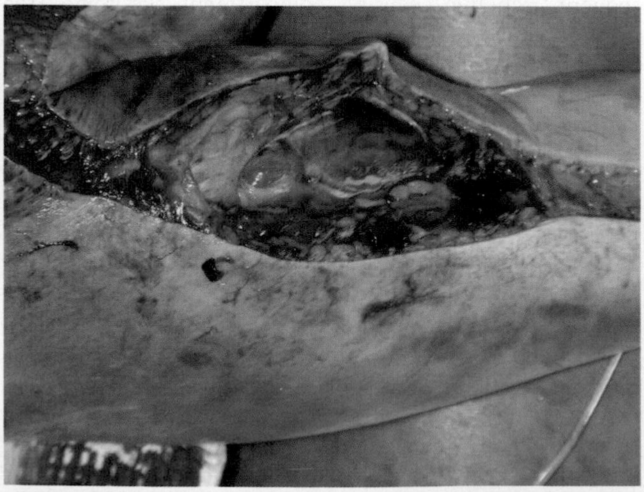

FIGURE 62-6 • Fasciotomy of upper arm. Used with permission from University of Texas Medical Branch, Galveston, TX.

Smoke (a colloid) contains airborne solid and liquid particles and gases that result when a substance undergoes combustion (Nuccio, 2011). The components of smoke that cause damage are (1) heat, (2) particulates, and (3) systemic toxins (Toon, Maybauer, Greenwood, et al., 2010). Because a common tendency in the presence of anxiety is tachypnea, victims of fire will then inhale more of these toxins. These effects were recognized in events such as the September 11, 2001, disaster in New York City (Nuccio, 2011).

Approximately 6% to 30% of patients admitted to burn centers have an inhalation injury (ABA, 2011a). The presence of inhalation injury is important to recognize. Studies of thermal burn injuries show a mortality rate of 13.9%; however, that mortality rate doubles, to 26.6%, with the presence of inhalation injury (Colohan, 2010). In one retrospective study of patients admitted to an intensive care unit with inhalation injury ($N = 105$), relative hypoxemia (a clinical finding consistent with inhalation injury), TBSA, and age were highly predictive of mortality (Hassan, Wong, Bush, et al., 2010). The extent of damage is directly related to the temperature and the concentration of toxic gases.

In smoke inhalation, the mechanisms of injury are thermal damage, asphyxiation, and irritation of the pulmonary tissues (Lafferty, 2010). It is important to note that respiratory failure may not always be in proportion to smoke exposure. This disproportion may be due to the toxin inhaled and/or patient response (Woodson, 2009). Indicators of possible inhalation injury include (1) injury occurring in an enclosed space; (2) burns of the face or neck; (3) singed nasal hair; (4) hoarseness, high-pitched voice change, stridor; (5) soot in sputum; (6) dyspnea or tachypnea and other signs of reduced oxygen levels (hypoxemia); and (7) erythema and blistering of the oral or pharyngeal mucosa (Kasten et al., 2011; Toon et al., 2010).

Inhalation injuries are categorized as upper airway injury (above the vocal cords) or lower airway injury (below the vocal cords). Injuries above the vocal cords can be thermal or chemical, whereas injuries below the vocal cords are usually chemical (ABA, 2011a).

Upper Airway Injury

Upper airway injury is the result of severe upper airway edema caused by direct thermal injury or face and neck burns, which can cause obstruction of the upper airway, including the pharynx and larynx, in the early hours post burn. Because of the cooling effect of rapid vaporization in the pulmonary tract, direct heat injury does not normally occur below the level of the glottis. However, with steam exposure, thermal injury to the lower airways is possible because the upper airway cannot effectively protect the lower airway from steam.

Lower Airway Injury

Inhalation injury below the vocal cords results from inhaling the products of incomplete combustion or noxious gases and is often the source of death at the scene of a fire. Smoke can contain more than 400 toxic compounds, including carbon monoxide, chlorine, cyanide, ammonia, aldehydes, acrolein, sulfur dioxide, phosgene, and isocyanates (Barillo, 2009; Kealey, 2009; Nuccio, 2011; Peck, 2011b). Smoke inhalation injuries cause loss of ciliary action, trigger an inflammatory response causing hypersecretion, and produce severe mucosal edema and possibly bronchospasm. Alveolar surfactant production is reduced, resulting in atelectasis (collapse of alveoli). Expectoration of carbon particles in the sputum is the cardinal sign of this injury.

Carbon monoxide poisoning is a factor in most fatalities at the scene of a fire. Carbon monoxide combines with hemoglobin to form **carboxyhemoglobin**. The affinity of hemoglobin for carbon monoxide is 200 times greater than that for oxygen, and if significant quantities of carbon monoxide are present, then tissue hypoxia will occur (ABA, 2011a). The brain and cardiovascular system are two organ systems that are particularly sensitive to hypoxic events secondary to the displacement of oxygen by carbon monoxide molecules (Cancio, 2009).

Pulmonary deterioration in severely burned patients can occur without obvious evidence of a smoke inhalation injury, and symptoms may be delayed as long as 24 to 36 hours after injury (Lafferty, 2010). Bronchoconstriction (caused by release of histamine, serotonin, and thromboxane [a powerful vasoconstrictor]) and chest constriction secondary to circumferential torso burns can contribute to this deterioration. Even without pulmonary injury, hypoxia may be present. Early in the postburn period, catecholamine release in response to the stress of the burn injury alters peripheral blood flow, thereby reducing oxygen delivery to the periphery. Later, hypermetabolism and continued catecholamine release lead to increased tissue oxygen consumption, which can lead to hypoxia. To ensure that adequate oxygen is available to the tissues, supplemental oxygen may be needed. Restrictive pulmonary excursion may occur with full-thickness burns encircling the thorax resulting in decreased tidal volume. In such situations, an escharotomy may be necessary to restore adequate chest excursion (see later discussion on escharotomy).

Late pulmonary complications secondary to inhalation injuries include mucosal sloughing of the airway, which can lead to obstruction, increased secretions, inflammation, atelectasis, airway ulceration, pulmonary edema, and tissue hypoxia.

Kidney Alterations

Kidney function may be altered as a result of decreased blood volume. Destruction of red blood cells at the injury site results in free hemoglobin in the urine. If muscle damage occurs (e.g., from electrical burns), myoglobin is released from the muscle cells and excreted by the kidneys. Adequate fluid volume replacement restores renal blood flow, increasing the glomerular filtration rate and urine volume. If there is inadequate blood flow through the kidneys, the hemoglobin and myoglobin occlude the renal tubules, resulting in acute tubular necrosis and kidney failure. Studies have indicated that as many as 25% to 50% of patients with severe burn injuries will develop acute kidney injury. Patients with acute kidney injury have significantly higher mortality than those without it (Brusselaers, Monstrey, Colpaert, et al., 2010; Palmieri, Lavrentieva, & Greenhalgh, 2010). (See Chapter 54 for discussion of kidney failure.)

Immunologic Alterations

The immunologic defenses of the body are greatly altered by a burn injury. The skin is the largest barrier to infection, and when it is compromised, the patient is continually exposed to the environment. The burn injury itself produces systemic release of cytokines and other substances that cause leukocyte and endothelial cell dysfunction. These alterations result in immunosuppression, and when combined with the required invasive procedures (line placements, surgical débridement, etc.) and continued loss of the skin barrier, there is increased risk for sepsis (Murphey, Sherwood, & Toliver-Kinsky, 2012). Burn centers provide an infection-controlled environment to protect the patient and minimize exposure to potentially harmful organisms.

Thermoregulatory Alterations

Loss of skin also results in an inability to regulate body temperature. Patients with burn injuries may therefore exhibit low body temperatures in the early hours after injury. Hypothermia on admission is associated with increases in mortality, ventilator days, length of stay, and infection rates (Emdad, Lee, Finnerty, et al., 2010). Most burn centers have heat panels at the bedside as additional heating sources to help maintain the patient's body temperature. The metabolic response of a burn injury may also cause core body temperature to be higher by as much as 2°C (3.6°F) compared to patients who do not have burn injuries (Williams, Jeschke, Chinkes, et al., 2009).

Gastrointestinal Alterations

Patients with large TBSA burns are at risk for life-threatening abdominal compartment syndrome (ACS) due to large volumes of fluid required for resuscitation, fluid shift causing edema formation, and decreased abdominal wall compliance due to eschar formation. ACS is defined as sustained intra-abdominal hypertension, which is intra-abdominal pressures greater than 30 cm H_2O (Kramer, 2012). ACS may result in multiple organ dysfunction syndrome (MODS) including impairment in hepatic, kidney, pulmonary, cardiac, and neurologic function. Early indicators include abdominal distension, oliguria, and difficulties with mechanical ventilation (Kasten et al., 2011).

Patients who are critically ill, including those with burns, are predisposed to altered gastrointestinal (GI) motility for many reasons, which may include impaired enteric nerve and smooth muscle function, inflammation, surgery, medications, and impaired tissue perfusion. Three of the most common GI alterations in burn-injured patients are paralytic ileus (absence of intestinal peristalsis), Curling's ulcer, and translocation of bacteria. Decreased peristalsis and bowel sounds are manifestations of paralytic ileus. Paralytic ileus and dilated bowel can lead to increased abdominal pressure, which can further increase ischemia (Fruhwald & Kainz, 2010; Ukleja, 2010). Gastric distention and nausea may lead to vomiting; therefore, gastric decompression is advised. Gastric bleeding secondary to massive physiologic stress may be signaled by occult blood in the stool, regurgitation of "coffee ground" material from the stomach, or bloody vomitus. These signs suggest gastric or duodenal erosion (Curling's ulcer). After a burn injury, the gastric mucosal barrier becomes permeable, allowing translocation of bacteria, which may result in severe sepsis and MODS (Nieves, Tobón, Rios, et al., 2011; Ukleja, 2010). (See Chapter 14 for further discussion.)

TABLE 62-3 Phases of Burn Care

Phase	Duration	Priorities
Emergent/resuscitative	From onset of injury to completion of fluid resuscitation	• Primary survey: A,B,C,D,E • Prevention of shock • Prevention of respiratory distress • Detection and treatment of concomitant injuries • Wound assessment and initial care
Acute/intermediate	From beginning of diuresis to near completion of wound closure	• Wound care and closure • Prevention or treatment of complications, including infection • Nutritional support
Rehabilitation	From major wound closure to return to individual's optimal level of physical and psychosocial adjustment	• Prevention and treatment of scars and contractures • Physical, occupational, and vocational rehabilitation • Functional and cosmetic reconstruction • Psychosocial counseling

Adapted from American Burn Association. (2011a). *Advanced burn life support (ABLS) course provider manual 2011*. Chicago: Author; Gallagher, J. J., Branski, L. K., Williams-Bouyer, N., et al. (2012). Treatment of infection in burns. In D. Herndon (Ed.). *Total burn care* (4th ed., pp. 137–156). Edinburgh, UK: Saunders Elsevier; Lee, J. O., Dibildox, M., Jimenez, C. J., et al. (2012). Operative wound management. In D. Herndon (Ed.). *Total burn care* (4th ed., pp. 157–172). Edinburgh, UK: Saunders Elsevier; Saffle, J. R., Graves C., & Cochran, A. (2012). Nutritional support of the burn patient. In D. Herndon (Ed.). *Total burn care* (4th ed., pp. 333–353). Edinburgh, UK: Saunders Elsevier; and Serghiou, M. A., Ott, S., Whitehead, C., et al. (2012). Comprehensive rehabilitation of the burn patient. In D. Herndon (Ed.). *Total burn care* (4th ed., pp. 517–550). Edinburgh, UK: Saunders Elsevier.

Management of Burn Injury

Burn care is typically categorized into three phases of care: emergent/resuscitative, acute/intermediate, and rehabilitation. Although priorities exist for each of the phases, assessment and management of problems and complications will overlap. Table 62-3 summarizes the priorities of care for each phase.

Emergent/Resuscitative Phase

On-the-Scene Care

The first step in management is to remove the patient from the source of injury and stop the burning process while preventing injury to the rescuer. Rescue workers' priorities include establishing an airway, supplying oxygen (100% oxygen if carbon monoxide poisoning is suspected), inserting at least one large-bore IV line, and covering the wound. Irrigation of chemical injury must begin immediately. Chart 62-4 describes the procedures and care required at the burn scene. The outward physical appearance of the person who has been burned is often distracting, but it is the internal systemic effects that pose greater threats to life.

An immediate primary survey of the patient is carried out to assess the ABCDEs: *a*irway (A) with consideration given to protecting the cervical spine, gas exchange or *b*reathing (B), *c*irculatory and cardiac status (C), *d*isability (D) including neurologic deficit, and *e*xpose and *e*xamine (E) while maintaining a warm environment (ABA, 2011a).

Quality and Safety Nursing Alert

Airway patency and breathing must be assessed during the initial minutes of emergency care. Immediate therapy is directed toward establishing an airway and administering humidified 100% oxygen. If qualified personnel and equipment are available and the victim has severe respiratory distress and/or airway edema, the rescuers must insert an endotracheal tube and initiate mechanical ventilation.

Quality and Safety Nursing Alert

No food or fluid is given by mouth, and the patient is placed in a position that will prevent aspiration of vomitus, because nausea and vomiting may occur and protection of the airway is always a priority.

The secondary survey focuses on obtaining a history, the completion of the total body system assessment, initial fluid resuscitation, and provision of psychosocial support of the conscious patient (ABA, 2011a). (See Chapter 72 for further discussion of primary and secondary surveys.)

Medical Management

Initially, the patient is transported to the nearest emergency department (ED) so that lifesaving measures can be initiated, and early referral to a burn center is made if indicated. Collaborative relationships between professionals from the prehospital emergency medical systems (EMS), EDs, and burn centers are critical for the promotion of positive outcomes for patients with burns (Zaletel, 2009).

Initial priorities in the ED remain airway, breathing, and circulation. For mild pulmonary injury, 100% humidified oxygen is administered, and the patient is encouraged to cough so that secretions can be expectorated or removed by suctioning. For more severe situations, it may be necessary to remove secretions by bronchial suctioning and to administer bronchodilators and mucolytic agents. It is essential to continuously monitor airway patency, because a previously stable airway may rapidly deteriorate as edema increases and toxic effects of smoke inhalation become apparent.

Once urgent respiratory needs are appropriately addressed, fluid resuscitation is initiated in burns greater than 20% TBSA. Fluids are necessary to support circulatory function and tissue perfusion (Zaletel, 2009). The kidney and GI tract are particularly susceptible to ischemia, organ dysfunction, and organ failure if fluid resuscitation is delayed or inadequate (Keck, Herndon, et al., 2009). Baseline weight and laboratory test results are obtained, and these parameters must be

Chart 62-4 Emergency Procedures at the Burn Scene

- **Extinguish the flames or remove from source.** When clothing catches fire, the flames can be extinguished if the person drops to the floor or ground and rolls ("stop, drop, and roll"); anything available to smother the flames, such as a blanket, rug, or coat, may be used. The older adult, or others with impaired mobility, could be instructed to "stop, sit, and pat" to prevent concomitant musculoskeletal injuries. Standing still forces the person to breathe flames and smoke, and running fans the flames. If the burn source is electrical, the electrical source must be disconnected safely before moving the patient.
- **Cool the burn.** After the flames are extinguished, the burned area and adherent clothing are soaked with *cool* water, *briefly,* to cool the wound and halt the burning process. However, *never* apply ice directly to the burn, *never* wrap the person in ice, and *never* use cold soaks or dressings for longer than several minutes; such procedures may worsen tissue damage and lead to hypothermia in people with large burns.
- **Remove restrictive objects.** If possible, remove clothing immediately. Adherent clothing may be left in place once cooled. Other clothing and all jewelry, including all piercings, should be removed to allow for assessment and to prevent constriction secondary to rapidly developing edema.
- **Cover the wound.** The burn should be covered as quickly as possible to minimize bacterial contamination, maintain body temperature, and decrease pain by preventing air currents from coming in contact with the injured surface. Any clean, dry cloth can be used as an emergency dressing. Ointments and salves should *not* be used. Other than the dressing, no medication or material should be applied to the burn wound at the scene.
- **Irrigate chemical burns.** Chemical burns resulting from contact with a corrosive material are irrigated immediately. Most chemical laboratories have a high-pressure shower for such emergencies. If such an injury occurs at home, brush off the chemical agent, remove clothes immediately, and rinse all areas of the body that have come in contact with the chemical. Rinsing can occur in the shower or any other source of continuous running water. If a chemical gets in or near the eyes, the eyes should be flushed with cool, clean water immediately. Outcomes for the patient with chemical burns are significantly improved by rapid, sustained flushing of the injury at the scene.

Adapted from American Burn Association. (2011a). *Advanced burn life support (ABLS) course provider manual 2011*. Chicago: Author.

monitored closely in the immediate postburn (resuscitation) period. Both underresuscitation and overresuscitation with fluids are associated with poor outcomes, such as shock, ischemic complications, and MODS for underresuscitation (see Chapter 14) and heart failure and pulmonary edema for overresuscitation (see Chapter 29).

In order to facilitate fluid administration, peripheral IV access may be obtained; however, in larger burns, central venous access is required (Kasten et al., 2011). TBSA is calculated and fluid resuscitation with lactated Ringers (LR) should be initiated using ABA fluid resuscitation formulas. LR is the crystalloid of choice because its composition and osmolality most closely resembles plasma and because the use of normal saline is associated with hyperchloremic acidosis (Zaletel, 2009).

The ABA (2011a) fluid resuscitation formula for adults within 24 hours post thermal or chemical burn is as follows:

$$2\text{ mL LR} \times \text{Patient's weight in kilograms} \times \%\text{TBSA 2nd-, 3rd-, and 4th-degree burns}$$

For adults with electrical burns:

$$4\text{ mL LR} \times \text{Patient's weight in kilograms} \times \%\text{TBSA 2nd-, 3rd-, and 4th-degree burns}$$

Timing is one of the most important considerations in calculating fluid needs in the first 24 hours post burn. The starting point is the time of injury—not the time of arrival to the treating facility (Williams, 2009). The infusion is regulated so that one half of the calculated volume is administered in the first 8 hours post burn injury. The second half of the calculated volume is administered over the next 16 hours. These formulas are only a guideline. It is imperative that the rate of infusion be titrated hourly as indicated by physiologic monitoring of the patient's response. For adults, a urine output of 30 to 50 mL per hour is used as an indication of appropriate resuscitation in thermal and chemical injuries, whereas in electrical injuries a urine output of 75 to 100 mL per hour is the goal (ABA, 2011a).

After adequate respiratory function and circulatory status have been established, the patient is assessed for cervical spine and/or head injuries if involved in a traumatic or electrical injury. All clothing and jewelry are removed because they may contain chemicals, retain heat, or become constrictive as edema rapidly develops. For chemical burns, flushing of the exposed areas is continued. The patient is checked for contact lenses. These are removed immediately if chemicals have come in contact with the eyes or if facial burns have occurred. In addition, the eyes are examined promptly for injury to the corneas. An ophthalmologist may be consulted for complete assessment via fluorescent staining. The patient's temperature must be monitored because hypothermia may develop rapidly and manipulation of the environment may be necessary. A temperature less than 35°C (95°F) causes vasoconstriction, which may increase tissue injury (ABA, 2011a).

It is important to validate an account of the burn scenario provided by the patient, witnesses at the scene, and first responders. Information needs to include the time and the source of the burn injury, the location (particularly if the patient was in an enclosed space), length of exposure, prior treatment at the scene, and any history of concomitant traumatic injury. A history of preexisting medical conditions, allergies, medications, and the use of drugs, alcohol, and tobacco is obtained to assist with the treatment plan.

An indwelling urinary catheter is inserted to permit accurate monitoring of urine output and as a measure of kidney function and fluid needs for patients with moderate to severe burns. If the burn exceeds 20% to 25% TBSA, a nasogastric tube is inserted and connected to low intermittent suction. All intubated patients should have a nasogastric tube inserted to decompress the stomach and prevent vomiting. Often, patients with large burns become nauseated as a result of the GI effects of the burn injury, such as paralytic ileus, and the effects of medications such as opioids.

Clean sheets are placed under and over the patient to protect the burn wound from contamination, maintain body temperature, and reduce pain caused by air currents passing over exposed nerve endings. Baseline height, weight, arterial blood gases, hematocrit, electrolyte values, blood alcohol level, drug panel, urinalysis, and chest x-rays may be obtained. Because poor tissue perfusion accompanies burn injuries, only IV analgesia is administered, which is paramount for pain relief in the emergent phase. If the patient has an electrical burn, a baseline electrocardiogram is also obtained and continuous monitoring is initiated. Because burns are contaminated wounds, tetanus prophylaxis is administered if the patient's immunization status is not current or is unknown.

Quality and Safety Nursing Alert

If necessary, a blood pressure cuff can be placed around a patient's burned extremity. The cuff must be of the correct size with accommodations made for edema.

Although the major focus of care during the emergent phase is physical stabilization, the nurse must also attend to the patient's and family's psychological needs. Anxiety accompanies burn injuries and must be addressed on an ongoing basis. Burn injury is a crisis—one that causes varying emotional responses that may result in conflicts that may cause ethical dilemmas (Chart 62-5). The patient's and family's coping abilities and available supports are considered. The nurse must consider circumstances surrounding the burn injury when providing care. Examples include cases of abuse, neglect, suicide attempts, and injury/death of other family members or friends from the same event.

Nursing Management

Nursing assessment in the emergent phase of burn injury focuses on the major priorities for any trauma patient; the burn wound is a secondary consideration to stabilization of airway, breathing, and circulation. Respiratory status is

Chart 62-5 ETHICAL DILEMMA
How Much Is Enough and What Is Comfort Care?

Case Scenario

A 71-year-old woman was flown to the nearest burn center, where you work as a staff nurse. She lives alone and was cooking on top of a gas stove. She reached over the burner and caught her robe on fire. The flame consumed most of her upper body. She sustained a full-thickness burn to 42% of her body (face, neck, both arms, and chest). In addition, she turned to remove herself from the room quickly, slipped, and fell, fracturing two ribs and spraining her right ankle. Due to the nature of the fire, she was enclosed in a burning area, so she also sustained an upper airway inhalation injury. She was awake and alert at the scene and indicated a past medical history of diabetes, hypertension, and kidney disease. She was in little pain at the time due to the depth of the burn injury; however, she was asking the team questions related to the survivability of her injury, "Am I going to die?"

Upon admission, she was intubated to treat her upper airway edema and suspected inhalation injury. Family arrived, and the burn surgeon explained the extent of injury. A decision was made to proceed with care and assess her progress post resuscitation. They were prepared for a future decision regarding withdrawal of care if survival becomes futile.

Initial resuscitation has occurred, and approximately 4 days post burn, the patient begins to show signs of kidney failure. She has had one surgical procedure for débridement and grafting to her chest, and although grafts are intact, they do not look healthy. She is showing increased signs of discomfort and pain due to the operative procedure and the addition of another wound (the new donor site). In a team meeting, the family decides that they do not want to initiate hemodialysis and therefore place the patient on comfort measures. Approximately 3 days later, she remains on comfort care and has increased signs of restlessness and pain. You are caring for the patient and request that the physician prescribe an increase in pain medication from 5 mg morphine to 10 mg IV. Within 2 hours after administration, the patient expired without discomfort. Six months later, the facility is contacted by an attorney and informed that the family is taking legal action against you and the physician for performing euthanasia on their mother.

Discussion

Clinicians on burn teams commonly face these types of ethical dilemmas. Whenever possible, the team should attempt to elicit the patient's wishes as early as possible, including any advance directives. The problem in this case is that due to absence of pain, it is often difficult for a patient to understand the severity of his or her burn injury. The patient begins to experience increased pain as care progresses, and increased medication is expected. Once the patient is unable to make decisions for herself, family members assume this responsibility. Clear and documented discussions with the family are essential in order to facilitate communication and decision making based on what is believed consistent with the patient's wishes.

Analysis

- Describe the ethical principles that are in conflict in this case (see Chart 3-3 in Chapter 3). Which principle should have preeminence as you proceeded to work with this family?
- Would you describe your role in this case as complicit with assisted suicide, active euthanasia, or providing palliation? What distinction, if any, might there be among these three acts? Do you believe that this was a survivable injury?
- Describe how the ethical principles of autonomy, beneficence, and nonmaleficence may intersect or be at odds with each other in this case (see Chart 3-3 in Chapter 3). As a member of the burn team when the patient arrived at your facility, what could you have done differently to ensure that the patient's autonomy was preserved before she was intubated?
- Note that the American Nurses Association (2001) *Code of Ethics* clearly states that nurses should not *intentionally* cause a patient's death. What were the "intentions" in this case? How might intentionality be instrumental in making this act of titrating medications that provided palliation morally defensible? Is this nursing intervention morally defensible according to the principle of double effect (see Chart 3-3 in Chapter 3)?

References

American Nurses Association. (2001). *Code of ethics for nurses with interpretive statements*. Washington, DC: American Nurses Publishing, American Nurses Foundation/American Nurses Association.

Resources

See Chapter 3, Chart 3-6 for ethics resources.

monitored closely, and pulses are evaluated, particularly in areas of circumferential burn injury to an extremity. Cardiac monitoring is indicated initially or if the patient has a history of cardiac disease, electrical injury, or respiratory conditions. Vital signs and hemodynamic status are monitored closely.

If all extremities are burned, determining blood pressure may be difficult. A clean dressing applied under the blood pressure cuff protects the wound from contamination. Because increasing edema makes blood pressure difficult to auscultate, a Doppler (ultrasound) device or a noninvasive electronic blood pressure device may be helpful. In patients with severe burns, an arterial catheter is used for blood pressure measurement and for collecting blood specimens. Peripheral pulses of burned extremities are checked frequently, either by palpation or the use of a Doppler if necessary. Elevation of burned extremities above the level of the heart is crucial to decrease edema. Large-bore IV catheters (e.g., 14 to 18 gauge) and an indwelling urinary catheter are inserted, if not already in place, and the nurse's assessment includes hourly assessment of intake and urine output.

Burgundy-colored urine suggests the presence of hemochromogens and myoglobin resulting from muscle damage. This is associated with deep burns caused by electrical injury or prolonged contact with flames. Glycosuria, a common finding in the early postburn hours, results from the release of liver glycogen stores in response to stress.

The nurse assists with calculating the patient's expected fluid requirements and monitoring the patient's response to fluid resuscitation. Nurse-driven resuscitation protocols are emerging as a mechanism to improve patient outcomes in the emergent/resuscitative phase (Endorf & Dries, 2011). Infusion pumps are used for the precise delivery of IV fluids. Nursing responsibilities include administration of fluids, strict monitoring of fluid intake and output, monitoring patient response, and notifying the treatment team of significant assessment findings and any abnormal laboratory values.

To help guide the treatment plan, the following are essential: documentation of body temperature, body weight, and preburn weight; history of allergies, tetanus immunization, past medical and surgical history, and current illnesses; and a list of current medications. A head-to-toe assessment is performed, focusing on signs and symptoms of concomitant illness, associated injury, or developing complications. Assessing the extent of the burn wound is facilitated with anatomic diagrams (described previously). In addition, the nurse works with the physician to assess the depth of the wound and areas of full- and partial-thickness injury. Psychosocial considerations of the patient and family and communication with the treatment team are important early in the course of care.

Nursing care of the patient during the emergent/resuscitative phase of burn injury is detailed in Chart 62-6.

Acute/Intermediate Phase

The acute/intermediate phase of burn care follows the emergent/resuscitative phase and begins 48 to 72 hours after the burn injury. During this phase, attention is directed toward continued assessment and maintenance of respiratory and circulatory status, fluid and electrolyte balance, and GI and kidney function. Infection prevention, burn wound care (e.g., wound cleaning and débridement, topical antibacterial therapy, application of dressings, and wound grafting), pain management, modulation of hypermetabolism, and early positioning/mobility are priorities at this stage and are discussed in detail in the following sections.

Medical Management

Pulmonary complications are common in burn injury. Airway obstruction caused by upper airway edema can take as long as 48 hours to develop. Changes detected by x-ray and arterial blood gas analysis may occur as the effects of resuscitative fluid and the chemical reactions of smoke ingredients with lung tissues become apparent. Diagnosis is largely based on history and clinical presentation, monitoring of arterial blood gases with carboxyhemoglobin levels, and direct observation of the airway by fiberoptic bronchoscopy (Woodson, 2009). In order to lessen the effects of edema, elevation of the patient's head of the bed may be helpful. Stridor and dyspnea are ominous because they are late signs of impending airway obstruction. Upper airway injury is treated by early intubation to maintain airway patency because obstruction may occur very rapidly (Nuccio, 2011; Purdue et al., 2011). However, intubation and mechanical ventilation are significant contributors to pulmonary infections. Ideally, the best practice is to remove the endotracheal tube as soon as possible so that a route for pathogens is not accessible to the lungs.

Mucosal sloughing of airway tissue can occur as late as 3 to 4 days postinjury; therefore, vigilant respiratory assessment during this period is crucial (Nuccio, 2011). Clots and casts may form and obstruct the airway, causing a life-threatening respiratory emergency (Cancio, 2009). Aerosolized heparin is commonly prescribed to inhibit airway fibrin clot formation, minimize barotrauma, and reduce pulmonary edema (Cancio, 2009; Zaletel, 2009). In one study, severe inhalation injuries were positively associated with an increased risk of ARDS and prolonged ventilator days (Mosier, Pham, Park, et al., 2012). The mortality related to ARDS is very high secondary to respiratory failure and hypoxia, MODS, or ventilator-associated pneumonia (VAP) (Palmieri, 2009). VAP is one complication of any patient who is hospitalized and mechanically ventilated and is particularly exacerbated in the patient with an inhalation injury. It affects as many as 10% to 20% of patients on mechanical ventilation longer than 48 hours. Strategies for the prevention of VAP in the patient with burns mirror those for other patients receiving mechanical ventilation (Mosier & Pham, 2009). (See Chart 21-13 for discussion of "bundled" strategies to prevent VAP and Chapter 23 for discussion of respiratory failure and ARDS.)

As capillaries regain integrity, 48 or more hours after the burn, fluid moves from the interstitial to the intravascular compartment and diuresis begins. If cardiac or kidney function is inadequate, fluid overload may occur and symptoms of congestive heart failure may result (see Chapter 29). Administration of fluids and electrolytes continues cautiously during this phase of burn care because of the

(text continues on page 1820)

Chart 62-6

PLAN OF NURSING CARE

Care of the Patient During the Emergent/Resuscitative Phase of Burn Injury

NURSING DIAGNOSIS: Impaired gas exchange related to carbon monoxide poisoning, smoke inhalation, and upper airway obstruction
GOAL: Maintenance of adequate tissue oxygenation

Nursing Interventions	Rationale	Expected Outcomes
1. Provide 100% humidified oxygen.	1. Humidification provides moisture to injured tissues; supplemental oxygen increases alveolar oxygenation.	• Absence of dyspnea • Arterial oxygen saturation >96% by pulse oximetry
2. Assess breath sounds, and respiratory rate, rhythm, depth, and symmetry of chest excursion. Monitor patient for signs of hypoxia. Report abnormalities to physician.	2. These factors provide baseline data of assessment and evidence of increasing respiratory compromise.	• Arterial blood gas levels within normal limits • Respiratory rate, pattern, and breath sounds normal
3. Observe for the following: a. Erythema or blistering of lips or buccal mucosa b. Singed nasal hairs c. Burns of face, neck, or chest d. Increasing hoarseness e. Soot in sputum or tracheal tissue in respiratory secretions	3. These signs indicate possible inhalation injury and risk of respiratory dysfunction.	
4. Monitor arterial blood gas values, pulse oximetry readings, and carboxyhemoglobin levels.	4. Increasing $PaCO_2$ and decreasing PaO_2 and O_2 saturation may indicate need for mechanical ventilation.	
5. Prepare to assist with intubation and escharotomies of chest.	5. Intubation allows airway protection and mechanical ventilation. Escharotomy enables adequate chest excursion in circumferential chest burns.	

NURSING DIAGNOSIS: Ineffective airway clearance related to edema and effects of smoke inhalation
GOAL: Maintain patent airway and adequate airway clearance

Nursing Interventions	Rationale	Expected Outcomes
1. Maintain patent airway through proper patient positioning, removal of secretions, and artificial airway if needed.	1. A patent airway is crucial to respiration.	• Patent airway • Respiratory secretions are minimal, colorless, and thin
2. Provide humidified oxygen.	2. Humidification liquefies secretions and facilitates expectoration.	
3. Encourage patient to turn, cough, and deep breathe. Encourage patient to use incentive spirometry. Suction as needed.	3. These activities promote mobilization and removal of secretions.	

NURSING DIAGNOSIS: Deficient fluid volume related to increased capillary permeability and evaporative losses from the burn wound
GOAL: Restoration of optimal fluid and electrolyte balance and perfusion of vital organs

Nursing Interventions	Rationale	Expected Outcomes
1. Monitor vital signs, hemodynamics, and urine output, as well as strict intake and output and daily weight.	1. Hypovolemia is a major risk immediately after the burn injury. Overresuscitation might cause fluid overload.	• Urine output between 0.5 and 1.0 mL/kg/h (30–50 mL/h; 75–100 mL/h if electrical burn injury)
2. Maintain IV lines and regulate fluids at appropriate rates, as prescribed.	2. Adequate fluids are necessary for perfusion of vital organs and maintenance of fluid and electrolyte balance.	• Mean arterial pressure ≥60 mm Hg • Voids clear yellow urine with specific gravity within normal limits • Serum electrolytes within normal limits
3. Observe for symptoms of deficiency or excess of serum sodium, potassium, calcium, phosphorus, and bicarbonate.	3. Rapid shifts in fluid and electrolyte status are possible in the postburn period.	
4. Elevate head of patient's bed, and elevate burned extremities.	4. Elevation promotes venous return.	
5. Notify physician immediately of decreased urine output and hemodynamic changes.	5. Rapid fluid shifts must be detected early to prevent complications.	

Chart 62-6

PLAN OF NURSING CARE

Care of the Patient During the Emergent/Resuscitative Phase of Burn Injury (continued)

NURSING DIAGNOSIS: Hypothermia related to loss of skin microcirculation and open wounds
GOAL: Maintenance of adequate body temperature

Nursing Interventions	Rationale	Expected Outcomes
1. Assess core body temperature frequently. **2.** Provide a warm environment by increasing room temperature or adjunct therapies as needed (overbed warmers, blankets, heat lamps, etc.) **3.** Work quickly when wounds must be exposed.	**1.** Frequent temperature assessments help detect hypothermia. **2.** Minimizes energy expenditure. **3.** Limiting exposure minimizes evaporative heat loss from wound.	• Body temperature remains >37°C (98.6°F) • Absence of chills or shivering

NURSING DIAGNOSIS: Acute pain related to tissue and nerve injury
GOAL: Control of pain

Nursing Interventions	Rationale	Expected Outcomes
1. Use pain intensity scale to assess pain level. Differentiate restlessness due to pain from restlessness due to hypoxia. **2.** Administer IV analgesics as prescribed and assess for effectiveness. **3.** Provide emotional support and reassurance.	**1.** Pain level provides baseline for evaluating effectiveness of pain relief measures. Hypoxia can cause similar signs and must be ruled out before analgesic medication is administered. **2.** IV administration is necessary because of altered tissue perfusion from burn injury. **3.** Fear and anxiety increase the perception of pain.	• States pain level is decreased acceptable to patient's pain goal • Absence of nonverbal cues of pain

NURSING DIAGNOSIS: Anxiety related to fear and the emotional impact of burn injury
GOAL: Minimization of patient's and family's anxiety

Nursing Interventions	Rationale	Expected Outcomes
1. Assess patient's and family's understanding of burn injury, coping skills, and family dynamics. **2.** Explain all procedures to the patient and the family in clear, simple terms. **3.** Consider administering prescribed antianxiety medications if the patient remains extremely anxious despite nonpharmacologic interventions.	**1.** Previous successful coping strategies can be fostered for use in the present crisis. Assessment allows planning of individualized interventions. **2.** Increased understanding alleviates fear of the unknown. High levels of anxiety may interfere with understanding of complex explanations. **3.** Anxiety levels during the emergent phase may exceed the patient's coping abilities.	• Patient and family verbalize understanding of emergent burn care • Patient and family's anxiety levels will be minimized

COLLABORATIVE PROBLEMS: Acute respiratory failure, distributive shock, acute kidney injury, compartment syndrome, paralytic ileus, Curling's ulcer
GOAL: Absence of complications

Nursing Interventions	Rationale	Expected Outcomes
Acute Respiratory Failure **1.** Assess for increasing dyspnea, stridor, changes in respiratory patterns. **2.** Monitor pulse oximetry, arterial blood gas values. **3.** Monitor chest x-ray results. **4.** Assess for restlessness, confusion, difficulty attending to questions, or decreasing level of consciousness. **5.** Report deteriorating respiratory status immediately to physician. **6.** Prepare to assist with intubation or escharotomies as indicated.	 **1.** Such signs reflect deteriorating respiratory status. **2.** Abnormal findings may indicate respiratory failure. **3.** X-ray may disclose pulmonary injury. **4.** Such manifestations may indicate cerebral hypoxia. **5.** Acute respiratory failure is life threatening, and immediate intervention is required. **6.** Intubation allows mechanical ventilation. Escharotomies allow adequate chest excursion with respirations.	 • Breathes spontaneously with adequate tidal volume • Arterial blood gas values within acceptable limits • Chest x-ray findings normal • Absence of cerebral signs of hypoxia

(continues on page 1820)

Chart 62-6

PLAN OF NURSING CARE

Care of the Patient During the Emergent/Resuscitative Phase of Burn Injury (continued)

Nursing Interventions	Rationale	Expected Outcomes
Distributive Shock		
1. Assess for decreasing urine output and alterations in vital signs and hemodynamics.	1. Such signs and symptoms may indicate distributive shock and inadequate intravascular volume.	• Urine output between 0.5 and 1.0 mL/kg/h (30–50 mL/h; 75–100 mL/h if electrical burn injury)
2. Assess for progressive edema as fluid shifts occur.	2. As fluid shifts into the interstitial spaces in burn shock, edema occurs and may compromise tissue perfusion.	• Blood pressure within patient's normal range • Hemodynamics remain within normal limits.
3. Adjust fluid resuscitation in collaboration with the physician in response to physiologic findings.	3. Optimal fluid resuscitation prevents distributive shock and improves patient outcomes.	• No signs or symptoms of impaired perfusion
Acute Renal Injury		
1. Monitor urine output, blood urea nitrogen (BUN), and serum creatinine levels.	1. These values reflect renal function.	• Adequate urine output • BUN and serum creatinine values remain normal.
2. Report decreased urine output or increased BUN and creatinine values to physician.	2. These laboratory values indicate possible kidney failure.	
3. Assess urine for hemoglobin or myoglobin. Administer increased fluids as prescribed.	3. Hemoglobin or myoglobin in the urine predisposes patient to increased risk of kidney failure. Fluids help to flush hemoglobin and myoglobin from renal tubules.	
Compartment Syndrome		
1. Assess peripheral pulses frequently with Doppler ultrasound device if needed.	1. Pulse assessments are crucial to assess for adequate perfusion.	• Peripheral pulses detectable • Signs of adequate peripheral perfusion
2. Assess warmth, capillary refill, sensation, and movement of extremity frequently. Compare affected with unaffected extremity.	2. These assessments may signal worsening peripheral perfusion.	
3. Remove blood pressure cuff after each reading.	3. Cuff may act as a tourniquet as extremities swell.	
4. Elevate burned extremities.	4. Elevation reduces edema formation.	
5. Report loss of pulse or sensation or presence of pain to physician immediately.	5. These signs and symptoms may indicate impaired tissue perfusion.	
6. Prepare to assist with escharotomies.	6. Escharotomies relieve the constriction caused by edema.	
Paralytic Ileus		
1. Auscultate for bowel sounds, abdominal distention.	1. Presence of bowel sounds indicate normal peristalsis. Abdominal distention reflects inadequate decompression.	• Normal bowel sounds • Absence of abdominal distention
2. Maintain nasogastric tube on low intermittent suction until bowel sounds resume.	2. This measure relieves gastric and abdominal distention.	
Curling's Ulcer		
1. Assess gastric aspirate and stools for blood.	1. Blood indicates possible gastric or duodenal ulcer bleeding.	• Gastric aspirate and stools do not contain blood.
2. Administer histamine-2 blockers and/or antacids as prescribed.	2. Such medications reduce gastric acidity and risk of ulceration.	

fluid shifts, loss of fluid from large burn wounds, and the patient's physiologic responses to the burn injury. Blood components are administered as needed to treat blood loss and anemia.

Hyperthermia is common in patients after burn shock resolves. A resetting of the core body temperature in patients who are severely burned results in a body temperature a few degrees higher than normal for several weeks after the burn. This can be compounded by body temperature increases due to bacteremia and septicemia.

Central venous, peripheral arterial, or specialty catheters may be required for monitoring hemodynamics. It is not uncommon for patients with major burns to have multiple invasive line sites due to the amount and frequency of fluid and medication that need to be administered. Whenever possible, burned areas of the body are avoided as sites for insertion of invasive lines.

One of the most important medical interventions for burn patients that has positively impacted mortality includes early surgical intervention. In surgical excision, the necrotic tissue is removed and underlying viable tissue is preserved. Early excision of dead tissue with wound closure and/or application of biologic or synthetic wound covering reduces the effects of inflammatory mediators that cause pathophysiologic abnormalities (Evers et al., 2010).

Infection Prevention

There are multifactorial reasons why patients with burns incur some of the highest risks for health care–associated infections. The first factor is the loss of the barrier function of the skin for protection against invasion of microorganisms. The second factor is that the necrotic tissue in burn eschar combined with serum proteins produces an environment conducive to bacterial growth. A third factor is that the effects of thermal injury compromise both local and systemic intrinsic immunity (Mayhall, 2012). It is therefore paramount that the nurse prioritizes infection prevention in the plan of care.

Causative agents of burn infections may include bacteria, fungi, or viruses. Common sources of potential contamination in a burn unit that require hypervigilance include **hydrotherapy** equipment (the use of bath, shower, shower cart table, or immersion to cleanse wounds), direct or indirect contamination from health care workers' hands, environmental surfaces, and translocation of microorganisms from other body systems—most notably, the GI tract. Risk factors that predispose the burn wound to infection include length of hospital stay, wound size, hyperglycemia, and increasing resistance of microorganisms to antimicrobial agents (Mayhall, 2012). Whether the burn wound is healing through spontaneous re-epithelialization or is being prepared for skin grafting, it must be protected from infection. Infection impedes burn wound healing by promoting excessive inflammation and damaging tissue. The pathologic diagnosis of burn wound infection is the presence of more than 10^5 bacteria/g of tissue (Gallagher, Branski, Williams-Bouyer, et al., 2012). Clinical signs of infection include progressive erythema, warmth, tenderness, and malodorous exudate.

A multiple-strategy approach is crucial in prevention and control of burn wound infections. Such strategies include (Mayhall, 2012):

- The use of barrier techniques (e.g., gowns, gloves, eye protection, and masks if needed)
- Environmental cleaning with periodic cultures of patient care equipment (with special attention to hydrotherapy equipment)
- Application of appropriate topical antimicrobial agents
- Appropriate use of systemic antibiotic and antifungal agents (Careful use and close monitoring of culture sensitivities are needed due to increasing challenges of antibiotic resistance in health care environments)
- Early excision and closure of the burn wound
- Control of hyperglycemia (with insulin as indicated, even in a patient without a prior diagnosis of diabetes)
- Management of the hypermetabolic response (see Chapter 14)

There are varying practices in the burn community regarding environmental and patient cultures for surveillance. In some burn centers, patients may be cultured on admission to screen for the presence of methicillin-resistant *Staphylococcus aureus* and vancomycin-resistant enterococci. Wounds are cultured on admission, with each surgical case, and for clinical suspicion of infection. Antimicrobial therapy is tailored to culture results.

A surgically excised burn wound is also at risk for infection. In one retrospective study of adults over a 5-year period with *at least* 20% TBSA burned (n = 62), 39% of patients developed a surgical wound infection. The most common pathogens identified were *Candida*, *Pseudomonas aeruginosa*, *Serratia marcescens*, *Staphylococcus aureus*, and *Aspergillus* species. In this study, a surgical wound infection that required regrafting was associated with less favorable outcomes, including increased length of hospital stay, additional surgeries, and increased area requiring autografting (Posluszny, Conrad, Halerz, et al., 2011).

Wound Cleaning

Proper management of burn wounds is required to prevent wound deterioration. The goal of wound care is débridement of nonviable tissue, removal of previously applied topical agents, and application of new topical agents (Purdue et al., 2011). Gentle cleaning with mild soap, water, and a washcloth can prevent infection by decreasing bacteria and debris on the wound surface. Hair in and around the burn area, except the eyebrows, should be clipped short or shaved. In addition, most burn experts recommend débridement of blisters larger than 0.5 cm to reduce risk of bacterial invasion (Kasten et al., 2011).

Various measures are used to clean the burn wound. Patients who are hemodynamically unstable may have their wounds washed at the bedside, whereas ambulatory patients may shower. The wounds of nonambulatory patients can be cleansed using shower carts—mobile stretchers made with removable sides, drainage holes, and positioning capabilities. Retractable shower hoses suspended from walls and ceilings provide the nurse with easy access to a water source for washing the wounds. Whatever method is used, the goal is to protect the wound from overwhelming proliferation of pathogenic organisms and invasion of deeper tissues until either spontaneous healing or skin grafting can be achieved. Strategies for the prevention of cross-contamination include the use of plastic liners, water filters, and thorough decontamination of equipment.

Patient comfort and ability to participate in the prescribed treatment are important considerations. During bathing, patient participation is encouraged to promote exercise of the extremities. At the time of wound cleaning, all skin is inspected for any signs of redness, breakdown, or local infection. This also provides the nurse an opportunity for patient education.

During treatment, the patient is continuously assessed for signs of hypothermia. The temperature of the water is maintained at 37.8°C (100°F), and the temperature of the room should be maintained between 26.6°C and 29.4°C (80°F and 85°F) to prevent hypothermia. Other assessment considerations include patient fatigue, changes in hemodynamic status, and pain unrelieved by analgesic medications or relaxation techniques.

Topical Antibacterial Therapy

Variations in topical wound care for nonsurgical burn wounds exist among burn centers across the country, and choices are made based on the individualized needs of each patient. The goal of topical therapy is to provide a dressing with the following characteristics:

- Is effective against gram-positive and gram-negative organisms and fungi
- Penetrates the eschar but is not systemically toxic
- Is cost-effective, available, and acceptable to the patient
- Is easy to apply and remove, and decreases the frequency of dressing changes, decreases pain, and minimizes nursing time

No single topical medication is universally effective, and the use of different agents at different times in the postburn period may be necessary.

Concept Mastery Alert

Prudent use and alternation of antimicrobial agents can result in reduction of resistant strains of bacteria, greater effectiveness of the agents, and a decreased risk of sepsis. Table 62-4 describes selected topical antimicrobial agents.

Wound Dressing

After the ordered topical agents are applied, the wound is covered with several layers of dressings. A lighter dressing is used over joints to allow for mobility. Dressings may also need to be modified to accommodate splints or other positioning devices. Circumferential dressings should always be applied distally to proximally in order to promote return of excess fluid to the central circulation. If the hand or foot is burned, the fingers and toes should be wrapped individually to promote function while healing.

Burns to the face may be left open to air once they have been cleaned and the topical agent has been applied. Careful attention is required to ensure that the topical agent does not come in contact with the eyes or mouth. A light dressing can be applied to the face to absorb excess exudates if needed.

Occlusive dressings, gauze, and a topical antimicrobial agent may be used over areas with new skin grafts to protect the new graft and promote an optimal condition for its adherence to the recipient site. Ideally, these surgical dressings remain in place for 3 to 5 days, at which time they are removed for examination of the graft. When occlusive dressings are applied, precautions are taken to prevent two body surfaces from touching, such as fingers or toes, ear and scalp, the areas under the breasts, any point of flexion, or between the genital folds. Functional body alignment positions are maintained by using splints or by regular repositioning of the patient.

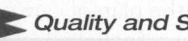

Quality and Safety Nursing Alert

Dressings can impede circulation if they are wrapped too tightly. The peripheral pulses must be checked frequently and burned extremities elevated. If the patient's pulse is diminished, this is a critical situation and must be addressed immediately.

TABLE 62-4 Overview of Selected Topical Antimicrobial Agents Used for Burn Wounds

Agent	Indication/Comment	Application	Nursing Implications
General			
Antimicrobial ointment	Antibacterial coverage and promotion of a moist wound environment	Apply 1/16-inch layer of ointment with a clean glove daily.	Ensure removal of old ointment at the time of wound cleaning prior to applying a new layer. Monitor closely for signs and symptoms of local infection.
Specific Agents			
Silver sulfadiazine 1% (Silvadene) water-soluble cream	Bactericidal agent for many gram-positive and gram-negative organisms, as well as yeast and *Candida albicans* Minimal penetration of eschar	Apply 1/16-inch layer of cream with a clean glove 1–3 times daily.	Anticipate formation of pseudo-eschar (proteinaceous gel), which can be removed.
Mafenide acetate 5%–10% (Sulfamylon) hydrophilic-based solution or cream	Antimicrobial agent for gram-positive and gram-negative organisms Diffuses through eschar and avascular tissue (e.g., cartilage)	Apply twice a day as prescribed.	Is a strong carbonic anhydrase inhibitor and may cause metabolic acidosis Application may cause considerable pain initially.
Silver nitrate 0.5% aqueous solution	Effective against most strains of *Staphylococcus* and *Pseudomonas* and many gram-negative organisms. Does not penetrate eschar	Apply solution to gauze dressing and place over wound. Keep the dressing wet but covered with dry gauze and dry blankets to decrease vaporization.	Monitor serum sodium (Na^+) and potassium (K^+) levels, and replace as prescribed. Silver nitrate solution is hypotonic and acts as a wick for sodium and potassium. Protect bed linens and clothing from contact with silver nitrate, which stains everything it touches.
Silver-impregnated dressings (sheets or mesh)	Broad antimicrobial effects (product specific) Delivers a uniform, antimicrobial concentration of silver ions to the burn wound.	Apply directly to wound. Cover with absorbent secondary dressing if needed.	May produce a pseudo-eschar from silver after application Can be left in place for several days (product specific)

Adapted from Micromedex 2.0. (2012). *Formulary Advisor, University of Texas–Medical Branch Galveston.* Available pass-protected at: www.micromedicsolutions.com/micromedex2/librarian

Dressings that adhere to the wound bed may be removed more comfortably and with less damage to healing tissue by moistening the dressing with water. The patient may participate in removing the dressings, providing some degree of control over this painful procedure. The wounds are then cleaned and débrided to remove any remaining topical agent, exudate, and nonviable tissue. Sterile scissors and forceps may be used to trim loose eschar and encourage separation of devitalized tissue. During this procedure, the wound and surrounding skin are carefully inspected. Documentation should include color, odor, size, exudate, signs of re-epithelialization, any changes from the previous dressing change, and other key characteristics.

Wound Débridement

The goals of **débridement** (the removal of devitalized tissue) are:

- Removal of devitalized tissue or burn eschar in preparation for grafting and wound healing
- Removal of tissue contaminated by bacteria and foreign bodies, thereby protecting the patient from invasion of bacteria

There are four types of débridement—natural, mechanical, chemical and surgical.

Natural Débridement

With natural débridement, the devitalized tissue separates from the underlying viable tissue spontaneously. This process may take weeks to months to occur. Bacteria present at the interface of the burned tissue and the viable tissue gradually liquefy the fibrils of **collagen** (a protein present in skin, tendon, bone, cartilage, and connective tissue) that hold the eschar in place. Proteolytic and other natural enzymes cause this phenomenon.

Mechanical Débridement

Mechanical débridement involves the use of surgical tools to separate and remove the eschar. This technique can be performed by skilled physicians, nurses, or physical therapists and is usually done with daily dressing changes. If bleeding occurs, hemostatic agents or pressure can be used to stop the bleeding from small vessels. Dressing changes and wound cleaning aid the removal of wound debris. Wet-to-dry dressings are not advocated in burn care because of the chance of removing viable cells along with necrotic tissue.

Chemical Débridement

Topical enzymatic agents are available to promote débridement of burn wounds. Because such agents usually do not have antimicrobial properties, they may be used together with topical antibacterial therapy to protect the patient from bacterial invasion. Heavy metals such as silver deactivate the débriding agent; therefore, caution is necessary to ensure that the topical antimicrobial agent does not interfere with the chemical débridement. Alternating topicals with dressing changes may be considered.

Surgical Débridement

Early surgical excision to remove devitalized tissue along with early burn wound closure is now recognized as one of the most important factors contributing to survival in a patient with a major burn injury. Surgical débridement is carried out before the natural separation of eschar is allowed to occur. The operative procedure involves either excision of the full thickness of the skin down to the fascia (tangential excision) or shaving of the burned skin layers gradually down to freely bleeding, viable tissue (Leon-Villapalos, Eldardiri, & Dziewulski, 2010). This may be performed as soon as possible after the burn, once the patient is hemodynamically stable and edema has decreased. Ideally, the wound is then covered immediately with a skin graft (if necessary) and a dressing. If the wound bed is not ready for a skin graft at the time of excision, a temporary biologic or synthetic dressing may be used until a skin graft can be applied during a subsequent surgery.

The use of surgical excision carries with it risks and complications, especially with large burns. The procedure creates a high risk of extensive blood loss and lengthy operating and anesthesia times. Furthermore, blood losses sustained during surgical procedures, wound care, and ongoing hemolysis exacerbate anemia. Blood transfusions may be required periodically to maintain adequate hemoglobin levels for oxygen delivery to the myocardium. (See Chapter 32 for discussion of blood component therapy.)

When conducted in a timely and efficient manner, surgical excision results in shorter lengths of hospital stay and possibly a decreased risk of complications from invasive burn wound sepsis. Once débrided, granulation tissue fills the space created by the wound, creates a barrier to bacteria, and serves as a bed for epithelial cell growth. A wound covering is then applied to keep the wound bed moist and promote the granulation process.

Wound Grafting

The patient with deep partial- or full-thickness burns may be a candidate for skin grafting to decrease the risk of infection; prevent further loss of protein, fluid, and electrolytes through the wound; and minimize evaporative heat loss. Special attention is warranted when grafting the face (for cosmetic, functional, and psychological reasons); functional areas, such as the hands and feet; and areas that involve joints. Grafting permits earlier functional ability and reduces wound **contractures** (shrinkage of burn scar through collagen maturation). When burns are very extensive, the order in which areas are grafted is chosen based on the ability to achieve wound closure as soon as possible; therefore, the chest and abdomen or back may be grafted first to reduce the total body burn surface area.

Autografts

Autografting remains the preferred autologous method for definitive burn wound closure after excision. **Autografts** are the ideal means of covering burn wounds because the grafts are the patient's own skin and therefore are not rejected by the patient's immune system. They can be split-thickness, full-thickness, or epithelial grafts. Because the **donor site** (the area from which skin is taken to provide the autograft) from a full-thickness graft includes both the epidermis and dermis, its use must be cautiously considered because it cannot heal spontaneously.

Split-thickness autografts are more commonly used and can be applied in sheets (Fig. 62-7), or they can be expanded by meshing so that they cover more than a given donor site

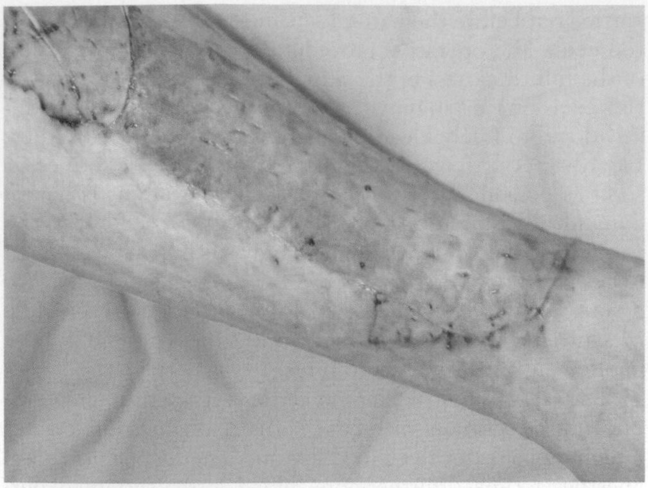

FIGURE 62-7 • Split-thickness sheet graft. Used with permission from University of Texas Medical Branch, Galveston, TX.

area (Fig. 62-8). Skin meshers enable the surgeon to cut tiny slits into a sheet of donor skin, making it possible to expand to cover larger areas with smaller amounts of donor skin. These expanded grafts adhere to the recipient site more easily than sheet grafts and prevent the accumulation of blood, serum, air, or purulent material under the graft. However, any kind of graft other than a sheet graft contributes to scar formation as it heals. The use of expanded grafts may be necessary in large wounds but should be viewed as a compromise in terms of cosmesis.

If blood, serum, air, fat, or necrotic tissue are present between the recipient site and the graft, there may be partial or total loss of the graft. Infection, mishandling of the graft, or trauma during dressing changes account for most other instances of graft loss. The use of split-thickness grafts allows the remaining donor site to retain sweat glands and hair follicles and minimizes donor site healing time.

Cultured epidermal autograft (CEA), an epithelial graft, has emerged as an important procedure in the management of massive burns. In burns that cover more than 90% TBSA, CEA may be the only option because the availability of nonburned skin as donor sites will not be sufficient for grafting. CEA involves obtaining full-thickness biopsies of the patient's unburned skin that are cultured to promote growth. The final product is available approximately 3 weeks later for grafting (Fig. 62-9). Meticulous attention is required in CEA applied to certain body surfaces because it is fragile and prone to graft loss. The use of CEA necessitates prolonged length of hospital stay and reconstructive surgeries (Lee, Dibildox, Jimenez, et al., 2012).

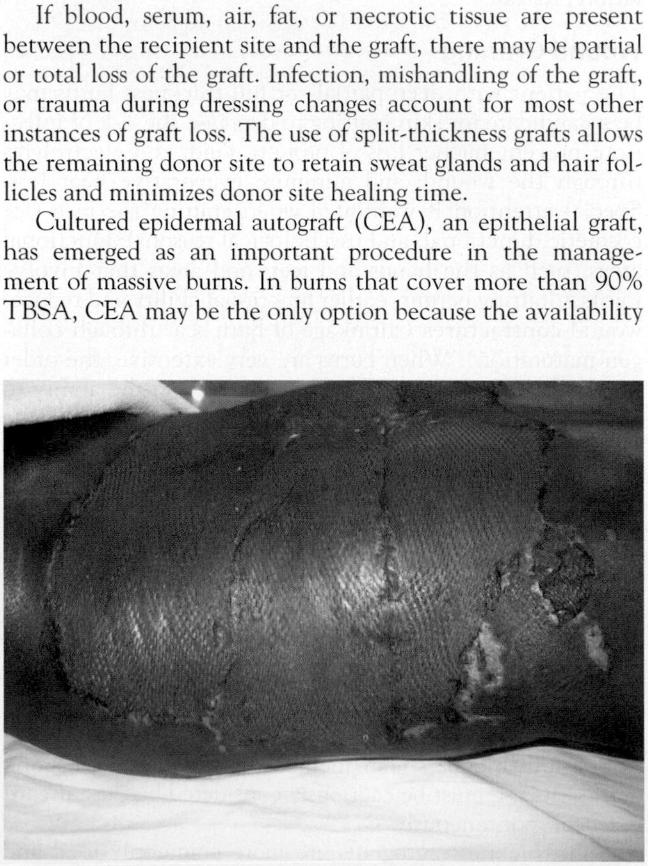

FIGURE 62-8 • Split-thickness meshed graft. Used with permission from University of Texas Medical Branch, Galveston, TX.

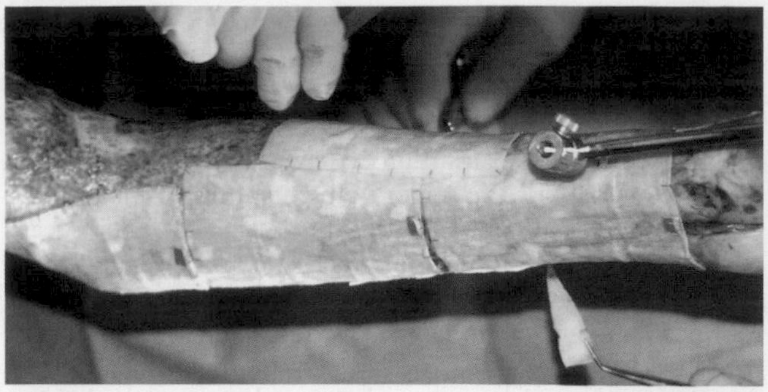

FIGURE 62-9 • Application of cultured epithelial autologous grafts. Used with permission from University of Texas Medical Branch, Galveston, TX.

Care of the Graft Site. Protection is the key goal of caring for skin grafts postoperatively. Occlusive dressings are commonly used initially after grafting to immobilize the graft. Occupational therapists may construct splints to immobilize newly grafted areas. Homograft, xenograft, or synthetic dressings (discussed later) may also be used to protect grafts.

The first dressing change is usually performed 2 to 5 days after surgery, or earlier in the case of clinical signs of infection or bleeding. Infection, bleeding beneath the graft, and shearing force are the most common reasons for graft loss in the early postoperative period. The patient is positioned and turned carefully to avoid disturbing the graft or putting pressure on the graft site. If an extremity has been grafted, it is elevated to minimize edema. The patient may begin exercising the grafted area 5 to 7 days after surgery. This may vary with individual burn center's protocols.

Care of the Donor Site. The donor site is a clean wound created in a surgical environment. After the donor skin is excised, a thrombostatic agent such as thrombin or epinephrine may be applied directly to the site to decrease bleeding. Multiple dressings are available to cover donor sites. Donor sites must remain clean, dry, and free from pressure. Because a donor site is usually a partial-thickness wound, it is very painful and an additional potential site of infection. With proper care, the donor site should heal spontaneously within 7 to 14 days.

Homografts and Xenografts

Homografts (or allografts) and xenografts (or heterografts) are also referred to as biologic dressings and are intended to be temporary wound coverage. **Homografts** are skin obtained from recently deceased or living humans other than the patient. **Xenografts** consist of skin taken from animals (usually pigs). Therefore, the body's immune response will eventually reject them as a foreign substance.

Biologic dressings have several uses. In extensive burns, they provide temporary wound coverage and protection of granulation tissue until autografting is possible. They can also be used as a test graft in preparation for the patient's own skin graft to determine if the bed will accept the graft. Once the biologic dressing appears to be "taking," or adhering to the granulating surface with minimal underlying exudation, the patient is ready for an autologous skin graft. As a temporary dressing, they also decrease the wound's evaporative water and protein loss, provide an effective barrier for entry of bacteria, and decrease pain by protecting nerve endings. Biologic materials can be left open or covered. They stay in place for varying lengths of time but are removed in instances of infection or rejection. Another advantage is that fewer dressing changes may be required.

Homografts tend to be the most expensive biologic dressings. They are available from skin banks in fresh and cryopreserved (frozen) forms. Homografts are thought to provide the best infection control of all biologic or biosynthetic dressings available. Revascularization occurs within 48 hours, and the graft may be left in place for several weeks.

Pigskin is available from commercial suppliers. It is available fresh, frozen, or lyophilized (freeze-dried) for longer shelf life. Pigskin is used for temporary covering of clean wounds such as superficial partial-thickness wounds and donor sites. Although pigskin does not vascularize, it adheres to clean superficial wounds, providing pain control and reducing evaporative fluid loss while the underlying wound re-epithelializes (Lee et al., 2012).

When grafting is not possible, skin substitutes have been created that surgically replace the epidermis and the dermis either temporarily or permanently. Each provides advantages and disadvantages that are important to consider in product selection. In addition, many of these dressings are institution specific and are likely utilized based on preferences, expertise of the treatment team, and commercial availability (Lou & Hickerson, 2009).

Biosynthetic and Synthetic Dressings

Problems with availability, sterility, and cost have prompted the search for biosynthetic and synthetic dressings, which may eventually replace biologic dressings as temporary wound coverings. One widely used synthetic dressing is Biobrane (UDL Laboratories, 2012), a dual-layer dressing of nylon and silicone. The material is porous, semitransparent, and sterile. It has an indefinite shelf life and is less costly than homograft or xenograft such as pigskin. Like biologic dressings, Biobrane protects the wound from fluid loss and bacterial invasion. It can remain in place until spontaneous re-epithelialization and wound healing occur. It can also be laid on top of a wide-meshed autograft to protect the wound until the autograft epithelium grows out to close the interstices. As the Biobrane gradually separates, it is trimmed away, leaving a healed wound (Fig. 62-10).

Pain Management

A burn injury is considered one of the most painful types of trauma that a patient can experience. The nature of the injury may expose nerve endings, and the patient may require multiple débridements, surgeries, and treatments. Moving, changing position, and receiving occupational and physical therapy cause additional discomfort. The most recent practice guidelines for pain management from the ABA (Faucher & Furukawa, 2006) emphasize an organized approach to pain management that addresses background, procedural, and breakthrough pain; the goal is for patients to be comfortable and alert, and pain needs to be differentiated from anxiety.

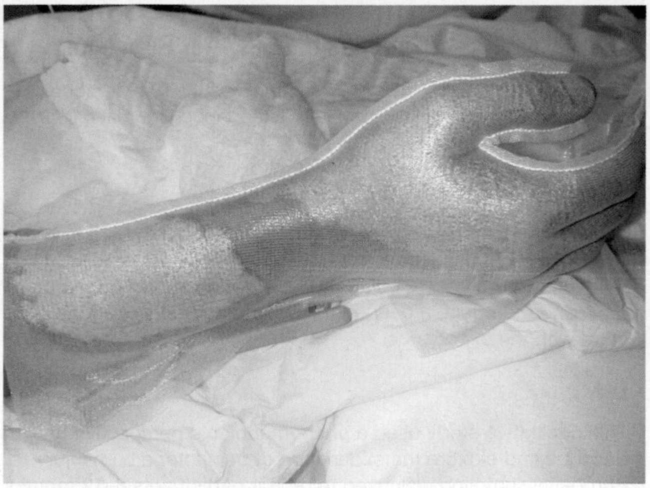

FIGURE 62-10 • Biobrane® applied to hand and forearm. Used with permission from University of Texas Medical Branch, Galveston, TX.

Background pain is a continuous level of discomfort even when the patient is inactive or not undergoing any procedures. The goal of treatment is to provide a long-acting analgesic agent that will provide even coverage for this long-term discomfort. It is helpful to use escalating doses when initiating the medication to reach the level of pain control that is acceptable to the patient. The use of patient-controlled analgesia gives control to the patient and achieves this goal. Breakthrough pain is described as acute, intense, and episodic. It is generally related to an activity or movement of the affected area. Short-acting agents are used for breakthrough pain to achieve pain control if needed. Procedural pain is discomfort that occurs with procedures such as daily wound treatments, invasive line insertions, and physical and occupational therapy. The goal is to plan proper analgesia that will place the patient in a state of comfort throughout the procedure.

Most severe burns are a combination of partial- and full-thickness burns that influence the amount of pain the patient experiences. Superficial and deep partial-thickness burns are very painful because the nerve endings are exposed, resulting in excruciating pain with exposure to temperature, pressure, air currents, and movement. In a full-thickness burn, the nerve endings are destroyed, and upon admission there is numbness and decreased sensation to that area. Thus, severe injuries are often underestimated by the patient because the pain at that time is minimal. Yet, memories of the pain patients experience may persist for a long time (Chart 62-7). Educating patients and their families about burn pain and its relationship to the depth of injury as well as the pain management plan is an important priority for the nurse.

Pharmacologic treatment for the management of burn pain includes the use of opioids, nonsteroidal anti-inflammatory

Chart 62-7
NURSING RESEARCH PROFILE
Memories of Pain After Burn Injury

Tengvall, O., Wickman, M., & Wengström, Y. (2010). Memories of pain after burn injury—the patient's experience. *Journal of Burn Care and Research*, *31*(2), 319–327.

Purpose

Pain is the most frequent complaint of patients with burn injuries. The purpose of this study was to describe the experiences of patients with burns, including their memories of pain during burn care, and to acquire a deeper understanding of how patients cope with the experience.

Design

This qualitative study used a phenomenologic perspective to describe and explore the experience of pain after a burn injury. Purposive sampling was used to recruit participants 6–12 months post burn injury. The researchers recruited participants of both genders (men = 8, women = 4), across a wide age range (19–72 years of age, with a mean age of 49.6 years), who had experienced different types of burns (flames = 7, scalds = 2, contact = 3). Length of hospital stay ranged from 8–45 days (mean 16 days), and burn size ranged from 3%–40% total body surface area (mean 10.6%). The interviews lasted approximately 60–90 minutes, were audiotaped, and were transcribed verbatim by one of the researchers.

Findings

Four themes emerged for participants' pain experience: (1) becoming aware of pain at the scene of the accident, (2) allowing oneself to feel pain on arrival to the hospital, (3) different pain experiences on the burn unit (this included two subthemes: procedural pain and background pain), and (4) fragile body surface after discharge.

Nursing Implications

The experience of surviving a severe burn injury is life altering. The pain associated with a burn injury is perceived as overwhelming, and in fact, one participant described procedural pain as being "torture." It is important to note that none of these patients expected to be pain free. Participants expressed a strong need to talk about their experiences and share their stories. Structured programs that encourage sharing of experiences may be invaluable in this population.

drugs, anxiolytics, and anesthetic agents. These and other pain management strategies are discussed in Chapter 12. Treatment of anxiety with benzodiazepines can be used along with opioids. The use of anesthetics in a nonoperative setting (i.e., moderate sedation) requires administration by qualified personnel. Recent advances include the use of agents with rapid onset and short duration, which have been very effective in pain control during a planned procedure. In the provision of holistic patient care, the use of nonpharmacologic pain and anxiety interventions must not be overlooked. Nonpharmacologic therapies include relaxation techniques, distraction, guided imagery, hypnosis, therapeutic touch, humor, music therapy, and more recently virtual reality techniques.

Modulation of Hypermetabolism

Burn injuries produce profound metabolic abnormalities fueled by the exaggerated stress response to the injury. The body's response has been classified as hyperdynamic, hypermetabolic, and hypercatabolic. Hypermetabolism can affect morbidity and mortality by increasing the risk of infection and slowing the healing rate. Persistent hypermetabolism may last up to 1 year after burn injury and well past the acute hospitalization (Williams et al., 2009).

In addition to the promotion of burn wound healing, early enteral nutrition is advocated to ameliorate the hypermetabolic response and prevent ileus and stress ulcerations (Mosier, Pham, Klein, et al., 2011). In one prospective multicenter study, administration of enteral nutrition within 24 hours of admission was associated with decreased wound infection and length of stay in the intensive care unit (Mosier et al., 2011). Nutrition should be provided upon arrival to the burn center and may require placement of a feeding tube. Patients who are critically ill may even have their feedings continued intraoperatively if the airway is protected. Effective nutrition management depends on delivery of appropriate amounts of micronutrients, carbohydrates, lipids, and protein (Purdue et al., 2011). Parenteral nutrition is associated with many complications and should only be provided to patients with intolerance to enteral feeding or prolonged ileus (Williams et al., 2009).

Several formulas exist for estimating the daily metabolic expenditure and caloric requirements of patients with burn injuries. A commonly used formula is the Harris-Benedict equation, or requirements may be measured by using indirect calorimetry to measure resting energy expenditure (REE) (Purdue et al., 2011). Once calculated, the caloric needs of the patient with burns can be met by providing 1.2 to 1.4 times the REE (Williams et al., 2009). Carbohydrates are the most important energy source for these patients to ensure proper wound healing. Fat, although a required nutrient, should be provided in more limited quantities. In addition, recommendations from recent literature advocate protein requirements of 1.5 to 2 g/kg/day (Saffle, Graves, & Cochran, 2012). When the oral route is used, high-protein, high-calorie meals and supplements are given. Dietary consultations are useful in helping patients meet their nutritional needs. Daily calorie counts aid in assessing the adequacy of nutritional intake.

A comprehensive review of strategies aimed at modulating the hypermetabolic response in patients with burns (Williams et al., 2009) discussed several other important interventions beyond providing nutritional support. Early excision and grafting of the burn wound is one of the most important factors in ameliorating hypermetabolism by removing eschar, thereby lessening the effects of inflammatory mediators. Appropriate manipulation of environmental temperatures may decrease energy expenditure by the patient. Insulin therapy in patients with burns is needed to treat the hyperglycemia that occurs from accelerated gluconeogenesis and is beneficial in muscle protein synthesis. Oxandrolone, an anabolic steroid, is commonly administered to patients with burns because it improves protein synthesis and metabolism. Administration of propranolol (Inderal) (a beta-blocker) not only decreases heart rate but also blocks harmful catecholamine effects. And lastly, exercise is described as an essential adjunct in the plan of care for patients with burns, because it attenuates the hypermetabolic response (in addition to the numerous other benefits of exercise). Most patients with large burns will require many of these strategies, because combination therapy is superior to any individual strategy in modulating the negative effects of hypermetabolism (Williams et al., 2009).

Nursing Management

Nursing management of the patient in the acute/intermediate phase is focused on the following priorities: restoring fluid balance, preventing infection, modulating hypermetabolism, promoting skin integrity, relieving pain and discomfort, promoting mobility, strengthening coping strategies, supporting patient and family processes, and monitoring and managing complications.

Restoring Normal Fluid Balance

To reduce the risk of fluid overload and consequent heart failure and pulmonary edema, daily weights and careful calculation of intake and output measurement are utilized to guide therapy. Changes in physical assessment and hemodynamic indicators are also useful in evaluating the patient's response to treatment.

Preventing Infection

The patient with burns is at risk for infection from multiple sources, which may include open wounds, lung injury, and indwelling catheters. Therapeutic interventions are complex due to the patient's altered physiology. Signs of infection including increased temperature, tachycardia, tachypnea, and leukocytosis may be present in the patient with burns without underlying infection, making infections more difficult to recognize. Furthermore, treatment with antipyretic agents (e.g., acetaminophen [Tylenol]) is not indicated (Kasten et al., 2011).

A major part of the nurse's role during the acute phase of burn care is detection and prevention of infection. The nurse is responsible for providing a clean environment, including the promotion of required isolation procedures. The nurse protects the patient from sources of contamination, including other patients, staff members, visitors, and equipment. Patients can inadvertently promote migration of microorganisms from one burned area to another by touching their wounds or dressings. Bed linens can also spread infection through either colonization with wound microorganisms or fecal contamination. Regular bathing of unburned areas and changing of linens can help prevent infection. Fresh flowers, plants, and fresh fruit baskets are not permitted in the patient's room because of the risk of microorganism growth. Invasive lines and tubing must be routinely changed according to CDC recommendations, and prompt removal of all indwelling catheters when no longer indicated provides the best prevention of health care–associated infections (Kasten et al., 2011).

Modulating Hypermetabolism

The nurse collaborates with the dietitian or nutrition support team to develop a plan that meets the needs of the patient. Family members may be encouraged to bring nutritious and favorite foods to the hospital. High-calorie nutritional supplements may be necessary. Nutritional intake must be accurately documented. Vitamin and mineral supplements may be prescribed.

If caloric goals cannot be met by oral feeding, a feeding tube is inserted and used for continuous or bolus feedings of specific formulas. The volume of residual gastric secretions should be checked to ensure absorption. The patient should be weighed each day and the results tracked to properly assess appropriate weight parameters and as an indicator of adequate nutrition administration.

Promoting Skin Integrity

Wound care is usually the single most time-consuming element of burn care after the emergent phase. The physician prescribes the desired topical antibacterial agents and specific biologic, biosynthetic, or synthetic wound coverings and plans for surgical excision and grafting. The nurse needs to make astute assessments of wound status, use creative approaches to wound dressing, and support the patient during the emotionally distressing and very painful experience of wound care.

Assessment of the burn wound requires an experienced eye, hand, and sense of smell. Important wound assessment features include size, color, odor, presence of eschar and exudate, epithelial buds (small pearl-like clusters of cells on the wound surface), bleeding, granulation tissue, the status of graft take, healing of the donor site, and the condition of the surrounding skin. Any significant changes in the wound are reported to the physician because they may indicate burn wound infection and require immediate intervention.

The nurse also assists the patient and family by providing instruction, support, and encouragement to take an active part in dressing changes and wound care when appropriate. Discharge planning needs for wound care must be anticipated early in the course of burn management, and the strengths of the patient and family are assessed and used in preparing for the patient's eventual discharge and home care needs.

Relieving Pain and Discomfort

Pain management continues to be a priority during the acute phase of burn recovery. Frequent assessment of pain is required, and analgesic and anxiolytic medications are administered as prescribed. To increase its effectiveness, analgesic medication is provided before the pain becomes severe. Nonpharmacologic interventions can be used to alter the patient's perceptions of and responses to pain. Frequent reassessment of responses to interventions, whether pharmacologic or nonpharmacologic, is essential.

Postburn pruritus (itching) affects almost all patients with burns and is one of the most distressing symptoms in the postburn period. Oral antipruritic agents, environmental conditions, frequent lubrication of the skin with water or silica-based lotion, and diversional activities all help to promote comfort in this phase. Recent research has demonstrated promising results with the use of gabapentin (Neurontin), an antiepileptic agent used to address neuropathic pain, with reported benefit in the treatment of itching (Ahuja, Gupta, Gupta, et al., 2011). The instructions "pat, don't scratch" must be reinforced with patients in order to prevent further discomfort and complications.

Lack of sleep and rest interfere with healing, comfort, and restoration of energy. If necessary, sleep aids are prescribed on a regular basis in addition to analgesic and anxiolytic agents.

Promoting Physical Mobility

An early priority is to prevent complications of immobility. Deep breathing, turning, and proper positioning are essential nursing practices that prevent atelectasis and pneumonia, control edema, and prevent pressure ulcers and contractures. Specialty beds may be useful, and early mobility is encouraged. If the lower extremities are burned, elastic pressure bandages should be applied before the patient is placed in an upright position to promote venous return and minimize edema formation. Prevention of venous thromboembolism (VTE) is an important factor in care. The most recently published practice guidelines for deep vein thrombosis (DVT) prophylaxis in patients with burns (Faucher & Conlon, 2007) report a 1% to 23% incidence of DVT in burn patients. Patients with burn injuries are at high risk because of their hypercoagulability, loss of vascular integrity, immobility, multiple invasive lines, and need for other operative procedures. In one small retrospective case control study, in-hospital risk factors for VTE formation included pneumonia, number of operations, and the presence of a central venous line (Pannucci, Osborne, Park, et al., 2012).

The burn wound is in a dynamic state for at least 1 year after wound closure. During this time, aggressive efforts must be made to prevent contracture and hypertrophic scarring. Both passive and active range-of-motion exercises are initiated from the day of admission and are continued after grafting within prescribed limitations. Splints or functional devices may be applied to the extremities for contracture control. The nurse monitors the splinted areas for signs of vascular insufficiency, nerve compression, and skin breakdown. Occupational and physical therapists are consulted to develop a patient-specific plan of care throughout hospitalization and recovery.

Strengthening Coping Strategies

Much of the patient's energy goes into maintaining vital physical functions and wound healing in the early postburn weeks, leaving little emotional energy for coping. In the acute phase of burn care, the patient is facing the reality of the burn injury. Grief, depression, anger, regression, and manipulative behavior are common responses of patients who have burn injuries. Withdrawal from participation in required treatments and regression must be viewed with an understanding that such behavior may help the patient cope with an enormously stressful event.

The patient may experience feelings of anger. At times, the anger may be directed inward because of a sense of guilt, perhaps for causing the fire or even for surviving when loved ones perished. The anger may also be directed outward toward those who escaped unharmed or those who are now providing care. One way to help the patient handle these emotions is to enlist someone to whom the patient can express feelings without fear of retaliation. A nurse, social worker, psychiatric liaison nurse, peer supporter, spiritual advisor, or counselor who is not involved in direct care activities may fill this role successfully.

Patients with burn injuries are very dependent on health care team members during the long period of treatment and recovery. However, even when physically unable to contribute much to self-care, they should be included in decisions regarding care and encouraged to assert their individuality in terms of preferences and recognition of their unique identities. As the patient improves in mobility and strength, the nurse works with the patient to set realistic expectations for self-care and planning for the future. Many patients respond positively to the use of contractual agreements and other strategies that recognize their independence, set expectations for behavior, and encourage positive communication. Application of Orem's self-care nursing model has been reported as a useful framework for delivery of burn nursing care throughout the continuum (Wilson & Gramling, 2009).

Supporting Patient and Family Processes

The burn injury has tremendous psychological, economic, and life-altering impact on the patient and family. The nurse is instrumental in providing support to the patient and family as they adapt to the burn injury. Referrals for social services or psychological counseling should be made as appropriate. This support continues into the rehabilitation phase. Some burn centers offer a structured peer support program in which a burn survivor visits the patient while hospitalized to provide support. Many patients appreciate the opportunity to be able to share their experience with another burn survivor.

Patients who experience major burns are commonly sent to burn centers far from home. Because burn injuries are sudden and unexpected, family roles are disrupted. If the primary provider in the family is injured, roles may change, which adds more stress to the family. Therefore, both the patient and the family need thorough information about the patient's burn care and expected course of treatment. Barriers to learning and preferred learning styles are assessed and considered. This information is used to tailor education activities. Patient and family education is a priority and is provided with a multimedia approach.

Monitoring and Managing Potential Complications

Acute Respiratory Failure and Acute Respiratory Distress Syndrome

The patient's respiratory status is monitored closely for increased difficulty in breathing, change in respiratory pattern, or onset of adventitious (abnormal) breath sounds. Typically, at this stage, signs and symptoms of injury to the respiratory tract become apparent. As described previously, signs of hypoxia (decreased oxygen to the tissues), decreased breath sounds, wheezing, tachypnea, stridor, and sputum tinged with soot (or in some cases containing sloughed tracheal tissue) are among the many possible findings. Medical management of the patient with acute respiratory failure requires intubation and mechanical ventilation (if not already in use). If ARDS has developed, higher oxygen levels, positive end-expiratory pressure, and pressure support are used with mechanical ventilation to promote gas exchange across the alveolar–capillary membrane (see Chapter 21).

TABLE 62-5 Complications in Rehabilitation Phase of Burn Care

Complications	Contributing Factors	Interventions
Neuropathies and nerve entrapment	Electrical injury, large deep burns, improper positioning, edema, scar tissue	Assess peripheral pulses and sensation (neurovascular checks). Prevent edema and pressure by elevation, positioning, and prevention of constricting dressings. Assess splints for proper fit and application. Consult OT and PT departments for positioning.
Wound breakdown and/or pressure ulcer formation	Shearing, pressure, inadequate nutrition	Protect wound from pressure and shearing forces. Educate patient about importance of good nutrition.
Hypertrophic scarring	Partial- and full-thickness burns	Keep skin pliable and soft by using emollients. Apply pressure garments as prescribed. Massage
Contractures	Partial- and full-thickness burns	Maintain position of joints in alignment. Perform gentle range-of-motion exercises. Consult OT and PT departments for exercises and positioning recommendations.
Joint instability	Burn wound, burn scar, and contractures	Maintain appropriate joint positioning through appropriate application of splints. Monitor joint pinning if indicated.
Complex pain	Trauma and burns	Provide adequate pain management. Consult OT and PT departments for exercises and desensitization. Promote gentle motion of affected extremities.

OT, occupational therapy; PT, physical therapy.

Heart Failure and Pulmonary Edema

If the cardiac and renal systems cannot compensate for the excess vascular volume as fluid shifts back to the intravascular space, heart failure and pulmonary edema may result. The patient is assessed for signs of heart failure, including decreased cardiac output, oliguria, jugular vein distention, edema, and the onset of an S_3 or S_4 heart sound. If invasive hemodynamic monitoring is used, increasing central venous, pulmonary artery, and pulmonary artery wedge pressures indicate increased fluid volume.

Crackles in the lungs and increased difficulty with respiration may indicate a fluid buildup in the lungs, which is reported promptly to the physician. In the meantime, the patient is positioned comfortably, with the head of the bed raised (if not contraindicated because of other treatments or injuries) to promote lung expansion and gas exchange. Management of this complication includes providing supplemental oxygen, administering IV diuretic agents, carefully assessing the patient's response, and providing vasoactive medications, if indicated (see Chapter 29 for further discussion).

Sepsis

Sepsis is a leading cause of morbidity and mortality in patients with burn injuries. The signs of early sepsis are subtle and require a high index of suspicion and very close monitoring of changes in the patient's status. One of the challenges in the recognition of sepsis is the lack of definition of sepsis in the burn population. Because patients with burns are hypermetabolic, this results in tachycardia, tachypnea, and elevated body temperature. These physiologic norms in patients with burns make the diagnosis of sepsis more challenging (Palmieri, 2009). (See Chapter 14 for treatment recommendations for sepsis.)

Rehabilitation Phase

Rehabilitation begins immediately after the burn has occurred and often extends for years after injury. For nurses who care for patients with burns, this can be one of the more physically demanding and challenging phases. One important focus of the burn team is to evaluate the patient carefully for late complications related to burn injuries as described in Table 62-5. There continue to be areas of rehabilitation best practices that are not yet defined; this represents an aspect of burn care that is ripe for research (Richard, Dewey, Forbes-Duchart, et al., 2009).

Burn rehabilitation is comprehensive and complex and requires a multidisciplinary approach to optimize the patient's physical and psychosocial recovery related to the injury. As patients begin to recover, they become more aware of the injuries and the challenges they face. Individualized plans of care that are specific to the severity and location of injury are developed and reevaluated frequently. The increased survival of patients with significant burn injuries has translated into the need for additional and comprehensive burn rehabilitation programs nationwide. The ultimate goal is to return patients to the highest level of function possible within the context of their injuries. Occupational and physical therapists are essential to optimize patient goals and outcomes.

Psychological Support

A patient's outlook, motivation, and support system are important to his or her overall well-being and ability to progress through the rehabilitation phase. First, psychiatric disorders may have contributed to the cause of the burn injury itself. Examples include self-inflicted burns and suicide attempts. In addition, burn injuries may be intentionally inflicted on one individual by another in cases of abuse and other situations involving psychiatric disorders. These are a few examples that illustrate the critical need for psychosocial resources in burn recovery. Although psychiatric disorders are not contributory to all burn injuries, the life-altering nature of burn injuries almost always causes temporary or permanent impairment of psychosocial adaptation.

In the acute phase of the injury, overwhelming shock, terror, disbelief, confusion, and anxiety are common. Patients may be at risk for delirium and may experience temporary psychoses. Patients may be confused from medications they are taking, but they have an underlying sense of fear, anxiety, and pain. Early consultation with mental health professionals will assist to best meet individual needs, which may include pharmaceutical interventions with concurrent counseling (Rosenberg, Lawrence, Rosenberg, et al., 2012).

The rehabilitation phase may present a new set of challenges. While the patient is physically recovering, the reality and impact of the injury begin to set in as patients recognize that survival is expected. Patients may experience devastating grief and loss. The sense of loss may originate from physical injury, loss of control from the forced dependency on others for care, or loss of family members/friends who may have perished in the accident. In residential fires, survivors may have lost their homes and all of their possessions. Emotional responses may include frustration, anger, depression, hopelessness, emotional lability, and helplessness (Rosenberg et al., 2012).

Posttraumatic stress disorder (PTSD) is the most common psychiatric disorder in burn survivors, with a prevalence that may be as high as 45% (Low, Meyer, Willebrand, et al., 2012). Other psychological disorders that may be experienced by patients with burns include anxiety, depression, and sleep disturbances (Low et al., 2012). The symptoms and psychological responses to traumatic events, including PTSD, are discussed further in Chapter 6.

As recovery progresses, discharge planning must include strategies to assist the patient in reintegrating into his or her home, community, workplace, and school. For many patients, issues regarding quality of life may become very real at this point in recovery. This is an emotional time as the patient and family begin to live with new physical limitations and challenges in relationships. In addition to preparing the patient's support system, the patient and family must also prepare for the reactions from strangers. Role-playing may be a helpful tool to assist the patient and family with what to say and how to respond to the reactions of others (Rosenberg et al., 2012).

Organizations such as the Phoenix Society for Burn Survivors, an international burn survivor support group, offer a myriad resources, education, opportunities for peer support, and strategies for reintegration. The Phoenix Society was founded in 1977 by Alan Breslau, a burn survivor, who recognized the importance of peer support in the psychosocial recovery of burn survivors (see the Resources section). Each year, the Phoenix Society hosts World Burn Congress, which is a conference for survivors, their families, caregivers, burn care professionals, and firefighters. This forum offers education as well as an opportunity for burn survivors and their families to connect with others who have experienced similar life-changing events. Interaction with other burn survivors allows the patient and family to see that adaptation to a burn injury is possible.

Organizations that provide support for reintegration are able to offer education and training geared specifically to patients with burn injuries. Workshops on application of makeup to reduce the appearance of scars can benefit those with obvious facial scarring. This is just one example of strategies available to assist burn survivors with body image disturbances. Cultural influences play a strong factor in this process, because some cultures are particularly sensitive to physical appearance, which is reinforced by current media. The role of the nurse is to encourage patients to voice their concerns, provide empathy, and provide them with resources.

Abnormal Wound Healing

Partial-thickness wounds involving the epidermis and superficial dermis tend to heal without scarring. Deeper wounds will likely develop scarring of variable degrees. As with other conditions, risk factors may be stratified as modifiable or nonmodifiable. For example, a nonmodifiable risk factor for scarring is heredity, because some patients are simply more prone to hypertrophic scar formation. The focus of patient education for nurses must be on how to best change or adapt risk factors that are modifiable. Patients should be strongly encouraged to follow occupational therapist recommendations for scar prevention and management.

Normal scarring occurs in a superficial tissue injury and begins forming within 7 to 10 days postinjury and progresses over the next 6 to 12 months. Abnormal scarring occurs after a longer period of wound healing and may form either hypertrophic or keloid scars.

Hypertrophic and Keloid Scars

Hypertrophic scars form within the boundaries of the initial wound and push outward on the perimeter of the wound. They are common in areas over joints and in the younger population. The scar becomes red (because of its hypervascular nature), raised, and hard.

Keloid scars are irregularly formed and extend beyond the margins of the original wound. They are large, nodular, and ropelike, often causing itching and tenderness. They are more common in dark-pigmented skin, uncommon in children and older adults, and have familial tendencies.

Prevention and Treatment of Scars

Preventive treatment modalities are used to prevent scar contractures and excess hypertrophic tissue. Compression is introduced early in burn wound treatment. Elastic bandage wraps are used initially to help promote adequate circulation, but they can also be used as the first form of compression for scar management, followed by elasticized tubular bandages until the patient can be measured for a customized garment. Application of elastic pressure garments loosens collagen bundles and encourages parallel orientation of the collagen to the skin surface. As pressure continues over time, collagen restructures and vascularity decreases. Although this therapy is somewhat controversial, pressure has shown to be beneficial in controlling scar formation over time. Garments are worn continuously (i.e., 23 hours/day).

Many areas of the body are difficult to compress due to the contours or location of the injury. Inserts, such as silicone sheets, are helpful for these small troublesome areas and are placed beneath the garment or compression dressing to enhance scar compression. Gentle superficial scar massage can be performed with a moisturizer several times a day.

Burn reconstruction is a treatment option after scars have matured and is discussed within the first few years after injury. This decision requires individualized planning, realistic expectations, and patience. The treatment team and the patient will ultimately decide on the best approach for long-term functionality and cosmesis.

NURSING PROCESS

Care of the Patient During the Rehabilitation Phase

Assessment

The nurse obtains information about the patient's education level, occupation, leisure activities, cultural background, religion, and family interactions. The patient's self-concept, mental status, emotional response to the injury and hospitalization, level of intellectual functioning, previous hospitalizations, response to pain and pain relief measures, and sleep pattern are also essential components of a comprehensive assessment. Information about the patient's general self-concept, self-esteem, and coping strategies in the past are valuable in addressing emotional needs.

Ongoing physical assessments related to rehabilitation goals include range of motion of affected joints, functional abilities in activities of daily living, early signs of skin breakdown, evidence of neuropathies (neurologic damage), activity tolerance, and quality or condition of healing skin. The patient's participation in care and ability to demonstrate self-care in such areas as ambulation, eating, wound cleaning, toileting, and applying pressure wraps are documented on a regular basis.

Diagnosis

Nursing Diagnoses

Based on the assessment data, nursing diagnoses may include the following:

- Activity intolerance related to pain with exercise, limited joint mobility, muscle wasting, and limited endurance
- Disturbed body image related to altered physical appearance and self-concept
- Impaired physical mobility due to contractures or hypertrophic scarring
- Deficient knowledge about postdischarge home care and recovery needs

Collaborative Problems/Potential Complications

Potential complications may include the following:

- Inadequate psychological adaptation to burn injury

Planning and Goals

The major goals for the patient include increased mobility and participation in activities of daily living; adaptation and adjustment to alterations in body image, self-concept, and lifestyle; increased understanding and knowledge of the injury, treatment, and planned follow-up care; and absence of complications.

Nursing Interventions

Promoting Activity Tolerance

The extensive physical rehabilitation can be painful and overwhelming for the patient. Strategies to maintain motivation and participation may be beneficial during this critical phase. The nurse incorporates physical therapy exercises in the patient's care to prevent muscle atrophy and to maintain the mobility required for daily activities. The patient's activity tolerance, strength, and endurance gradually increase if activity occurs over increasingly longer periods. Fatigue and pain tolerance are monitored and used to determine the amount of activity to be encouraged on a daily basis. Activities such as family visits and recreational or play therapy (e.g., video games, radio, television) can provide diversion, improve the patient's outlook, and increase tolerance for physical activity. In older adult patients and those with chronic illnesses and disabilities, rehabilitation must take into account preexisting functional abilities and limitations.

The nurse must schedule care in such a way that the patient has periods of rest and uninterrupted sleep. A good time for planned patient rest is after the stress of dressing changes and exercise, while pain interventions and sedatives are still effective. This plan must be communicated to family members and other care providers. The patient may have insomnia related to frequent nightmares about the burn injury or other fears and anxieties about the outcome of the injury. The nurse reassures the patient and administers agents, as prescribed, to promote sleep. Reducing metabolic stress by relieving pain, preventing chilling or fever, and promoting the physical integrity of all body systems help the patient conserve energy for therapeutic activities and wound healing.

Improving Body Image and Self-Concept

Patients who have survived burn injuries may lack the benefit of anticipatory grief often seen in a patient who is approaching surgery or dealing with the terminal illness of a loved one. As care progresses, the patient who is recovering from burns becomes aware of daily improvement and begins to exhibit basic concerns: Will I be disfigured or be disabled? How long will I be in the hospital? What about my job and family? Will I ever be independent again? How will this injury affect my sexual relationships? How can I pay for my care? Was my burn the result of my carelessness? Where will I live now?

When caring for a patient with a burn injury, the nurse needs to be aware that there are prejudices and misunderstandings in society about those who are viewed as different. Opportunities and accommodations available to others are often denied to those who are disfigured. These include social participation, employment, prestige, various roles, and status. The health care team must actively promote a healthy body image and self-concept in patients with burn injuries so that they can accept or challenge others' perceptions of those who are disfigured or disabled. Survivors themselves must show others who they are, how they function, and how they want to be treated.

Promoting Physical Mobility Through Preventing Contractures or Hypertrophic Scar Formation

With early and aggressive physical and occupational therapy, contractures or hypertrophic scars are rarely a long-term complication. However, surgical intervention is indicated if full range of motion in the burn patient is not achieved. (See Chapter 10 for a discussion of prevention of contractures and deformities.)

Monitoring and Managing Potential Complications

Impaired Psychological Adaptation to the Burn Injury. Some patients, particularly those with limited coping skills or psychological function or a history of psychiatric problems before the burn injury, may not achieve adequate psychological adaptation to the burn injury. Psychological counseling or psychiatric referral may be made to assess the patient's

Chart 62-8 HOME CARE CHECKLIST

The Patient With a Burn Injury

At the completion of home care education, the patient or caregiver will be able to:	PATIENT	CAREGIVER
Psychosocial Adaptation and Social Reintegration		
• Remember that changes in lifestyle and emotional adjustment to injury take time. It is not uncommon for patients to report nightmares or "flashbacks" of the accident. If this becomes disruptive, please discuss with your treatment team.	✔	✔
• Resume previous interests and activities gradually.	✔	
• Consider community and other resources such as burn support groups. Many books, videos, and Web sites are also available and may be invaluable in assisting with burn recovery.	✔	✔
• Modifications may be needed to the home care environment. If so, the social worker or care manager can provide instruction if needed.	✔	✔
• Programs are available to assist with school reintegration and return to work. Consult your treatment team if needed.	✔	✔
Burn Skin Precautions		
• Wear sunblock with the highest sun protection factor (SPF) possible to protect exposed burned skin from the sun. Light-colored clothing, long pants, and long-sleeved shirts may also be necessary to protect from the sun.	✔	
• Wear wide-brimmed hats if face or ears have been burned to protect the area from the sun.	✔	
• Avoid further trauma to burned skin; leave blisters that may form intact.	✔	
• Lubricate healed burned skin with mild lotion (as prescribed); avoid scratching.	✔	✔
• Use only mild soap and lotion (i.e., products without perfume) on burned areas. Keeping skin clean overall is important to support good hygiene and prevent infection.	✔	✔
• Avoid tight clothing over burned areas so as not to restrict movement or irritate newly healed areas.	✔	
• Select white cotton, loose-fitting clothing so that dyes in colored clothes do not irritate healing skin.	✔	
• Wear clothing and gloves to protect healing skin from unnecessary bruises, bumps, and scratches.	✔	✔
• Be aware that tolerance to extremes in temperatures may be affected.	✔	✔
Wound Care		
• Take prescribed pain medications if needed 30 minutes prior to wound care to achieve maximum effectiveness.	✔	
• Use mild soap, water, and a clean washcloth to clean wounds.	✔	✔
• Apply prescribed topical medications and dressings as instructed.	✔	✔
• Inspect wounds carefully with each dressing change for signs of infection, including increased redness, swelling, drainage, or foul odor.	✔	✔
Activities of Daily Living and Exercise		
• Perform as much of your own care as possible.	✔	
• Adhere to the exercise regimen given by the therapist.	✔	
• Although physical rehabilitation is tiring, daily participation is essential.	✔	
• Be sure to plan for adequate rest periods.	✔	
• When at rest, swollen limbs should be elevated.	✔	✔
Nutrition		
• Eat a healthy, nutritious diet. Consider nutrient rich foods, rather than empty calories.	✔	
• Drink adequate volume of fluids to prevent constipation associated with the use of analgesic medications.	✔	
Pain Management		
• Take analgesic medication as prescribed. Be aware of the side effects of your medications.	✔	
• Use nonmedication techniques such as meditation and distraction as much as possible.	✔	

Chart 62-8 HOME CARE CHECKLIST

The Patient With a Burn Injury (continued)

At the completion of home care education, the patient or caregiver will be able to:	PATIENT	CAREGIVER
Itching		
• Itching is a normal part of healing. Many patients describe this as one of the most uncomfortable aspects of burn recovery.	✔	
• Apply mild moisturizers to decrease itching from dryness. Medications can be discussed with your treatment team.	✔	✔
• Pat—do not scratch!	✔	
Management of Burn Scar		
• Massage with mild lotion or cream to stretch skin to maintain/increase its elasticity.	✔	✔
• Wear compression garments 23 hours a day if instructed.	✔	
• Expect discoloration of the skin for many months. This is a normal part of healing.	✔	✔
Medical Devices		
• Follow your occupational therapist's instructions for splint use and cleaning.	✔	✔
• Use your crutches, walkers, or other assistive devices as instructed.	✔	
• If devices such as shower chairs or grab bars are needed in the home, this should be arranged prior to discharge.	✔	✔
Resumption of Sexual Relations		
• Realize that resumption of sexual relationships is the rule rather than the exception. Resume at a pace that is comfortable for you.	✔	✔
• Expect sensitivity of and around the genital area for several months if these areas were burned.	✔	
Follow-Up Appointments		
• Keep a list of questions to ask the treatment team during your follow-up visit.	✔	✔
• Keeping scheduled appointments is critical to recovery.	✔	✔
• Bring your medications, or a list of current medications, to each visit for the team to review.	✔	✔
Financial Considerations		
• Burn care (dressings, medications, therapy) can be very expensive. Work with social worker or care manager to find funding assistance programs if needed.	✔	✔

emotional status, to help the patient develop coping skills, and to intervene if major psychological issues or ineffective coping is identified.

Promoting Home and Community-Based Care

Educating the Patient About Self-Care. The focus of rehabilitative interventions is directed toward outpatient care, home care, or care in a rehabilitation center. This includes wound care, dressing changes, pain management, nutrition, prevention of complications, and other care needs. Information and written instructions are provided about specific exercises and the use of pressure garments and splints. The patient and family are provided education to assist them in their continued care needs after discharge (see Chart 62-8).

Continuing Care. Follow-up care after discharge by a multidisciplinary treatment team is necessary. Some patients may require the services of an inpatient rehabilitation center before returning home. Patients should receive follow-up from a burn center when possible for periodic evaluation by the burn team, modification of outpatient treatment plan, and evaluation for reconstructive surgery. Many patients require outpatient physical or occupational therapy several times a week. The nurse is responsible for coordinating all aspects of care and ensuring that the patient's needs are met. Such coordination is an important aspect of assisting the patient to achieve independence.

Some patients may require home care referrals after discharge. The home care nurse assesses the patient's physical and psychological status as well as the adequacy of the home setting for safe and adequate care. The nurse monitors the patient's progress; assesses adherence to the plan of care; and assists the patient and family with wound care, exercises, and other physical needs. Patients experiencing difficulties with psychosocial adjustments are identified and appropriate referrals are made.

Evaluation

Expected patient outcomes may include:

1. Demonstrates adequate activity tolerance
   a. Has energy available to perform daily activities
   b. Shows gradual increased tolerance and endurance in physical activities
   c. Obtains adequate sleep and rest daily

2. Adapts to altered body image
   a. Verbalizes accurate description of alterations in body image and accepts physical appearance
   b. Demonstrates interest in resources that may improve function and perception of body appearance (examples include cosmetics, wigs, and prostheses as appropriate)
   c. Socializes with significant others, peers, and usual social group
   d. Seeks and achieves return to role in family, school, and community as a contributing member
3. Demonstrates physical mobility adequate to perform activities of daily living
   a. Demonstrates range of motion appropriate for injury
   b. Absence of complications from wound healing
4. Demonstrates knowledge of required self-care and follow-up care
   a. Verbalizes detailed plan for follow-up care
   b. Demonstrates ability to perform or direct wound care and prescribed exercises
   c. Returns for follow-up appointments as scheduled
   d. Identifies resource people and agencies to contact for specific problems
5. Exhibits no complications
   a. Exhibits psychosocial adaptation to burn injury
   b. Verbalizes understanding of diagnosis and treatment plan

Outpatient Burn Care

The increased availability of outpatient surgery and access to expert burn care in outpatient settings make this option possible for the treatment of minor burns as well as follow-up for the discharged patient with burns. The goals for treatment in an outpatient setting may include burn wound management, pain management, scar and reconstructive care, and rehabilitation. However, a number of factors must be considered when determining if outpatient care is appropriate for the patient: age, past medical history, extent and depth of the burn, location of the burn wounds, availability of family support systems and community resources, the patient's ability and willingness to adhere to a therapeutic regimen, distance from home, and availability of transportation from home to the outpatient setting.

The frequency of follow-up visits is individualized. The initial outpatient visit for a discharged patient with burn injuries is usually scheduled within 2 to 3 days after hospital discharge. Follow-up appointments will vary and over time decrease in frequency, based on the patient's needs. Patient and family education is very important and should include verbal and written instructions as well as return demonstration of the wound or scar care required. The importance of notifying the outpatient setting about early complications and of keeping follow-up appointments is emphasized to the patient and family. Physical therapy and occupational therapy are often provided in the outpatient burn setting. The rehabilitation goals are to increase range of motion and to strengthen and build the patient's endurance. This is accomplished with a specific plan of care and includes routine visits for up to 2 years following the injury. Adaptation to lifestyle changes and emotional status should be assessed during the outpatient visits and proper referrals made for counseling services. These assessments may be difficult to recognize due to the infrequent nature of the visits; therefore, it is helpful to incorporate family response and interactions into the assessment. The health care team must also be alert to issues of substance abuse, safety concerns, suicidal thoughts, depression, and PTSD.

Critical Thinking Exercises

1 pq A 34-year-old man working on a ship is injured when a pressurized steam pipe bursts. He sustains third-degree burns over 85% of his body and fractures to his right femur and pelvis. Inclement weather has delayed transportation to a burn center. What should be the priorities of the first responders on the scene? On arrival to the treating facility, the patient's temperature is 35.2°C (95.3°F), and his weight is 58 kg (128 lb). How should the treatment team at the burn center prioritize this patient's care?

2 ebp A 54-year-old man is deep-frying a turkey for Thanksgiving while enjoying some cocktails with friends. In the process of removing the turkey from the deep fryer, the pot overturns. The patient tries to grab it and sustains grease burns to his right forearm, his chest, and the anterior surface of both legs. He is transported by EMS to the hospital for care. Using the rule of nines, what is his estimated TBSA? Based on best evidence, what type of fluid should be used for resuscitation and why? What other information would you need to calculate his fluid requirements in the emergent/resuscitative phase?

3 ebp A 19-year-old woman is at a party when a burning candle is tipped into the drapes, causing a fire. She is knocked to the floor and requires the assistance of firefighters to finally extract her from the smoke-filled building. She does not sustain obvious cutaneous burns. Upon interview and examination, she is slightly tachypneic and exhibits hoarseness when speaking. You note singed nasal hair and lips. What is your immediate concern? What treatment should you expect the physician to prescribe first? Based on current best evidence, what tests or procedures would most likely be prescribed to assist in the diagnosis and management of this patient?

4 A 52-year-old man and his 24-year-old son are rewiring a historical home. When connecting to the power source, both sustain conductive electrical injuries. The son displays obvious contact points to his right hand and right foot. The father has no evidence of contact points but had been thrown across the room. When first responders arrive, he has no pulse or respirations, and advanced life support measures are initiated. Both men are transported to the ED. The father is pronounced dead after a prolonged, unsuccessful resuscitation attempt. What are the challenges associated with calculation of the son's fluid resuscitation needs? Does this patient meet the ABA criteria for referral to a burn center? What psychological and emotional needs should you anticipate that he will require immediately? What resources within the hospital may be beneficial to assist with these needs?

5 A 64-year-old woman is rescued from her burning home, where it has been reported that she was smoking in bed. The patient has a history of pulmonary fibrosis, congestive heart failure, and type 2 diabetes. She has been on home oxygen for several years. She presents with burn injuries to her face, ears, and neck. Sixteen months ago, she was treated at the burn center as an outpatient for right hand and forearm burns sustained while cooking. What assessment parameters must be closely monitored on this patient? What important factors need to be addressed as part of the discharge plan?

Brunner Suite Resources

Explore these additional resources to enhance learning for this chapter:

- **NCLEX-Style Questions and Other Resources on thePoint, http://thePoint.lww.com/Brunner13e**
- **Study Guide**
- **PrepU**
- **Clinical Handbook**
- **Handbook of Laboratory and Diagnostic Tests**

References

*Asterisk indicates nursing research.

Books

American Burn Association. (2011a). *Advanced burn life support (ABLS) course provider manual 2011*. Chicago: Author.

Branski, L. K., Herndon, D. N., & Barrow, R. E. (2012). A brief history of acute burn care management. In D. Herndon (Ed.): *Total burn care* (4th ed., pp. 1–7). Edinburgh, UK: Saunders Elsevier.

Gallagher, J. J., Branski, L. K., Williams-Bouyer, N., et al. (2012). Treatment of infection in burns. In D. Herndon (Ed.): *Total burn care* (4th ed., pp. 137–156). Edinburgh, UK: Saunders Elsevier.

Karter, M. (2011). *Fire loss in the United States during 2010*. Quincy, MA: National Fire Protection Association, Fire Analysis and Research Division.

Kramer, G. C. (2012). Pathophysiology of burn shock and burn edema. In D. Herndon (Ed.): *Total burn care* (4th ed., pp. 103–113). Edinburgh, UK: Saunders Elsevier.

Lee, J. O., Dibildox, M., Jimenez, C. J., et al. (2012). Operative wound management. In D. Herndon (Ed.): *Total burn care* (4th ed., pp. 157–172). Edinburgh, UK: Saunders Elsevier.

Low, J. F. A., Meyer, W. J., Willebrand, M., et al. (2012). Psychiatric disorders associated with burn injury. In D. Herndon (Ed.): *Total burn care* (4th ed., pp. 733–742). Edinburgh, UK: Saunders Elsevier.

Mayhall, C. G. (2012). Healthcare-associated burn wound infections. In C. G. Mayhall (Ed.): *Hospital epidemiology and infection control* (4th ed., pp. 338–351). Philadelphia: Lippincott Williams & Wilkins.

Milner, S. M., & Feldman, M. J. (2012). Radiation injuries and vesicant burns. In D. Herndon (Ed.): *Total burn care* (4th ed., pp. 461–469). Edinburgh, UK: Saunders Elsevier.

Murphey, E. D., Sherwood, E. R., & Toliver-Kinsky, T. (2012). The immunological response and strategies for intervention. In D. Herndon (Ed.): *Total burn care* (4th ed., pp. 265–276). Edinburgh, UK: Saunders Elsevier.

Rosenberg, L., Lawrence, J. W., Rosenberg, M., et al. (2012). Psychosocial recovery and reintegration of patients with burn injuries. In D. Herndon (Ed.): *Total burn care* (4th ed., pp. 743–753). Edinburgh, UK: Saunders Elsevier.

Saffle, J. R., Graves C., & Cochran, A. (2012). Nutritional support of the burn patient. In D. Herndon (Ed.): *Total burn care* (4th ed., pp. 333–353). Edinburgh, UK: Saunders Elsevier.

Journals and Electronic Sources

Ahuja, R. B., Gupta, R., Gupta, G., et al. (2011). A comparative analysis of cetirizine, gabapentin, and their combination in the relief of post-burn pruritis. *Burns, 37*(2), 203–207.

Albornoz, C. R., Villegas, J., Sylvester, M., et al. (2011). Burns are more aggressive in the elderly: Proportion of deep burn area/total burn area might have a role in mortality. *Burns, 37*(4), 1058–1061.

American Burn Association. (2005). *Burn center verification*. Available at: www.ameriburn.org/verification_verifiedcenters.php

American Burn Association. (2011b). *Burn incidence and treatment in the United States: 2011 fact sheet*. Available at: www.ameriburn.org/resources_factsheet.php

American Burn Association National Burn Repository. (2011). *Report of data from 2001–2010*. Available at: www.ameriburn.org/2011NBRAnnualReport.pdf

Barillo, D. J. (2009). Diagnosis and treatment of cyanide toxicity. *Journal of Burn Care and Research, 30*(1), 148–152.

Brusselaers, N., Monstrey, S., Colpaert, K., et al. (2010). Outcome of acute kidney injury in severe burns: A systematic review and meta-analysis. *Intensive Care Medicine, 36*(6), 915–925.

Brusselaers, N., Monstrey, S., Vogelaers, D., et al. (2010). Severe burn injury in Europe: A systematic review of the incidence, etiology, morbidity, and mortality. *Critical Care, 14*(5), 1–12.

Campbell, J. W., DeGolia, P. A., Fallon, W. F., et al. (2009). In harm's way: Moving the older trauma patient toward a better outcome. *Geriatrics, 64*(1), 8–28.

Cancio, L. (2009). Airway management and smoke inhalation injury in the burn patient. *Clinics in Plastic Surgery, 36*(4), 555–567.

Centers for Disease Control and Prevention. (2011). *Injury prevention and control: Data and statistics. Fatal injury reports 1999–2008*. Available at: www.cdc.gov/injury/wisqars/fatal_injury_reports.html

Colohan, S. M. (2010). Predicting prognosis in thermal burns with associated inhalation injury: A systematic review of prognostic factors in adult burn victims. *Journal of Burn Care and Research, 31*(4), 529–539.

Edgar, D. W., Hons, B., Fish, J. S., et al. (2011). Local and systemic treatments for acute edema after burn injury: A systematic review of the literature. *Journal of Burn Care and Research, 32*(2), 334–347.

Endlad, F., Lee, J. O., Finnerty, C. C., et al. (2010). Does hypothermia at admission affect morbidity and mortality post-burn? *Journal of Burn Care and Research, 31*(2), S97.

Endorf, F. W., & Dries, D. J. (2011). Burn resuscitation. *Scandinavian Journal of Trauma, Resuscitation and Emergency Medicine, 19*, 69.

Evers, L. H., Bhavsar, D., & Mailander, P. (2010). The biology of burn injury. *Experimental Dermatology, 19*(9), 777–783.

Faucher, L. D., & Conlon, K. M. (2007). Practice guidelines for deep vein thrombosis prophylaxis in burns. *Journal of Burn Care and Research, 28*(5), 661–663.

Faucher, L., & Furukawa, K. (2006). Practice guidelines for the management of pain. *Journal of Burn Care and Research, 27*(5), 659–668.

Fruhwald, S., & Kainz, J. (2010). Effect of ICU interventions on gastrointestinal motility. *Current Opinion in Critical Care, 16*(2), 159–164.

Geneva Association. (2012). *World fire statistics bulletin*. Available at: www.genevaassociation.org/PDF/WFSC/GA2012-FIRE28.pdf

Hassan, Z., Wong, J. K., Bush, J., et al. (2010). Assessing the severity of inhalation injuries in adults. *Burns, 36*(2), 212–216.

Kasten, K. R., Makley, A. T., & Kagan, R. J. (2011). Update on the critical care management of severe burns. *Journal of Intensive Care Medicine, 26*(4), 223–236.

Kealey, G. P. (2009). Carbon monoxide toxicity. *Journal of Burn Care and Research, 30*(1), 146–147.

Keck, M., Herndon, D. H., Kamolz, L. P., et al. (2009). Pathophysiology of burns. *Wiener Medizinische Wochenschrift, 159*(13–14), 327–336.

Keck, M., Lumenta, D. B., Andel, H., et al. (2009). Burn treatment in the elderly. *Burns, 35*(8), 1071–1079.

Lafferty, K. A. (2010). Smoke inhalation. *Medscape*. Available at: emedicine.medscape.com/article/771194-overview

Leon-Villapalos, J., Eldardiri, M., & Dziewulski, P. (2010). The use of human deceased donor skin allograft in burn care. *Cell Tissue Bank, 11*(1), 99–104.

Lou, R. B., & Hickerson, W. L. (2009). The use of skin substitutes in hand burns. *Hand Clinics, 25*(4), 497–509.

Lumenta, D. B., Vierhapper, M. F., Kamolz, L. P., et al. (2011). Train surfing and other high voltage trauma: Differences in injury-related mechanisms and operative outcomes after fasciotomy, amputation and soft-tissue coverage. *Burns, 37*(8), 1427–1434.

Luz, D. P., Millan, L. S., Alessi, M. S., et al. (2009). Electrical burns: A retrospective analysis across a 5-year period. *Burns, 35*(7), 1015–1019.

Mosier, M. J., & Pham, T. N. (2009). American Burn Association practice guidelines for prevention, diagnosis, and treatment of ventilator-associated pneumonia (VAP) in burn patients. *Journal of Burn Care and Research, 30*(6), 910–928.

Mosier, M. J., Pham, T. N., Klein, M. B., et al. (2011). Early enteral nutrition in burns: Compliance with guidelines and associated outcomes in a multicenter study. *Journal of Burn Care and Research, 32*(1), 104–109.

Mosier, M. J., Pham, T. N., Park, D. R., et al. (2012). Predictive value of bronchoscopy in assessing the severity of inhalation injury. *Journal of Burn Care and Research, 33*(1), 65–73.

National Oceanic and Atmospheric Administration. (2013). *Lightning safety.* Available at: www.lightningsafety.noaa.gov/index.htm

Nieves, E., Tobón, L. F., Rios, D. I., et al. (2011). Bacterial translocation in abdominal trauma and postoperative infections. *Journal of Trauma Injury, Infection, and Critical Care, 71*(5), 1258–1261.

Nuccio, P. F. (2011). Inhalation injuries. *Journal for Respiratory Care Practitioners, 24*(10), 34–36.

Orgill, D. P., & Piccolo, N. (2009). Escharotomy and decompressive therapies in burns. *Journal of Burn Care and Research, 30*(5), 759–768.

Palmieri, T. L. (2009). What's new in critical care of the burn-injured patient? *Clinical Plastic Surgery, 36*(4), 607–615.

Palmieri, T. L., Lavrentieva, A., & Greenhalgh, D. G. (2010). Acute kidney injury in critically ill burn patients. Risk factors, progression and impact on mortality. *Burns, 36*(7), 205–211.

Pannucci, C. J., Osborne, N. H., Jaber, R. M., et al. (2010). Early fasciotomy in electrically injured patients as a marker for injury severity and deep vein thrombosis risk: An analysis of the National Burn Repository. *Journal of Burn Care and Research, 31*(6), 882–887.

Pannucci, C. J., Osborne, N. H., Park, H. S., et al. (2012). Acquired inpatient risk factors for venous thromboembolism after thermal injury. *Journal of Burn Care and Research, 33*(1), 84–88.

Park, J. O., Shin, S. D., Kim, J., et al. (2009). Association between socioeconomic status and burn injury severity. *Burns, 35*(4), 482–490.

Peck, M. D. (2011a). Epidemiology of burns throughout the world. Part I: Distribution and risk factors. *Burns, 37*(7), 1087–1100.

Peck, M. D. (2011b). Structure fires, smoke production and smoke alarms. *Journal of Burn Care and Research, 32*(5), 511–518.

Posluszny, J. A., Conrad, P., Halerz, M., et al. (2011). Surgical burn wound infections and their clinical implications. *Journal of Burn Care and Research, 32*(2), 324–333.

Purdue, G. F., Arnoldo, B. D., & Hunt, J. L. (2011). Acute assessment and management of burn injuries. *Physical Medicine and Rehabilitation, 42*(12), 201–212.

Richard, R., Dewey, W. S., Forbes-Duchart, L., et al. (2009). Burn rehabilitation and research: Proceedings of a consensus summit. *Journal of Burn Care and Research, 30*(4), 543–573.

*Tengvall, O., Wickman, M., & Wengström, Y. (2010). Memories of pain after burn injury—the patient's experience. *Journal of Burn Care and Research, 31*(2), 319–327.

Tiengo, C., Castagnetti, M., Garolla, A., et al. (2011). High-voltage electrical burn of the genitalia, perineum, and upper extremities: The importance of a multidisciplinary approach. *Journal of Burn Care and Research, 32*(6), e168–e171.

Toon, M. H., Maybauer, M. O., Greenwood, J. E., et al. (2010). Management of acute smoke inhalation injury. *Critical Care and Resuscitation, 12*(1), 53–61.

UDL Laboratories. (2012). *Biobrane.* Available at: www.udllabs.com/burn_care/biobrane.aspx

Ukleja, A. (2010). Altered GI motility in critically ill patients: Current understanding of pathophysiology, clinical impact, and diagnostic approach. *Nutrition in Clinical Practice, 25*(1), 16–25.

Wikiel, K., Gemma, L. W., Yowler, C. J., et al. (2011). Takotsubo cardiomyopathy after minor burn injury. *Journal of Burn Care and Research, 32*(5), e161–e164.

Williams, C. (2009). Successful assessment and management of burn injuries. *Nursing Standard, 23*(32), 53–62.

Williams, F. N., Jeschke, M. G., Chinkes, D. L., et al. (2009). Modulation of the hypermetabolic response to trauma: Temperature, nutrition, and drugs. *Journal of the American College of Surgeons, 208*(4), 489–502.

Wilson, J., & Gramling, L. (2009). The application of Orem's self-care model to burn care. *Journal of Burn Care and Research, 30*(5), 852–858.

Woodson, L. C. (2009). Diagnosis and grading of inhalation injury. *Journal of Burn Care and Research, 28*(4), 143–145.

Zaletel, C. L. (2009). Factors affecting fluid resuscitation in the burn patient. The collaborative of the APN. *Advanced Emergency Nursing Journal, 31*(4), 309–320.

Resources

Alisa Ann Ruch Burn Foundation, www.aarbf.org
American Burn Association (ABA), www.ameriburn.org
American Red Cross, www.redcross.org
Burn Children Recovery Foundation, www.burnedchildrenrecovery.org
Burn Foundation, www.burnfoundation.org
Burn Institute, www.burninstitute.org
Burn Prevention Network, www.burnprevention.org
Firefighters Burn Institute, www.ffburn.org
International Association of Fire Fighters, www.iaff.org
International Society for Burn Injuries (ISBI), www.worldburn.org
Lund-Browder Classification, medical-dictionary.thefreedictionary.com/Lund-Browder+classification
Medline Plus, U.S. National Library of Medicine, www.nlm.nih.gov/medlineplus/tutorials/burns/htm/index.htm
National Fire Protection Association (NFPA), www.nfpa.org
Phoenix Society for Burn Survivors, www.phoenix-society.org
Texas Burn Survivor Society, www.texasburnsurvivors.org
Total Burn Care, www.totalburncare.com
U.S. Fire Administration, www.usfa.fema.gov

Unit 15 Sensory Function

Case Study

A PATIENT WITH IMPAIRED VISION AND DECREASED ATTENTION TO ONE SIDE OF THE BODY

Mr. Martin is a 60-year-old man who has had several strokes. Ophthalmologic testing reveals that he has homonymous hemianopsia of the left visual field and visual spatial neglect; as a result, he has limited vision in the left visual fields of both eyes. He has difficulty in many areas, such as bumping into objects and ignoring the left side of his body. He is admitted to the hospital from a skilled nursing facility with a urinary tract infection. The hospital has installed new stationary computers in all patient rooms that link to the electronic health records (EHRs). The nurse notes that the computer screen in Mr. Martin's room is on the left side of his bed.

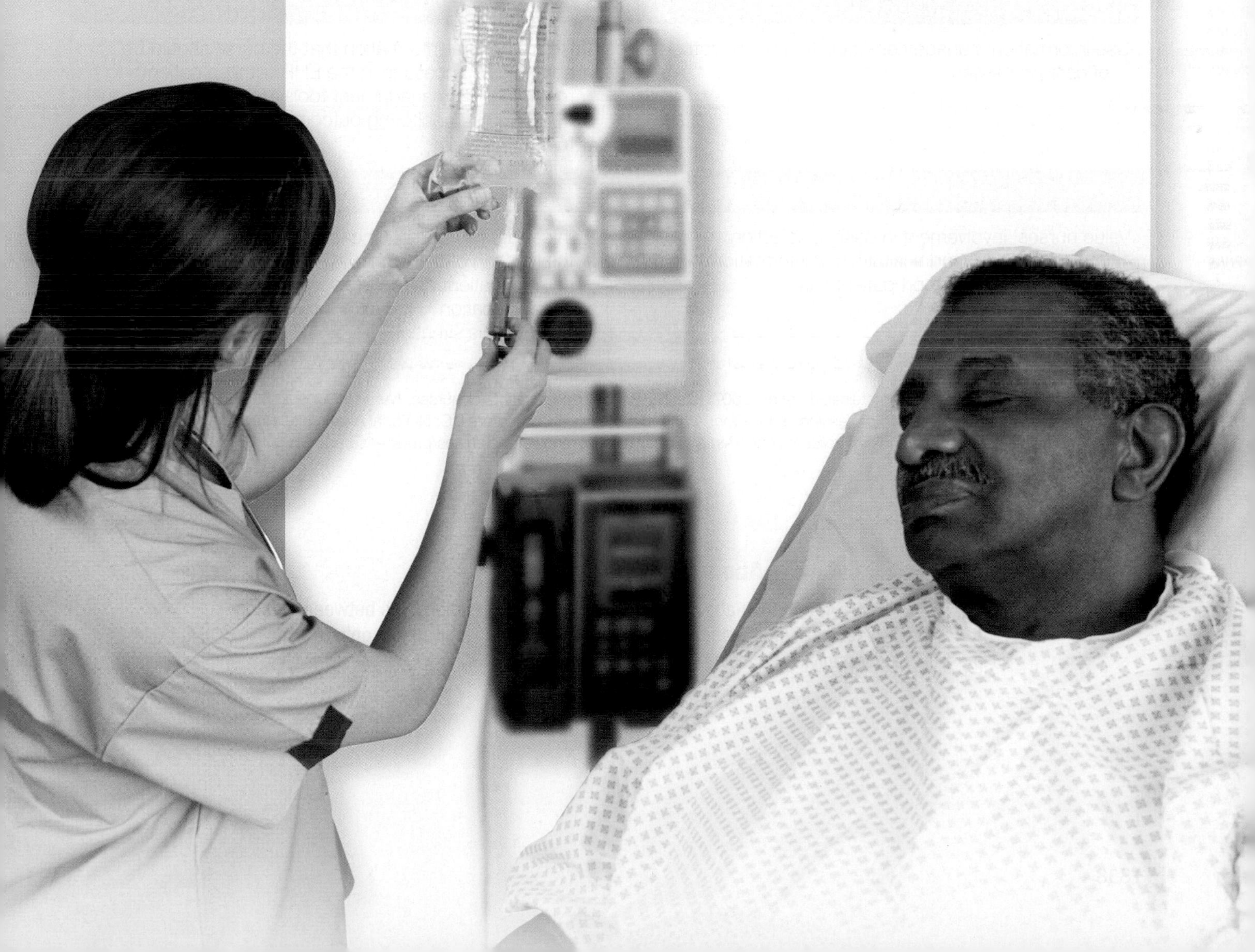

QSEN Competency Focus: *Informatics*

The complexities inherent in today's health care system challenge nurses to demonstrate integration of specific interdisciplinary core competencies. These competencies are aimed at ensuring the delivery of safe, quality patient care (Institute of Medicine, 2003). The concepts from the Quality and Safety Education for Nurses (QSEN) Institute (2012) provide a framework for the knowledge, skills, and attitudes (KSAs) required for nurses to demonstrate competency in these key areas, which include *patient-centered care, interdisciplinary teamwork and collaboration, evidence-based practice, quality improvement, safety,* and *informatics.*

Informatics Definition: Use information and technology to communicate, manage knowledge, mitigate error, and support decision making.

RELEVANT PRE-LICENSURE KSAs	APPLICATION AND REFLECTION
Knowledge	
Describe examples of how technology and information management are related to the quality and safety of patient care.	Identify how ready access to a patient's EHR within the patient's room can improve patient care processes, thus enhancing quality and safety of patient care.
Skills	
Use information management tools to monitor outcomes of care processes.	Describe general information that the nurse should be able to gain access to in the EHR within a patient's room. Identify management tools in the EHR that would be useful for monitoring outcomes of care processes for Mr. Martin.
Attitudes	
Value nurses' involvement in design, selection, implementation, and evaluation of information technologies to support patient care.	Think about this particular case scenario. How might the location of the computer screen potentially hamper patient care processes and nurse-to-patient communication? How might the nurse intervene to improve this situation?

Cronenwett, L., Sherwood, G., Barnsteiner, J., et al. (2007). Quality and safety education for nurses. *Nursing Outlook*, *55*(3), 122–131.
Institute of Medicine. (2003). *Health professions education: A bridge to quality*. Washington, DC: National Academies Press.
QSEN Institute. (2012). *Competencies: Prelicensure KSAs*. Available at: qsen.org/competencies/pre-licensure-ksas

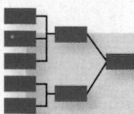

Read More About This Case

More information about this case study and the relationships between nursing diagnoses, interventions, and expected outcomes is available online. Visit thePoint for Applying Concepts From NANDA-I, NIC, and NOC.

Chapter 63

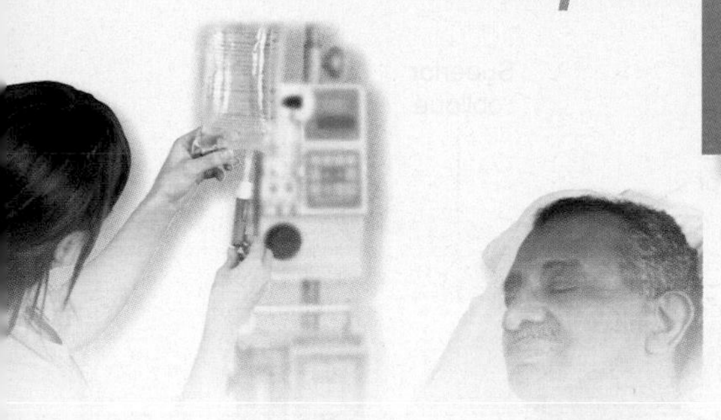

Assessment and Management of Patients With Eye and Vision Disorders

Learning Objectives

On completion of this chapter, the learner will be able to:

1 Identify significant eye structures and describe their functions.
2 Specify assessment and diagnostic findings used in the evaluation of ocular disorders.
3 Describe assessment and management strategies for patients with low vision and blindness.
4 Identify the pharmacologic actions and nursing management of common ophthalmic medications.
5 Discuss clinical features, assessment and diagnostic findings, and medical or surgical management of glaucoma, cataracts, and other ocular disorders.
6 Describe the nursing management of patients with glaucoma, cataracts, and ocular trauma.
7 Discuss general discharge education for patients after ocular surgery.

Glossary

accommodation: process by which the lens of the eye adjusts the focal length to focus a clear image on the retina

anterior chamber: aqueous-containing space in the eye between the posterior (endothelial) cornea and the anterior iris and pupil

aqueous humor: transparent nutrient-containing fluid that fills the anterior and posterior chambers of the eye

astigmatism: refractive error due to an irregularity in the curvature of the cornea

binocular vision: normal ability of both eyes to focus on one object and fuse the two images into one

blindness: inability to see, defined as corrected visual acuity of 20/400 or less, or a visual field of no more than 20 degrees in the better eye

bullous keratopathy: corneal edema with painful blisters in the epithelium due to corneal endothelial dysfunction

chemosis: edema of the conjunctiva

diplopia: seeing one object as two; double vision

ectropion: turning out of the lower eyelid

emmetropia: normal refractive condition resulting in clear focus on retina; no optical defects

endophthalmitis: intraocular infection

entropion: turning in of the lower eyelid

enucleation: removal of the eyeball and part of the optic nerve

evisceration: removal of the intraocular contents through a corneal or scleral incision; the optic nerve, sclera, extraocular muscles, and sometimes the cornea are left intact

exenteration: surgical removal of the entire contents of the orbit, surrounding soft tissue, and most or all of the eyelids

hyperemia: red eyes resulting from dilation of the vasculature of the conjunctiva

hyperopia: farsightedness; light rays focus behind the retina

hyphema: blood in the anterior chamber

hypopyon: collection of inflammatory cells in the anterior chamber of the eye

injection: congestion of blood vessels

keratoconus: cone-shaped deformity of the cornea

miotics: medications that cause constriction of the pupil

mydriatics: medications that cause dilation of the pupil

myopia: nearsightedness; light rays focus in front of the retina

neovascularization: growth of abnormal new blood vessels

nystagmus: involuntary oscillation of the eyeball

papilledema: swelling of the optic disc usually due to increased intracranial pressure

photophobia: ocular pain on exposure to light

proptosis or exophthalmos: abnormal protrusion of the eyeball

ptosis: drooping eyelid

refraction: determination of the refractive errors of the eye for the purpose of vision correction

scotomas: blind or partially blind areas in the visual field

sympathetic ophthalmia: an inflammatory condition created in the fellow eye by the affected eye

trachoma: an infectious disease caused by the bacterium *Chlamydia trachomatis*—the leading cause of preventable blindness in the world

trichiasis: turning in of the eyelashes

vitreous humor: transparent, colorless gelatinous material that fills the vitreous chamber behind the lens

Note: Common abbreviations related to vision and eye health are OD (oculus dexter, right eye), OS (oculus sinister, left eye), and OU (oculus uterque, both eyes).

The eye is a sensitive, highly specialized sense organ subject to various disorders, many of which can lead to impaired vision. Impaired vision may affect individuals in many ways, including their independence in self-care, sense of self-esteem, safety, and overall quality of life. Many of the leading causes of visual impairment are associated with aging (e.g., cataracts, glaucoma, and macular degeneration). Younger people are also at risk for eye disorders, particularly traumatic injuries.

Although most people with eye disorders are treated in an ambulatory care setting, various eye disorders are seen in many settings. In addition to understanding the prevention, treatment, and consequences of eye disorders, nurses in all settings assess visual acuity in patients at risk (e.g., older adults, those with diabetes or acquired immunodeficiency syndrome [AIDS]), refer patients to eye care specialists as appropriate, implement measures to prevent further visual loss, and help patients adapt to impaired vision.

ASSESSMENT OF THE EYE

Anatomic and Physiologic Overview

Unlike most organs of the body, the eye is available for external examination, and its anatomy is more easily assessed than other body parts (Fig. 63-1). The eyeball, or globe, sits in the bony protective orbit (Porth & Matfin, 2009). Lined with muscle and connective and adipose tissues, the orbit is shaped like a four-sided pyramid, surrounded on three sides by the sinuses: ethmoid (medially), frontal (superiorly), and maxillary (inferiorly). The optic nerve and the ophthalmic artery enter the orbit at its apex through the optic foramen. The eyeball is moved through all fields of gaze by the extraocular muscles. The four rectus muscles and two oblique muscles (Fig. 63-2) are innervated by cranial nerves (CNs) III, IV, and VI. Normally, the movements of the two eyes are coordinated and the brain perceives a single image.

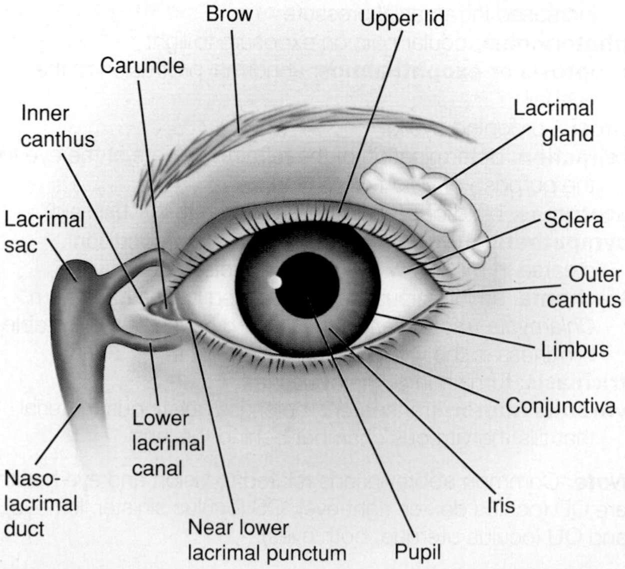
Brow
Upper lid
Caruncle
Inner canthus
Lacrimal gland
Lacrimal sac
Sclera
Outer canthus
Limbus
Conjunctiva
Lower lacrimal canal
Naso-lacrimal duct
Iris
Near lower lacrimal punctum
Pupil

FIGURE 63-1 • External structures of the eye and position of the lacrimal structures.

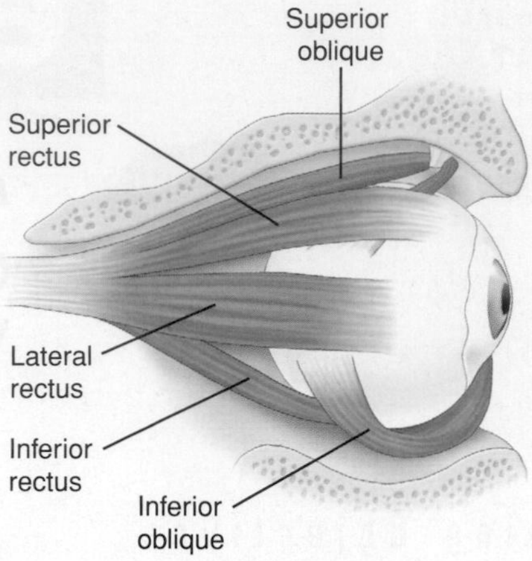
Superior oblique
Superior rectus
Lateral rectus
Inferior rectus
Inferior oblique

FIGURE 63-2 • The extraocular muscles responsible for eye movement. The medial rectus muscle (not shown) is responsible for opposing the movement of the lateral rectus muscle.

The eyelids are composed of thin, elastic skin that covers striated and smooth muscles and protect the anterior portion of the eye. The eyelids contain multiple glands (sebaceous, sweat, and lacrimal). The upper lid normally covers the uppermost portion of the iris and is innervated by the oculomotor nerve (CN III). The lid margins contain meibomian glands, the inferior and superior puncta, and the eyelashes. The triangular spaces formed by the junction of the eyelids are known as the inner or medial canthus and the outer or lateral canthus. With every blink, the eyelids wash the cornea and conjunctiva with tears.

Tears are vital to eye health. Formed by the lacrimal gland and the accessory lacrimal glands, tears are secreted in response to reflex or emotional stimuli. A healthy tear is composed of three layers: lipoid, aqueous, and mucoid. These layers nourish the cornea and create a smooth optical surface of the cornea and conjunctival epithelium. If there is a defect in the composition of any of these layers, the integrity of the cornea may be compromised.

The conjunctiva, a thin transparent mucous membrane, provides a barrier to the external environment extending under the eyelids (palpebral conjunctiva) and over the sclera (bulbar conjunctiva). The junction of the two portions is known as the fornix. The conjunctiva meets the cornea at the limbus on the outermost edge of the iris.

The eyeball is composed of the following three layers:

- The outer dense fibrous layer, including the sclera and transparent cornea
- The middle vascular layer, containing the iris, ciliary body, and choroid
- The inner neural layer, including the retina, optic nerve, and visual pathway

The eyeball is divided anatomically into two segments. The anterior segment is between the anterior cornea and posterior iris, including the anterior and posterior chamber. The posterior segment is between the posterior lens and the retina, including the vitreous chamber. The eyeball also has three fluid-containing chambers. The aqueous-filled **anterior chamber** lies between the posterior cornea

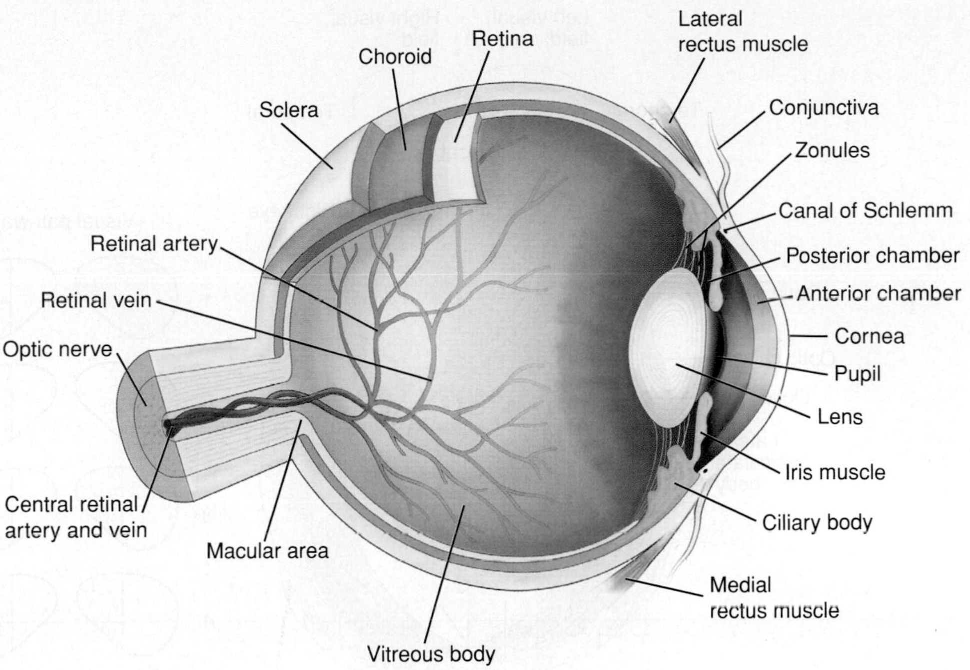
Lateral rectus muscle
Retina
Choroid
Sclera
Conjunctiva
Zonules
Canal of Schlemm
Retinal artery
Posterior chamber
Anterior chamber
Retinal vein
Cornea
Optic nerve
Pupil
Lens
Iris muscle
Central retinal artery and vein
Ciliary body
Macular area
Medial rectus muscle
Vitreous body

FIGURE 63-3 • Three-dimensional cross-section of the eye.

and the anterior iris and pupil. The posterior chamber is a small aqueous containing space between the posterior iris and pupil and anterior lens. The vitreous chamber, containing clear gelatinous vitreous fluid, is the largest chamber in the ocular fundus between the lens and retina (Fig. 63-3).

The **aqueous humor** (transparent nutrient-containing fluid that fills the anterior and posterior chambers of the eye) is produced in the posterior chamber by the ciliary body. It flows through the pupil into the anterior chamber and drains through the trabecular meshwork into the canal of Schlemm. Production of aqueous humor is related to the intraocular pressure (IOP). Normal IOP is 10 to 21 mm Hg (Gerstenblith & Rabinowitz, 2012; Riordan-Eva & Cunningham, 2011). **Vitreous humor**, which is composed mostly of water and encapsulated by a hyaloid membrane, helps maintain the shape of the eye. The vitreous is attached to the retina by scattered collagenous filaments. The vitreous shrinks and shifts with age. Through this degenerative process, the gel-like characteristics liquefy, causing stringy debris known as floaters.

The sclera is the white avascular dense fibrous structure that helps maintain the shape of the eyeball and protects the intraocular contents. Scleral thinning and changes of the scleral collagen fibers can cause the underlying uveal pigment to be seen, resulting in a blue or grey sclera. The episclera is vascularized loose elastic tissue that overlies the sclera supplying nutritional support and reacting to inflammation.

The cornea, a vulnerable transparent avascular domelike structure, forms the most anterior portion of the eyeball and is the main refracting surface of the eye. It is composed of five layers: the epithelium, Bowman's membrane, stroma, Descemet's membrane, and endothelium. It contains high concentrations of nerve fibers and is extremely sensitive to pain. The epithelium, the outermost protective layer, absorbs oxygen and nutrients from the tear film nourishing the cornea. The epithelial cells readily regenerate, unlike the innermost endothelial cells, which do not regenerate and result in corneal edema when injured.

The uveal tract is the vascular middle layer of the eye consisting of the iris, ciliary body, and the choroid.

The iris surrounding the pupil is a highly vascularized pigmented collection of fibers that give the eye color. The dilator and sphincter muscles of the iris control pupil size. The dilator muscles are controlled by the sympathetic nervous system. The sphincter muscles are controlled by the parasympathetic nervous system.

The ciliary body consists of ciliary processes, ciliary muscles, and zonular fibers (ligaments) that work together to form aqueous fluid and control focusing through the zonular fibers that suspend the crystalline lens.

The choroid lies between the retina and the sclera, supplying blood and oxygen to the outer retina. Pigmented cells containing melanocytes in the choroid assist to absorb scattered light.

Directly behind the pupil and iris lies the lens, an avascular and almost completely transparent biconvex structure held in position by zonular fibers in the ciliary body. The lens enables focusing for near and distance vision through **accommodation**, the process by which the lens of the eye adjusts the focal length to focus a clear image on the retina. With aging and certain conditions (e.g., diabetes or trauma), the lens loses its transparency and ability to focus due to formation of a cataract.

The retina—the innermost surface of the fundus composed of neural tissue—is an extension of the optic nerve. Viewed through the pupil, the landmarks of the retina are the optic disc, the retinal vessels, and the macula. The point of entrance of the optic nerve into the retina is the optic disc. The optic disc is pink, either oval or circular in shape, and has sharp margins. In the disc, a physiologic depression or cup is present centrally, with the retinal blood vessels emerging from it. The retinal tissues arise from the optic

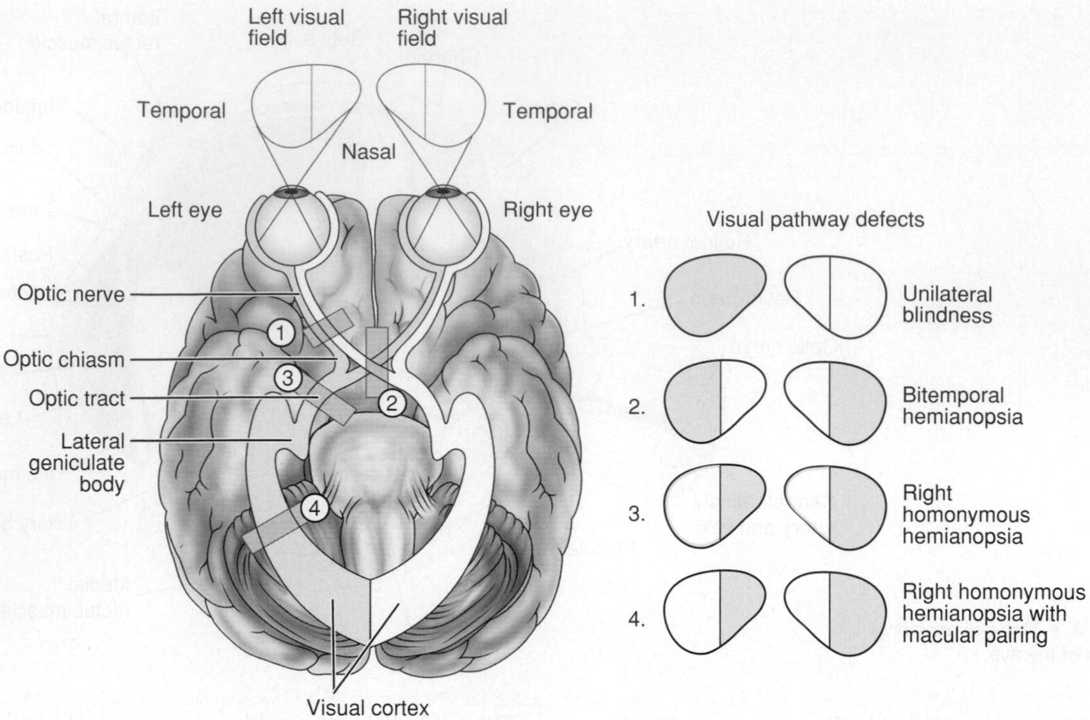

Left visual field
Right visual field
Temporal
Temporal
Nasal
Left eye
Right eye
Optic nerve
Optic chiasm
Optic tract
Lateral geniculate body
1
2
3
4
Visual cortex
Visual pathway defects
1.
Unilateral blindness
2.
Bitemporal hemianopsia
3.
Right homonymous hemianopsia
4.
Right homonymous hemianopsia with macular pairing

FIGURE 63-4 • The visual pathway. From Fuller, J., & Schaller-Ayers, J. (2000). *Health assessment: A nursing approach* (3rd ed.). Philadelphia: Lippincott Williams & Wilkins.

disc and line the inner surface of the vitreous chamber. The retinal vessels enter the eye through the optic disc, branching out through the retina and forming superior and inferior branches. The macula is the area of the retina responsible for central vision. The rest of the retina is responsible for peripheral vision. In the center of the macula is the most sensitive area—the fovea—which is avascular and surrounded by the superior and inferior vascular arcades. Two important layers of the retina are the retinal pigment epithelium (RPE) and the sensory retina. A single layer of cells constitutes the RPE. These cells have numerous functions, including the absorption of light. The sensory retina contains the photoreceptor cells: rods and cones. The rods are responsible for night or low light vision. The cones are retinal photoreceptor cells essential for visual acuity, color discrimination, and fine detail. Cones are distributed throughout the retina, with their greatest concentration in the fovea. Rods are absent in the fovea.

Visual acuity depends on a healthy functioning eye and an intact visual pathway. This pathway is made up of the retina, optic nerve, optic chiasm, optic tracks, lateral geniculate bodies, optic radiations, and the visual cortex area of the brain. The visual pathway is part of the central nervous system (Fig. 63-4).

The optic nerve (CN II) transmits impulses from the retina to the occipital lobe of the brain. The optic nerve head, or optic disc, is the physiologic blind spot in each eye. The optic nerve leaves the eye and then meets the optic nerve from the other eye at the optic chiasm. The chiasm is the anatomic point at which the nasal fibers from the nasal retina of each eye cross to the opposite side of the brain. The nerve fibers from the temporal retina of each eye remain uncrossed. Fibers from the right half of each eye, which would be the left visual field, carry impulses to the right occipital lobe. Fibers from the left half of each eye, or the right visual field, carry impulses to the left occipital lobe. Beyond the chiasm, these fibers are known as the optic tract. The optic tract continues on to the lateral geniculate body. The lateral geniculate body is connected by the optic radiations to the cortex of the occipital lobe of the brain.

Assessment

Ocular History

The nurse, through careful questioning, elicits the necessary information that can assist in diagnosis of an ophthalmic condition. Pertinent questions to ask when taking an ocular history are presented in Chart 63-1. Genetics plays a role in many eye and vision disorders (Bankert, 2011) (see Chart 63-2).

Visual Acuity

Following the health history, the patient's visual acuity is assessed. This is an essential part of the eye examination and a measure against which all therapeutic outcomes are based.

Visual acuity is tested for both near (14 inches away) and distance (20 feet away) vision and is performed on each eye separately with a standardized Snellen chart for distance and a Rosenbaum pocket screener for near vision. A tumbling "E," "illiterate E," number, or picture chart is used if the person is illiterate or unable to read the English alphabet (Riordan-Eva & Cunningham, 2011).

The Snellen chart is composed of a series of progressively smaller rows of letters; the person is asked to read the lowest

Chart 63-1

ASSESSMENT

Taking a History for Patients With Eye and Vision Disorders

- What does the patient perceive to be the problem?
- Is visual acuity diminished?
- Does the patient experience blurred, double, or distorted vision?
- Is there pain? Is it sharp or dull? Is it worse when blinking?
- Is the discomfort an itching sensation or more of a foreign body sensation?
- Are both eyes affected?
- Is there a history of discharge? If so, inquire about color, consistency, and odor.
- Describe the onset of the problem (sudden, gradual). Is it worsening?
- What is the duration of the problem?
- Is this a recurrence of a previous condition?
- How has the patient self-treated?
- What makes the symptoms improve or worsen?
- Has the condition affected performance of activities of daily living?
- Are there any systemic diseases? What medications are used in their treatment?
- What other eye conditions does the patient have?
- Is there a history of eye surgery?
- Have other family members had the same symptoms or condition?

Adapted from Weber, J., & Kelley, J. (2010). *Health assessment in nursing* (4th ed.). Philadelphia: Lippincott Williams & Wilkins.

line possible. The fraction 20/20 is considered the standard of normal vision. Most people can see the letters on the line designated as 20/20. The patient should be encouraged to read every letter possible. If 20/20 vision is not obtained with or without corrective lenses, then pinhole vision is tested. Pinhole vision is a quick method to assess best corrected vision. By looking through a pinhole (an occluder with small holes), the refractive errors of the peripheral cornea and crystalline lens of the eye are significantly reduced or eliminated, simulating corrected vision. No pinhole improvement indicates refractive error is less likely the cause of subnormal visual acuity.

Visual acuity is then recorded. The following would be an example of visual acuity documentation: A patient reads all five letters from the 20/20 line on the Snellen chart with the right eye (OD) and three of the five letters on the 20/30 line with the left eye (OS); the visual acuity is documented as OD: 20/20 and OS: 20/30.

If the patient cannot see the big "E" at the top of the Snellen chart, the examiner should determine next if the patient can count fingers ("CF" in documentation). Initially, the examiner stands 5 feet from the person, holds up a random number of fingers, and then asks the patient to count the number of fingers seen. If the patient is unable to count fingers at 5 feet, the examiner keeps moving a foot closer until either 1 foot away or the person can correctly count fingers. If the patient correctly counts the number of fingers at 3 feet, for example, the examiner would record vision as

Chart 63-2

GENETICS IN NURSING PRACTICE

Eye and Vision Disorders

Several eye and vision disorders are associated with genetic abnormalities. Some examples include:

- Albinism
- Aniridia
- Color blindness
- Glaucoma
- Homocystinuria
- Isolated familial congenital cataracts
- Leber hereditary optic neuropathy
- Leber congenital amaurosis
- Marfan syndrome
- Retinitis pigmentosa

Nursing Assessments

Family History Assessment

- Assess history of family members in the past three generations with glaucoma, cataracts, night blindness (retinitis pigmentosa), color blindness, or other vision impairment.
- Inquire about the age of onset of symptoms (the onset of Leber congenital amaurosis is in childhood, whereas the onset of Leber hereditary optic neuropathy is in young adulthood).
- Inquire about family members with other disorders that may include visual impairment, such as cutaneous, metabolic, or connective tissue disorders and hearing loss.

Patient Assessment

- Assess for other systemic and/or clinical features such as cutaneous or skeletal conditions, or hearing loss.

Management Issues Specific to Genetics

- Inquire whether deoxyribonucleic acid (DNA) gene mutation or other genetic testing has been performed on any affected family members.
- If indicated, refer for further genetic counseling and evaluation so that family members can discuss inheritance, risk to other family members, availability of genetic testing, and gene-based interventions.
- Offer appropriate genetic information and resources.
- Assess patient's understanding of genetic information.
- Provide support to families with newly diagnosed genetics-related sensorineural disorders.
- Participate in management and coordination of care of patients with genetics-related eye and vision disorders, patients with genetic conditions, and individuals predisposed to develop or pass on a genetic condition.

Genetics Resources

National Ophthalmic Disease Genotyping Network (eyeGene): a network created by the National Eye Institute to facilitate research into the genetic causes of ophthalmic disease, http://www.nei.nih.gov/eyegene/

See Chapter 8, Chart 8-6 for additional genetics resources.

CF/3ft (American Society of Ophthalmic Registered Nurses [ASORN], 2011).

If the patient cannot count fingers, the examiner raises one hand up and down or moves it side to side and asks in which direction the hand is moving. This level of vision is known as hand motion. A patient who can perceive only light is described as having light perception. The vision of a patient who cannot perceive light is described as no light perception (ASORN, 2011).

External Eye Examination

The nurse uses a systematic approach to perform an external eye examination by first assessing for symmetry and placement of eyelids, pupils, and muscles. CNs III, IV, and VI control movement and pupil size. The eyelids should rest just above and below the corneal limbus without exposure of the sclera. The nurse observes for **ptosis** (drooping of the eyelid), **ectropion** (turning out of the lower eyelid), or **entropion** (turning in of the lower eyelid). Entropion may involve **trichiasis** (turning in of the eyelashes). Eyelids and lashes should be free of drainage or scaling.

The room should be darkened so that the pupils can be examined. The pupillary response is checked with a penlight to determine if the pupils are equally reactive and regular. A normal pupil is black. An irregular pupil may result from trauma, previous surgery, or a disease process.

The patient's eyes are observed in primary or direct gaze, and any head tilt is noted. A head tilt may indicate cranial nerve palsy. The patient is asked to stare at a target; each eye is covered and uncovered quickly while the examiner looks for any shift in gaze. The examiner observes for **nystagmus** (oscillating movement of the eyeball). The extraocular movements of the eyes are tested by having the patient follow the examiner's finger, pencil, or a hand light through the six cardinal directions of gaze (i.e., up, down, right, left, and both diagonals). This is especially important when screening patients for ocular trauma or for neurologic disorders such as stroke or myasthenia gravis (Bader & Littlejohns, 2010).

Diagnostic Evaluation

A wide range of diagnostic studies may be performed in patients with eye disorders. The nurse should educate the patient about the purpose, what to expect, and any possible side effects related to these examinations prior to testing. The nurse should be aware of contraindications, potential complications, and trends in results. Trends provide information about disease progression as well as the patient's response to therapy.

Direct Ophthalmoscopy

A direct ophthalmoscope is a handheld instrument with various plus and minus lenses (Weber & Kelley, 2010). The lenses can be rotated into place, enabling the examiner to bring the cornea, lens, and retina into focus sequentially. The examiner holds the ophthalmoscope in the right hand and uses the right eye to examine the patient's right eye. The examiner switches to the left hand and left eye when examining the patient's left eye. During this examination, the room should be darkened, and the patient's eye should be on the same level as the examiner's eye. The patient and the examiner should be comfortable, and both should breathe normally. The patient is given a target to gaze at and is encouraged to keep both eyes open and steady.

When the fundus is examined, the vasculature comes into focus first. The veins are larger in diameter than the arteries. The examiner focuses on a large vessel and then follows it toward the midline of the body, which leads to the optic nerve. The central depression in the disc is known as the cup. The normal cup is about one third the size of the diameter of the disc. The size of the physiologic optic cup should be estimated and the disc margins described as sharp or blurred. A silvery or coppery appearance, which indicates arteriolosclerosis, should be noted. The periphery of the retina is examined by having the patient shift his or her gaze. The last area of the fundus to be examined is the macula, because this area is the most sensitive to light. The retina of a young person often has a glistening effect, sometimes referred to as a cellophane reflex.

The healthy fundus should be free of any lesions. The examiner looks for intraretinal hemorrhages, which may appear as red smudges, and, if the patient has hypertension, they may be somewhat flame shaped. Lipid with a yellowish appearance may be present in the retina of patients with hypercholesterolemia or diabetes. Soft exudates that have a fuzzy, white appearance (cotton-wool spots) should be noted. The examiner looks for microaneurysms, which look like little red dots, and nevi. Drusen (small, hyaline, globular deposits), commonly found in macular degeneration, appear as yellowish areas with indistinct edges. Small drusen have a more distinct edge. The examiner should sketch the fundus and document any abnormalities.

Indirect Ophthalmoscopy

The indirect ophthalmoscope is an instrument commonly used by the ophthalmologist to see larger areas of the retina, although in an unmagnified state. It produces a bright and intense light. The light source is affixed with a pair of binocular lenses mounted on the examiner's head. The ophthalmoscope is used with a handheld, 20-diopter lens.

Slit-Lamp Examination

The slit lamp is a binocular microscope mounted on a table. This instrument enables the user to examine the eye with magnification of 10 to 40 times the real image. The illumination can be varied from a broad to a narrow beam of light for different parts of the eye. For example, by varying the width and intensity of the light, the anterior chamber can be examined for signs of inflammation. Cataracts may be evaluated by changing the angle of the light. When a handheld contact lens, such as a three-mirror lens, is used with the slit lamp, the angle of the anterior chamber may be examined, as may the ocular fundus.

Tonometry

Tonometry is an essential part of a diagnostic evaluation; it measures IOP to screen for and manage glaucoma. The device used for measuring IOP is an accurately calibrated applanation tonometer, which measures the pressure needed to flatten the cornea. Because the probe or prism touches the highly sensitive cornea, a topical anesthetic is administered prior to measurement (ASORN, 2008).

Nursing Interventions

Providing patient education prior to tonometry helps avoid possible errors in IOP measurement. Patients are cautioned to avoid squeezing the eyelids, holding their breath, or performing a Valsalva maneuver, because these may result in abnormally increased IOP.

Color Vision Testing

The ability to differentiate colors has a dramatic effect on the activities of daily living (ADLs). For example, the inability to differentiate between red and green can compromise traffic safety. Some careers (e.g., commercial artist, [color] photographer, airline pilot, and electrician) may be closed to people with significant color deficiencies. The photoreceptor cells responsible for color vision are the cones, and the greatest area of color sensitivity is in the macula—the area of densest cone concentration.

A screening test, such as the polychromatic plates discussed in the next paragraph, can be used to establish whether a person's color vision is within normal range. Color vision deficits can be inherited. For example, red–green color deficiencies are inherited in an X-linked manner, affecting approximately 8% of men and 0.5% of women (Jaeger & Tasman, 2010). Acquired color vision losses may be caused by medications (e.g., digitalis) or pathology (e.g., cataracts). A simple test, such as asking a patient if the red top on a bottle of eye drops appears redder to one eye than the other, can be an effective tool. A difference in the perception of the intensity of the color red between the two eyes can be a symptom of a neurologic problem and may provide information about the location of the lesion.

Because alteration in color vision sometimes indicates conditions of the optic nerve, color vision testing is often performed in a neuro-ophthalmologic workup. The most common color vision test is performed using Ishihara polychromatic plates. These plates are bound together in a booklet. On each plate of this booklet are dots of primary colors that are integrated into a background of secondary colors. The dots are arranged in simple patterns, such as numbers or geometric shapes. Patients with diminished color vision may be unable to identify the hidden shapes. Patients with central vision conditions (e.g., macular degeneration) have more difficulty identifying colors than those with peripheral vision conditions (e.g., glaucoma) because central vision identifies color.

Amsler Grid

The Amsler grid is a test often used for patients with macular problems, such as macular degeneration. It consists of a geometric grid of identical squares with a central fixation point. The grid should be viewed by the patient wearing normal reading glasses. Each eye is tested separately. The patient is instructed to stare at the central fixation spot on the grid and report any distortion in the squares of the grid itself. For patients with macular problems, some of the squares may look faded, or the lines may be wavy. Patients with age-related macular degeneration (AMD) are commonly given these Amsler grids to take home. The patient is encouraged to check the grids frequently, as often as daily, to monitor macular function for early detection of changes requiring immediate attention (Gerstenblith & Rabinowitz, 2012).

Ultrasonography

Lesions in the globe or the orbit may not be directly visible and are evaluated by ultrasonography. Ultrasonography is a valuable diagnostic technique, especially when the view of the retina is obscured by opaque media such as cataract or hemorrhage. An ultrasonography B-scan identifies pathology such as orbital tumors, retinal detachment, and vitreous hemorrhage. Ultrasonography A-scans are used to measure the axial length for implants prior to cataract surgery (Gerstenblith & Rabinowitz, 2012).

Optical Coherence Tomography

Optical coherence tomography is a technology that involves low-coherence interferometry (Gerstenblith & Rabinowitz, 2012). Light is used to evaluate retinal and macular diseases as well as anterior segment conditions. This method is noninvasive and involves no physical contact with the eye.

Fundus Photography

Fundus photography is used to detect and document retinal lesions. The patient's pupils are usually widely dilated before the procedure. The resulting fundus photographs can be viewed stereoscopically so that elevations such as macular edema can be identified.

Laser Scanning

Various scanning techniques use laser light in the diagnostic evaluation of eye disorders. Confocal laser scanning ophthalmoscopy provides a three-dimensional image of the optic nerve topography and is used alone or in conjunction with fundus photography to provide comparative data for suspected optic nerve disease such as glaucoma and **papilledema** (swelling of the optic disc due to increased intracranial pressure) (Gerstenblith & Rabinowitz, 2012). Laser scanning polarimetry is used to measure nerve fiber layer thickness and is an important indicator of glaucoma progression.

Angiography

Angiography is done using fluorescein or indocyanine green as contrast agents. Fluorescein angiography is used to evaluate clinically significant macular edema, document macular capillary nonperfusion, and identify retinal and choroidal **neovascularization** (growth of abnormal new blood vessels) in AMD. It is an invasive procedure in which fluorescein dye is injected, usually into an antecubital vein. Within 10 to 15 seconds, this dye can be seen coursing through the retinal vessels. Over a 10-minute period, serial black-and-white photographs are taken of the retinal vasculature.

Indocyanine green angiography is used to evaluate abnormalities in the choroidal vasculature, which often are seen in macular degeneration. Indocyanine green dye is injected intravenously (IV), and multiple images are captured using digital video angiography over a period of 30 seconds to 20 minutes.

Nursing Interventions

Prior to the angiography, the patient's blood urea nitrogen and creatinine should be checked to ensure that the kidneys will excrete the contrast agent (Fischbach & Dunning, 2011). The patient should be well hydrated, and clear liquids are usually permitted up to the time of the test. The patient is instructed to remain immobile during the angiogram process and is told to expect a brief feeling of warmth in the face, behind the eyes, or in the jaw, teeth, tongue, and lips, and a metallic taste when the contrast agent is injected.

Nursing care after angiography includes observation of the injection site (usually the antecubital vein) for bleeding or hematoma formation (a localized collection of blood). Fluorescein may impart a gold tone to the skin in some patients, and urine may turn deep yellow or orange. This discoloration usually disappears in 24 hours. Indocyanine green dye is generally well tolerated, but some patients experience nausea and vomiting. Allergic reactions are rare; however, indocyanine green angiography is contraindicated in patients with a history of iodide reactions. Fluids are encouraged following the procedure to facilitate excretion of the contrast agent (Fischbach & Dunning, 2011).

Perimetry Testing

Perimetry testing evaluates the field of vision. Visual field testing (i.e., perimetry) helps identify which parts of the patient's central and peripheral visual fields have useful vision. It is most helpful in detecting central **scotomas** (blind or partially blind areas in the visual field) in macular degeneration and the peripheral field defects in glaucoma and retinitis pigmentosa. Visual field evaluation and optic nerve assessment are major components of monitoring and detecting glaucoma progression.

IMPAIRED VISION

Refractive Errors

In refractive errors, vision is impaired because a shortened or elongated eyeball prevents light rays from focusing sharply on the retina. Blurred vision from refractive error can be corrected with eyeglasses or contact lenses. Ophthalmic **refraction** is the determination of the refractive errors of the eye for the purpose of vision correction and consists of placing various types of lenses in front of the patient's eyes to determine which lens best improves the patient's vision.

The depth of the eyeball is important in determining refractive error (Fig. 63-5). Patients for whom the visual image focuses precisely on the macula and who do not need eyeglasses or contact lenses are said to have **emmetropia**, a normal refractive condition resulting in clear focus on retina with no optical defects (normal vision). Some people have deeper eyeballs, in which case the distant visual image focuses in front of, or short of, the retina; those with **myopia** are said to be nearsighted and have blurred distance vision. Other people have shallower eyeballs, in which case the visual image focuses beyond the retina; those with **hyperopia** are said to be farsighted and have excellent distance vision but blurry near vision (Weber & Kelley, 2010).

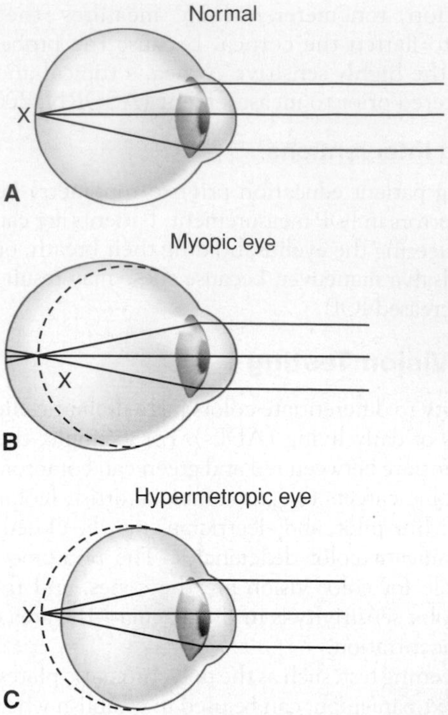
Normal
X
A
Myopic eye
X
B
Hypermetropic eye
X
C

FIGURE 63-5 • Eyeball shape determines visual acuity in refractive errors. **A.** Normal eye. **B.** Myopic eye. **C.** Hypermetropic eye.

Ophthalmology has entered the era of customized vision correction in its desire to achieve "super-normal vision." Wavefront technology to measure unique refractive imperfections of the cornea or higher aberrations (i.e., myopia, hyperopia, astigmatism) is currently used to customize laser-assisted in situ keratomileusis (LASIK) procedures. These procedures are described later in this chapter.

Vision Impairment and Blindness

Vision impairment is defined as having best corrected visual acuity of 20/40 or worse in the better-seeing eye. *Low vision* describes visual impairment that requires the use of devices and strategies to perform visual tasks.

Blindness is defined as having best corrected visual acuity that can range from 20/400 to no light perception. The clinical definition of absolute blindness is the absence of light perception. Legal blindness is a condition of impaired vision in which a person has best corrected visual acuity that does not exceed 20/200 in the better eye or whose widest visual field diameter is 20 degrees or less (Prevent Blindness America, 2012). This definition neither equates with functional ability nor classifies the degrees of visual impairment. Legal blindness ranges from an inability to perceive light to having some vision remaining. A person who meets the criteria for legal blindness may be eligible for government financial assistance through disability.

TABLE 63-1 Age-Related Changes in the Eye

External Eye	Structural Change	Functional Change	History and Physical Findings
Eyelids and lacrimal structures	Loss of skin elasticity and orbital fat, decreased muscle tone; wrinkles develop	Lid margins turn in, causing lashes to irritate cornea and conjunctiva (entropion); or lid margins may turn out, resulting in increased corneal exposure (ectropion).	Reports of burning, foreign body sensation, increased tearing (epiphora); injection, inflammation, and ulceration may occur.
Refractive changes; presbyopia	Loss of accommodative power in the lens with age	Reading materials must be held at increasing distance in order to focus.	Patient reports, "Arms are too short!"; need for increased light; reading glasses or bifocals needed
Cataract	Opacities in the normally crystalline lens	Interference with the focus of a sharp image on the retina	Patients report increased glare, decreased vision, changes in color values (blue and yellow especially affected).
Posterior vitreous detachment	Liquefaction and shrinkage of vitreous body	May lead to retinal tears and detachment	Reports light flashes, cobwebs, floaters
Age-related macular degeneration (AMD)	Drusen (yellowish aging spots in the retina) appear and coalesce in the macula. Abnormal choroidal blood vessels may lead to formation of fibrotic disciform scars in the macula.	Central vision is affected; onset is more gradual in dry AMD, more rapid in wet AMD; distortion and loss of central vision may occur.	Reading vision is affected; words may be missing letters, faded areas appear on the page, straight lines may appear wavy; drusen, pigmentary changes in retina; abnormal submacular choroidal vessels.

Impaired vision is often accompanied by difficulty in performing functional activities. People with visual acuity of 20/80 to 20/100 with a visual field restriction of 60 degrees to greater than 20 degrees can read at a nearly normal level with optical aids. Their visual orientation is near normal but requires increased scanning of the environment (i.e., systematic use of head and eye movements). In a visual acuity range of 20/200 to 20/400 with a 20-degree to greater than 10-degree visual field restriction, the person can read slowly with optical aids.

The most common causes of blindness and visual impairment among adults 40 years and older are diabetic retinopathy, macular degeneration, glaucoma, and cataracts (Prevent Blindness America, 2012). Macular degeneration is more prevalent among Caucasians, whereas glaucoma is more prevalent among African Americans. Age-related changes in the eye are summarized in Table 63-1.

Assessment and Diagnostic Testing

Assessment of vision impairment includes a thorough history and examination of distance and near visual acuity, visual field, contrast sensitivity, glare, color perception, and refraction. Specially designed, low vision visual acuity charts are used to evaluate patients.

Patient Interview

During history taking, the cause and duration of the patient's visual impairment are identified. Patients with retinitis pigmentosa, for example, have a genetic abnormality. Patients with diabetic macular edema typically have fluctuating visual acuity. Patients with macular degeneration have central acuity problems that cause difficulty in performing activities that require finer vision, such as reading. People with peripheral field defects have more difficulties with mobility. The patient's customary ADLs, medication regimen, habits (e.g., smoking), acceptance of the physical limitations brought about by the visual impairment, and realistic expectations of low vision aids are identified and included in the plan of care, as well as provision of guidelines for safety and referrals to social services.

Contrast-Sensitivity Testing and Glare Testing

Contrast-sensitivity testing measures visual acuity in different degrees of light and dark contrast to determine visual function. Glare testing is also used to determine visual function. Glare can reduce a person's ability to see, especially in patients with cataracts. Those affected by loss of contrast sensitivity and glare have difficulty functioning in low light, or driving at night or in foggy conditions. People with a loss of contrast sensitivity may benefit from better illumination.

Medical Management

Managing vision impairment involves magnification and image enhancement through the use of low vision aids and strategies, as well as referrals to social services and community agencies. The goals are to optimize the patient's remaining vision and assist the patient to perform customary activities. Table 63-2 presents low vision aids. Medications are prescribed for glaucoma. Ongoing research suggests that gene therapy may replace or be an adjunct to pharmacologic or surgical treatment for ocular disorders in the near future (Bankert, 2011).

A mobile telephone–based communication support system allows people with poor eyesight to communicate by mobile phone. The system consists of a low-power mobile phone (Personal Handy-phone System [PHS]) and a large screen. When the person telephones, a registered support personnel picture is displayed on the screen and the PHS telephone dials that person automatically (Ogawa, Yonezawa, Maki, et al., 2007).

Referrals to community agencies may be necessary for patients with low vision who live alone and cannot self-administer their medications. Community agencies, such as the Lighthouse International, offer services to patients with low vision that include training in independent living skills,

TABLE 63-2 Activities Affected by Visual Impairment and Suggestions for Low Vision Aids

Activity	Optical Aids	Nonoptical Aids
Shopping	Hand magnifier	Lighting, color cues
Fixing a snack	Bifocals	Color cues; consistent food storage plan
Eating out	Hand magnifier	Flashlight, portable lamp
Identifying money	Bifocals, hand magnifier	Arrange paper money in wallet compartments
Reading print	High-power spectacle, bifocals, hand magnifier, stand magnifier, closed-circuit television	Lighting, high-contrast print, large print, reading slit
Writing	Hand magnifier	Lighting, bold-tip pen, black ink
Using a telephone	Hand magnifier	Large-print buttons, hand-printed directory
Crossing streets	Lightweight handheld monoscopes/telescopes	Cane; ask directions.
Finding taxis and bus signs	Lightweight handheld monoscopes/telescopes	Ask for assistance.
Reading medication labels	Hand magnifier	Color codes, large print
Reading stove dials	Hand magnifier	Color codes, raised dots
Adjusting the thermostat	Hand magnifier	Enlarged-print model
Using a computer	Spectacles	High-contrast color, large-print program. Screen-reader programs convert text on the computer screen to synthesized speech (e.g., JAWS for Windows, Windows Eyes, Slimware Window Bridge, ProTalk 32, Hal Screen Reader, WinVision, WYNN, outSPOKEN for Windows).
Reading signs	Spectacles	Move closer.
Watching sporting event	Lightweight handheld monoscopes/telescopes	Sit in front rows.

Adapted from Riordan-Eva, P., & Cunningham, E. T. (2011) *Vaughan & Asbury's general ophthalmology* (18th ed.). New York: McGraw-Hill.

the provision of occupational and recreational activities, and a wide variety of assistive devices for vision enhancement and orientation and mobility.

Nursing Management

Research suggests that nurses need to increase their sensitivity to the challenges faced by patients with visual impairments (Sharts-Hopko, Smeltzer, Ott, et al., 2010). Coping with blindness involves emotional, physical, and social adaptation. The emotional adjustment to blindness or severe visual impairment determines the success of the physical and social adjustments of the patient. Successful emotional adjustment means acceptance of blindness or severe visual impairment.

Promoting Coping Efforts

Effective coping may not occur until the patient recognizes the permanence of the low vision or blindness. A patient who is newly visually impaired and his or her family members undergo the various steps of grieving: denial and shock, anger and protest, restitution, loss resolution, and acceptance. The ability to accept the changes that must come with visual loss and willingness to adapt to those changes influence the successful rehabilitation of the patient with vision loss. Additional aspects to consider are value changes, independence–dependence conflicts, coping with stigma, and learning to function in social settings without visual cues and landmarks (Chart 63-3).

Promoting Spatial Orientation and Mobility

A person who is blind or severely visually impaired requires strategies for adapting to the environment. ADLs, such as walking to a chair from a bed, require spatial concepts. The person needs to know where he or she is in relation to the rest of the room, to understand the changes that may occur, and to know how to approach the desired location safely. This requires a collaborative effort between the patient and the responsible adult who serves as the sighted guide. The nurse must assess the degree of physical assistance the person with vision loss requires and communicate this to other health care personnel.

The nurse should be aware of the importance of techniques in providing physical assistance, encouraging independence, and ensuring safety. Guidelines for interacting with the patient with vision loss are presented in Chart 63-4. Research supports the use of a validated protocol to educate staff to perform therapeutic communication, deliver services, and minimize communication barriers with patients who are blind (Reboucas, Pagliuca, Sawada, et al., 2012). The readiness of the patient and family to learn must be assessed before initiating orientation and mobility training.

Promoting Home and Community-Based Care

The nurse, social worker, family, and others collaborate to assess the patient's home condition and support system. A low vision specialist or occupational therapist should be consulted, particularly for patients for whom identifying and administering medications pose challenges. Referral for vision rehabilitation should be provided for appropriate patients (American Academy of Ophthalmology Vision Rehabilitation Committee, 2012).

Other interventions that are appropriate for some people with visual impairment or blindness include Braille and service animals. There has been an ever-increasing reliance on print magnification technology as well as computer-assisted speech output. However, although the use of Braille may be less important for adults who have already learned language

Chart 63-3

ETHICAL DILEMMA

When Might Someone's Commitment to an Independent Lifestyle Place Others at Risk?

Case Scenario

You work as a nurse manager in a family practice clinic in the neighborhood where you also live. It is early February, and there is a local outbreak of a variety of viral illnesses, including gastroenteritis and rhinosinusitis. The clinic is very busy seeing patients with these viral infections. An 82-year-old neighbor of yours presents to be seen with a 3-day history of a low-grade fever, cough, and runny nose. You note that no one has accompanied him to the clinic. This concerns you because you know that he has a history of "dry" macular degeneration, has been diagnosed as "legally blind," and has had his driver's license revoked. You think that you may have seen him driving to the local grocery store a couple of weeks ago and suspect that he may have driven himself to the clinic today. This patient lives alone and has no family close by. You wonder how he has managed to take care of his personal needs because he has refused links to local social support agencies.

Discussion

Patients newly diagnosed with disabilities, such as legal blindness, frequently experience threats to their lifestyle and sense of well-being when they face newly imposed limitations, such as restrictions on driving. And yet, by not adhering to these restrictions, they may place the lives of others at risk.

Analysis

- Describe the ethical principles that are in conflict in this case (see Chart 3-3 in Chapter 3). Which principle should have preeminence as you proceed to work with this patient?
- Is it possible to preserve this patient's autonomy? If so, what steps you might take to ensure that the patient's rights to "self-rule" are maintained?
- This patient has reputedly refused referrals to social service agencies in the past. Might you delve a bit more deeply into his background to find out why he has refused that support? What are his reasons? Are there other agencies or support services that you may be able to mobilize for him (e.g., if he is a veteran, there are veterans' benefits that he may be able to draw from; alternatively, if he is active in a church, there may be a parish nurse program that he could link into for some assistance)?

Resources

See Chapter 3, Chart 3-6 for ethics resources.

and grammar skills, educators and low vision specialists have continued to advocate that children who are legally blind be given the opportunity to learn Braille.

Guide dogs, also known as service dogs, are dogs that are specially bred, raised, and rigorously trained to assist people who are blind. The guide dog is a constant companion to the person who is blind (also referred to as the animal's handler) and is allowed on airplanes and in restaurants, stores, hotels, and other public places. With the assistance of the guide dog, the person who is blind can be extremely mobile and accomplish normal activities both within and outside of the home and workplace. A dog in harness is a working dog, not a pet. The dog should not be distracted from his job by well-intentioned strangers who want

Chart 63-4

Strategies for Interacting With People Who Are Blind or Have Low Vision

- Remember that the only difference between you and people who are blind or have low vision is that they are not able to see through their eyes what you are able to see through yours.
- Do not be uncomfortable when in the company of a person who is blind or has low vision. Talk with the person as you would talk with any other individual, honestly and without pity; do not be concerned about using words like "see" and "look." There is no need to raise your voice unless the person asks you to do so.
- Identify yourself as you approach the person and before you make physical contact. Tell the person your name and your role. If another person approaches, introduce him or her. When you leave the room, be sure to tell the person that you are leaving and if anyone else remains in the room.
- Keep in mind that it is often appropriate to touch the person's hand or arm lightly to indicate that you are about to speak.
- When talking, face the person and speak directly to him or her using a normal tone of voice.
- Be specific when communicating direction. Mention a specific distance or use clock cues when possible (e.g., walk left about 2 yards; walk about 20 feet to the right; the telephone is at 2 o'clock). Avoid using phrases such as "over there."
- When you offer to assist someone, allow the person to hold on to your arm just above the elbow and to walk a half-step behind you.
- When offering the person a seat, place the person's hand on the back or the arm of the seat.
- When you are about to go up or down a flight of stairs, tell the person and place his or her hand on the banister.
- Make sure that the environment is free of obstacles; close doors and cabinets so that they are not in the path.
- Offer to read written information, such as a menu.
- If you serve food to the person, use clock cues to specify where everything is on the plate.
- When the person who is blind or has low vision is a patient in a health care facility:
  - Make sure all objects the person will need are close at hand.
  - Identify the location of objects that the person may need (e.g., "The call light is near your right hand"; "The telephone is on the table on the left side of your bed.")
  - Remove obstacles that may be in the person's pathway and could cause a fall.
  - Place all assistive devices that the person uses close at hand; let the person feel the devices so that he or she knows their location.
- Do not distract a service animal unless the owner has given permission.
- Ask the person, "How can I help you?" At some times, the person needs help; at other times, help may not be needed.

Adapted from Reboucas, C., Pagliuca, L. M. F., Sawada, N. O., et al. (2012). Validation of a non-verbal communication protocol for nursing consultations with blind people. *Northeast Network Nursing Journal – Rev Rene (Revistarene)*, *13*(1), 125–139. Available at: http://www.revistarene.ufc.br/revista/index.php/revista/article/view/24/20

to pet, feed, or play with the animal. The dog's handler should always be consulted before approaching the working guide dog. Most health care facilities have a service animal policy that outlines the responsibilities of the handler with regard to the care of the animal.

OCULAR MEDICATION ADMINISTRATION

Because medications are often prescribed to treat ocular disorders, nurses must understand the actions of the commonly used medications and effective administration. The main objective of ocular medication delivery is to maximize the amount of medication that reaches the ocular site of action in sufficient concentration to produce a beneficial therapeutic effect. This is determined by the dynamics of ocular pharmacokinetics: absorption, distribution, metabolism, and excretion.

Ocular absorption involves the entry of a medication into the aqueous humor through the different routes of ocular medication administration. The rate and extent of aqueous humor absorption are determined by the characteristics of the medication and the anatomy and physiology of the eye. Natural barriers of absorption that diminish the efficacy of ocular medications include the following:

- *Limited size of the conjunctival sac*. The conjunctival sac can hold only 50 mcL, and any excess is wasted. The volume of one eye drop from commercial topical ocular solutions typically ranges from 20 to 35 mcL.
- *Corneal membrane barriers*. The epithelial, stromal, and endothelial layers are barriers to absorption.
- *Blood–ocular barriers*. Blood–ocular barriers prevent high ocular tissue concentration of most ophthalmic medications because they separate the bloodstream from the ocular tissues and keep foreign substances from entering the eye, thereby limiting a medication's efficacy.
- *Tearing, blinking, and drainage*. Increased tear production and drainage due to ocular irritation or an ocular condition may dilute or wash out an instilled eye drop; blinking expels an instilled eye drop from the conjunctival sac.

Distribution of an ocular medication into the various ocular tissues varies by tissue type—the conjunctiva, cornea, lens, iris, ciliary body, and choroids absorb medications to varying degrees. Medications penetrate the corneal epithelium by diffusion by passing through the cells (intracellular) or by passing between the cells (intercellular). Water-soluble (hydrophilic) medications diffuse through the intracellular route, and fat-soluble (lipophilic) medications diffuse through the intercellular route. Topical administration usually does not reach the retina in significant concentrations. Because the space between the ciliary process and the lens is small, medication diffusion in the vitreous is slow. When high concentrations of medication in the vitreous are required, intraocular injection is often chosen to bypass the natural ocular anatomic and physiologic barriers.

Aqueous solutions are most commonly used for the eye. They are the least expensive medications and interfere the least with vision. However, corneal contact time is brief because tears dilute the medication. Ophthalmic ointments have extended retention time in the conjunctival sac and provide a higher concentration than eye drops. The major disadvantage of ointments is the blurred vision that results after application. In general, eyelids and eyelid margins are best treated with ointments. The conjunctiva, limbus, cornea, and anterior chamber are treated most effectively with instilled solutions or suspensions. Subconjunctival injection may be necessary for better absorption in the anterior chamber. If high medication concentrations are required in the posterior chamber, intravitreal injections or systemically absorbed medications are considered. Contact lenses and collagen shields soaked in antibiotic agents are alternative delivery methods for treating corneal infections.

Of all these delivery methods, the topical route of administration—instilled eye drops and applied ointments—remains the most common. Topical instillation, which is the least invasive method, permits self-administration of medication and produces fewer side effects.

Preservatives are commonly used in ocular medications. Benzalkonium chloride, for example, prevents the growth of organisms and enhances the corneal permeability of most medications; however, some patients are allergic to this preservative. This may be suspected even if the patient had never before experienced an allergic reaction to systemic use of the medication in question. Eye drops without preservatives can be prepared by pharmacists.

Common Ocular Medications

Common ocular medications include topical anesthetic, mydriatic, and cycloplegic agents that reduce IOP; anti-infective medications; corticosteroids; nonsteroidal anti-inflammatory drugs (NSAIDs); antiallergy medications; eye irrigants; and lubricants.

Topical Anesthetic Agents

One or two drops of proparacaine hydrochloride (Ophthaine 0.5%) and tetracaine hydrochloride (Pontocaine 0.5%) are instilled before diagnostic procedures such as tonometry and minor ocular procedures such as removal of sutures or conjunctival or corneal scrapings. Topical anesthetic agents are also used for severe eye pain to allow the patient to open his or her eyes for examination or treatment (e.g., eye irrigation for chemical burns). Anesthesia occurs within 20 seconds to 1 minute and lasts 10 to 20 minutes. The nurse must educate the patient not to rub his or her eyes while anesthetized because this may result in corneal damage.

Most patients are not allowed to take topical anesthetics home because of the risk of overuse (ASORN, 2008). Patients with corneal abrasions and erosions experience severe pain and are often tempted to overuse topical anesthetic eye drops. The overuse of these drops results in softening of the cornea. Prolonged use of anesthetic drops can delay wound healing and can lead to permanent corneal opacification and scarring, resulting in visual loss.

Mydriatic and Cycloplegic Agents

Mydriasis, or pupil dilation, is the main objective of the administration of **mydriatics** and cycloplegics (Table 63-3).

TABLE 63-3 Mydriatics and Cycloplegics

Medication	Available Preparation/ Concentration	Indication/Dosage	Peak		Recovery Time	
			Mydriasis	Cycloplegia	Mydriasis	Cycloplegia
phenylephrine (Neo-Synephrine)	Solutions (2.5%, 10%)	Administered with cycloplegics in pupillary dilation for ophthalmoscopy and surgical procedures every 5–10 min × 3 or until the pupils are fully dilated	10–60 min	—	3–6 h	—
Isopto (Atropine Ophthalmic)	Ointment (0.5%–2%) Solutions (0.5%–3%)	In uveitis, or after surgery, 2–4 × daily	30–40 min	60–180 min	7–10 d	6–12 d
scopolamine (Isopto Hyoscine Ophthalmic)	Solution (0.25%)	Same as atropine	20–30 min	30–60 min	3–7 d	3–7 d
homatropine (Homatropine HBr)	Solution (5%–2.5%)	Same as atropine and scopolamine	40–60 min	30–60 min	1–3 d	1–3 d
cyclopentolate (AK-Pentolate)	Solution (0.5%–2%)	Administered with mydriatics every 5–10 min × 3 or until the pupils are fully dilated for ophthalmoscopy and surgical procedures	30–60 min	25–75 min	1 d	6–24 h
tropicamide (Myrdal)	Solution (0.25%–1%)		20–40 min	20–35 min	6 h	<6 h

Adapted from Bartlett, J. D., Bennett, E. S., & Fiscella, R. G. (Eds.). (2008). *Ophthalmic drug facts, 2008*. St. Louis: Facts & Comparisons; and Karch, A. (2012). *2012 Lippincott's nursing drug guide*. Philadelphia: Lippincott Williams & Wilkins.

These two types of medications function differently and are used in combination to achieve the maximal dilation that is needed during surgery and fundus examinations to give the ophthalmologist a better view of the internal eye structures. Mydriatics potentiate alpha-adrenergic sympathetic effects that result in the relaxation of the ciliary muscle. This causes the pupil to dilate. However, this sympathetic action alone is not enough to sustain mydriasis because of its short duration of action. The strong light used during an eye examination also stimulates miosis (i.e., pupillary contraction). Cycloplegic medications are administered to paralyze the iris sphincter.

The patient is educated about the temporary effects of mydriasis on vision, such as glare and the inability to focus properly. The patient may have difficulty reading. The effects of the various mydriatics and cycloplegics can last 3 hours to several days. The patient is advised to wear sunglasses (most eye clinics provide protective sunglasses). The ability to drive depends on the person's age, vision, and comfort level. Some patients can drive safely with the use of sunglasses, whereas others may need to be driven home.

Mydriatic and cycloplegic agents affect the central nervous system. Their effects are most prominent in younger and older adult patients; these patients must be assessed closely for symptoms, such as increased blood pressure, tachycardia, dizziness, ataxia, confusion, disorientation, incoherent speech, and hallucination. These medications are contraindicated in patients with narrow angles or shallow anterior chambers and in patients taking monoamine oxidase inhibitors or tricyclic antidepressants.

Medications Used to Treat Glaucoma

Therapeutic medications for glaucoma are used to lower IOP by decreasing aqueous production or increasing aqueous outflow. (See section on glaucoma later in this chapter.) Because glaucoma calls for lifetime therapy, the patient must be educated with regard to both the ocular and systemic side effects of the medications.

Anti-Infective Medications

Anti-infective medications include antibiotic, antifungal, and antiviral agents. Most are available as drops, ointments, or subconjunctival or intravitreal injections. Antibiotics include penicillin, cephalosporins, aminoglycosides, and fluoroquinolones. The main antifungal agent is amphotericin B. Side effects of amphotericin are serious and include severe pain, conjunctival necrosis, iritis, and retinal toxicity. Antiviral medications include acyclovir (Zovirax) and ganciclovir. They are used to treat ocular infections associated with herpes virus and cytomegalovirus (CMV). Patients receiving ocular anti-infective agents are subject to the same side effects and adverse reactions as those receiving oral or parenteral medications.

Corticosteroids and Nonsteroidal Anti-Inflammatory Drugs

Topical preparations of corticosteroids are commonly used in inflammatory conditions of the eyelids, conjunctiva, cornea, anterior chamber, lens, and uvea. In posterior segment diseases that involve the posterior sclera, retina, and optic nerve, topical agents are less effective and parenteral and oral routes are preferred. When a suspension is prescribed, the patient is instructed to shake the bottle several times to promote mixture of the medication and maximize its therapeutic effect. The most common ocular side effects of long-term topical corticosteroid administration are glaucoma, cataracts, susceptibility to infection, impaired wound healing, mydriasis, and ptosis. High IOP may develop, which is reversible after corticosteroid use is discontinued. To avoid the side effects of corticosteroids, NSAIDs are used as an alternative in controlling inflammatory eye conditions and postoperatively to reduce inflammation.

Antiallergy Medications

Ocular hypersensitivity reactions, such as allergic conjunctivitis, are extremely common. These conditions result primarily from responses to environmental allergens. Most allergens are airborne or carried to the eye by the hand or by other means, although allergic reactions may also be drug induced. Corticosteroids are commonly used as anti-inflammatory and immunosuppressive agents to control ocular hypersensitivity reactions.

Ocular Irrigants and Lubricants

Most irrigating solutions are used to cleanse the external lids to maintain lid hygiene, to irrigate the external corneal surface to regain normal pH (e.g., in chemical burns), to irrigate the corneal surface to eliminate debris, or to inflate the globe intraoperatively. These solutions have various compositions that include sodium, potassium, magnesium, calcium, bicarbonate, glucose, and glutathione (i.e., substance found in the aqueous humor). Sterile irrigating solutions, such as Dacriose, for lid hygiene are available. Irrigating solutions are safe to use with an intact corneal surface; however, the corneal surface should not be irrigated in cases of threatened corneal perforation. For patients with severe corneal ulcer, specific instructions from the ophthalmologist must be obtained regarding whether it is safe to irrigate the corneal surface or just to cleanse the external lids. Although it is good practice to promote hygiene, prevention of complications must be the primary concern. Normal saline solutions are commonly used to irrigate the corneal surface when chemical burns occur.

Lubricants, such as artificial tears, help alleviate corneal irritation, such as dry eye syndrome. Artificial tears are topical preparations of carboxymethylcellulose or hydroxypropyl methylcellulose that are prepared as eye drop solutions, ointments, or ocular inserts (inserted at the lower conjunctival cul-de-sac once each day). The eye drops can be instilled as often as every hour, depending on the severity of symptoms.

Nursing Management

The objectives in administering ocular medications are to ensure proper administration to maximize the therapeutic effects and to ensure the safety of the patient by monitoring for systemic and local side effects. Absorption of eye drops by the nasolacrimal duct is undesirable because of the potential systemic side effects of ocular medications. To diminish systemic absorption and minimize the side effects, it is important to occlude the puncta (Chart 63-5). This is especially important for patients who are most vulnerable to medication overdose, including older adults; women who are pregnant or lactating; and patients with cardiac, pulmonary, hepatic, or renal disease. A 1-minute interval between instillation of different types of ocular drops is recommended.

Before the administration of ocular medications, the nurse warns the patient that blurred vision, stinging, and a burning sensation are symptoms that ordinarily occur after instillation and are temporary. Risk for interactions of the ocular medication with other ocular and systemic medications must be emphasized; therefore, a careful patient interview regarding the medications being taken must be obtained.

Quality and Safety Nursing Alert

To prevent infection, meticulous hand hygiene before and after medication instillation is crucial. In addition, the tip of the eye drop bottle or the ointment tube must never touch any part of the eye, and the medication must be recapped immediately after each use.

If a patient who instills his or her own medications cannot feel the eye drops when they are instilled, the eye medication may be refrigerated, because a cold drop is easier to detect. A 5-minute interval between successive administrations allows adequate drug retention and absorption. The patient or the caregiver at home should be asked to demonstrate actual eye drop or ointment instillation and punctal occlusion.

Glaucoma

The term *glaucoma* is used to refer to a group of ocular conditions characterized by optic nerve damage. Increased IOP damages the optic nerve and nerve fiber layer, but the degree of harm is highly variable (Porth & Matfin 2009). The optic nerve damage is related to the IOP caused by congestion of aqueous humor in the eye. A range of IOPs are considered "normal," but these may also be associated with vision loss in some patients.

Glaucoma is estimated to affect 2.2 million Americans, and 3 to 6 million more are at risk for the disease (Porth & Matfin, 2009). Glaucoma is more prevalent in people older than 40 years, and it is the third most common age-related eye disease in the United States. Chart 63-6 presents the risk factors for glaucoma. There is no cure for glaucoma, but the disease can be controlled (Sharts-Hopko & Glynn-Milley, 2009).

Physiology

Aqueous humor flows between the iris and the lens, nourishing the cornea and lens. Most (90%) of the fluid then flows out of the anterior chamber, draining through the spongy trabecular meshwork into the canal of Schlemm and the episcleral veins (Fig. 63-6). About 10% of the aqueous fluid exits through the ciliary body into the suprachoroidal space and then drains into the venous circulation of the ciliary body, choroid, and sclera (Jaeger & Tasman, 2010). Unimpeded outflow of aqueous fluid depends on an intact drainage system and an open angle (about 45 degrees) between the iris and the cornea. A narrower angle places the iris closer to the trabecular meshwork, diminishing the angle. The amount of aqueous humor produced tends to decrease with age, in systemic diseases such as diabetes, and in ocular inflammatory conditions.

IOP is determined by the rate of aqueous production, the resistance encountered by the aqueous humor as it flows out of the passages, and the venous pressure of the episcleral veins that drain into the anterior ciliary vein. When aqueous fluid production and drainage are in balance, the IOP is between 10 and 21 mm Hg. When aqueous fluid is inhibited from flowing out, pressure builds up within the eye. Fluctuations in IOP occur with time of day, exertion, diet, and medications. IOP tends to increase with blinking, tight lid squeezing, and upward gazing. Systemic conditions such as diabetes and intraocular conditions such as uveitis and retinal detachment have been associated with elevated IOP. Glaucoma may not be recognized in people with thin corneas because measurement of the IOP may be falsely low as a result of this thinness.

Pathophysiology

There are two theories regarding how increased IOP damages the optic nerve in glaucoma. The direct mechanical theory suggests that high IOP damages the retinal layer as it passes through the optic nerve head. The indirect ischemic theory

Chart 63-5 PATIENT EDUCATION
Instilling Eye Medications

- Do not use solutions that have changed colors.
- Perform hand hygiene before and after the procedure.
- Ensure adequate lighting.
- Read the label of the eye medication to verify that it is the correct medication.
- Remove contact lens as needed.
- Assume a comfortable position.
- Do not touch the tip of the medication container to any part of the eye or face.
- Hold the lower lid down; do not press on the eyeball. Apply gentle pressure to the cheekbone to anchor the finger holding the lid.

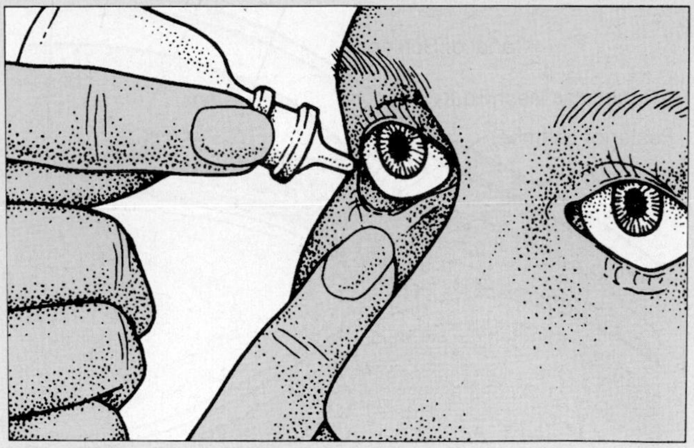

- Instill eye drops before applying ointments.
- Apply a 0.25–0.5-inch ribbon of ointment to the lower conjunctival sac.

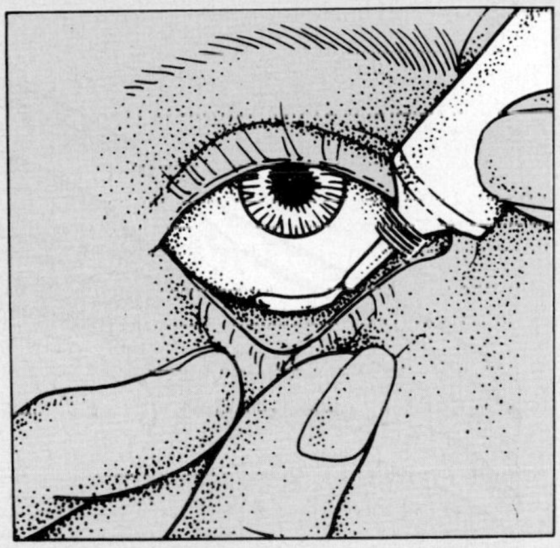

- Keep the eyelids closed, and apply gentle pressure on the inner canthus (punctal occlusion) near the bridge of the nose for 1 or 2 minutes immediately after instilling eye drops.
- Using a clean tissue, gently pat skin to absorb excess eye drops that run onto the cheeks.
- Wait 5 minutes before instilling another eye drop and 10 minutes before instilling another ointment.
- Reinsert contact lens if applicable.

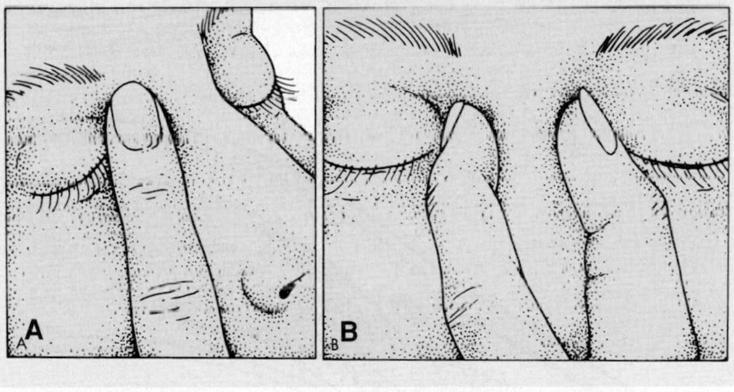
A
B

Adapted from Flach, A. J. (2009). Proposed mandate for instructions and labeling regarding the use of eye drops. *Archives of Ophthalmology*, *127*(9), 1207–1209; and Karch, A. (2012). *2012 Lippincott's nursing drug guide*. Philadelphia: Lippincott Williams & Wilkins.

Chart 63-6 RISK FACTORS
Glaucoma

- Family history of glaucoma
- Thin cornea
- African American race
- Older age
- Diabetes
- Cardiovascular disease
- Migraine syndromes
- Nearsightedness (myopia)
- Eye trauma
- Prolonged use of topical or systemic corticosteroids

suggests that high IOP compresses the microcirculation in the optic nerve head, resulting in cell injury and death. Some glaucomas appear as exclusively mechanical, and some are exclusively ischemic types. Typically, most cases are a combination of both.

Classification of Glaucoma

There are several types of glaucoma. Although glaucoma classification is changing as knowledge increases, current clinical forms of glaucoma are identified as open-angle glaucoma; angle-closure glaucoma (also called *pupillary block*); congenital glaucoma; and glaucoma associated with other conditions,

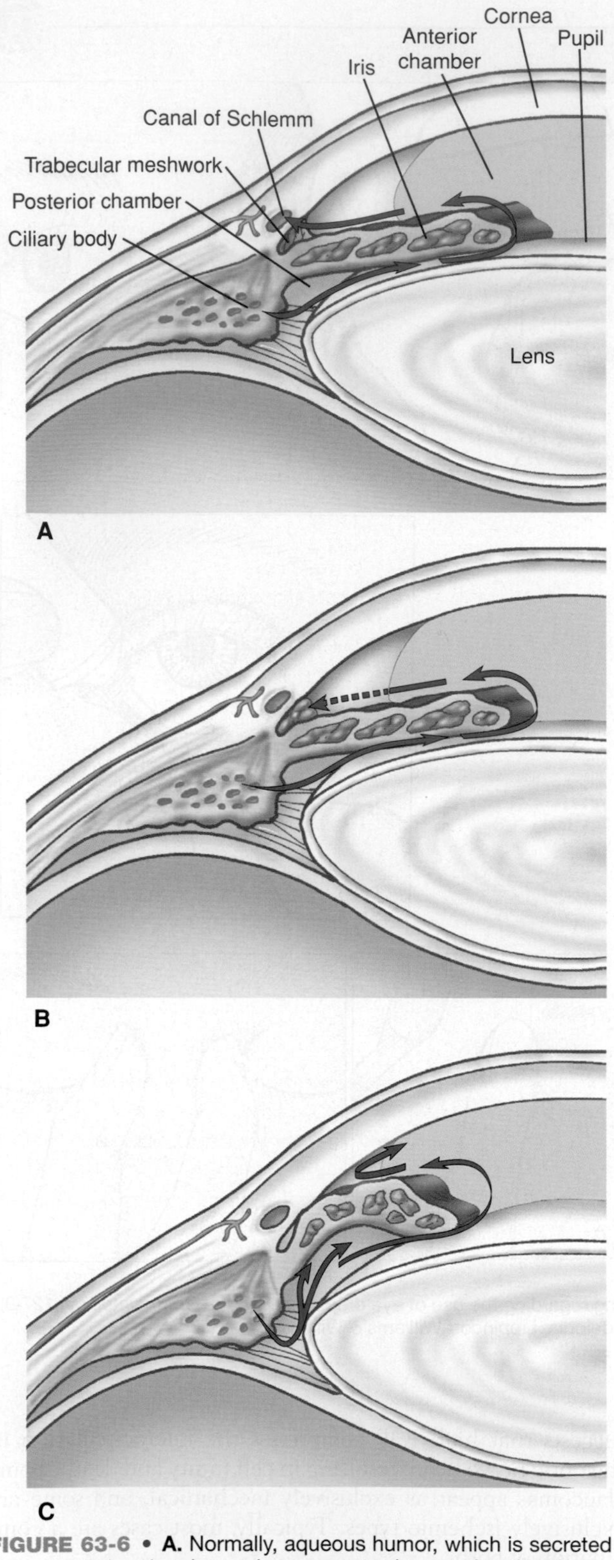
Cornea
Anterior chamber
Pupil
Iris
Canal of Schlemm
Trabecular meshwork
Posterior chamber
Ciliary body
Lens
A
B
C

FIGURE 63-6 • **A.** Normally, aqueous humor, which is secreted in the posterior chamber, gains access to the anterior chamber by flowing through the pupil. In the angle of the anterior chamber, it passes through the canal of Schlemm into the venous system. **B.** In open-angle glaucoma, the outflow of aqueous humor is obstructed at the trabecular meshwork. **C.** In angle-closure glaucoma, the aqueous humor encounters resistance to flow through the pupil. Increased pressure in the posterior chamber produces a forward bowing of the peripheral iris so that the iris blocks the trabecular meshwork. From Porth, C. M. (2010). *Essentials of pathophysiology: Concepts of altered health states* (3rd ed.). Philadelphia: Lippincott Williams & Wilkins.

such as developmental anomalies or corticosteroid use. Glaucoma can be primary or secondary, depending on whether associated factors contribute to the rise in IOP. The two common clinical forms of glaucoma encountered in adults are primary open-angle glaucoma (POAG) and angle-closure glaucoma, which are differentiated by the mechanisms that cause impaired aqueous outflow (Porth & Matfin, 2009). Table 63-4 summarizes the characteristics of the different types of adult glaucoma.

Clinical Manifestation

Glaucoma is often called the "silent thief of sight" because most patients are unaware that they have the disease until they have experienced visual changes and vision loss. The patient may not seek health care until he or she experiences blurred vision or "halos" around lights, difficulty focusing, difficulty adjusting eyes in low lighting, loss of peripheral vision, aching or discomfort around the eyes, and headache.

Assessment and Diagnostic Findings

The purpose of a glaucoma workup is to establish the diagnostic category, assess the optic nerve damage, and formulate a treatment plan. The patient's ocular and medical history must be detailed to investigate the history of predisposing factors. The types of examinations used in glaucoma include tonometry to measure the IOP, ophthalmoscopy to inspect the optic nerve, and central visual field testing (Porth & Matfin, 2009).

The changes in the optic nerve related to glaucoma are pallor and cupping of the optic nerve disc. The pallor of the optic nerve is caused by a lack of blood supply. Cupping is characterized by exaggerated bending of the blood vessels as they cross the optic disc, resulting in an enlarged optic cup that appears more basinlike compared with a normal cup. The progression of cupping in glaucoma is caused by the gradual loss of retinal nerve fibers and the loss of blood supply.

As the optic nerve damage increases, visual perception decreases. The localized areas of visual loss (i.e., scotomas) represent loss of retinal sensitivity and nerve fiber damage and are measured and mapped on a graph. In patients with glaucoma, the graph has a distinct pattern that is different from other ocular diseases and is useful in establishing the diagnosis. Figure 63-7 shows the visual changes caused by glaucoma.

Glaucoma

FIGURE 63-7 • Visual changes associated with glaucoma. Photo courtesy of the National Eye Institute, National Institutes of Health.

TABLE 63-4 Glaucoma Types, Clinical Manifestation, and Treatment

Types of Glaucoma	Clinical Manifestations	Treatment
Open Angle		
Usually bilateral, but one eye may be more severely affected than the other. In all three types of open-angle glaucoma, the anterior chamber angle is open and appears normal.		
Primary open-angle glaucoma (POAG)	Optic nerve damage, visual field defects, IOP >21 mm Hg. May have fluctuating IOPs. Usually no symptoms, but possible ocular pain, headache, and halos.	Decrease IOP 20%–50%. Additional topical and oral agents added as necessary.
Normal tension glaucoma	IOP ≤21 mm Hg. Optic nerve damage, visual field defects.	If medical treatment is unsuccessful, LT can decrease IOP by 20%. Glaucoma filtering surgery if continued optic nerve damage despite medication therapy and LT.
Ocular hypertension	Elevated IOP. Possible ocular pain or headache.	Treatment similar to POAG; however, the best management for normal tension glaucoma management is yet to be established. Goal is to lower the IOP by at least 30%.
Angle Closure (Pupillary Block)		
Obstruction in aqueous humor outflow due to the complete or partial closure of the angle from the forward shift of the peripheral iris to the trabecula. The obstruction results in an increased IOP.		
Acute angle-closure glaucoma	Rapidly progressive visual impairment, periocular pain, conjunctival hyperemia, and congestion. Pain may be associated with nausea, vomiting, bradycardia, and profuse sweating. Reduced central visual acuity, severely elevated IOP, corneal edema. Pupil is vertically oval, fixed in a semidilated position, and unreactive to light and accommodation.	Ocular emergency; administration of hyperosmotics, acetazolamide, and topical ocular hypotensive agents, such as pilocarpine and beta-blockers (betaxolol). Possible laser incision in the iris (iridotomy) to release blocked aqueous and reduce IOP. Other eye is also treated with pilocarpine eye drops and/or surgical management to avoid a similar spontaneous attack.
Subacute angle-closure glaucoma	Transient blurring of vision, halos around lights; temporal headaches and/or ocular pain; pupil may be semidilated.	Prophylactic peripheral laser iridotomy. Can lead to acute or chronic angle-closure glaucoma if untreated.
Chronic angle-closure glaucoma	Progression of glaucomatous cupping and significant visual field loss; IOP may be normal or elevated; ocular pain and headache.	Management similar to that for POAG, which includes laser iridotomy and medications

IOP, intraocular pressure; LT, laser trabeculoplasty.

Medical Management

The aim of all glaucoma treatment is prevention of optic nerve damage. Lifelong therapy is almost always necessary because glaucoma cannot be cured. Treatment focuses on pharmacologic therapy, laser procedures, surgery, or a combination of these approaches, all of which have potential complications and side effects. The object is to achieve the greatest benefit at the least risk, cost, and inconvenience to the patient. Although treatment cannot reverse optic nerve damage, further damage can be controlled. The goal is to maintain an IOP within a range unlikely to cause further damage.

The initial target for IOP among patients with elevated IOP and those with low-tension glaucoma with progressive visual field loss is typically set at 30% lower than the current pressure. The patient is monitored for changes in the appearance of the optic nerve. If there is evidence of progressive damage, the target IOP is again lowered until the optic nerve shows stability.

Pharmacologic Therapy

Medical management of glaucoma relies on systemic and topical ocular medications that lower IOP. Periodic follow-up examinations are essential to monitor IOP, the appearance of the optic nerve, the visual fields, and side effects of medications. Therapy takes into account the patient's health and stage of glaucoma. Comfort, affordability, convenience, lifestyle, and personality are factors to consider in the patient's adherence to the medical regimen.

The patient is usually started on the lowest dose of topical medication and then advanced to increased concentrations until the desired IOP level is reached and maintained. Because of their efficacy, minimal dosing (can be used once each day), and low cost, beta-blockers are the preferred initial topical medications. One eye is treated first, with the other eye used as a control in determining the efficacy of the medication; once efficacy has been established, treatment of the other eye is started. If the IOP is elevated in both eyes, both are treated. When results are not satisfactory, a new medication is substituted. The main markers of the efficacy of the medication in glaucoma control are lowering of the IOP to the target pressure, stable appearance of the optic nerve head, and the visual field.

Many ocular medications are used to treat glaucoma (Table 63-5), including **miotics** (medications that cause pupillary constriction), beta-blockers, alpha$_2$-agonists (i.e., adrenergic agents), carbonic anhydrase inhibitors, and prostaglandins. Cholinergics (i.e., miotics) increase the outflow of the aqueous humor by affecting ciliary muscle contraction and pupil constriction, allowing flow through a larger opening between the iris and the trabecular meshwork. Beta-blockers and carbonic anhydrase inhibitors decrease aqueous production. Prostaglandin analogues reduce IOP by increasing aqueous humor outflow (Karch, 2012).

Surgical Management

In laser trabeculoplasty for glaucoma, laser burns are applied to the inner surface of the trabecular meshwork to open the intratrabecular spaces and widen the canal of Schlemm, thereby promoting outflow of aqueous humor and decreasing IOP (Prevent Blindness America, 2011a). The procedure is indicated when IOP is inadequately controlled by medications, and it is contraindicated when the trabecular meshwork cannot be fully visualized because of a narrow

TABLE 63-5 Medications Used in the Management of Glaucoma

Medication	Action	Side Effects	Nursing Implications
Cholinergics (miotics) (pilocarpine, carbachol)	Increase aqueous fluid outflow by contracting the ciliary muscle and causing miosis (constriction of the pupil) and opening of trabecular meshwork	Periorbital pain, blurry vision, difficulty seeing in the dark	Caution patients about diminished vision in dimly lit areas. Pilocarpine can be stored at room temperature for up to 8 weeks and then should be discarded.
Beta-blockers (timolol maleate)	Decrease aqueous humor production	Can have systemic effects, including bradycardia, exacerbation of pulmonary disease, and hypotension	Contraindicated in patients with asthma, chronic obstructive pulmonary disease, 2nd- or 3rd-degree heart block, bradycardia, or cardiac failure; educate patients about punctal occlusion to limit systemic effects.
Alpha-adrenergic agonists (apraclonidine, brimonidine)	Decrease aqueous humor production	Eye redness, dry mouth and nasal passages	Educate patients about punctal occlusion to limit systemic effects.
Carbonic anhydrase inhibitors (acetazolamide, dorzolamide)	Decrease aqueous humor production	Oral medications (acetazolamide) are associated with serious side effects, including anaphylactic reactions, electrolyte loss, depression, lethargy, gastrointestinal upset, impotence, and weight loss; side effects of topical form (dorzolamide) include topical allergy.	Do not administer to patients with sulfa allergies; monitor electrolyte levels.
Prostaglandin analogues (latanoprost, bimatoprost)	Increase uveoscleral outflow	Darkening of the iris, conjunctival redness, possible rash	Instruct patients to report any side effects.

Adapted from Karch, A. (2012). *2012 Lippincott's nursing drug guide*. Philadelphia: Lippincott Williams & Wilkins.

angle. A potential complication of this procedure is a transient increase in IOP (usually 2 hours after surgery) that may become persistent. IOP assessment in the immediate postoperative period is essential.

In peripheral iridotomy for pupillary block glaucoma, an opening is made in the iris to eliminate the pupillary blockage (Prevent Blindness America, 2011a). Laser iridotomy is contraindicated in patients with corneal edema, which interferes with laser targeting and strength. Potential complications are burns to the cornea, lens, or retina; transient elevated IOP; closure of the iridotomy; uveitis; and blurring. Pilocarpine (Pilocar) is usually prescribed to prevent closure of the iridotomy.

Filtering procedures for chronic glaucoma are used to create an opening or fistula in the trabecular meshwork to drain aqueous humor from the anterior chamber to the subconjunctival space into a bleb (fluid collection on the outside of the eye), thereby bypassing the usual drainage structures. This allows the aqueous humor to flow and exit by different routes (i.e., absorption by the conjunctival vessels or mixing with tears). *Trabeculectomy* is the standard filtering technique used to remove part of the trabecular meshwork. Complications include hemorrhage, an extremely low (hypotony) or extremely elevated IOP, uveitis, cataracts, bleb failure, bleb leak, and **endophthalmitis** (i.e., intraocular infection). Unlike other surgical procedures, the goal of the filtering procedure is to achieve incomplete healing of the surgical wound. The outflow of aqueous humor in a newly created drainage fistula is circumvented by the granulation of fibrovascular tissue or scar tissue formation on the surgical site. Scarring is inhibited by using an antifibrosis agent such as mitomycin C (Mutamycin), but this medication is potent and administered only intraoperatively (ASORN, 2012). Like all antineoplastic agents, they require special handling procedures before, during, and after the procedure (see Chapter 15).

Drainage implants or *shunts* are tubes implanted in the anterior chamber to shunt aqueous humor to the episcleral plate in the conjunctival space. Implants are used when failure has occurred with one or more trabeculectomies in which antifibrotic agents were used. A fibrous capsule develops around the episcleral plate and filters the aqueous humor, thereby regulating the outflow and controlling IOP.

Trabectome surgery is reserved for patients in whom pharmacologic treatment or laser trabeculoplasty do not control the IOP sufficiently (Filippopoulos & Rhee, 2008). This minimally invasive procedure is specifically designed to improve fluid drainage from the eye to balance IOP. By restoring the eye's natural fluid balance, trabectome surgery stabilizes the optic nerve and minimizes further visual field damage. The surgery is performed through a small incision and does not require creation of a permanent hole in the eye wall or an external filtering bleb or an implant.

Nursing Management

Nurses in all settings encounter patients with glaucoma. Even patients with long-standing disease and those with glaucoma as a secondary diagnosis should be assessed for knowledge level and adherence to their prescribed therapeutic regimen.

Promoting Home and Community-Based Care

Educating Patients About Self-Care

The medical and surgical management of glaucoma slows the progression of glaucoma but does not cure it. The lifelong therapeutic regimen mandates patient education. The nature of the disease and the importance of strict adherence to the medication regimen must be included in an individualized education plan. A structured self-management program may increase adherence to the treatment regime (Amro, Cox, Waddington, et al., 2010). A thorough discussion of the medication program, particularly the interactions of glaucoma-control medications with other medications, is essential. For example, the diuretic effect of acetazolamide (Diamox) may have an additive effect on the diuretic effects of other antihypertensive medications (Karch, 2012). The effects of glaucoma-control medications on vision must also be explained. Miotics and sympathomimetics result in altered focus; therefore, patients need to be cautious in navigating their surroundings. Chart 63-5 presents patient education about instilling eye medications and preventing systemic absorption with punctal occlusion. Chart 63-7 contains additional educational information to review with patients with glaucoma.

Continuing Care

For the patient with severe glaucoma and impaired function, referral to services that assist the patient in performing ADLs may be needed. The loss of peripheral vision impairs mobility the most. These patients need to be referred for low vision and rehabilitation services. Patients who meet the criteria for legal blindness should be offered referrals to agencies that can assist them in obtaining federal assistance.

Chart 63-7 PATIENT EDUCATION
Managing Glaucoma

- Know your intraocular pressure measurement and the desired range.
- Be informed about the extent of your vision loss and optic nerve damage.
- Keep a record of your eye pressure measurements and visual field test results to monitor your own progress.
- Review all of your medications (including over-the-counter and herbal medications) with your ophthalmologist, and mention any side effects each time you visit.
- Ask about potential side effects and drug interactions of your eye medications.
- Ask whether generic or less costly forms of your eye medications are available.
- Review the dosing schedule with your ophthalmologist, and inform him or her if you have trouble following the schedule.
- Participate in the decision-making process. Let your primary provider know what dosing schedule works for you and other preferences regarding your eye care.
- Have the nurse observe you instilling eye medication to determine whether you are administering it properly (see Chart 63-5).
- Be aware that glaucoma medications can cause adverse effects if used inappropriately. Eye drops are to be administered as prescribed, not when eyes feel irritated.
- Ask your ophthalmologist to send a report to your primary provider after each appointment.
- Keep all follow-up appointments.

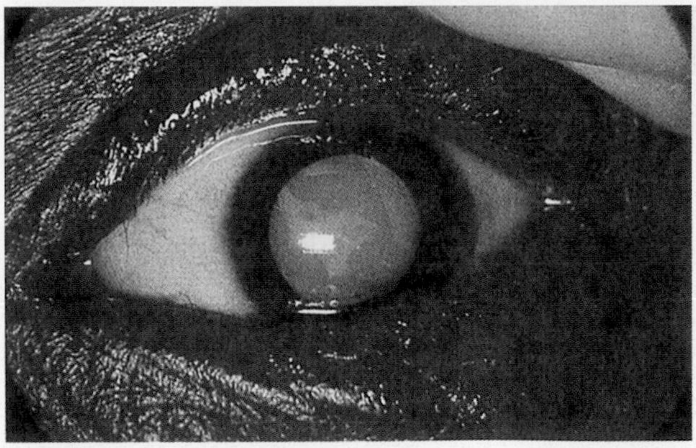

FIGURE 63-8 • A cataract is a cloudy or opaque lens. On visual inspection, the lens appears gray or milky. From Rubin, R., Strayer, D. S., & Rubin, E. (2012). *Rubin's pathology: Clinicopathologic foundations of medicine* (6th ed.). Philadelphia: Lippincott Williams & Wilkins.

Reassurance and emotional support are important aspects of care. A lifelong disease involving possible loss of sight has psychological, physical, social, and vocational ramifications. The family must be integrated into the plan of care, and because the disease has a familial tendency, family members should be encouraged to undergo examinations at least once every 2 years to detect glaucoma early.

Cataracts

A cataract is a lens opacity or cloudiness (Fig. 63-8). Cataracts affect nearly 24.4 million Americans who are 40 years or older, or about 1 in 6 people in this age range. By 80 years of age, more than half of all Americans have cataracts. Cataract is a leading cause of blindness in the world (Prevent Blindness America, 2012).

Pathophysiology

Cataracts can develop in one or both eyes at any age and have a variety of risk factors (Chart 63-8). The three most common types of cataracts are defined by their location in the lens: nuclear, cortical, and posterior subcapsular. The extent of visual impairment depends on their size, density, and location in the lens. More than one type can be present in the eye.

A nuclear cataract tends to have a substantial genetic component that causes a central opacity in the lens. It is associated with myopia (i.e., nearsightedness), which worsens when the cataract progresses.

A cortical cataract involves the anterior, posterior, or equatorial cortex of the lens. Cortical cataracts progress at a highly variable rate. Vision is worse in very bright light. Sunlight exposure is a risk factor in cortical cataract formation.

Posterior subcapsular cataracts occur in front of the posterior capsule. This type typically develops in younger people and is associated with prolonged corticosteroid use, diabetes, or ocular trauma. Near vision is diminished, and the eye is increasingly sensitive to glare from bright light. Caucasians are more likely to develop nuclear and posterior subcapsular cataracts, whereas cortical cataracts are more prevalent among African Americans.

Chart 63-8 RISK FACTORS
Cataract Formation

Aging

- Loss of lens transparency
- Clumping or aggregation of lens protein (which leads to light scattering)
- Accumulation of a yellow-brown pigment due to the breakdown of lens protein
- Decreased oxygen uptake
- Increase in sodium and calcium
- Decrease in levels of vitamin C, protein, and glutathione (an antioxidant)

Associated Ocular Conditions

- Retinitis pigmentosa
- Myopia
- Retinal detachment and retinal surgery
- Infection (e.g., herpes zoster, uveitis)

Toxic Factors

- Ionizing radiation
- Aspirin use
- Corticosteroids, especially at high doses and in long-term use
- Alkaline chemical eye burns, poisoning
- Cigarette smoking
- Calcium, copper, iron, gold, silver, and mercury, which tend to deposit in the pupillary area of the lens

Nutritional Factors

- Reduced levels of antioxidants
- Poor nutrition
- Obesity

Physical Factors

- Dehydration associated with chronic diarrhea, the use of purgatives in anorexia nervosa, and the use of hyperbaric oxygenation
- Blunt trauma, perforation of the lens with a sharp object or foreign body, electric shock
- Ultraviolet radiation in sunlight and x-ray

Systemic Diseases and Syndromes

- Diabetes
- Down syndrome
- Disorders related to lipid metabolism
- Renal disorders
- Musculoskeletal disorders

Adapted from American Academy of Ophthalmology Cataract and Anterior Segment Panel (2011). *Preferred Practice Pattern® guidelines, Cataract in the adult eye*. San Francisco: Author.

Clinical Manifestations

Painless, blurry vision is characteristic of cataracts. The person perceives that surroundings are dimmer, as if his or her glasses need cleaning. Light scattering is common, and the person experiences reduced contrast sensitivity, sensitivity to glare, and reduced visual acuity. Other effects include myopic shift (return of ability to do close work [e.g., reading fine print] without eyeglasses), **astigmatism** (refractive error due to an irregularity in the curvature of the cornea), monocular **diplopia** (double vision), and color changes as lens becomes more brown in color (Eliopoulos, 2010).

Assessment and Diagnostic Findings

Decreased visual acuity is directly proportionate to cataract density. The Snellen visual acuity test, ophthalmoscopy, and slit-lamp biomicroscopic examination are used to establish the degree of cataract formation. The degree of lens opacity does not always correlate with the patient's functional status. Some patients can perform normal activities despite clinically significant cataracts. Others with less lens opacification have a disproportionate decrease in visual acuity; hence, visual acuity is an imperfect measure of visual impairment.

Medical Management

No nonsurgical treatment (e.g., medications, eye drops, eyeglasses) cures cataracts or prevents age-related cataracts. Optimal medical management is prevention. Patients should be educated by primary providers about risk reduction strategies such as smoking cessation, weight reduction, and optimal blood sugar control for patients with diabetes, and should be advised to wear sunglasses outdoors to prevent early cataract formation (American Academy of Ophthalmology Cataract and Anterior Segment Panel, 2011).

Surgical Management

In general, if reduced vision from cataract does not interfere with normal activities, surgery may not be needed. In deciding when cataract surgery is to be performed, the patient's functional and visual status should be a primary consideration (Eliopoulos, 2010). Cataract removal is common, with more than 1 million such surgeries performed in the United States each year (Prevent Blindness America, 2012). Surgery is performed on an outpatient basis and usually takes less than 1 hour, with the patient being discharged in 30 minutes or less afterward. Although complications from cataract surgery are uncommon, they can have significant effects on vision (Table 63-6). Restoration of visual function through a safe and minimally invasive procedure is the surgical goal, which is achieved with advances in topical anesthesia, smaller wound incision (i.e., clear cornea incision), and lens design (i.e., foldable and more accurate intraocular lens [IOL] measurements).

Injection-free topical and intraocular anesthesia, such as 1% lidocaine gel applied to the surface of the eye, eliminates the hazards of regional (retrobulbar and peribulbar) anesthesia, such as ocular perforation, retrobulbar hemorrhage, optic injuries, diplopia, and ptosis, and is ideal for patients receiving anticoagulants. Furthermore, patients can communicate and cooperate during surgery. IV moderate sedation may be used to minimize anxiety and discomfort.

When both eyes have cataracts, one eye is treated first, with at least several weeks, preferably months, separating the two procedures. Because cataract surgery is performed to improve visual functioning, the delay for the other eye gives time for the patient and the surgeon to evaluate whether the results from the first surgery are adequate to preclude the need for a second operation. The delay also provides time for the first eye to recover; if there are any

TABLE 63-6 Potential Complications of Cataract Surgery

Complication	Effects	Management and Outcome
Immediate Preoperative		
Retrobulbar hemorrhage—can result from retrobulbar infiltration of anesthetic agents if the short ciliary artery is located by the injectia	Increased IOP, proptosis, lid tightness, and subconjunctival hemorrhage with or without edema	Emergent lateral canthotomy (slitting of the canthus) is performed to stop central retinal perfusion when the IOP is dangerously elevated. If this procedure fails to reduce IOP, a puncture of the anterior chamber with removal of fluid is considered. The patient must be closely monitored for at least a few hours. Postponement of cataract surgery for 2–4 weeks is advised. Complications such as iris prolapse, vitreous loss, and choroidal hemorrhage could result in a catastrophic visual outcome.
Intraoperative		
Rupture of the posterior capsule	May result in loss of vitreous	Anterior vitrectomy is required if vitreous loss occurs.
Suprachoroidal (expulsive) hemorrhage—profuse bleeding into the suprachoroidal space	Extrusion of intraocular contents from the eye or opposition of retinal surfaces	Closure of the incision and administration of a hyperosmotic agent to reduce IOP or corticosteroids to reduce intraocular inflammation. Vitrectomy is performed 1–2 weeks later. Visual prognosis is poor; some useful vision may be salvaged on rare occasions.
Early Postoperative		
Acute bacterial endophthalmitis—devastating complication that occurs in about 1 in 1,000 cases; the most common causative organisms are *Staphylococcus epidermidis*, *Staphylococcus aureus*, *Pseudomonas* and *Proteus* species	Characterized by marked visual loss, pain, lid edema, hypopyon, corneal haze, and chemosis	Managed by aggressive antibiotic therapy. Broad-spectrum antibiotics are administered while awaiting culture and sensitivity results. Once results are obtained, the appropriate antibiotics are administered via intravitreal injection. Corticosteroid therapy is also administered.
Toxic anterior segment syndrome—noninfectious inflammation that is a complication of anterior chamber surgery; caused by a toxic agent such as an agent used to sterilize surgical instruments	Corneal edema occurs <24 hours after surgery; symptoms include reduced visual acuity and pain.	If there is no growth of microorganisms, the treatment is topical steroids alone.
Late Postoperative		
Suture-related problems	Toxic reactions or mechanical injury from broken or loose sutures	Suture removal relieves the symptoms. Topical corticosteroids are used when the incision is not healed and sutures cannot be removed.
Malposition of the IOL	Results in astigmatism, sensitivity to glare, or appearance of halos	Miotics are used for mild cases, whereas IOL removal and replacement is necessary for severe cases.
Chronic endophthalmitis	Persistent, low-grade inflammation and granuloma	Corticosteroids and antibiotics are administered systemically. If the condition persists, removal of the IOL and capsular bag, vitrectomy, and intravitreal injection of antibiotics are required.
Opacification of the posterior capsule—most common late complication of extracapsular cataract extraction	Visual acuity is diminished.	Nd:YAG laser is used to create a hole in the posterior capsule. Blurred vision is cleared immediately.

IOP, intraocular pressure; IOL, intraocular lens; Nd:YAG, Neodymium: yttrium aluminum garnet.

complications, the surgeon may decide to perform the second procedure differently.

Phacoemulsification

In this method of extracapsular cataract surgery, a portion of the anterior capsule is removed, allowing extraction of the lens nucleus and cortex while the posterior capsule and zonular support are left intact. An ultrasonic device is used to liquefy the nucleus and cortex, which are then suctioned out through a tube. An intact zonular–capsular diaphragm provides the needed safe anchor for the posterior chamber IOL. The pupil is dilated to 7 mm or greater (Peng, Fong, Phaik, et al., 2012). The surgeon makes a small incision on the upper edge of the cornea and a viscoelastic substance (clear gel) is injected into the space between the cornea and the lens. This prevents the space from collapsing and facilitates insertion of the IOL. Because the incision is smaller than the manual extracapsular cataract extraction, the wound heals more rapidly, and there is early stabilization of refractive error and less astigmatism.

Lens Replacement

After removal of the crystalline lens, the patient is referred to as aphakic (i.e., without lens). The lens, which focuses light on the retina, must be replaced for the patient to see clearly. There are three lens replacement options: aphakic eyeglasses, contact lenses, and IOL implants.

Aphakic glasses, although effective, are rarely used. Objects are magnified by 25%, making them appear closer than they actually are. This magnification creates distortion. Peripheral vision is also limited, and **binocular vision** (i.e., ability of both eyes to focus on one object and fuse the two images into one) is impossible if the other eye is aphakic (without a natural lens).

Contact lenses provide patients with almost normal vision, but because contact lenses need to be removed occasionally, the patient also needs a pair of aphakic glasses. Contact lenses are not advised for patients who have difficulty inserting, removing, and cleaning them. Frequent handling and improper disinfection increase the risk of infection.

Insertion of IOLs during cataract surgery is the most common approach to lens replacement (Eliopoulos, 2010). After cataract extraction, or phacoemulsification, the surgeon implants an IOL. Cataract extraction and posterior chamber IOLs are associated with a relatively low incidence of complications (e.g., eye infection, loss of vitreous humor, and slipping of the implant). IOL implantation is contraindicated in patients with recurrent uveitis, proliferative diabetic retinopathy, neovascular glaucoma, or rubeosis iridis.

Nursing Management

Providing Preoperative Care

The patient with cataracts receives the usual preoperative care for ambulatory surgical patients undergoing eye surgery. The standard battery of preoperative tests (e.g., complete blood count, electrocardiogram, and urinalysis) commonly performed for most surgeries is prescribed only if indicated by the patient's medical history.

Alpha-antagonists (particularly tamsulosin [Flomax], which is used for treatment of enlarged prostate) are known to cause a condition called *intraoperative floppy iris syndrome*. Alpha-antagonists can interfere with pupil dilation during the surgical procedure, resulting in miosis and iris prolapse and leading to complications. Intraoperative floppy iris syndrome can occur even though a patient has stopped taking the drug. The nurse needs to ask patients about a history of taking alpha-antagonists. Surgical team members are then alerted to the risk of this complication (Chatziralli & Sergentanis, 2011).

Dilating drops are administered prior to surgery. Nurses in the ambulatory surgery setting begin patient education about eye medications (antibiotic, corticosteroid, and anti-inflammatory drops) that will need to be self-administered to prevent postoperative infection and inflammation.

Providing Postoperative Care

Before discharge, the patient receives verbal and written education regarding eye protection, administration of medications, recognition of complications, activities to avoid, and obtaining emergency care (Chart 63-9). An eye shield is usually worn at night for the first week to avoid injury. The nurse also explains that there should be minimal discomfort after surgery and educates the patient about taking a mild analgesic agent, such as acetaminophen, as needed. Antibiotic, anti-inflammatory, and corticosteroid eye drops or ointments are prescribed postoperatively.

Promoting Home and Community-Based Care

Educating Patients About Self-Care

To prevent accidental rubbing or poking of the eye, the patient wears a protective eye patch for about the first 24 hours after surgery, followed by eyeglasses worn during the day and an eye shield at night. The nurse educates the patient and family about applying and caring for the eye shield, if one is recommended. Sunglasses should be worn while outdoors during the day because the eye is sensitive to light.

Slight morning discharge, some redness, and a scratchy feeling may be expected for a few days. A clean, damp washcloth may be used to remove slight morning eye discharge. Because cataract surgery increases the risk of retinal detachment, the patient must know to notify the surgeon if new floaters (dots) in vision, flashing lights, decrease in vision, pain, or increase in redness occurs.

Continuing Care

If an eye patch is worn, it is removed after the first follow-up appointment, which should occur within 48 hours of surgery. Nurses should educate patients about the importance of keeping their follow-up appointments, because monitoring of visual status and prompt intervention of

Chart 63-9 HOME CARE CHECKLIST
Intraocular Lens Implant

At the completion of the home care education, the patient or caregiver will be able to:	PATIENT	CAREGIVER
• Wear glasses or eye shield following surgery as instructed.	✔	
• Always wash hands before touching or cleaning the postoperative eye.	✔	✔
• Clean postoperative eye with a clean tissue; wipe the closed eye with a single gesture from the inner canthus outward.	✔	✔
• When bathing or showering, shampoo hair cautiously or seek assistance.	✔	
• Avoid lying on the side of the affected eye the night after surgery.	✔	
• Keep activity light (e.g., walking, reading, watching television). Resume the following activities only as directed by the ophthalmologist: driving, sexual activity, unusually strenuous activity.	✔	
• Avoid lifting, pushing, or pulling objects heavier than 15 pounds.	✔	
• Avoid bending or stooping for an extended period.	✔	
• Be careful when climbing or descending stairs.	✔	
• Know when to contact the ophthalmologist.*	✔	✔

*Contact the ophthalmologist immediately if any of the following problems occur before the next appointment: (1) vision changes; (2) continuous flashing lights appear to the affected eye; (3) redness, swelling, or pain increase in the eye; (4) the amount or type of eye drainage changes; (5) the eye is injured in any way; (6) significant pain is not relieved by acetaminophen.

postoperative complications enhance good visual outcome. Vision is stabilized when the eye is completely healed, usually within 6 to 12 weeks, when final corrective prescription is completed. Visual correction may still be needed for any remaining refractive errors. Patients who choose multifocal IOLs should be aware that there may be increased night glare and contrast sensitivity.

CORNEAL DISORDERS

Corneal Dystrophies

Corneal dystrophies are inherited as autosomal dominant traits and manifest when the person is about 20 years of age. They are characterized by deposits in the corneal layers. Decreased vision is caused by the irregular corneal surface and corneal deposits. Corneal endothelial decomposition leads to corneal edema and blurring of vision. Persistent edema leads to **bullous keratopathy** (formation of blisters that cause pain and discomfort on rupturing). The two main types are **keratoconus** (cone-shaped deformity of the cornea) and Fuchs endothelial dystrophy.

Keratoconus

Keratoconus, the most common type of corneal dystrophy, is characterized by a conical protuberance of the cornea with progressive thinning on protrusion and irregular astigmatism. This hereditary condition has a higher incidence among women. Onset occurs at puberty; the condition may progress for more than 20 years and is bilateral. Corneal scarring occurs in severe cases. Blurred vision is a prominent symptom. Rigid, gas-permeable contact lenses correct irregular astigmatism and improve vision. Advances in contact lens design have reduced the need for surgery. Penetrating keratoplasty (PKP) is indicated when contact lens correction is no longer effective.

Fuchs Endothelial Dystrophy

Fuchs (pronounced *Fooks*) dystrophy is manifested by a slow death of cells in the endothelial cornea. It affects women more than men and is usually not noted until 50 years of age. Corneal endothelial cell death leads to corneal edema and blurring of vision. Persistent edema leads to bullous keratopathy. A bandage contact lens is used to flatten the bullae, protect the exposed corneal nerve endings, and relieve discomfort. Symptomatic treatments, such as hypertonic drops or ointment (5% sodium chloride), may reduce epithelial edema. Currently, the only cure is a corneal transplant (Kirkwood & Kirkwood, 2010).

Corneal Surgeries

Among the surgical procedures used to treat diseased corneal tissue are phototherapeutic keratectomy (PTK), PKP and corneal endothelial transplantation, and Descemet's stripping endothelial keratoplasty (DSEK).

Phototherapeutic Keratectomy

PTK is a laser procedure that is used to treat diseased corneal tissue by removing or reducing corneal opacities and smoothing the anterior corneal surface to improve functional vision. PTK is contraindicated in patients with active herpetic keratitis because the ultraviolet rays may reactivate latent virus. Common side effects are induced hyperopia and stromal haze. Complications are delayed re-epithelialization (particularly in patients with diabetes) and bacterial keratitis. Postoperative management consists of oral analgesic agents for eye pain. Re-epithelialization is promoted with a pressure patch or therapeutic soft contact lens. Antibiotic and corticosteroid ointments and NSAIDs are prescribed postoperatively. Follow-up examinations are required for up to 2 years.

Penetrating Keratoplasty

PKP (corneal transplantation or corneal grafting) involves replacing abnormal host tissue with healthy donor (cadaver) corneal tissue. Common indications are keratoconus, corneal dystrophy, corneal scarring from herpes simplex keratitis, and chemical burns.

Several factors affect the success of the graft: the condition of the ocular structures (e.g., lids, conjunctiva), quality of the tears, adequacy of blinking, and viability of the donor endothelium. Contraindications for the use of donor tissue are outlined in Chart 63-10.

In PKP, the surgeon determines the graft size before the procedure, and the appropriate size is marked on the surface of the cornea. The surgeon prepares the donor cornea and the recipient bed, removes the diseased cornea, places the donor cornea on the recipient bed, and sutures it in place.

Chart 63-10 Contraindications to the Use of Donor Tissue for Corneal Transplantation: Donor Characteristics

Systemic Disorders

- Death from unknown cause
- History or suspected history of the following:
  - Acquired immunodeficiency syndrome or high-risk for human immunodeficiency virus infection
  - Creutzfeldt-Jacob disease
  - Eye infection
  - Hepatitis
  - Rabies
  - Systemic infection

Intrinsic Eye Disease

- Disorders of the conjunctiva or corneal surface involving the optical zone of the cornea
- Malignant tumors of anterior segment
- Ocular inflammation
- Retinoblastoma

Other

- Corneal scars
- History of eye trauma
- Previous surgical eye procedures such as corneal graft or LASIK eye surgery

Sutures remain in place for 12 to 18 months and are then removed. Potential complications include early graft failure due to poor quality of donor tissue, surgical trauma, acute infection, and persistently increased IOP and late graft failure due to rejection.

Postoperatively, the patient receives mydriatics for 2 weeks and topical corticosteroids for 12 months (daily doses for 6 months and tapered doses thereafter). These mydriatics and corticosteroid drops should be preservative free to prevent a reactive inflammation. Patients typically describe a sensation of postoperative eye discomfort rather than acute pain.

Additional Keratoplasty Surgeries

For many years, PKP was the standard of care for patients with corneal endothelial failure but with poor refractive results. Recently, several newer techniques have been developed. Anterior lamellar keratoplasty (ALK) involves a partial-thickness graft for disorders such as anterior corneal dystrophies or scarring not involving the endothelial portion of the cornea. Deep anterior lamellar keratoplasty (DALK) involves replacement of only anterior corneal layers and not Descemet's membrane or endothelial layers for disorders such as keratoconus. Posterior lamellar keratoplasty (PLK or EK) and DSEK (which replaces only the endothelial layer of the cornea) are procedures indicated for conditions such as Fuchs endothelial dystrophy or bullous keratopathy (Kirkwood & Kirkwood, 2010). In DSEK, layers of the cornea are dissected and selectively replaced by donor cornea tissue. DSEK offers several advantages, such as less postoperative astigmatism, faster visual recovery, and stronger wound integrity. Theoretically, the risk of rejection is less because less of the patient's tissue is replaced.

Keratoprosthesis

An additional therapeutic option for patients with multiple graft failures or severe corneal disease is keratoprosthesis (artificial cornea; e.g., U.S. Food and Drug Administration [FDA]-approved Boston Keratoprosthesis [KPro] and AlphaCor). The general design of the keratoprosthesis consists of a central optic core and an outer rim that secures the prosthesis. This high-risk and controversial procedure has serious complications (e.g., glaucoma, endophthalmitis) and requires close follow-up and monitoring. However a report of long-term outcomes indicates that keratoprosthesis may be an option for corneal blindness (Pujari, Siddique, & Dohlman, 2011).

Nursing Management

For corneal surgeries, the nurse reinforces the surgeon's recommendations and instructions regarding visual rehabilitation and visual improvement. A technically successful graft may initially produce disappointing results because the procedure has produced a new optical surface. Only after several months do patients start seeing the natural and true colors of their environment. Correction of a resultant refractive error with eyeglasses or contact lenses determines the final visual outcome. The nurse assesses the patient's support system and his or her ability to comply with long-term follow-up, which includes frequent clinic visits for several months for tapering of topical corticosteroid therapy, selective suture removal, and ongoing evaluation of the graft site and visual acuity. The nurse also initiates appropriate referrals to community services when indicated.

Because graft failure is an ophthalmic emergency that can occur at any time, the primary goal of nursing care is to educate the patient to identify signs and symptoms of graft failure. The early symptoms are blurred vision, discomfort, tearing, or redness of the eye. Decreased vision results after graft destruction. The patient must contact the ophthalmologist as soon as symptoms occur. Treatment of graft rejection usually involves prompt administration of hourly topical corticosteroids and periocular corticosteroid injections. Systemic immunosuppressive agents may be necessary for severe, resistant cases.

Refractive Surgeries

Refractive surgeries are elective procedures performed to correct refractive errors (myopia or hyperopia) and astigmatism by reshaping the cornea. Laser vision correction alters the major optical function of the eye and carries certain surgical risks. The patient must fully understand the benefits, potential risks and complications, common side effects, and limitations of the procedure. Refractive surgery does not alter the normal aging process of the eye. If the reason for the procedure is to meet vision requirements for the patient's occupation, the results must satisfy both the patient and the employer. Precise visual outcome cannot be guaranteed. Typically, patients must be at least 18 years of age (American Academy of Ophthalmology Refractive Management/Intervention Preferred Practice Panel, 2012).

The corneal structure must be normal, and the refractive error must be stable. The patient is required to discontinue using contact lenses for a period before the procedure (2 to 3 weeks for soft lenses and 4 weeks for hard lenses). Patients with conditions that are likely to adversely affect corneal wound healing (e.g., corticosteroid use, immunosuppression, elevated IOP) are not good candidates for the procedure. Any superficial eye disease must be diagnosed and fully treated before a refractive procedure.

Patient satisfaction is the ultimate goal; therefore, patient education and counseling about potential risks, complications, and postoperative follow-up are critical. Minimal postoperative care includes topical corticosteroid or NSAID and antibiotic drops.

Laser Vision Correction Photorefractive Keratectomy

Photorefractive keratectomy (PRK) is used to treat myopia and hyperopia with or without astigmatism. The excimer laser is applied directly to the cornea according to carefully calculated measurements. For myopia, the relative curvature is decreased; for hyperopia, the relative curvature is increased. A bandage contact lens is placed over the cornea to promote epithelial healing and reduce pain, which is similar to that of a severe corneal abrasion. The major limitations of this procedure are postoperative pain, corneal haze, and prolonged recovery of vision, particularly night vision (American Academy of

Ophthalmology Refractive Management/Intervention Preferred Practice Panel, 2012).

Laser-Assisted In Situ Keratomileusis

An improvement over PRK, particularly for correcting high (severe) myopia, LASIK involves flattening the anterior curvature of the cornea by removing a stromal lamella or layer. The surgeon creates a corneal flap with a microkeratome, which is an automatic corneal shaper similar to a carpenter's plane. The surgeon retracts a flap of corneal tissue less than one third the thickness of a human hair to access the corneal stroma and then uses the excimer laser on the stromal bed to reshape the cornea according to calculated measurements (Fig. 63-9). LASIK causes less postoperative discomfort, has fewer side effects, and is safer than PRK. The patient has no corneal haze and requires less postoperative care. However, with LASIK, the cornea has been invaded at a deeper level, and any complications are more significant than those that can occur with PRK. With the increasing success and popularity of LASIK, PRK is now reserved for patients who are unsuitable for LASIK, such as people with very thin corneas.

Perioperative Complications

Surgically Induced Abnormalities

Corneal surface irregularities can occur after LASIK treatment. These include central islands (central areas of stiffness or elevation), decentered ablations resulting from misalignment of the laser treatment or from involuntary eye movement during laser treatment, and forms of irregular astigmatism. Symptoms of central islands and decentered ablations include monocular diplopia or ghost images, halos, glare, and decreased visual acuity.

Diffuse Lamellar Keratitis

As LASIK is performed more frequently, the vision-threatening complication known as diffuse lamellar keratitis (DLK) is reported more often. DLK is a peculiar, noninfectious, inflammatory reaction in the lamellar interface after LASIK. DLK seems to be strongly associated with a decrease of contrast sensitivity up to 6 months postoperatively (Han, Wee, Lee, et al., 2007). Depending on the severity of the condition, treatment methods range from administering corticosteroid drops to intervening surgically.

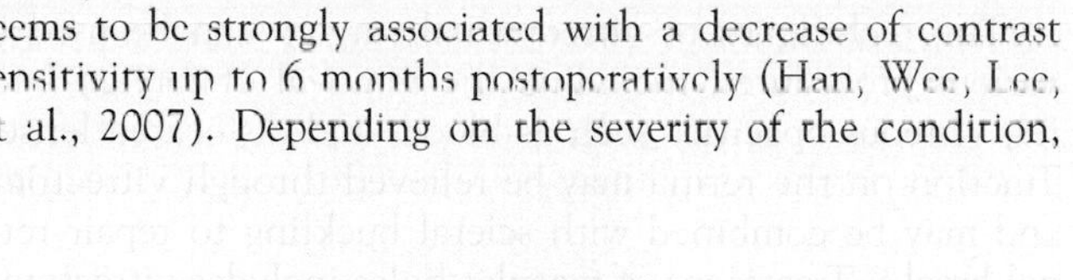
"Flap"
Laser pulse

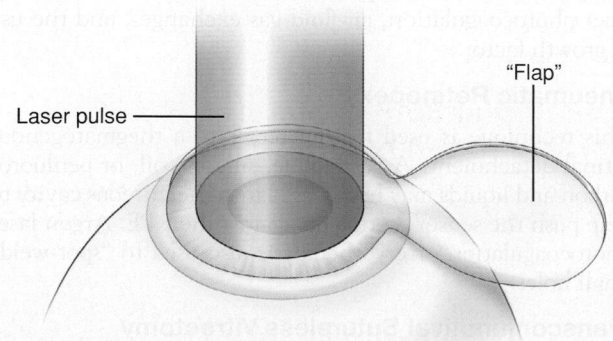

FIGURE 63-9 • LASIK combines delicate surgical procedures and laser treatment. A flap is surgically created and lifted to one side. A laser is then applied to the cornea to reshape it. With permission from the Wilmer Laser Vision Center, Lutherville, MD.

Phakic Intraocular Lenses

Because the results of refractive surgery on high (severe) myopia, hyperopia, and astigmatism are less predictable, there has been increasing interest in the use of phakic IOL implantation in patients who retain their natural lens. These phakic IOLs may be used in either the anterior or posterior chamber. The implantation of such devices is reversible because the natural lens is left in place and the normal architecture of the cornea is preserved. This procedure may provide more predictable refractive results than procedures that alter the corneal curvature. Potential complications include cataract, iritis or uveitis, endothelial cell loss, and increased IOP.

Although phakic IOL implantation provides a more predictable alternative to corneal refractive surgery, more controlled, longitudinal multicenter trials are needed to evaluate its long-term safety. The design of phakic IOLs continues to improve.

Conductive Keratoplasty

A recent innovation in refractive surgery for the correction of low to mild hyperopia uses the principles of thermal keratoplasty by applying radiofrequency current to the peripheral cornea using a thin, handheld probe. It does not involve the removal of cornea tissue. Clinical trials have shown that postprocedure visual acuity, predictability, and stability are as good as, if not better than, with other refractive procedures (Du, Fan, & Asbell, 2007).

RETINAL DISORDERS

Although the retina is composed of multiple microscopic layers, the two innermost layers—the sensory retina and the RPE—are most commonly implicated in retinal disorders.

Retinal Detachment

Retinal detachment refers to the separation of the RPE from the sensory layer. The four types of retinal detachment are rhegmatogenous, traction, a combination of rhegmatogenous and traction, and exudative. *Rhegmatogenous detachment* is the most common form. In this condition, a hole or tear develops in the sensory retina, allowing some of the liquid vitreous to seep through the sensory retina and detach it from the RPE (Fig. 63-10). People at risk for this type of detachment include those with high myopia or those who have aphakia (absence of the natural lens) after cataract surgery. Trauma may also play a role in rhegmatogenous retinal detachment. Between 5% and 10% of all rhegmatogenous retinal detachments are associated with proliferative retinopathy—a retinopathy associated with diabetic neovascularization (see Chapter 51).

Tension, or a pulling force, is responsible for *traction retinal detachment*. An ophthalmologist must ascertain all of the areas of retinal break and identify and release the

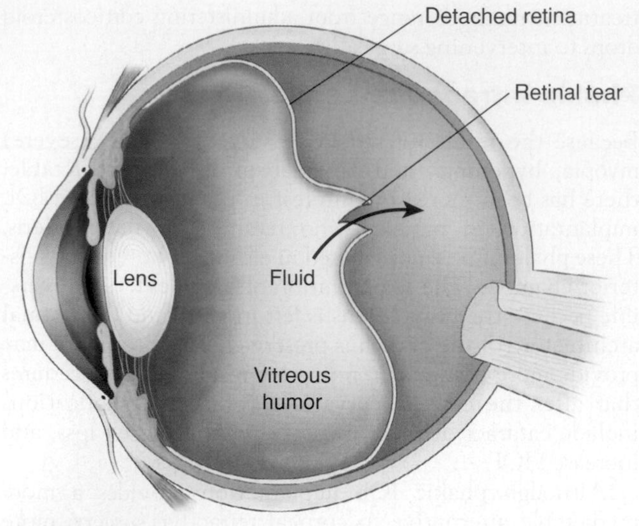
Detached retina
Retinal tear
Lens
Fluid
Vitreous
humor

FIGURE 63-10 • Retinal detachment.

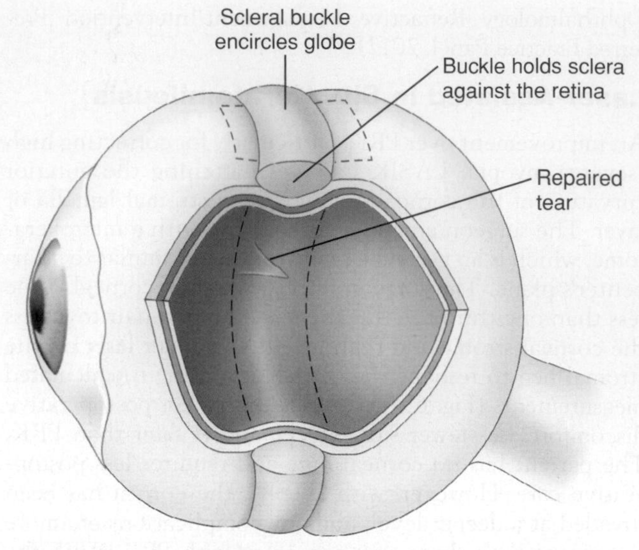
Scleral buckle
encircles globe
Buckle holds sclera
against the retina
Repaired
tear

FIGURE 63-11 • Scleral buckle.

scars or bands of fibrous material providing traction on the retina. Generally, patients with this condition have developed fibrous scar tissue from conditions such as diabetic retinopathy, vitreous hemorrhage, or the retinopathy of prematurity. The hemorrhages and fibrous proliferation associated with these conditions exert a pulling force on the delicate retina.

Patients can have both rhegmatogenous and traction retinal detachment. *Exudative retinal detachments* are the result of the production of a serous fluid under the retina from the choroid. Conditions such as uveitis and macular degeneration may cause the production of this serous fluid.

Clinical Manifestations

Patients may report the sensation of a shade or curtain coming across the vision of one eye, cobwebs, bright flashing lights, or the sudden onset of a great number of floaters. Patients do not complain of pain.

Assessment and Diagnostic Findings

After visual acuity is determined, the patient must have a dilated fundus examination using an indirect ophthalmoscope as well as slit-lamp biomicroscopy. Stereo fundus photography and fluorescein angiography are commonly used during the evaluation.

Increasingly, optical coherence tomography and ultrasound are used for the complete retinal assessment, especially if the view is obscured by a dense cataract or vitreal hemorrhage. All retinal breaks, all fibrous bands that may be causing traction on the retina, and all degenerative changes must be identified.

Surgical Management

In rhegmatogenous detachment, an attempt is made to surgically reattach the sensory retina to the RPE. In traction retinal detachment, the source of traction must be removed and the sensory retina reattached. New surgical techniques as well as advances in instrumentation have led to an increased rate of success of surgical reattachment and better visual outcomes. The most commonly used surgical interventions are the scleral buckle, the pars plana vitrectomy, and pneumatic retinopexy. Another procedure is the 25-gauge transconjunctival sutureless vitrectomy (Chong & Fuller, 2010).

Scleral Buckle

The retinal surgeon compresses the sclera (often with a scleral buckle [Fig. 63-11] or a silicone band) to indent the scleral wall from the outside of the eye and bring the two retinal layers in contact with each other.

Pars Plana Vitrectomy

A vitrectomy is an intraocular procedure that allows the introduction of a light source through an incision; a second incision serves as the portal for the vitrectomy instrument. The surgeon dissects preretinal membranes under direct visualization while the retina is stabilized by an intraoperative vitreous substitute.

The techniques of vitreoretinal surgery can be used in various procedures, including the removal of foreign bodies, vitreous opacities such as blood, and dislocated lenses. Traction on the retina may be relieved through vitrectomy and may be combined with scleral buckling to repair retinal breaks. Treatment of macular holes includes vitrectomy, laser photocoagulation, air-fluid-gas exchanges, and the use of growth factor.

Pneumatic Retinopexy

This technique is used for the repair of a rhegmatogenous retinal detachment. A gas bubble, silicone oil, or perfluorocarbon and liquids may be injected into the vitreous cavity to help push the sensory retina up against the RPE. Argon laser photocoagulation or cryotherapy is also used to "spot-weld" small holes.

Transconjunctival Sutureless Vitrectomy

The 25-gauge transconjunctival sutureless vitrectomy is a significant advancement in vitreoretinal surgery. Replacement of the larger 20-gauge approach with the less invasive 25-gauge technique allows for self-sealing transconjunctival

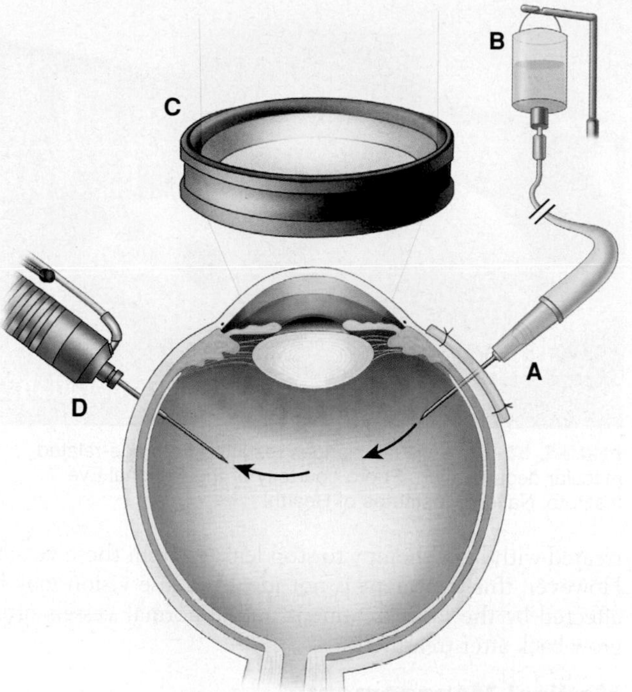
B
C
A
D

FIGURE 63-12 • The 25-gauge transconjunctival pars plana sutureless vitrectomy with simultaneous suction and infusion. **A.** Infusion of fluid through a 25-gauge needle secured by passing through a block of plastic sutured to the eye. **B.** Balanced salt solution is fed into the eye by gravity. **C.** Illumination and observation are provided with the indirect ophthalmoscope. **D.** The vitreous is aspirated into the port of the sharp-tipped vitreous cutter.

pars plana sclerotomies. As a result, postoperative inflammation is decreased, thus promoting rapid wound healing and patient recovery. The 25-gauge microcannula maintains the alignment between the entry site of the conjunctiva and the sclera (Chong & Fuller, 2010) (Fig 63-12).

Complications such as hypotony and endophthalmitis may be related to the unsutured sclerotomy. However, clinical experience has shown that this sutureless system is both safe and effective with decreased surgical times, reduced postoperative inflammation, and more rapid recovery (Chong & Fuller, 2010).

Nursing Management

For the most part, nursing interventions consist of educating the patient and providing supportive care. For pneumatic retinopexy, postoperative positioning of the patient is critical because the injected bubble must float into a position overlying the area of detachment, providing consistent pressure to reattach the sensory retina. The patient must maintain a prone position that would allow the gas bubble to act as a tamponade for the retinal break (Ross & Lavina, 2008). Patients and family members should be made aware of these special needs beforehand so that the patient can be made as comfortable as possible.

In many cases, vitreoretinal procedures are performed on an outpatient basis, and the patient is seen the next day for a follow-up examination and closely monitored thereafter as required. Postoperative complications may include increased IOP, endophthalmitis, other retinal detachments, and development of cataracts. Patients must be educated about the signs and symptoms of complications, particularly of increasing IOP and postoperative infection. Patients should be provided with telephone numbers of members of the ophthalmic team and encouraged to call immediately if discomfort escalates.

Retinal Vascular Disorders

Loss of vision can occur from occlusion of a retinal artery or vein. Such occlusions may result from atherosclerosis, valvular heart disease, venous stasis, hypertension, or increased blood viscosity. Associated risk factors include diabetes, glaucoma, and aging.

Central Retinal Vein Occlusion

Blood supply to and from the ocular fundus is provided by the central retinal artery and vein. Central retinal vein occlusions (CRVOs) are found most often in people older than 50 years. Patients who suffer a CRVO report decreased visual acuity ranging from mild blurring to severely limited.

Direct ophthalmoscopy of the retina shows optic disc swelling, venous dilation and tortuousness, retinal hemorrhages, cotton-wool spots, and a bloody appearance of the retina. The better the initial visual acuity, the better the general prognosis.

Fluorescein angiography may show extensive areas of capillary closure. The patient should be monitored carefully over the ensuing several months for signs of neovascularization and neovascular glaucoma. Laser panretinal photocoagulation may be necessary to treat the abnormal neovascularization. Neovascularization of the iris may cause neovascular glaucoma. Macular edema, macular nonperfusion, and vitreous hemorrhage from the neovascularization are among the potential complications of CRVO.

Branch Retinal Vein Occlusion

Some patients with branch retinal vein occlusion (BRVO) are symptom free, whereas others complain of a sudden loss of vision if the macular area is involved. A more gradual loss of vision may occur if macular edema associated with BRVO develops.

On examination, the ocular fundus appears similar to that found in CRVO. The occlusions generally occur at the arteriovenous crossings. The diagnostic evaluation and follow-up assessments and complications are the same as for CRVO. Associated conditions include glaucoma, systemic hypertension, diabetes, and hyperlipidemia.

Central Retinal Artery Occlusion

Patients with central retinal artery occlusion, a relatively rare disorder that accounts for approximately 1 in 10,000 ophthalmologic visits, present with a sudden loss of vision. Visual acuity is reduced to being able to count the examiner's fingers, or the field of vision is tremendously restricted. A relative afferent pupillary defect is present. Examination of the ocular fundus reveals a pale retina with a cherry-red spot at the fovea. The retinal arteries are thin, and emboli are occasionally seen in the central retinal artery or its branches.

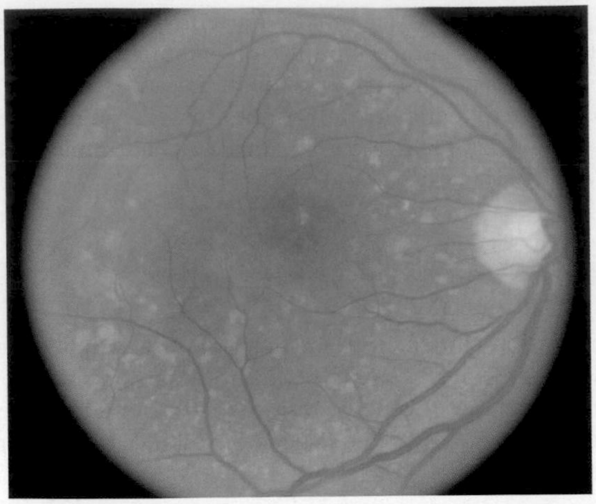

FIGURE 63-13 • Retina showing drusen and age-related macular degeneration.

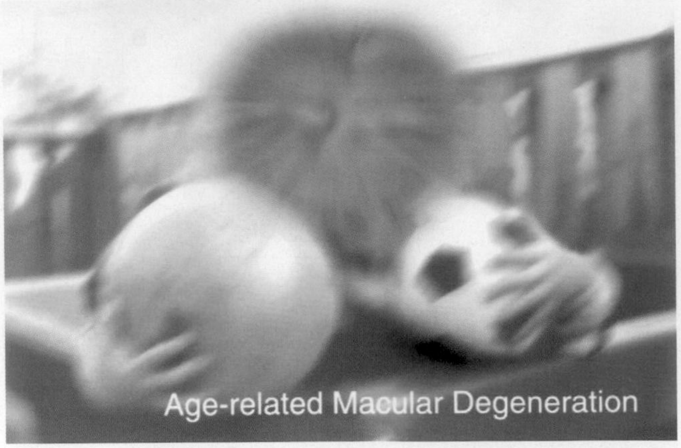
Age-related Macular Degeneration

FIGURE 63-14 • Visual changes resulting from age-related macular degeneration. Photo courtesy of the National Eye Institute, National Institutes of Health.

Central retinal artery occlusion is a true ocular emergency. Various treatments are used: ocular massage, anterior chamber paracentesis, IV administration of hyperosmotic agents such as acetazolamide (Diamox), and high concentrations of oxygen. An aggressive stepwise approach may be beneficial, depending on the underlying cause of the occlusion and the amount of time from onset of occlusion to treatment. Most visual loss associated with central retinal artery occlusion is severe and permanent.

Age-Related Macular Degeneration

AMD is the most common cause of visual loss in people older than 65 years in the United States (Prevent Blindness America, 2011b). AMD is characterized by tiny, yellowish spots called *drusen* beneath the retina (Fig. 63-13). Most people older than 60 years have at least a few small drusen, which are clusters of debris or waste material. When drusen are located in the macular area, they can affect vision. Patients with AMD have a wide range of visual loss, but only a small portion experience total blindness. Central vision is generally the most affected, with most patients retaining peripheral vision (Fig. 63-14). There are two types of AMD: the dry type and the wet type.

Between 85% and 90% of people with AMD have the dry (nonneovascular, nonexudative) type of the condition, in which the outer layers of the retina slowly break down (Fig. 63-15). With this breakdown comes the appearance of drusen. When the drusen occur outside of the macular area, patients generally have no symptoms. When the drusen occur within the macula, however, there is a gradual blurring of vision that patients may notice when they try to read.

The second type of AMD, the wet (neovascular, exudative) type, may have an abrupt onset. Patients report that straight lines appear crooked and distorted or that letters in words appear broken. This effect results from proliferation of abnormal blood vessels growing under the retina, within the choroid layer of the eye, a condition known as choroidal neovascularization. The affected vessels can leak fluid and blood, elevating the retina. Some patients can be treated with laser therapy to stop leakage from these vessels. However, this treatment is not ideal because vision may be affected by the laser treatment and abnormal vessels often grow back after treatment.

Medical Management

There is no known cure for the dry (nonexudative, nonneovascular) type of AMD. The National Eye Institute (NEI) is sponsoring an ongoing clinical trial, the Age-Related Eye Disease Study 2 (AREDS2), to study the effect of lutein and zeaxanthin (carotenoids) or fish oils in protecting the macula and preventing the progression of AMD (NEI, 2013).

An important component of treatment of neovascular (wet, exudative) AMD targets development and progression of angiogenesis (abnormal blood vessel formation). Studies continue toward identification of agents that can be used to inhibit angiogenesis (FDA, 2011).

Vasoproliferation in exudative AMD is believed to be caused by an underlying angiogenic stimulus known as vascular endothelial growth factor (VEGF). Research has resulted in the development of agents that inhibit the development of VEGF and therefore angiogenesis. Ranibizumab (Lucentis) is designed to bind and inactivate all isoenzymes of VEGF. Approved in 2006, it is administered by intravitreal injection once a month. Studies have shown that, on average, patients may gain one to two lines of vision on the Snellen chart after a year of treatment (Bressler, Quigley, & Schein, 2008). Another injectable medication, aflibercept (Eylea), received FDA approval in 2011; patients treated with this medication are reported to have visual improvement similar to ranibizumab (FDA, 2011).

Nursing Management

Amsler grids are given to patients to use in their homes to monitor for a sudden onset or distortion of vision. These may provide the earliest sign that macular degeneration is getting worse. Patients should be encouraged to look at these grids, one eye at a time, several times each week with glasses on if needed for corrected near vision. If there is a change in the grid (e.g., if the lines or squares appear distorted or faded), the patient should notify the ophthalmologist immediately and should arrange to be seen promptly.

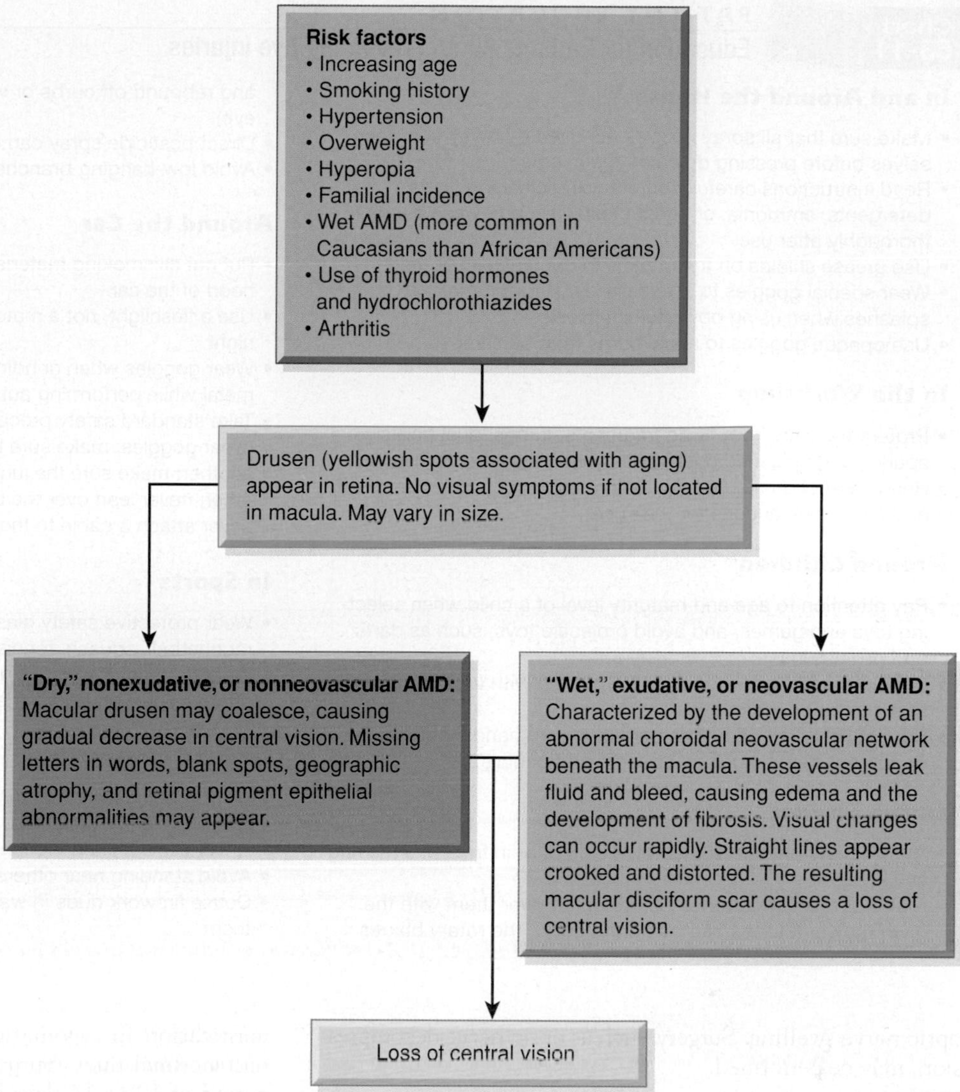
Risk factors
• Increasing age
• Smoking history
• Hypertension
• Overweight
• Hyperopia
• Familial incidence
• Wet AMD (more common in Caucasians than African Americans)
• Use of thyroid hormones and hydrochlorothiazides
• Arthritis
Drusen (yellowish spots associated with aging) appear in retina. No visual symptoms if not located in macula. May vary in size.
"Dry," nonexudative, or nonneovascular AMD: Macular drusen may coalesce, causing gradual decrease in central vision. Missing letters in words, blank spots, geographic atrophy, and retinal pigment epithelial abnormalities may appear.
"Wet," exudative, or neovascular AMD: Characterized by the development of an abnormal choroidal neovascular network beneath the macula. These vessels leak fluid and bleed, causing edema and the development of fibrosis. Visual changes can occur rapidly. Straight lines appear crooked and distorted. The resulting macular disciform scar causes a loss of central vision.
Loss of central vision

FIGURE 63-15 • Progression of age-related macular degeneration (AMD): pathways to vision loss.

ORBITAL AND OCULAR TRAUMA

Whether affecting the eye or the orbit, trauma to the eye and surrounding structures may have devastating consequences for vision. It is preferable to prevent injury rather than treat it. Chart 63-11 details measures to prevent eye injuries.

Orbital Trauma

Injury to the orbit is usually associated with a head injury; hence, the patient's general medical condition must first be stabilized before conducting an ocular examination. Only then is the globe assessed for soft tissue injury. During inspection, the face is meticulously assessed for underlying fractures, which should always be suspected in cases of blunt trauma. To establish the extent of ocular injury, visual acuity is assessed as soon as possible, even if it is only a rough estimate. Soft tissue orbital injuries often result in damage to the optic nerve. Major ocular injuries indicated by a soft globe, prolapsing tissue, ruptured globe, and hemorrhage require immediate surgical attention.

Soft Tissue Injury and Hemorrhage

The signs and symptoms of soft tissue injury from blunt or penetrating trauma include tenderness, ecchymosis, lid swelling, **proptosis or exophthalmos** (abnormal protrusion of the eyeball), and hemorrhage. Closed injuries lead to contusions with subconjunctival hemorrhage, commonly known as a black eye. Blood accumulates in the tissues of the conjunctiva. Hemorrhage may be caused by a soft tissue injury to the eyelid or by an underlying fracture.

Management of soft tissue hemorrhage that does not threaten vision is usually conservative and consists of thorough inspection, cleansing, and repair of wounds. Cold compresses are used in the early phase, followed by warm compresses. Hematomas that appear as swollen, fluctuating areas may be surgically drained or aspirated; if they are causing significant orbital pressure, they may be surgically evacuated.

Penetrating injuries or a severe blow to the head can result in severe optic nerve damage. Visual loss can be sudden or delayed and progressive. Immediate loss of vision after an ocular injury is usually irreversible. Delayed visual loss has a better prognosis. Corticosteroid therapy is indicated to reduce

Chart 63-11

PATIENT EDUCATION

Education for Patients About Preventing Eye Injuries

In and Around the House

- Make sure that all spray nozzles are directed away from themselves before pressing down on the handle.
- Read instructions carefully before using cleaning fluids, detergents, ammonia, or harsh chemicals, and wash hands thoroughly after use.
- Use grease shields on frying pans to decrease spattering.
- Wear special goggles to shield the eyes from fumes and splashes when using powerful chemicals.
- Use opaque goggles to avoid burns from sunlamps.

In the Workshop

- Protect the eyes from flying fragments, fumes, dust particles, sparks, and splashed chemicals by wearing safety glasses.
- Read instructions thoroughly before using tools and chemicals, and follow precautions for their use.

Around Children

- Pay attention to age and maturity level of a child when selecting toys and games, and avoid projectile toys, such as darts and pellet guns.
- Supervise children when they are playing with toys or games that can be dangerous.
- Educate children about the correct way to handle potentially dangerous items, such as scissors and pencils.

In the Garden

- Avoid letting anyone stand at the side of or in front of a moving lawn mower.
- Pick up rocks and stones before going over them with the lawn mower (stones can be hurled out of the rotary blades and rebound off curbs or walls, causing severe injury to the eye).
- Direct pesticide spray can nozzles away from the face.
- Avoid low-hanging branches.

Around the Car

- Put out all smoking materials and matches before opening the hood of the car.
- Use a flashlight, not a match or lighter, to look at the battery at night.
- Wear goggles when grinding metal or striking metal against metal while performing auto body repair.
- Take standard safety precautions when using jumper cables (wear goggles; make sure the cars are not touching one another; make sure the jumper cable clamps never touch each other; never lean over the battery when attaching cables; and never attach a cable to the negative terminal of a dead battery).

In Sports

- Wear protective safety glasses, especially for sports such as racquetball, squash, tennis, baseball, and basketball.
- Wear protective caps, helmets, or face protectors when appropriate, especially for sports such as ice hockey.

Around Fireworks

- Wear eyeglasses or safety goggles.
- Avoid explosive fireworks.
- Never allow children to ignite fireworks.
- Avoid standing near others when lighting fireworks.
- Douse firework duds in water instead of attempting to relight them.

optic nerve swelling. Surgery, such as optic nerve decompression, may be performed.

Orbital Fractures

Orbital fractures are detected by facial x-rays. Depending on the orbital structures involved, orbital fractures can be classified as blowout, zygomatic or tripod, maxillary, midfacial, orbital apex, and orbital roof fractures. Blowout fractures result from compression of soft tissue and the sudden increase in orbital pressure when the force is transmitted to the orbital floor, which is the area of least resistance.

The inferior rectus and inferior oblique muscles, with their fat and fascial attachments, or the nerve that courses along the inferior oblique muscle may become entrapped, and the globe may be displaced inward (i.e., enophthalmos). Computed tomography (CT) scanning can identify the muscle and its auxiliary structures that are entrapped. These fractures are usually caused by blunt small objects, such as a fist, knee, elbow, or tennis or golf ball.

Orbital roof fractures are dangerous because of potential complications to the brain. Surgical management of these fractures requires a neurosurgeon and an ophthalmologist. The most common indications for surgical intervention are displacement of bone fragments disfiguring the normal facial contours, interference with normal binocular vision caused by extraocular muscle entrapment, interference with mastication in zygomatic fracture, and obstruction of the nasolacrimal duct. Surgery is usually nonemergent, and a period of 10 to 14 days gives the ophthalmologist time to assess ocular function, especially the extraocular muscles and the nasolacrimal duct. Emergency surgical repair is usually not performed unless the globe is displaced into the maxillary sinus. Surgical repair is primarily directed at freeing the entrapped ocular structures and restoring the integrity of the orbital floor.

Foreign Bodies

Foreign bodies that enter the orbit are usually tolerated, except for copper, iron, and vegetable materials such as those from plants or trees, which may cause purulent infection. X-rays and CT scans are used to identify the foreign body. A careful history is important, especially if the foreign body has been in the orbit for a period of time and the incident forgotten. It is important to identify metallic foreign bodies because they prohibit the use of magnetic resonance imaging (MRI) as a diagnostic tool.

After the extent of the orbital damage is assessed, the decision to use conservative treatment or surgical removal is made. In general, orbital foreign bodies are removed if they are superficial and anterior in location; have sharp edges that may affect adjacent orbital structures; or are composed of copper, iron, or vegetable material. Surgical intervention is directed at preventing further ocular injury and maintaining the integrity of the affected areas. Cultures are usually obtained, and the patient is placed on prophylactic IV antibiotic medications that are later changed to ones in an oral form.

Ocular Trauma

Ocular trauma is a leading cause of blindness among children and young adults, especially male trauma victims. Ocular trauma occurs with occupational injuries (e.g., construction industry), contact sports, weapons (e.g., air guns, BB guns), assaults, motor vehicle crashes (e.g., broken windshields), and explosions (e.g., blast fragments). A study of military personnel in the Iraq and Afghanistan wars revealed closed-eye ocular injuries that included foreign bodies, angle recession, retinal tears, and optic nerve atrophy, among other ocular injuries (Cockerham, Rice, Hewes, et al., 2011).

There are two types of ocular trauma in which the first response is critical: chemical burn and foreign object in the eye. With a chemical burn, the eye should be immediately irrigated with tap water or normal saline. With a foreign body, no attempt should be made to remove the foreign object. The object should be protected from jarring or movement to prevent further ocular damage. No pressure or patch should be applied to the affected eye. All other traumatic eye injuries should be protected using a patch or shield if available or a stiff paper cup until medical treatment can be obtained (Fig. 63-16).

Assessment and Diagnostic Findings

A thorough history is obtained, particularly assessing the patient's ocular history, such as pre-injury vision in the affected eye or past ocular surgery. Details related to the injury that help in the diagnosis and assessment of the need for further tests include the nature of the ocular injury (i.e., blunt or penetrating trauma); the type of activity that caused the injury to determine the nature of the force striking the eye; and whether onset of vision loss was sudden, slow, or progressive. For chemical eye burns, the chemical agent must be identified and tested for pH if the agent is available. The corneal surface is examined for foreign bodies, wounds, and abrasions, after which the other external structures of the eye are examined. Pupil size, shape, and light reaction of the pupil of the affected eye are compared with the other eye. Ocular motility (ability of the eyes to move synchronously up, down, right, and left) is also assessed.

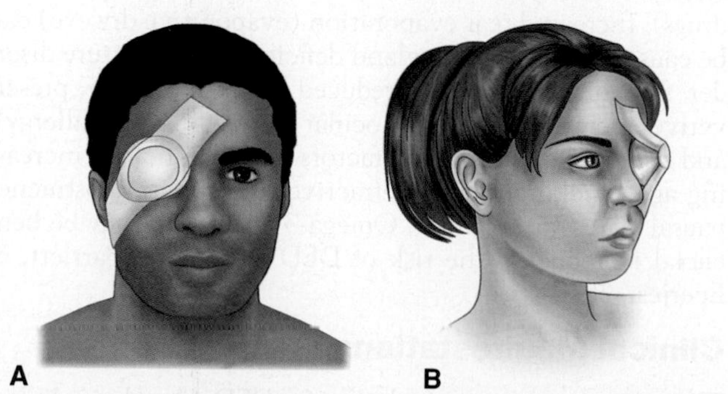
A
B

FIGURE 63-16 • Two kinds of eye patches. **A.** Aluminum shield. **B.** Stiff paper cup shield (innovative substitute when aluminum shield is unavailable).

Medical Management

Splash Injuries

Splash injuries are irrigated with normal saline solution before further evaluation occurs. In cases of a ruptured globe, cycloplegic agents (agents that paralyze the ciliary muscle) or topical antibiotics must be deferred because of potential toxicity to exposed intraocular tissues. Further manipulation of the eye must be avoided until the patient is under general anesthesia. Parenteral, broad-spectrum antibiotics are initiated. Tetanus antitoxin is administered, if indicated, as well as analgesic agents. (Tetanus prophylaxis is recommended for full-thickness ocular and skin wounds.) Any topical ophthalmic medication (e.g., anesthetic, dyes) must be sterile.

Foreign Bodies and Corneal Abrasions

After removal of a foreign body from the surface of the eye, an antibiotic ointment is applied and the eye is patched. The eye is examined daily for evidence of infection until the wound is completely healed.

Contact lens wear is a common cause of corneal abrasion. The patient experiences severe pain and **photophobia** (ocular pain on exposure to light). Corneal epithelial defects are treated with antibiotic ointment and, in some instances, a pressure patch to immobilize the eyelids. Topical anesthetic eye drops must not be given to the patient to take home for repeated use after corneal injury because their effects mask further damage, delay healing, and can lead to corneal scarring.

Penetrating Injuries and Contusions of the Eyeball

Sharp penetrating injury or blunt contusion force can rupture the eyeball. When the globe, cornea, and sclera rupture, rapid decompression or herniation of the orbital contents into adjacent sinuses can occur. Blunt traumatic injuries (with an increased incidence of retinal detachment, intraocular tissue avulsion, and herniation) have a worse prognosis than penetrating injuries. Most penetrating injuries result in marked loss of vision with the following signs: hemorrhagic **chemosis** (edema of the conjunctiva), conjunctival laceration, shallow anterior chamber with or without an eccentrically placed pupil, **hyphema** (blood within the anterior chamber), or vitreous hemorrhage.

Hyphema is caused by contusion forces that tear the vessels of the iris and damage the anterior chamber angle. Preventing rebleeding and prolonged increased IOP are the goals of treatment for hyphema. In severe cases, the patient is hospitalized with moderate activity restriction. An eye shield is applied. Topical corticosteroids are prescribed to reduce inflammation. An antifibrinolytic agent (aminocaproic acid [Amicar]) stabilizes clot formation at the site of hemorrhage. Aspirin is contraindicated. A ruptured globe and severe injuries with intraocular hemorrhage require surgical intervention. Vitrectomy is performed for traumatic retinal detachments. Primary **enucleation** (complete removal of the eyeball and part of the optic nerve) is considered only if the globe is irreparable and has no light perception. It is a general rule that enucleation is performed within 2 weeks of the initial injury (in an eye that has no useful vision after sustaining penetrating injury) to prevent the risk of **sympathetic ophthalmia** (an inflammation created in the uninjured eye by the affected eye that can result in blindness of the uninjured eye).

Intraocular Foreign Bodies

A patient who complains of blurred vision and discomfort should be questioned carefully about recent injuries and exposures. Patients may be injured in a number of different situations and experience an intraocular foreign body (IOFB). Precipitating circumstances can include working in construction, striking metal against metal, being involved in a motor vehicle crash with facial injury, a gunshot wound, grinding-wheel work, and explosions.

IOFB is diagnosed and localized by slit-lamp biomicroscopy and indirect ophthalmoscopy, as well as CT or ultrasonography scanning. MRI is contraindicated because most foreign bodies are metallic and magnetic. It is important to determine the composition, size, and location of the IOFB and affected eye structures. Every effort should be made to identify the type of IOFB and whether it is magnetic. Iron, steel, copper, and vegetable matter may cause intense inflammatory reactions. The incidence of endophthalmitis is high. Surgical excision of the foreign body depends on its location and composition and associated ocular injuries. Specially designed IOFB forceps and magnets are used to grasp and remove the foreign body. Any damaged area of the retina is treated to prevent retinal detachment.

Ocular Burns

Alkali, acid, and other chemically active organic substances, such as Mace and tear gas, cause chemical burns. Alkali burns (e.g., lye, ammonia) result in the most severe injury because they penetrate the ocular tissues rapidly and continue to cause long-term damage. Increased IOP also occurs. Acids (e.g., bleach, car batteries, refrigerant) generally cause less damage because the precipitated necrotic tissue proteins form a barrier to further penetration and damage. Chemical burns may appear as superficial punctate keratopathy (i.e., spotty damage to the cornea), subconjunctival hemorrhage, or complete marbleizing of the cornea.

In treating chemical burns, every minute counts. Immediate irrigation with tap water should be started on site before transport of the patient to an emergency department. A brief history and examination are performed. Critical information, if available, is the name of the substance that went into the eye (the actual container is best). Material safety data sheets (MSDS) should be accessed for reference (see Chapter 73). The corneal surfaces and conjunctival fornices are immediately and copiously irrigated with normal saline or any neutral solution. A local anesthetic is instilled, and a lid speculum is applied to overcome blepharospasm (i.e., spasms of the eyelid muscles that result in closure of the lids). Particulate matter must be removed from the fornices using moistened, cotton-tipped applicators and minimal pressure on the globe. Irrigation continues until the conjunctival pH normalizes (between 7.3 and 7.6). The pH of the corneal surface is checked by placing a pH paper strip in the fornix. Antibiotic agents are instilled, and the eye usually is patched.

The goal of intermediate treatment is to prevent tissue ulceration and promote re-epithelialization of the cornea. Intense lubrication using artificial tears without preservatives (to avoid allergic reactions) is essential. Patching or therapeutic soft lenses may also be used to promote corneal healing, and the patient is closely monitored. Prognosis depends on the type of injury and adequacy of the irrigation immediately after exposure. Long-term treatment consists of two phases: restoration of the ocular surface through grafting procedures and surgical restoration of corneal integrity and optical clarity.

Thermal injury is caused by exposure to a hot object (e.g., curling iron, tobacco, ash), whereas photochemical injury results from ultraviolet irradiation or infrared exposure (e.g., exposure to the reflections from snow, sun gazing, viewing an eclipse of the sun without an adequate filter). These injuries can cause corneal epithelial defect, corneal opacity, conjunctival chemosis and **injection** (congestion of blood vessels), and burns of the eyelids and periocular region. Antibiotic agents and a pressure patch for 24 hours constitute the treatment of mild injuries. Scarring of the eyelids may require oculoplastic surgery, whereas corneal scarring may require corneal surgery.

INFECTIOUS AND INFLAMMATORY CONDITIONS

Inflammation and infections of eye structures are common. Table 63-7 summarizes selected common infections and their treatment.

Dry Eye Disease

Formerly known as dry eye syndrome and keratoconjunctivitis sicca, dry eye disease (DED) of the tears and ocular surface is a multifactorial condition with an underlying inflammatory component (Davis, 2012). DED is caused by a decrease in tear production or increased tear evaporation, which can be episodic or chronic. Decrease in tear production (aqueous deficient) can be caused by systemic disease (Sjögren's, connective tissue disease), lacrimal gland obstruction, and systemic drugs (e.g., diuretics, antihistamines, psychotropic drugs). Increased tear evaporation (evaporative dry eye) can be caused by meibomian gland deficiency, lid aperture disorder, vitamin A deficiency, reduced lid blinking rate, preservatives from topical drugs, ocular surface disease (allergy), and contact lens wear. Risk factors for DED include increasing age, smoking, recent refractive surgery, and postmenopausal status (in women). Omega-3 fatty acids may be beneficial in reducing the risk of DED (Roncone, Bartlett, & Eperjesi, 2010).

Clinical Manifestations

The most common complaints in DED are photophobia, foreign body sensation, burning and stinging, redness, and decreased tearing.

TABLE 63-7 Common Infections and Inflammatory Disorders of Eye Structures

Disorder	Description	Management
Hordeolum (stye)	Acute suppurative infection of the glands of the eyelids caused by *Staphylococcus aureus*. The lid is red and edematous with a small collection of pus in the form of an abscess. There is considerable discomfort.	Warm compresses are applied directly to the affected lid area 3–4 times a day for 10–15 minutes. If the condition is not improved after 48 hours, incision and drainage may be indicated. Application of topical antibiotics may be prescribed thereafter.
Chalazion	Sterile inflammatory process involving chronic granulomatous inflammation of the meibomian glands; can appear as a single granuloma or multiple granulomas in the upper or lower eyelids	Warm compresses applied 3–4 times a day for 10–15 minutes may resolve the inflammation in the early stages. Most often, however, surgical excision is indicated. Corticosteroid injection to the chalazion lesion may be used for smaller lesions.
Blepharitis	Chronic bilateral inflammation of the eyelid margins. There are 2 types: staphylococcal and seborrheic. Staphylococcal blepharitis is usually ulcerative and is more serious due to the involvement of the base of hair follicles. Permanent scarring can result.	The seborrheic type is chronic and is usually resistant to treatment, but the milder cases may respond to lid hygiene. Staphylococcal blepharitis requires topical antibiotic treatment. Instructions on lid hygiene (to keep the lid margins clean and free of exudates) are given to the patient.
Bacterial keratitis	Infection of the cornea by *S. aureus*, *Streptococcus pneumoniae*, and *Pseudomonas aeruginosa*	Fortified (high-concentration) antibiotic eye drops are administered every 30 minutes around the clock for the first few days, then every 1–2 hours. Systemic antibiotics may be administered. Cycloplegics are administered to reduce pain caused by ciliary spasm. Corticosteroid therapy and subconjunctival injections of antibiotics are controversial.
Herpes simplex keratitis	Leading cause of corneal blindness in the United States. Symptoms are severe pain, tearing, and photophobia. The dendritic ulcer has a branching, linear pattern with feathery edges and terminal bulbs at its ends. Herpes simplex keratitis can lead to recurrent stromal keratitis and persist to 12 months with residual corneal scarring.	Many lesions heal without treatment and residual effects. The treatment goal is to minimize the damaging effect of the inflammatory response and eliminate viral replication within the cornea. Penetrating keratoplasty is indicated for corneal scarring and must be performed when the herpetic disease has been inactive for many months.

Assessment and Diagnostic Findings

Slit-lamp examination shows an absent or interrupted tear meniscus at the lower lid margin, and the conjunctiva is thickened, edematous, and hyperemic and has lost its luster (Davis, 2012). A tear meniscus is the crescent-shaped edge of the tear film in the lower lid margin. Chronic dry eyes may result in chronic conjunctival and corneal irritation that can lead to corneal erosion, scarring, ulceration, thinning, or perforation that can seriously threaten vision. Secondary bacterial infection can occur.

Management

Management of DED requires the cooperation of the patient with a regimen that needs to be followed at home for a long period; otherwise, complete relief of symptoms is unlikely. Instillation of artificial tears during the day and an ointment at night is the usual regimen to hydrate and lubricate the eye and preserve a moist ocular surface. Cyclosporine ophthalmic emulsion (Restasis) is an effective agent that increases tear production and is used once daily. Anti-inflammatory medications are also used, and moisture chambers (e.g., moisture chamber spectacles, swim goggles) may provide additional relief.

Patients may become hypersensitive to chemical preservatives such as benzalkonium chloride and thimerosal. For these patients, preservative-free ophthalmic solutions are used. Management of the dry eye syndrome also includes the concurrent treatment of infections, such as chronic blepharitis and acne rosacea, and treating the underlying systemic disease, such as Sjögren's syndrome (an autoimmune disease).

In advanced cases of DED, surgical treatment that includes punctal occlusion, grafting procedures, and lateral tarsorrhaphy (uniting the edges of the lids) are options. Punctal plugs are made of silicone material for the temporary or permanent occlusion of the puncta. These help to preserve the natural tears and prolong the effects of artificial tears. Short-term occlusion is performed by inserting punctal or silicone rods in all four puncta. If tearing is induced by the occlusion, the upper plugs are removed, and the remaining lower plugs are removed in another week. Permanent occlusion is performed only in severe cases in adults who do not develop tearing after partial occlusion and who have results on a repeated Schirmer's test of 2 mm or less (filter paper is used to measure tear production) (Davis, 2012).

Conjunctivitis

Conjunctivitis (inflammation of the conjunctiva) is a common ocular disorder worldwide. It is characterized by a pink appearance (hence the common term *pink eye*) because of subconjunctival blood vessel congestion.

Clinical Manifestations

General symptoms include foreign body sensation, scratching or burning sensation, itching, and photophobia. Conjunctivitis may be unilateral or bilateral, but the infection usually starts in one eye and then spreads to the other eye by hand contact.

Assessment and Diagnostic Findings

The four main clinical features important to evaluate are the type of discharge (watery, mucoid, purulent, or mucopurulent), type of conjunctival reaction (follicular or papillary), presence of pseudomembranes or true membranes, and presence or absence of lymphadenopathy (enlargement of the preauricular and submandibular lymph nodes where the eyelids drain). Pseudomembranes consist of coagulated exudate that adheres to the surface of the inflamed conjunctiva. True membranes form when the exudate adheres to the superficial layer of the conjunctiva, and removal results in bleeding. Follicles are multiple, slightly elevated lesions encircled by tiny blood vessels; they look like grains of rice. Papillae are hyperplastic conjunctival epithelium in numerous projections that are usually seen as a fine mosaic pattern under slit-lamp examination. Diagnosis is based on the distinctive characteristics of ocular signs, acute or chronic presentation, and identification of any precipitating events. Positive results of swab smear preparations and cultures confirm the diagnosis.

Types of Conjunctivitis

Conjunctivitis is classified according to its cause. The major causes are microbial infection, allergy, and irritating toxic stimuli. A wide spectrum of organisms can cause conjunctivitis, including bacteria (e.g., *Chlamydia*), viruses, fungus, and parasites. Conjunctivitis can also result from an existing ocular infection or can be a manifestation of a systemic disease.

Microbial Conjunctivitis

Bacterial Conjunctivitis

Bacterial conjunctivitis can be acute or chronic. The acute type can develop into a chronic condition. Signs and symptoms can vary from mild to severe. Chronic bacterial conjunctivitis is usually seen in patients with lacrimal duct obstruction, chronic dacryocystitis, and chronic blepharitis. The most common causative microorganisms are *Streptococcus pneumoniae*, *Haemophilus influenzae*, and *Staphylococcus aureus*.

Bacterial conjunctivitis manifests with an acute onset of redness, burning, and discharge. There is papillary formation, conjunctival irritation, and injection in the fornices. The exudates are variable but are usually present on waking in the morning. The eyes may be difficult to open because of adhesions caused by the exudate. Purulent discharge occurs in severe acute bacterial infections, whereas mucopurulent discharge appears in mild cases. In gonococcal conjunctivitis, the symptoms are more acute. The exudate is profuse and purulent, and there is lymphadenopathy. Pseudomembranes may be present.

Trachoma is an infectious disease caused by the bacterium *Chlamydia trachomatis*, an ancient disease and the leading cause of preventable blindness in the world. It is prevalent in areas with hot, dry, and dusty climates and in areas with poor living conditions. It is spread by direct contact or by carrier (e.g., insects such as flies and gnats). The onset of trachoma in children is usually insidious, but it can be acute or subacute in adults. The initial symptoms include red inflamed eyes, tearing, photophobia, ocular pain, purulent exudates, preauricular lymphadenopathy, and lid edema. Initial ocular signs include follicular and papillary formations. At the middle stage of the disease, there is an acute inflammation with papillary hypertrophy and follicular necrosis, after which trichiasis (turning inward of hair follicles) and entropion begin to develop. The lashes that are turned in rub against the cornea and, after prolonged irritation, cause corneal erosion and ulceration. The late stage of the disease is characterized by scarred conjunctiva, subepithelial keratitis, abnormal vascularization of the cornea (pannus), and residual scars from the follicles that look like depressions in the conjunctiva (Herbert's pits). Severe corneal ulceration can lead to perforation and blindness.

Inclusion conjunctivitis affects sexually active people who have genital chlamydial infection. Transmission is by oral–genital sex or hand-to-eye transmission. Indirect transmission can occur in inadequately chlorinated swimming pools. The eye lesions usually appear a week after exposure and may be associated with a nonspecific urethritis or cervicitis. The discharge is mucopurulent, follicles are present, and there is lymphadenopathy.

Viral Conjunctivitis

Viral conjunctivitis can be acute and chronic. The discharge is watery, and follicles are prominent. Severe cases include pseudomembranes. The common causative organisms are adenovirus and herpes simplex virus. Conjunctivitis caused by adenovirus is highly contagious. The condition is usually preceded by symptoms of upper respiratory infection. Corneal involvement causes extreme photophobia. Symptoms include tearing, redness, and foreign body sensation that can involve one or both eyes. There is lid edema, ptosis, and conjunctival **hyperemia** (red eyes caused by dilation of blood vessels) (Fig. 63-17). These signs and symptoms vary from mild to severe. Viral conjunctivitis, although self-limited, tends to last longer than bacterial conjunctivitis.

Epidemic keratoconjunctivitis (EKC) is a highly contagious viral conjunctivitis that is easily transmitted from one person to another among household members, schoolchildren, and health care workers. The outbreak of epidemics is seasonal, especially during the summer when people use swimming pools. EKC is most often accompanied by preauricular lymphadenopathy and occasionally periorbital pain. There are marked follicular and papillary formations. This type of conjunctivitis can lead to keratopathy.

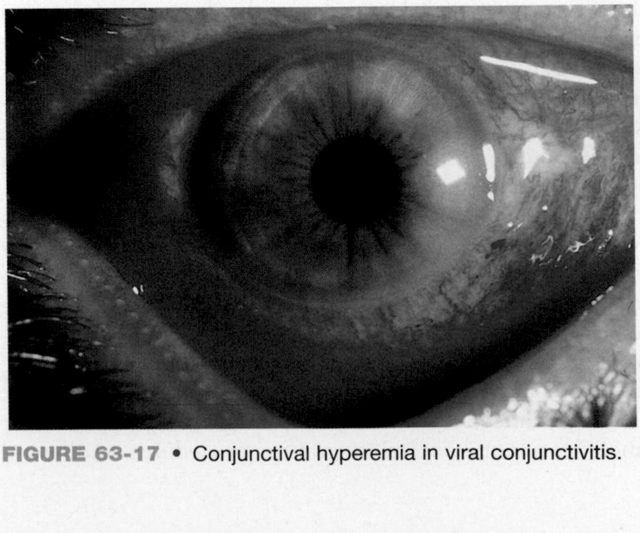

FIGURE 63-17 • Conjunctival hyperemia in viral conjunctivitis.

Allergic Conjunctivitis

Immunologic or allergic conjunctivitis is a hypersensitivity reaction that occurs as part of allergic rhinitis (hay fever), or it can be an independent allergic reaction. The patient usually has a history of an allergy to pollens and other environmental allergens. There is extreme pruritus, epiphora (excessive secretion of tears), injection, and usually severe photophobia. The stringlike mucoid discharge is usually associated with rubbing the eyes because of severe pruritus. Vernal conjunctivitis is also known as seasonal conjunctivitis because it appears mostly during warm weather. There may be large formations of papillae that have a cobblestone appearance. It is more common in children and young adults. Most affected people have a history of asthma or eczema.

Toxic Conjunctivitis

Chemical conjunctivitis can be the result of medications; chlorine from swimming pools; exposure to toxic fumes among industrial workers; or exposure to other irritants such as smoke, hair sprays, acids, and alkalis.

Medical Management

The management of conjunctivitis depends on the type. Most types of mild and viral conjunctivitis are self-limiting, benign conditions that may not require treatment and laboratory procedures. For more severe cases, topical antibiotic agents, eye drops, or ointments are prescribed. Patients with gonococcal conjunctivitis require urgent antibiotic therapy. If left untreated, this ocular disease can lead to corneal perforation and blindness. The systemic complications can include meningitis and generalized septicemia.

Bacterial Conjunctivitis

Acute bacterial conjunctivitis is almost always self-limiting, lasting 2 weeks if left untreated. If treated with antibiotics, it may last a few days, except for gonococcal and staphylococcal conjunctivitis.

For trachoma, usually broad-spectrum antibiotic agents are administered topically and systemically. Surgical management includes the correction of trichiasis (eyelashes growing inward toward the conjunctiva and cornea) to prevent conjunctival scarring.

Adult inclusion conjunctivitis requires 1 week of antibiotics. Prevention of reinfection is important, and affected people and their sexual partners must seek treatment for sexually transmitted infection, if indicated.

Viral Conjunctivitis

Viral conjunctivitis is not responsive to any treatment. Cold compresses may alleviate some symptoms. It is extremely important to remember that viral conjunctivitis, especially EKC, is highly contagious. Patients must be made aware of the contagious nature of the disease, and adequate education must be provided (Chart 63-12).

Proper steps must be taken to avoid health care–associated infections. Frequent hand hygiene and procedures for environmental cleaning and disinfection of equipment used for eye examination must be strictly followed at all times. To prevent spread during outbreaks of conjunctivitis caused by adenovirus, health care facilities must set aside specified areas for treating patients diagnosed with or suspected of having conjunctivitis caused by adenovirus. All forms of tonometry must be avoided unless medically indicated. All multidose ophthalmic medications must be discarded at the end of each day or when contaminated. Infected employees and others must not be allowed to work or attend school until symptoms have resolved, which can take 3 to 7 days.

Chart 63-12 PATIENT EDUCATION

Education for Patients With Viral Conjunctivitis

Viral conjunctivitis is a highly contagious eye infection that can easily spread from one person to another. The symptoms can be alarming but are not serious. The following information will help you understand this eye condition and how to take care of yourself and/or your family member at home.

- Your eyes will look red and will have watery discharge, and your lids will be swollen for about a week.
- You will experience eye pain, a sandy sensation in your eye, and sensitivity to light.
- Symptoms will resolve after about 1 week.
- You may use lightweight cold compresses over your eyes for about 10 minutes four to five times a day to soothe the pain.
- You may use artificial tears for the sandy sensation in your eye and mild pain medications such as acetaminophen (Tylenol).
- You need to stay at home and not go outside. You may return to work or school after 7 days, when the redness and discharge have cleared. You may obtain a note from your primary provider to return to work or school.
- Do not share towels, linens, makeup, or toys.
- Wash your hands thoroughly with soap and water frequently, including before and after you apply artificial tears or cold compresses.
- Use a new tissue every time you wipe the discharge from your eye. You may dampen the tissue with clean water to clean the outside of the eye.
- You may wash your face and take a shower as you normally do.
- Discard all of your makeup articles. You must not apply makeup until the infection has resolved.
- Wear dark glasses if bright lights bother you.
- If the discharge from your eye turns yellowish and purulent or you experience changes in your vision, you need to return to the primary provider for an examination.

Allergic Conjunctivitis

Patients with allergic conjunctivitis, especially recurrent vernal or seasonal conjunctivitis, are usually given corticosteroids in ophthalmic preparations. Depending on the severity of the disease, they may be given oral preparations. The use of vasoconstrictors, such as topical epinephrine solution, cold compresses, ice packs, and cool ventilation usually provide comfort by decreasing swelling.

Toxic Conjunctivitis

For conjunctivitis caused by chemical irritants, the eye must be irrigated immediately and profusely with saline or sterile water.

Uveitis

Inflammation of the uveal tract (uveitis) can affect the iris, the ciliary body, or the choroid. There are two types of uveitis: nongranulomatous and granulomatous.

The more common form of uveitis is the nongranulomatous type, which manifests as an acute condition with pain, photophobia, and a pattern of conjunctival injection, especially around the cornea. The pupil is small or irregular, and vision is blurred. There may be small, fine precipitates on the posterior corneal surface and cells in the aqueous humor (i.e., cell and flare). If the uveitis is severe, a **hypopyon** (accumulation of inflammatory cells in the anterior chamber of the eye) may develop. The condition may be unilateral or bilateral and may be recurrent. Repeated attacks of nongranulomatous anterior uveitis can cause anterior synechiae (peripheral iris adheres to the cornea and impedes outflow of aqueous humor). Posterior synechiae (adherence of the iris and lens) block aqueous outflow from the posterior chamber. Secondary glaucoma can result from either anterior or posterior synechiae. Cataracts may also occur as a sequela to uveitis.

Granulomatous uveitis can have a more insidious onset and can involve any portion of the uveal tract. It tends to be chronic. Symptoms such as photophobia and pain may be minimal. Vision is markedly and adversely affected. Conjunctival injection is diffuse, and there may be vitreous clouding. In a severe posterior uveitis, such as chorioretinitis, there may be retinal and choroidal hemorrhages.

Medical Management

Because photophobia is a common symptom, patients should wear dark glasses outdoors. Ciliary spasm and synechia are best avoided through mydriasis; cyclopentolate (Cyclogyl) and atropine are commonly used. Local corticosteroid drops, such as Pred Forte 1% and Flarex 0.1%, instilled four to six times a day are also used to decrease inflammation. In very severe cases, systemic corticosteroids, as well as intravitreal corticosteroids, may be used. Daclizumab (Zenapax), a monoclonal antibody, is designed to prevent a specific chemical interaction needed by immune cells, such as lymphocytes, to produce inflammation. The NEI of the National Institutes of Health (NIH) has conducted a preliminary clinical trial to examine the safety and effectiveness of treating uveitis with daclizumab with positive results. Larger-scale trials are planned (Yeh, Wroblewski, Buggage, et al., 2008). NSAID therapy in combination with topical and oral preparations is an important adjunct therapy in managing uveitis.

If the uveitis is recurrent, a careful history should be initiated to discover any underlying causes. This evaluation should include a complete history, physical examination, and diagnostic tests, including a complete blood count, erythrocyte sedimentation rate, antinuclear antibodies, and Venereal Disease Research Laboratory (VDRL) and Lyme disease titers. Underlying causes include autoimmune disorders such as ankylosing spondylitis and sarcoidosis as well as toxoplasmosis, herpes zoster virus, ocular candidiasis, histoplasmosis, herpes simplex virus, tuberculosis, and syphilis.

Orbital Cellulitis

Orbital cellulitis is inflammation of the tissues surrounding the eye that may result from bacterial, fungal, or viral inflammatory conditions of contiguous structures, such as the face, oropharynx, dental structures, or intracranial structures. It can also result from foreign bodies and preexisting ocular infection, such as dacryocystitis and panophthalmitis, or from generalized septicemia. Infection of the sinuses is the most frequent cause. Infection originating in the sinuses can spread easily to the orbit through the thin bony walls and foramina or by means of the interconnecting venous system of the orbit and sinuses. The most common causative organisms are staphylococci and streptococci in adults. The symptoms include pain, eyelid swelling, conjunctival edema, proptosis, and decreased ocular motility. With such edema, optic nerve compression can occur and IOP may increase.

The severe intraorbital tension caused by abscess formation and the impairment of optic nerve function in orbital cellulitis can result in permanent visual loss. Because of the orbit's proximity to the brain, orbital cellulitis can lead to life-threatening complications, such as intracranial abscess and cavernous sinus thrombosis.

Medical Management

Immediate administration of high-dose, broad-spectrum, systemic antibiotics is indicated. Cultures and Gram-stained smears are obtained. Monitoring changes in visual acuity, degree of proptosis, central nervous system function (e.g., nausea, vomiting, fever, cognitive changes), displacement of the globe, extraocular movements, pupillary signs, and the fundus is extremely important. Consultation with an otolaryngologist is necessary, especially when rhinosinusitis is suspected. In the event of abscess formation or progressive loss of vision, surgical drainage of the abscess or sinus is performed. Sinusotomy and antibiotic irrigation are also performed.

ORBITAL AND OCULAR TUMORS

Benign Tumors of the Orbit

Benign tumors can develop from infancy and grow rapidly or slowly and present in later life. Some benign tumors are superficial and are easily identifiable by external presentation, palpation, and x-rays, but some are deep and may require a CT scan to diagnose. There can be significant proptosis, and visual function may be jeopardized. Benign tumors are masses characterized by the lack of infiltration in the surrounding tissues. Examples are cystic dermoid cysts and mucocele, hemangiomas, lymphangiomas, lacrimal tumors, and neurofibromas.

To prevent recurrence, benign masses are excised completely when possible. Excision may be difficult because of the involvement of some portions of the orbital bones, such as deep dermoid cysts, in which dissection of the bone is required. Subtotal resection may be indicated in deep benign tumors that intertwine with other orbital structures, such as optic nerve meningiomas. Complete removal of the tumor may endanger visual function.

Benign Tumors of the Eyelids

Benign tumors include a wide variety of neoplasms and increase in frequency with age. Nevi may be unpigmented at birth and may enlarge and darken in adolescence or may never acquire any pigment at all. Hemangiomas are vascular capillary tumors that may be bright, superficial, strawberry-red lesions (i.e., strawberry nevi) or bluish and purplish deeper lesions. Milia are small, white, slightly elevated cysts of the eyelid that may occur in multiples. Xanthelasma are yellowish, lipoid deposits on both lids that commonly appear as a result of the aging of the skin or a lipid disorder. Molluscum contagiosum lesions are flat, symmetric growths along the lid margin caused by a virus that can result in conjunctivitis and keratitis if the lesion grows into the conjunctival sac.

Treatment of benign congenital lid lesions is rarely indicated except when visual function is affected. Corticosteroid injection to the hemangioma lesion is usually effective, but surgical excision may be performed. Benign lid lesions usually present aesthetic problems rather than visual function problems. Surgical excision, or electrocautery, is primarily performed for cosmetic reasons, except for cases of molluscum contagiosum, for which surgical intervention is performed to prevent an infectious process that may ensue.

Benign Tumors of the Conjunctiva

Conjunctival nevus, a congenital benign neoplasm, is a flat, slightly elevated, brown spot that becomes pigmented during late childhood or adolescence. This should be differentiated from the pigmented lesion melanosis, which is acquired at middle age and may become melanoma. Keratin- and sebum-containing dermoid cysts are congenital and can be found in the conjunctiva. Dermolipoma is a congenital tumor that manifests as a smooth, rounded growth in the conjunctiva near the lateral canthus. Papillomas are usually soft with irregular surfaces and appear on the lid margins. Treatment consists of surgical excision.

Malignant Tumors of the Orbit

Rhabdomyosarcoma is the most common malignant primary orbital tumor in childhood; it can also develop in older adults. The symptoms of rhabdomyosarcoma include sudden painless proptosis of one eye followed by lid swelling, conjunctival chemosis, and impairment of ocular motility. Imaging of these tumors establishes the size, configuration, location, and stage of the disease; delineates the degree of bone destruction; and is useful in estimating the field for radiation therapy. The most common site of metastasis is the lung.

Management of these primary malignant orbital tumors involves three major therapeutic modalities: surgery, radiation therapy, and adjuvant chemotherapy. The degree of orbital destruction is important in planning the surgical approach. Resection often involves removal of the eyeball. The psychological needs of the patient and family are paramount in planning the management approach.

Malignant Tumors of the Eyelid

Basal cell carcinoma is the most common malignant tumor of the eyelid. Squamous cell carcinoma occurs less frequently but is considered the second most common malignant tumor. Melanoma is rare. Malignant eyelid tumors occur more frequently among people with a fair complexion who have a history of chronic exposure to the sun (Porth & Matfin, 2009).

Basal cell carcinoma appears as a painless nodule that may ulcerate. The lesion is invasive, spreads to the surrounding tissues, and grows slowly but does not metastasize. It usually appears on the lower lid margin near the inner canthus with a pearly white margin. Squamous cell carcinoma of the eyelids may resemble basal cell carcinoma initially because it also grows slowly and painlessly. It tends to ulcerate and invade the surrounding tissues, but it can metastasize to the regional lymph nodes. Melanoma may not be pigmented and can arise from nevi. It spreads to the surrounding tissues and metastasizes to other organs.

Complete excision of these carcinomas is followed by reconstruction with skin grafting if the surgical excision is extensive. The ocular postoperative site and the graft donor site are monitored for bleeding. Donor graft sites may include the buccal mucosa, the thigh, or the abdomen. The patient is referred to an oncologist for evaluation of the need for radiation therapy and monitoring for metastasis. Early diagnosis and surgical management are the basis of a good prognosis. These conditions have life-threatening consequences, and surgical excisions may result in facial disfigurement. Emotional support is an extremely important aspect of nursing management.

Malignant Tumors of the Conjunctiva

Conjunctival carcinoma most often grows in the exposed areas of the conjunctiva. The typical lesions are usually gelatinous and whitish due to keratin formation. They grow gradually, and deep invasion and metastasis are rare. Melanoma is rare but may arise from a preexisting nevus or acquired melanosis during middle age. Squamous cell carcinoma is also rare but invasive.

The management is surgical incision. Some benign tumors and most malignant tumors recur. To avoid recurrences, patients usually undergo radiation therapy and cryotherapy after the excision of malignant tumors. Cosmetic disfigurement may result from extensive excision when deep invasion by the malignant tumor is involved.

Malignant Tumors of the Globe

Ocular melanoma is a rare malignant choroidal tumor sometimes discovered on a retinal examination. In its early stages, it could be mistaken for a nevus. Many ophthalmologists may practice for decades and never encounter this lesion. For this reason, any patient who is suspected of having ocular melanoma should be immediately referred to an ocular oncologist with experience in this disease.

Although many patients do not have symptoms in the early stages, some patients complain of blurred vision or a change in eye color. A number of such tumors have been found in people with blindness and painful eyes. In addition to a complete physical examination to discover any evidence of metastasis (to the liver, lung, and breast), retinal fundus photography, fluorescein angiography, and ultrasonography are performed. The diagnosis is confirmed at biopsy after enucleation.

Tumors are classified according to boundary lines (apical height and basal diameter) as small, medium, or large. Small tumors are generally monitored, whereas medium and large tumors require treatment. Treatment consists of radiation, enucleation, or both. Radiation therapy may be achieved by external beam performed in repeated episodes over several days or through the implantation of a small plaque that contains radioactive iodine (I-125) pellets over the tumor.

SURGICAL PROCEDURES AND ENUCLEATION

Orbital Surgeries

Orbital surgeries may be performed to repair fractures, remove a foreign body, or remove benign or malignant growths. Surgical procedures involving the orbit and lids affect facial appearance (cosmesis). The goals are to recover and preserve visual function and to maintain the anatomic relationship of the ocular structures to achieve cosmesis. During the repair of orbital fractures, the orbital bones are realigned to follow the anatomic positions of facial structures.

Orbital surgical procedures involve working around delicate structures of the eye, such as the optic nerve, retinal blood vessels, and ocular muscles. Complications of orbital surgical procedures may include blindness as a result of damage to the optic nerve and its blood supply. Sudden pain and loss of vision may indicate intraorbital hemorrhage or compression of the optic nerve. Ptosis and diplopia may result from trauma to the extraocular muscles during the surgical procedure, but these conditions typically resolve after a few weeks.

Prophylaxis with IV antibiotic agents is the usual postoperative regimen after orbital surgery, especially with repair of orbital fractures and intraorbital foreign body removal. IV corticosteroids may be used if there is concern about optic nerve swelling. Topical ocular antibiotics are typically instilled, and antibiotic ointments are applied externally to the skin suture sites.

For the first 24 to 48 hours postoperatively, ice compresses are applied over the periocular area to decrease periorbital swelling, facial swelling, and hematoma. The head of the patient's bed should be elevated to a comfortable position (30 to 45 degrees).

Discharge education should include information about oral antibiotic agents, instillation of ophthalmic medications, and application of ocular compresses.

Enucleation

Enucleation is removal of the eyeball (globe) from the orbit, leaving the muscles and orbital contents intact. It may be surgically performed for the following conditions:

- Injury resulting in prolapse of uveal tissue or loss of light perception
- A blind, painful, deformed, or disfigured eye, usually caused by glaucoma, retinal detachment, or chronic inflammation
- An eye without useful vision that is producing or has produced sympathetic ophthalmia in the other eye
- Intraocular tumors that are untreatable by other means

The procedure for enucleation involves the separation and cutting of each of the ocular muscles and surrounding soft tissue and cutting of the optic nerve from the eyeball. The insertion of an orbital implant typically follows, and the conjunctiva is closed. A large pressure dressing is applied over the area.

Evisceration involves the removal of the intraocular contents through an incision or opening in the cornea or sclera. Evisceration may be surgically performed to treat severe ocular trauma with ruptured globe, severe ocular inflammation, or severe ocular infection. The optic nerve, sclera, extraocular muscles, and sometimes the cornea are left intact. The main advantage of evisceration over enucleation is that the final cosmetic result and motility after fitting the ocular prosthesis are enhanced.

Exenteration is the surgical removal of the entire contents of the orbit, surrounding soft tissue, and most or all of the eyelids. This surgery is indicated in malignancies of the orbit that are life threatening or when more conservative modalities of treatment have failed or are inappropriate. An example is squamous cell carcinoma of the paranasal sinuses, skin, and conjunctiva with deep orbital involvement. In its most extensive form, exenteration may include the removal of all orbital tissues and resection of the orbital bones.

Ocular Prostheses

Orbital implants and conformers (ocular prostheses usually made of silicone rubber) maintain the shape of the eye after enucleation or evisceration to prevent a contracted, sunken appearance. The temporary conformer is placed over the conjunctival closure after the implantation of an orbital implant. A conformer is placed after the enucleation or evisceration procedure to protect the suture line, maintain the fornices, prevent contracture of the socket in preparation for the ocular prosthesis, and promote the integrity of the eyelids.

All ocular prosthetics have limitations in their motility. There are two designs of eye prostheses. The anophthalmic ocular prostheses are used in the absence of the globe. Scleral shells look just like the anophthalmic prosthesis (Fig. 63-18) but are thinner and fit over a globe with intact corneal

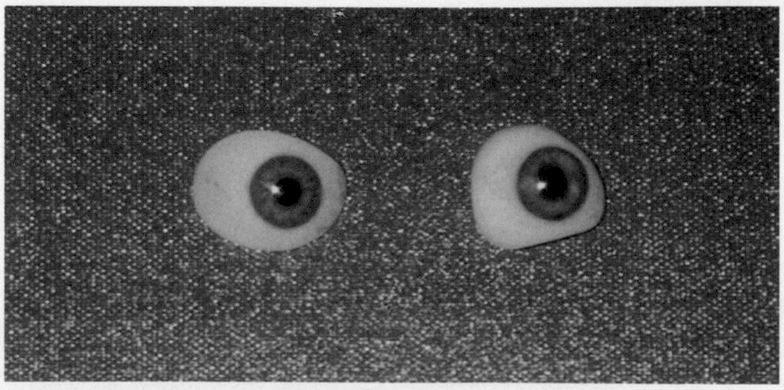

FIGURE 63-18 • Eye prostheses. (*Left*) Anophthalmic ocular prosthesis. (*Right*) Scleral shell.

sensation. An eye prosthesis usually lasts about 6 years, depending on the quality of fit, comfort, and cosmetic appearance. When the anophthalmic socket is completely healed, conformers are replaced with prosthetic eyes.

An ocularist is a specially trained and skilled professional who makes prosthetic eyes. After the ophthalmologist is satisfied that the anophthalmic socket is completely healed and is ready for prosthetic fitting, the patient is referred to an ocularist. The healing period is usually 6 to 8 weeks. It is advisable for the patient to have a consultation with the ocularist before the fitting. Obtaining accurate information and verbalizing concerns can lessen anxiety about wearing an ocular prosthesis.

Medical Management

Removal of an eye has physical, social, and psychological ramifications for any person. The significance of loss of the eye and vision must be addressed in the plan of care. The patient's preparation should include information about the surgical procedure and placement of orbital implants and conformers and the availability of ocular prosthetics to enhance cosmetic appearance. In some cases, patients may choose to see an ocularist before the surgery to discuss ocular prosthetics.

Nursing Management

Education About Postsurgical and Prosthetic Care

Patients who undergo eye removal need to know that they will usually have a large ocular pressure dressing, which is typically removed after a week, and that an ophthalmic topical antibiotic ointment is applied in the socket three times daily.

After the removal of an eye, there is a loss of depth perception. Patients must be advised to take extra caution in their ambulation and movement to avoid miscalculations that may result in injury. It may take some time to adjust to monocular vision.

The patient must be advised that conformers may accidentally fall out of the socket. If this happens, the conformer must be washed, wiped dry, and placed back in the socket.

When surgical eye removal is unexpected, such as in severe ocular trauma, leaving no time for the patient and family to prepare for the loss, the nurse's role in providing emotional support is crucial.

Promoting Home and Community-Based Care

Educating Patients About Self-Care

Patients need to be educated about how to insert, remove, and care for the prosthetic eye. Proper hand hygiene must be observed before inserting and removing an ocular prosthesis. A suction cup may be used if there are problems with manual dexterity. Precautions, such as draping a towel over the sink and closing the sink drain, must be taken to avoid loss of the prosthesis. When instructing patients or family members, a return demonstration is important to assess the level of understanding and ability to perform the procedure.

Before insertion, the inner punctal or outer lateral aspects and the superior and inferior aspects of the prosthesis must be identified by locating the identifying marks, such as a reddish color in the inner punctal area. For people with low vision, other forms of identifying markers, such as dots or notches, are used. The upper lid is raised high enough to create a space and then the patient learns to slide the prosthesis up, underneath, and behind the upper eyelid. Meanwhile, the patient pulls the lower eyelid down to help put the prosthesis in place and to have its inferior edge fall back gradually to the lower eyelid. The lower eyelid is checked for correct positioning.

To remove the prosthesis, the patient cups one hand on the cheek to catch the prosthesis, places the forefinger of the free hand against the midportion of the lower eyelid, and gazes upward. Gazing upward brings the inferior edge of the prosthesis nearer the inferior eyelid margin. With the finger pushing inward, downward, and laterally against the lower eyelid, the prosthesis slides out into the cupped hand.

Continuing Care

An eye prosthesis can be worn and left in place for several months. Hygiene and comfort are usually maintained with daily irrigation of the prosthesis in place with normal saline solution, hard contact lens solution, or artificial tears. In the case of dry eye symptoms, the use of ophthalmic ointment lubricants or oil-based drops, such as vitamin E and mineral oil, can be helpful. Removing crusting and mucus discharge that accumulate overnight is performed with the prosthesis in place. Malpositions may occur when wiping or rubbing the prosthesis in the socket. The prosthesis can be repositioned with the use of clean fingers. Proper wiping of the prosthesis should be a gentle temporal-to-nasal motion to avoid malpositions.

The prosthesis needs to be removed and cleaned when it becomes uncomfortable and when there is increased mucous discharge. The socket should also be rendered free of mucus and inspected for any signs of infection. Any unusual discomfort, irritation, or redness of the globe or eyelids may indicate excessive wear, debris under the shell, or lack of proper hygiene. Any infection or irritation that does not resolve needs medical attention.

OCULAR CONSEQUENCES OF SYSTEMIC DISEASE

Diabetic Retinopathy

One of the most serious complications of diabetes is retinopathy. In the United States today, diabetes is the leading cause of new cases of blindness in people between 20 and 74 years of age (Prevent Blindness America, 2012). With advancements in the treatment of diabetes, more and more patients are able to survive and enjoy relatively a normal lifespan, but they are confronted with the complications of long-term diabetes. (See Chapter 51 for a detailed discussion of diabetic retinopathy.)

Cytomegalovirus Retinitis

Many ophthalmic complications have been associated with AIDS, and CMV is the most common cause of retinal inflammation in patients with AIDS. Early symptoms of CMV retinitis vary from patient to patient. Some patients complain of floaters or a decrease in peripheral vision. Some have a paracentral or central scotoma, whereas others have fluctuations in vision from macular edema. The retina often becomes thin and atrophic and susceptible to retinal tears and breaks.

CMV retinitis generally takes one of three forms: hemorrhagic, brushfire, or granular. In the hemorrhagic type, large areas of white, necrotic retina may be associated with retinal hemorrhage. In the brushfire type, a yellow-white margin begins at the edge of burned-out atrophic retina. This retinitis expands and, if untreated, involves the entire retina. In the granular type, white granular lesions in the periphery of the retina gradually expand. The white, feathery infiltration of the retina destroys sensory retina and leads to necrosis, optic atrophy, and retinal detachment.

Medical Management

Management of CMV retinitis consists of prescribing the appropriate pharmacologic agent.

Pharmacologic Therapy

Pharmacologic agents available for treatment of CMV retinitis include ganciclovir (Cytovene), foscarnet (Foscavir), and cidofovir (Vistide).

Ganciclovir is administered IV, orally, or intravitreously in the acute stage of CMV retinitis. The intravitreous form is available as a 4-mm intraocular implant or insert containing the medication embedded in a polymer-based system that slowly releases the medication. The insert is surgically placed in the posterior segment of the eye, and the medication diffuses locally to the site of the infection over a period of 5 to 8 months before the insert must be replaced. When administered systemically, ganciclovir is a very potent medication; it can cause neutropenia, thrombocytopenia, anemia, and elevated serum creatinine levels. The surgically implanted sustained-release insert enables higher concentrations of ganciclovir to reach the CMV retinitis, but there are risks and complications associated with the inserts, including endophthalmitis, retinal detachment, and hypotony.

Foscarnet inhibits viral deoxyribonucleic acid (DNA) replication. It may be the medication of choice when ganciclovir is ineffective. It may be administered by IV or intravitreal injections. The combination of foscarnet and ganciclovir has been more effective than either medication alone. Nephrotoxicity may occur with systemic foscarnet, and renal function must be monitored carefully.

Cidofovir impedes CMV replication and is administered IV. Cidofovir has been shown to delay the progression of CMV retinitis significantly. Nephrotoxicity, proteinuria, and increased serum creatinine levels are significant side effects.

A nucleoside analogue such as zidovudine (Retrovir) administered in combination with one or more protease inhibitors such as ritonavir (Norvir) in the management of patients with AIDS has led to a suppression of human immunodeficiency virus replication for sustained periods. The immune system can then recover to a functional level. Several patients who had been treated for CMV retinitis have been able to discontinue secondary prophylaxis as their immune systems rebounded (AHRQ Guidelines, 2012). However, some patients develop immune recovery uveitis, characterized by intraocular inflammation, cystoid macular edema, and the formation of epiretinal membranes. Immune recovery uveitis is managed by corticosteroids or by injection of corticosteroids into the sub-Tenon's area of the eye.

Hypertension-Related Eye Changes

Long-standing hypertension is associated with atherosclerosis, and retinal changes are evidenced by the development of retinal arteriolar changes, such as tortuousness, narrowing, and a change in light reflex (Weber & Kelley, 2010). Funduscopic examination reveals a copper or silver coloration of the arterioles and venous compression (arteriovenous nicking) at the arteriolar and venous crossings. Intraretinal hemorrhages from hypertension appear flame shaped because they occur in the nerve fiber layer of the retina.

Hypertension can also occur as an acute consequence of conditions such as pheochromocytoma, acute kidney failure, and pregnancy-induced hypertension. The retinopathy associated with these crisis states is extensive, and the manifestations include cotton-wool spots, retinal hemorrhages, retinal edema, and retinal exudates, often clustered around the macula (Weber & Kelley, 2010).

The choroid is also affected by the profound and abrupt rise in blood pressure and resulting vasoconstriction, and ischemia may result in serious retinal detachments and infarction of the RPE. Ischemic optic neuropathy and papilledema may also result. Blood pressure in these more severe stages should be lowered in a controlled gradual fashion to avoid ischemia of the optic nerve and brain secondary to a too-rapid fall in blood pressure. (See Chapter 31 for further information about hypertension.)

Critical Thinking Exercises

1 A 56-year-old patient presents to the clinic with a new diagnosis of open-angle glaucoma. She lives alone and works full-time as a librarian. Identify the elements of a comprehensive ocular history. Describe your plan for educating this patient about how to manage her glaucoma. What skills will you have her return demonstrate? How will you modify your plan for a patient who has acute angle-closure glaucoma? For a patient who has had glaucoma for 10 years?

2 **ebp** You are working in an outpatient surgery center. A 76-year-old patient presents for cataract surgery. How would you find the evidence for best practices in cataract surgery? What databases and search terms would you use? What kind of research evidence is stronger than others?

3 **pq** A 38-year-old man presents to the emergency department with ocular trauma that he experienced at a construction site. The patient seems quite anxious and asks if he is going to be blind. How will you respond? What information will you give him to decrease his anxiety? What are the priority nursing assessments for ocular trauma and why? What are your priorities for the nursing management of this patient?

Brunner Suite Resources

Explore these additional resources to enhance learning for this chapter:

- NCLEX-Style Questions and Other Resources on thePoint, http://thePoint.lww.com/Brunner13e
- Study Guide
- PrepU
- Clinical Handbook
- Handbook of Laboratory and Diagnostic Tests

References

*Asterisk indicates nursing research.

Books

American Academy of Ophthalmology Cataract and Anterior Segment Panel. (2011). *Preferred Practice Pattern® guidelines, Cataract in the adult eye*. San Francisco: Author.

American Academy of Ophthalmology Refractive Management/Intervention Preferred Practice Panel. (2012). *Preferred Practice Pattern® guidelines, Refractive errors & refractive surgery*. San Francisco: Author.

American Academy of Ophthalmology Vision Rehabilitation Committee. (2012). *Preferred Practice Pattern® guidelines, Vision rehabilitation for adults*. San Francisco: Author.

American Society of Ophthalmic Registered Nurses. (2008). *Core curriculum for ophthalmic nursing* (3rd ed.). San Francisco: Author.

American Society of Ophthalmic Registered Nurses. (2011). *Ophthalmic procedures in the office and clinic* (3rd ed.). San Francisco: Author

Bader, M., & Littlejohns, L. R. (2010). *AANN core curriculum for neuroscience nursing* (5th ed.). Glenview, IL: American Association of Neuroscience Nurses.

Bressler, S. B., Quigley, H. A., & Schein, O. D. (2008). *Vision: The Johns Hopkins white papers*. New York: Medletter Associates.

Eliopoulos, C. (2010). *Gerontological nursing* (7th ed.). Philadelphia: Lippincott Williams & Wilkins.

Fischbach, F. T., & Dunning, M. B. (2011). *Nurse's quick reference to common laboratory and diagnostic tests* (4th ed.). Philadelphia: Lippincott Williams & Wilkins.

Gerstenblith, A. T., & Rabinowitz, M. P. (2012). *The Wills Eye manual: Office and emergency room diagnosis and treatment of eye disease* (6th ed.). Philadelphia: Lippincott Williams & Wilkins.

Jaeger, E. A., & Tasman, W. (2010). *Duane's ophthalmology*. Philadelphia: Lippincott Williams & Wilkins.

Karch, A. (2012). *2012 Lippincott's nursing drug guide*. Philadelphia: Lippincott Williams & Wilkins.

Porth, C. M., & Matfin, G. (2009). *Pathophysiology: Concepts of altered health states* (8th ed.). Philadelphia: Lippincott Williams & Wilkins.

Prevent Blindness America. (2012). *Vision problems in the U.S.: Update to the fifth edition*. Schaumberg, IL: Prevent Blindness America/National Eye Institute.

Riordan-Eva, P., & Cunningham, E. T. (2011) *Vaughan & Asbury's general ophthalmology*. (18th ed.). New York: McGraw-Hill.

Weber, J., & Kelley, J. (2010). *Health assessment in nursing* (4th ed.). Philadelphia: Lippincott Williams & Wilkins.

Journals and Electronic Documents

Agency for Healthcare Research and Quality (AHRQ) Guidelines. (2012). *Guidelines for prevention and treatment of opportunistic infections in HIV-infected adults and adolescents*. Available at: http://www.guideline.gov/content.aspx?id=14320

Amro, R., Cox, C. L., Waddington, K., et al. (2010). Glaucoma expert patient programme and ocular hypotensive treatment. *British Journal of Nursing, 19*(20), 1287–1292.

American Society of Ophthalmic Registered Nurses. (2012). *Ophthalmic Mitomycin C: Top tips for safe handling, use, and disposal*. Available at: asorn.org/client_data/files/2012/10_tipsheet_final.pdf

Bankert, J. (2011). Future of ophthalmic nursing care: Gene therapy. *Insight, 36*(3), 12–15.

Chatziralli, I. P., & Sergentanis, T. N. (2011). Risk factors for intraoperative floppy iris syndrome: A meta-analysis. *Ophthalmology, 118*(4), 730–735.

Chong D. Y., & Fuller, D. D. (2010). The declining use of scleral buckling with vitrectomy for primary retinal detachments. *Archives of Ophthalmology, 128*(9), 1206–1207.

Cockerham, G. C., Rice, T. A., Hewes, E. H., et al. (2011). Closed-eye ocular injuries in the Iraq and Afghanistan wars. *New England Journal of Medicine, 364*(22), 2172–2173.

Davis, C. P. (2012). *Dry eye syndrome*. Available at: www.medicinenet.com/dry_eyes/article.htm

Du, T. T., Fan, V. C., & Asbell, P. A. (2007). Conductive keratoplasty. *Current Opinion in Ophthalmology, 18*(4), 334–337.

Filippopoulos, T., & Rhee, D. J. (2008). Novel surgical procedures in glaucoma: Advances in penetrating glaucoma surgery. *Current Opinion in Ophthalmology, 19*(2), 149–154.

Han, E. S., Wee, W. R., Lee, J. H., et al. (2007). The effect of diffuse lamellar keratitis on visual acuity and contrast sensitivity following LASIK. *Korean Journal of Ophthalmology, 21*(1), 6–10.

Kirkwood, B. J., & Kirkwood, R. A. (2010). Penetrating and lamellar corneal transplantation: Current techniques and indications. *Insight, 35*(3), 6–12.

National Eye Institute. (2013). *Age-Related Eye Disease Study 2 (AREDS2)*. Available at: http://www.nei.nih.gov/areds2/

Ogawa, H., Yonezawa, Y., Maki, H., et al. (2007). A mobile phone-based Communication Support System for elderly persons. *Conference Proceedings: Annual International Conference of the IEEE. Engineering in Medicine & Biology Society, 2007*, 3798–3801.

Peng, L. H., Fong, A., Phaik, C. S., et al. (2012). Improving presurgical pupil dilation for cataract surgery patients. *Joint Commission Journal on Quality and Patient Safety, 38*(9), 420–425.

Prevent Blindness America. (2011a). *Glaucoma laser surgery*. Available at: www.preventblindness.org/glaucoma-laser-surgery

Prevent Blindness America. (2011b). *Age-related macular degeneration (AMD)*. Available at: www.preventblindness.org/age-related-macular-degeneration-amd

Pujari, S., Siddique, S. S., & Dohlman, C. H. (2011). The Boston keratoprosthesis type II: The Massachusetts eye and ear infirmary experience. *Cornea, 30*(12), 1298–1303.

*Reboucas, C., Pagliuca, L. M. F., Sawada, N. O., et al. (2012). Validation of a non-verbal communication protocol for nursing consultations with blind people. *Northeast Network Nursing Journal – Rev Rene (Revistarene), 13*(1), 125–139. Available at: www.revistarene.ufc.br/revista/index.php/revista/article/view/24/20

Roncone, M., Bartlett H., & Eperjesi, F. (2010). Essential fatty acids for dry eye: A review. *Contact Lens & Anterior Eye, 33*(2), 49–54.

Ross, W. H. & Lavina, A. (2008). Pneumatic retinopexy, scleral buckling, and vitrectomy surgery in the management of pseudophakic retinal detachments. *Canadian Journal of Ophthalmology, 43*(1), 65–72.

Sharts-Hopko, N. C., & Glynn-Milley, C. (2009). Primary open-angle glaucoma. Catching and treating the "sneak thief of sight." *American Journal of Nursing, 109*(2), 40–48.

*Sharts-Hopko, N. C., Smeltzer, S., Ott, B. B., et al. (2010). Healthcare experiences of women with visual impairment. *Clinical Nurse Specialist, 24*(3), 149–153.

Yeh, S., Wroblewski, K., Buggage, R., et al. (2008). High-dose humanized anti-IL-2 receptor alpha antibody (daclizumab) for the treatment of active, non-infectious uveitis. *Journal of Autoimmunity, 31*(2), 91–97.

U.S. Food and Drug Administration. (2011). *FDA approves Eylea for eye disorder in older people*. Available at: www.fda.gov/NewsEvents/Newsroom/PressAnnouncements/ucm280601.htm

Resources

American Academy of Ophthalmology, www.aao.org
American Academy of Ophthalmology EyeSmart, www.geteyesmart.org/eyesmart/
American Foundation for the Blind (AFB), www.afb.org
American Society of Ophthalmic Registered Nurses (ASORN), www.asorn.org
Association for Macular Diseases, www.macula.org
Foundation Fighting Blindness, www.blindness.org
Glaucoma Research Foundation, www.glaucoma.org
Lighthouse International, www.lighthouse.org
MAB Community Services, www.mabcommunity.org
Macular Degeneration Foundation, www.eyesight.org
National Association for Visually Handicapped, www.lighthouse.org/navh
National Diabetes Information Clearinghouse (NDIC), diabetes.niddk.nih.gov/dm/pubs/statistics/
National Eye Institute Information Office, www.nei.nih.gov
Prevent Blindness America, www.preventblindness.org
Research to Prevent Blindness, www.rpbusa.org

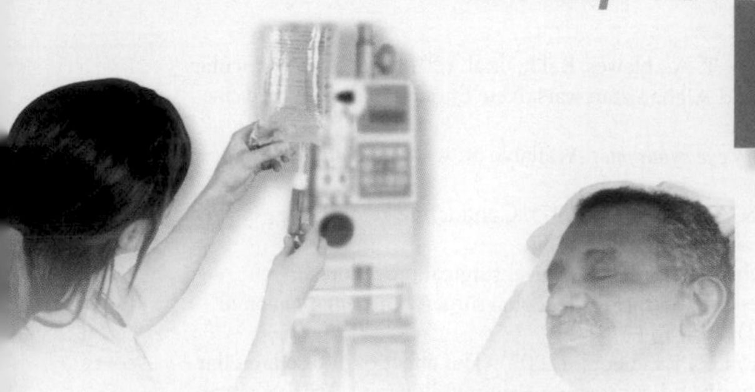

Chapter 64

Assessment and Management of Patients With Hearing and Balance Disorders

Learning Objectives

On completion of this chapter, the learner will be able to:

1 Describe the anatomy and physiology of the ear.
2 Describe methods used to assess hearing and to diagnose hearing and balance disorders.
3 List the manifestations that may be exhibited by a person with a hearing disorder.
4 Identify ways to communicate effectively with a person with a hearing disorder.
5 Differentiate problems of the external ear from those of the middle ear and inner ear.
6 Compare the various types of surgical procedures used for managing middle ear disorders and appropriate nursing care.
7 Use the nursing process as a framework of care for patients undergoing mastoid surgery.
8 Use the nursing process as a framework of care for patients with vertigo.
9 Describe the different types of inner ear disorders, including the clinical manifestations, diagnosis, and management.

Glossary

acute otitis media: inflammation in the middle ear lasting less than 6 weeks
cholesteatoma: tumor of the middle ear or mastoid, or both, that can destroy structures of the temporal bone
chronic otitis media: repeated episodes of acute otitis media causing irreversible tissue damage and persistent tympanic membrane perforation
conductive hearing loss: loss of hearing in which efficient sound transmission to the inner ear is interrupted by some obstruction or disease process
deafness: partial or complete loss of the ability to hear
dizziness: altered sensation of orientation in space
endolymphatic hydrops: dilation of the endolymphatic space of the inner ear; the pathologic correlate of Ménière's disease
exostoses: small, hard, bony protrusions in the lower posterior bony portion of the ear canal
external otitis (i.e., otitis externa): inflammation of the external auditory canal
labyrinthitis: inflammation of the labyrinth of the inner ear
Ménière's disease: condition of the inner ear characterized by a triad of symptoms: episodic vertigo, tinnitus, and fluctuating sensorineural hearing loss
middle ear effusion: fluid in the middle ear without evidence of infection
myringotomy (i.e., tympanotomy): incision in the tympanic membrane
nystagmus: involuntary rhythmic eye movement
ossiculoplasty: surgical reconstruction of the middle ear bones to restore hearing
otalgia: sensation of fullness or pain in the ear
otorrhea: drainage from the ear
otosclerosis: a condition characterized by abnormal spongy bone formation around the stapes
presbycusis: progressive hearing loss associated with aging
rhinorrhea: drainage from the nose
sensorineural hearing loss: loss of hearing related to damage to the end organ for hearing or cranial nerve VIII, or both
tinnitus: subjective perception of sound with internal origin; unwanted noises in the head or ear
tympanoplasty: surgical repair of the tympanic membrane
vertigo: illusion of movement in which the individual or the surroundings are sensed as moving

The ear is a delicate sensory organ with dual functions—hearing and balance. The sense of hearing is essential for normal development and maintenance of speech as well as the ability to communicate with others. Balance, or equilibrium, is essential for maintaining body movement, position, and coordination.

The early detection and accurate diagnosis of disorders is necessary for preservation of normal hearing and balance. The diagnosis and treatment of these disorders requires skilled health care professionals such as otolaryngologists, internists, and nurses. This chapter provides an overview of the anatomy and physiology of the ear and addresses the general assessment and management of hearing and balance disorders common to adults seen in many health care settings.

ASSESSMENT OF THE EAR

Anatomic and Physiologic Overview

The cranium encloses and protects the brain and surrounding structures, providing attachment for various muscles that control head and jaw movements. The ears are located on either side of the cranium at approximately eye level.

Anatomy of the External Ear

The external ear includes the auricle (pinna) and the external auditory canal (Fig. 64-1). The external ear is separated from the middle ear by a disk-shaped structure called the *tympanic membrane* (eardrum).

Auricle

The auricle, attached to the side of the head by skin, is composed mainly of cartilage, except for the fat and subcutaneous tissue in the earlobe. The auricle collects the sound waves and directs vibrations into the external auditory canal.

External Auditory Canal

The external auditory canal is approximately 2 to 3 cm long (Porth & Matfin, 2009). The lateral third is an elastic cartilaginous and dense fibrous framework to which thin skin is attached. The medial two thirds is bone lined with thin skin. The external auditory canal ends at the tympanic membrane.

The skin of the canal contains hair, sebaceous glands, and ceruminous glands, which secrete a brown, waxlike substance called *cerumen* (ear wax). The ear's self-cleaning mechanism moves old skin cells and cerumen to the outer part of the ear.

Just anterior to the external auditory canal is the temporomandibular joint. The head of the mandible can be felt by placing a fingertip in the external auditory canal while the patient opens and closes the mouth.

Anatomy of the Middle Ear

The middle ear, an air-filled cavity, includes the tympanic membrane laterally and the otic capsule medially. The middle ear cleft lies between the two. The middle ear is connected to the nasopharynx by the eustachian tube and is continuous with air-filled cells in the adjacent mastoid portion of the temporal bone.

The eustachian tube, which is approximately 1 mm wide and 35 mm long, connects the middle ear to the nasopharynx.

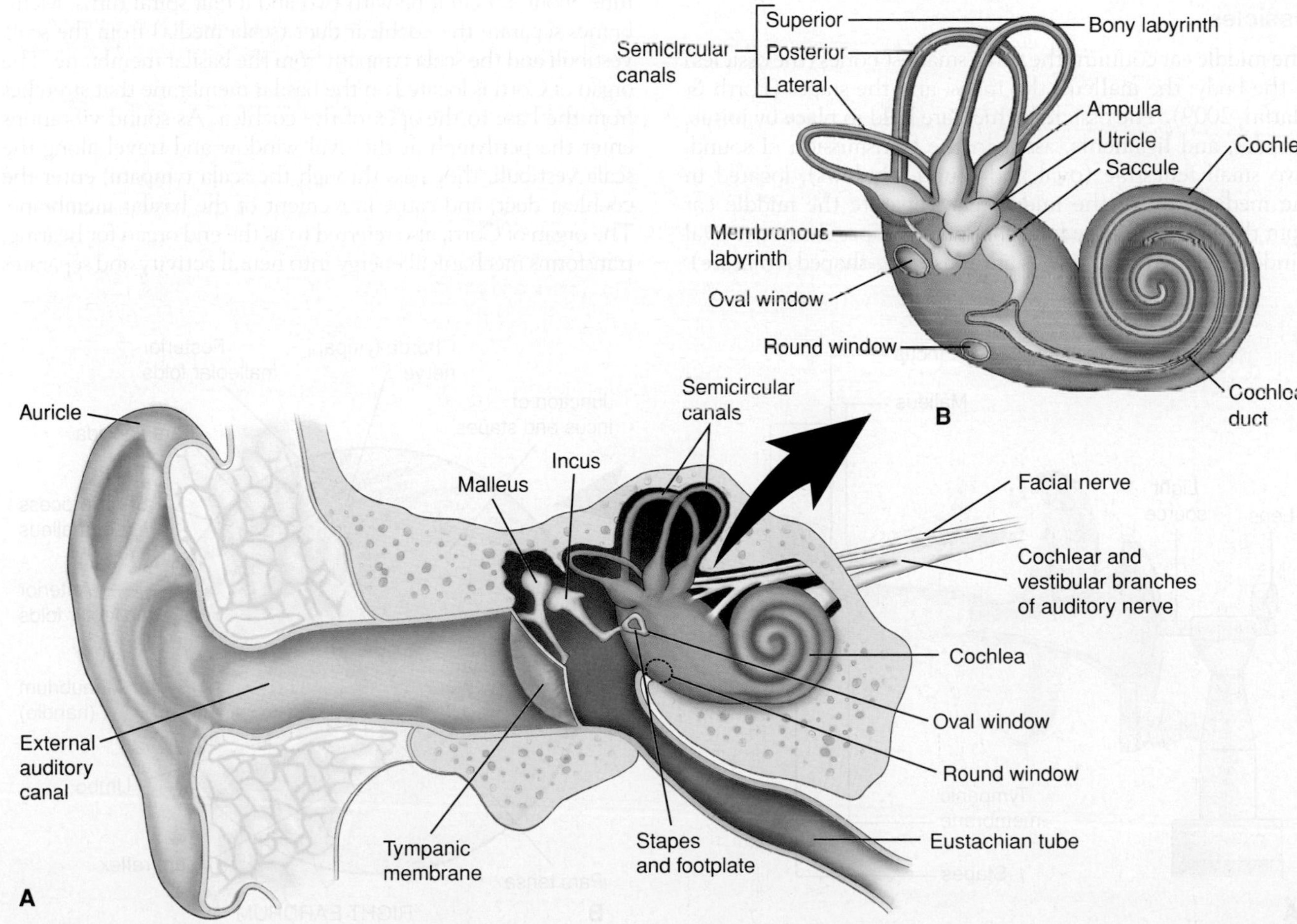
Superior
Posterior
Lateral
Semicircular canals
Bony labyrinth
Ampulla
Utricle
Saccule
Cochlea
Membranous labyrinth
Oval window
Round window
Cochlear duct
B
Auricle
Semicircular canals
Incus
Malleus
Facial nerve
Cochlear and vestibular branches of auditory nerve
Cochlea
Oval window
Round window
External auditory canal
Tympanic membrane
Stapes and footplate
Eustachian tube
A

FIGURE 64-1 • **A.** Anatomy of the ear. **B.** The inner ear.

Normally, the eustachian tube is closed, but it opens by action of the tensor veli palatini muscle when the person performs a Valsalva maneuver, yawns, or swallows. It drains normal and abnormal secretions of the middle ear and equalizes pressure in the middle ear with that of the atmosphere.

Tympanic Membrane

The tympanic membrane (eardrum), about 1 cm in diameter and very thin, is normally pearly gray and translucent. It consists of three layers of tissue: an outer layer, continuous with the skin of the ear canal; a fibrous middle layer; and an inner mucosal layer, continuous with the lining of the middle ear cavity. Approximately 80% of the tympanic membrane is composed of all three layers and is called the *pars tensa*. The remaining 20% lacks the middle layer and is called the *pars flaccida*. The absence of this fibrous middle layer makes the pars flaccida more vulnerable to pathologic disorders than the pars tensa. Distinguishing landmarks include the annulus, the fibrous border that attaches the eardrum to the temporal bone; the short process of the malleus; the long process of the malleus; the umbo of the malleus, which attaches to the tympanic membrane in the center; the pars flaccida; and the pars tensa (Fig. 64-2).

The tympanic membrane protects the middle ear and conducts sound vibrations from the external canal to the ossicles. The sound pressure is magnified 22 times as a result of transmission from a larger area to a smaller one.

Ossicles

The middle ear contains the three smallest bones (the ossicles) of the body: the malleus, the incus, and the stapes (Porth & Matfin, 2009). The ossicles, which are held in place by joints, muscles, and ligaments, assist in the transmission of sound. Two small fenestrae (oval and round windows), located in the medial wall of the middle ear, separate the middle ear from the inner ear. The footplate of the stapes sits in the oval window, secured by a fibrous annulus (ring-shaped structure). The footplate transmits sound to the inner ear. The round window, covered by a thin membrane, provides an exit for sound vibrations (see Fig. 64-1).

Anatomy of the Inner Ear

The inner ear is housed deep within the temporal bone. The organs for hearing (cochlea) and balance (semicircular canals), as well as cranial nerves VII (facial nerve) and VIII (vestibulocochlear nerve), are all part of this complex anatomy (see Fig. 64-1). The cochlea and semicircular canals are housed in the bony labyrinth. The bony labyrinth surrounds and protects the membranous labyrinth, which is bathed in a fluid called *perilymph*.

Membranous Labyrinth

The membranous labyrinth is composed of the utricle, the saccule, the cochlear duct, the semicircular canals, and the organ of Corti, all of which are surrounded by a fluid called *endolymph*. The three semicircular canals—posterior, superior, and lateral, which lie at 90-degree angles to one another—contain sensory receptor organs that are arranged to detect rotational movement. These receptor end organs are stimulated by changes in the rate or direction of a person's movement. The utricle and saccule are involved with linear movements.

Organ of Corti

The organ of Corti is housed in the cochlea, a snail-shaped, bony tube about 3.5 cm long with two and a half spiral turns. Membranes separate the cochlear duct (scala media) from the scala vestibuli and the scala tympani from the basilar membrane. The organ of Corti is located on the basilar membrane that stretches from the base to the apex of the cochlea. As sound vibrations enter the perilymph at the oval window and travel along the scala vestibuli, they pass through the scala tympani, enter the cochlear duct, and cause movement of the basilar membrane. The organ of Corti, also referred to as the end organ for hearing, transforms mechanical energy into neural activity and separates

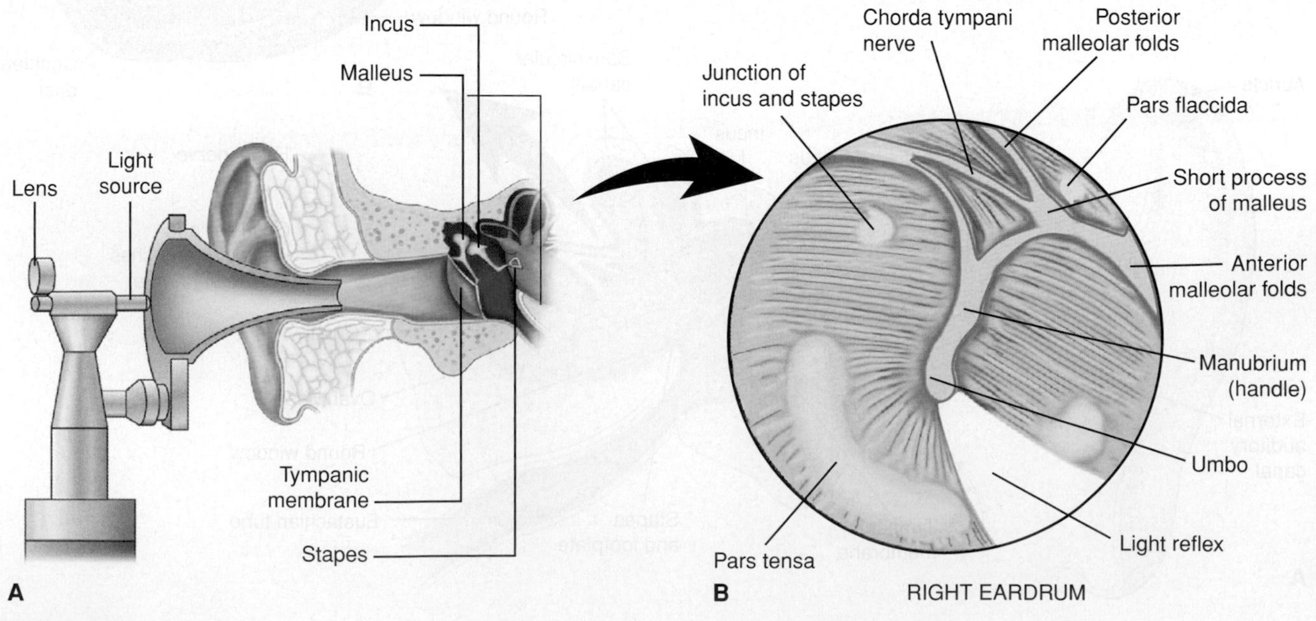
Incus
Malleus
Light source
Lens
Tympanic membrane
Stapes
A
Chorda tympani nerve
Posterior malleolar folds
Junction of incus and stapes
Pars flaccida
Short process of malleus
Anterior malleolar folds
Manubrium (handle)
Umbo
Light reflex
Pars tensa
B
RIGHT EARDRUM

FIGURE 64-2 • Technique for using the otoscope (**A**) to see the tympanic membrane (**B**).

sounds into different frequencies. This electrochemical impulse travels through the acoustic nerve to the temporal cortex of the brain to be interpreted as meaningful sound. In the internal auditory canal, the cochlear (acoustic) nerve, arising from the cochlea, joins the vestibular nerve, arising from the semicircular canals, utricle, and saccule, to become the vestibulocochlear nerve (cranial nerve VIII). This canal also houses the facial nerve and the blood supply from the ear to the brain.

Function of the Ears

Hearing

Hearing is conducted over two pathways: air and bone. Sounds transmitted by air conduction travel over the air-filled external and middle ear through vibration of the tympanic membrane and ossicles. Sounds transmitted by bone conduction travel directly through bone to the inner ear, bypassing the tympanic membrane and ossicles. Normally, air conduction is the more efficient pathway.

Sound Conduction and Transmission

Sound enters the ear through the external auditory canal and causes the tympanic membrane to vibrate. These vibrations transmit sound through the lever action of the ossicles to the oval window as mechanical energy. This mechanical energy is then transmitted through the inner ear fluids to the cochlea, stimulating the hair cells, and is subsequently converted to electrical energy. The electrical energy travels through the vestibulocochlear nerve to the central nervous system, where it is interpreted in its final form as sound.

Vibrations transmitted by the tympanic membrane to the ossicles of the middle ear are transmitted to the cochlea, located in the labyrinth of the inner ear. The stapes rocks, causing vibrations (waves) in fluids contained in the inner ear. These fluid waves cause movement of the basilar membrane, stimulating the hair cells of the organ of Corti in the cochlea to move in a wavelike manner. The movements of the tympanic membrane set up electrical currents that stimulate the various areas of the cochlea. The hair cells set up neural impulses that are encoded and then transferred to the auditory cortex in the brain, where they are decoded into a sound message.

The footplate of the stapes receives impulses transmitted by the incus and the malleus from the tympanic membrane. The round window, which opens on the opposite side of the cochlear duct, is protected from sound waves by the intact tympanic membrane, permitting motion of the inner ear fluids by sound wave stimulation. For example, in the normally intact tympanic membrane, sound waves stimulate the oval window first, and a lag occurs before the terminal effect of the stimulus reaches the round window. However, this lag phase is changed when a perforation of the tympanic membrane allows sound waves to impinge on the oval and round windows simultaneously. This effect cancels the lag and prevents the maximal effect of inner ear fluid motility and its subsequent effect in stimulating the hair cells in the organ of Corti. The result is a reduction in hearing ability (Fig. 64-3).

Balance and Equilibrium

Body balance is maintained by the cooperation of the muscles and joints of the body (proprioceptive system), the eyes (visual system), and the labyrinth (vestibular system). These areas send their information about equilibrium, or balance, to the brain (cerebellar system) for coordination and perception in the cerebral cortex. The brain obtains its blood supply from the heart and arterial system. A problem in any of these areas, such as arteriosclerosis or impaired vision, can cause a disturbance of balance. The vestibular apparatus of the inner ear provides feedback regarding the movements and the position of the head and body in space.

Assessment

Assessment of hearing and balance involves inspection of the external, middle, and inner ear. Evaluation of gross hearing acuity also is included in every physical examination.

Inspection of the External Ear

Inspection of the external ear is a simple procedure, but it is often overlooked. The external ear is examined by inspection and direct palpation; the auricle and surrounding tissues should be inspected for deformities, lesions, and discharge, as well as size, symmetry, and angle of attachment to the head. Manipulation of the auricle does not normally elicit pain. If this maneuver is painful, acute external otitis is suspected (Weber & Kelley, 2010). Tenderness on palpation in the area of the mastoid may indicate acute mastoiditis or inflammation of the posterior auricular node. Occasionally, sebaceous cysts and tophi (subcutaneous mineral deposits) are present on the pinna. A flaky scaliness on or behind the auricle usually indicates seborrheic dermatitis and can be present on the scalp and facial structures as well.

Otoscopic Examination

The tympanic membrane is inspected with an otoscope and indirect palpation with a pneumatic otoscope. To examine the external auditory canal and tympanic membrane, the otoscope should be held in the examiner's right hand, in a pencil-hold position, with the examiner's hand braced against the patient's face (Fig. 64-4). This position prevents the examiner from inserting the otoscope too far into the external canal. Using the opposite hand, the auricle is grasped and gently pulled back to straighten the canal in the adult.

The speculum is slowly inserted into the ear canal, with the examiner's eye held close to the magnifying lens of the otoscope to visualize the canal and tympanic membrane. The largest speculum that the canal can accommodate (usually 5 mm in an adult) is guided gently down into the canal and slightly forward. Because the distal portion of the canal is bony and covered by a sensitive layer of epithelium, only light pressure can be used without causing pain. The external auditory canal is examined for discharge, inflammation, or a foreign body.

The healthy tympanic membrane is pearly gray and is positioned obliquely at the base of the canal. The following landmarks are identified, if visible (see Fig. 64-2): the pars tensa, the umbo, the manubrium of the malleus, and its short process. A slow, circular movement of the speculum allows further visualization of the malleolar folds and periphery. The position and color of the membrane and any unusual

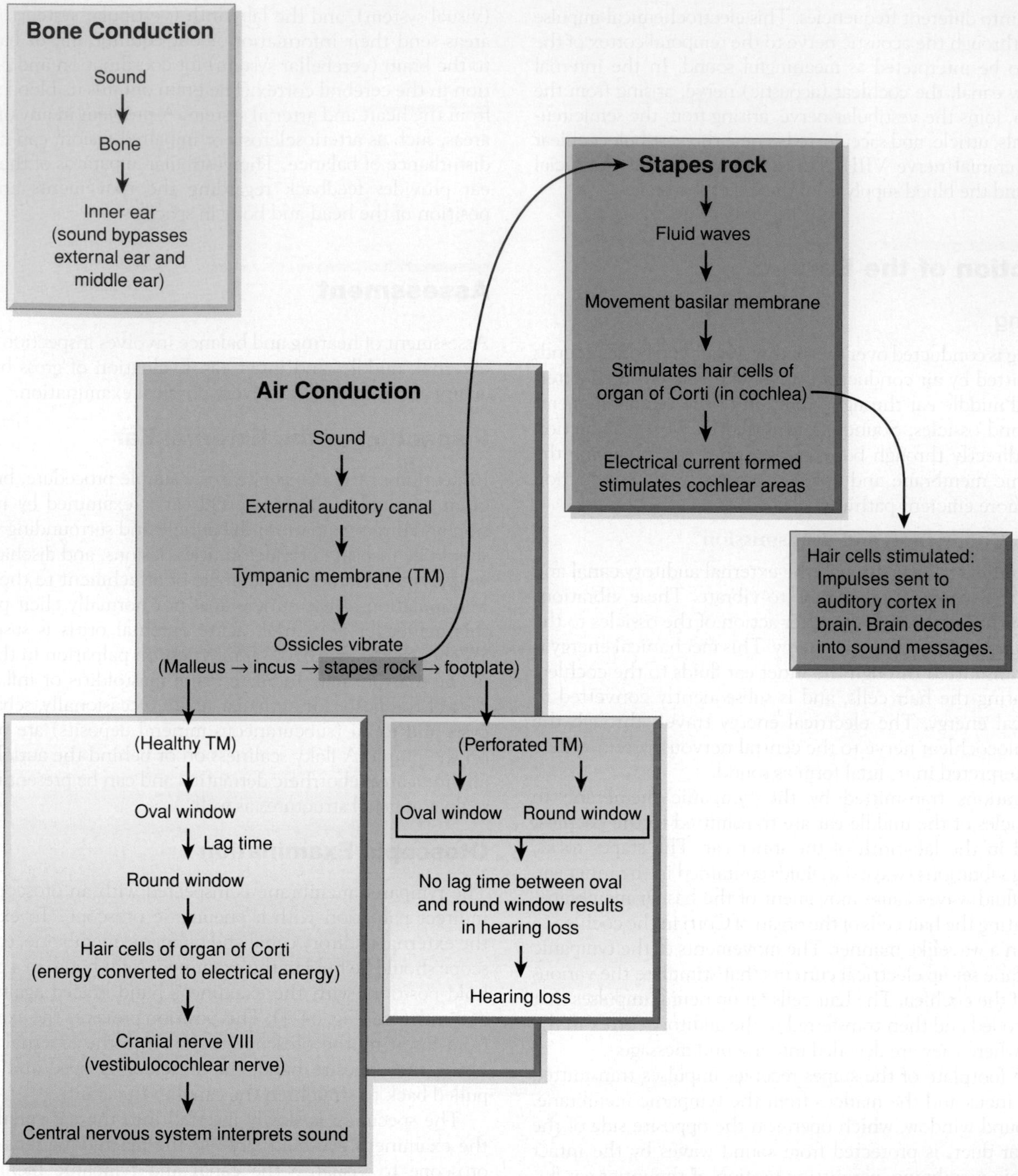
Bone Conduction
Sound
Bone
Inner ear
(sound bypasses
external ear and
middle ear)
Air Conduction
Sound
External auditory canal
Tympanic membrane (TM)
Ossicles vibrate
(Malleus → incus → stapes rock → footplate)
(Healthy TM)
Oval window
Lag time
Round window
Hair cells of organ of Corti
(energy converted to electrical energy)
Cranial nerve VIII
(vestibulocochlear nerve)
Central nervous system interprets sound
(Perforated TM)
Oval window
Round window
No lag time between oval
and round window results
in hearing loss
Hearing loss
Stapes rock
Fluid waves
Movement basilar membrane
Stimulates hair cells of
organ of Corti (in cochlea)
Electrical current formed
stimulates cochlear areas
Hair cells stimulated:
Impulses sent to
auditory cortex in
brain. Brain decodes
into sound messages.

FIGURE 64-3 • Bone conduction compared to air conduction.

markings or deviations from normal are documented. The presence of fluid, air bubbles, blood, or masses in the middle ear also should be noted.

Proper otoscopic examination of the external auditory canal and tympanic membrane requires that the canal be free of large amounts of cerumen. Cerumen is normally present in the external canal, and small amounts should not interfere with otoscopic examination. If the tympanic membrane cannot be visualized because of cerumen, the cerumen may be removed by gently irrigating the external canal with warm water (unless contraindicated). If adherent cerumen is present, a small amount of mineral oil or over-the-counter cerumen softener may be instilled into the ear canal and the patient instructed to return for subsequent removal of the cerumen and inspection of the ear. The use of instruments such as a cerumen curette for cerumen removal is reserved for otolaryngologists and nurses with specialized training because of the danger of perforating the tympanic membrane or excoriating the external auditory canal. Cerumen buildup is a common cause of hearing loss and local irritation (Porth & Matfin, 2009).

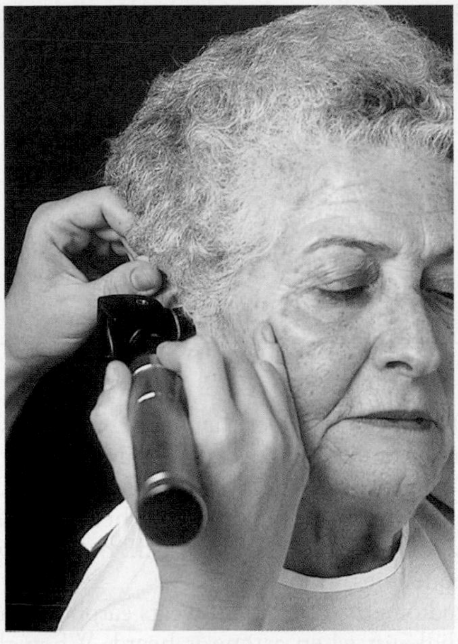

FIGURE 64-4 • Proper technique for examining the ear. Hold the otoscope in the right or left hand, in a "pencil-hold" position. Steady the hand against the patient's head to avoid inserting the otoscope too far into the external canal. From Bickley, L. S. (2009). *Bates' guide to physical examination and history taking* (10th ed., p. 225). Philadelphia: Lippincott Williams & Wilkins.

Evaluation of Gross Auditory Acuity

A general estimate of hearing can be made by assessing the patient's ability to hear a whispered phrase, testing one ear at a time. The Weber and Rinne tests may be used to distinguish conductive loss from sensorineural loss when hearing is impaired (Weber & Kelley, 2010). These tests are part of the usual screening physical examination and are useful if a more specific assessment is needed, if hearing loss is detected, or if confirmation of audiometric results is desired.

TABLE 64-1 Comparison of Weber and Rinne Tests

Hearing Status	Weber	Rinne
Normal hearing	Sound is heard equally in both ears.	Air conduction is audible longer than bone conduction.
Conductive hearing loss	Sound is heard best in affected ear (hearing loss).	Sound is heard as long or longer in affected ear (hearing loss).
Sensorineural hearing loss	Sound is heard best in normal hearing ear.	Air conduction is audible longer than bone conduction in affected ear.

Adapted from Weber, J., & Kelley, J. (2010). *Health assessment in nursing* (4th ed.). Philadelphia: Lippincott Williams & Wilkins.

Whisper Test

To exclude one ear from the testing, the examiner covers the untested ear with the palm of the hand. The examiner then whispers softly from a distance of 1 or 2 feet from the unoccluded ear and out of the patient's sight. The patient with normal acuity can correctly repeat what was whispered.

Weber Test

The Weber test uses bone conduction to test lateralization of sound. A tuning fork (ideally, 512 Hertz [Hz]), set in motion by grasping it firmly by its stem and tapping it on the examiner's knee or hand, is placed on the patient's head or forehead (Fig. 64-5A). A person with normal hearing hears the sound equally in both ears or describes the sound as centered in the middle of the head. A person with **conductive hearing loss**, such as from otosclerosis or otitis media, hears the sound better in the affected ear. A person with **sensorineural hearing loss**, resulting from damage to the cochlear or vestibulocochlear nerve, hears the sound in the better-hearing ear. The Weber test is useful for detecting unilateral hearing loss (Table 64-1).

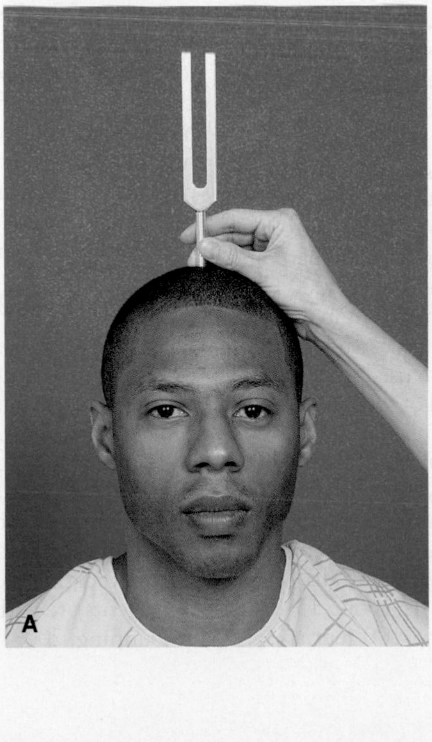

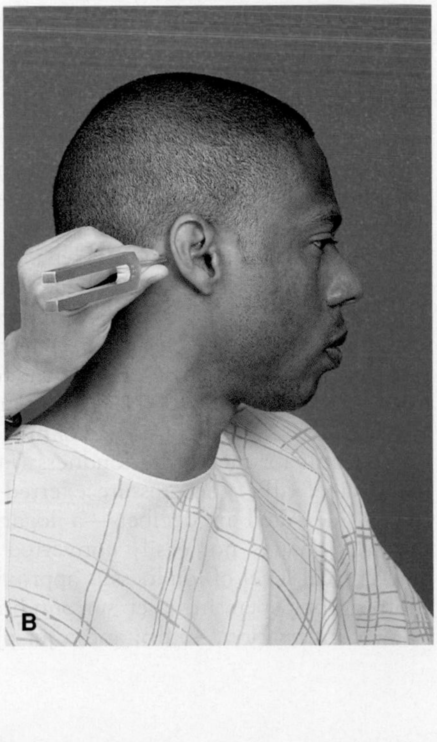

FIGURE 64-5 • **A.** The Weber test assesses bone conduction of sound. **B.** The Rinne test assesses both air and bone conduction of sound.

Concept Mastery Alert

Results of the Weber test are used to determine whether the patient has conductive hearing loss (sounds are heard better in the affected ear) or sensorineural hearing loss (sounds are heard better in the normal ear).

Rinne Test

In the Rinne test (pronounced *rin-ay*), the examiner shifts the stem of a vibrating tuning fork between two positions: 2 inches from the opening of the ear canal (for air conduction) and against the mastoid bone (for bone conduction) (Fig. 64-5B). As the position changes, the patient is asked to indicate which tone is louder or when the tone is no longer audible.

The Rinne test is useful for distinguishing between conductive and sensorineural hearing loss. A person with normal hearing reports that air-conducted sound is louder than bone-conducted sound. A person with a conductive hearing loss hears bone-conducted sound as long as or longer than air-conducted sound. A person with a sensorineural hearing loss hears air-conducted sound longer than bone-conducted sound.

Diagnostic Evaluation

Many diagnostic procedures are available to measure the auditory and vestibular systems indirectly. These tests are usually performed by a certified audiologist. The nurse educates the patient about the purpose, what to expect, and any possible side effects related to testing. The nurse notes trends in results because they provide information about disease progression as well as the patient's response to therapy.

Audiometry

In detecting hearing loss, audiometry is the single most important diagnostic instrument. Audiometric testing is of two kinds: pure-tone audiometry, in which the sound stimulus consists of a pure or musical tone (the louder the tone before the patient perceives it, the greater the hearing loss), and speech audiometry, in which the spoken word is used to determine the ability to hear and discriminate sounds and words.

When evaluating hearing, three characteristics are important: frequency, pitch, and intensity. *Frequency* refers to the number of sound waves emanating from a source per second, measured as cycles per second, or Hertz. The normal human ear perceives sounds ranging in frequency from 20 to 20,000 Hz. The frequencies from 500 to 2,000 Hz are important in understanding everyday speech and are referred to as the speech range or speech frequencies. *Pitch* is the term used to describe frequency; a tone with 100 Hz is considered of low pitch, and a tone of 10,000 Hz is considered of high pitch.

The unit for measuring loudness (*intensity* of sound) is the decibel (dB), the pressure exerted by sound. Hearing loss is measured in decibels—a logarithmic function of intensity that is not easily converted into a percentage. The critical level of loudness is approximately 30 dB. The shuffling of papers in quiet surroundings is about 15 dB; a low conversation, 40 dB; and a jet plane 100 feet away, about 150 dB. Sound louder than 80 dB is perceived by the human ear to be harsh and can be damaging to the inner ear. Table 64-2 classifies hearing loss based on decibel level. In surgical treatment of patients with hearing loss, the aim is to improve the hearing level to 30 dB or better within the speech frequencies.

TABLE 64-2 Severity of Hearing Loss

Loss in Decibels	Interpretation
0–15	Normal hearing
>15–25	Slight hearing loss
>25–40	Mild hearing loss
>40–55	Moderate hearing loss
>55–70	Moderate to severe hearing loss
>70–90	Severe hearing loss
>90	Profound hearing loss

With audiometry, the patient wears earphones and signals to the audiologist when a tone is heard. When the tone is applied directly over the external auditory canal, air conduction is measured. When the stimulus is applied to the mastoid bone, bypassing the conductive mechanism (i.e., the ossicles), nerve conduction is tested. For accuracy, testing is performed in a soundproof room. Responses are plotted on a graph known as an audiogram, which differentiates conductive from sensorineural hearing loss.

Tympanogram

A tympanogram, or impedance audiometry, measures middle ear muscle reflex to sound stimulation and compliance of the tympanic membrane by changing the air pressure in a sealed ear canal. Compliance is impaired with middle ear disease.

Auditory Brain Stem Response

The auditory brain stem response is a detectable electrical potential from cranial nerve VIII and the ascending auditory pathways of the brain stem in response to sound stimulation. Electrodes are placed on the patient's scalp and on each earlobe (Fischbach & Dunning, 2009). Acoustic stimuli (e.g., clicks) are made in the ear. The resulting electrophysiologic measurements can determine at which decibel level a patient hears and whether there are any impairments along the nerve pathways (e.g., tumor on cranial nerve VIII). Patients are instructed to wash and rinse their hair prior to this study but to avoid applying any other hair product.

Electronystagmography

Electronystagmography is the measurement and graphic recording of the changes in electrical potentials created by eye movements during spontaneous, positional, or calorically evoked nystagmus. It is also used to assess the oculomotor and vestibular systems and their corresponding interaction. It helps to diagnose causes of unilateral hearing loss of unknown origin, vertigo, or tinnitus. Any vestibular suppressants, such as caffeine and alcohol, are withheld for 84 hours before testing. Medications such as tranquilizers, stimulants, or antivertigo agents are withheld for 5 days before the test (Fischbach & Dunning, 2009).

Platform Posturography

Platform posturography is used to investigate postural control capabilities such as vertigo. It can be used to evaluate if a person's vertigo is becoming worse or to evaluate the person's response to treatment. The integration of visual, vestibular, and proprioceptive cues (i.e., sensory integration) with motor response output and coordination of the lower limbs is tested. The patient stands on a platform, surrounded by a screen, and different conditions such as a moving platform with a moving screen or a stationary platform with a moving screen are presented. The responses from the patient on six different conditions are measured and indicate which of the anatomic systems may be impaired. Preparation for the testing is the same as for electronystagmography.

Sinusoidal Harmonic Acceleration

Sinusoidal harmonic acceleration, or a rotary chair, is used to assess the vestibulo-ocular system by analyzing compensatory eye movements in response to the clockwise and counterclockwise rotation of the chair. Although such testing cannot identify the side of the lesion in unilateral disease, it helps to identify disease (e.g., Ménière's disease and tumors of the auditory canal) and evaluate the course of recovery. The same patient preparation is required as that for electronystagmography.

Middle Ear Endoscopy

Using endoscopes with very small diameters and acute angles, the ear can be examined by an endoscopist specializing in otolaryngology. Middle ear endoscopy is performed safely and effectively as an office procedure to evaluate suspected perilymphatic fistula and new-onset conductive hearing loss, the anatomy of the round window before transtympanic treatment of Ménière's disease, and the tympanic cavity before ear surgery to treat chronic middle ear and mastoid infections.

The tympanic membrane is anesthetized topically for about 10 minutes before the procedure. Then, the external auditory canal is irrigated with sterile normal saline solution. With the aid of a microscope, a tympanotomy is created with a laser beam or a myringotomy knife so that the endoscope can be inserted into the middle ear cavity. Video and photo documentation can be accomplished through the scope.

HEARING LOSS

In the United States, hearing impairment has been reported to occur in 3 of every 1,000 births, and approximately one half of the time it is related to genetic factors (U.S. Department of Health and Human Services [HHS], 2010). Genetic syndromes associated with hearing impairment include Waardenburg syndrome, Usher syndrome, Pendred syndrome, and Jervell and Lange-Nielsen syndrome (Antonio, 2012). Chart 64-1 contains more information about hearing disorders that have a genetic cause. Most hospitals or birthing centers offer universal hearing screenings for newborns after birth and prior to discharge.

Hearing loss is greater in men than in women. By the year 2050, about one of every five people in the United States, or almost 58 million people, will be 55 years or older. Of this population, almost half can expect to have a hearing impairment (HHS, 2010). Hearing loss is an important health issue, and as people age, hearing screening and treatment are indicated.

Many people are exposed on a daily basis to noise levels that produce high-frequency hearing loss. Occupations such as carpentry, plumbing, and coal mining have the highest risk of noise-induced hearing loss. Wise Ears was developed by the

Chart 64-1 GENETICS IN NURSING PRACTICE

Hearing Disorders

Several hearing disorders are associated with genetic abnormalities, such as the following:

- Autosomal dominant hearing loss
- Autosomal recessive hearing loss (e.g., connexin 26 gene)
- Otosclerosis
- Pendred syndrome
- Usher syndrome
- Waardenburg syndrome

Nursing Assessments

Family History Assessment

- Assess for other family members in several generations with hearing loss (autosomal dominant hearing loss).
- Inquire about genetic relatedness (e.g., individuals who are related, such as first cousins, have a higher chance to share the same recessive genes—autosomal recessive hearing loss).
- Inquire about age at onset of hearing loss.

Patient Assessment

- Assess for related genetic conditions, such as vision impairment (e.g., retinitis pigmentosa in Usher syndrome; thyroid disorder in Pendred syndrome).
- Assess for iris, pigment, and hair alterations (white forelock) seen in Waardenburg syndrome.

Management Issues Specific to Genetics

- Inquire whether deoxyribonucleic acid (DNA) gene mutation or other genetic testing has been performed on affected family members.
- If indicated, refer for further genetic counseling and evaluation so that family members can discuss inheritance, risk to other family members, availability of genetic testing, and gene-based interventions.
- Offer appropriate genetic information and resources.
- Assess patient's understanding of genetic information.
- Provide support to families with newly diagnosed genetics-related sensorineural disorders.
- Participate in management and coordination of care of patients with genetics-related hearing disorders, patients with genetic conditions, and individuals predisposed to develop or pass on a genetic condition.

Genetics Resources

See Chapter 8, Chart 8-6 for genetics resources.

National Institute on Deafness and Other Communication Disorders (NIDCD) and the National Institute for Occupational Safety and Health (NIOSH). It aims to educate the public about noise-induced hearing loss and ways to prevent this hearing loss (NIDCD, 2008b).

Conductive hearing loss usually results from an external ear disorder, such as impacted cerumen, or a middle ear disorder, such as otitis media or otosclerosis. In such instances, the efficient transmission of sound by air to the inner ear is interrupted. A sensorineural loss involves damage to the cochlea or vestibulocochlear nerve.

Mixed hearing loss and functional hearing loss also may occur. Patients with mixed hearing loss have conductive loss and sensorineural loss, resulting from dysfunction of air and bone conduction. A functional (or psychogenic) hearing loss is nonorganic and unrelated to detectable structural changes in the hearing mechanisms; it is usually a manifestation of an emotional reaction.

Clinical Manifestations

Deafness is the partial or complete loss of the ability to hear. Early manifestations may include tinnitus, increasing inability to hear when in a group, and a need to turn up the volume of the television. Hearing impairment can also trigger changes in attitude, the ability to communicate, the awareness of surroundings, and even the ability to protect oneself, thus affecting a person's quality of life. In a classroom, a student with impaired hearing may be uninterested and inattentive and have failing grades. A person at home may feel isolated because of an inability to hear the clock chime or to hear the telephone. A pedestrian who is hearing impaired may attempt to cross the street and fail to hear an approaching car. People with impaired hearing may miss parts of a conversation. Many people are unaware of their gradual hearing impairment. Often, it is not the person with the hearing loss but the people with whom he or she is communicating who recognizes the impairment first (Chart 64-2).

For various reasons, some people with hearing loss refuse to seek medical attention or wear a hearing aid. They may feel self-conscious about wearing a hearing aid. Other more insightful people generally ask those with whom they are trying to communicate to let them know whether difficulties in communication exist. The attitudes and behaviors of patients who need hearing assistance should be taken into account when counseling them. The decision to wear a hearing aid is a personal one that is affected by these attitudes and behaviors.

Prevention

Many environmental factors have an adverse effect on the auditory system and with time result in permanent sensorineural hearing loss. The most common is noise. Noise (unwanted and unavoidable sound) has been identified as one of today's environmental hazards. The volume of noise that surrounds us daily has increased into a potentially dangerous source of physical and psychological damage.

Loud, persistent noise has been found to cause constriction of peripheral blood vessels, increased blood pressure and heart rate (because of increased secretion of adrenalin), and increased gastrointestinal activity. Although research is needed to address the overall effects of noise on the human body, a quiet environment is more conducive to peace of mind. A person who is ill feels more at ease when noise is kept to a minimum.

Numerous factors contribute to hearing loss (Chart 64-3). *Noise-induced hearing loss* refers to hearing loss that follows a long period of exposure to loud noise (e.g., heavy machinery, engines, artillery, rock-band music). *Acoustic trauma* refers

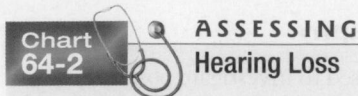

Chart 64-2 ASSESSING: Hearing Loss

The nurse should be alert to the following:

Speech deterioration: The person who slurs words, drops word endings, or produces flat-sounding speech may not be hearing correctly. The ears guide the voice, both in loudness and in pronunciation.

Fatigue: If a person tires easily when listening to conversation or to a speech, fatigue may be the result of straining to hear. Under these circumstances, the person may become irritable very easily.

Indifference: It is easy for the person who cannot hear what others say to become depressed and disinterested in life in general.

Social withdrawal: Not being able to hear what is going on causes the person who is hearing impaired to withdraw from situations that might prove embarrassing.

Insecurity: Lack of self-confidence and fear of mistakes create a feeling of insecurity in many people who are hearing impaired. No one likes to say the wrong thing or do anything that might appear foolish.

Indecision and procrastination: Loss of self-confidence makes it increasingly difficult for a person who is hearing impaired to make decisions.

Suspiciousness: The person who is hearing impaired—who often hears only part of what is being said—may suspect that others are talking about him or her, or that portions of the conversation are deliberately spoken softly so that he or she will not hear them.

False pride: The person who is hearing impaired wants to conceal the hearing loss and thus often pretends to be hearing when he or she actually is not.

Loneliness and unhappiness: Although everyone wishes for quiet now and then, *enforced* silence can be boring and even somewhat frightening. People with a hearing loss often feel isolated.

Tendency to dominate the conversation: Many people who are hearing impaired tend to dominate the conversation, knowing that as long as it is centered on them and they can control it, they are not so likely to be embarrassed by some mistake.

Chart 64-3 RISK FACTORS: Hearing Loss

- Family history of sensorineural impairment
- Congenital malformations of the cranial structure (ear)
- Low birth weight (<1,500 g)
- Use of ototoxic medications (e.g., gentamycin, loop diuretics)
- Recurrent ear infections
- Bacterial meningitis
- Chronic exposure to loud noises
- Perforation of the tympanic membrane

to hearing loss caused by a single exposure to an extremely intense noise, such as an explosion. Usually, noise-induced hearing loss occurs at a high frequency (about 4,000 Hz). However, with continued noise exposure, the hearing loss can become more severe and include adjacent frequencies. The minimum noise level known to cause noise-induced hearing loss, regardless of duration, is about 85 to 90 dB.

Noise exposure is inherent in many jobs (e.g., mechanics, printers, pilots, flight attendants, musicians) and in hobbies such as woodworking and hunting. Noise level regulations are based on the amount of noise a person is exposed to, with the maximum legal amount being exposure to noise over an average working day or week of 80 dB, with a peak sound pressure of 135 dB (NIDCD, 2008a). The Occupational Safety and Health Administration (OSHA) requires that workers wear ear protection to prevent noise-induced hearing loss when exposed to noise above the legal limits. Ear protection against noise is the most effective preventive measure available. Hearing loss due to noise is permanent because the hair cells in the organ of Corti are destroyed.

Gerontologic Considerations

With aging, changes occur in the ear that may eventually lead to hearing deficits. Although few changes occur in the external ear, cerumen tends to become harder and drier, posing a greater chance of impaction. In the middle ear, the tympanic membrane may atrophy or become sclerotic. In the inner ear, cells at the base of the cochlea degenerate. A familial predisposition to sensorineural hearing loss is also seen, manifested by inability to hear high-frequency sounds, followed in time by the loss of middle and lower frequencies. The term **presbycusis** is used to describe this progressive hearing loss (Eliopoulos, 2010).

In addition to age-related changes, other factors can affect hearing in the older adult population, such as lifelong exposure to loud noises. Certain medications, such as aminoglycosides, aspirin, loop diuretics and platinum-based antineoplastic medications have ototoxic effects when kidney changes result in delayed medication excretion and increased levels of the medications in the blood.

Many older people have taken quinine for treatment of leg cramps; this medication also can contribute to hearing loss. Psychogenic factors and other disease processes (e.g., diabetes) also may be partially responsible for sensorineural hearing loss. The Individuals with Disabilities Education Act (IDEA) was developed to ensure that children and adults, including older adults, receive the same opportunities in the educational system as those without hearing impairment (U.S. Department of Education, 2012).

Even with the best health care, people with hearing loss must learn to adjust to it. Care of older patients includes recognizing emotional reactions related to hearing loss, such as suspicion of others because of an inability to hear adequately; frustration and anger, with repeated statements such as "I didn't hear what you said!"; and feelings of insecurity because of the inability to hear the telephone or alarms. The Americans with Disabilities Act (ADA) of 1990 requires that all emergency services are accessible to people who have text message telephones (teletypewriters [TTYs]). In addition, all 911 centers in the United States must be accessible to people with TTYs.

Depression, isolation, and a decrease in cognitive function can have a negative impact on quality of life in the older adult with hearing loss. Feelings of social isolation, confusion, alterations in daily living activities, and increasing risk of falls have all been associated with hearing loss in older adults (Hardin, 2012). Hearing loss can also interfere with relationships due to a loss of communication. A hearing screening is recommended as one component of the physical examination for the older adult when joining Medicare for the first time. "Welcome to Medicare" examinations and annual screenings are also important.

Medical Management

If a hearing loss is permanent or untreatable or if the patient elects not to be treated, aural rehabilitation (discussed at the end of the chapter) may be beneficial.

Nursing Management

Early detection of hearing loss is one of the objectives of *Healthy People 2020*, and nurses are in a position to assist in meeting this goal (HHS, 2010). Another objective is for people diagnosed with hearing loss or deafness to use rehabilitation services and supplemental devices to improve communication with other people. Resources are available in workplaces and in schools. Questions used to assess for hearing loss may include:

- Have you experienced any hearing loss in the past?
- Are you experiencing any hearing loss now?
- Do your family members think that you are having difficulty hearing and/or experiencing any hearing loss?

These questions should be included in any routine nursing assessment and referrals made for further evaluation as needed.

Nurses who understand the different types of hearing loss are more successful in adopting a communication style to fit the needs and preferences of each patient. Trying to speak in a loud voice to a person who cannot hear high-frequency sounds only makes understanding more difficult. However, strategies such as talking into the less impaired ear and using gestures and facial expressions can help (Chart 64-4).

A major issue for many people who are deaf or hearing impaired is that they have other health problems that often do not receive attention, in large part because of communication barriers with their health care providers. To meet the health care needs of these patients, providers are legally obligated to make accommodations for a patient's inability to hear. Providing interpreters for those who can communicate through sign language is essential in many situations so that effective communication occurs.

During health care and screening procedures, the provider (e.g., dentist, physician, nurse) must be aware that patients who are deaf or hearing impaired are unable to read lips, see a signer, or read written materials in the dark rooms required during some diagnostic tests. The same situation exists if the provider is wearing a mask or is not in sight (e.g., x-ray studies, magnetic resonance imaging [MRI], colonoscopy).

Nurses and other health care providers must work with patients who are deaf or hearing impaired and their families to identify practical and effective means of communication. Nurses can serve as catalysts throughout the health care

Chart 64-4 Communicating With People Who Are Hearing Impaired

For the Person Who is Hearing Impaired Whose Speech Is Difficult to Understand

- Determine how the person prefers to communicate with others. Do not assume that writing, gestures, or other means are the best or preferred technique.
- Consider if the person uses sign language. Interpreters are available from American Sign Language Services, Inc. (ASLI). These specialists provide the best means of communication, providing accurate, professional services.
- Devote full attention to what the person is saying. Look and listen—do not try to attend to another task while listening.
- Engage the speaker in conversation when it is possible for you to anticipate the replies. This enables you to become accustomed to any peculiarities in speech patterns.
- Try to determine the essential context of what is being said; you can often fill in the details from context.
- Do not try to appear as if you understand if you do not.
- If you cannot understand at all or have serious doubt about your ability to understand what is being said, have the person write the message rather than risk misunderstanding. Having the person repeat the message in speech, after you know its content, also aids you in becoming accustomed to the person's pattern of speech.
- Written communication is an excellent resource. Written material should be written at a third-grade level so that the majority of people can understand it.

For the Person Who Is Hearing Impaired Who Speech Reads

- When speaking, always face the person as directly as possible.
- Make sure that your face is as clearly visible as possible. Locate yourself so that your face is well lighted; avoid being silhouetted against strong light. Do not obscure the person's view of your mouth in any way; avoid talking with any object held in your mouth.
- Be sure that the patient knows the topic or subject before going ahead with what you plan to say. This enables the person to use contextual clues in speech reading.
- Speak slowly and distinctly, pausing more frequently than you would normally.
- If you question whether some important direction or instruction has been understood, check to be certain that the patient has the full meaning of your message.
- If for any reason your mouth must be covered (as with a mask) and you must direct or instruct the patient, write the message.

system to ensure that accommodations are made to meet the communication needs of these patients.

CONDITIONS OF THE EXTERNAL EAR

Cerumen Impaction

Cerumen normally accumulates in the external canal in various amounts and colors. Although wax does not usually need to be removed, impaction occasionally occurs, causing **otalgia** (a sensation of fullness or pain in the ear) with or without a hearing loss. Accumulation of cerumen as a cause of hearing loss is especially significant in older adult patients (Eliopoulos, 2010). Attempts to clear the external auditory canal with matches, hairpins, and other implements are dangerous because trauma to the skin, infection, and damage to the tympanic membrane can occur.

Management

Cerumen can be removed by irrigation, suction, or instrumentation. Unless the patient has a perforated eardrum or an inflamed external ear (i.e., otitis externa), gentle irrigation with warm water usually helps remove impacted cerumen, particularly if it is not tightly packed in the external auditory canal. For successful removal, the water stream must flow behind the obstructing cerumen to move it first laterally and then out of the canal. To prevent injury, the lowest effective pressure should be used. However, if the eardrum behind the impaction is perforated, water can enter the middle ear, producing acute vertigo and infection. If irrigation is unsuccessful, direct visual, mechanical removal can be performed on a cooperative patient by a trained health care provider.

Quality and Safety Nursing Alert

Warm water (never cold or hot) and gentle (not forceful) irrigation should be used to remove cerumen. Irrigation that is too forceful can cause perforation of the tympanic membrane, and ice water causes vomiting.

Instilling a few drops of warmed glycerin, mineral oil, or half-strength hydrogen peroxide into the ear canal for 30 minutes prior to irrigation can soften cerumen before its removal. Ceruminolytic agents, such as peroxide in glyceryl (Debrox), are available. The use of any softening solution two or three times a day for several days is generally sufficient. If the cerumen cannot be dislodged by these methods, instruments, such as a cerumen curette, aural suction, and a binocular microscope for magnification, can be used.

Foreign Bodies

Some objects are inserted intentionally into the ear by adults who may have been trying to clean the external canal or relieve itching, or by children who introduce peas, beans, pebbles, toys, and beads. Insects may also enter the ear canal. In either case, the effects may range from no symptoms to profound pain and decreased hearing.

Management

Removing a foreign body from the external auditory canal can be quite challenging. The three standard methods for removing foreign bodies are the same as those for removing cerumen: irrigation, suction, and instrumentation. The contraindications for irrigation are also the same. Foreign vegetable bodies and insects tend to swell; thus, irrigation is contraindicated. Usually, an insect can be dislodged by instilling mineral oil, which will kill the insect and allow it to be removed.

Attempts to remove a foreign body from the external canal may be dangerous in unskilled hands. The object may be pushed completely into the bony portion of the canal, lacerating the skin and perforating the tympanic membrane. In rare circumstances, the foreign body may have to be extracted in the operating room with the patient under general anesthesia.

External Otitis (Otitis Externa)

External otitis (i.e., otitis externa), refers to an inflammation of the external auditory canal. Causes include water in the ear canal (swimmer's ear); trauma to the skin of the ear canal, permitting entrance of organisms into the tissues; and systemic conditions, such as vitamin deficiency and endocrine disorders. Bacterial or fungal infections are most frequently encountered. The most common bacterial pathogens associated with external otitis are *Staphylococcus aureus* and *Pseudomonas* species. The most common fungus isolated in both normal and infected ears is *Aspergillus* (Porth & Matfin, 2009). External otitis is often caused by a dermatosis such as psoriasis, eczema, or seborrheic dermatitis. Even allergic reactions to hair spray, hair dye, and permanent wave lotions can cause dermatitis, which clears when the offending agent is removed.

Clinical Manifestations

Patients usually report pain; discharge from the external auditory canal; aural tenderness (usually not present in middle ear infections); and occasionally fever, cellulitis, and lymphadenopathy. Other symptoms may include pruritus and hearing loss or a feeling of fullness. On otoscopic examination, the ear canal is erythematous and edematous. Discharge may be yellow or green and foul smelling. In fungal infections, hairlike black spores may even be visible.

Medical Management

The principles of therapy are aimed at relieving the discomfort, reducing the swelling of the ear canal, and eradicating the infection. Patients may require analgesic medications for the first 48 to 92 hours. Treatment most often includes antimicrobial or antifungal otic medications administered by dropper at room temperature. In bacterial infection, a combination antibiotic and corticosteroid agent may be used to soothe the inflamed tissues (Porth & Matfin, 2009).

Nursing Management

Nurses should instruct patients not to clean the external auditory canal with cotton-tipped applicators and to avoid events that traumatize the external canal, such as scratching the canal with the fingernail or other objects. Trauma may lead to infection of the canal. Patients should also avoid getting the canal wet when swimming or shampooing the hair. A cotton ball can be covered in a water-insoluble gel such as petrolatum jelly and placed in the ear as a barrier to water contamination. Infection can be prevented by using antiseptic otic preparations after swimming (e.g., Swim-Ear, Ear Dry), unless there is a history of tympanic membrane perforation or a current ear infection (Chart 64-5).

Chart 64-5 — PATIENT EDUCATION

Prevention of Otitis Externa

- Protect the external canal when swimming, showering, or washing hair. Ear plugs or a swim cap should be worn. Drying the external canal afterward with a hair dryer on low heat may be suggested.
- Alcohol drops may be placed in the external canal to act as an astringent to help prevent infection after water exposure.
- Prevent trauma to the external canal. Procedures, foreign objects (e.g., bobby pin), scratching, or any other trauma to the canal that breaks the skin integrity may cause infection.
- If otitis externa is diagnosed, refrain from any water sport activity for approximately 7–10 days to allow the canal to heal completely. Recurrence is highly likely unless you allow the external canal to heal completely.

Malignant External Otitis

A more serious, although rare, external ear infection is malignant external otitis (temporal bone osteomyelitis). This is a progressive, debilitating, and occasionally fatal infection of the external auditory canal, the surrounding tissue, and the base of the skull. *Pseudomonas aeruginosa* is usually the infecting organism in patients with low resistance to infection (e.g., patients with acquired immunodeficiency virus). Successful treatment includes administration of antibiotics (usually intravenously [IV]), and aggressive local wound care. Standard parenteral antibiotic treatment includes the combination of an antipseudomonal agent and an aminoglycoside, both of which have potentially serious side effects. Because aminoglycosides are nephrotoxic and ototoxic, serum aminoglycoside levels and kidney and auditory function must be monitored during therapy. Local wound care includes limited débridement of the infected tissue, including bone and cartilage, depending on the extent of the infection.

Masses of the External Ear

Exostoses are small, hard, bony protrusions found in the lower posterior bony portion of the ear canal; they usually occur bilaterally. The skin covering the exostosis is normal. It is believed that exostoses are caused by an exposure to cold water, as in scuba diving or surfing. The usual treatment, if any, is surgical excision.

Malignant tumors also may occur in the external ear. Most common are basal cell carcinomas on the pinna and squamous cell carcinomas in the ear canal. If untreated, squamous cell carcinoma may spread through the temporal bone, causing facial nerve paralysis and hearing loss. Carcinomas must be treated surgically.

CONDITIONS OF THE MIDDLE EAR

Tympanic Membrane Perforation

Perforation of the tympanic membrane is usually caused by infection or trauma. Sources of trauma include skull fracture, explosive injury, or a severe blow to the ear. Less frequently, perforation is caused by foreign objects (e.g., cotton-tipped applicators, hairpins, keys) that have been pushed too far into the external auditory canal. In addition to tympanic membrane perforation, injury to the ossicles and even the inner ear may result from this type of trauma. During infection, the tympanic membrane can rupture if the pressure in the middle ear exceeds the atmospheric pressure in the external auditory canal.

Medical Management

Although most tympanic membrane perforations heal spontaneously within weeks after rupture, some may take several months to heal. Some perforations persist because scar tissue grows over the edges of the perforation, preventing extension of the epithelial cells across the margins and final healing. In the case of a head injury or temporal bone fracture, a patient is observed for evidence of cerebrospinal fluid **otorrhea** or **rhinorrhea**—a clear, watery drainage from the ear or nose, respectively. While healing, the ear must be protected from water.

Surgical Management

Perforations that do not heal on their own may require surgery. The decision to perform a **tympanoplasty** (surgical repair of the tympanic membrane) is usually based on the need to prevent potential infection from water entering the ear or the desire to improve the patient's hearing. Performed on an outpatient basis, tympanoplasty may involve a variety of surgical techniques. In all techniques, tissue (commonly from the temporalis fascia) is placed across the perforation to allow healing. Surgery is usually successful in closing the perforation permanently and improving hearing.

Acute Otitis Media

Ear infections can occur at any age; however, they are most commonly seen in children. **Acute otitis media (AOM)** is an acute infection of the middle ear, lasting less than 6 weeks. Pathogens that cause AOM are usually bacterial or viral and enter the middle ear after eustachian tube dysfunction caused by obstruction related to upper respiratory infections, inflammation of surrounding structures (e.g., rhinosinusitis, adenoid hypertrophy), or allergic reactions (e.g., allergic rhinitis) (Porth & Matfin, 2009). Bacteria can enter the eustachian tube from contaminated secretions in the nasopharynx and the middle ear from a tympanic membrane perforation. A purulent exudate is usually present in the middle ear, resulting in a conductive hearing loss.

Clinical Manifestations

Symptoms of otitis media vary with the severity of the infection. The condition, usually unilateral in adults, may be accompanied by otalgia. The pain is relieved after spontaneous perforation or therapeutic incision of the tympanic membrane. Other symptoms may include drainage from the ear, fever, and hearing loss. Table 64-3 differentiates acute otitis externa from AOM. Risk factors for AOM include age (younger than 12 months), chronic upper respiratory infections, medical conditions that predispose to ear infections (Down syndrome, cystic fibrosis, cleft palate), and chronic exposure to secondhand cigarette smoke.

TABLE 64-3 Clinical Features of Otitis

Feature	Acute Otitis Externa	Acute Otitis Media
Otorrhea	May or may not be present	Present if tympanic membrane perforates; discharge is profuse.
Otalgia	Persistent; may awaken patient at night	Relieved if tympanic membrane ruptures
Aural tenderness	Present on palpation of auricle	Usually absent
Systemic symptoms	Absent	Fever, upper respiratory infection, rhinitis
Edema of external auditory canal	Present	Absent
Tympanic membrane	May appear normal	Erythema, bulging, may be perforated
Hearing loss	Conductive type	Conductive type

Medical Management

The outcome of AOM depends on the efficacy of therapy (the prescribed dose of an oral antibiotic and the duration of therapy), the virulence of the bacteria, and the physical status of the patient. With early and appropriate broad-spectrum antibiotic therapy, otitis media may resolve with no serious sequelae. If drainage occurs, an antibiotic otic preparation is usually prescribed. The condition may become subacute (lasting 2 weeks to 3 months), with persistent purulent discharge from the ear. Rarely does permanent hearing loss occur. Secondary complications involving the mastoid and other serious intracranial complications, such as meningitis or brain abscess, although rare, can occur.

Surgical Management

A **myringotomy (i.e., tympanotomy)** is an incision in the tympanic membrane. The tympanic membrane is numbed with a local anesthetic agent such as phenol or by iontophoresis (i.e., in which electrical current flows through a lidocaine and epinephrine solution to numb the ear canal and tympanic membrane). The procedure is painless and takes less than 15 minutes. Under microscopic guidance, an incision is made through the tympanic membrane to relieve pressure and to drain serous or purulent fluid from the middle ear.

Normally, this procedure is unnecessary for treating AOM, but it may be performed if pain persists. Myringotomy also allows the drainage to be analyzed (by culture and sensitivity testing) so that the infecting organism can be identified and appropriate antibiotic therapy prescribed. The incision heals within 24 to 72 hours.

If AOM recurs and there is no contraindication, a ventilating, or pressure-equalizing, tube may be inserted. The ventilating tube, which temporarily takes the place of the eustachian tube in equalizing pressure, is retained for 6 to 18 months. The ventilating tube is then extruded with normal skin migration of the tympanic membrane, with the hole healing in nearly every case. Ventilating tubes are used to treat recurrent episodes of AOM.

Serous Otitis Media

Serous otitis media (i.e., **middle ear effusion**) involves the presence of fluid, without evidence of active infection, in the middle ear. In theory, this fluid results from a negative pressure in the middle ear caused by eustachian tube obstruction. When this condition occurs in adults, an underlying cause for the eustachian tube dysfunction must be sought. Middle ear effusion is frequently seen in patients after radiation therapy or barotrauma and in patients with eustachian tube dysfunction from a concurrent upper respiratory infection or allergy. Barotrauma results from sudden pressure changes in the middle ear caused by changes in barometric pressure, as in scuba diving or airplane descent. A carcinoma (e.g., nasopharyngeal cancer) obstructing the eustachian tube should be ruled out in adults with persistent unilateral serous otitis media.

Clinical Manifestations

Patients may complain of hearing loss, fullness in the ear or a sensation of congestion, or popping and crackling noises that occur as the eustachian tube attempts to open. The tympanic membrane appears dull on otoscopy, and air bubbles may be visualized in the middle ear. Usually, the audiogram shows a conductive hearing loss.

Management

Serous otitis media need not be treated medically unless infection (i.e., AOM) occurs. If the hearing loss associated with middle ear effusion is significant, a myringotomy can be performed, and a tube may be placed to keep the middle ear ventilated. Corticosteroids in small doses may decrease the edema of the eustachian tube in cases of barotrauma. Decongestants have not proved to be effective. A Valsalva maneuver, which forcibly opens the eustachian tube by increasing nasopharyngeal pressure, may be cautiously performed; this maneuver may cause worsening pain or perforation of the tympanic membrane.

Chronic Otitis Media

Chronic otitis media is recurrent AOM that causes irreversible tissue pathology. Chronic infections of the middle ear damage the tympanic membrane, destroy the ossicles, and involve the mastoid. Before the discovery of antibiotic medications, infections of the mastoid were life threatening. Today, acute mastoiditis is rare in developed countries.

Clinical Manifestations

Symptoms may be minimal, with varying degrees of hearing loss and a persistent or intermittent, foul-smelling otorrhea. Pain is not usually experienced, except in cases of acute mastoiditis, when the postauricular area is tender and may be erythematous and edematous. Otoscopic examination may show a perforation, and cholesteatoma can be identified as a white mass behind the tympanic membrane or coming through to the external canal from a perforation.

Cholesteatoma is a tumor of the external layer of the eardrum into the middle ear. It is generally caused by a chronic retraction pocket of the tympanic membrane, creating a persistently high negative pressure of the middle ear. The skin forms a sac that fills with degenerated skin and sebaceous materials. The sac can attach to the structures of the middle ear, mastoid, or both.

Chronic otitis media can cause chronic mastoiditis and lead to the formation of cholesteatoma. It can occur in the middle ear, mastoid cavity, or both, often dictating the type of surgery to be performed. If untreated, cholesteatoma will continue to enlarge, possibly causing damage to the facial nerve and horizontal canal and destruction of other surrounding structures.

Cholesteatomas are cystlike lesions of the inner ear (Porth & Matfin, 2009). They usually do not cause pain; however, if treatment or surgery is delayed, they may burst or destroy the mastoid bone. These fast-growing lesions may cause severe sequelae such as hearing loss. Cholesteatomas found in older adult patients generally develop in the external canal.

Cholesteatomas may be asymptomatic, or they may cause hearing loss, facial pain and paralysis, tinnitus, or vertigo. Audiometric tests often show a conductive or mixed hearing loss. Based on presenting symptoms, diagnosis may be made by visual examination or by computed tomography (CT) or MRI. Therapy includes treatment of the acute infection and surgical removal of the mass to restore hearing.

Medical Management

Local treatment for chronic otitis media consists of careful suctioning of the ear under otoscopic guidance. Instillation of antibiotic drops or application of antibiotic powder is used to treat purulent discharge. Systemic antibiotic agents are prescribed only in cases of acute infection.

Surgical Management

Surgical procedures, including tympanoplasty, ossiculoplasty, and mastoidectomy, are used if medical treatments are ineffective.

Tympanoplasty

The most common surgical procedure for chronic otitis media is tympanoplasty, or surgical reconstruction of the tympanic membrane. Reconstruction of the ossicles may also be required. The purposes of a tympanoplasty are to reestablish middle ear function, close the perforation, prevent recurrent infection, and improve hearing.

There are five types of tympanoplasties. The simplest surgical procedure, type I (myringoplasty), is designed to close a perforation in the tympanic membrane. The other procedures, types II through V, involve more extensive repair of middle ear structures. The structures and the degree of involvement can differ, but all tympanoplasty procedures include restoring the continuity of the sound conduction mechanism.

Tympanoplasty is performed through the external auditory canal with a transcanal approach or through a postauricular incision. The contents of the middle ear are carefully inspected, and the ossicular chain (malleus and incus unit) is evaluated. Ossicular interruption is most frequent in chronic otitis media, but problems of reconstruction can also occur with malformations of the middle ear and ossicular dislocations due to head injuries. Dramatic improvement in hearing can result from closure of a perforation and reestablishment of the ossicles. Surgery is usually performed in an outpatient facility under moderate sedation or general anesthesia.

Ossiculoplasty

Ossiculoplasty is the surgical reconstruction of the middle ear bones to restore hearing. Prostheses made of materials such as Teflon, stainless steel, and hydroxyapatite are used to reconnect the ossicles, thereby reestablishing the sound conduction mechanism. However, the greater the damage, the lower the success rate for restoring normal hearing.

Mastoidectomy

The objectives of mastoid surgery are to remove the cholesteatoma, gain access to diseased structures, and create a dry (noninfected) and healthy ear. If possible, the ossicles are reconstructed during the initial surgical procedure. Occasionally, extensive disease or damage dictates that this be performed as part of a two-stage operation.

A mastoidectomy is usually performed through a postauricular incision. Infection is eliminated by removing the mastoid air cells. A second mastoidectomy may be necessary to check for recurrent or residual cholesteatoma. The hearing mechanism may be reconstructed at this time. The success rate for correcting this conductive hearing loss is approximately 75%. Surgery is usually performed in an outpatient setting. The patient has a mastoid pressure dressing, which can be removed 24 to 48 hours after surgery. Although infrequently injured, the facial nerve, which runs through the middle ear and mastoid, is at some risk for injury during mastoid surgery. As the patient awakens from anesthesia, any evidence of facial paresis should be reported to the physician.

NURSING PROCESS

The Patient Undergoing Mastoid Surgery

Although several otologic surgical procedures are performed under moderate sedation, mastoid surgery is performed using general anesthesia.

Assessment

The health history includes a complete description of the ear disorder, including infection, otalgia, otorrhea, hearing loss, and vertigo. Data are collected about the duration and intensity of the disorder, its causes, and previous treatments. Information is obtained about other health problems and all medications that the patient is taking. Medication allergies and family history of ear disease also should be obtained.

Physical assessment addresses erythema, edema, otorrhea, lesions, and characteristics such as odor and color of discharge. The results of the audiogram are reviewed.

Nursing Diagnoses

Based on the assessment data, major nursing diagnoses may include the following:

- Anxiety related to surgical procedure, potential loss of hearing, potential taste disturbance, and potential loss of facial movement
- Acute pain related to mastoid surgery
- Risk for infection related to mastoidectomy; placement of grafts, prostheses, and electrodes; and surgical trauma to surrounding tissues and structures
- Impaired verbal communication related to ear disorder, surgery, or packing
- Risk for injury related to impaired balance or vertigo during the immediate postoperative period, dislodgment of the graft or prosthesis, or injury to facial nerve (cranial nerve VII) and chorda tympani nerve
- Deficient knowledge about mastoid disease, surgical procedure, and postoperative care and expectations

Planning and Goals

The major goals of caring for a patient undergoing mastoidectomy include reduction of anxiety; freedom from pain and discomfort; prevention of infection; stable or improved hearing and communication; absence of vertigo and related injury; and increased knowledge regarding the disease, surgical procedure, and postoperative care.

Nursing Interventions

Reducing Anxiety

The nurse reinforces the information discussed by the otologic surgeon with the patient, including anesthesia, the location of the incision (postauricular), and expected surgical results (e.g., hearing, balance, taste, facial movement). The patient also is encouraged to discuss any anxieties and concerns about the surgery.

Relieving Pain

Although most patients complain very little about incisional pain after mastoid surgery, they do have some ear discomfort. Aural fullness or pressure after surgery is caused by residual blood or fluid in the middle ear. The prescribed analgesic medication may be taken for the first 24 hours after surgery and then only as needed.

A wick or external auditory canal packing is used if tympanoplasty was performed at the time of the mastoidectomy. For the next 2 to 3 weeks after surgery, the patient may experience sharp, shooting pains intermittently as the eustachian tube opens and allows air to enter the middle ear. Constant, throbbing pain accompanied by fever may indicate infection and should be reported to the primary provider.

Preventing Infection

Measures are initiated to prevent infection in the operated ear. The external auditory canal wick, or packing, may be impregnated with an antibiotic solution before instillation. Prophylactic antibiotic agents are administered as prescribed, and the patient is instructed to prevent water from entering the external auditory canal for 6 weeks. A cotton ball or lamb's wool covered with a water-insoluble substance (e.g., petrolatum jelly) and placed loosely in the ear canal usually prevents water contamination and should be used when the patient showers or washes his or her hair, or in any situations in which water may enter the ear. The postauricular incision should be kept dry for the first 2 days. Signs of infection such as an elevated temperature and purulent drainage should be reported. Some serosanguineous drainage from the external auditory canal is normal after surgery.

Improving Hearing and Communication

Hearing in the operated ear may be reduced for several weeks because of edema, accumulation of blood and tissue fluid in the middle ear, and dressings or packing. Measures are initiated to improve hearing and communication, such as reducing environmental noise, facing the patient when speaking, speaking clearly and distinctly without shouting, providing good lighting if the patient relies on speech reading, and using nonverbal clues (e.g., facial expression, pointing, gestures) and other forms of communication. Family members or significant others are instructed about effective ways to communicate with the patient. If the patient uses assistive hearing devices, one can be used in the unaffected ear.

Preventing Injury

Vertigo may occur after mastoid surgery if the semicircular canals or other areas of the inner ear are traumatized. Antiemetic or antivertiginous medications (e.g., antihistamines) can be prescribed if a balance disturbance or vertigo occurs. Safety measures such as assisted ambulation are implemented to prevent falls and injury. The patient is instructed to avoid heavy lifting, straining, exertion, and nose blowing for 2 to 3 weeks after surgery to prevent dislodging the tympanic membrane graft or ossicular prosthesis.

Facial nerve injury is a potential, although rare, complication of mastoid surgery. The patient is instructed to report immediately any evidence of facial nerve (cranial nerve VII) weakness, such as drooping of the mouth on the operated side, slurred speech, decreased sensation, and difficulty swallowing. A more frequent occurrence is a temporary disturbance in the chorda tympani nerve, which is a small branch of the facial nerve that runs through the middle ear. Patients experience a taste disturbance and dry mouth on the side of surgery for several months until the nerve regenerates.

Promoting Home and Community-Based Care

Educating Patients About Self-Care. Patients require education about medication therapy, such as analgesic and antivertiginous agents (e.g., antihistamines) prescribed for balance disturbance. Education includes information about the expected effects and potential side effects of the medication. Patients also need instruction about any activity restrictions. Possible complications such as infection, facial nerve weakness, or taste disturbances, including the signs and symptoms to report immediately, are included (Chart 64-6).

Continuing Care. Some patients, particularly older adult patients, who have had mastoid surgery may require the services of a home care nurse for a few days after returning home. However, most people find that assistance from a family member or a friend is sufficient. The caregiver and patient are cautioned that the patient may experience some vertigo and will therefore require help with ambulation to avoid falling. Any symptoms of complications are to be reported promptly to the primary provider. The importance of scheduling and keeping follow-up appointments is also stressed.

Evaluation

Expected patient outcomes may include:

1. Demonstrates reduced anxiety about surgical procedure
   a. Verbalizes and exhibits less stress, tension, and irritability
   b. Verbalizes acceptance of the results of surgery and adjustment to possible hearing impairment
2. Remains free of discomfort or pain
   a. Exhibits no facial grimacing, moaning, or crying, and reports absence of pain
   b. Uses analgesic agents appropriately
3. Demonstrates no signs or symptoms of infection
   a. Has normal vital signs, including temperature
   b. Demonstrates absence of purulent drainage from the external auditory canal
   c. Describes method for preventing water from contaminating packing
4. Exhibits signs that communication and hearing have stabilized or improved
   a. Describes surgical goal for hearing and judges whether the goal has been met
   b. Verbalizes that hearing has improved
5. Remains free of injury and trauma
   a. Reports absence of vertigo or balance disturbance
   b. Experiences no injury or fall
   c. Avoids activities that can cause dislodgement of graft or prosthesis
   d. Reports no taste disturbance, mouth dryness, or facial weakness
6. Verbalizes the reasons for and methods of care and treatment
   a. Discusses the discharge plan formulated with the nurse with regard to rest periods, medication, and activities permitted and restricted
   b. Lists symptoms that should be reported to health care personnel
   c. Keeps follow-up appointments

Chart 64-6 PATIENT EDUCATION

Self-Care After Middle Ear or Mastoid Surgery

Postoperative instructions for patients who have had middle ear and mastoid surgery may vary among otolaryngologists. General education guidelines for the patient may include the following:

- Take antibiotics and other medications as prescribed.
- Avoid nose blowing for 2 to 3 weeks after surgery.
- Sneeze and cough with the mouth open for a few weeks after surgery.
- Avoid heavy lifting (>10 lb), straining, and bending over for a few weeks after surgery.
- Popping and crackling sensations in the operative ear are normal for approximately 3 to 5 weeks after surgery.
- Temporary hearing loss is normal in the operative ear due to fluid, blood, or packing in the ear.
- Report excessive or purulent ear drainage to the physician.
- Avoid getting water in the operative ear for 2 weeks after surgery. You may shampoo the hair 2 to 3 days postoperatively if the ear is protected from water by saturating a cotton ball with petrolatum jelly (or some other water-insoluble substance) and loosely placing it in the ear. If the postauricular suture line becomes wet, pat (not rub) the area and cover it with a thin layer of antibiotic ointment.

Otosclerosis

Otosclerosis involves the stapes and is thought to result from the formation of new, abnormal spongy bone, especially around the oval window, with resulting fixation of the stapes (Porth & Matfin, 2009). The efficient transmission of sound is prevented because the stapes cannot vibrate and carry the sound as conducted from the malleus and incus to the inner ear. Otosclerosis is more common in women and frequently hereditary, and pregnancy may worsen it.

Clinical Manifestations

Otosclerosis may involve one or both ears and manifests as a progressive conductive or mixed hearing loss. The patient may or may not complain of tinnitus. Otoscopic examination usually reveals a normal tympanic membrane. Bone conduction is better than air conduction on Rinne testing. The audiogram confirms conductive hearing loss or mixed loss, especially in the low frequencies.

Medical Management

There is no evidence that any nonsurgical treatments are effective for treating otosclerosis. However, some physicians believe that the use of sodium fluoride can mature the abnormal spongy bone growth and prevent the breakdown of the bone tissue. Amplification with a hearing aid also may help (Porth & Matfin, 2009).

Surgical Management

One of two surgical procedures may be performed: the stapedectomy or the stapedotomy. A stapedectomy involves removing the stapes superstructure and part of the footplate and inserting a tissue graft and a suitable prosthesis (Fig. 64-6). The surgeon drills a small hole into the footplate to hold a prosthesis. The prosthesis bridges the gap between the incus and the inner ear, providing better sound conduction. The majority of patients experience resolution of conductive hearing loss following stapes surgery. Balance disturbance or true vertigo may occur during the postoperative period for several days. Long-term balance disorders are rare.

Middle Ear Masses

Other than cholesteatoma, masses in the middle ear are rare. Glomus tympanicum is a tumor that arises from Jacobson's nerve (in the temporal bone of the skull) and remains limited to the middle ear. On otoscopy, a red blemish on or behind the tympanic membrane is seen. Glomus jugulare tumors are rarely malignant; however, because of their location, treatment may be necessary to relieve symptoms. The treatment is

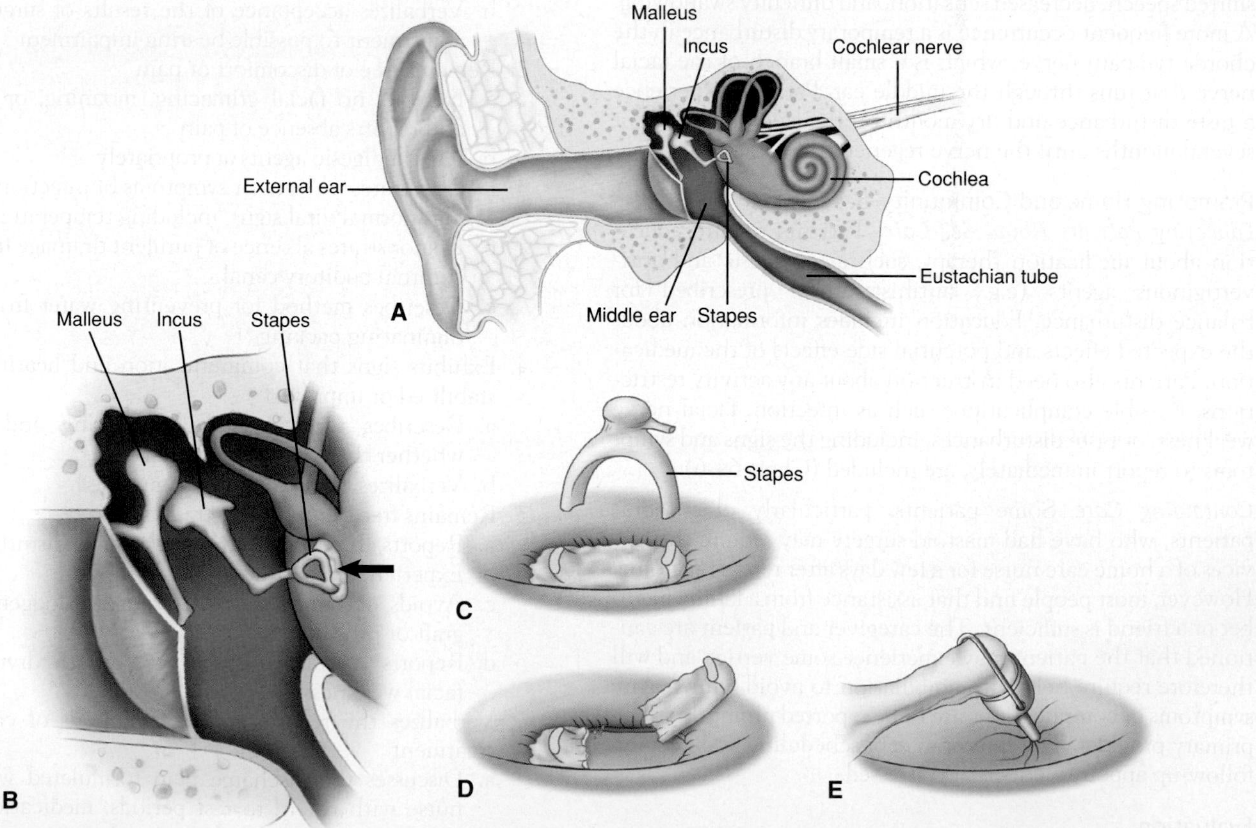

Malleus
Incus
Cochlear nerve
Cochlea
External ear
Eustachian tube
Middle ear
Stapes
A
Malleus
Incus
Stapes
B
Stapes
C
D
E

FIGURE 64-6 Stapedectomy for otosclerosis. **A.** Normal anatomy. **B.** *Arrow* points to sclerotic process at the foot of the stapes. **C.** Stapes broken away surgically from its diseased base. The hole in the footplate provides an area where an instrument can grasp the plate. **D.** The footplate is removed from its base. Some otosclerotic tissue may remain, and tissue is placed over it. **E.** Robinson stainless steel prosthesis in position.

surgical excision, except in poor surgical candidates, in whom radiation therapy is used.

A facial nerve neuroma is a tumor on cranial nerve VII. These types of tumors are usually not visible on otoscopic examination but are suspected when a patient presents with a facial nerve paresis. X-ray evaluation is used to identify the site of the tumor along the facial nerve. The treatment is surgical removal.

CONDITIONS OF THE INNER EAR

Disorders of balance are common (NIDCD, 2009), and dizziness is one of the most common reasons for patients presenting to the emergency department (Navi, Kamel, Shah, et al., 2012). The term **dizziness** is used frequently by patients and health care providers to describe any altered sensation of orientation in space and is more common in women and older adults (Kim, Fullerton, & Johnston, 2011). **Vertigo** is the misperception or illusion of motion of the person or the surroundings. Most people with vertigo describe a spinning sensation or say they feel as though objects are moving around them. Ataxia is a failure of muscular coordination and may be present in patients with vestibular disease. Syncope, fainting, and loss of consciousness are not forms of vertigo and usually indicate disease in the cardiovascular system.

Nystagmus is an involuntary rhythmic movement of the eyes. Nystagmus occurs normally when a person watches a rapidly moving object (e.g., through the side window of a moving car or train). However, pathologically, it is an ocular disorder associated with vestibular dysfunction. Nystagmus can be horizontal, vertical, or rotary and can be caused by a disorder in the central or peripheral nervous system.

Motion Sickness

Motion sickness is a disturbance of equilibrium caused by constant motion. For example, it can occur aboard a ship, while riding on a merry-go-round or swing, or in a car (Herron, 2010).

Clinical Manifestations

The syndrome manifests itself in sweating, pallor, nausea, and vomiting caused by vestibular overstimulation. These manifestations may persist for several hours after the stimulation stops.

Management

Over-the-counter antihistamines such as dimenhydrinate (Dramamine) or meclizine (Antivert) may provide some relief of nausea and vomiting by blocking the conduction of the vestibular pathway of the inner ear. Anticholinergic medications, such as scopolamine patches (Transderm Scop), may also be effective because they antagonize the histamine response. These must be replaced every 3 days (Karch, 2012). Side effects such as dry mouth and drowsiness may occur. Potentially hazardous activities such as driving a car or operating heavy machinery should be avoided if drowsiness occurs.

Ménière's Disease

Ménière's disease is an abnormality in inner ear fluid balance caused by a malabsorption in the endolymphatic sac or a blockage in the endolymphatic duct (Nelson & Viire, 2009; NIDCD, 2009). **Endolymphatic hydrops** (dilation of the endolymphatic space) develops, and either increased pressure in the system or rupture of the inner ear membrane occurs, producing symptoms of Ménière's disease.

Ménière's disease affects 10 to 12 of 1,000 people in the United States. It is estimated that there are 615,000 cases in the United States, with approximately 45,500 new cases diagnosed annually (NIDCD, 2010). More common in adults, onset is generally seen when adults reach their 40s, with symptoms usually beginning between the ages of 20 and 60 years. Ménière's disease appears to be equally common in men and women, and it occurs bilaterally in about 20% of patients. About 50% of the patients have a positive family history of the disease.

Clinical Manifestations

Ménière's disease is characterized by a triad of symptoms: episodic vertigo, **tinnitus** (unwanted noises in the head or ear), and fluctuating sensorineural hearing loss. It may also include a feeling of pressure or fullness in the ear and incapacitating vertigo, often accompanied by nausea and vomiting (Nelson & Viire, 2009; NIDCD, 2009). These symptoms range in severity from a minor nuisance to extreme disability, especially if the attacks of vertigo are severe. At the onset of the disease, usually only one or two of the symptoms are manifested.

Some characterize the disease into two subsets: cochlear and vestibular. Cochlear Ménière's disease is recognized as a fluctuating, progressive sensorineural hearing loss associated with tinnitus and aural pressure in the absence of vestibular symptoms or findings. Vestibular Ménière's disease is characterized as the occurrence of episodic vertigo associated with aural pressure but no cochlear symptoms. Patients may experience either cochlear or vestibular disease symptoms; however, eventually all of these symptoms develop.

Assessment and Diagnostic Findings

Vertigo is usually the most troublesome complaint related to Ménière's disease. A careful history is taken to determine the frequency, duration, severity, and character of the vertigo attacks. Vertigo may last minutes to hours, possibly accompanied by nausea or vomiting. Diaphoresis and a persistent feeling of imbalance or disequilibrium may awaken patients at night. Some patients report that these feelings last for days. However, they usually feel well between attacks. Hearing loss may fluctuate, with tinnitus and aural pressure waxing and waning with changes in hearing. These feelings may occur during or before attacks, or they may be constant.

Physical examination findings are usually normal, with the exception of those of cranial nerve VIII. Sounds from a tuning fork (Weber test) may lateralize to the ear opposite the hearing loss, the one affected with Ménière's disease. An audiogram typically reveals a sensorineural hearing loss in the affected ear. This can be in the form of a "Pike's Peak" pattern, which looks like a hill or mountain. A sensorineural loss in the low frequencies occurs as the disease progresses. The electronystagmogram may be normal or may show reduced vestibular response.

Medical Management

Most patients with Ménière's disease can be successfully treated with diet and medication. Many patients can control their symptoms by adhering to a low-sodium (1,000 to 1,500 mg/day or less) diet. Chart 64-7 describes dietary guidelines that may be useful in Ménière's disease. The amount of sodium is one of many factors that regulate the balance of fluid within the body. Sodium and fluid retention disrupts the delicate balance between endolymph and perilymph in the inner ear. Psychological evaluation and cognitive therapy may be indicated if a patient is anxious, uncertain, fearful, or depressed (NIDCD, 2010).

Pharmacologic Therapy

Pharmacologic therapy for Ménière's disease consists of antihistamines, such as meclizine, which shortens the attack (NIDCD, 2010). Tranquilizers such as diazepam (Valium) may be used in acute instances to help control vertigo. Antiemetic agents such as promethazine (Phenergan) suppositories help control the nausea and vomiting and the vertigo because of their antihistamine effect. Diuretic therapy (e.g., hydrochlorothiazide [Dyazide], triamterene [Dyrenium]) may relieve symptoms by lowering the pressure in the endolymphatic system. Intratympanic injection of gentamicin (Garamycin) is used to cause ablation of the vestibular hair cells; however, the risk of significant hearing loss is high (NIDCD, 2010).

Surgical Management

Although most patients respond well to conservative therapy, some continue to have disabling attacks of vertigo. If these attacks reduce their quality of life, patients may elect to undergo surgery for relief. Surgical procedures include endolymphatic sac procedures and vestibular nerve section (NIDCD, 2010). However, hearing loss, tinnitus, and aural fullness may continue, because the surgical treatment of Ménière's disease is aimed at eliminating the attacks of vertigo.

Endolymphatic Sac Decompression

Endolymphatic sac decompression, or shunting, theoretically equalizes the pressure in the endolymphatic space. A shunt or drain is inserted in the endolymphatic sac through a postauricular incision. This procedure is favored by many otolaryngologists as a first-line surgical approach to treat the vertigo of Ménière's disease because it is relatively simple and safe and can be performed on an outpatient basis.

Vestibular Nerve Sectioning

Vestibular nerve sectioning provides the greatest success rate (approximately 98%) in eliminating the attacks of vertigo. It can be performed by a translabyrinthine approach (i.e., through the hearing mechanism) or in a manner that can conserve hearing (i.e., suboccipital or middle cranial fossa), depending on the degree of hearing loss. Most patients with incapacitating Ménière's disease have little or no effective hearing. Cutting the nerve prevents the brain from receiving input from the semicircular canals. This procedure may require a brief hospital stay. A plan of nursing care for the patient with vertigo is presented in Chart 64-8.

Chart 64-7 **PATIENT EDUCATION**

Dietary Guidelines for Patients With Ménière's Disease

- Limit foods high in salt or sugar. Be aware of foods with hidden salts and sugars.
- Eat meals and snacks at regular intervals to stay hydrated. Missing meals or snacks may alter the fluid level in the inner ear.
- Eat fresh fruits, vegetables, and whole grains. Limit the amount of canned, frozen, or processed foods with high sodium content.
- Drink plenty of fluids daily. Water, milk, and low-sugar fruit juices are recommended. Limit intake of coffee, tea, and soft drinks. Avoid caffeine because of its diuretic effect.
- Limit alcohol intake. Alcohol may change the volume and concentration of the inner ear fluid and may worsen symptoms.
- Avoid monosodium glutamate (MSG), which may increase symptoms.
- Pay attention to the intake of foods containing potassium (e.g., bananas, tomatoes, oranges) if taking a diuretic that causes potassium loss.
- Avoid aspirin and aspirin-containing medications. Aspirin may increase tinnitus and dizziness.

Adapted from National Institute on Deafness and Other Communication Disorders. (2010). *Ménière's disease*. Available at: www.nidcd.nih.gov//health/balance/meniere.html

Benign Paroxysmal Positional Vertigo

Benign paroxysmal positional vertigo is a brief period of incapacitating vertigo that occurs when the position of the patient's head is changed with respect to gravity, typically by placing the head back with the affected ear turned down (Nelson & Viire, 2009; NIDCD, 2009). The onset is sudden and followed by a predisposition for positional vertigo, usually for hours to weeks but occasionally for months or years.

Benign paroxysmal positional vertigo is thought to be due to the disruption of debris within the semicircular canal. This debris is formed from small crystals of calcium carbonate from the inner ear structure (the utricle). This is frequently stimulated by head trauma, infection, or other events. In severe cases, vertigo may easily be induced by any head movement. The vertigo is usually accompanied by nausea and vomiting; however, hearing impairment does not generally occur.

Bed rest is recommended for patients with acute symptoms. Repositioning techniques can be used to treat vertigo. The canalith repositioning procedure, also known as the Epley maneuver, is commonly used. This noninvasive procedure, which involves quick movements of the body, rearranges the debris in the canal. The procedure is performed by placing the patient in a sitting position, turning the head to a 45-degree angle on the affected side, and then quickly moving the patient to the supine position. The procedure is safe, inexpensive, and easy to perform (Nelson & Viire, 2009).

Patients with acute vertigo may be treated with meclizine for 1 to 2 weeks. After this time, the meclizine is stopped and the patient is reassessed. Patients who continue to have severe positional vertigo may be premedicated with prochlorperazine

Chart 64-8

PLAN OF NURSING CARE

Care of the Patient With Vertigo

NURSING DIAGNOSIS: Risk for injury related to altered mobility because of gait disturbance and vertigo
GOAL: Remains free of any injuries associated with imbalance and/or falls

Nursing Interventions	Rationale	Expected Outcomes
1. Assess for vertigo, including history, onset, description of attacks, duration, frequency, and any associated ear symptoms (hearing loss, tinnitus, aural fullness).	**1.** History provides basis for interventions.	• Experiences no falls due to balance disturbance • Fear and anxiety are reduced • Performs exercises as prescribed • Takes prescribed medications appropriately • Assumes safe position when vertigo is present • Keeps head still when vertigo is present • Identifies a characteristic fullness or sense of pressure in the ear as occurring before a full-blown attack • Reports measures that help reduce vertigo
2. Assess extent of disability in relation to activities of daily living.	**2.** Extent of disability indicates risk of falling.	
3. Instruct or reinforce vestibular/balance therapy as prescribed.	**3.** Exercises hasten labyrinthine compensation, which may decrease vertigo and gait disturbance.	
4. Administer, or educate about administration of, antivertiginous medications and/or vestibular sedation medication; instruct patient about side effects.	**4.** Alleviates acute symptoms of vertigo.	
5. Encourage patient to sit down and restrict activity when dizzy.	**5.** Decreases possibility of falling and injury.	
6. Place pillow on each side of head to restrict movement.	**6.** Movement aggravates vertigo.	
7. Assist patient in identifying aura that suggests an impending attack.	**7.** Recognition of aura may trigger the need to take medication before an attack occurs, thereby minimizing the severity of effects.	
8. Recommend that the patient keep eyes open and stare straight ahead when lying down and experiencing vertigo.	**8.** Sensation of vertigo decreases and motion decelerates if eyes are kept in a fixed position.	

NURSING DIAGNOSIS: Risk-prone health behavior related to disability requiring change in lifestyle due to unpredictability of vertigo
GOAL: Modifies lifestyle to decrease disability and exert maximum control and independence within limits posed by chronic vertigo

Nursing Interventions	Rationale	Expected Outcomes
1. Encourage patient to identify personal strengths and roles that can still be fulfilled.	**1.** Maximizes sense of regaining control and independence.	• Exerts maximum control of environment and independence within limits imposed by vertigo • Is informed about condition • Family and significant others are included in rehabilitation process • Uses strengths and potentials to engage in the most independent and constructive lifestyle
2. Provide information about vertigo and what to expect.	**2.** Reduces fear and anxiety.	
3. Include family and significant others in rehabilitative process.	**3.** Perceived beliefs of significant others are important for patient's adherence to medical regimen.	
4. Encourage patient to maintain sense of control by making decisions and assuming more responsibility for care.	**4.** Reinforces positive psychological and social outcomes.	

NURSING DIAGNOSIS: Risk for deficient fluid volume related to increased fluid output, altered intake, and medications
GOAL: Maintains normal fluid and electrolyte balance

Nursing Interventions	Rationale	Expected Outcomes
1. Assess, or have patient assess, intake and output (including emesis, liquid stools, urine, and diaphoresis). Monitor laboratory values of electrolytes.	**1.** Accurate records provide basis for fluid replacement.	• Laboratory values within normal limits • Alert and oriented; vital signs within normal limits, skin turgor normal; electrolytes normal • Mucous membranes are moist • Vomiting or diarrhea has stopped; usual oral intake is resumed
2. Assess indicators of dehydration, including blood pressure (orthostasis), pulse, skin turgor, mucous membranes, and level of consciousness.	**2.** Prompt recognition of dehydration allows early intervention.	
3. Encourage oral fluids as tolerated; discourage beverages containing caffeine (a vestibular stimulant).	**3.** Oral replacement is begun as soon as possible to replace losses. Caffeine may increase diarrhea.	

(continues on page 1900)

Chart 64-8

PLAN OF NURSING CARE

Care of the Patient With Vertigo (continued)

Nursing Interventions	Rationale	Expected Outcomes
4. Administer, or educate about administration of, antiemetic and antidiarrheal medications as prescribed and needed. Educate patient about side effects.	**4.** Antiemetic medications reduce nausea and vomiting, reducing fluid losses and improving oral intake. Antidiarrheal medication reduces intestinal motility and fluid losses.	

NURSING DIAGNOSIS: Anxiety related to threat of, or change in, health status and disability effects of vertigo
GOAL: Experiences less or no anxiety

Nursing Interventions	Rationale	Expected Outcomes
1. Assess level of anxiety. Help patient identify coping skills used successfully in the past. **2.** Provide information about vertigo and its treatment. **3.** Encourage patient to discuss anxieties and explore concerns about vertigo attacks. **4.** Educate patient about stress management techniques or make appropriate referral. **5.** Provide comfort measures, and avoid stress-producing activities. **6.** Instruct patient in aspects of treatment regimen.	**1.** Guides therapeutic interventions and participation in self-care. Past coping skills can relieve anxiety. **2.** Increased knowledge helps to decrease anxiety. **3.** Promotes awareness and understanding of relationship between anxiety level and behavior. **4.** Improved stress management can reduce the frequency and severity of some vertiginous attacks. **5.** Stressful situations may exacerbate symptoms of the condition. **6.** Patient knowledge helps to decrease anxiety.	• Fear and anxiety about attacks of vertigo reduced or eliminated • Acquires knowledge and skills to deal with vertigo • Feels less tension, apprehension, and uncertainty • Uses stress management techniques when needed • Avoids upsetting encounters • Repeats instructions given and verbalizes understanding of treatments

NURSING DIAGNOSIS: Risk for trauma related to impaired balance
GOAL: Reduces the risk of trauma by adapting the home environment and by using assistive devices as necessary

Nursing Interventions	Rationale	Expected Outcomes
1. Assess for balance disturbance and/or vertigo by taking history and by examination for nystagmus, positive Romberg, and inability to perform tandem Romberg. **2.** Assist with ambulation when indicated. **3.** Assess for visual acuity and proprioceptive deficits. **4.** Encourage increased activity level with or without use of assistive devices. **5.** Help identify hazards in home environment.	**1.** Peripheral vestibular disorders cause these signs and symptoms. **2.** Abnormal gait can predispose patient to unsteadiness and falls. **3.** Balance depends on visual, vestibular, and proprioceptive systems. **4.** Increased activity may help retrain balance system. **5.** Adaptation of home environment can reduce risk of falls during rehabilitative process.	• Has adapted home environment or uses rehabilitative devices to reduce risk of falling • Ambulates with needed assistance • Visual and proprioceptive risks identified • Activity level increased • Home environment free of hazards

NURSING DIAGNOSIS: Self-care deficit: feeding, bathing/hygiene, dressing/grooming, toileting, related to labyrinth dysfunction and episodes of vertigo
GOAL: Able to care for self

Nursing Interventions	Rationale	Expected Outcomes
1. Administer, or educate about administration of, antiemetic and other prescribed medications to relieve nausea and vomiting associated with vertigo. **2.** Encourage patient to perform self-care when free of vertigo. **3.** Review diet with patient and caregivers. Offer fluids as necessary.	**1.** Antiemetic and sedative-type medications depress stimuli in the cerebellum. **2.** Spacing activities is important because episodes of vertigo vary in occurrence. **3.** Sodium restriction helps improve an inner ear fluid imbalance in some patients, thereby decreasing vertigo. Fluids help prevent dehydration.	• Carries out necessary functions during symptom-free periods and takes medications to relieve nausea, vomiting, or vertigo • Carries out daily activities • Accepts dietary plan and reports its effectiveness • Drinks fluids in sufficient amounts

Chart 64-8

PLAN OF NURSING CARE

Care of the Patient With Vertigo (continued)

NURSING DIAGNOSIS: Powerlessness related to illness regimen and being helpless in certain situations due to vertigo/balance disturbance
GOAL: Experiences increased sense of control over life and activities despite vertigo/balance disturbance

Nursing Interventions	Rationale	Expected Outcomes
1. Assess patient's needs, values, attitudes, and readiness to initiate activities.	1. Involving patient in planning activities and care enhances potential for mastery.	• Does not restrict activities unnecessarily due to vertigo
2. Provide opportunities for patient to express feelings about self and illness.	2. Expressing feelings increases understanding of individual coping styles and defense mechanisms.	• Verbalizes positive feelings about own ability to achieve a sense of power and control
3. Help patient identify previous coping behaviors that were successful.	3. Awareness increases understanding of stressors that trigger feeling of powerlessness. Awareness of past successes enhances self-confidence.	• Identifies previous successful coping behaviors

(Compazine) 1 hour before the canalith repositioning procedure is performed.

Vestibular rehabilitation can be used in the management of vestibular disorders. This strategy promotes active use of the vestibular system through an interdisciplinary team approach, including medical and nursing care, stress management, biofeedback, vocational rehabilitation, and physical therapy. A physical therapist prescribes balance exercises that help the brain compensate for the impairment to the balance system.

Tinnitus

Tinnitus is a symptom of an underlying disorder of the ear that is associated with hearing loss. This condition affects approximately 37 million people in the United States and is most prevalent between 40 and 70 years of age (Porth & Matfin, 2009). The severity of tinnitus may range from mild to severe. Patients describe tinnitus as a roaring, buzzing, or hissing sound in one or both ears. Numerous factors may contribute to the development of tinnitus, including several ototoxic substances (Chart 64-9). Underlying disorders that contribute to tinnitus may include thyroid disease, hyperlipidemia, vitamin B_{12} deficiency, psychological disorders (e.g., depression, anxiety), fibromyalgia, otologic disorders (Ménière's disease, acoustic neuroma), and neurologic disorders (head injury, multiple sclerosis).

A physical examination should be performed to determine the cause of tinnitus. Diagnostic testing determines if hearing loss is present. An audiograph speech discrimination test or a tympanogram may be used to help determine the cause. Some forms of tinnitus are irreversible; therefore, patients may need education and counseling about ways of adjusting to their treatment and dealing with tinnitus in the future. Research suggests that exposing patients to music that does not contain the frequency range that is disturbing to them may reduce their tinnitus (Anderson, 2009).

Labyrinthitis

Labyrinthitis, an inflammation of the labyrinth of the inner ear, can be bacterial or viral in origin. Bacterial labyrinthitis is rare because of antibiotic therapy, but it sometimes occurs as a complication of otitis media. The infection can spread to the inner ear by penetrating the membranes of the oval or round windows. Viral labyrinthitis is a common diagnosis, but little is known about this disorder, which affects hearing and balance. The most common viral causes are mumps, rubella, rubeola, and influenza. Viral illnesses of the upper respiratory tract and herpetiform disorders of the facial and acoustic nerves (i.e., Ramsay Hunt syndrome) also cause labyrinthitis.

Clinical Manifestations

Labyrinthitis is characterized by a sudden onset of incapacitating vertigo, usually with nausea and vomiting, various degrees of hearing loss, and possibly tinnitus. The first episode is usually the worst; subsequent attacks, which usually occur over a period of several weeks to months, are less severe.

Management

Treatment of bacterial labyrinthitis includes IV antibiotic therapy, fluid replacement, and administration of an antihistamine (e.g., meclizine) and antiemetic medications. Treatment of viral labyrinthitis is based on the patient's symptoms.

Ototoxicity

A variety of medications may have adverse effects on the cochlea, vestibular apparatus, or cranial nerve VIII. All but a few, such as aspirin and quinine, cause irreversible hearing loss. At high doses, aspirin toxicity can produce bilateral tinnitus. IV medications, especially the aminoglycosides, are a common cause of ototoxicity, because they destroy the hair cells in the organ of Corti (see Chart 64-9). Antineoplastic agents also cause hair cell death in the cochlea, which can lead to hearing loss (Mudd, Edmunds, Glatz, et al., 2012). These medications can be found in the body several months later; side effects are dose dependent, with higher doses causing increased ototoxicity. Therefore, hearing loss may occur at any time, even several months after the last dose of the medication was administered.

To prevent loss of hearing or balance, patients receiving potentially ototoxic medications should be counseled about

Chart 64-9 Selected Ototoxic Substances

- **Diuretic agents:** ethacrynic acid, furosemide, acetazolamide
- **Chemotherapeutic (antineoplastic) agents:** cisplatin, nitrogen mustard, carboplatin
- **Antimalarial agents:** quinine, chloroquine
- **Anti-inflammatory agents:** salicylates (aspirin), indomethacin
- **Chemicals:** alcohol, arsenic
- **Aminoglycoside antibiotic agents:** amikacin, gentamicin, kanamycin, netilmicin, neomycin, streptomycin, tobramycin
- **Other antibiotic agents:** erythromycin, minocycline, polymyxin B, vancomycin
- **Metals:** gold, mercury, lead

their side effects. These medications should be used with caution in patients who are at high risk for complications, such as children, older adults, pregnant patients, patients with kidney or liver problems, and patients with current hearing disorders. Blood levels of the medications should be monitored, and patients receiving long-term IV antibiotics should be monitored with an audiogram twice each week during therapy.

Acoustic Neuroma

Acoustic neuromas are slow-growing, benign tumors of cranial nerve VIII, usually arising from the Schwann cells of the vestibular portion of the nerve. Most acoustic tumors arise within the internal auditory canal and extend into the cerebellopontine angle to press on the brain stem, possibly destroying the vestibular nerve (Bader & Littlejohns, 2010). Most acoustic neuromas are unilateral, except in von Recklinghausen's disease (neurofibromatosis type 2), in which bilateral tumors occur (Porth & Matfin, 2009).

Acoustic neuromas develop in 1 of every 100,000 people per year. These neuromas account for 5% to 10% of all intracranial tumors and seem to occur with equal frequency in men and women at any age, although most occur during middle age (McDonald, 2011).

Assessment and Diagnostic Findings

The most common assessment findings of patients with acoustic neuromas are unilateral tinnitus and hearing loss with or without vertigo or balance disturbance. It is important to identify asymmetry in audiovestibular test results so that further workup can be performed to rule out an acoustic neuroma. Although conflicting data exists, the only known risk factor for acoustic neuroma is cell phone usage (Kundi, 2009). MRI with a contrast agent (i.e., gadolinium or gadopentetate [Magnevist]) is the imaging study of choice. If the patient is claustrophobic or cannot undergo an MRI for other reasons or if the scan is unavailable, a CT scan with contrast dye is performed. However, MRI is more sensitive than CT in delineating a small tumor.

Management

Conservative treatment is recommended for patients with tumors less than 1.5 cm and in those who are older. In addition, routine monitoring is recommended for these patients (McDonald, 2011). For low-risk patients, surgical removal of the acoustic tumor is the treatment of choice because these tumors do not respond well to radiation or chemotherapy. Because treatment of acoustic tumors crosses several specialties, the interdisciplinary treatment approach involves a neurologist and a neurosurgeon. The objective of the surgery is to remove the tumor while preserving facial nerve function. Most acoustic tumors have damaged the cochlear portion of cranial nerve VIII, and hearing is impaired. In these patients, the surgery is performed using a translabyrinthine approach, and the hearing mechanism is destroyed. If hearing is still good before surgery, a suboccipital or middle cranial fossa approach to removing the tumor may be used. This procedure exposes the lateral third of the internal auditory canal and preserves hearing.

Potential complications of surgery include facial nerve paralysis, cerebrospinal fluid leakage, meningitis, and cerebral edema. Death from acoustic neuroma surgery is rare.

AURAL REHABILITATION

If hearing loss is permanent or cannot be treated by medical or surgical means or if the patient elects not to undergo surgery, aural rehabilitation may be beneficial. The purpose of aural rehabilitation is to maximize the communication skills of the person with hearing impairment. Aural rehabilitation includes auditory training, speech reading, speech training, and the use of hearing aids and hearing guide dogs.

Auditory training emphasizes listening skills, so the person who is hearing impaired concentrates on the speaker. Speech reading (also known as lip reading) can help fill the gaps left by missed or misheard words. The goals of speech training are to conserve, develop, and prevent deterioration of current communication skills.

It is important to identify the type of hearing impairment a person has so that rehabilitative efforts can be directed at his or her particular need. Surgical correction may be all that is necessary to treat and improve a conductive hearing loss by eliminating the cause of the hearing loss. With advances in hearing aid technology, amplification for patients with sensorineural hearing loss is more helpful than ever.

Hearing Aids

A hearing aid is a device through which speech and environmental sounds are received by a microphone, converted to electrical signals, amplified, and reconverted to acoustic signals. Many aids available for sensorineural hearing loss depress the low frequencies, or tones, and enhance hearing for the high frequencies. A general guideline for assessing the patient's need for a hearing aid is a hearing loss exceeding 30 dB in the range of 500 to 2,000 Hz in the better-hearing ear.

A hearing aid makes sounds louder, but it does not improve a patient's ability to discriminate words or understand speech. People who have low discrimination scores (i.e., 20%) on audiograms may derive little benefit from a hearing aid. Hearing aids amplify all sounds, including background noise, which may be particularly disturbing to the first-time wearer. Chart 64-10 identifies additional problems associated with hearing aid use. Computerized hearing aids are available to

Chart 64-10 Hearing Aid Problems

Whistling Noise

Loose ear mold
Improperly made
Improperly worn
Worn out

Improper Aid Selection

Too much power required in aid, with inadequate separation between microphone and receiver
Open mold used inappropriately
Inadequate amplification
Dead batteries
Cerumen in ear
Cerumen or other material in mold
Wires or tubing disconnected from aid
Aid turned off or volume too low
Improper mold
Improper aid for degree of loss

Pain From Mold

Improperly fitted mold
Ear skin or cartilage infection
Middle ear infection
Ear tumor
Unrelated conditions of the temporomandibular joint, throat, or larynx

compensate for background noise or allow amplification at certain programmed frequencies rather than at all frequencies. Occasionally, depending on the type of hearing loss, binaural aids (i.e., one for each ear) may be indicated. Chart 64-11 provides tips for hearing aid care.

A hearing aid should be fitted according to the patient's needs (e.g., type of hearing loss, manual dexterity, and preferences), rather than the brand name, by a certified audiologist licensed to dispense hearing aids. Many states have consumer protection laws that allow the hearing aid to be returned after a trial use if the patient is not completely satisfied. In addition, to protect the health and safety of people with hearing impairments, the U.S. Food and Drug Administration (FDA) has established certain regulations. A medical evaluation of the impairment by a physician must be obtained within 6 months before the purchase of a hearing aid. However, the written statement from a physician may be waived if the patient (a fully informed adult 18 years or older) signs a document to this effect. Health care professionals who dispense hearing aids are required to refer prospective users to a physician if any of the following otologic conditions are evident:

- Visible congenital or traumatic deformity of the ear
- Active drainage from the ear within the previous 90 days
- Sudden or rapidly progressive hearing loss within the previous 90 days
- Complaints of dizziness or tinnitus
- Unilateral hearing loss that has occurred suddenly or within the previous 90 days
- Audiometric air–bone gap of 15 dB or more at 500, 1,000, and 2,000 Hz

Chart 64-11 PATIENT EDUCATION Tips for Hearing Aid Care

Cleaning

- The ear mold is the only part of the hearing aid that may be washed frequently.
- Wash ear mold daily with soap and water.
- Allow the ear mold to dry completely before it is snapped into the receiver.
- Clean the cannula with a small pipe cleaner–like device.
- Proper care of the ear device and keeping the ear canal clean and dry can prevent complications.

Malfunctioning

- Inadequate amplification, a whistling noise, or pain from the mold can occur when a hearing aid is not functioning properly.
- Check for malfunctions:
  - Is the switch on properly?
  - Are the batteries charged and positioned correctly?
  - Is the ear mold clogged with cerumen? Ear wax can be easily removed with pin, pipe cleaner, or wax loop.
- If the hearing aid is still not working properly, notify the hearing aid dealer.
- If the unit requires extended time for repair, the dealer may lend you a hearing aid until the repair can be accomplished.

Recognizing Complications

- Common medical complications include external otitis media and pressure ulcers in the external auditory canal. Signs and symptoms of these infections include painful ear, especially when the external ear is touched; canal swelling; redness; difficulty hearing; pain radiating to the jaw area; and fever.
- If any of these symptoms are present, notify your health care provider for evaluation. You may need medication to treat infection, pain, or both.

- Significant accumulation of cerumen or a foreign body in the external auditory canal
- Pain or discomfort in the ear

A user instruction brochure is provided with every hearing aid device. In this brochure, the following information is presented:

- Notification that good health practice requires a medical evaluation before purchasing a hearing aid
- Notification that any of the otologic conditions listed previously should be investigated by a physician before purchase of a hearing aid
- Instructions for proper use, maintenance, and care of the hearing aid, as well as instructions for replacing or recharging the batteries
- Repair service information
- Description of avoidable conditions that could damage the hearing aid
- List of any known side effects that may warrant physician consultation (e.g., skin irritation, accelerated cerumen accumulation)

The evolution in technology has led to the availability of many smaller and more effective devices as well as different options and features of hearing aids (FDA, 2009a) (Chart 64-12). The majority of hearing aids sold today are behind-the-ear, in-the-ear, or in-the-canal types (Table 64-4). One

Chart 64-12 Options and Features of Hearing Aids to Consider

- *T-Coil:* May improve hearing on the telephone by switching the settings from normal to telephone setting. This feature also assists in amplifying voices when the patient is in larger areas, such as theaters, auditoriums, and gymnasiums. Background noise may be toned down to adequately hear close conversation.
- *Directional Microphone:* Useful in environments with a lot of background noise and activity. The microphone may be directed toward the speaker and amplifies conversation while diminishing background noise.
- *Direct Audio Input:* Directly plugs into another device, such as a computer, television, or stereo, that is attached with an extension cord
- *Feedback Suppression:* Suppresses whistling feedback noise

Adapted from U.S. Food and Drug Administration. (2009a). *Medical devices: Types of hearing aids*. Available at: www.fda.gov/MedicalDevices/ProductsandMedicalProcedures/HomeHealthandConsumer/ConsumerProducts/HearingAids/ucm181470.htm

model is the Lyric, which is placed in the ear canal just 4 mm from the tympanic membrane. Its volume is controlled by a magnet, and when its batteries no longer function (1 to 4 months), a physician can remove it with the magnet and reinsert a new device. This device does not have many of the problems (e.g., feedback noise, overamplification of background noise) associated with other hearing aids, and it does not involve the expense and uncertainty of surgical procedures. However, it is not an option for a person whose ear canal is too narrow to accommodate it.

Implanted Hearing Devices

There are several types of implanted hearing devices, ranging from implantable to semi-implantable devices (FDA, 2009b).

Bone conduction devices, which transmit sound through the skull to the inner ear, are used in patients with a conductive hearing loss if a hearing aid is contraindicated (e.g., those with chronic infection). The device is implanted postauricularly under the skin into the skull, and an external device—worn above the ear, not in the canal—transmits the sound through the skin. There are two types of implantable hearing aids. The bone-anchored hearing aid (BAHA) is implanted behind the ear in the mastoid area. The middle ear implantation (MEI) is implanted in the middle ear cavity. The BAHA is used for conductive or mixed hearing loss, whereas the MEI is used for sensorineural hearing loss.

The implantable middle ear hearing device (IMEHD) comes in two styles: piezoelectric and electromagnetic, which are partially or totally implanted. Patients must be 18 years or older, must be diagnosed with mild to severe sensorineural loss, and must have tried other conventional devices but achieved poor results to be considered candidates for this type of device. The implantable device has several advantages—for example, it may eliminate feedback, achieve good cosmetic results, and allow the patient to perform most preferred leisure activities (e.g., dancing, swimming). Disadvantages are that this device is expensive, necessitates surgery, requires periodic recharging of batteries, and has unpredictable power output.

The FDA has also approved the semi-implantable Vibrant Soundbridge (electromagnetic) and the total implantable Envoy Esteem (piezoelectric) devices for use in the United States. The Vibrant Soundbridge has an external device attached to the postauricular bone that transmits sound to the magnet in the middle ear that is attached to the long process of the incus. The magnet surrounds the long axis of the stapes, which in turn vibrates and sound is heard. The Envoy Esteem works similarly to the natural ear. The piezoelectric transducer is located at the head of the incus, which sends a signal that is amplified, filtered, and then converted back to a vibration signal. This vibration is delivered by the driver (piezoelectric transducer) and attached to the stapes capitulum; then, via the stapes bone, the inner ear receives the signal and is converted into a nerve impulse and translated into sound by the brain. The incus is removed prior to insertion of this device to prevent feedback from the sensor. It is estimated that sound is amplified up to 110 dB with this device (Shohet & Myers, 2011).

A cochlear implant is an auditory prosthesis used for people with profound sensorineural hearing loss bilaterally who do not benefit from conventional hearing aids. The hearing loss may be congenital or acquired. An implant does not restore normal hearing; rather, it helps the person detect medium to loud environmental sounds and conversation. The implant provides stimulation directly to the auditory nerve, bypassing the nonfunctioning hair cells of the inner ear. The microphone

TABLE 64-4 Hearing Aids

Site (Range of Hearing Loss)	Advantages	Disadvantages
Body, usually on the trunk (mild–profound)	Separation of receiver and microphone prevents acoustic feedback, allowing high amplification; generally used in a school setting	Bulky; requires long wire, which may be cosmetically displeasing; some loss of high-frequency response
Behind the ear (mild–profound)	Economical; powerful, with no long wires; easily used by children—adapts easily as the child grows, with only the ear mold needing replacement	Large size
In the ear (mild–moderately severe)	One-piece custom fit to contour of ear; no tubes or cords; miniature microphone located in the ear, which is a more natural placement; more cosmetically appealing due to easy concealment	Smaller size limits output; patients who have arthritis or cannot perform tasks requiring good manual dexterity may have difficulty with the small size of aid and/or battery; can require more repair than the behind-the-ear aid.
In the canal (mild–moderately severe)	Same as in-the-ear aids; less visible, so more cosmetically pleasing	Even smaller than in-the-ear aids; requires good manual dexterity and good vision

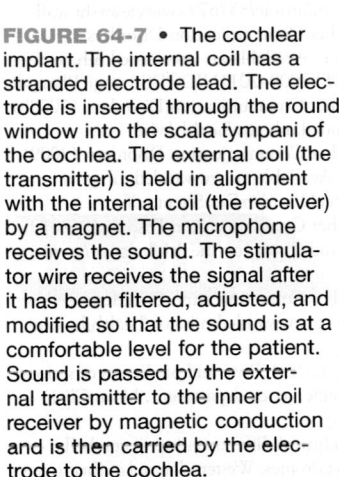

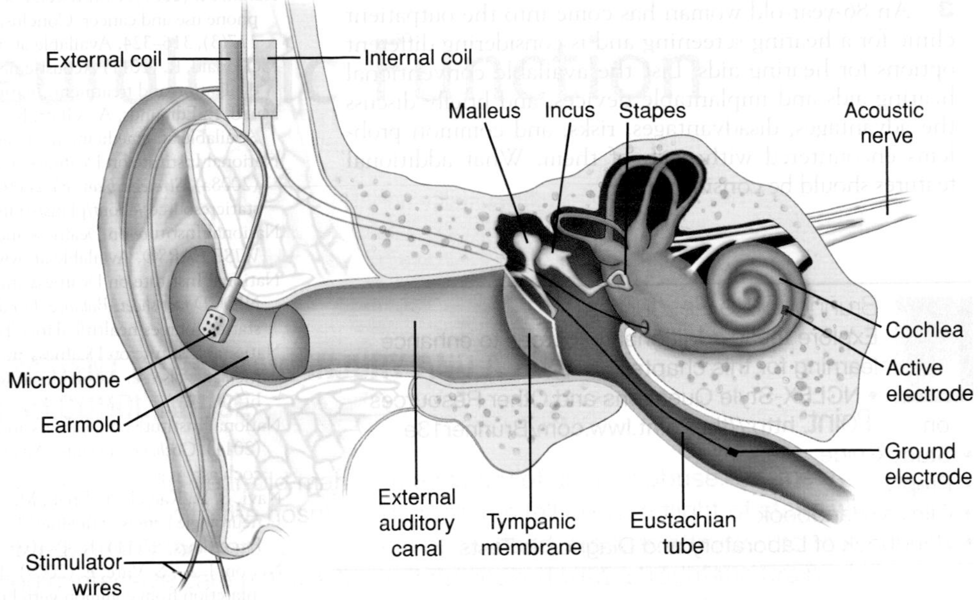
External coil
Internal coil
Malleus
Incus
Stapes
Acoustic nerve
Cochlea
Active electrode
Ground electrode
Microphone
Earmold
Stimulator wires
External auditory canal
Tympanic membrane
Eustachian tube

FIGURE 64-7 • The cochlear implant. The internal coil has a stranded electrode lead. The electrode is inserted through the round window into the scala tympani of the cochlea. The external coil (the transmitter) is held in alignment with the internal coil (the receiver) by a magnet. The microphone receives the sound. The stimulator wire receives the signal after it has been filtered, adjusted, and modified so that the sound is at a comfortable level for the patient. Sound is passed by the external transmitter to the inner coil receiver by magnetic conduction and is then carried by the electrode to the cochlea.

and signal processor, worn outside the body, transmit electrical stimuli to the implanted electrodes. The electrical signals stimulate the auditory nerve fibers and then the brain, where they are interpreted.

Worldwide, more than 219,000 people have received a cochlear implant. In the United States, more than 42,600 adults have cochlear implants (NIDCD, 2011). Candidates for a cochlear implant, who are usually at least 1 year of age, are selected after careful screening by otologic history, physical examination, audiologic testing, x-rays, and psychological testing. Criteria for choosing adults who may benefit from a cochlear implant include:

- Profound sensorineural hearing loss in both ears
- Inability to hear and recognize speech well with hearing aids
- No medical contraindication to a cochlear implant or general anesthesia
- Indications that being able to hear would enhance the patient's life

The surgery involves implanting a small receiver in the temporal bone through a postauricular incision and placing electrodes into the inner ear (Fig. 64-7). The microphone and transmitter are worn on an external unit. The patient undergoes extensive cochlear rehabilitation with the multidisciplinary team, which includes an audiologist and speech pathologist. Several months may be needed to learn to interpret the sounds heard. Children and adults who lost their hearing before they learned to speak take much longer to acquire speech. There are wide variations of success with cochlear implants, and there is also controversy about their use, especially among the deaf community. Patients who have had a cochlear implant are cautioned that an MRI will inactivate the implant; MRI should be used only when there is no other diagnostic option.

Hearing Guide Dogs

Specially trained dogs (service dogs) are available to assist the person with a hearing loss. People who live alone are eligible to apply for a dog trained by International Hearing Dog, Inc. The dog reacts to the sound of a telephone, a doorbell, an alarm clock, a baby's cry, a knock at the door, a smoke alarm, or an intruder. The dog alerts its master by physical contact; the dog then runs to the source of the noise. In public, the dog positions itself between the person with hearing impairment and any potential hazard that the person cannot hear, such as an oncoming vehicle or a loud, hostile person. In many states, a certified hearing guide dog is legally permitted access to public transportation, public eating places, and stores, including food markets.

Critical Thinking Exercises

1 **ebp** The charge nurse in the primary care clinic has asked you to develop a brief evidence-based questionnaire for patients to fill out while waiting to be seen. This clinic is focusing on health maintenance and meeting the *Healthy People 2020* objectives and would like to assess "hearing loss" in all individuals. Patients of all socioeconomic and educational levels are seen at this clinic. Develop an evidence-based questionnaire to assess for hearing loss that the patients can complete while in the clinic waiting room. Identify the criteria used to evaluate the strength of the evidence for the assessment of hearing loss.

2 **pq** You are assigned to care for an 85-year-old male patient who is being discharged from the hospital with the recent diagnosis of vertigo. Identify the priorities, approach, and techniques that you would incorporate into a comprehensive plan of care for his vertigo. How will your priorities, approach, and techniques differ if the patient lives alone? If the patient has a visual impairment or is hard of hearing? If the patient is from a culture with very different values from your own?

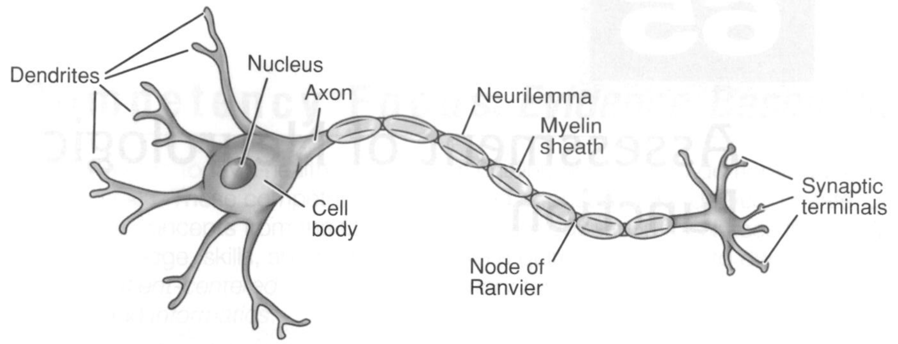
Dendrites
Nucleus
Axon
Neurilemma
Myelin sheath
Cell body
Node of Ranvier
Synaptic terminals

FIGURE 65-1 • Neuron.

activity through complex chemical and electrical messages (Klein & Stewart-Amidei, 2012).

Cells of the Nervous System

The basic functional unit of the brain is the neuron (Fig. 65-1). It is composed of dendrites, a cell body, and an axon. The **dendrites** are branch-type structures for receiving electrochemical messages. The **axon** is a long projection that carries electrical impulses away from the cell body. Some axons have a myelinated sheath that increases speed of conduction. Nerve cell bodies occurring in clusters are called *ganglia* or *nuclei*. A cluster of cell bodies with the same function is called a *center* (e.g., the respiratory center). Neurons are supported, protected, and nourished by glial cells, which are 50 times greater in number than neurons (Hickey, 2009).

Concept Mastery Alert

Axons carry impulses *away* from the cell body. Dendrites carry impulses *toward* the cell.

Neurotransmitters

Neurotransmitters communicate messages from one neuron to another or from a neuron to a target cell, such as muscle or endocrine cells. Neurotransmitters are manufactured and stored in synaptic vesicles. As an electrical action potential moves along the axon and reaches the nerve terminal, neurotransmitters are released into the synapse. The neurotransmitter is transported across the synapse, binding to receptors on the postsynaptic cell membrane. A neurotransmitter can either excite or inhibit activity of the target cell. Usually, multiple neurotransmitters are at work in the neural synapse. The source and action of major neurotransmitters are described in Table 65-1. Once released, enzymes either destroy the neurotransmitter or reabsorb it into the neuron for future use.

Many neurologic disorders are due, at least in part, to an imbalance in neurotransmitters. For example, Parkinson's disease develops from decreased availability of dopamine, whereas acetylcholine binding to muscle cells is impaired in myasthenia gravis (Porth, 2011). All brain functions are modulated through neurotransmitter receptor site activity, including memory and other cognitive processes (Hickey, 2009).

Ongoing research is evaluating diagnostic tests that can detect abnormal levels of neurotransmitters in the brain. Positron emission tomography (PET), for example, can detect dopamine, serotonin, and acetylcholine. Single-photon emission computed tomography (SPECT), similar to PET, can detect changes in some neurotransmitters, such as dopamine in Parkinson's disease (Bremner, 2005). Both PET and SPECT are discussed in more detail later in this chapter.

The Central Nervous System

The Brain

The brain accounts for approximately 2% of the total body weight; in an average young adult, the brain weighs approximately 1,400 g, whereas in an average older adult, the brain weighs approximately 1,200 g (Hickey, 2009). The brain is divided into three major areas: the cerebrum, the brain stem,

TABLE 65-1 Major Neurotransmitters

Neurotransmitter	Source	Action
Acetylcholine (major transmitter of the parasympathetic nervous system)	Many areas of the brain; autonomic nervous system	Usually excitatory; parasympathetic effects sometimes inhibitory (stimulation of heart by vagal nerve)
Serotonin	Brain stem, hypothalamus, dorsal horn of the spinal cord	Inhibitory; helps control mood and sleep, inhibits pain pathways
Dopamine	Substantia nigra and basal ganglia	Usually inhibitory; affects behavior (attention, emotions) and fine movement
Norepinephrine (major transmitter of the sympathetic nervous system)	Brain stem, hypothalamus, postganglionic neurons of the sympathetic nervous system	Usually excitatory; affects mood and overall activity
Gamma-aminobutyric acid	Spinal cord, cerebellum, basal ganglia, some cortical areas	Inhibitory
Enkephalin, endorphin	Nerve terminals in the spine, brain stem, thalamus and hypothalamus, pituitary gland	Excitatory; pleasurable sensation, inhibits pain transmission

Adapted from Porth, C. M. (2011). *Essentials of pathophysiology* (3rd ed.). Philadelphia: Lippincott Williams & Wilkins.

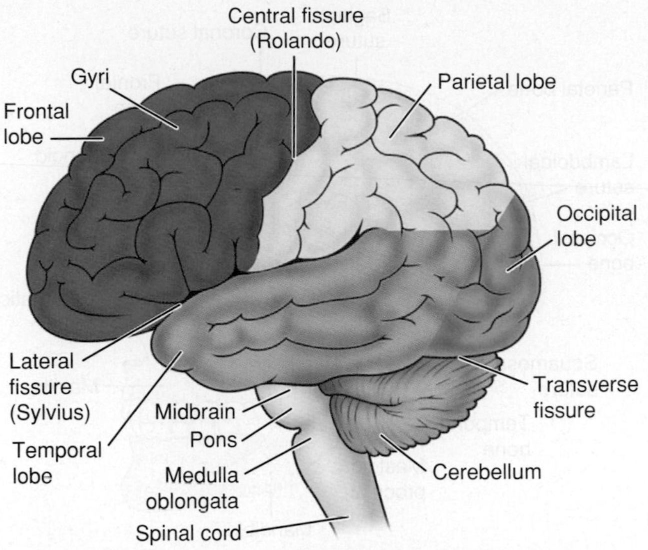
Central fissure
(Rolando)
Gyri
Parietal lobe
Frontal
lobe
Occipital
lobe
Lateral
fissure
(Sylvius)
Transverse
fissure
Midbrain
Pons
Temporal
lobe
Medulla
oblongata
Cerebellum
Spinal cord

FIGURE 65-2 • View of the external surface of the brain showing lobes, cerebellum, and brain stem.

and the cerebellum. The cerebrum is composed of two hemispheres, the thalamus, the hypothalamus, and the basal ganglia. The brain stem includes the midbrain, pons, and medulla. The cerebellum is located under the cerebrum and behind the brain stem (Fig. 65-2).

Cerebrum

The outside surface of the hemispheres has a wrinkled appearance that is the result of many folded layers or convolutions called *gyri,* which increase the surface area of the brain, accounting for the high level of activity carried out by such a small-appearing organ. Between each gyrus is a sulcus or fissure that serves as an anatomic division. In between the cerebral hemispheres is the great longitudinal fissure that separates the cerebrum into the right and left hemispheres. The two hemispheres are joined at the lower portion of the fissure by the corpus callosum. The external or outer portion of the hemispheres (the cerebral cortex) is made up of gray matter approximately 2 to 5 mm in depth; it contains billions of neuron cell bodies, giving it a gray appearance. White matter makes up the innermost layer and is composed of myelinated nerve fibers and neuroglia cells that form tracts or pathways connecting various parts of the brain with one another. These pathways also connect the cortex with lower portions of the brain and spinal cord. The cerebral hemispheres are divided into pairs of lobes as follows (see Fig. 65-2):

- *Frontal*—the largest lobe, located in the front of the brain. The major functions of this lobe are concentration, abstract thought, information storage or memory, and motor function. It contains Broca's area, which is located in the left hemisphere and is critical for motor control of speech. The frontal lobe is also responsible in large part for a person's affect, judgment, personality, and inhibitions (Hickey, 2009).
- *Parietal*—a predominantly sensory lobe posterior to the frontal lobe. This lobe analyzes sensory information and relays the interpretation of this information to other cortical areas and is essential to a person's awareness of body position in space, size and shape discrimination, and right–left orientation (Hickey, 2009).
- *Temporal*—located inferior to the frontal and parietal lobes, this lobe contains the auditory receptive areas and plays a role in memory of sound and understanding of language and music.
- *Occipital*—located posterior to the parietal lobe, this lobe is responsible for visual interpretation and memory.

The corpus callosum (Fig. 65-3), a thick collection of nerve fibers that connects the two hemispheres of the brain, is responsible for the transmission of information from one side of the brain to the other. Information transferred includes sensation, memory, and learned discrimination. Right-handed people and some left-handed people have cerebral dominance on the left side of the brain for verbal, linguistic, arithmetic, calculation, and analytic functions. The nondominant hemisphere is responsible for geometric, spatial, visual, pattern, and musical functions. Nuclei for cranial nerves I and II are also located in the cerebrum.

The thalami lie on either side of the third ventricle and act primarily as a relay station for all sensation except smell. All memory, sensation, and pain impulses pass through this section of the brain. The hypothalamus (see Fig. 65-3) is located anterior and inferior to the thalamus, and beneath and lateral to the third ventricle. The infundibulum of the hypothalamus connects it to the posterior pituitary gland. The hypothalamus plays an important role in the endocrine system because it regulates the pituitary secretion of hormones that influence metabolism, reproduction, stress response, and urine production. It works with the pituitary to maintain fluid balance through hormonal release and maintains temperature regulation by promoting vasoconstriction or vasodilatation. In addition, the hypothalamus is the site of the hunger center and is involved in appetite control. It contains centers that regulate the sleep–wake cycle, blood pressure, aggressive and sexual behavior, and emotional responses (i.e., blushing, rage, depression, panic, and fear). The hypothalamus also controls

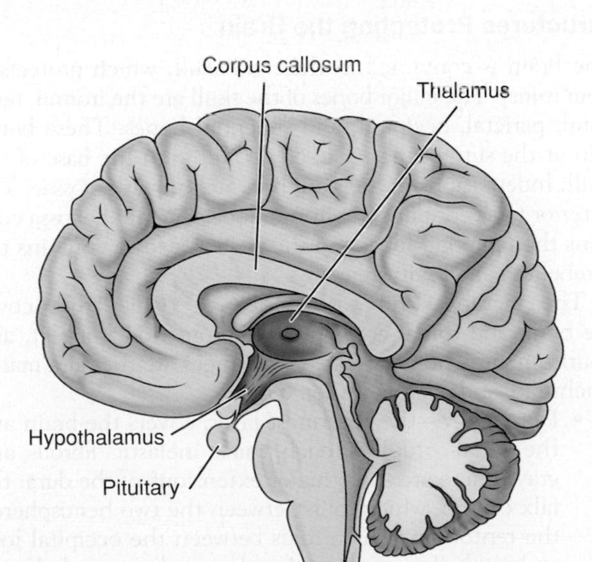
Corpus callosum
Thalamus
Hypothalamus
Pituitary

FIGURE 65-3 • Medial view of the brain.

and regulates the autonomic nervous system. The optic chiasm (the point at which the two optic tracts cross) and the mammillary bodies (involved in olfactory reflexes and emotional response to odors) are also found in this area.

The basal ganglia are masses of nuclei located deep in the cerebral hemispheres that are responsible for control of fine motor movements, including those of the hands and lower extremities.

Brain Stem

The brain stem consists of the midbrain, pons, and medulla oblongata (see Fig. 65-2). The midbrain connects the pons and the cerebellum with the cerebral hemispheres; it contains sensory and motor pathways and serves as the center for auditory and visual reflexes. Cranial nerves III and IV originate in the midbrain. The pons is situated in front of the cerebellum between the midbrain and the medulla and is a bridge between the two halves of the cerebellum, and between the medulla and the midbrain. Cranial nerves V through VIII originate in the pons. The pons also contains motor and sensory pathways. Portions of the pons help regulate respiration.

Motor fibers from the brain to the spinal cord and sensory fibers from the spinal cord to the brain are located in the medulla. Most of these fibers cross, or decussate, at this level. Cranial nerves IX through XII originate in the medulla. Reflex centers for respiration, blood pressure, heart rate, coughing, vomiting, swallowing, and sneezing are also located in the medulla. The reticular formation, responsible for arousal and the sleep–wake cycle, begins in the medulla and connects with numerous higher structures.

Cerebellum

The cerebellum is posterior to the midbrain and pons, and below the occipital lobe (see Fig. 65-2). The cerebellum integrates sensory information to provide smooth coordinated movement. It controls fine movement, balance, and **position (postural) sense** or proprioception (awareness of position of body parts without looking at them).

Structures Protecting the Brain

The brain is contained in the rigid skull, which protects it from injury. The major bones of the skull are the frontal, temporal, parietal, occipital, and sphenoid bones. These bones join at the suture lines (Fig. 65-4) and form the base of the skull. Indentations in the skull base are known as fossae. The anterior fossa contains the frontal lobe, the middle fossa contains the temporal lobe, and the posterior fossa contains the cerebellum and brain stem.

The meninges (fibrous connective tissues that cover the brain and spinal cord) provide protection, support, and nourishment. The layers of the meninges are the dura mater, arachnoid, and pia mater (Fig. 65-5):

- *Dura mater*—the outermost layer; covers the brain and the spinal cord. It is tough, thick, inelastic, fibrous, and gray. There are three major extensions of the dura: the falx cerebri, which folds between the two hemispheres; the tentorium, which folds between the occipital lobe and cerebellum to form a tough, membranous shelf; and the falx cerebelli, which is located between the right and left side of the cerebellum. When excess pressure occurs in the cranial cavity, brain tissue may be compressed against these dural folds or displaced around them, a process called *herniation*. A potential space exists between the dura and the skull, and between the periosteum and the dura in the vertebral column, known as the epidural space. Another potential space, the subdural space, also exists below the dura. Blood or an abscess can accumulate in these potential spaces.
- *Arachnoid*—the middle membrane; an extremely thin, delicate membrane that closely resembles a spider web (hence the name *arachnoid*). The arachnoid membrane has cerebrospinal fluid (CSF) in the space below it, known as the subarachnoid space. This membrane has arachnoid villi, which are unique fingerlike projections that absorb CSF into the venous system. When blood or bacteria enter the subarachnoid space, the villi become

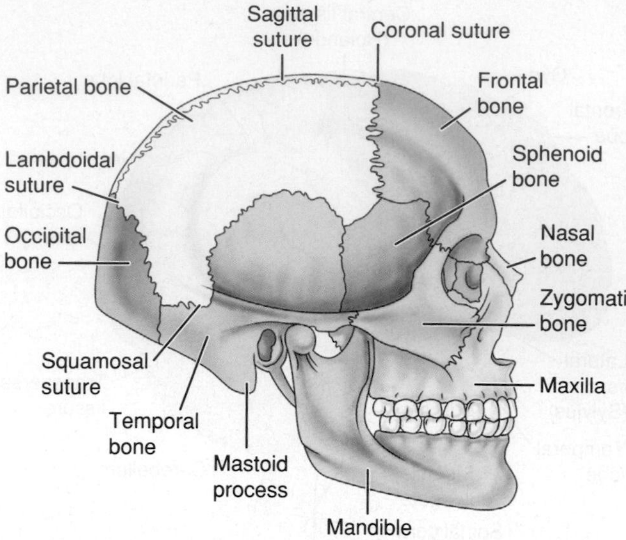

Sagittal suture
Coronal suture
Parietal bone
Frontal bone
Lambdoidal suture
Sphenoid bone
Occipital bone
Nasal bone
Zygomatic bone
Squamosal suture
Maxilla
Temporal bone
Mastoid process
Mandible

FIGURE 65-4 • Bones and sutures of the skull.

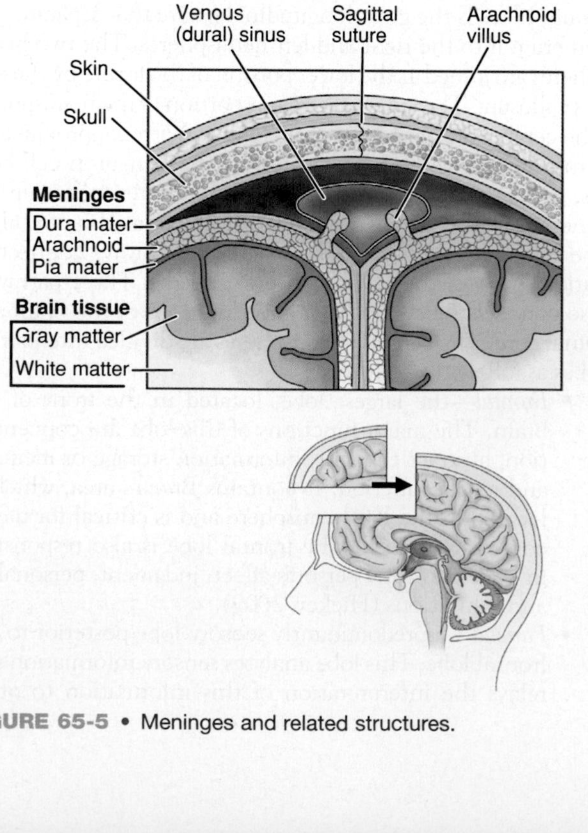

Venous (dural) sinus
Sagittal suture
Arachnoid villus
Skin
Skull
Meninges
Dura mater
Arachnoid
Pia mater
Brain tissue
Gray matter
White matter

FIGURE 65-5 • Meninges and related structures.

obstructed and *communicating* hydrocephalus (increased size of ventricles) may result.

- *Pia mater*—the innermost, thin, transparent layer that hugs the brain closely and extends into every fold of the brain's surface

Cerebrospinal Fluid

CSF is a clear and colorless fluid that is produced in the choroid plexus of the ventricles and circulates around the surface of the brain and the spinal cord. There are four ventricles: the right and left lateral and the third and fourth ventricles. The two lateral ventricles open into the third ventricle at the interventricular foramen (also known as the foramen of Monro). The third and fourth ventricles connect via the aqueduct of Sylvius. The fourth ventricle drains CSF into the subarachnoid space on the surface of the brain and spinal cord, where it is absorbed by the arachnoid villi. Blockage of the flow of CSF anywhere in the ventricular system produces *obstructive* hydrocephalus.

CSF is important in immune and metabolic functions in the brain. It is produced at a rate of about 500 mL/day; the ventricles and subarachnoid space contain approximately 150 mL of fluid (Hickey, 2009). The composition of CSF is similar to other extracellular fluids (such as blood plasma), but the concentrations of the various constituents differ. A laboratory analysis of CSF indicates color (clear), specific gravity (normal 1.007), protein count, cell count, glucose, and other electrolyte levels (see Table A-5 in Appendix A on thePoint). Normal CSF contains a minimal number of white blood cells and no red blood cells. The CSF may also be tested for immunoglobulins or the presence of bacteria.

Cerebral Circulation

The brain does not store nutrients and requires a constant supply of oxygen. These needs are met through cerebral circulation; the brain receives approximately 15% of the cardiac output, or 750 mL per minute of blood flow. Brain circulation is unique in several aspects. First, arterial and venous vessels are not parallel as in other organs in the body; this is due in part to the role the venous system plays in CSF absorption. Second, the brain has collateral circulation through the circle of Willis (see later discussion), allowing blood flow to be redirected on demand. Third, blood vessels in the brain have two rather than three layers, which may make them more prone to rupture when weakened or under pressure.

Arteries

Arterial blood supply to the anterior brain originates from the common carotid artery, which is the first bifurcation off the aorta. The internal carotid arteries arise at the bifurcation of the common carotid. Branches of the internal carotid arteries (the anterior and middle cerebral arteries) and their connections (the anterior and posterior communicating arteries) form the circle of Willis (Fig. 65-6).

The vertebral arteries branch from the subclavian arteries to supply most of the posterior circulation of the brain. At the level of the brain stem, the vertebral arteries join to form the basilar artery. The basilar artery divides to form the two branches of the posterior cerebral arteries. Functionally, the posterior and anterior portions of the circulation usually remain separate. However, the circle of Willis can provide collateral circulation through communicating arteries if one of the vessels supplying it becomes occluded or is ligated.

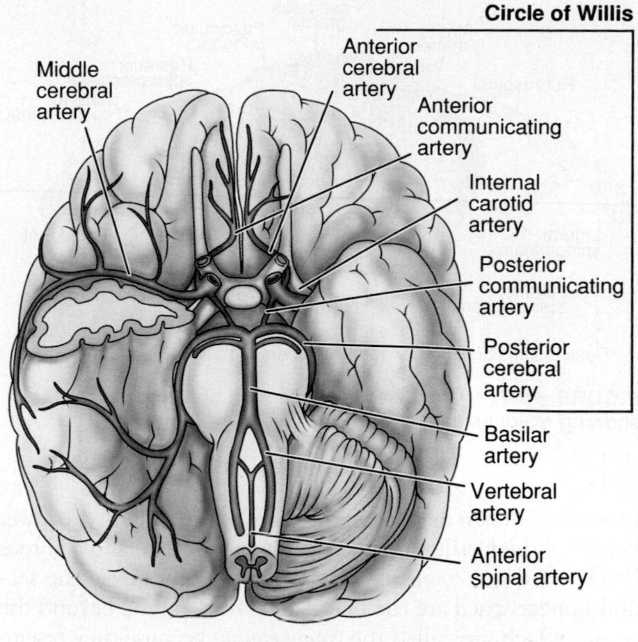
Circle of Willis
Middle cerebral artery
Anterior cerebral artery
Anterior communicating artery
Internal carotid artery
Posterior communicating artery
Posterior cerebral artery
Basilar artery
Vertebral artery
Anterior spinal artery

FIGURE 65-6 • Arterial blood supply of the brain, including the circle of Willis, as viewed from the ventral surface.

The bifurcations along the circle of Willis are frequent sites of aneurysm formation. Aneurysms are out-pouchings of the blood vessel due to vessel wall weakness. Aneurysms can rupture and cause a hemorrhagic stroke. (See Chapter 67 for a more detailed discussion of aneurysms.)

Veins

Venous drainage for the brain does not follow the arterial circulation as in other body structures. The veins reach the brain's surface, join larger veins, and then cross the subarachnoid space and empty into the dural sinuses, which are the vascular channels embedded in the dura (see Fig. 65-5). The network of the sinuses carries venous outflow from the brain and empties into the internal jugular veins, returning the blood to the heart. Cerebral veins are unique, because unlike other veins in the body, they do not have valves to prevent blood from flowing backward and depend on both gravity and blood pressure for flow.

Blood–Brain Barrier

The CNS is inaccessible to many substances that circulate in the blood plasma (e.g., dyes, medications, and antibiotic agents) because of the blood–brain barrier. This barrier is formed by the endothelial cells of the brain's capillaries, which form continuous tight junctions, creating a barrier to macromolecules and many compounds. All substances entering the CSF must filter through the capillary endothelial cells and astrocytes. The blood–brain barrier has a protective function but can be altered by trauma, cerebral edema, and cerebral hypoxemia; this has implications for treatment and selection of medications for CNS disorders (Hickey, 2009).

The Spinal Cord

The spinal cord is continuous with the medulla, extending from the cerebral hemispheres and serving as the connection between the brain and the periphery. Approximately 45 cm (18 inches) long and about the thickness of a finger, it extends

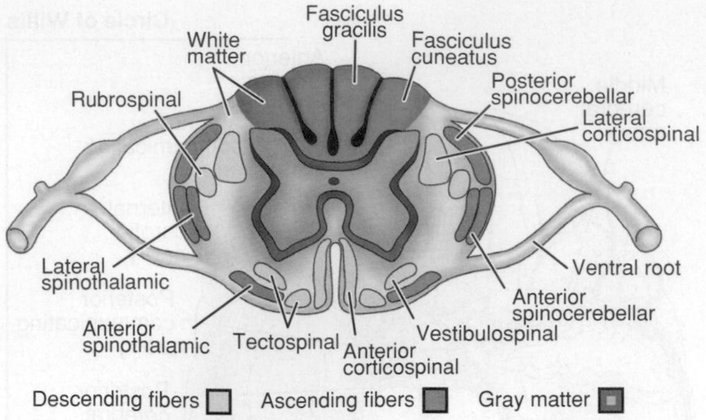
Fasciculus gracilis
White matter
Fasciculus cuneatus
Rubrospinal
Posterior spinocerebellar
Lateral corticospinal
Lateral spinothalamic
Ventral root
Anterior spinocerebellar
Anterior spinothalamic
Tectospinal
Vestibulospinal
Anterior corticospinal
Descending fibers
Ascending fibers
Gray matter

FIGURE 65-7 • Cross-sectional diagram of the spinal cord showing major spinal tracts.

from the foramen magnum at the base of the skull to the lower border of the first lumbar vertebra, where it tapers to a fibrous band called the *conus medullaris*. Continuing below the second lumbar space are the nerve roots that extend beyond the conus, which are called the *cauda equina* because they resemble a horse's tail. Meninges surround the spinal cord.

In a cross-sectional view, the spinal cord has an H-shaped central core of nerve cell bodies (gray matter) surrounded by ascending and descending tracts (white matter) (Fig. 65-7). The lower portion of the H is broader than the upper portion and corresponds to the anterior horns. The anterior horns contain cells with fibers that form the anterior (motor) root and are essential for the voluntary and reflex activity of the muscles they innervate. The thinner posterior (upper horns) portion contains cells with fibers that enter over the posterior (sensory) root and thus serve as a relay station in the sensory/reflex pathway.

The thoracic region of the spinal cord has a projection from each side at the crossbar of the H-shaped structure of gray matter called the *lateral horn*. It contains the cells that give rise to the autonomic fibers of the sympathetic division. The fibers leave the spinal cord through the anterior roots in the thoracic and upper lumbar segments.

The Spinal Tracts

The white matter of the spinal cord is composed of myelinated and unmyelinated nerve fibers. The fast-conducting myelinated fibers form bundles; fiber bundles with a common function are called *tracts*.

There are six ascending tracts (see Fig. 65-7). Two tracts, known as the fasciculus cuneatus and gracilis or the posterior columns, conduct sensations of deep touch, pressure, vibration, position, and passive motion from the same side of the body. Before reaching the cerebral cortex, these fibers cross to the opposite side in the medulla. The anterior and posterior spinocerebellar tracts conduct sensory impulses from muscle spindles, providing necessary input for coordinated muscle contraction. They ascend uncrossed and terminate in the cerebellum. The anterior and lateral spinothalamic tracts are responsible for conduction of pain, temperature, proprioception, fine touch, and vibratory sense from the upper body to the brain. They cross to the opposite side of the cord and then ascend to the brain, terminating in the thalamus (Klein & Stewart-Amidei, 2012).

There are eight descending tracts (see Fig. 65-7). The anterior and lateral corticospinal tracts conduct motor impulses to the anterior horn cells from the opposite side of the brain, cross in the medulla, and control voluntary muscle activity. The three vestibulospinal tracts descend uncrossed and are involved in some autonomic functions (sweating, pupil dilation, and circulation) and involuntary muscle control. The corticobulbar tract conducts impulses responsible for voluntary head and facial muscle movement and crosses at the level of the brain stem. The rubrospinal and reticulospinal tracts conduct impulses involved with involuntary muscle movement.

Vertebral Column

The bones of the vertebral column surround and protect the spinal cord and normally consist of 7 cervical, 12 thoracic, and 5 lumbar vertebrae, as well as the sacrum (a fused mass of 5 vertebrae), and terminate in the coccyx. Nerve roots exit from the vertebral column through the intervertebral foramina (openings). The vertebrae are separated by disks, except for the first and second cervical, the sacral, and the coccygeal vertebrae. Each vertebra has a ventral solid body and a dorsal segment or arch, which is posterior to the body. The arch is composed of two pedicles and two laminae supporting seven processes. The vertebral body, arch, pedicles, and laminae all encase and protect the spinal cord.

The Peripheral Nervous System

The peripheral nervous system includes the cranial nerves, the spinal nerves, and the autonomic nervous system.

Cranial Nerves

Twelve pairs of cranial nerves emerge from the lower surface of the brain and pass through openings in the base of the skull. Three cranial nerves are entirely sensory (I, II, VIII), five are motor (III, IV, VI, XI, and XII), and four are mixed sensory and motor (V, VII, IX, and X). The cranial nerves are numbered in the order in which they arise from the brain (Fig. 65-8). The cranial nerves innervate the head, neck, and

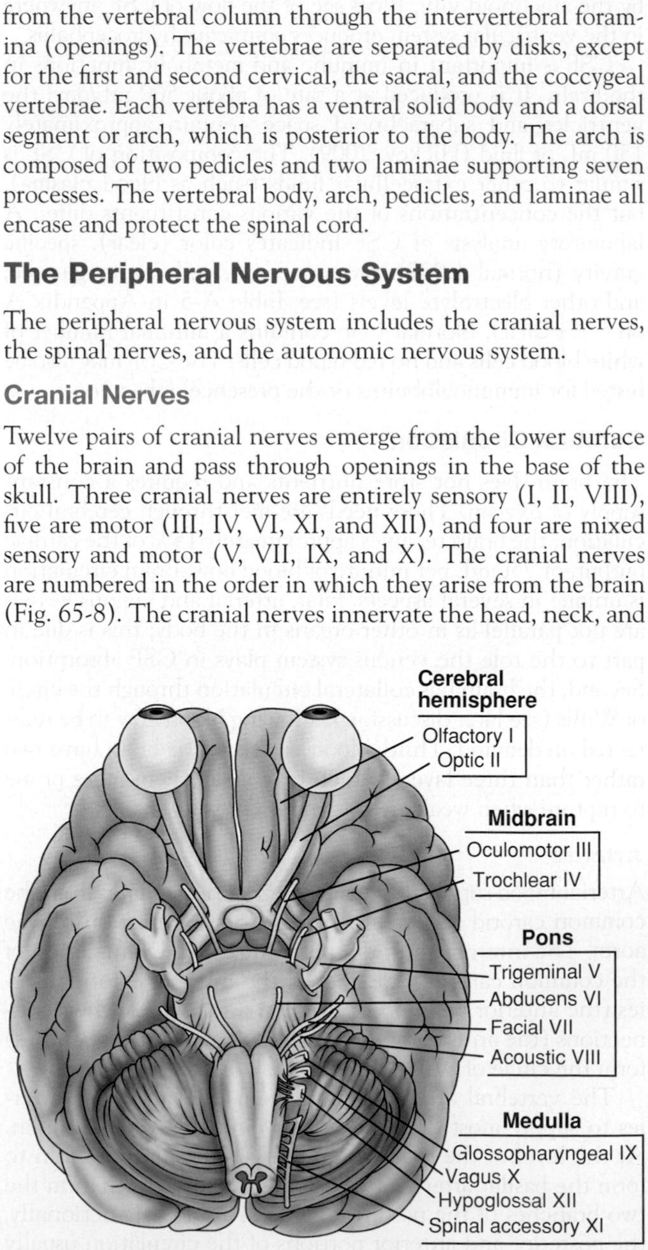
Cerebral hemisphere
Olfactory I
Optic II
Midbrain
Oculomotor III
Trochlear IV
Pons
Trigeminal V
Abducens VI
Facial VII
Acoustic VIII
Medulla
Glossopharyngeal IX
Vagus X
Hypoglossal XII
Spinal accessory XI

FIGURE 65-8 • Diagram of the base of the brain showing location of the cranial nerves.

TABLE 65-2 Cranial Nerves

Cranial Nerve	Type	Function
I (olfactory)	Sensory	Sense of smell
II (optic)	Sensory	Visual acuity and visual fields
III (oculomotor)	Motor	Muscles that move the eye and lid, pupillary constriction, lens accommodation
IV (trochlear)	Motor	Muscles that move the eye
V (trigeminal)	Mixed	Facial sensation, corneal reflex, mastication
VI (abducens)	Motor	Muscles that move the eye
VII (facial)	Mixed	Symmetry of facial expression and muscle movement in upper and lower face, salivation and tearing, taste, sensation in the ear
VIII (acoustic)	Sensory	Hearing and equilibrium
IX (glossopharyngeal)	Mixed	Taste, sensation in pharynx and tongue, pharyngeal muscles, swallowing
X (vagus)	Mixed	Muscles of pharynx, larynx, and soft palate; sensation in external ear, pharynx, larynx, thoracic and abdominal viscera; parasympathetic innervation of thoracic and abdominal organs
XI (spinal accessory)	Motor	Sternocleidomastoid and trapezius muscles
XII (hypoglossal)	Motor	Movement of the tongue

Adapted from Bader, M., & Littlejohns, L. R. (2010). *AANN core curriculum for neuroscience nursing* (5th ed.). Glenview, IL: American Association of Neuroscience Nurses.

special sense structures. Table 65-2 identifies primary functions of the cranial nerves.

Spinal Nerves

The spinal cord is composed of 31 pairs of spinal nerves: 8 cervical, 12 thoracic, 5 lumbar, 5 sacral, and 1 coccygeal. Each spinal nerve has a ventral root and a dorsal root. The dorsal roots are sensory and transmit sensory impulses from specific areas of the body known as dermatomes (Fig. 65-9) to the dorsal horn ganglia. The sensory fiber may be somatic, carrying information about pain, temperature, touch, and position sense (proprioception) from the tendons, joints, and body

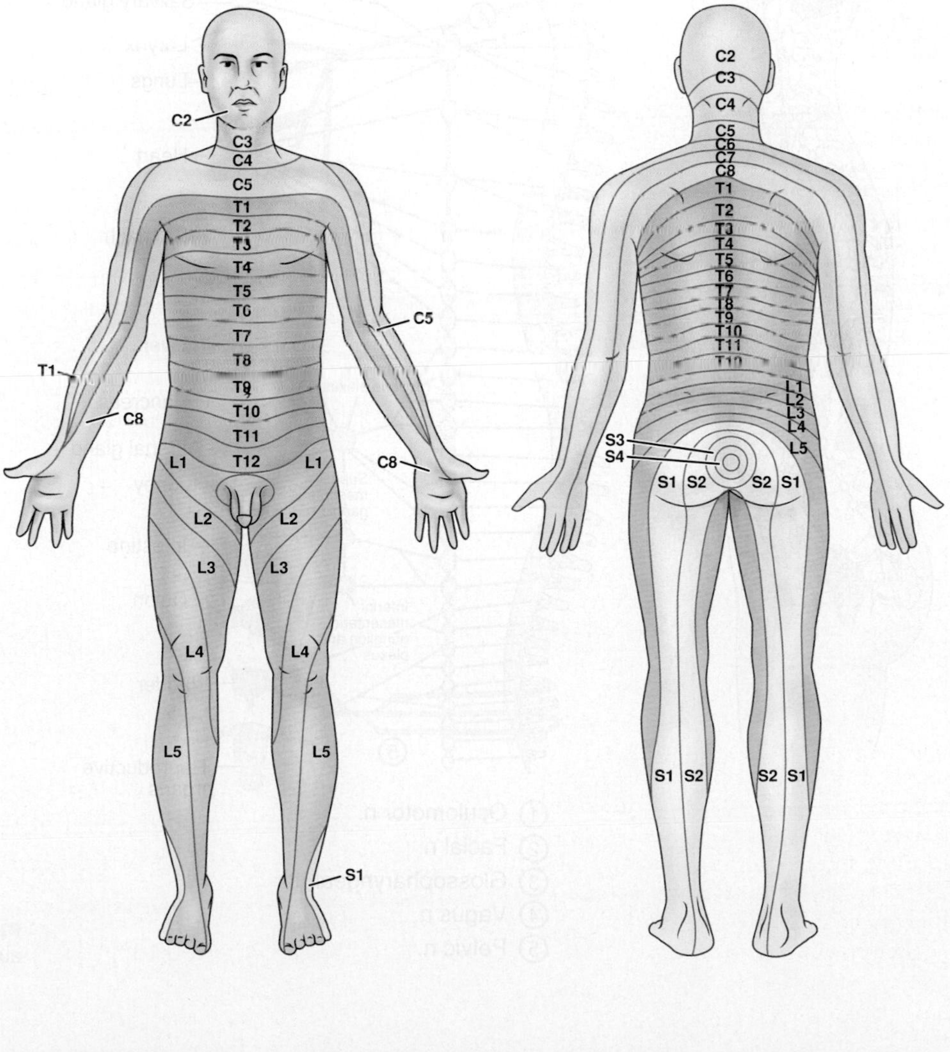

C2
C3
C4
C5
T1
T2
T3
T4
T5
T6
T7
T8
T9
T10
T11
T12
L1
L2
L3
L4
L5
S1
C8
C6
C7
S2
S3
S4

FIGURE 65-9 • Dermatome distribution.

surfaces; or visceral, carrying information from the internal organs.

The ventral roots are motor and transmit impulses from the spinal cord to the body; these fibers are also either somatic or visceral. The visceral fibers include autonomic fibers that control the cardiac muscles and glandular secretions.

Autonomic Nervous System

The **autonomic nervous system** regulates the activities of internal organs such as the heart, lungs, blood vessels, digestive organs, and glands (Fig. 65-10). Maintenance and restoration of internal homeostasis is largely the responsibility of the autonomic nervous system. There are two major divisions: the **sympathetic nervous system**, with predominantly excitatory responses (most notably the "fight-or-flight" response), and the **parasympathetic nervous system**, which controls mostly visceral functions.

The autonomic nervous system innervates most body organs. Although usually considered part of the peripheral nervous system, this system is regulated by centers in the spinal cord, brain stem, and hypothalamus.

The hypothalamus is the major subcortical center for the regulation of autonomic activities, serving an inhibitory–excitatory role. The hypothalamus has connections that link the autonomic system with the thalamus, the cortex, the olfactory apparatus, and the pituitary gland. Located here are the mechanisms for the control of visceral and somatic reactions that were originally important for defense or attack and are associated with emotional states (e.g., fear, anger, and anxiety); for the control of metabolic processes, including fat, carbohydrate, and water metabolism; for the regulation of body temperature, arterial pressure, and all muscular and glandular activities of the gastrointestinal tract; for control of genital functions; and for the sleep cycle.

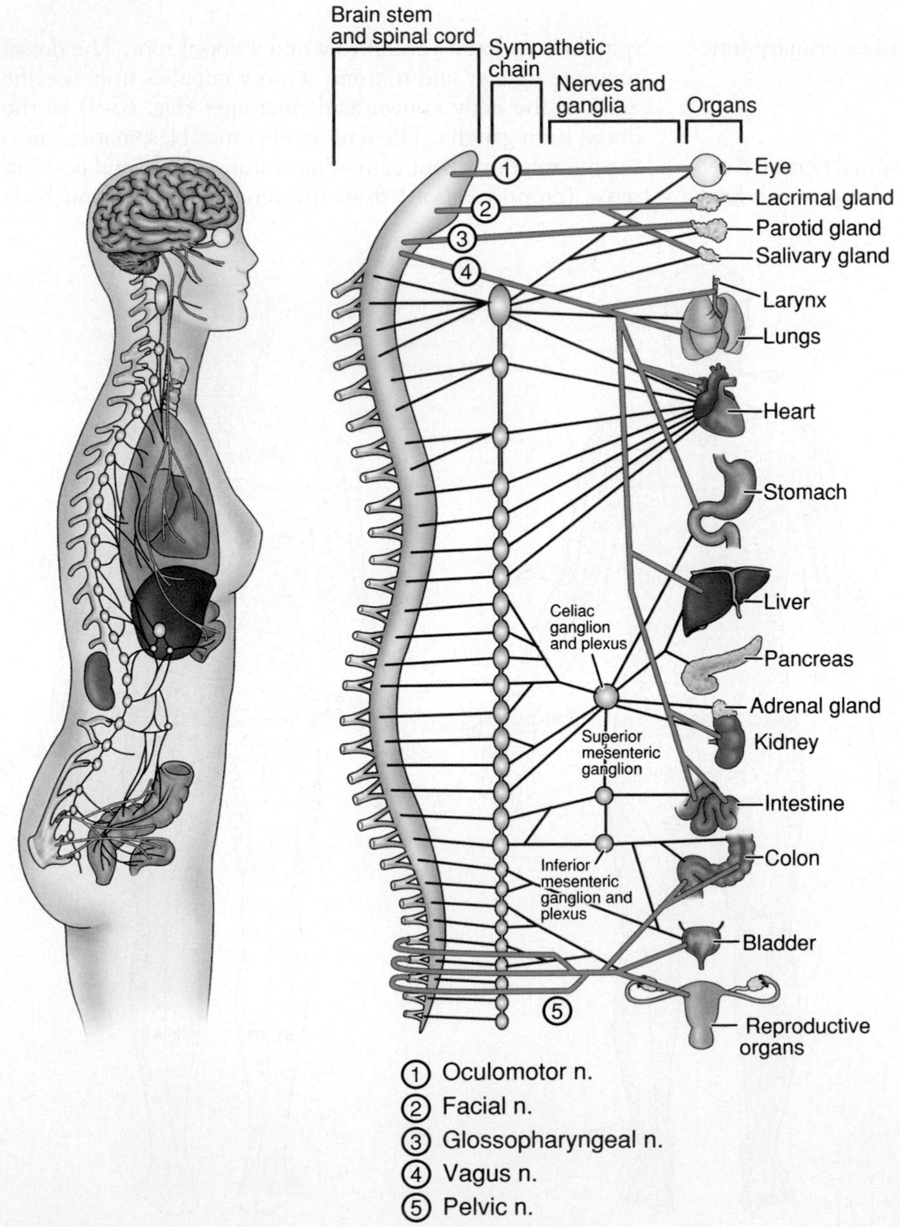
Brain stem and spinal cord
Sympathetic chain
Nerves and ganglia
Organs
Eye
Lacrimal gland
Parotid gland
Salivary gland
Larynx
Lungs
Heart
Stomach
Liver
Pancreas
Adrenal gland
Kidney
Intestine
Colon
Bladder
Reproductive organs
Celiac ganglion and plexus
Superior mesenteric ganglion
Inferior mesenteric ganglion and plexus
① Oculomotor n.
② Facial n.
③ Glossopharyngeal n.
④ Vagus n.
⑤ Pelvic n.

FIGURE 65-10 • Anatomy of the autonomic nervous system.

The autonomic nervous system is separated into the anatomically and functionally distinct sympathetic and parasympathetic divisions. Most of the tissues and the organs under autonomic control are innervated by both systems. For example, the parasympathetic division causes contraction (stimulation) of the urinary bladder muscles and a decrease (inhibition) in heart rate, whereas the sympathetic division produces relaxation (inhibition) of the urinary bladder and an increase (stimulation) in the rate and force of the heartbeat. Table 65-3 compares the sympathetic and the parasympathetic effects on the different systems of the body.

Sympathetic Nervous System

The sympathetic division of the autonomic nervous system is best known for its role in the body's fight-or-flight response.

TABLE 65-3 Effects of the Autonomic Nervous System

Structure or Activity	Parasympathetic Effects	Sympathetic Effects
Pupil of the Eye	Constricted	Dilated
Circulatory System		
Rate and force of heartbeat	Decreased	Increased
Blood vessels		
In heart muscle	Constricted	Dilated
In skeletal muscle	*	Dilated
In abdominal viscera and the skin	*	Constricted
Blood pressure	Decreased	Increased
Respiratory System		
Bronchioles	Constricted	Dilated
Rate of breathing	Decreased	Increased
Digestive System		
Peristaltic movements of digestive tube	Increased	Decreased
Muscular sphincters of digestive tube	Relaxed	Contracted
Secretion of salivary glands	Thin, watery saliva	Thick, viscid saliva
Secretions of stomach, intestine, and pancreas	Increased	*
Conversion of liver glycogen to glucose	*	Increased
Genitourinary System		
Urinary bladder		
Muscle walls	Contracted	Relaxed
Sphincters	Relaxed	Contracted
Muscles of the uterus	Relaxed; variable	Contracted under some conditions; varies with menstrual cycle and pregnancy
Blood vessels of external genitalia	Dilated	*
Integumentary System		
Secretion of sweat	*	Increased
Pilomotor muscles	*	Contracted (goose flesh)
Adrenal Medulla	*	Secretion of epinephrine and norepinephrine

*No direct effect.
Adapted from Hickey, J. (2009). *Clinical practice of neurological and neurosurgical nursing* (6th ed.). Philadelphia: Lippincott Williams & Wilkins.

Under stress from either physical or emotional causes, sympathetic impulses increase greatly. As a result, the bronchioles dilate for easier gas exchange; the heart's contractions are stronger and faster; the arteries to the heart and voluntary muscles dilate, carrying more blood to these organs; peripheral blood vessels constrict, making the skin feel cool but shunting blood to essential organs; the pupils dilate; the liver releases glucose for quick energy; peristalsis slows; hair stands on end; and perspiration increases. The main sympathetic neurotransmitter is norepinephrine (noradrenaline). A sympathetic discharge releases adrenalin (epinephrine)—hence, the term *adrenergic* is often used to refer to this division.

Sympathetic neurons are located primarily in the thoracic and lumbar segments of the spinal cord, and their axons, or the preganglionic fibers, emerge by way of anterior nerve roots from the eighth cervical or first thoracic segment to the second or third lumbar segment. A short distance from the cord, these fibers diverge to join a chain, composed of 22 linked ganglia, that extends the entire length of the spinal column, adjacent to the vertebral bodies on both sides. Some form multiple synapses with nerve cells within the chain. Others traverse the chain without making connections or losing continuity to join large "prevertebral" ganglia in the thorax, the abdomen, or the pelvis or one of the "terminal" ganglia in the vicinity of an organ, such as the bladder or the rectum at the end of the colon (see Fig. 65-10). Postganglionic nerve fibers originating in the sympathetic chain rejoin the spinal nerves that supply the extremities and are distributed to blood vessels, sweat glands, and smooth muscle tissue in the skin. Postganglionic fibers from the prevertebral plexuses (e.g., the cardiac, pulmonary, splanchnic, and pelvic plexuses) supply structures in the head and neck, thorax, abdomen, and pelvis, respectively, having been joined in these plexuses by fibers from the parasympathetic division.

The adrenal glands, kidneys, liver, spleen, stomach, and duodenum are under the control of the giant celiac plexus, commonly known as the solar plexus. This receives its sympathetic nerve components by way of the three splanchnic nerves, composed of preganglionic fibers from nine segments of the spinal cord (T4 to L1), and is joined by the vagus nerve, representing the parasympathetic division. From the celiac plexus, fibers of both divisions travel along the course of blood vessels to their target organs.

Certain syndromes are distinctive to the sympathetic nervous system. For example, sympathetic storm is a syndrome associated with changes in level of consciousness, altered vital signs, diaphoresis, and agitation that may result from hypothalamic stimulation of the sympathetic nervous system following traumatic brain injury (Fischbach & Dunning, 2011).

Parasympathetic Nervous System

The parasympathetic nervous system functions as the dominant controller for most visceral functions; the primary neurotransmitter is acetylcholine. During quiet, nonstressful conditions, impulses from parasympathetic fibers (cholinergic) predominate. The fibers of the parasympathetic system are located in two sections: one in the brain stem and the other from spinal segments below L2. Because of the location of these fibers, the parasympathetic system is referred to as the craniosacral division, as distinct from the thoracolumbar (sympathetic) division of the autonomic nervous system.

The parasympathetic nerves arise from the midbrain and the medulla oblongata. Fibers from cells in the midbrain travel with the third oculomotor nerve to the ciliary ganglia, where postganglionic fibers of this division are joined by those of the sympathetic system, creating controlled opposition, with a delicate balance maintained between the two at all times.

TABLE 65-4 Comparison of Upper Motor Neuron and Lower Motor Neuron Lesions

Upper Motor Neuron Lesions	Lower Motor Neuron Lesions
Loss of voluntary control	Loss of voluntary control
Increased muscle tone	Decreased muscle tone
Muscle spasticity	Flaccid muscle paralysis
No muscle atrophy	Muscle atrophy
Hyperactive and abnormal reflexes	Absent or decreased reflexes

Motor and Sensory Pathways of the Nervous System

Motor Pathways

The corticospinal tract begins in the motor cortex, a vertical band within each frontal lobe, and controls voluntary movements of the body. The exact locations within the brain at which the voluntary movements of the muscles of the face, thumb, hand, arm, trunk, and leg originate are known (Fig. 65-11). To initiate movement, these particular cells must send the stimulus along their fibers. Stimulation of these cells with an electric current also results in muscle contraction. En route to the pons, the motor fibers converge into a tight bundle known as the internal capsule. A comparatively small injury to the internal capsule results in a more severe paralysis than does a larger injury to the cortex itself.

At the medulla, the corticospinal tracts cross to the opposite side, continuing to the anterior horn of the spinal cord, in proximity to a motor nerve cell. Until this point, neurons are known as upper motor neurons. As they connect to motor fibers of the spinal nerves, they become lower motor neurons. The lower motor neurons receive the impulse in the posterior part of the cord and run to the myoneural junction located in the peripheral muscle.

Involuntary motor activity is also possible and is mediated through reflex arcs. Synaptic connections between anterior horn cells and sensory fibers that have entered adjacent or neighboring segments of the spinal cord serve as protective mechanisms. These connections are seen during deep tendon reflex testing.

Upper and Lower Motor Neurons

The voluntary motor system consists of two groups of neurons: upper motor neurons and lower motor neurons. Upper motor neurons originate in the cerebral cortex, the cerebellum, and the brain stem. Their fibers make up the descending motor pathways, are located entirely within the CNS, and modulate the activity of the lower motor neurons. Lower motor neurons are located either in the anterior horn of the spinal cord gray matter or within cranial nerve nuclei in the brain stem. Axons of lower motor neurons in both sites extend through peripheral nerves and terminate in skeletal muscle. Lower motor neurons are located in both the CNS and the peripheral nervous system.

The motor pathways from the brain to the spinal cord, as well as from the cerebrum to the brain stem, are formed by upper motor neurons. They begin in the cortex of one side of the brain, descend through the internal capsule, cross to the opposite side in the brain stem, descend through the corticospinal tract, and synapse with the lower motor neurons in the cord. The lower motor neurons receive the impulse in the posterior part of the cord and run to the myoneural junction located in the peripheral muscle. The clinical features of lesions of upper and lower motor neurons are discussed in the following sections and in Table 65-4.

Upper Motor Neuron Lesions. Upper motor neuron lesions can involve the motor cortex, the internal capsule, the spinal cord gray matter, and other structures of the brain through which the corticospinal tract descends. If the upper motor neurons are damaged or destroyed, as frequently occurs with stroke or spinal cord injury, paralysis (loss of voluntary movement) results. However, because the inhibitory influences of intact upper motor neurons are impaired, **reflex** (involuntary) movements are uninhibited, and hence hyperactive deep tendon reflexes, diminished or absent superficial reflexes, and pathologic reflexes such as a Babinski response occur. Severe leg spasms can occur as the result of an upper motor neuron lesion; the spasms result from the preserved reflex arc, which lacks inhibition along the spinal cord below the level of injury. There is little or no muscle atrophy, and muscles remain permanently tense, exhibiting spastic paralysis.

Paralysis associated with upper motor neuron lesions can affect a whole extremity, both extremities, or an entire half of the body. *Hemiplegia* (paralysis of an arm and leg on the same side of the body) can be the result of an upper motor neuron lesion. If hemorrhage, an embolus, or a thrombus destroys the fibers from the motor area in the internal capsule, the arm and the leg of the opposite side become stiff, weak, or paralyzed, and the reflexes are hyperactive (see further discussion of hemiplegia in Chapter 67). If both legs are paralyzed, the condition is called *paraplegia*. If all four extremities are paralyzed, the condition is called *tetraplegia*

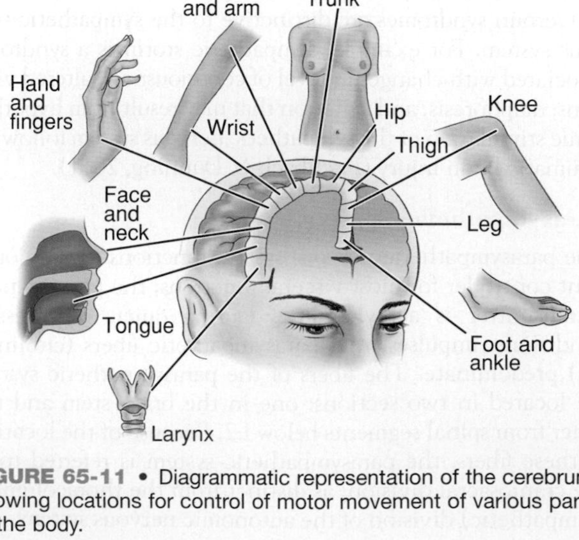
Elbow and arm
Trunk
Hand and fingers
Wrist
Hip
Knee
Thigh
Face and neck
Leg
Tongue
Foot and ankle
Larynx

FIGURE 65-11 • Diagrammatic representation of the cerebrum showing locations for control of motor movement of various parts of the body.

(quadriplegia). (See Chapter 68 for additional discussion of these disorders.)

Lower Motor Neuron Lesions. A patient is considered to have lower motor neuron damage if a motor nerve is damaged between the spinal cord and muscle. The result of lower motor neuron damage is muscle paralysis. Reflexes are lost, and the muscle becomes flaccid (limp) and atrophied from disuse. If the patient has injured the spinal trunk and it can heal, the use of muscles connected to that section of the spinal cord may be regained. However, if the anterior horn motor cells are destroyed, the nerves cannot regenerate and the muscles are never useful again.

Flaccid paralysis and atrophy of the affected muscles are the principal signs of lower motor neuron disease. Lower motor neuron lesions can be the result of trauma, infection (poliomyelitis), toxins, vascular disorders, congenital malformations, degenerative processes, and neoplasms. Compression of nerve roots by herniated intervertebral disks is a common cause of lower motor neuron dysfunction.

Coordination of Movement

The motor system is complex, and motor function depends not only on the integrity of the corticospinal tracts but also on other pathways from the basal ganglia and cerebellum that control and coordinate voluntary motor function. The smoothness, accuracy, and strength that characterize the muscular movements of a normal person are attributable to the influence of the cerebellum and the basal ganglia.

Through the action of the cerebellum, the contractions of opposing muscle groups are adjusted in relation to each other to maximal mechanical advantage; muscle contractions can be sustained evenly at the desired tension and without significant fluctuation, and reciprocal movements can be reproduced at high and constant speed, in stereotyped fashion and with relatively little effort.

The basal ganglia play an important role in planning and coordinating motor movements and posture. Complex neural connections link the basal ganglia with the cerebral cortex. The major effect of these structures is to inhibit unwanted muscular activity.

Impaired cerebellar function, which may occur as a result of an intracranial injury or some type of an expanding mass (e.g., a hemorrhage, an abscess, or a tumor), results in loss of muscle tone, weakness, and fatigue. Depending on the area of the brain affected, the patient has different motor symptoms or responses. The patient may demonstrate abnormal flexion, abnormal extension, or flaccid posturing. **Flaccidity** (lack of muscle tone) preceded by abnormal posturing in a patient with cerebral injury indicates severe neurologic impairment, which may herald brain death (Klein & Stewart-Amidei, 2012; Posner, Saper, Schiff, et al., 2007). (See Fig. 66-1 in Chapter 66 for further explanation of posturing.)

Destruction or dysfunction of the basal ganglia leads not to paralysis but to muscle rigidity, disturbances of posture, and difficulty initiating or changing movement. The patient tends to have involuntary movements. These may take the form of coarse tremors, most often in the upper extremities, particularly in the distal portions; athetosis, which is movement of a slow, squirming, writhing, twisting type; or chorea, marked by spasmodic, purposeless, irregular, uncoordinated motions of the trunk and the extremities, and facial grimacing. Disorders affecting basal ganglia activity include Parkinson's and Huntington diseases (see Chapter 70).

Sensory System Function

Receiving Sensory Impulses

Afferent impulses travel from their points of origin to their destinations in the cerebral cortex via the ascending pathways directly, or they may cross at the level of the spinal cord or in the medulla, depending on the type of sensation carried. Knowledge of these pathways is important for neurologic assessment and for understanding symptoms and their relationship to various lesions.

Sensory impulses convey sensations of heat, cold, and pain; position; and vibration. The axons enter the spinal cord by way of the posterior root, specifically in the posterior gray columns of the spinal cord, where they connect with the cells of secondary neurons. Pain and temperature fibers (located in the spinothalamic tract) cross immediately to the opposite side of the cord and course upward to the thalamus. Fibers carrying sensations of touch, light pressure, and localization do not connect immediately with the second neuron but ascend the cord for a variable distance before entering the gray matter and completing this connection. The axon of the secondary neuron traverses the cord, crosses in the medulla, and proceeds upward to the thalamus.

Position and vibratory sensations are produced by stimuli arising from muscles, joints, and bones. These stimuli are conveyed, uncrossed, all the way to the brain stem by the axon of the primary neuron. In the medulla, synaptic connections are made with cells of the secondary neurons, whose axons cross to the opposite side and then proceed to the thalamus.

Integrating Sensory Impulses

The thalamus integrates all sensory impulses except olfaction. It plays a role in the conscious awareness of pain and the recognition of variation in temperature and touch. The thalamus is responsible for the sense of movement and position as well as the ability to recognize the size, shape, and quality of objects. Sensory information is relayed from the thalamus to the parietal lobe for interpretation.

Sensory Losses

Destruction of a sensory nerve results in total loss of sensation in its area of distribution (see Fig. 65-9). Lesions affecting the posterior spinal nerve roots may impair tactile sensation, causing intermittent severe pain that is referred to their areas of distribution. Destruction of the spinal cord yields complete anesthesia below the level of injury. Selective destruction or degeneration of the posterior columns of the spinal cord is responsible for a loss of position and vibratory sense in segments distal to the lesion, without loss of touch, pain, or temperature perception. A cyst in the center of the spinal cord causes dissociation of sensation—loss of pain at the level of the lesion. This occurs because the fibers carrying pain and temperature cross within the cord immediately on entering; thus, any lesion that divides the cord longitudinally divides these fibers. Other sensory fibers ascend the cord for variable distances, some even to the medulla, before crossing, thereby bypassing the lesion and avoiding destruction. Lesions in the thalamus or parietal lobe result in impaired touch, pain, temperature, and proprioceptive sensations.

Assessment of the Nervous System

Health History

An important aspect of the neurologic assessment is the history of the present illness. The initial interview provides an excellent opportunity to systematically explore the patient's current condition and related events while simultaneously observing overall appearance, mental status, posture, movement, and affect. Depending on the patient's condition, the nurse may need to rely on yes-or-no answers to questions, a review of the medical record, input from witnesses or the family, or a combination of these.

Neurologic disorders may be stable or progressive, characterized by symptom-free periods as well as fluctuations in symptoms. The health history therefore includes details about the onset, character, severity, location, duration, and frequency of symptoms and signs; associated complaints; precipitating, aggravating, and relieving factors; progression, remission, and exacerbation; and the presence or absence of similar symptoms among family members.

Common Symptoms

The symptoms of neurologic disorders are as varied as the disease processes. Symptoms may be subtle or intense, fluctuating or permanent, inconvenient or devastating. This chapter discusses the most common signs and symptoms associated with neurologic disease; the relationship of specific signs and symptoms to a particular disorder is presented in later chapters in this unit.

Pain

Pain is considered an unpleasant sensory perception and emotional experience associated with actual or potential tissue damage or described in terms of such damage. Pain is therefore considered multidimensional and entirely subjective. Pain can be acute or chronic. In general, acute pain lasts for a relatively short period of time and remits as the pathology resolves. In neurologic disease, acute pain may be associated with brain hemorrhage, spinal disk disease (Jarvis, 2012), or trigeminal neuralgia. In contrast, chronic or persistent pain extends for long periods of time and may represent a broader pathology. This type of pain can occur with many degenerative and chronic neurologic conditions (e.g., multiple sclerosis). (See Chapter 12 for a more detailed discussion of pain.)

Seizures

Seizures are the result of abnormal electrical discharges in the cerebral cortex, which then manifest as an alteration in sensation, behavior, movement, perception, or consciousness. The alteration may be short, such as in a blank stare that lasts only a second, or of longer duration, such as a tonic–clonic grand mal seizure that can last several minutes. The seizure activity reflects the area of the brain affected. Seizures can occur as isolated events, such as when induced by a high fever, alcohol or drug withdrawal, or hypoglycemia. A seizure may also be the first obvious sign of a brain lesion (Hickey, 2009).

Dizziness and Vertigo

Dizziness is an abnormal sensation of imbalance or movement. It is fairly common in the older adult and a common complaint encountered by health professionals (Jarvis, 2012). Dizziness can have a variety of causes, including viral syndromes, hot weather, roller-coaster rides, and middle ear infections, to name a few. One difficulty confronting health care providers when assessing dizziness is the vague and varied terms patients that use to describe the sensation.

About 50% of all patients with dizziness have **vertigo**, or the illusion of movement in which the individual or the surroundings are sensed as moving, usually as rotation (Jarvis, 2012). Vertigo is usually a manifestation of vestibular dysfunction. It can be so severe as to result in spatial disorientation, lightheadedness, loss of equilibrium (staggering), and nausea and vomiting.

Visual Disturbances

Visual defects that cause people to seek health care can range from the decreased visual acuity associated with aging to sudden blindness caused by glaucoma. Normal vision depends on functioning visual pathways through the retina and optic chiasm and the radiations into the visual cortex in the occipital lobes. Lesions of the eye itself (e.g., cataract), lesions along the pathway (e.g., tumor), or lesions in the visual cortex (e.g., stroke) interfere with normal visual acuity. Abnormalities of eye movement (as in the nystagmus associated with multiple sclerosis) can also compromise vision by causing diplopia or double vision. (See Chapter 63 for a more detailed discussion of disorders that affect vision.)

Muscle Weakness

Muscle weakness is a common manifestation of neurologic disease. It frequently coexists with other symptoms of disease and can affect a variety of muscles, causing a wide range of disability. Weakness can be sudden and permanent, as in stroke, or progressive, as in neuromuscular diseases such as amyotrophic lateral sclerosis. Any muscle group can be affected.

Abnormal Sensation

Abnormal sensation is a neurologic manifestation of both central and peripheral nervous system disease. Altered sensation can affect small or large areas of the body. It is frequently associated with weakness or pain and is potentially disabling. Lack of sensation places a person at risk for falls and injury.

Past Health, Family, and Social History

The nurse may inquire about any family history of genetic diseases (Chart 65-1). A review of the medical history, including a system-by-system evaluation, is part of the health history. The nurse should be aware of any history of trauma or falls that may have involved the head or spinal cord. Questions regarding the use of alcohol, medications, and illicit drugs are also relevant. The history-taking portion of the neurologic assessment is critical and, in many cases of neurologic disease, leads to an accurate diagnosis.

Physical Assessment

The neurologic examination is a systematic process that includes a variety of clinical tests, observations, and assessments designed to evaluate the neurologic status of a complex system. Many neurologic rating scales exist (Herndon, 2006), and some of the more common ones are discussed in this chapter.

Chart 65-1

GENETICS IN NURSING PRACTICE
Neurologic Disorders

Several neurologic disorders are associated with genetic abnormalities. Some examples include:

- Alzheimer's disease
- Amyotrophic lateral sclerosis (ALS)
- Duchenne muscular dystrophy
- Epilepsy
- Friedrich ataxia
- Huntington disease
- Myotonic dystrophy
- Neurofibromatosis type 1
- Parkinson's disease
- Spina bifida
- Tourette syndrome

Nursing Assessments

Family History Assessment

- Assess for other similarly affected relatives with neurologic impairment.
- Inquire about age of onset (e.g., present at birth—spina bifida; developed in childhood—Duchenne muscular dystrophy; developed in adulthood—Huntington disease, Alzheimer's disease, ALS).
- Inquire about the presence of related conditions such as mental retardation and/or learning disabilities (neurofibromatosis type 1).

Patient Assessment

- Assess for the presence of other physical features suggestive of an underlying genetic condition, such as skin lesions seen in neurofibromatosis type 1 (*café-au-lait* spots).
- Assess for other congenital abnormalities (e.g., cardiac, ocular).

Management Issues Specific to Genetics

- Inquire whether deoxyribonucleic acid (DNA) mutation or other genetic testing has been performed on affected family members.
- If indicated, refer for further genetic counseling and evaluation so that family members can discuss inheritance, risk to other family members, availability of genetic testing, and gene-based interventions.
- Offer appropriate genetic information and resources.
- Assess patient's understanding of genetic information.
- Provide support to families with newly diagnosed genetics-related neurologic disorders.
- Participate in management and coordination of care of patients with genetics-related neurologic disorders, patients with genetic conditions and individuals predisposed to develop or pass on a genetic condition.

Genetics Resources

See Chapter 8, Chart 8-6 for genetics resources.

The brain and spinal cord cannot be examined as directly as other systems of the body. Therefore, much of the neurologic examination is an indirect evaluation that assesses the function of the specific body part or parts controlled by the nervous system. A neurologic assessment is divided into five components: consciousness and cognition, cranial nerves, motor system, sensory system, and reflexes. One or more components may become the priority assessment, depending on the patient's condition. For example, motor, sensory, and reflex assessments are the priority in patients with spinal injury, whereas in a comatose patient, the cranial nerves and level of consciousness become the priority.

Assessing Consciousness and Cognition

Cerebral abnormalities may cause disturbances in mental status, intellectual functioning, thought content, and emotional status. There may also be alterations in language abilities as well as lifestyle. The examiner must also be aware of the patient's overall level of consciousness and any changes over time (Posner et al., 2007).

The examiner records and reports specific observations regarding mental status, intellectual function, thought content, and emotional status, all of which permit comparison by others over time. Alterations should be described in specific and nonjudgmental terms. The use of terms such as "inappropriate" or "demented" are avoided, because they often mean different things to different people and are therefore not useful when describing behavior. Analysis and the conclusions that may be drawn from these findings usually depend on the examiner's knowledge of neuroanatomy, neurophysiology, and neuropathology.

Mental Status

An assessment of mental status begins by observing the patient's appearance and behavior, noting dress, grooming, and personal hygiene. Posture, gestures, movements, and facial expressions often provide important information about the patient. Does the patient appear to be aware of and interact with the surroundings?

Assessing orientation to time, place, and person assists in evaluating mental status. Does the patient know what day it is, what year it is, and the name of the president of the United States? Is the patient aware of where he or she is? Is the patient aware of who the examiner is and of his or her purpose for being in the room? Assessment of immediate and remote memory is also important. Is the capacity for immediate memory intact? (See Chapter 11.)

Intellectual Function

A person with an average intelligence quotient (IQ) can repeat seven digits without faltering and can recite five digits backward. The examiner might ask the patient to count backward from 100 or to subtract 7 from 100, then 7 from that, and so forth (referred to as serial 7s). The capacity to interpret well-known proverbs tests abstract reasoning, which is a higher intellectual function—for example, does the patient know what is meant by "a stitch in time saves nine"? The intellectual function of patients with damage to the frontal cortex appears intact until one or more tests of intellectual capacity are performed. Questions designed to assess this capacity might include the ability to recognize similarities—for example, how are a mouse and dog or pen and pencil alike? Can the patient make judgments about situations—for example, if the patient arrives home without a house key, what alternatives are there?

Thought Content

During the interview, it is important to assess the patient's thought content. Are the patient's thoughts spontaneous, natural, clear, relevant, and coherent? Does the patient have any fixed ideas, illusions, or preoccupations? What are his or her insights into these thoughts? Preoccupation with death or morbid events, hallucinations, and paranoid ideation are examples of unusual thoughts or perceptions that require further evaluation.

Emotional Status

An assessment of consciousness and cognition also includes the patient's emotional status. Is the patient's affect (external manifestation of mood) natural and even, or irritable and angry, anxious, apathetic or flat, or euphoric? Does his or her mood fluctuate normally, or does the patient unpredictably swing from joy to sadness during the interview? Is affect appropriate to words and thought content? Are verbal communications consistent with nonverbal cues?

Language Ability

The person with normal neurologic function can understand and communicate in spoken and written language. Does the patient answer questions appropriately? Can he or she read a sentence from a newspaper and explain its meaning? Can the patient write his or her name or copy a simple figure that the examiner has drawn? A deficiency in language function is called *aphasia*. Different types of aphasia result from injury to different parts of the brain (Table 65-5). (See Chapter 67 for a detailed discussion of aphasia.)

TABLE 65-5 Types of Aphasia and Region of Brain Involved

Type of Aphasia	Brain Area Involved
Auditory receptive	Temporal lobe
Visual receptive	Parietal & Occipital area
Expressive speaking	Inferior posterior frontal areas
Expressive writing	Posterior frontal area

Impact on Lifestyle

The nurse assesses the impact of any impairment on the patient's lifestyle. Issues to consider include the limitations imposed on the patient by any cognitive deficit and the patient's role in society, including family and community roles. The plan of care that the nurse develops needs to address and support adaptation to the neurologic deficit and continued function to the extent possible within the patient's support system.

Level of Consciousness

Consciousness is the patient's wakefulness and ability to respond to the environment. Level of consciousness is the most sensitive indicator of neurologic function. To assess level of consciousness, the examiner observes for alertness and ability to follow commands.

If the patient is not alert or able to follow commands, the examiner observes for eye opening; verbal response and motor response to stimuli, if any; and the type of stimuli needed to obtain a response (Chart 65-2). Noxious stimuli should be used first, then painful stimuli if no response is observed. In the patient with decreased level of consciousness, motor and cranial nerve function become the priority assessments, because abnormalities can indicate the area of involvement in the absence of responsiveness. (See Chapter 66 for further discussion of changes in level of consciousness.)

Examining the Cranial Nerves

Cranial nerves are assessed when level of consciousness is decreased, with brain stem pathology, or in the presence of peripheral nervous system disease (Caton-Richards, 2010). Chart 65-3 describes assessment of the cranial nerves and

Chart 65-2 NURSING RESEARCH PROFILE

Frequency of Neurologic Observations

Mavin, C. (2009). Does underpinning evidence influence the frequency of neurological observations? *British Journal of Neuroscience Nursing, 5*(10), 456–459.

Purpose

The purpose of neurologic assessment is to identify potentially life-threatening alterations in neurologic function so that interventions may be implemented in a timely manner. This review sought to identify whether evidence exists that supports the frequency of nurses performing neurologic assessments.

Design

A systematic review of five databases from 1979 to 2009 was conducted. Search terms included "Glasgow Coma Scale," "neurological assessment," "patient observation," "nursing assessment," and "brain injury." Fifty-two articles were examined for evidence of frequency of neurologic observations.

Findings

Evidence supporting frequency guidelines was largely anecdotal. Frequency of neurologic observations varied widely, and no data were found supporting specific frequencies of observations. This suggests that neurologic assessment practices are inconsistent among nurses. Vital signs, which often accompany neurologic assessment, have evidence supporting patterns of vital sign assessment from hourly for the first 24 hours to every 4 hours thereafter for patients with neurologic problems.

Nursing Implications

More research is needed; however, this review highlights the need to understand nursing practices. It may be that nurses make decisions independently about frequency of observations based on experience and education. Other factors that may influence how frequently nurses perform neurologic assessments, such as orders, patient condition, risk factors identified, and others, should be explored to facilitate development of guidelines for assessment frequency. Although frequency of assessment is often left to nursing discretion, nurses must be familiar with aspects of neurologic assessment.

Chart 65-3

GUIDELINES

Assessing Cranial Nerve Function

Equipment

• Tongue depressor • Flashlight • Sugar and salt samples • Watch • Cotton-tipped swab • Cotton swabs • Snellen chart
• Ophthalmoscope • Samples of familiar odors • Tuning fork • Tubes of hot and cold water

Implementation

Step	Rationale
1. Assess cranial nerve (CN) I (olfactory). With eyes closed, patient is asked to identify familiar odors (coffee, tobacco). Each nostril is tested separately.	The significant finding is loss of sense of smell (anosmia).
2. Assess CN II (optic). Assess vision using a Snellen eye chart. Assess visual fields. Perform ophthalmoscopic examination.	Significant findings include visual field defects (hemianopias) and decreased visual acuity or blindness.
3. Assess CN III (oculomotor). Test for eye movement toward the nose; inspect for conjugate movements and nystagmus. Evaluate papillary size, and test for pupillary reactivity to light; inspect ability to open eyelids.	Significant findings include dysconjugate gaze; gaze weakness or paralysis; double vision; dilated pupil, with or without impaired pupillary reaction to light; and inability to open the affected eyelid.
4. Assess CN IV (trochlear). Test for upward eye movement; inspect for conjugate movements and nystagmus.	Significant findings include dysconjugate gaze, gaze weakness or paralysis, and double vision.
5. Assess CN V (trigeminal). Have patient close the eyes. Touch cotton to forehead, cheeks, and jaw. Sensitivity to superficial pain is tested in these same three areas by using the sharp and dull ends of a broken tongue blade. Alternate between the sharp point and the dull end. Patient reports "sharp" or "dull" with each movement. If responses are incorrect, test for temperature sensation. Test tubes of cold and hot water are used alternately. While patient looks up, *lightly* touch a wisp of cotton against the temporal surface of each cornea. A blink and tearing are normal responses. Have patient clench and move the jaw from side to side. Palpate the masseter and temporal muscles, noting strength and equality.	Significant findings include impaired or absent corneal reflex, facial numbness, and jaw weakness.
6. Assess CN VI (abducens). Test for lateral eye movement; inspect for conjugate movement.	Significant findings include dysconjugate gaze, gaze weakness or paralysis, and double vision.
7. Assess CN VII (facial). Observe for symmetry while patient performs facial movements: smiles, whistles, elevates eyebrows, frowns, tightly closes eyelids against resistance (examiner attempts to open them). Observe face for flaccid paralysis (shallow nasolabial folds). Have patient extend tongue. Test ability to discriminate between sugar and salt.	Significant findings include facial weakness, inability to completely close the eyelid, and impaired taste.
8. Assess CN VIII (acoustic). Perform whisper or watch-tick test. Test for lateralization (Weber test). Test for air and bone conduction (Rinne test). Assess balance with eyes open and then closed for 20 seconds (Romberg test).	Significant findings include decreased hearing or deafness and impaired balance.
9. Assess CN IX (glossopharyngeal). Assess patient's ability to swallow and discriminate between sugar and salt on posterior third of the tongue.	Significant findings include difficulty swallowing (dysphagia) and impaired taste.
10. Assess CN X (vagus). Depress a tongue blade on posterior tongue, or stimulate posterior pharynx to elicit gag reflex. Note any hoarseness in voice. Check ability to swallow. Have patient say "ah." Observe for symmetric rise of uvula and soft palate.	Significant findings include weak or absent gag reflex, difficulty swallowing, aspiration, hoarseness, and slurred speech (dysarthria).
11. Assess CN XI (spinal accessory). While patient shrugs shoulders against resistance, palpate and note strength of trapezius muscles. As patient turns head against opposing pressure of the examiner's hand, palpate and note strength of each sternocleidomastoid muscle.	Significant findings include weak or absent shoulder shrug and inability to turn the head to the side.
12. Assess CN XII (hypoglossal). While patient protrudes the tongue, note any deviation or tremors. Test the strength of the tongue by having patient move the protruded tongue from side to side against a tongue depressor.	Significant findings include difficulty swallowing and slurred speech.

Adapted from Bader, M., & Littlejohns, L. R. (2010). *AANN core curriculum for neuroscience nursing* (5th ed.). Glenview, IL: American Association of Neuroscience Nurses; and Weber, J., & Kelley, J. (2010). *Health assessment in nursing* (4th ed.). Philadelphia: Lippincott Williams & Wilkins.

significant findings. Right and left cranial nerve functions are compared throughout the examination.

Examining the Motor System

Motor Ability

A thorough examination of the motor system includes an assessment of muscle size and tone as well as strength, coordination, and balance. The patient is instructed to walk across the room, if possible, while the examiner observes posture and gait. The muscles are inspected, and palpated if necessary, for their size and symmetry. Any evidence of atrophy or involuntary movements (tremors, tics) is noted. Muscle tone (the tension present in a muscle at rest) is evaluated by palpating various muscle groups at rest and during passive movement. Resistance to these movements is assessed and documented. Abnormalities in tone include **spasticity** (increased muscle tone), **rigidity** (resistance to passive stretch), and flaccidity.

Muscle Strength

Assessing the patient's ability to flex or extend the extremities against resistance tests muscle strength. The function of an individual muscle or group of muscles is evaluated by placing the muscle at a disadvantage. The quadriceps, for example, is a powerful muscle responsible for straightening the leg. Once the leg is straightened, it is exceedingly difficult for the examiner to flex the knee. If the knee is flexed and the patient is asked to straighten the leg against resistance, weakness can be elicited. The evaluation of muscle strength compares the sides of the body to each other. For example, the right upper extremity is compared to the left upper extremity. Subtle differences in strength may be evaluated by testing for drift. For example, both arms are out in front of the patient with palms up; drift is seen as pronation of the palm, indicating a subtle weakness that may not have been detected on the resistance examination.

Clinicians use a five-point scale to rate muscle strength. A 5 indicates full power of contraction against gravity and resistance or normal muscle strength; 4 indicates fair but not full strength against gravity and a moderate amount of resistance or slight weakness; 3 indicates just sufficient strength to overcome the force of gravity or moderate weakness; 2 indicates the ability to move but not to overcome the force of gravity or severe weakness; 1 indicates minimal contractile power (weak muscle contraction can be palpated but no movement is noted) or very severe weakness; and 0 indicates no movement (Jarvis, 2012).

Concept Mastery Alert

When recording muscle strength, a stick figure is used as a precise form to document findings. The five-point scale is used to rate and record distal and proximal strength in both upper and lower extremities. Figure 65-12 provides further details.

Assessment of muscle strength may be as detailed as necessary. One may quickly test the strength of the proximal muscles of the upper and lower extremities, always comparing both sides. The strength of the finer muscles that control the function of the hand (hand grasp) and the foot (dorsiflexion and plantar flexion) can then be assessed.

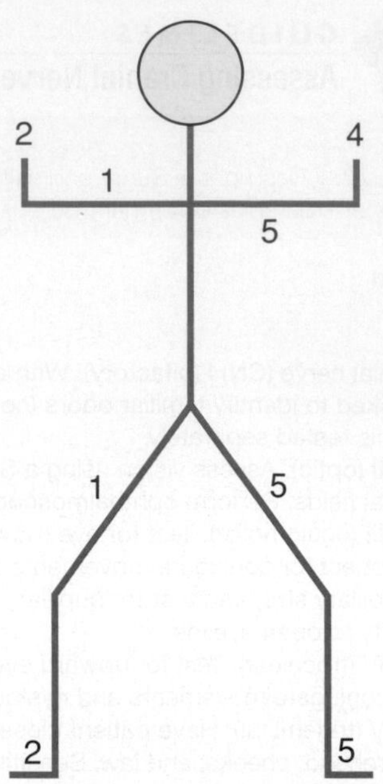
2
4
1
5
1
5
2
5

FIGURE 65-12 • A stick figure may be used to record muscle strength.

Balance and Coordination

Cerebellar and basal ganglia influence on the motor system is reflected in balance control and coordination. Coordination in the hands and upper extremities is tested by having the patient perform rapid, alternating movements and point-to-point testing. First, the patient is instructed to pat his or her thigh as fast as possible with each hand separately. Then, the patient is instructed to alternately pronate and supinate the hand as rapidly as possible. Last, the patient is asked to touch each of the fingers with the thumb in a consecutive motion. Speed, symmetry, and degree of difficulty are noted. Point-to-point testing is accomplished by having the patient touch the examiner's extended finger and then his or her own nose. This is repeated several times.

Coordination in the lower extremities is tested by having the patient run the heel down the anterior surface of the tibia of the other leg. Each leg is tested in turn. **Ataxia** is an incoordination of voluntary muscle action, particularly of the muscle groups used in activities such as walking or reaching for objects. Tremors (rhythmic, involuntary movements) noted at rest or during movement suggest a problem in the anatomic areas responsible for balance and coordination.

The **Romberg test** is a screening test for balance that can be done with the patient seated or standing The patient can be seated or stand with feet together and arms at the side, first with eyes open and then with both eyes closed for 20 seconds (Bader & Littlejohns, 2010; Weber & Kelley, 2010). The examiner stands close to support the standing patient if he or she begins to fall. Slight swaying is normal, but a loss of balance is abnormal and is considered a positive Romberg test. Additional cerebellar tests for balance in the ambulatory patient include hopping in place, alternating knee bends, and heel-to-toe walking (both forward and backward).

Examining the Sensory System

The sensory system is even more complex than the motor system, because sensory modalities are more widespread throughout the central and peripheral nervous systems. The sensory examination is largely subjective and requires the cooperation of the patient. The examiner should be familiar with dermatomes that represent the distribution of the peripheral nerves that arise from the spinal cord (Jarvis, 2012) (see Fig. 65-9).

Assessment of the sensory system involves tests for tactile sensation, superficial pain, temperature, vibration, and position sense (proprioception). During the sensory assessment, the patient's eyes are closed. Simple directions and reassurance that the examiner will not hurt or startle the patient encourage the cooperation of the patient.

Tactile sensation is assessed by lightly touching a cotton wisp or fingertip to corresponding areas on each side of the body. The sensitivity of proximal parts of the extremities is compared with that of distal parts, and the right and left sides are compared.

Pain and temperature sensations are transmitted together in the lateral part of the spinal cord, so it is unnecessary to test for temperature sense in most circumstances. Determining the patient's sensitivity to a sharp object can assess superficial pain perception. However, pain sensation is usually reserved for patients who do not respond to or cannot discriminate touch stimulation. The patient is asked to differentiate between the sharp and dull ends of a broken wooden cotton swab or tongue blade; using a safety pin is inadvisable because it breaks the integrity of the skin. Both the sharp and dull sides of the object are applied with equal intensity at all times, and the two sides are compared.

Vibration and proprioception are transmitted together in the posterior part of the cord. Vibration may be evaluated through the use of a low-frequency (128 or 256 Hertz [Hz]) tuning fork. The handle of the vibrating fork is placed against a bony prominence, and the patient is asked if he or she feels a sensation and is instructed to signal the examiner when the sensation ceases. Common locations used to test for vibratory sense include the distal joint of the great toe and the proximal thumb joint. If the patient does not perceive the vibrations at the distal bony prominences, the examiner progresses upward with the tuning fork until the patient perceives the vibrations. As with all measurements of sensation, a side-to-side comparison is made.

Position sense or proprioception may be determined by asking the patient to close both eyes and indicate, as the great toe or index finger is alternately moved up and down, in which direction movement has taken place. Vibration and position sense are often lost together, frequently in circumstances in which all other sensation remains intact.

Integration of sensation in the brain is evaluated by testing two-point discrimination. When the patient is touched with two sharp objects simultaneously, are they perceived as two or as one? If touched simultaneously on opposite sides of the body, the patient should normally report being touched in two places. If only one site is reported, the one not being recognized is said to demonstrate extinction. Another test of higher cortical sensory ability is tactile identification. The patient is instructed to close both eyes and identify an object (e.g., key, coin) that is placed in one hand by the examiner; inability to identify an object by touch is known as tactile agnosia or astereognosis. **Agnosia** is the general loss of ability to recognize objects through a particular sensory system. The patient can also be shown a familiar object and asked to identify it by name; inability to identify a visualized object is known as visual agnosia. Each of these dysfunctions implicates a different part of the brain (Table 65-6).

TABLE 65-6 Types of Agnosia and Corresponding Sites of Lesions

Type of Agnosia	Affected Cerebral Area
Visual	Occipital lobe
Auditory	Temporal lobe (lateral and superior portions)
Tactile	Parietal lobe
Body parts and relationships	Parietal lobe (posteroinferior regions)

Decreased or absent sensations occur with problems anywhere along the sensory pathway. Sensory deficits resulting from peripheral neuropathy or spinal cord injury follow anatomic dermatomes. Destructive lesions of the brain may affect sensation on an entire side of the body. Stroke affecting a portion of the sensory cortex will produce altered sensory discrimination.

Examining the Reflexes

Reflexes are involuntary contractions of muscles or muscle groups in response to a stimulus. Reflexes are classified as deep tendon, superficial, or pathologic. Testing reflexes enables the examiner to assess involuntary reflex arcs that depend on the presence of afferent stretch receptors, spinal or brain stem synapses, efferent motor fibers, and a variety of modifying influences from higher levels.

Deep Tendon Reflexes

A reflex hammer is used to elicit a deep tendon reflex. The handle of the hammer is held loosely between the thumb and index finger, allowing a full swinging motion. The wrist motion is similar to that used during percussion. The extremity is positioned so that the tendon is slightly stretched. This requires a sound knowledge of the location of muscles and their tendon attachments. The tendon is then struck briskly (Fig. 65-13), and the response is compared with that on the opposite side of the body. A wide variation in reflex response may be considered normal; however, it is more important that the reflexes be symmetrically equivalent. When the comparison is made, both sides should be equivalently relaxed and each tendon struck with equal force.

Valid findings depend on several factors: proper use of the reflex hammer, proper positioning of the extremity, and a relaxed patient (Jarvis, 2012). If the reflexes are symmetrically diminished or absent, the examiner may use isometric contraction of other muscle groups to increase reflex activity. For example, if lower extremity reflexes are diminished or absent, the patient is instructed to lock the fingers together and pull in opposite directions. Having the patient clench the jaw or press the heels against the floor or examining table may similarly elicit more reliable biceps, triceps, and brachioradialis reflexes.

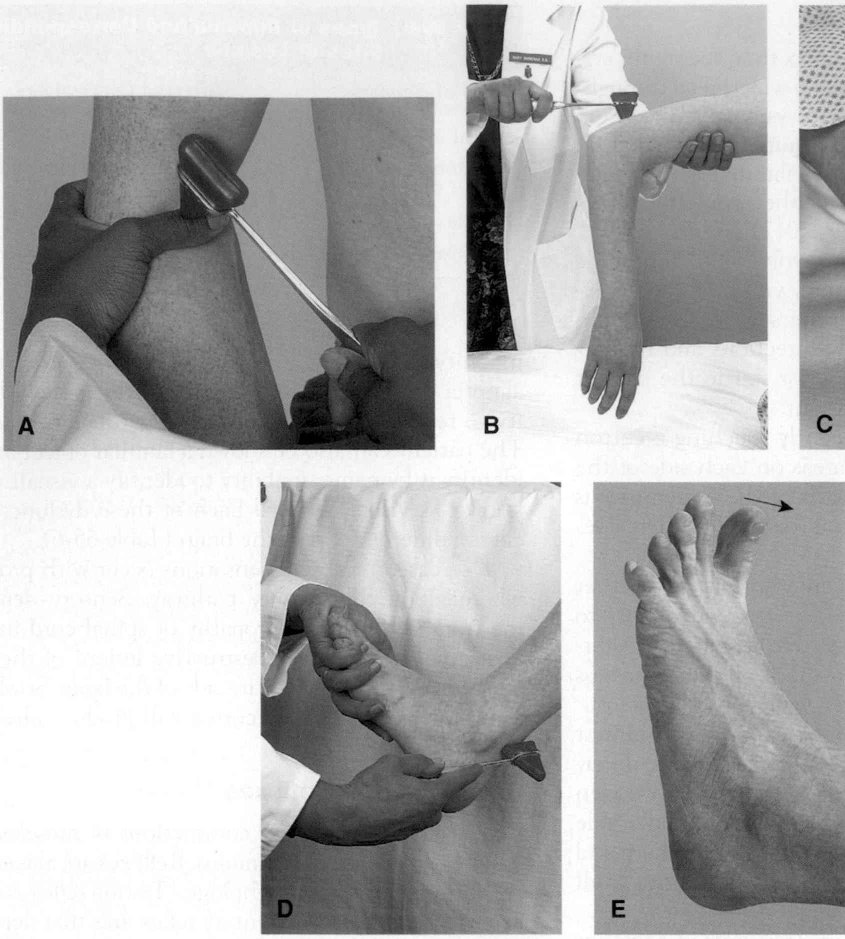
A
B
C
D
E

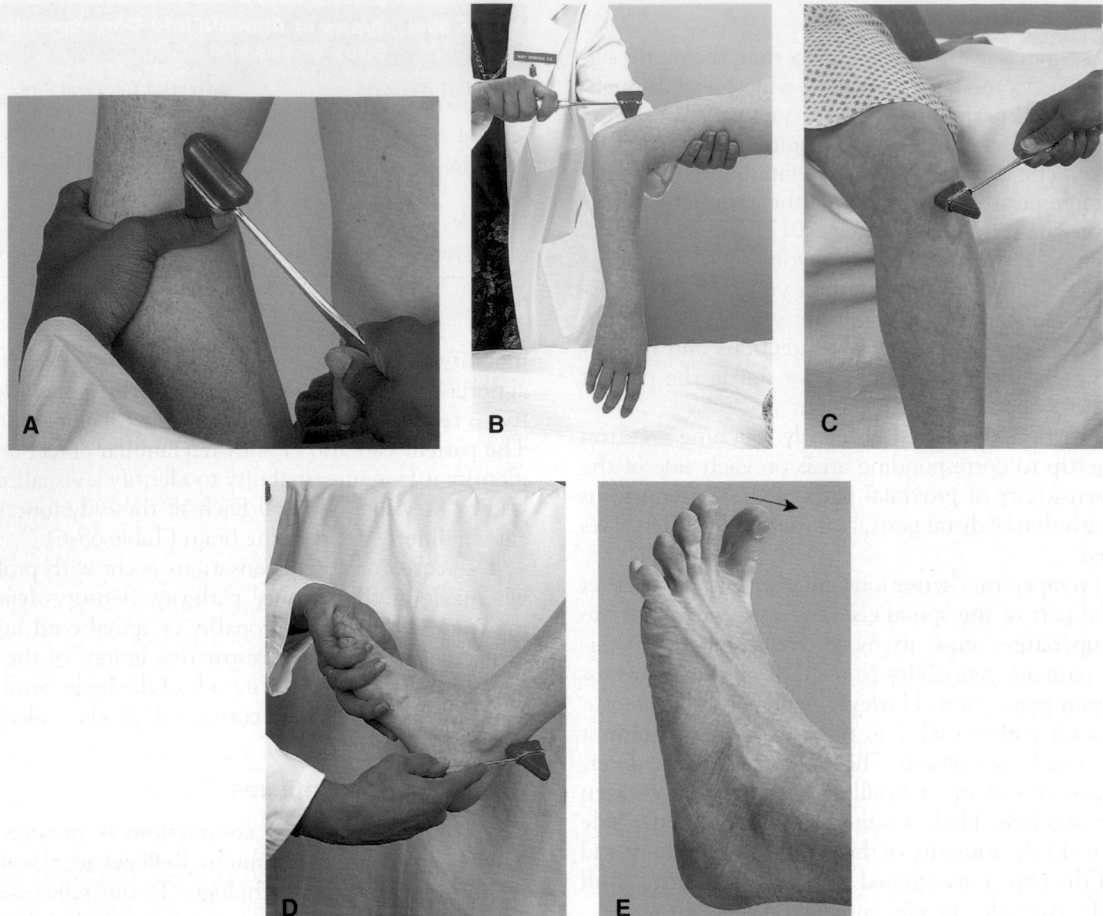

FIGURE 65-13 • Techniques for eliciting major reflexes. **A.** Biceps reflex. **B.** Triceps reflex. **C.** Patellar reflex. **D.** Ankle or Achilles reflex. **E.** Positive (abnormal) Babinski response. Parts A–D from Weber, J., & Kelley, J. (2010). *Health assessment in nursing* (4th ed.). Philadelphia: Lippincott Williams & Wilkins. Photos by B. Proud; Part E from Bickley, L. S. (2009). *Bates' guide to physical examination and history taking* (10th ed., p. 702). Philadelphia: Lippincott Williams & Wilkins.

The absence of reflexes is significant, although ankle jerks (Achilles reflex) may be normally absent in older people. Deep tendon reflex responses are often graded on a scale of 0 to 4+, with 2+ considered normal (Chart 65-4). As stated previously, scale ratings are highly subjective. Findings can be recorded as a fraction, indicating the scale range (e.g., 2/4). Some examiners prefer to use the terms *present*, *absent*, and *diminished* when describing reflexes. As with muscle strength recording, a stick figure may be used to record numerical findings.

Biceps Reflex. The biceps reflex is elicited by striking the biceps tendon over a slightly flexed elbow (see Fig. 65-13A). The examiner supports the forearm at the elbow with one arm while placing the thumb against the tendon and striking the thumb with the reflex hammer. The normal response is flexion at the elbow and contraction of the biceps.

Triceps Reflex. To elicit a triceps reflex, the patient's arm is flexed at the elbow and hanging freely at the side. The examiner supports the patient's arm and identifies the triceps tendon by palpating 2.5 to 5 cm (1 to 2 in) above the elbow. A direct blow on the tendon (see Fig. 65-13B) normally produces contraction of the triceps muscle and extension of the elbow.

Brachioradialis Reflex. With the patient's forearm resting on the lap or across the abdomen, the brachioradialis reflex is assessed. A gentle strike of the hammer 2.5 to 5 cm (1 to 2 in) above the wrist results in flexion and supination of the forearm (Jarvis, 2012).

Patellar Reflex. The patellar reflex is elicited by striking the patellar tendon just below the patella. The patient may be in a sitting or a lying position. If the patient is supine, the examiner supports the legs to facilitate relaxation of the muscles (see Fig. 65-13C). Contractions of the quadriceps and knee extension are normal responses.

Achilles Reflex. To elicit an Achilles reflex, the foot is dorsiflexed at the ankle and the hammer strikes the stretched Achilles tendon (see Fig. 65-13D). This reflex normally produces plantar flexion. If the examiner cannot elicit the ankle reflex and suspects that the patient cannot relax, the patient is instructed to kneel on a chair or similar elevated, flat surface. This position places the ankles in dorsiflexion and reduces any muscle tension in the gastrocnemius. The Achilles tendons are struck in turn, and plantar flexion is usually demonstrated (Jarvis, 2012).

Chart 65-4 Documenting Reflexes

Deep tendon reflexes are graded on a scale of 0–4:

0 No response
1+ Diminished (hypoactive)
2+ Normal
3+ Increased (may be interpreted as normal)
4+ Hyperactive (hyperreflexia)

The deep tendon responses and plantar reflexes are commonly recorded on stick figures. The arrow points downward if the plantar response is normal and upward if the response is abnormal.

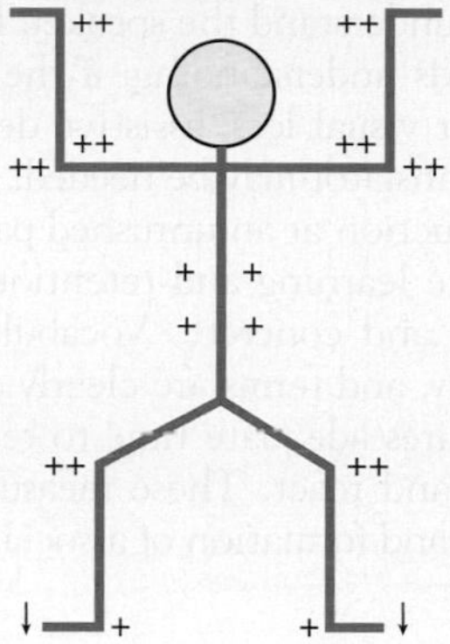

Clonus. When reflexes are hyperactive, a movement called **clonus** may be elicited. If the foot is abruptly dorsiflexed, it may continue to "beat" two or three times before it settles into a position of rest. Occasionally with CNS system disease, this activity persists, and the foot does not come to rest while the tendon is being stretched but persists in repetitive activity. The unsustained clonus associated with normal but hyperactive reflexes is not considered pathologic. Sustained clonus always indicates the presence of CNS disease and requires further evaluation.

Superficial Reflexes

The major superficial reflexes include corneal, palpebral, gag, upper/lower abdominal, cremasteric (men only), and perianal. These reflexes are graded differently than the motor reflexes and are noted to be present (+) or absent (–). Of these, only the corneal, gag, and plantar reflexes are commonly tested.

The corneal reflex is tested carefully using a clean wisp of cotton and lightly touching the outer corner of each eye on the sclera. The reflex is present if the action elicits a blink. A stroke or brain injury might result in loss of this reflex, either unilaterally or bilaterally. Loss of this reflex indicates the need for eye protection and possible lubrication to prevent corneal damage.

The gag reflex is elicited by gently touching the back of the pharynx with a cotton-tipped applicator, first on one side of the uvula and then the other. Positive response is an equal elevation of the uvula and "gag" with stimulation. Absent response on one or both sides can be seen following a stroke and requires careful evaluation and treatment of the resultant swallowing dysfunction to prevent aspiration of food and fluids.

Pathologic Reflexes

Pathologic reflexes are seen in the presence of neurologic disease; they often represent emergence of earlier reflexes that disappeared with maturity of the nervous system. A pathologic reflex indicative of CNS disease affecting the corticospinal tract is the **Babinski reflex (sign)**. In a person with an intact CNS, if the lateral aspect of the sole of the foot is stroked, the toes contract and draw together. However, in a person who has CNS disease of the motor system, the toes fan out and draw back (Jarvis, 2012) (see Fig. 65-13E). This is normal in newborns but represents a serious abnormality in adults. Other pathologic reflexes include the suck (sucking motions in response to touching the lips), snout (lip pursing in response to touching the lips), palmar (grasp in response to stroking the palm), and palmomental (contraction of the facial muscle in response to stimulation of the thenar eminence near the thumb) reflexes in adults. These reflexes often signify progressive nervous system degeneration (Klein & Stewart-Amidei, 2012).

Gerontologic Considerations

During the normal aging process, the nervous system undergoes many changes and is more vulnerable to illness. Age-related changes in the nervous system vary in degree and must be distinguished from those due to disease. It is important for clinicians not to attribute abnormality or dysfunction to aging without appropriate investigation (Kunz, Mylius, Schepelmann, et al., 2009). For example, although diminished strength and agility are a normal part of aging, localized weakness can only be attributed to disease.

Structural and Physiologic Changes

As the brain ages, neurons are lost, leading to a decrease in the number of synapses and neurotransmitters. This results in slowed nerve conduction and response time. Brain weight is decreased, and the ventricle size increases to maintain cranial volume. Cerebral blood flow and metabolism are reduced, leading to slower mental functions. Temperature regulation becomes less efficient. In the peripheral nervous system, myelin is lost, resulting in a decrease in conduction velocity in some nerves. Visual and auditory nerves degenerate, leading to loss of visual acuity and hearing. Taste buds atrophy, and nerve cell fibers in the olfactory bulb degenerate (Jarvis, 2012). Nerve cells in the vestibular system of the inner ear, cerebellum, and proprioceptive pathways also degenerate, leading to balance difficulties. Deep tendon reflexes can be decreased or in some cases absent. Hypothalamic function is modified such that stage IV sleep is reduced. There is an overall slowing of autonomic nervous system responses. Pupillary responses are reduced or may not appear at all in the presence of cataracts.

Motor Alterations

Reduced nerve input into muscle contributes to an overall reduction in muscle bulk, with atrophy most easily noted in the hands. Changes in motor function often result in decreased strength and agility, with increased reaction time. Gait is often slowed and wide based. These changes can create difficulties in maintaining balance, predisposing the older person to falls.

Sensory Alterations

Tactile sensation is dulled in the older adult due to a decrease in the number of sensory receptors. There may be difficulty in identifying objects by touch, because fewer tactile cues are received from the bottom of the feet and the person may become confused about body position and location (Wickremaratchi & Llewelyn, 2006).

Sensitivity to glare, decreased peripheral vision, and a constricted visual field occur due to degeneration of visual pathways, resulting in disorientation, especially at night when there is little or no light in the room. Because the older adult takes longer to recover visual sensitivity when moving from a light to dark area, nightlights and a safe and familiar arrangement of furniture are essential.

Loss of hearing can contribute to confusion, anxiety, disorientation, misinterpretation of the environment, feelings of inadequacy, and social isolation. A decreased sense of taste and smell may contribute to weight loss and disinterest in food. A decreased sense of smell may present a safety hazard, because older adults living alone may be unable to detect household gas leaks or fires. Smoke and carbon monoxide detectors—important in every residence—are critical for the older adult.

Temperature Regulation and Pain Perception

The older adult patient may feel cold more readily than heat and may require extra covering when in bed; a room temperature somewhat higher than usual may be desirable. Reaction to painful stimuli may be decreased with age (Kunz et al., 2009). Because pain is an important warning signal, caution must be used when hot or cold packs are used. The older patient may be burned or suffer frostbite before being aware of any discomfort. Complaints of pain such as abdominal discomfort or chest pain may be more serious than the patient's perception might indicate and thus require careful evaluation (Kunz et al., 2009). Two pain syndromes are common in the neurologic system in older adults: diabetic and postherpetic neuropathies.

Mental Status

Although mental processing time decreases with age, memory, language, and judgment capacities remain intact. Change in mental status should never be assumed to be a normal part of aging. **Delirium** (transient mental confusion, usually with delusions and hallucinations) is seen in older adult patients who have underlying CNS damage or are experiencing an acute condition such as infection, adverse medication reaction, or dehydration. Drug toxicity and depression may produce impairment of attention and memory and should be evaluated as a possible cause of mental status change. Delirium must be differentiated from dementia, which is a chronic and irreversible deterioration of cognitive status. (See Chapter 11 for further discussion of delirium and dementia.)

Nursing Implications

Nursing care for patients with age-related changes to the nervous system and for patients with long-term neurologic disability who are aging should include the modifications described previously. In addition, the consequences of any neurologic deficit and its impact on overall function such as activities of daily living, the use of assistive devices, and individual coping should be assessed and considered in planning patient care. Fall risk must be evaluated and fall prevention measures instituted for the patient who is hospitalized as well as in the home.

The nurse must understand the altered responses and the changing needs of the older adult patient before providing education. Visual and hearing deficits require adaptations in activities such as preoperative education, diet therapy, and education about new medications. When using visual materials for education or menu selection, adequate lighting without glare, contrasting colors, and large print are used to offset visual difficulties caused by rigidity and opacity of the lens in the eye and slower pupillary reaction. Procedures and preparations needed for diagnostic tests are explained, taking into account the possibility of impaired hearing and slowed responses in the older adult. Even with hearing loss, the older adult patient often hears adequately if the speaker uses a low-pitched, clear voice; shouting only makes it harder for the patient to understand the speaker. Providing auditory and visual cues aids understanding; if the patient has a significant hearing or visual loss, assistive devices, a signer, an interpreter, or a translator may be needed.

Providing instruction at an unrushed pace and using reinforcement enhance learning and retention. Material should be short, concise, and concrete. Vocabulary is matched to the patient's ability, and terms are clearly defined. The older adult patient requires adequate time to receive and respond to stimuli, learn, and react. These measures allow comprehension, memory, and formation of association and concepts.

Diagnostic Evaluation

A wide range of diagnostic studies may be performed in patients with altered neurologic function. The nurse should educate the patient about the purpose, what to expect, and any possible side effects related to these examinations prior to testing. Premenopausal women are advised to practice effective contraception before and for several days after any diagnostic procedure using contrast, and the woman who is breastfeeding is instructed to stop for the time period recommended by the nuclear medicine department (Pagana & Pagana, 2009). The nurse should note trends in results, because they provide information about disease progression as well as the patient's response to therapy.

Computed Tomography Scanning

Computed tomography (CT) scanning uses a narrow x-ray beam to scan body parts in successive layers. The images provide cross-sectional views of the brain, distinguishing differences in tissue densities of the skull, cortex, subcortical structures, and ventricles. An intravenous (IV) contrast agent may be used to highlight differences further. The brightness of each slice of the brain in the final image is proportional to the degree to which it absorbs x-rays. The image is displayed on an oscilloscope or television monitor and is photographed and stored digitally (Bremner, 2005). CT scanning is usually performed first without contrast material and then with IV contrast, if needed. The patient lies on an adjustable table with the head in a headrest while the scanning system rotates around the head and produces cross-sectional images. The patient must lie with the head held perfectly still without talking or moving the face, because head motion distorts the image. CT scanning is quick and painless and uses a small amount of radiation to produce images; it has a high degree of sensitivity for detecting lesions.

Brain lesions have a different tissue density from the surrounding normal brain tissue. Abnormalities detected on brain CT include tumor or other masses, infarction, hemorrhage, displacement of the ventricles, and cortical atrophy (Bremner, 2005). CT angiography allows visualization of blood vessels; in some situations, this eliminates the need for formal angiography. Whole-body CT scanners allow cross-sections of the spinal cord to be visualized. The injection of a water-soluble iodinated contrast agent into the subarachnoid space through lumbar puncture improves the visualization of the spinal and intracranial contents on these images. The CT scan, along with magnetic resonance imaging (MRI), has largely replaced myelography as a diagnostic procedure for the diagnosis of herniated lumbar disks.

Nursing Interventions

Essential nursing interventions include preparation for the procedure and patient monitoring. Preparation includes educating the patient about the need to lie quietly throughout the procedure. A review of relaxation techniques may be helpful for patients with claustrophobia. Sedation can be used if agitation, restlessness, or confusion interferes with a successful study. Ongoing patient monitoring during sedation is necessary. If a contrast agent is used, the patient must be assessed before the CT scan for an iodine/shellfish allergy, because the contrast agent used may be iodine based. Kidney function must also be evaluated because the contrast material is cleared through the kidneys. A suitable IV line for contrast injection and a period of fasting (usually 4 hours) are required prior to the study. Patients who receive an IV contrast agent are monitored during and after the procedure for allergic reactions and changes in kidney function (Karpoff & Labus, 2008). Fluid intake is also encouraged after IV contrast to facilitate contrast clearance through the kidney.

Magnetic Resonance Imaging

MRI uses a powerful magnetic field to obtain images of different areas of the body. The magnetic field causes the hydrogen nuclei (protons) within the body to align like small magnets in a magnetic field. In combination with radiofrequency pulses, the protons emit signals, which are converted to images. An MRI scan can be performed with or without a contrast agent and can identify a cerebral abnormality earlier and more clearly than other diagnostic tests (Bremner, 2005). It can provide information about the chemical changes within cells, allowing the clinician to monitor a tumor's response to treatment. It is particularly useful in the diagnosis of brain tumor, stroke, and multiple sclerosis and does not involve ionizing radiation. A MRI scan may take an hour or longer to complete, so its use in emergency situations is limited.

Several MRI applications allow imaging of brain blood flow and metabolism via special imaging techniques added to the MRI. Such techniques include diffusion-weighted imaging (DWI), perfusion-weighted imaging (PWI), magnetic resonance spectroscopy, and fluid attenuation inversion recovery (FLAIR) (Bremner, 2005). Magnetic resonance angiography (MRA) allows separate visualization of the cerebral vasculature without the administration of an arterial contrast agent. Both MRI and CT images are used as tools to plan and direct surgical intervention.

Nursing Interventions

Patient preparation includes providing education and obtaining an adequate history. Ferromagnetic substances in the body may become dislodged by the magnet, so history of working with metal fragments must be reviewed. The patient is questioned about any implants of any metal objects (e.g., aneurysm clips, orthopedic hardware, pacemakers, artificial heart valves, intrauterine devices). These objects could malfunction, be dislodged, or heat up as they absorb energy. Cochlear implants will be inactivated by MRI; therefore, other imaging procedures are considered. A complete list of metal compatibility may be found on MRI manufacturer Web sites.

Before the patient enters the room where the MRI is to be performed, all metal objects and credit cards (the magnetic field can erase them) must be removed. This includes medication patches that have a metal backing and metallic lead wires; these can cause burns if not removed (Bremner, 2005). No metal objects may be brought into the room where the MRI is located; this includes oxygen tanks, IV poles, ventilators, or even stethoscopes. The magnetic field generated by the unit is so strong that any metal-containing items will be strongly attracted and literally can be pulled away with such force that they fly like projectiles toward the magnet. There is a risk of severe injury and death. Further, damage to expensive equipment may occur.

Quality and Safety Nursing Alert

For patient safety, the nurse must make sure that no patient care equipment (e.g., portable oxygen tanks) that contains metal or metal parts enters the room where the MRI is located. In addition, the patient must be assessed for the presence of medication patches with foil backing (such as nicotine) that may cause a burn.

For the MRI, the patient lies with the head in a frame on a flat platform that is moved into a tube housing the magnet (Fig. 65-14). The tube is narrow; persons with a wide girth may not fit into the scanner. Patients who are unable to lie

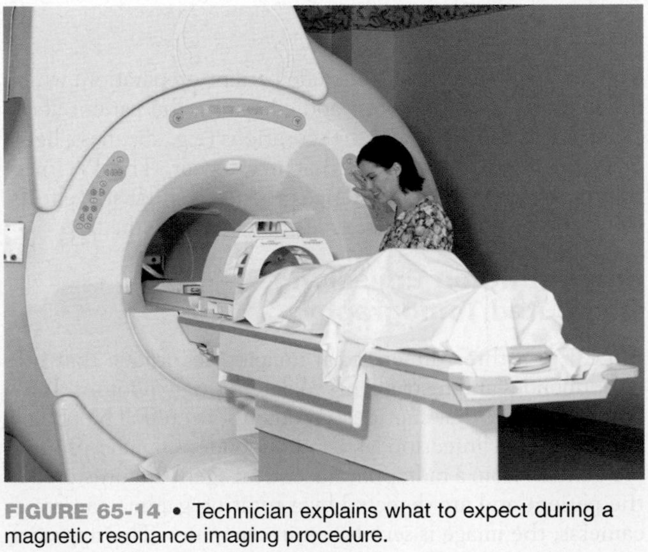

FIGURE 65-14 • Technician explains what to expect during a magnetic resonance imaging procedure.

flat will not be able to tolerate an MRI. The scanning process is painless, but the patient hears loud thumping of the magnetic coils as the magnetic field is being pulsed. Patients may experience claustrophobia while inside the narrow tube; sedation may be prescribed in these circumstances. "Open" MRI machines are less claustrophobic than the other devices and are available in many locations. However, the images produced on these machines are sometimes not as detailed, and traditional devices are preferred for accurate diagnosis. The patient may be educated about the use of relaxation techniques while in the scanner. The patient is informed that he or she will be able to talk to the staff during the scan through a microphone inside the scanner (Karpoff & Labus, 2008).

Positron Emission Tomography

PET is a computer-based nuclear imaging technique that produces images of actual organ functioning. The patient either inhales a radioactive gas or is injected with a radioactive substance that emits positively charged particles. When these positrons combine with negatively charged electrons (normally found in the body's cells), the resultant gamma rays can be detected by a scanning device that produces a series of two-dimensional views at various levels of the brain. This information is integrated by a computer and gives a composite picture of the brain at work.

PET permits the measurement of blood flow, tissue composition, and brain metabolism and thus indirectly evaluates brain function. The brain is one of the most metabolically active organs, consuming 80% of the glucose the body uses (Bader & Littlejohns, 2010). PET measures this activity in specific areas of the brain and can detect changes in glucose use.

PET is useful in showing metabolic changes in the brain (Alzheimer's disease), locating lesions (brain tumor, epileptogenic lesions), identifying blood flow and oxygen metabolism in patients with strokes, distinguishing tumor from areas of necrosis, and revealing biochemical abnormalities associated with mental illness. The isotopes used have a very short half-life and are expensive to produce, requiring specialized equipment for production. Improvement in the scanning procedure and production of isotopes, as well as the advent of reimbursement by third-party payers, has increased the clinical applications of PET studies.

Nursing Interventions

Key nursing interventions include patient preparation, which involves explaining the test and educating the patient about inhalation techniques and the sensations (e.g., dizziness, lightheadedness, and headache) that may occur. The IV injection of the radioactive substance produces similar side effects. Relaxation exercises may reduce anxiety during the test.

Single-Photon Emission Computed Tomography

SPECT is a three-dimensional imaging technique that uses radionuclides and instruments to detect single photons. It is a perfusion study that captures a moment of cerebral blood flow at the time of injection of a radionuclide. Gamma photons are emitted from a radiopharmaceutical agent administered to the patient and are detected by a rotating gamma camera or cameras; the image is sent to a minicomputer. This approach allows areas behind overlying structures or background to be viewed, greatly increasing the contrast between normal and abnormal tissue. It is relatively inexpensive, and the duration is similar to that of a CT scan.

SPECT is useful in detecting the extent and location of abnormally perfused areas of the brain, thus allowing detection, localization, and sizing of stroke (before it is visible by CT scan); localization of seizure foci in epilepsy; detection of tumor progression (Bremner, 2005); and evaluation of perfusion before and after neurosurgical procedures.

Nursing Interventions

The nursing interventions for SPECT primarily include patient preparation and patient monitoring. Providing education about what to expect before the test can allay anxiety and ensure patient cooperation during the test. Pregnancy and breast-feeding are contraindications to SPECT.

The nurse may need to accompany and monitor the patient during transport to the nuclear medicine department for the scan. Patients are monitored during and after the procedure for allergic reactions to the radiopharmaceutical agent.

Cerebral Angiography

Cerebral angiography is an x-ray study of the cerebral circulation with a contrast agent injected into a selected artery. A valuable tool in investigating vascular disease or anomalies, it is used to determine vessel patency, identify presence of collateral circulation, and provide detail on vascular anomalies that can be used in planning interventions. With the advent of additional imaging techniques, formal cerebral angiography is less frequently performed.

Cerebral angiograms are performed by threading a catheter through the femoral artery in the groin and up to the desired vessel. Alternatively, direct puncture of the carotid artery may be performed. X-ray images are obtained as the contrast agent flows through the vessels; the carotid and vertebral arterial systems are visualized, as well as venous drainage. Arterial access may also be used for interventional procedures, such as placing coils in an aneurysm or arteriovenous malformation.

Nursing Interventions

Prior to the angiography, the patient's blood urea nitrogen and creatinine should be checked to ensure the kidneys will be able to excrete the contrast agent. The patient should be well hydrated, and clear liquids are usually permitted up to the time of the test. The patient is instructed to void immediately before the test, and locations of the appropriate peripheral pulses are marked with a felt-tip pen. The patient is instructed to remain immobile during the angiogram process and is told to expect a brief feeling of warmth in the face, behind the eyes, or in the jaw, teeth, tongue, and lips, and a metallic taste when the contrast agent is injected.

After the groin is shaved and prepared, a local anesthetic agent is administered to minimize pain at the insertion site and to reduce arterial spasm. A catheter is introduced into the femoral artery, flushed with heparinized saline, and filled with contrast agent. Fluoroscopy is used to guide the catheter to the appropriate vessels. Neurologic assessment is conducted during and immediately following cerebral angiography to observe for embolism or arterial dissection that may occur during the test. Signs of these complications include new onset of alterations in the level of consciousness, weakness on one side of the body, motor or sensory deficits, and speech disturbances.

Nursing care after cerebral angiography includes observation of the injection site for bleeding or hematoma formation (a localized collection of blood). Because a hematoma at the puncture site or embolization to a distal artery affects peripheral pulses, the peripheral pulses that were marked prior to the test are monitored frequently. The color and temperature of the involved extremity are assessed to detect possible embolism (Karpoff & Labus, 2008). Fluids are encouraged to facilitate clearance of the contrast through the kidney. The nurse also monitors for an allergic reaction to the contrast agent.

Myelography

A myelogram is an x-ray of the spinal subarachnoid space taken after the injection of a contrast agent into the spinal subarachnoid space through a lumbar puncture. The water-based contrast agent disperses upward through the CSF to outline the spinal subarachnoid space and show any distortion of the spinal cord or spinal dural sac caused by tumors, cysts, herniated vertebral disks, or other lesions. Myelography is performed infrequently today because of the sensitivity of CT and MRI scanning (Bremner, 2005).

Nursing Interventions

The patient is educated about what to expect during the procedure and made aware that changes in position may be made during the procedure. After myelography, the patient lies in bed with the head of the bed elevated 30 to 45 degrees. The patient is advised to remain in bed in the recommended position for 3 hours or as prescribed. Drinking liberal amounts of fluid for rehydration and replacement of CSF may decrease the incidence of post–lumbar puncture headache. The blood pressure, pulse, respiratory rate, and temperature are monitored, as well as the patient's ability to void. Untoward signs include headache, fever, stiff neck, **photophobia** (sensitivity to light), seizures, and signs of chemical or bacterial meningitis (Hickey, 2009).

Noninvasive Carotid Flow Studies

Noninvasive carotid flow studies use ultrasound imagery and Doppler measurements of arterial blood flow to evaluate carotid and deep orbital circulation. The graph produced indicates blood velocity. Increased blood velocity can indicate stenosis or partial obstruction. These tests are often obtained before more invasive tests such as arteriography or are used as screening tools. Carotid Doppler, carotid ultrasonography, oculoplethysmography, and ophthalmodynamometry are four common noninvasive vascular techniques that permit evaluation of arterial blood flow and detection of arterial stenosis, occlusion, and plaques. These vascular studies allow noninvasive imaging of extra- and intracranial circulation (Fischbach & Dunning, 2011).

Transcranial Doppler

Transcranial Doppler uses the same noninvasive techniques as carotid flow studies but records the blood flow velocities of the intracranial vessels. Arterial flow velocities can be measured through thin areas of the temporal and occipital bones of the skull. A handheld Doppler probe emits a pulsed beam; the signal is reflected by the moving red blood cells within the blood vessels. Transcranial Doppler is a noninvasive technique that is helpful in assessing vasospasm (a complication following subarachnoid hemorrhage), altered cerebral blood flow found in occlusive vascular disease, other cerebral pathology, and brain death.

Nursing Interventions

When a carotid flow study or transcranial Doppler is scheduled, the procedure is described to the patient. The patient is informed that this is a noninvasive test, that a handheld transducer will be placed over the neck and the orbits of the eyes, and that a water-soluble jelly is used on the transducer. Either of these two low-risk tests can be performed at the patient's bedside (Littlejohns & Bader, 2009).

Electroencephalography

An electroencephalogram (EEG) represents a record of the electrical activity generated in the brain (Hickey, 2009). It is obtained through electrodes applied on the scalp or through microelectrodes placed within the brain tissue. It provides an assessment of cerebral electrical activity. It is useful for diagnosing and evaluating seizure disorders, coma, or organic brain syndrome. Tumors, brain abscesses, blood clots, and infection may cause abnormal patterns in electrical activity. The EEG is also used in making a determination of brain death.

Electrodes are applied to the scalp to record the electrical activity in various regions of the brain. The amplified activity of the neurons between any two of these electrodes is recorded on continuously moving paper; this record is called the *encephalogram*.

For a baseline recording, the patient lies quietly with both eyes closed. The patient may be asked to hyperventilate for 3 to 4 minutes or to look at a bright, flashing light for photic stimulation. These activation procedures are performed to evoke abnormal electrical discharges, such as seizure potentials. A sleep EEG may be recorded after sedation because some abnormal brain waves are seen only when the patient is asleep. If the epileptogenic area is inaccessible to conventional scalp electrodes, nasopharyngeal electrodes may be used.

Depth recording of EEG is performed by introducing electrodes stereotactically (radiologically placed using instrumentation) into a target area of the brain, as indicated by the patient's seizure pattern and scalp EEG. It is used to identify patients who may benefit from surgical excision of epileptogenic foci. Special transsphenoidal, mandibular, and nasopharyngeal electrodes can be used, and video recording combined with EEG monitoring and telemetry is used in hospital settings to capture epileptiform abnormalities and their sequelae. Some epilepsy centers provide long-term ambulatory EEG monitoring with portable recording devices (Bader & Littlejohns, 2010).

Nursing Interventions

To increase the chances of recording seizure activity, it is sometimes recommended that the patient be deprived of sleep the night before the EEG. Antiseizure agents, tranquilizers, stimulants, and depressants should be withheld 24 to 48 hours before an EEG, because these medications can alter the EEG wave patterns or mask the abnormal wave patterns of seizure disorders (Pagana & Pagana, 2009). Coffee, tea, chocolate, and cola drinks are omitted from the meal before the test because of their stimulating effect. However, the meal itself is not omitted, because an altered blood glucose level can cause changes in brain wave patterns.

The patient is informed that the standard EEG takes 45 to 60 minutes; a sleep EEG requires 12 hours. The patient is assured that the procedure does not cause an electric shock and that the EEG is a diagnostic test, not a form of treatment. An EEG requires the patient to lie quietly during the test. Sedation is not advisable, because it may lower the seizure threshold in patients with a seizure disorder and it alters brain wave activity in all patients. The nurse needs to check the prescription regarding the administration of antiseizure medication prior to testing.

Routine EEGs use a water-soluble lubricant for electrode contact, which can be wiped off and removed by shampooing later. Sleep EEGs involve the use of collodion glue for electrode contact, which requires acetone for removal.

Electromyography

An electromyogram (EMG) is obtained by inserting needle electrodes into the skeletal muscles to measure changes in the electrical potential of the muscles (Pagana & Pagana, 2009). The electrical potentials are shown on an oscilloscope and amplified so that both the sound and appearance of the waves can be analyzed and compared simultaneously.

An EMG is useful in determining the presence of neuromuscular disorders and myopathies. It helps distinguish weakness due to neuropathy (functional or pathologic changes in the peripheral nervous system) from weakness resulting from other causes.

Nursing Interventions

The procedure is explained, and the patient is warned to expect a sensation similar to that of an intramuscular injection as the needle is inserted into the muscle. The muscles examined may ache for a short time after the procedure.

Nerve Conduction Studies

Nerve conduction studies are performed by stimulating a peripheral nerve at several points along its course and recording the muscle action potential or the sensory action potential that results. Surface or needle electrodes are placed on the skin over the nerve to stimulate the nerve fibers. This test is useful in the study of peripheral neuropathies and is often included as part of the EMG.

Evoked Potential Studies

Evoked potential studies involve application of an external stimulus to specific peripheral sensory receptors with subsequent measurement of the electrical potential generated. Electrical changes are detected with the aid of computerized devices that extract the signal, display it on an oscilloscope, and store the data on magnetic tape or disk. In neurologic diagnosis, they reflect nerve conduction times in the peripheral nervous system. In clinical practice, the visual, auditory, and somatosensory systems are most often tested.

In visual evoked responses, the patient looks at a visual stimulus (flashing lights, a checkerboard pattern on a screen). The average of several hundred stimuli is recorded by EEG leads placed over the occipital lobe. The transit time from the retina to the occipital area is measured using computer-averaging methods.

Brainstem auditory evoked responses (BAERs) are measured by applying an auditory stimulus (repetitive auditory click) and measuring the transit time via the brain stem into the cortex. Specific lesions in the auditory pathway modify or delay the response. BAERs may be used in the diagnosis of brain stem abnormalities and in determination of brain death.

In somatosensory evoked responses (SERs), the peripheral nerves are stimulated (electrical stimulation through skin electrodes) and the transit time along the spinal cord to the cortex is measured and recorded from scalp electrodes. SERs are used to detect deficits in spinal cord or peripheral nerve conduction and to monitor spinal cord function during surgical procedures. It is also useful in the diagnosis of demyelinating diseases, such as multiple sclerosis and polyneuropathies, where nerve conduction is slowed.

Nursing Interventions

The nurse explains the procedure and reassures the patient and encourages him or her to relax. The patient is advised to remain perfectly still throughout the recording to prevent artifacts (signals not generated by the brain) that interfere with the recording and interpretation of the test.

Lumbar Puncture and Examination of Cerebrospinal Fluid

A lumbar puncture (spinal tap) is carried out by inserting a needle into the lumbar subarachnoid space to withdraw CSF. The test may be performed to obtain CSF for examination, to measure and reduce CSF pressure, to determine the presence or absence of blood in the CSF, and to administer medications intrathecally (into the spinal canal).

The needle is usually inserted into the subarachnoid space between the third and fourth or fourth and fifth lumbar vertebrae. Because the spinal cord ends at the first lumbar vertebra, insertion of the needle below the level of the third lumbar vertebra prevents puncture of the spinal cord.

A successful lumbar puncture requires that the patient be relaxed; an anxious patient is tense, and this may increase the pressure reading. CSF pressure with the patient in a lateral recumbent position is normally 50 to 200 mm H_2O (Karpoff & Labus, 2008).

A lumbar puncture may be risky in the presence of an intracranial mass lesion because intraspinal pressure is decreased by removal of CSF, and the brain may herniate downward through the foramen magnum. Chart 65-5 provides instructions for assisting with a lumbar puncture.

Cerebrospinal Fluid Analysis

The CSF should be clear and colorless. Pink, blood-tinged, or grossly bloody CSF may indicate a subarachnoid hemorrhage. The CSF may be bloody initially because of local trauma but becomes clearer as more fluid is drained. Specimens are obtained for cell count, culture, glucose, protein, and other tests as indicated. The specimens should be sent to the laboratory immediately because changes will take place and alter the result if the specimens are allowed to stand. (See Table A-5 in Appendix A on thePoint for the normal values of CSF.)

Post–Lumbar Puncture Headache

A post–lumbar puncture headache, ranging from mild to severe, may occur a few hours to several days after the procedure. It is a throbbing bifrontal or occipital headache that is dull and deep in character. It is particularly severe on sitting or standing but lessens or disappears when the patient lies down.

Chart 65-5 Assisting With a Lumbar Puncture

A needle is inserted into the subarachnoid space through the 3rd and 4th or 4th and 5th lumbar interface to withdraw spinal fluid.

Preprocedure

1. Determine whether written consent for the procedure has been obtained.
2. Explain the procedure to the patient, and describe sensations that are likely during the procedure (i.e., a sensation of cold as the site is cleansed with solution, a needle prick when local anesthetic agent is injected).
3. Determine whether the patient has any questions or misconceptions about the procedure; reassure the patient that the needle will not enter the spinal cord or cause paralysis.
4. Instruct the patient to void before the procedure.

Procedure

1. The patient is positioned on one side at the edge of the bed or examining table with the back toward the physician; the thighs and legs are flexed as much as possible to increase the space between the spinous processes of the vertebrae, for easier entry into the subarachnoid space.

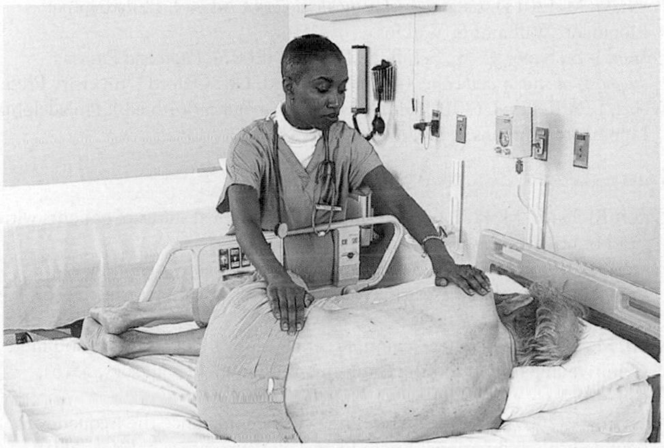

Photo by B. Proud.

2. A small pillow may be placed under the patient's head to maintain the spine in a horizontal position; a pillow may be placed between the legs to prevent the upper leg from rolling forward.
3. The nurse assists the patient to maintain the position to avoid sudden movement, which can produce a traumatic (bloody) tap.
4. The patient is encouraged to relax and is instructed to breathe normally, because hyperventilation may lower an elevated pressure.
5. The nurse describes the procedure step by step to the patient as it proceeds.
6. The physician cleanses the puncture site with an antiseptic agent solution and drapes the site.
7. The physician injects local anesthetic agent to numb the puncture site and then inserts a spinal needle into the subarachnoid space through the 3rd and 4th or 4th and 5th lumbar interspace. A pressure reading may be obtained.
8. A specimen of cerebrospinal fluid (CSF) is removed and usually collected in 3 test tubes, labeled in order of collection. The needle is withdrawn.
9. The physician applies a small dressing to the puncture site.
10. The tubes of CSF are sent to the laboratory immediately.

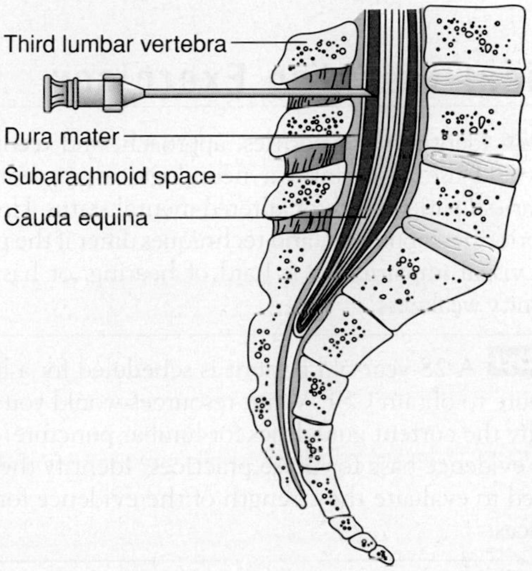
Third lumbar vertebra
Dura mater
Subarachnoid space
Cauda equina

Postprocedure

1. Instruct the patient to lie prone to separate the alignment of the dural and arachnoid needle punctures in the meninges, to reduce leakage of CSF.
2. Monitor the patient for complications of lumbar puncture; notify primary provider if complications occur.
3. Encourage increased fluid intake to reduce the risk of postprocedure headache.

The headache is caused by CSF leakage at the puncture site. The fluid continues to escape into the tissues by way of the needle track from the spinal canal. As a result of a leak, the supply of CSF in the cranium is depleted to a point at which it is insufficient to maintain proper mechanical stabilization of the brain. When the patient assumes an upright position, tension and stretching of the venous sinuses and pain-sensitive structures occur.

Post–lumbar puncture headache may be avoided if a small-gauge needle is used and if the patient remains prone after the procedure. When more than 20 mL of CSF is removed, the patient is positioned supine for 6 hours (Bader & Littlejohns, 2010). A postpuncture headache is usually managed by bed rest, analgesic agents, and hydration.

Other Complications of Lumbar Puncture

Herniation of the intracranial contents, spinal epidural abscess, spinal epidural hematoma, and meningitis are rare but serious complications of lumbar puncture. Other complications include temporary voiding problems, slight elevation of temperature, backache or spasms, and stiffness of the neck.

Promoting Home and Community-Based Care

Educating Patients About Self-Care

Many diagnostic tests that were once performed as part of a hospital stay are now carried out in short-procedure units or outpatient testing settings or units. As a result, family members often provide the postprocedure care. Therefore, the patient and family must receive adequate education about precautions to take after the procedure, complications to watch for, and steps to take if complications occur. Because many patients undergoing neurologic diagnostic studies are older adults or have neurologic deficits, provisions must be made to ensure that transportation, postprocedure care, and appropriate monitoring are available.

Continuing Care

Contacting the patient and family after diagnostic testing enables the nurse to determine whether they have any questions about the procedure or whether the patient had any untoward results. Education is reinforced and the patient and family are reminded to make and keep follow-up appointments. Patients, family members, and health care providers are focused on the immediate needs, issues, or deficits that necessitated the diagnostic testing.

Critical Thinking Exercises

1 Identify the priorities, approach, and techniques you would use to perform a neurologic assessment on a 65-year-old patient with an altered mental status. How will your priorities, approach, and techniques differ if the patient has a visual impairment, is hard of hearing, or has lower extremity weakness?

2 A 28-year-old patient is scheduled for a lumbar puncture to obtain CSF. What resources would you use to identify the current guidelines for lumbar puncture? What is the evidence base for these practices? Identify the criteria used to evaluate the strength of the evidence for these practices.

3 A 35-year-old female patient is scheduled for cerebral angiography. Identify the nursing interventions indicated before, during, and after the procedure. What patient education is indicated? How will your nursing interventions and patient education change if the patient is pregnant or lactating?

Brunner Suite Resources

Explore these additional resources to enhance learning for this chapter:

- NCLEX-Style Questions and Other Resources on thePoint, http://thePoint.lww.com/Brunner13e
- Study Guide
- PrepU
- Clinical Handbook
- Handbook of Laboratory and Diagnostic Tests

References

*Asterisk indicates nursing research.

Books

Bader, M., & Littlejohns, L. R. (2010). *AANN core curriculum for neuroscience nursing* (5th ed.). Glenview, IL: American Association of Neuroscience Nurses.

Bremner, J. D. (2005). *Brain imaging handbook*. New York: W. W. Norton.

Fischbach, F. T., & Dunning, M. B. (2011). *Nurse's quick reference to common laboratory and diagnostic tests* (4th ed.). Philadelphia: Lippincott Williams & Wilkins.

Herndon, R. M. (2006). *Handbook of neurologic rating scales* (2nd ed.). New York: Demos Medical Publishing.

Hickey, J. V. (2009). *The clinical practice of neurological & neurosurgical nursing* (6th ed.). Philadelphia: Lippincott Williams & Wilkins.

Jarvis, C. (2012). *Physical examination and health assessment* (6th ed.). Philadelphia: Saunders.

Karpoff, S., & Labus, D. (2008). *Portable diagnostic tests*. Philadelphia: Lippincott Williams & Wilkins.

Klein, D. G., & Stewart-Amidei, C. (2012). Nervous system alterations. In M. L. Sole, D. G. Klein, & M. J. Moseley (Eds.). *Introduction to critical care nursing* (6th ed.). St. Louis: Elsevier Saunders.

Littlejohns, L. R., & Bader, M. K. (2009). *AACN-AANN protocols for practice: Monitoring technologies in critically ill neuroscience patients*. Sudbury, MA: Jones & Bartlett.

Pagana, K. D., & Pagana, T. J. (2009). *Manual of diagnostic and laboratory tests* (4th ed.). St. Louis: Mosby Elsevier.

Porth, C. M. (2011). *Essentials of pathophysiology* (3rd ed.). Philadelphia: Lippincott Williams & Wilkins.

Posner, J. B., Saper, C. B., Schiff, N. D., et al. (2007). *Plum and Posner's diagnosis of stupor and coma* (4th ed.). Oxford, UK: Oxford University Press.

Weber, J., & Kelley, J. (2010). *Health assessment in nursing* (4th ed.). Philadelphia: Lippincott Williams & Wilkins.

Journals and Electronic Documents

Caton-Richards, M. (2010). Assessing the neurological status of patients with head injuries. *Emergency Nurse, 17*(10), 28–31.

*Doerksen, K., & Naimark, B. (2006). Nonspecific behaviors as early indicators of cerebral vasospasm. *Journal of Neuroscience Nursing, 38*(6), 409–415.

Kunz, M., Mylius, V., Schepelmann, K., et al. (2009). Effects of age and mild cognitive impairment on the pain response system. *Gerontology, 55*(6), 674–682.

*Mavin, C. (2009). Does underpinning evidence influence the frequency of neurological observations? *British Journal of Neuroscience Nursing, 5*(10), 456–459.

Wickremaratchi, M. M., & Llewelyn, J. G. (2006). Effects of aging on touch. *Postgraduate Medical Journal, 82*, 301–304.

Resources

American Headache Society, www.americanheadachesociety.org
Brain Injury Association of America, www.biausa.org
Brain Trauma Foundation, www.braintrauma.org
Epilepsy Foundation, www.epilepsyfoundation.org
Harvard Health Publications, Harvard Medical School Office of Public Affairs, www.health.harvard.edu/diagnostic-tests/#brain
National Headache Foundation, www.headaches.org

Chapter 66

Management of Patients With Neurologic Dysfunction

Learning Objectives

On completion of this chapter, the learner will be able to:

1. Describe the causes, clinical manifestations, and medical management of various neurologic dysfunctions.
2. Use the nursing process as a framework for care of the multiple needs of the patient with altered level of consciousness.
3. Identify the early and late clinical manifestations of increased intracranial pressure.
4. Use the nursing process as a framework for care of the patient with increased intracranial pressure.
5. Describe the indications for intracranial or transsphenoidal surgery.
6. Use the nursing process as a framework for care of the patient undergoing intracranial or transsphenoidal surgery.
7. Identify the various types and causes of seizures.
8. Use the nursing process to develop a plan of care for the patient experiencing seizures.
9. Identify the causes, clinical manifestations, and medical and nursing management of the patient experiencing headaches.

Glossary

akinetic mutism: unresponsiveness to the environment; the patient makes no movement or sound but sometimes opens the eyes

altered level of consciousness (LOC): when a patient is not oriented, does not follow commands, or needs persistent stimuli to achieve a state of alertness

autoregulation: in the context of neuroscience, describes the ability of cerebral blood vessels to dilate or constrict to maintain stable cerebral blood flow despite changes in systemic arterial blood pressure

brain death: irreversible loss of all functions of the entire brain, including the brain stem

coma: prolonged state of unconsciousness

cortical spreading depression: depolarization of neuronal cells; associated with migraine

craniectomy: a surgical procedure that involves removal of a portion of the skull

craniotomy: a surgical procedure that involves entry into the cranial vault

Cushing's response: the brain's attempt to restore blood flow by increasing arterial pressure to overcome the increased intracranial pressure

Cushing's triad: three classic signs—bradycardia, hypertension, and bradypnea—seen with pressure on the medulla as a result of brain stem herniation

decerebration: an abnormal body posture associated with a severe brain injury, characterized by extreme extension of the upper and lower extremities

decortication: an abnormal posture associated with severe brain injury, characterized by abnormal flexion of the upper extremities and extension of the lower extremities

epidural monitor: a sensor placed between the skull and the dura to monitor intracranial pressure

epilepsy: a group of syndromes characterized by paroxysmal transient disturbances of brain function

fiberoptic monitor: a system that uses light refraction to determine intracranial pressure

herniation: abnormal protrusion of tissue through a defect or natural opening

intracranial pressure (ICP): pressure exerted by the volume of the intracranial contents within the cranial vault

locked-in syndrome: condition resulting from a lesion in the pons in which the patient lacks all distal motor activity (paralysis) but cognition is intact

microdialysis: procedure in which an intracranial catheter is inserted near an injured area of brain to measure lactate, pyruvate, glutamate, and glucose levels

migraine headache: a severe, unrelenting headache often accompanied by symptoms such as nausea, vomiting, and visual disturbances

minimally conscious state: a state in which the patient demonstrates awareness but cannot communicate thoughts or feelings

Monro-Kellie hypothesis: theory that states that due to limited space for expansion within the skull, an increase in any one of the cranial contents—brain tissue, blood, or cerebrospinal fluid—causes a change in the volume of the others; also referred to as Monro-Kellie doctrine

persistent vegetative state: condition in which the patient is wakeful but devoid of conscious content, without cognitive or affective mental function

primary headache: a headache for which no specific organic cause can be found

secondary headache: headache identified as a symptom of another organic disorder (e.g., brain tumor, hypertension)

seizures: paroxysmal transient disturbance of the brain resulting from a discharge of abnormal electrical activity

(continues on page 1936)

status epilepticus: episode in which the patient experiences multiple seizure bursts with no recovery time in between

subarachnoid screw or bolt: device placed into the subarachnoid space to measure intracranial pressure

transsphenoidal: surgical approach to the pituitary via the sphenoid sinuses

ventriculostomy: a catheter placed in one of the lateral ventricles of the brain to measure intracranial pressure and allow for drainage of fluid

This chapter presents an overview of care of the patient with an altered level of consciousness (LOC); the patient with increased intracranial pressure (ICP); and the patient who is undergoing neurosurgical procedures, experiencing seizures, or experiencing headaches. Some of the disorders in this chapter, such as headaches and seizures, may be symptoms of dysfunction in another body system. Alternatively, headaches and seizures can be symptoms of a disruption of the neurologic system. These disorders can also be diagnosed at times as "idiopathic," or without an identifiable cause. The commonalities of these disorders are often the behaviors and needs of the patient and the approaches nurses use to support the patient.

The central nervous system (CNS) contains a vast network of neurons that control the body's vital functions. However, this system is vulnerable, and its optimal function depends on several key factors. First, the neurologic system relies on its structural integrity for support and homeostasis, but this integrity may be disrupted. Examples of structural disruption include head injury, brain tumor, intracranial hemorrhage, infection, and stroke. As brain tissue expands in the inflexible cranium, **intracranial pressure (ICP)** (pressure exerted by the volume of the intracranial contents within the cranial vault) rises, and cerebral perfusion is impaired. Further expansion places pressure on vital centers, which can cause permanent neurologic deficits or lead to brain death.

Second, the neurologic system relies on the body's ability to maintain a homeostatic environment. It requires the delivery of the essential elements of oxygen and glucose, as well as filtration of substrates that are toxic to the neurons. The functions of the neurologic system may be decreased or absent because of the effect of toxic substrates or the body's inability to provide essential substrates. Sepsis, hypovolemia, myocardial infarction, respiratory arrest, hypoglycemia, electrolyte imbalance, drug and/or alcohol overdose, encephalopathy, and ketoacidosis are examples of such circumstances. Some conditions can be treated and reversed; others result in permanent neurologic deficits and disabilities.

Although the specialty of neuroscience nursing requires an understanding of neuroanatomy, neurophysiology, neurodiagnostic testing, critical care nursing, and rehabilitation nursing, nurses in all settings care for patients with neurologic disorders (Hickey, 2009). Ongoing assessment of the patient's neurologic function and health needs, identification of problems, mutual goal setting, development and implementation of care plans (including education, counseling, and coordinating activities), and evaluation of the outcomes of care are nursing actions integral to the recovery of the patient. The nurse also collaborates with other members of the health care team to provide essential care, offer a variety of solutions to problems, help the patient and family gain control of their lives, and explore the educational and supportive resources available in the community. The goals are to achieve as high a level of function as possible and to enhance the quality of life for the patient with neurologic impairment and his or her family.

ALTERED LEVEL OF CONSCIOUSNESS

An **altered level of consciousness (LOC)** is present when the patient is not oriented, does not follow commands, or needs persistent stimuli to achieve a state of alertness. LOC is gauged on a continuum, with a normal state of alertness and full cognition (consciousness) on one end and coma on the other end. **Coma** is a clinical state of unarousable unresponsiveness in which there are no purposeful responses to internal or external stimuli, although nonpurposeful responses to painful stimuli and brain stem reflexes may be present (Laureys, Boly, Moonen, et al., 2009). The usual duration of coma is 2 to 4 weeks. **Akinetic mutism** is a state of unresponsiveness to the environment in which the patient makes no voluntary movement. **Persistent vegetative state** is a condition in which the unresponsive patient resumes sleep–wake cycles after coma but is devoid of cognitive or affective mental function. A **minimally conscious state** differs from persistent vegetative state in that the patient has inconsistent but reproducible signs of awareness (Cruse, Chennu, Chatelle, et al., 2012). **Locked-in syndrome** results from a lesion affecting the pons and results in paralysis and the inability to speak, but vertical eye movements and lid elevation remain intact and are used to indicate responsiveness (Maurer, 2008). The level of responsiveness and consciousness is the most important indicator of the patient's condition.

Pathophysiology

Altered LOC is not a disorder itself; rather, it is a result of multiple pathophysiologic phenomena. The cause may be neurologic (head injury, stroke), toxicologic (drug overdose, alcohol intoxication), or metabolic (hepatic or renal failure, diabetic ketoacidosis).

The underlying cause of neurologic dysfunction is disruption in the cells of the nervous system, neurotransmitters, or brain anatomy (see Chapter 65). Disruptions result from cellular edema or other mechanisms, such as disruption of chemical transmission at receptor sites by antibodies.

Intact anatomic structures of the brain are needed for normal function. The two hemispheres of the cerebrum must communicate, via an intact corpus callosum, and the lobes of the brain (frontal, parietal, temporal, and occipital) must communicate and coordinate their specific functions (see Chapter 65). Other anatomic structures of importance are the cerebellum and the brain stem. The cerebellum has both excitatory and inhibitory actions and is largely responsible for coordination of movement. The brain stem contains areas that control the heart, respiration, and blood pressure. Disruptions in the anatomic structures result from trauma, edema, pressure from tumors, or other mechanisms, such as an increase or decrease in the circulation of blood or cerebrospinal fluid (CSF).

Clinical Manifestations

Alterations in LOC occur along a continuum, and the clinical manifestations depend on where the patient is on this continuum. As the patient's state of alertness and consciousness decreases, changes occur in the pupillary response, eye opening response, verbal response, and motor response. However, initial alterations in LOC may be reflected by subtle behavioral changes, such as restlessness or increased anxiety. The pupils, normally round and quickly reactive to light, become sluggish (response is slower); as the patient becomes comatose, the pupils become fixed (no response to light). The patient in a coma does not open the eyes to voice or command, respond verbally, or move the extremities in response to a request to do so.

Assessment and Diagnostic Findings

The patient with an altered LOC is at risk for alterations in every body system. A complete assessment is performed, with particular attention to the neurologic system. The neurologic examination should be as complete as the LOC allows (American Association of Neuroscience Nurses [AANN], 2011). It includes an evaluation of mental status, cranial nerve function, cerebellar function (balance and coordination), reflexes, and motor and sensory function. LOC, a sensitive indicator of neurologic function, is assessed based on the criteria in the Glasgow Coma Scale: eye opening, verbal response, and motor response (Barlow, 2012). The patient's responses are rated on a scale from 3 to 15. A score of 3 indicates severe impairment of neurologic function, brain death, or pharmacologic inhibition of the neurologic response. A score of 15 indicates that the patient is fully responsive (see Chapter 68).

If the patient is comatose and has localized signs such as abnormal pupillary and motor responses, it is assumed that neurologic disease is present until proven otherwise. If the patient is comatose but pupillary light reflexes are preserved, a toxic or metabolic disorder is suspected. Common diagnostic procedures used to identify the cause of unconsciousness include computed tomography (CT) scanning, magnetic resonance imaging (MRI), and electroencephalography (EEG). Less common procedures include positron emission tomography (PET) and single-photon emission computed tomography (SPECT) (see Chapter 65.) Emerging research identifies EEG, MRI, and PET as important technologies in determining brain function through the evaluation of metabolic and electrical activity (Cruse et al., 2012). Laboratory tests include analysis of blood glucose, electrolytes, serum ammonia, and liver function tests; blood urea nitrogen (BUN) levels; serum osmolality; calcium level; and partial thromboplastin and prothrombin times. Other studies may be used to evaluate serum ketones, alcohol and drug concentrations, and arterial blood gases.

Medical Management

The first priority of treatment for the patient with altered LOC is to obtain and maintain a patent airway. The patient may be orally or nasally intubated, or a tracheostomy may be performed. Until the ability of the patient to breathe is determined, a mechanical ventilator is used to maintain adequate oxygenation and ventilation. The circulatory status (blood pressure, heart rate) is monitored to ensure adequate perfusion to the body and brain. An intravenous (IV) catheter is inserted to provide access for IV fluids and medications. Neurologic care focuses on the specific neurologic pathology, if known. Nutritional support, via a feeding tube or a gastrostomy tube, is initiated as soon as possible. In addition to measures designed to determine and treat the underlying causes of altered LOC, other medical interventions are aimed at pharmacologic management and prevention of complications.

NURSING PROCESS

The Patient With an Altered Level of Consciousness

Assessment

Assessment of the patient with an altered LOC often starts with assessing the verbal response through determining the patient's orientation to time, person, and place. Patients are asked to identify the day, date, or season of the year, as well as where they are or the clinicians, family members, or visitors present. Other questions such as "Who is the president?" or "What is the next holiday?" may be helpful in determining the patient's processing of information. (Verbal response cannot be evaluated if the patient is intubated or has a tracheostomy, and this should be clearly documented.)

Alertness is measured by the patient's ability to open the eyes spontaneously or in response to a vocal or noxious stimulus (pressure or pain). Patients with severe neurologic dysfunction cannot do this. The nurse assesses for periorbital edema (swelling around the eyes) or trauma, which may prevent the patient from opening the eyes, and documents any such condition that interferes with eye opening.

Motor response includes spontaneous, purposeful movement (e.g., the awake patient can move all four extremities with equal strength on command), movement only in response to painful stimuli, or abnormal posturing (Barlow, 2012). If the patient is not responding to commands, the motor response is tested by applying a painful stimulus (firm but gentle pressure) to the nail bed or by squeezing a muscle. If the patient attempts to push away or withdraw, the response is recorded as purposeful or appropriate ("Patient withdraws to painful stimulus"). This response is considered purposeful if the patient can cross the midline from one side of the body to the other in response to a painful stimulus. An inappropriate or nonpurposeful response is random and aimless. Posturing may be decorticate or decerebrate (Fig. 66-1; see also Chapter 65). The most severe neurologic impairment results in flaccidity. The motor response cannot be elicited or assessed when the patient has been administered pharmacologic paralyzing agents (i.e., neuromuscular blocking agents).

In addition to LOC, the nurse monitors parameters such as respiratory status, eye signs, and reflexes on an ongoing basis. Table 66-1 summarizes the assessment and the clinical significance of the findings. Body functions (circulation, respiration,

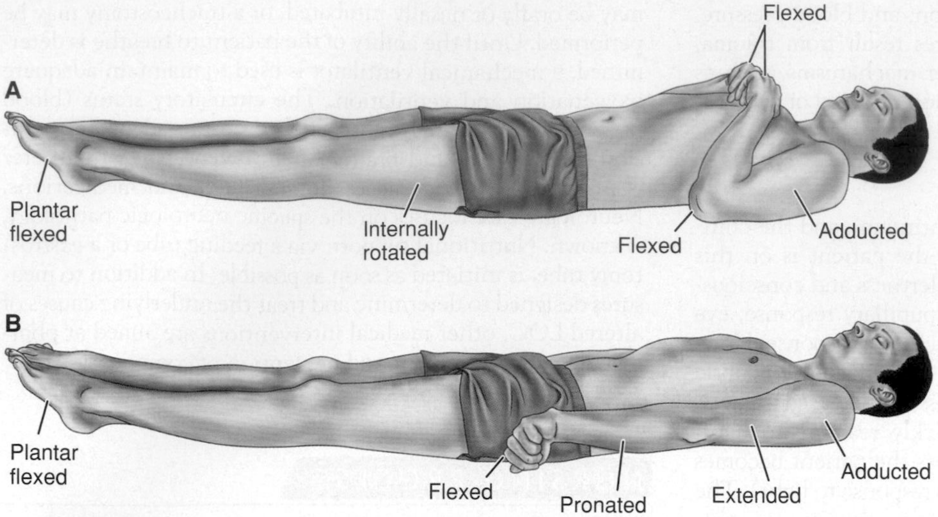
A
Flexed
Plantar flexed
Internally rotated
Flexed
Adducted
B
Plantar flexed
Flexed
Pronated
Extended
Adducted

FIGURE 66-1 • Abnormal posture response to stimuli. **A.** Decorticate posturing and flexion of the upper extremities, internal rotation of the lower extremities, and plantar flexion of the feet. **B.** Decerebrate posturing, involving extension and outward rotation of upper extremities and plantar flexion of the feet. Adapted from Posner, J. B., Saper, C. B., Schiff, N. D., et al. (2007). *Plum and Posner's diagnosis of stupor and coma* (4th ed.). Oxford, UK: Oxford University Press.

elimination, fluid and electrolyte balance) are examined in a systematic and ongoing manner.

Diagnosis

Nursing Diagnoses

Based on the assessment data, major nursing diagnoses may include the following:

- Ineffective airway clearance related to altered LOC
- Risk of injury related to decreased LOC
- Deficient fluid volume related to inability to take fluids by mouth
- Risk for imbalanced nutrition: less than body requirements related to inability to ingest nutrients to meet metabolic needs
- Impaired oral mucous membrane related to mouth breathing, absence of pharyngeal reflex, and altered fluid intake
- Risk for impaired skin integrity related to prolonged immobility
- Impaired tissue integrity of cornea related to diminished or absent corneal reflex
- Ineffective thermoregulation related to damage to hypothalamic center
- Impaired urinary elimination (incontinence or retention) related to impairment in neurologic sensing and control
- Bowel incontinence related to impairment in neurologic sensing and control and also related to changes in nutritional delivery methods
- Ineffective health maintenance related to neurologic impairment
- Interrupted family processes related to health crisis

Collaborative Problems/Potential Complications

Potential complications may include the following:

- Respiratory distress or failure
- Pneumonia
- Aspiration
- Pressure ulcer
- Venous thromboembolism (VTE)
- Contractures

Planning and Goals

The patient with altered LOC is subject to all of the complications associated with immobility. Therefore, the goals of care for the patient with altered LOC include maintenance of a clear airway, protection from injury, attainment of fluid volume balance, achievement of intact oral mucous membranes, maintenance of normal skin integrity, absence of corneal irritation, attainment of effective thermoregulation, and effective urinary elimination. Additional goals include bowel continence, restoration of health maintenance, maintenance of intact family or support system, and absence of complications.

Because the unconscious patient's protective reflexes are impaired, the quality of nursing care provided may mean the difference between life and death. The nurse must assume responsibility for the patient until the basic reflexes (coughing, blinking, and swallowing) return and the patient becomes conscious and oriented. Therefore, the major nursing goal is to compensate for the absence of these protective reflexes.

Nursing Interventions

Maintaining the Airway

The most important consideration in managing the patient with altered LOC is to establish an adequate airway and ensure ventilation. Obstruction of the airway is a risk because the epiglottis and tongue may relax, occluding the oropharynx, or the patient may aspirate vomitus or nasopharyngeal secretions.

The accumulation of secretions in the pharynx presents a serious problem. Because the patient cannot swallow and lacks pharyngeal reflexes, these secretions must be removed to eliminate the danger of aspiration. Elevating the head of the bed to 30 degrees helps prevent aspiration. Positioning the patient in a lateral or semiprone position also helps because it allows the jaw and tongue to fall forward, thus promoting drainage of secretions.

Positioning alone is not always adequate, however. Suctioning and oral hygiene may be required. Suctioning is performed to remove secretions from the posterior pharynx and upper trachea. Before and after suctioning, the patient is adequately ventilated to prevent hypoxia (Hickey, 2009). Chest physiotherapy and postural drainage may be initiated to promote pulmonary hygiene, unless contraindicated by

TABLE 66-1 Nursing Assessment of the Unconscious Patient

Examination	Clinical Assessment	Clinical Significance
Level of responsiveness or consciousness	Eye opening; verbal and motor responses; pupils (size, equality, reaction to light)	Obeying commands is a favorable response and demonstrates a return to consciousness.
Pattern of respiration	Respiratory pattern	Disturbances of respiratory center of brain may result in various respiratory patterns.
	Cheyne-Stokes respiration	Suggests lesions deep in both hemispheres; area of basal ganglia and upper brain stem
	Hyperventilation	Suggests onset of metabolic problem or brain stem damage
	Ataxic respiration with irregularity in depth/rate	Ominous sign of damage to medullary center
Eyes		
Pupils (size, equality, reaction to light)	Equal, normally reactive pupils	Suggests that coma is toxic or metabolic in origin
	Equal or unequal diameter	Helps determine location of lesion
	Progressive dilation	Indicates increasing intracranial pressure
	Fixed dilated pupils	Indicates injury at level of midbrain
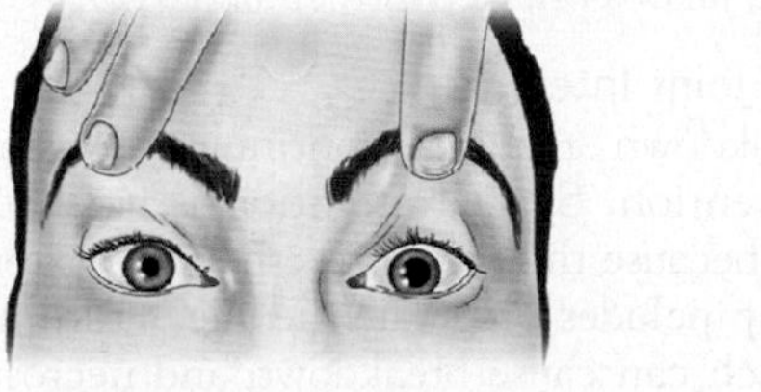		
Eye movements	Normally, eyes should move from side to side.	Functional and structural integrity of brain stem is assessed by inspection of extraocular movements; usually absent in deep coma.
Corneal reflex	When cornea is touched with a wisp of clean cotton, blink response is normal.	Tests cranial nerves V and VII; helps determine location of lesion if unilateral; absent in deep coma
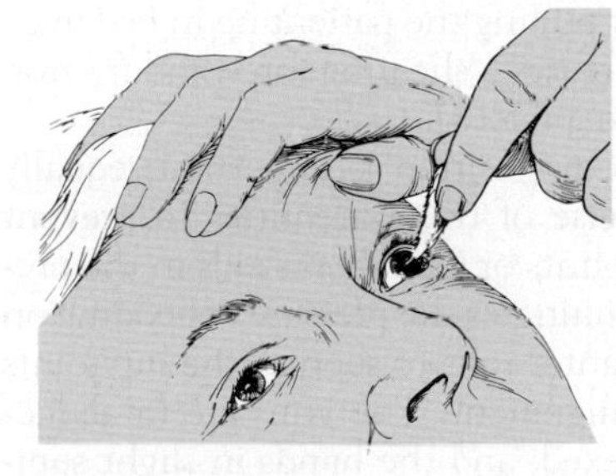		
Facial symmetry	Asymmetry (sagging, decrease in wrinkles)	Sign of paralysis
Swallowing reflex	Drooling versus spontaneous swallowing	Absent in coma
		Paralysis of cranial nerves X and XII
Neck	Stiff neck	Subarachnoid hemorrhage, meningitis
	Absence of spontaneous neck movement	Fracture or dislocation of cervical spine
Response of extremity to noxious stimuli	Firm pressure on a joint of the upper and lower extremity	Asymmetric response in paralysis
		Absent in deep coma
	Observe spontaneous movements.	
Deep tendon reflexes	Tap patellar and biceps tendons.	Brisk response may have localizing value.
		Asymmetric response in paralysis
		Absent in deep coma
Pathologic reflexes	Firm pressure with blunt object on sole of foot, moving along lateral margin and crossing to the ball of foot	Flexion of the toes, especially the great toe, is normal except in newborn.
		Dorsiflexion of toes (especially great toe) indicates contralateral pathology of corticospinal tract (Babinski reflex).
		Helps determine location of lesion in brain
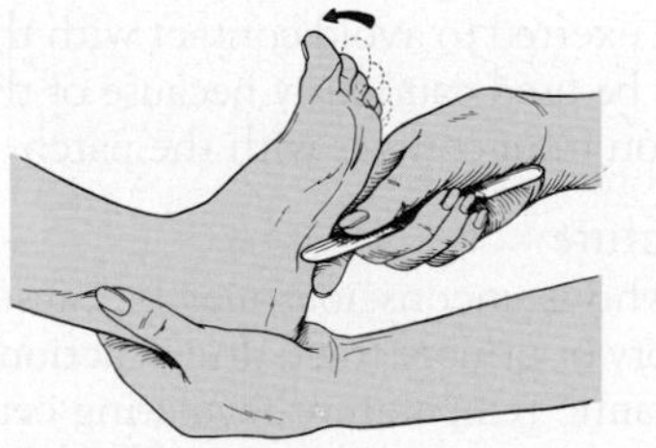		
Abnormal posture	Observation for posturing (spontaneous or in response to noxious stimuli)	Deep extensive brain lesion
	Flaccidity with absence of motor response	Seen with cerebral hemisphere pathology and in metabolic depression of brain function
	Decorticate posture (flexion and internal rotation of forearms and hands)	Decerebrate posturing indicates deeper and more severe dysfunction than does decorticate posturing; implies brain pathology; poor prognostic sign.
	Decerebrate posture (extension and external rotation)	

the patient's underlying condition. The chest should be auscultated at least every 8 hours to detect adventitious breath sounds or absence of breath sounds.

Despite these measures, or because of the severity of impairment, the patient with altered LOC often requires intubation and mechanical ventilation. Nursing actions for the mechanically ventilated patient include maintaining the patency of the endotracheal tube or tracheostomy, providing frequent oral care, monitoring arterial blood gas measurements, and maintaining ventilator settings (see Chapter 21).

Protecting the Patient

For the protection of the patient, side rails are padded. Two rails are kept in the raised position during the day and three at night; however, raising all four side rails is considered a restraint by the Joint Commission if the intent is to limit the patient's mobility. Care should be taken to prevent injury from invasive lines and equipment, and other potential sources of injury should be identified, such as restraints, tight dressings, environmental irritants, damp bedding or dressings, and tubes and drains.

Protection also includes ensuring the patient's dignity during altered LOC. Simple measures such as providing privacy and speaking to the patient during nursing care activities preserve the patient's dignity. Not speaking negatively about the patient's condition or prognosis is also important, because patients in a light coma may be able to hear. The comatose patient has an increased need for advocacy, and the nurse is responsible for seeing that these advocacy needs are met.

Quality and Safety Nursing Alert

If the patient begins to emerge from unconsciousness, every measure that is available and appropriate for calming and quieting the patient should be used. Any form of restraint is likely to be countered with resistance, leading to self-injury or to a dangerous increase in ICP. Therefore, physical restraints are avoided if possible; a written prescription must be obtained if their use is essential for the patient's well-being.

Maintaining Fluid Balance and Managing Nutritional Needs

Hydration status is assessed by examining tissue turgor and mucous membranes, assessing intake and output trends, and analyzing laboratory data. Fluid needs are met initially by administering the required IV fluids. However, IV solutions (and blood component therapy) for patients with intracranial conditions must be administered slowly. If they are administered too rapidly, they can increase ICP. The quantity of fluids administered may be restricted to minimize the possibility of cerebral edema.

If the patient does not recover quickly and sufficiently enough to take adequate fluids and calories by mouth, a feeding or gastrostomy tube will be inserted for the administration of fluids and enteral feedings. Research suggests that patients fed within 48 hours of injury have improved outcomes over those in whom nutrition is delayed (Chiang, Chao, Chu, et al., 2012).

Providing Mouth Care

The mouth is inspected for dryness, inflammation, and crusting. The unconscious patient requires careful oral care, because there is a risk of parotitis if the mouth is not kept scrupulously clean. The mouth is cleansed and rinsed carefully to remove secretions and crusts and to keep the mucous membranes moist. A thin coating of petrolatum on the lips prevents drying, cracking, and encrustations. If the patient has an endotracheal tube, the tube should be moved to the opposite side of the mouth daily to prevent ulceration of the mouth and lips. If the patient is intubated and mechanically ventilated, good oral care is also necessary. Research suggests that comprehensive mouth care with antiseptic such as chlorhexidine (Hunter, 2012) or using a tongue scraper, an electrical toothbrush, and pharmacologic moisturizers decreases ventilator-associated pneumonia and improves the oral health in patients who are intubated (Prendergast, Jakobsson, Renvert, et al., 2012).

Maintaining Skin and Joint Integrity

Preventing skin breakdown requires continuing nursing assessment and intervention. Special attention is given to unconscious patients, because they cannot respond to external stimuli. Assessment includes a regular schedule of turning to avoid pressure, which can cause breakdown and necrosis of the skin. Turning also provides kinesthetic (sensation of movement), proprioceptive (awareness of position), and vestibular (equilibrium) stimulation. After turning, the patient is carefully repositioned to prevent ischemic necrosis over pressure areas. Dragging or pulling the patient up in bed must be avoided, because this creates a shearing force and friction on the skin surface (see Chapter 10).

Maintaining correct body position is important; equally important is passive exercise of the extremities to prevent contractures. The use of splints or foam boots aids in the prevention of footdrop and eliminates the pressure of bedding on the toes. The use of trochanter rolls to support the hip joints keeps the legs in proper alignment. The arms are in abduction, the fingers lightly flexed, and the hands in slight supination. The heels of the feet are assessed for pressure areas. Specialty beds, such as fluidized or low-air-loss beds, may be used to decrease pressure on bony prominences (Hickey, 2009).

Preserving Corneal Integrity

Some patients who are unconscious have their eyes open and have inadequate or absent corneal reflexes. The cornea may become irritated, dry, or scratched, leading to ulceration. The eyes may be cleansed with cotton balls moistened with sterile normal saline to remove debris and discharge. If artificial tears are prescribed, they may be instilled every 2 hours. Periorbital edema often occurs after cranial surgery. If cold compresses are prescribed, care must be exerted to avoid contact with the cornea. Eye patches should be used cautiously because of the potential for corneal abrasion from contact with the patch.

Maintaining Body Temperature

High fever in the patient who is unconscious may be caused by infection of the respiratory or urinary tract, drug reactions, or damage to the hypothalamic temperature-regulating center. A slight elevation of temperature may be caused by dehydration. The environment can be adjusted, depending on the patient's condition, to promote a normal body temperature. If body temperature is elevated, a minimum amount of bedding is used. The room may be cooled to 18.3°C (65°F). However, if the patient is an older adult and does not have an elevated temperature, a warmer environment is needed.

Because of damage to the temperature-regulating center in the brain or severe intracranial infection, patients who are unconscious often develop very high temperatures. Such temperature elevations must be controlled, because the increased metabolic demands of the brain can exceed cerebral circulation and oxygen delivery, potentially resulting in cerebral deterioration (Hickey, 2009). Studies suggest that hyperthermia may contribute to poor outcome after brain injury but not through a decreased brain oxygen level (Spiotta, Stiefel, Heuer, et al., 2008). Persistent hyperthermia with no identified clinical source of infection indicates brain stem damage and a poor prognosis.

Quality and Safety Nursing Alert

The body temperature of a patient who is unconscious is never taken by mouth. Rectal or tympanic (if not contraindicated) temperature measurement is preferred to the less accurate axillary temperature.

Strategies for reducing fever include:

- Removing all bedding over the patient (with the possible exception of a light sheet, towel, or small drape)
- Administering acetaminophen as prescribed
- Giving cool sponge baths and allowing an electric fan to blow over the patient to increase surface cooling
- Using a hypothermia blanket
- Frequent temperature monitoring to assess the patient's response to the therapy and to prevent an excessive decrease in temperature and shivering

Preventing Urinary Retention

The patient with an altered LOC is often incontinent or has urinary retention. The bladder is palpated or scanned at intervals to determine whether urinary retention is present, because a full bladder may be an overlooked cause of overflow incontinence. A portable bladder ultrasound instrument is a useful tool in bladder management and retraining programs (Newman & Willson, 2011).

If the patient is not voiding, a program of intermittent catheterization should be devised in order to reduce the patient's risk of urinary tract infection. A catheter may be inserted during the acute phase of illness to monitor urinary output. Because catheters are a major cause of urinary tract infection, the patient is observed for fever and cloudy urine. The area around the urethral orifice is inspected for drainage and cleansed routinely. The urinary catheter is usually removed if the patient has a stable cardiovascular system and if no diuresis, sepsis, or voiding dysfunction existed before the onset of coma. Although many patients who are unconscious urinate spontaneously after catheter removal, the bladder should be palpated or scanned with a portable ultrasound device periodically for urinary retention (Newman & Willson, 2011) (see Fig. 53-8 in Chapter 53).

An external catheter (condom catheter) for the male patient and absorbent pads for the female patient can be used for patients who are unconscious and can urinate spontaneously, although involuntarily. As soon as consciousness is regained, a bladder training program is initiated (Hickey, 2009). The incontinent patient is monitored frequently for skin irritation and skin breakdown. Appropriate skin care is implemented to prevent these complications.

Promoting Bowel Function

The abdomen is assessed for distention by listening for bowel sounds and measuring the girth of the abdomen with a tape measure. There is a risk of diarrhea from infection, antibiotic agents, and hyperosmolar fluids. Frequent loose stools may also occur with fecal impaction. Commercial fecal collection bags are available for patients with fecal incontinence.

Immobility and lack of dietary fiber can cause constipation. The nurse monitors the number and consistency of bowel movements and performs a rectal examination for signs of fecal impaction. Stool softeners may be prescribed and can be administered with tube feedings. To facilitate bowel emptying, a glycerin suppository may be indicated. The patient may require an enema every other day to empty the lower colon.

Restoring Health Maintenance

Once increased ICP is not a problem, the nurse assists the patient and family to overcome the perceptual impairment of the unconscious patient. This involves using auditory, visual, olfactory, gustatory, tactile, and kinesthetic activities to stimulate the patient emerging from coma (Meyer, Megyesi, Meythaler, et al., 2010). Efforts are made to restore the sense of daily rhythm by maintaining usual day and night patterns for activity and sleep. The nurse touches and talks to the patient and encourages family members and friends to do so. Communication is extremely important and includes touching the patient and spending enough time with the patient to become sensitive to his or her needs. It is also important to avoid making any negative comments about the patient's status or prognosis in the patient's presence.

The nurse orients the patient to time and place at least once every 8 hours. Sounds from the patient's usual environment may be introduced using a tape recorder. Family members can read to the patient from a favorite book and may suggest radio and television programs that the patient previously enjoyed as a means of enriching the environment and providing familiar input.

When arousing from coma, many patients experience a period of agitation, indicating that they are becoming more aware of their surroundings but still cannot react or communicate in an appropriate fashion. Although this is disturbing for many family members, it is actually a positive clinical sign. At this time, it is necessary to minimize stimulation by limiting background noises, having only one person speak to the patient at a time, giving the patient a longer period of time to respond, and allowing for frequent rest or quiet times. After the patient has regained consciousness, videotaped family or social events may assist the patient in recognizing family and friends and allow him or her to experience missed events.

Programs of sensory stimulation for patients with brain injury have been developed in an effort to improve outcomes. Although these are controversial programs with inconsistent results, some research supports the concept of providing structured stimulation (Meyer et al., 2010).

Meeting the Family's Needs

The family of the patient with altered LOC may be thrown into a sudden state of crisis and go through the process of severe anxiety, denial, anger, remorse, grief, and reconciliation. Depending on the disorder that caused the altered LOC and the extent of the patient's recovery, the family may be unprepared for the changes in the cognitive and physical status of their loved one. If the patient has significant residual deficits, the family may require considerable time, assistance, and support to come to terms with these changes. To help family members mobilize resources and coping skills, the nurse reinforces and clarifies information about the patient's condition, encourages the family to be involved in care, and listens to and encourages ventilation of feelings and concerns while supporting decision making about management and placement after hospitalization. Families may benefit from participation in support groups offered through the hospital, rehabilitation facility, or community organizations.

The family may need to face the death of their loved one. The patient with a neurologic disorder is often pronounced brain dead before the heart stops beating. The term **brain death** describes irreversible loss of all functions of the entire brain, including the brain stem (Iltis & Cherry, 2010). The term may be misleading to the family because, although brain function has ceased, the patient appears to be alive, with the heart rate and blood pressure sustained by vasoactive medications and breathing continued by mechanical ventilation. When discussing a patient who is brain dead with family members, it is important to provide accurate, timely, understandable, and consistent information. (See Chapter 16 for discussion of end-of-life care.)

Monitoring and Managing Potential Complications

Pneumonia, aspiration, and respiratory failure are potential complications in any patient who has a depressed LOC and who cannot protect the airway or turn, cough, and take deep breaths. The longer the period of unconsciousness, the greater the risk of pulmonary complications.

Vital signs and respiratory function are monitored closely to detect any signs of respiratory failure or distress. Complete blood count and arterial blood gas measurements are assessed to determine whether there are adequate red blood cells to carry oxygen and whether ventilation is effective. Chest physiotherapy and suctioning are initiated to prevent respiratory complications such as pneumonia. Oral care interventions are performed for patients receiving mechanical ventilation to maintain oral health and decrease the incidence of pneumonia (Hunter, 2012; Prendergast, Hallberg, Jahnke, et al., 2009). If pneumonia develops, cultures are obtained to identify the organism so that appropriate antibiotic agents can be administered.

The patient with altered LOC is monitored closely for evidence of impaired skin integrity, and strategies to prevent skin breakdown and pressure ulcers are continued through all phases of care, including hospitalization, rehabilitation, and home care. Factors that contribute to impaired skin integrity (e.g., incontinence, inadequate dietary intake, pressure on bony prominences, edema) are addressed. If pressure ulcers develop, strategies to promote healing are undertaken. Care is taken to prevent bacterial contamination of pressure ulcers, which may lead to sepsis and septic shock. (See Chapter 10 for assessment and management of pressure ulcers.)

The patient should also be monitored for signs and symptoms of VTE, which may manifest as a deep vein thrombosis (DVT) or pulmonary embolism (PE). Prophylaxis with subcutaneous heparin or low-molecular-weight heparin (dalteparin [Fragmin], danaparoid [Orgaran]) as well as anti-embolism stockings or pneumatic compression devices are prescribed according to the patient's risk factors for thrombosis and bleeding (Kahn, Lim, Dunn, et al., 2012). The nurse observes for signs and symptoms of DVT or PE.

Patients with a prolonged decrease in LOC are at risk for developing contractures. During acute care, the patient is turned every 2 hours and passive range of motion performed at least twice a day (Bader & Littlejohns, 2010). Splints, provided by occupational therapy, are applied to the hands and feet in a rotating manner to maintain functional joint alignment. (See Chapter 10 for further information about management of contractures.)

Evaluation

Expected patient outcomes may include:

1. Maintains clear airway and demonstrates appropriate breath sounds
2. Experiences no injuries
3. Attains or maintains adequate fluid balance and nutritional status
   a. Has no clinical signs or symptoms of dehydration
   b. Demonstrates normal range of serum electrolytes
   c. Has no clinical signs or symptoms of overhydration or malnutrition
4. Achieves healthy oral mucous membranes
5. Maintains normal skin integrity
6. Has no corneal irritation
7. Attains or maintains thermoregulation
8. Has no urinary retention
9. Has no diarrhea or fecal impaction
10. Receives appropriate sensory stimulation
11. Has family members who cope with crisis
    a. Verbalize fears and concerns
    b. Participate in patient's care and provide sensory stimulation by talking and touching
12. Is free of complications
    a. Has arterial blood gas values or oxygen saturation levels within normal range
    b. Displays no signs or symptoms of pneumonia
    c. Exhibits intact skin over pressure areas
    d. Does not develop VTE such as DVT or PE

INCREASED INTRACRANIAL PRESSURE

The rigid cranial vault contains brain tissue (1,400 g), blood (75 mL), and CSF (75 mL). The volume and pressure of these three components are usually in a state of equilibrium and produce the ICP. ICP is usually measured in the lateral ventricles, with the normal pressure being 0 to 10 mm Hg, and 15 mm Hg being the upper limit of normal (Hickey, 2009).

The **Monro-Kellie hypothesis**, also known as the Monro-Kellie doctrine, explains the dynamic equilibrium of cranial contents. The hypothesis states that because of the limited space for expansion within the skull, an increase in any one of the components causes a change in the volume of the others. Because brain tissue has limited space to expand, compensation typically is accomplished by displacing or shifting CSF, increasing the absorption or diminishing the production of CSF, or decreasing cerebral blood volume. Without such changes, ICP begins to rise. Under normal circumstances, minor changes in blood volume and CSF volume occur constantly as a result of alterations in intrathoracic pressure (coughing, sneezing, straining), posture, blood pressure, and systemic oxygen and carbon dioxide levels (Bader & Littlejohns, 2010; Hickey, 2009).

Concept Mastery Alert

Nurses caring for patients with neurologic conditions need a clear understanding of the Monro-Kellie hypothesis, which states that because of limited space for expansion within the skull, an increase in the volume of either brain tissue, blood, or CSF causes a change in the volume of the other components.

Pathophysiology

Increased ICP affects many patients with acute neurologic conditions because pathologic conditions alter the relationship between intracranial volume and ICP. Although elevated ICP is most commonly associated with head injury, it also may be seen as a secondary effect in other conditions, such as brain tumors, subarachnoid hemorrhage, and toxic and viral encephalopathies. Increased ICP from any cause decreases cerebral perfusion, stimulates further swelling (edema), and may shift brain tissue, resulting in **herniation**—a dire and frequently fatal event.

Decreased Cerebral Blood Flow

Increased ICP may reduce cerebral blood flow, resulting in ischemia and cell death. In the early stages of cerebral ischemia, the vasomotor centers are stimulated and the systemic pressure rises to maintain cerebral blood flow. Usually, this is accompanied by a slow bounding pulse and respiratory irregularities. These changes in blood pressure, pulse, and respiration are important clinically because they suggest increased ICP.

The concentration of carbon dioxide in the blood and in the brain tissue also plays a role in the regulation of cerebral blood flow. An increase in the partial pressure of arterial carbon dioxide ($PaCO_2$) causes cerebral vasodilation, leading to increased cerebral blood flow and increased ICP. A decrease in $PaCO_2$ has a vasoconstrictive effect, limiting blood flow to the brain. Decreased venous outflow may also increase cerebral blood volume, thus raising ICP.

Cerebral Edema

Cerebral edema or swelling is defined as an abnormal accumulation of water or fluid in the intracellular space, extracellular space, or both, associated with an increase in the volume of brain tissue. Edema can occur in the gray, white, or interstitial matter. As brain tissue swells within the rigid skull, several mechanisms attempt to compensate for the increasing ICP. These compensatory mechanisms include autoregulation as well as decreased production and flow of CSF. **Autoregulation** refers to the brain's ability to change the diameter of its blood vessels to maintain a constant cerebral blood flow during alterations in systemic blood pressure. This mechanism can be impaired in patients who are experiencing a pathologic and sustained increase in ICP.

Cerebral Response to Increased Intracranial Pressure

As ICP rises, compensatory mechanisms in the brain work to maintain blood flow and prevent tissue damage. The brain can maintain a steady perfusion pressure if the arterial systolic blood pressure is 50 to 150 mm Hg and the ICP is less than 40 mm Hg. Changes in ICP are closely linked with cerebral perfusion pressure (CPP). The CPP is calculated by subtracting the ICP from the mean arterial pressure (MAP). For example, if the MAP is 100 mm Hg and the ICP is 15 mm Hg, then the CPP is 85 mm Hg. The normal CPP is 70 to 100 mm Hg (Hickey, 2009). As ICP rises and the autoregulatory mechanism of the brain is overwhelmed, the CPP can increase to greater than 100 mm Hg or decrease to less than 50 mm Hg. Patients with a CPP of less than 50 mm Hg experience irreversible neurologic damage. Therefore, the CPP must be maintained at 70 to 80 mm Hg to ensure adequate blood flow to the brain. If ICP is equal to MAP, cerebral circulation ceases.

A clinical phenomenon known as the **Cushing's response** (or Cushing's reflex) is seen when cerebral blood flow decreases significantly. When ischemic, the vasomotor center triggers an increase in arterial pressure in an effort to overcome the increased ICP. A sympathetically mediated response causes an increase in the systolic blood pressure with a widening of the pulse pressure and cardiac slowing. This response is seen clinically as an increase in systolic blood pressure, widening of the pulse pressure, and reflex slowing of the heart rate. It is a late sign requiring immediate intervention; however, perfusion may be recoverable if the Cushing's response is treated rapidly.

At a certain point, the brain's ability to autoregulate becomes ineffective and decompensation (ischemia and infarction) begins. When this occurs, the patient exhibits significant changes in mental status and vital signs. The bradycardia, hypertension, and bradypnea associated with this deterioration are known as **Cushing's triad**, which is a grave sign. At this point, herniation of the brain stem and occlusion of the cerebral blood flow occur if therapeutic intervention is not initiated. Herniation refers to the shifting of brain tissue from an area of high pressure to an area of lower pressure (Fig. 66-2). The herniated tissue exerts pressure on the brain area into which it has shifted, which interferes with the blood supply in that area. Cessation of cerebral blood flow results in cerebral ischemia, infarction, and brain death.

Clinical Manifestations

If ICP increases to the point at which the brain's ability to adjust has reached its limits, neural function is impaired; this may be manifested at first by clinical changes in LOC and later by abnormal respiratory and vasomotor responses.

Quality and Safety Nursing Alert

The earliest sign of increasing ICP is a change in LOC. Agitation, slowing of speech, and delay in response to verbal suggestions are other early indicators.

Any sudden change in the patient's condition, such as restlessness (without apparent cause), confusion, or increasing drowsiness, has neurologic significance. These signs may

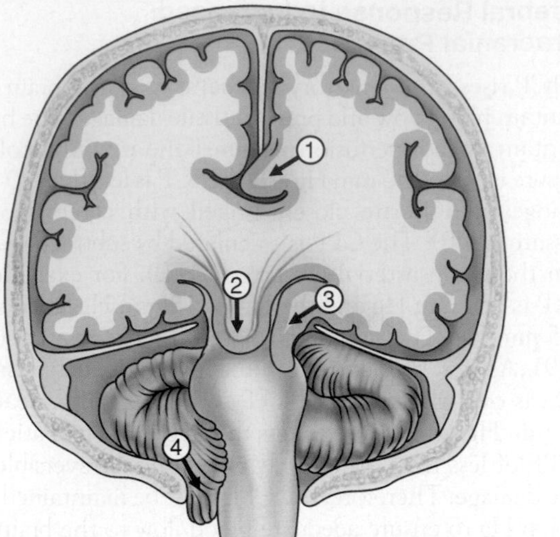
1
2
3
4
FIGURE 66-2 • Brain with intracranial shifts from supratentorial lesions. **1,** Herniation of the cingulate gyrus under the falx cerebri. **2,** Central transtentorial herniation. **3,** Uncal herniation of the temporal lobe into the tentorial notch. **4,** Infratentorial herniation of the cerebral tonsils. Adapted from Porth, C. M., & Matfin, G. (2009). *Pathophysiology: Concepts of altered health states* (8th ed.). Philadelphia: Lippincott Williams & Wilkins.

result from compression of the brain due to swelling from hemorrhage or edema, an expanding intracranial lesion (hematoma or tumor), or a combination of both.

As ICP increases, the patient becomes stuporous, reacting only to loud or painful stimuli. At this stage, serious impairment of brain circulation is probably taking place, and immediate intervention is required. As neurologic function deteriorates further, the patient becomes comatose and exhibits abnormal motor responses in the form of **decortication** (abnormal flexion of the upper extremities and extension of the lower extremities), **decerebration** (extreme extension of the upper and lower extremities), or flaccidity (see Fig. 66-1). If the coma is profound, with the pupils dilated and fixed and respirations impaired or absent, death is usually inevitable (Barlow, 2012).

Assessment and Diagnostic Findings

The diagnostic studies used to determine the underlying cause of increased ICP are discussed in detail in Chapter 65. The most common diagnostic tests are CT scanning and MRI. The patient may also undergo cerebral angiography, PET, or SPECT. Transcranial Doppler studies provide information about cerebral blood flow. The patient with increased ICP may also undergo electrophysiologic monitoring to observe cerebral blood flow indirectly. Evoked potential monitoring measures the electrical potentials produced by nerve tissue in response to external stimulation (auditory, visual, or sensory). Lumbar puncture is avoided in patients with increased ICP, because the sudden release of pressure in the lumbar area can cause the brain to herniate (AANN, 2011). (See Chapter 65 for further discussion of lumbar puncture and other diagnostic tests.)

Complications

Complications of increased ICP include brain stem herniation, diabetes insipidus, and syndrome of inappropriate antidiuretic hormone (SIADH).

Brain stem herniation results from an excessive increase in ICP in which the pressure builds in the cranial vault and the brain tissue presses down on the brain stem. This increasing pressure on the brain stem results in cessation of blood flow to the brain, leading to irreversible brain anoxia and brain death.

Diabetes insipidus is the result of decreased secretion of antidiuretic hormone (ADH). The patient has excessive urine output, decreased urine osmolality, and serum hyperosmolarity (Porth & Matfin, 2009). Therapy consists of administration of fluids, electrolyte replacement, and vasopressin (desmopressin [DDAVP]) therapy. (See Chapters 13 and 52 for a discussion of diabetes insipidus.)

SIADH is the result of increased secretion of ADH. The patient becomes volume overloaded, urine output diminishes, and serum sodium concentration becomes dilute. Treatment of SIADH includes fluid restriction (less than 800 mL/day with no free water), which is usually sufficient to correct the hyponatremia. In severe cases, careful administration of a 3% hypertonic saline solution may be therapeutic (Vaidya, Ho, & Freda, 2010). The change in serum sodium concentration should not exceed a correction rate of approximately 1.3 mEq/L/h. (See Chapters 13 and 52 for further discussion of SIADH.)

Medical Management

Increased ICP is a true emergency and must be treated promptly. Invasive monitoring of ICP is an important component of management. Immediate management to relieve increased ICP requires decreasing cerebral edema, lowering the volume of CSF, or decreasing cerebral blood volume while maintaining cerebral perfusion. These goals are accomplished by administering osmotic diuretics, restricting fluids, draining CSF, controlling fever, maintaining systemic blood pressure and oxygenation, and reducing cellular metabolic demands. (See Chapter 68 for a discussion of the management of increased ICP.)

Monitoring Intracranial Pressure and Cerebral Oxygenation

The purposes of ICP monitoring are to identify increased pressure early in its course (before cerebral damage occurs), to quantify the degree of elevation, to initiate appropriate treatment, to provide access to CSF for sampling and drainage, and to evaluate the effectiveness of treatment. ICP can be monitored with the use of an intraventricular catheter (ventriculostomy), a subarachnoid bolt, an epidural or subdural catheter, or a fiberoptic transducer-tipped catheter placed in the subdural space or in the ventricle (Fig. 66-3).

When a **ventriculostomy** or intraventricular catheter monitoring device is used for monitoring ICP, a fine-bore catheter is inserted into a lateral ventricle, preferably in the nondominant hemisphere of the brain (Hickey, 2009). The catheter is connected by a fluid-filled system to a transducer, which records the pressure in the form of an electrical impulse. In addition to obtaining continuous ICP recordings, the ventricular catheter allows CSF to drain, particularly during acute

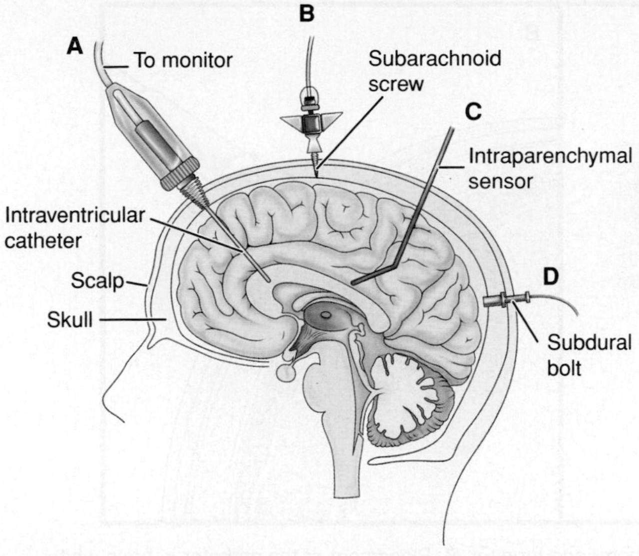
A
To monitor
B
Subarachnoid
screw
C
Intraparenchymal
sensor
Intraventricular
catheter
Scalp
Skull
D
Subdural
bolt

FIGURE 66-3 • Intracranial pressure monitoring. A device may be placed in the ventricle (**A**), the subarachnoid space (**B**), the intraparenchymal space (**C**), or the subdural space (**D**).

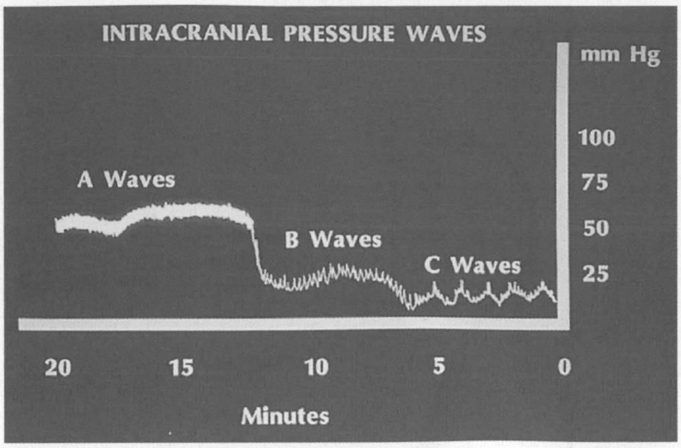
INTRACRANIAL PRESSURE WAVES
mm Hg
100
75
50
25
A Waves
B Waves
C Waves
20
15
10
5
0
Minutes

FIGURE 66-4 • Intracranial pressure waves. Composite diagram of A (plateau) waves, which indicate cerebral ischemia; B waves, which indicate intracranial hypertension and variations in the respiratory cycle; and C waves, which relate to variations in systemic arterial pressure and respirations.

increases in pressure. The ventriculostomy can also be used to drain blood from the ventricle. Continuous drainage of CSF under pressure control is an effective method of treating intracranial hypertension. Another advantage of a ventricular catheter is access for the intraventricular administration of medications and the occasional instillation of air or a contrast agent for ventriculography. Complications associated with its use include infection, meningitis, ventricular collapse, occlusion of the catheter by brain tissue or blood, and problems with the monitoring system.

The **subarachnoid screw or bolt** is a hollow device that is inserted through the skull and dura mater into the cranial subarachnoid space (Hickey, 2009). It has the advantage of not requiring a ventricular puncture. The subarachnoid screw is attached to a pressure transducer, and the output is recorded on an oscilloscope. The hollow screw technique also has the advantage of avoiding complications from brain shift and small ventricle size. Complications include infection and blockage of the screw by clot or brain tissue, which leads to a loss of pressure tracing and a decrease in accuracy at high ICP readings.

An **epidural monitor** uses a pneumatic flow sensor to detect ICP. The epidural ICP monitoring system has a low incidence of infection and complications and appears to read pressures accurately. Calibration of the system is maintained automatically, and abnormal pressure waves trigger an alarm system. One disadvantage of the epidural catheter is the inability to withdraw CSF for analysis.

A **fiberoptic monitor**, or transducer-tipped catheter, is an alternative to other intraventricular, subarachnoid, and subdural systems (Sandsmark, Kumar, Park, et al., 2012). The miniature transducer reflects pressure changes, which are converted to electrical signals in an amplifier and displayed on a digital monitor. The catheter can be inserted into the ventricle, subarachnoid space, subdural space, or brain parenchyma or under a bone flap. If inserted into the ventricle, it can also be used in conjunction with a CSF drainage device.

Interpreting Intracranial Pressure Waveforms

Waves of high pressure and troughs of relatively normal pressure indicate changes in ICP. Waveforms are captured and recorded on an oscilloscope. These waves have been classified as A waves (plateau waves), B waves, and C waves (Fig. 66-4). The plateau waves (A waves) are transient, paroxysmal, recurring elevations of ICP that may last 5 to 20 minutes and range in amplitude from 50 to 100 mm Hg (AANN, 2011). Plateau waves have clinical significance and indicate changes in vascular volume within the intracranial compartment that are beginning to compromise cerebral perfusion. The A waves may increase in amplitude and frequency, reflecting cerebral ischemia and brain damage that can occur before overt signs and symptoms of raised ICP are seen clinically. B waves are shorter (30 seconds to 2 minutes) and have smaller amplitude (up to 50 mm Hg). They have less clinical significance, but if seen in a series in a patient with depressed consciousness, they may precede the appearance of A waves. B waves may be seen in patients with intracranial hypertension and decreased intracranial compliance. C waves are small, rhythmic oscillations with frequencies of approximately six per minute. They appear to be related to rhythmic variations of the systemic arterial blood pressure and respirations. The clinical significance of C waves is unknown (Littlejohns & Bader, 2009).

Other Neurologic Monitoring Systems

Additional trends in neurologic monitoring include **microdialysis** of the patient with a brain injury (Sandsmark et al., 2012). Cortical probes are placed near the injured area and are used to measure levels of glutamate, lactate, pyruvate, and glucose, substances that reflect the metabolic function of the brain. Some researchers theorize that direct measurements of glucose and energy by-products in the brain will lead to better management of these patients and, ultimately, to improved outcomes.

An additional trend is monitoring of cerebral oxygenation through monitoring of the oxygen saturation in the jugular venous bulb ($SjvO_2$) or via a catheter in the brain. Cerebral oxygenation is thought to be important because changes in cerebral perfusion may reflect an increase in ICP. Readings

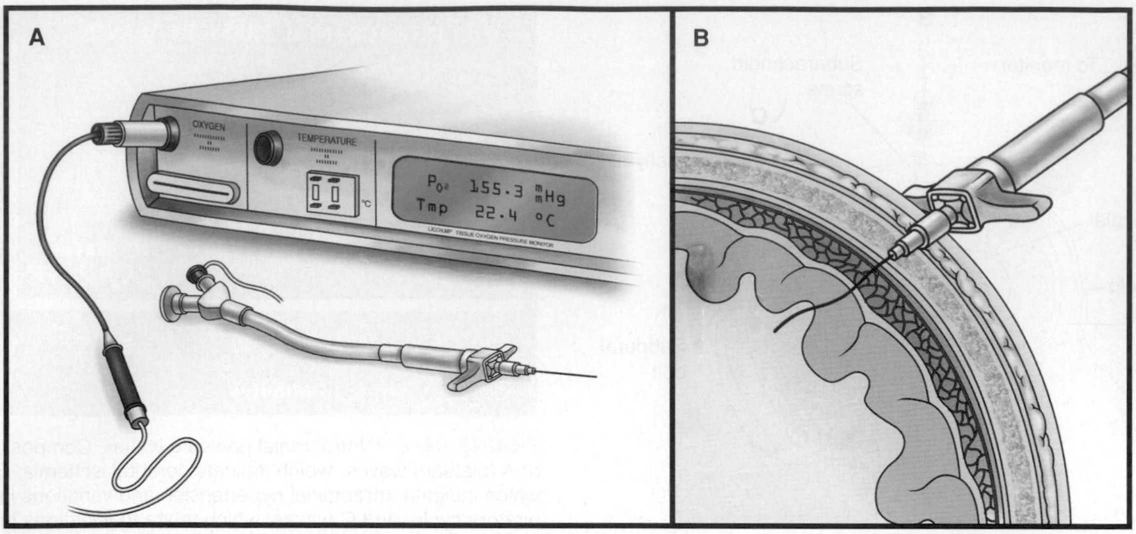
A
OXYGEN
TEMPERATURE
Po2 155.3 mmHg
Tmp 22.4 °C
LICOX MP TISSUE OXYGEN PRESSURE MONITOR
B

FIGURE 66-5 • Licox catheter system. **A.** The brain tissue oxygen catheter and monitor. **B.** Placement of the catheter in brain white matter. Redrawn with permission of Integra NeuroSciences, Plainsboro, NJ.

taken from a catheter residing in the jugular outflow tract allow for a comparison of arterial and venous oxygen saturation, and the balance of cerebral oxygen supply and demand is demonstrated. Venous jugular desaturations can reflect early cerebral ischemia, alerting the clinician before an increase in ICP occurs. Minimizing cerebral desaturations can potentially improve outcomes (Sandsmark et al., 2012). This type of monitoring is now widely available and has been successfully used to identify secondary brain insults. A limiting factor is that this saturation reflects overall perfusion of the brain rather than that of a specific injured area (Lescot, Abdennour, Boch, et al., 2008).

Another method of measuring cerebral oxygenation and temperature is by inserting a fiberoptic catheter into the brain matter (Sandsmark et al., 2012). The most common system is Licox (manufactured by Integra NeuroSciences, Plainsboro, NJ; Fig. 66-5). The system includes a monitor with a screen for the display of oxygen and temperature values and cables that connect to the monitoring probes in the brain (Hickey, 2009).

Decreasing Cerebral Edema

Osmotic diuretics such as mannitol may be administered to dehydrate the brain tissue and reduce cerebral edema. They act by drawing water across intact membranes, thereby reducing the volume of the brain and extracellular fluid. An indwelling urinary catheter is usually inserted to monitor urinary output and to manage the resulting diuresis. If the patient is receiving osmotic diuretics, serum osmolality should be determined to assess hydration status. If a brain tumor is the cause of the increased ICP, corticosteroids (e.g., dexamethasone [Decadron]) help reduce the edema surrounding the tumor.

Concept Mastery Alert

Typically, cerebral edema resulting from increased ICP is treated with osmotic diuretics. However, if the cerebral edema is due to the mass effect of a brain tumor, corticosteroids are the drug of choice.

Another method for decreasing cerebral edema is fluid restriction (Hickey, 2009). Limiting overall fluid intake leads to dehydration and hemoconcentration, which draws fluid across the osmotic gradient and decreases cerebral edema. Conversely, overhydration of the patient with increased ICP is avoided, because it increases cerebral edema.

Researchers have long hypothesized that lowering body temperature would decrease cerebral edema by reducing the oxygen and metabolic requirements of the brain, thus protecting the brain from continued ischemia. If body metabolism can be reduced by lowering the body temperature, the collateral circulation in the brain may be able to provide an adequate blood supply to the brain. The effect of hypothermia on ICP requires more study; thus far, induced hypothermia has not consistently been shown to be beneficial for patients with brain injury. Inducing and maintaining hypothermia is a major clinical treatment and requires knowledge and skilled nursing observation and management. The type and length of rewarming techniques after hypothermia may also be factors in the outcome of patients with neurologic injuries (Lescot et al., 2008).

Maintaining Cerebral Perfusion

Cardiac output may be manipulated to provide adequate perfusion to the brain. Improvements in cardiac output are made using fluid volume and inotropic agents such as dobutamine (Dobutrex) and norepinephrine (Levophed). The effectiveness of the cardiac output is reflected in the CPP, which is maintained at greater than 70 mm Hg (Lescot et al., 2008). A lower CPP indicates that the cardiac output is insufficient to maintain adequate cerebral perfusion. $SjvO_2$ and Licox, described earlier, assist in monitoring cerebral perfusion.

Reducing Cerebrospinal Fluid and Intracranial Blood Volume

CSF drainage is frequently performed, because the removal of CSF with a ventriculostomy drain can dramatically reduce ICP and restore CPP. Caution should be used in draining CSF, however, because excessive drainage may result in collapse of the ventricles and herniation. The reduction in $PaCO_2$ may result in hypoxia, ischemia, and an increase in cerebral lactate levels. Maintaining the $PaCO_2$ at greater than 30 mm Hg may prove beneficial (Hickey, 2009).

Controlling Fever

Preventing a temperature elevation is critical, because fever increases cerebral metabolism and the rate at which cerebral edema forms. Strategies to reduce body temperature include administration of antipyretic medications, as prescribed, and the use of a hypothermia blanket. Additional strategies for reducing fever were discussed previously in the Nursing Process section on altered LOC. The patient's temperature is monitored closely, and the patient is observed for shivering, which should be avoided because it is associated with increased oxygen consumption, increased levels of circulating catecholamines, and increased vasoconstriction. Shivering is associated with decreased levels of brain oxygenation; however, the association between shivering and neurologic outcome is unknown (Oddo, Frangos, Maloney-Wilensky, et al., 2010).

Maintaining Oxygenation and Reducing Metabolic Demands

Arterial blood gases and pulse oximetry are monitored to ensure that systemic oxygenation remains optimal. Metabolic demands may be reduced through the administration of high doses of barbiturates if the patient is unresponsive to conventional treatment. The mechanism by which barbiturates decrease ICP and protect the brain is uncertain, but the resultant comatose state is thought to reduce the metabolic requirements of the brain, thus providing cerebral protection (Blissit, 2012).

Another method of reducing cellular metabolic demand and improving oxygenation is the administration of medications causing sedation. The patient who receives these agents cannot move; this decreases the metabolic demands and results in a decrease in cerebral oxygen demand. The patient cannot respond to or report pain either. The most common agents used for sedation are pentobarbital (Nembutal), thiopental (Pentothal), propofol (Diprivan), and dexmedetomidine (Precedex) (Bader & Littlejohns, 2010; Blissit, 2012; Haddad & Arabi, 2012).

If sedative agents are used, the ability to perform serial neurologic assessments is lost. Therefore, other monitoring tools are needed to assess the patient's status and response to therapy. Important parameters that must be assessed include ICP, blood pressure, heart rate, respiratory rate, and the patient's response to ventilator therapy (e.g., "fighting or bucking the ventilator"). The level of pharmacologic paralysis is adjusted based on serum levels of the medications administered and the assessed parameters. Potential complications of these medications include hypotension caused by decreased sympathetic tone and myocardial depression.

Patients receiving high doses of barbiturates or pharmacologic sedatives require continuous cardiac monitoring, endotracheal intubation, mechanical ventilation, and arterial pressure monitoring, as well as ICP monitoring. In addition, serum barbiturate levels must be routinely monitored (Lescot et al., 2008).

NURSING PROCESS

The Patient With Increased Intracranial Pressure

PRACTICE & LEARN

Assessment

Initial assessment of the patient with increased ICP includes obtaining a history of events leading to the present illness and the pertinent past medical history. It is usually necessary to obtain this information from family or friends. The neurologic examination should be as complete as the patient's condition allows. It includes an evaluation of mental status, LOC, cranial nerve function, cerebellar function (balance and coordination), reflexes, and motor and sensory function. Because the patient is critically ill, ongoing assessment is more focused, including pupil checks, assessment of selected cranial nerves, frequent measurements of vital signs and ICP, and the use of the Glasgow Coma Scale (see Table 66-1).

Diagnosis

Nursing Diagnoses

Based on the assessment data, major nursing diagnoses include the following:

- Ineffective airway clearance related to diminished protective reflexes (cough, gag)
- Ineffective breathing patterns related to neurologic dysfunction (brain stem compression, structural displacement)
- Ineffective cerebral tissue perfusion related to the effects of increased ICP
- Deficient fluid volume related to fluid restriction
- Risk for infection related to ICP monitoring system (fiberoptic or intraventricular catheter)

Other relevant nursing diagnoses are included in the earlier section on altered LOC.

Collaborative Problems/Potential Complications

Potential complications may include the following:

- Brain stem herniation
- Diabetes insipidus
- SIADH

Planning and Goals

The goals for the patient include maintenance of a patent airway, normalization of respiration, adequate cerebral tissue perfusion through reduction in ICP, restoration of fluid balance, absence of infection, and absence of complications.

Nursing Interventions

Maintaining a Patent Airway

The patency of the airway is assessed. Secretions that are obstructing the airway must be suctioned with care, because transient elevations of ICP occur with suctioning (Hickey, 2009). Hypoxia caused by poor oxygenation leads to cerebral ischemia and edema. Coughing is discouraged because it increases ICP. The lung fields are auscultated at least every 8 hours to determine the presence of adventitious sounds or any areas of congestion. Elevating the head of the bed may aid in clearing secretions and improve venous drainage of the brain.

Achieving an Adequate Breathing Pattern

The patient must be monitored constantly for respiratory irregularities. Increased pressure on the frontal lobes or deep midline structures may result in Cheyne-Stokes respirations, whereas pressure in the midbrain can cause hyperventilation. If the lower portion of the brain stem (the pons and medulla) is involved, respirations become irregular and eventually cease.

Hyperventilation therapy is a controversial therapy in traumatic brain injury used in some centers to reduce ICP by causing cerebral vasoconstriction and a decrease in cerebral blood volume. The nurse collaborates with the respiratory therapist in monitoring the $PaCO_2$, which is usually maintained at less than 30 mm Hg. Patients undergoing hyperventilation therapy also benefit from multimodality monitoring to determine the overall effect of this therapy on brain perfusion (Carrera, Steiner, Castellani, et al., 2011).

A neurologic observation record (Fig. 66-6) is maintained, and all observations are made in relation to the patient's baseline condition. Repeated assessments of the patient are made (sometimes minute by minute) so that improvement or deterioration may be noted immediately. If the patient's condition deteriorates, preparations are made for surgical intervention.

Optimizing Cerebral Tissue Perfusion

In addition to ongoing nursing assessment, strategies are initiated to reduce factors contributing to the elevation of ICP (Table 66-2).

Proper positioning helps reduce ICP. The patient's head is kept in a neutral (midline) position, maintained with the use of a cervical collar if necessary, to promote venous drainage. Elevation of the head is maintained at 30 to 45 degrees unless contraindicated (Littlejohns & Bader, 2009). Extreme rotation of the neck and flexion of the neck are avoided, because compression or distortion of the jugular veins increases ICP. Extreme hip flexion is also avoided, because this position causes an increase in intra-abdominal and intrathoracic pressures, which can produce an increase in ICP. Relatively minor changes in position can significantly affect ICP. If monitoring reveals that turning the patient raises ICP, rotating beds, turning sheets, and holding the patient's head during turning may minimize the stimuli that increase ICP. Research suggests that patient response to position change is highly variable and requires close hemodynamic monitoring and individualized care (Ledwith, Bloom, Maloney-Wilensky, et al., 2010) (Chart 66-1).

The Valsalva maneuver, which can be produced by straining at defecation or even moving in bed, raises ICP and is to be avoided. Stool softeners may be prescribed. If the patient is alert and able to eat, a diet high in fiber may be indicated. Abdominal distention, which increases intra-abdominal and intrathoracic pressure and ICP, should be noted. Enemas and cathartics are avoided if possible. When moving or being turned in bed, the patient can be instructed to exhale (which opens the glottis) to avoid the Valsalva maneuver.

Mechanical ventilation presents unique problems for the patient with increased ICP. Before suctioning, the patient should be preoxygenated and briefly hyperventilated using 100% oxygen on the ventilator. Suctioning should not last longer than 15 seconds. High levels of positive end-expiratory pressure (PEEP) are avoided, because they may decrease venous return to the heart and decrease venous drainage from the brain through increased intrathoracic pressure (Littlejohns & Bader, 2009).

Activities that increase ICP, as indicated by changes in waveforms, should be avoided if possible. Spacing of nursing interventions may prevent transient increases in ICP. During

TABLE 66-2 Increased Intracranial Pressure and Interventions

Factor	Physiology	Interventions	Rationale
Cerebral edema	Can be caused by contusion, tumor, or abscess; water intoxication (hypo-osmolality); alteration in the blood–brain barrier (protein leaks into the tissue, causing water to follow)	Administer osmotic diuretics as prescribed (monitor serum osmolality). Maintain head of bed elevation at 30 degrees. Maintain alignment of the head.	Promotes venous return Prevents impairment of venous return through the jugular veins
Hypoxia	A decrease in PaO_2 to <60 mm Hg causes cerebral vasodilation.	Maintain PaO_2 >60 mm Hg. Maintain oxygen therapy. Monitor arterial blood gas values. Suction when needed. Maintain a patent airway.	Prevents hypoxia and vasodilation
Hypercapnia (elevated $PaCO_2$)	Causes vasodilation	Maintain $PaCO_2$ (normally 35–45 mm Hg) by establishing ventilation.	Normalizing $PaCO_2$ minimizes vasodilation and thus reduces the cerebral blood volume.
Impaired venous return	Increases the cerebral blood volume	Maintain head alignment. Elevate head of bed 30 degrees.	Hyperextension, rotation, or hyperflexion of the neck causes decreased venous return.
Increase in intrathoracic or abdominal pressure	An increase in these pressures due to coughing, PEEP, or Valsalva maneuver causes a decrease in venous return.	Monitor arterial blood gas values, and keep PEEP as low as possible. Provide humidified oxygen. Administer stool softeners as prescribed.	To keep secretions loose and easy to suction or expectorate Soft bowel movements will prevent straining or Valsalva maneuver.

PaO_2, partial pressure of arterial oxygen; $PaCO_2$, partial pressure of arterial carbon dioxide; PEEP, positive end-expiratory pressure.

Adapted from American Association of Neuroscience Nurses. (2011). *Guide to the care of the patient with intracranial pressure monitoring/external ventricular drainage or lumbar drainage: AANN reference series for clinical practice*. Glenview, IL: Author; Bader, M., & Littlejohns, L. R. (2010). *AANN core curriculum for neuroscience nursing* (5th ed.). Glenview, IL: American Association of Neuroscience Nurses; and Hickey, J. V. (2009). *The clinical practice of neurological & neurosurgical nursing* (6th ed.). Philadelphia: Lippincott Williams & Wilkins.

NURSING NEUROLOGICAL CRITICAL CARE FLOWSHEET

ADDRESSOGRAPH

	Date												
	Time												
	Initials												
Level of orientation (✓)	Person												
	Place												
	Date and time												
	No orientation												
Awakens to (✓)	Voice												
	Touch												
	Noxious stimuli												
	Painful stimuli												
	No response												
Best verbal response (✓)	Clear and appropriate												
	Clear and inappropriate												
	Difficulty speaking*												
	Perseveration												
	Aphasic expressive (non-fluent)												
	Aphasic receptive (fluent)												
	Sounds no speech												
	No verbal response												
	ETT/TRACH												
Best motor response (✓)	Moves all extremities purposetully												
	Withdraws and lifts to painful stimuli												
	Moves to painful stimuli												
	Decorticates (spinal reflex)												
	Decerebrates (spinal reflex)												
	No motor response												
Best motor strength upper extremities (✓)	No drifts (R/L)	R L	R L	R L	R L	R L	R L	R L	R L	R L	R L	R L	R L
	Drift (R/L)	R L	R L	R L	R L	R L	R L	R L	R L	R L	R L	R L	R L
	Can only lift forearm (R/L)	R L	R L	R L	R L	R L	R L	R L	R L	R L	R L	R L	R L
	Trace movement of hand or arm (R/L)	R L	R L	R L	R L	R L	R L	R L	R L	R L	R L	R L	R L
	Trace movement of fingers only (R/L)	R L	R L	R L	R L	R L	R L	R L	R L	R L	R L	R L	R L
	No motor response (R/L)	R L	R L	R L	R L	R L	R L	R L	R L	R L	R L	R L	R L
Best strength lower extremities (✓)	Raises leg off bed (R/L)	R L	R L	R L	R L	R L	R L	R L	R L	R L	R L	R L	R L
	Drags heel on bed and lifts knee (R/L)	R L	R L	R L	R L	R L	R L	R L	R L	R L	R L	R L	R L
	Trace movement of foot or leg (R/L)	R L	R L	R L	R L	R L	R L	R L	R L	R L	R L	R L	R L
	Trace movement of toes only (R/L)	R L	R L	R L	R L	R L	R L	R L	R L	R L	R L	R L	R L
	No response (R/L)	R L	R L	R L	R L	R L	R L	R L	R L	R L	R L	R L	R L
Seizure activity (✓)	No seizure activity												
	With loss of consciousness*												
	Without loss of consciousness*												
Ataxia (✓)	Gross ataxia												
	Fine motor ataxia												
	Does not apply												
ICP monitoring	Ventriculostomy mL												
	ICP mm Hg												
	Not applicable												

*= FURTHER DOCUMENTATION IS REQUIRED TO VALIDATE ASSESSMENT

FIGURE 66-6 • A neurologic assessment flow chart. (*continued*)

PUPIL GAUGE (mm)

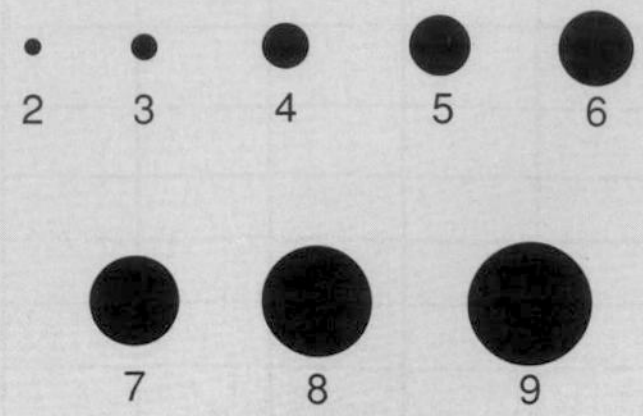
2
3
4
5
6
7
8
9

B=Brisk, S=Sluggish, F=Fixed

ADDRESSOGRAPH

	Date												
	Time												
	Initials												
Incision +/–	Dry and intact												
	Drainage												
Pupils: refer to above gauge (✓) (+)=Present (–)=Absent	Size (R/L)	R L	R L	R L	R L	R L	R L	R L	R L	R L	R L	R L	R L
	Regular (R/L)	R L	R L	R L	R L	R L	R L	R L	R L	R L	R L	R L	R L
	Irregular* (R/L)	R L	R L	R L	R L	R L	R L	R L	R L	R L	R L	R L	R L
	Reaction (R/L) (B) - (S) - (F)	R L	R L	R L	R L	R L	R L	R L	R L	R L	R L	R L	R L
	Ptosis (R/L) (+) (–)	R L	R L	R L	R L	R L	R L	R L	R L	R L	R L	R L	R L
	Gaze preference (R/L) (+)* (–)	R L	R L	R L	R L	R L	R L	R L	R L	R L	R L	R L	R L
Meningeal signs (+)=Present (–)=Absent	Headache												
	Nuchal rigidity												
	Photophobia												
Visual fields (+)=Present (–)=Absent* NA=Not applicable	Right upper outer												
	Right lower outer												
	Left upper outer												
	Left lower outer												
Nystagmus (+)=Present (–)=Absent	Lateral (R/L)	R L	R L	R L	R L	R L	R L	R L	R L	R L	R L	R L	R L
	Vertical (R/L)	R L	R L	R L	R L	R L	R L	R L	R L	R L	R L	R L	R L
Cranial nerves (+)=Present (–)=Absent	III, IV, VI, Extraocular movements												
	VII – Peripheral facial droop (R/L)	R L	R L	R L	R L	R L	R L	R L	R L	R L	R L	R L	R L
	XII – Tongue deviation (R/L)	R L	R L	R L	R L	R L	R L	R L	R L	R L	R L	R L	R L
	IX – Gag reflex												
	V, VII – Corneal reflex (R/L)	R L	R L	R L	R L	R L	R L	R L	R L	R L	R L	R L	R L
	X, IX – Cough reflex												
	Doll's eyes if appropriate												
Follows commands	Two step verbal command												
	One step verbal command												
	Unable to follow command												

*= **FURTHER DOCUMENTATION IS REQUIRED TO VALIDATE ASSESSMENT**

Initials	Signature	Title	Initials	Signature	Title

FIGURE 66-6 • (*Continued*)

Chart 66-1

NURSING RESEARCH PROFILE

Effects of Position Changes on Cerebral Oxygenation

Ledwith, M., Bloom, S., Maloney-Wilensky, E., et al. (2010). Effect of body position on cerebral oxygenation and physiologic parameters in patients with acute neurological conditions. *Journal of Neuroscience Nursing*, *42*(5), 280–287.

Purpose

There is not a good understanding of how body position changes influences brain tissue oxygen levels ($PbtO_2$) and intracranial pressure (ICP) in critically ill neurosurgical patients. The purpose of this study was to examine the effects of different body positions on $PbtO_2$, ICP, cerebral perfusion pressure (CPP), and mean arterial pressure (MAP) in patients who have had a severe brain injury.

Design

This prospective observational repeated measures study explored the effects of 12 different body positions on $PbtO_2$, ICP, CPP, and MAP. The sample included 33 patients who were receiving care in a neurocritical care unit, had a Glasgow Coma Scale score ≤8, and had a Licox monitoring system in place. Each patient was exposed to all 12 position changes in a random order. $PbtO_2$, ICP, CPP, and MAP were monitored at baseline and then repeated at 15 minutes following each position change.

Findings

The mean age of the patients was 48 years, 22 were men, and the main neurosurgical conditions included traumatic brain injury, subarachnoid hemorrhage, or post craniotomy for removal of a brain tumor. Changes from baseline to the 15-minute post position change assessment showed a downward trend in $PbtO_2$ levels in all patients, with a statistically significant decrease with supine head-of-bed (HOB) position at 30 and 45 degrees ($p < .01$) and right and left lateral positioning HOB at 30 degrees ($p < .05$). ICP decreased with supine HOB at 45 degrees ($p < .01$) and knee elevation with HOB at 30 and 45 degrees ($p < .05$). In contrast, the ICP increased with right and left lateral HOB at 15 degrees ($p < .05$).

Nursing Implications

Nurses need to be aware that position changes can positively or negatively affect $PbtO_2$ and ICP. Nurses must carefully consider the effects of position changes, carefully monitor hemodynamic changes in individual patients, and use the data to make critical decisions about optimal positions to maintain or improve cerebral oxygenation.

nursing interventions, the ICP should not increase above 25 mm Hg, and it should return to baseline levels within 5 minutes. Patients with increased ICP should not demonstrate a significant increase in pressure or change in the ICP waveform. Patients with the potential for a significant increase in ICP may need sedation before initiation of nursing activities (Bader & Littlejohns, 2010).

Emotional stress and frequent arousal from sleep are avoided. A calm atmosphere is maintained. Environmental stimuli (e.g., noise, conversation) should be minimal.

Maintaining Negative Fluid Balance

The administration of osmotic and loop diuretics is part of the treatment protocol to reduce ICP. Corticosteroids may be used to reduce cerebral edema (except when it results from trauma), and fluids may be restricted. All of these treatment modalities promote dehydration.

Skin turgor, mucous membranes, urine output, and serum and urine osmolality are monitored to assess fluid status. If IV fluids are prescribed, the nurse ensures that they are administered at a slow to moderate rate with an IV infusion pump, to prevent too-rapid administration and avoid overhydration. For the patient receiving mannitol, the nurse observes for the possible development of heart failure and pulmonary edema. The intent of treatment is to promote a shift of fluid from the intracellular to the intravascular compartment and to control cerebral edema. However, this shift of fluid volume to the intravascular compartment may overwhelm the ability of the myocardium to increase workload sufficient to meet these demands, which may cause failure and pulmonary edema.

For patients undergoing dehydrating procedures, vital signs, including blood pressure, must be monitored to assess fluid volume status. An indwelling urinary catheter is inserted to permit assessment of renal function and fluid status. During the acute phase, urine output is monitored hourly. An output greater than 200 mL per hour for 2 consecutive hours may indicate the onset of diabetes insipidus (Hickey, 2009). These patients need careful oral hygiene, because mouth dryness occurs with dehydration. Frequently rinsing the mouth with nondrying solutions, lubricating the lips, and removing encrustations relieve dryness and promote comfort.

Preventing Infection

The risk of infection is greatest when ICP is monitored with an intraventricular catheter and increases with the duration of the monitoring. Most health care facilities have written protocols for managing these systems and maintaining their sterility; strict adherence to the protocols is essential.

Aseptic technique must be used when managing the system and changing the ventricular drainage bag. The drainage system is also checked for loose connections, because they can cause leakage and contamination of the CSF as well as inaccurate readings of ICP. The nurse observes the character of the CSF drainage and reports increasing cloudiness or blood. The patient is monitored for signs and symptoms of meningitis: fever, chills, nuchal (neck) rigidity, and increasing or persistent headache. (See Chapter 69 for a discussion of meningitis.)

Monitoring and Managing Potential Complications

The primary complication of increased ICP is brain herniation resulting in death (see Fig. 66-2). Nursing management focuses on detecting early signs of increasing ICP, because medical interventions are usually ineffective once later signs develop (Bader & Littlejohns, 2010). Frequent neurologic assessments and documentation and analysis of trends will reveal the subtle changes that may indicate increasing ICP.

Detecting Indications of Increasing Intracranial Pressure. The nurse assesses for and immediately reports any signs or symptoms of increasing ICP, as noted in Chart 66-2.

Chart 66-2 Detecting Indications of Increasing Intracranial Pressure

Early Signs and Symptoms of Increasing Intracranial Pressure (ICP)

- *Disorientation, restlessness, increased respiratory effort, purposeless movements, and mental confusion.* These are early clinical indications of increasing ICP because the brain cells responsible for cognition are extremely sensitive to decreased oxygenation.
- *Pupillary changes and impaired extraocular movements.* These occur as the increasing pressure displaces the brain against the oculomotor and optic nerves (cranial nerves II, III, IV, and VI), which arise from the midbrain and brain stem (see Chapter 65).
- *Weakness in one extremity or on one side of the body.* This occurs as increasing ICP compresses the pyramidal tracts.
- *Headache that is constant, increasing in intensity, and aggravated by movement or straining.* This occurs as increasing ICP causes pressure and stretching of venous and arterial vessels in the base of the brain.

Later Signs and Symptoms of Increasing ICP

- The level of consciousness continues to deteriorate until the patient is comatose (Glasgow Coma Scale score ≤8).
- The pulse rate and respiratory rate decrease or become erratic, and the blood pressure and temperature increase. The pulse pressure (the difference between the systolic and the diastolic pressures) widens. The pulse fluctuates rapidly, varying from bradycardia to tachycardia.
- Altered respiratory patterns develop, including Cheyne-Stokes breathing (rhythmic waxing and waning of rate and depth of respirations alternating with brief periods of apnea) and ataxic breathing (irregular breathing with a random sequence of deep and shallow breaths).
- Projectile vomiting may occur with increased pressure on the reflex center in the medulla.
- Hemiplegia or decorticate or decerebrate posturing may develop as pressure on the brain stem increases; bilateral flaccidity occurs before death.
- Loss of brain stem reflexes, including pupillary, corneal, gag, and swallowing reflexes, is an ominous sign of approaching death.

Monitoring Intracranial Pressure. Because clinical assessment is not always a reliable guide in recognizing increased ICP, especially in comatose patients, monitoring of ICP and cerebral oxygenation is an essential part of management. ICP is monitored closely for continuous elevation or significant increase over baseline. The trend of ICP measurements over time is an important indication of the patient's underlying status. Vital signs are assessed when an increase in ICP is noted (Bader & Littlejohns, 2010).

Careful attention to aseptic technique is needed when handling any part of the monitoring system. The insertion site is inspected for signs of infection. Temperature, pulse, and respirations are closely monitored for systemic signs of infection. All connections and stopcocks are checked for leaks, because even small leaks can distort pressure readings and lead to infection (Littlejohns & Bader, 2009).

When ICP is monitored with a fluid system, the transducer is calibrated at a particular reference point, usually 2.5 cm

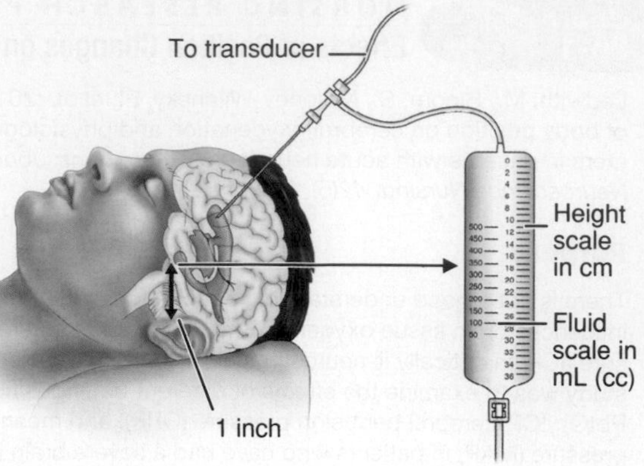
To transducer
Height scale in cm
Fluid scale in mL (cc)
1 inch

FIGURE 66-7 • Location of the foramen of Monro for calibration of the intracranial pressure monitoring system.

(1 in) above the ear with the patient in the supine position; this point corresponds to the level of the foramen of Monro (Fig. 66-7). CSF pressure readings depend on the patient's position. For subsequent pressure readings, the head should be in the same position relative to the transducer. Fiberoptic catheters are calibrated before insertion and do not require further referencing; they do not require the head of the bed to be at a specific position to obtain an accurate reading.

When technology is associated with patient management, the nurse must be certain that the technologic equipment is functioning properly. The most important concern must be the patient to whom equipment is attached. The patient and family must be informed about the technology and the goals of its use. The patient's response is monitored, and appropriate comfort measures are implemented to ensure that the patient's stress is minimized.

ICP measurement is only one parameter; repeated neurologic checks and clinical examinations remain important measures. Astute observation, comparison of findings with previous observations, and interventions can assist in preventing life-threatening ICP elevations.

Monitoring for Secondary Complications. The nurse also assesses for complications of increased ICP, including diabetes insipidus and SIADH (see Chapters 13 and 52). Urine output should be monitored closely. Diabetes insipidus requires fluid and electrolyte replacement, along with the administration of vasopressin, to replace and slow the urine output. Serum electrolyte levels are monitored for imbalances. SIADH requires fluid restriction and monitoring of serum electrolyte levels.

Evaluation

Expected patient outcomes may include:

1. Maintains patent airway
2. Attains optimal breathing pattern
   a. Breathes in a regular pattern
   b. Attains or maintains arterial blood gas values within acceptable range
3. Demonstrates optimal cerebral tissue perfusion
   a. Increasingly oriented to time, place, and person
   b. Follows verbal commands; answers questions correctly

4. Attains desired fluid balance
   a. Maintains fluid restriction
   b. Demonstrates serum and urine osmolality values within acceptable range
5. Has no signs or symptoms of infection
   a. Has no fever
   b. Shows no redness, swelling, or drainage at arterial, IV, and urinary catheter sites
   c. Has no redness, swelling, or purulent drainage from invasive intracranial monitoring device
6. Absence of complications
   a. Has ICP values that remain within normal limits
   b. Demonstrates urine output and serum electrolyte levels within acceptable limits

INTRACRANIAL SURGERY

A **craniotomy** involves opening the skull surgically to gain access to intracranial structures. This procedure is performed to remove a tumor, relieve elevated ICP, evacuate a blood clot, or control hemorrhage. The surgeon cuts the skull to create a bony flap, which can be repositioned after surgery and held in place by periosteal or wire sutures. One of two approaches through the skull is used: (1) above the tentorium (supratentorial craniotomy) into the supratentorial compartment, or (2) below the tentorium into the infratentorial (posterior fossa) compartment. A third approach, the **transsphenoidal** approach (through the mouth and nasal sinuses) is often used to gain access to the pituitary gland (Swearingen, 2012). Table 66-3 compares these three different surgical approaches.

Alternatively, intracranial structures may be approached through burr holes (Fig. 66-8), which are circular openings made in the skull by either a hand drill or an automatic craniotome (which has a self-controlled system to stop the drill when the bone is penetrated). Burr holes may be used to determine the presence of cerebral swelling and injury and the size and position of the ventricles. They are also a means of evacuating an intracranial hematoma or abscess and for making a bone flap in the skull that allows access to the ventricles for decompression, ventriculography, or shunting procedures. Other cranial procedures include **craniectomy** (excision of a portion of the skull) and cranioplasty (repair of a cranial defect using a plastic or metal plate).

TABLE 66-3 Comparison of Cranial Surgical Approaches

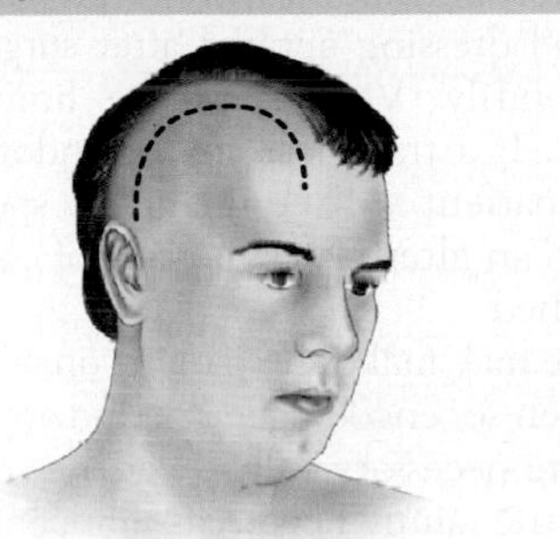
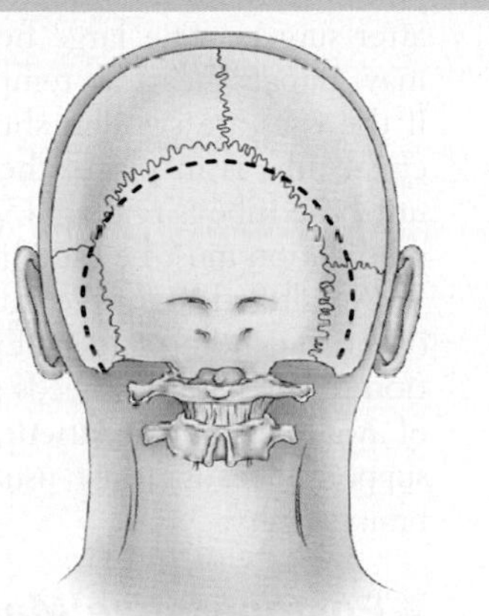
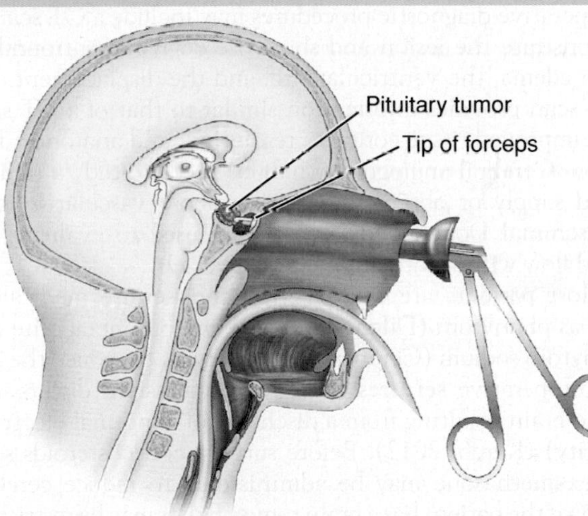
Pituitary tumor
Tip of forceps

Supratentorial	Infratentorial	Transsphenoidal
Site of Surgery		
Above the tentorium	Below the tentorium, brain stem	Sella turcica and pituitary region
Incision Location		
Incision is made above the area to be operated on; usually located behind the hairline.	Incision is made at the nape of the neck, around the occipital lobe.	Incision is made beneath the upper lip to gain access into the nasal cavity.
Selected Nursing Interventions		
Maintain head of bed elevated 30–45 degrees, with neck in neutral alignment. Position patient on either side or back. (Avoid positioning patient on operative side if a large tumor has been removed.)	Maintain neck in straight alignment. Avoid flexion of the neck to prevent possible tearing of the suture line. Position the patient on either side. (Check surgeon's preference for positioning of patient.)	Maintain nasal packing in place and reinforce as needed. Instruct patient to avoid blowing the nose. Provide oral care according to institutional procedure. Keep head of bed elevated to promote venous drainage and drainage from the surgical site.

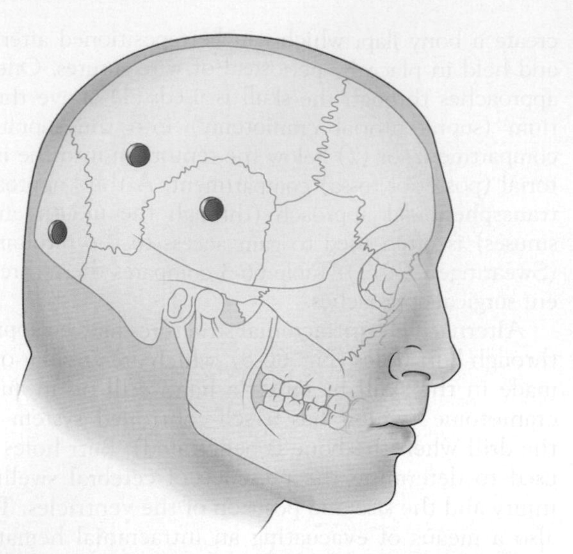

FIGURE 66-8 • Burr holes may be used in neurosurgical procedures to make a bone flap in the skull, to aspirate a brain abscess, or to evacuate a hematoma.

Supratentorial and Infratentorial Approaches

Preoperative Management

Medical Management

Preoperative diagnostic procedures may include a CT scan to demonstrate the lesion and show the degree of surrounding brain edema, the ventricular size, and the displacement. An MRI scan provides information similar to that of a CT scan with improved tissue contrast, resolution, and anatomic definition. Cerebral angiography may be used to study a tumor's blood supply or obtain information about vascular lesions. Transcranial Doppler flow studies are used to evaluate the blood flow within intracranial blood vessels.

Most patients are prescribed an antiseizure medication such as phenytoin (Dilantin) or a phenytoin metabolite fosphenytoin sodium (Cerebyx) before surgery to reduce the risk of postoperative **seizures** (paroxysmal transient disturbance of the brain resulting from a discharge of abnormal electrical activity) (Karch, 2012). Before surgery, corticosteroids such as dexamethasone may be administered to reduce cerebral edema if the patient has a brain tumor. Fluids may be restricted. A hyperosmotic agent (mannitol) and a diuretic agent such as furosemide (Lasix) may be administered IV immediately before and sometimes during surgery if the patient tends to retain fluid, as do many who have intracranial dysfunction. Antibiotic agents may be administered if there is a chance of cerebral contamination; diazepam (Valium) or lorazepam (Ativan) may be prescribed before surgery to allay anxiety.

Nursing Management

The preoperative assessment serves as a baseline against which postoperative status and recovery are compared. This assessment includes evaluating the LOC and responsiveness to stimuli and identifying any neurologic deficits, such as paralysis, visual dysfunction, alterations in personality or speech, and bladder and bowel disorders. Baseline distal and proximal motor strength in both upper and lower extremities is tested and recorded. (See Chapter 65 for a discussion of the testing of motor function.)

The patient's and family's understanding of and reactions to the anticipated surgical procedure and its possible sequelae are assessed, as is the availability of support systems for the patient and family. Adequate preparation for surgery, with attention to the patient's physical and emotional status, can reduce the risk of anxiety, fear, and postoperative complications. The patient is assessed for neurologic deficits and their potential impact after surgery. For motor deficits or weakness or paralysis of the arms or legs, trochanter rolls are applied to the extremities, and the feet are positioned against a footboard or the ankles are supported in a neutral position with orthotic boots. A patient who can ambulate is encouraged to do so. If the patient is aphasic, writing materials or picture and word cards showing the bedpan, glass of water, blanket, and other frequently used items may help improve communication.

Preparation of the patient and family includes providing education about what to expect during and after surgery. Hair is removed with the use of clippers and the surgical site prepared immediately before surgery (usually in the operating room) to decrease the chance of infection (Sebastian, 2012). An indwelling urinary catheter is inserted in the operating room to drain the bladder during the administration of diuretic agents and to permit monitoring of urinary output. The patient may have a central and arterial line placed for fluid administration and monitoring of pressures after surgery. The large head dressing applied after surgery may impair hearing temporarily. Vision may be limited if the eyes are swollen shut. If a tracheostomy or endotracheal tube is in place, the patient will be unable to speak until the tube is removed, so an alternative method of communication must be established.

An altered cognitive state may make the patient unaware of the impending surgery. Even so, encouragement and attention to the patient's needs are necessary. Whatever the state of awareness of the patient, the family need reassurance and support, because they usually recognize the seriousness of brain surgery.

Postoperative Management

Postoperatively, an arterial line and a central venous pressure line may be in place to monitor and manage blood pressure and central venous pressure. The patient may be intubated and may receive supplemental oxygen therapy. Ongoing postoperative management is aimed at detecting and reducing cerebral edema, relieving pain and preventing seizures, and monitoring ICP and neurologic status.

Reducing Cerebral Edema

Medications to reduce cerebral edema include mannitol, which increases serum osmolality and draws free water from areas of the brain (with an intact blood–brain barrier). The fluid is then excreted by osmotic diuresis. Dexamethasone may be administered IV every 6 hours for 24 to 72 hours; the route is changed to oral as soon as possible, and the dosage is tapered over 5 to 7 days (Karch, 2012).

Relieving Pain and Preventing Seizures

Acetaminophen is usually prescribed for temperatures exceeding 37.5°C (99.6°F) and for pain. The patient usually has a headache after a craniotomy as a result of stretching and irritation of nerves in the scalp during surgery. Codeine, administered IV, is often sufficient to relieve headache. Morphine sulfate may also be used in the management of postoperative pain in patients who have undergone a craniotomy (Hansen, Brennum, Moltke, et al., 2011).

Antiseizure medication (phenytoin, diazepam) is often prescribed prophylactically for patients who have undergone supratentorial craniotomy because of the high risk of seizures after these procedures (Hickey, 2009). Serum levels are monitored to check that the medication levels are within the therapeutic range.

Monitoring Intracranial Pressure

A patient undergoing intracranial surgery may have an ICP or cerebral oxygenation monitor inserted during surgery. Strict adherence to written protocols for managing these systems is essential, as discussed earlier, for preventing infection and managing ICP. The system is removed after the ICP or cerebral oxygenation is normal and stable. The neurosurgeon must be notified immediately if the system is not functioning.

NURSING PROCESS

The Patient Who Has Undergone Intracranial Surgery

Assessment

After surgery, the frequency of postoperative monitoring is based on the patient's clinical status. Assessing respiratory function is essential, because even a small degree of hypoxia can increase cerebral ischemia. The respiratory rate and pattern are monitored, and arterial blood gas values are assessed frequently. Fluctuations in vital signs are carefully monitored and documented, because they may indicate increased ICP. The patient's temperature is measured to assess for hyperthermia secondary to infection or damage to the hypothalamus. Neurologic checks are made frequently to detect increased ICP resulting from cerebral edema or bleeding. A change in LOC or response to stimuli may be the first sign of increasing ICP.

The surgical dressing is inspected for evidence of bleeding and CSF drainage. The incision is monitored for redness, tenderness, bulging, separation, or foul odor. Sodium retention may occur in the immediate postoperative period. Serum and urine electrolytes, BUN, blood glucose, weight, and clinical status are monitored. Intake and output are measured in view of losses associated with fever, respiration, and CSF drainage. The nurse must be alert to the development of complications; all assessments are carried out with these problems in mind. Seizures are a potential complication, and any seizure activity is carefully recorded and reported. Restlessness may occur as the patient becomes more responsive, or restlessness may be caused by pain, confusion, hypoxia, or other stimuli.

Diagnosis

Nursing Diagnoses

Based on the assessment data, major nursing diagnoses may include the following:

- Risk for ineffective cerebral tissue perfusion related to cerebral edema
- Risk for imbalanced body temperature related to damage to the hypothalamus, dehydration, and infection
- Potential for impaired gas exchange related to hypoventilation, aspiration, and immobility
- Ineffective coping related to sensory perception changes due to periorbital edema, head dressing, endotracheal tube, and effects of ICP
- Body image disturbance related to change in appearance or physical disabilities

Other nursing diagnoses may include impaired communication (aphasia) related to insult to brain tissue and high risk for impaired skin integrity related to immobility, pressure, and incontinence; impaired physical mobility related to a neurologic deficit secondary to the neurosurgical procedure or to the underlying disorder may also occur.

Collaborative Problems/Potential Complications

Potential complications may include the following:

- Increased ICP
- Bleeding and hypovolemic shock
- Fluid and electrolyte disturbances
- Infection
- Seizures

Planning and Goals

The major goals for the patient include maintaining or restoring neurologic homeostasis to improve cerebral tissue perfusion, adequate thermoregulation, normal ventilation and gas exchange, ability to cope with sensory deprivation, adaptation to changes in body image, and absence of complications.

Nursing Interventions

Maintaining Cerebral Tissue Perfusion

Attention to the patient's respiratory status is essential, because even slight decreases in the oxygen level (hypoxia) or slight increases in the carbon dioxide level (hypercarbia) can affect cerebral perfusion, the clinical course, and the patient's outcome. The endotracheal tube is left in place until the patient shows signs of awakening and has adequate spontaneous ventilation, as evaluated clinically and by arterial blood gas analysis. Secondary brain damage can result from impaired cerebral oxygenation.

Some degree of cerebral edema occurs after brain surgery; it tends to peak 24 to 36 hours after surgery, producing decreased responsiveness on the second postoperative day. The control of cerebral edema was discussed earlier. Nursing strategies used to control factors that may raise ICP were presented in the previous Nursing Process section on increased ICP. Intraventricular drainage is carefully monitored, using strict asepsis when any part of the system is handled.

Vital signs and neurologic status (LOC and responsiveness, pupillary and motor responses) are assessed every 15 to 60 minutes. Extreme head rotation is avoided, because this raises ICP. After supratentorial surgery, the patient is placed on his or her back or side (on the unoperated side if a large lesion was removed) with one pillow under the head. The head of the bed may be elevated 30 degrees, depending on the level of the ICP and the neurosurgeon's preference. After posterior fossa (infratentorial) surgery, the patient is kept flat on one side (off the back) with the head on a small, firm pillow. The patient may be turned on either side, keeping the neck in a neutral position. When the patient is being turned, the body should be turned as a unit to prevent placing strain on the incision and possibly tearing the sutures. The head of the bed may be elevated slowly as tolerated by the patient.

The patient's position is changed every 2 hours, and skin care is given frequently. During position changes, care is taken to prevent disruption of the ICP monitoring system. A turning sheet placed under the patient's head to midthigh makes it easier to move and turn the patient safely.

Regulating Temperature

Moderate temperature elevation can be expected after intracranial surgery because of the reaction to blood at the operative site or in the subarachnoid space. Injury to the hypothalamic centers that regulate body temperature can occur during surgery. Fever is treated vigorously to combat the effect of an elevated temperature on brain metabolism and function.

Nursing interventions include monitoring the patient's temperature and using the following measures to reduce body temperature: removing blankets, using a hypothermia blanket as prescribed, and administering prescribed antipyretics to reduce fever (McIlvoy, 2012).

Conversely, hypothermia may be seen after lengthy neurosurgical procedures. Therefore, frequent measurements of rectal temperatures are necessary. Rewarming should occur slowly to prevent shivering, which increases cellular oxygen demands.

Improving Gas Exchange

The patient undergoing neurosurgery is at risk for impaired gas exchange and pulmonary infections due to immobility, immunosuppression, decreased LOC, and fluid restriction. Immobility compromises the respiratory system by causing pooling and stasis of secretions in dependent areas and the development of atelectasis. The patient whose fluid intake is restricted may be more vulnerable to atelectasis as a result of inability to expectorate thickened secretions. Pneumonia can develop due to aspiration and restricted mobility.

Repositioning the patient every 2 hours helps to mobilize pulmonary secretions and prevent stasis. After the patient regains consciousness, additional measures to expand collapsed alveoli can be instituted, such as yawning, sighing, deep breathing, incentive spirometry, and coughing (unless contraindicated). If necessary, the oropharynx and trachea are suctioned to remove secretions that cannot be raised by coughing; however, coughing and suctioning increase ICP. Therefore, suctioning should be used cautiously. Increasing the humidity in the oxygen delivery system may help to loosen secretions. The nurse and the respiratory therapist work together to monitor the effects of chest physiotherapy.

Coping With Sensory Deprivation

Periorbital edema is a common consequence of intracranial surgery, because fluid drains into the dependent periorbital areas when the patient has been positioned in a prone position during surgery. A hematoma may form under the scalp and spread down to the orbit, producing an area of ecchymosis (black eye).

Before surgery, the patient and family should be informed that one or both eyes may be edematous temporarily after surgery. After surgery, elevating the head of the bed (if not contraindicated) and applying cold compresses over the eyes will help reduce the edema. The surgeon is notified if periorbital edema increases significantly, because this may indicate that a postoperative clot is developing or that there is increasing ICP and poor venous drainage. Health care personnel should announce their presence when entering the room to avoid startling the patient whose vision is impaired due to periorbital edema or neurologic deficits.

Additional factors that can affect sensation include a bulky head dressing, the presence of an endotracheal tube, and effects of increased ICP. The first postoperative dressing change is usually performed by the neurosurgeon. In the absence of bleeding or a CSF leak, every effort is made to minimize the size of the head dressing. If the patient requires an endotracheal tube for mechanical ventilation, every effort is made to extubate the patient as soon as clinical signs indicate it is possible. The patient is monitored closely for the effects of elevated ICP.

Enhancing Self-Image

The patient is encouraged to verbalize feelings and frustrations about any change in appearance. Nursing support is based on the patient's reactions and feelings. Factual information may need to be provided if the patient has misconceptions about puffiness about the face, periorbital bruising, and hair loss. Attention to grooming, the use of the patient's own clothing, and covering the head with a turban (and later a wig until hair growth occurs) are encouraged. Social interaction with close friends, family, and hospital personnel may increase the patient's sense of self-worth.

The family and social support system can be of assistance while the patient recovers from surgery.

Monitoring and Managing Potential Complications

The nurse must be vigilant for complications that may develop within hours of surgery and require close collaboration with the neurosurgeon. These include increased ICP, bleeding and hypovolemic shock, altered fluid and electrolyte balance (e.g., water intoxication and diabetes insipidus), infection, and seizures.

Monitoring for Increased Intracranial Pressure and Bleeding. Increased ICP and bleeding are life threatening to the patient who has undergone intracranial surgery. The following points must be kept in mind when caring for any patient who has undergone such surgery:

- An increase in blood pressure and decrease in pulse with respiratory failure may indicate increased ICP.
- An accumulation of blood under the bone flap (extradural, subdural, or intracerebral hematoma) may pose a threat to life. A clot must be suspected in any patient who does not awaken as expected or whose condition deteriorates. An intracranial hematoma is suspected if the patient has any new postoperative neurologic deficits (especially a dilated pupil on the operative side). In these circumstances, the patient is returned to the operating room immediately for evacuation of the clot, if indicated.
- Cerebral edema, infarction, metabolic disturbances, and hydrocephalus are conditions that may mimic the clinical manifestations of a clot.

The patient is monitored closely for indicators of complications, and early signs and trends in clinical status are reported to the surgeon. Treatments are initiated promptly, and the nurse assists in evaluating the patient's response to treatment. The nurse also provides support to the patient and family.

Quality and Safety Nursing Alert

If signs and symptoms of increased ICP occur, efforts to decrease the ICP are initiated: alignment of the head in a neutral position without flexion to promote venous drainage, elevation of the head of the bed to 30 degrees (when prescribed), administration of mannitol (an osmotic diuretic), and possible administration of pharmacologic paralyzing agents.

Managing Fluid and Electrolyte Disturbances. Fluid and electrolyte imbalances may occur because of the patient's underlying condition and its management or as complications of surgery. These disturbances can contribute to the development of cerebral edema.

The postoperative fluid regimen depends on the type of neurosurgical procedure and is determined on an individual basis. The volume and composition of fluids are adjusted based on daily serum electrolyte values, along with fluid intake and output. Fluids may have to be restricted in patients with cerebral edema.

Oral fluids are usually resumed after the first 24 hours. The presence of gag and swallowing reflexes must be checked before initiation of oral fluids. Some patients with posterior fossa tumors have impaired swallowing, so fluids may need to be administered by alternative routes. The patient should be observed for signs and symptoms of nausea and vomiting as the diet is progressed (Hickey, 2009).

Patients undergoing surgery for brain tumors often receive large doses of corticosteroids and are at risk for hyperglycemia. Serum glucose levels are measured every 4 to 6 hours, and sliding scale insulin is prescribed as needed (Bader & Littlejohns, 2010). These patients are prone to stress ulcers, so histamine-2 receptor antagonists (H_2 blockers) are prescribed to suppress the secretion of gastric acid. Patients also are monitored for bleeding and assessed for gastric pain.

If the surgical site is near to (or causes edema to) the pituitary gland and hypothalamus, the patient may develop symptoms of diabetes insipidus, which is characterized by excessive urinary output, elevated serum osmolality, decreased urine osmolality, hypernatremia, and a low urine specific gravity. The urine specific gravity is measured hourly, and fluid intake and output are monitored. Fluid replacement must compensate for urine output, and serum potassium levels must be monitored.

SIADH, which results in water retention with hyponatremia and serum hypo-osmolality, occurs in a wide variety of CNS disorders (e.g., brain tumor, head trauma) causing fluid disturbances. Nursing management includes careful intake and output measurements, specific gravity determinations of urine, and monitoring of serum and urine electrolyte levels while following directives for fluid restriction. SIADH is usually self-limited.

Preventing Infection. The patient undergoing neurosurgery is at risk for infection related to the neurosurgical procedure (brain exposure, bone exposure, wound hematomas) and the presence of IV and arterial lines for fluid administration and monitoring. Risk for infection is increased in patients who undergo lengthy intracranial operations, in those who have external ventricular drains in place longer than 5 days, and with those who have ventricular catheters placed outside of a controlled environment such as the operating room or intensive care unit (Sonabend, Korenfeld, Crisman, et al., 2011).

The dressing is often stained with blood in the immediate postoperative period. Because blood is an excellent culture medium for bacteria, the dressing is reinforced with sterile pads so that contamination and infection are avoided. A heavily stained or displaced dressing should be reported immediately. A drain is sometimes placed in the craniotomy incision to facilitate drainage.

After suboccipital surgical procedures, CSF may leak through the incision. This complication is dangerous because of the possibility of meningitis. After a craniotomy, the patient is instructed to avoid coughing, sneezing, or nose blowing, which can cause CSF leakage by creating pressure on the operative site.

Quality and Safety Nursing Alert

Any sudden discharge of fluid from a cranial incision is reported at once, because a large leak requires surgical repair. Attention should be paid to the patient who complains of a salty taste or "postnasal drip," because this can be caused by CSF trickling down the throat.

Aseptic technique is used when handling dressings, drainage systems, and IV and arterial lines. The patient is monitored carefully for signs and symptoms of infection, and cultures are obtained if infection is suspected. Appropriate antibiotic agents are administered as prescribed. Other causes of infection in the patient undergoing intracranial surgery, such as pneumonia and urinary tract infections, are similar to those in other postoperative patients.

Monitoring for Seizure Activity. Seizures may occur as complications after any intracranial neurosurgical procedure. Preventing seizures is essential to avoid further cerebral edema. Administering the prescribed antiseizure medication before and after surgery may prevent the development of seizures in subsequent months and years. **Status epilepticus** (prolonged seizures without recovery of consciousness in the intervals between seizures) may occur after craniotomy and also may be related to the development of complications (hematoma, ischemia). The management of status epilepticus is described later in this chapter.

Monitoring and Managing Other Complications. Other complications may occur during the first 2 weeks or later and may compromise the patient's recovery. The most important of these are VTE (DVT, PE), pulmonary and urinary tract infection, and pressure ulcers. Most of these complications may be avoided with frequent changes of position, adequate suctioning of secretions, thrombosis prophylaxis, early removal of indwelling urinary catheter, early ambulation, and skin care.

Promoting Home and Community-Based Care

Educating Patients About Self-Care. The recovery of a neurosurgical patient at home depends on the extent of the surgical procedure and its success. The patient's strengths as well as limitations are assessed and explained to the family, along with the family's part in promoting recovery. Because administration of antiseizure medication is a priority, the patient and family are educated to use a check-off system, pill boxes, and alarms to ensure that the medication is taken as prescribed.

The patient and family are educated about what to expect after surgery. Dietary restrictions usually are not required unless another health problem necessitates a special diet. Although showering or tub bathing is permitted, the scalp should be kept dry until all sutures have been removed. A clean scarf or cap may be worn until a wig or hairpiece is purchased. If skull bone has been removed, the neurosurgeon may prescribe a protective helmet. After a craniotomy, the patient may require rehabilitation, depending on the postoperative level of function. The patient may require physical therapy for residual weakness and mobility issues. An occupational therapist is consulted to assist with self-care issues. If the patient is aphasic, speech therapy may be necessary.

Continuing Care. Barring complications, patients are discharged from the hospital as soon as possible. Patients with severe motor deficits require extensive physical therapy and rehabilitation. Those with postoperative cognitive and speech impairments require psychological evaluation, speech therapy, and rehabilitation. The nurse collaborates with the physician and other health care professionals during hospitalization and home care to achieve as complete a rehabilitation as possible and to assist the patient in living with residual disability.

If tumor, injury, or disease makes the prognosis poor, care is directed toward making the patient as comfortable as possible. With return of the tumor or cerebral compression, the patient becomes less alert and aware. Other possible consequences include paralysis, blindness, and seizures. The home care nurse, hospice nurse, and social worker collaborate with the family to plan for additional home health care or hospice services or placement of the patient in an extended care facility (see the Cerebral Metastases section in Chapter 70). The patient and family are encouraged to discuss end-of-life preferences for care; the patient's end-of-life preferences must be respected (see Chapter 16). The nurse involved in home and continuing care of patients after cranial surgery also should remind patients and family members of the need for health promotion activities and recommended health screening.

Evaluation

Expected patient outcomes may include:

1. Achieves optimal cerebral tissue perfusion
   a. Opens eyes on request; uses recognizable words, progressing to normal speech
   b. Obeys commands with appropriate motor responses
2. Maintains normal body temperature
   a. Registers normal body temperature
3. Has normal gas exchange
   a. Has arterial blood gas values within normal ranges
   b. Breathes easily; lung sounds clear without adventitious sounds
   c. Takes deep breaths and changes position as directed
4. Copes with sensory deprivation
5. Demonstrates improving self-concept
   a. Pays attention to grooming
   b. Visits and interacts with others
6. Exhibits absence of complications
   a. Exhibits ICP within normal range
   b. Has minimal bleeding at surgical site; surgical incision healing without evidence of infection
   c. Shows fluid balance and electrolyte levels within desired ranges
   d. Exhibits no evidence of seizures

Transsphenoidal Approach

Tumors within the sella turcica and small adenomas of the pituitary can be removed through a transsphenoidal approach (see Table 66-3). Although an otorhinolaryngologist may make the initial opening, the neurosurgeon completes the opening into the sphenoidal sinus and exposes the floor of the sella. Microsurgical techniques provide improved illumination, magnification, and visualization so that nearby vital structures can be avoided.

The transsphenoidal approach offers direct access to the sella turcica with minimal risk of trauma and hemorrhage (Swearingen, 2012). It avoids many of the risks of craniotomy, and the postoperative discomfort is similar to that of other transnasal surgical procedures. It may also be used for pituitary ablation (destruction) in patients with disseminated breast or prostatic cancer.

Complications

Manipulation of the posterior pituitary gland during surgery may produce transient diabetes insipidus of several days' duration (Hickey, 2009). It is treated with vasopressin but occasionally persists. Other complications include CSF leakage, visual disturbances, postoperative meningitis, pneumocephalus (air in the intracranial cavity), and SIADH (see Chapter 52).

Preoperative Management

Medical Management

The preoperative workup includes a series of endocrine tests, rhinologic evaluation (to assess the status of the sinuses and nasal cavity), and neuroradiologic studies. Funduscopic examination and visual field determinations are performed, because the most serious effect of pituitary tumor is localized pressure on the optic nerve or chiasm. In addition, the nasopharyngeal secretions are cultured, because a sinus infection is a contraindication to an intracranial procedure using this approach. Corticosteroids may be administered before and after surgery, because the surgery involves removal of the pituitary, which is the source of adrenocorticotropic hormone (ACTH). Antibiotic agents may or may not be administered prophylactically.

Nursing Management

The patient is educated in deep breathing techniques before surgery. In addition, the patient is instructed that after the surgery, he or she will need to avoid vigorous coughing, blowing the nose, sucking through a straw, or sneezing, because these actions may place increased pressure at the surgical site and cause a CSF leak (Hickey, 2009).

Postoperative Management

Medical Management

Because the procedure disrupts the oral and nasal mucous membranes, management focuses on preventing infection and promoting healing. Medications include antimicrobial agents (which are continued until the nasal packing inserted at the time of surgery is removed), corticosteroids, analgesic agents for discomfort, and agents for the control of diabetes insipidus, if necessary (Hickey, 2009).

Nursing Management

Vital signs are measured to monitor hemodynamic, cardiac, and ventilatory status. Because of the anatomic proximity of the pituitary gland to the optic chiasm, visual acuity and visual fields are assessed at regular intervals. One method is to ask the patient to count the number of fingers held up by the nurse. Evidence of decreasing visual acuity suggests an expanding hematoma.

The head of the bed is raised to decrease pressure on the sella turcica and to promote normal drainage. The patient is cautioned against blowing the nose or engaging in any activity that raises ICP, such as bending over or straining during urination or defecation.

Intake and output are measured as a guide to fluid and electrolyte replacement and to assess for diabetes insipidus. The urine specific gravity is measured after each voiding. Daily weight is monitored. Fluids are usually given after nausea ceases, and the patient then progresses to a regular diet.

The nasal packing inserted during surgery is checked frequently for blood or CSF drainage. The major discomfort is related to the nasal packing and to mouth dryness and thirst caused by mouth breathing. Oral care is provided every 4 hours or more frequently. Usually, the teeth are not brushed until the incision above the teeth has healed. Warm saline mouth rinses and the use of a cool mist vaporizer are helpful. Petrolatum is soothing when applied to the lips. A room humidifier assists in keeping the mucous membranes moist. The packing is removed in 3 to 4 days, and only then can the area around the nares be cleaned with the prescribed solution to remove crusted blood and moisten the mucous membranes (Hickey, 2009).

Home care considerations include advising the patient to use a room humidifier to keep the mucous membranes moist and to soothe irritation. The head of the bed is elevated at 30 degrees for at least 2 weeks after surgery. The patient is cautioned against blowing the nose or sneezing for at least 1 month (Hickey, 2009).

SEIZURE DISORDERS

Seizures are episodes of abnormal motor, sensory, autonomic, or psychic activity (or a combination of these) that result from sudden excessive discharge from cerebral neurons (Hickey, 2009). A part or all of the brain may be involved. The International League Against Epilepsy (ILAE) differentiates between three main seizure types: generalized, focal, and unknown seizures (Chart 66-3). Generalized seizures occur in and rapidly engage bilaterally distributed networks (Berg, Berkovic, Brodie, et al., 2010). Focal seizures are thought to originate within one hemisphere in the brain. The unknown type includes epileptic spasms. Unclassified seizures are so termed because of incomplete data but are not considered a classification (Berg et al., 2010).

Chart 66-3 Classification of Seizures

Generalized Seizures

Tonic–Clonic (in any combination)
Absence
Typical
Atypical
Absence with special features
  Myoclonic absence
Eyelid myoclonia
Myoclonic
  Myoclonic
  Myoclonic atonic
  Myoclonic tonic
Tonic
Clonic
Atonic

Focal Seizures

Unknown

Epileptic spasms

From Berg, A. T., Berkovic, S. F., Brodie, M. J., et al. (2010). Revised terminology and concepts for organization of seizures and epilepsies: Report of the ILAE Commission on Classification and Terminology, 2005–2009. *Epilepsia*, *51*(4), 676–685.

Pathophysiology

The underlying cause is an electrical disturbance (dysrhythmia) in the nerve cells in one section of the brain; these cells emit abnormal, recurring, uncontrolled electrical discharges. The characteristic seizure is a manifestation of this excessive neuronal discharge. Associated loss of consciousness, excess movement or loss of muscle tone or movement, and disturbances of behavior, mood, sensation, and perception may also occur.

The specific causes of seizures are varied and can be categorized as genetic, due to a structural or metabolic condition, or the cause may be yet unknown (Berg et al., 2010). Causes of seizures include:

- Cerebrovascular disease
- Hypoxemia of any cause, including vascular insufficiency
- Fever (childhood)
- Head injury
- Hypertension
- CNS infections
- Metabolic and toxic conditions (e.g., renal failure, hyponatremia, hypocalcemia, hypoglycemia, pesticide exposure)
- Brain tumor
- Drug and alcohol withdrawal
- Allergies

Clinical Manifestations

Depending on the location of the discharging neurons, seizures may range from a simple staring episode (absence seizure) to prolonged convulsive movements with loss of consciousness.

The initial pattern of the seizures indicates the region of the brain in which the seizure originates (see Chart 66-3). Only a finger or hand may shake, or the mouth may jerk uncontrollably. The person may talk unintelligibly; may be dizzy; and may experience unusual or unpleasant sights, sounds, odors, or tastes, but without loss of consciousness (Hickey, 2009).

Generalized seizures often involve both hemispheres of the brain, causing both sides of the body to react (Berg et al., 2010). Intense rigidity of the entire body may occur, followed by alternating muscle relaxation and contraction (generalized tonic–clonic contraction). The simultaneous contractions of the diaphragm and chest muscles may produce a characteristic epileptic cry. The tongue is often chewed, and the patient is incontinent of urine and feces. After 1 or 2 minutes, the convulsive movements begin to subside; the patient relaxes and lies in deep coma, breathing noisily. The respirations at this point are chiefly abdominal. In the postictal state (after the seizure), the patient is often confused and hard to arouse and may sleep for hours. Many patients report headache, sore muscles, fatigue, and depression (AANN, 2009). Generalized seizures may also be asymptomatic, as with absence types of seizures (Berg et al., 2010).

Focal seizures have no natural classification. There may be an impairment of consciousness or awareness or other dyscognitive features, localization, and progression of ictal events (Berg et al., 2010).

Assessment and Diagnostic Findings

The diagnostic assessment is aimed at determining the type of seizures, their frequency and severity, and the factors that precipitate them. A developmental history is taken, including events of pregnancy and childbirth, to seek evidence of preexisting injury. The patient is also questioned about illnesses or head injuries that may have affected the brain. In addition to physical and neurologic evaluations, diagnostic examinations include biochemical, hematologic, and serologic studies. MRI is used to detect structural lesions such as focal abnormalities, cerebrovascular abnormalities, and cerebral degenerative changes (AANN, 2009).

The EEG furnishes diagnostic evidence for a substantial proportion of patients with epilepsy and assists in classifying the type of seizure (Karpoff & Labus, 2008). Abnormalities in the EEG usually continue between seizures or, if not apparent, may be elicited by hyperventilation or during sleep (Kotagal & Yardi, 2008). Microelectrodes (depth electrodes) can be inserted deep in the brain to probe the action of single brain cells. Some people with clinical seizures have normal EEGs, whereas others who have never had seizures have abnormal EEGs. Telemetry and computerized equipment are used to monitor electrical brain activity while the patient pursues his or her normal activities and to store the readings on computer tapes for analysis. Video recording of seizures taken simultaneously with EEG telemetry is useful in determining the type of seizure as well as its duration and magnitude. This type of intensive monitoring is changing the treatment of severe epilepsy (Dilorio, Reisinger, Yeager, et al., 2008).

SPECT is an additional tool that is sometimes used in the diagnostic workup. It is useful for identifying the epileptogenic zone so that the area in the brain giving rise to seizures can be removed surgically (AANN, 2009).

Nursing Management

During a Seizure

A major responsibility of the nurse is to observe and record the sequence of signs. The nature of the seizure usually indicates the type of treatment required (AANN, 2009). Before and during a seizure, the patient is assessed and the following items are documented:

- Circumstances before the seizure (visual, auditory, or olfactory stimuli; tactile stimuli; emotional or psychological disturbances; sleep; hyperventilation)
- Occurrence of an aura (a premonitory or warning sensation, which can be visual, auditory, or olfactory)
- First thing the patient does in the seizure—where the movements or the stiffness begins, conjugate gaze position, and the position of the head at the beginning of the seizure. This information gives clues to the location of the seizure origin in the brain. (In recording, it is important to state whether the beginning of the seizure was observed.)
- Type of movements in the part of the body involved
- Areas of the body involved (turn back bedding to expose patient)
- Size of both pupils and whether the eyes are open
- Whether the eyes or head are turned to one side
- Presence or absence of automatisms (involuntary motor activity, such as lip smacking or repeated swallowing)
- Incontinence of urine or stool
- Duration of each phase of the seizure
- Unconsciousness, if present, and its duration
- Any obvious paralysis or weakness of arms or legs after the seizure
- Inability to speak after the seizure
- Movements at the end of the seizure
- Whether or not the patient sleeps afterward
- Cognitive status (confused or not confused) after the seizure

In addition to providing data about the seizure, nursing care is directed at preventing injury and supporting the patient, not only physically but also psychologically. Consequences such as anxiety, embarrassment, fatigue, and depression can be devastating to the patient.

After a Seizure

After a patient has a seizure, the nurse's role is to document the events leading to and occurring during and after the seizure and to prevent complications (e.g., aspiration, injury). The patient is at risk for hypoxia, vomiting, and pulmonary aspiration. To prevent complications, the patient is placed in the side-lying position to facilitate drainage of oral secretions, and suctioning is performed, if needed, to maintain a patent airway and prevent aspiration (Chart 66-4). Seizure precautions are maintained, including having available functioning suction equipment with a suction catheter and oral airway.

Chart 66-4

GUIDELINES

Seizure Care

Nursing Care During a Seizure

- Provide privacy, and protect the patient from curious onlookers. (The patient who has an aura [warning of an impending seizure] may have time to seek a safe, private place.)
- Ease the patient to the floor, if possible.
- Protect the head with a pad to prevent injury (from striking a hard surface).
- Loosen constrictive clothing.
- Push aside any furniture that may injure the patient during the seizure.
- If the patient is in bed, remove pillows and raise side rails.
- If an aura precedes the seizure, insert an oral airway to reduce the possibility of the patient biting the tongue or cheek.
- *Do not attempt to pry open jaws that are clenched in a spasm or attempt to insert anything.* Broken teeth and injury to the lips and tongue may result from such an action.
- No attempt should be made to restrain the patient during the seizure, because muscular contractions are strong and restraint can produce injury.
- If possible, place the patient on one side with head flexed forward, which allows the tongue to fall forward and facilitates drainage of saliva and mucus. If suction is available, use it if necessary to clear secretions.

Nursing Care After the Seizure

- Keep the patient on one side to prevent aspiration. Make sure the airway is patent.
- There is usually a period of confusion after a grand mal seizure.
- A short apneic period may occur during or immediately after a generalized seizure.
- The patient, on awakening, should be reoriented to the environment.
- If the patient becomes agitated after a seizure (postictal), use persuasion and gentle restraint to assist him or her to stay calm.

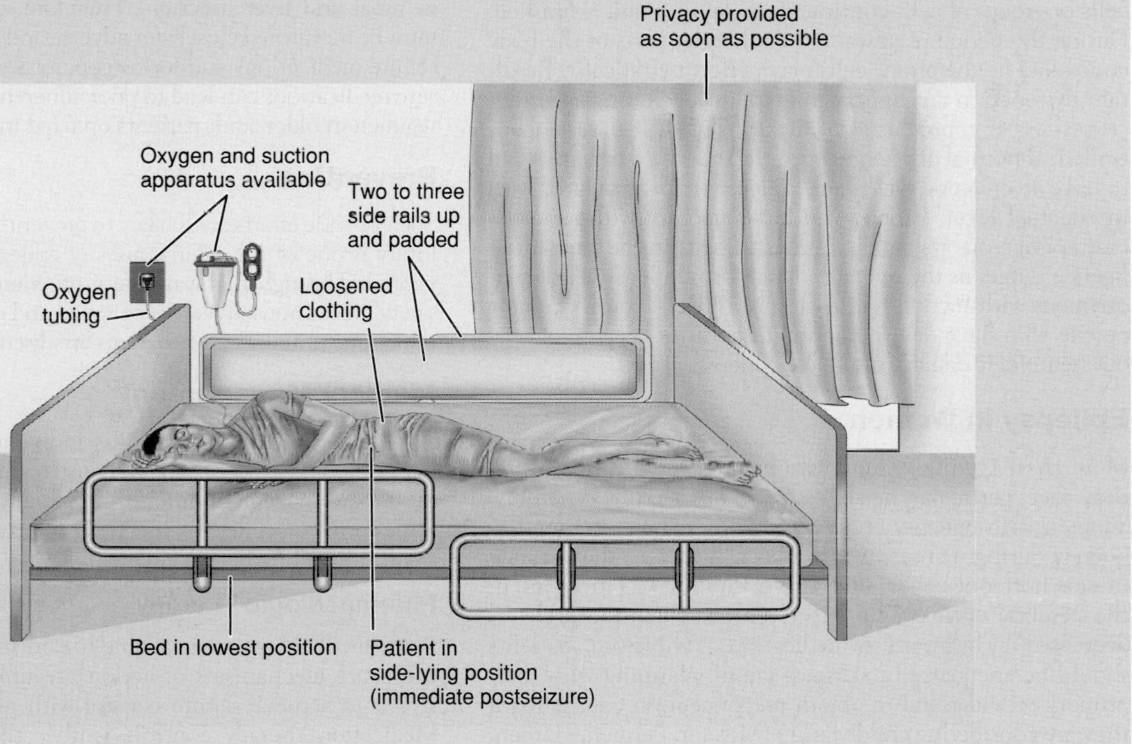

Privacy provided as soon as possible
Oxygen and suction apparatus available
Two to three side rails up and padded
Oxygen tubing
Loosened clothing
Bed in lowest position
Patient in side-lying position (immediate postseizure)

Adapted from American Association of Neuroscience Nurses. (2009). *Care of the patient with seizures: AANN clinical practice guidelines series.* Glenview, IL: Author.

The bed is placed in a low position with two to three side rails up and padded, if necessary, to prevent injury to the patient. The patient may be drowsy and may wish to sleep after the seizure; he or she may not remember events leading up to the seizure and for a short time thereafter.

The Epilepsies

Epilepsy is a group of syndromes characterized by unprovoked, recurring seizures (AANN, 2009). Epileptic syndromes are classified by specific patterns of clinical features, including age at onset, family history, and seizure type. The epilepsies include electroclinical syndromes (a complex of clinical features, signs, and symptoms) and other epilepsies (Berg et al., 2010; IOM, 2012). Epilepsy can be primary (idiopathic) or secondary (when the cause is known and the epilepsy is a symptom of another underlying condition, such as a brain tumor).

Epilepsy affects an estimated 3% of people during their lifetime, and most forms of epilepsy occur in childhood (Bader & Littlejohns, 2010). The improved treatment for cerebrovascular disorders, head injuries, brain tumors, meningitis, and encephalitis has increased the number of patients at risk for seizures after recovery from these conditions. In addition, advances in EEG have aided in the diagnosis of epilepsy. The general public has been educated about epilepsy, which has reduced the stigma associated with it; as a result, more people are willing to acknowledge that they have epilepsy.

Although some evidence suggests that susceptibility to some types of epilepsy may be inherited, the cause of seizures in many people is idiopathic (unknown). Epilepsy can follow birth trauma, asphyxia neonatorum, head injuries, some infectious diseases (bacterial, viral, parasitic), toxicity (carbon monoxide and lead poisoning), circulatory problems, fever, metabolic and nutritional disorders, or drug or alcohol intoxication. It is also associated with brain tumors, abscesses, and congenital malformations.

Pathophysiology

Messages from the body are carried by the neurons (nerve cells) of the brain by means of discharges of electrochemical energy that sweep along them. These impulses occur in bursts whenever a nerve cell has a task to perform. Sometimes, these cells or groups of cells continue firing after a task is finished. During the period of unwanted discharges, parts of the body controlled by the errant cells may perform erratically. Resultant dysfunction ranges from mild to incapacitating and often causes loss of consciousness (Hickey, 2009). If these uncontrolled, abnormal discharges occur repeatedly, a person is said to have an epileptic syndrome. Epilepsy is not associated with intellectual level. People who have epilepsy without other brain or nervous system disabilities fall within the same intelligence ranges as the overall population. Epilepsy is not synonymous with mental retardation or illness. However, many people who have developmental disabilities because of serious neurologic damage also have epilepsy.

Epilepsy in Women

More than 1 million American women have epilepsy, and they face particular needs associated with the syndrome. Women with epilepsy often note an increase in seizure frequency during menses; this has been linked to the increase in sex hormones that alter the excitability of neurons in the cerebral cortex. The effectiveness of contraceptives is decreased by antiseizure medications. Therefore, patients should be encouraged to discuss family planning with their primary provider and to obtain preconception counseling if they are considering childbearing (Meador, Pennell, Harden, et al., 2008).

Women of childbearing age who have epilepsy require special care and guidance before, during, and after pregnancy. Many women note a change in the pattern of seizure activity during pregnancy. The risk of congenital fetal anomaly is two to three times higher in women with epilepsy. Maternal seizures, antiseizure medications, and genetic predisposition all contribute to possible malformations. Women who take certain antiseizure medications for epilepsy are at risk and need careful monitoring, including blood studies to detect the level of antiseizure medications taken throughout pregnancy. High-risk mothers (teenagers, women with histories of difficult deliveries, women who use illicit drugs [e.g., "crack" cocaine], and women with diabetes or hypertension) should be identified and monitored closely during pregnancy, because damage to the fetus during pregnancy and delivery can increase the risk of epilepsy. All of these issues need further study (Meador et al., 2008).

Because of bone loss associated with the long-term use of antiseizure medications, patients receiving antiseizure agents should be assessed for low bone mass and osteoporosis. They should be instructed about strategies to reduce their risks of osteoporosis (AANN, 2009).

Gerontologic Considerations

Older adults have a high incidence of new-onset epilepsy (Hickey, 2009). Cerebrovascular disease is the leading cause of seizures in the older adult. The increased incidence is also associated with head injury, dementia, infection, alcoholism, and aging. Treatment depends on the underlying cause. Because many older adults have chronic health problems, they may be taking other medications that can interact with medications prescribed for seizure control. In addition, the absorption, distribution, metabolism, and excretion of medications are altered in the older adult as a result of age-related changes in renal and liver function. Therefore, older adult patients must be monitored closely for adverse and toxic effects of antiseizure medications and for osteoporosis. The cost of antiseizure medications can lead to poor adherence to the prescribed regimen in older adult patients on fixed incomes.

Prevention

Society-wide efforts are the key to prevention of epilepsy. Head injury is one of the main causes of epilepsy that can be prevented. Through highway safety programs and occupational safety precautions, lives can be saved and epilepsy due to head injury prevented; these programs are discussed in Chapter 68.

Medical Management

The management of epilepsy is individualized to meet the needs of each patient and not just to manage and prevent seizures. Management differs from patient to patient, because some forms of epilepsy arise from brain damage and others result from altered brain chemistry.

Pharmacologic Therapy

Many medications are available to control seizures, although the exact mechanisms of action are unknown. The objective is to achieve seizure control with minimal side effects. Medication therapy controls—rather than cures—seizures. Medications are selected on the basis of the type of seizure being treated and the effectiveness and safety of the medications. If properly prescribed and taken, medications control seizures in 70% to 80% of patients with seizures. However, 20% of patients with generalized seizures and 30% of those with focal seizures do not demonstrate improvement with any prescribed medication or may be unable to tolerate the side effects of medications (AANN, 2009). Table 66-4 lists selected antiseizure medications.

Treatment usually starts with a single medication. The starting dose and the rate at which the dosage is increased depend on the occurrence of side effects. The medication levels in the blood are monitored, because the rate of drug absorption varies among patients. Changing to another medication may be necessary if seizure control is not achieved or if toxicity makes it impossible to increase the dosage. The

TABLE 66-4 Selected Antiseizure Medications

Medication	Dose-Related Side Effects	Toxic Effects
carbamazepine (Tegretol)	Dizziness, drowsiness, unsteadiness, nausea and vomiting, diplopia, mild leukopenia	Severe skin rash, blood dyscrasias, hepatitis
clonazepam (Klonopin)	Drowsiness, behavior changes, headache, hirsutism, alopecia, palpitations	Hepatotoxicity, thrombocytopenia, bone marrow failure, ataxia
ethosuximide (Zarontin)	Nausea and vomiting, headache, gastric distress	Skin rash, blood dyscrasias, hepatitis, systemic lupus erythematosus
felbamate (Felbatol)	Cognitive impairments, insomnia, nausea, headache, fatigue	Aplastic anemia, hepatotoxicity
gabapentin (Neurontin)	Dizziness, drowsiness, somnolence, fatigue, ataxia, weight gain, nausea	Leukopenia, hepatotoxicity
lamotrigine (Lamictal)	Drowsiness, tremor, nausea, ataxia, dizziness, headache, weight gain	Severe rash (Stevens-Johnson syndrome)
levetiracetam (Keppra)	Somnolence, dizziness, fatigue	Unknown
oxcarbazepine (Trileptal)	Dizziness, somnolence, double vision, fatigue, nausea, vomiting, loss of coordination, abnormal vision, abdominal pain, tremor, abnormal gait	Hepatotoxicity
phenobarbital (Luminal)	Sedation, irritability, diplopia, ataxia	Skin rash, anemia
phenytoin (Dilantin)	Visual problems, hirsutism, gingival hyperplasia, dysrhythmias, dysarthria, nystagmus	Severe skin reaction, peripheral neuropathy, ataxia, drowsiness, blood dyscrasias
primidone (Mysoline)	Lethargy, irritability, diplopia, ataxia, impotence	Skin rash
tiagabine (Gabitril)	Dizziness, fatigue, nervousness, tremor, difficulty concentrating, dysarthria, weak or buckling knees, abdominal pain	Unknown
topiramate (Topamax)	Fatigue, somnolence, confusion, ataxia, anorexia, depression, weight loss	Nephrolithiasis
valproate (Depakote, Depakene)	Nausea and vomiting, weight gain, hair loss, tremor, menstrual irregularities	Hepatotoxicity, skin rash, blood dyscrasias, nephritis
zonisamide (Zonegran, Excegran)	Somnolence, dizziness, anorexia, headache, nausea, agitation, rash	Leukopenia, hepatotoxicity

Adapted from Karch, A. (2012). *2012 Lippincott's nursing drug guide*. Philadelphia: Lippincott Williams & Wilkins.

medication may need to be adjusted because of concurrent illness, weight changes, or increases in stress. Side effects of antiseizure medications may be divided into three groups: (1) idiosyncratic or allergic disorders, which manifest primarily as skin reactions; (2) acute toxicity, which may occur when the medication is initially prescribed; and (3) chronic toxicity, which occurs late in the course of therapy.

The manifestations of drug toxicity are variable, and any organ system may be involved. For example, gingival hyperplasia (swollen and tender gums) can be associated with long-term use of phenytoin (Dilantin) (Karch, 2012). Periodic physical and dental examinations and laboratory tests are performed for patients receiving medications that are known to have hematopoietic, genitourinary, or hepatic effects.

Surgical Management

Surgery is indicated for patients whose epilepsy results from intracranial tumors, abscesses, cysts, or vascular anomalies. Some patients have intractable seizure disorders that do not respond to medication. A focal atrophic process may occur secondary to trauma, inflammation, stroke, or anoxia. If the seizures originate in a reasonably well-circumscribed area of the brain that can be excised without producing significant neurologic deficits, the removal of the area generating the seizures may produce long-term control and improvement (AANN, 2009).

This type of neurosurgery has been aided by several advances, including microsurgical techniques, EEGs with depth electrodes, improved illumination and hemostasis, and the introduction of neuroleptanalgesic agents (droperidol and fentanyl). These techniques, combined with the use of local anesthetic agents, enable the neurosurgeon to perform surgery on an alert and cooperative patient. Using special testing devices, electrocortical mapping, and the patient's responses to stimulation, the boundaries of the epileptogenic focus (i.e., abnormal area of the brain) are determined. Any abnormal epileptogenic focus is then excised (AANN, 2009). Resection surgery significantly reduces the incidence of seizures in patients with refractory epilepsy.

When seizures are refractory to medication in adolescents and adults with focal seizures, a generator may be implanted under the clavicle. The device is connected to the vagus nerve in the cervical area, where it delivers electrical signals to the brain to control and reduce seizure activity (AANN, 2009). An external programming system is used by the physician to change stimulator settings. Patients can activate the stimulator with a magnet at the time of a seizure or aura. Some patients report that use of the vagal nerve stimulator diminishes the severity or duration of the seizure. Complications such as hoarseness, cough, and shortness of breath can occur with the use of this device (Noe, 2011).

More research is needed to determine the effects of the various surgical approaches on complication rates, quality of life, anxiety, and depression, all of which are issues for patients with epilepsy.

NURSING PROCESS

The Patient With Epilepsy

Assessment

The nurse elicits information about the patient's seizure history. The patient is asked about the factors or events that may precipitate the seizures. Alcohol intake is documented. The nurse determines whether the patient has an aura before an epileptic seizure, which may indicate the origin of the seizure (e.g., seeing a flashing light may indicate that the seizure originated in the occipital lobe). Observation and assessment during and after a seizure assist in identifying the type of seizure and its management.

The effects of epilepsy on the patient's lifestyle are assessed (AANN, 2009). What limitations are imposed by the seizure disorder? Does the patient participate in any recreational activities? Have any social contacts? Is the patient working, and is it a positive or stressful experience? What coping mechanisms are used?

Diagnosis

Nursing Diagnoses

Based on the assessment data, major nursing diagnoses may include the following:

- Risk for injury related to seizure activity
- Fear related to the possibility of seizures
- Ineffective individual coping related to stresses imposed by epilepsy
- Deficient knowledge related to epilepsy and its control

Collaborative Problems/Potential Complications

The major potential complications for patients with epilepsy are status epilepticus and medication side effects (toxicity).

Planning and Goals

The major goals for the patient may include prevention of injury, control of seizures, achievement of a satisfactory psychosocial adjustment, acquisition of knowledge and understanding about the condition, and absence of complications.

Nursing Interventions

Preventing Injury

Injury prevention for the patient with seizures is a priority. Patients for whom seizure precautions are instituted should have pads applied to the side rails while in bed. Steps to prevent or minimize injury are presented in Chart 66-4.

Reducing Fear of Seizures

Fear that a seizure may occur unexpectedly can be reduced by the patient's adherence to the prescribed treatment regimen. Cooperation of the patient and family and their trust in the prescribed regimen are essential for control of seizures. The nurse emphasizes that the prescribed antiseizure medication must be taken on a continuing basis and that drug dependence or addiction does not occur. Periodic monitoring is necessary to ensure the adequacy of the treatment regimen, to prevent side effects, and to monitor for drug resistance (Hickey, 2009).

In an effort to control seizures, factors that may precipitate them are identified, such as emotional disturbances, new environmental stressors, onset of menstruation in female patients, or fever (AANN, 2009). The patient is encouraged to follow a regular and moderate routine in lifestyle, diet (avoiding excessive stimulants), exercise, and rest (sleep deprivation may lower the seizure threshold). Moderate activity is therapeutic, but excessive exercise should be avoided. An additional dietary intervention, referred to as the ketogenic diet, may be helpful for control of seizures in some patients. This high-protein, low-carbohydrate, high-fat diet is most effective in children whose seizures have not been controlled with two antiseizure medications, but it is sometimes used for adults who have had poor seizure control (Mosek, Natour, Neufeld, et al., 2009).

Photic stimulation (e.g., bright flickering lights, television viewing) may precipitate seizures; wearing dark glasses or covering one eye may be preventive. Tension states (anxiety, frustration) induce seizures in some patients. Classes in stress management may be of value. Because seizures are known to occur with alcohol intake, alcoholic beverages should be avoided.

Improving Coping Mechanisms

The social, psychological, and behavioral problems that frequently accompany epilepsy can be more of a disability than the actual seizures. Epilepsy may be accompanied by feelings of stigmatization, alienation, depression, and uncertainty (IOM, 2012). The patient must cope with the constant fear of a seizure and the psychological consequences (AANN, 2009). Children with epilepsy may be ostracized and excluded from school and peer activities. These problems are compounded during adolescence and add to the challenges of dating, not being able to drive, and feeling different from other people. Adults face these problems in addition to the burden of finding employment, concerns about relationships and childbearing, insurance problems, and legal barriers. Alcohol abuse may complicate matters. Family reactions may vary from outright rejection of the person with epilepsy to overprotection.

Counseling assists the patient and family to understand the condition and the limitations it imposes. Social and recreational opportunities are necessary for good mental health. Nurses can improve the quality of life for patients with epilepsy by educating them and their families about symptoms and their management (AANN, 2009).

Providing Patient and Family Education

Perhaps the most valuable facets of care contributed by the nurse to the person with epilepsy are education and efforts to modify the attitudes of the patient and family toward the disorder. The person who experiences seizures may consider every seizure a potential source of humiliation and shame. This may result in anxiety, depression, hostility, and secrecy on the part of the patient and family. Ongoing education and encouragement should be given to patients to enable them to overcome these reactions. The patient with epilepsy should carry an emergency medical identification card or wear a medical information bracelet. The patient and family need to be educated about medications as well as care during a seizure.

Monitoring and Managing Potential Complications

Status epilepticus—the major complication—is described later in this chapter. Another complication is the toxicity of medications. The patient and family are instructed about side effects and are given specific guidelines to assess and report signs and symptoms that indicate medication overdose. Antiseizure medications require careful monitoring for

Chart 66-5 HOME CARE CHECKLIST
The Patient With Epilepsy

At the completion of home care education, the patient and caregiver will be able to:	PATIENT	CAREGIVER
• Take medications daily as prescribed to keep the drug level constant to prevent seizures. The patient should never discontinue medications, even if there is no seizure activity.	✔	
• Keep a medication and seizure record, noting when medications are taken and any seizure activity.	✔	✔
• Notify the primary provider if the patient cannot take medications due to illness.	✔	✔
• Have antiseizure medication serum levels checked regularly. When testing is prescribed, the patient should report to the laboratory for blood sampling before taking morning medication.	✔	
• Avoid activities that require alertness and coordination (driving, operating machinery) until after the effects of the medication have been evaluated.	✔	
• Report signs of toxicity so that dosage can be adjusted. Common signs include drowsiness, lethargy, dizziness, difficulty walking, hyperactivity, confusion, inappropriate sleep, and visual disturbances.	✔	✔
• Avoid over-the-counter medications unless approved by the primary provider.	✔	
• Carry a medical alert bracelet or identification card specifying the name of the patient's antiseizure medication and physician.	✔	
• Avoid seizure triggers, such as alcoholic beverages, electrical shocks, stress, caffeine, constipation, fever, hyperventilation, and hypoglycemia.	✔	
• Take showers rather than tub baths to avoid drowning if seizure occurs; never swim alone.	✔	
• Exercise in moderation in a temperature-controlled environment to avoid excessive heat.	✔	
• Develop regular sleep patterns to minimize fatigue and insomnia.	✔	✔
• Use the Epilepsy Foundation of America's special services, including help in obtaining medications, vocational rehabilitation, and coping with epilepsy.	✔	✔

therapeutic levels. The patient should plan to have serum drug levels assessed at regular intervals. Many known drug interactions occur with antiseizure medications. A complete pharmacologic profile should be reviewed with the patient to avoid interactions that either potentiate or inhibit the effectiveness of the medications.

Quality and Safety Nursing Alert

Patients with epilepsy are at risk for status epilepticus from having their medication regimen interrupted.

Promoting Home and Community-Based Care

Educating Patients About Self-Care. Thorough oral hygiene after each meal, gum massage, daily flossing, and regular dental care are essential to prevent or control gingival hyperplasia in patients receiving phenytoin. The patient is also educated to inform all health care providers of the medication being taken, because of the possibility of drug interactions. An individualized comprehensive education plan is needed to assist the patient and family to adjust to this chronic disorder. Written patient education materials must be appropriate for the patient's reading level and must be provided in alternative formats if warranted. See Chart 66-5 for home care education points.

Continuing Care. Because epilepsy can be lifelong, the use of costly medications can create a significant financial burden. The Epilepsy Foundation of America (EFA) offers a mail-order program to provide medications at minimal cost and access to life insurance. This organization also serves as a referral source for special services for people with epilepsy.

For many, overcoming employment problems is a challenge. State vocational rehabilitation agencies can provide information about job training. The EFA has a training and placement service. If seizures are not well controlled, information about sheltered workshops or home employment programs may be obtained. Federal and state agencies and federal legislation may be of assistance to people with epilepsy who experience job discrimination. As a result of the Americans with Disabilities Act, the number of employers who knowingly hire people with epilepsy is increasing, but barriers to employment still exist.

People who have uncontrollable seizures accompanied by psychological and social difficulties can be referred to comprehensive epilepsy centers where continuous audio-video and EEG monitoring, specialized treatment, and rehabilitation services are available (AANN, 2009). Patients and their families need to be reminded of the importance of following the prescribed treatment regimen and of keeping follow-up appointments. In addition, they are reminded of the importance of participating in health promotion activities and recommended health screenings to promote a healthy lifestyle. Genetic and preconception counseling is advised.

Evaluation

Expected patient outcomes may include:

1. Sustains no injury during seizure activity
   a. Complies with treatment regimen and identifies the hazards of stopping the medication

b. Can identify appropriate care during seizure; caregivers can do so as well
2. Indicates a decrease in fear
3. Displays effective individual coping
4. Exhibits knowledge and understanding of epilepsy
   a. Identifies the side effects of medications
   b. Avoids factors or situations that may precipitate seizures (e.g., flickering lights, hyperventilation, alcohol)
   c. Follows a healthy lifestyle by getting adequate sleep and eating meals at regular times to avoid hypoglycemia
5. Absence of complications

Status Epilepticus

Status epilepticus (acute prolonged seizure activity) is a series of generalized seizures that occur without full recovery of consciousness between attacks (Rabinstein, 2010). The term has been broadened to include continuous clinical or electrical seizures (on EEG) lasting at least 30 minutes, even without impairment of consciousness. It is considered a medical emergency. Status epilepticus produces cumulative effects. Vigorous muscular contractions impose a heavy metabolic demand and can interfere with respirations. Some respiratory arrest at the height of each seizure produces venous congestion and hypoxia of the brain. Repeated episodes of cerebral anoxia and edema may lead to irreversible and fatal brain damage. Factors that precipitate status epilepticus include withdrawal of antiseizure medication, fever, concurrent infection, or other illness.

Medical Management

The goals of treatment are to stop the seizures as quickly as possible, to ensure adequate cerebral oxygenation, and to maintain the patient in a seizure-free state. An airway and adequate oxygenation are established. If the patient remains unconscious and unresponsive, an endotracheal tube is inserted. IV diazepam (Valium), lorazepam (Ativan), or fosphenytoin is administered slowly in an attempt to halt seizures immediately. Other medications (phenytoin, phenobarbital) are administered later to maintain a seizure-free state.

An IV line is established, and blood samples are obtained to monitor serum electrolytes, glucose, and phenytoin levels. EEG monitoring may be useful in determining the nature of the seizure activity. Vital signs and neurologic signs are monitored on a continuing basis. An IV infusion of dextrose is administered if the seizure is caused by hypoglycemia. If initial treatment is unsuccessful, general anesthesia with a short-acting barbiturate may be used. The serum concentration of the antiseizure medication is measured, because a low level suggests that the patient was not taking the medication or that the dosage was too low. Cardiac involvement or respiratory depression may be life threatening. The potential for postictal cerebral edema also exists.

Nursing Management

The nurse initiates ongoing assessment and monitoring of respiratory and cardiac function because of the risk for delayed depression of respiration and blood pressure secondary to administration of antiseizure medications and sedatives to halt the seizures. Nursing assessment also includes monitoring and documenting the seizure activity and the patient's responsiveness.

The patient is turned to a side-lying position, if possible, to assist in draining pharyngeal secretions. Suction equipment must be available because of the risk of aspiration. The IV line is closely monitored, because it may become dislodged during seizures.

A person who has received long-term antiseizure therapy has a significant risk for fractures resulting from bone disease (osteoporosis, osteomalacia, and hyperparathyroidism), which is a side effect of therapy (Karch, 2012). Therefore, during seizures, the patient is protected from injury with the use of seizure precautions and is monitored closely. The patient having seizures can inadvertently injure nearby people, so nurses should protect themselves. Additional nursing interventions for the person having seizures are presented in Chart 66-4.

HEADACHE

Headache, or cephalalgia, is one of the most common of all human physical complaints. Headache is a symptom rather than a disease entity; it may indicate organic disease (neurologic or other disease), a stress response, vasodilation (migraine), skeletal muscle tension (tension headache), or a combination of factors. A **primary headache** is one for which no organic cause can be identified. This type of headache includes migraine, tension-type, and cluster headaches (Hickey, 2009). Cranial arteritis is another common cause of headache. A classification of headaches was issued first by the Headache Classification Committee of the International Headache Society in 1988. The International Headache Society revised the headache classification in 2004; an abbreviated list is shown in Chart 66-6.

Chart 66-6 International Headache Society Classification of Headache

1. Migraine
2. Tension-type headache
3. Cluster headache and other trigeminal autonomic cephalalgias
4. Other primary headaches
5. Headache attributed to head and/or neck trauma
6. Headache attributed to cranial or cervical vascular disorder
7. Headache attributed to nonvascular intracranial disorder
8. Headache attributed to a substance or its withdrawal
9. Headache attributed to infection
10. Headache attributed to disorder of homeostasis
11. Headache or facial pain attributed to disorder of cranium, neck, eyes, ears, nose, sinuses, teeth, mouth, or other facial or cranial structures
12. Headache attributed to psychiatric disorder
13. Cranial neuralgias and central causes of facial pain
14. Other headache

Adapted from Headache Classification Subcommittee of the International Headache Society. (2004). *The International Classification of Headache Disorders:* 2nd edition. *Cephalalgia*, *24*(Suppl. 1), 1–150.

Migraine is a complex of symptoms characterized by periodic and recurrent attacks of severe headache lasting from 4 to 72 hours in adults. The cause of migraine has not been clearly demonstrated, but it is primarily a vascular disturbance that has a strong familial tendency. The typical time of onset is at puberty, and the incidence is higher in women than men (Ward, 2012).

There are many subtypes of **migraine headache,** including migraine with and without aura. Most patients have migraine without an aura. Tension-type headaches tend to be chronic and less severe and are probably the most common type of headache. Cluster headaches are a severe form of vascular headache. They are seen five times more frequently in men than in women. Types of headaches not subsumed under these categories fall into the other primary headache group and include headaches triggered by cough, exertion, and sexual activity.

Cranial arteritis is a cause of headache in the older population, reaching its greatest incidence in those older than 70 years of age. Inflammation of the cranial arteries is characterized by a severe headache localized in the region of the temporal arteries. The inflammation may be generalized (in which case cranial arteritis is part of a vascular disease) or focal (in which case only the cranial arteries are involved).

A **secondary headache** is a symptom associated with another organic cause, such as a brain tumor or an aneurysm. Although most headaches do not indicate serious disease, persistent headaches require further investigation. Serious disorders related to headache include brain tumors, subarachnoid hemorrhage, stroke, severe hypertension, meningitis, and head injuries.

Pathophysiology

The cerebral signs and symptoms of migraine result from a hyperexcitable brain that is susceptible to a phenomenon known as **cortical spreading depression** (CSD). CSD can be described as a wave of depolarization over the cerebral cortex, cerebellum, and hippocampus. This depolarization activates inflammatory neuropeptides and other neurotransmitters (including serotonin), resulting in the stimulation of meningeal nociceptors. Vascular changes, inflammation, and a continuation of pain signal stimulation occur. The initial phase of this process is known as peripheral sensitization. If treatment is initiated at this point, the migraine may be fully terminated. As the attack progresses, central sensitization occurs, and the migraine becomes much harder to treat. Central sensitization can manifest as scalp, neck, face, or extremity pain (Ward, 2012).

Migraines are often hereditary and associated with low brain magnesium levels. Attacks can be triggered by hormonal changes associated with menstrual cycles, bright lights, stress, depression, sleep deprivation, fatigue, or odors. Certain foods containing tyramine (especially aged cheese), monosodium glutamate, nitrites, or milk products may be food triggers. The use of oral contraceptives may be associated with increased frequency and severity of attacks in some women.

Emotional or physical stress may cause contraction of the muscles in the neck and scalp, resulting in tension headache. The pathophysiology of cluster headache is not fully understood. One theory is that it is caused by dilation of orbital and nearby extracranial arteries. Cranial arteritis is thought to represent an immune vasculitis in which immune complexes are deposited within the walls of affected blood vessels, producing vascular injury and inflammation. A biopsy may be performed on the involved artery to make the diagnosis.

Clinical Manifestations

Migraine

The migraine with aura can be divided into four phases: prodrome, aura, the headache, and recovery (headache termination and postdrome).

Prodrome Phase

The prodrome phase is experienced by 60% of patients, with symptoms that occur hours to days before a migraine headache. Symptoms may include depression, irritability, feeling cold, food cravings, anorexia, change in activity level, increased urination, diarrhea, or constipation. Patients usually experience the same prodrome with each migraine headache.

Aura Phase

Aura occurs in a minority of patients who experience migraines (Ward, 2012). The aura usually lasts less than 1 hour and may provide enough time for the patient to take the prescribed medication to avert an attack (see later discussion). This period is characterized by focal neurologic symptoms. Visual disturbances (i.e., light flashes and bright spots) are most common and may be hemianopic (affecting only half of the visual field). Other symptoms that may follow include numbness and tingling of the lips, face, or hands; mild confusion; slight weakness of an extremity; drowsiness; and dizziness.

This period of aura corresponds to the phenomenon of cortical CSD that is associated with reduced metabolic demand in abnormally functioning neurons. This can be associated with decreased blood flow; however, cerebral blood flow studies performed during migraine headaches demonstrate that although changes in blood vessels occur during phases of migraine, cerebral blood flow is not the main abnormality (Ward, 2012).

Headache Phase

As a decline in serotonin levels occur, a throbbing headache (unilateral in 60% of patients) intensifies over several hours. This headache is severe and incapacitating and is often associated with photophobia (light and/or sound sensitivity), nausea, and vomiting; irritability; and fatigue (Meyer, 2009). Its duration varies, ranging from 4 to 72 hours (Meyer, 2009).

Recovery Phase

In the recovery phase (termination and postdrome), the pain gradually subsides and can last up to 25 hours (Meyer, 2009). Muscle contraction in the neck and scalp is common, with associated muscle ache and localized tenderness, exhaustion, and mood changes. Any physical exertion exacerbates the headache pain. During this postheadache phase, patients may sleep for extended periods.

Other Headache Types

The tension-type headache is characterized by a steady, constant feeling of pressure that usually begins in the forehead, temple, or back of the neck. It is often bandlike or may be described as "a weight on top of my head."

Cluster headaches are unilateral and come in clusters of one to eight daily, with excruciating pain localized to the eye and orbit and radiating to the facial and temporal regions. The pain is accompanied by watering of the eye and nasal congestion. Each attack lasts 15 minutes to 3 hours and may have a crescendo–decrescendo pattern (Hickey, 2009). The headache is often described as penetrating.

Cranial arteritis often begins with general manifestations, such as fatigue, malaise, weight loss, and fever. Clinical manifestations associated with inflammation (heat, redness, swelling, tenderness, or pain over the involved artery) usually are present. Sometimes a tender, swollen, or nodular temporal artery is visible. Visual problems are caused by ischemia of the involved structures.

Assessment and Diagnostic Findings

The diagnostic evaluation includes a detailed history, a physical assessment of the head and neck, and a complete neurologic examination. Headaches may manifest differently in the same person over the course of a lifetime, and the same type of headache may manifest differently from patient to patient. The health history focuses on assessing the headache itself, with emphasis on the factors that precipitate or provoke it. The patient is asked to describe the headache in his or her own words.

Because headache is often the presenting symptom of various physiologic and psychological disturbances, a general health history is an essential component of the patient database. Headache may be a symptom of endocrine, hematologic, gastrointestinal, infectious, renal, cardiovascular, or psychiatric disease. Therefore, questions addressed in the health history should cover major medical and surgical illness as well as a body systems review.

The medication history can provide insight into the patient's overall health status and indicate medications that may be provoking headaches. Antihypertensive agents, diuretic medications, anti-inflammatory agents, and monoamine oxidase (MAO) inhibitors are a few of the categories of medications that can provoke headaches. Daily use of over-the-counter or prescribed pain medications for 8 to 10 days out of a month can lead to a chronic headache due to medication overuse (Meyer, 2009). Emotional factors can play a role in precipitating headaches. Stress is thought to be a major initiating factor in migraine headaches; therefore, sleep patterns, level of stress, recreational interests, appetite, emotional problems, and family stressors are relevant. There is a strong familial tendency for headache disorders, and a positive family history may help in making a diagnosis.

A direct relationship may exist between exposure to toxic substances and headache. Careful questioning may uncover chemicals to which a worker has been exposed. Under the Right-to-know Law employees have access to the material safety data sheets (commonly referred to as MSDS) for all substances with which they come in contact in the workplace (see Chapter 72). The occupational history also includes assessment of the workplace as a possible source of stress and for a possible ergonomic basis of muscle strain and headache.

A complete description of the headache itself is crucial. The nurse reviews the age at onset of headache; the headache's frequency, location, and duration; the type of pain; factors that relieve and precipitate the event; and associated symptoms. The data obtained should include the patient's own words about the headache in response to the following questions:

- What is the location? Is it unilateral or bilateral? Does it radiate?
- What is the quality—dull, aching, steady, boring, burning, intermittent, continuous, paroxysmal?
- How many headaches occur during a given period of time?
- What are the precipitating factors, if any—environmental (e.g., sunlight, weather change), foods, exertion, other?
- What makes the headache worse (e.g., coughing, straining)?
- What time (day or night) does it occur?
- How long does a typical headache last?
- Are there any associated symptoms, such as facial pain, lacrimation (excessive tearing), or scotomas (blind spots in the field of vision)?
- What usually relieves the headache (aspirin, nonsteroidal anti-inflammatory drugs, ergot preparation, food, heat, rest, neck massage)?
- Does nausea, vomiting, weakness, or numbness in the extremities accompany the headache?
- Does the headache interfere with daily activities?
- Do you have any allergies?
- Do you have insomnia, poor appetite, loss of energy?
- Is there a family history of headache?
- What is the relationship of the headache to your lifestyle or physical or emotional stress?
- What medications are you taking?

Diagnostic testing often is not helpful in the investigation of headache, because usually there are few objective findings. In patients who demonstrate abnormalities on the neurologic examination, CT, cerebral angiography, or MRI may be used to detect underlying causes, such as tumor or aneurysm. Electromyography (EMG) may reveal a sustained contraction of the neck, scalp, or facial muscles. Laboratory tests may include complete blood count, erythrocyte sedimentation rate, electrolytes, glucose, creatinine, and thyroid hormone levels.

Prevention

Prevention begins by having the patient avoid specific triggers that are known to initiate the headache syndrome. Preventive medical management of migraine involves the daily use of one or more agents that are thought to block the physiologic events leading to an attack. Treatment regimens vary greatly, as do patient responses; therefore, close monitoring is indicated.

Alcohol, nitrites, vasodilators, and histamines may precipitate cluster headaches. Elimination of these factors helps prevent the headaches.

Medical Management

Therapy for migraine headache is divided into abortive (symptomatic) and preventive approaches. The abortive approach, best used in those patients who have less frequent attacks, is aimed at relieving or limiting a headache at the onset or while it is in progress. The preventive approach is used in patients who experience more frequent attacks at regular or predictable intervals and may have a medical condition that precludes the use of abortive therapies (Hickey, 2009).

The triptans, which are serotonin receptor agonists, are the most specific antimigraine agents available. These agents cause vasoconstriction, reduce inflammation, and may reduce pain transmission. The five triptans in routine clinical use include sumatriptan (Imitrex), naratriptan (Amerge), rizatriptan (Maxalt), zolmitriptan (Zomig), and almotriptan (Axert) (Mett & Tfelt-Hansen, 2008). Numerous serotonin receptor agonists are under study. Many of the triptan medications are available in a variety of formulations, such as nasal sprays, inhalers, suppositories, or injections. The nasal sprays are useful for patients experiencing nausea and vomiting (Hickey, 2009).

The triptans are considered first-line treatment of the management of moderate to severe migraine pain. Best results are achieved with early use of triptans; oral dosing takes effect within 20 to 60 minutes of taking the drug and if needed may be repeated in 2 to 4 hours. Triptans can cause chest pain and are contraindicated in patients with ischemic heart disease (Rizzoli, 2012). Careful administration and dosing instructions to patients are important to prevent adverse reactions such as increased blood pressure, drowsiness, muscle pain, sweating, and anxiety. Interactions are possible if the medication is taken in conjunction with St. John's wort (Karch, 2012).

Ergotamine preparations (taken orally, sublingually, subcutaneously, intramuscularly, by rectum, or by inhalation) may be effective in aborting the headache if taken early in the migraine process. They are low in cost. Ergotamine tartrate acts on smooth muscle, causing prolonged constriction of the cranial blood vessels. Each patient's dosage is based on individual needs. Side effects include aching muscles, paresthesias (numbness and tingling), nausea, and vomiting. Pretreatment with antiemetic agents may be required (Rizzoli, 2012). None of the triptan medications should be taken concurrently with medications containing ergotamine because of the potential for a prolonged vasoactive reaction (Karch, 2012).

Many widely used medications for the prevention of migraine are available. An analysis of research in the last decade reported the most effective medications for migraine treatment include antiseizure agents (divalproex sodium [Depakote], valproate [Depacon], topiramate [Topamax]), beta-blockers (metoprolol [Lopressor], propranolol [Inderal], timolol [Blocadren]), and triptans (frovatriptan [Frova]). Other medications prescribed for migraine prevention include antidepressant agents (amitriptyline [Elavil], venlafaxine [Effexor]) and additional beta-blockers (atenolol [Tenormin], nadolol [Corgard]) and triptans (naratriptan [Amerge], zolmitriptan [Zomig]) (Silberstein, Holland, Freitag, et al., 2012).

The medical management of an acute attack of cluster headaches may include 100% oxygen by facemask for 15 minutes, ergotamine tartrate, sumatriptan, corticosteroids, or a percutaneous sphenopalatine ganglion blockade (Hickey, 2009).

The medical management of cranial arteritis consists of early administration of a corticosteroid to prevent the possibility of loss of vision due to vascular occlusion or rupture of the involved artery. The patient is instructed not to stop the medication abruptly, because this can lead to relapse. Analgesic agents are prescribed for comfort.

Nursing Management

When migraine or the other types of headaches have been diagnosed, the goal of nursing management is pain relief. It is reasonable to try nonpharmacologic interventions first, but the use of medications should not be delayed. The first priority is to treat the acute event of the headache and the second is to prevent recurrent episodes. Prevention involves patient education regarding precipitating factors, possible lifestyle or habit changes that may be helpful, and pharmacologic measures.

Relieving Pain

Individualized treatment depends on the type of headache and differs for migraine, cluster headaches, cranial arteritis, and tension headache. Nursing care is directed toward treatment of the acute episode. A migraine or a cluster headache in the early phase requires abortive medication therapy instituted as soon as possible. Some headaches can be prevented if the appropriate medications are taken before the onset of pain. Nursing care during an attack includes comfort measures such as a quiet, dark environment; elevation of the head of the bed to 30 degrees; and symptomatic treatment (i.e., administration of antiemetic medication) (Hickey, 2009).

Symptomatic pain relief for tension headache may be obtained by application of local heat or massage. Additional strategies may include administration of analgesic agents, antidepressant medications, and muscle relaxants.

Promoting Home and Community-Based Care

Educating Patients About Self-Care

Headaches, especially migraines, are more likely to occur when the patient is ill, overly tired, or stressed. Nonpharmacologic therapies are important and include patient education about the type of headache, its mechanism (if known), and appropriate changes in lifestyle to avoid triggers. Regular sleep, meals, exercise, relaxation, and avoidance of dietary triggers may be helpful in avoiding headaches (Hickey, 2009).

The patient with tension headaches needs education and reassurance that the headache is not the result of a brain tumor or other intracranial disorder. Stress reduction techniques, such as biofeedback, exercise programs, and meditation, are examples of nonpharmacologic therapies that may prove helpful. The patient and family need to be educated about the importance of following the prescribed treatment regimen for headache and keeping follow-up appointments. In addition, the patient is reminded of the importance of participating in health promotion activities and recommended health screenings to promote a healthy lifestyle. Chart 66-7 presents a home care checklist for the patient with migraine headaches.

Continuing Care

The National Headache Foundation (see the Resources section) provides a list of clinics in the United States and the names of physicians who specialize in headache and who are members of the American Headache Society.

Chart 66-7 HOME CARE CHECKLIST

The Patient With Migraine Headaches

At the completion of home care education, the patient or caregiver will be able to:	PATIENT	CAREGIVER
• Define migraine headaches and describe characteristics and manifestations.	✔	✔
• Identify triggers of migraine headaches and how to avoid such triggers as:		
• Foods that contain tyramine, such as chocolate, cheese, coffee, dairy products	✔	✔
• Dietary habits that result in long periods between meals	✔	✔
• Menstruation and ovulation (caused by hormone fluctuation)	✔	✔
• Alcohol (causes vasodilation of blood vessels)	✔	✔
• Fatigue and fluctuations in sleep patterns	✔	✔
• State importance of developing and using a headache diary.	✔	✔
• State stress management and lifestyle changes to minimize the frequency of headaches.	✔	✔
• State pharmacologic management: acute therapy and prophylaxis, to include medication regimen and side effects.	✔	✔
• Identify comfort measures during headache attacks, such as resting in a quiet and dark environment, applying cold compresses to the painful area, and elevating the head.	✔	✔
• Identify resources for education and support, such as the National Headache Foundation.	✔	✔

Critical Thinking Exercises

1 ebp Your 65-year-old patient with a brain injury has signs of increased ICP. Describe the nursing measures that are indicated. How would you determine whether your interventions are effective in alleviating the increased ICP? What is the evidence base for practices to decrease ICP? Identify the criteria used to evaluate the strength of the evidence for these practices.

2 A patient is admitted to your unit for a supratentorial cranial procedure. Identify the nursing interventions indicated before, during, and after the procedure. What patient education is indicated? How will your nursing interventions and patient education change if a transsphenoidal approach is used?

3 pq Identify the priorities, approach, and techniques you would use to provide care to a 35-year-old patient with migraines. How will your priorities, approach, and techniques differ if the patient has tension-type or cluster headaches?

Brunner Suite Resources

Explore these additional resources to enhance learning for this chapter:

• NCLEX-Style Questions and Other Resources on thePoint, http://thePoint.lww.com/Brunner13e

• Study Guide
• PrepU
• Clinical Handbook
• Handbook of Laboratory and Diagnostic Tests

References

*Asterisk indicates nursing research.

Books

American Association of Neuroscience Nurses. (2009). *Care of the patient with seizures: AANN clinical practice guidelines series*. Glenview, IL: Author.

American Association of Neuroscience Nurses. (2011). *Guide to the care of the patient with intracranial pressure monitoring/ external ventricular drainage or lumbar drainage: AANN reference series for clinical practice*. Glenview, IL: Author.

Bader, M., & Littlejohns, L. R. (2010). *AANN core curriculum for neuroscience nursing* (5th ed.). Glenview, IL: American Association of Neuroscience Nurses.

Hickey, J. V. (2009). *The clinical practice of neurological & neurosurgical nursing* (6th ed.). Philadelphia: Lippincott Williams & Wilkins.

Karch, A. (2012). *2012 Lippincott's nursing drug guide*. Philadelphia: Lippincott Williams & Wilkins.

Karpoff, S., & Labus, D. M. (2008). *Portable diagnostic tests*. Philadelphia: Lippincott Williams & Wilkins.

Laureys, S., Boly, M., Moonen, G., et al. (2009). Coma. *Encyclopedia of Neuroscience*, *2*, 1133–1142.

Littlejohns, L. R., & Bader, M. K. (2009). *AACN-AANN protocols for practice: Monitoring technologies in critically ill neuroscience patients*. Sudbury, MA: Jones & Bartlett.

Porth, C. M., & Matfin, C. (2009). *Pathophysiology: Concepts of altered health states* (8th ed.). Philadelphia: Lippincott Williams & Wilkins.

Posner, J. B., Saper, C. B., Schiff, N. D., et al. (2007). *Plum and Posner's diagnosis of stupor and coma* (4th ed.). Oxford, UK: Oxford University Press.

Journals and Electronic Documents

Barlow, P. (2012). A practical review of the Glasgow Coma Scale and score. *Surgeon*, *10*(2), 114–119.

Berg, A. T., Berkovic, S. F., Brodie, M. J., et al. (2010). Revised terminology and concepts for organization of seizures and epilepsies: Report of the ILAE Commission on Classification and Terminology, 2005–2009. *Epilepsia*, *51*(4), 676–685.

Blissitt, P. (2012). Controversies in the management of adults with severe traumatic brain injury. *AACN Advanced Critical Care*, *23*(2), 186–302.

Carrera, E., Steiner, L., Castellani, G., et al. (2011). Changes in cerebral compartmental compliances during mild hypocapnia in patients with traumatic brain injury. *Journal of Neurotrauma*, *28*(6), 889–896.

Chiang, Y., Chao, D., Chu, S., et al. (2012). Early enteral nutrition and clinical outcomes of severe traumatic brain injury patients in acute stage: A multi-center cohort study. *Journal of Neurotrauma, 29*(1), 75–80.

Cruse, D., Chennu, S., Chatelle, C., et al. (2012). Relationship between etiology and covert cognition in the minimally conscious state. *Neurology, 78*(11), 816–822.

*Dilorio, C., Reisinger, E. L., Yeager, K., et al. (2008). A descriptive analysis of seizure events among adults who participated in a computer-based assessment. *Journal of Neuroscience Nursing, 40*(3), 134–141.

Haddad, S., & Arabi, Y. (2012). Critical care management of severe traumatic brain injury in adults. *Scandinavian Journal of Trauma, Resuscitation, and Emergency Medicine, 20*(3), 1–15.

Hansen, M., Brennum, J., Moltke, F., et al. (2011). Pain treatment after craniotomy: Where is the (procedure specific) evidence? A qualitative systematic review. *European Society of Anesthesiology, 28*(12), 821–829.

Hunter, J. D. (2012). Ventilator associated pneumonia. *British Medical Journal, 344*, 40–44.

Iltis, A., & Cherry, M. (2010). Death revisited: Rethinking death and the dead donor rule. *Journal of Medicine and Philosophy, 35*(3), 223–241

Institute of Medicine. (2012). Epilepsy across the spectrum: Promoting health and understanding. Available at: www.iom.edu/epilepsy.

Kahn, S. R., Lim, W., Dunn, A., et al. (2012). Prevention of VTE in nonsurgical patients. *Chest, 141*(2), e195S–e226S.

Kotagal, P., & Yardi, N. (2008). The relationship between sleep and epilepsy. *Seminars in Pediatric Neurology, 15*(2), 42–49.

*Ledwith, M., Bloom, S., Maloney-Wilensky, E., et al. (2010). Effect of body position on cerebral oxygenation and physiologic parameters in patients with acute neurological conditions. *Journal of Neuroscience Nursing, 42*(5), 280–287.

Lescot, T., Abdennour, L., Boch, A., et al. (2008). Treatment of intracranial hypertension. *Current Opinion in Critical Care, 14*(2), 129–134.

Maurer, B. T. (2008). Locked-in syndrome. *American Journal of Hospice and Palliative Care, 25*(2), 151.

McIlvoy, L. (2012). Fever management in patients with brain injury. *AACN Advanced Critical Care, 23*(2), 204–211.

Meador, K. J., Pennell, P. B., Harden, C. L., et al. (2008). Pregnancy registries in epilepsy: A consensus statement on health outcomes. *Neurology, 71*(14), 1109–1117.

Mett, A., & Tfelt-Hansen, P. (2008). Acute migraine therapy: Recent evidence from randomized comparative trials. *Current Opinions in Neurology, 21*(3), 331–337.

Meyer, H. (2009). Pharmacological management of the adult migraine sufferer. *Journal of Neuroscience Nursing, 41*(6), 348–352.

Meyer, M., Megyesi, J., Meythaler, J., et al. (2010). Acute management of acquired brain injury part II: An evidence-based review of interventions used to promote arousal from coma. *Brain Injury, 24*(5), 722–729.

Mosek, A., Natour, H., Neufeld, M., et al. (2009). Ketogenic diet treatment in adults with refractory epilepsy: A prospective pilot study. *Seizure, 18*(1), 30–33.

Newman, D. K., & Willson, M. M. (2011). Review of intermittent catheterization and current best practices. *Urologic Nursing, 31*(1), 12–28, 48.

Noe, K. (2011). Seizures: Diagnosis and management in the outpatient setting. *Seminars in Neurology, 31*(1), 54–64.

Oddo, M., Frangos, S., Maloney-Wilensky, E. K., et al. (2009). Effect of shivering on brain tissue oxygenation during induced normothermia in patients with severe brain injury. *Neurocritical Care, 12*(1), 10–16.

Prendergast, V., Hallberg, I. R., Jahnke, H., et al. (2009). Oral health, ventilator-associated pneumonia, and intracranial pressure in intubated patients in a neuroscience intensive care unit. *American Journal of Critical Care, 18*(4), 368–376.

Prendergast, V., Jakobsson, U., Renvert, S., et al. (2012). Effects of a standard versus comprehensive oral care protocol among intubated neuroscience ICU patients: Results of a randomized controlled trial. *Journal of Neuroscience Nursing, 44*(3), 134–146.

Rabinstein, A. (2010). Management of status epilepticus in adults. *Neurology Clinics, 28*(4), 853–862.

Rizzoli, P. (2012). Acute and preventive treatment of migraine. *Continuum of Lifelong Learning in Neurology, 18*(4), 764–782.

Sandsmark, D., Kumar, M., Park, S., et al. (2012). Multimodal monitoring in subarachnoid hemorrhage. *Stroke, 43*(5), 1440–1445.

*Sebastian, S. (2012). Does preoperative scalp shaving result in fewer postoperative wound infections when compared with no scalp shaving? A systematic review. *Journal of Neuroscience Nursing, 44*(3), 149–156.

Silberstein, S., Holland, S., Freitag, F. et al. (2012). Evidence-based guideline update: Pharmacologic treatment for episodic migraine prevention in adults *Neurology, 78*(17), 1337–1345.

Sonabend, A., Korenfeld, Y., Crisman, C., et al. (2011). Prevention of ventriculostomy-related infections with prophylactic antibiotics and antibiotic-coated external ventricular drains: A systematic review. *Neurosurgery, 68*(4), 996–1005.

Spiotta, A., Stiefel, M., Heuer, G., et al. (2008). Brain hyperthermia after traumatic brain injury does not reduce brain oxygen. *Neurosurgery, 62*(4), 664–872.

Swearingen, B. (2012). Update on pituitary surgery. *Journal of Clinical Endocrinology and Metabolism, 97*(4), 1073–1081.

Vaidya, C., Ho, W., & Freda, B. (2010). Management of hyponatremia: Providing treatment and avoiding harm. *Cleveland Clinic Journal of Medicine, 11*(10), 715–726.

Ward, T. (2012). Migraine diagnosis and pathophysiology. *Continuum of Lifelong Learning in Neurology, 18*(4), 753–763.

Resources

American Headache Society, www.americanheadachesociety.org
Brain Injury Association of America, www.biausa.org
Brain Trauma Foundation (BTF), www.braintrauma.org
Epilepsy Foundation of America, www.epilepsyfoundation.org
Hydrocephalus Association, www.hydroassoc.org
National Headache Foundation, www.headaches.org

Chapter 67

Management of Patients With Cerebrovascular Disorders

Learning Objectives

On completion of this chapter, the learner will be able to:

1 Describe the incidence and impact of cerebrovascular disorders.
2 Identify the risk factors for cerebrovascular disorders and related measures for prevention.
3 Compare the various types of cerebrovascular disorders: their causes, clinical manifestations, and medical management.
4 Apply the principles of nursing management to the care of a patient in the acute stage of an ischemic stroke.
5 Use the nursing process as a framework for care of a patient recovering from an ischemic stroke.
6 Use the nursing process as a framework for care of a patient with a hemorrhagic stroke.
7 Identify essential elements for family education and preparation for home care of the patient who has had a stroke.

Glossary

agnosia: loss of ability to recognize objects through a particular sensory system; may be visual, auditory, or tactile
aneurysm: a weakening or bulge in an arterial wall
aphasia: inability to express oneself or to understand language
apraxia: inability to perform previously learned purposeful motor acts on a voluntary basis
dysphagia: difficulty swallowing
dysarthria: defects of articulation due to neurologic causes
expressive aphasia: inability to express oneself; often associated with damage to the left frontal lobe area
hemianopsia: blindness of half of the field of vision in one or both eyes
hemiparesis: weakness of one side of the body, or part of it, due to an injury in the motor area of the brain
hemiplegia: paralysis of one side of the body, or part of it, due to an injury in the motor area of the brain
infarction: tissue necrosis in an area deprived of blood supply
penumbra region: area of low cerebral blood flow
receptive aphasia: inability to understand what someone else is saying; often associated with damage to the temporal lobe area

Cerebrovascular disorders is an umbrella term that refers to a functional abnormality of the central nervous system (CNS) that occurs when the blood supply to the brain is disrupted. Stroke is the primary cerebrovascular disorder in the United States, and it is the fourth leading cause of death after heart disease, cancer, and chronic lower respiratory diseases. Approximately 795,000 people experience a stroke each year in the United States. Approximately 610,000 of these are new strokes, and 185,000 are recurrent strokes (Roger, Go, Lloyd-Jones, et al., 2012). About 6.2 million noninstitutionalized stroke survivors are alive today (National Center for Health Statistics, 2012). Stroke is a leading cause of serious, long-term disability in the United States. The financial impact of stroke is profound, with estimated direct and indirect costs of $34.3 billion in 2008 (Roger et al., 2012).

Strokes can be divided into two major categories: ischemic (approximately 87%), in which vascular occlusion and significant hypoperfusion occur, and hemorrhagic (approximately 13%), in which there is extravasation of blood into the brain or subarachnoid space (Hickey, 2009). Although there are some similarities between the two types of stroke, differences exist in etiology, pathophysiology, medical management, surgical management, and nursing care. Table 67-1 compares ischemic and hemorrhagic strokes.

Ischemic Stroke

An ischemic stroke, also known as a cerebrovascular accident (CVA), or "brain attack" is a sudden loss of function resulting from disruption of the blood supply to a part of the brain. The term *brain attack* has been used to suggest to health care practitioners and the public that a stroke is an urgent health care issue similar to a heart attack. With the approval of thrombolytic therapy for the treatment of acute ischemic stroke in 1996 came a revolution in the care of patients after a stroke. Early treatment with thrombolytic therapy for ischemic stroke results in fewer stroke symptoms and less loss of function (National Institute of Neurologic Disorders and Stroke [NINDS], 1995). The only U.S. Food and Drug Administration (FDA)-approved

TABLE 67-1 Comparison of Major Types of Stroke

Types of Stroke	Causes	Main Presenting Symptoms	Functional Recovery
Ischemic	• Large artery thrombosis • Small penetrating artery thrombosis • Cardiogenic embolic • Cryptogenic (no known cause) • Other	Numbness or weakness of the face, arm, or leg, especially on one side of the body	Usually plateaus at 6 months
Hemorrhagic	• Intracerebral hemorrhage • Subarachnoid hemorrhage • Cerebral aneurysm • Arteriovenous malformation	• "Exploding headache" • Decreased level of consciousness	Slower, usually plateaus at about 18 months

thrombolytic therapy has a treatment window of 3 hours after the onset of a stroke, and scientific statements have endorsed its expanded use for up to 4.5 hours (Del Zoppo, Saver, Jauch, et al., 2009). Although the time frame for treatment has expanded in some centers, urgency is needed on the part of the public and health care practitioners for rapid transport of the patient to a hospital for assessment and administration of the medication.

Ischemic strokes are subdivided into five different types based on the cause: large artery thrombotic strokes (20%), small penetrating artery thrombotic strokes (25%), cardiogenic embolic strokes (20%), cryptogenic strokes (30%), and other (5%) (see Table 67-1). Large artery thrombotic strokes are caused by atherosclerotic plaques in the large blood vessels of the brain. Thrombus formation and occlusion at the site of the atherosclerosis result in ischemia and **infarction** (tissue necrosis in an area deprived of blood supply) (Hickey, 2009).

Small penetrating artery thrombotic strokes affect one or more vessels and are the most common type of ischemic stroke. Small artery thrombotic strokes are also called *lacunar strokes* because of the cavity that is created after the death of infarcted brain tissue (American Association of Neuroscience Nurses [AANN], 2011; Hickey, 2009).

Cardiogenic embolic strokes are associated with cardiac dysrhythmias, usually atrial fibrillation. Embolic strokes can also be associated with valvular heart disease and thrombi in the left ventricle. Emboli originate from the heart and circulate to the cerebral vasculature, most commonly the left middle cerebral artery, resulting in a stroke. Embolic strokes may be prevented by the use of anticoagulation therapy in patients with atrial fibrillation.

The last two classifications of ischemic strokes are cryptogenic strokes, which have no known cause, and strokes from other causes, such as illicit drug use, coagulopathies, migraine, and spontaneous dissection of the carotid or vertebral arteries.

Pathophysiology

In an ischemic brain attack, there is disruption of the cerebral blood flow due to obstruction of a blood vessel. This disruption in blood flow initiates a complex series of cellular metabolic events referred to as the ischemic cascade (Fig. 67-1).

The ischemic cascade begins when cerebral blood flow decreases to less than 25 mL per 100 g of blood per minute. At this point, neurons are no longer able to maintain aerobic respiration. The mitochondria must then switch to anaerobic respiration, which generates large amounts of lactic acid, causing a change in the pH. This switch to the less efficient anaerobic respiration also renders the neuron incapable of producing sufficient quantities of adenosine triphosphate (ATP) to fuel the depolarization processes. The membrane pumps that maintain electrolyte balances begin to fail, and the cells cease to function.

Early in the cascade, an area of low cerebral blood flow, referred to as the **penumbra region**, exists around the area of infarction. The penumbra region is ischemic brain tissue that may be salvaged with timely intervention. The ischemic cascade threatens cells in the penumbra because membrane depolarization of the cell wall leads to an increase in intracellular calcium and the release of glutamate. The influx of calcium and the release of glutamate, if continued, activate a number of damaging pathways that result in the destruction of the cell membrane, the release of more calcium and glutamate, vasoconstriction, and the generation of free radicals. These processes enlarge the area of infarction into the penumbra, extending the stroke. A person experiencing a stroke typically loses 1.9 million neurons each minute that a stroke is not treated, and the ischemic brain ages 3.6 years each hour without treatment (Saver, 2006).

Physiology Pathophysiology

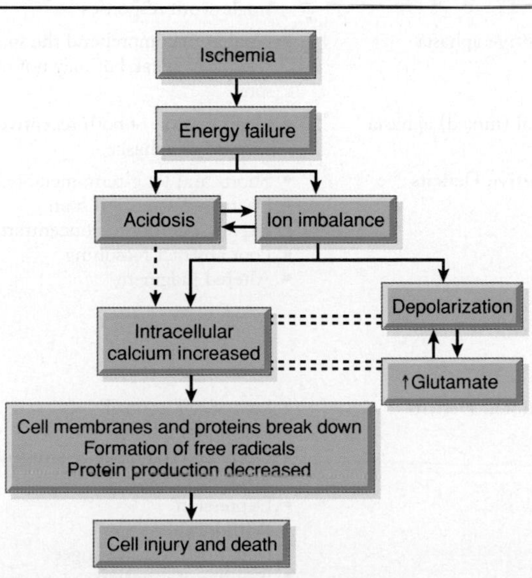
Ischemia
Energy failure
Acidosis
Ion imbalance
Depolarization
Intracellular calcium increased
↑Glutamate
Cell membranes and proteins break down
Formation of free radicals
Protein production decreased
Cell injury and death

FIGURE 67-1 • Some of the processes contributing to ischemic brain cell injury.

TABLE 67-2 Neurologic Deficits of Stroke: Manifestations and Nursing Implications

Neurologic Deficit	Manifestation	Nursing Implications/Patient Education Applications
Visual Field Deficits		
Homonymous hemianopsia (loss of half of the visual field)	• Unaware of persons or objects on side of visual loss • Neglect of one side of the body • Difficulty judging distances	Place objects within intact field of vision. Approach the patient from side of intact field of vision. Instruct/remind the patient to turn head in the direction of visual loss to compensate for loss of visual field. Encourage the use of eyeglasses if available. When educating the patient, do so within patient's intact visual field.
Loss of peripheral vision	• Difficulty seeing at night • Unaware of objects or the borders of objects	Place objects in center of patient's intact visual field. Encourage the use of a cane or other object to identify objects in the periphery of the visual field. Driving ability will need to be evaluated.
Diplopia	• Double vision	Explain to the patient the location of an object when placing it near the patient. Consistently place patient care items in the same location.
Motor Deficits		
Hemiparesis	• Weakness of the face, arm, and leg on the same side (due to a lesion in the opposite hemisphere)	Place objects within the patient's reach on the nonaffected side. Instruct the patient to exercise and increase the strength on the unaffected side.
Hemiplegia	• Paralysis of the face, arm, and leg on the same side (due to a lesion in the opposite hemisphere)	Encourage the patient to provide range-of-motion exercises to the affected side. Provide immobilization as needed to the affected side. Maintain body alignment in functional position. Exercise unaffected limb to increase mobility, strength, and use.
Ataxia	• Staggering, unsteady gait • Unable to keep feet together; needs a broad base to stand	Support patient during the initial ambulation phase. Provide supportive device for ambulation (walker, cane). Instruct the patient not to walk without assistance or supportive device.
Dysarthria	• Difficulty in forming words	Provide the patient with alternative methods of communicating. Allow the patient sufficient time to respond to verbal communication. Support patient and family to alleviate frustration related to difficulty in communicating.
Dysphagia	• Difficulty in swallowing	Test the patient's pharyngeal reflexes before offering food or fluids. Assist the patient with meals. Place food on the unaffected side of the mouth. Allow ample time to eat.
Sensory Deficits		
Paresthesia (occurs on the side opposite the lesion)	• Sensation of numbness, tingling, or a "pins and needles" sensation • Difficulty with proprioception	Instruct patient that sensation may be altered. Provide range of motion to affected areas and apply corrective devices as needed.
Verbal Deficits		
Expressive aphasia	• Unable to form words that are understandable; may be able to speak in single-word responses	Encourage patient to repeat sounds of the alphabet. Explore the patient's ability to write as an alternative means of communication.
Receptive aphasia	• Unable to comprehend the spoken word; can speak but may not make sense	Speak slowly and clearly to assist the patient in forming the sounds. Explore the patient's ability to read as an alternative means of communication.
Global (mixed) aphasia	• Combination of both receptive and expressive aphasia	Speak clearly and in simple sentences; use gestures or pictures when able. Establish alternative means of communication.
Cognitive Deficits	• Short- and long-term memory loss • Decreased attention span • Impaired ability to concentrate • Poor abstract reasoning • Altered judgment	Reorient patient to time, place, and situation frequently. Use verbal and auditory cues to orient patient. Provide familiar objects (family photographs, favorite objects). Use noncomplicated language. Match visual tasks with a verbal cue; holding a toothbrush, simulate brushing of teeth while saying, "I would like you to brush your teeth now." Minimize distracting noises and views when providing education to the patient. Repeat and reinforce instructions frequently.
Emotional Deficits	• Loss of self-control • Emotional lability • Decreased tolerance to stressful situations • Depression • Withdrawal • Fear, hostility, and anger • Feelings of isolation	Support patient during uncontrollable outbursts. Discuss with the patient and family that the outbursts are due to the disease process. Encourage patient to participate in group activity. Provide stimulation for the patient. Control stressful situations, if possible. Provide a safe environment. Encourage patient to express feelings and frustrations related to disease process.

Each step in the ischemic cascade represents an opportunity for intervention to limit the extent of secondary brain damage caused by a stroke. The penumbra area may be revitalized by administration of tissue plasminogen activator (t-PA). Medications that protect the brain from secondary injury are called *neuroprotectants*. A number of ongoing clinical trials focus on neuroprotective medications and strategies to improve stroke recovery and survival (Fisher, 2011).

Clinical Manifestations

An ischemic stroke can cause a wide variety of neurologic deficits, depending on the location of the lesion (which blood vessels are obstructed), the size of the area of inadequate perfusion, and the amount of collateral (secondary or accessory) blood flow (see Chapter 65 for discussion of anatomy and brain blood supply). The patient may present with any of the following signs or symptoms:

- Numbness or weakness of the face, arm, or leg, especially on one side of the body
- Confusion or change in mental status
- Trouble speaking or understanding speech
- Visual disturbances
- Difficulty walking, dizziness, or loss of balance or coordination
- Sudden severe headache

Motor, sensory, cranial nerve, cognitive, and other functions may be disrupted. Table 67-2 reviews the neurologic deficits frequently seen in patients with strokes. Table 67-3 compares the symptoms and behaviors seen in right hemispheric stroke with those seen in left hemispheric stroke.

Motor Loss

A stroke is an upper motor neuron lesion and results in loss of voluntary control over motor movements. Because the upper motor neurons decussate (cross), a disturbance of voluntary motor control on one side of the body may reflect damage to the upper motor neurons on the opposite side of the brain. The most common motor dysfunction is **hemiplegia** (paralysis of one side of the body, or part of it) caused by a lesion of the opposite side of the brain. **Hemiparesis**, or weakness of one side of the body, or part of it, is another sign. The concept of upper and lower motor neuron lesions is described in more detail in Table 65-4 in Chapter 65.

TABLE 67-3 Comparison of Left and Right Hemispheric Strokes

Left Hemispheric Stroke	Right Hemispheric Stroke
Paralysis or weakness on right side of body	Paralysis or weakness on left side of body
Right visual field deficit	Left visual field deficit
Aphasia (expressive, receptive, or global)	Spatial–perceptual deficits
Altered intellectual ability	Increased distractibility
Slow, cautious behavior	Impulsive behavior and poor judgment
	Lack of awareness of deficits

Adapted from Hickey, J. V. (2009). *The clinical practice of neurological and neurosurgical nursing* (6th ed., p. 600). Philadelphia: Lippincott Williams & Wilkins.

In the early stage of stroke, the initial clinical features may be flaccid paralysis and loss of or decrease in the deep tendon reflexes. When these deep reflexes reappear (usually by 48 hours), increased tone is observed along with spasticity (abnormal increase in muscle tone) of the extremities on the affected side.

Communication Loss

Other brain functions affected by stroke are language and communication. In fact, stroke is the most common cause of **aphasia** (inability to express oneself or to understand language). The following are dysfunctions of language and communication:

- **Dysarthria** (difficulty in speaking), caused by paralysis of the muscles responsible for producing speech
- Dysphasia (impaired speech) or aphasia, which can be **expressive aphasia** (inability to express oneself), **receptive aphasia** (inability to understand language), or global (mixed) aphasia (see Table 65-5 in Chapter 65)
- **Apraxia** (inability to perform a previously learned action), as may be seen when a patient makes verbal substitutions for desired syllables or words

Perceptual Disturbances

Perception is the ability to interpret sensation. Stroke can result in visual-perceptual dysfunctions, disturbances in visual-spatial relations, and sensory loss.

Visual-perceptual dysfunctions are caused by disturbances of the primary sensory pathways between the eye and visual cortex. Homonymous **hemianopsia** (blindness in half of the visual field) may occur from stroke and may be temporary or permanent. The affected side of vision corresponds to the paralyzed side of the body.

Disturbances in visual-spatial relations (perceiving the relationship of two or more objects in spatial areas) are frequently seen in patients with right hemispheric damage.

Sensory Loss

The sensory losses from stroke may take the form of slight impairment of touch, or it may be more severe, with loss of proprioception (ability to perceive the position and motion of body parts) as well as difficulty in interpreting visual, tactile, and auditory stimuli. **Agnosias** are deficits in the ability to recognize previously familiar objects perceived by one or more of the senses (see Table 65-6 in Chapter 65).

Cognitive Impairment and Psychological Effects

If damage has occurred to the frontal lobe, learning capacity, memory, or other higher cortical intellectual functions may be impaired. Such dysfunction may be reflected in a limited attention span, difficulties in comprehension, forgetfulness, and a lack of motivation. These changes can cause the patient to become easily frustrated during rehabilitation. Depression is common and may be exaggerated by the patient's natural response to this catastrophic event. Emotional lability, hostility, frustration, resentment, lack of cooperation, and other psychological problems may occur.

Assessment and Diagnostic Findings

Any patient with neurologic deficits needs a careful history and a complete physical and neurologic examination. Initial assessment focuses on airway patency, which may be compromised by loss of gag or cough reflexes and altered respiratory pattern; cardiovascular status (including blood pressure, cardiac rhythm and rate, carotid bruit); and gross neurologic deficits.

Patients may present to the acute care facility with temporary neurologic symptoms. A transient ischemic attack (TIA) is a neurologic deficit typically lasting less than 1 hour. A TIA is manifested by a sudden loss of motor, sensory, or visual function. The symptoms result from temporary ischemia (impairment of blood flow) to a specific region of the brain; however, when brain imaging is performed, there is no evidence of ischemia. A TIA may serve as a warning of impending stroke. Approximately 15% of all strokes are preceded by a TIA (Roger et al., 2012). Lack of evaluation and treatment of a patient who has experienced previous TIAs may result in a stroke and irreversible deficits.

The initial diagnostic test for a stroke is usually a noncontrast computed tomography (CT) scan. This should be performed within 25 minutes or less from the time the patient presents to the emergency department (ED) to determine if the event is ischemic or hemorrhagic (the category of stroke determines treatment). Further diagnostic workup for ischemic stroke involves attempting to identify the source of the thrombi or emboli. A 12-lead electrocardiogram (ECG) and a carotid ultrasound are standard tests. Other studies may include CT angiography or CT perfusion; magnetic resonance imaging (MRI) and magnetic resonance angiography of the brain and neck vessels; transcranial Doppler flow studies; transthoracic or transesophageal echocardiography; xenon-enhanced CT scan; and single-photon emission computed tomography scan (Jauch, Cucchiara, Adeoye, et al., 2010; Jauch, Saver, Adams, et al. 2013; Summers, Leonard, Wentworth, et al., 2009).

Prevention

Primary prevention of ischemic stroke remains the best approach. A healthy lifestyle including not smoking, maintaining a healthy weight, following a healthy diet (including modest alcohol consumption), and regular exercise can reduce the risk of having a stroke (Towfight, 2011). The risk of coronary heart disease and stroke decreases in both men and women on the Dietary Approaches to Stop Hypertension (DASH) diet. The DASH diet is high in fruits and vegetables, moderate in low-fat dairy products, and low in animal protein (has a substantial amount of plant protein from legumes and nuts). There is ample evidence for incorporating education about the DASH diet into stroke risk management programs (Satterfield, Anderson, & Moore, 2012).

Stroke risk screenings are an ideal opportunity to lower stroke risk by identifying people or groups of people who are at high risk for stroke and by educating patients and the community about recognition and prevention of stroke. Research findings suggest that low-dose aspirin may lower the risk of a first stroke for those who are at risk (Goldstein, Bushnell, Adams, et al., 2011).

Age, gender, and race are well-known nonmodifiable risk factors for stroke. High-risk groups include people older than 55 years, and the incidence of stroke more than doubles in each successive decade. Men have a higher age-adjusted rate of stroke than that of women. However, the magnitude of strokes in women should not be underplayed; 1 in 6 women die of stroke as compared to 1 in 25 who die of breast cancer (Ennen, 2013; Goldstein et al., 2011). Another high-risk group is African Americans; the incidence of first stroke in African Americans is almost twice that in Caucasian Americans (Roger et al., 2012).

Chart 67-1 MODIFIABLE RISK FACTORS
Ischemic Stroke

- Hypertension (Controlling hypertension, the major risk factor, is the key to preventing stroke.)
- Atrial fibrillation
- Dyslipidemia
- Diabetes (associated with accelerated atherogenesis)
- Smoking
- Asymptomatic carotid stenosis
- Obesity
- Sedentary lifestyle
- Sleep apnea
- Excessive alcohol consumption
- Periodontal disease

Adapted from Goldstein, L., Bushnell, C., Adams, R., et al. (2011). Guidelines for the primary prevention of stroke: A guideline for healthcare professionals from the American Heart Association/American Stroke Association. *Stroke, 42*(2), 517–584.

There are many risk factors for ischemic stroke (Chart 67-1). For people who are at high risk, interventions that alter modifiable factors, such as treating hypertension and hyperglycemia and stopping smoking, reduce stroke risk.

Additional treatable conditions that increase risk of stroke include sickle cell diseases and valvular heart disease (e.g., endocarditis, prosthetic heart valves). Periodontal disease has also been linked to stroke risk. The association between periodontal disease and stroke may result from the host inflammatory response and the chronic bacterial infection, but the exact mechanism is not fully understood. Periodontal disease is a treatable and preventable condition. Other chronic inflammatory conditions that have been associated with an increased risk of stroke are systemic lupus erythematosus and rheumatoid arthritis (Goldstein et al., 2011).

Several methods of preventing recurrent stroke have been identified for patients with TIAs or ischemic stroke. Patients with moderate to severe carotid stenosis are treated with carotid endarterectomy (CEA) or carotid angioplasty and stenting. In patients with atrial fibrillation, which increases the risk of emboli, administration of an anticoagulant that inhibits clot formation may prevent both thrombotic and embolic strokes (Furie, Kasner, Adams, et al., 2011).

Medical Management

Patients who have experienced a TIA or stroke should have medical management for secondary prevention. Those with atrial fibrillation (or cardioembolic strokes) are treated with dose-adjusted warfarin (Coumadin) with a target international normalized ratio (INR) of 2 to 3. Other, newer anticoagulants that may be prescribed as alternative drugs include dabigatran (Pradaxa) or rivaroxaban (Xarelto), unless they are contraindicated. If anticoagulants are contraindicated, aspirin is the best option, although other medications may be used if both are contraindicated (Furie, Goldstein, Albers, et al., 2012; Furie et al., 2011).

Platelet-inhibiting medications, including aspirin, extended-release dipyridamole plus aspirin (Aggrenox), and clopidogrel (Plavix) decrease the incidence of cerebral infarction in patients who have experienced TIAs and stroke from suspected embolic or thrombotic causes. The specific medication that is used is based on the patient's health history.

Research suggests that medications known as statins reduce coronary events and strokes, particulary in patient who also have diabetes (de Vries, Denig, Pauwels, et al., 2012). Benefits were independent of cholesterol levels, and these medications are now widely used for stroke prevention. The FDA has included indications for statin medications, such as simvastatin (Zocor), to include secondary stroke prevention. After the acute stroke period, antihypertensive medications are also used, if indicated, for secondary stroke prevention (Furie et al., 2011).

Ongoing research is focusing on several aspects of the medical management of acute ischemic stroke (Jauch, et al., 2013). The FDA has approved clot retrieval devices that open the blocked artery and restore blood flow to the brain. These devices include one shaped like a tiny corkscrew, one that uses suction, and one that uses a stent to compress and trap the clot.

Thrombolytic Therapy

Thrombolytic agents are used to treat ischemic stroke by dissolving the blood clot that is blocking blood flow to the brain. Recombinant t-PA is a genetically engineered form of t-PA (a thrombolytic substance made naturally by the body) (Karch, 2012). It works by binding to fibrin and converting plasminogen to plasmin, which stimulates fibrinolysis of the clot. Rapid diagnosis of stroke and initiation of thrombolytic therapy (within 3 hours) in patients with ischemic stroke leads to a decrease in the size of the stroke and an overall improvement in functional outcome after 3 months (Jauch, et al., 2013; NINDS, 1995). The goal is for intravenous (IV) t-PA to be given within 60 minutes of the patient arriving to the ED (Jauch et al., 2010). Intra-arterial delivery of t-PA is an alternative to IV administration. This type of administration can allow for higher concentrations of the drug to be given directly to the clot, and the time window for treatment may be extended up to 6 hours. Those not eligible for IV delivery may be eligible for intra-arterial delivery, and these methods may also be combined. Treatment using intra-arterial delivery must occur in specialized centers with access to emergent cerebral angiogram and interventional operating rooms/suites (Jauch, et al., 2013; Summers et al., 2009). Ongoing clinical trials continue to investigate the efficacy of other thrombolytic agents.

To realize the full potential of thrombolytic therapy, community education directed at recognizing the symptoms of stroke and obtaining appropriate emergency care is necessary to ensure rapid transport to a hospital and initiation of therapy within the recommended 3-hour period (which may be extended up to 4.5 hours in some stroke centers) (Del Zoppo et al., 2009). Delays make the patient ineligible for thrombolytic therapy, because revascularization of necrotic tissue (which develops after 3 hours) increases the risk of cerebral edema and hemorrhage.

Enhancing Prompt Diagnosis

After being notified by emergency medical service personnel, the ED contacts the appropriate staff (neurologist, neuroradiologist, radiology department, nursing staff, ECG, and laboratory technicians) and informs them of the patient's imminent arrival at the hospital. Many institutions have acute stroke teams that respond rapidly, ensuring that treatment occurs within the allotted period (AANN, 2011).

Initial management requires the definitive diagnosis of an ischemic stroke by brain imaging and a careful history to determine whether the patient meets the criteria for t-PA therapy (Chart 67-2). The goal is that diagnostic results from imaging are completed within 25 minutes of the patient's arrival to the ED (Jauch et al., 2010). Some of the contraindications for thrombolytic therapy include symptom onset greater than 3 hours before admission (expanded to 4.5 hours in some centers), a patient who is anticoagulated (with an INR above 1.7), or a patient who has recently had any type of intracranial pathology (e.g., previous stroke, head injury, trauma).

Before receiving t-PA, the patient is assessed using the National Institutes of Health Stroke Scale (NIHSS), a standardized assessment tool that helps evaluate stroke severity (Table 67-4). Total NIHSS scores range from 0 (normal) to 42 (severe stroke). Certification in the administration of the scale is recommended and is available for nurses and other health care professionals.

Dosage and Administration

The patient is weighed to determine the dose of t-PA. Typically, two or more IV sites are established prior to administration of t-PA (one for the t-PA and the other for administration of IV fluids). The dosage for t-PA is 0.9 mg/kg, with a maximum dose of 90 mg. Ten percent of the calculated dose is administered as an IV bolus over 1 minute. The remaining dose (90%) is administered IV over 1 hour via an infusion pump (Summers et al., 2009).

Chart 67-2 Eligibility Criteria for Tissue Plasminogen Activator Administration

- Age ≥18 years
- Clinical diagnosis of ischemic stroke
- Time of onset of stroke known and is within center guidelines
- Systolic blood pressure ≤185 mm Hg; diastolic ≤110 mm Hg
- No minor stroke or rapidly resolving stroke
- No seizure at onset of stroke
- Not taking warfarin (Coumadin)
- Prothrombin time ≤15 seconds or international normalized ratio ≤1.7
- Not received heparin during the past 48 hours with elevated partial thromboplastin time
- Platelet count ≥100,000/mm^3
- No prior intracranial hemorrhage, neoplasm, arteriovenous malformation, or aneurysm
- No major surgical procedures within 14 days
- No stroke, serious head injury, or intracranial surgery within 3 months
- No gastrointestinal or urinary bleeding within 21 days

Adapted from Adams, H. P., Zoppo, G., Alberts, M. J., et al. (2007). Guidelines for the early management of patients with ischemic stroke. A guideline from the American Heart Association/American Stroke Association Stroke Council, Clinical Cardiology Council, Cardiovascular Radiology and Intervention Council, and the Atherosclerotic Peripheral Vascular Disease and Quality of Care Outcomes in Research Interdisciplinary Working Groups. *Stroke, 38*(5), 1655–1711.

TABLE 67-4 Summary of National Institutes of Health Stroke Scale

Category	Description	Score
1a. LOC	Alert	0
	Arousable by minor stimulation	1
	Obtunded, strong stimulation to attend	2
	Unresponsive, or reflexic responses only	3
1b. LOC questions (month, age)	Answers both correctly	0
	Answers one correctly	1
	Both incorrect	2
1c. LOC commands (open, close eyes; make fist, let go)	Obeys both correctly	0
	Obeys one correctly	1
	Both incorrect	2
2. Best gaze (eyes open—patient follows examiner's finger or face)	Normal	0
	Partial gaze palsy	1
	Forced deviation	2
3. Visual (introduce visual stimulus/threat to patient's visual field quadrants)	No visual loss	0
	Partial hemianopsia	1
	Complete hemianopsia	2
	Bilateral hemianopsia	3
4. Facial palsy (show teeth, raise eyebrows and squeeze eyes shut)	Normal	0
	Minor	1
	Partial	2
	Complete	3
5a. Motor; arm—left (elevate extremity to 90 degrees and score drift/movement)	No drift	0
	Drift but maintains in air	1
	Unable to maintain in air	2
	No effort against gravity	3
	No movement	4
	Amputation, joint fusion (explain)	N/A
5b. Motor; arm—right (elevate extremity to 90 degrees and score drift/movement)	No drift	0
	Drift but maintains in air	1
	Unable to maintain in air	2
	No effort against gravity	3
	No movement	4
	Amputation, joint fusion (explain)	N/A
6a. Motor; leg—left (elevate extremity to 30 degrees and score drift/movement)	No drift	0
	Drift but maintains in air	1
	Unable to maintain in air	2
	No effort against gravity	3
	No movement	4
	Amputation, joint fusion (explain)	N/A
6b. Motor; leg—right (elevate extremity to 30 degrees and score drift/movement)	No drift	0
	Drift but maintains in air	1
	Unable to maintain in air	2
	No effort against gravity	3
	No movement	4
	Amputation, joint fusion (explain)	N/A
7. Limb ataxia (finger-to-nose and heel-to-shin testing)	Absent	0
	Present in one limb	1
	Present in two limbs	2
8. Sensory (pinprick to face, arm, trunk, and leg—compare side to side)	Normal	0
	Mild to moderate loss	1
	Severe to total loss	2
9. Best language (name items, describe a picture, and read sentences)	No aphasia	0
	Mild to moderate aphasia	1
	Severe aphasia	2
	Mute	3
10. Dysarthria (evaluate speech clarity by having patient repeat words)	Normal	0
	Mild to moderate dysarthria	1
	Severe dysarthria, mostly unintelligible or worse	2
	Intubated or other physical barrier	N/A
11. Extinction and inattention (use information from prior testing to score)	No abnormality	0
	Visual, tactile, auditory, or other extinction to bilateral simultaneous stimulation	1
	Profound hemiattention or extinction to more than one modality	2
Total score		

LOC, level of consciousness; N/A, not applicable.

Adapted from the version available at the National Institute of Neurological Disorders and Stroke, National Institutes of Health, Bethesda, MD 20892, www.ninds.nih.gov/doctors/NIH_Stroke_Scale.pdf. It is recommended that the full scale with all instructions be used.

The patient is admitted to the intensive care unit or an acute stroke unit, where continuous cardiac monitoring and frequent neurologic assessments are conducted. Vital signs are obtained frequently, with particular attention to blood pressure (with the goal of lowering the risk of intracranial hemorrhage). An example of a standard protocol would be to obtain vital signs every 15 minutes for the first 2 hours, every 30 minutes for the next 6 hours, then every hour until 24 hours after treatment. Blood pressure should be maintained with the systolic pressure less than 180 mm Hg and the diastolic pressure less than 105 mm Hg (Summers et al., 2009). Airway management is instituted based on the patient's clinical condition and arterial blood gas values.

Side Effects

Once it is determined that the patient is a candidate for t-PA therapy, no anticoagulant agents are administered for the next 24 hours. Bleeding is the most common side effect of t-PA administration, and the patient is closely monitored for any bleeding (IV insertion sites, urinary catheter site, endotracheal tube, nasogastric tube, urine, stool, emesis, other secretions). A 24-hour delay in placement of nasogastric tubes, urinary catheters, and intra-arterial pressure catheters is recommended. Intracranial bleeding is a major complication that occurred in approximately 6.4% of patients in the initial t-PA study (NINDS, 1995). A number of factors are associated with the occurrence of symptomatic intracranial bleeding: age greater than 70 years, baseline NIHSS score greater than 20, serum glucose concentration 300 mg/dL or higher, and edema or mass effect observed on the patient's initial CT scan (NINDS, 1995).

Therapy for Patients With Ischemic Stroke Not Receiving Tissue Plasminogen Activator

Not all patients are candidates for t-PA therapy. In some centers, other treatments may include anticoagulant administration (IV heparin or low-molecular-weight heparin). Because of the risks associated with anticoagulation, their general use is no longer recommended for patients with acute ischemic stroke (Jauch, et al., 2013).

Careful maintenance of cerebral hemodynamics to maintain cerebral perfusion is extremely important after a stroke. Increased intracranial pressure (ICP) from brain edema and associated complications may occur after a large ischemic stroke. Interventions during this period include measures to reduce ICP, such as administering an osmotic diuretic (e.g., mannitol), and maintaining the partial pressure of arterial carbon dioxide ($PaCO_2$) within a slightly lower range of 30 to 35 mm Hg. Other treatment measures include the following:

- Providing supplemental oxygen if oxygen saturation is below 92%
- Elevation of the head of the bed to 25 to 30 degrees to assist the patient in handling oral secretions and decrease intracranial pressure
- Possible hemicraniectomy for increased ICP from brain edema in a very large stroke
- Intubation with an endotracheal tube to establish a patent airway, if necessary
- Continuous hemodynamic monitoring (The goals for blood pressure remain controversial for a patient who has not received thrombolytic therapy; antihypertensive treatment may be withheld unless the systolic blood pressure exceeds 220 mm Hg or the diastolic blood pressure exceeds 120 mm Hg.)
- Frequent neurologic assessments to determine if the stroke is evolving and if other acute complications are developing (Such complications may include seizures, bleeding from anticoagulation, or medication-induced bradycardia, which can result in hypotension and subsequent decreases in cardiac output and cerebral perfusion pressure.)

Managing Potential Complications

Adequate cerebral blood flow is essential for cerebral oxygenation. If cerebral blood flow is inadequate, the amount of oxygen supplied to the brain will decrease, and tissue ischemia will result. Adequate oxygenation begins with pulmonary care, maintenance of a patent airway, and administration of supplemental oxygen as needed. The importance of adequate gas exchange in these patients cannot be overemphasized, because many patients are at risk for aspiration pneumonia.

Other potential complications after a stroke include urinary tract infections, cardiac dysrhythmias (ventricular ectopy, tachycardia, and heart blocks), and complications of immobility. Hyperglycemia has been associated with poor neurologic outcomes in acute stroke and should be treated if the blood glucose is above 140 mg/dL (Summers et al., 2009).

Surgical Prevention of Ischemic Stroke

The main surgical procedure for selected patients with TIAs and mild stroke is CEA, which currently is the most frequently performed noncardiac vascular procedure. A CEA is the removal of an atherosclerotic plaque or thrombus from the carotid artery to prevent stroke in patients with occlusive disease of the extracranial cerebral arteries (Fig. 67-2). This surgery is indicated for patients with symptoms of TIA or mild

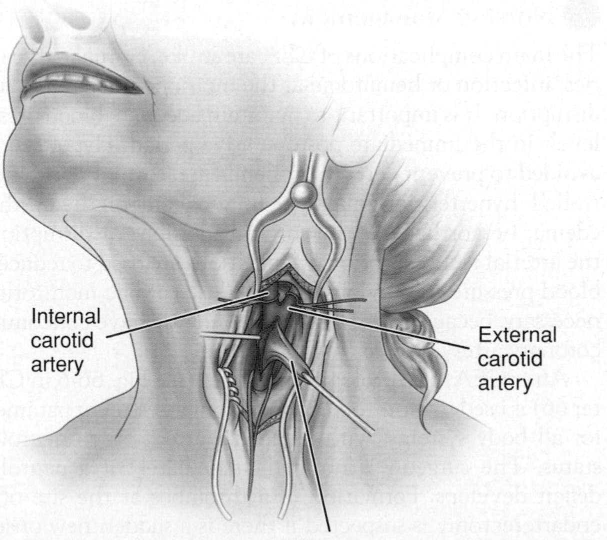
Internal carotid artery
External carotid artery
Plaque

FIGURE 67-2 • Plaque, a potential source of emboli in transient ischemic attack and stroke, is surgically removed from the carotid artery.

TABLE 67-5 Selected Complications of Carotid Endarterectomy (CEA) and Nursing Interventions

Complication	Characteristics	Nursing Interventions
Incision hematoma	Occurs in 5.5% of patients. Large or rapidly expanding hematomas require emergency treatment. If the airway is obstructed by the hematoma, the incision may be opened at the bedside.	Monitor neck discomfort and wound expansion. Report swelling, subjective feelings of pressure in the neck, difficulty breathing.
Hypertension	Poorly controlled hypertension increases the risk of postoperative complications, including hematoma and hyperperfusion syndrome. There is an increased incidence of neurologic impairment and death due to intracerebral hemorrhage. May be related to surgically induced abnormalities of carotid baroreceptor sensitivity.	Risk is highest in the first 48 hours after surgery. Check blood pressure frequently, and report deviations from baseline. Observe for and report new onset of neurologic deficits.
Postoperative hypotension	Occurs in approximately 5% of patients. Treated with fluids and low-dose phenylephrine infusion. Usually resolves in 24–48 hours. Patients with hypotension should have serial electrocardiograms to rule out myocardial infarction.	Monitor blood pressure, and observe for signs and symptoms of hypotension.
Hyperperfusion syndrome	Occurs when cerebral vessel autoregulation fails. Arteries accustomed to diminished blood flow may be permanently dilated; increased blood flow after endarterectomy coupled with insufficient vasoconstriction leads to capillary bed damage, edema, and hemorrhage.	Observe for severe unilateral headache improved by sitting upright or standing.
Intracerebral hemorrhage	Occurs infrequently but is often fatal (60%) or results in serious neurologic impairment. Can occur secondary to hyperperfusion syndrome. Increased risk with advanced age, hypertension, presence of high-grade stenosis, poor collateral flow, and slow flow in the region of the middle cerebral artery.	Monitor neurologic status, and report any changes in mental status or neurologic functioning immediately.

stroke (or those without symptoms) that are found to have severe (70% to 99%) carotid artery stenosis or moderate (50% to 69%) stenosis with other significant risk factors.

Carotid stenting, with or without angioplasty, is a less invasive procedure that is used for selected patients for severe stenosis. Outcomes from carotid artery stenting were recently compared to CEA in a study that included those with symptoms of TIA or stroke and those without symptoms (2,502 participants). This study demonstrated similar short- and long-term outcomes for both groups. Young patients had slightly better outcomes with carotid stenting, and older patients had better outcomes with CEA. It also showed that more strokes occurred after stenting and more myocardial infarctions occurred after CEA (Mantese, Timaran, Chiu, et al., 2010).

Nursing Management

The main complications of CEA are stroke, cranial nerve injuries, infection or hematoma at the incision, and carotid artery disruption. It is important to maintain adequate blood pressure levels in the immediate postoperative period. Hypotension is avoided to prevent cerebral ischemia and thrombosis. Uncontrolled hypertension may precipitate cerebral hemorrhage, edema, hemorrhage at the surgical incision, or disruption of the arterial reconstruction. Medications are used to reduce the blood pressure to previous levels. Close cardiac monitoring is necessary, because these patients frequently have concomitant coronary artery disease.

After CEA, a neurologic flow sheet (see Fig. 66-6 in Chapter 66) is used to monitor and document assessment parameters for all body systems, with particular attention to neurologic status. The surgeon is notified immediately if a neurologic deficit develops. Formation of a thrombus at the site of the endarterectomy is suspected if there is a sudden new onset of neurologic deficits, such as weakness on one side of the body. The patient should be prepared for repeat endarterectomy.

Difficulty in swallowing, hoarseness, or other signs of cranial nerve dysfunction must be assessed. The nurse focuses on assessment of the following cranial nerves: facial (VII), vagus (X), spinal accessory (XI), and hypoglossal (XII). Some edema in the neck after surgery is expected; however, extensive edema and hematoma formation can obstruct the airway. Emergency airway supplies, including those needed for a tracheostomy, must be available. Table 67-5 provides more information about potential complications of carotid surgery.

Management after carotid stenting also requires monitoring of neurologic status and evaluation for hematoma formation (at the catheterization site). Cardiac monitoring is necessary with assessment for bilateral pulses distal to the catheterization site. Typically, patients are discharged the day after stenting if there are no complications (Custer, 2009).

NURSING PROCESS

The Patient Recovering From an Ischemic Stroke

PRACTICE & LEARN

The acute phase of an ischemic stroke may last 1 to 3 days, but ongoing monitoring of all body systems is essential as long as the patient requires care. The patient who has had a stroke is at risk for multiple complications, including deconditioning and other musculoskeletal problems, swallowing difficulties, bowel and bladder dysfunction, inability to perform self-care, and skin breakdown. Nursing management focuses on the prompt initiation of rehabilitation for any deficits.

Assessment

During the acute phase, a neurologic flow sheet is maintained to provide data about the following important measures of the patient's clinical status:

- Change in level of consciousness or responsiveness as evidenced by movement, resistance to changes of position, and response to stimulation; orientation to time, place, and person

- Presence or absence of voluntary or involuntary movements of the extremities; muscle tone; body posture; and position of the head
- Eye opening, comparative size of pupils and pupillary reactions to light, and ocular position
- Color of the face and extremities; temperature and moisture of the skin
- Quality and rates of pulse and respiration; arterial blood gas values as indicated, body temperature, and arterial pressure
- Ability to speak
- Volume of fluids ingested or administered; volume of urine excreted each 24 hours
- Presence of bleeding
- Maintenance of blood pressure within the desired parameters

After the acute phase, the nurse assesses mental status (memory, attention span, perception, orientation, affect, speech/language), sensation/perception (the patient may have decreased awareness of pain and temperature), motor control (upper and lower extremity movement), swallowing ability, nutritional and hydration status, skin integrity, activity tolerance, and bowel and bladder function. Ongoing nursing assessment continues to focus on any impairment of function in the patient's daily activities, because the quality of life after stroke is closely related to the patient's functional status.

Diagnosis

Nursing Diagnoses

Based on the assessment data, major nursing diagnoses may include the following:

- Impaired physical mobility related to hemiparesis, loss of balance and coordination, spasticity, and brain injury
- Acute pain (painful shoulder) related to hemiplegia and disuse
- Self-care deficits (bathing, hygiene, toileting, dressing, grooming, and feeding) related to stroke sequelae
- Impaired physical comfort related to altered sensory reception, transmission, and/or integration
- Impaired swallowing
- Impaired urinary elimination related to flaccid bladder, detrusor instability, confusion, or difficulty in communicating
- Constipation related to change in mental status or difficulty communicating
- Acute confusion related to brain infarction
- Impaired verbal communication related to brain damage
- Risk for impaired skin integrity related to hemiparesis, hemiplegia, or decreased mobility
- Interrupted family processes related to catastrophic illness and caregiving burdens
- Sexual dysfunction related to neurologic deficits or fear of failure

Collaborative Problems/Potential Complications

Potential complications may include the following:

- Decreased cerebral blood flow due to increased ICP
- Inadequate oxygen delivery to the brain
- Pneumonia

Planning and Goals

Although rehabilitation begins on the day the patient has the stroke, the process is intensified during convalescence and requires a coordinated team effort. It is helpful for the team to know what the patient was like before the stroke: his or her illnesses, abilities, mental and emotional state, behavioral characteristics, and activities of daily living. It is also helpful for clinicians to be knowledgeable about the relative importance of predictors of stroke outcome (age, NIHSS score, and level of consciousness at time of admission) in order to provide stroke survivors and their families with realistic goals (Jauch, et al., 2013).

The major goals for the patient (and family) may include improved mobility, avoidance of shoulder pain, achievement of self-care, relief of discomfort, prevention of aspiration, continence of bowel and bladder, decreasing confusion, achieving a form of communication, maintaining skin integrity, restored family functioning, improved sexual function, and absence of complications.

Nursing Interventions

Nursing care has a significant impact on the patient's recovery. Often, many body systems are impaired as a result of the stroke, and conscientious care and timely interventions can prevent debilitating complications. During and after the acute phase, nursing interventions focus on the whole person. In addition to providing physical care, the nurse encourages and fosters recovery by listening to the patient and asking questions to elicit the meaning of the stroke experience.

Improving Mobility and Preventing Joint Deformities

A patient with hemiplegia has unilateral paralysis (paralysis on one side). When control of the voluntary muscles is lost, the strong flexor muscles exert control over the extensors. The arm tends to adduct (adductor muscles are stronger than abductors) and to rotate internally. The elbow and the wrist tend to flex, the affected leg tends to rotate externally at the hip joint and flex at the knee, and the foot at the ankle joint supinates and tends toward plantar flexion.

Correct positioning is important to prevent contractures; measures are used to relieve pressure, assist in maintaining good body alignment, and prevent compressive neuropathies, especially of the ulnar and peroneal nerves. Because flexor muscles are stronger than extensor muscles, a splint applied at night to the affected extremity may prevent flexion and maintain correct positioning during sleep. (See Chapter 10 for additional information.)

Preventing Shoulder Adduction. To prevent adduction of the affected shoulder while the patient is in bed, a pillow is placed in the axilla when there is limited external rotation; this keeps the arm away from the chest. A pillow is placed under the arm, and the arm is placed in a neutral (slightly flexed) position, with distal joints positioned higher than the more proximal joints (i.e., the elbow is positioned higher than the shoulder and the wrist higher than the elbow). This helps to prevent edema and the resultant joint fibrosis that will limit range of motion if the patient regains control of the arm (Fig. 67-3).

Positioning the Hand and Fingers. The fingers are positioned so that they are barely flexed. The hand is placed in slight supination (palm faces upward), which is its most functional

FIGURE 67-3 • Correct positioning to prevent shoulder adduction.

position. If the upper extremity is flaccid, a splint can be used to support the wrist and hand in a functional position. If the upper extremity is spastic, a hand roll is not used, because it stimulates the grasp reflex. In this instance, a dorsal wrist splint is useful in allowing the palm to be free of pressure. Every effort is made to prevent hand edema.

Spasticity, particularly in the hand, can be a disabling complication after stroke. Botulinum toxin type A injected intramuscularly into wrist and finger muscles has been shown to be effective in reducing this spasticity (although the effect is temporary, typically lasting 2 to 4 months) (Teasell, Foley, Pereira, et al., 2012). Other treatments for spasticity may include stretching, splinting, and oral medications such as baclofen (Lioresal).

Changing Positions. The patient's position should be changed every 2 hours. To place a patient in a lateral (side-lying) position, a pillow is placed between the legs before the patient is turned. To promote venous return and prevent edema, the upper thigh should not be acutely flexed. The patient may be turned from side to side, but if sensation is impaired, the amount of time spent on the affected side should be limited.

If possible, the patient is placed in a prone position for 15 to 30 minutes several times a day. A small pillow or a support is placed under the pelvis, extending from the level of the umbilicus to the upper third of the thigh (Fig. 67-4). This position helps promote hyperextension of the hip joints, which is essential for normal gait and helps prevent knee and hip flexion contractures. The prone position also helps drain bronchial secretions and prevents contractural deformities of the shoulders and knees. During positioning, it is important to reduce pressure and change position frequently to prevent pressure ulcers.

Establishing an Exercise Program. The affected extremities are exercised passively and put through a full range of motion four or five times a day to maintain joint mobility, regain motor control, prevent contractures in the paralyzed extremity, prevent further deterioration of the neuromuscular system, and enhance circulation. Exercise is helpful in preventing venous stasis, which may predispose the patient to thrombosis and pulmonary embolus.

Repetition of an activity forms new pathways in the CNS and therefore encourages new patterns of motion. At first, the extremities are usually flaccid. If tightness occurs in any area, the range-of-motion exercises should be performed more frequently (see Chapter 10).

The patient is observed for signs and symptoms that may indicate pulmonary embolus or excessive cardiac workload during exercise; these include shortness of breath, chest pain, cyanosis, and increasing pulse rate with exercise. Frequent short periods of exercise always are preferable to longer periods at infrequent intervals. Regularity in exercise is most important. Improvement in muscle strength and maintenance of range of motion can be achieved only through daily exercise.

The patient is encouraged and reminded to exercise the unaffected side at intervals throughout the day. It is helpful to develop a written schedule to remind the patient of the exercise activities. The nurse supervises and supports the patient during these activities. The patient can be instructed to put the unaffected leg under the affected one to assist in moving it when turning and exercising. Flexibility, strengthening, coordination, endurance, and balancing exercises prepare the patient for ambulation. Quadriceps muscle setting and gluteal setting exercises (see Chart 41-3 in Chapter 41) are started early to improve the muscle strength needed for walking; these are performed at least five times daily for 10 minutes at a time.

Preparing for Ambulation. As soon as possible, the patient is assisted out of bed and an active rehabilitation program is started. The patient is first educated to maintain balance while sitting and then to learn to balance while standing. If the patient has difficulty in achieving standing balance, a tilt table, which slowly brings the patient to an upright position, can be used. Tilt tables are especially helpful for patients who have been on bed rest for prolonged periods and have orthostatic blood pressure changes.

If the patient needs a wheelchair, the folding type with hand brakes is the most practical because it allows the patient to manipulate the chair. The chair should be low enough to allow the patient to propel it with the uninvolved foot and narrow enough to permit it to be used in the home. When the patient is transferred from the wheelchair, the brakes must be applied and locked on both sides of the chair.

The patient is usually ready to walk as soon as standing balance is achieved. Parallel bars are useful in these first efforts. A chair or wheelchair should be readily available in case the patient suddenly becomes fatigued or feels dizzy.

The training periods for ambulation should be short and frequent. As the patient gains strength and confidence, an adjustable cane can be used for support. Generally, a three- or four-pronged cane provides a stable support in the early phases of rehabilitation.

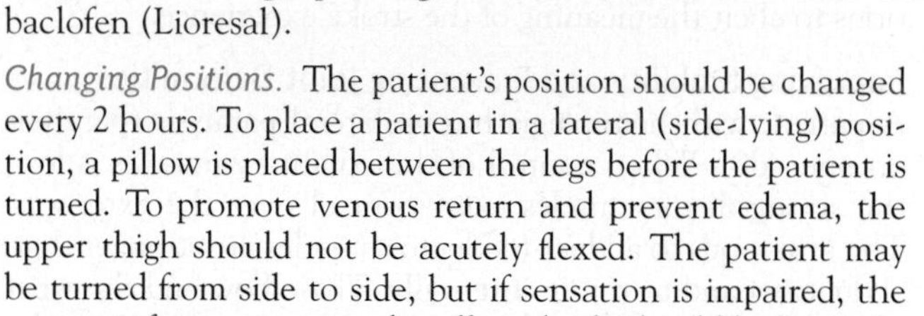

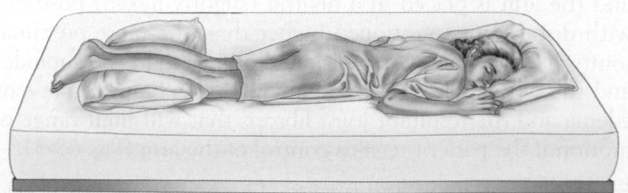

FIGURE 67-4 • Prone position with pillow support helps prevent hip flexion.

Preventing Shoulder Pain

As many as 84% of patients who have had a stroke have pain in the shoulder (Kalichman & Ratmansky, 2011). That pain may prevent them from learning new skills and affect their quality of life. Shoulder function is essential in achieving balance and performing transfers and self-care activities. Problems that can occur include rotator cuff disorders, spasticity of the shoulder muscles, painful shoulder, subluxation of the shoulder, and shoulder–hand syndrome. Development of a condition known as central pain syndrome may also contribute to the development of shoulder pain after a stroke.

A flaccid shoulder joint may be overstretched by the use of excessive force in turning the patient or from over strenuous arm and shoulder movement. To prevent shoulder pain, the nurse should never lift the patient by the flaccid shoulder or pull on the affected arm or shoulder. Overhead pulleys should also be avoided. If the arm is paralyzed, subluxation (incomplete dislocation) at the shoulder can occur as a result of overstretching of the joint capsule and musculature by the force of gravity when the patient sits or stands in the early stages after a stroke. This results in severe pain. Shoulder–hand syndrome (painful shoulder and generalized swelling of the hand) can cause a frozen shoulder and ultimately atrophy of subcutaneous tissues. When a shoulder becomes stiff, it is usually painful.

Many shoulder problems can be prevented by proper patient movement and positioning. The flaccid arm is positioned on a table or with pillows while the patient is seated. Some clinicians advocate the use of a properly worn sling when the patient first becomes ambulatory, to prevent the paralyzed upper extremity from dangling without support. Range-of-motion exercises are important in preventing painful shoulder. Overstrenuous arm movements are avoided. The patient is instructed to interlace the fingers, place the palms together, and push the clasped hands slowly forward to bring the scapulae forward; he or she then raises both hands above the head. This is repeated throughout the day. The patient is instructed to flex the affected wrist at intervals and move all the joints of the affected fingers. The patient is encouraged to touch, stroke, rub, and look at both hands. Pushing the heel of the hand firmly down on a surface is useful. Elevation of the arm and hand is also important in preventing dependent edema of the hand. Patients with continuing pain after attempted movement and positioning may require the addition of analgesia to their treatment program. Other treatments may include injections to the shoulder joint with corticosteroid medications or botulinum toxin type A, shoulder strapping, electrical stimulation, heat or ice, and soft tissue massage (Kumar, Kalita, Kumar, et al., 2009).

Medications are often helpful in the management of poststroke pain. Amitriptyline (Elavil) may be prescribed; in addition, antiseizure medications lamotrigine (Lamictal) and pregabalin (Lyrica) have been found to be effective for poststroke pain, and they may serve as alternatives for patients who cannot tolerate amitriptyline (Kim, Bashford, Murphy, et al., 2011; Kumar et al., 2009).

Enhancing Self-Care

As soon as the patient can sit up, he or she is encouraged to participate in personal hygiene activities. The patient is helped to set realistic goals; if feasible, a new task is added daily. The first step is to carry out all self-care activities on the unaffected side. Such activities as combing the hair, brushing the teeth, shaving with an electric razor, bathing, and eating can be carried out with one hand and should be encouraged. Although the patient may feel awkward at first, these motor skills can be learned by repetition, and the unaffected side will become stronger with use. The nurse must be sure that the patient does not neglect the affected side. Assistive devices will help make up for some of the patient's deficits (Chart 67-3). A small towel is easier to control while drying after bathing, and boxed paper tissues are easier to use than a roll of toilet tissue.

Return of functional ability is important to the patient recovering after a stroke. An early baseline assessment of functional ability with an instrument such as the Functional Independence Measure (FIM™) is important in team planning and goal setting for the patient. The FIM™ is a widely used instrument in stroke rehabilitation and provides valuable information about motor, social, and cognitive function. The patient's morale may improve if ambulatory activities are carried out in street clothes. The family is instructed to bring in clothing that is preferably a size larger than that usually worn. Clothing fitted with front or side fasteners or Velcro closures is the most suitable. The patient has better balance if most of the dressing activities are carried out while seated.

Perceptual problems may make it difficult for the patient to dress without assistance because of an inability to match the clothing to the body parts. To assist the patient, the nurse can take steps to keep the environment organized and uncluttered, because the patient with a perceptual problem is easily distracted. The clothing is placed on the affected side in the order in which the garments are to be put on. The use of a large mirror while dressing promotes the patient's awareness of what he or she is putting on the affected side. The patient has to make many compensatory movements when dressing; these can produce fatigue and painful twisting of the intercostal muscles. Support and encouragement are provided to

Chart 67-3 Assistive Devices to Enhance Self-Care After Stroke

Eating Devices
- Nonskid mats to stabilize plates
- Plate guards to prevent food from being pushed off plate
- Wide-grip utensils to accommodate a weak grasp

Bathing and Grooming Devices
- Long-handled bath sponge
- Grab bars, nonskid mats, handheld shower heads
- Electric razors with head at 90 degrees to handle
- Shower and tub seats, stationary or on wheels

Toileting Aids
- Raised toilet seat
- Grab bars next to toilet

Dressing Aids
- Velcro closures
- Elastic shoelaces
- Long-handled shoe horn

Mobility Aids
- Canes, walkers, wheelchairs
- Transfer devices such as transfer boards and belts

prevent the patient from becoming overly fatigued and discouraged. Even with intensive training, not all patients can achieve independence in dressing.

Adjusting to Physical Changes

Patients with a decreased field of vision should be approached on the side where visual perception is intact. All visual stimuli (e.g., clock, calendar, television) should be placed on this side. The patient can be educated to turn the head in the direction of the defective visual field to compensate for this loss. The nurse should make eye contact with the patient and draw his or her attention to the affected side by encouraging the patient to move the head. The nurse may also want to stand at a position that encourages the patient to move or turn to visualize who is in the room. Increasing the natural or artificial lighting in the room and providing eyeglasses are important aids to increasing vision.

The patient with homonymous hemianopsia turns away from the affected side of the body and tends to neglect that side and the space on that side; this is known as amorphosynthesis. In such instances, the patient cannot see food on half of the tray, and only half of the room is visible. It is important for the nurse to constantly remind the patient of the other side of the body; to maintain alignment of the extremities; and, if possible, to place the extremities where the patient can see them.

Assisting With Nutrition

Stroke can result in **dysphagia** (difficulty swallowing) due to impaired function of the mouth, tongue, palate, larynx, pharynx, or upper esophagus. Patients must be observed for paroxysms of coughing, food dribbling out of or pooling in one side of the mouth, food retained for long periods in the mouth, or nasal regurgitation when swallowing liquids. Swallowing difficulties place the patient at risk for aspiration, pneumonia, dehydration, and malnutrition.

A swallow assessment should be performed as soon as possible after the patient's arrival to the ED (preferably in the first 24 hours). This is done before allowing any oral intake. A speech therapist will evaluate the patient's swallowing ability, but an assessment may also be done by the nurse using a validated assessment tool (Summers et al., 2009).

If swallowing function is partially impaired, it may return over time, or the patient may be educated in alternative swallowing techniques, advised to take smaller boluses of food, and educated about types of foods that are easier to swallow. The patient may be started on a thick liquid or puréed diet, because these foods are easier to swallow than thin liquids. Having the patient sit upright, preferably out of bed in a chair, and instructing him or her to tuck the chin toward the chest as he or she swallows will help prevent aspiration. The diet may be advanced as the patient becomes more proficient at swallowing. If the patient cannot resume oral intake, a gastrointestinal feeding tube is placed for ongoing tube feedings and medication administration.

Enteral tubes can be either nasogastric (placed in the stomach) or nasoenteral (placed in the duodenum) to reduce the risk of aspiration. Nursing responsibilities in feeding include elevating the head of the bed at least 30 degrees to prevent aspiration, checking the position of the tube before feeding, ensuring that the cuff of the tracheostomy tube (if in place) is inflated, and giving the tube feeding slowly. The feeding tube is aspirated periodically to ensure that the feedings are passing through the gastrointestinal tract. Retained or residual feedings increase the risk of aspiration. Patients with retained feedings may benefit from the placement of a gastrostomy tube or a percutaneous endoscopic gastrostomy tube. In a patient with a nasogastric tube, the feeding tube should be placed in the duodenum to reduce the risk of aspiration. For long-term feedings, a gastrostomy tube is preferred. (See Chapter 45 for the management of patients with tube feedings.)

Attaining Bladder and Bowel Control

After a stroke, the patient may have transient urinary incontinence due to confusion, inability to communicate needs, and inability to use the urinal or bedpan because of impaired motor and postural control. Occasionally after a stroke, the bladder becomes atonic, with impaired sensation in response to bladder filling. Sometimes, control of the external urinary sphincter is lost or diminished. During this period, intermittent catheterization with sterile technique is carried out. After muscle tone increases and deep tendon reflexes return, bladder tone increases and spasticity of the bladder may develop. Because the patient's sense of awareness is clouded, persistent urinary incontinence or urinary retention may be symptomatic of bilateral brain damage. The voiding pattern is analyzed, and the urinal or bedpan is offered on this pattern or schedule. The upright posture and standing position are helpful for male patients during this aspect of rehabilitation.

Patients may have problems with bowel control, particularly constipation. Unless contraindicated, a high-fiber diet and adequate fluid intake (2 to 3 L/day) should be provided, and a regular time (usually after breakfast) should be established for toileting. (See Chapter 10 for additional information about bowel and bladder control.)

Improving Thought Processes

After a stroke, the patient may have problems with cognitive, behavioral, and emotional deficits related to brain damage. However, in many instances, a considerable degree of function can be recovered, because not all areas of the brain are equally damaged; some remain more intact and functional than others.

After assessment that delineates the patient's deficits, the neuropsychologist, in collaboration with the primary provider, psychiatrist, nurse, and other professionals, structures a training program using cognitive-perceptual retraining, visual imagery, reality orientation, and cueing procedures to compensate for losses.

The role of the nurse is supportive. The nurse reviews the results of neuropsychological testing; observes the patient's performance and progress; gives positive feedback; and, most importantly, conveys an attitude of confidence and hope. Interventions capitalize on the patient's strengths and remaining abilities while attempting to improve performance of affected functions. Other interventions are similar to those for improving cognitive functioning after a head injury (see Chapter 68).

Improving Communication

Aphasia, which impairs the patient's ability to express him- or herself and to understand what is being said, may become apparent in various ways. The cortical area that is responsible for integrating the myriad pathways required for the comprehension and formulation of language is called *Broca's area.*

It is located in a convolution adjoining the middle cerebral artery. This area is responsible for control of the combinations of muscular movements needed to speak each word. Broca's area is so close to the left motor area that a disturbance in the motor area often affects the speech area. This is why so many patients who are paralyzed on the right side (due to damage or injury to the left side of the brain) cannot speak, whereas those paralyzed on the left side are less likely to have speech disturbances.

The speech therapist assesses the communication needs of the stroke patient, describes the precise deficit, and suggests the best overall method of communication. Most language intervention strategies can be tailored for the individual patient. The patient is expected to take an active part in establishing goals.

A person with aphasia may become depressed. The inability to talk on the telephone, answer a question, or participate in conversation often causes anger, frustration, fear of the future, and hopelessness. Nursing interventions include strategies to make the atmosphere conducive to communication. This includes being sensitive to the patient's reactions and needs and responding to them in an appropriate manner while always treating the patient as an adult. The nurse provides strong emotional support and understanding to allay anxiety and frustration.

A common pitfall is for the nurse or other health care team member to complete the thoughts or sentences of the patient. This should be avoided, because it causes the patient to become more frustrated at not being allowed to speak and may deter efforts to practice putting thoughts together and completing sentences. A consistent schedule, routines, and repetition help the patient to function despite significant deficits. A written copy of the daily schedule, a folder of personal information (birth date, address, names of relatives), checklists, and an audiotaped list help improve the patient's memory and concentration. The patient may also benefit from a communication board, which has pictures of common needs and phrases. The board may be translated into any language.

When talking with the patient, it is important for the nurse to gain the patient's attention, speak slowly, and keep the language of instruction consistent. One instruction is given at a time, and time is allowed for the patient to process what has been said. The use of gestures may enhance comprehension. Speaking is thinking out loud, and the emphasis is on thinking. Listening and sorting out incoming messages requires mental effort; the patient must struggle against mental inertia and needs time to organize a response.

In working with the patient with aphasia, the nurse must remember to talk to the patient during care activities. This provides social contact for the patient. Chart 67-4 describes points to keep in mind when communicating with the patient with aphasia.

Chart 67-4 Communicating With the Patient With Aphasia

- Face the patient and establish eye contact.
- Speak in a usual manner and tone.
- Use short phrases, and pause between phrases to allow the patient time to understand what is being said.
- Limit conversation to practical and concrete matters.
- Use gestures, pictures, objects, and writing.
- As the patient uses and handles an object, say what the object is. It helps to match the words with the object or action.
- Be consistent in using the same words and gestures each time you give instructions or ask a question.
- Keep extraneous noises and sounds to a minimum. Too much background noise can distract the patient or make it difficult to sort out the message being spoken.

Maintaining Skin Integrity

The patient who has had a stroke may be at risk for skin and tissue breakdown because of altered sensation and inability to respond to pressure and discomfort by turning and moving. Preventing skin and tissue breakdown requires frequent assessment of the skin, with emphasis on bony areas and dependent parts of the body. During the acute phase, a specialty bed (e.g., low-air-loss bed) may be used until the patient can move independently or assist in moving.

A regular turning schedule (e.g., every 2 hours) is adhered to even if pressure-relieving devices are used to prevent tissue and skin breakdown. When the patient is positioned or turned, care must be used to minimize shear and friction forces, which cause damage to tissues and predispose the skin to breakdown.

The patient's skin must be kept clean and dry; gentle massage of healthy (nonreddened) skin and adequate nutrition are other factors that help to maintain skin and tissue integrity (see Chapter 10).

Improving Family Coping

Family members play an important role in the patient's recovery. Family members are encouraged to participate in counseling and to use support systems that will help with the emotional and physical stress of caring for the patient. Involving others in the patient's care and providing education about stress management techniques and methods for maintaining personal health also facilitate family coping.

The family may have difficulty accepting the patient's disability and may be unrealistic in their expectations. They are given information about the expected outcomes and are counseled to avoid doing activities for the patient that he or she can do. They are assured that their love and interest are part of the patient's therapy.

The family needs to be informed that the rehabilitation of the hemiplegic patient requires many months and that progress may be slow. The gains made by the patient in the hospital or rehabilitation unit must be maintained. All caregivers should approach the patient with a supportive and optimistic attitude, focusing on the patient's remaining abilities. The rehabilitation team, the medical and nursing team, the patient, and the family must all be involved in developing attainable goals for the patient at home.

Most relatives of patients with stroke handle the physical changes better than the emotional aspects of care. The family should be prepared to expect occasional episodes of emotional lability. The patient may laugh or cry easily and may be irritable and demanding or depressed and confused. The nurse can explain to the family that the patient's laughter does not necessarily connote happiness, nor does crying

Keeping the patient well informed of the plan of care provides reassurance and helps minimize anxiety. Appropriate reassurance also helps relieve the patient's fears and anxiety. The family also requires information and support.

Monitoring and Managing Potential Complications

Vasospasm. The patient is assessed for signs of possible vasospasm: intensified headaches, a decrease in level of responsiveness (confusion, disorientation, lethargy), or evidence of aphasia or partial paralysis. These signs may develop several days after surgery or on the initiation of treatment and must be reported immediately. The calcium channel blocker nimodipine should be administered for prevention of vasospasm, and fluid volume expanders may be prescribed as well (Connolly et al., 2012).

Seizures. Seizure precautions are maintained for every patient who may be at risk for seizure activity. Should a seizure occur, maintaining the airway and preventing injury are the primary goals. Medication therapy is initiated at this time.

Hydrocephalus. Blood in the subarachnoid space or ventricles impedes the circulation of CSF, resulting in hydrocephalus. A CT scan that indicates dilated ventricles confirms the diagnosis. Hydrocephalus can occur within the first 24 hours (acute) after subarachnoid hemorrhage or several days (subacute) to several weeks (delayed) later. Symptoms vary according to the time of onset and may be nonspecific. Acute hydrocephalus is characterized by sudden onset of stupor or coma and is managed with a ventriculostomy drain to decrease ICP. Symptoms of subacute and delayed hydrocephalus include gradual onset of drowsiness, behavioral changes, and ataxic gait. A ventriculoperitoneal shunt is surgically placed to treat chronic hydrocephalus. Changes in patient responsiveness are reported immediately.

Rebleeding. The rate of recurrent hemorrhage is approximately 2.1% to 3.7% per patient per year after a primary intracerebral hemorrhage (Morgenstern et al., 2010). Hypertension is the most serious and modifiable risk factor, which shows the importance of appropriate antihypertensive treatment.

Aneurysm rebleeding occurs most frequently during the first 2 weeks after the initial hemorrhage and is considered a major complication. Symptoms of rebleeding include sudden severe headache, nausea, vomiting, decreased level of consciousness, and neurologic deficit. Rebleeding is confirmed by CT scan. Blood pressure is carefully maintained with medications. The most effective preventive treatment is to secure the aneurysm if the patient is a candidate for surgery or endovascular treatment (Holmes, 2012).

Hyponatremia. After subarachnoid hemorrhage, hyponatremia is found in up to 50% of patients (AANN, 2009). Laboratory data must be checked frequently, and hyponatremia (defined as a serum sodium concentration less than 135 mEq/L) must be identified as early as possible. The patient's primary provider needs to be notified of a low serum sodium level that has persisted for 24 hours or longer. The patient is then evaluated for syndrome of inappropriate antidiuretic hormone (SIADH) or cerebral salt-wasting syndrome. (SIADH is described in Chapter 13.) Cerebral salt-wasting syndrome occurs when the kidneys are unable to conserve sodium and volume depletion results. The treatment most often is the use of IV hypertonic 3% saline.

Promoting Home and Community-Based Care

Educating Patients About Self-Care. The patient and family are provided with education that will enable them to cooperate with the care and restrictions required during the acute phase of hemorrhagic stroke and to prepare them to return home. Patient and family education includes information about the causes of hemorrhagic stroke and its possible consequences. In addition, the patient and family are informed about the medical treatments that are implemented, including surgical intervention if warranted, and the importance of interventions taken to prevent and detect complications (i.e., aneurysm precautions, close monitoring of the patient). Depending on the presence and severity of neurologic impairment and other complications resulting from the stroke, the patient may be transferred to a rehabilitation unit or center for additional patient and family education about strategies to regain self-care ability. Education addresses the use of assistive devices or modification of the home environment to help the patient live with the disability. Modifications of the home may be required to provide a safe environment.

Continuing Care. The acute and rehabilitation phase of care focuses on obvious needs, issues, and deficits for the patient with a hemorrhagic stroke. The patient and family are reminded of the importance of following recommendations to prevent further hemorrhagic stroke and keeping follow-up appointments with health care providers for monitoring of risk factors. Referral for home care may be warranted to assess the home environment and the ability of the patient and to ensure that the patient and family are able to manage at home. Home visits provide opportunities to monitor the physical and psychological status of the patient and the ability of the family to cope with any alterations in the patient's status. In addition, the home care nurse reminds the patient and family of the importance of continuing health promotion and screening practices. Chart 67-6 lists education for the patient recovering from a stroke.

Evaluation

Expected patient outcomes may include:

1. Demonstrates stable neurologic status and vital signs and respiratory patterns within normal limits
   a. Is alert and oriented to time, place, and person
   b. Demonstrates understandable speech patterns and stable cognitive processes
   c. Demonstrates usual and equal strength, movement, and sensation of all four extremities
   d. Exhibits deep tendon reflexes and pupillary responses within normal limits
2. Exhibits reduced anxiety level
   a. States rationale for aneurysm precautions
   b. Exhibits clear thought processes
   c. Is less restless
   d. Exhibits absence of physiologic indicators of anxiety (e.g., has vital signs within normal limits; usual respiratory rate; absence of excessive, fast speech)
3. Is free of complications
   a. Exhibits absence of vasospasm
   b. Exhibits vital signs within normal limits and is without seizures
   c. Verbalizes understanding of seizure precautions
   d. Exhibits intact mental, motor, and sensory status
   e. Reports no visual changes

Chart 67-6 HOME CARE CHECKLIST

The Patient Recovering From a Stroke

At the completion of the home care education, the patient or caregiver will be able to:	PATIENT	CAREGIVER
• Discuss measures to prevent subsequent strokes.	✔	✔
• Identify signs and symptoms of specific complications.	✔	✔
• Identify potential complications and discuss measures to prevent them (blood clots, aspiration, pneumonia, urinary tract infection, fecal impaction, skin breakdown, contracture).	✔	✔
• Identify psychosocial consequences of stroke and appropriate interventions.	✔	✔
• Identify safety measures to prevent falls.	✔	✔
• State names, doses, indications, and side effects of medications.	✔	✔
• Demonstrate adaptive techniques for accomplishing activities of daily living.	✔	✔
• Demonstrate swallowing techniques (for patients with dysphagia).	✔	
• Demonstrate care of enteral feeding tube, if applicable.	✔	✔
• Demonstrate home exercises, the use of splints or orthotics, proper positioning, and frequent repositioning.	✔	✔
• Describe procedures for maintaining skin integrity.	✔	✔
• Demonstrate indwelling catheter care, if applicable. Describe a bowel and bladder elimination program as appropriate.	✔	✔
• Identify appropriate recreational or diversional activities, support groups, and community resources.	✔	✔

Critical Thinking Exercises

1 pq Identify the priorities, approach, and techniques you would use to provide care to a patient just admitted to the ED for an ischemic stroke. How will your priorities, approach, and techniques differ if it is determined the patient has hemorrhagic stroke?

2 A 55-year-old patient with a history of hypertension and diabetes and is a current smoker is expected to be discharged to home today after a 4-day stay for a stroke. She has residual right-sided weakness and a visual field deficit. What education would be indicated to prevent another stroke?

3 ebp A patient is admitted to the hospital following a subarachnoid hemorrhage and has developed a fever. Should medical and nursing measures be implemented to reduce the temperature? Identify the evidence for and the criteria used to evaluate the strength of the evidence for the specific measures identified for fever.

Brunner Suite Resources

Explore these additional resources to enhance learning for this chapter:

• NCLEX-Style Questions and Other Resources on thePoint, http://thePoint.lww.com/Brunner13e

- Study Guide
- PrepU
- Clinical Handbook
- Handbook of Laboratory and Diagnostic Tests

References

*Asterisk indicates nursing research.
**Double asterisk indicates classic reference.

Books

American Association of Neuroscience Nurses. (2009). *Guide to the care of the patient with aneurysmal subarachnoid hemorrhage: AANN clinical practice guideline series*. Glenview, IL: Author.

American Association of Neuroscience Nurses. (2011). *Guide to the care of the hospitalized patient with ischemic stroke* (2nd ed.): *AANN clinical practice guideline series*. Glenview, IL: Author.

Bader, M., & Littlejohns, L. R. (2010). *AANN core curriculum for neuroscience nursing* (5th ed.). Glenview, IL: American Association of Neuroscience Nurses.

Hickey, J. V. (2009). *The clinical practice of neurological & neurosurgical nursing* (6th ed.). Philadelphia: Lippincott Williams & Wilkins.

Karch, A. (2012). *2012 Lippincott's nursing drug guide*. Philadelphia: Lippincott Williams & Wilkins.

Journals and Electronic Documents

Adams, H. P., Zoppo, G., Alberts, M. J., et al. (2007). Guidelines for the early management of patients with ischemic stroke. A guideline from the American Heart Association/American Stroke Association Stroke Council, Clinical Cardiology Council, Cardiovascular Radiology and Intervention Council, and the Atherosclerotic Peripheral Vascular Disease and Quality of Care Outcomes in Research Interdisciplinary Working Groups. *Stroke*, *38*(5), 1655–1711.

Adeoye, O., Ringer, A., Hornung, R., et al. (2010). Trends in surgical management and mortality of intracerebral hemorrhage in the United States before and after the STICH trial. *Neurocritical Care*, *13*(1), 82–86.

Connolly, E., Rabinstein, A., Carhuapoma, J., et al. (2012). Guidelines for the management of aneurysmal subarachnoid hemorrhage: A guideline for healthcare professionals from the American Heart Association/American Stroke Association. *Stroke*, *43*(6), 1711–1737.

Custer, N. (2009). What nurses should know about carotid stents. *Medsurg Nursing*, *18*(5), 277–282.

de Vries, F. M., Denig, P., Pauwels, K. B., et al. (2012). Primary prevention of major cardiovascular and cerebrovascular events with statins in diabetic patients: A meta-analysis. *Drugs*, *72*(18), 2365–2373.

Del Zoppo, G., Saver, J., Jauch, E., et al. (2009). Expansion of the time window for treatment of acute ischemic stroke with intravenous tissue plasminogen activator: A science advisory from the American Heart Association/American Stroke Association. *Stroke, 40*(8), 2945–2948.
Ennen, K. A. (2013). Taking a second look at stroke in women. *American Nurse Today, 8*(5), 12–15.
Fisher, M. (2011). New approaches to neuroprotective drug development. *Stroke, 42*(1 Suppl.), S24–S27.
Freeman, W., & Aguilar, M. (2012). Intracranial hemorrhage: Diagnosis and management. *Neurologic Clinics, 30*(1), 211–240.
Furie, K., Goldstein L., Albers, G., et al. (2012). AHA/ASA Science Advisory: Oral antithrombotic agents for the prevention of stroke in nonvalvular atrial fibrillation: A science advisory for healthcare professionals from the American Heart Association/American Stroke Association. *Stroke, 43*(8), 3422–3453.
Furie, K., Kasner, S., Adams, R., et al. (2011). Guidelines for the prevention of stroke in patients with stroke or transient ischemic attack: A guideline for healthcare professionals from the American Heart Association/American Stroke Association. *Stroke, 42*(1), 227–276.
Goldstein, L., Bushnell, C., Adams, R., et al. (2011). Guidelines for the primary prevention of stroke: A guideline for healthcare professionals from the American Heart Association/American Stroke Association. *Stroke, 42*(2), 517–584.
Holmes, T. D. (2012). Endovascular treatments/procedures: Are they efficacious and have they improved patient outcomes poststroke? *Journal of Neuroscience Nursing, 44*(5), 284–294.
Jauch, E., Cucchiara, B., Adeoye, O., et al. (2010). Part 11: Adult stroke: 2010 American Heart Association guidelines for cardiopulmonary resuscitation and emergency cardiovascular care. *Circulation, 122*(18 Suppl. 3), S818–S828.
Jauch, E. C., Saver, J. L., Adams, H. P., et al. (2013). Guidelines for the early management of patients with ischemic stroke: A guideline from the American Heart Association/American stroke Association. *Stroke, 44*(1), 870–947.
Kaatz, S., Kouides, P. A., Garcia, D. A., et al. (2012). Guidance on the emergent reversal of oral thrombin and factor Xa inhibitors. *American Journal of Hematology, 87*(Suppl. 1), S141–S145.
Kalichman, L., & Ratmansky, M. (2011). Underlying pathology and associated factors of hemiplegic shoulder pain. *American Journal of Physical Medicine & Rehabilitation, 90*(9), 768–780.
Kim, J. S., Bashford, G., Murphy, T. K., et al. (2011). Safety and efficacy of pregabalin in patients with central post-stroke pain. *Pain, 152*(5), 1018–1023.
*Klinedinst, J. N., Dunbar, S. B., & Clark, P. C. (2012). Stroke survivor and informal caregiver perceptions of poststroke depressive symptoms. *Journal of Neuroscience Nursing, 44*(2), 72–81.
Kumar, B., Kalita, J., Kumar, G., et al. (2009). Central poststroke pain: A review of pathophysiology and treatment. *Anesthesia & Analgesia, 108*(5), 1645–1657.
Mantese, V., Timaran, C., Chiu, D., et al. (2010). The Carotid Revascularization Endarterectomy versus Stenting Trial (CREST): Stenting versus carotid endarterectomy for carotid disease. *Stroke, 41*(10 Suppl.), S31–S34.
Miller, E., Murray, L., Richards, L., et al. (2010). Comprehensive overview of nursing and interdisciplinary rehabilitation care of the stroke patient: A scientific statement from the American Heart Association. *Stroke, 41*(10), 2402–2448.
Morgenstern, L., Hemphill, J., Anderson, C., et al. (2010). Guidelines for the management of spontaneous intracerebral hemorrhage: A guideline for healthcare professionals from the American Heart Association/American Stroke Association. *Stroke, 41*(9), 2108–2129.
National Center for Health Statistics. (2012). *Summary health statistics for U.S. adults: National Health Interview Survey, 2010* (Tables 1, 2). Washington, DC: U.S. Department of Health and Human Services. Available at: www.cdc.gov/nchs/data/series/sr_10/sr10_252.pdf
**National Institute of Neurologic Disorders and Stroke. (1995). Tissue plasminogen activator for acute ischemic stroke. The National Institute of Neurological Disorders and Stroke rt-PA Stroke Study Group. *New England Journal of Medicine, 333*(24), 1581–1587.
Roger, V., Go, A., Lloyd-Jones, D., et al. (2012). Heart disease and stroke statistics—2012 update: A report from the American Heart Association. *Circulation, 125*(1), e2–e220.
Satterfield, G., Anderson, J., & Moore, C. (2012). Evidence supporting the incorporation of the Dietary Approaches to Stop Hypertension (DASH) eating pattern into stroke self-management programs: A review. *Journal of Neuroscience Nursing, 44*(5), 244–252.
Saver, J. L. (2006). Time is brain quantified. *Stroke, 37*(1), 263–233.
Shaughnessy, M., Michael, K., & Resnick, B. (2012). Impact of treadmill exercise on efficacy expectations, physical activity, and stroke recovery. *Journal of Neuroscience Nursing, 44*(1), 27–35.
Summers, D., Leonard, A., Wentworth, D., et al. (2009). Comprehensive overview of nursing and interdisciplinary care of the acute ischemic stroke patient: A scientific statement from the American Heart Association. *Stroke, 40*(8), 2911–2944.
Teasell, R., Foley, N., Pereira, S., et al. (2012). Evidence to practice: Botulinum toxin in the treatment of spasticity post stroke. *Topics in Stroke Rehabilitation, 19*(2), 115–121.
Towfight, A. (2011). Insulin resistance, obesity, metabolic syndrome and lifestyle modification. *Neurology, 17*(6), 1293–1303.
*Wuchner, S., Bakas, T., Adams, G., et al. (2012). Nursing interventions and assessments for aneurysmal subarachnoid hemorrhage patients: A mixed methods study involving practicing nurses. *Journal of Neuroscience Nursing, 44*(4), 177–185.

Resources

American Association of Neuroscience Nurses (AANN), www.aann.org
American Stroke Association, a Division of the American Heart Association, www.strokeassociation.org
National Institute of Neurological Disorders and Stroke, www.ninds.nih.gov
National Stroke Association, www.stroke.org

Chapter 68

Management of Patients With Neurologic Trauma

Learning Objectives

On completion of this chapter, the learner will be able to:

1. Describe the mechanisms of injury, clinical signs and symptoms, diagnostic testing, and treatment options for patients with traumatic brain and spinal cord injuries.
2. Describe the nursing management of patients with brain injury.
3. Use the nursing process as a framework for care of patients with traumatic brain injury.
4. Identify the population at risk for spinal cord injury.
5. Describe the clinical features and management of the patient with neurogenic shock.
6. Discuss the pathophysiology of autonomic dysreflexia and describe the appropriate nursing interventions.
7. Use the nursing process as a framework for care of patients with spinal cord injury.

Glossary

autonomic dysreflexia: a life-threatening emergency in spinal cord injury patients that causes a hypertensive emergency; also called *autonomic hyperreflexia*

brain injury: an injury to the skull or brain that is severe enough to interfere with normal functioning

brain injury, closed (blunt): occurs when the head accelerates and then rapidly decelerates or collides with another object and brain tissue is damaged, but there is no opening through the skull and dura

brain injury, open: occurs when an object penetrates the skull, enters the brain, and damages the soft brain tissue in its path (penetrating injury), or when blunt trauma to the head is so severe that it opens the scalp, skull, and dura to expose the brain

complete spinal cord lesion: a condition that involves total loss of sensation and voluntary muscle control below the lesion

concussion: a temporary loss of neurologic function with no apparent structural damage to the brain

contusion: bruising of the brain surface

halo vest: a lightweight vest with an attached halo that stabilizes the cervical spine

head injury: an injury to the scalp, skull, and/or brain

incomplete spinal cord lesion: a condition in which there is preservation of the sensory or motor fibers, or both, below the lesion

neurogenic bladder: bladder dysfunction that results from a disorder or dysfunction of the nervous system; may result in either urinary retention or bladder overactivity

paraplegia: paralysis of the lower extremities with dysfunction of the bowel and bladder from a lesion in the thoracic, lumbar, or sacral region of the spinal cord

primary injury: initial damage to the brain that results from the traumatic event

secondary injury: an insult to the brain subsequent to the original traumatic event

spinal cord injury: an injury to the spinal cord, vertebral column, supporting soft tissue, or intervertebral disks caused by trauma

tetraplegia: paralysis of both arms and legs, with dysfunction of bowel and bladder from a lesion of the cervical segments of the spinal cord; formerly called *quadriplegia*

transection: severing of the spinal cord itself; transection can be complete (all the way through the cord) or incomplete (partially through)

Trauma involving the central nervous system can be life threatening. Even if not life threatening, brain and spinal cord injury may result in major physical and psychological dysfunction and can alter the patient's life completely. Neurologic trauma affects the patient, the family, the health care system, and society as a whole because of its major sequelae and the costs of acute and long-term care of patients with trauma to the brain and spinal cord.

Head Injuries

Head injury is a broad classification that includes injury to the scalp, skull, or brain. It is estimated that 1.7 million people sustain a head injury each year in the United States. Approximately 52,000 people die, 275,000 are hospitalized, and 80,000 to 90,000 will have long-term disabilities (Centers for Disease Control and Prevention [CDC], 2012;

Hickey, 2009; Kimbler, Murphy & Dhandapani, 2011). A head injury may lead to conditions ranging from mild concussion to coma and death; the most serious form is known as a traumatic brain injury (TBI). The most common causes of TBIs are falls (35.2%), motor vehicle crashes (17.3%), being struck by objects (16.5%), and assaults (10%). Children 0 to 4 years old, adolescents 15 to 19 years, and adults 65 years and older are most likely to sustain a TBI. In every age group, TBI rates are higher for males than females. An estimated 57 million people worldwide are currently living with a TBI-related disability, producing an annual economic impact of approximately $60 billion due to medical expenses and the cost of lost productivity (CDC, 2012; Kimbler et al., 2011). The best approach to head injury is prevention (Chart 68-1).

Pathophysiology

Research suggests that not all brain damage occurs at the moment of impact. Damage to the brain from traumatic injury takes two forms: primary injury and secondary injury. **Primary injury** is the initial damage to the brain that results from the traumatic event. This may include contusions, lacerations, and torn blood vessels due to impact; acceleration/deceleration; or foreign object penetration. **Secondary injury** evolves over the ensuing hours and days after the initial injury and results from inadequate delivery of nutrients and oxygen to the cells. These processes include intracranial hemorrhage, cerebral edema, increased intracranial pressure (ICP), hypoxic brain damage, and infection (Bader & Littlejohns, 2010; Hickey, 2009).

The Monro-Kellie hypothesis, also known as the Monro-Kellie doctrine, explains the dynamic equilibrium of cranial contents. The cranial vault contains three main components: brain, blood, and cerebrospinal fluid (CSF). According to the Monro-Kellie hypothesis, the cranial vault is a closed system, and if one of the three components increases in volume, at least one of the other two must decrease in volume or the pressure will increase. Any bleeding or swelling within the skull increases the volume of contents within the skull and therefore causes increased ICP (see Chapter 66). If the pressure increases enough, it can cause displacement of the brain through or against the rigid structures of the skull. This causes restriction of blood flow to the brain, decreasing oxygen delivery and waste removal. Cells within the brain become anoxic and cannot metabolize properly, producing ischemia, infarction, irreversible brain damage, and eventually brain death (Fig. 68-1).

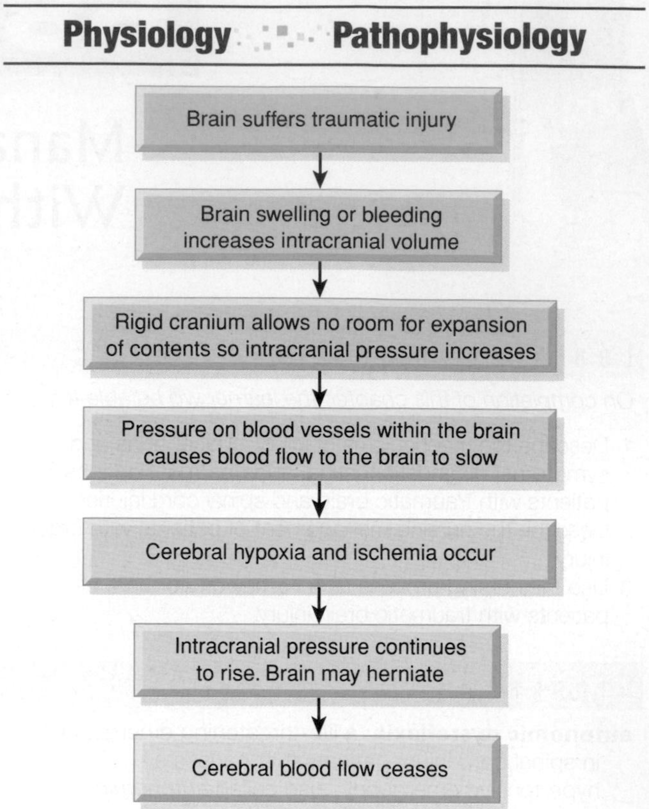
Physiology → Pathophysiology
Brain suffers traumatic injury
Brain swelling or bleeding increases intracranial volume
Rigid cranium allows no room for expansion of contents so intracranial pressure increases
Pressure on blood vessels within the brain causes blood flow to the brain to slow
Cerebral hypoxia and ischemia occur
Intracranial pressure continues to rise. Brain may herniate
Cerebral blood flow ceases

FIGURE 68-1 • Pathophysiology of traumatic brain injury.

Chart 68-1 HEALTH PROMOTION
Preventing Head and Spinal Cord Injuries

- Advise drivers to obey traffic laws and to avoid speeding or driving when under the influence of drugs or alcohol.
- Advise all drivers and passengers to wear seat belts and shoulder harnesses. Children younger than 12 years should be restrained in an age/size-appropriate system in the back seat.
- Caution passengers against riding in the back of pickup trucks.
- Advise motorcyclists, scooter riders, bicyclists, skateboarders, and roller skaters to wear helmets.
- Promote educational programs that are directed toward violence and suicide prevention in the community.
- Provide water safety instruction.
- Educate patients about steps that can be taken to prevent falls, particularly in older adults.
- Advise athletes to use protective devices. Recommend that coaches be educated in proper coaching techniques.
- Advise owners of firearms to keep them locked in a secure area where children cannot access them.

Scalp Injury

Isolated scalp trauma is generally classified as a minor injury. Because its many blood vessels constrict poorly, the scalp bleeds profusely when injured. Trauma may result in an abrasion (brush wound), contusion, laceration, or hematoma beneath the layers of tissue of the scalp (subgaleal hematoma). A large avulsion (tearing away) of the scalp may be potentially life threatening and is a true emergency. Diagnosis of a scalp injury is based on physical examination, inspection, and palpation. Scalp wounds are potential portals of entry for organisms that cause intracranial infections. Therefore, the area is irrigated before the laceration is sutured to remove foreign material and to reduce the risk for infection. Subgaleal hematomas (hematomas below the outer covering of the skull) usually reabsorb and do not require any specific treatment.

Skull Fractures

A skull fracture is a break in the continuity of the skull caused by forceful trauma. It may occur with or without damage to the brain. Skull fractures are classified by type and location. Types include linear, comminuted, and depressed skull fractures, whereas location fractures include frontal, temporal, and basilar skull fractures. A simple (linear) fracture is a break

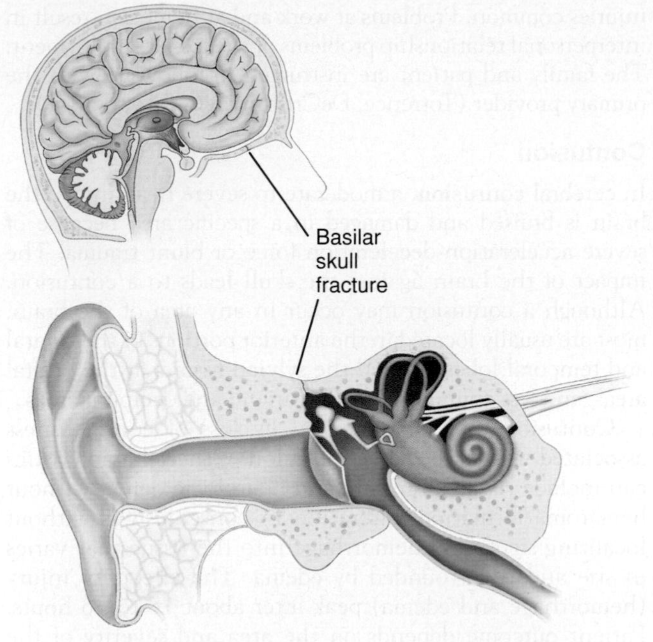
Basilar
skull
fracture

FIGURE 68-2 • Basilar fractures allow cerebrospinal fluid to leak from the nose and ears. Adapted from Hickey, J. V. (2009). *The clinical practice of neurological and neurosurgical nursing* (6th ed.). Philadelphia: Lippincott Williams & Wilkins.

in the continuity of the bone. A comminuted skull fracture refers to a splintered or multiple fracture line. Depressed skull fractures occur when the bones of the skull are forcefully displaced downward and can vary from a slight depression to bones of the skull being splintered and embedded within brain tissue. A fracture of the base of the skull is referred to as a basilar skull fracture (Hickey, 2009; Porth & Matfin, 2009) (Fig. 68-2). A fracture may be open, indicating a scalp laceration or tear in the dura (e.g., from a bullet or an ice pick), or closed, in which case the dura is intact.

Clinical Manifestations

Symptoms, apart from those of the local injury, depend on the severity and the anatomic location of the underlying brain injury. Persistent, localized pain usually suggests that a fracture is present. Fractures of the cranial vault may or may not produce swelling in the region of the fracture. Fractures of the base of the skull tend to traverse the paranasal sinus of the frontal bone or the middle ear located in the temporal bone (see Fig. 68-2). Therefore, they frequently produce hemorrhage from the nose, pharynx, or ears, and blood may appear under the conjunctiva. An area of ecchymosis (bruising) may be seen over the mastoid (Battle's sign). Basilar skull fractures are suspected when CSF escapes from the ears (CSF otorrhea) and the nose (CSF rhinorrhea). Drainage of CSF is a serious problem, because meningeal infection can occur if organisms gain access to the cranial contents via the nose, ear, or sinus through a tear in the dura.

Assessment and Diagnostic Findings

A computed tomography (CT) scan is used to diagnose a skull fracture. The ease with which a diagnosis of skull fracture is made depends on the site of the fracture. If a fracture is found on CT scan, there is always the question of associated brain injury, and magnetic resonance imaging (MRI) provides better resolution and clearer pictures of the injured area (Hickey, 2009).

Gerontologic Considerations

Older patients with head injuries differ from those who are younger in terms of etiology of injury, higher mortality rates, longer lengths of hospital stay, and poorer functional outcomes. The most common causes of injury in older adult patients are falls and motor vehicle crashes. Approximately 61% of all TBIs among adults aged 65 years and older result from falls (CDC, 2012). Physiologic changes related to aging may place the older adult at increased risk for injury, alter the type and severity of injury that occurs, or promote the development of complications. Two major factors place older adults at increased risk for hematomas. First, brain weight decreases, the dura becomes more adherent to the skull, and reaction times slow with increasing age (Eliopoulos, 2010). Second, many older adults take aspirin and anticoagulant agents as part of routine management of chronic conditions.

Medical Management

Nondepressed skull fractures generally do not require surgical treatment; however, close observation of the patient is essential. Nursing personnel may observe the patient in the hospital, but if no underlying brain injury is present, the patient may be allowed to return home. If the patient is discharged home, specific instructions must be given to the family (see later discussion of concussion).

Depressed skull fractures usually require surgery with elevation of the skull and débridement, usually within 24 hours of injury. Skull fractures can be a combination of open, compound, closed, or simple. Associated injuries include concurrent scalp laceration, dural tears, and brain injury directly below the fracture from compression of the tissue below the bony injury and from lacerations produced by the bony fragments (Hickey, 2009).

Brain Injury

The most important consideration in any head injury is whether the brain is injured. Even seemingly minor injury can cause significant brain damage secondary to obstructed blood flow and decreased tissue perfusion. The brain cannot store oxygen or glucose to any significant degree. Because the cerebral cells need an uninterrupted blood supply to obtain these nutrients, irreversible brain damage and cell death occur if the blood supply is interrupted for even a few minutes. Clinical manifestations of **brain injury** (injury to the skull or brain that is severe enough to interfere with normal functioning) are listed in Chart 68-2. **Closed (blunt) brain injury** occurs when the head accelerates and then rapidly decelerates or collides with another object (e.g., a wall, the dashboard of a car) and brain tissue is damaged but there is no opening through the skull and dura. **Open brain injury** occurs when an object penetrates the skull, enters the brain, and damages the soft brain tissue in its path (penetrating injury), or when blunt trauma to the head is so severe that it opens the scalp, skull,

Chart 68-2 ASSESSMENT
Assessing Traumatic Brain Injury

Be alert to the following signs and symptoms:

- Altered level of consciousness
- Confusion
- Pupillary abnormalities (changes in shape, size, and response to light)
- Altered or absent gag reflex
- Absent corneal reflex
- Sudden onset of neurologic deficits
- Changes in vital signs (altered respiratory pattern, widened pulse pressure, bradycardia, tachycardia, hypothermia, or hyperthermia)
- Vision and hearing impairment
- Sensory dysfunction
- Headache
- Seizures

and dura to expose the brain. Injuries to the brain can be focal or diffuse. Focal injuries include contusions and hematomas. Concussions and diffuse axonal injuries are the major diffuse injuries (Hickey, 2009).

Types of Brain Injury

Concussion

A **concussion** after head injury is a temporary loss of neurologic function with no apparent structural damage to the brain. A concussion (also referred to as a mild TBI) may or may not produce a brief loss of consciousness (Kimbler et al., 2011; West, Bergman, Biggins, et al., 2011). The mechanism of injury is usually blunt trauma from an acceleration-deceleration force, a direct blow, or a blast injury. If brain tissue in the frontal lobe is affected, the patient may exhibit bizarre irrational behavior, whereas involvement of the temporal lobe can produce temporary amnesia or disorientation.

There are three grades of concussion or mild TBI defined by the American Academy of Neurology when the injury is sports related (Ruff, Iverson, Barth, et al., 2009). A grade 1 concussion has symptoms of transient confusion, no loss of consciousness, and duration of mental status abnormalities on examination that resolve in less than 15 minutes. A grade 2 concussion also has symptoms of transient confusion and no loss of consciousness, but the concussion symptoms or mental status abnormalities on examination last more than 15 minutes. In a grade 3 concussion, there is any loss of consciousness lasting from seconds to minutes (Ruff et al., 2009).

A mild TBI is often overlooked in the emergency department (ED) because diagnostic studies may show no apparent structural sign of injury (Bay, 2011). The duration of mental status abnormalities is an indicator of the grade of the concussion. The patient is discharged from the hospital or ED once he or she returns to baseline after a concussion. Monitoring includes observing the patient for a decrease in level of consciousness (LOC), worsening headache, dizziness, seizures, abnormal pupil response, vomiting, irritability, slurred speech, and numbness or weakness in the arms or legs (West et al., 2011). The occurrence of these symptoms is a red flag indicating the need for further intervention. Recovery may appear complete, but long-term sequelae are possible and repeat injuries common. Problems at work and at home can result in interpersonal relationship problems or the loss of employment. The family and patient are instructed to follow up with the primary provider (Torrence, DeCristofaro, & Elliott, 2011).

Contusion

In cerebral **contusion**, a moderate to severe head injury, the brain is bruised and damaged in a specific area because of severe acceleration-deceleration force or blunt trauma. The impact of the brain against the skull leads to a contusion. Although a contusion may occur in any area of the brain, most are usually located in the anterior portions of the frontal and temporal lobes, around the sylvian fissure, at the orbital areas, and less commonly at the parietal and occipital areas.

Contusions are characterized by loss of consciousness associated with stupor and confusion. Other characteristics can include tissue alteration and neurologic deficit without hematoma formation, alteration in consciousness without localizing signs, and hemorrhage into the tissue that varies in size and is surrounded by edema. The effects of injury (hemorrhage and edema) peak after about 18 to 36 hours. Patient outcome depends on the area and severity of the injury. Temporal lobe contusions carry a greater risk of swelling, rapid deterioration, and brain herniation. Deep contusions are more often associated with hemorrhage and destruction of the reticular activating fibers altering arousal (Hickey, 2009).

Diffuse Axonal Injury

Diffuse axonal injury (DAI) results from widespread shearing and rotational forces that produce damage throughout the brain—to axons in the cerebral hemispheres, corpus callosum, and brain stem. The injured area may be diffuse with no identifiable focal lesion. DAI is associated with prolonged traumatic coma; it is more serious and is associated with a poorer prognosis than a focal lesion or ischemia. The patient with DAI in severe head trauma experiences no lucid interval, immediate coma, decorticate and decerebrate posturing (see Fig. 66-1 in Chapter 66), and global cerebral edema. Diagnosis is made by clinical signs in conjunction with a CT or MRI scan. Recovery depends on the severity of the axonal injury.

Intracranial Hemorrhage

Hematomas are collections of blood in the brain that may be epidural (above the dura), subdural (below the dura), or intracerebral (within the brain) (Fig. 68-3). Major symptoms are frequently delayed until the hematoma is large enough to cause distortion of the brain and increased ICP. The signs and symptoms of cerebral ischemia resulting from compression by a hematoma are variable and depend on the speed with which vital areas are affected and the area that is injured (American Association of Neuroscience Nurses, [AANN], 2011). In general, a rapidly developing hematoma, even if small, may be fatal, whereas a larger but slowly developing one may allow compensation for increases in ICP.

Epidural Hematoma

After a head injury, blood may collect in the epidural (extradural) space between the skull and the dura mater. This can result from a skull fracture that causes a rupture or laceration

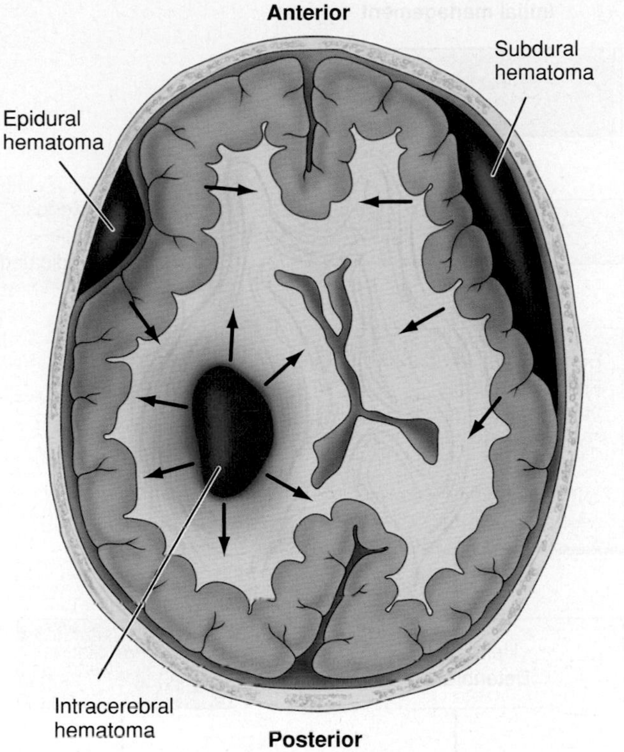

Anterior
Subdural hematoma
Epidural hematoma
Intracerebral hematoma
Posterior

FIGURE 68-3 • Location of epidural, subdural, and intracerebral hematomas.

of the middle meningeal artery, the artery that runs between the dura and the skull inferior to a thin portion of temporal bone. Hemorrhage from this artery causes rapid pressure on the brain (Hickey, 2009).

Symptoms are caused by the expanding hematoma. Epidural hematomas are often characterized by a brief loss of consciousness followed by a lucid interval in which the patient is awake and conversant. During this lucid interval, compensation for the expanding hematoma takes place by rapid absorption of CSF and decreased intravascular volume, both of which help to maintain the ICP within normal limits. When these mechanisms can no longer compensate, even a small increase in the volume of the blood clot produces a marked elevation in ICP. The patient then becomes increasingly restless, agitated, and confused as the condition progresses to coma. Then, often suddenly, signs of herniation appear (usually deterioration of consciousness and signs of focal neurologic deficits, such as dilation and fixation of a pupil or paralysis of an extremity), and the patient's condition deteriorates rapidly. The most common type of herniation syndrome associated with an epidural hematoma is uncal herniation (Hickey, 2009).

An epidural hematoma is considered an extreme emergency; marked neurologic deficit or even respiratory arrest can occur within minutes. Treatment consists of making openings through the skull (burr holes; see Fig. 66-8 in Chapter 66) to decrease ICP emergently, remove the clot, and control the bleeding. A craniotomy may be required to remove the clot and control the bleeding. A drain is usually inserted after creation of burr holes or a craniotomy to prevent reaccumulation of blood.

Subdural Hematoma

A subdural hematoma is a collection of blood between the dura and the brain, a space normally occupied by a thin cushion of fluid. The most common cause of subdural hematoma is trauma, but it can also occur as a result of coagulopathies or rupture of an aneurysm. A subdural hemorrhage is more frequently venous in origin and is caused by the rupture of small vessels that bridge the subdural space (Bader & Littlejohns, 2010). The subdural hematoma that results may be acute, subacute, or chronic, depending on the size of the involved vessel and the amount of bleeding.

Acute and Subacute Subdural Hematoma. Acute subdural hematomas are associated with major head injury involving contusion or laceration. Clinical symptoms develop over 24 to 48 hours. Signs and symptoms include changes in the LOC, pupillary signs, and hemiparesis. There may be minor or even no symptoms with small collections of blood. Coma, increasing blood pressure, decreasing heart rate, and slowing respiratory rate are all signs of a rapidly expanding mass requiring immediate intervention. Subacute subdural hematomas are the result of less severe contusions and head trauma. Clinical manifestations usually appear between 48 hours and 2 weeks after the injury. Signs and symptoms are similar to those of an acute subdural hematoma.

If the patient can be transported rapidly to the hospital, an immediate craniotomy is performed to open the dura, allowing the subdural clot to be evacuated. Successful outcome also depends on the control of ICP and careful monitoring of respiratory function (see the discussion of intracranial surgery in Chapter 66). The mortality rate for patients with acute or subacute subdural hematoma is high because of associated brain damage.

Chronic Subdural Hematoma. Chronic subdural hematomas can develop from seemingly minor head injuries and are seen most frequently in older adults who are prone to this type of head injury due to brain atrophy, which is a consequence of the aging process. Seemingly minor head trauma may produce enough impact to shift the brain contents abnormally. The time between injury and onset of symptoms can be lengthy (e.g., 3 weeks to months), so the actual injury may be forgotten.

A chronic subdural hematoma can resemble other conditions—for example, it may be mistaken for a stroke. The bleeding is less profuse, but compression of the intracranial contents still occurs. The blood within the brain changes in character in 2 to 4 days, becoming thicker and darker. In a few weeks, the clot breaks down and has the color and consistency of motor oil. Eventually, calcification or ossification of the clot takes place. The brain adapts to this foreign body invasion, and the clinical signs and symptoms fluctuate. Symptoms include severe headache, which tends to come and go; alternating focal neurologic signs; personality changes; mental deterioration; and focal seizures. The patient may be labeled neurotic or psychotic if the cause is overlooked.

The treatment for a chronic subdural hematoma consists of surgical evacuation of the clot. The procedure may be carried out through multiple burr holes, or a craniotomy may be performed for a sizable subdural mass that cannot be suctioned or drained through burr holes.

Intracerebral Hemorrhage and Hematoma

Intracerebral hemorrhage is bleeding into the parenchyma of the brain. It is commonly seen in head injuries when force is exerted to the head over a small area (e.g., missile injuries, bullet wounds, stab injuries). These hemorrhages within the brain may also result from the following:

- Systemic hypertension, which causes degeneration and rupture of a vessel
- Rupture of an aneurysm
- Vascular anomalies
- Intracranial tumors
- Bleeding disorders such as leukemia, hemophilia, aplastic anemia, and thrombocytopenia
- Complications of anticoagulant therapy

Nontraumatic causes of intracerebral hemorrhage are discussed in Chapter 67.

The onset may be insidious, beginning with the development of neurologic deficits followed by headache. Management includes supportive care; control of ICP; and careful administration of fluids, electrolytes, and antihypertensive medications. Surgical intervention by craniotomy or craniectomy permits removal of the blood clot and control of hemorrhage but may not be possible because of the inaccessible location of the bleeding or the lack of a clearly circumscribed area of blood that can be removed.

Management of Brain Injuries

Assessment and diagnosis of the extent of injury are accomplished by the initial physical and neurologic examinations. CT and MRI scans are the main neuroimaging diagnostic tools and are useful in evaluating the brain structure. Positron emission tomography (PET) is available in some trauma centers for assessing brain function. A flow chart developed by the Brain Trauma Foundation (2007) for the initial management of brain injury is presented in Figure 68-4.

Any patient with a head injury is presumed to have a cervical spine injury until proven otherwise. The patient is transported from the scene of the injury on a board with the head and neck maintained in alignment with the axis of the body. A cervical collar should be applied and maintained until cervical spine x-rays have been obtained and the absence of cervical spinal cord injury (SCI) documented.

All therapy is directed toward preserving brain homeostasis and preventing secondary brain injury, which is injury to the brain that occurs after the original traumatic event (Bader & Littlejohns, 2010). Common causes of secondary injury are cerebral edema, hypotension, and respiratory depression that may lead to hypoxemia and electrolyte imbalance. Treatments to prevent secondary injury include stabilization of cardiovascular and respiratory function to maintain adequate cerebral perfusion, control of hemorrhage and hypovolemia, and maintenance of optimal blood gas values.

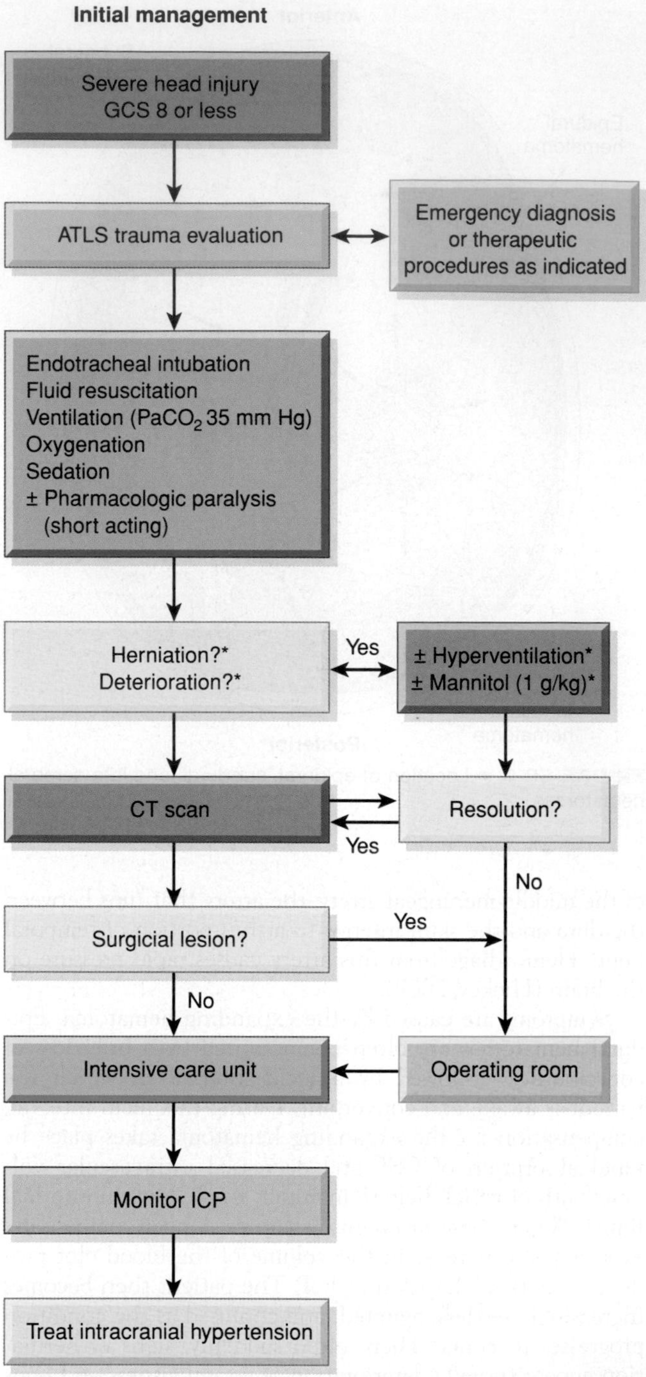
Initial management
Severe head injury
GCS 8 or less
ATLS trauma evaluation
Emergency diagnosis or therapeutic procedures as indicated
Endotracheal intubation
Fluid resuscitation
Ventilation ($PaCO_2$ 35 mm Hg)
Oxygenation
Sedation
± Pharmacologic paralysis (short acting)
Herniation?*
Deterioration?*
Yes
± Hyperventilation*
± Mannitol (1 g/kg)*
CT scan
Resolution?
Yes
No
Surgical lesion?
Yes
No
Intensive care unit
Operating room
Monitor ICP
Treat intracranial hypertension

FIGURE 68-4 • Initial management of the patient with traumatic brain injury (treatment option). Copyright © 2007 Brain Trauma Foundation. *Only in the presence of signs of herniation or progressive neurologic deterioration not attributable to extracranial factors. GCS, Glasgow Coma Scale; ATLS, Advanced Trauma Life Support; $PaCO_2$, partial pressure of arterial carbon dioxide; CT, computed tomography; ICP, intracranial pressure.

Treatment of Increased Intracranial Pressure

As the damaged brain swells with edema or as blood collects within the brain, an increase in ICP occurs; this requires aggressive treatment (see Chapter 66 for a discussion of the relationship between ICP and cerebral perfusion pressure [CPP]). If the ICP remains elevated, it can decrease the CPP. Initial management is based on the principle of preventing secondary injury and maintaining adequate cerebral oxygenation (see Fig. 68-4).

Surgery is required for evacuation of blood clots, débridement and elevation of depressed fractures of the skull, and suture of severe scalp lacerations. ICP is monitored closely; if increased, it is managed by maintaining adequate oxygenation, elevating the head of the bed, and maintaining normal blood volume. Devices to monitor ICP or drain CSF can be inserted during surgery or at the bedside using aseptic technique. The patient is cared for in the intensive care unit, where expert nursing care and medical treatment are readily available (AANN, 2011).

Supportive Measures

Treatment also includes ventilatory support, seizure prevention, fluid and electrolyte maintenance, nutritional support, and management of pain and anxiety. Patients who are comatose are intubated and mechanically ventilated to ensure adequate oxygenation and protect the airway.

Because seizures can occur after head injury and can cause secondary brain damage from hypoxia, antiseizure agents may be administered. If the patient is very agitated, benzodiazepines are the most commonly used sedative agents and do not affect cerebral blood flow or ICP. Lorazepam (Ativan) and midazolam (Versed) are frequently used but have active metabolites that my cause prolonged sedation, making it difficult to conduct a neurologic assessment. Propofol (Diprivan), on the other hand, a sedative-hypnotic agent that is supplied in an intralipid emulsion for intravenous (IV) use, is the sedative of choice. It is an ultra-short acting, rapid onset drug with elimination half-life of less than an hour. It has a major advantage of being titratable to its desired clinical effect but still provides the opportunity for an accurate neurologic assessment (Hickey, 2009). A nasogastric tube may be inserted, because reduced gastric motility and reverse peristalsis are associated with head injury, making regurgitation and aspiration common in the first few hours.

Brain Death

When a patient has sustained a severe head injury incompatible with life, the patient is a potential organ donor. The nurse may assist in the clinical examination for determination of brain death and in the process of organ procurement. All 50 states recognize the Uniform Determination of Death Act that indicates death will be determined by accepted medical standards and indicates irreversible loss of all brain function, including the brain stem (Morton, Fontaine, Hudak, et al., 2009). The three cardinal signs of brain death on clinical examination are coma, the absence of brain stem reflexes, and apnea. Adjunctive tests, such as cerebral blood flow studies, electroencephalogram (EEG), transcranial Doppler, and brain stem auditory evoked potential, are often used to confirm brain death (Hickey, 2009). The health care team provides information to the family and assists them with the decision-making process about end-of-life care.

Chart 68-3 ASSESSMENT
Glasgow Coma Scale

The Glasgow Coma Scale is a tool for assessing a patient's response to stimuli. Scores range from 3 (deep coma) to 15 (normal).

Eye opening response	Spontaneous	4
	To voice	3
	To pain	2
	None	1
Best verbal response	Oriented	5
	Confused	4
	Inappropriate words	3
	Incomprehensible sounds	2
	None	1
Best motor response	Obeys command	6
	Localizes pain	5
	Withdraws	4
	Flexion	3
	Extension	2
	None	1
Total		3 to 15

From Teasdale, G., & Jennett, B. (1974). Assessment of coma and impaired consciousness. A practical scale. *Lancet*, 2(7872), 81–84. Used with permission.

NURSING PROCESS

The Patient With a Traumatic Brain Injury

Assessment

Depending on the patient's neurologic status, the nurse may elicit information from the patient, from the family, or from witnesses or emergency rescue personnel. Although all usual baseline data may not be collected initially, the immediate health history should include the following questions:

- When did the injury occur?
- What caused the injury? A high-velocity missile? An object striking the head? A fall?
- What was the direction and force of the blow?

A history of unconsciousness or amnesia after a head injury indicates a significant degree of brain damage, and changes that occur minutes to hours after the initial injury can reflect recovery or indicate the development of secondary brain damage. The nurse should determine if there was a loss of consciousness, the duration of the unconscious period, and if the patient could be aroused.

In addition to asking questions that establish the nature of the injury and the patient's condition immediately after the injury, the nurse examines the patient thoroughly. This assessment includes determining the patient's LOC using the Glasgow Coma Scale (GCS) and assessing the patient's response to tactile stimuli (if unconscious), pupillary response to light, corneal and gag reflexes, and motor function (Teasdale & Jennett, 1974). The GCS (Chart 68-3) is based on the three criteria of eye opening, verbal responses, and motor responses to verbal commands or painful stimuli. It is particularly useful for monitoring changes during the acute phase, the first few days after a head injury. It does not take the place of an in-depth neurologic assessment. Additional detailed assessments are made initially and at frequent intervals throughout the acute phase of care (Hickey, 2009). Baseline and ongoing assessments are critical in nursing assessment of the patient with brain injury, whose condition can worsen dramatically and irrevocably if subtle signs are overlooked. More information on assessment is provided in the following sections and in Figure 68-5 and Table 68-1.

Maintaining the Airway

One of the most important nursing goals in the management of head injury is to establish and maintain an adequate airway. The brain is extremely sensitive to hypoxia, and a neurologic deficit can worsen if the patient is hypoxic. Therapy is directed toward maintaining optimal oxygenation to preserve cerebral function. An obstructed airway causes carbon dioxide retention and hypoventilation, which can produce cerebral vessel dilation and increased ICP.

Interventions to ensure an adequate exchange of air are discussed in Chapter 66 and include the following:

- Maintaining the unconscious patient in a position that facilitates drainage of oral secretions, with the head of the bed elevated about 30 degrees to decrease intracranial venous pressure (Brosche, 2011)
- Establishing effective suctioning procedures (pulmonary secretions produce coughing and straining, which increase ICP)
- Guarding against aspiration and respiratory insufficiency
- Closely monitoring arterial blood gas values to assess the adequacy of ventilation. The goal is to keep blood gas values within normal limits to ensure adequate cerebral blood flow.
- Monitoring the patient who is receiving mechanical ventilation for pulmonary complications such as acute respiratory distress syndrome and pneumonia (Hickey, 2009)

The patient who is intubated is at risk for deterioration of oral health with the development of dental plaque, xerostomia, and bacterial growth. Research suggests that providing oral care does not increase ICP (Prendergast, Hallberg, Jahnke, et al., 2009). Furthermore, a comprehensive oral care protocol that includes the use of a tongue scraper, an electric toothbrush, and pharmacologic moisturizers provides optimal oral hygiene throughout intubation and after extubation (Prendergast, Jakobsson, Renvert, et al., 2012) (see Chart 68-4).

Monitoring Neurologic Function

The importance of ongoing assessment and monitoring of the patient with brain injury cannot be overstated. Parameters are assessed initially and as frequently as the patient's condition requires. As soon as the initial assessment is made, the use of a neurologic flow chart is started and maintained.

Level of Consciousness. The GCS is used to assess LOC at regular intervals, because changes in the LOC precede all other changes in vital and neurologic signs. The patient's best responses to predetermined stimuli are recorded (see Chart 68-3). Each response is scored (the greater the number, the better the functioning), and the sum of these scores gives an indication of the severity of coma and a prediction of possible outcome. The lowest score is 3 (least responsive); the highest is 15 (most responsive). A GCS score between 3 and 8 is generally accepted as indicating a severe head injury (Bader & Littlejohns, 2010).

Concept Mastery Alert

The GCS is considered the most sensitive indicator of a lapse in neurologic functioning in patients with TBI and is often the earliest sign of acute change in ICP.

Vital Signs. Although a change in LOC is the most sensitive neurologic indication of deterioration of the patient's condition, vital signs also are monitored at frequent intervals to assess the intracranial status. Table 68-1 depicts the general assessment parameters for the patient with a head injury.

Signs of increasing ICP include slowing of the heart rate (bradycardia), increasing systolic blood pressure, and widening pulse pressure (Cushing's reflex). As brain compression increases, respirations become rapid, the blood pressure may decrease, and the pulse slows further. This is an ominous development, as is a rapid fluctuation of vital signs (Hickey, 2009). The temperature is maintained at less than 38°C (100.4°F). Tachycardia and arterial hypotension may indicate that bleeding is occurring elsewhere in the body.

Chart 68-4 NURSING RESEARCH PROFILE

Best Oral Hygiene Practices for Intubated Patients With Neuroscience Injuries

Prendergast, V., Jakobsson, U., Renvert, S., et al. (2012). Effects of a standard versus comprehensive oral care protocol among intubated neuroscience ICU patients: Results of a randomized controlled trial. *Journal of Neuroscience Nursing*, *44*(3), 134–146.

Purpose

The best oral hygiene practices for nurses to use for patients with neuroscience injuries who are intubated are not known. The purpose of this study was to compare the effects of two protocols on oral health.

Design

This was a randomized controlled trial that compared the effects a standard protocol of manual toothbrushing to a comprehensive protocol of tongue scraping, electric toothbrushing, and moisturizing on oral health. Oral health was evaluated using the Oral Assessment Guide (OAG) upon enrollment in the trial, before extubation, and 48 hours after extubation.

Findings

There were 31 patients in the standard protocol group and 25 patients in the comprehensive protocol group. Compliance with performing the oral care protocol was >91% in both groups. The OAG score significantly deteriorated ($p < .001$) in the standard oral care group and did not return to baseline after extubation. The OAG deteriorated during intubation within the comprehensive protocol group as well ($p < .004$) but returned to baseline after extubation.

Nursing Implications

Current published oral care protocols appear to be substandard in promoting and maintaining oral health in patients who are intubated. Nurses should use a comprehensive oral care protocol using a tongue scraper, an electric toothbrush, and pharmacologic moisturizers for optimal oral hygiene throughout intubation and after extubation for patients in intensive care units.

Concept Mastery Alert

In a patient with a head injury, a rapid increase in body temperature is regarded as unfavorable because hyperthermia increases the metabolic demands of the brain and may indicate brain stem damage—a poor prognostic sign.

Motor Function. Motor function is assessed frequently by observing spontaneous movements, asking the patient to raise and lower the extremities, and comparing the strength and equality of the upper and lower extremities at periodic intervals. To assess upper extremity strength, the nurse instructs the patient to squeeze the examiner's fingers tightly. The nurse assesses lower extremity motor strength by placing the hands on the soles of the patient's feet and asking the patient to push down against the examiner's hands. Examination of the motor system is discussed in more detail in Chapter 65. The presence or absence of spontaneous movement of each extremity is also noted, and speech and eye signs are assessed.

If the patient does not demonstrate spontaneous movement, responses to painful stimuli are assessed (Hickey, 2009). Motor response to pain is assessed by applying a central stimulus, such as pinching the pectoralis major muscle, to determine the patient's best response. Peripheral stimulation may provide inaccurate assessment data because it may result in a reflex movement rather than a voluntary motor response. Abnormal responses (lack of motor response; extension responses) are associated with a poorer prognosis.

Other Neurologic Signs. In addition to the patient's spontaneous eye opening, which is evaluated with the GCS, the size and equality of the pupils and their reaction to light are assessed. A unilaterally dilated and poorly responding pupil may indicate a developing hematoma, with subsequent pressure on the third cranial nerve due to shifting of the brain. If both pupils become fixed and dilated, this indicates overwhelming injury and intrinsic damage to the upper brain stem and is a poor prognostic sign (Bader & Littlejohns, 2010).

The patient with a head injury may develop deficits such as anosmia (lack of sense of smell), eye movement abnormalities, aphasia, memory deficits, and posttraumatic seizures or epilepsy. Patients may be left with residual psychological deficits (impulsiveness; emotional lability; or uninhibited, aggressive behaviors) and, as a consequence of the impairment, may lack insight into their emotional responses.

Monitoring Fluid and Electrolyte Balance

Brain damage can produce metabolic and hormonal dysfunctions. The monitoring of serum electrolyte levels is important, especially in patients receiving osmotic diuretics, those with syndrome of inappropriate antidiuretic hormone (SIADH) secretion, and those with posttraumatic diabetes insipidus.

Serial studies of blood and urine electrolytes and osmolality are carried out because head injuries may be accompanied by disorders of sodium regulation. Hyponatremia is common after head injury due to shifts in extracellular fluid, electrolytes, and volume. Hyperglycemia, for example, can cause an increase in extracellular fluid that lowers sodium (Butcavage, 2012). Hypernatremia may also occur as a result of sodium retention that may last several days, followed by sodium diuresis. Increasing lethargy, confusion, and seizures may be the result of electrolyte imbalance.

Endocrine function is evaluated by monitoring serum electrolytes, blood glucose values, and intake and output. Urine is tested regularly for acetone. A record of daily weights is maintained, especially if the patient has hypothalamic involvement and is at risk for the development of diabetes insipidus.

Promoting Adequate Nutrition

Head injury results in metabolic changes that increase calorie consumption and nitrogen excretion. Protein demand increases. Early initiation of nutritional therapy has been shown to improve outcomes in patients with head injury. Patients with brain injury are assumed to be catabolic, and nutritional support consultation should be considered as soon as the patient is admitted. Parenteral nutrition via a central line or enteral feedings administered via a nasogastric or nasojejunal feeding tube should be considered (Hickey, 2009). If CSF rhinorrhea occurs, an oral feeding tube should be inserted instead of a nasal tube.

Laboratory values should be monitored closely in patients receiving parenteral nutrition. Elevating the head of the bed and aspirating the enteral tube for evidence of residual feeding before administering additional feedings can help prevent distention, regurgitation, and aspiration. A continuous-drip infusion or pump may be used to regulate the feeding. Enteral or parenteral feedings are usually continued until the swallowing reflex returns and the patient can meet caloric requirements orally. (See Chapter 45 for the principles and technique of enteral feedings.)

Preventing Injury

Often, as the patient emerges from coma, a period of lethargy and stupor is followed by a period of agitation. Each phase is variable and depends on the person, the location of the injury, the depth and duration of coma, and the patient's age. Restlessness may be caused by hypoxia, fever, pain, or a full bladder. It may indicate injury to the brain but may also be a sign that the patient is regaining consciousness. (Some restlessness may be beneficial because the lungs and extremities are exercised.) Agitation may also be the result of discomfort from catheters, IV lines, restraints, and repeated neurologic checks. Alternatives to restraints must be used whenever possible.

Strategies to prevent injury include the following:

- Assessing the patient to ensure that oxygenation is adequate and the bladder is not distended. Dressings and casts are checked for constriction.
- Using padded side rails or wrapping the patient's hands in mitts to protect the patient from self-injury and dislodging of tubes. Restraints are avoided, because straining against them can increase ICP or cause other injury. Enclosed or floor-level specialty beds may be indicated.
- Avoiding opioids as a means of controlling restlessness, because they depress respiration, constrict the pupils, and alter responsiveness
- Reducing environmental stimuli by keeping the room quiet, limiting visitors, speaking calmly, and providing frequent orientation information (e.g., explaining where the patient is and what is being done)
- Providing adequate lighting to prevent visual hallucinations
- Minimizing disruption of the patient's sleep–wake cycles
- Lubricating the patient's skin with oil or emollient lotion to prevent irritation due to rubbing against the sheet

TABLE 68-2 Rancho Los Amigos Scale: Levels of Cognitive Function

Cognitive Level	Description	Nursing Management
For levels I–III, the key approach is to *provide stimulation*.		
I: No response	Completely unresponsive to all stimuli, including painful stimuli	Multiple modalities of sensory input should be used. Examples are listed here, but management should be individualized and expanded based on available materials and patient preferences (determined by obtaining information from the family).
II: Generalized response	Nonpurposeful response; responds to pain but in a nonpurposeful manner	*Olfactory:* Perfumes, flowers, shaving lotion *Visual:* Family pictures, card, personal items
III: Localized response	Responses more focused—withdraws to pain; turns toward sound; follows moving objects that pass within visual field; pulls on sources of discomfort (e.g., tubes, restraints); may follow simple commands but inconsistently and in a delayed manner	*Auditory:* Radio, television, tapes of family voices or favorite recordings, talking to patient (nurse, family members). The nurse should tell patient what is going to be done, discuss the environment, provide encouragement. *Tactile:* Touching of skin, rubbing various textures on skin *Movement:* Range-of-motion exercises, turning, repositioning, the use of water mattress
For levels IV–VI, the key approach is to *provide structure*.		
IV: Confused, agitated response	Alert, hyperactive state in which patient responds to internal confusion/agitation; behavior nonpurposeful in relation to the environment; aggressive, bizarre behavior common	For level IV, which lasts 2–4 weeks, interventions are directed at decreasing agitation, increasing environmental awareness, and promoting safety. • Approach patient in a calm manner, and use a soft voice. • Screen patient from environmental stimuli (e.g., sounds, sights); provide a quiet, controlled environment. • Remove devices that contribute to agitation (e.g., tubes), if possible. • Functional goals cannot be set because the patient is unable to cooperate.
V: Confused, inappropriate response	When agitation occurs, it is the result of external rather than internal stimuli; focused attention is difficult; memory is severely impaired; responses are fragmented and inappropriate to the situation; there is no carryover of learning from one situation to the other.	For levels V and VI, interventions are directed at decreasing confusion, improving cognitive function, and improving independence in performing ADLs. • Provide supervision. • Use repetition and cues to educate about ADLs. Focus the patient's attention, and help to increase his or her concentration. • Help the patient organize activity. • Clarify misinformation and reorient when confused. • Provide a consistent, predictable schedule (e.g., post daily schedule on large poster board).
VI: Confused, appropriate response	Follows simple directions consistently but is inconsistently oriented to time and place; short-term memory worse than long-term memory; can perform some ADLs	
For levels VII–X, the key approach is *integration into the community*.		
VII: Automatic, appropriate response	Appropriately responsive and oriented within the hospital setting; needs little supervision in ADLs; some carryover of learning; patient has superficial insight into disabilities; has decreased judgment and problem-solving abilities; lacks realistic planning for future	For levels VII–X, interventions are directed at increasing the patient's ability to function with minimal or no supervision in the community. • Reduce environmental structure. • Help the patient plan for adapting ADLs for self into the home environment. • Discuss and adapt home living skills (e.g., cleaning, cooking) to patient's ability.
VIII: Purposeful, appropriate	Alert, oriented, intact memory; has realistic goals for the future. Able to complete familiar tasks for 1 hour in a distracting environment; overestimates or underestimates abilities, argumentative, easily frustrated, self-centered; uncharacteristically dependent/independent	• Provide standby assistance as needed for ADLs and home living skills.
IX: Purposeful, appropriate	Independently shifts back and forth between tasks and completes them accurately for at least 2 consecutive hours; uses assistive memory devices to recall schedule and activities; aware of and acknowledges impairments and disabilities when they interfere with task completion; depression may continue; may be easily irritable and have a low frustration tolerance	• Provide assistance on request for adapting ADLs and home living skills.

Cognitive Level	Description	Nursing Management
X: Purposeful, appropriate	Able to handle multiple tasks simultaneously in all environments but may require periodic breaks; independently initiates and carries out familiar and unfamiliar tasks but may require more than usual amount of time and/or compensatory strategies to complete them; accurately estimates abilities and independently adjusts to task demands; periodic periods of depression may occur; irritability and low frustration tolerance when sick, fatigued, and/or under stress	• Monitor for signs and symptoms of depression. • Help the patient plan, anticipate concerns, and solve problems.

ADLs, activities of daily living.

- Using an external sheath catheter for a male patient if incontinence occurs. Because prolonged use of an indwelling catheter inevitably produces infection, the patient may be placed on an intermittent catheterization schedule.

Maintaining Body Temperature

Fever in the patient with a TBI can be the result of damage to the hypothalamus, cerebral irritation from hemorrhage, or infection. The nurse monitors the patient's temperature every 2 to 4 hours. If the temperature increases, efforts are made to identify the cause and to control it using acetaminophen and cooling blankets to maintain normothermia (Bader & Littlejohns, 2010; McIlvoy, 2012). Cooling blankets should be used with caution so as not to induce shivering, which increases ICP. If infection is suspected, potential sites of infection are cultured and antibiotic agents are prescribed and administered.

The use of mild hypothermia to 34°C to 35°C (94°F to 96°F) has been tested in small randomized controlled trials for at least 12 hours versus normothermia (control) in patients with closed head injury. Early research showed improvement in patient outcomes but needs to be repeated in larger trials. Because hypothermia increases the risk of pneumonia and has other side effects, this treatment is not currently recommended outside of controlled clinical trials (Brain Trauma Foundation, 2007).

Maintaining Skin Integrity

Patients with TBI often require assistance in turning and positioning because of immobility or unconsciousness. Prolonged pressure on the tissues decreases circulation and leads to tissue necrosis. Potential areas of breakdown need to be identified early to avoid the development of pressure ulcers. Specific nursing measures include the following:

- Assessing all body surfaces and documenting skin integrity every 8 hours
- Turning and repositioning the patient every 2 hours
- Providing skin care every 4 hours
- Assisting the patient to get out of bed to a chair three times a day

Improving Coping

Although many patients with head injury survive because of resuscitative and supportive technology, they frequently have ineffective coping due to cognitive sequelae. Cognitive impairment includes memory deficits; decreased ability to focus and sustain attention to a task (distractibility); reduced ability to process information; and slowness in thinking, perceiving, communicating, reading, and writing. Psychiatric, emotional, and relationship problems develop in many patients after head injury. Resulting psychosocial, behavioral, emotional, and cognitive impairments are devastating to the family as well as to the patient (Norup, Kristensen, Siert, et al., 2011).

These problems require collaboration among many disciplines. A neuropsychologist (specialist in evaluating and treating cognitive problems) plans a program and initiates therapy or counseling to help the patient reach maximal potential. Cognitive rehabilitation activities help the patient to devise new problem-solving strategies. The retraining is carried out over an extended period and may include the use of sensory stimulation and reinforcement, behavior modification, reality orientation, computer training programs, and video games. Assistance from many disciplines is necessary during this phase of recovery. Even if intellectual ability does not improve, social and behavioral abilities may improve.

The patient recovering from a TBI may experience fluctuations in the level of cognitive function, with orientation, attention, and memory frequently affected. Many types of sensory stimulation programs have been tried, and research on these programs is ongoing (Hickey, 2009). When pushed to a level greater than the impaired cortical functioning allows, the patient may show symptoms of fatigue, anger, and stress (headache, dizziness). The Rancho Levels of Cognitive Functioning scale is frequently used to assess cognitive function and evaluate ongoing recovery from head injury. Progress through the levels of cognitive function can vary widely for individual patients (Hagen, Malkmus, & Durham, 1972). Nursing management and a description of each level are included in Table 68-2.

Preventing Sleep Pattern Disturbance

Patients who require frequent monitoring of neurologic status may experience sleep deprivation as they are awakened hourly for assessment of LOC. To allow the patient longer times of uninterrupted sleep and rest, the nurse can group nursing care activities so that the patient is disturbed less frequently. Environmental noise is decreased, and the room lights are dimmed. Backrubs and other measures to increase comfort may promote sleep and rest.

Supporting Family Coping

Having a loved one sustain a TBI produces a great deal of stress in the family. This stress can result from the patient's physical and emotional deficits, the unpredictable outcome, and altered family relationships. Families report difficulties in coping with changes in the patient's temperament, behavior, and personality (Coco, Tossavainen, Jaaskelainen, et al., 2011). Such changes are associated with disruption in family cohesion, loss of leisure pursuits, and loss of work capacity, as well as social isolation of the caretaker. The family may experience marital disruption, anger, grief, guilt, and denial in recurring cycles.

To promote effective coping, the nurse can ask the family how the patient is different now, what has been lost, and what is most difficult about coping with this situation. Helpful interventions include providing family members with accurate and honest information and encouraging them to continue to set well-defined short-term goals. Family counseling helps address the family members' overwhelming feelings of loss and helplessness and gives them guidance for the management of inappropriate behaviors. Support groups help the family members share problems, develop insight, gain information, network, and gain assistance in maintaining realistic expectations, hope, and a good quality of life (Arango-Lasprilla, Nicolls, Villasenor Cabrera, et al., 2011).

The Brain Injury Association of America (see the Resources section) serves as a clearinghouse for information and resources for patients with head injuries and their families, including specific information on coma, rehabilitation, behavioral consequences of head injury, and family issues. This organization can provide names of facilities and professionals who work with patients with head injuries and can assist families in organizing local support groups.

Many patients with severe head injury die of their injuries, and many of those who survive experience long-term disabilities that prevent them from resuming their previous roles and functions. During the most acute phase of injury, family members need factual information and support from the health care team.

Many patients with severe head injuries that result in brain death are young and otherwise healthy and are therefore considered for organ donation. Family members of patients with such injuries need support during this extremely stressful time and assistance in making decisions to end life support and permit donation of organs. They need to know that the patient who is brain dead and whose respiratory and cardiovascular systems are maintained through life support is not going to survive and that the severe head injury, not the removal of the patient's organs or the removal of life support, is the cause of the patient's death. Bereavement counselors and members of the organ procurement team are often very helpful to family members in making decisions about organ donation and in helping them cope with stress.

Monitoring and Managing Potential Complications

Decreased Cerebral Perfusion Pressure. Maintenance of adequate CPP is important to prevent serious complications of head injury due to decreased cerebral perfusion. Adequate CPP is greater than 50 mm Hg. If CPP falls below a patient's threshold, a vasodilating cascade occurs, causing the volume of blood to increase inside the brain, which causes ICP to increase. Measures to maintain adequate CPP are essential because a decrease in CPP can impair cerebral perfusion and cause brain hypoxia and ischemia, leading to permanent brain damage. Once the threshold CPP is reached, vasoconstriction of the cerebral blood vessels occurs, causing ICP to decrease (Brosche, 2011). Therapy (e.g., elevation of the head of the bed and increased IV fluids) is directed toward decreasing cerebral edema and increasing venous outflow from the brain. Systemic hypotension, which causes vasoconstriction and a significant decrease in CPP, is treated with increased IV fluids or vasopressors.

Cerebral Edema and Herniation. The patient with a head injury is at risk for additional complications such as increased ICP and brain stem herniation. Cerebral edema is the most common cause of increased ICP in the patient with a head injury, with the swelling peaking approximately 48 to 72 hours after injury. Bleeding also may increase the volume of contents within the rigid, closed compartment of the skull, causing increased ICP and herniation of the brain stem and resulting in irreversible brain anoxia and brain death (Morton et al., 2009). Measures to control ICP are listed in Chart 68-5 and discussed in Chapter 66.

Impaired Oxygenation and Ventilation. Impaired oxygen and ventilation may require mechanical ventilatory support. The patient must be monitored for a patent airway, altered breathing patterns, and hypoxemia and pneumonia. Interventions may include endotracheal intubation, mechanical ventilation, and positive end-expiratory pressure. (See Chapters 21 and 66 for a detailed discussion of these topics.)

Impaired Fluid, Electrolyte, and Nutritional Balance. Fluid, electrolyte, and nutritional imbalances are common in the patient with a head injury. Common imbalances include hyponatremia, which is often associated with SIADH (see Chapters 13 and 52), hypokalemia, and hyperglycemia. Modifications in fluid intake with tube feedings or IV fluids, including hypertonic saline, may be necessary to treat these imbalances (Hickey, 2009). Insulin administration may be prescribed to treat hyperglycemia, but the intensity of glycemic control is an area of ongoing controversy and research (Butcavage, 2012).

Undernutrition is also a common problem in response to the increased metabolic needs associated with severe head injury. Decisions about early feeding should be individualized;

Chart 68-5

Controlling Intracranial Pressure in Patients With Severe Brain Injury

- Elevate the head of the bed as prescribed.
- Maintain the patient's head and neck in neutral alignment (no twisting or flexing the neck).
- Initiate measures to prevent the Valsalva maneuver (e.g., stool softeners).
- Maintain body temperature within normal limits.
- Administer oxygen (O_2) to maintain partial pressure of arterial oxygen (PaO_2) >90 mm Hg.
- Maintain fluid balance with normal saline solution.
- Avoid noxious stimuli (e.g., excessive suctioning, painful procedures).
- Administer sedation to reduce agitation.
- Maintain cerebral perfusion pressure of 50–70 mm Hg.

Adapted from Bader, M. K., & Littlejohns, L. R. (2010). *AANN core curriculum for neuroscience nursing*. Glenview, IL: American Association of Neuroscience Nurses.

options include IV hyperalimentation or placement of a feeding tube (jejunal or gastric). Caloric expenditure can increase up to 120% to 140% with TBI, requiring close monitoring of nutritional status. Feeding tubes should be placed 3 to 7 days after neurologic injury to replace energy and nitrogen losses, prevent increased mortality, and improve outcomes (Hickey, 2009).

Posttraumatic Seizures. Patients with head injury are at an increased risk for posttraumatic seizures. Posttraumatic seizures are classified as immediate (within 24 hours after injury), early (within 1 to 7 days after injury), or late (more than 7 days after injury) (Hickey, 2009). Seizure prophylaxis is the practice of administering antiseizure medications to patients with head injury to prevent seizures. It is important to prevent posttraumatic seizures, especially in the immediate and early phases of recovery, because seizures may increase ICP and decrease oxygenation. However, many antiseizure medications impair cognitive performance and can prolong the duration of rehabilitation. Therefore, the overall benefits of these medications must be weighed against their side effects. Research evidence supports the use of prophylactic antiseizure agents to prevent immediate and early seizures after head injury, but not for prevention of late seizures (Brain Trauma Foundation, 2007; Hickey, 2009). (See Chapter 66 for the nursing management of seizures.)

Promoting Home and Community-Based Care

Educating Patients About Self-Care. Education early in the course of head injury often focuses on reinforcing information given to the family about the patient's condition and prognosis. As the patient's status and expected outcome change over time, family education may focus on interpretation and explanation of changes in the patient's physical and psychological responses.

If the patient's physical status allows discharge to home, the patient and family are educated about limitations that can be expected and complications that may occur. The nurse explains to the patient and family, verbally and in writing, how to monitor for complications that merit contacting the primary provider. Depending on the patient's prognosis and physical and cognitive status, the patient may be included in education about self-care management strategies.

If the patient is at risk for late posttraumatic seizures, antiseizure medications may be prescribed at discharge. The patient and family require education about the side effects of these medications and the importance of continuing to take them as prescribed.

Continuing Care. The rehabilitation phase of care for the patient with a TBI begins at hospital admission. Admission to the rehabilitation unit is a milestone in a patient's recovery and requires intense work by the patient to complete the daily schedule of therapies. The goals of rehabilitation are to maximize the patient's ability to return to his or her highest level of functioning and to his or her home and the community, address concerns before discharge for a smooth transition to home or rehabilitation, and promote independence with adaptation to deficits (Phelan, Griffin, Hellerstedt, et al., 2011). The patient is encouraged to continue the rehabilitation program after discharge, because improvement in status may continue 3 or more years after injury. Changes in the patient with a TBI and the effects of long-term rehabilitation on the family and their coping abilities need ongoing assessment (Engstrom & Soderberg, 2011; Phelan et al., 2011; Schipper, Widdershoven, & Abma, 2011). Continued education and support of the patient and family are essential as their needs and the patient's status change. Education to address with the family of the patient who is about to return home is described in Chart 68-6.

Depending on status, the patient is encouraged to return to usual activities gradually. Referral to support groups and to the Brain Injury Association of America may be warranted (see the Resources section at the end of this chapter).

During the acute and rehabilitation phases of care, the focus of education is on obvious needs, issues, deficits, and complications. Complications after TBI include infections (e.g., pneumonia, urinary tract infection [UTI], septicemia, wound infection, osteomyelitis, meningitis, ventriculitis, brain abscess) and heterotopic ossification (painful bone overgrowth in weight-bearing joints).

Chart 68-6 HOME CARE CHECKLIST

The Patient With a Traumatic Brain Injury

At the completion of home care education, the patient or caregiver will be able to:	PATIENT	CAREGIVER
• Explain the need for monitoring for changes in neurologic status and for complications.	✔	✔
• Identify changes in neurologic status and signs and symptoms of complications that should be reported to the neurosurgeon or nurse.		✔
• Demonstrate safe techniques to assist patient with self-care, hygiene, and ambulation.		✔
• Demonstrate safe technique for eating, feeding patient, or assisting patient with eating.	✔	✔
• Explain rationale for taking medications as prescribed.	✔	✔
• Identify need for close monitoring of behavior due to changes in cognitive functioning.		✔
• Describe household modifications needed to ensure safe environment for the patient.		✔
• Describe strategies for reinforcing positive behaviors.		✔
• State importance of continuing follow-up by health care team.	✔	✔

The nurse needs to remind the patient and family of the need for continuing health promotion and screening practices after the initial phase of care. Patients who have not been involved in these practices in the past are educated about their importance and are referred to appropriate health care providers.

Evaluation

Expected patient outcomes may include:

1. Attains or maintains effective airway clearance, ventilation, and brain oxygenation
   a. Achieves blood gas values within normal limits and has breath sounds clear on auscultation
   b. Mobilizes and clears secretions
2. Achieves satisfactory fluid and electrolyte balance
   a. Demonstrates serum electrolytes within normal limits
   b. Has no clinical signs of dehydration or overhydration
3. Attains adequate nutritional status
   a. Has less than 50 mL of aspirate in stomach before each tube feeding
   b. Is free of gastric distention and vomiting
   c. Shows minimal weight loss
4. Avoids injury
   a. Shows lessening agitation and restlessness
   b. Is oriented to time, place, and person
5. Maintains body temperature within normal limits
   a. Absence of fever
   b. Absence of hypothermia
6. Demonstrates intact skin integrity
   a. Exhibits no redness or breaks in skin integrity
   b. Exhibits no pressure ulcers
7. Shows improvement in coping
8. Demonstrates usual sleep–wake cycle
9. Family demonstrates adaptive family processes
   a. Joins support group
   b. Shares feelings with appropriate health care personnel
   c. Makes end-of-life decisions, if needed
10. Demonstrates absence of complications
    a. Exhibits vital signs and body temperature within normal limits and increasing orientation to time, place, and person
    b. Demonstrates ICP within normal limits
11. Experiences no posttraumatic seizures
    a. Takes antiseizure medications as prescribed
    b. Identifies side effects/adverse effects of antiseizure medications
12. Participates in rehabilitation process as indicated for patient and family members
    a. Takes active role in identifying rehabilitation goals and participating in recommended patient care activities
    b. Prepares for discharge

Spinal Cord Injury

Spinal cord injury is a major health disorder. Almost 200,000 people in the United States live each day with a disability from SCI. Annually, 15 to 40 new cases per million people—or 12,000 to 20,000 new patients—are estimated to occur. The most common causes of SCIs are motor vehicle crashes (46%), falls (22%), violence (16%), and sports (12%). The average annual medical cost is $15,000 to $30,000 per patient per year. Estimated lifetime costs per patient range from $500,000 to more than $3 million depending on injury severity. Males account for 80% of patients with SCI. An estimated 50% to 70% of SCIs occur in those 15 to 35 years of age. Approximately 65% of patients with SCI are Caucasian, whereas 25% are African American, 8% are Hispanic, and 2% are from other racial groups (CDC, 2010). The predominant risk factors for SCI include young age, male gender, and alcohol and drug use. The frequency with which these risk factors are associated with SCI serves to emphasize the importance of primary prevention. The same interventions suggested earlier in this chapter for head injury prevention serve to decrease the incidence of SCI as well (see Chart 68-1). Life expectancy continues to increase for people with SCI because of improved health care but remains slightly lower than for those without SCI. The major causes of death are pneumonia, pulmonary embolism (PE), and sepsis (Hickey, 2009).

Paraplegia (paralysis of the lower body) and **tetraplegia** (paralysis of all four extremities; formerly called *quadriplegia*) can occur, with incomplete tetraplegia being the most frequently occurring injury followed by complete paraplegia, complete tetraplegia, and incomplete paraplegia.

Pathophysiology

Damage in SCI ranges from transient concussion (from which the patient fully recovers), to contusion, laceration, and compression of the spinal cord tissue (either alone or in combination), to complete **transection** (severing) of the spinal cord (which renders the patient paralyzed below the level of the injury). The vertebrae most frequently involved are the 5th, 6th, and 7th cervical vertebrae (C5 to C7), the 12th thoracic vertebra (T12), and the 1st lumbar vertebra (L1). These vertebrae are most susceptible because there is a greater range of mobility in the vertebral column in these areas (Lucas, 2012; National Institute of Neurological Disorders and Stroke [NINDS], 2012a; Stauber, 2011).

SCIs can be separated into two categories: primary injuries and secondary injuries. Primary injuries are the result of the initial insult or trauma and are usually permanent. Secondary injuries are usually the result of a contusion or tear injury, in which the nerve fibers begin to swell and disintegrate. A secondary chain of events produces ischemia, hypoxia, edema, and hemorrhagic lesions, which in turn result in destruction of myelin and axons. The secondary injury is a major concern for critical care nurses. Experts believe that secondary injury is the principal cause of spinal cord degeneration at the level of injury and that it is reversible during the first 4 to 6 hours after injury. Early treatment is essential to prevent partial damage from becoming total and permanent (NINDS, 2012a).

Clinical Manifestations

Manifestations of SCI depend on the type and level of injury (Chart 68-7). The type of injury refers to the extent of injury to the spinal cord itself. **Incomplete spinal cord lesions** (the sensory or motor fibers, or both, are preserved below the lesion) are classified according to the area of spinal cord damage: central,

Chart 68-7 Effects of Spinal Cord Injuries

Central Cord Syndrome

- *Characteristics:* Motor deficits (in the upper extremities compared to the lower extremities; sensory loss varies but is more pronounced in the upper extremities); bowel/bladder dysfunction is variable, or function may be completely preserved.
- *Cause:* Injury or edema of the central cord, usually of the cervical area. May be caused by hyperextension injuries.

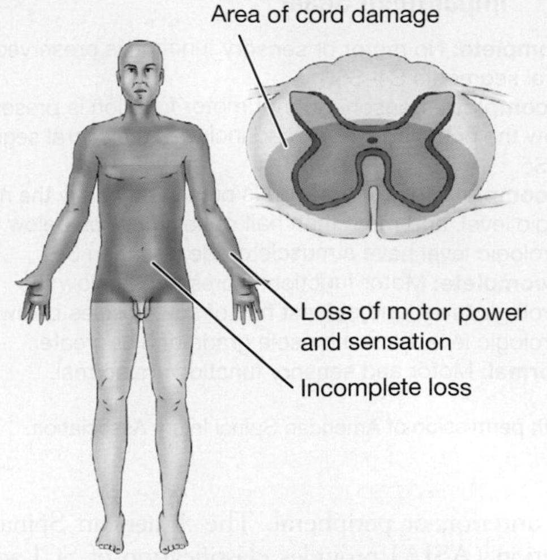
Area of cord damage
Loss of motor power and sensation
Incomplete loss

Central Cord Syndrome

Anterior Cord Syndrome

- *Characteristics:* Loss of pain, temperature, and motor function is noted below the level of the lesion; light touch, position, and vibration sensation remain intact.
- *Cause:* The syndrome may be caused by acute disk herniation or hyperflexion injuries associated with fracture/dislocation of vertebra. It also may occur as a result of injury to the anterior spinal artery, which supplies the anterior two thirds of the spinal cord.

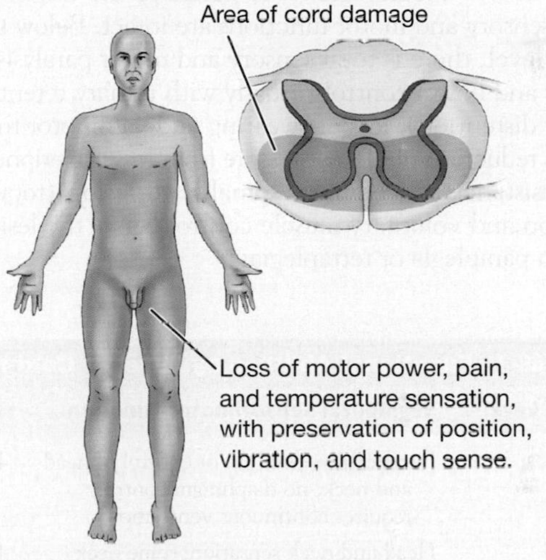
Area of cord damage
Loss of motor power, pain, and temperature sensation, with preservation of position, vibration, and touch sense.

Anterior Cord Syndrome

Lateral Cord Syndrome (Brown-Séquard Syndrome)

- *Characteristics:* Ipsilateral paralysis or paresis is noted, together with ipsilateral loss of touch, pressure, and vibration and contralateral loss of pain and temperature.
- *Cause:* The lesion is caused by a transverse hemisection of the cord (half of the cord is transected from north to south), usually as a result of a knife or missile injury, fracture/dislocation of a unilateral articular process, or possibly an acute ruptured disk.

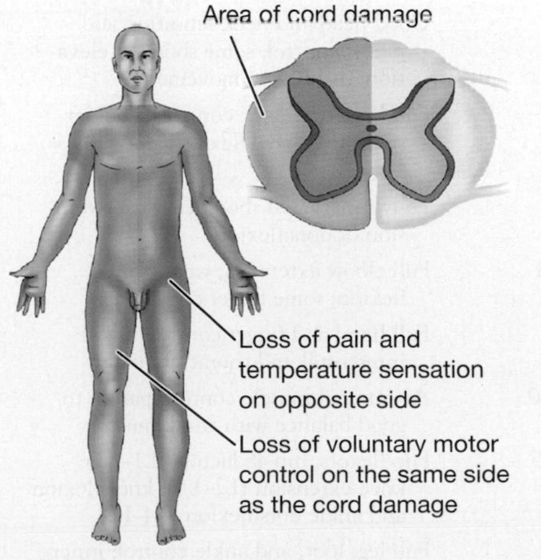
Area of cord damage
Loss of pain and temperature sensation on opposite side
Loss of voluntary motor control on the same side as the cord damage

Brown-Séquard Syndrome

Adapted from Hickey, J. (2009). *The clinical practice of neurological and neurosurgical nursing* (6th ed.). Philadelphia: Lippincott Williams & Wilkins.

Chart 68-8
American Spinal Injury Association Impairment Scale

A = Complete: No motor or sensory function is preserved in the sacral segments S4–S5.
B = Incomplete: Sensory but not motor function is preserved below the neurologic level, and includes the sacral segments S4–S5.
C = Incomplete: Motor function is preserved below the neurologic level, and more than half of key muscles below the neurologic level have a muscle grade less than 3.
D = Incomplete: Motor function is preserved below the neurologic level, and at least half of key muscles below the neurologic level have a muscle grade of 3 or greater.
E = Normal: Motor and sensory function are normal.

Used with permission of American Spinal Injury Association.

lateral, anterior, or peripheral. The American Spinal Injury Association (ASIA) provides classification of SCI according to the degree of sensory and motor function present after injury (Chart 68-8). "Neurologic level" refers to the lowest level at which sensory and motor functions are intact. Below the neurologic level, there is total sensory and motor paralysis, loss of bladder and bowel control (usually with urinary retention and bladder distention), loss of sweating and vasomotor tone, and marked reduction of blood pressure from loss of peripheral vascular resistance. A **complete spinal cord lesion** (total loss of sensation and voluntary muscle control below the lesion) can result in paraplegia or tetraplegia.

If conscious, the patient usually complains of acute pain in the back or neck, which may radiate along the involved nerve. However, absence of pain does not rule out spinal injury, and a careful assessment of the spine should be conducted if there has been a significant force and mechanism of injury (i.e., concomitant head injury). Often, the patient speaks of fear that the neck or back is broken.

Respiratory dysfunction is related to the level of injury. The muscles contributing to respiration are the abdominals and intercostals (T1 to T11) and the diaphragm (C4). In high cervical cord injury, acute respiratory failure is the leading cause of death. Functional abilities by level of injury are described in Table 68-3.

Assessment and Diagnostic Findings

A detailed neurologic examination is performed. Diagnostic x-rays (lateral cervical spine x-rays) and CT scanning are usually performed initially. An MRI scan may be ordered as a further workup if a ligamentous injury is suspected, because significant spinal cord damage may exist even in the absence of bony injury (Hickey, 2009). If an MRI scan is contraindicated, a myelogram may be used to visualize the spinal axis. An assessment is made for other injuries, because spinal trauma often is accompanied by concomitant injuries, commonly to the head and chest. Continuous electrocardiographic monitoring may be indicated if an SCI is suspected, because bradycardia (slow heart rate) and asystole (cardiac standstill) are common in patients with acute spinal cord injuries.

TABLE 68-3 Functional Abilities by Level of Cord Injury

Injury Level	Segmental Sensorimotor Function	Dressing, Eating	Elimination	Mobility*
C1	Little or no sensation or control of head and neck; no diaphragm control; requires continuous ventilation	Dependent	Dependent	Limited. Voice or sip-n-puff controlled electric wheelchair
C2–C3	Head and neck sensation; some neck control; independent of mechanical ventilation for short periods	Dependent	Dependent	Same as for C1
C4	Good head and neck sensation and motor control; some shoulder elevation; diaphragm movement	Dependent; may be able to eat with adaptive sling	Dependent	Limited to voice, mouth, head, chin, or shoulder-controlled electric wheelchair
C5	Full head and neck control; shoulder strength; elbow flexion	Independent with assistance	Maximal assistance	Electric or modified manual wheelchair, needs transfer assistance
C6	Fully innervated shoulder; wrist extension or dorsiflexion	Independent or with minimal assistance	Independent or with minimal assistance	Independent in transfers and wheelchair
C7–C8	Full elbow extension; wrist plantar flexion; some finger control	Independent	Independent	Independent; manual wheelchair
T1–T5	Full hand and finger control; use of intercostal and thoracic muscles	Independent	Independent	Independent; manual wheelchair
T6–T10	Abdominal muscle control, partial to good balance with trunk muscles	Independent	Independent	Independent; manual wheelchair
T11–L5	Hip flexors, hip abductors (L1–L3); knee extension (L2–L4); knee flexion and ankle dorsiflexion (L4–L5)	Independent	Independent	Short distance to full ambulation with assistance
S1–S5	Full leg, foot, and ankle control; innervation of perineal muscles for bowel, bladder, and sexual function (S2–S4)	Independent	Normal to impaired bowel and bladder function	Ambulate independently with or without assistance

*Assistance refers to adaptive equipment, setup, or physical assistance.
Adapted from Porth, C. M., & Matfin, G. (2009). *Pathophysiology: Concepts of altered health states* (8th ed.). Philadelphia: Lippincott Williams & Wilkins.

Emergency Management

The immediate management at the scene of the injury is critical, because improper handling of the patient can cause further damage and loss of neurologic function. Any patient who is involved in a motor vehicle crash, a diving or contact sports injury, a fall, or any direct trauma to the head and neck must be considered to have SCI until such an injury is ruled out. Initial care must include a rapid assessment, immobilization, extrication, and stabilization or control of life-threatening injuries, and transportation to the most appropriate medical facility. Immediate transportation to a trauma center with the capacity to manage major neurologic trauma is then necessary (Hickey, 2009).

At the scene of the injury, the patient must be immobilized on a spinal (back) board, with the head and neck maintained in a neutral position, to prevent an incomplete injury from becoming complete. One member of the team must assume control of the patient's head to prevent flexion, rotation, or extension; this is done by placing the hands on both sides of the patient's head at about ear level to limit movement and maintain alignment while a spinal board or cervical immobilizing device is applied. If possible, at least four people should slide the patient carefully onto a board for transfer to the hospital. Any twisting movement may irreversibly damage the spinal cord by causing a bony fragment of the vertebra to cut into, crush, or sever the cord completely.

The standard of care is that the patient is referred to a regional spinal injury or trauma center because of the multidisciplinary personnel and support services required to counteract the destructive changes that occur in the first 24 hours after injury. However, no randomized controlled trials have been conducted to confirm that referral results in better outcomes for the patient with SCI (NINDS, 2012b). During treatment in the emergency and x-ray departments, the patient is kept on the transfer board. The patient must always be maintained in an extended position. No part of the body should be twisted or turned, and the patient is not allowed to sit up. Once the extent of the injury has been determined, the patient may be placed on a rotating specialty bed or in a cervical collar (Fig. 68-6). Later, if SCI and bone instability have been ruled out, the patient may be moved to a conventional bed or the collar may be removed without harm. If a specialty bed is needed but not available, the patient should be placed in a cervical collar and on a firm mattress.

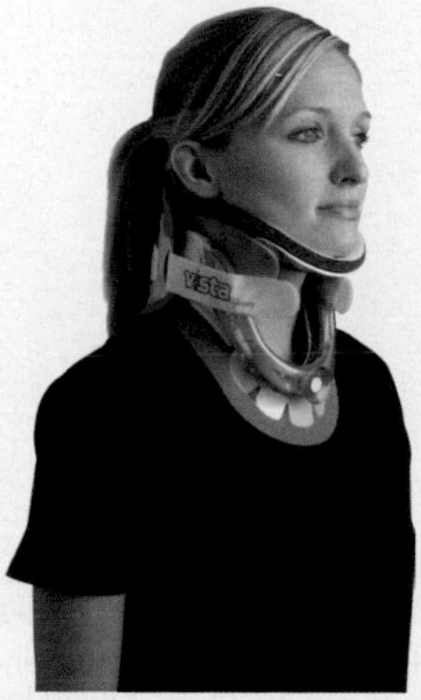
Vista

FIGURE 68-6 • Cervical collar. Used with permission from Aspen Medical Products.

Medical Management (Acute Phase)

The goals of management are to prevent secondary injury, to observe for symptoms of progressive neurologic deficits, and to prevent complications. The patient is resuscitated as necessary, and oxygenation and cardiovascular stability are maintained. SCI is a devastating event; new treatment methods and medications are continually being investigated for the acute and chronic phases of care (NINDS, 2012b).

Pharmacologic Therapy

Administration of high-dose IV corticosteroids (methylprednisolone sodium succinate [Solu-Medrol]) in the first 24 or 48 hours is controversial. The validity of studies has been questioned based on critical analysis of the original and additional data. As a result, there is now a consensus that corticosteroids may offer only a slight benefit. Corticosteroids are no longer considered the standard of care for acute SCI, although some centers continue to use corticosteroid protocols (Hickey, 2009).

Respiratory Therapy

Oxygen is administered to maintain a high partial pressure of arterial oxygen (PaO_2), because hypoxemia can create or worsen a neurologic deficit of the spinal cord. If endotracheal intubation is necessary, extreme care is taken to avoid flexing or extending the patient's neck, which can result in extension of a cervical injury.

In high cervical spine injuries, spinal cord innervation to the phrenic nerve, which stimulates the diaphragm, is lost. Diaphragmatic pacing (electrical stimulation of the phrenic nerve) attempts to stimulate the diaphragm to help the patient breathe (Sole, Klein, & Moseley, 2012). Intramuscular diaphragmatic pacing is currently in the clinical trial phase for the patient with a high cervical injury. This is implanted via laparoscopic surgery, usually after the acute phase.

Skeletal Fracture Reduction and Traction

Management of SCI requires immobilization and reduction of dislocations (restoration of pre-injury position) and stabilization of the vertebral column.

Cervical fractures are reduced, and the cervical spine is aligned with some form of skeletal traction, such as skeletal tongs or calipers, or with use of the halo device (Sole et al., 2012). A variety of skeletal tongs are available, all of which involve fixation in the skull in some manner. Gardner-Wells tongs require no predrilled holes in the skull. Crutchfield and Vinke tongs are inserted through holes made in the skull with a special drill under local anesthesia.

Traction is applied to the skeletal traction device by weights, the amount depending on the size of the patient and the degree of fracture displacement (Fig. 68-7). The traction force is exerted along the longitudinal axis of the vertebral bodies, with the patient's neck in a neutral position. The traction is then gradually increased by adding more weights. As the amount of traction is increased, the spaces between the intervertebral disks widen and the vertebrae are given a chance to slip back into position. Reduction usually occurs

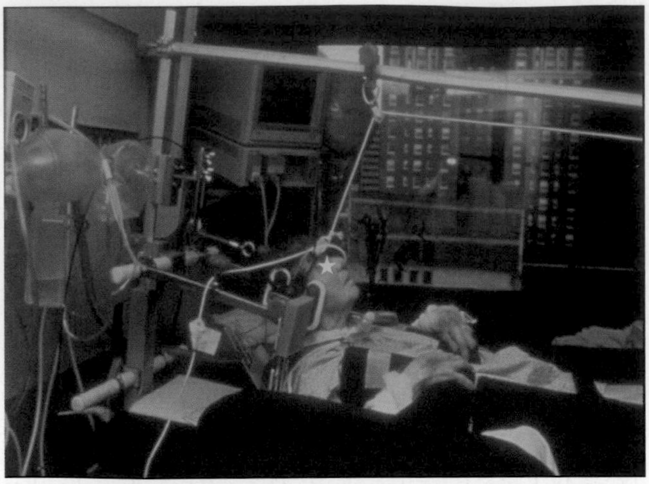

FIGURE 68-7 • Patient in skeletal traction in a rotating specialty bed. From Schwartz, E. D., Adam, E., & Flander, S. (2007). *Spinal trauma: Imaging, diagnosis, and management*. Philadelphia: Lippincott Williams & Wilkins.

after correct alignment has been restored. Once reduction is achieved, as verified by cervical spine x-rays and neurologic examination, the weights are gradually removed until the amount of weight needed to maintain the alignment is identified. The weights should hang freely so as not to interfere with the traction. Traction is sometimes supplemented with manual manipulation of the neck by a surgeon to help achieve realignment of the vertebral bodies. Cervical traction is used much less frequently with the advent of earlier and better surgical stabilization (Hickey, 2009).

A halo device may be used initially with traction or may be applied after removal of the tongs. It consists of a stainless steel halo ring that is fixed to the skull by four pins. The ring is attached to a removable **halo vest**, a device that suspends the weight of the unit circumferentially around the chest. A metal frame connects the ring to the chest. Halo devices provide immobilization of the cervical spine while allowing early ambulation (Fig. 68-8).

Thoracic and lumbar injuries are usually treated with surgical intervention followed by immobilization with a fitted brace. Traction is not indicated either before or after surgery, due to the relative stability of the spine in these regions.

Quality and Safety Nursing Alert

The patient's vital organ functions and body defenses must be supported and maintained until spinal and neurogenic shock abates and the neurologic system has recovered from the traumatic insult; this can take up to 4 months.

Surgical Management

Surgery is indicated in any of the following situations:

- Compression of the cord is evident.
- The injury results in a fragmented or unstable vertebral body.
- The injury involves a wound that penetrates the cord.
- Bony fragments are in the spinal canal.
- The patient's neurologic status is deteriorating.

Research indicates that early surgical stabilization improves the clinical outcome of patients compared to surgery performed later during the clinical course (NINDS, 2012b). The goals of surgical treatment are to preserve neurologic function by removing pressure from the spinal cord and to provide stability.

Management of Acute Complications of Spinal Cord Injury

Spinal and Neurogenic Shock

The spinal shock associated with SCI reflects a sudden depression of reflex activity in the spinal cord (areflexia) below the level of injury. The muscles innervated by the part of the spinal cord segment below the level of the lesion are without sensation, paralyzed, and flaccid, and the reflexes are absent. In particular, the reflexes that initiate bladder and bowel function are affected. Bowel distention and paralytic ileus can be caused by depression of the reflexes and are treated with intestinal decompression by insertion of a nasogastric tube. A paralytic ileus most often occurs within the first 2 to 3 days after SCI and resolves within 3 to 7 days (Bader & Littlejohns, 2010).

Neurogenic shock develops as a result of the loss of autonomic nervous system function below the level of the lesion. The vital organs are affected, causing decreases in blood pressure, heart rate, and cardiac output, as well as venous pooling in the extremities and peripheral vasodilation (Bader &

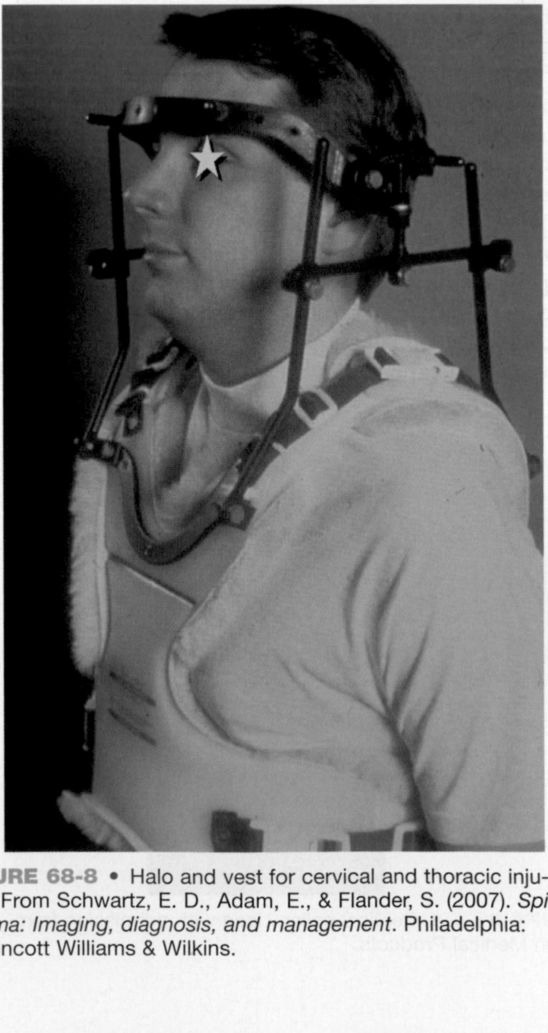

FIGURE 68-8 • Halo and vest for cervical and thoracic injuries. From Schwartz, E. D., Adam, E., & Flander, S. (2007). *Spinal trauma: Imaging, diagnosis, and management*. Philadelphia: Lippincott Williams & Wilkins.

Littlejohns, 2010). In addition, the patient does not perspire in the paralyzed portions of the body, because sympathetic activity is blocked; therefore, close observation is required for early detection of an abrupt onset of fever. (See Chapter 14 for further discussion of neurogenic shock.)

With injuries to the cervical and upper thoracic spinal cord, innervation to the major accessory muscles of respiration is lost and respiratory problems develop. These include decreased vital capacity, retention of secretions, increased partial pressure of arterial carbon dioxide ($PaCO_2$) levels and decreased oxygen levels, respiratory failure, and pulmonary edema.

Venous Thromboembolism

The risk of venous thromboembolism (VTE) is a potential complication of immobility and occurs in patients with SCI at the same rate as those having had other types of traumatic injuries (Bader & Littlejohns, 2010). Patients who develop VTE are at risk for both deep vein thrombosis (DVT) and PE. Manifestations of PE include pleuritic chest pain, anxiety, shortness of breath, and abnormal blood gas values (increased $PaCO_2$ and decreased PaO_2). Low-dose anticoagulation therapy usually is initiated to prevent DVT and PE, along with the use of anti-embolism stockings or pneumatic compression devices (Kahn, Lim, Dunn, et al., 2012). In some cases, permanent indwelling filters (see Chapter 23) may be placed prophylactically in the vena cava to prevent emboli (dislodged clots) from migrating to the lungs and causing PE.

Quality and Safety Nursing Alert

The calves or thighs of a patient who is immobile should never be massaged because of the danger of dislodging an undetected thromboemboli.

Other Complications

In addition to respiratory complications (respiratory failure, pneumonia) and autonomic dysreflexia, other complications that may occur include pressure ulcers and infection (urinary, respiratory, and local infection at the skeletal traction pin sites).

NURSING PROCESS

The Patient With Acute Spinal Cord Injury

PRACTICE & LEARN

Assessment

The patient's breathing pattern and the strength of the cough are assessed and the lungs are auscultated, because paralysis of abdominal and respiratory muscles diminishes coughing and makes clearing of bronchial and pharyngeal secretions difficult. Reduced excursion of the chest also results.

The patient is monitored closely for any changes in motor or sensory function and for symptoms of progressive neurologic damage. In the early stages of SCI, determining whether the cord has been severed may not be possible, because signs and symptoms of cord edema are indistinguishable from those of cord transection. Edema of the spinal cord may occur with any severe cord injury and may further compromise spinal cord function.

Motor and sensory functions are assessed through careful neurologic examination. These findings are recorded on a flow sheet so that changes in the baseline neurologic status can be monitored closely and accurately. The ASIA classification is commonly used to describe the level of function for patients with SCI (see Chart 68-8). Chart 68-7 gives examples of the effects of altered spinal cord function. At the minimum:

- Motor ability is tested by asking the patient to spread the fingers, squeeze the examiner's hand, and move the toes or turn the feet.
- Sensation is evaluated by gently pinching the skin or touching it lightly with an object such as a tongue blade, starting at shoulder level and working down both sides of the extremities. The patient should have both eyes closed so that the examination reveals true findings, not what the patient hopes to feel. The patient is asked where the sensation is felt.
- Any decrease in neurologic function is reported immediately.

The patient is also assessed for spinal shock, which is a complete loss of all reflex, motor, sensory, and autonomic activity below the level of the lesion that causes bladder paralysis and distention. The lower abdomen is palpated for signs of urinary retention and overdistention of the bladder. Further assessment is made for gastric dilation and paralytic ileus caused by an atonic bowel, a result of autonomic disruption.

Temperature is monitored, because the patient may have periods of hyperthermia as a result of alteration in temperature control due to autonomic disruption.

Diagnosis

Nursing Diagnoses

Based on the assessment data, major nursing diagnoses may include the following:

- Ineffective breathing patterns related to weakness or paralysis of abdominal and intercostal muscles and inability to clear secretions
- Ineffective airway clearance related to weakness of intercostal muscles
- Impaired bed and physical mobility related to motor and sensory impairments
- Risk for injury related to motor and sensory impairment
- Risk for impaired skin integrity related to immobility and sensory loss
- Impaired urinary elimination related to inability to void spontaneously
- Constipation related to presence of atonic bowel as a result of autonomic disruption
- Acute pain related to treatment and prolonged immobility
- Autonomic dysreflexia related to uninhibited sympathetic response of the nervous system following SCI

Collaborative Problems/Potential Complications

Potential complications may include the following:

- VTE
- Orthostatic hypotension

Planning and Goals

The goals for the patient may include improved breathing pattern and airway clearance, improved mobility, prevention of injury due to sensory impairment, maintenance of skin integrity, relief of urinary retention, improved bowel function, decreasing pain, early recognition of autonomic dysreflexia, and absence of complications.

Nursing Interventions

Promoting Adequate Breathing and Airway Clearance

Possible impending respiratory failure is detected by observing the patient, measuring vital capacity, monitoring oxygen saturation through pulse oximetry, and monitoring arterial blood gases. Early and vigorous attention to clearing bronchial and pharyngeal secretions can prevent retention of secretions and atelectasis. Suctioning may be indicated, but it should be used with caution to avoid stimulating the vagus nerve and producing bradycardia and cardiac arrest.

If the patient cannot cough effectively because of decreased inspiratory volume and inability to generate sufficient expiratory pressure, chest physiotherapy and assisted coughing may be indicated. Specific breathing exercises are supervised by the nurse to increase the strength and endurance of the inspiratory muscles, particularly the diaphragm. Assisted coughing promotes clearing of secretions from the upper respiratory tract and is similar to the use of abdominal thrusts to clear an airway (Bader & Littlejohns, 2010). Proper humidification and hydration are important to prevent secretions from becoming thick and difficult to remove even with coughing. The patient is assessed for signs of respiratory infection (e.g., cough, fever, dyspnea).

Ascending edema of the spinal cord in the acute phase may cause respiratory difficulty that requires immediate intervention. Therefore, the patient's respiratory status must be monitored closely.

Improving Mobility

Proper body alignment is maintained at all times. The patient is repositioned frequently and is assisted out of bed as soon as the spinal column is stabilized. The feet are prone to footdrop; therefore, various types of splints are used to prevent footdrop. When used, the splints are removed and reapplied every 2 hours. Trochanter rolls, applied from the crest of the ilium to the midthigh of both legs, help prevent external rotation of the hip joints.

Patients with lesions above the midthoracic level have loss of sympathetic control of peripheral vasoconstrictor activity, leading to hypotension. These patients may tolerate changes in position poorly and require monitoring of blood pressure when positions are changed. If not on a rotating specialty bed, the patient should not be turned unless the spine is stable and the physician has indicated that it is safe to do so.

Contractures can develop rapidly with immobility and muscle paralysis. A joint that is immobilized too long becomes fixed as a result of contractures of the tendon and joint capsule. Atrophy of the extremities results from disuse. Contractures and other complications may be prevented by range-of-motion exercises that help preserve joint motion and stimulate circulation. Passive range-of-motion exercises should be implemented as soon as possible after injury. Toes, metatarsals, ankles, knees, and hips should be put through a full range of motion at least four, and ideally five, times daily.

For most patients who have a cervical fracture without neurologic deficit, reduction in traction followed by rigid immobilization for 6 to 8 weeks restores skeletal integrity. These patients are allowed to move gradually to an erect position. A neck brace or molded collar is applied when the patient is mobilized after traction is removed (see Fig. 68-7).

Preventing Injury Due to Sensory and Perceptual Alterations

The nurse assists the patient to compensate for sensory and perceptual alterations that occur with SCI. The intact senses above the level of the injury are stimulated through touch, aromas, flavorful food and beverages, conversation, and music. Additional strategies include the following:

- Providing prism glasses to enable the patient to see from the supine position
- Encouraging the use of hearing aids, if indicated, to enable the patient to hear conversations and environmental sounds
- Providing emotional support to the patient
- Educating the patient and family about strategies to compensate for or cope with sensory deficits

Maintaining Skin Integrity

Pressure ulcers are a significant complication of SCI. The most common sites are over the ischial tuberosity, the greater trochanter, the sacrum, and the occiput (back of head). In the acute care setting, during the initial phase of hospitalization, it may be necessary to delay rehabilitation in 20% to 30% of patients because of pressure ulcers. Pressure ulcers may begin within hours of an acute SCI where pressure is continuous and where the peripheral circulation is inadequate as a result of spinal shock and a recumbent position. It is important to move the patient from the backboard as soon as possible and inspect the skin. In addition, patients who wear cervical collars for prolonged periods may develop breakdown from the pressure of the collar under the chin, on the shoulders, and at the occiput. Pressure ulcers can add substantially to the personal and economic costs of living with SCI. The prevalence of this complication ranges from 17% for people 2 months after injury to 33% for those living with SCI (Chart 68-9).

The most effective approach to addressing this costly complication of SCI is prevention. The patient's position is changed at least every 2 hours (Bader & Littlejohns, 2010). Turning not only assists in the prevention of pressure ulcers but also prevents pooling of blood and edema in the dependent areas. Careful inspection of the skin is made each time the patient is turned. The skin over the pressure points is assessed for redness or breaks; the perineum is checked for soilage, and the catheter is observed for adequate drainage. The patient's general body alignment and comfort are assessed. Special attention should be given to pressure areas in contact with the transfer board.

In addition, the patient's skin should be kept clean by washing with a mild soap, rinsing well, and blotting dry. Pressure-sensitive areas should be kept well lubricated and soft with oil or emollient lotion. The patient is educated about the danger of pressure ulcers and is encouraged to take control and make decisions about appropriate skin care. (See Chapter 10 for other aspects of the prevention of pressure ulcers.)

Maintaining Urinary Elimination

Immediately after SCI, the urinary bladder becomes atonic and cannot contract by reflex activity. Urinary retention is

Chart 68-9

ETHICAL DILEMMA

When Is it Appropriate to Offer Comfort Care?

Case Scenario

You work as a staff nurse on a medical-surgical unit of a community hospital. A 36-year-old man who has had paraplegia from a spinal cord injury that occurred when he was 18 years old is admitted with a large stage 4 pressure ulcer on his sacrum that is infected with methicillin-resistant *Staphylococcus aureus* (MRSA). You don your gown, glove, and mask and enter his room as the physician hospitalist leaves. The patient is crying softly with his face buried in his bed linens. When you approach him and ask him why he is crying, he tells you "That doctor that just left doesn't even know me. I've been independent my entire adult life and I have a graduate degree and a decent job. I have always taken good care of myself. This is the first time in years I've needed to come to a hospital. So this doctor walks in and tells me how bad this wound is, that getting treated for it could wipe out my savings, and that I might end up in a nursing home. And then he tells me that I have a right to refuse antibiotics. I know that if I don't get antibiotics that I will die. I guess he thinks my life is not worthwhile!"

Discussion

Patients with disabilities may face bias from health care providers, who may assume that they would prefer to not live than to undergo treatments that are expensive, painful, and could further exacerbate their disabilities (Peace, 2012).

Analysis

- Describe the ethical principles that are in conflict in this case (see Chart 3-3 in Chapter 3). Which principle should have preeminence when a treatment plan is devised for this patient?
- Although you did not directly hear the communication that transpired between the hospitalist and the patient, there may have been miscommunication between them. How would you respond to the patient? What are your professional obligations in this case? What steps might you take to ensure that the patient's autonomy is preserved?

Reference

Peace, W. J. (2012). Comfort care as denial of personhood. *Hastings Center Report*, *42*(4), 14–17.

Resources

See Chapter 3, Chart 3–6 for ethics resources.

the immediate result. Because the patient has no sensation of bladder distention, overstretching of the bladder and detrusor muscle may occur, delaying the return of bladder function.

Intermittent catheterization is carried out to avoid overdistention of the bladder and UTI. If this is not feasible, an indwelling catheter is inserted temporarily. At an early stage, family members are shown how to carry out intermittent catheterization and are encouraged to participate in this facet of care, because they will be involved in long-term follow-up and must be able to recognize complications so that treatment can be instituted.

The patient is educated to record fluid intake, voiding pattern, amounts of residual urine after catheterization, characteristics of urine, and any unusual sensations that may occur. The management of a **neurogenic bladder** (bladder dysfunction that results from a disorder or dysfunction of the nervous system) is discussed in detail in Chapter 10.

Improving Bowel Function

Immediately after SCI, a paralytic ileus usually develops as a result of neurogenic paralysis of the bowel; therefore, a nasogastric tube is often required to relieve distention and to prevent vomiting and aspiration (Bader & Littlejohns, 2010).

Bowel activity usually returns within the first week. As soon as bowel sounds are heard on auscultation, the patient is given a high-calorie, high-protein, high-fiber diet, with the amount of food gradually increased. The nurse administers prescribed stool softeners to counteract the effects of immobility and analgesic agents. A bowel program is instituted as early as possible (see Chapter 10).

Providing Comfort Measures: The Patient in Halo Traction

A patient who has had pins, tongs, or calipers placed for cervical stabilization may have a headache or discomfort for several days after the pins are inserted. Patients initially may be bothered by the rather startling appearance of these devices, but usually they readily adapt to it because the device provides comfort for the unstable neck (see Fig. 68-8). The patient may complain of being caged in and of noise created by any object coming in contact with the steel frame of a halo device, but he or she can be reassured that adaptation will occur.

The areas around the four pin sites of a halo device are cleaned daily and observed for redness, drainage, and pain. The pins are observed for loosening, which may contribute to infection. If one of the pins becomes detached, the head is stabilized in a neutral position by one person while another notifies the neurosurgeon. A torque screwdriver should be readily available in case the screws on the frame need tightening.

The skin under the halo vest is inspected for excessive perspiration, redness, and skin blistering, especially on the bony prominences. The vest is opened at the sides to allow the torso to be washed. The liner of the vest should not become wet, because dampness causes skin excoriation. Powder is not used inside the vest, because it may contribute to the development of pressure ulcers. The liner should be changed periodically to promote hygiene and good skin care. If the patient is to be discharged with the vest, detailed instructions must be given to the family, with time allowed for them to demonstrate the necessary skills of halo vest care (Chart 68-10).

Recognizing Autonomic Dysreflexia

Autonomic dysreflexia, also known as autonomic hyperreflexia, is an acute life-threatening emergency that occurs as a result of exaggerated autonomic responses to stimuli that are harmless in normal people. It occurs only after spinal shock has resolved. This syndrome is characterized by a severe, pounding headache with paroxysmal hypertension, profuse diaphoresis (most often of the forehead), nausea, nasal congestion, and bradycardia. It occurs among patients with cord lesions above T6 (the sympathetic visceral outflow level) after spinal shock has subsided (Bader & Littlejohns, 2010).

Chart 68-10 HOME CARE CHECKLIST

The Patient With a Halo Vest

At the completion of home care education, the patient or caregiver will be able to:	PATIENT	CAREGIVER
• Describe the rationale for the use of the halo vest.	✔	✔
• Demonstrate assessment of frame, traction, tongs, and pins.		✔
• Describe emergency measures if respiratory or other complications develop while the patient is in the halo vest or if the frame becomes dislodged.		✔
• Demonstrate pin care using correct technique.		✔
• Identify signs and symptoms of infection.	✔	✔
• Assess the skin for reddened or irritated areas and breakdown.		✔
• Demonstrate care of skin.		✔
• Explain the reasons for and the method for changing the vest liner.	✔	✔
• Demonstrate safe techniques to assist the patient with self-care, hygiene, and ambulation.		✔
• Identify signs and symptoms of complications (venous thromboembolism), respiratory impairment, urinary tract infection).	✔	✔

The sudden increase in blood pressure may cause a rupture of one or more cerebral blood vessels or lead to increased ICP. A number of stimuli may trigger this reflex: distended bladder (the most common cause); distention or contraction of the visceral organs, especially the bowel (from constipation, impaction); or stimulation of the skin (tactile, pain, thermal stimuli, pressure ulcer). Because this is an emergency situation, the objectives are to remove the triggering stimulus and to avoid the possibility of serious complications.

The following measures are carried out:

- The patient is placed immediately in a sitting position to lower blood pressure.
- Rapid assessment is performed to identify and alleviate the cause.
- The bladder is emptied immediately via a urinary catheter. If an indwelling catheter is not patent, it is irrigated or replaced with another catheter.
- The rectum is examined for a fecal mass. If one is present, a topical anesthetic agent is inserted 10 to 15 minutes before the mass is removed, because visceral distention or contraction can cause autonomic dysreflexia.
- The skin is examined for any areas of pressure, irritation, or broken skin.
- Any other stimulus that could be the triggering event, such as an object next to the skin or a draft of cold air, must be removed.
- If these measures do not relieve the hypertension and excruciating headache, a ganglionic blocking agent (hydralazine hydrochloride [Apresoline]) is prescribed and administered slowly by the IV route.
- The medical record is labeled with a clearly visible note about the risk of autonomic dysreflexia.
- The patient is instructed about prevention and management measures.
- Any patient with a lesion above the T6 segment is informed that such an episode is possible and may occur even many years after the initial injury (Bader & Littlejohns, 2010).

Monitoring and Managing Potential Complications

Patients are at high risk for VTE after SCI. The patient must be assessed for symptoms of VTE including DVT and PE. Chest pain, shortness of breath, and changes in arterial blood gas values must be reported promptly to the primary provider. The circumferences of the thighs and calves are measured and recorded daily; further diagnostic studies are performed if a significant increase is noted. Patients remain at high risk for thrombophlebitis for several months after the initial injury. Patients with paraplegia or tetraplegia are at increased risk for the rest of their lives. Immobilization and the associated venous stasis, as well as varying degrees of autonomic disruption, contribute to the high risk and susceptibility for DVT.

Anticoagulation is initiated once head injury and other systemic injuries have been ruled out and other risk factors assessed (Kahn et al., 2012). The use of low-molecular-weight heparin or low-dose unfractionated heparin may be followed by long-term oral anticoagulation (i.e., warfarin [Coumadin]). Additional measures such as range-of-motion exercises, anti-embolism stockings, and adequate hydration are important preventive measures. Pneumatic compression devices may also be used to reduce venous pooling and promote venous return. It is also important to avoid external pressure on the lower extremities that may result from flexion of the knees while the patient is in bed.

Orthostatic Hypotension. For the first 2 weeks after SCI, the blood pressure tends to be unstable and quite low. It gradually returns to pre-injury levels, but periodic episodes of severe orthostatic hypotension frequently interfere with efforts to mobilize the patient. Interruption in the reflex arcs that normally produce vasoconstriction in the upright position, coupled with vasodilation and pooling in abdominal and lower extremity vessels, can result in blood pressure readings of 40 mm Hg systolic and 0 mm Hg diastolic. Orthostatic hypotension is a particularly common problem for patients with lesions above T7. In some patients with tetraplegia, even slight elevations of the head can result in dramatic decreases in blood pressure.

A number of techniques can be used to reduce the frequency of hypotensive episodes. Close monitoring of vital signs before and during position changes is essential. Vasopressor medication can be used to treat the profound vasodilation. Anti-embolism stockings should be applied to improve venous return from the lower extremities. Abdominal binders may also be used to encourage venous return and provide diaphragmatic support when the patient is upright (Bader & Littlejohns, 2010). Activity should be planned in advance, and adequate time should be allowed for a slow progression of position changes from recumbent to sitting and upright. Tilt tables frequently are helpful in assisting patients to make this transition.

Promoting Home and Community-Based Care

Educating Patients About Self-Care. In most cases, patients with SCI (i.e., patients with tetraplegia or paraplegia) need long-term rehabilitation. The process begins during hospitalization as acute symptoms begin to subside or come under better control and the overall deficits and long-term effects of the injury become clear. The goals begin to shift from merely surviving the injury to learning strategies necessary to cope with the alterations that the injury imposes on activities of daily living (ADLs). The emphasis shifts from ensuring that the patient is stable and free of complications to specific assessment and planning designed to meet the patient's rehabilitation needs. Patient education may initially focus on the injury and its effects on mobility; dressing; and bowel, bladder, and sexual function. As the patient and family acknowledge the consequences of the injury and the resulting disability, the focus of education broadens to address issues necessary for carrying out the tasks of daily living and taking charge of their lives. Education must begin in the acute phase and continue throughout rehabilitation and the patient's entire life as changes occur, the patient ages, and problems arise (Rundquist, Gassaway, Bailey, et al., 2011).

Caring for the patient with SCI at home may at first seem a daunting task to the family. They will require dedicated nursing support to gradually assume full care of the patient. Although maintaining function and preventing complications will remain important, goals regarding self-care and preparation for discharge will assist in a smooth transition to rehabilitation and eventually to the community.

Continuing Care. The goal of the rehabilitation process is independence. The nurse becomes a support to both the patient and the family, assisting them to assume responsibility for increasing aspects of patient care and management. Care for the patient with SCI involves members of all health care disciplines, which may include nursing, medicine, rehabilitation, respiratory therapy, physical and occupational therapy, case management, and social services. The nurse often serves as coordinator of the management team and as a liaison with rehabilitation centers and home care agencies. The patient and family often require assistance in dealing with the psychological impact of the injury and its consequences; referral to a psychiatric clinical nurse specialist or other mental health care professional often is helpful. Therapeutic horseback riding may help increase balance, muscle strength, and self-esteem (Asselin, Ward, Penning, et al., 2012).

The nurse should reassure female patients with SCI that pregnancy is not contraindicated and fertility is relatively unaffected, but that pregnant women with acute or chronic SCI pose unique management challenges. The normal physiologic changes of pregnancy may predispose women with SCI to many potentially life-threatening complications, including autonomic dysreflexia, pyelonephritis, respiratory insufficiency, thrombophlebitis, PE, and unattended delivery. Preconception assessment and counseling are strongly recommended to ensure that the woman is in optimal health and to increase the likelihood of an uneventful pregnancy and healthy outcomes (Smeltzer & Wetzel-Effinger, 2009).

As more patients survive acute SCI, they face the changes associated with aging with a disability. Three common secondary health problems experienced by persons living with SCI include chronic pain, spasticity, and depression (Revell, 2011). Education in the home and community focuses on health promotion and addresses the need to minimize risk factors (e.g., smoking, alcohol and drug abuse, obesity). Routine health screening and preventive services are needed for the older adult with SCI for early detection of secondary health problems. Home care nurses and others who have contact with patients with SCI are in a position to educate patients about healthy lifestyles, remind them of the need for health screenings, and make referrals as appropriate. Assisting patients to identify accessible health care providers, clinical facilities, and imaging centers may increase the likelihood that they will participate in health screening.

Evaluation

Expected patient outcomes may include:

1. Demonstrates improvement in gas exchange and clearance of secretions, as evidenced by normal breath sounds on auscultation
   a. Breathes easily without shortness of breath
   b. Performs hourly deep-breathing exercises, coughs effectively, and clears pulmonary secretions
   c. Is free of respiratory infection (e.g., has temperature, respiratory rate, and pulse within normal limits; breath sounds clear to auscultation; absence of purulent sputum)
2. Moves within limits of the dysfunction and demonstrates completion of exercises within functional limitations
3. Avoids injury due to sensory and perceptual alterations
   a. Uses assistive devices (e.g., prism glasses, hearing aids, computers) as indicated
   b. Describes sensory and perceptual alterations as a consequence of injury
4. Demonstrates optimal skin integrity
   a. Exhibits normal skin turgor; skin is free of reddened areas or breaks
   b. Participates in skin care and monitoring procedures within functional limitations
5. Regains urinary bladder function
   a. Exhibits no signs of UTI (e.g., has temperature within normal limits; voids clear, dilute urine)
   b. Has adequate fluid intake
   c. Participates in bladder training program within functional limitations
6. Regains bowel function
   a. Reports regular pattern of bowel movement
   b. Consumes adequate dietary fiber and oral fluids
   c. Participates in bowel training program within functional limitations

7. Reports absence of pain and discomfort
8. Recognizes manifestations of autonomic dysreflexia if they occur (e.g., headache, diaphoresis, nasal congestion, bradycardia, or diaphoresis)
9. Is free of complications
   a. Demonstrates no signs of thrombophlebitis, DVT, or PE
   b. Maintains blood pressure within normal limits
   c. Reports no lightheadedness with position changes

Medical Management of Long-Term Complications of Spinal Cord Injury

The patient faces a lifetime of disability, requiring ongoing follow-up and care. The expertise of a number of health professionals, including physicians (specifically a physiatrist), rehabilitation nurses, occupational therapists, physical therapists, psychologists, social workers, rehabilitation engineers, and vocational counselors, is necessary at different times as the need arises.

As people with SCI age, they have the same medical problems as other people. In addition, they face the threat of complications associated with their disability (Hickey, 2009). Usually, patients are encouraged to attend a spine clinic when complications and other issues arise. Lifetime care includes assessment of the urinary tract at prescribed intervals, because there is the likelihood of continuing alteration in detrusor and sphincter function, and the patient is prone to UTI.

Long-term problems and complications of SCI include premature aging, disuse syndrome, autonomic dysreflexia (discussed earlier), bladder and kidney infections, spasticity, and depression (Revell, 2011). Pressure ulcers with potential complications of sepsis, osteomyelitis, and fistulas occur in about 10% of patients. Spasticity may be particularly disabling. Heterotopic ossification (overgrowth of bone) in the hips, knees, shoulders, and elbows occurs in many patients after SCI. Both of these complications are painful and can produce a loss of range of motion (Hickey, 2009). Management includes observing for and addressing any alteration in physiologic status and psychological outlook, as well as the prevention and treatment of long-term complications. The nursing role involves emphasizing the need for vigilance in self-assessment and care.

NURSING PROCESS

The Patient With Tetraplegia or Paraplegia

Assessment

Assessment focuses on the patient's general condition, complications, and how the patient is managing at that particular point in time. A head-to-toe assessment and review of systems should be part of the database, with emphasis on the areas that are prone to problems in this population. A thorough inspection of all areas of the skin for redness or breakdown is critical. The nurse reviews the established bowel and bladder program with the patient, because the program must continue uninterrupted. Patients with tetraplegia or paraplegia have varying degrees of loss of motor power, deep and superficial sensation, vasomotor control, bladder and bowel control, and sexual function. They are faced with potential complications related to immobility, skin breakdown and pressure ulcers, recurring UTIs, and contractures. Knowledge about these particular issues can further guide the assessment in any setting. Nurses in all settings, including home care, must be aware of these potential complications in the lifetime management of these patients.

An understanding of the emotional and psychological responses to tetraplegia or paraplegia is achieved by observing the responses and behaviors of the patient and family and by listening to their concerns (Chen, Lai, & Wu, 2011). Documenting these assessments and reviewing the plan with the entire team on a regular basis provide insight into how both the patient and the family are coping with the changes in lifestyle and body functioning. Additional information frequently can be gathered from the social worker or psychiatric/mental health worker.

It takes time for the patient and family to comprehend the magnitude of the disability. They may go through stages of grief, including shock, disbelief, denial, anger, depression, and acceptance. During the acute phase of the injury, denial can be a protective mechanism to shield the patient from the overwhelming reality of what has happened. As the patient realizes the permanent nature of paraplegia or tetraplegia, the grieving process may be prolonged and all-encompassing because of the recognition that long-held plans and expectations are interrupted or permanently altered. A period of depression often follows as the patient experiences a loss of self-esteem in areas of self-identity, sexual functioning, and social and emotional roles. Exploration and assessment of these issues can assist in developing a meaningful plan of care.

Diagnosis

Nursing Diagnoses

Based on the assessment data, major nursing diagnoses may include the following:

- Impaired bed and physical mobility related to loss of motor function
- Risk for disuse syndrome
- Risk for impaired skin integrity related to permanent sensory loss and immobility
- Impaired urinary elimination related to level of injury
- Constipation related to effects of spinal cord disruption
- Sexual dysfunction related to neurologic dysfunction
- Ineffective coping related to impact of disability on daily living
- Deficient knowledge about requirements for long-term management

Collaborative Problems/Potential Complications

Potential complications may include the following:

- Spasticity
- Infection and sepsis

Planning and Goals

The goals for the patient may include attainment of some form of mobility; maintenance of healthy, intact skin; achievement of bladder management without infection; achievement of bowel control; achievement of sexual expression; strengthening of coping mechanisms; knowledge of long-term management; and absence of complications.

Nursing Interventions

The patient requires extensive rehabilitation, which is less difficult if appropriate nursing management has been carried out during the acute phase of the injury or illness. Nursing care is one of the key factors determining the success of the rehabilitation program. The main objective is for the patient to live as independently as possible in the home and community.

Increasing Mobility

Exercise Programs. The unaffected parts of the body are built up to optimal strength to promote maximal self-care. The muscles of the hands, arms, shoulders, chest, spine, abdomen, and neck must be strengthened in the patient with paraplegia, because he or she must bear full weight on these muscles to ambulate. The triceps and the latissimus dorsi are important muscles used in crutch walking. The muscles of the abdomen and the back also are necessary for balance and for maintaining the upright position.

To strengthen these muscles, the patient can do push-ups when in a prone position and sit-ups when in a sitting position. Extending the arms while holding weights (traction weights can be used) also develops muscle strength. Squeezing rubber balls or crumbling newspaper promotes hand strength.

With encouragement from all members of the rehabilitation team, the patient with paraplegia can develop the increased exercise tolerance needed for gait training and ambulation activities. The importance of maintaining cardiovascular fitness is stressed to the patient. Alternative exercises to increase the heart rate to target levels must be designed within the patient's abilities.

Mobilization. After the spine is stable enough to allow the patient to assume an upright posture, mobilization activities are initiated. A brace or vest may be used, depending on the level of the lesion. A patient whose paralysis is a result of complete transection of the cord can begin weight bearing early, because no further damage can be incurred. The sooner muscles are used, the less chance there is of disuse atrophy. The earlier the patient is brought to a standing position, the less opportunity there is for osteoporotic changes to take place in the long bones. Weight bearing also reduces the possibility of renal calculi and enhances many other metabolic processes.

Braces and crutches enable some patients with paraplegia to ambulate for short distances. Ambulation using crutches requires a high expenditure of energy. Motorized wheelchairs and specially equipped vans can provide greater independence and mobility for patients with high-level SCI or other lesions. Every effort should be made to encourage the patient to be as mobile and active as possible.

Long-term risks include altered body composition, a decrease in lean body mass, decreased bone mineral density, and increased BMI. Patients are at high risk for obesity due to high fat intake combined with decreased physical activity (Perret & Stoffel-Kurt, 2011). This puts patients at higher risk of developing comorbid conditions such as diabetes and cardiovascular disease. Patients benefit from nutrition counseling to prevent these secondary complications. For those who are overweight or obese, weight loss programs must be designed to accommodate the dietary and physical activity barriers that are unique to this population (Magnuson, Peppard, & Flomenhoft, 2011).

Preventing Disuse Syndrome

Patients are at high risk for development of contractures as a result of disuse syndrome due to the musculoskeletal system changes (atrophy) brought about by the loss of motor and sensory functions below the level of injury. Range-of-motion exercises must be provided at least four times a day, and care is taken to stretch the Achilles tendon with exercises. The patient is repositioned frequently and is maintained in proper body alignment whether in bed or in a wheelchair.

Contractures can complicate day-to-day care, increasing the difficulty of positioning and decreasing mobility. A number of surgical procedures have been tried with varying degrees of success. These techniques are used if more conservative approaches fail, but the best treatment is prevention.

Promoting Skin Integrity

Because these patients spend a great portion of their lives in wheelchairs, pressure ulcers are an ever-present threat. Contributing factors are permanent sensory loss over pressure areas; immobility, which makes relief of pressure difficult; trauma from bumps (against the wheelchair, toilet, furniture, and so forth) that cause unnoticed abrasions and wounds; loss of protective function of the skin from excoriation and maceration due to excessive perspiration and possible incontinence; and poor general health (anemia, diabetes), leading to poor tissue perfusion. (See Chapter 10 for a detailed discussion of the prevention and management of pressure ulcers.)

The person with tetraplegia or paraplegia must take responsibility for monitoring (or directing monitoring) of his or her skin status. This involves relieving pressure and not remaining in any position for longer than 2 hours, in addition to ensuring that the skin receives meticulous attention and cleansing (Bader & Littlejohns, 2010). The patient is educated that ulcers develop over bony prominences that are exposed to unrelieved pressure in the lying and sitting positions. The most vulnerable areas are identified. The patient with paraplegia is instructed to use mirrors, if possible, to inspect these areas morning and night, observing for redness, slight edema, or any abrasions. While in bed, the patient should turn at 2-hour intervals and then inspect the skin again for redness that does not fade on pressure. The bottom sheet should be checked for wetness and for creases. The patient with tetraplegia or paraplegia who cannot perform these activities is encouraged to direct others to check these areas and prevent ulcers from developing.

The patient is educated to relieve pressure while in the wheelchair by doing push-ups, leaning from side to side to relieve ischial pressure, and tilting forward while leaning on a table. The caregiver for the patient with tetraplegia will need to perform these activities if the patient cannot do so independently. A wheelchair cushion is prescribed to meet individual needs, which may change in time with changes in posture, weight, and skin tolerance. A referral can be made to a rehabilitation engineer, who can measure pressure levels while the patient is sitting and then tailor the cushion and other necessary aids and assistive devices to the patient's needs.

The diet for the patient with tetraplegia or paraplegia should be high in protein, vitamins, and calories to ensure minimal wasting of muscle and the maintenance of healthy skin, and high in fluids to maintain well-functioning kidneys.

Excessive weight gain and obesity should be avoided, because they further limit mobility.

Improving Bladder Management

The effect of the spinal cord lesion on the bladder depends on the level of injury, the degree of cord damage, and the length of time after injury. A patient with tetraplegia or paraplegia usually has either a reflex or a nonreflex bladder (see Chapter 10). Both bladder types increase the risk of UTI.

The nurse emphasizes the importance of maintaining an adequate flow of urine by encouraging a fluid intake of about 2.5 L daily. The patient should empty the bladder frequently so that there is minimal residual urine and should pay attention to personal hygiene, because infection of the bladder and kidneys almost always occurs by the ascending route. The perineum must be kept clean and dry, and attention must be given to the perianal skin after defecation. Underwear should be cotton (which is more absorbent) and should be changed at least once a day.

If an external catheter (condom catheter) is used, the sheath is removed nightly; the penis is cleansed to remove urine and is dried carefully, because warm urine on the periurethral skin promotes the growth of bacteria. Attention also is given to the collection bag. The nurse emphasizes the importance of monitoring for signs of UTI: cloudy, foul-smelling urine or hematuria (blood in the urine); fever; or chills.

The female patient who cannot achieve reflex bladder control or self-catheterization may need to wear pads or waterproof undergarments. Surgical intervention may be indicated in some patients to create a urinary diversion.

Establishing Bowel Control

The objective of a bowel training program is to establish bowel evacuation through reflex conditioning, a technique described in Chapter 10. If the SCI occurs above the sacral segments or nerve roots and there is reflex activity, the anal sphincter may be massaged (digital stimulation) to stimulate defecation. If the cord lesion involves the sacral segment or nerve roots, anal massage is not performed because the anus may be relaxed and lack tone. Massage is also contraindicated if there is spasticity of the anal sphincter. The anal sphincter is massaged by inserting a gloved finger (which has been adequately lubricated) 2.5 to 3.7 cm (1 to 1.5 in) into the rectum and moving it in a circular motion or from side to side. It soon becomes apparent which area triggers the defecation response. This procedure should be performed at regular time intervals (usually every 48 hours), after a meal, and at a time that will be convenient for the patient at home. The patient also is educated about the symptoms of impaction (frequent loose stools; constipation) and is cautioned to watch for hemorrhoids. A diet with sufficient fluids and fiber is essential to developing a successful bowel training program, avoiding constipation, and decreasing the risk of autonomic dysreflexia.

Counseling on Sexual Expression

Many patients with tetraplegia and paraplegia can have some form of meaningful sexual relationship, although modifications are necessary. The patient and partner benefit from counseling about the range of sexual expression possible, special techniques and positions, exploration of body sensations offering sensual feelings, and urinary and bowel hygiene as related to sexual activity. For men with erectile failure, penile prostheses enable them to have and sustain an erection, and impotence drugs may be helpful. Sildenafil (Viagra), vardenafil (Levitra), and tadalafil (Cialis), for example, are oral smooth muscle relaxants that cause blood to flow into the penis, resulting in an erection (see Chapter 59).

Sexual education and counseling services are included in the rehabilitation services at spinal centers. Small group meetings in which patients can share their feelings, receive information, and discuss sexual concerns and practical aspects are helpful in producing effective attitudes and adjustments.

Enhancing Coping Mechanisms

The impact of the disability and loss becomes marked when the patient returns home. Each time something new enters the patient's life (e.g., a new relationship, going to work), the patient is reminded anew of his or her limitations. Grief reactions and depression are common.

To work through this depression, the patient must have some hope for relief in the future. The nurse can encourage the patient to feel confident in his or her ability to achieve self-care and relative independence. The role of the nurse ranges from caretaker during the acute phase to educator, counselor, and facilitator as the patient gains mobility and independence.

The patient's disability affects not only the patient but also the entire family. In many cases, family therapy is helpful in working through issues as they arise. Adjustment to the disability leads to the development of realistic goals for the future, making the best of the abilities that are left intact and reinvesting in other activities and relationships. Rejection of the disability causes self-destructive neglect and nonadherence to the therapeutic program, which leads to more frustration and depression. Crises for which interventions may be sought include social, psychological, marital, sexual, and psychiatric problems. The family usually requires counseling, social services, and other support systems to help them cope with the changes in their lifestyle and socioeconomic status.

A major goal of nursing management is to help the patient overcome his or her sense of futility and to encourage the patient in the emotional adjustment that must be made before he or she is willing to venture into the outside world. However, an excessively sympathetic attitude on the part of the nurse may cause the patient to develop an overdependence that defeats the purpose of the entire rehabilitation program. The patient is educated and assisted when necessary, but the nurse should avoid performing activities that the patient can do independently with a little effort. This approach to care more than repays itself in the satisfaction of seeing a completely demoralized and helpless patient become independent and find meaning in a newly emerging lifestyle.

Monitoring and Managing Potential Complications

Spasticity. Muscle spasticity can be a problematic complication of tetraplegia and paraplegia. Flexor or extensor spasms occur below the level of the spinal cord lesion and can interfere with the rehabilitation process, ADLs, and quality of life (Bhimani, Anderson, Henly, et al., 2011). Spasticity results from an imbalance between the facilitatory and inhibitory effects on neurons that exist normally. The area of the cord distal to the site of injury or lesion becomes disconnected from the higher inhibitory centers located in the brain, so facilitatory impulses, which originate from muscles, skin, and ligaments, predominate.

Spasticity is defined as a condition of increased muscle tone in a muscle that is weak. Initial resistance to stretching is quickly followed by sudden relaxation. The stimulus that precipitates spasm can be obvious, such as movement or a position change, or subtle, such as a slight jarring of the wheelchair. Most patients with tetraplegia or paraplegia have some degree of spasticity. Because it increases muscle tone, some degree of spasticity can be beneficial in weak patients (Bhimani et al., 2011). With SCI, the onset of spasticity usually occurs from a few weeks to 6 months after the injury. The same muscles that are flaccid during the period of spinal shock develop spasticity during recovery. The intensity of spasticity tends to peak approximately 2 years after the injury, after which the spasms tend to regress.

Management of spasticity is based on the severity of symptoms and the degree of incapacitation. Botulinum toxin (Botox) injections, as well as the antispasmodic medication baclofen (Lioresal), available in an oral and an intrathecal form, may be indicated (Bhimani et al., 2011). Oral medications such as diazepam (Valium), dantrolene (Dantrium), and tizanidine (Zanaflex) help control spasms by decreasing sympathetic outflow from the central nervous system. Newer forms of adjunctive therapy include oral and transdermal forms of clonidine (Catapres) (Karch, 2012). All of the antispasmodic medications cause drowsiness, weakness, and vertigo in some patients. Passive range-of-motion exercises and frequent turning and repositioning are helpful, because stiffness tends to increase spasticity. These activities also are essential in the prevention of contractures, pressure ulcers, and bowel and bladder dysfunction.

Infection and Sepsis. Patients with tetraplegia and paraplegia are at increased risk for infection and sepsis from a variety of sources: urinary tract, respiratory tract, and pressure ulcers. Sepsis remains a major cause of complications and death in these patients. Prevention of infection and sepsis is essential through maintenance of skin integrity, complete emptying of the bladder at regular intervals, and prevention of urinary and fecal incontinence. The risk of respiratory infection can be decreased by avoiding contact with people who have symptoms of respiratory infection, performing coughing and deep-breathing exercises to prevent pooling of respiratory secretions, receiving yearly influenza vaccines, and giving up smoking. A high-protein diet is important in maintaining an adequate immune system, as is avoiding factors that may reduce immune system function, such as excessive stress, drug abuse, and excessive alcohol intake.

If infection occurs, the patient requires thorough assessment and prompt treatment. Antibiotic therapy and adequate hydration, in addition to local measures (depending on the site of infection), are initiated immediately.

UTIs are minimized or prevented by aseptic technique in catheter management, adequate hydration, a bladder training program, and prevention of overdistention of the bladder and urinary stasis.

Skin breakdown and infection are prevented by maintenance of a turning schedule; frequent back care; regular assessment of all skin areas; regular cleaning and lubrication of the skin; passive range-of-motion exercise to prevent contractures; pressure relief over broken skin areas, bony prominences, and heels; and wrinkle-free bed linen.

Pulmonary infections are managed and prevented by frequent coughing, turning, and deep-breathing exercises and chest physiotherapy; aggressive respiratory care and suctioning of the airway if a tracheostomy is present; assisted coughing as needed; and adequate hydration.

Infections of any kind can be life threatening. Aggressive nursing interventions are key to prevention, detection, and early management.

Promoting Home and Community-Based Care

Educating Patients About Self-Care. Patients with tetraplegia or paraplegia are at risk for complications for the rest of their lives. Therefore, a major aspect of nursing care is educating the patient and family about these complications and about strategies to minimize risks. UTIs, contractures, infected pressure ulcers, and sepsis may necessitate hospitalization. Other late complications that may occur include lower extremity edema, joint contractures, respiratory dysfunction, and pain. To avoid these and other complications, the patient and family members are educated about skin care, catheter care, range-of-motion exercises, breathing exercises, and other care techniques. Education is initiated as soon as possible and extends into the rehabilitation or long-term care facility and home. In all aspects of care, it is important for the nurse and patient to set mutual goals and discuss the tasks the patient is capable of doing independently and which tasks the patient needs assistance to complete. (See Chapter 10 for a more detailed discussion of rehabilitation.)

Continuing Care. Referral for home care is often appropriate for assessment of the home setting, patient education, and evaluation of the patient's physical and emotional status. During visits by the home care nurse, education about strategies to prevent or minimize potential complications is reinforced. The home environment is assessed for adequacy of care and for safety. Environmental modifications are made, and specialized equipment is obtained, ideally before the patient goes home.

The home care nurse also assesses the patient's and the family's adherence to recommendations and their use of coping strategies. The use of inappropriate coping strategies (e.g., drug and alcohol use) is assessed, and referrals to counseling are made for the patient and family. Appropriate and effective coping strategies are reinforced. The nurse reviews previous education and determines the need for further physical or psychological assistance. The patient's self-esteem and body image may be very poor at this time. Because people with high levels of social support often report feelings of well-being despite major physical disability, it is beneficial for the nurse to assess and promote further development of the support system and effective coping strategies for each patient. Research suggests that caregivers play a critical role in helping patients feel less dependent, experiencing feelings of freedom, and reintegrating into the community (Gajraj-Singh, 2011; Gorzkowski, Kelly, Klaas, et al., 2011; Martinsen & Dreyer, 2012).

The patient requires continuing, lifelong follow-up by the primary provider, physical therapist, and other rehabilitation team members, because the neurologic deficit is usually permanent and new deficits, complications, and secondary conditions can develop. These require prompt attention before they take their toll in additional physical impairment, time, morale, and financial costs. One small study suggested that education and peer mentoring may decrease medical complications in the first year following SCI (Ljungberg, Kroll, Libin, et al., 2011). The local counselor for the Office of Vocational Rehabilitation works with the patient with respect to job placement or additional educational or vocational training. The nurse is in a good position to remind patients and family members of the need for continuing health promotion and screening practices. Referral to accessible health care providers and imaging centers is important in health promotion and health screening. (See Chapter 9 for more information on chronic illness and disability.)

Evaluation

Expected patient outcomes may include:

1. Attains maximum form of mobility
2. Contractures do not develop
3. Maintains healthy, intact skin
4. Achieves bladder control, absence of UTI
5. Achieves bowel control
6. Reports sexual satisfaction
7. Shows improved adaptation to environment and others
8. Exhibits reduction in spasticity
   a. Reports understanding of the precipitating factors
   b. Uses measures to reduce spasticity
9. Describes long-term management required
10. Exhibits absence of complications

Critical Thinking Exercises

1 **ebp** An 18-year-old man is brought to the ED by his family, who report that he was "dinged" in a football game 1 week ago. The patient does not recall the event. His family states that he is sleeping more than usual and seems forgetful. What is the evidence base for conducting a comprehensive and focused evaluation for a patient with a sports concussion injury? What evidence-based recommendations may guide return to sports for this patient? Identify the criteria used to evaluate the strength of the evidence for these practices.

2 A 19-year-old with an acute SCI at the T6 level complains of a severe headache. The patient is diaphoretic and flushed above the level of injury. Blood pressure is 230/110 mm Hg. What do you suspect is happening? What are the possible causes of this condition, and how would you intervene? What is included in your plan for educating the patient and family before discharge to rehabilitation?

3 **pq** A 66-year-old man was involved in a motor vehicle crash 2 days ago and sustained a C4 fracture with SCI. As a result, he has tetraplegia and is on a mechanical ventilator in a neurologic intensive care unit. Identify the priorities, approach, and techniques you would use to provide care for this patient during the acute phase of care. Identify how your priorities, approach, and techniques will differ once the patient reaches the rehabilitation phase of care.

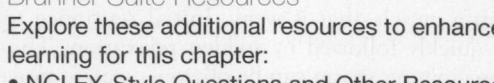

Brunner Suite Resources

Explore these additional resources to enhance learning for this chapter:

- NCLEX-Style Questions and Other Resources on thePoint, http://thePoint.lww.com/Brunner13e
- Study Guide
- PrepU
- Clinical Handbook
- Handbook of Laboratory and Diagnostic Tests

References

*Asterisk indicates nursing research.
**Double asterisk indicate classic reference.

Books

American Association of Neuroscience Nurses. (2011). *Guide to the care of the patient with intracranial pressure monitoring/external ventricular drainage or lumbar drainage: AANN reference series for clinical practice*. Glenview, IL: Author.

Bader, M. K., & Littlejohns, L. R. (2010). *AANN core curriculum for neuroscience nursing*. Glenview, IL: American Association of Neuroscience Nurses.

Eliopoulos, C. (2010). *Gerontological nursing* (7th ed.). Philadelphia: Lippincott Williams & Wilkins.

Hickey, J. V. (2009). *The clinical practice of neurological & neurosurgical nursing* (6th ed.). Philadelphia: Lippincott Williams & Wilkins.

Karch, A. M. (2012). *2012 Lippincott's nursing drug guide*. Philadelphia: Lippincott Williams & Wilkins.

Morton, P. G., Fontaine, D., Hudak, C., et al. (2009). *Critical care nursing a holistic approach* (9th ed.). Philadelphia: Lippincott Williams & Wilkins.

Porth, C. M., & Matfin, G. (2009). *Pathophysiology: Concepts of altered health status* (8th ed.). Philadelphia: Lippincott Williams & Wilkins.

Sole, M. L., Klein, D. G., & Moseley, M. J. (2012). *Introduction to critical care nursing* (6th ed.). St. Louis: Elsevier Saunders.

West, T. A., Bergman, A., Biggins, M.S., et al. (2011). *Care of the patient with mild traumatic brain injury: AANN and ARN clinical practice guidelines series*. Glenview, IL: American Association of Neuroscience Nurses.

Journals and Electronic Documents

Head Injury

Arango-Lasprilla, J. C., Nicolls, E., Villasenor Cabrera, T., et al. (2011). Health related quality of life in caregivers of individuals with traumatic brain injury from Guadalajara, Mexico. *Journal of Rehabilitation Medicine*, *43*(11), 983–986.

*Bay, E. (2011). Mild traumatic brain injury: A Midwest survey about the assessment and documentation practices of emergency department nurses. *Advanced Emergency Nursing Journal*, *33*(1), 71–83.

Brain Trauma Foundation. (2007). *Guidelines for the management of severe traumatic brain injury* (3rd ed.). New York: Author. Available at: www.braintrauma.org

Brosche, T. M. (2011). Intracranial pressure and cerebral perfusion pressure ranges. *Critical Care Nurse*, *31*(4), 18–19.

Butcavage, K. (2012). Glycemic control and intensive insulin protocols for neurologically injured patients. *Journal of Neuroscience Nursing*, *44*(4), E1–E9.

Centers for Disease Control and Prevention. (2012). *Injury prevention & control: Traumatic brain injury*. Available at: www.cdc.gov/traumaticbraininjury/statistics.html

Coco, K., Tossavainen, K., Jaaskelainen, J. E., et al. (2011). Support for traumatic brain injury patients' family members in neurosurgical nursing: A systematic review. *Journal of Neuroscience Nursing*, *43*(6), 337–348.

*Engstrom, A., & Soderberg, S. (2011) Transition as experience by close relatives of people with traumatic brain injury. *Journal of Neuroscience Nursing*, *43*(5), 253–260.

**Hagen, C., Malkmus, D., & Durham, P. (1972). *The Rancho Levels of Cognitive Functioning scale*. Communication Disorders Service, Rancho Los Amigos Hospital, Downey, CA. Revised 11/15/74 by Danese Malkmus and Kathryn Stenderup.

Kimbler, D. E., Murphy, M., & Dhandapani, K. M. (2011). Concussion and the adolescent athlete. *Journal of Neuroscience Nursing*, *43*(6), 286–290.

McIlvoy, L. (2012). Fever management in patients with brain injury. *AACN Advanced Critical Care, 23*(2), 204–211.

Norup, A., Kristensen, K. S., Siert, L., et al. (2011). Neuropsychological support to relatives of patients with severe traumatic brain injury in the subacute phase. *Neuropsychological Rehabilitation, 21*(3), 306–321.

Phelan, S. M., Griffin, J. M., Hellerstedt, W. L., et al. (2011). Perceived stigma, strain, and metal health among caregivers of veterans with traumatic brain injury. *Disability and Health Journal, 4*(3), 177–184.

Prendergast, V., Hallberg, I. R., Jahnke, H., et al. (2009). Oral health, ventilator-associated pneumonia, and intracranial pressure in intubated patients in a neuroscience intensive care unit. *American Journal of Critical Care, 18*(4), 368–376.

Prendergast, V., Jakobsson, U., Renvert, S., et al. (2012). Effects of a standard versus comprehensive oral care protocol among intubated neuroscience ICU patients: Results of a randomized controlled trial. *Journal of Neuroscience Nursing, 44*(3), 134–146.

Ruff, R. M., Iverson, G. L., Barth, J. T., et al. (2009). Recommendations for diagnosing a mild traumatic brain injury: A National Academy of Neuropsychology education paper. *Archives of Clinical Neuropsychology, 24*(1), 3–10.

Schipper, K., Widdershoven, G., & Abma, T. (2011). Citizenship and autonomy in acquired brain injury. *Nursing Ethics, 18*(4), 526–536.

**Teasdale, G., & Jennett, B. (1974). Assessment of coma and impaired consciousness. A practical scale. *Lancet, 2*(7872), 81–84.

Torrence, C. B., DeCristofaro, C., & Elliott, L. (2011). Empowering the primary care provider to optimally manage mild traumatic brain injury. *Journal of the American Academy of Nurse Practitioners, 23*(12), 638–647.

Spinal Cord Injury

Asselin, G., Ward, C., Penning, J. H., et al. (2012). Therapeutic horse back riding of a spinal cord injured veteran: A case study. *Rehabilitation Nursing, 37*(6), 270–276.

Bhimani, R. H., Anderson, L. C., Henly, S. J., et al. (2011). Clinical measurement of limb spasticity in adults: State of the science. *Journal of Neuroscience Nursing, 43*(2), 104–115.

Centers for Disease Control & Prevention. (2010). *Spinal Cord Injury (SCI): Fact Sheet-Traumatic Brain Injury-Injury Center.* Available at:www.cdc.gov/traumaticbraininjury/scifacts.html

Chen, H. Y., Lai, C. H., & Wu, T. J. (2011). A study of factors affecting moving forward behavior among people with spinal cord injury. *Rehabilitation Nursing, 36*(3), 91–97.

Gajraj-Singh, P. (2011). Psychological impact and the burden of caregiving for persons with spinal cord injury living in the community in Fiji. *Spinal Cord,* 49(8), 928–934.

Gorzkowski, J., Kelly, E. H., Klaas, S. J., et al. (2011). Obstacles to community participation among youth with spinal cord injury. *Journal of Spinal Cord Medicine, 34*(6), 576–585.

Kahn, S. R., Lim, W., Dunn, A., et al. (2012). Prevention of VTE in nonsurgical patients. *Chest, 141*(2), e195S–e226S.

Lucas, S. M. (2012). Malignant spinal cord compression. *Nursing, 42*(2), 72.

*Ljungberg, I., Kroll, T., Libin, A., et al. (2011). Using peer mentoring for people with spinal cord injury to enhance self-efficacy beliefs and prevent medical complications. *Journal of Clinical Nursing, 20*(3–4), 351–358.

Magnuson, B., Peppard, A., & Flomenhoft, D. A. (2011). Hypocaloric considerations in patients with potentially hypometabolic states. *Nutrition in Clinical Practice, 26*(3), 253–260.

*Martinsen B., & Dreyer, P. (2012). Dependence on care experienced by people living with Duchenne muscular dystrophy and spinal cord injury. *Journal of Neuroscience Nursing, 44*(2), 82–90.

National Institute of Neurological Disorders and Stroke. (2012a). *NINDS spinal cord information page.* Available at: www.ninds.nih.gov/disorders/sci/sci.htm

National Institute of Neurological Disorders and Stroke. (2012b). *Spinal cord injury: Hope through research.* Available at: www.ninds.nih.gov/disorders/sci/detail_sci.htm

Perret, C., & Stoffel-Kurt, N. (2011). Comparison of nutritional intake between individuals with acute and chronic spinal cord injury. *Journal of Spinal Cord Medicine, 34*(6), 569–575.

Revell, S. M. (2011). Symptom clusters in traumatic spinal cord injury: An exploratory literature review. *Journal of Neuroscience Nursing, 43*(2), 85–93.

Rundquist, J., Gassaway, J., Bailey, J., et al. (2011). The SCI rehab project: Treatment time spent in SCI rehabilitation. Nursing bedside education and care management time during inpatient spinal cord injury rehabilitation. *Journal of Spinal Cord Medicine, 34*(2), 205–215.

Smeltzer, S. C., & Wetzel-Effinger, L. (2009). Pregnancy in women with spinal cord injury. *Topics in Spinal Cord Injury Rehabilitation, 15*(1), 29–42.

Stauber, M. A. (2011). Not all spinal cord injuries involve a fracture. *Advanced Emergency Nursing Journal, 33*(3), 226–231.

Resources

American Association of Neuroscience Nurses (AANN), www.aann.org

American Academy of Spinal Cord Injury Professionals, Inc. (ASCIP; formerly known as the American Association of Spinal Cord Injury Nurses), www.academyscipro.org/Default.aspx

Association of Rehabilitation Nurses (ARN), www.rehabnurse.org

Brain Injury Association of America, www.biausa.org

Brain Trauma Foundation, www.braintrauma.org

Centers for Disease Control and Prevention (CDC), www.cdc.gov

Information Center for Individuals with Disabilities, www.disability.net

National Spinal Cord Injury Association (NSCIA), www.spinalcord.org

Paralyzed Veterans of America, www.pva.org

Chapter 69

Management of Patients With Neurologic Infections, Autoimmune Disorders, and Neuropathies

Learning Objectives

On completion of this chapter, the learner will be able to:

1. Differentiate among the infectious disorders of the nervous system according to causes, manifestations, medical care, and nursing management.
2. Describe the pathophysiology, clinical manifestations, and medical and nursing management of multiple sclerosis, myasthenia gravis, and Guillain-Barré syndrome.
3. Use the nursing process as a framework for care of patients with multiple sclerosis and Guillain-Barré syndrome.
4. Describe disorders of the cranial nerves, their manifestations, and indicated nursing interventions.
5. Use the nursing process as a framework for care of the patient with a cranial nerve disorder.

Glossary

ataxia: impaired coordination of movement during voluntary movement

bulbar paralysis: immobility of muscles innervated by cranial nerves with their cell bodies in the lower portion of the brain stem

diplopia: double vision, or the awareness of two images of the same object occurring in one or both eyes

dyskinesia: impaired ability to execute voluntary movements

dysphagia: difficulty swallowing

dysphonia: voice impairment or altered voice production

hemiparesis: weakness of one side of the body, or part of it, due to an injury in the motor area of the brain

hemiplegia: paralysis of one side of the body, or part of it, due to an injury in the motor area of the brain

neuropathy: general term indicating a disorder of the nervous system

paresthesia: numbness, tingling, or a "pins and needles" sensation

prion: a particle smaller than a virus that is resistant to standard sterilization procedures

ptosis: drooping of the eyelids

spasticity: muscular hypertonicity with increased resistance to stretch often associated with weakness, increased deep tendon reflexes, and diminished superficial reflexes

spongiform: having the appearance or quality of a sponge

The diverse group of neurologic disorders that make up infectious and autoimmune disorders and cranial and peripheral neuropathies presents unique challenges for nursing care. The nurse who cares for patients with these disorders must have a clear understanding of the pathologic processes and the clinical outcomes. Some of the issues nurses must help patients and families confront include adaptation to the effects of the disease, potential changes in family dynamics, and possibly end-of-life issues.

INFECTIOUS NEUROLOGIC DISORDERS

The infectious disorders of the nervous system include meningitis, brain abscesses, various types of encephalitis Creutzfeldt-Jakob disease (CJD), and variant Creutzfeldt-Jakob disease (vCJD). The clinical manifestations, assessment, and diagnostic findings, as well as the medical and nursing management, are related to the specific infectious process.

Meningitis

Meningitis is inflammation of the meninges, which cover and protect the brain and spinal cord. The three major causes of meningitis are bacterial, viral, and fungal infections (Bader & Littlejohns, 2010). Meningitis can be the main reason a patient is hospitalized, or it can develop during hospitalization; it is classified as septic or aseptic. Septic meningitis is caused by bacteria. The bacteria *Streptococcus pneumoniae* and *Neisseria meningitidis* are responsible for the majority of cases of bacterial meningitis in adults. In aseptic meningitis, the cause is viral or secondary to cancer or having a weakened immune system, such as in human immunodeficiency virus (HIV). The most common causative agents are the enteroviruses. Aseptic meningitis occurs more frequently in the summer and early fall.

Outbreaks of *N. meningitidis* infection are most likely to occur in dense community groups, such as college campuses and military installations. Although infections occur year-round, the peak incidence is in the winter and early spring. Factors that increase the risk of bacterial meningitis include tobacco use and viral upper respiratory infection, because they increase the amount of droplet production. Otitis media and mastoiditis increase the risk of bacterial meningitis, because the bacteria can cross the epithelial membrane and enter the subarachnoid space. People with immune system deficiencies are also at greater risk for development of bacterial meningitis (Bader & Littlejohns, 2010).

Pathophysiology

Meningeal infections generally originate in one of two ways: through the bloodstream as a consequence of other infections or by direct spread, such as might occur after a traumatic injury to the facial bones or secondary to invasive procedures.

N. meningitidis concentrates in the nasopharynx and is transmitted by secretion or aerosol contamination. Bacterial or meningococcal meningitis also occurs as an opportunistic infection in patients with acquired immunodeficiency syndrome (AIDS) and as a complication of Lyme disease (Chart 69-1).

Once the causative organism enters the bloodstream, it crosses the blood–brain barrier and proliferates in the cerebrospinal fluid (CSF). The host immune response stimulates the release of cell wall fragments and lipopolysaccharides, facilitating inflammation of the subarachnoid and pia mater. Because the cranial vault contains little room for expansion, the inflammation may cause increased intracranial pressure (ICP). CSF circulates through the subarachnoid space, where inflammatory cellular materials from the affected meningeal tissue enter and accumulate.

The prognosis for bacterial meningitis depends on the causative organism, the severity of the infection and illness, and the timeliness of treatment. Acute fulminant presentation may include adrenal damage, circulatory collapse, and widespread hemorrhages (Waterhouse-Friderichsen syndrome). This syndrome is the result of endothelial damage and vascular necrosis caused by the bacteria. Complications include visual impairment, deafness, seizures, paralysis, hydrocephalus, and septic shock.

Clinical Manifestations

Headache and fever are frequently the initial symptoms. Fever tends to remain high throughout the course of the illness. The headache is usually either steady or throbbing and very severe as a result of meningeal irritation (Bickley, 2009; Weber & Kelley, 2010). Meningeal irritation results in a number of other well-recognized signs common to all types of meningitis:

- *Neck immobility:* A stiff and painful neck (nuchal rigidity) can be an early sign, and any attempts at flexion of the head are difficult because of spasms in the muscles of the neck. Usually, the neck is supple, and the patient can easily bend the head and neck forward.
- *Positive Kernig's sign:* When the patient is lying with the thigh flexed on the abdomen, the leg cannot be completely extended (Fig. 69-1). When Kernig's sign is bilateral, meningeal irritation is suspected (Weber & Kelley, 2010).
- *Positive Brudzinski's sign:* When the patient's neck is flexed (after ruling out cervical trauma or injury), flexion of the knees and hips is produced; when the lower extremity of one side is passively flexed, a similar movement is seen in the opposite extremity (see Fig. 69-1). Brudzinski's sign is a more sensitive indicator of meningeal irritation than Kernig's sign.
- *Photophobia (extreme sensitivity to light):* This finding is common, although the cause is unclear.

A rash can be a striking feature of *N. meningitidis* infection, occurring in about half of patients with this type of meningitis. Skin lesions develop, ranging from a petechial rash with purpuric lesions to large areas of ecchymosis.

Chart 69-1 Meningitis in Specific Populations

Meningitis can occur as a complication of other diseases and is an opportunistic infection seen with greater frequency in patients who are immunocompromised.

Meningitis in Patients With Acquired Immunodeficiency Syndrome (AIDS)

- Cryptococcal meningitis is the most common fungal infection of the central nervous system in patients with AIDS. Patients may experience headache, nausea, vomiting, seizures, confusion, and lethargy. Treatment consists of IV administration of amphotericin B followed by fluconazole. Maintenance therapy with fluconazole may be necessary to prevent relapse.
- Some patients who are immunosuppressed develop few if any symptoms because of blunted inflammatory responses; others develop atypical features.
- Onset of fever, headache, nausea, and malaise most often occur over a few weeks. Only 25% of patients present with stiff neck and photophobia.

Meningitis in Patients With Lyme Disease

- Lyme disease is a multisystem inflammatory process caused by the tick-transmitted spirochete *Borrelia burgdorferi.*
- Neurologic deficits are seen in later stages (stages 2 or 3). Stage 2 occurs with the start of a characteristic rash or 1–6 months after the rash has disappeared.
- Neurologic comorbidities include aseptic meningitis, chronic lymphocytic meningitis, and encephalitis.
- Cranial nerve inflammation, including Bell's palsy and other peripheral neuropathies, is common.
- Stage 3 (the chronic form of the disease) begins years after the initial tick infection and is characterized by arthritis, skin lesions, and neurologic abnormalities.
- Most patients with stage 2 and 3 Lyme disease are treated with IV antibiotics, usually ceftriaxone or penicillin G.
- Meningeal and systemic symptoms begin to improve within days, although other symptoms, such as headache, may persist for weeks.

Adapted from Theroux, N., Phipps, M., Zimmerman, L., & Relf, M. V. (2013). *Journal of Neuroscience Nursing, 45*(1), 5–13.

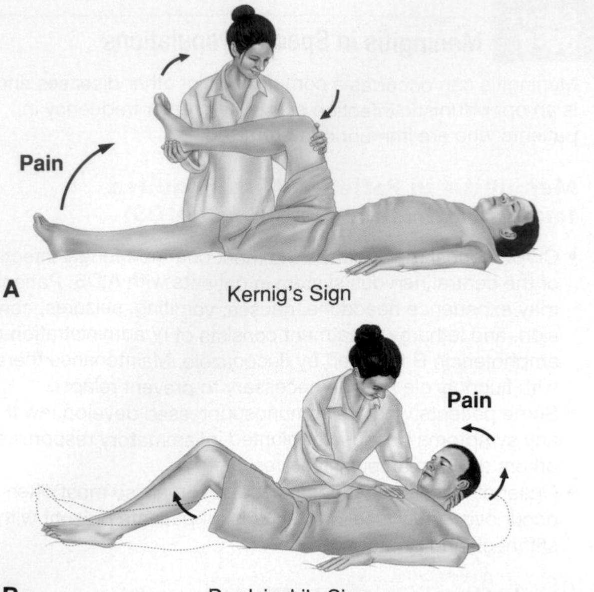
Pain
A
Kernig's Sign
Pain
B
Brudzinski's Sign

FIGURE 69-1 • Testing for meningeal irritation. **A.** Kernig's sign. Flex the patient's leg at both the hip and knee and then straighten the knee. **B.** Brudzinski's sign. As the neck is flexed, watch the hips and knees for a reaction.

Disorientation and memory impairment are common early in the course of the illness. The changes depend on the severity of the infection as well as the individual response to the physiologic processes. Behavioral manifestations are also common. As the illness progresses, lethargy, unresponsiveness, and coma may develop.

Seizures can occur and are the result of areas of irritability in the brain. ICP increases secondary to diffuse brain swelling or hydrocephalus (Bader & Littlejohns, 2010). The initial signs of increased ICP include decreased level of consciousness (LOC) and focal motor deficits. If ICP is not controlled, the uncus of the temporal lobe may herniate through the tentorium, causing pressure on the brain stem. Brain stem herniation is a life-threatening event that causes cranial nerve dysfunction and depresses the centers of vital functions, such as the medulla. (See Chapter 66 for discussion of the patient with a change in LOC or increased ICP.)

An acute fulminant infection occurs in about 10% of patients with meningococcal meningitis, producing signs of overwhelming septicemia: an abrupt onset of high fever, extensive purpuric lesions (over the face and extremities), shock, and signs of disseminated intravascular coagulation (see chapter 33). Death may occur within a few hours after onset of the infection.

Assessment and Diagnostic Findings

If the clinical presentation suggests meningitis, diagnostic testing is conducted to identify the causative organism. A computed tomography (CT) scan or magnetic resonance imaging (MRI) scan is used to detect a shift in brain contents (which may lead to herniation) prior to a lumbar puncture in patient with altered LOC, papilledema, neurologic deficits, new onset of seizure, immunocompromised state, or history of central nervous system (CNS) disease (Ziai & Lewin, 2009). Bacterial culture and Gram staining of CSF and blood are key diagnostic tests. An overview of CSF values and alterations in bacterial, viral, and fungal meningitis is presented in Table 69-1. Gram staining allows for rapid identification of the causative bacteria and initiation of appropriate antibiotic therapy.

A bedside risk score for use in adults with bacterial meningitis has been developed. Risks for an unfavorable outcome include older age, a heart rate of greater than 120 bpm, decreased Glasgow Coma Scale score, cranial nerve palsies, and a positive Gram stain 1 hour after presentation to the hospital (Weisfelt, van de Beek, Spanjaard, et al., 2007).

Prevention

The Advisory Committee on Immunization Practices of the Centers for Disease Control and Prevention (CDC, 2011) recommends that the meningococcal conjugated vaccine be given to youth at 11 to 12 years of age, with a booster dose at 16 years of age. First-year college students living in dormitories have a three times greater risk for developing meningococcal meningitis as compared with the general population and students living off campus. Most states mandate education addressing meningococcal meningitis and the availability of vaccination so that families can make informed decisions.

People in close contact with patients with meningococcal meningitis should be treated with antimicrobial chemoprophylaxis using rifampin (Rifadin), ciprofloxacin hydrochloride (Cipro), or ceftriaxone sodium (Rocephin). Therapy should be started within 24 hours after exposure because a delay in the initiation of therapy limits the effectiveness of the prophylaxis. Vaccination should also be considered as an adjunct to antibiotic chemoprophylaxis for anyone living with a person who develops meningococcal infection. Vaccination against

TABLE 69-1 Cerebrospinal Fluid Values Diagnostic for Meningitis

Parameter	Normal CSF	Bacterial Meningitis	Viral Meningitis	Fungal Meningitis
Opening pressure (mm H_2O)	100–180	200–500	≤250	>200
Leukocyte count (white blood cell/mm^3)	0–5	Increased 100–5,000	Increased 50–1,000	Increased >20
Neutrophils (%)	0	≥80	<40	
Protein (mg/dL)	18–45	Elevated 100–500	Elevated <200	Elevated >45
Glucose (mg/dL)	45–80; 0.6 times blood glucose level	5–40; <0.3 times blood glucose level	>45	<40

CSF, cerebrospinal fluid.
From Tunkel, A. R. (2001). *Bacterial meningitis*. Philadelphia: Lippincott Williams & Wilkins, p. 102.

Haemophilus influenzae and *S. pneumoniae* should be encouraged for children and at-risk adults (CDC, 2011).

Medical Management

Successful outcomes depend on the early administration of an antibiotic agent that crosses the blood–brain barrier into the subarachnoid space in sufficient concentration to halt the multiplication of bacteria. Penicillin G in combination with one of the cephalosporins (e.g., ceftriaxone sodium, cefotaxime sodium) is most often administered intravenously (IV), optimally within 30 minutes of hospital arrival (Bader & Littlejohns, 2010).

Dexamethasone (Decadron) has been shown to be beneficial as adjunct therapy in the treatment of acute bacterial meningitis and in pneumococcal meningitis if it is administered 15 to 20 minutes before the first dose of antibiotic and every 6 hours for the next 4 days. Research suggests that dexamethasone improves the outcome in adults and does not increase the risk of gastrointestinal bleeding (Bader & Littlejohns, 2010).

Dehydration and shock are treated with fluid volume expanders. Seizures, which may occur early in the course of the disease, are treated with phenytoin. Increased ICP is treated as necessary (see Chapter 66).

Nursing Management

The patient with meningitis is critically ill; therefore, many of the nursing interventions are collaborative with the physician, respiratory therapist, and other members of the health care team. The patient's safety and well-being depend on sound nursing judgment. Most patients will need the following nursing interventions:

- Instituting infection control precautions until 24 hours after initiation of antibiotic therapy (oral and nasal discharge is considered infectious)
- Assisting with pain management due to overall body aches and neck pain
- Assisting with getting rest in a quiet, darkened room
- Implementing interventions to treat the elevated temperature, such as antipyretic agents and cooling blankets
- Encouraging the patient to stay hydrated either orally or peripherally
- Ensuring close neurologic monitoring (see Chapter 66)

Neurologic status and vital signs are continually assessed. Pulse oximetry and arterial blood gas values are used to quickly identify the need for respiratory support if increasing ICP compromises the brain stem. Insertion of a cuffed endotracheal tube (or tracheotomy) and mechanical ventilation may be necessary to maintain adequate tissue oxygenation.

Blood pressure (usually monitored using an arterial line) is assessed for early manifestations of shock, which precedes cardiac or respiratory failure. Rapid IV fluid replacement may be prescribed, but care is taken to prevent fluid overload. Fever also increases the workload of the heart and cerebral metabolism. ICP will increase in response to increased cerebral metabolic demands. Therefore, measures are taken to reduce body temperature as quickly as possible.

Other important components of nursing care include the following measures:

- Protecting the patient from injury secondary to seizure activity or altered LOC
- Monitoring daily body weight; serum electrolytes; and urine volume, specific gravity, and osmolality, especially if syndrome of inappropriate antidiuretic hormone (SIADH) is suspected
- Preventing complications associated with immobility, such as pressure ulcers and pneumonia

Any sudden, critical illness can be devastating to the family. Because the patient's condition is often critical and the prognosis guarded, the family needs to be informed about the patient's condition. Periodic family visits are essential to facilitate coping of the patient and family. An important aspect of the nurse's role is to support the family and assist them in identifying others who can be supportive to them during the crisis.

Continuing Care in the Home and Community

After the patient has achieved physiologic homeostasis and has demonstrated achievement of major health care goals, rehabilitation continues either in a rehabilitation facility or at home. Continued support and evaluation by the home care nurse are essential.

Because patients may have been critically ill and focused on the most obvious needs and issues, the nurse reminds the patient and family about the importance of continuing health promotion and screening practices, such as regular physical examinations and appropriate diagnostic screening tests.

Brain Abscess

Brain abscesses account for less than 1% of space-occupying brain lesions in the United States (Bader & Littlejohns, 2010). Brain abscesses are rare in immunocompetent people; they are more frequently diagnosed in people who are immunosuppressed as a result of an underlying disease or the use of immunosuppressive medications.

Pathophysiology

A brain abscess is a collection of infectious material within the tissue of the brain. Bacteria are the most common causative organisms. The most common predisposing conditions for abscesses among adults who are immunocompetent are otitis media and rhinosinusitis. It is estimated that 50% of brain abscesses are otogenic in origin (Alaani, Couldon, McDermott, et al., 2010). An abscess can result from intracranial surgery, penetrating head injury, or tongue piercing. Organisms causing brain abscess may reach the brain by hematologic spread from the lungs, gums, tongue, or heart, or from a wound or intra-abdominal infection. Brain abscesses in those who are immunocompromised may result from various pathogens. To prevent brain abscess, otitis media, mastoiditis, rhinosinusitis, dental infections, and systemic infections should be treated promptly.

Clinical Manifestations

The clinical manifestations of a brain abscess result from alterations in intracranial dynamics (edema, brain shift), infection, or the location of the abscess.

Headache, usually worse in the morning, is the most prevalent symptom. Fever may or may not be present (Bader & Littlejohns, 2010). Vomiting and focal neurologic deficits

Chart 69-2 ASSESSMENT

Assessing for Brain Abscesses

Be alert to the following signs and symptoms of brain abscess:

Frontal Lobe

Hemiparesis (weakness on one side of the body)
Expressive aphasia (inability to express oneself)
Seizures
Frontal headache

Temporal Lobe

Localized headache
Changes in vision
Facial weakness
Receptive aphasia (inability to understand language)

Cerebellar Abscess

Occipital headache
Ataxia (inability to coordinate movements)
Nystagmus (rhythmic, involuntary movements of the eye)

occur as well. Focal deficits such as weakness and decreasing vision reflect the area of brain that is involved. As the abscess expands, symptoms of increased ICP such as decreasing LOC and seizures occur.

Assessment and Diagnostic Findings

The baseline neurological examination may reveal a variety of signs and symptoms (Chart 69-2). Neuroimaging with CT scanning is used most often to identify the size and location of the abscess (Alaani et al., 2010). Aspiration of the abscess, guided by CT or MRI, is often used to culture and identify the infectious organism. Blood cultures are obtained if the abscess is believed to arise from a distant source. Chest x-ray is performed to rule out predisposing lung infections, and electroencephalography (EEG) may help localize the lesion (Hickey, 2009).

Medical Management

Treatment is aimed at controlling increased ICP, draining the abscess, and providing antimicrobial therapy directed at the abscess and the main source of infection. Large IV doses of antibiotic agents are administered to penetrate the blood–brain barrier and reach the abscess. The choice of the specific antibiotic medication is based on culture and sensitivity testing and directed at the causative organism. Antibiotics should be started as soon as possible and thus the initial antibiotic started typically is ceftriaxone, which will be adjusted based on the culture and sensitivity results. A stereotactic CT-guided aspiration may be used to drain the abscess and identify the causative organism. Corticosteroids may be prescribed to help reduce the inflammatory cerebral edema if the patient shows evidence of an increasing neurologic deficit. Antiseizure medications may be prescribed to prevent or treat seizures (see Chapter 66).

Nursing Management

Nursing care focuses on continuing to assess the neurologic status, administering medications, assessing the response to treatment, and providing supportive care.

Ongoing neurologic assessment alerts the nurse to changes in ICP, which may indicate a need for more aggressive intervention. The nurse also assesses and documents the responses to medications. Blood laboratory test results, specifically blood glucose and serum potassium levels, need to be closely monitored when corticosteroids are prescribed (Karch, 2012). Administration of insulin or electrolyte replacement may be required to return these values to within normal limits.

Patient safety is another key nursing responsibility. Injury may result from decreased LOC or falls related to motor weakness or seizures.

The patient with a brain abscess is very ill, with neurologic deficits, such as **hemiplegia** (paralysis of one side of the body, or part of it), **hemiparesis** (weakness of one side of the body, or part of it), seizures, visual deficits, and cranial nerve palsies that may persist after treatment. The nurse must assess the family's ability to express distress at the patient's condition, cope with the patient's illness and deficits, and obtain support.

Herpes Simplex Virus Encephalitis

Encephalitis is an acute inflammatory process of the brain tissue. Herpes simplex virus (HSV) is the most common cause of acute encephalitis in the United States. There are two HSVs: HSV-1 and HSV-2. HSV-1 typically affects children and adults. HSV-2 most commonly affects neonates who acquire the disease from a mother who has an active genital herpes infection at the time of delivery (Steiner, 2011). HSV-2 is discussed in pediatric textbooks.

Pathophysiology

The pathology of encephalitis involves local necrotizing hemorrhage that becomes more generalized, followed by edema. There is also progressive deterioration of nerve cell bodies (Porth & Matfin, 2009).

Clinical Manifestations

The initial symptoms of HSV-1 encephalitis include fever, headache, confusion, and hallucinations. Focal neurologic symptoms reflect the areas of cerebral inflammation and necrosis and include fever, headache, behavioral changes, focal seizures, dysphasia, hemiparesis, and altered LOC (Porth & Matfin, 2009).

Assessment and Diagnostic Findings

Neuroimaging studies, such as EEG, and CSF examination are used to diagnose HSV encephalitis. MRI is the neuroimaging study of choice for detection of early changes caused by HSV-1; the study shows edema in the frontal and temporal lobes. The EEG shows diffuse slowing or focal changes in the temporal lobe in the majority of patients. Lumbar puncture often reveals a high opening pressure, glucose within normal limits, and high protein levels in CSF samples (see Table 69-1). Twenty percent of initial CSF studies have results within normal limits (Bader & Littlejohns, 2010). Viral cultures are almost always negative. The polymerase chain reaction (PCR) is the standard test for early diagnosis of HSV-1 encephalitis. PCR identifies the deoxyribonucleic acid (DNA) bands of

HSV-1 in the CSF. The validity of PCR is very high between the 3rd and 10th days after symptom onset.

Medical Management

Antiviral agents, either acyclovir (Zovirax) or ganciclovir (Cytovene), are the medications of choice in the treatment for HSV (Karch, 2012). Early administration of antiviral agents (usually well tolerated) improves the prognosis associated with HSV-1 encephalitis. The mode of action is inhibition of viral DNA replication. To prevent relapse, treatment should continue for up to 3 weeks. Slow IV administration over 1 hour prevents crystallization of the medication in the urine. The usual dose of acyclovir is decreased if the patient has a history of renal insufficiency.

Nursing Management

Assessment of neurologic function is key to monitoring the progression of disease. Comfort measures to reduce headache include dimming the lights, limiting noise and visitors, grouping nursing interventions, and administering analgesic agents. Opioid analgesic medications may mask neurologic symptoms; therefore, they are used cautiously. Seizures and altered LOC require care directed at injury prevention and safety. Nursing care addressing patient and family anxieties is ongoing throughout the illness. Monitoring of blood chemistry test results and urinary output alert the nurse to the presence of renal complications related to antiviral therapy.

Arthropod-Borne Virus Encephalitis

Arthropod-borne viruses, or arboviruses, are maintained in nature through biologic transmission between susceptible vertebrate hosts by blood feeding arthropods (mosquitoes, psychodids, ceratopogonids, and ticks) (CDC, 2009a). Arthropod vectors transmit several types of viruses that cause encephalitis. The main vector in North America is the mosquito. Arbovirus infection (transmitted by arthropod vectors) occurs in specific geographic areas during the summer and fall. In the United States, there are five main arboviral encephalitides: eastern equine encephalitis, western equine encephalitis, St. Louis encephalitis, La Crosse encephalitis, and West Nile virus encephalitis (Bader & Littlejohns, 2010). Powassan virus is one tick-transmitted cause of encephalitis seen in the United States.

Pathophysiology

Viral replication occurs at the site of the mosquito bite. The host immune response attempts to control viral replication. If the immune response is inadequate, viremia will ensue. The virus gains access to the CNS via the olfactory tract, resulting in encephalitis. It spreads from neuron to neuron, predominantly affecting the cortical gray matter, the brain stem, and the thalamus. Meningeal exudates compound the clinical presentation by irritating the meninges and increasing ICP.

Clinical Manifestations

St. Louis and West Nile virus encephalitis most commonly affect adults. Climate, an environment conducive to arthropod proliferation, and human behavior contribute to the occurrence of St. Louis and West Nile encephalitis. An arboviral encephalitis occurs along a continuum with some cases having only flulike symptoms (i.e., headache and fever) but others progressing to specific neurologic manifestations that vary depending on the viral type (CDC, 2009a). A unique clinical feature of St. Louis encephalitis is SIADH with hyponatremia. The incubation period in St. Louis encephalitis is from 5 to 15 days (CDC, 2009b). Onset of symptoms is abrupt with fever, headache, dizziness, nausea, and malaise. If the disease spreads to the CNS, symptoms include stiff neck, confusion, dizziness, and tremors. Coma can occur in severe cases, and mortality increases with age. Eastern equine encephalitis has the highest mortality rate at 30% to 80% (Bader & Littlejohns, 2010). Seizures, a poor prognostic indicator, are present in encephalitis and are more common in the St. Louis type.

Assessment and Diagnostic Findings

Preliminary diagnosis is based on clinical presentation and location and dates of recent travel because certain viruses are endemic to certain geographic locales. Neuroimaging and CSF evaluation are useful in the diagnosis of encephalitis. The MRI scan demonstrates inflammation of the basal ganglia in cases of St. Louis encephalitis and inflammation in the periventricular area in cases of West Nile encephalitis. Immunoglobulin M antibodies to West Nile virus are observed in serum and CSF. Serum cultures are not useful, because the viremia is brief. EEG can identify abnormal brain waves, which can help identify some viral infections.

Medical Management

No specific medication for arboviral encephalitis exists; therefore, symptom management is key. Controlling the seizures and the increased ICP are critical components of care in cases with neurologic manifestations. Interferon may be useful in treating St. Louis encephalitis, although no specific drug is indicated. Neuropsychiatric complications, such as emotional outbursts and other behavior changes, occur frequently.

Nursing Management

Many patients, particularly those with only fever and headache, are treated on an outpatient basis. If the patient is very ill, hospitalization may be required. The nurse carefully assesses neurologic status and identifies improvement or deterioration in the patient's condition. Injury prevention is key in light of the potential for falls or seizures. Any encephalitis that progresses may result in death or lifelong residual health issues such as neurologic deficits and seizures. The family will need support and education to cope with these outcomes.

Public education addressing the prevention of arboviral encephalitis is a key nursing role. Clothing that provides coverage and insect repellents containing 25% to 30% diethyltoluamide (DEET) should be used on exposed clothing and skin in high-risk areas to decrease mosquito and tick bites. Remaining indoors at dawn and dusk when mosquito activity is highest is recommended. Screens should be in good repair in the home, and standing water should be removed (Haley, 2012). All cases of arboviral encephalitis must be reported to the local health department (CDC, 2009a).

Fungal Encephalitis

Fungal infections of the CNS occur rarely in healthy people. The presentation of fungal encephalitis is related to geographic area or to an immune system that is compromised due to disease or immunosuppressive medication. Causes of fungal infections include *Cryptococcus neoformans*, *Blastomyces dermatitidis*, *Histoplasma capsulatum*, *Aspergillus fumigatus*, *Candida*, and *Coccidioides immitis* (Bader & Littlejohns, 2010). *C. immitis* is found mainly in California, Arizona, New Mexico, and Texas. *B. dermatitidis* exists in the southeastern United States and in the Ohio, St. Lawrence, and Mississippi River basins. It is a risk for coal miners, construction workers, and farmers. *C. neoformans* is associated with exposure to bird droppings and may be seen in bird handlers.

Pathophysiology

The fungal spores enter the body via inhalation. They initially infect the lungs, causing vague respiratory symptoms or pneumonitis. The fungi may enter the bloodstream, causing a fungemia. If the fungemia overwhelms the person's immune system, the fungus may spread to the CNS. The fungal invasion may cause meningitis, encephalitis, brain abscess, granuloma, or arterial thrombus.

Clinical Manifestations

The common symptoms of fungal encephalitis include fever, malaise, headache, meningeal signs, and change in LOC or cranial nerve dysfunction. Symptoms of increased ICP related to hydrocephalus often occur. *C. neoformans* and *C. immitis* are associated with specific skin lesions. *H. capsulatum* is associated with seizures, and *A. fumigatus* may cause ischemic or hemorrhagic strokes.

Assessment and Diagnostic Findings

A history of immunosuppression associated with AIDS or the use of immunosuppressive medications may indicate fungal disease of the brain (Theroux, Phipps, Zimmerman, et al., 2013). Occupational and travel history may point to a fungal cause of CNS infection. Infections caused by *H. capsulatum* and *C. immitis* will demonstrate fungal antibodies in serologic tests. The CSF usually demonstrates elevated white cell and protein levels; glucose levels are decreased. *C. neoformans* is easily identified in CSF fungal cultures. *Candida* may be cultured from the blood or CSF. To identify *B. dermatitidis*, cisternal or ventricular cultures of CSF may need to be obtained. *A. fumigatus* is difficult to isolate in CSF and is diagnosed by lung biopsy. Neuroimaging is used to identify CNS changes related to fungal infection. MRI is the study of choice; it demonstrates areas of hemorrhage, abscess, or enhanced meninges indicating inflammation.

Medical Management

Medical management is directed at the causative fungus and the neurologic consequences of the infection. Seizures are controlled by standard antiseizure medications. Increased ICP is controlled by repeated lumbar punctures or shunting of CSF.

Antifungal agents are administered for a specific period to cure the infection in patients with competent immune systems. Patients with compromised immune systems receive antifungal therapy until the infection is controlled, after which they receive a maintenance dose of the medication for an indefinite period. Amphotericin B is used for a progressive fungal infection that does not respond to conventional drug therapy (Karch, 2012). Dosing depends on the causative organism, and it is usually administered IV. Adverse reactions include fever, nausea and vomiting, anemia, uremia, and electrolyte abnormalities. Toxicities may occur, and renal insufficiency is a serious reaction to amphotericin B that can occur.

Fluconazole (Diflucan) or flucytosine (Ancobon) may be administered orally in conjunction with amphotericin B as maintenance therapy. Potential side effects of fluconazole include nausea, vomiting, and a transient increase in liver enzymes. The most common adverse reaction to flucytosine is bone marrow depression (Karch, 2012). Therefore, patients receiving flucytosine should have leukocyte and platelet counts monitored regularly.

Nursing Management

The ICP will increase if hydrocephalus develops and the inflammatory response progresses. Nursing assessment aimed at early identification of increased ICP is necessary to ensure early control and management. (See Chapter 66 for management of the patient with increased ICP.) Administering nonopioid analgesic agents, limiting environmental stimuli, and positioning may optimize patient comfort. Administering diphenhydramine (Benadryl) and acetaminophen (Tylenol) approximately 30 minutes before giving amphotericin B may prevent flulike side effects. If renal insufficiency develops, the dose may need to be reduced. Increasing levels of serum creatinine and blood urea nitrogen may alert the nurse to the development of renal insufficiency and the need to address the patient's renal status.

Providing support assists the patient and family to cope with the illness. Workup of the patient for immunodeficiency diseases such as AIDS may put additional stress on the family. The nurse may need to mobilize community support systems for the patient and family, because the recovery may be long.

Creutzfeldt-Jakob and Variant Creutzfeldt-Jakob Disease

Creutfeldt-Jakob Disease (CJD) and Variant Creutzfeldt-Jakob Disease (vCJD) belong to a group of degenerative, infectious neurologic disorders called *transmissible spongiform encephalopathies* (TSE). CJD is very rare and has no identifiable cause. vCJD, the human variation of bovine spongiform encephalopathy (commonly known as mad cow disease), results from the ingestion by humans of prions in infected meat. TSEs are caused by **prions**, which are particles smaller than a virus that are resistant to standard methods of sterilization (Porth & Matfin, 2009). Although CJD and vCJD have distinct clinical features, one characteristic they share is a lack of CNS inflammation. CJD may lie dormant for decades before causing neurologic degeneration. The incubation period of vCJD seems to be shorter (less than 10 years). In both diseases, the symptoms are progressive, there is no definitive treatment, and the outcome is fatal (Bader & Littlejohns, 2010).

Almost all cases of vCJD have been reported in the United Kingdom, with a smaller number of cases being identified in 10 other countries worldwide (Ironside, 2010). The risk of vCJD in the United States is thought to be low, because cattle are fed primarily with soy-derived feed as opposed to feed containing animal parts.

Pathophysiology

The prion is a unique pathogen because it lacks nucleic acid, which enables the organism to withstand conventional means of sterilization. How the prion replicates in the absence of nucleic acid is unknown (Porth & Matfin, 2009). In both CJD and vCJD, the prion crosses the blood–brain barrier and is deposited in brain tissue and causes degeneration of brain tissue. Cell death occurs, and **spongiform** changes (spongy vacuoles) are produced in the brain. The spongiform vacuoles are surrounded by amyloid plaque.

Approximately 85% of cases of CJD appear sporadically; the incidence is 1 case per 1 million people. Although it is not transmittable by typical human contact, some cases of sporadic CJD have resulted from contaminated neurosurgical instruments, cadaver-derived growth factor, or corneal transplants. Fifteen percent of cases appear to be familial (Bader & Littlejohns, 2010). There have been four confirmed cases of vCJD associated with blood transfusions (Dupiereux, Zorzi, Quadrio, et al., 2009).

The prion exists in lymphoid tissue and blood in both vCJD and CJD. Both prion diseases are believed to be bloodborne. No method is available to screen blood for infectivity. For this reason, the American Red Cross will not accept blood donation from anyone who has traveled to the United Kingdom or Europe for more than 3 to 6 months (Benjamin, 2010).

Clinical Manifestations

CJD and vCJD have several clinically distinct features. Psychiatric symptoms occur early in vCJD, whereas they are a late symptom in CJD. The mean age at onset of vCJD is 27 years, whereas the mean age for CJD onset is 65 years. The presenting symptoms of vCJD include affective symptoms (i.e., behavioral changes), sensory disturbance, and limb pain. Muscle spasms and rigidity, dysarthria, incoordination, cognitive impairment, and sleep disturbances follow. Patients with sporadic CJD present with mental deterioration, ataxia, and visual disturbance. Memory loss, involuntary movement, paralysis, and mutism occur as the disease progresses. After clinical presentation, people with vCJD survive an average of 22 months; those with CJD survive for less than 1 year (Bader & Littlejohns, 2010; Zieve & Eltz, 2011).

Assessment and Diagnostic Findings

Brain biopsy is not recommended to diagnose CJD. The three diagnostic tests currently used in suspicious clinical presentations to support the diagnosis of CJD are immunologic assessment, EEG, and MRI scanning. Immunologic assessment of CSF detects a protein kinase inhibitor referred to as 14–3–3. The presence of this inhibitor indicates neuronal cell death, which is not specific to CJD but does support the diagnosis. The EEG reveals a characteristic pattern over the duration of the disease. After initial slowing, the EEG shows periodic activity. Later in the course of the disease, the EEG shows burst suppressions characterized by periodic spikes alternating with slow periods. The MRI scan demonstrates symmetric or unilateral hyperintense signals arising from the basal ganglia (Bader & Littlejohns, 2010).

Medical Management

After the onset of specific neurologic symptoms, progression of disease occurs quickly. There is no effective treatment for CJD or vCJD. The care of the patient is supportive and palliative. Goals of care include prevention of injury related to immobility and dementia, promotion of patient comfort, and provision of support and education for the family.

Nursing Management

The nursing care of patients is primarily supportive and palliative. Psychological and emotional support of the patient and family throughout the course of the illness is needed. Care extends to providing for a dignified death and supporting the family through the processes of grief and loss. Hospice services are appropriate either at home or at an inpatient facility. (See Chapter 16 for an in-depth discussion of end-of-life issues.)

Prevention of disease transmission is an important part of nursing care. Although patient isolation is not necessary, the use of standard precautions is important. Institutional protocols are followed for blood and body fluid exposure and decontamination of equipment. In the operating room, it is recommended that disposable instruments be used and then incinerated, because conventional methods of sterilization do not destroy the prion (Bader & Littlejohns, 2010; Zieve & Eltz, 2011). The World Health Organization (WHO) has guidelines that outline the stringent sterilization methods that must be used to destroy prions on surfaces.

AUTOIMMUNE PROCESSES

Autoimmune nervous system disorders include multiple sclerosis (MS), myasthenia gravis, and Guillain-Barré syndrome.

Multiple Sclerosis

MS is an immune-mediated, progressive demyelinating disease of the CNS. Demyelination refers to the destruction of myelin—the fatty and protein material that surrounds certain nerve fibers in the brain and spinal cord; it results in impaired transmission of nerve impulses (Fig. 69-2). MS affects nearly 400,000 people in the United States (American Association of Neuroscience Nurses [AANN], 2011). MS may occur at any age, but the age of peak onset is between 25 and 35 years; it affects women more frequently than men (AANN, 2011).

The cause of MS is an area of ongoing research. Autoimmune activity results in demyelination, but the sensitized antigen has not been identified. Multiple factors play a role in the initiation of the immune process. Geographic prevalence is highest in Europe, New Zealand, southern Australia, the northern United States, and southern Canada. MS is less prevalent in Asians. There is a North–South gradient, with greater frequency in temperate regions and latitudes distancing from the equator (Bader & Littlejohns, 2010).

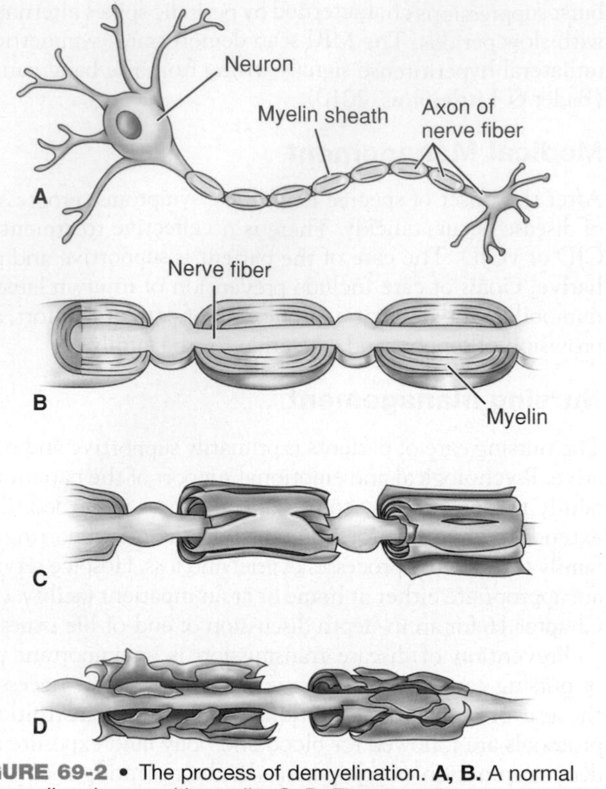
Neuron
Myelin sheath
Axon of
nerve fiber
A
Nerve fiber
B
Myelin
C
D

FIGURE 69-2 • The process of demyelination. **A, B.** A normal nerve cell and axon with myelin. **C, D.** The slow disintegration of myelin, resulting in a disruption in axon function.

MS is considered to have many risks, including genetic factors. However, it has not been found to be genetically transmitted (Bader & Littlejohns, 2010). Genetic predisposition is indicated by the presence of a specific cluster (haplotype) of human leukocyte antigens on the cell wall. Its presence may increase susceptibility to factors, such as viruses, that trigger the autoimmune response activated in MS. A specific virus capable of initiating the autoimmune response has not been identified. It is believed that DNA on the virus mimics the amino acid sequence of myelin, resulting in an immune system cross-reaction in the presence of a defective immune system. Environmental risks include smoking, lack of vitamin D exposure, and exposure to the Epstein-Barr virus (AANN, 2011).

Pathophysiology

Sensitized T- and B lymphocytes cross the blood–brain barrier; their function is to check the CNS for antigens and then leave. In MS, sensitized T cells remain in the CNS and promote the infiltration of other agents that damage the immune system. The immune system attack leads to inflammation that destroys myelin (which insulates the axon and speeds the conduction of impulses along the axon) and the oligodendroglial cells that produce myelin in the CNS.

Demyelination interrupts the flow of nerve impulses and results in a variety of manifestations, depending on the nerves affected. Plaques appear on demyelinated axons, further interrupting the transmission of impulses. Demyelinated axons are scattered irregularly throughout the CNS (Fig. 69-3). The

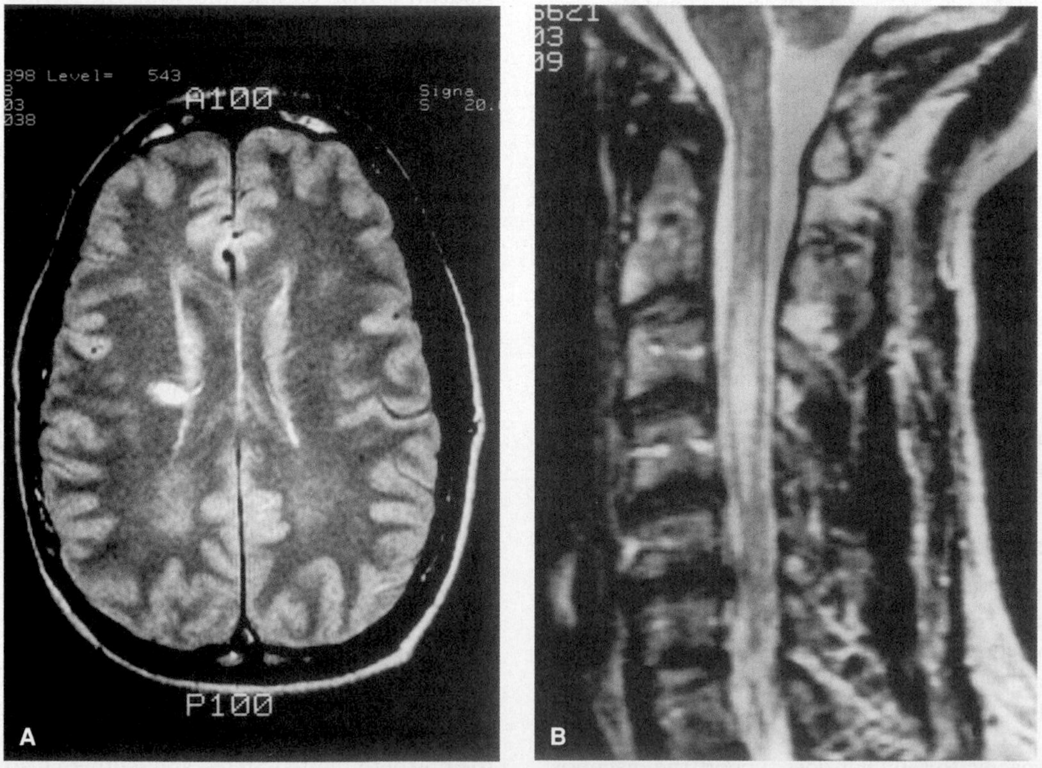
A100
P100
A
B

FIGURE 69-3 • Multiple sclerosis. **A.** A computed tomography scan of the brain demonstrates an area of demyelination in the periventricular white matter of the right frontal lobe. The plaque is perpendicular to the lateral ventricle, which is a typical finding in multiple sclerosis. **B.** A magnetic resonance image of the spinal cord in the same patient highlights another typical finding: a flame-shaped area of demyelination within the midcervical region of the spinal cord. Courtesy of the Danbury Hospital Department of Radiology.

areas most frequently affected are the optic nerves, chiasm, and tracts; the cerebrum; the brain stem and cerebellum; and the spinal cord. The axons themselves begin to degenerate, resulting in permanent and irreversible damage (Porth & Matfin, 2009).

Clinical Manifestations

The course of MS may assume many different patterns (Lublin & Reingold, 1996) (Fig. 69-4). In some patients, the disease follows a benign course, and symptoms are so mild that the patient does not seek health care or treatment. Approximately 85% of patients with MS have a *relapsing-remitting* (RR) course. With each relapse, recovery is usually complete; however, residual deficits may occur and accumulate over time, contributing to functional decline. Over time, most patients with the RR course of MS progress to a *secondary progressive* course, in which disease progression occurs with or without relapses. Approximately 15% of patients have a *primary progressive* course, in which disabling symptoms steadily increase, with rare plateaus and temporary minor improvement. Primary progressive MS may result in quadriparesis, cognitive dysfunction, visual loss, and brain stem syndromes. The least common presentation (about 5% of cases) is the *progressive-relapsing* course. It is characterized by relapses with continuous disabling progression between exacerbations (AANN, 2011).

The signs and symptoms of MS are varied and multiple, reflecting the location of the lesion (plaque) or combination

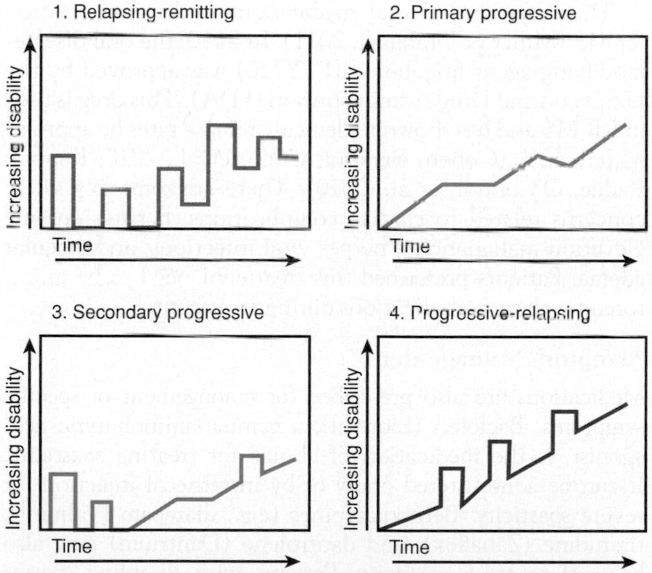
1. Relapsing-remitting
2. Primary progressive
3. Secondary progressive
4. Progressive-relapsing
Increasing disability
Time

FIGURE 69-4 • Types and courses of multiple sclerosis (MS). **1.** Relapsing-remitting (RR) MS is characterized by clearly acute attacks with full recovery or with sequelae and residual deficit upon recovery. Periods between disease relapses are characterized by lack of disease progression. **2.** Primary progressive MS is characterized by disease showing progression of disability from onset, without plateaus and temporary minor improvements. **3.** Secondary progressive MS begins with an initial RR course, followed by progression of variable rate, which may also include occasional relapses and minor remissions. **4.** Progressive-relapsing MS shows progression from onset but with clear acute relapses with or without recovery. From Lublin, F. D., & Reingold, S. C. (1996). Defining the clinical course of multiple sclerosis: Results of an international survey. *Neurology*, *46*(64), 907–911. Used with permission from Lippincott Williams & Wilkins.

of lesions. The main symptoms most commonly reported are fatigue, depression, weakness, numbness, difficulty in coordination, loss of balance, spasticity, and pain (AANN, 2011). Visual disturbances due to lesions in the optic nerves or their connections may include blurring of vision, **diplopia** (double vision), patchy blindness (scotoma), and total blindness.

Fatigue affects most people with MS and is often the most disabling symptom (Plow & Finlayson, 2012). Heat, depression, anemia, deconditioning, and medication may contribute to fatigue. Avoiding hot temperatures, effective treatment of depression and anemia, and occupational and physical therapies may help control fatigue. Additional strategies include a balance of rest and activities, good nutrition to avoid being overweight and obese, and a healthy lifestyle including avoidance of alcohol and cigarette smoking (AANN, 2011; Bader & Littlejohns, 2010).

Pain is another common symptom of MS that can contribute to social isolation. Lesions on the sensory pathways cause pain. Additional sensory manifestations include paresthesias, dysesthesias, and proprioception loss. Many people with MS need daily analgesic medications. Analgesics account for 30% of the medication used in treating MS (Solaro & Uccelli, 2011). In some cases, pain is managed with opioids, antiseizure medications, or antidepressants. Rarely, surgery may be needed to interrupt pain pathways.

Among perimenopausal women, those with MS are more likely to have pain related to osteoporosis. In addition to estrogen loss, immobility and corticosteroid therapy play a role in the development of osteoporosis among women with MS. Bone mineral density testing is recommended for this high-risk group. (See Chapter 42 for a discussion of the diagnosis of and treatment for osteoporosis.)

Spasticity (muscle hypertonicity) occurs in 90% of MS patients, most often in the lower extremities, and can include loss of the abdominal reflexes (AANN, 2011). Spasticity results from involvement of the main motor pathways (pyramidal tracts) of the spinal cord. Cognitive and psychosocial problems may reflect frontal or parietal lobe involvement. Some degree of cognitive change (e.g., memory loss, decreased concentration) occurs in about half of patients, but severe cognitive changes with dementia (progressive organic mental disorder) are rare.

Involvement of the cerebellum or basal ganglia can produce **ataxia** (impaired coordination of movements) and tremor. Loss of the control connections between the cortex and the basal ganglia may occur and cause emotional lability and euphoria. Bladder, bowel, and sexual dysfunctions are common.

Secondary complications of MS include urinary tract infections, constipation, pressure ulcers, contracture deformities, dependent pedal edema, pneumonia, reactive depression, and osteoporosis. Emotional, social, marital, economic, and vocational problems may also occur.

Exacerbations and remissions are characteristic of MS. During exacerbations, new symptoms appear and existing ones worsen; during remissions, symptoms decrease or disappear. Relapses may be associated with emotional and physical stress.

Gerontologic Considerations

The life expectancy for patients with MS is 5 to 7 years shorter than patients without MS (Bader & Littlejohns, 2010).

Those diagnosed with secondary progressive disease live an average of 35 years after onset. Older adult patients with MS have specific physical and psychosocial challenges. They may have chronic health problems, for which they may be taking additional medications that could interact with medications prescribed for MS (Awad & Stüve, 2010). The absorption, distribution, metabolism, and excretion of medications are altered in the older adult as a result of age-related changes in kidney and liver functions. Therefore, older adult patients must be monitored closely for adverse and toxic effects of MS medications and for osteoporosis (particularly with frequent corticosteroid use that may be needed to treat exacerbations). The cost of medications may lead to poor adherence to the prescribed regimen in older adult patients on fixed incomes.

Older adult patients with MS are particularly concerned about increasing disability, family burden, marital concern, and the possible future need for nursing home care. Immobility resulting in fewer social opportunities contributes to loneliness and depression. Along with functional loss, spasticity, pain and bladder dysfunction, impaired sleep, and an increased need for assistance with self-care contribute to the physical challenges experienced by the older adult patient with MS.

Assessment and Diagnostic Findings

The diagnosis of MS is based on the presence of multiple plaques in the CNS observed with MRI (Halper & Holland, 2011). Electrophoresis of CSF identifies the presence of oligoclonal banding (several bands of immunoglobulin G bonded together, indicating an immune system abnormality). Evoked potential studies can help define the extent of the disease process and monitor changes. Underlying bladder dysfunction is diagnosed by urodynamic studies. Neuropsychological testing may be indicated to assess cognitive impairment. A sexual history helps identify changes in sexual function.

Medical Management

No cure exists for MS. An individual treatment program is indicated to relieve the patient's symptoms and provide continuing support, particularly for patients with cognitive changes, who may need more structure and support. The goals of treatment are to delay the progression of the disease, manage chronic symptoms, and treat acute exacerbations. Many patients with MS have a stable disease course and require only intermittent treatment, whereas others experience steady progression of their disease. Symptoms requiring intervention include spasticity, fatigue, bladder dysfunction, and ataxia. Management strategies target the various motor and sensory symptoms and effects of immobility that can occur.

Pharmacologic Therapy

Medications prescribed for MS include those for disease modification and those for symptom management. The disease-modifying therapies available to treat MS include immunomodulating therapies and immunosuppressive agents (Miller & Umhauer, 2011). Multiple medications are used for symptom management in MS.

Disease-Modifying Therapies

The disease-modifying medications reduce the frequency of relapse, the duration of relapse, and the number and size of plaques observed on MRI.

Interferon beta-1a (Rebif) and interferon beta-1b (Betaseron) are administered subcutaneously every other day. Another preparation of interferon beta-1a, Avonex, is administered intramuscularly once a week. Side effects of all interferon-beta medications include flulike symptoms that can be managed with acetaminophen and ibuprofen and resolve after a few months. Additional side effects include potential liver damage, fetal abnormalities, and depression. For optimal control of disability, disease-modifying medications should be started early in the course of the disease (Karch, 2012).

Glatiramer acetate (Copaxone) reduces the rate of relapse in the RR course of MS. Copaxone is administered subcutaneously daily (Karch, 2012). Copaxone is an option for those with an RR course; however, it may take 6 months for evidence of an immune response to appear.

IV methylprednisolone, the key agent in treating acute relapse in the RR course, shortens the duration of relapse but has not been found to have long-term benefit (Bader & Littlejohns, 2010). It exerts anti-inflammatory effects by acting on T cells and cytokines. The medication is administered as 1g IV daily for 3 to 5 days, followed by an oral taper of prednisone. Side effects include mood swings, weight gain, and electrolyte imbalances (Halper & Holland, 2011; Karch, 2012).

The medication mitoxantrone (Novantrone) is administered via IV infusion every 3 months. Mitoxantrone can reduce the frequency of clinical relapses in patients with secondary progressive or worsening RR MS. Patients must be very closely monitored for side effects (i.e., cardiac toxicity), and there is a maximum lifetime dose that can be administered.

There are several oral medications under investigation for MS (Miller & Umhauer, 2011). In 2010, the oral disease-modifying agent fingolimod (FTY720) was approved by the U.S. Food and Drug Administration (FDA). This drug is used in RR MS and has shown to decrease relapse rates by approximately 50% (Cohen, Barkhof, Comi, et al., 2010; Kappos, Radue, O'Connor, et al., 2010). There are some key safety concerns related to cardiac complications, hepatic enzyme elevation malignancies, herpes viral infections, and macular edema. Patients prescribed this treatment need to be monitored for these complications during treatment.

Symptom Management

Medications are also prescribed for management of specific symptoms. Baclofen (Lioresal), a gamma-aminobutyric acid agonist, is the medication of choice for treating spasticity. It can be administered orally or by intrathecal injection for severe spasticity. Benzodiazepines (e.g., diazepam [Valium]), tizanidine (Zanaflex), and dantrolene (Dantrium) may also be used to treat spasticity. Patients with disabling spasms and contractures may require nerve blocks or surgical intervention. Fatigue that interferes with activities of daily living (ADLs) may be treated with amantadine (Symmetrel), pemoline (Cylert), or dalfampridine (Ampyra). Ataxia is a chronic problem most resistant to treatment. Medications used to treat ataxia include beta-adrenergic blockers (e.g., propranolol [Inderal]), the antiseizure agent gabapentin (Neurontin), and benzodiazepines (e.g., clonazepam [Klonopin]) (Halper & Holland, 2011).

Bladder and bowel problems are often among the most difficult ones for patients, and a variety of medications (anticholinergic agents, alpha-adrenergic blockers, antispasmodic agents) may be prescribed. Nonpharmacologic strategies also assist in establishing effective bowel and bladder elimination (see later discussion).

Urinary tract infection is often superimposed on the underlying neurologic dysfunction. Ascorbic acid (vitamin C) may be prescribed to acidify the urine, making bacterial growth less likely. Antibiotic agents are prescribed when appropriate.

NURSING PROCESS

The Patient With Multiple Sclerosis

Assessment

Nursing assessment addresses neurologic deficits, secondary complications, and the impact of the disease on the patient and family. The patient's mobility and balance are observed to determine whether there is risk of falling. Assessment of function is carried out both when the patient is well rested and when fatigued. The patient is assessed for weakness, spasticity, visual impairment, incontinence, and disorders of swallowing and speech. Additional areas of assessment include how MS has affected the patient's lifestyle, how the patient is coping, adherence to the prescribed medication regimen, and what the patient would like to improve.

Diagnosis

Nursing Diagnoses

Based on the assessment data, major nursing diagnoses may include the following:

- Impaired bed and physical mobility related to weakness, muscle paresis, spasticity, increased weight
- Risk for injury related to sensory and visual impairment
- Impaired urinary and bowel elimination (urgency, frequency, incontinence, constipation) related to nervous system dysfunction
- Impaired verbal communication and risk for aspiration related to cranial nerve involvement
- Chronic confusion related to cerebral dysfunction
- Ineffective individual coping related to uncertainty of course of MS
- Impaired home maintenance management related to physical, psychological, and social limits imposed by MS
- Potential for sexual dysfunction related to lesions or psychological reaction

Planning and Goals

The major goals for the patient may include promotion of physical mobility, avoidance of injury, achievement of bladder and bowel continence, promotion of speech and swallowing mechanisms, improvement of cognitive function, development of coping strengths, improved home maintenance management, and adaptation to sexual dysfunction.

Nursing Interventions

An individualized program of physical, occupational, and speech-language therapy, rehabilitation, and education is combined with emotional support. An educational plan of care is developed to enable the person with MS to deal with the physiologic, social, and psychological problems that accompany chronic disease. The presence of depression, pain, fatigue, and walking difficulty all decrease physical activity. Assisting patients with management of these symptoms may help increase the level of physical activity and overall sense of well-being (Halper & Holland, 2011).

Promoting Physical Mobility

Relaxation and coordination exercises promote muscle efficiency. Progressive resistive exercises are used to strengthen weak muscles, because diminishing muscle strength is often significant in MS.

Exercises. Walking improves the gait, particularly the problem of loss of position sense of the legs and feet. If certain muscle groups are irreversibly affected, other muscles can be trained to compensate. Instruction in the use of assistive devices may be needed to ensure their safe and correct use.

Minimizing Spasticity and Contractures. Muscle spasticity is common and, in its later stages, is characterized by severe adductor spasm of the hips with flexor spasm of the hips and knees. Without relief, fibrous contractures of these joints occur. Warm packs may be beneficial, but hot baths should be avoided because of risk of burn injury secondary to sensory loss and increasing symptoms that may occur with elevation of the body temperature.

Daily exercises for muscle stretching are prescribed to minimize joint contractures. Special attention is given to the hamstrings, gastrocnemius muscles, hip adductors, biceps, and wrist and finger flexors. Muscle spasticity is common and interferes with usual function. Application of prescribed orthotics may help maintain a functional position and reduce contractures. A stretch–hold–relax routine is helpful for relaxing and treating muscle spasticity. Swimming and stationary bicycling are useful, and progressive weight bearing can relieve spasticity in the legs. The patient should not be hurried in any of these activities, because this often increases spasticity.

Activity and Rest. The patient is encouraged to work and exercise to a point just short of fatigue. Very strenuous physical exercise is not advisable, because it raises the body temperature and may aggravate symptoms. The patient is advised to take frequent short rest periods, preferably lying down. Extreme fatigue may contribute to the exacerbation of symptoms.

Nutrition. Approximately half of all people with MS are overweight or obese (Marrie, Horwitz, Cutter, et al., 2011). Contributing factors include the use of corticosteroids for exacerbations of symptoms and mobility impairments as a result of the disease. Research suggests that people with MS have unique challenges to participating in nutritional behaviors (Chart 69-3). Interventions to promote healthy eating and weight reductions need to take into account that fatigue and mobility impairments are barriers to engagement in nutritional behaviors for people with MS. Nurses also need to be certain to include family members in interventions and nutrition education, because they are often the gatekeepers for food preparation and selection (Dudek, 2010; Plow & Finlayson, 2012).

Minimizing Effects of Immobility. Because of the decrease in physical activity that often occurs with MS, complications associated with immobility, including pressure ulcers, expiratory muscle weakness, and accumulation of bronchial

Chart 69-3

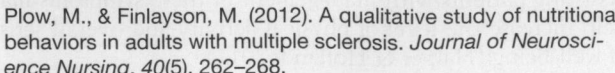

NURSING RESEARCH PROFILE

Nutritional Behaviors in People With Multiple Sclerosis

Plow, M., & Finlayson, M. (2012). A qualitative study of nutritional behaviors in adults with multiple sclerosis. *Journal of Neuroscience Nursing*, *40*(5), 262–268.

Purpose

Healthy eating habits are a challenge for people with multiple sclerosis (MS) because many are overweight or obese. This study explored how people with MS cope with impairment and disability to participant in nutritional behaviors.

Design

In this qualitative study, 8 people (6 women and 2 men) with MS who used a mobility aide participated in the 45-minute face-to-face interview. The purpose was to explore the subjective experience of engaging in nutritional behaviors defined as food selection and preparation as well as going to restaurants and grocery stores.

Findings

This study found 4 important themes. Overall themes were "it is a lot of work," "it is not up to me," sifting through nutritional information, and "why I eat what I eat." Participants reported fatigue and mobility impairments as barriers to engagement in nutritional behaviors and family members as the gatekeepers for food preparation and selection.

Nursing Implications

People with MS have unique challenges related to healthy eating. Nursing interventions to promote healthy eating need to take into account that fatigue and mobility impairments are barriers to engagement in nutritional behaviors for people with MS. Nurses also need to make every effort to include family members in interventions and nutrition education because they are often the gatekeepers for food preparation and selection.

secretions, need to be considered and steps taken to prevent them. Measures to prevent such complications include assessing and maintaining skin integrity and having the patient perform coughing and deep-breathing exercises.

Preventing Injury

If motor dysfunction causes problems of incoordination and clumsiness, or if ataxia is apparent, then the patient is at risk for falling. To overcome this disability, the patient is instructed to walk with feet apart to widen the base of support and to increase walking stability. If loss of position sense occurs, the patient is instructed to watch the feet while walking. Gait training may require assistive devices (walker, cane, braces, crutches, parallel bars) and instruction about their use by a physical therapist. If the gait remains inefficient, a wheelchair or motorized scooter may be the solution. The occupational therapist is a valuable resource person in suggesting and securing aids to promote independence. If incoordination is a problem and tremor of the upper extremities occurs when voluntary movement is attempted (intention tremor), weighted bracelets or wrist cuffs are helpful. The patient is trained in transfer and ADLs.

Because sensory loss may occur in addition to motor loss, pressure ulcers are a continuing threat to skin integrity. The need to use a wheelchair continuously increases the risk. (See Chapter 10 for a discussion of the prevention and treatment of pressure ulcers.)

Enhancing Bladder and Bowel Control

Generally, bladder symptoms fall into the following categories: (1) inability to store urine (hyperreflexic, uninhibited); (2) inability to empty the bladder (hyporeflexic, hypotonic); and (3) a mixture of both types. The patient with urinary frequency, urgency, or incontinence requires special support. The sensation of the need to void must be heeded immediately, so the bedpan or urinal should be readily available. A voiding time schedule is set up (every 1.5 to 2 hours initially, with gradual lengthening of the interval). The patient is instructed to drink a measured amount of fluid every 2 hours and then attempt to void 30 minutes after drinking. The use of a timer or wristwatch with an alarm may be helpful for the patient who does not have enough sensation to signal the need to empty the bladder. The nurse encourages the patient to take the prescribed medications to treat bladder spasticity, because this allows greater independence. Intermittent self-catheterization (see Chapter 10) has been successful in maintaining bladder control in patients with MS. If a female patient has permanent urinary incontinence, urinary diversion procedures may be considered. The male patient may wear a condom appliance for urine collection.

Bowel problems include constipation, fecal impaction, and incontinence. Adequate fluids, dietary fiber, and a bowel training program are frequently effective in solving these problems. (See Chapter 10 for a discussion of promoting bowel continence.)

Enhancing Communication and Managing Swallowing Difficulties

If the cranial nerves that control the mechanisms of speech and swallowing are affected, dysarthrias (defects of articulation) marked by slurring, low volume of speech, and difficulties in phonation may occur. **Dysphagia** (difficulty swallowing) may also occur. A speech-language pathologist evaluates speech and swallowing and instructs the patient, family, and health team members about strategies to compensate for speech and swallowing problems. The nurse reinforces this instruction and encourages the patient and family to adhere to the plan.

Quality and Safety Nursing Alert

Impaired swallowing increases the patient's risk of aspiration. To reduce that risk, the nurse implements strategies such as having suction apparatus available, ensuring careful feeding, confirming correct food and liquid consistency, and positioning properly for eating.

Improving Cognitive Function

Measures may be taken if visual defects or changes in cognitive status occur and result in chronic confusion.

Vision. The cranial nerves affecting vision may be affected by MS. An eye patch or a covered eyeglass lens may be used to block the visual impulses of one eye if the patient has diplopia.

Prism glasses may be helpful for patients who are confined to bed and have difficulty reading in the supine position. People who are unable to read regular-print materials are eligible for the free "talking book" services of the Library of Congress or may obtain large-print or audio books from local libraries.

Cognition and Emotional Responses. Cognitive impairment and emotional lability occur early in MS in some patients and may impose numerous stresses on the patient and family. Some patients with MS are forgetful and easily distracted and may exhibit emotional lability.

Patients adapt to illness in a variety of ways, including denial, depression, withdrawal, and hostility. Emotional support assists patients and their families to adapt to the changes and uncertainties associated with MS and to cope with the disruption in their lives. The family should be made aware of the nature and degree of cognitive impairment. Patients with MS have identified that the support of family and friends is a major need. The patient is assisted to set meaningful and realistic goals, to remain as active as possible, and to maintain interests and activities. Hobbies may help the patient's morale and provide satisfying interests if the disease progresses to the stage in which formerly enjoyed activities can no longer be pursued. The environment is kept structured, distractions are reduced, and lists and other memory aids are used to help the patient with cognitive changes maintain a daily routine. The occupational therapist can be helpful in formulating a structured daily routine.

Strengthening Coping Mechanisms

The diagnosis of MS is always distressing to the patient and family. They need to know that no two patients with MS have identical symptoms or courses of illness. Although some patients do experience significant disability early, others have a near-normal lifespan with minimal disability. Some families, however, face overwhelming frustrations and problems. MS affects people who are often in a productive stage of life and concerned about career and family responsibilities. Family conflict, disintegration, separation, and divorce are not uncommon. Often, very young family members assume the responsibility of caring for a parent with MS. Nursing interventions in this area include alleviating stress and making appropriate referrals for counseling and support to minimize the adverse effects of dealing with chronic illness.

The nurse, mindful of these complex problems, initiates home care and coordinates a network of services, including social services, speech therapy, physical therapy, and homemaker services. To strengthen the patient's coping skills, as much information as possible is provided. Patients need an updated list of available assistive devices, services, and resources.

Coping through problem solving involves helping the patient define the problem and develop alternatives for its management. Careful planning and maintaining flexibility and a hopeful attitude are useful for psychological and physical adaptation (Halper & Holland, 2011).

Improving Home Management

MS can affect every facet of daily living. Certain abilities are often impossible to regain after they are lost. Physical function may vary from day to day. Modifications that allow independence in home management should be implemented (e.g., assistive eating devices, raised toilet seat, bathing aids, telephone modifications, long-handled comb, tongs, modified clothing). Exposure to heat increases fatigue and muscle weakness, so air-conditioning is recommended in at least one room. Exposure to extreme cold may increase spasticity.

Promoting Sexual Functioning

Patients with MS and their partners face problems that interfere with sexual activity, both as a direct consequence of nerve damage and also from psychological reactions to the disease. Easy fatigability, conflicts arising from dependency and depression, emotional lability, and loss of self-esteem compound the problem. Erectile and ejaculatory disorders in men and orgasmic dysfunction and adductor spasms of the thigh muscles in women can make sexual intercourse difficult or impossible. Bladder and bowel incontinence and urinary tract infections add to the difficulties.

Collaboration between the patient, family, and primary provider is essential for supporting intimacy. A sexual counselor can help bring into focus the patient's or partner's sexual resources and suggest relevant information and supportive therapy. Sharing and communicating feelings, planning for sexual activity (to minimize the effects of fatigue), and exploring alternative methods of sexual expression may open up a wide range of sexual enjoyment and experiences.

Promoting Home and Community-Based Care

Educating Patients About Self-Care. As the disease progresses, the patient and family need to learn new strategies to maintain optimal independence. Educating about self-care techniques may be initiated in the hospital or clinic setting and reinforced in the home. Self-care education may address the use of assistive devices, self-catheterization, and administration of medications that affect the course of the disease or treat complications. An education plan that addresses intramuscular or subcutaneous administration of medications is developed for the patient and his or her family. The patient and family may be educated about exercises that enable the patient to continue some form of activity or that maintain or improve swallowing, speech, or respiratory function (Chart 69-4).

Continuing Care. After discharge, the home care nurse often provides education and reinforcement of new interventions in the patient's home. Nurses in the home setting assess for changes in the patient's physical and emotional status; provide physical care to the patient if required; coordinate outpatient services and resources; and encourage health promotion, appropriate health screenings, and adaptation. If changes in the disease or its course are noted, the home care nurse encourages the patient to contact the primary provider, because treatment of an acute exacerbation or new problem may be indicated. Continuing health care and follow-up are recommended.

The patient with MS is encouraged to contact the local chapter of the National Multiple Sclerosis Society for services, publications, and contact with others who have MS (see the Resources section). Local chapters also provide direct services to patients. Through group participation, the patient has an opportunity to meet others with similar problems, share experiences, and learn self-help methods in a social environment.

Chart 69-4 HOME CARE CHECKLIST
The Patient With Multiple Sclerosis

At the completion of the home care education, the patient or caregiver will be able to:	PATIENT	CAREGIVER
• State how to access the local chapter of the National Multiple Sclerosis (MS) Society and available resources.	✔	✔
• Discuss the clinical course of MS.	✔	✔
• Identify strategies to manage symptoms (pain, cognitive responses, dysphagia, tremors, visual disturbances).	✔	✔
• State how to prevent complications (pressure ulcers, pneumonia, depression).	✔	✔
• Identify coping strategies.	✔	✔
• Identify ways to minimize fatigue.	✔	✔
• Explain how to prevent injury.	✔	✔
• State ways to promote sexual function.	✔	✔
• Discuss ways to manage bowel and bladder function.	✔	✔
• Name benefits of exercise and physical activity.	✔	✔
• Identify ways to minimize immobility and spasticity.	✔	✔
• Describe medication regimen and potential adverse effects.	✔	✔
• Demonstrate correct techniques of administering injectable medications, if prescribed.	✔	✔

Evaluation

Expected patient outcomes may include:

1. Improves physical mobility
   a. Participates in gait training and rehabilitation program
   b. Establishes a balanced program of rest and exercise
   c. Uses assistive devices correctly and safely
2. Is free of injury
   a. Uses visual cues to compensate for decreased sense of touch or position
   b. Asks for assistance when necessary
3. Attains or maintains control of bladder and bowel patterns
   a. Monitors self for urine retention and employs intermittent self-catheterization technique, if indicated
   b. Identifies the signs and symptoms of urinary tract infection
   c. Maintains adequate fluid and fiber intake
   d. Identifies the foods that are constipating versus food that increase gastric motility
4. Participates in strategies to improve speech and swallowing
   a. Practices exercises recommended by speech therapist
   b. Maintains adequate nutritional intake without aspiration
5. Compensates for chronic confusion
   a. Uses lists and other aids to compensate for memory losses
   b. Discusses problems with trusted advisor or friend
   c. Substitutes new activities for those that are no longer possible
6. Demonstrates effective coping strategies
   a. Maintains sense of control
   b. Modifies lifestyle to fit goals and limitations
   c. Verbalizes desire to pursue goals and developmental tasks of adulthood
7. Adheres to plan for home maintenance management
   a. Uses appropriate techniques to maintain independence
   b. Engages in health promotion activities and health screenings as appropriate
8. Adapts to changes in sexual function
   a. Is able to discuss problem with partner and appropriate health professional
   b. Identifies alternative means of sexual expression

Myasthenia Gravis

Myasthenia gravis, an autoimmune disorder affecting the myoneural junction, is characterized by varying degrees of weakness of the voluntary muscles. It affects approximately 20 in 100,00 people in the United States. More women than men are affected, commonly in the second and third decades of life; however, after age 50, the gender distribution is more equal (AANN, 2013; Bader & Littlejohns, 2010).

Pathophysiology

Normally, a chemical impulse precipitates the release of acetylcholine from vesicles on the nerve terminal at the myoneural junction. The acetylcholine attaches to receptor sites on the motor endplate and stimulates muscle contraction. Continuous binding of acetylcholine to the receptor site is required for muscular contraction to be sustained.

In myasthenia gravis, antibodies directed at the acetylcholine receptor sites impair transmission of impulses across the myoneural junction. Therefore, fewer receptors are available for stimulation, resulting in voluntary muscle weakness

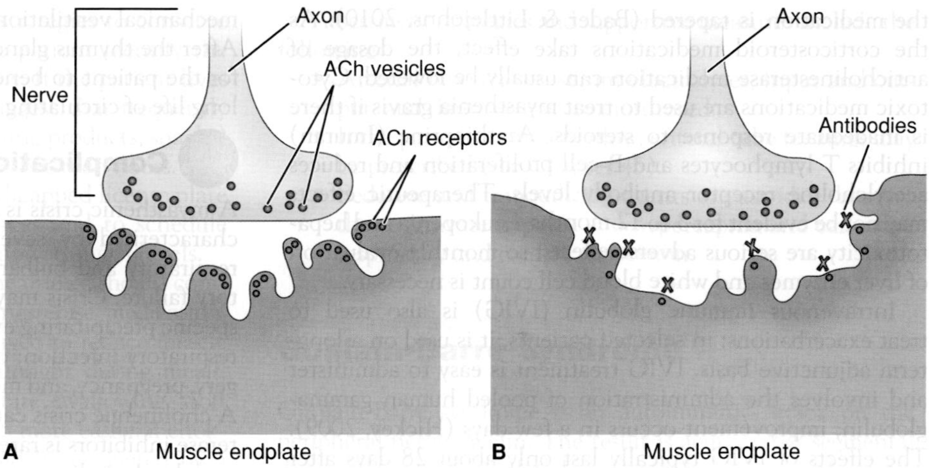

Axon
ACh vesicles
Nerve
ACh receptors
A
Muscle endplate
Axon
Antibodies
B
Muscle endplate

FIGURE 69-5 • Myasthenia gravis. **A.** Usual acetylcholine (ACh) receptor site. **B.** ACh receptor site in myasthenia gravis.

that escalates with continued activity (Fig. 69-5). These antibodies are found in 80% to 90% of people with myasthenia gravis (Hickey, 2009). Of people with myasthenia gravis, 75% have either thymic hyperplasia or a thymic tumor, and the thymus gland is believed to be the site of antibody production (Bader & Littlejohns, 2010). In patients who are antibody negative, researchers believe that the offending antibody is directed at a portion of the receptor site rather than the whole complex (AANN, 2013).

Clinical Manifestations

The initial manifestation of myasthenia gravis in 80% of patients involves the ocular muscles. Diplopia and **ptosis** (drooping of the eyelids) are common (Bader & Littlejohns, 2010). Many patients also experience weakness of the muscles of the face and throat (bulbar symptoms) and generalized weakness. Weakness of the facial muscles results in a bland facial expression. Laryngeal involvement produces **dysphonia** (voice impairment) and dysphagia, which increases the risk of choking and aspiration. Generalized weakness affects all extremities and the intercostal muscles, resulting in decreasing vital capacity and respiratory failure. Myasthenia gravis is purely a motor disorder with no effect on sensation or coordination.

Assessment and Diagnostic Findings

An acetylcholinesterase inhibitor test, called the *Tensilon test,* is used to diagnose myasthenia gravis. The acetylcholinesterase inhibitor stops the breakdown of acetylcholine, thereby increasing availability at the neuromuscular junction. Edrophonium chloride (Tensilon), a fast-acting acetylcholinesterase inhibitor, is administered IV to diagnose myasthenia gravis. Thirty seconds after injection, facial muscle weakness and ptosis should resolve for about 5 minutes (Hickey, 2009). Immediate improvement in muscle strength after administration of this agent represents a positive test and usually confirms the diagnosis. Atropine should be available to control the side effects of edrophonium, which include bradycardia, sweating, and cramping.

The ice test is a diagnostic test with comparable sensitivity to the Tensilon test. The ice test is indicated in patients who have cardiac conditions or asthma that may contraindicate the use of edrophonium. With the ice test, an ice pack is held over the eyes of the patient for 1 minute; the ptosis should temporarily resolve in a patient with myasthenia gravis (Browning, Wallace, Chana, et al., 2011).

The presence of acetylcholine receptor antibodies is identified in the serum (Bader & Littlejohns, 2010). Repetitive muscle stimulation demonstrates a decrease in successive action potentials. The thymus gland, a site of acetylcholine receptor antibody production, may be enlarged in myasthenia gravis and may be identified by MRI scan. A single-fiber electromyography (EMG) detects a delay or failure of neuromuscular transmission and is about 99% sensitive in confirming the diagnosis of myasthenia gravis (Hickey, 2009; Karpoff & Labus, 2008).

Medical Management

Management of myasthenia gravis is directed at improving function and reducing and removing circulating antibodies. Therapeutic modalities include administration of anticholinesterase medications and immunosuppressive therapy, plasmapheresis, and thymectomy. There is no cure for myasthenia gravis; treatments do not stop the production of the acetylcholine receptor antibodies.

Pharmacologic Therapy

Pyridostigmine bromide (Mestinon), an anticholinesterase medication, is the first line of therapy. It provides symptomatic relief by inhibiting the breakdown of acetylcholine and increasing the relative concentration of available acetylcholine at the neuromuscular junction. The dosage is gradually increased to a daily maximum and is administered in divided doses (usually four times a day). Adverse effects of anticholinesterase medications include fasciculations, abdominal pain, diarrhea, and increased oropharyngeal secretions (Karch, 2012). Pyridostigmine tends to have fewer side effects than other anticholinesterase medications.

If pyridostigmine bromide does not improve muscle strength and control fatigue, the next agents used are the immunomodulating drugs. The goal of immunosuppressive therapy is to reduce production of the antibody. Corticosteroids suppress the patient's immune response, decreasing the amount of antibody production, and this correlates with clinical improvement. An initial dose of prednisone is given daily and maintained for 1 to 2 months; as symptoms improve,

inflammation and destruction, interruption of nerve conduction, and axonal loss (Ho, Thakur, Gorson, et al., 2008).

Clinical Manifestations

Guillain-Barré syndrome typically begins with muscle weakness and diminished reflexes of the lower extremities. Hyporeflexia and weakness may progress to tetraplegia. Demyelination of the nerves that innervate the diaphragm and intercostal muscles results in neuromuscular respiratory failure. Sensory symptoms include paresthesias of the hands and feet and pain related to the demyelination of sensory fibers.

The antecedent event usually occurs 1 to 3 weeks before symptoms begin. Weakness usually begins in the legs and may progresses upward. Maximum weakness (the plateau) varies in length but usually includes neuromuscular respiratory failure and bulbar weakness. Guillain-Barré syndrome progresses to peak severity typically within 2 weeks and no longer than 4 weeks. If progression is longer, then the patient is classified as having chronic inflammatory demyelinating polyneuropathy (Hardy et al., 2011). Any residual symptoms are permanent and reflect axonal damage from demyelination.

Cranial nerve demyelination can result in a variety of clinical manifestations. Optic nerve demyelination may result in blindness. Bulbar muscle weakness related to demyelination of the glossopharyngeal and vagus nerves results in the inability to swallow or clear secretions. Vagus nerve demyelination results in autonomic dysfunction, manifested by instability of the cardiovascular system. The presentation is variable and may include tachycardia, bradycardia, hypertension, or orthostatic hypotension. The symptoms of autonomic dysfunction occur and resolve rapidly. Guillain-Barré syndrome does not affect cognitive function or LOC.

Although the classic clinical features include areflexia and ascending weakness, variation in presentation occurs. There may be a sensory presentation, with progressive sensory symptoms; an atypical axonal destruction; or the Miller-Fisher variant, which includes paralysis of the ocular muscles, ataxia, and areflexia.

Assessment and Diagnostic Findings

The patient presents with symmetric weakness, diminished reflexes, and upward progression of motor weakness. A history of a viral illness in the previous few weeks suggests the diagnosis. Changes in vital capacity and negative inspiratory force are assessed to identify impending neuromuscular respiratory failure. Serum laboratory tests are not useful in the diagnosis. However, elevated protein levels are detected in CSF evaluation, without an increase in other cells. Evoked potential studies demonstrate a progressive loss of nerve conduction velocity.

Medical Management

Because of the possibility of rapid progression and neuromuscular respiratory failure, Guillain-Barré syndrome is a medical emergency that may require management in an intensive care unit. After baseline values are identified, assessment of changes in muscle strength and respiratory function alert the clinician to the physical and respiratory needs of the patient. Respiratory therapy or mechanical ventilation may be necessary to support pulmonary function and adequate oxygenation. Some clinicians recommend elective intubation before the onset of extreme respiratory muscle fatigue. Emergent intubation may result in autonomic dysfunction, and mechanical ventilation may be required for an extended period. One study found the lack of foot flexion at the end of the therapy predicted a prolonged duration of mechanical ventilation (Fourrier, Robriquet, Hurtevent, et al., 2011). The patient is weaned from mechanical ventilation after the respiratory muscles can again support spontaneous respiration and maintain adequate tissue oxygenation.

Other interventions are aimed at preventing the complications of immobility. These may include the use of anticoagulant agents and sequential compression boots to prevent thrombosis and PE.

Plasmapheresis and IVIG are used to directly affect the peripheral nerve myelin antibody level. Both therapies decrease circulating antibody levels and reduce the amount of time the patient is immobilized and dependent on mechanical ventilation. Studies indicate that IVIG and plasmapheresis are equally effective in treating Guillain-Barré syndrome; however, IVIG is the therapy of choice because it is associated with fewer side effects when administered appropriately (Waterhouse, 2011). The cardiovascular risks posed by autonomic dysfunction require continuous electrocardiographic (ECG) monitoring. Tachycardia and hypertension are treated with short-acting medications such as alpha-adrenergic blocking agents. The use of short-acting agents is important, because autonomic dysfunction is very labile. Hypotension is managed by increasing the amount of IV fluid administered.

NURSING PROCESS

The Patient With Guillain-Barré Syndrome

Assessment

Ongoing assessment for disease progression is critical. The patient is monitored for life-threatening complications (respiratory failure, cardiac dysrhythmias, venous thromboembolism [VTE] including deep vein thrombosis [DVT] or PE) so that appropriate interventions can be initiated. Because of the threat to the patient in this sudden, potentially life-threatening disease, the nurse must assess the patient's and family's ability to cope and their use of coping strategies.

Diagnosis

Nursing Diagnoses

Based on the assessment data, major nursing diagnoses may include the following:

- Ineffective breathing pattern and impaired gas exchange related to rapidly progressive weakness and impending respiratory failure
- Impaired bed and physical mobility related to paralysis
- Imbalanced nutrition: less than body requirements related to inability to swallow
- Impaired verbal communication related to cranial nerve dysfunction
- Fear and anxiety related to loss of control and paralysis

Collaborative Problems/Potential Complications

Potential complications may include the following:

- Respiratory failure
- Autonomic dysfunction

Planning and Goals

The major goals for the patient may include improved respiratory function, increased mobility, improved nutritional status, effective communication, decreased fear and anxiety, and absence of complications.

Nursing Interventions

Maintaining Respiratory Function

Respiratory function can be maximized with incentive spirometry and chest physiotherapy. Monitoring for changes in vital capacity and negative inspiratory force is key to early intervention for neuromuscular respiratory failure. Mechanical ventilation is required if the vital capacity falls, making spontaneous breathing impossible and tissue oxygenation inadequate.

The potential need for mechanical ventilation should be discussed with the patient and family on admission to provide time for psychological preparation and decision making. Intubation and mechanical ventilation result in less anxiety if they are initiated on a nonemergency basis to a well-informed patient. The patient may require mechanical ventilation for a long period. (See Chapter 21 for the nursing management of the patient requiring mechanical ventilation.)

Bulbar weakness that impairs the ability to swallow and clear secretions is another factor in the development of respiratory failure in the patient with Guillain-Barré syndrome. Suctioning may be needed to maintain a clear airway.

The nurse assesses the blood pressure and heart rate frequently to identify autonomic dysfunction so that interventions can be initiated quickly if needed. Medications are administered or a temporary pacemaker placed for clinically significant bradycardia.

Enhancing Physical Mobility

Nursing interventions to enhance physical mobility and prevent the complications of immobility are key to the function and survival of patients. The paralyzed extremities are supported in functional positions, and passive range-of-motion exercises are performed at least twice daily. DVT and PE are threats to the patient who is paralyzed. Nursing interventions are aimed at preventing VTE. Range-of-motion exercises, position changes, anticoagulation, the use of anti-embolism stockings and sequential compression boots, and adequate hydration decrease the risk of VTE.

Padding may be placed over bony prominences, such as the elbows and heels, to reduce the risk of pressure ulcers. The need for consistent position changes every 2 hours cannot be overemphasized. The nurse evaluates laboratory test results that may indicate malnutrition or dehydration, both of which increase the risk of pressure ulcers and decrease mobility. The nurse collaborates with the physician and dietitian to develop a plan to meet the patient's nutritional and hydration needs.

Providing Adequate Nutrition

Paralytic ileus may result from insufficient parasympathetic activity. In this event, the nurse administers IV fluids and parenteral nutrition as a supplement and monitors for the return of bowel sounds. If the patient cannot swallow because of **bulbar paralysis** (immobility of muscles), a gastrostomy tube may be placed to administer nutrients. The nurse carefully assesses the return of the gag reflex and bowel sounds before resuming oral nutrition.

Improving Communication

Because of paralysis, the patient cannot talk, laugh, or cry and therefore has no method for communicating needs or expressing emotion. Although the patient may be unable to speak, cognition is completely intact. Establishing some form of communication with picture cards or an eye blink system provides a means of communication. Collaboration with the speech therapist may be helpful in developing a communication mechanism that is most effective for a specific patient.

Decreasing Fear and Anxiety

The patient and family are faced with a sudden, potentially life-threatening disease, and anxiety and fear are constant themes for them. The impact of disease on the family depends on the patient's role within the family. Referral to a support group may provide information and support to the patient and family.

The family may feel helpless in caring for the patient. Mechanical ventilation and monitoring devices may frighten and intimidate them. Family members often want to participate in physical care; with instruction and support by the nurse, they should be allowed and encouraged to do so.

In addition to fear, the patient may experience isolation, loneliness, and lack of control. Nursing interventions that increase the patient's sense of control include providing information about the condition, emphasizing a positive appraisal of coping resources, and providing instruction about relaxation exercises and distraction techniques. The positive attitude and atmosphere of the multidisciplinary team are important to promote a sense of well-being.

Diversional activities are encouraged to decrease loneliness and isolation. Encouraging visitors, engaging visitors or volunteers to read to the patient, listening to music or books on tape, and watching television are ways to alleviate the patient's sense of isolation.

Monitoring and Managing Potential Complications

Thorough assessment of respiratory function at regular and frequent intervals is essential, because respiratory insufficiency and subsequent failure due to weakness or paralysis of the intercostal muscles and diaphragm may develop quickly. Respiratory failure is the major cause of mortality. In addition to the respiratory rate and the quality of respirations, vital capacity is monitored frequently and at regular intervals so that respiratory insufficiency can be anticipated. Decreasing vital capacity with associated muscle weakness indicates impending respiratory failure. Signs and symptoms include breathlessness while speaking, shallow and irregular breathing, the use of accessory muscles, tachycardia, weak cough, and changes in respiratory pattern.

Other complications include cardiac dysrhythmias (which necessitate ECG monitoring), transient hypertension, orthostatic hypotension, DVT, PE, urinary retention, and other threats to any patient who is immobilized and paralyzed. These complications require monitoring and attention to prevent them and prompt treatment if indicated.

Promoting Home and Community-Based Care

Educating Patients About Self-Care. Patients with Guillain-Barré syndrome and their families are usually frightened by the sudden onset of life-threatening symptoms and their severity. Therefore, educating the patient and family about the disorder and its generally favorable prognosis is important (Chart 69-5).

Chart 69-5 HOME CARE CHECKLIST
The Patient With Guillain-Barré Syndrome

At the completion of the home care education, the patient or caregiver will be able to:	PATIENT	CAREGIVER
• Describe the disease process of Guillain-Barré syndrome.	✔	✔
• Manage respiratory needs—tracheostomy care, suctioning.		✔
• Demonstrate proper body mechanics regarding lifting and transfers.		✔
• Practice gait training and strength endurance.	✔	✔
• Perform range-of-motion exercises.	✔	✔
• Perform activities of daily living and manage self-care:		
• Nutrition	✔	✔
• Bowel and bladder management	✔	✔
• Skin care	✔	✔
• Adaptive equipment for bathing, hygiene, grooming, dressing	✔	✔
• Operate and explain function of medical equipment and mobility aids—walkers, wheelchairs, bedside commodes, tub transfer benches, adaptive devices.	✔	✔
• Use coping mechanisms and diversional activities appropriately.	✔	✔
• Implement safety measures in the home.	✔	✔
• Know how to contact and use community resources and the Guillain-Barré Syndrome Foundation International.	✔	✔

During the acute phase of the illness, the patient and family are educated about strategies that can be implemented to minimize the effects of immobility and other complications. As function begins to return, family members and other home care providers are educated about care of the patient and their role in the rehabilitation process. Preparation for discharge is an interdisciplinary effort requiring family or caregiver education by all team members, including the nurse, physician, occupational and physical therapists, speech therapist, and respiratory therapist.

Continuing Care. Most patients with Guillain-Barré syndrome experience complete recovery. Patients who have experienced total or prolonged paralysis require intensive rehabilitation; the extent depends on the patient's needs. Approaches include a comprehensive inpatient program if deficits are significant, an outpatient program if the patient can travel by car, or a home program of physical and occupational therapy. The recovery phase may be long and requires patience as well as involvement on the part of the patient and family.

During acute care, the focus is on immediate issues and deficits. The nurse needs to remind or educate patients and family members of the need for continuing health promotion and screening practices after this initial phase of care.

Evaluation

Expected patient outcomes may include the following:

1. Maintains effective respirations and airway clearance
   a. Has clear breath sounds on auscultation
   b. Demonstrates gradual improvement in respiratory function
   c. Breathes spontaneously
   d. Has vital capacity within normal range
   e. Exhibits arterial blood gases and pulse oximetry within normal limits
2. Shows increasing mobility
   a. Regains use of extremities
   b. Participates in rehabilitation program
   c. Demonstrates no contractures and minimal muscle atrophy
3. Receives adequate nutrition and hydration
   a. Consumes diet adequate to meet nutritional needs
   b. Swallows without aspiration
4. Demonstrates recovery of speech
   a. Communicates needs through alternative strategies
   b. Practices exercises recommended by the speech therapist
5. Shows lessening fear and anxiety
6. Has absence of complications
   a. Maintains intact skin integrity
   b. Does not develop VTE
   c. Voids without difficulty

CRANIAL NERVE DISORDERS

Because the brain stem and cranial nerves involve vital motor, sensory, and autonomic functions of the body, these nerves may be affected by conditions arising primarily within these structures or in secondary extension from adjacent disease processes. The cranial nerves are examined separately and in sequence (see Chapter 65). Some cranial nerve deficits can be detected by observing the patient's face, eye movements, speech, and swallowing. EMG is used to investigate motor and sensory dysfunction. An MRI scan is used to obtain images of the cranial nerves and brain stem. An overview of disorders that may affect each of the cranial nerves, including clinical manifestations and nursing interventions, is presented in Table 69-2. The following discussion centers on the most

TABLE 69-2 Disorders of Cranial Nerves

Disorder	Clinical Manifestations	Nursing Interventions
Olfactory Nerve—I		
Head trauma Intracranial tumor Intracranial surgery	Unilateral or bilateral anosmia (temporary or persistent) Diminished taste for food	Assess sense of smell. Assess for cerebrospinal fluid rhinorrhea if patient has sustained head trauma.
Optic Nerve—II		
Optic neuritis Increased intracranial pressure Pituitary tumor	Lesions of optic tract producing homonymous hemianopsia	Assess visual acuity. Restructure environment to prevent injuries. Educate patient to accommodate for visual loss.
Oculomotor Nerve—III		Assess extraocular movement and for nonreactive pupil.
Trochlear Nerve—IV		Assess extraocular movement and for nonreactive pupil.
Abducens Nerve—VI		Assess extraocular movement and for nonreactive pupil.
Vascular Brain stem ischemia Hemorrhage and infarction Neoplasm Trauma Infection	Dilation of pupil with loss of light reflex on one side Impairment of ocular movement Diplopia Gaze palsies Ptosis of eyelid	
Trigeminal Nerve—V		
Trigeminal neuralgia Head trauma Cerebellopontine lesion Sinus tract tumor and metastatic disease Compression of trigeminal root by tumor	Pain in face Diminished or loss of corneal reflex Chewing dysfunction	Assess for pain and triggering mechanisms for pain. Assess for difficulty in chewing. Discuss trigger zones and pain precipitants with patient. Protect cornea from abrasion. Ensure good oral hygiene. Educate patient about medication regimen.
Facial Nerve—VII		
Bell's palsy Facial nerve tumor Intracranial lesion Herpes zoster	Facial dysfunction; weakness and paralysis Hemifacial spasm Diminished or absent taste Pain	Recognize facial paralysis as emergency; refer for treatment as soon as possible. Discuss protective care for eyes. Select easily chewed foods; patient should eat and drink from unaffected side of mouth. Emphasize importance of oral hygiene. Provide emotional support for changed appearance of face.
Vestibulocochlear Nerve—VIII		
Tumors and acoustic neuroma Vascular compression of nerve Ménière's syndrome	Tinnitus Vertigo Hearing difficulties	Assess pattern of vertigo. Provide for safety measures to prevent falls. Ensure that patient can maintain balance before ambulating. Caution patient to change positions slowly. Assist with ambulation. Encourage the use of assistive devices.
Glossopharyngeal Nerve—IX		
Glossopharyngeal neuralgia from neurovascular compression of cranial nerves IX and X Trauma Inflammatory conditions Tumor Vertebral artery aneurysms	Pain at base of tongue Difficulty in swallowing Loss of gag reflex Palatal, pharyngeal, and laryngeal paralysis	Assess for paroxysmal pain in throat, decreased or absent swallowing, and gag and cough reflexes. Monitor for dysphagia, aspiration, and nasal dysarthric speech. Position patient upright for eating or tube feeding.
Vagus Nerve—X		
Spastic palsy of larynx; bulbar paralysis; high vagal paralysis Guillain-Barré syndrome Vagal body tumors Nerve paralysis from malignancy, surgical trauma such as carotid endarterectomy	Voice changes (temporary or permanent hoarseness) Vocal paralysis Dysphagia	Assess for airway obstruction/provide airway management. Prevent aspiration. Support patient having voice reconstruction procedures.
Spinal Accessory Nerve—XI		
Spinal cord disorder Amyotrophic lateral sclerosis Trauma Guillain-Barré syndrome	Drooping of affected shoulder with limited shoulder movement Weakness or paralysis of head rotation, flexion, extension; shoulder elevation	Support patient undergoing diagnostic tests.
Hypoglossal Nerve—XII		
Medullary lesions Amyotrophic lateral sclerosis Polio and motor system disease, which may destroy hypoglossal nuclei Multiple sclerosis Trauma	Abnormal movements of tongue Weakness or paralysis of tongue muscles Difficulty in talking, chewing, and swallowing	Observe swallowing ability. Observe speech pattern. Be aware of swallowing or vocal difficulties. Prepare for alternate feeding methods (tube feeding) to maintain nutrition.

common disorders of the cranial nerves: trigeminal neuralgia, a condition affecting the fifth cranial nerve, and Bell's palsy, caused by involvement of the seventh cranial nerve.

Trigeminal Neuralgia (Tic Douloureux)

Trigeminal neuralgia is a condition of the fifth cranial nerve that is characterized by paroxysms of sudden pain in the area innervated by any of the three branches of the nerve (Bader & Littlejohns, 2010) (Fig. 69-6). The pain ends as abruptly as it starts and is described as a unilateral shooting and stabbing or burning sensation. The unilateral nature of the pain is an important feature. Associated involuntary contraction of the facial muscles can cause sudden closing of the eye or twitching of the mouth, hence the former name *tic douloureux* (painful twitch). Although the cause is not certain, vascular compression and pressure are suggested causes. As the brain changes with age, a loop of a cerebral artery or vein may compress the nerve root entry point, and this can be identified on MRI scan (Gronseth, Cruccu, Alksne, et al., 2008).

Trigeminal neuralgia occurs most often as people age, most commonly between the fifth and sixth decade of life. It is more common in women and in people with MS compared to the general population (Bader & Littlejohns, 2010; Gronseth et al., 2008). Patients who develop trigeminal neuralgia before age 50 years should be evaluated for the coexistence of MS, because trigeminal neuralgia occurs in approximately 5% of patients with MS (Bader & Littlejohns, 2010). Pain-free intervals may be measured in terms of minutes, hours, days, or longer. With advancing years, the painful episodes tend to become more frequent and agonizing. The patient lives in constant fear of attacks.

Paroxysms can occur with any stimulation of the terminals of the affected nerve branches, such as washing the face, shaving, brushing the teeth, eating, and drinking. A draft of cold air or direct pressure against the nerve trunk may also cause pain. Certain areas are called *trigger points* because the slightest touch immediately starts a paroxysm or episode. To avoid stimulating these areas, patients with trigeminal neuralgia try not to touch or wash their faces, shave, chew, or do anything else that might cause an attack. These behaviors are a clue to the diagnosis.

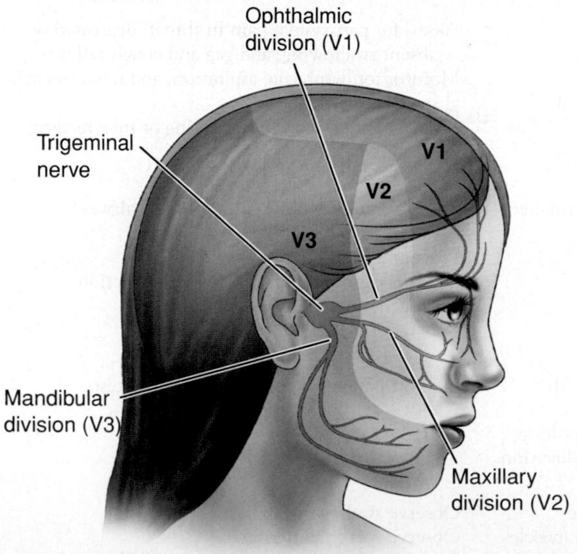
Ophthalmic division (V1)
Trigeminal nerve
V1
V2
V3
Mandibular division (V3)
Maxillary division (V2)

FIGURE 69-6 • Distribution of trigeminal nerve branches—the fifth cranial nerve.

Medical Management

Pharmacologic Therapy

Antiseizure agents, such as carbamazepine (Tegretol), relieve pain in most patients with trigeminal neuralgia by reducing the transmission of impulses at certain nerve terminals. Carbamazepine is taken with meals. Serum levels must be monitored to avoid toxicity in patients who require high doses to control the pain. Side effects include nausea, dizziness, drowsiness, and aplastic anemia. Serious drug reactions occur more often in people of Asian ethnicity (Bader & Littlejohns, 2010). The patient is monitored for bone marrow depression during long-term therapy. Gabapentin and baclofen are also used for pain control. If pain control is still not achieved, phenytoin may be used as adjunctive therapy.

Surgical Management

If pharmacologic management fails to relieve pain, a number of surgical options are available. Although these procedures may relieve facial pain for a few years, recurrence and complication rates are high (Hickey, 2009). The choice of procedure depends on the patient's preference and health status. The procedures are designed to either decompress the nerve and save nerve function or to damage the nerve and destroy nerve function to keep it from malfunctioning (Bader & Littlejohns, 2010).

Microvascular Decompression of the Trigeminal Nerve

An intracranial approach is used to relieve the contact between the cerebral vessel and the trigeminal nerve root entry. With the aid of an operating microscope, the artery loop is lifted from the nerve to relieve the pressure, and a small prosthetic device is inserted to prevent recurrence of impingement on the nerve. The postoperative management is the same as for other intracranial surgeries (see Chapter 66).

Radiofrequency Thermal Coagulation

Percutaneous radiofrequency produces a thermal lesion on the trigeminal nerve. Although immediate pain relief is experienced, dysesthesia of the face and loss of the corneal reflex may occur. The use of stereotactic MRI for identification of the trigeminal nerve followed by gamma knife radiosurgery is being used at some medical centers. Gamma knife radiosurgery is a noninvasive method of delivering focused radiation to the trigeminal nerve; it requires 6 to 8 weeks of treatment for maximal effect to occur (Zakrzewska & Akram, 2011).

Percutaneous Balloon Microcompression

Percutaneous balloon microcompression disrupts large myelinated fibers in all three branches of the trigeminal nerve. After its placement, the balloon is filled with a contrast material for fluoroscopic identification. The balloon compresses the nerve root for 1 minute and provides microvascular decompression (Bader & Littlejohns, 2010).

Nursing Management

Preventing Pain

Preoperative management of a patient with trigeminal neuralgia occurs mostly on an outpatient basis and includes recognizing factors that may aggravate excruciating facial pain, such as food that is too hot or too cold or jarring of the patient's bed or chair. Even washing the face, combing the hair, or brushing the teeth may produce acute pain. The nurse can assist the patient in preventing or reducing this pain by providing instructions about preventive strategies. Providing cotton pads and room temperature water for washing the face, instructing the patient to rinse with mouthwash after eating if toothbrushing causes pain, and performing personal hygiene during pain-free intervals are all effective strategies. The patient is instructed to take food and fluids at room temperature, to chew on the unaffected side, and to ingest soft foods. The nurse recognizes that anxiety, depression, and insomnia often accompany chronic painful conditions and uses appropriate interventions and referrals. (See Chapter 12 for management of patients with chronic pain.)

Providing Postoperative Care

Postoperative neurologic assessments are conducted to evaluate the patient for facial motor and sensory deficits in each of the three branches of the trigeminal nerve. If the surgery results in sensory deficits to the affected side of the face, the patient is instructed not to rub the eye because the pain of a resulting injury will not be detected. The eye is assessed for irritation or redness. Artificial tears may be prescribed to prevent dryness in the affected eye. The patient is cautioned not to chew on the affected side until numbness has diminished. The patient is observed carefully for any difficulty in eating or swallowing foods of different consistencies.

Bell's Palsy

Bell's palsy (facial paralysis) is caused by unilateral inflammation of the seventh cranial nerve, which results in weakness or paralysis of the facial muscles on the affected side (Fig. 69-7). Although the cause is unknown, theories about causes include vascular ischemia, viral disease (herpes simplex, herpes zoster), autoimmune disease, or a combination of all of these factors (Kennedy, 2010). Most adults with Bell's palsy are younger than 45 years.

Bell's palsy may be a type of pressure paralysis. The inflamed, edematous nerve becomes compressed to the point of damage, or its blood supply is occluded, producing ischemic necrosis of the nerve. The face is distorted from paralysis of the facial muscles; decreased lacrimation (tearing) occurs; and the patient experiences painful sensations in the face, behind the ear, and in the eye. The patient may also experience speech difficulties and may be unable to eat on the affected side because of weakness or paralysis of the facial muscles. Seventy percent of patients recover completely (Kennedy, 2010), and Bell's palsy rarely recurs.

Medical Management

The objectives of treatment are to maintain the muscle tone of the face and to prevent or minimize denervation. The patient should be reassured that no stroke has occurred and that spontaneous recovery occurs within 3 to 5 weeks in most patients.

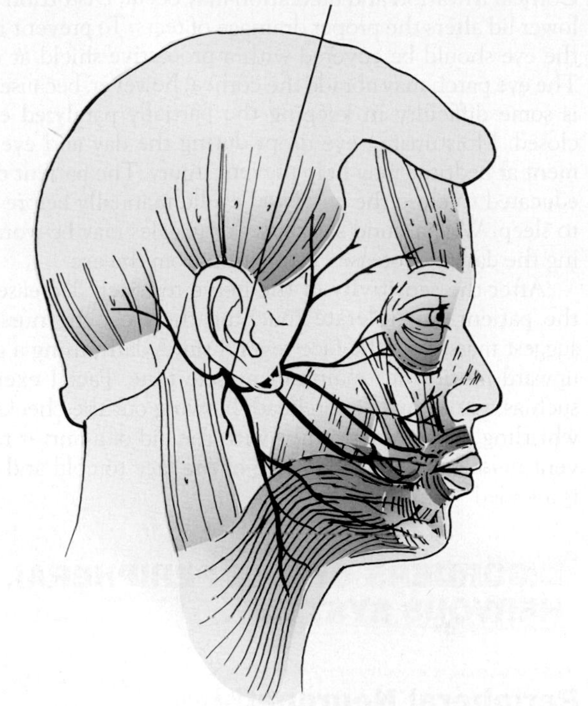
Facial
nerve

FIGURE 69-7 • Distribution of the facial nerve—the seventh cranial nerve.

Corticosteroid therapy (prednisone) may be prescribed to reduce inflammation and edema; this reduces vascular compression and permits restoration of blood circulation to the nerve. Early administration of corticosteroid therapy appears to diminish the severity of the disease, relieve the pain, and prevent or minimize denervation. There has been much research regarding the efficacy of antiviral agents prescribed in combination with corticosteroids (Kennedy, 2011; Lockhart, Daly, Pitkethly, et al., 2009; Numthavaj, Thakkinstian, Dejthevaporn, et al., 2011; Shahidullah, Haque, Islam, et al., 2011). Currently, the best evidence is for sole use of prednisone.

Facial pain is controlled with analgesic agents. Heat may be applied to the involved side of the face to promote comfort and blood flow through the muscles. Physical therapy with tailored exercises on the facial muscle can be prescribed, but true benefit has not been shown (Pereira, Obara, Dias, et al., 2011). Electrical stimulation may be applied to the face to prevent muscle atrophy. Although most patients recover with conservative treatment, surgical exploration of the facial nerve may be indicated if a tumor is suspected, for surgical decompression of the facial nerve, or for surgical treatment of a paralyzed face.

Nursing Management

While the paralysis is present, nursing care involves protection of the eye from injury. Frequently, the eyelid does not close completely and the blink reflex is diminished, so the eye is vulnerable to injury from dust and foreign particles. Corneal irritation and ulceration may occur. Distortion of the lower lid alters the proper drainage of tears. To prevent injury, the eye should be covered with a protective shield at night. The eye patch may abrade the cornea, however, because there is some difficulty in keeping the partially paralyzed eyelids closed. Moisturizing eye drops during the day and eye ointment at bedtime may help prevent injury. The patient can be educated to close the paralyzed eyelid manually before going to sleep. Wraparound sunglasses or goggles may be worn during the day to decrease evaporation from the eye.

After the sensitivity of the nerve to touch decreases and the patient can tolerate touching the face, the nurse can suggest massaging the face several times daily, using a gentle upward motion, to maintain muscle tone. Facial exercises, such as wrinkling the forehead, blowing out the cheeks, and whistling, may be performed with the aid of a mirror to prevent muscle atrophy. Exposure of the face to cold and drafts is avoided.

DISORDERS OF THE PERIPHERAL NERVOUS SYSTEM

Peripheral Neuropathies

A peripheral **neuropathy** (disorder of the nervous system) is a disorder affecting the peripheral motor and sensory nerves. Peripheral nerves connect the spinal cord and brain to all other organs. They transmit motor impulses from the brain and relay sensory impulses to the brain. Peripheral neuropathies are characterized by bilateral and symmetric disturbance of function, usually beginning in the feet and hands. The most common cause of peripheral neuropathy is diabetes with poor glycemic control. The major symptoms of peripheral nerve disorders are loss of sensation, muscle atrophy, weakness, diminished reflexes, pain, and paresthesia of the extremities.

Peripheral nerve disorders are diagnosed by history, physical examination, and electrodiagnostic studies such as EEG. The diagnosis of peripheral neuropathy in the geriatric population is challenging because many symptoms, such as decreased reflexes, can be associated with the normal aging process (Miller, 2012).

No specific treatment exists for peripheral neuropathy. Elimination or control of the cause may slow progression. Patients with peripheral neuropathy are at risk for falls, thermal injuries, and skin breakdown. The plan of care includes inspection of the lower extremities for skin breakdown. Assistive devices such as a walker or cane may decrease the risk of falls. Bathwater temperature is checked to avoid thermal injury. Footwear should be accurately sized. Driving may be limited or eliminated, thereby disrupting the patient's sense of independence.

Mononeuropathy

Mononeuropathy is limited to a single peripheral nerve and its branches. It arises when the trunk of the nerve is compressed or entrapped (as in carpal tunnel syndrome), traumatized (as when bruised by a blow), overstretched (as in joint dislocation), punctured by a needle used to inject a drug or damaged by the drugs thus injected, or inflamed because an adjacent infectious process extends to the nerve trunk. Mononeuropathy is frequently seen in patients with diabetes.

Pain is seldom a major symptom of mononeuropathy when the condition is due to trauma, but in patients with complicating inflammatory conditions such as arthritis, pain is prominent. Pain is increased with all body movements that tend to stretch, strain, or cause pressure on the injured nerve and sudden jarring of the body (e.g., from coughing or sneezing). The skin in the areas supplied by nerves that are injured or diseased may become reddened and glossy, the subcutaneous tissue may become edematous, and the nails and hair in this area are altered. Chemical injuries to a nerve trunk, such as those caused by drugs injected into or near it, are often permanent.

The objective of treatment of mononeuropathy is to remove the cause, if possible (e.g., freeing the compressed nerve). Local corticosteroid injections may reduce inflammation and the pressure on the nerve. Aspirin or codeine may be used to relieve pain. Chronic pain can be treated with neuropathic pain medications such as gabapentin (Bader & Littlejohns, 2010).

Nursing care involves protection of the affected limb or area from injury, as well as appropriate patient education about mononeuropathy and its treatment.

Critical Thinking Exercises

1 A 65-year-old patient is admitted with suspected meningitis. Describe the pathophysiology of meningitis. What diagnostic tests should be performed to confirm the diagnosis? What is the most common priority nursing diagnosis in the patient admitted with meningitis? What history and physical findings are common in the patient with meningitis?

2 ebp A 25-year-old woman is newly diagnosed with MS. Identify her treatment options. Describe the evidence base for the options that you have identified and the criteria used to evaluate the strength of the evidence for her various treatment options.

3 pq Identify the priorities, approach, and techniques you would use to develop a nursing plan of care for a patient with trigeminal neuralgia. How will your priorities, approach, and techniques differ if the patient has Bell's palsy?

Brunner Suite Resources

Explore these additional resources to enhance learning for this chapter:

- NCLEX-Style Questions and Other Resources on thePoint, http://thePoint.lww.com/Brunner13e
- Study Guide
- PrepU
- Clinical Handbook
- Handbook of Laboratory and Diagnostic Tests

References

*Asterisk indicates nursing research.
**Double asterisk indicates classic reference.

Books

American Association of Neuroscience Nurses. (2011). *Nursing management of the patient with multiple sclerosis: AANN and ARN clinical practice guideline series*. Glenview, IL: Author.

American Association of Neuroscience Nurses. (2013). *Care of the patient with Myathenia Gravis: AANN patient guideline series*. Glenview, Il: Author.

Bader, M., & Littlejohns, L. R. (2010). *AANN core curriculum for neuroscience nursing* (5th ed.). Glenview, IL: American Association of Neuroscience Nurses.

Bickley, L. S. (2009). *Bates' guide to physical examination and history taking* (10th ed.). Philadelphia: Lippincott Williams & Wilkins.

Dudek, S. G. (2010). *Nutrition essentials for nursing practice* (6th ed.). Philadelphia: Lippincott Williams & Wilkins.

Halper, J., & Holland, N. J. (2011). *Comprehensive nursing care in multiple sclerosis* (3rd ed.). New York: Springer.

Hickey, J. V. (2009). *The clinical practice of neurological & neurosurgical nursing* (6th ed.). Philadelphia: Lippincott Williams & Wilkins.

Karch, A. (2012). *2012 Lippincott's nursing drug guide*. Philadelphia: Lippincott Williams & Wilkins.

Karpoff, S., & Labus, D. M. (2008). *Portable diagnostic tests*. Philadelphia: Lippincott Williams & Wilkins.

Miller, C. A. (2012). *Nursing for wellness in older adults* (6th ed.). Philadelphia: Lippincott Williams & Wilkins.

Porth, C. M., & Matfin, C. (2009). *Pathophysiology: Concepts of altered health states* (8th ed.) Philadelphia: Lippincott Williams & Wilkins.

Weber, J., & Kelley, J. (2010). *Health assessment in nursing* (4th ed.). Philadelphia: Lippincott Williams & Wilkins.

Journals and Electronic Documents

Central Nervous System Infections

Alaani, A., Couldon, C., McDermott, A., et al. (2010). Transtemporal approach to otogenic brain abscesses. *Acta Oto-Laryngologica, 130*, 1214–1219.

Centers for Disease Control and Prevention. (2009a). *Fact sheet: Arboviral encephalitis*. Available at: www.cdc.gov/ncidod/dvbid/arbor/arbofact.htm

Centers for Disease Control and Prevention. (2009b). *Saint Louis encephalitis*. Available at: www.cdc.gov/sle/technical/symptoms.html

Centers for Disease Control and Prevention. (2011). Updated recommendations for use of meningococcal conjugate vaccines—Advisory Committee on Immunization Practices (ACIP), 2010. *Morbidity and Mortality Weekly Report, 60*(03), 72–76.

Haley, R. (2012). Controlling urban epidemics of West Nile virus infection. *Journal of the American Medical Association, 308*(13), 1325–1326.

Steiner, I. (2011). Herpes simplex virus encephalitis: New infection or reactivation? *Current Opinion in Neurology, 24*(3), 268–274.

Theroux, N., Phipps, M., Zimmerman, L., et al. (2013). *Journal of Neuroscience Nursing, 45*(1), 5–13.

Weisfelt, M., van de Beek, D., Spanjaard, L., et al. (2007). A risk score for unfavorable outcome in adults with bacterial meningitis. *Annals of Neurology, 63*(1), 90–97.

Ziai, W. C., & Lewin, J. J. (2009). Improving the role of intraventricular antimicrobial agents in the management of meningitis. *Current Opinion in Neurology, 22*(3), 277–282.

Creutzfeldt-Jakob Disease

Benjamin, R. (2010). Eligibility criteria by alphabetical listing. *American Red Cross*. Available at: www.redcrossblood.org/donating-blood/eligibility-requirements/eligibility-criteria-alphabetical-listing

Dupiereux, I., Zorzi, W., Quadrio, I., et al. (2009). Creutzfeldt-Jakob, Parkinson, Lewy body dementia and Alzheimer diseases: From diagnosis to therapy. *Central Nervous Systems Agents in Medicinal Chemistry, 9*(1), 2–11.

Ironside, J. (2010). Variant Creutzfeldt-Jakob disease. *Haemophilia, 16*(Suppl. 5), 175–180.

Zieve, D., & Eltz, D. R. (2011). *Creutzfeldt-Jakob disease—all information*. Available at: www.umm.edu/ency/article/000788all.htm

Multiple Sclerosis

Awad, A., & Stüve, O. (2010). Multiple sclerosis in the elderly patient. *Drugs & Aging, 27*(4), 283–294.

Cohen, J. A., Barkhof, F., Comi, G., et al. (2010). Oral fingolimod or intramuscular interferon for relapsing multiple sclerosis. *New England Journal of Medicine, 362*(5), 402–415.

Kappos, L., Radue, E., O'Connor, P., et al. (2010). A placebo-controlled trial of oral fingolimod in relapsing multiple sclerosis. *New England Journal of Medicine, 362*(5), 387–401.

**Lublin, F. D., & Reingold, S. C. (1996). Defining the clinical course of multiple sclerosis: Results of an international study. *Neurology, 46*(4), 907–911.

Marrie, R. A., Horwitz, R. I., Cutter, G. et al. (2011). Association between comorbidity and clinical characteristics of MS. *Acta Neurological Scandinavia, 124*(2), 134–141.

Miller, C. E., & Umhauer, M. A. (2011). Emerging oral therapies for multiple sclerosis. *Journal of Neuroscience Nursing, 43*(1), 3–14.

*Plow, M., & Finlayson, M. (2012). A qualitative study of nutritional behaviors in adults with multiple sclerosis. *Journal of Neuroscience Nursing, 40*(5), 262–268.

Solaro, C., & Uccelli, M. (2011). Management of pain in multiple sclerosis: A pharmacological approach. *Nature Reviews Neurology, 7*(9), 519–527.

Myasthenia Gravis and Guillain-Barré Syndrome

Browning, J., Wallace, M., Chana, J., et al. (2011). Bedside testing for myasthenia gravis: The ice-test. *Emergency Medicine Journal, 28*(8), 709–711.

Fourrier, F., Robriquet, L., Hurtevent, J. F., et al. (2011). A simple functional marker to predict the need for prolonged mechanical ventilation in patients with Guillain-Barré syndrome. *Critical Care, 15*(3), 426.

Hardy, T. A., Blum, S., McCombe, P. A., et al. (2011). Guillain-Barré syndrome: Modern theories of etiology. *Current Allergy and Asthma Report, 11*, 197–204.

Ho, D., Thakur, K., Gorson, K. C., et al. (2008). Influence of critical illness on axonal loss in Guillain-Barré syndrome. *Muscle & Nerve, 39*(1), 10–15.

Kumar, V., & Kaminski, H. (2011). Treatment of myasthenia gravis. *Current Neurology and Neuroscience Report, 11*, 89–96.

Lugg, J. (2010). Recognising and managing Guillain-Barré syndrome. *Emergency Nurse, 18*(3), 27–30.

Randell, D. J., Byars, A., Williams, F., et al. (2008). Glyconutrient supplementation in patients with myasthenia gravis. *Journal of Alternative and Complementary Medicine, 14*(9), 1–8.

Waterhouse, C. (2011). Promoting the safe administration of immunoglobulins in neuroscience. *British Journal of Neuroscience Nursing, 7*(3), 529–535.

Trigeminal Neuralgia and Neuropathies

Gronseth, G., Cruccu, G., Alksne, J., et al. (2008). Practice parameter: The diagnostic evaluation and treatment of trigeminal neuralgia. *Neurology, 71*(8), 1183–1190.

Kennedy, P. G. (2010). Herpes simplex virus type 1 and Bell's palsy—a current assessment of the controversy. *Journal of Neurovirology, 16*(1), 1–5.

Lockhart, P., Daly, F., Pitkethly, M., et al. (2009). Antiviral treatment for Bell's palsy (idiopathic facial paralysis). *Cochrane Database of Systematic Reviews*, (4), CD001869.

Numthavaj, P., Thakkinstian, A., Dejthevaporn, C., et al. (2011). Corticosteroid and antiviral therapy for Bell's palsy: A network meta-analysis. *BioMed Central Neurology, 11*(1), 1–10.

Pereira, L., Obara, K., Dias, J., et al. (2011). Facial exercise therapy for facial palsy: Systematic review and meta-analysis. *Clinical Rehabilitation, 25*(7), 649–658.

Shahidullah, M., Haque, A., Islam, M., et al. (2011). Comparative study between combination of famciclovir and prednisolone with prednisolone alone in acute Bell's palsy. *Mymensingh Medical Journal, 20*(4), 605–613.

Zakrzewska, J. M., & Akram, H. (2011). Neurosurgical interventions for the treatment of classical trigeminal neuralgia. *Cochrane Database of Systematic Reviews*, (9), CD007312.

Resources

Creutzfeldt-Jakob Disease Foundation, www.cjdfoundation.org
Guillain-Barré Syndrome Foundation International, gbs-cidp.org
Myasthenia Gravis Foundation of America (MGFA), www.myasthenia.org
National Multiple Sclerosis Society, www. nationalmssociety.org
Neuropathy Association, www.neuropathy.org

Chapter 70

Management of Patients With Oncologic or Degenerative Neurologic Disorders

Learning Objectives

On completion of this chapter, the learner will be able to:

1 Describe brain and spinal cord tumors: their classification, pathophysiology, clinical manifestations, diagnosis, and medical and nursing management.
2 Use the nursing process as a framework for care of patients with nervous system metastases or primary brain tumor.
3 Identify the pathophysiologic processes responsible for various degenerative neurologic disorders.
4 Use the nursing process as a framework for care of patients with Parkinson's disease.
5 Identify resources for patients and families with oncologic and degenerative neurologic disorders.
6 Use the nursing process as a framework for care of patients following a cervical diskectomy.

Glossary

akathisia: motor restlessness, urgent need to move around, and agitation
bradykinesia: very slow voluntary movements and speech
chorea: rapid, jerky, involuntary, purposeless movements of the extremities or facial muscles, including facial grimacing
dementia: a progressive organic mental disorder characterized by personality changes, confusion, disorientation, and deterioration of intellect associated with impaired memory and judgment
dyskinesia: impaired ability to execute voluntary movements
dysphonia: voice impairment or altered voice production
homonymous hemianopsia: visual loss in half of the visual field on the opposite side of a lesion
micrographia: small and often illegible handwriting
neurodegenerative: a disease, process, or condition leading to deterioration of cells or function of the nervous system
nystagmus: involuntary rhythmic eye movements
papilledema: edema of the optic nerve
paresthesia: numbness, tingling, or a "pins and needles" sensation
radiculopathy: disease of a spinal nerve root, often resulting in pain and extreme sensitivity to touch
sciatica: inflammation of the sciatic nerve, resulting in pain and tenderness along the nerve through the thigh and leg
spondylosis: degenerative changes occurring in a disk and adjacent vertebral bodies; can occur in the cervical or lumbar vertebrae

The occurrence of oncologic or degenerative disease processes in the neurologic system produces a unique set of nursing challenges. Care of the patient with oncologic or degenerative disease processes requires an understanding of the anatomy, physiology, diagnostic testing, nursing care, and rehabilitation of patients. Nurses care for patients with oncologic or degenerative disease processes in many settings, including inpatient hospital units, and outpatient cancer care centers, clinics, and home health agencies, to name a few. Oncologic disorders include brain and spinal cord tumors. Degenerative neurologic disorders include Parkinson's disease, Huntington disease, amyotrophic lateral sclerosis (ALS), muscular dystrophies, and degenerative disk disease. Post-polio syndrome is thought to be degenerative in nature and is included in this chapter.

ONCOLOGIC DISORDERS OF THE BRAIN AND SPINAL CORD

There are many types of brain and spinal cord tumors, each with its own biology, prognosis, and treatment options. Because of the unique anatomy and physiology, tumors of the central nervous system (CNS) are challenging to diagnose and treat.

Brain Tumors

A brain tumor occupies space within the skull, growing as a spherical mass or diffusely infiltrating tissue. The effects of brain tumors are caused by inflammation, compression, and infiltration of tissue. A variety of physiologic changes result, causing any or all of the following pathophysiologic events:

- Increased intracranial pressure (ICP) and cerebral edema
- Seizure activity and focal neurologic signs
- Hydrocephalus
- Altered pituitary function

Neoplastic lesions in the brain ultimately cause death by increasing ICP and impairing vital functions, such as respiration.

Brain tumors are classified as primary or secondary. Primary brain tumors originate from cells within the brain. In adults, most primary brain tumors originate from glial cells (cells that make up the structure and support system of the brain and spinal cord) and are supratentorial (located above the covering of the cerebellum). Primary tumors progress locally and rarely metastasize outside the CNS. The overall incidence of primary brain tumors is approximately 40 cases per 100,000 individuals (Central Brain Tumor Registry of the United States [CBTRUS], 2012). The cause of primary brain tumors is unknown, but specific genetic changes resulting in tumor development are implicated. Although many risk factors have been investigated, exposure to ionizing radiation is the only known modifiable risk factor (CBTRUS, 2012). Both glial and meningeal neoplasms have been linked to irradiation of the cranium, with a latency period of up to 20 years after exposure (CBTRUS, 2012). Genetic syndromes, such as neurofibromatosis, have been associated with brain tumor risk in families.

Secondary, or metastatic, brain tumors develop from structures outside the brain and are twice as common as primary brain tumors. Metastatic lesions to the brain can occur from the lung, breast, lower gastrointestinal tract, pancreas, kidney, and skin (melanomas) neoplasms. Single or multiple metastases may occur, and brain metastases may be found at any time during the disease course, even at initial diagnosis of the primary disease.

The highest incidence of brain tumors in adults occurs in the fifth through seventh decades of life. Brain tumors are slightly more common in females than males and approximately 85% occur in Caucasians (CBTRUS, 2012). Death rates for brain and other nervous system tumors have declined in both men and women in the past decade.

Types of Primary Brain Tumors

Brain tumors may be classified into several groups: those arising from the coverings of the brain (e.g., dural meningioma), those developing in or on the cranial nerves (e.g., acoustic neuroma), those originating within brain tissue (e.g., glioma), and metastatic lesions originating elsewhere in the body. Tumors of the pituitary and pineal glands and of cerebral blood vessels are also types of brain tumors. Relevant clinical considerations include the location and the histologic character of the tumor. Tumors may be benign or malignant. A benign tumor, such as a colloid cyst, can occur in a vital area and can grow large enough to have serious effects (Richards & Ballard, 2008). See Chart 70-1 for the classification of brain tumors.

Gliomas

Glial tumors, the most common type of intracerebral brain neoplasm, are divided into many categories (Wen & Kesari, 2008). Astrocytomas, arising from astrocytic cells, are the most common type of glioma and are graded from I to IV, indicating the degree of malignancy (Cahill & Armstrong, 2011). The grade is based on cellular density, cell mitosis, and degree of differentiation from the original cell type. Grades III and IV tumors are known as glioblastomas and have little resemblance to the original cell type. Astrocytomas infiltrate into the surrounding neural connective tissue and therefore cannot be totally removed without causing considerable damage to vital structures.

Oligodendroglial tumors, arising from oligodendroglial cells, represent 20% of gliomas and are categorized as low or high grade (anaplastic) (American Brain Tumor Association [ABTA] (2011). The histologic distinction between astrocytomas and oligodendrogliomas is difficult to make but is important, because oligodendrogliomas are more sensitive than astrocytomas to chemotherapy. Tumors originating from ependymal cells, another type of glial cell, are known as ependymomas and are more common in children than adults. Glial tumors may be treated with a combination of surgery, radiation therapy, and chemotherapy, depending on specific cell and patient characteristics.

Chart 70-1 Classification of Brain Tumors in Adults

I. Intracerebral Tumors
- A. Gliomas—infiltrate any portion of the brain; most common type of brain tumor
  1. Astrocytomas (grades I and II)
  2. Glioblastoma (astrocytoma grades III and IV)
  3. Oligodendroglioma (low and high grades)
  4. Ependymoma (grades I to IV)
  5. Medulloblastoma

II. Tumors Arising From Supporting Structures
- A. Meningiomas
- B. Neuromas (acoustic neuroma, schwannoma)
- C. Pituitary adenomas

III. Developmental Tumors
- A. Angiomas
- B. Dermoid, epidermoid, teratoma, craniopharyngioma

IV. Metastatic Lesions

Meningiomas

Meningiomas, which represent 15% of all primary brain tumors, are common benign encapsulated tumors of arachnoid cells on the meninges (CBTRUS, 2012). They are slow growing, occur most often in middle-aged adults, and are more common in women. Meningiomas most often occur in areas proximal to the venous sinuses. Manifestations depend on the area involved and are often the result of compression rather than invasion of brain tissue. Preferred treatment for symptomatic lesions is surgery with complete removal or partial dissection, although radiation therapy may be useful for some patients.

Acoustic Neuromas

An acoustic neuroma is a tumor of the eighth cranial nerve—the cranial nerve most responsible for hearing and balance. It usually arises just within the internal auditory meatus, where it frequently expands before filling the cerebellopontine recess. An acoustic neuroma may grow slowly and attain considerable size before it is diagnosed. The patient usually experiences loss of hearing, tinnitus, and episodes of vertigo and staggering gait. As the tumor becomes larger, painful sensations of the face may occur on the same side as a result of the tumor's compression of the fifth cranial nerve. Many acoustic neuromas are benign and can be managed conservatively. Many that continue to grow can be surgically removed and have a good prognosis (see Chapter 64). Some acoustic neuromas may be suitable for stereotactic radiotherapy rather than open craniotomy. Stereotactic radiotherapy is discussed later in this chapter.

Pituitary Adenomas

Pituitary tumors represent about 10% to 15% of all brain tumors and may be treated by surgery and occasionally radiation therapy (CBTRUS, 2012). Pituitary tumors cause symptoms as a result of pressure on adjacent structures or hormonal changes.

Pressure Effects of Pituitary Adenomas

Pressure from a pituitary adenoma may be exerted on the optic nerves, optic chiasm, or optic tracts or on the hypothalamus or the third ventricle if the tumor invades the cavernous sinuses or expands into the sphenoid bone. These pressure effects produce headache, visual dysfunction, hypothalamic disorders (disorders of sleep, appetite, temperature, and emotions), increased ICP, and enlargement and erosion of the sella turcica.

Hormonal Effects of Pituitary Adenomas

Functioning pituitary tumors can produce one or more hormones normally produced by the anterior pituitary. There are prolactin-secreting pituitary adenomas (prolactinomas), growth hormone–secreting pituitary adenomas that produce acromegaly in adults, and adrenocorticotropic hormone (ACTH)–producing pituitary adenomas that result in Cushing syndrome (Gordon, 2007). Adenomas that secrete thyroid-stimulating hormone or follicle-stimulating hormone and luteinizing hormone occur infrequently, whereas adenomas that produce both growth hormone and prolactin are relatively common.

The female patient whose pituitary gland is secreting excessive quantities of prolactin presents with amenorrhea or galactorrhea (excessive or spontaneous flow of milk). Male patients with prolactinomas may present with impotence and hypogonadism. Acromegaly, caused by excess growth hormone, produces enlargement of the hands and feet, distortion of the facial features, and pressure on peripheral nerves (entrapment syndromes). The clinical features of Cushing syndrome, a condition associated with prolonged overproduction of cortisol, occur with excessive production of ACTH. Manifestations include a form of obesity with redistribution of fat to the facial, supraclavicular, and abdominal areas; hypertension; purple striae and ecchymoses; osteoporosis; elevated blood glucose levels; and emotional disorders. (See Chapter 52 for a discussion of endocrine disorders resulting from these tumors.)

Gerontologic Considerations

The most frequent tumor types in the older adult are anaplastic astrocytoma, glioblastoma, and cerebral metastases from other sites. The incidence of primary brain tumors and the likelihood of malignancy increase with age. Intracranial tumors can produce personality changes, confusion, speech dysfunction, or disturbances of gait. In older adult patients, early signs and symptoms of intracranial tumors can be easily overlooked or incorrectly attributed to cognitive and neurologic changes associated with normal aging. Neurologic signs and symptoms in the older adult must be carefully evaluated, because brain metastases occur in patients with a history of prior cancer. Researchers are investigating patterns of care and clinical outcomes of older adult patients with primary brain tumors (Barnholtz-Sloan, Williams, Maldonado, et al., 2008).

Clinical Manifestations

Brain tumors can produce both focal or generalized neurologic signs and symptoms. Generalized symptoms reflect increased ICP, and the most common focal or specific signs and symptoms result from tumors that interfere with functions in specific brain regions. Figure 70-1 indicates common tumor sites in the brain.

Increased Intracranial Pressure

As discussed in Chapter 66, the skull is a rigid compartment containing essential noncompressible contents: brain matter, intravascular blood, and cerebrospinal fluid (CSF). The Monro-Kellie hypothesis, also known as the Monro-Kellie doctrine, explains the dynamic equilibrium of cranial contents. According to the Monro-Kellie hypothesis, if any one of these skull components increases in volume, ICP increases unless one of the other components decreases in volume. Consequently, any change in volume occupied by the brain (as occurs with disorders such as brain tumor or cerebral edema) produces signs and symptoms of increased ICP.

Symptoms of increased ICP result from a gradual compression of the brain by the enlarging tumor and associated edema. The effect is a disruption of the equilibrium that exists between the brain, the CSF, and the cerebral blood. As the tumor grows, compensatory adjustments may occur through compression of intracranial veins, reduction of CSF volume (by increased absorption or decreased production), a modest decrease in cerebral blood flow, or reduction of intra- and extracellular brain tissue mass. When these compensatory mechanisms fail, the patient develops signs and symptoms of increased ICP, most often including headache, nausea with or without vomiting, and **papilledema** (edema of the optic disc) (Cahill & Armstrong, 2011). Personality changes and a variety of focal deficits, including motor, sensory, and cranial nerve dysfunction, are common.

Headache

Headache, although not always present, is most common in the early morning and is made worse by coughing, straining, or sudden movement. It is thought to be caused by the tumor invading, compressing, or distorting the pain-sensitive structures or by edema that accompanies the tumor. Headaches are usually described as deep or expanding or as dull but unrelenting. Frontal tumors usually produce a bilateral frontal headache; pituitary gland tumors produce pain radiating between the two temples (bitemporal); in cerebellar tumors,

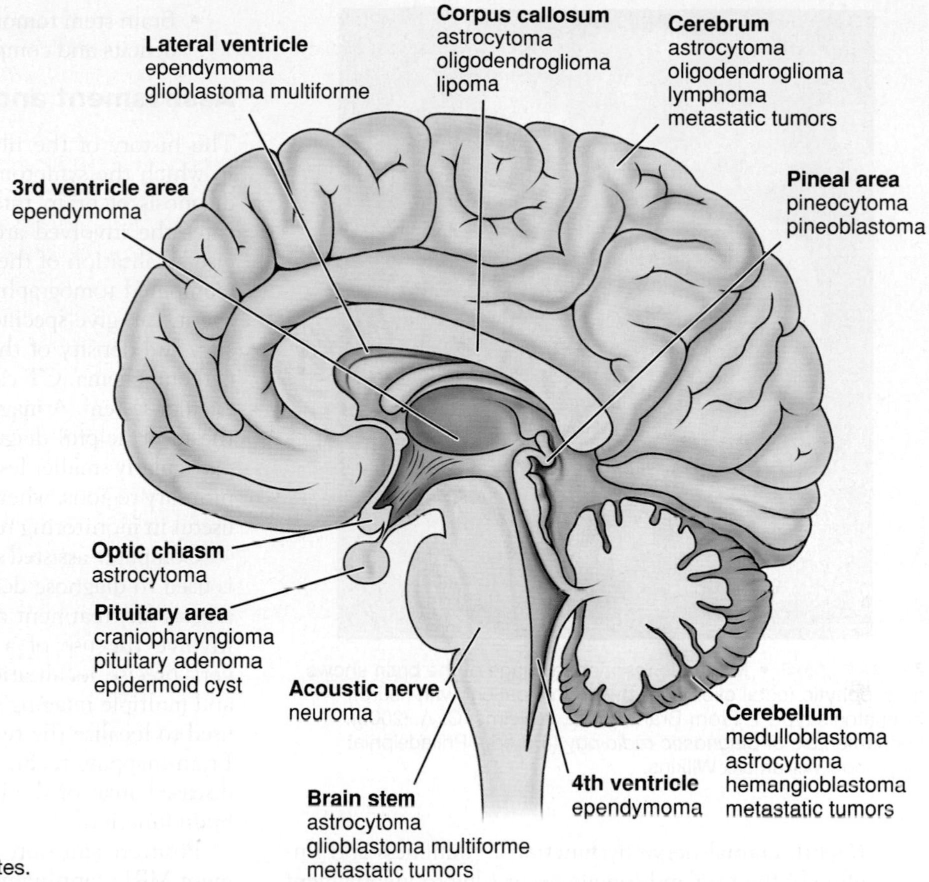

Lateral ventricle
ependymoma
glioblastoma multiforme
Corpus callosum
astrocytoma
oligodendroglioma
lipoma
Cerebrum
astrocytoma
oligodendroglioma
lymphoma
metastatic tumors
3rd ventricle area
ependymoma
Pineal area
pineocytoma
pineoblastoma
Optic chiasm
astrocytoma
Pituitary area
craniopharyngioma
pituitary adenoma
epidermoid cyst
Acoustic nerve
neuroma
Cerebellum
medulloblastoma
astrocytoma
hemangioblastoma
metastatic tumors
4th ventricle
ependymoma
Brain stem
astrocytoma
glioblastoma multiforme
metastatic tumors

FIGURE 70-1 • Common brain tumor sites.

the headache may be located in the suboccipital region at the back of the head.

Vomiting

Vomiting, seldom related to food intake, is usually the result of irritation of the vagal centers in the medulla. Forceful vomiting is described as projectile vomiting. Headache may be relieved by vomiting.

Visual Disturbances

Papilledema is present in 70% to 75% of patients and is associated with decreased visual acuity. Additional visual disturbances, such as diplopia (double vision), or visual field deficits may occur with tumors affecting visual pathways (Davis & Stoiber, 2011).

Seizures

Seizures are common, occurring in up to 70% of persons with brain tumors (Hickey, 2009). Seizures may be focal or generalized. Tumors of the frontal, parietal, and temporal lobes carry the greatest risk of seizures; seizures are unusual with brain stem or cerebellar tumors. (See Chapter 66 for discussion about seizures and related management.)

Localized Symptoms

When specific regions of the brain are affected, local signs and symptoms occur, such as sensory or motor abnormalities, visual alterations, alterations in cognition, or language disturbances (e.g., aphasia). Identifying the signs and symptoms is important, because it can help identify tumor location. Some tumors are not easily localized because they lie in so-called silent areas of the brain (i.e., areas in which functions are not well defined). Many tumors can be localized by correlating the signs and symptoms to specific areas in the brain, as follows:

- A tumor in the motor cortex of the frontal lobe produces hemiparesis and partial seizures on the opposite side of the body or generalized seizures A frontal lobe tumor may also produce changes in emotional state and behavior, as well as an apathetic mental attitude. The patient often becomes extremely untidy and careless and may use obscene language.
- A parietal lobe tumor may cause decreased sensation on the opposite side of the body and sensory or generalized seizures.
- A temporal lobe tumor may cause seizures as well as psychological disorders.
- An occipital lobe tumor produces visual manifestations: contralateral **homonymous hemianopsia** (visual loss in half of the visual field on the opposite side of the tumor) and visual hallucinations.
- A cerebellar tumor causes dizziness; an ataxic or staggering gait with a tendency to fall toward the side of the lesion; marked muscle incoordination; and **nystagmus** (involuntary rhythmic eye movements), usually in the horizontal direction.
- A cerebellopontine angle tumor usually originates in the sheath of the acoustic nerve and gives rise to a characteristic sequence of symptoms. Tinnitus and vertigo appear first, soon followed by progressive nerve deafness

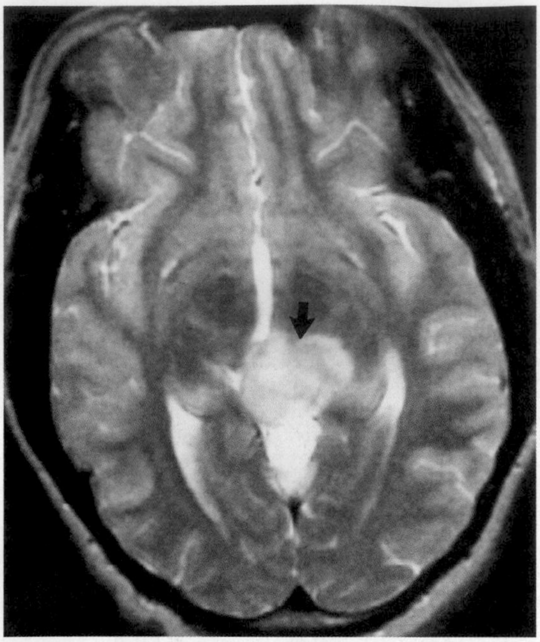

FIGURE 70-2 • Magnetic resonance image of the brain shows an exophytic tectal glioma (*arrow*) in this patient with neurofibromatosis type 1. From Brant, W. E., & Helms, C. A. (2006). *Fundamentals of diagnostic radiology* (3rd ed.). Philadelphia: Lippincott Williams & Wilkins.

(eighth cranial nerve dysfunction). Numbness and tingling of the face and tongue occur (due to involvement of the fifth cranial nerve). Later, weakness or paralysis of the face develops (seventh cranial nerve involvement). Finally, because the enlarging tumor presses on the cerebellum, abnormalities in motor function may be present.

- Brain stem tumors may be associated with cranial nerve deficits and complex motor and sensory function.

Assessment and Diagnostic Findings

The history of the illness and the manner and time frame in which the symptoms evolved are key components in the diagnosis of brain tumors. A neurologic examination indicates the involved areas of the CNS. To assist in the precise localization of the lesion, a battery of tests is performed. Computed tomography (CT) scans, enhanced by a contrast agent, can give specific information concerning the number, size, and density of the lesions and the extent of secondary cerebral edema. CT can provide information about the ventricular system. A magnetic resonance imaging (MRI) scan is the most helpful diagnostic tool for detecting brain tumors, particularly smaller lesions, and tumors in the brain stem and pituitary regions, where bone is thick (Fig. 70-2). MRI is also useful in monitoring response to treatment.

Computer-assisted stereotactic (three-dimensional) biopsy is used to diagnose deep-seated brain tumors and to provide a basis for treatment and prognosis. Stereotactic approaches involve the use of a three-dimensional frame that allows very precise localization of the tumor; a stereotactic frame and multiple imaging studies (x-rays, CT scans, or MRIs) are used to localize the tumor and verify its position (Fig. 70-3). Brain-mapping technology helps determine the proximity of diseased areas of the brain to structures essential for normal brain function.

Positron emission tomography (PET) is used to supplement MRI scanning in centers where it is available. On PET scans, low-grade tumors are associated with hypometabolism, and high-grade tumors show hypermetabolism. This information can be useful in making treatment decisions (ABTA, 2011). An electroencephalogram can detect abnormal brain

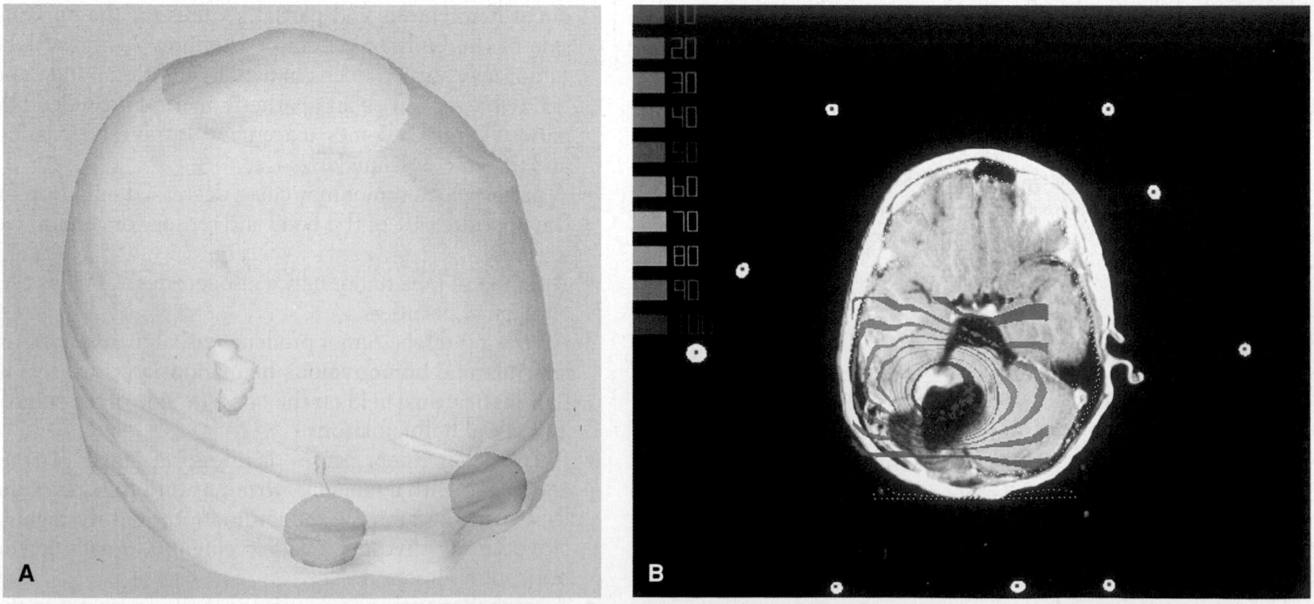
A
B

FIGURE 70-3 • **A.** Using stereotactic or "brain-mapping" guided approach, a three-dimensional computer image fuses the computed tomography image and magnetic resonance image to pinpoint the exact location of the brain tumor. This low-grade astrocytoma is localized adjacent to the brain stem, is inoperable, and is treated with radiation. Note the optic chasm and optic nerves. **B.** Computerized image of the prescribed radiation dose.

waves in regions occupied by or adjacent to tumor; it is used to evaluate temporal lobe seizures and to assist in ruling out other disorders. Cytologic studies of the CSF may be performed to detect malignant cells, because CNS tumors can shed cells into the CSF.

Medical Management

A variety of medical management approaches, including surgery, chemotherapy, and external-beam radiation therapy, are used alone or in combination (Wen & Kesari, 2008).

Secreting tumors may be treated with medications that suppress hormones. Nonfunctioning tumors may have no effect on pituitary function or may suppress hormone production and release. Hormone replacement may be necessary for these patients to restore normal endocrine function.

Surgical Management

The objective of surgical management is to remove as much tumor as possible without increasing the neurologic deficit (paralysis, blindness), or to relieve symptoms by partial removal (decompression). Surgery also provides tissue to establish a definitive diagnosis. A variety of surgical approaches may be used; the specific approach depends on the type of tumor, its location, and its accessibility. Conventional surgical approaches require a craniotomy (incision into the skull). (See Chapter 66 for a discussion of care of the patient who has undergone a craniotomy.) This approach is used in patients with meningiomas, acoustic neuromas, cystic astrocytomas of the cerebellum, colloid cysts of the third ventricle, congenital tumors such as dermoid cyst, and some of the granulomas. With improved imaging techniques and the availability of the operating microscope and microsurgical instrumentation, even large tumors can be removed through a relatively small craniotomy. For patients with malignant glioma, complete removal of the tumor and cure are not possible, but the rationale for resection includes relief of ICP, removal of any necrotic tissue, and reduction in the bulk of the tumor, which theoretically leaves behind fewer cells to become resistant to radiation or chemotherapy. Most pituitary adenomas are treated by transsphenoidal microsurgical removal (see Chapter 66), and the remainder of tumors that cannot be removed completely are treated by radiation (National Comprehensive Cancer Network [NCCN], 2012).

Radiation Therapy

Radiation therapy—the cornerstone of treatment for many brain tumors—decreases the incidence of recurrence of incompletely resected tumors (NCCN, 2012). Gamma radiation is delivered via an external beam to the tumor in multiple fractions. Brachytherapy (the surgical implantation of radiation sources to deliver high doses at a short distance) has had promising results for primary brain malignancies. It is usually used as an adjunct to conventional radiation therapy or as a rescue measure for recurrent disease. Radioisotopes such as iodine 131 (^{131}I) are used to minimize effects on surrounding brain tissue.

Stereotactic procedures may be performed using a linear accelerator or gamma knife to perform radiosurgery (NCCN, 2012). These procedures allow treatment of deep, inaccessible tumors, often in a single session. Precise localization of the tumor is accomplished by the stereotactic approach and by minute measurements and precise positioning of the patient. Multiple narrow beams then deliver a very high dose of radiation. An advantage of this method is that no surgical incision is needed; a disadvantage is the lag time between treatment and the desired result (Swinson & Friedman, 2008).

Chemotherapy

Chemotherapy may be used in conjunction with radiation therapy, or as the sole therapy, with the goal of increasing survival time. The greatest challenge in chemotherapy of brain tumors is that the blood–brain barrier prevents drugs from getting to the tumor in effective doses without causing systemic toxicity. Malignant glioma is usually treated with 6 weeks of oral temozolomide (Temodar) during radiation therapy, followed by 6 to 12 months of oral temozolomide. Low-grade gliomas may be treated with 6 months of oral temozolomide alone. Temozolomide is the first oral chemotherapy that crosses the blood–brain barrier (Hickey, 2009). Additional agents are under study (Eggert, 2010).

Autologous bone marrow transplantation is used in some patients who will receive chemotherapy or radiation therapy, because it can "rescue" the patient from the bone marrow toxicity associated with high doses of chemotherapy and radiation. A fraction of the patient's bone marrow is aspirated, usually from the iliac crest, and stored. The patient receives large doses of chemotherapy or radiation therapy to destroy large numbers of malignant cells. The marrow is then reinfused intravenously after treatment is completed (see Chapter 34 for discussion of bone marrow transplant). Experimental vaccines that enhance immune function are currently in clinical trials. Gene transfer therapy, also in clinical trials, uses retroviral vectors to carry genes to the tumor, reprogramming the tumor tissue to enhance susceptibility to treatment (Eggert, 2010).

Pharmacologic Therapy

Corticosteroids are useful in relieving headache and alterations in level of consciousness. Corticosteroids such as dexamethasone (Decadron) and prednisone are thought to reduce inflammation and edema around tumors (Karch, 2012). Other medications used include osmotic diuretics (e.g., mannitol [Osmitrol]) to decrease the fluid content of the brain, which leads to a decrease in ICP. Antiseizure medications are used to prevent and treat seizures (NCCN, 2012). Venous thromboembolic events, such as deep vein thrombosis and pulmonary embolism, occur in about 15% of patients and are associated with significant morbidity. Patients receiving anticoagulant agents must be closely monitored because of the risk of CNS hemorrhage (Mogensen, 2008).

Nursing Management

Headache characteristics should be assessed. Upright positioning and pain medications may be useful in managing pain; nurses should evaluate effectiveness of pain management interventions. Even if seizure history is absent, the patient and family should be educated about the possibility of seizure and the need to adhere to prophylactic seizure medications, if prescribed. The patient with a brain tumor may be at increased risk for aspiration as a result of cranial nerve dysfunction. Medications to alleviate nausea and prevent vomiting should be considered. Preoperatively, the gag reflex and ability to swallow are evaluated. In patients with diminished gag response, care includes educating the patient to direct food and fluids toward the unaffected side, having the patient sit upright to eat, offering a semisoft diet, and having suction readily available. The effects of increased ICP caused by the tumor mass are reviewed in Chapter 66. The nurse performs neurologic checks; monitors vital signs; maintains a neurologic flow chart; spaces nursing interventions to prevent rapid increase in ICP; and reorients the patient when necessary to person, time, and place. The use of corticosteroids to control headache and neurologic symptoms requires astute nursing assessment and intervention because many adverse effects can occur, including hyperglycemia, electrolyte abnormalities, and muscle weakness (see Table 52-5 in Chapter 52 for a summary of corticosteroid therapy and nursing implications). Patients with changes in cognition caused by their lesion require frequent reorientation and the use of orienting devices (e.g., personal possessions, photographs, lists, a clock), supervision of and assistance with self-care, and ongoing monitoring and intervention for prevention of injury. Patients with seizures are carefully monitored and protected from injury. Motor function is checked at intervals because specific motor deficits may occur, depending on the tumor's location. When muscle weakness is present, a multidisciplinary approach, including the nurse and physical and occupational therapists, can be used to preserve muscle strength, promote range of motion, and facilitate independence in self-care. Sensory disturbances are assessed and any area of numbness should be protected from injury. Speech is evaluated, and patients with speech deficits can be educated to use alternative forms of communication. Eye movement and pupillary size and reaction may be affected by cranial nerve involvement. Fatigue is common during therapy; efforts should be made to conserve energy and promote rest.

The psychosocial effects on family caregivers of a family member who has a primary malignant brain tumor may be significant (Chart 70-2). Caregiving family members should be included in the plan of care.

The nursing process for patients undergoing neurosurgery is discussed in Chapter 66. The patient's functional abilities should be reassessed postoperatively, because changes can occur.

Cerebral Metastases

A significant number of patients with cancer experience neurologic deficits caused by metastasis to the nervous system, which can include the brain, CSF, and meninges. Metastatic lesions to the brain are twice as common as primary brain tumors (CBTRUS, 2012). This high rate of occurrence is clinically important, as more patients with all forms of cancer live longer because of improved therapies. Neurologic signs and symptoms include headache, gait disturbances, visual impairment, personality changes, altered mentation (memory loss and confusion), focal weakness, paralysis, aphasia, and seizures. These signs and symptoms can be devastating to both patient and family. Metastases to the CSF and meninges, known as leptomeningeal metastases, can produce symptoms of headache and isolated cranial nerve deficits.

Medical Management

The treatment of metastatic nervous system cancer is palliative and involves eliminating or reducing serious symptoms. Even when palliation is the goal, distressing signs and symptoms can be relieved, thereby improving the quality of life for both patient and family. Patients with intracerebral metastases who are not treated have a steady downhill course with a limited survival time, whereas those who are treated may survive for slightly longer periods. However, for most patients, cure is not possible.

The therapeutic options include whole brain radiation therapy (the foundation of treatment) for multiple metastases or stereotactic radiosurgery for up to three sites of metastases. Surgery may be considered for a single symptomatic metastasis.

Chart 70-2 NURSING RESEARCH PROFILE

When a Family Member Has a Malignant Brain Tumor

Sherwood, P., Hricik, A., Donovan, H., et al. (2011). Changes in caregiver perceptions over time in response to providing care for a loved one with a primary malignant brain tumor. *Oncology Nursing Forum, 38*(2), 149–155.

Purpose

Persons with a malignant brain tumor require care from family members due to cognitive and neurologic deficits as well as fatigue. The purpose of this study was to examine caregiver's transition over time when caring for a family member with a malignant brain tumor.

Design

This qualitative descriptive study used interviews to examine caregiver perceptions over a 4-month period. Ten family caregivers participated in semistructured interviews within 1 month of diagnosis and 4 months after. Data were analyzed using content analysis to identify themes.

Findings

Data analysis uncovered three main themes. First, changes in the person with the brain tumor required switching of roles and routines over time, introducing a "new normal." Second, caregivers adjusted over time to changes in the relationship, valuing time together. Third, support was necessary but challenging, and caregivers perceived the need over time to receive more specific types of support from friends and family.

Nursing Implications

Family members of persons with brain tumors are placed in the caregiver role, and family member perceptions of that role change over time. Nurses should have frequent contact with caregivers within the first 4 months of diagnosis in order to facilitate caregiver needs, realizing that caregiver needs change over time. Nurses should also be prepared to offer direction on how to enlist the support of family and friends and access additional support services.

Systemic chemotherapy directed at the primary cancer may be ineffective in crossing the blood–brain barrier, but chemotherapy that crosses this barrier may be added. Intrathecal chemotherapy, with direct injection of chemotherapy agents into the CSF of the brain or spinal canal, may be useful in persons with leptomeningeal metastases (Aiello-Laws & Rutledge, 2008). Some combination of these treatments is usually the optimal method.

Pain can be a significant problem and is managed by means of a stepped progression in the doses and type of analgesic agents needed for relief. If the patient has severe pain, morphine can be infused into the epidural or subarachnoid space through a spinal needle and a catheter placed as near as possible to the spinal segment where the pain is projected. Small doses of morphine are administered at prescribed intervals (see Chapter 12).

NURSING PROCESS

The Patient With Nervous System Metastases or Primary Brain Tumor

Assessment

The nursing assessment includes a baseline neurologic examination and focuses on how the patient is functioning, moving, and walking; adapting to weakness or paralysis and to loss of vision and speech; and dealing with seizures. Assessment addresses symptoms that cause distress to the patient and affect the quality of life, including pain, respiratory problems, bowel and bladder disorders, and sleep disturbances, as well as impairment of skin integrity, fluid balance, and temperature regulation (Davis & Stoiber, 2011). Nutritional status is assessed, because cachexia and cancer-related anorexia-cachexia syndrome are common (see Chapter 15).

The nurse takes a dietary history to assess food intake, intolerance, and preferences. Calculation of body mass index can confirm the loss of subcutaneous fat and lean body mass (see Chapter 5). Biochemical measurements are reviewed to assess the degree of malnutrition, impaired cellular immunity, and electrolyte balance (please see Appendix A on thePoint for normal laboratory values). A dietitian assists in determining the caloric needs of the patient.

The nurse works with other members of the health care team to assess the impact of the illness on the family in terms of home care, altered relationships, financial problems, time pressures, and family problems. This information is important in helping family members cope with the diagnosis and the changes associated with it.

Diagnosis

Nursing Diagnoses

Based on the assessment data, major nursing diagnoses may include the following:

- Self-care deficit (feeding, bathing, and toileting) related to loss or impairment of motor and sensory function and decreased cognitive abilities
- Imbalanced nutrition: less than body requirements related to cachexia due to treatment and tumor effects, decreased nutritional intake, and malabsorption
- Imbalanced nutrition: more than body requirements related to increased nutritional intake and impaired metabolism
- Anxiety related to fear of dying, uncertainty, change in appearance, or altered lifestyle
- Interrupted family processes related to anticipatory grief and the burdens imposed by the care of the person with a terminal illness

Other nursing diagnoses of the patient with nervous system metastases may include acute pain related to tumor compression; impaired gas exchange related to dyspnea; constipation related to decreased fluid and dietary intake and medications; impaired urinary elimination related to reduced fluid intake, vomiting, and side effects of medications; sleep pattern disturbances related to discomfort and fear of dying; impairment of skin integrity related to cachexia, poor tissue perfusion, and decreased mobility; deficient fluid volume related to fever, vomiting, and low fluid intake; and ineffective thermoregulation related to hypothalamic involvement, fever, and chills. (See Chapter 15 for assessment and nursing interventions for the patient with cancer.)

Planning and Goals

The goals for the patient may include compensating for self-care deficits, improving nutrition, reducing anxiety, enhancing family coping skills, and absence of complications.

Nursing Interventions

Compensating for Self-Care Deficits

The patient may have difficulty participating in goal setting as the tumor metastasizes and affects cognitive function. The nurse should encourage the family to keep the patient as independent as possible for as long as possible. Increasing assistance with self-care activities is required. Because the patient with nervous system metastasis and the family live with uncertainty, they are encouraged to plan for each day and to make the most of each day. The tasks and challenges are to assist the patient to find useful coping mechanisms, adaptations, and compensations for solving problems that arise. This helps patients maintain some sense of control. An individualized exercise program helps maintain strength, endurance, and range of motion. Eventually, referral for home or hospice care may be necessary (see Chapter 16).

Improving Nutrition

Patients with nausea, vomiting, diarrhea, breathlessness, and pain are rarely interested in eating. These symptoms are managed or controlled through assessment, planning, and care. The nurse educates the family about how to position the patient for comfort during meals. Meals are planned for times when the patient is rested and in less distress from pain or the effects of treatment.

The patient needs to be clean, comfortable, and free of pain for meals, in an environment that is as attractive as possible. Oral hygiene before meals helps to improve appetite. Offensive sights, sounds, and odors are eliminated. Creative strategies may be required to make food more palatable, provide enough fluids, and increase opportunities for socialization during meals. The family may be asked to keep a daily weight chart and to record the quantity of food eaten to determine the daily calorie count. Dietary supplements, if acceptable to the patient, can be provided to meet increased caloric needs. If the patient is not interested in most usual foods, those foods preferred by the patient should be offered. When the patient shows marked deterioration as a result of tumor growth and effects, some other form of nutritional support (e.g., tube feeding, parenteral nutrition) may be indicated if consistent with the patient's end-of-life preferences (Dudek, 2010). Nursing interventions include assessing the patency of the central and intravenous lines or feeding tube, monitoring the insertion site for infection, checking the infusion rate, monitoring intake and output, and changing the intravenous tubing and dressing. Family members are instructed in these techniques if they will be providing care at home. Parenteral nutrition can be provided at home if indicated. The patient's quality of life may guide the selection, initiation, maintenance, and discontinuation of nutritional support. The nurse and family should not place too much emphasis on eating or on discussions about food, because the patient may not desire aggressive nutritional intervention. The subsequent course of action must be congruent with the wishes and choices of the patient and family.

Increased appetite may occur in patients receiving steroids and may result in weight gain as well as hyperglycemia. Patients and family members should be instructed to monitor weight and blood glucose (if appropriate), as well as maintain a healthy diet with caloric intake appropriate to patient needs.

Relieving Anxiety

Patients may be restless, with changing moods that may include intense depression, euphoria, paranoia, and severe anxiety. The response of patients to terminal illness reflects their pattern of reaction to other crisis situations. Serious illness imposes additional strains that often bring other unresolved problems to light. The patient's own coping strategies can help deal with anxious and depressed feelings. Health care providers need to be sensitive to the patient's concerns and fears.

Patients need the opportunity to exercise some control over their situation. A sense of mastery can be gained as they learn to understand the disease and its treatment and how to deal with their feelings. The presence of family, friends, a spiritual advisor, and health professionals may be supportive. Brain tumor support groups may provide a feeling of support and strength (see the Resources section).

Spending time with patients allows them time to talk and to communicate their fears and concerns. Open communication and acknowledgment of fears are often therapeutic. Touch is also a form of communication. These patients need reassurance that continuing care will be provided and that they will not be abandoned. The situation becomes more endurable when others share in the experience of dying. If a patient's emotional reactions are very intense or prolonged, additional help from a spiritual advisor, social worker, or mental health professional may be indicated.

Enhancing Family Processes

The family needs to be reassured that their loved one is receiving optimal care and that attention will be paid to the patient's changing symptoms and concerns. When the patient can no longer carry out self-care, the family and additional support systems (social worker, home health aide, home care nurse, hospice nurse) may be needed. Educating family members to care for their loved one may decrease the burden on the caregiver as well as improve the patient's quality of life (Sherwood, Hricik, Donovan, et al., 2011). End-of-life care is provided with respect, and reassurance is provided by communicating the plan of care to the family.

Promoting Home and Community-Based Care

Educating Patients About Self-Care. The patient and family often have major responsibility for care at home. Therefore, education includes pain management strategies, prevention of complications related to treatment strategies, and methods to ensure adequate fluid and food intake (Chart 70-3). Educational needs of the patient and family regarding care priorities are likely to change as the disease progresses. The nurse should assess the changing needs of the patient and the family and inform them about resources and services early to assist them in dealing with changes in the patient's condition (see the Resources section).

Continuing Care. Home care nursing and hospice services are valuable resources that should be made available to the patient and the family early in the course of a terminal illness. Anticipating needs before they occur can assist in smooth initiation of services. Home care needs and interventions focus on four major areas: palliation of symptoms and pain control, assistance in self-care, control of treatment complications, and administration of specific forms of treatment (e.g., parenteral nutrition). The home care nurse assesses pain management, respiratory status, complications of the disorder and its treatment, and the patient's cognitive and emotional status. In addition, the nurse assesses the family's ability to perform necessary care and notifies the primary provider about changing needs or complications, if indicated.

The patient and family who elect to care for the patient at home as the disease progresses benefit from the care and support provided through hospice and palliative care services (Sherwood et al., 2011). Steps to initiate hospice care, including discussion of hospice care as an option, should not be postponed until death is imminent. Exploration of hospice care as an option should be initiated at a time when hospice services can provide support and care to the patient and family consistent with their end-of-life decisions and can assist in allowing death with dignity. (See Chapter 16 for in-depth discussion of end-of-life care.)

Evaluation

Expected patient outcomes may include:

1. Engages in self-care activities as long as possible
   a. Uses assistive devices or accepts assistance as needed
   b. Schedules periodic rest periods to permit maximal participation in self-care
2. Maintains as optimal a nutritional status as possible
   a. Eats and accepts food within limits of condition and preferences
   b. Accepts alternative methods of providing nutrition if indicated
3. Reports being less anxious
   a. Is less restless and is sleeping better
   b. Verbalizes concerns and fears about death
   c. Participates in activities of personal importance as long as feasible

Chart 70-3 HOME CARE CHECKLIST

The Patient With Nervous System Metastases

At the completion of home care education, the patient or caregiver will be able to:	PATIENT	CAREGIVER
• State effects of the tumor according to its type and location in the brain.	✔	✔
• Describe side effects of treatment.	✔	✔
• Identify community resources, including:		
• Home health services	✔	✔
• Hospices	✔	✔
• Support groups	✔	✔
• American Brain Tumor Association	✔	✔
• Identify coping strategies, such as:		
• Taking control, setting daily goals, and staying positive	✔	✔
• Rehabilitation to improve self-care	✔	✔
• Relaxation techniques	✔	✔
• Family support		✔
• Verbalize an understanding of the treatment plan for:		
• Medications and pain control	✔	✔
• Nutritional needs	✔	✔
• Contacting the primary provider	✔	✔

4. Family members seek help as needed
   a. Demonstrate ability to bathe, feed, and care for the patient and participate in pain management and prevention of complications
   b. Express feelings and concerns to appropriate health professionals
   c. Discuss and seek hospice care as an option

Spinal Cord Tumors

Tumors within the spinal canal are classified according to their anatomic relation to the spinal cord (Aghayev, Vrionis, & Chamberlain, 2011). They include intramedullary lesions (within the spinal cord), extramedullary-intradural lesions (within or under the spinal dura), and extramedullary-extradural lesions (outside the dural membrane). Primary tumors are usually intramedullary, consisting of astrocytoma or ependymoma; meningiomas can occur as extramedullary-intradural lesions. Secondary tumors are far more common and are usually extramedullary-extradural lesions. Tumors that occur within the spinal canal or exert pressure on it cause symptoms ranging from localized or shooting pains and weakness and loss of reflexes below the tumor level to progressive loss of motor function and paralysis. Usually, sharp pain occurs in the area innervated by the spinal roots that arise from the cord in the region of the tumor. In addition, increasing sensory deficits develop below the level of the lesion. Loss of bowel and bladder function is common.

Assessment and Diagnostic Findings

Neurologic examination and diagnostic studies are used to make the diagnosis. Neurologic examination focuses on assessing pain and identifying loss of reflexes, sensation, or motor function. Helpful diagnostic studies include CT scans, MRI scans, and biopsy. The MRI scan is the most commonly used and the most sensitive diagnostic tool (Aghayev et al., 2011), and it is particularly helpful in detecting epidural spinal cord compression and metastases.

Medical Management

Treatment of specific intraspinal tumors depends on the type and location of the tumor and the presenting symptoms and physical status of the patient. Surgical intervention is the primary treatment for most spinal cord tumors. Other treatment modalities include chemotherapy and radiation therapy (including stereotactic radiosurgery), particularly for intramedullary tumors and metastatic lesions (Aghayev et al., 2011).

Extramedullary-extradural spinal cord compression occurs in 5% to 7% of patients who die of cancer and is considered a neurologic emergency. For the patient with spinal cord compression resulting from metastatic cancer (most commonly from breast, prostate, or lung), high-dose dexamethasone (Decadron) combined with radiation therapy is effective in relieving pain (see Chapter 15 for a discussion of care of the patient with spinal cord compression). Palliative care may be an option for the medical management of some patients. Chemotherapy specific to the tumor type may be considered (Eggert, 2010).

Surgical Management

Tumor removal is desirable but not always possible. The goal is to remove as much tumor as possible while sparing uninvolved portions of the spinal cord (Kim, Losina, Bono, et al., 2012). Microsurgical techniques have improved the prognosis for patients with intramedullary tumors. Extramedullary-intradural tumors may be completely resected. Prognosis is related to the degree of neurologic impairment at the time of surgery, the speed with which symptoms occurred, and the origin of the tumor. Patients with extensive neurologic deficits before surgery usually do not make significant functional recovery even after successful tumor removal.

Nursing Management

Providing Preoperative Care

The objectives of preoperative care include recognition of neurologic changes through ongoing assessments, pain control, and management of altered activities of daily living (ADLs) resulting from sensory and motor deficits and bowel and bladder dysfunction. The nurse assesses for weakness, muscle wasting, spasticity, sensory changes, bowel and bladder dysfunction, and potential respiratory problems, especially if a cervical tumor is present. The patient is also evaluated for coagulation deficiencies. A history of aspirin intake is obtained and reported, because the use of aspirin may impede hemostasis postoperatively. The patient is educated about breathing exercises, and they are demonstrated preoperatively. Postoperative pain management strategies are discussed with the patient before surgery.

Assessing the Patient After Surgery

The patient is monitored for deterioration in neurologic status. A sudden onset of neurologic deficit is an ominous sign and may be due to spinal cord ischemia or infarction. Frequent neurologic checks are carried out, with emphasis on movement, strength, and sensation of the upper and lower extremities. Assessment of sensory function involves pinching the skin of the arms, legs, and trunk to determine if there is loss of feeling and, if so, at what level. Vital signs are monitored at regular intervals.

Managing Pain

The prescribed pain medication should be administered in adequate amounts and at appropriate intervals to relieve pain and prevent its recurrence. Pain is the hallmark of spinal metastasis. Patients with sensory root involvement may suffer excruciating pain, which requires effective pain management.

The bed is usually kept flat initially. The nurse turns the patient as a unit, keeping shoulders and hips aligned and the back straight. The side-lying position is usually the most comfortable, because this position imposes the least pressure on the surgical site. Placement of a pillow between the knees of the patient in a side-lying position helps to prevent extreme knee flexion.

Monitoring and Managing Potential Complications

If the tumor was in the cervical area, respiratory compromise due to postoperative edema may occur. The nurse monitors the patient for asymmetric chest movement, abdominal breathing, and abnormal breath sounds. For a high cervical lesion, the endotracheal tube remains in place until adequate respiratory function is ensured. The patient is encouraged to perform deep-breathing and coughing exercises.

The area over the bladder is palpated or a bladder scan performed to assess for urinary retention. The nurse also monitors for incontinence, because urinary dysfunction usually implies significant decompensation of spinal cord function. An intake and output record is maintained. In addition, the abdomen is auscultated for bowel sounds.

Staining of the dressing may indicate leakage of CSF from the surgical site, which may lead to serious infection or to an inflammatory reaction in the surrounding tissues that can cause severe pain in the postoperative period. Bulging at the incision may indicate contained CSF leak. The site should be monitored for increasing bulging, known as pseudomeningocele, which may require surgical repair.

Promoting Home and Community-Based Care

Educating Patients About Self-Care

In preparation for discharge, the patient is assessed for the ability to function independently in the home and for the availability of resources such as family members to assist in caregiving. Patients with residual sensory involvement are cautioned about the dangers of extremes in temperature. They should be educated about the dangers of heating devices (e.g., hot water bottles, heating pads, space heaters). The patient is educated to check skin integrity daily. Patients with impaired motor function related to motor weakness or paralysis may require training in ADLs and safe use of assistive devices, such as a cane, walker, or wheelchair.

The patient and family members are educated about pain management strategies, bowel and bladder management, and assessment for signs and symptoms that should be reported promptly.

Continuing Care

Referral for inpatient or outpatient rehabilitation may be warranted to improve self-care abilities. A home care referral may be indicated and provides the home care nurse with the opportunity to assess the patient's physical and psychological status and the patient's and family's ability to adhere to recommended management strategies. During the home visit, the nurse determines whether changes in neurologic function have occurred. The patient's respiratory status and nutritional status are assessed. The adequacy of pain management is assessed, and modifications are made to ensure adequate pain relief. The need for hospice services or placement in an extended care facility is discussed with the patient and family if warranted, and the patient is asked about preferences for end-of-life care. In addition, social workers may be consulted to assist the patient and family members in identifying support groups and agencies that can provide help in coping with the disease process.

DEGENERATIVE DISORDERS

Disorders of the central and peripheral nervous system that are **neurodegenerative** (leading to deterioration of normal cells or function of the nervous system) are characterized by the slow onset of signs and symptoms. Patients are managed at home for as long as possible and are admitted to the acute care setting for exacerbations, treatments, and surgical interventions as needed.

Parkinson's Disease

Parkinson's disease is a slowly progressing neurologic movement disorder that eventually leads to disability. It is the fourth most common neurodegenerative disease, and 60,000 new cases are reported each year in the United States (Parkinson's Disease Foundation [PDF], 2012). The disease affects men more often than women. Symptoms usually first appear in the fifth decade of life; however, cases have been diagnosed as early as 30 years of age.

The degenerative or idiopathic form of Parkinson's disease is the most common; there is also a secondary form with a known or suspected cause. Although the cause of most cases is unknown, research suggests several causative factors, including genetics, atherosclerosis, excessive accumulation of oxygen free radicals, viral infections, brain trauma, chronic use of antipsychotic medications, and some environmental exposures.

Pathophysiology

Parkinson's disease is associated with decreased levels of dopamine resulting from degeneration of dopamine storage cells in the substantia nigra in the basal ganglia region of the brain (Fig. 70-4). Fibers or neuronal pathways project from the substantia nigra to the corpus striatum, where neurotransmitters are key to the control of complex body movements. Through the neurotransmitters acetylcholine (excitatory) and dopamine (inhibitory), striatal neurons relay messages to the higher motor centers that control and refine motor movements. The loss of dopamine stores in this area of the brain results in more excitatory neurotransmitters than inhibitory neurotransmitters, leading to an imbalance that affects voluntary movement.

Clinical symptoms do not appear until 60% of the pigmented neurons are lost and the striatal dopamine level is decreased by 80%. Cellular degeneration impairs the extrapyramidal tracts that control semiautomatic functions and coordinated movements; motor cells of the motor cortex and the pyramidal tracts are not affected. Researchers are working on uncovering the exact mechanism of neurodegeneration; current theories suggest that it results from oxidative stress that seeds axons with abnormal proteins that create inclusions in the cell, disrupting usual cell function and leading to cell death (Jancovic & Poewe, 2012).

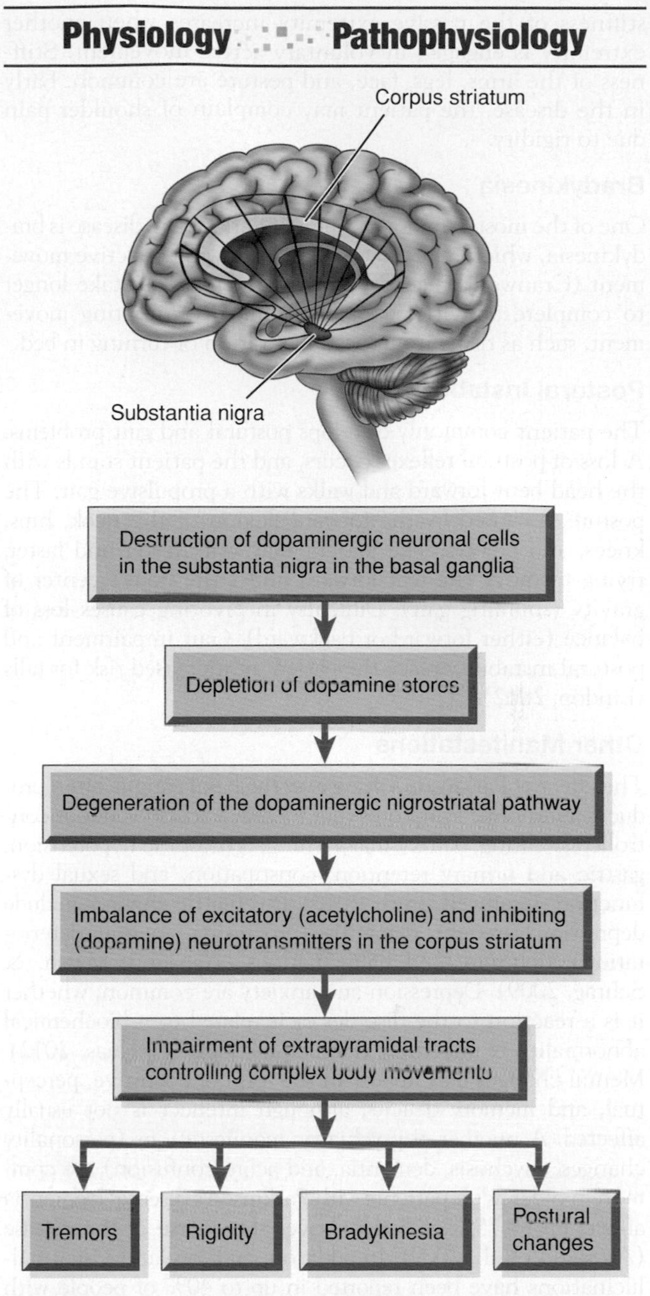
Physiology Pathophysiology
Corpus striatum
Substantia nigra
Destruction of dopaminergic neuronal cells in the substantia nigra in the basal ganglia
Depletion of dopamine stores
Degeneration of the dopaminergic nigrostriatal pathway
Imbalance of excitatory (acetylcholine) and inhibiting (dopamine) neurotransmitters in the corpus striatum
Impairment of extrapyramidal tracts controlling complex body movements
Tremors
Rigidity
Bradykinesia
Postural changes

FIGURE 70-4 • Pathophysiology of Parkinson's disease. The nuclei in the substantia nigra project fibers to the corpus striatum. The nerve fibers carry dopamine to the corpus striatum. The loss of dopamine nerve cells from the brain's substantia nigra is thought to be responsible for the symptoms of parkinsonism.

Clinical Manifestations

Parkinson's disease has a gradual onset, and symptoms progress slowly over a chronic, prolonged course. The cardinal signs are tremor, rigidity, **bradykinesia** (abnormally slow movements), and postural instability (Lindop, 2012; Cranwell-Bruce, 2010).

Tremor

Although symptoms are variable, a slow, unilateral resting tremor is present in the majority of patients at the time of diagnosis (Cranwell-Bruce, 2010). Resting tremor characteristically disappears with purposeful movement and during sleep but is evident when the extremities are motionless or at rest. The tremor may manifest as a rhythmic, slow turning motion (pronation–supination) of the forearm and the hand and a motion of the thumb against the fingers as if rolling a pill between the fingers.

Rigidity

Resistance to passive limb movement characterizes muscle rigidity. Passive movement of an extremity may cause the limb to move in jerky increments, referred to as lead-pipe or cogwheel movements (Cranwell-Bruce, 2010). Involuntary stiffness of the passive extremity increases when another extremity is engaged in voluntary active movement. Stiffness of the arms, legs, face, and posture are common. Early in the disease, the patient may complain of shoulder pain due to rigidity.

Bradykinesia

One of the most common features of Parkinson's disease is bradykinesia, which refers to the overall slowing of active movement (Cranwell-Bruce, 2010). Patients may also take longer to complete activities and have difficulty initiating movement, such as rising from a sitting position or turning in bed.

Postural Instability

The patient commonly develops postural and gait problems. A loss of postural reflexes occurs, and the patient stands with the head bent forward and walks with a propulsive gait. The posture is caused by the forward flexion of the neck, hips, knees, and elbows. The patient may walk faster and faster, trying to move the feet forward under the body's center of gravity (shuffling gait). Difficulty in pivoting causes loss of balance (either forward or backward). Gait impairment and postural instability place the patient at increased risk for falls (Lindop, 2012).

Other Manifestations

The effect of Parkinson's disease on the basal ganglia often produces autonomic symptoms that include excessive and uncontrolled sweating, paroxysmal flushing, orthostatic hypotension, gastric and urinary retention, constipation, and sexual dysfunction (Cranwell-Bruce, 2010). Psychiatric changes include depression, anxiety, **dementia** (progressive mental deterioration), delirium, and hallucinations (Aarsland, Marsh, & Schrag, 2009). Depression and anxiety are common; whether it is a reaction to the disorder or is related to a biochemical abnormality is uncertain (Hermanns, Deal, & Haas, 2012). Mental changes may appear in the form of cognitive, perceptual, and memory deficits, although intellect is not usually affected. A number of psychiatric manifestations (personality changes, psychosis, dementia, and acute confusion) are common in older adult patients with Parkinson's disease. Dementia affects up to 75% of patients over the course of the disease (Aarsland et al., 2009). In addition, auditory and visual hallucinations have been reported in up to 40% of people with Parkinson's disease and may be associated with depression, dementia, lack of sleep, or adverse effects of medications.

Hypokinesia (abnormally diminished movement) is also common and may appear after the tremor. The freezing phenomenon refers to a transient inability to perform active movement and is thought to be an extreme form of bradykinesia. The patient tends to shuffle and exhibits a decreased arm swing as well. As dexterity declines, **micrographia** (small handwriting) develops. The face becomes increasingly masklike and expressionless, and the frequency of blinking decreases. **Dysphonia** (voice impairment or altered voice production) may occur as a result of weakness and incoordination of the muscles responsible for speech. In many cases, the patient develops dysphagia, begins to drool, and is at risk for choking and aspiration.

Complications associated with Parkinson's disease are common and are typically related to disorders of movement. As the disease progresses, patients are at risk for respiratory and urinary tract infection, skin breakdown, and injury from falls. The adverse effects of medications used to treat the symptoms are associated with numerous complications such as dyskinesia or orthostatic hypotension (Cranwell-Bruce, 2010).

Assessment and Diagnostic Findings

Although laboratory tests and imaging studies are not helpful to the clinician in diagnosing Parkinson's disease, ongoing research with PET and single-photon emission computed tomography scanning has been helpful in understanding the disease and advancing treatment. Currently, the disease is diagnosed clinically from the patient's history and the presence of two of the four cardinal manifestations: tremor, rigidity, bradykinesia, and postural changes.

Early diagnosis can be challenging because patients rarely are able to pinpoint when the symptoms started. Often, a family member notices a change such as stooped posture; a stiff arm; a slight limp; tremor; or slow, small handwriting. The medical history, presenting symptoms, neurologic examination, and response to pharmacologic management are carefully evaluated when making the diagnosis. Diagnosis is often confirmed by a positive response to a levodopa (Larodopa) trial (Cranwell-Bruce, 2010).

Medical Management

Treatment is directed toward controlling symptoms and maintaining functional independence, because no medical or surgical approaches in current use prevent disease progression. Care is individualized for each patient based on presenting symptoms and social, occupational, and emotional needs. Pharmacologic management is the mainstay of treatment, although advances in research have led to increased surgical options. Patients are usually cared for at home and are admitted to the hospital only for complications or to initiate new treatments.

Pharmacologic Therapy

Antiparkinsonian medications act by (1) increasing striatal dopaminergic activity; (2) reducing the excessive influence of excitatory cholinergic neurons on the extrapyramidal tract, thereby restoring a balance between dopaminergic and cholinergic activities; or (3) acting on neurotransmitter pathways other than the dopaminergic pathway.

Levodopa is the most effective agent and the mainstay of treatment. Levodopa is converted to dopamine in the basal ganglia, producing symptom relief. Carbidopa is often added to levodopa to avoid metabolism of levodopa before it can reach the brain. The beneficial effects of levodopa therapy are most pronounced in the first year or two of treatment. Benefits begin to wane and adverse effects become more severe over time. Confusion, hallucinations, depression, and sleep alterations are associated with prolonged use. Within 5 to 10 years, most patients develop a response to the medication characterized by **dyskinesia** (impaired ability to execute voluntary movements), including facial grimacing, rhythmic jerking movements of the hands, head bobbing, chewing and smacking movements, and involuntary movements of the trunk and extremities. The patient may experience an on–off syndrome in which sudden periods of near immobility ("off effect") are followed by a sudden return of effectiveness of the medication ("on effect"). Changing the drug dosing regimen or switching to other drugs may be helpful in minimizing the on–off syndrome. Another

TABLE 70-1 Selected Medications Used to Treat Parkinson's Disease

Medications	Indications and Therapeutic Effects	Common Side Effects
Anticholinergic Agents		
trihexyphenidyl hydrochloride (Apo-Trihex) benztropine mesylate (Cogentin)	Control of tremor and rigidity Counteract the action of acetylcholine	Blurred vision, flushing, rash, constipation, urinary retention, and acute confusional states Contraindicated in patients with narrow-angle glaucoma
Antiviral Agent		
amantadine hydrochloride (Symmetrel)	Reduce rigidity, tremor, bradykinesia, and postural changes in early Parkinson's disease	Psychiatric disturbances (mood changes, confusion, depression, hallucinations), lower extremity edema, nausea, epigastric distress, urinary retention, headache, and visual impairment
Dopamine Agonists		
bromocriptine mesylate (Parlodel) pergolide (Permax)	Early Parkinson's disease as well as secondary drug therapy after carbidopa or levodopa loses effectiveness	Nausea, vomiting, diarrhea, lightheadedness, hypotension, impotence, and psychiatric effects
Nonergot Derivatives ropinirole hydrochloride (Requip) pramipexole (Mirapex)	Early stages of Parkinson's disease	May cause drowsiness or dizziness
Monoamine Oxidase Inhibitors		
selegiline (Eldepryl) rasagiline (Azilect)	Inhibit dopamine breakdown	Can cause hypertensive crisis
Catechol-O-Methyltransferase Inhibitors		
entacapone (Comtan) tolcapone (Tasmar)	Increase the duration of action of carbidopa or levodopa Reduce motor fluctuations in patients with advanced Parkinson's disease	Nausea, vomiting, diarrhea, lightheadedness, hypotension, confusion, psychiatric effects, liver failure, muscle weakness, rash
Antidepressants		
bupropion hydrochloride (Wellbutrin)		
Tricyclic Antidepressants amitriptyline hydrochloride (Elavil)	Anticholinergic and antidepressant	Hypertension, insomnia, dry mouth
Selective Serotonin Reuptake Inhibitors fluoxetine hydrochloride (Prozac)	Antidepressant	Clinical worsening and suicide risk
Antihistamines		
diphenhydramine hydrochloride (Benadryl) orphenadrine citrate (Banflex) phenylephrine hydrochloride (Neo-Synephrine)	May reduce tremors	Anticholinergic and sedative effects

Adapted from Cranwell-Bruce, L. (2010). Drugs for Parkinson's disease. *Medsurg Nursing*, 19(6), 347–355; and Karch, A. M. (2012). *2012 Lippincott's nursing drug guide*. Philadelphia: Lippincott Williams & Wilkins.

potential complication of long-term dopaminergic medication use is neuroleptic malignant syndrome, which is characterized by severe rigidity, stupor, and hyperthermia (Cranwell-Bruce, 2010). To minimize adverse effects of levodopa over time, current practice includes delaying use of levodopa-containing drugs as long as possible, with the use of other drugs for symptom control in the interim (Table 70-1).

Concept Mastery Alert

Nurses caring for patients with Parkinson's disease need a clear understanding of medications that are prescribed in addition to levodopa, which is the primary treatment agent. Table 70-1 provides a summary of the indications, therapeutic actions, and side effects of selected drugs.

Surgical Management

The limitations of levodopa therapy, improvements in surgical techniques, and new approaches in transplantation have renewed interest in the surgical treatment of Parkinson's disease. In patients with disabling tremor, rigidity, or severe levodopa-induced dyskinesia, surgery may be considered. Although surgery provides symptom relief in selected patients, it has not been shown to alter the course of the disease or to produce permanent improvement.

Stereotactic Procedures

Thalamotomy and pallidotomy are ablative procedures that were formerly used to relieve symptoms of Parkinson's disease (Jancovic & Poewe, 2012). However, these procedures permanently destroy brain tissue. Deep brain stimulation

(DBS) has largely replaced ablative procedures in the surgical treatment of Parkinson's disease (Jancovic & Poewe, 2012). DBS involves surgical implantation of an electrode into the brain in either the globus pallidus or subthalamic nucleus. Stimulation of these areas may increase dopamine release or block anticholinergic release, thereby improving tremor and rigidity. Levodopa medication dose may be able to be reduced, thus improving dyskinesias.

Patients eligible for DBS are those who have responded to levodopa but are impaired by dyskinesias, have had the disease for at least 5 years, and are disabled by tremor. Patients with dementia and atypical Parkinson's disease are usually not considered for surgical procedures. Parkinson's disease rating scales and specific neurologic tests are used to identify eligible patients.

A CT or MRI scan is used to localize the appropriate surgical site in the brain. Then, the patient's head is positioned in a stereotactic frame (Fig. 70-5). After the surgeon makes an incision in the skin and a burr hole, an electrode is passed through to the target area to the subthalamic nuclei or globus pallidus. The desired response of the patient to the electrical stimulation (i.e., a decrease in rigidity) is used to confirm electrode placement. Electrode placement is completed on one side of the brain at a time; bilateral electrodes are usually placed. Electrodes are then connected to a pulse generator that is implanted in a subcutaneous subclavicular or abdominal pouch (Fig. 70-6). The battery-powered pulse generator sends high-frequency electrical impulses through a wire placed under the skin to a lead anchored to the skull (see Fig. 70-6). These devices are not without complications that can result from both the surgical procedure needed for implantation (e.g., weakness, paresthesias, confusion, hemorrhage) and the device itself (e.g., infection lead leakage) (Jancovic & Poewe, 2012).

Neural Transplantation

Ongoing research is exploring transplantation of porcine neuronal cells, human fetal cells, and stem cells to replace degenerated striatal cells (Jancovic & Poewe, 2012). Legal, ethical, and political concerns surrounding the use of fetal brain cells and stem cells have limited the exploration of these procedures.

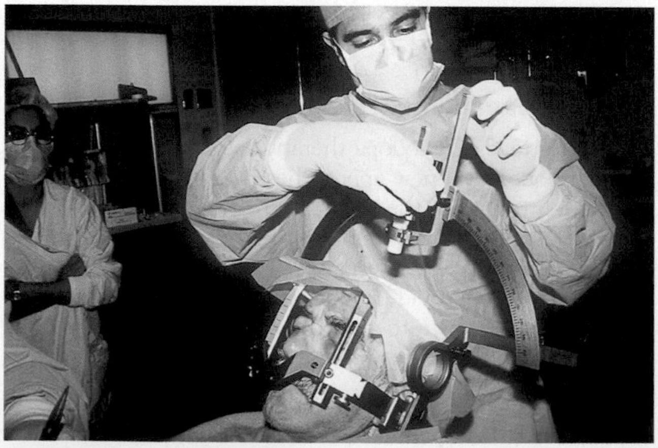

FIGURE 70-5 • A stereotactic frame is applied to a patient's head in preparation for deep brain stimulation. The frame provides external points of reference.

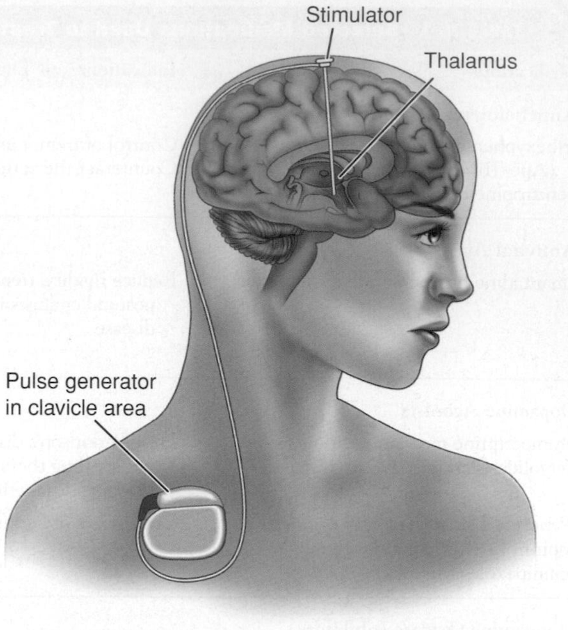
Stimulator
Thalamus
Pulse generator
in clavicle area

FIGURE 70-6 • Deep brain stimulation is provided by a pulse generator surgically implanted in a pouch beneath the clavicle. The generator sends high-frequency electrical impulses to the thalamus, thereby blocking the nerve pathways associated with tremors in Parkinson's disease.

NURSING PROCESS

The Patient With Parkinson's Disease

Assessment

Assessment focuses on how the disease has affected the patient's ADLs and functional abilities. The patient is observed for degree of disability and functional changes that occur throughout the day, such as responses to medication. Almost every patient with a movement disorder has some functional alteration and may have some type of behavioral dysfunction. The following questions may be useful to assess alterations:

- Do you have leg or arm stiffness?
- Have you experienced any irregular jerking of your arms or legs?
- Have you ever been "frozen" or rooted to the spot and unable to move?
- Does your mouth water excessively? Have you (or others) noticed yourself grimacing or making faces or chewing movements?
- What specific activities do you have difficulty doing?
- Have you had any recent falls?

During this assessment, the nurse observes the patient for quality of speech, loss of facial expression, swallowing deficits (drooling, poor head control, coughing), tremors, slowness of movement, weakness, forward posture, rigidity, evidence of mental slowness, and confusion. Parkinson's disease symptoms, as well as side effects of medications, put these patients at high risk for falls; therefore, a fall risk assessment should be conducted (Lindop, 2012).

Diagnosis

Nursing Diagnoses

Based on the assessment data, major nursing diagnoses may include the following:

- Impaired physical mobility related to muscle rigidity and postural impairment
- Self-care deficits (feeding, dressing, hygiene, and toileting) related to tremor and muscle rigidity
- Constipation related to medication and reduced activity
- Imbalanced nutrition: less than body requirements related to tremor, slowness in eating, difficulty in chewing and swallowing
- Impaired verbal communication related to decreased speech volume, slowness of speech, inability to move facial muscles
- Ineffective coping related to depression and dysfunction due to disease progression

Other nursing diagnoses may include sleep pattern disturbances, deficient knowledge, risk for injury, risk for activity intolerance, disturbed thought processes, and compromised family coping.

Planning and Goals

The goals for the patient may include improving functional mobility, maintaining independence in ADLs, achieving adequate bowel elimination, attaining and maintaining acceptable nutritional status, achieving effective communication, and developing positive coping mechanisms.

Nursing Interventions

Improving Mobility

A progressive program of daily exercise will increase muscle strength, improve coordination and dexterity, reduce muscular rigidity, and prevent contractures that occur when muscles are not used (Lindop, 2012). Walking, riding a stationary bicycle, swimming, and gardening are all exercises that help maintain joint mobility. Stretching (stretch–hold–relax) and range-of-motion exercises promote joint flexibility. Postural exercises are important to counter the tendency of the head and neck to be drawn forward and down. A physical therapist may be helpful in developing an individualized exercise program and can provide instruction to the patient and caregiver on exercising safely. Faithful adherence to an exercise and walking program helps delay the progress of the disease. Warm baths and massage, in addition to passive and active exercises, help relax muscles and relieve painful muscle spasms that accompany rigidity.

Balance may be adversely affected because of the rigidity of the arms (arm swinging is necessary in normal walking). Special walking techniques must be learned to offset the shuffling gait and the tendency to lean forward. The patient is educated to concentrate on walking erect, to watch the horizon, and to use a wide-based gait (i.e., walking with the feet separated). A conscious effort must be made to swing the arms, raise the feet while walking, and use a heel–toe placement of the feet with long strides. The patient is advised to practice walking to marching music or to the sound of a ticking metronome, because this provides sensory reinforcement. Performing breathing exercises while walking helps move the rib cage and aerate parts of the lungs. Frequent rest periods aid in preventing frustration and fatigue.

Enhancing Self-Care Activities

Encouraging, educating, and supporting the patient during ADLs promote self-care (Lindop, 2012). (See Chapter 10 for rehabilitation techniques.)

Environmental modifications are necessary to compensate for functional disabilities. Patients may have severe mobility problems that make normal activities impossible. Adaptive or assistive devices may be useful. A hospital bed at home with bedside rails, an overbed frame with a trapeze, or a rope tied to the foot of the bed can provide assistance in pulling up without help. An occupational therapist can evaluate the patient's needs in the home, make recommendations regarding adaptive devices, and educate the patient and caregiver how to improvise.

Improving Bowel Elimination

The patient may have severe problems with constipation. Among the factors causing constipation are weakness of the muscles used in defecation, lack of exercise, inadequate fluid intake, and decreased autonomic nervous system activity. The medications used for the treatment of the disease also inhibit normal intestinal secretions. A regular bowel routine may be established by encouraging the patient to follow a regular time pattern, consciously increase fluid intake, and eat foods with moderate fiber content. Laxatives should be avoided. Psyllium (Metamucil), for example, decreases constipation but carries the risk of bowel obstruction (Karch, 2012). A raised toilet seat is useful because the patient has difficulty in moving from a standing to a sitting position.

Improving Nutrition

Patients may have difficulty maintaining their weight. Eating becomes a very slow process, requiring concentration due to a dry mouth from medications and difficulty chewing and swallowing. These patients are at risk for aspiration because of impaired swallowing and the accumulation of saliva. They may be unaware that they are aspirating; subsequently, bronchopneumonia may develop.

Monitoring weight on a weekly basis indicates whether caloric intake is adequate. Supplemental feedings increase caloric intake. As the disease progresses, a nasogastric or percutaneous endoscopic gastrostomy (PEG) tube may be necessary to maintain adequate nutrition. A dietitian can be consulted regarding nutritional needs.

Enhancing Swallowing

Swallowing difficulties and choking are common in Parkinson's disease (Lindop, 2012). These can lead to problems with poor head control, tongue tremor, hesitancy in initiating swallowing, difficulty shaping food into a bolus, and disturbances in pharyngeal motility. To offset these problems, the patient should sit in an upright position during mealtime. A semisolid diet with thick liquids is easier to swallow than solids; thin liquids should be avoided. Thinking through the swallowing sequence is helpful. The patient is instructed to place the food on the tongue, close the lips and teeth, lift the tongue up and then back, and swallow. The patient is encouraged to chew first on one side of the mouth and then on the other. To control the buildup of saliva, the patient is reminded to hold the head upright and make a conscious effort to swallow.

Massaging the facial and neck muscles before meals may be beneficial.

Encouraging the Use of Assistive Devices

An electric warming tray keeps food hot and allows the patient to rest during the prolonged time that it may take to eat. Special utensils also assist at mealtime. A plate that is stabilized, a nonspill cup, and eating utensils with built-up handles are useful self-help devices. The occupational therapist can assist in identifying appropriate adaptive devices.

Improving Communication

Speech disorders are present in most patients with Parkinson's disease. The low-pitched, monotonous, soft speech of patients requires that they make a conscious effort to speak slowly, with deliberate attention to what they are saying. The patient is reminded to face the listener, exaggerate the pronunciation of words, speak in short sentences, and take a few deep breaths before speaking. A speech therapist may be helpful in designing speech improvement exercises and assisting the family and health care personnel to develop and use a method of communication that meets the patient's needs. A small electronic amplifier is helpful if the patient has difficulty being heard.

Supporting Coping Abilities

Support can be given by encouraging the patient and pointing out that activities will be maintained through active participation. A combination of physiotherapy, psychotherapy, medication therapy, and support group participation may help reduce the depression that often occurs (Lindop, 2012). Patients with Parkinson's disease can become withdrawn. It is best if patients are active participants in their therapeutic program, including social and recreational events. A planned program of activity throughout the day prevents too much daytime sleeping as well as disinterest and apathy.

Patients often feel embarrassed, apathetic, inadequate, bored, and lonely. In part, these feelings may result from physical slowness and the great effort that even small tasks require. The patient is assisted and encouraged to set achievable goals (e.g., improvement of mobility). Every effort should be made to encourage patients to carry out the tasks involved in meeting their own daily needs and to remain independent. Doing things for the patient merely to save time undermines the basic goal of improving coping abilities and promoting a positive self-concept.

Promoting Home and Community-Based Care

Educating Patients About Self-Care. Patient and family education is important in the management of Parkinson's disease. The needs depend on the severity of symptoms and the stage of the disease. Care must be taken not to overwhelm the patient and family with too much information early in the disease process. The patient's and family's need for education is ongoing as adaptations become necessary. The education plan should include a clear explanation of the disease and the goal of assisting the patient to remain functionally independent as long as possible. Every effort is made to explain the nature of the disease and its management to offset disabling anxieties and fears. The patient and family must be educated about the effects and side effects of medications and about the importance of reporting side effects to the primary provider (Chart 70-4).

Continuing Care. Family members often serve as caregivers, with home care or community services available to assist in meeting health care needs as the disease progresses. The

Chart 70-4 HOME CARE CHECKLIST
The Patient With Parkinson's Disease

At the completion of home care education, the patient or caregiver will be able to:	PATIENT	CAREGIVER
• Define Parkinson's disease and discuss its long-term effects.	✔	✔
• Identify the medication regimen and name adverse effects as well as precautions.	✔	✔
• Discuss the risk of injury; prevent falls; implement adaptive measures in the home.	✔	✔
• Describe nutritional needs, dietary restrictions, dysphagia management, and ways to prevent aspiration.	✔	✔
• Manage constipation: fluid intake, bowel routine.	✔	✔
• Manage urinary problems: functional incontinence, retention (indwelling urinary catheter care, suprapubic catheter care).	✔	✔
• Explain effects of immobility and define preventive care: skin breakdown (frequent turning, pressure release, skin care), pneumonia (deep breathing, movement), contractures (range-of-motion exercises).	✔	✔
• Define benefits of daily exercise program.	✔	✔
• Walk and balance safely.	✔	
• Demonstrate speech and communication skills: speech exercises, communication techniques, breathing exercises.	✔	
• Name signs and symptoms of infection (urinary and respiratory), and state when primary provider should be notified.	✔	✔
• Describe strategies to promote self-care activities and independence.	✔	✔
• Identify resources: American Parkinson Disease Association, National Parkinson Foundation, and local support groups.	✔	✔

family caregiver may be under considerable stress from living with and caring for a person with a significant disability. Providing information about treatment and care prevents many unnecessary problems. The caregiver is included in the plan and may be advised to learn stress reduction techniques, to include others in the caregiving process, to obtain periodic relief from responsibilities, and to have a yearly health assessment. Allowing family members to express feelings of frustration, anger, and guilt is often helpful to them.

The patient should be evaluated in the home for adaptation and safety needs and compliance with the plan of care. In the advanced stages, patients usually enter long-term care facilities if family support is absent. Periodically, admission to an acute care facility may be necessary for changes in medical management or treatment of complications. Nurses provide support, education, and monitoring of patients over the course of the illness.

The nurse involved in home and continuing care educates the patient and family members about the importance of addressing health promotion needs such as screening for hypertension and stroke risk assessments in this predominantly older adult population. Patients are referred to appropriate health care providers. Information is available from both the PDF and the American Parkinson's Disease Association (see the Resources section).

Evaluation

Expected patient outcomes may include:

1. Strives toward improved mobility
   a. Participates in exercise program daily
   b. Walks with wide base of support; exaggerates arm swinging when walking
   c. Takes medications as prescribed
2. Progresses toward self-care
   a. Allows time for self-care activities
   b. Uses self-help devices
3. Maintains bowel function
   a. Consumes adequate fluid
   b. Increases dietary intake of fiber
   c. Reports regular pattern of bowel function
4. Attains improved nutritional status
   a. Swallows without aspiration
   b. Takes time while eating
5. Achieves a method of communication
   a. Communicates needs
   b. Practices speech exercises
6. Copes with effects of Parkinson's disease
   a. Sets realistic goals
   b. Demonstrates persistence in meaningful activities
   c. Verbalizes feelings to appropriate person

Huntington Disease

Huntington disease is a chronic, progressive, hereditary disease of the nervous system that results in progressive involuntary choreiform movement and dementia. The disease affects approximately 1 in 10,000 men or women of all races at midlife. It is transmitted as an autosomal dominant genetic disorder; therefore, each child of a parent with Huntington disease has a 50% risk of inheriting the disorder (Ha & Fung, 2012).

A genetic mutation in Huntington disease, the presence of a repeat in the Huntington gene (*HTT*), has been identified. Genetic testing can identify people who will develop this disease but cannot predict timing of disease onset. Although the gene was mapped in 1983 and presymptomatic testing has been offered since 1986, many patients choose not to be tested because of concerns about employment and health care discrimination. People of childbearing age with a family history of Huntington disease often seek information about their risk of disease transmission. Genetic counseling is crucial, and patients and their families may require long-term psychological counseling and emotional, financial, and legal support (see Chapter 8).

Pathophysiology

The basic pathology involves premature death of cells in the striatum (caudate and putamen) of the basal ganglia, the region deep within the brain that is involved in the control of movement. Cells also are lost in the cortex, the region of the brain associated with thinking, memory, perception, judgment, and behavior, and in the cerebellum, the area that coordinates voluntary muscle activity. Why the protein destroys only certain brain cells is unknown, but several theories have been proposed to explain the phenomenon (Ha & Fung, 2012). One possible theory is that glutamine, a building block for protein, abnormally collects in the cell nucleus, causing cell death (Ha & Fung, 2012).

Clinical Manifestations

The most prominent clinical features of the disease are **chorea** (rapid, jerky, involuntary, purposeless movements), impaired voluntary movement, intellectual decline, and often personality changes (Aubeeluck & Wilson, 2008). As the disease progresses, a constant writhing, twisting, uncontrollable movement may involve the entire body. These motions are devoid of purpose or rhythm, although patients may try to turn them into purposeful movement. All of the body musculature is involved. Facial movements produce tics and grimaces. Speech becomes slurred, hesitant, often explosive, and eventually unintelligible. Chewing and swallowing are difficult, and there is a constant danger of choking and aspiration. Choreiform movements persist during sleep but are diminished.

As with speech, the gait becomes disorganized to the point that ambulation eventually is impossible. Although independent ambulation should be encouraged for as long as possible, a wheelchair usually becomes necessary. Eventually, the patient is confined to bed when the chorea interferes with walking, sitting, and all other activities. Bladder and bowel control is lost. Cognitive function is usually affected, and in later stages, marked dementia is present. Initially, the patient is aware that the disease is responsible for the myriad dysfunctions that are occurring. The mental and emotional changes may be more devastating to the patient and family than the abnormal movements. Personality changes may result in nervous, irritable, or impatient behaviors. In the early stages, patients are particularly subject to uncontrollable fits of anger, profound and often suicidal depression, apathy, anxiety, psychosis, or euphoria. Judgment and memory are impaired, and dementia eventually ensues. Hallucinations, delusions, and paranoid thinking may precede the appearance of disjointed movements. Emotional and cognitive symptoms often become less acute as the disease progresses (Aubeeluck, Wilson, & Stupple, 2011).

Onset usually occurs between 35 and 45 years of age, although about 10% of patients are children. The disease progresses slowly. Despite a ravenous appetite, patients usually become emaciated and exhausted. Patients succumb in 10 to 20 years to heart failure, pneumonia, or infection, or as a result of a fall or choking.

Assessment and Diagnostic Findings

The diagnosis is made based on the clinical presentation of characteristic symptoms, a positive family history, the known presence of a genetic marker, and exclusion of other causes.

Management

Although no treatment halts or reverses the underlying process, medications may reduce chorea. Tetrabenazine (Xenazine) is the only approved drug for the treatment of the chorea, although thiothixene hydrochloride (Navane) and haloperidol decanoate (Haldol), which predominantly block dopamine receptors, have been used in the past. Benzodiazepines and neuroleptic drugs have also been reported to control chorea. Motor signs must be assessed and evaluated on an ongoing basis so that optimal therapeutic drug levels can be reached. **Akathisia** (motor restlessness) in the patient who is overmedicated is dangerous because it may be mistaken for the restless fidgeting of the illness and consequently may be overlooked. In certain types of the disease, hypokinetic motor impairment resembles Parkinson's disease. In patients who present with rigidity, some temporary benefit may be obtained from antiparkinson medications, such as levodopa.

Selective serotonin reuptake inhibitors and tricyclic antidepressants have been recommended for control of psychiatric symptoms (Aubeeluck et al., 2011). The threat of suicide is present particularly early in the course of the disease (Aubeeleck & Wilson, 2008). Psychotic symptoms usually respond to antipsychotic medications. Psychotherapy aimed at allaying anxiety and reducing stress may be beneficial. Nurses must look beyond the disease to focus on the patient's needs and capabilities (Chart 70-5).

No disease-modifying therapy has been found to be effective. Studies involving high doses of creatine, which may improve intracellular protein metabolism, are in progress (Ha & Fung, 2012).

Promoting Home and Community-Based Care

Educating Patients About Self-Care

The needs of the patient and family for education depend on the nature and severity of the physical, cognitive, and psychological changes experienced by the patient. The patient and family members are educated about the medications prescribed and about signs indicating a need for change in medication or dosage. The educational plan addresses strategies to manage symptoms such as chorea, swallowing problems, limitations in ambulation, memory loss, irritability, depression, and loss of bowel and bladder function. Consultation with a speech therapist may be indicated to assist in identifying alternative communication strategies if speech is affected. A PEG tube may be considered for nutritional support later in the disease.

Continuing Care

A program combining medical, nursing, psychological, social, occupational, speech, and physical rehabilitation services and palliative care is needed to help the patient and family cope with this severely disabling illness (Aubeeluck & Wilson, 2008). Huntington disease exacts enormous emotional, physical, social, and financial tolls on every member of the family. The family needs supportive care as they adjust to the impact of the illness. Regular follow-up visits help allay the fear of abandonment.

Home care assistance, day care centers, respite care, and eventually skilled long-term care can assist the patient and family in coping with the constant strain of the illness. Although the relentless progression of the disease cannot be halted, families can benefit from supportive care. Planning for end-of-life care should occur early in the disease when possible (see Chapter 16).

Voluntary organizations can be major aids to families and have been largely responsible for bringing the illness to national attention. The Huntington's Disease Society of America helps patients and families by providing information, referrals, family and public education, and support for research (see the Resources section).

Amyotrophic Lateral Sclerosis

ALS is a disease of unknown cause in which there is a loss of motor neurons (nerve cells controlling muscles) in the anterior horns of the spinal cord and the motor nuclei of the lower brain stem. It is often referred to as Lou Gehrig's disease, after the famous baseball player who suffered from the disease. As motor neuron cells die, the muscle fibers that they supply undergo atrophic changes. Neuronal degeneration may occur in both the upper and lower motor neuron systems (see Chapter 65). The leading theory held by researchers is that overexcitation of nerve cells by the neurotransmitter glutamate results in cell injury and neuronal degeneration. Possible causes of ALS include autoimmune disease, free radical damage, and oxidative stress. Although most cases of ALS arise sporadically, 5% to 10% of cases are familial (Corcia & Meininger, 2008). Several risk factors have been identified, but the exact cause is still unknown (Corcia & Meininger, 2008).

Clinical Manifestations

Clinical manifestations depend on the location of the affected motor neurons, because specific neurons activate specific muscle fibers. The chief symptoms are fatigue, progressive muscle weakness, cramps, fasciculations (twitching), and incoordination. Loss of motor neurons in the anterior horns of the spinal cord results in progressive weakness and atrophy of the muscles of the arms, trunk, or legs. Spasticity usually is present, and the deep tendon stretch reflexes become brisk and overactive. Usually, the function of the anal and bladder sphincters remains intact, because the spinal nerves that control muscles of the rectum and urinary bladder are not affected.

Chart 70-5 **Care of the Patient With Huntington Disease**

Nursing Diagnosis: Risk for injury from falls and possible skin breakdown (pressure ulcers, abrasions), resulting from constant movement

Nursing Interventions

Pad the sides and head of the bed; ensure that the patient can see over the sides of bed.
Use padded heel and elbow protectors.
Keep the skin meticulously clean.
Apply emollient cleansing agent and skin lotion as needed.
Use soft sheets and bedding.
Have patient wear football padding or other forms of padding.
Encourage ambulation with assistance to maintain muscle tone.
Secure the patient (only if necessary) in bed or chair with padded protective devices, making sure that they are loosened frequently.

Nursing Diagnosis: Imbalanced nutrition: less than body requirements due to inadequate intake and dehydration resulting from swallowing or chewing disorders and danger of choking or aspirating food

Nursing Interventions

Administer phenothiazines (chlorpromazine) as prescribed before meals (calms some patients) (Karch, 2012).
Talk to the patient before mealtime to promote relaxation; use mealtime for social interaction. Provide undivided attention and help the patient enjoy the mealtime experience.
Use a warming tray to keep food warm.
Learn the position that is best for *this* patient. Keep patient as close to upright as possible while feeding. Stabilize patient's head gently with one hand while feeding.
Show the food, explain what the foods are, and temperature (e.g., whether hot or cold).
Encircle the patient with one arm and get as close as possible to provide stability and support while feeding. Use pillows and wedges for additional support.
Do not interpret stiffness, turning away, or sudden turning of the head as rejection; these are uncontrollable choreiform movements.
For feeding, use a long-handled spoon (iced tea spoon). Place spoon on middle of tongue and exert slight pressure.
Place bite-sized food between patient's teeth. Serve stews, casseroles, and thick liquids.
Disregard messiness, and treat the person with dignity.
Wait for the patient to chew and swallow before introducing another spoonful. Make sure that bite-sized food is small.
Give between-meal feedings. Constant movement expends more calories. Patients often have voracious appetites, particularly for sweets.
Use blenderized meals if patient cannot chew; do not repeatedly give the same strained baby foods. Gradually introduce increased textures and consistencies to the diet.
For swallowing difficulties:
- Apply gentle deep pressure around the patient's mouth.
- Rub fingers in circles on the patient's cheeks and then down each side of the patient's throat.

Develop skill in the Heimlich maneuver (to be used in the event of choking).

Nursing Diagnosis: Anxiety and impaired communication from excessive grimacing and unintelligible speech

Nursing Interventions

Read to the patient.
Employ biofeedback and relaxation therapy to reduce stress.
Consult with speech therapist to help maintain and prolong communication abilities.
Try to devise a communication system, perhaps using cards with words or pictures of familiar objects, before verbal communication becomes too difficult. Patients can indicate correct card by hitting it with hand, grunting, or blinking the eyes.
Learn how this particular patient expresses needs and wants—particularly nonverbal messages (widening of eyes, responses).
Patients can understand even if unable to speak. Do not isolate patients by ceasing to communicate with them.

Nursing Diagnosis: Confusion and impaired social interaction

Nursing Interventions

Reorient the patient after awakening.
Have clock, calendar, and wall posters in view to assist in orientation.
Use every opportunity for one-to-one contact.
Use music for relaxation.
Have the patient wear a medical identification bracelet.
Keep the patient in the social mainstream.
Recruit and train volunteers for social interaction. Role-model appropriate and creative interactions.
Do not abandon a patient because the disease is terminal. Patients are *living* until the end.

In about 25% of patients, weakness starts in the muscles supplied by the cranial nerves, and difficulty in talking, swallowing, and ultimately breathing occurs. When the patient ingests liquids, soft palate and upper esophageal weakness causes the liquid to be regurgitated through the nose. Weakness of the posterior tongue and palate impairs the ability to laugh, cough, or even blow the nose. If bulbar muscles are impaired, speaking and swallowing are progressively difficult, and aspiration becomes a risk. The voice assumes a nasal sound, and articulation becomes so disrupted that speech is unintelligible. Some emotional liability may be present. It was traditionally believed that ALS spared cognitive function, but it is now recognized that some patients experience cognitive impairment.

The prognosis generally is based on the area of CNS involvement and the speed with which the disease progresses. Eventually, respiratory function is compromised. Death usually occurs as a result of infection, respiratory insufficiency, or aspiration.

Assessment and Diagnostic Findings

ALS is diagnosed on the basis of the signs and symptoms, because no clinical or laboratory tests are specific to this disease. Electromyography and muscle biopsy studies of the affected muscles indicate reduction in the number of functioning motor units. An MRI scan may show high signal intensity in the corticospinal tracts; this differentiates ALS from a multifocal motor neuropathy. Neuropsychological testing can assist in assessment and diagnosis (Corcia & Meininger, 2008).

Management

No cure exists for ALS. The main focus of medical and nursing management is on interventions to maintain or improve function, well-being, and quality of life. The average survival time is 3 to 5 years with death most commonly due to respiratory insufficiency.

Riluzole (Rilutek), a glutamate antagonist, has been shown to prolong survival for persons with ALS for 3 to 6 months (Kiernan, Vucic, Cheah, et al., 2011). The action of riluzole is not clear, but its pharmacologic properties suggest that it may have a neuroprotective effect in the early stages of ALS. Symptomatic treatment and rehabilitative measures are used to support the patient and improve the quality of life. Baclofen (Lioresal), dantrolene sodium (Dantrium), or diazepam (Valium) may be useful for patients troubled by spasticity, which causes pain and interferes with self-care. Modafinil (Provigil) may be used for fatigue, and additional medications may be added to manage the pain, depression, drooling, and constipation that often accompany the disease (Corcia & Meininger, 2008). Research suggests that greater functional impairment is associated with greater depressive symptoms (Oh, Sin, Schepp, et al., 2012). It is important to treat depression to maintain a good quality of life.

Most patients with ALS are managed at home and in the community, with hospitalization for acute problems. The most common reasons for hospitalization are dehydration and malnutrition, pneumonia, and respiratory failure; recognizing these problems at an early stage in the illness allows for the development of preventive strategies. End-of-life issues include pain, dyspnea, and delirium (Clarke & Levine, 2011).

Mechanical ventilation (using negative-pressure ventilators) is an option if alveolar hypoventilation develops. Noninvasive positive-pressure ventilation is also an option. The use of noninvasive positive-pressure ventilation is particularly helpful at night and postpones the decision about whether to undergo a tracheotomy for long-term mechanical ventilation.

A patient experiencing aspiration and swallowing difficulties may require enteral feeding. A PEG tube is inserted before the forced vital capacity drops below 50% of the predicted value. The tube can be safely placed in patients who are using noninvasive positive-pressure ventilation for ventilatory support (Hickey, 2009).

Decisions about life support measures are made by the patient and family and should be based on a thorough understanding of the disease, the prognosis, and the implications of initiating such therapy. Patients are encouraged to complete an advance directive ("living will") to preserve their autonomy in decision making. (See Chapter 16 for additional discussion of end-of-life care.)

The ALS Association has broad programs of research funding, patient and clinical services, patient information and support, and medical and public information (see the Resources section). The *ALS Association Newsletter* is a source of practical information.

Muscular Dystrophies

The muscular dystrophies are a group of incurable muscle disorders characterized by progressive weakening and wasting of the skeletal or voluntary muscles. Most of these diseases are inherited. Duchenne muscular dystrophy, the most common and severe type, occurs in 1 of every 3,500 male births (Spurney, 2011). The pathologic features include degeneration and loss of muscle fibers, variation in muscle fiber size, phagocytosis and regeneration, and replacement of muscle tissue by connective tissue. The common characteristics of these diseases include varying degrees of muscle wasting and weakness and abnormal elevation in serum levels of muscle enzymes (Barker & Barasi, 2008). Differences among these diseases center on the genetic pattern of inheritance, the muscles involved, the age at onset, and the rate of disease progression. The unique needs of these patients, who in the past did not live to adulthood, must be addressed, considering they live longer because of better supportive care.

Medical Management

Treatment of the muscular dystrophies focuses on supportive care and prevention of complications in the absence of a cure or specific pharmacologic interventions (Partridge, 2011). The goal of supportive management is to keep the patient active and functioning as normally as possible and to minimize functional deterioration. An individualized therapeutic exercise program is prescribed to prevent muscle tightness, contractures, and disuse atrophy. Night splints and stretching exercises are used to delay contractures of the joints, especially the ankles, knees, and hips. Braces may compensate for muscle weakness.

Spinal deformity is a severe problem. Weakness of trunk muscles and spinal collapse occur almost routinely in patients with severe neuromuscular disease. To help prevent spinal deformity, the patient is fitted with an orthotic jacket to improve sitting stability and reduce trunk deformity. This measure also supports cardiovascular status. In time, spinal fusion is performed to maintain spinal stability. Other procedures may be carried out to correct deformities.

Compromised pulmonary function may result either from progression of the disease or from deformity of the thorax secondary to severe scoliosis. Upper respiratory infections and fractures from falls must be vigorously treated in a way that minimizes immobilization because joint contractures become worse when the patient's activities are more restricted than usual.

Other difficulties may manifest in relation to the underlying disease. Weakness of the facial muscles makes it difficult to attend to dental hygiene and to speak clearly. Gastrointestinal tract problems may include gastric dilation, rectal prolapse, and fecal impaction. Finally, cardiomyopathy appears to be a common complication in all forms of muscular dystrophy (Spurney, 2011).

Several genetic therapies are under study to prevent or ameliorate the genetic mutations that cause symptoms (Verhaart & Aartsma-Rus, 2012). Genetic counseling is advised for parents and siblings of the patient because of the genetic nature of this disease. MDA (the Muscular Dystrophy Association) works to combat neuromuscular disease through research, programs of patient services and clinical care, and professional and public education (see the Resources section).

Nursing Management

The goals of the patient and the nurse are to maintain function at optimal levels and to enhance the quality of life.

Therefore, the patient's physical requirements, which are considerable, are addressed without losing sight of emotional and developmental needs. The patient and family are actively involved in decision making, including end-of-life decisions.

During hospitalization for treatment of complications, the knowledge and expertise of the patient and family members responsible for caregiving in the home are assessed. Because the patient and family caregivers often have developed caregiving strategies that work effectively for them, these strategies need to be acknowledged and accepted, and provisions must be made to ensure that they are maintained during hospitalization.

Families of adolescents and young adults with muscular dystrophy need assistance to shift the focus of care from pediatric to adult care and to understand the usual disease course. Nursing goals include assisting the adolescent to make the transition to adult values and expectations while providing age-appropriate ongoing care. The nurse may need to help build the confidence of an older adolescent or adult patient by encouraging him or her to pursue job training to become economically independent. Other nursing interventions might include guidance in accessing adult health care and finding appropriate programs in sex education.

Promoting Home and Community-Based Care

Educating Patients About Self-Care

Management goals are addressed in special rehabilitation programs or in the patient's home and community. Therefore, the patient and family require information and instruction about the disorder, its anticipated course, and care and management strategies that will optimize the patient's growth and development and physical and psychological status. Members of a variety of health-related disciplines are involved in patient and family education; recommendations are communicated to all members of the health care team so that they may work toward common goals (Partridge, 2011).

Continuing Care

Both the neuromuscular disease and the associated deformities may progress in adolescence and adulthood. Self-help and assistive devices can aid in maintaining maximum independence. These devices, recommended by physical and occupational therapists, often become necessary as more muscle groups are affected.

The family is instructed to monitor the patient for respiratory problems, because respiratory infection and cardiac failure are the most common causes of death (Bader & Littlejohns, 2010). As respiratory difficulties develop, patients and their families need information regarding respiratory support. Options currently exist that can provide ventilatory support (e.g., negative-pressure devices, positive-pressure ventilators) while allowing mobility. Patients can remain relatively independent in a wheelchair, for example, while being maintained on a ventilator at home for many years.

The patient is encouraged to continue with range-of-motion exercises to prevent contractures, which are particularly disabling. Practical adaptations must be made, however, to cope with the effects of chronic neuromuscular disability. The patient at various stages of the disease may require a manual or an electric wheelchair, gait aids, upper and lower extremity and spinal orthoses, seating systems, bathroom equipment, lifts, ramps, and additional assistive devices, all of which require a team approach. The home care nurse assesses how the patient and family are managing, makes referrals, and coordinates the activities of the physical therapist, occupational therapist, and social services.

The patient is greatly concerned about issues surrounding the threat of increasing disability and dependence on others, accompanied by a significant deterioration in health-related quality of life. The patient is faced with a progressive loss of function, leading eventually to death. Feelings of helplessness and powerlessness are common. Each functional loss is accompanied by grief and mourning. The patient and family are assessed for depression, anger, or denial. The patient and family are assisted and encouraged to address decisions about end-of-life options before their need arises.

A psychiatric nurse clinician or other mental health professional may assist the patient to cope and adapt to the disease. By understanding and addressing the physical and psychological needs of the patient and family, the nurse provides a hopeful, supportive, and nurturing environment.

Degenerative Disk Disease

Low back pain is the second most common neurologic disorder in the United States. It is estimated that there were almost 500,000 spinal surgeries for herniated disks, stenosis, and degenerative changes in 2010 in the United States (Agency for Healthcare Research and Quality, 2011). There are significant economic costs to patients, their families, and society. Acute low back pain lasts less than 3 months, whereas chronic or degenerative disease has a duration of 3 months or longer. Most back problems are related to disk disease.

Pathophysiology

The intervertebral disk is a cartilaginous plate that forms a cushion between the vertebral bodies (Fig. 70-7A). This tough, fibrous material is incorporated in a capsule. A ball-like cushion in the center of the disk is called the *nucleus pulposus*. In herniation of the intervertebral disk (ruptured disk), the nucleus of the disk protrudes into the annulus (the fibrous ring around the disk), with subsequent nerve compression. Protrusion or rupture of the nucleus pulposus usually is preceded by degenerative changes that occur with aging. Loss of protein polysaccharides in the disk decreases the water content of the nucleus pulposus. The development of radiating cracks in the annulus weakens resistance to nucleus herniation. After trauma (falls and repeated minor stresses such as lifting incorrectly), the cartilage may be injured.

For most patients, the immediate symptoms of trauma are short-lived, and those resulting from injury to the disk do not appear for months or years. Then, with degeneration in the disk, the capsule pushes back into the spinal canal, or it may rupture and allow the nucleus pulposus to be pushed back against the dural sac or against a spinal nerve as it emerges from the spinal column (see Fig. 70-7B). This disease of the spinal root, produces pain and extreme sensitivity to touch due to **radiculopathy** (pressure in the area of distribution of the involved nerve endings). Continued pressure may produce degenerative changes in the involved nerve, such as changes in sensation and deep tendon reflexes.

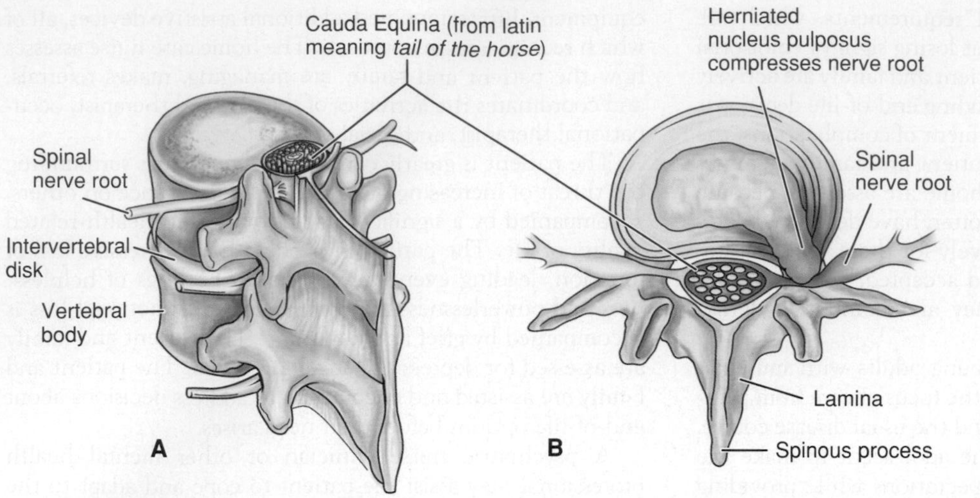
Cauda Equina (from latin meaning *tail of the horse*)
Spinal nerve root
Intervertebral disk
Vertebral body
A
Herniated nucleus pulposus compresses nerve root
Spinal nerve root
Lamina
Spinous process
B

FIGURE 70-7 • **A.** Normal lumbar spine vertebrae, invertebral disks, and spinal nerve root. **B.** Ruptured vertebral disk.

Clinical Manifestations

A herniated disk with accompanying pain may occur in any portion of the spine: cervical, thoracic (rare), or lumbar. The clinical manifestations depend on the location, the rate of development (acute or chronic), and the effect on the surrounding structures.

Assessment and Diagnostic Findings

A thorough health history and physical examination are important to rule out potentially serious conditions that may manifest as low back pain, including fracture, tumor, infection, or cauda equina syndrome (Hickey, 2009).

The MRI scan has become the diagnostic tool of choice for localizing even small disk protrusions, particularly for lumbar spine disease. If the clinical symptoms are not consistent with the pathology seen on MRI, CT scanning and myelography are performed. A neurologic examination is carried out to determine whether reflex, sensory, or motor impairment from root compression is present and to provide a baseline for future assessment. Electromyography may be used to localize the specific spinal nerve roots involved.

Medical Management

Herniations of the cervical and the lumbar disks occur most commonly and are usually managed conservatively with medication and physical therapy or exercise (Hickey, 2009). Surgery is sometimes necessary.

Surgical Management

Surgical excision of a herniated disk is performed if there is evidence of a progressing neurologic deficit (muscle weakness and atrophy, loss of sensory and motor function, loss of sphincter control) and radicular pain (pain that follows the dermatomal distribution [see Fig. 65-9 in Chapter 65] of the compressed nerve) that are unresponsive to conservative management (American Association of Neuroscience Nurses [AANN], 2012). The goal of surgical treatment is to reduce the pressure on the nerve root to relieve pain and reverse neurologic deficits. Microsurgical techniques make it possible to remove only the amount of tissue that is necessary, which preserves the integrity of normal tissue better and imposes less trauma on the body. During these procedures, spinal cord function can be monitored electrophysiologically.

To achieve the goal of pain relief, several surgical techniques are used, depending on the type and location of disk herniation, surgical morbidity, and results of previous surgery. Some of the surgical techniques available include (AANN, 2012):

- *Microdiskectomy:* Removal of herniated or extruded fragments of intervertebral disk material
- *Laminectomy:* Removal of the bone between the spinal process and facet pedicle junction to expose the neural elements in the spinal canal (Hickey, 2009); this allows the surgeon to inspect the spinal canal, identify and remove pathologic tissue, and relieve compression of the cord and roots
- *Hemilaminectomy:* Removal of part of the lamina and part of the posterior arch of the vertebra
- *Partial laminectomy or laminotomy:* Creation of a hole in the lamina of a vertebra
- *Diskectomy with fusion:* Fusion of the vertebral spinous process with a bone graft (from iliac crest or bone bank), with the object of spinal fusion being to bridge over the defective disk to stabilize the spine and reduce the rate of recurrence
- *Foraminotomy:* Removal of the intervertebral foramen to increase the space for exit of a spinal nerve, resulting in reduced pain, compression, and edema

Herniation of a Cervical Intervertebral Disk

The cervical spine is subjected to stresses that result from disk degeneration (due to aging, occupational stresses) and **spondylosis** (degenerative changes occurring in a disk and adjacent vertebral bodies). Cervical disk degeneration may lead to lesions that can cause damage to the spinal cord and its roots.

Clinical Manifestations

A cervical disk herniation usually occurs at the C5–C6 or C6–C7 interspaces and compresses a unilateral nerve root. Pain and stiffness may occur in the neck, the top of the shoulders, and the region of the scapulae. Patients sometimes interpret these signs as symptoms of heart trouble or bursitis. Pain may also occur in the arm and hand, accompanied by **paresthesia** (numbness, tingling, or a "pins and needles" sensation) of the upper extremity. Cervical MRI usually confirms the diagnosis. Occasionally, the disk herniates centrally onto the spinal cord, causing Lhermitte's syndrome, an electriclike shock sensation in the extremities or spine with neck flexion or straining, and bilateral arm and leg weakness (myelopathy).

Medical Management

The goals of treatment are to rest and immobilize the cervical spine to give the soft tissues time to heal and to reduce inflammation in the supporting tissues and the affected nerve roots in the cervical spine. It also reduces inflammation and edema in soft tissues around the disk, relieving pressure on the nerve roots. Proper positioning on a firm mattress may bring dramatic relief from pain.

The cervical spine may be rested and immobilized by a cervical collar, cervical traction, or a brace. A collar allows maximal opening of the intervertebral foramina and holds the head in a neutral or slightly flexed position. The patient may have to wear the collar 24 hours a day during the acute phase. The skin under the collar is inspected for irritation. After the patient is free of pain, cervical isometric exercises are started to strengthen the neck muscles.

Pharmacologic Therapy

Analgesic agents (nonsteroidal anti-inflammatory agents [NSAIDs], acetaminophen/oxycodone [Tylox], or acetaminophen/hydrocodone [Vicodin]) are prescribed during the acute phase to relieve pain, and sedative agents may be administered to control the anxiety that is often associated with cervical disk disease. Muscle relaxants (cyclobenzaprine [Flexeril], methocarbamol [Robaxin], metaxalone [Skelaxin]) are administered to interrupt muscle spasm and to promote comfort. NSAIDs (aspirin, ibuprofen [Motrin, Advil], naproxen [Naprosyn, Anaprox]) or corticosteroids are prescribed to treat the inflammation and swelling that usually occurs in the affected nerve roots and supporting tissues. Occasionally, a corticosteroid is injected into the epidural space for relief of radicular (spinal nerve root) pain. NSAIDs are administered with food and antacids to prevent gastrointestinal irritation (Karch, 2012). Hot, moist compresses (for 10 to 20 minutes) applied to the back of the neck several times daily increase blood flow to the muscles and help relax the patient and reduce muscle spasm.

Surgical Management

Surgical excision of the herniated disk may be necessary if there is a significant neurologic deficit, progression of the deficit, evidence of cord compression, or pain that either worsens or fails to improve. A cervical diskectomy, with or without fusion, may be performed to alleviate symptoms. An anterior surgical approach may be used through a transverse incision to remove disk material that has herniated into the spinal canal and foramina, or a posterior approach may be used at the appropriate level of the cervical spine. Potential complications with the anterior approach include carotid or vertebral artery injury, recurrent laryngeal nerve dysfunction, esophageal perforation, and airway obstruction. Complications of the posterior approach include damage to the nerve root or the spinal cord due to retraction or contusion of either of these structures, resulting in weakness of muscles supplied by the nerve root or cord.

Microsurgery, such as endoscopic microdiskectomy, may be performed in selected patients through a small incision, using magnification techniques. This usually results in less tissue trauma and pain, and patients consequently have a shorter length of hospital stay compared with those who have conventional surgery. Artificial cervical disk replacement is a recent development (Cepoiu-Martin, Faris, Lorenzetti, et al., 2011).

NURSING PROCESS

The Patient Undergoing a Cervical Diskectomy

Assessment

The patient is asked about past injuries to the neck (whiplash), because unresolved trauma can cause persistent discomfort, pain and tenderness, and symptoms of arthritis in the injured joint of the cervical spine. Assessment includes determining the onset, location, and radiation of pain and assessing for paresthesias, limited movement, and diminished function of the neck, shoulders, and upper extremities. It is important to determine whether the symptoms are bilateral; with large herniations, bilateral symptoms may be caused by cord compression. The area around the cervical spine is palpated to assess muscle tone and tenderness. Range of motion in the neck and shoulders is evaluated.

The patient is asked about any health issues that may influence the postoperative course and quality of life. It is also important to assess mood and stress levels. The nurse determines the patient's need for information about the surgical procedure and reinforces what the primary provider has explained. Strategies for pain management are discussed with the patient.

Diagnosis

Nursing Diagnoses

Based on the assessment data, major nursing diagnoses may include the following:

- Acute pain related to the surgical procedure
- Impaired physical mobility related to the postoperative surgical regimen
- Deficient knowledge about the postoperative course and home care management

Other nursing diagnoses may include preoperative anxiety, postoperative constipation, urinary retention related to the surgical procedure, self-care deficits related to use of a neck orthosis, and sleep pattern disturbance related to disruption in lifestyle.

Collaborative Problems/Potential Complications

Potential complications may include the following:

- Hematoma at the surgical site, resulting in cord compression and neurologic deficit
- Recurrent or persistent pain after surgery

Planning and Goals

The goals for the patient may include relief of pain, improved mobility, increased knowledge and self-care ability, and prevention of complications.

Nursing Interventions

Relieving Pain

Incisional pain is expected. Radicular pain improves over time as the nerve recovers. If the patient has had a bone fusion with bone removed from the iliac crest, considerable pain may be experienced at the donor site. Interventions consist of monitoring the donor site for hematoma formation, administering the prescribed postoperative analgesic agent, positioning for comfort, and reassuring the patient that the pain can be relieved. If the patient experiences a sudden increase in pain, extrusion of the graft may have occurred, requiring reoperation. A sudden increase in pain should be promptly reported to the surgeon.

The patient may experience a sore throat, hoarseness, and dysphagia due to temporary edema. These symptoms are relieved by throat lozenges, voice rest, and humidification. A puréed diet may be given if the patient has dysphagia.

Improving Mobility

Postoperatively, a cervical collar (neck orthosis) may be worn, which contributes to limited neck motion and altered mobility. The patient is instructed to turn the body instead of the neck when looking from side to side. The neck should be kept in a neutral (midline) position. The patient is assisted during position changes to make sure that the head, shoulders, and thorax are kept aligned. When assisting the patient to a sitting position, the nurse supports the patient's neck and shoulders. To increase stability, the patient should wear shoes when ambulating.

Monitoring and Managing Potential Complications

The patient is evaluated for bleeding and hematoma formation by assessing for swelling, excessive pressure in the neck, or severe pain in the incision area. The dressing is inspected for serosanguineous drainage, which suggests a dural leak. If this occurs, meningitis is a threat. A complaint of headache requires careful evaluation. Neurologic checks are made for swallowing deficits and upper and lower extremity weakness, because cord compression may produce rapid or delayed onset of paralysis. The patient who has had an anterior cervical diskectomy is also assessed for a sudden return of radicular (spinal nerve root) pain, which may indicate instability of the spine.

Throughout the postoperative course, the patient is monitored frequently to detect any signs of respiratory difficulty, because retractors used during surgery may injure the recurrent laryngeal nerve, resulting in hoarseness and the inability to cough effectively and clear pulmonary secretions. In addition, the blood pressure and pulse are monitored to evaluate cardiovascular status.

Bleeding at the surgical site and subsequent hematoma formation may occur. Severe localized pain not relieved by analgesic agents should be reported. A change in neurologic status (motor or sensory function) should be reported promptly, because it suggests hematoma formation that may necessitate surgery to prevent irreversible motor and sensory deficits (Hickey, 2009).

Promoting Home and Community-Based Care

Educating Patients About Self-Care. The patient's length of hospital stay is likely to be short; therefore, the patient and family should understand the care that is important for a smooth recovery. A cervical collar, if prescribed, is usually worn for about 6 weeks. The patient is educated in the use and care of the cervical collar. The patient will need to alternate tasks that involve minimal body movement (e.g., reading) with tasks that require greater body movement.

The patient is educated about strategies to manage the incision, for pain management, and about signs and symptoms that may indicate complications that should be reported to the primary provider. The nurse assesses the patient's understanding of these management strategies, limitations, and recommendations. In addition, the nurse assists the patient in identifying strategies to cope with ADLs (e.g., self-care, child care) and minimize risks to the surgical site (Chart 70-6). A discharge educational plan is developed collaboratively by members of the health care team to decrease the risk of recurrent disk herniation. Topics include those previously discussed as well as proper body mechanics, maintenance of optimal weight, proper exercise techniques, and modifications in activity.

Continuing Care. The patient is instructed to see the primary provider at prescribed intervals so that the provider can document the disappearance of old symptoms and assess the range of motion of the neck. Recurrent or persistent pain may occur despite removal of the offending disk or disk fragments. Patients who undergo diskectomy usually have consented to surgery after prolonged pain; they have often undergone repeated courses of ineffective conservative management and previous surgeries to relieve the pain. Therefore, the recurrence or persistence of symptoms postoperatively, including pain and sensory deficits, is often discouraging for the patient and family. The patient who experiences recurrence of symptoms requires emotional support and understanding. In addition, the patient is assisted in modifying activities and in considering options for subsequent treatment. The nurse educates the patient and family members about the need to participate in health promotion and health screening practices.

Evaluation

Expected patient outcomes may include:

1. Reports decreasing frequency and severity of pain
2. Demonstrates improved mobility
   a. Demonstrates progressive participation in self-care activities
   b. Identifies prescribed activity limitations and restrictions
   c. Demonstrates proper body mechanics
3. Is knowledgeable about postoperative course, medications, and home care management
   a. Lists the signs and symptoms to be reported postoperatively
   b. Identifies dose, action, and potential side effects of medications
   c. Identifies appropriate home care management activities and any restrictions
4. Has absence of complications
   a. Reports no increase in incision pain or sensory symptoms
   b. Demonstrates normal findings on neurologic assessment

Chart 70-6 HOME CARE CHECKLIST

The Patient With Cervical Diskectomy and Cervical Collar

At the completion of home care education, the patient or caregiver will be able to:	PATIENT	CAREGIVER
• Care for the surgical incision site.		
• Keep staples or sutures clean and dry and cover with dry dressing.		✔
• Notify primary provider if any signs or symptoms of infection occur, such as fever, redness or irritation, drainage, increased pain.	✔	✔
• Demonstrate proper body mechanics and prescribed exercise techniques.	✔	
• Modify activity.		
• Avoid sitting or standing for more than 30 minutes.	✔	
• Avoid twisting, flexing, extending, or rotating the neck.	✔	
• Avoid long automobile rides.	✔	
• Avoid sleeping in a prone position or the use of pillows to minimize neck flexion in bed; keep head in a neutral position.	✔	
• Use adequate mattress and chair support.	✔	
• Wear low-heeled shoes.	✔	
• Follow instructions regarding lifting, climbing stairs, driving a car, sexual activity, sports, exercise, and return to work.	✔	
• Practice stress reduction and relaxation techniques.	✔	
• Care of the cervical collar:		
• Wear the collar at all times until directed otherwise by the provider.	✔	
• Wash the neck twice a day with mild soap.	✔	✔
• Keep the neck still while the collar is open.	✔	
• With the assistance of a helper, wash the neck in steps.		
• Lie flat and supine.	✔	
• Open the Velcro tabs on each side of the collar and remove its front portion.		✔
• Gently wash and dry the neck.		✔
• Replace the front part of the collar and refasten the tabs.		✔
• Turn to one side with a thin pillow under the head.	✔	
• Open one tab.	✔	
• Gently wash and dry the back of the neck. Refasten the tab.		✔
• Turn to the other side and wash and dry this side. Refasten the tab.		✔
• Place a wrinkle-free silk scarf under the collar to increase comfort.	✔	✔
• ***For men:*** Shave without twisting or moving the neck. This may be done with help while lying flat or sitting. Remove only the front part of the collar for shaving.	✔	✔

Herniation of a Lumbar Disk

Approximately 90% to 95% of lumbar disk herniations occur at the L5–S1 region (AANN, 2012). A herniated lumbar disk produces low back pain accompanied by varying degrees of sensory and motor impairment.

Clinical Manifestations

The patient complains of low back pain with muscle spasm with radicular pain (radiation of the pain into one hip and down into the leg ([sciatica]). Pain is aggravated by actions that increase intraspinal fluid pressure, such as bending, lifting, or straining (as in sneezing or coughing), and usually is relieved by bed rest. There is often some type of postural deformity, because pain causes an alteration of the normal spinal mechanics. If the patient lies on the back and attempts to raise a leg in a straight position, pain radiates into the leg; this maneuver, called the *straight-leg raising test*, stretches the sciatic nerve. Additional signs include muscle weakness, alterations in tendon reflexes, and sensory loss.

Assessment and Diagnostic Findings

The diagnosis of lumbar disk disease is based on the history and physical findings and the use of imaging techniques such as MRI, CT, and myelography.

Medical Management

The objectives of treatment are to relieve pain, slow disease progression, and increase the patient's functional ability (Kose & Hatipoglu, 2012). Bed rest is discouraged because it may weaken muscles, but activities that exacerbate pain should be avoided (AANN, 2012).

Because muscle spasm is prominent during the acute phase, muscle relaxants are used. NSAIDs and systemic corticosteroids may be administered to counter the inflammation and swelling that usually occurs in the supporting tissues and the affected nerve roots. Moist heat and massage help relax muscles. Strategies for increasing the patient's functional ability include weight reduction, physical therapy, and biofeedback. Exercises, prescribed by physical therapists, can help strengthen back muscles and decrease pain (AANN, 2012). (See Chapter 12 for descriptions of nursing interventions for the patient with pain.)

Surgical Management

In the lumbar region, surgical treatment includes lumbar disk excision through a posterolateral laminotomy and the newer techniques of microdiskectomy and percutaneous diskectomy. In microdiskectomy, an operating microscope is used to visualize the offending disk and compressed nerve roots; it permits a small incision (2.5 cm [1 in]) and minimal blood loss and takes about 30 minutes of operating time. Generally, the length of hospital stay is short, and the patient makes a rapid recovery. Several minimally invasive techniques in spinal surgery have led to improved patient outcomes and lower hospital costs, and research on these techniques is ongoing (Starkweather, Witek-Janusek, Nockels, et al., 2008). Fusion is considered when vertebral instability is present and contributes to pain. Fusion procedures are longer and cause more postoperative pain. Research suggests that intraoperative wound infiltration with bupivacaine hydrochloride solution decreases pain and the need for opioids postoperatively (Ersayli, Gurget, Bekar, et al., 2006). Recent studies suggest that gel substances for disk replacement may improve pain and spinal function (Leckie & Kang, 2009).

A patient undergoing a disk procedure at one level of the vertebral column may have a degenerative process at other levels. A herniation relapse may occur at the same level or elsewhere, so the patient may become a candidate for another disk procedure. Arachnoiditis (inflammation of the arachnoid membrane) may occur after surgery (and after myelography); it involves an insidious onset of diffuse, frequently burning pain in the lower back, radiating into the buttocks. Disk excision can leave adhesions and scarring around the spinal nerves and dura, which then produce inflammatory changes that create chronic neuritis and neurofibrosis. Disk surgery may relieve pressure on the spinal nerves, but it does not reverse the effects of neural injury and scarring and the pain that results. Failed disk syndrome (recurrence of sciatica after lumbar diskectomy) remains a cause of disability (Hickey, 2009).

Nursing Management

Providing Preoperative Care

Most patients fear surgery on any part of the spine and therefore need explanations about the surgery and reassurance that it will not weaken the back. When data are being collected for the health history, any reports of pain, paresthesia, or muscle spasm are recorded to provide a baseline for comparison after surgery. Health issues that may influence the postoperative course and quality of life (e.g., fatigue, mood, stress, patient expectations) are important to assess (Starkweather et al., 2008). Preoperative assessment also includes an evaluation of movement of the extremities as well as bladder and bowel function. To facilitate the postoperative turning procedure, the patient is instructed to turn as a unit (called *logrolling*) as part of the preoperative preparation (Fig. 70-8). Before surgery, the patient is also encouraged to take deep breaths, cough, and perform muscle setting exercises to maintain muscle tone.

Assessing the Patient After Surgery

After lumbar disk excision, vital signs are checked frequently and the wound is inspected for hemorrhage, because vascular injury is a complication of disk surgery. Because postoperative neurologic deficits may occur from nerve root injury, the sensation and motor strength of the lower extremities are evaluated at specified intervals, along with the color and temperature of the legs and sensation of the toes. It is important to assess for urinary retention, which is another sign of neurologic deterioration.

In diskectomy with fusion, the patient has an additional surgical incision if bone fragments were taken from the iliac crest or fibula to serve as wedges in the spine. The recovery period is longer than for those patients who underwent diskectomy with spinal fusion, because bony union must take place.

Positioning the Patient

To position the patient, a pillow is placed under the head, and the knee rest is elevated slightly to relax the back muscles.

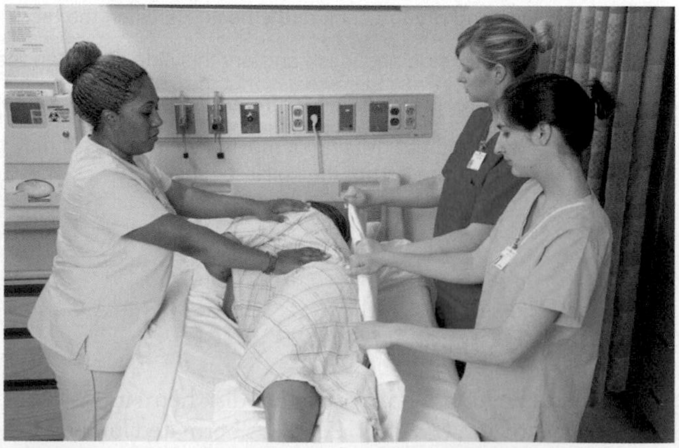

FIGURE 70-8 • Before the patient undergoes laminectomy surgery, the logrolling technique that will be used for turning the patient should be demonstrated. The patient's arms will be crossed and the spine aligned. To avoid twisting the spine, the head, shoulders, knees, and hips are turned at the same time so that the patient rolls over like a log. When in a side-lying position, the patient's back, buttocks, and legs are supported with pillows.

When the patient is lying on one side, however, extreme knee flexion must be avoided. The patient is encouraged to move from side to side to relieve pressure and is reassured that no injury will result from moving. When the patient is ready to turn, the bed is placed in a flat position and a pillow is placed between the patient's legs. The patient turns as a unit (logrolls) without twisting the back.

To get out of bed, the patient lies on one side while pushing up to a sitting position. At the same time, the nurse or family member eases the patient's legs over the side of the bed. Coming to a sitting or standing posture is accomplished in one long, smooth motion. Most patients walk to the bathroom on the same day as the surgery. Sitting is discouraged except for defecation.

Promoting Home and Community-Based Care

Educating Patients About Self-Care

The patient is instructed to increase activity gradually, as tolerated, because it takes up to 6 weeks for the ligaments to heal. Excessive activity may result in spasm of the paraspinal muscles.

Activities that produce flexion strain on the spine (e.g., driving a car) should be avoided until healing has taken place. Heat may be applied to the back to relax muscle spasms. Scheduled rest periods are important, and the patient is advised to avoid heavy work for 2 to 3 months after surgery. Exercises are prescribed to strengthen the abdominal and erector spinal muscles. A back brace or corset may be necessary if back pain persists.

Continuing Care

Referral for inpatient or outpatient rehabilitation may be warranted to improve self-care abilities after medical or surgical treatment for herniation of a lumbar disk. A home care referral may be indicated and provides the home care nurse with the opportunity to assess the patient's physical and psychological status, as well as his or her ability to adhere to recommended management strategies. During the home visit, the nurse determines whether changes in neurologic function have occurred. The adequacy of pain management is assessed, and modifications are made to ensure adequate pain relief.

Post-polio Syndrome

People who survived the polio epidemic of the 1950s, many of whom are now older adults, are developing new symptoms of weakness, fatigue, and musculoskeletal pain. Researchers estimate that as many as 80% of the 1,000,000 polio survivors are experiencing the phenomenon known as post-polio syndrome. Men and women appear to be equally at risk for this condition (LaRocco, 2011).

Pathophysiology

The exact cause of post-polio syndrome is not known, but researchers suspect that with aging or muscle overuse, the neurons not destroyed by the poliovirus continue generating axon sprouts (Werhagen & Borg, 2011). These new terminal axon sprouts reinnervate the affected muscles after the initial insult but may become more vulnerable as the body ages.

Assessment and Diagnostic Findings

No specific diagnostic test exists for post-polio syndrome. Clinical diagnosis is made on the basis of the history and physical examination and exclusion of other medical conditions that could be causing the new symptoms. Patients report a history of paralytic poliomyelitis followed by partial or complete recovery of function, with a plateau of function and then the recurrence of symptoms. Signs and symptoms may occur decades after the original onset of poliomyelitis (LaRocco, 2011).

Management

No specific medical or surgical treatment is available for this syndrome, and therefore nurses play a pivotal role in the team approach to assisting patients and families in dealing with the symptoms of progressive loss of muscle strength and significant fatigue. Pain and weakness may be improved with infusion of intravenous immunoglobulin (Werhagen & Borg, 2011). Other health care professionals who may assist in patient care include physical, occupational, speech, and respiratory therapists. Nursing interventions are aimed at maintaining the patient's strength as well as physical, psychological, and social well-being.

The patient needs to plan and coordinate activities to conserve energy and reduce fatigue. Rest periods should be planned and assistive devices used to reduce weakness and fatigue. Important activities should be planned for the morning, because fatigue often increases in the afternoon and evening.

One study proposed an explanatory model of health promotion and quality of life in patients with post-polio syndrome (Stuifbergen, Seraphine, Harrison, et al., 2005). The results suggested that quality of life is the result of a complex interplay among contextual factors such as severity of impairment, antecedent variables, and health-promoting behaviors.

Pain in muscles and joints may be a problem. Nonpharmacologic techniques such as the application of heat and cold are most appropriate, because older patients may not tolerate or may have strong reactions to medications.

Maintaining a balance between adequate nutritional intake and avoiding excess calories that can lead to obesity in this sedentary group of patients is a challenge. Pulmonary hygiene and adequate fluid intake can help with airway management. Several interventions can improve sleep, including limiting caffeine intake before bedtime and assessing for nocturia. If nocturia is an issue, the patient needs to be evaluated for obstructive sleep apnea. Supportive ventilation may be appropriate, with continuous positive airway pressure if sleep apnea is a problem.

Bone density testing in patients with post-polio syndrome has revealed low bone mass and osteoporosis (Smeltzer, Zimmerman, & Capriotti, 2005). Therefore, the importance of identifying risks, preventing falls, and treating osteoporosis must be discussed with patients and families. Families also need to be made aware of the possibility of changes in individual and family relationships due to the many symptoms of post-polio syndrome (LaRocco, 2011). The nurse also needs to remind patients and family members of the need for health promotion activities and health screening.

Critical Thinking Exercises

1 A 68-year-old woman has been newly diagnosed with a metastatic brain tumor. Assess and prioritize the patient's physiologic and psychosocial needs. What nursing interventions and actions would you suggest to assist in the management of the brain tumor?

2 A 35-year-old man newly diagnosed with Huntington disease asks what type of medication he will be given. What are the possible medication regimens that may be used to treat his disease and the efficacy of those regimens? What patient education is appropriate?

3 pq Identify the priorities, approach, and techniques you would use to develop a comprehensive plan of care for a 45-year-old female patient with ALS. How will your priorities, approach, and techniques differ if the patient lives alone, is hearing impaired, or is from a culture that is different from your own?

4 ebp A 28-year-old patient with low back pain is seen in the clinic. Identify the evidence-based practices for the management of low back pain. Describe the evidence base for the practices that you have identified and the criteria used to evaluate the strength of that evidence. Identify the health promotion activities you would recommend to this patient and the rationale for your recommendations.

Brunner Suite Resources

Explore these additional resources to enhance learning for this chapter:

- NCLEX-Style Questions and Other Resources on thePoint, http://thePoint.lww.com/Brunner13e
- Study Guide
- PrepU
- Clinical Handbook
- Handbook of Laboratory and Diagnostic Tests

References

*Asterisk indicates nursing research.

Books

American Association of Neuroscience Nurses. (2012). *Thoracolumbar spine surgery: A guide to preoperative and postoperative care: AANN clinical practice guideline series*. Glenview, IL: Author.

Bader, M., & Littlejohns, L. R. (2010). *AANN core curriculum for neuroscience nursing* (5th ed.). Glenview, IL: American Association of Neuroscience Nurses.

Barker, R. A., & Barasi, S. (2008). *Neuroscience at a glance* (3rd ed.). Oxford: Blackwell.

Dudek, S. G. (2010). *Nutrition essentials for nursing practice* (6th ed.). Philadelphia: Lippincott Williams & Wilkins.

Eggert, J. (Ed.). (2010). *Cancer basics*. Pittsburgh: Oncology Nursing Society.

Hickey, J. V. (2009). *The clinical practice of neurological & neurosurgical nursing* (6th ed.). Philadelphia: Lippincott Williams & Wilkins.

Karch, A. M. (2012). *2012 Lippincott's nursing drug guide*. Philadelphia: Lippincott Williams & Wilkins.

Weber, J., & Kelley, J. (2010). *Health assessment in nursing* (4th ed.). Philadelphia: Lippincott Williams & Wilkins.

Journals and Electronic Documents

Amyotrophic Lateral Sclerosis

Clarke, K., & Levine, T. (2011). Clinical recognition and management of amyotrophic lateral sclerosis: The nurse's role. *Journal of Neuroscience Nursing, 43*(4), 205–214.

Corcia P., & Meininger, V. (2008). Management of amyotrophic lateral sclerosis. *Drugs*, 68(8), 1037–1048.

Kiernan, M., Vucic, S., Cheah, B., et al. (2011). Amyotrophic lateral sclerosis. *Lancet, 377*, 942–955.

Oh, H., Sin, M., Schepp, J., et al. (2012). Depressive symptoms and functional impairment among amyotrophic lateral sclerosis patients in South Korea. *Rehabilitation Nursing, 37* (3), 136–144.

Degenerative Disk Disease

Agency for Healthcare Research and Quality. (2011). *Healthcare cost and utilization project*. Available at: www.ahrq.gov/data/hcup/

Cepoiu-Martin, M., Faris, P., Lorenzetti, D., et al. (2011). Artificial cervical disc arthroplasty: A systematic review. *Spine*, 36(25), E1623–E1633.

Ersayli, D. T., Gurget, A., Bekar, A., et al. (2006). Effects of perioperatively administered bupivacaine and bupivacaine-methylprednisolone on pain after lumbar diskectomy. *Spine*, 31(19), 2221–2226.

Kose, G., & Hatipoglu, S. (2012). The effect of low back pain on the daily activities of patients with lumbar disc herniation: A Turkish military hospital experience. *Journal of Neuroscience Nursing*, 44(2), 98–104.

Leckie, S., & Kang, J. (2009). Recent advances in nucleus pulposus replacement technology. *Current Orthopedic Practice*, 20(3), 222–226.

*Starkweather, A. R., Witek-Janusek, L., Nockels, R. P., et al. (2008). The multiple benefits of minimally invasive spinal surgery: Results comparing transformational lumbar underbody fusion and posterior lumbar fusion. *Journal of Neuroscience Nursing*, 40(1), 32–39.

Huntington Disease

Aubeeluck, A., & Wilson, E. (2008). Huntington's disease. Part 1: Essential background and management. *British Journal of Nursing*, 17(3), 146–151.

Aubeeluck, A., Wilson, E., & Stupple, E. (2011). Obtaining quality of life for Huntington's disease patients and their families. *British Journal of Neuroscience Nursing*, 7(5), 634–638.

Ha, A. D., & Fung, C. (2012). Huntington's disease. *Current Opinion in Neurology*, 25(4), 491–498.

Muscular Dystrophy

Partridge, T. A. (2011). Impending therapies for Duchenne muscular dystrophy. *Current Opinion in Neurology*, 24(5), 415–422.

Spurney, C. F. (2011). Cardiomyopathy of Duchenne muscular dystrophy: Current understanding and future directions. *Muscle & Nerve*, 44(1), 8–19.

Verhaart, I. E. C., & Aartsma-Rus, A. (2012). Gene therapy for Duchenne muscular dystrophy. *Current Opinion in Neurology*, 25(5), 588–596.

Oncologic Disorders

Aghayev, K., Vrionis, F., & Chamberlain, M. C. (2011). Adult intradural primary spinal cord tumors. *Journal of the National Comprehensive Cancer Network*, 9(4), 434–447.

Aiello-Laws, L., & Rutledge, D. (2008). Management of adult patients receiving intraventricular chemotherapy for the treatment of leptomeningela metastasis. *Clinical Journal of Oncology Nursing*, 12(3), 429–435.

American Brain Tumor Association. (2011). *A primer of brain tumors: A patient's reference manual* (9th ed.). Available at: www.abta.org

Barnholtz-Sloan, J. S., Williams, V. L., Maldonado, J. L., et al. (2008). Patterns of care and outcomes among elderly individuals with primary malignant astrocytoma. *Journal of Neurosurgery*, 108(4), 642–648.

Cahill, J., & Armstrong, T. (2011). Caring for an adult with a malignant primary brain tumor. *Nursing 2011*, 41(6), 28–33.

Central Brain Tumor Registry of the United States. (2012). *CBTRUS statistical report: Primary brain and central nervous system tumors diagnosed in the United States in 2004–2008*. Available at: www.cbtrus.org/2012-NPCR-SEER/CBTRUS_Report_2004-2008_3-23-2012.pdf

Davis, M., & Stoiber, A. (2011). Glioblastoma multiforme: Enhancing survival and quality of life. *Clinical Journal of Oncology Nursing*, 15(3), 291–297.

Gordon, B. M. (2007). Pharmacological management of secreting pituitary tumors. *Journal of Neuroscience Nursing*, 39(1), 52–57.

*Kim, J. M., Losina, E., Bono, C. M., et al. (2012). Clinical outcome of metastatic spinal cord compression treated with surgical excision ± radiation versus radiation therapy alone: A systematic review of literature. *Spine*, 37(1), 78–84.

Mogensen, K. (2008). The whole picture: Addressing the diverse needs of the patient treated for a brain tumor. *Clinical Journal of Oncology Nursing, 12*(5), 817–819.

National Comprehensive Cancer Network. (2012). *Central nervous system cancers: Version 2.2012.* Available at: www.nccn.com

Richards, J., & Ballard, N. (2008). Colloid cyst: A case study. *Journal of Neuroscience Nursing, 40*(2), 103–105.

*Sherwood, P., Hricik, A., Donovan, H., et al. (2011). Changes in caregiver perceptions over time in response to providing care for a loved one with a primary malignant brain tumor. *Oncology Nursing Forum, 38*(2), 149–155.

Swinson, B., & Friedman, W. (2008). Linear accelerator stereotactic radiosurgery for metastatic brain tumors: Years of experience at the University of Florida. *Neurosurgery, 62*(5), 1018–1032.

Wen, P. Y., & Kesari, S. (2008). Malignant gliomas in adults. *New England Journal of Medicine, 359*(5), 492–507.

Parkinson's Disease

Aarsland, D., Marsh, L., & Schrag, A. (2009). Neuropsychiatric symptoms in Parkinson's disease. *Movement Disorders, 24*(15), 2175–2186.

Cranwell-Bruce, L. (2010). Drugs for Parkinson's disease. *Medsurg Nursing, 19*(6), 347–355.

Hermanns, M., Deal, B., & Haas, B. (2012). Biopsychosocial and spiritual aspects of Parkinson disease: An integrative review. *Journal of Neuroscience Nursing, 44*(4), 194–205.

Jancovic, J., & Poewe, W. (2012). Therapies in Parkinson's disease. *Current Opinion in Neurology, 25*(4), 433–447.

Lindop, F. (2012). A multidisciplinary approach to Parkinson's disease. *British Journal of Neuroscience Nursing, 8*(2), 61–64.

Parkinson's Disease Foundation. (2012). *Statistics on Parkinson's.* Available at: www.pdf.org/en/parkinson_statistics

Post-polio Syndrome

LaRocco, S. A. (2011). Post-polio syndrome: Unraveling the mystery. *Nursing, 41*(2), 26–30.

*Smeltzer, S. C., Zimmerman, V., & Capriotti, T. (2005). Osteoporosis risk and bone mineral density in women with physical disabilities. *Archives of Physical Medicine and Rehabilitation, 86*(3), 582–586.

*Stuifbergen, A., Seraphine, A., Harrison, T., et al. (2005). An explanatory model of health promotion and quality of life for persons with post-polio syndrome. *Social Science and Medicine, 60*(2), 383–393.

Werhagen, L., & Borg, K. (2011). Effect of intravenous immunoglobulin on pain in patients with post-polio syndrome. *Journal of Rehabilitation Medicine, 43*(11), 1038–1040.

Resources

American Brain Tumor Association (ABTA), www.abta.org
American Cancer Society, www.cancer.org
American Parkinson Disease Association (APDA), www.apdaparkinson.org
ALS Association, www.alsa.org
Huntington's Disease Society of America, www.hdsa.org
MDA (Muscular Dystrophy Association), www.mda.org
Michael J. Fox Foundation for Parkinson's Research, Church Street Station, www.michaeljfox.org
National Brain Tumor Society, www.braintumor.org
Parkinson's Disease Foundation (PDF), www.pdf.org

Unit 17 Acute Community-Based Challenges

Case Study

A PATIENT WITH MULTIPLE TRAUMA RESULTING IN HEMORRHAGE AND RISK FOR SHOCK

Rescue personnel bring 19-year-old Anna Woo to the emergency department after a motorcycle crash. She has facial contusions and lacerations, a fractured sternum, three fractured ribs, a hemothorax, a dislocated hip, a fractured pelvis, and multiple minor lacerations. She is moaning but responds to her name. Initial findings include an unobstructed airway with absent breath sounds in the right basilar lung field, tachypnea with 32 shallow respirations per minute, and arterial blood gas values as follows: pH 7.29; pCO_2 48; SaO_2 90%; HCO_3^- 24. Blood pressure is 94/70; skin is cool and clammy; heart rate is 100 bpm; and peripheral pulses are intact. After a preliminary survey and placement of intravenous access lines, a chest tube is inserted. Immediately, 300 mL of bloody fluid drains into the collection chamber. Grossly bloody fluid continues to drain from the chest tube at a rate of 20 mL every 15 minutes.

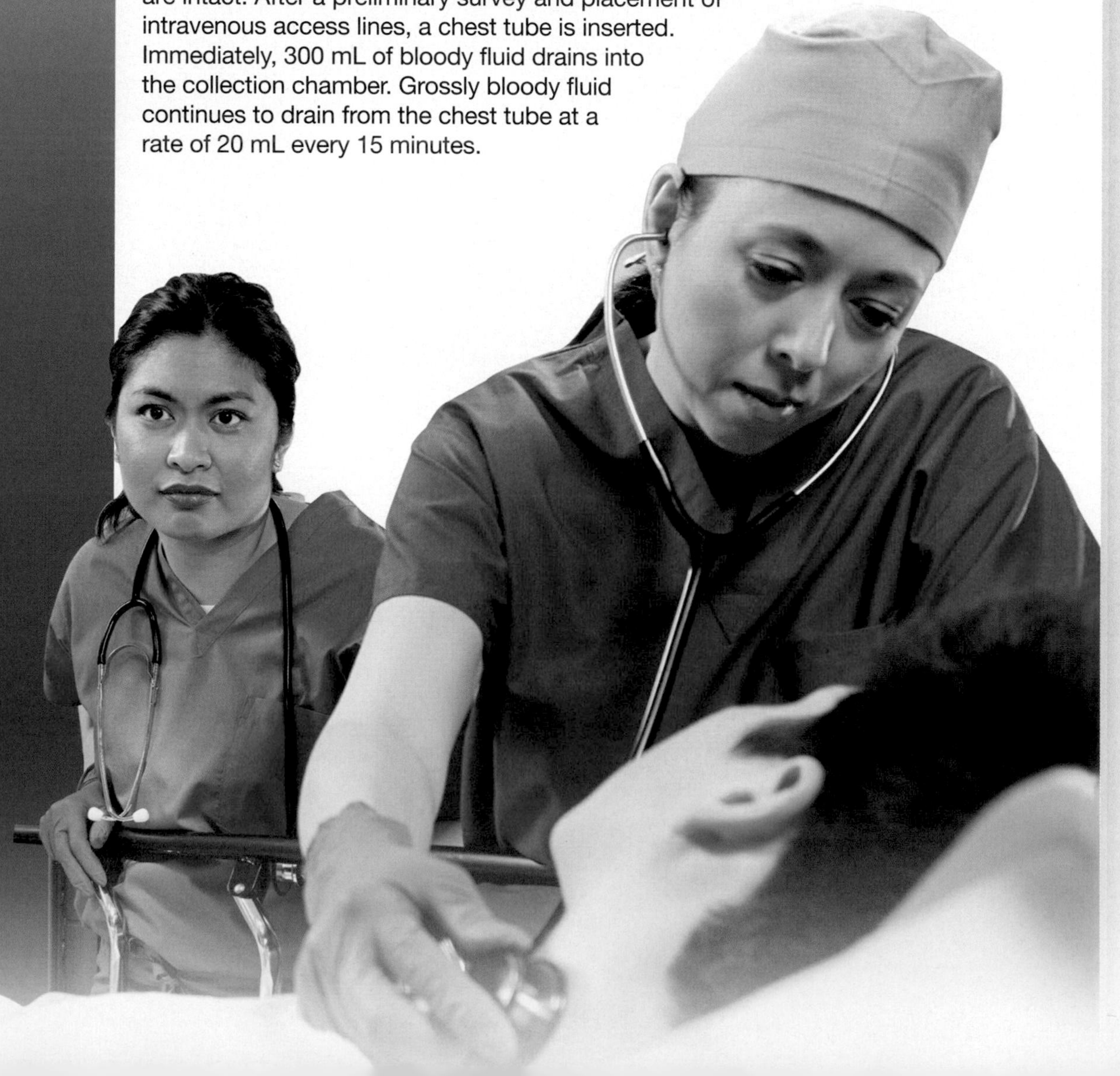

QSEN Competency Focus: *Teamwork and Collaboration*

The complexities inherent in today's health care system challenge nurses to demonstrate integration of specific interdisciplinary core competencies. These competencies are aimed at ensuring the delivery of safe, quality patient care (Institute of Medicine, 2003). The concepts from the Quality and Safety Education for Nurses (QSEN) Institute (2012) provide a framework for the knowledge, skills, and attitudes (KSAs) required for nurses to demonstrate competency in these key areas, which include *patient-centered care, interdisciplinary teamwork and collaboration, evidence-based practice, quality improvement, safety,* and *informatics*.

Teamwork and Collaboration Definition: Function effectively within nursing and inter-professional teams, fostering open communication, mutual respect, and shared decision-making to achieve quality patient care.

RELEVANT PRE-LICENSURE KSAs	APPLICATION AND REFLECTION
Knowledge	
Describe own strengths, limitations, and values in functioning as a member of a team.	Describe the nurse's priorities of care for Ms. Woo. How would these intersect with those of other members of the interdisciplinary trauma team? Describe the professional health care team members involved in her care, their respective roles, and how these roles interplay with each other.
Skills	
Demonstrate awareness of own strengths and limitations as a team member. Initiate plan for self-development as a team member.	Identify your priorities of care for Ms. Woo if you were the nurse assigned to her care. As a novice in the trauma setting, on whom would you rely to help you develop your skill set? Identify your professional development growth goals so that you would be an effective member of the trauma health care team.
Attitudes	
Acknowledge own potential to contribute to effective team functioning.	Reflect upon how you might feel threatened being responsible for caring for a multi-trauma patient, such as Ms. Woo, as a new graduate nurse. How might you learn the skills that you need to be most effective in this setting from seasoned nurses and from other members of the interdisciplinary health care team? What else might you do to hone your skill set and knowledge base?

Cronenwett, L., Sherwood, G., Barnsteiner, J., et al. (2007). Quality and safety education for nurses. *Nursing Outlook*, *55*(3), 122–131.
Institute of Medicine. (2003). *Health professions education: A bridge to quality*. Washington, DC: National Academies Press.
QSEN Institute. (2012). *Competencies: Prelicensure KSAs*. Available at: qsen.org/competencies/pre-licensure-ksas

Read More About This Case

More information about this case study and the relationships between nursing diagnoses, interventions, and expected outcomes is available online. Visit thePoint for Applying Concepts From NANDA-I, NIC, and NOC.

Chapter

71

Management of Patients With Infectious Diseases

Learning Objectives

On completion of this chapter, the learner will be able to:

1 Differentiate between colonization, infection, and disease.
2 Identify federal and local resources available to the nurse seeking information about infectious diseases.
3 Identify the benefits of vaccines recommended for health care workers.
4 Identify standard and transmission-based precautions and discuss the elements of these standards.
5 Describe the concept and the nursing management of emerging infectious diseases.
6 Use the nursing process as a framework for care of patients with sexually transmitted infections.
7 Describe home health care measures that reduce the risk of infection.
8 Use the nursing process as a framework for care of patients with infectious diseases.

Glossary

bacteremia: laboratory-confirmed presence of bacteria in the bloodstream

carrier: person who has an organism without apparent signs and symptoms; one who is able to transmit an infection to others

colonization: microorganisms present in or on a host, without host interference or interaction and without eliciting symptoms in the host

community-associated methicillin-resistant *Staphylococcus aureus* (CA-MRSA): a strain of MRSA infecting persons who have not been treated in a health care setting

emerging infectious diseases: human infectious diseases with incidence increased within the past two decades or potential increase in the near future

fungemia: a bloodstream infection caused by a fungal organism

health care–associated infection (HAI): an infection not present or incubating at the time of admission to the health care setting; this term has replaced the term *nosocomial infection*

host: an organism that provides living conditions to support a microorganism

immune: person with protection from a previous infection or vaccination who resists reinfection when re-exposed to the same agent

incubation period: time between contact and onset of signs and symptoms

infection: condition in which the host interacts physiologically and immunologically with a microorganism

infectious disease: the consequences that result from invasion of the body by microorganisms that can produce harm to the body and potentially death

latency: time interval after primary infection when a microorganism lives within the host without producing clinical evidence

methicillin-resistant *Staphylococcus aureus* (MRSA): *Staphylococcus aureus* bacterium that is not susceptible to extended-penicillin antibiotic formulas, such as methicillin, oxacillin, or nafcillin; MRSA may occur in a health care or community setting

normal flora: persistent nonpathogenic organisms colonizing a host

reservoir: any person, plant, animal, substance, or location that provides living conditions for microorganisms and that enables further dispersal of the organism

standard precautions: strategy of assuming all patients may carry infectious agents and using appropriate barrier precautions for all health care worker–patient interactions

susceptible: not possessing immunity to a particular pathogen

transient flora: organisms that have been recently acquired and are likely to be shed in a relatively short period

transmission-based precautions: precautions used in addition to standard precautions when contagious or epidemiologically significant organisms are recognized; the three types of transmission-based precautions are airborne, droplet, and contact precautions

vancomycin-resistant *Enterococcus* (VRE): *Enterococcus* bacterium that is resistant to the antibiotic vancomycin

vancomycin-resistant *Staphylococcus aureus* (VRSA): *Staphylococcus aureus* bacterium that is not susceptible to vancomycin

virulence: degree of pathogenicity of an organism

An infectious disease is any disease caused by the growth of pathogenic microbes in the body. It may or may not be communicable (i.e., contagious). Modern science has controlled, eradicated, or decreased the incidence of many infectious diseases. However, increases in the incidence of other infections, such as those caused by antibiotic-resistant organisms and emerging infectious diseases, are of growing concern. Examples of these infectious diseases are presented in this chapter. Other infectious diseases are discussed in the appropriate chapters (e.g., see Chapter 23 for information on tuberculosis [TB]). It is important to understand infectious causes and the treatment of contagious, serious, and common infections. Table 71-1 presents an overview of many infectious diseases, their causative organisms, mode of transmission, and usual **incubation periods** (time between contact and development of the first signs and symptoms).

The nurse plays an important role in infection control and prevention. Educating patients may decrease their risk of becoming infected or may decrease the sequelae of infection. Using appropriate barrier precautions, observing prudent hand hygiene, and ensuring aseptic care of intravenous (IV) catheters and other invasive equipment also assists in reducing infections.

The Infectious Process

The Chain of Infection

A complete chain of events is necessary for infection to occur. Six elements are necessary, including a causative organism, a reservoir of available organisms, a portal of exit from the reservoir, a mode of transmission from the reservoir to the host and a mode of entry into a susceptible host.

Concept Mastery Alert

Nurses must clearly understand the elements of the chain of infection in order to identify points at which they can intervene to interrupt the chain, thus protecting patients, themselves, and others from infectious disease. Figure 71-1 illustrates these concepts.

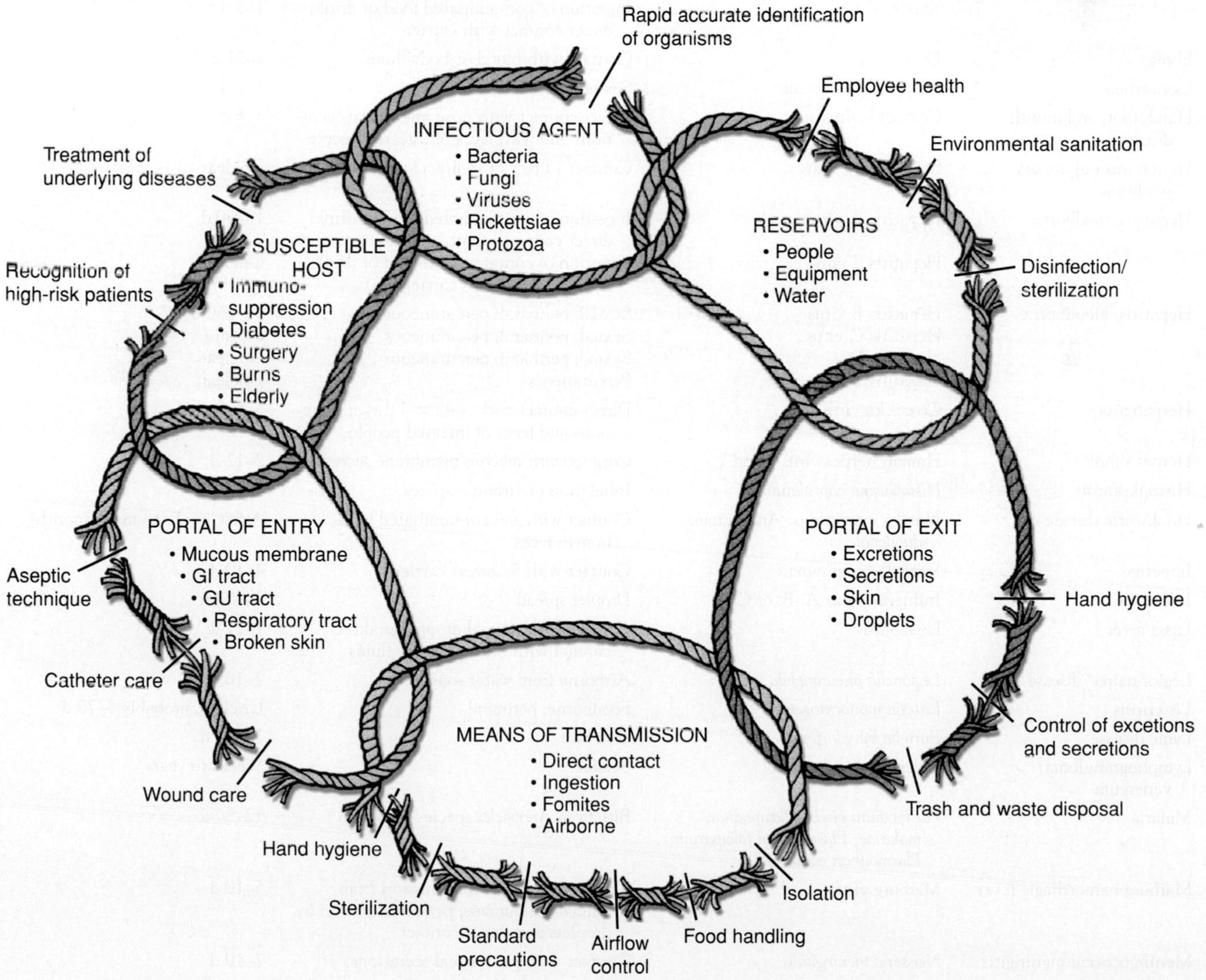
Rapid accurate identification of organisms
Employee health
Environmental sanitation
Disinfection/ sterilization
Hand hygiene
Control of excetions and secretions
Trash and waste disposal
Isolation
Food handling
Airflow control
Standard precautions
Sterilization
Hand hygiene
Wound care
Catheter care
Aseptic technique
Recognition of high-risk patients
Treatment of underlying diseases
INFECTIOUS AGENT
• Bacteria
• Fungi
• Viruses
• Rickettsiae
• Protozoa
RESERVOIRS
• People
• Equipment
• Water
PORTAL OF EXIT
• Excretions
• Secretions
• Skin
• Droplets
MEANS OF TRANSMISSION
• Direct contact
• Ingestion
• Fomites
• Airborne
PORTAL OF ENTRY
• Mucous membrane
• GI tract
• GU tract
• Respiratory tract
• Broken skin
SUSCEPTIBLE HOST
• Immuno-suppression
• Diabetes
• Surgery
• Burns
• Elderly

FIGURE 71-1 • Health care workers' interventions used to break the chain of infection transmission.

TABLE 71-1 Infectious Diseases, Causative Organisms, Modes of Transmission, and Usual Incubation Periods

Disease or Condition	Organism	Usual Mode of Transmission	Usual Incubation Period (Infection to First Symptom)
Acquired immunodeficiency syndrome	Human immunodeficiency virus	Sexual; percutaneous; perinatal	Median of 10 y
Amebiasis	*Entamoeba histolytica*	Contaminated water	2–4 wk
Anthrax	*Bacillus anthracis*	Airborne or contact	2–60 d
Chancroid	*Haemophilus ducreyi*	Sexual	3–5 d
Chickenpox	Varicella zoster	Airborne or contact	~14 d
Cholera	*Vibrio cholerae*	Ingestion of water contaminated with human waste	A few hours to 5 d
Cryptococcosis	*Cryptococcus neoformans*	Probably by inhalation	Unknown
Cryptosporidiosis	*Cryptosporidium* species	Ingestion of contaminated water; direct contact with carrier	Probably 1–12 d
Cytomegalovirus infection	Cytomegalovirus	Transfusion and transplantation; sexual; perinatal	Highly variable: 3–8 wk after transfusion, 3–12 wk after delivery of newborn
Diarrheal disease (common causes)	*Campylobacter* species	Ingestion of contaminated food	3–5 d
	Clostridium difficile	Fecal–oral	Variable; in part related to the influence of antibiotic agents
	Salmonella species	Ingestion of contaminated food or drink	12–36 h
	Shigella species	Ingestion of contaminated food or drink; direct contact with carrier	1–3 d
	Yersinia species	Ingestion of contaminated food or drink; direct contact with carrier	1–3 d
Ebola	Ebola virus	Contact with blood or body fluids	2–21 d
Gonorrhea	*Neisseria gonorrhoeae*	Sexual; perinatal	2–7 d
Hand, foot, and mouth disease	Coxsackievirus	Direct contact with nose and throat secretions and with feces of infected people	3–5 d
Hantavirus pulmonary syndrome	Sin Nombre virus	Contact (direct or indirect) with rodents	Unclear
Hepatitis, foodborne	Hepatitis A virus	Ingestion of contaminated food or drink; direct contact with carrier	15–50 d
	Hepatitis E virus	Ingestion of contaminated food or drink; direct contact with carrier	Unclear
Hepatitis, bloodborne	Hepatitis B virus	Sexual; perinatal; percutaneous	45–160 d
	Hepatitis C virus	Sexual; perinatal; percutaneous	6–9 mo
	Hepatitis D	Sexual; perinatal; percutaneous	Unclear
	Hepatitis G	Percutaneous	Unclear
Herpangina	Coxsackievirus	Direct contact with nose and throat secretions and feces of infected people	3–5 d
Herpes simplex	Human herpes virus 1 and 2	Contact with mucous membrane secretions	2–12 d
Histoplasmosis	*Histoplasma capsulatum*	Inhalation of airborne spores	5–18 d
Hookworm disease	*Necator americanus; Ancylostoma duodenale*	Contact with soil contaminated with human feces	A few weeks to many months
Impetigo	*Staphylococcus aureus*	Contact with *S. aureus* carrier	4–10 d
Influenza	Influenza virus A, B, or C	Droplet spread	24–72 h
Lassa fever	Lassa virus	Contact with animal droppings; direct contact with blood or body fluids	7–21 d
Legionnaires' disease	*Legionella pneumophila*	Airborne from water source	2–10 d
Listeriosis	*Listeria monocytogenes*	Foodborne; perinatal	Unclear; probably 3–70 d
Lyme disease	*Borrelia burgdorferi*	Tick bite	14–23 d
Lymphogranuloma venereum	*Chlamydia trachomatis*	Sexual	Weeks to years
Malaria	*Plasmodium vivax; Plasmodium malariae; Plasmodium falciparum; Plasmodium ovale*	Bite from *Anopheles* species mosquito	12–30 d
Marburg hemorrhagic fever	Marburg virus	Unknown route of transmission from animals to humans; person-to-person by droplets and direct contact	5–10 d
Meningococcal meningitis or bacteremia	*Neisseria meningitidis*	Contact with pharyngeal secretions; perhaps airborne	2–10 d

Disease or Condition	Organism	Usual Mode of Transmission	Usual Incubation Period (Infection to First Symptom)
Mononucleosis	Epstein-Barr virus	Contact with pharyngeal secretions	4–6 wk
Mycobacterial diseases (nontuberculosis *Mycobacterium* species)	*Mycobacterium avium; Mycobacterium kansasii; Mycobacterium fortuitum; Mycobacterium gordonae;* other *Mycobacterium* species	Variable; probably contact with soil, water, or other environmental source; none is person-to-person transmissible	Variable
Mycoplasmal pneumonia	*Mycoplasma pneumoniae*	Droplet inhalation	14–21 d
Norovirus	*Norovirus*	Fecal–oral by food or water or by person-to-person spread	24–48 h
Pediculosis	*Pediculus humanus capitis* (head louse); *Pthirus pubis* (crab louse)	Direct contact	1–2 wk
Pertussis (whooping cough)	*Bordetella pertussis*	Contact with respiratory droplets	7–10 d
Pinworm disease	*Enterobius vermicularis*	Direct contact with egg-contaminated articles	4- to 6-wk life cycle; often takes months of infection before recognition
Pneumocystis jiroveci pneumonia	*Pneumocystis jiroveci*	Unknown; not transmitted person to person	Infants: 1–2 mo; adults: unclear
Pneumococcal pneumonia	*Streptococcus pneumoniae*	Droplet spread	Probably 1–3 d
Rabies	Rabies virus	Bite from rabid animal	2–8 wk
Respiratory syncytial disease	Respiratory syncytial virus	Self-inoculation by mouth or nose after contact with infectious respiratory secretions	3–7 d
Ringworm	*Microsporum* species; *Trichophyton* species	Direct and indirect contact with lesions	4–10 d
Rocky Mountain spotted fever	*Rickettsia rickettsii*	Bite from infected tick	3–14 d
Roseola infantum	Human herpes virus 6	Saliva	10–15 d
Rotavirus gastroenteritis	*Rotavirus*	Fecal–oral route	~48 h
Rubella	Rubella virus	Droplet spread; direct contact	14–21 d
Scabies	*Sarcoptes scabei*	Direct skin contact	2–6 wk
Severe acute respiratory syndrome (SARS)	SARS-associated coronavirus (SARS-CoV)	Droplet; direct contact; occasionally airborne	2–10 d
Smallpox	*Variola major*	Airborne and contact	7–14 d
Syphilis	*Treponema pallidum*	Sexual; perinatal	10 d to 10 wk
Tetanus	*Clostridium tetani*	Puncture wound	4–21 d
Trichinosis	*Trichinella spiralis*	Ingestion of insufficiently cooked foods, especially pork and beef	10–14 d
Tuberculosis	*Mycobacterium tuberculosis*	Airborne	4–12 wk to the formation of primary lesion
West Nile virus	West Nile virus	Bite of infected mosquitoes; from transfusions and transplants; perinatal	3–14 d

Causative Organism

The types of microorganisms that cause infections are bacteria, rickettsiae, viruses, protozoa, fungi, and helminths.

Reservoir

Reservoir is the term used for any person, plant, animal, substance, or location that provides nourishment for microorganisms and enables further dispersal of the organism. Infections may be prevented by eliminating the causative organisms from the reservoir.

Mode of Exit

The organism must have a mode of exit from a reservoir. An infected host must shed organisms to another or to the environment for transmission to occur. Organisms exit through the respiratory tract, the gastrointestinal tract, the genitourinary tract, or the blood.

Route of Transmission

A route of transmission is necessary to connect the infectious source with its new host. Organisms may be transmitted through sexual contact, skin-to-skin contact, percutaneous injection, or infectious particles carried in the air. A person who carries or transmits an organism but does not have apparent signs and symptoms of infection is called a **carrier**.

Specific organisms require specific routes of transmission for infection to occur. For example, *Mycobacterium tuberculosis* is almost always transmitted by the airborne route. Health care providers do not "carry" *M. tuberculosis* bacteria on their hands or clothing. In contrast, bacteria such as *Staphylococcus aureus* are easily transmitted from patient to patient on the hands of health care providers. When appropriate, the nurse should explain routes of disease transmission to patients.

Susceptible Host

For infection to occur, the **host** must be **susceptible** (not possessing immunity to a particular pathogen). Previous infection or vaccine administration may render the host **immune** (not susceptible) to further infection with an agent. Although exposure to potentially infectious microorganisms occurs essentially on a constant basis, people have elaborate immune systems that generally prevent infection from occurring. A person who is immunosuppressed has much greater susceptibility to infection than a healthy person.

Portal of Entry

A portal of entry is needed for the organism to gain access to the host. Again, specific organisms may require specific portals of entry for infection to occur. For example, airborne *M. tuberculosis* does not cause disease when it settles on the skin of an exposed host; the only entry route for *M. tuberculosis* is through the respiratory tract.

Colonization, Infection, and Infectious Disease

Relatively few anatomic sites (e.g., brain, blood, bone, heart, vascular system) are sterile. Bacteria found throughout the body usually provide beneficial **normal flora** to compete with potential pathogens, to facilitate digestion, or to work in other ways symbiotically with the host.

Colonization

The term **colonization** is used to describe microorganisms present without host interference or interaction. Organisms reported in microbiology test results often reflect colonization rather than infection. The patient's health care team must interpret microbiology test results accurately to ensure appropriate treatment.

Infection

Infection indicates a host interaction with an organism. A patient colonized with *S. aureus* may have staphylococci on the skin without any skin interruption or irritation. However, if the patient has an incision, *S. aureus* could enter the wound, resulting in an immune system reaction of local inflammation and migration of white cells to the site. Clinical evidence of redness, heat, and pain and laboratory evidence of white blood cells on the wound specimen smear suggest infection. In this situation, the host identifies the staphylococci as *foreign*. Infection is recognized by the host reaction (manifested by signs and symptoms) and by laboratory-based evidence of white blood cell reaction and microbiologic organism identification.

Infectious Disease

It is important to recognize the difference between infection and infectious disease. **Infectious disease** is the state in which the infected host displays a decline in wellness due to the infection. When the host interacts immunologically with an organism but remains symptom free, the definition of infectious disease has not been met. For example, when a person is first infected with *M. tuberculosis*, infection can be detected by a positive tuberculin skin test (TST), which demonstrates immunologic recognition. Most people who are infected with *M. tuberculosis* have latent infection, but some become ill and demonstrate symptoms of TB such as fever, weight loss, and advancing pneumonia (Porth & Matfin, 2009). Figure 71-2 depicts response to bacterial infection at the cellular level and at the host level.

The primary source of information about most bacterial infections is the microbiology laboratory report, which should be viewed as a tool to be used along with clinical indicators to determine if a patient is colonized, infected, or diseased. Microbiology reports from clinical specimens usually show three components: the smear and stain, the culture and organism identification, and the antimicrobial susceptibility (i.e., sensitivity). As a marker for the likelihood of infection, the smear and stain generally provide the most helpful information

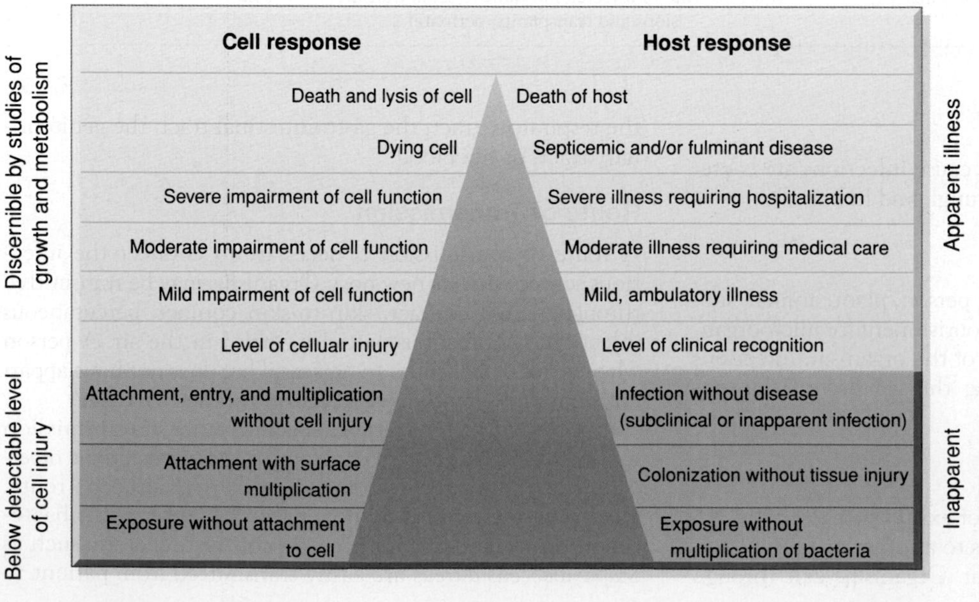
Cell response
Host response
Discernible by studies of growth and metabolism
Below detectable level of cell injury
Apparent illness
Inapparent
Death and lysis of cell
Death of host
Dying cell
Septicemic and/or fulminant disease
Severe impairment of cell function
Severe illness requiring hospitalization
Moderate impairment of cell function
Moderate illness requiring medical care
Mild impairment of cell function
Mild, ambulatory illness
Level of cellualr injury
Level of clinical recognition
Attachment, entry, and multiplication without cell injury
Infection without disease (subclinical or inapparent infection)
Attachment with surface multiplication
Colonization without tissue injury
Exposure without attachment to cell
Exposure without multiplication of bacteria

FIGURE 71-2 • Biologic spectrum of response to bacterial infection at the cellular level (*left*) and of the intact host (*right*). Redrawn from Evans, A. S., & Brachman, P. S. (1998). *Bacterial infections in humans* (3rd ed., p. 40). New York: Plenum.

because they describe the mix of cells present at the anatomic site at the time of specimen collection. Culture and sensitivity results specify which organisms are recognized and which antibiotic agents actively affect the bacteria.

Infection Control and Prevention

The World Health Organization (WHO) and the Centers for Disease Control and Prevention (CDC) are the principal agencies involved in setting guidelines about infection prevention. In recent years, the WHO and CDC have focused more attention on **health care–associated infections (HAIs)**, which are infections not present or incubating at the time of admission to the health care setting. Previously known as nosocomial infections, HAIs have also received increased attention from the Joint Commission, the Institute for Healthcare Improvement (IHI), and Medicare.

The impact of infectious diseases changes over time as microorganisms mutate, as human behavior patterns shift, or as therapeutic options change. The CDC provides timely recommendations about many of the situations that a nurse may face when caring for or educating a patient with an infectious disease. The CDC routinely publishes recommendations, guidelines, and summaries. Through its Internet site and its weekly journal, the *Morbidity and Mortality Weekly Report (MMWR)*, the CDC reports significant cases, outbreaks, environmental hazards, or other public health problems. Examples of important CDC guidelines and summaries are *Sexually Transmitted Diseases Treatment Guidelines* (2010a) and *Immunization Schedule* (CDC, 2013a).

This chapter summarizes several aspects of infectious diseases. However, the field of infection control and prevention changes rapidly. Nurses should seek the most current information when they address patient care concerns or develop infection control policies (see the Resources section at the end of this chapter).

Preventing Infection in the Hospital

Isolation Precautions

Isolation precautions are guidelines created to prevent transmission of microorganisms in hospitals. The Healthcare Infection Control Practices Advisory Committee (HICPAC) of the CDC recommends two tiers of isolation precautions. The first tier, called *standard precautions*, is designed for the care of *all* patients in the hospital and is the primary strategy for preventing HAIs. The second tier, called **transmission-based precautions**, is designed for care of patients with known or suspected infectious diseases spread by airborne, droplet, or contact routes.

Standard Precautions

The premise of **standard precautions** is that all patients are colonized or infected with microorganisms, whether or not there are signs or symptoms, and that a uniform level of caution should be used in the care of all patients. The health care worker should use additional barriers in the form of personal protective equipment (PPE), including masks, eye protection, and cover gowns, depending on the expected degree of exposure to patient excretions or secretions. The elements of standard precautions include hand hygiene, the use of PPE, proper handling of patient care equipment and linen, environmental control, prevention of injury from sharps devices, and patients' room assignments within health care facilities. Hand hygiene, glove use, needlestick prevention, and avoidance of splash or spray of body fluids are discussed in the following sections. (See Chart 37-4 in Chapter 37 for a detailed description of standard precautions.)

Hand Hygiene. The most frequent cause of bacterial transmission in health care institutions is spread of microorganisms by the hands of health care workers (Mortell, 2012). Hands should be washed or decontaminated frequently during patient care. Chart 71-1 describes the recommended hand hygiene methods.

When hands are visibly dirty or contaminated with biologic material from patient care, the hands should be washed with soap and water. In intensive care units and other locations in which virulent or resistant organisms are likely to be present, antimicrobial agents (e.g., chlorhexidine gluconate, iodophor, chloroxylenol) may be used. Effective handwashing requires at least *15 seconds of vigorous scrubbing,* with special attention to the area around nail beds and between fingers, where there is a high bacterial load. Hands should be thoroughly rinsed after washing (CDC, 2002).

If hands are not visibly soiled, health care providers are strongly encouraged to use alcohol-based, waterless antiseptic agents for routine hand decontamination. These solutions are superior to soap or antimicrobial handwashing agents in their speed of action and effectiveness against most microorganisms. Because they are formulated with emollients, they are usually better tolerated than other agents, and because they can be used without sinks and towels, health care workers may be more

Chart 71-1 Hand Hygiene Methods

Hand Decontamination With Alcohol-Based Product

- After contact with body fluids, excretions, mucous membranes, nonintact skin, or wound dressings as long as hands are not visibly soiled
- After contact with a patient's intact skin (e.g., after taking pulse or blood pressure or lifting a patient)
- In patient care, when moving from a contaminated body site to a clean body site
- After contact with inanimate objects in the patient's immediate vicinity
- Before caring for patients with severe neutropenia or other forms of severe immune suppression
- Before donning sterile gloves when inserting central catheters
- Before inserting urinary catheters or other devices that do not require a surgical procedure
- After removing gloves

Handwashing

- When hands are visibly dirty or contaminated with biologic material from patient care
- When health care workers do not tolerate waterless alcohol product

Adapted from Centers for Disease Control and Prevention. (2002). Guideline for hand hygiene in health care settings. *MMWR: Morbidity and Mortality Weekly Report, 51*(RR 16), 1–56.

adherent with their use. Nurses working in home health care or other settings where they are relatively mobile should carry pocket-sized containers of alcohol-based solutions. The spore form of the bacterium *Clostridium difficile* is resistant to alcohol and other hand disinfectants; therefore, the use of gloves and handwashing (soap and water for physical removal) are required when C. *difficile* has been identified (Cohen, Gerding, Johnson, et al., 2010; Dubberke & Gerding, 2011).

Normal skin flora usually consist of coagulase-negative staphylococci or diphtheroids. In the health care setting, workers may temporarily carry other bacteria (i.e., **transient flora**) such as *S. aureus*, *Pseudomonas aeruginosa*, or other organisms with increased pathogenic potential. Generally, transient flora are superficially attached and are shed with hand hygiene and skin regeneration.

Hand hygiene decreases the risk of transfer of bacteria to vulnerable patients by reducing bacterial load on health care workers' hands. The Joint Commission has included hand hygiene as one of the National Patient Safety Goals and focuses on this behavior in all surveys of health care facilities (Joint Commission, 2013). All health care settings should have mechanisms to evaluate compliance with hand hygiene by all personnel who care for patients.

When providing patient care, nurses should not wear artificial fingernails or nail extenders because they have been epidemiologically linked to several significant outbreaks of infections. Natural nails should be kept less than 0.6 cm (0.25 in) long, and nail polish should be removed when chipped because it can support increased bacterial growth (CDC, 2002).

Glove Use. Gloves provide an effective barrier for hands from the microflora associated with patient care. Gloves should be worn when a health care worker has contact with any patient secretions or excretions and must be discarded after each patient care contact. Because microbial organisms colonizing health care workers' hands can proliferate in the warm, moist environment provided by gloves, hands must be washed or disinfected after gloves are removed. As patient advocates, nurses have an important role in promoting hand hygiene and glove use by other hospital workers, such as laboratory personnel, technicians, physicians, and others who have contact with patients.

Compared with vinyl gloves, latex or nitrile gloves are preferred because they resist puncture better and provide greater comfort and fit. Latex gloves have been improved to reduce the incidence of latex hypersensitivity, but some workers continue to experience local skin irritation or more severe reactions, including generalized dermatitis, conjunctivitis, asthma, angioedema, and anaphylaxis (see Chapter 38). The nurse who experiences irritation or an allergic reaction associated with exposure to latex should report symptoms to an occupational health specialist or a physician.

Needlestick Prevention. The most important aspect of reducing the risk of bloodborne infection is avoidance of percutaneous injury. Extreme care is essential in all situations in which needles, scalpels, and other sharp objects are handled. Used needles should not be recapped. Instead, they are placed directly into puncture-resistant containers near the place where they are used. If a situation dictates that a needle must be recapped, the nurse must use a mechanical device to hold the cap or use a one-handed approach to decrease the likelihood of skin puncture. Since 2001, the Occupational Safety and Health Administration (OSHA) has required use of needleless devices and other instruments designed to prevent injury from sharps when appropriate.

Avoidance of Splash and Spray. When the health care professional is involved in an activity in which body fluids may be sprayed or splashed, appropriate barriers must be used. If a splash to the face may occur, goggles and a facemask are warranted. If the health care worker is involved in a procedure in which clothing may be contaminated with biologic material, a cover gown should be worn (CDC, 2002).

Transmission-Based Precautions

Some microbes are so contagious or epidemiologically significant that in addition to standard precautions, a second tier of precautions—transmission-based precautions—should be used when such organisms have been identified. The isolation categories are airborne, droplet, and contact precautions (Siegel, Rhinehart, Jackson, et al., 2007).

Airborne precautions are required for patients with presumed or proven pulmonary TB, varicella, or other airborne pathogens. When hospitalized, patients should be in airborne infection isolation rooms, engineered to provide negative air pressure, rapid turnover of air, and air either highly filtered or exhausted directly to the outside. Health care providers should wear an N95 respirator (i.e., protective mask) at all times while in the patient's room. The nurse should be able to validate negative pressure by reading a pressure manometer placed outside the room and/or by witnessing that a tissue held at the gap between the door and the floor will be pulled toward the room.

Droplet precautions are used for organisms such as influenza or meningococcus that can be transmitted by close contact with respiratory or pharyngeal secretions. When caring for a patient requiring droplet precautions, the nurse should wear a facemask within 3 to 6 feet of the patient; however, because the risk of transmission is limited to close contact, the door may remain open.

Contact precautions are used for organisms that are spread by skin-to-skin contact, such as antibiotic-resistant organisms or C. *difficile*. Contact precautions are designed to emphasize cautious technique and the use of barriers for organisms that have serious epidemiologic consequences or those easily transmitted by contact between health care worker and patient. When possible, the patient requiring contact isolation is placed in a private room to facilitate hand hygiene and decreased environmental contamination. Masks are not needed, and doors do not need to be closed (Chart 71-2).

Specific Organisms With Health Care–Associated Infection Potential

Clostridium difficile

This spore-forming bacterium has significant HAI potential. An especially virulent strain has affected health care facilities throughout North America in the past several years, and rates of C. *difficile* infection increased significantly in the United States since 2000 (CDC, 2012a). Infection is usually preceded by antibiotic agents that disrupt normal intestinal flora and allow the antibiotic-resistant C. *difficile* spores to

Chart 71-2 **Summary of Types of Precautions and Patients Requiring the Precautions**

Standard Precautions

Use standard precautions for the care of all patients.

Airborne Precautions

In addition to standard precautions, use airborne precautions for patients known or suspected to have serious illnesses transmitted by airborne droplet nuclei. Examples of such illnesses include:

- Measles
- Varicella (including disseminated zoster)*
- Tuberculosis

Droplet Precautions

In addition to standard precautions, use droplet precautions for patients known or suspected to have serious illnesses transmitted by large particle droplets. Examples of such illnesses include:

- Invasive *Haemophilus influenzae* type b disease, including meningitis, pneumonia, epiglottitis, and sepsis
- Invasive *Neisseria meningitidis* disease, including meningitis, pneumonia, and sepsis
- Other serious bacterial respiratory infections spread by droplet transmission, including:
  - Diphtheria (pharyngeal)
  - Primary atypical pneumonia (*Mycoplasma pneumoniae*)
  - Pertussis
  - Pneumonic plague
  - Streptococcal (group A) pharyngitis, pneumonia, or scarlet fever in infants and young children
- Serious viral infections spread by droplet transmission, including:
  - Adenovirus*
  - Influenza
  - Mumps
  - Parvovirus B19
  - Rubella

Contact Precautions

In addition to standard precautions, use contact precautions for patients known or suspected to have serious illnesses easily transmitted by direct patient contact or by contact with items in the patient's environment. Examples of such illnesses include:

- Gastrointestinal, respiratory, skin, or wound infections or colonization with multidrug-resistant bacteria judged by the infection control program, based on current state, regional, or national recommendations, to be of special clinical and epidemiologic significance
- Enteric infections with a low infectious dose or prolonged environmental survival, including:
  - *Clostridium difficile*
  - For diapered or incontinent patients: enterohemorrhagic *Escherichia coli* O157:H7, *Shigella* species, hepatitis A, or rotavirus
- Respiratory syncytial virus, parainfluenza virus, or enteroviral infections in infants and young children
- Skin infections that are highly contagious or that may occur on dry skin, including:
  - Diphtheria (cutaneous)
  - Herpes simplex virus (neonatal or mucocutaneous)
  - Impetigo
  - Major (noncontained) abscesses, cellulitis, or pressure ulcers
  - Pediculosis
  - Scabies
  - Staphylococcal furunculosis in infants and young children
  - Zoster (disseminated or in the immunocompromised host)*
- Viral and hemorrhagic conjunctivitis
- Viral hemorrhagic infections (Ebola, Lassa, or Marburg)

*Certain infections require more than one type of precaution.
Adapted from Siegel, J. D., Rhinehart, E., Jackson, M., et al. (2007). *Guideline for isolation precautions: Preventing transmission of infectious agents in healthcare settings*. Available at: www.cdc.gov/hicpac/2007ip/2007isolationprecautions.html

proliferate within the intestine. The organism causes pathology by releasing toxins into the lumen of the bowel. In pseudomembranous colitis (the most extreme form of *C. difficile* infection), debris from the injured lumen of the bowel and from white blood cells accumulates in the form of pseudomembranes or studded areas of the colon. The destruction of such a large anatomic area can cause severe sepsis.

Because antibiotic medications are used so extensively in healthcare settings, many patients are at risk for infection with *C. difficile*. The potential for healthcare–associated acquisition is increased because the spore is relatively resistant to disinfectants and can be spread on the hands of health care providers after contact with equipment previously contaminated with *C. difficile*. Control is best achieved by using contact precautions for infected patients, with use of gowns and gloves for all patient contact. Because the spores are resistant to alcohol, waterless hand products are not as effective as handwashing with soap and water for use in hand hygiene. Bleach-based cleaning products are optimal because bleach can kill spores, whereas other cleaning agents often do not. Frequently touched equipment (such as the overbed table and side rails) should be cleaned daily and whenever visibly soiled. IV poles and other peripheral items should be cleaned when the patient is discharged (CDC, 2012a).

Methicillin-Resistant *Staphylococcus aureus*

Methicillin-resistant ***Staphylococcus aureus*** **(MRSA)**, a common human pathogen, refers to *S. aureus* that is resistant to methicillin or its comparable pharmaceutical agents, oxacillin and nafcillin. Soon after penicillin was discovered in the 1940s, *S. aureus* became all but universally penicillin resistant. Alternative therapies in the form of cephalosporins and synthetic penicillin solutions such as methicillin were introduced. During the late 1970s, MRSA became increasingly more prevalent, as transmission within hospitals and nursing homes was well documented. However, since 2006, the incidence of serious infections due to MRSA has declined (Kallens, Mu, Bulens, et al., 2010; CDC, 2012b).

Healthcare–Associated MRSA. Healthcare providers transmit MRSA to patients easily because *S. aureus* has an affinity for skin colonization. The patient who is colonized with MRSA has an increased probability of developing health care–associated methicillin-resistant *Staphylococcus aureus* (HA-MRSA), especially when invasive procedures, such as

IV therapy, respiratory therapy, or surgery, are performed. The colonized patient also serves as a reservoir for MRSA transmission to others. HA-MRSA may persist as normal flora in the patient for an extended time. The CDC (2012b) reports decreasing incidence of invasive HAIs with MRSA. Although the exact reasons for the decline are unknown, it is likely that infection control efforts (especially those focused on reduction of bloodstream infections) and declining lengths of hospital stay (therefore decreasing exposure within health care agencies) are important factors (Society for Healthcare Epidemiology of America, 2013).

Community-Associated MRSA. In recent decades, new strains of MRSA caused significant infections and outbreaks of infection in children, members of sports teams, and prison inmates, and in other people who had no apparent healthcare exposure. These **community-associated methicillin-resistant *Staphylococcus aureus* (CA-MRSA)** infections are typically caused by strains of *S. aureus* that are molecularly distinct from HA-MRSA. The CA-MRSA strains typically produce more toxins than HA-MRSA, and localized skin symptoms can lead to necrotizing fasciitis or bacteremia. Often, skin symptoms are initially mistaken for spider or bug bites. CA-MRSA infections have resulted in serious skin and soft tissue infections; pneumonia; and, in rare cases, death. Once introduced into a healthcare facility, the virulent CA-MRSA strains can be transmitted to other patients in a manner typically associated with HA-MRSA (Chua, Laurent, Coombs, et al., 2011).

Control of MRSA in Healthcare Facilities. The CDC recommends contact precautions for patients with MRSA colonization or infection (CDC, 2006). In recent years, there has been a growing debate about the benefits of active surveillance cultures to determine unrecognized carriers. In active surveillance, asymptomatic patients are cultured routinely (on admission and/or on a regular basis) to determine unrecognized carriers. Although the epidemiologic evidence does not clearly indicate whether this strategy reduces patient risk, a growing number of states have mandated this approach (Milstone & Perl, 2008). Nurses who participate in active surveillance programs must be prepared to educate patients and their families about the definitions of colonization and infection and the reason for the vigorous isolation efforts. Vancomycin (Vancocin) and linezolid (Zyvox) are typically the preferred treatment options for serious MRSA infection. However, there is concern that MRSA will eventually become resistant to even these medications because they are used so frequently. A small number of patients have been diagnosed with *S. aureus* infections that are intermediately sensitive to vancomycin (i.e., vancomycin-intermediate *Staphylococcus aureus* [VISA]) or completely resistant to vancomycin (i.e., **vancomycin-resistant *Staphylococcus aureus* [VRSA]**). The threat of VRSA is considered a very serious public health concern because of the incidence and pathogenicity of *S. aureus* infections. With the control of MRSA, the emergence of VRSA strains should decrease.

Vancomycin-Resistant *Enterococcus*

Vancomycin-resistant *Enterococcus* (VRE) is the second most frequently isolated source of HAIs in the United States. This gram-positive bacterium, which is part of the normal flora of the gastrointestinal tract, can produce significant disease when it infects blood, wounds, or the urinary tract.

Enterococcus has several traits that make it an easily transmittable HAI organism. It is a normal part of the gastrointestinal flora of the host; it is bile resistant and able to withstand harsh anatomic sites, such as the intestine; and it persists well on the hands of healthcare providers and on environmental objects.

As a relatively resistant organism at baseline, therapy for *Enterococcus* is limited to penicillin formulations (e.g., ampicillin), vancomycin in combination with an aminoglycoside (e.g., gentamicin), or linezolid. Because many strains of VRE are resistant to all other antimicrobial therapies, clinicians are left with few choices for effective therapy. Equally important, VRE colonization and infection may serve as a reservoir of vancomycin-resistant coded genes that may be transferred to the more virulent *S. aureus*. The two most frequently cultured enterococcal species, *Enterococcus faecalis* and *Enterococcus faecium*, are both increasingly found to be resistant. Between 2009 and 2010, more than 42% of central line–associated bloodstream infections (CLABSI) were due to vancomycin-resistant strains of *Enterococcus* (Sievert, Ricks, Edwards, et al., 2013).

Multidrug-Resistant Gram-Negative Organisms

During the past several decades, the incidence of infections caused by multidrug-resistant gram-negative organisms has also increased significantly. Extensive use of antibiotic agents in agriculture and healthcare has led to a growing prevalence of organisms with few, if any, effective antibiotics. The bacteria that most commonly develop resistance include *P. aeruginosa* (resistant to fluoroquinolone antibiotics and/or carbapenems), *Acinetobacter* species (resistant to many antibiotics, including carbapenems), and both *Klebsiella pneumoniae* and *Escherichia coli* (resistant to extended-spectrum beta-lactam antibiotics). These pathogens are also associated with outbreaks in healthcare facilities. For example, an outbreak in 2011 of carbapenem-resistant *Klebsiella* with a high mortality rate at the Clinical Center of the National Institutes of Health progressed for months despite extremely aggressive control methods (Sandora & Goldmann, 2012).

Preventing Healthcare–Associated Bloodstream Infections (Bacteremia and Fungemia)

Reducing the risk of healthcare–associated bloodstream infections requires preventive activities in addition to implementing standard precautions. If a healthcare–associated bloodstream infection occurs, early diagnosis is important to prevent complications, such as endocarditis and brain abscess. Mortality rates associated with infection by some organisms are estimated to average 18%. The estimated average cost attributed to catheter-related bloodstream infections is $16,500 (CDC, 2011a).

Bacteremia is defined as laboratory-confirmed presence of bacteria in the bloodstream. **Fungemia** is a bloodstream infection caused by a fungal organism. Any vascular catheter can serve as the source for a bloodstream infection. Vascular catheters are used for most patients who are hospitalized, and increasingly, long-term central catheters are used to provide IV therapy to outpatients in clinic or home settings. In all instances, the nurse must use appropriate care to reduce the risk of bacteremia and to be alert to signs of bacteremia. Chart 71-3 identifies conditions that suggest the presence of health care–associated catheter-related bacteremia or fungemia.

Chart 71-3 Conditions That Suggest the Presence of Healthcare–Associated Vascular Catheter-Related Bacteremia or Fungemia

- The patient has catheter in place, appears septic, but has no obvious reason to suggest predisposition to sepsis.
- There is no infection at another body site to indicate probable source of sepsis.
- The site of vascular line insertion is red, swollen, or draining (especially purulent drainage).
- The patient has a central vascular line in place at the onset of sepsis.
- The bloodstream infection is caused by *Candida* species or by common skin organisms such as coagulase-negative staphylococci, *Bacillus* species, or *Corynebacterium* species.
- The patient remains septic after appropriate therapy without removal of the vascular catheter-related device.

Prevention of CLABSI is important in preventing sepsis (see Chapter 14). A bundle approach is recommended, including (1) hand hygiene; (2) maximal barrier precautions; (3) chlorhexidine skin antisepsis; (4) optimal catheter site selection, with avoidance of the femoral vein for central venous access in adult patients; and (5) daily review of line necessity with prompt removal of unnecessary lines (see Chart 14–2 in Chapter 14 to review the CLABSI bundle) (CDC, 2011b; IHI, 2012).

Preventing Infection in the Community

The CDC and state and local public health departments share responsibility for prevention and control of infection in the community. Methods of infection prevention include sanitation techniques (e.g., water purification, disposal of sewage and other potentially infectious materials), regulated health practices (e.g., the handling, storage, packaging, and preparation of food by institutions), and immunization programs. In the United States, immunization programs have markedly decreased the incidence of infectious diseases.

Vaccination Programs

The goal of vaccination programs is to use wide-scale efforts to prevent specific infectious diseases from occurring in a population. Public health decisions about vaccination efforts are complex. Risks and benefits for the person and the community must be evaluated in terms of morbidity, mortality, and financial cost and benefit. Successful vaccine programs have reduced the incidence of many infectious diseases in the United States (see Tables 71-2 and 71-3).

More than 50 vaccines are currently licensed in the United States (CDC, 2012c). Vaccines are suspensions of antigen preparations that are intended to produce a human immune response to protect the host from future encounters with the organism. Because no vaccine is completely safe for all recipients, contraindications on package inserts of a vaccine and the CDC-produced "Vaccine Information Statements" must be heeded. These documents provide details about studied experiences with allergy and other complications and provide crucial information about refrigeration, storage, dosage, and administration. The most common adverse effects are allergic reaction to the antigen or carrier solution and the occurrence of the actual disease (often in modified form) when live vaccine is used (Stratton, Ford, Rusch, et al., 2012).

TABLE 71-2 Impact of Vaccines in the Twentieth and Twenty-First Centuries

Disease	20th-Century Annual Morbidity	2010 Total	Decrease (%)
Smallpox	29,005	0	100
Diphtheria	21,053	0	100
Pertussis	200,752	21,291	89
Tetanus	580	8	99
Polio (paralytic)	16,316	0	100
Measles	530,217	61	>99
Mumps	162,344	2,528	98
Rubella	47,745	6	>99.9
Congenital rubella	152	0	100
Haemophilus influenzae (<5 y of age)	20,000 (estimated)	270 (serotype b or unknown)	99

Adapted from Centers for Disease Control and Prevention. (2012c). *Epidemiology and prevention of vaccine-preventable diseases*. Washington, DC: Public Health Foundation.

The standard recommended immunization schedules are revised by the CDC (2013a) as epidemiologic evidence warrants. Variations to the recommended immunization schedule should be made on a case-by-case basis, depending on the patient's risk factors as well as likely exposures. An annual influenza vaccine is recommended for all people 6 months or older, unless contraindicated. In addition, adults who are immunosuppressed (including those who have had a splenectomy) should be vaccinated for pneumococcus (*Streptococcus pneumoniae*) and meningococcus (*Neisseria meningitidis*). Healthcare workers should be immune to measles, mumps, rubella, pertussis, tetanus, hepatitis B, and varicella.

The CDC (2012c) gives information about individual vaccines and vaccine-preventable diseases. It also provides a 24-hour telephone hotline (see the Resources section) for routine pediatric or adult vaccine advice. Advice about optimal

TABLE 71-3 Comparison of Pre-Vaccine Era Estimated Annual Morbidity With Current Estimate

Disease	Pre-Vaccine Era Annual Estimate	2008 Estimate	Decrease (%)
Hepatitis A	117,333	11,049	91
Hepatitis B (acute)	66,232	11,269	83
Pneumococcus (invasive) All ages	63,067	44,000	30
<5 y of age	16,069	4,167	74
Rotavirus (hospitalizations, age <5 y)	62,500	7,500	88
Varicella	4,085,120	449,363	89

Adapted from Centers for Disease Control and Prevention. (2012c). *Epidemiology and prevention of vaccine-preventable diseases*. Washington, DC: Public Health Foundation.

vaccinations for travelers is also available at the CDC (2009) Web site and by phone (see the Resources section).

The incidence of vaccine-preventable diseases, such as measles, mumps, rubella, and diphtheria, is affected by immigration from developing countries. Vaccine campaigns in developing countries are often financially and logistically constrained, and immigrants from such areas may be more likely than U.S. residents to be unprotected. Individual risk and epidemic risk are reduced when vaccination campaigns reach all communities, including those with a high proportion of immigrants.

Reporting Problems With Vaccines

Nurses should ask adult vaccine recipients to provide information about any problems encountered after vaccination. As mandated by law, a Vaccine Adverse Event Reporting System (VAERS) form must be completed with the following information: type of vaccine received, timing of vaccination, onset of the adverse event, current illnesses or medication, history of adverse events after vaccination, and demographic information about the recipient. Forms are obtained by telephone or via the Internet (see the Resources section) and can be submitted online.

Contraindications to Vaccines

As a general rule, all vaccines may be given at the same visit, and separating the dosing by time is not indicated. If they must be given on separate visits, it is then wise to separate by at least 4 weeks so that there is no immune reaction interfering with the response to the second vaccine (CDC, 2012c). Patients who have developed anaphylaxis or other moderate or severe sequelae after a previous dose or those who have developed encephalopathy within 7 days of a previous pertussis vaccine dose should not receive further doses. Some live vaccines (e.g., varicella, MMR [against measles, mumps, and rubella], yellow fever) are contraindicated for people who are severely immunosuppressed or pregnant. All decisions about vaccination should be made by the patient's primary provider after careful review of vaccine-specific contraindications.

Common Vaccines

Measles, Mumps, and Rubella Vaccine. Since the time of licensing of the MMR vaccines, endemic rubella has been eliminated in the United States (CDC, 2005a), and mumps and rubella have decreased by more than 99%. To maintain this effective public health strategy, routine MMR vaccination should be administered to children at 12 to 15 months of age, with repeat dosing at 4 to 6 years of age (CDC, 2013a). Adults who have not received the MMR vaccine should receive one to two doses (CDC, 2013a).

Patients should be advised that fever, transient lymphadenopathy, or hypersensitivity reaction might occur following an MMR vaccination. The risk of side effects is greater in vaccine recipients who have not previously received the vaccine than in those who have received repeat doses. Antipyretics may be used to decrease the risk of fever.

Varicella (Chickenpox) Vaccine and Zoster (Shingles) Vaccine. Varicella zoster is the virus that causes chickenpox and herpes zoster. In its natural state, the varicella virus often attacks children, causing disseminated disease in the form of chickenpox. Although the incidence of varicella is lower in adults, the severity of chickenpox and possible sequelae, including death, is substantially greater (CDC, 2012d). Transmission occurs by the airborne and contact routes. With rare exception, varicella infects a person only once. The incubation period is about 2 weeks (range, 10 to 21 days). During a prodrome of general malaise (often noticed about 2 days before the rash develops), the newly infected host is capable of transmitting the virus to other susceptible contacts. Typically, the vesicular, pustular rash spreads rapidly from few to many lesions in a matter of hours. New lesions continue to form for 2 to 3 days and appear at different stages throughout this time. By the fourth symptomatic day, the lesions begin to dry, and new lesions usually do not develop. Fever is common during the 4 to 6 days of rash progression. When the lesions have crusted, the patient is no longer contagious.

Between 1995, when the varicella vaccine was first licensed in the United States, and 2008, the incidence of chickenpox decreased by 89%. The vaccine is effective in preventing chickenpox in approximately 98% of people who receive two doses of vaccine, and it is almost 100% effective in preventing severe varicella (CDC, 2012d). The vaccine should not be given to those who have depressed immune function, are pregnant, or have demonstrated allergy to varicella vaccine.

Herpes zoster, also known as shingles, is a painful, localized rash caused by recurrent varicella. Vesicles are restricted to areas supplied by single associated nerve groups. Varicella may be transmitted from the rash of those with shingles to people who are susceptible to varicella; the new varicella infections are manifested as chickenpox. It is estimated that more than 50% of unvaccinated people who had a primary varicella infection and live to 85 years of age will develop shingles (CDC, 2012e). Zostavax, a vaccine to reduce the risk of shingles, is recommended for people older than 60 years of age because it reduces the risk of shingles by approximately 50% (CDC, 2012c).

Influenza Vaccine. Influenza is an acute viral respiratory disease that predictably and periodically causes worldwide epidemics known as pandemics. Epidemics occur every 2 to 3 years, with a highly variable degree of severity. An estimated 23,000 deaths per year are associated with influenza or its sequelae (i.e., pneumonia, cardiopulmonary collapse). Older people are more susceptible to influenza, and the incidence of the disease in the United States is increasing as the proportion of older adults increases (CDC, 2012c).

Each year, a new vaccine is composed of the three virus strains (two type A influenza strains and one type B influenza strain) considered most likely to occur in the coming season. If the correct influenza agents have been included in that year's vaccine, the vaccine offers approximately 70% to 90% protection for healthy children and adults younger than 65 years. Although less effective in preventing disease in older adults, it decreases the severity of illness in those who become infected. In extended care facilities, the vaccine is approximately 80% effective in preventing death and 50% to 60% effective in preventing hospitalization (CDC, 2012c). The vaccine is administered as an injection with inactivated virus or as a nasal spray with live attenuated virus.

Until 2013, all influenza vaccines were manufactured by the process of growing the influenza virus in eggs. In January 2013, the U.S. Food and Drug Administration (FDA) approved the manufacture of influenza vaccine using a process called *Flublok*. The process for Flublok does not use eggs or live influenza virus but rapidly replicates large quantities of a protein component of influenza called *hemagglutinin*—the active ingredient in all inactivated influenza vaccines that is essential for entry of the virus into cells in the body. This improvement will be especially important in years when a new strain of influenza is circulating and a vaccine must be produced quickly. Flublok is currently approved for the prevention of seasonal influenza in adults 18 to 49 years of age (FDA, 2013).

Human Papillomavirus Vaccine. Two vaccines that reduce the risk of human papillomavirus (HPV), a sexually transmitted infection (STI), have been approved since 2006. Gardasil (Merck), which is approved for females and males 9 to 26 of age, contains proteins from four HPVs. Cervarix (GlaxoSmithKline), which is approved only for females 10 to 25 years of age, contains proteins from two HPVs. HPV is the most prevalent of all sexually transmitted viruses and is the principal cause of cervical cancer (CDC, 2012c) (see Chapter 57).

Planning for a Pandemic

A pandemic is a global outbreak of a disease. For example, influenza caused three pandemics in the 20th century. In the United States, pandemic influenza caused more than 500,000 deaths in the 1918 Spanish flu pandemic, about 70,000 deaths in 1957 in the Asian flu pandemic, and about 34,000 deaths in 1968 in the Hong Kong flu pandemic. This century, the H1N1 (named for the characteristics of the viral surface proteins hemagglutinin and neuraminidase) pandemic of 2009 caused approximately 12,500 deaths and 270,000 hospitalizations among the more than 60 million infected in the United States. Unlike typical years, when more than 90% of deaths occur in those older than 65 years, during the pandemic period, 90% of deaths occurred in those younger than 65 years (CDC, 2012c).

Subtypes of influenza viruses have pandemic potential because they constantly change within animals and secondarily within humans. As a result of these changes, essentially new viruses can "emerge" and can expose entire populations who are immunologically unprotected.

Influenza pandemics are likely to be more catastrophic than other anticipated public health problems because they last longer than other emergency events, often occur in "waves," deplete the available healthcare workforce, and reduce the supply of medical equipment because of their widespread nature. The frequency and severity of pandemics cannot be accurately predicted, but models suggest that a medium-intensity pandemic could cause more than 200,000 deaths in the United States and quickly overwhelm the existing healthcare infrastructure. The CDC encourages all healthcare institutions to have a pandemic plan and to test the components of the plan regularly.

Avian influenza (bird flu) is an infection caused by influenza viruses that chiefly infect birds and poultry. The H5N1 strain is of particular concern. It has caused a number of outbreaks in poultry since 2003, and the problem is ongoing; flocks of migratory birds have rapidly disseminated the virus throughout much of the world. Although many avian influenza viruses are natural and nonpathogenic in birds, H5N1 is unusual because of its high mortality rate in birds and because it has shown a limited ability to be transmitted from a bird source to mammals, including humans. The human mortality rate from H5N1 avian influenza has been more than 60% (WHO, 2013). The majority of human cases of H5N1 are attributed to direct contact with poultry, but there are rare instances that suggest occasional human-to-human transmission.

Scientists are especially concerned that avian influenza H5N1 may change, either through mutation or reassortment, to become easily transmitted from human to human. If H5N1 were easily transmissible to humans, it would be likely to cause a severe pandemic because the human population has no immunity to the virus and because development of an appropriate vaccine would take too long to effectively stop the pandemic (CDC, 2012f).

The symptoms associated with H5N1 avian influenza in humans have ranged from the symptoms typically seen with seasonal influenza (cough, fever, and muscle aches) to severe pneumonia and multiorgan failure. Influenza antiviral therapy (i.e., oseltamivir [Tamiflu] and zanamivir [Relenza]) is stockpiled as part of a national strategy, with hopes that such therapy would be effective in treating humans infected with the pandemic strain of H5N1. Simple infection control strategies using careful hand hygiene and masks will be especially important in an avian influenza pandemic. Health departments throughout the country are also planning how to implement isolation programs to keep symptomatic people away from schools and workplaces (CDC, 2012f).

Home-Based Care of the Patient With an Infectious Disease

The nurse who cares for the patient with an infectious disease in the home should provide information about infection risk prevention to the patient, the family, and the caregiver (Chart 71-4). Recognizing that a health history may not identify all active or latent infections, the caregiver should carefully follow standard precautions in the home. The nurse should establish a work environment that facilitates hand hygiene and aseptic technique.

Family caregivers should receive an annual influenza vaccine. This is especially true if the caregiver or the patient is older than 50 years, has underlying cardiac or pulmonary disease, or has underlying immunosuppression.

Patients requiring home care are often people with immunosuppression from underlying conditions, such as human immunodeficiency virus (HIV) infection or cancer, or those who have treatment-induced immunosuppression, as occurs with many antineoplastic agents. Careful assessment for signs of infection is important.

Reducing Risk to the Patient

Equipment Care

All caregivers must pay careful attention to disinfection and aseptic technique while using medical equipment. Catheter-related sepsis should be suspected in a patient who has unexplained fever, redness, swelling, and drainage around a vascular catheter insertion site. Indwelling urinary catheters should be discontinued whenever possible, because each day of use increases the risk of infection. The nurse should

Chart 71-4 HOME CARE CHECKLIST
Prevention of Infection in the Home Care Setting

At the completion of home care education, the patient or caregiver will be able to:	PATIENT	CAREGIVER
• Demonstrate aseptic technique in the care of technical equipment such as IV catheter and indwelling urinary catheter.	✔	✔
• Demonstrate thorough hand hygiene after patient care. (Use alcohol-based disinfectant *or* handwashing.)	✔	✔
• Adhere to antibiotic regimen (patient) or completion of vaccination series (patient and caregiver).	✔	✔
• State the rationale for thoroughly cooking all foods and storing meat products separate from other food groups.	✔	✔
• Use separate eating utensils and towels.	✔	✔
• Avoid contact with someone who has a known infectious disease.	✔	✔

promptly report signs of urinary tract infection or generalized sepsis to the patient's primary provider.

Patient Education

When assessing the risk of infection in the home environment of the patient who is immunosuppressed, it is important to realize that intrinsic colonizing bacteria and latent viral infections present a greater risk than do extrinsic environmental contaminants. The nurse should reassure the patient and family that their home needs to be clean but not sterile. Commonsense approaches to cleanliness and risk reduction are helpful.

For patients with neutropenia or T-cell dysfunction (e.g., patients with acquired immunodeficiency syndrome [AIDS]), it is wise to restrict visits of people with potentially contagious illnesses. The patient who is immunosuppressed is vulnerable to acquiring bacterial infection with enteric pathogens from food; therefore, family members should be reminded about the need to follow recommendations for hygiene and safe cooking times and temperatures.

Reducing Risk to Household Members

Establishing reasonable barriers to infection transmission in the household is an important part of home care. The route of transmission of the organism in question must first be determined. The nurse can then educate household members about strategies to reduce their risk of becoming infected. If the patient has active pulmonary TB, the public health department should be contacted to provide screening and treatment for family members. If the patient has shingles (herpes zoster), family members who have had varicella vaccine or who have previously had chickenpox are considered immune and need no precautions. However, if a family member is immunosuppressed or otherwise susceptible to varicella, maintaining physical separation may be an important strategy during the time when the patient has draining lesions. When the patient is infected with *Shigella*, *Salmonella*, *C. difficile*, hepatitis A, or other enteric organisms, the family should be reassured that common household disinfectants are effective in controlling environmental contamination.

Family members who assist in the care of a patient with a bloodborne infection such as HIV or hepatitis C can prevent transmission by carefully handling any sharp objects that are contaminated with blood. Family education may include discussion about the need for caution when shaving the patient; performing dressing changes; or administering any IV, intramuscular, or subcutaneous medication. To collect and dispose of used needles, syringes, and vascular access equipment, the family should use containers designed for sharps disposal. With the exception of TB, the opportunistic infections associated with AIDS do not usually pose a risk to the healthy family member. Family members should be reassured that dishes are safe to use after being washed with hot water and that linens and clothing are safe to use after being washed in a hot water cycle.

Nursing Management

Assessment

Symptoms of infectious diseases vary significantly between and within diseases. For some infections, visible symptoms such as rash, redness, or swelling provide early warnings of infection. In other infections, such as TB and HIV, asymptomatic latency is prolonged, and infection must be determined through diagnostic procedures.

The history is obtained to establish the likelihood and probable source of infection as well as the degree of associated pathology and symptoms. The patient's previous medical record is reviewed when possible. Chart 71-5 outlines questions to ask when obtaining a health history.

Because infection may occur in any body system, physical examination may reveal signs of infection at any body site. Generalized signs of chronic infection may include significant weight loss or pallor associated with anemia of chronic diseases. Acute infection may manifest with fever, chills, lymphadenopathy, or rash. Localized signs vary by source of infection. Purulent drainage, pain, edema, and redness are strongly associated with localized infection. Cough and shortness of breath may be caused by influenza, pneumonia, or TB, as well as many noninfectious causes.

Nursing Interventions

Preventing Infection Transmission

Preventing the spread of infection requires an understanding of the usual routes of transmission of the organism. The patient who is hospitalized may pose a contagious risk to others if the disease is easily spread (such as *C. difficile*) or is spread through an airborne route (such as TB). In these situations,

Chart 71-5

ASSESSMENT

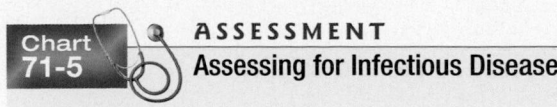

Assessing for Infectious Disease

Ask the following:

- Does the patient have a history of previous or recurrent infections?
- Has there been fever? How high has the patient's temperature been? Is the temperature constant, or does it rise and fall? Has fever been associated with chills? Has the patient taken any medication to relieve fever?
- Is there cough? Is the cough chronic or acute? Is it associated with shortness of breath? Does the cough produce sputum? Is the sputum bloody? Has the patient had a tuberculin skin test recently? If so, what were the results? Has the patient been given isoniazid (INH) prophylaxis for tuberculosis (TB) infection? Has the patient been treated for TB in the past?
- Is there pain? Where is the pain? What is the nature of the pain? Does the patient have a sore throat, headache, myalgias, or arthralgias? Is there pain on urination or other activity?
- Is there edema? Is there drainage associated with the edema? Is the edematous area warm to the touch?
- Is there a draining lesion? Is the drainage associated with trauma or a previous procedure? Is the drainage purulent or clear?
- Does the patient have diarrhea, vomiting, or abdominal pain?
- Is there a rash? What is the nature of the rash—is it flat, raised, red, crusted, purulent, or lacelike? Has the patient taken medications that could induce rash? Has there been exposure to another person who has an identified infectious disease or rash?
- What is the patient's vaccination history? Are immunizations up to date?
- Has there been an insect or animal bite? Has there been an animal scratch or other exposure to pets, farm animals, or experimental animals?
- What medications are used? Have antibiotic agents been taken recently or long term? Is the patient being treated with corticosteroids, immunosuppressive agents, or chemotherapy?
- Has the patient been treated for other infectious diseases in the past? Has the patient been hospitalized for infectious diseases?
- If sexual history is pertinent, has there been sexual exposure to another person with a known sexually transmitted infection (STI)? Has the patient been treated for STIs in the past? Is the patient pregnant, or has she recently been pregnant? Has the patient been tested for human immunodeficiency virus?
- Has the patient traveled abroad, including developing countries? What was the immunization or antimicrobial prophylaxis used for protection while traveling?
- What is the patient's occupation? What are the patient's recreational activities? Hobbies?

strict adherence to isolation measures is important in reducing the opportunity for spread. Preventing transmission of organisms from patient to patient requires participation of all members of the healthcare team. Transmission of organisms on the hands and gloves of healthcare workers remains a common source of cross-infection in the hospital or clinic setting.

Nurses serve an important role in preventing the transfer of organisms in two ways (Mortell, 2012). First, nurses have many opportunities for spreading organisms, given their frequent encounters with patients and families. For example, the nurse who has performed endotracheal suctioning should remove the gloves, wash the hands, and put on a new pair of gloves before performing wound care on the same patient. Second, nurses can reduce hand-to-hand spread of organisms by serving as patient advocates. The nurse should observe the hand hygiene activities of other professionals and alert them to lapses in technique that are observed.

Quality and Safety Nursing Alert

It is imperative that nurses disinfect their hands before and after contact with patients and after performing a potentially hand-contaminating activity. Hands must be disinfected each time gloves are removed.

Educating About the Infectious Process

Interruption of transmission requires diagnosis and patient compliance with the treatment regimen. The nurse's role is to educate the patient and, in some situations, to report the case to public health officials for contact tracing and verification of follow-up.

The nurse must stress the importance of immunization to parents of young children and to others for whom vaccines are recommended, such as patients who are older, are immunosuppressed, or have chronic illnesses or disabilities. Nurses should recognize their personal responsibility to receive the hepatitis B vaccine and an annual influenza vaccine to reduce potential transmission to themselves and vulnerable patient groups.

Infectious diseases often seem mysterious and frequently are socially stigmatizing. Patient education requires empathy and sensitivity. For example, the patient may have negative feelings about a MRSA diagnosis and feel that isolation practices are unjust (Chart 71-6).

Controlling Fever and Accompanying Discomforts

Fever must always be investigated to determine whether infection is the source. Evidence indicates that fever, mediated by the hypothalamus, may potentiate beneficial functions in the syndrome of reactions known as *acute-phase reaction*. These reactions include changes in liver protein synthesis; alterations in serum metals, such as iron; and increased production of certain classes of white blood cells and other cells of the immune system (Porth & Matfin, 2009). Most fevers are physiologically controlled so that the temperature remains below 41°C (105.8°F). However, severe fever, as often occurs with meningococcal meningitis, may cause complications. Even milder fevers accompanied by fatigue, chills, and diaphoresis are often uncomfortable for the patient. Whether fever is treated or untreated, adequate fluid intake is important during febrile episodes.

Quality and Safety Nursing Alert

Because fever offers clues about infection severity and the success of antibiotic therapy, outpatients with fever should be instructed to obtain accurate temperature readings. Frequently, family caregivers know that a patient has warm skin but do not take a temperature reading. Body temperature information can be very helpful in adjusting therapy or in reevaluating a preliminary diagnosis.

Monitoring and Managing Potential Complications

The patient with a rapidly progressive infectious disease should have vital signs and level of consciousness closely monitored.

Chart 71-6

NURSING RESEARCH PROFILE

Patient Experience of Methicillin-Resistant *Staphylococcus aureus* and Isolation

Webber, K., Macpherson, S., Meagher, A., et al. (2012). The impact of strict isolation on MRSA positive patients: An action-based study undertaken in a rehabilitation center. *Rehabilitation Nursing*, *37*(1), 43–50.

Purpose

Being in long-term isolation can have a negative impact on emotional well-being of patients. The purpose of this research was to seek insights from rehabilitation inpatients to better understand the experience of methicillin-resistant *Staphylococcus aureus* (MRSA) and strict isolation.

Design

In this action-based, qualitative project, nine rehabilitation inpatients with MRSA who had been in isolation for at least 1 month participated in semistructured interviews until thematic saturation was reached. A follow-up focus group validated themes from the interviews and shared ideas for improving isolations practices.

Findings

This study found six important themes. The themes were as follows: "MRSA is a nuisance," "MRSA is not serious," "negative feelings most often result with having an MRSA diagnosis and being placed isolation," "strict isolation policy is unjust and illogical," "Why should we care if you don't?" and "It's not up to me to educate people about MRSA." Following the focus group, there was immediate implementation of four action plans and ongoing development of four additional action plans.

Nursing Implications

Nurses need to consider the psychosocial effects of isolation. People with MRSA have many insights and negative feelings when placed in isolation. Nurses need to take these feelings into account when providing education and support services for patients with MRSA.

X-ray findings and microbiologic, immunologic, hematologic, cytologic, and parasitologic laboratory values must be interpreted in the context of other clinical findings to assess the course of the infectious disease.

Antibiotic therapy is frequently complex, and modifications are necessary because of sensitivity test results and disease progression. To rapidly ensure therapeutic blood levels antibiotic therapy should be initiated as soon as it is prescribed rather than waiting until routine medication scheduling times. Chart 71-7 describes nursing interventions for specific complications of infection.

Diarrheal Diseases

In the United States, the epidemiology of diarrheal diseases changes constantly. Water disinfection, pasteurization, and appropriate food packaging have decreased the incidence of diseases such as typhoid and cholera. However, importation of foreign foods, environmental and ecologic changes, and changes in diagnostic test modalities have led to recognition of important new trends and outbreaks.

Transmission

The portal of entry of all diarrheal pathogens is oral ingestion. Although food is far from sterile, the high acidity of the stomach and the antibody-producing cells of the small bowel generally serve to decrease the potential of pathogens. Infection can occur when the infectious dose is high enough or if the food neutralizes the acidic environment. Decreased gastric acidity with disruption of normal bowel flora (as occurs after surgery), the use of antimicrobial agents, and AIDS all decrease intestinal defenses.

Causes

There are many bacterial, viral, and parasitic causes of diarrheal diseases. Common causes of bacterial infection include *E. coli* and *Salmonella, Shigella, Campylobacter,* and *Yersinia* species. The most significant viral causes of diarrhea are *Rotavirus,* which commonly results in diarrhea in young children, and *Calicivirus* (often called *Norovirus*), a virus associated with outbreaks in long-term care facilities and cruise ships (CDC, 2011d). Parasitic infections of importance include *Giardia* and *Cryptosporidium* species and *Entamoeba histolytica*.

Campylobacter Infections

Campylobacter species are among the most frequent causes of diarrheal disease in the United States. The bacterium, which is abundant in animal foods, is especially common in poultry but can also be found in beef and pork. Direct person-to-person transmission appears to be less common than it is for other enteric pathogens, such as *Shigella*. Guillain-Barré syndrome, a serious neurologic disorder characterized by temporary paralysis, is a rare complication of approximately 1 in 1,000 cases of *Campylobacter* infection (Bowyer & Glover, 2010).

Cooking and storing food at appropriate temperatures protects against *Campylobacter*. It is important that kitchen utensils used in meat preparation be kept away from other food to prevent *Campylobacter* transmission.

After a person is infected, the bacterium directly attacks the lumen of the intestine and may cause disease through enterotoxin release. Symptoms can range from mild abdominal cramping and minimal diarrhea to severe disease with profuse watery bloody diarrhea and debilitating abdominal cramping. Antimicrobial therapy is recommended only for patients who are seriously ill (CDC, 2012g).

Salmonella Infection

Salmonella is a gram-negative bacillus with many species, including the very pathogenic *Salmonella typhi* (i.e., typhoid fever). Of the nontyphi species, most organisms are prevalent in animal food sources. Annually in the United States, *Salmonella* species contaminate approximately 2.2 million eggs (1 in 20,000 eggs) and one in eight chickens raised as meat (CDC, 2013b). *Salmonella* outbreaks have also been associated with contaminated sprouts, fruits and vegetables.

(*text continues on page 2104*)

Chart 71-7

PLAN OF NURSING CARE

Care of the Patient With an Infectious Disease

NURSING DIAGNOSIS: Risk for infection transmission
GOAL: Prevention of transmission of infectious agents

Nursing Interventions	Rationale	Expected Outcomes
1. Prevent patient-to-patient infection spread.	**1.** Organisms that are spread through an airborne route or are very contagious through direct contact can be transmitted in a healthcare setting.	• No evidence of patient-to-patient transmission of infection • No evidence of transmission via health care workers • No occupationally acquired infections in nurses and other healthcare workers • No evidence of transmission due to contaminated equipment • Absence of bacteremia, septicemia, and sepsis • Absence of urinary tract infections • Absence of pneumonia
a. Provide isolation according to Centers for Disease Control and Prevention (CDC) guidelines and standard precautions.	**a.** CDC isolation strategies are developed to reduce the likelihood of transmission from patient to patient.	
b. Ensure that patients with airborne infections remain in private rooms during hospital stay. If they must leave their rooms, arrangements should be made to decrease the likelihood of contact with other patients. Rooms should be ventilated according to CDC criteria. Personal protective equipment in the form of N95 respirators should be worn as indicated.	**b.** Engineering controls are important in the prevention of airborne diseases. Influenza vaccine safely reduces risk of illness associated with this highly communicable, and frequently virulent, condition. The N95 respirator is the minimal level of personal protection for tuberculosis (TB) control. The "N" indicates the filter resistance to oil aerosols; the "95" indicates that the respirator has 95% effectiveness in filtering test particles.	
c. Ensure that patients with highly transmissible, nonairborne organisms such as *Clostridium difficile* and *Shigella* species are physically separated from other patients if hygiene or institutional policy dictates.	**c.** Increased prevention strategies are needed when the organism has high epidemic potential.	
2. Prevent healthcare workers' transfer of organisms from patient to patient.	**2.** Transfer of organisms on the hands of healthcare workers is a common route of transmission. Hospital organisms colonizing the hands of healthcare workers may be virulent.	
a. Perform hand hygiene (by handwashing or by using alcohol-based solution) consistently and thoroughly, disinfecting hands before and after each patient contact, and after procedures that offer contamination risk while caring for an individual patient.	**a.** Hand hygiene is important in reducing transient flora on outer epidermal layers of skin. Alcohol-based hand disinfectants are effective methods to reduce transient flora.	
b. Use gloves when handling any body fluid from any patient. Change gloves between patient care activities, and disinfect hands after gloves are removed.	**b.** Gloves provide effective barrier protection. Gloves quickly become contaminated and then become a potential vehicle for the transfer of organisms between patients. Microflora on hands are likely to proliferate while gloves are worn.	
c. Avoid wearing artificial fingernails or extenders when providing patient care. Keep natural nails less than ¼ inch long.	**c.** Artificial fingernails and extenders harbor microorganisms.	
d. Monitor the hand hygiene and glove use behaviors of healthcare professionals caring for the patient.	**d.** Poor adherence to hand hygiene among healthcare workers has been well documented and should be anticipated. It is important for the nurse (as the patient's advocate) to communicate protective behavior.	

(continues on page 2100)

Chart 71-7

PLAN OF NURSING CARE

Care of the Patient With an Infectious Disease (continued)

Nursing Interventions	Rationale	Expected Outcomes
3. Prevent transmission of infection from patient to healthcare worker.	**3.** Healthcare workers may acquire infections occupationally due to close contact with patients.	
a. Avoid risk of infection with TB.	**a.** The most important element in the reduction of TB is early identification. Many of the symptoms of TB are subtle and may be first observed by the nurse who has prolonged contact with the patient.	
(1) Participate in the early identification of patients with active disease. Patients will be asked about risk factors, symptoms, previous exposure, and status of tuberculin skin test or other rapid tests.	**(1)** Identification of patients at risk can help to prevent exposure.	
(2) Expedite diagnostic workup with chest x-ray, sputum analysis for organisms, and administration of TB testing as appropriate.	**(2)** Confirmation of diagnosis facilitates development of an appropriate treatment plan, including prevention of spread of infection.	
(3) Maintain engineering controls. Keep the patient in a private room with a closed door.	**(3)** Confining airflow to the immediate vicinity of the patient and exhausting air to the outside reduces the likelihood of transmission to healthcare workers in areas outside of the patient room.	
(4) Use protection in isolation room or when participating in procedures that are likely to generate cough, such as suctioning, intubation, or administering nebulized medications.	**(4)** N95 respirators are designed to reduce healthcare workers' risk.	
b. Avoid risk of transmission of bloodborne diseases such as hepatitis B, hepatitis C, and the human immunodeficiency virus.	**b.** Healthcare workers can contract bloodborne diseases via percutaneous injury such as needlestick or by contact with blood or body fluids to mucous membranes, such as eyes and mouth.	
(1) Get hepatitis B vaccination.	**(1)** Hepatitis B vaccine should be administered to reduce risk from this contagious bloodborne virus.	
(2) Use standard precautions as defined by the CDC (see Chart 37-4 in Chapter 37).	**(2)** Standard precautions are based on the recognition that most patients are not identified as infected by physical assessment or history taking. Health care workers must assume that all patients may be infected with bloodborne or other infection and must use barrier precautions appropriately for *all* patients.	
(3) Use "needleless" syringes and other injury-preventing devices.	**(3)** The use of injury-preventing devices decreases risk of transmission of bloodborne diseases.	
c. Avoid risk of airborne diseases. **(1)** Receive influenza vaccination annually. **(2)** Get vaccinated or produce proof of immunity to measles, mumps, rubella, and varicella.	**c.** Influenza vaccine is recommended for healthcare workers to reduce the likelihood of transmission in health care settings where immunocompromised patients can be exposed.	

Chart 71-7

PLAN OF NURSING CARE

Care of the Patient With an Infectious Disease (continued)

Nursing Interventions	Rationale	Expected Outcomes
4. Prevent patient exposure to contaminated medical equipment.	**4.** Technologic advances offer increased opportunity for invasive procedures. Equipment may be complex and difficult to clean.	
a. Ensure that equipment being inserted through intact skin is sterilized between patient uses.	**a.** Sterilization renders equipment free of all microorganisms.	
b. Ensure that equipment that has contact with mucous membranes is sterilized or receives "high-level disinfection" between patient uses.	**b.** High-level disinfection renders an object free of all microorganisms with the possible exception of spore-producing organisms.	
c. Ensure that equipment used against intact skin is thoroughly cleaned and receives "low-level disinfection" between patient uses.	**c.** The disinfection goal for low-level disinfection is to reduce the load of microorganisms to a level that is not threatening to the host with intact skin.	
5. Follow established guidelines for the routine removal and replacement of IV devices.	**5.** Indwelling IV devices can serve as a conduit for organisms to migrate into the bloodstream.	
6. Remove urinary catheters at the earliest time possible.	**6.** The risk of urinary tract infections is directly proportional to the length of time that a urinary catheter remains in place.	
7. Remove endotracheal and nasogastric tubes as soon as possible.	**7.** The risk for pneumonia is increased as the use of indwelling equipment increases.	

NURSING DIAGNOSIS: Knowledge deficit about disease, cause of infection, and preventive measures
GOAL: Acquisition of knowledge about the infectious process

Nursing Interventions	Rationale	Expected Outcomes
1. Listen carefully to what the patient says about illness and previous treatment.	**1.** Listening facilitates detection of misunderstanding and misinformation and provides opportunity for education.	• Patient actively participates in treatment • Patient adheres to infection control measures
2. Provide pertinent explanations about: **a.** Organism and route of transmission **b.** Treatment goals **c.** Follow-up schedule **d.** Prevention of transmission to others	**2.** Knowledge about specific diagnoses and treatments may promote adherence.	
3. Allow opportunities for questions and discussions.	**3.** The patient's questions indicate issues that need clarification.	
4. Educate the patient and family about: **a.** Prophylaxis or immunization, if recommended **b.** Community resources, if necessary **c.** Means of preventing transmission within the home	**4.** Understanding of the risks and precautions associated with an infectious disease may reduce the opportunity for further spread.	

NURSING DIAGNOSIS: Risk for imbalanced body temperature (fever) related to the presence of infection
GOAL: Patient comfort and return of normal temperature

Nursing Interventions	Rationale	Expected Outcomes
1. Monitor temperature, pulse, and respirations at regular intervals.	**1.** Graph fever curve to help evaluate when fever occurs, how long it lasts, and whether it responds to therapy.	• Body temperature within normal limits • Maintenance of fluid and electrolyte balance
2. Administer antipyretic agents as prescribed.	**2.** Prompt treatment will improve outcomes.	• Patient comfortable

(continues on page 2102)

Chart 71-7

PLAN OF NURSING CARE

Care of the Patient With an Infectious Disease (continued)

COLLABORATIVE PROBLEMS: Among potential complications are septicemia, bacteremia, or sepsis, septic shock, dehydration, abscess formation, endocarditis, infectious disease–related cancers, and infertility
GOAL: Absence of complications

Nursing Interventions	Rationale	Expected Outcomes
Septicemia, Bacteremia, Sepsis		
1. Monitor patient for evidence of infection at any location.	**1.** Vigilance for bacterial or fungal infection at any site promotes early recognition and treatment and reduces the likelihood of secondary infections.	• No episode of infection • Effective treatment for identified bacterial and fungal infections without progression to bloodstream infection
2. Assess treatment effectiveness of all identified infections.	**2.** The natural course of some infections may be rapid unless antibiotic agents are administered promptly.	• Early improvement in septic course
3. Administer antibiotic agents as prescribed with first dose given at the earliest time possible.	**3.** Prompt treatment will improve outcomes.	
Septic Shock		
1. Routinely, and as warranted, monitor vital signs for patients with recognized infections and severely immunosuppressed patients at risk for shock. In particular, be alert to signs of: **a.** Fever **b.** Tachycardia (>90 bpm) **c.** Tachypnea (>20 breaths/min) **d.** Evidence of decreased perfusion or dysfunction of vital organs in the form of: **(1)** Change in mental status **(2)** Hypoxemia as measured by arterial blood gases **(3)** Elevated lactate levels **(4)** Urine output (<0.5 mL/kg/h)	**1.** Early recognition of the signs and prompt treatment of impending shock may reduce the associated severity or mortality.	• Absence of symptoms of septic shock • Hemodynamic and respiratory status within normal range
2. Administer antibiotic agents, fluid replacement, vasopressors, and oxygen as prescribed.	**2.** Therapeutic maintenance of hemodynamic and respiratory status is necessary until infection is effectively treated with an antimicrobial regimen.	
Dehydration		
1. Assess for dehydration (thirst, dryness of mucous membranes, loss of skin turgor, reduced peripheral pulses, urine output <0.5 mL/kg/h).	**1.** Signs of dehydration provide a basis for fluid replacement and suggest possible further complications of circulatory collapse.	• Attains fluid balance (output approximates intake: body weight unchanged) • Mucous membranes appear moist; normal skin turgor • Serum electrolytes within normal limits
2. Monitor weight.	**2.** Rapid changes in weight indicate fluid volume changes.	
3. Monitor intake and output and serum electrolyte levels.	**3.** Dehydration produces a deficit in some electrolytes. Decreased urine production may indicate hypovolemia and decreased renal perfusion.	
4. Replace fluids as needed. If the patient can tolerate oral fluids, offer fluids every 2–4 hours. Administer IV fluids as prescribed.	**4.** When possible, oral hydration is preferable because the patient can select the beverage, control the rate and interval of replacement, and care for self at home. Additionally, the risks associated with vascular devices are avoided. If IV fluid is required, IV solutions are selected to facilitate intestinal reabsorption of fluid and electrolytes.	

Chart 71-7

PLAN OF NURSING CARE

Care of the Patient With an Infectious Disease (continued)

Nursing Interventions	Rationale	Expected Outcomes
Abscess Formation		
1. Assess vascular access sites, wound sites, pressure ulcers, and other appropriate sites for apparent collections of purulent material.	1. Collections of purulent material often require drainage before antimicrobial therapy is effective.	• Absence of abscess • Takes antibiotic agents as prescribed
2. Assess the patient who has had abdominal surgery or trauma to abdominal area for localized signs of intra-abdominal abscess. These signs include: a. Low-grade fever b. Elevated peripheral white blood cell count c. Localized pain d. Abdominal tenderness e. Visible or palpable mass f. Postoperative diarrhea g. Gastrointestinal (GI) bleeding	2. Intra-abdominal abscess formation is most common following traumatic or surgical disruption of the GI tract. Signs are often initially subtle.	
3. Assess patient who has had percutaneous abscess drainage to determine whether drainage has been successful. Be alert to all of the previously mentioned signs and symptoms.	3. After percutaneous drainage, recurrent or persistent signs of abscess may indicate the need for surgical treatment.	
4. Administer antibiotic agents as prescribed.	4. Antibiotic agents, along with drainage, are the most important elements of intra-abdominal abscess management.	
Endocarditis		
Prevention		
1. Educate patients with the following conditions about the importance of antibiotic prophylaxis for events and procedures that may introduce the risk of endocarditis: a. Valvular disease b. Congenital heart disease c. Intracardiac prosthesis d. Previous endocarditis	1. Patients with underlying valvular disease and other cardiac abnormalities are at increased risk for "seeding" of the cardiac valves during procedures that can cause bacteremia.	• Informs healthcare professionals of cardiac conditions that require antibiotic prophylaxis before invasive procedures • Takes prophylactic antibiotics as prescribed
Management		
1. Obtain blood cultures as prescribed; carefully record results. Note persistent bloodstream infections with a particular organism.	1. A definitive diagnosis of endocarditis requires blood culture confirmation.	• Endocarditis is diagnosed, treated, and cured
2. Obtain a detailed history about the duration of fever in the absence of well-recognized cause.	2. Endocarditis should be suspected in patients who report an unexplained fever of more than 1 week's duration.	
3. Administer IV antibiotic therapy at prescribed time schedule.	3. IV therapy is usually required for cure. The goal of therapy is complete eradication of all organisms. Careful adherence to following the scheduled administration is therefore essential.	

Infectious Disease–Related Cancers and Infertility

These potential complications of infectious diseases are prevented by primary avoidance of infection. Management of them is directed toward treating each of them as a noninfectious entity. For example, the management of cancer secondary to hepatitis B is handled as an oncology issue, not as an infectious disease issue.

Variable symptoms are associated with *Salmonella* species infection, including an asymptomatic carrier state, gastroenteritis, and systemic infection. Diarrhea with gastroenteritis is common. Disseminated disease and bacteremia, sometimes accompanied by diarrhea, occur less often.

The person with *Salmonella*-caused diarrhea can on rare occasions be a source of transmission to others. The importance of good hygiene should be emphasized, and healthcare workers should use special care when handling bedpans, stool specimens, or other objects that may be contaminated with feces. Hand hygiene is imperative after any contact with a person with *Salmonella* diarrhea. Although patients with systemic salmonellosis require antimicrobial therapy, those with gastroenteritis only are not usually treated, because antibiotic use may increase the period of time that the patient carries the bacteria while not improving the clinical outcome.

Shigella Infection

The *Shigella* species is a gram-negative organism that invades the lumen of the intestine and causes disease and severe watery (possibly bloody) diarrhea. *Shigella* species are spread through the fecal–oral route, with easy transmission from one person to another. *Shigella* exhibits high levels of **virulence** (degree of pathogenicity of an organism); infection with a very small number of organisms can cause disease. Because transmission occurs easily with improper hygiene, it is not surprising that *Shigella* organisms disproportionately affect pediatric populations. Disease in the very young may infrequently be complicated by pulmonary or neurologic symptoms.

Antimicrobial therapy should be instituted early. Frequently, initial therapy choices must be altered when final microbiologic testing reveals the organism's sensitivity.

Escherichia coli

E. coli is the most common aerobic organism colonizing the large bowel. When *E. coli* bacteria are cultured from fecal specimens, the results usually reflect normal flora. However, certain strains of *E. coli* with increased virulence have been responsible for significant outbreaks of diarrheal disease in recent years. These stronger pathologic strains are subgrouped as Shiga toxin–producing *Escherichia coli* (STEC) because of their production of enterotoxins. STEC strains often cause choleralike disease, with rapid, severe dehydration and an increased risk of death.

Several outbreaks of an *E. coli* STEC species, 0157:H7, have been linked to the ingestion of undercooked beef and to vegetables that have been contaminated by animal wastewater. This bacterium lives in the intestines of cattle and can be introduced into meat at the time of slaughter. Prevention of disease from *STEC* strains is aimed at educating the public to cook ground beef thoroughly (i.e., until the juices run clear) (CDC, 2012L).

Calicivirus (Norwalklike Virus; *Norovirus*)

Calicivirus, which is often referred to as the Norwalklike virus or the *Norovirus*, is a the most common cause of foodborne illness and gastroenteritis in the United States. Onset of illness is usually acute, with vomiting and watery diarrhea that generally last for approximately 2 days. Most outbreaks occur between November and April. Dehydration is the most common complication. This agent has been associated with important diarrheal outbreaks in schools, day care centers, cruise ships, long-term care facilities, and hospitals.

Calicivirus is transmitted easily from person to person by direct contact and by ingesting contaminated food. Waterborne outbreaks have been associated with sewage-contaminated wells and contaminated swimming pools. Although people with *Calicivirus* infection typically recover within 2 to 3 days, they may continue to transmit the virus to others for approximately 2 more weeks.

Caliciviruses can withstand environmental extremes of heat or cold and are resistant to chemical disinfection, which are significant reasons for their epidemic potential. Control of *Calicivirus* in healthcare facilities requires a coordinated program with decisions about isolation, environmental disinfection, diagnosis, and coordination with public health officials. Contact precautions should be used when caring for patients with incontinence and during outbreaks of the virus. Workers should wear masks if they are cleaning heavily soiled areas or caring for a patient who is actively vomiting. A number of disinfectants have U.S. Environmental Protection Agency (EPA) approval based on specific claims about effectiveness for *Calicivirus*. However, the CDC recommends that surface disinfection be accomplished with a freshly prepared solution of 1 part bleach to 50 parts water or with the peroxygen compound Virkon-S, which has been approved for feline *Calicivirus* and may be similarly effective for human viruses (CDC, 2012h).

Giardia lamblia

Transmission of the protozoan *Giardia lamblia* occurs when food or drink is contaminated with viable cysts of the organism. People often become infected while traveling to endemic areas or by drinking contaminated water from mountain streams within the United States. The organism can be transmitted by close contact, such as occurs in day care settings. Transmission by sexual contact has also been documented.

Frequently, the infection goes unnoticed. Infection is often recognized more easily in children than in adults. In extreme cases, the patient may experience abdominal pain and chronic diarrhea, usually described as containing mucus and fat but not blood. Microscopic examination of stool specimens reveals the trophozoite or cyst stages of the parasitic life cycle.

Metronidazole (Flagyl) is recommended by the CDC to treat *Giardia* (Karch, 2012). Patients with *Giardia* infections should be instructed that the organism can be easily transmitted in family or group settings. Personal hygiene measures should be reinforced, and those who travel or camp where water is not treated and filtered should be advised to avoid local water supplies unless water is purified before drinking or using it in cooking.

Vibrio cholerae

Although reported cases of cholera are rare in the United States, the incidence has increased worldwide in recent years. Historically, epidemics of cholera have influenced all aspects of life—from medical to political—and infection rates have been significant enough to destroy governments and armies. Ten months after a devastating earthquake in January 2010, cholera was introduced to Haiti after decades without incidence in the Caribbean. After likely introduction of the bacteria by relief workers, cholera has been epidemic with more than 600,000 cases and more than 7,400 deaths in the first

2 years (Barzilay, Schaad, Magloire, et al., 2013; Tappero & Tauxe, 2011).

V. cholerae is a gram-negative organism with several different serotypes. The type usually associated with epidemics is toxigenic *V. cholerae* 01. The organism is transmitted by contaminated food or water. Most recent cases in the United States have been from contaminated shellfish found in the Gulf of Mexico or from contaminated shellfish brought into the United States by visitors. Cholera causes disease with a very rapid onset of copious diarrhea in which up to 1 L of fluid per hour can be lost. Dehydration, with subsequent cardiopulmonary collapse, may cause rapid progression from onset of signs and symptoms to death. Rehydration efforts should be vigorous and sustained. If oral rehydration cannot be accomplished, the patient needs IV therapy.

In the United States, cholera should be suspected in patients who have watery diarrhea after eating shellfish harvested from the Gulf of Mexico. Confirmation of the causative organism can be made by stool culture. It is imperative that all cases are reported to local and state public health authorities. People traveling to areas where cholera occurs regularly should remember the simple rule of thumb: "Boil it, cook it, peel it, or forget it."

NURSING PROCESS

The Patient With Infectious Diarrhea

Assessment

The most important element of assessment in the patient with diarrhea is to determine hydration status. The goal of rehydration is to correct the dehydration. Assessment includes evaluation for thirst, dryness of oral mucous membranes, sunken eyes, a weakened pulse, and loss of skin turgor. Careful observation for these signs is especially important in cases of rapidly dehydrating diseases (most notably cholera) and in younger children.

Intake and output measurements are crucial in determining fluid balance. Liquid stool should be measured and recorded, along with the frequency of stools. It is important to note the consistency and appearance of stool as key indicators of the type and severity of the diarrheal disease. The presence of mucus or blood should also be documented.

When conducting a health history, the nurse asks if the patient has recently traveled, if the patient is being treated with antibiotic medications, if the patient has been in contact with anyone who has recently had diarrheal disease, and what the patient has recently eaten. Frequently, patients attribute the most recent meal eaten as the cause of symptoms. However, the incubation period for most diarrheal conditions is longer than the time interval between meals, and the nurse needs to get detailed information about the meal preceding the illness and about all food intake in the previous 3 to 4 days. When eliciting this kind of history, it is helpful to ask the patient to list every food tasted. The nurse also asks the patient if he or she is employed in a food preparation service, because the local public health departments should be notified about any person with infectious diarrhea who works in the food industry.

Diagnosis

Nursing Diagnoses

Based on the assessment data, major nursing diagnoses may include the following:

- Deficient fluid volume related to fluid lost through diarrhea
- Deficient knowledge about the infection and the risk of transmission to others

Collaborative Problems/Potential Complications

Potential complications may include the following:

- Bacteremia
- Hypovolemic shock

Planning and Goals

The most important goals are maintenance of fluid and electrolyte balance, increased knowledge about the disease and risk of transmission, and absence of complications.

Nursing Interventions

Correcting Dehydration Associated With Diarrhea

The patient is assessed to determine the degree of dehydration and the amount and route of rehydration needed. Oral rehydration therapy is a strategy used to reduce the severe complications of diarrheal disease regardless of causative agent. It is inexpensive and effective for most patients, but it is often underused because of cultural beliefs discouraging oral intake during episodes of diarrhea. The WHO and the United Nations Children's Fund (UNICEF) recommend zinc replacement and oral rehydration salts (ORS) solution for the treatment of children and adults with dehydration and electrolyte imbalance associated with cholera and other forms of diarrheal disease. The ORS formula contains (in grams per liter) sodium chloride, 2.6; glucose (anhydrous), 13.5; potassium chloride, 1.5; and trisodium citrate (dihydrate), 2.9 (WHO, 2006). Sports drinks do not replace fluid losses correctly and should not be used.

Mild Dehydration. The patient exhibits dry oral mucous membranes of the mouth and increased thirst. The rehydration goal at this level of dehydration is to deliver about 50 mL of ORS per 1 kg of weight over a 4-hour interval.

Moderate Dehydration. Common findings are sunken eyes, loss of skin turgor, increased thirst, and dry oral mucous membranes. The rehydration goal at this level of dehydration is to deliver about 100 mL/kg of ORS over 4 hours.

Severe Dehydration. The patient with severe dehydration shows signs of shock (i.e., rapid thready pulse, cyanosis, cold extremities, rapid breathing, lethargy, or coma) and should receive IV replacement until hemodynamic and mental status return to normal. When improvement is evident, the patient can be treated with ORS.

Administering Rehydration Therapy

Because diarrheal episodes are often accompanied by vomiting, rehydration and refeeding can be difficult. Oral rehydration therapy should be delivered frequently in small amounts. When patients are persistently vomiting, they often require frequent administration of fluids by spoonfuls. IV therapy is necessary for the patient who is severely dehydrated or in shock.

It is important for children and adults with acute diarrheal symptoms to maintain caloric intake. As soon as dehydration has been corrected, an age-appropriate, unrestricted diet is allowed. Recommended foods include starches, cereals, yogurt, fruits, and vegetables. Foods that are high in simple sugars, such as undiluted apple juice or gelatin, should be avoided.

Increasing Knowledge and Preventing Spread of Infection

Public health nurses, school nurses, and others who are involved in patient education should emphasize principles of safe food preparation, with special attention to meat preparation and cooking. Ground beef should be thoroughly cooked and all meat should be maintained at temperatures below 40°F or above 140°F (CDC, 2012L). In planning events for groups of people, adequate provision for storage and reheating to temperature thresholds is important. When preparing food, it is important to use different surfaces, knives, and other equipment for meat and nonmeat items.

Diarrheal diseases discussed in this section must be reported to local or state health departments. The goal of reporting is to provide information for determining incidence trends and promptly identifying any restaurants or other food preparation establishments that have served contaminated food.

In both homes and healthcare delivery settings, good hygiene and principles of standard precautions should be emphasized.

Monitoring and Managing Potential Complications

Bacteremia. *E. coli*, *Salmonella*, and *Shigella* are organisms that can enter the bloodstream and disseminate to other organs. Blood cultures are necessary in the acutely febrile patient with diarrhea. If initial smear results reveal gram-negative organisms, antibiotic therapy is instituted.

Hypovolemic Shock. Shock associated with diarrheal diseases demands accurate intake and output assessment and vigorous fluid replacement. In rare instances, patients with severe fluid imbalance require intensive care nursing support with aggressive hemodynamic monitoring. (See Chapter 14 for further information.)

Evaluation

Expected patient outcomes may include:

1. Attains fluid balance
   a. Output approximates intake.
   b. Mucous membranes appear moist.
   c. Skin turgor is normal.
   d. Adequate amounts of fluids and calories ingested
   e. Absence of vomiting
   f. Stools are of normal color and consistency.
2. Acquires knowledge and understanding about infectious diarrhea and transmission potential
   a. Takes proper precautions to prevent spread of infection to others
   b. Describes principles and techniques of safe food storage, preparation, and cooking
3. Absence of complications
   a. Temperature is within normal range.
   b. Blood culture reports are negative.
   c. Fluid balance is achieved.

TABLE 71-4 Sexually Transmitted Infections and Their Routes of Transmission

Disease	Route(s) of Transmission
Chancroid, *Lymphogranuloma venereum*, and *Granuloma inguinale*	Sexual
Chlamydia	Sexual
Cytomegalovirus	Sexual, less intimate contact
Gonorrhea	Sexual, perinatal
Hepatitis B	Sexual, percutaneous, perinatal
Hepatitis C	Percutaneous, probably sexual, probably perinatal
Herpes simplex	Sexual
HIV infection/AIDS	Sexual, percutaneous, perinatal
Human papillomavirus	Sexual
Syphilis	Sexual, perinatal

HIV, human immunodeficiency virus; AIDS, acquired immunodeficiency virus.

Sexually Transmitted Infections

An STI is a disease acquired through sexual contact with an infected person. Table 71-4 identifies diseases that can be classified as STIs. Infections caused by organisms not generally considered STIs can also be transmitted during sexual contact—for example, *G. lamblia*, usually associated with contaminated water, can be transmitted through sexual exposure.

STIs are the most common infectious diseases in the United States and are epidemic in most parts of the world. Portals of entry of STI-causing microorganisms and sites of infection include the skin and mucosal linings of the urethra, cervix, vagina, rectum, and oropharynx.

Approximately 19 million Americans become infected with STIs annually. STIs have severe health consequences. In addition, they represent a financial burden estimated to be as high as $17 billion per year (CDC, 2013c).

Education about prevention of STIs includes information about risk factors and behaviors that can lead to infection. Using straightforward language and personal testimonials for targeted audiences (e.g., people who want information about protecting themselves) and conducting presentations in trusted establishments (e.g., churches, healthcare facilities) are recommended educational strategies (CDC, 2004a, 2004b). Included in this education is information about the relative value of condoms in reducing the risk of infection. The use of condoms to provide a protective barrier from transmission of STI-related organisms has been broadly promoted, especially since the recognition of HIV/AIDS. At first referred to as a method to ensure *safe sex*, the use of condoms has been shown to reduce but not eliminate the risk of transmission of HIV and other STIs. Thus, the term *safer sex* more appropriately connotes the public health message to be used when promoting the use of condoms. (See Chapter 37 for most information about AIDS and HIV.)

STIs provide a unique set of challenges for nurses, physicians, and public health officials. Because of perceived stigma and possible threat to emotional relationships, people with symptoms of STIs are often reluctant to seek healthcare in a timely fashion. STIs may progress without symptoms, and a delay in diagnosis and treatment is potentially harmful because the risk of complications for the infected person and the risk of transmission to others increase over time.

Infection with one STI suggests the possibility of infection with other diseases as well. After one STI is identified, diagnostic evaluation for others should be conducted. The possibility of HIV infection should be pursued when any STI is diagnosed.

Syphilis

Syphilis is an acute and chronic infectious disease caused by the spirochete *Treponema pallidum*. It is acquired through sexual contact or may be congenital in origin. The rates of primary and secondary syphilis have been on the rise in men since 2001 and in women since 2005 (Porth & Matfin, 2009).

Stages of Syphilis

In the untreated person, the course of syphilis can be divided into three stages: primary, secondary, and tertiary. These stages reflect the time from infection and the clinical manifestations observed in that period and are the basis for treatment decisions.

Primary syphilis occurs 2 to 3 weeks after initial inoculation with the organism. A painless lesion at the site of infection is called a *chancre*. These lesions usually resolve spontaneously within 3 to 12 weeks, with or without treatment (Porth & Matfin, 2009).

Secondary syphilis occurs when the hematogenous spread of organisms from the original chancre leads to generalized infection. The rash of secondary syphilis occurs about 2 to 8 weeks after the chancre and involves the trunk and the extremities, including the palms of the hands and the soles of the feet. Transmission of the organism can occur through contact with these lesions. Generalized signs of infection may include lymphadenopathy, arthritis, meningitis, hair loss, fever, malaise, and weight loss.

After the secondary stage, there is a period of **latency,** when the infected person has no signs or symptoms of syphilis. Latency can be interrupted by a recurrence of secondary syphilis symptoms.

Tertiary syphilis is the final stage in the natural history of the disease. It is estimated that between 20% and 40% of those infected do not exhibit signs and symptoms in this final stage. Tertiary syphilis presents as a slowly progressive inflammatory disease with the potential to affect multiple organs. The most common manifestations at this level are aortitis and neurosyphilis, as evidenced by dementia, psychosis, paresis, stroke, or meningitis.

Assessment and Diagnostic Findings

Because syphilis shares symptoms with many diseases, clinical history and laboratory evaluation are important. The conclusive diagnosis of syphilis can be made by direct identification of the spirochete obtained from the chancre lesions of primary syphilis. Serologic tests used in the diagnosis of secondary and tertiary syphilis require clinical correlation in interpretation. The serologic tests are summarized as follows:

- *Nontreponemal* or *reagin tests*, such as the Venereal Disease Research Laboratory (VDRL) or the rapid plasma reagin circle test (RPR-CT), are generally used for screening and diagnosis. After adequate therapy, the test result is expected to decrease quantitatively until it is read as negative, usually about 2 years after therapy is completed.
- *Treponemal tests*, such as the fluorescent treponemal antibody absorption test (FTA-ABS) and the microhemagglutination test for *Treponema pallidum* (MHA-TP), are used to verify that the screening test did not represent a false-positive result. Positive results usually are positive for life and therefore are not appropriate to determine therapeutic effectiveness.

Medical Management

Treatment of all stages of syphilis is administration of antibiotic medications. Penicillin G benzathine is the medication of choice for early syphilis or early latent syphilis of less than 1 year's duration. It is administered by intramuscular injection at a single session. Patients with late latent or latent syphilis of unknown duration should receive three injections at 1-week intervals. Patients who are allergic to penicillin are usually treated with doxycycline (Adoxa). The patient treated with penicillin is monitored for 30 minutes after the injection to observe for a possible allergic reaction.

Treatment guidelines established by the CDC are updated on a regular basis. Recommendations provide special guidelines for treatment in the setting of pregnancy, allergy, HIV infection, pediatric infection, congenital infection, and neurosyphilis (CDC, 2010a).

Nursing Management

Syphilis is a reportable communicable disease. In any health care facility, a mechanism must be in place to ensure that all diagnosed patients are reported to the state or local public health department to ensure community follow-up. The public health department is responsible for identification of sexual contacts, contact notification, and contact screening.

Lesions of primary and secondary syphilis may be highly infective. Gloves are worn when direct contact with lesions is likely, and hand hygiene is performed after gloves are removed. Isolation in a private room is not required (Chart 71-8).

Chlamydia trachomatis *and* Neisseria gonorrhoeae *Infections*

Chlamydia trachomatis and *Neisseria gonorrhoeae* are the most commonly reported infectious diseases in the United States.

Chart 71-8 PATIENT EDUCATION

Preventing the Spread of Syphilis

- If multiple penicillin injections are required, complete the full course of therapy.
- Refrain from sexual contact with previous or current partners until the partners have been treated.
- If you have primary or secondary syphilis, be aware that with proper treatment, skin lesions and other sequelae of infection will improve, and serology eventually will reflect cure.
- Know that condoms significantly reduce the risk of transmission of syphilis and other sexually transmitted infections.
- Be aware that having multiple sexual partners increases the risk of acquiring syphilis and other sexually transmitted infections.

Coinfection with *C. trachomatis* often occurs in patients infected with *N. gonorrhoeae*. The greatest risk of *C. trachomatis* infection occurs in young women between 15 and 19 years of age (CDC, 2010a).

Clinical Manifestations

Women

Both *C. trachomatis* and *N. gonorrhoeae* infections frequently do not cause symptoms in women. When symptoms are present, mucopurulent cervicitis with exudates in the endocervical canal is the most frequent finding. Women with gonorrhea can also present with symptoms of urinary tract infection or vaginitis. (See Chapter 57 for more in-depth coverage of STIs in women.)

Men

Although men are more likely than women to have symptoms when infected, infection with *N. gonorrhoeae* or *C. trachomatis* can be asymptomatic. When symptoms are present, they may include burning during urination and penile discharge. Patients with *N. gonorrhoeae* infection may also report painful, swollen testicles.

Complications

In women, pelvic inflammatory disease (PID), ectopic pregnancy, endometritis, and infertility are possible complications of either *N. gonorrhoeae* or *C. trachomatis* infection. In men, epididymitis, a painful disease that may lead to infertility, may result from infection with either bacterium. In people of either gender, arthritis or bloodstream infection may be caused by *N. gonorrhoeae*.

Assessment and Diagnostic Findings

The patient is assessed for fever, discharge (urethral, vaginal, or rectal), and signs of arthritis. Diagnostic methods used in *N. gonorrhoeae* infection include Gram stain (appropriate only for male urethral samples), culture, and nucleic acid amplification tests (NAATs). Gram stain and the direct fluorescent antibody test can be used in chlamydia. NAATs are also available for *C. trachomatis* but demand strict attention to laboratory procedures to ensure test reliability. In the female patient, samples are obtained from the endocervix, anal canal, and pharynx. In the male patient, specimens are obtained from the urethra, anal canal, and pharynx. Because *N. gonorrhoeae* organisms are susceptible to environmental changes, specimens for culture must be delivered to the laboratory immediately after they are obtained.

Because as many as 70% of chlamydial infections are asymptomatic, the CDC recommends annual *Chlamydia* testing for all pregnant women, sexually active women younger than 25 years, and older women with a new sexual partner or multiple partners (CDC, 2010a).

Medical Management

Because patients are often coinfected with both gonorrhea and chlamydia, the CDC recommends dual therapy even if only gonorrhea has been laboratory proven. CDC guidelines should be used to determine alternative therapy for the patient who is pregnant or allergic or who has a complicated chlamydial infection. The CDC updates STI therapy recommendations regularly because of growing problems with bacterial antibiotic resistance patterns and drug shortages.

Concept Mastery Alert

Although the number of resistant strains of gonorrhea has increased, that is not the reason for the use of combination antibiotic therapy. Such therapy is prescribed in order to treat both gonorrhea and chlamydia, because many patients with gonorrhea have a coexisting chlamydial infection.

Patients with uncomplicated gonorrhea who are treated with CDC-recommended therapy do not routinely need to return for a proof-of-cure visit. If the patient reports a new episode of symptoms or tests are positive for gonorrhea again, the most likely explanation is reinfection rather than treatment failure. Serologic testing for syphilis and HIV should be offered to patients with gonorrhea or chlamydia, because any STI increases the risk of other STIs.

Nursing Management

Gonorrhea and chlamydia are reportable communicable diseases. In any healthcare facility, a mechanism should be in place to ensure that all diagnosed patients are reported to the local public health department to ensure follow-up of the patient. The public health department also is responsible for interviewing the patient to identify sexual contacts so that contact notification and screening can be initiated.

The target group for preventive patient education about gonorrhea and chlamydia is the adolescent and young adult population. Along with reinforcing the importance of abstinence, when appropriate, education should address postponing the age of initial sexual exposure, limiting the number of sexual partners, and using condoms for barrier protection. Young women and pregnant women should also be instructed about the importance of routine screening for chlamydia.

NURSING PROCESS

The Patient With a Sexually Transmitted Infection

Assessment

The patient should be asked to describe the onset and progression of symptoms and to characterize any lesions by location and by describing drainage, if present. Brief explanations of why the information is needed are often helpful. Clarification of terms may be necessary if either the patient or nurse uses words that are unfamiliar to the other.

Protecting confidentiality is important when discussing sexual issues. When a detailed sexual history is necessary, it is important to respect the patient's right to privacy. When obtaining a sexual history, the CDC recommends the following systematic interview of key areas, the "five Ps": *p*artners, *p*revention of pregnancy, *p*rotection from STIs, *p*ractices, *p*ast history of STIs.

Asking specific information about sexual contacts usually should be done only when the nurse is part of a team that will conduct partner notification. The nurse should describe to the patient the public health notification process and resources that are available to assist sexual partners or infants and children.

During the physical examination, the examiner looks for rashes, lesions, drainage, discharge, or swelling. Inguinal nodes are palpated to elicit tenderness and to assess swelling. Women are examined for abdominal or uterine tenderness. The mouth and throat are examined for signs of inflammation or exudate. The nurse wears gloves while examining the mucous membranes, and gloves are changed and replaced after vaginal or rectal examination.

Diagnosis

Nursing Diagnoses

Based on assessment data, major nursing diagnoses may include the following:

- Knowledge deficit about the disease and risk for spread of infection and reinfection
- Anxiety related to anticipated stigmatization and to prognosis and complications
- Nonadherence with treatment

Collaborative Problems/Potential Complications

Potential complications may include the following:

- Ectopic pregnancy
- Infertility
- Transmission of infection to fetus, resulting in congenital abnormalities and other outcomes
- Neurosyphilis
- Gonococcal meningitis
- Gonococcal arthritis
- Syphilitic aortitis
- HIV-related complications

Planning and Goals

Major goals are increased patient understanding of the natural history and treatment of the infection, reduction in anxiety, increased compliance with therapeutic and preventive goals, and absence of complications.

Nursing Interventions

Increasing Knowledge and Preventing Spread of Disease

Education about STIs and prevention of the spread to others is often accomplished simultaneously. The infected patient should be told what the causative organism is and should receive an explanation of the usual course of the infection (including the interval of potential communicability to others) and possible complications. The nurse should stress the importance of following therapy as prescribed and the need to report any side effects or symptom progression.

Discussion should emphasize that the same behaviors that led to infection with one STI increase the risk of any other STI, including HIV. Methods used to contact sexual partners should be discussed. The patient should understand that until the partner has been treated, continued sexual exposure to the same person may lead to reinfection. The relative value of condoms in reducing the risk for infection with STIs should be addressed. When appropriate, the patient should be encouraged to discuss any reasons for resistance to condom use to promote thoughtful decision making about this preventive method.

Reducing Anxiety

When appropriate, the patient is encouraged to discuss anxieties and fear associated with the diagnosis, treatment, or prognosis. By individualizing education, factual information applied to specific needs may offer reassurance. Patients may need help in planning discussion with partners. If the patient is especially apprehensive about this aspect, referral to a social worker or other specialist may be appropriate. For example, such support is especially important when the patient has newly diagnosed HIV infection. Patients with HIV may benefit from programs that combine support, education, counseling, and therapeutic goals. Such programs are designed to offer coordinated care throughout the course of disease progression.

Increasing Adherence

In group settings (e.g., an outpatient obstetric setting) or in a one-to-one setting, open discussion about STI information facilitates patient education. Discomfort can be reduced by factual explanation of causes, consequences, treatments, prevention, and responsibilities. Because most communities have expanded STI prevention resources, referrals to appropriate agencies can complement individual educational efforts and ensure that later questions or uncertainties can be addressed by experts.

Monitoring and Managing Potential Complications

Infertility and Increased Risk of Ectopic Pregnancy. STIs may lead to PID and, with it, increased risk of ectopic pregnancy and infertility. (See Chapters 56 and 57 for additional information.)

Congenital Infections. All STIs can be transmitted to infants in utero or at the time of birth. Complications of congenital infection can range from localized infection (e.g., throat infection with *N. gonorrhoeae*), to congenital abnormalities (e.g., stunting of growth or deafness from congenital syphilis), to life-threatening disease (e.g., congenital herpes simplex virus).

Neurosyphilis, Gonococcal Meningitis, Gonococcal Arthritis, and Syphilitic Aortitis. STIs can cause disseminated infection. The central nervous system may be infected, as seen in cases of neurosyphilis or gonococcal meningitis. Gonorrhea that infects the skeletal system may result in gonococcal arthritis. Syphilis can infect the cardiovascular system by forming vegetative lesions on the mitral or aortic valves.

Human Immunodeficiency Virus–Related Complications. HIV infection leads to the profound immunosuppression that is characteristic of AIDS. Complications of HIV infection include many opportunistic infections, including those due to *Pneumocystis jiroveci*, *Cryptococcus neoformans*, cytomegalovirus, and *Mycobacterium avium* (see Chapter 37).

Evaluation

Expected patient outcomes may include:

1. Exhibits knowledge about STIs and their transmission
2. Demonstrates a less anxious demeanor
   a. Discusses anxieties and goals for treatment
   b. Inspects self for lesions, rashes, and discharge
   c. Accepts support, education, and counseling when indicated
   d. Assists with sharing information about infection with sexual partners
   e. Discusses risk reduction behaviors and safer sex practices

3. Complies with treatment
4. Achieves effective treatment
5. Reports for follow-up examinations if necessary
6. Absence of complications

Emerging Infectious Diseases

As defined by the CDC, **emerging infectious diseases** are human diseases of infectious origin that have increased within the past two decades or that are likely to increase in the near future. Examples of emerging infectious diseases presented here include West Nile virus, Legionnaires' disease, pertussis, hantavirus pulmonary syndrome, and viral hemorrhagic fevers. Some multiple resistant strains of common bacteria such as CA-MRSA and gram-negative organisms that are resistant to extended-spectrum beta-lactam antibiotics or carbapenem-based antibiotics are also often considered causes of emerging diseases (Chua et al., 2011). The newest is carbapenem-resistant Enterobacteriaceae; approximately 4% of U.S. hospitals and 18% of nursing homes have treated at least one patient with the bacteria (Reinberg, 2013).

Many factors contribute to newly emerging or re-emerging infectious diseases. These include travel, globalization of food supply and central processing of food, population growth, increased urban crowding, population movements (e.g., those that result from war, famine, or man-made or natural disasters), ecologic changes, human behavior (e.g., risky sexual behavior, IV/injection drug use), antimicrobial resistance, and breakdown in public health measures.

These diseases are important from an epidemiologic standpoint because their incidence has not yet stabilized. When the pattern of disease in a community is not well understood in the medical-scientific community, patients, families, and others in the community often become alarmed about these diseases. During times of increased concern about bioterrorism, whether triggered by actual events or by hoaxes, nurses have responsibility to rationally separate facts from fears. In discussions with patients and other caregivers, it is important to keep the focus on what is known and to clarify the plan for diagnosis, treatment, and containment.

West Nile Virus

The West Nile virus was first recognized in the 1930s in Africa and was first seen in humans in the United States in 1999. Although most human infections are mild or asymptomatic, a range of presentations is possible. Approximately 20% of infected people have a mild disease called *West Nile fever*. These patients usually experience headache, fever, and a persistent fatigue that may continue for several months. In these patients, fewer than 1 in 150 infections develop into more serious disease, which is characterized by severe neuroinvasive illness, meningitis, encephalitis, and paralysis or poliomyelitis. The mortality rate for people with the milder West Nile fever and meningitis is less than 1%, but this rate increases to approximately 20% for those with encephalitis (Peterson & Fischer, 2012). Although age-related factors do not appear to affect a person's chances of acquiring West Nile virus, the risk of neuroinvasive disease is greater in those older than 50 years.

The incubation period (i.e., from mosquito bite to onset of symptoms) is between 3 and 14 days. Currently, there is no treatment for West Nile virus infection. Medical and nursing management consists of fluid replacement, airway management, and supportive nursing care when meningitis or symptoms are present.

Birds are the natural reservoir for the virus, and since 1999, the population of infected birds in the United States has increased steadily. Mosquitoes become infected when feeding on birds and can transmit the virus to animals and humans. Although human-to-human transmission of West Nile virus is very rare, transmission has occurred as the result of occupational exposure in laboratory workers, infant exposure transplacentally and from breast-feeding, and blood transfusion or organ transplant from infected donors (WHO, 2007).

Legionnaires' Disease

Legionnaires' disease is a multisystem illness that usually includes pneumonia and is caused by the gram-negative bacterium *Legionella pneumophila*. Named after an outbreak among people attending a convention of the American Legion in 1976, its potential to cause outbreaks has been demonstrated repeatedly in hospitals and other settings. It continues to be considered an emerging infectious disease because there are new patterns in recent years. There has been an approximate 70% increase in new cases since 2003 compared with earlier time periods, and a disproportionate increase in the eastern United States (Neil & Berkeman, 2008).

Legionella organisms are found in many man-made and naturally occurring water sources. Although the organisms may initially be introduced to the plumbing system in low numbers, growth is enhanced by water storage, sediment, temperatures ranging from 25°C to 42°C (77°F to 107°F), and certain amoebae frequently present in water that can support intracellular growth of legionellae. Because incidence appears to increase in the summer and autumn months, vacation-related exposure to hotel or cruise ship plumbing and air-conditioning systems, whirlpool spas, and decorative fountains may be the causative risk.

Pathophysiology

L. pneumophila is transmitted by the aerosolized route from an environmental source to a person's respiratory tract. It is not transmitted from person to person. In hospitals, patients may be exposed to aerosols created by cooling towers, water exposure from in-room plumbing, and respiratory therapy equipment. Because underlying medical conditions can increase host susceptibility and subsequent severity of disease and because hospital plumbing systems are often very complex, outbreaks occur in hospitals more frequently than at other centers within the community. The mortality rate for Legionnaires' disease may be as high as 30% in some populations (CDC, 2011c).

Risk Factors

Risk factors for *Legionella* infection include diseases that lead to severe immunosuppression, such as AIDS, hematologic malignancy, end-stage kidney disease, or the use of immunosuppressive agents. Other factors associated with increased risk include diabetes, smoking, exposure to whirlpool spas, and recent travel.

Clinical Manifestations

The lungs are the principal organs of infection; however, other organs may also be involved. The incubation period ranges from 2 to 10 days. Early symptoms may include malaise, myalgias, headache, and dry cough. The patient develops increasing pulmonary symptoms, including productive cough, dyspnea, and chest pain. Patients are usually febrile, and body temperatures may reach or exceed 39.4°C (103°F). Diarrhea and other gastrointestinal symptoms are common. In severe cases, multiorgan involvement and failure may follow.

Assessment and Diagnostic Findings

The diagnostic approach generally involves using information obtained from the history, physical examination, x-rays, laboratory findings, and assessment of therapeutic effectiveness. Chest x-ray abnormalities may vary in severity and in location within the lungs. Laboratory tests available for the diagnosis of *Legionella* include culture or tests that detect either antigen or antibody. The most frequently used test is the urinary antigen. The greatest limitation of the test is that it detects only one subgroup of one of the several species of *Legionella*. The CDC recommends using multiple tests when Legionnaires' disease is suspected because none of the tests is completely accurate (CDC, 2011c).

Medical Management

The antibiotic agents of choice are azithromycin (Zithromax) or a fluoroquinolone such as moxifloxacin (Avelox) (Mandell, Wunderink, Anzueto, et al., 2007). The antibiotic doxycycline may also be used.

Nursing Management

The nursing management described for the patient with any pneumonia (see Chapter 23) should form the basis of care for the patient with *Legionella* pneumonia. Isolation is not required because *Legionella* is not transmitted between humans. When the patient has acquired the infection in a healthcare facility, water cultures should be performed to determine if the water supply is contaminated.

Pertussis

Pertussis, also known as whooping cough, a common childhood disease in the pre–vaccine era, is an example of a disease that has re-emerged. Incidence rates declined until the 1980s, when rates for all age groups began to increase steadily for the next three decades. In the short period from 2001 to 2003, the incidence of cases in adolescents and adults more than doubled compared with the period from 1990 to 1993 (CDC, 2012i) (Fig. 71-3).

Pertussis is highly contagious, and patients usually present to healthcare professionals with a sudden (paroxysmal) cough that is accompanied by a characteristic whoop—a high-pitched noise heard when inhaling. Whooping cough is caused by the bacterium *Bordetella pertussis*.

Pathophysiology

B. pertussis is transmitted by droplets. The bacteria easily attach to pharyngeal epithelial cells, where they release a number of antigens, toxins, and other substances that trigger the immune system. Because most of the disease manifestations are caused by this immune reaction, patients are usually contagious only early in the disease (when the bacteria are still present) and not during the protracted period of cough (when the immune reaction is causing the pathology).

Clinical Manifestations

Pertussis causes a range of respiratory symptoms, with cough being the most frequent. It is generally most severe for infants who have not yet been vaccinated. Pneumonia is the most common consequence of infection, but the disease can also lead to seizures, encephalopathy, and rarely death. People who have been vaccinated seldom have severe disease.

Assessment and Diagnostic Findings

Most diagnoses of pertussis are made, at least initially, without laboratory confirmation. The clinical case definition, unless there is a preexisting condition to explain the symptoms, is a new cough lasting at least 2 weeks with inspiratory whoop or vomiting after cough. Laboratory confirmation can be made

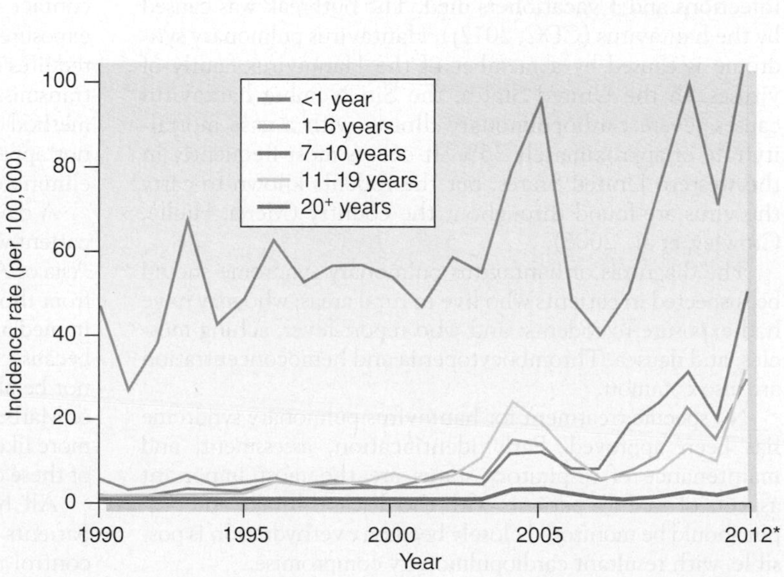

100
80
60
40
20
0
Incidence rate (per 100,000)
<1 year
1–6 years
7–10 years
11–19 years
20+ years
1990
1995
2000
2005
2012*
Year

FIGURE 71-3 • Reported pertussis incidence (per 100,000 persons) by age group in the United States from 1990–2012. *2012 data are provisional. From CDC, National Notifiable Diseases Surveillance System and Supplemental Pertussis Surveillance System. *Pertussis (whooping cough)*. Available at: http://www.cdc.gov/pertussis/surv-reporting.html

by clinical culture or by polymerase chain reaction assay for *B. pertussis*. Serologic testing, although less reliable, can also strengthen the diagnostic suspicion. The best source for a culture is a nasopharyngeal specimen.

Medical Management

The antibiotic agents of choice are azithromycin, erythromycin (Erythrocin), or clarithromycin (Biaxin). The antibiotic trimethoprim sulfate and co-trimoxazole (TMP-SMZ) (Bactrim) may also be used (Mandell et al., 2007). Close contacts of a patient with proven or suspected pertussis should receive prophylaxis with one of these agents to reduce the risk of disease.

Nursing Management

Patients who are hospitalized with pertussis should be isolated in droplet precautions until they have received 5 days of appropriate therapy. Household members should receive antimicrobial prophylaxis and should be advised to report any symptoms of an upper respiratory infection.

Because pertussis is most severe for infants, it is important to ensure that all pregnant women are vaccinated with Tdap (tetanus, diphtheria, acellular pertussis vaccine). The best time for Tdap for women who have not been vaccinated before becoming pregnant is late in the second trimester or in the third trimester. If a woman has not been vaccinated prior to delivery of the infant, Tdap should be administered immediately postpartum. Fathers, siblings, grandparents, and others who are around infants under 12 months should also be vaccinated to reduce the risk of transmitting pertussis.

For the protection of both worker and patient, all health care workers should receive a single dose of Tdap vaccine. Workers who are unsure if they have ever had Tdap should receive the vaccine regardless of the possible interval since the previous vaccination. No redosing for pertussis will be necessary through life, but boosters for tetanus and diphtheria (Td vaccine) should be given every 10 years (CDC, 2012c).

Hantavirus Pulmonary Syndrome

In the summer of 2012, 10 vacationers who stayed in tent cabins in Yosemite National Park developed acute respiratory infections and 3 vacationers died. The outbreak was caused by the hantavirus (CDC, 2012j). Hantavirus pulmonary syndrome is caused by a member of the Hantavirus family of viruses. In the United States, the Sin Nombre hantavirus causes severe cardiopulmonary illness, with a case mortality rate of approximately 35%. It occurs most frequently in the western United States, but the rodents known to carry the virus are found throughout the country (Mertz, Hjelle, Crowley, et al., 2006).

The diagnosis of hantavirus pulmonary syndrome should be suspected in patients who live in rural areas; who may have had exposure to rodents; and who report fever, aching muscles, and nausea. Thrombocytopenia and hemoconcentration are also common.

No specific treatment for hantavirus pulmonary syndrome has been approved. Early identification, assessment, and maintenance of respiratory status are the most important aspects of care for patients with the disease. Intake and output should be monitored closely because overhydration is possible, with resultant cardiopulmonary compromise.

Prevention requires strategies to reduce human contact with rodents and their droppings. Public health programs and clinics in rural areas should regularly educate people to eliminate rodent food sources that are in areas close to humans. Openings in walls or cabinets should be sealed. Traps should be used in areas such as sheds and barns in which humans work and rodents may enter. Gloves should be worn when removing an animal from a trap, and the trap should be disinfected with a 1:10 bleach solution. People entering such areas should be educated to avoid stirring up dust or breathing potentially contaminated dust. Brooms and vacuum cleaners should be used with caution; areas that may emit dust while being cleaned should be first dampened with a bleach solution to reduce viral contaminants and the potential for dust dispersion.

Viral Hemorrhagic Fevers

Viral hemorrhagic fevers are a group of illnesses caused by several families of viruses (the arenaviruses, filoviruses, bunyaviruses, and flaviviruses). These viruses cause a syndrome characterized by multisystem involvement, resulting in a damaged vascular system. Although most cases of hemorrhagic fever are severe, some cases are less acute. The viruses as a whole can be found throughout the world; however, each virus usually causes disease only in its own limited geographic area.

The Ebola and Marburg viruses, both belonging to the filovirus family, are the best-known viral hemorrhagic fever viruses. Since the 1960s, they have been the source of an irregular pattern of sporadic outbreaks. The clinical course differs among patients but often includes fever, hemorrhage, vomiting, diarrhea, cough, and jaundice. Symptoms usually occur rapidly, and the course of the illness often progresses rapidly to profound hemorrhage, organ destruction, and shock. The mortality rate is often as high as 90% in outbreaks. When patients survive, the recovery period is often prolonged, and weakness, malaise, and cachexia are common (WHO, 2012b).

Nonhuman animals or insects appear to be the natural reservoirs of the viruses. Humans usually become infected when exposed to the natural reservoir (e.g., after exposure to an unrecognized host or an insect bite). However, human-to-human transmission occurs occasionally; it involves close contact and usually occurs via the bloodborne route after exposure to blood or other body fluid. Percutaneous exposure requires only a very low inoculum of contaminated blood for transmission to occur. Mucous membrane exposure is another method of transmission. Although airborne transmission does not appear to be likely, the possibility has not been entirely eliminated.

A diagnosis of Ebola or Marburg should be considered in a patient who has a febrile, hemorrhagic illness after traveling to Asia or Africa or who has handled animals or animal carcasses from those parts of the world. The CDC should be contacted immediately when Ebola and Marburg viruses are suspected, because hospital and local public health laboratories would not be able to confirm a diagnosis. Because no cases of Ebola or Marburg have been diagnosed in the United States to date, more likely diagnoses should also be considered whenever one of these diseases is a diagnostic possibility (CDC, 2005b).

All healthcare workers who are involved in caring for patients with filoviruses must adhere to strict infection control measures. Systems must be set up to have objective observers ensure that each worker wears complete protective equipment in the form of cap, goggles, mask, gown, gloves, and shoe covers.

Treatment is largely supportive maintenance of the circulatory system and respiratory systems. It is likely that the patient who is infected will need ventilator and dialysis support during the acute phases of illness (CDC, 2005b). Supportive care for a patient with such a devastating disease requires psychological support for the patient and family. The patient, family, healthcare workers, and others in the community need substantial, coordinated education about the known elements and approach. Intervention may be required from those trained to provide psychological support for national emergencies or crises.

Travel and Immigration

Travel, trade, migration, and wars have led to many epidemics throughout history. The potential for epidemics is greatest when travelers and immigrants introduce microorganisms to which the host population has little or no immunity. Examples of important epidemics in the Western Hemisphere have included yellow fever, malaria, hookworm, leprosy, smallpox, measles, mumps, and syphilis. The HIV epidemic demonstrates the way that travel and immigration allow a disease to spread undetected worldwide. The 2003 severe acute respiratory syndrome (SARS) outbreak demonstrates how global travel contributes to a rapidly occurring epidemic involving an unrecognized pathogen.

In the United States, an infrastructure with enforced vaccination, clean water, and insect and rodent control decreases the risk that epidemics will progress even when travel may introduce exotic microorganisms. However, the recent experience with West Nile virus has reinforced knowledge that insect transmission can lead to significant human outbreaks. Thus, the concern grows that vector-borne diseases such as dengue and malaria may increase in the United States as mosquitoes can transmit disease locally when a reservoir of infected humans is established. The CDC maintains an active surveillance system to monitor and halt the incidence of many diseases prospectively.

Immigration and Acquired Immunodeficiency Disease Syndrome

The fact that AIDS reached pandemic proportions less than a decade after its recognition attests to the efficiency of world travel in spreading disease. Such rapid transmission rates are especially dramatic because HIV essentially requires intimate contact between two people through sexual activity or sharing blood through needles (see Chapter 37).

Immigration and Tuberculosis

Immigration has always been an important influence in the dynamic epidemiology of TB in the United States. In 2011, the incidence of TB in the United States was 11.5 times greater in foreign-born people than in native-born people (CDC, 2012k; WHO, 2012a).

The association between immigration and transmission risk is greatest in urban areas, because these locations are frequently heavily populated and frequently visited by foreign-born people. These locales are also often the epicenter of the HIV epidemic. Because HIV infection depletes T cells, which are necessary for TB protection, the geographic closeness of these two microorganisms potentiates increased rates of both infections.

A positive TST establishes that TB infection has occurred at some time in a person's life but does not provide information about current infectivity. The reliability of TST interpretation is decreased among foreign-born people because the bacille Calmette-Guérin (BCG) vaccine is used in many countries. After receiving BCG, people often have some degree of TST reactivity for a prolonged time.

The QuantiFERON-TB Gold (QFT-G) test is an enzyme-linked immunosorbent assay (ELISA) that detects the release of interferon-gamma by white blood cells when the blood of a patient with TB is incubated with peptides similar to those in M. *tuberculosis*. The results of the QFT-G test are available in less than 24 hours and are not affected by prior vaccination with BCG. Additional rapid tests for TB include the QuantiFERON-TB Gold In-Tube test (QFT-GIT); T-SPOT TB test (T-Spot); and Xpert MTB/RIF, which was endorsed by WHO in 2011 (CDC, 2010b). (See Chapter 23 for more detailed discussion.)

Immigration and Vector-Borne Diseases

Malaria and dengue are diseases that cause significant morbidity and mortality throughout the developing world. These diseases may be "imported" to the United States via travel, immigration, or commerce. They are caused by microorganisms that can be spread to humans by mosquitoes in the United States that thrive in tropical zones and breed in stagnant water sources. Although malaria was eradicated in the United States in the 1950s, limited local outbreaks have occurred regularly when mosquitoes acquire the bacteria from a person recently traveling from an area in which malaria is endemic and transmit it to a small number of people. Similarly, an increase of dengue virus in the Caribbean has caused concern that outbreaks may occur in the United States.

Critical Thinking Exercises

1 Several patients on your unit have recently been diagnosed with C. *difficile*. Identify how this organism causes pathology. What nursing interventions should be used to reduce the risk of new infections? What strategies are used to control C. *difficile* in healthcare facilities?

2 ebp You are the nurse on a cruise ship where an outbreak of *Norovirus* occurs. What is the evidence base to assist in intervening and containing this outbreak? Identify the criteria used to evaluate the strength of the evidence for the practices.

3 pq Identify the priorities, approach, and techniques you would use to perform a comprehensive assessment on a 25-year-old woman who presents to the local public health clinic with an STI. How would your priorities, approach, and techniques differ if the patient is a 25-year-old man? If the patient has AIDS? If the patient is from a culture with very different values from your own?

Brunner Suite Resources

Explore these additional resources to enhance learning for this chapter:

- NCLEX-Style Questions and Other Resources on thePoint, http://thePoint.lww.com/Brunner13e
- Study Guide
- PrepU
- Clinical Handbook
- Handbook of Laboratory and Diagnostic Tests

References

*Asterisk indicates nursing research.

**Double asterisk indicates classic reference.

Books

Centers for Disease Control and Prevention. (2012c). *Epidemiology and prevention of vaccine-preventable diseases*. Washington, DC: Public Health Foundation.

Karch, A. (2012). *2012 Lippincott's nursing drug guide*. Philadelphia: Lippincott Williams & Wilkins.

Porth, C. M., & Matfin, G. (2009). *Pathophysiology: Concepts of altered health states* (8th ed.). Philadelphia: Lippincott Williams & Wilkins.

Stratton, K., Ford, A., Rusch, E., & Clayton E. W. (Eds.). (2012). *Adverse effects of vaccines: Evidence and causality*. Washington, DC: National Academies Press.

Journals and Electronic Documents

Barzilay, E. J., Schaad, N., Magloire, R., et al. (2013). Cholera surveillance during the Haiti epidemic: The first two years. *New England Journal of Medicine, 368*(6), 599–609.

Bowyer, H. R., & Glover, M. (2010). Guillain-Barré syndrome: Management and treatment options for patients with moderate to severe progression. *Journal of Neuroscience Nursing, 42*(5), 288–293.

**Centers for Disease Control and Prevention. (2002). Guideline for hand hygiene in health care settings. *MMWR: Morbidity and Mortality Weekly Report, 51*(RR 16), 1–56.

Centers for Disease Control and Prevention. (2004a). National Nosocomial Infections Surveillance System Report, data summary from January 1992 through June 2004, issued October 2004. *American Journal of Infection Control, 32*(8), 470–485.

Centers for Disease Control and Prevention. (2004b). *Executive summary*. Available at: www.cdc.gov/std/HealthComm/ExecSumHPVGenPub2004.pdf

Centers for Disease Control and Prevention. (2005a). Achievements in public health: Elimination of rubella and congenital rubella syndrome—United States, 1969–2004. *MMWR: Morbidity and Mortality Weekly Report, 54*(11), 279–282.

Centers for Disease Control and Prevention. (2005b). Brief report: Outbreak of Marburg virus hemorrhagic fever—Angola, October 1, 2004–March 29, 2005. *MMWR: Morbidity and Mortality Weekly Report, 54*(Dispatch), 1–2.

Centers for Disease Control and Prevention. (2006). *Management of multidrug-resistant organisms in healthcare settings, 2006*. Available at: http://www.cdc.gov/hicpac/pdf/guidelines/MDROGuideline2006.pdf

Centers for Disease Control and Prevention. (2009). *Travelers health: Vaccinations*. Available at: www.cdc.gov/travel/page/vaccinations.htm

Centers for Disease Control and Prevention. (2010a). *2010 STD treatment guidelines*. Available at: www.cdc.gov/std/treatment/2010/

Centers for Disease Control and Prevention. (2010b). Updated guidelines for using interferon gamma release assays to detect *Mycobacterium tuberculosis* infection—United States, 2010. *MMWR: Morbidity Mortality Weekly Report, 59*(RR 5), 1–13.

Centers for Disease Control and Prevention. (2011a). *Vital signs: Central line–associated bloodstream infections—United States, 2001, 2008, and 2009*. Available at: www.cdc.gov/mmwr/preview/mmwrhtml/mm6008a4.htm?s_cid=mm6008a4_w

Centers for Disease Control and Prevention. (2011b). *2011 guidelines for the prevention of intravascular catheter-related infections*. Available at: www.cdc.gov/hicpac/bsi/bsi-guidelines-2011.html

Centers for Disease Control and Prevention. (2011c). *Top 10 things every clinician needs to know about legionellosis*. Available at: http://www.cdc.gov/legionella/clinicians.html

Centers for Disease Control and Prevention. (2011d). *Updated Norovirus outbreak management and disease prevention guidelines*. Available at: www.cdc.gov/mmwr/preview/mmwrhtml/rr6003a1.htm

Centers for Disease Control and Prevention. (2012a). *Vital signs: Preventing* Clostridium difficile *infections, 2012*. Available at: www.cdc.gov/mmwr/preview/mmwrhtml/mm6109a3.htm?s_cid=mm6109a3_w

Centers for Disease Control and Prevention. (2012b). *Active Bacterial Core surveillance (ABCs). Surveillance reports: Methicillin-resistant* Staphylococcus aureus *(MRSA)*. Available at: http://www.cdc.gov/abcs/reports-findings/surv-reports.html

Centers for Disease Control and Prevention. (2012d). *Chickenpox (varicella)*. Available at: www.cdc.gov/chickenpox/hcp/clinical-overview.html

Centers for Disease Control and Prevention. (2012e). *Shingles (herpes zoster)*. Available at: www.cdc.gov/shingles/hcp/clinical-overview.html

Centers for Disease Control and Prevention. (2012f). *Highly pathogenic avian influenza A (H5N1) in people*. Available at: www.cdc.gov/flu/avianflu/h5n1-people.htm

Centers for Disease Control and Prevention. (2012g). *Division of Foodborne, Waterborne and Environmental Diseases: Enteric diseases epidemiology branch*. Available at: www.cdc.gov/ncezid/dfwed/edeb/

Centers for Disease Control and Prevention. (2012h). *Preventing Norovirus infection*. Available at: www.cdc.gov/norovirus/preventing-infection.html

Centers for Disease Control and Prevention. (2012i). *Pertussis (whooping cough)*. Available at: www.cdc.gov/pertussis

Centers for Disease Control and Prevention. (2012j). *Hantavirus: Outbreak of Hantavirus infection at Yosemite National Park*. Available at: www.cdc.gov/hantavirus/outbreaks/yosemite-national-park-2012.html

Centers for Disease Control and Prevention. (2012k). *Tuberculosis (TB): Fact sheet—trends in tuberculosis, 2011*. Available at: www.cdc.gov/tb/publications/factsheets/statistics/TBTrends.htm

Centers for Disease Control and Prevention. (2012L). *Studies on restaurant food handling and food safety practices*. Available at: www.cdc.gov/nceh/ehs/ehsnet/Restaurant_Policies_Practices.htm#beef-handling

Centers for Disease Control and Prevention. (2013a). *Immunization schedules*. Available at: www.cdc.gov/vaccines/schedules

Centers for Disease Control and Prevention. (2013b). *Salmonella*. Available at: www.cdc.gov/salmonella/general/technical.html

Centers for Disease Control and Prevention. (2013c). *2011 sexually transmitted diseases surveillance*. Available at: www.cdc.gov/std/stats11/default.htm

Chua, K., Laurent, F., Coombs, G., et al. (2011). Antimicrobial resistance: Not community-associated methicillin-resistant *Staphylococcus aureus* (CA-MRSA)! A clinician's guide to community MRSA—its evolving antimicrobial resistance and implications for therapy. *Clinical Infectious Disease, 52*(1), 99–114.

Cohen, S. H., Gerding, D., Johnson, N., et al. (2010). Clinical practice guidelines for *Clostridium difficile* infection in adults: 2010 update by the Society of Healthcare Epidemiology of America (SHEA) and the Infectious Disease Society of America (IDSA). *Infection Control and Hospital Epidemiology, 31*(5), 431–455.

Dubberke E., & Gerding, D. (2011). *Rationale for hand hygiene: Recommendations after caring for a patient with Clostridium difficile infection. A compendium of strategies to prevent healthcare-associated infections in acute care hospitals—fall 2011 update*. Society of Healthcare Epidemiology of America. Available at: www.shea-online.org/Portals/0/CDI%20hand%20hygiene%20Update.pdf

Institute for Healthcare Improvement. (2012). *How-to guide: Prevent central line–associated bloodstream infections*. Available at: www.ihi.org

Joint Commission. (2013). *2013 national patient safety goals*. Available at: www.jointcommission.org/2013_npsgs_slides/

Kallens, A. J., Mu, Y., Bulens, S., et al. (2010). Health care–associated invasive MRSA infections, 2005–2008. *Journal of the American Medical Association, 304*(6), 641–647.

Mandell, L. A., Wunderink, R. G., Anzueto, A., et al. (2007). Infectious Disease Society of America/American Thoracic Society consensus guidelines on the management of community-acquired pneumonia in adults. *Clinical Infectious Diseases, 44*(Suppl. 2), S27–S72.

Mertz, G. J., Hjelle, B., Crowley, M., et al. (2006). Diagnosis and treatment of new world hantavirus infections. *Current Opinions Infectious Diseases, 19*(5), 457–442.

Milstone, A. M., & Perl, T. M. (2008). Fact, fiction, or no data: What does surveillance for methicillin-resistant *Staphylococcus aureus* prevent in the intensive care unit? *Clinical Infectious Diseases, 46*(11), 1726–1728.

Mortell, M. (2012). Hand hygiene compliance: Is there a theory-practice-ethics gap? *British Journal of Nursing, 21*(17), 1011–1014.

Neil, K., & Berkeman, R. (2008). Increasing incidence of legionellosis in the United States, 1990–2005: Changing epidemiologic trends. *Clinical Infectious Diseases, 47*(5), 591–599.

Peterson, L. R., & Fischer, M. (2012). Unpredictable and difficult to control—the adolescence of West Nile virus. *New England Journal of Medicine, 367*(14), 1281–1284.

Reinberg, S. (2013). *'Nightmare' bacteria spreading in U.S. hospitals, nursing homes: CDC*. Available at: consumer.healthday.com/Article.asp?AID=674130

Sandora, T. J., & Goldmann, D. A. (2012). Preventing lethal hospital outbreaks of antibiotic resistant bacteria. *New England Journal of Medicine, 367*(23), 2168–2170.

Siegel, J. D., Rhinehart, E., Jackson, M., et al. (2007). *Guideline for isolation precautions: Preventing transmission of infectious agents in healthcare settings*. Available at: www.cdc.gov/hicpac/2007ip/2007isolationprecautions.html

Sievert, D. M., Ricks, P., Edwards, J. R., et al.; National Healthcare Safety Network (NHSN) Team and Participating NHSN Facilities. (2013). Antimicrobial-resistant pathogens associated with healthcare-associated infections: Summary of data reported to the National Healthcare Safety Network at the Centers for Disease Control and Prevention, 2009–2010. *Infection Control and Hospital Epidemiology, 34*(1), 1–14

Society for Healthcare Epidemiology of America. (2013). *FAQs (frequently asked questions) about "MRSA" (methicillin-resistant Staphylococcus aureus)*. Available at: www.shea-online.org/Assets/files/patient%20guides/NNL_MRSA.pdf

Tappero, J. W., & Tauxe, R. V. (2011). Lessons learned during public health response to cholera epidemic in Haiti and the Dominican Republic. *Emergency Infectious Disease*. Available at: dx.doi.org/10.3201/eid1711.110827

U.S. Food and Drug Administration. (2013). *FDA news release, January 16, 2013: FDA approves new seasonal influenza vaccine made using novel technology*. Available at: www.fda.gov/NewsEvents/Newsroom/PressAnnouncements/ucm335891.htm

World Health Organization. (2006). *Oral rehydration salts. Production of the new ORS*. Available at: www.who.int/maternal_child_adolescent/documents/fch_cah_06_1/en/index.html

World Health Organization. (2007). *West Nile virus*. Available at: www.who.int/mediacentre/factsheets/fs354/en/index.html

World Health Organization. (2012a). *Global tuberculosis report 2012*. Available at: www.who.int/tb/publications/global_report/en/

World Health Organization. (2012b). *Ebola haemorrhagic fever*. Available at: www.who.int/mediacentre/factsheets/fs103/en/

World Health Organization. (2013). *Monthly risk assessment summary: Influenza at the human–animal interface. Summary and Assessment as of 4 January 2013*. Available at: www.who.int/influenza/human_animal_interface/HAI_Risk_Assessment/en/index.html

Resources

American Lung Association, www.lungusa.org

American Public Health Association (APHA), www.apha.org

Association for Professionals in Infection Control and Epidemiology (APIC), www.apic.org

Centers for Disease Control and Prevention (CDC), www.cdc.gov

Centers for Disease Control and Prevention (CDC), adult or pediatric vaccine advice hotline, 1-800-232-4636

Centers for Disease Control and Prevention (CDC), patient education about MRSA, www.cdc.gov/mrsa/groups/index.html

Centers for Disease Control and Prevention (CDC), traveler's vaccination advice, 1-877-FYI-TRIP (1-877-394-8747)

Infectious Diseases Society of America (IDSA), www.idsociety.org

National Foundation for Infectious Diseases (NFID), www.nfid.org

National Institute of Allergy and Infectious Diseases (NIAID), www.niaid.nih.gov/

Occupational Safety and Health Administration (OSHA), www.osha.gov

Society for Healthcare Epidemiology of America (SHEA), www.shea-online.org

Vaccine Adverse Event Reporting System (VAERS), 1-800-822-7967, www.vaers.hhs.gov

World Health Organization (WHO), www.who.int/home-page

Chapter 72

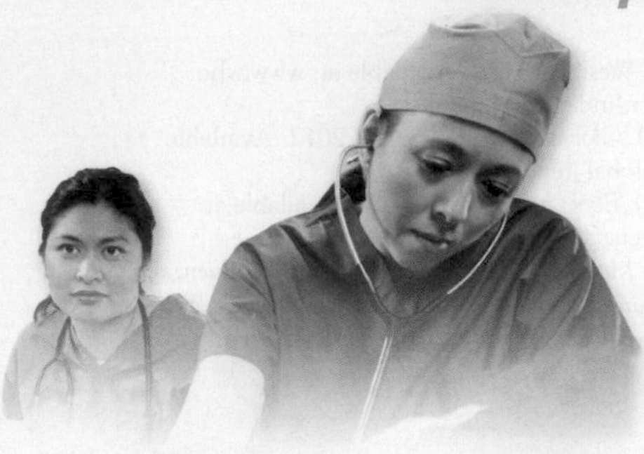

Emergency Nursing

Learning Objectives

On completion of this chapter, the learner will be able to:

1. Describe emergency care as a collaborative, holistic approach that includes the patient, the family, and significant others.
2. Discuss priority emergency measures instituted for the patient with an emergency condition.
3. Identify the priorities of care for the patient with intra-abdominal injuries and multiple injuries.
4. Compare and contrast the emergency management of patients with heat stroke, frostbite, and hypothermia.
5. Specify the similarities and differences of the emergency management of patients with swallowed or inhaled poisons, skin contamination, and food poisoning.
6. Discuss the emergency management of patients with drug overdose, those with acute alcohol intoxication, and those who have been sexually assaulted.
7. Differentiate between the emergency care of patients who are overactive, those who are violent, those who are depressed, and those who are suicidal.

Glossary

antivenin: antitoxin manufactured from venom of poisonous snakes to assist the patient's immune system response to an envenomation

carboxyhemoglobin: hemoglobin that is bound to carbon monoxide and therefore is unable to bind with oxygen, resulting in hypoxemia

corrosive poison: alkaline or acidic agent; causes tissue destruction after contact

cricothyroidotomy: surgical opening of the cricothyroid membrane to obtain an airway that is maintained with a tracheostomy or endotracheal tube

diagnostic peritoneal lavage: instillation of lactated Ringer's or normal saline solution into the abdominal cavity to detect red blood cells, white blood cells, bile, bacteria, amylase, or gastrointestinal contents

envenomation: injection of a poisonous material by sting, spine, bite, or other means

fasciotomy: surgical incision of the extremity to the level of the fascia to relieve pressure and restore neurovascular function to the extremity

Hare traction: portable in-line traction applied to the lower extremity to manage femur or hip fractures or dislocations

multiple trauma: trauma caused by a single catastrophic event that causes life-threatening injuries to at least two distinct organs or organ systems

primary survey: an assessment of the patient triaged to the emergent or resuscitation category that focuses on stabilizing life-threatening conditions; uses the mnemonic ABCDE, which stands for *a*irway, *b*reathing, *c*irculation, *d*isability, and *e*xposure

secondary survey: an assessment of the patient triaged to the emergent or resuscitation category that commences after the primary survey is completed and life-threatening insults addressed; includes obtaining vital signs, completing a head-to-toe examination, and obtaining the patient's pertinent medical-surgical history, including the history of the current event

triage: process of assessing patients to determine management priorities

The term *emergency management* traditionally refers to care given to patients with urgent and critical needs. However, because many people lack access to health care, the emergency department (ED) is increasingly used for nonurgent problems. Therefore, the philosophy of emergency management has broadened to include the concept that an emergency is whatever the patient or the family considers it to be.

The emergency nurse has had specialized education, training, experience, and expertise in assessing and identifying patients' health care problems in crisis situations. In addition, the emergency nurse establishes priorities, monitors, and continuously assesses acutely ill and injured patients, supports and attends to families, supervises allied health personnel, and educates patients and families within a time-limited, high-pressured care environment. Nursing interventions are accomplished interdependently, in consultation with or under the direction of a physician or advanced practice registered nurse. The roles of nursing and medicine are complementary in an emergency situation. Appropriate nursing and medical interventions are anticipated based on assessment data. The emergency health care staff members work as a team in performing the highly technical, hands-on skills required to

Chart 72-1 Facts About Emergency Department Visits—2009

In 2009, there were 136.1 million visits to emergency departments (EDs), a 10% increase from 2005. There was a concomitant increase in injury cases to 45.4 million patients. Of these, 77% were seen for urgent reasons, whereas only 12% were immediate or emergent in nature. There was an equal distribution of patients seen in the ED by race, gender, and ethnicity that mirrors the general population.

- The most common reasons for ED visits were abdominal pain, fever, and chest pain.
- Approximately 16% of patients arrived at the ED by ambulance.
- Patients with private health insurance totaled 39%, whereas Medicare/Medicaid was the second most common insurance at 29%.
- Injuries accounted for 25% of all ED visits.
- The leading causes of injuries were unintentional, totaling 65% of injury admissions, with falls and motor vehicle collisions making up 31% of these. The most common injuries were orthopedic (32%), followed by head injuries (13%). The age range of those most frequently injured was 25 to 44 years (28%), with 22% 45 to 64 years of age, and another 25% older than 64 years; 75% of all injuries necessitating ED care occurred in persons older than 24 years.
- The median ED waiting time before being seen by a health care provider for definitive treatment was 33 minutes, with 39% of patients waiting <1 hour to be seen and 50% spending 2–4 hours in the ED.
- 80% of patients were seen by an emergency physician.
- 63% of patients were discharged to home from the ED.

Adapted from Centers for Disease Control and Prevention. (2009). *NHAMCS: 2009 emergency department summary.* Available at: www.cdc.gov/nchs/ahcd/web_tables.htm#2009

care for patients in emergency situations (Emergency Nurses Association [ENA], 2013).

Large numbers of people seek emergency care for serious life-threatening conditions, such as cardiac dysrhythmias, acute coronary syndrome, acute heart failure, pulmonary edema, and stroke. Priorities for managing these cardiac and other conditions are discussed in Chapters 26, 27, 29, and 67. Emergency management of trauma and conditions not found elsewhere in this book are discussed in this chapter. Facts about ED visits in the United States are presented in Chart 72-1.

ISSUES IN EMERGENCY NURSING CARE

Emergency nursing is demanding because of the diversity of conditions and situations that present unique challenges. These challenges include legal issues, occupational health and safety risks for ED staff, and the challenge of providing holistic care in the context of a fast-paced, technology-driven environment in which serious illness and death are encountered on a daily basis. Another dimension of emergency nursing is nursing in disasters. With the increasing use of weapons of terror and mass destruction, both internationally and at home, the emergency nurse must recognize and treat patients exposed to biologic and other weapons anticipating nursing care in the event of a mass casualty incident from natural causes or a terrorist event (see Chapter 73).

Documentation of Consent and Privacy

Consent to examine and treat the patient is part of the ED record. The patient needs to give consent for invasive procedures (e.g., angiography, lumbar puncture) unless he or she is unconscious or in a critical condition and unable to make decisions. If the patient is unconscious and brought to the ED without family or friends, this fact must be documented. Monitoring of the patient's condition, as well as all instituted treatments and the times at which they were performed, must be documented. After treatment, a notation is made on the record about the patient's condition, response to the treatment, and condition at discharge or transfer and about instructions given to the patient and family for follow-up care.

The patient is also provided with a statement of the privacy policy of the health care agency, according to federal law. Patients involved in violent events can be provided with an alias, and access to the electronic or paper medical record is limited to protect the privacy of the patient. A patient may also request extra privacy by limiting access to his or her room and by choosing not to receive phone calls, mail, flowers, other gifts, or certain visitors. These practices relate to the federally mandated privacy policy stipulated in the Health Insurance Portability and Accountability Act (HIPAA).

According to the Emergency Medical Treatment and Active Labor Act (EMTALA), every ED with a Medicare provider agreement must perform a medical screening examination on all patients arriving with an emergency medical complaint if their acute signs and symptoms could result in serious injury or death if left untreated. EDs are also required to provide treatment aimed at stabilizing each patient's condition. If the patient must be transferred to another facility, the patient's consent for transfer should be obtained, if possible. In addition, acceptance by the receiving facility and physician must be obtained, and an appropriate method of transfer for the patient should be secured. Documentation of assessment and treatment must be sent with the patient upon transfer (ENA, 2013).

Limiting Exposure to Health Risks

Because of the increasing numbers of people infected with hepatitis, human immunodeficiency virus (HIV), tuberculosis, and other infectious diseases, health care providers are at an increased risk for exposure to communicable diseases through blood, respiratory droplets, or other body fluids. This risk is further compounded in the ED because of the common use of invasive treatments in patients who may have a wide range of conditions and are unable to provide a comprehensive medical history. All emergency health care providers must adhere strictly to standard precautions for minimizing exposure.

The re-emergence of tuberculosis as a major health problem is complicated by multidrug-resistant tuberculosis and by tuberculosis concomitant with HIV infection. Nurses in the ED are usually fitted with personal high-efficiency particulate air (HEPA) filter masks to use when treating patients with airborne diseases.

> ***Quality and Safety Nursing Alert***
>
> **To limit the risk of exposure to airborne diseases, early identification and strict adherence to transmission-based precautions for patients who are potentially infectious is crucial.**

The potential for exposure to highly contagious organisms, hazardous chemicals or gases, and radiation related to acts of terrorism or natural or man-made disasters presents additional risks to ED staff (see Chapter 73 for information about decontamination procedures).

Violence in the Emergency Department

Not only do ED staff members encounter patients who may be violent because of the effects of substance abuse, injury, or other emergencies, they may also encounter other violent situations. Patients and families waiting for assistance may be emotionally volatile. Often, waiting rooms are the sites where feelings of dissatisfaction, fear, and anger are channeled violently. Some EDs assign security officers to the area and have installed silent alarm systems or metal detectors to identify weapons in order to protect patients, families, and staff. Safety is the first priority.

Although not a typical occurrence, a patient or family member may to come to the ED armed. To avoid angry confrontations, members of gangs and feuding families need to be separated in the ED, in the waiting room, and later in the inpatient nursing unit. Nurses and other personnel must be prepared to deal with these circumstances. The ED should be locked against entry if security is questionable.

Violent or potentially violent patients must be vigilantly monitored by the ED staff. Care must be taken to avoid injury. Patients from prison and those who are under guard need to be handcuffed to the bed and appropriately assessed to ensure the safety of hospital staff and other patients. Precautions to be taken to avoid injury include the following:

- For prisoners, the hand or ankle restraint (handcuff) is never released, and a guard is always present in the room.
- A mask can be placed on the patient to prevent spitting or biting.
- Physical restraints are used on any violent patient only as needed and, if used, should be humanely and professionally administered (American College of Emergency Physicians [ACEP], 2007); nonetheless, the staff should be cognizant that the patient could head-butt if restrained.
- Distance should be maintained from the patient to avoid grabbing; staff should not wear items that can be grabbed by the patient, such as dangling jewelry and stethoscopes. Furthermore, distance should be maintained between the patient and the door so that an escape route for the staff member is preserved.
- Objects should not be left within patient reach; even an intravenous (IV) line spike can become a tool of violence if the patient is determined.
- Courses on safety (de-escalation and physical restraint techniques) assist the staff with preparing for various violent situations (Greenlund, 2011).
- Chemical restraints (i.e., medications) may be administered only as necessary to control violent behavior until definitive treatment can be obtained (ACEP, 2007).

In the case of gunfire in the ED, self-protection is a priority. There is no advantage to protecting others if health care providers are injured. Security officers and police must gain control of the situation first and then care is provided to the injured.

Providing Holistic Care

Patients and families experiencing sudden injury or illness are often overwhelmed by anxiety because they have not had time to adapt to the crisis. They experience real and terrifying fear of death, mutilation, immobilization, and other assaults on their personal identity and body integrity. When confronted with trauma, severe disfigurement, severe illness, or sudden death, the family experiences several stages of crisis. The stages begin with anxiety and progress through denial, remorse and guilt, anger, grief, and reconciliation. The initial goal for the patient and family is anxiety reduction, a prerequisite to effective and appropriate coping. During this stressful time, safety is of prime importance. Close observation and preplanning are essential and security personnel are stationed nearby in the event that a patient or family member responds to stress with physical violence.

Assessment of the patient and family's psychological function includes evaluating emotional expression, degree of anxiety, and cognitive functioning. Possible nursing diagnoses include:

- Anxiety or death anxiety related to uncertain potential outcomes of the illness or trauma
- Ineffective coping related to acute situational crisis

Possible nursing diagnoses for the family include:

- Grieving
- Interrupted family processes
- Compromised or disabled family coping related to acute situational crises

Patient-Focused Interventions

Clinicians caring for the patient should act confidently and competently to relieve anxiety and promote a sense of security. Explanations should be given that the patient can understand. Human contact and reassuring words reduce the panic of the severely injured or ill person and aid in dispelling fear of the unknown.

The unconscious patient should be treated as if conscious—that is, the patient should be touched, called by name, and given an explanation of every procedure that is performed. As the patient regains consciousness, the nurse should orient the patient by stating his or her name, the date, and the location. This basic information should be provided repeatedly, as needed, in a reassuring way.

Ensuring patient safety is a major focus in clinical practice settings, including within EDs. Despite this focus, ED visits accounted for 4.4% of sentinel events (unanticipated events that result in patient harm, including serious injury or death) in 2010. Some of the most common of these included delays to care and medication errors. Common root causes for these sentinel events revolved around nurse staffing patterns, patient volume, and specialty availability. Solutions to patient safety issues in the ED include ensuring optimal nurse staffing, pharmacy presence, and rapid diagnostics turnaround times to minimize wait time to diagnosis and fostering teamwork and support by leadership. All errors should be reported and investigated even if a patient was not harmed. In this way, future injury or death may be avoided (ENA, 2013).

Family-Focused Interventions

The family is kept informed about where the patient is, how he or she is doing, and the care that is being given. Encouraging family members to stay with the patient, when possible, also helps allay their anxieties. In many facilities, family presence during resuscitation is permitted to assist the family to cope through this difficult time. Many family members respond very well to this approach. One study found that 88% of surveyed providers felt that family presence did not interfere with the care being provided, nor did the presence of the family adversely affect team communication (Oman & Duran, 2010). The presence of a family facilitator, who is trained to provide support to family members, is vital to the success of a family presence program. Additional interventions are based on the assessment of the stage of crisis that the family is experiencing. Measures to help family members cope with sudden death are presented in Chart 72-2.

Anxiety and Denial

During these crises, family members are encouraged to recognize and talk about their feelings of anxiety. Asking questions is encouraged. Honest answers given at the level of the family's understanding must be provided. Although denial is an ego-defense mechanism that protects one from recognizing painful and disturbing aspects of reality, prolonged denial is not encouraged or supported. The family must be prepared for the reality of what has happened and what may come.

Remorse and Guilt

Expressions of remorse and guilt are common, with family members accusing themselves (or each other) of negligence or minor omissions. Family members are urged to verbalize their feelings to help them cope appropriately.

Anger

Expressions of anger, common in crisis situations, are a way of handling anxiety and fear. Anger is frequently directed by the family at the patient, but it is also often expressed toward the physician, the nurse, or admitting personnel. The therapeutic approach is to allow the anger to be expressed and to assist the family members to identify their feelings of frustration.

Grief

Grief is a complex emotional response to anticipated or actual loss. The key nursing intervention is to help family members work through their grief and to support their coping mechanisms, letting them know that it is normal and acceptable for them to cry, feel pain, and express loss. The hospital chaplain and social services staff serve as invaluable members of the team when assisting families to work through their grief (see Chapter 16).

Caring for Emergency Personnel

Concerted efforts have been made to focus on the needs of the ED staff, especially after serious and stressful events (ENA, 2013). Events can range from a local trauma case involving children; to treating someone known to the emergency worker, such as a colleague or family member; to a more complex natural disaster or multicasualty situation. It is important to remember that all staff members may not necessarily respond in the same way; an event that is stressful to one person may not be as stressful to another. In addition, because stress is a daily occurrence in the ED, the staff may not recognize the personal effect of any one event or the cumulative effect of day-to-day crisis interventions. ED leadership should be aware of staff coping patterns and support systems and appropriately assist with identifying behaviors caused by workplace stress. The availability of nonjudgmental counseling is essential to promoting a healthy staff.

After serious events, critical incident stress management (CISM) is necessary to critique individual and group performance and to facilitate healthy coping. Optimally, this may consist of three steps: defusing, debriefing, and follow-up. Defusing occurs immediately after the critical incident. During this session, affected staff are encouraged to discuss their feelings about the incident and are given contact information so that they may talk to someone if they have disturbing symptoms (e.g., sleeplessness, excessive worry). Debriefing typically occurs 1 to 10 days after the critical incident. Debriefing sessions follow a format similar to the initial defusing session; however, during these sessions, participating staff are encouraged to discuss their feelings about the incident and are reassured that their negative reactions and feelings are normal and that their negative feelings will diminish over time. At the end of these sessions, participants should have a feeling of closure and be able to resume their professional roles at an

Chart 72-2 Helping Family Members Cope With Sudden Death

- Take the family to a private place.
- Talk to the family together so that they can grieve together and hear the information given together.
- Reassure the family that everything possible was done; inform them of the treatment rendered.
- Avoid using euphemisms such as "passed on." Show the family that you care by touching, offering coffee, water, and the services of a chaplain.
- Encourage family members to support each other and to express emotions freely (grief, loss, anger, helplessness, tears, disbelief).
- Avoid giving sedation to family members; this may mask or delay the grieving process, which is necessary to achieve emotional equilibrium and to prevent prolonged depression.
- Encourage the family to view the body if they wish; this action helps to integrate the loss. Cover disfigured and injured areas before the family sees the body. Go with the family and do not leave them alone. Show acceptance by touching the body to give the family "permission" to touch.
- Spend time with the family, listening to them and identifying any needs that they may have for which the nursing staff can be helpful.
- Allow family members to talk about the deceased and what he or she meant to them; this permits ventilation of feelings of loss. Encourage the family to talk about events preceding admission to the emergency department. Do not challenge initial feelings of anger or denial.
- Avoid volunteering unnecessary information (e.g., the patient was drinking).

emotional level commensurate to that prior to the critical incident. Some staff may require further professional follow-up, however. Follow-up may occur after the debriefing session is completed for those participants who have persistent negative symptoms, which may consist of continued individual or group counseling and therapy (Everly & Mitchell, 2010).

EMERGENCY NURSING AND THE CONTINUUM OF CARE

A key principle underlying emergency care is that the patient is rapidly assessed, treated, and referred to the appropriate setting for ongoing care. This makes the ED a temporary point on the continuum of care. As noted in Chart 72-1, most patients who receive emergency care are discharged directly from the ED to their homes (i.e., 63%), and emergency nurses must plan and facilitate the patient's safe discharge and follow-up care in the home and the community (Centers for Disease Control and Prevention [CDC], 2009).

Discharge Planning

Before discharge, verbal and written instructions for continuing care are given to the patient and the family or significant others. Many EDs have preprinted standard instruction sheets for the more common conditions (i.e., concussion), which can then be individualized. Discharge instructions should be available in a variety of languages. A language interpreter should be used as necessary to provide both written and verbal instructions.

Instructions should include information about prescribed medications, treatments, diet, activity, and when to contact a health care provider or schedule follow-up appointments. Discharge planning is the "teachable moment" for the patient, providing the opportunity to present injury prevention or smoking cessation strategies, alcohol counseling opportunities, and more. It is imperative that instructions are written legibly, use simple language, and are clear in their important points. When providing discharge instructions, the nurse also considers any special needs the patient may have related to hearing or visual impairments. Alternate formats of instruction (e.g., large print, Braille, audiotape) should be available to meet the needs of patients with hearing or visual impairments.

Community Services

Before discharge, some patients require the services of a social worker to help them meet continuing health care needs. Home care resources may be contacted before discharge to arrange services. This is particularly important for patients who are older adults or disabled and who need assistance. Identifying continuing health care needs and making arrangements for meeting these needs can prevent return visits to the ED or readmission to the hospital.

For patients who are returning to long-term care facilities and for those who already rely on community agencies for continuing health care, communication about the patient's condition and any changes in health care needs that have occurred must be provided to the appropriate facilities or agencies. This communication is essential to promote continuity of care and to ensure ongoing care that meets the patient's changing health care needs.

Gerontologic Considerations

The ED is a common point of entry into the health care system for patients 65 years and older. Findings from one study reported that 65% of ED visits by patients older than 74 years resulted in hospital admission (LaMantia, Platts-Mills, Biese, et al., 2010). Older adult patients typically arrive with one or more presenting conditions. Nonspecific symptoms, such as weakness and fatigue, episodes of falling, incontinence, and change in mental status, may be manifestations of acute, potentially life-threatening illness in the older person. Emergencies in this age group may be more difficult to manage because older adult patients may have an atypical presentation, an altered response to treatment, a greater risk of developing complications, or a combination of these factors.

The older adult patient may perceive the emergency as a crisis signaling the end of an independent lifestyle or even resulting in death. The nurse should give attention to the patient's feelings of anxiety and fear.

The older patient may have limited sources of social and financial support during these times of crises. The nurse should assess the psychosocial resources of the patient (and of the caregiver, if necessary) and anticipate discharge needs. Referrals for support services (e.g., to the social service department or a gerontologic nurse specialist) may be necessary.

Obesity Considerations

The growing rates of obesity in the United States has implications for treating patients who are obese within the ED, in terms of stocking appropriately sized equipment, gowns, and stretchers; ensuring that equipment is able to handle a greater weight capacity (e.g., computed tomography [CT] scanners); and recognizing specific disorders and complications that may occur in these patients. For instance, research suggests that increased morbidity is associated in patients with traumatic injuries who have a body mass index (BMI) in excess of 40 kg/m^2 (Newell, Bard, Goettler, et al., 2007). Complications that patients with obesity are more prone to experience during hospitalization include respiratory failure, acute kidney injury, pneumonia, deep vein thrombosis, and pressure ulcers. Although many of these complications do not occur until later in the hospital stay, some preventive measures (e.g., encouraging turning, coughing, and deep-breathing exercises to prevent atelectasis) may be initiated in the ED. Other considerations in ED management of patients who are obese include an understanding that it is generally more challenging to insert IV lines and airways.

PRINCIPLES OF EMERGENCY CARE

By definition, emergency care is care that must be rendered without delay. In an ED, several patients with diverse health problems—some life threatening, some not—may present to the ED simultaneously. One of the first principles of emergency care is triage.

Triage

The word **triage** comes from the French word *trier*, meaning "to sort." In the daily routine of the ED, triage is used to sort patients into groups based on the severity of their health problems and the immediacy with which these problems must be treated.

A basic and widely used triage system that had been in use for many years utilized three categories: emergent, urgent, and nonurgent. In this system, emergent patients had the highest priority, urgent patients had serious health problems but not immediately life-threatening ones, and nonurgent patients had episodic illnesses.

A comprehensive, valid, and reliable five-level triage severity rating system that recognizes that EDs are used for both emergency and routine health care is now endorsed by the ENA and ACEP. The increased number of triage levels assists the triage nurse to more precisely determine the needs of the patient and the urgency for treatment. Systems that meet these criteria for validity and reliability that are commonly used in the United States are the Emergency Severity Index (ESI) and the Canadian Triage and Acuity Scale (CTAS). The ESI assigns patients into five levels, from level 1 (most urgent) to level 5 (least urgent). With the ESI, patients are assigned to triage levels based on both their acuity and their anticipated resource needs (Gilboy, Tanabe, Travers, et al., 2012). The CTAS system's five levels include time parameters that guide how frequently patients must be reassessed by either a nurse or provider. Patients assigned to the *resuscitation* category must receive continuous nursing surveillance, those in the *emergent* category must be reassessed at least every 15 minutes, patients in the *urgent* category must be reassessed at least every 30 minutes, patients in the *less urgent* category must be reassessed at least every 60 minutes, and those in the *nonurgent* category must be reassessed at least every 120 minutes (ENA, 2013).

Although the ESI and the CTAS are valid and reliable triage severity rating systems, many EDs in the United States have problems with patient volume and flow. Patients arriving at the ED may experience bottlenecks at triage. To further refine the system, triage bypass moves an incoming patient directly to a bed if open beds are available in the department. This eliminates the patient waiting and the receiving nurse performs the initial assessment and vital signs. In team triage, the triage nurse works with the physician or advanced practitioner within the triage area itself. Team triage can move the patients to diagnostics and possibly discharge without full admission to the department. Both of these additional concepts added to triage decrease wait time for treatment (ENA, 2013).

Triage is an advanced skill. Emergency nurses spend many hours learning to classify different illnesses and injuries to ensure that patients most in need of care do not needlessly wait. Protocols may be followed to initiate laboratory or x-ray studies while the patient is in the triage area. Collaborative protocols are developed and used by the triage nurse based on his or her level of experience. Nurses in the triage area collect additional crucial baseline data: full vital signs including pain assessment, history of the current event and past medical history, neurologic assessment findings, weight, allergies (especially to latex and medications), domestic violence screening, and necessary diagnostic data. Some facilities collect these data in a computerized system, which helps guide the nurse through assessment and documentation. The following questions reflect the minimum information that should be obtained from the patient or from the person who accompanied the patient to the ED and then are documented:

- What were the circumstances, precipitating events, location, and time of the injury or illness?
- When did the symptoms appear?
- Was the patient unconscious after the injury or onset of illness?
- How did the patient get to the ED?
- What was the health status of the patient before the injury or illness?
- Is there a history of medical illness or previous surgeries? A history of admissions to the hospital?
- Is the patient currently taking any medications, especially hormones, insulin, digitalis, or anticoagulants? Is the patient using any complementary or alternative therapies such as herbology, naturopathy, Reiki, massage, or acupuncture?
- Does the patient have any allergies, especially to latex, medications, eggs, or nuts?
- Does the patient smoke or use recreational drugs? How frequently? What type? When was the last time they were used?
- Does the patient have any fears? Does the patient feel that he or she is in danger or in an unsafe situation?
- When was the last meal eaten? (This is important if general anesthesia is to be given or if the patient is unconscious.)
- When was the last menstrual period?
- Is the patient under a physician's care? What are the name and contact information for the physician?
- What was the date of the patient's most recent tetanus immunization?

In addition to the collection of initial vital signs and medical history, triage consists of providing basic first aid, which may include application of ice, bleeding control, and basic wound care, as well as initiating protocol-based orders (e.g., x-rays, administering antipyretic or mild analgesic agents, obtaining an electrocardiogram [ECG] or urinalysis, removing sutures). The triage nurse also is responsible for and monitors the waiting area, maintains a safe environment, reassesses waiting patients, and is the initial liaison to the families of patients.

Routine ED triage protocols differ significantly from the triage protocols used in disasters and mass casualty incidents (field triage). Routine triage directs all available resources to the patients who are most critically ill, regardless of potential outcome. In field triage (or hospital triage during a disaster), scarce resources must be used to benefit the most people possible. This distinction affects triage decisions (see Chapter 73).

Assess and Intervene

For the patient assigned to an urgent or higher triage category, stabilization, provision of critical treatments, and prompt transfer to the appropriate setting (intensive care unit, operating room, general care unit) are the priorities of emergency care. Although treatment is initiated in the ED, ongoing definitive treatment of the underlying problem is provided in other settings, and the sooner the patient is stabilized and moved to that area, the better the outcome.

A systematic approach to effectively establishing and treating health priorities is the primary survey/secondary survey approach. The **primary survey** focuses on stabilizing life-threatening conditions. The ED staff work collaboratively and follow the ABCDE (*a*irway, *b*reathing, *c*irculation, *d*isability, *e*xposure) method:

- Establish a patent airway.
- Provide adequate ventilation, employing resuscitation measures when necessary. (Trauma patients must have the cervical spine protected and chest injuries assessed first, immediately after the airway is established.)
- Evaluate and restore cardiac output by controlling hemorrhage, preventing and treating shock, and maintaining or restoring effective circulation. This includes the prevention and management of hypothermia. In addition, peripheral pulses are examined, and any immediate closed reductions of fractures or dislocations are performed if an extremity is pulseless.
- Determine neurologic disability by assessing neurologic function using the Glasgow Coma Scale and a motor and sensory evaluation of the spine (see Chapter 65). A quick neurologic assessment may be performed using the AVPU mnemonic:
  - *A*—alert. Is the patient alert and responsive?
  - *V*—verbal. Does the patient respond to verbal stimuli?
  - *P*—pain. Does the patient respond only to painful stimuli?
  - *U*—unresponsive. Is the patient unresponsive to all stimuli, including pain?
- Undress the patient quickly but gently so that any wounds or areas of injury are identified; this may entail cutting away articles of clothing (American College of Surgeons [ACS], 2008).

After these priorities have been addressed, the ED team proceeds with the **secondary survey**. This includes the following:

- Complete health history, including the history of the current event
- Head-to-toe assessment (includes a reassessment of airway and breathing parameters and vital signs)
- Diagnostic and laboratory testing
- Insertion or application of monitoring devices such as ECG electrodes, arterial lines, or urinary catheters
- Splinting of suspected fractures
- Cleansing, closure, and dressing of wounds
- Performance of other necessary interventions based on the patient's condition

Once the patient has been assessed, stabilized, and tested, appropriate medical and nursing diagnoses are formulated, initial important treatment is started, and plans for the proper disposition of the patient are made. Many emergent and urgent conditions and priority emergency interventions are discussed in detail in the remaining sections of this chapter.

In addition to the management of the illness or injury, the ED nurse must also focus on providing comfort and emotional support to the patient and family. Included in this is pain management. Effective pain management must be instituted early and should include rapid-acting agents that result in minimal sedation so that the patient can continue to interact with the staff for ongoing assessment. Moderate sedation can help facilitate short procedures in the ED; the patient will not remember the procedure later. The patient is closely monitored during the procedure and then rapidly awakens when it is complete (see Chapter 18).

It is essential that family crisis intervention services are available for families of ED patients. Even if a patient's condition is not emergent, the situation may be perceived as such by the family. Every family needs attention and support. The chaplain and social worker may be available to assist with interventions.

AIRWAY OBSTRUCTION

Acute upper airway obstruction is a life-threatening medical emergency.

Pathophysiology

The airway may be partially or completely occluded. Partial obstruction of the airway can lead to progressive hypoxia, hypercarbia, and respiratory and cardiac arrest. If the airway is completely obstructed, permanent brain injury or death will occur within 3 to 5 minutes secondary to hypoxia. Air movement is absent in the presence of complete airway obstruction. Oxygen saturation of the blood decreases rapidly because obstruction of the airway prevents entry of air into the lungs. Oxygen deficit occurs in the brain, resulting in unconsciousness, with death following rapidly.

Upper airway obstruction has a number of causes, including aspiration of foreign bodies, anaphylaxis, viral or bacterial infection, trauma, and inhalation or chemical burns. For older adult patients, especially those in extended care facilities, sedative and hypnotic medications, diseases affecting motor coordination (e.g., Parkinson's disease), and mental dysfunction (e.g., dementia, mental retardation) are risk factors for asphyxiation by food. In adults, aspiration of a bolus of meat is the most common cause of airway obstruction. Peritonsillar abscesses, epiglottitis, and other acute infectious processes of the posterior pharynx can also result in airway obstruction (Sinz & Navarro, 2011).

Clinical Manifestations

Typically, a person with a foreign body airway obstruction cannot speak, breathe, or cough. The patient may clutch the neck between the thumb and fingers (i.e., universal distress signal). Other common signs and symptoms include choking, apprehensive appearance, refusing to lie flat, inspiratory and expiratory stridor, labored breathing, the use of accessory muscles (suprasternal and intercostal retraction), flaring nostrils, increasing anxiety, restlessness, and confusion. Cyanosis and loss of consciousness develop as hypoxia worsens. Cyanosis and loss of consciousness are late signs. Action must be taken before these manifestations develop, if possible, or immediately if the patient has already exhibited these signs.

Assessment and Diagnostic Findings

Assessment of the patient who has a foreign object occluding the airway may involve simply asking the person whether he or she is choking and requires help. If the person is unconscious, inspection of the oropharynx may reveal the offending object. X-rays, laryngoscopy, or bronchoscopy also may be performed. Oxygen supplementation should be considered immediately.

Management

If the patient can breathe and cough spontaneously, a partial obstruction should be suspected. The victim is encouraged to cough forcefully and to persist with spontaneous coughing and breathing efforts as long as good air exchange exists. There may be some wheezing between coughs. If the patient demonstrates a weak, ineffective cough, high-pitched noise while inhaling, increased respiratory difficulty, or cyanosis, the patient should be managed as if there were complete airway obstruction.

After the obstruction is removed, rescue breathing is initiated. If the patient has no pulse, cardiac compressions are instituted. These measures provide oxygen to the brain, heart, and other vital organs until definitive medical treatment can restore and support normal heart and ventilatory activity (Travers, Rhea, Bobrow, et al., 2010). (See Chapter 29 for review of current CPR guidelines.)

Establishing an Airway

Establishing an airway may be as simple as repositioning the patient's head to prevent the tongue from obstructing the pharynx. Alternatively, other maneuvers, such as the head-tilt/chin-lift maneuver, the jaw-thrust maneuver, or insertion of specialized equipment, may be needed to open the airway, remove a foreign body, or maintain the airway. In all maneuvers, the cervical spine must be protected from injury. After these maneuvers are performed, the patient is assessed for breathing by watching for chest movement and listening and feeling for air movement. In such a case, nursing diagnoses would include ineffective airway clearance related to obstruction of the airway by the tongue, an object, or fluids (blood, saliva) and ineffective breathing pattern related to airway obstruction or injury. (See Chart 21-6 in Chapter 21 for information regarding how to clear an upper airway obstruction.)

Oropharyngeal/Nasopharyngeal Airway Insertion

An oropharyngeal airway is a semicircular tube or tubelike plastic device that is inserted over the back of the tongue into the lower posterior pharynx in a patient who is breathing spontaneously but who is unconscious (Chart 72-3). This type of airway prevents the tongue from falling back against the posterior pharynx and obstructing the airway. It also allows health care providers to suction secretions. The nasopharyngeal airway provides the same airway access but is inserted through the nares.

Quality and Safety Nursing Alert

In the case of potential facial trauma or basilar skull fracture, the nasopharyngeal airway should not be used because it could enter the brain cavity instead of the pharynx.

Concept Mastery Alert

Inserting an oropharyngeal airway is a crucial intervention for the patient with an airway obstruction. Chart 72-3 describes the steps for this procedure.

Endotracheal Intubation

The purpose of endotracheal intubation is to establish and maintain the airway in patients with respiratory insufficiency or hypoxia. Endotracheal intubation is indicated to establish

Chart 72-3 Inserting an Oropharyngeal Airway

1. Measure the oral airway alongside the head. The airway should reach from lip to ear.
2. Extend the patient's head by placing one hand under the bony chin (*only if the cervical spine is uninjured*). With the other hand, tilt the head backward by applying pressure to the forehead while simultaneously lifting the chin forward.
3. Open the patient's mouth.
4. **A.** Insert the oropharyngeal airway with the tip facing up toward the roof of the mouth until it passes the uvula. **B.** Rotate the tip 180 degrees so that the tip is pointed down toward the pharynx. This displaces the tongue anteriorly, and the patient then breathes through and around the airway. An alternate method is to use a tongue blade to hold the tongue and insert the oropharyngeal airway directly without rotation.
5. The distal end of the oropharyngeal airway is in the hypopharynx, and the flange is approximately at the patient's lips. Make sure that the tongue has not been pushed into the airway.

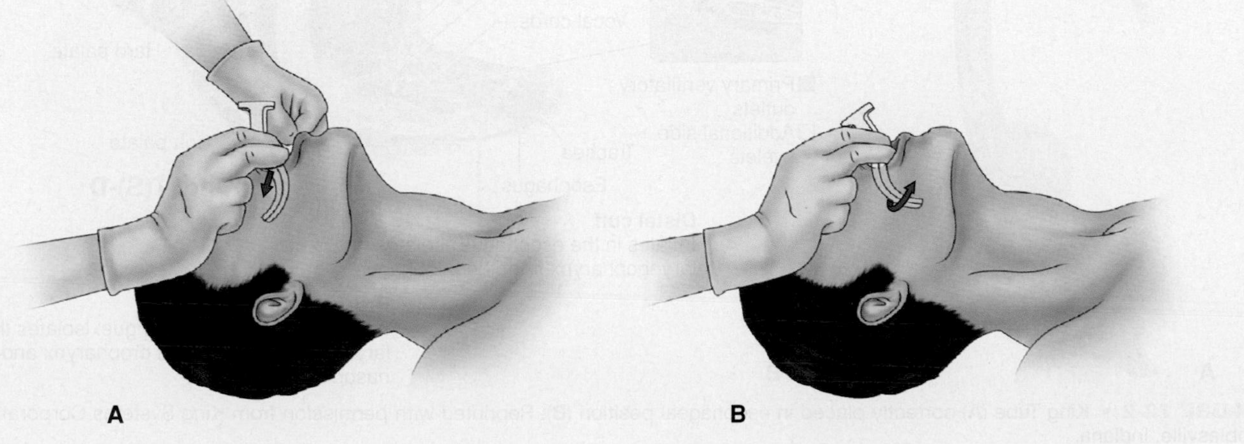

A B

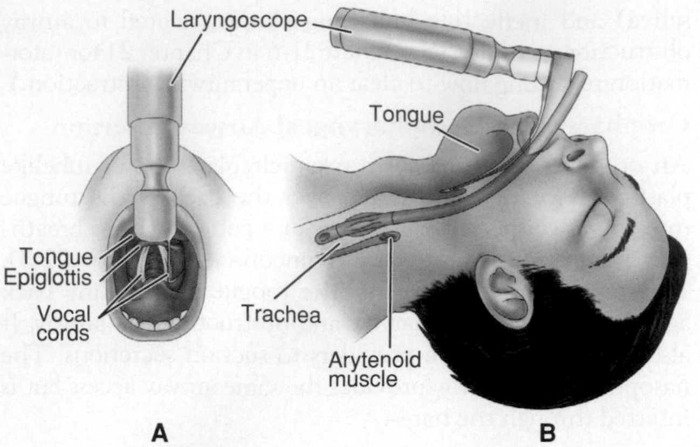
Laryngoscope
Tongue
Tongue
Epiglottis
Vocal cords
Trachea
Arytenoid muscle
A
B

FIGURE 72-1 • Endotracheal intubation in a patient without a cervical spine injury. **A.** The primary glottic landmarks for tracheal intubation as visualized with proper placement of the laryngoscope. **B.** Positioning the endotracheal tube.

an airway for a patient who cannot be adequately ventilated with an oropharyngeal airway, bypass an upper airway obstruction, prevent aspiration, permit connection of the patient to a resuscitation bag or mechanical ventilator, or facilitate the removal of tracheobronchial secretions (Fig. 72-1). Because the procedure requires skill, endotracheal intubation is performed only by those who have had extensive training. These may include physicians, nurse anesthetists, respiratory therapists, flight nurses, and nurse practitioners. However, the emergency nurse commonly assists with intubation.

Rapid sequence intubation may be indicated, which provides management of the patient in a situation similar to that in the operating room. Medications used to facilitate rapid sequence intubation include a sedative, an analgesic, and a neuromuscular blockade agent; these are usually administered by the practitioner performing the intubation.

Intubation With a King Tube or Laryngeal Mask Airway

If the patient is not hospitalized and cannot be intubated in the field, emergency medical personnel may insert a King Tube, which rapidly provides pharyngeal ventilation. When the tube is inserted into the trachea, it functions like an endotracheal tube (Fig. 72-2).

The two balloons that surround the tube are inflated after the tube is inserted. One balloon is large and occludes the oropharynx. This permits ventilation by forcing air through the larynx. The smaller balloon is inflated with air and occludes the esophagus at a site distal to the glottis. Breath sounds are auscultated after balloon inflation to make sure that the oropharyngeal balloon (or cuff) does not obstruct the glottis. One variant type of King Tube is designed so that a gastric tube may also be passed for suction. If it is difficult to establish an airway, a laryngeal mask airway (LMA) may be inserted as an interim airway device. The design of the LMA provides a "mask" in the subglottic airway with a cuff inflated within the esophagus. It allows easy insertion for rapid airway control until a more definitive airway can be placed. Some LMAs also permit removal of secretions from the esophagus (see Fig. 18-3A in Chapter 18).

Cricothyroidotomy (Cricothyroid Membrane Puncture)

Cricothyroidotomy is the opening of the cricothyroid membrane to establish an airway. This procedure is used in emergency situations in which endotracheal intubation is either not possible or contraindicated, as in airway obstruction from extensive maxillofacial trauma, cervical spine injuries, laryngospasm, laryngeal edema (after an allergic reaction or extubation), hemorrhage into neck tissue, or obstruction of the larynx. A cricothyroidotomy is replaced with a formal tracheostomy when the patient is able to tolerate this procedure.

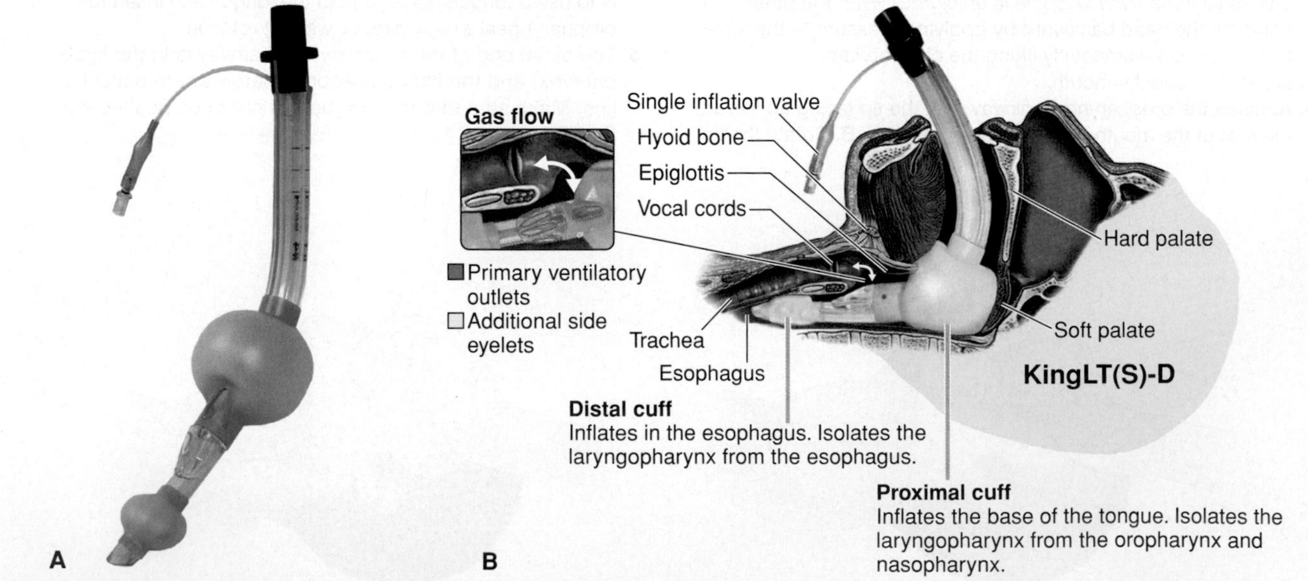
Gas flow
Primary ventilatory outlets
Additional side eyelets
Single inflation valve
Hyoid bone
Epiglottis
Vocal cords
Hard palate
Soft palate
Trachea
Esophagus
KingLT(S)-D
Distal cuff
Inflates in the esophagus. Isolates the laryngopharynx from the esophagus.
Proximal cuff
Inflates the base of the tongue. Isolates the laryngopharynx from the oropharynx and nasopharynx.
A
B

FIGURE 72-2 • King Tube (**A**) correctly placed in esophageal position (**B**). Reprinted with permission from King Systems Corporation, Noblesville, Indiana.

Maintaining Ventilation

After the airway is determined to be unobstructed, the nurse must ensure that ventilation is adequate by checking for equal bilateral breath sounds. Satisfactory management of ventilations may prevent hypoxia and hypercapnia. The nurse must quickly assess for absent or diminished breath sounds, open chest wounds, and difficulty delivering artificial breaths for the patient. The nurse should monitor pulse oximetry, capnography, and arterial blood gases if the patient requires airway or ventilatory assistance. A tension pneumothorax can mimic hypovolemia, so ventilatory assessment precedes assessment for hemorrhage. A pneumothorax (both simple and tension) or sucking (open) chest wound is managed with a chest tube and occlusion of the sucking wound; immediate relief of increasing positive intrathoracic pressure and maintenance of adequate ventilation should occur.

HEMORRHAGE

Stopping bleeding is essential to the care and survival of patients in an emergency or disaster situation. Hemorrhage that results in the reduction of circulating blood volume is a main cause of shock. Minor bleeding, which is usually venous, generally stops spontaneously unless the patient has a bleeding disorder or has been taking anticoagulant agents. Internal hemorrhage can hide in many anatomic spaces and compartments, resulting in shock without external evidence of hemorrhage. The internal spaces and compartments that are capable of housing large amounts of blood include the retroperitoneum, pelvis, chest, and thighs (ACS, 2010).

The patient is assessed for signs and symptoms of shock: cool, moist skin (resulting from poor peripheral perfusion), decreasing blood pressure, increasing heart rate, delayed capillary refill, and decreasing urine volume (see Chapter 14). The goals of emergency management are to control the bleeding, maintain adequate circulating blood volume for tissue oxygenation, and prevent shock. Patients who hemorrhage are at risk for cardiac arrest caused by hypovolemia with secondary anoxia. Nursing interventions are carried out collaboratively with other members of the emergency health care team.

Management

Fluid Replacement

Whenever a patient is hemorrhaging—whether externally or internally—a loss of circulating blood results in a fluid volume deficit and decreased cardiac output. Therefore, fluid replacement is imperative to maintain circulation. Typically, two large-gauge IV catheters are inserted, preferably in an uninjured extremity, to provide a means for fluid and blood replacement. Blood samples are obtained for analysis, typing, and cross-matching. Replacement fluids are administered as prescribed, depending on clinical estimates of the type and volume of fluid lost. Replacement fluids may include isotonic electrolyte solutions (e.g., lactated Ringer's, normal saline), colloids, and blood component therapy.

Packed red blood cells are infused when there is massive blood loss, which may also necessitate transfusion of other blood components, including platelets and clotting factors. (See Chapter 32 for full discussion of blood component therapy indications and treatment.)

> ***Quality and Safety Nursing Alert***
>
> **The infusion rate is determined by the severity of the blood loss and the clinical evidence of hypovolemia. If massive blood replacement is necessary, it should be administered via warmer when possible, because administration of large amounts of blood that has been refrigerated has a core cooling effect that may lead to cardiac arrest and coagulopathy.**

Control of External Hemorrhage

If a patient is hemorrhaging externally (e.g., from a wound), a rapid physical assessment is performed as the patient's clothing is cut away in an attempt to identify the area of hemorrhage. Direct, firm pressure is applied over the bleeding area or the involved artery at a site that is proximal to the wound (Fig. 72-3). Most bleeding can be stopped or at least controlled by application of direct pressure. Otherwise, unchecked arterial bleeding results in death. A firm pressure dressing is applied, and the injured part is elevated to stop venous and capillary bleeding, if possible. If the injured area is an extremity, the extremity is immobilized to control blood loss.

A tourniquet is applied to an extremity only as a *last resort* when the external hemorrhage cannot be controlled in any other way and immediate surgery is not feasible. Care must be taken when applying a tourniquet because of the risk of loss of the extremity. The tourniquet is applied just proximal to the wound and tied tightly enough to control arterial blood flow. The patient is tagged with a skin-marking pencil or on adhesive tape on the forehead with a "T," stating the location of the tourniquet and the time applied. If there is no arterial bleeding, the tourniquet is removed and a pressure dressing is applied. If the patient has suffered a traumatic amputation with uncontrollable hemorrhage, the tourniquet remains in place until the patient is in the operating room. Time of tourniquet application and removal should be documented. Tourniquet placement among military personnel with battle-associated trauma has demonstrated clear mortality reduction, although it occasionally has led to amputation or fasciotomy (Kragh, Walters, Baer, et al., 2008).

Control of Internal Bleeding

If the patient shows no external signs of bleeding but exhibits tachycardia, falling blood pressure, thirst, apprehension, cool and moist skin, or delayed capillary refill, internal hemorrhage is suspected. Typically, packed red blood cells are administered at a rapid rate, and the patient is prepared for more definitive treatment (e.g., surgery, pharmacologic therapy). In addition, arterial blood gas specimens are obtained to evaluate pulmonary function and tissue perfusion and to establish baseline hemodynamic parameters, which are then used as an index for determining the amount of fluid replacement the patient can tolerate and the response to therapy. The patient is maintained in the supine position and monitored closely until hemodynamic or circulatory parameters improve, or until he or she is transported to the operating room or intensive care unit.

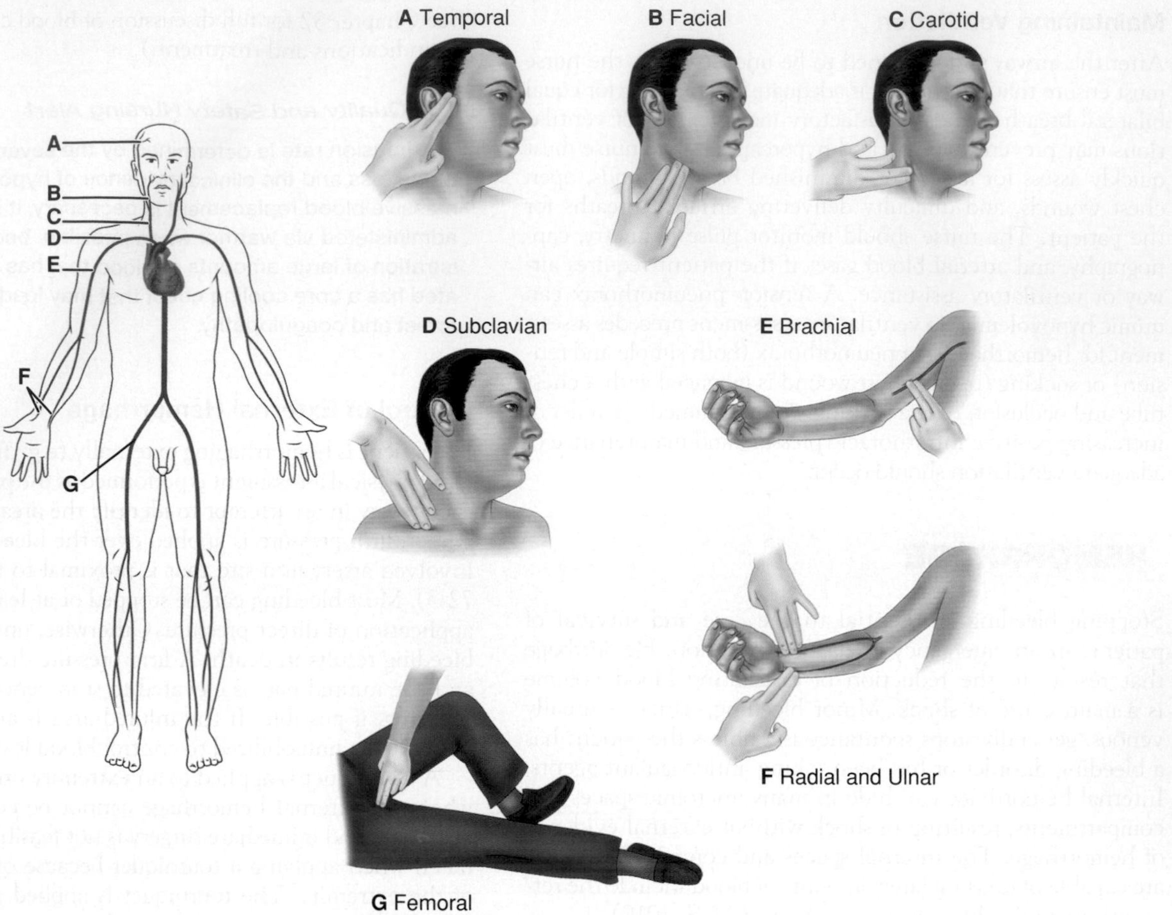
A Temporal
B Facial
C Carotid
D Subclavian
E Brachial
F Radial and Ulnar
G Femoral
A
B
C
D
E
F
G

FIGURE 72-3 • Pressure points for control of hemorrhage.

HYPOVOLEMIC SHOCK

Shock is a condition in which there is loss of effective circulating blood volume. Inadequate organ and tissue perfusion follows, ultimately resulting in cellular metabolic derangements. In any emergency situation, the onset of shock should be anticipated by assessing all injured people immediately. The underlying cause of shock (hypovolemic, cardiogenic, neurogenic, anaphylactic, or septic) must be determined. Of these, hypovolemia is the most common cause (see Chapter 14 for further discussion of management of hypovolemic shock).

WOUNDS

Wounds involving injury to soft tissues can vary from minor tears to severe crushing injuries. The types of wounds that may occur are defined in Chart 72-4. The main goal of treatment is to restore the physical integrity and function of the injured tissue while minimizing scarring and preventing infection. Proper documentation of the characteristics of the wound, using precise descriptions and correct terminology, is essential. Such information may be needed in the future for forensic evidence. Photographs are helpful because they provide an accurate, visible depiction of the wound. Photographs also become important for exigent wounds (i.e., wounds that will eventually heal). Patients involved in domestic violence or trauma may need the photographs later to visually describe the extent of injury.

Determining *when* and *how* the wound occurred is important because a treatment delay increases infection risk. Using aseptic technique, the clinician inspects the wound to determine the extent of damage to underlying structures or the presence of a foreign body. Sensory, motor, and vascular function is evaluated for changes that might indicate complications.

Chart 72-4 Definition of Terms: Wounds

Laceration: skin tear with irregular edges and vein bridging
Avulsion: tearing away of tissue from supporting structures
Abrasion: denuded skin
Ecchymosis/contusion: blood trapped under the surface of the skin
Hematoma: tumorlike mass of blood trapped under the skin
Stab: incision of the skin with well-defined edges, usually caused by a sharp instrument; a stab wound is typically deeper than long
Cut: incision of the skin with well-defined edges, usually longer than deep
Patterned: wound representing the outline of the object (e.g., steering wheel) causing the wound

Management

Wound Cleansing

Hair around the wound may be clipped (only as directed) if it is anticipated that the hair will interfere with wound closure. Typically, the area around the wound is cleansed with normal saline solution or a polymer agent (e.g., Shur-Clens). Antibacterial agents, such as povidone-iodine (Betadine) or hydrogen peroxide, should not be allowed to get deep into the wound without thorough rinsing. These agents are used only for the initial cleansing because they injure exposed and healthy tissue, resulting in further tissue damage (ENA, 2013).

If indicated, the area is infiltrated with a local intradermal anesthetic through the wound margins or by regional block. Patients with soft tissue injuries usually have localized pain at the site of injury. The nurse then assists with cleaning and débriding the wound. The wound is irrigated gently and copiously with sterile isotonic saline solution to remove surface dirt. Devitalized tissue and foreign matter are removed because they impede healing and may promote infection. Any small bleeding vessels are clamped, tied, or cauterized. After wound treatment, a nonadherent dressing is applied to protect the wound and to serve as a splint and as a reminder to the patient that the area is injured.

Primary Closure

The decision to suture a wound depends on the nature of the wound, the time since the injury was sustained, the degree of contamination, and the vascularity of tissues. If primary closure is indicated, the wound is sutured or stapled, usually by the physician, with the patient receiving either local anesthesia or moderate sedation (see Chapter 18). Wound closure begins when subcutaneous fat is brought together loosely with a few sutures to close off the dead space. The subcuticular layer is then closed, and finally the epidermis is closed. Sutures are placed near the wound edge, with the skin edges leveled carefully to promote optimal healing. Instead of sutures, sterile strips of reinforced microporous tape or a bonding agent (skin glue) may be used to close clean, superficial wounds.

Delayed Primary Closure

Delayed primary closure may be indicated if tissue has been lost or there is a high potential for infection. A thin layer of gauze (to ensure drainage and prevent pooling of exudate), covered by an occlusive dressing, may be used. The wound is splinted in a functional position to prevent motion and decrease the possibility of contracture.

If there are no signs of suppuration (formation of purulent drainage), the wound may be sutured (with the patient receiving a local anesthetic). The use of antibiotic agents to prevent infection depends on factors such as how the injury occurred, the age of the wound, and the risk of contamination. The site is immobilized and elevated to limit accumulation of fluid in the interstitial spaces of the wound.

Tetanus prophylaxis is administered as prescribed, based on the condition of the wound and the patient's immunization status. If the patient's last tetanus booster was administered more than 5 years ago, or if the patient's immunization status is unknown, a tetanus booster must be given (ACS, 2010). The patient is instructed about signs and symptoms of infection and is instructed to contact the primary provider or clinic if there is sudden or persistent pain, fever or chills, bleeding, rapid swelling, foul odor, drainage, or redness surrounding the wound.

TRAUMA

Trauma (an unintentional or intentional wound or injury inflicted on the body from a mechanism against which the body cannot protect itself) is the fourth leading cause of death in the United States. Trauma is the leading cause of death in children and in adults younger than 44 years. The incidence is increasing in adults older than 44 years. Alcohol and drug abuse are often implicated as factors in both blunt and penetrating trauma (McQuillan, VonReuden, Hartsock, et al., 2008).

Collection of Forensic Evidence

In assessing and managing any patient with an emergency condition, but especially the patient experiencing trauma, meticulous documentation is essential. Included in documentation are descriptions of all wounds, mechanism of injury, time of events, and collection of evidence. In trauma care, the nurse must be exceedingly careful with all potential evidence, handling and documenting it properly.

The basics of care management for patients with traumatic injury include an understanding that trauma in any patient (living or dead) has potential legal or forensic science implications if criminal activity is suspected. Hence, proper management from both a medical and forensic evidence perspective is essential.

When clothing is removed from the patient who has experienced trauma, the nurse must be careful not to cut through or disrupt any tears, holes, blood stains, or dirt present on the clothing if criminal activity is suspected. Each piece of clothing should be placed in an individual paper bag. Plastic bags are not used because they retain moisture; moisture may promote mold and mildew formation, which can destroy evidence. If the clothing is wet, it should be hung to dry. Clothing should not be given to families. Valuables should be inventoried and either placed in the hospital safe or clearly documented to which family member they were given. If a police officer is present to collect clothing or any other items from the patient, each item is labeled and the transfer of custody to the officer, the officer's name, the date, and the time are documented. Evidence cannot be left unattended in the room; a formal chain of custody must be maintained for the evidence to be valid and useful for legal purposes (Nayduch, 2009).

If suicide or homicide is suspected in a deceased trauma patient, the medical examiner examines the body on site or has the body moved to the coroner's office for autopsy. All tubes and lines must remain in place. The patient's hands must be covered with paper bags to protect evidence on the hands or under the fingernails. In the surviving patient, tissue specimens may be swabbed from the hands and nails as potential evidence. Photographs of wounds or clothing are essential and should include a reference ruler in one photo and another without the ruler.

Documentation should also include any statements made by the patient in the patient's own words and surrounded by quotation marks. A chain of evidence is essential. If the patient's case is reviewed in a court of law in the future, clear documentation assists the judicial process and helps to identify the activities that occurred in the ED.

Injury Prevention

Any discussion of trauma management must address injury prevention. A component of the emergency nurse's daily role is to provide injury prevention information to every patient with whom there is contact, including patients admitted for reasons other than injury. The only way to reduce the incidence of trauma is through prevention.

There are three components of injury prevention. The first is education. Providing information and materials to help prevent violence and to maintain safety at home and in vehicles is important. Involvement in local injury prevention organizations, nursing organizations, and health fairs promotes wellness and safety. In practice, nursing and other health care professionals should avoid using the word *accident*, because trauma events are *preventable* and should be viewed as such rather than as "fate" or "happenstance." Responsibility and accountability must be assigned to traumatic incidents, particularly because of the high rate of trauma recidivism (repeated trauma). People who are at risk for trauma and trauma recidivism should be identified and provided with education and counseling directed toward altering risky behaviors and preventing further trauma (ENA, 2013).

The second component of injury prevention is legislation. Nurses should be actively involved in safety legislation at the local, state, and federal levels. Such legislation is meant to provide universal safety measures, not to infringe on rights.

The third component is automatic protection. Airbags and automotive design are included in this category. These mechanisms provide for safety without requiring personal intervention.

Emergency nurses may develop injury prevention programs using a focus similar to the ABCDE approach used in the primary survey in trauma care. In this case, however (ACS, 2010):

- *A*—describes *a*ssessment of the community for common injury mechanisms
- *B*— is used to describe *b*uilding a coalition of key community members
- *C*—refers to *c*ommunicating awareness of the trauma mechanisms and risks prevalent in the local community
- *D*—stands for *d*eveloping and implementing interventions, which may be educational or legislative
- *E*—refers to *e*valuating the injury prevention program soon after it is launched, which may result in either continuation or revision of the program.

Multiple Trauma

Multiple trauma is caused by a single catastrophic event that causes life-threatening injuries to at least two distinct organs or organ systems. Patients with single-system trauma still receive full assessment, because even single-system injuries can be life threatening or more severe than they initially appear. Mortality in patients with multiple trauma is related to the severity of the injuries, the number of systems and organs involved, and the severity of each injury alone and in combination. Immediately after injury, the body is hypermetabolic, hypercoagulable, and severely stressed.

Care of the patient with multiple injuries requires a team approach, with one person responsible for coordinating the treatment. The nursing staff assumes responsibility for assessing and monitoring the patient, ensuring/maintaining airway and IV access, administering prescribed medications, collecting laboratory specimens, and documenting activities and the patient's subsequent responses.

Assessment and Diagnostic Findings

External evidence of trauma may be sparse or absent. Patients with multiple trauma should be assumed to have a spinal cord injury until it is proven otherwise. The injury regarded as the least significant in appearance may be the most lethal. For example, the pelvic fracture not identified until an x-ray is obtained may cause rapid and massive hemorrhage into the pelvic cavity, but an obvious amputation of the arm may have already stopped bleeding from the body's normal response of vasoconstriction.

Management

The goals of treatment are to determine the extent of injuries and to establish priorities of treatment. Any injury interfering with a vital physiologic function (e.g., airway, breathing, circulation) is an immediate threat to life and has the highest priority for immediate treatment. Essential lifesaving procedures are performed simultaneously by the emergency team. As soon as the patient is resuscitated, clothes are removed or cut off and a rapid physical assessment is performed. Transfer from field management to the ED must be orderly and controlled, with attention given to the verbal report from emergency medical services. Treatment in a trauma center is appropriate for patients experiencing major trauma. Treatment priorities are presented in Chart 72-5.

Intra-Abdominal Injuries

Intra-abdominal injuries are categorized as penetrating or blunt trauma. *Penetrating* abdominal injuries (i.e., gunshot wounds, stab wounds) are serious and usually require surgery. Penetrating abdominal trauma results in a high incidence of injury to hollow organs, particularly the small bowel. The liver is the most frequently injured solid organ due to its size and anterior placement in the right upper quadrant of the abdomen. In gunshot wounds, the most important prognostic factor is the velocity at which the missile enters the body. High-velocity missiles (bullets) produce extensive tissue damage. All abdominal gunshot wounds that cross the peritoneum or are associated with peritoneal signs require surgical exploration. On the other hand, some stab wounds may be managed nonoperatively due to low velocity and less penetration of the implement (i.e., weapon) (ACS, 2008).

Blunt trauma to the abdomen may result from motor vehicle crashes, falls, blows, or explosions. Blunt trauma is commonly associated with extra-abdominal injuries to the chest,

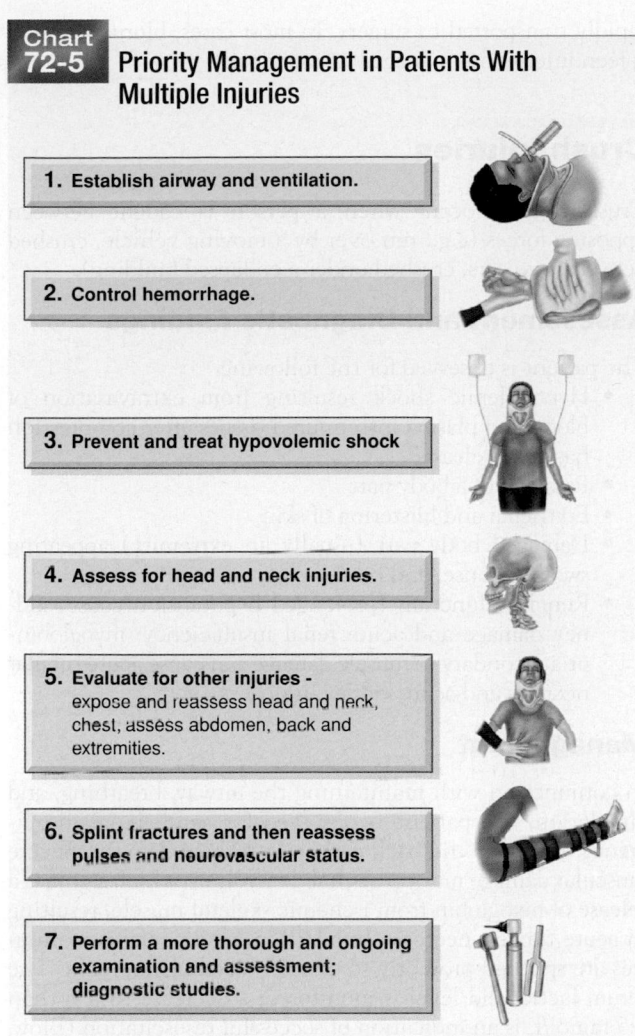

Chart 72-5 **Priority Management in Patients With Multiple Injuries**

1. Establish airway and ventilation.

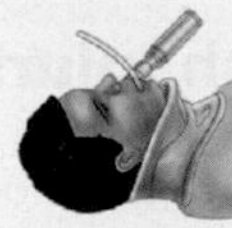

2. Control hemorrhage.

3. Prevent and treat hypovolemic shock

4. Assess for head and neck injuries.

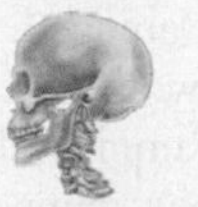

5. Evaluate for other injuries - expose and reassess head and neck, chest, assess abdomen, back and extremities.

6. Splint fractures and then reassess pulses and neurovascular status.

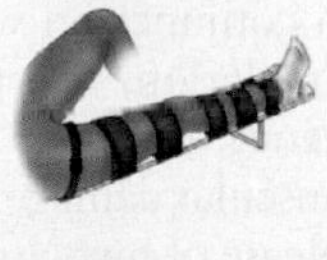

7. Perform a more thorough and ongoing examination and assessment; diagnostic studies.

Adapted from American College of Surgeons. (2008). *Advanced trauma life support* (8th ed.). Chicago: Author.

head, or extremities. Patients with blunt trauma are a challenge because injuries may be difficult to detect. The incidence of delayed and trauma-related complications is greater than for penetrating injuries. This is especially true of blunt injuries involving the liver, kidneys, spleen, or blood vessels, which can lead to massive blood loss into the peritoneal cavity (ACS, 2008).

Assessment and Diagnostic Findings

As the history of the traumatic event is obtained, the abdomen is inspected as a part of the secondary survey for obvious signs of injury, including penetrating injuries, bruises, and abrasions. Abdominal assessment continues with auscultation of bowel sounds to provide baseline data from which changes can be noted. Absence of bowel sounds may be an early sign of intraperitoneal involvement, although stress can also decrease or halt peristalsis and thus bowel sounds. Further abdominal assessment may reveal progressive abdominal distention, involuntary guarding, tenderness, pain, muscular rigidity, or rebound tenderness along with changes in bowel sounds, all of which are signs of peritoneal irritation. Hypotension and signs and symptoms of shock may also be noted. In addition, the chest and other body systems are assessed for injuries that frequently accompany intra-abdominal injuries (ACS, 2008).

Laboratory studies that aid in assessment include the following:

- Urinalysis to detect hematuria (indicative of a urinary tract injury)
- Serial hemoglobin and hematocrit levels to evaluate trends reflecting the presence or absence of bleeding
- White blood cell (WBC) count to detect elevation (generally associated with trauma)
- Serum amylase/lipase analysis to detect increasing levels, which suggest pancreatic injury or perforation of the gastrointestinal tract

Internal Bleeding

Hemorrhage frequently accompanies abdominal injury, especially if the liver or spleen has been traumatized. Therefore, the patient is assessed continuously for signs and symptoms of external and internal bleeding. The front of the body, flanks, and back are inspected for bluish discoloration, asymmetry, abrasion, and contusion. Abdominal CT scans permit detailed evaluation of abdominal contents and retroperitoneal examination. Abdominal ultrasounds can rapidly assess hemodynamically unstable patients to detect intraperitoneal bleeding. This is referred to as the focused assessment with sonography for trauma (FAST) examination (ACS, 2008). During the resuscitation period, pain is managed using administration of small dosages of opioids (ACS, 2008).

Concept Mastery Alert

The location of pain can indicate certain types of intra-abdominal injuries. Pain in the left shoulder is common in a patient with bleeding from a ruptured spleen, whereas pain in the right shoulder can result from laceration of the liver.

Intraperitoneal Injury

The abdomen is assessed for tenderness, rebound tenderness, guarding, rigidity, spasm, increasing distention, and pain. Referred pain is a significant finding because it suggests intraperitoneal injury. To determine if there is intraperitoneal injury and bleeding, the patient is usually prepared for diagnostic procedures, such as peritoneal lavage, abdominal ultrasonography, or abdominal CT scanning. **Diagnostic peritoneal lavage** (DPL), although no longer the standard diagnostic study used to evaluate a traumatized abdomen, remains a backup procedure that is easily performed and is very useful during mass casualty situations when CT scanners may not be readily available. DPL involves the instillation of 1 L of warmed lactated Ringer's or normal saline solution into the abdominal cavity. After a minimum of 400 mL has been returned, a fluid specimen is sent to the laboratory for analysis. Positive laboratory findings include a red blood cell count greater than 100,000/mm^3; a WBC count greater than 500/mm^3; or the presence of bile, feces, or food (ACS, 2008).

In patients with stab wounds, sinography may be performed to detect peritoneal penetration; a purse-string suture is placed around the wound and a small catheter is introduced through the wound. A contrast agent is then introduced through the catheter, and x-rays are taken to identify any peritoneal penetration.

Genitourinary Injury

A focused genitourinary examination, which typically includes a rectal and/or vaginal examination, is performed to determine any injury to the pelvis, bladder, urethra, vagina, or intestinal wall. In the male patient, a high-riding prostate gland (abnormal position) discovered during a rectal examination indicates a potential urethral injury. A digital vaginal examination is performed on female patients to determine if there is an open pelvic fracture that has torn the vagina.

Quality and Safety Nursing Alert

To decompress the bladder and monitor urine output in a patient with a genitourinary injury, an indwelling catheter is inserted after a rectal examination has been completed, not before. In addition, urethral catheter insertion when a possible urethral injury is present is contraindicated; a urology consultation and further evaluation of the urethra are required.

Management

As indicated by the patient's condition, resuscitation procedures (restoration of airway, breathing, and circulation) are initiated as described previously.

With blunt trauma, the patient is kept on a stretcher to immobilize the spine. A backboard may be used for transporting the patient to the x-ray department, to the operating room, or to the intensive care unit. If the patient has been placed on a backboard, it should be removed as early as possible to prevent skin breakdown. Cervical spine immobilization is maintained until cervical x-rays have been obtained and cervical spine injury has been ruled out. Likewise, once the patient has arrived at the definitive destination, the backboard is removed, and logrolling can be used to protect the spine until x-rays are obtained and confirm that there is no evidence of injuries.

Knowing the mechanism of injury (e.g., penetrating force from a gunshot or knife, blunt force from a blow) is essential to determining the type of management needed. All wounds are located, counted, and documented. If abdominal viscera protrude, the area is covered with sterile, moist saline dressings to keep the viscera from drying.

Typically, oral fluids are withheld in anticipation of surgery, and the stomach contents are aspirated with a nasogastric tube to reduce the risk of aspiration and to decompress the stomach in preparation for diagnostic procedures.

Trauma predisposes the patient to infection by disruption of mechanical barriers, exposure to exogenous bacteria from the environment at the time of injury, aspiration of vomitus, and diagnostic and therapeutic procedures (hospital-acquired infection). Tetanus prophylaxis and broad-spectrum antibiotics are administered as prescribed.

Throughout the stay in the ED, the patient's condition is continuously monitored for changes. If there is continuing evidence of shock, blood loss, free air under the diaphragm, evisceration, hematuria, severe head injury, musculoskeletal injury, or suspected or known abdominal injury, the patient is rapidly transported to surgery. In most cases, blunt liver and spleen injuries are managed nonsurgically.

Crush Injuries

Crush injuries occur when a person is caught between opposing forces (e.g., run over by a moving vehicle, crushed between two cars, crushed under a collapsed building).

Assessment and Diagnostic Findings

The patient is observed for the following

- Hypovolemic shock resulting from extravasation of blood and plasma into injured tissues after compression has been released
- Paralysis of a body part
- Erythema and blistering of skin
- Damaged body part (usually an extremity) appearing swollen, tense, and hard
- Renal dysfunction (prolonged hypotension causes kidney damage and acute renal insufficiency; myoglobinuria secondary to muscle damage can cause acute tubular necrosis and acute kidney injury)

Management

In conjunction with maintaining the airway, breathing, and circulation, the patient is observed for acute renal insufficiency. Injury to the back can cause kidney damage. Severe muscular damage may cause rhabdomyolysis, which signifies a release of myoglobin from ischemic skeletal muscle, resulting in acute tubular necrosis. In addition, major soft tissue injuries are splinted promptly to control bleeding and pain. The serum lactic acid level is monitored; a decrease to less than 2.5 mmol/L is an indication of successful resuscitation (Blow, Magliore, Claridge, et al., 1999).

If an extremity is injured, it is elevated to relieve swelling and pressure. If compartment syndrome develops, the physician may perform a **fasciotomy** (i.e., surgical incision to the level of the fascia) to restore neurovascular function (see Chapter 43). Medications for pain and anxiety are then administered as prescribed, and the patient is quickly transported to the operating suite for wound débridement and fracture repair. A hyperbaric oxygen chamber (if available) may be used to hyperoxygenate crushed tissue, if indicated.

Fractures

Immediate appropriate management of a fracture may determine the patient's eventual outcome and may mean the difference between recovery and disability. When the patient is being examined for fracture, the body part is handled gently and as little as possible. Clothing is cut off to visualize the affected body part. Assessment is conducted for pain over or near a bone, swelling (from blood, lymph, and exudate infiltrating the tissue), and circulatory disturbance. The patient is assessed for ecchymosis, tenderness, and crepitation (see Chapter 43). The nurse must remember that the patient may have multiple fractures accompanied by head, chest, spine, or abdominal injuries.

Management

Immediate attention is given to the patient's general condition. Assessment of airway, breathing, and circulation (which includes pulses in the extremities) is conducted. The patient is also evaluated for neurologic or abdominal injuries before the extremity is treated, unless a pulseless extremity is detected.

If a pulseless extremity is identified, repositioning of the extremity to proper alignment is required. If the pulseless extremity involves a fractured hip or femur, **Hare traction** (a portable in-line traction device) may be applied to assist with alignment. If repositioning is ineffective in restoring the pulse, a rapid total-body assessment must be completed, followed by transfer of the patient to the operating room for arteriography and possible arterial repair.

After the initial evaluation has been completed, all injuries identified are evaluated and treated. The fractured body part is inspected. Using a systematic head-to-toe approach, the nurse inspects the entire body, observing for lacerations, swelling, and deformities, including angulation (bending), shortening, rotation, and asymmetry. All peripheral pulses, especially those distal to the fractured extremity, are palpated. The extremity is also assessed for coolness, blanching, and decreased sensation and motor function, which are indicative of injury to the extremity's neurovascular supply.

A splint is applied before the patient is moved. Splinting immobilizes the joint at a site distal and proximal to the fracture, relieves pain, restores or improves circulation, prevents further tissue injury, and prevents a closed fracture from becoming an open one. To splint an extremity, one hand is placed distal to the fracture and some traction is applied while the other hand is placed beneath the fracture for support. The splints should extend beyond the joints adjacent to the fracture. Upper extremities must be splinted in a functional position. If the fracture is open, a moist, sterile dressing is applied.

After splinting, the vascular status of the extremity is checked by assessing color, temperature, pulse, and blanching of the nail bed. In addition, the patient is assessed for neurovascular compromise if pain or pressure is reported. (See Chapter 43 for a complete description of fracture management.)

ENVIRONMENTAL EMERGENCIES

Heat-Induced Illnesses

Heat-induced illnesses may range in severity from mild and self-limiting to life-threatening emergencies. The most serious of these—heat stroke—is an acute medical emergency caused by failure of the heat-regulating mechanisms of the body. The most common cause of heat stroke is nonexertional, prolonged exposure to an environmental temperature of greater than 39.2°C (102.5°F), although a heat index of greater than 35°C (95°F) is associated with increased mortality (Becker & Stewart, 2011). It usually occurs during extended heat waves, especially when they are accompanied by high humidity. Exertional heat stroke is caused by strenuous physical activity that occurs in a hot environment (Cline, Ma, Cydulka, et al., 2012; ENA, 2013).

Chart 72-6 HEALTH PROMOTION

Preventing Heat-Induced Illnesses

- Advise the patient treated for heat-induced illness to avoid immediate re-exposure to high temperatures; hypersensitivity to high temperatures may remain for a considerable time.
- Emphasize the importance of maintaining adequate fluid intake, wearing loose clothing, and reducing activity in hot weather.
- Advise athletes to monitor fluid losses and weight loss during workout activities or exercise and to replace fluids and electrolytes.
- Advise the patient to use a gradual approach to physical conditioning, allowing sufficient time for return to baseline temperature.
- Direct older adult patients living in urban settings with high environmental temperatures to places where air-conditioning is available (e.g., shopping mall, library, church). Fans alone are not adequate to prevent heat-induced illness.
- Advise patients to plan outdoor activities to avoid the hottest part of the day (between 10 AM and 2 PM).

People at risk for nonexertional heat stroke are those not acclimatized to heat, those who are older or very young, those unable to care for themselves, those with chronic and debilitating diseases, and those taking certain medications (e.g., major tranquilizers, anticholinergics, diuretics, beta-blockers) (Cline et al., 2012; ENA, 2013). Exertional heat stroke occurs in healthy individuals during sports or work activities (e.g., exercising in extreme heat and humidity). Hyperthermia results because of inadequate heat loss. Strategies used to prevent heat stroke are reviewed in Chart 72-6.

Less severe forms of heat-induced illnesses include heat exhaustion and heat cramps. The causes of heat exhaustion are the same as for heat stroke. Heat cramps are caused by a loss of electrolytes, typically during strenuous physical activity in a hot environment (ENA, 2013).

Gerontologic Considerations

Most heat-related deaths occur in older adults because their circulatory systems are unable to compensate for stress imposed by heat. Older adults have a decreased ability to perspire as well as a decreased ability to vasodilate and vasoconstrict. They have less subcutaneous tissue, a decreased thirst mechanism, and a diminished ability to concentrate urine to compensate for heat. Many older adults do not drink adequate amounts of fluid, partly because of fear of incontinence, and thus have a greater risk of heat stroke. In addition, many older adults fear being victims of crime, so they tend to keep windows closed even when the temperature and humidity levels are high (ENA, 2013).

Assessment and Diagnostic Findings

Heat stroke, whether the cause is exertional or nonexertional, causes thermal injury at the cellular level, resulting in coagulopathies and widespread damage to the heart, liver, and kidneys. Recent patient history reveals exposure to elevated ambient temperature or excessive exercise during extreme heat. When assessing the patient, the nurse notes the following symptoms: profound central nervous system (CNS) dysfunction (manifested by confusion, delirium, bizarre behavior, coma, seizures); elevated body temperature (40.6°C [105°F] or higher); hot, dry skin; and usually anhidrosis (absence of sweating), tachypnea, hypotension, and tachycardia. The patient with heat exhaustion, on the other hand, may exhibit similarly high body temperatures accompanied by headaches, anxiety, syncope, profuse diaphoresis, gooseflesh, and orthostasis. The cardinal manifestations of heat cramps include muscle cramps, particularly in the shoulders, abdomen, and lower extremities; profound diaphoresis; and profound thirst (ENA, 2013).

Management

The main goal is to reduce the high body temperature as quickly as possible, because mortality in heat stroke or morbid progression to heat stroke with less serious forms of heat-induced illnesses is directly related to the duration of hyperthermia. For the patient with heat stroke, simultaneous treatment focuses on stabilizing oxygenation using the CABs (circulation, *a*irway, and *b*reathing) (formerly referred to as the ABCs) of basic life support. This includes establishing IV access for fluid administration.

After the patient's clothing is removed, the core (internal) temperature is reduced to 39°C (102°F) as rapidly as possible, preferably within 1 hour (Becker & Stewart, 2011). One or more of the following methods may be used as prescribed (Auerbach, 2012):

- Cool sheets and towels or continuous sponging with cool water
- Ice applied to the neck, groin, chest, and axillae while spraying with tepid water
- Cooling blankets
- Immersion of the patient in a cold water bath is the optimal method for cooling (if available).

During cooling procedures, an electric fan is positioned so that it blows on the patient to augment heat dissipation by convection and evaporation. The patient's temperature is constantly monitored with a thermistor placed in the rectum, bladder, or esophagus to evaluate core temperature. Caution is used to avoid hypothermia and to prevent hyperthermia, which may recur spontaneously within 3 to 4 hours. The cooling process should stop at 38°C (100.4°F) in order to avoid iatrogenic hypothermia (ENA, 2013).

Throughout treatment, the patient's status is monitored carefully, including vital signs, ECG findings (for possible myocardial ischemia, myocardial infarction, and dysrhythmias), central venous pressure (CVP), and level of responsiveness, all of which may change with rapid alterations in body temperature. A seizure may be followed by recurrence of hyperthermia. To meet tissue needs exaggerated by the hypermetabolic condition, 100% oxygen is administered. Endotracheal intubation and mechanical ventilation to support failing cardiopulmonary systems may be required.

IV infusion therapy of normal saline or lactated Ringer's solution is initiated as directed to replace fluid losses and maintain adequate circulation. Fluids are administered carefully because of the dangers of myocardial injury from high body temperature and poor kidney function. Cooling redistributes fluid volume from the periphery to the core.

Urine output is also measured frequently, because acute tubular necrosis may occur as a complication of heat stroke from rhabdomyolysis (myoglobin in the urine). Blood specimens are obtained for serial testing to detect bleeding disorders, such as disseminated intravascular coagulation, and for serial enzyme studies to estimate thermal hypoxic injury to the liver, heart, and muscle tissue. Permanent liver, cardiac, and CNS damage may occur.

Additional supportive care may include dialysis for acute kidney injury, antiseizure medications to control seizures, potassium for hypokalemia, and sodium bicarbonate to correct metabolic acidosis. Benzodiazepines such as diazepam (Valium) may be prescribed to suppress seizure activity while a phenothiazine such as chlorpromazine (Thorazine) may be prescribed to suppress shivering (Cline et al., 2012).

Patients with heat exhaustion or heat cramps may be managed less aggressively. These patients should lie supine in a cool environment. Patients with heat exhaustion may require IV fluids but may also take oral fluids, if they are tolerated. Patients with heat cramps are given oral sodium supplements and oral electrolyte solutions (ENA, 2013). Patients who have experienced a heat-induced illness should receive education to prevent another heat-related illness (see Chart 72-6).

Frostbite

Frostbite is trauma from exposure to freezing temperatures and freezing of the intracellular fluid and fluids in the intercellular spaces. It results in cellular and vascular damage. Frostbite can result in venous stasis and thrombosis. Body parts most frequently affected by frostbite include the feet, hands, nose, and ears. Frostbite ranges from first degree (redness and erythema) to fourth degree (full-depth tissue destruction).

Assessment and Diagnostic Findings

A frozen extremity may be hard, cold, and insensitive to touch and may appear white or mottled blue-white. The extent of injury from exposure to cold is not always initially known. The patient history should include environmental temperature, duration of exposure, clothing worn, humidity, and the presence of wet conditions. Protective clothing may partially prevent exposure to cold environments; however, wearing wet socks and exercise/movement may diminish the protective effects of insulation by 45% (Flatt, 2010).

Management

The goal of management is to restore normal body temperature. Constrictive clothing and jewelry that could impair circulation are removed. Wet clothing is removed as rapidly as possible. If the lower extremities are involved, the patient should not be allowed to ambulate.

Controlled yet rapid rewarming is instituted. Frozen extremities are usually placed in a 37°C to 40°C (98.6°F to 104°F) circulating bath for 30- to 40-minute spans. This treatment is repeated until circulation is effectively restored. Early rewarming appears to decrease the amount of ultimate tissue loss. During rewarming, an analgesic for pain is administered as prescribed, because the rewarming process may be very painful. To avoid further mechanical injury, the body part is not handled. Massage is contraindicated.

Once rewarmed, the part is protected from further injury and is elevated to help control swelling. Sterile gauze or cotton is placed between affected fingers or toes to prevent maceration, and a bulky dressing is placed on the extremity. A foot cradle may be used to prevent contact with bedclothes if the feet are involved. Hemorrhagic blebs, which may develop 1 hour to a few days after rewarming, are left intact and not ruptured. Nonhemorrhagic blisters are débrided to decrease the inflammatory mediators found in the blister fluid.

A physical assessment is conducted with rewarming to observe for concomitant injury, such as soft tissue injury, dehydration, alcohol intoxication, or fat embolism. Problems such as hyperkalemia (e.g., from release of potassium in the damaged cells) and hypovolemia, which occur frequently in people with frostbite, are corrected. Risk of infection is also great; therefore, aseptic technique is used during dressing changes, and tetanus prophylaxis is administered as indicated. Nonsteroidal anti-inflammatory drugs (NSAIDs) are prescribed for their anti-inflammatory effects and to control pain.

Additional measures that may be carried out when appropriate after emergency stabilization measures have been instituted include the following:

- Whirlpool bath for the affected body parts to aid circulation and débridement of necrotic tissue to help prevent infection
- Escharotomy (incision through the eschar) to prevent further tissue damage, to allow for normal circulation, and to permit joint motion
- Fasciotomy to treat compartment syndrome

After rewarming, hourly active motion of any affected digits is encouraged to promote maximal restoration of function and to prevent contractures. Discharge instructions also include encouraging the patient to avoid tobacco, alcohol, and caffeine because of their vasoconstrictive effects, which further reduce the already deficient blood supply to injured tissues.

Hypothermia

Hypothermia is a condition in which the core (internal) temperature is 35°C (95°F) or less as a result of exposure to cold or an inability to maintain body temperature in the absence of low ambient temperatures. Urban hypothermia (extreme exposure to cold in an urban setting) is associated with a high mortality rate; older adults, infants, people with concurrent illnesses, and those who are homeless are particularly susceptible. Alcohol ingestion increases susceptibility because it causes systemic vasodilation. Some medications (e.g., phenothiazines) or medical conditions (e.g., hypothyroidism, spinal cord injury) decrease the ability to shiver, hampering the body's innate ability to generate body heat. Heat loss of 2% is normal but increases with exposure. Wet clothing accelerates heat loss, and immersion in cold water increases heat loss by 25% (ENA, 2013). Trauma victims are also at risk for hypothermia resulting from treatment with cold fluids, unwarmed oxygen, and exposure during examination. The patient may also have frostbite, but hypothermia takes precedence in treatment.

Assessment and Diagnostic Findings

Hypothermia leads to physiologic changes in all organ systems. There is progressive deterioration, with apathy, poor judgment, ataxia, dysarthria, drowsiness, pulmonary edema, acid–base abnormalities, coagulopathy, and eventual coma. Shivering may be suppressed at a temperature of less than 32.2°C (90°F), because the body's self-warming mechanisms become ineffective. The heartbeat and blood pressure may be so weak that peripheral pulses become undetectable. Cardiac dysrhythmias may also occur. Other physiologic abnormalities include hypoxemia and acidosis.

Management

Management consists of removal of wet clothing, continuous monitoring, rewarming, and supportive care.

Monitoring

The CABs of basic life support are a priority. The patient's vital signs, CVP, urine output, arterial blood gas levels, blood chemistry determinations (blood urea nitrogen, creatinine, glucose, electrolytes), and chest x-rays are evaluated frequently. Core body temperature is monitored with an esophageal, bladder, or rectal thermistor. Continuous ECG monitoring is performed, because cold-induced myocardial irritability leads to conduction disturbances, especially ventricular fibrillation. An arterial line is inserted and maintained to record blood pressure and to facilitate blood sampling.

Rewarming

Rewarming methods include active internal (core) rewarming and passive (spontaneous) or active external rewarming.

Active internal (core) rewarming methods are used for moderate to severe hypothermia (less than 28°C to 32.2°C [82.5°F to 90°F]) and include cardiopulmonary bypass, warm fluid administration, warm humidified oxygen by ventilator, and warmed peritoneal lavage. Monitoring for ventricular fibrillation as the patient's temperature increases from 31°C to 32°C (88°F to 90°F) is essential.

Passive or active external rewarming is used for mild hypothermia (32.2°C to 35°C [90°F to 95°F]). Passive external rewarming uses over-the-bed heaters to the extremities and increases blood flow to the acidotic, anaerobic extremities. The cold blood from peripheral tissues has high lactic acid levels. As this blood returns to the core, it causes a significant drop in the core temperature (i.e., core temperature afterdrop) and can potentially cause cardiac dysrhythmias and electrolyte disturbances. Active external rewarming uses forced-air warming blankets. Care must be taken to prevent extremity burn from these devices, because the patient may not have effective sensation to feel the burn.

Supportive Care

Supportive care during rewarming includes the following as directed:

- External cardiac compression (typically performed only as directed in patients with temperatures higher than 31°C [88°F])
- Defibrillation of ventricular fibrillation. A patient whose temperature is less than 32°C [90°F] experiences spontaneous ventricular fibrillation if moved or touched. Defibrillation is ineffective in patients with temperatures lower than 31°C (88°F); therefore, the patient must be rewarmed first.

- Mechanical ventilation with positive end-expiratory pressure (PEEP) and heated humidified oxygen to maintain tissue oxygenation
- Administration of warmed IV fluids to correct hypotension and to maintain urine output and core rewarming, as described previously
- Administration of sodium bicarbonate to correct metabolic acidosis if necessary
- Administration of antiarrhythmic medications
- Insertion of an indwelling urinary catheter to monitor urinary output and kidney function

Nonfatal Drowning

Nonfatal drowning is defined as survival for at least 24 hours after submersion that caused a respiratory arrest. The most common consequence is hypoxemia. Drowning is the leading cause of unintentional death in boys 5 through 14 years of age and all children younger than 4 years. An estimated 500,000 drownings occur throughout the world, and, for every death, approximately 4 nonfatal drownings occur (Szpilman, Bierens, Handley, et al., 2012). Drowning and nonfatal drowning can be prevented by avoiding rip currents offshore; approximately 85% of shore drownings involve a rip current. Pool drownings can be prevented by surrounding the pool with fencing and a self-latching/closing gate. It is estimated that with the enforcement of these safety measures, the incidence of drowning would decrease by 50% (Szpilman et al., 2012).

Factors associated with drowning and nonfatal drowning include alcohol ingestion, inability to swim, diving injuries, hypothermia, and exhaustion. The majority of drowning events occur in pools, lakes, and bathtubs. Suicide by drowning rarely occurs in pools and rarely involves alcohol (Auerbach, 2012).

Efforts to save the patient should not be abandoned prematurely. Successful resuscitation with full neurologic recovery has occurred in nonfatal drowning patients after prolonged submersion in cold water. This is possible because of a decrease in metabolic demands and/or the diving reflex. The nonfatal drowning process involves the onset of hypoxia, hypercapnia, bradycardia, and dysrhythmias. If there is a violent struggle associated with the nonfatal drowning episode, exercise-induced acidosis and tachypnea can result in aspiration. Hypoxia and acidosis cause eventual apnea and loss of consciousness. When the victim loses consciousness and makes a final effort to breathe, the terminal gasp occurs. Water then moves passively into the airways prior to death.

After resuscitation, hypoxia and acidosis are the major complications experienced by a person who has experienced nonfatal drowning; immediate intervention in the ED is essential. Resultant pathophysiologic changes and pulmonary injury depend on the type of fluid (fresh or salt water) and the volume aspirated. Freshwater aspiration results in a loss of surfactant and, therefore, an inability to expand the lungs. Saltwater aspiration leads to pulmonary edema from the osmotic effects of the salt within the lungs. If a person survives submersion, acute respiratory distress syndrome (ARDS), resulting in hypoxia, hypercarbia, and respiratory or metabolic acidosis, can occur.

Management

Therapeutic goals include maintaining cerebral perfusion and adequate oxygenation to prevent further damage to vital organs. Cardiopulmonary resuscitation is the factor with the greatest influence on survival. The most important priority in resuscitation is to manage the hypoxia, acidosis, and hypothermia. Prevention and management of hypoxia are accomplished by ensuring an adequate airway and respiration, thus improving ventilation (which helps correct respiratory acidosis) and oxygenation. Arterial blood gases are monitored to evaluate oxygen, carbon dioxide, bicarbonate levels, and pH. These parameters determine the type of ventilatory support needed. The use of endotracheal intubation with PEEP improves oxygenation, prevents aspiration, and corrects intrapulmonary shunting and ventilation–perfusion abnormalities (caused by aspiration of water). If the patient is breathing spontaneously, supplemental oxygen may be administered by mask. However, an endotracheal tube is necessary if the patient does not breathe spontaneously.

Because of submersion, the patient is usually hypothermic. A rectal probe or other core measurement device is used to determine the degree of hypothermia. Prescribed rewarming procedures (e.g., extracorporeal warming, warmed peritoneal dialysis, inhalation of warm aerosolized oxygen, torso warming) are started during resuscitation. The choice of warming method is determined by the severity and duration of hypothermia and available resources. Intravascular volume expansion and inotropic agents are used to treat hypotension and impaired tissue perfusion. ECG monitoring is initiated, because dysrhythmias frequently occur. An indwelling urinary catheter is inserted to measure urine output. Hypothermia and accompanying metabolic acidosis may compromise kidney function. Nasogastric intubation is used to decompress the stomach and to prevent the patient from aspirating gastric contents.

Even if the patient appears healthy, close monitoring continues with serial vital signs, serial arterial blood gas values, ECG monitoring, intracranial pressure assessments, serum electrolyte levels, intake and output, and serial chest x-rays. After a nonfatal drowning, the patient is at risk for complications such as hypoxic or ischemic cerebral injury, ARDS, and life-threatening cardiac arrest. The patient is also at heightened risk for aspiration; vomiting occurs in more than 65% of patients requiring rescue breathing and in 86% of patients requiring CPR (Szpilman et al., 2012).

Decompression Sickness

Decompression sickness, also known as "the bends," occurs in patients who have engaged in diving (lake/ocean diving), high-altitude flying, or flying in commercial aircraft within 24 hours after diving. It occurs relatively infrequently in the United States, but its effects can be hazardous. Being aware of decompression sickness and assessing the patient properly ensures appropriate management and results in decreased morbidity.

Decompression sickness results from formation of nitrogen bubbles that occur with rapid changes in atmospheric pressure. They may occur in joint or muscle spaces, resulting in musculoskeletal pain, numbness, or hypesthesia. More significantly, nitrogen bubbles can become air emboli in the bloodstream and thereby produce stroke, paralysis, or death. Taking a rapid history about the events preceding the symptoms is essential.

Assessment and Diagnostic Findings

To identify decompression sickness, a detailed history is obtained from the patient or diving partner. Evidence of rapid ascent, loss of air in the tank, buddy breathing, recent alcohol intake or lack of sleep, or a flight within 24 hours after diving suggests possible decompression sickness. Some patients describe a perfect dive yet still have the signs and symptoms of decompression sickness, in which case they must receive treatment for the condition.

Signs and symptoms include joint or extremity pain, numbness, hypesthesia, and loss of range of motion. Neurologic symptoms mimicking those of a stroke or spinal cord injury can indicate an air embolus. Cardiopulmonary arrest can also occur in severe cases and is usually fatal. Any neurologic symptoms should be rapidly assessed. All patients with decompression sickness need rapid transfer to a hyperbaric chamber.

Management

A patent airway and adequate ventilation are established, as described previously, and 100% oxygen is administered throughout treatment and transport. A chest x-ray is obtained to identify aspiration, and at least one IV line is started with lactated Ringer's or normal saline solution.

The cardiopulmonary and neurologic systems are supported as needed. If an air embolus is suspected, the head of the bed should be lowered. If the patient's wet clothing is still present, it is removed. The patient is kept warm. Transfer to the closest hyperbaric chamber for treatment is initiated. If air transport is necessary, low-altitude flight (below 1,000 feet) is required. However, the patient who is awake and alert without central neurologic deficits may be able to travel by ground ambulance or by automobile, depending on the severity of symptoms. Throughout treatment, the patient is continually assessed, and changes are documented. If aspiration is suspected, antibiotic agents and other treatment may be prescribed. The addition of NSAIDs and/or the use of heliox in the hyperbaric chamber may decrease the number of recompression treatments needed for the more severely affected patients (Bennett, Lehm, Mitchell, et al., 2010).

Animal and Human Bites

Bites are a common reason for visits to the ED. Dog bites constitute 80% to 90% of these bites and are responsible for the majority of deaths from bites by a nonvenomous animal (Cline et al., 2012). Cat bites have a high risk of infection because of the presence of *Pasteurella* in their saliva. All animal bites must be reported to public health authorities, which must provide follow-up screening of the offending animal for rabies. If the animal cannot be located and rabies vaccination verified, rabies prophylaxis for the person who has been bitten must be instituted (ENA, 2013).

Human bites are frequently associated with rapes, sexual assaults, or other forms of battery. The human mouth contains more bacteria than that of most other animals, so a high risk of bite-related infection exists. Depending on the circumstances surrounding the event, the victim may delay seeking treatment. The ED nurse should inspect any bitten tissue for pus, erythema, or necrosis. A health care provider should take photographs, which can be used as evidence in criminal and legal proceedings. Guidelines for collecting forensic evidence for photographing with and without a measuring device should be followed. Cleansing with soap and water is then necessary, followed by the administration of antibiotics and tetanus toxoid as prescribed (Cline et al., 2012).

Snakebites

Venomous (poisonous) snakes caused more than 2,800 bites in the United States in 2008 (Moriarty, Dryer, Repolgie, et al., 2012). Across the globe, more than 50,000 snakebites occur each year; approximately 7,000 of these are venomous (Ahmed, Ahmed, Nadeem, et al., 2008). Children between 1 and 9 years of age are the most likely victims. The greatest number of bites occurs during the daylight hours and early evening of the summer months. The most frequent poisonous snakebite in the United States occurs from pit vipers (Crotalidae). The most common site is the upper extremity (ENA, 2013). Of these bites, only 20% to 25% result in **envenomation** (injection of a poisonous material by sting, spine, bite, or other means). Typically, even a venomous snake will produce a dry bite (little to no venom) most of the time. Venomous snakebites are medical emergencies.

Nineteen different species of venomous snakes are found in various regions within the United States. Nurses should be familiar with the types of snakes common to the geographic region in which they practice. However, the exotic pet industry sells atypical snakes as "pets." Because of this, venomous snakes such as cobras and asps may be found outside of their typical place of residence (Lubich & Krenzelok, 2009).

Clinical Manifestations

Snake venom consists primarily of proteins and has a broad range of physiologic effects. It may affect multiple organ systems, especially the neurologic, cardiovascular, and respiratory systems.

Classic clinical signs of envenomation are edema, ecchymosis, and hemorrhagic bullae, leading to necrosis at the site of envenomation. Symptoms include lymph node tenderness, nausea, vomiting, numbness, and a metallic taste in the mouth. Without decisive treatment, these clinical manifestations may progress to include fasciculations, hypotension, paresthesias, seizures, and coma (Moriarty et al., 2012).

Management

Initial first aid at the site of the snakebite includes having the person lie down, removing constrictive items such as rings, providing warmth, cleansing the wound, covering the wound with a light sterile dressing, and immobilizing the injured body part below the level of the heart. Airway, breathing, and circulation are the priorities of care. Ice, incision and suction, or a tourniquet is *not* applied. Tetanus and analgesia should be administered as necessary. Initial evaluation in the

ED is performed quickly and includes information about the following:

- Whether the snake was venomous or nonvenomous; discourage bringing the snake for identification—even a dead snake's venom is poisonous). Do *not* handle any snake brought to the ED. If the snake is transported to the ED, caution should be taken because the snake is frequently in a stunned, not dead, state.
- Where and when the bite occurred and the circumstances of the bite
- Sequence of events, signs and symptoms (fang punctures, pain, edema, and erythema of the bite and nearby tissues)
- Severity of poisonous effects. Call the local poison control phone number to gain access to information about an exotic snakebite presentation and management, as necessary. The poison control center may also be able to assist with retrieving antivenin for these particular species (Lubich & Krenzelok, 2009).
- Vital signs
- Circumference of the bitten extremity or area at several points. The circumference of the extremity that was bitten is compared with the circumference of the opposite extremity.
- Laboratory data (complete blood count, urinalysis, and coagulation studies)

The course and prognosis of snakebite injuries depend on the kind and amount of venom injected; where on the body the bite occurred; and the general health, age, and size of the patient. There is no one specific protocol for treatment of snakebites. Generally, ice, tourniquets, heparin, and corticosteroids are not used during the acute stage. Corticosteroids are contraindicated in the first 6 to 8 hours after the bite because they may depress antibody production and hinder the action of **antivenin** (antitoxin manufactured from the snake venom and used to treat snakebites).

Parenteral fluids may be used to treat hypotension. If vasopressors are used to treat hypotension, their use should be short term. Surgical exploration of the bite is rarely indicated. Typically, the patient is observed closely for at least 6 hours. The patient is *never* left unattended.

Administration of Antivenin (Antitoxin)

Although envenomation is rare, it can occur with snakebites. An assessment of progressive signs and symptoms is essential before considering administration of antivenin, which is most effective if administered within 4 hours and no greater than 12 hours after the snakebite. The decision to administer antivenin depends on worsening tissue injury and evidence of systemic and coagulopathic symptoms. Rattlesnakes are more likely to cause coagulation abnormalities as well as more systemic effects. Coagulation abnormalities are not restricted to severe envenomation (Moriarty et al., 2012).

The most readily available antivenin in the United States is Crotalidae polyvalent immune Fab antivenom (FabAV or CroFab) (Auerbach, 2012). The dose depends on the type of snake and the estimated severity of the bite. Indications for antivenin depend on the progression of symptoms, including coagulopathy and systemic reaction.

Crotalidae polyvalent immune Fab antivenom does not require pretesting (i.e., skin sensitivity screening for an allergic reaction; see Chapter 38), albeit monitoring for a hypersensitivity reaction is still necessary. Outside of the United States, however, other antivenin formulas that may be commercially available may still result in severe serum sickness. If the dose exceeds 10 vials, serum sickness will most likely occur. Serum sickness is a type of hypersensitivity response that results in fever, arthralgias, pruritus, lymphadenopathy, and proteinuria and can progress to neuropathies (Auerbach, 2012). However, FabAV must be administered cautiously to patients receiving anticoagulation therapy. Administration of FabAV may result in a recurring coagulopathy. The dosage and administration of FabAV are different from previously manufactured types of antivenin and should be reviewed carefully before the medication is given.

Before administering antivenin and every 15 minutes thereafter, the circumference of the affected part is measured. Premedication with diphenhydramine (Benadryl) or cimetidine (Tagamet) may be indicated, because these antihistamines may decrease the allergic response to antivenin. Antivenin is administered as an IV infusion whenever possible, although intramuscular administration can be used.

Depending on the severity of the snakebite, the antivenin is diluted in 500 to 1,000 mL of normal saline solution. The infusion is started slowly, and the rate is increased after 10 minutes if there is no reaction. The total dose should be infused during the first 4 to 6 hours after the bite. The initial dose is repeated until symptoms decrease, after which time the circumference of the affected part should be measured every 30 to 60 minutes for the next 48 hours to detect symptoms of compartment syndrome (swelling, loss of pulse, increased pain, and paresthesias).

The most common cause of allergic reaction to the antivenin is too-rapid infusion, although a recent study demonstrated that the rate of infusion did not affect the degree of severity of systemic hypersensitivity (Isbister, Shahmy, Mohamed, et al., 2012). Reactions may consist of a feeling of fullness in the face, urticaria, pruritus, malaise, and apprehension. These symptoms may be followed by tachycardia, shortness of breath, hypotension, and shock. In this situation, the infusion should be stopped immediately and IV diphenhydramine administered. Vasopressors are used for patients in shock, and resuscitation equipment must be on standby while antivenin is infusing.

Spider Bites

There are two venomous spiders found in the United States that may interact with humans: the brown recluse and the black widow. Both are usually found in dark places such as closets, woodpiles, and attics, as well as in shoes (ENA, 2013).

Brown recluse spider bites are painless. Systemic effects such as fever and chills, nausea and vomiting, malaise, and joint pain develop within 24 to 72 hours. The site of the bite may appear reddish to purple in color within 2 to 8 hours after the bite. Necrosis occurs in the next 2 to 4 days in approximately 10% of cases. The center of the bite may become necrotic, and surgical débridement may be necessary. Wound care consists of cleansing with soap and water, and hyperbaric oxygen treatments may be helpful. Most wounds heal within 2 to 3 months (Cline et al. 2012; ENA, 2013).

Black widow spider bites feel like pinpricks. Systemic effects usually occur within 30 minutes—much more rapidly than with brown recluse spider bites. Signs and symptoms include abdominal rigidity, nausea and vomiting, hypertension, tachycardia, and paresthesias. Severe pain also develops within 60 minutes and increases over 1 to 2 days. Treatment involves application of ice to the site to decrease swelling and discomfort, along with elevation and assessment of tetanus immunization status. Analgesic agents and benzodiazepines may relieve muscle spasms. Cardiopulmonary monitoring is essential. Antivenin is effective for black widow spider bites. This antivenin is horse serum based; therefore, testing for sensitivity must be performed prior to administration (Auerbach, 2012; Cline et al., 2012).

Tick Bites

Tick bites are common in many areas of the United States, and they usually occur in grassy or wooded areas. It is important to learn the place where the bite occurred as well as the location of the bite on the body. The tick bite itself is not usually the problem; rather, it is the pathogen transmitted by the tick that can cause serious disease. Ticks can carry diseases such as Rocky Mountain spotted fever, tularemia, and Lyme disease.

Ticks transmit pathogens through their saliva; therefore, the earlier the tick is removed, the better the prognosis. The tick should be removed with tweezers using a straight upward pull (Fig. 72-4), and the patient should be informed of the signs and symptoms of diseases carried by ticks, especially if the patient lives in an area endemic for tick-related diseases (e.g., Lyme disease) (Cline et al., 2012; ENA, 2013).

Lyme disease has three stages. Stage I presents with a classic "bull's-eye" rash (i.e., erythema migrans) that typically can be found in the axilla, groin, or thigh area and that appears within 4 weeks after the tick bite, with a peak manifestation time of 7 days after the bite. Classically, this rash is at least 5 cm in diameter with bright red borders. It is accompanied by flulike signs and symptoms that may include chills, fever, myalgia, fatigue, and headache. Without treatment, the rash subsides within 3 to 4 weeks. However, the rash and flulike manifestations can be significantly reduced within days if prompt treatment with antibiotic agents (e.g., doxycycline [Vibramycin]) is initiated. If antibiotics are not administered, stage II Lyme disease may present within 4 to 10 weeks following the tick bite and may manifest with joint pain, memory loss, poor motor coordination, and meningitis. Stage III

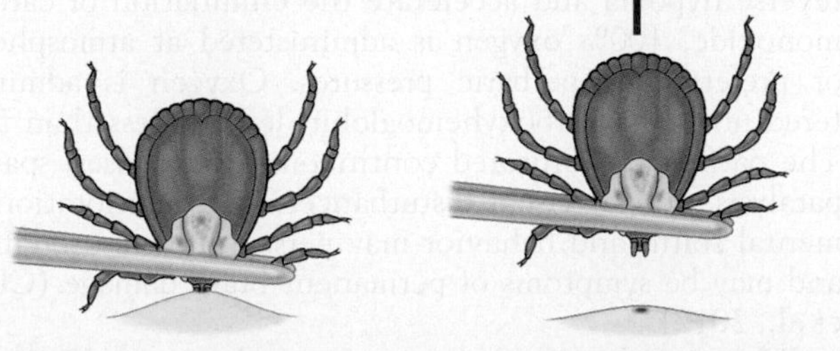

FIGURE 72-4 • Removal of tick with tweezers using a straight upward pull.

can begin anywhere from weeks to more than a year after the bite and has serious long-term chronic sequelae, including arthritis, neuropathy, myalgia, and myocarditis.

POISONING

A poison is any substance that, when ingested, inhaled, absorbed, applied to the skin, or produced within the body in relatively small amounts, injures the body by its chemical action. Poisoning from inhalation and ingestion of toxic materials, both intentional and unintentional, constitutes a major health hazard and an emergency situation. Emergency treatment is initiated with the following goals:

- Removal or inactivation of the poison before it is absorbed
- Provision of supportive care in maintaining vital organ function
- Administration of a specific antidote to neutralize a specific poison
- Implementation of treatment that hastens the elimination of the absorbed poison

Ingested (Swallowed) Poisons

Swallowed poisons may be corrosive. **Corrosive poisons** include alkaline and acid agents that can cause tissue destruction after coming in contact with mucous membranes. Alkaline products include lye, drain cleaners, toilet bowl cleaners, bleach, nonphosphate detergents, oven cleaners, and button batteries (batteries used to power watches, calculators, hearing aids, or cameras). Acid products include toilet bowl cleaners, pool cleaners, metal cleaners, rust removers, and battery acid.

Control of the airway, ventilation, and oxygenation are essential. In the absence of cerebral or renal damage, the patient's prognosis depends largely on successful management of respiration and circulation. Measures are instituted to stabilize cardiovascular and other body functions. ECG, vital signs, and neurologic status are monitored closely for changes. Shock may result from the cardiodepressant action of the substance ingested, from venous pooling in the lower extremities, or from reduced circulating blood volume resulting from increased capillary permeability (see Chapter 14 for further discussion of management of shock). An indwelling urinary catheter is inserted to monitor kidney function. Blood specimens are obtained to determine the concentration of drug or poison.

Efforts are made to determine what substance was ingested; the amount; the time since ingestion; signs and symptoms, such as pain or burning sensations, any evidence of redness or burn in the mouth or throat, pain on swallowing or an inability to swallow, vomiting, or drooling; age and weight of the patient; and pertinent health history.

Quality and Safety Nursing Alert

The local poison control center should be called if an unknown toxic agent has been taken or if it is necessary to identify an antidote for a known toxic agent.

Measures are instituted to remove the toxin or decrease its absorption. If there is a specific chemical or physiologic antagonist (antidote), it is administered as early as possible to reverse or diminish the effects of the toxin. If this measure is ineffective, procedures may be initiated to remove or dilute the ingested substance. These procedures include administration of multiple doses of activated charcoal, dialysis, or hemoperfusion. Hemoperfusion involves detoxification of the blood by processing it through an extracorporeal circuit and an adsorbent cartridge containing charcoal or resin, after which the cleansed blood is returned to the patient.

The patient who has ingested a corrosive poison, which can be a strong acid or alkaline substance, is given water or milk to drink for dilution. However, dilution is not attempted if the patient has acute airway edema or obstruction; potential for vomiting; or if there is clinical evidence of esophageal, gastric, or intestinal burn or perforation. The following gastric emptying procedures may be used as prescribed:

- Gastric lavage for the obtunded patient is only useful within 1 hour of ingestion, for sustained release substances, or massive life-threatening amounts of a substance; however, complications of aspiration and stomach or esophageal perforation outweigh its usefulness. If performed, gastric aspirate is saved and sent to the laboratory for testing (toxicology screens).
- Activated charcoal administration if the poison is one that is absorbed by charcoal; given orally or by nasogastric tube, it is effective in small intermittent doses to decrease vomiting. It should be diluted as a slurry so that it is easier to drink or pass through the nasogastric tube. Activated charcoal absorbs most commonly ingested poisons except corrosives, heavy metals and hydrocarbons, iron, and lithium.

Cathartics, which had traditionally accompanied the use of activated charcoal, are now rarely indicated because they can result in severe electrolyte imbalances, diarrhea, and hypovolemia (ENA, 2013). Furthermore, syrup of ipecac to induce vomiting in the alert patient is no longer recommended due to the risk of aspiration and should *never* be used with corrosive poisons or with petroleum distillates (e.g., lubricating oil, fuel oil).

Quality and Safety Nursing Alert

Vomiting is never induced after ingestion of caustic substances (acid or alkaline) or petroleum distillates.

Throughout detoxification, the patient's vital signs, CVP, and fluid and electrolyte balance are monitored closely. Hypotension and cardiac dysrhythmias are possible. Seizures are also possible because of CNS stimulation from the poison or from oxygen deprivation. If the patient complains of pain, analgesic agents are administered cautiously. Severe pain causes vasomotor collapse and reflex inhibition of normal physiologic functions.

After the patient's condition has stabilized and discharge is imminent, written material should be given to the patient indicating the signs and symptoms of potential problems related to the poison ingested and signs or symptoms requiring evaluation by a health care provider. If poisoning was determined to be a suicide or self-harm attempt, a psychiatric consultation should be requested before the patient is discharged. In cases of inadvertent poison ingestion, poison prevention and home poison-proofing instructions should be provided to the patient and family.

Carbon Monoxide Poisoning

Carbon monoxide poisoning may occur as a result of industrial or household incidents or attempted suicide. It is implicated in more deaths than any other toxin except alcohol. Carbon monoxide exerts its toxic effect by binding to circulating hemoglobin and thereby reducing the oxygen-carrying capacity of the blood. Hemoglobin absorbs carbon monoxide 200 times more readily than it absorbs oxygen. Carbon monoxide–bound hemoglobin, called **carboxyhemoglobin**, does not transport oxygen.

Clinical Manifestations

Because the CNS has a critical need for oxygen, CNS symptoms predominate with carbon monoxide toxicity. A person with carbon monoxide poisoning may appear intoxicated (from cerebral hypoxia). Other signs and symptoms include headache, muscular weakness, palpitation, dizziness, and confusion, which can progress rapidly to coma. Skin color, which can range from pink or cherry-red to cyanotic and pale, is not a reliable sign. Pulse oximetry may reveal a high hemoglobin saturation, which may be deceiving, since the hemoglobin molecule is saturated with carbon monoxide rather than oxygen (ENA, 2013).

Management

Exposure to carbon monoxide requires immediate treatment. Goals of management are to reverse cerebral and myocardial hypoxia and to hasten elimination of carbon monoxide. Whenever a patient inhales a poison, the following general measures apply:

- Carry the patient to fresh air immediately; open all doors and windows.
- Loosen all tight clothing.
- Initiate traditional cardiopulmonary resuscitation, if required; administer 100% oxygen.
- Prevent chilling; wrap the patient in blankets.
- Keep the patient as quiet as possible.
- Do not give alcohol in any form or permit the patient to smoke.

In addition, for the patient with carbon monoxide poisoning, carboxyhemoglobin levels are analyzed on arrival at the ED and before treatment with oxygen if possible. To reverse hypoxia and accelerate the elimination of carbon monoxide, 100% oxygen is administered at atmospheric or preferably hyperbaric pressures. Oxygen is administered until the carboxyhemoglobin level is less than 5%. The patient is monitored continuously. Psychoses, spastic paralysis, ataxia, visual disturbances, and deterioration of mental status and behavior may persist after resuscitation and may be symptoms of permanent brain damage (Cline et al., 2012).

When unintentional carbon monoxide poisoning occurs, the health department should be contacted so that the dwelling or building in question can be inspected. A psychiatric consultation is warranted if it has been determined that the poisoning was a suicide attempt.

Skin Contamination Poisoning (Chemical Burns)

Skin contamination injuries from exposure to chemicals are challenging because of the large number of possible offending agents with diverse actions and metabolic effects. The severity of a chemical burn is determined by the mechanism of action, the penetrating strength and concentration, and the amount and duration of exposure of the skin to the chemical.

The skin should be drenched immediately with running water from a shower, hose, or faucet, except in the case of lye and white phosphorus, which should be brushed off the skin dry (Cline et al., 2012).

> ***Quality and Safety Nursing Alert***
>
> **Water should not be applied to burns from lye or white phosphorus because of the potential for an explosion or for deepening of the burn. All evidence of these chemicals should be brushed off the patient before any flushing occurs.**

The skin should be flushed with a constant stream of water as the patient's clothing is removed. The skin of health care personnel assisting the patient should be appropriately protected if the burn is extensive or if the agent is significantly toxic or still present. Prolonged lavage with generous amounts of tepid water is important.

Attempts to determine the identity and characteristics of the chemical agent are necessary in order to specify future treatment. The standard burn treatment appropriate for the size and location of the wound (antimicrobial treatment, débridement, tetanus prophylaxis, antidote administration as prescribed) is instituted (Cline et al., 2012) (see Chapter 62). The patient may require plastic surgery for further wound management. The patient is instructed to have the affected area reexamined at 24 and 72 hours and in 7 days because of the risk of underestimating the extent and depth of these types of injuries.

Food Poisoning

Food poisoning is a sudden illness that occurs after ingestion of contaminated food or drink. Botulism is a serious form of food poisoning that requires continual surveillance (see Table 73-6 in Chapter 73). Assessment questions for patients with food poisoning are discussed in Chart 72-7.

The key to treatment is determining the source and type of food poisoning. If possible, the suspected food should be brought to the medical facility and a history obtained from the patient or family.

Food, gastric contents, vomitus, serum, and feces are collected for examination. The patient's respirations, blood pressure, level of consciousness, CVP (if indicated), and muscular activity are monitored closely. Measures are instituted to support the respiratory system. Death from respiratory paralysis can occur with botulism, fish poisoning, and some other food poisonings.

Because large volumes of electrolytes and water are lost by vomiting and diarrhea, fluid and electrolyte status should be assessed. Severe vomiting produces alkalosis, and severe diarrhea produces acidosis. Hypovolemic shock may also occur from severe fluid and electrolyte losses. The patient is assessed for signs and symptoms of fluid and electrolyte imbalances, including lethargy, rapid pulse rate, fever, oliguria, anuria, hypotension, and delirium. Baseline weight and serum electrolyte levels are obtained for future comparisons.

Measures to control nausea are also important to prevent vomiting, which could exacerbate fluid and electrolyte imbalances. An antiemetic medication is administered parenterally as prescribed if the patient cannot tolerate fluids or medications by mouth (Cline et al., 2012). For mild nausea, the patient is encouraged to take sips of weak tea, carbonated drinks, or tap water. After nausea and vomiting subside, clear liquids are usually prescribed for 12 to 24 hours, and the diet is gradually progressed to a low-residue, bland diet.

Chart 72-7 ASSESSMENT
Food Poisoning

Use the following questions to elicit information about the circumstances surrounding the possibility of food poisoning:

- How soon after eating did the symptoms occur? (Immediate onset suggests chemical, plant, or animal poisoning.)
- What was eaten in the previous meal? Did the food have an unusual odor or taste? (Most foods causing bacterial poisoning *do not* have unusual odor or taste.)
- Did anyone else become ill from eating the same food?
- Did vomiting occur? What was the appearance of the vomitus?
- Did diarrhea occur? (Diarrhea is usually absent with botulism and with shellfish or other fish poisoning.)
- Are any neurologic symptoms present? (These occur in botulism and in chemical, plant, and animal poisoning.)
- Does the patient have a fever? (Fever is characteristic in salmonella, ingestion of fava beans, and some fish poisoning.)

SUBSTANCE ABUSE

Substance abuse is the misuse of specific substances, such as drugs or alcohol, to alter mood or behavior. Drug abuse is the use of drugs for other than legitimate medical purposes. People who abuse drugs often take a variety of drugs simultaneously (such as alcohol, barbiturates, opioids, and tranquilizers), and the combination may have additive and addictive effects.

"Rave parties" are large-scale parties attended by hundreds of people involved in illicit drug use. At these events, one of the most commonly used drugs is 3,4-methylenedioxymethamphetamine (MDMA), or Ecstasy, a methamphetamine-based drug that users believe produces a "harmless high." ED nurses should be aware of rave parties in their geographic area so that they can prepare for a potential influx of patients who abuse this drug. Others may combine Ecstasy with sildenafil (Viagra); this drug combination is nicknamed "sextasy."

Spice is a synthetic cannabinoid sold commercially as a smoking mixture under the names "spice," "incense," or "K2." Its chemical structure and effects are similar to marijuana, targeting the same receptor sites in the brain. Spice is sold with variable concentrations and unregulated potency (Jerry, Collins, & Streem, 2012).

Bath salts are synthetic stimulants similar to Ecstasy known as "mephedrone," "drone," or "MCAT." Their effects are similar to amphetamines, MDMA, and cocaine. This designer drug had been legally sold in the United States until 2010, under the premise that it was "not for human consumption." Although bath salts are most commonly swallowed or snorted, they may also be smoked or injected; the method of intake affects the severity and duration of effects (Table 72-1). The structural formula 3,4-methylenedioxypyrovalerone (MDPV) is the most common type of bath salt abused (Jerry et al., 2012).

Abuse of various inhalants (see Table 72-1) has also increased in popularity; these products generally result more often in cravings than withdrawal when their use is stopped. The method of inhalation varies with the product chosen and requires several deep inhalations to reach euphoria. Methods include sniffing or snorting by directly inhaling the fumes. "Bagging" (from a bag) or "huffing" (from a rag or cloth) provide the greatest concentration; "dusting" is another method that delivers the inhalant by directly spraying it into the nostrils. Long-term use results in cortical atrophy and brain stem dysfunction, in addition to cardiomyopathy and emphysemalike abnormalities of the lung. Significant others or parents may report that the patient has had poor school or work performance or attendance, weight loss, poor hygiene, fatigue, nosebleeds, and decreased appetite (Baydala, 2010; Kumar, Grover, Kulhara, et al., 2008).

Clinical manifestations vary with the substance used, but the underlying principles of management are essentially the same. Table 72-1 identifies commonly abused drugs, listing their clinical manifestations and therapeutic management. Treatment goals for a patient with a drug overdose are to support the respiratory and cardiovascular functions, to enhance clearance of the agent, and to provide for safety of the patient and staff. People who abuse IV/injection drugs are at increased risk for HIV infection, acquired immunodeficiency syndrome, hepatitis B and C, and tetanus.

Acute Alcohol Intoxication

Alcohol is a psychotropic drug that affects mood, judgment, behavior, concentration, and consciousness. Many heavy drinkers are young adults or people older than 60 years. There is a high prevalence of alcoholism among ED patients. Of the patients visiting EDs in the United States in 2010, 6.5% of these visits were related to alcohol use (ENA, 2013). Because patients who abuse alcohol return frequently to the ED, they often frustrate and tax the patience of the health care professionals who care for them. The CDC advocates routine screening for alcohol abuse in all outpatient settings, including EDs. Therefore, screenings, brief interventions, and referral to treatment (SBIRT) for patients presenting with suspected alcohol abuse are recommended. All level I and II verified trauma centers are required to provide this service. SBIRT is considered cost-effective in saving quality of life-years lost and preventing the morbid consequences of continued alcohol abuse (ENA, 2013).

Alcohol, or ethanol, is a multisystem toxin and CNS depressant that causes drowsiness, impaired coordination, slurring of speech, sudden mood changes, aggression, belligerence, grandiosity, and uninhibited behavior. In excess, it can also cause stupor and eventually coma and death (i.e., alcohol poisoning). Increasingly, underage minors and college students arrive at the ED with alcohol poisoning from binge drinking.

In the ED, the patient who is intoxicated with alcohol or who presents with alcohol poisoning is assessed for head injury, hypoglycemia (which mimics intoxication), and other health problems. Possible nursing diagnoses include ineffective breathing pattern related to CNS depression and ineffective impulse control related to severe intoxication from alcohol.

Treatment involves detoxification of the acute poisoning, recovery, and rehabilitation. Commonly, the patient uses mechanisms of denial and defensiveness. The nurse should approach the patient in a nonjudgmental manner, using a firm, consistent, accepting, and reasonable attitude. Speaking in a calm and slow manner is helpful because alcohol interferes with thought processes. If the patient appears intoxicated, hypoxia, hypovolemia, and neurologic impairment must be ruled out before it is assumed that the patient is intoxicated. Typically, a blood specimen is obtained for analysis of the blood alcohol level.

If drowsy, the patient should be allowed to sleep off the state of alcoholic intoxication. During this time, maintenance of a patent airway and observation for symptoms of CNS depression are essential. The patient should be undressed and kept warm with blankets. On the other hand, if the patient is noisy or belligerent, sedation may be necessary. If sedation is used, the patient should be monitored carefully for hypotension and decreased level of consciousness.

In addition, the patient is examined for alcohol withdrawal delirium and also for injuries and organic disease (such as head injury, seizures, pulmonary infections, hypoglycemia, and nutritional deficiencies) that may be masked by alcoholic intoxication. People with alcoholism suffer more injuries than the general population. Acute alcohol intoxication is the cause of trauma for many nonalcoholic patients as well. Pulmonary infections are also more common in patients with alcoholism, resulting from respiratory depression, an impaired defense system, and a tendency toward aspiration of gastric contents. The patient may show little increase in temperature or WBC count. The patient may be hospitalized or admitted to a detoxification center in an effort to examine problems underlying the substance abuse (ENA, 2013).

Alcohol Withdrawal Syndrome/ Delirium Tremens

Alcohol withdrawal syndrome is an acute toxic state that occurs as a result of sudden cessation of alcohol intake after a bout of heavy drinking or, more typically, after prolonged intake of alcohol. Severity of symptoms depends on how much

(continues on page 2145)

TABLE 72-1 Emergency Management of Patients With Drug Overdose

Drug	Clinical Manifestations	Therapeutic Management
Cocaine *Routes may include:* • Intranasally ("snorting")—inhaled into nostrils through straws • By smoking ("freebasing")—cocaine hydrochloride dissolved in ether to yield a pure cocaine alkaloid base ("crack," "rocks"); smoking in a small pipe delivers large quantities of cocaine to lungs. • IV • Polysubstance (cocaine and heroin)	Cocaine is a CNS stimulant that can increase heart rate and blood pressure and cause hyperpyrexia, seizures, increased energy, agitation, aggression, and ventricular dysrhythmias. It produces intense euphoria, then anxiety, sadness, insomnia, and sexual indifference; cocaine hallucinations with delusions; psychosis with extreme paranoia and ideas of persecution; and hypervigilance. Chronic psychotic symptoms may persist. Overall psychotic symptoms are short-lived compared to methamphetamines.	1. Maintain airway and provide respiratory support. 2. Control seizures. 3. Monitor cardiovascular effects; have antiarrhythmic drugs and defibrillator available. 4. Treat for hyperthermia. 5. If cocaine was ingested, evacuate stomach contents and use activated charcoal to treat. Whole bowel irrigation may be necessary to treat body packers ("mules"). 6. Refer for psychiatric evaluation and treatment in an inpatient unit that eliminates access to the drug. Include drug rehabilitation counseling.
Opioids Heroin Opium or paregoric Morphine, codeine, semisynthetic derivatives: oxycodone (OxyContin), methadone, meperidine (Demerol), tramadol (Ultram), fentanyl (Sublimaze)	Acute intoxication (overdose) Pinpoint pupils (may be dilated with severe hypoxia); decreased blood pressure Marked respiratory depression/arrest Pulmonary edema Stupor → coma Seizures Fresh needle marks along course of any superficial vein; skin abscesses (from "popping")	1. Support respiratory and cardiovascular functions. 2. Establish an IV line; obtain blood for chemical and toxicologic analysis. Patient may be given bolus of glucose to eliminate possibility of hypoglycemia. 3. Administer narcotic antagonist (naloxone hydrochloride IV, IM [Narcan]) as prescribed to reverse severe respiratory depression and coma. 4. Continue to monitor level of responsiveness and respirations, pulse, and blood pressure. Duration of action of naloxone hydrochloride is shorter than that of heroin; repeated dosages may be necessary. 5. Send urine for analysis; opioids can be detected in urine. 6. Obtain an ECG. 7. Do not leave patient unattended; he or she may lapse back into coma rapidly. Clinical status may change from minute to minute. Hemodialysis may be indicated for severe drug intoxication. Activated charcoal may be considered if opioids were taken orally and if the patient is alert. 8. Monitor for pulmonary edema, which is frequently seen in patients who abuse/overdose on narcotics. 9. Refer patient for psychiatric and drug rehabilitation evaluation before discharge.
Barbiturates Pentobarbital (Nembutal), secobarbital (Seconal), amobarbital (Amytal), gamma-hydroxybutyrate (GHB, "liquid Ecstasy")	*Acute intoxication (may mimic alcohol intoxication):* • Respiratory depression • Flushed face • Decreased pulse rate; decreased blood pressure • Increasing nystagmus • Depressed deep tendon reflexes • Decreasing mental alertness • Difficulty in speaking • Poor motor coordination • Coma, death *GHB:* • Sexual disinhibition • Amnesia, myoclonus, agitation • Overdoses when mixed with alcohol	1. Maintain airway and provide respiratory support. 2. Endotracheal intubation or tracheostomy is considered if there is any doubt about the adequacy of airway exchange. a. Check airway frequently. b. Perform suctioning as necessary. 3. Support cardiovascular and respiratory functions; most deaths result from respiratory depression or shock. 4. Start infusion through large-gauge needle or IV catheter to support blood pressure; coma and dehydration result in hypotension and respond to infusion of IV fluids with elevation of blood pressure. 5. Evacuate stomach contents or lavage if within 1 hour of ingestion to prevent absorption; repeated doses of activated charcoal may be administered. 6. Assist with hemodialysis for severely overdosed patient. 7. Maintain neurologic and vital sign flow sheet. 8. Patient awakening from overdose may demonstrate combative behavior. 9. Refer for psychiatric and drug rehabilitation consultation to evaluate suicide potential and drug abuse.

(continues on page 2142)

TABLE 72-1 Emergency Management of Patients With Drug Overdose (continued)

Drug	Clinical Manifestations	Therapeutic Management
Inhalants Amyl nitrate Freon Propane Trichloroethylene Gasoline Perchloroethylene Toluene (metallic paint spray) Helium Canned air Hand sanitizer *Routes may include:* • Sniffing/snorting—direct inhalation of fumes • "Bagging"—sniff from a bag • "Huffing"—sniff from a rag/cloth • "Dusting"—direct spray into the nostrils	Effects mimic those of alcohol, with dizziness and imbalance Euphoria, headache, disinhibition, altered level of consciousness to coma Renal, hepatic, and cardiac toxicity Aplastic anemia Fetal growth retardation Respiratory depression, arrest from CNS depression Vasodilation Nosebleeding Circumoral red spots/rash Air embolus	1. Provide airway support, ventilation, and oxygen. 2. Treat cardiac dysrhythmias and hypotension. 3. Provide advanced cardiac life support as needed. 4. Monitor for profound hypotension when amyl nitrate is combined with MDMA and sildenafil.
Amphetamine-Type Drugs (pep pills, "uppers," "speed," "crystal meth") amphetamine (Benzedrine) dextroamphetamine (Dexedrine) methamphetamine (Desoxyn, "speed") 3,4-methylenedioxymethamphetamine (MDMA) ("Ecstasy," "Adam")* 3,4-methylenedioxy-*N*-ethylamphetamine (MDEA) ("Eve") 3,4-methylenedioxyamphetamine (MDA); methylphenidate (Ritalin) "ice," "rocks," "crystal meth" 3,4-methylenedioxypyrovalerone (MDPV) or 4-methylmethcathinone (mephedrone); "Bath salts" (synthetic stimulant)	Nausea, vomiting, anorexia, palpitations, tachycardia, increased blood pressure, tachypnea, anxiety, nervousness, diaphoresis, mydriasis Repetitive or stereotyped behavior Irritability, insomnia, agitation Visual misperceptions, auditory hallucinations Fearfulness, anxiety, depression, hostility, paranoia Hyperactivity, rapid speech, euphoria, hyperalertness Decreased inhibition Seizures, coma, hyperthermia, cardiovascular collapse, rhabdomyolysis MDMA is both a hallucinogenic and stimulant. MDPV and mephedrone effects last >24 hours.	1. Provide airway support, ventilation, cardiac monitoring; insert IV line. 2. Use GI evacuation in cases of oral overdose; activated charcoal, gastric lavage if within 1 hour of ingestion. 3. Keep in calm, cool, quiet environment; elevated temperature potentiates amphetamine toxicity. Maintain normothermia, cooling the patient as necessary. 4. Small doses of diazepam (Valium) (IV) or haloperidol (Haldol) as prescribed for CNS and muscular hyperactivity 5. Administer appropriate pharmacologic therapy as prescribed for severe hypertension and ventricular dysrhythmias. 6. Treat seizures with benzodiazepines (e.g., diazepam) as prescribed. 7. Treat sympathetic stimulation with beta-blocker agents as prescribed. 8. Try to communicate with patient if delusions or hallucinations are present. 9. Place in a protective environment (preferably psychiatric security room with video monitoring) to observe for suicide attempt. 10. Refer for psychiatric and drug rehabilitation evaluation.
Hallucinogens or Psychedelic-Type Drugs Lysergic acid diethylamide (LSD) Phencyclidine HCl (PCP, "angel dust") Mescaline, psilocybin Cannabinoids (marijuana) Ketamine ("special K") Synthetic cannabinoids ("spice," "incense," "K2")	Nystagmus Mild hypertension Marked confusion bordering on panic Incoherence, hyperactivity Withdrawn Combative behavior; delirium, mania, self-injury (lasts 6–12 hours) Hallucinations, body image distortion Hypertension, hyperthermia, acute kidney injury Flashback—recurrence of LSD-like state without having taken the drug; may occur weeks or months after drug was taken Ketamine—"out-of-body" experience; increased aggressiveness Synthetic cannabinoids—euphoria, increased sensory experience, relaxation	1. Evaluate and maintain patient's circulation, airway, and breathing. 2. Determine by urine or serum drug screen whether the patient has ingested hallucinogenic drug or has a toxic psychosis. 3. Try to communicate with and reassure the patient. a. "Talking down" involves understanding the process through which the patient is proceeding and helping him overcome his fears while establishing contact with reality. b. Remind the patient that fear is common with this problem. c. Reassure the patient that he is not losing his mind but is experiencing the effect of drugs and that this will wear off. d. Instruct the patient to keep the eyes open; this reduces the intensity of reaction. e. Reduce sensory stimuli by minimizing noise, lights, movement, tactile stimulation. 4. Sedate the patient as prescribed if hyperactivity cannot be controlled; diazepam or a barbiturate may be prescribed.

Drug	Clinical Manifestations	Therapeutic Management
		5. Search for evidence of trauma; hallucinogen users have a tendency to "act out" their hallucinations. 6. Manage seizures with benzodiazepines (e.g., diazepam) as necessary. 7. Observe patient closely; patient's behavior may become hazardous. Have safety officers stationed near the patient's room. 8. Monitor for hypertensive crisis if patient has prolonged psychosis due to drug ingestion. 9. Place patient in a protected environment under proper medical supervision to prevent self-inflicted bodily harm. **Management for Phencyclidine Abusers** 1. Place patient in a calm, supportive environment to minimize stimuli; protect from self-injury. 2. Avoid talking down. 3. Do not leave patient unobserved. Treat symptoms as they occur. a. Drug effects are unpredictable and prolonged. b. Symptoms are likely to exacerbate; patient becomes out of control. 4. Refer all patients in this category for psychiatric and drug evaluation/rehabilitation.
Drugs Producing Sedation, Intoxication, or Psychological and Physical Dependence (nonbarbiturate sedatives)		
diazepam (Valium) chlordiazepoxide (Librium) oxazepam (Serax) lorazepam (Ativan) midazolam (Versed) flunitrazepam (Rohypnol, "roofies," "date rape drug")*	Seizures, coma, circulatory collapse, death *Acute intoxication:* • Respiratory depression • Decreasing mental alertness • Confusion • Slurred speech, decreased blood pressure • Ataxia • Pulmonary edema • Coma, death **Flunitrazepam:** • Disinhibition with antegrade amnesia • Weakness and unsteadiness with impaired judgment • Powerlessness	1. Endotracheal tube is inserted as a precaution; use assisted ventilation to stabilize and correct respiratory depression. Observe for sudden apnea and laryngeal spasm. 2. Assess for hypotension. a. Insert indwelling urinary catheter for comatose patient; decreased urinary volume is an index of reduced renal flow associated with reduced intravascular volume or vascular collapse. b. Start volume expansion with saline or dextrose as prescribed. 3. Evacuate stomach contents; lavage (if within 1 hour of ingestion); activated charcoal. 4. Start ECG monitoring. Observe for dysrhythmias. 5. Administer flumazenil (Romazicon), a benzodiazepine antagonist (reversal agent). 6. Refer patient for psychiatric evaluation (potential suicide intent).
Salicylate Poisoning		
Aspirin (present in compound analgesic tablets) • Toxic levels (150–200 mg/kg body weight) • Chronic toxicity (occurs in older adults due to decreased kidney function) • Long-term intoxication (>100 mg/kg/day for more than 2 days)	Restlessness, tinnitus, deafness, blurring of vision Hyperpnea, hyperpyrexia, sweating Epigastric pain, vomiting, dehydration Respiratory alkalosis and metabolic acidosis Disorientation, coma, cardiovascular collapse Coagulopathy	1. Treat respiratory depression. 2. Induce gastric emptying by lavage (if within 1 hour after ingestion). 3. Give activated charcoal to adsorb aspirin. 4. Support patient with IV infusions as prescribed to establish hydration and correct electrolyte imbalances, including administration of sodium bicarbonate. 5. Enhance elimination of salicylates as directed by forced diuresis, alkalinization of urine, peritoneal dialysis, or hemodialysis, according to severity of intoxication. 6. Monitor serum salicylate level for efficacy of treatment. 7. Administer specific prescribed pharmacologic agent for bleeding and other problems. 8. Recognize that concretions formed in the gut may result in prolonged exposure as they are digested. 9. Refer patient for psychiatric evaluation (potential suicide intent).

(continues on page 2144)

TABLE 72-1 Emergency Management of Patients With Drug Overdose (continued)

Drug	Clinical Manifestations	Therapeutic Management
Acetaminophen (present in prescription and nonprescription analgesics, antipyretics, and cold remedies)	Lethargy to encephalopathy and death GI upset, diaphoresis Right upper quadrant pain Abnormal liver function tests, prolonged prothrombin time, increased bilirubin, disseminated intravascular coagulation Hepatomegaly leading to liver failure Metabolic acidosis Hypoglycemia Stage I—within 24 hours; GI irritation, possible metabolic acidosis and coma if severe ingestion Stage II—24–48 hours; monitor liver and coagulation studies. Stage III—after 48 hours; hepatic encephalopathy/jaundice, vomiting, right upper quadrant pain, coagulopathy, hypoglycemia, acute kidney injury	1. Maintain airway. 2. Obtain acetaminophen level. Levels ≥140 mg/kg are toxic. 3. Laboratory studies—liver function tests, prothrombin time/partial thromboplastin time, complete blood count, blood urea nitrogen, creatinine 4. Lavage (if within 1 hour after ingestion); activated charcoal 5. Prepare for possible hemodialysis, which clears acetaminophen but does not halt liver damage. 6. Administer N-acetylcysteine (Mucomyst) as soon as possible. N-acetylcysteine replenishes essential liver enzymes and requires a total of 18 doses every 4 hours. Charcoal absorbs N-acetylcysteine; do not administer together. Repeat N-acetylcysteine dose if patient vomits. 7. Refer patient for psychiatric evaluation (potential suicide intent).
Tricyclic Antidepressants amitriptyline (Elavil) doxepin (Sinequan) nortriptyline (Aventyl) imipramine (Tofranil)	Dysrhythmia: ventricular fibrillation/tachycardia, sinus tachycardia Hypotension Pulmonary edema, hypoxemia, acidosis Confusion, agitation, coma Visual hallucinations Clonus, tremors, hyperactive reflexes, nystagmus, myoclonic jerking Seizures Blurred vision, flushing, hyperthermia	1. Provide airway support, ventilation, cardiac monitoring; insert IV line with normal saline solution. 2. If within 1–2 hours after overdose, insert a nasogastric tube and instill activated charcoal every 4 hours × 3. 3. Administer a sodium bicarbonate drip to decrease dysrhythmias; the alkaline environment increases the protein binding of the metabolite. Synchronized cardioversion may be indicated with some dysrhythmias refractory to sodium bicarbonate. Torsades de pointes should be treated with IV magnesium sulfate. 4. Administer vasopressors. 5. Manage seizure activity with benzodiazepines (e.g., diazepam) as necessary. 6. Refer patient for psychiatric evaluation for potential suicide intent and evaluation of medication regimen for effectiveness.
Selective Serotonin Reuptake Inhibitors and Other Antidepressants trazodone (Desyrel) fluoxetine (Prozac) paroxetine (Paxil) sertraline (Zoloft) venlafaxine (Effexor) escitalopram (Lexapro) bupropion (Wellbutrin)	Decreased level of consciousness, confusion Respiratory depression Increased heart rate **Serotonin syndrome** may occur if the SSRI was taken in conjunction with dextromethorphan or meperidine Agitation, seizures Hyperthermia, diaphoresis Hypertension, headache, shivering, "goose flesh," cardiac dysrhythmias, loss of consciousness	1. Administer activated charcoal with possibly whole bowel irrigation if a sustained-release medication was taken. 2. Use seizure precautions and administer benzodiazepines (e.g., diazepam) as ordered.
Anabolic Steroids "roids," "juice," methandrostenolone, stanozolol, nandrolone Synthetic testosterone	Increase in LDL, decrease in HDL Alter carbohydrate metabolism Hyponatremia, Hypokalemia Hypocalcemia/osteoporosis Mood swings/violent behaviors Invincibility, depression, potential for suicide attempts Memory loss, cognitive disabilities Immunosuppression Used to bulk up muscles, so skeletal muscle hypertrophy is a common manifestation.	1. Provide supportive therapy appropriate to patient's emotional manifestations. 2. Protect the patient from self harm/harming others. 3. Encourage the patient to stop use; refer patient for psychiatric evaluation.

CNS, central nervous system; IV, intravenous; IM, intramuscular; ECG, electrocardiogram; GI, gastrointestinal; NAC, N-acetylcysteine; SSRI, selective serotonin reuptake inhibitor; LDL, low-density lipoprotein; HDL, high-density lipoprotein.

*Polydrug use at "rave clubs" frequently involves MDMA, alcohol, amphetamines, LSD, and sometimes dextromethorphan. Terms such as "Ecstasy" may refer to flunitrazepam (Rohypnol), GHB, ephedrine, and/or caffeine, in addition to MDMA. The Web site www.drugabuse.gov/drugs-abuse/club-drugs provides more information about possible drug abuse and how it may relate to emergency nursing care.

Adapted from Cline, D. M., Ma, O. J., Cydulka, R. K., et al. (2012). *Tintinalli's emergency medicine manual* (7th ed.). New York: McGraw-Hill Medical; Emergency Nurses Association. (2013). *Sheehy's manual of emergency care* (7th ed.). St. Louis: Mosby; and Jerry, J., Collins, G., & Streem, O. (2012). Synthetic legal intoxicating drugs: The emerging 'incense' and 'bath salt' phenomenon. *Cleveland Clinic Journal of Medicine*, 79(4), 258–264.

alcohol was ingested and for how long. Delirium tremens may be precipitated by acute injury or infection (pneumonia, pancreatitis, hepatitis) and is the most severe form of alcohol withdrawal syndrome (Larson, 2008).

Patients with alcohol withdrawal syndrome show signs of anxiety, uncontrollable fear, tremor, irritability, agitation, insomnia, and incontinence. They are talkative and preoccupied and experience visual, tactile, olfactory, and auditory hallucinations that often are terrifying. Autonomic overactivity occurs and is evidenced by tachycardia, dilated pupils, and profuse perspiration. Usually, all vital signs are elevated in the alcoholic toxic state. Delirium tremens is a life-threatening condition and carries a high mortality rate if untreated (Burns, Chiang, Huff, et al., 2011).

The goals of management are to give adequate sedation and support to allow the patient to rest and recover without danger of injury or peripheral vascular collapse. A physical examination is performed to identify preexisting or contributing illnesses or injuries (e.g., head injury, pneumonia). A drug history is obtained to elicit information that may facilitate adjustment of any sedative requirements. Baseline blood pressure is determined, because the patient's subsequent treatment may depend on blood pressure changes.

Usually, the patient is sedated as directed with a sufficient dosage of benzodiazepines to establish and maintain sedation, which reduces agitation, prevents exhaustion, prevents seizures, and promotes sleep. The patient should be calm, able to respond, and able to maintain an airway safely on his or her own. A variety of medications and combinations of medications are used (e.g., chlordiazepoxide [Librium], lorazepam [Ativan], and clonidine [Catapres]). Haloperidol (Haldol) or esmolol (Brevibloc) or midazolam (Versed) may be administered for severe acute alcohol withdrawal syndrome. Dosages are adjusted according to the patient's symptoms (agitation, anxiety) and blood pressure response (Burns et al., 2011).

The patient is placed in a calm, nonstressful environment (usually a private room) and observed closely. The room remains lighted to minimize the potential for illusions (visual misrepresentations) and hallucinations. Homicidal or suicidal responses may result from hallucinations. Closet and bathroom doors are closed to eliminate shadows. Someone is designated to stay with the patient as much as possible. The presence of another person has a reassuring and calming effect, which helps the patient maintain contact with reality. To orient the patient to reality, any illusions are explained.

Quality and Safety Nursing Alert

Restraints are used as prescribed, if necessary, if the client is aggressive or violent, but only when other alternatives have been unsuccessful. The least restrictive device that will prevent the patient from injuring him- or herself or others is used. Caution is taken to ensure that restraints are applied properly and that they are not impairing circulation to any part of the body or interfering with respirations. Restraints should be used in tandem with verbal intervention to calm the patient and promote compliance. Restraints must be released according to protocol. Physical observation (e.g., skin integrity, circulatory status, respiratory status) is ongoing, and the patient's response is documented.

Fluid losses may result from gastrointestinal losses (vomiting), profuse perspiration, and hyperventilation. In addition, the patient may be dehydrated as a result of alcohol's effect of decreasing antidiuretic hormone. The oral or IV route is used to restore fluid and electrolyte balance.

Temperature, pulse, respiration, and blood pressure are recorded frequently (every 30 minutes in severe forms of delirium) to monitor for peripheral circulatory collapse or hyperthermia (the two most serious complications). Phenytoin (Dilantin) or other antiseizure medications may be prescribed to prevent repeated withdrawal seizures.

Frequently seen complications include infections (e.g., pneumonia), trauma, hepatic failure, hypoglycemia, and cardiovascular problems. Hypoglycemia may accompany alcohol withdrawal, because alcohol depletes liver glycogen stores and impairs gluconeogenesis; many patients with alcoholism also are malnourished. Parenteral dextrose may be prescribed if the liver glycogen level is depleted. Orange juice, sports drinks, or other sources of carbohydrates are given to stabilize the blood glucose level and counteract tremulousness. Supplemental vitamin therapy and a high-protein diet are provided as prescribed to counteract nutritional deficits. The patient should be referred to an alcoholic treatment center for follow-up care and rehabilitation.

VIOLENCE, ABUSE, AND NEGLECT

Family Violence, Abuse, and Neglect

EDs are often the first place where victims of family violence, abuse, or neglect go to seek help. In the United States, 18% of women and 1.4% of men have been raped at sometime in their lives; however, the percentages are equal by gender if rape was perpetrated by an intimate partner. Most female victims experience rape before the age of 25 years, with 42% of those before age 18 years; yet for men, 28% experience rape much younger, before 10 years of age. Intimate partner violence (IPV) many times involves violations beyond rape, including physical assault and stalking. Approximately 35% of women and 28% of men have experienced some component of IPV (CDC, 2011).

It is estimated that up to one third of all patients in the ED have experienced IPV at some point in their lives (Daugherty & Houry, 2008).IPV involves fear of one partner by another and control by threats, intimidation, and physical abuse. Approximately 95% to 99% of IPV is committed by men against women and can be emotional, physical, sexual, or economic in nature (Wong, Tiwari, Fong et al, 2011). ED nurses must be vigilant in their assessments of both women and men who present with injuries that may be consistent with IPV. In addition, ED nurses must be aware that men and women with disabilities are at higher risk of domestic violence and abuse than nondisabled people and should include questions that screen for IPV in their evaluations.

It is estimated that between 1 million and 2 million adults older than 65 years have been abused or neglected, a type of IPV called *elder abuse* (National Center on Elder Abuse, 2005). Elder abuse takes many forms, including physical, emotional,

Chart 72-8 ASSESSMENT

Assessing for Abuse, Maltreatment, and Neglect

The following questions may be helpful when assessing a patient for abuse, maltreatment, and neglect:

- I noticed that you have a number of bruises. Can you tell me how they happened? Has anyone hurt you?
- You seem frightened. Has anyone ever hurt you?
- Patients sometimes tell me that they have been hurt by someone at home or at work. Could this be happening to you?
- Are you afraid of anyone at home or work, or of anyone with whom you come in contact?
- Has anyone failed to help you to take care of yourself when you needed help?
- Has anyone prevented you from seeing friends or other people whom you wish to see?
- Have you signed any papers that you did not understand or did not wish to sign?
- Has anyone forced you to sign papers against your will?
- Has anyone forced you to engage in sexual activities within the past year?
- Has anyone prevented you from using an assistive device (e.g., wheelchair, walker) within the past year?
- Has anyone you depend on refused to help you take your medicine, bathe, groom, or eat within the past year?

and verbal abuse; neglect; violation of personal rights; and financial abuse (see Chapters 5 and 56).

Clinical Manifestations

When people who have been abused seek treatment, they may present with physical injuries or health problems such as anxiety, insomnia, or gastrointestinal symptoms related to stress. The possibility of abuse should be investigated whenever a person presents with multiple injuries that are in various stages of healing, when injuries are unexplained, and when the explanation does not fit the physical picture (Chart 72-8). The possibility of neglect should be investigated whenever a dependent person shows evidence of inattention to hygiene, to nutrition, or to known medical needs (e.g., unfilled medication prescriptions, missed appointments with health care providers). In the ED, the most common physical injuries seen are unexplained bruises, lacerations, abrasions, head injuries, or fractures. The most common clinical manifestations of neglect are malnutrition and dehydration.

Assessment and Diagnostic Findings

Nurses in EDs are in an ideal position to provide early detection and interventions for victims of IPV. This requires an acute awareness of the signs of possible abuse, maltreatment, and neglect. Nurses must be skilled in interviewing techniques that are likely to elicit accurate information. A careful history is crucial in the screening process. Asking questions in private—away from others—may be helpful in eliciting information about abuse, maltreatment, and neglect.

Whenever evidence leads one to suspect abuse or neglect, an evaluation with careful documentation of descriptions of events and drawings or photographs of injuries is important, because the medical record may be used as part of a legal proceeding. Assessment of the patient's general appearance and interactions with significant others, an examination of the entire surface area of the body, and a mental status examination are crucial.

Management

Whenever abuse, maltreatment, or neglect is suspected, the health care provider's main concern should be the safety and welfare of the patient. Treatment focuses on the consequences of the abuse, violence, or neglect and on prevention of further injury. Protocols of most EDs require that a multidisciplinary approach be used. Nurses, physicians, social workers, and community agencies work collaboratively to develop and implement a plan for meeting the patient's needs.

If the patient is in immediate danger, he or she should be separated from the abusing or neglecting person whenever possible. Referral to a shelter may be the most appropriate action, but many shelters are inaccessible to people with mobility limitations.

When abuse or neglect is the result of stress experienced by a caregiver who is no longer able to cope with the burden of caring for an older adult or a person with chronic disease or a disability, respite services may be necessary. Support groups may be helpful to these caregivers. When mental illness of the abuser or neglecter is responsible for the situation, alternative living arrangements may be required.

Nurses must be mindful that competent adults are free to accept or refuse the help that is offered to them. Some patients insist on remaining in the home environment where the abuse or neglect is occurring. The wishes of patients who are competent and not cognitively impaired should be respected. However, all possible alternatives, available resources, and safety plans should be explored with the patient.

Mandatory reporting laws in most states require health care workers to report *suspected* child abuse or abuse of older adults to an official agency, usually Adult (or Child) Protective Services. All that is required for reporting is the suspicion of abuse; the health care worker is not required to prove abuse or neglect. Likewise, health care workers who report suspected abuse are immune from civil or criminal liability if the report is made in good faith. Subsequent home visits resulting from the report of suspected abuse are a part of gathering information about the patient in the home environment. In addition, many states have resource hotlines for use by health care workers and by patients who seek answers to questions about abuse and neglect.

Sexual Assault

The definition of *rape* is forced sexual acts, especially if these acts involve vaginal or anal penetration. Perpetrators and victims may be either male or female. Rape crisis centers offer support and education and help people who have been sexually assaulted through the subsequent police investigation and courtroom experience.

The manner in which the patient is received and treated in the ED is important to his or her future psychological well-being. Crisis intervention should begin when the patient enters the health care facility. The patient should be seen immediately. Most hospitals have a written protocol that addresses the patient's physical and emotional needs as well as collection of forensic evidence.

In many states, the emergency nurse has the opportunity to become trained as a sexual assault nurse examiner (SANE). Preparing for this role requires specific training in forensic evidence collection, history taking, documentation, and ways to approach the patient and family. Specialized training also includes learning proper photographic methods and the use of colposcopy. Colposcopy facilitates assessment by magnifying tissues and looking for evidence of microtrauma. Evidence is collected through photography, videography, and analysis of specimens. Another tool useful to the SANE is the light-staining microscope, which enables the examiner to identify motile and nonmotile sperm and infectious organisms. This tool saves time and also enhances assessment. The SANE complements the ED staff and can spend more time with both the patient and police officers investigating the incident (ENA, 2013).

Assessment and Diagnostic Findings

The patient's reaction to rape has been termed *rape trauma syndrome* and is seen as an acute stress reaction to a life-threatening situation. The nurse performing the assessment is aware that the patient may go through several phases of psychological reactions, which have been described as follows:

- An acute disorganization phase, which may manifest as an expressed state in which shock, disbelief, fear, guilt, humiliation, anger, and other such emotions are encountered or as a controlled state in which feelings are masked or hidden and the victim appears composed
- A phase of denial and unwillingness to talk about the incident, followed by a phase of heightened anxiety, fear, flashbacks, sleep disturbances, hyperalertness, and psychosomatic reactions that is consistent with post-traumatic stress disorder (PTSD) (see later discussion)
- A phase of reorganization, in which the incident is put into perspective. Some victims never fully recover and go on to develop chronic stress disorders and phobias.

Management

The goals of management are to provide support, to reduce the patient's emotional trauma, and to gather available evidence for possible legal proceedings. All of the interventions are aimed at encouraging the patient to gain a sense of control over his or her life.

Throughout the patient's stay in the ED, the patient's privacy and sensitivity must be respected. The patient may exhibit a wide range of emotional reactions, such as hysteria, stoicism, or feelings of being overwhelmed. Support and caring are crucial. The patient should be reassured that anxiety is natural and asked whether a support person may be called. Appropriate support is available from professional and community resources. The National Sexual Assault Hotline (see the Resources section) will automatically route the patient to the nearest assault or crisis intervention center for services, as needed. The patient should never be left alone.

Physical Examination

A written, witnessed informed consent must be obtained from the patient (or parent or guardian if the patient is a minor) for examination, for taking of photographs, and for release of findings to police. A history is obtained only if the patient has not already talked to a police officer, social worker, or crisis intervention worker. The patient should not be asked to repeat the history. Any history of the event that is obtained should be recorded in the patient's own words. The patient is asked whether he or she has bathed, douched, brushed his or her teeth, changed clothes, urinated, or defecated since the attack, because these actions may alter interpretation of subsequent findings. The time of admission, time of examination, date and time of the alleged rape, and the patient's emotional state and general appearance (including any evidence of trauma, such as discoloration, bruises, lacerations, secretions, or torn and bloody clothing) are documented. If the patient has no recollection of the event, drugs that induce retrograde amnesia may have been involved, such as alcohol, ketamine, gamma-hydroxybutyrate, benzodiazepines, or flunitrazepam (Rohypnol) (Luce, Schrager, Gilchrist, et al., 2010).

For the physical examination, the patient is helped to undress and is draped properly. Each item of clothing is placed in a separate paper bag. As noted earlier, plastic bags are not used in order to avoid possible destruction of evidence from mold or mildew formation. The bags are labeled and given to appropriate law enforcement authorities.

The patient is examined (from head to toe) for injuries, especially injuries to the head, neck, breasts, thighs, back, and buttocks. Body diagrams and photographs aid in documenting the evidence of trauma. The physical examination focuses on the following:

- External evidence of trauma (bruises, contusions, lacerations, stab wounds)
- Dried semen stains (appearing as crusted, flaking areas) on the patient's body or clothes
- Broken fingernails and body tissue and foreign materials under nails (if found, samples are taken)
- Oral examination, including a specimen of saliva and cultures of gum and tooth areas

Pelvic and rectal examinations are also performed. The perineum and other areas are examined with a Wood's lamp or other filtered ultraviolet light. Areas that appear fluorescent may indicate semen stains. The color and consistency of any discharge present is noted. A water-moistened, rather than lubricated, vaginal speculum is used for the examination. Lubricant contains chemicals that may interfere with later forensic testing of specimens and acid phosphatase determinations. The rectum is examined for signs of trauma, blood, and semen. During the examination, the patient should be advised of the nature and necessity of each procedure and given the rationale for each (Cline et al., 2012; ENA, 2013).

Specimen Collection

During the physical examination, numerous laboratory specimens may be collected, including the following:

- Vaginal aspirate, examined for presence or absence of motile and nonmotile sperm
- Secretions (obtained with a sterile swab) from the vaginal pool for acid phosphatase, blood group antigen of semen, and precipitin test against human sperm and blood
- Separate smears from the oral, vaginal, and anal areas
- Culture of body orifices for gonorrhea

- Blood serum for syphilis and HIV testing and deoxyribonucleic acid (DNA) analysis. A sample of serum for syphilis may be frozen and saved for future testing.
- Pregnancy test if there is a possibility that the female patient may be pregnant
- Any foreign material (leaves, grass, dirt), which is placed in a clean envelope
- Pubic hair samples obtained by combing or trimming. Several pubic hairs with follicles are placed in separate containers and identified as the patient's hair.

To preserve the chain of evidence, each specimen is labeled with the name of the patient, the date and time of collection, the body area from which the specimen was obtained, and the names of personnel collecting specimens. The specimens are then given to a designated person (e.g., crime laboratory technician), and an itemized receipt is obtained (Cline et al., 2012; ENA, 2013).

Treating Potential Consequences of Rape

After the initial physical examination is completed and specimens have been obtained, any associated injuries are treated as indicated. The patient is given the option of prophylaxis against sexually transmitted infections (STIs) (also referred to as sexually transmitted disease [STDs]). Ceftriaxone (Rocephin), administered intramuscularly with 1% lidocaine (Xylocaine), may be prescribed as prophylaxis for gonorrhea. In addition, a single oral dose of metronidazole (Flagyl) and either a single oral dose of azithromycin (Zithromax) or a 7-day oral regimen of doxycycline may be prescribed as prophylaxis for syphilis and chlamydia (Luce et al., 2010).

Antipregnancy measures may be considered if the patient is a female of childbearing age. A postcoital contraceptive medication, such as an oral contraceptive medication that contains levonorgestrel and ethinyl estradiol (Alesse, Seasonique), may be prescribed after a pregnancy test. To promote effectiveness, the contraceptive medication should be administered within 12 to 24 hours and no later than 72 hours after intercourse. The 21-day package is prescribed so that the patient does not mistakenly take the inert tablets included in the 28-day package. An antiemetic agent may be administered as prescribed to decrease discomfort from side effects. A cleansing douche, mouthwash, and fresh clothing are usually offered.

Follow-Up Care

The patient is informed of counseling services to prevent long-term psychological effects. Counseling services should be made available to both the patient and the family. A referral is made to the National Sexual Assault Hotline (see the Resources section) or directly to a local crisis intervention center. Appointments for follow-up surveillance for pregnancy and for STI and HIV testing also are made (Luce et al., 2010).

The patient is encouraged to return to his or her previous level of functioning as soon as possible. When leaving the ED, the patient should be accompanied by a family member or friend.

PSYCHIATRIC EMERGENCIES

A psychiatric emergency is an urgent, serious disturbance of behavior, affect, or thought that makes the patient unable to cope with life situations and interpersonal relationships. A patient presenting with a psychiatric emergency may display overactive or violent, underactive or depressed, or suicidal behaviors.

The most important concern of the ED personnel is determining whether the patient is at risk for injuring self or others. The aim is to try to maintain the patient's self-esteem (and life, if necessary) while providing care. Determining whether the patient is under psychiatric care is important so that contact can be made with the therapist or physician who works with the patient.

Overactive Patients

Patients who display disturbed, uncooperative, and paranoid behavior and those who feel anxious and panicky may be prone to assaultive and destructive impulses and abnormal social behavior. Intense nervousness, depression, and crying are evident in some patients. Disturbed and noisy behavior may be exacerbated or compounded by alcohol or drug intoxication.

A reliable source for obtaining an accurate history is needed to identify events leading to the crisis. Past mental illness, hospitalizations, injuries, serious illnesses, the use of alcohol or drugs, crises in interpersonal relationships, or intrapsychic conflicts are explored. Because abnormal thoughts and behavior may be manifestations of an underlying physical disorder, such as hypoglycemia, drug or alcohol toxicity, a stroke, a seizure disorder, or head injury, a physical assessment is also performed.

The immediate goal is to gain control of the situation. If the patient is potentially violent, security or police should be nearby. Restraints are used as a *last* resort and only as prescribed. Approaching the patient with a calm, confident, and firm manner is therapeutic and has a calming effect. Helpful interventions include the following:

- Introduce yourself by name.
- Tell the patient, "I am here to help you."
- Repeat the patient's name from time to time.
- Speak in one-thought sentences, and be consistent.
- Give the patient space and time to slow down.
- Show interest in, listen to, and encourage the patient to talk about personal thoughts and feelings.
- Offer appropriate and honest explanations.

A psychotropic agent (e.g., one that exerts an effect on the mind) may be prescribed for emergency management of functional psychosis. However, a patient with a personality disorder should not be treated with psychotropic medications, and psychotropic medications should not be used if the patient's behavior results from the use of hallucinogens (e.g., lysergic acid diethylamide [LSD]).

Agents such as chlorpromazine and haloperidol act specifically against psychotic symptoms of thought fragmentation and perceptual and behavioral aberrations. The initial dose depends on the patient's body weight and the severity of the symptoms. After administration of the initial dose, the patient is observed closely to determine the degree of change in psychotic behavior. Subsequent doses depend on the patient's response. Typically, after stabilization, the patient is transferred to an inpatient psychiatric unit, or psychiatric outpatient treatment is arranged.

Posttraumatic Stress Disorder

PTSD is the development of characteristic symptoms after a psychologically stressful event that is considered outside the range of normal human experience (e.g., rape, combat, motor vehicle crash, natural catastrophe, terrorist attack). Symptoms of this disorder include intrusive thoughts and dreams, phobic avoidance reaction (avoidance of activities that arouse recollection of the traumatic event), heightened vigilance, exaggerated startle reaction, generalized anxiety, and societal withdrawal. PTSD may be acute, chronic, or delayed. PTSD often presents as multiple readmissions to the ED for minor or recurring complaints without evidence of injury.

Underactive or Depressed Patients

In the ED, depression may be seen as the main condition bringing the patient to the health care facility, or it may be masked by anxiety and somatic complaints. The depressed person has a mood disturbance.

Any patient who is depressed may be at risk for suicide. Attempts are made to find out whether the patient has thought about or attempted suicide. Questions such as "Have you ever thought about taking your own life?" may be helpful. Generally, the patient is relieved to have an opportunity to discuss personal feelings. If the patient is seriously depressed, relatives should be notified. The patient should never be left alone, because suicide is usually committed in solitude.

Suicidal Patients

Attempted suicide is an act that stems from depression (e.g., loss of a loved one, loss of body integrity or status, poor self-image) and can be viewed as a cry for help and intervention. Males are at greater risk than females. Others at risk are older adults; young adults; people who are enduring unusual loss or stress; those who are unemployed, divorced, widowed, or living alone; those showing signs of significant depression (e.g., weight loss, sleep disturbances, somatic complaints, suicidal preoccupation); and those with a history of a previous suicide attempt, suicide in the family, or psychiatric illness.

Being aware of people at risk and assessing for specific factors that predispose a person to suicide are key management strategies. Specific signs and symptoms of potential suicide include the following:

- Communication of *suicidal intent*, such as preoccupation with death or speaking of someone else's suicide (e.g., "I'm tired of living. I've put my affairs in order. I'm better off dead. I'm a burden to my family.")
- History of a previous suicide attempt, with risk being much greater in these cases
- Family history of suicide
- Loss of a parent at an early age
- Specific plan for suicide
- A means to carry out the plan

Emergency management focuses on treating the consequences of the suicide attempt (e.g., gunshot wound, drug overdose) and preventing further self-injury. A patient who has made a suicidal attempt may do so again. Crisis intervention is used to determine suicidal potential, to discover areas of depression and conflict, to find out about the patient's support system, and to determine whether hospitalization or psychiatric referral is necessary. Depending on the patient's potential for suicide, the patient may be admitted to the intensive care unit, referred for follow-up care, or admitted to the psychiatric unit.

Critical Thinking Exercises

1 pq An 80-year-old woman arrives at the ED by car after a fall down the stairs, colliding with the railing at the bottom. There is a bruise across her abdomen and lower chest. She is groaning loudly and seems confused. Although she complains of distinct abdominal and left shoulder pain, she cannot tell you what has happened or why she fell, where she is at present, or what day of the week it is, although she can tell you her full name. You note a shallow, rapid breathing pattern of 26 breaths/min; her systolic BP is 90 mm Hg; and her pulse rate is 70 bpm and regular. How would you prioritize the patient's needs? Develop an assessment strategy and identify diagnostic studies that will benefit the patient. In light of the injury pattern, respiratory rate, and blood pressure, can you give a plausible explanation why her pulse rate is not elevated?

2 ebp pq Two men choose to go ice fishing toward the end of the winter season. They are dressed in light clothing, including rubber fishing boots, as the weather has been warming and they expected the cabin to be warm. They did not notice the weather report of high humidity and wind. They spend the day fishing on the ice without moving much. On exiting the ice, one man splashes into the lake up to his hip. The other man now presents to the ED for treatment of frostbite of his feet. His friend has been massaging them en route to the ED, and the patient insists that this makes his feet feel better. You tell them to cease massage. You also notice that the second man is wet up to the level of his hips and shivering. Describe the explanation you would give to these patients and the evidence base that guides your response regarding massage in frostbite cases. Describe how you would proceed with managing the care of the patient with frostbite. What would you do for the second patient regarding registration in the ED and management? What are your priority assessments for him?

3 pq The following four patients present to the triage desk of the ED within minutes of each other. How would you prioritize and categorize each of these patients? Which ones need immediate attention? What initial care would you provide at triage? Which patient could wait or be sent to the clinic for management?

a A college student runs to the triage desk and screams, "My buddy is in the car outside—I think he stopped breathing!" The student appears disheveled and reeks of alcohol; he appears to have splattered vomitus across his chest.

b A woman presents cradling her left forearm in her right hand. Her husband tells you that she fell down the steps at home and he thinks she broke her arm. Her eyes are downcast, she does not speak for herself, and she is crying softly. She has a fresh bruise on her left eye that appears to be swollen shut, and there is blood trickling from her tightly closed, quivering mouth. Her husband insists that he wants to be with her during her assessment.

c A middle-aged woman experiences sudden dyspnea and midscapular back pain while making dinner. Instead of calling 911, she takes an aspirin and her husband drives her to the ED. She is now complaining of left scapular pain and tingling in her left arm, her skin appears ashen, and she is diaphoretic.

d A middle-aged man who is obese and has a known history of diabetes presents with complaints of a high fever that has waxed and waned over the course of the past 48 hours. His vital signs are normal, but he is diaphoretic and appears weak.

4 **pq** A 72-year-old woman is brought to the ED by ambulance. Her daughter called the ambulance after her mother told her she took an overdose of aspirin. The patient presents with the following vital signs: blood pressure 88/58; heart rate 118 bpm and irregular; respiratory rate 32 breaths/min. She appears highly anxious and is disoriented to person, place, and time. She is profusely diaphoretic. Describe your priority nursing diagnoses and interventions for this patient. Her daughter arrives at the ED shortly after the ambulance and blurts out to you tearfully, "I never should have agreed with Mom that she could live alone!" Describe how you might therapeutically interact with the daughter, what critical information you may wish to gain from her, and what type of support systems you might consider mobilizing for this family.

Brunner Suite Resources

Explore these additional resources to enhance learning for this chapter:

- NCLEX-Style Questions and Other Resources on thePoint, http://thePoint.lww.com/Brunner13e
- Study Guide
- PrepU
- Clinical Handbook
- Handbook of Laboratory and Diagnostic Tests

References

*Asterisk indicates nursing research.
**Double asterisk indicates classic reference.

Books

American College of Surgeons. (2008). *Advanced trauma life support* (8th ed.). Chicago: Author.

American College of Surgeons. (2010). *Trauma evaluation and management* (3rd ed.). Chicago: Author.

Auerbach, P. S. (2012). *Wilderness medicine* (6th ed.). St. Louis: Elsevier Mosby.

Cline, D. M., Ma, O. J, Cydulka, R. K., et al., (2012). *Tintinalli's emergency medicine manual* (7th ed.). New York: McGraw-Hill Medical.

Emergency Nurses Association. (2013). *Sheehy's manual of emergency care* (7th ed.). St. Louis: Mosby.

Gilboy, N., Tanabe, P., Travers, D., et al. (2012). *Emergency severity index (ESI): A triage tool for emergency department care, version 4. Implementation handbook*. Publication No. 12–0014. Rockville, MD: Agency for Healthcare Research and Quality, Publication.

McQuillan, K., VonReuden, K., Hartsock, R., et al. (2008). *Trauma nursing: Resuscitation through rehabilitation* (4th ed.). Philadelphia: Saunders.

Nayduch, D. (2009). *Nurse to nurse trauma care*. New York: McGraw-Hill.

Sinz, E., & Navarro, K. (2011). *Advanced cardiovascular life support provider manual*. Dallas: American Heart Association.

Journals and Electronic Documents

American College of Emergency Physicians. (2007). Use of patient restraints. *Clinical & Practice Management*. Available at: www.acep.org/Clinical—Practice-Management/Use-of-Patient-Restraints/

Ahmed, S. M., Ahmed, M., Nadeem, A., et al. (2008). Emergency treatment of a snakebite: Pearls from literature. *Journal of Emergencies, Trauma, and Shock*, *1*(2), 97–105.

Baydala, L. (2010). Inhalant abuse. *Journal of Pediatrics and Child Health*, *15*(7), 443–448.

Becker, J. A., & Stewart, L. K. (2011). Heat-related illness. *American Family Physician*, *83*(11), 1325–1330.

Bennett, M. H., Lehm, J. P., Mitchell, S. J., et al. (2010). Recompression and adjunctive therapy for decompression illness: A systematic review of randomized controlled trials. *Anesthesia and Analgesia*, *11*(3), 757–762.

**Blow, O., Magliore, L., Claridge, J. A., et al. (1999). The golden hour and the silver day: Detection and correction of occult hypoperfusion within 24 hours improves outcome from major trauma. *Journal of Trauma*, *47*(5), 964–969.

Burns, M. J., Chiang, W. K., Huff, J. S., et al. (2011). Delirium tremens (DTs). *Medscape*. Available at: emedicine.medscape.com/article/166032-overview

Centers for Disease Control and Prevention. (2009). *NHAMCS: 2009 emergency department summary*. Available at: www.cdc.gov/nchs/ahcd/web_tables.htm#2009

Centers for Disease Control and Prevention. (2011). *National Intimate Partner and Sexual Violence Survey: 2010 summary report*. Available at: www.cdc.gov/violenceprevention/pdf/NISVS_Report2010-a.pdf

Daugherty, J. D., & Houry, D. E. (2008). Intimate partner violence screening in the emergency department. *Journal of Postgraduate Medicine*, *54*(4), 301–305.

Everly, G. S., & Mitchell, J. T. (2010). *A primer on critical incident stress management (CISM)*. Available at: www.icisf.org/who-we-are/what-is-cism

Flatt, A. E. (2010). Frostbite. *Baylor University Medical Center Proceedings*, *23*(3), 261–262.

Greenlund, L. (2011). ED violence: Occupational hazard? *Nursing Management*, *42*(7), 28–32.

Isbister, G. K., Shahmy, S., Mohamed, F., et al. (2012). A randomised controlled trial of 2 infusion rates to decrease reactions to antivenom. *PLoS One*, *7*(6), e38739; doi:10.1371/journal.pone.0038739.

Jerry, J., Collins, G., & Streem, O. (2012). Synthetic legal intoxicating drugs: The emerging 'incense' and 'bath salt' phenomenon. *Cleveland Clinic Journal of Medicine*, *79*(4), 258–264.

Kragh, J. F., Walters, T. J., Baer, D. G., et al. (2008). Practical use of emergency tourniquets to stop bleeding in major limb trauma. *Journal of Trauma*, *64*(2), S38–S50.

Kumar, S., Grover, S., Kulhara, P., et al. (2008). Inhalant abuse: A clinic-based study. *Indian Journal of Psychiatry*, *50*(2), 117–120.

Larson, M. (2008). Alcohol-related psychosis. *Medscape*. Available at: emedicine.medscape.com/article/289848-overview

LaMantia, M. A., Platts-Mills, T. F., Biese, K., et al. (2010). Predicting hospital admission and return to the ED for elderly patients. *Academic Emergency Medicine*, *17*(3), 252–259.

Lubich, C., & Krenzelok, E. P. (2009). Exotic snakes are not always found in exotic places: How poison centers can assist emergency departments. *BMJ Case Reports*, doi:10.1136/bcr.11.2008.1229. Epub 2009 May 10.

Luce, H., Schrager, S., & Gilchrist, V. (2010). Sexual assault of women. *American Family Physician*, *81*(4), 489–495.

Moriarty, R. S., Dryer, S., Repolgie, W., et al. (2012). The role for coagulation markers in mild snakebite envenomations. *Western Journal of Emergency Medicine*, *13*(1), 68–74.

National Center on Elder Abuse. (2005). Fact sheet. *Statistics at a Glance*. Available at:www.ncea.aoa.gov/Resources/Publication/docs/FinalStatistics050331.pdf

Newell, M. A., Bard, M. R., Goettler, C. E., et al. (2007). Body mass index and outcomes in critically injured blunt trauma patients: Weighing the impact. *Journal of the American College of Surgeons, 204*(5), 1056–1061.
*Oman, K. S., & Duran, C. R. (2010). Health care providers evaluation of family presence during resuscitation. *Journal of Emergency Nursing, 36*(6), 524–533.
Szpilman, D., Bierens, J. J., Handley, A. J., et al. (2012). Current concepts: Drowning. *New England Journal of Medicine, 366*(22), 2102–2110.
Travers, A. H., Rea, T. D., Bobrow, B. J., et al. (2010). Part 4: CPR overview: 2010 American Heart Association guidelines for cardiopulmonary resuscitation and emergency cardiovascular care. *Circulation, 122*(Suppl. 3), S676–S684.
*Wong, J. Y., Tiwari, A., Fong, S. Y. et al. (2011). Depression among women experiencing intimate partner violence in a Chinese community. *Nursing Research, 60*(1), 58–65.

Resources

American Association of Poison Control Centers (AAPCC), www.aapcc.org
American College of Surgeons (ACS), Committee on Trauma, www.facs.org
American Heart Association, www.heart.org
American Trauma Society (ATS), www.amtrauma.org
Centers for Disease Control and Prevention (CDC), www.cdc.gov
Divers Alert Network (DAN), www.diversalertnetwork.org
Emergency Nurses Association (ENA), www.ena.org
National Institute on Drug Abuse, www.drugabuse.gov/drugs-abuse/club-drugs
National Safety Council, www.nsc.org
Rape, Abuse, and Incest National Network (RAINN), National Sexual Assault Hotline, www.rainn.org, 1-800-656-HOPE (1-800-656-4673)
Society of Trauma Nurses (STN), www.traumanurses.org

Chapter 73

Terrorism, Mass Casualty, and Disaster Nursing

Learning Objectives

On completion of this chapter, the learner will be able to:

1 Identify the necessary components of an emergency operations plan.
2 Discuss how triage in a disaster differs from triage in an emergency department.
3 Evaluate the different levels of personal protection and decontamination procedures that may be necessary during an event involving mass casualties or weapons of mass destruction.
4 Describe the physical injuries that may occur after blast events and isolation precautions necessary for bioterrorism agents.
5 Identify the differences among the various chemical agents used in terrorist events, their effects, and the decontamination and treatment procedures that are necessary.
6 Determine the injuries associated with varying levels of radiation or chemical exposure and the associated decontamination processes.

Glossary

biologic weapon: a biologic agent, such as anthrax, that is used to spread disease among the general population or the military

chemical weapon: a chemical agent, such as chlorine, that is used to cause disability and mortality in the general population or the military

decontamination: process of removing, or rendering harmless, contaminants that have accumulated on personnel, patients, and equipment

mass casualty incident (MCI): situation in which the number of casualties exceeds the number of available resources

material safety data sheet (MSDS): provides information to employees and health care providers regarding specific chemical agents; includes chemical name, physical data, chemical ingredients, fire and explosive hazard data, health and reactive data, spill or leak procedures, special protection information, and special precautions; also known as the Worker's Right to Know

personal protective equipment (PPE): equipment beyond standard precautions; may include different levels of equipment to provide complete protection, depending on the nature of the suspected biologic, chemical or radiologic agent

radiologic weapon: by-products of radiation contamination that are used to cause morbidity and mortality in the general population or the military

terrorism: unlawful use of violence or threats of violence against people in order to coerce or intimidate

weapons of mass destruction (WMD): weapons used to cause widespread death and destruction

The possibility and reality of mass casualties associated with disasters, terrorism, and biologic warfare are not new to human history, nor is the concept of using **weapons of mass destruction (WMD)** (weapons used to cause widespread death and destruction). In fact, the use of WMD dates as far back as the 6th century BCE (for the use of biologic weapons) and the year 436 BCE (for the use of chemical weapons) (U.S. Army Medical Research Institute of Chemical Defense, 2012). However, geopolitical forces and interests and the availability of destructive technology have brought the possibility of more terrorist events to our doorstep. **Terrorism** systematically involves using violence to create feelings of fear. Examples include the total destruction of the World Trade Center towers and the damage to the Pentagon on September 11, 2001; the London subway bombing of 2005; and the Boston Marathon bombing in 2013. Terrorists have become increasingly sophisticated, organized, and therefore effective. It is no longer a question of *whether* a terrorist event will again lead to mass casualties, but *when* such an event will occur.

The U.S. Department of Homeland Security was created after the attacks of September 11, 2001, to coordinate federal and state efforts to combat terrorist activity. Every facility receiving preparedness funds must have a plan that adheres to guidelines devised by the National Incident Management System (NIMS), which is directed by the Federal Emergency Management Agency (FEMA). The preparation based on the NIMS guidelines is effective for terrorist events, as well as any disaster situation, including natural disasters. Floods, tornadoes, hurricanes (e.g., Hurricane Katrina, the category 5 storm that struck New Orleans, Louisiana, and Biloxi, Mississippi, in 2006), fires, and earthquakes kill and injure hundreds of thousands of people worldwide each year. Notably, 2011 was the worst year for deaths by tornado in the United States and also significant for the earthquake and tsunami in Japan followed by the nuclear meltdown (Veenema, 2013).

Warfare, terrorism, and natural disasters are just some of the reasons nurses must plan for mass casualties. Airplane crashes, train crashes, toxic substance spills, and infectious disease outbreaks are other disasters that can result in mass casualties and tax the resources of health care facilities and their communities. Acute care facilities must be prepared for any and all of these disasters. This chapter focuses on disaster preparedness, including information about possible terrorist-related injuries and illnesses that can occur after biologic, chemical, and nuclear or radiation attacks. Information about the process that nurses should follow in order to respond to these emergencies is applicable to other types of mass disasters as well.

Federal, State, and Local Responses to Emergencies

Many resources are available at the federal, state, and local levels to assist in the management of **mass casualty incidents (MCIs)**, disasters, and emergencies. An MCI is defined as any incident that causes a large numbers of casualties to the extent that necessary resources become too scarce (American College of Surgeons [ACS], 2010). When resources to care for casualties become scarce, the greatest good for the greatest number of patients becomes the mode of operation (Chart 73-1). Local communities must be prepared to act in isolation (i.e., called *sustainability planning*) and provide competent care for up to 5 days before federal or other state resources may become available (ACS, 2010).

Disasters are often categorized by level to indicate the anticipated level of response (Chart 73-2). A list of local resources with specific instructions about how and when to contact these agencies or organizations should be readily available to local disaster planning committees and frequently reviewed by those committees for necessary updates.

Chart 73-2 Disaster Levels

Disasters are often classified by the resultant anticipated necessary response:

Level I: Local emergency response personnel and organizations can contain and effectively manage the disaster and its aftermath.

Level II: Regional efforts and aid from surrounding communities are sufficient to manage the effects of the disaster.

Level III: Local and regional assets are overwhelmed; statewide or federal assistance is required.

Adapted from Goolsby, C. A., & Mothershead, J. L. (2011). Disaster planning. *Medscape*. Available at: emedicine.medscape.com/article/765495-overview

A disaster response strategy cannot succeed without appropriate physical assets and a staff trained and prepared to carry out the plan. Assets such as increased security; stockpiles of equipment and medications; and planning, drills, and training are essential (Veenema, 2013). Hazard vulnerability assessments should be performed to identify potential and actual threats that involve a particular facility and community. Mutual aid agreements among various communities must take these vulnerabilities into account. Successful execution of a response plan is based on knowledge, confidence, and readiness.

Federal Agencies

In 2009, the National Health Security Strategy was designed to protect the health of all citizens of the United States in the event of any large-scale incident (Veenema, 2013). This is a national strategy targeted at prioritizing the use of limited resources and ensuring a rapid coordinated response by the entire affected community so that the maximum number of

Chart 73-1 ETHICAL DILEMMA

Who Receives Care First During a Mass Casualty?

Case Scenario

You work as a staff nurse in an emergency department (ED) of a community hospital. The community high school and its athletic complexes are across the street from the hospital. One Friday evening at 9 PM, you are at work and hear a loud boom. The floor and equipment violently shake for more than a minute, and the ambulance door shatters. You hear widespread screams and see a black cloud of smoke rise from the football field across the street. Within minutes, dozens of patients, some aided by family members and friends, begin to arrive to the ED with lacerations, traumatic amputations, obviously displaced fractures, and other traumatic injuries. A young woman runs screaming into the ED through the blown-out ambulance door and grabs your arm, shrieking, "Please come with me—my little girl is crushed! She's over there! She's not moving! Please, dear God, come with me and save my child!"

Discussion

One of the most difficult and morally distressing situations that ED nurses may face is how to triage and treat patients when resources are limited. Participating in disaster drills may prepare ED nurses in how to respond if an emergency operations plan needs to be executed for a true mass casualty situation. The key principle of triage in these situations is to do the greatest good for the greatest number of people.

Analysis

- Describe the ethical principles that are in conflict in this case (see Chart 3-3 in Chapter 3). Which principle should have preeminence as you respond to this mother?
- Reflect upon your own desire to do good (i.e., practice beneficence) and how this desire to help a mother and young girl may affect your triage decision. How would you respond to the mother? What are your professional obligations to the patients within the ED? Your fellow ED clinician colleagues?

Resources

See Chapter 3, Chart 3-6 for ethics resources.

lives are saved, property is preserved and protected, and basic health care needs are met in the aftermath of any incident that results in mass casualties (U.S. Department of Health and Human Services [HHS], 2011). State authorities must request federal assistance with resources through appropriate government channels. A request for federal resources generally is made when local resources have become or are expected to become depleted.

Federal agencies that may provide resources in response to an MCI or a disaster include the HHS, the Department of Justice, the Department of Defense, and the Department of Homeland Security. Each of these federal departments oversees hundreds of agencies that may respond to MCIs. For example, personnel from the Federal Bureau of Investigation (FBI) (structured under the Department of Justice) may be used for scene control and collection of forensic evidence. FEMA, which is overseen by the Department of Homeland Security, can activate teams such as the Urban Search and Rescue Teams (USRTs). The HHS administers the Centers for Disease Control and Prevention (CDC) and the National Disaster Medical System. The National Disaster Medical System has many medical support teams, such as Disaster Medical Assistance Teams (DMATs), Disaster Mortuary Operational Response Teams (DMORTs), Veterinary Medical Assistance Teams (VMATs), and National Medical Response Teams for Weapons of Mass Destruction (NMRTs).

The DMAT organizes voluntary medical personnel who can set up and staff a field hospital; DMATs are located across the country. There are four NMRTs: the mobile California, North Carolina, and Colorado teams, and the Washington, DC, team, which is stationary. These specialty teams were developed to respond to situations involving WMD. They consist of specially trained medical and technical personnel. The National Guard Civil Support Teams are specially trained units that may respond to MCIs or disasters. In addition, the Department of Homeland Security designates a level of security threat that is intended to alert the country to credible threats (Table 73-1).

Additional federal resources that may be activated include teams from the CDC. This is the main federal agency for disease prevention, and it controls activities and provides backup support to state and local health departments. Additional support is available from the American Red Cross, which provides many support systems and shelters as needed.

TABLE 73-1 Department of Homeland Security Threat Levels

Level of Security Threat	Actions
Imminent (formerly Level Red): Severe, usually site is specified; credible impending threat	Lockdown for security Prevention of parking close to the site Activation of the emergency operations plan and Incident Command System
Elevated (formerly Level Orange or Yellow): High/elevated risk of attack but a specific site or timing may not yet be identified	Increased security and screenings, awareness and monitoring Continued training for the Emergency Operations Plan
Sunset Provision	Specific threats will have an expiration date/time instead of continuous alert

From U.S. Department of Homeland Security. (2012). *NTAS public guide: National terrorism advisory system*. Available at: www.dhs.gov/ntas-public-guide#0

State and Local Agencies

Some state and local agencies may be branches of the same agencies already listed (e.g., local CDC and FBI). Other state and local resources may include the American Red Cross, poison control centers, and other local volunteer organizations. The Metropolitan Medical Response Teams Systems are local teams of health care providers who are located in cities considered possible terrorist targets and are funded for specialty response to WMD. Many state and federal task forces have been developed to assist in the development and improvement of civilian medical response to chemical and biologic terrorism.

Most cities and all states have an Office of Emergency Management (OEM) or Office of Emergency Services (OES). The OEM/OES coordinates the disaster relief efforts at the state and local levels. The OEM/OES is responsible for providing interagency coordination during an emergency. It maintains a corps of emergency management personnel, including a leader, responders, planners, and administrative and support staff.

The Incident Command System

The Incident Command System (ICS) is a federally mandated command structure that coordinates personnel, facilities, equipment, and communication in any emergency situation. The ICS is the center of operations for organization, planning, and transport of patients in the event of a specific local MCI. Successful incident management requires equipment compatibility, effective communication, adequate distribution of resources, and clear differentiation of members' roles. The ICS ensures that any hazardous substances used during an MCI are identified promptly and that appropriate personal protection equipment is distributed. In addition to all of these responsibilities, the ICS is also responsible for determining when an MCI has ended.

The Hospital Incident Command System (HICS) is a modification of the ICS that is used by both hospitals and law enforcement agencies. The HICS incident commander is the hospital emergency preparedness coordinator who oversees and coordinates all efforts surrounding the event. The HICS team includes a safety officer, public information officer, liaison officer, operations chief, logistics chief, planning chief, and finance chief. Each team member has a specific responsibility and communicates directly back to the incident commander (ACS, 2008; ACS, 2010).

Hospital Emergency Preparedness Plans

Health care facilities are required by the Joint Commission to create a plan for emergency preparedness and to practice this plan at least twice a year (Joint Commission, 2011). Generally, these plans are developed by the facility's safety/disaster management committee and are overseen by an administrative liaison.

Before the basic emergency operations plan (EOP) can be developed, the planning committee of the health care facility first evaluates characteristics of the community to identify the likely types of natural and man-made disasters that might occur. This hazard vulnerability analysis process is the responsibility of the local health care facility and its safety committee, safety officer, or emergency department (ED) manager. This information can be gathered by questioning local law enforcement, fire departments, and emergency medical systems and assessing the patterns of local train traffic, automobile traffic, and flood, earthquake, tornado, or hurricane activity. Consideration is also given to possible mass casualties that could arise because of the community's proximity to chemical plants, nuclear facilities, or military bases. Federal, judicial, or financial buildings, schools, and any places where large groups of people gather can be considered high-risk areas.

The emergency preparedness planning committee must have a realistic understanding of its resources. It must determine, for example, whether the facility has or needs a pharmaceutical stockpile (e.g., vaccinations, antibiotics) available to treat specific chemical or biologic agents (ACS, 2010). Another scenario that might be anticipated may include the dispersal of a pulmonary intoxicant or choking agent, which would require that emergency operations planners determine how many ventilators are available within the facility and throughout the greater community. The committee might also outline how staff would triage and assign priority to patients when the number of ventilators is limited. Multiple factors influence a facility's ability to respond effectively to a sudden influx of injured patients, and the committee must anticipate various scenarios to improve its preparedness.

Components of the Emergency Operations Plan

The principles of emergency management must be a part of the EOP design and include a comprehensive plan for tackling all potential and actual hazards. The main goal is protection of the community. The EOP should be integrated with local, state, and federal government plans and coordinated with the private sector and volunteers. It must be coordinated in advance to achieve a single common purpose, yet the plan must be flexible enough to adapt to any likely situation at hand. Predetermined organization is essential to minimize confusion, ensure that all key operations are directed, identify and correct flaws in the plan, and promote a well-coordinated response. Essential components of the EOP include the following (ACS, 2010):

- *Activation response:* The EOP activation response of a health care facility defines where, how, and when the response is initiated.
- *Internal/external communication plan:* Communication is critical for all parties involved, including communication to and from the prehospital arena.
- *Plan for coordinated patient care:* A response is planned for organized patient care into and out of the facility, including transfers from within the hospital to other facilities. The site of the disaster can determine where the greater number of patients may self-refer.
- *Security plans:* A coordinated security plan involving facility and community agencies is key to the control of an otherwise chaotic situation.
- *Identification of external resources:* Resources outside of the facility are identified, including local, state, and federal resources and information about how to activate these resources.
- *Plan for people management and traffic flow:* "People management" includes strategies to manage the patients, the public, the media, and personnel. Specific areas are assigned and a designated person is delegated to manage each of these groups
- *Data management strategy:* A data management plan for every aspect of the disaster will save time at every step. A backup system for charting, tracking, and staffing is developed if the facility has a computer system.
- *Demobilization response:* Deactivation of the response is as important as activation; resources should not be unnecessarily exhausted. The person who decides when the facility resumes daily activities is clearly identified. Any possible residual effects of a disaster must be considered before this decision is made.
- *After-action report or corrective plan:* Facilities often see increased volumes of patients 3 months or more after an incident. Postincident response must include a critique and a debriefing for all parties involved, immediately and again at a later date.
- *Plan for practice drills:* Practice drills that include community participation allow for troubleshooting any issues before a real-life incident occurs.
- *Anticipated resources:* Food and water must be available for staff, families, and others who may be at the facility for an extended period.
- *MCI planning:* MCI planning includes such issues as planning for mass fatalities and morgue readiness.
- *Education plan for all of the above:* A strong education plan for all personnel regarding each step of the plan allows for improved readiness and additional input for fine-tuning of the EOP.

As noted earlier, hospitals are required to periodically hold disaster drills. Results from these drills can identify flaws within the EOP as well as unanticipated needs prior to any real disaster situation. Full-scale regional exercises that coordinate responses from both hospitals and emergency medical services (EMS) have been demonstrated as the most effective drills because they clearly identify breakdowns in communication (Klima, Seiler, Peterson, et al., 2012).

Initiating the Emergency Operations Plan

Notification of a disaster situation to a health care facility varies with each situation. Generally, the notification to the facility comes from outside sources unless the initial incident occurred at the facility. The disaster activation plan should clearly state how the EOP is to be initiated. If communication is functioning, field incident command will give notice of the approximate number of arriving patients, although the number of self-referring patients will not be known.

Identifying Patients and Documenting Patient Information

Patient tracking is a critical component of casualty management. Disaster tags, which are numbered and include triage priority, name, address, age, location and description of injuries, and treatments or medications given, are used to communicate patient information. The tag should be securely placed on the patient and remain with the patient at all times. The tag number and the patient's name, if known, are recorded in a disaster log. The log is used by the command center to track patients, assign beds, and provide families with information.

Triage

Triage is the sorting of patients to determine the priority of their health care needs and the proper site for treatment. In nondisaster situations, health care workers assign a high priority and allocate the most resources to those who are the most critically ill (see Chapter 72). For example, a young adult who has a chest injury and is in full cardiac arrest would receive advanced cardiopulmonary resuscitation, including medications, chest tubes, intravenous (IV) fluids, blood, and possibly emergency surgery in an effort to restore life. However, in a disaster, when health care providers are faced with a large number of casualties, the fundamental principle guiding resource allocation is to do the greatest good for the greatest number of people. Decisions are based on the likelihood of survival and consumption of available resources. Therefore, this same patient, and others with conditions associated with a high mortality rate, will be assigned a low triage priority in a disaster situation, even if the person is conscious. Although this may sound uncaring, from an ethical standpoint the expenditure of limited resources on people with a low chance of survival, and denial of those resources to others with serious but treatable conditions, cannot be justified.

The triage officer rapidly assesses those injured at the disaster scene. Patients are immediately tagged and transported or given lifesaving interventions. One person performs the initial triage while other EMS personnel perform immediate lifesaving measures (e.g., intubation) and transport patients. Although EMS personnel carry out initial field triage, secondary and continuous triage at all subsequent levels of care is essential.

Staff should control all entrances to the acute care facility so that incoming patients are directed to the triage area first. Traffic control within the facility is one of the most important components of managing the disaster and resources (ACS, 2010). The triage area may be outside the entry or just at the door of the ED. This facilitates the triage of all patients—those arriving by medical transport and those who walk into the ED. Some patients who have already been seen in the field may be reclassified in the triage area based on their current presentation.

Triage categories separate patients according to severity of injury. A special color-coded tagging system is used during an MCI so that the triage category is immediately obvious. There are several triage systems in use across the country, and every nurse should be aware of the system used by his or her facility and community. The North Atlantic Treaty Organization (NATO) triage system is one that is widely used. It consists of four colors: red, yellow, green, and black. Each color signifies a different level of priority. Table 73-2 describes each category and gives examples of how different injuries would be classified. (See Chapter 72 for discussion of triage in non-MCI situations.)

Managing Internal Problems

The Red Cross has developed a basic survival/shelter resource kit; however, each facility must determine its supply lists based on its own needs assessment. The EOP committee should determine the top 10 critical medications used during normal day-to-day operations and then anticipate which other medications may be required in a disaster or an MCI. For example, the health care facility might plan to have available a stockpile of antidotes (e.g., cyanide kits) or antibiotics used in treating biologic agents. Information should be available about stocking or restocking any of the basic and special supplies, how those supplies are requested, and the time required to receive those supplies.

Communicating With the Media and Family

Communication is a key component of disaster management. Communication within the vast team of disaster responders is

TABLE 73-2 Triage Categories During a Mass Casualty Incident

Triage Category	Priority	Color	Typical Conditions
Immediate: Injuries are life threatening but survivable with minimal intervention. Individuals in this group can progress rapidly to expectant if treatment is delayed.	1	Red	Sucking chest wound, airway obstruction secondary to mechanical cause, shock, hemothorax, tension pneumothorax, asphyxia, unstable chest and abdominal wounds, incomplete amputations, open fractures of long bones, and 2nd/3rd degree burns of 15%–40% total body surface area
Delayed: Injuries are significant and require medical care but can wait hours without threat to life or limb. Individuals in this group receive treatment only after immediate casualties are treated.	2	Yellow	Stable abdominal wounds without evidence of significant hemorrhage; soft tissue injuries; maxillofacial wounds without airway compromise; vascular injuries with adequate collateral circulation; genitourinary tract disruption; fractures requiring open reduction, débridement, and external fixation; most eye and central nervous system injuries
Minimal: Injuries are minor, and treatment can be delayed hours to days. Individuals in this group should be moved away from the main triage area.	3	Green	Upper extremity fractures, minor burns, sprains, small lacerations without significant bleeding, behavioral disorders or psychological disturbances
Expectant: Injuries are extensive, and chances of survival are unlikely even with definitive care. Persons in this group should be separated from other casualties, but not abandoned. Comfort measures should be provided when possible.	4	Black	Unresponsive patients with penetrating head wounds, high spinal cord injuries, wounds involving multiple anatomic sites and organs, 2nd/3rd degree burns in excess of 60% of body surface area, seizures or vomiting within 24 hours after radiation exposure, profound shock with multiple injuries, agonal respirations; no pulse, no blood pressure, pupils fixed and dilated

Adapted from American College of Surgeons. (2010). *Disaster management and emergency preparedness course*. Chicago: Author.

paramount; however, effective, informative communication with the media and worried family members is also crucial.

Managing Media Requests for Information

Although the media have an obligation to report the news and can play a significant positive role in communication, the number of reporters and newscasters and their support teams can be overwhelming, possibly compromising operations and patient confidentiality. The disaster plan should include a clearly defined process for managing media requests, including a designated spokesperson, the public information officer, a site for the dissemination of information (away from patient care areas), and a regular schedule for providing updates.

The EOP helps prevent the release of contradictory or inaccurate information. Initial statements should focus on current efforts and what is being done to better understand the scope and impact of the situation. Information about casualties should not be released. Security staff should not allow media personnel access to patient care areas. However, media resources may be mobilized to notify the general population when disease containment is needed in case of an epidemic or the potential for one, including the location of shelters, necessity of quarantines, and point of dispensing units in the case of bioterrorism (ACS, 2010).

Caring for Families

Friends and family members converging on the scene must be cared for by the facility. The public information officer's role is to provide direction for the families and provide them with information as it becomes available. They may be feeling intense anxiety, shock, or grief and should be provided with information and updates about their loved ones as soon as possible and regularly thereafter. They should not be in the triage or treatment areas but in a designated area staffed by available social workers, counselors, therapists, or clergy. Access to this area should be controlled to prevent families from being disturbed. Information regarding loved ones can also be obtained at this time, which assists in identification of both the injured and deceased. Chart 73-3 discusses cultural variables to consider when coping with disaster-related injuries and death.

Chart 73-3 Cultural Considerations

Any disaster or mass casualty incident can be expected to involve members of diverse religious, ethnic, and cultural groups or may be targeted at and predominately affect a specific religious or ethnic group. Health care providers likewise include members of all religious, ethnic, and cultural backgrounds and should bear in mind that victims may have needs relating to the following:

- Language difficulties that increase fears and frustrations
- Specific religious practices related to medical treatment, hygiene, or diet
- Specific places/times for prayer
- Rituals about handling the dead
- Timing of funeral services
- Family roles and extended family importance
- Privacy

Some religious communities have plans for emergencies and disasters, and local hospitals should integrate these plans to the extent possible into their emergency operations plans.

The Nurse's Role in Disaster Response Plans

Nurses make up the largest component of the health care workforce across the globe (Veenema, 2013). The role of the nurse during a disaster varies. Nurses may be asked to perform duties outside their areas of expertise and may take on responsibilities normally held by physicians or advanced practice nurses. For example, a critical care nurse may intubate a patient or even insert a chest tube. A nurse may perform wound débridement or suturing. A nurse may serve as the triage officer. In these situations, it is imperative that nurses strive to maximize patient safety and be aware of state regulations related to nursing practice (American Nurses Association [ANA], 2008).

Although the exact role of a nurse in disaster management depends on the specific needs of the facility at the time, it should be clear which nurse or physician is in charge of a given patient care area and which procedures each individual nurse may or may not perform. Assistance can be obtained through the HICS, and nonmedical personnel can provide services where possible. For example, family members can provide nonskilled interventions for their loved ones. Nurses should remember that nursing care in a disaster focuses on essential care from a perspective of what is best for all patients. In addition, acquiring knowledge of the hospital disaster plan, participating in drills, and honing competencies relating to disaster management are essential (ANA, 2008). Nurse leaders/administrators should be cognizant of potential security issues and assess and plan for surge capacity capability and in-house resources such as water, supplies, pharmaceuticals, and generator power. They also should predetermine means to evacuate the hospital if necessary, including an ultimate destination or destinations. The hospital must take into account in its plans shortages of all kinds—staff, medications, water/food, and equipment (Hick, Davries, Fink-Kocken, et al., 2012).

New settings and atypical roles for nurses arise during a disaster—for example, the nurse may provide shelter care in a temporary housing area or bereavement support and assistance with identification of deceased loved ones. People may require crisis intervention, or the nurse may participate in counseling other staff members and in Critical Incident Stress Management (CISM). Special care may be warranted for at-risk populations during a disaster (Chart 73-4).

Considering Ethical Conflicts

Disasters can present a disparity between the resources of the health care agency and the needs of the victims. This generates ethical dilemmas for nurses and other health care providers. Issues include conflicts related to the following:

- Rationing care
- Futile therapy
- Consent
- Duty
- Confidentiality
- Resuscitation
- Assisted suicide

Nurses may find it difficult to not provide care to the dying or to withhold information to avoid spreading fear and panic. Clinical scenarios that are unimaginable in normal circumstances confront the nurse in extreme instances. Other ethical dilemmas may arise out of health care providers' instincts for self-protection and protection of their families. For example, what should a pregnant nurse do when incoming disaster

Chart 73-4 Caring for People With Disabilities During a Disaster

When a disaster occurs, the multiple agencies involved attempt to provide food, water, and shelter to all those affected. People with disabilities have specific needs that require attention. It is recommended that people with disabilities have a personal support network to check on them after a disaster and to provide needed assistance. They should also have a backup system and an evacuation plan. Agencies need to be aware that service animals are also affected during a disaster and may be brought to shelters with their companions.

Evacuation assistance is imperative for people with disabilities. Directions to personal equipment (e.g., communication aids, medications, oxygen) should be available to rescue personnel. In a rapid evacuation, mobility devices, oxygen, suction, and medications will be needed at the shelters. Special efforts to keep those with vision or hearing impairment informed should be implemented. People skilled in sign language are also valuable resources during a disaster. On their Web sites, the American Red Cross and the National Organization on Disability provide information for disaster preparedness for people with disabilities (see the Resources section).

Adapted from American Red Cross Disaster Services. (2012). *Preparing for disaster for people with disabilities and other special needs*. Available at: www.redcross.org/prepare/location/home-family/disabilities; and National Organization on Disability. (2009). *Guide for emergency planners, managers, and responders*. Available at: www.nod.org/research_publications/emergency_preparedness_materials/

victims have been exposed to radiation yet too few nurses are available?

Nurses can plan for the ethical dilemmas they will face during disasters by establishing a framework for evaluating ethical questions before they arise and by identifying and exploring possible responses to difficult clinical situations. They can consider how the fundamental ethical principles of nonmaleficence, beneficence, and distributive justice will influence their decisions and care in disaster response (see Chapter 3).

Managing Behavioral Issues

Although most people pull together and function well during a disaster, both people and communities suffer immediate and sometimes long-term psychological trauma (Emergency Nurses Association [ENA], 2013). Common responses to disaster include:

- Depression
- Anxiety
- Somatization (fatigue, general malaise, headaches, gastrointestinal disturbances, skin rashes)
- Posttraumatic stress disorder
- Substance abuse
- Interpersonal conflicts
- Impaired performance

Factors that influence a person's response to disaster include the degree and nature of the exposure to the disaster, loss of friends and loved ones, existing coping strategies, available resources and support, and the personal meaning attached to the event. Other factors, such as loss of home and valued possessions, extended exposure to danger, and exposure to toxic contamination, also influence response and increase the risk of adjustment problems. Those exposed to the dead and injured, those endangered by the event, older adults, children, emergency first responders, and health care personnel caring for victims are considered to be at higher risk for emotional sequelae. A person's normal response to stress and bereavement will also affect the response after a disaster. A bioterrorism attack can have psychological effects, resulting in psychogenic illnesses such as panic, fear, and/or anger (ENA, 2013).

Nurses can assist disaster victims through active listening and providing emotional support, giving information, and referring patients to therapists or social workers. Health care workers must refer people to mental health care services because experience has shown that few disaster victims seek these services, and early intervention minimizes psychological consequences. Nurses can also discourage victims from subjecting themselves to repeated exposure to the event through media replays and news articles, as well as encourage them to return to normal activities and social roles when appropriate (ENA, 2013).

Critical Incident Stress Management

CISM is an approach to preventing and treating the emotional trauma that can affect emergency responders as a consequence of their jobs and that can also occur to anyone involved in a disaster or MCI. CISM is handled by teams, which are available to the OEM. There are 350 such teams in the United States. All branches of emergency services have CISM teams, as do the military and other industries (e.g., airline industry).

Components of a management plan include education (preparedness) before an incident occurs about critical incident stress and coping strategies; field support (ensuring that staff get adequate rest, food, and fluids, and rotating workloads) during an incident; and defusings, debriefings, demobilization, supportive services to the family, and follow-up care after the incident (Veenema, 2013).

Defusing is a process by which the person receives education about recognition of stress reactions and management strategies for handling stress. Debriefing is a more complicated intervention; it involves a 2- to 3-hour process during which participants are asked about their emotional reactions to the incident, what symptoms they may be experiencing (e.g., flashbacks, difficulty sleeping, intrusive thoughts), and other psychological ramifications. In follow-up, members of the CISM team contact the participants of a debriefing and schedule a follow-up meeting if necessary. People with ongoing stress reactions are referred to mental health specialists.

Preparedness and Response

Recognition and Awareness

Preparedness for natural disasters and acts of terrorism includes awareness of the potential for covert use of WMD, self-protection, and early detection, containment, or decontamination of substances and agents that may affect others by secondary exposure. The strength of many toxins, mobility of many members of society, and long incubation periods for some organisms and diseases can result in an epidemic that can quickly and silently spread across the entire country. For example, a formerly healthy person with a rapid onset of flu-like symptoms can have an ominous illness, such as anthrax or severe acute respiratory syndrome (SARS), both of which are discussed later in the chapter.

Nurses should have a heightened awareness of trends that may suggest deliberate dispersal of toxic or infectious agents or pandemic onset that may include the following:

- An unusual increase in the number of people seeking care for fever, respiratory, or gastrointestinal symptoms
- Clusters of patients who present with the same unusual illness from a single location. For example, clusters can be from a specific geographic location, such as a city, or from a single sporting or entertainment event.
- A large number of fatalities, especially when death occurs within 72 hours after hospital admission
- Any increase in disease incidence in a normally healthy population. These cases should be reported to the state health department and to the CDC.

If any of these trends are noted, an extensive patient history is taken in an attempt to identify the possible agent involved. This history includes an occupational, work, and environmental assessment, in addition to the regular admission history. An exposure history contains, at a minimum, information about current and past exposures to possible hazards and an assessment of the patient's typical day and any deviations in routines. The work history includes, at a minimum, a description of all previous jobs, including short-term, seasonal, and part-time employment and any military service. The environmental history includes assessment of present and previous home locations, water supply, and any hobbies, to name a few factors. The admission history should include such information as recent travel and contact with others who have been ill or have recently died of a fatal illness. This is just a brief review of the extensive history that may need to be obtained to identify an exposure agent (Agency for Toxic Substances and Disease Registry, 2012).

Suspicions or findings are reported to the appropriate resources in the facility and to proper authorities in the community. Resources can include the infection control department, the state health department, the CDC, the local poison control center, various Internet sites, and **material safety data sheets (MSDS)** or the Chemtrac database (see the Resource section). The MSDS provides information to employees and health care providers regarding specific chemical agents; it includes the chemical name, physical data, chemical ingredients, fire and explosive hazard data, health and reactive data, spill or leak procedures, special protection information, and special precautions. Reporting furnishes data elements to those agencies responsible for epidemiology and response. Reporting also allows for sharing of information among facilities and jurisdictions and can help determine the source of infections or exposure and prevent further exposures and even deaths.

Personal Protective Equipment

Another component of preparedness and response involves the protection of the health care provider by additional **personal protective equipment (PPE)**. Chemical or biologic agents and radiation are silent killers and are generally colorless and odorless. The purpose of PPE is to shield health care workers from the chemical, physical, biologic, and radiologic hazards that may exist when caring for contaminated patients. The U.S. Environmental Protection Agency (EPA) has divided protective clothing and respiratory protection into the following four categories, levels A through D:

- *Level A* protection is worn when the highest level of respiratory, skin, eye, and mucous membrane protection is required. This includes a self-contained breathing apparatus (SCBA) and a fully encapsulating, vapor-tight, chemical-resistant suit with chemical-resistant gloves and boots.
- *Level B* protection requires the highest level of respiratory protection but a lesser level of skin and eye protection than with level A situations. This level of protection includes the SCBA and a chemical-resistant suit, but the suit is not vapor tight.
- *Level C* protection requires the air-purified respirator, which uses filters or sorbent materials to remove harmful substances from the air. A chemical-resistant coverall with splash hood, chemical-resistant gloves, and boots are included in level C protection.
- *Level D* protection is the typical work uniform.

Levels C and D PPE are the levels most often used in hospital facilities. Protective equipment must be donned before contact with a contaminated patient. The acute care facility's standard precaution PPE (level D) generally is not adequate for protection from a chemically, biologically, or radiologically contaminated patient. Level C PPE is adequate for the average patient exposure. The health care provider must use equipment that is capable of providing protection against the agent involved. This may mean using a splash suit along with a full-face positive- or negative-pressure respirator (a filter-type gas mask) or even an SCBA for medical personnel in the field.

No single PPE is capable of protecting against all hazards. Under no circumstances should responders wear any PPE without proper training, practice, and fit testing of respirator masks as necessary.

Decontamination

Decontamination, the process of removing accumulated contaminants or rendering them harmless, is critical to the health and safety of health care providers by preventing secondary contamination. The decontamination plan should establish procedures and educate employees about decontamination procedures, identify the equipment needed and methods to be used, and establish methods for disposal of contaminated materials (Veenema, 2013).

Although many principles and theories surround decontamination of a patient, authorities agree that to be effective, decontamination must include a minimum of two steps. The first step is removal of the patient's clothing and jewelry and then rinsing the patient with water. Depending on the type of exposure, this step alone can remove a large amount of the contamination and decrease secondary contamination. The second step consists of a thorough soap-and-water wash and rinse. When patients arrive at the facility after being assessed and treated by a prehospital provider, it should not be assumed that they have been thoroughly decontaminated. The hospital must be prepared to perform additional decontamination prior to entry into the facility. The hospital personnel may also treat "walking wounded" who did not receive any decontamination at the scene.

Natural Disasters

Natural disasters may result in mass casualties. Natural disasters can occur anywhere at any time and include events such as tornadoes, hurricanes, floods, avalanches, tidal waves (e.g., tsunamis), earthquakes, and volcanic eruptions (Table 73-3). In the event of a natural disaster, loss of communications, potable water, and electricity is usually the greatest obstacle to a well-coordinated emergency response, and preparatory planning is essential. Even wireless technology (e.g., cellular phones, computers, other communication devices) may not be functional.

The majority of the immediate casualties are trauma related. These mass casualties tax the trauma system to provide triage, transport of patients (in poor weather and road conditions), and management within the trauma centers. Most patients usually begin arriving within an hour of the event. However, the "walking wounded" may not seek care for 5 days to 2 weeks after the event or may seek care for injuries received during cleanup activities. Casualties arrive at hospitals in three waves. The first wave consists of minimally (generally) injured people who arrive of their own accord. The second wave consists of severely injured patients. The third wave consists of injured patients who arrive after they are discovered by rescuers. For example, in the event of earthquakes, buildings collapse and cause the majority of fatalities from injuries primarily involving the head and chest (Veenema, 2013).

TABLE 73-3 Natural Disasters

Event	Issues and Injuries
Hurricanes	Causes flooding and tornadoes (see later discussion) Failure to evacuate Food and water safety *Injuries:* From recovery activities (e.g., chainsaws), stress-related disorders, and GI and other vector-borne disease; physical injury; bites from traumatized pets
Earthquakes	Associated with multiple aftershocks, tsunami Buildings require tethering in earthquake-prone areas *Injuries:* Physical injury; dehydration; pulmonary problems
Flooding	Can accompany other natural disasters Results in home and community destruction *Injuries:* Nonfatal drowning/drowning (e.g., people swept away in currents); waterborne and vector-borne disease (e.g., shigellosis, *Escherichia coli* infection, hepatitis A, giardiasis, leptospirosis, malaria, plague, dengue fever); physical injury from debris
Tsunami	As with flooding (above) but with much more rapid onset resulting in immediate large volume of water on land *Injuries:* Physical injury from debris; vector-borne disease (see the Flooding section); cholera
Tornadoes	Minimal warning, fast moving (approximately 30 mph and travel approximately 20 km) Massive destruction, shelter loss *Injuries:* Physical injury; blastlike effects from pressure (see discussion of blast injuries in text)
Volcanic eruptions	Hazards from lava, openings in ground, gases, ash up to a 20 mile radius *Injuries:* Acid rain; toxic gases result in inhalation injury; physical injury
Exposure—heat/cold	*Cold:* Frostbite, hypothermia, (see Chapter 72) *Heat:* Burns, heat exhaustion/stroke (see Chapter 72)
Epidemic/Pandemic	Foodborne poisonings Acute outbreak, unexpected; local or distant transmission due to travel *Risk:* Determined by population density, travel, method of transmission *Illness:* Depends on organism involved (see Chapter 71)

Adapted from American College of Surgeons. (2010). *Disaster management and emergency preparedness course*. Chicago: Author; and Veenema, T. G. (2013). *Disaster nursing and emergency preparedness* (3rd ed.). New York: Springer.

Excessive exposure to the natural elements and the need for food and water (by both patients and emergency responders) are critical issues. Without cover (e.g., buildings may be unsafe or destroyed) or potable water (e.g., water may be either contaminated or unavailable), injuries from exposure to heat, cold, or contaminated food or water can occur. Safety equipment that protects rescue workers from injury, exposure, and potentially dangerous animals (e.g., snakes, alligators, spiders) must be readily available. Rescue workers may also injure themselves in the process of extrication or cleanup (e.g., chain saws, building collapse). Hypothermia can occur rapidly in workers who are exposed to water at temperatures of 23.9°C (75°F) or less. As is true during all disasters, mental health workers and shelters are needed throughout the community. Veterinary assistance is also essential because pets are frequently abandoned and injured. In addition, emergency response workers must be prepared to treat the most common ailments experienced after exposure to a specific natural disaster. For instance, pulmonary problems peak with earthquakes and volcanic eruptions because of the increased particulate matter in the air. Most volcano-related deaths are from suffocation and exposure to noxious gases. After floods or water disasters, waterborne transmission of agents such as *Escherichia coli*, salmonella, shigella, typhoid, leptospirosis, malaria, and tularemia are common and cause widespread disease.

In some instances, early warning systems have assisted in decreasing the number of deaths from tornadoes and hurricanes. However, even with the advent of early warning systems, some people are unable or unwilling to leave prior to the occurrence of the natural disaster. When buildings collapse, rapid response to identify and remove trapped victims is the only means of improving survivability. There is a direct relationship between time trapped and survival; fewer than 50 percent of people survive if they are trapped more than 2 to 6 hours (ACS, 2010). Water-damaged buildings are not safe and require extensive examination before experts can ensure safe occupancy. Larger-scale issues that can cause significant later morbidity and mortality include the absence of water purification, waste removal, removal of human and animal remains, and vector control. Removal or disposal of biologic, chemical, and nuclear agents must also be considered.

Weapons of Terror

Although biologic, chemical, and radiologic events are not everyday events, they can occur at any facility, and every nurse needs to know the basics of caring for affected patients. An explosion may spread dangerous material. Therefore, treatment of blast injuries must be anticipated and planned.

Blast Injury

A blast may result from terrorism but can also occur anywhere at any time if the right (or wrong) circumstances come together (e.g., welding inside of a tank that formerly contained tar but was not properly cleaned can result in an explosion as well as severe tar burns to the worker).

Types of Explosive Devices

The bomb most commonly utilized by terrorists is the pipe bomb, which contains low-velocity explosives and may also contain nails or other implements that cause more damage when the explosive ignites. Another type of commonly used explosive device is the Molotov cocktail, which uses a common flammable liquid such as gasoline in a glass bottle and a source of ignition, such as a rag. This forms a simple yet effective incendiary device. Other types of explosive devices include fertilizer bombs and dirty bombs, which include a radioactive source that spreads radiation after the initial blast.

Hazards following a bombing include secondary devices (set to explode at a predetermined time, typically after the arrival of rescue personnel); building collapse; contamination from biologic, chemical, or radiologic weapons; and the presence of terrorists among the patients and bystanders. The entire scene of the bombing is a crime scene and is treated as such. Triage of patients involved in a bombing is the same as for all disasters, with a heightened awareness that serious internal injuries from the blast wave may not be immediately evident.

Physical Injuries

Distance from the blast, whether the blast space was enclosed, composition of the explosive, whether a building collapsed, and the efficiency of medical resources available after the blast all affect patient outcomes after a blast injury. The actual blast that occurs during the initial seconds of the bombing causes a pressure wave or primary blast wave. Injuries can result from the impact of the explosion, the primary blast wave, or shrapnel (i.e., debris from the bomb). The majority of injuries are caused by the primary blast wave (ACS, 2010). A blast wave has four effects. These include spalling, which refers to the pressure wave itself; implosion, which refers to rupture of organs from entrapped gases; shearing, which refers to the blast response of different body tissues, dependent on their density; and irreversible work, which refers to the presence of forces that exceed the tensile strength of an organ or tissue. If the blast occurs in an enclosed space, the wave has the opportunity to be reflected and thus amplified (Veenema, 2013).

Concept Mastery Alert

It is important for the nurse to understand that different phases of the blast may result in common injuries, including blast lung, tympanic membrane (TM) rupture, and head and abdominal injuries. The various phases and related injuries are detailed in Table 73-4.

Although approximately 50% to 70% of deaths result from head injuries, most head injuries are not life threatening (ACS, 2010).

Blast Lung

Blast lung results from the blast wave as it passes through air-filled lungs. The result is hemorrhage and tearing of the lung, ventilation–perfusion mismatch, and possible air emboli. Typical signs and symptoms include dyspnea, hypoxia, tachypnea or apnea (depending on severity), cough, chest pain, and hemodynamic instability. Management involves providing respiratory support that includes administration of supplemental oxygen with a nonrebreathing mask but may also require endotracheal intubation and mechanical ventilation. If a hemothorax or pneumothorax is present, a chest tube must be inserted to re-expand the lung. In the event of an air embolus, the patient should be immediately placed in the prone left lateral position to prevent migration of the embolus and will require emergent treatment in a hyperbaric chamber (CDC, 2013). Complications following blast lung can include respiratory failure as well as acute respiratory distress syndrome (see Chapters 21 and 23 for further discussion).

Tympanic Membrane Rupture

TM rupture is the most frequent injury after subjection to a pressure wave because the TM is the body's most sensitive organ to pressure. There is an increased incidence of TM

TABLE 73-4 Phases of Blasts and Associated Common Injuries

Phase of Blast Injury	Common Injuries
Primary: Results from pressure wave	Pulmonary barotraumas, including pulmonary contusions Head injuries, including concussion, other severe brain injuries Tympanic membrane rupture, middle ear injury Abdominal hollow organ perforation, hemorrhage
Secondary: Results from debris from the scene or shrapnel from the bomb	Penetrating trunk, skin, and soft tissue injuries Fractures, traumatic amputations
Tertiary: Results from pressure wave that causes the victim to be thrown	Head injuries Fractures, including skull
Quaternary: Results from preexisting conditions exacerbated by the force of the blast or by postblast injury complications	Severe injuries with complex injury patterns—burns, crush injuries, head injuries Common preexisting conditions that become exacerbated—COPD, asthma, cardiac conditions, diabetes, and hypertension
Quinary: Thought to result from a hyperinflammatory state commonly seen in bystanders near to the blast and due to toxic substances or uncommon explosives	Hyperpyrexia Diaphoresis

COPD, chronic obstructive pulmonary disease.
Adapted from Kluger, Y., Nimrod, A., Biderman, P., et al. (2006). The quinary pattern of blast injury. *Journal of Emergency Management*, 4(1), 51–55.

rupture when a blast occurs in close proximity to the patient and when it occurs in an enclosed space. Signs and symptoms include hearing loss, tinnitus, pain, dizziness, and otorrhea (CDC, 2013). The majority of TM ruptures heal spontaneously. Approximately 5% of patients with TM rupture from a blast will require hearing aids, whereas the majority will suffer only mild high-frequency hearing loss (Ritenour, Wickley, Ritenour, et al., 2008). Other ear injuries may include ossicular disruption and impaction of foreign bodies.

Abdominal and Head Injuries

Blast abdomen may be evidenced by abdominal hemorrhage and internal organ injury. The typical signs and symptoms of internal abdominal injury can include pain, guarding, rebound tenderness, rectal bleeding, nausea, and vomiting (CDC, 2013) (see Chapter 72 for further discussion of abdominal trauma).

Head injuries are typically minor, but those that are severe result in the majority of postblast deaths. These injuries can occur without a direct blow to the head and may result from the blast itself, building collapse, or flying debris. Concussions commonly occur postblast, and the usual follow-up evaluation and treatment for postconcussive syndrome is indicated (see Chapter 68 for further discussion of head injuries). Approximately 30% of head injuries involve vascular structures (e.g., arteriovenous fistula, pseudoaneurysm, or dissection) (Veenema, 2013).

Special Populations

Special populations may have different blast-associated risks. For instance, older adults are particularly susceptible to bone fractures because they tend to have decreased bone density. They also tend to have more preexisting morbid conditions that may be exacerbated by the explosion. Pregnant patients' abdomens are particularly susceptible to placental shear forces that may result in abruptio placentae. People with mobility disabilities may have difficulty extricating themselves from the site of the blast (CDC, 2013).

Biologic Weapons

Biologic weapons are weapons that spread disease among the general population or the military. They can be used for sabotage, such as food or water contamination with a small target area, or may be used by global terrorists with intentions to enable global objectives.

Effects of Biologic Weapons

Biologic weapons are easily obtained and easily disseminated and can result in significant mortality and morbidity (Table 73-5). The potential use of biologic weapons calls for continuous increased surveillance by health departments and an increased index of suspicion by clinicians. Many biologic weapons result in signs and symptoms similar to those of common disease processes. Appropriate management of a biologic threat includes rapid recognition of the potential weapon; the use of proper PPE; decontamination, isolation, or quarantine of infected patients when appropriate; and the administration of appropriate vaccinations, antidotes, or medications to people at risk.

Biologic weapons are delivered in either a liquid or dry state, applied to foods or water, or vaporized for inhalation or direct contact. Vaporization may be accomplished through spray or explosives loaded with the weapon. Because of increases in business and pleasure travel by people in industrialized nations, a biologic weapon could be released in one city and affect people in other cities thousands of miles away. The vector can be an insect, animal, or person, or there may be direct contact with the weapon itself.

Two of the biologic agents most likely to be used or weaponized are discussed in the next section. Table 73-6 describes other easily weaponized biologic agents.

TABLE 73-5 Categories of Biologic Weapons

Category	Mortality and Morbidity	Example of Biologic Agent That May Be Weaponized
Category A	High mortality	*Bacillus anthracis* (anthrax) *Clostridium botulinum* (botulism) *Francisella tularensis* (tularemia) Viral hemorrhagic fevers (e.g., dengue, Ebola) Variola (i.e., smallpox) *Yersinia pestis* (plague)
Category B	Low mortality, moderate morbidity	*Brucella* species (brucellosis) *Coxiella burnetii* (Q fever) *Staphylococcus aureus*, *Vibrio* species (food poisoning) *Rickettsia typhi* (typhus) Arboviruses (viral encephalitis) *Cryptosporidium parvum* (Cryptosporidiosis)
Category C	Low mortality, low morbidity	Hantavirus

Adapted from Veenema, T. G. (2013). *Disaster nursing and emergency preparedness* (3rd ed.). New York: Springer.

Types of Biologic Weapons

Anthrax

Anthrax is recognized as the most likely weaponized biologic agent available and has been recognized as a highly debilitating agent for centuries. *Bacillus anthracis* is a naturally occurring gram-positive, encapsulated rod that lives in the soil in the spore state throughout the world. The bacterium sporulates (i.e., is liberated) when exposed to air and is infective only in the spore form. Contact with infected animal products (raw meat) or inhalation of the spores results in infection. Cattle and other herbivores are vaccinated against anthrax to prevent transmission through contaminated meat. As an aerosol, anthrax is odorless and invisible and can travel a great distance before disseminating; hence, the site of release and the site of infection can be miles apart.

Clinical Manifestations. Anthrax is caused by replicating bacteria that release toxin, resulting in hemorrhage, edema, and necrosis. The incubation period is 1 to 6 days. There are three main methods of infection: skin contact, gastrointestinal ingestion, and inhalation. Skin lesions (the most common infection) cause edema with pruritus and macule or papule formation, resulting in ulceration with 1- to 3-mm vesicles. A painless eschar develops, which falls off in 1 to 2 weeks (Auerbach, 2012).

Ingestion of anthrax results in fever, nausea and vomiting, abdominal pain, bloody diarrhea, and occasionally ascites.

TABLE 73-6 Examples of Biologic Agents That Can Be Used As Weapons

Agent/Organism	Contagion	Decontamination and Protective Equipment	Signs and Symptoms	Treatment (Mortality Rate)
Tularemia—*Francisella tularensis*: Gram-negative coccobacillus, one of the most infectious bacteria known	Direct contact with infected animals or aerosolized as a bioterror weapon; bites Not contagious through human-to-human contact	Standard barrier precautions Clothing and linens should be laundered under the usual hospital protocol.	*Initial:* Abrupt onset of fever, fatigue, chills, headache, lower backache, malaise, rigor, coryza, dry cough, and sore throat without adenopathy. Nausea and vomiting or diarrhea possible. *As disease progresses:* Sweating, fever, progressive weakness, anorexia, and weight loss demonstrate continued illness. *Mortality secondary to:* Pneumonitis (if inhalation is the source) with copious watery or purulent sputum, hemoptysis, respiratory insufficiency, sepsis, and shock	Streptomycin or gentamicin/aminoglycoside for 10–14 days Inhalation tularemia must be treated within 48 hours of onset. In mass casualty situations, doxycycline or ciprofloxacin is recommended. For persons exposed to tularemia, tetracycline or doxycycline is recommended for 14 days. (Mortality rate = 2%)
Botulism—*Clostridium botulinum*: Botulinum blocks acetylcholine-containing vesicles from fusing with the terminal membranes of the motor neuron end-plate, resulting in a flaccid paralysis	Direct contact Not contagious through human-to-human contact	Any skin exposure to the botulism toxin can be treated with soap and water or a 0.1% hypochlorite solution. Standard precautions are used when treating patients with botulism.	*Gastrointestinal botulism:* Abdominal cramps, nausea, vomiting, and diarrhea *Inhalation botulism:* Fever; symmetric descending flaccid paralysis with multiple cranial nerve palsies. Classic signs and symptoms include diplopia, dysphagia, dry mouth, lack of fever, and alert mental status. Other possible symptoms include ptosis of the eyelids, blurred vision, enlarged sluggish pupils, dysarthria, and dysphonia. *Mortality secondary to:* Airway obstruction and inadequate tidal volume	Supportive ventilatory therapy is necessary if respiratory infection occurs. Aminoglycosides and clindamycin are contraindicated because they exacerbate neuromuscular blockage. Equine antitoxin is used to minimize subsequent nerve damage. There is a 2% rate of anaphylaxis to the antitoxin; therefore, diphenhydramine (Benadryl) and epinephrine must be immediately available for use. Supportive care—mechanical ventilation, nutrition, fluids, prevention of complications (Mortality rate = 5%)
Plague—*Yersinia pestis:* non-sporulating gram-negative coccobacillus. The bacterium causes destruction and necrosis of the lymph nodes.	Contagious *Bubonic plague:* Transmitted through flea bites with no person-to-person transmission *Pneumonic plague:* Transmitted through respiratory droplet contact	Isolation barrier precautions with full face respirators; the patient should wear a mask. Rooms should receive a terminal cleaning. Clothing and linens with body fluids on them should be cleaned with the usual disinfectant. Routine precautions should be used in the case of death.	*Bubonic plague:* Sudden fever and chills, weakness, a swollen and tender lymph node (bubo) in the groin, axilla, or cervical area. The resultant bacteremia progresses to septicemia from the endotoxin and, finally, shock and death. *Primary septicemic plague:* Disseminated intravascular coagulation, necrosis of small vessels, purpura, and gangrene of the digits and nose (black death) *Pneumonic plague:* Severe bronchospasm, chest pain, dyspnea, cough, and hemoptysis. There is a 100% mortality associated with pneumonic plague if not treated within the first 24 hours.	Streptomycin or gentamicin for 10–14 days. Tetracycline or doxycycline is an acceptable alternative if an aminoglycoside cannot be given. People with close contact exposure (<2 m) require prophylaxis with doxycycline for 7 days. (Mortality rate = 50%)

Adapted from Dire, D. J. (2011). CBRNE—Biological warfare agents. *Medscape*. Available at: emedicine.medscape.com/article/829613-overview; Emergency Nurses Association. (2013). *Sheehy's manual of emergency care* (7th ed). St. Louis: Elsevier Mosby; and Veenema, T. G. (2013). *Disaster nursing and emergency preparedness* (3rd ed.). New York: Springer.

If severe diarrhea develops, decreased intravascular volume becomes the major treatment concern. The bacterium targets the terminal ileum and cecum. Sepsis can occur.

Inhaling anthrax results in the most severe clinical manifestations. Its symptoms mimic those of the flu, and usually treatment is sought only when the second stage of severe respiratory distress occurs. Current antibiotic therapy does not halt the progress of the disease. Inhaled anthrax can incubate for up to 60 days, making it difficult to identify its source. Initial signs and symptoms include cough, headache, fever, vomiting, chills, weakness, mild chest discomfort, dyspnea, and syncope, without rhinorrhea or nasal congestion.

Most patients have a brief recovery period followed by the second stage within 1 to 3 days, characterized by fever, severe respiratory distress, stridor, hypoxia, cyanosis, diaphoresis, hypotension, and shock. These patients require optimization of oxygenation, correction of electrolyte imbalances, and ventilatory and hemodynamic support. More than 50% of these patients have hemorrhagic mediastinitis on a chest x-ray (a hallmark sign). The disease can also progress to include meningitis with subarachnoid hemorrhage. Death results approximately 24 to 36 hours after the onset of severe respiratory distress. The mortality rate approaches 100% (Auerbach, 2012).

Treatment. At present, anthrax is penicillin sensitive; however, strains of penicillin-resistant anthrax are thought to exist. Recommended treatment includes penicillin (Penicillin V), erythromycin (Erythrocin), gentamicin (Garamycin), or doxycycline (Vibramycin). If antibiotic treatment begins within 24 hours after exposure, death can be prevented. In a mass casualty situation, treatment with ciprofloxacin (Cipro) or doxycycline is recommended, because these easily administered oral antibiotic agents are stockpiled and there should be sufficient dosages to fully treat many anthrax-exposed patients. Treatment is continued for 60 days. For patients who have been directly exposed to anthrax but have no signs and symptoms of disease, ciprofloxacin or doxycycline is used for prophylaxis for 60 days (Auerbach, 2012).

Standard precautions are the only ones indicated to protect the caregiver exposed to a patient infected with anthrax. The patient is not contagious, and the disease cannot spread from person to person. Equipment should be cleaned using standard hospital disinfectant. After death, cremation is recommended because the spores can survive for decades and represent a threat to morticians and forensic medicine personnel. There is a vaccine for anthrax that has been used in the military; however, it is not yet widely used because it requires multiple time-interval–sensitive boosters.

Smallpox

Smallpox (variola) is classified as a deoxyribonucleic acid (DNA) virus. It has an incubation period of approximately 12 days. It is extremely contagious and is spread by direct contact, by contact with clothing or linens, or by droplets from person to person only after the fever has decreased and the rash phase has begun. Aerosolization of the virus would result in widespread dissemination.

The World Health Organization (WHO) declared smallpox eradicated in 1977 and stopped worldwide vaccination in 1980. In the United States, the last child was vaccinated in 1972. Therefore, a large portion of the current population has no immunity to the virus. A smallpox vaccination plan introduced in 2003 proposed that a designated number of ED staff receive the first vaccinations to ensure that ED staff would be immunized in the event of a smallpox outbreak. The government estimated that 0.1% of those people receiving the vaccine would have serious side effects. Of these, approximately 4% would have life-threatening complications, and 0.1% would die. Currently, only people with a high likelihood of exposure to smallpox are encouraged to receive the vaccination (CDC, 2012b).

Clinical Manifestations. Signs and symptoms of smallpox infection include high fever, malaise, headache, backache, and prostration. After 1 to 2 days, a maculopapular rash appears, evolving at the same rate, beginning on the face, mouth, pharynx, and forearms (Fig. 73-1). Only then does the rash progress to the trunk and also become vesicular to pustular (Veenema, 2013). There is a large amount of the virus in the saliva and pustules. Smallpox is contagious only after the appearance of the rash. There are two forms of smallpox: variola major and variola minor. Variola major is more common, results in a higher fever and more extensive rash, and has a 30% case fatality rate (i.e., the likelihood of fatality per case diagnosed). Hemorrhagic smallpox, a subtype of variola major, includes all of the above signs and symptoms plus a dusky erythema and petechiae leading to frank hemorrhage of the skin and mucous membranes, and it results in death by day 5 or 6 (Veneema, 2013).

Treatment. Treatment includes supportive care with antibiotic agents for any additional infection. The patient must be isolated with the use of transmission precautions. Laundry and biologic wastes should be autoclaved before being washed with hot water and bleach. Standard decontamination of the room is effective. All people who have household or face-to-face contact with the patient after the fever begins should be vaccinated within 4 days to prevent infection and death. A patient with a temperature of 38°C (101°F) or higher within 17 days after exposure must be placed in isolation. Cremation is preferred for all deaths, because the virus can survive in scabs for up to 13 years (Veneema, 2013).

Severe Acute Respiratory Syndrome

Not all mass casualty biologic events are terrorist based. Air travel and worldwide trade have increased the possibility that any contagious disease may spread rapidly. SARS, for example, is caused by a virus, officially named SARS coronavirus

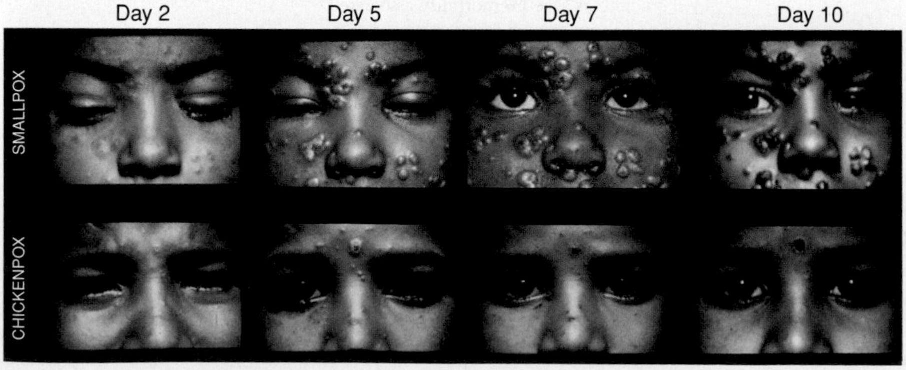
Day 2
Day 5
Day 7
Day 10
SMALLPOX
CHICKENPOX

FIGURE 73-1 • Comparison of progression of smallpox rash and chickenpox rash. From World Health Organization. (2001). *WHO slide set on the diagnosis of smallpox*. Reproduced by permission of the World Health Organization. Available at www.who.int/emc/diseases/smallpox/slideset/index.htm

TABLE 73-7 Common Chemical Agents

Agent	Action	Signs and Symptoms	Decontamination and Treatment
Nerve Agents			
Sarin Soman	Inhibition of cholinesterase	Increased secretions, gastrointestinal motility, diarrhea, bronchospasm	Soap and water Supportive care Benzodiazepines Pralidoxime Atropine
Blood Agent			
Cyanide	Inhibition of aerobic metabolism	Inhalation—tachypnea, tachycardia, coma, seizures; can progress to respiratory arrest, respiratory failure, cardiac arrest, death	Sodium nitrite Sodium thiocyanate Amyl nitrate Hydroxocobalamin
Vesicant Agents			
Lewisite Sulfur mustard Nitrogen mustard Phosgene	Blistering agents	Superficial to partial-thickness burn with vesicles that coalesce	Soap and water Blot; do not rub dry.
Pulmonary Agents			
Phosgene Chlorine	Separation of alveoli from capillary bed	Pulmonary edema, bronchospasm	Airway management Ventilatory support Bronchoscopy

Adapted from Ciottone, G. R., & Arnold, J. L. (2008). Chemical warfare agents. *Medscape*. Available at: emedicine.medscape.com/article/829454-overview; Emergency Nurses Association. (2013). *Sheehy's manual of emergency care* (7th ed). St Louis: Elsevier Mosby; and Veenema, T. G. (2013). *Disaster nursing and emergency preparedness* (3rd ed.). New York: Springer.

(SARS-CoV). Its incubation period is 2 to 10 days. People at risk include health care workers who have had unprotected exposure to SARS-CoV (CDC, 2012a).

Chemical Weapons

Chemical weapons may be used as agents in warfare or for terrorist purposes; they are overt agents in that the effects are more apparent and occur more quickly than those caused by biologic weapons (Table 73-7). Poisonous exposure to everyday chemicals also may occur; the same management principles discussed later apply when patients are exposed to these chemical agents even when not used as weapons. Typical exposures in these instances include industrial chemicals, gasoline, turpentine, kerosene, and insecticides (ENA, 2013).

Characteristics of Chemicals

Volatility

Volatility is the tendency for a chemical to become a vapor. The most common volatile agents are phosgene and cyanide. Most chemicals are heavier than air, except for hydrogen cyanide. Therefore, in the presence of most chemicals, people should stand up to avoid heavy exposure (because the chemical will sink toward the floor or ground).

Persistence

Persistence means that the chemical is less likely to vaporize and disperse. More volatile chemicals do not evaporate very quickly. Most industrial chemicals (e.g., cyanide) are not very persistent. Weaponized agents (chemicals developed as weapons by the military or terrorists [e.g., mustard gas]) are more likely than industrial chemicals to penetrate skin and mucous membranes and also cause secondary exposure.

Toxicity

Toxicity is the potential of an agent to cause injury to the body. The median lethal dose (LD_{50}) is the amount of the chemical that will cause death in 50% of those who are exposed. For example, cyanogen chloride has the highest LD_{50}, twice that of hydrogen cyanide and eight times higher than sulfur mustard (ACS, 2010). The median effective dose (ED_{50}) is the amount of the chemical that will cause signs and symptoms in 50% of those who are exposed. The concentration time (CT) is the concentration released multiplied by the time exposed (in milligrams per minute). For example, if 1,000 mg of a chemical is released and the time a person is exposed to this amount of chemical is 10 minutes, then the CT would be 10,000 mg/min.

Latency

Latency is the time from absorption to the appearance of signs and symptoms. Sulfur mustards and pulmonary agents have the longest latency, whereas other vesicants, nerve agents, and cyanide produce signs and symptoms within seconds.

Limiting Exposure

Evacuation is essential, as is removal of the person's clothing and decontamination as close to the scene as possible and before transport of the exposed person. Soap and water are effective means of decontamination in most cases. Staff involved in decontamination efforts must wear PPE and contain and dispose of the runoff after decontamination procedures (ENA, 2013).

Types of Chemicals

Vesicants

Vesicants are chemicals that cause blistering and result in burning, conjunctivitis, bronchitis, pneumonia, hematopoietic suppression, and death. Examples of vesicants include lewisite, phosgene, nitrogen mustard, and sulfur mustard. In World War I and in the Iran–Iraq conflict of 1980 to 1988, vesicants were used to disable opponents. Vesicants were the main incapacitating agents, resulting in minimal (less than

5%) death but large numbers of injuries (Veenema, 2013). Liquid sulfur mustard was the most frequently used vesicant in these conflicts.

Clinical Manifestations. The initial presentation after exposure to a vesicant is similar to that of a large superficial to partial-thickness burn in the warm and moist areas of the body (i.e., perineum, axillae, antecubital spaces). There is stinging and erythema for approximately 24 hours, followed by pruritus, painful burning, and small vesicle formation after 2 to 18 hours. These vesicles can coalesce into large, fluid-filled bullae. Lewisite and phosgene result in immediate pain after exposure. Tissue damage occurs within minutes.

Concept Mastery Alert

Different types of skin lesions can occur with exposure to chemical agents. Bullae are serous fluid-filled circumscribed areas often resembling a dome, whereas papules are elevated, solid circumscribed masses.

If the eye is exposed, there is pain, photophobia, lacrimation, and decreased vision. This progresses to conjunctivitis, blepharospasm, corneal ulcer, and corneal edema.

Respiratory effects are more serious and often are the cause of mortality with vesicant exposure. Purulent fibrinous pseudomembrane discharge may cause obstruction of the airways. Bacterial pneumonia may be the cause of death within approximately a week of pulmonary exposure (ACS, 2010). Gastrointestinal exposure may cause nausea and vomiting, leukopenia, and upper gastrointestinal bleeding.

Treatment. Appropriate decontamination includes soap and water. Scrubbing and the use of hypochlorite solutions should be avoided because they increase penetration. Once the substance has penetrated, it cannot be removed. Eye exposure requires copious irrigation. For respiratory exposure, intubation and bronchoscopy to remove necrotic tissue are essential. With lewisite exposure, dimercaprol (BAL In Oil) is administered IV for systemic toxicity and topically for skin lesions. All persons with sulfur mustard exposures should be monitored for 24 hours for delayed (latent) effects (ENA, 2013).

Nerve Agents

The most toxic agents in existence are the nerve agents such as sarin, soman, tabun. They are inexpensive, effective in small quantities, and easily dispersed. In the liquid form, nerve agents evaporate into a colorless, odorless vapor. Organophosphates (pesticides) are similar in nature to the nerve agents used in warfare and are readily available in the farming industry.

Nerve agents can be inhaled or absorbed percutaneously or subcutaneously. These agents bond with acetylcholinesterase so that acetylcholine is not inactivated; the adverse result is continuous stimulation (hyperstimulation) of the nerve endings. Carbamates, which are insecticides originally extracted from the Calabar bean, are derivatives of carbamic acid; they are nerve agents that specifically inhibit acetylcholinesterase for several hours and then spontaneously become unbound from the acetylcholinesterase. However, organophosphates require the formation of new enzyme (acetylcholinesterase) before nervous system function can be restored.

A very small drop of a nerve agent is enough to result in sweating and twitching at the site of exposure. A larger amount results in more systemic symptoms. Effects can begin anywhere from 30 minutes up to 18 hours after exposure. The more common organophosphates and carbamates (e.g., Sevin and malathion) that are used in agriculture result in less severe symptoms than do those used in warfare or in terrorist attacks. In an ordinary situation (e.g., nonwarfare, nonterrorist attack situation), a patient could arrive at the ED having been unintentionally exposed to organophosphates or intentionally exposed to these agents in a suicidal attempt (ENA, 2013).

Clinical Manifestations. Signs and symptoms of nerve gas exposure are those of cholinergic crisis and include bilateral miosis, visual disturbances, increased gastrointestinal motility, nausea and vomiting, diarrhea, substernal spasm, indigestion, bradycardia and atrioventricular block, bronchoconstriction, laryngeal spasm, weakness, fasciculations, and incontinence. The patient must be examined in a dark area to truly identify miosis. Neurologic responses include insomnia, forgetfulness, impaired judgment, depression, and irritability. A lethal dose results in loss of consciousness, seizures, copious secretions, fasciculations, flaccid muscles, and apnea (ENA, 2013).

Treatment. Decontamination with copious amounts of soap and water or saline solution for 8 to 20 minutes is essential. The water is blotted off, not wiped off, the skin. Wiping may have the unintended effect of rubbing more of the agent into the skin. Fresh 0.5% hypochlorite solution (bleach) can also be used. The airway is maintained, and suctioning is frequently required. Plastic airway equipment should not be used, because plastic will absorb sarin gas and may result in continued exposure to the agent.

Atropine 2 to 4 mg is administered by IV, followed by 2 mg every 3 to 8 minutes for up to 24 hours of treatment. Alternatively, IV atropine 1 to 2 mg per hour may be administered until clear signs of anticholinergic activity have returned (decreased secretions, tachycardia, and decreased gastrointestinal motility). Another medication that may serve as an antidote is pralidoxime (Protopam), which allows cholinesterase to become active against acetylcholine. Pralidoxime 1 to 2 g in 100 to 150 mL of normal saline solution is administered over 15 to 30 minutes. Pralidoxime has no effect on secretions and may have any of the following side effects: hypertension, tachycardia, weakness, dizziness, blurred vision, and diplopia. Diazepam (Valium) or other benzodiazepines are used to control seizures, to decrease fasciculations, and to alleviate apprehension and agitation (ENA, 2013).

Military personnel believed to be at risk for chemical attack are provided with Mark I automatic injectors, which contain 2 mg atropine and 600 mg pralidoxime chloride. Diazepam may be administered by a partner.

Blood Agents

Blood agents such as hydrogen cyanide and cyanogen chloride have a direct effect on cellular metabolism, resulting in asphyxiation through alterations in hemoglobin. Cyanide is an agent that has profound systemic effects. It is commonly used in the mining of gold and silver and in the plastics and dye industries.

A cyanide release is often associated with the odor of bitter almonds. In house fires, cyanide is released during the combustion of plastics, rugs, silk, furniture, and other construction materials. There is a significant correlation between blood cyanide and carbon monoxide levels in patients who survive fires, and in some cases, the cause of death is cyanide poisoning.

Clinical Manifestations. Cyanide can be ingested, inhaled, or absorbed through the skin and mucous membranes. Cyanide is protein bound and inhibits aerobic metabolism, leading to respiratory muscle failure, respiratory arrest, cardiac arrest, and death. Its inhalation results in flushing, tachypnea, tachycardia, nonspecific neurologic symptoms, stupor, coma, and seizure preceding respiratory arrest (Cline, Ma, Cydulka, et al., 2012).

Treatment. Rapid administration of amyl nitrate, sodium nitrite, and sodium thiosulfate is essential to the successful management of cyanide exposure. First, the patient is intubated and placed on a ventilator. Next, amyl nitrate pearls are crushed and placed in the ventilator reservoir to induce methemoglobinemia. Cyanide has a 20% to 25% higher affinity for methemoglobin than it does for hemoglobin; it binds methemoglobin to form either cyanomethemoglobin or sulfmethemoglobin. The cyanomethemoglobin is then detoxified in the liver by the enzyme rhodanese. Next, IV sodium nitrite is administered to induce the rapid formation of methemoglobin. IV sodium thiosulfate is then administered; it has a higher affinity for cyanide than methemoglobin and stimulates the conversion of cyanide to sodium thiocyanate, which can be excreted by the kidneys (Cline et al., 2012). Although they may be lifesaving, these emergency medications do have side effects—sodium nitrite can result in severe hypotension, and thiocyanate can cause vomiting, psychosis, arthralgia, and myalgia.

The production of methemoglobin is contraindicated in patients with smoke inhalation, because they already have decreased oxygen-carrying capacity secondary to the carboxyhemoglobin produced by smoke inhalation. In facilities where a hyperbaric chamber is available, it may be used to provide oxygenation while the previously discussed therapies are initiated. An alternative suggested treatment for cyanide poisoning is hydroxocobalamin (vitamin B_{12}a). Hydroxocobalamin binds cyanide to form cyanocobalamin (vitamin B_{12}). It must be administered IV in large doses (Cline et al., 2012). Administration of vitamin B_{12} can result in a transient pink discoloration of mucous membranes, skin, and urine. In high doses, tachycardia and hypertension can occur, but they usually resolve within 48 hours.

Pulmonary Agents

Pulmonary agents such as phosgene and chlorine destroy the pulmonary membrane that separates the alveolus from the capillary bed, disrupting alveolar–capillary oxygen transport mechanisms. Capillary leakage results in fluid-filled alveoli. Phosgene and chlorine both vaporize, rapidly causing this pulmonary injury. Phosgene has the odor of freshly mown hay.

Signs and symptoms include pulmonary edema with shortness of breath, especially during exertion. An initial hacking cough is followed by frothy sputum production. A particulate air filter mask is the only protection required to protect health care personnel. Phosgene does not injure the eyes.

Nuclear Radiation Exposure

The threat of nuclear warfare or exposure to a **radiologic weapon** is very real with the availability of nuclear material and easily concealed simple devices, such as the so-called dirty bomb, for dispersal. A dirty bomb is a conventional explosive (e.g., dynamite) that is packaged with radioactive material that scatters when the bomb is detonated. It disperses radioactive material and may be called a *radiologic* weapon, but it is not a *nuclear* weapon, which uses a complex nuclear fission reaction that is thousands of times more devastating than the dirty bomb.

Sources of radioactive material include not only nuclear weapons but also reactors and simple radioactive samples, such as weapons-grade plutonium or uranium, freshly spent nuclear fuel, or medical supplies (e.g., radium, certain cesium isotopes) used in cancer treatments and radiology. Exposure to a large number of people can be accomplished by placing a radioactive sample in a public place. Thousands may be exposed this way; some may be immediately affected, and others may require health monitoring for many years to assess long-term effects.

Nuclear reactor incidents have occurred in nuclear facilities in Japan (2011), Chernobyl in the former Soviet Union (1986) and Three Mile Island, Pennsylvania (1979). There were 31 official deaths on the day of the Chernobyl incident, which involved a core meltdown and explosion, releasing radiation throughout the community. The long-term effects of this incident, including increased incidence of thyroid cancers and leukemia, continue to be evaluated. The town of Chernobyl remains a ghost town.

Any terrorist act or unintentional radiation release can be sizable and may require the entire hospital and prehospital staff to be prepared, recognize signs and symptoms of exposure, and rapidly treat victims without contamination of personnel, visitors, patients, or the facility itself.

Types of Radiation

Atoms consist of protons, neutrons, and electrons. The protons and neutrons are in balance in the nucleus. The protons repel each other because they are all positively charged. The number of protons is specific for each element in the periodic table. There is a specific ratio of protons and neutrons for each different atom, and the result is element stability. When an element is radioactive, there is an imbalance in the nucleus, resulting from an excess of neutrons.

To achieve stability, a radioactive nuclide can eject particles until the most stable number (an even number) of protons and neutrons exists. A proton can become a neutron by ejecting a positron; conversely, a neutron can become a proton by ejecting a negative electron. An alpha particle is released when two protons and two electrons are ejected (beta particles are electrons).

Alpha particles cannot penetrate the skin. A thin layer of paper or clothing is all that is necessary to protect the skin from alpha radiation. However, this low-level radiation can enter the body through inhalation, ingestion, or injection (open wound). Only localized damage occurs.

Beta particles have the ability to moderately penetrate the skin to the layer in which skin cells are being produced. This high-energy radiation can cause skin damage if the skin is exposed for a prolonged period and can cause injury if beta particles penetrate the skin.

Gamma radiation is a short-wavelength electromagnetic energy that is emitted when there is excess core nucleus energy. Gamma particles are penetrating. Therefore, it is difficult to shield against gamma radiation. X-rays are an example of gamma radiation. Gamma radiation often accompanies both alpha particle and beta particle emission.

Measurement and Detection

Radiation is measured in several different units. The *rad* is the basic unit of measurement. A rad is equivalent to 0.01 J of energy per kilogram of tissue. To determine the damaging effect of the rad, a conversion to the *rem* (roentgen equivalents man) is necessary. The rem reflects the type of radiation absorbed and the potential for damage. For example, 200,000 mrem results in mild radiation sickness (1 rem = 1,000 mrem) (ACS, 2010). Typical natural yearly exposure for a person is 360 mrem. Another important concept is *half-life*. The half-life of a radioactive product is the time it takes to lose half of its radioactivity.

The only way to detect radiation is through a device that determines the exposure per minute. There are various devices for this purpose. The Geiger counter (or Geiger-Mueller survey meter) can measure background radiation quickly through detection of gamma radiation and some beta radiation. With high-level radiation, the Geiger counter may underestimate exposure. Other devices include the ionization chamber survey meter, alpha monitors, and dose rate meters. Personal dosimeters are simple tools that identify radiation exposure and are worn by radiology personnel every day.

Exposure

Exposure is affected by time, distance, and shielding. The longer a person is within the radiation area, the higher the exposure. In addition, the larger the amount of radioactive material in the area, the greater the exposure. The farther away the person is from the radiation source, the lower the exposure. Shielding from the radiation source also decreases exposure.

Three types of radiation-induced injury can occur: external irradiation, contamination with radioactive materials, and incorporation of radioactive material into body cells, tissues, or organs:

- *External irradiation* exposure occurs when all or part of the body is exposed to radiation that penetrates or passes completely through the body. In this type of exposure, the person is not radioactive and does not require special isolation or decontamination measures. Irradiation does not necessarily constitute a medical emergency.
- *Contamination* occurs when the body is exposed to radioactive gases, liquids, or solids either externally or internally. If internal, the contaminant can be deposited within the body. Contamination requires immediate medical management to prevent incorporation.
- *Incorporation* is the actual uptake of radioactive material into the cells, tissues, and susceptible organs. The organs involved are usually the kidneys, bones, liver, and thyroid.

Sequelae of contamination and incorporation can occur days to years later. The thyroid gland can be largely protected from radiation exposure by administration of stable iodine (potassium iodide, or KI) before or promptly after the intake of radioactive iodine. Priorities in the treatment of any type of radiation exposure are always treatment for life-threatening injuries and illnesses first, followed by measures to limit exposure, contamination control, and finally decontamination.

Decontamination

Hospital and community disaster plans should be in effect when managing a radiation disaster. Access restriction is essential to prevent contamination of other areas of the hospital. Triage outside the hospital is the most effective means of preventing contamination of the facility itself. Floors are covered to prevent tracking of contaminants throughout the treatment areas. Strict isolation precautions should be in effect. All air ducts and vents must be sealed to prevent spread. Waste is controlled through double-bagging and the use of plastic-lined containers outside of the facility. All radiation-contaminated waste must be disposed in appropriate color-coded yellow and magenta canisters.

Staff are required to wear protective clothing, such as water-resistant gowns, two pairs of gloves, masks, caps, goggles, and booties. Dosimetry devices should be worn by all staff members participating in patient care. The radiation safety officer in the hospital should be notified immediately to assist with surveys (using a radiation survey meter) of the incoming patients and to provide dosimeters to all staff personnel involved in direct care of exposed patients. There is minimal risk to staff if the patients are properly surveyed and decontaminated.

Each patient arriving at the hospital should first be surveyed with the radiation survey meter for external contamination and then directed toward the decontamination area as needed. The majority of patients can be safely decontaminated with soap and water. Decontamination occurs outside of the ED with a shower, collection pool, tarp, and collection containers for patient belongings, as well as soap, towels, and disposable paper gowns for patients. Water runoff needs to be contained. Patients who are uninjured can perform self-decontamination with handheld showers. After the patient has showered, a resurvey is conducted to determine whether the radioactive contaminants have been removed. Additional washings should occur until the patient is free of contamination. It is important to ensure that during showers, previously clean areas are not contaminated with runoff from the washed contaminated areas (e.g., hair should be washed in a position that protects the body from contamination). Wounds are irrigated and then covered with a water-resistant dressing prior to total body decontamination.

Internal contamination or incorporation requires decontamination through catharsis, gastric lavage with chelating agents (agents that bind with radioactive substances and are then excreted), or both. Samples of urine, feces, and vomitus are surveyed to determine internal contamination levels. Biologic samples are taken through nasal and throat swabs, and a complete blood count with differential is obtained.

Acute Radiation Syndrome

Acute radiation syndrome (ARS) can occur after exposure to radiation. It is the dose, rather than the source, that determines whether ARS develops. Factors that determine whether the patient's response to exposure will result in ARS include a high dose (minimum 100 rad) and rate of radiation with total body exposure and penetrating-type radiation. Age, medical history, and genetics also affect the outcome after exposure. The course is predictable. Table 73-8 identifies the phases of ARS.

TABLE 73-8 Phases of Effects of Radiation Exposure

Phase	Time of Occurrence	Signs and Symptoms
Prodromal phase (presenting symptoms)	48–72 hours after exposure	Nausea, vomiting, loss of appetite, diarrhea, fatigue High-dose radiation—fever, respiratory distress, and increased excitability
Latent phase (a symptom-free period)	After resolution of prodromal phase; can last up to 3 weeks With high-dose radiation, latent period is shorter.	Decreasing lymphocytes, leukocytes, thrombocytes, red blood cells
Illness phase	After latent period phase	Infection, fluid and electrolyte imbalance, bleeding, diarrhea, shock, and altered level of consciousness
Recovery phase OR	After illness phase	Can take weeks to months for full recovery
Death	After illness phase	Increased intracranial pressure is a sign of impending death.

Adapted from Centers for Disease Control and Prevention. (2006). *Acute radiation syndrome: A fact sheet for physicians*. Available at: www.emergency.cdc.gov/radiation/arsphysicianfactsheet.asp

Each body system is affected differently in ARS. Systems with cells that rapidly reproduce are most commonly affected. The hematopoietic system is the first system affected and serves as an indicator of the severity of radiation exposure (Veenema, 2013). A predictor of outcome is the absolute lymphocyte count at 48 hours after exposure. A significant exposure would be indicated by blood lymphocyte counts of 300 to 1,200/mm^3. Barrier precautions should be implemented to protect the patient from infection. Neutrophils decrease within 1 week, platelets decrease within 2 weeks, and red blood cells decrease within 3 weeks. Hemorrhagic complications including fever and sepsis are common.

The gastrointestinal system, with its rapidly reproducing cells, is also readily affected by radiation. Doses of radiation required to produce symptoms are approximately 600 rad or higher. The gastrointestinal symptoms usually occur at the same time as the changes in the hematopoietic system. Nausea and vomiting occur within 2 hours after exposure. Sepsis, fluid and electrolyte imbalance, and opportunistic infections can occur as complications. An ominous sign is the presence of high fever and bloody diarrhea; these typically appear on day 10 after exposure (CDC, 2006).

The central nervous system is affected when the dose exceeds 1,000 rad (CDC, 2006). The symptoms occur when damage to the blood vessels of the brain results in fluid leakage. Signs and symptoms include cerebral edema; nausea; vomiting; headache; and increased intracranial pressure, which heralds a poor outcome and imminent death. Central nervous system injury with this amount of exposure is irreversible and occurs before hematopoietic or gastrointestinal system symptoms appear. Cardiovascular collapse is usually seen in conjunction with these injuries.

Skin effects can also indicate the dose of radiation exposure. With exposure of 600 to 1,000 rad, erythema occurs; it can disappear within hours and then reappear. The exposed patient must be evaluated hourly for the presence of erythema. With exposures greater than 1,000 rad, desquamation (radiation dermatitis) of the skin occurs. Necrosis becomes evident within a few days to months at doses greater than 5,000 rad.

Secondary injury can occur when the radiation exposure occurs during a traumatic event such as a blast or burn. Trauma in addition to radiation exposure increases patient mortality. Attention must first be directed toward the primary assessment for trauma. Airway, breathing, circulation, and fracture reduction require immediate attention. All definitive treatments must occur within the first 48 hours. Thereafter, all surgical procedures should be delayed for 2 to 3 months because of the potential for delayed wound healing and the possible development of opportunistic infections several weeks after exposure.

Survival

There are three categories of predicted survival after radiation exposure: probable, possible, and improbable. Triage of victims at the scene, after decontamination, is conducted using the routine system for disaster triage. Presenting signs and symptoms determine the potential for survival and therefore the category of predicted survival during triage.

Probable survivors have either no initial symptoms or only minimal symptoms (e.g., nausea and vomiting), or these symptoms resolve within a few hours. These patients should have a complete blood count drawn and may be discharged with instructions to return if any symptoms recur.

Possible survivors present with nausea and vomiting that persist for 24 to 48 hours. They experience a latent period, during which leukopenia, thrombocytopenia, and lymphocytopenia occur. Barrier precautions and protective isolation are implemented if the patient's lymphocyte count is less than 1,200/mm^3. Supportive treatment includes administration of blood products, prevention of infection, and provision of enhanced nutrition.

Improbable survivors have received more than 800 rad of total-body penetrating irradiation. People in this group demonstrate an acute onset of vomiting, bloody diarrhea, and shock. Any neurologic symptoms suggest a lethal dose of radiation (CDC, 2006). These patients still require decontamination to prevent further contamination of the area and of others. Personal protection is essential, because it is virtually impossible to fully decontaminate these patients; all of their internal organs have been irradiated. The survival time is variable; however, death usually occurs swiftly due to shock. If there are no neurologic symptoms, patients may be alert and oriented, similar to a patient with extensive burns. In a mass casualty situation, these patients would be triaged into the black category, where they will receive comfort measures and emotional support. If it is not a mass casualty situation, aggressive fluid and electrolyte therapies are essential.

Critical Thinking Exercises

1 pq You are the triage nurse at the receiving facility for casualties after an earthquake. As you prepare to set up a triage station, "walking wounded" begin to arrive on their own. Among them is a crying 19-year-old woman with an arm laceration who is visibly shaking and keeps shrieking, "Why?"; a woman who is 26 weeks pregnant and carrying a 3-year-old child who has a Glasgow Coma Scale score of 6 and is apneic; and a 20-year-old man who believes he has broken his ankle and is leaning on a friend who is supporting him and appears to be without injury. As you quickly assess these patients, an ambulance arrives with a 90-year-old man with a head injury who is apneic and with a pulse barely palpable only at the carotids. The EMS personnel have been unable to intubate him. In addition, a 25-year-old woman who is 32 weeks pregnant arrives with blunt abdominal trauma. How would you initially triage these patients to provide optimal use of resources, remembering that this is only the first wave of injuries from the scene? Family members, members of the press, and city officials also begin to arrive at the hospital in large numbers, and all attempt hospital entry via your triage area. You have assistance of nonclinical personnel available to you. How should the family members be managed? How should the other people be managed?

2 A patient arrives at the triage desk of the ED complaining of a sudden onset of a high fever and respiratory flulike symptoms. Three other patients with similar complaints arrived to the ED within the past 2 hours. The CDC has recently issued an alert that the swine flu may be returning this year and to be prepared for patients with these symptoms. What signs and symptoms should you identify if you suspect a biologic weapon may be the culprit? Which agents cause pneumonialike or flulike signs and symptoms? What precautions should be taken to protect staff and patients? What type of isolation would be typical if you suspect a contagion that is transmitted by the respiratory tract? What other questions would you ask the patient to determine his exposure risks?

3 ebp You are a member of the safety committee at your hospital. The committee is charged with updating the hospital's EOP and initiating a mass casualty subcommittee. What types of specific disasters are more likely to occur in and around your community? How do you determine the hazards and vulnerability of your facility? How does your facility plan comply with the NIMS requirements? What plan for triage will be included in the EOP? What types of PPE will you recommend? What is the strength of the evidence that supports your recommendation for the use of a particular type of PPE?

Brunner Suite Resources

Explore these additional resources to enhance learning for this chapter:

- **NCLEX-Style Questions and Other Resources on thePoint, http://thePoint.lww.com/Brunner13e**
- Study Guide
- PrepU
- Clinical Handbook
- Handbook of Laboratory and Diagnostic Tests

References

Books

American College of Surgeons. (2008). *Advanced trauma life support* (8th ed.). Chicago: Author.

American College of Surgeons. (2010). *Disaster management and emergency preparedness course*. Chicago: Author.

American Nurses Association. (2008). *Adapting standards of care under extreme conditions: Guidance for professionals during disasters, pandemics, and other extreme emergencies*. Washington, DC: American Nurses Publishing

Auerbach, P. S. (2012). *Wilderness medicine* (6th ed.). Philadelphia: Mosby Elsevier.

Cline, D. M., Ma, O. J., Cydulka, R. K., et al. (2012). *Tintinalli's emergency medicine manual* (7th ed.). New York: McGraw-Hill Medical.

Emergency Nurses Association. (2013). *Sheehy's manual of emergency care* (7th ed.). St. Louis: Elsevier Mosby.

Veenema, T. G. (2013). *Disaster nursing and emergency preparedness* (3rd ed.). New York: Springer.

Journals and Electronic Documents

Agency for Toxic Substances and Disease Registry. (2012). *HazMat emergency preparedness training and tools for responders*. Available at: www.atsdr.cdc.gov/hazmat-emergency-preparedness.html

Centers for Disease Control and Prevention. (2006). *Acute radiation syndrome: A fact sheet for physicians*. Available at: www.emergency.cdc.gov/radiation/arsphysicianfactsheet.asp

Centers for Disease Control and Prevention. (2012a). *Severe acute respiratory syndrome (SARS)*. Available at: www.cdc.gov/sars

Centers for Disease Control and Prevention. (2012b). *Smallpox vaccination*. Available at: www.emergency.cdc.gov/agent/smallpox/vaccination

Centers for Disease Control and Prevention. (2013). *Blast injury: Fact sheets for professionals* Available at: www.emergency.cdc.gov/BlastInjuries

Hick, J. L., Davries, A. S., Fink-Kocken, P., et al. (2012). Allocating resources during a crisis: You can't always get what you want. *Minnesota Medicine*, *95*(4), 46–50.

Joint Commission. (2011). Hospital accreditation standards for emergency management planning. *Joint Commission Perspectives*, *31*(10), 2–21.

Klima, D. A., Seiler, S. H., Peterson, J. B., et al. (2012). Full-scale regional exercises: Closing the gaps in disaster preparedness. *Journal of Trauma*, *73*(3), 592–598.

Ritenour, A. E., Wickley, A., Ritenour, J. S., et al. (2008). Tympanic membrane perforation and hearing loss from blast overpressure in Operation Enduring Freedom and Operation Iraqi Freedom wounded. *Journal of Trauma*, *64*(2), S174–S178.

U.S. Army Medical Research Institute of Chemical Defense. (2012). Available at: chemdef.apgea.army.mil

U.S. Department of Health and Human Services. (2011). *Public health emergency: National health security strategy*. Available at: www.phe.gov/preparedness/planning/authority/nhss/Pages/default.aspx

Resources

Agency for Toxic Substances and Disease Registry, www.atsdr.cdc.gov

American Red Cross, www.redcross.org/news/article/Beat-Disaster-with-Preparedness

Centers for Disease Control and Prevention (CDC), www.cdc.gov

Chemtrac, www.chemtrac.co.uk/

Federal Emergency Management Agency (FEMA), www.fema.gov

National Organization on Disability, www.nod.org/research_publications/emergency_preparedness_materials/

U.S. Department of Homeland Security, www.dhs.gov

Index

Page numbers followed by c indicate charts; those followed by f indicate figures; those followed by t indicate tables.

M

Z

Features Index

ASSESSMENT

ETHICAL DILEMMA

GENETICS IN NURSING PRACTICE

GERONTOLOGIC CONSIDERATIONS

GUIDELINES

HEALTH PROMOTION

HOME CARE CHECKLIST

NURSING RESEARCH PROFILE